Basic Curves

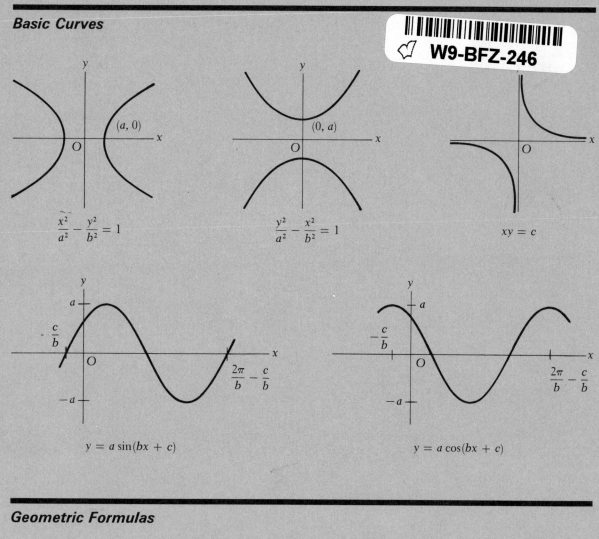

$$\frac{x^2}{a^2} - \frac{y^2}{b^2} = 1$$

$$\frac{y^2}{a^2} - \frac{x^2}{b^2} = 1$$

$$xy = c$$

$$y = a \sin(bx + c)$$

$$y = a \cos(bx + c)$$

Geometric Formulas

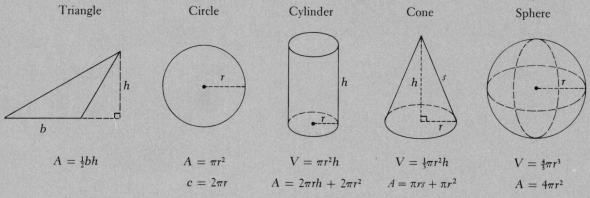

Triangle	Circle	Cylinder	Cone	Sphere
$A = \frac{1}{2}bh$	$A = \pi r^2$	$V = \pi r^2 h$	$V = \frac{1}{3}\pi r^2 h$	$V = \frac{4}{3}\pi r^3$
	$c = 2\pi r$	$A = 2\pi rh + 2\pi r^2$	$A = \pi rs + \pi r^2$	$A = 4\pi r^2$

FIFTH EDITION

BASIC TECHNICAL MATHEMATICS WITH CALCULUS

Other Benjamin/Cummings Titles of Related Interest

Basic Technical Mathematics, Fifth Edition, by Allyn J. Washington

Basic Technical Mathematics with Calculus, Metric Version, Fifth Edition, by Allyn J. Washington

Student Solutions Manual for the Washington Series in Basic Technical Mathematics, Fifth Edition, by Frances Willbanks and Anne Zeigler

Introduction to Technical Mathematics, Fourth Edition, by Allyn J. Washington and Mario F. Triola

Technical Calculus with Analytic Geometry, Third Edition, by Allyn J. Washington

Introduction to Geometry (module), by Allyn J. Washington

Arithmetic and Beginning Algebra, by Allyn J. Washington

Essentials of Basic Mathematics, Third Edition, by Allyn J. Washington, Samuel H. Plotkin, and Carolyn E. Edmond

FIFTH EDITION

BASIC TECHNICAL MATHEMATICS WITH CALCULUS

Allyn J. Washington
Dutchess Community College

The Benjamin/Cummings Publishing Company
Redwood City, California • Fort Collins, Colorado
Menlo Park, California • Reading, Massachusetts
New York • Don Mills, Ontario • Wokingham, U.K. • Amsterdam
Bonn • Sydney • Singapore • Tokyo • Madrid • San Juan

Sponsoring editor: Lisa Moller

Production supervisor: Sharon Montooth

Production service: Greg Hubit Bookworks

Book and cover designer: John Edeen

Cover photo: Peter Menzel

Copyeditor: Patricia Harris

Library of Congress Cataloging-in-Publication Data

Washington, Allyn J.
   Basic technical mathematics with calculus / Allyn J. Washington. — 5th ed.
    p.   cm.

   Includes index.
   ISBN 0-8053-8889-3
   1. Mathematics. I. Title.
QA37.2.W37 1989
510—dc20   89-6999
                                       CIP

ABCDEFGHIJK-VH-8932109

The Benjamin/Cummings Publishing Company, Inc.
390 Bridge Parkway
Redwood City, California 94065

Preface

Scope of the Book

This book is intended primarily for students in technical and pre-engineering technology programs or other programs for which a coverage of basic mathematics is required.

Chapters 1 through 19 provide the necessary background in algebra and trigonometry for analytic geometry and calculus courses, and Chapters 1 through 20 provide the background necessary for calculus courses. There is an integrated treatment of mathematical topics, from algebra to calculus, which are necessary for a sound mathematical background for the technician. Numerous applications from many fields of technology are included, primarily to indicate where and how mathematical techniques are used. However, it is not necessary that the student have a specific knowledge of the technical area from which any given problem is taken.

It is assumed that students using this text will have a background including algebra and geometry. However, the material is presented in sufficient detail for use by those whose background is possibly deficient to some extent in these areas. The material presented here is sufficient for three or four semesters.

One of the primary reasons for the arrangement of topics in this text is to present material in an order that allows a student to take courses in allied technical areas, such as physics and electricity, concurrently. These allied courses normally require a student to know certain mathematical topics by certain definite times; yet the traditional order of topics in mathematics makes it difficult to attain this coverage without loss of continuity. However, the material in this book can be rearranged to fit any appropriate sequence of topics if this is deemed necessary. Another feature of this text is that certain topics that are traditionally included primarily for mathematical completeness have been covered only briefly or have been omitted.

The approach used here is basically an intuitive one. It is not unduly rigorous mathematically, although all appropriate terms and concepts are introduced as needed and given an intuitive or algebraic foundation. The book's aim is to help the student develop a feeling for mathematical methods, and not simply to provide a collection of formulas. The text emphasizes that it is essential for the student to be fluent in algebra and trigonometry in order to understand and succeed in any subsequent work in mathematics.

New Features

This fifth edition of *Basic Technical Mathematics with Calculus* includes all the basic features of the earlier editions. However, many sections have been rewritten to some degree to include additional or revised explanatory material, examples, and exercises. Some sections have been extensively revised. Specifically, among the new features of this edition are the following:

■ **Special Explanatory Comments**
Throughout the book, special explanatory comments **in color** have been used in the examples to emphasize and clarify certain important points. Arrows are used to clearly indicate the part of the example to which reference is made.

■ **Chapter Introductions**
Each chapter is introduced by identifying the concepts to be developed and some of the important areas of technical applications. A particular type of application is shown in a photograph, and in Chapters 2 through 29 a problem related to this application is solved in an example later in the chapter.

■ **Chapter Equations**
At the end of each chapter, all important chapter equations are listed together for easy reference.

■ **Chapter Practice Tests**
At the end of each chapter there is a practice test which a student may use to check on his or her understanding of the material. Solutions to all problems in each test are given in the Appendix.

■ **Special Note Margin Indicator**
A special margin indicator (as shown) is used to clearly identify points with which students commonly make errors or which they tend to have more difficulty in handling. The particular part of the text to which the symbol refers

NOTE▷ is shown in ***boldface italic print.***

■ **Key Terms in Margin**
Certain important key terms and topics are noted in the margin for emphasis and easy reference.

■ **New Sections**
New sections are Section 2-2, which introduces the concepts of *domain* and *range* of a function, Section 22-9, which introduces *higher derivatives*, and Section 28-1, which covers concepts of *infinite series* used later in Chapter 28.

■ **Revised Coverage**
A full section is no longer devoted to computations with logarithms. Section 12-4 on *logarithms to the base 10* now includes examples of computations illustrating the properties of logarithms and computations which possibly cannot be done directly on a calculator. Also, Section 16-2 is now devoted to the solution of *linear inequalities*, rather than the graphical solution of inequalities. This method is illustrated in Section 16-3. Then in Chapter 19, Section

19-6 combines the two sections of the fourth edition on *inverse trigonometric functions.*

■ **Supplementary Topics**

In order to respond to the needs of certain programs, eight new additional sections of text material are included after Chapter 29. The topics covered are *Gaussian elimination, rotation of axes, functions of two variables, curves and surfaces in three dimensions, partial derivatives, double integrals,* and two sections on *integration by partial fractions.*

■ **Exercises**

There are now over 2000 exercises illustrating technical applications. Of these, about 68% are new to the fifth edition. This also represents an increase of over 25% from the fourth edition. In all, there are now over 9600 exercises, of which about 28% are new to this edition.

■ **Examples**

There are now over 260 worked examples illustrating technical applications. Of these, about 45% are new to the fifth edition. This also represents an increase of about 50% from the fourth edition. In all, there are now over 1350 examples, an increase of over 10%.

■ **Figures**

Many new figures have been included to assist the student in the examples and exercises. There are now over 970 figures (an increase of about 68%) in the text, and over 600 figures (an increase of over 10%) in the answer section.

Additional Features

■ **Applications**

The text material and exercises illustrate the application of mathematics to all fields of technology. Many of the new exercises relate to modern technology in areas such as computer design, computer-assisted design (CAD), electronics, solar energy, lasers, fiber optics, holography, and space technology. A special *Index of Applications* in the exercises is included near the end of the book.

■ **Calculator Use**

It is assumed that students will use a standard scientific calculator for most of their calculating work. Calculator material is integrated throughout the text, and some of the exercises are specifically designed for solution using a calculator.

■ **Examples**

Many of the examples are used advantageously to introduce concepts, as well as to clarify and illustrate points made in the text.

■ **Graphical Methods**

In addition to specific topics from analytic geometry, graphical techniques and interpretations are included throughout the text.

■ **Stated Problems**
There is a greater emphasis on the solution of stated problems in this edition. They are introduced in Chapter 1, and about 120 examples and 900 exercises appear throughout the text. Including them a few at a time, but regularly, allows students to better develop techniques of solution.

■ **Review Exercises**
Each chapter is followed by review exercises. These can be used either as additional problems or as review assignments.

■ **Important Formulas**
These formulas are set off and displayed so that they are easily located and used.

■ **Computer Programs in BASIC**
Appendix E gives 25 selected computer programs in BASIC, of which 10 are new in this edition. These programs are keyed to text material by use of a computer symbol (shown to the left).

■ **Units of Measurement**
Both U.S. customary units and SI metric units are used. Appendix B includes a discussion of units.

■ **Answers to Exercises**
The answers to all the odd-numbered exercises are given at the back of the book. Also, the *Student Solutions Manual* contains solutions for every other odd-numbered exercise. The answers to the even-numbered exercises are given in the *Instructor's Guide*.

■ **Use of Color**
A second color is used extensively throughout the text to assist in highlighting important points and features. It is used for special symbols, special explanatory comments, margin terms, application exercises, graphs and figures, and other display features.

■ **Flexibility of Material Coverage**
The order of material coverage can be changed in many places, and certain sections may be omitted without loss of continuity of coverage. Users of earlier editions have indicated the successful use of numerous variations in coverage. Any omissions or changes will depend on the type of course and completeness required. Several of the possible variations in coverage are discussed in the *Instructor's Guide*.

■ **Computer Supplement**
TECHDISK is a software package which allows students to have additional practice on the concepts presented in the text. It does not require previous computer experience. Additional information on *TECHDISK* is presented in the *Instructor's Guide*.

■ **Other Supplements**
Other supplements to this text include an *Instructor's Guide* and a *Student Solutions Manual*. The *Instructor's Guide* presents information on the use of

the text, some of which has been previously noted in this preface. The *Student Solutions Manual* contains solutions for every other odd-numbered exercise, as was previously noted.

Acknowledgments

The author gratefully acknowledges the contributions of the following reviewers, each of whom read all or most of the fifth edition manuscript. Their detailed comments and many valuable suggestions were of great assistance.

Kathleen Davidoff
University of Cincinnati – Clermont
    College

Dewey Furness
Ricks College

Janet K. Hertzberg
New Hampshire Technical
    Institute

The author also sincerely thanks the following individuals for their helpful responses and comments in completing a detailed questionnaire for the fifth edition.

Dennis Adams
State Technical Institute at
    Knoxville

Gary Adams
Thames Valley State Technical
    College

Haya Adner
Queensborough Community
    College

W. Alderson
Conestoga College

George Allan
Lambton College

Alfred Amatangelo
Central Maine Vocational
    Technical Institute

Bill Anderson
St. Lawrence College

Larry Badois
County College of Morris

William J. Ballock
College of Lake County

L. Edward Bednar
Purdue University North Central

Walter Bell
Orange County Community
    College

Don Benbow
Marshalltown Community College

Donald Bennett
Murray State University

Louise Bernauer
County College of Morris

Robert C. Blanchard
Embry-Riddle Aeronautical
    University

Edward L. Bloxom
Coastal Carolina Community
    College

Dwin Bohn
Delaware Technical Community
    College

Glenn R. Boston
Catawba Valley Technical College

Richard Boyles
Pennsylvania State University

Doris Bratcher
Northwest Iowa Technical College

William D. Brower
New Jersey Institute of Technology

Dick Buhman
University of Nebraska

Malcolm Bull
Sheridan College

Arthur Caisse
Waterbury State Technical College

Frank Calwell
York Technical College

Jim Carney
Lorain County Community College

William Casolara
Tompkins-Cortland Community
    College

Ed Champy Jr.
Northern Essex Community
    College

Herb Cheeley
Spokane Community College

Alan S. Chepen
Miami University

Jane L. Colapietro
Broome Community College

Joseph E. Collins
Kirkwood Community College

Bonnie Connell
Monroe Community College

Sharron Cook
University of Nebraska – Omaha

Michael T. Crane
Bristol Community College

Barbara Cribbs
Stark Technical College

Bill Cunning
Humber College

Joe DeBlassio
Allegheny Community College

Eugene Decaire
Durham College

Joseph M. DeGuilmo
Hudson County Community
College

Larry DeLonais
National Education Center

George DeRise
Thomas Nelson Community
College

Ron DiMenna
St. Clair College

Tom Divver
Spartanburg Technical College

Alexander Donahue
Eastern Community College

Mona Fabricant
Queensborough Community
College

David Finlay
Seneca College

Alexander Gerardo
University of Alabama

Francis Graham
Beaver Community College

Yvette Grimmond
Seneca College

Joan Gundrum
Embry-Riddle Aeronautical
University

W. A. Habkirk
Seneca College

Ronald L. Harris
Weber State College

Alleyne Hartnett
Corning Community College

Sally Haws
Northwest Vocational Technical
School

John A. Heublein
Kansas Technical Institute

David E. Heyd
Pennsylvania State University

G. R. Heyland
Northern College

Walton B. Hill
Pennsylvania Institute of
Technology

Tommy Hinson
Forsyth Technical Institute

Margie Hobbs
State Technical Institute at
Memphis

Edward Horan
College of Aeronautics

Arla M. Huber
Pennsylvania State University

Tim S. Hungate
Illinois Central College

John Jenkins
Embry-Riddle Aeronautical
University

Arnold A. John
Humber College

J. B. Jones Jr.
Central Piedmont Community
College

Joe Jordan
John Tyler Community College

W. J. Keun
Conestoga College

William A. Kinnel
Mount Aloysius Junior College

Robert Kokernak
Fitchburg State College

John S. Kostoff
Delta College

Ralph Krauss
Prairie State College

Linda Kuroski
Erie Community College

Pat Lechherter
State Technical Institute at
Memphis

Martin Levine
Virginia Western Community
College

Larry Liddle
Wytheville Community College

Alice A. Lindsey
Greenville Technical College

James E. Martin
Wake Technical College

Raymond Maruca
Pennsylvania State University

Salvatore Mascia
Pennsylvania State University

Ralph McGrew
Iowa Western Community College

Owen McKinnie
New River Community College

Joseph Meeks
Tompkins-Cortland Community
College

P. J. Melroy
Kirkwood Community College

Alvin S. Metts
Forsyth Technical Institute

Wayne R. Michie
Virginia Western Community
College

John J. Mikalauskas
Ulster County Community College

P. M. Moanfeldt
Greater New Haven State
Technical College

Theodore F. Moore
Mohawk Valley Community
College

Donald Moser
St. Louis Community College

Beverly Mugrage
University of Akron

Neal Nettler
Holyoke Community College

Robert B. Niedbala
Thames Valley State Technical
College

Kylene Norman
Clark Technical College

Harold Oxsen
Diablo Valley College

Kathy E. Page
Stanly Technical College

Robert Parker
Weber State College

Marilyn L. Peacock
Tidewater Community College

Edward Pecce
Kettering College of Medical Arts

Carl B. Petersen
Northeast Metropolitan Technical
Institute

Maxine D. Reed
State Technical Institute at
Memphis

Patrick L. Reilly
Southern Illinois University

Steven D. Rice
Missoula Vocational Technical
Institute

L. Rosenman
DeVry Technical Institute

Robert Rouse
Agricultural Technical Institute at
Morrisville

Thomas M. Royal
Coastal Carolina Community
College

Edith M. Ruben
New River Community College

Gordon O. Schlafmann
Western Wisconsin Technical
Institute

James Sedivel
Triton College

M. C. Sexton Jr.
Wake Technical College

Craig R. Sibley
Union County College

Nancy Siemor
Tompkins-Cortland Community
College

Edith Silver
Mercer County College

Renald Simmons
Vincennes University

Kath Soderbom
Massasoit Community College

Marjorie O. South
Community College of Beaver
County

Wai Stam
Seneca College

E. Stedman
Mississippi County Community
College

Debra M. Swedberg
Casper College

Thomas W. Taylor
San Antonio College

Linda Tully
University of Pittsburgh

Wilfred Turner
Craven Community College

Ronald Virden
Lehigh County Community College

Dick Wagenknecht
St. Paul Technical Vocational
Institute

Alfred E. Wandrei
Quinsigamond Community College

Eugene R. Waschbusch
St. Paul Technical Vocational
Institute

Richard Watkins
Tidewater Community College

Lee Webster
Greenfield Community College

Suzanne Welsch
Sierra Nevada College

Bob Wheeler
Weber State College

Charles Wilson
St. Clair College

J. D. Wilson
Murray State University

John L. Wistoff
Anne Arundel Community College

Ben F. Zirkle
Virginia Western Community
College

I again wish to thank members of the Mathematics Department of Dutchess Community College for their help and suggestions for the earlier editions. Also, I wish to thank William J. Flynn of Dutchess Community College for reviewing all of the computer programs.

Special thanks go to Anne Ziegler of Florence Darlington Technical College and Frances Willbanks of Southern Bell for again preparing the *Student Solutions Manual*, and to Robert Seaver of Lorain County Community College and William Thomas of the University of Toledo for their diligent efforts in preparing *TECHDISK*.

My thanks and gratitude also go to Gloria Langer for assisting in the tedious task of reading proofs and checking answers. I also wish to thank Carolyn

Edmond for her assistance in checking answers.

I gratefully acknowledge my appreciation for the assistance, cooperation, and support of my editor, Lisa Moller, the production supervisor, Sharon Montooth, the production editor, Greg Hubit, and the many others of the Benjamin/Cummings staff during the production of this text. Finally, special mention is due my wife, Millie, who helped with reading proof and checking answers, as well as with her patience and support all through the preparation of this edition, and all of the earlier editions.

A. J. W.

Chapter Photograph Acknowledgments

The author wishes to thank the following for use of the chapter introduction photographs:

Chapter 1: William Felgar / Grant Heilman Photography, Inc.
Chapters 2, 5, 8, 12, 14, 24: John Baron
Chapters 3, 23: Vance Houghton
Chapters 4, 27: Comstock, Inc. / Russ Kinne
Chapters 6, 16, 18, 26, 29: Comstock, Inc.
Chapters 7, 17: NASA / Grant Heilman Photography, Inc.
Chapter 9: Runk / Schoenberger / Grant Heilman Photography, Inc.
Chapters 10, 22: Comstock, Inc. / George Lepp
Chapter 11: Barry L. Runk / Grant Heilman Photography, Inc.
Chapter 15: Comstock, Inc. / Marvin Coner
Chapter 19: John Colwell / Grant Heilman Photography, Inc.
Chapter 20: Comstock, Inc. / Mark Daniel Ippolito
Chapter 21: Comstock, Inc. / Michael Stuckey
Chapter 25: Grant Heilman Photography, Inc.
Chapter 28: Comstock, Inc. / Mike and Carol Werner

Contents

1 Fundamental Concepts and Operations

Mathematics has played a most important role in the development and understanding of the various fields of technology and in the endless chain of technological and scientific advances of our time. This has resulted in a continually increased use of mathematics by technicians in all fields.

With the mathematics we shall develop in this text, many kinds of applied problems can and will be solved. Simply to give one illustration, consider the microwave receiver "dish" shown at the left. A few of the technical areas of application associated with the receiver include

- surveying and construction (site preparation)
- physics, mechanical design, and drafting (dish design)
- mechanical design and machine design (support structure)
- electricity and electronics (circuitry)

as well as many others. To support the use of these technical areas, others would also be required. For example, in surveying the site lasers might well be used, and in designing the receiver and support structure a computer would undoubtedly be used. To solve the applied problems presented in this text will require a knowledge of the relevant mathematics, but will *not* require any prior knowledge of the field of application.

Of course, we cannot solve the more advanced types of problems which arise, but we can form a foundation for the more advanced mathematics which is used to solve such problems. Therefore, the development of a real understanding of the mathematics presented in this text will be of great value to you in your future work.

A thorough understanding of algebra is essential in the study of any of the fields of mathematics. It is important for you to learn and understand the basic concepts and operations given here; otherwise the result will be a weak foundation for further work in mathematics and in other fields where mathematics is applied, which include all technical areas, at least to some degree. Development of this understanding will require a serious commitment on your part. The author sincerely wishes you the best success.

We shall begin our study of mathematics by reviewing some of the basic concepts and operations that deal with numbers and symbols. With these we shall be able to develop the topics in algebra which are necessary for progress into other fields of mathematics, such as trigonometry and calculus.

1-1   Numbers and Literal Symbols

The way we represent numbers today has been evolving for thousands of years. *The first numbers used were those which stand for whole quantities, and these we call the* **positive integers** *(or* **natural numbers**). The positive integers are represented by the symbols 1, 2, 3, 4, and so forth.

Of course, it is necessary to have numbers to represent parts of certain quantities, and for this purpose fractional quantities are introduced. *The name* **positive rational number** *is given to any number that we can represent by the division of one positive integer by another. Numbers that cannot be designated by the division of one integer by another are termed* **irrational.**

EXAMPLE A ⎤ The numbers 5, 18, and 19 are positive integers. They are also rational numbers, since they may be written as $\frac{5}{1}$, $\frac{18}{1}$, and $\frac{19}{1}$. Normally we do not write the 1's in the denominators.

The numbers $\frac{1}{2}$, $\frac{5}{8}$, $\frac{11}{3}$, and $\frac{106}{17}$ are positive rational numbers, since the numerator and the denominator of each are integers.

The numbers $\sqrt{2}$, $\sqrt{3}$, and π are irrational. It is not possible to find two integers which represent these numbers when one of the integers is divided by the other. For example, $\frac{22}{7}$ is not an *exact* representation of π; it is only an *approximation.*

The number $\frac{2}{\sqrt{3}}$ is irrational. The numerator is an integer, but the denominator is irrational and $\frac{2}{\sqrt{3}}$ cannot be written as one integer divided by another integer.

In addition to the positive numbers, it is necessary to introduce **negative numbers,** not only because we need to have a numerical answer to problems such as $5 - 8$, but also because the negative sign is used to designate direction. *Thus,* $-1, -2, -3,$ *and so on are the* **negative integers.** *The number* **zero** *is an integer, but it is neither positive nor negative. This means that the* **integers** *are the numbers* . . . , $-3, -2, -1, 0, 1, 2, 3, \ldots,$ *and so on.*

The integers, the rational numbers, and the irrational numbers, which include all such numbers which are zero, positive, or negative, make up what we call the **real number system.** We shall use real numbers throughout this text, with one important exception. In Chapter 11 we shall be using **imaginary numbers,** *which is the name given to square roots of negative numbers.* (The symbol j is used to

real number system

designate $\sqrt{-1}$, which is not part of the real number system.) However, until Chapter 11, when we discuss the operations on imaginary numbers in detail, it will be necessary only to recognize them if they occur.

EXAMPLE B

The number 7 is an integer. It is also a rational number since $7 = \frac{7}{1}$, and it is a real number since the real numbers include all of the rational numbers.

The number 3π is irrational, and it is real since the real numbers include all of the irrational numbers.

The numbers $\sqrt{-10}$ and $-\sqrt{-7}$ are imaginary.

The number $\frac{1}{8}$ is rational and real. The number $\sqrt{5}$ is irrational and real.

The number $\frac{-3}{7}$ is rational and real. The number $-\sqrt{7}$ is irrational and real.

The number $\frac{\pi}{6}$ is irrational and real. The number $\frac{\sqrt{-3}}{2}$ is imaginary.

A **fraction** *may contain any number or symbol representing a number in its numerator or in its denominator. Thus, a fraction may be rational, irrational, or imaginary.*

EXAMPLE C

The numbers $\frac{2}{7}$ and $\frac{-3}{2}$ are fractions, and they are also rational.

The numbers $\frac{\sqrt{2}}{9}$ and $\frac{6}{\pi}$ are fractions, but they are not rational numbers. It is not possible to express either as the ratio of one integer to another.

The number $\frac{\sqrt{-3}}{2}$ is a fraction, and it is also imaginary.

The real numbers may be represented as points on a line. We draw a horizontal line and designate some point on it by O, which we call the **origin** (see Fig. 1-1). The number zero, which is an integer, is located at this point. Then equal intervals are marked off from this point toward the right, and the positive integers are placed at these positions. The other positive rational numbers are located between the positions of the integers.

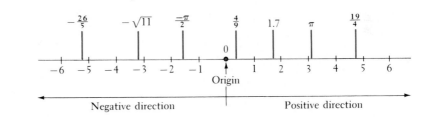

Fig. 1-1

Now we can give the direction interpretation to negative numbers. By starting at the origin and proceeding to the left, in the **negative direction,** we locate all the negative numbers. *As shown in Fig. 1-1, the positive numbers are to the right of the origin and the negative numbers are to the left.* Representing numbers in this way will be especially useful when we study graphical methods.

It will not be proved here, but the rational numbers do not take up all the positions on the line; the remaining points represent irrational numbers.

Another important mathematical concept we use in dealing with numbers is the **absolute value** of a number. By definition, *the absolute value of a positive number is the number itself, and the absolute value of a negative number is the corresponding positive number (obtained by changing its sign).* We may interpret the absolute value as being the number of units a given number is from the origin, regardless of direction. The absolute value is designated by ‖ placed around the number.

EXAMPLE D

The absolute value of 6 is 6, and the absolute value of -7 is 7. We designate these by $|6| = 6$ and $|-7| = 7$. See Fig. 1-2.

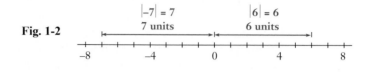

Fig. 1-2

Other examples are: $|-\pi| = \pi,$ $\left|\dfrac{7}{5}\right| = \dfrac{7}{5},$ $|-\sqrt{2}| = \sqrt{2},$ $|0| = 0$

On the number scale, *if a first number is to the right of a second number, then the first number is said to be* **greater than** *the second. If the first number is to the left of the second, it is* **less than** *the second number.* The symbol $>$ is used to designate "is greater than," and the symbol $<$ is used to designate "is less than." These are called **signs of inequality.**

EXAMPLE E

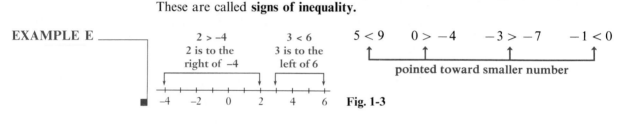

Fig. 1-3

Every number, except zero, has a **reciprocal.** *The reciprocal of any number is 1 divided by that number.*

EXAMPLE F

The reciprocal of 7 is $\frac{1}{7}$. The reciprocal of $\frac{2}{3}$ is

$$\frac{1}{\frac{2}{3}} = 1 \times \frac{3}{2} = \frac{3}{2}$$

invert and multiply (from arithmetic)

The reciprocal of π is $1/\pi$. The reciprocal of -5 is $-\frac{1}{5}$. Notice that the negative sign is retained in the reciprocal.

In applications, numbers are often used to represent some type of measurement and therefore have certain units of measurement associated with them. *Such numbers are referred to as* **denominate numbers.** For a discussion of units of measurement and the symbols which are used, see Appendix B. The following example illustrates the use of units and the symbols which represent them.

EXAMPLE G

To indicate that a wire is 10 feet long, we represent the length as 10 ft.

To indicate that the speed of a projectile is 150 meters per second, we represent the speed as 150 m/s. (Note the use of s for second. We use s rather than sec, which is also common.)

To indicate that the volume of a container is 500 cubic centimeters, we represent the volume by 500 cm^3. (Also common are cc and cu cm.)

literal numbers

To this point we have been dealing with numbers in their explicit form. However, it is normally more convenient to state definitions and operations on numbers in a generalized form. *To do this we represent the numbers by letters, often referred to as* **literal numbers.**

For example, we can say, "If a is to the right of b on the number scale, then a is greater than b, or $a > b$." This is more convenient than saying, "If a first number is to the right of a second number, then the first number is greater than the second number." The statement "the reciprocal of a number a is $1/a$" is another example of using letters to stand for numbers in general.

In an algebraic discussion, certain letters are sometimes allowed to take on any value, while other letters represent the same number throughout the discussion. *Those literal numbers which may vary in a given problem are called* **variables,** *and those which are held fixed are called* **constants.**

Common usage today normally designates the letters near the end of the alphabet as variables and letters near the beginning of the alphabet as constants. There are exceptions, but these are specifically noted. Letters in the middle of the alphabet are also used, but their meaning in any problem is specified.

EXAMPLE H _____ The resistance of an electric resistor is R. The current I in the resistor is equal to the voltage V divided by R. We write this as $I = V/R$, and for this resistor I and V may take on various values and R is fixed. This means that I and V are variables and R is a constant. For a different resistor the value of R may change. ■

EXAMPLE I _____ The fixed costs for a microprocessor manufacturer to operate a certain plant is b dollars per day, and it costs a dollars to produce each microprocessor. The total daily cost C to produce n microprocessors is

$$C = an + b$$

Here C and n are variables, and a and b are constants. For another plant, a and b might differ. ■

NOTE ▷ Special note: In the following exercises, and in the exercises and examples throughout the text, if an explanation of the units of measurement and their symbols is needed, see Appendix B.

Exercises 1-1

In Exercises 1 through 4, designate each of the given numbers as being an integer, rational, irrational, real, or imaginary. (More than one designation may be correct.)

1. $3, -\pi$ **2.** $\dfrac{5}{4}, \sqrt{-4}$ **3.** $-\sqrt{-6}, \dfrac{\sqrt{7}}{3}$ **4.** $-\dfrac{7}{3}, \dfrac{\pi}{6}$

In Exercises 5 through 8, find the absolute value of each of the given numbers.

5. $3, \dfrac{7}{2}$ **6.** $-4, \sqrt{2}$ **7.** $-\dfrac{6}{7}, -\sqrt{3}$ **8.** $-\dfrac{\pi}{2}, -\dfrac{19}{4}$

In Exercises 9 through 16, insert the correct sign of inequality ($>$ or $<$) between the given pairs of numbers.

9. 6   8   **10.** 7   5   **11.** π -1 **12.** -4 0

13. -4 -3 **14.** $-\sqrt{2}$ -9 **15.** $-\dfrac{1}{3}$ $-\dfrac{1}{2}$ **16.** -0.6 0.2

In Exercises 17 through 20, find the reciprocals of the given numbers.

17. 3, -2 **18.** $\frac{1}{6}$, $-\frac{7}{4}$ **19.** $-\frac{5}{\pi}$, x **20.** $-\frac{8}{3}$, $\frac{y}{b}$

In Exercises 21 through 24, locate the given numbers on a number line as in Fig. 1-1.

21. 2.5, $-\frac{1}{2}$ **22.** $\sqrt{3}$, $-\frac{12}{5}$ **23.** $-\frac{\sqrt{2}}{2}$, 2π **24.** $\frac{123}{19}$, $-\frac{\pi}{6}$

In Exercises 25 through 36, solve the given problems.

25. List the following numbers in numerical order, starting with the smallest: -1, 9, π, $\sqrt{5}$, $|-8|$, $-|-3|$, -18.

26. List the following numbers in numerical order, starting with the smallest: $\frac{1}{5}$, $-\sqrt{10}$, $-|-6|$, -4, 0.25, $|-\pi|$.

27. If a and b are positive integers and $b > a$, what type of number is represented by (a) $b - a$, (b) $a - b$, (c) $\frac{b - a}{b + a}$?

28. If a and b represent positive integers, what kind of number is represented by (a) $a + b$, (b) a/b, (c) $a \times b$?

29. Describe the location of a number x on the number line when (a) $x > 0$, (b) $x < -4$.

30. Describe the location of a number x on the number line when (a) $|x| < 1$, (b) $|x| > 2$.

31. The heat loss L through a certain type of insulation of thickness t is given by $L = a/t$, where a has a fixed value for this type of insulation. Identify the variables and constants.

32. The velocity v of a device dropped from a weather balloon is related to the distance fallen s by the equation $v = \sqrt{2gs}$, where g is the acceleration of gravity near the surface of the earth. Identify the variables and constants.

33. The memory of a certain computer has a bits in each byte. Express the number N of bits in n kilobytes in an equation. (A *bit* is a single digit, and bits are grouped in *bytes* in order to represent special characters. Commonly there are 6 or 8 bits per byte. If necessary, see Appendix B for the meaning of "kilo.")

34. A piece y inches long is cut from a board x feet long. Give an equation for the length L, in inches, of the remaining piece.

35. In writing a laboratory report, a student wrote "$-20°C > -30°C$." Is this statement correct?

36. After 5 s, the pressure on a valve is less than 60 lb/in.^{2} (pounds per square inch). Using t to represent time and p to represent pressure, this statement can be written "for $t > 5$ s, $p < 60$ lb/in.^{2}." In this way write the statement "when the current I in a circuit is less than 4 A, the voltage V is greater than 12 V."

1-2   Fundamental Laws of Algebra and Order of Operations ■

In performing the basic operations with numbers, we know that certain basic laws are valid. *These basic statements are called the* **fundamental laws of algebra.**

For example, we know that if two numbers are to be added, it does not matter in which order they are added. Thus $5 + 3 = 8$, as well as $3 + 5 = 8$. For this case we can say that $5 + 3 = 3 + 5$. This statement, generalized and assumed correct for all possible combinations of numbers to be added, is called the **commutative law** for addition. The law states that *the sum of two numbers is the same, regardless of the order in which they are added.* We make no attempt to prove this in general, but accept its validity.

commutative and associative laws

In the same way we have the **associative law** for addition, which states that *the sum of three or more numbers is the same, regardless of the manner in which they are grouped for addition.* For example,

$$3 + (5 + 6) = (3 + 5) + 6$$

The laws which we have just stated for addition are also true for multiplication. Therefore, *the product of two numbers is the same, regardless of the order in which they are multiplied,* and *the product of three or more numbers is the same, regardless of the manner in which they are grouped for multiplication.* For example, $2 \times 5 = 5 \times 2$ and $5 \times (4 \times 2) = (5 \times 4) \times 2$.

distributive law

There is one more very important law, called the **distributive law.** It states that *the product of one number and the sum of two or more other numbers is equal to the sum of the products of the first number and each of the other numbers of the sum.* For example,

$$4(3 + 5) = 4 \times 3 + 4 \times 5$$

In practice these laws are used intuitively, except perhaps for the distributive law. However, it is necessary to state them and to accept their validity, so that we may build our later results with them.

Not all operations are associative and commutative. For example, division is not commutative, since the indicated order of division of two numbers does matter. For example, $\frac{6}{5} \neq \frac{5}{6}$ ($\neq$ is read "does not equal").

Using literal symbols, the fundamental laws of algebra are as follows:

Commutative law of addition: $a + b = b + a$

Associative law of addition: $a + (b + c) = (a + b) + c$

Commutative law of multiplication: $ab = ba$

Associative law of multiplication: $a(bc) = (ab)c$

Distributive law: $a(b + c) = ab + ac$

Having identified the fundamental laws of algebra, we shall state the laws which govern the operations of addition, subtraction, multiplication, and division of signed numbers. These laws will be of primary and direct use in all of our work.

1. *To add two real numbers with like signs, add their absolute values and affix their common sign to the result.*

EXAMPLE A

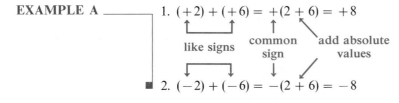

1. $(+2) + (+6) = +(2 + 6) = +8$

 like signs common sign add absolute values

2. $(-2) + (-6) = -(2 + 6) = -8$

2. *To add two real numbers with unlike signs, subtract the smaller absolute value from the larger absolute value and affix the sign of the number with the larger absolute value to the result.*

EXAMPLE B

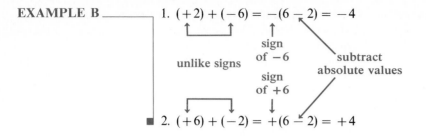

1. $(+2) + (-6) = -(6 - 2) = -4$

unlike signs

sign of -6

subtract absolute values

sign of $+6$

2. $(+6) + (-2) = +(6 - 2) = +4$

3. *To subtract one real number from another, change the sign of the number to be subtracted, and then proceed as in addition.*

EXAMPLE C

subtraction   addition

change sign

1. $(+2) - (+6) = +2 + (-6) = -(6 - 2) = -4$
2. $(-a) - (-a) = -a + (+a) = -a + a = 0$

The second part of Example C shows that subtracting a number from itself results in zero, even if the number is negative. Therefore, we see that *subtracting a negative number is equivalent to adding a positive number of the same absolute value.* This reasoning is the basis of the rule which states, "The negative of a negative number is a positive number."

4. *The product (or quotient) of two real numbers of like signs is the product (or quotient) of their absolute values. The product (or quotient) of two real numbers of unlike signs is the negative of the product (or quotient) of their absolute values.*

EXAMPLE D

1. $(+3)(+5) = +(3 \times 5) = +15$ $\dfrac{+3}{+5} = +\left(\dfrac{3}{5}\right) = +\dfrac{3}{5}$

like signs

2. $(-3)(+5) = -(3 \times 5) = -15$ $\dfrac{-3}{+5} = -\left(\dfrac{3}{5}\right) = -\dfrac{3}{5}$

unlike signs

3. $(-3)(-5) = +(3 \times 5) = +15$ $\dfrac{-3}{-5} = +\left(\dfrac{3}{5}\right) = +\dfrac{3}{5}$

like signs

When we have an expression in which there is a combination of the basic operations, we must be careful to perform them in the proper order. Generally it is clear by the grouping of numbers as to the proper order of performing these operations. Numbers are grouped by the use of symbols such as **parentheses, ( )**, and the **bar,** ——, between the numerator and the denominator of a fraction. However, if the order of at least some of the operations is not defined by specific groupings, we use the following **order of operations** related to additions, subtractions, multiplications, and divisions.

order of operations

5. (a) *Operations within specific groupings are done first.*

In completing the operations within specific groupings and otherwise where the order of operations is not indicated by specific grouping,

(b) *multiplications and divisions are performed first, and*

(c) *then additions and subtractions are performed.*

In performing additions and subtractions, or multiplications and divisions, we work from left to right.

EXAMPLE E

The expression $20 \div (2 + 3)$ is evaluated by first adding $2 + 3$ and then dividing 20 by 5 to obtain the result 4. Here, the grouping of $2 + 3$ is clearly shown by the parentheses.

The expression $20 \div 2 + 3$ is evaluated by first dividing 20 by 2 and adding this quotient of 10 to 3 in order to obtain the result of 13. Here no specific grouping is shown, and therefore the division is performed before the addition.

NOTE ▷

The expression $16 - 2 \times 3$ is evaluated by *first multiplying 2 by 3* and subtracting the product from 16 in order to obtain the result of 10. We do *not* first subtract 2 from 16.

When evaluating expressions, it is generally more convenient to change the operations and numbers so that the result is obtained by the addition and subtraction of unsigned numbers (equivalent to positive numbers). When this is done we must remember that adding a negative number is equivalent to subtracting an unsigned number, and subtracting a negative number is equivalent to adding an unsigned number.

EXAMPLE F

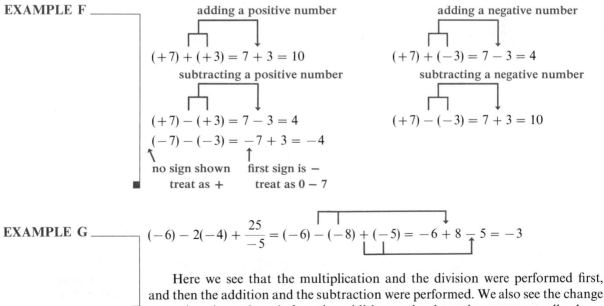

adding a positive number

$(+7) + (+3) = 7 + 3 = 10$

adding a negative number

$(+7) + (-3) = 7 - 3 = 4$

subtracting a positive number

$(+7) - (+3) = 7 - 3 = 4$

$(-7) - (-3) = -7 + 3 = -4$

no sign shown   first sign is $-$
treat as $+$ treat as $0 - 7$

subtracting a negative number

$(+7) - (-3) = 7 + 3 = 10$

EXAMPLE G

$$(-6) - 2(-4) + \frac{25}{-5} = (-6) - (-8) + (-5) = -6 + 8 - 5 = -3$$

Here we see that the multiplication and the division were performed first, and then the addition and the subtraction were performed. We also see the change to unsigned numbers before the addition and subtraction were actually done.

EXAMPLE H _____ An investigator determined that the velocity, in miles per hour, of an automobile immediately following a collision with another vehicle can be found by performing the following evaluation:

$$\frac{3000(40) + (2000)(-20)}{3000 + 2000} = \frac{120,000 + (-40,000)}{3000 + 2000} = \frac{120,000 - 40,000}{5000}$$

$$= \frac{80,000}{5000} = 16 \text{ mi/h}$$

The numerator and the denominator must be evaluated before the division is performed. Therefore, the multiplications in the numerator are performed first, followed by the addition in the denominator and the subtraction in the numerator. We also see the use of unsigned numbers. ■

1-3   Operations with Zero

Since the basic operations with zero tend to cause some difficulty, we shall demonstrate them separately in this section.

If a represents any real number, the operations of addition, subtraction, multiplication, and division with zero are defined as follows:

$$a + 0 = a$$

$$a - 0 = a \qquad 0 - a = -a$$

$$a \times 0 = 0$$

$$0 \div a = \frac{0}{a} = 0 \qquad \text{if} \qquad a \neq 0 \qquad (\neq \text{ means "is not equal to")}$$

Note that there is no result defined for division by zero. To understand the reason for this, consider the results for $\frac{6}{2}$ and $\frac{6}{0}$.

$$\frac{6}{2} = 3 \qquad \text{since} \qquad 2 \times 3 = 6$$

If $\frac{6}{0} = b$, then $0 \times b = 6$. This cannot be true because $0 \times b = 0$ for any value of b. Thus,

division by zero is undefined

(The special case of $\frac{0}{0}$ is termed *indeterminate*. Setting $\frac{0}{0} = b$, we see that $0 = 0 \times b$, which is true for any value of b. This means that no specific value of b can be determined.)

EXAMPLE A _____ $5 + 0 = 5, \qquad 7 - 0 = 7, \qquad 0 - 4 = -4,$

$$\frac{0}{6} = 0, \qquad \frac{0}{-3} = 0, \qquad \frac{5 \times 0}{7} = 0$$

EXAMPLE B $\dfrac{2}{5} \div 0$ is undefined $\dfrac{8}{0}$ is undefined $\dfrac{7}{0 \times 6}$ is undefined

$$\left(\dfrac{7 \times 0}{0 \times 6} \text{ is indeterminate} \right)$$

There is no need for confusion in the operations with zero. They will not cause any difficulty if we remember to

NOTE ▷   *never divide by 0*

Division by zero is the only undefined basic operation. All of the other operations with zero may be performed as for any other number.

Exercises 1-2, 1-3

In Exercises 1 through 36, evaluate each of the given expressions by performing the indicated operations.

1. $6 + (+5)$

2. $8 + (-4)$

3. $(-4) + (-7)$

4. $(-3) + (+9)$

5. $16 - 7$

6. $(+18) - (+21)$

7. $-19 - (-16)$

8. $8 - (-4)$

9. $(8)(-3)$

10. $(-9)(+3)$

11. $(-7)(-5)$

12. $(+5)(-8)$

13. $\dfrac{-9}{+3}$

14. $\dfrac{-18}{-6}$

15. $\dfrac{-60}{-3}$

16. $\dfrac{+28}{-7}$

17. $(-2)(+4)(-5)$

18. $(+3)(-4)(+6)$

19. $\dfrac{(+2)(-5)}{10}$

20. $\dfrac{-64}{(2)(-4)}$

21. $9 - 0$

22. $\dfrac{0}{-6}$

23. $\dfrac{+17}{0}$

24. $\dfrac{2 - (-5)}{0}$

25. $8 - 3(-4)$

26. $20 + 8 \div 4$

27. $3 - 2(6) + \dfrac{8}{2}$

28. $0 - (-6)(-8) + (-10)$

29. $\dfrac{(+3)(-6)(-2)}{0 - 4}$

30. $\dfrac{7 - (-5)}{(-1)(-2)}$

31. $\dfrac{24}{3 + (-5)} - 4(-9)$

32. $\dfrac{-18}{+3} - \dfrac{4 - 6}{-1}$

33. $(-7) - \dfrac{-14}{2} - 3(2)$

34. $-7(-3) + \dfrac{+6}{-3} - (-9)$

35. $\dfrac{(+3)(-9) - 2(-3)}{3 - 10}$

36. $\dfrac{(+2)(-7) - 4(-2)}{-9 - (-9)}$

In Exercises 37 through 44, determine which of the fundamental laws of algebra is demonstrated.

37. $(6)(7) = (7)(6)$

38. $6 + 8 = 8 + 6$

39. $6(3 + 1) = 6(3) + 6(1)$

40. $4(5 \times 7) = (4 \times 5)(7)$

41. $3 + (5 + 9) = (3 + 5) + 9$

42. $8(3 - 2) = 8(3) - 8(2)$

43. $(2 \times 3) \times 9 = 2 \times (3 \times 9)$

44. $(3 \times 6) \times 7 = 7 \times (3 \times 6)$

In Exercises 45 through 52, answer the given questions.

45. What is the sign of the product of an even number of negative numbers?

46. What is the sign of the product of an odd number of negative numbers?

47. Using the division of 4 by 2 and then of 2 by 4, show that division is not commutative.

48. Is subtraction commutative? Illustrate.

49. One oil well drilling rig drills 100 m deep the first day and 200 m deeper the second day. A second rig drills 200 m deep the first day and 100 m deeper the second day. Set up the equation showing that the total depth drilled by each was the same, and state what fundamental law is illustrated.

50. Set up the equation showing that the total force of five 7-lb forces is the same as that for seven 5-lb forces. What fundamental law is illustrated?

51. In a month, a personal computer dealer sold eight computers, which cost the dealer $2000 each. On each a profit of $1000 was made. Set up the expression which would be used to show the total amount the dealer received for these computers. What fundamental law is illustrated?

52. A jet travels 600 mi/h relative to the air. The wind is blowing at 50 mi/h. If the jet travels with the wind for 3 h, set up the expression which would be used to evaluate the distance traveled. What fundamental law is illustrated?

1-4 The Calculator

In the preceding sections, a number of basic calculations have been used. Throughout the remainder of the book, there will be numerous calculations to do. Many of them can be done mentally, but many others will require the use of a calculator.

A scientific calculator can be used for any of the calculations found in the book, and it is assumed that you will use a scientific calculator for most such calculations. The more advanced models can perform many types of calculations and operations beyond those shown in this text. However, a calculator that performs only the basic arithmetic operations is not sufficient for most of our needs.

In Appendix D there is a discussion of the use of the basic keys on a scientific calculator. Using Appendix D as a reference, we shall show how the calculator can be used to do calculations wherever it may be necessary. In many examples in the remainder of the text, as basic types of calculation are encountered, we shall show the sequence of keys that is to be used.

If you are not sufficiently familiar with your calculator, you should review your calculator manual or Appendix D, for *you must be familiar with the order in which the keys must be used on your particular calculator.* The order in which the operations are performed by the calculator, which determines the order in which the keys are used, is called the **logic** of the calculator. The sequence of keys we show in the examples of this book is that of a calculator using *algebraic logic* and it is proper for many calculators, but it may require a change for use on other calculators.

The following examples show the use of a calculator in making calculations of the types shown in the previous sections.

EXAMPLE A ——— Find the reciprocal of 25.

By using the definition of a reciprocal, we know that we can find the reciprocal of 25 by dividing 1 by 25. The sequence of calculator keys for this is

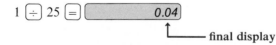

final display

The result is 0.04, as the final display shows.

However, most scientific calculators have a key for reciprocals, the $1/x$ key, as shown in Appendix D. By using it, the sequence of keys is simply

■ Again, the result is 0.04.

EXAMPLE B _____ Calculate the value of $38.3 - 12.9(-3.58)$.

Recalling the proper order of operations, we know that the multiplication $12.9(-3.58)$ is to be done before the subtraction. However, if your calculator uses algebraic logic, the numbers can be entered as shown, and the calculator will do the multiplication first. Also, the sign of -3.58 can be entered by use of the $\boxed{+/-}$ key. Therefore, the sequence of keys is

$$38.3 \; \boxed{-} \; 12.9 \; \boxed{\times} \; 3.58 \; \boxed{+/-} \; \boxed{=} \boxed{84.482}$$

which means the result is 84.5 (rounded off).

If your calculator does not use algebraic logic, it may be necessary to perform the multiplication first.

NOTE ▷ In Example B the result we found was 84.482. If the numbers being used were _approximate_, and that generally is the case in most calculations, the result should be _rounded off_ to 84.5. **The final result should not be expressed with any more significant digits _than is proper_,** even if the calculator displays more.

A discussion of the terms **rounding off, significant digits,** and **approximate numbers** can be found in Appendix B.

EXAMPLE C _____ In designing a thermostatic control, the temperature, in degrees Celsius, at which it is to be set is found by performing the following calculation:

$$\frac{726.721}{0.00689910 + 0.00523680} - (105.55)(567.15)$$

The proper order of operations tells us first to find the value of the denominator of the first expression, then to do the division in the first expression and the multiplication in the second expression, and finally to subtract. A calculator with algebraic logic and parentheses will perform these operations in the proper order. Noting that the fraction in the first expression can be written as $726.721 \div (0.0068991 + 0.0052368)$, we have the following sequence of keys:

$$726.721 \; \boxed{\div} \; \boxed{(} \; .0068991 \; \boxed{+} \; .0052368 \; \boxed{)} \; \boxed{-} \; 105.55$$
$$\boxed{\times} \; 567.15 \; \boxed{=} \boxed{19.23808}$$

■ Therefore, the result is 19.2°C (rounded off).

After this section we will often show the calculator key sequences in the left-hand margin of the page.

Exercises 1-4

In all of the following exercises, perform the required calculations on a scientific calculator.

In Exercises 1 through 8, perform the indicated operations. In Exercises 1 through 4, round off results to three significant digits and in Exercises 5 through 8, round off results to four significant digits.

1. $3.16 + 53.9 \div 117.8$

2. $807 - 245 \times 3.19$

3. $0.702 - 0.0886 \div 0.108$

4. $46.7 \times 0.923 + 39.8 \times 0.362$

5. $\dfrac{23.962 \times 0.01537}{10.965 - 8.249}$

6. $\dfrac{0.69378 + 0.04997}{257.4 \times 3.216}$

7. $\dfrac{3872}{503.1} - \dfrac{2.056 \times 309.6}{395.2}$

8. $\dfrac{1}{0.5926} + \dfrac{3.6957}{2.935 - 1.054}$

In Exercises 9 through 12, verify the basic laws of algebra by calculating the value of each side of the equal sign. Do not round off results. Indicate which law is illustrated.

9. $37.962 + 5.049 = 5.049 + 37.962$ **10.** $0.3526 \times 2.4953 = 2.4953 \times 0.3526$

11. $68.572(5.0496 + 1.9256) = 68.572 \times 5.0496 + 68.572 \times 1.9256$

12. $0.0273(2.56 \times 8.99) = (0.0273 \times 2.56) \times 8.99$

In Exercises 13 through 16, find the products and quotients of the indicated signed numbers. Round off results to three significant digits.

13. -1.46×5.62 **14.** $(39.6) \times (-2.57)$ **15.** $(-0.3627) \times (-2.69)$ **16.** $(-7.064) \times (-0.0962)$

17. $37.2 \div 2.49$ **18.** $8968 \div 478$ **19.** $\dfrac{0.0307}{0.9295}$ **20.** $\dfrac{0.1209}{-3.66}$

In Exercises 21 through 32, perform the indicated calculations.

21. Find the reciprocal of 2.8 by (a) using the definition of a reciprocal, (b) using the $\boxed{1/x}$ key. Round off the result, assuming that 2.8 is approximate.

22. Find the reciprocal of 0.48 by (a) using the definition of a reciprocal, (b) using the $\boxed{1/x}$ key. Round off the result, assuming that 0.48 is approximate.

23. Is $\pi > 3.1416$? Determine the answer by using the key sequence $\boxed{\pi}$ $\boxed{-}$ 3.1416.

24. Show that the value of π is not exactly the same as $\frac{22}{7}$ by using the key $\boxed{\pi}$ and by dividing 22 by 7.

25. If we find the decimal equivalent of a rational number, at some point some sequence of digits will start repeating endlessly. For an irrational number there is never such an endlessly repeating sequence of digits. Find the decimal equivalents of (a) $\frac{8}{33}$, (b) π. Note the repetition for $\frac{8}{33}$ and that no such repetition occurs for π.

26. Following Exercise 25, show that the decimal equivalent of the fraction $\frac{124}{990}$ indicates that it is rational.

27. Divide 2 by 0. What is the calculator display?

28. Divide 0 by 0. What is the calculator display?

29. In an experiment, a biologist found that a plant added 460 g of weight in 180 days. Find the average weight added each day, and round off the result to two significant digits.

30. In order to find the length of belting between two pulleys, it is necessary to calculate the value of

$$2 \times 136 + \frac{3.25(36.0 + 28.5)}{2}$$

Find the length of the belting (in inches) by performing the calculation and rounding off the result to three significant digits.

31. The percent of alcohol needed in a certain automobile engine coolant is found by performing the calculation $\dfrac{100(40.63 + 52.96)}{105.30 + 52.96}$. Find the percent, and round off the result to four significant digits.

32. The evaporation rate, in gallons per day, of a wastewater holding pond is found by calculating $145(1.05 + \frac{1}{236})$. Determine this rate, and round off the result to three significant digits.

1-5   Exponents

We have introduced numbers and the fundamental laws which are used with them in the fundamental operations. Then we showed the use of the calculator in performing these operations. Also, we have shown the use of literal numbers to represent numbers. In this section we shall introduce some basic terminology and notation which are important to the basic algebraic operations developed in the following sections.

In multiplication we often encounter a number which is to be multiplied by itself several times. Rather than writing this number over and over repeatedly, we use the notation a^n, where a is the number being considered and n is the number of times it appears in the product. *In the expression a^n, the number a is called the **base**, the number n is called the **exponent**, and in words, the expression is read as the "nth power of a."*

exponent

EXAMPLE A

5 factors of 4

1. $4 \times 4 \times 4 \times 4 \times 4 = 4^5$ (the fifth power of 4)

4 factors of -2

2. $(-2)(-2)(-2)(-2) = (-2)^4$ (the fourth power of -2)
3. $a \times a = a^2$ (the second power of a, called "a squared")
4. $(\frac{1}{5})(\frac{1}{5})(\frac{1}{5}) = (\frac{1}{5})^3$ (the third power of $\frac{1}{5}$, called "$\frac{1}{5}$ cubed")
5. $8 \times 8 \times 8 \times 8 \times 8 \times 8 \times 8 \times 8 \times 8 = 8^9$ (the ninth power of 8)

The basic operations with exponents will now be stated symbolically. We first state them for positive integers as exponents. Therefore, if m and n are positive integers, we have the following important operations for exponents.

$$a^m \times a^n = a^{m+n} \tag{1-1}$$

$$\frac{a^m}{a^n} = a^{m-n} \quad (m > n, a \neq 0), \qquad \frac{a^m}{a^n} = \frac{1}{a^{n-m}} \quad (m < n, a \neq 0) \tag{1-2}$$

$$(a^m)^n = a^{mn} \tag{1-3}$$

$$(ab)^n = a^n b^n, \qquad \left(\frac{a}{b}\right)^n = \frac{a^n}{b^n} \quad (b \neq 0) \tag{1-4}$$

Two forms are shown for Eq. (1-2) in order that the resulting exponent is a positive integer. This is generally, but not always, the best form of the result. After the following three examples we will discuss zero and negative exponents. This will cover the case for $m = n$, and also show that the first form of Eq. (1-2) is its basic form.

In applying Eqs. (1-1) and (1-2), the base a must be the same for the exponents to be added or subtracted. When a problem involves a product of different bases, *only exponents of the same base may be combined.* In the following three examples, Eqs. (1-1) to (1-4) are verified and illustrated.

EXAMPLE B _____ Applying Eq. (1-1), we have

add exponents

$$a^3 \times a^5 = a^{3+5} = a^8$$

We see that this result is correct since we can also write

8 factors of a

(3 factors of a)(5 factors of a) ————

$$a^3 \times a^5 = (a \times a \times a)(a \times a \times a \times a \times a) = a^8$$

Applying the first form of Eqs. (1-2), we have

$5 > 3$

$$\frac{a^5}{a^3} = a^{5-3} = a^2, \qquad \frac{a^5}{a^3} = \frac{\overset{1}{a} \times \overset{1}{a} \times \overset{1}{a} \times a \times a}{\underset{1}{a} \times \underset{1}{a} \times \underset{1}{a}} = a^2$$

Applying the second form of Eqs. (1-2), we have

$$\frac{a^3}{a^5} = \frac{1}{a^{5-3}} = \frac{1}{a^2}, \qquad \frac{a^3}{a^5} = \frac{\overset{1}{a} \times \overset{1}{a} \times \overset{1}{a}}{\underset{1}{a} \times \underset{1}{a} \times \underset{1}{a} \times a \times a} = \frac{1}{a^2}$$

$5 > 3$

■

EXAMPLE C _____ Applying Eq. (1-3), we have

multiply exponents

$$(a^5)^3 = a^{5(3)} = a^{15}, \qquad (a^5)^3 = (a^5)(a^5)(a^5) = a^{5+5+5} = a^{15}$$

Applying the first form of Eqs. (1-4), we have

$$(ab)^3 = a^3b^3, \qquad (ab)^3 = (ab)(ab)(ab) = a^3b^3$$

Applying the second form of Eqs. (1-4), we have

$$\left(\frac{a}{b}\right)^3 = \frac{a^3}{b^3}, \qquad \left(\frac{a}{b}\right)^3 = \left(\frac{a}{b}\right)\left(\frac{a}{b}\right)\left(\frac{a}{b}\right) = \frac{a^3}{b^3}$$

■

EXAMPLE D _____ Other illustrations of the use of Eqs. (1-1) to (1-4) are as follows:

1. $$\frac{(3 \times 2)^4}{(3 \times 5)^3} = \frac{3^4 2^4}{3^3 5^3} = \frac{3 \times 2^4}{5^3}$$

2. $$(-x^2)^3 = [(-1)x^2]^3 = (-1)^3(x^2)^3 = -x^6$$

3. $ax^2(ax)^3 = ax^2(a^3x^3) = a^4x^5$ ← add exponents of x

exponent of 1 ↓   add exponents of a ↓

4. $\dfrac{(ry^3)^2}{r(y^2)^4} = \dfrac{r^2y^6}{ry^8} = \dfrac{r}{y^2}$

NOTE ▷ In the third illustration, note that **ax^2 means a times the square of x, and does not mean a^2x^2**, whereas $(ax)^3$ *does* mean a^3x^3. ■

EXAMPLE E ─── In the analysis of the deflection of a beam, the following expression and simplification are used.

$$\frac{1}{2}\left(\frac{PL}{4EI}\right)\left(\frac{2}{3}\right)\left(\frac{L}{2}\right)^2 = \frac{1}{2}\left(\frac{PL}{4EI}\right)\left(\frac{2}{3}\right)\left(\frac{L^2}{2^2}\right) = \frac{\overset{1}{\cancel{2}}PL(L^2)}{\underset{1}{\cancel{2}}(3)(4)(4)EI} = \frac{PL^3}{48EI}$$

Here L is the length of the beam and P is the force applied to it. E and I are other constants related to the beam. ■

As we previously pointed out, Eqs. (1-1) to (1-4) were developed for use with positive integers as exponents. We shall now show how their use can be extended to include zero and negative integers as exponents.

In Eq. (1-2), if $n = m$, we would have $a^m/a^m = a^{m-m} = a^0$. Also $a^m/a^m = 1$, since any nonzero quantity divided by itself equals 1. Therefore, for Eqs. (1-2) to hold when $m = n$, we have

$$a^0 = 1 \qquad (a \neq 0) \tag{1-5}$$

Equation (1-5) gives the definition of zero as an exponent. Since a has not been specified, this equation states that *any nonzero algebraic expression raised to the zero power is 1*. Also, the other laws of exponents are valid for this definition.

EXAMPLE F ─── Equation (1-1) states that $a^m \times a^n = a^{m+n}$. If $n = 0$, we thus have $a^m \times a^0 = a^{m+0} = a^m$. Since $a^0 = 1$, this equation could be written as $a^m(1) = a^m$. This provides further verification for the validity of Eq. (1-5). ■

EXAMPLE G ─── $5^0 = 1, \qquad (2x)^0 = 1, \qquad (ax + b)^0 = 1$

$(a^2xb^4)^0 = 1, \qquad (a^2b^0c)^2 = a^4b^0c^2 = a^4c^2$

$\qquad\qquad\qquad\qquad\qquad\qquad\qquad\qquad b^0 = 1$

$2t^0 = 2(1) = 2$

NOTE ▷ We note in the last illustration that **only t is raised to the zero power.** If the quantity $2t$ were raised to the zero power, it would be written as $(2t)^0$. ■

If we apply the first form of Eqs. (1-2) to the case where $n > m$, the resulting exponent is negative. This leads us to the definition of a negative exponent.

EXAMPLE H

Applying the first form of Eqs. (1-2) to a^2/a^7, we have

$$\frac{a^2}{a^7} = a^{2-7} = a^{-5}$$

Applying the second form of Eqs. (1-2) to the same fraction leads to

$$\frac{a^2}{a^7} = \frac{1}{a^{7-2}} = \frac{1}{a^5}$$

In order that these results can be consistent, it must be true that

$$a^{-5} = \frac{1}{a^5}$$

Following the reasoning in Example H, if we define

$$a^{-n} = \frac{1}{a^n} \qquad (a \neq 0) \tag{1-6}$$

then all of the laws of exponents will hold for negative integers.

EXAMPLE I

EXAMPLE J

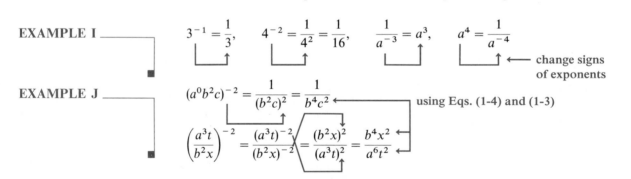

When we discussed the operations with signed numbers in Section 1–2, we noted that it is necessary to follow a particular order when performing the basic operations. Since raising a number to a power is a form of multiplication, this operation is also done before additions and subtractions. In fact it is done before divisions and other multiplications. Therefore, *the order of operations is*

1. *specific groupings,*
2. *powers,*
3. *multiplications and divisions, and*
4. *additions and subtractions.*

EXAMPLE K

$$8 - (-1)^2 - 2(-3)^2 = 8 - (+1) - 2(+9) = 8 - 1 - 18 = -11$$

NOTE ▷

Note that we squared -1 and -3 as the first operation. The next operation was finding the product in the last term. Finally the subtractions were performed. *We did not change the sign of -1 before we squared it,* for this would have been incorrect. ∎

Frequently an algebraic expression is known, and it is necessary to **evaluate** the expression for given values of the literal symbols. *The evaluation is done by* **substituting** *the numbers for the literal symbols and then calculating the value of the expression.*

If the evaluation is done on a calculator, we use the $\boxed{x^2}$ key to square numbers and the $\boxed{x^y}$ key for other powers. (See Appendix D for the use of the calculator keys.) On some calculators, the $\boxed{x^y}$ key can be used only if x is positive (or zero). For raising a negative number to a power, we must either determine the sign of the result by inspection or use successive multiplications.

EXAMPLE L

In order to find the surface area of a ball bearing, we must evaluate the expression $4\pi r^2$, where r is the radius of the bearing.

For a ball bearing of radius 2.55 mm, we find its surface area by substituting the values of π and r into $4\pi r^2$, and we get

$$\overset{\pi}{\underset{\downarrow}{}} \quad \overset{r\,=\,2.55\text{ mm}\;\leftarrow\;\text{substituting}}{\underset{\downarrow}{}}$$

$$4(3.14)(2.55)^2$$

4 $\boxed{\times}$ $\boxed{\pi}$ $\boxed{\times}$
2.55 $\boxed{x^2}$ $\boxed{=}$
$\boxed{\;81.712825\;}$

as the expression. On a calculator, it is not necessary to substitute 3.14 (or any other decimal equivalent), for π, but we simply use the $\boxed{\pi}$ key. Also, we use the $\boxed{x^2}$ key for 2.55^2. A calculator which uses algebraic logic will square the 2.55 before performing the multiplication. The sequence of keys is shown at the left. Rounding off the result to three significant digits (this is the accuracy of r) we find that the surface area is 81.7 mm^2. (See Appendix B for units of measurement.) ∎

It is usually possible to *estimate* a rough *approximation* to a calculation. In Example L, if we *mentally* round off the values as $4(3)(3^2) = 12(9) = 108$, we know that the answer should be roughly 100. We know 108 is high since we treated 2.55 as 3. Therefore, the answer of 81.7 mm^2 is reasonable. Such estimations will often prevent our accepting an incorrect result after using an incorrect calculator sequence, particularly when the estimation is far from the value in the final display.

Exercises 1-5

In Exercises 1 through 48, simplify the given expressions. Express results with positive exponents only.

1. x^3x^4 **2.** y^2y^7 **3.** $2b^4b^2$ **4.** $3k(k^5)$ **5.** $\dfrac{m^5}{m^3}$ **6.** $\dfrac{x^6}{x}$

7. $\dfrac{n^5}{n^9}$ **8.** $\dfrac{s}{s^4}$ **9.** $(a^2)^4$ **10.** $(x^8)^3$ **11.** $(t^5)^4$ **12.** $(n^3)^7$

13. $(2n)^3$ **14.** $(ax)^5$ **15.** $(ax^4)^2$ **16.** $(3a^2)^3$ **17.** $\left(\dfrac{2}{b}\right)^3$ **18.** $\left(\dfrac{x}{y}\right)^7$

19. $\left(\dfrac{x^2}{2}\right)^4$ **20.** $\left(\dfrac{3}{n^3}\right)^3$ **21.** 7^0 **22.** $(8a)^0$ **23.** $3x^0$ **24.** $6v^0$

25. 6^{-1} **26.** 10^{-3} **27.** $\dfrac{1}{s^{-2}}$ **28.** $\dfrac{1}{t^{-5}}$ **29.** $(-t^2)^7$ **30.** $(-y^3)^5$

31. $(2x^2)^6$ **32.** $(-c^4)^4$ **33.** $(4xa^{-2})^0$ **34.** $3(ab^{-1})^0$ **35.** b^5b^{-3} **36.** $2c^4c^{-7}$

37. $\dfrac{2a^4}{(2a)^4}$ **38.** $\dfrac{x^2x^3}{(x^2)^3}$ **39.** $\dfrac{(n^2)^4}{(n^4)^2}$ **40.** $\dfrac{(3t)^{-1}}{3t^0}$ **41.** $(5^0x^2a^{-1})^{-1}$ **42.** $(3m^{-2}n^4)^{-2}$

43. $\left(\dfrac{4a}{x}\right)^{-3}$ **44.** $\left(\dfrac{2b^2}{y^5}\right)^{-2}$ **45.** $(-8gs^3)^2$ **46.** $ax^2(-a^2x)^2$ **47.** $\dfrac{15a^2n^5}{3an^6}$ **48.** $\dfrac{(ab^2)^3}{a^2b^8}$

In Exercises 49 through 52, evaluate the given expressions.

49. $7(-4) - (-5)^2$ **50.** $\dfrac{12}{-3} - (-1)^3$ **51.** $6 + (-2)^5 - (-2)(8)$ **52.** $9 - 2(-3)^4 - (-7)$

In Exercises 53 through 58, perform the indicated operations on a calculator. In each case round off the results to three significant digits.

53. $2.38(60.7)^2 - 2540$ **54.** $(0.513)(-2.778) - (-3.67)^3$

55. $\dfrac{3.07(-1.86)}{(-1.86)^4 + 1.596}$ **56.** $\dfrac{(15.66)^2 - (14.07)^3}{-3.68}$

57. In designing a cam for a pump, the expression $\pi\left(\dfrac{r}{2}\right)^3\left(\dfrac{4}{3\pi r^2}\right)$ is used. Simplify this expression.

58. For an integrated electric circuit it might be necessary to simplify the expression $\dfrac{gM}{2\pi f C(2\pi f M)^2}$. Perform this simplification.

In Exercises 59 and 60, evaluate the given expressions.

59. In order to find the electric power (in watts) consumed by an electric light, the expression i^2R must be evaluated, where i is the current in amperes and R is the resistance in ohms. Find the power consumed if $i = 0.525$ A and $R = 250\ \Omega$.

60. In designing a building, it was determined that the forces acting on an I beam would deflect the beam an amount, in centimeters, given by $\dfrac{x(1000 - 20x^2 + x^3)}{1850}$, where x is the distance, in meters, from one end of the beam. Find the deflection for $x = 6.85$ m.

1-6   Scientific Notation

In technical and scientific work we often encounter numbers which are either very large or very small in magnitude. Illustrations of such numbers are given in the following example.

EXAMPLE A ——————— Television signals travel at about 30,000,000,000 cm/s. The mass of the earth is about 6,600,000,000,000,000,000,000 tons. A typical individual fiber in a fiber-optic communications cable has a diameter of 0.000005 m. The wavelength of ■ some X rays is about 0.000000095 cm.

Writing numbers such as these is inconvenient in ordinary notation, as shown in Example A, particularly when the numbers of zeros needed for the proper location of the decimal point are excessive. Also, limitations of calculators and computers require a more efficient way of expressing such numbers in order to work with them. Therefore, a convenient and useful notation, known as **scientific**
scientific notation **notation,** is normally used to represent such numbers.

A number written in scientific notation is expressed as the product of a number greater than or equal to 1 and less than 10, and a power of 10. Symbolically this can be written as

$$P \times 10^k$$

where $1 \le P < 10$, and k is an integer. The following example illustrates how numbers are written in scientific notation.

EXAMPLE B ———————
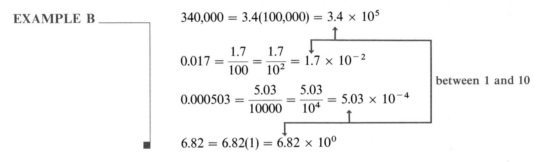

$$340{,}000 = 3.4(100{,}000) = 3.4 \times 10^5$$

$$0.017 = \frac{1.7}{100} = \frac{1.7}{10^2} = 1.7 \times 10^{-2}$$

$$0.000503 = \frac{5.03}{10000} = \frac{5.03}{10^4} = 5.03 \times 10^{-4}$$

between 1 and 10

$$6.82 = 6.82(1) = 6.82 \times 10^0$$

From Example B we can establish a method for changing numbers from ordinary notation to scientific notation. *The decimal point is moved so that only one nonzero digit is to its left. The number of places moved is the value of k. It is positive if the decimal point is moved to the left, and it is negative if it is moved to the right.* Consider the illustrations in the following example.

EXAMPLE C ———————
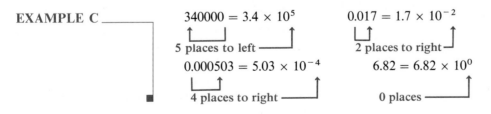

$$340000 = 3.4 \times 10^5 \qquad\qquad 0.017 = 1.7 \times 10^{-2}$$

5 places to left ——————— 2 places to right ——

$$0.000503 = 5.03 \times 10^{-4} \qquad\qquad 6.82 = 6.82 \times 10^0$$

4 places to right ——————— 0 places ———————

To change a number from scientific notation to ordinary notation, the procedure above is reversed. The following example illustrates the procedure.

EXAMPLE D
To change 5.83×10^6 to ordinary notation, we must move the decimal point 6 places to the right. Therefore, additional zeros must be included for the proper location of the decimal point. This means we write

$$5.83 \times 10^6 = 5,830,000$$

6 places to right

To change 8.06×10^{-3} to ordinary notation, we must move the decimal point 3 places to the left. Again, additional zeros must be included. Thus,

$$8.06 \times 10^{-3} = 0.00806$$

3 places to left

An illustration of the importance of scientific notation is demonstrated by the metric system use of prefixes on units to denote certain powers of 10. These are illustrated in Appendix B.

As we see from the previous examples, scientific notation provides an important application of the use of exponents, positive and negative. Also, scientific notation provides a practical way to handle calculations involving numbers of very large or very small magnitudes.

The calculation can be made by first expressing all numbers in scientific notation. Then the actual calculation can be performed on numbers between 1 and 10, with the laws of exponents used to find the proper power of 10 for the result. It is proper to leave the result in scientific notation.

EXAMPLE E
In designing a computer, it was determined that it would be able to process 803,000 bits of data in 0.00000525 s. The rate of processing the data may be set up and calculated as

$$5 - (-6) = 11$$

$$\frac{803,000}{0.00000525} = \frac{8.03 \times 10^5}{5.25 \times 10^{-6}} = \left(\frac{8.03}{5.25}\right) \times 10^{11} = 1.53 \times 10^{11} \text{ bits/s}$$

The power of 10 here is sufficiently large that we would normally leave the result in this form. (See Exercise 33 in Exercises 1-1 for a brief discussion of computer data.)

EXAMPLE F
In evaluating the product $(7.50 \times 10^{11})(6.44 \times 10^{-3})$, we obtain

$$(7.50 \times 10^{11})(6.44 \times 10^{-3}) = 48.3 \times 10^8$$

However, since the answer is to be given in scientific notation, we should express it as the product of a number between 1 and 10 and a power of 10, which means we should rewrite it as

$$48.3 \times 10^8 = (4.83 \times 10)(10^8)$$

not between 1 and 10 $= 4.83 \times 10^9$

In Example F the product 7.50×6.44 can be found by use of a calculator, but it would not be possible even to enter 7.50×10^{11} in the form 750,000,000,000 on most calculators. However, most scientific calculators have the feature of scientific notation, and it is possible to enter numbers in scientific notation form as well as to have answers given automatically in scientific notation form.

To enter a number into a calculator in scientific notation form, we use the $\boxed{\text{EE}}$ key (see Appendix D). If scientific notation form has not been used and the display cannot show the answer in standard form, the calculator will automatically use scientific notation form to display the answer.

EXAMPLE G

To perform the calculation of Example E by use of the scientific notation feature of a calculator, we first express each number in scientific notation, as shown in Example E.

8.03 $\boxed{\text{EE}}$ 5 $\boxed{\div}$
5.25 $\boxed{\text{EE}}$ 6 $\boxed{+/-}$
$\boxed{=}$ $\boxed{1.5295238 \ 11}$

$$\frac{803,000}{0.00000525} = \frac{8.03 \times 10^5}{5.25 \times 10^{-6}}$$

The sequence of keys used for this calculation is shown at the left. The display means that the quotient is equal to 1.53×10^{11}, rounded off.

Exercises 1-6

In Exercises 1 through 8, change the numbers from scientific notation to ordinary notation.

1. 4.5×10^4
2. 6.8×10^7
3. 2.01×10^{-3}
4. 9.61×10^{-5}
5. 3.23×10^0
6. 8.40×10^0
7. 1.86×10
8. 1×10^{-1}

In Exercises 9 through 16, change the given numbers from ordinary notation to scientific notation.

9. 40000
10. 5600000
11. 0.0087
12. 0.7
13. 6
14. 1.09
15. 0.063
16. 0.0000908

In Exercises 17 through 20, perform the indicated calculations without the use of a calculator and by first expressing all numbers in scientific notation. (See Example E.)

17. $(28,000)(2,000,000,000)$
18. $(50,000)(0.006)$
19. $\dfrac{88,000}{0.0004}$
20. $\dfrac{0.00003}{6,000,000}$

In Exercises 21 through 28, perform the indicated calculations by use of a scientific calculator which has scientific notation. In each case, round off results to the number of significant digits used in each of the numbers.

21. $(1280)(865,000)(43.8)$
22. $(0.0000659)(0.00486)(3,190,000,000)$
23. $\dfrac{(0.0732)(6710)}{0.00134}$
24. $\dfrac{(2430)(97,100)}{0.00452}$
25. $(3.642 \times 10^{-8})(2.736 \times 10^5)$
26. $\dfrac{9.368 \times 10^{-12}}{4.651 \times 10^4}$
27. $\dfrac{(3.1075 \times 10^{-5})(1.0772 \times 10^{14})}{6.6483 \times 10^7}$
28. $\dfrac{7.3009 \times 10^{-2}}{(5.9843)(2.5036 \times 10^{-20})}$

In Exercises 29 through 40, change any numbers in ordinary notation to scientific notation or change any numbers in scientific notation to ordinary notation. (See Appendix B for an explanation of symbols used.)

29. The power plant at Grand Coulee Dam produces 6,500,000 kW of power.

30. The maximum pressure exerted by the human heart is 16,000 Pa.

31. The power of a radio signal received from a lunar probe is 1.6×10^{-12} W.

32. Some computers can perform an addition in 4.5×10^{-12} s.

33. In a given FM receiver the output voltage is 2×10^5 times the signal voltage.

34. To attain an energy density of that in some laser beams, an object would have to be heated to about 10^{30}°C.

35. A fiber-optic system requires 0.000003 W of power.

36. A plutonium isotope is radioactive, and 0.0000079% of it disintegrates in one day.

37. The electrical force between two electrons is about 2.4×10^{-43} times the gravitational force between them.

38. Among the stars nearest the earth, Centaurus A is about 2.53×10^{13} mi away.

39. The altitude of a communications satellite is 36,000 km.

40. A reforestation machine can plant about 250,000 seedlings in one day.

In Exercises 41 through 44, perform the indicated calculations by first expressing all numbers in scientific notation.

41. There are about 161,000 cm in one mile. The area of Lake Erie is 9930 mi². What is the area of Lake Erie in square centimeters?

42. A particular virus is a sphere 0.0000175 cm in diameter. What volume does the virus occupy?

43. In a microwave receiver circuit the resistance R of a wire 1 m long is given by $R = k/d^2$, where d is the diameter of the wire. Find R if $k = 0.00000002196$ $\Omega \cdot$ m² and $d = 0.00007998$ m.

44. At 0°C, the refrigerant Freon is a vapor at a pressure of $P = 1.378 \times 10^5$ Pa. If the volume of vapor is $V = 1.185 \times 10^3$ cm³, find the value of PV.

1-7   Roots and Radicals

A problem that is often encountered is: What number multiplied by itself n times gives another specified number? For example, we may ask: What number squared is 9? We can see that either $+3$ or -3 is a proper answer to this question. *We therefore call either $+3$ or -3 a* **square root** *of* 9, since $(+3)^2 = 9$ or $(-3)^2 = 9$.

In order to have a general notation for the square root of a number, one which is not ambiguous and does not represent more than one number, *we define the* **principal square root** *of a to be positive if a is positive and represent it by* $\sqrt{a}$. This means that $\sqrt{9} = 3$ and not -3.

principal *n*th root

The general notation for the **principal nth root** of a is $\sqrt[n]{a}$. (When $n = 2$, it is common practice not to put the 2 where n appears.) *The* $\sqrt{}$ *sign is called a* **radical sign.** Unless we state otherwise, when we refer to the root of a number, we refer to the principal root. (In Chapter 11 we consider roots other than principal roots.)

EXAMPLE A

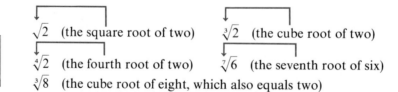

$\sqrt{2}$ (the square root of two) $\sqrt[3]{2}$ (the cube root of two)

$\sqrt[4]{2}$ (the fourth root of two) $\sqrt[7]{6}$ (the seventh root of six)

$\sqrt[3]{8}$ (the cube root of eight, which also equals two)

In considering the question of what number squared is 9, we saw that either $+3$ or -3 gives a proper result. Therefore, in order to avoid ambiguity, we defined the principal square root of a positive number to be positive. (Again, $\sqrt{9} = 3$

and not -3.) In order to have a single defined value for other roots, and to consider only real number roots, *we define the **principal nth root** of a to be positive if a is positive, and the principal nth root of a to be negative if a is negative and n is odd*. (If a is negative and n is even, the roots are not real.)

EXAMPLE B

$$\sqrt{4} = 2 \quad (\sqrt{4} \neq -2), \qquad \sqrt{169} = 13 \quad (\sqrt{169} \neq -13),$$
$$-\sqrt{64} = -8, \qquad -\sqrt{81} = -9, \qquad -\sqrt[4]{256} = -4,$$

$$\sqrt[3]{27} = 3, \qquad \overset{\text{odd}}{\sqrt[3]{-27}} = -3, \qquad -\sqrt[3]{27} = -(+3) = -3$$

Another important property of radicals is developed by noting illustrations such as

$$\sqrt{36} = 6 \quad \text{and} \quad \sqrt{36} = \sqrt{4 \times 9} = \sqrt{4} \times \sqrt{9} = 2 \times 3 = 6$$

In general, this property states that *the square root of a product of positive numbers is the product of the square roots*. That is,

$$\boxed{\sqrt{ab} = \sqrt{a}\,\sqrt{b} \qquad \textbf{(a and b positive real numbers)}} \qquad (1\text{-}7)$$

This property is useful in simplifying radicals. It is most useful if either a or b is a **perfect square,** *which is the square of a rational number.* Consider the following example.

EXAMPLE C

1. $\sqrt{8} = \sqrt{(4)(2)} = \sqrt{4}\sqrt{2} = 2\sqrt{2}$

2. $\sqrt{75} = \sqrt{(25)(3)} = \sqrt{25}\sqrt{3} = 5\sqrt{3}$

perfect squares

3. $\sqrt{80} = \sqrt{(16)(5)} = \sqrt{16}\sqrt{5} = 4\sqrt{5}$

When we simplify a radical and do not give its *approximate* decimal equivalent, we are generally interested in an *exact* value. In some cases the exact value is preferred, although a decimal value can easily be found on a calculator.

If a decimal value for a square root is acceptable, we can use the $\boxed{\sqrt{x}}$ key on a calculator. The method for finding other roots on a calculator is discussed in Chapter 10.

EXAMPLE D

.25 $\boxed{\times}$ 1260

$\boxed{\sqrt{x}}$ $\boxed{=}$

$\boxed{8.8741197}$

After reaching its greatest height, the time, in seconds, for a rocket to fall h feet due to gravity is found by evaluating $0.25\sqrt{h}$. Find the time, to the nearest 0.1 s, for a rocket to fall 1260 ft.

We must evaluate $0.25\sqrt{1260}$, and the calculator sequence for this is shown at the left. From the display we know that it takes 8.9 s, to the nearest 0.1 s.

Another important point regarding the simplification of a radical is that *all operations under a radical must be done before finding the root or using Eq. (1-7)*. That is, the horizontal bar of a radical sign is a grouping symbol which groups the numbers and symbols under it.

EXAMPLE E

1. $\sqrt{16 + 9} = \sqrt{25} = 5$,   but

add first

NOTE ▷ $\sqrt{16 + 9}$ *is* **not** $\sqrt{16} + \sqrt{9} = 4 + 3 = 7$

2. $\sqrt{2^2 + 6^2} = \sqrt{4 + 36} = \sqrt{40} = \sqrt{4}\sqrt{10} = 2\sqrt{10}$,   but

$\sqrt{2^2 + 6^2}$ is **not** $\sqrt{2^2} + \sqrt{6^2} = 2 + 6 = 8$

In defining the principal square root, we did not define the square root of a negative number. Earlier *we defined the square root of a negative number to be an* **imaginary number.** There are certain very important places where it is necessary to recognize when imaginary numbers occur, but until Chapter 11 we will not use them otherwise.

It should be emphasized that although the square root of a negative number gives an imaginary number, the cube root of a negative number gives a negative real number. More generally, *the even root of a negative number is imaginary, and the odd root of a negative number is real.* A more detailed discussion of exponents, radicals, and imaginary numbers is found in Chapters 10 and 11.

EXAMPLE F

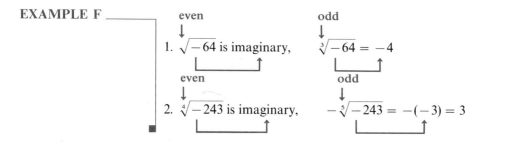

even   odd

1. $\sqrt{-64}$ is imaginary, $\sqrt[3]{-64} = -4$

even   odd

2. $\sqrt[4]{-243}$ is imaginary, $-\sqrt[5]{-243} = -(-3) = 3$

Exercises 1-7

In Exercises 1 through 32, simplify the given expressions. In each of 1 through 16, the result is an integer.

1. $\sqrt{25}$	**2.** $\sqrt{81}$	**3.** $-\sqrt{121}$	**4.** $-\sqrt{36}$	**5.** $-\sqrt{49}$	**6.** $\sqrt{225}$
7. $\sqrt{400}$	**8.** $-\sqrt{900}$	**9.** $\sqrt[3]{125}$	**10.** $\sqrt[4]{16}$	**11.** $\sqrt[3]{-216}$	**12.** $\sqrt[5]{-32}$
13. $(\sqrt{5})^2$	**14.** $(\sqrt{19})^2$	**15.** $(\sqrt[3]{31})^3$	**16.** $(\sqrt[4]{53})^4$	**17.** $\sqrt{18}$	**18.** $\sqrt{32}$
19. $\sqrt{12}$	**20.** $\sqrt{50}$	**21.** $\sqrt{44}$	**22.** $\sqrt{54}$	**23.** $\sqrt{6300}$	**24.** $\sqrt{160}$

25. $2\sqrt{84}$ **26.** $4\sqrt{108}$ **27.** $\dfrac{7^2\sqrt{81}}{3^2\sqrt{49}}$ **28.** $\dfrac{2^5\sqrt[5]{243}}{3\sqrt{144}}$ **29.** $\sqrt{36 + 64}$ **30.** $\sqrt{25 + 144}$

31. $\sqrt{3^2 + 9^2}$ **32.** $\sqrt{8^2 - 4^2}$

In Exercises 33 through 36, find the value of each square root by use of a calculator. Round off the result to the accuracy of the given number.

33. $\sqrt{85.4}$ **34.** $\sqrt{3762}$ **35.** $\sqrt{0.4729}$ **36.** $\sqrt{0.0627}$

In Exercises 37 through 44, solve the given problems.

37. The period, in seconds, of a pendulum whose length is L feet can be found by evaluating $1.11\sqrt{L}$. Find the period, to the nearest 0.01 s, of a pendulum for which $L = 1.75$ ft.

38. The resistance in an amplifier circuit is found by evaluating $\sqrt{Z^2 - X^2}$. Find the resistance for $Z = 5.362\ \Omega$ and $X = 2.875\ \Omega$.

39. In order to satisfy zoning requirements, a store plans a square parking lot with an area of 1250 m². Find the length of a side of the parking lot to the nearest 0.1 m.

40. A protective coating seal completely covers a cubical container of medicine. If the total area of the coating is 25.6 in.², what is the edge of the cube to the nearest 0.01 in.?

41. A rectangular television screen is 15.2 in. wide and 11.4 in. high. What is the length of the diagonal of the screen (the measurement normally used to describe it)? (*Hint:* Use the Pythagorean theorem—see Appendix C.)

42. A cubical wastewater holding tank has a capacity of 3375 m³. What is the length of an edge of the tank? (*Hint:* See Table 1 in Appendix F.)

43. Is it always true that $\sqrt{a^2} = a$?

44. If $0 < x < 1$ (x between 0 and 1), is $x > \sqrt{x}$?

1-8 Addition and Subtraction of Algebraic Expressions ▬▬

It is the basic characteristic of algebra that letters are used to represent numbers. Since we have used literal symbols to represent numbers, even if in a general sense, we may conclude that all operations valid for numbers are valid for these literal symbols. In this section we shall discuss the terminology and methods for combining literal symbols.

Addition, subtraction, multiplication, division, and taking of roots are known as **algebraic operations.** *Any combination of numbers and literal symbols which results from algebraic operations is known as an* **algebraic expression.**

When an algebraic expression consists of several parts connected by plus signs and minus signs, each part (along with its sign) is known as a **term** *of the expression.*

terms and factors

If a given expression is made up of the product of a number of quantities, each of these quantities, or any product of them, is called a **factor** *of the expression.*

NOTE▷   It is important to ***distinguish clearly between terms and factors*** since some operations that are valid for terms are not valid for factors, and conversely.

EXAMPLE A

$$3xy + 6x^2 - 7x\sqrt{y}$$

is an algebraic expression with three terms. As indicated, they are $3xy$, $6x^2$, and $-7x\sqrt{y}$.

The first term, $3xy$, has individual factors of 3, x, and y. Also, any product of these factors is also a factor of $3xy$. Thus, additional expressions which are factors of $3xy$ are $3x$, $3y$, xy, and $3xy$ itself.

EXAMPLE B

$7x(y^2 + x) - \dfrac{x + y}{6x}$ is an algebraic expression with terms $7x(y^2 + x)$ and

$\dfrac{-(x + y)}{6x}$

The term $7x(y^2 + x)$ has individual factors of 7, x, and $(y^2 + x)$, as well as products of these factors. The factor $y^2 + x$ has two terms, y^2 and x.

The numerator of the term $-\dfrac{x + y}{6x}$ has two terms, and the denominator has

■ individual factors of 2, 3, and x. The minus sign can be treated as a factor of -1.

An algebraic expression containing only one term is called a **monomial.** *An expression containing two terms is a* **binomial,** *and one containing three terms is a* **trinomial.** *Any expression containing two or more terms is called a* **multinomial.** *Thus, any binomial or trinomial expression can also be considered as a multinomial.*

coefficient

In any given term, the numbers and literal symbols multiplying any given factor constitute the **coefficient** *of that factor. The product of all the numbers in explicit form is known as the* **numerical coefficient** *of the term. All terms which differ at most in their numerical coefficients are known as* **similar** *or* **like** *terms. That is, similar terms have the same variables with the same exponents.*

EXAMPLE C

$7x^3\sqrt{y}$ is a monomial. It has a numerical coefficient of 7. The coefficient of $\sqrt{y}$
■ is $7x^3$, and the coefficient of x^3 is $7\sqrt{y}$.

EXAMPLE D

like terms   ┌─── unlike other terms due to factor of a

1. $4 \times 2b + 81b - 6ab$

   is a multinomial of three terms (a trinomial). The first term has a numerical coefficient of 8 ($= 4 \times 2$), the second has a numerical coefficient of 81, and the third has a numerical coefficient of -6 (the sign of the term is attached to the numerical coefficient). The first and second terms are similar, since they differ only in their numerical coefficient. The third term is not similar to either of the others, for it has the factor a.

2. The commutative law tells us that $x^2y^3 = y^3x^2$. Therefore, the two terms of
■ the expression $3x^2y^3 + 5y^3x^2$ are similar.

In adding and subtracting algebraic expressions, we combine similar (or like) terms. In doing so, we are combining quantities which are alike. All of the similar terms may be combined into a single term, and the final simplified expression will be made up entirely of terms which are not similar.

EXAMPLE E

1. $3x + 2x - 5y = 5x - 5y$

   Since there are two similar terms in the original expression they are added together, so the simplified result has two unlike terms.

2. $6a^2 - 7a + 8ax$

   cannot be simplified, since there are no like terms.

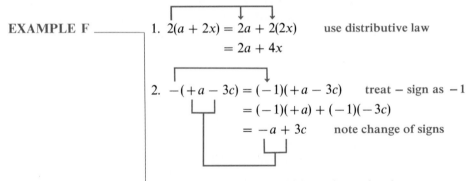

3. $6a + 5c + 2a - c = 6a + 2a + 5c - c$ commutative law
$$= 8a + 4c$$ add like terms

symbols of grouping

In writing algebraic expressions, it is often necessary to group certain terms together. For this purpose we use **symbols of grouping.** In this text we use **parentheses ( )**, **brackets [ ]**, and **braces { }**. The **bar,** which is used with radicals and fractions, also groups terms. The bar attached to the radical sign groups the terms below it, and the bar separating the numerator and denominator of a fraction groups the terms above and below it. In earlier sections, when we discussed the order of operations, we used parentheses and the bar for grouping.

When adding and subtracting algebraic expressions, it is often necessary to remove symbols of grouping. To do so we must *change the sign of **every term** within the symbols if the grouping is preceded by a minus sign. If the symbols of grouping are preceded by a plus sign, each term within the symbols retains its original sign.* This is a result of the distributive law, $a(b + c) = ab + ac$.

EXAMPLE F

1. $2(a + 2x) = 2a + 2(2x)$ use distributive law
$$= 2a + 4x$$

2. $-(+a - 3c) = (-1)(+a - 3c)$ treat $-$ sign as -1
$$= (-1)(+a) + (-1)(-3c)$$
$$= -a + 3c$$ note change of signs

■ Normally, the $+a$ would be written simply as a.

EXAMPLE G

$+$ sign before parentheses

1. $3c + (2b - c) = 3c + 2b - c = 2b + 2c$
signs retained
$2b = +2b$

$-$ sign before parentheses

2. $3c - (2b - c) = 3c - 2b + c = -2b + 4c$
signs changed
$2b = +2b$

3. $3c - (-2b + c) = 3c + 2b - c = 2b + 2c$
signs changed

Note in each case that the parentheses are removed and the sign before the parentheses is also removed. Also, in the first two illustrations, $2b = +2b$.

EXAMPLE H

1. $-(2x - 3c) + (c - x) = -2x + 3c + c - x = 4c - 3x$

2. $3 - 2(m^2 - 2) = 3 - 2m^2 + 4$

$$= 7 - 2m^2$$

3. $4(t - 3 - 2t^2) - (6t + t^2 - 4) = 4t - 12 - 8t^2 - 6t - t^2 + 4$

$$= -9t^2 - 2t - 8$$

EXAMPLE I

In designing a machine part, it is necessary to simplify the expression

$$16(8 - x) - 2(8x - x^2) - (64 - 16x + x^2)$$

This simplification is performed as follows:

$$16(8 - x) - 2(8x - x^2) - (64 - 16x + x^2) = 128 - 16x - 16x + 2x^2 - 64 + 16x - x^2$$

$$= 64 - 16x + x^2$$

It is fairly common to have expressions in which more than one symbol of grouping is to be removed in the simplification. Normally, *when several symbols of grouping are to be removed, it is more convenient to remove the innermost symbols first.* This is illustrated in the following example.

EXAMPLE J

1. $3ax - [ax - (5s - 2ax)] = 3ax - [ax - 5s + 2ax]$ remove parentheses

$$= 3ax - ax + 5s - 2ax$$ remove brackets

$$= 5s$$

2. $[a^2b - ab + (ab - 2a^2b)] - \{[(3a^2b + b) - (4ab - 2a^2b)] - b\}$

$$= [a^2b - ab + ab - 2a^2b] - \{[3a^2b + b - 4ab + 2a^2b] - b\}$$

remove parentheses

$$= a^2b - ab + ab - 2a^2b - \{3a^2b + b - 4ab + 2a^2b - b\}$$

remove brackets

$$= a^2b - ab + ab - 2a^2b - 3a^2b - b + 4ab - 2a^2b + b$$ remove braces

$$= -6a^2b + 4ab$$

Calculators and computers generally use only parentheses for grouping symbols. Therefore, in writing expressions in which only parentheses are used to group terms it is frequently necessary to use one set of parentheses within another set to group more than one set of symbols. These are called **nested parentheses.** The following example illustrates the simplification of an algebraic expression with nested parentheses. Note that innermost parentheses are removed first.

EXAMPLE K

$$2 - (3x - 2(5 - (7 - x))) = 2 - (3x - 2(5 - 7 + x))$$

$$= 2 - (3x - 10 + 14 - 2x)$$

$$= 2 - 3x + 10 - 14 + 2x$$

$$= -x - 2$$

One of the most common errors made by beginning students is changing the sign of only the first term when removing symbols of grouping preceded by a minus sign. *Remember, if the symbols are preceded by a minus sign, we must change the sign of all terms.*

NOTE ▷

Exercises 1-8

In the following exercises, simplify the given algebraic expressions.

1. $5x + 7x - 4x$

2. $6t - 3t - 4t$

3. $2y - y + 4x$

4. $4c + d - 6c$

5. $2a - 2c - 2 + 3c - a$

6. $x - 2y + 3x - y + z$

7. $a^2b - a^2b^2 - 2a^2b$

8. $xy^2 - 3x^2y^2 + 2xy^2$

9. $s + (4 + 3s)$

10. $5 + (3 - 4n + p)$

11. $v - (4 - 5x + 2v)$

12. $2a - (b - a)$

13. $2 - 3 - (4 - 5a)$

14. $\sqrt{x} + (y - 2\sqrt{x}) - 3\sqrt{x}$

15. $(a - 3) + (5 - 6a)$

16. $(4x - y) - (-2x - 4y)$

17. $-(t - 2u) + (3u - t)$

18. $2(x - 2y) + (5x - y)$

19. $3(2r + s) - (-5s - r)$

20. $3(a - b) - 2(a - 2b)$

21. $-7(6 - 3c) - 2(c + 4)$

22. $-(5t + a^2) - 2(3a^2 - 2st)$

23. $-[(6 - n) - (2n - 3)]$

24. $-[(a - b) - (b - a)]$

25. $2[4 - (t^2 - 5)]$

26. $3[-3 - (a - 4)]$

27. $-2[-x - 2a - (a - x)]$

28. $-2[-3(x - 2y) + 4x]$

29. $a\sqrt{xy} - [3 - (a\sqrt{xy} + 4)]$

30. $9v - [6 - (v - 4) + 4v]$

31. $8c - \{5 - [2 - (3 + 4c)]\}$

32. $7y - \{y - [2y - (x - y)]\}$

33. $5p - (q - 2p) - [3q - (p - q)]$

34. $-(4 - x) - [(5x - 7) - (6x + 2)]$

35. $-2\{-(4 - x^2) - [3 + (4 - x^2)]\}$

36. $-\{-[-(x - 2a) - b] - a\}$

37. $3a - (6 - (a + 3))$

38. $-2x + 2((2x - 1) - 5)$

39. $-(3t - (7 + 2t - (5t - 6)))$

40. $a^2 - 2(x - 5 - (7 - 2(a^2 - 2x) - 3x))$

41. In determining the size of a V belt to be used with an engine, the expression $3D - (D - d)$ is used. Simplify this expression.

42. When finding the current in a transistor circuit, the expression $i_1 - (2 - 3i_2) + i_2$ is used. Simplify this expression. (The numbers below the i's are *subscripts*. Different subscripts denote different variables.)

43. A company analyzing production costs of three products uses the expression $3x + (4 - 4x) - (9x - 18)$. Simplify this expression.

44. Research in the development of a plastic building material uses the expression $[(B + \frac{4}{3}\alpha) + 2(B - \frac{2}{3}\alpha)] - [(B + \frac{4}{3}\alpha) - (B - \frac{2}{3}\alpha)]$. Simplify this expression.

1-9 Multiplication of Algebraic Expressions

To find the product of two or more monomials, we use the laws of exponents as given in Section 1-5 and the laws for multiplying signed numbers as stated in Section 1-2. We first multiply the numerical coefficients to determine the numerical coefficient of the product. Then we multiply the literal factors, remembering that *the exponents may be combined only if the base is the same.*

EXAMPLE A

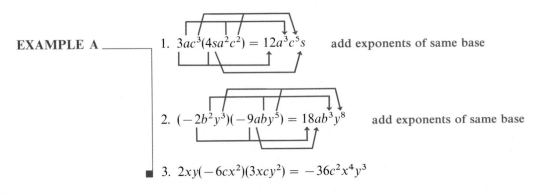

1. $3ac^3(4sa^2c^2) = 12a^3c^5s$ **add exponents of same base**

2. $(-2b^2y^3)(-9aby^5) = 18ab^3y^8$ **add exponents of same base**

3. $2xy(-6cx^2)(3xcy^2) = -36c^2x^4y^3$

If a product contains a monomial which itself is raised to a power, *we must first raise it to the indicated power* before proceeding with the remainder of the multiplication.

EXAMPLE B

1. $3(2a^2x)^3(-ax) = 3(8a^6x^3)(-ax) = -24a^7x^4$

2. $2s^3(-st^4)^2(4s^2t) = 2s^3(s^2t^8)(4s^2t) = 8s^7t^9$

We find the product of a monomial and a multinomial by using the distributive law, which states that we *multiply each term of the multinomial by the monomial.* We must be careful to assign the correct sign to each term of the result, using the rules for multiplication of signed numbers. Also, we must properly combine literal factors in each term of the result.

EXAMPLE C

1. $2ax(3ax^2 - 4yz) = 2ax(3ax^2) + (2ax)(-4yz) = 6a^2x^3 - 8axyz$

2. $5cy^2(-7cx - ac) = (5cy^2)(-7cx) + (5cy^2)(-ac)$
   $$= -35c^2xy^2 - 5ac^2y^2$$

In practice, it is generally not necessary to write out the middle step as it appears in the example above. We can generally write the answer directly. For example, the first part of Example C would usually appear as

$$2ax(3ax^2 - 4yz) = 6a^2x^3 - 8axyz$$

We find the product of two or more multinomials by using the distributive law and the laws of exponents. The result is that we *multiply each term of one multinomial by each term of the other, and add the results.*

EXAMPLE D

1. $(x - 2)(x + 3) = x(x) + x(3) + (-2)(x) + (-2)(3)$
   $$= x^2 + 3x - 2x - 6$$
   $$= x^2 + x - 6$$

2. $(x - 2y)(x^2 + 2xy + 4y^2) = x^3 + 2x^2y + 4xy^2 - 2x^2y - 4xy^2 - 8y^3$
   $$= x^3 - 8y^3$$

Finding the power of an algebraic expression is equivalent to using the expression as a factor the number of times indicated by the exponent. In practice, it is often convenient to write the power of an algebraic expression in this form before multiplying. Consider the following example.

EXAMPLE E

1. $(x + 5)^2 = (x + 5)(x + 5) = x^2 + 5x + 5x + 25$

$$= x^2 + 10x + 25$$

2. $2(3 - 2x)^2 = 2(3 - 2x)(3 - 2x)$

$$= 2(9 - 6x - 6x + 4x^2)$$
$$= 2(9 - 12x + 4x^2)$$
$$= 18 - 24x + 8x^2 = 8x^2 - 24x + 18$$

3. $(2a - b)^3 = (2a - b)(2a - b)(2a - b)$

$$= (2a - b)(4a^2 - 4ab + b^2)$$
$$= 8a^3 - 8a^2b + 2ab^2 - 4a^2b + 4ab^2 - b^3$$
$$= 8a^3 - 12a^2b + 6ab^2 - b^3$$

We should note in the first illustration that

NOTE ▷ $(x + 5)^2$ *is not equal to* $x^2 + 25$

since the term $10x$ is not included. We must follow the proper procedure for multiplication in squaring an expression and not simply square each of the terms within the parentheses. ∎

EXAMPLE F

An expression used with a lens of a certain type of telescope is

$$a(a + b)^2 + a^3 - (a + b)(2a^2 - s^2)$$

Simplifying this expression, we have the following:

$$a(a + b)^2 + a^3 - (a + b)(2a^2 - s^2)$$
$$= a(a + b)(a + b) + a^3 - (2a^3 - as^2 + 2a^2b - bs^2)$$
$$= a(a^2 + ab + ab + b^2) + a^3 - 2a^3 + as^2 - 2a^2b + bs^2$$
$$= a^3 + a^2b + a^2b + ab^2 - a^3 + as^2 - 2a^2b + bs^2$$
$$= ab^2 + as^2 + bs^2$$
∎

Exercises 1-9

In the following exercises, perform the indicated multiplications.

1. $(a^2)(ax)$ 　　　**2.** $(2xy)(x^2y^3)$ 　　　**3.** $-ac^2(acx^3)$ 　　　**4.** $-2s^2(-4cs)^2$

5. $(2ax^2)^2(-2ax)$ 　　**6.** $6pq^3(3pq^2)^2$ 　　**7.** $a(-a^2x)^3(-2a)$ 　　**8.** $-2m^2(-3mn)(m^2n)^2$

9. $a^2(x + y)$ 　　　**10.** $2x(p - q)$ 　　　**11.** $-3s(s^2 - 5t)$ 　　　**12.** $-3b(2b^2 - b)$

13. $5m(m^2n + 3mn)$ 　**14.** $a^2bc(2ac - 3a^2b)$ 　**15.** $3x(-x - y + 2)$ 　**16.** $b^2x^2(x^2 - 2x + 1)$

17. $ab^2c^4(ac - bc - ab)$ 　**18.** $-4c^2(-9gc - 2c + g^2)$ 　**19.** $ax(cx^2)(x + y^3)$ 　**20.** $-2(-3st^3)(3s - 4t)$

21. $(x - 3)(x + 5)$ 　**22.** $(a + 7)(a + 1)$ 　**23.** $(x + 5)(2x - 1)$ 　**24.** $(4t + s)(2t - 3s)$

25. $(2a - b)(3a - 2b)$ 　**26.** $(4x - 3)(3x - 1)$ 　**27.** $(2s + 7t)(3s - 5t)$ 　**28.** $(5p - 2q)(p + 8q)$

29. $(x^2 - 1)(2x + 5)$ 　　**30.** $(3y^2 + 2)(2y - 9)$ 　　**31.** $(x^2 - 2x)(x + 4)$

32. $(2ab^2 - 5t)(-ab^2 - 6t)$ 　　**33.** $(x + 1)(x^2 - 3x + 2)$ 　　**34.** $(2x + 3)(x^2 - x - 5)$

35. $(4x - x^3)(2 + x - x^2)$ **36.** $(5a - 3c)(a^2 + ac - c^2)$ **37.** $2(a + 1)(a - 9)$

38. $-5(y - 3)(y + 6)$ **39.** $2x(x - 1)(x + 4)$ **40.** $ax(x + 4)(7 - x^2)$

41. $(2x - 5)^2$ **42.** $(x - 3)^2$ **43.** $(x + 3a)^2$ **44.** $(2m + 1)^2$

45. $(xyz - 2)^2$ **46.** $(b - 2x^2)^2$ **47.** $2(x + 8)^2$ **48.** $3(a + 4)^2$

49. $(2 + x)(3 - x)(x - 1)$ **50.** $(3x - c^2)^3$ **51.** $3x(x + 2)^2(2x - 1)$ **52.** $[(x - 2)^2(x + 2)]^2$

53. A square microprocessor chip of side x millimeters is redesigned into a rectangular chip. One dimension is increased by 2 mm and the other is decreased by 1 mm. Find an expression for the area of the new chip.

54. An expression found in chemical thermodynamics is $p(c - 1) + 2 - c(p - 1)$. Multiply and simplify.

55. In finding the maximum power in part of a microwave transmitter circuit, the expression $(R + r)^2 - 2r(R + r)$ is used. Multiply and simplify.

56. In determining the deflection of a certain steel beam, the expression $27x^2 - 24(x - 6)^2 - (x - 12)^3$ is used. Multiply and simplify.

1-10 Division of Algebraic Expressions

To find the quotient of one monomial divided by another, we use the laws of exponents as given in Section 1-5 and the laws for dividing signed numbers as stated in Section 1-2. Again, the exponents may be combined only if the base is the same.

EXAMPLE A

1. $\dfrac{16x^3y^5}{4xy^2} = \dfrac{16}{4}(x^{3-1})(y^{5-2}) = 4x^2y^3$

2. $\dfrac{-6a^2xy^2}{2axy^4} = -\left(\dfrac{6}{2}\right)\dfrac{a^{2-1}x^{1-1}}{y^{4-2}} = -\dfrac{3a}{y^2}$

As noted in the second illustration, we use only positive exponents in the final result, unless there are specific instructions otherwise.

From arithmetic we may show how a multinomial is to be divided by a monomial. When adding fractions, say $\frac{2}{7}$ and $\frac{3}{7}$, we have

$$\frac{2}{7} + \frac{3}{7} = \frac{2 + 3}{7}$$

Looking at this equation now from *right to left,* we see that we may divide each term of the numerator separately by the denominator and have an equal result. Therefore, we see that *the quotient of a multinomial divided by a monomial is found by dividing each term of the multinomial by the monomial and adding the results.* This process can be shown as

$$\frac{a + b}{c} = \frac{a}{c} + \frac{b}{c}$$

NOTE ▷ $\left(\text{Be careful: Although } \dfrac{a + b}{c} = \dfrac{a}{c} + \dfrac{b}{c}, \text{ we must note that } \dfrac{c}{a + b} \text{ is } \textit{not } \dfrac{c}{a} + \dfrac{c}{b}.\right)$

EXAMPLE B

1. $\dfrac{4a^2 + 8a}{2a} = \dfrac{4a^2}{2a} + \dfrac{8a}{2a} = 2a + 4$

each term of numerator divided by denominator

2. $\dfrac{4x^3y - 8x^3y^2 + 2x^2y}{2x^2y} = \dfrac{4x^3y}{2x^2y} - \dfrac{8x^3y^2}{2x^2y} + \dfrac{2x^2y}{2x^2y}$

$= 2x - 4xy + 1$

3. $\dfrac{a^3bc^4 - 6abc + 9a^2b^3c - 3}{3ab^2c^3} = \dfrac{a^2c}{3b} - \dfrac{2}{bc^2} + \dfrac{3ab}{c^2} - \dfrac{1}{ab^2c^3}$

EXAMPLE C

Related to the operation of an irrigation pump, the expression

$$\frac{2p + v^2d + 2ydg}{2dg}$$

is used. Performing the indicated division, we have

$$\frac{2p + v^2d + 2ydg}{2dg} = \frac{p}{dg} + \frac{v^2}{2g} + y$$

In practice we would usually not write the middle step shown in the first two illustrations of Example B. The divisions are done by inspection, and the expression would appear as shown in the third illustration. However, we must remember that each term in the numerator is divided by the monomial in the denominator.

polynomials

If each term of an algebraic sum is a number or is of the form ax^n where n is a nonnegative integer, we call the expression a **polynomial** *in* x. The distinction between a multinomial and a polynomial is that a polynomial does not contain terms like $\sqrt{x}$ or $1/x^2$, whereas a multinomial may contain such terms. Also, a polynomial may consist of only one term, whereas a multinomial must have at least two terms. In a polynomial, *the greatest value of n which appears is the* **degree** *of the polynomial*.

EXAMPLE D

$3 + 2x^2 - x^3$ is a polynomial of degree 3

$x^4 - 3x^2 - \sqrt{x}$ is not a polynomial

$4x^5$ is a polynomial of degree 5

$\dfrac{1}{4x^5}$ is not a polynomial

The first two expressions are also multinomials since each contains more than one term. The second illustration is not a polynomial due to the presence of the $\sqrt{x}$ term. The third illustration is a polynomial since the exponent is a positive integer. It is also a monomial. The fourth illustration is not a polynomial since it can be written as $\frac{1}{4}x^{-5}$, which means that when it is written in the form ax^n, n is not positive.

At times it is necessary to divide one polynomial by another. To do this, we first arrange the dividend (the polynomial to be divided) and the divisor in descending powers of the variable. Then we divide the first term of the dividend by the first term of the divisor. The result gives the first term of the quotient. Next, we multiply the entire divisor by the first term of the quotient and subtract the product from the dividend. We divide the first term of this difference by the first term of the divisor. This gives the second term of the quotient. We multiply this term by each of the terms of the divisor and subtract this result from the first difference. We repeat this process until the remainder is either zero or a term which is of lower degree than the divisor. This process is similar to that of long division of numbers.

EXAMPLE E

Divide $6x^2 + x - 2$ by $2x - 1$. $\left(\text{This division can also be indicated by the form}\right.$

$(6x^2 + x - 2) \div (2x - 1)$ or in the fractional form $\left.\dfrac{6x^2 + x - 2}{2x - 1}.\right)$

We set up the division in the same way we would for long division in arithmetic. Then, following the procedure outlined above, we have the following:

NOTE ▷

$$
\begin{array}{r}
3x + 2 \qquad \dfrac{6x^2}{2x} \\
2x - 1 \overline{\big)\, 6x^2 + x - 2} \\
\end{array}
$$

$3x(2x - 1)$

$6x^2 - 3x$ $\qquad$ *subtract*

$6x^2 - 6x^2 = 0$

$x - (-3x) = 4x$

$4x - 2$

$4x - 2$ $\quad$ subtract

0

The remainder is zero and the quotient is $3x + 2$. Note that when we subtracted $-3x$ from x, we obtained $4x$.

EXAMPLE F

Divide $4x^3 + 6x^2 + 1$ by $2x - 1$. Since there is no x-term in the dividend, it is advisable to leave space for any x-terms which might arise.

$$
\begin{array}{r}
2x^2 + 4x + 2 \\
\text{divisor} \quad 2x - 1 \overline{\big)\, 4x^3 + 6x^2 \qquad + 1} \quad \text{dividend} \\
4x^3 - 2x^2 \\
\end{array}
$$

$6x^2 - (-2x^2) = 8x^2 \longrightarrow 8x^2 \qquad + 1$

$8x^2 - 4x$

$0 - (-4x) = 4x \longrightarrow 4x + 1$

$4x - 2$

$3 \qquad$ remainder

The quotient in this case is written as

$$2x^2 + 4x + 2 + \frac{3}{2x - 1}$$

Note how the remainder is expressed as part of the quotient.

Exercises 1-10

In the following exercises, perform the indicated divisions.

1. $\dfrac{8x^3y^2}{-2xy}$

2. $\dfrac{-18b^7c^3}{bc^2}$

3. $\dfrac{-16r^3t^5}{-4r^5t}$

4. $\dfrac{51mn^5}{17m^2n^2}$

5. $\dfrac{(15x^2)(4bx)(2y)}{30bxy}$

6. $\dfrac{(5st)(8s^2t^3)}{10s^3t^2}$

7. $\dfrac{6(ax)^2}{-ax^2}$

8. $\dfrac{12a^2b}{(3ab^2)^2}$

9. $\dfrac{a^2x + 4xy}{x}$

10. $\dfrac{2m^2n - 6mn}{2m}$

11. $\dfrac{3rst - 6r^2st^2}{3rs}$

12. $\dfrac{-5a^2n - 10an^2}{5an}$

13. $\dfrac{4pq^3 + 8p^2q^2 - 16pq^5}{4pq^2}$

14. $\dfrac{a^2xy^2 + ax^3 - ax}{ax}$

15. $\dfrac{2\pi f L - \pi f R^2}{\pi f R}$

16. $\dfrac{2(ab)^4 - a^3b^4}{3(ab)^3}$

17. $\dfrac{3ab^2 - 6ab^3 + 9a^2b^2}{9a^2b^2}$

18. $\dfrac{2x^2y^2 + 8xy - 12x^2y^4}{2x^2y^2}$

19. $\dfrac{x^{n+2} + ax^n}{x^n}$

20. $\dfrac{3a(x + y)b^2 - (x + y)}{a(x + y)}$

21. $(2x^2 + 7x + 3) \div (x + 3)$

22. $(3x^2 - 11x - 4) \div (x - 4)$

23. $\dfrac{x^2 - 3x + 2}{x - 2}$

24. $\dfrac{2x^2 - 5x - 7}{x + 1}$

25. $\dfrac{x - 14x^2 + 8x^3}{2x - 3}$

26. $\dfrac{6x^2 + 6 + 7x}{2x + 1}$

27. $(4x^2 + 23x + 15) \div (4x + 3)$

28. $(6x^2 - 20x + 16) \div (3x - 4)$

29. $\dfrac{x^3 + 3x^2 - 4x - 12}{x + 2}$

30. $\dfrac{3x^3 + 19x^2 + 16x - 20}{3x - 2}$

31. $\dfrac{2x^4 + 4x^3 + 2}{x^2 - 1}$

32. $\dfrac{2x^3 - 3x^2 + 8x - 2}{x^2 - x + 2}$

33. $\dfrac{x^3 + 8}{x + 2}$

34. $\dfrac{x^3 - 1}{x - 1}$

35. $\dfrac{x^2 - 2xy + y^2}{x - y}$

36. $\dfrac{3a^2 - 5ab + 2b^2}{a - 3b}$

37. In the optical theory dealing with lasers, the following expression arises:

$$\dfrac{8A^5 + 4A^3\mu^2E^2 - A\mu^4E^4}{8A^4}$$

Perform the indicated division.

38. The reciprocal of the total resistance of parallel resistors of 6 Ω, R_1 and R_2, is

$$\dfrac{6R_1 + 6R_2 + R_1R_2}{6R_1R_2}$$

Perform the indicated division. See Fig. 1-4.

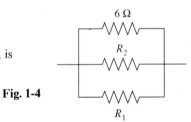

6 Ω

R_2

Fig. 1-4

R_1

39. In the development of a new adhesive, it is found to cover a rectangular area given by $10x^2 + 17x - 48$. Find the width of the area if its length is $2x - 3$.

40. In analyzing the displacement of a certain valve, the following expression is used:

$$\dfrac{s^2 - 2s - 2}{s^4 + 4}$$

Find the reciprocal of this expression, and perform the indicated division.

1-11   Equations

The basic operations for algebraic expressions that we have developed are used in the important process of solving equations. In this section we show how the basic algebraic operations are used in solving equations, and in the following sections we demonstrate some of the important applications of equations.

An **equation** *is an algebraic statement that two algebraic expressions are equal.* It is possible that many values of the letter representing the **unknown** will **satisfy** the equation; that is, many values may produce equality when **substituted** in the equation. It is also possible that only one value for the unknown will satisfy the equation (and this will be true of nearly all of the equations we solve in this section). Or possibly there may be no values which satisfy the equation, although the statement is still an equation.

EXAMPLE A

The equation $x^2 - 4 = (x - 2)(x + 2)$ is true for all values of x. For example, if we substitute $x = 3$ we have $9 - 4 = (3 - 2)(3 + 2)$ or $5 = 5$. If we let $x = -1$, we have $-3 = -3$. *An equation that is true for all values of the unknown is termed an* **identity.**

The equation $x^2 - 2 = x$ is true if $x = 2$ or if $x = -1$, but it is not true for any other values of x. If $x = 2$ we obtain $2 = 2$, and if $x = -1$ we obtain $-1 = -1$. However, if we let $x = 4$, we obtain $14 = 4$, which of course is not correct. *An equation valid only for certain values of the unknown is termed a* **conditional equation.** These equations are those which are generally encountered.

EXAMPLE B

The equation $3x - 5 = x + 1$ is true only for $x = 3$. When $x = 3$ we obtain $4 = 4$; if we let $x = 2$, we obtain $1 = 3$, which is not correct.

The equation $x + 5 = x + 1$ is not true for any value of x. For any value of x we try, we will find that the left side is 4 greater than the right side. *Such an equation is called a* **contradiction.**

To **solve** *an equation we find the values of the unknown which satisfy it.* There is one basic rule to follow when solving an equation:

Perform the same operation on both sides of the equation.

We do this to isolate the unknown and thus to find its values.

By performing the same operation on both sides of an equation, the two sides remain equal. Thus,

we may add the same number to both sides, subtract the same number from both sides, multiply both sides by the same number, or divide both sides by the same number (not zero).

Although we may multiply both sides of an equation by zero, this produces $0 = 0$, which is not useful in finding the solution.

EXAMPLE C _____ In solving the following equations, we note that we may isolate x, and thereby solve the equation, by performing the indicated operation.

$x - 3 = 12$	$x + 3 = 12$	$\dfrac{x}{3} = 12$	$3x = 12$
add 3 to both sides	subtract 3 from both sides	multiply both sides by 3	divide both sides by 3
$x - 3 + 3 = 12 + 3$	$x + 3 - 3 = 12 - 3$	$3\left(\dfrac{x}{3}\right) = 3(12)$	$\dfrac{3x}{3} = \dfrac{12}{3}$
$x = 15$	$x = 9$	$x = 36$	$x = 4$

Each can be checked by substitution in the original equation. (The term *transposing* is often used to denote the result of adding or subtracting a term from both sides of the equation. In transposing, a term is moved from one side of the equation to the other, and its sign is changed.)

The solution of an equation generally requires a combination of the basic operations. The following examples illustrate the solution of such equations.

EXAMPLE D _____ Solve the equation $2t - 7 = 9$.

We are to perform basic operations to both sides of the equation to finally isolate t on one side. The steps to be followed are suggested by the form of the equation, and in this case are as follows.

$2t - 7 = 9$	original equation
$2t - 7 + 7 = 9 + 7$	add 7 to both sides
$2t = 16$	combine like terms
$\dfrac{2t}{2} = \dfrac{16}{2}$	divide both sides by 2
$t = 8$	simplify

Therefore, we conclude that $t = 8$. Checking *in the original equation,* we have

$$2(8) - 7 = 9, \qquad 16 - 7 = 9, \quad \text{or} \quad 9 = 9$$

Therefore, the solution checks.

When simpler numbers are involved, the step of adding or subtracting a term, or multiplying or dividing by a factor, is usually done by inspection and not actually written down. This is done in the later examples, when applicable.

EXAMPLE E _____ Solve the equation $3n + 4 = n - 6$.

$2n + 4 = -6$	n subtracted from both sides—by inspection
$2n = -10$	4 subtracted from both sides—by inspection
$n = -5$	both sides divided by 2—by inspection

Checking *in the original equation,* we have $-11 = -11$.

EXAMPLE F ⎯⎯⎯⎯⎯ Solve the equation $x - 7 = 3x - (6x - 8)$.

$$x - 7 = 3x - 6x + 8 \qquad \text{parentheses removed}$$
$$x - 7 = -3x + 8 \qquad x\text{-terms combined on right}$$
$$4x - 7 = 8 \qquad 3x \text{ added to both sides}$$
$$4x = 15 \qquad 7 \text{ added to both sides}$$
$$x = \tfrac{15}{4} \qquad \text{both sides divided by 4}$$

■ Checking in the original equation, we obtain (after simplifying) $-\frac{13}{4} = -\frac{13}{4}$.

NOTE ▷   Note that we **always check in the original equation.** This is done since errors may have been made in finding the later equations.

If an equation contains decimals, the best procedure is to set up the solution first, before actually performing the calculations. Once the unknown is isolated, a calculator can be used to perform the calculations.

EXAMPLE G ⎯⎯⎯⎯⎯ When finding the electric current i, in amperes, in a circuit in a radio, the following equation and solution are used:

$$0.0595 - 0.525i - 8.85(i + 0.00316) = 0$$
$$0.0595 - 0.525i - 8.85i - 8.85(0.00316) = 0$$
$$(-0.525 - 8.85)i = 8.85(0.00316) - 0.0595$$
$$i = \frac{8.85(0.00316) - 0.0595}{-0.525 - 8.85}$$
$$= 0.00336 \text{ A} \quad \text{(rounded off)}$$

8.85 ⨯ .00316
⊟ .0595 ⊜ ⨸
⓪ .525 ⊞∕⊟ ⊟
8.85 ⦆ ⊜
[3.3636266 −03]

In checking, we find the left side of the original equation to be 3.4×10^{-5}. If we use the unrounded calculator value, we get 0. The calculator sequence for ■ this solution is shown at the left.

solving equations   From the discussion and examples of this section, we can see the following steps are used in solving the type of equation we have encountered.

1. *Remove grouping symbols (distributive law).*
2. *Combine any like terms on each side.*
3. *Perform the same operations on both sides until $x = $ (result) is obtained.*
4. *Check the solution in the original equation.*

Exercises 1-11

In Exercises 1 through 28, solve the given equations.

1. $x - 2 = 7$ **2.** $x - 4 = 1$ **3.** $x + 5 = 4$ **4.** $s + 6 = -3$

5. $\dfrac{t}{2} = 5$ **6.** $\dfrac{x}{4} = -2$ **7.** $4x = -20$ **8.** $2x = 12$

9. $3t + 5 = -4$ **10.** $5x - 2 = 13$ **11.** $5 - 2y = 3$ **12.** $8 - 5t = 18$

13. $3x + 7 = x$ **14.** $6 + 8y = 5 - y$ **15.** $2(s - 4) = s$ **16.** $3(n - 2) = -n$

17. $6 - (r - 4) = 2r$

18. $5 - (x + 2) = 5x$

19. $2(x - 3) - 5x = 7$

20. $4(x + 7) - x = -7$

21. $x - 5(x - 2) = 2$

22. $5x - 2(x - 5) = 4x$

23. $7 - 3(1 - 2p) = 4 + 2p$

24. $3 - 6(2 - 3t) = t - 5$

25. $5.8 - 0.3(x - 6.0) = 0.5x$

26. $1.9t = 0.5(4.0 - t) - 0.8$

27. $0.15 - 0.24(y - 0.50) = 0.63$

28. $27.5(5.17 - 1.44x) = 73.4$

In Exercises 29 through 32, what conclusion can be made about each of the equations?

29. $2(x - 3) + 1 = 2x - 5$ **30.** $3(x + 2) = 3x + 4$ **31.** $7 - (2 - x) = x + 2$ **32.** $1 - (3 - x) = x - 2$

In Exercises 33 through 36, solve the indicated equations.

33. In finding the rate v, in miles per hour, at which a polluted stream is flowing, the equation $15(5.5 + v) = 24(5.5 - v)$ is used. Find v to the nearest 0.1 mi/h.

34. In order to find the voltage in a circuit in a television remote control unit, the equation $1.12V - 0.67(10.5 - V) = 0$ is used. Find the voltage V, to the nearest 0.01 V.

35. In blending a gasohol fuel mixture, in order to find the number of gallons of one fuel needed, it is necessary to solve the equation $0.14a + 0.06(2000 - a) = 0.09(2000)$. Solve for a.

36. In order to find the distance x such that the weights are balanced on the lever shown in Fig. 1-5, the equation $210(3x) = 55.3x + 38.5(8.25 - 3x)$ must be solved. Find x.

Fig. 1-5

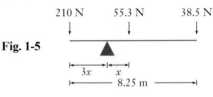

1-12   Formulas and Literal Equations

Equations and their solutions are of great importance in most fields of technology and science. They are used to attain, study, and confirm information of all kinds. One of the most important applications occurs in the use of formulas in mathematics, physics, engineering, and other fields. A **formula** *is an equation which expresses a rule and uses letters to represent certain quantities.* For example, the formula for the area of a circle is $A = \pi r^2$. The symbol A stands for the area as does the expression πr^2, and the formula states that the area is found by multiplying the square of the radius by π.

Often it is necessary to solve a formula or any equation containing more than one literal symbol for a particular letter or symbol which appears in it. We do this in the same manner as we solve any equation: We isolate the letter or symbol desired by use of the basic algebraic operations.

EXAMPLE A

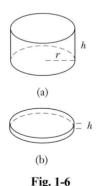

(a)

(b)

Fig. 1-6

The volume V of a right circular cylinder is given by the formula $V = \pi r^2 h$. See Fig. 1-6(a). This formula is of importance in numerous applications, including that of minting coins. In the case of a coin, the height h is its thickness. See Fig. 1-6(b). In the formula, solve for h.

$$V = \pi r^2 h \qquad \text{original equation}$$

$$\frac{V}{\pi r^2} = h \qquad \text{divide both sides by } \pi r^2$$

$$h = \frac{V}{\pi r^2} \qquad \begin{array}{l}\text{since each side equals the other, it makes}\\ \text{no difference which appears on the left}\end{array}$$

Since the symbol for which we are solving is usually shown on the left of the equal sign, the sides were switched, as shown. ∎

EXAMPLE B

A formula relating acceleration a, velocity v, initial velocity v_0, and time t is $v = v_0 + at$. Solve for t.

$$v - v_0 = at \qquad v_0 \text{ subtracted from both sides}$$

$$t = \frac{v - v_0}{a} \qquad \text{both sides divided by } a \text{ and then sides switched}$$

As we can see from Examples A and B, we can solve for the indicated literal number just as we solved for the unknown in the previous section. That is, we perform the basic algebraic operations on the various literal numbers which appear in the same way we perform them on explicit numbers. Other illustrations appear in the following examples.

EXAMPLE C

In the study of the forces on a certain beam, the equation

$$M = \frac{L(wL + 2P)}{8}$$

is used. Solve for P.

$$8M = \frac{8L(wL + 2P)}{8} \qquad \text{multiply both sides by 8}$$

$$8M = L(wL + 2P) \qquad \text{simplify right side}$$

$$8M = wL^2 + 2LP \qquad \text{remove parentheses}$$

$$8M - wL^2 = 2LP \qquad \text{subtract } wL^2 \text{ from both sides}$$

$$P = \frac{8M - wL^2}{2L} \qquad \text{divide both sides by } 2L \text{ and switch sides}$$

EXAMPLE D

The effect of temperature is important when accurate instrumentation is required. The volume V of a precision container at temperature T in terms of the volume V_0 at temperature T_0 is given by

$$V = V_0[1 + b(T - T_0)]$$

where b depends on the material of which the container is made. Solve for T.

Since we are to solve for T, we must isolate the term containing T. This can be done by first removing the grouping symbols, and then isolate the term with T.

$$V = V_0[1 + b(T - T_0)] \qquad \text{original equation}$$

$$V = V_0[1 + bT - bT_0] \qquad \text{remove parentheses}$$

$$V = V_0 + bTV_0 - bT_0V_0 \qquad \text{remove brackets}$$

$$V - V_0 + bT_0V_0 = bTV_0 \qquad \text{subtract } V_0 \text{ and add } bT_0V_0 \text{ to both sides}$$

$$T = \frac{V - V_0 + bT_0V_0}{bV_0} \qquad \text{divide both sides by } bV_0 \text{ and switch sides}$$

If we wish to determine the value of any literal number in an expression for which we know values of the other literal numbers, we should *first solve for the required symbol and then substitute the given values.* This is illustrated in the following example.

EXAMPLE E ———— The electric resistance R, in ohms, of a resistor changes with temperature T according to the formula $R = R_0(1 + \alpha T)$, where R_0 is the resistance at $0°C$. For a given resistor $R_0 = 712\ \Omega$ and $\alpha = 0.00455/°C$. Determine the value of T for $R = 825\ \Omega$.

Following the procedure given above, we first solve for T and then substitute the given values.

$$R = R_0 + R_0\alpha T$$

$$R - R_0 = R_0\alpha T$$

$$T = \frac{R - R_0}{\alpha R_0}$$

Now substituting, we have

$$T = \frac{825 - 712}{(0.00455)(712)}$$

$$= 34.9°C$$

Here a calculator is used to perform the calculation, and the result is rounded to three significant digits.

Exercises 1-12

In Exercises 1 through 8, solve for the indicated letter.

1. $ax = b$, for x

2. $cy + d = 0$, for y

3. $4n + 1 = 4m$, for n

4. $bt - 3 = a$, for t

5. $ax + 6 = 2ax - c$, for x

6. $s - 6n^2 = 3s + 4$, for s

7. $\frac{1}{2}t - (4 - a) = 2a$, for t

8. $7 - (p - \frac{1}{3}x) = 3p$, for x

In Exercises 9 through 32, each of the given formulas arises in the technical or scientific area of study listed. Solve for the indicated letter.

9. $\theta = kA + \lambda$, for λ (robotics)

10. $C = a + bx$, for a (economics)

11. $E = IR$, for R (electricity)

12. $F = pDL$, for p (mechanics)

13. $P = 2\pi Tf$, for T (mechanics)

14. $PV = nRT$, for T (chemistry)

15. $p = p_a + dgh$, for h (hydrodynamics)

16. $2Q = 2I + A + S$, for I (nuclear physics)

17. $P = \dfrac{\pi^2 EI}{L^2}$, for E (mechanics)

18. $u = -\dfrac{eL}{2m}$, for L (spectroscopy)

19. $s = vt - 16t^2$, for v (physics: motion)

20. $FL = P_1L - P_1d + P_2L$, for d (construction)

21. $C_0^2 = C_1^2(1 + 2V)$, for V (electronics)

22. $A_1 = A(M + 1)$, for M (photography)

23. $a = V(k - PV)$, for k (biology)

24. $T = 3(T_2 - T_1)$, for T_1 (oil drilling)

25. $Q_1 = P(Q_2 - Q_1)$, for Q_2 (refrigeration)

26. $p - p_a = dg(y_2 - y_1)$, for y_2 (pressure gauges)

27. $N = N_1 T - N_2(1 - T)$, for N_1 (machine design)

28. $t_a = t_c + (1 - h)t_m$, for h (computer access time)

29. $L = \pi(r_1 + r_2) + 2x_1 + x_2$, for r_1 (pulleys)

30. $r_e + r_c(1 - a) = \dfrac{1}{h}$, for r_e (electronics: transistors)

31. $P = \dfrac{V_1(V_2 - V_1)}{gJ}$, for V_2 (jet engine power)

32. $W = T(S_1 - S_2) - Q$, for S_2 (refrigeration)

In Exercises 33 through 36, determine the value of the indicated symbol for the given values of the other symbols.

33. The formula for the perimeter of the window as designed in Fig. 1-7 is
$p = \pi r + 2r + 2h$. Determine the value of h if $r = 22.5$ in. and $p = 172$ in.

34. A formula used in determining the total transmitted power P_t in an AM radio
signal is $P_t = P_c(1 + 0.5m^2)$. Find P_c if $P_t = 680$ W and $m = 0.925$.

Fig. 1-7

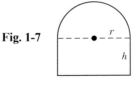

35. A formula relating the Fahrenheit temperature F and the Celsius temperature C is
$F = \frac{9}{5}C + 32$. Find the Celsius temperature which corresponds to $90.2°$F.

36. In forestry, a formula used to determine the volume V of a log is $V = \frac{1}{2}L(B + b)$, where L is the length of the
log and B and b are the areas of the ends. Find b, in square feet, if $V = 38.6$ ft^3, $L = 16.1$ ft, and $B = 2.63$ ft^2.

1-13 Applied Verbal Problems

Mathematics is very useful in most technical areas, because with it we can solve many kinds of applied problems. Some of these problems are in formula form and can therefore be solved directly. However, in practice it is often necessary to set up equations to be solved by using known formulas and given conditions. Such problems are first formulated as verbal problems, and it is necessary to translate them into mathematical terms for solution.

NOTE ▷ Usually the most difficult part in solving a stated problem is identifying the information that leads to the equation. Often this is due to the fact that *some of the information is implied, but not explicitly stated,* in the problem. A careful reading of the problem and an understanding of all terms and expressions are very important to being able to set up the equation properly for solution.

Since a careful reading and analysis are important to the solution of stated problems, it is possible only to give a general guideline to follow. Thus,

solving verbal problems

1. *read* the statement of the problem carefully;

2. clearly *identify* the unknown quantities, *assign* an appropriate letter to represent one of them, and specify the others in terms of this unknown;

3. *analyze* the statement clearly to *establish* the necessary equation;

4. *solve* the equation; and

5. *check* the solution in the original statement of the problem.

Carefully read the following examples.

EXAMPLE A

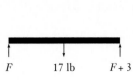

F 17 lb $F + 3$

Fig. 1-8

A 17-lb beam is supported at each end. The supporting force at one end is 3 lb more than that at the other end. Find the forces. See Fig. 1-8.

Since the force at each end is required, we write

let $F =$ the smaller force

as a way of establishing the unknown for the equation. Any appropriate letter could be used, and we could have let it represent the larger force.

Also, since the other force is 3 lb more, we write

$F + 3 =$ the larger force

Since the forces at each end of the beam support the weight of the beam, we have the equation

$$F + (F + 3) = 17$$

This equation can now be solved.

$$2F + 3 = 17$$
$$2F = 14$$
$$F = 7 \text{ lb}$$

Thus, the smaller force is 7 lb and the larger force is 10 lb. This checks with the original statement of the problem. ∎

EXAMPLE B

In designing an electric circuit, it is found that 34 resistors, with a total resistance of 56 Ω, of two resistances, 1.5 Ω and 2.0 Ω, are required. How many of each are in the circuit?

Since we want to find the number of each resistance, we

let $x =$ number of 1.5-Ω resistors

Also, since there are 34 resistors in all,

$34 - x =$ number of 2.0-Ω resistors

We also know that the total resistance of all resistors is 56 Ω. This means that

$$\underbrace{1.5x}_{\substack{\text{total resistance} \\ \text{of 1.5-}\Omega\text{ resistors}}} + \underbrace{2.0(34 - x)}_{\substack{\text{total resistance} \\ \text{of 2.0-}\Omega\text{ resistors}}} = 56 \leftarrow \text{total resistance of all resistors}$$

where the $1.5\,\Omega$ each and number give $1.5x$, and the $2.0\,\Omega$ each and number give $2.0(34 - x)$.

$$1.5x + 68 - 2.0x = 56$$
$$-0.5x = -12$$
$$x = 24$$

Therefore, there are 24 1.5-Ω resistors and 10 2.0-Ω resistors. The total resistance of these is $24(1.5) + 10(2.0) = 36 + 20 = 56\ \Omega$. We see that this checks with the statement of the problem. ∎

EXAMPLE C ——————

A medical researcher finds that a given sample of an experimental drug can be divided into 4 more slides with 5 mg each than with 6 mg each. How many slides with 5 mg each can be made up?

We are asked to find the number of slides with 5 mg, and therefore we

let x = number of slides with 5 mg

Since the sample may be divided into 4 more slides with 5 mg each than of 6 mg each, we know that

$x - 4$ = number of slides with 6 mg

Since *it is the same sample which is to be divided,* the total weight of the drug on each type is the same. This means

$$
\underset{\substack{\text{total weight}\\\text{5-mg slides}}}{\underset{5 \text{ mg}\\ \text{each} \searrow \downarrow}{5x}} \quad = \quad \underset{\substack{\text{total weight}\\\text{6-mg slides}}}{\underset{6 \text{ mg}\\ \text{each} \searrow \downarrow}{6(x - 4)}}
$$

$$5x = 6x - 24$$

$$-x = -24 \quad \text{or} \quad x = 24$$

Therefore, the sample can be divided into 24 slides with 5 mg each, or 20 slides with 6 mg each. Since the total weight, 120 mg, is the same for each set of slides, the solution checks with the statement of the problem.

EXAMPLE D ——————

A certain microprocessor chip is rectangular, and its length is 2.0 mm more than its width. Find the dimensions of the chip if its perimeter is 26.4 mm.

Since the dimensions, the length and the width, are required, we

let w = the width of the chip

Since the length is 2.0 mm more than its width, we know that

$w + 2.0$ = the length of the chip

The perimeter (distance around) of a rectangle is twice the length plus twice the width. This gives us the equation

$$2(w + 2.0) + 2w = 26.4$$

since the perimeter is given as 26.4 mm. This is the equation we need.

Solving this equation, we have

$$2w + 4.0 + 2w = 26.4$$

$$4w = 22.4$$

$$w = 5.6 \text{ mm}$$

and

$$w + 2 = 7.6 \text{ mm}$$

Thus, the length is 7.6 mm and the width is 5.6 mm. We see that these values check with the statements in the original problem.

EXAMPLE E _____ A space shuttle maneuvers into the same orbit as an already orbiting satellite which is 1200 km ahead in the orbit. If the satellite is traveling at 24,800 km/h and the shuttle is traveling at 25,300 km/h, how long will it take the shuttle to reach the satellite?

First we let t = the time for the shuttle to reach the satellite. Next, we use the facts that (1) the shuttle must go 1200 km farther in the same time and (2) _distance = rate × time_. This leads to the following equation and solution.

speed of
shuttle ⌐ time   speed of
satellite ⌐ time

$$25,300t = 1200 + 24,800t$$

distance traveled   distance between   distance traveled
by shuttle   at beginning   by satellite

$$500t = 1200$$
$$t = 2.4 \text{ h}$$

This means that it will take the shuttle 2.4 h to reach the satellite. In 2.4 h, the shuttle will travel 60,720 km and the satellite will travel 59,520 km. We see that the solution checks with the statement of the problem.

EXAMPLE F _____ An environmentalist checking a stream travels 8 mi in 1 h when going downstream. Then, with the motor set for twice its downstream speed, the environmentalist is able to travel only 7 mi upstream in 1 h. Find the speed of the boat relative to the water and the rate of the stream.

Here we let x = the downstream speed of the boat relative to the water. Since 8 mi were covered in going downstream in 1 h, the rate of the stream plus that of the boat is 8 mi/h, which in turn tells us that

$$8 - x = \text{rate of the stream}$$

In going upstream, the stream moves against the boat, and

$$2x - (8 - x) = \text{actual speed upstream}$$

since we subtract the rate of the stream from that of the boat. Now, using distance = rate × time, we have

$$7 = [2x - (8 - x)](1)$$

upstream   upstream   time
distance   speed

Solving for x, we have

$$7 = 2x - (8 - x)$$
$$7 = 2x - 8 + x$$
$$15 = 3x \qquad x = 5 \text{ mi/h}$$

Therefore, the downstream speed of the boat relative to the water is 5 mi/h, and the stream flows at the rate of 3 mi/h. We see that this checks, in that the distance covered in 1 h going downstream is $(5 + 3)(1) = 8$ mi, and the distance covered in 1 h going upstream is $[(2)(5) - 3](1) = 7$ mi.

NOTE ▷   Again we carefully note that *the solution is checked with the* **statement** *of the problem.* This is done because of the possibility of an error in setting up the equation.

EXAMPLE G _____   An alcohol and water mixture of 8 L is 25% alcohol. How much pure alcohol must be added to this solution so that the resulting solution is 40% alcohol?

First we let x = the number of liters of alcohol to be added. From the statement of the problem, we want the volume of alcohol to be 40% of the final mixture, and we know that this volume of alcohol is 25% of the original mixture plus that which is added. This leads to the equation

$$
\underset{\substack{\text{alcohol in}\\\text{original mixture}}}{\underbrace{\underset{25\%\text{—}}{\overset{\substack{\text{original}\\\text{volume}}}{0.25(8)}}}} \quad + \quad \underset{\substack{\text{alcohol}\\\text{added}}}{x} \quad = \quad \underset{\substack{\text{alcohol in}\\\text{final mixture}}}{\underbrace{\underset{40\%\text{—}}{\overset{\text{final volume}}{0.40(8 + x)}}}}
$$

Now, completing the solution, we have

$$2 + x = 0.40(8) + 0.40x$$

$$0.60x = 3.2 - 2$$

$$x = \frac{1.2}{0.60} = 2 \text{ L}$$

Therefore, 2 L of alcohol are to be added to the solution. Note that this result checks, since there would be 4 L of alcohol of a total volume of 10 L when 2 L of pure alcohol are added to the original solution.   ∎

Exercises 1-13

Solve the following problems by first setting up an appropriate equation.

1. Two computer software programs cost $390 together. If one costs $114 more than the other, what is the cost of each?

2. Two pipes drain an oil tank. One pipe releases 50 L/min more than the other. If they release 3200 L in 10 min together, what is the drainage rate of each?

3. Three meshed spur gears have a total of 107 teeth. If the second gear has 13 more teeth than the first, and the third has 15 more teeth than the second, how many teeth does each have?

4. In order to produce equilibrium on a particular beam, the sum of two forces must equal a third force. If the second of the two forces is 6.4 N more than the first and the third force is four times the first, what are the forces?

5. A developer purchased 70 acres of land for $900,000. If part of the land cost $20,000 per acre and the remainder cost $10,000 per acre, how much did the developer buy at each price?

6. A vial contains 2000 mg, which is to be used for two dosages. One patient is to be administered 660 mg more than the other. How much should be administered to each?

7. In the design of a bridge, an engineer determines that four fewer 18-m girders are needed for the span than 15-m girders. How many 18-m girders are needed?

8. A fuel oil storage depot had an 8-weeks supply on hand. However, cold weather caused the supply to be used in 6 weeks when 5000 gal extra were used each week. How many gallons were in the original supply?

9. The sum of three electric currents which come together at a point in an integrated circuit is zero. If the second current is double the first, and the third current is 9.2 μA more than the first, what are the currents? (The sign of a current indicates the direction of flow.)

10. A trucking firm uses two fleets of trucks. One fleet is used on round-trip delivery routes of 8 h, and the second fleet, with five more trucks than the first, is used on round-trip delivery routes of 6 h. Budget allotments allow for 198 h of delivery time in a week. How many trucks are in each fleet?

11. An architect designs a rectangular window such that the width of the window is 18 in. less than the height. If the perimeter of the window is 180 in., what are its dimensions?

12. It takes 300 cm of trim to go around a rectangular solar panel. The length of the panel is 90 cm more than the width. What are the dimensions of the panel?

13. A primary natural gas pipeline feeds into three smaller pipelines, each of which is 2.6 km longer than the main pipeline. If the total length of the four pipelines is 35.4 km, what is the length of each section of the line?

14. A rectangular security area is enclosed on one side by a wall, and the other three sides are fenced. The length of the wall is twice the width of the area, and the total cost of building the wall and fence is $13,200. If the wall costs $50/m and the fence costs $5/m, find the dimensions of the area.

15. In a race, one race car stalls at the beginning and starts 30 s after a second car. However, the first car travels at 260 ft/s and the second car travels at 240 ft/s. How long will it take the first car to overtake the second car? If the race is 20 mi is length, who will come in first?

16. A supersonic jet made one trip by averaging 100 mi/h less than the speed of sound for 1 h and then averaging 400 mi/h more than the speed of sound for 3 h. If the trip covered 3980 mi, what is the speed of sound?

17. Two supersonic jets, originally 5400 mi apart, start at the same time and travel toward each other. Find the speed of each if one travels 400 mi/h faster than the other and they pass each other in 1.5 h.

18. A corporate executive leaves the manufacturing plant and travels on an interstate highway at 55 mi/h to the corporate headquarters. The executive later returns in a helicopter, which travels at 125 mi/h on a route parallel to the highway. If the total travel time is 1.8 h, how far is it from the manufacturing plant to the corporate headquarters?

19. A ski lift takes a skier up a slope at 50 m/min. The skier then skis down the slope at 150 m/min. If one round-trip takes 24 min, how long is the slope?

20. A computer part manufacturer produces two types of parts. In testing a total of 6100 parts of both types, it was found that 0.5% of one type and 0.8% of the other type were defective. If a total of 38 defective parts were found, how many of each type were tested?

21. A certain type of engine uses a fuel mixture of 15 parts of gasoline to 1 part of oil. How much gasoline must be mixed with a gasoline-oil mixture, which is 75% gasoline, to make 8 L of the required mixture for the engine?

22. How many grams of solder which is 15% tin must be mixed with 90 g of solder which is 45% tin in order to have solder which is 25% tin?

23. By weight, a certain roadbed material is 75% crushed rock, and another is 30% crushed rock. How many tons of each must be mixed in order to have 250 tons of material which is 50% crushed rock?

24. The air in a rectangular room 5.60 m long, 4.67 m wide, and 2.44 m high contains 0.042% carbon dioxide. How much air in the room must a ventilating system replace with air containing 0.012% carbon dioxide in order that the room have air with 0.020% carbon dioxide?

1-14   Chapter Equations, Review Exercises, and Practice Test

Chapter Equations ▬▬▬▬▬▬▬▬

Commutative law of addition: $a + b = b + a$

Associative law of addition: $a + (b + c) = (a + b) + c$

Commutative law of multiplication: $ab = ba$

Associative law of multiplication: $a(bc) = (ab)c$

Distributive law: $a(b + c) = ab + ac$

$$a^m \times a^n = a^{m+n} \tag{1-1}$$

$$\frac{a^m}{a^n} = a^{m-n} \quad (m > n, a \neq 0), \qquad \frac{a^m}{a^n} = \frac{1}{a^{n-m}} \quad (m < n, a \neq 0) \tag{1-2}$$

$$(a^m)^n = a^{mn} \tag{1-3}$$

$$(ab)^n = a^n b^n, \qquad \left(\frac{a}{b}\right)^n = \frac{a^n}{b^n} \quad (b \neq 0) \tag{1-4}$$

$$a^0 = 1 \quad (a \neq 0) \tag{1-5}$$

$$a^{-n} = \frac{1}{a^n} \quad (a \neq 0) \tag{1-6}$$

$$\sqrt{ab} = \sqrt{a}\sqrt{b} \quad (a \text{ and } b \text{ positive real numbers}) \tag{1-7}$$

Review Exercises ▬▬▬▬▬▬▬▬

In Exercises 1 through 12, simplify the given expressions.

1. $(-2) + (-5) - (+3)$

2. $(+6) - (+8) - (-4)$

3. $\dfrac{(-5)(+6)(-4)}{(-2)(+3)}$

4. $\dfrac{(-9)(-12)(-4)}{24}$

5. $-5 - 2(-6) + \dfrac{-15}{+3}$

6. $3 - 5(-2) - \dfrac{12}{-4}$

7. $\dfrac{18}{3-5} - (-4)^2$

8. $-(-3)^2 - \dfrac{-8}{(-2) - (-4)}$

9. $\sqrt{16} - \sqrt{64}$

10. $-\sqrt{144} + \sqrt{49}$

11. $(\sqrt{7})^2 - \sqrt[3]{8}$

12. $-\sqrt[4]{16} + (\sqrt{6})^2$

In Exercises 13 through 24, simplify the given expressions. Where appropriate, express results with positive exponents only.

13. $(-2rt^2)^2$

14. $(3x^4 y)^3$

15. $\dfrac{18m^3 n^4 t}{-3mn^5 t^3}$

16. $\dfrac{15p^4 q^2 r}{5pq^5 r}$

17. $(x^0 y^{-1} z^3)^2$

18. $(3a^0 b^{-2})^3$

19. $\dfrac{-16s^{-2}(st^2)}{-2st^{-1}}$

20. $\dfrac{-35x^{-1} y(x^2 y)}{5xy^{-1}}$

21. $\sqrt{45}$

22. $\sqrt{68}$

23. $\sqrt{4 + 16}$

24. $\sqrt{9 + 36}$

In Exercises 25 through 28, perform the indicated operations on a calculator. In each case round off the results to four significant digits.

25. $37.38 - 16.92(1.067)^2$

26. $\dfrac{8.896 \times 10^{-12}}{3.5954 + 6.0449}$

27. $(0.6723)^3 - \sqrt{0.1958} + 2.844$

28. $\dfrac{1}{0.03568} + \dfrac{37{,}466}{29.63^2}$

In Exercises 29 through 60, perform the indicated operations.

29. $a - 3ab - 2a + ab$

30. $xy - y - 5y - 4xy$

31. $6xy - (xy - 3)$

32. $-(2x - b) - 3(-x - 5b)$

33. $(2x - 1)(x + 5)$

34. $(x - 4y)(2x + y)$

35. $(x + 8)^2$

36. $(2x + 3y)^2$

37. $\dfrac{2h^3k^2 - 6h^4k^5}{2h^2k}$

38. $\dfrac{4a^2x^3 - 8ax^4}{2ax^2}$

39. $4a - [2b - (3a - 4b)]$

40. $3b - [2b + 3a - (2a - 3b)] + 4a$

41. $2xy - \{3z - [5xy - (7z - 6xy)]\}$

42. $x^2 + 3b + [(b - y) - 3(2b - y + z)]$

43. $(2x + 1)(x^2 - x - 3)$

44. $(x - 3)(2x^2 - 3x + 1)$

45. $-3y(x - 4y)^2$

46. $-s(4s - 3t)^2$

47. $3p[(q - p) - 2p(1 - 3q)]$

48. $3x[2y - r - 4(s - 2r)]$

49. $\dfrac{12p^3q^2 - 4p^4q + 6pq^5}{2p^4q}$

50. $\dfrac{27s^3t^2 - 18s^4t + 9s^2t}{9s^2t}$

51. $(2x^2 + 7x - 30) \div (x + 6)$

52. $(4x^2 + 15x - 21) \div (2x + 7)$

53. $\dfrac{3x^3 - 7x^2 + 11x - 3}{3x - 1}$

54. $\dfrac{x^3 - 4x^2 + 7x - 12}{x - 3}$

55. $\dfrac{4x^4 + 10x^3 + 18x - 1}{x + 3}$

56. $\dfrac{8x^3 - 14x + 3}{2x + 3}$

57. $-3\{(r + s - t) - 2[(3r - 2s) - (t - 2s)]\}$

58. $(1 - 2x)(x - 3) - (x + 4)(4 - 3x)$

59. $\dfrac{2y^3 + 9y^2 - 7y + 5}{2y - 1}$

60. $\dfrac{6x^2 + 5xy - 4y^2}{2x - y}$

In Exercises 61 through 72, solve the given equations.

61. $3s + 8 = 5s$

62. $6n = 14 - n$

63. $3x + 1 = x - 8$

64. $4y - 3 = 5y + 7$

65. $6x - 5 = 3(x - 4)$

66. $-2(-4 - y) = 3y$

67. $2s + 4(3 - s) = 6$

68. $-(4 + v) = 2(2v - 5)$

69. $3t - 2(7 - t) = 5(2t + 1)$

70. $6 - 3x - (8 - x) = x - 2(2 - x)$

71. $2.7 + 2.0(2.1x - 3.4) = 0.1$

72. $0.250(6.721 - 2.44x) = 2.08$

In Exercises 73 through 80, change any numbers in ordinary notation to scientific notation or change any numbers in scientific notation to ordinary notation. (See Appendix B for an explanation of symbols which are used.)

73. The escape velocity (the velocity required to leave the earth's gravitational field) of a rocket is in excess of 25,000 mi/h.

74. When the first pictures of the surface of Mars were transmitted to Earth, Mars was 213,000,000 mi from Earth.

75. Police radar operates at a frequency of 1,020,000,000 Hz.

76. A computer can retrieve 15,600,000 units of data in one second.

77. A biological cell has a surface area of 0.0000012 cm^2.

78. An optical coating on glass to reduce reflections is 0.00000015 m thick.

79. The density of helium is 1.8×10^{-4} kg/m^3.

80. One kilogram will compress a certain coil spring 7.2×10^{-3} m.

In Exercises 81 through 96, solve for the indicated letter. Where noted the given formula arises in the technical or scientific area of study listed.

81. $3s + 2 = 5a$, for s

82. $5 - 7t = 6b$, for t

83. $3(4 - x) = 8 + 2n$, for x

84. $6 - 3b = 5(7 + 2v)$, for v

85. $R = n^2Z$, for Z (electricity)

86. $D = \dfrac{KI^2t}{A}$, for t (medicine)

87. $I = P + Prt$, for t (business)

88. $V = IR + Ir$, for R (electricity)

89. $m = dV(1 - e)$, for e (solar heating)

90. $mu = (m + M)v$, for M (physics: momentum)

91. $C = \dfrac{m(N_1 + N_2)}{2}$, for N_1 (mechanics: gears)

92. $2(J + 1) = \dfrac{f}{B}$, for J (spectroscopy)

93. $E = \dfrac{J - K}{1 - S}$, for K (chemistry)

94. $Z^2\left(1 - \dfrac{\lambda}{2a}\right) = k$, for λ (radar design)

95. $d = kx^2[3(a + b) - x]$, for a (mechanics: beams)

96. $V = V_0[1 + 3a(T_2 - T_1)]$, for T_2 (physics: thermal expansion)

In Exercises 97 through 100, evaluate the given expressions. Perform all calculations by use of a scientific calculator and round results to three significant digits.

97. The change in length of a steel girder can be found by evaluating the expression $aL(T_2 - T_1)$. Find the change in length, in feet, of a girder for which $a = 0.670 \times 10^{-5}/°\text{F}$, $L = 75.0$ ft, $T_2 = 75.6°\text{F}$, and $T_1 = 39.8°\text{F}$.

98. The combined resistance R of two electric resistors in parallel is found from the formula $R(R_1 + R_2) = R_1 R_2$. Find R, in ohms, if $R_1 = 3.78 \times 10^3\ \Omega$ and $R_2 = 8.75 \times 10^3\ \Omega$.

99. Three forces on a certain beam are related by the equation $3.50F_1 = 1.80F_2 + 3.25F_3$. Find F_3 if $F_1 = 2.63$ N and $F_2 = 2.50$ N.

100. A formula which gives the height h of an object in terms of its velocity v, initial velocity v_0, and the acceleration due to gravity is $v^2 = v_0^2 + 2gh$. Find h, in feet, if $v_0 = 55.0$ ft/s, $v = 25.5$ ft/s, and $g = -32.2$ ft/s^2.

In Exercises 101 through 104, perform the indicated operations.

101. When analyzing the load on a leaf spring, the expression $1 - r^2 - 4r(1 - r)$ is used. Simplify this expression.

102. In studying the electric potential of a conductor, the expression $2V(r - a) - V(b - a)$ is used. Simplify this expression.

103. When determining the illuminance from a light source, the expression $8(100 - x)^2 + x^2$ is used. Simplify this expression.

104. In finding the value of an annuity, the expression $(Ai - R)(1 + i)^2$ is used. Multiply out this expression.

In Exercises 105 through 116, solve the stated problems by first setting up an appropriate equation.

105. One computer has four times the storage capacity of another computer. If their combined storage capacity is 81,920 bytes, what is the storage capacity of each?

106. Three chemical reactions each produce oxygen. If the first produces twice that of the second, the third produces twice that of the first, and the combined total is 560 cm^3, what volume is produced by each reaction?

107. The voltage across a resistor equals the current times the resistance. In a microprocessor circuit, one resistor is 1200 Ω greater than another. The sum of the voltages across them is 12.0 mV. Find the resistances if the current is 2.4 μA in each.

108. An air sample contains 4 ppm (parts per million) of two harmful pollutants. The concentration of one is four times the other. What is the concentration of each?

109. A protective screen is designed to be placed in front of a fan blade. The screen is made of equally spaced straight wire strands and is 95 cm wide and 135 cm high. There are 30 more horizontal strands than vertical ones. How many horizontal and vertical strands are there if a total of 14,810 cm of wire is used in making the screen?

110. A rectangular field is enclosed with a fence of four strands of barbed wire. The length is 20 m more than the width, and a total of 1360 m of barbed wire is used for the fencing. What are the dimensions of the field?

111. Two motorboats, 55.3 mi apart, start toward each other. One travels at the rate of 22.5 mi/h and the other at 17.0 mi/h. When will they meet?

112. A helicopter used in fighting a forest fire travels 105 mi/h from the fire to a pond and 70 mi/h with water from the pond to the fire. If a round-trip takes 30 min, how long does it take from the pond to the fire?

113. One grade of oil has 0.50% of a certain additive and a second grade has 0.75% of the same additive. How many liters of the first grade of oil must be added to the second grade in order to have 1000 L with 0.65% of the additive?

114. Fifty pounds of a cement-sand mixture is 40% sand. How many pounds of sand must be added for the resulting mixture to be 60% sand?

115. An architect intends to have 25% of the floor area of a house in ceramic tile. In all but the kitchen and entry, the plans have 2200 ft^2 of floor area, of which 15% is tile. What area can the architect plan for the kitchen and entry if each has an all-tile floor?

116. A *karat* equals $\frac{1}{24}$ part of gold in an alloy (e.g., 9-karat gold is $\frac{9}{24}$ gold). How many grams of 9-karat gold must be mixed with 18-karat gold to get 200 g of 14-karat gold?

Practice Test

In Problems 1 through 5, evaluate the given expressions.

1. $\sqrt{9 + 16}$

2. $\dfrac{(7)(-3)(-2)}{(-6)(0)}$

3. $\dfrac{3.372 \times 10^{-3}}{7.526 \times 10^{12}}$

4. $\dfrac{(+6)(-2) - 3(-1)}{5 - 2}$

5. $\dfrac{523.0}{207.7} - \dfrac{396.4 - 23.5}{249.8}$

In Problems 6 through 12, perform the indicated operations and simplify. Use only positive exponents when exponents are used in the final result.

6. (a) $2x^0$ (b) $(2a^{-2}b^3)^{-3}$

7. $(3s^2y)^2(-2s)$

8. $\dfrac{8a^3x^2 - 4a^2x^4}{-2ax^2}$

9. $3m^2(am - 2m^3)$

10. $(2x - 3)(x + 7)$

11. $\dfrac{6x^2 - 13x + 7}{2x - 1}$

12. $(2x + 3)^2$

In Problems 13 and 14, solve for x.

13. $3(x - 3) = x - d$

14. $5x - 2(x - 4) = 7$

15. Express 0.0000036 in scientific notation.

16. List the numbers $-3, |-4|, -\pi, \sqrt{2}$, and 0.3 in numerical order.

17. What fundamental law is illustrated by $3(5 + 8) = 3(5) + 3(8)$?

18. The following equation is used in physics. Solve for t_2. $L = L_0[1 + a(t_2 - t_1)]$

19. A square tract of land is enclosed with fencing and then divided in half by additional fencing parallel to two of the sides. If 75 m of fencing is used, what is the length of one side of the tract? Solve by first setting up an appropriate equation.

20. An alloy weighing 20 lb is 30% copper. How many pounds of another alloy, which is 80% copper, must be added in order for the final alloy to be 60% copper? Solve by first setting up an appropriate equation.

2 Functions and Graphs

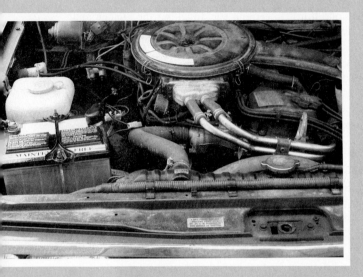

The electric power produced in a circuit depends on the resistance in the circuit. In Section 2-4 we draw a graph to see this type of relationship.

In technology and science, as well as in everyday life, we see that one quantity depends on one or more other quantities. Plant growth depends on sunlight and rainfall; traffic flow depends on the roadway design; the sales tax on an item depends on the cost of the item. These are but a few of the innumerable possible examples.

Determining how one quantity depends on other quantities is one of the primary goals of science. A rule that relates such quantities is of great importance and usefulness in science and technology. In mathematics such a rule is called a *function*, and we will start this chapter with a discussion of functions.

A way of actually seeing how one quantity depends on another is by means of a *graph*. The basic method of drawing and using graphs is also taken up in this chapter.

2-1 Introduction to Functions

In the later part of Chapter 1, we discussed the solution of equations, with applications to formulas. In most of the formulas, one quantity was given in terms of one or more other quantities. It is obvious, then, that the various quantities are related by means of the formula. One important method of finding such formulas is through scientific observation and experimentation.

If we were to perform an experiment to determine whether or not a relationship exists between the distance an object drops and the time it falls, observation of the results would indicate (approximately, at least) that $s = 16t^2$, where s is the distance in feet and t is the time in seconds. We would therefore see that distance and time for a falling object are related.

A similar study of the pressure and the volume of a gas at constant temperature would show that as pressure increases, volume decreases according to the formula $PV = k$, where k is a constant. Electrical measurements of current and voltage with respect to a particular resistor would show that $V = kI$, where V is the voltage, I is the current, and k is a constant.

Considerations such as these lead us to one of the most important and basic concepts in mathematics.

definition of a function

*Whenever a relationship exists between two variables such that for every permissible value of the first, there is only one corresponding value of the second, we say that the second variable is a **function** of the first variable.*

*The first variable is called the **independent variable**,* since permissible values can be assigned to it arbitrarily. *The second variable is called the **dependent variable**,* since its value is determined by the choice of the independent variable. *Values of the independent variable and dependent variable are to be **real numbers**.* Therefore, it is possible that there are restrictions on the possible values of the variables. This is discussed in the following section.

There are many ways to express functions. Formulas such as those we have discussed define functions. Other ways to express functions are by means of tables, charts, and graphs.

EXAMPLE A
In the equation $y = 2x$, we see that y is a function of x, since for each value of x there is only one value of y. For example, if $x = 3$, $y = 6$ and no other value. By arbitrarily assigning values to x, we make it the independent variable and y the dependent variable. ∎

EXAMPLE B
The power P developed in a certain resistor by a current I is given by $P = 4I^2$. Here P is a function of I. The dependent variable is P, and the independent variable is I. ∎

EXAMPLE C
For the equation $y = 2x^2 - 6x$, we have y as a function of x. The dependent variable is y, and the independent variable is x. Some of the values of y corresponding to chosen values of x are given in the following table.

x	-2	-1	0	$\frac{1}{2}$	1	2	3	π	10
y	20	8	0	$-\frac{5}{2}$	-4	-4	0	$2\pi^2 - 6\pi$	140

∎

In the following example we see how basic geometric information is used to establish the formula for a function.

EXAMPLE D

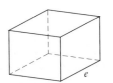

Figure 2-1 shows a cube of edge e. In order to express the volume V as a function of the edge, we recall from geometry that the volume is the cube of the edge. The formula is $V = e^3$. Here V is a function of e, since for each value of e there is only one value of V. The dependent variable is V, and the independent variable is e.

If the equation relating the volume and edge of a cube were written as $e = \sqrt[3]{V}$, that is, if the edge of a cube were expressed in terms of its volume, we would say that e is a function of V. In this case e would be the dependent variable and V the independent variable. ∎

Fig. 2-1

For convenience of notation,

the phrase "function of x" is written as f(x).

functional notation

This, in turn, means that the statement "*y* is a function of *x*" may be written as $y = f(x)$. In the form $y = f(x)$, *y* and *x* may represent measurable quantities, such as volume and edge, or power and current. However, the letter *f* denotes *dependence* and is not a measurable quantity. Used in this way,

NOTE▷   *f(x)* **does** not **mean f times x.**

EXAMPLE E _____ If $y = 6x^3 - 5x$, we may say that *y* is a function of *x*, where this function $f(x)$ is $6x^3 - 5x$. We may also write $f(x) = 6x^3 - 5x$ to denote the dependence. It is common to write functions in this form rather than in the form $y = 6x^3 - 5x$. However, *y* and $f(x)$ represent the same expression. Using *y*, the quantities are ■ shown, and using $f(x)$, the functional dependence is shown.

One of the most important uses of this notation is to designate the value of a function for a particular value of the independent variable. That is,

the value of the function f(x) when x = a is written as f(a).

EXAMPLE F _____ For the function $f(x) = 3x^2 - 5$, the value of $f(x)$ for $x = 2$ may be represented as $f(2)$. Thus, substituting 2 for *x*, we have

$$f(2) = 3(2^2) - 5 = 7 \qquad \text{substitute 2 for } x$$

In the same way, the value of $f(x)$ for $x = -1$ is

$$f(-1) = 3(-1)^2 - 5 = -2 \qquad \text{substitute } -1 \text{ for } x$$

The value of $f(x)$ for $x = -2.73$ is

$$f(-2.73) = 3(-2.73)^2 - 5 = 17.4 \qquad \text{substitute } -2.73 \text{ for } x$$

■ where a calculator was used and the result is rounded off.

In certain instances we need to define more than one function of *x*. Then we use different symbols to denote the functions. For example, $f(x)$ and $g(x)$ may represent different functions of *x*, such as $f(x) = 5x^2 - 3$ and $g(x) = 6x - 7$. Special functions are represented by particular symbols. For example, in trigonometry we shall come across the "sine of the angle θ," where the sine is a function of θ. This is designated by $\sin \theta$.

EXAMPLE G _____ If $f(x) = \sqrt{3x} + x$ and $g(x) = ax^4 - 5x$, then

$$f(3) = \sqrt{3(3)} + 3 = 3 + 3 = 6 \qquad \text{substitute 3 for } x \text{ in } f(x)$$

and

$$g(3) = a(3^4) - 5(3) = 81a - 15 \qquad \text{substitute 3 for } x \text{ in } g(x)$$

There are occasions when we wish to evaluate a function in terms of a literal number rather than an explicit number. However, whatever number a represents in $f(a)$, we substitute a for x in $f(x)$.

EXAMPLE H — If $g(t) = 4t^2 - 5t$, to find $g(a^3)$ we substitute a^3 for t in the function. Thus,

$$g(a^3) = 4(a^3)^2 - 5(a^3) = 4a^6 - 5a^3 \qquad \text{substitute } a^3 \text{ for } t$$

For the same function,

$$
\begin{aligned}
g(b + 1) &= 4(b + 1)^2 - 5(b + 1) \qquad \text{substitute } b + 1 \text{ for } t \\
&= 4(b^2 + 2b + 1) - 5(b + 1) \\
&= 4b^2 + 8b + 4 - 5b - 5 \\
&= 4b^2 + 3b - 1
\end{aligned}
$$

EXAMPLE I — The resistance of a particular resistor as a function of temperature is given by $R = 10.0 + 0.10T + 0.001\ T^2$. If a given temperature T is increased by $10°C$, what is the value of R for the increased temperature as a function of the temperature T?

We are to determine R for a temperature of $T + 10$. Since

$$f(T) = 10.0 + 0.10T + 0.001 T^2$$

we know that

$$
\begin{aligned}
f(T + 10) &= 10.0 + 0.10(T + 10) + 0.001(T + 10)^2 \qquad \text{substitute } T + 10 \text{ for } T \\
&= 10.0 + 0.10T + 1.0 + 0.001 T^2 + 0.02T + 0.1 \\
&= 11.1 + 0.12T + 0.001 T^2
\end{aligned}
$$

A function may be looked upon as a set of instructions. These instructions tell us how to obtain the value of the dependent variable for a particular value of the independent variable, even if the set of instructions is expressed in literal symbols.

EXAMPLE J — The function $f(x) = x^2 - 3x$ tells us to "square the value of the independent variable, multiply the value of the independent variable by 3, and subtract the second result from the first." An analogy would be a computer which was programmed so that when a number was entered into the computer, it would square the number, then multiply the number by 3, and finally subtract the second result from the first. This is represented in diagram form in Fig. 2-2.

The functions $f(t) = t^2 - 3t$ and $f(n) = n^2 - 3n$ are the same as the function $f(x) = x^2 - 3x$, since the operations performed on the independent variable are the same. ***Although different literal symbols appear, this does not change the function.***

NOTE ▷

Fig. 2-2

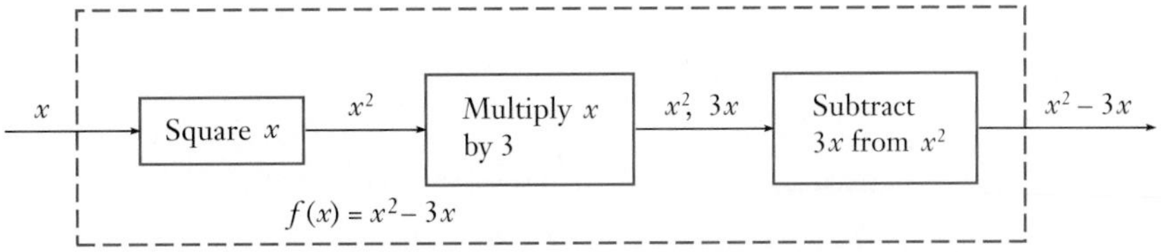

$f(x) = x^2 - 3x$

EXAMPLE K ——————— If we define the meanings of y and x by the statement "y equals the cube root of x," we know that $y = \sqrt[3]{x}$. Here, we may note that y is a function of x by writing $y = f(x)$, and we may specifically show that function as

$$f(x) = \sqrt[3]{x}$$

In Example D, we saw that the edge e of a cube in terms of its volume V is $e = \sqrt[3]{V}$. Here, $e = f(V)$ with

$$f(V) = \sqrt[3]{V}$$

Again we note that $f(x) = \sqrt[3]{x}$ is the same function as $f(V) = \sqrt[3]{V}$. The operation in each is finding the cube root, and the use of different symbols does not change the function. ∎

Exercises 2-1

In Exercises 1 through 8, determine the appropriate functions.

1. Express the area A of a circle as a function of its radius r.

2. Express the area A of a circle as a function of its diameter d.

3. Express the circumference c of a circle as a function of its radius r.

4. Express the circumference c of a circle as a function of its diameter d.

5. Express the area A of a rectangle of width 5 as a function of its length l.

6. Express the volume V of a right circular cone of height 8 as a function of the radius r of the base.

7. Express the area A of a square as a function of its side s; express the side s of a square as a function of its area A.

8. Express the perimeter p of a square as a function of its side s; express the side s of a square as a function of its perimeter p.

In Exercises 9 through 20, evaluate the given functions.

9. Given $f(x) = 2x + 1$, find $f(1)$ and $f(-1)$.

10. Given $f(x) = 5x - 9$, find $f(2)$ and $f(-2)$.

11. Given $f(x) = 5 - 3x$, find $f(-2)$ and $f(0.4)$.

12. Given $f(x) = 7 - 2x$, find $f(2.6)$ and $f(-4)$.

13. Given $f(n) = n^2 - 9n$, find $f(3)$ and $f(-5)$.

14. Given $f(v) = 2v^3 - 7v$, find $f(1)$ and $f(\frac{1}{2})$.

15. Given $\phi(x) = \dfrac{6 - x^2}{2x}$, find $\phi(1)$ and $\phi(-2)$.

See Appendix E for a computer program for evaluating a function.

16. Given $H(q) = \dfrac{8}{q} + 2\sqrt{q}$, find $H(4)$ and $H(0.16)$.

17. Given $g(t) = at^2 - a^2t$, find $g(-\frac{1}{2})$ and $g(a)$.

18. Given $s(y) = 6\sqrt{y} - 3$, find $s(9)$ and $s(a^2)$.

19. Given $K(s) = 3s^2 - s + 6$, find $K(-s)$ and $K(2s)$.

20. Given $T(t) = 5t + 7$, find $T(-2t)$ and $T(t + 1)$.

In Exercises 21 through 24, evaluate the given functions by use of a calculator. Round off results to the accuracy of the value of the independent variable used.

21. Given $f(x) = 5x^2 - 3x$, find $f(3.86)$ and $f(-6.92)$.

22. Given $g(t) = \sqrt{t + 1.0604} - 6t^3$, find $g(0.9261)$ and $g(-0.3256)$.

23. Given $F(y) = \dfrac{2y^2}{y + 0.03685}$, find $F(0.02474)$ and $F(-0.08466)$.

24. Given $f(x) = \dfrac{x^4 - 2.0965}{6x}$, find $f(1.9654)$ and $f(-2.3865)$.

In Exercises 25 through 28, state the instructions of the function in words as in Example J.

25. $f(x) = x^2 + 2$ **26.** $f(x) = 2x - 6$ **27.** $g(y) = 6y - y^3$ **28.** $\phi(s) = 8 - 5s + s^5$

In Exercises 29 through 32, following Example K, show that the dependent variable is a function of the independent variable in $y = f(x)$ form. Then write the indicated function using functional notation.

29. y is equal to the square of x.   **30.** s is equal to the square root of $t + 2$.

31. The net profit P made on selling 40 items, if each costs \$24, is equal to the product of 40 and $p - 24$, where p is the price charged.

32. The electrical resistance R of a certain ammeter, in which the resistance of the coil is R_c, is the product of 10 and R_c divided by the sum of 10 and R_c.

In Exercises 33 through 36, solve the given problems.

33. A demolition ball is used to tear down a building. Its distance s, in meters, above the ground as a function of the time t, in seconds, after it is dropped is given by $s = 17.5 - 9.8t^2$. Since $s = f(t)$, find $f(1.2)$.

34. The change C, in inches, in the length of a 100-ft steel bridge girder from its length at $40°F$, as a function of the temperature T, is given by $C = 0.014(T - 40)$. Since $C = f(T)$, find $f(15)$.

35. The stopping distance d, in feet, of a car going v miles per hour is given by the function $d = v + 0.05v^2$. Since $d = f(v)$, find $f(30)$, $f(2v)$, and $f(60)$ using both $f(v)$ and $f(2v)$.

36. The electric power P, in watts, dissipated in a resistor of resistance R, in ohms, is given by the function $P = \dfrac{200R}{(100 + R)^2}$. Since $P = f(R)$, find $f(R + 10)$.

2-2   More About Functions

As we mentioned in the previous section, using only real numbers may result in restrictions as to the permissible values of the independent and dependent variables. *The complete set of possible values of the independent variable is called the* **domain** *of the function,* and *the complete set of all possible resulting values of the dependent variable is called the* **range** *of the function.* Therefore, using real numbers in the domain and range of a function,

domain and range

> *values which lead to division by zero or*
> *to imaginary values may not be included.*

(Only in very specific cases, mostly in Chapter 11, will we include imaginary numbers in our discussion.) Consider the following examples.

EXAMPLE A

The function $f(x) = x^2 + 2$ is defined for all real values of x. This means its domain is written as *all real numbers*. However, since x^2 is never negative, $x^2 + 2$ is never less than 2. We then write the range as *all real numbers $f(x) \geq 2$*, where the symbol $\geq$ means "is greater than or equal to."

The function $f(t) = \frac{1}{t+2}$ is not defined for $t = -2$, for this value would require division by zero. Also, $f(t)$ will never equal zero, for no matter how large t becomes, $f(t)$ will never exactly equal zero. Therefore, the domain of this function is *all real numbers except -2*, and the range is *all real numbers except 0*.

EXAMPLE B _____ The function $g(s) = \sqrt{3 - s}$ is not defined for real numbers greater than 3, since such values make $3 - s$ negative and would result in imaginary values for $g(s)$. This means that the domain of this function is *all real numbers $s \leq 3$*, where the symbol $\leq$ means "is less than or equal to."

Also, since $\sqrt{3 - s}$ means the principal square root of $3 - s$ (see Section 1-7), we know that $g(s)$ cannot be negative. This tells us that the range of the function is *all real numbers $g(s) \geq 0$*.

In Examples A and B we determined the domain of each function by looking for those values of the independent variable which cannot be used. The range of each was found through an inspection of the function. This is normally the procedure, except that more advanced methods are often necessary to find the range. Even the second illustration in Example A required a special look at the function. For this reason, we will look for only the domain for some functions.

EXAMPLE C _____ Find the domain of the function $f(x) = 16\sqrt{x} + \dfrac{1}{x}$.

From the term $16\sqrt{x}$ we see that x must be greater than or equal to zero in order to have real values. The term $\dfrac{1}{x}$ indicates that x cannot be zero, because of division by zero. Thus, putting these together, the domain is *all real numbers $x > 0$*.

As for the range, it is *all real numbers $f(x) \geq 12$*. More advanced methods are needed to determine this.

We have seen that the domains of some functions are restricted to particular values. It can also happen that the domain of a function is restricted by definition or by practical considerations in an application. Consider the illustrations in the following example.

EXAMPLE D _____ A function defined as

$$f(x) = x^2 + 4 \quad \text{for } x > 2$$

has a domain restricted to real numbers greater than 2 by definition. Thus, $f(5) = 29$, but $f(1)$ is not defined, since 1 is not in the domain. Also, the range is all real numbers greater than 8.

The height h, in meters, of a certain projectile as a function of the time t, in seconds, is

$$h = 20t - 4.9t^2$$

Generally, negative values of time do not have meaning in such an application. This leads us to state the domain as values of $t \geq 0$. Of course, the projectile will not continue in flight indefinitely, and there is some upper limit on the value of t. These restrictions are not usually stated unless there is a particular reason which affects the solution.

The following example illustrates a function which is defined differently for different intervals of the domain.

EXAMPLE E — For the function $f(x) = \begin{cases} 2x - 1 & \text{for } x < 2 \\ 5 & \text{for } x \geq 2 \end{cases}$ find $f(-1)$ and $f(3)$.

We see that values of this function are found differently for values of x less than 2 than for values of x greater than or equal to 2. Since -1 is less than 2,

$$f(-1) = 2(-1) - 1 = -3$$

Since 3 is greater than 2,

$$f(3) = 5$$

■ We note that $f(x) = 5$ for all values of x which are 2 or greater.

In the previous section we wrote functions in mathematical form from given statements and by using geometric information. It is often necessary to determine a mathematical function from a given statement. The function is formed using methods similar to those we used in establishing equations from statements in Chapter 1. In the following examples we set up such functions.

EXAMPLE F — The fixed cost for a company to operate a certain plant is $3000 per day. It also costs $4 for each unit produced in the plant. Express the daily total cost C of operating the plant as a function of the number n of units produced.

The daily total cost C equals the fixed cost of $3000 plus the cost of producing n units. Since the cost of producing one unit is $4, the cost of producing n units is $4n$. Thus, the total cost C, where $C = f(n)$, is

$$C = 3000 + 4n$$

Here we know that the domain is all values of $n \geq 0$, with some upper limit on ■ n based on the production capacity of the plant.

EXAMPLE G — A metallurgist melts and mixes m grams of solder which is 40% tin with n grams of another solder which is 20% tin to get a final solder mixture which contains 200 g of tin. Express n as a function of m.

The statement leads to the following equation.

tin in first solder		tin in second solder		total amount of tin
$0.40m$	$+$	$0.20n$	$=$	200

Since we want $n = f(m)$, we now solve for n.

$$0.20n = 200 - 0.40m$$
$$n = 1000 - 2m$$

This is the required function. Since neither m nor n can be negative, the domain is all values $0 \leq m \leq 500$ g, which means that m is greater than or equal to 0 g ■ and less than or equal to 500 g. The range is all values $0 \leq n \leq 1000$ g.

EXAMPLE H

$\frac{1}{2}(2\pi r)$

r

$2r + 10$

$2r$

Fig. 2-3

An architect designs a window such that it has the shape of a rectangle with a semicircle on top. See Fig. 2-3. The base of the window is 10 cm less than the height of the rectangular part. Express the perimeter p of the window as a function of the radius r of the circular part.

We know that the perimeter is the distance around the window. Since the top part is a semicircle, and the circumference of a circle is $2\pi r$, the length of the top circular part is $\frac{1}{2}(2\pi r)$. This also tells us that the dashed line, and therefore the base of the window, is $2r$. Finally, the fact that the base is 10 cm less than the height of the rectangular part tells us that each vertical part is $2r + 10$. This means that the perimeter p, where $p = f(r)$, is

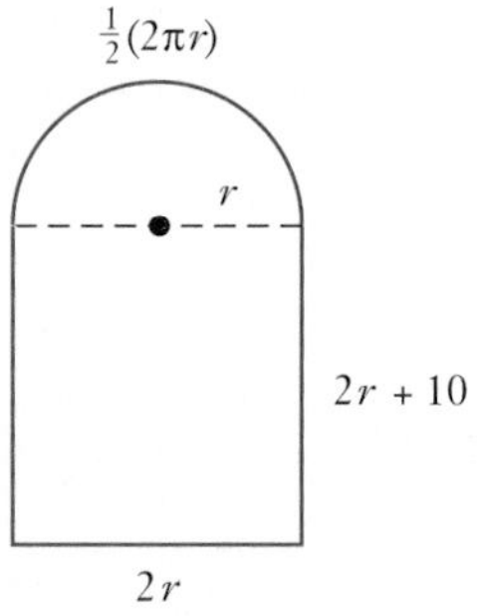

$$p = \frac{1}{2}(2\pi r) + 2r + 2(2r + 10)$$

$$= \pi r + 2r + 4r + 20$$

$$= \pi r + 6r + 20$$

We see that the required function is $p = \pi r + 6r + 20$. Since the radius cannot be negative, and there would be no window if $r = 0$, the domain of the function is all values $0 < r \leq R$, where R is a maximum possible value of r determined by design considerations. ∎

In the definition of a function, it was stipulated that any value for the independent variable must yield only a single value of the dependent variable. This requirement is stressed in more advanced courses, and we will use it again in Chapter 19. *If a value of the independent variable yields more than one value of the dependent variable, the relationship is called a* **relation** *instead of a function.* A relation involves two variables related so that values of the second variable can be determined from values of the first variable. A function is a relation in which each value of the first variable yields only one value of the second. A function is therefore a special type of relation. However, there are relations that are not functions.

EXAMPLE I For $y^2 = 4x^2$, if $x = 2$, then y can be either $+4$ or -4. Since a value of x yields more than one single value for y, we have a relation, not a function. ∎

Exercises 2-2

In Exercises 1 through 8, determine the domain and the range of the given functions.

1. $f(x) = x + 5$ **2.** $g(u) = 3 - u^2$ **3.** $G(z) = \dfrac{3}{z}$ **4.** $F(r) = \sqrt{r + 4}$

5. $f(s) = \dfrac{2}{s^2}$ **6.** $f(x) = \dfrac{6}{\sqrt{2 - x}}$ **7.** $H(h) = 2h + \sqrt{h + 1}$ **8.** $T(t) = 2t^4 + t^2 - 1$

In Exercises 9 through 12, determine the domain of the given functions.

9. $Y(y) = \dfrac{y + 1}{\sqrt{y - 2}}$ **10.** $f(n) = \dfrac{n}{6 - 2n}$ **11.** $f(u) = \dfrac{u}{u - 2} + \dfrac{4}{u + 4}$ **12.** $g(x) = \dfrac{\sqrt{x - 2}}{x - 3}$

In Exercises 13 through 20, evaluate the following functions as indicated.

$$F(t) = 3t - t^2 \quad \text{for } t \le 2$$

$$h(s) = \begin{cases} 2s & \text{for } s < -1 \\ s + 1 & \text{for } s \ge -1 \end{cases}$$

$$f(x) = \begin{cases} x + 1 & \text{for } x < 1 \\ \sqrt{x + 3} & \text{for } x \ge 1 \end{cases}$$

—"not equal to"

$$g(x) = \begin{cases} \dfrac{1}{x} & \text{for } x \ne 0 \\ 0 & \text{for } x = 0 \end{cases}$$

13. Find $F(-2)$ and $F\left(\dfrac{1}{3}\right)$.

14. Find $F(2)$ and $F(3)$.

15. Find $h(-8)$ and $h\left(-\dfrac{1}{2}\right)$.

16. Find $h(-1)$ and $h(3)$.

17. Find $f(-2)$ and $f(2)$.

18. Find $f(1)$ and $f\left(-\dfrac{1}{4}\right)$.

19. Find $g(2)$ and $g(0)$.

20. Find $g(-2)$ and $g\left(\dfrac{1}{5}\right)$.

In Exercises 21 through 32, determine the appropriate functions.

21. A motorist travels at 40 mi/h for 2 h and then at 55 mi/h for t hours. Express the distance d traveled as a function of t.

22. Express the cost C of fencing a rectangular field 100 m long and w meters wide as a function of w if fencing cost $5 per meter.

23. A rocket burns up at the rate of 2 tons/min after falling out of orbit into the atmosphere. If the rocket weighed 5500 tons before reentry, express its weight w as a function of the time t, in minutes, of reentry.

24. A computer part costs $3 to produce and distribute. Express the profit p made by selling 100 of these parts as a function of the price of c dollars each.

25. Upon ascending, a weather balloon ices up at the rate of 0.5 kg/m after reaching an altitude of 1000 m. If the weight of the balloon below 1000 m is 110 kg, express its weight w as a function of its altitude h if $h > 1000$ m.

26. A chemist adds x liters of a solution which is 50% alcohol to 100 L of a solution which is 70% alcohol. Express the number n of liters of alcohol in the final solution as a function of x.

27. A company installs underground cable at a cost of $500 for the first 50 ft (or up to 50 ft) and $5 for each foot thereafter. Express the cost C as a function of the length ℓ of underground cable if $\ell > 50$ ft.

28. A state income tax rate is 4% of all taxable income over $12,000. If a person's taxable income I is greater than $12,000, express the tax T as a function of I.

29. The capacities (in L) of two oil storage tanks are x and y. The tanks are initially full; 1200 L of oil is removed from them by taking 10% of the contents of the first tank and 40% of the contents of the second tank. Express y as a function of x.

30. A manufacturer finds that it earns a profit of $2 on each box of computer disks and a profit of $4 on each box of computer paper. If x boxes of disks and y boxes of paper are produced, the profit is $5000. Express y as a function of x.

31. In studying the electric current which is induced in wire rotating through a magnetic field, a piece of wire 60 cm long is cut into two pieces. One of these is bent into a circle, and the other is bent into a square. Express the total area A of the two figures as a function of the perimeter p of the square.

32. The cross section of an air-conditioning duct is in the shape of a square with semicircles on each side. See Fig. 2-4. Express the area A of this cross section as a function of the diameter d (in cm) of the circular part.

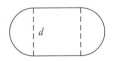

Fig. 2-4

In Exercises 33 through 36, solve the given problems.

33. The resonant frequency f (in Hz) in a certain electric circuit as a function of the capacitance C (in farads) in the circuit is $f = \dfrac{1}{2\pi\sqrt{C}}$. Describe the domain of this function. See Examples D and F.

34. A jet is traveling directly between cities A and B, which are 2400 km apart. If the jet is x km from city A and y km from city B, find the domain of $y = f(x)$.

35. Express the weight w of the weather balloon in Exercise 25 as a function of any height h in the same manner as the functions in Example E and Exercises 13 through 20 were expressed.

36. Express the cost C of installing any length ℓ of the underground cable in Exercise 27 in the same manner as the functions in Example E and Exercises 13 through 20 were expressed.

2-3 Rectangular Coordinates

One of the most valuable ways of representing functions is by graphical representation. By using graphs we are able to obtain a "picture" of the function, and by using this picture we can learn a great deal about the function.

To make a graphical representation, we recall that numbers can be represented by points on a line. For a function we have values of the independent variable and the corresponding values of the dependent variable. Therefore, it is necessary to use two lines to represent the values from these sets of numbers. We do this most conveniently by placing the lines perpendicular to each other.

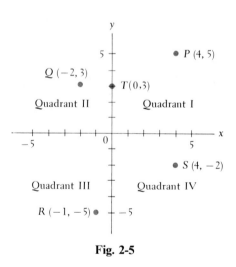

Fig. 2-5

We place one line horizontally and label it the x-axis. The numbers of the set for the independent variable are normally placed on this axis. *The other line we place vertically, and label the y-axis.* Normally the y-axis is used for values of the dependent variable. *The point of intersection is called the origin.* This is the **rectangular coordinate system.**

On the x-axis, positive values are to the right of the origin, and negative values are to the left of the origin. On the y-axis, positive values are above the origin, and negative values are below it. *The four parts into which the plane is divided are called* **quadrants,** which are numbered as in Fig. 2-5.

A point P on the plane is designated by the pair of numbers (x, y), where x is the value of the independent variable and y is the corresponding value of the dependent variable. *The x-value, called the* **abscissa,** *is the perpendicular distance of P from the y-axis. The y-value, called the* **ordinate,** *is the perpendicular distance of P from the x-axis.* The values x and y together, written as (x, y), are the **coordinates** of the point P.

EXAMPLE A

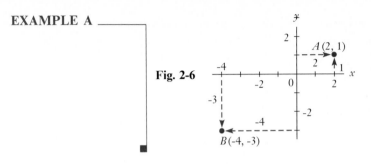

Fig. 2-6

Locate the points $A(2, 1)$ and $B(-4, -3)$ on the rectangular coordinate system.

The coordinates $(2, 1)$ for A mean that the point is 2 units to the *right* of the y-axis and 1 unit *above* the x-axis, as shown in Fig. 2-6. The coordinates $(-4, -3)$ for B mean that the point is 4 units to the *left* of the y-axis and 3 units *below* the x-axis, as shown.

EXAMPLE B

The positions of points $P(4, 5)$, $Q(-2, 3)$, $R(-1, -5)$, $S(4, -2)$, and $T(0, 3)$ are shown in Fig. 2-5, which is located on the previous page. We see that this representation allows for *one point for any pair of values* (x, y). Also, we note that the point $T(0, 3)$ is on the y-axis. Any such point which is on either axis is not *in* any of the quadrants.

EXAMPLE C

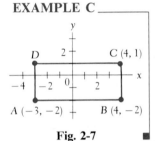

Fig. 2-7

Three vertices of the rectangle in Fig. 2-7 are $A(-3, -2)$, $B(4, -2)$, and $C(4, 1)$. What is the fourth vertex?

We use the fact that opposite sides of a rectangle are equal and parallel to find the solution. Since both vertices of the base AB of the rectangle have a y-coordinate of -2, the base is parallel to the x-axis. Therefore, the top of the rectangle must also be parallel to the x-axis. Thus, the vertices of the top must both have a y-coordinate of 1, since one of them has a y-coordinate of 1. In the same way the x-coordinates of the left side must both be -3. Therefore, the fourth vertex is $D(-3, 1)$.

EXAMPLE D

Where are all the points whose ordinates are 2?

Since the ordinate is the y-value, we can see that the question could be stated as: "Where are all points for which $y = 2$?" Since all such points are 2 units above the x-axis, the answer can be stated as "on a line 2 units above the x-axis." See Fig. 2-8.

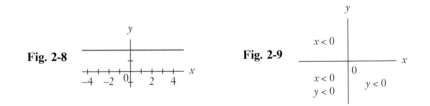

Fig. 2-8   **Fig. 2-9**

EXAMPLE E

Where are all points (x, y) for which $x < 0$ and $y < 0$?

Noting that $x < 0$ means "x is less than zero," or "x is negative," and that $y < 0$ means the same for y, we want to determine where both x and y are negative. Our answer is "in the third quadrant," since both coordinates are negative for all points in the third quadrant, and this is the only quadrant for which this is true. See Fig. 2-9.

Exercises 2-3

In Exercises 1 and 2, determine (at least approximately) the coordinates of the points specified in Fig. 2-10.

1. A, B, C **2.** D, E, F

In Exercises 3 and 4, plot (at least approximately) the given points.

3. $A(2, 7)$, $B(-1, -2)$, $C(-4, 2)$ **4.** $A(3, \frac{1}{2})$, $B(-6, 0)$, $C(-\frac{5}{2}, -5)$

In Exercises 5 through 8, plot the given points and then join these points, in the order given by straight-line segments. Name the geometric figure formed.

5. $A(-1, 4)$, $B(3, 4)$, $C(1, -2)$ **6.** $A(0, 3)$, $B(0, -1)$, $C(4, -1)$

7. $A(-2, -1)$, $B(3, -1)$, $C(3, 5)$, $D(-2, 5)$ **8.** $A(-5, -2)$, $B(4, -2)$, $C(6, 3)$, $D(-3, 3)$

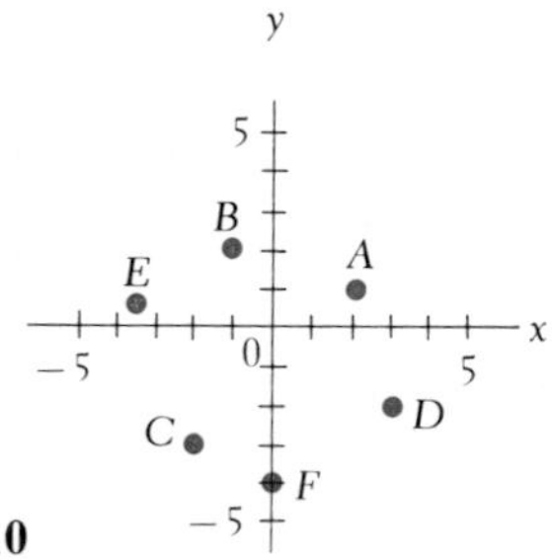

In Exercises 9 through 12, find the indicated coordinates.

Fig. 2-10

9. Three vertices of a rectangle are $(5, 2)$, $(-1, 2)$, and $(-1, 4)$. What are the coordinates of the fourth vertex?

10. Two vertices of an equilateral triangle are $(7, 1)$ and $(2, 1)$. What is the abscissa of the third vertex?

11. P is the point $(3, 2)$. Locate point Q such that the x-axis is the perpendicular bisector of the line segment joining P and Q.

12. P is the point $(-4, 1)$. Locate point Q such that the line segment joining P and Q is bisected by the origin.

In Exercises 13 through 28, answer the given questions.

13. Where are all the points whose abscissas are 1?   **14.** Where are all the points whose ordinates are -3?

15. Where are all points such that $y = 3$?   **16.** Where are all points such that $x = -2$?

17. Where are all the points whose abscissas equal their ordinates?

18. Where are all the points whose abscissas equal the negative of their ordinates?

19. What is the abscissa of all points on the y-axis?   **20.** What is the ordinate of all points on the x-axis?

21. Where are all the points for which $x > 0$?   **22.** Where are all the points for which $y < 0$?

23. Where are all the points for which $x < -1$?   **24.** Where are all the points for which $y > 4$?

25. Where are all points (x, y) for which $x = 0$ and $y < 0$?   **26.** Where are all points (x, y) for which $x < 0$ and $y > 1$?

27. In which quadrants is the ratio y/x positive?   **28.** In which quadrants is the ratio y/x negative?

2-4 The Graph of a Function

Now that we have introduced the concepts of a function and the rectangular coordinate system, we are in a position to determine the graph of a function. In this way we shall obtain a visual representation of a function.

The graph of a function is the set of all points whose coordinates (x, y) satisfy the functional relationship $y = f(x)$. Since $y = f(x)$, we can write the coordinates of the points on the graph as $[x, f(x)]$. Writing the coordinates in this manner tells us exactly how to find them. *We assume a certain value for x and then find the value of the function of x. These two numbers are the coordinates of the point.*

Since there is no limit to the possible number of points which can be chosen, we normally select a few values of x, obtain the corresponding values of the function, and plot these points. These points are then connected by a *smooth* curve (not short straight lines from one point to the next), and are connected from left to right. In this section we shall develop only the method of plotting points to obtain the graph of a function. We shall develop other methods for some functions in later chapters.

EXAMPLE A _____ Graph the function $f(x) = 3x - 5$.

For purposes of graphing, we let $y = f(x)$, or $y = 3x - 5$. We then let x take on various values and determine the corresponding values of y. Note that once we choose a given value of x, we have no choice about the corresponding y-value, as it is determined by evaluating the function. If $x = 0$, we find that $y = -5$. This means that the point $(0, -5)$ is on the graph of the function $3x - 5$. Choosing another value of x, for example, 1, we find that $y = -2$. This means that the point $(1, -2)$ is on the graph of the function $3x - 5$. Continuing to choose a few other values of x, we tabulate the results, as shown in Fig. 2-11. It is best to arrange the table so that the values of x increase; then there is no doubt how they are to be connected, for they are then connected in the order shown. Finally, we connect the points as shown in Fig. 2-11, and see that the graph of the function $3x - 5$ is a straight line.

Fig. 2-11

$f(x) = 3x - 5$

$f(-1) = 3(-1) - 5 = -8$

$f(0) = 3(0) - 5 = -5$

$f(1) = 3(1) - 5 = -2$

$f(2) = 3(2) - 5 = 1$

$f(3) = 3(3) - 5 = 4$

x	y
−1	−8
0	−5
1	−2
2	1
3	4

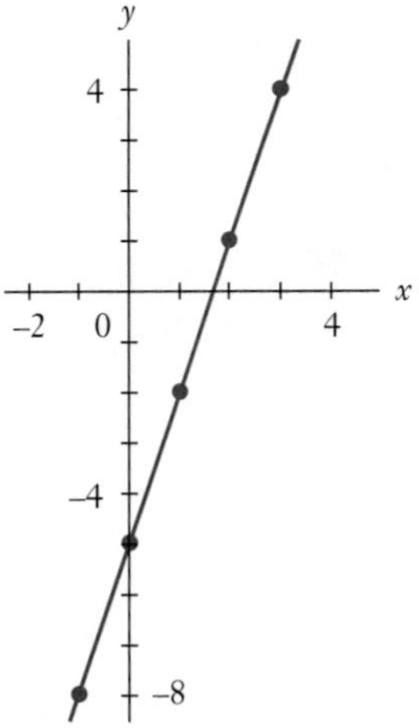

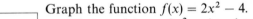

EXAMPLE B _____ Graph the function $f(x) = 2x^2 - 4$.

First, we let $y = 2x^2 - 4$ and tabulate the values as shown in Fig. 2-12. In determining the values in the table, we must take particular care to obtain the

NOTE ▷ correct values of y for negative values of x. **Mistakes are relatively common when dealing with negative numbers.** We must carefully use the laws for signed numbers. For example, if $x = -2$, we have $y = 2(-2)^2 - 4 = 2(4) - 4 = 8 - 4 = 4$. Once the values are obtained, we plot and connect the points with a smooth curve, as shown in Fig. 2-12.

Fig. 2-12

$f(x) = 2x^2 - 4$

$f(-2) = 2(-2)^2 - 4 = 4$

$f(-1) = 2(-1)^2 - 4 = -2$

$f(0) = 2(0)^2 - 4 = -4$

$f(1) = 2(1)^2 - 4 = -2$

$f(2) = 2(2)^2 - 4 = 4$

x	y
−2	4
−1	−2
0	−4
1	−2
2	4

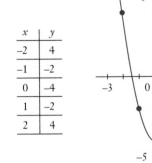

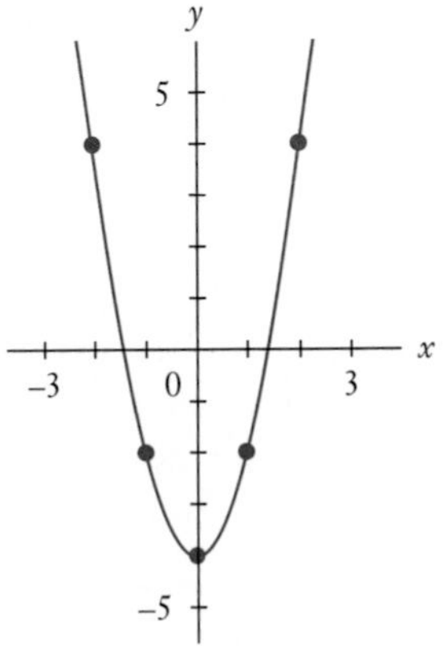

There are some special points which should be noted. Since the graphs of most common functions are smooth, any irregularities in the graph should be carefully checked. In these cases it usually helps to take values of x between those values where the question arises. Also, the domain of the function may not include

restrictions on a graph

all values of x (remember, *division by zero is not defined, and only real values of the variables are permissible*). Finally, in applications, we must be careful to plot values that are meaningful. As we have seen, negative values for quantities, such as time, are not generally meaningful. The following examples illustrate these points.

EXAMPLE C _____

NOTE ▷

Graph the function $y = x - x^2$.

First we determine the values in the table, as shown with Fig. 2-13. Again, we must be careful with negative values of x. For the value $x = -1$, we have $y = (-1) - (-1)^2 = -1 - (+1) = -1 - 1 = -2$. Once all the values have been found and plotted, we note that **$y = 0$ for both $x = 0$ and $x = 1$.** The question arises—***what happens between these values?*** Trying $x = \frac{1}{2}$, we find that $y = \frac{1}{4}$. Using this point completes the necessary information. Note that in plotting these graphs we do not stop the graph with the last point determined, but indicate that the curve continues.

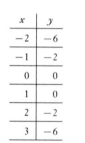

x	y
-2	-6
-1	-2
0	0
1	0
2	-2
3	-6

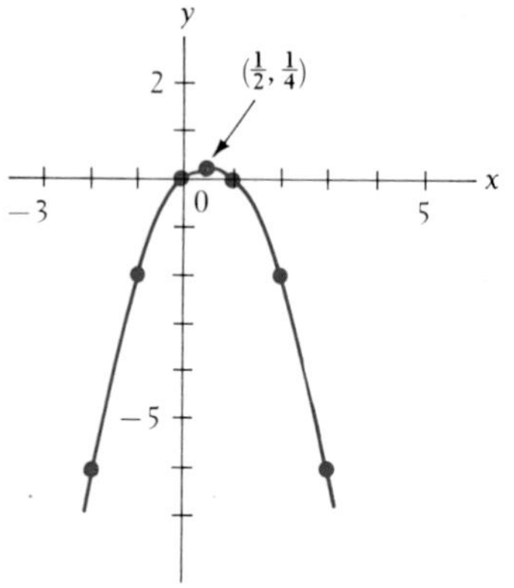

Fig. 2-13

x	y
-4	$3/4$
-3	$2/3$
-2	$1/2$
-1	0
$-1/2$	-1
$-1/3$	-2
$1/3$	4
$1/2$	3
1	2
2	$3/2$
3	$4/3$
4	$5/4$

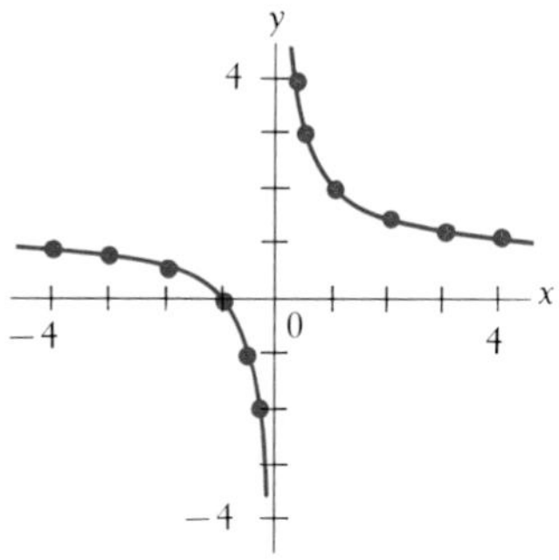

Fig. 2-14

EXAMPLE D _____

NOTE ▷

Graph the function $y = 1 + \dfrac{1}{x}$.

In finding the points on this graph, as shown in Fig. 2-14, we note that y is not defined for $x = 0$, due to division by zero. Thus, $x = 0$ is not in the domain and **we *must be careful not to have any part of the curve cross the y-axis* ($x = 0$).** Although we cannot let $x = 0$, we can choose other values of x between -1 and 1 which are close to zero. In doing so, we find that as x gets closer to zero, the points get closer and closer to the y-axis, although they do not reach or touch it. In this case the y-axis is called an **asymptote** of the curve.

EXAMPLE E

NOTE ▷

Graph the function $y = \sqrt{x + 1}$.

When finding the points for the graph, we may not let x take on any value less than -1, for *all such values would lead to imaginary values for y* and are not in the domain. Also, since we have the positive square root indicated, the range consists of all values of y which are positive or zero ($y \geq 0$). See Fig. 2-15. Note that the graph starts at $(-1, 0)$.

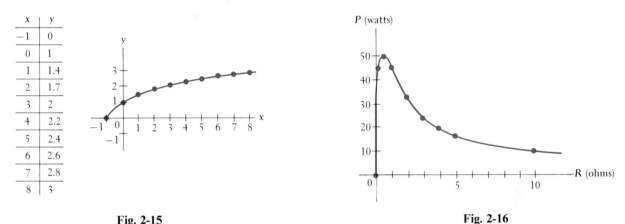

x	y
−1	0
0	1
1	1.4
2	1.7
3	2
4	2.2
5	2.4
6	2.6
7	2.8
8	3

Fig. 2-15

Fig. 2-16

EXAMPLE F

See the chapter introduction.

100 ⊗ .25 ÷
.75 x^2 =

    44.444444

The electric power delivered by a certain battery as a function of the resistance in the circuit is given by

$$P = \frac{100R}{(0.50 + R)^2}$$

where P is measured in watts and R in ohms. Plot the power as a function of the resistance.

Since a negative value for the resistance has no physical significance, we need not plot P for negative values of R. The following table of values is obtained. The calculator sequence for $R = 0.25 \, \Omega$, with the addition done *mentally*, is shown at the left.

R(ohms)	0	0.25	0.50	1.0	2.0	3.0	4.0	5.0	10.0
P(watts)	0.0	44.4	50.0	44.4	32.0	24.5	19.8	16.5	9.1

The values 0.25 and 0.50 are used for R when it is found that P is less for $R = 2$ than for $R = 1$. In this way a smoother curve is obtained (see Fig. 2-16).

Also note that the scale on the P-axis is different from that on the R-axis. This reflects the different magnitudes and ranges of values used for each of the variables. Different scales are normally used in such cases.

We can make various conclusions from the graph. For example, we see that the maximum power of 50 W occurs for $R = 0.5 \, \Omega$. Also, P continually decreases as R increases beyond $0.5 \, \Omega$. We will consider further the information which can be read from a graph in the next section.

The following example illustrates the graph of a function which is defined differently for different intervals of the domain.

EXAMPLE G ——— Graph the function $f(x) = \begin{cases} 2x + 1 & \text{for } x \le 1 \\ 6 - x^2 & \text{for } x > 1 \end{cases}$.

First we let $y = f(x)$ and then tabulate the necessary values, as shown in Fig. 2-17. In evaluating $f(x)$ we must be careful to use the proper part of the definition. To see where to start the curve for $x > 1$, we evaluate $6 - x^2$ for $x = 1$, but we

NOTE ▷ must realize that ***the curve does not include this point*** $(1, 5)$ and starts immediately to its right. To show that this point is not part of the curve, we draw it as an open ■ circle. Such a function is called *discontinuous* at $x = 1$.

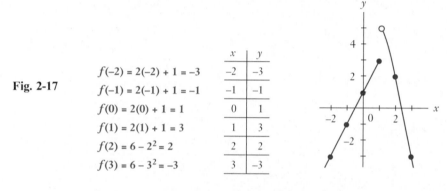

Fig. 2-17

	x	y
$f(-2) = 2(-2) + 1 = -3$	-2	-3
$f(-1) = 2(-1) + 1 = -1$	-1	-1
$f(0) = 2(0) + 1 = 1$	0	1
$f(1) = 2(1) + 1 = 3$	1	3
$f(2) = 6 - 2^2 = 2$	2	2
$f(3) = 6 - 3^2 = -3$	3	-3

In Section 2-1 we defined a function such that it has only one value of the dependent variable for each value of the independent variable. At the end of Section 2-2 we stated that a relation may have more than one such value of the dependent variable. The following example illustrates how to use the graph to determine whether or not a relation is a function.

EXAMPLE H ——— In Example I of Section 2-2 we noted that $y^2 = 4x^2$ is a relation, but not a function. Since $y = 4$ or $y = -4$, normally written as $y = \pm 4$, for $x = 2$, we have two values of y for $x = 2$. Therefore, it is not a function. Constructing the following table of values, we then draw the graph in Fig. 2-18.

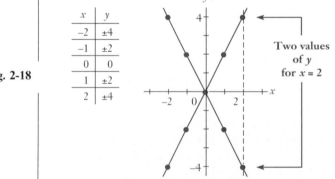

Fig. 2-18

x	y
-2	± 4
-1	± 2
0	0
1	± 2
2	± 4

Two values
of y
for $x = 2$

We note, in fact, that there are two values for all values of x except $x = 0$.

If any vertical line which crosses the x-axis in the domain intersects the graph in more than one point, it is the graph of a relation which is not a function. The dashed line in Fig. 2-18 shows that $y^2 = 4x^2$ is not a function. Any such vertical line in any of the previous graphs of this section would intersect the graph in only one point, which shows that they are graphs of functions.

In addition to computers, there are calculators that can be used to display graphs of functions. Such calculators have rectangular display windows capable of producing graphs as well as other data. We could use such a calculator to display the graph of $y = 1 + \dfrac{1}{x}$, and the result is shown in Fig. 2-19. It is essentially the same as the graph in Fig. 2-14 with the numbers and axis labels excluded.

Functions of a particular type have graphs which are of a specific form, and many of them have been named. We noted that the graph of the function in Example A is a **straight line.** A **parabola** is illustrated in Example B and in Example C. The graph of the function in Example D is a **hyperbola.** Other types of graphs are found in the exercises and in later chapters. The straight line is considered again in Chapter 4 and the parabola in Chapter 6. A more detailed analysis of several of these curves is found in Chapter 20. The use of graphs is extensive in mathematics and in nearly all areas of application. For example, in Section 2-6 we see how graphs can be used to solve equations. Many other uses of graphical methods appear in the later chapters.

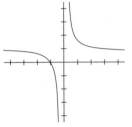

Fig. 2-19

Exercises 2-4

In Exercises 1 through 32, graph the given functions.

1. $y = 3x$

2. $y = -2x$

3. $y = 2x - 4$

4. $y = 3x + 5$

5. $y = 7 - 2x$

6. $y = 5 - 3x$

7. $y = \dfrac{1}{2}x - 2$

8. $y = 6 - \dfrac{1}{3}x$

9. $y = x^2$

10. $y = -2x^2$

11. $y = 3 - x^2$

12. $y = x^2 - 3$

13. $y = \dfrac{1}{2}x^2 + 2$

14. $y = 2x^2 + 1$

15. $y = x^2 + 2x$

16. $y = 2x - x^2$

17. $y = x^2 - 3x + 1$

18. $y = 2 + 3x + x^2$

19. $y = x^3$

20. $y = -2x^3$

21. $y = x^3 - x^2$

22. $y = 3x - x^3$

23. $y = x^4 - 4x^2$

24. $y = x^3 - x^4$

25. $y = \dfrac{1}{x}$

26. $y = \dfrac{2}{x + 2}$

27. $y = \dfrac{4}{x^2}$

28. $y = \dfrac{1}{x^2 + 1}$

29. $y = \sqrt{x}$

30. $y = \sqrt{4 - x}$

31. $y = \sqrt{16 - x^2}$

32. $y = \sqrt{x^2 - 16}$

See Appendix E for a computer program for plotting the graph of a function.

In Exercises 33 through 48, graph the indicated functions. In each case, plot the independent variable as the abscissa and the dependent variable as the ordinate.

33. For a certain model of truck, its resale value V, in dollars, as a function of its mileage m, is $V = 50{,}000 - 0.2m$. Plot V as a function of m for $m \leq 100{,}000$ mi.

34. The resistance R, in ohms, of a resistor as a function of the temperature T, in degrees Celsius, is given by $R = 250(1 + 0.0032T)$. Plot R as a function of T.

35. The consumption c of fuel, in liters per hour (L/h), of a certain engine is determined as a function of the number r, in revolutions per minute (r/min) of the engine to be $c = 0.011r + 4.0$. This formula is valid from 500 r/min to 3000 r/min. Plot c as a function of r.

36. The profit P, in dollars, a manufacturer makes in producing x units is given by $P = 3.5x - 120$. Plot P as a function of x for $x \leq 100$.

37. The maximum speed v, in miles per hour, at which a car can safely travel around a circular turn of radius r, in feet, is given by $r = 0.42v^2$. Plot r as a function of v.

38. The height h, in meters, of a rocket as a function of the time t, in seconds, is given by the function $h = 1500t - 4.9t^2$. Plot h as a function of t, assuming level terrain.

39. A formula used to determine the number N of board feet of lumber which can be cut from a 4-ft section of a log of diameter d, in inches, is $N = 0.22d^2 - 0.71d$. Plot N as a function of d for values of d from 10 in. to 40 in.

40. A copper electrode weighing 25.0 g is placed in a solution of copper sulfate. An electric current is passed through the solution and 1.6 g of copper is deposited on the electrode each hour. Express the total weight w of the electrode as a function of the time t and plot the graph.

41. The perimeter of a rectangular tarpaulin is 60 m. Express the area A of the tarpaulin as a function of its width w and plot the graph.

42. The distance p, in meters, from a camera with a 50-mm lens to the object being photographed is a function of the magnification m of the camera given by $p = \dfrac{0.05(1 + m)}{m}$. Plot the graph for positive values of m up to 0.50.

43. A measure of the light beam which can be passed through an optic fiber is its numerical aperture N. For a particular optic fiber, N is a function of the index of refraction n of the glass in the fiber, given by $N = \sqrt{n^2 - 1.69}$. Plot the graph for values of n up to 2.00.

44. The force F, in newtons, exerted by a cam on the arm of a robot, is a function of the distance x shown in Fig. 2-20. The function is $F = x^4 - 12x^3 + 46x^2 - 60x + 25$. Plot the graph. (Note that x varies from 1 cm to 5 cm.)

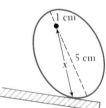

45. Plot the graphs of $y = x$ and $y = |x|$ on the same coordinate system. Note how the graphs differ.

46. Plot the graphs of $y = 2 - x$ and $y = |2 - x|$ on the same coordinate system. Note how the graphs differ.

Fig. 2-20

47. Plot the graph of $f(x) = \begin{cases} 3 - x & \text{for } x < 1 \\ x^2 + 1 & \text{for } x \geq 1 \end{cases}$.

48. Plot the graph of $f(x) = \begin{cases} \dfrac{1}{x - 1} & \text{for } x < 0 \\ \sqrt{x + 1} & \text{for } x \geq 0 \end{cases}$.

In Exercises 49 through 52, determine whether or not the indicated graph is that of a relation which is a function.

49. Fig. 2-21(a)
50. Fig. 2-21(b)
51. Fig. 2-21(c)
52. Fig. 2-21(d)

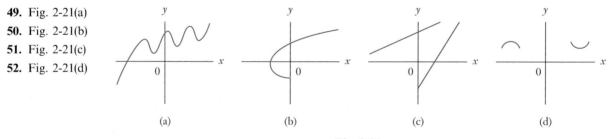

    (a)   (b)   (c)   (d)

Fig. 2-21

2-5 Graphs of Functions Defined by Tables of Data ▬▬▬▬

In the previous section, we showed how to construct the graph of a function, where the function was defined by a formula or equation. However, as we stated in Section 2-1, there are other ways of expressing functions. One of the most important ways of showing the relationship between variables is by use of a table of values obtained by observation or experimentation.

Often the data from an experiment indicate that the variables could have a formula which relates them, although the formula may not be known. Data from experiments from the various fields of science and technology generally have variables which are related in such a way. In this case, when we are plotting the graph, the points should be connected by a smooth curve.

Statistical data in tabular form often give values which are taken only for certain intervals or are averaged over specified intervals. In such cases, no real meaning can be given to the intervals between the points on the graph. On these graphs, the points should be connected by straight-line segments, where this is done only to make the points stand out better and make the graph easier to read. Example A illustrates this type of graph.

EXAMPLE A ———— The electric energy usage, in kilowatt-hours, for a particular all-electric house for each month of a certain year is given in the following table. Plot these data.

Month	Jan	Feb	Mar	Apr	May	Jun
Energy usage	2626	3090	2542	1875	1207	892

Month	July	Aug	Sep	Oct	Nov	Dec
Energy usage	637	722	825	1437	1825	2427

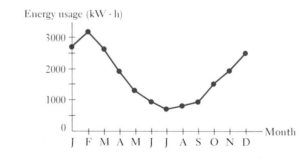

Fig. 2-22

We know that there is no meaning to the intervals between the months, since we have the total number of kilowatt-hours for each month. Therefore, we use straight-line segments, but only to make the points stand out better. ■

In the next example we illustrate data that give a graph in which the points are connected by a smooth curve and not by straight-line segments.

EXAMPLE B _____   Steam in a boiler was heated to 150°C. Its temperature was then recorded each
minute, giving the values in the following table. Plot the graph.

Time (minutes)	0.0	1.0	2.0	3.0	4.0	5.0
Temperature (°C)	150.0	142.8	138.5	135.2	132.7	130.8

Since the temperature changes in a continuous way, there is meaning to the
values in the intervals between points. Therefore these points are joined by a
smooth curve, as in Fig. 2-23. Also note that most of the vertical scale was used
for the required values, with the indicated break in the scale between 0 and 130. ■

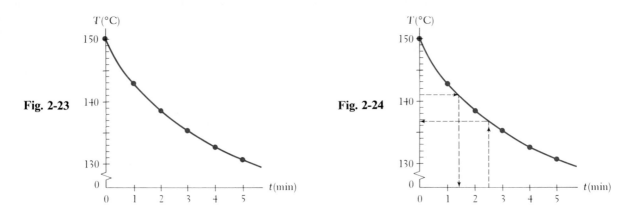

Fig. 2-23   **Fig. 2-24**

If the graph relating two variables is known, values can be obtained directly
from the graph. In finding such a value through the inspection of the graph, we
are *reading the graph.*

EXAMPLE C _____   For the cooling steam in Example B, we can estimate values of one variable for
given values of the other.

If we want to know the temperature after 2.5 min, we estimate 0.5 of the
interval between 2 and 3 on the t-axis and mark it. See Fig. 2-24. Then we draw
a line vertically from this point to the graph. From the point where it intersects
the graph, we draw a horizontal line to the T-axis. We now estimate that the
line crosses at $T = 136.7$°C. Obviously, the number of tenths is a rough estimate.

In the same way, if we want to determine how long the object took to cool
down to 141.0°, we go from 141.0 on the T-axis to the graph and then to the
t-axis. This crosses at about $t = 1.4$ min. ■

We can see from Example C that we can estimate values from a graph.
However, unless a very accurate graph is drawn with expanded scales for both
variables, only quite approximate values can be found. There is a method, using
the table of values itself, which generally gives reasonably accurate results. *The
method is called* **linear interpolation.**

linear interpolation   *Linear interpolation assumes that if a particular value of one variable lies between two of those listed in the table, then the corresponding value of the other
variable is at the same proportional distance between its listed values.* From the
point of view of the graph, linear interpolation assumes that the two points de-

fined in the table are connected by a straight line. Although this is generally not correct, it is a very good approximation if the values in the table are sufficiently close together.

EXAMPLE D —————— For the cooling steam in Example B, we can use interpolation to find the temperature of the object after 1.4 min. Since 1.4 min is $\frac{4}{10}$ of the way from 1.0 min to 2.0 min, we shall assume that the value of T which we want is $\frac{4}{10}$ of the way between 142.8 and 138.5, the values of T for 1.0 min and 2.0 min, respectively. The difference between these values is 4.3, and $\frac{4}{10}$ of 4.3 is 1.7 (rounded off to tenths). Subtracting (the values of T are decreasing) 1.7 from 142.8, we obtain 141.1. Thus, the required value of T is about 141.1°C. (Note that this agrees well with the result in Example C.)

Another method of indicating the interpolation is shown in Fig. 2-25. From the figure we see that

$$\frac{x}{4.3} = \frac{0.4}{1.0}$$

$$x = 1.7 \quad \text{(rounded off)}$$

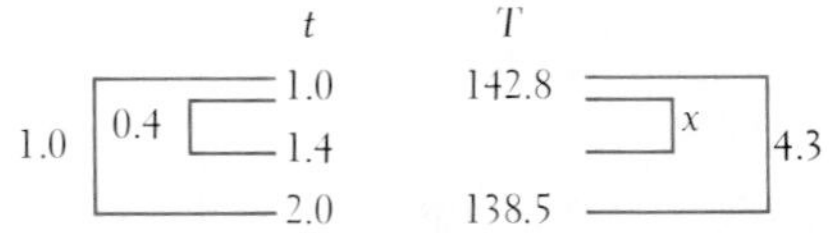

Therefore,

$$142.8 - 1.7 = 141.1°C$$

Fig. 2-25

is the required value of T. If the values of T had been increasing, we would have added 1.7 to the value of T for 1.0 min.

Exercises 2-5

In Exercises 1 through 8, represent the data graphically.

1. The diesel fuel production, in thousands of gallons, at a certain refinery during a 10-week period was as follows.

Week	1	2	3	4	5	6	7	8	9	10
Production	765	780	840	850	880	840	845	820	760	810

2. The number of isotopes of some of the heavier elements are in the following table.

Atomic number	82	84	86	88	90	92	94
Number of isotopes	16	19	12	11	12	14	11

These elements, by atomic number, are: 82, lead; 84, polonium; 86, radon; 88, radium; 90, thorium; 92, uranium; 94, plutonium.

3. The amount of material necessary to make a cylindrical gallon container depends on the diameter, as showm in the following table.

Diameter (in.)	3.0	4.0	5.0	6.0	7.0	8.0	9.0
Material (in.²)	322	256	224	211	209	216	230

4. An oil burner propels air which has been heated to 90°C. The temperature then drops as the distance from the burner increases, as shown in the following table.

Distance (m)	0.0	1.0	2.0	3.0	4.0	5.0	6.0
Temperature (°C)	90	84	76	66	54	46	41

5. A changing electric current in a coil of wire will induce a voltage in a nearby coil. Important in the design of transformers, the effect is called *mutual inductance*. For two coils, the mutual inductance, in henrys, as a function of the distance between them is given in the following table.

Distance (cm)	0.0	2.0	4.0	6.0	8.0	10.0	12.0
M. inductance (H)	0.77	0.75	0.61	0.49	0.38	0.25	0.17

6. The boiling point of water, in degrees Fahrenheit, was recorded at the atmospheric pressures, in kilopascals, as shown in the following table.

Pressure (kPa)	1.0	10	20	40	60	100
Temperature (°F)	58	115	138	168	192	212

7. The time required for a sum of money to double in value, when compounded annually, is given as a function of the interest rate in the following table.

Rate (%)	4	5	6	7	8	9	10	11	12
Time (years)	17.7	14.2	11.9	10.2	9.0	8.0	7.3	6.6	6.1

8. The force required to break a metal rod, as a function of its diameter, is given in the following table.

Diameter (cm)	1.0	1.5	2.0	2.5	3.0	3.5	4.0
Force (kg)	5.0	12	20	30	42	58	80

In Exercises 9 and 10, use the graph in Fig. 2-24, which relates the temperature of the cooling steam and time. Find the indicated values by reading the graph.

9. (a) For $t = 4.3$ min, find T. (b) For $T = 145.0°C$, find t.

10. (a) For $t = 1.8$ min, find T. (b) For $T = 133.5°C$, find t.

In Exercises 11 and 12, use the following table, which gives the voltage produced by a certain thermocouple as a function of the temperature of the thermocouple. Plot the graph. Find the indicated values by reading the graph.

Temperature (°C)	0	10	20	30	40	50
Voltage (volts)	0.0	2.9	5.9	9.0	12.3	15.8

11. (a) For $T = 26°C$, find V. (b) For $V = 13.5$ V, find T.

12. (a) For $T = 32°C$, find V. (b) For $V = 4.8$ V, find T.

In Exercises 13 through 16, find the indicated values by means of linear interpolation.

13. Using the table with Exercise 5, find the inductance for $d = 9.2$ cm.

14. Using the table with Exercise 6, find the temperature for $p = 34$ kPa.

15. Using the table with Exercise 7, find the rate for $t = 10$ years.

16. Using the table with Exercise 8, find the diameter for $F = 65$ kg.

In Exercises 17 through 20, use the following table, which gives the rate of discharge from a tank of water as a function of the height of water in the tank. For Exercises 17 and 18, plot the graph and find the values from the graph. For Exercises 19 and 20, find the indicated values by means of linear interpolation.

Height (ft)	0	1.0	2.0	4.0	6.0	8.0	12
Rate (ft³/s)	0	10	15	22	27	31	35

17. (a) For $R = 20$ ft^3/s, find H. (b) For $H = 2.5$ ft, find R.

18. (a) For $R = 34$ ft^3/s, find H. (b) For $H = 6.4$ ft, find R.

19. Find R for $H = 1.7$ ft.   **20.** Find H for $R = 25$ ft^3/s.

In Exercises 21 through 24, use the following table, which gives the fraction f (as a decimal) of the total heating load of a certain system that will be supplied by a solar collector of area A, in square meters. Find the indicated values by means of linear interpolation.

f	0.22	0.30	0.37	0.44	0.50	0.56	0.61
A (m²)	20	30	40	50	60	70	80

21. For $A = 36$ m², find f.   **22.** For $A = 52$ m², find f.

23. For $f = 0.59$, find A.   **24.** For $f = 0.27$, find A.

In Exercises 25 through 28, a method of finding values beyond those given is considered. By using a straight line segment to extend a graph beyond the last known point we can estimate values from the extension of the graph. The method is known as **linear extrapolation.** *Use this method to estimate the required values from the given graphs.*

25. Using Fig. 2-24, estimate T for $t = 5.3$ min.

26. Using the graph for Exercises 11 and 12, estimate V for $T = 55°C$.

27. Using the graph for Exercises 17 through 20, estimate R for $H = 13$ ft.

28. Using the graph for Exercises 17 through 20, estimate R for $H = 16$ ft.

2-6 Solving Equations Graphically

It is possible to solve equations by the use of graphs. This is particularly useful when algebraic methods cannot be applied conveniently to the equation or as a check of the algebraic solution. Before taking up graphical solutions, however, we shall briefly introduce the related concept of the zero of a function.

zeros of a function

Those values of the independent variable for which the function equals zero are known as the **zeros** *of the function. To find the zeros of a given function, we must set the function equal to zero and solve the resulting equation.* Using functional notation, we are looking for the solutions of $f(x) = 0$.

EXAMPLE A

Find any zeros of the function $f(x) = 3x - 9$.

To find the zeros we let $f(x) = 0$, which means that $3x - 9 = 0$. Thus we obtain $x = 3$, which means that 3 is a zero of the function $f(x) = 3x - 9$. ∎

Graphically, the zeros of a function are found where the curve crosses the *x*-axis. *These points are called the* **x-intercepts** *of the curve.* The function is zero at these points since the *x*-axis represents all points for which $y = 0$, or $f(x) = 0$.

EXAMPLE B

Graphically determine any zeros of the function $f(x) = x^2 - 2x - 1$.

First we set $y = x^2 - 2x - 1$ and then find the points for the graph. After plotting the graph in Fig. 2-26, we see that the curve crosses the *x*-axis at approximately $x = -0.4$ and $x = 2.4$ (estimating to the nearest tenth). Thus the zeros of this function are approximately -0.4 and 2.4. Checking these values in the function, we have $f(-0.4) = -0.04$ and $f(2.4) = -0.04$, which shows that -0.4 and 2.4 are very close to the exact values.

x	y
-2	7
-1	2
0	-1
1	-2
2	-1
3	2
4	7

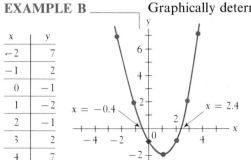

Fig. 2-26

EXAMPLE C _____ Graphically determine any zeros of the function $f(x) = \dfrac{6}{x^2 + 2}$.

Graphing the function $y = \dfrac{6}{x^2 + 2}$ in Fig. 2-27, we see that it does not cross the x-axis anywhere. Therefore, this function does not have any real zeros. We see, therefore, that not all functions have real zeros.

Examining the function, we note that x^2 is never negative, which means that $x^2 + 2$ is never less than 2. Thus, the range of $f(x)$ is positive and is 3 or less, which we can write as $0 < f(x) \leq 3$. This shows that the function cannot have any zeros.

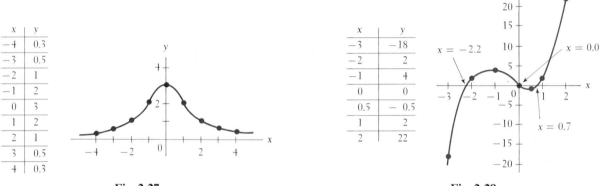

x	y
-4	0.3
-3	0.5
-2	1
-1	2
0	3
1	2
2	1
3	0.5
4	0.3

Fig. 2-27

x	y
-3	-18
-2	2
-1	4
0	0
0.5	-0.5
1	2
2	22

Fig. 2-28

We can now see how to solve equations graphically and how this is related to the zeros of a function. To solve an equation, we collect all terms on one side of the equal sign, giving the equation $f(x) = 0$. This equation is solved graphically by first setting $y = f(x)$ and graphing this function. We then find those values of x for which $y = 0$. This means that we are finding the zeros of this function. The following examples illustrate the method.

EXAMPLE D _____ Solve the equation $x^2(2x + 3) = 3x$ graphically.

We first collect all terms of the left side of the equal sign. This leads to the equation $2x^3 + 3x^2 - 3x = 0$. We then let $y = 2x^3 + 3x^2 - 3x$ and graph this function, as shown in Fig. 2-28. From the graph we see that $y = 0$ (which is equivalent to $2x^3 + 3x^2 - 3x = 0$) for the approximate values $x = -2.2$, $x = 0.0$, and $x = 0.7$. Therefore, these values are approximate solutions to the original equation. (Actually, $x = 0.0$ is exact, since we found this value when obtaining the table.) Checking these values in the original equation, we obtain $-6.776 = -6.6$ for $x = -2.2$, $0 = 0$ for $x = 0.0$, and $2.156 = 2.1$ for $x = 0.7$, which shows that the approximate solutions we obtained are very close to the exact solutions. We also note that we were able to find the solution $x = 0.7$ by being careful with values around $x = 0$, where we noticed that $f(x)$ was positive for both $x = -1$ and $x = 1$.

EXAMPLE E

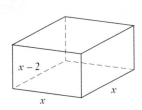

$x - 2$

x

Fig. 2-29

A rectangular box whose volume is 30 cm³ is made with a square base and a height which is 2 cm less than the length of a side of the base. Find the dimensions of the box by first setting up the necessary equation and then solving it graphically.

Let x = the length of a side of the base (see Fig. 2-29); the height is then $x - 2$. This means the volume is $(x)(x)(x - 2)$, or $x^2(x - 2)$. Since the volume is 30 cm³, we have the equation

$$x^2(x - 2) = 30$$

Now, solving the equation graphically, we write it as $x^3 - 2x^2 - 30 = 0$. Now we set $y = x^3 - 2x^2 - 30$ and graph this equation as shown in Fig. 2-30. Only positive values of x are used, since negative values of x have no meaning to the problem. From the graph we see that $x = 3.9$ is the approximate solution. Therefore, the approximate dimensions are 3.9 cm, 3.9 cm, and 1.9 cm. Checking, these dimensions give a volume of 29 cm³. The difference from the given value is due to the approximate value of x found from the graph.

Fig. 2-30

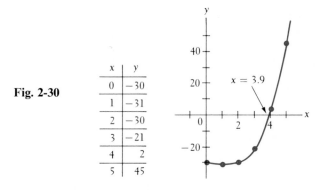

x	y
0	-30
1	-31
2	-30
3	-21
4	2
5	45

Exercises 2-6

In Exercises 1 through 4, find the zeros of the given functions algebraically, as in Example A.

1. $5x - 10$ **2.** $7x - 4$ **3.** $4x + (9 - 2x)$ **4.** $2 - 3(x - 5)$

In Exercises 5 through 12, find the zeros of the given functions graphically. Check the solutions in Exercises 5 through 8 algebraically. Check the solutions in Exercises 9 through 12 by substituting in the function.

5. $2x - 7$ **6.** $3x - 2$ **7.** $5x - (3 - x)$ **8.** $3 - 2(2x - 7)$

9. $x^2 + x$ **10.** $2x^2 - x$ **11.** $x^3 - x + 3$ **12.** $x^4 + 3x^2 - 7$

In Exercises 13 through 28, solve the given equations graphically.

13. $7x - 5 = 0$ **14.** $8x + 3 = 0$ **15.** $6x = 15$ **16.** $7x = -18$

17. $x^2 + x - 5 = 0$ **18.** $x^2 - 2x - 4 = 0$ **19.** $2x^2 - x = 7$ **20.** $x(x - 4) = 9$

21. $x^3 - 4x = 0$ **22.** $x^3 - 3x - 3 = 0$ **23.** $x^4 - 2x = 0$ **24.** $2x^5 - 5x = 0$

25. $\sqrt{2x + 2} = 3$ **26.** $\sqrt{x} + 3x = 7$ **27.** $\dfrac{1}{x^2 + 1} = 0$ **28.** $x - 2 = \dfrac{1}{x}$

In Exercises 29 through 36, solve the indicated equations graphically.

29. In an electric circuit, the current i, in amperes, as a function of the voltage v is given by $i = 0.01v - 0.06$. Find v for $i = 0$.

30. For tax purposes a corporation assumes that one of its computers depreciates in value according to the equation $V = 90{,}000 - 12{,}000t$, where V is the value, in dollars, of the computer after t years. According to this formula, when will the computer be fully depreciated (no value)?

31. A box is suspended by two ropes, as shown in Fig. 2-31. In analyzing the tensions in the supporting ropes, it was found that they satisfied the following equation: $0.60T_1 + 0.87T_2 = 10.0$. Find T_1 if $T_2 = 8.0$ N.

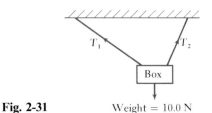

Fig. 2-31   Weight $= 10.0$ N

32. One industrial cleaner contains 30% of a certain solvent and another contains 10% of the solvent. To get a mixture containing 50 gal of the solvent, 120 gal of the first cleaner is used. How much of the second must be used?

33. The length of a rectangular solar panel is 12 cm more than its width. If its area is 520 cm², find its dimensions.

34. The height h, in feet, of a rocket as a function of the time t, in seconds, of flight is given by $h = 50 + 280t - 16t^2$. Determine when the rocket is at ground level.

35. In finding the illumination at a point x feet from one of two light sources which are 100 ft apart, it is necessary to solve the equation $9x^3 - 2400x^2 + 240{,}000x - 8{,}000{,}000 = 0$. Find x.

36. A rectangular storage bin is to be made from a rectangular piece of sheet metal 12 in. by 10 in., by cutting out equal corners of side x and bending up the sides. See Fig. 2-32. Find x if the storage bin is to hold 90 in.³.

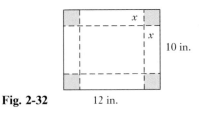

Fig. 2-32   12 in.

2-7   Review Exercises and Practice Test

Review Exercises

In Exercises 1 through 4, determine the appropriate functions.

1. Find the volume of a right circular cylinder of height 8 ft as a function of the radius of the base.

2. Find the surface area A of a cube as a function of one of the edges e.

3. One computer printer prints at the rate of 2000 lines/min for x minutes, and a second printer prints at the rate of 1800 lines/min for y minutes. Together they print 50,000 lines. Find y as a function of x.

4. Fencing around a rectangular storage depot area costs twice as much along the front as along the other three sides. The back costs $10 per foot. Express the cost C of the fencing as a function of the width if the length (along the front) is 20 ft more than the width.

In Exercises 5 through 12, evaluate the given functions.

5. Given $f(x) = 7x - 5$, find $f(3)$ and $f(-6)$.

6. Given $g(x) = 8 - 3x$, find $g\left(\dfrac{1}{6}\right)$ and $g(-4)$.

7. Given $F(u) = 3u - 2u^2$, find $F(-1)$ and $F(-0.3)$.

8. Given $h(y) = 2y^2 - y - 2$, find $h(5)$ and $h(-3)$.

9. Given $H(h) = \sqrt{1 - 2h}$, find $H(-4)$ and $H(2h)$.

10. Given $\phi(v) = \dfrac{3v - 2}{v + 1}$, find $\phi(-2)$ and $\phi(v + 1)$.

11. Given $f(x) = 3x^2 - 2x + 4$, find $f(x + h) - f(x)$.

12. Given $F(x) = x^3 + 2x^2 - 3x$, find $F(3 + h) - F(3)$.

In Exercises 13 through 16, evaluate the given functions by use of a calculator. Round off results to the accuracy of the value of the independent variable used.

13. Given $f(x) = 8.07 - 2x$, find $f(5.87)$ and $f(-4.29)$.

14. Given $g(x) = 7x - x^2$, find $g(45.81)$ and $g(-21.85)$.

15. Given $G(y) = \dfrac{y - 0.087629}{3y}$, find $G(0.17427)$ and $G(0.053206)$.

16. Given $h(t) = \dfrac{t^2 - 4t}{t^3 + 564}$, find $h(8.91)$ and $h(-4.91)$.

In Exercises 17 through 20, determine the domain and range of the given functions.

17. $f(x) = x^4 + 1$ **18.** $G(z) = \dfrac{4}{z^3}$ **19.** $g(t) = \dfrac{2}{\sqrt{t+4}}$ **20.** $F(y) = 1 - 2\sqrt{y}$

In Exercises 21 through 32, graph the given functions.

21. $y = 4x + 2$ **22.** $y = 5x - 10$ **23.** $y = 4x - x^2$ **24.** $y = x^2 - 8x - 5$

25. $y = 3 - x - 2x^2$ **26.** $y = 6 + 4x + x^2$ **27.** $y = x^3 - 6x$ **28.** $y = 3 - x^3$

29. $y = 2 - x^4$ **30.** $y = x^4 - 4x$ **31.** $y = \dfrac{x}{x+1}$ **32.** $y = \sqrt{25 - x^2}$

In Exercises 33 through 44, find any real zeros of the indicated functions by examining their graphs. Estimate the answer to the nearest tenth where necessary. Use the functions from Exercises 21 through 32, as indicated.

33. Exercise 21 **34.** Exercise 22 **35.** Exercise 23 **36.** Exercise 24

37. Exercise 25 **38.** Exercise 26 **39.** Exercise 27 **40.** Exercise 28

41. Exercise 29 **42.** Exercise 30 **43.** Exercise 31 **44.** Exercise 32

In Exercises 45 through 52, solve the given equations graphically.

45. $7x - 3 = 0$ **46.** $3x + 11 = 0$ **47.** $x^2 + 1 = 6x$ **48.** $3x - 2 = x^2$

49. $x^3 - x^2 = 2 - x$ **50.** $5 - x^3 = 2x^2$ **51.** $\dfrac{1}{x} = 2x$ **52.** $\sqrt{x} = 2x - 1$

In Exercises 53 through 56, answer the given questions.

53. Explain how $A(a, b)$ and $B(b, a)$ may be in different quadrants.

54. The points $(1, 2)$ and $(1, -3)$ are two adjacent vertices of a square. Find the other vertices.

55. An equation used in electronics with a transformer antenna is $I = 12.5\sqrt{1 + 0.5m^2}$. For $I = f(m)$, find $f(0.55)$.

56. The percent p of wood lost in cutting it into boards 1.5 in. thick due to the thickness t, in inches, of the saw blade is $p = \dfrac{100t}{t + 1.5}$. Find p if $t = 0.4$ in. That is, since $p = f(t)$, find $f(0.4)$.

In Exercises 57 through 68, plot the graphs of the indicated functions.

57. In determining the forces which act on a certain structure, it was found that two of them satisfied the equation $F_2 = 0.75F_1 + 45$, where the forces are in newtons. Plot the graph of F_2 as a function of F_1.

58. A company which makes digital clocks determines that in order to sell x clocks, the price must be $p = 50 - 0.25x$. Plot the graph of p as a function of x, for $x = 1$ to $x = 60$. (Although only the points for integral values of x have real meaning, join the points.)

59. For a given temperature, five times the Fahrenheit reading less nine times the Celsius reading is 160. Plot the graph of $C = f(F)$.

60. There are 5000 L of oil in a tank which has a capacity of 100,000 L. It is filled at the rate of 7000 L/h. Determine the function relating the number of liters N and the time t while the tank is being filled. Plot N as a function of t.

61. For a certain laser device, the laser output power P, in milliwatts, is negligible if the drive current i, in milliamperes, is less than 80 mA. From 80 mA to 140 mA, $P = 1.5 \times 10^{-6}i^3 - 0.77$. Plot the graph of $P = f(i)$.

62. It is determined that a good approximation for the cost C, in cents per mile, of operating a certain car at a constant speed v, in miles per hour, is given by $C = 0.025v^2 - 1.4v + 35$. Plot C as a function of v for $v = 10$ mi/h to $v = 60$ mi/h.

63. A medical researcher exposed a virus culture to an experimental vaccine. It was observed that the number of live cells N in the culture as a function of the time t, in hours, after exposure was given by $N = \dfrac{1000}{\sqrt{t+1}}$. Plot the graph of $N = f(t)$.

64. The electric field E, in volts per meter, from a certain electric charge is given by the function $E = 25/r^2$, where r is the distance, in meters, from the charge. Plot the graph of $E = f(r)$ for values of r up to 10 cm.

65. To draw the approximate shape of an irregular shoreline, a surveyor measured the distances from a straight wall to the shoreline at 20-ft intervals along the wall. The measurements are in the following table. Plot the graph of the distance d to the shoreline as a function of the distance D along the wall.

D (ft)	0	20	40	60	80	100	120	140	160
d (ft)	15	32	56	33	29	47	68	31	52

66. The percent of a computer network which is in use during a particular loading cycle as a function of the time, in seconds, is given in the following table. Plot the graph of the function.

Time (s)	0.0	0.2	0.4	0.6	0.8	1.0	1.2	1.4	1.6
Percent (%)	0	45	85	90	85	85	60	10	0

67. In an experiment measuring the pressure p, in kilopascals, at a given depth d, in meters, of seawater, the results in the following table were found. Plot the graph of $p = f(d)$, and from the graph determine $f(10)$.

Depth (m)	0.0	3.0	6.0	9.0	12	15
Pressure (kPa)	101	131	161	193	225	256

68. The vertical sag s, in feet, at the middle of an 800-ft power line, as a function of the temperature T, in degrees Fahrenheit, is given in the following table. For $s = f(T)$, find $f(47)$ by linear interpolation.

T (°F)	0	20	40	60
S (ft)	10.2	10.6	11.1	11.7

In Exercises 69 through 72, solve the indicated equations graphically.

69. The solubility s, in kilograms per cubic meter of water, of a certain type of fertilizer is given by $s = 135 + 4.9T + 0.19T^2$, where T is the temperature in degrees Celsius. Find T for $s = 500$ kg/m^3.

70. The rectangular cross section of a beam has a perimeter of 42 in. Express the cross-sectional area A as a function of the width w of the beam. Then find w for $A = 96$ in.^{2}.

71. A computer, using data from a refrigeration plant, estimates that in the event of a power failure, the temperature, in degrees Celsius, in the freezers would be given by $T = \dfrac{4t^2}{t+2} - 20$, where t is the number of hours after the power failure. How long would it take for the temperature to reach 0°C?

72. Two electrical resistors in parallel have a combined resistance R_T, in ohms, given by $R_T = \dfrac{R_1 R_2}{R_1 + R_2}$. If $R_2 = R_1 + 2.0$, express R_T as a function of R_1 and find R_1 if $R_T = 6.0$ Ω.

Practice Test

1. Given $f(x) = 2x - x^2 + \dfrac{8}{x}$, find $f(-4)$ and $f(2.385)$.

2. A rocket weighs 2000 Mg at lift-off. If the first-stage engines burn fuel at the rate of 10 Mg/s, find the weight w of the rocket as a function of the time t while the first-stage engines operate.

3. Plot the graph of the function $f(x) = 4 - 2x$.

4. Find the zeros of the function $f(x) = 2x^2 - 3x - 3$ graphically.

5. Plot the graph of the function $y = \sqrt{4 + 2x}$.

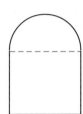

6. Locate all points (x, y) for which $x < 0$ and $y = 0$.

7. A window has the shape of a semicircle over a square, as shown in Fig. 2-33. Express the area of the window as a function of the radius of the circular part.

8. Find the domain and the range of the function $f(x) = \sqrt{6 - x}$.

Fig. 2-33

9. The voltage V and current i, in milliamperes, for a certain electrical experiment were measured as shown in the following table. Plot the graph of $i = f(V)$, and from the graph find $f(45.0)$.

Voltage (V)	10.0	20.0	30.0	40.0	50.0	60.0
Current (mA)	145	188	220	255	285	315

10. From the table in Problem 9, find the voltage for $i = 200$ mA.

3 The Trigonometric Functions

Many applied problems in science and technology require the use of triangles, especially right triangles, for their solution. Included among these are problems in air navigation, surveying, the motion of rockets and missiles, and optics. Problems involving forces acting on objects and the measurement of distances between various parts of the universe can also be solved. Even certain types of electric circuits are analyzed by the use of triangles.

In **trigonometry** we develop methods for measuring sides and angles of triangles as well as for solving related applied problems. Because of the extensive use of these concepts, trigonometry is considered one of the most practical and relevant branches of mathematics.

The angle through which STOP is seen by the driver must be large enough for easy vision. In Section 3-5 we show how this angle may be measured.

In this chapter we shall introduce the basic trigonometric functions and some of their applications. Later chapters will consider additional topics in trigonometry.

3-1 Angles

*An **angle** is defined as being **generated** by rotating a **half-line**, or **ray**, about its endpoint from an initial position to a terminal position.* A half-line is that portion of a line to one side of a fixed point on the line. *We call the initial position of the half-line the **initial side**, and the terminal position of the half-line the **terminal side**. The fixed point is the **vertex**.* The angle itself is the amount of rotation from the initial side to the terminal side.

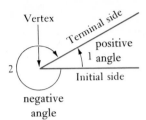

Vertex

negative
angle

Fig. 3-1

*If the rotation of the terminal side from the initial side is counterclockwise, the angle is said to be **positive**. If the rotation is clockwise the angle is **negative**.* In Fig. 3-1, angle 1 is positive and angle 2 is negative.

There are many symbols used to designate angles. Among the most widely used are certain Greek letters such as θ (theta), ϕ (phi), α (alpha), and β (beta). Capital letters representing the vertex (e.g. $\angle A$ or simply A) and other literal symbols, such as x and y, are also used commonly.

Two measurements of angles are widely used. These are the **degree** and the **radian.** In our work in this chapter, we will use degrees. However, since degrees and radians are both used on calculators (and computers), we will briefly show the relationship between them in this section. In Chapter 7 we will define and discuss the use of radians.

A *degree is defined as* $\frac{1}{360}$ *of one complete rotation.* The symbol $^\circ$ is used to designate degrees.

EXAMPLE A

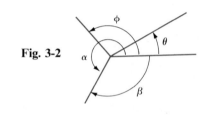

Fig. 3-2

The angles $\theta = 30°$, $\phi = 140°$, $\alpha = 240°$, and $\beta = -120°$ are shown in Fig. 3-2. Note again that β is drawn in a clockwise direction to show that it is negative.

One method of subdividing a degree is to use decimal parts of a degree. Most calculators use degrees in decimal form. Another traditional way is to divide a degree into 60 equal parts called **minutes;** each minute is divided into 60 equal parts called **seconds.** The symbols $'$ and $''$ are used to designate minutes and seconds, respectively.

In Fig. 3-2 we note that angles α and β have the same initial and terminal sides. *Such angles are called* **coterminal angles.** An understanding of coterminal angles is important in certain concepts in trigonometry.

EXAMPLE B

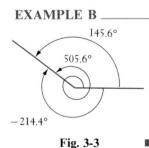

Fig. 3-3

Determine the values of two angles which are coterminal with an angle of 145.6°.

Since there are 360° in a complete rotation, we can find one coterminal angle by considering the angle which is 360° larger than the given angle. This gives us an angle of 505.6°. Another method of finding a coterminal angle is to subtract 145.6° from 360°, and then consider the resulting angle to be negative. This means that the original angle and the derived angle would make up one complete rotation, when put together. This method leads us to the angle of $-214.4°$ (see Fig. 3-3). These methods could be employed repeatedly to find other coterminal angles.

Although we will only use degrees in this chapter, when using a computer it may be necessary to change an angle expressed in radians to one expressed in degrees. Therefore, at this point we briefly note that this can be done on a calculator by use of the [DRG] key (see Appendix D). Also, we can use the fact that 1 radian is approximately 57.30°, or 1 rad = 57.30°.

EXAMPLE C

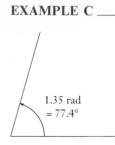

1.35 rad
= 77.4°

Fig. 3-4

Express 1.35 rad in degrees.

In order to do this we multiply 1.35 by 57.30. This gives 77.4. Thus,

1.35 rad = 77.4°

This angle is shown in Fig. 3-4. We must carefully note that degrees and radians ■ are simply two different ways of measuring an angle.

As we stated previously, the radian is defined and discussed in detail in Chapter 7. It might also be noted here that the "G" on the DRG key of a calculator represents *grads*, where 100 grad = 90°. We will not use grads in this text.

Traditionally it has been very common to use degrees and minutes in tables, and most calculators use degrees and decimal parts. Therefore, it may also be necessary to change from one form to the other. On many calculators this can be done directly by use of the DMS (or °′″) key (see Appendix D). It can also be done by using the meanings of minutes and seconds, as shown in the following examples.

EXAMPLE D

43 ⊞ 24 ⊟ 60 ⊨

43.4

217°53′ = 217.9°

Fig. 3-5

To change the angle 43°24′ to decimal form, we use the fact that $1' = (\frac{1}{60})°$. Therefore, $24' = (\frac{24}{60})° = 0.4°$. Thus,

43°24′ = 43.4°

The calculator sequence for this is shown at the left.

Also, 217°53′ is changed to decimal form as follows:

$$53' = \left(\frac{53}{60}\right)° = 0.88°$$

Thus, 217°53′ = 217.88° (to the nearest hundredth), or 217°53′ = 217.9° (to the ■ nearest tenth). This angle is shown in Fig. 3-5.

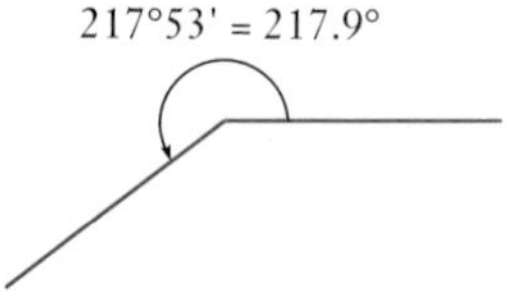

EXAMPLE E

154.36° = 154°22′

Fig. 3-6

To change 154.36° to an angle measured to the nearest minute, we use the fact that 1° = 60′. Therefore,

0.36° = 0.36(60′) = 21.6′

■ To the nearest minute, we have 154.36° = 154°22′. See Fig. 3-6.

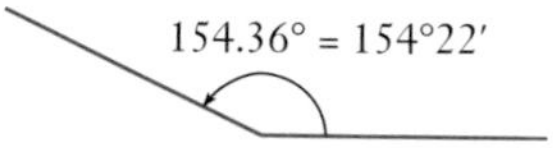

standard position of angle

If the initial side of the angle is the positive x-axis and the vertex is at the origin, the angle is said to be in **standard position.** *The angle is then determined by the position of the terminal side. If the terminal side is in the first quadrant, the angle is called a *first-quadrant angle*. Similar terms are used when the terminal side is in the other quadrants. If the terminal side coincides with one of the axes, the angle is a* **quadrantal angle.** *When an angle is in standard position, the terminal side can be determined if we know any point, other than the origin, on the terminal side.*

EXAMPLE F

To draw a third-quadrant angle of 205°, we simply measure 205° from the positive ↓ x-axis and draw the terminal side. See angle α in Fig. 3-7.

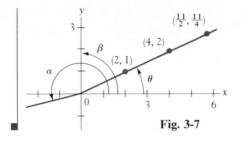

Fig. 3-7

Also in Fig. 3-7, θ is in standard position and the terminal side is uniquely determined by knowing that it passes through the point (2, 1). The same terminal side passes through (4, 2) and $(\frac{11}{2}, \frac{11}{4})$, among other points. Knowing that the terminal side passes through any one of these points makes it possible to determine the terminal side.

Angle β in Fig. 3-7 is a quadrantal angle since its terminal side is the positive y-axis.

Exercises 3-1

In Exercises 1 through 4, draw the given angles.

1. $60°$, $120°$, $-90°$ **2.** $330°$, $-150°$, $450°$ **3.** $50°$, $-360°$, $-30°$ **4.** $45°$, $225°$, $-250°$

In Exercises 5 through 12, determine one positive and one negative coterminal angle for each angle given.

5. $45°$ **6.** $73°$ **7.** $-150°$ **8.** $162°$

9. $70°30'$ **10.** $153°47'$ **11.** $278.1°$ **12.** $-197.6°$

In Exercises 13 through 16, change the given angles in radians to equal angles expressed in degrees. Round off results.

13. 0.265 rad **14.** 0.838 rad **15.** 1.447 rad **16.** 3.642 rad

In Exercises 17 through 24, change the given angles to equal angles expressed in decimal form. In Exercises 21 through 24, round off results to hundredths.

17. $15°12'$ **18.** $246°48'$ **19.** $86°3'$ **20.** $157°39'$

21. $301°16'$ **22.** $-4°47'$ **23.** $-96°8'$ **24.** $38°28'$

In Exercises 25 through 32, change the given angles to equal angles expressed to the nearest minute.

25. $47.5°$ **26.** $315.8°$ **27.** $19.75°$ **28.** $-84.55°$

29. $-5.62°$ **30.** $238.21°$ **31.** $24.92°$ **32.** $142.87°$

In Exercises 33 through 40, draw angles in standard position such that the terminal side passes through the given point.

33. (4, 2) **34.** $(-3, 8)$ **35.** $(-3, -5)$ **36.** $(6, -1)$

37. $(-7, 5)$ **38.** $(-4, -2)$ **39.** $(2, -5)$ **40.** (1, 6)

In Exercises 41 and 42, change the given angles to equal angles expressed in decimal form to the nearest thousandth. In Exercises 43 and 44, change the given angles to equal angles expressed to the nearest second. Use a calculator.

41. $21°42'36''$ **42.** $7°16'23''$ **43.** $86.274°$ **44.** $57.019°$

3-2 Defining the Trigonometric Functions

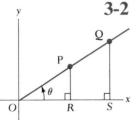

Fig. 3-8

Let us place an angle θ in standard position and drop perpendicular lines from points on the terminal side to the x-axis, as shown in Fig. 3-8. In doing this we set up similar triangles, each with one vertex at the origin and one side along the x-axis. Using the basic fact from geometry that *corresponding sides of similar triangles are proportional,* we may set up equal ratios of corresponding sides.

EXAMPLE A

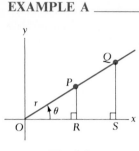

Fig. 3-8

In Fig. 3-8 we can see that triangles ORP and OSQ are similar triangles (they have exactly the same shape), since their corresponding angles are equal (each has the same angle at O, a right angle, and equal angles at P and Q). This means that the ratios of the lengths of corresponding sides are equal.

For example, sides RP and SQ are corresponding sides, and sides OR and OS are corresponding sides. The ratio of RP to OR, which is $\dfrac{RP}{OR}$, must equal the ratio of SQ to OS, which is $\dfrac{SQ}{OS}$. This means that $\dfrac{RP}{OR} = \dfrac{SQ}{OS}$.

This proportion holds for any positions (except at O) which P or Q may have on the terminal side of θ. (Here we have used the terms *ratio* and *proportion*. A detailed discussion of these terms is found in Chapter 17.)

There are three important distances associated with a given point on the terminal side of an angle. *They are the* **abscissa** *(x-coordinate), the* **ordinate** *(y-coordinate), and the* **radius vector** *(the distance from the origin to the point).* The length of the radius vector is denoted by r.

EXAMPLE B

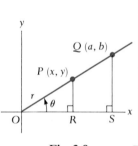

Fig. 3-9

In Fig. 3-9 we show the same angle θ as in Fig. 3-8, except that we have given coordinates to points P and Q and have shown the distance r. The value of r would be the distance OP for point P, or the distance OQ for point Q.

In Example A we showed that triangles ORP and OSQ are similar. We also gave the proportion which showed the equality of the ratios of two pairs of corresponding sides. Since RP is the ordinate of P, OR is the abscissa of P, SQ is the ordinate of Q, and OS is the abscissa of Q,

$$\frac{RP}{OR} = \frac{SQ}{OS} \qquad \text{is the same as} \qquad \frac{y}{x} = \frac{b}{a}$$

For any position (except at O) of Q on the terminal side of θ, the ratio of its ordinate to its abscissa will be equal to y/x.

For any angle θ in standard position, there are six different ratios which may be set up. Because of the property of similar triangles, each of these ratios is the same, regardless of the point on the terminal side which is chosen. For a different angle, with a different terminal side, the ratios have different values. Thus, the ratios depend on the position of the terminal side, which means that *the ratios depend on the size of the angle, and there is only one value of each ratio for a given angle*. Recalling the meaning of a function, we see that *the ratios are functions of the* **angle.** *These functions are called the* **trigonometric functions** and are defined as follows (see Fig. 3-10):

definition of
trigonometric
functions

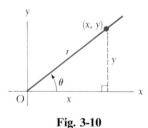

Fig. 3-10

$$
\begin{array}{cc}
\text{sine } \theta = \dfrac{y}{r} & \text{cosine } \theta = \dfrac{x}{r} \\[2mm]
\text{tangent } \theta = \dfrac{y}{x} & \text{cotangent } \theta = \dfrac{x}{y} \\[2mm]
\text{secant } \theta = \dfrac{r}{x} & \text{cosecant } \theta = \dfrac{r}{y}
\end{array}
$$

(3-1)

In this chapter we shall restrict our attention to the trigonometric functions of acute angles (angles between $0°$ and $90°$). However, it should be emphasized that the definitions in Eqs. (3-1) are general and may be used for angles of any magnitude. Discussion of the trigonometric functions of angles in general, along with additional important properties, is found in Chapters 7 and 19.

For convenience, the names of the various functions are usually abbreviated to

$$\sin \theta, \quad \cos \theta, \quad \tan \theta, \quad \cot \theta, \quad \sec \theta, \quad \text{and} \quad \csc \theta$$

Note that a given function is not defined if the denominator is zero. The denominator is zero in $\tan \theta$ and $\sec \theta$ for $x = 0$, and in $\cot \theta$ and $\csc \theta$ for $y = 0$. In all cases we will assume that $r > 0$. If $r = 0$ there would be no terminal side and therefore no angle. These restrictions affect the domains of these functions. We will discuss the domains and ranges of the trigonometric functions in Chapter 9, when we consider the graphs of these functions.

When finding values of the trigonometric functions, it is very often necessary to use the **Pythagorean theorem.** From geometry we recall that the Pythagorean theorem states that

in a right triangle, the square of the length of the hypotenuse equals the sum of the squares of the lengths of the other two sides.

For the right triangle in Fig. 3-11, with hypotenuse c and legs a and b, we therefore see that

$$c^2 = a^2 + b^2 \tag{3-2}$$

Fig. 3-11

As shown, *the **hypotenuse** is the side opposite the right angle.*

EXAMPLE C

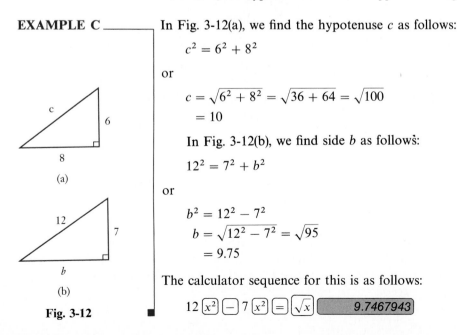

(a)

(b)

Fig. 3-12

In Fig. 3-12(a), we find the hypotenuse c as follows:

$$c^2 = 6^2 + 8^2$$

or

$$c = \sqrt{6^2 + 8^2} = \sqrt{36 + 64} = \sqrt{100}$$
$$= 10$$

In Fig. 3-12(b), we find side b as follows:

$$12^2 = 7^2 + b^2$$

or

$$b^2 = 12^2 - 7^2$$
$$b = \sqrt{12^2 - 7^2} = \sqrt{95}$$
$$= 9.75$$

The calculator sequence for this is as follows:

$12\ \boxed{x^2}\ \boxed{-}\ 7\ \boxed{x^2}\ \boxed{=}\ \boxed{\sqrt{x}}\ \boxed{9.7467943}$

EXAMPLE D

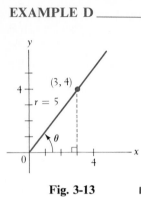

Fig. 3-13

Determine the trigonometric functions of the angle θ with a terminal side passing through the point (3, 4).

By placing the angle in standard position, as shown in Fig. 3-13, and drawing the terminal side through (3, 4), we find by use of the Pythagorean theorem that

$$r = \sqrt{3^2 + 4^2} = \sqrt{25} = 5$$

Using the values $x = 3$, $y = 4$, and $r = 5$, we find that

$$\sin \theta = \frac{4}{5} \qquad \cos \theta = \frac{3}{5} \qquad \tan \theta = \frac{4}{3}$$

$$\cot \theta = \frac{3}{4} \qquad \sec \theta = \frac{5}{3} \qquad \csc \theta = \frac{5}{4}$$

∎

EXAMPLE E

Find the trigonometric functions of the angle whose terminal side passes through (7.27, 4.49). Round off the results to three significant digits.

We show the angle and the given point in Fig. 3-14. From the Pythagorean theorem, we have

$$r = \sqrt{7.27^2 + 4.49^2} = 8.545$$

(Here we keep one extra significant digit since r is an intermediate result; see Appendix B.) Therefore, we have the following values for the trigonometric functions of θ.

$$\sin \theta = \frac{4.49}{8.545} = 0.525 \qquad \cos \theta = \frac{7.27}{8.545} = 0.851$$

$$\tan \theta = \frac{4.49}{7.27} = 0.618 \qquad \cot \theta = \frac{7.27}{4.49} = 1.62$$

$$\sec \theta = \frac{8.545}{7.27} = 1.18 \qquad \csc \theta = \frac{8.545}{4.49} = 1.90$$

Fig. 3-14

In finding these values on a calculator, it would not actually be necessary to round off the value of r, as its value can be kept in memory. The reason for showing the value of r here is to show the values used in the calculation of each of the trigonometric functions. A sequence of steps on a calculator which can be used to find r and place its value in memory is

7.27 $\boxed{x^2}$ $\boxed{+}$ 4.49 $\boxed{x^2}$ $\boxed{=}$ $\boxed{\sqrt{x}}$ $\boxed{\text{STO}}$

At this point the necessary divisions can be performed using the $\boxed{\text{RCL}}$ key for r. For example, to find $\sin \theta$ we continue the calculator sequence with

4.49 $\boxed{\div}$ $\boxed{\text{RCL}}$ $\boxed{=}$ $\boxed{\qquad 0.525468}$

NOTE ▷

In Example E, we expressed the result for $\sin \theta$ as $\sin \theta = 0.525$. It is a common error to omit the angle and give the value as $\sin = 0.525$. This is a meaningless expression, for *we must show the angle* for which we have the value of the function.

If one of the trigonometric functions is known, we can determine the other functions of the angle using the Pythagorean theorem and the definitions of the functions. The following example illustrates the method.

EXAMPLE F If we know that $\sin \theta = \frac{3}{7}$ and that θ is a first-quadrant angle, we know that the ratio of the ordinate to the radius vector (of y to r) is 3 to 7. Therefore, the point on the terminal side for which y is 3 can be determined by use of the Pythagorean theorem. The x-value for this point is

$$x = \sqrt{7^2 - 3^2} = \sqrt{49 - 9} = \sqrt{40} = 2\sqrt{10}$$

Therefore, the point $(2\sqrt{10}, 3)$ is on the terminal side, as shown in Fig. 3-15.
Therefore, using the values $x = 2\sqrt{10}$, $y = 3$, and $r = 7$, we have the other trigonometric functions of θ. They are

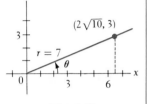

Fig. 3-15

$$\cos \theta = \frac{2\sqrt{10}}{7}, \qquad \tan \theta = \frac{3}{2\sqrt{10}}, \qquad \cot \theta = \frac{2\sqrt{10}}{3},$$

$$\sec \theta = \frac{7}{2\sqrt{10}}, \qquad \csc \theta = \frac{7}{3}$$

The values shown here are *exact*. *Approximate* decimal values can be found by the use of a calculator.

Exercises 3-2

In Exercises 1 through 12, determine the values of the trigonometric functions of the angles (in standard position) whose terminal sides pass through the given points. Give answers in exact form, as in Examples D and F.

1. (6, 8)   **2.** (5, 12)   **3.** (15, 8)   **4.** (24, 7)   **5.** (9, 40)   **6.** (16, 30)

7. $(1, \sqrt{15})$ **8.** $(\sqrt{3}, 2)$ **9.** (1, 1)   **10.** (6, 5)   **11.** (5, 2)   **12.** $\left(1, \frac{1}{2}\right)$

In Exercises 13 through 16, determine the values of the trigonometric functions of the angles (in standard position) whose terminal sides pass through the given points. Give answers in decimal form. Round off the results to the accuracy of the coordinates given.

13. (3.25, 5.15)   **14.** (0.687, 0.943)   **15.** (0.08623, 0.01327)   **16.** (37.65, 21.87)

In Exercises 17 through 24, use the given trigonometric functions to find the indicated trigonometric functions. In Exercises 17 through 20, give answers in exact form. In Exercises 21 through 24, give answers in decimal form, rounded off to three significant digits.

17. Given $\cos \theta = \dfrac{12}{13}$, find $\sin \theta$ and $\cot \theta$.

18. Given $\sin \theta = \dfrac{1}{2}$, find $\cos \theta$ and $\csc \theta$.

19. Given $\tan \theta = 1$, find $\sin \theta$ and $\sec \theta$.

20. Given $\sec \theta = \dfrac{4}{3}$, find $\tan \theta$ and $\cos \theta$.

21. Given $\sin \theta = 0.750$, find $\cot \theta$ and $\csc \theta$.

22. Given $\cos \theta = 0.326$, find $\sin \theta$ and $\tan \theta$.

23. Given $\cot \theta = 0.254$, find $\cos \theta$ and $\tan \theta$.

24. Given $\csc \theta = 1.20$, find $\sec \theta$ and $\cot \theta$.

In Exercises 25 through 28, each of the listed points is on the terminal side of an angle. Show that each of the indicated functions is the same for each of the points.

25. (3, 4), (6, 8), (4.5, 6), $\sin \theta$ and $\tan \theta$

26. (5, 12), (15, 36), (7.5, 18), $\cos \theta$ and $\cot \theta$

27. (2, 1), (4, 2), (8, 4), $\tan \theta$ and $\sec \theta$

28. (3, 2), (6, 4), (9, 6), $\csc \theta$ and $\cos \theta$

In Exercises 29 through 32, answer the given questions.

29. From the definitions of the trigonometric functions, it can be seen that csc θ is the reciprocal of sin θ. What function is the reciprocal of cos θ?

30. Following Exercise 29, what function is the reciprocal of tan θ?

31. Multiply the expression for cot θ by that for sin θ. Is the result the expression for any of the other functions?

32. Divide the expression for sin θ by that for cos θ. Is the result the expression for any of the other functions?

3-3   Values of the Trigonometric Functions

We have been able to calculate the trigonometric functions if we knew one point on the terminal side of the angle. However, in practice it is more common to know the angle in degrees, for example, and to be required to find the functions of this angle. Therefore, we must be able to determine the trigonometric functions of angles in degrees.

One way to determine the functions of a given angle is to make a scale drawing. That is, we draw the angle in standard position using a protractor and then measure the lengths of the values of x, y, and r for some point on the terminal side. By using the proper ratios we may determine the functions of this angle.

We may also use certain geometric facts to determine the functions of some particular angles. The following two examples illustrate this procedure.

EXAMPLE A

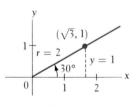

Fig. 3-16

From geometry we know that the side opposite a 30° angle in a right triangle is one-half the hypotenuse. By using this fact and letting $y = 1$ and $r = 2$ (see Fig. 3-16), we calculate that $x = \sqrt{2^2 - 1^2} = \sqrt{3}$ from the Pythagorean theorem. Therefore, with $x = \sqrt{3}$, $y = 1$, and $r = 2$, we have

$$\sin 30° = \frac{1}{2}, \qquad \cos 30° = \frac{\sqrt{3}}{2}, \quad \text{and} \quad \tan 30° = \frac{1}{\sqrt{3}}$$

In a similar way we may determine the values of the functions of 60° to be

$$\sin 60° = \frac{\sqrt{3}}{2}, \qquad \cos 60° = \frac{1}{2}, \quad \text{and} \quad \tan 60° = \sqrt{3}$$

EXAMPLE B

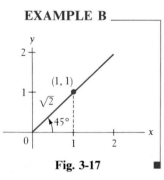

Fig. 3-17

Determine sin 45°, cos 45°, and tan 45°.

From geometry we know that in an isosceles right triangle the angles are 45°, 45°, and 90°. We know that the sides are in proportion 1, 1, $\sqrt{2}$, respectively. Putting the 45° angle in standard position, we find $x = 1$, $y = 1$, and $r = \sqrt{2}$ (see Fig. 3-17). From this we determine

$$\sin 45° = \frac{1}{\sqrt{2}}, \qquad \cos 45° = \frac{1}{\sqrt{2}}, \quad \text{and} \quad \tan 45° = 1$$

As in Example A, we have given the exact values. Decimal approximations are given in the table which follows.

Summarizing the results for 30°, 45°, and 60°, we have the following table.

θ	30°	45°	60°	(decimal approximations) 30°	45°	60°
$\sin \theta$	$\dfrac{1}{2}$	$\dfrac{1}{\sqrt{2}}$	$\dfrac{\sqrt{3}}{2}$	0.500	0.707	0.866
$\cos \theta$	$\dfrac{\sqrt{3}}{2}$	$\dfrac{1}{\sqrt{2}}$	$\dfrac{1}{2}$	0.866	0.707	0.500
$\tan \theta$	$\dfrac{1}{\sqrt{3}}$	1	$\sqrt{3}$	0.577	1.000	1.732

It is helpful to be familiar with these values, as they are used at times in later sections.

The scale-drawing method is only approximate, and the geometric methods work only for certain angles. However, the values of the functions may be determined through more advanced methods (using calculus and what are known as *power series*).

All the values of the trigonometric functions have been programmed into the scientific calculator. Therefore, we now show how the calculator is used to find these values.

NOTE ▷ To find the values of sin θ, cos θ, and tan θ, we use the $\boxed{\text{SIN}}$, $\boxed{\text{COS}}$, and $\boxed{\text{TAN}}$ keys, respectively. The angle is entered, and then the appropriate function key is used. *If your calculator has a degree-radian key or switch, **be sure that it is set for degrees.***

EXAMPLE C Using a calculator to find sin 37°, we use the sequence

37 $\boxed{\text{SIN}}$ $\boxed{ 0.601815}$

Therefore, sin 37° = 0.601815.

To find tan 67.36°, we use the sequence

67.36 $\boxed{\text{TAN}}$ $\boxed{ 2.3976264}$

■ Therefore, tan 67.36° = 2.3976264.

Not only are we able to find values of the functions if we know the angle, but also we can find the angle if we know the value of a function. This requires that we reverse the procedures mentioned above. In doing this, we are actually using another important type of mathematical function, an **inverse trigonometric function.** These are discussed in some detail in Chapter 19. For calculator purposes at this point, it is sufficient to recognize the notation which is used. The notation for "the angle whose sine is x" is $\text{Sin}^{-1} x$, or Arcsin x. Similar meanings are given to $\text{Cos}^{-1} x$, Arccos x, $\text{Tan}^{-1} x$, and Arctan x. (The -1 in $\text{Sin}^{-1} x$ used with the *function* in this way indicates the inverse function and is *not* a negative exponent.) On a calculator the $\boxed{\text{ARCSIN}}$, $\boxed{\text{SIN}^{-1}}$, or the sequence of $\boxed{\text{INV}}$ $\boxed{\text{SIN}}$ keys are used to find Arcsin x. Since these functions are usually second functions for a given key on a calculator, we will use the sequence $\boxed{\text{INV}}$ $\boxed{\text{SIN}}$ in our examples.

EXAMPLE D _____   If we know that the cosine of an angle is 0.3527, we can find the angle by using the calculator sequence

.3527 $\boxed{\text{INV}}$ $\boxed{\text{COS}}$ $\boxed{\textit{69.347452}}$

Thus, if cos $\theta = 0.3527$, then $\theta = 69.35°$, where the result has been rounded off according to the guidelines in Appendix B for the accuracy of angles and trigonometric functions. ∎

It is generally possible to set up the solution of a problem in terms of the sine, cosine, or tangent. However, if a value of the cotangent, secant, or cosecant of an angle is needed, it can also be found on the calculator.

From the definitions of the trigonometric functions we see that csc $\theta = r/y$ and sin $\theta = y/r$. This means that *the value of csc θ is the reciprocal of the value of* sin θ. In the same way, sec θ *is the reciprocal of* cos θ *and* cot θ *is the reciprocal of* tan θ. Thus, by use of the reciprocal key, $\boxed{1/x}$, the values of these functions can be found.

EXAMPLE E _____   To find the value sec 27.82°, we use the fact that

$$\sec 27.82° = \frac{1}{\cos 27.82°}$$

This indicates that we first find cos 27.82° and then find the reciprocal of this value. This leads to the calculator sequence

27.82 $\boxed{\text{COS}}$ $\boxed{1/x}$ $\boxed{\textit{1.1306869}}$

Therefore, sec 27.82° = 1.131, with the result rounded off according to the guidelines in Appendix B. ∎

EXAMPLE F _____   To find the value of θ if cot $\theta = 0.354$, we use the fact that

$$\tan \theta = \frac{1}{\cot \theta} = \frac{1}{0.354}$$

The calculator sequence is

.354 $\boxed{1/x}$ $\boxed{\text{INV}}$ $\boxed{\text{TAN}}$ $\boxed{\textit{70.506037}}$

This means that $\theta = 70.5°$, with the result rounded off according to the guidelines in Appendix B. ∎

In the following example, we see how to find the value of one function if we know the value of another function of the same angle.

EXAMPLE G _____   Find sin θ if sec $\theta = 2.504$.

Since we know sec θ, we find θ by use of the $\boxed{1/x}$ and $\boxed{\text{COS}}$ keys. Then we find sin θ. This is done with the following calculator sequence:

$$\begin{array}{cccc} & \text{to get} & & \text{to get} \\ & \cos \theta & \text{to get } \theta & \sin \theta \end{array}$$

2.504 $\boxed{1/x}$ $\boxed{\text{INV}}$ $\boxed{\text{COS}}$ $\boxed{\text{SIN}}$ $\boxed{\textit{0.9167937}}$

∎ Thus, sin $\theta = 0.9168$, rounded off to the same accuracy as sec θ.

Before the extensive use of calculators, the values of the trigonometric functions were generally obtained from tables. Although a calculator may be used to find these values, we briefly outline the use of a table here in case it becomes necessary to use such a table. Table 3 in Appendix F gives the values of the trigonometric functions for each degree.

To obtain a value of a function from Table 3, we note that the angles from 0° to 45° are listed in the left-hand column and are read down. The angles from 45° to 90° are listed on the right-hand side and are read up. The functions for angles from 0° to 45° are listed along the top, and those for the angles from 45° to 90° are listed along the bottom.

EXAMPLE H — Using Table 3, find tan 42° and sin 64°.

Tan 42° is found under tan θ (at the top) and to the right of 42°. Tan 42° = 0.900.

Sin 64° is found over sin θ (at the bottom) and to the left of 64°. ■ Sin 64° = 0.899.

EXAMPLE I — Given that cos θ = 0.293, find θ to the nearest degree.

We look for 0.293, or the number nearest to it, in the columns for cos θ in Table 3. Since the nearest number which appears is 0.292, and this appears over cos θ, we conclude that θ = 73°, to the nearest degree. ■

Tables with precision to at least the nearest 10′ or 0.1° are available in many standard sources. We can get reasonably accurate values to these precisions by using Table 3 and linear interpolation (see Exercises 49 through 56).

We can see that the calculator is easier to use than a table, and it can also give values to a much greater degree of accuracy. However, tables were used extensively until the late 1970s.

The following example illustrates the use of a trigonometric value in an applied problem. We will consider various types of applications later in the chapter.

EXAMPLE J — When a rocket is launched, its horizontal velocity v_x is related to the velocity v with which it is fired by the relation $v_x = v \cos \theta$ [which means, $v(\cos \theta)$]. Here θ is the angle between the horizontal and the direction in which it is fired (see Fig. 3-18). Find v_x if v = 1250 m/s and θ = 36.0°.

Substituting the given values of v and θ in $v_x = v \cos \theta$, we have

$$v_x = 1250 \cos 36.0°$$

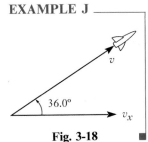

Fig. 3-18

Using the calculator sequence 1250 $\boxed{\times}$ 36 $\boxed{\cos}$ $\boxed{=}$ $\boxed{1011.2712}$, we find ■ that v_x = 1010 m/s.

Exercises 3-3

In Exercises 1 through 4, use a protractor to draw the given angle. Measure off 10 units (centimeters are convenient) along the radius vector. Then measure the corresponding values of x and y. From these values determine the trigonometric functions of the angle.

1. 40°   **2.** 75°   **3.** 15°   **4.** 53°

In Exercises 5 through 20, find the values of the trigonometric functions by use of a calculator. Round off results according to the guidelines in Appendix B.

5. sin 22.4°   **6.** cos 72.5°   **7.** tan 57.6°   **8.** sin 36.0°

9. cos 15.71°   **10.** tan 8.653°   **11.** sin 84°   **12.** cos 47°

13. cot 67.78°   **14.** csc 22.81°   **15.** sec 50.4°   **16.** cot 41.8°

17. csc 49.3°   **18.** sec 7.8°   **19.** cot 85.96°   **20.** csc 76.30°

In Exercises 21 through 36, find θ for each of the given trigonometric functions by use of a calculator. Round off the results according to the guidelines in Appendix B.

21. $\cos \theta = 0.3261$ **22.** $\tan \theta = 2.470$ **23.** $\sin \theta = 0.9114$ **24.** $\cos \theta = 0.0427$

25. $\tan \theta = 0.207$ **26.** $\sin \theta = 0.109$ **27.** $\cos \theta = 0.65007$ **28.** $\tan \theta = 5.7706$

29. $\csc \theta = 1.245$ **30.** $\sec \theta = 2.045$ **31.** $\cot \theta = 0.1443$ **32.** $\csc \theta = 1.012$

33. $\sec \theta = 3.65$ **34.** $\cot \theta = 2.08$ **35.** $\csc \theta = 3.262$ **36.** $\cot \theta = 0.1519$

In Exercises 37 through 40, find the values of the indicated trigonometric functions.

37. Find $\sin \theta$, given $\tan \theta = 1.936$.

38. Find $\cos \theta$, given $\sin \theta = 0.6725$.

39. Find $\tan \theta$, given $\sec \theta = 1.3698$.

40. Find $\csc \theta$, given $\cos \theta = 0.1063$.

In Exercises 41 through 44, find the value of each of the trigonometric functions from Table 3 in Appendix F.

41. sin 19°   **42.** cos 43°   **43.** tan 67°   **44.** cot 76°

In Exercises 45 through 48, use Table 3 to find θ to the nearest degree for each of the given functions.

45. $\tan \theta = 0.844$ **46.** $\sin \theta = 0.918$ **47.** $\cos \theta = 0.126$ **48.** $\tan \theta = 1.52$

Linear interpolation, as described in Section 2-5, can be used to find values of the trigonometric functions for angles expressed to the nearest 0.1° or 10′ from Table 3. In Exercises 49 through 52, use interpolation to find the values of each of the given functions. Use interpolation to find θ to the nearest 0.1° in Exercises 53 and 54, and to the nearest 10′ in Exercises 55 and 56.

49. tan 28.8°   **50.** cos 48.7°   **51.** sin 61° 40′   **52.** tan 53° 10′

53. $\cos \theta = 0.296$ **54.** $\tan \theta = 0.109$ **55.** $\sin \theta = 0.576$ **56.** $\cot \theta = 0.807$

In Exercises 57 through 60, solve the given problems.

57. The sound produced by a jet engine was measured at a distance of 100 m in all directions. The loudness of the sound (in decibels) was found to be $d = 70 + 30 \cos \theta$, where the 0° line was directed in front of the engine. Calculate d for $\theta = 54.5°$.

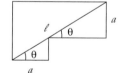

58. A brace is used in the structure shown in Fig. 3-19. Its length is $\ell = a(\sec \theta + \csc \theta)$. Find ℓ if $a = 28.0$ cm and $\theta = 34.5°$.

Fig. 3-19

59. The voltage e at any instant in a coil of wire which is turning in a magnetic field is given by $e = E \cos \alpha$, where E is the maximum voltage and α is the angle the coil makes with the field. Find the acute angle α if $e = 56.9$ V and $E = 170$ V.

60. A submarine dives such that the horizontal distance h and vertical distance v traveled are related by $v = h \tan \theta$. Here θ is the angle of the dive, as shown in Fig. 3-20. Find θ if $h = 2.35$ mi and $v = 1.52$ mi.

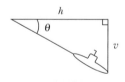

Fig. 3-20

3-4   The Right Triangle

From geometry we know that a triangle, by definition, consists of three sides and has three angles. If one side and any other two of these six parts of the triangle are known, it is possible to determine the other three parts. One of the three known parts must be a side, for if we know only the three angles, we can conclude only that an entire set of similar triangles has those particular angles.

EXAMPLE A

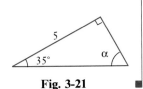

Fig. 3-21

Assume that one side and two angles are known, such as the side of 5 and the angles of 35° and 90° (the meaning of the symbol ⌐) in the triangle in Fig. 3-21. Then we may determine the third angle α by the fact that the sum of the angles of a triangle is always 180°. Of all the possible similar triangles having these three angles [35°, 90°, and 55° (α)], we have the one with the particular side of 5 between angles of 35° and 90°. Only one triangle with these parts is possible (in the sense that all triangles with the given parts are congruent and have equal corresponding angles and sides).

Fig. 3-22

To **solve a triangle** *means that, when we are given three parts of a triangle (at least one a side), we are to find the other three parts.* In this section we are going to demonstrate the method of solving a right triangle. *Since one angle of the triangle will be* 90°, *it is necessary to know one side and one other part.* Also, we know that the sum of the three angles of a triangle is 180°, and this in turn tells us that *the sum of the other two angles, both acute, is* 90°. *Any two acute angles whose sum is* 90° *are said to be* **complementary.**

For consistency, when we are labeling the parts of the right triangle *we shall use the letters A and B to denote the acute angles, and C to denote the right angle. The letters a, b, and c will denote the sides opposite these angles, respectively. Thus, side c is the hypotenuse of the right triangle.* See Fig. 3-22.

In solving right triangles we shall find it convenient to express the trigonometric functions of the acute angles in terms of the sides. By placing the vertex of angle A at the origin and the vertex of right angle C on the positive x-axis, as shown in Fig. 3-23, we have the following ratios for angle A in terms of the sides of the triangle.

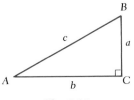

Fig. 3-23

$$\sin A = \frac{a}{c} \qquad \cos A = \frac{b}{c} \qquad \tan A = \frac{a}{b}$$

$$\cot A = \frac{b}{a} \qquad \sec A = \frac{c}{b} \qquad \csc A = \frac{c}{a}$$

(3-3)

If we should place the vertex of B at the origin, instead of the vertex of angle A, we would obtain the following ratios for the functions of angle B (see Fig. 3-24):

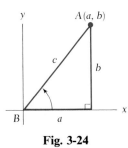

Fig. 3-24

$$\sin B = \frac{b}{c} \qquad \cos B = \frac{a}{c} \qquad \tan B = \frac{b}{a}$$

$$\cot B = \frac{a}{b} \qquad \sec B = \frac{c}{a} \qquad \csc B = \frac{c}{b}$$

(3-4)

trigonometric functions of acute angle of right triangle

Inspecting these results, we may generalize our definitions of the trigonometric functions of any acute angle α of a right triangle to be as follows (see Fig. 3-25):

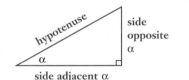

Fig. 3-25

$$\sin \alpha = \frac{\text{side opposite } \alpha}{\text{hypotenuse}} \qquad \cos \alpha = \frac{\text{side adjacent } \alpha}{\text{hypotenuse}}$$

$$\tan \alpha = \frac{\text{side opposite } \alpha}{\text{side adjacent } \alpha} \qquad \cot \alpha = \frac{\text{side adjacent } \alpha}{\text{side opposite } \alpha} \qquad (3\text{-}5)$$

$$\sec \alpha = \frac{\text{hypotenuse}}{\text{side adjacent } \alpha} \qquad \csc \alpha = \frac{\text{hypotenuse}}{\text{side opposite } \alpha}$$

Using the definitions in this form, we can solve right triangles without placing the angle in standard position. The angle need only be a part of any right triangle.

We note from the above discussion that $\sin A = \cos B$, $\tan A = \cot B$, and $\sec A = \csc B$. From this we conclude that *cofunctions of acute complementary angles are equal.* The sine function and cosine function are cofunctions, the tangent function and cotangent function are cofunctions, and the secant function and cosecant function are cofunctions. From this we can see how the tables of trigonometric functions are constructed. Since $\sin A = \cos (90° - A)$, the number representing either of these need appear only once in the tables.

EXAMPLE B

Give $a = 4$, $b = 7$, and $c = \sqrt{65}$ $(C = 90°)$, find $\sin A$, $\cos A$, and $\tan A$ (see Fig. 3-26).

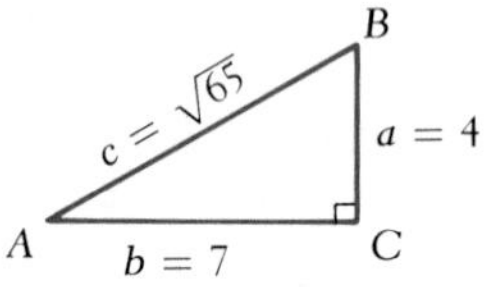

Fig. 3-26

$$\sin A = \frac{\text{side opposite angle } A}{\text{hypotenuse}} = \frac{4}{\sqrt{65}} = 0.496$$

$$\cos A = \frac{\text{side adjacent angle } A}{\text{hypotenuse}} = \frac{7}{\sqrt{65}} = 0.868$$

$$\tan A = \frac{\text{side opposite angle } A}{\text{side adjacent angle } A} = \frac{4}{7} = 0.571$$ ■

EXAMPLE C

In Fig. 3-26, we have

$$\sin B = \frac{\text{side opposite angle } B}{\text{hypotenuse}} = \frac{7}{\sqrt{65}} = 0.868$$

$$\cos B = \frac{\text{side adjacent angle } B}{\text{hypotenuse}} = \frac{4}{\sqrt{65}} = 0.496$$

$$\tan B = \frac{\text{side opposite angle } B}{\text{side adjacent angle } B} = \frac{7}{4} = 1.75$$

■ We also note that $\sin A = \cos B$ and $\cos A = \sin B$.

We are now ready to solve right triangles. We do this by first expressing the unknown parts in terms of the known parts. Results should be rounded off to the accuracy of the given parts. (Again, discussions of rounding and significant digits are given in Appendix B.)

EXAMPLE D

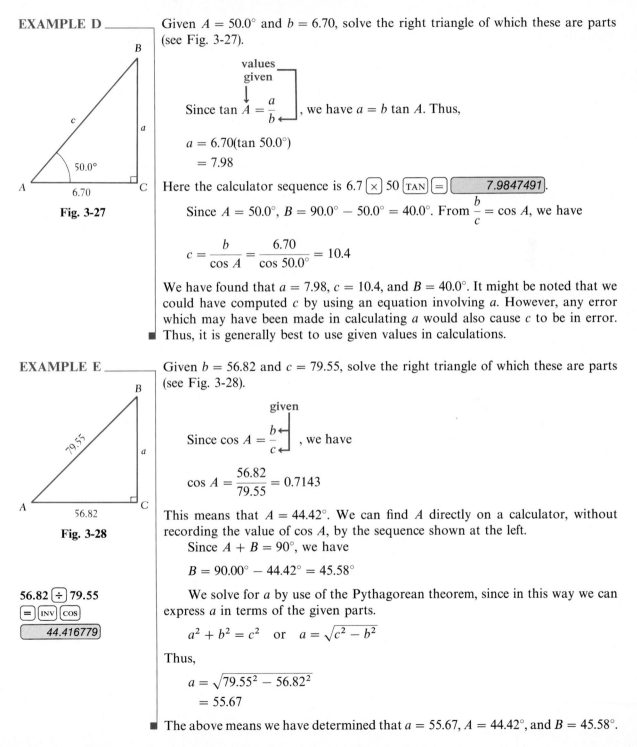

Fig. 3-27

Given $A = 50.0°$ and $b = 6.70$, solve the right triangle of which these are parts (see Fig. 3-27).

Since $\tan A = \dfrac{a}{b}$, we have $a = b \tan A$. Thus,

(values given)

$a = 6.70(\tan 50.0°)$

$= 7.98$

Here the calculator sequence is 6.7 $\boxed{\times}$ 50 $\boxed{\text{TAN}}$ $\boxed{=}$ $\boxed{7.9847491}$.

Since $A = 50.0°$, $B = 90.0° - 50.0° = 40.0°$. From $\dfrac{b}{c} = \cos A$, we have

$$c = \frac{b}{\cos A} = \frac{6.70}{\cos 50.0°} = 10.4$$

We have found that $a = 7.98$, $c = 10.4$, and $B = 40.0°$. It might be noted that we could have computed c by using an equation involving a. However, any error which may have been made in calculating a would also cause c to be in error. Thus, it is generally best to use given values in calculations.

EXAMPLE E

Given $b = 56.82$ and $c = 79.55$, solve the right triangle of which these are parts (see Fig. 3-28).

Fig. 3-28

Since $\cos A = \dfrac{b}{c}$ (given), we have

$$\cos A = \frac{56.82}{79.55} = 0.7143$$

This means that $A = 44.42°$. We can find A directly on a calculator, without recording the value of $\cos A$, by the sequence shown at the left.

56.82 $\boxed{\div}$ 79.55
$\boxed{=}$ $\boxed{\text{INV}}$ $\boxed{\text{COS}}$
$\boxed{44.416779}$

Since $A + B = 90°$, we have

$$B = 90.00° - 44.42° = 45.58°$$

We solve for a by use of the Pythagorean theorem, since in this way we can express a in terms of the given parts.

$$a^2 + b^2 = c^2 \quad \text{or} \quad a = \sqrt{c^2 - b^2}$$

Thus,

$$a = \sqrt{79.55^2 - 56.82^2}$$
$$= 55.67$$

The above means we have determined that $a = 55.67$, $A = 44.42°$, and $B = 45.58°$.

In Example E we expressed the result for *a* to four significant digits and the results for *A* and *B* to the nearest 0.01°. This is consistent with the guidelines for significant digits with angles discussed in Appendix B.

EXAMPLE F

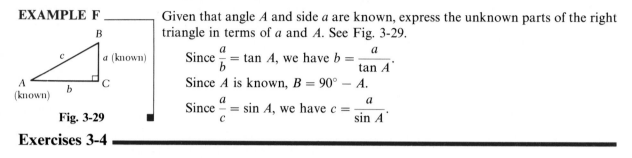

Fig. 3-29 ∎

Given that angle *A* and side *a* are known, express the unknown parts of the right triangle in terms of *a* and *A*. See Fig. 3-29.

Since $\frac{a}{b} = \tan A$, we have $b = \frac{a}{\tan A}$.

Since *A* is known, $B = 90° - A$.

Since $\frac{a}{c} = \sin A$, we have $c = \frac{a}{\sin A}$.

Exercises 3-4

In Exercises 1 through 4, draw appropriate figures and verify through observation that only one triangle may contain the given parts (that is, any others which may be drawn will be congruent—have equal corresponding sides and angles).

1. A 60° angle included between sides of 3 in. and 6 in.

2. A side of 4 in. included between angles of 40° and 50°

3. A right triangle with a hypotenuse of 5 cm and a leg of 3 cm

4. A right triangle with a 70° angle between the hypotenuse and a leg of 5 cm

In Exercises 5 through 24, solve the right triangles which have the given parts. Round off results. Refer to Fig. 3-30.

Fig. 3-30

5. $A = 77.8°, a = 6700$

6. $A = 18.4°, c = 0.0897$

7. $a = 150, c = 345$

8. $a = 93.2, c = 124$

9. $B = 32.1°, c = 23.8$

10. $B = 64.3°, b = 0.652$

11. $b = 82, c = 88$

12. $a = 5920, b = 4110$

13. $A = 32.10°, c = 56.85$

14. $B = 12.60°, c = 18.42$

15. $a = 56.73, b = 44.09$

16. $a = 9.908, c = 12.63$

17. $B = 37.5°, a = 0.862$

18. $A = 52°, b = 8.4$

19. $B = 74.18°, b = 1.849$

20. $A = 51.36°, a = 369.2$

21. $a = 591.87, b = 264.93$

22. $b = 2.9507, c = 5.0864$

23. $A = 12.975°, b = 14.592$

24. $B = 84.942°, a = 7413.5$

See Appendix E for a computer program for solving a right triangle.

In Exercises 25 through 28, find the part of the indicated triangle labeled either x or A.

25. The triangle in Fig. 3-31(a)

26. The triangle in Fig. 3-31(b)

27. The triangle in Fig. 3-31(c)

28. The triangle in Fig. 3-31(d)

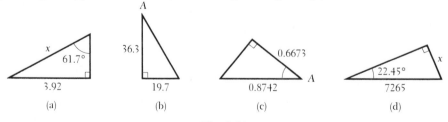

Fig. 3-31

In Exercises 29 through 32, solve the right triangles which have the given parts. Express angles to the nearest 10′ by use of Table 3 in Appendix F. Refer to Fig. 3-30.

29. $B = 37°40′, a = 0.886$

30. $A = 70°10′, a = 137$

31. $b = 86.7, c = 167$

32. $a = 6.85, b = 2.12$

In Exercises 33 through 36, the parts listed refer to those in Fig. 3-30 and are assumed to be known. Express the other parts in terms of these known parts.

33. A, c **34.** a, b **35.** B, a **36.** b, c

3-5   Applications of Right Triangles ▬▬▬▬

Many applied problems can be solved by setting up the solutions in terms of right triangles. These applications are essentially the same as solving right triangles, although it is usually one specific part of the triangle that we wish to determine. The following examples illustrate some of the basic applications of right triangles.

EXAMPLE A

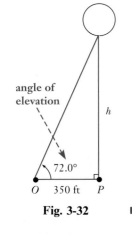

angle of elevation

h

72.0°

O 350 ft P

Fig. 3-32

A weather balloon is seen directly above a point on the ground known to be 350 ft from an observer on the ground. The **angle of elevation** (*the angle between the horizontal and the line of sight, when the object is above the horizontal*) from the observer to the balloon is measured to be 72.0°. How high is the balloon?

By drawing an appropriate figure, as shown in Fig. 3-32, we note the given information and that which is required. Here we have let h be the height of the balloon, O the position of the observer, and P the point under the balloon. From the figure, we see that

$$\frac{h}{350} = \tan 72.0° \qquad \frac{\text{required opposite side}}{\text{given adjacent side}} = \text{tangent of given angle}$$

or

$$h = 350 \tan 72.0°$$
$$= 1080 \text{ ft}$$

EXAMPLE B

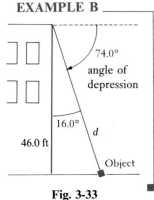

74.0°

angle of depression

16.0°

d

46.0 ft

Object

Fig. 3-33

From the roof of a building 46.0 ft high, the **angle of depression** (the angle between the horizontal and the line of sight, when the object is below the horizontal) of an object in the street is 74.0°. What is the distance of the observer from the object?

Again we draw an appropriate figure (Fig. 3-33). Here we let d represent the required distance. From the figure we see that

$$\frac{46.0}{d} = \cos 16.0° \qquad \frac{\text{given adjacent side}}{\text{required hypotenuse}} = \text{cosine of known angle}$$

$$d = \frac{46.0}{\cos 16.0°}$$
$$= 47.9 \text{ ft}$$

EXAMPLE C

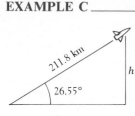

Fig. 3-34

A missile is launched at an angle of 26.55° with respect to the horizontal. If it travels in a straight line over level terrain for 2.000 min and its average speed is 6355 km/h, what is its altitude at this time?

In Fig. 3-34 we have let h represent the altitude of the missile after 2.000 min (altitude is measured on a perpendicular). Also, we determine that the missile has flown 211.8 km in a direct line from the launching site. This is found from the fact that it travels at 6355 km/h for $\dfrac{1}{30.00}$ h (2.000 min) and from the fact that $(6355 \text{ km/h})\left(\dfrac{1}{30.00}\text{ h}\right) = 211.8$ km. Therefore,

$$\frac{h}{211.8} = \sin 26.55° \qquad \frac{\text{required opposite side}}{\text{known hypotenuse}} = \text{sine of given angle}$$

$$h = 211.8(\sin 26.55°)$$
$$= 94.67 \text{ km}$$

∎

EXAMPLE D

See the chapter introduction.

A driver coming to an intersection sees the word STOP in the roadway. From the measurements shown in Fig. 3-35, determine the angle θ which the letters make at the driver's eye.

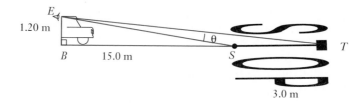

Fig. 3-35

From the figure we know sides BS and BE in triangle BES, and sides BT and BE in triangle BET. This means we can find $\angle TEB$ and $\angle SEB$ by use of the tangent. We then find θ from the fact that $\theta = \angle TEB - \angle SEB$.

$$\tan \angle TEB = \frac{18.0}{1.20}, \qquad \angle TEB = 86.2°$$

$$\tan \angle SEB = \frac{15.0}{1.20}, \qquad \angle SEB = 85.4°$$

$$\theta = 86.2° - 85.4° = 0.8°$$

∎

EXAMPLE E

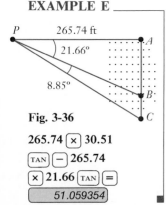

Fig. 3-36

265.74 ⊗ 30.51
TAN ⊖ 265.74
⊗ 21.66 TAN ⊜
51.059354

Using lasers, a surveyor measured the distances and angles shown in Fig. 3-36, where points B and C are in a marsh. Find the distance between B and C.

Since the distance $BC = AC - AB$, BC is found by finding AC and AB and subtracting.

$$\frac{AB}{265.74} = \tan 21.66° \quad \text{or} \quad AB = 265.74 \tan 21.66°$$

$$\frac{AC}{265.74} = \tan(21.66° + 8.85°) \quad \text{or} \quad AC = 265.74 \tan 30.51°$$

$$BC = AC - AB = 265.74 \tan 30.51° - 265.74 \tan 21.66°$$
$$= 51.06 \text{ ft}$$

∎ The calculator sequence for this calculation is shown at the left.

Exercises 3-5

In the following exercises, solve the given problems. Draw an appropriate figure unless the figure is given.

1. A straight 120-ft culvert is built down a hillside which makes an angle of 54.0° with the horizontal. Find the height of the hill.

2. A plumb line is dropped from a window of an old building. The plumb line makes an angle of 3.2° with the building, and the plumb at the bottom is 4.62 ft from the base of the building. How long is the plumb line?

3. A tree has a shadow 22.8 ft long when the angle of elevation of the sun is 62.6°. How tall is the tree?

4. The straight arm of a robot is 1.25 m long and makes an angle of 13.0° above a horizontal conveyor belt. How high above the belt is the end of the arm? See Fig. 3-37.

5. The headlights of an automobile are set such that the beam drops 2.00 in. for each 25.0 ft in front of the car. What is the angle between the beam and the road?

6. A bullet was fired such that it just grazed the top of a table. It entered a wall, which was 12.60 ft from the graze in the table, at a point 4.63 ft above the table top. At what angle was the bullet fired above the horizontal? See Fig. 3-38.

7. A robot is on the surface of Mars. The angle of depression from a camera in the robot to a rock on the surface of Mars is 13.33°. The camera is 196.0 cm above the surface. How far is the camera from the rock?

8. The observation level of a tower is 160 ft above a level plain. A brush fire is observed at an angle of depression of 1.9° from the observation level. How far from the base of the tower is the fire?

9. In designing a new building, a doorway is 2.65 ft above the ground. A ramp for the disabled, at an angle of 6.0° with the ground, is to be built to the doorway. How long will the ramp be?

10. On a test flight, during the landing of the space shuttle, the ship was 325 ft above the end of the landing strip. It then came in on a constant angle of 7.5° with the landing strip. How far from the end of the landing strip did it first touch?

11. A rectangular piece of plywood 4.00 ft by 8.00 ft is cut from one corner to an opposite corner. What are the angles between edges of the resulting pieces?

12. A rectangular solar panel is 125 cm long. It is supported by a vertical rod 97.5 cm long (see Fig. 3-39). What is the angle between the panel and the horizontal?

13. The approach to a bridge is designed to rise 15.3 ft at an angle of 4.5° to the horizontal. How long is the approach?

14. A tabletop is in the shape of a regular octagon. What is the greatest distance across the table if one edge is 0.750 m?

15. A guardrail is to be constructed around the top of a circular observation tower. The diameter of the observation area is 12.3 m. If the railing is constructed with 30 equal straight sections, what should be the length of each section?

16. A street light is designed as shown in Fig. 3-40. How high above the street is the light?

17. A straight driveway is 85.0 ft long, and the top is 12.0 ft above the bottom. What angle does the driveway make with the horizontal?

18. A level drawbridge is 250 ft long. When each half is raised, the distance between them is 80.0 ft. What angle does each make with the horizontal?

Fig. 3-37   **Fig. 3-38**

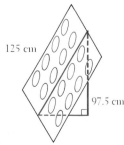

Fig. 3-39   **Fig. 3-40**

19. From a space probe circling Io, one of Jupiter's moons, at an altitude of 550 km, it was observed that the angle of depression of the horizon was 39.7°. What is the radius of Io?

20. A square wire loop is rotating in the magnetic field between two poles of a magnet in order to induce an electric current. The axis of rotation passes through the center of the loop and is midway between the poles, as shown in the side view in Fig. 3-41. How far is the edge of the loop from either pole if the side of the square is 7.30 cm and the poles are 7.66 cm apart when the angle between the loop and the vertical is 78.0°?

21. A manufacturing plant is designed to be in the shape of a regular pentagon with 92.5 ft on each side. A security fence surrounds the building to form a circle, and each corner of the building is to be 25.0 ft from the closest point on the fence. How much fencing is required?

22. A surveyor wishes to determine the height of a cliff on the other side of a river. The measurements made are shown in Fig. 3-42. How high is the cliff? (In the figure, the triangle containing the height h is vertical and perpendicular to the river.)

23. Find the angle θ in the taper shown in Fig. 3-43. (The front face is an isosceles trapezoid.)

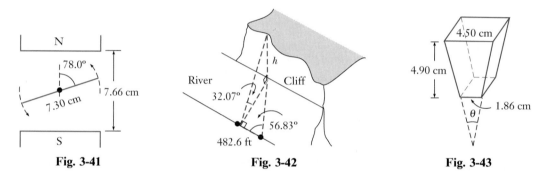

Fig. 3-41   Fig. 3-42   Fig. 3-43

24. A light beam 3.87 in. in diameter strikes the floor at an angle of 26.0° with the floor. What is the longest dimension of the area lit by the beam on the floor?

25. A stairway 1.0 m wide goes from the bottom of a cylindrical storage tank to the top at a point halfway around the tank. The handrail on the outside of the stairway makes an angle of 31.8° with the horizontal, and the radius of the tank is 11.8 m. Find the length of the handrail. See Fig. 3-44.

26. A television antenna is on the roof of a building. From a point on the ground 36.0 ft from the building, the angles of elevation of the top and the bottom of the antenna are 51.0° and 42.5°, respectively. How tall is the antenna?

27. Some of the streets in a certain city intersect as shown in Fig. 3-45. The distances between intersections A and E, D and E, and B and C are shown in the figure. How far is it between intersections B and D?

28. A supporting girder structure is shown in Fig. 3-46. Find the length x.

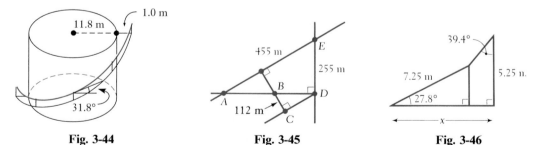

Fig. 3-44   Fig. 3-45   Fig. 3-46

3-6 Chapter Equations, Review Exercises, and Practice Test

Chapter Equations

$$\text{sine } \theta = \frac{y}{r} \qquad \cos \text{ine } \theta = \frac{x}{r}$$

$$\text{tangent } \theta = \frac{y}{x} \qquad \text{cotangent } \theta = \frac{x}{y} \tag{3-1}$$

$$\text{secant } \theta = \frac{r}{x} \qquad \text{cosecant } \theta = \frac{r}{y}$$

Pythagorean theorem
$$c^2 = a^2 + b^2 \tag{3-2}$$

$$\sin \alpha = \frac{\text{side opposite } \alpha}{\text{hypotenuse}} \qquad \cos \alpha = \frac{\text{side adjacent } \alpha}{\text{hypotenuse}}$$

$$\tan \alpha = \frac{\text{side opposite } \alpha}{\text{side adjacent } \alpha} \qquad \cot \alpha = \frac{\text{side adjacent } \alpha}{\text{side opposite } \alpha} \tag{3-5}$$

$$\sec \alpha = \frac{\text{hypotenuse}}{\text{side adjacent } \alpha} \qquad \csc \alpha = \frac{\text{hypotenuse}}{\text{side opposite } \alpha}$$

Review Exercises

In Exercises 1 through 4, find the smallest positive angle and the smallest negative angle (numerically) coterminal with, but not equal to, the given angles.

1. $17.0°$ **2.** $248.3°$ **3.** $-217.5°$ **4.** $-7.6°$

In Exercises 5 through 8, change the given angles to equal angles expressed in decimal form.

5. $31°54'$ **6.** $174°45'$ **7.** $38°6'$ **8.** $321°27'$

In Exercises 9 through 12, change the given angles to equal angles expressed to the nearest minute.

9. $17.5°$ **10.** $65.4°$ **11.** $49.7°$ **12.** $126.25°$

In Exercises 13 through 16, determine the trigonometric functions of the angles (in standard position) whose terminal side passes through the given points. Give the answers in exact form.

13. $(24, 7)$ **14.** $(5, 4)$ **15.** $(4, 4)$ **16.** $(1.2, 0.5)$

In Exercises 17 through 20, use the given trigonometric functions to find the indicated trigonometric functions. Give answers in decimal form, rounded off to three significant digits.

17. Given $\sin \theta = \frac{5}{13}$, find $\cos \theta$ and $\cot \theta$. **18.** Given $\cos \theta = \frac{3}{8}$, find $\sin \theta$ and $\tan \theta$.

19. Given $\tan \theta = 2$, find $\cos \theta$ and $\csc \theta$. **20.** Given $\cot \theta = 4$, find $\sin \theta$ and $\sec \theta$.

In Exercises 21 through 28, find the values of the trigonometric functions by use of a calculator. Round off the results.

21. $\sin 72.1°$ **22.** $\cos 40.3°$ **23.** $\tan 61.64°$ **24.** $\sin 49.09°$

25. $\sec 18.4°$ **26.** $\csc 82.4°$ **27.** $\cot 7.06°$ **28.** $\sec 79.36°$

In Exercises 29 through 36, find θ for each of the given trigonometric functions by use of a calculator. Round off the results.

29. $\cos \theta = 0.950$ **30.** $\sin \theta = 0.63052$ **31.** $\tan \theta = 1.574$ **32.** $\cos \theta = 0.1345$

33. $\csc \theta = 4.713$ **34.** $\cot \theta = 0.7561$ **35.** $\sec \theta = 2.54$ **36.** $\csc \theta = 1.92$

In Exercises 37 through 48, solve the right triangles which have the given parts. Round off the results. Refer to Fig. 3-47.

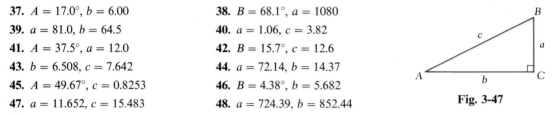

37. $A = 17.0°$, $b = 6.00$ **38.** $B = 68.1°$, $a = 1080$

39. $a = 81.0$, $b = 64.5$ **40.** $a = 1.06$, $c = 3.82$

41. $A = 37.5°$, $a = 12.0$ **42.** $B = 15.7°$, $c = 12.6$

43. $b = 6.508$, $c = 7.642$ **44.** $a = 72.14$, $b = 14.37$

45. $A = 49.67°$, $c = 0.8253$ **46.** $B = 4.38°$, $b = 5.682$

47. $a = 11.652$, $c = 15.483$ **48.** $a = 724.39$, $b = 852.44$

Fig. 3-47

In Exercises 49 through 76, solve the given applied problems.

49. In analyzing the forces on a certain hinge, one of the vertical forces F_y was found from the equation $F_y = F \cos \theta$, where θ is the angle between one part of the hinge and the vertical. Find the value of F_y if $F = 56.0$ N and $\theta = 37.5°$.

50. The velocity v, in feet per second, of an object which slides s feet down an inclined plane is given by $v = \sqrt{2gs} \sin \theta$, where g is the acceleration due to gravity and θ is the angle which the plane makes with the horizontal. Find v if $g = 32.2$ ft/s², $s = 35.5$ ft, and $\theta = 25.2°$.

51. In finding the area A of a triangular tract of land, a surveyor uses the formula $A = \frac{1}{2}ab \sin C$, where a and b are two sides of the triangle and C is the angle included between them. Find A if $a = 31.96$ m, $b = 47.25$ m, and $C = 64.09°$.

52. A water channel has the cross section of an isosceles trapezoid. See Fig. 3-48. The area of the cross section is $A = bh + h^2 \cot \theta$. Find A if $b = 12.6$ ft, $h = 4.75$ ft, and $\theta = 37.2°$.

53. For a car rounding a curve, the road should be banked at an angle θ according to the equation $\tan \theta = \dfrac{v^2}{gr}$. Here v is the speed of the car, g is the acceleration due to gravity, and r is the radius of the curve in the road. Find θ for $v = 80.7$ ft/s (55.0 mi/h), $g = 32.2$ ft/s², and $r = 950$ ft.

54. The *apparent power S* in an electric circuit in which the power is P and the impedance phase angle is θ is given by $S = P \sec \theta$. Given $P = 12.0$ V·A and $\theta = 29.4°$, find S.

55. In tracking an airplane on radar, it is found that the plane is 27.5 km on a direct line from the control tower, with an angle of elevation of 10.3°. What is the altitude of the plane?

56. A conveyor belt 75.0 ft long is inclined at 34.7° above the horizontal. Through what height can the belt lift objects placed on it?

57. The window of a house is shaded as shown in Fig. 3-49. What percent of the window is shaded when the angle of elevation θ of the sun is 65.0°?

58. The windshield on an automobile in inclined 42.5° with respect to the horizontal. Assuming that the windshield is flat and rectangular, what is its area if it is 4.80 ft wide and the bottom is 1.50 ft in front of the top?

Fig. 3-48

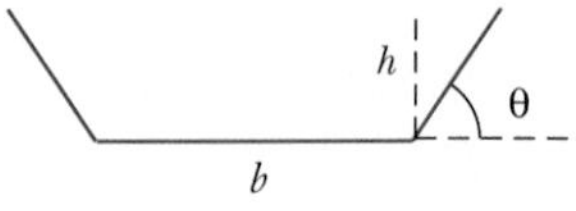

Fig. 3-49

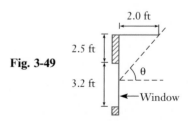

59. The bottom of a hanging lamp is 82.0 cm from the ceiling. It is then pulled 15.0 cm to the side. How much nearer the ceiling is the bottom of the lamp?

60. A theater stage inclines 3.5° upward from front to back. If it measures 35.0 ft from front to back along the stage, how much higher is the back of the stage than the front?

61. The vertical cross section of an attic room in a house is shown in Fig. 3-50. Find the distance d across the floor.

62. The impedance Z and resistance R in an alternating-current circuit may be represented by letting the impedance be the hypotenuse of a right triangle and the resistance be the side adjacent to the phase angle θ. If $R = 1750\ \Omega$ and $\theta = 17.38°$, find Z.

63. A certain straight section of a natural gas pipeline is 3.25 km long, and it rises at an angle of 3.25°. How much higher is one end of the section than the other end?

64. A Coast Guard boat which is 2.75 km from a straight beach can travel at 37.5 km/h. By traveling along a line which is at 69.0° with the beach, how long will it take to reach the beach? See Fig. 3-51.

65. In the structural support shown in Fig. 3-52, find the length x.

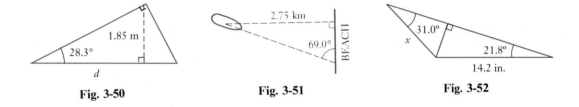

Fig. 3-50   Fig. 3-51   Fig. 3-52

66. A straight emergency chute for an airplane is 16.0 ft long. In being tested, the end at the plane is 8.5 ft above the ground. What angle does the chute make with the level ground?

67. A bridge 880 m long is sighted from a helicopter. The angle made (subtended) by the bridge at the eye of the observer is 2.2°. Show that the distance from the helicopter to the bridge is calculated as approximately the same if the line of sight is perpendicular to the end or to the middle of the bridge.

68. The top and height of the trellis shown in Fig. 3-53 are each 2.25 m. Each side piece makes an angle of 80.0° with the ground. Find the length of each side piece and the area covered by the trellis.

69. The distance from ground level to the underside of a cloud is called the *ceiling*. A ground observer 1200 m from a searchlight aimed vertically notes that the angle of elevation of the spot of light on a cloud is 76°. What is the ceiling?

70. Through what angle θ must the rectangular crate shown in Fig. 3-54 be tipped in order that its center C is directly above the pivot point P?

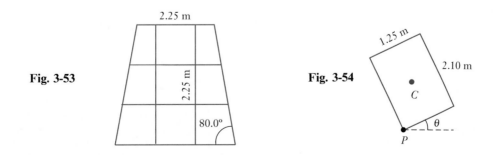

Fig. 3-53   Fig. 3-54

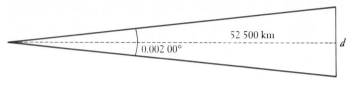

Fig. 3-55

71. A laser beam is transmitted with a "width" of 0.00200°. What is the diameter of a spot of the beam on an object 52,500 km distant? See Fig. 3-55.

72. Find the gear angle θ in Fig. 3-56 if $t = 0.180$ in.

73. A hang glider is directly above the shore of a small lake. An observer on a hill is 375 m along a straight line from the shore. From the observer, the angle of elevation of the hang glider is 42.0°, and the angle of depression of the shore is 25.0°. How far above the shore is the hang glider?

74. A crop-dusting plane flies over a field at a height of 25 ft. If the dust leaves the plane through an angle of 30° and hits the ground after the plane has traveled 75 ft, how wide a strip is dusted? See Fig. 3-57.

75. A uniform strip of wood 5.0 cm wide frames a trapezoidal window as shown in Fig. 3-58. Find the left dimension ℓ of the outside of the frame.

76. A ground observer sights a weather balloon to the east at an angle of elevation of 15.0°. A second observer 2.35 mi to the east of the first also sights the balloon to the east at an angle of elevation of 24.0°. How high is the balloon?

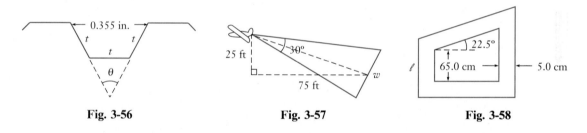

Fig. 3-56   **Fig. 3-57**   **Fig. 3-58**

Practice Test

1. Express 37°39′ in decimal form.

2. Find θ to the nearest 0.01° if $\cos \theta = 0.3726$.

3. A ship's captain, desiring to travel due south, discovers that due to an improperly functioning instrument, the ship has gone 22.62 km in a direction 4.05° east of south. How far from its course (to the east) is the ship?

4. Find $\tan \theta$ in fractional form if $\sin \theta = \frac{2}{3}$.

5. Find $\csc \theta$ if $\tan \theta = 1.294$.

6. Solve the right triangle in Fig. 3-59 if $A = 37.4°$ and $b = 52.8$.

7. Solve the right triangle in Fig. 3-59 if $a = 2.49$ and $c = 3.88$.   **Fig. 3-59**

8. In finding the wavelength λ (the Greek lambda) of light, the equation $\lambda = d \sin \theta$ is used. Find λ if $d = 30.05$ μm and $\theta = 1.167°$ (μ is the prefix for 10^{-6}).

9. Determine the trigonometric functions of an angle in standard position if its terminal side passes through (5, 2). Give answers in exact and decimal forms.

10. A surveyor sights two points directly ahead. Both are at an elevation 18.525 m lower than the observation point. How far apart are the points if the angles of depression are 13.500° and 21.375°, respectively?

4 Systems of Linear Equations; Determinants

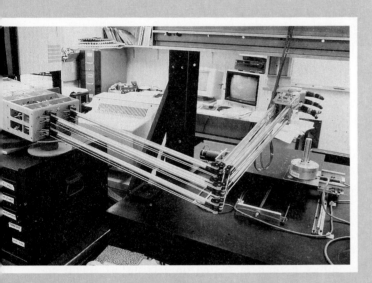

In designing an industrial robot, the forces acting on each link must be carefully analyzed. In Section 4-6 we show how the equations involving these forces are solved.

The solution of most technical and scientific problems requires that we consider several quantities that may be related in a number of ways. This can lead to more than one equation relating the quantities. For example, when determining the price at which to sell a computer, the costs for research, development, and production must be considered as well as the costs of the various components.

More than one equation relating quantities is also found for the various forces acting on an object, the electric currents in different parts of a given circuit, and the different components of a chemical or of a medical dosage. Applications exist in all technical areas.

In this chapter we will consider the solution of systems of two equations with two unknowns and of three equations with three unknowns.

4-1 Linear Equations

In general, *an equation is termed* **linear** *in a given set of variables if each term contains only one variable, to the first power, or is a constant.*

EXAMPLE A

$5x - t + 6 = 0$ is linear in x and t, but $5x^2 - t + 6 = 0$ is not, due to the x^2.

The equation $4x + y = 8$ is linear in x and y, but $4xy + y = 8$ is not, due to the presence of xy.

The equation $x - 6y + z - 4w = 7$ is linear in x, y, z, and w, but the equation $x - \frac{6}{y} + z - 4w = 7$ is not, due to the presence of $\frac{6}{y}$, where y appears in the denominator.

An equation which can be written in the form

$$ax + b = 0 \qquad (4\text{-}1)$$

is known as a **linear equation in one unknown.** We have already discussed the solution to this type of equation in Section 1–11. In general, *the* **solution,** *or* **root,** *of the equation is* $x = -b/a$. Also, it will be noted that finding the solution is equivalent to finding the zero of the **linear function** $f(x) = ax + b$.

EXAMPLE B _____ The equation $2x + 7 = 0$ is a linear equation of the form of Eq. (4-1) with $a = 2$ and $b = 7$.

From Chapter 2 we know that the linear function $2x + 7$ can be shown as $f(x) = 2x + 7$. To find the zero of this function, we set $f(x) = 0$ and get $2x + 7 = 0$.

The solution to the equation $2x + 7 = 0$, or the zero of the function $2x + 7$, is $-\frac{7}{2}$. ∎

As we noted in the chapter introduction, there are a great many applied problems which involve more than one unknown quantity. In the following example, two specific illustrations are given.

EXAMPLE C _____
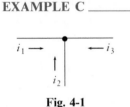
Fig. 4-1

1. A basic law of direct-current electricity, known as Kirchhoff's first law, may be stated as "The algebraic sum of the currents entering any junction in a circuit is zero." If three wires are joined at a junction, this law leads to the linear equation

    $$i_1 + i_2 + i_3 = 0$$

    where i_1, i_2, and i_3 are the currents in each of the wires. See Fig. 4-1.

2. When determining two forces F_1 and F_2 acting on a beam, we might encounter an equation such as

    $$2F_1 + 4F_2 = 200$$
    ∎

An equation which can be written in the form

$$ax + by = c \qquad (4\text{-}2)$$

is known as a **linear equation in two unknowns.** In Chapter 2 we considered many equations which can be written in this form. We found that for each value of x, there is a corresponding value for y. Each of these pairs of numbers is a **solution** to the equation, although we did not call it that at the time. *A solution is any set of numbers, one for each variable, which satisfies the equation.* When we represent the solutions in the form of a graph, we see that the graph of any linear equation in two unknowns is a straight line. Also, graphs of linear equations in one unknown, those for which $a = 0$ or $b = 0$, are also straight lines. Thus we see the significance of the name *linear*. In the next section we shall further consider the graph of the linear equation.

EXAMPLE D

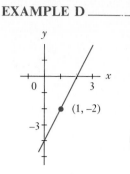

Fig. 4-2

The equation $2x - y - 4 = 0$ is a linear equation in two unknowns, x and y, since it can be written in the form of Eq. (4-2) as $2x - y = 4$. Following the methods of Chapter 2, to graph this equation we would write it in the more convenient form of $y = 2x - 4$. We see from Fig. 4-2 that the graph is a straight line. The coordinates of any point on the line give us a solution of this equation. For example, the point $(1, -2)$ is a point on the line. This means that $x = 1$, $y = -2$ is a solution of the equation. In the same way, $x = 0$, $y = -4$ is a solution since the point $(0, -4)$ is on the line. This means that there is an unlimited number of solutions to the equation. ∎

Two linear equations, each containing the same two unknowns,

**system of
linear equations**

$$a_1x + b_1y = c_1$$
$$a_2x + b_2y = c_2$$

(4-3)

are said to form a **system of simultaneous linear equations.** *A* **solution of the system** *is any pair of values (x, y) which satisfies both equations.* Methods of finding the solutions to such systems are the principal concern of this chapter.

EXAMPLE E

A piece of computer paper (rectangular) has a perimeter of 104 cm. The length is 4 cm more than the width.

From these two statements, we can set up two equations in the two unknown quantities, the length and the width. Letting ℓ = the length and w = the width, we have

$$2\ell + 2w = 104$$
$$\ell - w = 4$$

as a system of simultaneous linear equations. The solution of this system is $\ell = 28$ cm and $w = 24$ cm. These values satisfy both equations since

$$2(28) + 2(24) = 104$$

and
$$28 - 24 = 4$$

This is the only pair of values which satisfies *both* equations. Methods for finding such solutions are taken up in later sections of this chapter. ∎

Exercises 4-1

In Exercises 1 through 4, determine whether or not the given pairs of values are solutions of the given linear equations in two unknowns.

1. $2x + 3y = 9$; $(3, 1)$, $(5, \frac{1}{3})$

2. $5x + 2y = 1$; $(2, -4)$, $(1, -2)$

3. $-3x + 5y = 13$; $(-1, 2)$, $(4, 5)$

4. $x - 4y = 10$; $(2, -2)$, $(2, 2)$

In Exercises 5 through 8, for each given value of x, determine the value of y which gives a solution to the given linear equations in two unknowns.

5. $5x - y = 6$; $x = 1$, $x = -2$

6. $2x + 7y = 8$; $x = -3$, $x = 2$

7. $x - 4y = 2$; $x = 3$, $x = -0.4$

8. $3x - 2y = 9$; $x = \frac{2}{3}$, $x = -3$

In Exercises 9 through 16, determine whether or not the given pair of values is a solution of the given system of simultaneous linear equations.

9. $x - y = 5$ $x = 4, y = -1$
$2x + y = 7$

10. $2x + y = 8$ $x = -1, y = 10$
$3x - y = -13$

11. $x + 5y = -7$ $x = -2, y = 1$
$3x - 4y = -10$

12. $-3x + y = 1$ $x = \frac{1}{3}, y = 2$
$6x - 3y = -4$

13. $2x - 5y = 0$ $x = \frac{1}{2}, y = -\frac{1}{5}$
$4x + 10y = 4$

14. $6x + y = 5$ $x = 1, y = -1$
$3x - 4y = -1$

15. $3x - 2y = 2.2$ $x = 0.6, y = -0.2$
$5x + y = 2.8$

16. $x - 7y = -3.2$ $x = -1.1, y = 0.3$
$2x + y = 2.5$

In Exercises 17 through 20, answer the given questions.

17. An airplane is flying at p miles per hour relative to a wind blowing at w miles per hour. Traveling with the wind, the ground speed of the plane is 300 mi/h, and traveling against the wind, the ground speed is 220 mi/h. This leads to the two equations

$$p + w = 300$$
$$p - w = 220$$

Are the speeds 260 mi/h and 40 mi/h?

18. The electric resistance R of a certain resistor is a function of the temperature T given by the equation $R = aT + b$, where a and b are constants. If $R = 1200\ \Omega$ when $T = 10°C$, and $R = 1280\ \Omega$ when $T = 50°C$, we can find the constants a and b by substituting and obtaining the equations

$$1200 = 10a + b$$
$$1280 = 50a + b$$

Are the constants $a = 4\ \Omega/°C$ and $b = 1160\ \Omega$?

19. The forces acting on part of a structure are shown in Fig. 4-3. An analysis of the forces leads to the equations

$$0.8F_1 + 0.5F_2 = 50$$
$$0.6F_1 - 0.9F_2 = 12$$

Are the forces 45 N and 28 N?

Fig. 4-3

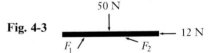

20. Using the data that fuel consumption for transportation contributes a percent p_1 of pollution which is 16% less than the percent p_2 of all other sources combined, the equations

$$p_1 + p_2 = 100$$
$$p_2 - p_1 = 16$$

can be set up. Are the percents $p_1 = 58$ and $p_2 = 42$?

4-2 Graphs of Linear Equations

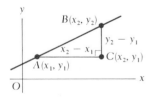

Fig. 4-4

As we mentioned in Section 4-1, the main topic of this chapter is finding solutions of systems of linear equations. Before we take up the first method, which is a graphical method, we shall develop additional ways of graphing a linear equation.

Consider the line which passes through points A and B in Fig. 4-4. Point A has coordinates (x_1, y_1) and point B has coordinates (x_2, y_2). Point C is directly horizontal from point A and directly vertical from point B. Therefore, point C has coordinates (x_2, y_1), and there is a right angle at C.

One way of measuring the steepness of a line is to determine the vertical distance between two points in relation to the horizontal distance between them. Therefore, *we define the* **slope** *of a line through two points as the difference in the* *y-coordinates divided by the difference in the x-coordinates.* For points *A* and *B*, this means

slope

$$m = \frac{y_2 - y_1}{x_2 - x_1}$$

(4-4)

where *m* is the slope of the line. Note that the slope of a vertical line, for which $x_2 = x_1$, is undefined.

EXAMPLE A

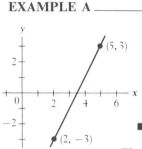

Fig. 4-5

Find the slope of the line which passes through the points $(2, -3)$ and $(5, 3)$.

In Fig. 4-5, we draw the line through the two given points. By taking (5, 3) as (x_2, y_2), then (x_1, y_1) is $(2, -3)$. Although we may choose either point as (x_2, y_2), *once the choice is made the order must be maintained.* Using Eq. (4-4), the slope is

$$m = \frac{3 - (-3)}{5 - 2} = \frac{6}{3} = 2$$

■ This means that the line rises 2 units for each unit it moves from left to right.

EXAMPLE B

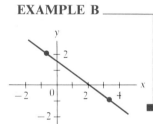

Fig. 4-6

Find the slope of the line which passes through the points $(-1, 2)$ and $(3, -1)$.

In Fig. 4-6, we draw the line through these two points. Taking (x_2, y_2) as $(3, -1)$ and (x_1, y_1) as $(-1, 2)$, the slope is

$$m = \frac{-1 - 2}{3 - (-1)} = \frac{-3}{3 + 1} = -\frac{3}{4}$$

■ This means that the line *falls* 3 units for each 4 units it moves from left to right.

We note in Example A that *as x increases, y increases and that the slope is positive.* In Example B we see that *as x increases, y decreases and that the slope is negative.* We can also see that *the larger the numerical value of the slope, the more nearly vertical is the line.* Also, if the slope is numerically small, the line rises or falls slowly. This is further illustrated in Example C.

EXAMPLE C

In Fig. 4-7(a), a line with a slope of $+5$ is illustrated. It rises sharply.
In Fig. 4-7(b), a line with a slope of $+\frac{1}{2}$ is illustrated. It rises slowly.
In Fig. 4-7(c), a line with a slope of -5 is shown. It falls sharply.
In Fig. 4-7(d), a line with a slope of $-\frac{1}{2}$ is shown. It falls slowly.

For each of these lines, we have shown the difference in the *y*-coordinates and the difference in the *x*-coordinates between two points. The ratio of these differences in each case gives the slope of each line.

Fig. 4-7

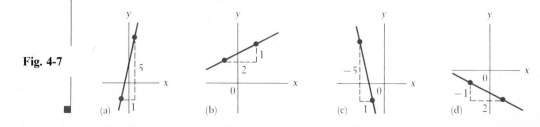

(a)   (b)   (c)   (d)

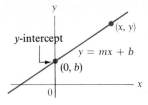

Fig. 4-8

We shall now show how the slope is related to the equation of a straight line. In Fig. 4-8, if we have two points, $(0, b)$ and a general point (x, y), the slope of this line is

$$m = \frac{y - b}{x - 0}$$

Simplifying this, we have $mx = y - b$, or

$$y = mx + b \tag{4-5}$$

intercepts

In Eq. (4-5), m is the slope of the line and b is the y-coordinate of the point where it crosses the y-axis. *This point is called the* **y-intercept** *of the line,* and its co-ordinates are $(0, b)$. *Equation (4-5) is known as the* **slope-intercept** *form of the equation of a straight line. If an equation is written in the form of Eq. (4-5), the coefficient of x is the slope and the constant is the ordinate of the y-intercept of the line.* The point $(0, b)$ and simply the value of b are both referred to as the y-intercept. Since the x-coordinate of the y-intercept must be 0, this should cause no confusion.

EXAMPLE D

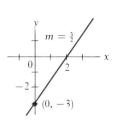

Fig. 4-9

Find the slope and the y-intercept of the line $y = \frac{3}{2}x - 3$.

Since the equation is written as y as a function of x, it is in the form of Eq. (4-5). Therefore, since the coefficient of x is $\frac{3}{2}$, the slope of the line is $\frac{3}{2}$. Also, since we can write the equation as

$$y = \overset{\text{slope}}{\frac{3}{2}} x + \overset{\text{y-intercept ordinate}}{(-3)}$$

we see that the constant is -3, which means that the y-intercept is the point $(0, -3)$. The line is shown in Fig. 4-9. ■

EXAMPLE E

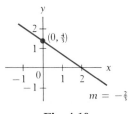

Fig. 4-10

Find the slope and y-intercept of the line $2x + 3y = 4$.

Here, *we must first write the equation in the slope-intercept form.* Solving for y gives us

$$y = \overset{\text{slope}}{-\frac{2}{3}} x + \overset{\text{y-intercept ordinate}}{\frac{4}{3}}$$

Therefore, the slope is $-\frac{2}{3}$ and the y-intercept is the point $(0, \frac{4}{3})$. The line is shown in Fig. 4-10. ■

EXAMPLE F

In analyzing an electric circuit, two of the currents, i_1 and i_2, were found to be related by the equation $i_2 = 2i_1 + 1$. Graph this equation by use of the slope-intercept form.

Since the equation is solved for i_2, we treat i_1 as the independent variable and i_2 as the dependent variable. This means that the slope of the line is 2. Also, since the b-term is 1, we can see that the intercept is $(0, 1)$.

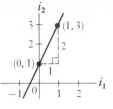

Fig. 4-11

We can use this information to sketch the line, as shown in Fig. 4-11. Since the slope is 2, we know that i_2 increases 2 units for each unit of increase of i_1. Thus, starting at the i_2-intercept $(0, 1)$, if i_1 increases by 1, i_2 increases by 2, and we are at the point $(1, 3)$. The line must pass through $(1, 3)$, as well as through $(0, 1)$. Therefore, we draw the line through these points. (Negative values may be used for electric currents since the sign depends on the direction of flow of the current.)

Another way of sketching the graph of a straight line is to find two points on the line and then draw the line through them. Two points which are easily determined are those where the line crosses the y-axis and the x-axis. We already know that the point where it crosses the y-axis is the y-intercept. In the same way, *the point where it crosses the x-axis is called the* **x-intercept,** and the coordinates of the x-intercept are $(a, 0)$. These points are easily found because in each case one of the coordinates is zero. By setting $x = 0$ and $y = 0$, in turn, and determining the corresponding value of the other unknown, we obtain the coordinates of the intercepts. A third point should be found as a check. This method is sufficient unless the line passes through the origin. Then both intercepts are at the origin and one more point must be determined, or we must use the slope-intercept method. Example G illustrates how a line is sketched by finding its intercepts.

EXAMPLE G _____ Plot the graph of $2x - 3y = 6$ by finding its intercepts and one check point (see Fig. 4-12).

First let $x = 0$. This gives $-3y = 6$, or $y = -2$. Thus, the point $(0, -2)$ is on the graph. Next we let $y = 0$, and this gives $2x = 6$, or $x = 3$. Thus, the point $(3, 0)$ is on the graph. The point $(0, -2)$ is the y-intercept, and $(3, 0)$ is the x-intercept. These two points are sufficient to plot the line, but we should find another point as a check. Choosing $x = 1$, we find $y = -\frac{4}{3}$. Thus, the point $(1, -\frac{4}{3})$ should be on the line. From Fig. 4-12 we see that it is on the line.

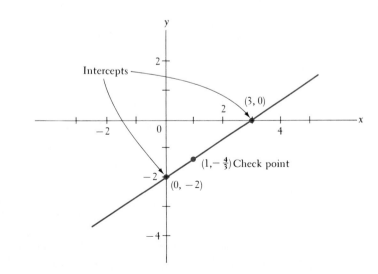

Fig. 4-12

Further details and discussion of the concept of slope and of the graphing of linear equations are found in Chapter 20.

Exercises 4-2

In Exercises 1 through 8, find the slopes of the lines which pass through the given points.

1. $(1, 0), (3, 8)$ **2.** $(3, 1), (2, 7)$ **3.** $(-1, 2), (-4, 17)$ **4.** $(-1, -2), (2, 10)$

5. $(5, -3), (-2, -5)$ **6.** $(3, -4), (-7, 1)$ **7.** $(0.4, 0.5), (-0.2, 0.2)$ **8.** $(-2.8, 3.4), (1.2, 4.2)$

In Exercises 9 through 16, sketch the lines with the given slopes and y-intercepts.

9. $m = 2, (0, -1)$ **10.** $m = 3, (0, 1)$ **11.** $m = -3, (0, 2)$ **12.** $m = -4, (0, -2)$

13. $m = \dfrac{1}{2}, (0, 0)$ **14.** $m = \dfrac{2}{3}, (0, -1)$ **15.** $m = -9, (0, 20)$ **16.** $m = -0.3, (0, -1.4)$

In Exercises 17 through 24, find the slopes and y-intercepts of the lines with the given equations and sketch the graphs.

17. $y = -2x + 1$ **18.** $y = -4x$ **19.** $y = x + 4$ **20.** $y = \dfrac{4}{5}x + 2$

21. $5x - 2y = 4$ **22.** $6x - 2y = 7$ **23.** $x + 3y = 3$ **24.** $2x + 6y = 3$

In Exercises 25 through 32, find the x-intercepts and the y-intercepts of the lines with the given equations and sketch the lines.

25. $x + 2y = 4$ **26.** $3x + y = 3$ **27.** $4x - 3y = 12$ **28.** $x - 5y = 5$

29. $y = 3x + 6$ **30.** $y = -2x - 4$ **31.** $y = -x + 3$ **32.** $y = 2x + 3$

In Exercises 33 through 36, sketch the indicated lines.

33. The diameter of the large end, d (in inches), of a certain type of machine tool can be found from the equation $d = 0.2\ell + 1.2$, where ℓ is the length of the tool. Sketch d as a function of ℓ, for values of ℓ to 10 in.

34. In testing an anticholesterol drug, it was found that each gram of drug administered reduced a person's blood cholesterol level by 2 units. Set up the function relating the cholesterol level C as a function of the dosage d for a person whose cholesterol level is 310 before taking the drug. Sketch the graph.

35. Two forces, F_1 and F_2 (in newtons), act on a supporting brace. The equation relating the forces is $0.5F_1 + 0.6F_2 = 30$. Sketch the graph of F_2 as a function of F_1.

36. Two electric currents, I_1 and I_2 (in amperes), in part of a circuit in a microcomputer are related by the equation $4I_1 - 5I_2 = 2$. Sketch I_2 as a function of I_1. These currents can be considered to be negative.

4-3 Solving Systems of Two Linear Equations in Two Unknowns Graphically

We shall now take up the problem of solving for the unknowns when we have a system of two simultaneous linear equations in two unknowns. In this section we shall show how the solution may be found graphically. The sections which follow will discuss other basic methods of solution.

Since a solution of a system of simultaneous linear equations in two unknowns is any pair of values (x, y) which satisfies both equations, graphically, *the solution would be the coordinates of the point of intersection of the two lines.* This must be the case, for the coordinates of this point constitute the only pair of values to satisfy *both* equations. (In some special cases there may be no solution; in others there may be many solutions. See Examples E and F.)

Therefore, when we solve two simultaneous linear equations in two unknowns graphically, *we must plot the graph of each line and determine the point of intersection.* This may, of course, lead to approximate results if the lines cross at values between those chosen to determine the graph.

EXAMPLE A

Solve the system of equations

$$y = x - 3$$
$$y = -2x + 1$$

Since each of the equations is in slope-intercept form, we see that $m = 1$ and $b = -3$ for the first line and that $m = -2$ and $b = 1$ for the second line. Using these values we sketch the lines, as shown in Fig. 4-13.

From the figure we see that

the lines cross at about $(1.3, -1.7)$

This means that the solution of the system is approximately $x = 1.3$, $y = -1.7$. (The actual, exact solution is $x = \frac{4}{3}$, $y = -\frac{5}{3}$.) Checking this solution in both equations, we get

$$-1.7 = 1.3 - 3 \qquad \text{and} \qquad -1.7 = -2(1.3) + 1$$
$$= -1.7 \qquad\qquad\qquad = -1.6$$

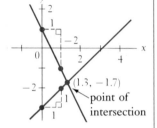

Fig. 4-13

The slight discrepancy in the second equation occurs because the solution is not exact. ∎

EXAMPLE B

Solve the system of equations

$$2x + 5y = 10$$
$$3x - y = 6$$

We could write each equation in slope-intercept form in order to sketch the lines. Also, we could use the form in which they are written to find the intercepts. Choosing to find the intercepts and draw lines through them, we let $y = 0$; then $x = 0$. Therefore, we find that the intercepts of the first line are the points $(5, 0)$ and $(0, 2)$. A third point is $(-1, \frac{12}{5})$. The intercepts of the second line are $(2, 0)$ and $(0, -6)$. A third point is $(1, -3)$. Plotting these points and drawing the proper straight lines, we see that the lines cross at about $(2.3, 1.1)$. [The exact values are $(\frac{40}{17}, \frac{18}{17})$.] The solution of the system of equations is approximately $x = 2.3$, $y = 1.1$ (see Fig. 4-14).

Checking, we have

$$2(2.3) + 5(1.1) = 10 \qquad \text{and} \qquad 3(2.3) - 1.1 = 6$$
$$10.1 = 10 \qquad\qquad\qquad 5.8 = 6$$

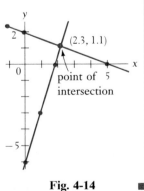

Fig. 4-14

This verifies that the solution is correct to the accuracy obtainable from the graph. ∎

EXAMPLE C

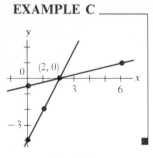

Fig. 4-15

Solve the system of equations

$$x - 4y = 2$$
$$-2x + y = -4$$

The intercepts and a third point for the first line are $(2, 0)$, $(0, -0.5)$, and $(6, 1)$. For the second line they are $(2, 0)$, $(0, -4)$, and $(1, -2)$. Since they have one point in common, the point $(2, 0)$, we conclude that the exact solution to the system is $x = 2$, $y = 0$ (see Fig. 4-15). This solution can be verified by substituting ■ into each of the equations.

Linear equations in two unknowns are often useful in solving stated problems. In such problems we must read the statement carefully in order to identify the unknowns and the method of setting up the proper equations. Example E of Section 4-1 illustrates the procedure, and the following example gives another complete illustration of the method.

EXAMPLE D

v_c (mi/h)

$v_c = -3v_h + 200$

$m = -3$, $b = 200$

$v_c = v_h - 20$

$m = 1$, $b = -20$

$(55, 35)$

v_h (mi/h)

Fig. 4-16

A driver traveled for 1.5 h at a constant speed along a highway. Then, through a construction zone, the driver reduced the car's speed by 20 mi/h for 30 min. If a total of 100 mi was covered in the 2 h, what were the two speeds?

First we let $v_h =$ the highway speed and $v_c =$ the speed through the construction zone. Two equations are found by using (1) distance = rate × time, and (2) the fact that "the car's speed was reduced by 20 mi/h." The equations are as follows:

time on   highway   30 min.   speed in
highway   speed   = 0.5 h   construction zone

$$1.5\, v_h \qquad + \qquad 0.5\, v_c \qquad = 100 \qquad \text{total distance}$$

distance   distance in
on highway   construction

$$v_c = v_h - 20$$
$$\uparrow$$
speed reduced by 20 mi/h

Using v_h as the independent variable and v_c as the dependent variable, the graphs are drawn in Fig. 4-16. (The slope-intercept form of the first equation is $v_c = -3v_h + 200$. Part of the graph is dashed because only positive values have meaning.)

We see that the point of intersection is $(55, 35)$, which means the solution is $v_h = 55$ mi/h and $v_c = 35$ mi/h. Substitution in both equations verifies this ■ solution.

EXAMPLE E

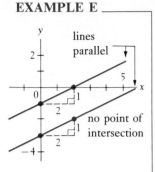

Fig. 4-17

Solve the system of equations

$$x = 2y + 6$$
$$6y = 3x - 6$$

Writing each of these equations in slope-intercept form, we have

$$y = \frac{1}{2}x - 3$$

$$y = \frac{1}{2}x - 1$$

We note that each line has a slope of $\frac{1}{2}$ and that the y-intercepts are $(0, -3)$ and $(0, -1)$. This means that the y-intercepts are different, but the slopes are the same. Since the slope indicates that each line rises $\frac{1}{2}$ unit for y for each unit of x,

NOTE ▷ *the lines are parallel and do not intersect* (see Fig. 4-17). Therefore, ***there are no solutions.** Such a system is called* **inconsistent.**

EXAMPLE F

Fig. 4-18

Solve the system of equations

$$x - 3y = 9$$
$$-2x + 6y = -18$$

The intercepts and a third point for the first line are $(9, 0)$, $(0, -3)$, and $(3, -2)$. In determining the intercepts for the second line, we find that they are $(9, 0)$ and $(0, -3)$, which are also the intercepts of the first line (see Fig. 4-18). As a check we note that $(3, -2)$ also satisfies the equation of the second line. This means the two lines are really the same line, and the *coordinates of any point*

NOTE ▷ *on this common line constitute a solution of the system.* Since ***no unique solution may be determined,*** *such a system is called* **dependent.**

Exercises 4-3

In the following exercises, solve each system of equations graphically. Estimate the answer to the nearest tenth of a unit if necessary.

1. $y = -x + 4$
$y = x - 2$

2. $y = \frac{1}{2}x - 1$
$y = -x + 8$

3. $y = 2x - 6$
$y = -\frac{1}{3}x + 1$

4. $y = \frac{1}{2}x - 4$
$y = 2x + 2$

5. $3x + 2y = 6$
$x - 3y = 3$

6. $4x - 3y = -8$
$6x + y = 6$

7. $2x - 5y = 10$
$3x + 4y = -12$

8. $-5x + 3y = 15$
$2x + 7y = 14$

9. $x - 4y = 8$
$2x + 5y = 10$

10. $y = 4x - 6$
$y = 2x + 4$

11. $y = -x + 3$
$y = -2x + 3$

12. $x - 6 = 6y$
$y = 3 - 3x$

13. $x - 4y = 6$
$2y = x + 4$

14. $x + y = 3$
$3x - 2y = 14$

15. $-2x + 2y = 7$
$4x - 2y = 1$

16. $2x - 3y = -5$
$3x + 2y = 12$

17. $x = 4y + 2$
$3y = 2x + 3$

18. $1.2x - 2.4y = 4.8$
$3.0x = -2.0y + 7.2$

19. $4.0x - 3.5y = 1.5$
$0.7y + 0.1x = 0.7$

20. $5x - 2y = 7$
$3x + 4y = 8$

21. $x - 5y = 10$
$2x - 10y = 20$

22. $18x - 3y = 7$
$2y = 1 + 12x$

23. $y = 3x$
$x - 2y = 6$

24. $4x - y = 3$
$2x + 3y = 0$

25. $5x = y + 3$
$4x = 2y - 3$

26. $5x + 7y = 5$
$2x - 3y = 4$

27. $3x = 8y + 12$
$-6x + 16y = 6$

28. $y = 6x + 2$
$12x - 2y = -4$

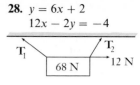

29. Chains support a crate as shown in Fig. 4-19. The equations relating tensions T_1 and T_2 are given below. Estimate T_1 and T_2 to the nearest 1 N from the graph.

$$0.8T_1 - 0.6T_2 = 12$$
$$0.6T_1 + 0.8T_2 = 68$$

Fig. 4-19

30. The equations relating the currents i_1 and i_2 shown in Fig. 4-20 are given below. Estimate the currents to the nearest 0.1 A.

$$2i_1 + 6(i_1 + i_2) = 12$$
$$4i_2 + 6(i_1 + i_2) = 12$$

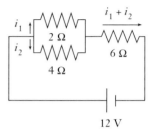

31. A computer requires 12 s to do two series of calculations. The first series requires twice the time as the second series. Set up the appropriate equations and solve graphically for the time required for each series.

32. A total of 40 tons of two types of ore is to be loaded into a smelter. The first type contains 6% copper, and the second contains 2% copper. Set up the appropriate equations relating the necessary amounts of each ore to produce 2 tons of copper. Solve graphically.

Fig. 4-20

4-4   Solving Systems of Two Linear Equations in Two Unknowns Algebraically

We have just seen how a system of two linear equations can be solved graphically. This technique is good for obtaining a "picture" of the solution. Finding the solution of systems of equations by graphical methods has one difficulty: The results are usually approximate. If exact solutions are required, we must turn to other methods. In this section we shall present two algebraic methods of solution.

solution by substitution

The first method involves the *elimination of one variable by* **substitution.** To follow this method, *we first solve one of the equations for one of the unknowns. This solution is then substituted into the other equation,* resulting in one linear equation in one unknown. This linear equation can then be solved for the unknown it contains. By substituting this value into one of the original equations, we can find the corresponding value of the other unknown. The following two examples illustrate the method.

EXAMPLE A

Use the method of elimination by substitution to solve the system of equations

$$x - 3y = 6$$
$$2x + 3y = 3$$

The first step is to solve one of the equations for one of the unknowns. The choice of which equation and which unknown depends on ease of algebraic manipulation. In this system, it is somewhat easier to solve the first equation for x. Therefore, performing this operation we have

$$x = 3y + 6 \tag{A1}$$

We then substitute this expression into the second equation in place of x, giving

in second equation, x replaced by $3y + 6$

$$2(3y + 6) + 3y = 3$$

Solving this equation for y, we obtain

$$6y + 12 + 3y = 3$$
$$9y = -9$$
$$y = -1$$

We now put the value $y = -1$ into the first of the original equations. Since we have already solved this equation for x in terms of y, Eq. (A1), we obtain

$$x = 3(-1) + 6 = 3$$

Therefore, the solution of the system is $x = 3$, $y = -1$. As a check, we substitute these values into each of the original equations. We obtain $3 - 3(-1) = 6$ and $2(3) + 3(-1) = 3$, which verifies the solution.

EXAMPLE B ———— Use the method of elimination by substitution to solve the system of equations

$$-5x + 2y = -4$$
$$10x + 6y = 3$$

It makes little difference which equation or which unknown is chosen. Therefore, choosing to solve the first equation for y, we obtain

$$2y = 5x - 4$$
$$y = \frac{5x - 4}{2} \tag{B1}$$

Substituting this expression into the second equation, we have

$$10x + 6\left(\frac{5x - 4}{2}\right) = 3 \qquad \text{in second equation, } y \text{ replaced by } \frac{5x - 4}{2}$$

We now proceed to solve this equation for x.

$$10x + 3(5x - 4) = 3$$
$$10x + 15x - 12 = 3$$
$$25x = 15$$
$$x = \frac{3}{5}$$

Substituting this value into the expression for y, Eq. (B1), we obtain

$$y = \frac{5(3/5) - 4}{2} = \frac{3 - 4}{2} = -\frac{1}{2}$$

Therefore, the solution of this system is $x = \frac{3}{5}$, $y = -\frac{1}{2}$. Substituting these values in both original equations shows that the solution checks.

The method of elimination by substitution is useful if one equation can easily be solved for one of the unknowns. However, the numerical coefficients often make this method somewhat cumbersome. So we shall now develop another algebraic method of solving a system of linear equations.

solution by addition or subtraction

The second method is the *elimination of a variable by means of* **addition or subtraction.** To use this method *we multiply each equation by a number chosen so that the coefficients for one of the unknowns will be numerically the same in* **both** *equations.* If these numerically equal coefficients are opposite in sign, we **add** the two equations. If the numerically equal coefficients have the same signs, we subtract one equation from the other. That is, we **subtract** the left side of one equation from the left side of the other equation, and also do the same to the right sides. After adding or subtracting, we have a simple linear equation in one unknown, which we then solve for the unknown. We then substitute this value into one of the original equations to obtain the value of the other unknown.

EXAMPLE C _____ Use the method of elimination by addition or subtraction to solve the system of equations

$$x - 3y = 6$$
$$2x + 3y = 3$$

We look at the coefficients to determine the best way to eliminate one of the unknowns. In this case, since the coefficients of the y-terms are numerically the same and are opposite in sign, we may immediately add the two equations to eliminate y. Adding the left sides together and the right sides together, we obtain

$$x + 2x - 3y + 3y = 6 + 3$$
$$3x = 9$$
$$x = 3$$

Substituting this value into the first equation, we obtain

$$3 - 3y = 6$$
$$-3y = 3$$
$$y = -1$$

The solution $x = 3$, $y = -1$ agrees with the results obtained for the same problem illustrated in Example A of this section.

EXAMPLE D _____ Use the method of elimination by addition or subtraction to solve the system of equations

$$3x - 2y = 4$$
$$x + 3y = 2$$

Looking at the coefficients of x and y, we see that it is necessary to multiply the second equation by 3 in order to make the coefficients of x the same. In order to make the coefficients of y numerically the same, we have to multiply the first equation by 3 and the second equation by 2. Therefore, the more convenient method is probably to multiply the second equation by 3 and eliminate x. Doing

NOTE ▷ this (be careful to multiply the terms on *both* sides; ***a common error is to forget to multiply the value on the right***) and then subtracting the second equation from the first, we have

$$3x - 2y = 4$$
$$3x + 9y = 6 \qquad \text{each term of second equation multiplied by 3}$$

$3x - 3x = 0 \longrightarrow -11y = -2$ subtract

$-2y - (+9y) = -11y \longrightarrow y = \dfrac{2}{11}$ $4 - 6 = -2$

In order to find the value of x, we substitute $y = \frac{2}{11}$ into one of the original equations. Choosing the second equation (its form is somewhat simpler), we have

$$x + 3\left(\frac{2}{11}\right) = 2$$

$$11x + 6 = 22 \qquad \text{multiply each term by 11}$$

$$x = \frac{16}{11}$$

We arrive at the solution $x = \frac{16}{11}$, $y = \frac{2}{11}$. Substituting these values into both of the original equations shows that the solution checks. ∎

EXAMPLE E ———— As we noted in Example D, we can solve the system of equations by first multiplying the first equation by 3 and the second equation by 2, thereby eliminating y. Doing this, we have

$$9x - 6y = 12 \qquad \text{each term of first equation multiplied by 3}$$
$$2x + 6y = 4 \qquad \text{each term of second equation multiplied by 2}$$
$$11x = 16 \qquad \text{add}$$

$9x + 2x = 11x \longrightarrow \quad x = \dfrac{16}{11}$ $12 + 4 = 16$

$-6y + 6y = 0$

At this point we can find the value of y by substituting $x = \frac{16}{11}$ into one of the original equations, or we could eliminate x as is done in Example D. Substitution in the first of the original equations gives us

$$3\left(\frac{16}{11}\right) - 2y = 4$$

$$48 - 22y = 44 \qquad \text{multiply each term by 11}$$

$$-22y = -4$$

$$y = \frac{2}{11}$$

∎ Therefore, the solution is $x = \frac{16}{11}$, $y = \frac{2}{11}$, as we obtained in Example D.

The best form in which to have the equations for solution of the system by the addition or subtraction method is the form shown in Examples C and D. That is, the x-term and then the y-term are on the left (the reverse order in both equations would also be acceptable) and the constant is on the right. If the equations are not written in this form, they both should be written this way before proceeding with the solution.

EXAMPLE F _____ In solving the system of equations

$$3x = 2y + 4$$
$$3y + x - 2 = 0$$

we should first rewrite the equations in the form noted above. Doing this, we have

$$3x - 2y = 4$$
$$x + 3y = 2$$

We then proceed with the solution. This system is the same as in Examples D and E. ∎

EXAMPLE G _____ Use the method of elimination by addition or subtraction to solve the system of equations

$$4x - 2y = 3$$
$$2x - \ y = 2$$

When we multiply the second equation by 2 and subtract, we obtain

$$
\begin{aligned}
4x - 2y &= \quad 3 \\
4x - 2y &= \quad 4 \\
\hline
\end{aligned}
$$

$$4x - 4x = 0 \longrightarrow 0 = -1 \longleftarrow 3 - 4 = -1$$

$$-2y - (-2y) = 0$$

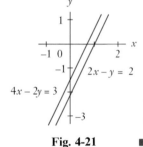

Fig. 4-21

Since we know 0 does not equal -1, we conclude that there is no solution. When we obtain a result of $0 = a$ ($a \neq 0$), the system of equations is *inconsistent*. As we discussed in the previous section, the lines which represent the equations are parallel, as shown in Fig. 4-21. ∎

In Example G we showed that a result of $0 = a$ ($a \neq 0$) indicates the system of equations is inconsistent. If we obtain the result $0 = 0$, the system is *dependent*. As shown in the previous section, this means that there is an unlimited number of solutions and the lines which represent the equations are the same line.

After solving systems of equations by these methods, it is always a good policy to check the results by substituting the values of the two unknowns into both of the original equations to see that the values satisfy them. Remember, we are finding the one pair of values which satisfies both equations.

The following example gives another complete illustration of solving a stated problem by first setting up the proper equations.

EXAMPLE H _____ By weight, one alloy is 70% copper and 30% zinc. Another alloy is 40% copper and 60% zinc. How many grams of each of these would be required to make 300 g of an alloy which is 60% copper and 40% zinc?

Let A = required number of grams of first alloy, and B = required number of grams of second alloy. We know that the total weight of the final alloy is 300 g, which leads us to the equation $A + B = 300$. We also know that the final alloy will contain 180 g of copper (60% of 300). The weight of copper from the first alloy is $0.70A$ and that from the second is $0.40B$. This leads to the equation $0.70A + 0.40B = 180$. These two equations can now be solved simultaneously.

$$A + \quad B = 300 \qquad \text{sum of weights is 300 g}$$

copper $\longrightarrow$ $\quad 0.70A + 0.40B = 180 \longleftarrow \quad$ 60% of 300 g

70% weight of first alloy \qquad 40% weight of second alloy

$$4A + 4B = 1200 \qquad \text{multiply each term of first equation by 4}$$
$$7A + 4B = 1800 \qquad \text{multiply each term of second equation by 10}$$
$$\overline{3A \qquad\quad = \quad 600} \qquad \text{subtract first equation from second equation}$$
$$A = 200 \text{ g}$$
$$B = 100 \text{ g} \qquad \text{by substituting into the first equation}$$

Substitution shows that this solution checks with the given information.

Exercises 4-4

In Exercises 1 through 12, solve the given systems of equations by the method of elimination by substitution.

1. $x = y + 3$
$x - 2y = 5$

2. $x = 2y + 1$
$2x - 3y = 4$

3. $y = x - 4$
$x + y = 10$

4. $y = 2x + 10$
$2x + y = -2$

5. $x + y = -5$
$2x - y = 2$

6. $3x + y = 1$
$3x - 2y = 16$

7. $2x + 3y = 7$
$6x - y = 1$

8. $2x + 2y = 1$
$4x - 2y = 17$

9. $3x + 2y = 7$
$2y = 9x + 11$

10. $3x + 3y = -1$
$5x = -6y - 1$

11. $0.4x - 0.3y = 0.6$
$0.2x + 0.4y = -0.5$

12. $6.0x + 4.8y = -8.4$
$4.8x - 6.5y = -7.8$

In Exercises 13 through 24, solve the given systems of equations by the method of elimination by addition or subtraction.

13. $x + 2y = 5$
$x - 2y = 1$

14. $x + 3y = 7$
$2x + 3y = 5$

15. $2x - 3y = 4$
$2x + y = -4$

16. $x - 4y = 17$
$3x + 4y = 3$

17. $2x + 3y = 8$
$x = 2y - 3$

18. $3x - y = 3$
$4x = 3y + 14$

19. $x + 2y = 7$
$2x + 4y = 9$

20. $3x - y = 5$
$-9x + 3y = -15$

21. $2x - 3y - 4 = 0$
$3x + 2 = 2y$

22. $3x + 5 = -4y$
$3y = 5x - 2$

23. $0.3x = 0.7y + 0.4$
$0.5y = 0.7 - 0.2x$

24. $2.50x + 2.25y = 4.00$
$3.75x - 6.75y = 3.25$

In Exercises 25 through 32, solve the given systems of equations by either method of this section.

25. $2x - y = 5$
$6x + 2y = -5$

26. $3x + 2y = 4$
$6x - 6y = 13$

27. $6x + 3y + 4 = 0$
$5y = -9x - 6$

28. $1 + 6y = 5x$
$3x - 4y = 7$

29. $3x - 6y = 15$
$4x - 8y = 20$

30. $2x + 6y = -3$
$-6x - 18y = 5$

31. $1.2y + 10.8 = -8.4x$
$3.6x + 4.8y + 13.2 = 0$

32. $0.66x + 0.66y = -0.77$
$0.33x - 1.32y = 1.43$

In Exercises 33 through 36, solve the given systems of equations by an appropriate algebraic method.

33. Find the voltages V_1 and V_2 of the batteries shown in Fig. 4-22. The terminals are aligned in the same direction in Fig. 4-22(a) and in opposite directions in Fig. 4-22(b).

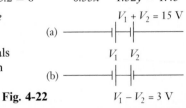

(a) $V_1 + V_2 = 15 \text{ V}$

$V_1 \quad V_2$

(b) $V_1 - V_2 = 3 \text{ V}$

Fig. 4-22

34. In determining the dimensions of a rectangular solar panel, the following equations are used:

$$2\ell + 2w = 460$$
$$\ell = w + 100$$

Find the length ℓ and width w, in centimeters, of the panel.

35. Two grades of gasoline are mixed in order to get a mixture which has 1.5% of a certain special additive. By mixing x liters of a grade with 1.8% of the additive to y liters of a grade with 1.0% of the additive, 10,000 L of the mixture are produced. The equations used to find x and y are

$$x + y = 10,000$$
$$0.018x + 0.010y = 0.015(10,000)$$

Find x and y. (Verify the equations.)

36. While a pulley is making one complete revolution, one of the pulley wheels makes 6 revolutions and the other makes 15 revolutions. The circumference of one wheel is 2 ft more than twice the circumference of the other wheel. The circumferences c_1 and c_2 can be found by solving the following equations:

$$6c_1 = 15c_2$$
$$c_1 - 2c_2 = 2$$

In Exercises 37 through 44, set up appropriate systems of two linear equations in two unknowns and solve the systems algebraically.

37. In a test of a heat-seeking rocket, a first rocket is launched at 2000 ft/s and the heat-seeking rocket is launched along the same flight path 12 s later at a speed of 3200 ft/s. Find the times t_1 and t_2 of flight of the rockets until the heat-seeking rocket destroys the first rocket.

38. The *torque* of a force is the product of the force and the perpendicular distance from a specified point. If a lever is supported at only one point, and is in balance, the sum of the torques (about the support) of forces acting on one side of the support must equal the sum of the torques of the forces acting on the other side. Find the forces F_1 and F_2 which are in the positions shown in Fig. 4-23(a) and then move to the positions in Fig. 4-23(b). The lever weighs 20 N and is in balance.

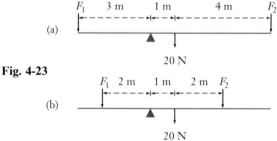

Fig. 4-23

39. One personal computer can perform a calculations per second, and a second personal computer can perform b calculations per second. If the first calculates for 3 s and the second calculates for 2 s, 17.0 million calculations are performed. If the times are reversed, 15.5 million calculations are performed. Find the rates a and b.

40. An underwater (but near the surface) explosion is detected by sonar on a ship 30 s before it is heard on the deck. If sound travels at 5000 ft/s in water and 1100 ft/s in air, how far is the ship from the explosion?

41. For proper dosage a drug must be a 10% solution. How many milliliters of a 5% solution and a 25% solution should be mixed to obtain 1000 mL of the required solution?

42. In mixing a weed-killing chemical, a 40% solution of the chemical is mixed with an 85% solution to get 20 L of a 60% solution. How much of each is needed?

43. What conclusion can be drawn from a sales report which states that "sales this month were $8000 more than last month, which means that total sales for both months are $4000 more than twice the sales last month"?

44. For the circuit shown in Fig. 4-24, a report stated that current i_1 is twice current i_2, and twice the sum of the two currents minus 6 times i_2 is 6. What values of the current are found from this report?

Fig. 4-24

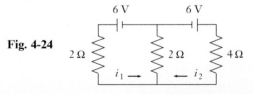

4-5   Solving Systems of Two Linear Equations in Two Unknowns by Determinants

Consider two linear equations in two unknowns, as given in Eq. (4-3):

$$a_1 x + b_1 y = c_1$$
$$a_2 x + b_2 y = c_2$$

(4-3)

If we multiply the first of these equations by b_2 and the second by b_1, we obtain

$$a_1 b_2 x + b_1 b_2 y = c_1 b_2$$
(4-6)
$$a_2 b_1 x + b_2 b_1 y = c_2 b_1$$

If we now subtract the second equation of (4-6) from the first, we obtain

$$a_1 b_2 x - a_2 b_1 x = c_1 b_2 - c_2 b_1$$

which by use of the distributive law may be written as

$$(a_1 b_2 - a_2 b_1)x = c_1 b_2 - c_2 b_1$$

(4-7)

Solving Eq. (4-7) for x, we obtain

$$x = \frac{c_1 b_2 - c_2 b_1}{a_1 b_2 - a_2 b_1}$$

(4-8)

In the same manner, we may show that

$$y = \frac{a_1 c_2 - a_2 c_1}{a_1 b_2 - a_2 b_1}$$

(4-9)

If the denominator $a_1 b_2 - a_2 b_1 = 0$, there is no solution for Eqs. (4-8) and (4-9), since division by zero is not defined.

The expression $a_1 b_2 - a_2 b_1$, which appears in each of the denominators of Eqs. (4-8) and (4-9), is an example of a special kind of expression called a **determinant of the second order**. The determinant $a_1 b_2 - a_2 b_1$ is denoted by the symbol

$$\begin{vmatrix} a_1 & b_1 \\ a_2 & b_2 \end{vmatrix}$$

Thus, by definition, a determinant of the second order is given by

definition of
second-order
determinant

$$\begin{vmatrix} a_1 & b_1 \\ a_2 & b_2 \end{vmatrix} = a_1 b_2 - a_2 b_1$$

(4-10)

The numbers a_1 and b_1 are called the **elements** *of the first* **row** *of the determinant. The numbers a_1 and a_2 are the elements of the first* **column** *of the determinant. In the same manner, the numbers a_2 and b_2 are the elements of the second row,*

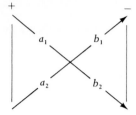

Fig. 4-25

and the numbers b_1 and b_2 are the elements of the second column. *The numbers a_1 and b_2 are the elements of the* **principal diagonal,** *and the numbers a_2 and b_1 are the elements of the* **secondary diagonal.** Thus, one way of stating the definition indicated in Eq. (4-10) is that *the value of a determinant of the second order is found by taking the product of the elements of the principal diagonal and subtracting the product of the elements of the secondary diagonal.*

A diagram which is often helpful for remembering the expansion of a second-order determinant is shown in Fig. 4-25. The following examples illustrate how we carry out the evaluation of determinants.

EXAMPLE A ⎯⎯⎯

$$\begin{vmatrix} 1 & 4 \\ 3 & 2 \end{vmatrix} = 1(2) - 3(4) = 2 - 12 = -10$$

EXAMPLE B ⎯⎯⎯

$$\begin{vmatrix} -5 & 8 \\ 3 & 7 \end{vmatrix} = (-5)(7) - 3(8) = -35 - 24 = -59$$

EXAMPLE C ⎯⎯⎯

1. $\begin{vmatrix} 4 & 6 \\ 3 & 17 \end{vmatrix} = 4(17) - (3)(6) = 68 - 18 = 50$

2. $\begin{vmatrix} 4 & 6 \\ -3 & 17 \end{vmatrix} = 4(17) - (-3)(6) = 68 + 18 = 86$

3. $\begin{vmatrix} 4 & 6 \\ -3 & -17 \end{vmatrix} = 4(-17) - (-3)(6) = -68 + 18 = -50$

▪ Note the signs of the terms being combined.

We note that the numerators of Eqs. (4-8) and (4-9) may also be written as determinants. The numerators of Eqs. (4-8) and (4-9) are

$$\begin{vmatrix} c_1 & b_1 \\ c_2 & b_2 \end{vmatrix} \quad \text{and} \quad \begin{vmatrix} a_1 & c_1 \\ a_2 & c_2 \end{vmatrix} \tag{4-11}$$

Therefore, the solutions for x and y of the system of equations (4-3) may be written directly in terms of determinants, without any algebraic operations, as

$$x = \frac{\begin{vmatrix} c_1 & b_1 \\ c_2 & b_2 \end{vmatrix}}{\begin{vmatrix} a_1 & b_1 \\ a_2 & b_2 \end{vmatrix}} \quad \text{and} \quad y = \frac{\begin{vmatrix} a_1 & c_1 \\ a_2 & c_2 \end{vmatrix}}{\begin{vmatrix} a_1 & b_1 \\ a_2 & b_2 \end{vmatrix}} \tag{4-12}$$

For this reason determinants provide a very quick and easy method of solution of systems of equations. Here again, the denominator of each of Eqs. (4-12) is the same. *The determinant of the denominator is made up of the coefficients of x and y.* Also, we can see that *the determinant of the numerator of the solution for x may be obtained by replacing the column of a's by the column of c's. The numerator of the solution for y may be obtained from the determinant of the denominator by*

Cramer's rule

replacing the column of b's by the column of c's. This result is often referred to as **Cramer's rule.**

The following examples illustrate the method of solving systems of equations by determinants.

EXAMPLE D — Solve the following system of equations by determinants:

$$2x + y = 1$$
$$5x - 2y = -11$$

First we set up the determinant for the denominator. This consists of the four coefficients in the system written as shown. Therefore, the determinant of the denominator is

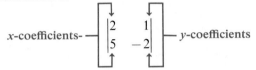

x-coefficients- $\begin{vmatrix} 2 & 1 \\ 5 & -2 \end{vmatrix}$ - y-coefficients

For finding x, the determinant in the numerator is obtained from this determinant by replacing the first column by the constants which appear on the right sides of the equations. Thus, *the numerator for the solution for x is*

$\begin{vmatrix} 1 & 1 \\ -11 & -2 \end{vmatrix}$ — replace x-coefficients with the constants

For finding y, the determinant in the numerator is obtained from the determinant of the denominator by replacing the second column by the constants which appear on the right sides of the equations. Thus, *the numerator for the solution for y is*

$\begin{vmatrix} 2 & 1 \\ 5 & -11 \end{vmatrix}$ — replace y-coefficients with the constants

Now we set up the solutions for x and y using the determinants above.

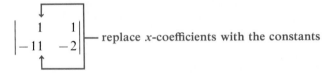

$$x = \frac{\begin{vmatrix} 1 & 1 \\ -11 & -2 \end{vmatrix}}{\begin{vmatrix} 2 & 1 \\ 5 & -2 \end{vmatrix}} = \frac{1(-2) - (-11)(1)}{2(-2) - (5)(1)} = \frac{-2 + 11}{-4 - 5} = \frac{9}{-9} = -1$$

$$y = \frac{\begin{vmatrix} 2 & 1 \\ 5 & -11 \end{vmatrix}}{\begin{vmatrix} 2 & 1 \\ 5 & -2 \end{vmatrix}} = \frac{2(-11) - (5)(1)}{-9} = \frac{-22 - 5}{-9} = 3$$

Therefore, the solution to the system of equations is $x = -1$, $y = 3$.

Since the determinant in the denominators is the same, it needs to be evaluated only once. This means that three determinants are to be evaluated in order to solve the system.

EXAMPLE E _____ Solve the following system of equations by determinants:

$$x = y + 4$$
$$y + 2x - 11 = 0$$

Since these equations are not written in the form of Eqs. (4-3), we must first write them in that form so that we can properly determine the elements of the determinants. Doing this, we proceed with the solution.

$$x - y = 4$$
$$2x + y = 11$$

constants —————→

$$x = \frac{\begin{vmatrix} 4 & -1 \\ 11 & 1 \end{vmatrix}}{\begin{vmatrix} 1 & -1 \\ 2 & 1 \end{vmatrix}} = \frac{4 - (-11)}{1 - (-2)} = 5$$

coefficients ————→

constants

$$y = \frac{\begin{vmatrix} 1 & 4 \\ 2 & 11 \end{vmatrix}}{\begin{vmatrix} 1 & -1 \\ 2 & 1 \end{vmatrix}} = \frac{11 - 8}{3} = 1$$

coefficients ————→

■ Therefore, the solution is $x = 5$, $y = 1$.

EXAMPLE F _____ Solve the following system of equations by determinants:

$$5x + 7y = 4$$
$$3x - 6y = 5$$

constants —————→

$$x = \frac{\begin{vmatrix} 4 & 7 \\ 5 & -6 \end{vmatrix}}{\begin{vmatrix} 5 & 7 \\ 3 & -6 \end{vmatrix}} = \frac{-24 - 35}{-30 - 21} = \frac{59}{51}$$

coefficients ————→

constants

$$y = \frac{\begin{vmatrix} 5 & 4 \\ 3 & 5 \end{vmatrix}}{\begin{vmatrix} 5 & 7 \\ 3 & -6 \end{vmatrix}} = \frac{25 - 12}{-51} = -\frac{13}{51}$$

coefficients ————→

■ Therefore, the solution is $x = \frac{59}{51}$, $y = -\frac{13}{51}$.

EXAMPLE G _____ Two investments totaling $18,000 yield an annual income of $700. If the first investment has an interest rate of 5.5% and the second a rate of 3.0%, what is the value of each of the investments?

Let $x =$ the value of the first investment, and $y =$ the value of the second investment. We know that the total of the two investments is $18,000. This leads to the equation $x + y = 18,000$. The first investment yields $0.055x$ dollars annually, and the second yields $0.03y$ dollars annually. This leads to the equation $0.055x + 0.03y = 700$. These two equations are then solved simultaneously.

$$x + y = 18,000 \quad \text{sum of investments}$$

$$0.055x \quad + 0.030y \quad = 700 \longleftarrow \text{income}$$

$$\uparrow \uparrow \qquad \uparrow \uparrow$$
5.5% value   3.0% value

$$x = \frac{\begin{vmatrix} 18,000 & 1 \\ 700 & 0.03 \end{vmatrix}}{\begin{vmatrix} 1 & 1 \\ 0.055 & 0.03 \end{vmatrix}} = \frac{540 - 700}{0.03 - 0.055} = \frac{160}{0.025} = 6400$$

The value of y can be found most easily by substituting this value of x into the first equation.

$$y = 18,000 - x = 18,000 - 6400 = 11,600$$

■ Therefore, the values invested are $6400 and $11,600, respectively.

Certain points should be made here. As we noted in Example E, *the equations must be in the form of Eqs. (4-3) before the determinants are set up.* This is because the equations for the solutions in terms of determinants are based on that form of writing the system. Also, if either of the unknowns is missing from either equation, its coefficient is taken as zero, and zero is put in the appropriate position in that determinant. Finally, if the determinant of the denominator is zero and that of the numerator is not zero, the system is inconsistent. If determinants of both numerator and denominator are zero, the system is dependent.

Exercises 4-5

In Exercises 1 through 12, evaluate the given determinants.

1. $\begin{vmatrix} 2 & 4 \\ 3 & 1 \end{vmatrix}$

2. $\begin{vmatrix} -1 & 3 \\ 2 & 6 \end{vmatrix}$

3. $\begin{vmatrix} 3 & -5 \\ 7 & -2 \end{vmatrix}$

4. $\begin{vmatrix} -4 & 7 \\ 1 & -3 \end{vmatrix}$

5. $\begin{vmatrix} 8 & -10 \\ 0 & 4 \end{vmatrix}$

6. $\begin{vmatrix} -4 & -3 \\ -8 & -6 \end{vmatrix}$

7. $\begin{vmatrix} -2 & 11 \\ -7 & -8 \end{vmatrix}$

8. $\begin{vmatrix} -6 & 12 \\ -15 & 3 \end{vmatrix}$

9. $\begin{vmatrix} 0.7 & -1.3 \\ 0.1 & 1.1 \end{vmatrix}$

10. $\begin{vmatrix} 0.20 & -0.05 \\ 0.28 & 0.09 \end{vmatrix}$

11. $\begin{vmatrix} 16 & -8 \\ 42 & -15 \end{vmatrix}$

12. $\begin{vmatrix} 43 & -7 \\ -81 & 16 \end{vmatrix}$

In Exercises 13 through 32, solve the given systems of equations by use of determinants. (These systems are the same as those for Exercises 13 through 32 of Section 4-4.)

13. $x + 2y = 5$
$x - 2y = 1$

14. $x + 3y = 7$
$2x + 3y = 5$

15. $2x - 3y = 4$
$2x + y = -4$

16. $x - 4y = 17$
$3x + 4y = 3$

17. $2x + 3y = 8$
$x = 2y - 3$

18. $3x - y = 3$
$4x = 3y + 14$

19. $x + 2y = 7$
$2x + 4y = 9$

20. $3x - y = 5$
$-9x + 3y = -15$

21. $2x - 3y - 4 = 0$
$3x + 2 = 2y$

22. $3x + 5 = -4y$
$3y = 5x - 2$

23. $0.3x = 0.7y + 0.4$
$0.5y = 0.7 - 0.2x$

24. $2.50x + 2.25y = 4.00$
$3.75x - 6.75y = 3.25$

See Appendix E for a computer program for solving a system of equations by determinants.

25. $2x - y = 5$
$6x + 2y = -5$

26. $3x + 2y = 4$
$6x - 6y = 13$

27. $6x + 3y + 4 = 0$
$5y = -9x - 6$

28. $1 + 6y = 5x$
$3x - 4y = 7$

29. $3x - 6y = 15$
$4x - 8y = 20$

30. $2x + 6y = -3$
$-6x - 18y = 5$

31. $1.2y + 10.8 = -8.4x$
$3.6x + 4.8y + 13.2 = 0$

32. $0.66x + 0.66y = -0.77$
$0.33x - 1.32y = 1.43$

In Exercises 33 through 36, solve the given systems of equations by use of determinants.

33. The forces acting on a link of an industrial robot are shown in Fig. 4-26. The equations for finding forces F_1 and F_2 are

$$F_1 + F_2 = 21$$
$$2F_1 = 5F_2$$

Find F_1 and F_2.

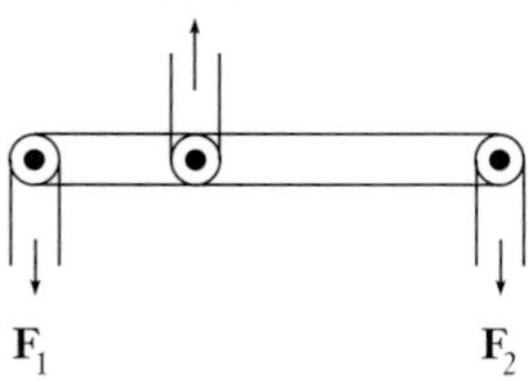

Fig. 4-26

21 lb

F_1 F_2

34. An agricultural expert estimated that a certain organism spreads over an area A, in square kilometers, after t years according to the equation $A = at + b$, where a and b are constants. If $A = 6.0 \text{ km}^2$ when $t = 2$ years and $A = 7.5 \text{ km}^2$ when $t = 3$ years, then a and b can be determined by solving the system of equations

$$6.0 = 2a + b$$
$$7.5 = 3a + b$$

Solve for a and b, and find A as a function of t.

35. If $6000 is invested, part at 7% and part at 12%, with a total annual income of $560, the amounts A and B invested at these rates can be found by solving the system of equations

$$A + B = 6000$$
$$0.07A + 0.12B = 560$$

Find A and B.

36. A certain amount of fuel contains 150,000 Btu of potential heat. Part is burned at 80% efficiency and the remainder is burned at 70% efficiency such that the total amount of heat actually delivered is 114,000 Btu. To find the amounts x and y burned with each efficiency, the following equations must be solved. Find x and y.

$$x + y = 150{,}000$$
$$0.80x + 0.70y = 114{,}000$$

In Exercises 37 through 44, set up appropriate systems of two linear equations in two unknowns and then solve the system by use of determinants.

37. A boat carrying illegal drugs leaves a port and travels at 42 mi/h. A Coast Guard cutter leaves the port 24 min later and travels at 50 mi/h in pursuit of the boat. Find the times each has traveled when the cutter overtakes the boat with drugs.

38. In framing the wall of a house, each horizontal plate is 3 ft 3 in. longer than each vertical stud. Find the lengths of each plate and each of the 9 studs if a total of 89 ft of lumber is used. See Fig. 4-27.

39. A computer manufacturer received two types of computer components in a shipment of 5600 components. Type A components cost $0.75 each and type B components cost $0.45 each; the total cost of the shipment was $3780. How many of each were in the shipment?

Fig. 4-27

40. Two types of electromechanical carburetors are being assembled and tested. Each of the first type requires 15 min of assembly time and 2 min of testing time. Each of the second type requires 12 min of assembly time and 3 min of testing time. If there are 222 min of assembly time and 45 min of testing time available, how many of each can be assembled and tested if all the time is used?

41. A pharmacist is mixing a 3% saline solution and an 8% saline solution to get 2 L of a 6% saline solution. How much of each solution is needed?

42. The velocity of sound in steel is 15,900 ft/s faster than the velocity of sound in air. One end of a long steel bar is struck and an observer at the other end measures the time it takes for the sound to reach him. He finds that the sound through the bar takes 0.012 s to reach him and that the sound through the air takes 0.180 s. What are the velocities of sound in air and in steel?

43. In an experiment, a variable voltage V is in a circuit with a fixed voltage V_0 and a resistance R. The voltage V is related to the current i in the circuit by $V = Ri - V_0$. If $V = 5.8$ V for $i = 2.0$ A, and $V = 24.7$ V for $i = 6.2$ A, find V as a function of i.

44. The distance s that a metal block travels down an inclined plane is given by $s = v_0 t + \frac{1}{2} at^2$, where v_0 is the initial velocity, a is the acceleration, and t is the time of descent. If $s = 16.4$ m for $t = 1.50$ s, and $s = 88.8$ m for $t = 5.30$ s, find v_0 and a.

4-6   Solving Systems of Three Linear Equations in Three Unknowns Algebraically

Many problems involve the solution of systems of linear equations which involve three, four, and occasionally even more unknowns. Solving such systems algebraically or by determinants is essentially the same as solving systems of two linear equations in two unknowns. In this section we shall discuss the algebraic method of solving a system of three linear equations in three unknowns. Graphical solutions are not used, since a linear equation in three unknowns represents a plane in space. We shall, however, briefly show graphical interpretations of systems of three linear equations in three unknowns at the end of this section.

A system of three linear equations in three unknowns written in the form

$$
\begin{aligned}
a_1 x + b_1 y + c_1 z &= d_1 \\
a_2 x + b_2 y + c_2 z &= d_2 \\
a_3 x + b_3 y + c_3 z &= d_3
\end{aligned}
\tag{4-13}
$$

has as its solution the set of values x, y, and z which satisfy all three equations simultaneously. The method of solution involves multiplying *two* of the equations by the proper numbers to eliminate *one* of the unknowns between these equations. We then repeat this process, using a *different pair* of the original equations, being sure that we eliminate the same unknown as we did between the first pair of equations. At this point we have two linear equations in two unknowns which can be solved by any of the methods previously discussed. The unknown originally eliminated may then be found by substitution into one of the original equations. The solution should then be checked in the original system of equations.

EXAMPLE A _____ Solve the following system of equations:

$$(1)\qquad x + 2y - z = -5$$
$$(2)\qquad 2x - y + 2z = 8$$
$$(3)\qquad 3x + 3y + 4z = 5$$

(4)	$2x + 4y - 2z = -10$	(1) multiplied by 2
	$2x - y + 2z = 8$	(2)
(5)	$4x + 3y = -2$	adding (4) and (2)
(6)	$4x - 2y + 4z = 16$	(2) multiplied by 2
	$3x + 3y + 4z = 5$	(3)
(7)	$x - 5y = 11$	subtracting
	$4x + 3y = -2$	(5)
(8)	$4x - 20y = 44$	(7) multiplied by 4
(9)	$23y = -46$	subtracting
(10)	$y = -2$	
(11)	$x - 5(-2) = 11$	substituting (10) in (7)
(12)	$x = 1$	
(13)	$1 + 2(-2) - z = -5$	substituting (12) and (10) in (1)
(14)	$z = 2$	

Checking, we substitute the solution $x = 1$, $y = -2$, $z = 2$ in the original system.

$$1 + 2(-2) - 2 = -5 \qquad 2(1) - (-2) + 2(2) = 8 \qquad 3(1) + 3(-2) + 4(2) = 5$$
$$-5 = -5 \qquad\qquad\qquad 8 = 8 \qquad\qquad\qquad\qquad 5 = 5$$

We see that the solution checks.

It should be noted that Eqs. (1), (2), and (3) could be solved just as well by eliminating y between (1) and (2), and then again between (1) and (3). We would then have two equations to solve in x and z. Also, z could have been eliminated between Eqs. (1) and (3) to obtain the second equation in x and y.

EXAMPLE B _____ Solve the following system of equations:

$$(1)\qquad 4x + y + 3z = 1$$
$$(2)\qquad 2x - 2y + 6z = 11$$
$$(3)\qquad -6x + 3y + 12z = -4$$

(4)	$8x + 2y + 6z = 2$	(1) multiplied by 2
	$2x - 2y + 6z = 11$	(2)
(5)	$10x + 12z = 13$	adding
(6)	$12x + 3y + 9z = 3$	(1) multiplied by 3
	$-6x + 3y + 12z = -4$	(3)
(7)	$18x - 3z = 7$	subtracting
	$10x + 12z = 13$	(5)
(8)	$72x - 12z = 28$	(7) multiplied by 4
(9)	$82x = 41$	adding

$$(10) \qquad\qquad x = \frac{1}{2}$$

$$(11) \qquad 18\left(\frac{1}{2}\right) - 3z = 7 \qquad\qquad \text{substituting (10) in (7)}$$

$$(12) \qquad\qquad -3z = -2$$

$$(13) \qquad\qquad z = \frac{2}{3}$$

$$(14) \quad 4\left(\frac{1}{2}\right) + y + 3\left(\frac{2}{3}\right) = 1 \qquad \text{substituting (13) and (10) in (1)}$$

$$(15) \qquad\qquad 2 + y + 2 = 1$$

$$(16) \qquad\qquad y = -3$$

Therefore, the solution is $x = \frac{1}{2}$, $y = -3$, $z = \frac{2}{3}$. This solution checks when substituted into the original system of equations.

EXAMPLE C

See the chapter
introduction.

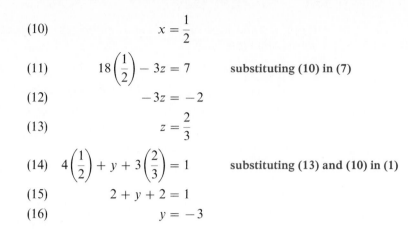

Fig. 4-28

An industrial robot and the forces acting on its main link are shown in Fig. 4-28.
An analysis of the forces leads to the following equations relating the forces.
Determine the forces.

$$
\begin{array}{llll}
(1) & A + 60 = 0.8T & \\
(2) & B = 0.6T & \\
(3) & 8A + 6B + 80 = \quad 5T & \\
(4) & 5A \qquad\quad - 4T = -300 & \text{(1) mult. by 5 and rewritten} \\
(5) & \qquad 5B - 3T = \quad 0 & \text{(2) mult. by 5 and rewritten} \\
(6) & 8A + 6B - 5T = -80 & \text{(3) rewritten} \\
(7) & 40A \qquad\quad - 32T = -2400 & \text{(4) multiplied by 8} \\
(8) & 40A + 30B - 25T = -400 & \text{(6) multiplied by 5} \\
(9) & \quad -30B - \; 7T = -2000 & \text{subtracting} \\
(10) & \quad 30B - 18T = \quad 0 & \text{(5) multiplied by 6} \\
(11) & \qquad -25T = -2000 & \text{adding} \\
(12) & \qquad T = 80 \text{ N} & \\
(13) & A + 60 = 64 & \text{(12) substituted in (1)} \\
(14) & A = 4 \text{ N} & \\
(15) & B = 48 \text{ N} & \text{(12) substituted in (2)} \\
\end{array}
$$

Therefore, the forces are $A = 4$ N, $B = 48$ N, and $T = 80$ N. The solution checks
when substituted into the original system of equations.

EXAMPLE D

A triangular brace has a perimeter of 37 in. The longest side is 3 in. longer than
the next longest, which in turn is 8 in. longer than the shortest side. Find the
length of each side.

Let a = length of the longest side, b = length of the next-longest side, and
c = length of the shortest side. Since the perimeter is 37 in., we have the equation $a + b + c = 37$. The statement of the problem also leads to the equations

(Continued on next page)

$a = b + 3$ and $b = c + 8$. These equations are then put into standard form and solved simultaneously.

(1)	$a + b + c = 37$	
(2)	$a - b \quad = 3$	rewriting second equation
(3)	$\underline{b - c = 8}$	rewriting third equation
(4)	$a + 2b \quad = 45$	adding (1) and (3)
	$\underline{a - b \quad = 3}$	(2)
(5)	$3b \quad = 42$	subtracting
(6)	$b = 14$	
(7)	$a - 14 = 3$	substituting (6) in (2)
(8)	$a = 17$	
(9)	$14 - c = 8$	substituting (6) in (3)
(10)	$c = 6$	

Therefore, the three sides of the triangle are 17 in., 14 in., and 6 in.

Checking the solution, the sum of the lengths of the three sides of the brace is 17 in. + 14 in. + 6 in. = 37 in., and the perimeter was given to be 37 in.

The systems of equations we have solved in this section have had unique solutions. However, linear systems with more than two unknowns may also have an unlimited number of solutions or may be inconsistent. After eliminating unknowns, if we obtain $0 = 0$, there is an unlimited number of solutions. If we obtain $0 = a$ $(a \neq 0)$, the system is inconsistent and there is no solution. (See Exercises 25 through 28.)

In the introduction to this section we noted that a linear equation in three unknowns represents a plane in space. For a system of three linear equations in three unknowns, if the planes intersect at a point, there is a unique solution [Fig. 4-29(a)]. If the three planes intersect in a line, there is an unlimited number of solutions [Fig. 4-29(b)]. If the planes do not have a common intersection, the system is inconsistent. The planes can be parallel [Fig. 4-30(a)], two of the planes can be parallel [Fig. 4-30(b)], or they can intersect in pairs in three parallel lines [Fig. 4-30(c)]. If any plane is coincident with another plane, the system is inconsistent or there is an unlimited number of solutions, depending on whether or not any of the planes are parallel.

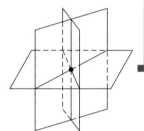
(a)

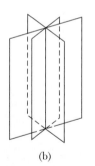
(b)

Fig. 4-29

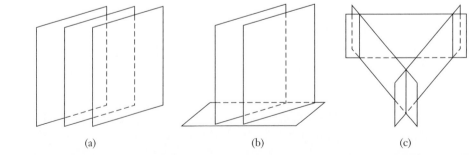

Fig. 4-30

(a)　　　(b)　　　(c)

For systems of equations with more than three unknowns, the solution is found in a manner similar to that used with three unknowns. For example, with four unknowns one of the unknowns is eliminated between three different pairs of equations. The result is three equations in the remaining three unknowns. The solution then follows the procedure used with three unknowns.

Exercises 4-6

In Exercises 1 through 20, solve the given systems of equations.

1. $x + y + z = 2$
$x - z = 1$
$x + y = 1$

2. $x + y - z = -3$
$x + z = 2$
$2x - y + 2z = 3$

3. $2x + 3y + z = 2$
$-x + 2y + 3z = -1$
$-3x - 3y + z = 0$

4. $2x + y - z = 4$
$4x - 3y - 2z = -2$
$8x - 2y - 3z = 3$

5. $5x + 6y - 3z = 6$
$4x - 7y - 2z = -3$
$3x + y - 7z = 1$

6. $3r + s - t = 2$
$r - 2s + t = 0$
$4r - s + t = 3$

7. $2x - 2y + 3z = 5$
$2x + y - 2z = -1$
$4x - y - 3z = 0$

8. $2u + 2v + 3w = 0$
$3u + v + 4w = 21$
$-u - 3v + 7w = 15$

9. $3x - 7y + 3z = 6$
$3x + 3y + 6z = 1$
$5x - 5y + 2z = 5$

10. $8x + y + z = 1$
$7x - 2y + 9z = -3$
$4x - 6y + 8z = -5$

11. $x + 2y + 2z = 0$
$2x + 6y - 3z = -1$
$4x - 3y + 6z = -8$

12. $3x + 3y + z = 6$
$2x + 2y - z = 9$
$4x + 2y - 3z = 16$

13. $2x + 3y - 5z = 7$
$4x - 3y - 2z = 1$
$8x - y + 4z = 3$

14. $2x - 4y - 4z = 3$
$3x + 8y + 2z = -11$
$4x + 6y - z = -8$

15. $r - s - 3t - u = 1$
$2r + 4s - 2u = 2$
$3r + 4s - 2t = 0$
$r + 2t - 3u = 3$

16. $3x + 2y - 4z + 2t = 3$
$5x - 3y - 5z + 6t = 8$
$2x - y + 3z - 2t = 1$
$-2x + 3y + 2z - 3t = -2$

17. A medical supply company has 1150 worker-hours for production, maintenance, and inspection. Using this and other factors, the number of hours used for each operation, P, M, and I, respectively, is found by solving the following system of equations:

$$P + M + I = 1150$$
$$P = 4I - 100$$
$$P = 6M + 50$$

18. Three oil pumps fill three different tanks. The pumping rates of the pumps in liters per hour are r_1, r_2, and r_3, respectively. Because of malfunctions, they do not operate at capacity each time. Their rates can be found by solving the system of equations

$$r_1 + r_2 + r_3 = 14{,}000$$
$$r_1 + 2r_2 = 13{,}000$$
$$3r_1 + 3r_2 + 2r_3 = 36{,}000$$

Find the pumping rates.

19. The forces acting on a certain girder, as shown in Fig. 4-31, can be found by solving the following system of equations:

$$0.707F_1 - 0.800F_2 = 0$$
$$0.707F_1 + 0.600F_2 - F_3 = 10.0$$
$$3.00F_2 - 3.00F_3 = 20.0$$

Find the forces, in newtons.

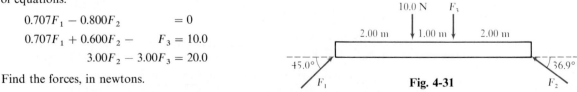

Fig. 4-31

20. In applying Kirchhoff's laws (e.g., see Beiser, *Modern Technical Physics,* 5th ed., p. 542) to the electric circuit shown in Fig. 4-32, the following equations are found. Determine the indicated currents. In Fig. 4-32, I signifies current, in amperes.

$$I_A + \quad I_B + \quad I_C = 0$$
$$4I_A - 10I_B \qquad = 3$$
$$- 10I_B + 5I_C = 6$$

Fig. 4-32

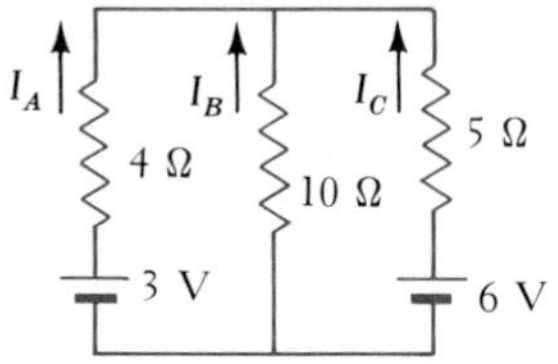

In Exercises 21 through 24, set up systems of three linear equations and solve for the indicated quantities.

21. Find angles A, B, and C in the roof truss shown in Fig. 4-33.

22. Three computer line printers print 18,500 lines when operating together for 2 min. They print 17,500 lines if they operate for 3 min, 2 min, and 1 min, respectively. The first and third printers print 6000 lines when they operate together for 1 min. What is the rate, in lines per minute, at which each prints?

23. By weight, one fertilizer is 20% potassium, 30% nitrogen, and 50% phosphorus. A second fertilizer has percents of 10, 20, and 70, respectively, and a third fertilizer has percents of 0, 30, and 70, respectively. How much of each must be mixed together to get 200 lb of fertilizer with percents of 12, 25, and 63, respectively?

24. The average traffic flow (number of vehicles) from noon until 1 P.M. in a certain section of one-way streets in a city is shown in Fig. 4-34. Show that an analysis of the flow through intersections is not sufficient to obtain unique values for x, y, and z.

Fig. 4-33

Fig. 4-34

In Exercises 25 through 28, show that the given systems of equations have either an unlimited number of solutions or no solution. If there is an unlimited number of solutions, find one of them.

25.
$$x - 2y - 3z = 2$$
$$x - 4y - 13z = 14$$
$$-3x + 5y + 4z = 0$$

26.
$$x - 2y - 3z = 2$$
$$x - 4y - 13z = 14$$
$$-3x + 5y + 4z = 2$$

27.
$$3x + 3y - 2z = 2$$
$$2x - y + z = 1$$
$$x - 5y + 4z = -3$$

28.
$$3x + y - z = -3$$
$$x + y - 3z = -5$$
$$-5x - 2y + 3z = 7$$

4-7   Solving Systems of Three Linear Equations in Three Unknowns by Determinants

Just as systems of two linear equations in two unknowns can be solved by the use of determinants, so can systems of three linear equations in three unknowns. The system as given in Eqs. (4-13) can be solved in general terms by the method of elimination by addition or subtraction. This leads to the following solutions for x, y, and z:

$$x = \frac{d_1b_2c_3 + d_3b_1c_2 + d_2b_3c_1 - d_3b_2c_1 - d_1b_3c_2 - d_2b_1c_3}{a_1b_2c_3 + a_3b_1c_2 + a_2b_3c_1 - a_3b_2c_1 - a_1b_3c_2 - a_2b_1c_3}$$

$$y = \frac{a_1d_2c_3 + a_3d_1c_2 + a_2d_3c_1 - a_3d_2c_1 - a_1d_3c_2 - a_2d_1c_3}{a_1b_2c_3 + a_3b_1c_2 + a_2b_3c_1 - a_3b_2c_1 - a_1b_3c_2 - a_2b_1c_3} \qquad (4\text{-}14)$$

$$z = \frac{a_1b_2d_3 + a_3b_1d_2 + a_2b_3d_1 - a_3b_2d_1 - a_1b_3d_2 - a_2b_1d_3}{a_1b_2c_3 + a_3b_1c_2 + a_2b_3c_1 - a_3b_2c_1 - a_1b_3c_2 - a_2b_1c_3}$$

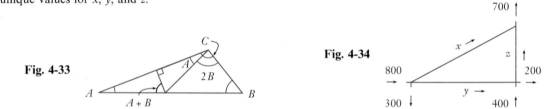

The expression that appears in each of the denominators of Eqs. (4-14) is an example of a **determinant of the third order.** This determinant is denoted by

$$\begin{vmatrix} a_1 & b_1 & c_1 \\ a_2 & b_2 & c_2 \\ a_3 & b_3 & c_3 \end{vmatrix}$$

Therefore, a determinant of the third order is defined by the equation

definition of third-order determinant

$$\begin{vmatrix} a_1 & b_1 & c_1 \\ a_2 & b_2 & c_2 \\ a_3 & b_3 & c_3 \end{vmatrix} = a_1 b_2 c_3 + a_3 b_1 c_2 + a_2 b_3 c_1 - a_3 b_2 c_1 - a_1 b_3 c_2 - a_2 b_1 c_3 \qquad (4\text{-}15)$$

The elements, rows, columns, and diagonals of a third-order determinant are defined just as are those of a second-order determinant. For example, the principal diagonal is made up of the elements a_1, b_2, and c_3.

Probably the easiest way of remembering the method of determining the value of a third-order determinant is as follows (this method does *not* work for determinants of order higher than three): *Rewrite the first and second columns to the right of the determinant. The products of the elements of the principal diagonal and the two parallel diagonals to the right of it are then added. The products of the elements of the secondary diagonal and the two parallel diagonals to the right of it are subtracted from the first sum. The algebraic sum of these six products gives the value of the determinant.* These products are indicated in Fig. 4-35.

Fig. 4-35

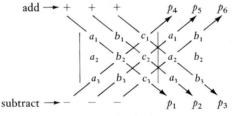

Examples A and B illustrate the evaluation of third-order determinants.

EXAMPLE A

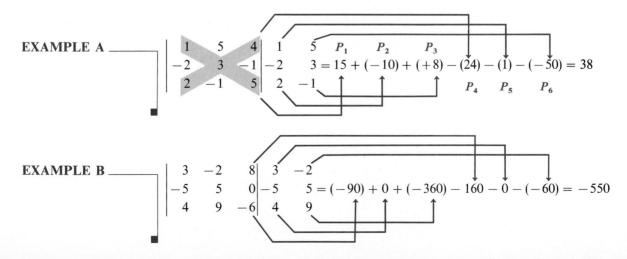

$$\begin{vmatrix} 1 & 5 & 4 \\ -2 & 3 & -1 \\ 2 & -1 & 5 \end{vmatrix} \begin{matrix} 1 & 5 \\ -2 & 3 \\ 2 & -1 \end{matrix} = 15 + (-10) + (+8) - (24) - (1) - (-50) = 38$$

EXAMPLE B

$$\begin{vmatrix} 3 & -2 & 8 \\ -5 & 5 & 0 \\ 4 & 9 & -6 \end{vmatrix} \begin{matrix} 3 & -2 \\ -5 & 5 \\ 4 & 9 \end{matrix} = (-90) + 0 + (-360) - 160 - 0 - (-60) = -550$$

Inspection of Eqs. (4-14) reveals that the numerators of these solutions may also be written in terms of determinants. Thus, we may write the general solution to a system of three equations in three unknowns as

$$
x = \frac{\begin{vmatrix} d_1 & b_1 & c_1 \\ d_2 & b_2 & c_2 \\ d_3 & b_3 & c_3 \end{vmatrix}}{\begin{vmatrix} a_1 & b_1 & c_1 \\ a_2 & b_2 & c_2 \\ a_3 & b_3 & c_3 \end{vmatrix}} \qquad
y = \frac{\begin{vmatrix} a_1 & d_1 & c_1 \\ a_2 & d_2 & c_2 \\ a_3 & d_3 & c_3 \end{vmatrix}}{\begin{vmatrix} a_1 & b_1 & c_1 \\ a_2 & b_2 & c_2 \\ a_3 & b_3 & c_3 \end{vmatrix}} \qquad
z = \frac{\begin{vmatrix} a_1 & b_1 & d_1 \\ a_2 & b_2 & d_2 \\ a_3 & b_3 & d_3 \end{vmatrix}}{\begin{vmatrix} a_1 & b_1 & c_1 \\ a_2 & b_2 & c_2 \\ a_3 & b_3 & c_3 \end{vmatrix}} \qquad (4\text{-}16)
$$

If the determinant of the denominator is not zero, there is a unique solution to the system of equations. (If all determinants are zero, there is an *unlimited number of solutions.* If the determinant of the denominator is zero, and any of the determinants of the numerators is not zero, the system is *inconsistent,* and there is *no solution.*)

An analysis of Eqs. (4-16) shows that the situation is precisely the same as it was when we were using determinants to solve systems of two linear equations. That is, the determinants in the denominators in the expressions for x, y, and z are the same. They consist of elements which are the coefficients of the unknowns. The determinant of the numerator of the solution for x is the same as that of the denominator, except that the column of d's replaces the column of a's. The determinant in the numerator of the solution for y is the same as that of the denominator, except that the column of d's replaces the column of b's. The determinant of the numerator of the solution for z is the same as the determinant of the denominator, except that the column of d's replaces the column of c's. To summarize, we can state that *the determinants in the numerators are the same as those in the denominators, except that the column of d's replaces the column of coefficients of* Cramer's rule *the unknown for which we are solving.* This again is **Cramer's rule.** Remember that the equations must be written in the standard form shown in Eqs. (4-13) before the determinants are formed.

EXAMPLE C _____ Solve the following system of equations by determinants:

$$x + 2y + 2z = 1$$
$$2x - \ y + \ z = 3$$
$$4x + \ y + 2z = 0$$

constants

$$
x = \frac{\begin{array}{ccc|cc} 1 & 2 & 2 & 1 & 2 \\ 3 & -1 & 1 & 3 & -1 \\ 0 & 1 & 2 & 0 & 1 \end{array}}{\begin{array}{ccc|cc} 1 & 2 & 2 & 1 & 2 \\ 2 & -1 & 1 & 2 & -1 \\ 4 & 1 & 2 & 4 & 1 \end{array}} = \frac{-2 + 0 + 6 - 0 - 1 - 12}{-2 + 8 + 4 - (-8) - 1 - 8}
$$

coefficients

$$= \frac{-9}{+9} = -1$$

Since the value of the denominator is already determined, there is no need to write the denominator in determinant form when solving for y and z.

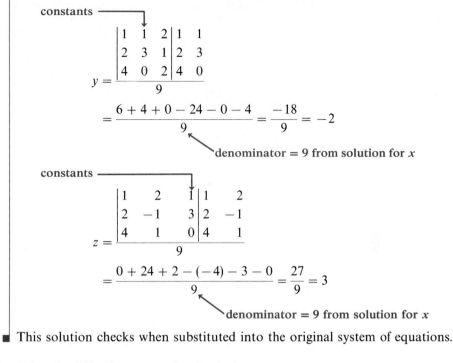

constants

$$y = \frac{\begin{vmatrix} 1 & 1 & 2 \\ 2 & 3 & 1 \\ 4 & 0 & 2 \end{vmatrix} \begin{matrix} 1 & 1 \\ 2 & 3 \\ 4 & 0 \end{matrix}}{9}$$

$$= \frac{6 + 4 + 0 - 24 - 0 - 4}{9} = \frac{-18}{9} = -2$$

denominator = 9 from solution for x

constants

$$z = \frac{\begin{vmatrix} 1 & 2 & 1 \\ 2 & -1 & 3 \\ 4 & 1 & 0 \end{vmatrix} \begin{matrix} 1 & 2 \\ 2 & -1 \\ 4 & 1 \end{matrix}}{9}$$

$$= \frac{0 + 24 + 2 - (-4) - 3 - 0}{9} = \frac{27}{9} = 3$$

denominator = 9 from solution for x

■ This solution checks when substituted into the original system of equations.

EXAMPLE D ———— Solve the following system by determinants:

$$3x + 2y - 5z = -1$$
$$2x - 3y - z = 11$$
$$5x - 2y + 7z = 9$$

constants

$$x = \frac{\begin{vmatrix} -1 & 2 & -5 \\ 11 & -3 & -1 \\ 9 & -2 & 7 \end{vmatrix} \begin{matrix} -1 & 2 \\ 11 & -3 \\ 9 & -2 \end{matrix}}{\begin{vmatrix} 3 & 2 & -5 \\ 2 & -3 & -1 \\ 5 & -2 & 7 \end{vmatrix} \begin{matrix} 3 & 2 \\ 2 & -3 \\ 5 & -2 \end{matrix}}$$

coefficients ⟶

$$= \frac{21 - 18 + 110 - 135 + 2 - 154}{-63 - 10 + 20 - 75 - 6 - 28} = \frac{-174}{-162} = \frac{29}{27}$$

(*Continued on next page*)

constants

$$y = \frac{\begin{vmatrix} 3 & -1 & -5 \\ 2 & 11 & -1 \\ 5 & 9 & 7 \end{vmatrix} \begin{matrix} 3 & -1 \\ 2 & 11 \\ 5 & 9 \end{matrix}}{-162} = \frac{231 + 5 - 90 + 275 + 27 + 14}{-162}$$

denominator $= -162$ from solution for x

$$= \frac{462}{-162} = -\frac{77}{27}$$

constants

$$z = \frac{\begin{vmatrix} 3 & 2 & -1 \\ 2 & -3 & 11 \\ 5 & -2 & 9 \end{vmatrix} \begin{matrix} 3 & 2 \\ 2 & -3 \\ 5 & -2 \end{matrix}}{-162} = \frac{-81 + 110 + 4 - 15 + 66 - 36}{-162}$$

denominator $= -162$ from solution for x

$$= \frac{48}{-162} = -\frac{8}{27}$$

Substituting in each of the original equations shows that the solution checks.

$$3(\tfrac{29}{27}) + 2(-\tfrac{77}{27}) - 5(-\tfrac{8}{27}) = \frac{87 - 154 + 40}{27} = \frac{-27}{27} = -1$$

$$2(\tfrac{29}{27}) - 3(-\tfrac{77}{27}) - (-\tfrac{8}{27}) = \frac{58 + 231 + 8}{27} = \frac{297}{27} = 11$$

$$5(\tfrac{29}{27}) - 2(-\tfrac{77}{27}) + 7(-\tfrac{8}{27}) = \frac{145 + 154 - 56}{27} = \frac{243}{27} = 9$$

After the values of x and y were determined, we could have evaluated z by substituting the values of x and y into one of the original equations. ■

EXAMPLE E   An 8% solution, a 10% solution, and a 20% solution of nitric acid are to be mixed in order to get 100 mL of a 12% solution. If the volume of acid from the 8% solution equals half the volume of acid from the other two solutions, how much of each is needed?

Let $x =$ volume of 8% solution needed, $y =$ volume of 10% solution needed, and $z =$ volume of 20% solution needed.

We first use the fact that the sum of the volumes of the three solutions is 100 mL. This leads to the equation $x + y + z = 100$. Next we note that there are $0.8x$ milliliters of pure nitric acid from the first solution, $0.10y$ milliliters from the second, $0.20z$ milliliters from the third solution, and $0.12(100)$ mL in the final solution. This leads to the equation $0.08x + 0.10y + 0.20z = 12$. Finally, using the last stated condition, we have $0.08x = \tfrac{1}{2}(0.10y + 0.20z)$. These equations are then rewritten in standard form, simplified, and solved.

$$x + y + z = 100 \qquad \text{sum of volumes}$$

$$0.08x + 0.10y + 0.20z = 12 \qquad \text{volumes of pure acid}$$

acid in
8% solution

$$\underline{0.08x = 0.05y + 0.10z} \qquad \begin{array}{l}\text{one-half of}\\ \text{acid in others}\end{array}$$

$$x + y + z = 100$$

$$4x + 5y + 10z = 600 \qquad \text{rewriting second equation}$$

$$8x - 5y - 10z = 0 \qquad \text{rewriting third equation}$$

$$x = \frac{\begin{vmatrix} 100 & 1 & 1 \\ 600 & 5 & 10 \\ 0 & -5 & -10 \end{vmatrix} \begin{matrix} 100 & 1 \\ 600 & 5 \\ 0 & -5 \end{matrix}}{\begin{vmatrix} 1 & 1 & 1 \\ 4 & 5 & 10 \\ 8 & -5 & -10 \end{vmatrix} \begin{matrix} 1 & 1 \\ 4 & 5 \\ 8 & -5 \end{matrix}}$$

$$= \frac{-5000 + 0 - 3000 - 0 + 5000 + 6000}{-50 + 80 - 20 - 40 + 50 + 40} = \frac{3000}{60} = 50$$

$$y = \frac{\begin{vmatrix} 1 & 100 & 1 \\ 4 & 600 & 10 \\ 8 & 0 & -10 \end{vmatrix} \begin{matrix} 1 & 100 \\ 4 & 600 \\ 8 & 0 \end{matrix}}{60}$$

$$= \frac{-6000 + 8000 + 0 - 4800 - 0 + 4000}{60} = \frac{1200}{60} = 20$$

$$z = \frac{\begin{vmatrix} 1 & 1 & 100 \\ 4 & 5 & 600 \\ 8 & -5 & 0 \end{vmatrix} \begin{matrix} 1 & 1 \\ 4 & 5 \\ 8 & -5 \end{matrix}}{60}$$

$$= \frac{0 + 4800 - 2000 - 4000 + 3000 - 0}{60} = \frac{1800}{60} = 30$$

Therefore, 50 mL of the 8% solution, 20 mL of the 10% solution, and 30 mL of the 20% solution are required to make the 12% solution. Checking with the statement of the problem verifies the solution.

Additional techniques which are useful in solving systems of equations are taken up in Chapter 15. Methods of evaluating determinants which are particularly useful for higher order determinants are also discussed.

Exercises 4-7

In Exercises 1 through 12, evaluate the given third-order determinants.

1. $\begin{vmatrix} 5 & 4 & -1 \\ -2 & -6 & 8 \\ 7 & 1 & 1 \end{vmatrix}$

2. $\begin{vmatrix} -7 & 0 & 0 \\ 2 & 4 & 5 \\ 1 & 4 & 2 \end{vmatrix}$

3. $\begin{vmatrix} 8 & 9 & -6 \\ -3 & 7 & 2 \\ 4 & -2 & 5 \end{vmatrix}$

4. $\begin{vmatrix} -2 & 4 & -1 \\ 5 & -1 & 4 \\ 4 & -8 & 2 \end{vmatrix}$

5. $\begin{vmatrix} -3 & -4 & -8 \\ 5 & -1 & 0 \\ 2 & 10 & -1 \end{vmatrix}$ **6.** $\begin{vmatrix} 10 & 2 & -7 \\ -2 & -3 & 6 \\ 6 & 5 & -2 \end{vmatrix}$ **7.** $\begin{vmatrix} 4 & -3 & -11 \\ -9 & 2 & -2 \\ 0 & 1 & -5 \end{vmatrix}$ **8.** $\begin{vmatrix} 9 & -2 & 0 \\ -1 & 3 & -6 \\ -4 & -6 & -2 \end{vmatrix}$

9. $\begin{vmatrix} 5 & 4 & -5 \\ -3 & 2 & -1 \\ 7 & 1 & 3 \end{vmatrix}$ **10.** $\begin{vmatrix} 20 & 0 & -15 \\ -4 & 30 & 1 \\ 6 & -1 & 40 \end{vmatrix}$ **11.** $\begin{vmatrix} 0.1 & -0.2 & 0 \\ -0.5 & 1 & 0.4 \\ -2 & 0.8 & 2 \end{vmatrix}$ **12.** $\begin{vmatrix} 0.2 & -0.5 & -0.4 \\ 1.2 & 0.3 & 0.2 \\ -0.5 & 0.1 & -0.4 \end{vmatrix}$

In Exercises 13 through 28, solve the given systems of equations by use of determinants. (Exercises 15 through 26 are the same as Exercises 1 through 12 of Section 4-6.)

13. $2x + 3y + z = 4$
$3x - z = -3$
$x - 2y + 2z = -5$

14. $4x + y + z = 2$
$2x - y - z = 4$
$3y + z = 2$

15. $x + y + z = 2$
$x - z = 1$
$x + y = 1$

16. $x + y - z = -3$
$x + z = 2$
$2x - y + 2z = 3$

17. $2x + 3y + z = 2$
$-x + 2y + 3z = -1$
$-3x - 3y + z = 0$

18. $2x + y - z = 4$
$4x - 3y - 2z = -2$
$8x - 2y - 3z = 3$

19. $5x + 6y - 3z = 6$
$4x - 7y - 2z = -3$
$3x + y - 7z = 1$

20. $3r + s - t = 2$
$r - 2s + t = 0$
$4r - s + t = 3$

21. $2x - 2y + 3z = 5$
$2x + y - 2z = -1$
$4x - y - 3z = 0$

22. $2u + 2v + 3w = 0$
$3u + v + 4w = 21$
$-u - 3v + 7w = 15$

23. $3x - 7y + 3z = 6$
$3x + 3y + 6z = 1$
$5x - 5y + 2z = 5$

24. $8x + y + z = 1$
$7x - 2y + 9z = -3$
$4x - 6y + 8z = -5$

25. $x + 2y + 2z = 0$
$2x + 6y - 3z = -1$
$4x - 3y + 6z = -8$

26. $3x + 3y + z = 6$
$2x + 2y - z = 9$
$4x + 2y - 3z = 16$

27. $3.0x + 4.5y - 7.5z = 10.5$
$4.8x - 3.6y - 2.4z = 1.2$
$4.0x - 0.5y + 2.0z = 1.5$

28. $0.26x - 0.52y - 0.52z = 0.39$
$0.45x + 0.96y + 0.40z = -0.80$
$0.55x + 0.62y - 0.11z = -0.48$

In Exercises 29 through 32, solve the given problems by determinants. In Exercises 31 and 32, set up appropriate equations and then solve them.

29. In analyzing the forces on the bell-crank mechanism shown in Fig. 4-36, the following equations are obtained. Find the indicated forces.

$A \quad - 0.6F = 80$
$B - 0.8F = 0$
$6A \quad - 10F = 0$

Fig. 4-36

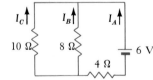

30. In applying Kirchhoff's laws (see Exercise 20 of Section 4-6) to the given electric circuit, these equations result.

$I_A + I_B + I_C = 0$
$-8I_B + 10I_C = 0$
$4I_A - 8I_B = 6$

Fig. 4-37

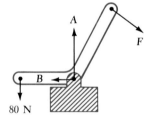

Determine the indicated currents, in amperes. (See Fig. 4-37.)

31. A person spent 1 h in a car while going to an airport, 2 h flying in a jet, and 0.5 h in a taxi to reach the final destination. The speed of the jet averaged 12 times that of the car, which averaged 15 mi/h more than the taxi. What was the average speed of each if the total trip covered 1140 mi?

32. An intravenous aqueous solution is to be made from three existing mixtures to get 500 mL with 6% of one medication, 8% of a second medication, and 86% water. The percentages in the three existing mixtures are, respectively, 5, 20, 75 in the first, 0, 5, 95 in the second, and 10, 5, 85 in the third. How much of each is needed?

4-8   Chapter Equations, Review Exercises, and Practice Test

Chapter Equations

Linear equation in one unknown	$ax + b = 0$	(4-1)
Linear equation in two unknowns	$ax + by = c$	(4-2)
System of two linear equations	$a_1x + b_1y = c_1$ $a_2x + b_2y = c_2$	(4-3)
Definition of slope	$m = \dfrac{y_2 - y_1}{x_2 - x_1}$	(4-4)
Slope-intercept form	$y = mx + b$	(4-5)
Second-order determinant	$\begin{vmatrix} a_1 & b_1 \\ a_2 & b_2 \end{vmatrix} = a_1b_2 - a_2b_1$	(4-10)
Cramer's rule	$x = \dfrac{\begin{vmatrix} c_1 & b_1 \\ c_2 & b_2 \end{vmatrix}}{\begin{vmatrix} a_1 & b_1 \\ a_2 & b_2 \end{vmatrix}}$ and $y = \dfrac{\begin{vmatrix} a_1 & c_1 \\ a_2 & c_2 \end{vmatrix}}{\begin{vmatrix} a_1 & b_1 \\ a_2 & b_2 \end{vmatrix}}$	(4-12)
System of three linear equations	$a_1x + b_1y + c_1z = d_1$ $a_2x + b_2y + c_2z = d_2$ $a_3x + b_3y + c_3z = d_3$	(4-13)
Third-order determinant	$\begin{vmatrix} a_1 & b_1 & c_1 \\ a_2 & b_2 & c_2 \\ a_3 & b_3 & c_3 \end{vmatrix} = a_1b_2c_3 + a_3b_1c_2 + a_2b_3c_1 - a_3b_2c_1 - a_1b_3c_2 - a_2b_1c_3$	(4-15)
Cramer's rule	$x = \dfrac{\begin{vmatrix} d_1 & b_1 & c_1 \\ d_2 & b_2 & c_2 \\ d_3 & b_3 & c_3 \end{vmatrix}}{\begin{vmatrix} a_1 & b_1 & c_1 \\ a_2 & b_2 & c_2 \\ a_3 & b_3 & c_3 \end{vmatrix}}$ $y = \dfrac{\begin{vmatrix} a_1 & d_1 & c_1 \\ a_2 & d_2 & c_2 \\ a_3 & d_3 & c_3 \end{vmatrix}}{\begin{vmatrix} a_1 & b_1 & c_1 \\ a_2 & b_2 & c_2 \\ a_3 & b_3 & c_3 \end{vmatrix}}$ $z = \dfrac{\begin{vmatrix} a_1 & b_1 & d_1 \\ a_2 & b_2 & d_2 \\ a_3 & b_3 & d_3 \end{vmatrix}}{\begin{vmatrix} a_1 & b_1 & c_1 \\ a_2 & b_2 & c_2 \\ a_3 & b_3 & c_3 \end{vmatrix}}$	(4-16)

Review Exercises

In Exercises 1 through 4, evaluate the given determinants.

1. $\begin{vmatrix} -2 & 5 \\ 3 & 1 \end{vmatrix}$ **2.** $\begin{vmatrix} 4 & 0 \\ -2 & -6 \end{vmatrix}$ **3.** $\begin{vmatrix} -18 & -33 \\ -21 & 44 \end{vmatrix}$ **4.** $\begin{vmatrix} 0.91 & -1.2 \\ 0.73 & -5.0 \end{vmatrix}$

In Exercises 5 through 8, find the slopes of the lines which pass through the given points.

5. $(2, 0), (4, -8)$ **6.** $(-1, -5), (-4, 4)$ **7.** $(4, -2), (-3, -4)$ **8.** $(-6, \frac{1}{2}), (1, -\frac{7}{2})$

In Exercises 9 through 12, find the slopes and y-intercepts of the lines with the given equations, and sketch the graphs.

9. $y = -2x + 4$ **10.** $y = \dfrac{2}{3}x - 3$ **11.** $8x - 2y = 5$ **12.** $3x = 8 + 3y$

In Exercises 13 through 20, solve the given systems of equations graphically.

13. $y = 2x - 4$
$y = -\dfrac{3}{2}x + 3$

14. $y = -3x + 3$
$y = 2x - 6$

15. $4x - y = 6$
$3x + 2y = 12$

16. $2x - 5y = 10$
$3x + y = 6$

17. $7x = 2y + 14$
$y = -4x + 4$

18. $5x = 15 - 3y$
$y = 6x - 12$

19. $3x + 4y = 6$
$2x - 3y = 2$

20. $5x + 2y = 5$
$2x - 4y = 3$

In Exercises 21 through 32, solve the given systems of equations algebraically.

21. $x + 2y = 5$
$x + 3y = 7$

22. $2x - y = 7$
$x + y = 2$

23. $4x + 3y = -4$
$y = 2x - 3$

24. $x = -3y - 2$
$-2x - 9y = 2$

25. $3x + 4y = 6$
$9x + 8y = 11$

26. $3x - 6y = 5$
$7x + 2y = 4$

27. $2x - 5y = 8$
$5x - 3y = 7$

28. $3x + 4y = 8$
$2x - 3y = 9$

29. $7x = 2y - 6$
$7y = 12 - 4x$

30. $3y = 8 - 5x$
$6x = 8y + 11$

31. $0.9x - 1.1y = 0.4$
$0.6x - 0.3y = 0.5$

32. $0.42x - 0.56y = 1.26$
$0.98x - 1.40y = -0.28$

In Exercises 33 through 44, solve the given systems of equations by use of determinants. (These systems are the same as for Exercises 21 through 32.)

33. $x + 2y = 5$
$x + 3y = 7$

34. $2x - y = 7$
$x + y = 2$

35. $4x + 3y = -4$
$y = 2x - 3$

36. $x = -3y - 2$
$-2x - 9y = 2$

37. $3x + 4y = 6$
$9x + 8y = 11$

38. $3x - 6y = 5$
$7x + 2y = 4$

39. $2x - 5y = 8$
$5x - 3y = 7$

40. $3x + 4y = 8$
$2x - 3y = 9$

41. $7x = 2y - 6$
$7y = 12 - 4x$

42. $3y = 8 - 5x$
$6x = 8y + 11$

43. $0.9x - 1.1y = 0.4$
$0.6x - 0.3y = 0.5$

44. $0.42x - 0.56y = 1.26$
$0.98x - 1.40y = -0.28$

In Exercises 45 through 48, evaluate the given determinants.

45. $\begin{vmatrix} 4 & -1 & 8 \\ -1 & 6 & -2 \\ 2 & 1 & -1 \end{vmatrix}$

46. $\begin{vmatrix} -5 & 0 & -5 \\ 2 & 3 & -1 \\ -3 & 2 & 2 \end{vmatrix}$

47. $\begin{vmatrix} -2.2 & -4.1 & 7.0 \\ 1.2 & 6.4 & -3.5 \\ -7.2 & 2.4 & -1.0 \end{vmatrix}$

48. $\begin{vmatrix} 30 & 22 & -12 \\ 0 & -34 & 44 \\ 35 & -41 & -27 \end{vmatrix}$

In Exercises 49 through 56, solve the given systems of equations algebraically.

49. $2x + y + z = 4$
$x - 2y - z = 3$
$3x + 3y - 2z = 1$

50. $x + 2y + z = 2$
$3x - 6y + 2z = 2$
$2x - z = 8$

51. $3x + 2y + z = 1$
$9x - 4y + 2z = 8$
$12x - 18y = 17$

52. $2x + 2y - z = 2$
$3x + 4y + z = -4$
$5x - 2y - 3z = 5$

53. $2r + s + 2t = 8$
$3r - 2s - 4t = 5$
$-2r + 3s + 4t = -3$

54. $2u + 2v - w = -2$
$4u - 3v + 2w = -2$
$8u - 4v - 3w = 13$

55. $4x + 6y - z = -2$
$3x - 5y + 4z = 8$
$6x + 4y + 2z = 5$

56. $3t + 2u + 6v = 3$
$4t - 3u + 2v = 13$
$5t + u + v = 0$

In Exercises 57 through 64, solve the given systems of equations by use of determinants. (These systems are the same as for Exercises 49 through 56.)

57. $2x + y + z = 4$
$x - 2y - z = 3$
$3x + 3y - 2z = 1$

58. $x + 2y + z = 2$
$3x - 6y + 2z = 2$
$2x - z = 8$

59. $3x + 2y + z = 1$
$9x - 4y + 2z = 8$
$12x - 18y = 17$

60. $2x + 2y - z = 2$
$3x + 4y + z = -4$
$5x - 2y - 3z = 5$

61. $2r + s + 2t = 8$
$3r - 2s - 4t = 5$
$-2r + 3s + 4t = -3$

62. $2u + 2v - w = -2$
$4u - 3v + 2w = -2$
$8u - 4v - 3w = 13$

63. $4x + 6y - z = -2$
$3x - 5y + 4z = 8$
$6x + 4y + 2z = 5$

64. $3t + 2u + 6v = 3$
$4t - 3u + 2v = 13$
$5t + u + v = 0$

In Exercises 65 through 68, let $1/x = u$ and $1/y = v$. Solve for u and v, and then solve for x and y. In this way we will see how to solve systems of equations involving reciprocals.

65. $\dfrac{1}{x} - \dfrac{1}{y} = \dfrac{1}{2}$
$\dfrac{1}{x} + \dfrac{1}{y} = \dfrac{1}{4}$

66. $\dfrac{1}{x} + \dfrac{1}{y} = 3$
$\dfrac{2}{x} + \dfrac{1}{y} = 1$

67. $\dfrac{2}{x} + \dfrac{3}{y} = 3$
$\dfrac{5}{x} - \dfrac{6}{y} = 3$

68. $\dfrac{3}{x} - \dfrac{2}{y} = 4$
$\dfrac{2}{x} + \dfrac{4}{y} = 1$

In Exercises 69 and 70, determine the value of k which makes the system dependent. In Exercises 71 and 72, determine the value of k which makes the system inconsistent.

69. $3x - ky = 6$
$x + 2y = 2$

70. $5x + 20y = 15$
$2x + ky = 6$

71. $kx - 2y = 5$
$4x + 6y = 1$

72. $2x - 5y = 7$
$kx + 10y = 2$

In Exercises 73 and 74, solve the given systems of equations by any appropriate method.

73. A 20-ft crane arm with a supporting cable and with a 100-lb box suspended from its end has forces acting on it as shown in Fig. 4-38. Equations which can be used to find the forces are

$$F_1 + 2F_2 = 280$$
$$0.866F_1 \qquad - F_3 = 0$$
$$3F_1 - 4F_2 = 600$$

Find the forces, in pounds.

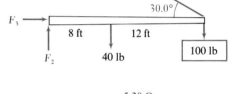

Fig. 4-38

74. In applying Kirchhoff's laws (see Exercise 20 of Section 4-6) to the electric circuit shown in Fig. 4-39, the following equations result. Determine the indicated currents, in amperes.

$$i_1 + i_2 + i_3 = 0$$
$$5.20i_1 - 3.25i_2 = 8.33 - 6.45$$
$$3.25i_2 - 2.62i_3 = 6.45 - 9.80$$

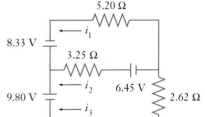

Fig. 4-39

In Exercises 75 through 84, set up systems of equations and solve by any appropriate method.

75. A computer analysis showed that the temperature T of the ocean water within 1000 m of a nuclear plant discharge pipe was given by $T = \dfrac{a}{x + 100} + b$, where x is the distance from the pipe and a and b are constants. If $T = 14°C$ for $x = 0$, and $T = 10°C$ for $x = 900$ m, find a and b.

76. A sales representative receives a fixed amount plus a percentage of sales commissions each month. If the representative received $3260 on sales of $22,000 in one month and $4380 on sales of $36,000 in the following month, what is the fixed amount and the commission percent?

77. A satellite is to be launched from a space shuttle. It is calculated that the satellite's speed will be 24,200 km/h if launched directly ahead of the shuttle, or 21,400 km/h if launched directly to the rear of the shuttle. What is the speed of the shuttle and the launching speed of the satellite relative to the shuttle?

78. The velocity v of sound is a function of the temperature T according to the function $v = aT + b$, where a and b are constants. If $v = 337.5$ m/s for $T = 10.0°C$, and $v = 346.6$ m/s for $T = 25.0°C$, find v as a function of T.

79. The power, in watts, dissipated in an electric resistance, in ohms, equals the resistance times the square of the current, in amperes. If 1 A flows through resistance R_1 and 3 A flows through resistance R_2, the total power dissipated is 14 W. If 3 A flows through R_1 and 1 A flows through R_2, the total power dissipated is 6 W. Find R_1 and R_2.

80. Twelve equal rectangular ceiling panels are used as shown in Fig. 4-40. If each panel is 6.0 in. longer than it is wide, and a total of 132.0 ft of edge and middle strips are used, what are the dimensions of the room?

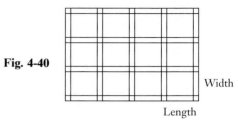

Fig. 4-40

81. The weight of a lever may be considered to be at its center. A 20-ft lever of weight w is balanced on a fulcrum 8 ft from one end by a load L at that end. If 4 times the load is placed at that end, it requires a 20-lb weight at the other end to balance the lever. What is the initial load L and the weight w of the lever? (See Exercise 38 of Section 4-4.)

82. Two fuel mixtures, one of 2% oil and 98% gasoline and another of 8% oil and 92% gasoline, are to be mixed to make 10 L of a fuel which is 4% oil and 96% gasoline which is to be used in a chain saw. How much of each is needed?

83. A manufacturer of televisons, videocassette recorders, and compact disc players makes 1750 total units each month. The number of CDs is twice that of TVs, and four times the number of CDs is 700 more than three times the number of TVs and VCRs combined. How many of each are produced?

84. One ampere of electric current is passed through a solution of sulfuric acid, silver nitrate, and cupric sulfate, releasing hydrogen gas, silver, and copper. A total weight of 1.750 g is released. The weight of silver deposited is 3.40 times the weight of copper deposited, and the weight of copper and 70 times the weight of hydrogen combined equals the weight of silver deposited less 0.037 g. How much of each is released?

Practice Test

1. Find the slope of the line through $(2, -5)$ and $(-1, 4)$.

2. Solve by substitution:

    $x + 2y = 5$
    $4y = 3 - 2x$

3. Solve by determinants:

    $3x - 2y = 4$
    $2x + 5y = -1$

4. Sketch the graph of $2x + y = 4$ by finding its slope and y-intercept.

5. The perimeter of a rectangular ranch is 24 km, and the length is 6 km more than the width. Set up equations relating the length x and the width y, and then solve for x and y.

6. Solve graphically by first finding both intercepts of each line. Estimate values to the nearest tenth.

    $2x - 3y = 6$
    $4x + y = 4$

7. By volume, one alloy is 60% copper, 30% zinc, and 10% nickel. A second alloy has percents of 50, 30, and 20, respectively, of the three metals. A third alloy is 30% copper and 70% nickel. How much of each must be used so that 100 cm^3 of the resulting alloy has percents of 40, 15, and 45, respectively?

8. Solve for y by determinants:

    $3x + 2y - z = 4$
    $2x - y + 3z = -2$
    $x + 4z = 5$

5 Factoring and Fractions

Time for processing data by a computer system may be reduced by adding components to the system. In Section 5-7 we show how this time may be calculated.

In Chapter 1 we introduced certain fundamental algebraic operations. These have been sufficient for our purposes to this point. However, material which we shall encounter requires algebraic methods beyond those we have developed to this point. Therefore, in this chapter we shall develop certain additional basic algebraic topics, which in turn will allow us to develop other topics having technical and scientific applications.

Although the primary purpose in this chapter is to develop additional algebraic methods, we will show many of the applied areas in which these algebraic methods are used. Also, certain direct applications will be demonstrated.

5-1 Special Products

In working with algebraic expressions, we encounter certain types of products so frequently that we should become extremely familiar with them. These products are obtained by the methods of multiplication of algebraic expressions developed in Chapter 1 and are stated here in general form.

$$a(x + y) = ax + ay \tag{5-1}$$
$$(x + y)(x - y) = x^2 - y^2 \tag{5-2}$$
$$(x + y)^2 = x^2 + 2xy + y^2 \tag{5-3}$$
$$(x - y)^2 = x^2 - 2xy + y^2 \tag{5-4}$$
$$(x + a)(x + b) = x^2 + (a + b)x + ab \tag{5-5}$$
$$(ax + b)(cx + d) = acx^2 + (ad + bc)x + bd \tag{5-6}$$

These **special products** *should be known* ***thoroughly*** so that the multiplications represented are easily and clearly recognized. When this is the case, they allow us to perform many multiplications quickly and easily, often by inspection. We must realize that they are written in their most concise form. Any of the literal numbers appearing in these products may represent an expression which in turn represents a number.

EXAMPLE A _____ Using Eq. (5-1) in the following product, we have

$$6(3r + 2s) = 6(3r) + 6(2s) = 18r + 12s$$

Using Eq. (5-2), we have

$$(3r + 2s)(3r - 2s) = (3r)^2 - (2s)^2 = 9r^2 - 4s^2$$

sum of 3r and 2s   difference of 3r and 2s   difference of squares

When we use Eq. (5-1) in the first illustration, we see that $a = 6$. In both illustrations $3r = x$ and $2s = y$. ∎

EXAMPLE B _____ Using Eqs. (5-3) and (5-4) in the following products, we have

$$(5a + 2)^2 = (5a)^2 + 2(5a)(2) + 2^2 = 25a^2 + 20a + 4$$

square   twice product   square

$$(5a - 2)^2 = (5a)^2 - 2(5a)(2) + 2^2 = 25a^2 - 20a + 4$$

NOTE ▷ In these illustrations, we have let $x = 5a$ and $y = 2$. *It should also be emphasized that* ***(5a + 2)²*** *is* **not** ***(5a)² + 2²***, *or* $25a^2 + 4$. We must be careful to follow the correct form of Eqs. (5-3) and (5-4) properly and include the middle term, $20a$. (See Example E of Section 1-9.) ∎

EXAMPLE C _____ Using Eqs. (5-5) and (5-6) in the following products, we have

$$(x + 5)(x - 3) = x^2 + [5 + (-3)]x + (5)(-3) = x^2 + 2x - 15$$

$$(4x + 5)(2x - 3) = (4x)(2x) + [(4)(-3) + (5)(2)]x + (5)(-3)$$
$$= 8x^2 - 2x - 15$$

∎

Generally, when we use these special products, we do the middle step as shown in each of the examples above mentally and write down the result directly. This is indicated in the following example.

EXAMPLE D

$$2(x - 6) = 2x - 12$$
$$(y - 5)(y + 5) = y^2 - 25$$
$$(3x - 2)^2 = 9x^2 - 12x + 4$$
$$(x - 4)(x + 7) = x^2 + 3x - 28$$

At times these special products may appear in combinations. When this happens, it may be necessary to indicate an intermediate step.

EXAMPLE E

1. In analyzing the forces on a beam, the expression $Fa(L - a)(L + a)$ occurs. In expanding this expression, we first multiply $L - a$ by $L + a$ by use of Eq. (5-2). The expansion is then completed by use of Eq. (5-1), which is the distributive law.

$$Fa(L - a)(L + a) = Fa(L^2 - a^2) = FaL^2 - Fa^3$$

2. In electricity, the expression $R(i_1 + i_2)^2$ is used. In expanding this expression, we first perform the square by use of Eq. (5-3), then complete the expansion by use of Eq. (5-1).

$$R(i_1 + i_2)^2 = R(i_1^2 + 2i_1i_2 + i_2^2) = Ri_1^2 + 2Ri_1i_2 + Ri_2^2$$

EXAMPLE F

In determining the product $(x + y - 2)^2$, we may group the quantity $(x + y)$ in an intermediate step. This leads to

$$(x + y - 2)^2 = [(x + y) - 2]^2 = (x + y)^2 - 2(x + y)(2) + 2^2$$
$$= x^2 + 2xy + y^2 - 4x - 4y + 4$$

In this example we used Eqs. (5-3) and (5-4).

There are four other special products which occur less frequently. However, they are sufficiently important that they should be readily recognized. They are shown in Eqs. (5-7) through (5-10).

$$(x + y)^3 = x^3 + 3x^2y + 3xy^2 + y^3 \tag{5-7}$$
$$(x - y)^3 = x^3 - 3x^2y + 3xy^2 - y^3 \tag{5-8}$$
$$(x + y)(x^2 - xy + y^2) = x^3 + y^3 \tag{5-9}$$
$$(x - y)(x^2 + xy + y^2) = x^3 - y^3 \tag{5-10}$$

The following examples illustrate the use of Eqs. (5-7) through (5-10).

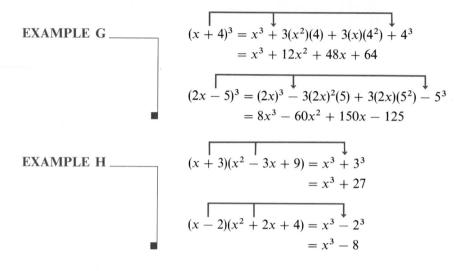

EXAMPLE G

$$(x + 4)^3 = x^3 + 3(x^2)(4) + 3(x)(4^2) + 4^3$$
$$= x^3 + 12x^2 + 48x + 64$$

$$(2x - 5)^3 = (2x)^3 - 3(2x)^2(5) + 3(2x)(5^2) - 5^3$$
$$= 8x^3 - 60x^2 + 150x - 125$$

EXAMPLE H

$$(x + 3)(x^2 - 3x + 9) = x^3 + 3^3$$
$$= x^3 + 27$$

$$(x - 2)(x^2 + 2x + 4) = x^3 - 2^3$$
$$= x^3 - 8$$

Exercises 5-1

In Exercises 1 through 36, find the indicated products directly by inspection. It should not be necessary to write down intermediate steps.

1. $40(x - y)$ **2.** $2x(a - 3)$ **3.** $2x^2(x - 4)$ **4.** $3a^2(2a + 7)$

5. $(y + 6)(y - 6)$ **6.** $(s + 2t)(s - 2t)$ **7.** $(3v - 2)(3v + 2)$ **8.** $(ab - c)(ab + c)$

9. $(4x - 5y)(4x + 5y)$ **10.** $(7s + 2t)(7s - 2t)$ **11.** $(12 + 5ab)(12 - 5ab)$ **12.** $(2xy - 11)(2xy + 11)$

13. $(5f + 4)^2$ **14.** $(i_1 + 3)^2$ **15.** $(2x + 7)^2$ **16.** $(5a + 2b)^2$

17. $(x - 1)^2$ **18.** $(y - 6)^2$ **19.** $(4a + 7xy)^2$ **20.** $(3x + 10y)^2$

21. $(4x - 2y)^2$ **22.** $(a - 5p)^2$ **23.** $(6s - t)^2$ **24.** $(3p - 4q)^2$

25. $(x + 1)(x + 5)$ **26.** $(y - 8)(y + 5)$ **27.** $(3 + c)(6 + c)$ **28.** $(1 - t)(7 - t)$

29. $(3x - 1)(2x + 5)$ **30.** $(2x - 7)(2x + 1)$ **31.** $(4x - 5)(5x + 1)$ **32.** $(2y - 1)(3y - 1)$

33. $(5v - 3)(4v + 5)$ **34.** $(7s + 6)(2s + 5)$ **35.** $(3x + 7y)(2x - 9y)$ **36.** $(8x - y)(3x + 4y)$

Use the special products of this section to determine the products of Exercises 37 through 60. You may need to write down one or two intermediate steps.

37. $2(x - 2)(x + 2)$ **38.** $5(n - 5)(n + 5)$ **39.** $2a(2a - 1)(2a + 1)$ **40.** $4c(2c - 3)(2c + 3)$

41. $6a(x + 2b)^2$ **42.** $7r(5r + 2b)^2$ **43.** $5n^2(2n + 5)^2$ **44.** $8p(p - 7)^2$

45. $4a(2a - 3)^2$ **46.** $6t^2(5t - 3s)^2$ **47.** $(x + y + 1)^2$ **48.** $(x + 2 + 3y)^2$

49. $(3 - x - y)^2$ **50.** $2(x - y + 1)^2$ **51.** $(5 - t)^3$ **52.** $(2s + 3)^3$

53. $(2x + 5t)^3$ **54.** $(x - 5y)^3$ **55.** $(x + y - 1)(x + y + 1)$

56. $(2a - c + 2)(2a - c - 2)$ **57.** $(x + 2)(x^2 - 2x + 4)$ **58.** $(a - 3)(a^2 + 3a + 9)$

59. $(4 - 3x)(16 + 12x + 9x^2)$ **60.** $(2x + 3a)(4x^2 - 6ax + 9a^2)$

Use the special products of this section to determine the products in Exercises 61 through 68. Each comes from the technical area indicated.

61. $P_1(P_0c + G)$ (computers)   **62.** $h^2L(L + 1)$ (spectroscopy)

63. $4(p + DA)^2$ (photography)

64. $(2J + 3)(2J - 1)$ (lasers)

65. $\frac{1}{2}\pi(R + r)(R - r)$ (architecture)

66. $w(1 - h)(4 - h^2)$ (hydrodynamics)

67. $\dfrac{L}{6}(x - a)^3$ (mechanics: beams)

68. $(1 - z)^2(1 + z)$ (motion: gyroscope)

5-2 Factoring: Common Factor and Difference of Squares ▬▬

We often find that we want to determine what expressions can be multiplied together to equal a given algebraic expression. We know from Section 1-8 that when an algebraic expression is the product of two or more quantities, each of these quantities is called a **factor** of the expression. *Therefore, determining these factors, which is essentially reversing the process of finding a product, is called* **factoring.**

In our work on factoring we shall consider only the factoring of polynomials (see Section 1-10) which have integers as coefficients for all terms. Also, all factors will have integral coefficients. *A polynomial or a factor is called* **prime** *if it contains no factors other than* $+1$ *or* -1 *and plus or minus itself. We say that an expression is* **factored completely** *if it is expressed as a product of its prime factors.*

EXAMPLE A — When we factor the expression $12x + 6x^2$ as

$$12x + 6x^2 = 2(6x + 3x^2)$$

we see that it has not been factored completely. The factor $6x + 3x^2$ is not prime, for it may be factored as

$$6x + 3x^2 = 3x(2 + x)$$

Therefore, the expression $12x + 6x^2$ is factored completely as

$$12x + 6x^2 = 6x(2 + x)$$

Here the factors x and $2 + x$ are prime. The numerical coefficient, 6, could be ■ factored into $(2)(3)$, but it is normal practice not to factor numerical coefficients.

NOTE ▷ To factor expressions easily, we must be familiar with algebraic multiplication, particularly the special products of the preceding section. *The solution of factoring problems is heavily dependent on the recognition of special products.* The special products also provide methods of checking answers and deciding whether or not a given factor is prime.

Often an algebraic expression contains a monomial that is common to each term of the expression. Therefore, *the first step in factoring any expression should be to factor out any* **common monomial factor** *that may exist.* To do this, we note the common factor by inspection and then use the reverse of the distributive law, Eq. (5-1), to show the factored form.

For reference,
Eq. (5-1) is
$a(x + y) = ax + ay.$

In Example A there is a common monomial factor of $6x$. However, the purpose of the example is to show the meaning of factors and complete factoring. The following three examples illustrate how a common monomial factor is identified and factored out of an expression.

EXAMPLE B In factoring the expression $6x - 2y$, we note that each term contains the factor 2. Therefore,

$$6x - 2y = 2(3x) - 2y = 2(3x - y)$$

Here, 2 is the common monomial factor, and $2(3x - y)$ is the required factored form of $6x - 2y$. Once the common factor has been identified, it is not actually necessary to write a term like $6x$ as $2(3x)$. The result can be written directly.

We can check our result by multiplication. In this case,

$$2(3x - y) = 6x - 2y$$

which is the original expression.

In Example B we determined the common factor of 2 by inspection. This is normally the way in which the common factor is found. Once the common factor has been found, the other factor can be determined by dividing the original expression by the common factor.

EXAMPLE C Factor: $4ax^2 + 2ax$.

The numerical factor 2 and the literal factors a and x are common to each term. Therefore, the common monomial factor of $4ax^2 + 2ax$ is $2ax$. This means that

$$4ax^2 + 2ax = 2ax(2x + 1)$$

Note the presence of the 1 in the factored form. When we divide $4ax^2 + 2ax$ by $2ax$ we get

$$\frac{4ax^2 + 2ax}{2ax} = \frac{4ax^2}{2ax} + \frac{2ax}{2ax} = 2x + 1$$

NOTE ▷ *We must include the **1** in the factor **2x + 1**,* although it is a common error to omit it. Without the 1, when the factored form is multiplied out, we would not obtain the proper expression.

EXAMPLE D Factor: $6a^5x^2 - 9a^3x^3 + 3a^3x^2$.

After inspecting each term, we determine that each contains a factor of 3, a^3, and x^2. Thus, the common monomial factor is $3a^3x^2$. This means that

$$6a^5x^2 - 9a^3x^3 + 3a^3x^2 = 3a^3x^2(2a^2 - 3x + 1)$$

It is often necessary to use factoring in solving an equation. This is illustrated in the following example.

EXAMPLE E _____ An equation used in the analysis of FM reception is $R_F = \alpha(2R_A + R_F)$. Solve for R_F.

The steps in the solution are as follows:

$$R_F = \alpha(2R_A + R_F) \qquad \text{original equation}$$

$$R_F = 2\alpha R_A + \alpha R_F \qquad \text{use distributive law}$$

$$R_F - \alpha R_F = 2\alpha R_A \qquad \text{subtract } \alpha R_F \text{ from both sides}$$

$$R_F(1 - \alpha) = 2\alpha R_A \qquad \text{factor out } R_F \text{ on left}$$

$$R_F = \frac{2\alpha R_A}{1 - \alpha} \qquad \text{divide both sides by } 1 - \alpha$$

We see that we collected both terms containing R_F on the left in order that we
■ could factor and thereby solve for R_F.

For reference,   Another important form for factoring is based on the special product of Eq.
Eq. (5-2) is   (5-2). In Eq. (5-2) we see that the product of the sum and the difference of two
$(x + y)(x - y) = x^2 - y^2.$ numbers results in the difference between the squares of the numbers. Therefore,
_factoring the difference of two squares gives factors which are the sum and the dif-
ference of the numbers._

EXAMPLE F _____ In factoring $x^2 - 16$, we note that x^2 is the square of x and that 16 is the square of 4. Therefore,

$$\text{squares}$$
$$x^2 - 16 = \quad x^2 - 4^2 = (x + 4) \ (x - 4)$$
$$\text{difference} \qquad \text{sum} \quad \text{difference}$$

Usually in factoring an expression of this type we do not actually write out the
■ middle step as shown.

EXAMPLE G _____ Since $4x^2$ is the square of $2x$ and 9 is the square of 3, we may factor $4x^2 - 9$ as

$$4x^2 - 9 = (2x + 3)(2x - 3)$$

In the same way,

$$(y - 3)^2 - 16x^4 = (y - 3 + 4x^2)(y - 3 - 4x^2)$$

■ where we note that $16x^4 = (4x^2)^2$.

NOTE ▷   As indicated previously, _if it is possible to factor out a common monomial
factor, **this factoring should be done first.**_ We should then inspect the resulting
factors to see if further factoring can be done. It is possible, for example, that the
complete factoring resulting factor is a difference of squares. Thus, complete factoring often requires
more than one step. Be sure to include all prime factors in writing the result.

EXAMPLE H _____ In factoring $20x^2 - 45$, we note that there is a common factor of 5 in each term. Therefore, $20x^2 - 45 = 5(4x^2 - 9)$. However, the factor $4x^2 - 9$ itself is the difference of squares. Therefore, $20x^2 - 45$ is completely factored as

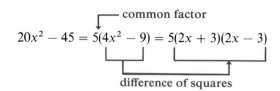

In factoring $x^4 - y^4$, we note that we have the difference of squares. Therefore, $x^4 - y^4 = (x^2 + y^2)(x^2 - y^2)$. However, the factor $x^2 - y^2$ is also the difference of squares. This means that

$$x^4 - y^4 = (x^2 + y^2)(x^2 - y^2) = (x^2 + y^2)(x + y)(x - y)$$

NOTE ▷ **The factor $x^2 + y^2$ is prime.** It is not equal to $(x + y)^2$. (See Example B of Section 5-1.)

■

The terms of a polynomial can sometimes be grouped such that the polynomial can then be factored by the methods of this section. The following example illustrates this method of **factoring by grouping.**

EXAMPLE I _____ Factor: $2x - 2y + ax - ay$.
We note that each of the first two terms contains a factor of 2, and each of the third and fourth terms contains a factor of a. Grouping the terms in this way, and then factoring each group, we have

$$2x - 2y + ax - ay = (2x - 2y) + (ax - ay)$$
$$= 2(x - y) + a(x - y)$$

Each of the two terms we now have contains a **common binomial factor** of $x - y$. Since this is a common factor, we have

$$2(x - y) + a(x - y) = (x - y)(2 + a)$$

This means that

$$2x - 2y + ax - ay = (x - y)(2 + a)$$

■ where the expression on the right is the factored form of the polynomial.

The general method of factoring by grouping can be used with several types of groupings. We will discuss another type in the following section.

Exercises 5-2 ━━━━━━━━━━━━━━━━━

In Exercises 1 through 40, factor the given expressions completely.

1. $6x + 6y$

2. $3a - 3b$

3. $5a - 5$

4. $2x^2 + 2$

5. $3x^2 - 9x$

6. $4s^2 + 20s$

7. $7b^2y - 28b$

8. $5a^2 - 20ax$

9. $12n^2 + 6n$

10. $18p^3 - 3p^2$

11. $2x + 4y - 8z$

12. $10a - 5b + 15c$

13. $3ab^2 - 6ab + 12ab^3$ **14.** $4pq - 14q^2 - 16pq^2$ **15.** $12pq^2 - 8pq - 28pq^3$

16. $27a^2b - 24ab - 9a$ **17.** $2a^2 - 2b^2 + 4c^2 - 6d^2$ **18.** $5a + 10ax - 5ay + 20az$

19. $x^2 - 4$ **20.** $r^2 - 25$ **21.** $100 - y^2$ **22.** $49 - z^2$ **23.** $36a^2 - 1$

24. $81z^2 - 1$ **25.** $81s^2 - 25t^2$ **26.** $36s^2 - 121t^2$ **27.** $144n^2 - 169p^4$ **28.** $36a^2b^2 - 169c^2$

29. $(x + y)^2 - 9$ **30.** $(a - b)^2 - 1$ **31.** $2x^2 - 8$ **32.** $5a^2 - 125$

33. $3x^2 - 27z^2$ **34.** $4x^2 - 100y^2$ **35.** $2(a - 3)^2 - 8$ **36.** $a(x + 2)^2 - ay^2$

37. $x^4 - 16$ **38.** $y^4 - 81$ **39.** $x^8 - 1$ **40.** $2x^4 - 8y^4$

In Exercises 41 through 44, solve for the indicated letter.

41. $2a - b = ab + 3$, for a **42.** $n(x + 1) = 5 - x$, for x

43. $3 - 2s = 2(3 - st)$, for s **44.** $k(2 - y) = y(2k - 1)$, for y

In Exercises 45 through 52, factor the given expressions by grouping as illustrated in Example 1.

45. $3x - 3y + bx - by$ **46.** $am + an + cn + cm$ **47.** $a^2 + ax - ab - bx$

48. $2y - y^2 - 6y^4 + 12y^3$ **49.** $x^3 + 3x^2 - 4x - 12$ **50.** $x^3 - 5x^2 - x + 5$

51. $x^2 - y^2 + x - y$ **52.** $4p^2 - q^2 + 2p + q$

In Exercises 53 through 58, factor the given expressions. In Exercises 59 and 60, solve for the indicated letter. Each comes from the technical area indicated.

53. $Rv + Rv^2 + Rv^3$ (business)   **54.** $4d^2D^2 - 4d^3D - d^4$ (machine design)

55. $aD_1^2 - aD_2^2$ (surveying)   **56.** $K - Ka^2$ (chemistry)

57. $PbL^2 - Pb^3$ (architecture)   **58.** $4RI^2 - 9Ri^2$ (electricity)

59. $3BY + 5Y = 9BS$ (Solve for B.)   (physics: elasticity)

60. $R = kT_2^4 - kT_1^4$ (Solve for k and factor the resulting denominator.)   (energy: radiation)

5-3   Factoring Trinomials

In the previous section we introduced the concept of factoring and considered factoring based on special products of Eqs. (5-1) and (5-2). We now note that the special products formed from Eqs. (5-3) through (5-6) all result in trinomial (three-term) polynomials. Thus, trinomials of the types formed from these products are important expressions to be factored, and this section is devoted to them.

When factoring an expression based on Eq. (5-5), we start with the expression on the right and then find the factors which are at the left. Therefore, writing Eq. (5-5) with sides reversed, we have

For reference,
Eq. (5-5) is
$(x + a)(x + b) =$
$x^2 + (a + b)x + ab$.

$$\text{coefficient} \atop =1 \qquad x^2 + (a + b)x + ab = (x + a)(x + b)$$

We are to find integers a and b, and they are found by noting that

1. *the coefficient of x^2 is 1,*

2. *the product of a and b is the final constant ab,* and

3. *the sum of a and b is the coefficient of x.*

As in Section 5-2, we shall consider only factors in which all terms have integral coefficients.

EXAMPLE A _____   In factoring $x^2 + 3x + 2$, we set it up as

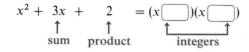

The constant 2 tells us that the product of the required integers is 2. Thus, the only possibilities are 2 and 1 (or 1 and 2). The plus sign before the 2 indicates that the sign before the 1 and 2 in the factors must be the same, either plus or minus. Since the coefficient of x, 3, is the sum of the integers, the plus sign before the 3 tells us that both signs are positive. Therefore,

$$x^2 + 3x + 2 = (x + 2)(x + 1)$$

In factoring $x^2 - 3x + 2$, the analysis is the same until we note that the middle term is negative. This tells us that both integers are negative in this case. Therefore,

$$x^2 - 3x + 2 = (x - 2)(x - 1)$$

For a trinomial containing x^2 and 2 to be factorable, the middle term must be $+3x$ or $-3x$. No other combination of integers gives the proper middle term. Therefore, the expression

$$x^2 + 4x + 2$$

cannot be factored. The integers would have to be 2 and 1, but the middle term would not be $4x$.

EXAMPLE B _____   In order to factor $x^2 + 7x - 8$, _we must find two integers whose **product** is_ -8 _and whose **sum** is_ $+7$. The possible factors of -8 are -8 and $+1$, $+8$ and -1, -4 and $+2$, and $+4$ and -2. Inspecting these we see that only $+8$ and -1 have the sum of $+7$. Therefore,

$$x^2 + 7x - 8 = (x + 8)(x - 1)$$

In the same way, we have

$$x^2 - x - 12 = (x - 4)(x + 3)$$

since -4 and $+3$ is the only pair of integers whose product is -12 and whose sum is -1.
   Also,

$$x^2 - 5xy + 6y^2 = (x - 3y)(x - 2y)$$

since -3 and -2 is the only pair of integers whose product is $+6$ and whose sum is -5. Here we note that we were to find second terms of each factor such that their product was $6y^2$ and their sum was $-5xy$. This means that each term must contain a factor of y.

For reference,
Eqs. (5-3) and (5-4) are
$(x + y)^2 = x^2 + 2xy + y^2$
$(x - y)^2 = x^2 - 2xy + y^2$.

In factoring a trinomial in which the second power term is x^2, we may find that the expression fits the form of Eq. (5-3) or Eq. (5-4), as well as Eq. (5-5). The following example illustrates this case.

EXAMPLE C In order to factor $x^2 + 10x + 25$, we must find two integers whose sum is $+10$ and whose product is $+25$. Since $5^2 = 25$ we note that this expression may fit the form of Eq. (5-3). This could be the case only if the first and third terms were perfect squares. We see that the sum of $+5$ and $+5$ is $+10$, which means

$$x^2 + 10x + 25 = (x + 5)(x + 5)$$

or

$$x^2 + 10x + 25 = (x + 5)^2$$

For reference,
Eq. (5-6) is
$(ax + b)(cx + d) =$
$acx^2 + (ad + bc)x + bd.$

Factoring expressions based upon the special product of Eq. (5-6) often requires some trial and error. However, the amount of trial and error can be kept to a minimum with a careful analysis of the coefficients of x^2 and the constant. Rewriting Eq. (5-6), with sides reversed, we have

This tells us that the coefficient of x^2 gives the possibilities for the coefficients a and c in the factors, and that the constant bd gives the possibilities for the integers b and d in the factors. It is then necessary to try possible combinations to determine which combination provides the proper outer and inner products for the middle term of the given expression.

EXAMPLE D When factoring the expression $2x^2 + 11x + 5$, we take the factors of 2 to be $+2$ and $+1$ (we will use only positive coefficients a and c when the coefficient of x^2 is positive). We now set up the factoring as

$$2x^2 + 11x + 5 = (2x\boxed{})(x\boxed{})$$

NOTE▷ Since the product of the integers to be found is $+5$, only integers of the same sign need be considered. Also, since *the sum of the outer and inner products is* **$+11$**, the integers are positive. The factors of $+5$ are $+1$ and $+5$, and -1 and -5, which means that $+1$ and $+5$ is the only possible pair. Now, trying the factors

$$+2x + 5x = +7x$$

we see that $7x$ is not the correct middle term.

(Continued on next page)

Next, trying

$$(2x + 1)(x + 5)$$

$+x$
$+10x$ $+ x + 10x = +11x$

we have the correct sum of $+11x$. Therefore,

$$2x^2 + 11x + 5 = (2x + 1)(x + 5)$$

According to this analysis, the expression $2x^2 + 10x + 5$ is not factorable, but the following expression is:

$$2x^2 + 7x + 5 = (2x + 5)(x + 1)$$

EXAMPLE E

In factoring $4x^2 + 4x - 3$, the coefficient 4 in $4x^2$ indicates that 4 and 1, or 2 and 2, are possible coefficients of x in the factors. The 3 indicates that only 3 and 1 are possible integers, and the minus sign with the 3 shows that the integers have different signs. The plus sign with the $4x$ tells us that the larger of the outer and inner products is positive. This gives us possible combinations of

$$(4x - 3)(x + 1), \qquad (4x - 1)(x + 3), \quad \text{or} \quad (2x + 3)(2x - 1)$$

The only one of these which has a middle term of $+4x$ is $(2x + 3)(2x - 1)$. Therefore,

$$4x^2 + 4x - 3 = (2x + 3)(2x - 1)$$

$+6x$
$-2x$
$-2x + 6x = +4x$

EXAMPLE F

Other examples of factoring based upon the special product of Eq. (5-6) are:

1. $3x^2 - 13x + 12 = (3x - 4)(x - 3)$

$-9x - 4x = -13x$

2. $6s^2 + 19st - 20t^2 = (6s - 5t)(s + 4t)$

$+24st - 5st = +19st$

In the first illustration, other possible factorizations of 12, such as 6×2 and 12×1, can be shown to give improper middle terms. In the second illustration, there are numerous possibilities for the combination of 6 and 20. *We must remember to* **check carefully that the** **middle term** *of the expression is the proper result* **NOTE** ▷ *of the factors* we have chosen.

EXAMPLE G

In factoring $9x^2 - 6x + 1$, we note that $9x^2$ is the square of $3x$ and 1 is the square of 1. Therefore, we recognize that this expression might fit the perfect square form of Eq. (5-4). This leads us to factor it tentatively as

$$9x^2 - 6x + 1 = (3x - 1)^2$$

However, before we can be certain that this factorization is correct, we must check to see if the middle term of the expansion of $(3x - 1)^2$ is $-6x$. Since $-6x$ properly fits the form of Eq. (5-4), the factorization is correct.

In the same way, we have

$$36x^2 + 84xy + 49y^2 = (6x + 7y)^2$$

NOTE ▷ As we pointed out in Section 5-2, we must be careful to see that we have factored an expression completely. *We look for common monomial factors first,* and then check each resulting factor. This check of each factor should be made every time we complete a step in factoring.

EXAMPLE H — When factoring $2x^2 + 6x - 8$, we first note the common monomial factor of 2. This leads to

$$2x^2 + 6x - 8 = 2(x^2 + 3x - 4)$$

We now notice that $x^2 + 3x - 4$ is also factorable. Therefore,

$$2x^2 + 6x - 8 = 2(x + 4)(x - 1)$$

Now each factor is prime.

Having noted the common factor of 2 prevents our having to check factors of 2 and 8. If we had not noted the common factor, we might have arrived at factorizations of

$$(2x + 8)(x - 1) \quad \text{or} \quad (2x - 2)(x + 4)$$

(possibly after a number of trials). Each is correct as far as it goes, but it is **not complete.** Since $2x + 8 = 2(x + 4)$, or $2x - 2 = 2(x - 1)$, we would eventually arrive at the proper result shown above.

EXAMPLE I — A study of the path of a rocket leads to the expression $16t^2 + 240t - 1600$, where t is the time of flight. Factor this expression.

An inspection shows that there is a common factor of 16. (This might be found by noting successive factors of 2 or 4.) Factoring out 16 leads to

$$16t^2 + 240t - 1600 = 16(t^2 + 15t - 100)$$
$$= 16(t + 20)(t - 5)$$

Here, factors of 100 need to be checked for sums equal to 15. This might take a little time, but it is much simpler than looking for factors of 16 and 1600 with sums equal to 240.

EXAMPLE J — When factoring $4x^3 + 18x^2 - 10x$, we must first see the factor $2x$ in each term. Factoring out this $2x$, we have

$$4x^3 + 18x^2 - 10x = 2x(2x^2 + 9x - 5)$$

Now we factor $2x^2 + 9x - 5$ as

$$2x^2 + 9x - 5 = (2x - 1)(x + 5)$$

Therefore,

$$4x^3 + 18x^2 - 10x = 2x(2x - 1)(x + 5)$$

In the previous section, we introduced factoring by grouping for an expression which can be written with a common binomial factor. The following example illustrates the method for an expression which uses factoring of a trinomial to write it as the difference of squares. Although the method can be used for several types of grouping, we will restrict our attention to these two types.

EXAMPLE K _____ Factor: $x^2 - 4xy + 4y^2 - 9$.

We see that the first three terms of this expression represent $(x - 2y)^2$. Thus, grouping these terms, we have the following solution:

$$x^2 - 4xy + 4y^2 - 9$$
$$= (x^2 - 4xy + 4y^2) - 9 \qquad \text{group terms}$$
$$= (x - 2y)^2 - 9 \qquad \text{factor grouping (note difference of squares)}$$
$$= [(x - 2y) + 3][(x - 2y) - 3] \quad \text{factor difference of squares}$$
$$= (x - 2y + 3)(x - 2y - 3)$$

The solution was completed after noting that the grouping gave us the difference of squares.

It should be noted that not all groupings work. If we had seen the combination $4y^2 - 9$, which is factorable, and then grouped the first two terms and the last two terms, this would not have led to the factorization.

Exercises 5-3

In Exercises 1 through 44, factor the given expressions completely.

1. $x^2 + 5x + 4$ **2.** $x^2 - 5x - 6$ **3.** $s^2 - s - 42$ **4.** $a^2 + 14a - 32$

5. $t^2 + 5t - 24$ **6.** $r^2 - 11r + 18$ **7.** $x^2 + 2x + 1$ **8.** $y^2 + 8y + 16$

9. $x^2 - 4xy + 4y^2$ **10.** $b^2 - 12bc + 36c^2$ **11.** $3x^2 - 5x - 2$ **12.** $2n^2 - 13n - 7$

13. $3y^2 - 8y - 3$ **14.** $5x^2 + 9x - 2$ **15.** $2s^2 + 13s + 11$ **16.** $7y^2 - 12y + 5$

17. $3f^2 - 16f + 5$ **18.** $5x^2 - 3x - 2$ **19.** $2t^2 + 7t - 15$ **20.** $3n^2 - 20n + 20$

21. $3t^2 - 7tu + 4u^2$ **22.** $3x^2 + xy - 14y^2$ **23.** $4x^2 - 3x - 7$ **24.** $2z^2 + 13z - 5$

25. $9x^2 + 7xy - 2y^2$ **26.** $4r^2 + 11rs - 3s^2$ **27.** $4m^2 + 20m + 25$ **28.** $16q^2 + 24q + 9$

29. $4x^2 - 12x + 9$ **30.** $a^2c^2 - 2ac + 1$ **31.** $9t^2 - 15t + 4$ **32.** $6x^2 + x - 12$

33. $8b^2 + 31b - 4$ **34.** $12n^2 + 8n - 15$ **35.** $4p^2 - 25pq + 6q^2$ **36.** $12x^2 + 4xy - 5y^2$

37. $12x^2 + 47xy - 4y^2$ **38.** $8r^2 - 14rs - 9s^2$ **39.** $2x^2 - 14x + 12$ **40.** $6y^2 - 33y - 18$

41. $4x^2 + 14x - 8$ **42.** $12x^2 + 22xy - 4y^2$ **43.** $ax^3 + 4a^2x^2 - 12a^3x$ **44.** $6x^4 - 13x^3 + 5x^2$

In Exercises 45 through 48, factor the given expressions by grouping, as illustrated in Example K.

45. $a^2 + 2ab + b^2 - 4$ **46.** $x^2 - 6xy + 9y^2 - 4z^2$

47. $25a^2 - 25x^2 - 10xy - y^2$ **48.** $r^2 - s^2 + 2st - t^2$

In Exercises 49 through 52, factor the given expression by referring directly to the special products in Eqs. (5-7) through (5-10), respectively. These expressions are not trinomials, but their factorization depends on the proper recognition of their forms, as with Eqs. (5-3) and (5-4).

49. $x^3 + 3x^2 + 3x + 1$ **50.** $x^3 - 6x^2 + 12x - 8$ **51.** $8x^3 + 1$ **52.** $x^3 - 27$

Factor the expressions given in Exercises 53 through 60. Each comes from the technical area indicated.

53. $4s^2 + 16s + 12$ (electricity)

54. $3p^2 + 9p - 54$ (business)

55. $200n^2 - 2100n - 3600$ (biology)

56. $2x^3 - 28x^2 + 98x$ (container design)

57. $wx^4 - 5wLx^3 + 6wL^2x^2$ (beam design)

58. $bT^2 - 40bT + 400b$ (thermodynamics)

59. $3Adu^2 - 4Aduv + Adv^2$ (water power)

60. $D^4 - d^3D$ (machine design)

5-4 Equivalent Fractions

When we deal with algebraic expressions, we must be able to work effectively with fractions. Since algebraic expressions are representations of numbers, the basic operations on fractions from arithmetic will form the basis of our algebraic operations. In this section we shall demonstrate a very important property of fractions, and in the following two sections we shall establish the basic algebraic operations with fractions.

fundamental principle of fractions

This important property of fractions, often referred to as the **fundamental principle of fractions,** is that *the value of a fraction is unchanged if both numerator and denominator are multiplied or divided by the same number, provided this number is not zero.* Two fractions are said to be **equivalent** if one can be obtained from the other by use of the fundamental theorem.

EXAMPLE A

If we multiply the numerator and denominator of the fraction $\frac{6}{8}$ by 2, we obtain the equivalent fraction $\frac{12}{16}$. If we divide the numerator and denominator of $\frac{6}{8}$ by 2, we obtain the equivalent fraction $\frac{3}{4}$. Therefore, the fractions $\frac{6}{8}, \frac{3}{4}$, and $\frac{12}{16}$ are equivalent. ∎

EXAMPLE B

We may write

$$\frac{ax}{2} = \frac{3a^2x}{6a}$$

since the fraction on the right is obtained from the fraction on the left by multiplying the numerator and the denominator by $3a$. Therefore, the fractions are equivalent. ∎

simplest form of fraction

One of the most important operations to be performed on a fraction is that of reducing it to its **simplest form,** or **lowest terms.** *A fraction is said to be in its simplest form if the numerator and the denominator have no common integral factors other than $+1$ or -1.* In reducing a fraction to its simplest form, we use the fundamental theorem by dividing both the numerator and the denominator by all factors which are common to each. (It will be assumed throughout this text that if any of the literal symbols were to be evaluated, numerical values would be restricted so that none of the denominators would be zero. Thereby, we avoid the undefined operation of division by zero.)

EXAMPLE C ———   In order to reduce the fraction

$$\frac{16ab^3c^2}{24ab^2c^5}$$

to its lowest terms, we note that both the numerator and the denominator contain the factor $8ab^2c^2$. Therefore, we may write

$$\frac{16ab^3c^2}{24ab^2c^5} = \frac{2b(8ab^2c^2)}{3c^3(8ab^2c^2)} = \frac{2b}{3c^3}$$ — common factor

Here we divided out the common factor. The resulting fraction is in lowest terms, since there are no common factors in the numerator and the denominator other ■ than $+1$ or -1.

NOTE ▷

cancellation

We must note very carefully that in simplifying fractions, *we divide both the numerator and the denominator* **by the common** *factor. This process is called* **cancellation.** However, many students are tempted to try to remove *any* expression which appears in both the numerator and the denominator. If a *term* is removed in this way, it is an incorrect application of the cancellation process. The following example illustrates this common error in the simplification of fractions.

EXAMPLE D ———   When simplifying the expression

$$\frac{x^2(x-2)}{x^2-4}$$ a term, but not a factor, of the denominator

NOTE ▷

many students would "cancel" the x^2 from the numerator and the denominator. This is *incorrect*, since x^2 *is a term only* of the denominator.

In order to simplify the above fraction properly, we should factor the denominator. We obtain

$$\frac{x^2(x-2)}{(x-2)(x+2)} = \frac{x^2}{x+2}$$

■ Here, the common *factor* $x-2$ has been divided out.

The following examples illustrate the proper simplification of fractions.

EXAMPLE E ——— $$\frac{2a}{2ax} = \frac{1}{x}$$ — 2a is a factor of the numerator and the denominator

We divide out the common factor of $2a$.

$$\frac{2a}{2a + x} \quad 2a \text{ is a term, but not a factor, of the denominator}$$

NOTE ▷ **This cannot be reduced,** since *there are no common **factors** in the numerator and the denominator.*

EXAMPLE F

$$\frac{2x^2 + 8x}{x + 4} = \frac{2x(x + 4)}{x + 4} = \frac{2x}{1} = 2x$$

In this simplification, the numerator and the denominator were each divided by $x + 4$ after the numerator was factored. Since the only remaining factor in the denominator after the division is 1, it generally is not written in the final answer. Another way of writing the denominator is $1(x + 4)$, which shows the **factor** of 1 more clearly.

EXAMPLE G

$$\frac{x^2 - 4x + 4}{x^2 - 4} = \frac{(x - 2)(x - 2)}{(x + 2)(x - 2)} = \frac{x - 2}{x + 2} \quad \begin{array}{l} x \text{ is a term, but} \\ \text{not a factor} \end{array}$$

NOTE ▷ In this simplification, the numerator and the denominator have each been **factored first and then the common factor $x - 2$ has been divided out.** In the final form, neither the x's nor 2s may be canceled, since they are not common **factors.**

EXAMPLE H

In the mathematical analysis of the vibrations in a certain mechanical system, the following expression and simplification are used:

$$\frac{8s + 12}{4s^2 + 26s + 30} = \frac{4(2s + 3)}{2(2s^2 + 13s + 15)} = \frac{4(2s + 3)}{2(2s + 3)(s + 5)}$$

$$= \frac{2}{s + 5}$$

Here, the factors common to the numerator and the denominator are 2 and $(2s + 3)$.

In simplifying fractions we must be able to distinguish between factors which differ only in **sign.** Since $-(y - x) = -y + x = x - y$, we have

$$\boxed{x - y = -(y - x)} \tag{5-11}$$

Here we see that **factors $x - y$ and $y - x$ differ only in sign.** The following examples illustrate the simplification of fractions where a change of signs is necessary.

EXAMPLE I

$$\frac{x^2 - 1}{1 - x} = \frac{(x - 1)(x + 1)}{-(x - 1)} = \frac{x + 1}{-1} = -(x + 1)$$

NOTE ▷   In the second fraction, *we replaced 1 − x with the equal expression −(x − 1).* In the third fraction, the common factor $x - 1$ was divided out. Finally, we expressed the result in the more convenient form by dividing $x + 1$ by -1, which makes the quantity $x + 1$ negative.

EXAMPLE J

$$\frac{2x^3 - 32x}{20 + 7x - 3x^2} = \frac{2x(x^2 - 16)}{(4 - x)(5 + 3x)} = \frac{2x(x - 4)(x + 4)}{-(x - 4)(3x + 5)}$$

$$= -\frac{2x(x + 4)}{3x + 5}$$

Again, the factor $4 - x$ has been replaced by the equal expression $-(x - 4)$. This allows us to recognize the common factor of $x - 4$. Notice also that the order of the terms of the factor $5 + 3x$ has been changed to $3x + 5$. This is merely an application of the commutative law of addition.

Exercises 5-4

In Exercises 1 through 8, multiply the numerator and the denominator of each of the given fractions by the given factor and obtain an equivalent fraction.

1. $\frac{2}{3}$ (by 7)

2. $\frac{7}{5}$ (by 9)

3. $\frac{ax}{y}$ (by 2x)

4. $\frac{2x^2 y}{3n}$ (by $2xn^2$)

5. $\frac{2}{x + 3}$ (by $x - 2$)

6. $\frac{7}{a - 1}$ (by $a + 2$)

7. $\frac{a(x - y)}{x - 2y}$ (by $x + y$)

8. $\frac{x - 1}{x + 1}$ (by $x - 1$)

In Exercises 9 through 16, divide the numerator and the denominator of each of the given fractions by the given factor and obtain an equivalent fraction.

9. $\frac{28}{44}$ (by 4)

10. $\frac{25}{65}$ (by 5)

11. $\frac{4x^2 y}{8xy^2}$ (by 2x)

12. $\frac{6a^3 b^2}{9a^5 b^4}$ (by $3a^2 b^2$)

13. $\frac{2(x - 1)}{(x - 1)(x + 1)}$ (by $x - 1$)

14. $\frac{(x + 5)(x - 3)}{3(x + 5)}$ (by $x + 5$)

15. $\frac{x^2 - 3x - 10}{2x^2 + 3x - 2}$ (by $x + 2$)

16. $\frac{6x^2 + 13x - 5}{6x^3 - 2x^2}$ (by $3x - 1$)

In Exercises 17 through 48, reduce each fraction to its simplest form.

17. $\frac{2a}{8a}$

18. $\frac{6x}{15x}$

19. $\frac{18x^2 y}{24xy}$

20. $\frac{2a^2 xy}{6axyz^2}$

21. $\frac{a + b}{5a^2 + 5ab}$

22. $\frac{t - a}{t^2 - a^2}$

23. $\frac{6a - 4b}{4a - 2b}$

24. $\frac{5r - 20s}{10r - 5s}$

25. $\dfrac{4x^2 + 1}{4x^2 - 1}$

26. $\dfrac{x^2 - y^2}{x^2 + y^2}$

27. $\dfrac{3x^2 - 6x}{x - 2}$

28. $\dfrac{10x^2 + 15x}{2x + 3}$

29. $\dfrac{2y + 3}{4y^3 + 6y^2}$

30. $\dfrac{3t - 6}{4t^3 - 8t^2}$

31. $\dfrac{x^2 - 8x + 16}{x^2 - 16}$

32. $\dfrac{4a^2 + 12ab + 9b^2}{4a^2 + 6ab}$

33. $\dfrac{2x^2 + 5x - 3}{x^2 + 11x + 24}$

34. $\dfrac{3y^3 + 7y^2 + 4y}{y^2 + 5y + 4}$

35. $\dfrac{5x^2 - 6x - 8}{x^3 + x^2 - 6x}$

36. $\dfrac{4r^2 - 8rs - 5s^2}{6r^2 - 17rs + 5s^2}$

37. $\dfrac{x^4 - 16}{x + 2}$

38. $\dfrac{2x^2 - 8}{4x + 8}$

39. $\dfrac{x^2y^4 - x^4y^2}{y^2 - 2xy + x^2}$

40. $\dfrac{8x^3 + 8x^2 + 2x}{4x + 2}$

41. $\dfrac{(x - 1)(3 + x)}{(3 - x)(1 - x)}$

42. $\dfrac{(2x - 1)(x + 6)}{(x - 3)(1 - 2x)}$

43. $\dfrac{y - x}{2x - 2y}$

44. $\dfrac{x^2 - y^2}{y - x}$

45. $\dfrac{2x^2 - 9x + 4}{4x - x^2}$

46. $\dfrac{3a^2 - 13a - 10}{5 + 4a - a^2}$

47. $\dfrac{(x + 5)(x - 2)(x + 2)(3 - x)}{(2 - x)(5 - x)(3 + x)(2 + x)}$

48. $\dfrac{(2x - 3)(3 - x)(x - 7)(3x + 1)}{(3x + 2)(3 - 2x)(x - 3)(7 + x)}$

In Exercises 49 through 52, reduce each fraction to its simplest form. This will require the use of Eqs. (5-7) through (5-10).

49. $\dfrac{x^3 + y^3}{2x + 2y}$

50. $\dfrac{x^3 - 8}{x^2 + 2x + 4}$

51. $\dfrac{x^3 + 3x^2 + 3x + 1}{x^3 + 1}$

52. $\dfrac{a^3 - 6a^2 + 12a - 8}{a^2 - 4a + 4}$

In Exercises 53 through 56, determine which fractions are in simplest form.

53. (a) $\dfrac{x^2(x + 2)}{x^2 + 4}$ (b) $\dfrac{x^4 + 4x^2}{x^4 - 16}$

54. (a) $\dfrac{2x + 3}{2x + 6}$ (b) $\dfrac{2(x + 6)}{2x + 6}$

55. (a) $\dfrac{x^2 - x - 2}{x^2 - x}$ (b) $\dfrac{x^2 - x - 2}{x^2 + x}$

56. (a) $\dfrac{x^3 - x}{1 - x}$ (b) $\dfrac{2x^2 + 4x}{2x^2 + 4}$

In Exercises 57 through 60, reduce the indicated fractions to simplest form. Each occurs in the indicated area of application.

57. $\dfrac{8\pi r^3}{6\pi^2 r^2}$ (hydrostatics)

58. $\dfrac{16(t^2 - 2tt_0 + t_0^2)(t - t_0 - 3)}{3t - 3t_0}$ (rocket motion)

59. $\dfrac{E^2 R^2 - E^2 r^2}{(R^2 + 2Rr + r^2)^2}$ (electricity)

60. $\dfrac{r_0^3 - r_i^3}{r_0^2 - r_i^2}$ (machine design)

5-5   Multiplication and Division of Fractions ▬▬▬

From arithmetic we recall that *the product of two fractions is a fraction whose numerator is the product of the numerators and whose denominator is the product of the denominators of the given fractions.* Also, we recall that *we can find the quotient of two fractions by inverting the divisor and proceeding as in multiplication.* Symbolically, multiplication of fractions is indicated by

multiplication of fractions

$$\frac{a}{b} \times \frac{c}{d} = \frac{ac}{bd}$$

and division is indicated by

division of fractions

$$\frac{\dfrac{a}{b}}{\dfrac{c}{d}} = \frac{a}{b} \times \frac{d}{c} = \frac{ad}{bc}$$

The rule for division may be verified by use of the fundamental principle of fractions. By multiplying the numerator and denominator of the fraction

$$\frac{\dfrac{a}{b}}{\dfrac{c}{d}} \quad \text{by} \quad \frac{d}{c} \quad \text{we obtain} \quad \frac{\dfrac{a}{b} \times \dfrac{d}{c}}{\dfrac{c}{d} \times \dfrac{d}{c}}$$

In the resulting denominator,

$$\frac{c}{d} \times \frac{d}{c}$$

becomes 1, and therefore the fraction is written as ad/bc.

EXAMPLE A

$$\frac{3}{5} \times \frac{2}{7} = \frac{(3)(2)}{(5)(7)} = \frac{6}{35} \longleftarrow \text{multiply numerators} \atop \longleftarrow \text{multiply denominators}$$

$$\frac{3a}{5b} \times \frac{15b^2}{a} = \frac{(3a)(15b^2)}{(5b)(a)} = \frac{45ab^2}{5ab} = \frac{9b}{1} = 9b$$

In the second illustration, we have divided out the common factor of $5ab$ to reduce the resulting fraction to its simplest form.

We shall usually want to express the result in its simplest form, which is generally its most useful form. Since all factors in the numerators and all factors in the denominators are to be multiplied, we should *first only* **indicate** *the multiplication, but not actually perform it, and then factor the numerator and the denominator.* In this way we can easily identify any factors common to both. If we were to multiply out the numerator and the denominator before factoring, it is very possible that we would be unable to factor the result to simplify it. The following example illustrates this point.

EXAMPLE B

In performing the multiplication

$$\frac{3(x - y)}{(x - y)^2} \times \frac{(x^2 - y^2)}{6x + 9y}$$

if we multiplied out the numerators and the denominators before performing any factoring, we would have to simplify the fraction

$$\frac{3x^3 - 3x^2y - 3xy^2 + 3y^3}{6x^3 - 3x^2y - 12xy^2 + 9y^3}$$

It is possible to factor the resulting numerator and denominator, but finding any common factors this way is very difficult. If we first indicate the multiplications, but do not actually perform them, and then factor completely, we have

$$\frac{3(x - y)}{(x - y)^2} \times \frac{(x^2 - y^2)}{6x + 9y} = \frac{3(x - y)(x^2 - y^2)}{(x - y)^2(6x + 9y)} = \frac{3(x - y)(x + y)(x - y)}{(x - y)^2(3)(2x + 3y)}$$

$$= \frac{3(x-y)^2(x+y)}{3(x-y)^2(2x+3y)}$$

$$= \frac{x+y}{2x+3y}$$

■ The common factor of $3(x-y)^2$ is readily recognized using this procedure.

EXAMPLE C

$$\frac{2x-4}{4x+12} \times \frac{2x^2+x-15}{3x-1} = \frac{2(x-2)(2x-5)(x+3)}{4(x+3)(3x-1)} \overset{\text{multiplications}}{\underset{\text{indicated}}{\leftarrow}}$$

$$= \frac{(x-2)(2x-5)}{2(3x-1)}$$

Here the common factor is $2(x+3)$. It is permissible to multiply out the final form of the numerator and the denominator, but it is often preferable to ■ leave the numerator and denominator in factored form, as indicated.

The following examples illustrate the division of fractions.

EXAMPLE D

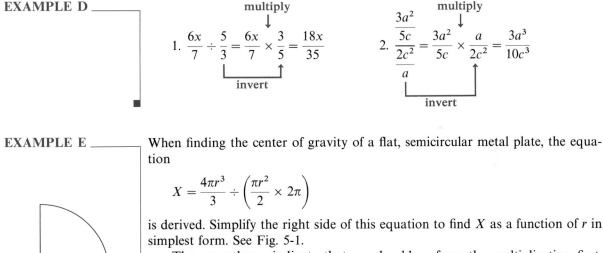

■ 1.
$$\frac{6x}{7} \div \frac{5}{3} = \frac{6x}{7} \times \frac{3}{5} = \frac{18x}{35}$$

2.
$$\frac{\frac{3a^2}{5c}}{\frac{2c^2}{a}} = \frac{3a^2}{5c} \times \frac{a}{2c^2} = \frac{3a^3}{10c^3}$$

EXAMPLE E

When finding the center of gravity of a flat, semicircular metal plate, the equation

$$X = \frac{4\pi r^3}{3} \div \left(\frac{\pi r^2}{2} \times 2\pi\right)$$

is derived. Simplify the right side of this equation to find X as a function of r in simplest form. See Fig. 5-1.

The parentheses indicate that we should perform the multiplication first. Proceeding, we have

$$X = \frac{4\pi r^3}{3} \div \left(\frac{\pi r^2}{2} \times 2\pi\right) = \frac{4\pi r^3}{3} \div \left(\frac{2\pi^2 r^2}{2}\right)$$

$$= \frac{4\pi r^3}{3} \div (\pi^2 r^2) = \frac{4\pi r^3}{3} \times \frac{1}{\pi^2 r^2}$$

$$= \frac{4\pi r^3}{3\pi^2 r^2} = \frac{4r}{3\pi}$$

Fig. 5-1

■ This is the exact solution. Approximately, $X = 0.424r$.

EXAMPLE F

$$\frac{x+y}{3} \div \frac{2x+2y}{6x+15y} = \frac{x+y}{3} \times \frac{6x+15y}{2x+2y} = \frac{(x+y)(3)(2x+5y)}{3(2)(x+y)}$$ indicate multiplications

invert

$$= \frac{2x+5y}{2}$$ simplify

EXAMPLE G

$$\frac{\dfrac{4-x^2}{x^2-3x+2}}{\dfrac{x+2}{x^2-9}} = \frac{4-x^2}{x^2-3x+2} \times \frac{x^2-9}{x+2}$$ invert

$$= \frac{(2-x)(2+x)(x-3)(x+3)}{(x-2)(x-1)(x+2)}$$ factor and indicate multiplications

$$= \frac{-(x-2)(x+2)(x-3)(x+3)}{(x-2)(x-1)(x+2)}$$ replace $(2-x)$ with $-(x-2)$ and $(2+x)$ with $(x+2)$

$$= -\frac{(x-3)(x+3)}{x-1} \quad \text{or} \quad \frac{(x-3)(x+3)}{1-x}$$ simplify

Note the use of Eq. (5-11) in the simplification and in expressing an alternate form of the result. The factor $2-x$ was replaced by its equivalent $-(x-2)$, and then $x-1$ was replaced by $-(1-x)$.

Exercises 5-5

In Exercises 1 through 32, perform the indicated operations and simplify.

1. $\dfrac{3}{8} \times \dfrac{2}{7}$

2. $\dfrac{11}{5} \times \dfrac{13}{33}$

3. $\dfrac{4x}{3y} \times \dfrac{9y^2}{2}$

4. $\dfrac{18sy^3}{ax^2} \times \dfrac{(ax)^2}{3s}$

5. $\dfrac{2}{9} \div \dfrac{4}{7}$

6. $\dfrac{5}{16} \div \dfrac{25}{13}$

7. $\dfrac{xy}{az} \div \dfrac{bz}{ay}$

8. $\dfrac{sr^2}{2t} \div \dfrac{st}{4}$

9. $\dfrac{4x+12}{5} \times \dfrac{15t}{3x+9}$

10. $\dfrac{y^2+2y}{6z} \times \dfrac{z^3}{y^2-4}$

11. $\dfrac{u^2-v^2}{u+2v}(3u+6v)$

12. $(x-y)\dfrac{x+2y}{x^2-y^2}$

13. $\dfrac{2a+8}{15} \div \dfrac{a^2+8a+16}{25}$

14. $\dfrac{a^2-a}{3a+9} \div \dfrac{a^2-2a+1}{a^2-9}$

15. $\dfrac{x^2-9}{x} \div (x+3)^2$

16. $\dfrac{9x^2-16}{x+1} \div (4-3x)$

17. $\dfrac{3ax^2-9ax}{10x^2+5x} \times \dfrac{2x^2+x}{a^2x-3a^2}$

18. $\dfrac{2x^2-18}{x^3-25x} \times \dfrac{3x-15}{2x^2+6x}$

19. $\dfrac{x^4-1}{8x+16} \times \dfrac{2x^2-8x}{x^3+x}$

20. $\dfrac{2x^2-4x-6}{x^2-3x} \times \dfrac{x^3-4x^2}{4x^2-4x-8}$

21. $\dfrac{ax+x^2}{2b-cx} \div \dfrac{a^2+2ax+x^2}{2bx-cx^2}$

22. $\dfrac{x^2-11x+28}{x+3} \div \dfrac{x-4}{2x+3}$

23. $\dfrac{35a+25}{12a+33} \div \dfrac{28a+20}{36a+99}$

24. $\dfrac{2a^3+a^2}{2b^3+b^2} \div \dfrac{2ab+a}{2ab+b}$

25. $\dfrac{x^2 - 6x + 5}{4x^2 - 17x - 15} \times \dfrac{6x + 21}{2x^2 + 5x - 7}$

26. $\dfrac{x^2 + 5x}{3x^2 + 8x - 4} \times \dfrac{2x^2 - 8}{x^3 + 2x^2 - 15x}$

27. $\dfrac{7x^2 + 27x - 4}{6x^2 + x - 15} \div \dfrac{4x^2 + 17x + 4}{8x^2 - 10x - 3}$

28. $\dfrac{4x^3 - 9x}{8x^2 + 10x - 3} \div \dfrac{2x^3 - 3x^2}{8x^2 + 18x - 5}$

29. $\dfrac{7x^2}{3a} \div \left(\dfrac{a}{x} \times \dfrac{a^2 x}{x^2}\right)$

30. $\left(\dfrac{3u}{8v^2} \div \dfrac{9u^2}{2w^2}\right) \times \dfrac{2u^4}{15vw}$

31. $\left(\dfrac{4t^2 - 1}{t - 5} \div \dfrac{2t + 1}{2t}\right) \times \dfrac{2t^2 - 50}{4t^2 + 4t + 1}$

32. $\dfrac{2x^2 - 5x - 3}{x - 4} \div \left(\dfrac{x - 3}{x^2 - 16} \times \dfrac{1}{3 - x}\right)$

In Exercises 33 through 36, perform the indicated operations and simplify. Exercises 33 and 34 require the use of Eqs. (5-7) through (5-10), and Exercises 35 and 36 require the use of factoring by grouping.

33. $\dfrac{x^3 - y^3}{2x^2 - 2y^2} \times \dfrac{x^2 + 2xy + y^2}{x^2 + xy + y^2}$

34. $\dfrac{x^3 + 3x^2 + 3x + 1}{6x - 6} \div \dfrac{5x + 5}{x^2 - 1}$

35. $\dfrac{ax + bx + ay + by}{p - q} \times \dfrac{3p^2 + 4pq - 7q^2}{a + b}$

36. $\dfrac{x^4 + x^5 - 1 - x}{x - 1} \div \dfrac{x + 1}{x}$

In Exercises 37 through 40, simplify the given expressions. The technical application of each is indicated.

37. $\dfrac{n^2 a^2}{v^2} \div \dfrac{n - an}{v}$ (chemistry)

38. $\dfrac{w\pi rbt}{2btg} \left(\dfrac{2r}{12\pi}\right)\left(\dfrac{144v^2}{r^2}\right)$ (stress on a rotating hoop)

39. $\dfrac{2\pi}{\lambda}\left(\dfrac{a + b}{2ab}\right)\left(\dfrac{ab\lambda}{2a + 2b}\right)$ (optics)

40. $(p_1 - p_2) \div \left(\dfrac{\pi a^4 p_1 - \pi a^4 p_2}{8lu}\right)$ (hydrodynamics)

5-6 Addition and Subtraction of Fractions

From arithmetic we recall that *the sum of a set of fractions that all have the same denominator is the sum of the numerators divided by the common denominator.* Since algebraic expressions represent numbers, this fact is also true in algebra. Addition and subtraction of such fractions are illustrated in the following example.

EXAMPLE A

$\dfrac{5}{9} + \dfrac{2}{9} - \dfrac{4}{9} = \dfrac{5 + 2 - 4}{9}$ ⟵ sum of numerators
    ⟵ same denominator

$= \dfrac{3}{9} = \dfrac{1}{3}$ ⟵ final result in lowest terms

use parentheses to show subtraction of both terms

$\dfrac{b}{ax} + \dfrac{1}{ax} - \dfrac{2b - 1}{ax} = \dfrac{b + 1 - (2b - 1)}{ax} = \dfrac{b + 1 - 2b + 1}{ax}$

$= \dfrac{2 - b}{ax}$

If the fractions to be combined do not all have the same denominator, we must first change each to an equivalent fraction so that the resulting fractions

lowest common denominator

do have the same denominator. Normally the denominator which is most convenient and useful is the **lowest common denominator** (abbreviated as **LCD**). *This is the product of all of the prime factors which appear in the denominators, with each factor raised to the **highest power** to which it appears **in any one** of the denominators.* This means that the lowest common denominator is the *simplest* algebraic expression into which all given denominators will divide exactly. The following two examples illustrate the method used in finding the lowest common denominator of a set of fractions.

EXAMPLE B

Find the lowest common denominator of the fractions

$$\frac{3}{4a^2b}, \qquad \frac{5}{6ab^3}, \quad \text{and} \quad \frac{1}{4ab^2}$$

Expressing each of the denominators in terms of powers of the prime factors, we have

$$\overbrace{}^{\text{highest powers}} \qquad \qquad \text{already seen to be highest power of 2}$$

$$4a^2b = 2^2a^2b, \qquad 6ab^3 = 2 \times 3 \times ab^3, \quad \text{and} \quad 4ab^2 = 2^2ab^2$$

The prime factors to be considered are 2, 3, a, and b. The largest exponent of 2 which appears is 2. This means that 2^2 is a factor of the lowest common denominator.

NOTE ▷

*What matters is that **the highest power of 2 which appears is 2,** not the fact that 2 appears in all three denominators with a total of five factors.*

The largest exponent of 3 which appears is 1 (understood in the second denominator). Therefore, 3 is a factor of the lowest common denominator. The largest exponent of a which appears is 2, and the largest exponent of b which appears is 3. Thus, a^2 and b^3 are factors of the lowest common denominator. Therefore, the lowest common denominator of the fractions is

$$2^2 \times 3 \times a^2b^3 = 12a^2b^3$$

This is the simplest expression into which *each* of the denominators above will divide exactly.

EXAMPLE C

Find the lowest common denominator of the following fractions:

$$\frac{x-4}{x^2-2x+1}, \qquad \frac{1}{x^2-1}, \qquad \frac{x+3}{x^2-x}$$

Factoring each of the denominators, we find that the fractions are

$$\frac{x-4}{(x-1)^2}, \qquad \frac{1}{(x-1)(x+1)}, \quad \text{and} \quad \frac{x+3}{x(x-1)}$$

The factor $(x-1)$ appears in all of the denominators. It is squared in the denominator of the first fraction and appears to the first power only in the other two fractions. Thus, we must have $(x-1)^2$ as a factor of the common denominator. We do not need a higher power of $x-1$ since, as far as this factor is concerned, each denominator will divide into it exactly. Next, the second denominator contains a factor of $(x + 1)$. Therefore, the common denominator must also contain a factor of $(x + 1)$; otherwise, the second denominator would not divide into it exactly. Finally, the third denominator indicates that a factor of x is also required in the common denominator. The lowest common denominator is, therefore, $x(x + 1)(x - 1)^2$. All three denominators will divide exactly into this expression, and there is no simpler expression for which this is true. ∎

addition and subtraction of fractions

Once we have found the lowest common denominator for the fractions, we multiply the numerator and denominator of each fraction by the proper quantity to make the resulting denominator in each case the lowest common denominator. After this step, it is necessary only to add the numerators, place this result over the common denominator, and simplify.

EXAMPLE D

Combine $\dfrac{2}{3r^2} + \dfrac{4}{rs^3} - \dfrac{5}{3s}$.

By looking at the denominators, we see that the factors necessary in the lowest common denominator are 3, r, and s. The 3 appears only to the first power, the largest exponent of r is 2, and the largest exponent of s is 3. Therefore, the lowest common denominator is $3r^2s^3$. We now wish to write each fraction with this quantity as the denominator. Since the denominator of the first fraction already contains factors of 3 and r^2, **it is necessary to introduce the factor of s^3.** In other words, we must multiply the numerator and denominator of this fraction by s^3. For similar reasons, we must multiply the numerators and the denominators of the second and third fractions by $3r$ and r^2s^2, respectively. This leads to

NOTE ▷

$$\frac{2}{3r^2} + \frac{4}{rs^3} - \frac{5}{3s} = \frac{2(s^3)}{(3r^2)(s^3)} + \frac{4(3r)}{(rs^3)(3r)} - \frac{5(r^2s^2)}{(3s)(r^2s^2)}$$ change to equivalent fractions with LCD

factors needed in each

$$= \frac{2s^3}{3r^2s^3} + \frac{12r}{3r^2s^3} - \frac{5r^2s^2}{3r^2s^3}$$

$$= \frac{2s^3 + 12r - 5r^2s^2}{3r^2s^3}$$ combine numerators over LCD

∎

EXAMPLE E

$$\frac{a}{x - 1} + \frac{a}{x + 1}$$

$$= \frac{a(x + 1)}{(x - 1)(x + 1)} + \frac{a(x - 1)}{(x + 1)(x - 1)}$$ change to equivalent fractions with LCD

factors needed

$$= \frac{ax + a + ax - a}{(x + 1)(x - 1)}$$ combine numerators over LCD

$$= \frac{2ax}{(x + 1)(x - 1)}$$ simplify

(*Continued on next page*)

When we multiply each fraction by the quantity required to obtain the proper denominator, we do not actually have to write the common denominator under each numerator. Placing all the products which appear in the numerators over the common denominator is sufficient. Hence the illustration in this example would appear as

$$\frac{a}{x-1} + \frac{a}{x+1} = \frac{a(x+1) + a(x-1)}{(x-1)(x+1)} = \frac{ax + a + ax - a}{(x-1)(x+1)}$$

$$= \frac{2ax}{(x-1)(x+1)}$$

EXAMPLE F ——— The following expression is found in the analysis of the dynamics of missile firing. The indicated addition is performed as shown:

$$\frac{1}{s} - \frac{1}{s+4} + \frac{8}{s^2 + 8s + 16}$$

$$= \frac{1}{s} - \frac{1}{s+4} + \frac{8}{(s+4)^2} \qquad \text{factor third denominator}$$

$$= \frac{1(s+4)^2 - 1(s)(s+4) + 8s}{s(s+4)^2} \qquad \begin{array}{l}\text{LCD has one factor of } s \\ \text{and two factors of } (s+4)\end{array}$$

$$= \frac{s^2 + 8s + 16 - s^2 - 4s + 8s}{s(s+4)^2} \qquad \begin{array}{l}\text{expand terms of} \\ \text{numerator}\end{array}$$

$$= \frac{12s + 16}{s(s+4)^2} = \frac{4(3s+4)}{s(s+4)^2} \qquad \text{simplify/factor}$$

We factored the numerator in the final result to see whether or not there were any factors common to the numerator and the denominator. Since there are none, either form of the result is acceptable.

EXAMPLE G ——— $\dfrac{3x}{x^2 - x - 12} - \dfrac{x-1}{x^2 - 8x + 16} - \dfrac{6-x}{2x-8}$

$$= \frac{3x}{(x-4)(x+3)} - \frac{x-1}{(x-4)^2} - \frac{6-x}{2(x-4)} \qquad \text{factor denominators}$$

$$= \frac{3x(2)(x-4) - (x-1)(2)(x+3) - (6-x)(x-4)(x+3)}{2(x-4)^2(x+3)} \qquad \begin{array}{l}\text{change to equivalent} \\ \text{fraction with LCD}\end{array}$$

$$= \frac{6x^2 - 24x - 2x^2 - 4x + 6 + x^3 - 7x^2 - 6x + 72}{2(x-4)^2(x+3)} \qquad \text{expand in numerator}$$

$$= \frac{x^3 - 3x^2 - 34x + 78}{2(x-4)^2(x+3)} \qquad \text{simplify}$$

One note of caution must be sounded here. In doing this kind of problem, many errors may arise in the use of the minus sign. Remember, if a minus sign

NOTE ▷ precedes a given expression, the *signs of all terms in that expression must be changed* before they can be combined with the other terms.

EXAMPLE H Simplify the fraction

$$\frac{1 + \dfrac{2}{x - 1}}{\dfrac{x^2 + x}{x^2 + x - 2}}$$

Before performing the indicated division, we *first perform the indicated addition in the numerator.* The numerator becomes

$$\frac{(x - 1) + 2}{x - 1} \quad \text{or} \quad \frac{x + 1}{x - 1}$$

This expression now replaces the numerator of the original fraction. Making this substitution and inverting the divisor, we then proceed with the simplification:

$$\frac{x + 1}{x - 1} \times \frac{x^2 + x - 2}{x^2 + x} = \frac{(x + 1)(x + 2)(x - 1)}{(x - 1)(x)(x + 1)} = \frac{x + 2}{x}$$

complex fractions

This is an example of what is known as a **complex fraction.** *In a complex fraction the numerator, the denominator, or both the numerator and denominator contain fractions.*

EXAMPLE I In simplifying the following fraction, we first perform the additions in the numerator and the denominator. We then divide the resulting fraction in the numerator by that in the denominator.

$$\frac{\dfrac{1}{x} + \dfrac{1}{x^2 - 4x}}{\dfrac{3}{x^2 - 16} + \dfrac{2}{x^2 + 4x}} = \frac{\dfrac{1}{x} + \dfrac{1}{x(x - 4)}}{\dfrac{3}{(x - 4)(x + 4)} + \dfrac{2}{x(x + 4)}}$$ ← factor denominators

$$= \frac{\dfrac{x - 4 + 1}{x(x - 4)}}{\dfrac{3x + 2(x - 4)}{x(x - 4)(x + 4)}}$$ ← change to equivalent fractions with LCDs

$$= \frac{\dfrac{x - 3}{x(x - 4)}}{\dfrac{5x - 8}{x(x - 4)(x + 4)}}$$ ← simplify numerators

$$= \frac{x - 3}{x(x - 4)} \times \frac{x(x - 4)(x + 4)}{5x - 8}$$ invert divisor and multiply

$$= \frac{x(x - 3)(x - 4)(x + 4)}{x(x - 4)(5x - 8)}$$ indicate multiplication

$$= \frac{(x - 3)(x + 4)}{5x - 8}$$ simplify

Exercises 5-6

In Exercises 1 through 40, perform the indicated operations and simplify.

1. $\dfrac{3}{5} + \dfrac{6}{5}$

2. $\dfrac{2}{13} + \dfrac{6}{13}$

3. $\dfrac{1}{x} + \dfrac{7}{x}$

4. $\dfrac{2}{a} + \dfrac{3}{a}$

5. $\dfrac{1}{2} + \dfrac{3}{4}$

6. $\dfrac{5}{9} - \dfrac{1}{3}$

7. $\dfrac{3}{4x} + \dfrac{7a}{4}$

8. $\dfrac{t-3}{a} - \dfrac{t}{2a}$

9. $\dfrac{a}{x} - \dfrac{b}{x^2}$

10. $\dfrac{2}{s^2} + \dfrac{3}{s}$

11. $\dfrac{6}{5x^3} + \dfrac{a}{25x}$

12. $\dfrac{a}{6y} - \dfrac{2b}{3y^4}$

13. $\dfrac{2}{5a} + \dfrac{1}{a} - \dfrac{a}{10}$

14. $\dfrac{1}{2a} - \dfrac{6}{b} - \dfrac{9}{4c}$

15. $\dfrac{x+1}{x} - \dfrac{x-3}{y} - \dfrac{2-x}{xy}$

16. $5 + \dfrac{1-x}{2} - \dfrac{3+x}{4}$

17. $\dfrac{3}{2x-1} + \dfrac{1}{4x-2}$

18. $\dfrac{5}{6y+3} - \dfrac{a}{8y+4}$

19. $\dfrac{4}{x(x+1)} - \dfrac{3}{2x}$

20. $\dfrac{3}{ax+ay} - \dfrac{1}{a^2}$

21. $\dfrac{s}{2s-6} + \dfrac{1}{4} - \dfrac{3s}{4s-12}$

22. $\dfrac{2}{x+2} - \dfrac{3-x}{x^2+2x} + \dfrac{1}{x}$

23. $\dfrac{3x}{x^2-9} - \dfrac{2}{3x+9}$

24. $\dfrac{2}{x^2+4x+4} - \dfrac{3}{x+2}$

25. $\dfrac{3}{x^2-8x+16} - \dfrac{2}{4-x}$

26. $\dfrac{1}{a^2-1} - \dfrac{2}{3-3a}$

27. $\dfrac{3}{x^2-11x+30} - \dfrac{2}{x^2-25}$

28. $\dfrac{x-1}{2x^3-4x^2} + \dfrac{5}{x-2}$

29. $\dfrac{x-1}{3x^2-13x+4} - \dfrac{3x+1}{4-x}$

30. $\dfrac{x}{4x^2-12x+5} + \dfrac{2x-1}{4x^2-4x-15}$

31. $\dfrac{t}{t^2-t-6} - \dfrac{2t}{t^2+6t+9} + \dfrac{t}{t^2-9}$

32. $\dfrac{5}{2x^3-3x^2+x} - \dfrac{x}{x^4-x^2} + \dfrac{2-x}{2x^2+x-1}$

33. $\dfrac{1+\dfrac{1}{x}}{1-\dfrac{1}{x}}$

34. $\dfrac{x-\dfrac{1}{x}}{1-\dfrac{1}{x}}$

35. $\dfrac{2-\dfrac{1}{x}-\dfrac{2}{x+1}}{\dfrac{1}{x^2+2x+1}-1}$

36. $\dfrac{\dfrac{2}{a}-\dfrac{1}{4}-\dfrac{3}{4a-4b}}{\dfrac{1}{4a^2-4b^2}-\dfrac{2}{b}}$

37. $\dfrac{\dfrac{3}{x}+\dfrac{1}{x^2+x}}{\dfrac{1}{x+1}-\dfrac{1}{x-1}}$

38. $\dfrac{\dfrac{1}{2x}-\dfrac{1}{4x^2-2x}}{\dfrac{6}{4x^2-1}-\dfrac{2}{2x+1}}$

39. $\dfrac{\dfrac{r+s}{r-s}-\dfrac{r-s}{r+s}}{1+\dfrac{r-s}{r+s}}$

40. $\dfrac{\dfrac{1}{u-v}+\dfrac{1}{2u+2v}}{\dfrac{2u}{2u^2-3uv+v^2}+\dfrac{2}{2u-v}}$

The expression $f(x+h) - f(x)$ is frequently used in the study of calculus. In Exercises 41 through 44, determine and then simplify this expression for the given functions.

41. $f(x) = \dfrac{x}{x+1}$

42. $f(x) = \dfrac{3}{2x-1}$

43. $f(x) = \dfrac{1}{x^2}$

44. $f(x) = \dfrac{2}{x^2+4}$

In Exercises 45 through 48, simplify the given expressions.

45. Using the definitions of the trigonometric functions given in Section 3-2, find an expression equivalent to $(\tan \theta)(\cot \theta) + (\sin \theta)^2 - \cos \theta$ in terms of x, y, and r.

46. Using the definitions of the trigonometric functions given in Section 3-2, find an expression equivalent to $\sec \theta - (\cot \theta)^2 + \csc \theta$ in terms of x, y, and r.

47. If $f(x) = 2x - x^2$, find $f(\frac{1}{a})$.

48. If $f(x) = x^2 + x$, find $f(a + \frac{1}{a})$.

In Exercises 49 through 52, perform the indicated operations. Each expression occurs in the indicated area of application.

49. $\dfrac{3}{4\pi} - \dfrac{3H_0}{4\pi H}$ (transistor theory)

50. $1 + \dfrac{9}{128T} - \dfrac{27P}{64T^3}$ (thermodynamics)

51. $\dfrac{2n^2 - n - 4}{2n^2 + 2n - 4} + \dfrac{1}{n - 1}$ (optics)

52. $\dfrac{\dfrac{L}{C} + \dfrac{R}{sC}}{sL + R + \dfrac{1}{sC}}$ (electricity)

5-7 Equations Involving Fractions

Many important equations in science and technology have fractions in them. Although the solution of these equations will still involve the use of the basic operations stated in Section 1-11, an additional procedure can be used to eliminate the fractions and thereby help lead to the solution. The method is to *multiply each term of the equation by the lowest common denominator.*

The resulting equation will not involve fractions and can be solved by methods previously discussed. The following examples illustrate how to solve equations involving fractions.

EXAMPLE A

Solve for x: $\dfrac{x}{12} - \dfrac{1}{8} = \dfrac{x + 2}{6}$.

We first note that the lowest common denominator of the terms of the equation is 24. Therefore, we multiply each term by 24. This gives

$$\frac{24(x)}{12} - \frac{24(1)}{8} = \frac{24(x + 2)}{6} \qquad \textbf{each term multiplied by LCD}$$

We reduce each term to its lowest terms and solve the resulting equation.

$$2x - 3 = 4(x + 2) \qquad \textbf{each term reduced}$$
$$2x - 3 = 4x + 8$$
$$-2x = 11$$
$$x = -\frac{11}{2}$$

When we check this solution in the original equation, we obtain $-\frac{7}{12}$ on each side of the equal sign. Therefore, the solution is correct.

EXAMPLE B

Solve for x: $\dfrac{x}{2} - \dfrac{1}{b^2} = \dfrac{x}{2b}$.

We first determine that the lowest common denominator of the terms of the equation is $2b^2$. We then multiply each term by $2b^2$ and continue with the solution.

$$\frac{2b^2(x)}{2} - \frac{2b^2(1)}{b^2} = \frac{2b^2(x)}{2b} \qquad \text{each term multiplied by LCD}$$

$$b^2x - 2 = bx \qquad \text{each term reduced}$$

$$b^2x - bx = 2$$

$$x(b^2 - b) = 2 \qquad \text{factor}$$

$$x = \frac{2}{b^2 - b}$$

Note the use of factoring in arriving at the final result. Checking shows that each side of the original equation is equal to $\dfrac{1}{b^2(b-1)}$.

■

EXAMPLE C

An equation relating the focal length f of a lens with the object distance p and image distance q is given below. See Fig. 5-2. Solve for q.

$$f = \frac{pq}{p+q} \qquad \text{given equation}$$

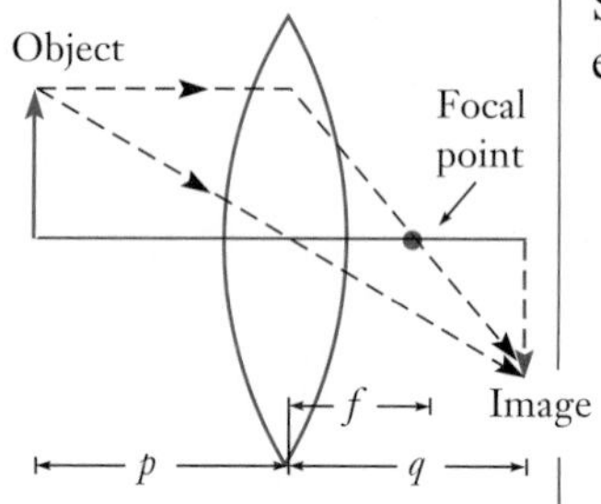

Since the only denominator is $p + q$, the LCD is also $p + q$. By first multiplying each term by $p + q$, the solution is completed as follows:

$$f(p+q) = \frac{pq(p+q)}{p+q} \qquad \text{each term multiplied by LCD}$$

$$fp + fq = pq \qquad \text{reduce term on right}$$

$$fq - pq = -fp$$

$$q(f - p) = -fp \qquad \text{factor}$$

$$q = \frac{-fp}{f - p}$$

$$= -\frac{fp}{-(p-f)} = \frac{fp}{p-f} \qquad \text{use Eq. (5-11)}$$

Object

Focal point

f

p — q

Image

Fig. 5-2

For reference, Eq. (5-11) is

$$x - y = -(y - x)$$

■

The last form is preferred since there is no minus sign before the fraction. However, either form of the result is correct.

EXAMPLE D

When developing the equations which describe the motion of the planets, the equation

$$\frac{1}{2}v^2 - \frac{GM}{r} = -\frac{GM}{2a}$$

is found. Solve for M.

We first determine that the lowest common denominator of the terms of the equation is $2ar$. Multiplying each term by $2ar$ and proceeding, we have

$$\frac{2ar(v^2)}{2} - \frac{2ar(GM)}{r} = -\frac{2ar(GM)}{2a} \qquad \text{each term multiplied by LCD}$$

$$arv^2 - 2aGM = -rGM \qquad \text{each term reduced}$$

$$rGM - 2aGM = -arv^2$$

$$M(rG - 2aG) = -arv^2 \qquad \text{factor}$$

$$M = -\frac{arv^2}{rG - 2aG} \quad \text{or} \quad \frac{arv^2}{2aG - rG}$$

The second form of the result is obtained by using Eq. (5-11). Again, note the use of factoring to arrive at the final result.

EXAMPLE E ——— Solve for x: $\dfrac{2}{x + 1} - \dfrac{1}{x} = -\dfrac{2}{x^2 + x}$.

Multiplying each term by the lowest common denominator $x(x + 1)$, we have

$$\frac{2(x)(x + 1)}{x + 1} - \frac{x(x + 1)}{x} = -\frac{2x(x + 1)}{x(x + 1)}$$

Now, simplifying each fraction, we have

$$2x - (x + 1) = -2$$

We now complete the solution.

$$2x - x - 1 = -2$$
$$x = -1$$

NOTE ▷ Checking this solution *in the original equation,* we see that we have zero in the denominators of the first and third terms of the equation. Since division by zero is undefined (see Section 1-3), $x = -1$ cannot be a solution. *Thus **there is** no solution to this equation.* This example points out clearly why it is necessary to check solutions in the original equation. It also shows that *whenever we multiply each term by a common denominator which **contains the unknown,** it is possible to obtain a value which is not a solution of the original equation. Such a value is termed an* extraneous solutions **extraneous solution.** Only certain equations will lead to extraneous solutions, but we must be careful to identify them when they occur.

A number of stated problems give rise to equations involving fractions. The following example illustrates the solution of such a problem.

EXAMPLE F ———

See the chapter introduction.

An industrial firm uses a computer system which processes and prints out its data for an average day in 20 h. In order to process this data more rapidly and to handle increased future computer needs, the firm plans to add new components to the system. One set of new components can process the data in 12 h, without the present system. How long would it take the new system, combining the present system and the new components, to process the data?

(Continued on next page)

First, we let x = the number of hours for the new system to process the data. Next, we know that it takes the present system 20 h to do it. This means that it processes $\frac{1}{20}$ of the data in one hour, or $\frac{1}{20}x$ of the data in x hours. In the same way, the new components can process $\frac{1}{12}x$ of the data in x hours. When x hours have passed, the new system will have processed all of the data. Therefore, the solution is as follows:

$$\underset{\substack{\text{part of data processed} \\ \text{by present system}}}{\frac{x}{20}} + \underset{\substack{\text{part of data processed} \\ \text{by new components}}}{\frac{x}{12}} = \underset{\substack{\text{one complete processing} \\ \text{(all of data)}}}{1}$$

$$\frac{60x}{20} + \frac{60x}{12} = 60(1) \qquad \text{each term multiplied by LCD of 60}$$

$$3x + 5x = 60 \qquad \text{each term reduced}$$

$$8x = 60$$

$$x = \frac{60}{8} = 7.5 \text{ h}$$

■ Therefore, the new system should take about 7.5 h to process the data.

Exercises 5-7

In Exercises 1 through 28, solve the given equations and check the results.

1. $\dfrac{x}{2} + 6 = 2x$

2. $\dfrac{x}{5} + 2 = \dfrac{15 + x}{10}$

3. $\dfrac{x}{6} - \dfrac{1}{2} = \dfrac{x}{3}$

4. $\dfrac{3x}{8} - \dfrac{3}{4} = \dfrac{x - 4}{2}$

5. $\dfrac{1}{2} - \dfrac{t - 5}{6} = \dfrac{3}{4}$

6. $\dfrac{2x - 7}{3} + 5 = \dfrac{1}{5}$

7. $\dfrac{3x}{7} - \dfrac{5}{21} = \dfrac{2 - x}{14}$

8. $\dfrac{x - 3}{12} - \dfrac{2}{3} = \dfrac{1 - 3x}{2}$

9. $\dfrac{3}{x} + 2 = \dfrac{5}{3}$

10. $\dfrac{1}{2y} - \dfrac{1}{2} = 4$

11. $3 - \dfrac{x - 2}{5x} = \dfrac{1}{5}$

12. $\dfrac{1}{2x} - \dfrac{1}{3} = \dfrac{2}{3x}$

13. $\dfrac{2y}{y - 1} = 5$

14. $\dfrac{x}{2x - 3} = 4$

15. $\dfrac{2}{s} = \dfrac{3}{s - 1}$

16. $\dfrac{5}{n + 2} = \dfrac{3}{2n}$

17. $\dfrac{5}{2x + 4} + \dfrac{3}{x + 2} = 2$

18. $\dfrac{3}{4x - 6} + \dfrac{1}{4} = \dfrac{5}{2x - 3}$

19. $\dfrac{4}{4 - x} + 2 - \dfrac{2}{12 - 3x} = \dfrac{1}{3}$

20. $\dfrac{2}{z - 5} - \dfrac{3}{10 - 2z} = 3$

21. $\dfrac{1}{x} + \dfrac{3}{2x} = \dfrac{2}{x + 1}$

22. $\dfrac{3}{t + 3} - \dfrac{1}{t} = \dfrac{5}{2t + 6}$

23. $\dfrac{1}{2x + 3} = \dfrac{5}{2x} - \dfrac{4}{2x^2 + 3x}$

24. $\dfrac{7}{y} = \dfrac{3}{y - 4} + \dfrac{7}{2y^2 - 8y}$

25. $\dfrac{1}{x^2 - x} - \dfrac{1}{x} = \dfrac{1}{x - 1}$

26. $\dfrac{2}{x^2 - 1} - \dfrac{2}{x + 1} = \dfrac{1}{x - 1}$

27. $\dfrac{2}{x^2 - 4} - \dfrac{1}{x - 2} = \dfrac{1}{2x + 4}$

28. $\dfrac{2}{2x^2 + 5x - 3} - \dfrac{1}{4x - 2} + \dfrac{3}{2x + 6} = 0$

In Exercises 29 through 32, solve for the indicated letter.

29. $2 - \dfrac{1}{b} + \dfrac{3}{c} = 0$, for c

30. $\dfrac{2}{3} - \dfrac{h}{x} = \dfrac{1}{6x}$, for x

31. $\dfrac{t - 3}{b} - \dfrac{t}{2b - 1} = \dfrac{1}{2}$, for t

32. $\dfrac{1}{a^2 + 2a} - \dfrac{y}{2a} = \dfrac{2y}{a + 2}$, for y

In Exercises 33 through 44, each of the given formulas arises in the technical or scientific area of study listed. Solve for the indicated letter.

33. $n = n_1 - \dfrac{n_1 v}{V}$, for v (acoustics)

34. $S = \dfrac{P}{A} + \dfrac{Mc}{I}$, for P (machine design)

35. $V_0 = \dfrac{V_r A}{1 + \beta A}$, for β (electricity)

36. $K = \dfrac{ax}{x + b}$, for x (medicine)

37. $z = \dfrac{1}{g_m} - \dfrac{jX}{g_m R}$, for R (FM transmission)

38. $A = \dfrac{1}{2} wp - \dfrac{1}{2} w^2 - \dfrac{\pi}{8} w^2$, for p (architecture)

39. $P = \dfrac{RT}{V - b} - \dfrac{a}{V^2}$, for T (thermodynamics)

40. $\dfrac{1}{x} + \dfrac{1}{nx} = \dfrac{1}{f}$, for n (photography)

41. $R = \dfrac{L_1}{kA_1} + \dfrac{L_2}{kA_2}$, for L_1 (air conditioning)

42. $D = \dfrac{wx^4}{24EI} - \dfrac{wLx^3}{6EI} + \dfrac{wL^2 x^2}{4EI}$, for w (beam design)

43. $\dfrac{1}{f} = (n - 1)\left(\dfrac{1}{R_1} + \dfrac{1}{R_2}\right)$, for R_1 (optics)

44. $P = \dfrac{\dfrac{1}{1 + i}}{1 - \dfrac{1}{1 + i}}$, for i (business)

In Exercises 45 through 52, set up appropriate equations and solve the given stated problems.

45. One pipe can fill a certain oil storage tank in 4.0 h, while a second pipe can fill it in 6.0 h. How long will it take to fill the tank if both pipes operate together?

46. One company determines that it will take its crew 450 h to clean up a chemical dump site, and a second company determines that it will take its crew 600 h to clean up the site. How long will it take the two crews working together?

47. One automatic packaging machine can package 100 boxes of machine parts in 12 min, and a second machine can do it in 10 min. A newer model machine can do it in 8.0 min. How long would it take the three machines working together?

48. A painting crew can paint a structure in 12 h and can paint the structure in 7.2 h when working with a second crew. How long would it take the second crew to do the job if working alone?

49. A jet takes the same time to travel 2580 km with the wind as it does to travel 1800 km against the wind. If its speed relative to the air is 450 km/h, what is the speed of the wind?

50. In making deliveries, a truck averaged 15 mi/gal for 65 mi more than for the distance it averaged 20 mi/gal. If 16 gal of gas were consumed, how long was the delivery route?

51. An industrial cleaning solution is to be $\frac{2}{5}$ acid. How many quarts of pure acid must be added to 10 qt of a solution that is $\frac{1}{4}$ acid to get the proper mixture?

52. How many grams of an alloy which is $\frac{3}{8}$ copper must be mixed with an alloy which is $\frac{2}{3}$ copper in order to get 100 g of alloy which is $\frac{1}{2}$ copper?

5-8   Chapter Equations, Review Exercises, and Practice Test

Chapter Equations

$$a(x + y) = ax + ay \tag{5-1}$$

$$(x + y)(x - y) = x^2 - y^2 \tag{5-2}$$

$$(x + y)^2 = x^2 + 2xy + y^2 \tag{5-3}$$

$$(x - y)^2 = x^2 - 2xy + y^2 \tag{5-4}$$

$$(x + a)(x + b) = x^2 + (a + b)x + ab \tag{5-5}$$

$$(ax + b)(cx + d) = acx^2 + (ad + bc)x + bd \tag{5-6}$$

$$(x + y)^3 = x^3 + 3x^2y + 3xy^2 + y^3 \tag{5-7}$$

$$(x - y)^3 = x^3 - 3x^2y + 3xy^2 - y^3 \tag{5-8}$$

$$(x + y)(x^2 - xy + y^2) = x^3 + y^3 \tag{5-9}$$

$$(x - y)(x^2 + xy + y^2) = x^3 - y^3 \tag{5-10}$$

$$x - y = -(y - x) \tag{5-11}$$

Review Exercises

In Exercises 1 through 12, find the products by inspection. No intermediate steps should be necessary.

1. $3a(4x + 5a)$ **2.** $-7xy(4x^2 - 7y)$ **3.** $(2a + 7b)(2a - 7b)$ **4.** $(x - 4z)(x + 4z)$

5. $(2a + 1)^2$ **6.** $(4x - 3y)^2$ **7.** $(b - 4)(b + 7)$ **8.** $(y - 5)(y - 7)$

9. $(2x + 5)(x - 9)$ **10.** $(4ax - 3)(5ax + 7)$ **11.** $(2c + d)(8c - d)$ **12.** $(3s - 2t)(8s + 3t)$

In Exercises 13 through 44, factor the given expressions completely. Exercises 37 through 40 illustrate Eqs. (5-7) through (5-10), and Exercises 41 through 44 illustrate factoring by grouping.

13. $3s + 9t$ **14.** $7x - 28y$ **15.** $a^2x^2 + a^2$ **16.** $3ax - 6ax^4 - 9a$

17. $x^2 - 144$ **18.** $900 - n^2$ **19.** $16(x + 2)^2 - t^4$ **20.** $25s^4 - 36t^2$

21. $9t^2 - 6t + 1$ **22.** $4x^2 - 12x + 9$ **23.** $25t^2 + 10t + 1$ **24.** $4x^2 + 36xy + 81y^2$

25. $x^2 + x - 56$ **26.** $x^2 - 4x - 45$ **27.** $t^2 - 5t - 36$ **28.** $n^2 - 11n + 10$

29. $2x^2 - x - 36$ **30.** $5x^2 + 2x - 3$ **31.** $4x^2 - 4x - 35$ **32.** $9x^2 + 7x - 16$

33. $10b^2 + 23b - 5$ **34.** $12x^2 - 7xy - 12y^2$ **35.** $4x^2 - 64y^2$ **36.** $4a^2x^2 + 26a^2x + 36a^2$

37. $x^3 + 9x^2 + 27x + 27$ **38.** $x^3 - 3x^2 + 3x - 1$ **39.** $8x^3 + 27$ **40.** $x^3 - 125a^3$

41. $ab^2 - 3b^2 + a - 3$ **42.** $axy - ay + ax - a$ **43.** $nx + 5n - x^2 + 25$ **44.** $ty - 4t + y^2 - 16$

In Exercises 45 through 68, perform the indicated operations and express results in simplest form.

45. $\dfrac{48ax^3y^6}{9a^3xy^6}$ **46.** $\dfrac{-39r^2s^4t^8}{52rs^5t}$ **47.** $\dfrac{6x^2 - 7x - 3}{4x^2 - 8x + 3}$ **48.** $\dfrac{x^2 - 3x - 4}{x^2 - x - 12}$

49. $\dfrac{4x + 4y}{35x^2} \times \dfrac{28x}{x^2 - y^2}$ **50.** $\left(\dfrac{6x - 3}{x^2}\right)\left(\dfrac{4x^2 - 12x}{12x - 6}\right)$

51. $\dfrac{18 - 6x}{x^2 - 6x + 9} \div \dfrac{x^2 - 2x - 15}{x^2 - 9}$ **52.** $\dfrac{6x^2 - xy - y^2}{2x^2 + xy - y^2} \div \dfrac{4x^2 - 16y^2}{x^2 + 3xy + 2y^2}$

53. $\dfrac{\dfrac{7x^2 + 13x - 2}{6x^2}}{\dfrac{3x}{x^2 + 4x + 4}}$ **54.** $\dfrac{\dfrac{3x - 3y}{2x^2 + 3xy - 2y^2}}{\dfrac{3x^2 - 3y^2}{x^2 + 4xy + 4y^2}}$ **55.** $\dfrac{x + \dfrac{1}{x} + 1}{x^2 - \dfrac{1}{x}}$ **56.** $\dfrac{\dfrac{4}{y} - 4y}{2 - \dfrac{2}{y}}$

57. $\dfrac{4}{9x} - \dfrac{5}{12x^2}$ **58.** $\dfrac{3}{10a^2} + \dfrac{1}{4a^3}$ **59.** $\dfrac{6}{x} - \dfrac{7}{2x} + \dfrac{3}{xy}$ **60.** $\dfrac{4}{a^2b} - \dfrac{5}{2ab} + \dfrac{1}{2b}$

61. $\dfrac{a + 1}{a + 2} - \dfrac{a + 3}{a}$ **62.** $\dfrac{y}{y + 2} - \dfrac{1}{y^2 + 2y}$ **63.** $\dfrac{2x}{x^2 + 2x - 3} - \dfrac{1}{x^2 + 3x}$

64. $\dfrac{x}{4x^2 + 4x - 3} - \dfrac{3}{4x^2 - 9}$ **65.** $\dfrac{3x}{2x^2 - 2} - \dfrac{2}{4x^2 - 5x + 1}$ **66.** $\dfrac{2x - 1}{4 - x} + \dfrac{x + 2}{5x - 20}$

67. $\dfrac{3x}{x^2 + 2x - 3} - \dfrac{2}{x^2 + 3x} + \dfrac{x}{x - 1}$

68. $\dfrac{3}{y^4 - 2y^3 - 8y^2} + \dfrac{y - 1}{y^2 + 2y} - \dfrac{y - 3}{y^2 - 4y}$

In Exercises 69 through 76, solve the given equations.

69. $\dfrac{x}{2} - 3 = \dfrac{x - 10}{4}$

70. $\dfrac{x}{6} - \dfrac{1}{2} = \dfrac{3 - x}{12}$

71. $\dfrac{2x}{c} - \dfrac{1}{2c} = \dfrac{3}{c} - x$, for x

72. $\dfrac{x}{2a} - b + \dfrac{x}{2c} = \dfrac{a}{b} - c$, for x

73. $\dfrac{2}{t} - \dfrac{1}{at} = 2 + \dfrac{a}{t}$, for t

74. $\dfrac{3}{a^2 y} - \dfrac{1}{ay} = \dfrac{9}{a}$, for y

75. $\dfrac{2x}{x^2 - 3x} - \dfrac{3}{x} = \dfrac{1}{2x - 6}$

76. $\dfrac{3}{x^2 + 3x} - \dfrac{1}{x} = \dfrac{1}{x + 3}$

In Exercises 77 through 88, perform the given operations. Where indicated, the expression is found in the technical area of application stated.

77. Show that $xy = \dfrac{1}{4}[(x + y)^2 - (x - y)^2]$.

78. Show that $x^2 + y^2 = \dfrac{1}{2}[(x + y)^2 + (x - y)^2]$.

79. Multiply: $2zS(S + 1)$.   (solid state physics)

80. Expand: $[2b + (n - 1)\lambda]^2$.   (optics)

81. Factor: $\pi r_1^2 \ell - \pi r_2^2 \ell$.   (jet plane fuel supply)

82. Factor: $4x^3 - 20x^2 + 25x$.   (mechanics: center of mass)

83. Express in factored form: $(t + 1)^2 - 2t(t + 1)$.   (velocity)

84. Express in factored form: $2R(R + r) - (R + r)^2$.   (electricity: power)

85. Expand and simplify: $(W^2 - 2L^2)^2 + 4L^2(W^2 + k^2 - 2L^2)$.
(mechanical vibrations)

86. Expand the third term and then factor by grouping:
$pa^2 + (1 - p)b^2 - [pa + (1 - p)b]^2$.   (nuclear physics)

87. A metal cube of edge x is heated and each edge increases by 4 mm. Express the increase in volume in factored form.

88. A machine part is made from a rectangular metal plate (see Fig. 5-3) by cutting out a square piece and a rectangular piece (shaded). Express the area of the face of the machine part in factored form.

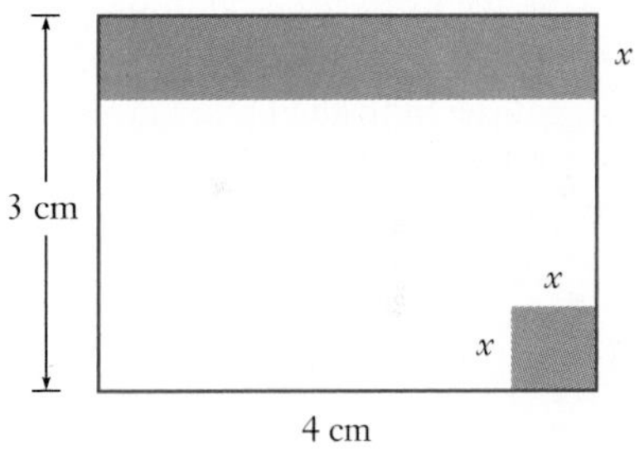

Fig. 5-3

In Exercises 89 through 98, perform the indicated operations and simplify the given expressions. Each comes from the indicated technical area of application.

89. $\left(\dfrac{2wtv^2}{Dg}\right)\left(\dfrac{b\pi^2 D^2}{n^2}\right)\left(\dfrac{6}{bt^2}\right)$ (machine design)

90. $\dfrac{m}{c} \div \left[1 - \left(\dfrac{p}{c}\right)^2\right]$ (airfoil design)

91. $10t + \dfrac{3}{2}t^2 + \dfrac{1}{3}t^3$ (business)

92. $\dfrac{V}{kp} - \dfrac{RT}{k^2 p^2}$ (electric motors)

93. $1 - \dfrac{d^2}{2} + \dfrac{d^4}{24} - \dfrac{d^6}{120}$ (aircraft emergency locator transmitter)

94. $\dfrac{5}{3} + \dfrac{3L}{8Cr} - \dfrac{L^3}{8C^3 r^3}$ (steel column safety factor)

95. $\dfrac{N + n}{2} + \dfrac{(N - n)^2}{4\pi^2 C}$ (machine design)

96. $\dfrac{Am}{k} - \dfrac{g}{2}\left(\dfrac{m}{k}\right)^2 + \dfrac{AML}{k}$ (rocket fuel)

97. $1 - \dfrac{3a}{4r} - \dfrac{a^3}{4r^3}$ (hydrodynamics)

98. $\dfrac{V}{\dfrac{1}{2R} + \dfrac{1}{2R + 2}}$ (electricity)

In Exericses 99 through 104, solve for the indicated letter. Each equation comes from the indicated technical area of application.

99. $\dfrac{q_2 - q_1}{d} = \dfrac{f + q_1}{D}$, for q_1 (photography)

100. $\dfrac{110}{R} = \dfrac{180}{R + 5}$, for R (electricity)

101. $R = \dfrac{wL}{H(w + L)}$, for L (architecture)

102. $y = \dfrac{1000a - bx}{x + a}$, for x (production of medication)

103. $s^2 + \dfrac{cs}{m} + \dfrac{kL^2}{mb^2} = 0$, for c (mechanical vibrations)

104. $I = \dfrac{A}{x^2} + \dfrac{B}{(10 - x)^2}$, for A (optics)

In Exercises 105 through 112, set up appropriate equations and solve the given stated problems.

105. If a certain car's lights are left on, the battery will be dead in 4 h. If only the radio is left on, the battery will be dead in 24 h. How long will the battery last if both the lights and the radio are left on?

106. Two pumps are being used to fight a fire. One pumps 5000 gal in 20 min, and the other pumps 5000 gal in 25 min. How long will it take the two pumps together to pump 5000 gal?

107. One computer can solve a certain problem in 30 s. With the aid of a second computer, the problem is solved in 10 s. How long would the second computer take to solve the problem alone?

108. An auto mechanic can do a certain motor job in 3.0 h, and with an assistant he can do it in 2.1 h. How long would it take the assistant to do the job alone?

109. The *relative density* of an object may be defined as its weight in air w_a divided by the difference of its weight in air and its weight when submerged in water, w_w. For a lead weight, $w_a = 1.097 w_w$. Find the relative density of lead.

110. A car travels halfway to its destination at 80.0 km/h and the remainder of the distance at 60.0 km/h. What is the average speed of the car for the trip?

111. For electric resistors in parallel, the reciprocal of the combined resistance equals the sum of the reciprocals of the individual resistances. For three resistors of 12 Ω, R ohms, and $2R$ ohms in parallel, the combined resistance is 6 Ω. Find R.

112. An ambulance averaged 36 mi/h going to an accident and 48 mi/h on its return to the hospital. If the total time for the round trip was 40 min, including 5 min at the accident scene, how far from the hospital was the accident?

Practice Test

1. Find the product: $2x(2x - 3)^2$.

2. The following equation is used in electricity. Solve for R_1: $\dfrac{1}{R} = \dfrac{1}{R_1 + r} + \dfrac{1}{R_2}$.

3. Reduce to simplest form: $\dfrac{2x^2 + 5x - 3}{2x^2 + 12x + 18}$.

4. Factor: $4x^2 - 16y^2$.

In Problems 5 through 7, perform the indicated operations and simplify.

5. $\dfrac{3}{4x^2} - \dfrac{2}{x^2 - x} - \dfrac{x}{2x - 2}$

6. $\dfrac{x^2 + x}{2 - x} \div \dfrac{x^2}{x^2 - 4x + 4}$

7. $\dfrac{1 - \dfrac{3}{2x + 2}}{\dfrac{x}{5} - \dfrac{1}{2}}$

8. If one riveter can do a job in 12 days, and a second riveter can do it in 16 days, how long would it take for them to do it together?

6 Quadratic Equations

The frame around the panels of a door affects the strength and appearance of the door. In Section 6-3 we see how a frame design may involve the solution of a quadratic equation.

The solution of basic equations was introduced in Chapter 1. Then, in Chapter 4, we extended the solution of equations to systems of linear equations. There are many other types of equations that we shall discuss. Among these is the important quadratic equation. To develop the methods of solving a quadratic equation, we use the algebraic operations of Chapter 5.

There are many applications of the use of quadratic equations. These include projectile motion, electric circuits, mechanical systems, and product design.

In the first three sections of this chapter we present algebraic methods of solving quadratic equations. In Section 6-4 we discuss the graphical solution.

6-1 Quadratic Equations; Solution by Factoring

Given that a, b, and c are constants, the equation

$$ax^2 + bx + c = 0 \tag{6-1}$$

is called the **general quadratic equation in x.** From Eq. (6-1) we can see that the left side of the equation is a polynomial function of degree 2. *This function, $ax^2 + bx + c$, is known as the* **quadratic function.**

Among the applications of quadratic equations and functions we have the following examples: in describing projectile motion, the equation $s_0 + v_0 t - 16t^2 = 0$ is found; in analyzing electric power, the function $EI - RI^2$ is found; and in determining the forces on beams, the function $ax^2 + bLx + cL^2$ is used.

Since it is the x^2-term that distinguishes the quadratic equation from other types of equations, the equation is not quadratic if $a = 0$. However, b or c or both may be zero, and the equation is quadratic. We should recognize a quadratic equation even when it does not initially appear in the form of Eq. (6-1). The following examples illustrate the recognition of quadratic equations.

EXAMPLE A

The following are quadratic equations.

$$\underset{\substack{\uparrow \\ a = 1}}{x^2} - \underset{\substack{\uparrow \\ b = -4}}{4x} - \underset{\substack{\uparrow \\ c = -5}}{5} = 0$$

To show this equation in the form of Eq. (6-1), it can be written as $1x^2 + (-4)x + (-5) = 0$.

$$\underset{\substack{\uparrow \\ a = 3}}{3x^2} - \underset{\substack{\uparrow \\ c = -6}}{6} = 0$$

Since there is no x-term, $b = 0$.

$$\underset{\substack{\uparrow \\ a = 2}}{2x^2} + \underset{\substack{\uparrow \\ b = 7}}{7x} = 0$$

Since no constant appears, $c = 0$.

$(a - 3)x^2 - ax + 7 = 0$

The constants in Eq. (6-1) may include literal expressions. In this case, $a - 3$ takes the place of a, $-a$ takes the place of b, and $c = 7$.

$4x^2 - 2x = x^2$

After all nonzero terms have been collected on the left side, the equation becomes $3x^2 - 2x = 0$.

$(x + 1)^2 = 4$

Expanding the left side and collecting all nonzero terms on the left, we have $x^2 + 2x - 3 = 0$.

EXAMPLE B

The following are not quadratic equations.

$bx - 6 = 0$

There is no x^2-term.

$x^3 - x^2 - 5 = 0$

There should be no term of degree higher than 2. Thus there can be no x^3-term in a quadratic equation.

$x^2 + x - 7 = x^2$

When terms are collected, there will be no x^2-term.

From our previous work, we recall that *the solution of an equation consists of all numbers which, when substituted in the equation, produce equality. There are* **two** *such roots for a quadratic equation. Occasionally these roots are equal,* as is shown in Example C, and only one number is actually a solution. Also, due to the presence of the x^2-term, the roots may be imaginary numbers.

solutions (roots)
of a quadratic
equation

EXAMPLE C

The quadratic equation

$$3x^2 - 7x + 2 = 0$$

has the roots $x = \frac{1}{3}$ and $x = 2$. This can be seen by substituting these values into the equation.

$$3\left(\frac{1}{3}\right)^2 - 7\left(\frac{1}{3}\right) + 2 = 3\left(\frac{1}{9}\right) - \frac{7}{3} + 2 = \frac{1}{3} - \frac{7}{3} + 2 = \frac{0}{3} = 0$$

$$3(2)^2 - 7(2) + 2 = 3(4) - 14 + 2 = 12 - 14 + 2 = 0$$

The quadratic equation

$$4x^2 - 4x + 1 = 0$$

has the **double root** *(both roots are the same)* of $x = \frac{1}{2}$. This can be seen to be a solution by substitution:

$$4\left(\frac{1}{2}\right)^2 - 4\left(\frac{1}{2}\right) + 1 = 4\left(\frac{1}{4}\right) - 2 + 1 = 1 - 2 + 1 = 0$$

The quadratic equation

$$x^2 + 9 = 0$$

has the imaginary roots of $\sqrt{-3}$ and $-\sqrt{-3}$. (At this point all we wish to do is to recognize imaginary roots when they occur.)

In this section we shall deal only with those quadratic equations whose quadratic expression is factorable. Therefore, all roots will be rational. To solve a quadratic equation by factoring, use the following steps:

1. *Collect all terms on the left* [the equation will then be in the form of Eq. (6-1)].
2. *Factor the quadratic expression.*
3. *Set each factor equal to zero.*
4. *Solve the resulting linear equations.*

Here we are using the fact that

a product is zero if any of its factors is zero.

The solutions of the resulting linear equations make up the solution of the quadratic equation.

EXAMPLE D

$$x^2 - x - 12 = 0$$
$$(x - 4)(x + 3) = 0 \qquad \text{factor}$$

$$x - 4 = 0 \qquad\qquad x + 3 = 0 \qquad \text{set each factor equal to zero}$$
$$x = 4 \qquad\qquad\quad x = -3 \qquad \text{solve}$$

The roots are $x = 4$ and $x = -3$. We can check them in the original equation by substitution. For the root $x = 4$, we have

$$(4)^2 - (4) - 12 \overset{?}{=} 0$$
$$0 = 0$$

For the root $x = -3$, we have

$$(-3)^2 - (-3) - 12 \overset{?}{=} 0$$
$$0 = 0$$

Both roots satisfy the original equation.

EXAMPLE E

$$2x^2 + 7x - 4 = 0$$

$$(2x - 1)(x + 4) = 0 \qquad \text{factor}$$

$$2x - 1 = 0, \quad \text{or} \quad x = \frac{1}{2} \qquad \text{set each factor equal to zero and solve}$$

$$x + 4 = 0, \quad \text{or} \quad x = -4$$

Therefore, the roots are $x = \frac{1}{2}$ and $x = -4$. These roots can be checked by the same procedure used in Example D. ■

EXAMPLE F

$$x^2 + 4 = 4x \qquad \text{equation not in form of Eq. (6-1)}$$

$$x^2 - 4x + 4 = 0 \qquad \text{subtract } 4x \text{ from both sides}$$

$$(x - 2)^2 = 0 \qquad \text{factor}$$

$$x - 2 = 0, \quad \text{or} \quad x = 2 \qquad \text{solve}$$

Since $(x - 2)^2 = (x - 2)(x - 2)$, both factors are the same. This means there is a double root of $x = 2$. Substitution shows that $x = 2$ satisfies the original equation. ■

NOTE ▷   *It is essential for the expression on the left to be equal to zero,* because if a product equals a nonzero number, it is probable that neither factor will give us a correct root. Again, the first step must be to write the equation in the form of Eq. (6-1).

A number of equations involving fractions lead to quadratic equations after the fractions are eliminated. The following two examples, the second being a stated problem, illustrate the process of solving such equations with fractions.

EXAMPLE G   Solve for x:

$$\frac{1}{x} + 3 = \frac{2}{x + 2}$$

$$\frac{x(x + 2)}{x} + 3x(x + 2) = \frac{2x(x + 2)}{x + 2} \qquad \text{multiply each term by the LCD, } x(x + 2)$$

$$x + 2 + 3x^2 + 6x = 2x \qquad \text{reduce each term}$$

$$3x^2 + 5x + 2 = 0 \qquad \text{collect terms on left}$$

$$(3x + 2)(x + 1) = 0 \qquad \text{factor}$$

$$3x + 2 = 0, \quad \text{or} \quad x = -\frac{2}{3} \qquad \text{set each factor equal to zero}$$

$$x + 1 = 0, \quad \text{or} \quad x = -1 \qquad \text{and solve}$$

Checking in the original equation, we have

$$\frac{1}{-\frac{2}{3}} + 3 \overset{?}{=} \frac{2}{-\frac{2}{3} + 2} \qquad\qquad \frac{1}{-1} + 3 \overset{?}{=} \frac{2}{-1 + 2}$$

$$-\frac{3}{2} + 3 \overset{?}{=} \frac{2}{\frac{4}{3}} \qquad\qquad -1 + 3 \overset{?}{=} \frac{2}{1}$$

$$\frac{3}{2} = \frac{6}{4} \qquad\qquad 2 = 2$$

We see that the roots check. Remember, if either value gives division by zero, the root is extraneous. ■

EXAMPLE H _____

A car travels to and from a city 180 mi distant in 8.5 h. If the average speed on the return trip is 5 mi/h less than on the trip to the city, what was the average speed of the car when it was going toward the city?

Let x = average speed of car going to the city, and t = time to travel to the city. By our choice of unknowns, we may state that $xt = 180$ (speed times time equals distance). Also, we know that the speed on the return trip was $x - 5$ and that the required time for the return trip was $8.5 - t$. Since the distance traveled returning was also 180 mi, we may state that $(x - 5)(8.5 - t) = 180$. Because we wish to find x, we can eliminate t between the equations by substitution.

$$(x - 5)\left(8.5 - \frac{180}{x}\right) = 180$$

$$(x - 5)(17x - 360) = 360x \qquad \text{multiply each side by}$$
$$17x^2 - 360x - 85x + 1800 = 360x \qquad 2x; \ (8.5 = \tfrac{17}{2})$$

$$17x^2 - 805x + 1800 = 0$$

$$(17x - 40)(x - 45) = 0$$

$$17x - 40 = 0, \quad \text{or} \quad x = \frac{40}{17} \qquad \text{set each factor equal to}$$
$$\text{zero and solve}$$

$$x - 45 = 0, \quad \text{or} \quad x = 45$$

The factors lead to two possible solutions, but only one of them has meaning for this problem. The solution $x = \frac{40}{17}$ cannot be the solution, since the return rate of 5 mi/h less would then be negative. Therefore, the solution is $x = 45$ mi/h. By substitution it is found that this solution satisfies the given conditions.

Exercises 6-1

In Exercises 1 through 8, determine whether or not the given equations are quadratic by performing algebraic operations which could put each in the form of Eq. (6-1). If the resulting form is quadratic, identify a, b, and c, with a > 0.

1. $x^2 + 5 = 8x$ **2.** $5x^2 = 9 - x$ **3.** $x(x - 2) = 4$ **4.** $(3x - 2)^2 = 2$

5. $x^2 = (x + 2)^2$ **6.** $x(2x + 5) = 7 + 2x^2$ **7.** $x(x^2 + x - 1) = x^3$ **8.** $(x - 7)^2 = (2x + 3)^2$

In Exercises 9 through 44, solve the given quadratic equations by factoring.

9. $x^2 - 4 = 0$ **10.** $x^2 - 400 = 0$ **11.** $4y^2 - 9 = 0$ **12.** $x^2 - 0.16 = 0$

13. $x^2 - 8x - 9 = 0$ **14.** $s^2 + s - 6 = 0$ **15.** $x^2 - 7x + 12 = 0$ **16.** $x^2 - 11x + 30 = 0$

17. $x^2 = -2x$ **18.** $x^2 = 7x$ **19.** $27m^2 = 3$ **20.** $5p^2 = 80$

21. $3x^2 - 13x + 4 = 0$ **22.** $7x^2 + 3x - 4 = 0$ **23.** $x^2 + 8x + 16 = 0$ **24.** $4x^2 - 20x + 25 = 0$

25. $6x^2 = 13x - 6$ **26.** $6z^2 = 6 + 5z$ **27.** $4x(x + 1) = 3$ **28.** $9t^2 = 9 - t(43 + t)$

29. $x^2 - x - 1 = 1$ **30.** $2x^2 - 7x + 6 = 3$ **31.** $x^2 - 4b^2 = 0$ **32.** $a^2x^2 - 1 = 0$

33. $40x - 16x^2 = 0$ **34.** $15x = 20x^2$ **35.** $8s^2 + 16s = 90$ **36.** $18t^2 - 48t + 32 = 0$

37. $(x + 2)^3 = x^3 + 8$ **38.** $x(x^2 - 4) = x^2(x - 1)$

39. $(x + a)^2 - b^2 = 0$ **40.** $x^2(a^2 + 2ab + b^2) - x(a + b) = 0$

41. Under experimental conditions, the partial pressure P, in pascals, of a gas can be found by solving the equation $P^2 - 3P = 70$. Find P.

42. The weight w, in megagrams, of the fuel supply in the first-stage booster of a rocket is $w = 135 - 6t - t^2$, where t is the time, in seconds, after launch. When does the booster run out of fuel?

43. The power, in megawatts, produced between midnight and noon by a nuclear power plant is given by $P = 4h^2 - 48h + 744$, where h is the hour of the day. At what time is the power 664 MW?

44. In determining the speed s, in miles per hour, of a car while studying its fuel economy, the equation $s^2 - 16s = 3072$ is used. Find s.

In Exercises 45 through 48, solve the given equations involving fractions.

45. $\dfrac{1}{x-3} + \dfrac{4}{x} = 2$

46. $2 - \dfrac{1}{x} = \dfrac{3}{x+2}$

47. $\dfrac{1}{2x} - \dfrac{3}{4} = \dfrac{1}{2x+3}$

48. $\dfrac{x}{2} + \dfrac{1}{x-3} = 3$

In Exercises 49 through 52, set up the appropriate quadratic equations and solve.

49. The spring constant k is the force F on a spring divided by the amount x it stretches ($k = F/x$). See Fig. 6-1(a). For two springs in series [see Fig. 6-1(b)], the reciprocal of the spring constant k_c for the combination equals the sum of the reciprocals of the individual spring constants. Find the spring constants for each of two springs in series if $k_c = 2$ N/cm and one spring constant is 3 N/cm more than the other.

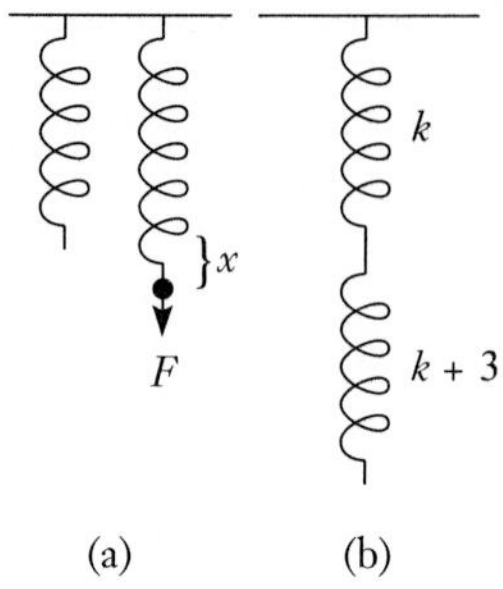

50. An equation used in analyzing a chemical solution is $x = \dfrac{k}{T_1} + \dfrac{k}{T}$. Find T_1 if $T_1 = T - 20$ and $x/k = 1/24$.

51. On a delivery route, a truck gets 4 mi/gal less for an 80-mi part of the route than for a 240-mi part of the route. If 30 gal of fuel are used, find the number of miles traveled per gallon for the two parts of the route.

52. A rectangular solar panel is 20 cm by 30 cm. By adding the same amount to each dimension, the area is doubled. How much is added?

(a)   (b)

Fig. 6-1

6-2   Completing the Square

Many quadratic equations cannot be solved by factoring. This is true of most quadratic equations which arise in applied situations. In this section, therefore, we develop a method which can be used to solve any quadratic equation. *The method is called* **completing the square.** In the following section we shall use completing the square to develop a formula which also may be used to solve any quadratic equation.

In the first example which follows we show the solution of a type of quadratic equation which arises while using the method of completing the square. In the examples which follow it, the method itself is used and described.

EXAMPLE A ⎯⎯⎯⎯ In solving $x^2 = 16$, we may write it as $x^2 - 16 = 0$ and then complete the solution by factoring. This gives us solutions of $x = 4$ and $x = -4$. Therefore, we see that the principal square root of 16 and its negative both satisfy the original equation. Thus, we may solve $x^2 = 16$ by equating x to the principal square root of 16 and to its negative. The roots of $x^2 = 16$ are 4 and -4.

We may solve $(x - 3)^2 = 16$ in a similar way by equating $x - 3$ to 4 and to -4. Thus,

$$x - 3 = 4 \quad \text{or} \quad x - 3 = -4$$

Solving these equations, we obtain the roots 7 and -1.

We may solve $(x - 3)^2 = 17$ in the same way. Thus,

$$x - 3 = \sqrt{17} \quad \text{or} \quad x - 3 = -\sqrt{17}$$

The roots are therefore $3 + \sqrt{17}$ and $3 - \sqrt{17}$. Decimal approximations of these roots are 7.123 and -1.123.

EXAMPLE B _____

We wish to find the roots of the quadratic equation

$$x^2 - 6x - 8 = 0$$

First we note that this equation is not factorable. However, we do recognize that $x^2 - 6x$ is part of one of the special products. If 9 were added to this expression, we would have $x^2 - 6x + 9$, which is $(x - 3)^2$. Therefore, we rewrite the original equation as

$$x^2 - 6x = 8$$

NOTE ▷ and then **add 9 to** **both** **sides** of the equation. The result is

$$x^2 - 6x + 9 = 17$$

The left side of this equation may be rewritten, giving

$$(x - 3)^2 = 17$$

Now, as in the third illustration of Example A, we have

$$x - 3 = \pm\sqrt{17}$$

The $\pm$ sign means that $x - 3 = \sqrt{17}$ or $x - 3 = -\sqrt{17}$.

By adding 3 to each side, we obtain

$$x = 3 \pm \sqrt{17}$$

which means that $x = 3 + \sqrt{17}$ and $x = 3 - \sqrt{17}$ are the two roots of the equation.

Therefore, we see that by creating an expression which is a perfect square and then using the principal square root and its negative, we were finally able to solve the equation as two linear equations.

How do we determine the number which must be added to complete the square? The answer to this question is based on the special products in Eqs. (5-3) and (5-4). We rewrite these in the form

$$(x + a)^2 = x^2 + 2ax + a^2 \tag{6-2}$$

and

$$(x - a)^2 = x^2 - 2ax + a^2 \tag{6-3}$$

We must be certain that the coefficient of the x^2-term is 1 before we start to complete the square. The coefficient of x in each case is numerically $2a$, and the

NOTE ▷ number added to complete the square is a^2. Thus *if we* **take half the coefficient** **of the x-term and square this result,** *we have the number which completes the* *square.* In our example, the numerical coefficient of the x-term was 6, and 9 was added to complete the square. The following example outlines the steps necessary to complete the square.

EXAMPLE C _____ Solve the following quadratic equation by the method of completing the square:

$$2x^2 + 16x - 9 = 0$$

First we divide each term by 2 so that the coefficient of the x^2-term becomes 1.

$$x^2 + 8x - \frac{9}{2} = 0$$

Now we put the constant term on the right-hand side by adding $\frac{9}{2}$ to both sides of the equation.

$$x^2 + 8x = \frac{9}{2}$$

Next we divide the coefficient of the x-term, 8, by 2, which gives us 4. We square 4 and obtain 16, which is the number to be added to both sides of the equation.

$$\tfrac{1}{2}(8) = 4; \ 4^2 = 16$$

$$x^2 + 8x + 16 = \frac{9}{2} + 16 = \frac{41}{2}$$

We write the left side as the square of $(x + 4)$.

$$(x + 4)^2 = \frac{41}{2}$$

Equating $x + 4$ to the principal square root of $\frac{41}{2}$ and its negative, we have

$$x + 4 = \pm\sqrt{\tfrac{41}{2}}$$

Solving for x, we have

$$x = -4 \pm \sqrt{\tfrac{41}{2}}$$

4 $\boxed{+/-}$ $\boxed{+}$ $\boxed{(}$
41 $\boxed{÷}$ 2 $\boxed{)}$ $\boxed{\sqrt{x}}$ $\boxed{=}$

$\boxed{0.5276926}$

Therefore the roots are $-4 + \sqrt{\frac{41}{2}}$ and $-4 - \sqrt{\frac{41}{2}}$. Using a calculator to approximate these roots, we get 0.5277 and -8.528. The sequence for the first root is shown at the left.

EXAMPLE D _____ Solve $4x^2 - 12x + 5 = 0$ by completing the square.

$$4x^2 - 12x + 5 = 0 \qquad \text{divide by 4 to make coefficient of } x^2 \text{ equal to 1}$$

$$x^2 - 3x + \frac{5}{4} = 0$$

$$x^2 - 3x = -\frac{5}{4} \qquad \text{subtract } \frac{5}{4} \text{ from both sides}$$

$$x^2 - 3x + \frac{9}{4} = -\frac{5}{4} + \frac{9}{4} \qquad \tfrac{1}{2}(-3) = -\frac{3}{2}; \left(-\frac{3}{2}\right)^2 = \frac{9}{4} \text{ added to both sides}$$

$$\left(x - \frac{3}{2}\right)^2 = \frac{4}{4} = 1 \qquad \text{recognize perfect square on left}$$

$$x - \frac{3}{2} = \pm 1 \qquad \text{square roots}$$

$$x = \frac{3}{2} \pm 1$$

$$x = \frac{5}{2}, x = \frac{1}{2} \qquad \frac{3}{2} + 1 = \frac{5}{2}, \frac{3}{2} - 1 = \frac{1}{2}$$

This equation could have been solved by factoring. However, at this point, we wanted to illustrate the *method* of completing the square.

Exercises 6-2

In Exercises 1 through 8, solve the given quadratic equations by finding the appropriate square roots as in Example A.

1. $x^2 = 25$ **2.** $x^2 = 100$ **3.** $x^2 = 7$ **4.** $x^2 = 15$

5. $(x - 2)^2 = 25$ **6.** $(x + 2)^2 = 100$ **7.** $(x + 3)^2 = 7$ **8.** $(x - 4)^2 = 10$

In Exercises 9 through 24, solve the given quadratic equations by completing the square. Exercises 9 through 12 and 15 through 18 may be checked by factoring.

9. $x^2 + 2x - 8 = 0$ **10.** $x^2 - x - 6 = 0$ **11.** $x^2 + 3x + 2 = 0$ **12.** $t^2 + 5t - 6 = 0$

13. $x^2 - 4x + 2 = 0$ **14.** $x^2 + 10x - 4 = 0$ **15.** $v^2 + 2v - 15 = 0$ **16.** $x^2 - 8x + 12 = 0$

17. $2s^2 + 5s = 3$ **18.** $4x^2 + x = 3$ **19.** $3y^2 = 3y + 2$ **20.** $3x^2 = 3 - 4x$

21. $2y^2 - y - 2 = 0$ **22.** $9v^2 - 6v - 2 = 0$ **23.** $x^2 + 2bx + c = 0$ **24.** $px^2 + qx + r = 0$

6-3   The Quadratic Formula

We shall now use the method of completing the square to derive a general formula which may be used for the solution of any quadratic equation.

Consider Eq. (6-1), the general quadratic equation

$$ax^2 + bx + c = 0$$

with $a > 0$. When we divide through by a, we obtain

$$x^2 + \frac{b}{a}x + \frac{c}{a} = 0$$

Subtracting c/a from each side, we have

$$x^2 + \frac{b}{a}x = -\frac{c}{a}$$

Half of b/a is $b/2a$, which squared is $b^2/4a^2$. Adding $b^2/4a^2$ to each side gives us

$$x^2 + \frac{b}{a}x + \frac{b^2}{4a^2} = -\frac{c}{a} + \frac{b^2}{4a^2}$$

Writing the left side as a perfect square, and combining fractions on the right side, we have

$$\left(x + \frac{b}{2a}\right)^2 = \frac{b^2 - 4ac}{4a^2}$$

Equating $x + \dfrac{b}{2a}$ to the principal square root of the right side and its negative,

$$x + \frac{b}{2a} = \frac{\pm\sqrt{b^2 - 4ac}}{2a}$$

When we subtract $b/2a$ from each side and simplify the resulting expression, we obtain the **quadratic formula:**

quadratic formula

$$x = \frac{-b \pm \sqrt{b^2 - 4ac}}{2a}$$

(6-4)

To solve a quadratic equation by using the quadratic formula, we need only write the equation in standard form (see Eq. 6-1), identify a, b, and c, and substitute these numbers directly into the formula. We shall use the quadratic formula to solve the quadratic equations in the following examples.

EXAMPLE A

Solve: $x^2 - 5x + 6 = 0$.

$$a = 1 \quad b = -5 \quad c = 6$$

Here, using the indicated values of a, b, and c in the quadratic equation, we have

$$x = \frac{-(-5) \pm \sqrt{(-5)^2 - 4(1)(6)}}{2(1)} = \frac{5 \pm \sqrt{25 - 24}}{2} = \frac{5 \pm 1}{2}$$

$$x = \frac{5 + 1}{2} = 3, \quad \text{or} \quad x = \frac{5 - 1}{2} = 2$$

The roots are $x = 3$ and $x = 2$. These values check when substituted in the original equation.

EXAMPLE B

Solve: $2x^2 - 7x - 5 = 0$.

$$a = 2 \quad b = -7 \quad c = -5$$

Substituting the values for a, b, and c in the quadratic formula, we have

$$x = \frac{-(-7) \pm \sqrt{(-7)^2 - 4(2)(-5)}}{2(2)} = \frac{7 \pm \sqrt{49 + 40}}{4} = \frac{7 \pm \sqrt{89}}{4}$$

$$x = \frac{7 + \sqrt{89}}{4} = 4.108, \quad \text{or} \quad x = \frac{7 - \sqrt{89}}{4} = -0.6085$$

$\boxed{(\,)\,7\,\boxed{+}\,89}$

$\boxed{\sqrt{x}}\,\boxed{)}\,\boxed{\div}\,4\,\boxed{=}$

$\boxed{4.1084953}$

The exact roots are $x = \dfrac{7 \pm \sqrt{89}}{4}$ (this form is commonly used when the roots are irrational). The approximate decimal values are $x = 4.108$ and $x = -0.6085$. The calculator sequence for the first root is shown at the left.

EXAMPLE C

Solve: $9x^2 + 24x + 16 = 0$.

In this example, $a = 9$, $b = 24$, and $c = 16$. Thus,

$$x = \frac{-24 \pm \sqrt{24^2 - 4(9)(16)}}{2(9)} = \frac{-24 \pm \sqrt{576 - 576}}{18} = \frac{-24 \pm 0}{18} = -\frac{4}{3}$$

Here both roots are $-\frac{4}{3}$, and we write the result as $x = -\frac{4}{3}$ and $x = -\frac{4}{3}$. We will get a double root when $b^2 = 4ac$, as in this case. ∎

EXAMPLE D

Solve: $3x^2 - 5x + 4 = 0$.

In this example, $a = 3$, $b = -5$, and $c = 4$. Therefore,

$$x = \frac{-(-5) \pm \sqrt{(-5)^2 - 4(3)(4)}}{2(3)} = \frac{5 \pm \sqrt{25 - 48}}{6} = \frac{5 \pm \sqrt{-23}}{6}$$

∎ We see that the roots contain imaginary numbers. This happens if $b^2 < 4ac$.

We can generalize on the results of these examples as to the character of the roots of a quadratic equation. This is done by noting the value of $b^2 - 4ac$, which is called the **discriminant.** If a, b, and c are rational numbers (see Section 1-1), we have the following:

1. If $b^2 - 4ac$ is positive and a perfect square (see Section 1-7), the roots are real, rational, and unequal. (See Example A, where $b^2 - 4ac = 1$.)
2. If $b^2 - 4ac$ is positive but not a perfect square, the roots are real, irrational, and unequal. (See Example B, where $b^2 - 4ac = 89$.)
3. If $b^2 - 4ac = 0$, the roots are real, rational, and equal. (See Example C, where $b^2 - 4ac = 0$.)
4. If $b^2 - 4ac < 0$, the roots contain imaginary numbers and are unequal. (See Example D, where $b^2 - 4ac = -23$.)

EXAMPLE E

Solve: $2x^2 = 4x + 3$.

First we must put the equation in the proper form. This is

$$2x^2 - 4x - 3 = 0$$

Now we identify $a = 2$, $b = -4$, and $c = -3$, which leads to the solution

$$x = \frac{-(-4) \pm \sqrt{(-4)^2 - 4(2)(-3)}}{2(2)} = \frac{4 \pm \sqrt{16 + 24}}{4}$$

$$= \frac{4 \pm \sqrt{40}}{4} = \frac{4 \pm 2\sqrt{10}}{4} = \frac{2(2 \pm \sqrt{10})}{4}$$

$$= \frac{2 \pm \sqrt{10}}{2}$$

Approximate decimal results are $x = 2.581$ and $x = -0.581$. The radical form of the answer is obtained by simplifying the radicals, as shown in Section 1-7.

For this quadratic equation $b^2 - 4ac = 40$, and we see that the roots are real, irrational, and unequal. ∎

EXAMPLE F

Solve for x: $dx^2 - 3x = dx - 4$

Putting this equation in the form of Eq. (6-1), we have

$$dx^2 - 3x - dx + 4 = 0$$
$$dx^2 - (3 + d)x + 4 = 0$$

Note the use of factoring to get the coefficient of x. We now see that $a = d$, $b = -(3 + d)$, and $c = 4$. Using the quadratic formula, we complete the solution.

$$x = \frac{-[-(3 + d)] \pm \sqrt{[-(3 + d)]^2 - 4(d)(4)}}{2d}$$

$$= \frac{3 + d \pm \sqrt{9 + 6d + d^2 - 16d}}{2d} = \frac{3 + d \pm \sqrt{9 - 10d + d^2}}{2d}$$

We see that we can use the quadratic formula to solve quadratic equations in which there are literal coefficients. ∎

EXAMPLE G

See the chapter introduction.

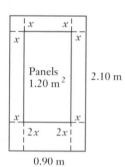

Panels 1.20 m^2

2.10 m

0.90 m

Fig. 6-2

In designing a custom door for a home, an architect plans it to be 0.90 m wide and 2.10 m high. The door is to have a frame of uniform width on the top and sides and twice this width at the bottom. The area within the frame must be 1.20 m^2 to allow for decorative panels. How wide a frame is the door to have?

First we let x be the width of the frame at the top and sides. This means the width of the frame at the bottom is $2x$. Knowing that the area within the frame is 1.20 m^2, we find x as follows:

$$\underset{\substack{\text{interior}\\\text{width}}}{(0.90 - 2x)} \underset{\substack{\text{interior}\\\text{height}}}{(2.10 - 3x)} = \underset{\substack{\text{interior}\\\text{area}}}{1.20}$$

$$1.89 - 4.20x - 2.70x + 6x^2 = 1.20$$

$$6x^2 - 6.90x + 0.69 = 0$$

$$x = \frac{-(-6.90) \pm \sqrt{(-6.90)^2 - 4(6)(0.69)}}{2(6)}$$

$$= \frac{6.90 \pm 5.572}{12}$$

Checking the roots, $\frac{6.90 + 5.572}{12} = 1.04$ m. This cannot be the required result, since the width of the frame would be greater than the width of the door. The other root is $\frac{6.90 - 5.572}{12} = 0.11$ m $= 11$ cm. Checking this root, the interior area is $(0.90 - 0.22)(2.10 - 0.33) = 1.20$ m^2. Thus, $x = 11$ cm, which means the frame is 11 cm wide at the top and sides and 22 cm wide at the bottom. ∎

NOTE ▷

In using the quadratic formula it must be emphasized that the entire expression $-b \pm \sqrt{b^2 - 4ac}$ is divided by $2a$. **It is a relatively common error to divide only the radical $\sqrt{b^2 - 4ac}$.**

The quadratic formula provides a quick general method for solving quadratic equations. Proper recognition and substitution of the coefficients a, b, and c is all that is required to complete the solution, regardless of the nature of the roots.

Exercises 6-3

In Exercises 1 through 36, solve the given quadratic equations using the quadratic formula. Exercises 1 through 14 are the same as Exercises 9 through 22 of Section 6-2.

1. $x^2 + 2x - 8 = 0$

2. $x^2 - x - 6 = 0$

3. $x^2 + 3x + 2 = 0$

4. $t^2 + 5t - 6 = 0$

5. $x^2 - 4x + 2 = 0$

6. $x^2 + 10x - 4 = 0$

7. $v^2 + 2v - 15 = 0$

8. $x^2 - 8x + 12 = 0$

9. $2s^2 + 5s = 3$

10. $4x^2 + x = 3$

11. $3y^2 = 3y + 2$

12. $3x^2 = 3 - 4x$

13. $2y^2 - y - 2 = 0$

14. $9v^2 - 6v - 2 = 0$

15. $30y^2 + 23y - 40 = 0$

16. $40x^2 - 62x - 63 = 0$

17. $2t^2 + 10t = -15$

18. $2d(d - 2) = -7$

19. $s^2 = 9 + s(1 - 2s)$

20. $6r^2 = 6r + 1$

21. $4x^2 = 9$

22. $6x = x^2$

23. $15 + 4z = 32z^2$

24. $4x^2 - 12x = 7$

25. $x^2 - 0.20x - 0.40 = 0$

26. $3.2x^2 = 2.5x + 7.6$

27. $0.29x^2 - 0.18 = 0.63x$

28. $12.5x^2 + 13.2x = 15.5$

29. $x^2 + 2cx - 1 = 0$

30. $x^2 - 7x + (6 + a) = 0$

31. $b^2x^2 - (b + 1)x + (1 - a) = 0$

32. $c^2x^2 - x - 1 = x^2$

See Appendix E for a computer program for solving a quadratic equation by use of the quadratic formula.

33. In machine design, in finding the outside diameter D_0 of a hollow shaft, the equation $D_0^2 - DD_0 - 0.25D^2 = 0$ is used. Solve for D_0 if $D = 3.625$ cm.

34. A missile is fired vertically into the air. The distance (in feet) above the ground as a function of time (in seconds) is given by the formula $s = 300 + 500t - 16t^2$. (a) When will the missile hit the ground? (b) When will the missile be 1000 ft above the ground?

35. In calculating the current in an electric circuit with an inductance L, a resistance R, and a capacitance C, it is necessary to solve the equation $Lm^2 + Rm + \dfrac{1}{C} = 0$. Solve for m in terms of L, R, and C.

36. A medicine tablet 2.50 mm thick is coated with 52.5 mm² of a special coating. What is the radius of the tablet? (The surface area A of a cylinder of radius r and height h is $A = 2\pi r^2 + 2\pi rh$.)

In Exercises 37 through 40, set up appropriate equations and solve the stated problems.

37. The height of a rectangular sign is 2.0 ft more than its width. The area of the sign is 29 ft². What are the dimensions of the sign?

38. Two circular oil spills are tangent to each other. If the distance between centers is 800 m, and they cover a combined area of 1.02×10^6 m², what is the radius of each?

39. A security fence is to be built around a rectangular parking area of 20,000 ft². If the front side of the fence costs \$20 per foot and the other three sides cost \$10 per foot, what are the dimensions of the parking area if the fence is to cost \$7500?

40. Two pipes together drain a wastewater holding tank in 6.00 h. If used alone to empty the tank, one takes 2.00 h longer than the other. How long does each take to empty the tank if used alone?

6-4   The Graph of the Quadratic Function

We have developed the basic algebraic methods of solving a quadratic equation. In this section we consider the graph of the quadratic function. Then, following the method developed in Section 2-6, we show the graphical solution of a quadratic equation.

In Section 6-1 we noted that $ax^2 + bx + c$ is the quadratic function. As in Chapter 2, by letting $y = ax^2 + bx + c$ we can graph this function. Example A briefly reviews the graph of a quadratic function done in this way.

EXAMPLE A

Graph the quadratic function $f(x) = x^2 + 2x - 3$.

First we let $y = x^2 + 2x - 3$, then we set up a table of values. The graph is shown in Fig. 6-3.

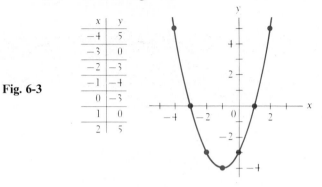

x	y
-4	5
-3	0
-2	-3
-1	-4
0	-3
1	0
2	5

Fig. 6-3

The graph of any quadratic function $y = ax^2 + bx + c$ *will have the same general* **shape** *as that shown in Fig. 6-3, and the graph is called a* **parabola.** (In Section 2-4 we briefly noted that the graphs for Examples B and C were parabolas.) A parabola defined by a quadratic function can open upward (as in Example A) or downward. The exact location of the graph and how it opens depend on the values of a, b, and c.

From Fig. 6-3 we see that the parabola has a minimum point at $(-1, -4)$ and that the curve opens upward. *All parabolas have an* **extreme point** *of this type. If $a > 0$, the parabola will have a* **minimum point** *and it will open* **upward.** *If $a < 0$, the parabola will have a* **maximum point** *and it will open* **downward.**

extreme points

EXAMPLE B

The graph of $y = 2x^2 - 8x + 6$ is shown in Fig. 6-4(a). Here $a = 2(a > 0)$ and the graph opens upward. The minimum point is $(2, -2)$.

The graph of $y = -2x^2 + 8x - 6$ is shown in Fig. 6-4(b). Here $a = -2$ ($a < 0$) and the graph opens downward. The maximum point is $(2, 2)$.

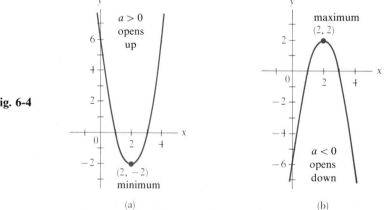

Fig. 6-4

(a)   (b)

We can quickly sketch the graph of a parabola by using its basic shape and by knowing the location of two or three points, including the extreme point. The location of the extreme point can be found by using the method of completing the square, in a way similar to that used in deriving the quadratic formula.

In order to find the coordinates of the extreme point we start with the quadratic function

$$y = ax^2 + bx + c$$

Then we factor a from the two terms containing x, obtaining

$$y = a\left(x^2 + \frac{b}{a}x\right) + c$$

Now, completing the square of the terms within parentheses, we have

$$y = a\left(x^2 + \frac{b}{a}x + \frac{b^2}{4a^2}\right) + c - \frac{b^2}{4a}$$

$$= a\left(x + \frac{b}{2a}\right)^2 + c - \frac{b^2}{4a}$$

We now look at the factor $\left(x + \frac{b}{2a}\right)^2$. If $x = -b/2a$, the term is zero. If x is any other value, $\left(x + \frac{b}{2a}\right)^2$ is positive. Thus, if $a > 0$, the value of y increases from that we have for $x = -b/2a$, and if $a < 0$, the value of y decreases from that we have for $x = -b/2a$. *This means that $x = -b/2a$ is the x-coordinate of the extreme point. The y-coordinate can be found by substituting in the function.*

Another easily found point is the y-intercept. As with a linear equation, we find the y-intercept where $x = 0$. For $y = ax^2 + bx + c$, if $x = 0$, then $y = c$. *This means that the point $(0, c)$ is the y-intercept.*

EXAMPLE C

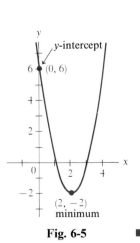

Fig. 6-5

For the function $y = 2x^2 - 8x + 6$, find the extreme point and the y-intercept and sketch the graph. (This function is also used in Example B.)

First, $a = 2$ and $b = -8$. This means that the x-coordinate of the extreme point is

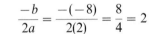

$$\frac{-b}{2a} = \frac{-(-8)}{2(2)} = \frac{8}{4} = 2$$

and the y-coordinate is

$$y = 2(2^2) - 8(2) + 6 = -2$$

Thus, the extreme point is $(2, -2)$. Since $a > 0$, it is a minimum point.

Since $c = 6$, the y-intercept is $(0, 6)$.

We can use the minimum point $(2, -2)$ and the y-intercept $(0, 6)$, along with the fact that the graph is a parabola, to get an approximate sketch of the graph. Noting that a parabola increases (or decreases) away from the extreme point in the same way on each side of it (it is *symmetric* to a vertical line through the extreme point), we sketch the graph in Fig. 6-5. We see that it is the same as that shown in Fig. 6-4(a). ∎

We may need one or two additional points to get a reasonable sketch of a parabola. This would be true if the y-intercept is close to the extreme point. Two points we can find are the x-intercepts, if the parabola crosses the x-axis (one point if the extreme point is on the x-axis). They are found by setting $y = 0$ and solving the quadratic equation $ax^2 + bx + c = 0$. Also, we may simply find one or two points other than the extreme point and the y-intercept.

EXAMPLE D

Sketch the graph of $y = -x^2 + x + 6$.

We first note that $a = -1$ and $b = 1$. Therefore, the x-coordinate of the maximum point ($a < 0$) is $-\frac{1}{2(-1)} = \frac{1}{2}$. The y-coordinate is $-(\frac{1}{2})^2 + \frac{1}{2} + 6 = \frac{25}{4}$. This means that the maximum point is $(\frac{1}{2}, \frac{25}{4})$.

The y-intercept is $(0, 6)$.

Using these points in Fig. 6-6, we see that they are close together and do not give a good idea of how wide the parabola opens. Therefore, setting $y = 0$, we solve the equation

$$-x^2 + x + 6 = 0$$

or

$$x^2 - x - 6 = 0$$

This equation is factorable. Thus,

$$(x - 3)(x + 2) = 0$$
$$x = 3, -2$$

This means that the x-intercepts are $(3, 0)$ and $(-2, 0)$, as shown in Fig. 6-6.

Also, rather than finding the x-intercepts, we can let $x = 2$ and then use the point $(2, 4)$.

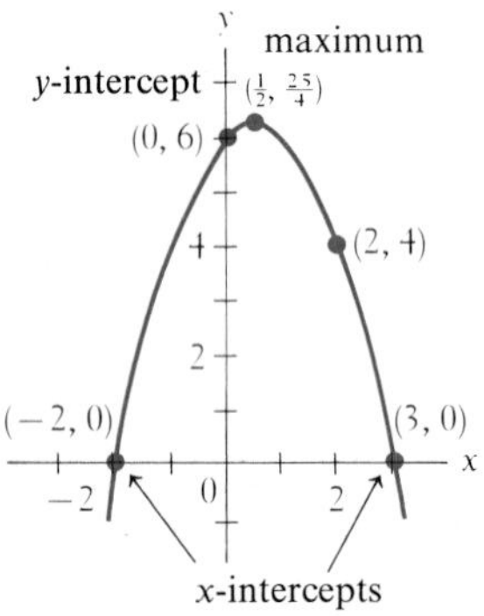

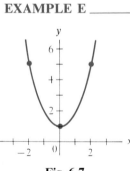

Fig. 6-6

EXAMPLE E

Sketch the graph of $y = x^2 + 1$.

Since there is no x-term, $b = 0$. This means that the x-coordinate of the minimum point ($a > 0$) is 0 and that the minimum point and the y-intercept are both $(0, 1)$. We know that the graph opens upward, since $a > 0$, which in turn means that it does not cross the x-axis. Now, letting $x = 2$ and $x = -2$, we find the points $(2, 5)$ and $(-2, 5)$ on the graph, which is shown in Fig. 6-7.

We can see that the domain of the quadratic function is all x. Knowing that the graph of the quadratic function has either a maximum or a minimum point, we see that the range must be restricted. In Example C, the range is $f(x) \geq -2$; in Example D, the range is $f(x) \leq \frac{25}{4}$; and in Example E, the range is $f(x) \geq 1$.

Fig. 6-7

The x-intercepts of the quadratic function $y = ax^2 + bx + c$ are the same as the zeros of the function, and we can find them by letting $y = 0$, as we have already noted. We can also use this idea to solve a quadratic equation graphically. This is illustrated in the following example.

EXAMPLE F

Solve the equation $3x = x(2 - x) + 3$ graphically.

We first algebraically collect all terms on the left side of the equal sign. This leads to the equation $x^2 + x - 3 = 0$. We then let $y = x^2 + x - 3$ and graph the function, as shown in Fig. 6-8. We use the minimum point $(-\frac{1}{2}, -\frac{13}{4})$, the y-intercept $(0, -3)$, and the points $(-3, 3)$ and $(2, 3)$ to sketch the graph. The points $(-2, -1)$ and $(1, -1)$ were then added for extra accuracy when it was determined approximately where the curve crossed the x-axis. Since the x-intercepts are approximately $x = -2.3$ and $x = 1.3$, these are also the solutions to the equation.

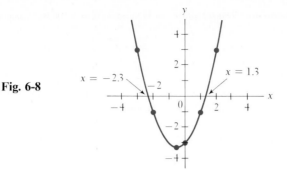

Fig. 6-8

If we were to try to solve the equation $x^2 + 1 = 0$ graphically, we would have the graph shown in Fig. 6-7. *Since it does not cross the x-axis, there are no **real** solutions to the equation.*

EXAMPLE G _____

A projectile is fired vertically upward from the ground with a velocity of 38 m/s. Its distance above the ground is given by $s = -4.9t^2 + 38t$, where s is the distance in meters and t is the time in seconds. Graph the function, and from the graph determine (1) when the projectile will hit the ground, (2) how high it will go, and (3) how long it takes to reach 45 m above the ground.

Since $c = 0$, the s-intercept is at the origin. Using $-b/2a$, we find the maximum point at about (3.9, 74). The t-intercepts are (0, 0) and about (7.8, 0). Using these points, we graph the function, as shown in Fig. 6-9.

1. The projectile will hit the ground when $s = 0$, which is shown by the right t-intercept. Thus, it takes about 7.8 s to hit the ground.

2. Its maximum height is the s-coordinate of the maximum point. Therefore, it goes to about 74 m above the ground.

3. We find the time it takes to get 45 m above the ground by drawing a horizontal line through 45 on the s-axis. We see that this line intersects the curve twice—when $t = 1.5$ s and when $t = 6.3$ s. (Two additional points, for $t = 2$ and $t = 6$, are helpful to get better accuracy for this part.)

Fig. 6-9

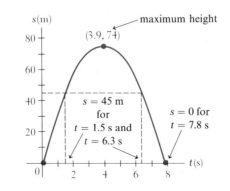

Exercises 6-4

In Exercises 1 through 8, sketch the graphs of the given parabolas by using only the extreme point and the y-intercept.

1. $y = x^2 - 6x + 5$ **2.** $y = -x^2 - 4x - 3$ **3.** $y = -3x^2 + 10x - 4$ **4.** $y = 2x^2 + 8x - 5$

5. $y = x^2 - 4x$ **6.** $y = -2x^2 - 5x$ **7.** $y = -2x^2 - 4x - 3$ **8.** $y = x^2 - 3x + 4$

In Exercises 9 through 12, sketch the graphs of the given parabolas by using the extreme point, the y-intercept, and the x-intercepts.

9. $y = x^2 - 4$ **10.** $y = x^2 + 3x$ **11.** $y = -2x^2 - 6x + 8$ **12.** $y = -3x^2 + 12x - 5$

In Exercises 13 through 16, sketch the graphs of the given parabolas by using the extreme point, the y-intercept, and two other points, not including the x-intercepts.

13. $y = 2x^2 + 3$ **14.** $y = x^2 + 2x + 2$ **15.** $y = -2x^2 - 2x - 6$ **16.** $y = -3x^2 - x$

In Exercises 17 through 24, solve the given quadratic equations graphically. If there are no real roots, simply state this as the answer.

17. $2x^2 - 3 = 0$ **18.** $5 - x^2 = 0$ **19.** $-3x^2 + 11x - 5 = 0$ **20.** $2x^2 = 7x + 4$

21. $x(2x - 1) = -3$ **22.** $2x - 5 = x^2$ **23.** $6x^2 = 18 - 7x$ **24.** $3x^2 - 25 = 20x$

In Exercises 25 through 32, solve the given applied problems.

25. The area A of a certain rectangular parcel of land is given by $A = w(7 - 2w)$, where w is the width of the parcel. Sketch the graph of A as a function of w.

26. Under specified conditions, the pressure loss L, in pounds per square inch, in the flow of water through a water line in which the flow is q gallons per minute is given by $L = 0.0002q^2 + 0.005q$. Sketch the graph of L as a function of q, for $q < 100$ gal/min.

27. When analyzing the power P, in watts, dissipated in an electric circuit, the equation $P = 50i - 3i^2$ results. Here i is the current in amperes. Sketch the graph of $P = f(i)$.

28. The vertical distance d, in centimeters, of the end of a robot arm above a conveyor belt in its 8-s cycle is given by $d = 2t^2 - 16t + 47$. Sketch the graph of $d = f(t)$.

29. A missile is fired vertically upward such that its distance s, in feet, above the ground is given by $s = 150 + 250t - 16t^2$, where t is the time in seconds. Sketch the graph, and then determine from the graph (a) when the missile will hit the ground, (b) how high it will go, and (c) how long it takes to reach 800 ft above the ground.

30. In a certain electric circuit, the resistance R, in ohms, that gives resonance is found by solving the equation $25R = 3(R^2 + 4)$. Solve this equation graphically.

31. In remodeling a house, an architect finds that by adding the same amount to each dimension of a 12-ft by 16-ft rectangular room the area would be increased by 80 ft². Set up the appropriate equation and solve it graphically for the amount added to each dimension.

32. An airplane pilot could decrease the time needed to cover 350 mi by 1 h if the plane's speed is increased by 50 mi/h. Set up the appropriate equation and solve graphically for the plane's speed.

6-5   Chapter Equations, Review Exercises, and Practice Test

Chapter Equations

Quadratic equation $ax^2 + bx + c = 0$ (6-1)

Quadratic formula $x = \dfrac{-b \pm \sqrt{b^2 - 4ac}}{2a}$ (6-4)

Review Exercises

In Exercises 1 through 12, solve the given quadratic equations by factoring.

1. $x^2 + 3x - 4 = 0$

2. $x^2 + 3x - 10 = 0$

3. $x^2 - 10x + 16 = 0$

4. $x^2 - 6x - 27 = 0$

5. $3x^2 + 11x = 4$

6. $6y^2 = 11y - 3$

7. $6t^2 = 13t - 5$

8. $3x^2 + 5x + 2 = 0$

9. $6s^2 = 25s$

10. $6n^2 - 23n - 35 = 0$

11. $4x^2 - 8x = 21$

12. $6x^2 = 8 - 47x$

In Exercises 13 through 24, solve the given quadratic equations by using the quadratic formula.

13. $x^2 - x - 110 = 0$

14. $x^2 + 3x - 18 = 0$

15. $x^2 + 2x - 5 = 0$

16. $x^2 - 7x - 1 = 0$

17. $2x^2 - x = 36$

18. $3x^2 + x = 14$

19. $4x^2 - 3x - 2 = 0$

20. $5x^2 + 7x - 2 = 0$

21. $2.1x^2 + 2.3x + 5.5 = 0$

22. $0.30x^2 - 0.42x = 0.15$

23. $6x^2 = 9 - 4x$

24. $24x^2 = 25x + 20$

In Exercises 25 through 36, solve the given quadratic equations by any appropriate method.

25. $x^2 + 4x - 4 = 0$

26. $x^2 + 3x + 1 = 0$

27. $3x^2 + 8x + 2 = 0$

28. $3p^2 = 28 - 5p$

29. $4v^2 = v + 5$

30. $6x^2 - x + 2 = 0$

31. $2x^2 + 3x + 7 = 0$

32. $4y^2 - 5y = 8$

33. $a^2x^2 + 2ax + 2 = 0$

34. $16r^2 - 8r + 1 = 0$

35. $ax^2 = a^2 - 3x$

36. $2bx = x^2 - 3b$

In Exercises 37 through 40, solve the given quadratic equations by completing the square.

37. $x^2 - x - 30 = 0$

38. $x^2 - 2x - 5 = 0$

39. $2x^2 - x - 4 = 0$

40. $4x^2 - 8x - 3 = 0$

In Exercises 41 through 44, solve the equations involving fractions.

41. $\dfrac{x-4}{x-1} = \dfrac{2}{x}$

42. $\dfrac{x-1}{3} = \dfrac{5}{x} + 1$

43. $\dfrac{x^2 - 3x}{x - 3} = \dfrac{x^2}{x + 2}$

44. $\dfrac{x-2}{x-5} = \dfrac{15}{x^2 - 5x}$

In Exercises 45 through 48, sketch the graphs of the given functions by using the extreme point, the y-intercept, and one or two other points.

45. $y = 2x^2 - x - 1$

46. $y = -4x^2 - 1$

47. $y = x - 3x^2$

48. $y = 2x^2 + 8x - 10$

In Exercises 49 through 52, solve the given quadratic equations graphically. If there are no real roots, simply state this as the answer.

49. $2x^2 + x - 4 = 0$

50. $-4x^2 - x - 1 = 0$

51. $3x^2 = -x - 2$

52. $x(15x - 12) = 8$

In Exercises 53 through 60, solve the given quadratic equations by any appropriate method.

53. In a natural gas pipeline, the velocity v, in meters per second, of the gas as a function of the distance x, in centimeters, from the wall of the pipe is given by $v = 5.2x - x^2$. Determine x for $v = 4.8$ m/s.

54. At an altitude h feet above sea level, the boiling point of water is lower by T degrees Fahrenheit than the boiling point at sea level, which is $212°F$. The difference can be approximated by solving the equation $T^2 + 520T - h = 0$. What is the boiling point at an altitude of 5300 ft?

55. The height h of an object ejected at an angle θ from a vehicle moving with velocity v is given by $h = vt \sin \theta - 16t^2$, where t is the time of flight. Find t if $v = 44$ ft/s, $\theta = 65°$, and $h = 18$ ft.

56. In studying the emission of light, in order to determine the angle at which the intensity is a given value, the equation $\sin^2 A - 4 \sin A + 1 = 0$ must be solved. Find angle A. $[\sin^2 A = (\sin A)^2.]$

57. A computer analysis shows that the number n of electronic components a company should produce for supply to equal demand is found by solving $\dfrac{n^2}{500{,}000} = 144 - \dfrac{n}{500}$. Find n.

58. To determine the resistances of two resistors which are to be in parallel in an electric circuit, it is necessary to solve the equation $\dfrac{20}{R} + \dfrac{20}{R + 10} = \dfrac{1}{5}$. Find R, in ohms.

59. In the study of ionization in chemistry, the equation $a^2 c = k(1 - a)$ occurs. Solve for a in terms of c and k.

60. In the analysis of mechanical vibrations, the equation $s^2 + \dfrac{c}{m} s + \dfrac{k\ell^2}{mb^2} = 0$ is found. Solve for s in terms of b, c, k, ℓ, and m.

In Exercises 61 through 72, set up the necessary equation where appropriate and solve the given problems.

61. In testing the effects of a drug, the percent of the drug in the blood was given by $p = 0.090t - 0.015t^2$, where t is the time, in hours, after the drug was administered. Sketch the graph of $p = f(t)$.

62. In an electric circuit, the voltage V as a function of the time t is given by $V = 9.8 - 9.2t + 2.3t^2$. Sketch the graph of $V = f(t)$, for $t \le 5$ min.

63. By adding the same amount to its length and its width, a developer increased the area of a rectangular lot by 3000 m^2 to make it 80 m by 100 m. What were the dimensions of the lot before the change?

64. A machinery pedestal is made of two concrete cubes, one on top of the other. The pedestal is 8.00 ft high and contains 152 ft^3 of concrete. Find the edge of each cube.

65. A metal cube expands when heated. If the volume changes by 6.00 mm^3 and each edge is 0.20 mm longer after being heated, what was the original length of an edge of the cube?

66. A military jet flies directly over and at right angles to the straight course of a commercial jet. The military jet is flying at 200 mi/h faster than four times the speed of the commercial jet. How fast is each going if they are 2050 mi apart (on a direct line) after 1 h?

67. A rectangular table is designed such that the length is 1.5 ft more than the width, and the diagonal is 6.0 ft. Find the width.

68. A given length of wire weighs 8.50 lb. It is stretched to a length 1.00 ft longer, and then each foot of wire weighs 0.050 lb less. Find the original length of the wire.

69. An electric utility company is placing utility poles along a road. It is determined that five fewer poles per kilometer would be necessary if the distance between poles were increased by 10 m. How many poles are being placed each kilometer?

70. A rectangular duct in a building's ventilating system is made of sheet metal 7.0 ft wide and has a cross-sectional area of 3.0 ft^2. What are the cross-sectional dimensions of the duct?

71. A testing station found that the parts p per million of sulfur dioxide in the air as a function of the hour h of the day was given by $p = 0.00174(10 + 24h - h^2)$. Sketch the graph of $p = f(h)$ and, from the graph, find the time when $p = 0.200$ parts per million.

72. A hole is drilled in the center of a circular plate such that it is 10.0 cm from the edge of the hole to the edge of the plate. Find the radius of the hole if 10% of the material of the plate is removed in drilling the hole.

Practice Test

In Problems 1 through 4, solve for x.

1. $x^2 - 3x - 5 = 0$

2. $2x^2 = 9x - 4$

3. $\dfrac{3}{x} - \dfrac{2}{x+2} = 1$

4. $2x^2 - x = 6 - 2x(3 - x)$

5. Sketch the graph of $y = 2x^2 + 8x + 5$ using the extreme point and the y-intercept.

6. In electricity the formula $P = EI - RI^2$ is used. Solve for I in terms of E, P, and R.

7. Solve by completing the square: $x^2 - 6x - 9 = 0$.

8. The perimeter of a rectangular window is 8.4 m, and its area is 3.8 m^2. Find its dimensions.

7 Trigonometric Functions of Any Angle

In Section 7-4 we see how the speed of a satellite orbiting the earth may be calculated.

When we were dealing with the trigonometric functions in Chapter 3, we restricted ourselves primarily to right triangles and functions of acute angles measured in degrees. Since we did define the functions in general, we can use these same definitions for finding the functions of any possible angle. In this chapter we shall not only find the trigonometric functions of angles measured in degrees, but we shall develop the use of radian measure as well.

We will find angles larger than 90° and triangles which are not right triangles important in many types of applied problems. Radian measure is also used in numerous applications, including mechanical vibrations, electric currents, and rotational motion.

7-1 Signs of the Trigonometric Functions

We recall the definitions of the trigonometric functions which were given in Section 3-2: *Here the point (x, y) is a point on the terminal side of angle θ, and r is the radius vector.* See Fig. 7-1.

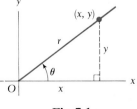

Fig. 7-1

$$\sin \theta = \frac{y}{r} \qquad \cos \theta = \frac{x}{r} \qquad \tan \theta = \frac{y}{x}$$

$$\cot \theta = \frac{x}{y} \qquad \sec \theta = \frac{r}{x} \qquad \csc \theta = \frac{r}{y}$$

(7-1)

We see that we can find the functions if we know the values of the coordinates (x, y) on the terminal side of θ and the radius vector r. Of course, if either x or y is zero in the denominator, the function is undefined, and we will consider this further in the next section. *Remembering that r is always considered to be positive, we can see that the various functions will vary in sign in each of the quadrants, depending on the signs of x and y.*

If the terminal side of the angle is in the first or second quadrant, the value of $\sin \theta$ will be positive, but if the terminal side is in the third or fourth quadrant, $\sin \theta$ is negative. This is because *the sign of sin θ depends on the sign of the y coordinate* of the point on the terminal side, and y is positive if the point is above the x-axis, and y is negative if this point is below the x-axis. See Fig. 7-2.

Quadrant			
I	II	III	IV
$\sin \theta$ +	+	−	−

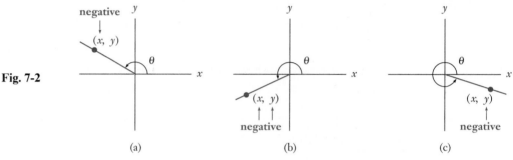

Fig. 7-2

(a)   (b)   (c)

EXAMPLE A — The value of $\sin 20°$ is positive, since the terminal side of $20°$ is in the first quadrant. The value of $\sin 160°$ is positive, since the terminal side of $160°$ is in the second quadrant. The values of $\sin 200°$ and $\sin 340°$ are negative, since the terminal sides of these angles are in the third and fourth quadrants, respectively. ∎

Quadrant			
I	II	III	IV
$\tan \theta$ +	−	+	−

The sign of tan θ depends upon the ratio of y to x. In the first quadrant both x and y are positive, and therefore the ratio y/x is positive. In the third quadrant both x and y are negative, and therefore the ratio y/x is positive. In the second and fourth quadrants either x or y is positive and the other negative, and so the ratio of y/x is negative. See Fig. 7-2.

EXAMPLE B — The values of $\tan 20°$ and $\tan 200°$ are positive, since the terminal sides of these angles are in the first and third quadrants, respectively. The values of $\tan 160°$ and $\tan 340°$ are negative, since the terminal sides of these angles are in the second and fourth quadrants, respectively. ∎

Quadrant			
I	II	III	IV
$\cos \theta$ +	−	−	+

The sign of cos θ depends upon the sign of x. Since x is positive in the first and fourth quadrants, $\cos \theta$ is positive in these quadrants. In the same way, $\cos \theta$ is negative in the second and third quadrants. See Fig. 7-2.

EXAMPLE C — The values of $\cos 20°$ and $\cos 340°$ are positive, since these angles are first- and fourth-quadrant angles, respectively. The values of $\cos 160°$ and $\cos 200°$ are negative, since these angles are second- and third-quadrant angles, respectively. ∎

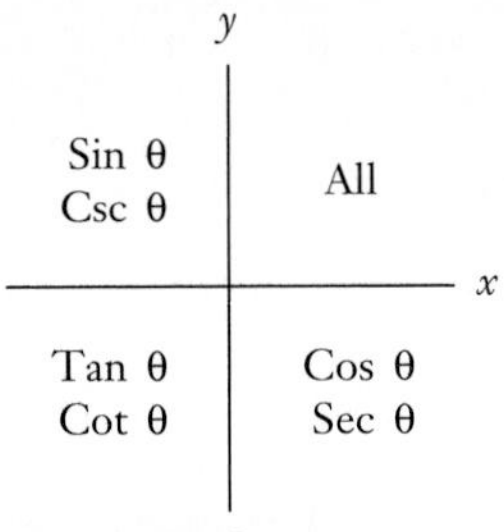

Sin θ
Csc θ

All

Tan θ
Cot θ

Cos θ
Sec θ

Positive functions

Fig. 7-3

Since csc θ is defined in terms of r and y, as is sin θ, the sign of csc θ is the same as that of sin θ. For the same reason, cot θ has the same sign as tan θ, and sec θ has the same sign as cos θ. A method for remembering the signs of the functions in the four quadrants is as follows:

All functions of first-quadrant angles are positive. Sin θ and csc θ are positive for second-quadrant angles. Tan θ and cot θ are positive for third-quadrant angles. Cos θ and sec θ are positive for fourth-quadrant angles. All others are negative.

This is shown in Fig. 7-3.

This discussion does not include the quadrantal angles, those angles with terminal sides on one of the axes. They will be discussed in the following section.

EXAMPLE D ___ sin 50°, sin 150°, sin (−200°), cos 8°, cos 300°, cos (−40°), tan 220°, tan (−100°), cot 260°, cot (−310°), sec 280°, sec (−37°), csc 140°, and csc (−190°) are all positive.

EXAMPLE E ___ sin 190°, sin 325°, cos 100°, cos (−95°), tan 172°, tan 295°, cot 105°, cot (−6°), sec 135°, sec (−135°), csc 240°, and csc 355° are all negative.

EXAMPLE F ___

Fig. 7-4

Determine the trigonometric functions of θ if the terminal side of θ passes through $(-1, \sqrt{3})$. See Fig. 7-4.

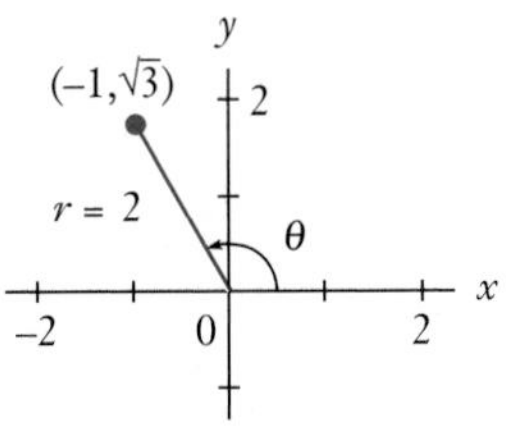

We know that $x = -1$, $y = +\sqrt{3}$, and from the Pythagorean theorem we find that $r = 2$. Therefore, the trigonometric functions of θ are

$$\sin \theta = +\frac{\sqrt{3}}{2} \qquad \cos \theta = -\frac{1}{2} \qquad \tan \theta = -\sqrt{3}$$

$$\cot \theta = -\frac{1}{\sqrt{3}} \qquad \sec \theta = -2 \qquad \csc \theta = +\frac{2}{\sqrt{3}}$$

We note that the point $(-1, \sqrt{3})$ is on the terminal side of a second-quadrant angle, and that the signs of the functons of θ are those of a second-quadrant angle.

Exercises 7-1

In Exercises 1 through 8, determine the algebraic sign of the given trigonometric functions.

1. sin 60°, cos 120°, tan 320°

2. tan 185°, sec 115°, sin (−36°)

3. cos 300°, csc 97°, cot (−35°)

4. sin 100°, sec (−15°), cos 188°

5. cot 186°, sec 280°, sin 470°

6. tan (−91°), csc 87°, cot 103°

7. cos 700°, tan (−560°), csc 530°

8. sin 256°, tan 321°, cos (−370°)

In Exercises 9 through 16, find the trigonometric functions of θ, where the terminal side of θ passes through the given point.

9. (2, 1)

10. (−1, 1)

11. (−2, −3)

12. (4, −3)

13. (−5, 12)

14. (−3, −4)

15. (50, −20)

16. (9, 40)

In Exercises 17 through 24, determine the quadrant in which the terminal side of θ lies, subject to the given conditions.

17. $\sin \theta$ positive, $\cos \theta$ negative

18. $\tan \theta$ positive, $\cos \theta$ negative

19. $\sec \theta$ negative, $\cot \theta$ negative

20. $\cos \theta$ positive, $\csc \theta$ negative

21. $\csc \theta$ negative, $\tan \theta$ negative

22. $\sec \theta$ positive, $\csc \theta$ positive

23. $\sin \theta$ negative, $\tan \theta$ positive

24. $\cot \theta$ negative, $\sin \theta$ negative

7-2   Trigonometric Functions of Any Angle

The trigonometric functions of acute angles were discussed in Section 3-3, and in the last section we determined the signs of the trigonometric functions in each of the four quadrants. In this section we shall show how we can find the trigonometric functions of an angle of any magnitude. This information will be very important in Chapter 8, when we discuss vectors and oblique triangles, and in Chapter 9, when we graph the trigonometric functions. Even *a calculator will not always give the required angle for a given value of a function.*

Any angle in standard position is coterminal with some positive angle less than 360°. Since the terminal sides of coterminal angles are the same, the trigonometric functions of coterminal angles are the same. Therefore, we need consider only the problem of finding the values of the trigonometric functions of positive angles less than 360°.

EXAMPLE A

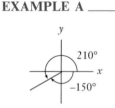

Fig. 7-5

The following pairs of angles are coterminal.

$$390° \quad \text{and} \quad 30°, \qquad -60° \quad \text{and} \quad 300°$$
$$900° \quad \text{and} \quad 180°, \qquad -150° \quad \text{and} \quad 210°$$

From this we conclude that the trigonometric functions of both angles in these pairs are equal. That is, for example, $\sin 390° = \sin 30°$ and $\tan(-150°) = \tan 210°$. See Fig. 7-5. ■

Considering the definitions of the functions, we see that the values of the functions depend only on the values of x, y, and r. The values of the functions of second-quadrant angles are numerically equal to the functions of corresponding first-quadrant angles. For example, considering the angles shown in Fig. 7-6, for angle θ_2 with terminal side passing through $(-3, 4)$, $\tan \theta_2 = -\frac{4}{3}$, and for angle θ_1 with terminal side passing through $(3, 4)$, $\tan \theta_1 = \frac{4}{3}$. In Fig. 7-6, we see that the triangle containing angles θ_1 and α are congruent, which means that θ_1 and α are equal. We know that the trigonometric functions of θ_1 and θ_2 are numerically equal. This means that

$$|F(\theta_2)| = |F(\theta_1)| = |F(\alpha)| \tag{7-2}$$

where F represents any of the trigonometric functions.

reference angle   The angle labeled α is called the **reference angle.** *The reference angle of a given angle is the acute angle formed by the terminal side of the angle and the x-axis.*

Using Eq. (7-2) and the fact that $\alpha = 180° - \theta_2$, we may conclude that the value of any trigonometric function of any second-quadrant angle is found from

$$F(\theta_2) = \pm F(180° - \theta_2) = \pm F(\alpha) \tag{7-3}$$

The sign to be used depends on whether the *function* is positive or negative in the second quadrant.

EXAMPLE B In Fig. 7-6, the trigonometric functions of θ_2 are as follows:

$$\sin \theta_2 = +\sin (180° - \theta_2) = +\sin \alpha = +\sin \theta_1 = \frac{4}{5}$$

$$\cos \theta_2 = -\cos \theta_1 = -\frac{3}{5}$$

$$\tan \theta_2 = -\frac{4}{3}, \qquad \cot \theta_2 = -\frac{3}{4}$$

$$\sec \theta_2 = -\frac{5}{3}, \qquad \csc \theta_2 = +\frac{5}{4}$$

In the same way we may derive the formulas for finding the trigonometric functions of any third- or fourth-quadrant angle. Considering the angles shown in Fig. 7-7, we see that the reference angle α is found by subtracting 180° from θ_3 and that functions of α and θ_1 are numerically equal. Considering the angles shown in Fig. 7-8, we see that the reference angle α is found by subtracting θ_4 from 360°. Therefore, we have

$$F(\theta_3) = \pm F(\theta_3 - 180°) = \pm F(\alpha) \tag{7-4}$$

$$F(\theta_4) = \pm F(360° - \theta_4) = \pm F(\alpha) \tag{7-5}$$

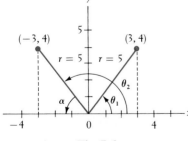

Fig. 7-6

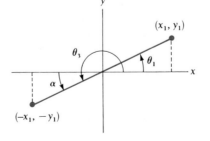

Fig. 7-7

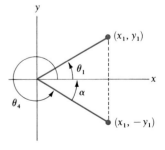

Fig. 7-8

EXAMPLE C _____

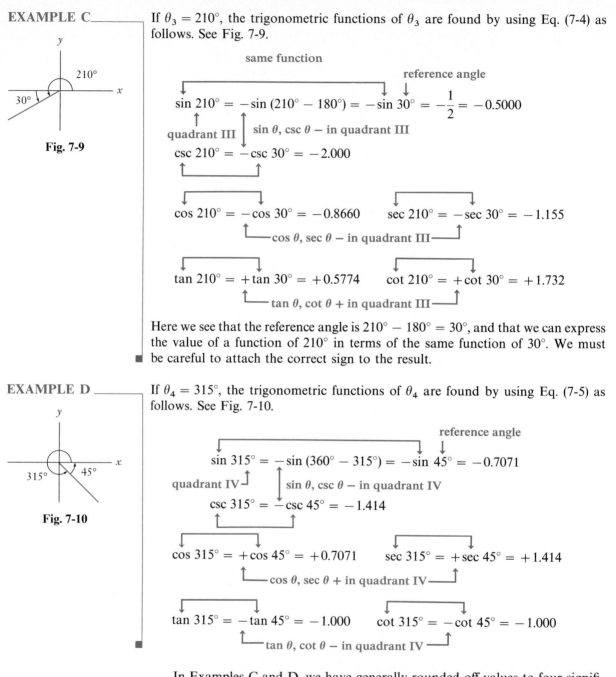

If $\theta_3 = 210°$, the trigonometric functions of θ_3 are found by using Eq. (7-4) as follows. See Fig. 7-9.

same function

reference angle

$$\sin 210° = -\sin(210° - 180°) = -\sin 30° = -\frac{1}{2} = -0.5000$$

quadrant III $\sin \theta, \csc \theta$ — in quadrant III

$$\csc 210° = -\csc 30° = -2.000$$

$$\cos 210° = -\cos 30° = -0.8660 \qquad \sec 210° = -\sec 30° = -1.155$$

$\cos \theta, \sec \theta$ — in quadrant III

$$\tan 210° = +\tan 30° = +0.5774 \qquad \cot 210° = +\cot 30° = +1.732$$

$\tan \theta, \cot \theta$ + in quadrant III

Fig. 7-9

Here we see that the reference angle is $210° - 180° = 30°$, and that we can express the value of a function of $210°$ in terms of the same function of $30°$. We must be careful to attach the correct sign to the result. ∎

EXAMPLE D _____

If $\theta_4 = 315°$, the trigonometric functions of θ_4 are found by using Eq. (7-5) as follows. See Fig. 7-10.

reference angle

$$\sin 315° = -\sin(360° - 315°) = -\sin 45° = -0.7071$$

quadrant IV $\sin \theta, \csc \theta$ — in quadrant IV

$$\csc 315° = -\csc 45° = -1.414$$

$$\cos 315° = +\cos 45° = +0.7071 \qquad \sec 315° = +\sec 45° = +1.414$$

$\cos \theta, \sec \theta$ + in quadrant IV

$$\tan 315° = -\tan 45° = -1.000 \qquad \cot 315° = -\cot 45° = -1.000$$

$\tan \theta, \cot \theta$ — in quadrant IV

Fig. 7-10

In Examples C and D, we have generally rounded off values to four significant digits. If we know that an angle is approximate, we will use the guidelines given in Appendix B for rounding off values of a function. Also, calculator displays will generally show an eight-digit value.

EXAMPLE E —————— Other illustrations of the use of Eqs. (7-3), (7-4), and (7-5) are as follows:

$$\sin 160° = +\sin (180° - 160°) = \sin 20° = 0.3420$$
$$\tan 110° = -\tan (180° - 110°) = -\tan 70° = -2.747$$
$$\cos 225° = -\cos (225° - 180°) = -\cos 45° = -0.7071$$
$$\cot 260° = +\cot (260° - 180°) = \cot 80° = 0.1763$$
$$\sec 304° = +\sec (360° - 304°) = \sec 56° = 1.788$$
$$\sin 357° = -\sin (360° - 357°) = -\sin 3° = -0.0523$$

same function → reference angle ↓

↑ determines quadrant   ↑ proper sign for function in quadrant

A calculator can be used directly to find values like those in Examples C, D, and E. We simply enter the angle and then the required function (and use the reciprocal key for cot θ, sec θ, and csc θ, as shown in Section 3-3). The calculator will give us the value of the function of any angle, with the proper sign. This is illustrated in the following example.

EXAMPLE F —————— A formula for finding the area of a triangle, knowing sides a and b and the included angle C, is $A = \frac{1}{2}ab \sin C$. A surveyor uses this formula to find the area of a triangular tract of land for which $a = 173.2$ m, $b = 156.3$ m, and $C = 112.51°$. See Fig. 7-11.

To find the area, we substitute the given values into the formula, which gives us

$$A = \tfrac{1}{2}(173.2)(156.3)\sin 112.51°$$

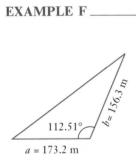

112.51°

$a = 173.2$ m

$b = 156.3$ m

Fig. 7-11

The calculator sequence for this is

173.2 $\boxed{\times}$ 156.3 $\boxed{\times}$ 112.51 $\boxed{\text{SIN}}$ $\boxed{\div}$ 2 $\boxed{=}$ | **12504.341** |

which means that the area is 12,500 m².

Examples C, D, and E show how the reference angle is used, and this is important when using a calculator in finding the angle for a given value of a function. As we noted earlier, *if we have the value of the function and want to find the angle, **the calculator will not necessarily give us directly the required angle.*** It will give us an angle we can use, but whether or not it is the required angle for the problem will depend on the problem being solved.

NOTE ▷ If a value of a trigonometric function is entered into a calculator, depending on the sign of the value of the function, it is programmed to give the angle as follows. *For positive values of a function, the calculator displays positive acute angles. For negative values of sin θ and tan θ, the calculator displays negative acute angles for* $\boxed{\text{INV}}$ $\boxed{\text{SIN}}$ *and* $\boxed{\text{INV}}$ $\boxed{\text{TAN}}$. *For negative values of cos θ, the calculator displays positive angles between 90° and 180° for* $\boxed{\text{INV}}$ $\boxed{\text{COS}}$. The reason for this type of display is explained in Chapter 19, when these functions are discussed in more detail.

EXAMPLE G _____ For $\sin \theta = 0.2250$, the calculator sequence to find θ is (with the calculator in degree mode)

.225 $\boxed{\text{INV}}\boxed{\text{SIN}}$ $\boxed{13.002878}$

which means that $\theta = 13.00°$ (rounded off). (Depending on your calculator, $\boxed{\text{SIN}^{-1}}$ or $\boxed{\text{ARCSIN}}$ may be used for $\boxed{\text{INV}}\boxed{\text{SIN}}$.)

This result is correct, but we must remember that

$$\sin (180° - 13.00°) = \sin 167.00° = 0.2250$$

also. If we are dealing only with acute angles, the answer $\theta = 13.00°$ is correct. However, if the problem requires a second-quadrant angle for the answer, then $\theta = 167.00°$ is the required angle. See Fig. 7-12.

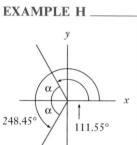

Fig. 7-12

EXAMPLE H _____ For $\sec \theta = -2.722$ and $0° < \theta < 360°$, the calculator sequence to find θ is

2.722 $\boxed{+/-}\boxed{1/x}\boxed{\text{INV}}\boxed{\text{COS}}$ $\boxed{111.55394}$

where we use the reciprocal key since $\sec \theta = 1/\cos \theta$. The final display indicates that $\theta = 111.55°$.

This is the correct second-quadrant angle, but we must remember that $\sec \theta < 0$ in the third quadrant as well. We then find the reference angle $\alpha = 180° - 111.55° = 68.45°$. The third-quadrant angle is then $180° + 68.45° = 248.45°$. Therefore, the two angles between $0°$ and $360°$ for which $\sec \theta = -2.722$ are $111.55°$ and $248.45°$. See Fig. 7-13.

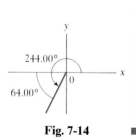

Fig. 7-13

EXAMPLE I _____ Given that $\tan \theta = 2.050$ and $\cos \theta < 0$, find θ when $0° \le \theta < 360°$.

Since $\tan \theta$ is positive and $\cos \theta$ is negative, θ must be a third-quadrant angle. Using the calculator sequence 2.050 $\boxed{\text{INV}}\boxed{\text{TAN}}$ $\boxed{63.996654}$, we find that $\tan 64.00° = 2.050$. However, we require a third-quadrant angle, which means that we must add $64.00°$ to $180°$. Thus, the required angle is $244.00°$. See Fig. 7-14.

If we are given that $\tan \theta = -2.050$ and $\cos \theta < 0$, we use the calculator sequence 2.050 $\boxed{+/-}\boxed{\text{INV}}\boxed{\text{TAN}}$ $\boxed{-63.996654}$. We would have to recognize that the reference angle is $64.00°$ and subtract it from $180°$ to get $116.00°$, the required second-quadrant angle.

Fig. 7-14

Considering the type of calculator displays described earlier, we see that the calculator gives the reference angle (disregarding any minus signs) in all cases except when $\cos \theta$ is negative. To avoid confusion from the angle displayed by the calculator, *a good procedure is to find the reference angle first*. Then it can be used to determine the angle required by the problem.

We can find the reference angle by entering the absolute value of the function. The angle displayed will be the reference angle. Then the required angle θ ($0° \leq \theta < 360°$) is found by use of the reference angle α as described previously, and as shown in Eqs. (7-6) for θ in the indicated quadrant.

$$
\begin{array}{ll}
\theta = \alpha & \text{(first quadrant)} \\
\theta = 180° - \alpha & \text{(second quadrant)} \\
\theta = 180° + \alpha & \text{(third quadrant)} \\
\theta = 360° - \alpha & \text{(fourth quadrant)}
\end{array}
\tag{7-6}
$$

EXAMPLE J

Given that $\cos \theta = -0.1298$, find θ for $0° \leq \theta < 360°$.

Since $\cos \theta$ is negative, θ is either a second-quadrant angle or a third-quadrant angle. Using the calculator sequence

.1298 $\boxed{\text{INV}}$ $\boxed{\text{COS}}$ $\boxed{82.541965}$

tells us that the reference angle is 82.54°.

To get the required second-quadrant angle, we subtract 82.54° from 180° and obtain 97.46°. To get the required third-quadrant angle, we add 82.54° to 180° to obtain 262.54°. See Fig. 7-15.

If we were to use the calculator sequence

.1298 $\boxed{+/-}$ $\boxed{\text{INV}}$ $\boxed{\text{COS}}$ $\boxed{97.458035}$

we get the required second-quadrant angle of 97.46° directly. However, to get the third-quadrant angle we must then subtract 97.46° from 180° to get the reference angle of 82.54°. The reference angle is then added to 180° to obtain the result of 262.54°. We see that it is better to store the reference angle in memory rather than recalculate it.

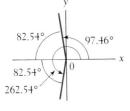

Fig. 7-15

With the use of Eqs. (7-3) through (7-5) we may find the value of any function, as long as the terminal side of the angle lies *in* one of the quadrants. This problem reduces to finding the function of an acute angle. We are left with *the angle for which the terminal side is along one of the axes, a* **quadrantal angle.** Using the definitions of the functions, and remembering that $r > 0$, we arrive at the values in the following table.

θ	$\sin \theta$	$\cos \theta$	$\tan \theta$	$\cot \theta$	$\sec \theta$	$\csc \theta$
0°	0.000	1.000	0.000	undef.	1.000	undef.
90°	1.000	0.000	undef.	0.000	undef.	1.000
180°	0.000	−1.000	0.000	undef.	−1.000	undef.
270°	−1.000	0.000	undef.	0.000	undef.	−1.000
360°	Same as the functions of 0° (same terminal side)					

The values in the table may be verified by referring to the figures in Fig. 7-16.

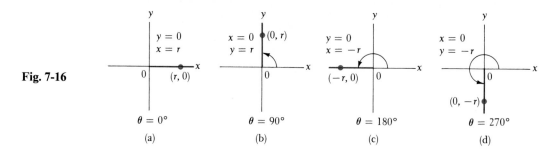

Fig. 7-16

$\theta = 0°$ (a) $\theta = 90°$ (b) $\theta = 180°$ (c) $\theta = 270°$ (d)

EXAMPLE K

Since $\sin \theta = y/r$, from Fig. 7-16(a) we see that $\sin 0° = 0/r = 0$.

Since $\tan \theta = y/x$, from Fig. 7-16(b) we see that $\tan 90° = r/0$, which is undefined due to the division by zero. If we used a calculator to find $\tan 90°$, the display would indicate an error (due to division by zero).

Since $\cos \theta = x/r$, from Fig. 7-16(c) we see that $\cos 180° = -r/r = -1$.

Since $\cot \theta = x/y$, from Fig. 7-16(d) we see that $\cot 270° = 0/-r = 0$. ∎

Exercises 7-2

In Exercises 1 through 8, express the given trigonometric function in terms of the same function of a positive acute angle.

1. $\sin 160°$, $\cos 220°$ **2.** $\tan 91°$, $\sec 345°$ **3.** $\tan 105°$, $\csc 302°$ **4.** $\cos 190°$, $\cot 290°$

5. $\sin (-123°)$, $\cot 174°$ **6.** $\sin 98°$, $\sec (-315°)$ **7.** $\cos 400°$, $\tan (-400°)$ **8.** $\tan 920°$, $\csc (-550°)$

In Exercises 9 through 44, assume that the angles are approximate. In Exercises 9 through 16, find the values of the given trigonometric functions by finding the reference angle, using a calculator, and attaching the proper sign.

9. $\sin 195°$ **10.** $\tan 311°$ **11.** $\cos 106.3°$ **12.** $\sin 103.4°$

13. $\tan 219.15°$ **14.** $\cos 198.82°$ **15.** $\sec 328.33°$ **16.** $\cot 136.53°$

In Exercises 17 through 24, find the values of the given trigonometric functions directly, using a calculator.

17. $\tan 152.4°$ **18.** $\cos 341.4°$ **19.** $\sin 310.36°$ **20.** $\tan 242.68°$

21. $\cos 110°$ **22.** $\sin 163°$ **23.** $\csc 194.82°$ **24.** $\sec 261.08°$

In Exercises 25 through 32, find θ for $0° \leq \theta < 360°$. Use a calculator.

25. $\sin \theta = -0.8480$ **26.** $\tan \theta = -1.830$ **27.** $\cos \theta = 0.4003$ **28.** $\sin \theta = 0.6374$

29. $\tan \theta = 0.283$ **30.** $\cos \theta = -0.928$ **31.** $\cot \theta = -0.212$ **32.** $\csc \theta = -1.09$

In Exercises 33 through 40, find θ for $0° \leq \theta < 360°$. Use a calculator.

33. $\sin \theta = 0.870$, $\cos \theta < 0$ **34.** $\tan \theta = 0.932$, $\sin \theta < 0$ **35.** $\cos \theta = -0.12$, $\tan \theta > 0$

36. $\sin \theta = -0.192$, $\tan \theta < 0$ **37.** $\tan \theta = -1.366$, $\cos \theta > 0$ **38.** $\cos \theta = 0.5726$, $\sin \theta < 0$

39. $\sec \theta = 2.047$, $\tan \theta < 0$ **40.** $\cot \theta = -0.3256$, $\sin \theta > 0$

In Exercises 41 through 44, determine the function which satisfies the given conditions. Use a calculator.

41. Find $\tan \theta$ when $\sin \theta = -0.5736$ and $\cos \theta > 0$.   **42.** Find $\sin \theta$ when $\cos \theta = 0.422$ and $\tan \theta < 0$.

43. Find $\cos \theta$ when $\tan \theta = -0.809$ and $\csc \theta > 0$.   **44.** Find $\cot \theta$ when $\sec \theta = 1.122$ and $\sin \theta < 0$.

In Exercises 45 through 48, insert the proper sign, > or < or =, between the given expressions.

45. sin 90°   2 sin 45°   **46.** cos 360°   2 cos 180°   **47.** tan 180°   tan 0°   **48.** sin 270°   3 sin 90°

In Exercises 49 through 52, evaluate the given expressions. Use a calculator.

49. The current i in an alternating-current circuit is given by $i = i_m \sin \theta$, where i_m is the maximum current. Find i if $i_m = 0.0259$ A and $\theta = 495.2°$

50. A force F is related to force F_x directed along the x-axis by $F = F_x \sec \theta$, where θ is the standard position angle for F. Find F if $F_x = -29.2$ N and $\theta = 127.6°$. See Fig. 7-17.

51. For the shelf support shown in Fig. 7-18, $y \sin \alpha = x \sin \beta$. Find y if $x = 14.2$ in., $\alpha = 35.0°$, and $\beta = 42.5°$.

52. A laser follows the path shown in Fig. 7-19. The angle θ is related to the distances a, b, and c by $2ab \cos \theta = a^2 + b^2 - c^2$. Find θ if $a = 15.3$ cm, $b = 12.9$ cm, and $c = 24.5$ cm.

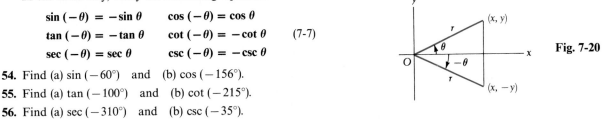

Fig. 7-17   **Fig. 7-18**   **Fig. 7-19**

In Exercises 53 through 56, the trigonometric functions of negative angles are considered. In Exercises 54, 55, and 56, use the equations derived in Exercise 53.

53. From Fig. 7-20 we see that $\sin \theta = y/r$ and $\sin(-\theta) = -y/r$. From this we conclude that $\sin(-\theta) = -\sin \theta$. In the same way, verify the remaining equations.

$$\sin(-\theta) = -\sin \theta \qquad \cos(-\theta) = \cos \theta$$
$$\tan(-\theta) = -\tan \theta \qquad \cot(-\theta) = -\cot \theta \qquad (7\text{-}7)$$
$$\sec(-\theta) = \sec \theta \qquad \csc(-\theta) = -\csc \theta$$

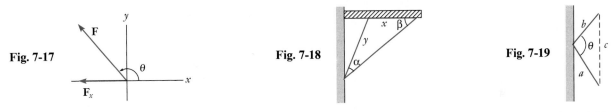

Fig. 7-20

54. Find (a) $\sin(-60°)$ and (b) $\cos(-156°)$.

55. Find (a) $\tan(-100°)$ and (b) $\cot(-215°)$.

56. Find (a) $\sec(-310°)$ and (b) $\csc(-35°)$.

7-3   Radians

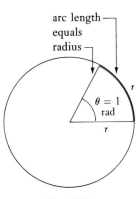

Fig. 7-21

For many problems in which trigonometric functions are used, particularly those involving the solution of triangles, degree measurements of angles are convenient and quite sufficient. However, division of a circle into 360 equal parts is by definition, and it is arbitrary and artificial. This definition comes from the ancient Babylonians and their use of a number system based on 60 rather than on 10, as is the system we use today. (Historians are uncertain as to the precise reason for the choice of 360° in a circle.) The *grad*, where 90° = 100 grad, is also arbitrary and is defined to fit base 10 numbers better.

In numerous other types of applications and in more theoretical discussions, another way of expressing the measure of angle is more meaningful and convenient. This unit measurement is the **radian,** which we briefly introduced in Chapter 3. *A radian is the measure of an angle with its vertex at the center of a circle and with an intercepted arc on the circle equal in length to the radius of the circle.* See Fig. 7-21.

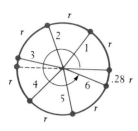

Fig. 7-22

Since the circumference of any circle in terms of its radius is given by $c = 2\pi r$, the ratio of the circumference to the radius is 2π. This means that the radius may be laid off 2π (about 6.28) times along the circumference, regardless of the length of the radius. Therefore, we see that radian measure is independent of the radius of the circle. The definition of a radian is based on an important property of a circle and is therefore a more natural measure of an angle. In Fig. 7-22 the numbers on each of the radii indicate the number of radians in the angle measured in standard position. The circular arrow shows an angle of 6 radians.

Since the radius may be laid off 2π times along the circumference, it follows that there are 2π radians in one complete rotation. Also, there are 360° in one complete rotation. Therefore, 360° is *equivalent* to 2π radians. It then follows that the relation between degrees and radians is 2π rad = 360°, or

radians and
degrees

$$\pi \text{ rad} = 180° \tag{7-8}$$

From this relation we find that

$$1° = \frac{\pi}{180} \text{ rad} = 0.01745 \text{ rad} \tag{7-9}$$

and that

$$1 \text{ rad} = \frac{180°}{\pi} = 57.30° \tag{7-10}$$

We see from Eqs. (7-8) through (7-10) that (1) *to convert an angle measured in degrees to the same angle measured in radians, we multiply the number of degrees by $\pi/180$, and* (2) *to convert an angle measured in radians to the same angle measured in degrees, we multiply the number of radians by $180/\pi$.*

EXAMPLE A

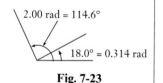

Fig. 7-23

┌ converting degrees to radians

$$18.0° = \left(\frac{\pi}{180}\right)(18.0) = \frac{\pi}{10.0} = \frac{3.14}{10.0} = 0.314 \text{ rad} \qquad \text{(See Fig. 7-23.)}$$

$$120° = \left(\frac{\pi}{180}\right)(120) = \frac{6.28}{3.00} = 2.09 \text{ rad}$$

┌ converting radians to degrees

$$0.400 \text{ rad} = \left(\frac{180°}{\pi}\right)(0.400) = \frac{72.0°}{3.14} = 22.9°$$

$$2.00 \text{ rad} = \left(\frac{180°}{\pi}\right)(2.00) = \frac{360°}{3.14} = 114.6° \qquad \text{(See Fig. 7-23.)}$$

Due to the nature of the definition of the radian, it is very common to express radians in terms of π. Expressing angles in terms of π is illustrated in the following example.

EXAMPLE B

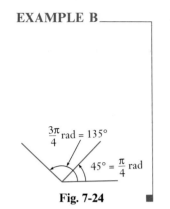

$\frac{3\pi}{4}$ rad = 135°

45° = $\frac{\pi}{4}$ rad

Fig. 7-24

┌─ converting degrees to radians

$$30° = \left(\frac{\pi}{180}\right)(30) = \frac{\pi}{6} \text{ rad}$$

$$45° = \left(\frac{\pi}{180}\right)(45) = \frac{\pi}{4} \text{ rad} \qquad \text{(See Fig. 7-24.)}$$

┌─ converting radians to degrees

$$\frac{\pi}{2} \text{ rad} = \left(\frac{180°}{\pi}\right)\left(\frac{\pi}{2}\right) = 90°$$

$$\frac{3\pi}{4} \text{ rad} = \left(\frac{180°}{\pi}\right)\left(\frac{3\pi}{4}\right) = 135° \qquad \text{(See Fig. 7-24.)}$$

We wish now to make a very important point. Since π is a special way of writing the number (slightly greater than 3) that is the ratio of the circumference of a circle to its diameter, it is the ratio of one distance to another. Thus radians really have no units and *radian measure amounts to measuring angles in terms of real numbers.* It is this property of radians that makes them useful in many situations. Therefore, when radians are being used, it is customary that no units are indicated for the angle. ***When no units are indicated, the radian is understood to be the unit of angle measurement.***

NOTE ▷

EXAMPLE C

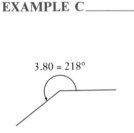

3.80 = 218°

Fig. 7-25

$$60.0° = \left(\frac{\pi}{180}\right)(60.0) = \frac{\pi}{3.00} = 1.05$$

↑

no units indicates radian measure

↓

$$3.80 = \left(\frac{180°}{\pi}\right)(3.80) = 218°$$

Since no units are indicated for 1.05 and 3.80 in this example, they are known to be in radian measure. Here we must know that 3.80 is an angle measure. See Fig. 7-25.

Many calculators have a key which can be used directly to change an angle expressed in degrees to an angle expressed in radians, or from an angle expressed in radians to one expressed in degrees. Generally the key is a second-function key.

We can use a calculator to find the value of a function of an angle in radians. Some calculators use a switch to indicate that either degrees or radians are being used. Other calculators are normally in degree mode, unless the mode is changed by a second-function key. If the calculator is in radian mode, then it uses values in radians directly. ***Always be careful to use the proper mode.*** The angle in radians should have the same number of significant digits as the angle in degrees in decimal form.

NOTE ▷

EXAMPLE D ———— To find the value of sin 0.7538, put the calculator in radian mode, and then use the sequence .7538 [SIN] [0.6844142], which means that

$$\sin 0.7538 = 0.6844$$

no units indicates radian measure

In the same way, using the sequence .9977 [TAN] [1.5495571],

$$\tan 0.9977 = 1.550$$

and using the sequence 2.074 [COS] [−0.4822346],

$$\cos 2.074 = -0.4822$$

EXAMPLE E ———— The velocity v of an object undergoing simple harmonic motion at the end of a spring is given by

$$v = A\sqrt{\frac{k}{m}}\cos\sqrt{\frac{k}{m}}\,t$$

Here m is the mass of the object (in grams), k is a constant depending on the spring, A is the maximum distance the object moves, and t is the time (in seconds). Find the velocity (in centimeters per second) after 0.100 s of a 36.0-g object at the end of a spring for which $k = 400$ g/s^2, if $A = 5.00$ cm.
Substituting, we have

5 [×] [(] 400 [÷]
36 [)] [√x] [STO]
[×] [(] [RCL] [×] .1
[)] [COS] [=]
[15.749282]

$$v = 5.00\sqrt{\frac{400}{36.0}}\cos\sqrt{\frac{400}{36.0}}\,(0.100)$$

Using calculator memory for $\sqrt{\frac{400}{36.0}}$, and with the calculator in radian mode, we use the calculator sequence shown at the left. Thus,

$$v = 15.7 \text{ cm/s}$$

For certain special situations, we may need to know a reference angle in radians. In order to determine the proper quadrant, we should remember that $\frac{1}{2}\pi = 90°$, $\pi = 180°$, $\frac{3}{2}\pi = 270°$, and $2\pi = 360°$. These are shown in Table 7-1 along with the decimal values for angles in radians. See Fig. 7-26.

TABLE 7-1   Quadrantal Angles

Degrees	Radians	Radians (decimal)
90°	$\frac{1}{2}\pi$	1.571
180°	π	3.142
270°	$\frac{3}{2}\pi$	4.712
360°	2π	6.283

Fig. 7-26

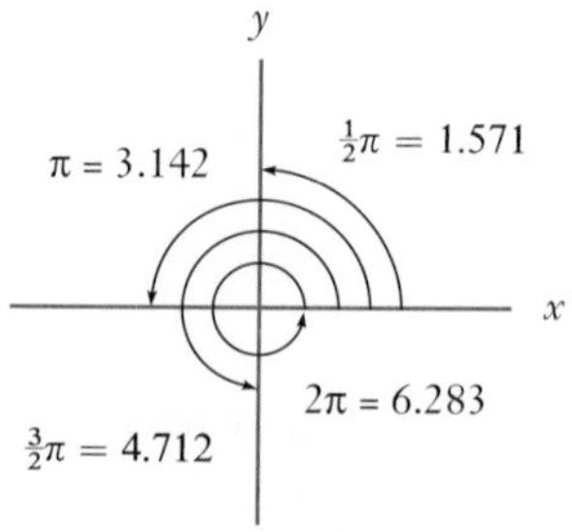

EXAMPLE F

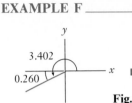

Fig. 7-27

An angle of 3.402 is greater than 3.142, but less than 4.712. This means that it is a third-quadrant angle, and that the reference angle is $3.402 - \pi = 0.260$. The $\boxed{\pi}$ key can be used for this. See Fig. 7-27.

An angle of 5.210 is between 4.712 and 6.283. Therefore, it is in the fourth quadrant and the reference angle is $2\pi - 5.210 = 1.073$. ∎

EXAMPLE G

Express θ in radians, such that $\cos\theta = 0.8829$ and $0 < \theta < 2\pi$.

We are to find θ in radians for the given value of the $\cos\theta$. Also, since θ is restricted to values between 0 and 2π, we must find a first-quadrant angle and a fourth-quadrant angle ($\cos\theta$ is positive in the first and fourth quadrants). With the calculator in radian mode, we use the sequence

$$.8829 \boxed{\text{INV}} \boxed{\text{COS}} \qquad \boxed{0.4887936}$$

and find that

$$\cos 0.4888 = 0.8829$$

Therefore, for the fourth-quadrant angle,

$$\cos(2\pi - 0.4888) = \cos 5.794$$

This means

$$\theta = 0.4888 \quad \text{or} \quad \theta = 5.794$$

Fig. 7-28

∎ See Fig. 7-28.

NOTE ▷ Often when one first encounters radian measure, *expressions such as* **sin 1** *and* **sin $\theta = 1$** *are confused.* The first is equivalent to sin 57.30°, since 57.30° = 1 (radian). The second means that θ is the angle for which the sine is 1. Since sin 90° = 1, we can say that $\theta = 90°$ or that $\theta = \pi/2$. The following example gives additional illustrations of evaluating expressions involving radians.

EXAMPLE H

$$\sin\frac{\pi}{3} = \frac{\sqrt{3}}{2} \qquad \text{since } \frac{\pi}{3} = 60°$$

$$\sin 0.6050 = 0.5688 \qquad \text{(We note that } 0.6050 = 34.66°.)$$

$$\tan\theta = 1.709 \qquad \text{means that } \theta = 59.67° \text{ (smallest positive } \theta)$$

∎ Since 59.67° = 1.041, we can state that tan 1.041 = 1.708.

Exercises 7-3

In Exercises 1 through 8, express the given angle measurements in radian measure in terms of π.

1. 15°, 150°   **2.** 12°, 225°   **3.** 75°, 330°
4. 36°, 315°   **5.** 210°, 270°   **6.** 240°, 300°
7. 160°, 260°   **8.** 66°, 350°

See Appendix E for a computer program for converting angles between radians and degrees.

In Exercises 9 through 16, the given numbers express angle measure. Express the measure of each angle in terms of degrees.

9. $\dfrac{2\pi}{5}, \dfrac{3\pi}{2}$

10. $\dfrac{3\pi}{10}, \dfrac{5\pi}{6}$

11. $\dfrac{\pi}{18}, \dfrac{7\pi}{4}$

12. $\dfrac{7\pi}{15}, \dfrac{4\pi}{3}$

13. $\dfrac{17\pi}{18}, \dfrac{5\pi}{3}$

14. $\dfrac{11\pi}{36}, \dfrac{5\pi}{4}$

15. $\dfrac{\pi}{12}, \dfrac{3\pi}{20}$

16. $\dfrac{7\pi}{30}, \dfrac{4\pi}{15}$

In Exercises 17 through 24, express the given angles in radian measure. Round off results to the number of significant digits in the given angle.

17. $23.0°$

18. $54.3°$

19. $252°$

20. $104°$

21. $333.5°$

22. $168.7°$

23. $178.5°$

24. $86.1°$

In Exercises 25 through 32, the given numbers express angle measure. Express the measure of each angle in terms of degrees, with the same accuracy as in the given value.

25. 0.750

26. 0.240

27. 3.407

28. 1.703

29. 2.45

30. 34.4

31. 16.42

32. 100.0

In Exercises 33 through 40, evaluate the given trigonometric functions by first changing the radian measure to degree measure. Round off results to four significant digits.

33. $\sin \dfrac{\pi}{4}$

34. $\cos \dfrac{\pi}{6}$

35. $\tan \dfrac{5\pi}{12}$

36. $\sin \dfrac{7\pi}{18}$

37. $\cos \dfrac{5\pi}{6}$

38. $\tan \dfrac{4\pi}{3}$

39. $\sec 4.5920$

40. $\cot 3.2732$

In Exercises 41 through 48, evaluate the given trigonometric functions directly, without first changing the radian measure to degree measure. Use a calculator in radian mode.

41. $\tan 0.7359$

42. $\cos 0.9308$

43. $\sin 4.24$

44. $\tan 3.47$

45. $\cos 2.07$

46. $\sin 2.34$

47. $\cot 4.86$

48. $\csc 6.19$

In Exercises 49 through 56, find θ to four significant digits for $0 \le \theta < 2\pi$.

49. $\sin \theta = 0.3090$

50. $\cos \theta = -0.9135$

51. $\tan \theta = -0.2126$

52. $\sin \theta = -0.0436$

53. $\cos \theta = 0.6742$

54. $\tan \theta = 1.860$

55. $\sec \theta = -1.307$

56. $\csc \theta = 3.940$

In Exercises 57 through 60, evaluate the given expressions.

57. A flat plate of weight W is attached at the top and is oscillating as shown in Fig. 7-29. Its potential energy V is given by $V = \frac{1}{2}Wb\theta^2$, where θ is measured in radians. Find V if $W = 8.75$ lb, $b = 0.75$ ft, and $\theta = 5.5°$.

58. The charge q, in coulombs, on a capacitor as a function of time is $q = 0.0130 \sin 50t - 0.0065 \sin 100t$. Find q for $t = 0.010$ s.

59. The height h of a rocket which is launched 1000 m from an observer is found to be $h = 1000 \tan \dfrac{5t}{3t + 10}$ for $t < 10$ s, where t is the time after launch. Find h for $t = 8.0$ s.

60. The electric intensity I, in W/m², from the two radio antennas shown in Fig. 7-30 is a function of the angle θ given by $I = 0.023 \cos^2(\pi \sin \theta)$. Find I for $\theta = 40.0°$. [$\cos^2 \alpha = (\cos \alpha)^2$.]

Fig. 7-29

Fig. 7-30

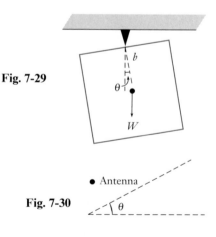

7-4   **Applications of the Use of Radian Measure** ▰▰▰

Radian measure has numerous applications in mathematics and technology, some of which were indicated in the last four exercises of the previous section. In this section we shall illustrate the usefulness of radian measure in certain specific geometric and physical applications.

arc length

From geometry we know that *the length of an arc on a circle is proportional to the central angle* and that the length of the arc of a complete circle is the circumference. Letting s stand for the length of arc, we may state that $s = 2\pi r$ for a complete circle. Since 2π is the central angle (in radians) of the complete circle, *we have as the length of arc*

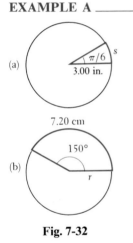

Fig. 7-31

$$s = \theta r \qquad\qquad (7\text{-}11)$$

for any circular arc with central angle. θ. If we know the central angle in radians and the radius of a circle, we can find the length of a circular arc directly by using Eq. (7-11). See Fig. 7-31.

EXAMPLE A

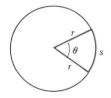

(a)

If $\theta = \pi/6$ and $r = 3.00$ in.,

$\ulcorner\theta$ in radians

$$s = \left(\frac{\pi}{6}\right)(3.00) = \frac{\pi}{2.00} = 1.57 \text{ in.}$$

See Fig. 7-32(a).

If the arc length is 7.20 cm for a central angle of 150° on a certain circle, we may find the radius of the circle by solving $s = \theta r$ for r and then substituting. Thus, $r = s/\theta$. Substituting, we have

(b)

$$r = \frac{7.20}{150\left(\dfrac{\pi}{180}\right)} = \frac{(7.20)(180)}{150\pi} = 2.75 \text{ cm}$$

θ in radians

Fig. 7-32

■ See Fig. 7-32(b).

area of sector

Another geometric application of radians is in finding the area of a sector of a circle (see Fig. 7-33). We recall from geometry that areas of sectors of circles are proportional to their central angles. The area of circle is given by $A = \pi r^2$. This can be written as $A = \frac{1}{2}(2\pi)r^2$. We now note that the angle for a complete circle is 2π, and therefore *the area of any sector of a circle in terms of the radius and the central angle is*

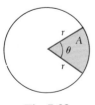

Fig. 7-33

$$A = \frac{1}{2}\theta r^2 \qquad\qquad (7\text{-}12)$$

EXAMPLE B

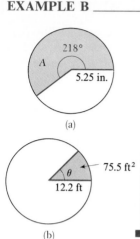

(a)

(b)

Fig. 7-34

The area of a sector of a circle with central angle 218° and a radius of 5.25 in. is

$$A = \frac{1}{2}(218)\overset{\displaystyle\frown \theta \text{ in radians}}{\left(\frac{\pi}{180}\right)}(5.25)^2 = 52.4 \text{ in.}^2$$

See Fig. 7-34(a).

Given that the area of a sector is 75.5 ft² and the radius is 12.2 ft, we can find the central angle by solving for θ and then substituting.

$$\theta = \frac{2A}{r^2}$$

$$= \frac{2(75.5)}{(12.2)^2} \overset{\displaystyle\frown \text{no units indicates radian measure}}{= 1.01}$$

■ This means that the central angle is 1.01 rad, or 57.9°. See Fig. 7-34(b).

NOTE ▷   We should note again that the equations of this section require that the angle θ is *expressed in radians.* A relatively common error is to use θ in degrees.

The next illustration deals with velocity. We know that average velocity is defined by the equation $v = s/t$, where v is the average velocity, s is the distance traveled, and t is the elapsed time. If an object is moving around a circular path with constant speed, the actual distance traveled is the length of arc traversed. Therefore, if we divide both sides of Eq. (7-11) by t, we obtain

$$\frac{s}{t} = \frac{\theta}{t} r$$

or

$$\boxed{v = \omega r} \qquad (7\text{-}13)$$

*Equation (7-13) expresses the relationship between the **linear velocity** v and the **angular velocity** ω of an object moving around a circle of radius r.* See Fig. 7-35. In the figure, v is shown directed tangent to the circle, for that is its direction for the position shown. The direction of v changes constantly.

The most convenient units for ω are radians per unit of time. In this way the formula can be used directly. However, in practice, ω is often given in revolutions per minute, or in some similar unit. In cases like these, it is necessary to convert the units of ω to radians per unit of time before substituting in Eq. (7-13).

velocity

Velocity is
tangent to circle

Fig. 7-35

EXAMPLE C

A woman on a hang glider is moving in a horizontal circular arc of radius 90.0 m with an angular velocity of 0.125 rad/s. Her linear velocity is

$$v = (0.125)(90.0) = 11.3 \text{ m/s}$$

(Remember that radians are numbers and are not included in the final set of units.) This means that she is moving along the circumference of the arc at
■ 11.3 m/s (40.7 km/h).

EXAMPLE D

See the chapter introduction.

A communications satellite remains at an altitude of 22,320 mi above a point on the equator. If the radius of the earth is 3960 mi, what is the velocity of the satellite?

In order for the satellite to remain over a point on the equator, it must rotate exactly once each day around the center of the earth. Since there are 2π radians in each revolution, the angular velocity is

$$\omega = \frac{1 \text{ r}}{1 \text{ day}} = \frac{2\pi \text{ rad}}{24 \text{ h}} = 0.2618 \text{ rad/h}$$

The radius of the circle through which the satellite moves is $22,320 + 3960 = 26,280$ mi. Thus, the velocity is

$$v = 0.2618(26,280) = 6880 \text{ mi/h}$$

EXAMPLE E

Point on rim

Fig. 7-36

A pulley belt 10.0 ft long takes 2.00 s to make one complete revolution. The radius of the pulley is 6.00 in. What is the angular velocity (in revolutions per minute) of a point on the rim of the pulley? See Fig. 7-36.

Since the linear velocity of a point on the rim of the pulley is the same as the velocity of the belt, $v = 10.0/2.00 = 5.00$ ft/s. The radius of the pulley is $r = 6.00$ in. $= 0.500$ ft, and we can find ω by substituting into Eq. (7-13). This gives us

$$5.00 = \omega(0.500)$$

or

$$\omega = 10.0 \text{ rad/s}$$
$$= 600 \text{ rad/min}$$
$$= 95.5 \text{ r/min}$$

As shown in Appendix B, the change of units can be handled algebraically as

$$10.0 \frac{\text{rad}}{\text{s}} \times 60 \frac{\text{s}}{\text{min}} = 600 \frac{\text{rad}}{\text{min}}, \qquad \frac{600 \text{ rad/min}}{2\pi \text{ rad/r}} = 600 \frac{\text{rad}}{\text{min}} \times \frac{1}{2\pi} \frac{\text{r}}{\text{rad}}$$

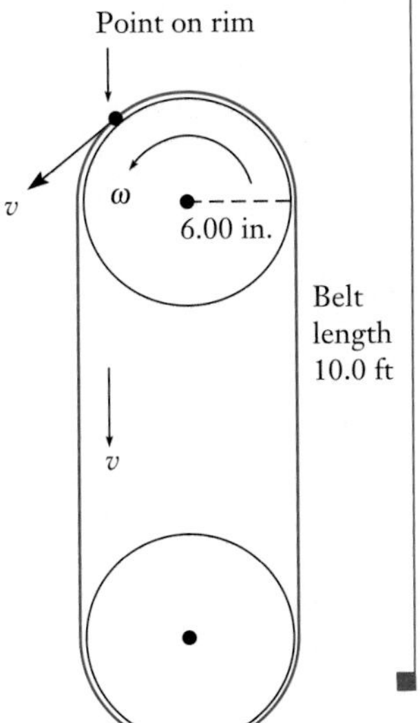

EXAMPLE F

The current at any time in a certain alternating-current electric circuit is given by $i = I \sin 120\pi t$, where I is the maximum current and t is the time in seconds. Given that $I = 0.0685$ A, find i for $t = 0.00500$ s.

Substituting, we get

$$i = 0.0685 \sin (120\pi)(0.00500)$$
$$= 0.0651 \text{ A}$$

With a calculator in radian mode, the sequence of keys is

.0685 ⨯ (120 ⨯ π ⨯ .005) SIN =   0.0651474

Exercises 7-4

In Exercises 1 through 36, solve the given problems.

1. Part of the center line of a highway follows a circular arc of which the radius is 320 ft and the central angle is 62.0°. Find the length of this part of the center line.

2. While playing, the left spool of a videocassette recorder turns through 820°. For this part of the tape, it is 3.30 cm from the center of the spool to the tape. What length of tape is played?

3. A door 76.2 cm wide is opened through an angle of 110.0°. What floor area does the door pass over?

4. A section of sidewalk is a circular sector of radius 3.00 ft and central angle 50.6°. What is the area of this section of sidewalk?

5. A section of a natural gas pipeline 3.25 km long is part of a circular arc. If the radius of the circle is 8.50 km, what is the central angle of the arc?

6. A cam is in the shape of a circular sector, as shown in Fig. 7-37. What is the perimeter of (distance around) the cam?

7. A lawn sprinkler can water up to a distance of 65.0 ft. It turns through an angle of 115°. What area can the sprinkler water?

8. A beam of light from a spotlight sweeps through a horizontal angle of 75.0°. If the range of the spotlight is 250 ft, how large an area can it cover?

9. If a car makes a U-turn in 6 s, what is its average angular velocity in the turn?

10. The roller on a computer printer makes 2000 r/min. What is its angular velocity?

11. What is the floor area of the hallway shown in Fig. 7-38? The outside and inside of the hallway are circular arcs.

12. The arm of a car windshield wiper is 12.75 in. long and is attached at the middle of a 15.00 in. blade. (Assume that the arm and blade are in line.) What area of the windshield is cleaned by the wiper if it swings through 110.0° arcs?

13. Part of a railroad track follows a circular arc with a central angle of 28.0°. If the radius of the arc of the inner rail is 93.67 ft and the rails are 4.71 ft apart, how much longer is the outer rail than the inner rail?

14. A weight on a string is dropped as shown in Fig. 7-39. Its velocity at the bottom of its swing is $v = \sqrt{2gh}$, where g is the acceleration due to gravity. What is its angular velocity at the bottom if $g = 9.80$ m/s² and $h = 1.20$ m?

15. Part of a security fence is built 2.50 m from a cylindrical storage tank 11.2 m in diameter. What is the area between the tank and this part of the fence if the central angle of the fence is 75.5°?

16. Through what angle does the drum in Fig. 7-40 turn in order to lower the crate 18.5 ft?

17. A section of road follows a circular arc with a central angle of 15.6°. The radius of the inside of the curve is 285.0 m, and the road is 15.2 m wide. What is the volume of the concrete in the road if it is 0.305 m thick?

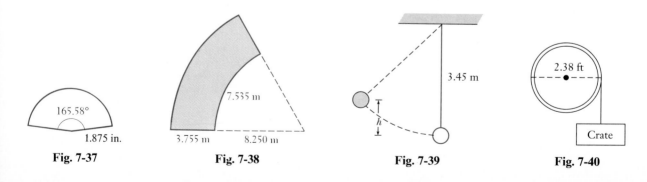

| Fig. 7-37 | Fig. 7-38 | Fig. 7-39 | Fig. 7-40 |

18. The propeller of the motor on a motor boat is rotating at 130 rad/s. What is the linear velocity of a point on the tip of a blade if it is 22.5 cm long?

19. A storm causes a pilot to follow a circular-arc route, with a central angle of 12.8°, from city A to city B rather than the straight-line route of 185.0 km. How much farther does the plane fly due to the storm?

20. An interstate route exit is a circular arc 330 m long with a central angle of 79.4°. What is the radius of curvature of the exit?

21. A special vehicle for traveling on glacial ice has tires that are 12.0 ft in diameter. If the vehicle travels at 3.5 mi/h, what is the angular velocity of the tire in revolutions per minute?

22. The sweep second hand of a watch is 15.0 mm long. What is the linear velocity of the tip?

23. A floppy disk used in a personal computer has a diameter of 5.25 in. and rotates at 360 r/min. What is the linear velocity of a point on the outer edge?

24. A man walking at 3.0 ft/s pushes a lawn roller of diameter 2.0 ft. How many revolutions per minute does the roller make?

25. Two streets meet at an angle of 82.0°. What is the length of the piece of curved curbing at the intersection if it is constructed along the arc of a circle 15.0 ft in radius? See Fig. 7-41.

26. An ammeter needle is deflected 52.00° by a current of 0.2500 A. The needle is 3.750 in. long, and a circular scale is used. How long is the scale for a maximum current of 1.500 A?

27. A drill bit $\frac{3}{8}$ in. in diameter rotates at 1200 r/min. What is the linear velocity of a point on its circumference?

28. A helicopter blade is 2.75 m long and is rotating at 420 r/min. What is the linear velocity of the tip of the blade?

29. A waterwheel used to generate electricity has paddles 3.75 m long. The speed of the end of a paddle is one-fourth that of the water. If the water is flowing at the rate of 6.50 m/s, what is the angular velocity of the waterwheel?

30. A jet is traveling westward with the sun directly overhead (the jet is on a line between the sun and the center of the earth). How fast must the jet fly in order to keep the sun directly overhead? (Assume that the earth's radius is 3960 mi, the altitude of the jet is low, and the earth rotates about its axis once in 24.0 h.)

31. A 1500-kW wind turbine (windmill) rotates at 40.0 r/min. What is the linear velocity of a point on the end of a blade, if the blade is 100 ft. long (from the center of rotation)?

32. What is the linear velocity of a point in New York City which is at a latitude of 40°45′ N? The radius of the earth is 3960 mi.

33. Through what total angle does the drive shaft of a car rotate in one second when the tachometer reads 2000 r/min?

34. A patio is in the shape of a circular sector with a central angle of 160.0°. It is enclosed by a railing of which the circular part is 11.6 m long. What is the area of the patio?

35. An oil storage tank 4.25 m long has a flat bottom as shown in Fig. 7-42. The radius of the circular part is 1.10 m. What volume of oil does the tank hold?

36. Two equal beams of light illuminate the area shown in Fig. 7-43. What area is lit by both beams?

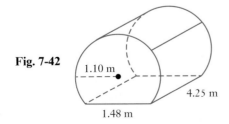

Fig. 7-42   **Fig. 7-43**

In Exercises 37 through 40, another use of radians is illustrated.

37. It can be shown through advanced mathematics that an excellent approximate method of evaluating $\sin \theta$ or $\tan \theta$ is given by

$$\sin \theta = \tan \theta = \theta \qquad (7\text{-}14)$$

for small values of θ (the equivalent of a few degrees or less), if θ is expressed in radians. Equation (7-14) can be used for very small values of θ—even some calculators cannot adequately handle very small angles. Using Eq. (7-14), express $1''$ in radians and evaluate $\sin 1''$ and $\tan 1''$.

38. Using Eq. (7-14), evaluate $\tan 0.001°$. Compare with a calculator value.

39. An astronomer observes that a star 12.5 light years away moves through an angle of $0.2''$ in one year. Assuming it moved in a straight line perpendicular to the initial line of observation, how many miles did the star move? (One light year $= 5.88 \times 10^{12}$ mi.) Use Eq. (7-14).

40. In calculating a back line of a lot, a surveyor discovers an error of $0.05°$ in an angle measurement. If the lot is 136.0 m deep, by how much is the back line calculation in error? See Fig. 7-44. Use Eq. (7-14).

Fig. 7-44

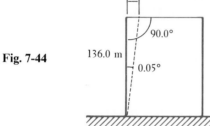

x

$90.0°$

136.0 m

$0.05°$

7-5 Chapter Equations, Review Exercises, and Practice Test

Chapter Equations

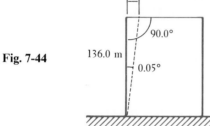

$$\sin \theta = \frac{y}{r} \qquad \cos \theta = \frac{x}{r} \qquad \tan \theta = \frac{y}{x}$$

$$\cot \theta = \frac{x}{y} \qquad \sec \theta = \frac{r}{x} \qquad \csc \theta = \frac{r}{y} \qquad (7\text{-}1)$$

α is
reference angle

$$F(\theta_2) = \pm F(180° - \theta_2) = \pm F(\alpha) \qquad (7\text{-}3)$$

$$F(\theta_3) = \pm F(\theta_3 - 180°) = \pm F(\alpha) \qquad (7\text{-}4)$$

$$F(\theta_4) = \pm F(360° - \theta_4) = \pm F(\alpha) \qquad (7\text{-}5)$$

$$
\begin{array}{ll}
\theta = \alpha & \text{(first quadrant)} \\
\theta = 180° - \alpha & \text{(second quadrant)} \\
\theta = 180° + \alpha & \text{(third quadrant)} \\
\theta = 360° - \alpha & \text{(fourth quadrant)}
\end{array}
\qquad (7\text{-}6)
$$

Functions of
negative angles

$$
\begin{array}{ll}
\sin(-\theta) = -\sin \theta & \cos(-\theta) = \cos \theta \\
\tan(-\theta) = -\tan \theta & \cot(-\theta) = -\cot \theta \\
\sec(-\theta) = \sec \theta & \csc(-\theta) = -\csc \theta
\end{array}
\qquad (7\text{-}7)
$$

$$\pi \text{ rad} = 180° \tag{7-8}$$

Radian-degree
conversions

$$1° = \frac{\pi}{180} \text{ rad} = 0.01745 \text{ rad} \tag{7-9}$$

$$1 \text{ rad} = \frac{180°}{\pi} = 57.30° \tag{7-10}$$

Circular arc length
$$s = \theta r \tag{7-11}$$

Circular sector area
$$A = \frac{1}{2}\theta r^2 \tag{7-12}$$

Linear and angular velocity
$$v = \omega r \tag{7-13}$$

For small θ
$$\sin \theta = \tan \theta = \theta \tag{7-14}$$

Review Exercises

In Exercises 1 through 4, find the trigonometric functions of θ given that the terminal side of θ passes through the given point.

1. (6, 8) **2.** (−12, 5) **3.** (7, −2) **4.** (−2, −3)

In Exercises 5 through 8, express the given trigonometric functions in terms of the same function of a positive acute angle.

5. cos 132°, tan 194° **6.** sin 243°, cot 318° **7.** sin 289°, sec(−15°) **8.** cos 103°, csc (−100°)

In Exercises 9 through 12, express the given angle measurements in terms of π.

9. 40°, 153° **10.** 22.5°, 324° **11.** 48°, 202.5° **12.** 27°, 162°

In Exercises 13 through 20, the given numbers represent angle measure. Express the measure of each angle in terms of degrees.

13. $\dfrac{7\pi}{5}, \dfrac{13\pi}{18}$ **14.** $\dfrac{3\pi}{8}, \dfrac{7\pi}{20}$ **15.** $\dfrac{\pi}{15}, \dfrac{11\pi}{6}$ **16.** $\dfrac{17\pi}{10}, \dfrac{5\pi}{4}$

17. 0.560 **18.** 1.354 **19.** 3.607 **20.** 14.5

In Exercises 21 through 28, express the given angles in radians. (Do not answer in terms of π.)

21. 102° **22.** 305° **23.** 20.25° **24.** 148.38°

25. 262.05° **26.** 18.72° **27.** 136.2° **28.** 385.4°

In Exercises 29 through 48, determine the values of the given trigonometric functions directly on a calculator. Assume that the angles are approximate. Express answers to Exercises 41 through 44 to four significant digits.

29. cos 245.5° **30.** sin 141.3° **31.** cot 295° **32.** tan 184°

33. csc 247.82° **34.** sec 96.17° **35.** sin 205.24° **36.** cos 326.72°

37. tan 301.4° **38.** sin 103.9° **39.** tan 256.42° **40.** cos 162.32°

41. $\sin \dfrac{9\pi}{5}$ **42.** $\sec \dfrac{5\pi}{8}$ **43.** $\cos \dfrac{7\pi}{6}$ **44.** $\tan \dfrac{23\pi}{12}$

45. sin 0.5906 **46.** tan 0.8035 **47.** csc 2.153 **48.** cot 5.190

In Exercises 49 through 52, find θ to the nearest 0.01° for $0° \le \theta < 360°$. Use a calculator.

49. tan θ = 0.1817 **50.** sin θ = −0.9323 **51.** cos θ = −0.4730 **52.** cot θ = 1.196

In Exercises 53 through 56, find θ for 0 ≤ θ < 2π. Use a calculator.

53. $\cos \theta = 0.8387$ **54.** $\sin \theta = 0.1045$ **55.** $\sin \theta = -0.8650$ **56.** $\tan \theta = 2.840$

In Exercises 57 through 60, find θ to the nearest 0.01° for 0° ≤ θ < 360°. Use a calculator.

57. $\cos \theta = -0.7222$, $\sin \theta < 0$ **58.** $\tan \theta = -1.683$, $\cos \theta < 0$

59. $\cot \theta = 0.4291$, $\cos \theta < 0$ **60.** $\sin \theta = 0.2626$, $\tan \theta < 0$

In Exercises 61 through 76, solve the given problems.

61. The instantaneous power p, in watts, input to a resistor in an alternating-current circuit is given by $p = p_m \sin^2 377t$, where p_m is the maximum power input and t is the time in seconds. Find p for $p_m = 0.120$ W and $t = 2.00$ ms. [$\sin^2 \theta = (\sin \theta)^2$.]

62. The horizontal distance x through which a pendulum moves is given by $x = a(\theta + \sin \theta)$, where a is a constant and θ is the angle between the vertical and the pendulum. Find x for $a = 45.0$ cm and $\theta = 0.175$.

63. A sector gear with a pitch radius of 8.25 in. and a 6.60-in. arc of contact is shown in Fig. 7-45. What is the sector angle θ?

64. Two pulleys have radii of 10.0 in. and 6.00 in., and their centers are 40.0 in. apart. If the pulley belt is uncrossed, what must be the length of the belt?

65. A thermometer needle passes through 55.25° for a temperature change of 40.00°C. If the needle is 5.250 cm long and the scale is circular, how long must the scale be for a maximum temperature change of 150.00°C?

66. A piece of circular filter paper 15.0 cm in diameter is folded such that its effective filtering area is the same as that of a sector with central angle of 220°. What is the filtering area?

67. A circular hood is to be used over a piece of machinery. It is to be made from a circular piece of sheet metal 3.25 ft in radius. A hole 0.75 ft in radius and a sector of central angle 80.0° are to be removed to make the hood. What is the area of the top of the hood?

68. Find the area of the decorative glass panel shown in Fig. 7-46. The panel is made of two equal circular sectors and an isosceles triangle.

69. To produce an electric current, a circular loop of wire of diameter 25.0 cm is rotating about its diameter at 60.0 r/s in a magnetic field. What is the greatest linear velocity of any point on the loop?

70. The chain on a chain saw is driven by a sprocket 7.50 cm in diameter. If the chain is 108 cm long and makes one revolution in 0.250 s, what is the angular velocity, in revolutions per second, of the sprocket?

71. An *ultracentrifuge*, used to observe the sedimentation of particles such as proteins, may rotate as fast as 80,000 r/min. If it rotates at this rate and is 7.20 cm in diameter, what is the linear velocity of a particle at the outer edge?

72. A horizontal water pipe has a radius of 6.00 ft. If the depth of water in the pipe is 3.00 ft, what percentage of the volume of the pipe is filled?

73. Part of a parking lot is shown in Fig. 7-47. The widths of the parking spaces are equal and the lengths of the dividing lines are equal. The circular section has a radius of 24.0 ft and a central angle of 26.0°. What is the total length of the lines and circular section shown?

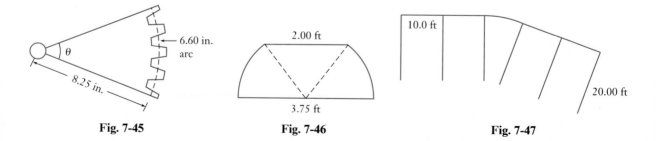

Fig. 7-45   Fig. 7-46   Fig. 7-47

74. An antique fan opens to the shape of a circular sector. If the arc length is 45.0 cm and the radius is 15.0 cm, what is the central angle (in degrees) of the fan?

75. A laser beam is transmitted with a "width" of 0.0008° and makes a circular spot of radius 2.50 km on a distant object. How far is the object from the source of the laser beam? Use Eq. (7-14).

76. The planet Venus subtends an angle of 15″ to an observer on Earth. If the distance between Venus and Earth is 1.04×10^8 mi, what is the diameter of Venus? Use Eq. (7-14).

Practice Test

1. Change 150° to radians in terms of π.

2. Express sin 205° in terms of the sine of a positive acute angle. Do not evaluate.

3. Find sin θ and sec θ if θ is in standard position and the terminal side passes through $(-9, 12)$.

4. The armature of a dynamo is 1.38 ft in diameter and is rotating at 1200 r/min. What is the linear velocity of a point on the rim of the armature?

5. Given that 3.572 is the measure of an angle, express the angle in degrees.

6. If tan $\theta = 0.2396$, find θ, in degrees, for $0° \leq \theta < 360°$.

7. If cos $\theta = -0.8244$ and csc $\theta < 0$, find θ in radians for $0 \leq \theta < 2\pi$.

8. The floor of a sunroom is in the shape of a circular sector of arc length 32.0 ft and radius 8.50 ft. What is the area of the floor?

8 Vectors and Oblique Triangles

The wind must be considered to find the proper heading for an aircraft. In Section 8-5 we see how this may be done.

We have been dealing with many types of applications in technology. However, to this point we have considered only the magnitude of various quantities. In this chapter we shall consider *vectors,* for which it is necessary to specify the magnitude and the direction for a complete and meaningful description. Vectors are of the greatest importance in many fields of science and technology, including physics and navigation, as the examples and exercises of the chapter will show.

After developing the concept of a vector, we shall take up the solution of oblique triangles. We will develop methods of solving any triangle and will not be limited only to right triangles. As with right triangles, the applications using oblique triangles are numerous.

8-1 Introduction to Vectors

A great many quantities with which we deal may be described by specifying their magnitudes. Generally, one can describe lengths of objects, areas, time intervals, monetary amounts, temperatures, and numerous other quantities by specifying a number: the magnitude of the quantity. *Such quantities are known as **scalar** quantities.*

scalars

*Many other quantities are fully described only when both their magnitude and direction are specified. Such quantities are known as **vectors.** Examples of*

vectors

230

vectors are velocity, force, and momentum. The following example illustrates a vector quantity and the distinction between scalars and vectors.

EXAMPLE A A jet flies over a certain point traveling at 600 mi/h. From this statement alone we know only the *speed* of the jet. *Speed is a scalar quantity,* and it designates only the *magnitude* of the rate.

If we were to add the phrase "in a direction 10° south of west" to the sentence above about the jet, we would be specifying the direction of travel as well as the speed. We then know the *velocity* of the jet; that is, we know the *direction* of travel as well as the *magnitude* of the rate at which the jet is traveling. *Velocity is a vector quantity.*

Let us analyze an example of the action of two vectors: Consider a boat moving in a stream. For purposes of this example, we shall assume that the boat is driven by a motor which can move it at 4 mi/h in still water. We shall assume the current is moving downstream at 3 mi/h. We immediately see that the motion of the boat depends on the direction in which it is headed. If the boat heads downstream, it can travel at 7 mi/h, for the current is going 3 mi/h and the boat moves at 4 mi/h with respect to the water. If the boat heads upstream, however, it progresses at the rate of only 1 mi/h, since the action of the boat and that of the stream are counter to each other. If the boat heads across the stream, the point which it reaches on the other side will not be directly opposite the point from which it started. We can see that this is so because we know that as the boat heads across the stream, the stream is moving the boat downstream *at the same time.*

This last case should be investigated further. Let us assume that the stream is $\frac{1}{2}$ mi wide where the boat is crossing. It will then take the boat $\frac{1}{8}$ h to cross. In $\frac{1}{8}$ h the stream will carry the boat $\frac{3}{8}$ mi downstream. Therefore, when the boat reaches the other side it will be $\frac{3}{8}$ mi downstream. From the Pythagorean theorem, we find that the boat traveled $\frac{5}{8}$ mi from its starting point to its finishing point.

$$d^2 = \left(\frac{4}{8}\right)^2 + \left(\frac{3}{8}\right)^2 = \frac{16+9}{64} = \frac{25}{64}; \qquad d = \frac{5}{8} \text{ mi}$$

Since this $\frac{5}{8}$ mi was traveled in $\frac{1}{8}$ h, the magnitude of the velocity of the boat was actually

$$v = \frac{d}{t} = \frac{\frac{5}{8}}{\frac{1}{8}} = \frac{5}{8} \times \frac{8}{1} = 5 \text{ mi/h}$$

Also, we see that the direction of this velocity can be represented along a line making an angle θ with the line directed across the stream (see Fig. 8-1).

We have just seen two velocity vectors being *added.* Note that these vectors are not added the way numbers are added. We have to take into account their direction as well as their magnitude. Reasoning along these lines, let us now define the sum of two vectors.

We will represent a vector quantity by a letter printed in boldface type. The same letter in italic (lightface) type represents the magnitude only. Thus, **A** is a

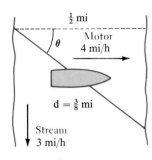

Fig. 8-1

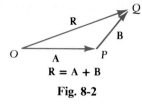

$R = A + B$

Fig. 8-2

vector of magnitude A. In handwriting, one usually places an arrow over the letter to represent a vector, such as $\vec{A}$.

Let **A** and **B** represent vectors directed from O to P and P to Q, respectively (see Fig. 8-2). *The vector sum* **A + B** *is the vector* **R**, *from the* **initial point** O *to the* **terminal point** Q. *Here, vector* **R** *is called the* **resultant.** *In general, a resultant is a single vector which can replace any number of other vectors and still produce the same physical effect.*

There are two common methods of adding vectors by means of a diagram. The first is illustrated in Fig. 8-3. To add **B** to **A**, we shift **B** parallel to itself until its tail touches the head of **A**. In doing so we are using the meaning of a vector, and we may move it for addition as long as *we **keep its magnitude and direction unchanged.** The vector sum* **A + B** *is the vector* **R**, *which is drawn from the tail of* **A** *to the head of* **B**. Clearly, when using a diagram to add vectors, the diagram must be drawn with reasonable accuracy.

NOTE ▷

addition of vectors

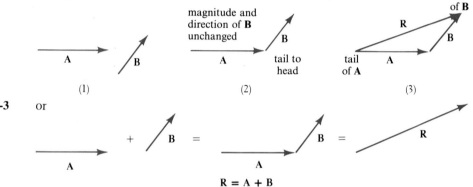

Fig. 8-3   or

$R = A + B$

Three or more vectors may also be added in the same general manner. We place the initial point of the second vector at the terminal point of the first vector, the initial point of the third vector at the terminal point of the second vector, and so forth. The resultant is the vector from the initial point of the first vector to the terminal point of the last vector. The order in which they are placed together does not matter, as shown in the following example.

EXAMPLE B

Fig. 8-4

The addition of vectors **A**, **B**, and **C** is shown in Fig. 8-4.

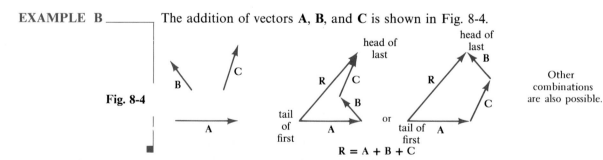

Other combinations are also possible.

$R = A + B + C$

Another method which is convenient when two vectors are being added is to *let the two vectors being added be the sides of a parallelogram. The resultant is then the diagonal of the parallelogram.* The initial point of the resultant is the **common initial point** of the vectors being added. In using this method the vectors are first placed tail to tail. See the following example.

EXAMPLE C ____ Using the parallelogram method, add vectors **A** and **B** of Fig. 8-3. See Fig. 8-5.

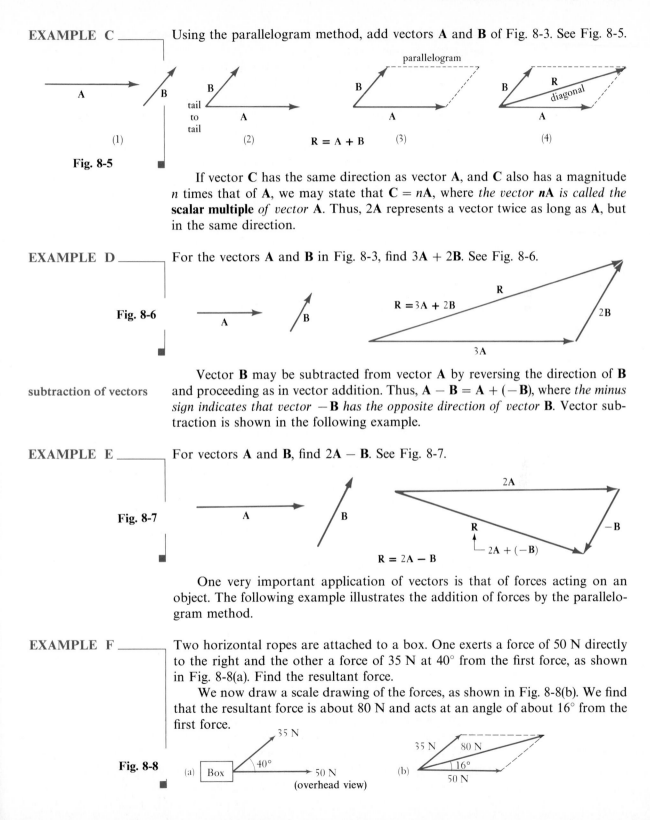

Fig. 8-5

If vector **C** has the same direction as vector **A**, and **C** also has a magnitude *n* times that of **A**, we may state that **C** = *n***A**, where *the vector n***A** *is called the* **scalar multiple** *of vector* **A**. Thus, 2**A** represents a vector twice as long as **A**, but in the same direction.

EXAMPLE D ____ For the vectors **A** and **B** in Fig. 8-3, find 3**A** + 2**B**. See Fig. 8-6.

Fig. 8-6

subtraction of vectors

Vector **B** may be subtracted from vector **A** by reversing the direction of **B** and proceeding as in vector addition. Thus, **A** − **B** = **A** + (−**B**), where *the minus sign indicates that vector* −**B** *has the opposite direction of vector* **B**. Vector subtraction is shown in the following example.

EXAMPLE E ____ For vectors **A** and **B**, find 2**A** − **B**. See Fig. 8-7.

Fig. 8-7

One very important application of vectors is that of forces acting on an object. The following example illustrates the addition of forces by the parallelogram method.

EXAMPLE F ____ Two horizontal ropes are attached to a box. One exerts a force of 50 N directly to the right and the other a force of 35 N at 40° from the first force, as shown in Fig. 8-8(a). Find the resultant force.

We now draw a scale drawing of the forces, as shown in Fig. 8-8(b). We find that the resultant force is about 80 N and acts at an angle of about 16° from the first force.

Fig. 8-8

Two other important vector quantities are *velocity* and *displacement*. Velocity as a vector is illustrated in Example A. *The* **displacement** *of an object is given by the distance from a point of reference and the angle from a reference direction.* Displacement is illustrated in the following example.

EXAMPLE G

A jet travels from city A due east for 100 mi and then turns 30° north of east, and travels another 50 mi. Find its displacement from city A.

We draw a diagram to illustrate the position of the jet, as shown in Fig. 8-9. From the figure we see that the plane is about 145 mi from city A, at an angle of about 10° north of east of the city. By giving both the magnitude and direction, we have given its displacement from the city.

Fig. 8-9

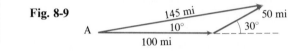

Exercises 8-1

In Exercises 1 through 4, determine whether a scalar or a vector is described in part (a) and part (b). Explain your answers.

1. (a) A person traveled 300 km to the southwest. (b) A person traveled 300 km.

2. (a) A small craft warning reports winds of 25 mi/h. (b) A small craft warning reports winds out of the north at 25 mi/h.

3. (a) An arm of an industrial robot pushes with a 10-lb force downward on a part. (b) A part is being pushed with a 10-lb force by an arm of an industrial robot.

4. (a) A ballistics test shows that a bullet hit a wall at a speed of 400 ft/s. (b) A ballistics test shows that a bullet hit a wall at a speed of 400 ft/s perpendicular to the wall.

In Exercises 5 through 8, add the given vectors by drawing the appropriate resultant. Use the parallelogram method in Exercises 7 and 8.

In Exercises 9 through 28, find the indicated vector sums and differences with the given vectors by means of diagrams. (You might find it helpful to work on graph paper.)

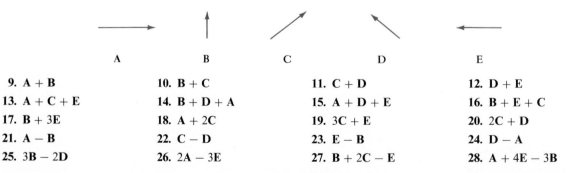

9. $\mathbf{A} + \mathbf{B}$	10. $\mathbf{B} + \mathbf{C}$	11. $\mathbf{C} + \mathbf{D}$	12. $\mathbf{D} + \mathbf{E}$
13. $\mathbf{A} + \mathbf{C} + \mathbf{E}$	14. $\mathbf{B} + \mathbf{D} + \mathbf{A}$	15. $\mathbf{A} + \mathbf{D} + \mathbf{E}$	16. $\mathbf{B} + \mathbf{E} + \mathbf{C}$
17. $\mathbf{B} + 3\mathbf{E}$	18. $\mathbf{A} + 2\mathbf{C}$	19. $3\mathbf{C} + \mathbf{E}$	20. $2\mathbf{C} + \mathbf{D}$
21. $\mathbf{A} - \mathbf{B}$	22. $\mathbf{C} - \mathbf{D}$	23. $\mathbf{E} - \mathbf{B}$	24. $\mathbf{D} - \mathbf{A}$
25. $3\mathbf{B} - 2\mathbf{D}$	26. $2\mathbf{A} - 3\mathbf{E}$	27. $\mathbf{B} + 2\mathbf{C} - \mathbf{E}$	28. $\mathbf{A} + 4\mathbf{E} - 3\mathbf{B}$

In Exercises 29 through 36, solve the given problems by use of an appropriate diagram.

29. Two forces act on a bolt. One is 10 lb and is directed vertically upward, and the other is 12 lb and is directed to the right. Find the resultant force.

30. Two electric charges create an electric field intensity, a vector quantity, at a given point. The field intensity is 30 kN/C to the right and 60 kN/C at an angle of 45° above the horizontal to the right. Find the resultant electric field intensity at this point.

31. A rocket takes off, moving at 2000 ft/s horizontally and 1500 ft/s vertically. Find its resultant velocity.

32. A small plane travels at 120 mi/h in still air. It is traveling due south in a wind of 30 mi/h from the northeast. What is the resultant velocity of the plane?

33. A driver takes the wrong road at an intersection and travels 4 mi north, then 6 mi east, and finally 10 mi to the southeast to reach the home of a friend. What is the displacement of the friend's home from the intersection?

34. A ship travels 20 km in a direction of 30° south of east and then turns due south for another 40 km. What is the ship's displacement from its initial position?

35. Three forces act on a small ring in a vertical plane. One force of 20 N acts vertically downward. A second force of 15 N acts horizontally to the right, and the third force of 25 N acts at an angle of 53° above the horizontal to the left. What is the resultant force on the ring?

36. A crate which has a weight of 100 N is suspended by two ropes. The force in one rope is 70 N and is directed to the left at an angle of 60° above the horizontal. What must be the other force in order that the resultant force (including the weight) on the crate is zero?

8-2   Components of Vectors

Adding vectors by means of diagrams is very useful in developing an understanding of vector quantities. However, unless the diagrams are drawn with great care and accuracy, the results we can obtain are quite approximate. Therefore, it is necessary to develop other methods to obtain results of sufficient accuracy.

In this section we show how a given vector can be considered to be the sum of two other vectors, with any required degree of accuracy. In the following section we show how this will allow us to add vectors in order to obtain the sum of vectors with the required accuracy in the result.

Two vectors which, when added together, give the original vector are called **components** *of the original vector.* In the illustration of the boat in Section 8-1, the velocities of 4 mi/h across the stream and 3 mi/h downstream are components of the 5 mi/h vector directed at the angle θ.

In practice, there are certain components of a vector which are of particular importance. If a vector is placed so that its initial point is at the origin of a rectangular coordinate system and its direction is indicated by an angle in standard position, we may find its **x-** and **y-components.** *These components are vectors directed along the coordinate axes which, when added together, result in the given vector.* The initial points of these vectors are at the origin, and the terminal points are located at the points where perpendicular lines from the terminal point of the vector cross the axes. *Finding these component vectors is called* **resolving** *the vector into its components.*

resolving a vector
into components

EXAMPLE A

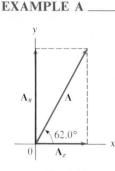

Fig. 8-10

Find the *x*- and *y*-components of the vector **A** shown in Fig. 8-10. The magnitude of **A** is 7.25.

From the figure we see that A_x, the magnitude of the *x*-component $\mathbf{A}_x$, is related to **A** by

$$\frac{A_x}{A} = \cos 62.0°$$

or

$$A_x = A \cos 62.0°$$

In the same way, A_y, the magnitude of the *y*-component $\mathbf{A}_y$, is related to **A** ($\mathbf{A}_y$ could be placed along the vertical dashed line) by

$$\frac{A_y}{A} = \sin 62.0°$$

or

$$A_y = A \sin 62.0°$$

From these relations, knowing that $A = 7.25$, we have

$$A_x = 7.25 \cos 62.0° = 3.40$$
$$A_y = 7.25 \sin 62.0° = 6.40$$

This means that the *x*-component is directed along the *x*-axis to the right and has a magnitude of 3.40. Also, the *y*-component is directed along the *y*-axis upward and its magnitude is 6.40. The calculator sequences for finding A_x and A_y are ■ shown at the left.

EXAMPLE B

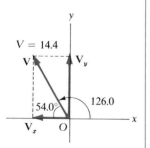

Fig. 8-11

Resolve a vector 14.4 units long and directed at an angle of 126.0° into its *x*- and *y*-components (see Fig. 8-11).

Placing the initial point of the vector at the origin and putting the angle in standard position, we see that the vector directed along the *x*-axis, $\mathbf{V}_x$, is related to the vector **V** of magnitude *V* by

$$V_x = V \cos 126.0°$$

    ⌐magnitude of vector

    └standard position angle

or in terms of the reference angle by

$$V_x = -V \cos 54.0°$$

    ⌐reference angle

    └directed along negative *x*-axis

since $\cos 126.0° = -\cos 54.0°$. We see that the minus sign shows that the *x*-component is directed in the negative direction, that is, to the left.

Since the vector directed along the *y*-axis, $\mathbf{V}_y$, could also be placed along the vertical dashed line, it is related to the vector **V** by

$$V_y = V \sin 126.0° = V \sin 54.0°$$

Thus, the vectors V_x and V_y have the magnitudes

$$V_x = 14.4 \cos 126.0° = -8.46, \qquad V_y = 14.4 \sin 126.0° = 11.6$$

Therefore, we have resolved the given vector into two components: one, directed along the negative x-axis, of magnitude 8.46, and the other, directed along the positive y-axis, of magnitude 11.6.

EXAMPLE C

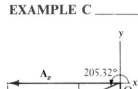

Fig. 8-12

Resolve vector **A**, of magnitude 375.4 and direction $\theta = 205.32°$, into its x- and y-components. See Fig. 8-12.

By placing **A** such that θ is in standard position, we see that

$$A_x = A \cos 205.32° = 375.4 \cos 205.32° = -339.3$$

and directed along negative axis

$$A_y = A \sin 205.32° = 375.4 \sin 205.32° = -160.5$$

Thus, **A** has two components: one, directed along the negative x-axis, of magnitude 339.3, and the other, directed along the negative y-axis, of magnitude 160.5.

EXAMPLE D

Fig. 8-13

The tension in a cable supporting a sign, as shown in Fig. 8-13(a), is 85.0 lb. Find the horizontal and vertical components of the tension.

The tension in the cable is the force that the cable exerts on the sign. Showing the tension in Fig. 8-13(b), we see that

$$T_x = T \cos 53.5° = 85.0 \cos 53.5° = 50.6 \text{ lb}$$
$$T_y = T \sin 53.5° = 85.0 \sin 53.5° = 68.3 \text{ lb}$$

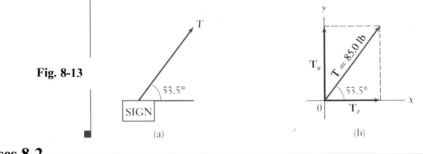

(a)　　　　(b)

Exercises 8-2

In Exercises 1 through 4, find the horizontal and vertical components of the vectors shown in the indicated figures. In each the magnitude of the vector is 750.

1.　　　**2.**　　　**3.**　　　**4.**

In Exercises 5 through 16, find the x- and y-components of the given vectors by use of the trigonometric functions.

5. Magnitude 8.60, $\theta = 68.0°$　　**6.** Magnitude 9750, $\theta = 243.0°$　　**7.** Magnitude 76.8, $\theta = 145.0°$

8. Magnitude 0.0998, $\theta = 296.0°$　　**9.** Magnitude 9.04, $\theta = 283.3°$　　**10.** Magnitude 16,400, $\theta = 156.5°$

11. Magnitude 2.65, $\theta = 197.3°$

12. Magnitude 67.8, $\theta = 22.5°$

13. Magnitude 0.8734, $\theta = 157.83°$

14. Magnitude 509.4, $\theta = 221.87°$

15. Magnitude 89,760, $\theta = 7.84°$

16. Magnitude 1.806, $\theta = 301.83°$

In Exercises 17 through 24, find the required horizontal and vertical components of the given vectors.

17. A nuclear submarine approaches the surface of the ocean at 25.0 km/h and at an angle of 17.3° with the surface. What are the components of its velocity?

18. Water is flowing downhill at 18.0 ft/s through a pipe which is at an angle of 66.4° with the horizontal. What are the components of its velocity?

19. The tension in a rope attached to a boat is 55.0 lb. The rope is attached to the boat 12.0 ft below the level at which it is being drawn in. At the point where there are 36.0 ft of rope out, what force tends to bring the boat toward the wharf, and what force tends to raise the boat?

20. A magnetic force of 0.250 N acts on a steel ball. It is directed to the left at an angle 13.6° above the horizontal. What are the components of the force?

21. A jet is 145 km at a position 37.5° north of east of a city. What are the components of the jet's displacement from the city?

22. The end of a robot arm is 3.50 ft on a line 78.6° above the horizontal from the point where it does a weld. What are the components of the displacement from the end of the robot arm to the welding point?

23. At one point the Pioneer space probe was entering the gravitational field of Jupiter at an angle of 2.55° below the horizontal with a velocity of 18,550 mi/h. What were the components of its velocity?

24. Two upward forces are acting on a bolt. One force of 60.5 lb acts at an angle of 82.4° above the horizontal and the other force of 37.2 lb acts at an angle of 50.5° below the first force. What is the total upward force on the bolt?

8-3   Vector Addition by Components

Now that we have developed the meaning of the components of a vector, we are in a position to add vectors to any degree of required accuracy. To do this we use the vector components, the Pythagorean theorem, and the tangent of the standard position angle of the resultant. In the following example, we show how two vectors at right angles are added.

EXAMPLE A

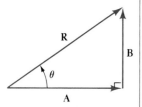

Fig. 8-14

Add vectors **A** and **B**, with $A = 14.5$ and $B = 9.10$. The vectors are at right angles, as shown in Fig. 8-14.

We can find the magnitude R of the resultant vector **R** by use of the Pythagorean theorem. This leads to

$$R = \sqrt{A^2 + B^2} = \sqrt{(14.5)^2 + (9.10)^2}$$
$$= 17.1$$

We shall now determine the direction of **R** by specifying its direction as the angle θ, that is, the angle **R** makes with **A**. Therefore,

$$\tan \theta = \frac{B}{A} = \frac{9.10}{14.5}, \qquad \theta = 32.1°$$

The following calculator sequence can be used to find R and θ.

$14.5 \; \boxed{x^2} \; \boxed{+} \; 9.1 \; \boxed{x^2} \; \boxed{=} \; \boxed{\sqrt{x}} \quad \boxed{17.118995}$

and

9.1 $\boxed{\div}$ 14.5 $\boxed{=}$ $\boxed{\text{INV}}$ $\boxed{\text{TAN}}$ $\boxed{\quad 32.111815 \quad}$

Therefore, we see that **R** is a vector of magnitude $R = 17.1$ and in a direction ■ 32.1° from vector **A**.

If vectors are to be added and they are not at right angles, we first place each with tail at the origin. Next, we resolve each vector into its x- and y-components. We then add all the x-components and add the y-components to determine the x- and y-components of the resultant. Then by use of the Pythagorean theorem and the tangent, as in Example A, we find the magnitude and directon of the resultant. *Remember, a vector is not completely specified unless*
NOTE▷ *both its magnitude **and its direction** are given.*

EXAMPLE B ———— Find the resultant of two vectors **A** and **B** such that $A = 1200$, $\theta_A = 270.0°$, $B = 1750$, and $\theta_B = 115.0°$.

We first place the vectors on a coordinate system with the tail of each at the origin as shown in Fig. 8-15(a). We then resolve each into its x- and y-components, as shown in Fig. 8-15(b) and as calculated below. (Note that **A** is vertical and has no horizontal component.) Next, the components are combined, as in Fig. 8-15(c) and as calculated. Finally, the magnitude and direction of the resultant, as shown in Fig. 8-15(d), are calculated.

$$A_x = A \cos 270.0° = 1200 \cos 270.0° = 0 \qquad A_y = A \sin 270.0° = 1200 \sin 270.0° = -1200$$
$$B_x = B \cos 115.0° = 1750 \cos 115.0° = -739.6 \qquad B_y = B \sin 115.0° = 1750 \sin 115.0° = 1586$$

$$R_x = A_x + B_x = 0 - 739.6 = -739.6$$
$$R_y = A_y + B_y = -1200 + 1586 = 386$$
$$R = \sqrt{R_x^2 + R_y^2} = \sqrt{(-739.6)^2 + 386^2} = 834$$

$$\overset{\displaystyle\quad\quad\quad\quad\quad 180° - 27.6°}{\tan \theta = \frac{R_y}{R_x} = \frac{386}{-739.6}, \qquad \theta = 152.4°}$$

Thus, the resultant has a magnitude of 834 and is directed at an angle of 152.4°. NOTE▷ In finding θ from the calculator, *the display shows an angle of $-27.6°$. However, we know that θ is a second-quadrant angle, since R_x is negative and R_y is positive.*

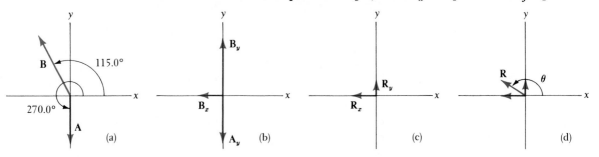

Fig. 8-15

(Continued on next page)

Therefore, we must use 27.6° as a reference angle. For this reason it may be advisable to find the reference angle first by disregarding the signs of R_x and R_y when finding θ. Thus,

$$\tan \theta_{\text{ref}} = \left|\frac{R_y}{R_x}\right| = \frac{386}{739.6}, \qquad \theta_{\text{ref}} = 27.6°$$

The above values have been rounded off. (In order not to round off intermediate results too soon, one extra digit was carried until R and θ were found.) If, in using the calculator, R_x and R_y are calculated in continuous steps, the values of R and θ may vary slightly. The calculator sequence for finding a resultant is shown in the next example. In this example we wished to show individual steps and results.

EXAMPLE C

Find the resultant **R** of the two vectors shown in Fig. 8-16(a), **A** of magnitude 8.075 and direction 57.26° and **B** of magnitude 5.437 and direction 322.15°.

$$A_x = 8.075 \cos 57.26° = 4.367 \qquad A_y = 8.075 \sin 57.26° = 6.792$$
$$B_x = 5.437 \cos 322.15° = 4.293 \qquad B_y = 5.437 \sin 322.15° = -3.336$$
$$R_x = A_x + B_x = 4.367 + 4.293 = 8.660$$
$$R_y = A_y + B_y = 6.792 - 3.336 = 3.456$$
$$R = \sqrt{R_x^2 + R_y^2} = \sqrt{(8.660)^2 + (3.456)^2} = 9.324$$

$$\tan \theta = \frac{R_y}{R_x} = \frac{3.456}{8.660} = 0.3991, \qquad \theta = 21.76° \longleftarrow \begin{array}{l}\text{don't forget} \\ \text{the direction}\end{array}$$

The resultant vector is 9.324 units long and is directed at an angle of 21.76°, as shown in Fig. 8-16(c). We know that the resultant is in the first quadrant since both R_x and R_y are positive.

The above results are rounded off from the calculator displays. In using a calculator we may calculate first R_x, then R_y, then R, and finally θ in a continuous sequence of calculator steps. If a calculator has more than one memory, R_x and R_y can be stored and need not actually be written down. Using a calculator with only one memory, by writing down R_y we can use the following sequence:

8.075 ⨯ 57.26 cos + 5.437 ⨯ 322.15 cos = STO ⟵ R_x

8.075 ⨯ 57.26 sin + 5.437 ⨯ 322.15 sin = [3.4560289] ⟵ R_y

x^2 + RCL x^2 = $\sqrt{x}$ [9.3244699] ⟵ R

3.4560289 ÷ RCL = INV TAN [21.755142] ⟵ θ

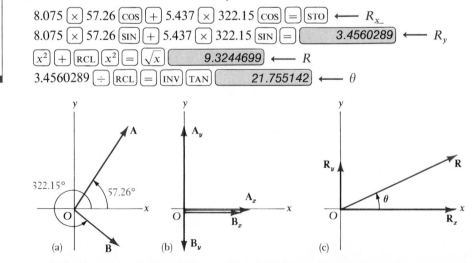

Fig. 8-16

Some general formulas can be derived from the previous examples. For a given vector **A**, directed at an angle θ, of magnitude A, and with components A_x and A_y, we have the following relations:

$$A_x = A \cos \theta, \qquad A_y = A \sin \theta \qquad (8\text{-}1)$$

$$A = \sqrt{A_x^2 + A_y^2} \qquad (8\text{-}2)$$

$$\tan \theta = \frac{A_y}{A_x} \qquad (8\text{-}3)$$

From the previous examples we can also see that the following procedure is used for adding vectors.

1. *Resolve the given vectors into their x- and y-components. Use Eqs. (8-1).*

2. *Add the x-components to obtain* **R**$_x$*, and add the y-components to obtain* **R**$_y$*.*

3. *Find the magnitude and the direction of the resultant* **R** *from the components* **R**$_x$ *and* **R**$_y$*. Use Eqs. (8-2) and (8-3) in the forms*

$$R = \sqrt{R_x^2 + R_y^2} \quad \text{and} \quad \tan \theta = \frac{R_y}{R_x}$$

EXAMPLE D ———— Find the resultant of the three given vectors with $A = 422$, $B = 405$, and $C = 210$, as shown in Fig. 8-17.

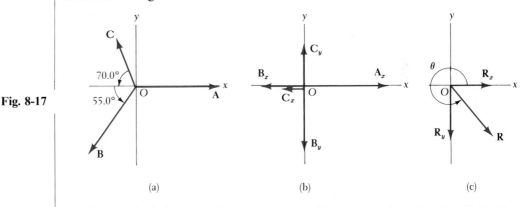

Fig. 8-17

(a)   (b)   (c)

We can find the x- and y-components of the vectors by using Eq. (8-1). The following table is helpful for determining the necessary values.

Vector	x-component	y-component
A	$422 \cos 0° = +422$	$422 \sin 0° = \quad 0$
B	$-405 \cos 55.0° = -232$	$-405 \sin 55.0° = -332$
C	$-210 \cos 70.0° = \quad -72$	$+210 \sin 70.0° = +197$
R	$+118$	-135

(Continued on next page)

From this table it is possible to compute R and θ:

$$R = \sqrt{(118)^2 + (-135)^2} = 179, \qquad \tan \theta_{\text{ref}} = \frac{135}{118}$$

$$\theta_{\text{ref}} = 48.8°, \qquad \theta = 311.2°$$

NOTE ▷   We note that we could have changed the given angles to standard position angles before finding the components. Also, we know that **θ is a fourth-quadrant angle since R_x is positive and R_y is negative.**

Exercises 8-3

In Exercises 1 through 4, vectors **A** and **B** are at right angles. Find the magnitude and direction (the angle from vector A) of the resultant.

1. $A = 14.7$
$B = 19.2$

2. $A = 592$
$B = 195$

3. $A = 3.086$
$B = 7.143$

4. $A = 1734$
$B = 3297$

In Exercises 5 through 12, with the given sets of components, find R and θ.

5. $R_x = 5.18, R_y = 8.56$

6. $R_x = 89.6, R_y = -52.0$

7. $R_x = -0.982, R_y = 2.56$

8. $R_x = -729, R_y = -209$

9. $R_x = -646, R_y = 2030$

10. $R_x = -31.2, R_y = -41.2$

11. $R_x = 0.6941, R_y = -1.246$

12. $R_x = 7.627, R_y = -6.353$

See Appendix E for a computer program for adding vectors.

In Exercises 13 through 24, add the given vectors by using the trigonometric functions and the Pythagorean theorem.

13. $A = 18.0, \theta_A = 0.0°$
$B = 12.0, \theta_B = 27.0°$

14. $A = 154, \theta_A = 90.0°$
$B = 128, \theta_B = 43.0°$

15. $A = 56.0, \theta_A = 76.0°$
$B = 24.0, \theta_B = 200.0°$

16. $A = 6.89, \theta_A = 123.0°$
$B = 29.0, \theta_B = 260.0°$

17. $A = 9.821, \theta_A = 34.27°$
$B = 17.45, \theta_B = 752.50°$

18. $A = 1.653. \theta_A = 36.37°$
$B = 0.9807, \theta_B = 253.06°$

19. $A = 12.653, \theta_A = 98.472°$
$B = 15.147, \theta_B = 332.092°$

20. $A = 121.36, \theta_A = 292.362°$
$B = 112.98, \theta_B = 197.892°$

21. $A = 21.9, \theta_A = 236.2°$
$B = 96.7, \theta_B = 11.5°$
$C = 62.9, \theta_C = 143.4°$

22. $A = 6300, \theta_A = 189.6°$
$B = 1760, \theta_B = 320.1°$
$C = 3240, \theta_C = 75.4°$

23. The vectors shown in Fig. 8-18

24. The vectors shown in Fig. 8-19

Fig. 8-18

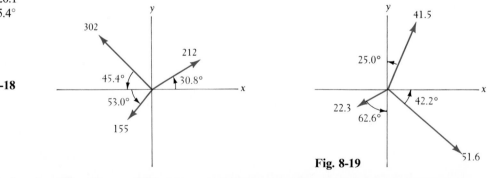

Fig. 8-19

8-4 Application of Vectors

In Section 8-1 we introduced the important vector quantities of force, velocity, and displacement, and we found vector sums by use of diagrams. Now we can use the method of Section 8-3 to find sums of these kinds of vectors and others and to use them in various types of applications.

EXAMPLE A

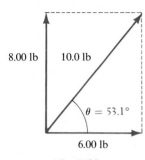

8.00 lb   10.0 lb

$\theta = 53.1°$

6.00 lb

Fig. 8-20

An object on a horizontal table is acted on by two horizontal forces. The two forces have magnitudes of 6.00 and 8.00 lb, and the angle between their lines of action is 90.0°. What is the resultant of these forces?

By means of an appropriate diagram (Fig. 8-20), we may better visualize the actual situation. We note that a good choice of axes (unless specified, it is often convenient to choose the x- and y-axes to fit the problem) is to have the x-axis in the direction of the 6.00-lb force and the y-axis in the direction of the 8.00-lb force. (This is possible since the angle between them is 90°.) With this choice we note that the two given forces will be the x- and y-components of the resultant. Therefore, we arrive at the following results:

$$F_x = 6.00 \text{ lb}, \qquad F_y = 8.00 \text{ lb}$$
$$F = \sqrt{(6.00)^2 + (8.00)^2} = 10.0 \text{ lb}$$
$$\tan \theta = \frac{F_y}{F_x} = \frac{8.00}{6.00}, \qquad \theta = 53.1°$$

We would state that the resultant has a magnitude of 10.0 lb and acts at an angle of 53.1° from the 6.00-lb force.

EXAMPLE B

A ship sails 32.50 mi due east and then turns 41.25° north of east. After sailing an additional 16.18 mi, where is it with reference to the starting point?

In this problem we are to determine the resultant displacement of the ship from the two given displacements. The problem is diagrammed in Fig. 8-21, where the first displacement has been labeled vector **A** and the second as vector **B**.

Since east corresponds to the positive x-direction, we see that the x-component of the resultant is $\mathbf{A} + \mathbf{B}_x$, and the y-component of the resultant is $\mathbf{B}_y$. Therefore, we have the following results:

$$R_x = A + B_x = 32.50 + 16.18 \cos 41.25°$$
$$= 32.50 + 12.16$$
$$= 44.66 \text{ mi}$$
$$R_y = 16.18 \sin 41.25° = 10.67 \text{ mi}$$
$$R = \sqrt{(44.66)^2 + (10.67)^2} = 45.92 \text{ mi}$$
$$\tan \theta = \frac{10.67}{44.66}, \qquad \theta = 13.44°$$

Therefore, the ship is 45.92 mi from the starting point, in a direction 13.44° north of east.

Fig. 8-21

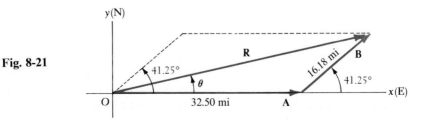

EXAMPLE C ———————— An airplane headed due east is in a wind blowing from the southeast. What is the resultant velocity of the plane with respect to the surface of the earth if the plane's velocity with respect to the air is 600 km/h and that of the wind is 100 km/h (see Fig. 8-22)?

Fig. 8-22

Let v_{px} be the velocity of the plane in the x-direction (east), v_{py} the velocity of the plane in the y-direction, v_{wx} the x-component of the velocity of the wind, v_{wy} the y-component of the velocity of the wind, and v_{pa} the velocity of the plane with respect to the air. Therefore,

$$v_{px} = v_{pa} - v_{wx} = 600 - 100(\cos 45.0°) = 529 \text{ km/h}$$
$$v_{py} = v_{wy} = 100(\sin 45.0°) = 70.7 \text{ km/h}$$
$$v = \sqrt{(529)^2 + (70.7)^2} = 534 \text{ km/h}$$
$$\tan \theta = \frac{v_{py}}{v_{px}} = \frac{70.7}{529}, \qquad \theta = 7.6°$$

We have determined that the plane is traveling 534 km/h and is flying in a direction 7.6° north of east. From this we observe that a plane does not necessarily head in the direction of its desired destination. ■

As we have seen, an important vector quantity is the force acting on an object. One of the most important applications of vectors involves forces which are in **equilibrium.** *For an object to be in equilibrium, the **net force** acting on it in any direction must be zero.* This condition is satisfied if the sum of the x-components of the force is zero and the sum of the y-components of the force is zero. The following two examples illustrate forces in equilibrium.

equilibrium of forces

EXAMPLE D ———————— A block is resting on an inclined plane which makes an angle of 30.0° with the horizontal. If the block weighs 100 lb, what is the force of friction between the block and the plane?

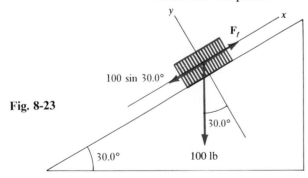

Fig. 8-23

All forces are assumed to act at the center of the block.

The weight of the block is the force exerted on the block due to gravity. Therefore, the weight is directed vertically downward. The frictional force tends to oppose the motion of the block and is directed upward along the plane. The frictional force must be sufficient to counterbalance that component of the weight of the block which is directed down the plane for the block to be at rest. The plane itself "holds up" that component of the weight which is perpendicular to the plane. A convenient set of coordinates (Fig. 8-23) would be one with the x-axis directed up the plane and y-axis perpendicular to the plane. The magnitude of the frictional force $\mathbf{F}_f$ is given by

$$F_f = 100 \sin 30.0° = 100(0.5000) = 50.0 \text{ lb}$$

$\qquad\qquad$ └─component of weight down
$\qquad\qquad\qquad$ plane equals frictional force

Here we have assumed that the block is small enough that all forces act at the center of the block.

EXAMPLE E

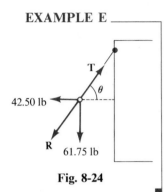

42.50 lb

Fig. 8-24

A 61.75-lb object hangs from a hook on a wall. If a horizontal force of 42.50 lb pulls the object away from the wall so that the object is in equilibrium (no resultant force in any direction), what is the tension T in the rope attached to the wall?

$\qquad$For the object to be in equilibrium, the tension in the rope must be equal and opposite to the resultant of the 42.50-lb force and the 61.75-lb force which is the weight of the object (see Fig. 8-24). Thus, the magnitude of the x-component of the tension is 42.50 lb and the magnitude of the y-component is 61.75 lb.

$$T = \sqrt{(42.50)^2 + (61.75)^2} = 74.96 \text{ lb}$$

$$\tan \theta = \frac{61.75}{42.50} = 1.453, \qquad \theta = 55.46°$$

EXAMPLE F

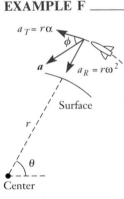

Fig. 8-25

For a spacecraft moving in a circular path around the earth, its tangential component of acceleration $\mathbf{a}_T$ and its centripetal component of acceleration $\mathbf{a}_R$ are given by the expressions shown in Fig. 8-25. The radius of the circle through which it is moving is r, the angular velocity at any instant is ω, and α is the angular acceleration, which is the rate at which the angular velocity ω is changing.

$\qquad$While going into orbit, at one point a spacecraft is moving in a circular path 230 km above the surface of the earth. At this point $r = 6.60 \times 10^6$ m, $\omega = 1.10 \times 10^{-3}$ rad/s, and $\alpha = 0.420 \times 10^{-6}$ rad/s². Calculate the magnitude of the resultant acceleration and the angle it makes with the tangential component.

$$a_R = r\omega^2 = 6.60 \times 10^6 (1.10 \times 10^{-3})^2 = 7.99 \text{ m/s}^2$$

$$a_T = r\alpha = 6.60 \times 10^6 (0.420 \times 10^{-6}) = 2.77 \text{ m/s}^2$$

Since a tangent line to a circle is perpendicular to the radius at the point of tangency, $\mathbf{a}_T$ is perpendicular to $\mathbf{a}_R$. Thus,

$$a = \sqrt{a_T^2 + a_R^2} = \sqrt{7.99^2 + 2.77^2} = 8.46 \text{ m/s}^2$$

$$\tan \phi = \frac{a_R}{a_T} = \frac{7.99}{2.77}, \qquad \phi = 70.9°$$

Exercises 8-4

In Exercises 1 through 24, solve the given problems.

1. Two forces, one of 5.75 lb and the other of 3.25 lb, act at right angles to each other on a shearing pin. Find the resultant of these forces.

2. Two ropes attached to a tree exert perpendicular horizontal forces of 18.5 lb and 23.0 lb. Find the resultant of these forces.

3. In lifting a heavy piece of equipment from the mud, a cable from a crane exerts a vertical force of 6500 N and a cable from a truck exerts a force of 8300 N at 10.0° above the horizontal. Find the resultant of these forces.

4. At a point in the plane, two electric charges create an electric field (a vector quantity) of 25.9 kN/C at 10.8° above the horizontal to the right and 12.6 kN/C at 83.4° below the horizontal to the right. Find the resultant electric field.

5. A motorboat leaves a dock and travels 1580 ft due west, then turns 35.0° to the south and travels another 1640 ft to a second dock. What is the displacement of the second dock from the first dock?

6. Toronto is 650 km at 19.0° north of east from Chicago. Cincinnati is 390 km at 48.0° south of east from Chicago. What is the displacement of Cincinnati from Toronto?

7. From a fixed point, a surveyor locates a pole at 215.6 ft due east and a building corner at 358.2 ft at 37.72° north of east. What is the displacement of the building from the pole?

8. A rocket is launched with a vertical component of velocity of 2840 km/h and a horizontal component of velocity of 1520 km/h. What is its resultant velocity?

9. A storm front is moving east at 22.0 km/h and south at 12.5 km/h. Find the resultant velocity of the front.

10. In an accident, a truck with momentum (a vector quantity) of 22,000 kg · m/s strikes a car with momentum of 17,800 kg · m/s from the rear. The angle between their directions of motion is 25.0°. What is the resultant momentum?

11. In an automobile safety test, a shoulder and seat belt exerts a force of 95.0 lb directly backward and a force of 83.0 lb backward at an angle of 20.0° below the horizontal on a dummy. If the belt holds the dummy from moving farther forward, what force did the dummy exert on the belt?

12. Two forces act on a ring at the end of a rope which passes over a (frictionless) pulley and holds weight w. See Fig. 8-26. Find w.

13. A plane flies at 550 km/h into a head wind of 60 km/h at 78.0° with the direction of the plane. Find the resultant velocity of the plane with respect to the ground.

14. A ship's navigator determines that the ship is moving through the water at 17.5 mi/h with a heading of 26.3° north of east, but that the ship is actually moving at 19.3 mi/h in a direction of 33.7° north of east. What is the velocity of the current?

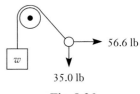

Fig. 8-26

15. A space shuttle is moving in orbit at 18,250 mi/h. A satellite is launched to the rear at 120 mi/h at an angle of 5.2° from the direction of the shuttle. Find the velocity of the satellite.

16. A block slides down an inclined (frictionless) plane with an acceleration of 5.3 m/s². If the angle of the plane is 32.7°, find the acceleration due to gravity.

17. While starting up, a circular saw blade 8.20 in. in diameter is rotating at 210 rad/min and has an angular acceleration of 320 rad/min². What is the acceleration of the tip of one of the teeth? (See Example F.)

18. A boat travels across a river, reaching the opposite bank at a point directly opposite that from which it left. If the boat travels 6.00 km/h in still water, and the current of the river flows at 3.00 km/h, what was the velocity of the boat in the water?

19. In searching for a boat lost at sea, a Coast Guard cutter leaves a port and travels 75.0 mi due east. It then turns 65.0° north of east and travels another 75.0 mi, and finally turns another 65.0° toward the west and travels another 75.0 mi. What is its displacement from the port?

20. In Fig. 8-23, if a horizontal 10.0-lb force is exerted to the right on the 100-lb object, what would the force of friction have to be so that the object does not move down the plane?

21. A plane is moving at 75.0 m/s, and a package with weather instruments is ejected from the plane at 15.0 m/s, perpendicular to the direction of the plane. If the vertical velocity v_v (in m/s), as a function of time of fall, is given by $v_v = 9.80t$, what is the velocity of the package after 2.00 s (before its parachute opens)?

22. A flat rectangular barge, 48.0 m long and 20.0 m wide, is headed directly across a stream at 4.5 km/h. The stream flows at 3.8 km/h. What is the velocity, relative to the riverbed, of a person walking diagonally across the barge at 5.0 km/h while facing the opposite upstream bank?

23. In Fig. 8-27, a long, straight conductor perpendicular to the plane of the paper carries an electric current i. A bar magnet having poles of strength m lies in the plane of the paper. The vectors $\mathbf{H}_i$, $\mathbf{H}_N$, and $\mathbf{H}_S$ represent the components of the magnetic intensity $\mathbf{H}$ due to the current and to the N and S poles of the magnet, respectively. The magnitude of the components of $\mathbf{H}$ are given by

$$H_i = \frac{1}{2\pi}\frac{i}{a}, \quad H_N = \frac{1}{4\pi}\frac{m}{b^2}, \quad \text{and} \quad H_S = \frac{1}{4\pi}\frac{m}{c^2}$$

Given that $a = 0.300$ m, $b = 0.400$ m, $c = 0.300$ m, the length of the magnet is 0.500 m, $i = 4.00$ A, and $m = 2.00$ A $\cdot$ m, calculate the resultant magnetic intensity $\mathbf{H}$. The component $\mathbf{H}_i$ is parallel to the magnet.

24. Solve the problem of Exercise 23 if $\mathbf{H}_i$ is directed away from the magnet, making an angle of 10.0° with the direction of the magnet.

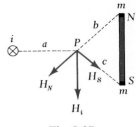

Fig. 8-27

8-5   Oblique Triangles, the Law of Sines

To this point we have limited our study of triangle solution to right triangles. However, *many triangles which require solution do not contain a right angle. Such triangles are called* **oblique triangles.** Let us now discuss the solutions of oblique triangles.

In Section 3-4 we stated that we need three parts, at least one of them a side, in order to solve any triangle. With this in mind we may determine that there are four possible combinations of parts from which we may solve a triangle. These combinations are as follows:

Case 1.   Two angles and one side.

Case 2.   Two sides and the angle opposite one of them.

Case 3.   Two sides and the included angle.

Case 4.   Three sides.

There are several ways in which oblique triangles may be solved, but we shall restrict our attention to the two most useful methods, the **law of sines** and the **law of cosines.** In this section we shall discuss the law of sines and show that it may be used to solve Case 1 and Case 2.

Let ABC be an oblique triangle with sides a, b, and c opposite angles A, B, and C, respectively. By drawing a perpendicular h from B to side b, or its extension, we see from Fig. 8-28(a) that

$$h = c \sin A \quad \text{or} \quad h = a \sin C \tag{8-4}$$

and from Fig. 8-28(b)

$$h = c \sin A \quad \text{or} \quad h = a \sin(180° - C) = a \sin C \tag{8-5}$$

We note that the results are precisely the same in Eqs. (8-4) and (8-5). Setting the results for h equal to each other, we have

$$c \sin A = a \sin C \quad \text{or} \quad \frac{a}{\sin A} = \frac{c}{\sin C} \tag{8-6}$$

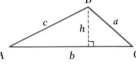

(a)

(b)

Fig. 8-28

By dropping a perpendicular from A to a, we also derive the result

$$c \sin B = b \sin C \quad \text{or} \quad \frac{b}{\sin B} = \frac{c}{\sin C} \tag{8-7}$$

Combining Eqs. (8-6) and (8-7) we have the **law of sines:**

law of sines

$$\boxed{\frac{a}{\sin A} = \frac{b}{\sin B} = \frac{c}{\sin C}} \tag{8-8}$$

Another form of the law of sines can be obtained by equating the reciprocals of each of the fractions in Eq. (8-8). The law of sines is a statement of proportionality between the sides of a triangle and the sines of the angles opposite them. We should note that there are actually three equations combined in Eq. (8-8). Of these, we use the one with three known parts of the triangle and we find the fourth part. In finding the complete solution of a triangle, it may be necessary to use two of the three equations.

Now we may see how the law of sines is applied to the solution of a triangle in which two angles and one side are known (Case 1). If two angles are known, the third may be found from the fact that the sum of the angles in a triangle is 180°. At this point we must be able to determine the ratio between the given side and the sine of the angle opposite it. Then, by use of the law of sines, we may find the other sides.

EXAMPLE A

Given that $c = 6.00$, $A = 60.0°$, and $B = 40.0°$, find a, b, and C.
First we can see that

$$C = 180.0° - (60.0° + 40.0°) = 80.0°$$

We now know side c and angle C, which allows us to use Eq. (8-8). Therefore, using the equation relating a, A, c, and C, we have

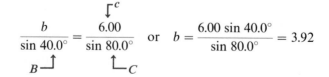

$$\frac{a}{\sin 60.0°} = \frac{6.00}{\sin 80.0°} \quad \text{or} \quad a = \frac{6.00 \sin 60.0°}{\sin 80.0°} = 5.28$$

Now, using the equation relating b, B, c, and C, we have

$$\frac{b}{\sin 40.0°} = \frac{6.00}{\sin 80.0°} \quad \text{or} \quad b = \frac{6.00 \sin 40.0°}{\sin 80.0°} = 3.92$$

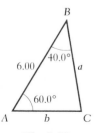

Fig. 8-29

Thus, $a = 5.28$, $b = 3.92$, and $C = 80.0°$. See Fig. 8-29. We could also have used the form of Eq. (8-8) relating a, A, b, and B in order to find b, but any error in calculating a would make b in error as well. Of course, any error in calculating C would make both a and b in error.

Since the quotient 6.00/sin 80.0° is used in the solution for both a and b, we can use the memory in a calculator as follows:

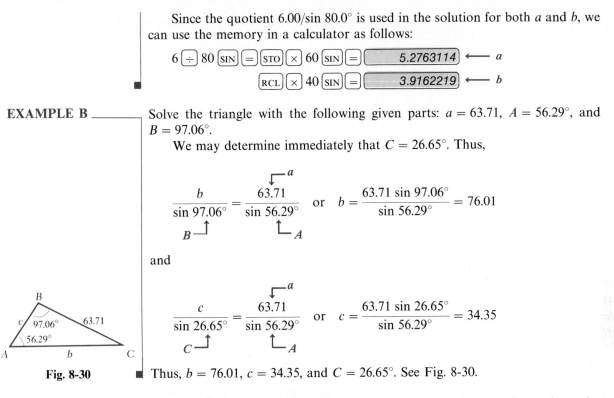

6 $\div$ 80 $\boxed{\text{SIN}}$ $\boxed{=}$ $\boxed{\text{STO}}$ $\times$ 60 $\boxed{\text{SIN}}$ $\boxed{=}$ $\boxed{5.2763114}$ $\longleftarrow$ a

$\boxed{\text{RCL}}$ $\times$ 40 $\boxed{\text{SIN}}$ $\boxed{=}$ $\boxed{3.9162219}$ $\longleftarrow$ b

EXAMPLE B ___ Solve the triangle with the following given parts: $a = 63.71$, $A = 56.29°$, and $B = 97.06°$.

We may determine immediately that $C = 26.65°$. Thus,

$$\frac{b}{\sin 97.06°} = \frac{63.71}{\sin 56.29°} \quad \text{or} \quad b = \frac{63.71 \sin 97.06°}{\sin 56.29°} = 76.01$$

with B and A labeled beneath, and a labeled above 63.71.

and

$$\frac{c}{\sin 26.65°} = \frac{63.71}{\sin 56.29°} \quad \text{or} \quad c = \frac{63.71 \sin 26.65°}{\sin 56.29°} = 34.35$$

with C and A labeled beneath, and a labeled above 63.71.

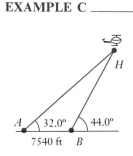

Fig. 8-30

Thus, $b = 76.01$, $c = 34.35$, and $C = 26.65°$. See Fig. 8-30.

If the given information is appropriate, the law of sines may be used to solve applied problems. The following example illustrates the use of the law of sines in such a problem.

EXAMPLE C ___ Two observers A and B sight a helicopter due east. The observers are 7540 ft apart, and the angles of elevation they each measure to the helicopter are 32.0° and 44.0°, respectively. How far is observer A from the helicopter? See Fig. 8-31.

Letting H represent the position of the helicopter, we see that angle B within the triangle ABH is $180° - 44.0° = 136.0°$. This means that the angle at H within the triangle is

$$H = 180° - (32.0° + 136.0°) = 12.0°$$

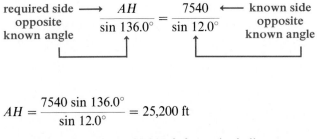

Fig. 8-31

Now, using the law of sines to find required side AH, we have

$$\begin{matrix} \text{required side} \longrightarrow & \dfrac{AH}{\sin 136.0°} = \dfrac{7540}{\sin 12.0°} & \longleftarrow \text{known side} \\ \text{opposite} & & \text{opposite} \\ \text{known angle} & & \text{known angle} \end{matrix}$$

or

$$AH = \frac{7540 \sin 136.0°}{\sin 12.0°} = 25{,}200 \text{ ft}$$

Thus, observer A is about 25,200 ft from the helicopter.

If we have information equivalent to Case 2 (two sides and the angle oppo-site one of them), we may find that there are *two* triangles which satisfy the given information. The following example illustrates this point.

EXAMPLE D

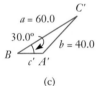

Side b reaches A at either of two points

(a)

C
$a = 60.0$ $b = 40.0$
B $30.0°$ A
c

(b)

C'
$a = 60.0$
$30.0°$ $b = 40.0$
B c' A'

(c)

Fig. 8-32

Solve the triangle with the following given parts: $a = 60.0$, $b = 40.0$, and $B = 30.0°$.

By making a good scale drawing, Fig. 8-32(a), we note that the angle opposite a may be either at position A or A'. Both positions of this angle satisfy the given parts. Therefore, there are two triangles which result. Using the law of sines, we solve the case in which A, opposite side a, is an acute angle.

$$\frac{60.0}{\sin A} = \frac{40.0}{\sin 30.0°} \quad \text{or} \quad \sin A = \frac{60.0 \sin 30.0°}{40.0}, \quad A = 48.6°$$

Therefore, $A = 48.6°$ and $C = 101.4°$. Using the law of sines again to find c, we have

$$\frac{c}{\sin 101.4°} = \frac{40.0}{\sin 30.0°} \quad \text{or} \quad c = \frac{40.0 \sin 101.4°}{\sin 30.0°} = 78.4$$

Thus we have $A = 48.6°$, $C = 101.4°$, and $c = 78.4$. See Fig. 8-32(b).

The other solution is the case in which A', opposite side a, is an obtuse angle. Here we have

$$\frac{60.0}{\sin A'} = \frac{40.0}{\sin 30.0°}, \quad \sin A = 0.7500$$

which leads to $A' = 180° - 48.6° = 131.4°$. For this case we have C' (the angle opposite c when $A' = 131.4°$) as $18.6°$.

Using the law of sines to find c', we have

$$\frac{c'}{\sin 18.6°} = \frac{40.0}{\sin 30.0°} \quad \text{or} \quad c' = \frac{40.0 \sin 18.6°}{\sin 30.0°} = 25.5$$

This means that the second solution is $A' = 131.4°$, $C' = 18.6°$, and $c' = 25.5$. See Fig. 8-32(c). ■

EXAMPLE E

In Example D, if $b > 60.0$, only one solution would result. In this case, side b would intercept side c at A. It also intercepts the extension of side c, but this would require that angle B not be included in the triangle (see Fig. 8-33). Thus only one solution may result if $b > a$.

In Example D, there would be no solution if side b were not at least 30.0. For if this were the case, side b would not be sufficiently long to even touch side c. It can be seen that b must at least equal $a \sin B$. If it is just equal to $a \sin B$, there is one solution, a right triangle (Fig. 8-34). ■

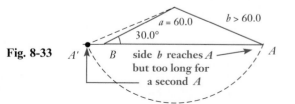

Fig. 8-33

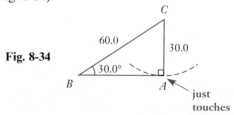

Fig. 8-34

Summarizing the results for Case 2 as illustrated in Examples D and E, we make the following conclusions. Given sides a and b and angle A (assuming here that a and A $[A < 90°]$ are the given corresponding parts), we have

ambiguous case

1. *no solution if $a < b \sin A$;*
2. *a right triangle solution if $a = b \sin A$;*
3. *two solutions if $b \sin A < a < b$;*
4. *one solution if $a > b$.*

See Fig. 8-35, parts (a) through (d), respectively.

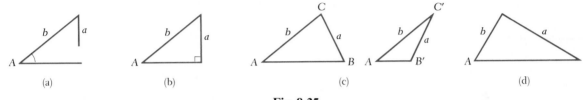

(a)   (b)   (c)   (d)

Fig. 8-35

For the reason that two solutions may result from it, Case 2 is often referred to as the **ambiguous case.** The following example illustrates Case 2 in an applied problem.

EXAMPLE F

See the chapter introduction.

City B is $43.2°$ south of east of city A. Find the direction in which a pilot should head a plane in flying from A to B if the wind is from the west at 40.0 km/h and the plane's velocity with respect to the air is 300 km/h.

The plane's heading should be set so that the resultant of the plane's velocity with respect to the air $\mathbf{v}_{pa}$ and the velocity of the wind $\mathbf{v}_w$ will be in the direction from city A to city B. This means that the resultant velocity $\mathbf{v}_{pg}$, representing the velocity of the plane with respect to the ground, must be at an angle of $43.2°$ south of east from city A.

With this analysis, we use the given information and draw the vector triangle shown in Fig. 8-36. In triangle ABC, we know that $\angle ABC = 43.2°$ by noting the alternate interior angles (see Appendix C). By finding angle θ, the required heading can be determined. There can be only one solution, since $v_{pa} > v_w$. Using the law of sines, we have

$$\underset{\substack{\text{known side}\\\text{opposite}\\\text{required angle}}}{} \rightarrow \frac{40.0}{\sin \theta} = \frac{300}{\sin 43.2°} \leftarrow \underset{\substack{\text{known side}\\\text{opposite}\\\text{known angle}}}{}$$

from which we get

$$\sin \theta = \frac{40.0 \sin 43.2°}{300}, \qquad \theta = 5.2°$$

Therefore the heading should be $43.2° + 5.2° = 48.4°$ south of east.

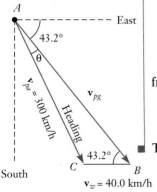

Fig. 8-36

If we attempt to use the law of sines for the solution of Case 3 or Case 4, we find that we do not have sufficient information to complete one of the ratios. These cases can, however, be solved by the law of cosines, which we shall consider in the next section.

EXAMPLE G — Given the three sides, $a = 5$, $b = 6$, $c = 7$, we would set up the ratios

$$\frac{5}{\sin A} = \frac{6}{\sin B} = \frac{7}{\sin C}$$

However, since there is no way to determine a complete ratio from these equations, we cannot find the solution of the triangle in this manner. ∎

Exercises 8-5

In Exercises 1 through 20, solve the triangles with the given parts.

1. $a = 45.7$, $A = 65.0°$, $B = 49.0°$

2. $b = 3.07$, $A = 26.0°$, $C = 120.0°$

3. $c = 4380$, $A = 37.4°$, $B = 34.6°$

4. $a = 93.2$, $B = 17.9°$, $C = 82.6°$

5. $a = 4.601$, $b = 3.107$, $A = 18.23°$

6. $b = 3.625$, $c = 2.946$, $B = 69.37°$

7. $b = 77.51$, $c = 36.42$, $B = 20.73°$

8. $a = 150.4$, $c = 250.9$, $C = 76.43°$

9. $b = 0.0742$, $B = 51.0°$, $C = 3.4°$

10. $c = 729$, $B = 121.0°$, $C = 44.2°$

11. $a = 63.8$, $B = 58.4°$, $C = 22.2°$

12. $a = 13.0$, $A = 55.2°$, $B = 67.5°$

13. $b = 4384$, $B = 47.43°$, $C = 64.56°$

14. $b = 283.2$, $B = 13.79°$, $C = 76.38°$

15. $a = 5.240$, $b = 4.446$, $B = 48.13°$

16. $a = 89.45$, $c = 37.36$, $C = 15.62°$

17. $b = 2880$, $c = 3650$, $B = 31.4°$

18. $a = 0.841$, $b = 0.965$, $A = 57.1°$

19. $a = 45.0$, $b = 126$, $A = 64.8°$

20. $a = 20$, $c = 10$, $C = 30°$

In Exercises 21 through 32, use the law of sines to solve the given problems.

21. A tabletop is in the shape of a regular pentagon (5 sides). A diagonal measures 4.00 ft. What is the length of each of the sides?

22. Two ropes hold a 175-lb crate as shown in Fig. 8-37. Find the tensions $\mathbf{T}_1$ and $\mathbf{T}_2$ in the ropes. (*Hint:* Move vectors so that they are tail to head to form a triangle. The vector sum $\mathbf{T}_1 + \mathbf{T}_2$ must equal 175 lb for equilibrium. See page 244.)

23. Find the tension $\mathbf{T}$ in the left guy wire attached to the top of the tower shown in Fig. 8-38. (*Hint:* The horizontal components of the tensions must be equal and opposite for equilibrium. See page 244. Thus, move the tension vectors tail to head to form a triangle with a vertical resultant. This resultant equals the upward force at the top by the tower for equilibrium. This last force is not shown and does not have to be calculated.)

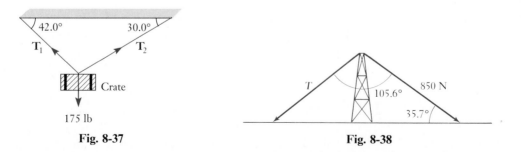

Fig. 8-37   Fig. 8-38

24. Find the distance from city A to city B from the diagram shown in Fig. 8-39.

25. Find the distance x between street intersections shown in Fig. 8-40.

26. When an airplane is landing at an 8250-ft runway, the angles of depression to the ends of the runway are 10.0° and 13.5°. How far is the plane from the near end of the runway?

27. Find the total length of the laser path shown in Fig. 8-41.

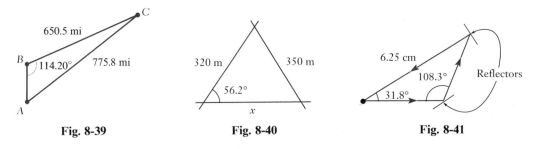

Fig. 8-39   Fig. 8-40   Fig. 8-41

28. In widening a highway, it is necessary for a construction crew to cut into the bank along the highway. The present angle of elevation of the straight slope of the bank is 23.0° and the new angle is to be 38.5°, leaving the top of the slope at its present position. If the slope of the present bank is 220 ft long, how far horizontally into the bank at its base must they dig?

29. A communications satellite is directly above the extension of a line between receiving towers A and B. It is determined from radio signals that the angle of elevation of the satellite from tower A is 89.2°, and the angle of elevation from tower B is 86.5°. If A and B are 1290 km apart, how far is the satellite from A? (Neglect the curvature of the earth.)

30. Point P on the mechanism shown in Fig. 8-42 is driven back and forth horizontally. If the minimum value of angle θ is 32.0°, what is the distance between extreme positions of P? What is the maximum possible value of angle θ?

Fig. 8-42

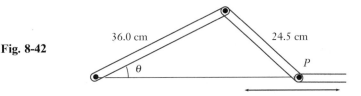

31. A boat owner wishes to cross a river 2.60 km wide and go directly to a point on the opposite side 1.75 km downstream. The boat goes 8.00 km/h in still water, and the stream flows at 3.50 km/h. What should the boat's heading be?

32. A triangular support was measured to have a side of 25.3 in., a second side of 14.0 in., and an angle of 36.5° opposite the second side. Find the length of the third side.

8-6   The Law of Cosines ▰▰▰▰▰▰▰▰▰▰

As we noted at the end of the preceding section, the law of sines cannot be used if the only information given is that of Case 3 or Case 4. Therefore, it is necessary to develop a method of finding at least one more part of the triangle. Here we can use the law of cosines. After obtaining another part by the law of cosines, we can then use the law of sines to complete the solution. We do this because the law of sines generally provides a simpler method of solution than the law of cosines.

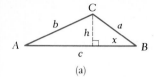

(a)

Consider any oblique triangle, for example either of the ones in Fig. 8-43. For each we obtain $h = b \sin A$. By using the Pythagorean theorem, we obtain $a^2 = h^2 + x^2$ for each. Thus (with $(\sin A)^2 = \sin^2 A$),

$$a^2 = b^2 \sin^2 A + x^2 \qquad (8\text{-}9)$$

In Fig. 8-43(a), we have $c - x = b \cos A$, or $x = c - b \cos A$. In Fig. 8-43(b), we have $c + x = b \cos A$, or $x = b \cos A - c$. Substituting these relations into Eq. (8-9), we obtain

$$a^2 = b^2 \sin^2 A + (c - b \cos A)^2$$

and $\qquad (8\text{-}10)$

$$a^2 = b^2 \sin^2 A + (b \cos A - c)^2$$

respectively. When expanded, these give

$$a^2 = b^2 \sin^2 A + b^2 \cos^2 A + c^2 - 2bc \cos A$$

and

$$a^2 = b^2 (\sin^2 A + \cos^2 A) + c^2 - 2bc \cos A \qquad (8\text{-}11)$$

Recalling the definitions of the trigonometric functions, we know that $\sin \theta = y/r$ and $\cos \theta = x/r$. Thus, $\sin^2 \theta + \cos^2 \theta = (y^2 + x^2)/r^2$. However, $x^2 + y^2 = r^2$, which means that

$$\sin^2 \theta + \cos^2 \theta = 1 \qquad (8\text{-}12)$$

This equation holds for any angle θ, since we made no assumptions as to the properties of θ. By substituting Eq. (8-12) into Eq. (8-11), we arrive at the **law of cosines:**

law of cosines

$$\boxed{a^2 = b^2 + c^2 - 2bc \cos A} \qquad (8\text{-}13)$$

Using the method above, we may also show that

$$\boxed{b^2 = a^2 + c^2 - 2ac \cos B}$$

and

$$\boxed{c^2 = a^2 + b^2 - 2ab \cos C}$$

Therefore, if we know two sides and the included angle (Case 3) we may directly solve for the side opposite the given angle. Then, by using the law of sines, we may complete the solution. If we are given all three sides (Case 4), we may solve for the angle opposite one of these sides by use of the law of cosines. Again we use the law of sines to complete the solution.

EXAMPLE A _____   Solve the triangle with $a = 45.0$, $b = 67.0$, and $C = 35.0°$ (see Fig. 8-44). Using the law of cosines, we have

Fig. 8-43

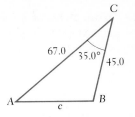

Fig. 8-44

$$\overbrace{c^2 = (45.0)^2 + (67.0)^2 - 2(45.0)(67.0) \cos 35.0^\circ}$$

unknown side opposite known angle

known sides

$$c = \sqrt{45.0^2 + 67.0^2 - 2(45.0)(67.0) \cos 35.0^\circ} = 39.7$$

The calculator sequence for this calculation is

$$45 \boxed{x^2} \boxed{+} 67 \boxed{x^2} \boxed{-} 2 \boxed{\times} 45 \boxed{\times} 67 \boxed{\times} 35 \boxed{\cos} \boxed{=} \boxed{\sqrt{x}} \quad \boxed{39.680136}$$

From the law of sines, we now have

$$\frac{45.0}{\sin A} = \frac{67.0}{\sin B} = \frac{39.7}{\sin 35.0^\circ} \quad \leftarrow \text{sides} \atop \leftarrow \text{opposite} \atop \leftarrow \text{angles}$$

which leads to

$$\sin A = \frac{45.0 \sin 35.0^\circ}{39.7}, \qquad A = 40.6^\circ$$

We could solve for B from the above relation or by use of the fact that the sum of all three angles is 180°. Thus,

$$B = 104.4^\circ$$

EXAMPLE B

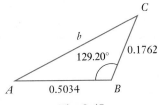

Fig. 8-45

Solve the triangle with $a = 0.1762$, $c = 0.5034$, and $B = 129.20^\circ$ (see Fig. 8-45).

Again, the given parts are two sides and the included angle, or Case 3. Since B is the known angle, we use the form of the law of cosines which includes angle B. This means we will find b first.

$$b^2 = 0.1762^2 + 0.5034^2 - 2(0.1762)(0.5034) \cos 129.20^\circ$$
$$b = 0.6297$$

From the law of sines, we have

$$\frac{0.1762}{\sin A} = \frac{0.6297}{\sin 129.20^\circ} = \frac{0.5034}{\sin C}$$

from which we get $A = 12.52^\circ$ and $C = 38.28^\circ$.

EXAMPLE C

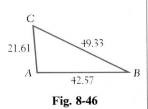

Fig. 8-46

Solve the triangle for which $a = 49.33$, $b = 21.61$, and $c = 42.57$ (see Fig. 8-46).

In this example the three sides are known, which is Case 4. Therefore, we must solve the law of cosines for the cosine of one of the angles. *The best procedure is to find the **largest angle**, for this will avoid the ambiguous case when we switch to the law of sines, if there is an obtuse angle in the triangle. The largest angle is always opposite the largest side, which means that in this example we should solve for cos A, which will then give us A.

$$\cos A = \frac{b^2 + c^2 - a^2}{2bc} = \frac{(21.61)^2 + (42.57)^2 - (49.33)^2}{2(21.61)(42.57)}$$
$$= -0.08384$$
$$A = 94.81^\circ$$

(Continued on next page)

NOTE ▷

We note that *the calculator shows an obtuse angle,* which is consistent with the fact that cos A is negative.

From the law of sines, we now have

$$\frac{49.33}{\sin 94.81°} = \frac{21.61}{\sin B} = \frac{42.57}{\sin C}$$

■ which gives us $B = 25.88°$ and $C = 59.31°$.

EXAMPLE D

Two forces are acting on a bolt. One is a 78.0-N force acting horizontally to the right, and the other is a 45.0-N force acting upward to the right, 15.0° from the vertical. Find the resultant force **F**. See Fig. 8-47.

Moving the 45.0-N vector to the right, and using the lower right triangle with the marked 105.0° angle, we find the magnitude of the resultant force **F** as

$$F = \sqrt{78.0^2 + 45.0^2 - 2(78.0)(45.0)\cos 105.0°}$$
$$= 99.6 \text{ N}$$

To find θ, we use the law of sines:

$$\frac{45.0}{\sin \theta} = \frac{99.6}{\sin 105.0°}, \qquad \sin \theta = \frac{45.0 \sin 105.0°}{99.6}$$

This gives us $\theta = 25.9°$.

This problem can also be solved by using vector components, as we did
■ earlier in the chapter.

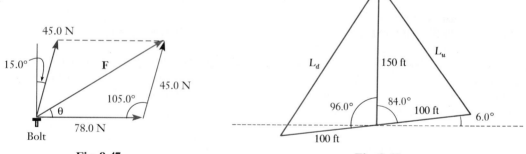

Fig. 8-47   Fig. 8-48

EXAMPLE E

A vertical radio antenna is to be built on a hill which makes an angle of 6.0° with the horizontal. If guy wires are to be attached at a point that is 150 ft up on the antenna and at points 100 ft from the base of the antenna, what will be the lengths of guy wires which are positioned directly up and directly down the hill?

Making an appropriate figure, such as Fig. 8-48, we are able to establish the equations necessary for the solution:

$$L_u^2 = 100^2 + 150^2 - 2(100)(150)\cos 84.0°$$
$$L_u = 171 \text{ ft}$$
$$L_d^2 = 100^2 + 150^2 - 2(100)(150)\cos 96.0°$$
$$L_d = 189 \text{ ft}$$

Exercises 8-6

In Exercises 1 through 20, solve the triangles with the given parts.

1. $a = 6.00$, $b = 7.56$, $C = 54.0°$

2. $b = 87.3$, $c = 34.0$, $A = 130.0°$

3. $a = 4530$, $b = 924$, $C = 98.0°$

4. $a = 0.0845$, $c = 0.116$, $B = 85.0°$

5. $a = 39.53$, $b = 45.22$, $c = 67.15$

6. $a = 23.31$, $b = 27.26$, $c = 29.17$

7. $a = 385.4$, $b = 467.7$, $c = 800.9$

8. $a = 0.2433$, $b = 0.2635$, $c = 0.1538$

9. $a = 320$, $b = 847$, $C = 158.0°$

10. $b = 18.3$, $c = 27.1$, $A = 58.7°$

11. $a = 21.4$, $c = 4.28$, $B = 86.3°$

12. $a = 11.3$, $b = 5.10$, $C = 77.6°$

13. $b = 103.7$, $c = 159.1$, $C = 104.67°$

14. $a = 49.32$, $b = 54.55$, $B = 114.36°$

15. $a = 0.4937$, $b = 0.5956$, $c = 0.6398$

16. $a = 69.72$, $b = 49.30$, $c = 56.29$

17. $a = 723$, $b = 598$, $c = 158$

18. $a = 1.78$, $b = 6.04$, $c = 4.80$

19. $a = 15$, $A = 15°$, $B = 140°$

20. $a = 17$, $b = 24$, $c = 37$

See Appendix E for a computer program for solving a triangle given three sides.

In Exercises 21 through 32, use the law of cosines to solve the given problems.

21. A nuclear submarine leaves its base and travels at 23.5 mi/h. For two hours it travels along a course 32.1° north of west. It then turns an additional 21.5° north of west and travels for another hour. How far from its base is it?

22. The robot arm shown in Fig. 8-49 places packages on a conveyor belt. What is the distance x?

23. In order to get around an obstruction, an oil pipeline is constructed in two straight sections, one 3.756 km long and the other 4.675 km long, with an angle of 168.85° between the sections where they are joined. How much more pipeline was necessary due to the obstruction?

24. A tabletop in the shape of an isosceles trapezoid is shown in Fig. 8-50. What are the angles between the edges?

25. A plane leaves an airport and travels 624 km due east. It then turns toward the north and travels another 326 km. It then turns again less than 180° and travels another 846 km directly back to the airport. Through what angles did it turn?

26. The apparent depth of an object submerged in water is less than its actual depth. A coin is actually 5.00 in. from an observer's eye just above the surface, but the distance appears to be only 4.25 in. The real light ray from the coin makes an angle with the surface which is 8.1° greater than the angle that the apparent ray makes. How much deeper is the coin than it appears to be? See Fig. 8-51.

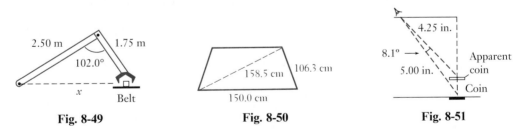

Fig. 8-49   Fig. 8-50   Fig. 8-51

27. A room is in the shape of a regular hexagon (6 sides). If each side is 10.5 ft, what is the length of the shortest diagonal of the room?

28. Two ropes support a 78.3 lb crate from above. The tensions in the ropes are 50.6 lb and 37.5 lb. What is the angle between the ropes? (See Exercise 22 of Section 8-5.)

29. A ferryboat travels at 11.5 km/h with respect to the water. Because of the river current, it is traveling at 12.7 km/h with respect to the land in the direction of its destination. If the ferryboat's heading is 23.6° from the direction of its destination, what is the velocity of the current?

30. The airline distance from Denver to Dallas is 660 mi. It is 800 mi from Denver to St. Louis and 550 mi from Dallas to St. Louis. Find the angle between the routes from Dallas.

31. A guy wire is attached to a utility pole 6.38 m from the base of the pole. It is 5.27 m from the base of the pole to the foot of the guy wire. If the guy wire is 8.60 m long, what is the angle the pole makes with the ground?

32. An air traffic controller sights two planes which are due east from the control tower and which are headed toward each other. One is 15.8 mi from the tower at an angle of elevation of 26.4°, and the other is 32.7 mi from the tower at an angle of elevation of 12.4°. How far apart are the planes?

8-7 Chapter Equations, Review Exercises, and Practice Test

Chapter Equations

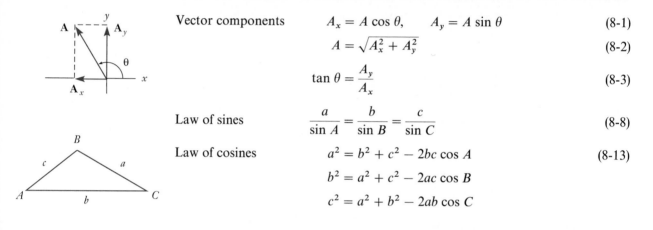

Vector components	$A_x = A \cos \theta, \qquad A_y = A \sin \theta$	(8-1)
	$A = \sqrt{A_x^2 + A_y^2}$	(8-2)
	$\tan \theta = \dfrac{A_y}{A_x}$	(8-3)
Law of sines	$\dfrac{a}{\sin A} = \dfrac{b}{\sin B} = \dfrac{c}{\sin C}$	(8-8)
Law of cosines	$a^2 = b^2 + c^2 - 2bc \cos A$	(8-13)
	$b^2 = a^2 + c^2 - 2ac \cos B$	
	$c^2 = a^2 + b^2 - 2ab \cos C$	

Review Exercises

In Exercises 1 through 4, find the x- and y-components of the given vectors by use of the trigonometric functions.

1. $A = 65.0$, $\theta_A = 28.0°$

2. $A = 8.05$, $\theta_A = 149.0°$

3. $A = 0.9204$, $\theta_A = 215.59°$

4. $A = 657.1$, $\theta_A = 343.74°$

In Exercises 5 through 8, vectors A and B are at right angles. Find the magnitude and direction of the resultant.

5. $A = 327$
$B = 505$

6. $A = 68$
$B = 29$

7. $A = 4964$
$B = 3298$

8. $A = 26.52$
$B = 89.86$

In Exercises 9 through 16, add the given vectors by use of the trigonometric functions and the Pythagorean theorem.

9. $A = 780$, $\theta_A = 28.0°$
$B = 346$, $\theta_B = 320.0°$

10. $A = 0.0120$, $\theta_A = 10.5°$
$B = 0.00781$, $\theta_B = 260.0°$

11. $A = 22.51$, $\theta_A = 130.16°$
$B = 7.604$, $\theta_B = 200.09°$

12. $A = 18,760$, $\theta_A = 110.43°$
$B = 4835$, $\theta_B = 350.20°$

13. $A = 51.33$, $\theta_A = 12.25°$
$B = 42.61$, $\theta_B = 291.77°$

14. $A = 70.31$, $\theta_A = 122.54°$
$B = 30.29$, $\theta_B = 214.82°$

15. $A = 75.0$, $\theta_A = 15.0°$
$B = 26.5$, $\theta_B = 192.4°$
$C = 54.8$, $\theta_C = 344.7°$

16. $A = 8120$, $\theta_A = 141.9°$
$B = 1540$, $\theta_B = 165.2°$
$C = 3470$, $\theta_C = 296.0°$

In Exercises 17 through 36, solve the triangles with the given parts.

17. $A = 48.0°$, $B = 68.0°$, $a = 14.5$

18. $A = 132.0°$, $b = 7.50$, $C = 32.0°$

19. $a = 22.8$, $B = 33.5°$, $C = 125.3°$

20. $A = 71.0°$, $B = 48.5°$, $c = 8.42$

21. $A = 17.85°, B = 154.16°, c = 7863$

22. $a = 1.985, b = 4.189, c = 3.652$

23. $b = 76.07, c = 40.53, B = 110.09°$

24. $A = 77.06°, a = 12.07, c = 5.104$

25. $b = 14.5, c = 13.0, C = 56.6°$

26. $B = 40.6°, b = 7.00, c = 18.0$

27. $a = 186, B = 130.0°, c = 106$

28. $b = 750, c = 1100, A = 56°$

29. $a = 7.86, b = 2.45, C = 22.0°$

30. $a = 0.208, c = 0.697, B = 105.4°$

31. $A = 67.16°, B = 96.84°, c = 532.9$

32. $A = 43.12°, a = 7.893, b = 4.113$

33. $a = 17, b = 12, c = 25$

34. $a = 9064, b = 9953, c = 1106$

35. $a = 5.30, b = 8.75, c = 12.5$

36. $a = 47.4, b = 40.0, c = 45.5$

In Exercises 37 through 56, solve the given problems.

37. Find the horizontal and vertical components of the force shown in Fig. 8-52.

38. Find the horizontal and vertical components of the velocity shown in Fig. 8-53.

39. In a ballistics test, a bullet was fired into a block of wood with a velocity of 2200 ft/s and at an angle of 71.3° with the surface of the block. What was the component of the velocity perpendicular to the surface?

40. A storm cloud is moving at 15 mi/h from the northwest. A television tower is 60° south of east of the cloud. What is the component of the cloud's velocity toward the tower?

41. In Fig. 8-54, force **F** represents the total surface tension force around the circumference on the liquid in the capillary tube. The vertical component of **F** holds up the liquid in the tube above the liquid surface outside the tube. What is this vertical component of **F**?

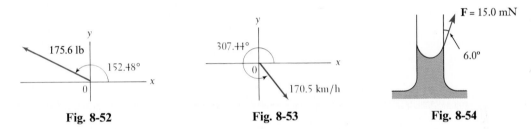

Fig. 8-52   Fig. 8-53   Fig. 8-54

42. During a three-minute period after taking off, the supersonic jet Concorde traveled at 480 km/h at an angle of 24.0° above the horizontal. What was its gain in altitude during the three minutes?

43. An ice cube is sliding down a (frictionless) plane inclined at 22.0° with the horizontal. Given that the acceleration due to gravity is 32.2 ft/s², what is the acceleration of the ice cube?

44. A crater on the moon is 150 mi in diameter. If the distance to the moon (to each side of the crater) from the earth is 240,000 mi, what angle is subtended by the crater at an observer's position on the earth?

45. In Fig. 8-55 a damper mechanism in an air-conditioning system is shown. If $\theta = 27.5°$ when the spring is at its shortest and longest lengths, what are these lengths?

46. Sixteen equally spaced holes are drilled in a metal plate on the circumference of a circle 28.5 cm in radius. What is the center-to-center distance between adjacent holes?

47. Two satellites are being observed at the same observing station. One is 22,500 mi from the station, and the other is 18,700 mi away. The angle between their lines of observation is 105.4°. How far apart are the satellites?

48. Find the side x in the truss shown in Fig. 8-56.

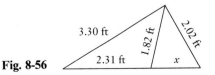

Fig. 8-55   Fig. 8-56

49. The angle of depression of a fire noticed west of a fire tower is 6.2°. The angle of depression of a pond, also west of the tower, is 13.5°. If it is known that the tower is 2.25 km from the pond on a direct line to the pond, how far is the fire from the pond?

50. A surveyor wishes to find the distance between two points between which there is a security-restricted area. The surveyor measures the distance from each of these points to a third point and finds them to be 226.73 m and 185.12 m. If the angle between the lines of sight from the third point to the other points is 126.724°, how far apart are the two points?

51. Atlanta is 290 mi and 51.0° south of east from Nashville. The pilot of an airplane due north of Atlanta radios Nashville and finds the plane is on a line 10.5° south of east from Nashville. How far is the plane from Nashville?

52. In going around a storm, a plane flies 125 mi south, then 140 mi at 30.0° south of west, and finally 225 mi at 15.0° north of west. What is the displacement of the plane from its original position?

53. A sailboat is headed due north, and its sail is set perpendicular to the wind, which is from south of west. The component of the force of the wind in the direction of the heading is 480 N, and the component perpendicular to the heading (the *drift* component) is 650 N. What is the force exerted by the wind, and what is the direction of the wind? See Fig. 8-57.

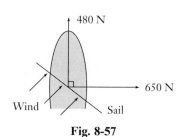

Fig. 8-57

54. Boston is 650 km and 21.0° south of west from Halifax, Nova Scotia. Radio signals locate a ship 10.5° east of south from Halifax and 5.6° north of east from Boston. How far is the ship from each city?

55. One end of a 1450-ft bridge is sighted from a distance of 3250 ft. The angle between the lines of sight of the ends of the bridge is 25.2°. From these data, how far is the observer from the other end of the bridge?

56. A plane is traveling horizontally at 1200 ft/s. A missile is fired horizontally from it 30.0° from the direction in which the plane is traveling. If the missile leaves the plane at 2000 ft/s, what is its velocity 10.0 s later if the vertical component is given by $v_V = -32.0\,t$ (in feet per second)?

Practice Test

In all triangle solutions, sides a, b, c are opposite angles A, B, C, respectively.

1. By use of a diagram, find the vector sum 2**A** + **B** for the given vectors.

2. For the triangle in which $a = 22.5$, $B = 78.6°$, and $c = 30.9$, find b.

3. A surveyor locates a tree 36.50 m to the northeast of a set position. The tree is 21.38 m north of a utility pole. What is the displacement of the utility pole from the set position?

4. For the triangle in which $A = 18.9°$, $B = 104.2°$, and $a = 426$, find c.

5. Solve the triangle in which $a = 9.84$, $b = 3.29$, and $c = 8.44$.

6. Find the horizontal and vertical components of a vector of magnitude 870 which is directed at a standard position angle of 284.3°.

7. A ship leaves a port and travels due west. At a certain point it turns 31.5° north of west and travels an additional 42.0 mi to a point 63.0 mi on a direct line from the port. How far from the port is the point where the ship turned?

8. Find the sum of the vectors for which the magnitudes and standard position angles are given as follows: $A = 450$, $B = 285$, $\theta_A = 74.2°$, $\theta_B = 208.9°$. Use the trigonometric functions and the Pythagorean theorem.

9 Graphs of the Trigonometric Functions

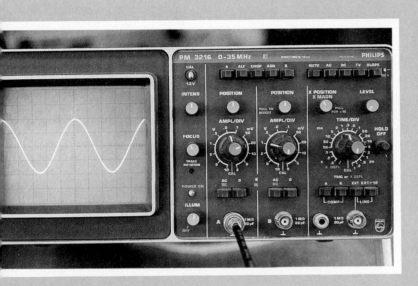

One of the clearest ways to show the properties of the trigonometric functions is by means of their graphs. Also, the graphs are valuable in a number of important applications.

We will see that the graphs of the trigonometric functions are particularly useful in applications which involve periodic values, values which repeat on a regular basis. Such applications are found in electronics, in mechanical vibrations, and in many areas of physics.

In Section 9-6 we show the resulting curve when an oscilloscope is used to combine and display electric signals.

9-1   Graphs of $y = a \sin x$ and $y = a \cos x$

The graphs of the trigonometric functions are constructed on the regular rectangular coordinate system. In plotting the trigonometric functions, *it is normal to express the angle in radians. In this way **x and the function of x are expressed as real numbers,** and these numbers may have any desired unit of measurement. Therefore, in order to determine the graphs, it is necessary to be able to readily use angles expressed in radians. If necessary, Section 7-3 should be reviewed for this purpose.

NOTE ▷

In this section the graphs of the sine and cosine functions are demonstrated. We begin by constructing a table of values of x and y for the function $y = \sin x$:

x	0	$\frac{\pi}{6}$	$\frac{\pi}{3}$	$\frac{\pi}{2}$	$\frac{2\pi}{3}$	$\frac{5\pi}{6}$	π	$\frac{7\pi}{6}$	$\frac{4\pi}{3}$	$\frac{3\pi}{2}$	$\frac{5\pi}{3}$	$\frac{11\pi}{6}$	2π
y	0	0.5	0.87	1	0.87	0.5	0	-0.5	-0.87	-1	-0.87	-0.5	0

Plotting these values, we obtain the graph shown in Fig. 9-1.

Fig. 9-1

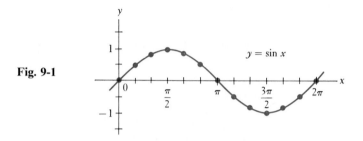

The graph of $y = \cos x$ may be constructed in the same manner. The following table gives the proper values for the graph of $y = \cos x$. The graph is shown in Fig. 9-2.

x	0	$\frac{\pi}{6}$	$\frac{\pi}{3}$	$\frac{\pi}{2}$	$\frac{2\pi}{3}$	$\frac{5\pi}{6}$	π	$\frac{7\pi}{6}$	$\frac{4\pi}{3}$	$\frac{3\pi}{2}$	$\frac{5\pi}{3}$	$\frac{11\pi}{6}$	2π
y	1	0.87	0.5	0	-0.5	-0.87	-1	-0.87	-0.5	0	0.5	0.87	1

Fig. 9-2

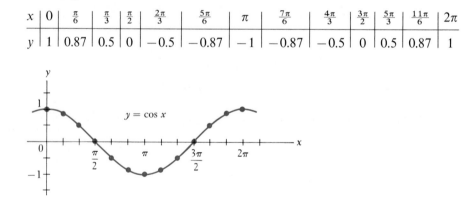

The graphs are continued beyond the values shown in the table to indicate that they continue on indefinitely in each direction. From the values and the graphs, *it can be seen that the two graphs are of exactly the same shape, with the cosine curve displaced $\pi/2$ units to the left of the sine curve.* The shape of these curves should be recognized readily, with special note as to the points at which they cross the axes. This information will be especially valuable in "sketching" similar curves, since the basic shape always remains the same. We shall find it unnecessary to plot numerous points every time we wish to sketch such a curve.

To obtain the graph of $y = a \sin x$, we note that all the y-values obtained for the graph of $y = \sin x$ are to be multiplied by the number a. In this case the greatest value of the sine function is $|a|$. *The number $|a|$ is called the* **amplitude** *of the curve and represents the greatest y-value of the curve.* Also, the curve will have no value less than $-|a|$. This is true for $y = a \cos x$ as well as for $y = a \sin x$.

amplitude

EXAMPLE A

Plot the curve of $y = 2 \sin x$.

Since $a = 2$, the amplitude of this curve is $|2| = 2$. This means that the maximum value of y is 2 and the minimum value of y is -2. The table of values follows, and the curve is shown in Fig. 9-3.

x	0	$\frac{\pi}{6}$	$\frac{\pi}{3}$	$\frac{\pi}{2}$	$\frac{2\pi}{3}$	$\frac{5\pi}{6}$	π	$\frac{7\pi}{6}$	$\frac{4\pi}{3}$	$\frac{3\pi}{2}$	$\frac{5\pi}{3}$	$\frac{11\pi}{6}$	2π
y	0	1	1.73	2	1.73	1	0	-1	-1.73	-2	-1.73	-1	0

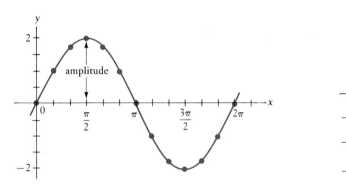

Fig. 9-3   **Fig. 9-4**

EXAMPLE B

Plot the curve of $y = -3 \cos x$.

In this case $a = -3$, which means that the amplitude is $|-3| = 3$. Therefore the maximum value of y is 3, and the minimum value of y is -3. The table of values follows, and the curve is shown in Fig. 9-4.

x	0	$\frac{\pi}{6}$	$\frac{\pi}{3}$	$\frac{\pi}{2}$	$\frac{2\pi}{3}$	$\frac{5\pi}{6}$	π	$\frac{7\pi}{6}$	$\frac{4\pi}{3}$	$\frac{3\pi}{2}$	$\frac{5\pi}{3}$	$\frac{11\pi}{6}$	2π
y	-3	-2.6	-1.5	0	1.5	2.6	3	2.6	1.5	0	-1.5	-2.6	-3

TABLE 9-1

	$x = 0, \pi, 2\pi$	$\frac{\pi}{2}, \frac{3\pi}{2}$
$y = a \sin x$	zeros	max. or min.
$y = a \cos x$	max. or min.	zeros

Note from Example B that *the effect of the minus sign before the number a is to invert the curve.* The effect of the number a can also be seen readily from these examples.

From the previous examples we see that the function $y = a \sin x$ has zeros for $x = 0$, π, 2π and that it has its maximum or minimum values for $x = \pi/2$, $3\pi/2$. The function $y = a \cos x$ has its zeros for $x = \pi/2$, $3\pi/2$ and its maximum or minimum values for $x = 0$, π, 2π. This is summarized in Table 9-1 at the left. Therefore, by knowing the general shape of the sine curve, where it has its zeros, and what its amplitude is, *we can rapidly **sketch** curves of the form $y = a \sin x$ and $y = a \cos x$.*

Since the graphs of $y = a \sin x$ and $y = a \cos x$ can extend indefinitely to the right and to the left, we see that the domain of each is all real numbers. We should note that the key values of $x = 0$, $\pi/2$, π, $3\pi/2$, and 2π are those only for x from 0 to 2π. Corresponding values ($x = 5\pi/2$, 3π, and their negatives) could also be used. Also from the graphs we can readily see that the range of these functions is $-|a| \leq f(x) \leq |a|$.

EXAMPLE C ———— Sketch the graph of $y = 4 \cos x$.

First we set up a table of values for the points where the curve has its zeros, maximum points, and minimum points.

x	0	$\frac{\pi}{2}$	π	$\frac{3\pi}{2}$	2π
y	4	0	-4	0	4
	max.		min.		max.

Now we plot the above points and join them, knowing the basic shape of the curve. See Fig. 9-5.

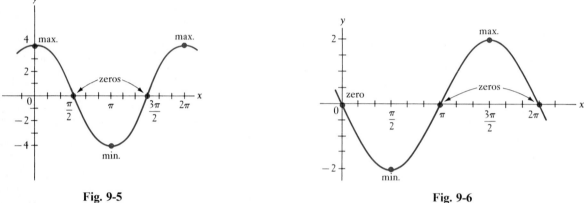

Fig. 9-5   Fig. 9-6

EXAMPLE D ———— Sketch the curve of $y = -2 \sin x$.

We list here the important values associated with this curve.

x	0	$\frac{\pi}{2}$	π	$\frac{3\pi}{2}$	2π
y	0	-2	0	2	0
		min.		max.	

Since we know the general shape of the sine curve, we can now sketch the graph, as shown in Fig. 9-6. Note the inversion of the curve due to the minus sign.

Exercises 9-1

In Exercises 1 through 4, complete the following table for the given functions, and then plot the resulting graph.

x	$-\pi$	$-\frac{3\pi}{4}$	$-\frac{\pi}{2}$	$-\frac{\pi}{4}$	0	$\frac{\pi}{4}$	$\frac{\pi}{2}$	$\frac{3\pi}{4}$	π	$\frac{5\pi}{4}$	$\frac{3\pi}{2}$	$\frac{7\pi}{4}$	2π	$\frac{9\pi}{4}$	$\frac{5\pi}{2}$	$\frac{11\pi}{4}$	3π
y																	

1. $y = \sin x$ **2.** $y = \cos x$ **3.** $y = 3 \cos x$ **4.** $y = -4 \sin x$

In Exercises 5 through 20, sketch the curves of the indicated functions.

5. $y = 3 \sin x$ **6.** $y = 5 \sin x$ **7.** $y = \frac{5}{2} \sin x$ **8.** $y = 0.5 \sin x$

9. $y = 2 \cos x$ **10.** $y = 3 \cos x$ **11.** $y = 0.8 \cos x$ **12.** $y = \frac{3}{2} \cos x$

13. $y = -\sin x$ **14.** $y = -3 \sin x.$ **15.** $y = -1.5 \sin x$ **16.** $y = -0.2 \sin x$

17. $y = -\cos x$ **18.** $y = -8 \cos x$ **19.** $y = -2.5 \cos x$ **20.** $y = -0.4 \cos x$

Although units of π are often convenient, we must remember that π is really only a number. Numbers which are not multiples of π may be used as well. In Exercises 21 through 24, plot the indicated graphs by finding the values of y corresponding to the values of 0, 1, 2, 3, 4, 5, 6, and 7 for x by use of a calculator. (Remember, the numbers 0, 1, 2, and so forth represent radian measure.)

21. $y = \sin x$ **22.** $y = -3 \sin x$ **23.** $y = \cos x$ **24.** $y = 2 \cos x$

9-2   Graphs of $y = a \sin bx$ and $y = a \cos bx$ ▬▬▬▬▬▬

period of a function

In graphing the curve of $y = \sin x$, we note that the values of y start repeating every 2π units of x. This is because $\sin x = \sin(x + 2\pi) = \sin(x + 4\pi)$, and so forth. For any trigonometric function F, we say that it has a **period** P if $F(x) = F(x + P)$. For functions which are periodic, such as the sine and cosine, *the period refers to the x-distance between any point and the next corresponding point for which the values of y start repeating.*

Let us now plot the curve $y = \sin 2x$. This means that we choose a value for x, multiply this value by two, and find the sine of the result. This leads to the following table of values for this function.

x	0	$\frac{\pi}{8}$	$\frac{\pi}{4}$	$\frac{3\pi}{8}$	$\frac{\pi}{2}$	$\frac{5\pi}{8}$	$\frac{3\pi}{4}$	$\frac{7\pi}{8}$	π	$\frac{9\pi}{8}$	$\frac{5\pi}{4}$
$2x$	0	$\frac{\pi}{4}$	$\frac{\pi}{2}$	$\frac{3\pi}{4}$	π	$\frac{5\pi}{4}$	$\frac{3\pi}{2}$	$\frac{7\pi}{4}$	2π	$\frac{9\pi}{4}$	$\frac{5\pi}{2}$
y	0	0.7	1	0.7	0	-0.7	-1	-0.7	0	0.7	1

Plotting these values, we have the curve shown in Fig. 9-7.

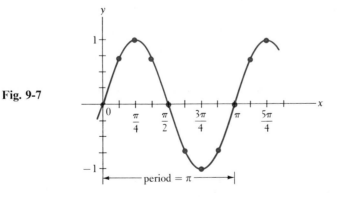

Fig. 9-7

period = π

From the table and Fig. 9-7, we note that the function $y = \sin 2x$ starts repeating after π units of x. The effect of the 2 before the x has been to make the period of this curve half the period of the curve of $\sin x$. This leads us to the following conclusion: If the period of the trigonometric function $F(x)$ is P, then the period of $F(bx)$ is P/b. Since each of the functions $\sin x$ and $\cos x$ has a period of 2π, *each of the functions $\sin bx$ and $\cos bx$ has a period of $2\pi/b$.*

EXAMPLE A

1. The period of sin $3x$ is $\dfrac{2\pi}{3}$, which means that the curve of the function $y =$ sin $3x$ will repeat every $\frac{2\pi}{3}$ (approximately 2.09) units of x.

2. The period of cos $4x$ is $\dfrac{2\pi}{4} = \dfrac{\pi}{2}$.

3. The period of sin $\frac{1}{2}x$ is $\dfrac{2\pi}{\frac{1}{2}} = 4\pi$. In this case we see that the period is longer than that of the basic sine curve. ∎

EXAMPLE B

The period of sin πx is $\frac{2\pi}{\pi} = 2$. That is, the curve of the function sin πx will repeat every 2 units. It is then noted that the periods of sin $3x$ and sin πx are nearly equal. This is to be expected, since π is only slightly greater than 3.

The period of cos $3\pi x$ is $\frac{2\pi}{3\pi} = \frac{2}{3}$.

The period of sin $\frac{\pi}{4}x$ is $2\pi/\frac{\pi}{4} = 8$. ∎

Combining the result for the period with the results of Section 9-1, we conclude that *each of the functions $y = a \sin bx$ and $y = a \cos bx$ has an amplitude of $|a|$ and a period of $2\pi/b$.* These properties are very useful in sketching these functions, as is shown in the following examples.

EXAMPLE C

Sketch the graph of $y = 3 \sin 4x$ for $0 \le x \le \pi$.

We immediately conclude that the amplitude is 3 and the period is $\frac{2\pi}{4} = \frac{\pi}{2}$. Therefore, we know that $y = 0$ when $x = 0$ and $y = 0$ when $x = \frac{\pi}{2}$. Also, we recall that the sine function is zero halfway between these values, which means that $y = 0$ when $x = \frac{\pi}{4}$. The function reaches its maximum or minimum values halfway between the zeros. Therefore, $y = 3$ for $x = \frac{\pi}{8}$ and $y = -3$ for $x = \frac{3\pi}{8}$. A table for these important values of the function $y = 3 \sin 4x$ follows.

x	0	$\frac{\pi}{8}$	$\frac{\pi}{4}$	$\frac{3\pi}{8}$	$\frac{\pi}{2}$	$\frac{5\pi}{8}$	$\frac{3\pi}{4}$	$\frac{7\pi}{8}$	π
y	0	3	0	-3	0	3	0	-3	0

Using this table and the knowledge of the form of the sine curve, we sketch the function (Fig. 9-8). ∎

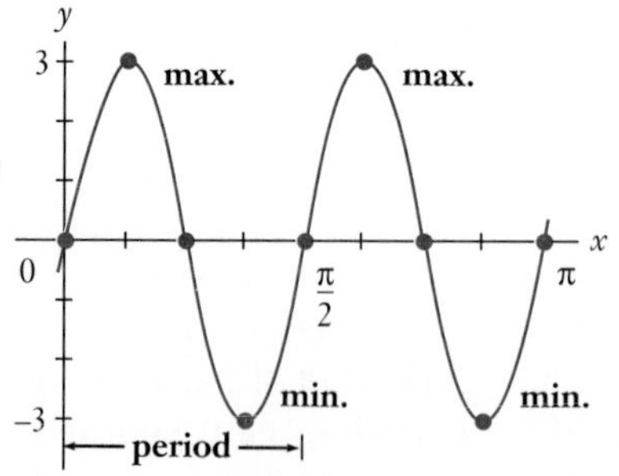

Fig. 9-8

We see from Example C that an important distance in sketching a sine curve or a cosine curve is one-fourth of the period. For $y = a \sin bx$, it is one-fourth of the period from the origin to the first value of x where y is at its maximum (or minimum) value. Then we proceed another one-fourth period to a zero, another one-fourth period to the next minimum (or maximum) value, another to the next zero (this is where one period is completed), and so on. Thus, *by finding one-fourth of the period, we can easily find the important values for sketching the curve.* Similarly, one-fourth of the period can be used in sketching $y = a \cos bx$.

The table of important values in sketching $y = a \sin bx$ or $y = a \cos bx$ can be found by

1. *finding the amplitude, $|a|$,*
2. *finding the period, $2\pi/b$, and*
3. *finding values of the function for each one-fourth period.*

EXAMPLE D

Sketch the graph of $y = -2 \cos 3x$ for $0 \le x \le 2\pi$.

We note that the amplitude is 2 and the period is $\frac{2\pi}{3}$. This means that one-fourth of the period is $\frac{1}{4} \times \frac{2\pi}{3} = \frac{\pi}{6}$. Since the cosine curve is at a maximum or minimum for $x = 0$, we find that $y = -2$ for $x = 0$ (the negative value is due to the minus sign before the function), which means it is a minimum point. The curve then has a zero at $x = \frac{\pi}{6}$, a maximum value of 2 at $x = 2(\frac{\pi}{6}) = \frac{\pi}{3}$, a zero at $x = 3(\frac{\pi}{6}) = \frac{\pi}{2}$, and its next value of -2 at $x = 4(\frac{\pi}{6}) = \frac{2\pi}{3}$, and so on. Therefore, we have the table of important values.

x	0	$\frac{\pi}{6}$	$\frac{\pi}{3}$	$\frac{\pi}{2}$	$\frac{2\pi}{3}$	$\frac{5\pi}{6}$	π	$\frac{7\pi}{6}$	$\frac{4\pi}{3}$	$\frac{3\pi}{2}$	$\frac{5\pi}{3}$	$\frac{11\pi}{6}$	2π
y	-2	0	2	0	-2	0	2	0	-2	0	2	0	-2

Using this table and the knowledge of the form of the cosine curve, we sketch the function as shown in Fig. 9-9.

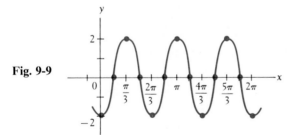

Fig. 9-9

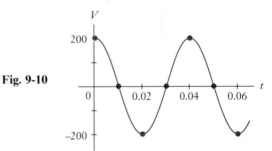

Fig. 9-10

EXAMPLE E

A generator produces a voltage given by $V = 200 \cos 50\pi t$, where t is the time in seconds. Sketch the graph of V as a function of t for $0 \le t \le 0.06$ s.

For this function the amplitude is 200 V and the period is $\frac{2\pi}{50\pi} = 0.04$ s. Since the value of the period is not in terms of π, it is more convenient to use regular decimal units for t when sketching rather than to use units in terms of π as in the previous graphs. Therefore we have the following table.

t (seconds)	0	0.01	0.02	0.03	0.04	0.05	0.06
V (volts)	200	0	-200	0	200	0	-200

The graph of this function is shown in Fig. 9-10.

Exercises 9-2

In Exercises 1 through 20, find the period of each of the given functions.

1. $y = 2 \sin 6x$

2. $y = 4 \sin 2x$

3. $y = 3 \cos 8x$

4. $y = \cos 10x$

5. $y = -2 \sin 12x$

6. $y = -\sin 5x$

7. $y = -\cos 16x$

8. $y = -4 \cos 2x$

9. $y = 5 \sin 2\pi x$

10. $y = 2 \sin 3\pi x$

11. $y = 3 \cos 4\pi x$

12. $y = 4 \cos 10\pi x$

13. $y = 3 \sin \frac{1}{3} x$

14. $y = -2 \sin \frac{2}{5} x$

15. $y = -\frac{1}{2} \cos \frac{2}{3} x$

16. $y = \frac{1}{3} \cos \frac{1}{4} x$

17. $y = 0.4 \sin \frac{2\pi x}{3}$

18. $y = 1.5 \cos \frac{\pi x}{10}$

19. $y = 3.3 \cos \pi^2 x$

20. $y = 2.5 \sin \frac{2x}{\pi}$

See Appendix E for a computer program for sketching the graphs of $y = a \sin bx$ and $y = \sin x$.

In Exercises 21 through 40, sketch the graphs of the given functions. For this, use the functions given for Exercises 1 through 20.

In Exercises 41 through 44, the period is given for a function of the form $y = \sin bx$. Write the function corresponding to the given value of the period.

41. $\frac{\pi}{3}$

42. $\frac{2\pi}{5}$

43. 2

44. 6

In Exercises 45 through 48, sketch the indicated graphs.

45. The standard electric voltage in a 60-Hz alternating-current circuit is given by $V = 170 \sin 120\pi t$, where t is the time in seconds. Sketch the graph of V as a function of t for $0 \le t \le 0.05$ s.

46. The end of a tuning fork moves with a displacement given by $y = 1.60 \cos 460\pi t$, where y is in millimeters and t is in seconds. Sketch the graph of y as a function of t for $0 \le t \le 0.02$ s.

47. The velocity of a piston is given by $v = 450 \cos 3600t$, where v is in inches per second and t is in seconds. Sketch the graph of v as a function of t for $0 \le t \le 0.006$ s.

48. The displacement y of the end of a robot arm for welding is given by $y = 12.75 \sin 0.419t$, where y is in meters and t is in seconds. Sketch the graph of y as a function of t for $0 \le t \le 15$ s.

9-3   Graphs of $y = a \sin (bx + c)$ and $y = a \cos (bx + c)$ ▬▬▬

There is one more important quantity to be discussed in relation to graphing the sine and cosine functions. *This quantity is the* **phase angle** *of the function. In the function* $y = a \sin (bx + c)$, *c represents this phase angle.* Its meaning is illustrated in the following example.

EXAMPLE A ———— Sketch the graph of $y = \sin (2x + \frac{\pi}{4})$.

Note that $c = \frac{\pi}{4}$. This means that in order to obtain the values for the table, we must assume a value of x, multiply it by two, add $\frac{\pi}{4}$ to this value, and then find the sine of this result. In this manner we arrive at the following table.

x	$-\frac{\pi}{8}$	0	$\frac{\pi}{8}$	$\frac{\pi}{4}$	$\frac{3\pi}{8}$	$\frac{\pi}{2}$	$\frac{5\pi}{8}$	$\frac{3\pi}{4}$	$\frac{7\pi}{8}$	π
y	0	0.7	1	0.7	0	-0.7	-1	-0.7	0	0.7

We use the value of $x = -\frac{\pi}{8}$ in the table, for we note that it corresponds to finding $\sin 0$. Using the values listed in the table, we plot the graph of $y = \sin (2x + \frac{\pi}{4})$.
■ See Fig. 9-11.

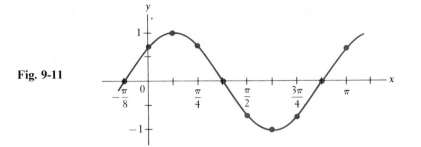

Fig. 9-11

We can see from the table and from the graph in Example A that the curve of

$$y = \sin \left(2x + \frac{\pi}{4} \right)$$

is precisely the same as that of $y = \sin 2x$, except that it is shifted $\frac{\pi}{8}$ units to the left. The effect of c in the equation of $y = a \sin (bx + c)$ is to shift the curve of $y = a \sin bx$ to the left if $c > 0$, and to shift the curve to the right if $c < 0$. Therefore, the amount of this shift is given by $-c/b$. Due to its importance in sketching curves, *the quantity $-c/b$ is called the* **displacement** (*or* **phase shift**).

displacement

Therefore, the results above, combined with the results of Section 9-2, may be used to sketch curves of the functions $y = a \sin (bx + c)$ and $y = a \cos (bx + c)$ where $b > 0$. These are the important quantities to determine:

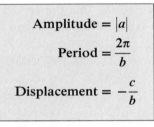

$$\textbf{Amplitude} = |a|$$

$$\textbf{Period} = \frac{2\pi}{b}$$ (9-1)

$$\textbf{Displacement} = -\frac{c}{b}$$

By use of these quantities and the one-fourth period distance, the curves of the sine and cosine functions can be readily sketched. A general illustration of the curve of $y = a \sin (bx + c)$ is shown in Fig. 9-12. Note that **the displacement is negative (to the left) for $c > 0$**, as in Fig. 9-12(a), and that **the displacement is positive (to the right) for $c < 0$**, as in Fig. 9-12(b).

NOTE ▷

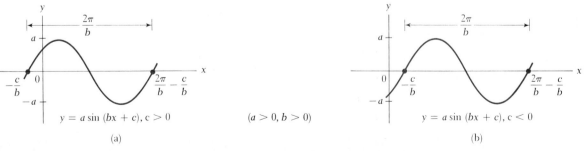

$y = a \sin (bx + c), c > 0$ $(a > 0, b > 0)$ $y = a \sin (bx + c), c < 0$

(a)   (b)

Fig. 9-12

EXAMPLE B ——— Sketch the graph of $y = 2 \sin(3x - \pi)$.

First we note that $a = 2$, $b = 3$, and $c = -\pi$. Therefore, the amplitude is 2, the period is $\frac{2\pi}{3}$, and the displacement is $-(\frac{-\pi}{3}) = \frac{\pi}{3}$.

With this information we can tell that the curve "starts" at $x = \frac{\pi}{3}$ and starts repeating $\frac{2\pi}{3}$ units to the right of this point. (Be sure to grasp this point well.

NOTE ▷ *The period tells how many units there are along the x-axis between such corresponding points.*) One-fourth of the period is $\frac{1}{4}(\frac{2\pi}{3}) = \frac{\pi}{6}$. Therefore, the important values are at $\frac{\pi}{3}, \frac{\pi}{3} + \frac{\pi}{6} = \frac{\pi}{2}, \frac{\pi}{3} + 2(\frac{\pi}{6}) = \frac{2\pi}{3}$, and so on. Therefore, we obtain the table of important values and sketch the graph shown in Fig. 9-13. Extending the curve to the left, we note that—since the period is $\frac{2\pi}{3}$—the curve passes ■ through the origin.

x	0	$\frac{\pi}{6}$	$\frac{\pi}{3}$	$\frac{\pi}{2}$	$\frac{2\pi}{3}$	$\frac{5\pi}{6}$	π
y	0	-2	0	2	0	-2	0

Fig. 9-13

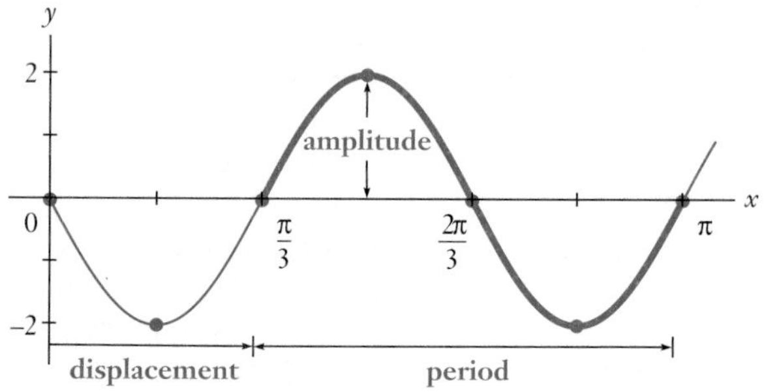

EXAMPLE C ——— Sketch the graph of the function $y = -\cos(2x + \frac{\pi}{6})$.

First we determine that the amplitude is 1, the period is $\frac{2\pi}{2} = \pi$, and the displacement is $-\frac{\pi}{6} \div 2 = -\frac{\pi}{12}$ (to the left, $c > 0$). From these values we construct the following table, remembering that the curve starts repeating π units to the right of $-\frac{\pi}{12}$. From this table we sketch the graph, as shown in Fig. 9-14.

x	$-\frac{\pi}{12}$	$\frac{\pi}{6}$	$\frac{5\pi}{12}$	$\frac{2\pi}{3}$	$\frac{11\pi}{12}$
y	-1	0	1	0	-1

Fig. 9-14

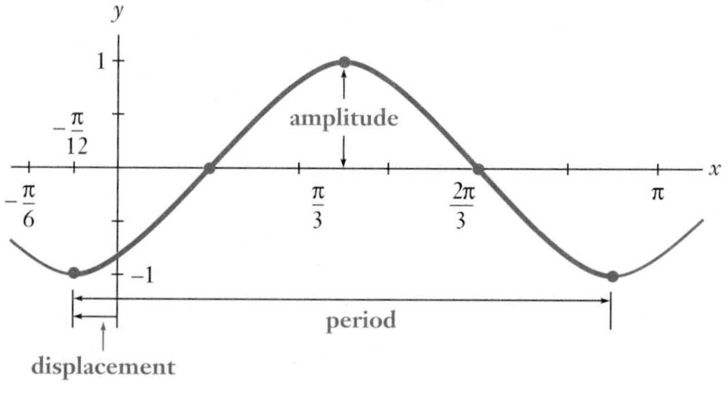

EXAMPLE D ——— Sketch the graph of the function $y = 2 \cos(\frac{1}{2}x - \frac{\pi}{6})$.

From the values $a = 2$, $b = \frac{1}{2}$, and $c = -\frac{\pi}{6}$, we determine that the amplitude is 2, the period is $2\pi \div \frac{1}{2} = 4\pi$, and the displacement is $-(-\frac{\pi}{6}) \div \frac{1}{2} = \frac{\pi}{3}$. From these values we construct the table of values.

x	$\frac{\pi}{3}$	$\frac{4\pi}{3}$	$\frac{7\pi}{3}$	$\frac{10\pi}{3}$	$\frac{13\pi}{3}$
y	2	0	-2	0	2

The graph is shown in Fig. 9-15. We note again that when the coefficient of x ■ is less than 1, the period is greater than 2π.

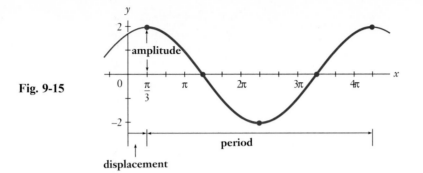

Fig. 9-15

Each of the heavy portions of the graphs in Figs. 9-13, 9-14, and 9-15 is called a **cycle** of the curve. *A cycle is the shortest section of the graph which includes one period.*

The following example illustrates the use of the graph of a trigonometric function in an applied problem.

EXAMPLE E ————— The cross section of a particular water wave is $y = 0.7 \sin (\frac{\pi}{2} x + \frac{\pi}{4})$, where x and y are measured in feet. Sketch two cycles of y vs. x.

From the values $a = 0.7$ ft, $b = \frac{\pi}{2}$ ft^{-1} (this means 1/ft or per foot), and $c = \frac{\pi}{4}$, we can find the amplitude, period, and displacement.

$$\text{amplitude} = 0.7 \text{ ft}, \quad \text{period} = \frac{2\pi}{\frac{\pi}{2}} = 4 \text{ ft}, \quad \text{displacement} = -\frac{\frac{\pi}{4}}{\frac{\pi}{2}} = -0.5 \text{ ft}$$

From these we construct the necessary table of values.

x (ft)	−0.5	0.5	1.5	2.5	3.5	4.5	5.5	6.5	7.5
y (ft)	0.0	0.7	0.0	−0.7	0.0	0.7	0.0	−0.7	0.0

Fig. 9-16

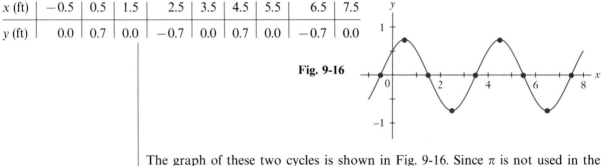

The graph of these two cycles is shown in Fig. 9-16. Since π is not used in the values of x, it is more convenient to use decimal number units for the graph. The negative values of x have the significance of giving points on the wave to the left of the origin. When time is the independent variable, no actual physical meaning is generally given to negative values of t. ∎

Exercises 9-3

In Exercises 1 through 24, determine the amplitude, period, and displacement for each of the functions. Then sketch the graphs of the functions.

1. $y = \sin \left(x - \dfrac{\pi}{6} \right)$ **2.** $y = 3 \sin \left(x + \dfrac{\pi}{4} \right)$ **3.** $y = \cos \left(x + \dfrac{\pi}{6} \right)$ **4.** $y = 2 \cos \left(x - \dfrac{\pi}{8} \right)$

5. $y = 2 \sin \left(2x + \dfrac{\pi}{2} \right)$ **6.** $y = -\sin \left(3x - \dfrac{\pi}{2} \right)$ **7.** $y = -\cos (2x - \pi)$ **8.** $y = 4 \cos \left(3x + \dfrac{\pi}{3} \right)$

9. $y = \dfrac{1}{2} \sin \left(\dfrac{1}{2}x - \dfrac{\pi}{4} \right)$ **10.** $y = 2 \sin \left(\dfrac{1}{4}x + \dfrac{\pi}{2} \right)$ **11.** $y = 3 \cos \left(\dfrac{1}{3}x + \dfrac{\pi}{3} \right)$ **12.** $y = \dfrac{1}{3} \cos \left(\dfrac{1}{2}x - \dfrac{\pi}{8} \right)$

13. $y = \sin \left(\pi x + \dfrac{\pi}{8} \right)$ **14.** $y = -2 \sin (2\pi x - \pi)$ **15.** $y = \dfrac{3}{4} \cos \left(4\pi x - \dfrac{\pi}{5} \right)$

16. $y = 25 \cos \left(3\pi x + \dfrac{\pi}{2} \right)$ **17.** $y = -0.6 \sin (2\pi x - 1)$ **18.** $y = 1.8 \sin \left(\pi x + \dfrac{1}{3} \right)$

19. $y = 40 \cos (3\pi x + 2)$ **20.** $y = 3 \cos (6\pi x - 1)$ **21.** $y = \sin (\pi^2 x - \pi)$

22. $y = -\dfrac{1}{2} \sin \left(2x - \dfrac{1}{\pi} \right)$ **23.** $y = -\dfrac{3}{2} \cos \left(\pi x + \dfrac{\pi^2}{6} \right)$ **24.** $y = \pi \cos \left(\dfrac{1}{\pi}x + \dfrac{1}{3} \right)$

In Exercises 25 through 28, sketch the indicated curves.

25. A wave traveling in a string may be represented by the equation

$$y = A \sin 2\pi \left(\dfrac{t}{T} - \dfrac{x}{\lambda} \right)$$

Here A is the amplitude, t is the time the wave has traveled, x is the distance from the origin, T is the time required for the wave to travel one *wavelength* λ (the Greek letter lambda). Sketch three cycles of the wave for which $A = 2.00$ cm, $T = 0.100$ s, $\lambda = 20.0$ cm, and $x = 5.00$ cm.

26. The electric current i, in microamperes, in a certain circuit is given by $i = 3.8 \cos 2\pi(t + 0.20)$, where t is the time in seconds. Sketch three cycles of this function.

27. A certain satellite circles the earth such that its distance y, in miles north or south (altitude is not considered), from the equator is $y = 4500 \cos (0.025t - 0.25)$, where t is the number of minutes after launch. Sketch two cycles of the graph.

28. In performing a test on a patient, a medical technician used an ultrasonic signal given by the equation $I = A \sin (\omega t + \theta)$. Sketch two cycles of the graph of this function if $A = 5$ nW/m^2, $\omega = 2 \times 10^5$ rad/s, and $\theta = 0.4$.

9-4   Graphs of $y = \tan x$, $y = \cot x$, $y = \sec x$, $y = \csc x$ ▬▬▬

In this section we shall briefly consider the graphs of the other trigonometric functions. We shall establish the basic form of each curve, and from these we shall be able to sketch other curves for these functions.

Considering the values and signs of the trigonometric functions as established in Chapter 7, we set up the following table for the function $y = \tan x$. The graph is shown in Fig. 9-17.

x	0	$\frac{\pi}{6}$	$\frac{\pi}{3}$	$\frac{\pi}{2}$	$\frac{2\pi}{3}$	$\frac{5\pi}{6}$	π
y	0	0.6	1.7	*	-1.7	-0.6	0

x	$\frac{7\pi}{6}$	$\frac{4\pi}{3}$	$\frac{3\pi}{2}$	$\frac{5\pi}{3}$	$\frac{11\pi}{6}$	2π
y	0.6	1.7	*	-1.7	-0.6	0

* Undefined.

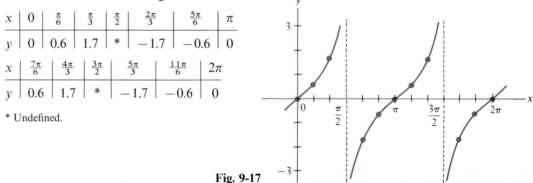

Fig. 9-17

Since the curve is not defined for $x = \frac{\pi}{2}$, $x = \frac{3\pi}{2}$, and so forth, we look at the table and note that the value of $\tan x$ becomes very large when x approaches the value $\frac{\pi}{2}$. We must keep in mind, however, that there is no point on the curve corresponding to $x = \frac{\pi}{2}$. We note that *the period of the tangent curve is π.* This differs from the period of the sine and cosine functions.

By following the same procedure, we can set up tables for the graphs of the other functions. In Figs. 9-18 through 9-21, we present the graphs of $y = \tan x$, $y = \cot x$, $y = \sec x$, and $y = \csc x$ (the graph of $y = \tan x$ is shown again to show it more completely). We see from these graphs that the period of $y = \tan x$ and $y = \cot x$ is π, and that of $y = \sec x$ and $y = \csc x$ is 2π. *The dashed lines in these figures are* **asymptotes** (see Sections 2-4 and 20-6). The functions are not defined for these values of x, which means that the domains are all real numbers, except for these values of x.

We see that the ranges of $y = \tan x$ and $y = \cot x$ are all real numbers, but that the ranges of $y = \sec x$ and $y = \csc x$ do not include real numbers between -1 and $+1$.

To sketch functions such as $y = a \sec x$, we first sketch $y = \sec x$, then multiply each y-value by a. *Here a is not an amplitude,* since the ranges of these functions are not limited in the way they are for the sine and cosine functions.

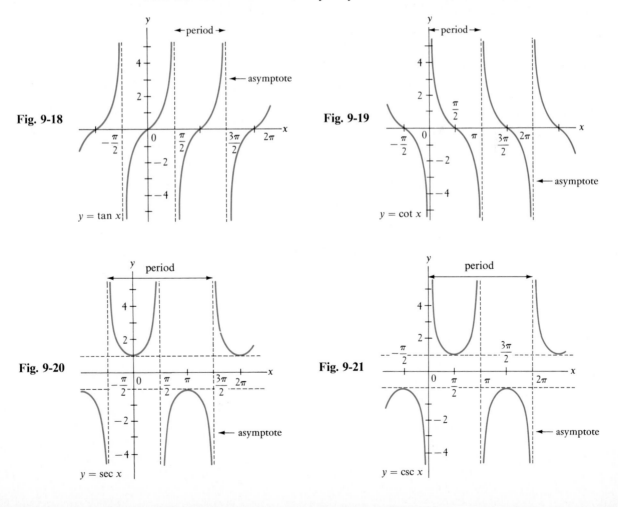

Fig. 9-18 $y = \tan x$

Fig. 9-19 $y = \cot x$

Fig. 9-20 $y = \sec x$

Fig. 9-21 $y = \csc x$

EXAMPLE A _____ Sketch the graph of $y = 2 \sec x$.

First we sketch in $y = \sec x$, shown as the light curve in Fig. 9-22. Now we multiply the y-values of the secant function by 2 (approximately, of course). In ■ this way we obtain the desired curve, shown as the curve in color in Fig. 9-22.

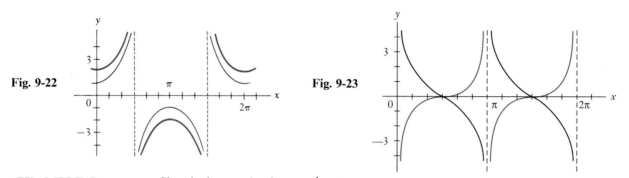

Fig. 9-22

Fig. 9-23

EXAMPLE B _____ Sketch the graph of $y = -\frac{1}{2} \cot x$.

We sketch in $y = \cot x$, shown as the light curve in Fig. 9-23. Now we multiply each y-value by $-\frac{1}{2}$. The effect of the negative sign is to invert the curve. ■ The resulting curve is shown as the curve in color in Fig. 9-23.

By knowing the graphs of the sine, cosine, and tangent functions, it is possible to graph the other three functions. This is due to the reciprocal relationships among the functions. In Section 3-3, when we were discussing how to find the values of the cotangent, secant, and cosecant functions on a calculator, we noted that $\csc x$ and $\sin x$ are reciprocals, $\sec x$ and $\cos x$ are reciprocals, and $\cot x$ and $\tan x$ are reciprocals. These relationships come as a result of the definitions of the various functions. We can show these reciprocal relationships by

$$\csc x = \frac{1}{\sin x} \qquad \sec x = \frac{1}{\cos x} \qquad \cot x = \frac{1}{\tan x} \tag{9-2}$$

Thus, to sketch $y = \cot x$, $y = \sec x$, or $y = \csc x$, we sketch the corresponding reciprocal function, and from this graph determine the necessary values.

EXAMPLE C _____ Sketch the graph of $y = \csc x$.

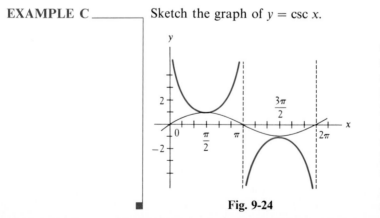

Fig. 9-24

We first sketch in the graph of $y = \sin x$ (light curve). Where $\sin x$ is 1, $\csc x$ will also be 1, since $1/1 = 1$. Where $\sin x$ is 0, $\csc x$ is undefined, since $1/0$ is undefined. Where $\sin x$ is 0.5, $\csc x$ is 2, since $1/0.5 = 2$. Thus, as $\sin x$ becomes larger, $\csc x$ becomes smaller, and as $\sin x$ becomes smaller, $\csc x$ becomes larger. The two functions always have the same sign. We sketch the graph of $y = \csc x$ from   this   information,   as shown by the curve in color in Fig. 9-24.

We can use the reciprocal relationships to help in making out a table of key values in order to plot the graphs of the tangent, cotangent, secant, and cosecant functions. This is illustrated in the following example.

EXAMPLE D ———— Plot the graph of $y = 2 \sec (2x - \frac{\pi}{4})$.

Since the secant function is the reciprocal of the cosine function, we can find the period and displacement of $y = \cos (2x - \frac{\pi}{4})$ to determine key values for the graph. The period is $\frac{2\pi}{2} = \pi$ and the displacement is $-\frac{-\pi/4}{2} = \frac{\pi}{8}$. Thus, making a table for $y = \cos (2x - \frac{\pi}{4})$, then multiplying the reciprocals of these values by 2, we get key values for $y = 2 \sec (2x - \frac{\pi}{4})$.

x	$\frac{\pi}{8}$	$\frac{3\pi}{8}$	$\frac{5\pi}{8}$	$\frac{7\pi}{8}$	$\frac{9\pi}{8}$
$\cos (2x - \frac{\pi}{4})$	1	0	-1	0	1
$2 \sec (2x - \frac{\pi}{4})$	2	undef.	-2	undef.	2

Knowing the basic shape of the curve of the secant function and these values, we get the graph in Fig. 9-25.

Fig. 9-25

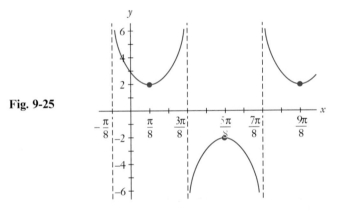

Exercises 9-4

In Exercises 1 through 4, fill in the following table for each function and then plot the curve from these points.

x	$-\frac{\pi}{2}$	$-\frac{\pi}{3}$	$-\frac{\pi}{4}$	$-\frac{\pi}{6}$	0	$\frac{\pi}{6}$	$\frac{\pi}{4}$	$\frac{\pi}{3}$	$\frac{\pi}{2}$	$\frac{2\pi}{3}$	$\frac{3\pi}{4}$	$\frac{5\pi}{6}$	π
y													

1. $y = \tan x$ **2.** $y = \cot x$ **3.** $y = \sec x$ **4.** $y = \csc x$

In Exercises 5 through 12, sketch the curves of the given functions by use of the basic curve forms (Figs. 9-18, 9-19, 9-20, 9-21). See Examples A and B.

5. $y = 2 \tan x$ **6.** $y = 3 \cot x$ **7.** $y = \frac{1}{2} \sec x$ **8.** $y = \frac{3}{2} \csc x$

9. $y = -2 \cot x$ **10.** $y = -\tan x$ **11.** $y = -3 \csc x$ **12.** $y = -\frac{1}{2} \sec x$

In Exercises 13 through 20, plot the graphs by first making an appropriate table for $0 \le x \le \pi$.

13. $y = \tan 2x$ **14.** $y = 2 \cot 3x$ **15.** $y = \frac{1}{2} \sec 3x$ **16.** $y = 4 \csc 2x$

17. $y = 2 \cot \left(2x + \frac{\pi}{6}\right)$ **18.** $y = \tan \left(3x - \frac{\pi}{2}\right)$ **19.** $y = \csc \left(3x - \frac{\pi}{3}\right)$ **20.** $y = 3 \sec \left(2x + \frac{\pi}{4}\right)$

In Exercises 21 through 24, sketch the given curves by first sketching the appropriate reciprocal function. See Example C.

21. $y = \sec x$ **22.** $y = \cot x$ **23.** $y = \csc 2x$ **24.** $y = \sec \pi x$

In Exercises 25 through 28, construct the appropriate graphs.

25. A drafting student draws a circle through the three vertices of a right triangle. The hypotenuse of the triangle is the diameter d of the circle, and from Fig. 9-26, we see that $d = a \sec \theta$. Sketch the graph of d as a function of θ for $a = 3.00$ in.

26. At a distance x from the base of a building 200 m high, the angle of elevation θ of the top of the building can be found from the equation $x = 200 \cot \theta$. Sketch x as a function of θ.

27. A mechanism with two springs is shown in Fig. 9-27, where point A is restricted to move horizontally. From the law of sines we see that $b = (a \sin B) \csc A$. Sketch the graph of b as a function of A for $a = 4.00$ cm and $B = \frac{\pi}{4}$.

28. In a laser experiment, two mirrors move horizontally in equal and opposite distances from point A. The laser path from and to point B is shown in Fig. 9-28. From the figure we see that $x = a \tan \theta$. Sketch the graph of x as a function of θ for $a = 5.00$ cm.

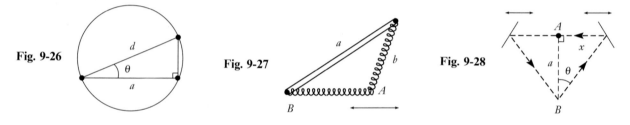

Fig. 9-26   **Fig. 9-27**   **Fig. 9-28**

9-5   Applications of the Trigonometric Graphs

There are a great many applications of the trigonometric functions and their graphs, a few of which have been indicated in the exercises of the previous sections. In this section we shall introduce an important physical concept and indicate some of the technical applications.

In Section 7-4 we discussed the velocity of an object moving in a circular path. When this object moves with constant velocity, its *projection* on a diameter moves with what is known as **simple harmonic motion.** For example, this could be the motion of the shadow of an object which is moving around a circle. Another example is the vertical position of the end of a spoke of a wheel in motion. A different illustration of simple harmonic motion is that of the displacement of a weight moving up and down at the end of a spring.

The following example illustrates the simple harmonic motion of a projection on the diameter of a particle moving in a circular path.

EXAMPLE A _____ When we consider Fig. 9-29, let us assume that motion starts with the end of the radius at $(R, 0)$ and that it is moving with constant angular velocity ω. This means that the length of the projection of the radius along the y-axis is given by $d = R \sin \theta$. The length of this projection is shown for a few different positions of the end of the radius. Since $\theta/t = \omega$, or $\theta = \omega t$, we have

$$d = R \sin \omega t$$

(9-3)

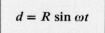

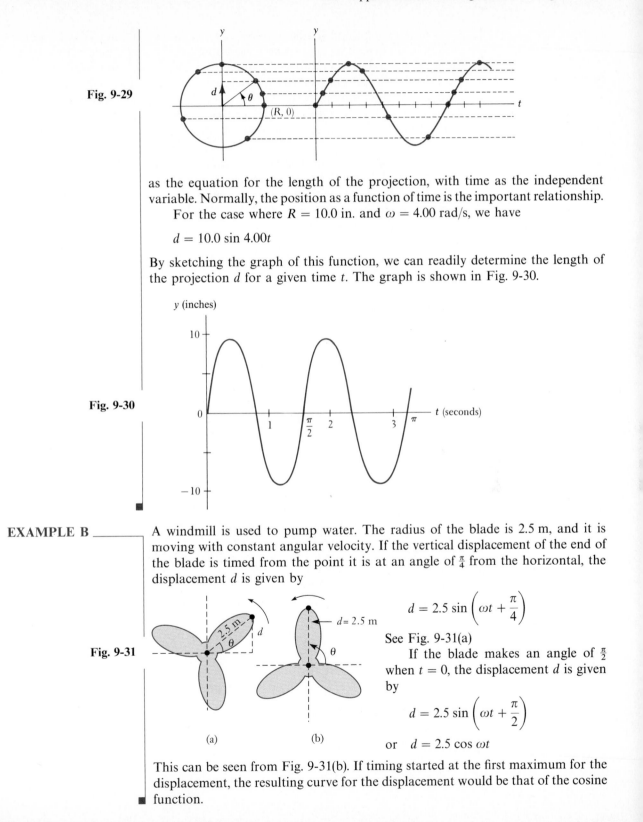

Fig. 9-29

as the equation for the length of the projection, with time as the independent variable. Normally, the position as a function of time is the important relationship.

For the case where $R = 10.0$ in. and $\omega = 4.00$ rad/s, we have

$$d = 10.0 \sin 4.00t$$

By sketching the graph of this function, we can readily determine the length of the projection d for a given time t. The graph is shown in Fig. 9-30.

Fig. 9-30

EXAMPLE B

A windmill is used to pump water. The radius of the blade is 2.5 m, and it is moving with constant angular velocity. If the vertical displacement of the end of the blade is timed from the point it is at an angle of $\frac{\pi}{4}$ from the horizontal, the displacement d is given by

$$d = 2.5 \sin\left(\omega t + \frac{\pi}{4}\right)$$

See Fig. 9-31(a)

If the blade makes an angle of $\frac{\pi}{2}$ when $t = 0$, the displacement d is given by

$$d = 2.5 \sin\left(\omega t + \frac{\pi}{2}\right)$$

or $d = 2.5 \cos \omega t$

Fig. 9-31

(a)   (b)

This can be seen from Fig. 9-31(b). If timing started at the first maximum for the displacement, the resulting curve for the displacement would be that of the cosine function.

Other examples of simple harmonic motion are (1) the movement of a pendulum bob through its arc (a very close approximation to simple harmonic motion), (2) the motion of an object "bobbing" in the water, and (3) the movement of the end of a vibrating rod (which we hear as sound). Other phenomena which give rise to equations just like those for simple harmonic motion are found in the fields of optics, sound, and electricity. Such phenomena have the same mathematical form because they result from vibratory motion or motion in a circle.

EXAMPLE C　A very important use of the trigonometric curves arises in the study of alternating current, which is caused by the motion of a wire passing through a magnetic field. If this wire is moving in a circular path, with angular velocity ω, the current i in the wire at time t is given by an equation of the form

$$i = I_m \sin (\omega t + \alpha)$$

where I_m is the maximum current attainable and α is the phase angle. The current may be represented by a sine wave, as in the following example. ∎

EXAMPLE D　In Example C, given that $I_m = 6.00$ A, $\omega = 120\pi$ rad/s, and $\alpha = \pi/6$, we have the equation

$$i = 6.00 \sin \left(120\pi t + \frac{\pi}{6} \right)$$

From this equation we see that the amplitude is 6.00 A, the period is $\frac{1}{60}$ s, and the displacement is $-\frac{1}{720}$ s. From these values we draw the graph as shown in Fig. 9-32. Since the current takes on both positive and negative values, we conclude that it moves alternately in one direction and then the other.

Fig. 9-32

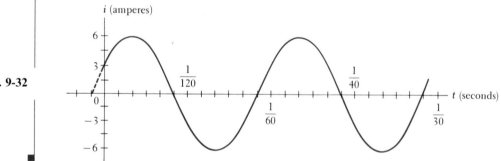

It is a common practice to express the rate of rotation in terms of *the frequency f, the number of cycles per second,* rather than directly in terms of the angular velocity ω, the number of radians per second. *The unit for frequency is the hertz (Hz), and 1 Hz = 1 cycle/s.* Since there are 2π rad in one cycle, we have

$$\omega = 2\pi f$$

(9-4)

EXAMPLE E ⎯⎯⎯⎯⎯ In Example D, we showed the angular velocity ω to be 120π rad/s. The corresponding frequency f is

$$f = \frac{120\pi}{2\pi} = 60 \text{ Hz}$$

■ This means that 120π rad/s corresponds to 60 cycles/s.

Exercises 9-5

In Exercises 1 and 2, draw two cycles of the curve of the projection of Example A as a function of time for the given values.

1. $R = 2.40$ cm, $\omega = 2.00$ rad/s

2. $R = 1.80$ ft, $f = 0.250$ Hz

In Exercises 3 and 4, a point on the edge of a cam is 8.30 cm from the center of rotation. The cam is rotating with constant angular velocity, and the vertical displacement $d = 8.30$ cm for $t = 0$ s. Draw two cycles of d as a function of t for the given values.

3. $f = 3.20$ Hz

4. $\omega = 3.20$ rad/s

In Exercises 5 and 6, a satellite is orbiting the earth such that its displacement D north of the equator (or south if D is negative) is given by $D = A \sin(\omega t + \alpha)$. Sketch two cycles of D as a function of t for the given values.

5. $A = 500$ mi, $\omega = 3.60$ rad/h, $\alpha = 0$

6. $A = 850$ km, $f = 1.6 \times 10^{-4}$ Hz, $\alpha = \frac{\pi}{3}$

In Exercises 7 and 8, for an alternating-current circuit in which the voltage e is given by $e = E \cos(\omega t + \alpha)$, draw two cycles of the voltage as a function of time for the given values.

7. $E = 170$ V, $f = 60.0$ Hz, $\alpha = -\frac{\pi}{3}$

8. $E = 80$ V, $\omega = 377$ rad/s, $\alpha = \frac{\pi}{2}$

In Exercises 9 and 10, refer to the wave in the string described in Exercise 25 of Section 9-3. The displacement y of a point on the string in given by

$$y = A \sin 2\pi \left(\frac{t}{T} - \frac{x}{\lambda} \right)$$

We see that each point on the string moves with simple harmonic motion. Draw two cycles of y as a function of t for the given values.

9. $A = 3.20$ cm, $T = 0.050$ s, $\lambda = 40.0$ cm, $x = 5.00$ cm

10. $A = 1.45$ in., $T = 0.250$ s, $\lambda = 24.0$ in., $x = 20.0$ in.

In Exercises 11 and 12, the air pressure within a plastic container changes above and below the external atmospheric pressure by $p = p_0 \sin 2\pi ft$. Draw two cycles of p as a function of t for the given values.

11. $p_0 = 2.80$ lb/in.², $f = 2.30$ Hz

12. $p_0 = 45.0$ kPa, $f = 0.450$ Hz

In Exercises 13 through 16, draw the required curves.

13. The displacement of the end of a vibrating rod is given by $y = 1.50 \cos 200\pi t$. Sketch two cycles of y (in cm) as a function of t (in seconds).

14. A simple pendulum is started by giving it a velocity from its equilibrium position. The angle θ between the vertical and the pendulum is given by $\theta = \theta_0 \sin \sqrt{g/\ell}\, t$, where θ_0 is the amplitude, g is the acceleration due to gravity, ℓ is the length of the pendulum, and t is the time of motion. Draw two cycles of θ as a function of t for the values $\theta_0 = 0.100$ rad, $g = 32.0$ ft/s², and $\ell = 2.00$ ft.

15. The signal received by a radio is given by $e = 0.014 \cos(2\pi ft + \frac{\pi}{4})$, where e is in volts and f is in hertz. Draw two cycles of e as a function of t for a station which is broadcasting on a frequency of $f = 950$ kHz ("95" on the AM radio dial).

16. The acoustical intensity of a sound wave is given by $I = A \cos(2\pi ft - \alpha)$, where f is the frequency of the sound. Draw two cycles of I as a function of t if $A = 0.027$ W/cm², $f = 240$ Hz, and $\alpha = 0.80$.

9-6 Composite Trigonometric Curves

Many applications involve functions which in themselves are a combination of two or more simpler functions. In this section we discuss methods by which the curve of such a function can be found by combining values from the simpler functions.

EXAMPLE A ———

Sketch the graph of $y = 2 + \sin 2x$.

This function is the sum of the simpler functions $y_1 = 2$ and $y_2 = \sin 2x$. We may find key values for y by adding 2 to each of the key values of $y_2 = \sin 2x$.

For $y_2 = \sin 2x$, the amplitude is 1 and the period is $\frac{2\pi}{2} = \pi$. Since $y = y_1 + y_2 = 2 + \sin 2x$, we obtain the values in the following table and sketch the graph in Fig. 9-33.

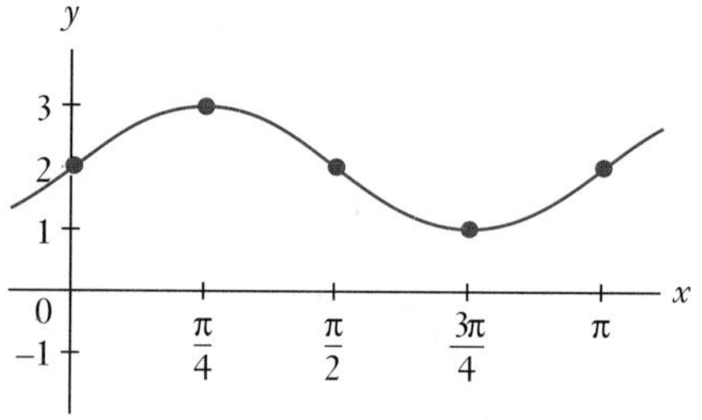

Fig. 9-33

x	0	$\frac{\pi}{4}$	$\frac{\pi}{2}$	$\frac{3\pi}{4}$	π
$\sin 2x$	0	1	0	-1	0
$2 + \sin 2x$	2	3	2	1	2

addition of ordinates

Another way to find the resulting graph is to *first sketch the two simpler functions and then add the y-values graphically. This method is called* **addition of ordinates** *and is illustrated in the following examples.*

EXAMPLE B ———

Sketch the graph of $y = 2 \cos x + \sin 2x$.

On the same set of coordinate axes we sketch the curves $y = 2 \cos x$ and $y = \sin 2x$. These are shown as dashed and solid light curves in Fig. 9-34. For various values of x, we determine the distance above or below the x-axis of each curve and add these distances, noting that those above the axis are positive and those below the axis are negative. We thereby graphically **add** the y-values of these two curves for these values of x to obtain the points on the resulting curve shown as a curve in color in Fig. 9-34.

For example, for the x-value at A we add the two lengths shown (side-by-side for clarity) to get the length for y. At B we see that both lengths are negative, and the value for y is the sum of these two negative values. At C one length is positive and one is negative, and we must subtract the lower length from the upper one to get the length for y.

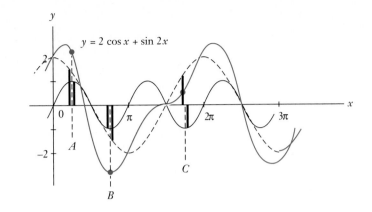

Fig. 9-34

We add (or subtract) these lengths at a sufficient number of x-values to obtain the proper representation. Some points are easily found. Where one curve crosses the x-axis, its y-value is zero, and therefore the resulting curve has its point on the other curve for this value of x. In this example, $\sin 2x$ is zero at $x = 0$, $\frac{\pi}{2}$, π, and so forth. We see that points on the resulting curve lie on the curve of $2 \cos x$. We should also add the values where each curve is at its maximum or minimum values. In this case, $\sin 2x$ equals 1 at $\frac{\pi}{4}$, and the two y-values should be added together here to get a point on the resulting curve. At $x = \frac{5\pi}{4}$, we must

NOTE ▷ take care in adding the values, since **$\sin 2x$ is positive and $2 \cos x$ is negative.** ■ Reasonable care and accuracy are necessary to obtain a proper resulting curve.

EXAMPLE C — Sketch the graph of $y = \frac{x}{2} - \cos x$.

The method of addition of ordinates is applicable regardless of the kinds of functions being added. Here we note that $y = \frac{x}{2}$ is a straight line and that it is to be combined with a trigonometric curve.

We could graph the functions $y = \frac{x}{2}$ and $y = \cos x$ and then subtract the

NOTE ▷ ordinates of $y = \cos x$ from the ordinates of $y = \frac{x}{2}$. But **it is easier and far less confusing to add values,** so we shall sketch $y = \frac{x}{2}$ and $y = -\cos x$ and add the ordinates to obtain points on the resulting curve. These graphs are shown as dashed curves in Fig. 9-35. The important points on the resulting curve are obtained by using the values of x corresponding to the zeros and amplitude values of $y = -\cos x$.

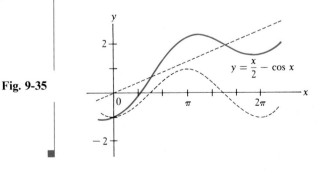

Fig. 9-35

EXAMPLE D _____ Sketch the graph of $y = \cos \pi x - 2 \sin 2x$.

The curves of $y = \cos \pi x$ and $y = -2 \sin 2x$ are shown as solid light and dashed curves in Fig. 9-36. Points for the resulting curve are found primarily at the x-values where each of the curves has its zero or amplitude values. Again, special care should be taken where one of the curves is negative and the other is positive.

Fig. 9-36

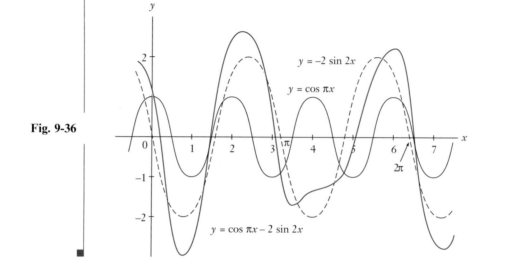

Another important application of trigonometric curves is made when they are added at **right angles.** The methods for doing this are shown in the following examples.

EXAMPLE E _____ Plot the graph for which the values of x and y are given by the equations $y = \sin 2\pi t$ and $x = 2 \cos \pi t$. *(Equations given in this form, x and y in terms of a third variable, are called* **parametric equations.***)*

Since both x and y are given in terms of t, by assuming values for t we may find corresponding values of x and y, and use these values to plot the resulting points (see Fig. 9-37).

t	0	$\frac{1}{4}$	$\frac{1}{2}$	$\frac{3}{4}$	1	$\frac{5}{4}$	$\frac{3}{2}$	$\frac{7}{4}$	2	$\frac{9}{4}$
x	2	1.4	0	-1.4	-2	-1.4	0	1.4	2	1.4
y	0	1	0	-1	0	1	0	-1	0	1
Point number	1	2	3	4	5	6	7	8	9	10

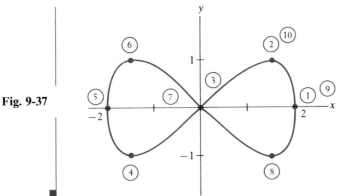

Fig. 9-37

Since x and y are trigonometric functions of a third variable t and since the x- and y-axes are at right angles, values of x and y obtained in this manner result in a combination of two trigonometric curves at right angles. *Figures obtained in this manner are called* **Lissajous figures.**

Lissajous figures

In practice, Lissajous figures can be shown by applying different voltages to an ***oscilloscope*** and displaying the electric signals on a screen similar to that on a television set.

EXAMPLE F

See the chapter introduction.

On an oscilloscope, the curve would result when two electric signals are used. The first would have twice the amplitude and one-half the frequency of the other.

If we place a circle on the x-axis and another on the y-axis, we may represent the coordinates (x, y) for the curve of Example E by the lengths of the projections (see Example A of Section 9-5) of a point moving around each circle. A careful study of Fig. 9-38 will clarify this. We note that the radius of the circle giving the x-values is 2, whereas the radius of the other is 1. This is due to the manner in which x and y are defined. Also, due to the definitions, the point revolves around the y-circle twice as fast as the corresponding point revolves around the x-circle.

Fig. 9-38

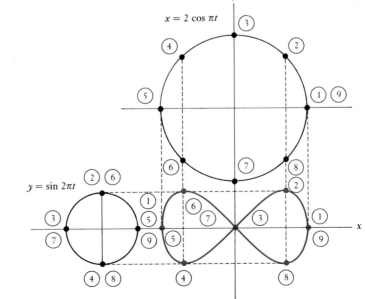

EXAMPLE G

Plot the Lissajous figure for which the x- and y-values are given by the equations $x = 2 \sin 3t$ and $y = 3 \sin \left(t + \frac{\pi}{3}\right)$.

Since values of t which are multiples of π give convenient values of x and y, the table is constructed with these values of t. Fig. 9-39 shows the graph.

t	x	y	Point Number
0	0	2.6	1
$\frac{\pi}{6}$	2	3	2
$\frac{\pi}{3}$	0	2.6	3
$\frac{\pi}{2}$	-2	1.5	4
$\frac{2\pi}{3}$	0	0	5
$\frac{5\pi}{6}$	2	-1.5	6
π	0	-2.6	7
$\frac{7\pi}{6}$	-2	-3	8
$\frac{4\pi}{3}$	0	-2.6	9
$\frac{3\pi}{2}$	2	-1.5	10
$\frac{5\pi}{3}$	0	0	11
$\frac{11\pi}{6}$	-2	1.5	12
2π	0	2.6	13

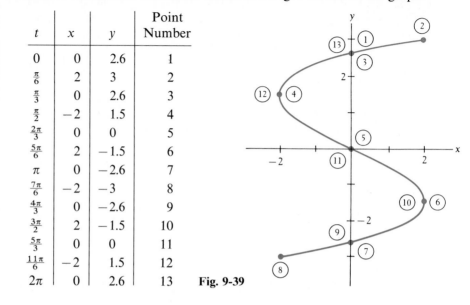

Fig. 9-39

Exercises 9-6

In Exercises 1 through 20, sketch the curves of the given functions.

1. $y = 1 + \sin x$

2. $y = 3 - 2 \cos x$

3. $y = \frac{1}{2} - \cos x$

4. $y = x - \sin x$

5. $y = \frac{1}{3}x + \sin 2x$

6. $y = \frac{1}{4}x + \cos 3x$

7. $y = \frac{1}{10}x^2 - \sin \pi x$

8. $y = \frac{1}{x^2 + 1} - \cos \pi x$

9. $y = \sin x + \cos x$

10. $y = \sin x + \sin 2x$

11. $y = \sin x - \sin 2x$

12. $y = \cos 3x - \sin x$

13. $y = 2 \cos 2x + 3 \sin x$

14. $y = \frac{1}{2} \sin 4x + \cos 2x$

15. $y = 2 \sin x - \cos x$

16. $y = \sin \frac{x}{2} - \sin x$

17. $y = 2 \cos 4x - \cos \left(x - \frac{\pi}{4}\right)$

18. $y = \sin \pi x - \cos 2x$

19. $y = 2 \sin \left(2x - \frac{\pi}{6}\right) + \cos \left(2x + \frac{\pi}{3}\right)$

20. $y = 3 \cos 2\pi x + \sin \frac{\pi}{2} x$

In Exercises 21 through 28, plot the Lissajous figures.

21. $y = \sin t, \; x = \sin t$

22. $y = 2 \cos t, \; x = \cos t$

23. $y = \sin \pi t, \; x = \cos \pi t$

24. $y = \sin 2t, \; x = \cos \left(t + \frac{\pi}{4}\right)$

25. $y = 2 \sin \pi t, \; x = \cos \pi \left(t + \frac{1}{6}\right)$

26. $y = \sin^2 \pi t, \; x = \cos 2\pi t$

27. $y = \cos 2t, \; x = 2 \cos 3t$

28. $y = 3 \sin 3\pi t, \; x = 2 \sin \pi t$

In Exercises 29 through 36, sketch the appropriate figures.

29. A computer analysis showed that the average daily temperature, in degrees Celsius, throughout the year at a certain weather station was given by $T = 18 - 10 \cos 2\pi (t - 0.10)$, where t is measured in years. Sketch the graph of T as a function of t for a two-year period, starting at $t = 0$.

30. Data showed that the bird population in a certain remote area was given by $P = 8000 + 1500 \sin 3t$, where t is measured in years. Sketch the graph of P as a function of t for the first five years for which the data were taken.

31. The vertical displacement of water in a wave is given by $y = 3.0 \cos 0.2t + 1.0 \sin 0.4t$, where y is measured in feet and t is measured in seconds. Sketch the graph of y as a function of t for the first 40 s.

32. In optics, two waves are said to interfere destructively if, when they pass through the same medium, the amplitude of the resulting wave is zero. Sketch the curve of $y = \sin x + \cos (x + \frac{\pi}{2})$, and determine whether or not it would represent destructive interference of two waves.

33. The electric current in a certain circuit is given by $i = 0.32 + 0.50 \sin t - 0.20 \cos 2t$, where i is in milliamperes and t is in milliseconds. Sketch two cycles of i as a function of t.

34. The available solar energy depends on the amount of sunlight, and the available time in a day for sunlight depends on the time of the year. An approximate correction factor, in minutes, to standard time is $C = 10 \sin \frac{1}{29}(n - 80) - 7.5 \cos \frac{1}{58}(n - 80)$, where n is the number of the day of the year. Sketch C as a function of n.

35. Two signals are sent to an oscilloscope and are seen on the oscilloscope as being at right angles. The equations for the displacements of these signals are $x = 4 \cos \pi t$ and $y = 2 \sin 3\pi t$. Sketch the figure which appears on the oscilloscope.

36. In the study of optics, light is said to be elliptically polarized if certain optic vibrations are out of phase. These may be represented by Lissajous figures. Determine the Lissajous figure for two waves of light given by $w_1 = \sin \omega t$, $w_2 = \sin (\omega t + \frac{\pi}{4})$.

9-7   Chapter Equations, Review Exercises, and Practice Test

Chapter Equations

For the graphs of $y = a \sin (bx + c)$ and $y = a \cos (bx + c)$

$$\text{Amplitude} = |a|$$

$$\text{Period} = \frac{2\pi}{b} \tag{9-1}$$

$$\text{Displacement} = -\frac{c}{b}$$

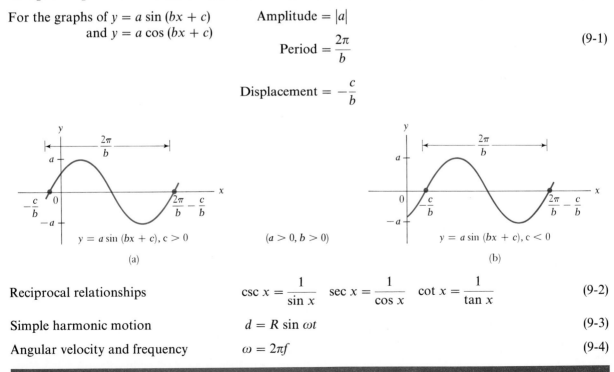

$y = a \sin (bx + c), c > 0$

(a)

$(a > 0, b > 0)$

$y = a \sin (bx + c), c < 0$

(b)

| Reciprocal relationships | $\csc x = \dfrac{1}{\sin x}$ $\sec x = \dfrac{1}{\cos x}$ $\cot x = \dfrac{1}{\tan x}$ | (9-2) |

Simple harmonic motion $d = R \sin \omega t$ $\qquad\qquad$ (9-3)

Angular velocity and frequency $\omega = 2\pi f$ $\qquad\qquad$ (9-4)

Review Exercises

In Exercises 1 through 28, sketch the curves of the given trigonometric functions.

1. $y = \dfrac{2}{3} \sin x$

2. $y = -4 \sin x$

3. $y = -2 \cos x$

4. $y = 2.3 \cos x$

5. $y = 2 \sin 3x$

6. $y = 4.5 \sin 12x$

7. $y = 2 \cos 2x$

8. $y = 24 \cos 6x$

9. $y = 3 \cos \dfrac{1}{3} x$

10. $y = 3 \sin \dfrac{1}{2} x$

11. $y = \sin \pi x$

12. $y = 3 \sin 4\pi x$

13. $y = 5 \cos 2\pi x$

14. $y = -\cos 6\pi x$

15. $y = -0.5 \sin \dfrac{\pi}{6} x$

16. $y = 8 \sin \dfrac{\pi}{4} x$

17. $y = 2 \sin \left(3x - \dfrac{\pi}{2}\right)$

18. $y = 3 \sin \left(\dfrac{x}{2} + \dfrac{\pi}{2}\right)$

19. $y = -2 \cos (4x + \pi)$

20. $y = 0.8 \cos \left(\dfrac{x}{6} - \dfrac{\pi}{2}\right)$

21. $y = -\sin \left(\pi x + \dfrac{\pi}{6}\right)$

22. $y = 2 \sin (3\pi x - \pi)$

23. $y = 8 \cos \left(4\pi x - \dfrac{\pi}{2}\right)$

24. $y = 3 \cos (2\pi x + \pi)$

25. $y = 3 \tan x$

26. $y = \dfrac{1}{4} \sec x$

27. $y = -\dfrac{1}{3} \csc x$

28. $y = -5 \cot x$

In Exercises 29 through 36, sketch the given curves by the method of addition of ordinates.

29. $y = 2 + \dfrac{1}{2} \sin 2x$

30. $y = \dfrac{1}{2} x - \cos \dfrac{1}{3} x$

31. $y = \sin 2x + 3 \cos x$

32. $y = \sin 3x + 2 \cos 2x$

33. $y = 2 \sin x - \cos 2x$

34. $y = \sin 3x - 2 \cos x$

35. $y = \cos \left(x + \dfrac{\pi}{4}\right) - 2 \sin 2x$

36. $y = 2 \cos \pi x + \cos (2\pi x - \pi)$

In Exercises 37 through 40, give the specific form of the indicated equation by evaluating a, b, and c through an inspection of the indicated curve

37. $y = a \sin (bx + c)$
(Figure 9-40)

38. $y = a \cos (bx + c)$
(Figure 9-40)

39. $y = a \cos (bx + c)$
(Figure 9-41)

40. $y = a \sin (bx + c)$
(Figure 9-41)

Fig. 9-40

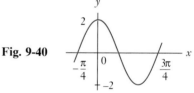

In Exercises 41 through 44, plot the Lissajous figures.

41. $y = 2 \sin \pi t, \ x = -\cos 2\pi t$

42. $y = \sin t, \ x = \sin \left(t + \dfrac{\pi}{6}\right)$

43. $y = \cos \pi t, \ x = \cos \left(2\pi t + \dfrac{\pi}{4}\right)$

44. $y = \cos \left(2t + \dfrac{\pi}{3}\right), \ x = \cos \left(t - \dfrac{\pi}{6}\right)$

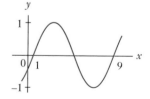

Fig. 9-41

In Exercises 45 through 60, sketch the appropriate figures.

45. The range R of a rocket is given by

$$R = \dfrac{v_0^2 \sin 2\theta}{g}$$

Fig. 9-42

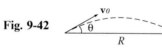

See Fig. 9-42. Sketch R as a function of θ for $v_0 = 1000$ m/s and $g = 9.8$ m/s^2.

46. A certain object is oscillating at the end of a spring. The displacement as a function of time is given by the relation $y = 0.200 \cos 8t$, where y is measured in meters and t in seconds. Plot the graph of y vs. t.

47. The velocity v, in centimeters per second, of a piston in a certain engine is given by $v = \omega D \cos \omega t$, where ω is the angular velocity of the crankshaft in radians per second and t is the time in seconds. Sketch the graph of v vs. t if the engine is at 3000 r/min and $D = 3.6$ cm.

48. A light wave for the color yellow can be represented by the equation $y = A \sin 3.4 \times 10^{15} t$. With A as a constant, sketch two cycles of y as a function of t, in seconds.

49. The displacement y of the end of a gear tooth is given by $y = A \sin (6t + 0.5)$. Sketch the graph of two cycles of y as a function of t, in seconds, for $A = 5.2$ cm.

50. A circular disk suspended by a thin wire attached to the center at one of its flat faces is twisted through an angle θ. Torsion in the wire tends to turn the disk back in the opposite direction (thus the name *torsion pendulum* is given to this device). The angular displacement as a function of time is given by $\theta = \theta_0 \cos (\omega t + \alpha)$, where θ_0 is the maximum angular displacement, ω is a constant which depends on the properties of the disk and wire, and α is the phase angle. Plot the graph of θ vs. t if $\theta_0 = 0.100$ rad, $\omega = 2.50$ rad/s and $\alpha = \frac{\pi}{4}$.

51. The charge q on a certain capacitor as a function of time is given by $q = 0.001 (1 - \cos 100t)$. Sketch two cycles of q as a function of t. Charge is measured in coulombs, and time is measured in seconds.

52. The weekly profit p, in thousands of dollars, for a seasonal business is given by $p = 35 + 15 \sin 0.24n$, where n is the number of the week of the year. Sketch p as a function of n for one year.

53. If the upper end of a spring is not fixed and is being moved with a sinusoidal motion, the motion of the bob at the end of the spring is affected. Plot the curve if the motion of the upper end of a spring is being moved by an external force and the bob moves according to the equation $y = 4 \sin 2t - 2 \cos 2t$.

54. Referring to Exercise 32 of Section 9-6, determine whether or not the curve of $y = 2 \sin (x - \frac{\pi}{4}) + 2 \cos (x + \frac{\pi}{4})$ represents the destructive interference of two waves.

55. The path of a roller mechanism used in an assembly line process is given by the equations $x = \theta - \sin \theta$ and $y = 1 - \cos \theta$. Sketch the path for $0 \leq \theta \leq 2\pi$.

56. Two voltage signals give a resulting curve on an oscilloscope. The equations for these signals are $x = 6 \sin \pi t$ and $y = 4 \cos 4\pi t$. Sketch the graph of the curve displayed by the oscilloscope.

57. The impedance Z, in ohms, and resistance R, in ohms, for an alternating-current circuit are related by $Z = R \sec \theta$, where θ is known as the *phase angle*. Sketch the graph for Z as a function of θ for $-\frac{\pi}{2} < \theta < \frac{\pi}{2}$.

58. An equation relating the components of a force $\mathbf{F}$ is $F_x = F_y \cot \theta$, where θ is the angle between $\mathbf{F}_x$ and $\mathbf{F}$. Sketch F_x as a function of θ if $F_y = 420$ lb.

59. The length ℓ of a conveyor belt and the height h it rises are related by $\ell = h \csc \theta$, where θ is the angle of inclination of the belt. Sketch ℓ as a function of θ for $h = 3.2$ m.

60. The instantaneous power in an electric circuit is defined as the product of the instantaneous voltage e and the instantaneous current i. If we have $e = 100 \cos 200t$ and $i = 2 \cos (200t + \frac{\pi}{4})$, plot the graph of the voltage and the graph of the current (in amperes), on the same coordinate system, vs. the time (in seconds). Then sketch the power (in watts) vs. time by multiplying appropriate values of e and i.

Practice Test

In Problems 1 through 4, sketch the graphs of the given functions by first finding the appropriate amplitude, period, and displacement.

1. $y = 0.5 \cos \frac{\pi}{2} x$ **2.** $y = 2 + 3 \sin x$ **3.** $y = 2 \sin x + \cos 2x$ **4.** $y = 2 \sin (2x - \frac{\pi}{3})$

5. A wave is traveling in a string. The displacement, as a function of time, from its equilibrium position is given by $y = A \cos (2\pi/T)t$. T is the period (measured in seconds) of the motion. If $A = 0.200$ in. and $T = 0.100$ s, draw two cycles of the displacement as a function of time.

6. Sketch the graph of $y = 3 \sec x$ for $0 \leq x \leq 2\pi$.

7. Plot the Lissajous figure for which $x = \sin \pi t$ and $y = 2 \cos 2\pi t$.

8. Draw two cycles of the curve of a projection on the end of a radius on the y-axis. The radius is of length R, and it is rotating counterclockwise about the origin at 2.00 rad/s. It starts at an angle of $\pi/6$ with the positive x-axis.

10 Exponents and Radicals

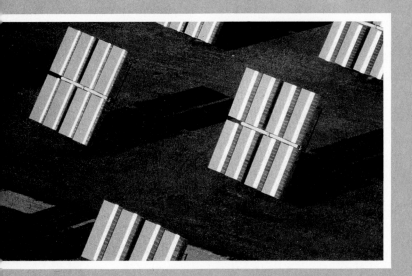

In Chapter 1 we introduced exponents and radicals. To this point, only a basic understanding of the meaning and elementary operations with them has been necessary. However, in our future work a more detailed understanding of exponents and radicals, and operations on them, will be required. Therefore, in this chapter we shall develop the necessary operations.

We will show the relationship between exponents and radicals when we introduce fractional exponents. In many algebraic operations and the applications which use them, it is common and more convenient to use fractional exponents rather than radicals.

In finding the rate at which solar radiation changes at a solar energy collector, the expression

$$\frac{(t^4 + 100)^{1/2} - 2t^3(t + 6)(t^4 + 100)^{-1/2}}{[(t^4 + 100)^{1/2}]^2}$$

is found. In Section 10-2 we show that this expression can be written in a much simpler form.

10-1 Integral Exponents

The laws of exponents were given in Section 1-5. For reference, they are:

$$a^m \times a^n = a^{m+n} \tag{10-1}$$

$$\frac{a^m}{a^n} = a^{m-n} \quad \text{or} \quad \frac{a^m}{a^n} = \frac{1}{a^{n-m}}, \qquad a \neq 0 \tag{10-2}$$

$$(a^m)^n = a^{mn} \tag{10-3}$$

$$(ab)^n = a^n b^n, \qquad \left(\frac{a}{b}\right)^n = \frac{a^n}{b^n}, \qquad b \neq 0 \tag{10-4}$$

$$a^0 = 1, \qquad a \neq 0 \tag{10-5}$$

$$a^{-n} = \frac{1}{a^n}, \qquad a \neq 0 \tag{10-6}$$

Although Eqs. (10-1) through (10-4) were originally defined for positive integers as exponents, we showed in Section 1-5 that with the definitions given in Eqs. (10-5) and (10-6), they are valid for all integral exponents. Later in this chapter we shall show how fractions may be used as exponents. Since the equations above are very important to the development of the topics in this chapter, they should again be reviewed—and learned thoroughly.

In this section we review the use of exponents in using Eqs. (10-1) through (10-6). Then we show how they are used and handled in somewhat more involved expressions.

EXAMPLE A

Applying Eq. (10-1), we have

$$a^5 \times a^{-3} = a^{5+(-3)} = a^{5-3} = a^2$$

Applying Eq. (10-1) and then Eq. (10-6), we have

$$a^3 \times a^{-5} = a^{3-5} = a^{-2} = \frac{1}{a^2}$$

We note that *negative exponents are not used in the expression of a final result*, unless specified otherwise. They are, however, used in intermediate steps in many operations.

EXAMPLE B

Applying Eq. (10-1), then (10-6), and then (10-4), we have

$$(2^3 \times 2^{-4})^2 = (2^{3-4})^2 = (2^{-1})^2 = \left(\frac{1}{2}\right)^2 = \frac{1}{2^2} = \frac{1}{4}$$

Often there are several combinations of the laws that can be used to simplify an expression. For example, this expression can be simplified by using Eq. (10-1), then (10-3), and then (10-6) as follows:

$$(2^3 \times 2^{-4})^2 = (2^{3-4})^2 = (2^{-1})^2 = 2^{-2} = \frac{1}{2^2} = \frac{1}{4}$$

The result is in a proper form as either $1/2^2$ or $1/4$. If the exponent is large, then it is common to leave the exponent in the answer.

EXAMPLE C

Applying Eqs. (10-2) and (10-5), we have

$$\frac{a^2 b^3 c^0}{ab^7} = \frac{a^{2-1}(1)}{b^{7-3}} = \frac{a}{b^4}$$

Applying Eqs. (10-4) and (10-3), we have

$$(x^{-2}y)^3 = (x^{-2})^3(y^3) = x^{-6}y^3 = \frac{y^3}{x^6}$$

Here, the simplification was completed by the use of Eq. (10-6).

EXAMPLE D

$$\left(\frac{4}{a^2}\right)^{-3} = \frac{1}{\left(\frac{4}{a^2}\right)^3} = \frac{1}{\frac{4^3}{a^6}} = \frac{a^6}{4^3}$$

or

$$\left(\frac{4}{a^2}\right)^{-3} = \frac{4^{-3}}{a^{-6}} = \frac{a^6}{4^3}$$

In the first method we used Eq. (10-6) first, then Eq. (10-4), and finally, we inverted the divisor. In the second method we first used Eq. (10-4) and then Eq. (10-6).

EXAMPLE E

$$(x^2y)^2\left(\frac{2}{x}\right)^{-2} = \frac{(x^4y^2)}{\left(\frac{2}{x}\right)^2} = \frac{x^4y^2}{\frac{4}{x^2}} = \frac{x^4y^2}{1} \times \frac{x^2}{4} = \frac{x^6y^2}{4}$$

or

$$(x^2y)^2\left(\frac{2}{x}\right)^{-2} = (x^4y^2)\left(\frac{2^{-2}}{x^{-2}}\right) = (x^4y^2)\left(\frac{x^2}{2^2}\right) = \frac{x^6y^2}{4}$$

The physical units associated with a denominate number can be expressed in terms of negative exponents. This is illustrated in the following example.

EXAMPLE F

The metric unit for pressure is the *pascal, where* 1 Pa = 1 N/m². Using a negative exponent, this can be expressed as

$$1 \text{ Pa} = 1 \text{ N/m}^2 = 1 \text{ N} \cdot \text{m}^{-2}$$

where $1/\text{m}^2 = \text{m}^{-2}$.

The metric unit for energy is the *joule,* where 1 J = 1 kg·(m·s⁻¹)²,

or

$$1 \text{ J} = 1 \text{ kg} \cdot \text{m}^2 \cdot \text{s}^{-2} = 1 \text{ kg} \cdot \text{m}^2/\text{s}^2.$$

In simplifying expressions, care must be taken to apply the laws of exponents properly. Certain relatively common problems are pointed out in the following examples.

EXAMPLE G

The expression $(-5x)^0$ equals 1, whereas the expression $-5x^0$ equals -5. For $(-5x)^0$ the parentheses indicate that the expression $-5x$ is raised to the zero power, whereas for $-5x^0$ only x is raised to the zero power and we have

$$-5x^0 = -5(1) = -5$$

NOTE ▷   Also, $(-5)^0 = 1$, whereas $-5^0 = -1$. Again note the use of parentheses. For $(-5)^0$ it is -5 which is raised to the zero power, whereas for -5^0 only 5 is raised to the zero power.

Similarly,

$$(-2)^2 = 4 \quad \text{and} \quad -2^2 = -4$$

For the same basic reasons,

NOTE ▷ $2x^{-1} = \dfrac{2}{x}$ whereas $(2x)^{-1} = \dfrac{1}{2x}$

EXAMPLE H ──── Simplify $(2a + b^{-1})^{-2}$.

In simplifying an expression we should give the result with positive exponents. For this expression, we may use various orders of operations. One is as follows:

$$(2a + b^{-1})^{-2} = \frac{1}{(2a + b^{-1})^2} = \frac{1}{\left(2a + \dfrac{1}{b}\right)^2} = \frac{1}{\left(\dfrac{2ab + 1}{b}\right)^2}$$

$$= \frac{1}{\dfrac{(2ab + 1)^2}{b^2}} = \frac{b^2}{(2ab + 1)^2}$$

Another order of operations for simplifying this expression is

$$(2a + b^{-1})^{-2} = \left(2a + \frac{1}{b}\right)^{-2} = \left(\frac{2ab + 1}{b}\right)^{-2}$$

$$= \frac{(2ab + 1)^{-2}}{b^{-2}} = \frac{b^2}{(2ab + 1)^2}$$

■ It is not necessary to expand the denominator of this result.

EXAMPLE I ──── There is an error which is commonly made in simplifying the type of expression in Example H. We must be careful to see that

NOTE ▷ $(2a + b^{-1})^{-2}$ is **not** equal to $(2a)^{-2} + (b^{-1})^{-2}$, or $\dfrac{1}{4a^2} + b^2$

Remember: As noted in Section 5-1, when raising a binomial (or any multinomial) to a power, we cannot simply raise each term to the power to obtain the result.

However, when raising a product of factors to a power, we use Eq. (10-4). Thus

$$(2ab^{-1})^{-2} = (2a)^{-2}(b^{-1})^{-2} = \frac{b^2}{(2a)^2} = \frac{b^2}{4a^2}$$

We see that we must be careful to distinguish between the power of a sum of
■ terms and the power of a product of factors.

From the above examples, we see that *when a factor is moved from the denominator to the numerator of a fraction, or conversely, the sign of the exponent is changed*. We should heed carefully the word *factor*; this rule does not apply to moving *terms* in the numerator or denominator.

EXAMPLE J

$$3a^{-1} - (2a)^{-2} = \frac{3}{a} - \frac{1}{(2a)^2} = \frac{3}{a} - \frac{1}{4a^2}$$

$$= \frac{12a - 1}{4a^2}$$

EXAMPLE K

$$3^{-1}\left(\frac{4^{-2}}{3 - 3^{-1}}\right) = \frac{1}{3}\left(\frac{1}{4^2}\right)\left(\frac{1}{3 - \frac{1}{3}}\right) = \frac{1}{3 \times 4^2}\left(\frac{1}{\frac{9 - 1}{3}}\right)$$

$$= \frac{1}{3 \times 4^2}\left(\frac{3}{8}\right) = \frac{1}{128}$$

EXAMPLE L

terms

$$\frac{1}{x^{-1}}\left(\frac{x^{-1} - y^{-1}}{x^2 - y^2}\right) = \frac{x}{1}\left(\frac{\frac{1}{x} - \frac{1}{y}}{x^2 - y^2}\right) = x\left(\frac{\frac{y - x}{xy}}{x^2 - y^2}\right)$$

$$= \frac{\frac{x(y - x)}{xy}}{(x - y)(x + y)} = \frac{x(y - x)}{xy} \times \frac{1}{(x - y)(x + y)}$$

$$= \frac{x(y - x)}{xy(x - y)(x + y)} = \frac{-(x - y)}{y(x - y)(x + y)}$$

$$= -\frac{1}{y(x + y)}$$

NOTE ▷ Note that in this example **the x^{-1} and y^{-1} in the numerator could not be moved directly to the denominator with positive exponents** because they are only terms of the original numerator.

EXAMPLE M

$$3(x + 4)^2(x - 3)^{-2} - 2(x - 3)^{-3}(x + 4)^3$$

$$= \frac{3(x + 4)^2}{(x - 3)^2} - \frac{2(x + 4)^3}{(x - 3)^3} = \frac{3(x - 3)(x + 4)^2 - 2(x + 4)^3}{(x - 3)^3}$$

$$= \frac{(x + 4)^2[3(x - 3) - 2(x + 4)]}{(x - 3)^3} = \frac{(x + 4)^2(x - 17)}{(x - 3)^3}$$

Expressions such as the one in this example are commonly found in problems in calculus.

Exercises 10-1

In Exercises 1 through 56, express each of the given expressions in the simplest form which contains only positive exponents.

1. x^7x^{-4}

2. y^9y^{-2}

3. a^2a^{-6}

4. ss^{-5}

5. 5×5^{-3}

6. $2 \times 7^4 \times 7^{-2}$

7. $(2^{-1} \times 5)^2$

8. $(3^2 \times 4^{-3})^3$

9. $(2ax^{-1})^2$

10. $(3xy^{-2})^3$

11. $(5an^{-2})^{-1}$

12. $(6s^2t^{-1})^{-2}$

13. $(-4)^0$

14. -4^0

15. $-7x^0$

16. $(-7x)^0$

17. $3x^{-2}$

18. $(3x)^{-2}$

19. $(7ax)^{-3}$

20. $7ax^{-3}$

21. $\left(\dfrac{2}{n^3}\right)^{-1}$

22. $\left(\dfrac{3}{x^3}\right)^{-2}$

23. $\left(\dfrac{a}{b^{-2}}\right)^{-3}$

24. $\left(\dfrac{2n^{-2}}{m^{-1}}\right)^{-2}$

25. $(a + b)^{-1}$

26. $a^{-1} + b^{-1}$

27. $3x^{-2} + 2y^{-2}$

28. $(3x + 2y)^{-2}$

29. $(2 \times 3^{-2})^2 \left(\dfrac{3}{2}\right)^{-1}$

30. $(3^{-1} \times 7^2)^{-1} \left(\dfrac{3}{7}\right)^2$

31. $ab^2(3a^{-1})^{-1}$

32. $(3st^{-1})^3 \left(\dfrac{s}{t}\right)^{-2}$

33. $\left(\dfrac{3a^2}{4b}\right)^{-3} \left(\dfrac{4}{a}\right)^{-5}$

34. $(2np^{-2})^{-2}(4^{-1}p^2)^{-1}$

35. $\left(\dfrac{v^{-1}}{2t}\right)^{-2} \left(\dfrac{t^2}{v^{-2}}\right)^{-3}$

36. $\left(\dfrac{a^{-2}}{b^2}\right)^{-3} \left(\dfrac{a^{-3}}{b^5}\right)^2$

37. $(x^2y^{-1})^2 - x^{-4}$

38. $s^2s^{-4} - (s^2)^{-4}$

39. $2a^{-2} + (2a^{-2})^4$

40. $3(a^{-1}z^2)^{-3} + c^{-2}z^{-1}$

41. $2 \times 3^{-1} + 4 \times 3^{-2}$

42. $5 \times 2^{-2} - 3^{-1} \times 2^3$

43. $(a^{-1} + b^{-1})^{-1}$

44. $(2a - b^{-2})^{-1}$

45. $(n^{-2} - 2n^{-1})^2$

46. $(2^{-3} - 4^{-1})^{-2}$

47. $\dfrac{3 - 2^{-1}}{3^{-2}}$

48. $\dfrac{6^{-1}}{4^{-2} + 2}$

49. $\dfrac{x - y^{-1}}{x^{-1} - y}$

50. $\dfrac{x^{-2} - y^{-2}}{x^{-1} - y^{-1}}$

51. $\dfrac{ax^{-2} + a^{-2}x}{a^{-1} + x^{-1}}$

52. $\dfrac{2x^{-2} - 2y^{-2}}{(xy)^{-3}}$

53. $2t^{-2} + t^{-1}(t + 1)$

54. $3x^{-1} - x^{-3}(y + 2)$

55. $(x - 1)^{-1} + (x + 1)^{-1}$

56. $4(2x - 1)(x + 2)^{-1} - (2x - 1)^2(x + 2)^{-2}$

In Exercises 57 through 64, perform the indicated operations.

57. Express $4^2 \times 64$ (a) as a power of 4, and (b) as a power of 2.

58. Express $1/81$ (a) as a power of 9, and (b) as a power of 3.

59. By use of Eqs. (10-4) and (10-6), show that

$$\left(\dfrac{a}{b}\right)^{-n} = \left(\dfrac{b}{a}\right)^n$$

60. Using a calculator, verify the equation in Exercise 59 by evaluating the expression on each side with $a = 3.576$, $b = 8.091$, and $n = 7$.

61. When studying the internal energy of an ideal gas, the units encountered are

$$\dfrac{\text{Pa} \cdot \text{m}^6 \cdot \text{mol}^{-2}}{\text{m}^3 \cdot \text{mol}^{-1}}$$

(see Appendix B). Simplify this form, using only positive exponents.

62. An expression encountered in the mathematics of finance is

$$\dfrac{p(1 + i)^{-1}[(1 + i)^{-n} - 1]}{(1 + i)^{-1} - 1}$$

where n is an integer. Simplify this expression.

63. In analyzing the tuning of an electronic circuit, the expression $[\omega\omega_0^{-1} - \omega_0\omega^{-1}]^2$ is used. Expand and simplify this expression.

64. In optics, the combined focal length F of two lenses is given by $F = [f_1^{-1} + f_2^{-1} + d(f_1 f_2)^{-1}]^{-1}$, where f_1 and f_2 are the focal lengths of the lenses and d is the distance between them. Simplify the right side of this equation.

10-2 Fractional Exponents

In Section 10-1 we reviewed the use of integral exponents, including exponents which are negative integers and zero. We now show how rational numbers may be used as exponents. With the appropriate definitions, all the laws of exponents are valid for all rational numbers as exponents.

Equation (10-3) states that $(a^m)^n = a^{mn}$. If we were to let $m = \frac{1}{2}$ and $n = 2$, we would have $(a^{1/2})^2 = a^1$. However, we already have a way of writing a quantity which when squared equals a. This is written as $\sqrt{a}$. To be consistent with previous definitions and to allow the laws of exponents to hold, we define

$$a^{1/n} = \sqrt[n]{a} \tag{10-7}$$

In order that Eqs. (10-3) and (10-7) may hold at the same time, we define

meaning of a fractional exponent

$$a^{m/n} = \sqrt[n]{a^m} = (\sqrt[n]{a})^m \tag{10-8}$$

It can be shown that these definitions are valid for all the laws of exponents.

EXAMPLE A

We shall verify here that Eq. (10-1) holds for the above definitions:

$$a^{1/4}a^{1/4}a^{1/4}a^{1/4} = a^{(1/4)+(1/4)+(1/4)+(1/4)} = a^1$$

Now $a^{1/4} = \sqrt[4]{a}$, by definition. Also, by definition $\sqrt[4]{a}\sqrt[4]{a}\sqrt[4]{a}\sqrt[4]{a} = a$. Equation (10-1) is thereby verified for $n = 4$. Equation (10-3) is verified by the following:

$$a^{1/4}a^{1/4}a^{1/4}a^{1/4} = a^{4(1/4)} = a = \sqrt[4]{a^4}$$

We note that we may interpret $a^{m/n}$ in Eq. (10-8) as the mth power of the nth root of a, as well as the nth root of the mth power of a. This is illustrated in the following example.

EXAMPLE B

1. $(\sqrt[3]{a})^2 = \sqrt[3]{a^2} = a^{2/3}$

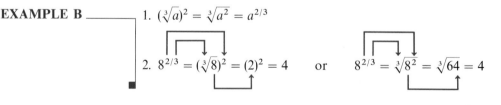

2. $8^{2/3} = (\sqrt[3]{8})^2 = (2)^2 = 4$ or $8^{2/3} = \sqrt[3]{8^2} = \sqrt[3]{64} = 4$

We must note that Eq. (10-8) is valid as long as $\sqrt[n]{a}$ does not involve the even root of a negative number. Such numbers are imaginary and are considered in Chapter 11.

Although both interpretations of Eq. (10-8) are possible, as indicated in Example B, in evaluating numerical expressions involving fractional exponents without a calculator, it is almost always best to *find the root first, as indicated by the denominator* of the fractional exponent. This will allow us to find the root of the smaller number, which is normally easier to find.

EXAMPLE C

To evaluate $(64)^{5/2}$, we should proceed as follows:

$$(64)^{5/2} = [(64)^{1/2}]^5 = 8^5 = 32{,}768$$

If we raised 64 to the fifth power first, we would have

$$(64)^{5/2} = (64^5)^{1/2} = (1{,}073{,}741{,}824)^{1/2}$$

We would now have to evaluate the indicated square root. This demonstrates why it is preferable to find the indicated root first. ∎

EXAMPLE D

1. $(16)^{3/4} = (16^{1/4})^3 = 2^3 = 8$

2. $4^{-1/2} = \dfrac{1}{4^{1/2}} = \dfrac{1}{2}, \qquad 9^{3/2} = (9^{1/2})^3 = 3^3 = 27$

We note in the illustration of $4^{-1/2}$ that Eq. (10-6) must also hold for negative rational exponents. The only change in the exponent which is made in writing it as $1/4^{1/2}$ is that the sign of the exponent is changed. ∎

Fractional exponents allow us to find roots of numbers on a calculator. By use of the $\boxed{x^y}$ (or $\boxed{y^x}$) key, we may raise any positive number to any power. For fractional exponents, the decimal form is used.

EXAMPLE E

The thermodynamic temperature T, in kelvins (K), is related to the pressure P, in kilopascals, of a gas by the equation $T = 80.5P^{2/7}$. Find T for $P = 750$ kPa.

Substituting, we have

$$T = 80.5(750)^{2/7}$$

Using a calculator, one way to find this value is first to find the decimal form of $\frac{2}{7}$ and put its value in memory. Doing this, we have the following calculator sequence of keys:

$2 \boxed{\div} 7 \boxed{=} \boxed{\text{STO}}\ 80.5 \boxed{\times}\ 750 \boxed{x^y} \boxed{\text{RCL}} \boxed{=}$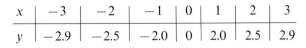

This means that $T = 534$ K.

Another way to perform this evaluation on a calculator is to use parentheses. This sequence is $80.5 \boxed{\times}\ 750 \boxed{x^y} \boxed{(}\ 2 \boxed{\div} 7 \boxed{)} \boxed{=}$ 533.63211. ∎

In Example E we evaluated the expression on a calculator by using $\frac{2}{7}$ in decimal form. When the exponent is known exactly, such as $x^{1/3}$ to indicate a cube root, the fractional exponent is generally used, although it is evaluated using the decimal form. However, if the exponent is approximate, the decimal form may be used directly, such as $x^{0.35}$.

EXAMPLE F

Graph the function $y = 2x^{1/3}$.

Here we use the fractional exponent in expressing the function. In obtaining points for the graph we can use a calculator, but it will use $\frac{1}{3}$ in approximate decimal form. With $y = f(x)$, we now get the following table and the graph shown in Fig. 10-1.

x	-3	-2	-1	0	1	2	3
y	-2.9	-2.5	-2.0	0	2.0	2.5	2.9

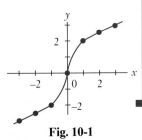

Fig. 10-1

When finding powers of negative numbers, some calculators will show an error. In Example F we must remember that $x^{1/3} = \sqrt[3]{x}$ and that the domain of this function is all real numbers and we must show negative values of y for negative values of x.

Another reason for developing fractional exponents is that they are often easier to use in more complex expressions involving roots. Therefore, any expression involving radicals can also be expressed with fractional exponents and then simplified. We now show some additional examples with fractional exponents.

EXAMPLE G

1. $(8a^2b^4)^{1/3} = [(8^{1/3})(a^2)^{1/3}(b^4)^{1/3}]$ using Eq. (10-4)

             $= 2a^{2/3}b^{4/3}$ using Eqs. (10-7) and (10-3)

2. $a^{3/4}a^{4/5} = a^{3/4+4/5} = a^{31/20}$ using Eq. (10-1)

EXAMPLE H

$$\left(\frac{4^{-3/2}x^{2/3}y^{-7/4}}{2^{3/2}x^{-1/3}y^{3/4}}\right)^{2/3} = \left(\frac{x^{2/3}x^{1/3}}{2^{3/2}4^{3/2}y^{3/4}y^{7/4}}\right)^{2/3}$$ using Eq. (10-6)

$$= \left(\frac{x^{2/3+1/3}}{2^{3/2}4^{3/2}y^{3/4+7/4}}\right)^{2/3}$$ using Eq. (10-1)

$$= \frac{x^{(1)(2/3)}}{2^{(3/2)(2/3)}4^{(3/2)(2/3)}y^{(10/4)(2/3)}}$$ using Eq. (10-4)

$$= \frac{x^{2/3}}{8y^{5/3}}$$

EXAMPLE I

$$(4x^4)^{-1/2} - 3x^{-3} = \frac{1}{(4x^4)^{1/2}} - \frac{3}{x^3}$$ using Eq. (10-6)

$$= \frac{1}{2x^2} - \frac{3}{x^3}$$ using Eq. (10-7)

$$= \frac{x-6}{2x^3}$$ common denominator

EXAMPLE J

See the chapter introduction.

The rate R at which solar radiation changes at a solar energy collector during a day is given by the equation

$$R = \frac{(t^4 + 100)^{1/2} - 2t^3(t + 6)(t^4 + 100)^{-1/2}}{[(t^4 + 100)^{1/2}]^2}$$

Here R is measured in kW/m²·h, t is the number of hours from noon, and $-6\text{ h} \le t \le 8\text{ h}$. Express the right side of this equation in simpler form, and find R for $t = 0$ (noon) and for $t = 4$ h (4 P.M.).

Performing the simplification, we have the following steps:

$$R = \frac{(t^4 + 100)^{1/2} - \dfrac{2t^3(t + 6)}{(t^4 + 100)^{1/2}}}{(t^4 + 100)}$$

       ← using Eq. (10-6)

       ← using Eq. (10-3)

$$= \frac{\dfrac{(t^4 + 100)^{1/2}(t^4 + 100)^{1/2} - 2t^3(t + 6)}{(t^4 + 100)^{1/2}}}{(t^4 + 100)} \quad \longleftarrow \text{common denominator}$$

$$= \frac{(t^4 + 100) - 2t^3(t + 6)}{(t^4 + 100)^{1/2}} \times \frac{1}{t^4 + 100} \quad \longleftarrow \text{invert divisor and multiply}$$

$$= \frac{100 - 12t^3 - t^4}{(t^4 + 100)^{1/2}(t^4 + 100)} = \frac{100 - 12t^3 - t^4}{(t^4 + 100)^{3/2}} \quad \longleftarrow \text{using Eq. (10-1)}$$

For $t = 0$: $\quad R = \dfrac{100 - 12(0^3) - 0^4}{(0^4 + 100)^{3/2}} = 0.10 \text{ kW/m}^2 \cdot \text{h}$

For $t = 4$ h: $\quad R = \dfrac{100 - 12(4^3) - 4^4}{(4^4 + 100)^{3/2}} = -0.14 \text{ kW/m}^2 \cdot \text{h}$

We see that the radiation is increasing at noon, and the negative sign tells us that it is decreasing at 4 P.M. ■

Exercises 10-2

In Exercises 1 through 28, evaluate the given expressions.

1. $(25)^{1/2}$ **2.** $(49)^{1/2}$ **3.** $(27)^{1/3}$ **4.** $(81)^{1/4}$

5. $8^{4/3}$ **6.** $(125)^{2/3}$ **7.** $(100)^{25/2}$ **8.** $(16)^{5/4}$

9. $8^{-1/3}$ **10.** $16^{-1/4}$ **11.** $(64)^{-2/3}$ **12.** $(32)^{-4/5}$

13. $5^{1/2}5^{3/2}$ **14.** $8^{1/3}4^{-1/2}$ **15.** $(4^4)^{3/2}$ **16.** $(3^6)^{2/3}$

17. $\dfrac{121^{-1/2}}{100^{1/2}}$ **18.** $\dfrac{1000^{1/3}}{400^{-1/2}}$ **19.** $\dfrac{7^{-1/2}}{6^{-1}7^{1/2}}$ **20.** $\dfrac{15^{2/3}}{5^2 15^{-1/3}}$

21. $\dfrac{(-27)^{1/3}}{6}$ **22.** $\dfrac{(-8)^{2/3}}{-2}$ **23.** $\dfrac{-8}{(-27)^{-1/3}}$ **24.** $\dfrac{-4}{(-64)^{-2/3}}$

25. $(125)^{-2/3} - (100)^{-3/2}$ **26.** $32^{0.4} + 25^{-0.5}$ **27.** $\dfrac{16^{-0.25}}{5} + \dfrac{2^{-0.6}}{2^{0.4}}$ **28.** $\dfrac{4^{-1}}{(36)^{-1/2}} - \dfrac{5^{-1/2}}{5^{1/2}}$

In Exercises 29 through 32, evaluate the given expressions by use of a calculator.

29. $(17.98)^{1/4}$ **30.** $(750.81)^{2/3}$ **31.** $(4.0187)^{-4/9}$ **32.** $(0.1863)^{-1/6}$

In Exercises 33 through 60, use the laws of exponents to simplify the given expressions. Express all answers with positive exponents.

33. $a^{2/3}a^{1/2}$ **34.** $x^{5/6}x^{-1/3}$ **35.** $\dfrac{y^{-1/2}}{y^{2/5}}$ **36.** $\dfrac{2r^{4/5}}{r^{-1}}$

37. $\dfrac{s^{1/4}s^{2/3}}{s^{-1}}$ **38.** $\dfrac{x^{3/10}}{x^{-1/5}x^2}$ **39.** $\dfrac{y^{-1}}{y^{1/3}y^{-1/4}}$ **40.** $\dfrac{a^{-2/5}a^2}{a^{-3/10}}$

41. $(8a^3b^6)^{1/3}$ **42.** $(8b^{-4}c^2)^{2/3}$ **43.** $(16a^4b^3)^{-3/4}$ **44.** $(32x^5y^4)^{-2/5}$

45. $\dfrac{1}{2}(4x^2 + 1)^{-1/2}(8x)$ **46.** $\dfrac{2}{3}(x^3 + 1)^{-1/3}(3x^2)$ **47.** $\left(\dfrac{9t^{-2}}{16}\right)^{3/2}$ **48.** $\left(\dfrac{a^{5/7}}{a^{2/3}}\right)^{7/4}$

49. $\left(\dfrac{4a^{5/6}b^{-1/5}}{a^{2/3}b^2}\right)^{-1/2}$ **50.** $\left(\dfrac{a^0b^8c^{-1/8}}{ab^{63/64}}\right)^{32/3}$ **51.** $\left(\dfrac{6x^{-1/2}y^{2/3}}{18x^{-1}}\right)\left(\dfrac{2y^{1/4}}{x^{1/3}}\right)$ **52.** $\dfrac{3^{-1}a^{1/2}}{4^{-1/2}b} \div \dfrac{9^{1/2}a^{-1/3}}{2b^{-1/4}}$

53. $(x^{-1} + 2x^{-2})^{-1/2}$ **54.** $(a^{-2} - a^{-4})^{-1/4}$ **55.** $(a^3)^{-4/3} + a^{-2}$ **56.** $(4x^6)^{-1/2} - 2x^{-1}$

57. $[(a^{1/2} - a^{-1/2})^2 + 4]^{1/2}$

58. $4x^{1/2} + \dfrac{1}{2}x^{-1/2}(4x + 1)$

59. $x^2(2x - 1)^{-1/2} + 2x(2x - 1)^{1/2}$

60. $(3x - 1)^{-2/3}(1 - x) - (3x - 1)^{1/3}$

In Exercises 61 through 64, graph the given functions.

61. $f(x) = 3x^{1/2}$ **62.** $f(x) = 2x^{2/3}$ **63.** $f(x) = x^{4/5}$ **64.** $f(x) = 4x^{3/2}$

In Exercises 65 through 68, perform the indicated operations.

65. A factor used in determining the performance of a solar energy storage system is $(A/S)^{-1/4}$, where A is the actual storage capacity and S is a standard storage capacity. Find the value of this factor if $A = 2S$.

66. A factor used in measuring the loudness sensed by the human ear is $(I/I_0)^{0.3}$, where I is the intensity of the sound and I_0 is a reference intensity. Evaluate this factor for $I = 3.2 \times 10^{-6}$ W/m² (ordinary conversation) and $I_0 = 10^{-12}$ W/m².

67. The electric current i, in amperes, in a circuit containing a battery of voltage E, a resistance R, and an inductance L is given by $i = \dfrac{E}{R}(1 - e^{-Rt/L})$, where t is the time after the circuit is closed. See Fig. 10-2. Find i for $E = 6.20$ V, $R = 120$ Ω, $L = 3.24$ H, and $t = 0.00100$ s. (The number e is irrational and can be found from the calculator by the sequence $1\,\boxed{e^x}$.)

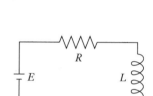

Fig. 10-2

68. For a heat-seeking rocket in pursuit of an aircraft, the distance from the rocket to the aircraft is

$$d = \frac{500(\sin \theta)^{1/2}}{(1 - \cos \theta)^{3/2}}$$

where θ is shown in Fig. 10-3. Find d, in km, for $\theta = 125.0°$.

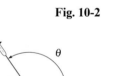

Fig. 10-3

10-3   Simplest Radical Form

Radicals were first introduced in Section 1-7, and we used them again in developing the concept of a fractional exponent. As we mentioned in the preceding section, it is possible to use fractional exponents for any operation required with radicals. For operations involving multiplication and division, this method has certain advantages. But for adding and subtracting radicals, there is normally little advantage in changing form.

We shall now define the operations with radicals so that these definitions are consistent with the laws of exponents. This will enable us from now on to use either fractional exponents or radicals, whichever is more convenient.

$$\sqrt[n]{a^n} = (\sqrt[n]{a})^n = a \tag{10-9}$$

$$\sqrt[n]{a}\,\sqrt[n]{b} = \sqrt[n]{ab} \tag{10-10}$$

$$\sqrt[m]{\sqrt[n]{a}} = \sqrt[mn]{a} \tag{10-11}$$

$$\frac{\sqrt[n]{a}}{\sqrt[n]{b}} = \sqrt[n]{\frac{a}{b}}, \quad b \neq 0 \tag{10-12}$$

The number under the radical is called the **radicand,** *and the number indicating the root being taken is called the* **order** *of the radical. To avoid difficulties with* imaginary numbers (which are considered in the next chapter), we shall assume that all letters represent positive numbers.

EXAMPLE A

Following are illustrations of each of Eqs. (10-9) through (10-12).

1. $\sqrt[5]{4^5} = (\sqrt[5]{4})^5 = 4$ using Eq. (10-9)
2. $\sqrt[3]{2}\sqrt[3]{3} = \sqrt[3]{2 \times 3} = \sqrt[3]{6}$ using Eq. (10-10)
3. $\sqrt[3]{\sqrt{5}} = \sqrt[3 \times 2]{5} = \sqrt[6]{5}$ using Eq. (10-11)
4. $\dfrac{\sqrt{7}}{\sqrt{3}} = \sqrt{\dfrac{7}{3}}$ using Eq. (10-12)

EXAMPLE B

In Example E of Section 1-7, it was seen that

$$\sqrt{16 + 9} \quad \text{is } \textbf{\textit{not}} \text{ equal to} \quad \sqrt{16} + \sqrt{9}$$

However, using Eq. (10-10),

$$\sqrt{16 \times 9} = \sqrt{16} \times \sqrt{9} = 4 \times 3 = 12$$

Therefore, we must be careful to distinguish between the root of a sum of terms and the root of a product of factors. This is the same as with powers of sums and powers of products, as shown in Example I of Section 10-1. It should be the same, as a root can be interpreted as a fractional exponent.

There are certain operations which are performed on radicals in order to put them in their simplest form. The following two examples illustrate one of these operations.

EXAMPLE C

To simplify $\sqrt{75}$, we recall that $75 = (25)(3)$ and that $\sqrt{25} = 5$. Using Eq. (10-10), we write

$$\sqrt{75} = \sqrt{25 \times 3} = \sqrt{25}\sqrt{3} = 5\sqrt{3}$$

↳ perfect square

This illustrates one step which should always be carried out in simplifying radicals. *Always remove all perfect nth-power factors from the radicand of a radical of order n.*

EXAMPLE D

1. $\sqrt{72} = \sqrt{(36)(2)} = \sqrt{36}\sqrt{2} = 6\sqrt{2}$

    ↳ perfect square

2. $\sqrt{a^3 b^2} = \sqrt{(a^2)(a)(b^2)} = \sqrt{a^2}\sqrt{a}\sqrt{b^2} = ab\sqrt{a}$

    ↳ perfect square

3. cube root ⟶ $\sqrt[3]{40} = \sqrt[3]{8 \times 5} = \sqrt[3]{8}\sqrt[3]{5} = 2\sqrt[3]{5}$

    ↳ perfect cube

4. fifth root ⟶ $\sqrt[5]{64x^8 y^{12}} = \sqrt[5]{(32)(2)(x^5)(x^3)(y^{10})(y^2)}$

    ↑   ↑   ↑ ── perfect fifth powers

$$= \sqrt[5]{(32)(x^5)(y^{10})}\sqrt[5]{2x^3 y^2}$$
$$= 2xy^2\sqrt[5]{2x^3 y^2}$$

The following examples illustrate another procedure which can be used to simplify certain radicals. The procedure is to *reduce the order of the radical,* when possible.

EXAMPLE E

$$\sqrt[6]{8} = \sqrt[6]{2^3} = 2^{3/6} = 2^{1/2} = \sqrt{2}$$

In this example we started with a sixth root and ended with a square root. Thus, the order of the radical was reduced. Fractional exponents are often helpful when we perform this operation.

EXAMPLE F

1. $\sqrt[8]{16} = \sqrt[8]{2^4} = 2^{4/8} = 2^{1/2} = \sqrt{2}$

2. $\dfrac{\sqrt[4]{9}}{\sqrt{3}} = \dfrac{\sqrt[4]{3^2}}{\sqrt{3}} = \dfrac{3^{2/4}}{3^{1/2}} = 1$

3. $\dfrac{\sqrt[6]{8}}{\sqrt{7}} = \dfrac{\sqrt[6]{2^3}}{\sqrt{7}} = \dfrac{2^{1/2}}{7^{1/2}} = \sqrt{\dfrac{2}{7}}$

4. $\sqrt[9]{27x^6 y^{12}} = \sqrt[9]{3^3 x^6 y^9 y^3} = 3^{3/9} x^{6/9} y^{9/9} y^{3/9} = 3^{1/3} x^{2/3} y y^{1/3}$
   $= y\sqrt[3]{3x^2 y}$

If a radical is to be written in *simplest form,* the two operations illustrated in the last four examples must be performed. They are:

simplest form of a radical

1. *all perfect nth-power factors are removed from a radical of order n, and*

2. *if possible, the order of the radical is reduced.*

When working with fractions, it has traditionally been the practice to write a fraction with radicals in a form in which the denominator contains no radicals. Such a fraction was not considered to be in simplest form unless this was done. This step of simplification was performed primarily for ease of calculation, but with a calculator it does not matter to any extent that there is a radical in the denominator. However, the procedure of writing a radical in this form, called **rationalizing the denominator,** is at times useful for other purposes. Therefore, the following examples show how the process of rationalizing the denominator is carried out.

EXAMPLE G

To write $\sqrt{\tfrac{2}{5}}$ in an equivalent form in which the denominator is not included under the radical sign, we create a perfect square in the denominator by multiplying the numerator and the denominator under the radical by 5. This give us $\sqrt{\tfrac{10}{25}}$, which may be written as $\tfrac{1}{5}\sqrt{10}$ or $\tfrac{\sqrt{10}}{5}$. These steps are written as follows:

$$\sqrt{\frac{2}{5}} = \sqrt{\frac{2 \times 5}{5 \times 5}} = \sqrt{\frac{10}{25}} = \frac{\sqrt{10}}{\sqrt{25}} = \frac{\sqrt{10}}{5}$$

perfect square

EXAMPLE H

$$\sqrt{\frac{5}{7}} = \sqrt{\frac{5 \times 7}{7 \times 7}} = \frac{\sqrt{35}}{\sqrt{49}} = \frac{\sqrt{35}}{7}$$

$\uparrow$ perfect square

$$\frac{3}{\sqrt{8}} = \frac{3\sqrt{2}}{\sqrt{8 \times 2}} = \frac{3\sqrt{2}}{\sqrt{16}} = \frac{3\sqrt{2}}{4}$$

$\uparrow$ perfect square

cube root $\longrightarrow$ $\sqrt[3]{\frac{2}{3}} = \sqrt[3]{\frac{2 \times 9}{3 \times 9}} = \sqrt[3]{\frac{18}{27}} = \frac{\sqrt[3]{18}}{\sqrt[3]{27}} = \frac{\sqrt[3]{18}}{3}$

$\uparrow$ perfect cube

NOTE ▷ In the second illustration, a perfect square was made by multiplying by $\sqrt{2}$. We should try to *choose the smallest possible number which can be used.* In the third illustration we wanted a perfect cube in the denominator, since a cube root is being found.

EXAMPLE I

The period T, in seconds, for one cycle of a simple pendulum is given by $T = 2\pi\sqrt{L/g}$, where L is the length of the pendulum and g is the acceleration due to gravity. Rationalize the denominator of the right side of this equation if $L = 3$ ft and $g = 32$ ft/s².

Substituting, and then rationalizing, we have

$$T = 2\pi\sqrt{\frac{3}{32}} = 2\pi\sqrt{\frac{3 \times 2}{32 \times 2}} = 2\pi\sqrt{\frac{6}{64}} = \frac{2\pi}{8}\sqrt{6} = \frac{\pi}{4}\sqrt{6} = 1.9 \text{ s}$$

EXAMPLE J

Simplify the radical $\sqrt{\frac{3a}{4b} - 2 + \frac{4b}{3a}}$, for $3a \geq 4b$, and rationalize the denominator.

$$\sqrt{\frac{3a}{4b} - 2 + \frac{4b}{3a}} = \sqrt{\frac{(3a)(3a) - 2(3a)(4b) + (4b)(4b)}{(3a)(4b)}}$$

$$= \sqrt{\frac{(3a - 4b)^2}{4(3ab)}} = \frac{3a - 4b}{2}\sqrt{\frac{1}{3ab}}$$

Rationalizing the denominator, we have

$$\frac{3a - 4b}{2}\sqrt{\frac{1}{3ab}} = \frac{3a - 4b}{2}\sqrt{\frac{1(3ab)}{3ab(3ab)}} = \frac{(3a - 4b)\sqrt{3ab}}{6ab}$$

Exercises 10-3

In Exercises 1 through 60, write each expression in simplest radical form. Where a radical appears in a denominator, rationalize the denominator.

1. $\sqrt{24}$ **2.** $\sqrt{150}$ **3.** $\sqrt{45}$ **4.** $\sqrt{98}$

5. $\sqrt{x^2y^5}$ **6.** $\sqrt{s^3t^6}$ **7.** $\sqrt{pq^2r^7}$ **8.** $\sqrt{x^2y^4z^3}$

9. $\sqrt{5x^2}$ **10.** $\sqrt{12ab^2}$ **11.** $\sqrt{18a^3bc^4}$ **12.** $\sqrt{54m^5n^3}$

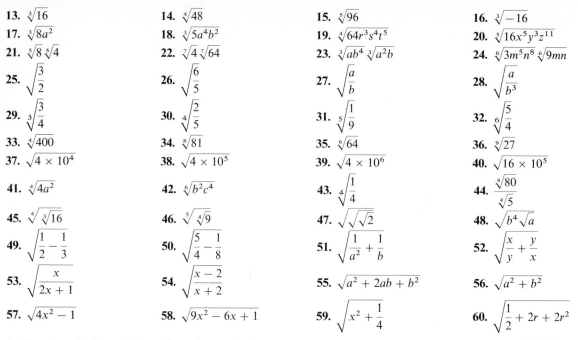

13. $\sqrt[3]{16}$ 14. $\sqrt[4]{48}$ 15. $\sqrt[5]{96}$ 16. $\sqrt[3]{-16}$

17. $\sqrt[3]{8a^2}$ 18. $\sqrt[3]{5a^4b^2}$ 19. $\sqrt[4]{64r^3s^4t^5}$ 20. $\sqrt[5]{16x^5y^3z^{11}}$

21. $\sqrt[5]{8}\sqrt[5]{4}$ 22. $\sqrt[7]{4}\sqrt[7]{64}$ 23. $\sqrt[3]{ab^4}\sqrt[3]{a^2b}$ 24. $\sqrt[6]{3m^5n^8}\sqrt[6]{9mn}$

25. $\sqrt{\dfrac{3}{2}}$ 26. $\sqrt{\dfrac{6}{5}}$ 27. $\sqrt{\dfrac{a}{b}}$ 28. $\sqrt{\dfrac{a}{b^3}}$

29. $\sqrt[3]{\dfrac{3}{4}}$ 30. $\sqrt[4]{\dfrac{2}{5}}$ 31. $\sqrt[5]{\dfrac{1}{9}}$ 32. $\sqrt[6]{\dfrac{5}{4}}$

33. $\sqrt[4]{400}$ 34. $\sqrt[8]{81}$ 35. $\sqrt[6]{64}$ 36. $\sqrt[9]{27}$

37. $\sqrt{4 \times 10^4}$ 38. $\sqrt{4 \times 10^5}$ 39. $\sqrt{4 \times 10^6}$ 40. $\sqrt{16 \times 10^5}$

41. $\sqrt[4]{4a^2}$ 42. $\sqrt[6]{b^2c^4}$ 43. $\sqrt[4]{\dfrac{1}{4}}$ 44. $\dfrac{\sqrt[4]{80}}{\sqrt[4]{5}}$

45. $\sqrt[4]{\sqrt[3]{16}}$ 46. $\sqrt[5]{\sqrt[4]{9}}$ 47. $\sqrt{\sqrt{\sqrt{2}}}$ 48. $\sqrt{b^4\sqrt{a}}$

49. $\sqrt{\dfrac{1}{2} - \dfrac{1}{3}}$ 50. $\sqrt{\dfrac{5}{4} - \dfrac{1}{8}}$ 51. $\sqrt{\dfrac{1}{a^2} + \dfrac{1}{b}}$ 52. $\sqrt{\dfrac{x}{y} + \dfrac{y}{x}}$

53. $\sqrt{\dfrac{x}{2x+1}}$ 54. $\sqrt{\dfrac{x-2}{x+2}}$ 55. $\sqrt{a^2 + 2ab + b^2}$ 56. $\sqrt{a^2 + b^2}$

57. $\sqrt{4x^2 - 1}$ 58. $\sqrt{9x^2 - 6x + 1}$ 59. $\sqrt{x^2 + \dfrac{1}{4}}$ 60. $\sqrt{\dfrac{1}{2} + 2r + 2r^2}$

In Exercises 61 through 64, perform the required operation.

61. An approximate equation for the efficiency E, in percent, of an engine is given by $E = 100(1 - 1/\sqrt[5]{R^2})$, where R is the compression ratio. Express this equation with a fractional exponent and find E for $R = 7.35$.

62. When analyzing the velocity of an object which falls through a very great distance, the expression $a\sqrt{2g/a}$ is derived. Show by rationalizing the denominator that this expression takes on a simpler form.

63. A formula for the angular velocity of a disc is $\omega = \sqrt{\dfrac{576EIg}{WL^3}}$. Rationalize the denominator on the right side of this equation.

64. In analyzing an electronic filter circuit, the expression $\dfrac{8A}{\pi^2\sqrt{1 + (f_0/f)^2}}$ is used. Rationalize the denominator, expressing the answer without the fraction f_0/f.

10-4   Addition and Subtraction of Radicals

When we first introduced the concept of adding algebraic expressions, we found that it was possible to combine similar terms, that is, those which differed only in numerical coefficients. The same is true of adding radicals. *We must have similar radicals in order to perform the addition*, rather than simply to be able to indicate addition. *By similar radicals we mean radicals which differ only in their numerical coefficients*, and which must therefore be of the same order and have the same radicand.

In order to add radicals, we first express each radical in its simplest form, rationalize any denominators, and then add those which are similar. For those which are not similar, we can only indicate the addition.

EXAMPLE A ⎯⎯⎯⎯

1. $2\sqrt{7} - 5\sqrt{7} + \sqrt{7} = -2\sqrt{7}$ all similar radicals
2. $\sqrt[5]{6} + 4\sqrt[5]{6} - 2\sqrt[5]{6} = 3\sqrt[5]{6}$ all similar radicals
3. $\sqrt{5} + 2\sqrt{3} - 5\sqrt{5} = 2\sqrt{3} - 4\sqrt{5}$ answer contains two terms

similar radicals

We note that in the last illustration we are able only to indicate the final subtraction.

EXAMPLE B ⎯⎯⎯⎯

1. $\sqrt{2} + \sqrt{8} = \sqrt{2} + \sqrt{4 \times 2} = \sqrt{2} + \sqrt{4}\sqrt{2}$
$= \sqrt{2} + 2\sqrt{2} = 3\sqrt{2}$
2. $\sqrt[3]{24} + \sqrt[3]{81} = \sqrt[3]{8 \times 3} + \sqrt[3]{27 \times 3} = \sqrt[3]{8}\sqrt[3]{3} + \sqrt[3]{27}\sqrt[3]{3}$
$= 2\sqrt[3]{3} + 3\sqrt[3]{3} = 5\sqrt[3]{3}$

NOTE ▷

Notice that $\sqrt{8}$, $\sqrt[3]{24}$, and $\sqrt[3]{81}$ were simplified before performing the addition. We also note that $\sqrt{2} + \sqrt{8}$ **is not** *equal to* $\sqrt{2 + 8}$.

We note in the illustrations of Example B that the radicals do not initially appear to be similar. However, after each is simplified we are able to recognize the similar radicals.

EXAMPLE C ⎯⎯⎯⎯

1. $6\sqrt{7} - \sqrt{28} + 3\sqrt{63} = 6\sqrt{7} - \sqrt{4 \times 7} + 3\sqrt{9 \times 7}$
$= 6\sqrt{7} - 2\sqrt{7} + 3(3\sqrt{7})$ all similar radicals
$= 6\sqrt{7} - 2\sqrt{7} + 9\sqrt{7}$
$= 13\sqrt{7}$
2. $3\sqrt{125} - \sqrt{20} + \sqrt{27} = 3\sqrt{25 \times 5} - \sqrt{4 \times 5} + \sqrt{9 \times 3}$
$= 3(5\sqrt{5}) - 2\sqrt{5} + 3\sqrt{3}$
$= 13\sqrt{5} + 3\sqrt{3}$ not similar to others

EXAMPLE D ⎯⎯⎯⎯

$\sqrt{24} + \sqrt{\dfrac{3}{2}} = \sqrt{4 \times 6} + \sqrt{\dfrac{3 \times 2}{2 \times 2}} = \sqrt{4}\sqrt{6} + \sqrt{\dfrac{6}{4}}$

$= 2\sqrt{6} + \dfrac{\sqrt{6}}{2} = \dfrac{4\sqrt{6} + \sqrt{6}}{2} = \dfrac{5}{2}\sqrt{6}$

One radical was simplified by removing the perfect square factor and in the other we rationalized the denominator. Note that we would not be able to combine the radicals if we did not rationalize the denominator of the second radical.

Our main purpose in this section is to add radicals in radical form. However, a decimal value can be obtained by use of a calculator, and in the following example we use decimal values to verify the result of the addition.

EXAMPLE E _____

Perform the addition $\sqrt{32} + 7\sqrt{18} - 2\sqrt{200}$ and use a calculator to verify the result.

$$
\begin{aligned}
\sqrt{32} + 7\sqrt{18} - 2\sqrt{200} &= \sqrt{16 \times 2} + 7\sqrt{9 \times 2} - 2\sqrt{100 \times 2} \\
&= 4\sqrt{2} + 7(3\sqrt{2}) - 2(10\sqrt{2}) \\
&= 4\sqrt{2} + 21\sqrt{2} - 20\sqrt{2} \\
&= 5\sqrt{2}
\end{aligned}
$$

Using the calculator sequence

$$32 \boxed{\sqrt{x}} \boxed{+} 7 \boxed{\times} 18 \boxed{\sqrt{x}} \boxed{-} 2 \boxed{\times} 200 \boxed{\sqrt{x}} \boxed{=} \boxed{7.0710678}$$

we find that the decimal value of the original expression is 7.0710678, which is the same as the value of $5\sqrt{2}$.

We now show two examples of adding radical expressions which contain literal symbols.

EXAMPLE F _____

$$
\begin{aligned}
\sqrt{\frac{2}{3a}} - 2\sqrt{\frac{3}{2a}} &= \sqrt{\frac{2(3a)}{3a(3a)}} - 2\sqrt{\frac{3(2a)}{2a(2a)}} = \sqrt{\frac{6a}{9a^2}} - 2\sqrt{\frac{6a}{4a^2}} \\
&= \frac{1}{3a}\sqrt{6a} - \frac{2}{2a}\sqrt{6a} = \frac{1}{3a}\sqrt{6a} - \frac{1}{a}\sqrt{6a} \\
&= \frac{\sqrt{6a} - 3\sqrt{6a}}{3a} = \frac{-2\sqrt{6a}}{3a} = -\frac{2}{3a}\sqrt{6a}
\end{aligned}
$$

EXAMPLE G _____

$$
\begin{aligned}
\sqrt{\frac{4}{a} - 4 + a} + \sqrt{\frac{1}{a} - \sqrt{16a^3}} &= \sqrt{\frac{4 - 4a + a^2}{a}} + \sqrt{\frac{1}{a} - 4a\sqrt{a}} \\
&= \sqrt{\frac{(2-a)^2 \times a}{a \times a}} + \sqrt{\frac{1 \times a}{a \times a}} - 4a\sqrt{a} \\
&= \frac{2-a}{a}\sqrt{a} + \frac{1}{a}\sqrt{a} - 4a\sqrt{a} \\
&= \sqrt{a}\left(\frac{2-a}{a} + \frac{1}{a} - 4a\right) = \sqrt{a}\left(\frac{2 - a + 1 - 4a^2}{a}\right) \\
&= \frac{(3 - a - 4a^2)\sqrt{a}}{a}
\end{aligned}
$$

This simplification is valid for $a \leq 2$, since we let $\sqrt{(2-a)^2} = 2 - a$.

Exercises 10-4

In Exercises 1 through 36, express each radical in simplest form, rationalize denominators, and perform the indicated operations.

1. $2\sqrt{3} + 5\sqrt{3}$

2. $8\sqrt{11} - 3\sqrt{11}$

3. $2\sqrt{7} + \sqrt{5} - 3\sqrt{7}$

4. $8\sqrt{6} - 2\sqrt{3} - 5\sqrt{6}$

5. $\sqrt{5} + \sqrt{20}$

6. $\sqrt{7} + \sqrt{63}$

7. $2\sqrt{3} - 3\sqrt{12}$

8. $4\sqrt{2} - \sqrt{50}$

9. $\sqrt{8a} - \sqrt{32a}$

10. $\sqrt{27x} + 2\sqrt{18x}$

11. $2\sqrt{28} + 3\sqrt{175}$

12. $5\sqrt{300} - 7\sqrt{48}$

13. $2\sqrt{20} - \sqrt{125} - \sqrt{45}$

14. $2\sqrt{44} - \sqrt{99} + \sqrt{2}\sqrt{88}$

15. $3\sqrt{75} + 2\sqrt{48} - 2\sqrt{18}$

16. $2\sqrt{28} - \sqrt{108} - 2\sqrt{175}$

17. $\sqrt{60} + \sqrt{\dfrac{5}{3}}$

18. $\sqrt{84} - \sqrt{\dfrac{3}{7}}$

19. $\sqrt{\dfrac{1}{2}} + \sqrt{\dfrac{25}{2}} - \sqrt{18}$

20. $\sqrt{6} - \sqrt{\dfrac{2}{3}} - \sqrt{18}$

21. $\sqrt[3]{81} + \sqrt[3]{3000}$

22. $\sqrt[3]{-16} + \sqrt[3]{54}$

23. $\sqrt[4]{32} - \sqrt[8]{4}$

24. $\sqrt[6]{\sqrt{2}} - \sqrt[12]{2^{13}}$

25. $\sqrt{a^3 b} - \sqrt{4ab^5}$

26. $\sqrt{2x^2 y} + \sqrt{8}\sqrt{y^5}$

27. $\sqrt{6}\sqrt{5}\sqrt{3} - \sqrt{40a^2}$

28. $\sqrt{60b^2 n} - b\sqrt{135n}$

29. $\sqrt[3]{24a^2 b^4} - \sqrt[3]{3a^5 b}$

30. $\sqrt[5]{32a^6 b^4} + 3a\sqrt[5]{243ab^9}$

31. $\sqrt{\dfrac{a}{c^5}} - \sqrt{\dfrac{c}{a^3}}$

32. $\sqrt{\dfrac{2x}{3y}} + \sqrt{\dfrac{27y}{8x}}$

33. $\sqrt[3]{\dfrac{a}{b}} - \sqrt[3]{\dfrac{8b^2}{a^2}}$

34. $\sqrt[4]{\dfrac{c}{b}} - \sqrt[4]{bc}$

35. $\sqrt{\dfrac{a-b}{a+b}} - \sqrt{\dfrac{a+b}{a-b}}$

36. $\sqrt{\dfrac{16}{x} + 8 + x} - \sqrt{1 - \dfrac{1}{x}}$

In Exercises 37 through 40, express each radical in simplified form, rationalize denominators, and perform the indicated operations. Then use a calculator to verify the result.

37. $3\sqrt{45} + 3\sqrt{75} - 2\sqrt{500}$

38. $2\sqrt{40} + 3\sqrt{90} - 5\sqrt{250}$

39. $2\sqrt{\tfrac{2}{3}} + \sqrt{24} - 5\sqrt{\tfrac{3}{2}}$

40. $\sqrt{\tfrac{2}{7}} - 2\sqrt{\tfrac{7}{2}} + 5\sqrt{56}$

In Exercises 41 through 44, solve the given problems.

41. Find the exact sum of the positive roots of $x^2 - 2x - 2 = 0$ and $x^2 + 2x - 11 = 0$.

42. Find the sum of the two roots of the quadratic equation $ax^2 + bx + c = 0$.

43. A rectangular piece of plywood 4 ft by 8 ft has corners cut from it as shown in Fig. 10-4. Find the perimeter of the remaining piece in exact form and in decimal form.

44. In the study of the kinetic theory of gases, the expression

$$a\sqrt{\dfrac{h^3 m^5}{\pi}} + b\sqrt{\dfrac{h^3 m^3}{\pi}}$$

is found. Perform the indicated addition.

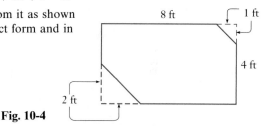

Fig. 10-4

10-5 Multiplication of Radicals

When multiplying expressions containing radicals, we use Eq. (10-10), along with the normal procedures of algebraic multiplication. Note that the orders of the radicals being multiplied in Eq. (10-10) are the same. The following examples illustrate the method.

EXAMPLE A

1. $\sqrt{5}\sqrt{2} = \sqrt{5 \times 2} = \sqrt{10}$

2. $\sqrt{33}\sqrt{3} = \sqrt{33 \times 3} = \sqrt{99} = \sqrt{9 \times 11} = \sqrt{9}\sqrt{11}$
   $= 3\sqrt{11}$

We note that we must be careful to express the resulting radical in simplest form.

EXAMPLE B

1. $\sqrt[3]{6}\sqrt[3]{4} = \sqrt[3]{6(4)} = \sqrt[3]{24} = \sqrt[3]{8}\sqrt[3]{3}$
$$= 2\sqrt[3]{3}$$

2. $\sqrt[5]{8a^3b^4}\sqrt[5]{8a^2b^3} = \sqrt[5]{(8a^3b^4)(8a^2b^3)} = \sqrt[5]{64a^5b^7} = \sqrt[5]{32a^5b^5}\sqrt[5]{2b^2}$
$$= 2ab\sqrt[5]{2b^2}$$

EXAMPLE C

$\sqrt{2}(3\sqrt{5} - 4\sqrt{2}) = 3\sqrt{2}\sqrt{5} - 4\sqrt{2}\sqrt{2} = 3\sqrt{10} - 4\sqrt{4}$
$$= 3\sqrt{10} - 4(2) = 3\sqrt{10} - 8$$

EXAMPLE D

$(5\sqrt{7} - 2\sqrt{3})(4\sqrt{7} + 3\sqrt{3})$
$$= (5)(4)\sqrt{7}\sqrt{7} + (5)(3)\sqrt{7}\sqrt{3} - (2)(4)\sqrt{3}\sqrt{7} - (2)(3)\sqrt{3}\sqrt{3}$$
$$= 20(7) + 15\sqrt{21} - 8\sqrt{21} - 6(3)$$
$$= 140 + 7\sqrt{21} - 18$$
$$= 122 + 7\sqrt{21}$$

Using a calculator, we find that the decimal value of the original expression is 154.07803, which is the same as that for $122 + 7\sqrt{21}$.

For reference, Eqs. (5-3) and (5-4) are $(x + y)^2 = x^2 + 2xy + y^2$ and $(x - y)^2 = x^2 - 2xy + y^2$.

When raising a single-term radical expression to a power, we use the basic meaning of the power. When raising a binomial to a power, we proceed as with any binomial. We use Eqs. (5-3) and (5-4), along with Eq. (10-10) if the binomial is squared. These are illustrated in the following two examples.

EXAMPLE E

1. $(2\sqrt{7})^2 = 2^2(\sqrt{7})^2 = 4(7) = 28$
2. $(2\sqrt{7})^3 = 2^3(\sqrt{7})^3 = 8(\sqrt{7})^2(\sqrt{7}) = 8(7)\sqrt{7}$
$$= 56\sqrt{7}$$

EXAMPLE F

1. $(3 + \sqrt{5})^2 = 3^2 + 2(3)\sqrt{5} + (\sqrt{5})^2 = 9 + 6\sqrt{5} + 5$
$$= 14 + 6\sqrt{5}$$

2. $(\sqrt{a} - \sqrt{b})^2 = (\sqrt{a})^2 - 2\sqrt{a}\sqrt{b} + (\sqrt{b})^2$
$$= a + b - 2\sqrt{ab}$$

EXAMPLE G

$(\sqrt{6} - \sqrt{2} - \sqrt{3})(\sqrt{6} + \sqrt{2}) = (\sqrt{6} - \sqrt{2})(\sqrt{6} + \sqrt{2}) - \sqrt{3}(\sqrt{6} + \sqrt{2})$
$$= (6 - 2) - \sqrt{18} - \sqrt{6}$$
$$= 4 - 3\sqrt{2} - \sqrt{6}$$

Note that by grouping $\sqrt{6} - \sqrt{2}$ together for multiplication we reduced the number of steps required.

EXAMPLE H

$\left(3\sqrt{\dfrac{a}{b}} - \sqrt{ab}\right)\left(2\sqrt{\dfrac{a}{b}} - \sqrt{ab}\right)$
$$= 6\sqrt{\dfrac{a}{b}}\sqrt{\dfrac{a}{b}} - 3\sqrt{\dfrac{a}{b}}\sqrt{ab} - 2\sqrt{ab}\sqrt{\dfrac{a}{b}} + \sqrt{ab}\sqrt{ab}$$

$$= 6\sqrt{\frac{a^2}{b^2}} - 3\sqrt{\frac{a^2b}{b}} - 2\sqrt{\frac{a^2b}{b}} + \sqrt{a^2b^2}$$

$$= 6\frac{a}{b} - 5\sqrt{\frac{a^2b}{b}} + ab = \frac{6a}{b} - 5a + ab$$

$$= \frac{6a - 5ab + ab^2}{b} = \frac{a(6 - 5b + b^2)}{b}$$

$$= \frac{a(3 - b)(2 - b)}{b}$$

NOTE $\triangleright$ Again, we note that *to multiply radicals and combine them under one radical sign, **it is necessary that the order of the radicals be the same.*** If necessary we can make the order of each radical the same by appropriate operations on each radical separately. Fractional exponents are frequently useful for this purpose.

EXAMPLE I _____

1. $\sqrt[3]{2}\sqrt{5} = 2^{1/3}5^{1/2} = 2^{2/6}5^{3/6} = (2^25^3)^{1/6} = \sqrt[6]{500}$

2. $\sqrt[3]{4a^2b}\sqrt[4]{8a^3b^2} = (2^2a^2b)^{1/3}(2^3a^3b^2)^{1/4} = (2^2a^2b)^{4/12}(2^3a^3b^2)^{3/12}$

$$= (2^8a^8b^4)^{1/12}(2^9a^9b^6)^{1/12} = (2^{17}a^{17}b^{10})^{1/12}$$

$$= 2a(2^5a^5b^{10})^{1/12}$$

$$= 2a\sqrt[12]{32a^5b^{10}}$$

Exercises 10-5 ──────────

In Exercises 1 through 48, perform the indicated multiplications, expressing answers in simplest form with rationalized denominators.

1. $\sqrt{3}\sqrt{10}$ **2.** $\sqrt{2}\sqrt{51}$ **3.** $\sqrt{6}\sqrt{2}$ **4.** $\sqrt{7}\sqrt{14}$

5. $\sqrt[3]{4}\sqrt[3]{2}$ **6.** $\sqrt[4]{3}\sqrt[4]{27}$ **7.** $\sqrt[5]{4}\sqrt[5]{16}$ **8.** $\sqrt[3]{25}\sqrt[3]{50}$

9. $(5\sqrt{2})^2$ **10.** $(3\sqrt{3})^2$ **11.** $(2\sqrt[3]{2})^3$ **12.** $(3\sqrt{5})^3$

13. $\sqrt{\frac{2}{3}}\sqrt{5}$ **14.** $\sqrt{8}\sqrt{\frac{5}{2}}$ **15.** $\sqrt{\frac{5}{6}}\sqrt{\frac{2}{11}}$ **16.** $\sqrt{\frac{6}{7}}\sqrt{\frac{2}{3}}$

17. $\sqrt{3}(\sqrt{2} - \sqrt{5})$ **18.** $\sqrt{5}(\sqrt{7} + \sqrt{2})$ **19.** $2\sqrt{2}(\sqrt{8} - 3\sqrt{6})$ **20.** $3\sqrt{5}(\sqrt{15} - 2\sqrt{5})$

21. $(2 - \sqrt{5})(2 + \sqrt{5})$ **22.** $(6 - \sqrt{3})(6 + \sqrt{3})$ **23.** $(6 - \sqrt{3})^2$ **24.** $(2 - \sqrt{5})^2$

25. $(3\sqrt{5} - 2\sqrt{3})(6\sqrt{5} + 7\sqrt{3})$ **26.** $(3\sqrt{7a} - \sqrt{8})(\sqrt{7a} + \sqrt{2})$

27. $(3\sqrt{11} - \sqrt{x})(2\sqrt{11} + 5\sqrt{x})$ **28.** $(2\sqrt{10} + 3\sqrt{15})(\sqrt{10} - 7\sqrt{15})$

29. $\sqrt{a}(\sqrt{ab} + \sqrt{c^3})$ **30.** $\sqrt{3x}(\sqrt{3x} - \sqrt{xy})$

31. $\sqrt{5n}(\sqrt{15n} + \sqrt{20m})$ **32.** $\sqrt{2x}(\sqrt{8xy} - 3\sqrt{y})$

33. $(\sqrt{2a} - \sqrt{b})(\sqrt{2a} + 3\sqrt{b})$ **34.** $(2\sqrt{mn} + 3\sqrt{n})^2$

35. $(\sqrt{2} + \sqrt{3} + \sqrt{5})(\sqrt{3} - \sqrt{5})$ **36.** $(2\sqrt{7} - \sqrt{5})(\sqrt{14} - 2\sqrt{5} + \sqrt{7})$

37. $\sqrt{2}\sqrt[3]{3}$ **38.** $\sqrt[3]{16}\sqrt[3]{8}$

39. $\sqrt[4]{ab}\sqrt[3]{bc^4}$ **40.** $\sqrt{2x}\sqrt[3]{16x}$

41. $(\sqrt[5]{\sqrt{6}} - \sqrt{5})(\sqrt[5]{\sqrt{6}} + \sqrt{5})$ **42.** $\sqrt[3]{5} - \sqrt{17}\sqrt[3]{5} + \sqrt{17}$

43. $(\sqrt{2x^2} - \sqrt[3]{y})(\sqrt{4x} + \sqrt[3]{y^2})$ **44.** $(\sqrt{a} - \sqrt[3]{b})(2\sqrt{a} - \sqrt[3]{b})$

45. $\left(\sqrt{\dfrac{2}{a}} + \sqrt{\dfrac{a}{2}} \right)\left(\sqrt{\dfrac{2}{a}} - 2\sqrt{\dfrac{a}{2}} \right)$

46. $\left(\sqrt{\dfrac{x}{y}} - \sqrt{xy} \right)\left(\sqrt{\dfrac{y}{x}} + \sqrt{xy} - 1 \right)$

47. $(2x - \sqrt{x - 2y})^2$

48. $(3 + \sqrt{6 - 2a})(2 - \sqrt{6 - 2a})$

In Exercises 49 through 52, perform the indicated multiplications, expressing answers in simplest form. Then use a calculator to verify the result.

49. $(\sqrt{11} + \sqrt{6})(\sqrt{11} - 2\sqrt{6})$

50. $(2\sqrt{5} - \sqrt{7})(3\sqrt{5} + \sqrt{7})$

51. $(5\sqrt{13} + 2\sqrt{14})(2\sqrt{13} + \sqrt{14})$

52. $(3\sqrt{3} - 5\sqrt{6})(8\sqrt{3} - 7\sqrt{6})$

In Exercises 53 through 56, combine the terms into a single fraction, but do not rationalize the denominators.

53. $2\sqrt{x} + \dfrac{1}{\sqrt{x}}$

54. $\dfrac{3}{2\sqrt{3x - 4}} - \sqrt{3x - 4}$

55. $\dfrac{x^2}{\sqrt{2x + 1}} + 2x\sqrt{2x + 1}$

56. $4\sqrt{x^2 + 1} - \dfrac{4x}{\sqrt{x^2 + 1}}$

In Exercises 57 through 60, solve the given problems.

57. By substitution, show that $x = 1 - \sqrt{2}$ is a solution to the equation $x^2 - 2x - 1 = 0$.

58. Determine the product of the two roots of the quadratic equation $ax^2 + bx + c = 0$.

59. For an object oscillating at the end of a spring, and on which there is a force retarding the motion, the equation $m^2 + bm + k^2 = 0$ must be solved. Here, b is a constant related to the retarding force and k is the spring constant. By substitution, show that $m = \frac{1}{2}(\sqrt{b^2 - 4k^2} - b)$ is a solution.

60. A square plastic sheet of side x is stretched by an amount equal to $\sqrt{x}$ in each direction. Find the expression for the percent increase in area of the sheet.

10-6 Division of Radicals

We have already dealt with some cases of division of radicals in the previous sections. When we have had the indicated division of one radical by another, we have generally expressed the answer with no radicals in the denominator by rationalizing the denominator. Therefore, if a change is to be made in an expression involving division by a radical, rationalization of the denominator is the principal step to be carried out. When we rationalize the denominator, we change the fraction to an equivalent form in which the denominator is free of radicals. In doing so, multiplication of the numerator and the denominator by the appropriate quantity is a primary step.

EXAMPLE A

1. $\dfrac{\sqrt{3}}{\sqrt{5}} = \dfrac{\sqrt{3}\sqrt{5}}{\sqrt{5}\sqrt{5}} = \dfrac{\sqrt{15}}{5}$ ⟵ no radical in denominator

2. $\dfrac{\sqrt{a}}{\sqrt[3]{b}} = \dfrac{\sqrt{a}}{\sqrt[3]{b}} \times \dfrac{\sqrt[3]{b^2}}{\sqrt[3]{b^2}} = \dfrac{\sqrt{a}\,\sqrt[3]{b^2}}{b} = \dfrac{a^{3/6}b^{4/6}}{b} = \dfrac{\sqrt[6]{a^3 b^4}}{b}$ ⟵

Notice in the second illustration that the denominator was rationalized and that the factors of the numerator were written in terms of fractional exponents so they could be combined under one radical.

If the denominator is the sum (or difference) of two terms, at least one of which is a radical, the fraction can be rationalized by multiplying both the numerator and the denominator by the difference (or sum) of the same two terms, if the radicals are square roots.

EXAMPLE B _____ The fraction

$$\frac{1}{\sqrt{3} - \sqrt{2}}$$

can be rationalized by multiplying the numerator and the denominator by $\sqrt{3} + \sqrt{2}$. In this way the radicals will be removed from the denominator.

$$\frac{1}{\sqrt{3} - \sqrt{2}} \times \frac{\sqrt{3} + \sqrt{2}}{\sqrt{3} + \sqrt{2}} = \frac{\sqrt{3} + \sqrt{2}}{(\sqrt{3})^2 - (\sqrt{2})^2} = \frac{\sqrt{3} + \sqrt{2}}{3 - 2} = \sqrt{3} + \sqrt{2}$$

NOTE ▷   change sign

The reason this technique works is that an expression of the form $a^2 - b^2$ is created in the denominator, where a or b (or both) is a radical. We see that ■ the result is a denominator free of radicals.

EXAMPLE C _____

$$\frac{\sqrt{2}}{2\sqrt{5} + \sqrt{3}} = \frac{\sqrt{2}}{2\sqrt{5} + \sqrt{3}} \times \frac{2\sqrt{5} - \sqrt{3}}{2\sqrt{5} - \sqrt{3}} = \frac{2\sqrt{2}\sqrt{5} - \sqrt{2}\sqrt{3}}{(2\sqrt{5})^2 - (\sqrt{3})^2}$$

change sign

$$= \frac{2\sqrt{10} - \sqrt{6}}{2^2(\sqrt{5})^2 - (\sqrt{3})^2} = \frac{2\sqrt{10} - \sqrt{6}}{20 - 3} = \frac{2\sqrt{10} - \sqrt{6}}{17}$$

Calculating the decimal value of the original expression and that of the answer, ■ we find that they are both 0.2279450.

EXAMPLE D _____

$$\frac{1 + \dfrac{\sqrt{3}}{2}}{1 - \dfrac{\sqrt{3}}{2}} = \frac{\dfrac{2 + \sqrt{3}}{2}}{\dfrac{2 - \sqrt{3}}{2}} = \frac{2 + \sqrt{3}}{2} \times \frac{2}{2 - \sqrt{3}} = \frac{2 + \sqrt{3}}{2 - \sqrt{3}}$$

$$= \frac{(2 + \sqrt{3})(2 + \sqrt{3})}{(2 - \sqrt{3})(2 + \sqrt{3})} = \frac{4 + 4\sqrt{3} + 3}{4 - 3} = 7 + 4\sqrt{3}$$

We note that in this particular case, the result has a much simpler form after ra- ■ tionalizing the denominator than the original expression.

EXAMPLE E

In studying the properties of a semiconductor, the expression

$$\frac{\sqrt{1 + 2V}}{C_1 + C_2\sqrt{1 + 2V}}$$

is used. Here C_1 and C_2 are constants and V is the voltage across a junction of the semiconductor. Rationalize the denominator of this expression.

Multiplying the numerator and the denominator by $C_1 - C_2\sqrt{1 + 2V}$, we have

$$\frac{\sqrt{1 + 2V}}{C_1 + C_2\sqrt{1 + 2V}} = \frac{\sqrt{1 + 2V}(C_1 - C_2\sqrt{1 + 2V})}{(C_1 + C_2\sqrt{1 + 2V})(C_1 - C_2\sqrt{1 + 2V})}$$

$$= \frac{C_1\sqrt{1 + 2V} + C_2\sqrt{1 + 2V}\sqrt{1 + 2V}}{C_1^2 - C_2^2(\sqrt{1 + 2V})^2}$$

$$= \frac{C_1\sqrt{1 + 2V} + C_2(1 + 2V)}{C_1^2 - C_2^2(1 + 2V)}$$

In certain types of algebraic operations it may be necessary to rationalize the numerator of an expression. This procedure is illustrated in the following example.

EXAMPLE F

Rationalize the numerator of the expression $\dfrac{\sqrt{2x + 3} - \sqrt{2x}}{6}$.

We follow the same basic procedure as in rationalizing the denominator of an expression. In this case we multiply the numerator and the denominator of this fraction by $\sqrt{2x + 3} + \sqrt{2x}$. This gives us the following solution:

$$\frac{\sqrt{2x + 3} - \sqrt{2x}}{6} = \frac{(\sqrt{2x + 3} - \sqrt{2x})(\sqrt{2x + 3} + \sqrt{2x})}{6(\sqrt{2x + 3} + \sqrt{2x})}$$

$$= \frac{(\sqrt{2x + 3})^2 - (\sqrt{2x})^2}{6(\sqrt{2x + 3} + \sqrt{2x})} = \frac{(2x + 3) - 2x}{6(\sqrt{2x + 3} + \sqrt{2x})}$$

$$= \frac{3}{6(\sqrt{2x + 3} + \sqrt{2x})} = \frac{1}{2(\sqrt{2x + 3} + \sqrt{2x})}$$

■ Note that the resulting *numerator* does not contain any radicals.

Exercises 10-6

In Exercises 1 through 40, perform the indicated operations and express the answers in simplest form with rationalized denominators.

1. $\dfrac{\sqrt{21}}{\sqrt{3}}$

2. $\dfrac{\sqrt{105}}{\sqrt{5}}$

3. $\sqrt{7} \div \sqrt{2}$

4. $3\sqrt{2} \div 2\sqrt{3}$

5. $\dfrac{\sqrt[3]{x^2}}{\sqrt[3]{24}}$

6. $\dfrac{\sqrt[4]{3}}{\sqrt[4]{2}}$

7. $\dfrac{\sqrt[3]{5}}{\sqrt{2}}$

8. $\dfrac{\sqrt{6}}{\sqrt[3]{2}}$

9. $\dfrac{\sqrt{a}}{\sqrt[3]{4}}$

10. $\dfrac{\sqrt[4]{32}}{\sqrt[5]{b^3}}$

11. $\dfrac{\sqrt{6}-3}{\sqrt{6}}$

12. $\dfrac{5-\sqrt{10}}{\sqrt{10}}$

13. $\dfrac{\sqrt{2a}-b}{\sqrt{a}}$

14. $\dfrac{\sqrt{8x}+\sqrt{2}}{\sqrt{2}}$

15. $\dfrac{\sqrt{3a}-\sqrt{b}}{\sqrt{3}}$

16. $\dfrac{\sqrt{7x}-\sqrt{14}}{\sqrt{7}}$

17. $\dfrac{1}{\sqrt{7}+\sqrt{3}}$

18. $\dfrac{4}{6+\sqrt{2}}$

19. $\dfrac{\sqrt{7}}{5-\sqrt{2}}$

20. $\dfrac{\sqrt{8}}{2\sqrt{3}-\sqrt{5}}$

21. $\dfrac{3}{2\sqrt{5}-6}$

22. $\dfrac{\sqrt{7}}{4-2\sqrt{7}}$

23. $\dfrac{2\sqrt{3}}{3\sqrt{3}-1}$

24. $\dfrac{6\sqrt{5}}{5-2\sqrt{5}}$

25. $\dfrac{\sqrt{2}-1}{\sqrt{7}-3\sqrt{2}}$

26. $\dfrac{3-\sqrt{5}}{2\sqrt{2}+\sqrt{5}}$

27. $\dfrac{2-\sqrt{3}}{5-2\sqrt{3}}$

28. $\dfrac{2\sqrt{15}-3}{\sqrt{15}+4}$

29. $\dfrac{2\sqrt{3}-5\sqrt{5}}{\sqrt{3}+2\sqrt{5}}$

30. $\dfrac{2\sqrt{6}+\sqrt{11}}{\sqrt{6}-3\sqrt{11}}$

31. $\dfrac{\sqrt{7}-\sqrt{14}}{2\sqrt{7}-3\sqrt{14}}$

32. $\dfrac{\sqrt{15}-3\sqrt{5}}{2\sqrt{15}-\sqrt{5}}$

33. $\dfrac{2\sqrt{x}}{\sqrt{x}-\sqrt{y}}$

34. $\dfrac{6\sqrt{a}}{2\sqrt{a}-b}$

35. $\dfrac{8}{3\sqrt{a}-2\sqrt{b}}$

36. $\dfrac{6}{1+2\sqrt{x}}$

37. $\dfrac{\sqrt{2c}+3d}{\sqrt{2c}-d}$

38. $\dfrac{3\sqrt{2x}+2\sqrt{x}}{\sqrt{2x}-\sqrt{x}}$

39. $\dfrac{\sqrt{x+y}}{\sqrt{x-y}-\sqrt{x}}$

40. $\dfrac{\sqrt{1+a}}{a-\sqrt{1-a}}$

In Exercises 41 through 44, perform the indicated operations and express the answers in simplest form with rationalized denominators. Then use a calculator to verify the result.

41. $\dfrac{2\sqrt{7}}{\sqrt{7}+\sqrt{6}}$

42. $\dfrac{3\sqrt{13}}{\sqrt{13}-\sqrt{11}}$

43. $\dfrac{2\sqrt{6}-\sqrt{5}}{3\sqrt{6}-4\sqrt{5}}$

44. $\dfrac{\sqrt{7}-4\sqrt{2}}{5\sqrt{7}-4\sqrt{2}}$

In Exercises 45 through 48, rationalize the numerator of each of the given fractions. See Example F.

45. $\dfrac{\sqrt{5}+\sqrt{2}}{3\sqrt{6}}$

46. $\dfrac{\sqrt{19}-3}{5}$

47. $\dfrac{\sqrt{x+h}-\sqrt{x}}{h}$

48. $\dfrac{\sqrt{3x+4}+\sqrt{3x}}{8}$

In Exercises 49 through 52, rationalize the given expressions.

49. The rim velocity of a flywheel is given by the equation $v=\dfrac{\sqrt{gs}}{\sqrt{12w}}$. Rationalize the denominator of the right side of this equation.

50. An expression used in determining the characteristics of a spur gear is $\dfrac{50}{50+\sqrt{V}}$. Write this expression in rationalized form.

51. In the hydrodynamics of fluid flow, the expression $\dfrac{1}{\sqrt{2g}(\sqrt{h_2}-\sqrt{h_1})}$ is used. Rationalize the denominator.

52. In analyzing a tuned electronic amplifier circuit, the expression $\dfrac{2Q}{\sqrt{\sqrt{2}-1}}$ is used. Rationalize the denominator.

10-7   Chapter Equations, Review Exercises, and Practice Test

Chapter Equations

Exponents
$$a^m \times a^n = a^{m+n} \tag{10-1}$$

$$\frac{a^m}{a^n} = a^{m-n} \quad \text{or} \quad \frac{a^m}{a^n} = \frac{1}{a^{n-m}}, \quad a \neq 0 \tag{10-2}$$

$$(a^m)^n = a^{mn} \tag{10-3}$$

$$(ab)^n = a^n b^n, \qquad \left(\frac{a}{b}\right)^n = \frac{a^n}{b^n}, \quad b \neq 0 \tag{10-4}$$

$$a^0 = 1, \quad a \neq 0 \tag{10-5}$$

$$a^{-n} = \frac{1}{a^n}, \quad a \neq 0 \tag{10-6}$$

Fractional exponents
$$a^{1/n} = \sqrt[n]{a} \tag{10-7}$$

$$a^{m/n} = \sqrt[n]{a^m} = (\sqrt[n]{a})^m \tag{10-8}$$

Radicals
$$\sqrt[n]{a^n} = (\sqrt[n]{a})^n = a \tag{10-9}$$

$$\sqrt[n]{a}\,\sqrt[n]{b} = \sqrt[n]{ab} \tag{10-10}$$

$$\sqrt[m]{\sqrt[n]{a}} = \sqrt[mn]{a} \tag{10-11}$$

$$\frac{\sqrt[n]{a}}{\sqrt[n]{b}} = \sqrt[n]{\frac{a}{b}}, \quad b \neq 0 \tag{10-12}$$

Review Exercises

In Exercises 1 through 28, express each of the given expressions in the simplest form which contains only positive exponents.

1. $2a^{-2}b^0$ **2.** $(2c)^{-1}z^{-2}$ **3.** $\dfrac{2c^{-1}}{d^{-3}}$ **4.** $\dfrac{-5x^0}{3y^{-1}}$

5. $3(25)^{3/2}$ **6.** $32^{2/5}$ **7.** $400^{-3/2}$ **8.** $1000^{-2/3}$

9. $\left(\dfrac{3}{t^2}\right)^{-2}$ **10.** $\left(\dfrac{2x^3}{3}\right)^{-3}$ **11.** $\dfrac{-8^{2/3}}{49^{-1/2}}$ **12.** $\dfrac{81^{-0.75}}{6^{-3}}$

13. $(2a^{1/3}b^{5/6})^6$ **14.** $(ax^{-1/2}y^{1/4})^8$ **15.** $(-32m^{15}n^{10})^{3/5}$ **16.** $(27x^{-6}y^9)^{2/3}$

17. $2x^{-2} - y^{-1}$ **18.** $a^4a^{-1} + (a^4)^{-1}$ **19.** $\dfrac{2x^{-1}}{2x^{-1} + y^{-1}}$ **20.** $\dfrac{3a}{(2a)^{-1} - a}$

21. $(a - 3b^{-1})^{-1}$ **22.** $(2s^{-2} + t)^{-2}$ **23.** $(x^3 - y^{-3})^{1/3}$ **24.** $(x^2 + 2xy + y^2)^{-1/2}$

25. $(8a^3)^{2/3}(4a^{-2} + 1)^{1/2}$

26. $\left[\dfrac{(9a)^0(4x^2)^{1/3}(3b^{1/2})}{(2b^0)^2}\right]^{-6}$

27. $2x(x - 1)^{-2} - 2(x^2 + 1)(x - 1)^{-3}$

28. $4(1 - x^2)^{1/2} - (1 - x^2)^{-1/2}$

In Exercises 29 through 72, perform the indicated operations and express the answer in simplest radical form with ratio-nalized denominators.

29. $\sqrt{68}$ **30.** $\sqrt{96}$ **31.** $\sqrt{ab^5c^2}$ **32.** $\sqrt{x^3y^4z^6}$

33. $\sqrt{9a^3b^4}$ **34.** $\sqrt{8x^5y^2}$ **35.** $\sqrt{84st^3u^2}$ **36.** $\sqrt{52x^2y^5}$

37. $\dfrac{5}{\sqrt{2s}}$ **38.** $\dfrac{3a}{\sqrt{5x}}$ **39.** $\sqrt{\dfrac{11}{27}}$ **40.** $\sqrt{\dfrac{7}{8}}$

41. $\sqrt[4]{8m^6n^9}$ **42.** $\sqrt[3]{9a^7b^{-3}}$ **43.** $\sqrt[4]{\sqrt[3]{64}}$ **44.** $\sqrt{a^{-3}\sqrt[5]{b^{12}}}$

45. $\sqrt{200} + \sqrt{32}$ **46.** $2\sqrt{68x} - \sqrt{153x}$ **47.** $\sqrt{63} - 2\sqrt{112} - \sqrt{28}$

48. $2\sqrt{20} - \sqrt{80} - 2\sqrt{125}$ **49.** $a\sqrt{2x^3} + \sqrt{8a^2x^3}$ **50.** $2\sqrt{m^2n^3} - \sqrt{n^5}$

51. $\sqrt[3]{8a^4} + b\sqrt[3]{a}$ **52.** $\sqrt[4]{2xy^5} - \sqrt[4]{32xy}$ **53.** $\sqrt{5}(2\sqrt{5} - \sqrt{11})$

54. $2\sqrt{8}(5\sqrt{2} - \sqrt{6})$ **55.** $2\sqrt{2}(\sqrt{6} - \sqrt{10})$ **56.** $3\sqrt{5}(\sqrt{15} + 2\sqrt{35})$

57. $(2 - 3\sqrt{17})(3 + \sqrt{17})$ **58.** $(5\sqrt{6} - 4)(3\sqrt{6} + 5)$

59. $(2\sqrt{7} - 3\sqrt{a})(3\sqrt{7} + \sqrt{a})$ **60.** $(3\sqrt{2} - \sqrt{13})(5\sqrt{2} + 3\sqrt{13})$

61. $\dfrac{\sqrt{3x}}{2\sqrt{3x} - \sqrt{y}}$ **62.** $\dfrac{5\sqrt{a}}{2\sqrt{a} - c}$ **63.** $\dfrac{\sqrt{2}}{\sqrt{3} - 4\sqrt{2}}$ **64.** $\dfrac{4}{3 - 2\sqrt{7}}$

65. $\dfrac{\sqrt{7} - \sqrt{5}}{\sqrt{5} + 3\sqrt{7}}$ **66.** $\dfrac{4 - 2\sqrt{6}}{3 + 2\sqrt{6}}$ **67.** $\dfrac{2\sqrt{x} - a}{3\sqrt{x} + 5a}$ **68.** $\dfrac{3\sqrt{y} - \sqrt{z}}{2\sqrt{y} + 5\sqrt{z}}$

69. $\sqrt{4b^2 + 1}$ **70.** $\sqrt{a^{-2} + \dfrac{1}{b^2}}$

71. $\left(\dfrac{2 - \sqrt{15}}{2}\right)^2 - \left(\dfrac{2 - \sqrt{15}}{2}\right)$ **72.** $\sqrt{2 + \dfrac{b}{a} + \dfrac{a}{b}} + \sqrt{a^4b^2 + 2a^3b^2 + a^2b^2}$

In Exercises 73 through 76, perform the indicated operations and express the answer in simplest radical form with ratio-nalized denominators. Then verify the results with a calculator.

73. $\sqrt{52} + 4\sqrt{24} - \sqrt{54}$ **74.** $2\sqrt{5}(6\sqrt{5} - 5\sqrt{6})$

75. $(\sqrt{7} - 2\sqrt{15})(3\sqrt{7} - \sqrt{15})$ **76.** $\dfrac{2\sqrt{3} - 7\sqrt{14}}{3\sqrt{3} + 2\sqrt{14}}$

In Exercises 77 through 88, perform the indicated operations.

77. The average annual increase i, in percent, of the cost of living over n years is given by $i = 100[(C_2/C_1)^{1/n} - 1]$, where C_1 is the cost of living index for the first year of the period and C_2 is the cost of living index for the last year of the period. Evaluate i if $C_1 = 247.0$ and $C_2 = 442.3$ are the values for 1980 and 1990, respectively.

78. Kepler's third law of planetary motion may be given as $T = kr^{3/2}$, where T is the time for one revolution of a planet around the sun, r is its mean radius from the sun, and $k = 1.115 \times 10^{-12}$ year/mi$^{3/2}$. Find the time for one revolution of Venus about the sun if $r = 6.73 \times 10^7$ mi.

79. The speed v of a ship of weight W whose engines produce power P is given by $v = k\sqrt[3]{P/W}$. Express this equation (a) with a fractional exponent and (b) as a radical with the denominator rationalized.

80. In the theory of semiconductors, the expression $km^{3/2}(E - E_1)^{1/2}$ is found. Write this expression in simplified radical form.

81. When studying atomic structure, the expression $\dfrac{v}{n_2^{-2} - n_1^{-2}}$ is used. Express this in simplest form with only positive exponents.

82. In hydrodynamics, the expression $v(4 + 3ar^{-1} - a^3r^{-3})^{-1}$ arises. Express this in simplest form with only positive exponents.

83. The period T of the motion of a torsion pendulum (see Exercise 50 of Section 9-7) is given by $T = 2\pi\sqrt{\dfrac{2\ell I}{\pi\mu a^4}}$, where ℓ is the length of the pendulum, r is the radius of the wire, and I and μ are constants. Express this with the denominator rationalized.

84. In thermodynamics, the expression $\left(\dfrac{1}{a^3} + \dfrac{1}{b^3}\right)^{-1/3}$ is used. Combine the fractions and simplify, expressing the answer without negative exponents.

85. In an experiment, a laser follows the path shown in Fig. 10-5. Express the length of the path in simplest radical form.

86. The frequency of a certain electric circuit is given by

$$\dfrac{1}{2\pi\sqrt{\dfrac{LC_1C_2}{C_1 + C_2}}}$$

Express this in simplest rationalized radical form.

Fig. 10-5

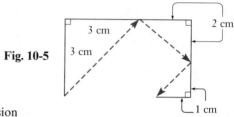

87. In determining the deflection of a certain type of beam, the expression

$$\ell\sqrt{\dfrac{a}{2\ell + a}}$$

is used. Express this in simplest rationalized radical form.

88. A computer analysis of an experiment showed that the fraction f of viruses surviving X-ray dosages was given by $f = \dfrac{20}{d + \sqrt{3d + 400}}$, where d is the dosage. Express this with the denominator rationalized.

Practice Test

In Problems 1 through 12, simplify the given expressions. For those with exponents, express each answer with only positive exponents. For the radicals, rationalize the denominator where applicable.

1. $2\sqrt{20} - \sqrt{125}$

2. $\dfrac{100^{3/2}}{8^{-2/3}}$

3. $\sqrt{x^2 + \dfrac{1}{9}}$

4. $(2x^{-1} + y^{-2})^{-1}$

5. $(\sqrt{2x} - 3\sqrt{y})^2$

6. $\sqrt[3]{\sqrt[4]{4}}$

7. $\dfrac{3 - 2\sqrt{2}}{2\sqrt{x}}$

8. $\sqrt{27a^4b^3}$

9. $(2x + 3)^{1/2} + (x + 1)(2x + 3)^{-1/2}$

10. $2\sqrt{2}(3\sqrt{10} - \sqrt{6})$

11. $2\sqrt{\dfrac{5}{3a}} - \sqrt{\dfrac{3}{5a}}$

12. $\dfrac{2\sqrt{15} + \sqrt{3}}{\sqrt{15} - 2\sqrt{3}}$

13. Express $\dfrac{3^{-1/2}}{2}$ in simplest radical form with a rationalized denominator.

14. In the study of fluid flow in pipes, the expression $0.220N^{-1/6}$ is found. Evaluate this expression for $N = 64 \times 10^6$.

11 Complex Numbers

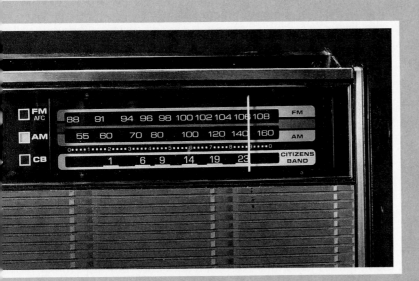

In Section 11-7 we see a basic application of complex numbers to electricity, including the tuning of a radio.

In Chapter 1, when we were introducing the topic of numbers, imaginary numbers were mentioned. Again, when we considered quadratic equations and their solutions in Chapter 6, we briefly came across this type of number. However, until now we have purposely avoided any extended discussion of imaginary numbers. In this chapter we shall discuss *complex numbers,* which include both the real numbers and the imaginary numbers.

Despite their names, complex numbers and imaginary numbers have very real and useful applications in many technical areas. One of the most important is in electricity and electronics, where these numbers are used extensively. Other applications are found in mechanical vibrations and in optics.

11-1 Basic Definitions

When we defined radicals we were able to define square roots of positive numbers easily, since any positive or negative number squared equals a positive number. For this reason we can see that it is impossible to square any real number and have the product equal a negative number. We must define a new number system if we wish to include square roots of negative numbers. With the proper definitions, we shall find that these numbers can be used to great advantage in certain applications.

If the radicand in a square root is negative, we can express the indicated root as the product of $\sqrt{-1}$ and the square root of a positive number. *The symbol $\sqrt{-1}$ is defined as the* **imaginary unit** *and is denoted by the symbol j.* In keeping with the definition of j, we have

$$j^2 = -1 \qquad (11\text{-}1)$$

Generally, mathematicians use the symbol i for $\sqrt{-1}$, and therefore most non-technical textbooks use i. However, one of the major technical applications of complex numbers is in electronics, where i represents electric current. Therefore, we shall use j for $\sqrt{-1}$, which is also the standard symbol in electronics textbooks.

EXAMPLE A

1. $\sqrt{-9} = \sqrt{(9)(-1)} = \sqrt{9}\sqrt{-1} = 3j$
2. $\sqrt{-16} = \sqrt{16}\sqrt{-1} = 4j$
3. $\sqrt{-0.25} = \sqrt{0.25}\sqrt{-1} = 0.5j$
4. $\sqrt{-5} = \sqrt{5}\sqrt{-1} = \sqrt{5}j = j\sqrt{5}$ **this form is better**

When a radical appears in the final result, we will write this result in a form with the j before the radical as in the last illustration. This clearly shows that the j is not *under* the radical. ∎

EXAMPLE B

$$(\sqrt{-4})^2 = (j\sqrt{4})^2 = 4j^2 = -4$$

We note that the simplification of this expression does not follow Eq. (10-10), which states that $\sqrt{ab} = \sqrt{a}\sqrt{b}$ for square roots. This is the reason it was noted as being valid only if a and b are positive. In fact, *Eq. (10-10) does not necessarily hold in general for negative values for a or b.* If $(\sqrt{-4})^2$ did follow Eq. (10-10) we would have

$$(\sqrt{-4})^2 = \sqrt{-4}\sqrt{-4} = \sqrt{(-4)(-4)} = \sqrt{16} = 4$$

NOTE ▷

this step is incorrect if both are negative, as in this case

We note that we obtain 4 and do not obtain the correct result of -4. ∎

EXAMPLE C

To further illustrate the method of handling square roots of negative numbers, consider the difference between $\sqrt{-3}\sqrt{-12}$ and $\sqrt{(-3)(-12)}$. For these expressions we have

$$\sqrt{-3}\sqrt{-12} = (j\sqrt{3})(j\sqrt{12}) = (\sqrt{3}\sqrt{12})j^2 = (\sqrt{36})j^2$$
$$= 6(-1) = -6$$

and

$$\sqrt{(-3)(-12)} = \sqrt{36} = 6$$

For $\sqrt{-3}\sqrt{-12}$ we have the product of square roots of negative numbers, whereas for $\sqrt{(-3)(-12)}$ we have the product of negative numbers under the radical. We must be careful to note the difference. ∎

From Examples B and C we see that *when we are dealing with the square roots of negative numbers,* **each should be expressed in terms of j before proceeding.** To do this, for any positive real number *a* we write

$$\sqrt{-a} = j\sqrt{a}, \quad (a > 0) \tag{11-2}$$

EXAMPLE D —— 1. $\sqrt{-6} = \sqrt{(6)(-1)} = \sqrt{6}\sqrt{-1} = j\sqrt{6}$

this step is correct if only
one is negative, as in this case

2. $-\sqrt{-75} = -\sqrt{(25)(3)(-1)} = -\sqrt{(25)(3)}\sqrt{-1} = -5j\sqrt{3}$

■ We note that $-\sqrt{-75}$ *is* **not** *equal to* $\sqrt{75}$. The minus signs are handled as shown.

In working with imaginary numbers, we often need to be able to raise these numbers to some power. Therefore, using the definitions of exponents and of *j*, we have the following results:

$$j = j, \qquad\qquad j^4 = j^2j^2 = (-1)(-1) = 1$$
$$j^2 = -1, \qquad\qquad j^5 = j^4j = j$$
$$j^3 = j^2j = -j, \qquad j^6 = j^4j^2 = (1)(-1) = -1$$

The powers of *j* go through the cycle of *j*, -1, $-j$, 1, *j*, -1, $-j$, 1, and so forth. Noting this and the fact that *j* raised to a power which is a multiple of 4 equals 1 allows us to raise *j* to any integral power almost on sight.

EXAMPLE E —— 1. $j^{10} = j^8j^2 = (1)(-1) = -1$
2. $j^{45} = j^{44}j = (1)(j) = j$
3. $j^{531} = j^{528}j^3 = (1)(-j) = -j$

⌐ exponents are multiples of 4

Using real numbers and the imaginary unit *j*, we define a new kind of number. *A* **complex number** *is any number which can be written in the form a + bj, where a and b are real numbers. If a = 0 and b ≠ 0, we have a number of the form bj, which is a* **pure imaginary number.** *If b = 0, then a + bj is a real number. The* **rectangular form** *of a complex number, where a is* **of a complex** *known as the* **real part** *and b is known as the* **imaginary part.** We can see that **number** complex numbers include all the real numbers and all of the pure imaginary numbers.

A comment here about the words *imaginary* and *complex* is in order. The choice of the names of these numbers is historical in nature, and unfortunately it leads to some misconceptions about the numbers. The use of imaginary does not imply that the numbers do not exist. Imaginary numbers do in fact exist, as they are defined above. In the same way, the use of complex does not imply that the numbers are complicated and therefore difficult to understand. With the appropriate definitions and operations, we can work with complex numbers, just as with any type of number.

In the following section we define the basic operations for complex numbers. Before doing this it is necessary to define the equality of complex numbers. Since a complex number is generally the sum of a real number and an imaginary number, it is not positive or negative in the usual sense, but each of the real part and the imaginary part is positive or negative. Therefore, *we define two complex numbers to be equal if the real parts are equal and the imaginary parts are equal.* That

NOTE ▷ is, ***two complex numbers, a + bj and x + yj, are equal if a = x and b = y.***

EXAMPLE F

1. $a + bj = 3 + 4j$ if $a = 3$ and $b = 4$
2. $x + yj = 5 - 3j$ if $x = 5$ and $y = -3$

↑ └─ imaginary parts

■ real parts

EXAMPLE G

What values of x and y satisfy the equation $4 - 6j - x = j + jy$?

One way to solve this is to rearrange the terms so that all the known terms are on the right and all the terms containing the unknowns x and y are on the left. This leads to $-x - jy = -4 + 7j$. From the definition of equality of complex numbers, $-x = -4$ and $-y = 7$, or $x = 4$ and $y = -7$.

EXAMPLE H

What values of x and y satisfy the equation

$$x + 3(xj + y) = 5 - j - jy$$

Rearranging the terms so that the known terms are on the right and the terms containing x and y are on the left, we have

$$x + 3y + 3jx + jy = 5 - j$$

Next, factoring j from the two terms on the left will put the expression on the left into proper form. This leads to

$$(x + 3y) + (3x + y)j = 5 - j$$

Using the definition of equality, we have

$$x + 3y = 5 \quad \text{and} \quad 3x + y = -1$$

We now solve this system of equations. The solution is $x = -1$ and $y = 2$. Actually, the solution can be obtained at any point by writing each side of the equation in the form $a + bj$ and then equating first the real parts and then the imaginary parts.

The **conjugate** *of the complex number* $a + bj$ *is the complex number* $a - bj$. We see that the sign of the imaginary part of a complex number is changed to obtain its conjugate.

EXAMPLE I —— $3 - 2j$ is the conjugate of $3 + 2j$. We may also say that $3 + 2j$ is the conjugate of $3 - 2j$. Thus each is the conjugate of the other.

$-2 - 5j$ and $-2 + 5j$ are conjugates.

$6j$ and $-6j$ are conjugates.

3 is the conjugate of 3 (imaginary part is zero).

Exercises 11-1

In Exercises 1 through 12, express each number in terms of j.

1. $\sqrt{-81}$ **2.** $\sqrt{-121}$ **3.** $-\sqrt{-4}$ **4.** $-\sqrt{-49}$

5. $\sqrt{-0.36}$ **6.** $-\sqrt{-0.01}$ **7.** $\sqrt{-8}$ **8.** $\sqrt{-48}$

9. $\sqrt{-\frac{7}{4}}$ **10.** $-\sqrt{-\frac{5}{9}}$ **11.** $-\sqrt{-\frac{2}{5}}$ **12.** $\sqrt{-\frac{5}{3}}$

In Exercises 13 through 16, simplify each of the given expressions.

13. $(\sqrt{-7})^2$; $\sqrt{(-7)^2}$ **14.** $\sqrt{(-15)^2}$; $(\sqrt{-15})^2$

15. $\sqrt{(-2)(-8)}$; $\sqrt{-2}\sqrt{-8}$ **16.** $\sqrt{-9}\sqrt{-16}$; $\sqrt{(-9)(-16)}$

In Exercises 17 through 24, simplify the given expressions.

17. j^7 **18.** j^{49} **19.** $-j^{22}$ **20.** j^{408}

21. $j^2 - j^6$ **22.** $2j^5 - j^7$ **23.** $j^{15} - j^{13}$ **24.** $3j^{48} + j^{200}$

In Exercises 25 through 36, perform the indicated operations and simplify each complex number to its rectangular form $a + bj$.

25. $2 + \sqrt{-9}$ **26.** $-6 + \sqrt{-64}$ **27.** $3j - \sqrt{-100}$ **28.** $-\sqrt{64} - \sqrt{-400}$

29. $8 - \sqrt{4} + \sqrt{-4}$ **30.** $5 - j + 2\sqrt{-25}$ **31.** $2j^2 + 3j$ **32.** $j^3 - 6$

33. $\sqrt{18} - \sqrt{-8}$ **34.** $\sqrt{-27} + \sqrt{12}$ **35.** $(\sqrt{-2})^2 + j^4$ **36.** $(2\sqrt{2})^2 - (\sqrt{-1})^2$

In Exercises 37 through 40, find the conjugate of each complex number.

37. $6 - 7j$ **38.** $-3 + 2j$ **39.** $2j$ **40.** 6

In Exercises 41 through 48, find the values of x and y which satisfy the given equations.

41. $7x - 2yj = 14 + 4j$ **42.** $2x + 3jy = -6 + 12j$ **43.** $6j - 7 = 3 - x - yj$

44. $9 - j = xj + 1 - y$ **45.** $x - y = 1 - xj - yj - j$ **46.** $2x - 2j = 4 - 2xj - yj$

47. $x + 2 + 7j = yj - 2xj$ **48.** $2x + 6xj + 3 = yj - y + 7j$

In Exercises 49 through 52, answer the given questions.

49. Are $8j$ and $-8j$ the solutions to the equation $x^2 + 64 = 0$?

50. Are $2j\sqrt{5}$ and $-2j\sqrt{5}$ the solutions to equation $x^2 + 20 = 0$?

51. What condition must be satisfied if a complex number and its conjugate are to be equal?

52. What type of number is a complex number if it is equal to the negative of its conjugate?

11-2   **Basic Operations with Complex Numbers**

The definitions of the operations of addition, subtraction, multiplication, and division for complex numbers are based on the operations for binomials with real coefficients. (See Chapters 1 and 5.) These operations are performed without regard for the fact that j has a special meaning. However, *we must be careful to express all complex numbers in terms of j before performing these operations.* Once this is done, we may proceed as with real numbers. We have the following definitions for these operations on complex numbers.

> **Addition (and subtraction):**
>
> $$(a + bj) + (c + dj) = (a + c) + (b + d)j \qquad (11\text{-}3)$$
>
> **Multiplication:**
>
> $$(a + bj)(c + dj) = (ac - bd) + (ad + bc)j \qquad (11\text{-}4)$$
>
> **Division:**
>
> $$\frac{a + bj}{c + dj} = \frac{(a + bj)(c - dj)}{(c + dj)(c - dj)} = \frac{(ac + bd) + (bc - ad)j}{c^2 + d^2} \qquad (11\text{-}5)$$

[Compare Eq. (11-4) with Eq. (5-6).]

We note from Eq. (11-3) that *the addition or subtraction of complex numbers is accomplished by combining the real parts and combining the imaginary parts.* Consider the following examples.

EXAMPLE A

1. $(3 - 2j) + (-5 + 7j) = (3 - 5) + (-2 + 7)j$
$$= -2 + 5j$$

2. $(7 + 9j) - (6 - 4j) = 7 + 9j - 6 + 4j$
$$= 1 + 13j$$

EXAMPLE B

$(3\sqrt{-4} - 4) - (6 - 2\sqrt{-25}) - \sqrt{-81}$

$= [3(2j) - 4] - [6 - 2(5j)] - 9j \qquad$ write in terms of j

$= [6j - 4] - [6 - 10j] - 9j$

$= 6j - 4 - 6 + 10j - 9j$

$= -10 + 7j$

When complex numbers are multiplied, Eq. (11-4) indicates that *we proceed as in any algebraic multiplication,* properly expressing numbers in terms of j and evaluating powers of j. This is illustrated in Examples C and D.

EXAMPLE C

$(6 - \sqrt{-4})(\sqrt{-9}) = (6 - 2j)(3j) \qquad$ write in terms of j

$= 18j - 6j^2 = 18j - 6(-1)$

$= 6 + 18j$

EXAMPLE D

$$(-9.4 - 6.2j)(2.5 + 1.5j) = -23.5 - 14.1j - 15.5j - 9.3j^2$$
$$= -23.5 - 29.6j - 9.3(-1)$$
$$= -14.2 - 29.6j$$

We note that our procedure in dividing two complex numbers is the same procedure that we used for rationalizing the denominator of a fraction with a radical in the denominator. We use this procedure so that we can express any answer in the form of a complex number. We need merely to *multiply numerator and denominator by the conjugate of the denominator* in order to perform this operation.

EXAMPLE E

$$\frac{7 - 2j}{3 + 4j} = \frac{(7 - 2j)(3 - 4j)}{(3 + 4j)(3 - 4j)} \longleftarrow \text{multiply by conjugate of denominator}$$
$$= \frac{21 - 28j - 6j + 8j^2}{9 - 16j^2} = \frac{21 - 34j + 8(-1)}{9 - 16(-1)}$$
$$= \frac{13 - 34j}{25}$$

This could be written in the form $a + bj$ as $\frac{13}{25} - \frac{34}{25}j$, but is generally left as a single fraction. In decimal form, the result could be expressed as $0.52 - 1.36j$.

EXAMPLE F

1. $\dfrac{6 + j}{2j} = \dfrac{6 + j}{2j} \times \dfrac{-2j}{-2j} = \dfrac{-12j - 2j^2}{4} = \dfrac{2 - 12j}{4} = \dfrac{1 - 6j}{2}$

2. $\dfrac{j^3 + 2j}{1 - j^5} = \dfrac{-j + 2j}{1 - j} = \dfrac{j}{1 - j} \times \dfrac{1 + j}{1 + j} = \dfrac{-1 + j}{2}$

EXAMPLE G

In an alternating-current circuit, the voltage E is given by $E = IZ$, where I is the current in amperes and Z is the impedance in ohms. Each of these can be represented by complex numbers. Find the complex number representation for I if $E = 4.20 - 3.00j$ volts and $Z = 5.30 + 2.65j$ ohms. (This type of circuit is discussed in more detail in Section 11-7.)

Since $I = E/Z$, we have

$$I = \frac{4.20 - 3.00j}{5.30 + 2.65j} = \frac{(4.20 - 3.00j)(5.30 - 2.65j)}{(5.30 + 2.65j)(5.30 - 2.65j)}$$
$$= \frac{22.26 - 11.13j - 15.90j + 7.95j^2}{5.30^2 - 2.65^2j^2} = \frac{22.26 - 7.95 - 27.03j}{5.30^2 + 2.65^2}$$
$$= \frac{14.31 - 27.03j}{35.11} = 0.408 - 0.770j \text{ amperes}$$

On a calculator, the result can be found directly from the expression on the first line, and the intermediate steps need not be written. The calculator sequence for this is shown below.

5.3 $\boxed{x^2}$ $\boxed{+}$ 2.65 $\boxed{x^2}$ $\boxed{=}$ $\boxed{\text{STO}}$ $\boxed{(}$ $\boxed{(}$ 4.2 $\boxed{\times}$ 5.3 $\boxed{-}$ 3 $\boxed{\times}$ 2.65 $\boxed{)}$ $\boxed{\div}$ $\boxed{\text{RCL}}$ $\boxed{=}$ $\boxed{0.4075472}$

$\boxed{(}$ 4.2 $\boxed{\times}$ 2.65 $\boxed{+/-}$ $\boxed{-}$ 3 $\boxed{\times}$ 5.3 $\boxed{)}$ $\boxed{\div}$ $\boxed{\text{RCL}}$ $\boxed{=}$ $\boxed{-0.7698113}$

Exercises 11-2

In Exercises 1 through 48, perform the indicated operations, expressing all answers in the form a + bj.

1. $(3 - 7j) + (2 - j)$

2. $(-4 - j) + (-7 - 4j)$

3. $(7j - 6) - (3 + j)$

4. $(0.23 + 0.67j) - (0.46 - 0.19j)$

5. $(4 + \sqrt{-16}) + (3 - \sqrt{-81})$

6. $(-1 + 3\sqrt{-4}) + (8 - 4\sqrt{-49})$

7. $(5 - \sqrt{-9}) - (\sqrt{-4} + 5)$

8. $(\sqrt{-25} - 1) - \sqrt{-9}$

9. $j - (j - 7) - 8$

10. $(7 - j) - (4 - 4j) + (6 - j)$

11. $(2\sqrt{-25} - 3) - (5 - 3\sqrt{-36}) - (\sqrt{-49})$

12. $(6 - 2\sqrt{-64}) - \sqrt{-100} - (\sqrt{-81} - 5)$

13. $(7 - j)(7j)$

14. $(-2.2j)(1.5j - 4.0)$

15. $\sqrt{-16}(2.8\sqrt{-4.0} + 1.6)$

16. $(\sqrt{-4} - 1)(\sqrt{-9})$

17. $(4 - j)(5 + 2j)$

18. $(3 - 5j)(6 + 7j)$

19. $(2\sqrt{-9} - 3)(3\sqrt{-4} + 2)$

20. $(5\sqrt{-64} - 5)(7 + \sqrt{-16})$

21. $(\sqrt{-18}\sqrt{-4})(3j)$

22. $\sqrt{-6}\sqrt{-12}\sqrt{3}$

23. $(\sqrt{-5})^5$

24. $(\sqrt{-36})^4$

25. $\sqrt{-108} - \sqrt{-27}$

26. $2\sqrt{-54} + \sqrt{-24}$

27. $3\sqrt{-28} - 2\sqrt{12}$

28. $5\sqrt{24} - 3\sqrt{-45}$

29. $7j^3 - 7\sqrt{-9}$

30. $6j - 5j^2\sqrt{-63}$

31. $j\sqrt{-7} - j^6\sqrt{112} + 3j$

32. $j^2\sqrt{-7} - \sqrt{-28} + 8$

33. $(3 - 7j)^2$

34. $(4j + 5)^2$

35. $(1 - j)^3$

36. $(1 + j)(1 - j)^2$

37. $\dfrac{6j}{2 - 5j}$

38. $\dfrac{4}{3 + 7j}$

39. $\dfrac{0.25}{3.0 - \sqrt{-1.0}}$

40. $\dfrac{\sqrt{-4}}{2 + \sqrt{-9}}$

41. $\dfrac{1 - j}{3j}$

42. $\dfrac{9 - 8j}{-4j}$

43. $\dfrac{\sqrt{-2} - 5}{\sqrt{-2} + 3}$

44. $\dfrac{2 + 3\sqrt{-3}}{5 - \sqrt{-3}}$

45. $\dfrac{\sqrt{-16} - \sqrt{2}}{\sqrt{2} + j}$

46. $\dfrac{1 - \sqrt{-4}}{2 + 9j}$

47. $\dfrac{j^2 - j}{2j - j^8}$

48. $\dfrac{j^5 - j^3}{3 + j}$

In Exercises 49 through 56, solve the given problems. In Exercises 55 and 56, refer to Example G.

49. Show that $-1 + j$ is a solution to the equation $x^2 + 2x + 2 = 0$.

50. Show that $-1 - j$ is a solution to the equation $x^2 + 2x + 2 = 0$.

51. Multiply $2 - 3j$ by its conjugate.

52. Multiply $-3 + j$ by its conjugate.

53. Divide $2 - 3j$ by its conjugate.

54. Divide $-3 + j$ by its conjugate.

55. If $I = 0.835 - 0.427j$ amperes and $Z = 250 + 170j$ ohms, find the complex number representation for E.

56. In an alternating-current circuit, two impedances Z_1 and Z_2 have a total impedance Z_T given by $Z_T = \dfrac{Z_1 Z_2}{Z_1 + Z_2}$. Find Z_T for $Z_1 = 2 + 3j$ ohms and $Z_2 = 3 - 4j$ ohms.

In Exercises 57 through 60, demonstrate the indicated properties.

57. Show that the sum of a complex number and its conjugate is a real number.

58. Show that the product of a complex number and its conjugate is a real number.

59. Show that the difference between a complex number and its conjugate is an imaginary number ($b \neq 0$).

60. Show that the reciprocal of the imaginary unit is the negative of the imaginary unit.

11-3   Graphical Representation of Complex Numbers

We showed in Section 1-1 how we could represent real numbers as points on a line. Because complex numbers include all real numbers as well as imaginary numbers, it is necessary to represent them graphically in a different way. Since there are two numbers associated with each complex number (the real part and the imaginary part), we find that we can represent complex numbers by representing the real parts by the *x*-values of the rectangular coordinate system, and the imaginary parts by the *y*-values. In this way *each complex number is represented as a point in the plane,* the point being designated as *a + bj. When the rectangular coordinate system is used in this manner, it is called the* **complex plane.** *The horizontal axis is called the* **real axis,** *and the vertical axis is called the* **imaginary axis.**

complex plane

EXAMPLE A

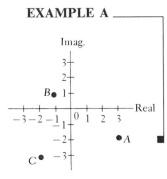

Fig. 11-1

In Fig. 11-1, the point *A* represents the complex number $3 - 2j$; point *B* represents $-1 + j$; point *C* represents $-2 - 3j$. We note that these complex numbers are represented by the points $(3, -2)$, $(-1, 1)$, and $(-2, -3)$ of the standard rectangular coordinate system.

We must keep in mind that the meaning given to the points representing complex numbers in the complex plane is different from the meaning given to the points in the standard rectangular coordinate system. A point in the complex plane represents a single complex number, whereas a point in the rectangular coordinate system represents a pair of real numbers. ∎

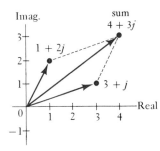

Fig. 11-2

Let us represent two complex numbers and their sum in the complex plane. Consider, for example, the two complex numbers $1 + 2j$ and $3 + j$. By algebraic addition the sum is $4 + 3j$. When we draw lines from the origin to these points (see Fig. 11-2), we note that if we think of the complex numbers as being vectors, their sum is the vector sum. Because complex numbers can be used to represent vectors, these numbers are particularly important. *Any complex number can be thought of as representing a vector from the origin to its point in the complex plane.* To add two complex numbers graphically, we find the point corresponding to one of them and draw a line from the origin to this point. We repeat this process for the second point. Next we complete a parallelogram with the lines drawn as adjacent sides. The resulting fourth vertex is the point representing the sum of the two complex numbers. Note that *this is equivalent to adding vectors by graphical means.*

EXAMPLE B

Add the complex numbers $5 - 2j$ and $-2 - j$ graphically.

The solution is indicated in Fig. 11-3. We can see that the fourth vertex of the parallelogram is at $3 - 3j$, which is, of course, the algebraic sum.

Fig. 11-3

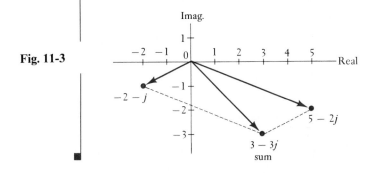

EXAMPLE C _____   Add the complex numbers -3 and $1 + 4j$ graphically.

First we note that $-3 = -3 + 0j$, which means that the point representing -3 is on the negative real axis. In Fig. 11-4, we show the numbers -3 and $1 + 4j$ on the graph and complete the parallelogram. From the graph we see ■ that the sum is $-2 + 4j$.

EXAMPLE D _____   Subtract $4 - 2j$ from $2 - 3j$ graphically.

Subtracting $4 - 2j$ is equivalent to adding $-4 + 2j$. Thus, we complete the ■ solution by adding $-4 + 2j$ and $2 - 3j$ (see Fig. 11-5). The result is $-2 - j$.

EXAMPLE E _____   Show graphically that the sum of a complex number and its conjugate is a real number.

If we choose the complex number $a + bj$, we know that its conjugate is $a - bj$. The imaginary coordinate for the conjugate is as far below the real axis as the imaginary coordinate of $a + bj$ is above it. Therefore, the sum of the imaginary parts must be zero and the sum of the two numbers must therefore lie on the real axis, as shown in Fig. 11-6. Therefore we have shown that the sum of $a + bj$ ■ and $a - bj$ is real.

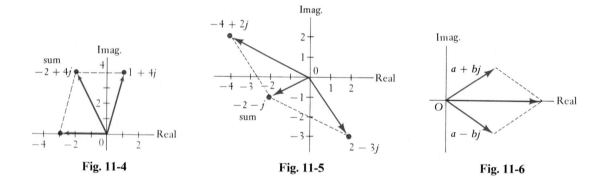

Fig. 11-4   Fig. 11-5   Fig. 11-6

Exercises 11-3

In Exercises 1 through 4, locate the given complex numbers in the complex plane.

1. $2 + 6j$ **2.** $-5 + j$ **3.** $-4 - 3j$ **4.** $3 - 4j$

In Exercises 5 through 24, perform the indicated operations graphically; check them algebraically.

5. $2 + (3 + 4j)$ **6.** $2j + (-2 + 3j)$ **7.** $(5 - j) + (3 + 2j)$

8. $(3 - 2j) + (-1 - j)$ **9.** $5 - (1 - 4j)$ **10.** $(2 - j) - j$

11. $(2 - 4j) + (-2 + j)$ **12.** $(-1 - 2j) + (6 - j)$ **13.** $(3 - 2j) - (4 - 6j)$

14. $(-2 - 4j) - (2 - 5j)$ **15.** $(1 + 4j) - (3 + j)$ **16.** $(-j - 2) - (-1 - 3j)$

17. $(1.5 - 0.5j) + (3.0 + 2.5j)$ **18.** $(3.5 + 2.0j) - (-4.0 - 1.5j)$ **19.** $(3 - 6j) - (-1 + 5j)$

20. $(-6 - 3j) + (2 - 7j)$ **21.** $(2j + 1) - 3j - (j + 1)$ **22.** $(6 - j) - 9 - (2j - 3)$

23. $(j - 6) - j + (j - 7)$ **24.** $j - (1 - j) + (3 + 2j)$

In Exercises 25 through 28, show the given number, its negative, and its conjugate on the same coordinate system.

25. $3 + 2j$ **26.** $-2 + 4j$ **27.** $-3 - 5j$ **28.** $5 - j$

In Exercises 29 through 32, show the given number $a + bj$, $3(a + bj)$, and $-3(a + bj)$ on the same coordinate system. The multiplication of a complex number by a real number is called the **scalar multiplication** of the complex number.

29. $-2 + j$ **30.** $-1 - 3j$ **31.** $3 - j$ **32.** $2 + j$

11-4 Polar Form of a Complex Number

We have just seen the relationship between complex numbers and vectors. Since one can be used to represent the other, we shall use this fact to write complex numbers in another way. The new form has certain advantages when basic operations are performed on complex numbers.

By drawing a vector from the origin to the point in the complex plane which represents the number $x + yj$, we see the relation between vectors and complex numbers. Further observation indicates an angle in standard position has been formed. Also, the point $x + yj$ is r units from the origin. In fact, *we can find any point in the complex plane by knowing this angle θ and the value of r.* We have already developed the relations between x, y, r, and θ in Eqs. (8-1) to (8-3). Let us rewrite these equations in a slightly different form. By referring to Eqs. (8-1) through (8-3) and to Fig. 11-7, we see that

$$x = r\cos\theta \qquad y = r\sin\theta \qquad (11\text{-}6)$$

$$r^2 = x^2 + y^2 \qquad \tan\theta = \frac{y}{x} \qquad (11\text{-}7)$$

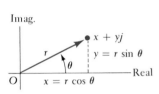

Fig. 11-7

Substituting Eqs. (11-6) into the rectangular form $x + yj$ of a complex number, we have

$$x + yj = r\cos\theta + j(r\sin\theta)$$

or

polar form

$$x + yj = r(\cos\theta + j\sin\theta) \qquad (11\text{-}8)$$

The right side of Eq. (11-8) is called the **polar form** of a complex number. Sometimes it is referred to as the **trigonometric form**. An abbreviated form of writing the polar form which is sometimes used is r cis θ. The length r is called the **absolute value,** or the **modulus,** and the angle θ is called the **argument** of the complex number. Therefore, Eq. (11-8), along with Eqs. (11-7), defines the polar form of a complex number.

EXAMPLE A

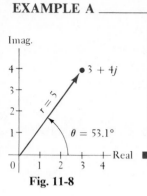

Fig. 11-8

Represent the complex number $3 + 4j$ graphically, and give its polar form.

From the rectangular form $3 + 4j$, we see that $x = 3$ and $y = 4$. Using Eqs. (11-7), we have

$$r = \sqrt{3^2 + 4^2} = 5, \qquad \tan \theta = \frac{4}{3}, \qquad \theta = 53.1°$$

Thus, the polar form is

$$5(\cos 53.1° + j \sin 53.1°)$$

The graphical representation is shown in Fig. 11-8.

A note on significant digits is in order here. In writing a complex number as $3 + 4j$ in Example A, no approximate values are intended. However, in expressing the polar form as $5(\cos 53.1° + j \sin 53.1°)$, we rounded off the angle to the nearest 0.1°, as it is not possible to express the result exactly in degrees. Thus, in dealing with nonexact numbers, we shall express angles to the nearest 0.1°. Other results, when approximate, will be expressed to three significant digits, unless a different accuracy is given in the problem. Of course, in applied situations most numbers used are derived through measurement and are therefore approximate.

Another convenient and widely used notation for the polar form is $r/\underline{\theta}$. We must remember in using this form that it represents a complex number and is simply a shorthand way of writing $r(\cos \theta + j \sin \theta)$. Therefore,

$$\boxed{r\,\underline{/\theta} = r(\cos \theta + j \sin \theta)} \qquad\qquad (11\text{-}9)$$

EXAMPLE B

$$3(\cos 40° + j \sin 40°) = 3\,\underline{/40°}$$
$$6.26(\cos 217.3° + j \sin 217.3°) = 6.26\,\underline{/217.3°}$$
$$5\,\underline{/120°} = 5(\cos 120° + j \sin 120°)$$
$$14.5\,\underline{/306.2°} = 14.5(\cos 306.2° + j \sin 306.2°)$$

EXAMPLE C

Represent the complex number $-2.08 - 3.12j$ graphically, and give its polar forms.

From Eqs. (11-7), we have

$$r = \sqrt{(-2.08)^2 + (-3.12)^2} = 3.75$$
$$\tan \theta_{\text{ref}} = \frac{3.12}{2.08}, \qquad \theta_{\text{ref}} = 56.3°, \qquad \theta = 180° + 56.3° = 236.3°$$

Since both the real and imaginary parts are negative, we know that θ is a third-quadrant angle. Therefore, we found the reference angle before finding θ. This means the polar forms are

$$3.75(\cos 236.3° + j \sin 236.3°) = 3.75\,\underline{/236.3°}$$

See Fig. 11-9.

The calculator sequence for this is

2.08 [x^2] [+] 3.12 [x^2] [=] [$\sqrt{x}$] 3.7497733

3.12 [÷] 2.08 [=] [INV] [TAN] [+] 180 [=] 236.30993

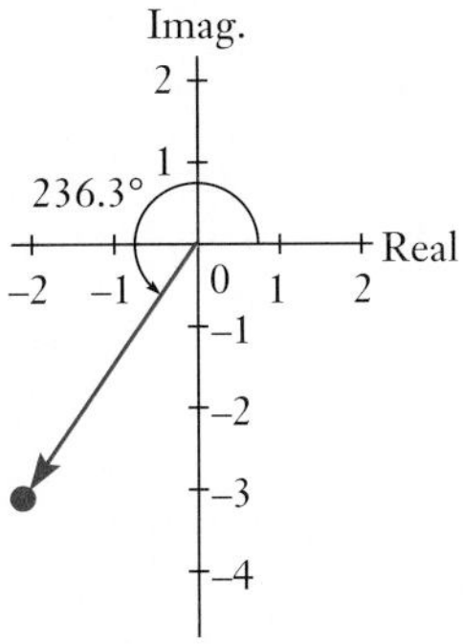

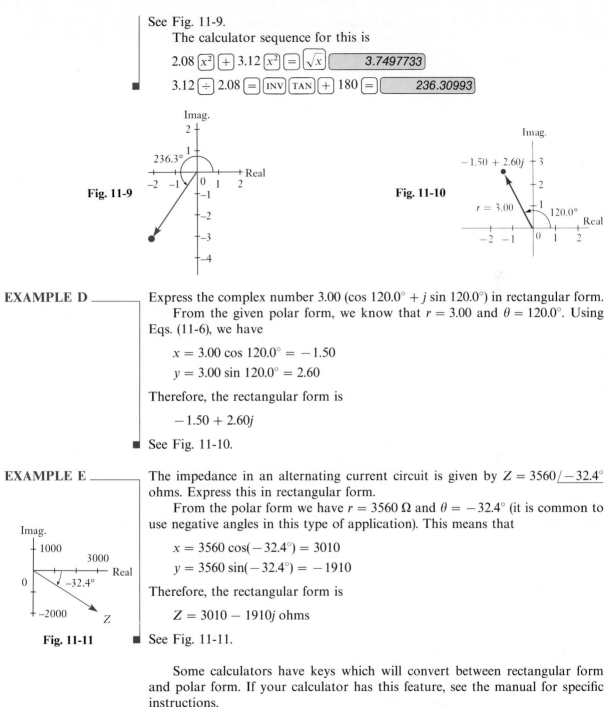

Fig. 11-9

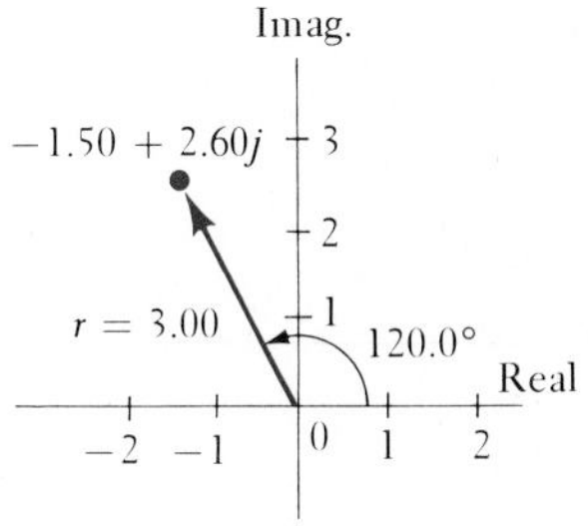

Fig. 11-10

EXAMPLE D

Express the complex number 3.00 (cos 120.0° + j sin 120.0°) in rectangular form.

From the given polar form, we know that $r = 3.00$ and $\theta = 120.0°$. Using Eqs. (11-6), we have

$$x = 3.00 \cos 120.0° = -1.50$$
$$y = 3.00 \sin 120.0° = 2.60$$

Therefore, the rectangular form is

$$-1.50 + 2.60j$$

See Fig. 11-10.

EXAMPLE E

The impedance in an alternating current circuit is given by $Z = 3560\underline{/-32.4°}$ ohms. Express this in rectangular form.

From the polar form we have $r = 3560\ \Omega$ and $\theta = -32.4°$ (it is common to use negative angles in this type of application). This means that

$$x = 3560 \cos(-32.4°) = 3010$$
$$y = 3560 \sin(-32.4°) = -1910$$

Therefore, the rectangular form is

$$Z = 3010 - 1910j \text{ ohms}$$

See Fig. 11-11.

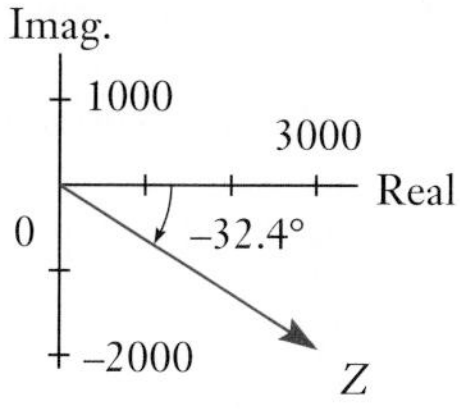

Fig. 11-11

Some calculators have keys which will convert between rectangular form and polar form. If your calculator has this feature, see the manual for specific instructions.

EXAMPLE F ——————— Represent the numbers 5, -5, $7j$, and $-7j$ in polar form.

Since any positive real number lies on the positive real axis in the complex plane, it is expressed in polar form by

$$a = a(\cos 0° + j \sin 0°) = a\,\underline{/0°}$$

Negative real numbers, being on the negative real axis, are written as

$$a = |a|(\cos 180° + j \sin 180°) = |a|\,\underline{/180°}$$

Thus,

$$5 = 5(\cos 0° + j \sin 0°) = 5\,\underline{/0°}$$

and

$$-5 = 5(\cos 180° + j \sin 180°) = 5\,\underline{/180°}$$

Positive pure imaginary numbers lie on the positive imaginary axis and are expressed in polar form by

$$bj = b(\cos 90° + j \sin 90°) = b\,\underline{/90°}$$

Similarly, negative pure imaginary numbers, being on the negative imaginary axis, are written as

$$bj = |b|(\cos 270° + j \sin 270°) = |b|\,\underline{/270°}$$

Thus,

$$7j = 7(\cos 90° + j \sin 90°) = 7\,\underline{/90°}$$

and

$$-7j = 7(\cos 270° + j \sin 270°) = 7\,\underline{/270°}$$

The graphical representations of the *complex numbers* 5, -5, $7j$, and $-7j$ are shown in Fig. 11-12.

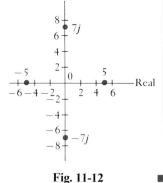

Fig. 11-12

Exercises 11-4

In Exercises 1 through 16, represent each of the complex numbers graphically, and give the polar form of each number.

1. $8 + 6j$

2. $-8 - 15j$

3. $3 - 4j$

4. $-5 + 12j$

5. $-2.00 + 3.00j$

6. $7.00 - 5.00j$

7. $-5.50 - 2.40j$

8. $4.60 - 4.60j$

9. $1 + j\sqrt{3}$

10. $\sqrt{2} - j\sqrt{2}$

11. $3.514 - 7.256j$

12. $6.231 + 9.527j$

13. -3

14. 6

15. $9j$

16. $-2j$

See Appendix E for a computer program for changing a complex number to polar form.

In Exercises 17 through 36, represent each of the complex numbers graphically, and give the rectangular form of each number.

17. $5.00(\cos 54.0° + j \sin 54.0°)$

18. $3.00(\cos 232.0° + j \sin 232.0°)$

19. $1.60(\cos 150.0° + j \sin 150.0°)$

20. $2.50(\cos 315.0° + j \sin 315.0°)$

21. $6(\cos 180° + j \sin 180°)$

22. $12(\cos 270° + j \sin 270°)$

23. $8(\cos 360° + j \sin 360°)$

24. $15(\cos 0° + j \sin 0°)$

25. $12.36(\cos 345.56° + j \sin 345.56°)$

26. $220.8(\cos 155.13° + j \sin 155.13°)$

27. $\cos 240.0° + j \sin 240.0°$

28. $\cos 299.0° + j \sin 299.0°$

29. $4.75\underline{/172.8°}$ **30.** $1.50\underline{/62.3°}$

31. $0.9326\underline{/229.54°}$ **32.** $277.8\underline{/-342.63°}$

33. $7.32\underline{/-270°}$ **34.** $18.3\underline{/180.0°}$

35. $86.42\underline{/94.62°}$ **36.** $4629\underline{/182.44°}$

In Exercises 37 through 40, solve the given problems.

37. Considering the relationship between complex numbers and vectors, what are the magnitude and direction of a force vector which is represented by $25.6 - 34.2j$ newtons?

38. What are the magnitude and direction of the displacement of a weight at the end of a spring which can be described (at a given time) by $y = 0.395 + 0.148j$ meters? (See Exercise 37.)

39. The electric field intensity of a light wave can be described by the complex number $12.4\underline{/78.3°}$ V/m. Write this in rectangular form.

40. The current in a certain microprocessor circuit is represented by the complex number $3.75\underline{/15.0°}$ μA. Write this in rectangular form.

11-5 Exponential Form of a Complex Number ▬▬▬▬

Another important form of a complex number is known as the **exponential form.** It is commonly used in electronics, engineering, and physics applications. As we will see in the next section, it is also convenient for multiplication and division of complex numbers, as the rectangular form is convenient for addition and subtraction.

The exponential form is written as $re^{j\theta}$, where r and θ have the same meanings as given in the last section, although in $re^{j\theta}$, θ is expressed in radians. *The number e represents a very special irrational number, where an approximate value of e is*

$$e = 2.7182818$$

This number e is very important in mathematics, and we shall see it again in the next chapter. At this point it is necessary to accept the value for e, although in calculus its basic meaning is shown along with the reason it has the above value.

We now define

exponential form

$$re^{j\theta} = r(\cos \theta + j \sin \theta) \qquad (11\text{-}10)$$

NOTE▷ By expressing θ in radians, the expression $j\theta$ is an exponent, and it can be shown to obey all of the laws of exponents which we discussed in Chapter 10. Therefore, *we shall always express θ in radians when using the exponential form.* The following examples show how complex numbers can be changed to and from exponential form.

EXAMPLE A ———— Express the number $3 + 4j$ in exponential form.

From Example A of Section 11-4, we know that this complex number may be written in polar form as $5(\cos 53.1° + j \sin 53.1°)$. Therefore, we know that $r = 5$. We now express $53.1°$ in terms of radians as

$$\frac{53.1\pi}{180} = 0.927 \text{ rad}$$

Thus, the exponential form is $5e^{0.927j}$. This means that

degrees to radians

$$3 + 4j = 5(\cos 53.1° + j \sin 53.1°) = 5e^{0.927j}$$

value of r

EXAMPLE B ———— Express the number $8.50\underline{/136.3°}$ in exponential form.

Since this complex number is in polar form we note that $r = 8.50$ and that we must express $136.3°$ in radians. Changing $136.3°$ to radians, we have

$$\frac{136.3\pi}{180} = 2.38 \text{ rad}$$

Therefore, the required exponential form is $8.50e^{2.38j}$. This means that

$$8.50\underline{/136.3°} = 8.50e^{2.38j}$$

We see that the principal step in changing from polar form to exponential form is to change θ from degrees to radians.

EXAMPLE C ———— Express the number $3.07 - 7.43j$ in exponential form.

From the rectangular form of the number, we have $x = 3.07$ and $y = -7.43$. Therefore,

$$r = \sqrt{(3.07)^2 + (-7.43)^2} = 8.04$$

$$\tan \theta = \frac{-7.43}{3.07}, \qquad \theta = 292.5°$$

Changing $292.5°$ to radians, we have $292.5° = 5.10$ rad. Therefore, the exponential form is $8.04e^{5.10j}$. This means that

$$3.07 - 7.43j = 8.04e^{5.10j}$$

EXAMPLE D ———— Express the complex number $2.00e^{4.80j}$ in polar and rectangular forms.

We first express 4.80 rad as $275.0°$. From the exponential form we know that $r = 2.00$. Thus, the polar form is

$$2.00(\cos 275.0° + j \sin 275.0°)$$

Next, by use of the distributive law we rewrite the polar form and then evaluate. Thus,

$$2.00e^{4.80j} = 2.00(\cos 275.0° + j \sin 275.0°)$$

$$= 2.00 \cos 275.0° + (2.00 \sin 275.0°)j = 0.174 - 1.99j$$

EXAMPLE E ———

Express the complex number $3.408e^{2.457j}$ in polar and rectangular forms.

We first express 2.457 rad as 140.78°. From the exponential form, we know that $r = 3.408$. Thus, the polar form is $3.408\underline{/140.78°}$. Thus,

$$3.408e^{2.457j} = 3.408(\cos 140.78° + j \sin 140.78°)$$
$$= -2.640 + 2.155j$$

∎

As we noted earlier, an important application of the use of complex numbers is in alternating-current analysis. When an alternating current flows through a given circuit, usually the current and voltage have different phases. That is, they do not reach their peak values at the same time. Therefore, one way of accounting for the magnitude as well as the phase of an electric current or voltage is to write it as a complex number. Here the modulus is the actual value of the current or voltage, and the argument is a measure of the phase.

EXAMPLE F ———

A current of $2.00 - 4.00j$ amperes flows through a given circuit. Write this current in exponential form and determine the magnitude of the current in the circuit.

From the rectangular form, we have $x = 2.00$ and $y = -4.00$. Therefore,

$$r = \sqrt{(2.00)^2 + (-4.00)^2} = 4.47$$

Also,

$$\tan \theta = -\tfrac{4.00}{2.00}$$

This means that $\theta = -63.4°$ (it is normal to express the phase in terms of negative angles). Changing 63.4° to radians, we have $63.4° = 1.11$ rad. Therefore, the exponential form of the current is $4.47e^{-1.11j}$. The modulus is 4.47, meaning the ∎ magnitude of the current is 4.47 A.

At this point we shall summarize the three important forms of a complex number. See Fig. 11-13 for the graphical representation.

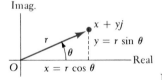

Fig. 11-13

> **Rectangular:** $x + yj$
> **Polar:** $r(\cos \theta + j \sin \theta) = r\underline{/\theta}$
> **Exponential:** $re^{j\theta}$

It follows that

$$x + yj = r(\cos \theta + j \sin \theta) = r\underline{/\theta} = re^{j\theta} \qquad (11\text{-}11)$$

where

$$r^2 = x^2 + y^2 \qquad \tan \theta = \frac{y}{x} \qquad (11\text{-}7)$$

Exercises 11-5

In Exercises 1 through 20, express the given complex numbers in exponential form.

1. $3.00(\cos 60.0° + j \sin 60.0°)$

2. $5.00(\cos 135.0° + j \sin 135.0°)$

3. $4.50(\cos 282.3° + j \sin 282.3°)$

4. $2.10(\cos 228.7° + j \sin 228.7°)$

5. $375.5(\cos 95.46° + j \sin 95.46°)$

6. $16.72[\cos(-7.14°) + j \sin(-7.14°)]$

7. $0.515/198.3°$

8. $4650/326.5°$

9. $4.06/-61.4°$

10. $0.0192/76.7°$

11. $9245/296.32°$

12. $827.6/110.09°$

13. $3 - 4j$

14. $-1 - 5j$

15. $-3 + 2j$

16. $6 + j$

17. $5.90 + 2.40j$

18. $47.3 - 10.9j$

19. $-634.6 - 528.2j$

20. $-8573 + 5477j$

In Exercises 21 through 28, express the given complex numbers in polar and rectangular forms.

21. $3.00e^{0.500j}$

22. $2.00e^{1.00j}$

23. $4.64e^{1.85j}$

24. $2.50e^{3.84j}$

25. $3.20e^{5.41j}$

26. $0.800e^{3.00j}$

27. $0.1724e^{2.391j}$

28. $820.7e^{3.492j}$

In Exercises 29 through 32, perform the indicated operations.

29. The impedance in an antenna circuit is $375 + 110j$ ohms. Write this in exponential form and find the magnitude of the impedance.

30. The intensity of the signal from a radar microwave signal is $37.0[\cos(-65.3°) + j \sin(-65.3°)]$ V/m. Write this in exponential form.

31. The displacement of the end of a vibrating rod is $5.83e^{-1.20j}$ cm. Write this in rectangular form.

32. In an electric circuit, the *admittance* is the reciprocal of the impedance. In a transistor circuit, the impedance is $2800 - 1450j$ ohms. Find the exponential form of the admittance.

11-6   Products, Quotients, Powers, and Roots of Complex Numbers

We have previously performed products and quotients using the rectangular forms of the given numbers. However, these operations can also be performed with complex numbers in polar and exponential forms. We find that these operations are convenient, and also useful for purposes of finding powers and roots of complex numbers.

We may find the product of two complex numbers by using the exponential form and the laws of exponents. Multiplying $r_1 e^{j\theta_1}$ by $r_2 e^{j\theta_2}$, we have

$$r_1 e^{j\theta_1} \times r_2 e^{j\theta_2} = r_1 r_2 e^{j\theta_1 + j\theta_2} = r_1 r_2 e^{j(\theta_1 + \theta_2)}$$

We use this equation to express the product of two complex numbers in polar form:

$$r_1 e^{j\theta_1} \times r_2 e^{j\theta_2} = r_1(\cos \theta_1 + j \sin \theta_1) \times r_2(\cos \theta_2 + j \sin \theta_2)$$

and

$$r_1 r_2 e^{j(\theta_1 + \theta_2)} = r_1 r_2 [\cos(\theta_1 + \theta_2) + j \sin(\theta_1 + \theta_2)]$$

Therefore, the polar expressions are equal, which means that *the product of two complex numbers is*

$$r_1(\cos\theta_1 + j\sin\theta_1)r_2(\cos\theta_2 + j\sin\theta_2)$$
$$= r_1r_2[\cos(\theta_1 + \theta_2) + j\sin(\theta_1 + \theta_2)]$$
$$(r_1\underline{/\theta_1})(r_2\underline{/\theta_2}) = r_1r_2\underline{/\theta_1 + \theta_2}$$

(11-12)

EXAMPLE A

Multiply the complex numbers $2 + 3j$ and $1 - j$ by using the polar form of each.

For $2 + 3j$: $r_1 = \sqrt{2^2 + 3^2} = 3.61,$ $\tan\theta_1 = \dfrac{3}{2},$ $\theta_1 = 56.3°$

For $1 - j$: $r_2 = \sqrt{1^2 + (-1)^2} = 1.41,$ $\tan\theta_2 = \dfrac{-1}{1},$ $\theta_2 = 315.0°$

$(3.61)(\cos 56.3° + j\sin 56.3°)(1.41)(\cos 315.0° + j\sin 315.0°)$

— sum —

$= (3.61)(1.41)[\cos(56.3° + 315.0°) + j\sin(56.3° + 315.0°)]$
$= 5.09(\cos 371.3° + j\sin 371.3°)$
$= 5.09(\cos 11.3° + j\sin 11.3°)$ ∎

EXAMPLE B

When we use the $r\underline{/\theta}$ polar form to multiply the two complex numbers in Example A, we have

$r_1 = 3.61$ $\theta_1 = 56.3°$
$r_2 = 1.41$ $\theta_2 = 315.0°$

$(3.61\underline{/56.3°})(1.41\underline{/315.0°}) = (3.61)(1.41)\underline{/56.3° + 315.0°}$

$= 5.09\underline{/371.3°}$
$= 5.09\underline{/11.3°}$ ∎

If we wish to *divide* one complex number in exponential form by another, we arrive at the following result:

$$r_1e^{j\theta_1} \div r_2e^{j\theta_2} = \frac{r_1}{r_2}e^{j(\theta_1 - \theta_2)}$$

(11-13)

Therefore, *the result of dividing one number in polar form by another is given by*

$$\frac{r_1(\cos\theta_1 + j\sin\theta_1)}{r_2(\cos\theta_2 + j\sin\theta_2)} = \frac{r_1}{r_2}[\cos(\theta_1 - \theta_2) + j\sin(\theta_1 - \theta_2)]$$

$$\frac{r_1\underline{/\theta_1}}{r_2\underline{/\theta_2}} = \frac{r_1}{r_2}\underline{/\theta_1 - \theta_2}$$

(11-14)

EXAMPLE C Divide the first complex number of Example A by the second. Using polar form, we have the following:

$$\frac{3.61(\cos 56.3° + j \sin 56.3°)}{1.41(\cos 315.0° + j \sin 315.0°)} = \frac{3.61}{1.41}[\cos(56.3° - 315.0°) + j \sin(56.3° - 315.0°)]$$

$$= 2.56[\cos(-258.7°) + j \sin(-258.7°)]$$
$$= 2.56(\cos 101.3° + j \sin 101.3°)$$

EXAMPLE D Repeating Example C using $r\underline{/\theta}$ polar form, we have

$$\frac{3.61\,\underline{/56.3°}}{1.41\,\underline{/315.0°}} = \frac{3.61}{1.41}\,\underline{/56.3° - 315.0°} = 2.56\,\underline{/-258.7°}$$

$$= 2.56\,\underline{/101.3°}$$

We have just seen that multiplying and dividing numbers in polar form can be very readily performed directly. However, if we are to add or subtract num-

NOTE ▷ bers in polar form, *we must do the addition or subtraction by using rectangular form.* This is illustrated in the following example.

EXAMPLE E Perform the addition $1.563\,\underline{/37.56°} + 3.827\,\underline{/146.23°}$.

In order to do this addition, we must change each number to rectangular form.

$$1.563\,\underline{/37.56°} + 3.827\,\underline{/146.23°}$$

$$= 1.563(\cos 37.56° + j \sin 37.56°) + 3.827(\cos 146.23° + j \sin 146.23°)$$
$$= 1.2390 + 0.9528j - 3.1813 + 2.1273j$$
$$= -1.9423 + 3.0801j$$

Now we change this to polar form.

$$r = \sqrt{(-1.9423)^2 + (3.0801)^2} = 3.641$$
$$\tan \theta = \frac{3.0801}{-1.9423}, \qquad \theta = 122.24°$$

Therefore,

$$1.563\,\underline{/37.56°} + 3.827\,\underline{/146.23°} = 3.641\,\underline{/122.24°}$$

To raise a complex number to a power, we simply multiply one complex number by itself the required number of times. For example, squaring a number in exponential form, we have

$$(re^{j\theta})^2 = r^2 e^{j2\theta} \tag{11-15}$$

Multiplying the expression in Eq. (11-15) by $re^{j\theta}$ gives $r^3 e^{j3\theta}$. This leads to the general expression for raising a complex number to the nth power,

$$(re^{j\theta})^n = r^n e^{jn\theta} \tag{11-16}$$

Extending this to polar form, we have

DeMoivre's theorem

$$[r(\cos\theta + j\sin\theta)]^n = r^n(\cos n\theta + j\sin n\theta)$$
$$(r\,\underline{/\theta}\,)^n = r^n\,\underline{/n\theta}$$

(11-17)

*Equation (11-17) is known as **DeMoivre's theorem.** It is valid for all real values of n and may also be used for finding the roots of complex numbers if n is a fractional exponent.*

EXAMPLE F

Using DeMoivre's theorem, find $(2 + 3j)^3$.

From Example A of this section, we know $r = 3.61$ and $\theta = 56.3°$. Thus, we have

$$[3.61(\cos 56.3° + j\sin 56.3°)]^3 = (3.61)^3[\cos (3 \times 56.3°) + j\sin (3 \times 56.3°)]$$
$$= 47.0(\cos 168.9° + j\sin 168.9°)$$
$$= 47.0\,\underline{/168.9°}$$

Expressing θ in radians, we have $\theta = 0.983$ rad. Thus, in exponential form,

$$(3.61e^{0.983j})^3 = (3.61)^3 e^{3 \times 0.983j} = 47.0e^{2.95j}$$

Therefore,

$$(2 + 3j)^3 = 47.0(\cos 168.9° + j\sin 168.9°)$$
$$= 47.0\,\underline{/168.9°} = 47.0e^{2.95j}$$

EXAMPLE G

Find the cube root of -1.

Since we know that -1 is a real number, we can find its cube root by means of the definition. That is, $(-1)^3 = -1$. We shall check this by DeMoivre's theorem. Writing -1 in polar form, we have

$$-1 = 1(\cos 180° + j\sin 180°)$$

Applying DeMoivre's theorem, with $n = \frac{1}{3}$, we obtain

$$(-1)^{1/3} = 1^{1/3}(\cos \tfrac{1}{3} 180° + j\sin \tfrac{1}{3} 180°) = \cos 60° + j\sin 60°$$
$$= 0.5000 + 0.8660j$$

Observe that we did not obtain -1 as an answer. If we check the answer which was obtained, in the form $\frac{1}{2}(1 + \sqrt{3}j)$, by actually cubing it, we obtain -1! Thus it is a correct answer.

We should note that it is possible to take $\frac{1}{3}$ of any angle up to $1080°$ and still have an angle less than $360°$. Since $180°$ and $540°$ have the same terminal side, let us try writing -1 as $1(\cos 540° + j\sin 540°)$. Using DeMoivre's theorem, we

(Continued on next page)

have

$$(-1)^{1/3} = 1^{1/3}(\cos \tfrac{1}{3}\,540° + j \sin \tfrac{1}{3}\,540°)$$
$$= \cos 180° + j \sin 180° = -1$$

We have found the answer we originally anticipated.

Angles of 180° and 900° also have the same terminal side, so we try

$$(-1)^{1/3} = 1^{1/3}(\cos \tfrac{1}{3}\,900° + j \sin \tfrac{1}{3}\,\sin 900°) = \cos 300° + j \sin 300°$$
$$= 0.5000 - 0.8660j$$

Checking this, we find that it is also a correct root. We may try 1260°, but $\tfrac{1}{3}(1260°) = 420°$, which has the same functional values as 60°, and would give us the answer $0.5000 + 0.8660j$ again.

We have found, therefore, *three cube roots* of -1. They are

$$-1, \qquad 0.5000 + 0.8660j, \quad \text{and} \quad 0.5000 - 0.8660j$$

These roots are graphed in Fig. 11-14. Note that they are equally spaced on the circumference of a circle of radius 1.

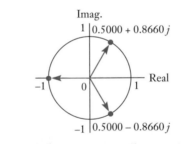

Fig. 11-14

When we generalize on the results of Example G, it can be proved that

there are n nth roots of a complex number

NOTE ▷ The method for finding the n roots is to use θ to find one root and **then add 360° to θ, $n - 1$ times,** in order to find the other roots.

EXAMPLE H _____ Find the two square roots of $2j$.

We must first write $2j$ in polar form so that we may use DeMoivre's theorem to find the roots. In polar form, $2j$ is

$$2j = 2(\cos 90° + j \sin 90°)$$

To find the square roots, we apply DeMoivre's theorem with $n = \tfrac{1}{2}$.

$$(2j)^{1/2} = 2^{1/2}\left(\cos \frac{90°}{2} + j \sin \frac{90°}{2}\right) = \sqrt{2}(\cos 45° + j \sin 45°) = 1 + j$$

To find the other square root using DeMoivre's theorem, we must write $2j$ in polar form as

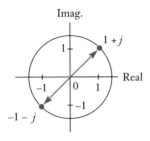

Fig. 11-15

for second root

$$2j = 2[\cos (90° + 360°) + j \sin (90° + 360°)] = 2(\cos 450° + j \sin 450°)$$

Applying DeMoivre's theorem to $2j$ in this form, we have

$$(2j)^{1/2} = 2^{1/2}\left(\cos \frac{450°}{2} + j \sin \frac{450°}{2}\right) = \sqrt{2}(\cos 225° + j \sin 225°) = -1 - j$$

Thus, the two square roots of $2j$ are $1 + j$ and $-1 - j$. These roots are graphed in Fig. 11-15. We see that they are on the circumference of a circle of radius $\sqrt{2}$ and are 180° apart. ∎

EXAMPLE I

Find the six sixth roots of 64.

Here we shall directly use the method for finding the roots of a number as outlined at the end of Example G.

$$64 = 64(\cos 0° + j \sin 0°)$$

First root: $64^{1/6} = 64^{1/6}\left(\cos \dfrac{0°}{6} + j \sin \dfrac{0°}{6}\right) = 2(\cos 0° + j \sin 0°) = 2$

add 360°

Second root: $64^{1/6} = 64^{1/6}\left(\cos \dfrac{0° + 360°}{6} + j \sin \dfrac{0° + 360°}{6}\right)$

$$= 2(\cos 60° + j \sin 60°) = 1 + j\sqrt{3}$$

add 2 × 360°

Third root: $64^{1/6} = 64^{1/6}\left(\cos \dfrac{0° + 720°}{6} + j \sin \dfrac{0° + 720°}{6}\right)$

$$= 2(\cos 120° + j \sin 120°) = -1 + j\sqrt{3}$$

add 3 × 360°

Fourth root: $64^{1/6} = 64^{1/6}\left(\cos \dfrac{0° + 1080°}{6} + j \sin \dfrac{0° + 1080°}{6}\right)$

$$= 2(\cos 180° + j \sin 180°) = -2$$

add 4 × 360°

Fifth root: $64^{1/6} = 64^{1/6}\left(\cos \dfrac{0° + 1440°}{6} + j \sin \dfrac{0° + 1440°}{6}\right)$

$$= 2(\cos 240° + j \sin 240°) = -1 - j\sqrt{3}$$

add 5 × 360°

Sixth root: $64^{1/6} = 64^{1/6}\left(\cos \dfrac{0° + 1800°}{6} + j \sin \dfrac{0° + 1800°}{6}\right)$

$$= 2(\cos 300° + j \sin 300°) = 1 - j\sqrt{3}$$

These roots are graphed in Fig. 11-16. Note that they are equally spaced 60° apart on the circumference of a circle of radius 2. ∎

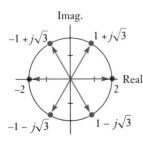

Fig. 11-16

At this point we can see advantages for the various forms of writing complex numbers. Rectangular form lends itself best to addition and subtraction. Polar form is generally used for multiplying, dividing, raising to powers, and finding roots. Exponential form is used for theoretical purposes (e.g., deriving DeMoivre's theorem).

Exercises 11-6

In Exercises 1 through 16, perform the indicated operations. Leave the result in polar form.

1. $[4(\cos 60° + j \sin 60°)][2(\cos 20° + j \sin 20°)]$

2. $[3(\cos 120° + j \sin 120°)][5(\cos 45° + j \sin 45°)]$

3. $(0.5\underline{/140°})(6\underline{/110°})$

4. $(0.4\underline{/320°})(5.5\underline{/150°})$

5. $\dfrac{8(\cos 100° + j \sin 100°)}{4(\cos 65° + j \sin 65°)}$

6. $\dfrac{9(\cos 230° + j \sin 230°)}{3(\cos 80° + j \sin 80°)}$

7. $\dfrac{12\underline{/320°}}{5\underline{/210°}}$

8. $\dfrac{2\underline{/90°}}{4\underline{/75°}}$

9. $[2(\cos 35° + j \sin 35°)]^3$

10. $[3(\cos 120° + j \sin 120°)]^4$

11. $(2\underline{/135°})^8$

12. $(1\underline{/142°})^{10}$

13. $2.78\underline{/56.8°} + 1.37\underline{/207.3°}$

14. $15.9\underline{/142.6°} - 18.5\underline{/71.4°}$

15. $7085(\cos 115.62° + j \sin 115.62°) - 4667(\cos 296.34° + j \sin 296.34°)$

16. $307.5(\cos 326.54° + j \sin 326.54°) + 726.3(\cos 96.41° + j \sin 96.41°)$

In Exercises 17 through 28, change each number to polar form and then perform the indicated operations. Express the final result in rectangular and polar forms. Check by performing the same operation in rectangular form.

17. $(3 + 4j)(5 - 12j)$

18. $(-2 + 5j)(-1 - j)$

19. $(7 - 3j)(8 + j)$

20. $(1 + 5j)(4 + 2j)$

21. $\dfrac{7}{1 - 3j}$

22. $\dfrac{8j}{7 + 2j}$

23. $\dfrac{3 + 4j}{5 - 12j}$

24. $\dfrac{-2 + 5j}{-1 - j}$

25. $(3 + 4j)^4$

26. $(-1 - j)^8$

27. $(2 + 3j)^5$

28. $(1 - 2j)^6$

In Exercises 29 through 36, use DeMoivre's theorem to find the indicated roots. Be sure to find all roots.

29. The two square roots of $4(\cos 60° + j \sin 60°)$.

30. The three cube roots of $27(\cos 120° + j \sin 120°)$.

31. The three cube roots of $3 - 4j$.

32. The two square roots of $-5 + 12j$.

33. The fourth roots of 1.   **34.** The cube roots of 8.

35. The cube roots of $-27j$.   **36.** The fourth roots of j.

In Exercises 37 through 40, perform the indicated operations.

37. In Example G we showed that one cube root of -1 is $0.500 - 0.866j$. The exact form of this root is $\frac{1}{2}(1 - j\sqrt{3})$. Cube this expression in rectangular form and show that the result is -1.

38. In Example H we showed that one of the square roots of $2j$ is equal to $1 + j$. Square this expression and show that the result is $2j$.

39. The power p, in watts, supplied to an element in an electric circuit is the product of the voltage e and the current i, in amperes. Find the expression for the power supplied if $e = 6.80\underline{/56.3°}$ volts and $i = 7.05\underline{/-15.8°}$ amperes.

40. The displacement d of a weight suspended on a system of two springs is $d = 6.03\underline{/22.5°} + 3.26\underline{/76.0°}$ inches. Perform the addition and express the answer in polar form.

11-7   An Application to Alternating Current (ac) Circuits

We shall complete our study of complex numbers by showing their use in one aspect of alternating-current circuit theory. This application will be made to measuring voltage between any two points in a simple ac circuit, similar to the

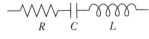

$R \qquad C \qquad L$

Fig. 11-17

application mentioned in some of the examples and exercises of earlier sections of this chapter. We shall consider a circuit containing a resistance, a capacitance, and an inductance.

Briefly, a *resistance* is any part of a circuit which tends to obstruct the flow of electric current through the circuit. It is denoted by R (units in ohms, Ω) and in diagrams by ⌇, as shown in Fig. 11-17. In essence, a *capacitance* is two nonconnected plates in a circuit; no current actually flows across the gap between them. In an ac circuit, an electric charge is continually going to and from each plate and, therefore, the current in the circuit is not effectively stopped. It is denoted by C (units in farads, F) and in diagrams by ⊣⊢ (see Fig. 11-17). An *inductance* is basically a coil of wire in which current is induced because the current is continually changing in the circuit. It is denoted by L (units in henrys, H) and in diagrams by ⏥ (see Fig. 11-17). All these elements affect the voltage in an alternating-current circuit. We shall state here the relation each has to the voltage and current in the circuit.

In Chapter 9, when we were discussing the graphs of the trigonometric functions, we noted that the current and voltage in an ac circuit could be represented by a sine or cosine curve. Therefore, each reaches peak values periodically. *If they reach their respective peak values at the same time, we say they are **in phase**. If the voltage reaches its peak before the current, we say that the voltage **leads** the current. If the voltage reaches its peak after the current, we say that the voltage **lags** the current.*

In the study of electricity, it is shown that the voltage across a resistance is in phase with the current. The voltage across a capacitor lags the current by 90°, and the voltage across an inductance leads the current by 90°. This is shown in Fig. 11-18, where, in a given circuit, I represents the current, V_R is the voltage across a resistor, V_C is the voltage across a capacitor, V_L is the voltage across an inductor, and t represents time.

Each element in an ac circuit tends to offer a type of resistance to the flow of current. *The effective resistance of any part of the circuit is called the **reactance**,* and it is denoted by X. The voltage across any part of the circuit whose reactance is X is given by $V = IX$, where I is the current (in amperes) and V is the voltage (in volts). Therefore, the voltages across a resistor, a capacitor, and an inductor are, respectively,

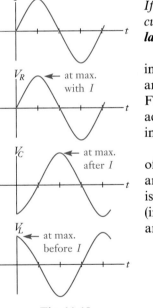

I

V_R ← at max. with I

V_C ← at max. after I

V_L ← at max. before I

Fig. 11-18

$$V_R = IX_R \qquad V_C = IX_C \qquad V_L = IX_L \qquad (11\text{-}18)$$

To determine the voltage across a combination of these elements of a circuit, we must account for the reactance, as well as the phase of the voltage across the individual elements. Since the voltage across a resistor is in phase with the current, we represent V_R along the positive real axis as a real number. Since the voltage across an inductance leads the current by 90°, we shall represent this voltage as a positive, pure imaginary number. In the same way, by representing the voltage across a capacitor as a negative, pure imaginary number, we show that the voltage *lags* the current by 90°. These representations are meaningful since the positive imaginary axis is $+90°$ from the positive real axis, and the negative imaginary axis is $-90°$ from the positive real axis. See Fig. 11-19.

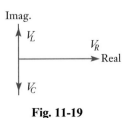

Imag.

V_L

V_R → Real

V_C

Fig. 11-19

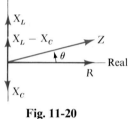

R C L

Fig. 11-17

The circuit elements shown in Fig. 11-17 are in *series,* and all circuits we shall consider are series circuits. The total voltage across a series of all three elements is given by $V_R + V_L + V_C$, which we shall represent by V_{RLC}. Therefore,

$$V_{RLC} = IR + IX_Lj - IX_cj = I[R + j(X_L - X_C)]$$

This expression is also written as

impedance

$$\boxed{V_{RLC} = IZ}$$ (11-19)

where the symbol Z is called the **impedance** *of the circuit. It is the total effective resistance to the flow of current by a combination of the elements in the circuit,* taking into account the phase of the voltage in each element. From its definition, we see that Z is a complex number.

$$\boxed{Z = R + j(X_L - X_C)}$$ (11-20)

with a magnitude

$$\boxed{|Z| = \sqrt{R^2 + (X_L - X_C)^2}}$$ (11-21)

Also, as a complex number, it makes an angle θ with the x-axis, given by

phase angle

$$\boxed{\tan \theta = \frac{X_L - X_C}{R}}$$ (11-22)

All these equations are based on phase relations of voltages with respect to the current. Therefore, *the angle θ represents the phase angle between the current and the voltage* (see Fig. 11-20).

In the examples and exercises of this section, the commonly used units and symbols for them are used. For a summary of these units and symbols, including prefixes, see Appendix B.

EXAMPLE A _____

In the series circuit shown in Fig. 11-21(a), $R = 12.0\ \Omega$ and $X_L = 5.00\ \Omega$. A current of 2.00 A is in the circuit. Find the voltage across each element, the impedance, the voltage across the combination, and the phase angle between the current and voltage.

Since the voltage across any element is the product of the current and reactance, we have the voltage across the resistor (between points a and b) as $V_R = (2.00)(12.0) = 24.0$ V. The voltage across the inductor (between points b and c) is $V_L = (2.00)(5.00) = 10.0$V. To find the voltage across the combination, between points a and c, we must first find the magnitude of the impedance. **The**

NOTE ▷

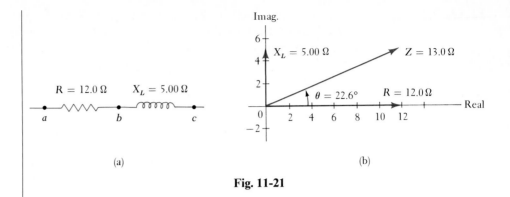

(a)   (b)

Fig. 11-21

*voltage is **not** the arithmetic sum of V_R and V_L;* we must account for the phase. By Eq. (11-20), the impedance is

$$Z = 12.0 + 5.00j$$

with magnitude

$$|Z| = \sqrt{R^2 + X_L^2} = \sqrt{(12.0)^2 + (5.00)^2} = 13.0 \ \Omega$$

Therefore, the voltage across the combination is

$$V_{RL} = (2.00)(13.0) = 26.0 \text{ V}$$

The phase angle between the voltage and current is found by Eq. (11-22). This gives

$$\tan \theta = \frac{5.00}{12.0}, \qquad \theta = 22.6°$$

■ The voltage leads the current by 22.6°, as shown in Fig. 11-21(b).

EXAMPLE B

For a circuit in which $R = 8.00 \ \Omega$, $X_L = 7.00 \ \Omega$, and $X_C = 13.0 \ \Omega$, find the impedance and the phase angle between the current and the voltage.

By the definition of impedance, Eq. (11-20), we have

$$Z = 8.00 + (7.00 - 13.0)j = 8.00 - 6.00j$$

where the magnitude of the impedance is

$$|Z| = \sqrt{(8.00)^2 + (-6.00)^2} = 10.0 \ \Omega$$

The phase angle is found by

$$\tan \theta = \frac{-6.00}{8.00}, \qquad \theta = -36.9°$$

We see that negative angles have a useful purpose in this type of problem. This means that the voltage lags the current by 36.9°. See Fig. 11-22.

From the values above, we can write the impedance in polar form as

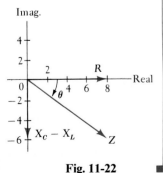

Fig. 11-22   ■ $Z = 10.0 \underline{/-36.9°}$ ohms.

EXAMPLE C____   Let $R = 6145 \, \Omega$, $X_L = 8304 \, \Omega$, and $X_C = 4018 \, \Omega$. Find the impedance and the phase angle between the current and the voltage.

$$Z = 6145 + (8304 - 4018)j = 6145 + 4286j$$
$$|Z| = \sqrt{6145^2 + 4286^2} = 7492 \, \Omega$$
$$\tan \theta = \frac{4286}{6145}, \qquad \theta = 34.89°$$

■   The voltage leads the current by $34.89°$. See Fig. 11-23.

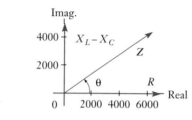

Fig. 11-23

Note that the resistance is represented in the same way as a vector along the positive x-axis. Actually resistance is not a vector quantity, but is represented in this manner in order to assign an angle as the phase of the current. The important concept in this analysis is that *the phase **difference** between the current and voltage is constant,* and therefore any direction may be chosen arbitrarily for one of them. Once this choice is made, other phase angles are measured with respect to this direction. A common choice, as above, is to make the phase angle of the current zero. If an arbitrary angle is chosen, it is necessary to treat the current, voltage, and impedance as complex numbers.

EXAMPLE D____   In a particular circuit, the current is $2.00 - 3.00j$ amperes and the impedance is $6.00 + 2.00j$ ohms. The voltage across this part of the circuit is

$$V = (2.00 - 3.00j)(6.00 + 2.00j) = 12.0 - 14.0j - 6.00j^2$$
$$= 12.0 - 14.0j + 6.00$$
$$= 18.0 - 14.0j \text{ volts}$$

The magnitude of the voltage is

$$|V| = \sqrt{(18.0)^2 + (-14.0)^2} = 22.8 \text{ V}$$

Since the voltage across a resistor is in phase with the current, this voltage can be represented as having a phase difference of zero with respect to the current. Therefore, the resistance is indicated as an arrow in the positive real direction, denoting the fact that the current and voltage are in phase, *Such a representation is called a* **phasor.** The arrow denoted by R, as in Fig. 11-20, is actually the phasor representing the voltage across the resistor. Remember, the positive real axis is arbitrarily chosen as the direction of the phase of the current.

To show properly that the voltage across an inductance leads the current by $90°$, its reactance (effective resistance) is multiplied by j. We know that there is a positive $90°$ angle between a positive real number and a positive imaginary number. In the same way, by multiplying the capacitive reactance by $-j$, we show the 90° difference in phase between the voltage and current in a capacitor, with the current leading. Therefore, jX_L represents the phasor for the voltage across an inductor and $-jX_C$ is the phasor for the voltage across the capacitor. The phasor for the voltage across the combination of the resistance, inductance, and capacitance is Z, where the phase difference between the voltage and current for the combination is the angle θ.

From this we see that *multiplying a phasor by j means to perform the operation of rotating it through 90°.* For this reason, j is also called the *j-operator.*

EXAMPLE E

Imag.

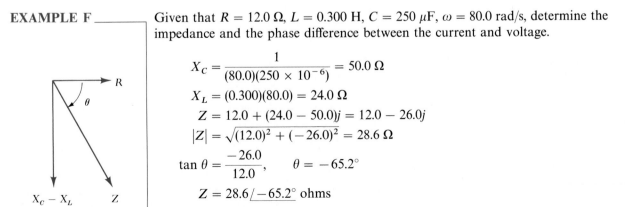

Fig. 11-24

Multiplying a positive real number A by j, we have $A \times j = Aj$, which is a positive imaginary number. In the complex plane, Aj is 90° from A, which means that by multiplying A by j we rotated it by 90°. Similarly, $Aj \times j = Aj^2 = -A$, which is a negative real number, rotated 90° from Aj. Thus, successive multiplications of A by j give us

$$A \times j = Aj \qquad \text{positive imaginary number}$$
$$Aj \times j = Aj^2 = -A \qquad \text{negative real number}$$
$$-A \times j = -Aj \qquad \text{negative imaginary number}$$
$$-Aj \times j = -Aj^2 = A \qquad \text{positive real number}$$

■ See Fig. 11-24.

An alternating current is produced by a coil of wire rotating through a magnetic field. If the angular velocity of this wire is ω, the capacitive and inductive reactances are given by the relations

$$X_C = \frac{1}{\omega C} \quad \text{and} \quad X_L = \omega L \tag{11-23}$$

Therefore, if ω, C, and L are known, the reactance of the circuit may be determined.

EXAMPLE F

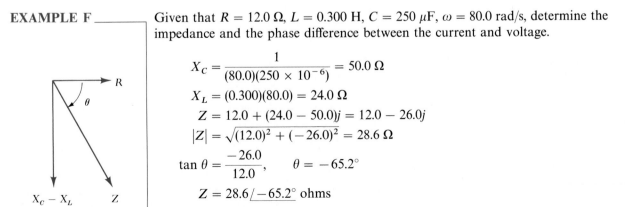

Fig. 11-25

Given that $R = 12.0 \ \Omega$, $L = 0.300$ H, $C = 250 \ \mu$F, $\omega = 80.0$ rad/s, determine the impedance and the phase difference between the current and voltage.

$$X_C = \frac{1}{(80.0)(250 \times 10^{-6})} = 50.0 \ \Omega$$
$$X_L = (0.300)(80.0) = 24.0 \ \Omega$$
$$Z = 12.0 + (24.0 - 50.0)j = 12.0 - 26.0j$$
$$|Z| = \sqrt{(12.0)^2 + (-26.0)^2} = 28.6 \ \Omega$$
$$\tan \theta = \frac{-26.0}{12.0}, \qquad \theta = -65.2°$$
$$Z = 28.6\underline{/-65.2°} \text{ ohms}$$

■ The voltage lags the current (see Fig. 11-25).

We recall from Section 9-5 that the angular velocity ω is related to the frequency f by the relation $\omega = 2\pi f$. It is very common to use frequency when discussing alternating current.

An important concept in the application of this theory is that of **resonance.** *For resonance, the impedance of any circuit is a minimum, or the total impedance is R.* Thus, $X_L - X_C = 0$. Also, it can be seen that the current and voltage are in phase under these conditions. Resonance is required for the tuning of radio and television receivers.

EXAMPLE G

See the chapter introduction.

In the antenna circuit of a radio, the inductance is 4.20 mH and the capacitance can be varied. What range of values of capacitance are necessary for the radio to receive the AM band of radio stations, with frequencies from 530 kHz to 1600 kHz?

For proper tuning, the circuit should be in resonance, or $X_L = X_C$. This means that

$$2\pi f L = \frac{1}{2\pi f C} \quad \text{or} \quad C = \frac{1}{(2\pi f)^2 L}$$

For $f_1 = 530 \text{ kHz} = 5.30 \times 10^5$ Hz and $L = 4.20 \text{ mH} = 4.20 \times 10^{-3}$ H, we have

$$C_1 = \frac{1}{(2\pi)^2 (5.30 \times 10^5)^2 (4.20 \times 10^{-3})} = 2.15 \times 10^{-11} \text{ F} = 21.5 \text{ pF}$$

and for $f_2 = 1600 \text{ kHz} = 1.60 \times 10^6$ Hz and $L = 4.20 \times 10^{-3}$ H, we have

$$C_2 = \frac{1}{(2\pi)^2 (1.60 \times 10^6)^2 (4.20 \times 10^{-3})} = 2.36 \times 10^{-12} \text{ F} = 2.36 \text{ pF}$$

Therefore, the capacitor should be capable of having capacitances from 2.36 pF to 21.5 pF.

Exercises 11-7

For Exercises 1 through 4, use the circuit shown in Fig. 11-26. The current in the circuit is 5.75 mA. Determine the indicated quantities.

1. Find the voltage across the resistor (between points *a* and *b*).

2. Find the voltage across the inductor (between points *b* and *c*).

3. (a) Find the magnitude of the impedance across the resistor and inductor (between points *a* and *c*).
   (b) Find the phase angle between the current and voltage for this combination.
   (c) Find the voltage across this combination.

4. (a) Find the magnitude of the impedance across the resistor, inductor, and capacitor (between points *a* and *d*).
   (b) Find the phase angle between the current and voltage for this combination.
   (c) Find the voltage across this combination.

Fig. 11-26

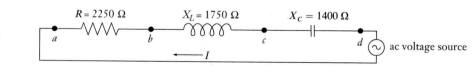

$R = 2250 \, \Omega$ $X_L = 1750 \, \Omega$ $X_C = 1400 \, \Omega$

a   *b*   *c*   *d*   ac voltage source

$\longleftarrow I$

In Exercises 5 through 8, an ac circuit contains the given combinations of circuit elements from among a resistor ($R = 45.0 \; \Omega$), *a capacitor* ($C = 86.2 \; \mu F$), *and an inductor* ($L = 42.9 \; mH$). *If the frequency in the circuit is* $f = 60.0 \; Hz$, *find (a) the magnitude of the impedance and (b) the phase angle between the current and voltage.*

5. The circuit has the inductor and the capacitor (an *LC* circuit).

6. The circuit has the resistor and the capacitor (an *RC* circuit).

7. The circuit has the resistor and the inductor (an *RL* circuit).

8. The circuit has the resistor, the inductor, and the capacitor (an *RLC* circuit).

In Exercises 9 through 20, find the required quantities.

9. Given that the current in a given circuit is $3.90 - 6.04j$ mA and the impedance is $5.16 + 1.14j$ kΩ, find the magnitude of the voltage.

10. Given that the voltage in a given circuit is $8.375 - 3.140j$ V and the impedance is $2.146 - 1.114j \; \Omega$, find the magnitude of the current.

11. A resistance ($R = 25.3 \; \Omega$) and a capacitance ($C = 2.75 \; nF$) are in an AM radio circuit. If $f = 1200 \; kHz$, find the impedance across the resistor and capacitor.

12. A resistance ($R = 64.5 \; \Omega$) and an inductance ($L = 1.08 \; mH$) are in a telephone circuit. If $f = 8.53 \; kHz$, find the impedance across the resistor and inductor.

13. The reactance of an inductor is $1200 \; \Omega$ for $f = 280 \; Hz$. What is its inductance?

14. A resistor, an inductor, and a capacitor are connected in series across an ac voltage source. A voltmeter measures 12.0 V, 15.5 V, and 10.5 V, respectively, when placed across each element separately. What is the voltage of the source?

15. An inductance ($L = 25.0 \; mH$) and a capacitance ($C = 7.18 \; \mu F$) are in series. Find the frequency for resonance.

16. A capacitance ($C = 95.2 \; nF$) and an inductance are in series in the circuit of a receiver for navigation signals. Find the inductance if the frequency for resonance is 50.0 kHz.

17. In Example G, what should be the capacitance in order to receive a 680-kHz radio signal?

18. A 220-V source with $f = 60.0 \; Hz$ is connected in series to an inductance ($L = 2.05 \; H$) and a resistance R. Find R if the current is 0.250 A.

19. The power supplied to a series combination of elements in an ac circuit is given by the relation $P = VI \cos \theta$, where P is the power (in watts), V is the effective voltage, I is the effective current, and θ is the phase angle between the current and voltage. Assuming that the effective voltage across the resistor, capacitor, and inductor combination in Exercise 8 is 225 mV, determine the power supplied to these elements.

20. Find the power supplied to a resistor of $120 \; \Omega$ and a capacitor of $0.725 \; \mu F$ if $\omega = 980 \; rad/s$ and the effective voltage is 110 V.

11-8 Chapter Equations, Review Exercises, and Practice Test

Chapter Equations for Complex Numbers

Imaginary unit	$j^2 = -1$	(11-1)
	$\sqrt{-a} = j\sqrt{a}, \quad (a > 0)$	(11-2)
Basic operations	Addition (and subtraction):	
	$(a + bj) + (c + dj) = (a + c) + (b + d)j$	(11-3)
	Multiplication:	
	$(a + bj)(c + dj) = (ac - bd) + (ad + bc)j$	(11-4)

Division:

$$\frac{a + bj}{c + dj} = \frac{(a + bj)(c - dj)}{(c + dj)(c - dj)} = \frac{(ac + bd) + (bc - ad)j}{c^2 + d^2} \qquad (11\text{-}5)$$

Complex number forms   Rectangular: $x + yj$

Polar: $r(\cos \theta + j \sin \theta) = r\underline{/\theta}$

Exponential: $re^{j\theta}$

$$x = r \cos \theta \qquad\qquad y = r \sin \theta \qquad (11\text{-}6)$$

$$r^2 = x^2 + y^2 \qquad \tan \theta = \frac{y}{x} \qquad (11\text{-}7)$$

$$x + yj = r(\cos \theta + j \sin \theta) = r\underline{/\theta} = re^{j\theta} \qquad (11\text{-}11)$$

Product in polar form $r_1(\cos \theta_1 + j \sin \theta_1)r_2(\cos \theta_2 + j \sin \theta_2)$

$$= r_1 r_2[\cos(\theta_1 + \theta_2) + j \sin(\theta_1 + \theta_2)] \qquad (11\text{-}12)$$

$$(r_1\underline{/\theta_1})(r_2\underline{/\theta_2}) = r_1 r_2\underline{/\theta_1 + \theta_2}$$

Quotient in polar form $\dfrac{r_1(\cos \theta_1 + j \sin \theta_1)}{r_2(\cos \theta_2 + j \sin \theta_2)} = \dfrac{r_1}{r_2}[\cos(\theta_1 - \theta_2) + j \sin(\theta_1 - \theta_2)]$

$$\frac{r_1\underline{/\theta_1}}{r_2\underline{/\theta_2}} = \frac{r_1}{r_2}\underline{/\theta_1 - \theta_2} \qquad (11\text{-}14)$$

DeMoivre's theorem $[r(\cos \theta + j \sin \theta)]^n = r^n(\cos n\theta + j \sin n\theta)$

$$(r\underline{/\theta})^n = r^n\underline{/n\theta} \qquad (11\text{-}17)$$

Chapter Equations for Alternating-Current Series Circuits ▬▬▬▬

Voltage, current, reactance $V_R = IX_R \qquad V_C = IX_C \qquad V_L = IX_L \qquad (11\text{-}18)$

Impedance $V_{RLC} = IZ \qquad\qquad\qquad\qquad\qquad\qquad (11\text{-}19)$

$$Z = R + j(X_L - X_C) \qquad (11\text{-}20)$$

$$|Z| = \sqrt{R^2 + (X_L - X_C)^2} \qquad (11\text{-}21)$$

Phase angle $\tan \theta = \dfrac{X_L - X_C}{R} \qquad (11\text{-}22)$

Capacitive reactance
and inductive reactance $X_C = \dfrac{1}{\omega C} \quad \text{and} \quad X_L = \omega L \qquad (11\text{-}23)$

Review Exercises ▬▬▬▬

In Exercises 1 through 16, perform the indicated operations, expressing all answers in the simplest rectangular form.

1. $(6 - 2j) + (4 + j)$ **2.** $(12 + 7j) + (-8 + 6j)$

3. $(18 - 3j) - (12 - 5j)$ **4.** $(-4 - 2j) - (-6 - \sqrt{-49})$

5. $(2 + j)(4 - j)$

6. $(-5 + 3j)(8 - 4j)$

7. $(2j^5)(6 - 3j)(4 + 3j)$

8. $j(3 - 2j) - (j^3)(5 + j)$

9. $\dfrac{3}{7 - 6j}$

10. $\dfrac{4j}{2 + 9j}$

11. $\dfrac{6 - \sqrt{-16}}{\sqrt{-4}}$

12. $\dfrac{3 + \sqrt{-4}}{4 - j}$

13. $\dfrac{5j - (3 - j)}{4 - 2j}$

14. $\dfrac{2 + (j - 6)}{1 - 2j}$

15. $\dfrac{j(7 - 3j)}{2 + j}$

16. $\dfrac{(2 - j)(3 + 2j)}{4 - 3j}$

In Exercises 17 through 20, find the values of x and y for which the equations are valid.

17. $3x - 2j = yj - 2$

18. $2xj - 2y = (y + 3)j - 3$

19. $2x - j + 4 = 6y + 2xj$

20. $3yj + xj = 6 + 3x + y$

In Exercises 21 through 24, perform the indicated operations graphically; check them algebraically.

21. $(-1 + 5j) + (4 + 6j)$

22. $(7 - 2j) + (-5 + 4j)$

23. $(9 + 2j) - (5 - 6j)$

24. $(1 + 4j) - (-3 - 3j)$

In Exercises 25 through 32, give the polar and exponential forms of each of the complex numbers.

25. $1 - j$

26. $4 + 3j$

27. $-2 - 7j$

28. $6 - 2j$

29. $1.07 + 4.55j$

30. $-327 + 158j$

31. 10

32. $-4j$

In Exercises 33 through 44, give the rectangular form of each of the complex numbers.

33. $2(\cos 225° + j \sin 225°)$

34. $4(\cos 60° + j \sin 60°)$

35. $5.011(\cos 123.82° + j \sin 123.82°)$

36. $2.417(\cos 296.26° + j \sin 296.26°)$

37. $0.62\underline{/-72°}$

38. $20\underline{/160°}$

39. $27.08\underline{/346.27°}$

40. $1.689\underline{/194.36°}$

41. $2.00e^{0.25j}$

42. $e^{3.62j}$

43. $25.37e^{1.906j}$

44. $44.47e^{6.046j}$

In Exercises 45 through 56, perform the indicated operations. Leave the result in polar form.

45. $3(\cos 32° + j \sin 32°)][5(\cos 52° + j \sin 52°)]$

46. $[2.5(\cos 162° + j \sin 162°)][8(\cos 115° + j \sin 115°)]$

47. $(40\underline{/18°})(0.5\underline{/245°})$

48. $(0.1254\underline{/172.38°})(27.17\underline{/204.34°})$

49. $\dfrac{24(\cos 165° + j \sin 165°)}{3(\cos 106° + j \sin 106°)}$

50. $\dfrac{18(\cos 403° + j \sin 403°)}{4(\cos 192° + j \sin 192°)}$

51. $\dfrac{245.6\underline{/326.44°}}{17.19\underline{/192.83°}}$

52. $\dfrac{100\underline{/206°}}{4\underline{/320°}}$

53. $0.983\underline{/47.2°} + 0.366\underline{/95.1°}$

54. $17.8\underline{/110.4°} - 14.9\underline{/226.3°}$

55. $7644\underline{/294.36°} - 6871\underline{/17.86°}$

56. $4.944\underline{/327.49°} + 8.009\underline{/7.37°}$

57. $[2(\cos 16° + j \sin 16°)]^{10}$

58. $[3(\cos 36° + j \sin 36°)]^6$

59. $(3\underline{/110.5°})^3$

60. $(5.36\underline{/220.3°})^4$

In Exercises 61 through 64, change each number to polar form and then perform the indicated operations. Express the final result in rectangular and polar forms. Check by performing the same operation in rectangular form.

61. $(1 - j)^{10}$

62. $(\sqrt{3} + j)^8(1 + j)^5$

63. $\dfrac{(5 + 5j)^4}{(-1 - j)^6}$

64. $(\sqrt{3} - j)^{-8}$

In Exercises 65 through 68, use DeMoivre's theorem to find the indicated roots. Be sure to find all roots.

65. The cube roots of -8

66. The cube roots of 1

67. The fourth roots of $-j$

68. The fifth roots of -32

In Exercises 69 through 80, find the required quantities.

69. A 60-V ac voltage source is connected in series across a resistor, an inductor, and a capacitor. The voltage across the inductor is 60 V, and the voltage across the capacitor is 60 V. What is the voltage across the resistor?

70. In a series ac circuit with a resistor, an inductor, and a capacitor, $R = 6.50\ \Omega$, $X_C = 3.74\ \Omega$, and $Z = 7.50\ \Omega$. Find X_L.

71. In a series ac circuit with a resistor, an inductor, and a capacitor, $R = 6250\ \Omega$, $Z = 6720\ \Omega$, and $X_L = 1320\ \Omega$. Find the phase angle θ.

72. A coil of wire rotates at 120.0 r/s. If the coil generates a current in a circuit containing a resistance of 12.07 Ω, an inductance of 0.1405 H and an impedance of 22.35 Ω, what must be the value of a capacitor (in farads) in the circuit?

73. What is the frequency f for resonance in a circuit for which $L = 2.65$ H and $C = 18.3\ \mu F$?

74. The displacement of an electromagnetic wave is given by $d = A(\cos \omega t + j \sin \omega t) + B(\cos \omega t - j \sin \omega t)$. Find the expressions for the magnitude and phase angle of the displacement.

75. What are the magnitude and direction of a force which is represented by $600 - 550j$ pounds?

76. What are the magnitude and direction of a velocity which is represented by $2500 + 1500j$ kilometers per hour?

77. In the study of shearing effects in the spinal column, the expression

$$\frac{1}{u + j\omega n}$$

is found. Express this in rectangular form.

78. In the theory of light reflection on metals, the expression

$$\frac{\mu(1 - kj) - 1}{\mu(1 - kj) + 1}$$

is encountered. Simplify this expression.

79. Show that $e^{j\pi} = -1$.   80. Show that $(e^{j\pi})^{1/2} = j$.

Practice Test

1. Add, expressing the result in rectangular form: $(3 - \sqrt{-4}) + (5\sqrt{-9} - 1)$.

2. Multiply, expressing the result in polar form: $(2\underline{/130°})(3\underline{/45°})$.

3. Express $2 - 7j$ in polar form.

4. Express in terms of j: (a) $-\sqrt{-64}$, (b) $-j^{15}$.

5. Add graphically: $(4 - 3j) + (-1 + 4j)$.

6. Simplify, expressing the result in rectangular form: $\dfrac{2 - 4j}{5 + 3j}$.

7. Express $2.56(\cos 125.2° + j \sin 125.2°)$ in exponential form.

8. For an ac circuit in which $R = 3.50\ \Omega$, $X_L = 6.20\ \Omega$, and $X_C = 7.35\ \Omega$, find the impedance and the phase angle between the current and the voltage.

9. Express $3.47 - 2.81j$ in exponential form.

10. Find the values of x and y: $x + 2j - y = yj - 3xj$.

11. What is the capacitance of a radio which has an inductance of 8.75 mH if it is to receive a station with frequency 600 kHz?

12. Find the cube roots of j.

12 Exponential and Logarithmic Functions

The intensity level of sound is measured on a logarithmic scale. This is shown in Section 12-6.

In this chapter we introduce two more important types of functions, the *exponential function* and the *logarithmic function*. Historically, logarithms were first used for computational work, but with the extensive use of calculators and computers, they are not directly used for this purpose to any extent today. However, these functions are very important in many scientific and technical applications, as well as in many areas of advanced mathematics.

To illustrate the importance of logarithms, many applications may be cited. The basic units used to measure the intensity of sound and those used to measure the intensity of earthquakes are defined in terms of logarithms. In chemistry, the distinction between a base and an acid is defined in terms of logarithms. In electrical transmission lines, power gains and losses are measured in terms of logarithmic units. In electronics and in mechanical systems, the use of exponential functions, which are closely related to logarithms, is extensive. Many of these applications are illustrated throughout the chapter.

12-1 The Exponential and Logarithmic Functions

Chapter 10 dealt with exponents in expressions of the form x^n, where we showed that n could be any rational number. Here we shall deal with expressions of the form b^x, where x is any real number. When we look at these expressions, we note the primary difference is that in the second expression *the exponent is variable*. We have not previously dealt with variable exponents. *Thus let us define*

349

the **exponential function** *to be*

exponential function

$$y = b^x$$

(12-1)

In Eq. (12-1), x *is called the* **logarithm** *of the number* y *to the base* b. In our work with logarithms we shall restrict all numbers to the real number system. *This leads us to choose the base as a positive number other than 1.* We know that 1 raised to any power will result in 1, which would make y a constant regardless of the value of x. Negative numbers for b would result in imaginary values for y if x were any fractional exponent with an even integer for its denominator.

EXAMPLE A

$y = 2^x$ is an exponential function, where x is the logarithm of y to the base 2. This means that 2 raised to a given power gives us the corresponding value of y.

If $x = 2$, $y = 2^2 = 4$; this means that 2 is the logarithm of 4 to the base 2. If $x = 4$, $y = 2^4 = 16$; this means that 4 is the logarithm of 16 to the base 2. If $x = \frac{1}{2}$, $y = 2^{1/2} = 1.41$; this means that $\frac{1}{2}$ is the logarithm of 1.41 to the base 2.

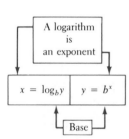

Fig. 12-1

Using the definition of a logarithm, we may express x in terms of y in the form

$$x = \log_b y$$

(12-2)

NOTE ▷

This equation is read in accordance with the definition of x in Eq. (12-1): **x equals the logarithm of y to the base b.** This means that x is the power to which the base b must be raised in order to equal the number y; that is, x is a logarithm, and *a logarithm is an exponent.* Note that Eqs. (12-1) and (12-2) state the same relationship, but in a different manner. Equation (12-1) is the **exponential form,** and Eq. (12-2) is the **logarithmic form.** See Fig. 12-1.

EXAMPLE B

The equation $y = 2^x$ would be written as $x = \log_2 y$ if we put it in logarithmic form. When we choose values of y to find the corresponding values of x from this equation, we ask ourselves, "2 raised to what power gives y?" Hence if $y = 4$, we know that 2^2 is 4, and x would be 2. If $y = 8$, $2^3 = 8$, or $x = 3$.

EXAMPLE C
NOTE ▷

$3^2 = 9$ in logarithmic form is $2 = \log_3 9$; $4^{-1} = \frac{1}{4}$ in logarithmic form is $-1 = \log_4 (\frac{1}{4})$. Remember, *the exponent may be negative.* The base must be positive.

EXAMPLE D

1. $(64)^{1/3} = 4$ in logarithmic form is $\frac{1}{3} = \log_{64} 4$,
2. $(32)^{3/5} = 8$ in logarithmic form is $\frac{3}{5} = \log_{32} 8$

EXAMPLE E

1. $\log_2 32 = 5$ in exponential form is $32 = 2^5$,
2. $\log_6 (\frac{1}{36}) = -2$ in exponential form is $\frac{1}{36} = 6^{-2}$

EXAMPLE F

Find b, given that $-4 = \log_b (\frac{1}{81})$.

Writing this in exponential form, we have $\frac{1}{81} = b^{-4}$. Thus, $\frac{1}{81} = \frac{1}{b^4}$ or $\frac{1}{3^4} = \frac{1}{b^4}$. Therefore, $b = 3$.

EXAMPLE G _____ Find y, given that $\log_4 y = \frac{1}{2}$.

In exponential form we have $y = 4^{1/2}$, or $y = 2$. ∎

We see that exponential form is very useful for determining values written in logarithmic form. For this reason it is important that you learn to transform readily from one form to the other.

EXAMPLE H _____ The power supply P, in watts, of a certain satellite is given by $P = 75e^{-0.005t}$, where t is the time in days after launch. By writing this equation in logarithmic form, solve for t.

In order to have the equation in the exponential form of Eq. (12-1), we must have only $e^{-0.005t}$ on the right. Therefore, by dividing by 75, we have

$$\frac{P}{75} = e^{-0.005t}$$

Writing this in logarithmic form, we have

$$\log_e \left(\frac{P}{75}\right) = -0.005t$$

or

$$t = \frac{\log_e \left(\dfrac{P}{75}\right)}{-0.005} = -200 \log_e \left(\frac{P}{75}\right)$$ ∎

In Example H we saw that in order to change a function of the form $y = ab^x$ into logarithmic form, we must first write it as $y/a = b^x$. We must have the coefficient of b^x equal to 1, which is the form of Eq. (12-1). In the same way, the coefficient of $\log_b y$ must be 1 in order to change it into exponential form.

When we are working with functions, we must keep in mind that a function is defined by the operation being performed on the independent variable, and not by the letter chosen to represent it. However, for consistency, it is standard practice to let y represent the dependent variable and x represent the independent variable. Therefore, *the* **logarithmic function** *is*

logarithmic function

$$\boxed{y = \log_b x}$$

(12-3)

As with the exponential function, $b > 0$ and $b \neq 1$.

Equations (12-2) and (12-3) express the same *function*, the logarithmic function. They do not represent different functions, due to the difference in location of the variables, since they represent the same operation on the independent variable which appears in each. However, Eq. (12-3) expresses the function with the standard dependent and independent variables.

EXAMPLE I _____ For the logarithmic function $y = \log_2 x$, we have the standard independent variable x and the standard dependent variable y.

If $x = 16$, $y = \log_2 16$, which means that $y = 4$, since $2^4 = 16$.

If $x = \frac{1}{16}$, $y = \log_2 \left(\frac{1}{16}\right)$, which means that $y = -4$, since $2^{-4} = \frac{1}{16}$. ∎

We note that for the exponential function $y = b^x$ and the logarithmic function $y = \log_b x$, if we solve for the independent variable in one by changing the form, then interchange the variables x and y, we obtain the other function. *Such functions are called* **inverse functions.** (This was essentially the procedure we used in defining the logarithmic function.)

EXAMPLE J ————— The functions $y = 2^x$ and $y = \log_2 x$ are inverse functions. We show this by solving $y = 2^x$ for x, then interchanging x and y.

$$y = 2^x \quad \text{in logarithmic form is} \quad x = \log_2 y$$

■ Interchanging x and y, we have $y = \log_2 x$, which is the inverse function.

For a function, there is exactly one value of y in the range for each value of x in the domain. This must also hold for the inverse function. Thus, for a function to have an inverse there must be only one x for each y. This holds for $y = b^x$ and $y = \log_b x$, and we will show this in the following section.

Exercises 12-1

In Exercises 1 through 4, evaluate the exponential function $y = 9^x$ for the given values of x.

1. $x = 0.5$ **2.** $x = 4$ **3.** $x = -2$ **4.** $x = -0.5$

In Exercises 5 through 16, express the given equations in logarithmic form.

5. $3^3 = 27$ **6.** $5^2 = 25$ **7.** $4^4 = 256$ **8.** $8^2 = 64$

9. $4^{-2} = \frac{1}{16}$ **10.** $3^{-2} = \frac{1}{9}$ **11.** $2^{-6} = \frac{1}{64}$ **12.** $(12)^0 = 1$

13. $8^{1/3} = 2$ **14.** $(81)^{3/4} = 27$ **15.** $(\frac{1}{4})^2 = \frac{1}{16}$ **16.** $(\frac{1}{2})^{-2} = 4$

In Exercises 17 through 28, express the given equations in exponential form.

17. $\log_3 81 = 4$ **18.** $\log_{11} 121 = 2$ **19.** $\log_9 9 = 1$ **20.** $\log_{15} 1 = 0$

21. $\log_{25} 5 = \frac{1}{2}$ **22.** $\log_8 16 = \frac{4}{3}$ **23.** $\log_{243} 3 = \frac{1}{5}$ **24.** $\log_{32} (\frac{1}{8}) = -\frac{3}{5}$

25. $\log_{10} 0.1 = -1$ **26.** $\log_7 (\frac{1}{49}) = -2$ **27.** $\log_{0.5} 16 = -4$ **28.** $\log_{1/3} 3 = -1$

In Exercises 29 through 44, determine the value of the unknown.

29. $\log_4 16 = x$ **30.** $\log_5 125 = x$ **31.** $\log_{10} 0.01 = x$ **32.** $\log_{16} (\frac{1}{4}) = x$

33. $\log_7 y = 3$ **34.** $\log_8 N = 3$ **35.** $\log_8 y = -\frac{2}{3}$ **36.** $\log_7 y = -2$

37. $\log_b 81 = 2$ **38.** $\log_b 625 = 4$ **39.** $\log_b 4 = -\frac{1}{3}$ **40.** $\log_b 4 = \frac{2}{3}$

41. $\log_{10} 10^{0.2} = x$ **42.** $\log_5 5^{1.3} = x$ **43.** $\log_3 27^{-1} = x$ **44.** $\log_b (\frac{1}{4}) = -\frac{1}{2}$

In Exercises 45 through 48, evaluate the logarithmic function $y = \log_4 x$ for the given values of x.

45. $x = 64$ **46.** $x = \frac{1}{64}$ **47.** $x = \frac{1}{2}$ **48.** $x = 2$

In Exercises 49 through 56, perform the indicated operations.

49. The value V of a bank account in which A dollars is invested at 10% interest, compounded annually, is given by $V = A(1.1)^t$, where t is the time in years. Solve for t.

50. The intensity I of an earthquarke is given by $I = I_0(10)^R$, where I_0 is a minimum intensity for comparison and R is the Richter scale magnitude of the earthquake. Solve for R.

51. In the theory dealing with the optical brightness of objects, the equation $D = \log_{10} (I_0/I)$ is used. Solve for I.

52. The velocity v of a rocket at the point at which its fuel is completely burned is given by $v = u \log_e (w_0/w)$, where u is the exhaust velocity, w_0 is the lift-off weight, and w is the burnout weight. Solve for w.

53. An equation relating the number N of atoms of radium at any time t in terms of the numbers of atoms at $t = 0$, N_0, is $\log_e (N/N_0) = -kt$, where k is a constant. Solve for N.

54. The charge q on a capacitor is given by $q = q_0(1 - e^{-at})$, where q_0 is the initial charge, a is a constant, and t is the time. Solve for t.

55. Show that $y = \log_3 x$ and $y = 3^x$ are inverse functions.

56. Show that $y = 2 \log_5 x$ and $y = 5^{x/2}$ are inverse functions.

12-2   Graphs of $y = b^x$ and $y = \log_b x$

Graphical representation of functions is often valuable when we wish to demonstrate their properties. We shall now show representative graphs of the exponential function $y = b^x$ and the logarithmic function $y = \log_b x$.

EXAMPLE A

Plot the graph of $y = 2^x$.

Assuming values for x and then finding the corresponding values for y, we obtain the following table.

x	-3	-2	-1	0	1	2	3
y	$\frac{1}{8}$	$\frac{1}{4}$	$\frac{1}{2}$	1	2	4	8

$\;\;\;\;\;\sqcup 2^{-3} = \frac{1}{8}$ $\;\;\;\;\;\sqcup 2^0 = 1$ $\;\;\;\;\;\sqcup 2^3 = 8$

From these values we plot the curve, as shown in Fig. 12-2. We note that the x-axis is an asymptote of the curve.

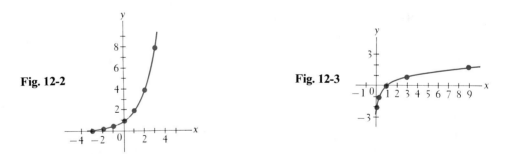

Fig. 12-2

Fig. 12-3

EXAMPLE B

Plot the graph of $y = \log_3 x$.

We can find the points for this graph more easily if we first put the equation in exponential form: $x = 3^y$. By assuming values for y, we can find the corresponding values for x.

$\;\;\;\;\;\sqcap 3^{-2} = \frac{1}{9}$ $\;\;\;\;\;\sqcap 3^2 = 9$

x	$\frac{1}{9}$	$\frac{1}{3}$	1	3	9
y	-2	-1	0	1	2

Using these values, we construct the graph seen in Fig. 12-3.

Any exponential or logarithmic curve, where $b > 1$, will be similar in shape to those of Examples A and B. From these curves we can draw certain conclusions:

1. *If $0 < x < 1$, $\log_b x < 0$; if $x = 1$, $\log_b 1 = 0$; if $x > 1$, $\log_b x > 0$.*
2. *If $x > 1$, x increases more rapidly than $\log_b x$.*
3. *For all values of x, $b^x > 0$.*
4. *If $x > 1$, b^x increases more rapidly than x.*

From the figures and analysis above, we note that the domain of the function $y = b^x$ is all real numbers and that its range is $y > 0$. For the function $y = \log_b x$, the domain is $x > 0$ and the range is all real numbers. Referring to the last paragraph of the previous section, we note that there is only one value of x for each value of y and that the values of the domain and the range are interchanged for these two inverse functions.

Although the bases important to applications are greater than 1, to understand how the curve of the exponential function differs somewhat if $b < 1$, let us consider the following example.

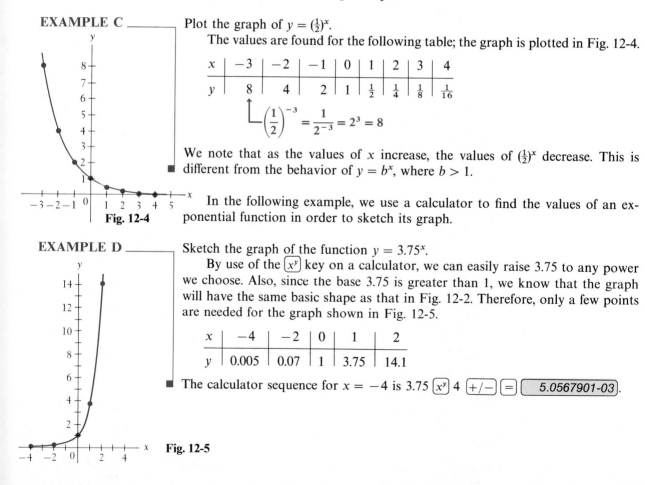

EXAMPLE C

Plot the graph of $y = (\tfrac{1}{2})^x$.

The values are found for the following table; the graph is plotted in Fig. 12-4.

x	-3	-2	-1	0	1	2	3	4
y	8	4	2	1	$\tfrac{1}{2}$	$\tfrac{1}{4}$	$\tfrac{1}{8}$	$\tfrac{1}{16}$

$$\left(\frac{1}{2}\right)^{-3} = \frac{1}{2^{-3}} = 2^3 = 8$$

We note that as the values of x increase, the values of $(\tfrac{1}{2})^x$ decrease. This is different from the behavior of $y = b^x$, where $b > 1$.

In the following example, we use a calculator to find the values of an exponential function in order to sketch its graph.

Fig. 12-4

EXAMPLE D

Sketch the graph of the function $y = 3.75^x$.

By use of the $\boxed{x^y}$ key on a calculator, we can easily raise 3.75 to any power we choose. Also, since the base 3.75 is greater than 1, we know that the graph will have the same basic shape as that in Fig. 12-2. Therefore, only a few points are needed for the graph shown in Fig. 12-5.

x	-4	-2	0	1	2
y	0.005	0.07	1	3.75	14.1

The calculator sequence for $x = -4$ is 3.75 $\boxed{x^y}$ 4 $\boxed{+/-}$ $\boxed{=}$ $\boxed{5.0567901\text{-}03}$.

Fig. 12-5

EXAMPLE E

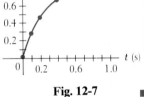

Fig. 12-6

Plot the graph of $y = 2 \log_2 x$.

As in Example B, it is generally more convenient to work with the exponential form. In order to change this equation to exponential form, we first divide each side by 2. Thus, we have

$$\frac{y}{2} = \log_2 x, \qquad x = 2^{y/2}$$

Choosing values of y, we then calculate corresponding values of x and obtain the following table.

x	0.5	0.7	1	1.4	2	2.8	4	5.7	8
y	-2	-1	0	1	2	3	4	5	6

■ See Fig. 12-6.

The following example illustrates an application in which there is an exponential function and for which we show its graph.

EXAMPLE F

Fig. 12-7

In an electric circuit in which there is a battery, an inductor, and a resistor, the current i, in amperes—as a function of the time t, in seconds—is given by $i = 0.8(1 - e^{-4t})$. Here e is the same irrational number which was used in Section 11-5. Its value is approximately $e = 2.718$.

Here we see that values of the exponential function e^{-4t} are to be subtracted from 1 and the result is then multiplied by 0.8. Also, e^{-4t} becomes very small after a short period of time, and this means that i quickly becomes nearly 0.8 A, although theoretically it never quite reaches this value. We shall calculate values for i using the $\boxed{e^x}$ key on the calculator and values of t from 0 to 1 s.

t	0	0.1	0.2	0.4	0.6	0.8	1.0
i	0	0.26	0.44	0.64	0.73	0.77	0.79

■ See Fig. 12-7.

Exercises 12-2

In Exercises 1 through 12, plot graphs of the given functions. Values of x from -3 or -2 through 2 or 3 are appropriate.

1. $y = 3^x$
2. $y = 4^x$
3. $y = 6^x$
4. $y = 10^x$
5. $y = (1.65)^x$
6. $y = (2.72)^x$
7. $y = (\frac{1}{3})^x$
8. $y = (\frac{1}{4})^x$
9. $y = 2(2^x)$
10. $y = 1.5(4.15)^x$
11. $y = 0.5(3.06)^{-x}$
12. $y = 0.1(10^{-x})$

In Exercises 13 through 24, plot graphs of the given functions. Values of y from -3 or -2 through 2 or 3 are appropriate for Exercises 13 through 20. Care should be taken in selecting appropriate values used in Exercises 21 through 24.

13. $y = \log_2 x$
14. $y = \log_4 x$
15. $y = \log_6 x$
16. $y = \log_{10} x$
17. $y = \log_{1.65} x$
18. $y = \log_{2.72} x$
19. $y = \log_{32} x$
20. $y = \log_{0.5} x$
21. $y = 2 \log_3 x$
22. $y = 3 \log_2 x$
23. $y = 0.2 \log_4 x$
24. $y = 5 \log_{10} x$

In Exercises 25 through 32, plot the indicated graphs.

25. If an amount of P dollars is invested at an annual interest rate r (expressed as a decimal), the value V of the investment after t years is given by $V = P(1 + r/n)^{nt}$, if interest is compounded n times a year. If \$1000 is invested at an annual interest rate of 12%, compounded semiannually, express V as a function of t and plot the graph for $0 \leq t \leq 8$ years.

26. If \$5000 is invested at an annual interest rate of 9%, compounded monthly (see Exercise 25), express V as a function of t and plot the graph for $0 \leq t \leq 10$ years.

27. Considering air resistance and other conditions, the velocity v, in meters per second, of a certain falling object is given by $v = 95(1 - e^{-0.1t})$, where t is the time of fall in seconds. Sketch the graph of this function.

28. The current i, in amperes, in a certain electric circuit is given by $i = 16(1 - e^{-250t})$, where t is the time in seconds. Using appropriate values of t, sketch the graph of this function.

29. The time t, in picoseconds, required for N calculations by a certain computer design is $t = N + \log_2 N$. Sketch the graph of this function.

30. An original amount of 100 mg of radium radioactively decomposes such that N mg remain after t years. The function relating t and N is $t = 2350 (\log_e 100 - \log_e N)$. Sketch the graph of this function.

31. Plot the graph of $y = 2^{-x}$ and compare it with the graph of Example C.

32. In Exercise 20, the graph of $y = \log_{0.5} x$ is plotted. By inspecting the graph and noting the properties of $\log_{0.5} x$, determine some of the differences of logarithms to a base less than 1 from those to a base greater than 1.

12-3 Properties of Logarithms

Since a logarithm is an exponent, it must follow the laws of exponents. Those laws which are of the greatest importance at this time are listed here for reference.

$$b^u b^v = b^{u+v} \tag{12-4}$$

$$\frac{b^u}{b^v} = b^{u-v} \tag{12-5}$$

$$(b^u)^n = b^{nu} \tag{12-6}$$

We will use these laws of exponents to derive certain useful properties of logarithms. The following example illustrates the reasoning used in deriving these properties.

EXAMPLE A

We know that $8 \times 16 = 128$. Writing these numbers as powers of 2, we have

$$8 = 2^3, \qquad 16 = 2^4, \qquad 128 = 2^7 = 2^{3+4}$$

The logarithmic forms can be written as

$$3 = \log_2 8, \qquad 4 = \log_2 16, \qquad 3 + 4 = \log_2 128$$

This means that

$$\log_2 8 + \log_2 16 = \log_2 128$$

where

$$8 \times 16 = 128$$

Following Example A, if we let $u = \log_b x$ and $v = \log_b y$ and write these equations in exponential form, we have $x = b^u$ and $y = b^v$. Therefore, forming the product of x and y, we obtain

$$xy = b^u b^v = b^{u+v} \quad \text{or} \quad xy = b^{u+v}$$

Writing this last equation in logarithmic form yields

$$u + v = \log_b xy$$

or

logarithm of a product

$$\boxed{\log_b x + \log_b y = \log_b xy} \tag{12-7}$$

Equation (12-7) states the property that *the logarithm of the product of two numbers is equal to the sum of the logarithms of the numbers.*

Using the same definitions of u and v to form the quotient of x and y, we then have

$$\frac{x}{y} = \frac{b^u}{b^v} = b^{u-v} \quad \text{or} \quad \frac{x}{y} = b^{u-v}$$

Writing this last equation in logarithmic form, we have

$$u - v = \log_b \left(\frac{x}{y}\right)$$

or

logarithm of a quotient

$$\boxed{\log_b x - \log_b y = \log_b \left(\frac{x}{y}\right)} \tag{12-8}$$

Equation (12-8) states the property that *the logarithm of the quotient of two numbers is equal to the logarithm of the numerator minus the logarithm of the denominator.*

If we again let $u = \log_b x$ and write this in exponential form, we have $x = b^u$. To find the nth power of x, we write

$$x^n = (b^u)^n = b^{nu}$$

Expressing this equation in logarithmic form yields

$$nu = \log_b (x^n)$$

or

logarithm of a power

$$\boxed{n \log_b x = \log_b (x^n)} \tag{12-9}$$

This last equation states that *the logarithm of the nth power of a number is equal to n times the logarithm of the number.* The exponent n may be integral or fractional and, therefore, we may use Eq. (12-9) for finding powers and roots of numbers.

We now recall from Section 12-1 that we showed that the base b of logarithms must be a positive number. Since $x = b^u$ and $y = b^v$, this means that x and y are also positive numbers. Therefore, *the properties of logarithms which have just been derived are valid only for positive values of x and y.*

EXAMPLE B

Using Eq. (12-7) we may express $\log_4 15$ as a sum of logarithms as follows:

$$\log_4 15 = \log_4 (3 \times 5) = \log_4 3 + \log_4 5 \qquad \text{logarithm of product} \atop \text{sum of logarithms}$$

Using Eq. (12-8) we may express $\log_4 \left(\frac{5}{3}\right)$ as the difference of logarithms as follows:

$$\log_4 \left(\frac{5}{3}\right) = \log_4 5 - \log_4 3 \qquad \text{logarithm of quotient} \atop \text{difference of logarithms}$$

Using Eq. (12-9) we may express $\log_4 9$ as twice $\log_4 3$ as follows:

$$\log_4 9 = \log_4 (3^2) = 2 \log_4 3 \qquad \text{logarithm of power} \atop \text{multiple of logarithm}$$

EXAMPLE C

Using Eqs. (12-7) through (12-9), we may express a sum or difference of logarithms as the logarithm of a single quantity.

1. $\log_4 3 + \log_4 x = \log_4 (3 \times x) = \log_4 3x$ using Eq. (12-7)

2. $\log_4 3 - \log_4 x = \log_4 \left(\frac{3}{x}\right)$ using Eq. (12-8)

3. $\log_4 3 + 2 \log_4 x = \log_4 3 + \log_4 (x^2) = \log_4 3x^2$ using Eqs. (12-7) and (12-9)

4. $\log_4 3 + 2 \log_4 x - \log_4 y = \log_4 \left(\frac{3x^2}{y}\right)$ using Eqs. (12-7), (12-8), and (12-9)

In Section 12-2 we noted another important property of logarithms, $\log_b 1 = 0$. Also, since $b = b^1$ in logarithmic form is $\log_b b = 1$, we have

$$\log_b (b^n) = n \log_b b = n(1) = n$$

Summarizing these properties, we have

$$\boxed{\log_b 1 = 0, \qquad \log_b b = 1} \qquad (12\text{-}10)$$

and

$$\boxed{\log_b (b^n) = n} \qquad (12\text{-}11)$$

These equations may be used to evaluate certain logarithms when the exact values may be determined.

EXAMPLE D

We may evaluate $\log_3 9$ in the following manner: Using Eq. (12-11), we have

$$\log_3 9 = \log_3 (3^2) = 2$$

We can establish the exact value since the base of logarithms and the number being raised to the power are the same. Of course, this could have been evaluated directly from the definition of a logarithm.

However, we cannot establish an exact value of $\log_4 9$ in this way. We have

$$\log_4 9 = \log_4 (3^2) = 2 \log_4 3$$

We must leave it in this form until we are able to evaluate $\log_4 3$. This can be done, and we shall consider this type of expression later in the chapter. ∎

EXAMPLE E

Calculate $\log_3 (3^{0.4})$.

Using Eq. (12-11) we may calculate this logarithm. This gives us

$$\log_3 (3^{0.4}) = 0.4$$

Although we did not calculate $3^{0.4}$ at this point, we were able to calculate $\log_3 3^{0.4}$. ∎

EXAMPLE F

We may also use Eqs. (12-10) and (12-11) in combination with other properties of logarithms. The following illustration shows how they can be used alone or in combination with Eq. (12-8) to evaluate the indicated logarithm.

$$\log_5 \left(\tfrac{1}{25}\right) = \log_5 1 - \log_5 25 = 0 - \log_5 (5^2) = -2$$
$$\log_5 \left(\tfrac{1}{25}\right) = \log_5 (5^{-2}) = -2$$

Either method is appropriate. ∎

In the following examples, the properties of logarithms of Eqs. (12-7) through (12-11) are further illustrated.

EXAMPLE G

1. $\log_2 6 = \log_2 (2 \times 3) = \log_2 2 + \log_2 3 = 1 + \log_2 3$
2. $\log_3 \left(\tfrac{2}{9}\right) = \log_3 2 - \log_3 9 = \log_3 2 - \log_3 (3^2) = \log_3 2 - 2$
   $\qquad = -2 + \log_3 2$ ∎

EXAMPLE H

$$\log_{10} \sqrt{7} = \log_{10} (7^{1/2}) = \tfrac{1}{2} \log_{10} 7$$

This demonstrates the property which is especially useful for finding roots of numbers. We see that we need merely to multiply the logarithm of the number by the fractional exponent representing the root to obtain the logarithm of the root.

Similarly,

$$\log_{10} \sqrt[3]{10x} = \log_{10} (10x)^{1/3} = \tfrac{1}{3} \log_{10}(10x)$$
$$= \tfrac{1}{3}(\log_{10} 10 + \log_{10} x)$$
$$= \tfrac{1}{3}(1 + \log_{10} x)$$

EXAMPLE I _____ Use the basic properties of logarithms to solve for y in terms of x: $\log_b y = 2 \log_b x + \log_b a$.

Using Eq. (12-9) and then Eq. (12-7), we have

$$\log_b y = \log_b (x^2) + \log_b a = \log_b (ax^2)$$

Now, since we have the logarithm to the base b of different expressions on each ■ side of the resulting equation, the expressions must be equal. Therefore, $y = ax^2$.

EXAMPLE J _____ An equation relating the current i and the time t in an electric circuit containing a resistance R and a capacitance C is $\log_e i - \log_e I = -t/RC$. Here I is the current for $t = 0$. Solve for i as a function of t.

Using Eq. (12-8), we rewrite the left side of this equation, obtaining

$$\log_e \left(\frac{i}{I} \right) = -\frac{t}{RC}$$

Rewriting this in exponential form, we have

$$\frac{i}{I} = e^{-t/RC}$$

or

$$i = Ie^{-t/RC}$$

Exercises 12-3

In Exercises 1 through 12, express each as a sum, difference, or multiple of logarithms. See Example B.

1. $\log_5 xy$

2. $\log_3 7y$

3. $\log_7 \left(\dfrac{5}{a} \right)$

4. $\log_3 \left(\dfrac{r}{s} \right)$

5. $\log_2 (a^3)$

6. $\log_8 (n^5)$

7. $\log_6 abc$

8. $\log_2 \left(\dfrac{xy}{z^2} \right)$

9. $\log_5 \sqrt[4]{y}$

10. $\log_4 \sqrt[7]{x}$

11. $\log_2 \left(\dfrac{\sqrt{x}}{a^2} \right)$

12. $\log_3 \left(\dfrac{\sqrt[3]{y}}{8} \right)$

See Appendix E for a computer program for the properties of logarithms.

In Exercises 13 through 20, express each as the logarithm of a single quantity. See Example C.

13. $\log_b a + \log_b c$

14. $\log_2 3 + \log_2 x$

15. $\log_5 9 - \log_5 3$

16. $\log_8 6 - \log_8 a$

17. $\log_b x^2 - \log_b \sqrt{x}$

18. $\log_4 3^3 + \log_4 9$

19. $2 \log_e 2 + 3 \log_e n$

20. $\dfrac{1}{2} \log_b a - 2 \log_b 5$

In Exercises 21 through 28, determine the exact value of each of the given logarithms.

21. $\log_2 \left(\dfrac{1}{32} \right)$

22. $\log_3 \left(\dfrac{1}{81} \right)$

23. $\log_2 (2^{2.5})$

24. $\log_5 (5^{0.1})$

25. $\log_7 \sqrt{7}$

26. $\log_6 \sqrt[3]{6}$

27. $\log_3 \sqrt[4]{27}$

28. $\log_5 \sqrt[3]{25}$

In Exercises 29 through 40, express each as a sum, difference, or multiple of logarithms. In each case part of the logarithm may be determined exactly. See Examples G and H.

29. $\log_3 18$ **30.** $\log_5 75$ **31.** $\log_2\left(\dfrac{1}{6}\right)$ **32.** $\log_{10}(0.05)$

33. $\log_3 \sqrt{6}$ **34.** $\log_2 \sqrt[3]{24}$ **35.** $\log_2 (4^2)(3^3)$ **36.** $\log_7 (7^4)(3^5)$

37. $\log_{10} 3000$ **38.** $\log_{10} (40^2)$ **39.** $\log_{10}\left(\dfrac{27}{100}\right)$ **40.** $\log_5\left(\dfrac{4}{125}\right)$

In Exercises 41 through 52, solve for y in terms of x.

41. $\log_b y = \log_b 2 + \log_b x$ **42.** $\log_b y = \log_b 6 + \log_b x$

43. $\log_4 y = \log_4 x - \log_4 5 + \log_4 3$ **44.** $\log_3 y = \log_3 7 - 2 \log_3 x$

45. $\log_{10} y = 2 \log_{10} 7 - 3 \log_{10} x$ **46.** $\log_b y = 3 \log_b \sqrt{x} + 2 \log_b 10$

47. $5 \log_2 y - \log_2 x = 3 \log_2 4 + \log_2 a$ **48.** $4 \log_2 x - 3 \log_2 y = \log_2 27$

49. $\log_2 x + \log_2 y = 1$ **50.** $3 \log_4 x + \log_4 y = 1$

51. $2 \log_5 x - \log_5 y = 2$ **52.** $\log_8 x - 2 \log_8 y = 4$

In Exercises 53 through 56, evaluate each of the given expressions using $\log_{10} 2 = 0.301$. (Hint: Each number can be expressed in terms of 2 or 10 or both.)

53. $\log_{10} 4$ **54.** $\log_{10} 20$ **55.** $\log_{10} (0.5)$ **56.** $\log_{10} 8000$

In Exercises 57 through 60, plot the indicated graphs and perform the indicated operations.

57. Plot the graphs of $y = 2 \log_2 x$ and $y = \log_2 x^2$, and show that they are the same.

58. Plot the graphs of $y = \log_2 4x$ and $y = 2 + \log_2 x$, and show that they are the same.

59. Under certain conditions, the temperature T, in degrees Celsius, of a cooling object is related to the time t, in minutes, by the equation $\log_e T = \log_e 65.0 - 0.41t$. Solve for T as a function of t.

60. In analyzing the power gain in an electric circuit, the equation

$$N = 10(2 \log_{10} I_1 - 2 \log_{10} I_2 + \log_{10} R_1 - \log_{10} R_2)$$

is used. Express this with a single logarithm on the right side.

12-4   Logarithms to the Base 10

In Section 12-1 we stated that a base of logarithms must be a positive number, not equal to one. In the examples and exercises of the previous sections we used a number of different bases. There are, however, only two bases which are generally used. They are 10 and e, where e is the irrational number approximately equal to 2.718 that we introduced in Section 11-5 and used again in Section 12-2.

Base 10 logarithms were developed for calculational purposes and were used a great deal for making calculations until the 1970s, when the modern electronic calculator became widely available. However, there are a number of scientific measurements in which base 10 logarithms are still used, and a need therefore still exists for them. Base e logarithms are used extensively in technical and scientific work, and they are considered in Section 12-5. In this section we discuss how the base 10 logarithm of a number is determined.

Logarithms to the base 10 are called **common logarithms.** They may be found directly by use of a calculator. The [LOG] key is used for this purpose. This calculator key indicates the common notation: *When no base is shown, it is assumed to be base 10.*

EXAMPLE A

Find log 426 by use of a calculator.

By using the sequence 426 [LOG] [2.6294096], we find that

$$\log 426 = 2.629$$

└── no base shown means base is 10

when the result is rounded off. The decimal part of a logarithm is normally expressed to the same accuracy as that of the number of which it is the logarithm, although showing one additional digit in the logarithm is generally acceptable.

EXAMPLE B

Find log 0.03654 by use of a calculator.

In this case the calculator display is -1.4372315, which means that

$$\log 0.03654 = -1.4372$$

We note that the logarithm here is negative. This should be the case when we recall the meaning of a logarithm. Raising 10 to a negative power gives us a number between 0 and 1, and here we have

$$10^{-1.4372} = 0.03654$$

EXAMPLE C

To find log sin 76.4°, we use the sequence 76.4 [SIN] [LOG] [-0.0123512]. Thus log sin 76.4° = -0.0124.

To find log $\sqrt{80.9}$, we may use the sequence

80.9 [$\sqrt{x}$] [LOG] [0.9539743].

Also, since

$$\log \sqrt{80.9} = \log (80.9)^{1/2} = \tfrac{1}{2} \log 80.9$$

we may use the sequence 80.9 [LOG] [÷] 2 [=] [0.9539743]. Thus,

$$\log \sqrt{80.9} = 0.954.$$

We may also use the calculator to find a number N if we know log N. *In this case we refer to N as the **antilogarithm** of log N.* On the calculator we use either [INV] [LOG] key sequence or the [10^x] key, depending on the calculator. Note that the [INV] [LOG] sequence uses the fact that the exponential and logarithmic functions are inverse functions, as we stated in Section 12-2. The [10^x] key uses the basic definition of a logarithm.

antilogarithms

EXAMPLE D

Given log $N = 2.1854$, find N.

Using either the sequence 2.1854 [INV] [LOG] [153.24983] or the sequence 2.1854 [10^x] [153.24983], we find that

$$N = 153.2$$

where the result has been rounded off.

We can also use a table of logarithms to find the common logarithm of a number or to find an antilogarithm. Table 2 in Appendix F is such a table. The use of the table is shown along with the table in Appendix F.

The following example illustrates an application in which a measurement requires the direct use of the value of a logarithm.

EXAMPLE E —— The power gain G, in decibels, of an electronic device is given by the equation $G = 10 \log (P_0/P_i)$, where P_0 is the output power, in watts, and P_i is the input power. Determine the power gain for an amplifier for which $P_0 = 15.8$ W and $P_i = 0.625$ W.

Substituting the given values, we have

$$G = 10 \log \frac{15.8}{0.625}$$

$$= 10 \log 25.28$$

$$= 14.0 \text{ dB}$$

where the result has been rounded off to three significant digits, the accuracy of the given data.

As we have mentioned previously, logarithms were developed for calculational purposes. They were first used in the seventeenth century for making tedious and complicated calculations which arose in astronomy and navigation. Logarithms were used to make these more complicated calculations by means of basic additions, subtractions, multiplications, and divisions. Performing calculations in this way provides an opportunity to understand better the meaning and properties of logarithms. Also, there are certain calculations which cannot be done directly on a calculator but can be done by logarithms. The following examples illustrate this use of logarithms.

EXAMPLE F —— By the use of logarithms, calculate the value of (42.8)(215).

Equation (12-7) tells us that $\log xy = \log x + \log y$. If we find log 42.8 and log 215 and add them, we shall have the logarithm of the product. Using this result, we look up the antilogarithm, which is the desired product.

$$\log 42.8 = 1.6314$$
$$\log 215 = 2.3324 \qquad \text{add}$$
$$\log [(42.8)(215)] = \overline{3.9638}$$
$$\log 9200 = 3.9638 \qquad \text{9200 is the antilogarithm}$$

Thus, (42.8)(215) = 9200.

EXAMPLE G —— A certain computer design has 64 different sequences of 10 binary digits so that the total number of possible states is $(2^{10})^{64} = 1024^{64}$. Evaluate 1024^{64} using logarithms.

Using Eq. (12-9) we know that $\log x^n = n \log x$. This means that $\log 1024^{64} = 64 \log 1024$. While most calculators will not directly evaluate 1024^{64}, we can use

(Continued on next page)

one to determine the value of 64 log 1024. We therefore can evaluate 1024^{64} as follows:

Let $N = 1024^{64}$

$\log N = \log 1024^{64} = 64 \log 1024$ **using Eq. (12-9)**

$= 64(3.0103000)$

characteristic (see Table 2)

$= 192.65920$

mantissa (see Table 2)

$N = 10^{192.65920} = 10^{0.65920} \times 10^{192}$

$= 4.5625 \times 10^{192}$ **antilogarithm of 0.65920 is 4.5625**

$1024^{64} = 4.5625 \times 10^{192}$

In the above steps we used a calculator to find the value 192.65920 and the antilogarithm of 0.65920. (We could also have used a table to find that the antilog of 0.6592 is 4.56.) However, the calculation was done essentially by logarithms. ∎

These examples show that calculations by logarithms are based on Eqs. (12-7), (12-8), and (12-9). Multiplication is performed by the addition of logarithms, division is performed by the subtraction of logarithms, and a power is found by a multiple of a logarithm. A root of a number is found by using the fractional exponent form of the power.

Exercises 12-4

In Exercises 1 through 16, find the common logarithm of each of the given numbers by use of a calculator.

1. 567
2. 60.5
3. 0.0640
4. 0.000566

5. 9.24×10^6
6. 3.19×10^{15}
7. 1.172×10^{-4}
8. 8.043×10^{-8}

9. 73.27
10. 0.008726
11. 0.18643
12. 164,850

13. $\cos 12.5°$
14. $\tan 50.8°$
15. $\sqrt{274}$
16. $\sqrt[3]{0.1275}$

In Exercises 17 through 28, find the antilogarithm from the given logarithms by use of a calculator.

17. 4.437
18. 0.929
19. -1.3045
20. -6.9788

21. 3.30112
22. 8.82436
23. $0.8594 - 1$
24. $0.4412 - 3$

25. 0.15485
26. 10.27562
27. -2.23746
28. -10.33577

In Exercises 29 through 36, use logarithms to perform the indicated calculations.

29. $(5.98)(14.3)$
30. $\dfrac{790}{8.02}$
31. $(6.75)^6$
32. $\sqrt[8]{308}$

33. $(47.3)(22.8)^{250}$
34. $\dfrac{895}{73.4^{86}}$
35. $(\sqrt[10]{7.32})(2470)^{30}$
36. $\dfrac{126,000^{20}}{2.63^{2.5}}$

In Exercises 37 through 40, find the logarithms of the given numbers.

37. A certain radar signal has a frequency of 1.15×10^9 Hz.

38. The bending moment of a particular concrete column is 4.60×10^6 lb·in.

39. A typical X-ray tube operates with a voltage of 150,000 V.

40. In an air sample taken in an urban area, $5/10^6$ of the air was carbon monoxide.

In Exercises 41 and 42, solve the given problems by finding the appropriate logarithms.

41. A stereo amplifier has an input power of 0.750 W and an output power of 25.0 W. What is the power gain? (See Example E.)

42. Measured on the Richter scale, the magnitude of an earthquake of intensity I is defined as $R = \log (I/I_0)$, where I_0 is a minimum level for comparison. What is the Richter scale reading for an earthquake for which $I = 75{,}000I_0$?

In Exercises 43 and 44, use logarithms to perform the indicated calculations.

43. Carbon 14 is radioactive such that one-half of it decays in 5600 years. This means that the amount N remaining after t years is given by $N = N_0(0.5)^{t/5600}$, where N_0 is the initial amount present. Find N for $t = 1400$ years if $N_0 = 1500$ g.

44. The peak current I_m, in amperes, in an alternating-current circuit is given by $I_m = \sqrt{\dfrac{2P}{Z \cos \theta}}$, where P is the power developed, Z is the magnitude of the impedance, and θ is the phase angle between the current and voltage. Evaluate I_m for $P = 5.25$ W, $Z = 320$ Ω, and $\theta = 35.4°$.

12-5   Natural Logarithms

As we mentioned earlier, another number which is important as a base of logarithms is the number e. *Logarithms to the base e are called* **natural logarithms.** Since e is an irrational number equal to approximately 2.718, it may appear to be a very unnatural choice as a base of logarithms. However, in the development of the calculus, the reason for its choice and the fact that it is a very natural number for a base of logarithms are shown.

Just as the notation $\log x$ refers to logarithms to the base 10, the notation $\ln x$ is used to denote logarithms to the base e. Due to the extensive use of natural logarithms, the notation $\boldsymbol{ln\ x}$ is more convenient than $\boldsymbol{log_e\ x,}$ although they mean the same thing.

Since more than one base is important, there are times when it is useful to be able to change a logarithm in one base to another base. If $u = \log_b x$, then $b^u = x$. Taking logarithms of both sides of this last expression to the base a, we have

$$\log_a b^u = \log_a x$$
$$u \log_a b = \log_a x$$
$$u = \frac{\log_a x}{\log_a b}$$

However, $u = \log_b x$, which means that

$$\log_b x = \frac{\log_a x}{\log_a b} \tag{12-12}$$

Equation (12-12) allows us to change a logarithm in one base to a logarithm in another base. The following examples illustrate the method of performing this operation.

EXAMPLE A _____ Change $\log 20 = 1.3010$ to a logarithm with base e; that is, find $\ln 20$.

In Eq. (12-12), if we let $b = e$ and $a = 10$, we have

$$\log_e x = \frac{\log_{10} x}{\log_{10} e}$$

or

$$\ln x = \frac{\log x}{\log e}$$

↑
└─ means base is e

In this example, $x = 20$. Therefore,

$$\ln 20 = \frac{\log 20}{\log e} = 2.996$$

The calculator sequence for this is

$$20 \boxed{\text{LOG}} \boxed{\div} 1 \boxed{e^x} \boxed{\text{LOG}} \boxed{=} \boxed{\qquad 2.9957323 \qquad}$$

└──────┘
↑
└──────── to obtain $\log e$

■

EXAMPLE B _____ Find $\log_5 560$.

In Eq. (12-12), if we let $b = 5$ and $a = 10$, we have

$$\log_5 x = \frac{\log x}{\log 5}$$

In this example, $x = 560$. Therefore, we have

$$\log_5 560 = \frac{\log 560}{\log 5} = 3.932$$

From the definition of a logarithm, this means that

$$5^{3.932} = 560$$

■

Since natural logarithms are used extensively, it is often convenient to have Eq. (12-12) written specifically for use with logarithms to the base 10 and natural logarithms. Using Eq. (12-12) in the forms $\ln x = \log x/\log e$ and $\log x = (\log e)(\ln x)$, where $\log e = 0.4343$, we have

$$\boxed{\ln x = 2.3026 \log x} \qquad\qquad (12\text{-}13)$$

and

$$\boxed{\log x = 0.4343 \ln x} \qquad\qquad (12\text{-}14)$$

EXAMPLE C ____ Using Eq. (12-13) to find ln 0.811, we have

$$\ln 0.811 = 2.3026 \log 0.811$$
$$= -0.2095$$

∎

Values of natural logarithms can be found directly on a scientific calculator. The $\boxed{\text{LN}}$ key is used for this purpose. In order to find the antilogarithm of a natural logarithm we use the $\boxed{e^x}$ key or the key sequence $\boxed{\text{INV}}\boxed{\text{LN}}$, depending on the calculator.

EXAMPLE D ____ Find ln 236.5 by use of a calculator.
The calculator key sequence of 236.5 $\boxed{\text{LN}}$ $\boxed{5.4659482}$ tells us that

∎ $\ln 236.5 = 5.4659$

EXAMPLE E ____ Find N if ln $N = -0.8729$ by use of a calculator.
The calculator key sequence .8729 $\boxed{+/-}\boxed{e^x}$ $\boxed{0.4177383}$ or the sequence .8729 $\boxed{+/-}\boxed{\text{INV}}\boxed{\text{LN}}$ $\boxed{0.4177383}$ tells us that the natural antilogarithm of -0.8729 is 0.4177. Thus,

∎ $N = 0.4177$

Natural logarithms can be found by use of tables such as Table 4 in Appendix F. The use of the table is shown along with the table in Appendix F.
Applications of natural logarithms are found in many fields of technology. One such application is shown in the following example, and others are found in the exercises.

EXAMPLE F ____ Under certain conditions, the electric current i in a circuit containing a resistance and an inductance (see Section 11–7) is given by

$$\ln \frac{i}{I} = -\frac{Rt}{L}$$

where I is the current at $t = 0$, R is the resistance, t is the time, and L is the inductance. Calculate how long (in seconds) it takes i to reach 0.430 A, if $I = 0.750$ A, $R = 7.50 \ \Omega$, and $L = 1.25$ H.
Solving for t, we have

$$t = -\frac{L \ln (i/I)}{R} = -\frac{L(\ln i - \ln I)}{R}$$ either form can be used

Thus, for the given values, we have

$$t = -\frac{1.25(\ln 0.430 - \ln 0.750)}{7.50}$$

$$= 0.0927 \text{ s}$$

∎ Therefore, the current changes from 0.750 A to 0.430 A in 0.0927 s.

Exercises 12-5

In Exercises 1 through 8, use logarithms to the base 10 to find the natural logarithms of the given numbers.

1. 26.0   **2.** 631   **3.** 1.562   **4.** 45.73

5. 0.5017   **6.** 0.05294   **7.** 0.0073267   **8.** 0.00044348

In Exercises 9 through 16, use logarithms to the base 10 to find the indicated logarithms.

9. $\log_7 42$ **10.** $\log_2 86$ **11.** $\log_5 245$ **12.** $\log_3 706$

13. $\log_{12} 122$ **14.** $\log_{20} 86$ **15.** $\log_{40} 750$ **16.** $\log_{100} 3720$

In Exercises 17 through 24, use Eq. (12-13) to find the natural logarithms of the indicated numbers.

17. 51.4   **18.** 293   **19.** 1.394   **20.** 65.62

21. 0.9917   **22.** 0.002086   **23.** 0.012937   **24.** 0.000060808

In Exercises 25 through 28, use a calculator to find the natural logarithms of the given numbers.

25. 45.17   **26.** 8765   **27.** 0.68528   **28.** 0.0014298

In Exercises 29 through 36, use a calculator to find the natural antilogarithms of the given logarithms.

29. 2.190   **30.** 3.420   **31.** 0.0084210   **32.** 0.632

33. -0.7429 **34.** -2.94218 **35.** -23.504 **36.** -0.00804

In Exercises 37 through 44, solve the given problems.

37. Solve for y in terms of x: $\ln y - \ln x = 1.0986$.

38. Solve for y in terms of x: $\ln y + 2 \ln x = 1 + \ln 5$.

39. If interest is compounded continuously (daily compounded interest closely approximates this), a bank account can double in t years according to the equation $i = \dfrac{\ln 2}{t}$, where i is the interest rate. What interest rate is required for an account to double in 8.5 years?

40. One approximate formula for world population growth is $T = 50.0 \ln 2$, where T is the number of years for the population to double. According to this formula, how long does it take for the population to double?

41. For the electric circuit of Example F, find how long it takes the current to reach 0.1 of the initial value of 0.750 A.

42. The intensity I of light decreases from its value I_0 at the edge of a medium as it passes through the medium. The distance x through which the light passes is a function of I given by $x = k(\ln I_0 - \ln I)$, where k is a constant depending on the medium. Calculate x for $I = 0.850 I_0$ and $k = 5.00$ cm.

43. The distance x traveled by a motorboat in t seconds after the engine is cut off is given by $x = \dfrac{1}{k} \ln (kv_0 t + 1)$, where v_0 is the velocity of the boat at the time the engine is cut and k is a constant. Find how long it takes a boat to go 150 m if $v_0 = 12.0$ m/s and $k = 6.80 \times 10^{-3}$/m.

44. The electric current i in a circuit containing a 1-H inductor, a 10-Ω resistor, and a 6-V battery is a function of time given by $i = 0.6(1 - e^{-10t})$. Solve for t as a function of i.

12-6   Exponential and Logarithmic Equations

In solving equations in which there is an unknown exponent, it is often advantageous to take logarithms of both sides of the equation and then proceed. In solving logarithmic equations, one should keep in mind the basic properties of logarithms, since these can often help transform the equation into a solvable form. There is, however, no general algebraic method for solving such equations, and here we shall solve only some special cases.

EXAMPLE A _____ Solve the equation $2^x = 8$.
By writing this in logarithmic form, we have

$$x = \log_2 8 = 3$$

Since 2^x and 8 are equal, the logarithms of 2^x and 8 are also equal. Thus, we can also solve this equation by taking logarithms (to any proper base) of both sides and equating these logarithms. This gives us

$$\log 2^x = \log 8 \qquad \text{or} \qquad \ln 2^x = \ln 8$$
$$x \log 2 = \log 8 \qquad\qquad x \ln 2 = \ln 8 \qquad \text{using Eq. (12-9)}$$
$$x = \frac{\log 8}{\log 2} = 3 \qquad\qquad x = \frac{\ln 8}{\ln 2} = 3$$

This last method is more generally applicable, since the first method is good only ■ if we can directly evaluate the logarithm which results.

EXAMPLE B _____ Solve the equation $3^{x-2} = 5$.
Taking logarithms of both sides and equating them, we have

$$\log 3^{x-2} = \log 5$$

or

$$(x - 2) \log 3 = \log 5 \qquad \text{using Eq. (12-9)}$$

Solving this last equation for x, we have

$$x = 2 + \frac{\log 5}{\log 3} = 3.465$$

3 $\boxed{x^y}$ 1.465 $\boxed{=}$ This solution means that
$\boxed{\quad 5.0001455 \quad}$

$$3^{3.465-2} = 3^{1.465} = 5$$

■ which can be checked by the calculator sequence shown at the left.

EXAMPLE C _____ Solve the equation $2(4^{x-1}) = 17^x$.
By taking logarithms of both sides, we have the following:

$$\log 2 + (x - 1) \log 4 = x \log 17 \qquad \text{using Eqs. (12-7) and (12-9)}$$
$$x \log 4 - x \log 17 = \log 4 - \log 2$$
$$x(\log 4 - \log 17) = \log 4 - \log 2$$
$$x = \frac{\log 4 - \log 2}{\log 4 - \log 17} = \frac{\log (4/2)}{\log 4 - \log 17} \qquad \text{using Eq. (12-8)}$$
$$= \frac{\log 2}{\log 4 - \log 17}$$
$$= -0.479$$

The equations in these examples could also have been solved by using natural logarithms. The following example illustrates an applied problem which is most easily solved by using natural logarithms.

EXAMPLE D — Under the condition of constant temperature, the atmospheric pressure p, in pascals, at an altitude h, in meters, is given by $p = p_0 e^{kh}$, where p_0 is the pressure where $h = 0$ (usually taken as sea level). Given that $p_0 = 101.3$ kPa (atmospheric pressure at sea level) and $p = 68.9$ kPa for $h = 3050$ m, find the value of k.

Since the equation is defined in terms of e, we can solve it most easily by taking natural logarithms of both sides. By doing this we have the following solution.

$$\ln p = \ln (p_0 e^{kh}) = \ln p_0 + \ln e^{kh} \qquad \text{using Eq. (12-7)}$$

$$= \ln p_0 + kh \ln e = \ln p_0 + kh \qquad \text{using Eq. (12-9), } \ln e = 1$$

$$k = \frac{\ln p - \ln p_0}{h}$$

Substituting the given values, we have

$$k = \frac{\ln(68.9 \times 10^3) - \ln (101.3 \times 10^3)}{3050} = -0.000126/\text{m}$$

Some of the important measurements in scientific and technical work are defined in terms of logarithms. We shall consider here some of the important applications which are basic logarithmic formulas. In using them we shall solve some equations in which logarithms are involved. The following example illustrates one such area of application, and others are found in the exercises.

EXAMPLE E —

See the chapter introduction.

It has been found that the human ear responds to sound on a scale which is approximately proportional to the logarithm of the intensity of the sound. Thus, the loudness of sound, measured in decibels, is defined by the equation $b = 10 \log (I/I_0)$, where I is the intensity of the sound and I_0 is the minimum intensity detectable.

A busy city street has a loudness of 70 dB, and riveting has a loudness of 100 dB. How many times greater is the intensity of the sound of riveting I_r than the sound of the city street I_c?

First, we substitute the decibel readings into the above definition. This gives us

$$70 = 10 \log \left(\frac{I_c}{I_0}\right) \quad \text{and} \quad 100 = 10 \log \left(\frac{I_r}{I_0}\right)$$

To solve these equations for I_c and I_r, we divide each side by 10 and then use the exponential form. Thus, we have

$$7.0 = \log \left(\frac{I_c}{I_0}\right) \quad \text{and} \quad 10 = \log \left(\frac{I_r}{I_0}\right)$$

$$\frac{I_c}{I_0} = 10^{7.0} \qquad\qquad \frac{I_r}{I_0} = 10^{10}$$

$$I_c = I_0(10^{7.0}) \qquad\qquad I_r = I_0(10^{10})$$

Since we want the number of times I_r is greater than I_c, we divide I_r by I_c. This gives us

$$\frac{I_r}{I_c} = \frac{I_0(10^{10})}{I_0(10^{7.0})} = \frac{10^{10}}{10^{7.0}} = 10^{3.0} \qquad \text{or} \qquad I_r = 10^{3.0}I_c = 1000I_c$$

Thus, the sound of riveting is 1000 times as intense as the sound of the city street. This demonstrates that sound intensity levels are considerably greater than loudness levels. (See Exercise 42.) ∎

The following examples illustrate the solution of other logarithmic equations.

EXAMPLE F ——— Solve the equation $\log_2 7 - \log_2 14 = x$.

Using the basic properties of logarithms, we arrive at the following result:

$$\log_2 \left(\tfrac{7}{14}\right) = x \qquad\qquad \text{using Eq. (12-8)}$$
$$\log_2 \left(\tfrac{1}{2}\right) = x \quad \text{or} \quad \tfrac{1}{2} = 2^x \qquad \text{exponential form}$$

∎ Thus, since $2^{-1} = \tfrac{1}{2}$, we have the result $x = -1$.

EXAMPLE G ——— Solve the equation $2 \ln 2 + \ln x = \ln 3$.

Using the properties of logarithms, we have the following solution.

$$2 \ln 2 + \ln x = \ln 3$$
$$\ln 2^2 + \ln x - \ln 3 = 0 \qquad\qquad \text{using Eq. (12-9)}$$
$$\ln \frac{4x}{3} = 0 \qquad\qquad \text{using Eqs. (12-7), (12-8)}$$
$$\frac{4x}{3} = e^0 = 1 \qquad \text{exponential form}$$
$$4x = 3$$
$$x = \frac{3}{4}$$

∎

EXAMPLE H ——— Solve the equation $2 \log x - 1 = \log (1 - 2x)$.

$$\log x^2 - \log (1 - 2x) = 1$$
$$\log \frac{x^2}{1 - 2x} = 1 \qquad\qquad \text{using Eq. (12-8)}$$
$$\frac{x^2}{1 - 2x} = 10^1 \qquad\qquad \text{exponential form}$$
$$x^2 = 10 - 20x$$
$$x^2 + 20x - 10 = 0$$
$$x = \frac{-20 \pm \sqrt{400 + 40}}{2} = -10 \pm \sqrt{110}$$

Since logarithms of negative numbers are not defined, and $-10 - \sqrt{110}$ is negative and cannot be used in the first term of the original equation, we have

$$x = -10 + \sqrt{110} = 0.488$$

Exercises 12-6

In Exercises 1 through 36, solve the given equations.

1. $2^x = 16$

2. $3^x = \frac{1}{81}$

3. $5^x = 0.3$

4. $6^x = 15$

5. $3^{-x} = 0.525$

6. $15^{-x} = 1.326$

7. $e^{2x} = 3.625$

8. $e^{-x} = 17.54$

9. $6^{x+1} = 10$

10. $5^{x-1} = 2$

11. $4(3^x) = 5$

12. $3(14^x) = 40$

13. $0.8^x = 0.4$

14. $0.6^x = 100$

15. $(15.6)^{x+2} = 23^x$

16. $5^{x+2} = e^{2x}$

17. $3 \log_8 x = -2$

18. $5 \log_{32} x = -3$

19. $2 \ln x = 1$

20. $3 \ln 2x = 2$

21. $\log_2 x + \log_2 7 = \log_2 21$

22. $2 \log_2 3 - \log_2 x = \log_2 45$

23. $2 \log (3 - x) = 1$

24. $3 \log (2x - 1) = 1$

25. $\log 4x + \log x = 2$

26. $\log 12x^2 - \log 3x = 3$

27. $\ln x + \ln 3 = 1$

28. $2 \ln 2 - \ln x = -1$

29. $3 \ln 2 + \ln (x - 1) = \ln 24$

30. $\ln (2x - 1) - 2 \ln 4 = 3 \ln 2$

31. $\frac{1}{2} \log (x + 2) + \log 5 = 1$

32. $\frac{1}{2} \log (x - 1) - \log x = 0$

33. $\log_5 (x - 3) + \log_5 x = \log_5 4$

34. $\log_7 x + \log_7 (2x - 5) = \log_7 3$

35. $\log (2x - 1) + \log (x + 4) = 1$

36. $\log_2 x + \log_2 (x + 2) = 3$

In Exercises 37 through 48, determine the indicated quantities.

37. In computer design, the number N of bits of memory is often expressed as a power of 2, or $N = 2^x$. Find x if $N = 2.68 \times 10^8$ bits.

38. If half of a radioactive substance decays in a year, the amount N remaining after t years is given by $N = N_0(0.5)^t$, where N_0 is the original amount. Find t if $N = 0.10 N_0$.

39. The temperature T, in degrees Celsius, of a certain cooling object is given by $T = T_0 + 70(0.40)^{0.20t}$, where T_0 is the temperature of the surroundings of the object and t is the time in minutes. Determine how long it takes the object to cool to 50°C if $T_0 = 30°C$.

40. The electric current i, in amperes, in a circuit containing a resistance R, an inductor L, and a voltage source E is given by $i = \dfrac{E}{R}(1 - e^{-Rt/L})$, where t is the time in seconds. Find t if $i = 0.750$ A, $E = 6.00$ V, $R = 4.50 \, \Omega$, and $L = 2.50$ H.

41. In chemistry, the pH value of a solution is a measure of its acidity. The pH value is defined by the relation $pH = -\log (H^+)$, where H^+ is the hydrogen ion concentration. If the pH of a certain wine is 3.4065, find the hydrogen ion concentration. (If the pH value is less than 7, the solution is acid. If the pH value is above 7, the solution is basic.)

42. Referring to Example E, show that if the difference in loudness of two sounds is d decibels, the louder sound is $10^{d/10}$ more intense than the quieter sound.

43. Measured on the Richter scale, the magnitude of an earthquake of intensity I is defined as $R = \log (I/I_0)$, where I_0 is a minimum level for comparison. How many times I_0 was the 1906 San Francisco earthquake whose magnitude was 8.25 on the Richter scale?

44. How many times more intense was the 1964 Alaska earthquake, $R = 7.5$, than the 1971 Los Angeles earthquake, $R = 6.7$? (See Exercise 43.)

45. Pure water is running into a certain brine solution, and the same amount of solution is running out. The number n of kilograms of salt in the solution after t minutes is found by solving the equation $\ln n = -0.04t + \ln 20$. Solve for n as a function of t.

46. In an electric circuit containing a resistor and a capacitor with an initial charge q_0, the charge q on the capacitor at any time t after closing the switch can be found by solving the equation $\ln q = -\dfrac{t}{RC} + \ln q_0$. Here R is the resistance and C is the capacitance. Solve for q as a function of t.

47. An earth satellite loses 0.1% of its remaining power each week. An equation relating the power P, the initial power P_0, and the time t in weeks is $\ln P = t \ln 0.999 + \ln P_0$. Solve for P as a function of t.

48. An equation relating the distance s through which a falling object moves and its velocity v is $\log s + \log 2g = 2 \log v$, where g is the acceleration due to gravity. Solve for s as a function of v.

To solve more complicated problems, we may use graphical methods. For example, if we wish to solve the equation $2^x + 3^x = 50$, we can set up the function $y = 2^x + 3^x - 50$ and then determine its zeros graphically. Note that the given equation can be written as $2^x + 3^x - 50 = 0$, and therefore the zeros of the function which has been set up will give the desired solution. In Exercises 49 through 52, solve the given equations in this way.

49. $2^x + 3^x = 50$ **50.** $4^x + x^2 = 25$

51. The curve in which a uniform wire hangs under its own weight is called a *catenary*. For a particular wire, the equation of the catenary is $y = 2(e^{x/4} + e^{-x/4})$, where (x, y) is a point on the curve. Find x for $y = 5.8$ m.

52. In finding the current i in a certain electric circuit, the equation $i = 2 \ln (t + 2) - t$ relates the current and the time t in seconds. Find t for $i = 0.05$ A.

12-7 Graphs on Logarithmic and Semilogarithmic Paper ▬▬▬

When constructing the graphs of some functions, one of the variables changes much more rapidly than the other. This was the case in graphing the exponential and logarithmic functions in Section 12-2. The following example illustrates this point.

EXAMPLE A _____ Graph $y = 4(3^x)$.

Constructing the following table of values

Fig. 12-8

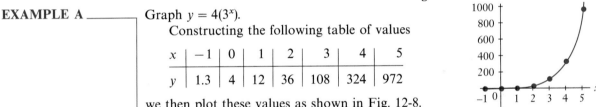

x	-1	0	1	2	3	4	5
y	1.3	4	12	36	108	324	972

we then plot these values as shown in Fig. 12-8.

We see that as x changes from -1 to 5, y changes much more rapidly, taking on values from near 1 to nearly 1000. Also, because of the scale which must be used, we see that it is not possible to show accurately the differences in the y-values of 1.3 to 36, or to read accurately any of the y-values from the graph.

It is possible to graph a function with a large change in values, for one or both variables, more accurately than can be done on the standard rectangular coordinate system. This is done by using a scale which is marked off in distances proportional to the logarithms of the values being represented. *Such a scale is called a* **logarithmic scale.** For example, $\log 1 = 0$, $\log 2 = 0.301$, and $\log 10 = 1$. Thus, on a logarithmic scale the 2 is placed 0.301 units of distance from the 1 to the 10. Figure 12-9 shows a logarithmic scale with the numbers represented and the distance used for each.

On a logarithmic scale the distances between numbers are not even, but this scale does allow for a much greater range of values and much greater accuracy for many of the numbers. There is another advantage to using logarithmic scales. Many equations which would have more complex curves on the standard coordinate system will work out as straight lines using logarithmic scales. In many cases this makes the analysis of the curve easier.

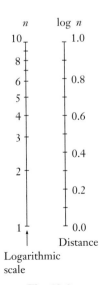

Logarithmic scale

Fig. 12-9

If we wish to use a large range of values for only one of the variables, we use what is known as **semilogarithmic,** or **semilog,** graph paper. On this graph paper, only one axis (usually the y-axis) uses a logarithmic scale. If we wish to use a large range of values for both variables, we use **logarithmic,** or **log-log,** graph paper. Both axes are marked with logarithmic scales.

The following examples illustrate the use of semilog and log-log graph paper.

EXAMPLE B _____

Construct the graph of $y = 4(3^x)$ on semilogarithmic graph paper.

Since this is the same function as in Example A, we repeat the table of values from Example A.

x	-1	0	1	2	3	4	5
y	1.3	4	12	36	108	324	972

Again, we see that the range of y-values is large. When we plotted this curve on the regular coordinate system in Example A, we had to use large units along the y-axis. This made the values of 1.3, 4, 12, and 36 appear at practically the same level. However, when we use semilog graph paper, we can label each axis such that all y-values are accurately plotted as well as the x-values.

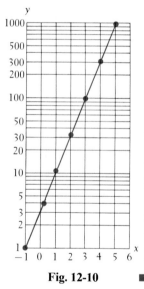

Fig. 12-10

The logarithmic scale is shown in **cycles,** and we must label the base line of the first cycle as 1 times a power of ten (0.01, 0.1, 1, 10, 100, and so on) with the following cycle labeled with the next power of ten. The lines between are labeled with 2, 3, 4, and so on, times the proper power of ten. See the vertical scale in Fig. 12-10. We now plot the points in the table on the graph. The resulting graph is a straight line, as we see in Fig. 12-10. Taking logarithms of both sides of the equation, we have

$$\log y = \log\left[4(3^x)\right] = \log 4 + \log 3^x \qquad \text{using Eq. (12-7)}$$
$$= \log 4 + x \log 3 \qquad \text{using Eq. (12-9)}$$

However, since log y was plotted automatically (because we used semilogarithmic paper), the graph really represents

$$u = \log 4 + x \log 3$$

where $u = \log y$; log 3 and log 4 are constants, and therefore this equation is of the form $u = mx + b$, which is a straight line (see Section 4-2).

The logarithmic scale in Fig. 12-10 has _three cycles,_ since all values of three powers of ten are represented.

We note that zero and negative numbers do not appear on the logarithmic scale. In fact, all numbers used on the logarithmic scale must be positive, since the domain of the logarithmic function includes only positive real numbers. Thus, the bottom scale must start at some number greater than zero. This number is a power of ten and can be very small, say $10^{-6} = 0.000001$, but is positive.

EXAMPLE C _____

Construct the graph of $x^4 y^2 = 1$ on logarithmic paper.

First we solve for y and then construct a table of values. Considering positive values of x and y, we have

$$y = \sqrt{\frac{1}{x^4}} = \frac{1}{x^2}$$

x	0.5	1	2	8	20
y	4	1	0.25	0.0156	0.0025

We now plot these values on log-log paper on which both scales are logarithmic, as shown in Fig. 12-11. We again note that we have a straight line. Taking logarithms of both sides of the equation, we have

$$\log (x^4 y^2) = \log 1$$

$$\log x^4 + \log y^2 = 0 \qquad \text{using Eq. (12-7)}$$

$$4 \log x + 2 \log y = 0 \qquad \text{using Eq. (12-9)}$$

If we let $u = \log y$ and $v = \log x$, we then have

$$4v + 2u = 0 \qquad \text{or} \qquad u = -2v$$

which is the equation of a straight line, as shown in Fig. 12-11. It should be pointed out that not all graphs on logarithmic paper are straight lines.

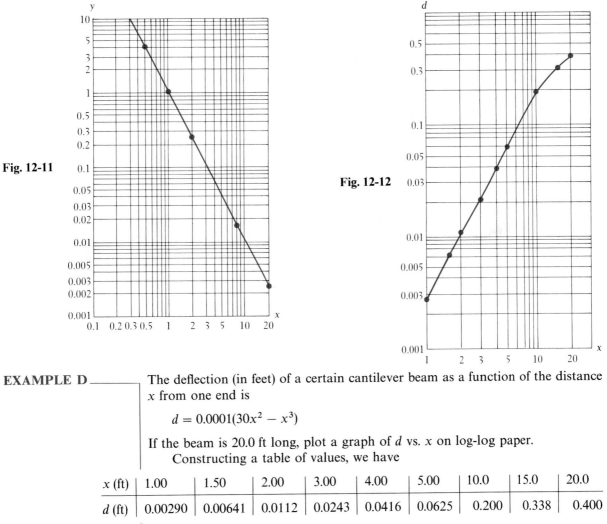

Fig. 12-11

Fig. 12-12

EXAMPLE D — The deflection (in feet) of a certain cantilever beam as a function of the distance x from one end is

$$d = 0.0001(30x^2 - x^3)$$

If the beam is 20.0 ft long, plot a graph of d vs. x on log-log paper. Constructing a table of values, we have

x (ft)	1.00	1.50	2.00	3.00	4.00	5.00	10.0	15.0	20.0
d (ft)	0.00290	0.00641	0.0112	0.0243	0.0416	0.0625	0.200	0.338	0.400

■ The graph is shown in Fig. 12-12.

Logarithmic and semilogarithmic paper are often useful for plotting data derived from experimentation. Often the data cover too large a range of values to be plotted on ordinary graph paper. The following example illustrates how we use semilogarithmic paper to plot data.

EXAMPLE E _____ The vapor pressure of water depends on the temperature. The following table gives the vapor pressure (in kilopascals) for corresponding values of temperature (in degrees Celsius).

Temp.	10	20	40	60	80	100	120	140	160
Pressure	1.19	2.33	7.34	19.9	47.3	101	199	361	617

These data are then plotted on semilogarithmic paper, as shown in Fig. 12-13. Intermediate values of temperature and pressure can then be read directly from ■ the graph.

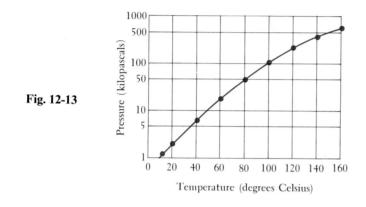

Fig. 12-13

Exercises 12-7

In Exercises 1 through 12, plot the graphs of the given functions on semilogarithmic paper.

1. $y = 2^x$ **2.** $y = 5^x$ **3.** $y = 2(4^x)$ **4.** $y = 5(10^x)$

5. $y = 3^{-x}$ **6.** $y = 2^{-x}$ **7.** $y = x^3$ **8.** $y = x^5$

9. $y = 3x^2$ **10.** $y = 2x^4$ **11.** $y = 2x^3 + 4x$ **12.** $y = 4x^3 + 2x^2$

In Exercises 13 through 24, plot the graphs of the given functions on log-log paper.

13. $y = 0.01x^4$ **14.** $y = 0.02x^3$ **15.** $y = \sqrt{x}$ **16.** $y = x^{2/3}$

17. $y = x^2 + 2x$ **18.** $y = x + \sqrt{x}$ **19.** $xy = 4$ **20.** $xy^2 = 10$

21. $y^2x = 1$ **22.** $x^2y^3 = 1$ **23.** $x^2y^2 = 25$ **24.** $x^3y = 8$

In Exercises 25 through 34, plot the indicated graphs.

25. The distance s through which a lead weight falls due to gravity is given by $s = \frac{1}{2}gt^2$, where g is the acceleration due to gravity and t is the time, in seconds, of fall. Given that $g = 9.80$ m/s^2, plot the graph of s as a function of t for $0 \le t \le 10$ s on (a) a regular rectangular coordinate system and (b) a semilogarithmic coordinate system.

26. By pumping, the air pressure within a tank is reduced by 18% each second. Thus, the pressure p, in kilopascals, in the tank is given by $p = 101(0.82)^t$, where t is the time in seconds. Plot the graph of p as a function of t for $0 \le t \le 30$ s on (a) a regular rectangular coordinate system and (b) a semilogarithmic coordinate system.

27. Strontium 90 decays according to the equation $N = N_0 e^{-0.028t}$, where N is the amount present after t years and N_0 is the original amount. Plot N as a function of t on semilog paper.

28. The electric power delivered by a certain battery as a function of the resistance in the circuit is given by
$$P = \frac{100R}{(0.50 + R)^2},$$ where P is measured in watts and R is in ohms. Plot P as a function of R on semilog paper, using the logarithmic scale for R and values of R from 0.01 Ω to 10 Ω. Compare the graph with that in Fig. 2-16 of Section 2-4.

29. The acceleration g produced by the gravitational force of the earth on a spacecraft is given by $g = 3.99 \times 10^{14}/r^2$, where r is the distance from the center of the earth to the spacecraft. On log-log paper, graph g (in m/s^2) as a function of r from $r = 6.37 \times 10^6$ m (the earth's surface) to $r = 3.91 \times 10^8$ m (the distance to the moon).

30. In undergoing an adiabatic (no *heat* gained or lost) expansion of a gas, the relation between the pressure p and the volume v is $p^2v^3 = 850$. On log-log paper, graph p (in kPa) as a function of v from $v = 0.10$ m^3 to $v = 10$ m^3.

31. The intensity level B, in decibels, and the frequency f, in hertz, for a sound of constant loudness were measured as follows:

f (Hz)	100	200	500	1000	2000	5000	10,000
B (dB)	40	30	22	20	18	24	30

Plot the data for B as a function of f on semilog paper, using the logarithmic scale for f.

32. The atmospheric pressure p at a given altitude h is given in the following table:

h (km)	0	10	20	30	40
p (kPa)	101	25	6.3	2.0	0.53

On semilog paper, plot p as a function of h.

33. One end of a very hot steel bar is sprayed with a stream of cool water. The rate of cooling R as a function of the distance d from the end of the bar is then measured. The following results are obtained:

d (in.)	0.063	0.13	0.19	0.25	0.38	0.50	0.75	1.0	1.5
R (°F/s)	600	190	100	72	46	29	17	10	6.0

On log-log paper, plot R as a function of d. Such experiments are made to determine the hardenability of steel.

34. The magnetic intensity H (in A/m) and flux density B (in teslas) of annealed iron are given in the following table.

B (T)	0.0042	0.043	0.67	1.01	1.18	1.44	1.58	1.72
H (A/m)	10	50	100	150	200	500	1000	10,000

Plot H as a function of B on logarithmic paper.

In Exercises 35 and 36, plot the indicated semilogarithmic graphs for the following application.

In a particular electric circuit, called a low-pass filter, the input voltage V_i is across a resistor and a capacitor, and the output voltage V_0 is across the capacitor (see Fig. 12-14). The voltage gain G, in decibels, in such a circuit is given by

$$G = 20 \log \frac{1}{\sqrt{1 + (\omega T)^2}} \quad \text{where} \quad \tan \phi = -\omega T$$

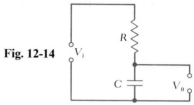

Fig. 12-14

Here ϕ is the phase angle of V_0/V_i. For values of ωT of 0.01, 0.1, 0.3, 1.0, 3.0, 10.0, 30.0, and 100, plot the indicated graphs. These graphs are called a Bode diagram for the circuit.

35. Calculate values of G for the given values of ωT and plot a semilogarithmic graph of G vs. ωT.

36. Calculate values of ϕ (as negative angles) for the given values of ωT, and plot a semilogarithmic graph of ϕ vs. ωT.

12-8   Chapter Equations, Review Exercises, and Practice Test

Chapter Equations

Exponential function	$y = b^x$	(12-1)
Logarithmic form	$x = \log_b y$	(12-2)
Logarithmic function	$y = \log_b x$	(12-3)
Laws of exponents	$b^u b^v = b^{u+v}$	(12-4)
	$\dfrac{b^u}{b^v} = b^{u-v}$	(12-5)
	$(b^u)^n = b^{nu}$	(12-6)
Properties of logarithms	$\log_b x + \log_b y = \log_b xy$	(12-7)
	$\log_b x - \log_b y = \log_b \left(\dfrac{x}{y}\right)$	(12-8)
	$n \log_b x = \log_b (x^n)$	(12-9)
	$\log_b 1 = 0, \quad \log_b b = 1$	(12-10)
	$\log_b (b^n) = n$	(12-11)
Changing base of logarithms	$\log_b x = \dfrac{\log_a x}{\log_a b}$	(12-12)
	$\ln x = 2.3026 \log x$	(12-13)
	$\log x = 0.4343 \ln x$	(12-14)

Review Exercises

In Exercises 1 through 12, determine the value of x.

1. $\log_{10} x = 4$

2. $\log_9 x = 3$

3. $\log_5 x = -1$

4. $\log_4 x = -\frac{1}{2}$

5. $\log_2 64 = x$

6. $\log_{12} 144 = x$

7. $\log_8 32 = x$

8. $\log_9 27 = x$

9. $\log_x 36 = 2$

10. $\log_x 243 = 5$

11. $\log_x 10 = \frac{1}{2}$

12. $\log_x 8 = \frac{3}{4}$

In Exercises 13 through 24, express each as a sum, difference, or multiple of logarithms. Wherever possible, evaluate logarithms of the result.

13. $\log_3 2x$

14. $\log_5 \left(\dfrac{7}{a}\right)$

15. $\log_3(t^2)$

16. $\log_6 \sqrt{5}$

17. $\log_2 28$

18. $\log_7 98$

19. $\log_3 \left(\dfrac{9}{x}\right)$

20. $\log_6 \left(\dfrac{5}{36}\right)$

21. $\log_4 \sqrt{48}$

22. $\log_6 \sqrt{72y}$

23. $\log_{10} (1000x^4)$

24. $\log_3 (9^2 \times 6^3)$

In Exercises 25 through 32, solve for y in terms of x.

25. $\log_6 y = \log_6 4 - \log_6 x$

26. $\log_3 y = \frac{1}{2} \log_3 7 + \frac{1}{2} \log_3 x$

27. $\log_2 y + \log_2 x = 3$

28. $6 \log_4 y = 8 \log_4 4 - 3 \log_4 x$

29. $\log_5 x + \log_5 y = \log_5 3 + 1$

30. $\log_7 y = 2 \log_7 5 + \log_7 x + 2$

31. $3(\log_8 y - \log_8 x) = 1$

32. $2(\log_9 y + 2 \log_9 x) = 1$

In Exercises 33 through 40, graph the given functions.

33. $y = 0.5(5^x)$

34. $y = 3(2^{-x})$

35. $y = 0.5 \log_4 x$

36. $y = 10 \log_{16} x$

37. $y = \log_{3.15} x$

38. $y = 0.1 \log_{4.65} x$

39. $y = 1 - e^{-x}$

40. $y = 2(1 - e^{-0.2x})$

In Exercises 41 through 44, use logarithms to perform the indicated calculations.

41. $(13.6)(0.693)$

42. $(0.00624)^{0.2}$

43. $\dfrac{195^{195}}{\sqrt{86.4}}$

44. $45.1(60.7)^{64}$

In Exercises 45 through 48, by using logarithms to the base 10, find the natural logarithms of the given numbers.

45. 8.86

46. 33.0

47. 2.07

48. 0.542

In Exercises 49 through 56, solve the given equations.

49. $e^{2x} = 5$

50. $2(5^x) = 15$

51. $3^{x+2} = 5^x$

52. $6^{x+2} = 12^{x-1}$

53. $\log_4 x + \log_4 6 = \log_4 12$

54. $2 \log_3 2 - \log_3 (x + 1) = \log_3 5$

55. $\log_8 (x + 2) + \log_8 2 = 2$

56. $\log (x + 2) + \log x = 0.4771$

In Exercises 57 and 58, plot the graphs of the given functions on semilogarithmic paper. In Exercises 59 and 60, plot the graphs of the given functions on log-log paper.

57. $y = 6^x$

58. $y = 5x^3$

59. $y = \sqrt[3]{x}$

60. $xy^4 = 16$

If x is eliminated between Eqs. (12-1) and (12-2), we have

$$y = b^{\log_b y} \tag{12-15}$$

In Exercises 61 through 64, evaluate the given expressions using Eq. (12-15).

61. $10^{\log 4}$

62. $2e^{\ln 7.5}$

63. $3e^{2 \ln 2}$

64. $5(10^{2 \log 3})$

In Exercises 65 through 80, solve the given problems.

65. A \$1000 certificate of deposit earns 8.3% annual interest, compounded continuously. Its value after t years is $V = 1000e^{0.083t}$. Solve for t.

66. An equation used in hydrodynamics when analyzing an incompressible fluid is $P = wa^2 \ln r + D$. Solve for r.

67. In a certain electric circuit, the current i is given by $i = i_0 e^{-5t}$, where i_0 is the current for $t = 0$ and t is the time. Solve for t.

68. If \$500 is invested at an annual interest rate of 10%, compounded quarterly (see Exercise 25 of Section 12-2), express V as a function of t and plot the graph for $0 \le t \le 8$ years.

69. The bacterial population in a certain culture is given by $N = 1000(1.5)^t$, where t is in hours. Sketch the graph of N as a function of t for $0 \le t \le 6$ h.

70. A computer analysis of the luminous efficiency E, in lumens per watt, of a tungsten lamp as a function of its input power P, in watts, is given by $E = 22.0(1 - 0.65e^{-0.008P})$. Sketch the graph of E as a function of P for $0 \le P \le 1000$ W.

71. An equation which may be used for the angular velocity ω of the slider mechanism in Fig. 12-15 is $2 \ln \omega = \ln 3g + \ln \sin \theta - \ln \ell$, where g is the acceleration due to gravity. Solve for $\sin \theta$.

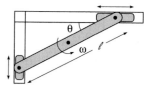

72. Taking into account the weight loss of fuel, the maximum velocity v_m of a rocket is given by the equation

$$v_m = u(\ln m_0 - \ln m_s) - gt_f$$

where m_0 is the initial mass of the rocket and fuel, m_s is the mass of the rocket shell, t_f is the time during which fuel is expended, u is the velocity of the expelled fuel, and g is the acceleration due to gravity. Solve for m_0.

Fig. 12-15

73. An equation used to calculate the capacity C, in bits per second, of a telephone channel is $C = B \log_2 (1 + R)$. Solve for R.

74. An equation used in studying the action of a protein molecule is $\ln A = \ln \theta - \ln (1 - \theta)$. Solve for θ.

75. For first-order chemical reactions, concentration of a reacting chemical species is related to time by the expression $\log (x_0/x) = kt$, where x_0 is the initial concentration and x is the concentration after time t. Determine the concentration of sucrose remaining after 3 h if the initial concentration is 9.00 g·mol/L and $k = 0.00158$ per minute.

76. The power gain of an electronic device such as an amplifier is defined as $n = 10 \log (P_0/P_i)$, where n is measured in decibels, P_0 is the power output, and P_i is the power input. If $P_0 = 10.0$ W and $P_i = 0.125$ W, calculate the power gain. (See Example E of Section 12-6.)

77. In studying the frictional effects on a flywheel, the revolutions per minute R that it makes as a function of the time t, in minutes, is given by $R = 4500(0.750)^{2.50t}$. Find t for $R = 2000$ r/min.

78. The efficiency e of a gasoline engine as a function of its compression ratio r is given by $e = 1 - r^{1-\gamma}$, where γ is a constant. Find γ for $e = 0.55$ and $r = 7.5$.

79. For a particular solar energy system, the collector area A required to supply a fraction F of the total energy is given by $A = 480 F^{2.2}$. Plot A, in m², as a function of F, from $F = 0.1$ to $F = 0.9$, on semilog paper.

80. The current I and resistance R were measured as follows in a certain microcomputer circuit.

R (Ω)	100	200	500	1000	2000	5000	10,000
I (μA)	81	41	16	8.2	4.0	1.6	0.8

Plot I as a function of R on log-log paper.

Practice Test

In Problems 1 through 4, determine the value of x.

1. $\log_9 x = -\frac{1}{2}$ **2.** $\log_3 x - \log_3 2 = 2$ **3.** $\log_x 64 = 3$ **4.** $3^{3x+1} = 8$

5. Graph the function $y = 2 \log_4 x$.

6. Graph the function $y = 2(3^x)$ on semilog paper.

7. Express as a combination of a sum, difference, and multiple of logarithms, including $\log_5 2$: $\log_5 \left(\dfrac{4a^3}{7} \right)$.

8. Solve for y in terms of x: $3 \log_7 x - \log_7 y = 2$.

9. An equation used for a certain electric circuit is $\ln i - \ln I = -t/RC$. Solve for i.

10. Evaluate: $\dfrac{2 \ln 0.9523}{\log 6066}$.

11. Evaluate: $\log_5 732$.

12. If A_0 dollars are invested at 8%, compounded continuously for t years, the value A of the investment is given by $A = A_0 e^{0.08t}$. Determine how long it takes for the investment to double in value.

13 Additional Types of Equations and Systems of Equations

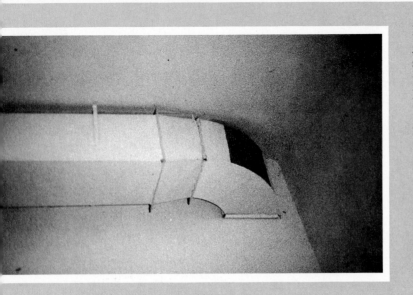

The dimensions of a heating system vent are important to its design. In Section 13-1 we discuss a problem in vent design.

In Chapter 2 we determined how to graph a function, as well as how to solve equations graphically. Since then we have discussed the graphs of linear and quadratic functions and we have dealt with methods for solving quadratic equations and systems of linear equations. Also, we have graphed the trigonometric, exponential, and logarithmic functions. In the first section of this chapter, we shall introduce a more general type of quadratic equation and then discuss graphical solutions of systems of equations involving these quadratic equations, as well as other types of equations.

In the second section we shall discuss the algebraic solution of certain systems of equations involving quadratic equations. In the final sections we shall consider the solution of equations which are not quadratic, but which can be solved by methods developed for quadratic equations, and equations which involve radicals.

Applications of these equations and systems of equations are found in various fields of science and technology. These include physics, electricity, business, and structural design.

13-1 Graphical Solution of Systems of Equations

As we noted, in this section we introduce another general type of equation: the general quadratic equation. Then we consider graphical solutions of systems of equations.

An equation of the form

$$ax^2 + bxy + cy^2 + dx + ey + f = 0 \qquad (13\text{-}1)$$

is called a **general quadratic equation in x and y.** We shall be interested primarily in some special cases of this equation. *The graphs of the various possible forms of this equation result in curves known as* **conic sections.** These curves are the circle, parabola, ellipse, and hyperbola. We were previously introduced to the parabola when we discussed the graph of the quadratic function in Chapter 6. The following examples illustrate these curves. A more complete discussion is found in Chapter 20.

EXAMPLE A

Graph the equation $y = 3x^2 - 6x$.

We graphed equations of this form in Section 2-4 and in Section 6-4. Using the method of Section 6-4 (and the constants a, b, and c as defined there) we see that $a = 3$, $b = -6$, and $c = 0$. This means that $-b/2a = -(-6)/2(3) = 1$, which tells us that the x-coordinate of the extreme point is 1. For $x = 1$, $y = -3$, which means that the extreme point is $(1, -3)$. Also, we know that it is a minimum point, since $a > 0$.

Since $c = 0$, the y-intercept is $(0, 0)$. Setting $y = 0$, we have $3x^2 - 6x = 0$, or $3x(x - 2) = 0$. This means that $y = 0$ for $x = 0$ and $x = 2$ and the points $(0, 0)$ and $(2, 0)$ are on the curve. (We have found the point $(0, 0)$ in two ways.) To define the curve a little better, we find the points $(-1, 9)$ and $(3, 9)$. These values are summarized in the following table, and the graph is shown in Fig. 13-1.

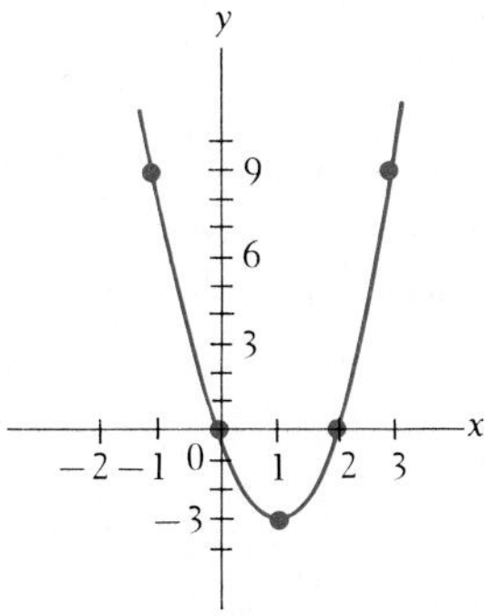

x	-1	0	1	2	3
y	9	0	-3	0	9

Fig. 13-1

As we showed in Section 6-4, the curve is a *parabola*. Using the constants of the general quadratic equation in Eq. (13-1), a parabola always results if the equation is of the form $y = ax^2 + dx + f$.

EXAMPLE B

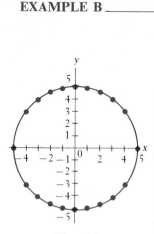

Fig. 13-2

Plot the graph of the equation $x^2 + y^2 = 25$.

We first solve this equation for y, and we obtain $y = \sqrt{25 - x^2}$, or $y = -\sqrt{25 - x^2}$, which we write as $y = \pm\sqrt{25 - x^2}$. We now assume values for x and find the corresponding values for y.

x	0	± 1	± 2	± 3	± 4	± 5
y	± 5	± 4.9	± 4.6	± 4	± 3	0

If we try values greater than 5, we have imaginary numbers. These cannot be plotted, for we assume that both x and y are real. (The complex plane is only for *numbers* of the form $a + bj$ and does not represent pairs of numbers representing two variables.) When we give the value $x = \pm 4$ when $y = \pm 3$, this is simply a short way of representing 4 points. These points are $(4, 3)$, $(4, -3)$, $(-4, 3)$, $(-4, -3)$. We note in Fig. 13-2 that *the resulting curve is a* **circle.** A circle always results from an equation of the form $x^2 + y^2 = r^2$, and r is the radius of the circle with its center at the origin.

EXAMPLE C

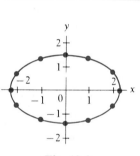

Fig. 13-3

Plot the graph of the equation $2x^2 + 5y^2 = 10$.

We first solve for y, then construct the table of values.

$$y = \pm\sqrt{\frac{10 - 2x^2}{5}}$$

x	0	± 1	± 2	$\pm\sqrt{5}(= \pm 2.2)$
y	± 1.4	± 1.3	± 0.6	0

Values of x greater than $\sqrt{5}$ result in imaginary values of y. *The curve* (Fig. 13-3) *is an* **ellipse.** An ellipse results from an equation that can be written in the form $ax^2 + cy^2 = k$. (Constants a, c, and k must all have the same sign, and $a \neq c$.)

EXAMPLE D

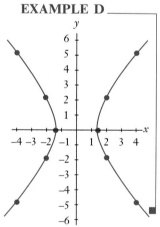

Fig. 13-4

Plot the graph of the equation $2x^2 - y^2 = 4$.

Solving for y, we get

$$y = \pm\sqrt{2x^2 - 4}$$

Now, constructing the table of values, we have the following points:

x	± 1.4	± 2	± 4
y	0	± 2	± 5.3

We note that values $-\sqrt{2} < x < \sqrt{2}$ are not in the domain of either $y = \sqrt{2x^2 - 4}$ or $y = -\sqrt{2x^2 - 4}$, since such values would lead to imaginary values of y.

Plotting these points, we obtain the curve in Fig. 13-4. This curve is called a **hyperbola.** A hyperbola results if the equation is of the form $ax^2 + cy^2 = k$, if a and c have different signs.

As in solving systems of linear equations, we obtain the desired solutions of any system by finding the values of x and y which satisfy both equations at the same time. To solve a system of equations graphically, we find all points which the graphs have in common. This means we need only graph the equations and then locate the points of intersection. If the curves do not intersect, the system has no real solutions. The following example illustrates solving a system of equations graphically in an applied situation.

EXAMPLE E

See the chapter introduction.

For proper ventilation, the vent for a natural gas heating system is to have a rectangular cross-sectional area of 6.0 ft², and it is to be made from sheet metal which is 12.0 ft wide. What are to be the length and width of this cross-sectional area of the vent?

In Fig. 13-5, we have let ℓ be the length and w be the width of the area. Since the area is 6.0 ft², we have $\ell w = 6.0$. Also, since the sheet metal was 12.0 ft wide, this is the perimeter of the area. This gives us $2\ell + 2w = 12.0$, or $\ell + w = 6.0$. This means that the system of equations to be solved is

$$\ell w = 6.0$$
$$\ell + w = 6.0$$

Solving the first equation for ℓ, we get $\ell = 6.0/w$. Constructing a table of values, we have

w	1.0	2.0	3.0	4.0	5.0	6.0
ℓ	6.0	3.0	2.0	1.5	1.2	1.0

We do not use negative values since they have no meaning to the solution.

Solving the second equation for ℓ, we get $\ell = -w + 6.0$, which we recognize as the equation for a straight line. Using coordinates (w, ℓ), we see that the intercepts for the line are (0.0, 6.0) and (6.0, 0.0). The slope of the line is -1.

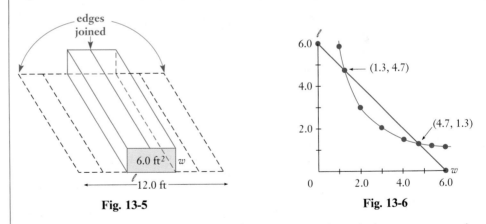

Fig. 13-5   Fig. 13-6

Graphing both curves, we see in Fig. 13-6 that the solutions are approximately (1.3, 4.7) and (4.7, 1.3). Using the length as the longer dimension, we have the solution $\ell = 4.7$ ft and $w = 1.3$ ft. We see that this solution checks with the statement of the problem.

If negative values had been used, there would have been another branch of the curve, $\ell w = 6.0$, in the third quadrant, and the straight line would have extended into the second and fourth quadrants. The curve of $\ell w = 6.0$ is another hyperbola. Another form of the equation of a hyperbola is $xy = k$.

The following two examples illustrate the graphical solution of a system of equations in which one of the equations is not a general quadratic type (Example F) and the graphical solution of a system which has no real solution (Example G).

EXAMPLE F

Graphically solve the system of equations

$$9x^2 + 4y^2 = 36$$
$$y = 3^x$$

The first equation is of the form represented by an ellipse, as indicated in Example C. The second equation is an exponential function, as discussed in Chapter 12. Solving the first equation for y, we have $y = \pm\frac{1}{2}\sqrt{36 - 9x^2}$. Substituting values for x, we obtain the following table:

x	0	± 1	± 2
y	± 3	± 2.6	0

For the exponential function, we obtain the following table:

x	-3	-2	-1	0	1	2
y	$\frac{1}{27}$	$\frac{1}{9}$	$\frac{1}{3}$	1	3	9

We plot these curves as shown in Fig. 13-7. The points of intersection are approximately $x = -1.9$, $y = 0.1$ and $x = 0.9$, $y = 2.7$.

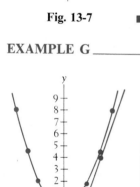

$(-1.9, 0.1)$ $(0.9, 2.7)$

Fig. 13-7

EXAMPLE G

Graphically solve the system of equations

$$x^2 = 2y$$
$$3x - y = 5$$

We note that the two curves in this system are a parabola and a straight line. Solving the equation of the parabola for y, we obtain $y = \frac{1}{2}x^2$. We construct the following table:

x	0	± 1	± 2	± 3	± 4
y	0	$\frac{1}{2}$	2	$\frac{9}{2}$	8

For the straight line, we have the following points:

x	0	$\frac{5}{3}$	3
y	-5	0	4

We plot these curves and see in Fig. 13-8 that they do not intersect, so we conclude that there are no real solutions to the system.

Fig. 13-8

Exercises 13-1

In Exercises 1 through 24, solve the given systems of equations graphically.

1. $y = 2x$
   $x^2 + y^2 = 16$

2. $3x - y = 4$
   $y = 6 - 2x^2$

3. $x^2 + 2y^2 = 8$
   $x - 2y = 4$

4. $y = 3x - 6$
   $xy = 6$

5. $y = x^2 - 2$
$4y = 12x - 17$

6. $x^2 + 4y^2 = 4$
$2y = 12 - x$

7. $y = x^2$
$xy = 4$

8. $y = -2x^2$
$y = x^2 - 6$

9. $y = -x^2 + 4$
$x^2 + y^2 = 9$

10. $y = 2x^2 - 1$
$x^2 + 2y^2 = 16$

11. $x^2 - 4y^2 = 16$
$x^2 + y^2 = 1$

12. $y = 2x^2 - 4x$
$xy = -4$

13. $2x^2 + 3y^2 = 19$
$x^2 + y^2 = 9$

14. $x^2 - y^2 = 4$
$2x^2 + y^2 = 16$

15. $x^2 + y^2 = 1$
$xy = \frac{1}{2}$

16. $x^2 + y^2 = 25$
$x^2 - y^2 = 7$

17. $y = x^2$
$y = \sin x$

18. $y = 4x - x^2$
$y = 2 \cos x$

19. $y = e^{-x}$
$x + y = 2$

20. $y = 2^x$
$x^2 + y^2 = 4$

21. $x^2 - y^2 = 1$
$y = \log_2 x$

22. $x^2 + 4y^2 = 16$
$y = 2 \ln x$

23. $y = \ln(x - 1)$
$y = \sin \frac{1}{2}x$

24. $y = \cos x$
$y = \log_3 x$

In Exercises 25 through 28, set up the indicated systems of equations and solve them graphically.

25. A helicopter is located 5.2 mi north of east of a radio tower such that it is three times as far north as it is east of the tower. Find the northern and eastern components of the displacement from the tower.

26. A rectangular field has a perimeter of 1140 m and an area of 75,600 m². Find the length and width of the field.

27. The power developed in an electric resistor is i^2R, where i is the current. If a first current passes through a 2-Ω resistor and a second current passes through a 3-Ω resistor, the total power produced is 12 W. If the resistors are reversed, the total power produced is 16 W. Find the currents ($i > 0$).

28. Two circular holes are drilled tangent to each other in a metal plate. The center-to-center distance is 4.9 cm, and the total area drilled out is 41 cm². Find the radius of each hole.

13-2 Algebraic Solution of Systems of Equations ▬▬▬▬

Often the graphical method is the easiest way to solve a system of equations. However, this method does not usually give an exact answer. Using algebraic methods to find exact solutions for some systems of equations is either not possible or quite involved. There are some systems, however, which do lend themselves to relatively simple solutions by algebraic means. In this section we shall consider two useful methods, both of which we discussed before when we were studying systems of linear equations.

solution by substitution

The first method is *substitution*. If we can solve one equation for one of its variables, we can substitute this solution into the other equation. We then have only one unknown in the resulting equation, and we can solve this equation by methods discussed in earlier chapters.

EXAMPLE A ————— Solve, by substitution, the system of equations

$$2x - y = 4$$
$$x^2 - y^2 = 4$$

We solve the first equation for y, obtaining $y = 2x - 4$. We now substitute $2x - 4$ for y in the second equation, getting

———— in second equation, y replaced by $2x - 4$

$$x^2 - (2x - 4)^2 = 4$$

(Continued on next page)

When simplified, this gives a quadratic equation.

$$x^2 - (4x^2 - 16x + 16) = 4$$

$$-3x^2 + 16x - 20 = 0$$

$$x = \frac{-16 \pm \sqrt{256 - 4(-3)(-20)}}{-6} = \frac{-16 \pm \sqrt{16}}{-6} = \frac{-16 \pm 4}{-6} = \frac{10}{3}, 2$$

We now find the corresponding values of y by substituting into $y = 2x - 4$. Thus, we have the solutions $x = \frac{10}{3}$, $y = \frac{8}{3}$, and $x = 2$, $y = 0$. As a check, we find that these values also satisfy the equation $x^2 - y^2 = 4$. Compare these solutions ■ with those which would be obtained from Fig. 13-9.

Fig. 13-9

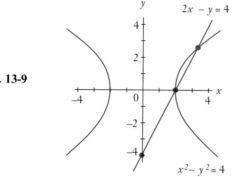

EXAMPLE B

Solve, by substitution, the system of equations

$$xy = -2$$

$$2x + y = 2$$

From the first equation we have $y = -2/x$. Substituting this into the second equation, we have

in second equation, y replaced by $-\dfrac{2}{x}$

$$2x + \left(-\frac{2}{x}\right) = 2$$

$$2x^2 - 2 = 2x$$

$$x^2 - x - 1 = 0$$

$$x = \frac{1 \pm \sqrt{1 + 4}}{2} = \frac{1 \pm \sqrt{5}}{2}$$

By substituting these values for x into either of the original equations, we find the corresponding values of y, and we have the solutions

$$x = \frac{1 + \sqrt{5}}{2}, y = 1 - \sqrt{5} \quad \text{and} \quad x = \frac{1 - \sqrt{5}}{2}, y = 1 + \sqrt{5}$$

These can be checked by substitution in the original equations.

Starting with the solutions for x of $x = \frac{1}{2}(1 \pm \sqrt{5})$, we can get the decimal results for x and then get the results for y by substituting in $y = -2/x$. The calculator sequence for the first solution is shown at the left. These results are

$$x = 1.618, \quad y = -1.236 \quad \text{and} \quad x = -0.618, \quad y = 3.236$$

Compare these solutions with those which would be obtained from Fig. 13-10.

The solutions could have been found by first solving the second equation for y, or either equation for x, then making the proper substitution in the other equation. ∎

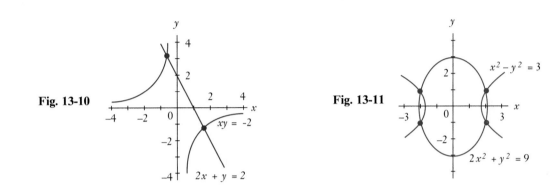

Fig. 13-10

Fig. 13-11

solution by addition or subtraction

Now let us use the other algebraic method of solution, that of *addition or subtraction*. This method can be used to great advantage if both equations have only squared terms and constants.

EXAMPLE C _____

Solve, by addition or subtraction, the system of equations

$$2x^2 + y^2 = 9$$
$$x^2 - y^2 = 3$$

We note that if we add the corresponding sides of each equation, y^2 is eliminated. This leads to the following solution:

$$
\begin{array}{rl}
2x^2 + y^2 &= 9 \\
x^2 - y^2 &= 3 \\
\hline
3x^2 \qquad &= 12 \qquad \text{add} \\
x^2 &= 4 \\
x &= \pm 2
\end{array}
$$

For $x = 2$, we have two corresponding y-values, $y = \pm 1$. Also for $x = -2$, we have two corresponding y-values, $y = \pm 1$. Thus, we have four solutions:

$$x = 2, y = 1; \qquad x = 2, y = -1; \qquad x = -2, y = 1; \quad \text{and} \quad x = -2, y = -1$$

Each of these solutions checks in the original equations. Compare these solutions with those which would be obtained from Fig. 13-11. ∎

(calculator display at left:)

(1 + 5 $\sqrt{x}$
) ÷ 2 =
 1.6180340
STO 2 +/− ÷
RCL =
 − 1.2360680

EXAMPLE D

Solve, by addition or subtraction, the system of equations

$$3x^2 - 2y^2 = 5$$
$$x^2 + y^2 = 5$$

If we multiply the second equation by 2 and then add the two resulting equations, we get

$$3x^2 - 2y^2 = 5$$
$$2x^2 + 2y^2 = 10 \qquad \text{each term of second equation multiplied by 2}$$
$$\overline{5x^2 \qquad\quad = 15} \qquad \text{add}$$
$$x^2 = 3 \qquad x = \pm\sqrt{3}$$

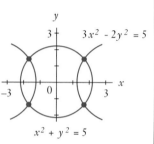

Fig. 13-12

The corresponding values of y for each value of x are $y = \pm\sqrt{2}$. Again we have four solutions:

$$x = \sqrt{3}, y = \sqrt{2}; x = \sqrt{3}, y = -\sqrt{2}; x = -\sqrt{3}, y = \sqrt{2}; x = -\sqrt{3}, y = -\sqrt{2}$$

Each of these solutions checks when substituted in the original equations. Compare these solutions with those which would be obtained from Fig. 13-12. ∎

EXAMPLE E

A certain number of machine parts cost $1000. If they cost $5 less per part, ten additional parts could be purchased for the same amount of money. What is the cost of each?

Since the cost of each part is required, we let $c =$ the cost per part. Also, we let $n =$ the number of parts. From the first statement of the problem, we see that $cn = 1000$. Also, from the second statement, we have $(c - 5)(n + 10) = 1000$. Therefore, we are to solve the system of equations

$$cn = 1000$$
$$(c - 5)(n + 10) = 1000$$

Solving the first equation for n, and multiplying out the second equation, we have $n = \dfrac{1000}{c}$ and $cn + 10c - 5n - 50 = 1000$. Now, substituting the expression for n into the second equation, we solve for c.

$$c\left(\frac{1000}{c}\right) + 10c - 5\left(\frac{1000}{c}\right) - 50 = 1000$$

$$1000 + 10c - \frac{5000}{c} - 50 = 1000$$

$$10c - \frac{5000}{c} - 50 = 0$$

$$c^2 - 5c - 500 = 0$$

$$(c + 20)(c - 25) = 0$$

$$c = -20, 25$$

Since a negative answer has no significance in this particular situation, we see that the solution is $c = \$25$ per part. Checking with the original statement of the problem, we see that this is correct. ∎

Exercises 13-2

In Exercises 1 through 24, solve the given systems of equations algebraically.

1. $y = x + 1$
$y = x^2 + 1$

2. $y = 2x - 1$
$y = 2x^2 + 2x - 3$

3. $x + 2y = 3$
$x^2 + y^2 = 26$

4. $y = x + 1$
$x^2 + y^2 = 25$

5. $x + y = 1$
$x^2 - y^2 = 1$

6. $x + y = 2$
$2x^2 - y^2 = 1$

7. $2x - y = 2$
$2x^2 + 3y^2 = 4$

8. $6y - x = 6$
$x^2 + 3y^2 = 36$

9. $xy = 1$
$x + y = 2$

10. $xy = 2$
$x + y = 3$

11. $xy = 3$
$3x - 2y = -7$

12. $xy = -4$
$2x + y = -2$

13. $y = x^2$
$y = 3x^2 - 8$

14. $y = x^2 - 1$
$2x^2 - y^2 = 2$

15. $x^2 - y = -1$
$x^2 + y^2 = 5$

16. $x^2 + y = 5$
$x^2 + y^2 = 25$

17. $x^2 - 1 = y$
$x^2 - 2y^2 = 1$

18. $2y^2 - 4x = 7$
$y^2 + 2x^2 = 3$

19. $x^2 + y^2 = 25$
$x^2 - 2y^2 = 7$

20. $3x^2 - y^2 = 4$
$x^2 + 4y^2 = 10$

21. $y^2 - 2x^2 = 6$
$5x^2 + 3y^2 = 20$

22. $y^2 - 2x^2 = 17$
$2y^2 + x^2 = 54$

23. $x^2 + 3y^2 = 37$
$2x^2 - 9y^2 = 14$

24. $5x^2 - 4y^2 = 15$
$3y^2 + 4x^2 = 12$

In Exercises 25 through 32, solve the indicated systems of equations algebraically. In Exercises 27 through 32, it is necessary to set up the systems of equations properly.

25. A 2-kg block collides with an 8-kg block. Using the physical laws of conservation of energy and conservation of momentum, along with given conditions, the following equations involving the velocities are established:

$$v_1^2 + 4v_2^2 = 41$$
$$2v_1 + 8v_2 = 12$$

Find these velocities (in m/s) if $v_2 > 0$.

26. A rocket is fired from behind a ship and follows the path given by $h = 3x - 0.05x^2$, where h is its altitude, x is the horizontal distance traveled, and distances are measured in miles. A missile fired from the ship follows the path given by $h = 0.8x - 15$. For $h > 0$ and $x > 0$, determine where the paths of the rocket and missile cross.

27. A rectangular computer chip has a surface area of 2.1 cm^2 and a perimeter of 5.8 cm. Find the length and width of the chip.

28. The impedance Z in an alternating-current circuit is 2.00 Ω. If the resistance R is numerically equal to the square of the reactance X, find R and X. See Section 11-7.

29. A right triangular lawn area has a hypotenuse of 40.0 ft and a perimeter of 90.0 ft. Find the sides of the area.

30. In a certain roller mechanism, the radius of one steel ball is 2.00 cm greater than the radius of a second steel ball. If the difference in their weights is 7100 g, find the radii of the balls. The density of steel is 7.70 g/cm^3.

31. A plane travels at 400 mi/h relative to the air. It takes the plane 2.5 h longer to travel 2400 mi against a certain wind than it does with the wind. Find the velocity of the wind.

32. In a marketing survey, a company found that the total gross income for selling t tables at a price of p dollars each was $35,000. It then increased the price of each table by $100 and found that the total income was only $27,000 because 40 fewer tables were sold. Find p and t.

13-3   Equations in Quadratic Form

NOTE ▷

Often we encounter equations which can be solved by methods applicable to quadratic equations, even though these equations are not actually quadratic. They do have the property, however, that *with a proper substitution **they may be written in the form of a quadratic equation.*** All that is necessary is that the equation have terms including some variable quantity, its square, and perhaps a constant term. The following example illustrates these types of equations.

EXAMPLE A   The equation $x - 2\sqrt{x} - 5 = 0$ is an equation in quadratic form, because if we let $y = \sqrt{x}$, we have $x = (\sqrt{x})^2 = y^2$, and the resulting equation is $y^2 - 2y - 5 = 0$. Other examples of equations in quadratic form are as follows:

$t^{-4} - 5t^{-2} + 3 = 0$

By letting $y = t^{-2}$, we have $\overset{(t^{-2})^2}{y^2} - 5y + 3 = 0.$

$t^3 - 3t^{3/2} - 7 = 0$

By letting $y = t^{3/2}$, we have $\overset{(t^{3/2})^2}{y^2} - 3y - 7 = 0.$

$(x + 1)^4 - (x + 1)^2 - 1 = 0$

By letting $y = (x + 1)^2$, we have $\overset{[(x+1)^2]^2}{y^2} - y - 1 = 0.$

$x^{10} - 2x^5 + 1 = 0$

By letting $y = x^5$, we have $\overset{(x^5)^2}{y^2} - 2y + 1 = 0.$

$(x - 3) + \sqrt{x - 3} - 6 = 0$

By letting $y = \sqrt{x - 3}$, we have $\overset{(\sqrt{x-3})^2}{y^2} + y - 6 = 0.$

The following examples illustrate the method of solving equations in quadratic form.

EXAMPLE B   Solve the equation $x^4 - 5x^2 + 4 = 0$.
We first let $y = x^2$ and obtain the resulting equivalent equation

$$y^2 - 5y + 4 = 0$$

This may be factored and solved as

$$(y - 4)(y - 1) = 0$$
$$y = 4 \quad \text{or} \quad y = 1$$

Since we want values for x, and since $y = x^2$, we have

$$x^2 = 4 \quad \text{or} \quad x^2 = 1$$
$$x = \pm 2 \quad \text{or} \quad x = \pm 1$$

Substitution into the original equation verifies that each of these values is a solution.

EXAMPLE C   Solve the equation $2x^4 + 7x^2 = 4$.
As in Example B, we let $y = x^2$ and then write the resulting equation in quadratic form and solve.

$$2y^2 + 7y - 4 = 0$$
$$(2y - 1)(y + 4) = 0$$
$$y = \frac{1}{2} \quad \text{or} \quad y = -4$$
$$x^2 = \frac{1}{2} \quad \text{or} \quad x^2 = -4 \qquad y = x^2$$
$$x = \pm\frac{1}{\sqrt{2}} \quad \text{or} \quad x = \pm 2j$$

Substitution into the original equation shows that each value is a solution.

Two of the solutions in Example C are complex numbers. We were able to find these solutions directly from the definition of the square root of a negative number. In some cases (see Exercises 23 and 24 of this section) it is necessary to use the method of Section 11-6 to find such complex number solutions.

EXAMPLE D _____ Solve the equation $x - \sqrt{x} - 2 = 0$.

By letting $y = \sqrt{x}$, we have

$$y^2 - y - 2 = 0$$
$$(y - 2)(y + 1) = 0$$
$$y = 2 \quad \text{or} \quad y = -1$$

Since $y = \sqrt{x}$, we note that y cannot be negative, and this in turn tells us that $y = -1$ cannot lead to a solution. For $y = 2$ we have $x = 4$. Checking, we find that $x = 4$ satisfies the original equation. Thus, the only solution is $x = 4$.

Example D illustrates a very important point: *Whenever any operation involving the unknown is performed on an equation, this operation may introduce roots into a subsequent equation which are not roots of the original equation.*

NOTE ▷ *Therefore, we must* **check all answers in the original equation.** Only operations involving constants—that is, adding, subtracting, multiplying by, or dividing by constants—are certain not to introduce these **extraneous roots.** Squaring both

extraneous roots

sides of an equation is a common way of introducing extraneous roots. We first encountered the concept of an extraneous root in Section 5-7, when we were discussing equations involving fractions.

EXAMPLE E _____ Solve the equation $x^{-2} + 3x^{-1} + 1 = 0$.

By substituting $y = x^{-1}$, we have $y^2 + 3y + 1 = 0$. To solve this equation we may use the quadratic formula:

$$y = \frac{-3 \pm \sqrt{9 - 4}}{2} = \frac{-3 \pm \sqrt{5}}{2}$$

Thus, since $x = 1/y$,

$$x = \frac{2}{-3 + \sqrt{5}} \quad \text{or} \quad x = \frac{2}{-3 - \sqrt{5}}$$

These answers in decimal form are

$$x = -2.618 \quad \text{or} \quad x = -0.382$$

The calculator sequence for the first is shown at the left.

In checking these solutions in the original equation, it is more accurate to use the unrounded calculator values. This is done by storing the unrounded value in memory and then using the following calculator sequence:

$$\boxed{\text{RCL}}\,\boxed{x^2}\,\boxed{1/x}\,\boxed{+}\,3\,\boxed{\times}\,\boxed{\text{RCL}}\,\boxed{1/x}\,\boxed{+}\,1\,\boxed{=}\,\boxed{\qquad\qquad 0.}$$

This shows that these answers check.

EXAMPLE F _____

Solve the equation $(x^2 - x)^2 - 8(x^2 - x) + 12 = 0$.

By substituting $y = x^2 - x$, we have

$$y^2 - 8y + 12 = 0$$
$$(y - 2)(y - 6) = 0$$
$$y = 2 \quad \text{or} \quad y = 6$$

This means that

$$x^2 - x = 2 \quad \text{or} \quad x^2 - x = 6$$

Solving each of these, we have

$$x^2 - x - 2 = 0 \qquad\qquad x^2 - x - 6 = 0$$
$$(x - 2)(x + 1) = 0 \qquad (x - 3)(x + 2) = 0$$
$$x = 2 \quad \text{or} \quad x = -1 \qquad x = 3 \quad \text{or} \quad x = -2$$

Substituting each of these four values in the original equations shows that all

∎ are solutions.

Example G illustrates a problem that leads to an equation in quadratic form.

EXAMPLE G _____

A rectangular solar cell has an area of 60 cm². The diagonal of the cell is 13 cm. Find the length and width of the cell.

Since the required quantities are the length and width, let ℓ = the length of the cell and w = the width of the cell. Since the area is 60 cm², $\ell w = 60$. Also, using the Pythagorean theorem and the fact that the diagonal is 13 cm, we have $\ell^2 + w^2 = 169$. Therefore, we are to solve the system of equations

$$\ell w = 60, \qquad \ell^2 + w^2 = 169$$

Solving the first equation for ℓ, we have $\ell = 60/w$. Substituting this expression into the second equation, we have

$$\left(\frac{60}{w}\right)^2 + w^2 = 169$$

We now solve for w as follows:

$$\frac{3600}{w^2} + w^2 = 169$$
$$3600 + w^4 = 169w^2$$
$$w^4 - 169w^2 + 3600 = 0$$

Let $x = w^2$.

$$x^2 - 169x + 3600 = 0$$
$$(x - 144)(x - 25) = 0$$
$$x = 144 \quad \text{or} \quad x = 25$$

Therefore,

$$w^2 = 144 \quad \text{or} \quad w^2 = 25$$

Solving for w, we obtain $w = \pm 12$ or $w = \pm 5$. Only the positive values of w are meaningful in this problem. Therefore, if $w = 5$ cm, then $\ell = 12$ cm. Normally, we designate the length as the longer dimension, although by letting $w = 12$ we get $\ell = 5$ cm. Checking with the statement of the problem, we see that these dimensions for the solar cell give an area of 60 cm^2 and a diagonal of 13 cm.

Exercises 13-3

In Exercises 1 through 24, solve the given equations.

1. $x^4 - 13x^2 + 36 = 0$

2. $x^4 - 20x^2 + 64 = 0$

3. $4x^4 - 5x^2 = 9$

4. $4x^4 + 15x^2 = 4$

5. $x^{-2} - 2x^{-1} - 8 = 0$

6. $10x^{-2} + 3x^{-1} - 1 = 0$

7. $x^{-4} + 2x^{-2} = 24$

8. $x^{-4} + 1 = 2x^{-2}$

9. $2x - 7\sqrt{x} + 5 = 0$

10. $4x + 3\sqrt{x} = 1$

11. $3\sqrt[3]{x} - 5\sqrt[6]{x} + 2 = 0$

12. $\sqrt{x} + 3\sqrt[4]{x} = 28$

13. $x^{2/3} - 2x^{1/3} - 15 = 0$

14. $x^3 + 2x^{3/2} - 80 = 0$

15. $x^{1/2} + x^{1/4} = 20$

16. $4x^{4/3} + 9 = 13x^{2/3}$

17. $(x - 1) - \sqrt{x - 1} - 2 = 0$

18. $(x + 1)^{-2/3} + 5(x + 1)^{-1/3} - 6 = 0$

19. $(x^2 - 2x)^2 - 11(x^2 - 2x) + 24 = 0$

20. $3(x^2 + 3x)^2 - 2(x^2 + 3x) - 5 = 0$

21. $x - 3\sqrt{x - 2} = 6$ (Let $y = \sqrt{x - 2}$.)

22. $(x^2 - 1)^2 + (x^2 - 1)^{-2} = 2$

23. $x^6 + 7x^3 - 8 = 0$

24. $x^6 - 19x^3 = 216.$

In Exercises 25 through 28, solve the given problems.

25. The equivalent resistance R_T of two resistors R_1 and R_2 in parallel is given by $R_T^{-1} = R_1^{-1} + R_2^{-1}$. If $R_T = 1.00\ \Omega$ and $R_2 = \sqrt{R_1}$, find R_1 and R_2.

26. An equation used in the study of the dispersion of light is $\mu = A + B\lambda^{-2} + C\lambda^{-4}$. Solve for λ.

27. A rectangular floor has an area of 192 ft^2 and a diagonal of 20 ft. Find the dimensions of the floor.

28. A metal plate is in the shape of an isosceles triangle. The length of the base equals the square root of one of the equal sides. Determine the lengths of the sides if the perimeter of the plate is 55 cm.

13-4   Equations with Radicals

Equations with radicals in them are normally solved by squaring both sides of the equation if the radical represents a square root, or by a similar operation for the other roots. However, when we do this, we often introduce *extraneous roots*. Thus, it is very important that all solutions be checked in the original equation.

EXAMPLE A

Solve the equation $\sqrt{x - 4} = 2$.

By squaring both sides of the equation, we have

$$(\sqrt{x - 4})^2 = 2^2$$
$$x - 4 = 4$$
$$x = 8$$

This solution checks when put into the original equation.

EXAMPLE B _____ Solve the equation $2\sqrt{3x-1} = 3x$.

Squaring both sides of the equation gives us

$$(2\sqrt{3x-1})^2 = (3x)^2 \quad \longleftarrow \text{ don't forget to square the 2}$$

$$4(3x-1) = 9x^2$$

$$12x - 4 = 9x^2$$

$$9x^2 - 12x + 4 = 0$$

$$(3x-2)^2 = 0$$

$$x = \frac{2}{3} \quad \text{(double root)}$$

Checking this solution in the original equation, we have

$$2\sqrt{3(\tfrac{2}{3})-1} = 3(\tfrac{2}{3}), \qquad 2\sqrt{2-1} = 2, \qquad 2 = 2$$

■ Therefore, the solution $x = \frac{2}{3}$ checks.

EXAMPLE C _____ Solve the equation $\sqrt[3]{x-8} = 2$.

Cubing both sides of the equation, we have $x - 8 = 8$. Thus $x = 16$, which ■ checks.

If one side of the equation contains a radical as well as other terms, we *first isolate the radical.* That is, we rewrite the equation with the radical on one side and collect the other terms on the other side.

EXAMPLE D _____ Solve the equation $\sqrt{x-1} + 3 = x$.

We first isolate the radical by subtracting 3 from each side. This gives us

$$\sqrt{x-1} = x - 3$$

We now square both sides and proceed with the solution.

$$\text{NOTE} \triangleright \qquad (\sqrt{x-1})^2 = (x-3)^2 \qquad \begin{array}{l}\text{square the expression on each side,}\\ \text{not just the terms separately}\end{array}$$

$$x - 1 = x^2 - 6x + 9$$

$$x^2 - 7x + 10 = 0$$

$$(x-5)(x-2) = 0$$

$$x = 5 \quad \text{or} \quad x = 2$$

The solution $x = 5$ checks, but the solution $x = 2$ gives $4 = 2$. Thus, the solu-
■ tion is $x = 5$. The value $x = 2$ is an extraneous root.

EXAMPLE E _____ Solve the equation $\sqrt{x+1} + \sqrt{x-4} = 5$.

This is most easily solved by first isolating one of the radicals by placing the other radical on the right side of the equation. We then square both sides of the resulting equation.

$$\sqrt{x+1} = 5 - \sqrt{x-4} \quad \longleftarrow \text{ two terms}$$

$$\text{NOTE} \triangleright \qquad (\sqrt{x+1})^2 = (5 - \sqrt{x-4})^2$$

$$x + 1 = 25 - 10\sqrt{x-4} + (\sqrt{x-4})^2 \quad \longleftarrow \text{ be careful!}$$

Now, isolating the radical on one side of the equation and squaring again, we have

$$10\sqrt{x-4} = 20$$
$$\sqrt{x-4} = 2 \qquad \text{divide by 10}$$
$$x - 4 = 4 \qquad \text{square both sides}$$
$$x = 8$$

This solution checks.

We note again that in squaring $5 - \sqrt{x-4}$, we do not simply square 5 and $\sqrt{x-4}$. This is similar to $(5a-2)^2$ in Example B of Section 5-1.

EXAMPLE F ———— Solve the equation $\sqrt{x} - \sqrt[4]{x} = 2$.

We can solve this most easily by handling it as an equation in quadratic form. By letting $y = \sqrt[4]{x}$, we have

$$y^2 - y - 2 = 0$$
$$(y - 2)(y + 1) = 0$$
$$y = 2 \quad \text{or} \quad y = -1$$

Since $y = \sqrt[4]{x}$, we know that a negative value of y does not lead to a solution of the original equation. Therefore, we see that $y = -1$ cannot give us a solution. For the other value, $y = 2$, we have

$$\sqrt[4]{x} = 2 \quad \text{or} \quad x = 16$$

This checks, because $\sqrt{16} - \sqrt[4]{16} = 4 - 2 = 2$. Therefore, the only solution of the original equation is $x = 16$.

EXAMPLE G ———— A piece of sheet metal is being cut into the shape of a right triangle. Its perimeter is 60 cm and its area is 120 cm². Find the lengths of its three sides.

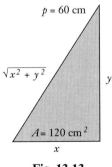

$p = 60$ cm

$\sqrt{x^2 + y^2}$

y

$A = 120$ cm²

x

Fig. 13-13

If we let the two legs of the triangle be x and y, as shown in Fig. 13-13, from the formulas for the perimeter p and the area A of a triangle, we have

$$p = x + y + \sqrt{x^2 + y^2} \quad \text{and} \quad A = \tfrac{1}{2}xy$$

where the hypotenuse was found by use of the Pythagorean theorem. Using the information given in the statement of the problem, we arrive at the equations

$$x + y + \sqrt{x^2 + y^2} = 60 \quad \text{and} \quad xy = 240$$

Isolating the radical in the first equation and then squaring both sides, we have

$$\sqrt{x^2 + y^2} = 60 - x - y$$
$$x^2 + y^2 = 3600 - 120x - 120y + x^2 + 2xy + y^2$$
$$0 = 3600 - 120x - 120y + 2xy$$

Solving the second of the original equations for y, we have $y = 240/x$. Substituting, we have

$$0 = 3600 - 120x - 120\left(\frac{240}{x}\right) + 480$$

(Continued on next page)

$$0 = 3600x - 120x^2 - 120(240) + 480x \qquad \text{multiply by } x$$
$$0 = 30x - x^2 - 240 + 4x \qquad \text{divide by } 120$$
$$x^2 - 34x + 240 = 0 \qquad \text{collect terms on left}$$
$$(x - 10)(x - 24) = 0$$
$$x = 10 \text{ cm} \quad \text{or} \quad x = 24 \text{ cm}$$

If $x = 10$ cm, then $y = 24$ cm, or if $x = 24$ cm, then $y = 10$ cm. Thus, the two legs are 10 cm and 24 cm and the hypotenuse is 26 cm. These sides give a perimeter of 60 cm and an area of 120 cm², which checks with the statement of the problem. ∎

Exercises 13-4

In Exercises 1 through 32, solve the given equations.

1. $\sqrt{x - 8} = 2$

2. $\sqrt{x + 4} = 3$

3. $\sqrt{8 - 2x} = x$

4. $\sqrt{3x + 4} = x$

5. $\sqrt{3x + 2} = 3x$

6. $\sqrt{2x + 6} = 2x$

7. $\sqrt{x - 2} + 3 = x$

8. $\sqrt{5x - 1} + 3 = x$

9. $2\sqrt{3 - x} - x = 5$

10. $x - 3\sqrt{2x + 1} = -5$

11. $\sqrt[3]{y - 5} = 3$

12. $\sqrt[4]{5 - x} = 2$

13. $\sqrt{x + 12} = x$

14. $\sqrt{x + 3} = 4x$

15. $5\sqrt{x + 3} = 2x$

16. $4\sqrt{x} = x + 3$

17. $\sqrt{x + 4} + 8 = x$

18. $\sqrt{x + 15} + 5 = x$

19. $\sqrt{5 + \sqrt{x}} = \sqrt{x} - 1$

20. $\sqrt{13 + \sqrt{x}} = \sqrt{x} + 1$

21. $3\sqrt{1 - 2t} + 1 = 2t$

22. $1 - 2\sqrt{y + 4} = y$

23. $2\sqrt{x + 2} - \sqrt{3x + 4} = 1$

24. $\sqrt{x - 1} + \sqrt{x + 2} = 3$

25. $\sqrt{5x + 1} - 1 = 3\sqrt{x}$

26. $\sqrt{2x + 1} + \sqrt{3x} = 11$

27. $\sqrt{2x - 1} - \sqrt{x + 11} = -1$

28. $\sqrt{5x - 4} - \sqrt{x} = 2$

29. $\sqrt[3]{2x - 1} = \sqrt[3]{x + 5}$

30. $\sqrt[4]{x + 10} = \sqrt{x - 2}$

31. $\sqrt{x - 2} = \sqrt[4]{x - 2} + 12$

32. $\sqrt{3x + \sqrt{3x + 4}} = 4$

In Exercises 33 through 36, solve for the indicated letter.

33. The resonant frequency f in an electric circuit with an inductance L and a capacitance C is given by $f = \dfrac{1}{2\pi\sqrt{LC}}$. Solve for L.

34. In the study of atomic spectra, the equation $v = \sqrt{\dfrac{2(nf - \phi)}{m}}$ is used. Solve for ϕ.

35. An equation used in analyzing a certain type of concrete beam is $k = \sqrt{2np + (np)^2} - np$. Solve for p.

36. In the study of spur gears in contact, the equation $kC = \sqrt{R_1^2 - R_2^2} + \sqrt{r_1^2 - r_2^2} - A$ is used. Solve for r_1^2.

In Exercises 37 through 40, set up the proper equations and solve them.

37. A triangular shelf support has a perimeter of 6.0 ft and sides of x, $x + 1$, and $\sqrt{2x + 1}$. Find the lengths of the sides of the support.

38. The velocity v of an object which falls through a distance h is given by $v = \sqrt{2gh}$, where g is the acceleration due to gravity. Two objects are dropped from heights which differ by 10.0 m such that the sum of their velocities when they strike the ground is 20.0 m/s. Find the heights from which they are dropped if $g = 9.80$ m/s.

39. An island is 3.0 mi offshore from the nearest point P on a straight beach. A person in a motorboat travels straight from the island to a point on the beach x miles from P, and then travels x miles farther along the beach away from P. Find x if the person traveled a total of 8.0 mi.

40. The length of the roller belt in Fig. 13-14 is 28.0 ft. Find x. **Fig. 13-14**

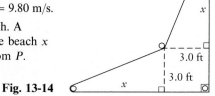

13-5   Chapter Equation, Review Exercises, and Practice Test

Chapter Equation

General quadratic equation $ax^2 + bxy + cy^2 + dx + ey + f = 0$ (13-1)

Review Exercises

In Exercises 1 through 10, solve the given systems of equations graphically.

1. $x + 2y = 6$
  $y = 4x^2$

2. $x + y = 3$
  $x^2 + y^2 = 25$

3. $3x + 2y = 6$
  $x^2 + 4y^2 = 4$

4. $x^2 - 2y = 0$
  $y = 3x - 5$

5. $y = x^2 + 1$
  $2x^2 + y^2 = 4$

6. $\dfrac{x^2}{4} + y^2 = 1$
  $x^2 - y^2 = 1$

7. $y = 4 - x^2$
  $y = 2x^2$

8. $xy = -2$
  $y = 1 - 2x^2$

9. $y = x^2 - 2x$
  $y = 1 - e^{-x}$

10. $y = \ln x$
  $y = \sin x$

In Exercises 11 through 20, solve the given systems of equations algebraically.

11. $y = 4x^2$
  $y = 8x$

12. $x + y = 2$
  $xy = 1$

13. $2y = x^2$
  $x^2 + y^2 = 3$

14. $y = x^2$
  $2x^2 - y^2 = 1$

15. $4x^2 + y = 3$
  $2x + 3y = 1$

16. $2x^2 + y^2 = 3$
  $x + 2y = 1$

17. $4x^2 - 7y^2 = 21$
  $x^2 + 2y^2 = 99$

18. $3x^2 + 2y^2 = 11$
  $2x^2 - y^2 = 30$

19. $4x^2 + 3xy = 4$
  $x + 3y = 4$

20. $\dfrac{6}{x} + \dfrac{3}{y} = 4$
  $\dfrac{36}{x^2} + \dfrac{36}{y^2} = 13$

In Exercises 21 through 40, solve the given equations.

21. $x^4 - 20x^2 + 64 = 0$

22. $x^6 - 26x^3 - 27 = 0$

23. $x^{3/2} - 9x^{3/4} + 8 = 0$

24. $x^{1/2} + 3x^{1/4} - 28 = 0$

25. $x^{-2} + 4x^{-1} - 21 = 0$

26. $4x^{-4} + 35x^{-2} = 9$

27. $2x - 3\sqrt{x} - 5 = 0$

28. $x^{-1} + x^{-1/2} = 6$

29. $\left(\dfrac{1}{x+1}\right)^2 - \dfrac{1}{x+1} = 2$

30. $(x^2 + 5x)^2 - 5(x^2 + 5x) = 6$

31. $\sqrt{x + 5} = 4$

32. $\sqrt[3]{x - 2} = 3$

33. $\sqrt{5x - 4} = x$

34. $\sqrt{6x + 8} = 3x$

35. $\sqrt{5x + 9} + 1 = x$

36. $2\sqrt{5x - 3} - 1 = 2x$

37. $\sqrt{x + 1} + \sqrt{x} = 2$

38. $\sqrt{3 + x} + \sqrt{3x - 2} = 1$

39. $\sqrt{x + 4} + 2\sqrt{x + 2} = 3$

40. $\sqrt{3x - 2} - \sqrt{x + 7} = 1$

In Exercises 41 through 46, solve for the indicated quantities.

41. In the study of atomic structure, the equation $L = \dfrac{h}{2\pi}\sqrt{\ell(\ell + 1)}$ is used. Solve for $\ell(\ell > 0)$.

42. Under certain conditions, the frequency ω of an *RLC* circuit is given by

$$\omega = \frac{\sqrt{R^2 + 4(L/C)} + R}{2L}$$ Solve for *C*.

43. In the theory dealing with a suspended cable, the equation $y = \sqrt{s^2 - m^2} - m$ is used. Solve for *m*.

44. The equation $V = e^2cr^{-2} - e^2Zr^{-1}$ is used in spectroscopy. Solve for *r*.

45. In an experiment, an object is allowed to fall, stops, and then falls for twice the initial time. The total distance the object falls is 45 ft. The equations relating the times t_1 and t_2, in seconds, of fall are $16t_1^2 + 16t_2^2 = 45$, and $t_2 = 2t_1$. Find the times of fall.

46. If two objects collide and the kinetic energy remains constant, the collision is termed perfectly elastic. Under these conditions, if an object of mass m_1 and initial velocity u_1 strikes a second object (initially at rest) of mass m_2, such that the velocities after collision are v_1 and v_2, the following equations are found:

$$m_1u_1 = m_1v_1 + m_2v_2$$
$$\tfrac{1}{2}m_1u_1^2 = \tfrac{1}{2}m_1v_1^2 + \tfrac{1}{2}m_2v_2^2$$

Solve these equations for m_2 in terms of u_1, v_1, and m_1.

In Exercises 47 through 56, set up the appropriate equations and solve them.

47. A wrench is dropped by a worker at a construction site. Four seconds later the worker hears it hit the ground below. How high is the worker above the ground? (The velocity of sound is 1100 ft/s, and the distance the wrench falls as a function of time is $s = 16t^2$.)

48. A rectangular door has an area of 19.5 ft^2 and a perimeter of 19.0 ft. Find the width and height of the door.

49. In a certain electric circuit the impedance Z is twice the square of the reactance X, and the resistance R is 0.800 Ω. Find the impedance and reactance. See Section 11-7.

50. The perimeter of the machine part shown in Fig. 13-15 is 10.0 cm. Find x.

51. A rectangular television screen has an area of 173 in.^{2} and a diagonal of 19.0 in. Find the width and height of the screen.

52. The circular solar cell and square solar cell shown in Fig. 13-16 have a combined surface area of 40.0 cm^2. Find the radius of the circular cell and the side of the square cell.

53. A trough is made from a piece of sheet metal 12.0 in. wide. The cross section of the trough is shown in Fig. 13-17. Find x.

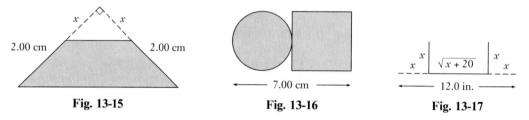

Fig. 13-15   **Fig. 13-16**   **Fig. 13-17**

54. A plastic band 19.0 cm long is bent into the shape of a triangle with sides of $\sqrt{x-1}$, $\sqrt{5x-1}$, and 9. Find x.

55. A Coast Guard ship travels from port A to port B, and later it returns to port A at a speed which is 6.0 mi/h faster. If port A is 420 mi from port B, and the total travel time is 29 h, find the speed of the ship in each direction.

56. Two trains are approaching the same crossing on tracks which are at right angles to each other. Each is traveling at 60.0 km/h. If one is 6.00 km from the crossing when the other is 3.00 km from it, how much later will they be 4.00 km apart?

Practice Test

1. Solve for x: $x^{1/2} - 2x^{1/4} = 3$.
2. Solve for x: $3\sqrt{x-2} - \sqrt{x+1} = 1$.
3. Solve for x and y algebraically:

$$x^2 - 2y = 5$$
$$2x + 6y = 1$$

4. The velocity v of an object falling under the influence of gravity in terms of its initial velocity v_0, the acceleration due to gravity g, and the height h fallen is given by $v = \sqrt{v_0^2 + 2gh}$. Solve for h.
5. Solve for x and y graphically:

$$x^2 - y^2 = 4$$
$$xy = 2$$

6. A rectangular desktop has a perimeter of 14.0 ft and an area of 10.0 ft^2. Find the length and width of the desktop.

14 Equations of Higher Degree

In Section 14-4 we solve a higher-degree equation relating the force and displacement of a spring system.

In previous chapters we have discussed methods of solving many kinds of equations. Except for special cases, however, we have not solved polynomial equations of degree higher than two (a polynomial equation of the first degree is a linear equation, and a polynomial equation of the second degree is a quadratic equation). In this chapter we shall develop certain methods for solving polynomial equations, especially the higher-degree equations. Since we shall be discussing equations involving only polynomials, in this chapter $f(x)$ will be assumed to be a polynomial.

Applications of higher-degree equations arise in a number of technical areas. Included among these are electricity, business analysis, beam structure, and mechanical oscillating systems.

14-1 The Remainder Theorem and the Factor Theorem

Any polynomial is a function of the form

$$f(x) = a_0 x^n + a_1 x^{n-1} + \cdots + a_n \tag{14-1}$$

where $a_0 \neq 0$ and n is a positive integer or zero. We will consider only polynomials in which the coefficients $a_0, a_1, \ldots a_n$ are real numbers.

If we divide a polynomial by $x - r$, we find a result of the following form:

$$f(x) = (x - r)q(x) + R \qquad (14\text{-}2)$$

where $q(x)$ is the quotient and R is the remainder.

EXAMPLE A Divide $f(x) = 3x^2 + 5x - 8$ by $x - 2$.

$$
\begin{array}{r}
3x \; + 11 \\
x - 2\overline{\smash{)}\,3x^2 + \; 5x - \; 8} \\
3x^2 - \; 6x \\
\hline
11x - \; 8 \\
11x - 22 \\
\hline
14
\end{array}
$$

Thus,

$$3x^2 + 5x - 8 = (x - 2)(3x + 11) + 14$$

where, for this function $f(x)$ with $r = 2$, we identify $q(x)$ and R as

$$q(x) = 3x + 11, \qquad R = 14$$

If we now set $x = r$ in Eq. (14-2), we have $f(r) = q(r)(r - r) + R$, or

$$f(r) = R \qquad (14\text{-}3)$$

remainder theorem

The equation above states that the remainder equals the value of the function of x at $x = r$. This leads us to the **remainder theorem,** which states that *if a polynomial $f(x)$ is divided by $x - r$ until a constant remainder (R) is obtained, then $f(r) = R$.*

EXAMPLE B In Example A, $f(x) = 3x^2 + 5x - 8$, $R = 14$, $r = 2$
We find that

$$f(2) = 3(4) + 5(2) - 8 = 14$$

Thus, $f(2) = 14$ verifies that $f(r) = R$.

EXAMPLE C By using the remainder theorem, determine the remainder when we divide $3x^3 - x^2 - 20x + 5$ by $x + 4$.

In using the remainder theorem, we determine the remainder when the function is divided by $x - r$ by evaluating the function for $x = r$. To have $x + 4$ in the proper form to identify r, we write it as $x - (-4)$. This means that $r = -4$, and we therefore are to evaluate the function $f(x) = 3x^3 - x^2 - 20x + 5$ for $x = -4$, or find $f(-4)$. Thus,

$$f(-4) = 3(-4)^3 - (-4)^2 - 20(-4) + 5 = -192 - 16 + 80 + 5$$
$$= -123$$

Thus, the remainder when $3x^3 - x^2 - 20x + 5$ is divided by $x + 4$ is -123.

In using the remainder theorem, it is convenient to use a calculator to evaluate the function. In doing so, we use the $\boxed{x^y}$ key to evaluate the higher powers. On some calculators, use of this key will show an error if the base, x, is negative, as in Example C. If your calculator does this, use x as a positive value and change the sign of the value if the power, y, is odd.

factor theorem

The remainder theorem leads immediately to another important theorem known as the **factor theorem.** The factor theorem states that *if $f(r) = R = 0$, then $x - r$ is a factor of $f(x)$.* Inspection of Eq. (14-2) justifies this theorem. Recalling material from earlier chapters, for the function $f(x)$ we see that

$$if\ f(r) = 0,\ then\ x = r\ is\ a\ \pmb{zero}\ of f(x),$$
$$x - r\ is\ a\ \pmb{factor}\ of\ f(x),\ and$$
$$x = r\ is\ a\ \pmb{root}\ of\ the\ equation\ f(x) = 0$$

EXAMPLE D

Is $x + 1$ a factor of $f(x) = x^3 + 2x^2 - 5x - 6$?
Here $r = -1$, and thus

$$f(-1) = -1 + 2 + 5 - 6 = 0$$

■ Therefore, since $f(-1) = 0$, $x + 1$ is a factor of $f(x)$.

EXAMPLE E

The expression $x + 2$ is not a factor of the function in Example D, since

$$f(-2) = -8 + 8 + 10 - 6 = 4$$

■ But $x - 2$ is a factor, since $f(2) = 8 + 8 - 10 - 6 = 0$.

EXAMPLE F

Determine if $\frac{2}{3}$ is a zero of the function $f(x) = 3x^3 + 4x^2 - 16x + 8$.
Evaluating $f(\frac{2}{3})$ we have

$$f\left(\frac{2}{3}\right) = 3\left(\frac{2}{3}\right)^3 + 4\left(\frac{2}{3}\right)^2 - 16\left(\frac{2}{3}\right) + 8$$

$$= \frac{8}{9} + \frac{16}{9} - \frac{32}{3} + 8 = \frac{8 + 16 - 96 + 72}{9}$$

$$= 0$$

■ Since $f(\frac{2}{3}) = 0$, $\frac{2}{3}$ is a zero of the function.

We now have one way of determining whether or not an expression of the form $x - r$ is a factor of a function $f(x)$. By finding $f(r)$, we can determine whether or not $x - r$ is a factor and whether or not r is a zero of the function.

Exercises 14-1

In Exercises 1 through 8, find the remainder R by long division and by the remainder theorem.

1. $(x^3 + 2x^2 - x - 2) \div (x - 1)$

2. $(x^3 - 3x^2 - x + 2) \div (x - 2)$

3. $(x^3 + 2x + 3) \div (x + 1)$

4. $(x^4 - 4x^3 - x^2 + x - 100) \div (x + 3)$

5. $(2x^5 - x^2 + 8x + 44) \div (x + 2)$

6. $(x^3 + 4x^2 - 25x - 98) \div (x - 5)$

7. $(3x^4 - 9x^3 - x^2 + 5x - 10) \div (x - 3)$

8. $(2x^4 - 10x^2 + 30x - 60) \div (x + 4)$

In Exercises 9 through 16, find the remainder using the remainder theorem.

9. $(x^3 + 2x^2 - 3x + 4) \div (x + 1)$

10. $(2x^3 - 4x^2 + x - 1) \div (x + 2)$

11. $(x^4 + x^3 - 2x^2 - 5x + 3) \div (x + 4)$

12. $(4x^4 - x^2 + 5x - 7) \div (x - 3)$

13. $(2x^4 - 7x^3 - x^2 + 8) \div (x - 3)$

14. $(x^4 - 5x^3 + x^2 - 2x + 6) \div (x + 4)$

15. $(x^5 - 3x^3 + 5x^2 - 10x + 6) \div (x - 2)$

16. $(3x^4 - 12x^3 - 60x + 4) \div (x - 5)$

In Exercises 17 through 24, use the factor theorem to determine whether or not the second expression is a factor of the first.

17. $x^2 - 2x - 3, \; x - 3$

18. $3x^3 + 2x^2 - 3x - 2, \; x + 2$

19. $4x^3 + x^2 - 16x - 4, \; x - 2$

20. $3x^3 + 14x^2 + 7x - 4, \; x + 4$

21. $5x^3 - 3x^2 + 4, \; x - 2$

22. $x^5 - 2x^4 + 3x^3 - 6x^2 - 4x + 8, \; x - 2$

23. $x^6 + 1, \; x + 1$

24. $x^7 - 128, \; x + 2$

In Exercises 25 through 28, determine whether or not the given numbers are zeros of the given functions.

25. $f(x) = x^3 - 2x^2 - 9x + 18; \quad 2$

26. $f(x) = 2x^3 + 3x^2 - 8x - 12; \quad -\frac{3}{2}$

27. $f(x) = 4x^4 - 4x^3 + 23x^2 + x - 6; \quad \frac{1}{2}$

28. $f(x) = 2x^4 + 3x^3 - 12x^2 - 7x + 6; \quad -3$

In Exercises 29 through 32, answer the given questions.

29. By division, show that $2x - 1$ is a factor of $f(x) = 4x^3 + 8x^2 - x - 2$. May we therefore conclude that $f(1) = 0$?

30. By division, show that $x^2 + 2$ is a factor of $f(x) = 3x^3 - x^2 + 6x - 2$. May we therefore conclude that $f(-2) = 0$?

31. For what value of k is $x - 2$ a factor of $f(x) = 2x^3 + kx^2 - x + 14$?

32. For what value of k is $x + 1$ a factor of $f(x) = 3x^4 + 3x^3 + 2x^2 + kx - 4$?

14-2   Synthetic Division

We shall now develop a method which greatly simplifies the procedure for dividing a polynomial by an expression of the form $x - r$. Using **synthetic division**, which is an abbreviated form of long division, we can determine the coefficients of the quotient as well as the remainder. Of course, for some values of r, we can easily calculate $f(r)$ directly. However, if the degree of the equation is high, this requires finding and combining higher powers of r. Synthetic division therefore allows us to find $f(r)$ easily by finding the remainder. The method is developed in the following example.

EXAMPLE A

Divide $x^4 + 4x^3 - x^2 - 16x - 14$ by $x - 2$.

We shall first perform this division in the usual manner.

$$
\begin{array}{r}
x^3 + 6x^2 + 11x + 6 \\
x - 2 \overline{\smash{\big)}\ x^4 + 4x^3 - x^2 - 16x - 14} \\
\underline{x^4 - 2x^3} \\
6x^3 - x^2 \\
\underline{6x^3 - 12x^2} \\
11x^2 - 16x \\
\underline{11x^2 - 22x} \\
6x - 14 \\
\underline{6x - 12} \\
-2
\end{array}
$$

We now note that, when we performed this division, we repeated many terms. Also, the only quantities of importance in the function being divided are the coefficients. There is no real need to put in the powers of x all the time. Therefore, we shall now write the above example without any x's and also eliminate the writing of identical terms:

$$
\begin{array}{r}
\;1 \quad\;\; 6 \quad\;\; 11 \quad\;\; 6 \\
\hline
-2\,\big|\;1 \quad\;\; 4 \quad -1 \quad -16 \quad -14 \\
\;-2 \\
\hline
\;6 \\
\;-12 \\
\hline
\;11 \\
\;-22 \\
\hline
\;6 \\
\;-12 \\
\hline
\;-2
\end{array}
$$

All but the first of the numbers which represent coefficients of the quotient are repeated below. Also, all the numbers below the dividend may be written in two lines. Thus, we have the following form:

$$
\begin{array}{r|rrrr}
-2 & 1 & 4 & -1 & -16 & -14 \\
 & & -2 & -12 & -22 & -12 \\
\hline
 & 6 & 11 & 6 & -2
\end{array}
$$

All the coefficients of the actual quotient appear except the first, so we shall now repeat the 1 in the bottom line. Also, we shall change -2 to 2, which is the actual value of r. Then, writing r to the right, we have the following form:

$$
\begin{array}{rrrrr|r}
1 & 4 & -1 & -16 & -14 & \underline{2} \\
 & -2 & -12 & -22 & -12 \\
\hline
1 & 6 & 11 & 6 & -2
\end{array}
$$

In this form the 1, 6, 11, and 6 represent the coefficients of the x^3, x^2, x, and constant terms of the quotient. The -2 is the remainder. Finally, we find it easier to use addition rather than subtraction in the process, so we change the signs of the numbers in the middle row. Remember that originally the bottom line was found by subtraction. Thus, we have

$$
\begin{array}{rrrrr|r}
1 & 4 & -1 & -16 & -14 & \underline{2} \\
 & 2 & 12 & 22 & 12 \\
\hline
1 & 6 & 11 & 6 & -2
\end{array}
$$

When we inspect this form we find the following: The 1 multiplied by the 2 (r) gives 2, which is the first number of the middle row. The 4 and 2 (of the middle row) added is 6, which is the second number in the bottom row. The 6 multiplied by $2(r)$ is 12, which is the second number in the middle row. This 12 and the -1 give 11. The 11 multiplied by 2 is 22. The 22 added to -16 is 6. This 6 multiplied by 2 gives 12. This 12 added to -14 is -2. When this process is followed in general, *the method is called **synthetic division**.*

**division by
synthetic division**

Generalizing on this last example, we have the following procedure: We write down the coefficients of $f(x)$, being certain that the powers are in descending order and that zeros are inserted for missing powers. We write the value of r to the right. We carry down the left coefficient, multiply this number by r, and place this product under the second coefficient of the top line. We add the two numbers in this second column and place the result below; then we multiply this number by r and place the result under the third coefficient of the top line. We continue until the bottom row has as many numbers as the top row. The last number in the bottom row is the remainder, the other numbers being the respective coefficients of the quotient. The first term of the quotient is of degree one less than the dividend.

EXAMPLE B

By synthetic division, divide $x^5 + 2x^4 - 4x^2 + 3x - 4$ by $x + 3$.

Since the powers of x are in descending order, we write down the coefficients of x. In doing so we must be certain to include a zero for the missing x^3 term. Next we note that the divisor is $x + 3$, which means that $r = -3$. The -3 is placed to the right. This gives us a top line of

$$\text{coefficients} \longrightarrow 1 \quad 2 \quad 0 \quad -4 \quad 3 \quad -4 \quad \underline{-3} \longleftarrow r$$

Next we carry the left coefficient, 1, to the bottom line and multiply it by r, -3, placing the product, -3, in the middle line under the second coefficient, 2. We then add the 2 and the -3 and place the result, -1, below. This gives

Now we multiply the -1 by r, -3, and place the result, 3, in the middle line under the zero. We now add, and continue the process, obtaining the following result:

$$
\begin{array}{rrrrrr|r}
1 & 2 & 0 & -4 & 3 & -4 & \underline{-3} \\
 & -3 & 3 & -9 & 39 & -126 & \\
\hline
1 & -1 & 3 & -13 & 42 & -130 &
\end{array}
$$

coefficients and constant of quotient ⟶ remainder

Since the degree of the dividend is 5, the degree of the quotient is 4. Thus, the quotient is $x^4 - x^3 + 3x^2 - 13x + 42$ and the remainder is -130.

EXAMPLE C

By synthetic division, divide $3x^4 - 5x + 6$ by $x - 4$.

$$
\begin{array}{rrrrr|r}
3 & 0 & 0 & -5 & 6 & \underline{4} \\
 & 12 & 48 & 192 & 748 & \\
\hline
3 & 12 & 48 & 187 & 754 &
\end{array}
$$

The quotient is $3x^3 + 12x^2 + 48x + 187$ and the remainder is 754.

EXAMPLE D By synthetic division, determine whether or not $x - 4$ is a factor of $x^4 + 2x^3 - 15x^2 - 32x - 16$.

$$
\begin{array}{rrrrr|r}
1 & 2 & -15 & -32 & -16 & \underline{4} \\
& 4 & 24 & 36 & 16 & \\
\hline
1 & 6 & 9 & 4 & 0 &
\end{array}
$$

Since the remainder is zero, $x - 4$ is a factor. We may also conclude that $f(x) = (x - 4)(x^3 + 6x^2 + 9x + 4)$, since the bottom line gives us the coefficients in the quotient. ∎

EXAMPLE E By using synthetic division, determine whether $2x - 3$ is a factor of $2x^3 - 3x^2 + 8x - 12$.

We first note that the coefficient of x in the possible factor is not 1. Thus,

NOTE▷ *we cannot use $r = 3$, since the factor is not of the form $x - r$.* However, $2x - 3 = 2(x - \frac{3}{2})$, which means that if $2(x - \frac{3}{2})$ is a factor of the function, $2x - 3$ is a factor. If we use $r = \frac{3}{2}$, and find that the remainder is zero, then $x - \frac{3}{2}$ is a factor.

$$
\begin{array}{rrrr|r}
2 & -3 & 8 & -12 & \dfrac{3}{2} \\
& 3 & 0 & 12 & \\
\hline
2 & 0 & 8 & 0 &
\end{array}
$$

Since the remainder is zero, $x - \frac{3}{2}$ is a factor. Also, the quotient is $2x^2 + 8$, which may be factored into $2(x^2 + 4)$. Thus, 2 is also a factor of the function. This means that $2(x - \frac{3}{2})$ is a factor of the function, and this in turn means that $2x - 3$ is a factor. This tells us that

$$2x^3 - 3x^2 + 8x - 12 = (2x - 3)(x^2 + 4)$$
∎

EXAMPLE F By synthetic division, determine whether or not $\frac{1}{3}$ is a zero of the function $3x^3 + 2x^2 - 4x + 1$.

This problem is equivalent to dividing the function by $x - \frac{1}{3}$. If the remainder is zero, $\frac{1}{3}$ is a zero of the function.

$$
\begin{array}{rrrr|r}
3 & 2 & -4 & 1 & \dfrac{1}{3} \\
& 1 & 1 & -1 & \\
\hline
3 & 3 & -3 & 0 &
\end{array}
$$

Since the remainder is zero, we conclude that $\frac{1}{3}$ is a zero of the function.

$$3x^3 + 2x^2 - 4x + 1 = \left(x - \frac{1}{3}\right)(3x^2 + 3x - 3)$$

$$= 3\left(x - \frac{1}{3}\right)(x^2 + x - 1)$$

EXAMPLE G _____ | Determine whether or not -12.5 is a zero of the function
$f(x) = 6x^3 + 61x^2 - 171x + 100$.

If -12.5 is a zero of $f(x)$, then $x - (-12.5)$, or $x + 12.5$, is a factor of $f(x)$, and $f(-12.5) = 0$. We can find the remainder by direct use of the remainder theorem or by synthetic division.

Using synthetic division and a calculator to make the calculations, we have the following setup and calculator sequence:

$$
\begin{array}{ccccc}
6 & 61 & -171 & 100 & \underline{\,|\,-12.5} \\
 & -75 & 175 & -50 & \\
\hline
6 & -14 & 4 & 50 &
\end{array}
$$

12.5 $\boxed{+/-}$ $\boxed{\text{STO}}$ $\boxed{\times}$ 6 $\boxed{+}$ 61 $\boxed{=}$ $\boxed{\times}$ $\boxed{\text{RCL}}$ $\boxed{+}$ 171 $\boxed{+/-}$ $\boxed{=}$ $\boxed{\times}$ $\boxed{\text{RCL}}$ $\boxed{+}$ 100 $\boxed{=}$ $\boxed{50.}$

The arrows indicate the calculator steps which give the displays for the given values. Since the remainder is 50, not zero, -12.5 is not a zero of $f(x)$.

Exercises 14-2

In Exercises 1 through 20, perform the required divisions by synthetic division. Exercises 1 through 16 are the same as those of Section 14-1.

1. $(x^3 + 2x^2 - x - 2) \div (x - 1)$

2. $(x^3 - 3x^2 - x + 2) \div (x - 2)$

3. $(x^3 + 2x + 3) \div (x + 1)$

4. $(x^4 - 4x^3 - x^2 + x - 100) \div (x + 3)$

5. $(2x^5 - x^2 + 8x + 44) \div (x + 2)$

6. $(x^3 + 4x^2 - 25x - 98) \div (x - 5)$

7. $(3x^4 - 9x^3 - x^2 + 5x - 10) \div (x - 3)$

8. $(2x^4 - 10x^2 + 30x - 60) \div (x + 4)$

9. $(x^3 + 2x^2 - 3x + 4) \div (x + 1)$

10. $(2x^3 - 4x^2 + x - 1) \div (x + 2)$

11. $(x^4 + x^3 - 2x^2 - 5x + 3) \div (x + 4)$

12. $(4x^4 - x^2 + 5x - 7) \div (x - 3)$

13. $(2x^4 - 7x^3 - x^2 + 8) \div (x - 3)$

14. $(x^4 - 5x^3 + x^2 - 2x + 6) \div (x + 4)$

15. $(x^5 - 3x^3 + 5x^2 - 10x + 6) \div (x - 2)$

16. $(3x^4 - 12x^3 - 60x + 4) \div (x - 5)$

17. $(x^6 + 2x^2 - 6) \div (x - 2)$

18. $(x^5 + 4x^4 - 8) \div (x + 1)$

19. $(x^7 - 128) \div (x - 2)$

20. $(20x^4 + 11x^3 - 89x^2 + 60x - 77) \div (x + 2.75)$

See Appendix E for a computer program for synthetic division.

In Exercises 21 through 32, use the factor theorem and synthetic division to determine whether or not the second expression is a factor of the first.

21. $x^3 + x^2 - x + 2;\quad x + 2$

22. $x^3 + 6x^2 + 10x + 6;\quad x + 3$

23. $x^4 - 6x^2 - 3x - 2;\quad x - 3$

24. $2x^4 - 5x^3 - 24x^2 + 5;\quad x - 5$

25. $2x^5 - x^3 + 3x^2 - 4;\quad x + 1$

26. $x^5 - 3x^4 - x^2 - 6;\quad x - 3$

27. $4x^3 - 6x^2 + 2x - 2;\quad x - \frac{1}{2}$

28. $3x^3 - 5x^2 + x + 1;\quad x + \frac{1}{3}$

29. $2x^4 - x^3 + 2x^2 - 3x + 1;\quad 2x - 1$

30. $6x^4 + 5x^3 - x^2 + 6x - 2;\quad 3x - 1$

31. $4x^4 + 2x^3 - 8x^2 + 3x + 12;\quad 2x + 3$

32. $3x^4 - 2x^3 + x^2 + 15x + 4;\quad 3x + 4$

In Exercises 33 through 36, use synthetic division to determine whether or not the given numbers are zeros of the given functions.

33. $x^4 - 5x^3 - 15x^2 + 5x + 14;\quad 7$

34. $x^4 + 7x^3 + 12x^2 + x + 4;\quad -4$

35. $85x^3 + 348x^2 - 263x + 120;\quad -4.8$

36. $2x^3 + 13x^2 + 10x - 4;\quad \frac{1}{2}$

14-3   The Roots of an Equation

In this section we shall present certain theorems which are useful in determining the number of roots in the equation $f(x) = 0$, and the nature of some of these roots. In dealing with polynomial equations of higher degree, it is helpful to have as much of this kind of information as is readily obtainable before proceeding to solve for the roots.

The first of these theorems is so important that is is called **the fundamental theorem of algebra.** It states that *every polynomial equation has at least one (real or complex) root.* The proof of this theorem is of an advanced nature, and therefore we must accept its validity at this time. However, using the fundamental theorem, we can show the validity of other useful theorems.

Let us now assume that we have a polynomial equation $f(x) = 0$, and that we are looking for its roots. By the fundamental theorem, we know that it has at least one root. Assuming that we can find this root by some means (the factor theorem, for example), we shall call this root r_1. Thus,

$$f(x) = (x - r_1)f_1(x)$$

where $f_1(x)$ is the polynomial quotient found by dividing $f(x)$ by $(x - r_1)$. However, since the fundamental theorem states that any polynomial equation has at least one root, this must apply to $f_1(x) = 0$ as well. Let us assume that $f_1(x) = 0$ has the root r_2. Therefore, this means that $f(x) = (x - r_1)(x - r_2)f_2(x)$. Continuing this process until one of the quotients is a constant a, we have

$$f(x) = a(x - r_1)(x - r_2) \cdots (x - r_n)$$

Note that one linear factor appears each time a root is found, and that the degree of the quotient is one less each time. Thus there are n factors, if the degree of $f(x)$ is n. This leads us to two theorems. The first of these states that

each polynomial of the nth degree can be factored into n linear factors

The second theorem states that

each polynomial equation of degree n has exactly n roots

EXAMPLE A

Consider the equation $f(x) = 2x^4 - 3x^3 - 12x^2 + 7x + 6 = 0$.

$$2x^4 - 3x^3 - 12x^2 + 7x + 6 = (x - 3)(2x^3 + 3x^2 - 3x - 2)$$
$$2x^3 + 3x^2 - 3x - 2 = (x + 2)(2x^2 - x - 1)$$
$$2x^2 - x - 1 = (x - 1)(2x + 1)$$
$$2x + 1 = 2\left(x + \frac{1}{2}\right)$$

Therefore,

$$2x^4 - 3x^3 - 12x^2 + 7x + 6 = 2(x - 3)(x + 2)(x - 1)\left(x + \frac{1}{2}\right) = 0$$

The degree of $f(x)$ is 4. There are 4 linear factors: $(x - 3)$, $(x + 2)$, $(x - 1)$, and $(x + \frac{1}{2})$. There are 4 roots of the equation: 3, -2, 1, and $-\frac{1}{2}$. Thus, we have verified each of the theorems above for this example.

It is not necessary for each root of an equation to be different from the others. For example, the equation $(x - 1)^2 = 0$ has two roots, both of which are 1. Such roots are referred to as multiple roots.

When we solve the equation $x^2 + 1 = 0$, we get two roots, j and $-j$. In fact, if we have any equation (with real coefficients) for which the roots are complex, for every root of the form $a + bj$ $(b \neq 0)$ there is also a root of the form $a - bj$. This is so because any quadratic equation can be solved by the quadratic formula. The solutions from the quadratic formula (for an equation of the form $ax^2 + bx + c = 0$) are

$$\frac{-b + \sqrt{b^2 - 4ac}}{2a} \quad \text{and} \quad \frac{-b - \sqrt{b^2 - 4ac}}{2a}$$

and the only difference between these roots is the sign before the radical. Thus we have the following theorem.

If the coefficients of the equation $f(x) = 0$ are real and $a + bj$ $(b \neq 0)$ is a complex root, then its conjugate, $a - bj$, is also a root.

EXAMPLE B

Consider the equation $f(x) = (x - 1)^3(x^2 + x + 1) = 0$.

We observe directly (since three factors of $x - 1$ are already indicated) that there is a triple root of 1. To find the other two roots, we use the quadratic formula on the *factor* $(x^2 + x + 1)$. This is permissible, because what we are actually finding are those values of x which make

$$x^2 + x + 1 = 0$$

For this we have

$$x = \frac{-1 \pm \sqrt{1 - 4}}{2}$$

Thus,

$$x = \frac{-1 + j\sqrt{3}}{2} \quad \text{and} \quad x = \frac{-1 - j\sqrt{3}}{2}$$

Therefore, the roots of $f(x)$ are

$$1, \quad 1, \quad 1, \quad \frac{-1 + j\sqrt{3}}{2}, \quad \frac{-1 - j\sqrt{3}}{2}$$

One further observation can be made from Example B. *Whenever enough roots are known so that the remaining factor is quadratic, it is always possible to find the remaining roots from the quadratic formula.* This is true for finding real or complex roots. If there are n roots and if we find $n - 2$ of these roots, the solution may be completed by using the quadratic formula.

EXAMPLE C

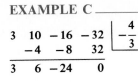

$$3 \quad 10 \quad -16 \quad -32 \quad \underline{\left|-\dfrac{4}{3}\right.}$$
$$\underline{\quad\;\; -4 \quad -8 \quad 32\;}$$
$$3 \quad\;\; 6 \quad -24 \quad\;\; 0$$

Solve the equation $3x^3 + 10x^2 - 16x - 32 = 0$ given that $-\frac{4}{3}$ is a root.

   Using synthetic division and the given root we have the setup shown at the left. From this we see that

$$3x^3 + 10x^2 - 16x - 32 = (x + \tfrac{4}{3})(3x^2 + 6x - 24)$$

We know that $x + \frac{4}{3}$ is a factor from the given root, and that $3x^2 + 6x - 24$ is a factor found from the synthetic division. This second factor itself can be factored as follows:

$$3x^2 + 6x - 24 = 3(x^2 + 2x - 8)$$
$$= 3(x + 4)(x - 2)$$

Therefore we have

$$3x^3 + 10x^2 - 16x - 32 = 3\left(x + \frac{4}{3}\right)(x + 4)(x - 2)$$

■ This means the roots are $-\frac{4}{3}$, -4, and 2.

EXAMPLE D

$$1 \quad 3 \quad -4 \quad -10 \quad -4 \quad \underline{\left|-1\right.}$$
$$\underline{\quad\; -1 \quad -2 \quad\;\; 6 \quad\;\; 4\;}$$
$$1 \quad 2 \quad -6 \quad -4 \quad\;\; 0$$

$$1 \quad 2 \quad -6 \quad -4 \quad \underline{\left|2\right.}$$
$$\underline{\quad\;\; 2 \quad\;\; 8 \quad\;\; 4\;}$$
$$1 \quad 4 \quad\;\; 2 \quad\;\; 0$$

Solve the equation $x^4 + 3x^3 - 4x^2 - 10x - 4 = 0$, given that -1 and 2 are roots.

   Using synthetic division and the root -1, we have the first setup shown at the left. This tells us that

$$x^4 + 3x^3 - 4x^2 - 10x - 4 = (x + 1)(x^3 + 2x^2 - 6x - 4)$$

We now know that $x - 2$ must be a factor of $x^3 + 2x^2 - 6x - 4$, since it is a factor of the original function. Again, using synthetic division and this time the root 2, we have the second setup at the left. Thus,

$$x^4 + 3x^3 - 4x^2 - 10x - 4 = (x + 1)(x - 2)(x^2 + 4x + 2)$$

Since the original equation can now be written as

$$(x + 1)(x - 2)(x^2 + 4x + 2) = 0$$

the remaining two roots are found by solving

$$x^2 + 4x + 2 = 0$$

by the quadratic formula. This gives us

$$x = \frac{-4 \pm \sqrt{16 - 8}}{2} = \frac{-4 \pm 2\sqrt{2}}{2} = -2 \pm \sqrt{2}$$

■ Therefore, the roots are -1, 2, $-2 + \sqrt{2}$, and $-2 - \sqrt{2}$.

EXAMPLE E

$$3 \;-26 \quad 63 \;-36 \;-20 \quad \underline{\left|2\right.}$$
$$\underline{\quad\quad 6 \;-40 \quad 46 \quad 20\;}$$
$$3 \;-20 \quad 23 \quad 10 \quad\;\; 0$$

$$3 \;-20 \quad 23 \quad 10 \quad \underline{\left|2\right.}$$
$$\underline{\quad\quad 6 \;-28 \;-10\;}$$
$$3 \;-14 \quad -5 \quad\;\; 0$$

Solve the equation $3x^4 - 26x^3 + 63x^2 - 36x - 20 = 0$, given that 2 is a double root.

   Using synthetic division, we have the first setup at the left. It tells us that

$$3x^4 - 26x^3 + 63x^2 - 36x - 20 = (x - 2)(3x^3 - 20x^2 + 23x + 10)$$

Also, since 2 is a double root, it must be a root of $3x^3 - 20x^2 + 23x + 10$. Using synthetic division again, we have the second setup at the left. From that division, the quotient $3x^2 - 14x - 5$ factors into $(3x + 1)(x - 5)$. Therefore, the roots of the equation are 2, 2, $-\frac{1}{3}$, and 5.

(*Continued on next page*)

Since the quotient of the first division is the dividend for the second division, it is not necessary to rewrite the coefficients of the first quotient. Both divisions can be done as follows:

$$
\begin{array}{rrrrr|r}
3 & -26 & 63 & -36 & -20 & \underline{2} \\
  & 6 & -40 & 46 & 20 & \\
\hline
3 & -20 & 23 & 10 & 0 & \underline{2} \\
  & 6 & -28 & -10 & & \\
\hline
3 & -14 & -5 & 0 & &
\end{array}
$$

first division → (first fraction row)
second division → (second fraction row)

EXAMPLE F

Solve the equation $2x^4 - 5x^3 + 11x^2 - 3x - 5 = 0$, given that $1 + 2j$ is a root.

Since $1 + 2j$ is a root, we know that $1 - 2j$ is also a root. Using synthetic division twice, we can then reduce the remaining factor to a quadratic function.

$$
\begin{array}{rrrrr|r}
2 & -5 & 11 & -3 & -5 & \underline{1 + 2j} \\
  & 2 + 4j & -11 - 2j & 4 - 2j & 5 & \\
\hline
2 & -3 + 4j & -2j & 1 - 2j & 0 & \underline{1 - 2j} \\
  & 2 - 4j & -1 + 2j & -1 + 2j & & \\
\hline
2 & -1 & -1 & 0 & &
\end{array}
$$

The quadratic factor $2x^2 - x - 1$ factors into $(2x + 1)(x - 1)$. Therefore, the roots of the equation are $1 + 2j$, $1 - 2j$, 1, and $-\frac{1}{2}$.

We now briefly discuss the graphical meaning of the roots of a polynomial equation. As we have seen, a polynomial equation $f(x) = 0$ of degree n has exactly n roots. If the roots are all real and different, the graph of $f(x)$ will cross the x-axis n times. See Fig. 14-1, where $n = 4$ and the graph crosses the x-axis four times.

The number of times the graph of $f(x)$ crosses the x-axis is reduced from n by 2 for each pair of complex roots (see Fig. 14-2). Also, the graph of $f(x)$ crosses the x-axis only once for a multiple root if the multiple is odd, or it is tangent to the x-axis if the multiple is even (see Fig. 14-3). The graph must cross at least once if the degree of n is odd (see Figs. 14-2 and 14-3), although it may not cross at all if n is even (see Fig. 14-4).

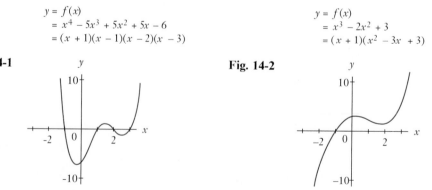

$y = f(x)$
$= x^4 - 5x^3 + 5x^2 + 5x - 6$
$= (x + 1)(x - 1)(x - 2)(x - 3)$

Fig. 14-1

$y = f(x)$
$= x^3 - 2x^2 + 3$
$= (x + 1)(x^2 - 3x + 3)$

Fig. 14-2

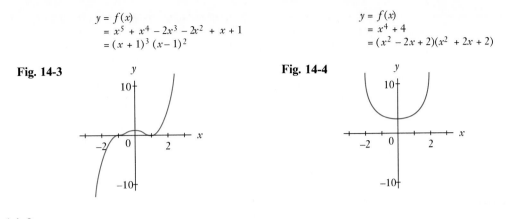

$$y = f(x)$$
$$= x^5 + x^4 - 2x^3 - 2x^2 + x + 1$$
$$= (x + 1)^3 (x - 1)^2$$

Fig. 14-3

$$y = f(x)$$
$$= x^4 + 4$$
$$= (x^2 - 2x + 2)(x^2 + 2x + 2)$$

Fig. 14-4

Exercises 14-3

In Exercises 1 through 24, solve the given equations using synthetic division, given the roots indicated.

1. $x^3 + 2x^2 - x - 2 = 0$ $(r_1 = 1)$

2. $x^3 + 2x^2 + x + 2 = 0$ $(r_1 = -2)$

3. $x^3 + x^2 - 8x - 12 = 0$ $(r_1 = -2)$

4. $x^3 - 1 = 0$ $(r_1 = 1)$

5. $2x^3 + 11x^2 + 20x + 12 = 0$ $(r_1 = -\frac{3}{2})$

6. $4x^3 - 20x^2 - x + 5 = 0$ $(r_1 = \frac{1}{2})$

7. $3x^3 + 2x^2 + 3x + 2 = 0$ $(r_1 = j)$

8. $x^3 + 5x^2 + 9x + 5 = 0$ $(r_1 = -2 + j)$

9. $x^4 + x^3 - 2x^2 + 4x - 24 = 0$ $(r_1 = 2, r_2 = -3)$

10. $x^4 + 2x^3 - 4x^2 - 5x + 6 = 0$ $(r_1 = 1, r_2 = -2)$

11. $x^4 - 9x^2 + 4x + 12 = 0$ (2 is a double root)

12. $4x^4 + 28x^3 + 61x^2 + 42x + 9 = 0$ $(-3$ is a double root)

13. $6x^4 + 5x^3 - 15x^2 + 4 = 0$ $(r_1 = -\frac{1}{2}, r_2 = \frac{2}{3})$

14. $6x^4 - 5x^3 - 14x^2 + 14x - 3 = 0$ $(r_1 = \frac{1}{3}, r_2 = \frac{3}{2})$

15. $2x^4 - x^3 - 4x^2 + 10x - 4 = 0$ $(r_1 = 1 + j)$

16. $x^4 - 8x^3 - 72x - 81 = 0$ $(r_1 = 3j)$

17. $2x^5 + 11x^4 + 16x^3 - 8x^2 - 32x - 16 = 0$ $(-2$ is a triple root)

18. $x^5 - 3x^4 + 4x^3 - 4x^2 + 3x - 1 = 0$ (1 is a triple root)

19. $2x^5 + x^4 - 15x^3 + 5x^2 + 13x - 6 = 0$ $(r_1 = 1, r_2 = -1, r_3 = \frac{1}{2})$

20. $12x^5 - 7x^4 + 41x^3 - 26x^2 - 28x + 8 = 0$ $(r_1 = 1, r_2 = \frac{1}{4}, r_3 = -\frac{2}{3})$

21. $x^5 - 3x^4 - x + 3 = 0$ $(r_1 = 3, r_2 = j)$

22. $4x^5 + x^3 - 4x^2 - 1 = 0$ $(r_1 = 1, r_2 = \frac{1}{2}j)$

23. $x^6 + 2x^5 - 4x^4 - 10x^3 - 41x^2 - 72x - 36 = 0$ $(-1$ is a double root; $2j$ is a root)

24. $x^6 - x^5 - 2x^3 - 3x^2 - x - 2 = 0$ $(j$ is a double root)

14-4 **Rational and Irrational Roots**

If we form the product of the factors $(x + 2)(x - 4)(x + 3)$, we obtain $x^3 + x^2 - 14x - 24$. In forming this product, we find that the constant 24 which results is determined only by the numbers 2, 4, and 3. We note that these numbers represent the roots of the equation if the given function is set equal to zero. In fact, if we found all the integral roots of an equation, and represented the equation in the form

$$f(x) = (x - r_1)(x - r_2) \cdots (x - r_k)f_{k+1}(x) = 0$$

where all the roots indicated are integers, the constant term of $f(x)$ must have factors of $r_1, r_2, \ldots, r_k$. This leads us to the theorem which states that

in a polynomial equation $f(x) = 0$, if the coefficient of the highest power is 1, then any integral roots are factors of the constant term of $f(x)$.

EXAMPLE A ⎤ The equation $f(x) = x^5 - 4x^4 - 7x^3 + 14x^2 - 44x + 120 = 0$ can be written as

$$(x - 5)(x + 3)(x - 2)(x^2 + 4) = 0$$

We now note that $5(3)(2)(4) = 120$. Thus, the roots 5, -3, and 2 are numerical
■ factors of $|120|$. The theorem states nothing in regard to the signs involved.

If the coefficient a_0 of the highest-power term of $f(x)$ is an integer not equal to 1, the polynomial equation $f(x) = 0$ may have rational roots which are not integers. This coefficient a_0 can be factored from every term of $f(x)$. Thus any polynomial equation $f(x) = a_0x^n + a_1x^{n-1} + \cdots + a_n = 0$ with integral coefficients can be written in the form

$$f(x) = a_0\left(x^n + \frac{a_1}{a_0}x^{n-1} + \cdots + \frac{a_n}{a_0}\right) = 0$$

Since a_n and a_0 are integers, a_n/a_0 is a rational number. Using the same reasoning as with integral roots applied to the factor within the parentheses, we see that any rational roots are factors of a_n/a_0. This leads to the following theorem:

Any rational root of a polynomial equation

$$f(x) = a_0x^n + a_1x^{n-1} + \cdots + a_n = 0$$

is an integral factor of a_n divided by an integral factor of a_0

We may show this rational root r_r as

$$r_r = \frac{\textbf{integral factor of } a_n}{\textbf{integral factor of } a_0} \qquad\qquad (14\text{-}4)$$

EXAMPLE B ⎤ If $f(x) = 4x^3 - 3x^2 - 25x - 6 = 0$, any rational roots, if they exist, must be integral factors of 6 divided by integral factors of 4. The integral factors of 6 are 1, 2, 3, and 6 and the integral factors of 4 are 1, 2, and 4. Forming all possible positive and negative quotients, any rational roots that exist will be found in the following list: $\pm 1, \pm\frac{1}{2}, \pm\frac{1}{4}, \pm 2, \pm 3, \pm\frac{3}{2}, \pm\frac{3}{4}, \pm 6$.
■ The roots of this equation are -2, 3, and $-\frac{1}{4}$.

There are 16 different possible rational roots in Example B. Since we have no way of telling which of these are the actual roots, we now present a rule which will help us to find these roots. This rule is known as **Descartes' rule of signs.** It states that *the number of positive roots of a polynomial equation $f(x) = 0$ cannot exceed the number of changes in sign in $f(x)$ in going from one term to the next in $f(x)$. The number of negative roots cannot exceed the number of sign changes in $f(-x)$.*

Descartes' rule of signs

We can reason this way: If $f(x)$ has all positive terms, then any positive number substituted in $f(x)$ must give a positive value for the function. This indicates that the number substituted in the function is not a root. Thus, there must be at least one negative and one positive term in the function for any positive number to be a root. This is not a proof, but does indicate the type of reasoning which is used in developing the theorem.

EXAMPLE C

By Descartes' rule of signs, determine the maximum number of positive and negative roots of $3x^3 - x^2 - x + 4 = 0$.

Here $f(x) = 3x^3 - x^2 - x + 4$. The first term is positive and the second is negative, which indicates a change of sign. The third term is also negative; there is no change of sign from the second to the third term. The fourth term is positive, thus giving us a second change of sign, from the third to the fourth term. Hence there are two changes in sign, which we can show as follows:

$$f(x) = 3x^3 - x^2 - x + 4$$

$\underset{1}{\llcorner\!\uparrow} \quad \underset{2}{\llcorner\!\uparrow}$ ⟵ two sign changes

Since there are *two* changes of sign in $f(x)$, there are *no more than two* positive roots of $f(x) = 0$.

To find the maximum possible number of negative roots, we must find the number of sign changes in $f(-x)$. Thus,

$$f(-x) = 3(-x)^3 - (-x)^2 - (-x) + 4$$
$$= -3x^3 - x^2 + x + 4$$

$\llcorner\!\uparrow$ ⟵ one sign change

There is only one change of sign in $f(-x)$; therefore, there is one negative root. *When there is just one change of sign in $f(x)$, there is a positive root, and when there is just one change of sign in $f(-x)$, there is a negative root.* ∎

EXAMPLE D

For the equation $4x^5 - x^4 - 4x^3 + x^2 - 5x - 6 = 0$, we write

$$f(x) = 4x^5 - x^4 - 4x^3 + x^2 - 5x - 6$$

⟵ three sign changes

and

$$f(-x) = -4x^5 - x^4 + 4x^3 + x^2 + 5x - 6$$

⟵ two sign changes

Thus, there are no more than three positive and two negative roots. ∎

roots of a
polynomial
equation

At this point let us summarize the information we can determine about the roots of a polynomial equation $f(x) = 0$ of degree n and with real coefficients:

1. There are n roots.

2. Complex roots appear in conjugate pairs.

3. Any rational roots must be factors of the constant term divided by factors of the coefficient of the highest-power term.

4. The maximum number of positive roots is the number of sign changes in $f(x)$, and the maximum number of negative roots is the number of sign changes in $f(-x)$.

5. Once we determine $n - 2$ of the roots, the remaining roots can be found by the quadratic formula.

Synthetic division is normally used to try possible roots. This is because synthetic division is relatively easy to perform, and when a root is found we have the quotient factor, which is of degree one less than the degree of the dividend. Each root we find makes the ensuing work simpler. The following examples indicate the complete method, as well as two other helpful rules.

EXAMPLE E

Determine the roots of the equation $2x^3 + x^2 + 5x - 3 = 0$.

Since $n = 3$, there are three roots. If we can find one of these roots, we can use the quadratic formula to find the other two. We have

$$f(x) = 2x^3 + x^2 + 5x - 3 \quad \text{and} \quad f(-x) = -2x^3 + x^2 - 5x - 3$$

which tells us that there is one positive root and no more than two negative roots. These roots may or may not be rational. The *possible* rational roots are ± 1, $\pm\frac{1}{2}$, $\pm\frac{3}{2}$, ± 3. Thus, using synthetic division, we shall try these.

```
2   1   5  -3  | 1
    2   3   8
2   3   8   5
```

First, trying the root 1 (always a possibility if there are positive roots), we have the synthetic division shown at the left. The remainder of 5 tells us that 1 is not a root, but we have gained some additional information, if we observe closely. If we try any positive number larger than 1, the results in the last row will be larger positive numbers than we now have. The products will be larger, and therefore the sums will also be larger. Thus there is no positive root larger than 1. This leads to the following rule: *When we are trying a positive root, if the bottom row contains all positive numbers, then there are no roots larger than the value tried.* This rule tells us that there is no reason to try $+\frac{3}{2}$ and $+3$ as roots.

```
2   1   5  -3  | 1/2
    1   1   3
2   2   6   0
```

Now let us try $+\frac{1}{2}$, as shown at the left. The zero remainder tells us that $+\frac{1}{2}$ is a root, and the remaining factor is $2x^2 + 2x + 6$, which itself factors to $2(x^2 + x + 3)$. By the quadratic formula we find the remaining roots by solving the equation $x^2 + x + 3 = 0$. This gives us

$$x = \frac{-1 \pm \sqrt{1 - 12}}{2} = \frac{-1 \pm j\sqrt{11}}{2}$$

The three roots are

$$\frac{1}{2}, \quad \frac{-1 + j\sqrt{11}}{2}, \quad \text{and} \quad \frac{-1 - j\sqrt{11}}{2}$$

We note that there were actually no negative roots, because the nonpositive roots are complex. Also, in proceeding in this way, we never found it necessary to try any negative roots. It must be admitted, however, that the solutions to all problems may not be so easily determined.

EXAMPLE F

See the chapter introduction.

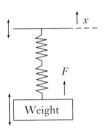

Fig. 14-5

During a cycle of the movement of the weight on the double spring shown in Fig. 14-5, the force F (in newtons) exerted on the weight by the spring is

$$F = x^4 - 7x^3 + 12x^2 + 4x$$

where x is the displacement (in cm) of the top of the double spring. For what values of x is $F = 16$ N?

Substituting 16 for F, we see that we are to solve the equation

$$x^4 - 7x^3 + 12x^2 + 4x - 16 = 0$$

To solve this equation, we write

$$f(x) = x^4 - 7x^3 + 12x^2 + 4x - 16$$

$$f(-x) = x^4 + 7x^3 + 12x^2 - 4x - 16$$

We see that there are four roots; there are no more than three positive roots, and there is one negative root. Since the coefficient of x^4 is 1, any possible rational roots must be integers. These possible rational roots are ±1, ±2, ±4, ±8, and ±16. Since there is only one negative root, we shall look for this one first. Trying -2, we have

$$
\begin{array}{rrrrr|r}
1 & -7 & +12 & +4 & -16 & \underline{-2} \\
  & -2 & +18 & -60 & +112 & \\
\hline
1 & -9 & +30 & -56 & +96 &
\end{array}
$$

$$
\begin{array}{rrrrr|r}
1 & -7 & +12 & +4 & -16 & \underline{-1} \\
  & -1 & 8 & -20 & 16 & \\
\hline
1 & -8 & 20 & -16 & 0 &
\end{array}
$$

$$
\begin{array}{rrrr|r}
1 & -8 & 20 & -16 & \underline{1} \\
  & 1 & -7 & 13 & \\
\hline
1 & -7 & 13 & -3 &
\end{array}
$$

$$
\begin{array}{rrrr|r}
1 & -8 & 20 & -16 & \underline{2} \\
  & 2 & -12 & 16 & \\
\hline
1 & -6 & 8 & 0 &
\end{array}
$$

If we were to try any negative roots less than -2 (remember, -3 is less than -2), we would find that the numbers would still alternate from term to term in the quotient. Thus, we have this rule: *When we are trying a negative root, if the signs alternate in the bottom row, then there are no roots less than the value tried.* In this case it means that we now know that -4, -8, *and* -16 cannot be roots.

Next we try -1, as shown at the left. The remainder of zero tells us that -1 is the negative root.

Now that we have found the one negative root, we look for the positive roots. Trying $+1$, we have the setup at the left. The remainder of -3 tells us that $+1$ is not a root. Next we try $+2$, as shown at the left. We see that $+2$ is a root.

It is not necessary to find any more roots by trial and error. We may now use the quadratic formula or factoring on the equation $x^2 - 6x + 8 = 0$. The remaining roots are 2 and 4. Thus, the roots are -1, 2, 2, and 4. (Note that 2 is a double root.)

These roots now indicate that $F = 16$ N for displacements of -1 cm, 2 cm, and 4 cm.

In concluding this discussion of the roots of an equation, we note that when a polynomial equation has more than two irrational roots, we cannot generally find these roots by the methods just presented. These irrational roots can be approximated by graphical methods, such as those discussed in Section 2-6. A calculator can be used to determine the values of these roots to the desired accuracy. The use of a calculator is illustrated in the following examples.

EXAMPLE G

Find the irrational root of the equation $x^3 + 2x^2 + 8x - 2 = 0$ which lies between 0 and 1.

We know that whenever a curve crosses the x-axis, that value of x is a root. We are told here that the root we want is between 0 and 1. We would then expect to find the function either positive or negative when $x = 0$, and to have the opposite sign when $x = 1$. To check this, the remainder theorem may be used. We find that $f(0) = -2$ and $f(1) = 9$.

We shall now use a method of *successive approximations* to approximate the value of the root. We shall evaluate $f(x)$ at values halfway, or nearly halfway, between those values for which $f(x)$ has different signs.

The halfway point between $x = 0$ and $x = 1$ is $x = 0.5$. We could use 0.5, but it is likely that the root is nearer $x = 0$ than $x = 1$, since $f(0) = -2$ and $f(1) = 9$. This can be seen by approximating the curve by a straight line between points $(0, -2)$ and $(1, 9)$, as shown in Fig. 14-6. Therefore, we will first try $x = 0.4$. Using the calculator sequence

$$.4 \; \boxed{x^y} \; 3 \; \boxed{+} \; 2 \; \boxed{\times} \; .4 \; \boxed{x^2} \; \boxed{+} \; 8 \; \boxed{\times} \; .4 \; \boxed{-} \; 2 \; \boxed{=} \; \boxed{\qquad 1.584}$$

we find that $f(0.4) = 1.584$. Since $f(0.4)$ is positive, we know that the root is between 0 and 0.4.

Since $f(0)$ and $f(0.4)$ are nearly equal in absolute value, we now evaluate $f(0.2)$. This value is -0.312.

We now know that the root is between 0.2 and 0.4. Therefore, we find that $f(0.3) = 0.607$. This tells us that the root is between 0.2 and 0.3, and that it is likely to be closer to 0.2.

Next we find that $f(0.24) = 0.049024$, $f(0.22) = -0.132552$, and $f(0.23) = -0.042033$. Therefore, the root is between 0.23 and 0.24. Since $f(0.235) = 0.0034279$, we know that the value of the root is $x = 0.23$ to two decimal places.

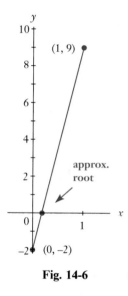

Fig. 14-6

EXAMPLE H

A shelf of area 2.50 m² is constructed in the shape of a right triangle in which the hypotenuse is 1.00 m longer than one of the legs. Find the lengths of the edges of the shelf.

If we let x equal the length of the leg, then $x + 1$ is the length of the hypotenuse. (Although for convenience we will use only one or two significant digits in developing the equation, the values are valid to three significant digits.) From the Pythagorean theorem, the other leg is

$$\sqrt{(x + 1)^2 - x^2} = \sqrt{2x + 1}$$

Since the area is one-half the product of the length of the legs, we have

$$\frac{1}{2} x \sqrt{2x + 1} = 2.5 \qquad \text{area} = 2.50 \text{ m}^2$$

$$x\sqrt{2x+1} = 5 \qquad \text{multiply each side by 2}$$
$$x^2(2x+1) = 25 \qquad \text{square each side}$$
$$2x^3 + x^2 - 25 = 0 \qquad \text{simplify with terms on left}$$

With $f(x) = 2x^3 + x^2 - 25$ we calculate values of $f(x)$, including the following, which give us the solution.

$$f(2) = -5 \qquad \rfloor\_ \quad \text{different signs mean at least}$$
$$f(3) = 38 \qquad \qquad \text{one root between 2 and 3}$$
$$f(2.2) = 1.136 \qquad \text{root between 2.0 and 2.2}$$
$$f(2.16) = -0.179008 \qquad \text{root between 2.16 and 2.20}$$
$$f(2.17) = 0.145526 \qquad \text{root between 2.16 and 2.17}$$

From the values of $f(2.16)$ and $f(2.17)$ we see that $x = 2.17$ m, to three significant digits. Thus, the edges of the shelf are

$$2.17 \text{ m}, \qquad 2.31 \text{ m}, \quad \text{and} \quad 3.17 \text{ m}$$

With legs of 2.17 m and 2.31 m, the area is $\frac{1}{2}(2.17)(2.31) = 2.51$ m^2, which checks closely with the given area.

Exercises 14-4

In Exercises 1 through 20, solve the given equations.

1. $x^3 + 2x^2 - x - 2 = 0$

2. $x^3 + x^2 - 5x + 3 = 0$

3. $x^3 + 2x^2 - 5x - 6 = 0$

4. $x^3 + 1 = 0$

5. $2x^3 - 5x^2 - 28x + 15 = 0$

6. $2x^3 - x^2 - 3x - 1 = 0$

7. $3x^3 + 11x^2 + 5x - 3 = 0$

8. $4x^3 - 5x^2 - 23x + 6 = 0$

9. $x^4 - 11x^2 - 12x + 4 = 0$

10. $x^4 + x^3 - 2x^2 - 4x - 8 = 0$

11. $x^4 - 2x^3 - 13x^2 + 14x + 24 = 0$

12. $x^4 - x^3 + 2x^2 - 4x - 8 = 0$

13. $2x^4 - 5x^3 - 3x^2 + 4x + 2 = 0$

14. $2x^4 + 7x^3 + 9x^2 + 5x + 1 = 0$

15. $12x^4 + 44x^3 + 21x^2 - 11x - 6 = 0$

16. $9x^4 - 3x^3 + 34x^2 - 12x - 8 = 0$

17. $x^5 + x^4 - 9x^3 - 5x^2 + 16x + 12 = 0$

18. $x^6 - x^4 - 14x^2 + 24 = 0$

19. $2x^5 - 5x^4 + 6x^3 - 6x^2 + 4x - 1 = 0$

20. $2x^5 + 5x^4 - 4x^3 - 19x^2 - 16x - 4 = 0$

In Exercises 21 through 24, use a calculator to find the irrational root, to two decimal places, which lies between the listed values.

21. $x^3 - 6x^2 + 10x - 4 = 0$ (0 and 1)

22. $x^4 - x^3 - 3x^2 - x - 4 = 0$ (2 and 3)

23. $3x^3 + 13x^2 + 3x - 4 = 0$ (−1 and 0)

24. $3x^4 - 3x^3 - 11x^2 - x - 4 = 0$ (−2 and −1)

In Exercises 25 through 36, solve the given problems.

25. In finding the volume V of a certain gas in equilibrium with a liquid, it is necessary to solve the equation $V^3 - 6V^2 + 12V - 8 = 0$. Find V (in cm^3).

26. In determining one of the dimensions d of the support columns of a building, the equation $3d^3 + 5d^2 - 400d - 18000 = 0$ is found. Determine this dimension (in inches).

27. The deflection y of a beam at a horizontal distance x from one end is given by $y = k(x^4 - 2Lx^3 + L^3x)$, where L is the length of the beam and k is a constant. For what values of x is the deflection zero?

28. In the theory of the motion of a sphere moving through a fluid, the expression $4r^3 - 3ar^2 - a^3$ is found. In terms of a, solve for r if this expression is zero.

29. A company found that the cost C, in dollars, of producing x pounds of a sealant for a space vehicle is given by $C = 8x^3 - 36x^2 + 90$. Find x for $C = \$36$.

30. The angle θ of a robot arm with the horizontal as a function of time is given by $\theta = 15 + 20t^2 - 4t^3$ for $0 \le t \le 5$ s. Find t for $\theta = 40°$.

31. For electrical resistors connected in parallel, the reciprocal of the combined resistance equals the sum of the reciprocals of the individual resistances. If three resistors are connected in parallel such that the second resistance is $1\ \Omega$ more than the first and the third is $4\ \Omega$ more than the first, find the resistances for a combined resistance of $1\ \Omega$.

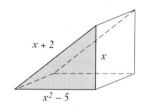

32. The right triangular end of a glass prism used in optics has the dimensions shown in Fig. 14-7. Find the lengths of the sides (in cm) of the triangle.

Fig. 14-7

33. The edge of one cubical block of steel is 1.0 in. longer than the edge of another block. Find the length of the edge of each if the sum of their volumes is 91.0 in³.

34. A rectangular tray is made from a square piece of sheet metal 10.0 cm on a side by cutting equal squares from each corner, bending up the sides, and then welding them together. How long is the side of the square which must be cut out if the volume of the tray is 70.0 cm³?

35. Find all of the real roots of the equation $x^3 - 2x^2 - 5x + 4 = 0$ to two decimal places. Graph the function to locate the roots approximately, and then use a calculator to complete the solution.

36. Find all of the real roots of the equation $x^4 - x^3 - 2x^2 - x - 3 = 0$ by following the method outlined in Exercise 35.

14-5   Chapter Equations, Review Exercises, and Practice Test

Chapter Equations

Polynomial function	$f(x) = a_0 x^n + a_1 x^{n-1} + \cdots + a_n$	(14-1)
Remainder theorem	$f(x) = (x - r)q(x) + R$	(14-2)
	$f(r) = R$	(14-3)
Rational roots	$r_r = \dfrac{\text{integral factor of } a_n}{\text{integral factor of } a_0}$	(14-4)

Review Exercises

In Exercises 1 through 4, find the remainder of the indicated division by the remainder theorem.

1. $(2x^3 - 4x^2 - x + 4) \div (x - 1)$ **2.** $(x^3 - 2x^2 + 9) \div (x + 2)$

3. $(4x^3 + x + 4) \div (x + 3)$ **4.** $(x^4 - 5x^3 + 8x^2 + 15x - 2) \div (x - 3)$

In Exercises 5 through 8, use the factor theorem to determine whether or not the second expression is a factor of the first.

5. $x^4 + x^3 + x^2 - 2x - 3;\quad x + 1$ **6.** $2x^3 - 2x^2 - 3x - 2;\quad x - 2$

7. $x^4 + 4x^3 + 5x^2 + 5x - 6;\quad x + 3$ **8.** $9x^3 + 6x^2 + 4x + 2;\quad 3x + 1$

In Exercises 9 through 16, use synthetic division to perform the indicated divisions.

9. $(x^3 + 3x^2 + 6x + 1) \div (x - 1)$ **10.** $(3x^3 - 2x^2 + 7) \div (x - 3)$

11. $(2x^3 - 3x^2 - 4x + 3) \div (x + 2)$ **12.** $(3x^3 - 5x^2 + 7x - 6) \div (x + 4)$

13. $(x^4 - 2x^3 - 3x^2 - 4x - 8) \div (x + 1)$

14. $(x^4 - 6x^3 + x - 8) \div (x - 3)$

15. $(2x^5 - 46x^3 + x^2 - 9) \div (x - 5)$

16. $(x^6 + 63x^3 + 5x^2 - 9x - 8) \div (x + 4)$

In Exercises 17 through 20, use synthetic division to determine whether or not the given numbers are zeros of the given functions.

17. $x^3 + 8x^2 + 17x - 6$; -3

18. $2x^3 + x^2 - 4x + 4$; -2

19. $2x^4 - x^3 + 2x^2 + x - 1$; $\frac{1}{2}$

20. $6x^4 - 7x^3 + 2x^2 - 9x - 6$; $-\frac{2}{3}$

In Exercises 21 through 32, find all the roots of the given equations, with the aid of synthetic division and with the roots indicated.

21. $x^3 + 8x^2 + 17x + 6 = 0$ $(r_1 = -3)$

22. $2x^3 + 7x^2 - 6x - 8 = 0$ $(r_1 = -4)$

23. $3x^4 + 5x^3 + x^2 + x - 10 = 0$ $(r_1 = 1, r_2 = -2)$

24. $x^4 - x^3 - 5x^2 - x - 6 = 0$ $(r_1 = 3, r_2 = -2)$

25. $2x^4 + x^3 - 29x^2 - 34x + 24 = 0$ $(r_1 = -2, r_2 = \frac{1}{2})$

26. $x^4 + x^3 - 11x^2 - 9x + 18 = 0$ $(r_1 = -3, r_2 = 1)$

27. $4x^4 + 4x^3 + x^2 + 4x - 3 = 0$ $(r_1 = j)$

28. $x^4 + 2x^3 - 4x - 4 = 0$ $(r_1 = -1 + j)$

29. $x^5 + 3x^4 - x^3 - 11x^2 - 12x - 4 = 0$ $(-1$ is a triple root$)$

30. $24x^5 + 10x^4 + 7x^2 - 6x + 1 = 0$ $(r_1 = -1, r_2 = \frac{1}{4}, r_3 = \frac{1}{3})$

31. $x^5 + 4x^4 + 5x^3 - x^2 - 4x - 5 = 0$ $(r_1 = 1, r_2 = -2 + j)$

32. $2x^5 - x^4 + 8x - 4 = 0$ $(r_1 = \frac{1}{2}, r_2 = 1 + j)$

In Exercises 33 through 40, solve the given equations.

33. $x^3 + x^2 - 10x + 8 = 0$

34. $x^3 - 8x^2 + 20x - 16 = 0$

35. $2x^3 - x^2 - 8x - 5 = 0$

36. $2x^3 - 3x^2 - 11x + 6 = 0$

37. $6x^3 - x^2 - 12x - 5 = 0$

38. $6x^3 + 19x^2 + 2x - 3 = 0$

39. $2x^4 + x^3 + 3x^2 + 2x - 2 = 0$

40. $2x^4 + 5x^3 - 14x^2 - 23x + 30 = 0$

In Exercises 41 through 56, determine the required quantities. Where appropriate, set up the required equations.

41. For what value of k is $x + 2$ a factor of $f(x) = 3x^3 + kx^2 - 8x - 8$?

42. For what value of k is $x - 3$ a factor of $f(x) = kx^4 - 15x^2 - 5x - 12$?

43. Where does the graph of the function $f(x) = 6x^4 - 14x^3 + 5x^2 + 5x - 2$ cross the x-axis?

44. Where does the graph of the function $f(x) = 2x^4 - 7x^3 + 11x^2 - 28x + 12$ cross the x-axis?

45. Find the irrational root of the equation $3x^3 - x^2 - 8x - 2 = 0$ which lies between 1 and 2.

46. Find the irrational root of the equation $x^4 + 3x^3 + 6x + 4 = 0$ which lies between -1 and 0.

47. A computer analysis of the number of crimes committed each month in a certain city for the first 10 months of a year showed that $n = x^3 - 9x^2 + 15x + 600$. Here n is the number of monthly crimes and x is the number of the month (as of the last day). In what month were 580 crimes committed?

48. A company determined that the number s (in thousands) of computer chips that it could supply at a price p of less than \$5 is given by $s = 4p^2 - 25$, whereas the demand d (in thousands) for the chips is given by $d = p^3 - 22p + 50$. For what price is the supply equal to the demand?

49. In order to find the diameter d (in cm) of a helical spring subject to given forces, it is necessary to solve the equation $64d^3 - 144d^2 + 108d - 27 = 0$. Solve for d.

50. A cubical tablet for purifying water is wrapped in a sheet of foil 0.500 mm thick. The total volume of tablet and foil is 33.1% greater than the volume of the tablet alone. Find the length of the edge of the tablet.

51. For the mirror shown in Fig. 14-8, the reciprocal of the focal distance f equals the sum of the reciprocals of the object distance p and image distance q. If $q = p + 4$ and $f = \dfrac{p+1}{p}$, find p. Distances are in inches.

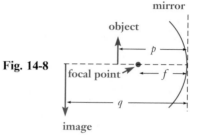

Fig. 14-8

52. Three electric capacitors are connected in series. The capacitance of the second is 1 μF more than that of the first, and the third is 2 μF more than that of the second. The capacitance of the combination is 1.33 μF. The equation used to determine C, the capacitance of the first capacitor, is

$$\frac{1}{C} + \frac{1}{C+1} + \frac{1}{C+3} = \frac{3}{4}$$

Find the values of the capacitances.

53. A metal support frame is in the shape of a triangle with sides $\sqrt{x+1}$, $\sqrt{2x+3}$, and $x+1$. If the perimeter of the frame is 9 cm, find the lengths of the sides.

54. A grain storage bin has a square base, each side of which is 5.5 m longer than the height of the bin. If the bin holds 160 m³ of grain, find its dimensions.

55. A rectangular slab of concrete has a diagonal which is 2.0 ft longer than one of the sides. If the area of the slab is 120 ft², find the lengths of the sides.

56. The radius of one ball bearing is 1.0 mm greater than the radius of a second ball bearing. If the sum of their volumes is 100 mm³, find the radius of each.

Practice Test

1. Is -3 a zero for the function $2x^3 + 3x^2 + 7x - 6$?

2. Find the remaining roots of the equation $x^4 - 2x^3 - 7x^2 + 20x - 12 = 0$, given that 2 is a double root.

3. Use synthetic division to perform the division $(x^3 - 5x^2 + 4x - 9) \div (x - 3)$.

4. Use the factor theorem and synthetic division to determine whether or not $(2x + 1)$ is a factor of $2x^4 + 15x^3 + 23x^2 - 16$.

5. Use the remainder theorem to find the remainder of the division $(x^3 + 4x^2 + 7x - 9) \div (x + 4)$.

6. Solve: $2x^4 - x^3 + 5x^2 - 4x - 12 = 0$.

7. The ends of a 10-ft beam are supported at different levels. The deflection y of the beam is given by $y = kx^2(x^3 + 436x - 4000)$, where x is the horizontal distance from one end and k is a constant. Determine the values of x for which the deflection is zero.

8. A cubical metal block is heated such that its edge increases by 1.0 mm and its volume is doubled. Find the edge of the cube to tenths.

15 Determinants and Matrices

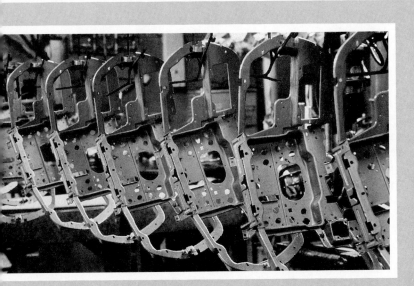

In Chapter 4 we first met the concept of a determinant and saw how it is used to solve systems of linear equations. However, at that time we limited our discussion to second- and third-order determinants. In the first two sections of this chapter we shall show methods of evaluating higher-order determinants. In the remainder of the chapter we shall develop the related concept of a matrix and its use in solving systems of linear equations.

Methods developed in this chapter are used in solving applied problems in areas such as electric circuits and analysis of forces. Also, these methods are used extensively in business and industry in making appropriate decisions for research, development, and production.

In Section 15-4 we see how material amounts and worker time may be determined for the production of machine parts.

15-1 Determinants: Expansion by Minors

From Section 4-7 we recall that *a third-order* **determinant** *is defined by the equation*

$$\begin{vmatrix} a_1 & b_1 & c_1 \\ a_2 & b_2 & c_2 \\ a_3 & b_3 & c_3 \end{vmatrix} = a_1 b_2 c_3 + a_3 b_1 c_2 + a_2 b_3 c_1 - a_3 b_2 c_1 - a_1 b_3 c_2 - a_2 b_1 c_3 \qquad (15\text{-}1)$$

If we rearrange the terms on the right and factor a_1, a_2, and a_3 from the terms in which they are contained, we have

$$\begin{vmatrix} a_1 & b_1 & c_1 \\ a_2 & b_2 & c_2 \\ a_3 & b_3 & c_3 \end{vmatrix} = a_1(b_2c_3 - b_3c_2) - a_2(b_1c_3 - b_3c_1) + a_3(b_1c_2 - b_2c_1) \qquad (15\text{-}2)$$

Recalling the definition of a second-order determinant, we have

$$\begin{vmatrix} a_1 & b_1 & c_1 \\ a_2 & b_2 & c_2 \\ a_3 & b_3 & c_3 \end{vmatrix} = a_1 \begin{vmatrix} b_2 & c_2 \\ b_3 & c_3 \end{vmatrix} - a_2 \begin{vmatrix} b_1 & c_1 \\ b_3 & c_3 \end{vmatrix} + a_3 \begin{vmatrix} b_1 & c_1 \\ b_2 & c_2 \end{vmatrix} \qquad (15\text{-}3)$$

In Eq. (15-3) we note that the third-order determinant is expanded with the terms of the expansion as products of the elements of the first column and specific second-order determinants. In each case the elements of the second-order determinant are those elements which are in neither the same row nor the same column as the element from the first column. *These determinants are called* **minors.**

In general, *the minor of a given element of a determinant is the determinant which results by deleting the row and the column in which the element lies.* Consider the following example.

EXAMPLE A

Consider the determinant $\begin{vmatrix} 1 & 2 & 3 \\ 4 & 5 & 6 \\ 7 & 8 & 9 \end{vmatrix}$

We find the minor of the element 1 by deleting the elements in the first row and first column because the element 1 is located in the first row and in the first column. The minor for the element 6 is formed by deleting the elements in the second row and in the third column, for this is the location of the 6. These minors are shown below.

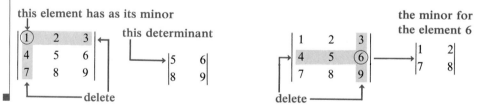

We now see that Eq. (15-3) expresses the expansion of a third-order determinant as the sum of the products of the elements of the first column and their minors, with the second term assigned a minus sign. Actually this is only one of several ways of expressing the expansion. However, it does lead to a general theorem regarding the expansion of a determinant of any order. The foregoing provides a basis for this theorem, although it cannot be considered as a proof. The theorem is as follows:

expansion by minors

The value of a determinant of order n may be found by forming the n products of the elements of any column (or row) and their minors. A product is given a plus sign if the sum of the number of the column and the number of the row in which the element lies is even, and a minus sign if this sum is odd. The algebraic sum of the terms thus obtained is the value of the determinant.

The following two examples illustrate the expansion of a third-order determinant and a fourth-order determinant by minors using the theorem above.

EXAMPLE B

Evaluate $\begin{vmatrix} 1 & -3 & -2 \\ 4 & -1 & 0 \\ 4 & 3 & -5 \end{vmatrix}$ by expansion by minors.

Since we may expand by any column or row, using the first row, we have:

$$\begin{vmatrix} 1 & -3 & -2 \\ 4 & -1 & 0 \\ 4 & 3 & -5 \end{vmatrix} = +(1)\begin{vmatrix} -1 & 0 \\ 3 & -5 \end{vmatrix} - (-3)\begin{vmatrix} 4 & 0 \\ 4 & -5 \end{vmatrix} + (-2)\begin{vmatrix} 4 & -1 \\ 4 & 3 \end{vmatrix}$$

column 1   column 2   column 3
row 1   row 1   row 1
2 (even)   3 (odd)   4 (even)

The first term of the expansion is assigned a plus sign since the element 1 is in column 1 and row 1, and $1 + 1 = 2$ (even). The second term is assigned a minus sign since the element -3 is in column 2 and row 1, and $2 + 1 = 3$ (odd). The third term is assigned a plus sign since the element -2 is in column 3 and row 1, and $3 + 1 = 4$ (even). Actually, once the first sign has been properly determined, the others are known since the signs alternate from term to term. Using the definition of a second-order determinant, Eq. (4-10), we complete the evaluation.

For reference, Eq. (4-10) is $\begin{vmatrix} a_1 & b_1 \\ a_2 & b_2 \end{vmatrix} = a_1b_2 - a_2b_1.$

$$\begin{vmatrix} 1 & -3 & -2 \\ 4 & -1 & 0 \\ 4 & 3 & -5 \end{vmatrix} = +(1)(5 - 0) - (-3)(-20 - 0) + (-2)[12 - (-4)]$$

$$= 1(5) + 3(-20) - 2(16) = 5 - 60 - 32 = -87$$

Expansion of this same determinant by the third column is as follows:

$$\begin{vmatrix} 1 & -3 & -2 \\ 4 & -1 & 0 \\ 4 & 3 & -5 \end{vmatrix} = +(-2)\begin{vmatrix} 4 & -1 \\ 4 & 3 \end{vmatrix} - (0)\begin{vmatrix} 1 & -3 \\ 4 & 3 \end{vmatrix} + (-5)\begin{vmatrix} 1 & -3 \\ 4 & -1 \end{vmatrix}$$

column 3
row 1
4 (even)   signs alternate

$$= -2(12 + 4) + 0 - 5(-1 + 12) = -87$$

This expansion has one advantage: since one of the elements of the third column is zero, its minor does not have to be evaluated because zero times whatever the value of the determinant will give the product of zero.

EXAMPLE C

$$\text{Evaluate} \quad \begin{vmatrix} 3 & -2 & 0 & 2 \\ 1 & 0 & -1 & 4 \\ -3 & 1 & 2 & -2 \\ 2 & -1 & 0 & -1 \end{vmatrix}$$

Expanding by the third column, we have

$$\begin{vmatrix} 3 & -2 & 0 & 2 \\ 1 & 0 & -1 & 4 \\ -3 & 1 & 2 & -2 \\ 2 & -1 & 0 & -1 \end{vmatrix} = +(0)\begin{vmatrix} 1 & 0 & 4 \\ -3 & 1 & -2 \\ 2 & -1 & -1 \end{vmatrix} -(-1)\begin{vmatrix} 3 & -2 & 2 \\ -3 & 1 & -2 \\ 2 & -1 & -1 \end{vmatrix} +(2)\begin{vmatrix} 3 & -2 & 2 \\ 1 & 0 & 4 \\ 2 & -1 & -1 \end{vmatrix} -(0)\begin{vmatrix} 3 & -2 & 2 \\ 1 & 0 & 4 \\ -3 & 1 & -2 \end{vmatrix}$$

It is not necessary to expand the minors in the first and fourth terms, since the element in each case is zero. The minors in the second and third terms can be expanded as third-order determinants or by minors. This illustrates well that expansion by minors effectively reduces the order of the determinant to be evaluated by one. Completing the evaluation by minors, we have

$$\begin{vmatrix} 3 & -2 & 0 & 2 \\ 1 & 0 & -1 & 4 \\ -3 & 1 & 2 & -2 \\ 2 & -1 & 0 & -1 \end{vmatrix} = \begin{vmatrix} 3 & -2 & 2 \\ -3 & 1 & -2 \\ 2 & -1 & -1 \end{vmatrix} + 2\begin{vmatrix} 3 & -2 & 2 \\ 1 & 0 & 4 \\ 2 & -1 & -1 \end{vmatrix}$$

$$= \left[3\begin{vmatrix} 1 & -2 \\ -1 & -1 \end{vmatrix} -(-3)\begin{vmatrix} -2 & 2 \\ -1 & -1 \end{vmatrix} + 2\begin{vmatrix} -2 & 2 \\ 1 & -2 \end{vmatrix}\right] + 2\left[-(-2)\begin{vmatrix} 1 & 4 \\ 2 & -1 \end{vmatrix} + 0\begin{vmatrix} 3 & 2 \\ 2 & -1 \end{vmatrix} -(-1)\begin{vmatrix} 3 & 2 \\ 1 & 4 \end{vmatrix}\right]$$

expanding first determinant by first column   expanding second determinant by
second column

$$= [3(-1-2) + 3(2+2) + 2(4-2)] + 2[2(-1-8) + (12-2)] = [-9 + 12 + 4] + 2[-18 + 10]$$

$$= +7 + 2(-8) = -9$$

We can use the expansion of determinants by minors to solve systems of linear equations. **Cramer's rule** *for solving systems of equations, as stated in Section 4-7, is valid for any system of n equations in n unknowns.* The following examples illustrate the solution of systems of equations.

EXAMPLE D

The production of a particular computer component is done in three stages, taking a total of 7 h. The second stage is 1 h less than the first, and the third stage is twice as long as the second stage. How long is each stage of production?

First, we let $a =$ the number of hours of the first production stage, $b =$ the number of hours of the second stage, and $c =$ the number of hours of the third stage.

The total production time of 7 h gives us $a + b + c = 7$. Since the second stage is 1 h less than the first, we then have $b = a - 1$. The fact that the third stage is twice as long as the second gives us $c = 2b$. Stating these equations in standard form for solution, we have

$$a + b + c = 7$$
$$a - b \quad\quad = 1$$
$$\quad\quad 2b - c = 0$$

Using Cramer's rule, we have

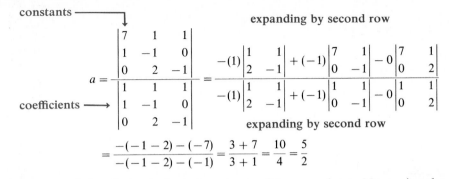

constants

$$a = \frac{\begin{vmatrix} 7 & 1 & 1 \\ 1 & -1 & 0 \\ 0 & 2 & -1 \end{vmatrix}}{\begin{vmatrix} 1 & 1 & 1 \\ 1 & -1 & 0 \\ 0 & 2 & -1 \end{vmatrix}} = \frac{-(1)\begin{vmatrix} 1 & 1 \\ 2 & -1 \end{vmatrix} + (-1)\begin{vmatrix} 7 & 1 \\ 0 & -1 \end{vmatrix} - 0\begin{vmatrix} 7 & 1 \\ 0 & 2 \end{vmatrix}}{-(1)\begin{vmatrix} 1 & 1 \\ 2 & -1 \end{vmatrix} + (-1)\begin{vmatrix} 1 & 1 \\ 0 & -1 \end{vmatrix} - 0\begin{vmatrix} 1 & 1 \\ 0 & 2 \end{vmatrix}}$$

expanding by second row

coefficients

expanding by second row

$$= \frac{-(-1-2)-(-7)}{-(-1-2)-(-1)} = \frac{3+7}{3+1} = \frac{10}{4} = \frac{5}{2}$$

Here we expanded each determinant by minors of the second row. Now using the second equation we have $\frac{5}{2} - b = 1$, or $b = \frac{3}{2}$. Using the third equation we have $2(\frac{3}{2}) - c = 0$, or $c = 3$. Thus,

$$a = \frac{5}{2}\,h, \qquad b = \frac{3}{2}\,h, \quad \text{and} \quad c = 3\,h$$

This means that the first stage takes 2.5 h, the second stage takes 1.5 h, and the third stage takes 3 h. We see that these times agree with the given information.

EXAMPLE E Solve the following system of equations.

$$\begin{aligned} x + 2y + z &= 5 \\ 2x \quad + z + 2t &= 1 \\ x - y + 3z + 4t &= -6 \\ 4x - y \quad - 2t &= 0 \end{aligned}$$

constants

$$x = \frac{\begin{vmatrix} 5 & 2 & 1 & 0 \\ 1 & 0 & 1 & 2 \\ -6 & -1 & 3 & 4 \\ 0 & -1 & 0 & -2 \end{vmatrix}}{\begin{vmatrix} 1 & 2 & 1 & 0 \\ 2 & 0 & 1 & 2 \\ 1 & -1 & 3 & 4 \\ 4 & -1 & 0 & -2 \end{vmatrix}}$$

expanding by fourth row

$$= \frac{-(0)\begin{vmatrix} 2 & 1 & 0 \\ 0 & 1 & 2 \\ -1 & 3 & 4 \end{vmatrix} + (-1)\begin{vmatrix} 5 & 1 & 0 \\ 1 & 1 & 2 \\ -6 & 3 & 4 \end{vmatrix} - (0)\begin{vmatrix} 5 & 2 & 0 \\ 1 & 0 & 2 \\ -6 & -1 & 4 \end{vmatrix} + (-2)\begin{vmatrix} 5 & 2 & 1 \\ 1 & 0 & 1 \\ -6 & -1 & 3 \end{vmatrix}}{(1)\begin{vmatrix} 0 & 1 & 2 \\ -1 & 3 & 4 \\ -1 & 0 & -2 \end{vmatrix} - 2\begin{vmatrix} 2 & 1 & 2 \\ 1 & 3 & 4 \\ 4 & 0 & -2 \end{vmatrix} + (1)\begin{vmatrix} 2 & 0 & 2 \\ 1 & -1 & 4 \\ 4 & -1 & -2 \end{vmatrix} - (0)\begin{vmatrix} 2 & 0 & 1 \\ 1 & -1 & 3 \\ 4 & -1 & 0 \end{vmatrix}}$$

expanding by first row

$$= \frac{-(-26) - 2(-14)}{1(0) - 2(-18) + 1(18)} = \frac{26 + 28}{36 + 18} = \frac{54}{54} = 1$$

(Continued on next page)

In solving for x the determinant in the numerator was evaluated by expanding by the minors of the fourth row, since it contained two zeros. The determinant in the denominator was evaluated by expanding by the minors of the first row. Now we solve for y, and we again note two zeros in the fourth row of the determinant of the numerator.

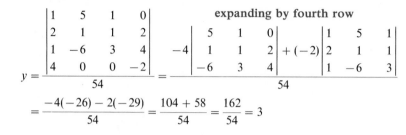

$$y = \frac{\begin{vmatrix} 1 & 5 & 1 & 0 \\ 2 & 1 & 1 & 2 \\ 1 & -6 & 3 & 4 \\ 4 & 0 & 0 & -2 \end{vmatrix}}{54} = \frac{-4\begin{vmatrix} 5 & 1 & 0 \\ 1 & 1 & 2 \\ -6 & 3 & 4 \end{vmatrix} + (-2)\begin{vmatrix} 1 & 5 & 1 \\ 2 & 1 & 1 \\ 1 & -6 & 3 \end{vmatrix}}{54}$$

$$= \frac{-4(-26) - 2(-29)}{54} = \frac{104 + 58}{54} = \frac{162}{54} = 3$$

Substituting these values for x and y into the first equation, we can solve for z. This gives $z = -2$. Again, substituting the values for x and y into the fourth equation, we find $t = \frac{1}{2}$. Thus the required solution is $x = 1$, $y = 3$, $z = -2$, $t = \frac{1}{2}$. The solution can be checked by substituting these values into either the second or third equation.

Exercises 15-1

In Exercises 1 through 16, evaluate the given determinants by expansion by minors.

1. $\begin{vmatrix} 3 & 0 & 0 \\ -2 & 1 & 4 \\ 4 & -2 & 5 \end{vmatrix}$
2. $\begin{vmatrix} 10 & 0 & -3 \\ -2 & -4 & 1 \\ 3 & 0 & 2 \end{vmatrix}$
3. $\begin{vmatrix} -2 & -4 & 2 \\ 1 & 3 & 0 \\ -4 & 5 & 2 \end{vmatrix}$
4. $\begin{vmatrix} 5 & -1 & 2 \\ 8 & 3 & -4 \\ 0 & 2 & -6 \end{vmatrix}$

5. $\begin{vmatrix} -6 & -1 & 3 \\ 2 & -2 & -3 \\ 10 & 1 & -2 \end{vmatrix}$
6. $\begin{vmatrix} 9 & -3 & 1 \\ -1 & 2 & -1 \\ 2 & -1 & 3 \end{vmatrix}$
7. $\begin{vmatrix} -30 & -25 & 54 \\ 12 & 21 & -14 \\ 37 & -46 & 24 \end{vmatrix}$
8. $\begin{vmatrix} 4.8 & -3.7 & 3.6 \\ -3.8 & 5.7 & 6.5 \\ 2.1 & -1.8 & 2.3 \end{vmatrix}$

9. $\begin{vmatrix} 1 & 0 & 1 & 0 \\ 2 & 4 & -3 & 1 \\ 1 & 1 & 1 & 1 \\ 3 & 5 & 0 & 2 \end{vmatrix}$
10. $\begin{vmatrix} 2 & 0 & 3 & 1 \\ -1 & -1 & 4 & 0 \\ 1 & 2 & 1 & 2 \\ 3 & 3 & -2 & -1 \end{vmatrix}$
11. $\begin{vmatrix} 2 & -1 & 1 & -4 \\ 2 & 1 & 3 & -5 \\ 3 & -1 & -1 & 0 \\ 1 & 2 & 2 & 6 \end{vmatrix}$

12. $\begin{vmatrix} 3 & 6 & -2 & 4 \\ 2 & -5 & 2 & 6 \\ 5 & 3 & 4 & 0 \\ 1 & 2 & 0 & -1 \end{vmatrix}$
13. $\begin{vmatrix} 1 & 2 & -1 & -2 \\ 3 & 1 & 2 & 1 \\ -1 & 3 & -1 & 2 \\ 2 & 1 & 3 & -3 \end{vmatrix}$
14. $\begin{vmatrix} 3 & -1 & 2 & -5 \\ 1 & 4 & 2 & 5 \\ -1 & 1 & 1 & 3 \\ 1 & 2 & -1 & -2 \end{vmatrix}$

15. $\begin{vmatrix} 1 & 2 & 1 & 2 & 1 \\ 1 & 0 & 0 & 1 & 0 \\ 0 & 1 & 1 & 0 & 1 \\ 1 & 1 & 2 & 2 & 1 \\ 0 & 1 & 1 & 0 & 2 \end{vmatrix}$
16. $\begin{vmatrix} 3 & 1 & 1 & 1 & 2 \\ 1 & 1 & 0 & 0 & 1 \\ 1 & 1 & 2 & 2 & 3 \\ 0 & 2 & 1 & 0 & 3 \\ 1 & 1 & 0 & 1 & 0 \end{vmatrix}$

In Exercises 17 through 24, solve the given systems of equations by determinants. Evaluate the determinants by expansion by minors.

17. $2x + y + z = 6$
$x - 2y + 2z = 10$
$3x - y - z = 4$

18. $2x + y = -1$
$4x - 2y - z = 5$
$2x + 3y + 3z = -2$

19. $3x + 6y + 2z = -2$
$x + 3y - 4z = 2$
$2x - 3y - 2z = -2$

20. $x + 3y + z = 4$
$2x - 6y - 3z = 10$
$4x - 9y + 3z = 4$

21. $x + t = 0$
$3x + y + z = -1$
$2y - z + 3t = 1$
$2z - 3t = 1$

22. $2x + y + z = 4$
$2y - 2z - t = 3$
$3y - 3z + 2t = 1$
$6x - y + t = 0$

23. $x + 2y - z = 6$
$y - 2z - 3t = -5$
$3x - 2y + t = 2$
$2x + y + z - t = 0$

24. $2x + 3y + z = 4$
$x - 2y - 3z + 4t = -1$
$3x + y + z - 5t = 3$
$-x + 2y + z + 3t = 2$

In Exercises 25 through 28, solve the given problems by use of determinants, using methods of this section.

25. In applying Kirchhoff's laws (see Exercise 20 of Section 4–6) to the given electric circuit, the following equations are found. Determine the indicated currents in amperes (see Fig. 15-1).

$$I_A + I_B + I_C + I_D = 0$$
$$2I_A - I_B = -2$$
$$3I_C - 2I_D = 0$$
$$I_B - 3I_C = 6$$

Fig. 15-1

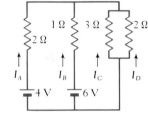

26. In analyzing the motion of four equal particles which are equally spaced along a string, the equation shown below is found. Here C depends on the string and the mass of each object. Solve for C ($C > 0$).

$$\begin{vmatrix} C & -1 & 0 & 0 \\ -1 & C & -1 & 0 \\ 0 & -1 & C & -1 \\ 0 & 0 & -1 & C \end{vmatrix} = 0$$

27. A firm plans to produce three different appliances, A, B, and C. Total production is to be 1500 appliances each week. A total of 3500 worker-hours are available for production, and 950 worker-hours are available for inspection. Each type A appliance requires 3 worker-hours for production and 1 worker-hour for inspection. Each type B appliance requires 2 worker-hours for production and 20 worker-minutes for inspection, and each type C appliance requires 2 worker-hours for production and 30 worker-minutes for inspection. How many of each are to be produced each week?

28. An alloy is to be made from four other alloys containing copper (Cu), nickel (Ni), zinc (Zn), and iron (Fe). The first is 80% Cu and 20% Ni. The second is 60% Cu, 20% Ni, and 20% Zn. The third is 30% Cu, 60% Ni, and 10% Fe. The fourth is 20% Ni, 40% Zn, and 40% Fe. How much is needed such that the final alloy has 56 g of copper, 28 g of nickel, 10 g of zinc, and 6 g of iron?

15-2 Some Properties of Determinants

Expansion of determinants by minors allows us to evaluate a determinant of any order. However, even for a fourth-order determinant, the amount of work necessary for the evaluation is usually excessive. There are a number of basic properties of determinants which allow us to perform the evaluation with considerably less work. We will present these properties here without proof, although each will be illustrated.

1. *If each element above or each element below the principal diagonal of a determinant is zero, then the product of the elements of the principal diagonal is the value of the determinant.*

EXAMPLE A

Following Property 1, we have

$$\begin{vmatrix} 2 & 1 & 5 & 8 \\ 0 & -5 & 7 & 9 \\ 0 & 0 & 4 & -6 \\ 0 & 0 & 0 & 3 \end{vmatrix} = 2(-5)(4)(3) = -120$$

Since all of the elements below the principal diagonal are zero, there is no need to expand the determinant. It will be noted, however, that if the determinant is expanded by the first column, and successive determinants are expanded by their first columns, the same value is found. Performing the expansion, we have

$$\begin{vmatrix} 2 & 1 & 5 & 8 \\ 0 & -5 & 7 & 9 \\ 0 & 0 & 4 & -6 \\ 0 & 0 & 0 & 3 \end{vmatrix} = 2\begin{vmatrix} -5 & 7 & 9 \\ 0 & 4 & -6 \\ 0 & 0 & 3 \end{vmatrix} = 2(-5)\begin{vmatrix} 4 & -6 \\ 0 & 3 \end{vmatrix}$$

$$= 2(-5)(4)(3) = -120$$

2. *If the corresponding rows and columns of a determinant are interchanged, the value of the determinant is unchanged.*

EXAMPLE B

If we interchange the first row and first column, the second row and second column, and the third row and third column of the determinant,

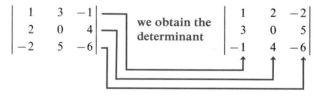

By expanding, we can show that the value of each is the same. We obtain very similar expansions which have equal values if we expand the first by the first column and the second by the first row. These expansions are

$$\begin{vmatrix} 1 & 3 & -1 \\ 2 & 0 & 4 \\ -2 & 5 & -6 \end{vmatrix} = (1)\begin{vmatrix} 0 & 4 \\ 5 & -6 \end{vmatrix} - 2\begin{vmatrix} 3 & -1 \\ 5 & -6 \end{vmatrix} + (-2)\begin{vmatrix} 3 & -1 \\ 0 & 4 \end{vmatrix}$$

$$= (-20) - 2(-13) - 2(12) = -18$$

$$\begin{vmatrix} 1 & 2 & -2 \\ 3 & 0 & 5 \\ -1 & 4 & -6 \end{vmatrix} = (1)\begin{vmatrix} 0 & 5 \\ 4 & -6 \end{vmatrix} - 2\begin{vmatrix} 3 & 5 \\ -1 & -6 \end{vmatrix} + (-2)\begin{vmatrix} 3 & 0 \\ -1 & 4 \end{vmatrix}$$

$$= (-20) - 2(-13) - 2(12) = -18$$

Therefore, we see that

$$\begin{vmatrix} 1 & 3 & -1 \\ 2 & 0 & 4 \\ -2 & 5 & -6 \end{vmatrix} = \begin{vmatrix} 1 & 2 & -2 \\ 3 & 0 & 5 \\ -1 & 4 & -6 \end{vmatrix}$$

■

3. *If two columns (or rows) of a determinant are identical, the value of the determinant is zero.*

EXAMPLE C

Property 3 tells us that

$$\text{identical} \begin{array}{c} \rightarrow \\ \rightarrow \end{array} \begin{vmatrix} 3 & 5 & 2 \\ -4 & 6 & 9 \\ -4 & 6 & 9 \end{vmatrix} = 0$$

since the second and third rows are identical. This is easily verified by expanding by the first row.

$$\begin{vmatrix} 3 & 5 & 2 \\ -4 & 6 & 9 \\ -4 & 6 & 9 \end{vmatrix} = 3\begin{vmatrix} 6 & 9 \\ 6 & 9 \end{vmatrix} - 5\begin{vmatrix} -4 & 9 \\ -4 & 9 \end{vmatrix} + 2\begin{vmatrix} -4 & 6 \\ -4 & 6 \end{vmatrix}$$

$$= 3(54 - 54) - 5(-36 + 36) + 2(-24 + 24)$$

$$= 3(0) - 5(0) + 2(0) = 0$$

■

4. *If two columns (or rows) of a determinant are interchanged, the value of the determinant is changed in sign.*

EXAMPLE D

The values of the determinants

$$\begin{vmatrix} 3 & 0 & 2 \\ 1 & 1 & 5 \\ 2 & 1 & 3 \end{vmatrix} \quad \text{and} \quad \begin{vmatrix} 2 & 0 & 3 \\ 5 & 1 & 1 \\ 3 & 1 & 2 \end{vmatrix}$$

differ in sign, since the first and third columns are interchanged. We shall verify this by expanding each by the second column.

$$\begin{vmatrix} 3 & 0 & 2 \\ 1 & 1 & 5 \\ 2 & 1 & 3 \end{vmatrix} = -(0)\begin{vmatrix} 1 & 5 \\ 2 & 3 \end{vmatrix} + (1)\begin{vmatrix} 3 & 2 \\ 2 & 3 \end{vmatrix} - (1)\begin{vmatrix} 3 & 2 \\ 1 & 5 \end{vmatrix}$$

$$= 0 + (9 - 4) - (15 - 2) = 5 - 13 = -8$$

$$\begin{vmatrix} 2 & 0 & 3 \\ 5 & 1 & 1 \\ 3 & 1 & 2 \end{vmatrix} = -(0)\begin{vmatrix} 5 & 1 \\ 3 & 2 \end{vmatrix} + (1)\begin{vmatrix} 2 & 3 \\ 3 & 2 \end{vmatrix} - (1)\begin{vmatrix} 2 & 3 \\ 5 & 1 \end{vmatrix}$$

$$= 0 + (4 - 9) - (2 - 15) = -5 + 13 = 8$$

Therefore,

$$\begin{vmatrix} 3 & 0 & 2 \\ 1 & 1 & 5 \\ 2 & 1 & 3 \end{vmatrix} = -\begin{vmatrix} 2 & 0 & 3 \\ 5 & 1 & 1 \\ 3 & 1 & 2 \end{vmatrix}$$

■

5. *If all elements of a column (or row) are multiplied by the same number k, the value of the determinant is multiplied by k.*

EXAMPLE E ———— From Property 5 we know that

$$
\begin{vmatrix} -1 & 0 & 6 \\ 6 & 3 & -6 \\ 0 & 5 & 3 \end{vmatrix} = 3 \begin{vmatrix} -1 & 0 & 6 \\ 2 & 1 & -2 \\ 0 & 5 & 3 \end{vmatrix}
$$

since each element of the second row in the left determinant is three times the corresponding element of the second row in the right determinant, and *the other rows are unchanged.*

By expansion, we can show that

$$
\begin{vmatrix} -1 & 0 & 6 \\ 6 & 3 & -6 \\ 0 & 5 & 3 \end{vmatrix} = 141 \quad \text{and} \quad \begin{vmatrix} -1 & 0 & 6 \\ 2 & 1 & -2 \\ 0 & 5 & 3 \end{vmatrix} = 47
$$

which verifies the original determinant equation, since $141 = 3(47)$.

The validity of this property can be seen by expanding each determinant by the second row. In each case one element is three times the other corresponding element, but the minors are the same. Look at the element in the second row, first column and its minor for each determinant.

$$
2 \begin{vmatrix} 0 & 6 \\ 5 & 3 \end{vmatrix} \quad \text{and} \quad 6 \begin{vmatrix} 0 & 6 \\ 5 & 3 \end{vmatrix}
$$

The element 6 in the second determinant is three times the corresponding element 2 in the first determinant, and the minors are the same. ■

EXAMPLE F ————

$$
\begin{vmatrix} 2 & 5 & -1 \\ -4 & 1 & 2 \\ 8 & 7 & -4 \end{vmatrix} = (-2) \begin{vmatrix} -1 & 5 & -1 \\ 2 & 1 & 2 \\ -4 & 7 & -4 \end{vmatrix} = 0
$$

The first step follows since each element of the first column of the first determinant is -2 times the corresponding element of the first column of the second determinant. Since the first and third columns of the second determinant are identical, the value is zero, following Property 3. ■

6. *If all the elements of any column (or row) are multiplied by the same number k, and the resulting numbers are added to the corresponding elements of another column (or row), the value of the determinant is unchanged.*

EXAMPLE G ———— The value of the following determinant at the left is unchanged if we multiply each element of the first row by 2 and add these numbers to the corresponding elements of the second row. That is,

$$
4 \times 2 = 8 \quad -1 \times 2 = -2 \quad 3 \times 2 = 6
$$

$$
\begin{vmatrix} 4 & -1 & 3 \\ 2 & 2 & 1 \\ 1 & 0 & -3 \end{vmatrix} = \begin{vmatrix} 4 & -1 & 3 \\ 2+8 & 2+(-2) & 1+6 \\ 1 & 0 & -3 \end{vmatrix} = \begin{vmatrix} 4 & -1 & 3 \\ 10 & 0 & 7 \\ 1 & 0 & -3 \end{vmatrix} \quad \text{or} \quad \begin{vmatrix} 4 & -1 & 3 \\ 10 & 0 & 7 \\ 1 & 0 & -3 \end{vmatrix} = \begin{vmatrix} 4 & -1 & 3 \\ 2 & 2 & 1 \\ 1 & 0 & -3 \end{vmatrix}
$$

When each determinant is expanded, the value -37 is obtained. The great value in property 6 is that by its use zeros can be purposely placed in the resulting determinant. ■

With the use of the properties above, determinants of higher order can be evaluated much more easily. The technique is to

obtain zeros in a given column (or row) in all positions except one.

We can then expand by this column (or row), thereby reducing the order of the determinant. Property 6 is probably the most valuable for obtaining the zeros. The following example illustrates the method.

EXAMPLE H ———— Using the properties of determinants, we have the following evaluation.

$$
\begin{vmatrix} 3 & 2 & -1 & 1 \\ -1 & 1 & 2 & 3 \\ 2 & 2 & 1 & 4 \\ 0 & -1 & -2 & 2 \end{vmatrix} = \begin{vmatrix} 0 & 5 & 5 & 10 \\ -1 & 1 & 2 & 3 \\ 2 & 2 & 1 & 4 \\ 0 & -1 & -2 & 2 \end{vmatrix}
$$

Each element of the second row is multiplied by 3, and the resulting numbers are added to the corresponding elements of the first row. Here we have used Property 6. In this way a zero has been placed in column 1, row 1.

$$
= \begin{vmatrix} 0 & 5 & 5 & 10 \\ -1 & 1 & 2 & 3 \\ 0 & 4 & 5 & 10 \\ 0 & -1 & -2 & 2 \end{vmatrix}
$$

Each element of the second row is multiplied by 2, and the resulting numbers are added to the corresponding elements of the third row. Again, we have used Property 6. Also, a zero has been placed in the first column, third row. We now have three zeros in the first column.

$$
= -(-1) \begin{vmatrix} 5 & 5 & 10 \\ 4 & 5 & 10 \\ -1 & -2 & 2 \end{vmatrix}
$$

Expand the determinant by the first column. We have now reduced the determinant to a third-order determinant.

$$
= 5 \begin{vmatrix} 1 & 1 & 2 \\ 4 & 5 & 10 \\ -1 & -2 & 2 \end{vmatrix}
$$

Factor 5 from each element of the first row. Here we are using Property 5.

$$
= 5(2) \begin{vmatrix} 1 & 1 & 1 \\ 4 & 5 & 5 \\ -1 & -2 & 1 \end{vmatrix}
$$

Factor 2 from each element of the third column. Again we are using Property 5. Also, by doing this we have reduced the size of the numbers, and the resulting numbers are somewhat easier to work with.

$$
= 10 \begin{vmatrix} 1 & 1 & 1 \\ 0 & 1 & 1 \\ -1 & -2 & 1 \end{vmatrix}
$$

Each element of the first row is multiplied by -4, and the resulting numbers are added to the corresponding elements of the second row. Here we are using Property 6. We have placed a zero in the first column, second row.

$$
= 10 \begin{vmatrix} 1 & 1 & 1 \\ 0 & 1 & 1 \\ 0 & -1 & 2 \end{vmatrix}
$$

Each element of the first row is added to the corresponding element of the third row. Again, we have used Property 6. A zero has been placed in the first column, third row. We now have two zeros in the first column.

$$
= 10(1) \begin{vmatrix} 1 & 1 \\ -1 & 2 \end{vmatrix}
$$

Expand the determinant by the first column.

$$
= 10(2 + 1) = 30
$$

Expand the second-order determinant.

A somewhat more systematic method is to place zeros below the principal diagonal and then use Property 1. However, all such techniques are essentially equivalent. ■

Exercises 15-2

In Exercises 1 through 8, evaluate each of the determinants by inspection. Careful observation will allow evaluation by the use of one or more of the basic properties of this section.

1. $\begin{vmatrix} 4 & -5 & 8 \\ 0 & 3 & -8 \\ 0 & 0 & -5 \end{vmatrix}$

2. $\begin{vmatrix} 6 & 4 & 0 \\ 0 & -2 & 3 \\ 0 & 0 & -6 \end{vmatrix}$

3. $\begin{vmatrix} -2 & 0 & 0 \\ 15 & 4 & 0 \\ 2 & -7 & 7 \end{vmatrix}$

4. $\begin{vmatrix} 3 & 0 & 0 \\ 0 & 10 & 0 \\ -9 & -1 & -5 \end{vmatrix}$

5. $\begin{vmatrix} -2 & 0 & -1 \\ 5 & 0 & 3 \\ 3 & 0 & -4 \end{vmatrix}$

6. $\begin{vmatrix} -6 & -3 & 1 \\ 1 & 2 & -5 \\ 0 & 0 & 0 \end{vmatrix}$

7. $\begin{vmatrix} 3 & -2 & 4 & 2 \\ 5 & -1 & 2 & -1 \\ 3 & -2 & 4 & 2 \\ 0 & 3 & -6 & 0 \end{vmatrix}$

8. $\begin{vmatrix} -12 & -24 & -24 & 15 \\ 12 & 32 & 32 & -35 \\ -22 & 18 & 18 & 18 \\ 44 & 0 & 0 & -26 \end{vmatrix}$

In Exercises 9 through 20, evaluate the determinants using the properties given in this section. Do not evaluate directly more than one second-order determinant for each.

9. $\begin{vmatrix} 3 & 1 & 0 \\ -2 & 3 & -1 \\ 4 & 2 & 5 \end{vmatrix}$

10. $\begin{vmatrix} 6 & -1 & 3 \\ 0 & 2 & -2 \\ -1 & 4 & 3 \end{vmatrix}$

11. $\begin{vmatrix} 5 & -1 & -2 \\ 3 & -5 & -2 \\ 1 & 4 & 6 \end{vmatrix}$

12. $\begin{vmatrix} -4 & 3 & -2 \\ -2 & 2 & 4 \\ -1 & 5 & -3 \end{vmatrix}$

13. $\begin{vmatrix} 4 & 3 & 6 & 0 \\ 3 & 0 & 0 & 4 \\ 5 & 0 & 1 & 2 \\ 2 & 1 & 1 & 7 \end{vmatrix}$

14. $\begin{vmatrix} -2 & 1 & 3 & 0 \\ 1 & 3 & 0 & 0 \\ 0 & 2 & -3 & -1 \\ 4 & -1 & 2 & 1 \end{vmatrix}$

15. $\begin{vmatrix} 3 & 1 & 2 & -1 \\ 2 & -1 & 3 & -1 \\ 1 & 2 & 1 & 3 \\ 1 & -2 & -3 & 2 \end{vmatrix}$

16. $\begin{vmatrix} 6 & -3 & -6 & 3 \\ -2 & 1 & 2 & -1 \\ 18 & 7 & -1 & 5 \\ 0 & -1 & 10 & 10 \end{vmatrix}$

17. $\begin{vmatrix} 1 & 3 & -3 & 5 \\ 4 & 2 & 1 & 2 \\ 3 & 2 & -2 & 2 \\ 0 & 1 & 2 & -1 \end{vmatrix}$

18. $\begin{vmatrix} -2 & 2 & 1 & 3 \\ 1 & 4 & 3 & 1 \\ 4 & 3 & -2 & -2 \\ 3 & -2 & 1 & 5 \end{vmatrix}$

19. $\begin{vmatrix} 1 & 2 & 0 & 1 & 0 \\ 0 & 2 & 1 & 0 & 1 \\ 1 & 0 & -1 & 1 & -1 \\ -2 & 0 & -1 & 2 & 1 \\ 1 & 0 & 2 & -1 & -2 \end{vmatrix}$

20. $\begin{vmatrix} -1 & 3 & 5 & 0 & -5 \\ 0 & 1 & 7 & 3 & -2 \\ 5 & -2 & -1 & 0 & 3 \\ -3 & 0 & 2 & -1 & 3 \\ 6 & 2 & 1 & -4 & 2 \end{vmatrix}$

In Exercises 21 through 28, solve the given systems of equations by determinants. Evaluate the determinants by the properties of determinants given in this section.

21. $2x - y + z = 5$
$x + 2y + 3z = 10$
$3x + 3y + 2z = 5$

22. $2x + y + z = 5$
$x + 3y - 3z = -13$
$3x + 2y - z = -1$

23. $3x + 2y + z = 1$
$9x + 2z = 5$
$6x - 4y - z = 3$

24. $3x + y + 2z = 4$
$x - y + 4z = 2$
$6x + 3y - 2z = 10$

25. $2x + y + z = 2$
$3y - z + 2t = 4$
$y + 2z + t = 0$
$3x + 2z = 4$

26. $2x + y + z = 0$
$x - y + 2t = 2$
$2y + z + 4t = 2$
$5x + 2z + 2t = 4$

27. $x + y + 2z = 1$
$2x - y + t = -2$
$x - y - z - 2t = 4$
$2x - y + 2z - t = 0$

28. $3x + y + t = 0$
$3z + 2t = 8$
$6x + 2y + 2z + t = 3$
$3x - y - z - t = 0$

In Exercises 29 through 32, solve the given problems by determinants.

29. In applying Kirchhoff's laws (see Exercise 20 of Section 4-6) to the circuit shown in Fig. 15-2, the following equations are found. Determine the indicated currents, in amperes.

$$I_A + I_B + I_C + I_D + I_E = 0$$
$$-2I_A + 3I_B = 0$$
$$3I_B - 3I_C = 6$$
$$-3I_C + I_D = 0$$
$$-I_D + 2I_E = 0$$

Fig. 15-2

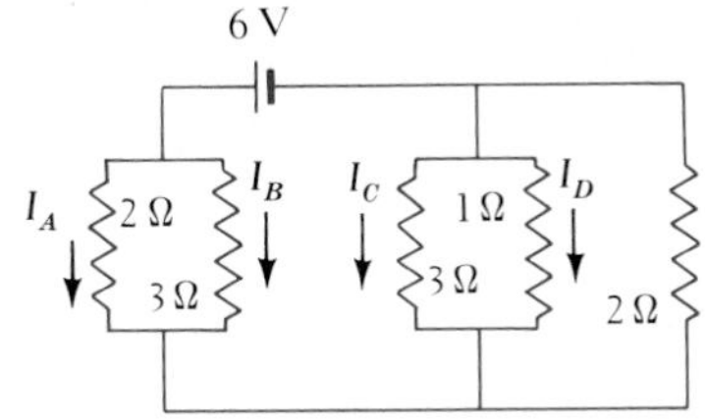

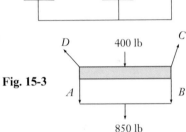

30. In analyzing the forces *A*, *B*, *C*, and *D* shown on the beam in Fig. 15-3, the following equations are used. Find these forces.

$$A + B = 850$$
$$A + B + 400 = 0.8C + 0.6D$$
$$0.6C = 0.8D$$
$$5A - 5B + 4C - 3D = 0$$

Fig. 15-3

31. In testing for air pollution, a given air sample contained a total of 6 parts per million (ppm) of four pollutants, sulfur dioxide (SO_2), nitric oxide (NO), nitrogen dioxide (NO_2), and carbon monoxide (CO). The ppm of CO was ten times that of SO_2, which in turn equaled those of NO and NO_2. There was a total of 0.8 ppm of SO_2 and NO. How many ppm of each were present in the air sample?

32. A business firm installed a system of 32 microcomputers for $24,000. Included were four different models costing $500, $600, $1000, and $1500 each, respectively. There are as many $600 models as $1000 models and $1500 models combined, and twice as many $500 models as $1000 models. How many of each model are in the computer system?

15-3   Matrices: Definitions and Basic Operations

Systems of linear equations occur in several areas of important technical and scientific applications. We indicated a few of these in Chapter 4 and in the first two sections of this chapter. Since a considerable amount of work is generally required to solve a system of equations, numerous methods have been developed for their solution.

Since the use of computers has been rapidly increasing in importance over the last several years, another mathematical concept which can be used to solve systems of equations is becoming used much more widely than in previous years. It is also used in numerous applications other than systems of equations, in such fields as business, economics, and psychology, as well as the scientific and technical areas. Since it is readily adaptable to use on a computer and is applicable to numerous areas, its importance will increase for some time to come. At this point, however, we shall only be able to introduce its definitions and basic operations.

A **matrix** *is an ordered rectangular array of numbers.* To distinguish such an array from a determinant, we shall enclose it within parentheses. As with a determinant, *the individual numbers are called* **elements** *of the matrix.*

EXAMPLE A _____

Some examples of matrices are as follows:

$$\begin{pmatrix} 2 & 8 \\ 1 & 0 \end{pmatrix} \quad \begin{pmatrix} 2 & -4 & 6 \\ -1 & 0 & 5 \end{pmatrix} \quad \begin{pmatrix} 4 & 6 \\ 0 & -1 \\ -2 & 5 \\ 3 & 0 \end{pmatrix}$$

$$\begin{pmatrix} -1 & 8 & 6 & 7 & 9 \\ 2 & 6 & 0 & 4 & 3 \\ 5 & -1 & 8 & 10 & 2 \end{pmatrix} \quad (-1 \quad 2 \quad 0 \quad 9)$$

As we can see, it is not necessary for the number of columns and number of rows to be the same, although such is the case for a determinant. However, *if the number of rows does equal the number of columns, the matrix is called a* **square matrix.** We shall find that square matrices are of some special importance. *If all the elements of a matrix are zero, the matrix is called a* **zero matrix.** We shall find it convenient to designate a given matrix by a capital letter.

NOTE ▷   We must be careful to distinguish between a matrix and a determinant. *A matrix is simply any* **rectangular array** *of numbers, whereas a determinant is a specific value which is associated with a* **square matrix.**

EXAMPLE B _____

Consider the following matrices:

$$A = \begin{pmatrix} 5 & 0 & -1 \\ 1 & 2 & 6 \\ 0 & -4 & -5 \end{pmatrix} \quad B = \begin{pmatrix} 9 \\ 8 \\ 1 \\ 5 \end{pmatrix}, \quad C = (-1 \quad 6 \quad 8 \quad 9), \quad O = \begin{pmatrix} 0 & 0 \\ 0 & 0 \end{pmatrix}$$

Matrix A is an example of a square matrix, matrix B is an example of a matrix with four rows and one column, matrix C is an example of a matrix with one row and four columns, and matrix O is an example of a zero matrix.

To be able to refer to specific elements of a matrix and to give a general representation, a double-subscript notation is usually employed. That is,

$$A = \begin{pmatrix} a_{11} & a_{12} & a_{13} \\ a_{21} & a_{22} & a_{23} \\ a_{31} & a_{32} & a_{33} \end{pmatrix}$$

row $\overset{\uparrow\nwarrow}{}$ column

We see that the first subscript refers to the row in which the element lies, and the second subscript refers to the column in which the element lies.

Two matrices are said to be **equal** *if and only if they are identical.* That is, they must have the same number of columns, the same number of rows, and the elements must respectively be equal. If these conditions are not satisfied, the matrices are not equal.

EXAMPLE C

$$\begin{pmatrix} a_{11} & a_{12} & a_{13} \\ a_{21} & a_{22} & a_{23} \end{pmatrix} = \begin{pmatrix} 1 & -5 & 0 \\ 4 & 6 & -3 \end{pmatrix}$$

if and only if $a_{11} = 1$, $a_{12} = -5$, $a_{13} = 0$, $a_{21} = 4$, $a_{22} = 6$, and $a_{23} = -3$.
The matrices

$$\begin{pmatrix} 1 & 2 & 3 \\ -1 & -2 & -5 \end{pmatrix} \quad \text{and} \quad \begin{pmatrix} 1 & 2 & -5 \\ -1 & -2 & 3 \end{pmatrix}$$

are not equal, since the elements in the third column are reversed.
The matrices

$$\begin{pmatrix} 2 & 3 \\ -1 & 5 \end{pmatrix} \quad \text{and} \quad \begin{pmatrix} 2 & 3 & 0 \\ -1 & 5 & 0 \end{pmatrix}$$

are not equal, since the number of columns is different. This is true despite the
fact that both elements of the third column are zeros. ∎

EXAMPLE D

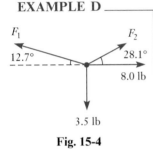

Fig. 15-4

The forces acting on a bolt are in equlibrium, as shown in Fig. 15-4. Analyzing
the horizontal and vertical components as in Section 8-4, we find the following
matrix equation. Find forces F_1 and F_2.

$$\begin{pmatrix} 0.98F_1 - 0.88F_2 \\ 0.22F_1 + 0.47F_2 \end{pmatrix} = \begin{pmatrix} 8.0 \\ 3.5 \end{pmatrix}$$

From the equality of matrices we know that $0.98F_1 - 0.88F_2 = 8.0$ and
$0.22F_1 + 0.47F_2 = 3.5$. Therefore, to find the forces F_1 and F_2 we must solve the
system of equations

$$0.98F_1 - 0.88F_2 = 8.0$$
$$0.22F_1 + 0.47F_2 = 3.5$$

Using determinants, we have

$$F_1 = \frac{\begin{vmatrix} 8.0 & -0.88 \\ 3.5 & 0.47 \end{vmatrix}}{\begin{vmatrix} 0.98 & -0.88 \\ 0.22 & 0.47 \end{vmatrix}} = \frac{8.0(0.47) - 3.5(-0.88)}{0.98(0.47) - 0.22(-0.88)} = 10.5 \text{ lb}$$

Using determinants again, or by substituting this value into either equation, we
find that $F_2 = 2.6$ lb. These values check when substituted into the original matrix equation. ∎

sum of matrices

*If two matrices have the same number of rows and the same number of columns,
their* **sum** *is defined as the matrix consisting of the sums of the corresponding elements.* If the number of rows or the number of columns of the two matrices is
not equal, they cannot be added.

EXAMPLE E

$$\begin{pmatrix} 8 & 1 & -5 & 9 \\ 0 & -2 & 3 & 7 \end{pmatrix} + \begin{pmatrix} -3 & 4 & 6 & 0 \\ 6 & -2 & 6 & 5 \end{pmatrix} = \begin{pmatrix} 8 + (-3) & 1+4 & -5+6 & 9+0 \\ 0+6 & -2+(-2) & 3+6 & 7+5 \end{pmatrix}$$

$$= \begin{pmatrix} 5 & 5 & 1 & 9 \\ 6 & -4 & 9 & 12 \end{pmatrix}$$

The matrices

$$\begin{pmatrix} 3 & -5 & 8 \\ 2 & 9 & 0 \\ 4 & -2 & 3 \end{pmatrix} \quad \text{and} \quad \begin{pmatrix} 3 & -5 & 8 & 0 \\ 2 & 9 & 0 & 0 \\ 4 & -2 & 3 & 0 \end{pmatrix}$$

cannot be added since the second matrix has one more column than the first matrix. This conclusion is not changed by the fact that the extra column contains only zeros. ∎

scalar multiplication

The product of a number and a matrix (known as **scalar multiplication** *of a matrix) is defined as the matrix whose elements are obtained by multiplying each element of the given matrix by the given number.* That is, kA is the matrix obtained by multiplying the elements of matrix A by k. In this way $A + A$ and $2A$ will result in the same matrix.

EXAMPLE F

For the matrix

$$A = \begin{pmatrix} -5 & 7 \\ 3 & 0 \end{pmatrix}$$

we have

$$2A = \begin{pmatrix} 2(-5) & 2(7) \\ 2(3) & 2(0) \end{pmatrix} = \begin{pmatrix} -10 & 14 \\ 6 & 0 \end{pmatrix}$$

Also,

$$5A = \begin{pmatrix} -25 & 35 \\ 15 & 0 \end{pmatrix} \quad \text{and} \quad -A = \begin{pmatrix} 5 & -7 \\ -3 & 0 \end{pmatrix}$$ ∎

By combining the definitions for the addition of matrices and for the scalar multiplication of a matrix, we can define the subtraction of matrices. That is, *the* **difference** *of matrices A and B is given by* $A - B = A + (-B)$. Therefore, we would change the sign of each element of B, and proceed as in addition.

By the preceding definitions we can see that the operations of addition, subtraction, and multiplication by a number of matrices are like those for real numbers. For these operations, we say that the algebra of matrices is like the algebra of real numbers. Although it is not our primary purpose to develop the algebra of matrices, we can see that the following laws hold for matrices.

$A + B = B + A$	(commutative law)	(15-4)
$A + (B + C) = (A + B) + C$	(associative law)	(15-5)
$k(A + B) = kA + kB$		(15-6)
$A + O = A$		(15-7)

Here we have let O represent the zero matrix. We shall find in the next section that not all laws for matrix operations are similar to those for real numbers.

Exercises 15-3

In Exercises 1 through 8, determine the value of the literal symbols for the following, in which the equality is properly defined.

1. $\begin{pmatrix} a & b \\ c & d \end{pmatrix} = \begin{pmatrix} 1 & -3 \\ 4 & 7 \end{pmatrix}$

2. $\begin{pmatrix} x & y & z \\ r & -s & -t \end{pmatrix} = \begin{pmatrix} -2 & 7 & -9 \\ 4 & -4 & 5 \end{pmatrix}$

3. $\begin{pmatrix} x \\ x + y \end{pmatrix} = \begin{pmatrix} 2 \\ 5 \end{pmatrix}$

4. $(a + bj \quad 2c - dj \quad -3e + fj) = (5j \quad a + 6 \quad 3b + c)$
$(j = \sqrt{-1})$

5. $\begin{vmatrix} x & x + y \\ x - z & y + z \\ x + t & y - t \end{vmatrix} = \begin{pmatrix} 2 & 3 \\ 4 & -1 \end{pmatrix}$

6. $\begin{pmatrix} 2x - 3y \\ x + 4y \end{pmatrix} = \begin{pmatrix} 13 \\ 1 \end{pmatrix}$

7. $\begin{vmatrix} x \\ x + 2 \\ 2y - 3 \end{vmatrix} = \begin{vmatrix} 4 \\ y \\ z \end{vmatrix}$

8. $\begin{pmatrix} x & y & z \\ x + y & 2x - y & x + 2 \end{pmatrix} = \begin{pmatrix} 2 & -3 \\ z & t \end{pmatrix}$

In Exercises 9 through 12, find the indicated sums of matrices.

9. $\begin{pmatrix} 2 & 3 \\ -5 & 4 \end{pmatrix} + \begin{pmatrix} -1 & 7 \\ 5 & -2 \end{pmatrix}$

10. $\begin{pmatrix} 1 & 0 & 9 \\ 3 & -5 & -2 \end{pmatrix} + \begin{pmatrix} 4 & -1 & 7 \\ 2 & 0 & -3 \end{pmatrix}$

11. $\begin{pmatrix} 50 & -82 \\ -34 & 57 \\ -15 & 62 \end{pmatrix} + \begin{pmatrix} -55 & 82 \\ 45 & 14 \\ 26 & -67 \end{pmatrix}$

12. $\begin{pmatrix} 4.7 & 2.1 & -9.6 \\ -6.8 & 4.8 & 7.4 \\ -1.9 & 0.7 & 5.9 \end{pmatrix} + \begin{pmatrix} -4.9 & -9.6 & -2.1 \\ 3.4 & 0.7 & 0.0 \\ 5.6 & 10.1 & -1.6 \end{pmatrix}$

In Exercises 13 through 20, use the following matrices to determine the indicated matrices.

$$A = \begin{pmatrix} -1 & 4 & -7 & 0 \\ 2 & -6 & -1 & 2 \end{pmatrix}, \quad B = \begin{pmatrix} 1 & 5 & -6 & 3 \\ 4 & -1 & 8 & -2 \end{pmatrix}, \quad C = \begin{pmatrix} 3 & -6 & 9 \\ -4 & 1 & 2 \end{pmatrix}$$

13. $A + B$

14. $A - B$

15. $A + C$

16. $B + C$

17. $2A + B$

18. $2B + A$

19. $A - 2B$

20. $3A - B$

In Exercises 21 through 24, use the given matrices to verify the indicated laws.

$$A = \begin{pmatrix} -1 & 2 & 3 & 7 \\ 0 & -3 & -1 & 4 \\ 9 & -1 & 0 & -2 \end{pmatrix}, \quad B = \begin{pmatrix} 4 & -1 & -3 & 0 \\ 5 & 0 & -1 & 1 \\ 1 & 11 & 8 & 2 \end{pmatrix}$$

21. $A + B = B + A$

22. $A + 0 = A$

23. $-(A - B) = B - A$

24. $3(A + B) = 3A + 3B$

In Exercises 25 and 26, find the unknown quantities in the given matrix equalities.

25. The magnitudes v_1 and v_2 of the components of the velocity vector shown in Fig. 15-5 can be found by solving for them in the following matrix equation:

$$\begin{pmatrix} v_1 \cos 35.0° - v_2 \cos 51.0° \\ v_1 \sin 35.0° + v_2 \sin 51.0° \end{pmatrix} = \begin{pmatrix} 71.8 \cos 75.0° \\ 71.8 \sin 75.0° \end{pmatrix}$$

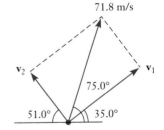

Fig. 15-5

26. The electric currents shown in Fig. 15-6 can be found by solving for them in the following matrix equation:

$$\begin{pmatrix} I_1 + I_2 + I_3 \\ -2I_1 + 3I_2 \\ -3I_2 + 6I_3 \end{pmatrix} = \begin{pmatrix} 0 \\ 24 \\ 0 \end{pmatrix}$$

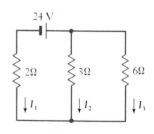

In Exercises 27 and 28, perform the indicated matrix operations.

27. The contractor of a housing development constructs four different types of houses, with either a carport, a one-car garage, or a two-car garage. The following matrix shows the number of houses of each type, and the type of garage.

	Type A	Type B	Type C	Type D
Carport	8	6	0	0
1-car garage	5	4	3	0
2-car garage	0	3	5	6

Fig. 15-6

If the contractor builds two additional identical developments, find the matrix showing the total number of each house-garage type built.

28. The inventory of a drug supply company shows that the following numbers of cases of bottles of vitamins C and E are in stock: Vitamin C—25 cases of 100-mg bottles, 10 cases of 250-mg bottles, and 32 cases of 500-mg bottles; vitamin E—30 cases of 100-mg bottles, 18 cases of 250-mg bottles, and 40 cases of 500-mg bottles. This can be represented by matrix A below. After sending two shipments, each of which can be represented by matrix B below, find the matrix representing the remaining inventory.

$$A = \begin{pmatrix} 25 & 10 & 32 \\ 30 & 18 & 40 \end{pmatrix} \qquad B = \begin{pmatrix} 10 & 5 & 6 \\ 12 & 4 & 8 \end{pmatrix}$$

15-4 Multiplication of Matrices

The definition for the multiplication of matrices does not have an intuitive basis. However, through the solution of a system of linear equations we can, at least in part, show why multiplication is defined as it is. Consider Example A.

EXAMPLE A

If we solve the system of equations

$$2x + y = 1$$
$$7x + 3y = 5$$

we obtain the solution $x = 2$, $y = -3$. In checking this solution in each of the equations, we obtain

$$2(2) + 1(-3) = 1$$
$$7(2) + 3(-3) = 5$$

Let us represent the coefficients of the equations by the matrix $\begin{pmatrix} 2 & 1 \\ 7 & 3 \end{pmatrix}$ and the solutions by the matrix $\begin{pmatrix} 2 \\ -3 \end{pmatrix}$. If we now indicate the multiplication of these matrices, and perform it as shown

$$\begin{pmatrix} 2 & 1 \\ 7 & 3 \end{pmatrix}\begin{pmatrix} 2 \\ -3 \end{pmatrix} = \begin{pmatrix} 2(2) + 1(-3) \\ 7(2) + 3(-3) \end{pmatrix} = \begin{pmatrix} 1 \\ 5 \end{pmatrix}$$

we note that we obtain a matrix which properly represents the right-side values of the equations. (Note carefully the products and sums in the resulting matrix.)

Following reasons along the lines indicated in Example A, we shall now define the **multiplication of matrices.** If the number of columns in a first matrix equals the number of rows in a second matrix, the product of these matrices is formed as follows: *The element in a specified row and a specified column of the product matrix is the sum of the products formed by multiplying each element in the specified row of the first matrix by the corresponding element in the specific column of the second matrix.* The product matrix will have the same number of rows as the first matrix and the same number of columns as the second matrix. Consider the following examples.

product of matrices

EXAMPLE B ———— Find the product AB, where

$$A = \begin{pmatrix} 2 & 1 \\ -3 & 0 \\ 1 & 2 \end{pmatrix} \quad \text{and} \quad B = \begin{pmatrix} -1 & 6 & 5 & -2 \\ 3 & 0 & 1 & -4 \end{pmatrix}$$

Since there are two columns in matrix A and two rows in matrix B, the product can be formed. To find the element in the first row and first column of the product, we find the sum of the products of corresponding elements of the first row of A and first column of B. To find the element in the first row and second column of the product, we find the sum of the products of corresponding elements in the first row of A and the second column of B. We continue this process until we have found the three rows (the number of rows in A) and the four columns (the number of columns in B) of the product. The product is formed as follows:

$$\begin{pmatrix} 2 & 1 \\ -3 & 0 \\ 1 & 2 \end{pmatrix}\begin{pmatrix} -1 & 6 & 5 & -2 \\ 3 & 0 & 1 & -4 \end{pmatrix} = \begin{pmatrix} 2(-1)+1(3) & 2(6)+1(0) & 2(5)+1(1) & 2(-2)+1(-4) \\ -3(-1)+0(3) & -3(6)+0(0) & -3(5)+0(1) & -3(-2)+0(-4) \\ 1(-1)+2(3) & 1(6)+2(0) & 1(5)+2(1) & 1(-2)+2(-4) \end{pmatrix}$$

$$= \begin{pmatrix} 1 & 12 & 11 & -8 \\ 3 & -18 & -15 & 6 \\ 5 & 6 & 7 & -10 \end{pmatrix}$$

The specific combination of elements used to form the element in the first row and first column and the element in the third row and second column of the product are outlined in color.

If we attempt to form the product BA, we find that B has four columns and A has three rows. Since the number of columns in B does not equal the number of rows in A, the product BA cannot be formed. In this way we see that $AB \neq BA$, which means that *matrix multiplication is not commutative (except in special cases).* Therefore, matrix multiplication differs from the multiplication of real numbers.

EXAMPLE C _____ The product of the two matrices below may be formed because the first matrix has four columns and the second matrix has four rows. The matrix is formed as shown:

$$\begin{pmatrix} -1 & 9 & 3 & -2 \\ 2 & 0 & -7 & 1 \end{pmatrix} \begin{pmatrix} 6 & -2 \\ 1 & 0 \\ 3 & -5 \\ 3 & 9 \end{pmatrix} = \begin{pmatrix} -1(6) + 9(1) + 3(3) + (-2)(3) & -1(-2) + 9(0) + 3(-5) + (-2)(9) \\ 2(6) + 0(1) + (-7)(3) + 1(3) & 2(-2) + 0(0) + (-7)(-5) + 1(9) \end{pmatrix}$$

$$= \begin{pmatrix} -6 + 9 + 9 - 6 & 2 + 0 - 15 - 18 \\ 12 + 0 - 21 + 3 & -4 + 0 + 35 + 9 \end{pmatrix} = \begin{pmatrix} 6 & -31 \\ -6 & 40 \end{pmatrix}$$

∎

identity matrix

There are two special matrices of particular importance in the multiplication of matrices. The first of these is the **identity matrix I,** *which is a square matrix with 1's for elements on the principal diagonal, with all other elements zero.* (The principal diagonal starts with element a_{11}.) It has the property that if it is multiplied by another square matrix with the same number of rows and columns, then the second matrix equals the product matrix.

EXAMPLE D _____ Show that $AI = IA = A$ for the matrix

$$A = \begin{pmatrix} 2 & -3 \\ 4 & 1 \end{pmatrix}$$

Since A has two rows and two columns, we choose I with two rows and two columns. Therefore, for this case

$$I = \begin{pmatrix} 1 & 0 \\ 0 & 1 \end{pmatrix} \longleftarrow \text{ elements of principal diagonal are 1's}$$

Forming the indicated products, we have results as follows:

$$AI = \begin{pmatrix} 2 & -3 \\ 4 & 1 \end{pmatrix} \begin{pmatrix} 1 & 0 \\ 0 & 1 \end{pmatrix}$$

$$= \begin{pmatrix} 2(1) + (-3)(0) & 2(0) + (-3)(1) \\ 4(1) + 1(0) & 4(0) + 1(1) \end{pmatrix} = \begin{pmatrix} 2 & -3 \\ 4 & 1 \end{pmatrix}$$

$$IA = \begin{pmatrix} 1 & 0 \\ 0 & 1 \end{pmatrix} \begin{pmatrix} 2 & -3 \\ 4 & 1 \end{pmatrix}$$

$$= \begin{pmatrix} 1(2) + 0(4) & 1(-3) + 0(1) \\ 0(2) + 1(4) & 0(-3) + 1(1) \end{pmatrix} = \begin{pmatrix} 2 & -3 \\ 4 & 1 \end{pmatrix}$$

∎ Therefore, we see that $AI = IA = A$.

inverse of matrix

For a given square matrix A, its **inverse A^{-1}** is the other important special matrix. *The matrix A and its inverse have the property that*

$$\boxed{AA^{-1} = A^{-1}A = I} \tag{15-8}$$

If the product of two square matrices equals the identity matrix, the matrices are called inverses of each other. Under certain conditions the inverse of a given square matrix may not exist, although for most square matrices the inverse does exist. In the next section we shall develop the procedure for finding the inverse of a square matrix, and the section which follows shows how the inverse is used in the solution of systems of equations. At this point we shall simply show that the product of certain matrices equals the identity matrix, and that therefore these matrices are inverses of each other.

EXAMPLE E

For the given matrices A and B, show that $AB = BA = I$, and therefore that $B = A^{-1}$.

$$A = \begin{pmatrix} 1 & -3 \\ -2 & 7 \end{pmatrix}, \quad B = \begin{pmatrix} 7 & 3 \\ 2 & 1 \end{pmatrix}$$

Forming the products AB and BA, we have the following:

$$AB = \begin{pmatrix} 1 & -3 \\ -2 & 7 \end{pmatrix}\begin{pmatrix} 7 & 3 \\ 2 & 1 \end{pmatrix} = \begin{pmatrix} 7-6 & 3-3 \\ -14+14 & -6+7 \end{pmatrix} = \begin{pmatrix} 1 & 0 \\ 0 & 1 \end{pmatrix}$$

$$BA = \begin{pmatrix} 7 & 3 \\ 2 & 1 \end{pmatrix}\begin{pmatrix} 1 & -3 \\ -2 & 7 \end{pmatrix} = \begin{pmatrix} 7-6 & -21+21 \\ 2-2 & -6+7 \end{pmatrix} = \begin{pmatrix} 1 & 0 \\ 0 & 1 \end{pmatrix}$$

■ Since $AB = I$ and $BA = I$, $B = A^{-1}$ and $A = B^{-1}$.

The following example illustrates one kind of application of the multiplication of matrices.

EXAMPLE F

See the chapter introduction.

A particular firm produces three types of machine parts. On a given day it produces 40 of type X, 50 of type Y, and 80 of type Z. Each of type X requires 4 units of material and 1 worker-hour to produce; each of type Y requires 5 units of material and 2 worker-hours to produce; each of type Z requires 3 units of material and 2 worker-hours to produce. By representing the number of each type produced as matrix A, and the material and time requirements as matrix B, we have

$$\begin{array}{c} \text{type X type Y type Z} \\ \downarrow \quad\quad \downarrow \quad\quad \downarrow \\ A = (40 \quad\quad 50 \quad\quad 80) \\ \\ \text{number of each type} \\ \text{produced} \end{array} \qquad \begin{array}{c} \text{units of   worker-hours} \\ \text{material} \quad\quad \downarrow \\ \downarrow \quad\quad \downarrow \\ B = \begin{pmatrix} 4 & 1 \\ 5 & 2 \\ 3 & 2 \end{pmatrix} \begin{array}{l} \leftarrow \text{type X} \\ \leftarrow \text{type Y} \\ \leftarrow \text{type Z} \end{array} \\ \\ \text{material and time} \\ \text{required for each} \end{array}$$

The product AB gives the total number of units of material and the total number of worker-hours needed for the day's production in a one-row, two-column matrix.

$$AB = (40 \quad 50 \quad 80)\begin{pmatrix} 4 & 1 \\ 5 & 2 \\ 3 & 2 \end{pmatrix} \qquad \begin{array}{c} \text{total} \quad\quad \text{total} \\ \text{units of worker-hours} \\ \text{material} \\ \downarrow \end{array}$$

$$= (160 + 250 + 240 \quad 40 + 100 + 160) = (650 \quad 300)$$

■ Therefore, 650 units of material and 300 worker-hours are required.

We now have seen how multiplication is defined for matrices. We see that *matrix multiplication is not commutative;* that is, $AB \neq BA$ in general. This is a major difference from the multiplication of real numbers. Another difference is that it is possible that $AB = O$, even though neither A nor B is O. There are some similarities, however, in that $AI = A$, where we make I and the number 1 equivalent for the two types of multiplication. Also, *the distributive property $A(B + C) = AB + AC$ holds for matrix multiplication.* This points out some more of the properties of the algebra of matrices.

Exercises 15-4

In Exercises 1 through 12, perform the indicated matrix multiplications.

1. $(4 \quad -2)\begin{pmatrix} -1 & 0 \\ 2 & 6 \end{pmatrix}$

2. $(-1 \quad 5 \quad -2)\begin{pmatrix} 6 & 3 \\ 2 & -1 \\ 0 & 2 \end{pmatrix}$

3. $\begin{pmatrix} 2 & -3 \\ 5 & -1 \end{pmatrix}\begin{pmatrix} 3 & 0 & -1 \\ 7 & -5 & 8 \end{pmatrix}$

4. $\begin{pmatrix} -7 & 8 \\ 5 & 0 \end{pmatrix}\begin{pmatrix} -9 & 10 \\ 1 & 4 \end{pmatrix}$

5. $\begin{pmatrix} 2 & -3 & 1 \\ 0 & 7 & -3 \end{pmatrix}\begin{pmatrix} 9 \\ -2 \\ 5 \end{pmatrix}$

6. $\begin{pmatrix} 0 & -1 & 2 \\ 4 & 11 & 2 \end{pmatrix}\begin{pmatrix} 3 & -1 \\ 1 & 2 \\ 6 & 1 \end{pmatrix}$

7. $\begin{pmatrix} -1.7 & -5.6 \\ 4.6 & 0.0 \\ 2.5 & 0.4 \end{pmatrix}\begin{pmatrix} 2.5 & 0.0 \\ 1.6 & 1.7 \end{pmatrix}$

8. $\begin{pmatrix} 12 & -47 \\ 43 & -18 \\ 36 & -22 \end{pmatrix}\begin{pmatrix} 25 & -10 & 32 \\ 66 & -37 & 92 \end{pmatrix}$

9. $\begin{pmatrix} -1 & 7 \\ 3 & 5 \\ 10 & -1 \\ -5 & 12 \end{pmatrix}\begin{pmatrix} 2 & 1 & 0 \\ 5 & -3 & 1 \end{pmatrix}$

10. $\begin{pmatrix} 3 & -1 & 8 \\ 0 & 2 & -4 \\ -1 & 6 & 7 \end{pmatrix}\begin{pmatrix} 7 & -1 \\ 0 & 3 \\ 1 & -2 \end{pmatrix}$

11. $\begin{pmatrix} -9 & -1 & 4 \\ 6 & 9 & -1 \end{pmatrix}\begin{pmatrix} 6 & -5 \\ 4 & 1 \\ -1 & 6 \end{pmatrix}$

12. $\begin{pmatrix} 1 & 2 & -6 & 6 & 1 \\ -2 & 4 & 0 & 1 & 2 \end{pmatrix}\begin{pmatrix} 1 \\ -1 \\ 0 \\ 5 \\ 2 \end{pmatrix}$

In Exercises 13 through 16, find, if possible, AB and BA.

13. $A = (1 \quad -3 \quad 8), \quad B = \begin{pmatrix} -1 \\ 5 \\ 7 \end{pmatrix}$

14. $A = \begin{pmatrix} -3 & 2 & 0 \\ 1 & -4 & 5 \end{pmatrix}, \quad B = \begin{pmatrix} -2 & 0 \\ 4 & -6 \\ 5 & 1 \end{pmatrix}$

15. $A = \begin{pmatrix} -1 & 2 & 3 \\ 5 & -1 & 0 \end{pmatrix}, \quad B = \begin{pmatrix} 1 \\ -5 \\ 2 \end{pmatrix}$

16. $A = \begin{pmatrix} -2 & 1 & 7 \\ 3 & -1 & 0 \\ 0 & 2 & -1 \end{pmatrix}, \quad B = (4 \quad -1 \quad 5)$

In Exercises 17 through 20, show that $AI = IA = A$.

17. $A = \begin{pmatrix} 1 & 8 \\ -2 & 2 \end{pmatrix}$

18. $A = \begin{pmatrix} -3 & 4 \\ 1 & 2 \end{pmatrix}$

19. $A = \begin{pmatrix} 1 & 3 & -5 \\ 2 & 0 & 1 \\ 1 & -2 & 4 \end{pmatrix}$

20. $A = \begin{pmatrix} -1 & 2 & 0 \\ 4 & -3 & 1 \\ 2 & 1 & 3 \end{pmatrix}$

In Exercises 21 through 24, determine whether or not $B = A^{-1}$.

21. $A = \begin{pmatrix} 5 & -2 \\ -2 & 1 \end{pmatrix}$, $B = \begin{pmatrix} 1 & 2 \\ 2 & 5 \end{pmatrix}$ **22.** $A = \begin{pmatrix} 3 & -4 \\ 5 & -7 \end{pmatrix}$, $B = \begin{pmatrix} 7 & -4 \\ 5 & -2 \end{pmatrix}$

23. $A = \begin{pmatrix} 1 & -2 & 3 \\ 2 & -5 & 7 \\ -1 & 3 & -5 \end{pmatrix}$, $B = \begin{pmatrix} 4 & -1 & 1 \\ 3 & -2 & -1 \\ 1 & -1 & -1 \end{pmatrix}$ **24.** $A = \begin{pmatrix} 1 & -1 & 3 \\ 3 & -4 & 8 \\ -2 & 3 & -4 \end{pmatrix}$, $B = \begin{pmatrix} 8 & -5 & -4 \\ 4 & -2 & -1 \\ -1 & 1 & 1 \end{pmatrix}$

In Exercises 25 through 28, determine by matrix multiplication whether or not A is the proper matrix of solution values.

25. $3x - 2y = -1$
$4x + y = 6$ $A = \begin{pmatrix} 1 \\ 2 \end{pmatrix}$

26. $4x + y = -5$
$3x + 4y = 6$ $A = \begin{pmatrix} -2 \\ 3 \end{pmatrix}$

27. $3x + y + 2z = 1$
$x - 3y + 4z = -3$
$2x + 2y + z = 1$ $A = \begin{pmatrix} -1 \\ 2 \\ 1 \end{pmatrix}$

28. $2x - y + z = 7$
$x - 3y + 2z = 6$
$3x + y - z = 8$ $A = \begin{pmatrix} 3 \\ -2 \\ -1 \end{pmatrix}$

In Exercise 29 through 36, perform the indicated matrix multiplications.

29. For the identity matrix with two rows and two columns, show that $(-I)^2 = I$.

30. Show that $N^2 = -I$, where $N = \begin{pmatrix} 0 & -1 \\ 1 & 0 \end{pmatrix}$.   **31.** Show that $A^2 - I = (A + I)(A - I)$ for $A = \begin{pmatrix} 2 & 4 \\ 3 & 5 \end{pmatrix}$.

32. For $J = \begin{pmatrix} j & 0 \\ 0 & j \end{pmatrix}$, where $j = \sqrt{-1}$, show that $J^2 = -I$, $J^3 = -J$, and $J^4 = I$. Note the similarity with j^2, j^3, and j^4, and compare J^2 with N^2 of Exercise 30.

33. In studying the motion of electrons, one of the Pauli spin matrices used is $s_y = \begin{pmatrix} 0 & -j \\ j & 0 \end{pmatrix}$, where $j = \sqrt{-1}$. Show that $s_y^2 = I$.

34. In analyzing the motion of a robotic mechanism, the following matrix multiplication is used. Perform the multiplication and evaluate each element.

$$\begin{pmatrix} \cos 60° & -\sin 60° & 0 \\ \sin 60° & \cos 60° & 0 \\ 0 & 0 & 1 \end{pmatrix}\begin{pmatrix} 2 \\ 4 \\ 0 \end{pmatrix}$$

35. An *ammeter* is used to measure the current in an electric circuit. Nearly all of the current passes through a low-resistance *shunt*, and the remaining known fraction of current is measured by the meter. See Fig. 15-7. The voltages, currents, and resistance of the meter are given by the matrix equation below. Find the individual equations for v_2 and i_2 in terms of whichever of v_1, i_1, and R may be applicable.

$$\begin{pmatrix} v_2 \\ i_2 \end{pmatrix} = \begin{pmatrix} 1 & 0 \\ -\dfrac{1}{R} & 1 \end{pmatrix}\begin{pmatrix} v_1 \\ i_1 \end{pmatrix}$$

Fig. 15-7

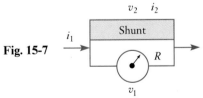

36. In the theory related to the reproduction of color photography, the equations

$$\begin{pmatrix} X \\ Y \\ Z \end{pmatrix} = \begin{pmatrix} 1.0 & 0.1 & 0 \\ 0.5 & 1.0 & 0.1 \\ 0.3 & 0.4 & 1.0 \end{pmatrix}\begin{pmatrix} x \\ y \\ z \end{pmatrix}$$

are found. The X, Y, and Z represent the red, green, and blue densities of the reproductions, respectively, and the x, y, and z represent the red, green, and blue densities, respectively, of the subject. Give the equations relating X, Y, and Z and x, y, and z.

15-5   Finding the Inverse of a Matrix

In the last section we introduced the concept of the inverse of a matrix. In this section we shall show how the inverse is found, and in the following section we shall show how this inverse is used in the solution of a system of linear equations.

inverse of a
2 × 2 matrix

We shall first show two methods of finding the inverse of a two-row, two-column (2 × 2) matrix. The first method is as follows:

1. *Interchange the elements on the principal diagonal.*

2. *Change the signs of the off-diagonal elements.*

3. *Divide each resulting element by the determinant of the given matrix.*

This method, which can be used with second-order square matrices *but not with higher-order matrices,* is illustrated in the following example.

EXAMPLE A

Find the inverse of the matrix

$$A = \begin{pmatrix} 2 & -3 \\ 4 & -7 \end{pmatrix}$$

First we interchange the elements on the principal diagonal and change the signs of the off-diagonal elements. This gives us the matrix

$$\begin{pmatrix} -7 & 3 \\ -4 & 2 \end{pmatrix}$$ ——— signs changed
——— elements interchanged

Now we find the determinant of the original matrix, which means we evaluate

$$\begin{vmatrix} 2 & -3 \\ 4 & -7 \end{vmatrix} = -14 - (-12) = -2$$

(Note again that the matrix is the array of numbers, whereas the determinant of the matrix has a value associated with it.) We now divide each element of the second matrix by -2. This gives

$$A^{-1} = \frac{1}{-2}\begin{pmatrix} -7 & 3 \\ -4 & 2 \end{pmatrix} = \begin{pmatrix} \dfrac{-7}{-2} & \dfrac{3}{-2} \\ \dfrac{-4}{-2} & \dfrac{2}{-2} \end{pmatrix} = \begin{pmatrix} \dfrac{7}{2} & -\dfrac{3}{2} \\ 2 & -1 \end{pmatrix}$$

This last matrix is the inverse of matrix A.

Checking by multiplication gives

$$AA^{-1} = \begin{pmatrix} 2 & -3 \\ 4 & -7 \end{pmatrix}\begin{pmatrix} \frac{7}{2} & -\frac{3}{2} \\ 2 & -1 \end{pmatrix} = \begin{pmatrix} 7 - 6 & -3 + 3 \\ 14 - 14 & -6 + 7 \end{pmatrix} = \begin{pmatrix} 1 & 0 \\ 0 & 1 \end{pmatrix} = I$$

■ Since $AA^{-1} = I$, the matrix A^{-1} is the proper inverse matrix.

Gauss-Jordan
method

NOTE ▷

The second method, called the *Gauss-Jordan method,* involves transforming the given matrix into the identity matrix while at the same time **transforming the identity matrix into the inverse.** There are three types of steps allowable in making these transformations:

1. *Any two rows may be interchanged.*

2. *Every element in any row may be multiplied by any given number other than zero.*

3. *Any row may be replaced by a row whose elements are the sum of a non-zero multiple of itself and a nonzero multiple of another row.*

Note carefully that these are row operations and not column operations.

Some reflection shows that these **row operations** are those which are performed in solving a system of equations by addition or subtraction. The following example illustrates the method.

EXAMPLE B Find the inverse of the matrix

$$A = \begin{pmatrix} 2 & -3 \\ 4 & -7 \end{pmatrix}$$

First we set up the given matrix along with the identity matrix in the following manner:

$$\begin{pmatrix} 2 & -3 & | & 1 & 0 \\ 4 & -7 & | & 0 & 1 \end{pmatrix}$$

The vertical line simply shows the separation of the two matrices.

We wish to transform the left matrix into the identity matrix. Therefore, the first requirement is a 1 for element a_{11}. Therefore, we divide all elements of the first row by 2. This gives the following setup:

$$\begin{pmatrix} 1 & -\frac{3}{2} & | & \frac{1}{2} & 0 \\ 4 & -7 & | & 0 & 1 \end{pmatrix}$$

Next we want to have a zero for element a_{21}. Therefore, we shall subtract 4 times each element of row 1 from the corresponding element in row 2, replacing the elements of row 2. This gives us the following setup:

$$\begin{pmatrix} 1 & -\frac{3}{2} & | & \frac{1}{2} & 0 \\ 4-4(1) & -7-4(-\frac{3}{2}) & | & 0-4(\frac{1}{2}) & 1-4(0) \end{pmatrix} \quad \text{or} \quad \begin{pmatrix} 1 & -\frac{3}{2} & | & \frac{1}{2} & 0 \\ 0 & -1 & | & -2 & 1 \end{pmatrix}$$

Next, we want to have 1, not -1, for element a_{22}. Therefore, we multiply each element of row 2 by -1. This gives

$$\begin{pmatrix} 1 & -\frac{3}{2} & | & \frac{1}{2} & 0 \\ 0 & 1 & | & 2 & -1 \end{pmatrix}$$

Finally, we want zero for element a_{12}. Therefore, we add $\frac{3}{2}$ times each element of row 2 to the corresponding elements of row 1, replacing row 1. This gives

$$\begin{pmatrix} 1+\frac{3}{2}(0) & -\frac{3}{2}+\frac{3}{2}(1) & | & \frac{1}{2}+\frac{3}{2}(2) & 0+\frac{3}{2}(-1) \\ 0 & 1 & | & 2 & -1 \end{pmatrix} \quad \text{or} \quad \begin{pmatrix} 1 & 0 & | & \frac{7}{2} & -\frac{3}{2} \\ 0 & 1 & | & 2 & -1 \end{pmatrix}$$

At this point, we have transformed the given matrix into the identity matrix, and the identity matrix into the inverse. Therefore, the matrix to the right of the vertical bar in the last setup is the required inverse. Thus,

$$A^{-1} = \begin{pmatrix} \frac{7}{2} & -\frac{3}{2} \\ 2 & -1 \end{pmatrix}$$

This is the same matrix and inverse as illustrated in Example A.

The idea to be noted most carefully in Example B is the order in which the zeros and ones were placed in transforming the given matrix to the identity matrix. We shall now give another example of finding the inverse for a 2×2 matrix, and then we shall find the inverse for a 3×3 matrix with the same method. This method is applicable for a square matrix of any number of rows or columns.

EXAMPLE C   Find the inverse of the matrix $\begin{pmatrix} -3 & 6 \\ 4 & 5 \end{pmatrix}$.

original setup

$$\begin{pmatrix} -3 & 6 & | & 1 & 0 \\ 4 & 5 & | & 0 & 1 \end{pmatrix} \rightarrow \begin{pmatrix} 1 & -2 & | & -\frac{1}{3} & 0 \\ 0 & 13 & | & \frac{4}{3} & 1 \end{pmatrix} \rightarrow \begin{pmatrix} 1 & 0 & | & -\frac{5}{39} & \frac{2}{13} \\ 0 & 1 & | & \frac{4}{39} & \frac{1}{13} \end{pmatrix}$$

row 1 divided by -3 row 2 divided by 13 I A^{-1}

$$\begin{pmatrix} 1 & -2 & | & -\frac{1}{3} & 0 \\ 4 & 5 & | & 0 & 1 \end{pmatrix} \quad \begin{pmatrix} 1 & -2 & | & -\frac{1}{3} & 0 \\ 0 & 1 & | & \frac{4}{39} & \frac{1}{13} \end{pmatrix}$$

-4 times row 1   2 times row 2
added to row 2   added to row 1

Therefore, $A^{-1} = \begin{pmatrix} -\frac{5}{39} & \frac{2}{13} \\ \frac{4}{39} & \frac{1}{13} \end{pmatrix}$, which can be checked by multiplication. ■

EXAMPLE D   Find the inverse of the matrix $\begin{pmatrix} 1 & 2 & -1 \\ 3 & 5 & -1 \\ -2 & -1 & -2 \end{pmatrix}$.

original setup

$$\begin{pmatrix} 1 & 2 & -1 & | & 1 & 0 & 0 \\ 3 & 5 & -1 & | & 0 & 1 & 0 \\ -2 & -1 & -2 & | & 0 & 0 & 1 \end{pmatrix} \rightarrow \begin{pmatrix} 1 & 2 & -1 & | & 1 & 0 & 0 \\ 0 & 1 & -2 & | & 3 & -1 & 0 \\ 0 & 3 & -4 & | & 2 & 0 & 1 \end{pmatrix} \rightarrow \begin{pmatrix} 1 & 0 & 3 & | & -5 & 2 & 0 \\ 0 & 1 & -2 & | & 3 & -1 & 0 \\ 0 & 0 & 1 & | & -\frac{7}{2} & \frac{3}{2} & \frac{1}{2} \end{pmatrix}$$

-3 times row 1 -2 times row 2   2 times row 3
added to row 2   added to row 1   added to row 2

$$\begin{pmatrix} 1 & 2 & -1 & | & 1 & 0 & 0 \\ 0 & -1 & 2 & | & -3 & 1 & 0 \\ -2 & -1 & -2 & | & 0 & 0 & 1 \end{pmatrix} \quad \begin{pmatrix} 1 & 0 & 3 & | & -5 & 2 & 0 \\ 0 & 1 & -2 & | & 3 & -1 & 0 \\ 0 & 3 & -4 & | & 2 & 0 & 1 \end{pmatrix} \quad \begin{pmatrix} 1 & 0 & 3 & | & -5 & 2 & 0 \\ 0 & 1 & 0 & | & -4 & 2 & 1 \\ 0 & 0 & 1 & | & -\frac{7}{2} & \frac{3}{2} & \frac{1}{2} \end{pmatrix}$$

2 times row 1 -3 times row 2 -3 times row 3
added to row 3   added to row 3   added to row 1

$$\begin{pmatrix} 1 & 2 & -1 & | & 1 & 0 & 0 \\ 0 & -1 & 2 & | & -3 & 1 & 0 \\ 0 & 3 & -4 & | & 2 & 0 & 1 \end{pmatrix} \quad \begin{pmatrix} 1 & 0 & 3 & | & -5 & 2 & 0 \\ 0 & 1 & -2 & | & 3 & -1 & 0 \\ 0 & 0 & 2 & | & -7 & 3 & 1 \end{pmatrix} \quad \begin{pmatrix} 1 & 0 & 0 & | & \frac{11}{2} & -\frac{5}{2} & -\frac{3}{2} \\ 0 & 1 & 0 & | & -4 & 2 & 1 \\ 0 & 0 & 1 & | & -\frac{7}{2} & \frac{3}{2} & \frac{1}{2} \end{pmatrix}$$

row 2 multiplied by -1 row 3 divided by 2 I A^{-1}

Therefore, the required inverse matrix is

$$\begin{pmatrix} \frac{11}{2} & -\frac{5}{2} & -\frac{3}{2} \\ -4 & 2 & 1 \\ -\frac{7}{2} & \frac{3}{2} & \frac{1}{2} \end{pmatrix}$$

■ which may be checked by multiplication.

In transforming a matrix into the identity matrix, we work on one column at a time, transforming the columns in order from left to right. It is generally wisest to make the element on the principal diagonal for the column 1 first, and then to make all the other elements in the column 0. Looking back to Example D, we see that this procedure has been systematically followed, first on column 1, then on column 2, and finally on column 3.

There are other methods of finding the inverse of a matrix. One of these other methods is shown in Exercises 25 through 28, which follow.

Exercises 15-5

In Exercises 1 through 8, find the inverse of each of the given matrices by the method of Example A of this section.

1. $\begin{pmatrix} 2 & -5 \\ -2 & 4 \end{pmatrix}$

2. $\begin{pmatrix} -6 & 3 \\ 3 & -2 \end{pmatrix}$

3. $\begin{pmatrix} -1 & 5 \\ 4 & 10 \end{pmatrix}$

4. $\begin{pmatrix} 8 & -1 \\ -4 & -5 \end{pmatrix}$

5. $\begin{pmatrix} 0 & -4 \\ 2 & 6 \end{pmatrix}$

6. $\begin{pmatrix} 7 & -2 \\ -6 & 2 \end{pmatrix}$

7. $\begin{pmatrix} -50 & -45 \\ 26 & 80 \end{pmatrix}$

8. $\begin{pmatrix} 7.2 & -3.6 \\ -1.3 & -5.7 \end{pmatrix}$

See Appendix E for a computer program for finding the inverse of a 2 × 2 matrix.

In Exercises 9 through 24, find the inverse of each of the given matrices by transforming the identity matrix, as in Examples B through D.

9. $\begin{pmatrix} 1 & 2 \\ 2 & 3 \end{pmatrix}$

10. $\begin{pmatrix} 1 & 5 \\ -1 & -4 \end{pmatrix}$

11. $\begin{pmatrix} 2 & 4 \\ -1 & -1 \end{pmatrix}$

12. $\begin{pmatrix} -2 & 6 \\ 3 & -4 \end{pmatrix}$

13. $\begin{pmatrix} 2 & 5 \\ -1 & 2 \end{pmatrix}$

14. $\begin{pmatrix} -2 & 3 \\ -3 & 5 \end{pmatrix}$

15. $\begin{pmatrix} 2 & -1 \\ 4 & 6 \end{pmatrix}$

16. $\begin{pmatrix} 1 & -3 \\ 7 & -5 \end{pmatrix}$

17. $\begin{pmatrix} 1 & -3 & -2 \\ -2 & 7 & 3 \\ 1 & -1 & -3 \end{pmatrix}$

18. $\begin{pmatrix} 1 & 2 & -1 \\ 3 & 7 & -5 \\ -1 & -2 & 0 \end{pmatrix}$

19. $\begin{pmatrix} 1 & -1 & -3 \\ 0 & -1 & -2 \\ 2 & 1 & -1 \end{pmatrix}$

20. $\begin{pmatrix} 1 & 4 & 1 \\ -3 & -13 & -1 \\ 0 & -2 & 5 \end{pmatrix}$

21. $\begin{pmatrix} 1 & 3 & 2 \\ -2 & -5 & -1 \\ 2 & 4 & 0 \end{pmatrix}$

22. $\begin{pmatrix} 1 & 3 & 4 \\ -1 & -4 & -2 \\ 4 & 9 & 20 \end{pmatrix}$

23. $\begin{pmatrix} 2 & 4 & 0 \\ 3 & 4 & -2 \\ -1 & 1 & 2 \end{pmatrix}$

24. $\begin{pmatrix} -2 & 6 & 1 \\ 0 & 3 & -3 \\ 4 & -7 & 3 \end{pmatrix}$

In Exercises 25 through 28, find the inverse of each of the given matrices (same as those for Exercises 21 through 24) by use of the following information. For matrix A, its inverse A^{-1} is found from

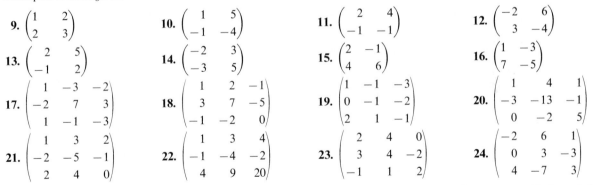

$$A = \begin{pmatrix} a_{11} & a_{12} & a_{13} \\ a_{21} & a_{22} & a_{23} \\ a_{31} & a_{32} & a_{33} \end{pmatrix}$$

$$A^{-1} = \frac{1}{|A|} \begin{pmatrix} \begin{vmatrix} a_{22} & a_{23} \\ a_{32} & a_{33} \end{vmatrix} & -\begin{vmatrix} a_{12} & a_{13} \\ a_{32} & a_{33} \end{vmatrix} & \begin{vmatrix} a_{12} & a_{13} \\ a_{22} & a_{23} \end{vmatrix} \\ -\begin{vmatrix} a_{21} & a_{23} \\ a_{31} & a_{33} \end{vmatrix} & \begin{vmatrix} a_{11} & a_{13} \\ a_{31} & a_{33} \end{vmatrix} & -\begin{vmatrix} a_{11} & a_{13} \\ a_{21} & a_{23} \end{vmatrix} \\ \begin{vmatrix} a_{21} & a_{22} \\ a_{31} & a_{32} \end{vmatrix} & -\begin{vmatrix} a_{11} & a_{12} \\ a_{31} & a_{32} \end{vmatrix} & \begin{vmatrix} a_{11} & a_{12} \\ a_{21} & a_{22} \end{vmatrix} \end{pmatrix}$$

25. $\begin{pmatrix} 1 & 3 & 2 \\ -2 & -5 & -1 \\ 2 & 4 & 0 \end{pmatrix}$

26. $\begin{pmatrix} 1 & 3 & 4 \\ -1 & -4 & -2 \\ 4 & 9 & 20 \end{pmatrix}$

27. $\begin{pmatrix} 2 & 4 & 0 \\ 3 & 4 & -2 \\ -1 & 1 & 2 \end{pmatrix}$

28. $\begin{pmatrix} -2 & 6 & 1 \\ 0 & 3 & -3 \\ 4 & -7 & 3 \end{pmatrix}$

In Exercises 29 and 30, perform the indicated matrix operations. They verify the validity of the method of Example A.

29. For the matrix $A = \begin{pmatrix} a & b \\ c & d \end{pmatrix}$, show that $\dfrac{1}{ad - bc} \begin{pmatrix} a & b \\ c & d \end{pmatrix} \begin{pmatrix} d & -b \\ -c & a \end{pmatrix} = \begin{pmatrix} 1 & 0 \\ 0 & 1 \end{pmatrix}$.

30. Find the inverse of matrix A of Exercise 29 by the method of Examples B through D.

In Exercises 31 and 32, solve the given problems.

31. For the *four-terminal network* shown in Fig. 15-8, it can be shown that the voltage matrix V is related to the coefficient matrix A and the current matrix I by $V = A^{-1}I$, where

$$V = \begin{pmatrix} v_1 \\ v_2 \end{pmatrix}, \qquad A = \begin{pmatrix} a_{11} & a_{12} \\ a_{21} & a_{22} \end{pmatrix}, \qquad I = \begin{pmatrix} i_1 \\ i_2 \end{pmatrix}$$

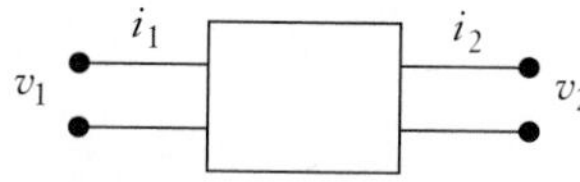

Fig. 15-8

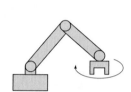

Find the individual equations for v_1 and v_2 which give each in terms of i_1 and i_2.

32. The rotations of a robot arm such as that shown in Fig. 15-9 are often represented by matrices. The values represent trigonometric functions of the angles of rotation. For the following rotation matrix R, find R^{-1}.

$$R = \begin{pmatrix} 0.8 & 0.0 & -0.6 \\ 0.0 & 1.0 & 0.0 \\ 0.6 & 0.0 & 0.8 \end{pmatrix}$$

Fig. 15-9

15-6 Matrices and Linear Equations

As we stated earlier, matrices can be used to solve systems of equations. In this section we shall show one of the methods by which this is done.

Let us consider the system of equations

$$a_1 x + b_1 y = c_1$$
$$a_2 x + b_2 y = c_2$$

Recalling the definition of equality of matrices, we can write this system as

$$\begin{pmatrix} a_1 x + b_1 y \\ a_2 x + b_2 y \end{pmatrix} = \begin{pmatrix} c_1 \\ c_2 \end{pmatrix} \tag{15-9}$$

If we let

$$A = \begin{pmatrix} a_1 & b_1 \\ a_2 & b_2 \end{pmatrix}, \qquad X = \begin{pmatrix} x \\ y \end{pmatrix}, \qquad C = \begin{pmatrix} c_1 \\ c_2 \end{pmatrix} \tag{15-10}$$

then the left side of Eq. (15-9) can be written as the product of matrices A and X, and we have

$$AX = C \tag{15-11}$$

If we now multiply (on the left) each side of this matrix equation by A^{-1}, we have

$$A^{-1}AX = A^{-1}C$$

Since $A^{-1}A = I$, we have

$$IX = A^{-1}C$$

However, $IX = X$. Therefore,

$$\boxed{X = A^{-1}C} \tag{15-12}$$

Equation (15-12) states that *we can solve a system of linear equations by multiplying the one-column matrix of the constants on the right by the inverse of the matrix of the coefficients.* The result is a one-column matrix whose elements are the required values. Note carefully that $X = A^{-1}C$ and **not** CA^{-1}. The following examples illustrate the method.

EXAMPLE A — Use matrices to solve the system of equations

$$2x - y = 7$$
$$5x - 3y = 18$$

We set up the matrix of coefficients and the matrix of constants as

$$A = \begin{pmatrix} 2 & -1 \\ 5 & -3 \end{pmatrix} \quad \text{and} \quad C = \begin{pmatrix} 7 \\ 18 \end{pmatrix}$$

By either of the methods of the previous section, we can determine the inverse of matrix A to be

$$A^{-1} = \begin{pmatrix} 3 & -1 \\ 5 & -2 \end{pmatrix}$$

We now form the matrix product $A^{-1}C$.

$$A^{-1}C = \begin{pmatrix} 3 & -1 \\ 5 & -2 \end{pmatrix}\begin{pmatrix} 7 \\ 18 \end{pmatrix} = \begin{pmatrix} 21 - 18 \\ 35 - 36 \end{pmatrix} = \begin{pmatrix} 3 \\ -1 \end{pmatrix}$$

Since $X = A^{-1}C$, this means that

$$\begin{pmatrix} x \\ y \end{pmatrix} = \begin{pmatrix} 3 \\ -1 \end{pmatrix}$$

Therefore, the required solution is $x = 3$ and $y = -1$, which checks when these values are substituted into the original equations. ∎

EXAMPLE B — For the electric circuit shown in Fig. 15-10, the equations used to find the currents (in amperes) i_1 and i_2 are

$$2.30i_1 + 6.45(i_1 + i_2) = 15.0 \qquad 8.75i_1 + 6.45i_2 = 15.0$$
$$\text{or}$$
$$1.25i_2 + 6.45(i_1 + i_2) = 12.5 \qquad 6.45i_1 + 7.70i_2 = 12.5$$

Using matrices to solve this system of equations, we set up the matrix A of

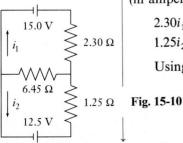

Fig. 15-10

(*Continued on next page*)

coefficients, the matrix C of constants, and the matrix X of currents as

$$A = \begin{pmatrix} 8.75 & 6.45 \\ 6.45 & 7.70 \end{pmatrix}, \qquad C = \begin{pmatrix} 15.0 \\ 12.5 \end{pmatrix}, \qquad \text{and} \qquad X = \begin{pmatrix} i_1 \\ i_2 \end{pmatrix}$$

We now find the inverse of A as

$$A^{-1} = \frac{1}{8.75(7.70) - 6.45(6.45)} \begin{pmatrix} 7.70 & -6.45 \\ -6.45 & 8.75 \end{pmatrix} = \begin{pmatrix} 0.2988 & -0.2503 \\ -0.2503 & 0.3395 \end{pmatrix}$$

Therefore,

$$X = A^{-1}C = \begin{pmatrix} 0.2988 & -0.2503 \\ -0.2503 & 0.3395 \end{pmatrix}\begin{pmatrix} 15.0 \\ 12.5 \end{pmatrix} = \begin{pmatrix} 0.2988(15.0) - 0.2503(12.5) \\ -0.2503(15.0) + 0.3395(12.5) \end{pmatrix} = \begin{pmatrix} 1.35 \\ 0.49 \end{pmatrix}$$

Therefore, the required currents are $i_1 = 1.35$ A and $i_2 = 0.49$ A. These values check when substituted into the original equations. ∎

EXAMPLE C _____ Use matrices to solve the system of equations

$$x + 4y - z = 4$$
$$x + 3y + z = 8$$
$$2x + 6y + z = 13$$

Setting up matrices A, C, and X, we have

$$A = \begin{pmatrix} 1 & 4 & -1 \\ 1 & 3 & 1 \\ 2 & 6 & 1 \end{pmatrix}, \qquad C = \begin{pmatrix} 4 \\ 8 \\ 13 \end{pmatrix}, \qquad X = \begin{pmatrix} x \\ y \\ z \end{pmatrix}$$

To give another example of finding the inverse of a 3×3 matrix, we shall briefly show the steps for finding A^{-1}.

$$\left[\begin{array}{ccc|ccc} 1 & 4 & -1 & 1 & 0 & 0 \\ 1 & 3 & 1 & 0 & 1 & 0 \\ 2 & 6 & 1 & 0 & 0 & 1 \end{array}\right] \rightarrow \left[\begin{array}{ccc|ccc} 1 & 4 & -1 & 1 & 0 & 0 \\ 0 & 1 & -2 & 1 & -1 & 0 \\ 0 & -2 & 3 & -2 & 0 & 1 \end{array}\right] \rightarrow \left[\begin{array}{ccc|ccc} 1 & 0 & 7 & -3 & 4 & 0 \\ 0 & 1 & -2 & 1 & -1 & 0 \\ 0 & 0 & 1 & 0 & 2 & -1 \end{array}\right]$$

$$\rightarrow \left[\begin{array}{ccc|ccc} 1 & 4 & -1 & 1 & 0 & 0 \\ 0 & -1 & 2 & -1 & 1 & 0 \\ 2 & 6 & 1 & 0 & 0 & 1 \end{array}\right] \rightarrow \left[\begin{array}{ccc|ccc} 1 & 0 & 7 & -3 & 4 & 0 \\ 0 & 1 & -2 & 1 & -1 & 0 \\ 0 & -2 & 3 & -2 & 0 & 1 \end{array}\right] \rightarrow \left[\begin{array}{ccc|ccc} 1 & 0 & 7 & -3 & 4 & 0 \\ 0 & 1 & 0 & 1 & 3 & -2 \\ 0 & 0 & 1 & 0 & 2 & -1 \end{array}\right]$$

$$\rightarrow \left[\begin{array}{ccc|ccc} 1 & 4 & -1 & 1 & 0 & 0 \\ 0 & -1 & 2 & -1 & 1 & 0 \\ 0 & -2 & 3 & -2 & 0 & 1 \end{array}\right] \rightarrow \left[\begin{array}{ccc|ccc} 1 & 0 & 7 & -3 & 4 & 0 \\ 0 & 1 & -2 & 1 & -1 & 0 \\ 0 & 0 & -1 & 0 & -2 & 1 \end{array}\right] \rightarrow \left[\begin{array}{ccc|ccc} 1 & 0 & 0 & -3 & -10 & 7 \\ 0 & 1 & 0 & 1 & 3 & -2 \\ 0 & 0 & 1 & 0 & 2 & -1 \end{array}\right]$$

Thus, $A^{-1} = \begin{pmatrix} -3 & -10 & 7 \\ 1 & 3 & -2 \\ 0 & 2 & -1 \end{pmatrix}$ and

$$X = A^{-1}C = \begin{pmatrix} -3 & -10 & 7 \\ 1 & 3 & -2 \\ 0 & 2 & -1 \end{pmatrix}\begin{pmatrix} 4 \\ 8 \\ 13 \end{pmatrix} = \begin{pmatrix} -12 - 80 + 91 \\ 4 + 24 - 26 \\ 0 + 16 - 13 \end{pmatrix} = \begin{pmatrix} -1 \\ 2 \\ 3 \end{pmatrix}$$

∎ This means that $x = -1$, $y = 2$, and $z = 3$.

EXAMPLE D   Use matrices to solve the system of equations

$$x + 2y - z = -4$$
$$3x + 5y - z = -5$$
$$-2x - y - 2z = -5$$

Setting up matrices A, C, and X, we have

$$A = \begin{pmatrix} 1 & 2 & -1 \\ 3 & 5 & -1 \\ -2 & -1 & -2 \end{pmatrix}, \qquad C = \begin{pmatrix} -4 \\ -5 \\ -5 \end{pmatrix}, \qquad X = \begin{pmatrix} x \\ y \\ z \end{pmatrix}$$

We now find the inverse of A to be

$$A^{-1} = \begin{pmatrix} \frac{11}{2} & -\frac{5}{2} & -\frac{3}{2} \\ -4 & 2 & 1 \\ -\frac{7}{2} & \frac{3}{2} & \frac{1}{2} \end{pmatrix}$$

(see Example D of Section 15-5). Therefore,

$$X = A^{-1}C = \begin{pmatrix} \frac{11}{2} & -\frac{5}{2} & -\frac{3}{2} \\ -4 & 2 & 1 \\ -\frac{7}{2} & \frac{3}{2} & \frac{1}{2} \end{pmatrix} \begin{pmatrix} -4 \\ -5 \\ -5 \end{pmatrix} = \begin{pmatrix} -2 \\ 1 \\ 4 \end{pmatrix}$$

This means that the solution is $x = -2$, $y = 1$, $z = 4$.

NOTE ▷   After solving systems of equations in this manner, the reader may feel that the method is much longer and more tedious than previously developed techniques. ***The principal problem with this method is that a great deal of numerical computation is generally required.*** However, methods such as this one are easily programmed for use on a computer, which can do the arithmetic work very rapidly. Therefore it is the *method* of solving the system of equations which is of primary importance here.

Exercises 15-6

In Exercises 1 through 8, solve the given systems of equations by using the inverse of the coefficient matrix. The numbers in parentheses refer to exercises from Section 15-5, where the inverses may be checked.

1. $2x - 5y = -14$ (1)
    $-2x + 4y = 11$

2. $-x + 5y = 4$ (3)
    $4x + 10y = -4$

3. $2x + 4y = -9$ (11)
    $-x - y = 2$

4. $2x + 5y = -6$ (13)
    $-x + 2y = -6$

5. $x - 3y - 2z = -8$ (17)
    $-2x + 7y + 3z = 19$
    $x - y - 3z = -3$

6. $x - y - 3z = -1$ (19)
    $-y - 2z = -2$
    $2x + y - z = 2$

7. $x + 3y + 2z = 5$ (21)
    $-2x - 5y - z = -1$
    $2x + 4y = -2$

8. $2x + 4y = -2$ (23)
    $3x + 4y - 2z = -6$
    $-x + y + 2z = 5$

In Exercises 9 through 20, solve the given systems of equations by using the inverse of the coefficient matrix.

9. $2x + 7y = 16$
    $x + 4y = 9$

10. $4x - 3y = -13$
    $-3x + 2y = 9$

11. $2x - 3y = 3$
    $4x - 5y = 4$

12. $x + 2y = 3$
$3x + 4y = 11$

13. $5x - 2y = -14$
$3x + 4y = -11$

14. $4x - 3y = -1$
$8x + 3y = 4$

15. $2.5x + 2.8y = -3.0$
$3.5x - 1.6y = 9.6$

16. $12x - 5y = -400$
$31x + 25y = 180$

17. $x + 2y + 2z = -4$
$4x + 9y + 10z = -18$
$-x + 3y + 7z = -7$

18. $x - 4y - 2z = -7$
$-x + 5y + 5z = 18$
$3x - 7y + 10z = 38$

19. $2x + 4y + z = 5$
$-2x - 2y - z = -6$
$-x + 2y + z = 0$

20. $4x + y = 2$
$-2x - y + 3z = -18$
$2x + y - z = 8$

In Exercises 21 through 24, solve the indicated systems of equations by using the inverse of the coefficient matrix. In Exercises 23 and 24, it is necessary to set up the appropriate equations.

21. Forces A and B hold up a beam that weighs 254 N, as shown in Fig. 15-11. The equations used to find the forces are

$$A \sin 47.2° + B \sin 64.4° = 254$$
$$A \cos 47.2° - B \cos 64.4° = 0$$

Find the forces.

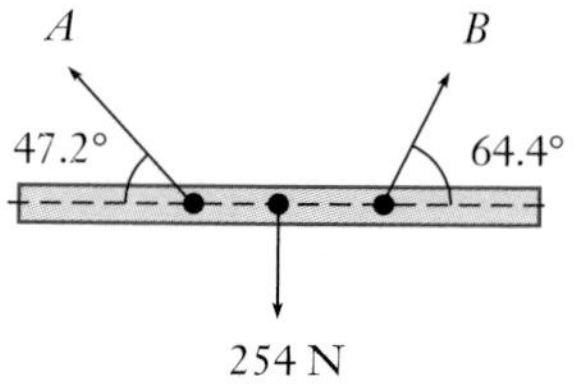

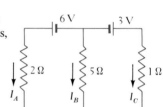

Fig. 15-11

254 N

22. In applying Kirchhoff's laws (see Exercise 20 of Section 4-6) to the circuit shown in Fig. 15-12, the following equations are found. Determine the indicated currents, in amperes.

$$I_A + I_B + I_C = 0$$
$$2I_A - 5I_B = 6$$
$$5I_B - I_C = -3$$

Fig. 15-12

23. A research chemist wants to make 10.0 L of gasoline containing 2.0% of a new experimental additive. Gasoline without additive and two mixtures of gasoline with additive, one with 5.0% and the other with 6.0%, are to be used. If four times as much gasoline without additive than the 5.0% mixture are to be used, how much of each is needed?

24. A person buys two certificates of deposit, one yielding 6.0% annual interest and the other yielding 7.5% annual interest, for $13,700. If the total annual income is $915, how much did each certificate cost?

15-7   Chapter Equations, Review Exercises, and Practice Test

Chapter Equations

Determinants

$$\begin{vmatrix} a_1 & b_1 & c_1 \\ a_2 & b_2 & c_2 \\ a_3 & b_3 & c_3 \end{vmatrix} = a_1b_2c_3 + a_3b_1c_2 + a_2b_3c_1 - a_3b_2c_1 - a_1b_3c_2 - a_2b_1c_3 \qquad (15\text{-}1)$$

$$\begin{vmatrix} a_1 & b_1 & c_1 \\ a_2 & b_2 & c_2 \\ a_3 & b_3 & c_3 \end{vmatrix} = a_1(b_2c_3 - b_3c_2) - a_2(b_1c_3 - b_3c_1) + a_3(b_1c_2 - b_2c_1) \qquad (15\text{-}2)$$

Minors

$$\begin{vmatrix} a_1 & b_1 & c_1 \\ a_2 & b_2 & c_2 \\ a_3 & b_3 & c_3 \end{vmatrix} = a_1\begin{vmatrix} b_2 & c_2 \\ b_3 & c_3 \end{vmatrix} - a_2\begin{vmatrix} b_1 & c_1 \\ b_3 & c_3 \end{vmatrix} + a_3\begin{vmatrix} b_1 & c_1 \\ b_2 & c_2 \end{vmatrix} \qquad (15\text{-}3)$$

Basic laws for matrices

$$A + B = B + A \qquad \text{(commutative law)} \qquad (15\text{-}4)$$

$$A + (B + C) = (A + B) + C \qquad \text{(associative law)} \qquad (15\text{-}5)$$

$$k(A + B) = kA + kB \tag{15-6}$$

$$A + O = A \tag{15-7}$$

Inverse matrix $AA^{-1} = A^{-1}A = I$ \hfill (15-8)

Solving systems of $\begin{pmatrix} a_1x + b_1y \\ a_2x + b_2y \end{pmatrix} = \begin{pmatrix} c_1 \\ c_2 \end{pmatrix}$ \hfill (15-9)
equations by matrices

$$A = \begin{pmatrix} a_1 & b_1 \\ a_2 & b_2 \end{pmatrix}, \qquad X = \begin{pmatrix} x \\ y \end{pmatrix}, \qquad C = \begin{pmatrix} c_1 \\ c_2 \end{pmatrix} \tag{15-10}$$

$$AX = C \tag{15-11}$$

$$X = A^{-1}C \tag{15-12}$$

Review Exercises

In Exercises 1 through 8, evaluate the given determinants by expansion by minors.

1. $\begin{vmatrix} 1 & 2 & -1 \\ 4 & 1 & -3 \\ -3 & -5 & 2 \end{vmatrix}$
    2. $\begin{vmatrix} 3 & -1 & 2 \\ 7 & -1 & 4 \\ 2 & 1 & -3 \end{vmatrix}$
    3. $\begin{vmatrix} -1 & 3 & -7 \\ 0 & 5 & 4 \\ 4 & -3 & -2 \end{vmatrix}$

4. $\begin{vmatrix} 60 & -54 & -76 \\ -10 & 24 & 40 \\ 25 & -37 & 18 \end{vmatrix}$
    5. $\begin{vmatrix} 2 & 6 & 2 & 5 \\ 2 & 0 & 4 & -1 \\ 4 & -3 & 6 & 1 \\ 3 & -1 & 0 & -2 \end{vmatrix}$
    6. $\begin{vmatrix} 1 & -2 & 2 & 4 \\ 0 & 1 & 2 & 3 \\ 3 & 2 & 2 & 5 \\ 2 & 1 & -2 & 0 \end{vmatrix}$

7. $\begin{vmatrix} 1 & 3 & -2 & 4 \\ 2 & 0 & 3 & -2 \\ 5 & -1 & 5 & -3 \\ -6 & 4 & -1 & 2 \end{vmatrix}$
    8. $\begin{vmatrix} 2 & 3 & -1 & -1 \\ -3 & -2 & 5 & -6 \\ 2 & 1 & -3 & 2 \\ 4 & 0 & -2 & 1 \end{vmatrix}$

In Exercises 9 through 16, evaluate the determinants of Exercises 1 through 8 by using the basic properties of determinants.

In Exercises 17 through 20, evaluate the given determinants by using the basic properties of determinants.

17. $\begin{vmatrix} 1 & 0 & -3 & -2 \\ 1 & -1 & 2 & 0 \\ -1 & 1 & 1 & 1 \\ 5 & -1 & 2 & -1 \end{vmatrix}$
    18. $\begin{vmatrix} 2 & 6 & -2 & 4 \\ -2 & 2 & -3 & 3 \\ 3 & 2 & 2 & -2 \\ 2 & -6 & 4 & 1 \end{vmatrix}$

19. $\begin{vmatrix} 1 & -1 & 3 & 0 & 2 \\ 4 & 0 & 4 & -2 & 2 \\ 0 & 4 & 0 & -1 & -1 \\ -2 & 2 & -1 & 4 & 0 \\ 1 & -1 & 2 & 0 & 1 \end{vmatrix}$
    20. $\begin{vmatrix} 1 & 4 & -3 & 3 & 0 \\ 3 & 1 & -1 & 2 & 2 \\ 1 & 2 & 1 & 1 & 1 \\ -3 & -5 & -5 & 0 & -6 \\ 2 & 2 & -2 & 3 & -2 \end{vmatrix}$

In Exercises 21 through 24, determine the values of the literal symbols.

21. $\begin{pmatrix} 2a \\ a - b \end{pmatrix} = \begin{pmatrix} 8 \\ 5 \end{pmatrix}$
    22. $\begin{pmatrix} a + bj & b \\ aj & b - aj \end{pmatrix} = \begin{pmatrix} 6j & 2d \\ 2cj & ej^2 \end{pmatrix} \qquad (j = \sqrt{-1})$

23. $\begin{pmatrix} 2x & 3y & 2z \\ x + y & 2y + z & z - x \end{pmatrix} = \begin{pmatrix} 4 & -9 & 5 \\ a & b & c \end{pmatrix}$
    24. $\begin{pmatrix} x - y \\ 2x + 2z \\ 4y + z \end{pmatrix} = \begin{pmatrix} 1 \\ 3 \\ -1 \end{pmatrix}$

In Exercises 25 through 32, use the given matrices and perform the indicated operations.

$$A = \begin{pmatrix} 2 & -3 \\ 4 & 1 \\ -5 & 0 \\ 2 & -3 \end{pmatrix}, \qquad B = \begin{pmatrix} -1 & 0 \\ 4 & -6 \\ -3 & -2 \\ 1 & -7 \end{pmatrix}, \qquad C = \begin{pmatrix} 5 & -6 \\ 2 & 8 \\ 0 & -2 \end{pmatrix}$$

25. $A + B$
26. $2C$
27. $-3B$
28. $B - A$
29. $A - C$
30. $2C - B$
31. $2A - 3B$
32. $2(A - B)$

In Exercises 33 through 36, perform the indicated matrix multiplications.

33. $\begin{pmatrix} 5 & -1 \\ 3 & 2 \end{pmatrix}\begin{pmatrix} 1 \\ -8 \end{pmatrix}$

34. $\begin{pmatrix} 6 & -4 & 1 & 0 \\ 2 & 0 & -4 & 3 \end{pmatrix}\begin{pmatrix} 7 & -1 & 6 \\ 4 & 0 & 1 \\ 3 & -2 & 5 \\ 9 & 1 & 0 \end{pmatrix}$

35. $\begin{pmatrix} -1 & 7 \\ 2 & 0 \\ 4 & -1 \end{pmatrix}\begin{pmatrix} 1 & -4 & 5 \\ 5 & 1 & 0 \end{pmatrix}$

36. $\begin{pmatrix} 0 & -1 & 6 \\ 8 & 1 & 4 \\ 7 & -2 & -1 \end{pmatrix}\begin{pmatrix} 5 & -1 & 7 & 1 & 5 \\ 0 & 1 & 0 & 4 & 1 \\ 1 & -2 & 3 & 0 & 1 \end{pmatrix}$

In Exercises 37 through 44, find the inverses of the given matrices.

37. $\begin{pmatrix} 2 & -5 \\ 2 & -4 \end{pmatrix}$

38. $\begin{pmatrix} -1 & -6 \\ 2 & 10 \end{pmatrix}$

39. $\begin{pmatrix} 7 & -1 \\ 4 & 8 \end{pmatrix}$

40. $\begin{pmatrix} 50 & -12 \\ 42 & -80 \end{pmatrix}$

41. $\begin{pmatrix} 1 & 1 & -2 \\ -1 & -2 & 1 \\ 0 & 3 & 4 \end{pmatrix}$

42. $\begin{pmatrix} -1 & -1 & 2 \\ 2 & 3 & 0 \\ 1 & 4 & 1 \end{pmatrix}$

43. $\begin{pmatrix} 2 & -4 & 3 \\ 4 & -6 & 5 \\ -2 & 1 & -1 \end{pmatrix}$

44. $\begin{pmatrix} 3 & 1 & -4 \\ -3 & 1 & -2 \\ -6 & 0 & 3 \end{pmatrix}$

In Exercises 45 through 52, solve the given systems of equations using the inverse of the coefficient matrix.

45. $2x - 3y = -9$
$\quad\; 4x - y = -13$

46. $5x - 7y = 62$
$\quad\; 6x + 5y = -6$

47. $33x + 52y = -450$
$\quad\; 45x - 62y = 1380$

48. $0.24x - 0.26y = -3.1$
$\quad\; 0.40x + 0.34y = -1.3$

49. $2x - 3y + 2z = 7$
$\quad\; 3x + y - 3z = -6$
$\quad\; x + 4y + z = -13$

50. $2x + 2y - z = 8$
$\quad\; x + 4y + 2z = 5$
$\quad\; 3x - 2y + z = 17$

51. $x + 2y + 3z = 1$
$\quad\; 3x - 4y - 3z = 2$
$\quad\; 7x - 6y + 6z = 2$

52. $3x + 2y + z = 2$
$\quad\; 2x + 3y - 6z = 3$
$\quad\; x + 3y + 3z = 1$

In Exercises 53 through 56, solve the given systems of equations by determinants. Use the basic properties of determinants.

53. $3x - 2y + z = 6$
$\quad\; 2x + 3z = 3$
$\quad\; 4x - y + 5z = 6$

54. $7x + y + 2z = 3$
$\quad\; 4x - 2y + 4z = -2$
$\quad\; 2x + 3y - 6z = 3$

55. $2x - 3y + z - t = -8$
$\quad\; 4x + 3z + 2t = -3$
$\quad\; 2y - 3z - t = 12$
$\quad\; x - y - z + t = 3$

56. $3x + 2y - 2z - 2t = 0$
$\quad\; 5y + 3z + 4t = 3$
$\quad\; 6y - 3z + 4t = 9$
$\quad\; 6x - y + 2z - 2t = -3$

In Exercises 57 and 58, evaluate the given determinants by minors.

57. $\begin{vmatrix} 1 + \sqrt{2} & 2 - \sqrt{3} & 0 \\ 3 + \sqrt{5} & 7 + \sqrt{6} & \sqrt{2} \\ 2 + \sqrt{3} & 1 - \sqrt{2} & 0 \end{vmatrix}$

58. $\begin{vmatrix} \cos\dfrac{\pi}{3} & \sin\dfrac{\pi}{6} & \cos 0 \\ \cos\pi & \sin\pi & \tan\dfrac{\pi}{4} \\ \sin\dfrac{\pi}{2} & \cos\dfrac{\pi}{2} & \sin 0 \end{vmatrix}$

In Exercises 59 and 60, evaluate the given determinants by use of the properties of determinants of Section 15-2.

59.
$$\begin{vmatrix} \ln e & \log_3 1 & \frac{1}{2}\log 100 \\ \ln \frac{1}{e} & \log_2 8 & \log 0.1 \\ \ln \sqrt{e} & \log 10 & \log_4 2 \end{vmatrix}$$

60.
$$\begin{vmatrix} j & 1 & 1+j \\ -j & -1+j & -j \\ 2j & 2 & 2+3j \end{vmatrix} \quad (j = \sqrt{-1})$$

In Exercises 61 and 62, use the matrix

$$N = \begin{pmatrix} 0 & -1 \\ 1 & 0 \end{pmatrix}$$

61. Show that $N^{-1} = -N$.

62. Show that $N^3 = -N$.

In Exercises 63 through 66, use the matrices

$$A = \begin{pmatrix} 1 & -2 \\ 0 & 3 \end{pmatrix} \quad \text{and} \quad B = \begin{pmatrix} -3 & 1 \\ 2 & -1 \end{pmatrix}$$

63. Show that $(A + B)(A - B) \neq A^2 - B^2$.

64. Show that $(A + B)^2 \neq A^2 + 2AB + B^2$.

65. Show that the inverse of $2A$ is $A^{-1}/2$.

66. Show that the inverse of B^2 is $(B^{-1})^2$.

In Exercises 67 through 70, solve the given systems of equations by any appropriate method of this chapter.

67. Two electric resistors, R_1 and R_2, are tested with currents and voltages such that the following equations are found.

$$2R_1 + 3R_2 = 26 \qquad 3R_1 + 2R_2 = 24$$

Find the resistances R_1 and R_2 (in ohms).

68. A company produces two products, each of which is processed in two departments. Considering the worker-time available, the numbers x and y of each product produced each week can be found by solving the system of equations

$$4.0x + 2.5y = 1200 \qquad 3.2x + 4.0y = 1200$$

Find x and y.

69. A steel beam is supported as shown in Fig. 15-13. Find the force F and the tension T by solving the system of equations

$$0.500F = 0.866T \qquad 0.866F + 0.500T = 350$$

70. To find the electric currents (in amperes) indicated in Fig. 15-14, it is necessary to solve the following equations.

$$I_A + I_B + I_C = 0$$
$$5I_A - 2I_B = -4$$
$$2I_B - I_C = 0$$

Fig. 15-14

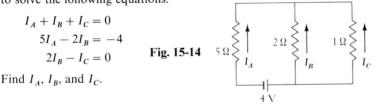

Find I_A, I_B, and I_C.

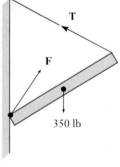

Fig. 15-13

In Exercises 71 through 76, set up systems of linear equations and solve them by any appropriate method developed in this chapter.

71. A crime suspect passes an intersection in a car traveling at 110 mi/h. The police pass the intersection 3.0 min later in a car traveling at 135 mi/h. How long is it before the police overtake the suspect?

72. The distance s that a crate moves down an inclined chute is given by $s = v_0 t + \frac{1}{2}at^2$, where v_0 is its inital velocity, a is its acceleration, and t is the time of descent. Find v_0 and a if $s = 90.0$ cm for $t = 2.00$ s, and $s = 280$ cm for $t = 4.00$ s.

73. By mass, three alloys have the following percentages of lead, zinc, and copper.

	Lead	Zinc	Copper
Alloy A	60%	30%	10%
Alloy B	40%	30%	30%
Alloy C	30%	70%	

How many grams of each of alloys A, B, and C must be mixed to get 100 g of an alloy which is 44% lead, 38% zinc, and 18% copper?

74. Three computer line printers can print a total of 8200 lines/min when printing at the same time. With the first printing for 2 min and the second printing for 3 min, a total of 12,200 lines can be printed. With the first printing for 1 min, the second for 2 min, and the third for 3 min, a total of 17,600 lines can be printed. How many lines per minute can each print?

75. On a 750-mi trip which took a total of 5.5 h, a person took a limousine to the airport, then a plane, and finally a car to reach the final destination. The limousine took as long as the final car trip and the time for connections. The limousine averaged 55 mi/h, the plane averaged 400 mi/h, and the car averaged 40 mi/h. The plane traveled four times as far as the limousine and car combined. How long did each part of the trip and the connections take?

76. A department store sells a certain type of sweater in four sizes—small, medium, large, and extra large. The cost of a sweater of each size is $22, $24, $25, and $28, respectively. In a certain week 13 sweaters were sold, with a gross income of $320. The number of large size sold equaled the total of the small and extra large sold, and the income from sales of the large size equaled the income from the small size and extra large size combined. How many of each were sold?

In Exercises 77 through 80, perform the indicated matrix operations.

77. An automobile maker has two assembly plants at which cars with 4 cylinders,·6 cylinders, or 8 cylinders, and with either standard transmission or automatic transmission are assembled. The annual production at the first plant of cars with number of cylinders-transmission type (standard-automatic) is as follows: 4—15,000, 10,000; 6—20,000, 18,000; 8—8000, 30,000. At the second plant the production is 4—18,000, 12,000; 6—30,000, 22,000; 8—12,000, 40,000. Set up matrices for this information, and by matrix addition find the matrix for total production by the number of cylinders and type of transmission.

78. Set up a matrix representing the information given in Exercise 73. A given shipment contains 500 g of alloy A, 800 g of alloy B, and 700 g of alloy C. Set up a matrix for this information. By multiplying these matrices, obtain a matrix which gives the total weight of lead, zinc, and copper in the shipment.

79. The matrix equation

$$\left(\begin{pmatrix} R_1 & -R_2 \\ -R_2 & R_1 \end{pmatrix} + R_2 \begin{pmatrix} 1 & 0 \\ 0 & 1 \end{pmatrix} \right) \begin{pmatrix} i_1 \\ i_2 \end{pmatrix} = \begin{pmatrix} 6 \\ 0 \end{pmatrix}$$

may be used to represent the system of equations relating the currents and resistances of the circuit in Fig. 15-15. Find this system of equations by performing the indicated matrix operations.

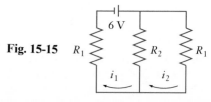

Fig. 15-15

80. In taking inventory, a firm finds that it has in one warehouse 6 pieces of 20-ft brass pipe, 8 pieces of 30-ft brass pipe, 11 pieces of 40-ft brass pipe, 5 pieces of 20-ft steel pipe, 10 pieces of 30-ft steel pipe, and 15 pieces of 40-ft steel pipe. This inventory can be represented by the matrix A below. In each of two other warehouses, the inventory of the same items is represented by the matrix B below

$$A = \begin{pmatrix} 6 & 8 & 11 \\ 5 & 10 & 15 \end{pmatrix} \qquad B = \begin{pmatrix} 8 & 3 & 4 \\ 6 & 10 & 5 \end{pmatrix}$$

By matrix addition and scalar multiplication, find the matrix which represents the total number of each item in the three warehouses.

Practice Test

1. For matrices A and B, find $A - 2B$.

$$A = \begin{pmatrix} 3 & -1 & 4 \\ 2 & 0 & -2 \end{pmatrix} \qquad B = \begin{pmatrix} 1 & 4 & 5 \\ -1 & -2 & 3 \end{pmatrix}$$

2. Evaluate using minors:

$$\begin{vmatrix} 4 & 0 & -2 \\ 3 & -3 & 2 \\ -4 & 1 & -1 \end{vmatrix}$$

3. For matrices C and D, find CD and DC.

$$C = \begin{pmatrix} 1 & 0 & 4 \\ 2 & -2 & 1 \\ -1 & 3 & 2 \end{pmatrix} \qquad D = \begin{pmatrix} 2 & -2 \\ 4 & -5 \\ 6 & 1 \end{pmatrix}$$

5. For matrix C of Problem 3, find C^{-1}.

4. Evaluate using the properties of determinants:

$$\begin{vmatrix} 1 & 0 & 4 & -2 \\ -2 & -1 & 3 & 0 \\ 3 & 2 & -1 & 2 \\ 1 & 1 & -1 & -2 \end{vmatrix}$$

6. Fifty shares of stock A and 30 shares of stock B cost $2600. Thirty shares of stock A and 40 shares of stock B cost $2000. What is the price per share of each stock? Solve by setting up the appropriate equations and then using the inverse of the coefficient matrix.

16 Inequalities

In Section 16-5 we use inequalities to show how a company can maximize profit in making products such as speaker systems.

Until now we have devoted a great deal of time to the solution of equations. Equation-solving does play an extremely important role in mathematics, but there are also times in mathematics and in applications when we must solve inequalities. In mathematics this occurs, for example, in finding the domain of a function by determining the values of the independent variable which give real numbers for the function. Inequalities are also useful in areas of advanced mathematics such as calculus.

Applications of inequalities occur in many technical areas, such as the analyses of projectile motion, beam deflections, and safe operating levels for equipment. Another important application occurs in business, where inequalities may be used to set production levels for maximizing profits or minimizing costs.

16-1 Properties of Inequalities

In Chapter 1 we first introduced the signs of inequality. To this point only a basic understanding of their meanings has been necessary. In this section we review the meanings and develop certain basic properties of inequalities.

The expression $a < b$ is read as "a is less than b," and the expression $a > b$ is read as "a is greater than b." *These signs define what is known as the **sense** (indicated by the direction of the sign) of the inequality.* Two inequalities are said to have the same sense if the signs of inequality point in the same direction. They are said to have the opposite sense if the signs of inequality point in opposite directions. *The two sides of the inequality are called **members** of the inequality.*

EXAMPLE A ——— The inequalities $x + 3 > 2$ and $x + 1 > 0$ have the same sense, as do the inequalities $3x - 1 < 4$ and $x^2 - 1 < 3$.
The inequalities $x - 4 < 0$ and $x > -4$ have the opposite sense, as do the inequalities $2x + 4 > 1$ and $3x^2 - 7 < 1$. ∎

The **solution** *of an inequality consists of those values of the variable for which the inequality is satisfied.* Most inequalities with which we shall deal are known as **conditional inequalities.** That is, *there are some values of the variable which satisfy the inequality, and also there are some values which do not satisfy it.* Also, *some inequalities are satisfied for all values of the variable. These are called* **absolute inequalities.** A solution of an inequality consists of only real numbers, as the terms "greater than" and "less than" have not been defined for complex numbers.

EXAMPLE B ——— The inequality $x + 1 > 0$ is satisfied by all values of x greater than -1. Thus, the values of x which satisfy this inequality are written as $x > -1$. This illustrates the difference between the solution of an equation and the solution of an inequality. The solution of an equation normally consists of a few specific numbers, whereas *the solution to an inequality normally consists of an interval of values of the variable.* Any and all values within this interval are termed solutions of the inequality. Since the inequality $x + 1 > 0$ is satisfied only by the values of x in the interval $x > -1$, it is a *conditional inequality.*
The inequality $x^2 + 1 > 0$ is true for all values of x, since x^2 is never negative. This is an *absolute inequality.* ∎

There are occasions when it is convenient to combine an inequality with an equality. For such purposes, the symbols $\leq$, read "less than or equal to," and $\geq$, read "greater than or equal to," are used.

EXAMPLE C ——— If we wish to state that x is positive, we can write $x > 0$. However, the value zero is not included in the solution. If we wish to state that x is not negative, that is, that zero is included as part of the solution, we can write $x \geq 0$. In order to state that x is less than or equal to -5, we write $x \leq -5$. ∎

We shall now present the basic operations performed on inequalities. These operations are the same as those performed on equations, but in certain cases the results take on a different form. *The following are referred to as the* **properties of inequalities:**

properties of
inequalities

    1. *The sense of an inequality is not changed when the same number is added to—or subtracted from—both members of the inequality.* Symbolically this may be stated as "if $a > b$, then $a + c > b + c$ and $a - c > b - c$."

EXAMPLE D ——— Using Property 1 on the inequality $9 > 6$, we have the following results:

$$9 > 6 \qquad\qquad\qquad 9 > 6$$

add 4 to each member   subtract 12 from each member

$$9 + 4 > 6 + 4 \qquad\qquad 9 - 12 > 6 - 12$$

$$13 > 10 \qquad\qquad\qquad -3 > -6$$

∎

2. *The sense of an inequality is not changed if both members are multiplied or divided by the same positive number.* Symbolically this is stated as "if $a > b$, then $ac > bc$, and $a/c > b/c$, provided that $c > 0$."

EXAMPLE E

Using Property 2 on the inequality $8 < 15$, we have the following results:

$$8 < 15 \qquad\qquad 8 < 15$$

multiply both members by 2 $\qquad$ divide both members by 2

$$2(8) < 2(15) \qquad\qquad \frac{8}{2} < \frac{15}{2}$$

$$16 < 30 \qquad\qquad\qquad 4 < \frac{15}{2}$$

NOTE ▷

3. *The sense of an inequality is reversed if both members are multiplied or divided by the same negative number.* Symbolically this is stated as "if $a > b$, then $ac < bc$, and $a/c < b/c$, provided that $c < 0$." Be very careful to note that *we obtain different results depending on whether both members are multiplied by a positive number or by a negative number.*

EXAMPLE F

Using Property 3 on the inequality $4 > -2$, we have the following results:

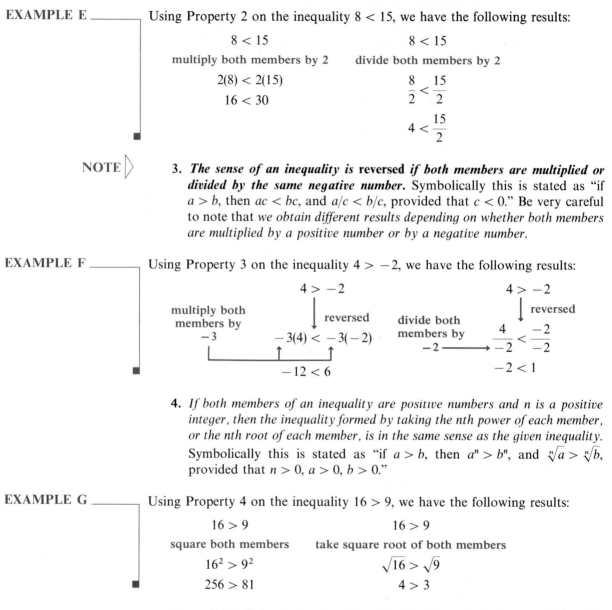

$$4 > -2 \qquad\qquad\qquad\qquad 4 > -2$$

multiply both members by -3 $\qquad$ reversed $\qquad$ divide both members by -2 $\qquad$ reversed

$$-3(4) < -3(-2) \qquad\qquad \frac{4}{-2} < \frac{-2}{-2}$$

$$-12 < 6 \qquad\qquad\qquad\qquad -2 < 1$$

4. *If both members of an inequality are positive numbers and n is a positive integer, then the inequality formed by taking the nth power of each member, or the nth root of each member, is in the same sense as the given inequality.* Symbolically this is stated as "if $a > b$, then $a^n > b^n$, and $\sqrt[n]{a} > \sqrt[n]{b}$, provided that $n > 0$, $a > 0$, $b > 0$."

EXAMPLE G

Using Property 4 on the inequality $16 > 9$, we have the following results:

$$16 > 9 \qquad\qquad 16 > 9$$

square both members $\qquad$ take square root of both members

$$16^2 > 9^2 \qquad\qquad \sqrt{16} > \sqrt{9}$$

$$256 > 81 \qquad\qquad 4 > 3$$

Many inequalities have more than two members. In fact, inequalities with three members are very common. All the operations stated above hold for inequalities with more than two members. Some care must be used, however, in stating inequalities with more than two members.

EXAMPLE H _____ In order to state that 5 is less than 6, and also greater than 2, which says that 5 is between 2 and 6, we may write $2 < 5 < 6$, or $6 > 5 > 2$. However, generally the form with the *less than* inequality signs is preferred.

In order to state that a number x may be equal to or greater than 2, *and* also less than 6, we write $2 \leq x < 6$.

By writing $2 \leq x \leq 6$ we are stating that x is greater than or equal to 2, and at the same time less than or equal to 6.

NOTE ▷ By writing $x \leq -5$, $x > 7$ we are stating that x is less than or equal to -5, *or greater than 7*. ***This may not be stated as $7 < x \leq -5$,*** for this shows x as being less than -5, while at the same time greater than 7, and no such numbers exist.

Notice the use of the words *and* and *or* in Example H. In stating inequalities, *and* is used when the solution consists of values which satisfy ***both*** statements. The word *or* is used when the solution consists of values which make ***either*** statement true. (In everyday speech, *or* can sometimes also mean that either one statement is true or another statement is true, but *not* that both statements are true.)

EXAMPLE I _____ The inequality $x^2 - 3x + 2 > 0$ is satisfied if x is either greater than 2 *or* less than 1. This would be written as $x > 2$ or $x < 1$, but it would be incorrect to state it as $1 > x > 2$. (If we wrote it this way, we would be saying that the same value of x is less than 1 *and* at the same time greater than 2. Of course, as we noted for this type of situation in Example H, no such number exists.) Any inequality must be valid for all values satisfying it. However, we could say that the inequality is not satisfied for $1 \leq x \leq 2$, which means those values of x greater than *or* equal to 1 *and* less than *or* equal to 2 (between or equal to 1 and 2).

EXAMPLE J _____ The design of a rectangular solar panel shows that its length l is to be between 80 cm and 90 cm and that its width w is to be between 40 cm and 80 cm. See Fig. 16-1. Find the values of area the panel may have.

80 cm < l < 90 cm

40 cm < w < 80 cm

Fig. 16-1

Since the maximum length is less than 90 cm and the maximum width is less than 80 cm, the area must be less than $(90)(80) = 7200$ cm². Also, since the minimum length is greater than 80 cm and the minimum width is greater than 40 cm, the area must be greater than $(80)(40) = 3200$ cm². Thus, the area A may be represented as

$$3200 \text{ cm}^2 < A < 7200 \text{ cm}^2$$

This means that the area is greater than 3200 cm² *and* less than 7200 cm².

solution on a number line

In the sections which follow, we will be solving inequalities. It is often useful to show the solution on a number line. The following example illustrates how this is done.

EXAMPLE K

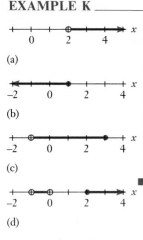

Fig. 16-2

To graph $x > 2$, we draw a small open circle at 2 on the number line (which is equivalent to the x-axis). Then we draw a solid line to the right of this point with an arrowhead pointing to the right, indicating all values greater than 2. See Fig. 16-2(a). The open circle shows that the point is not part of the indicated solution.

To graph $x \leq 1$, we follow the same basic procedure as above, except that we use a solid circle and the arrowhead points to the left. See Fig. 16-2(b). The solid circle shows that the point is part of the indicated solution.

To graph $-1 < x \leq 3$, we draw the solid line between -1 and 3. We use an open circle at -1 to show that it is not part of the solution, and we use a solid circle at 3 to show that it is part of the solution. See Fig. 16-2(c).

The graph of $-1 < x < 0$ or $x \geq 2$ is shown in Fig. 16-2(d).

In Section 2-2 we illustrated functions which were defined differently for different intervals of the domain, and we used basic inequalities to denote these intervals. The following example further illustrates this use of inequalities and includes a graphical representation.

EXAMPLE L

A semiconductor diode, an electronic device, has the property that an electric current can flow through it in only one direction. Thus, if a diode is in a circuit with an alternating-current source, the current in the circuit exists only during the half-cycle when the direction is correct for the diode. If a source of current given by $i = 2 \sin 50\pi t$ milliamperes is connected in series with a diode, write the inequalities which are appropriate for the first four half-cycles, assuming that the diode allows a positive current to flow. Also, graph the resulting current as a function of time, and graph the values of t for which the current is positive.

We are to find the values of t which correspond to $i > 0$ from the source. From the properties of the sine curve, we know that

$$\sin \theta = 0 \quad \text{for} \quad \theta = 0, \pi, 2\pi, 3\pi, 4\pi, \ldots$$

and

$$\sin \theta > 0 \quad \text{for} \quad 0 < \theta < \pi, \qquad 2\pi < \theta < 3\pi, \qquad \ldots$$

Since $i = 2 \sin 50\pi t$, and $\theta = 50\pi t$, we know that

$$i > 0 \quad \text{for} \quad 0 < 50\pi t < \pi, \qquad 2\pi < 50\pi t < 3\pi, \qquad \ldots$$
$$\text{or} \quad 0 < t < 0.02 \text{ s}, \qquad 0.04 \text{ s} < t < 0.06 \text{ s}, \qquad \ldots$$

From the statement of the problem, we know that $i = 0$ in the circuit for other values of t because of the diode. This means that

$$i = 0 \quad \text{for} \quad t = 0, \qquad 0.02 \text{ s} \leq t \leq 0.04 \text{ s}, \qquad 0.06 \text{ s} \leq t \leq 0.08 \text{ s}, \qquad \ldots$$

A graph of the current in the circuit as a function of time is shown in Fig. 16-3. In Fig. 16-4 the values of t for which $i > 0$ are shown.

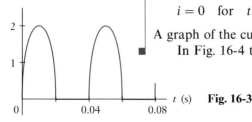

Fig. 16-3

Fig. 16-4

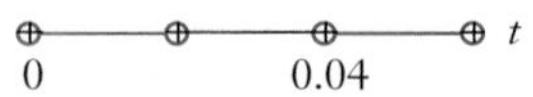

Exercises 16-1

In Exercises 1 through 8, for the inequality $4 < 9$, state the inequality resulting when the operations given are performed on both members.

1. Add 3.   **2.** Subtract 6.   **3.** Multiply by 5.   **4.** Multiply by -2.

5. Divide by -1.   **6.** Divide by 2.   **7.** Square both.   **8.** Take square roots.

In Exercises 9 through 20, give the inequalities which are equivalent to the following statements about a number x.

9. greater than -2 **10.** less than 7

11. less than or equal to 4   **12.** greater than or equal to -6

13. greater than 1 and less than 7   **14.** greater than or equal to -2 and less than 6

15. less than -9, or greater than or equal to -4 **16.** less than or equal to 8, or greater than or equal to 12

17. less than 1, or greater than 3 and less than or equal to 5

18. greater than or equal to 0 and less than or equal to 2, or greater than 5

19. greater than -2 and less than 2, or greater than or equal to 3 and less than 4

20. less than -4, or greater than or equal to 0 and less than or equal to 1, or greater than or equal to 5

In Exercises 21 through 24, give statements which are equivalent to the given inequalities involving a number x.

21. $0 < x \le 2$ **22.** $x < 5$ or $x > 7$ **23.** $x < -1$ or $1 \le x < 2$ **24.** $-1 \le x < 3$ or $5 < x < 7$

In Exercises 25 through 36, graph the given inequalities on a number line.

25. $x < 3$ **26.** $x \ge -1$ **27.** $x \le 1$ or $x > 3$

28. $x < -3$ or $x \ge 0$ **29.** $0 \le x < 5$ **30.** $-2 < x < 4$

31. $x < -1$ or $1 \le x < 4$ **32.** $-1 < x < 2$ or $x > 3$ **33.** $-3 < x < -1$ or $1 < x \le 3$

34. $1 < x \le 2$ or $3 \le x < 4$ **35.** $x < -3$ or $x > -3$ **36.** $x < 1$ or $1 < x \le 4$

In Exercises 37 through 44, some applications of inequalities are shown.

37. When Pioneer II left the solar system in 1989, it was 3×10^{12} mi from the sun. Assuming it never returns, express its distance d from the sun in the future as an inequality. Graph these values of d.

38. The temperature T within a certain refrigeration unit must be at least 36°F and no more than 40°F. Express the temperatures which should not exist in this refrigeration unit by use of inequalities. Graph these values of T.

39. An earth satellite put into orbit near the earth's surface will have an elliptic orbit if its velocity v is between 18,000 mi/h and 25,000 mi/h. State this as an inequality, and graph these values of v.

40. A geologist reported that the layers of soil in a certain region were formed between 25,000 years ago and 40,000 years ago. Write this as an inequality, with t representing past time in years. Graph these values of t.

41. In executing a program, a computer must perform a set of calculations. Any one of the calculations takes no more than 2565 steps. Express the number n of steps required for a given calculation by an inequality. (Note that n is a positive *integer*.)

42. A surveyor measures the side of a parcel of land and reports that its length ℓ is $\ell = 72.37$ m ± 0.05 m, where the ± 0.05 m gives the possible error in the measurement. Express the length ℓ by an inequality.

43. The electric intensity E within a charged spherical conductor is zero. The intensity on the surface and outside the sphere equals a constant k divided by the square of the distance r from the center of the sphere. State these relations for a sphere of radius a by using inequalities, and graph E as a function of r.

44. If the current from the source in Example L is $i = 5\cos 4\pi t$ and the diode allows only a negative current to flow, write the inequalities and draw the graph for the current in the circuit as a function of time for $0 \le t \le 1$ s.

16-2 Solving Linear Inequalities

Using the properties and definitions discussed in Section 16-1, we can now proceed to solve inequalities. In this section we solve linear inequalities in one variable. Similar to linear functions as defined in Chapter 4, *a* **linear inequality** *is one in which each term contains only one variable and the exponent of each variable is 1.* We will consider linear inequalities in two variables in Section 16-5.

The procedure for solving a linear inequality in one variable is like that we used in solving basic equations in Chapter 1. In order to isolate the variable, we perform the same operation on each member of the inequality. These operations are based on the properties of inequalities given in Section 16-1.

EXAMPLE A ____

In each of the following inequalities, by performing the indicated operation we isolate x and thereby solve the inequality.

$x + 2 < 4$	$\dfrac{x}{2} > 4$	$2x \le 4$
Subtract 2 from each member.	Multiply each member by 2.	Divide each member by 2.
$x < 2$	$x > 8$	$x \le 2$

Each solution can be checked by substituting any number in the indicated interval into the original inequality. For example, any value less than 2 will satisfy the first inequality, whereas 2 or any number less than 2 will satisfy the third inequality.

EXAMPLE B ____

Solve the inequality $3 - 2x \ge 15$.

We have the following solution:

$$3 - 2x \ge 15 \quad \text{original inequality}$$
$$-2x \ge 12 \quad \text{subtract 3 from each member}$$

inequality reversed $\longrightarrow$

$$x \le -6 \quad \text{divide each member by } -2$$

NOTE ▷

Again, carefully note that *the sign of inequality was reversed when each number was divided by* -2**.** We check the solution by substituting -7 in the original inequality, obtaining $17 \ge 15$.

EXAMPLE C ____

Solve the inequality $2x \le 3 - x$.

The solution proceeds as follows:

$$2x \le 3 - x \quad \text{original inequality}$$
$$3x \le 3 \quad \text{add } x \text{ to each member}$$
$$x \le 1 \quad \text{divide each member by 3}$$

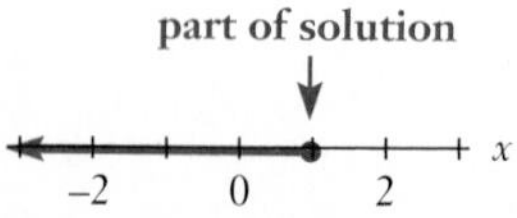

part of solution

Fig. 16-5

This solution checks and is represented in Fig. 16-5, as we showed in the last section.

This inequality could have been solved by combining x-terms on the right. In doing so, we would obtain $1 \ge x$. Since this might be misread, it is best to combine the variable terms on the left, as we did above.

EXAMPLE D _____ Solve the inequality $\frac{3}{2}(1 - x) > \frac{1}{4} - x$.

$$\frac{3}{2}(1 - x) > \frac{1}{4} - x \qquad \text{original inequality}$$

$$6(1 - x) > 1 - 4x \qquad \text{multiply each member by 4}$$

$$6 - 6x > 1 - 4x \qquad \text{remove parentheses}$$

$$-6x > -5 - 4x \qquad \text{subtract 6 from each member}$$

$$-2x > -5 \qquad \text{add } 4x \text{ to each member}$$

$$\downarrow$$

$$x < \frac{5}{2}$$

$$\text{divide each member by } -2$$

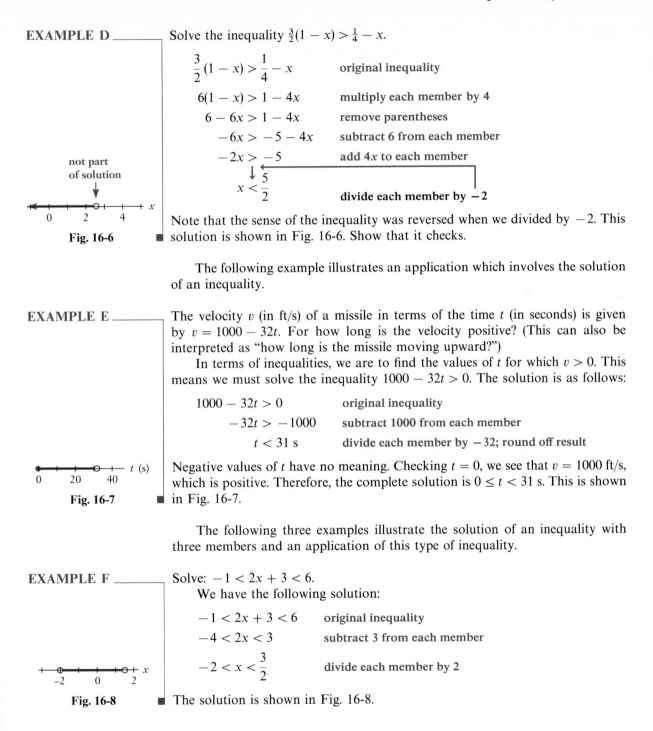

not part
of solution

Fig. 16-6

Note that the sense of the inequality was reversed when we divided by -2. This solution is shown in Fig. 16-6. Show that it checks. ■

The following example illustrates an application which involves the solution of an inequality.

EXAMPLE E _____ The velocity v (in ft/s) of a missile in terms of the time t (in seconds) is given by $v = 1000 - 32t$. For how long is the velocity positive? (This can also be interpreted as "how long is the missile moving upward?")

In terms of inequalities, we are to find the values of t for which $v > 0$. This means we must solve the inequality $1000 - 32t > 0$. The solution is as follows:

$$1000 - 32t > 0 \qquad \text{original inequality}$$

$$-32t > -1000 \qquad \text{subtract 1000 from each member}$$

$$t < 31 \text{ s} \qquad \text{divide each member by } -32; \text{ round off result}$$

Fig. 16-7

Negative values of t have no meaning. Checking $t = 0$, we see that $v = 1000$ ft/s, which is positive. Therefore, the complete solution is $0 \le t < 31$ s. This is shown in Fig. 16-7. ■

The following three examples illustrate the solution of an inequality with three members and an application of this type of inequality.

EXAMPLE F _____ Solve: $-1 < 2x + 3 < 6$.

We have the following solution:

$$-1 < 2x + 3 < 6 \qquad \text{original inequality}$$

$$-4 < 2x < 3 \qquad \text{subtract 3 from each member}$$

$$-2 < x < \frac{3}{2} \qquad \text{divide each member by 2}$$

Fig. 16-8

The solution is shown in Fig. 16-8. ■

EXAMPLE G _____ | Solve: $2x < x - 4 \le 3x + 8$.

Since we cannot isolate x in the middle member (or in any member), we rewrite the inequality as

$$2x < x - 4 \quad \text{and} \quad x - 4 \le 3x + 8$$

We then solve each of these inequalities, keeping in mind that the solution must satisfy both of them. Therefore, we have

$$2x < x - 4 \quad \text{and} \quad x - 4 \le 3x + 8$$
$$-2x \le 12$$
$$x < -4 \quad \text{and} \quad x \ge -6$$

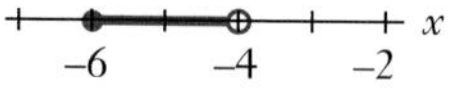

Fig. 16-9

■ This solution can be written as $-6 \le x < -4$, which is shown in Fig. 16-9.

EXAMPLE H _____ | In emptying a wastewater tank, one pump can remove no more than 40 L/min. If it operates for 8.0 min and a second pump operates for 5.0 min, what must be the pumping rate of the second pump if 480 L are to be removed?

Let x = the pumping rate of the first pump and y = the pumping rate of the second pump. Since the first operates for 8.0 min and the second for 5.0 min to remove 480 L, we have

$$\begin{array}{ccc} \text{first} & \text{second} \\ \text{pump} & \text{pump} & \text{total} \longleftarrow \text{amounts pumped} \\ 8.0x & + \quad 5.0y & = 480 \end{array}$$

Since we know that the first pump can remove no more than 40 L/min, which means that $0 \le x \le 40$ L/min, we solve for x, then substitute in this inequality.

Fig. 16-10

$x = 60 - 0.625y$	solve for x
$0 \le 60 - 0.625y \le 40$	substitute in inequality
$-60 \le -0.625y \le -20$	subtract 60 from each member
$96 \ge y \ge 32$	divide each member by -0.625
$32 \le y \le 96$ L/min	use < symbol (optional step)

This means that the second pump must be able to pump at least 32 L/min and no more than 96 L/min. See Fig. 16-10.

We note that although this was a three-member inequality and it was combined with equalities, the solution was performed in the same way as with a two-member inequality.

Exercises 16-2

In Exercises 1 through 24, solve the given inequalities. Graph each solution.

1. $x - 3 > -4$ **2.** $x + 2 \le 6$ **3.** $\dfrac{1}{2}x < 3$ **4.** $4x > -12$

5. $3x - 5 \le -11$ **6.** $\dfrac{1}{3}x + 2 \ge 1$ **7.** $6 - x > 4$ **8.** $3 - 3x < -1$

9. $4x - 5 \leq 2x$

10. $2x \geq 6 - x$

11. $2 - (x + 1) > x + 3$

12. $-2(x + 4) > 1 - 5x$

13. $x + 4 \geq 3(x - 3)$

14. $2x - 7 \leq 4 - (x + 2)$

15. $\dfrac{1}{3} - \dfrac{x}{2} < x + \dfrac{3}{2}$

16. $\dfrac{x}{5} - 2 > \dfrac{2}{3}(x + 3)$

17. $-1 < 2x + 1 < 3$

18. $2 < 3x + 1 \leq 8$

19. $-4 \leq 1 - x < -1$

20. $0 \leq 3 - 2x \leq 6$

21. $2x < x - 1 \leq 3x + 5$

22. $x + 1 \leq 7 - x < 2x$

23. $2x - 3 < x - 5 < 3x - 3$

24. $x - 1 < 2x + 2 < 3x + 1$

In Exercises 25 through 32, solve the given problems by setting up and solving appropriate inequalities. Graph each solution.

25. The value V, in dollars, of each building lot in a development is estimated as $V = 40{,}000 + 4000t$, where t is the time in years. For how long is the value of each lot no more than $64{,}000$?

26. A beam is supported at each end as shown in Fig. 16-11. Analyzing the forces on the beam leads to the equation $F_1 = 13 - 3d$. For what values of d is F_1 more than 6 N?

27. The velocity v (in ft/s) of a projectile as a function of the time t (in seconds) is $v = 120 - 32t$. For what values of t is the object ascending ($v > 0$)?

28. The relationship between Fahrenheit degrees F and Celsius degrees C is $5F = 9C + 160$. For what values of C is $F \geq 98.6$? (98.6°F is normal body temperature.)

29. During a given rush hour, the numbers of vehicles shown in Fig. 16-12 go in the indicated directions in a one-way-street section of a city. By finding the possible values of x and the equation relating x and y, find the possible values of y.

30. One computer line printer can print no more than 2500 lines/min. If it operates for 3.0 min and a second printer operates for 2.0 min, what printing rates must the second printer have in order to print 9000 lines?

31. In obtaining a solution of hydrochloric acid containing 2 L of pure acid, a chemist mixes x liters of a 5% solution with y liters of a 10% solution. If y is no more than 8 L, what are the possible values of x?

32. An oil company plans to install eight storage tanks, each with a capacity of x liters, and five additional tanks, each with a capacity of y liters, such that the total capacity of all tanks is $440{,}000$ L. If capacity y will be at least $40{,}000$ L, what are the values of capacity x?

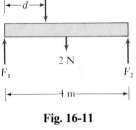

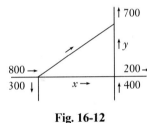

Fig. 16-11

Fig. 16-12

16-3   Solving Nonlinear Inequalities

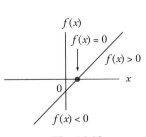

Fig. 16-13

In this section we will show how to solve inequalities which involve polynomial and fractional expressions. We will also introduce a graphical method which can be used with nonfactorable and nonalgebraic inequalities. In order to develop the method for polynomial inequalities, we first look again at a linear inequality.

If we graph the linear function $f(x) = ax + b$ ($a \neq 0$), we see that all values of $f(x)$ are positive for all values of x on one side of the point where $f(x) = 0$, and all values of $f(x)$ are negative on the other side of this point. See Fig. 16-13. This leads us to another method of solving a linear inequality. This method is to *express the given inequality with **zero** on the right side and then determine the **sign** of the resulting function on either side of the zero of the function.*

EXAMPLE A _____

Solve the inequality $2x - 5 > 1$.

We first find the equivalent inequality with zero on the right. This is done by subtracting 1 from each member. Thus, we have $2x - 6 > 0$. We now set the left member equal to zero. Thus,

$$2x - 6 = 0 \quad \text{for} \quad x = 3$$

which means that 3 is the zero of the function $f(x) = 2x - 6$. We know that the function $f(x)$ has one sign for $x < 3$ and has the opposite sign for $x > 3$. Testing values in these intervals, we find, for example, that

$$f(x) = -2 \quad \text{for} \quad x = 2 \quad \text{and} \quad f(x) = +2 \quad \text{for} \quad x = 4$$

This means that for $x > 3$, $2x - 6 > 0$, which tells us that the solution to the original inequality is $x > 3$. See Fig. 16-14.

Of course, we could have solved this inequality by the methods of Section 16-2. However, the idea of using the sign of the function on the left, with zero on the right, is the important concept here. ∎

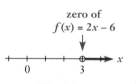

zero of
$f(x) = 2x - 6$

Fig. 16-14

This method can be extended to solving inequalities involving higher-degree polynomials or inequalities involving functions with fractions having x in the denominator as well as in the numerator, and it is especially useful in solving such inequalities. When the equivalent inequality with zero on the right has been found, the function on the left is then factored into linear factors and those quadratic factors which lead to complex roots. These quadratic factors will not change sign, but each linear factor can change sign only at the value for which it is zero, as all possible values of x are considered.

If a linear factor occurs in the numerator, the function is zero at the value of x for which the linear factor is zero. If such a factor occurs in the denominator, the function is undefined where the factor is zero. *The values of x for which a function is zero or undefined are called the **critical values** of the function.* The function can change sign only at a critical value. Therefore, we

critical values

> *find all of the critical values and then determine the sign of the function to the left of the leftmost critical value, between the critical values, and to the right of the rightmost critical value. Those intervals in which we have the proper sign will satisfy the given inequality.*

EXAMPLE B _____

Solve the inequality $x^2 - 3 > 2x$.

We first find the equivalent inequality with zero on the right. Thus, we have $x^2 - 2x - 3 > 0$. We then factor the left member, and have

$$(x - 3)(x + 1) > 0$$

NOTE ▷

We find the critical value for each of the factors, for these are the only values for which $f(x) = x^2 - 2x - 3$ is zero. The left critical value is -1, and the right critical value is 3. All values of x to the left of -1 give the same sign for the function. All values of x between -1 and 3 give the function the same sign. All values of x to the right of 3 give the same sign to the function. Therefore, *we must determine the **sign of $f(x)$ for each of the intervals $x < -1$, $-1 < x < 3$, and $x > 3$.*** For the interval $x < -1$, we find that each of the factors is negative. However, the product of two negative numbers is a positive number. Therefore, if $x < -1$, then $(x - 3)(x + 1) > 0$.

For the interval $-1 < x < 3$, we find that the left factor is negative, but the right factor is positive. The product of a negative and positive number gives a negative number. Thus, for the interval $-1 < x < 3$, $(x - 3)(x + 1) < 0$. For the interval $x > 3$, both factors are positive, making $(x - 3)(x + 1) > 0$. We tabulate the results.

$$\begin{array}{lll} \text{If} & x < -1 & (x - 3)(x + 1) > 0 \\ \text{If} & -1 < x < 3 & (x - 3)(x + 1) < 0 \\ \text{If} & x > 3 & (x - 3)(x + 1) > 0 \end{array}$$

■ Thus, the inequality is satisfied for $x < -1$ or $x > 3$. See Fig. 16-15.

Fig. 16-15

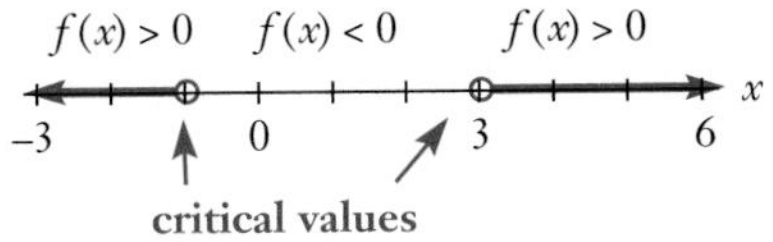

EXAMPLE C _____ Solve the inequality $x^3 - 4x^2 + x + 6 < 0$.

By methods developed in Chapter 14, we factor the function on the left and obtain $(x + 1)(x - 2)(x - 3) < 0$. The critical values are $-1, 2, 3$. We wish to determine the sign of the left member for the intervals $x < -1$, $-1 < x < 2$, $2 < x < 3$, and $x > 3$. The following table shows each interval, the sign of each factor in each interval, and the resulting sign of the function

$$f(x) = (x + 1)(x - 2)(x - 3)$$

Interval	$(x + 1)(x - 2)(x - 3)$			Sign of $f(x)$
$x < -1$	$-$	$-$	$-$	$-$
$-1 < x < 2$	$+$	$-$	$-$	$+$
$2 < x < 3$	$+$	$+$	$-$	$-$
$x > 3$	$+$	$+$	$+$	$+$

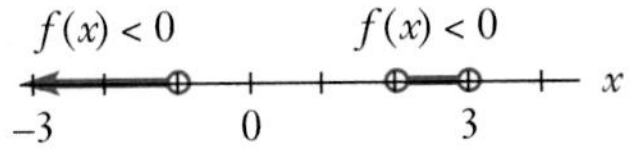

Fig. 16-16

Since we want values of $f(x)$ which are less than zero, the inequality is satisfied for $x < -1$ or $2 < x < 3$. See Fig. 16-16.

EXAMPLE D _____ The force F, in newtons, acting on a cam varies according to the time t, in seconds, and is given by the function $F = 2t^2 - 12t + 20$. For what values of t, $0 \le t \le 6$ s, is the force at least 4 N?

To have a force of at least 4 N, we know that $F \ge 4$ N, or $2t^2 - 12t + 20 \ge 4$. This means we are to solve the inequality $2t^2 - 12t + 16 \ge 0$. We have the solution shown on the next page.

(Continued on next page)

$$2t^2 - 12t + 16 \geq 0$$
$$t^2 - 6t + 8 \geq 0$$
$$(t - 2)(t - 4) = 0$$

The critical values are $t = 2$ and $t = 4$, which lead to the following table:

Interval	$(t - 2)(t - 4)$		Sign of $(t - 2)(t - 4)$
$0 \leq t < 2$	−	−	+
$2 < t < 4$	+	−	−
$4 < t \leq 6$	+	+	+

We see that the values of t which satisfy the *greater than* part of the problem are $0 \leq t < 2$ and $4 < t \leq 6$. Since we know that $(t - 2)(t - 4) = 0$ for $t = 2$ and $t = 4$, the solution is

$$0 \leq t \leq 2 \text{ s}, \qquad 4 \text{ s} \leq t \leq 6 \text{ s}$$

See Fig. 16-17.

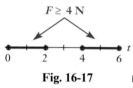

$F \geq 4\,\text{N}$

Fig. 16-17

EXAMPLE E Determine the values of x for which $\sqrt{\dfrac{x - 3}{x + 4}}$ represents a real number.

In order that the expression may represent a real number, the algebraic fraction under the radical must be greater than or equal to zero. This means that we must solve the inequality

$$\frac{x - 3}{x + 4} \geq 0$$

The critical values are found from the factors, whether they are in the numerator or in the denominator. Thus, the critical values are -4 and 3. Considering now the *greater than* part of the $\geq$ sign, we set up the following table:

Interval	$\dfrac{x - 3}{x + 4}$	Sign of $\dfrac{x - 3}{x + 4}$
$x < -4$	$\dfrac{-}{-}$	+
$-4 < x < 3$	$\dfrac{-}{+}$	−
$x > 3$	$\dfrac{+}{+}$	+

NOTE ▷ Thus, the values which satisfy the *greater than* part of the problem are those for which $x < -4$ or for which $x > 3$. Now considering the equality part of the $\geq$ sign, we note that $x = 3$ is valid, for the fraction is zero. However, ***if $x = -4$, we have division by zero, and thus x may not equal -4.*** Therefore, the inequality is satisfied for $x < -4$ or $x \geq 3$, and these are the values of x for which the original expression represents a real number. See Fig. 16-18.

Fig. 16-18

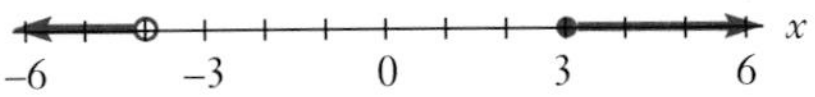

Fig. 16-19

EXAMPLE F ___ Solve the inequality $x^3 - x^2 + x - 1 > 0$.

This leads to $(x^2 + 1)(x - 1) > 0$. There is only one linear factor with a critical value. The factor $x^2 + 1$ is never negative. The inequality is satisfied for $x > 1$. See Fig. 16-19.

EXAMPLE G ___ Solve the inequality $\dfrac{(x - 2)^2(x + 3)}{4 - x} < 0$

The critical values are -3, 2, and 4. Thus, we have the following table:

Interval	$\dfrac{(x-2)^2(x+3)}{4-x}$	Sign of $\dfrac{(x-2)^2(x+3)}{4-x}$
$x < -3$	$\dfrac{+\quad-}{+}$	$-$
$-3 < x < 2$	$\dfrac{+\quad+}{+}$	$+$
$2 < x < 4$	$\dfrac{+\quad+}{+}$	$+$
$x > 4$	$\dfrac{+\quad+}{-}$	$-$

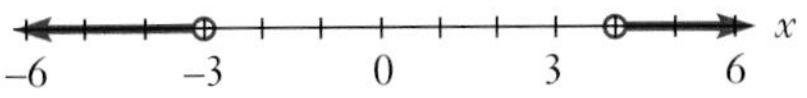

Fig. 16-20

The inequality is satisfied for $x < -3$ or $x > 4$. See Fig. 16-20.

To this point we have considered only inequalities in which the function on the left can be factored. Inequalities in which the function is not factorable, including nonalgebraic functions, can be solved approximately by graphing the function.

In solving an inequality by graphing, we first write the inequality in an equivalent form with zero on the right. Next we set y equal to the left member and graph the resulting function. Those values of x corresponding to the proper values of y (either above or below the x-axis) are those values which satisfy the inequality.

EXAMPLE H

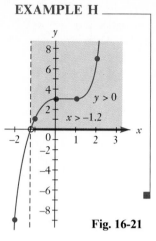

Fig. 16-21

Graphically solve the inequality $x^3 > x^2 - 3$.

Finding the equivalent inequality with 0 on the right, we then have $x^3 - x^2 + 3 > 0$. By letting $y = x^3 - x^2 + 3$, we may solve the inequality by finding those values of x for which y is positive. To graph this function, we use the following points:

x	-2	-1	0	1	2
y	-9	1	3	3	7

The graph is shown in Fig. 16-21. We see that $y > 0$, which corresponds to $x^3 - x^2 + 3 > 0$, occurs for $x > -1.2$ (which is approximated from the graph). Therefore, the solution to the original inequality is $x > -1.2$. If greater accuracy is required, a calculator may be used to find the root. ∎

Exercises 16-3

In Exercises 1 through 28, solve the given inequalities. Graph each solution.

1. $x^2 - 1 < 0$

2. $x^2 + 3x \geq 0$

3. $2x^2 \leq 4x$

4. $x^2 - 4x > 5$

5. $3x^2 + 5x \geq 2$

6. $2x^2 - 12 \leq -5x$

7. $6x^2 + 1 < 5x$

8. $12x^2 + x > 1$

9. $x^2 + 4x \leq -4$

10. $9x^2 + 6x > -1$

11. $x^2 + 4 > 0$

12. $x^4 + 2 < 1$

13. $x^3 + x^2 - 2x > 0$

14. $x^3 - 2x^2 + x \geq 0$

15. $x^3 + 2x^2 - x - 2 \geq 0$

See Appendix E for a computer program for finding the sign of a function.

16. $x^4 - 2x^3 - 7x^2 + 8x + 12 < 0$

17. $\dfrac{x - 8}{3 - x} < 0$

18. $\dfrac{x + 5}{x - 1} > 0$

19. $\dfrac{2x - 3}{x + 6} \leq 0$

20. $\dfrac{3x + 1}{x + 3} \geq 0$

21. $\dfrac{2}{x^2 - x - 2} < 0$

22. $\dfrac{-5}{2x^2 + 3x - 2} < 0$

23. $\dfrac{x^2 - 6x - 7}{x + 5} > 0$

24. $\dfrac{4 - x}{3 + 2x - x^2} > 0$

25. $\dfrac{6 - x}{3 - x - 4x^2} \geq 0$

26. $\dfrac{(x - 2)^2(5 - x)}{(4 - x)^3} \leq 0$

27. $\dfrac{x^4(9 - x)(x - 5)(2 - x)}{(4 - x)^5} > 0$

28. $\dfrac{x^3(1 - x)(x - 2)(3 - x)(4 - x)}{(5 - x)^2(x - 6)^3} < 0$

In Exercises 29 through 32, determine the values of x for which the given radicals represent real numbers.

29. $\sqrt{(x - 1)(x + 2)}$

30. $\sqrt{x^2 - 3x}$

31. $\sqrt{-x - x^2}$

32. $\sqrt{\dfrac{x^3 + 6x^2 + 8x}{3 - x}}$

In Exercises 33 through 40, solve the given inequalities graphically.

33. $x^3 > 2$

34. $x^3 > x + 4$

35. $x^4 < x^2 - 2x - 1$

36. $\log x > 0.5$

37. $2^x > 3$

38. $x^4 - 6x^3 + 7x^2 > 18 - 12x$

39. $\sin x < 0.1$ $(0 \leq x \leq 2\pi)$

40. $\cos 2x > 0.6$ $(0 \leq x \leq 4)$

In Exercises 41 through 48, answer the given questions by solving the appropriate inequalities.

41. The electric power p delivered to part of a circuit is given by $p = 6i - 4i^2$, where i is the current in amperes. For what positive values of i is the power greater than 2 W?

42. The weight w, in tons, of fuel in a rocket after launch is $w = 2000 - t^2 - 140t$, where t is measured in minutes. During what period of time is the weight of fuel greater than 500 tons?

43. The power p, in watts, dissipated in an electric circuit containing a resistance and an inductance is $p = 15e^{-4.0t}$, where t is measured in seconds. For what period of time is $p > 1.0$ W?

44. The object distance p and image distance q for a camera of focal length 3.00 cm is given by $p = 3.00q/(q - 3.00)$. For what values of q is $p > 12.0$ cm?

45. The length of a microprocessor chip is 2.0 mm more than its width. If its area is less than 35 mm^2, what values are possible for the width if it must be at least 3.0 mm?

46. A laser source is 2.0 in. from the nearest point P on a flat mirror, and the laser beam is directed at a point Q which is on the mirror and is x in. from P. The beam is then reflected to the receiver, which is x in. from Q. What is x if the total length of the beam is greater than 6.5 in.?

47. A plane takes off from an airport and flies due north at 400 mi/h. A second plane takes off an hour later from the same airport and flies due east at 400 mi/h. For how long after the second plane leaves are the planes less than 1000 mi apart?

48. An open box (no top) is formed from a piece of cardboard 8.00 in. square by cutting equal squares from the corners and turning up the resulting sides. Find the edges of the squares which are cut out in order that the volume of the box is greater than 32.0 in.^{3}.

16-4 Inequalities Involving Absolute Values

If we wish to write the inequality $|x| > 1$ without absolute-value signs, we must note that we are considering values of x which are *numerically* larger than 1. Thus we may write this inequality in the equivalent form $x < -1$ or $x > 1$. We now note that *the original inequality, with an absolute-value sign, can be written in terms of two equivalent inequalities, neither involving absolute values.* If we are asked to write the inequality $|x| < 1$ without the absolute-value signs, we write $-1 < x < 1$ since we are considering values of x numerically less than 1.

Following reasoning similar to the above, whenever absolute values are involved in inequalities, the following two relations allow us to write equivalent inequalities without absolute values.

> **If $|f(x)| > n$, then $f(x) < -n$ or $f(x) > n$.**   (16-1)
>
> **If $|f(x)| < n$, then $-n < f(x) < n$.**   (16-2)

The use of these relations is indicated in the following examples.

EXAMPLE A   Solve the inequality $|x - 3| < 2$.

Inspection of this inequality shows that we wish to find the values of x which are within 2 units of $x = 3$. Of course, such values are given by $1 < x < 5$. Let us now see how Eq. (16-2) gives us this result.

By using Eq. (16-2), we have

$$-2 < x - 3 < 2$$

By adding 3 to all three members of this inequality, we have

$$1 < x < 5$$

which is the proper interval. See Fig. 16-22.

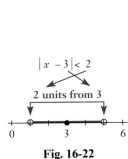

Fig. 16-22

EXAMPLE B   Solve the inequality $|2x - 1| > 5$.
By using Eq. (16-1), we have

$$2x - 1 < -5 \quad \text{or} \quad 2x - 1 > 5$$

Completing the solution, we have

$2x < -4$	or $2x > 6$	add 1 to each member
$x < -2$	or $x > 3$	divide each member by 2

This means that the given inequality is satisfied for $x < -2$ or for $x > 3$ (which *cannot* be written as $3 < x < -2$). See Fig. 16-23.

The meaning of the inequality is that the numerical value of $2x - 1$ is more than 5. We see that values of x less than -2 or greater than 3 make this true.

Fig. 16-23   **Fig. 16-24**

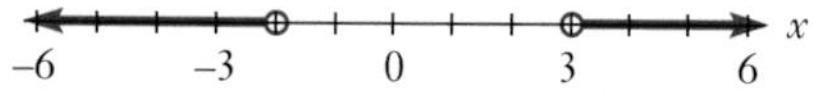

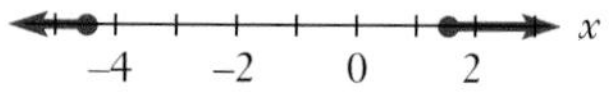

EXAMPLE C   Solve the inequality $2\left|\dfrac{2x}{3} + 1\right| \geq 4$.

The solution is as follows:

$2\left\|\dfrac{2x}{3} + 1\right\| \geq 4$	original inequality
$\left\|\dfrac{2x}{3} + 1\right\| \geq 2$	divide each member by 2
$\dfrac{2x}{3} + 1 \leq -2 \quad \text{or} \quad \dfrac{2x}{3} + 1 \geq 2$	using Eq. (16-1)
$2x + 3 \leq -6 \quad \text{or} \quad 2x + 3 \geq 6$	
$2x \leq -9 \quad \text{or} \quad 2x \geq 3$	
$x \leq -\dfrac{9}{2} \quad \text{or} \quad x \geq \dfrac{3}{2}$	solution

This solution is shown in Fig. 16-24. Note that the sign of equality does not change the method of solution. It simply indicates that $-\frac{9}{2}$ and $\frac{3}{2}$ are included in the solution.

EXAMPLE D   Solve the inequality $|3 - 2x| < 3$.
We have the following solution:

$\|3 - 2x\| < 3$	original inequality
$-3 < 3 - 2x < 3$	using Eq. (16-2)
$-6 < -2x < 0$	
$3 > x > 0$	divide by -2 and reverse
$0 < x < 3$	signs of inequality

Fig. 16-25   See Fig. 16-25.

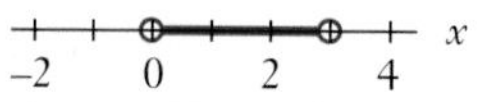

EXAMPLE E ———— A technician measures an electric current and reports that it is 0.036 A with a possible error of ± 0.002 A. Write this result for the current i using an inequality with absolute values.

The statement of the problem tells us that the current is no less than 0.034 A and no more than 0.038 A. Another way of stating this is that the numerical difference between the true value of i (unknown exactly) and the measured value, 0.036 A, is less than or equal to 0.002 A. Using an absolute-value inequality, this is written as

$$|i - 0.036| \le 0.002$$

where values are in amperes.

We can see that this inequality is correct by using Eq. (16-2).

$$-0.002 \le i - 0.036 \le 0.002$$
$$0.034 \le i \le 0.038 \qquad \textbf{add 0.036 to each member}$$

This verifies that i is no less than 0.034 A and no more than 0.038 A. See Fig. 16-26.

Fig. 16-26

Exercises 16-4

In Exercises 1 through 20, solve the given inequalities. Graph each solution.

1. $|x - 4| < 1$

2. $|x + 1| < 3$

3. $|5x + 4| > 6$

4. $|2x - 1| > 1$

5. $|6x - 5| \le 4$

6. $|5 - x| \le 2$

7. $|3 - 4x| > 3$

8. $|3x + 1| \ge 2$

9. $\left|\dfrac{x + 1}{5}\right| < 3$

10. $\left|\dfrac{2x - 9}{4}\right| < 1$

11. $|20x + 85| \le 43$

12. $|2.6x - 9.1| > 10.4$

13. $2|x - 4| > 8$

14. $3|4 - 3x| \le 10$

15. $4|2 - 5x| \ge 6$

16. $2|7 - 2x| < 12$

17. $\left|\dfrac{x}{2} + 1\right| < 8$

18. $\left|\dfrac{4x}{3} - 5\right| \ge 7$

19. $\left|6.5 - \dfrac{x}{2}\right| \ge 2.3$

20. $\left|27 - \dfrac{2x}{3}\right| > 17$

In Exercises 21 through 24, solve the given quadratic inequalities.

21. $|x^2 + x - 4| > 2$

[After using Eq. (16-1), you will have two inequalities. The solution includes the values of x which satisfy *either* of the inequalities.]

22. $|x^2 + 3x - 1| > 3$ (See Exercise 21.)

23. $|x^2 + x - 4| < 2$

[Use Eq. (16-2), then treat the resulting inequality as two inequalities of the form $f(x) > -n$ and $f(x) < n$. The solution includes the values of x which satisfy *both* of the inequalities.]

24. $|x^2 + 3x - 1| < 3$ (See Exercise 23.)

In Exercises 25 through 28, use inequalities involving absolute values to solve the given problems.

25. Using a vernier micrometer, a technician measured the diameter d of a metal rod and wrote down the result as 0.2537 ± 0.0003 in. Express the possible values of d using an inequality with absolute values.

26. The production p (in barrels) of an oil refinery for the coming month is estimated at $|p - 2{,}000{,}000| < 200{,}000$. By solving this inequality, determine the production which is anticipated.

27. A straight bridge support is 10.0 ft below the water level at the shoreline, and it rises 3.0 ft for each foot measured horizontally from the shoreline. Set up the vertical distance d from the water level to the support as a function of the horizontal distance x from the shoreline, then find the values of x for which the support is within 6.0 ft of water level.

28. A rocket is fired from a plane which is flying horizontally at 9000 ft. The height h, in feet, of the rocket above the plane is given by $h = 560t - 16t^2$, where t is the time of flight of the rocket in seconds. When is the rocket more than 4000 ft above or below the plane?

16-5   Graphical Solution of Inequalities with Two Variables ▬▬

To this point we have considered inequalities with one variable and certain methods of solving them. We may also graphically solve inequalities involving two variables, such as x and y. In this section we consider the solution of such inequalities, as well as one important type of application.

Let us consider the function $y = f(x)$. We know that the coordinates of points on the graph satisfy the equation $y = f(x)$. However, for points above the graph of the function, we have $y > f(x)$, and for points below the graph of the function we have $y < f(x)$. Consider the following example.

EXAMPLE A _____ Consider the linear function $y = 2x - 1$. This equation is satisfied for points on the line. For example, the point $(2, 3)$ is on the line and we have $3 = 2(2) - 1 = 3$. Therefore, for points on the line we have $y = 2x - 1$, or $y - 2x + 1 = 0$.

The point $(2, 4)$ is above the line, since we have $4 > 2(2) - 1$, or $4 > 3$. Therefore, for points above the line we have $y > 2x - 1$, or $y - 2x + 1 > 0$. In the same way, for points below the line, $y < 2x - 1$ or $y - 2x + 1 < 0$. We note this is true for the point $(2, 1)$, since $1 < 2(2) - 1$, or $1 < 3$.

The line for which $y = 2x - 1$, and the regions for which $y > 2x - 1$, and for which $y < 2x - 1$ are shown in Fig. 16-27.

Summarizing,

$y > 2x - 1$ for points *above* the line
$y = 2x - 1$ for points *on* the line
$y < 2x - 1$ for points *below* the line

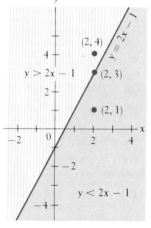

Fig. 16-27

The illustration of Example A leads us to the graphical method of indicating the points which satisfy an inequality with two variables. First we solve the inequality for y and then determine the graph of the function $y = f(x)$. *If we wish to solve the inequality $y > f(x)$, we indicate the appropriate points by shading in the region above the curve. For the inequality $y < f(x)$, we indicate the appropriate points by shading in the region below the curve.* We note that the complete solution to the inequality consists of all points in an entire region of the plane.

EXAMPLE B _____ Draw a sketch of the graph of the inequality $y < x + 3$.

First we graph the function $y = x + 3$, as shown in Fig. 16-28. Since we wish to find the points which satisfy the inequality $y < x + 3$, we show these points by shading in the region below the line. We show the line as a *dashed line* to indicate that points on it do not satisfy the inequality.

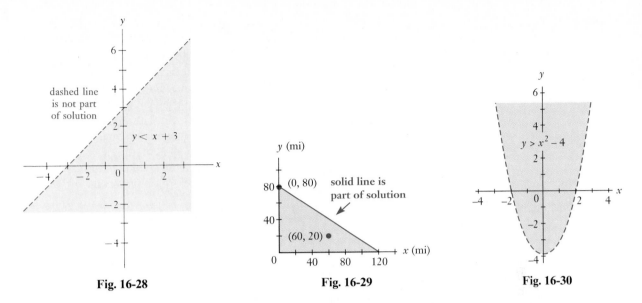

Fig. 16-28   **Fig. 16-29**   **Fig. 16-30**

EXAMPLE C

After a snowstorm, it is estimated that it will take 30 min to plow each mile of Route 15 and 45 min to plow each mile of Route 80. If no more than 60 plowing-hours are available, what combinations of Route 15 and Route 80 can be plowed?

Let x = miles of Route 15 which can be plowed and y = miles of Route 80 which can be plowed. The time to plow along each route is the product of the time for each mile and the number of miles to be plowed. This gives us

time to plow Rt. 15   time to plow Rt. 80   max. available time

$$(0.50 \text{ h/mi})(x \text{ mi}) \; + \; (0.75 \text{ h/mi})(y \text{ mi}) \; \leq \; 60 \text{ h}$$

30 min⌐   ⌐45 min

$$0.50x + 0.75y \leq 60$$
$$y \leq 80 - 0.67x$$

Noting that negative values of x and y do not have meaning, we have the graph in Fig. 16-29, shading in the region below the line since we have $y < 80 - 0.67x$ for that region. Any point in the shaded region, or on the axes or the line around the shaded region, gives a solution. The *solid line* indicates that points on it are part of the solution.

The point $(0, 80)$, for example, is a solution and tells us that 80 mi of Route 80 can be plowed if none of Route 15 is plowed. In this case, all 60 h of plowing time are used for Route 80. Another possibility is shown by the point $(60, 20)$, which indicates that 60 mi of Route 15 and 20 mi of Route 80 can be plowed. In this case, not all of the 60 plowing hours are used.

EXAMPLE D

Draw a sketch of the graph of the inequality $y > x^2 - 4$.

Although the graph of $y = x^2 - 4$ is not a straight line, the method of solution is the same. We graph the function $y = x^2 - 4$ as a dashed curve, since it is not part of the solution, as shown in Fig. 16-30. We then shade in the region above the curve to indicate the points which satisfy the inequality.

EXAMPLE E

Draw a sketch of the region which is defined by the system of inequalities $y \geq -x - 2$ and $y + x^2 < 0$.

In this case we sketch the graph of both inequalities, then determine the region common to both graphs. First we draw the graph of $y = -x - 2$ (the straight line), and shade in the region above the line. See Fig. 16-31. Next we have $y < -x^2$ and draw the graph of $y = -x^2$, shading the region below it. The sketch of the region which is defined by this system of inequalities is the darkly shaded region below the parabola which is above and on the line.

If we had wanted the region defined by the system of inequalities $y \geq -x - 2$ or $y + x^2 < 0$, it would consist of all points in all shaded areas and on the line. This is consistent with the discussion of the words *or* and *and* in Section 16-1.

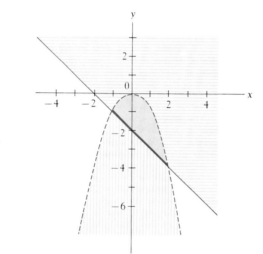

Fig. 16-31

linear programming

An important area in which graphs of inequalities with two or more variables are used is the branch of mathematics known as **linear programming** (in this context, "programming" has no relation to computer programming). This subject is widely applied in industry, business, economics, and technology. The analysis of many social problems can also be made by the use of linear programming.

Linear programming is used to analyze problems such as those related to maximizing profit, minimizing costs, or the use of materials, with certain constraints of production. The following serves as an example of the use of linear programming.

EXAMPLE F

See the chapter introduction.

A company makes two types of stereo speaker systems, their good-quality system and their highest-quality system. The production of the systems requires assembly of the speaker system itself and the production of the cabinets in which they are installed. The good-quality system requires 3 worker-hours for speaker assembly and 2 worker-hours for cabinet production for each complete system. The highest-quality system requires 4 worker-hours for speaker assembly and 6 worker-hours for cabinet production for each complete system. Available skilled labor allows for a maximum of 480 worker-hours per week for speaker assembly and a maximum of 540 worker-hours per week for cabinet production. It is anticipated that all systems will be sold and that the profit will be $10 for each good-quality system and $25 for each highest-quality system. How many of each should be produced to provide the greatest profit?

First, let x = the number of good-quality systems and y = the number of highest-quality systems made in one week. Thus, the profit p is given by

$$p = 10x + 25y$$

We know that negative numbers are not valid for either x or y, and therefore we have $x \geq 0$ and $y \geq 0$. Also, the number of available worker-hours per week for each part of the production restricts the number of systems which can be made. Both speaker assembly and cabinet production are required for all systems. The number of worker-hours needed to produce the x good-quality systems is $3x$ in the speaker assembly shop. Also, $4y$ worker-hours are required in the speaker assembly shop for the highest-quality systems. Thus,

$$3x + 4y \leq 480$$

since no more than 480 worker-hours are available in the speaker assembly shop. In the cabinet shop, we have

$$2x + 6y \leq 540$$

since no more than 540 worker-hours are available in the cabinet shop.

Therefore, we wish to maximize the profit p under the **constraints**

$x \geq 0$,	$y \geq 0$	number of systems produced cannot be negative
$3x + 4y \leq 480$		worker-hours for speaker assembly
$2x + 6y \leq 540$		worker-hours for cabinet production

In order to do this we sketch the region of points which satisfy this system of inequalities. From the previous examples, we see that the appropriate region is in the first quadrant (since $x \geq 0$ and $y \geq 0$) and under both lines. See Fig. 16-32.

Fig. 16-32

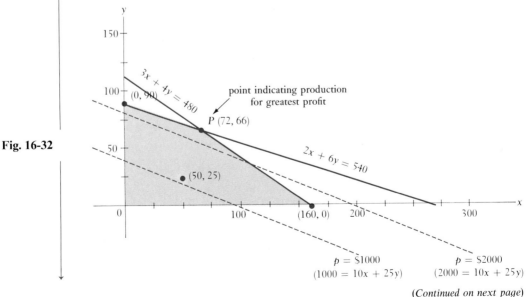

(Continued on next page)

Any point in the shaded region which is defined by the preceding system of in-equalities is known as a **feasible point.** In this case it means that it is possible to produce the number of systems of each type according to the coordinates of the point. For example, the point (50, 25) is in the region, which means that it is pos-sible to produce 50 good-quality systems and 25 highest-quality systems under the given constraints of available skilled labor. However, we wish to find the point which indicates the number of each kind of system which produces the greatest profit.

If we assume values for the profit, the resulting equations are straight lines. Thus, by finding the greatest value of p for which the line passes through a feasible point, we may solve the given problem. If $p = \$1000$, or if $p = \$2000$, we have the lines shown. Both are possible with various combinations of speaker systems being produced. However, we note the line for $p = \$2000$ passes through feasible points farther from the origin. It is also clear, since these lines are parallel, that the greatest profit attainable is given by the line passing through P, where $3x + 4y = 480$ and $2x + 6y = 540$ intersect. The coordinates of P are (72, 66). Thus, the production should be 72 good-quality systems and 66 highest-quality systems to produce a weekly profit of $p = 10(72) + 25(66) = \$2370$.

For this type of problem, the solution will be given by one of the vertices of the region. However, it could be any one of them, which means it is possible that only one type of product should be produced. (See Exercise 38.) Thus, we can solve the problem by finding the appropriate region and then testing the coordi-nates of the vertex points. Here the vertex points are (160, 0), which indicates a profit of $1600; (0, 90), which indicates a profit of $2250; and (72, 66), which in-dicates a profit of $2370.

Exercises 16-5

In Exercises 1 through 24, draw a sketch of the graph of the given inequality.

1. $y > x - 1$ **2.** $y < 3x - 2$ **3.** $y \geq 2x + 5$ **4.** $y \leq 3 - x$

5. $2x + y < 5$ **6.** $4x - y > 1$ **7.** $3x + 2y + 6 > 0$ **8.** $x + 4y - 8 < 0$

9. $y < x^2$ **10.** $y > -2x^2$ **11.** $y \geq 1 - x^2$ **12.** $y \leq 2x^2 - 3$

13. $x^2 + 2x + y < 0$ **14.** $2x^2 - 4x - y > 0$ **15.** $y \leq x^3$ **16.** $y \geq 3x - x^3$

17. $y > x^4 - 8$ **18.** $y < 32x - x^4$ **19.** $y \leq \sqrt{2x + 5}$ **20.** $y > \dfrac{1}{x^2 + 1}$

21. $y < \ln x$ **22.** $y > \sin 2x$ **23.** $y \geq 2 \cos \pi x$ **24.** $y \leq 1 - e^{-x}$

In Exercises 25 through 32, draw a sketch of the graph of the region in which the points satisfy the given system of inequalities.

25. $y > x$
    $y > 1 - x$

26. $y \leq 2x$
    $y \geq x - 1$

27. $y \leq 2x^2$
    $y > x - 2$

28. $y > x^2$
    $y < x + 4$

29. $y > \frac{1}{2}x^2$
    $y \leq 4x - x^2$

30. $y > 4 - x$
    $y < \sqrt{16 - x^2}$

31. $y \geq 0$
    $y \leq \sin x$
    $0 \leq x \leq 3\pi$

32. $y > 0$
    $y > 1 - x$
    $y < e^x$

In Exercises 33 through 36, set up the necessary inequalities and sketch the graph of the region in which the points satisfy the indicated systems of inequalities.

33. A telephone company is installing two types of fiber-optic cable in an area. It is estimated that no more than 300 m of type A cable, and at least 200 m but no more than 400 m of type B cable, are needed. Graph the possible lengths of cable which are needed.

34. A refinery can produce gasoline and diesel fuel, in amounts of any combination, except that equipment restricts maximum total production to 2500 gal per day. Graph the different possible production combinations of the two fuels.

35. The elements of an electric circuit dissipate p watts of power. The power p_R dissipated by a resistor in the circuit is given by $p_R = Ri^2$, where R is the resistance and i is the current, in amperes. Graph the possible values of p and i for $p > p_R$ and $R = 0.5\ \Omega$.

36. A person wishes to invest no more than \$25,000 in two stocks which pay 8% and 4% annual dividends, respectively. Graph the possible values invested in each if more than \$800 is to be received in annual dividends.

In Exercises 37 through 40, solve the given linear programming problems.

37. A manufacturer makes two types of calculators, a business model and a scientific model. Each model is assembled in two sets of operations, where each operation is in production 8 h each day. The average time required for a business model in the first operation is 3 min, and 6 min is required in the second operation. The scientific model averages 6 min in the first operation and 4 min in the second operation. All calculators can be sold; the profit for a business model is \$8, and the profit for a scientific model is \$10. How many of each model should be made each day in order to maximize profit?

38. Using the information of Example F, with the single exception that the profit on each good-quality system is \$20, how many of each system should be made?

39. A company makes brands A and B of breakfast cereal, both of which are enriched with vitamins P and Q. The necessary information about these cereals is given in the following table.

	Cereal A	Cereal B	Recommended Daily Allowance (RDA)
Vitamin P	1 unit/oz	2 units/oz	10 units
Vitamin Q	5 units/oz	3 units/oz	30 units
Cost per ounce	2¢	3¢	

Find the number of ounces of each cereal which together satisfies the recommended daily allowance of vitamins P and Q at the lowest cost. (*Note:* We wish to *minimize* cost; be careful in determining the feasible region.)

40. A computer company can produce two different computer parts, A and B, in each of two different production plants. It costs \$4000 per day to operate the first plant and \$5000 per day to operate the second plant. Each day the first plant produces 100 of A and 200 of B, while the second plant produces 250 of A and 100 of B. How many days should each plant operate to produce at least 2000 of each part and keep operating costs at a minimum?

16-6   Chapter Equations, Review Exercises, and Practice Test

Chapter Equations ▄▄▄▄▄▄▄▄▄▄▄▄▄▄▄▄▄▄▄▄▄▄▄▄▄▄▄▄▄▄▄▄▄▄▄

If $|f(x)| > n$, then $f(x) < -n$ or $f(x) > n$.   (16-1)

If $|f(x)| < n$, then $-n < f(x) < n$.   (16-2)

Review Exercises

In Exercises 1 through 32, solve the given inequalities algebraically. Graph each solution.

1. $2x - 12 > 0$

2. $5 - 3x < 0$

3. $3x + 5 \le 0$

4. $\frac{1}{4}x - 2 \ge 3x$

5. $3(x - 7) \ge 5x + 8$

6. $2x + 6 < 7(x - 3)$

7. $4 < 2x - 1 < 11$

8. $6 \le 4x - 2 < 9$

9. $2x < x + 1 < 4x + 7$

10. $3x < 2x + 1 < x - 5$

11. $5x^2 + 9x < 2$

12. $x^2 - 7x \ge 8$

13. $x^2 + 2x > 63$

14. $6x^2 - x > 35$

15. $x^3 + 4x^2 - x > 4$

16. $2x^3 + 4 \le x^2 + 8x$

17. $\dfrac{x - 8}{2x + 1} \le 0$

18. $\dfrac{3x + 2}{x - 3} > 0$

19. $\dfrac{(2x - 1)(3 - x)}{x + 4} > 0$

20. $\dfrac{(3 - x)^2}{2x + 7} \le 0$

21. $x^4 + x^2 \le 0$

22. $3x^3 + 7x^2 - 20x < 0$

23. $\dfrac{1}{x} < 2$

24. $\dfrac{1}{x - 2} < \dfrac{1}{4}$

25. $|x - 2| > 3$

26. $|2x + 1| < 5$

27. $|3x + 2| \le 4$

28. $|4 - 3x| \ge 1$

29. $|3 - 5x| > 7$

30. $2|2x - 9| < 8$

31. $\left|2 - \dfrac{x}{2}\right| \le 5$

32. $\left|\dfrac{5x + 1}{5}\right| \ge 4$

In Exercises 33 through 36, solve the given inequalities graphically.

33. $x^3 + x + 1 < 0$

34. $\dfrac{1}{x} > 2$

35. $e^{-x} > 0.5$

36. $\sin 2x < 0.8 \qquad (0 < x < 4)$

In Exercises 37 through 40, determine the values of x for which the given radicals represent real numbers.

37. $\sqrt{3 - x}$

38. $\sqrt{x + 5}$

39. $\sqrt{x^2 + 4x}$

40. $\sqrt{\dfrac{x - 1}{x + 2}}$

In Exercises 41 through 48, draw a sketch of the graph of the given inequality.

41. $y > 4 - x$

42. $y < \dfrac{1}{2}x + 2$

43. $2y - 3x - 4 \le 0$

44. $3y - x + 6 \ge 0$

45. $y > x^2 + 1$

46. $y \le \dfrac{1}{x^2 - 4}$

47. $y - x^3 + 1 < 0$

48. $2y + 2x^3 + 6x - 3 > 0$

In Exercises 49 through 52, draw a sketch of the region in which the points satisfy the given systems of inequalities.

49. $y > x + 1$
   $y < 4 - x^2$

50. $y > 2x - x^2$
   $y \ge -2$

51. $y \le \dfrac{1}{x^2 + 1}$
   $y < x - 1$

52. $y < \cos\dfrac{1}{2}x$
   $y > \dfrac{1}{2}e^x$
   $-\pi < x < \pi$

In Exercises 53 through 64, solve the given problems using inequalities.

53. The length L, in centimeters, of a certain metal bar is given by $L = 150 + 0.00025T$, where T is the temperature in degrees Celsius. For what values of T is $L \ge 151$ cm?

54. After conducting tests, it was determined that the stopping distance x (in feet) of a car traveling 60 mi/h was $|x - 290| \le 35$. Express this inequality without absolute values, and determine the interval of stopping distances which were found in the tests.

55. A heating unit with 80% efficiency and a second unit with 90% efficiency deliver 360,000 Btu of heat to an office complex. If the first unit consumes an amount of fuel which contains no more than 261,000 Btu, what is the Btu content of the fuel consumed by the second unit?

56. A rectangular parking lot is to have a perimeter of 100 m and an area no greater than 600 m². What are the possible dimensions of the lot?

57. The electric power p dissipated in a resistor is given by $p = Ri^2$, where R is the resistance and i is the current in amperes. For a given resistor, $R = 12.0\ \Omega$, and the power varies between 2.50 W and 8.00 W. Find the values of the current.

58. The reciprocal of the total resistance of two resistances in parallel equals the sum of the reciprocals of the resistances. If a 2.0 Ω resistance is in parallel with a resistance R, with a total resistance greater than 0.5 Ω, find R.

59. The displacement y of a weight oscillating at the end of a spring is given by $y = 4.0 \sin t$. For what values of t during the first oscillation is the weight within 2.0 cm of the equilibrium position?

60. A rocket is fired such that its height h is given by $h = 41t - t^2$. For what values of t, in minutes, is the height greater than 400 mi?

61. In developing a new product, a company estimates that it will take no more than 1200 min of computer time for research and no more than 1000 min of computer time for development. Graph the possible combinations of the computer times which are needed.

62. A natural gas supplier has a maximum of 120 worker-hours per week for delivery and for customer service. Graph the possible combinations of times available for these two services.

63. A company produces two types of cameras, the regular model and the deluxe model. For each regular model produced there is a profit of $8, and for each deluxe model the profit is $15. The same amount of materials is used to make each model, but the supply is sufficient only for 450 cameras per day. The deluxe model requires twice the time to produce as the regular model. If only regular models were made, there would be time enough to produce 600 per day. Assuming all models will be sold, how many of each model should be produced if the profit is to be a maximum?

64. A company which manufactures compact disc players gets two different parts, A and B, from two different suppliers. Each package of parts from the first supplier costs $2.00 and contains 6 of each type of part. Each package of parts from the second supplier costs $1.50 and contains 4 of A and 8 of B. How many packages should be bought from each supplier to keep the total cost to a minimum, if production requirements are 600 of A and 900 of B?

Practice Test

1. State conditions on x and y in terms of inequalities if the point (x, y) is in the second quadrant.

For Problems 2 through 6, solve the given inequalities algebraically and show the solution on a number line.

2. $\dfrac{-x}{2} \geq 3$　　　　　3. $3x + 1 < -5$　　　　4. $-1 < 1 - 2x < 5$　　　5. $\dfrac{x^2 + x}{x - 2} \leq 0$　　　　6. $|2x + 1| \geq 3$

7. Sketch the region in which the points satisfy the system of inequalities

$$y < x^2$$
$$y \geq x + 1$$

8. Determine the values of x for which $\sqrt{x^2 - x - 6}$ represents a real number.

9. The length of a rectangular lot is 20 m more than its width. If the area is to be at least 4800 m^2, what values may the width be?

10. Type A wire costs $0.10 per foot, and type B wire costs $0.20 per foot. Show the possible combinations of lengths of wire which can be purchased for less than $5.00.

17 Variation

Newton's universal law of gravitation is expressed in the language of variation. See Section 17-2 for a space-age application.

Through experimentation and observation, important relationships among quantities being measured may be found. In studying the measured values, it is often possible to see how one quantity changes as other, related quantities change. In this chapter we shall see how such information can be used to set up functions which relate these quantities.

We will begin this chapter by studying the meanings of *ratio* and *proportion*, which were first introduced in Chapter 3, in more detail. Then we will show how *variation* is used to set up many important functional relationships.

Applications of the topics of this chapter are found in all areas of science and technology. They are frequently used in acoustics, biology, chemistry, computer technology, economics, electronics, environmental technology, hydrodynamics, mechanics, navigation, optics, physics, space technology, thermodynamics, and other fields.

17-1 Ratio and Proportion

ratio

When we first introduced the trigonometric functions in Chapter 3, we used ratios which had a specific meaning. Considering now the general definition, *the* **ratio** *of a number a to a number b ($b \neq 0$) is the quotient a/b*. Thus, a fraction is a ratio.

Any measurement made is the ratio of the measured magnitude to an accepted unit of measurement. For example, when we say that an object is 5 ft long, we are saying that the length of that object is five times as long as an accepted unit of length, the foot. Other examples of ratios are density (weight/volume), relative density (density of object/density of water), and pressure (force/area). As these examples illustrate, ratios may compare quantities of the same kind, or they may express a division of magnitudes of different quantities (such a ratio is also called a **rate**).

EXAMPLE A ——— The approximate airline distance from New York to San Francisco is 2500 mi, and the approximate airline distance from New York to Minneapolis is 1000 mi. The ratio of these distances is

$$\frac{2500 \text{ mi}}{1000 \text{ mi}} = \frac{5}{2}$$

Since the units in both are miles, the resulting ratio is a dimensionless number.
   If a jet travels from New York to San Francisco in 4 h, its average speed is

$$\frac{2500 \text{ mi}}{4 \text{ h}} = 625 \text{ mi/h}$$

■ In this case we must attach the proper units to the resulting ratio.

As we noted in Example A, we must be careful to attach the proper units to the resulting ratio. Generally, the ratio of measurements of the same kind should be expressed as a dimensionless number. Consider the following example.

EXAMPLE B ——— The length of a certain room is 24 ft, and the width of the room is 18 ft. Therefore, the ratio of the length to the width is $\frac{24}{18}$, or $\frac{4}{3}$.
   If the width of the room is expressed as 6 yd, we have the ratio 24 ft/6 yd = 4 ft/1 yd. However, this does not clearly show the ratio. It is better and more meaningful first to change the units of one of the measurements to the units of the other measurement. Changing the length from 6 yd to 18 ft, we express the ratio as $\frac{4}{3}$, as we saw above. From this ratio we can easily see that
■ the length is $\frac{4}{3}$ as long as the width.

proportion   *A statement of equality between two ratios is called a* **proportion.** By this definition

$$\boxed{\frac{a}{b} = \frac{c}{d}} \tag{17-1}$$

is a proportion. (Another way of showing a proportion is $a:b = c:d$, which is read "*a* is to *b* as *c* is to *d*.") From Eq. (17-1) we see that a proportion is an equation, which means that any operation applicable to an equation is also applicable to a proportion.

EXAMPLE C _____ On a certain map 1 in. represents 10 mi. Thus on this map we have a ratio of 1 in./10 mi. To find the distance represented by 3.5 in., we can set up the proportion

$$\overbrace{\frac{3.5 \text{ in.}}{x} = \frac{1 \text{ in.}}{10 \text{ mi}}}^{\text{map distances}}$$

land distances

$$(10x)\left(\frac{3.5}{x}\right) = 10x\left(\frac{1}{10}\right) \qquad \text{multiply each side by LCD} = 10x$$

$$35 = x \quad \text{or} \quad x = 35 \text{ mi}$$

The ratio 1 in./10 mi is the scale of the map and has a special meaning, relating map distances in inches to land distances in miles. In this case we should ■ not change either unit to the other, even though they are both units of length.

EXAMPLE D _____ Given that 1 in. = 2.54 cm, what length in centimeters is 15.0 in.?

If we equate the ratio of known lengths to the ratio of the given length to the required length, we can find the required length. This gives us

$$\frac{1 \text{ in.}}{2.54 \text{ cm}} = \frac{15.0 \text{ in.}}{x \text{ cm}}$$

$$x = (15.0)(2.54) \qquad \text{multiply both sides by LCD}$$

$$= 38.1 \text{ cm}$$

■ Additional illustrations of changing units can be found in Appendix B.

EXAMPLE E _____ The magnitude of an electric field E is defined as the ratio between the force F on a charge q and the magnitude of q. This can be written as $E = F/q$. If we know the force exerted on a particular charge at some point in the field, we can determine the force which would be exerted on another charge placed at the same point. For example, if we know that a force of 10 nN is exerted on a charge of 4.0 nC, we can then determine the force which would be exerted on a charge of 6.0 nC by the proportion

$$\overbrace{\frac{10 \times 10^{-9}}{4.0 \times 10^{-9}} = \frac{F}{6.0 \times 10^{-9}}}^{\text{forces at point}}$$

charges at point

$$F = \frac{(6.0 \times 10^{-9})(10 \times 10^{-9})}{4.0 \times 10^{-9}}$$

$$= 15 \times 10^{-9} = 15 \text{ nN}$$

EXAMPLE F _____ A certain alloy is 5 parts tin and 3 parts lead. How many grams of each are there in 40 g of the alloy?

First, we let x = the number of grams of tin in the given amount of the alloy.

Next we note that there are 8 total parts of alloy, of which 5 are tin. Thus, 5 is to 8 as x is to 40. This gives the equation

$$\text{parts tin} \longrightarrow \frac{5}{8} = \frac{x}{40} \longleftarrow \text{grams of tin}$$
$$\text{total parts} \longrightarrow 8 \qquad 40 \longleftarrow \text{total grams}$$

$$x = 40\left(\frac{5}{8}\right) = 25 \text{ g}$$

Therefore, there are 25 g of tin and 15 g of lead. The ratio 25 to 15 is the same as 5 to 3. ∎

Exercises 17-1

In Exercises 1 through 8, express the ratios in the simplest form.

1. 18 V to 3 V

2. 27 ft to 18 ft

3. 48 in. to 3 ft

4. 120 s to 4 min

5. 20 qt to 2.5 gal

6. 6500 cL to 2.6 L

7. 0.14 kg to 3500 mg

8. 2000 μm to 6 mm

In Exercises 9 through 16, find the required ratios.

9. The *Mach number* of a moving object is the ratio of its velocity to the velocity of sound (740 mi/h). Find the Mach number of a supersonic jet traveling at 1250 mi/h.

10. A virus 3.0×10^{-5} cm long appears to be 1.2 cm long through a microscope. What is the *magnification* (ratio of image length to object length) of the microscope?

11. The *coefficient of friction* for two contacting surfaces is the ratio of the frictional force between them to the perpendicular force which presses them together. If it takes 45 N to overcome friction to move a 110-N crate along the floor, what is the coefficient of friction between the crate and the floor?

12. The *atomic mass* of an atom of carbon is defined to be 12 u. The ratio of the atomic mass of an atom of oxygen to that of an atom of carbon is $\frac{4}{3}$. What is the atomic mass of an atom of oxygen? (The symbol u represents the *unified atomic mass unit*, where $1 \text{ u} = 1.66 \times 10^{-27}$ kg.)

13. The *percent error* in a measurement is the ratio of the error in the measurement to the measurement itself, expressed as a percent. When writing a computer program, the memory remaining is determined as 2450 bytes, and then it is correctly found to be 2540 bytes. What is the percent error in the first reading?

14. The electric *current* in a given circuit is the ratio of the voltage to the resistance. What is the current $(1 \text{ V}/1 \Omega = 1 \text{ A})$ for a circuit where the voltage is 24.0 V and the resistance is 10.0 Ω?

15. The *mass* of an object is the ratio of its weight to the acceleration g due to gravity. If a space probe weighs 8460 N on earth, where $g = 9.80$ m/s^2, find its mass. (See Appendix B.)

16. *Power* is defined as the ratio of work done to the time required to do the work. If an engine performs 3650 J of work in 15.0 s, find the power developed by the engine. (See Appendix B.)

In Exercises 17 through 20, find the required quantities from the given proportions.

17. In an electric instrument called a "Wheatstone bridge," electric resistances are related by

$$\frac{R_1}{R_2} = \frac{R_3}{R_4}$$

Find R_2 if $R_1 = 6.00 \Omega$, $R_3 = 62.5 \Omega$, and $R_4 = 15.0 \Omega$. See Fig. 17-1.

18. For two connected gears, the relation

$$\frac{d_1}{d_2} = \frac{N_1}{N_2}$$

holds, where d is the diameter of the gear and N is the number of teeth. Find N_1 if $d_1 = 2.60$ in., $d_2 = 11.7$ in., and $N_2 = 45$.

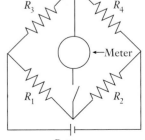

Fig. 17-1

19. According to Boyle's law, the relation

$$\frac{p_1}{p_2} = \frac{V_2}{V_1}$$

holds for pressures p_1 and p_2 and volumes V_1 and V_2 of a gas at constant temperature. Find V_1 if $p_1 = 36.6 \text{ kPa}$, $p_2 = 84.4 \text{ kPa}$, and $V_2 = 0.0447 \text{ m}^3$.

20. In a transformer, an electric current in one coil of wire induces a current in a second coil. For a transformer

$$\frac{i_1}{i_2} = \frac{t_2}{t_1}$$

where i is the current and t is the number of windings in each coil, find i_2 for $i_1 = 0.0350 \text{ A}$, $t_1 = 560$, and $t_2 = 1500$.

In Exercises 21 through 36, answer the given questions by setting up and solving the appropriate proportions.

21. Given that $1.00 \text{ in.}^2 = 6.45 \text{ cm}^2$, what area in square inches is 36.3 cm^2?

22. Given that $1.000 \text{ kg} = 2.205 \text{ lb}$, what weight in kilograms is 175.5 lb?

23. Given that $1.00 \text{ hp} = 746 \text{ W}$, what power in horsepower is 250 W?

24. Given that $1.50 \text{ L} = 1.59 \text{ qt}$, what capacity in quarts is 2.75 L?

25. Given that $2.00 \text{ km} = 1.24 \text{ mi}$, what distance in kilometers is 5.00 mi?

26. Given that $10^4 \text{ cm}^2 = 10^6 \text{ mm}^2$, what area in square centimeters is $2.50 \times 10^5 \text{ mm}^2$?

27. How many meters per second are equivalent to 45.0 km/h?

28. How many gallons per hour are equivalent to 540 L/min?

29. A particular type of automobile engine produces $60,000 \text{ cm}^3$ of carbon monoxide in 2.00 min. How much carbon monoxide is produced in 45.0 s?

30. A person 1.80 m tall is photographed, and the film image is 20.0 mm high. Under the same conditions, how tall is a person whose film image is 14.5 mm high?

31. By weight, the ratio of chlorine to sodium in table salt is 35.46 to 23.00. How much sodium is contained in 50.00 kg of salt?

32. Ten clicks on an adjustment screw cause an inlet valve opening to change by 0.035 cm. How many clicks are required for a valve adjustment of 0.049 cm?

33. A physician has 220 mg of medication, which is just sufficient for dosages for two particular patients. The amount given to each patient should be proportional to the patient's weight. If these patients weigh 60 kg and 72 kg, what should be the dosages?

34. An electric current of 0.772 mA passes into two wires in which it is divided into currents in the ratio of 2.83 to 1.09. What are the currents in the two wires?

35. One computer line printer can print 2000 lines/min, and a second can print 2800 lines/min. If they print a total of 8400 lines while printing together, how many lines does each print?

36. A person pays $\$4500$ in state and federal income taxes. Find the amount paid for each if the state taxes were 30% of the taxes paid.

17-2   Variable

Scientific laws are often stated in terms of ratios and proportions. For example, Charles' law can be stated as "for a perfect gas under constant pressure, the ratio of any two volumes this gas may occupy equals the ratio of the absolute temperatures." Symbolically this could be stated as $V_1/V_2 = T_1/T_2$. Thus, if the ratio of the volumes and one of the values of the temperature are known, we can easily find the other temperature.

By multiplying both sides of the proportion of Charles' law by V_2/T_1, we can change the form of the proportion to $V_1/T_1 = V_2/T_2$. This statement says that the ratio of the volume to the temperature (for constant pressure) is constant. Thus, if any pair of values of volume and temperature is known, this ratio of V_1/T_1 can be calculated. This ratio of V_1/T_1 can be called a constant k, which means that Charles' law can be written as $V/T = k$. We now have the statement that the ratio of the volume to temperature is always constant; or, as it is normally stated, "the volume is proportional to the temperature." Therefore, we write $V = kT$, the clearest and most informative statement of Charles' law.

Thus, *for any two quantities always in the same proportion, we say that one is* **proportional to** (*or* **varies directly as**) *the second. To show that y is proportional to x (or varies directly as x), we write*

direct variation

$$y = kx$$

(17-2)

where k is the **constant of proportionality.** This type of relationship is known as **direct variation.**

EXAMPLE A ——— The circumference c of a circle is proportional to (varies directly as) the radius r. We write this as $c = kr$. Since we know that $c = 2\pi r$ for a circle, we know in this case that $k = 2\pi$. ∎

EXAMPLE B ——— The fact that the electric resistance R of a wire varies directly as (is proportional to) its length ℓ is written as $R = k\ell$. As the length of the wire increases (or decreases), this equation tells us that the resistance increases (or decreases) proportionally. ∎

It is very common that, when two quantities are related, the product of the two quantities remains constant. In such a case $yx = k$, or

inverse variation

$$y = \frac{k}{x}$$

(17-3)

This is read as "y **varies inversely as** *x" or "y* **is inversely proportional to** *x."* This type of relationship is known as **inverse variation.**

EXAMPLE C ——— Boyle's law states that "at a given temperature, the pressure p of a gas varies inversely as the volume V." We write this as $p = k/V$. In this case, as the volume increases, the pressure decreases. ∎

In Fig. 17-2(a) the graph of the equation for direct variation $y = kx$ $(x \geq 0)$ is shown. It is a straight line, with slope of k $(k > 0)$ and a y-intercept of 0. We see that y increases as x increases. In Fig. 17-2(b) the graph of the equation for inverse variation $y = k/x$ $(k > 0, x > 0)$ is shown. It is a hyperbola (see Example E of Section 13-1), and we see that y decreases as x increases.

Fig. 17-2

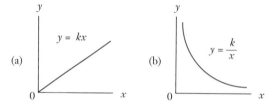

For many relationships, one quantity varies as a specified power of another quantity. The terms *varies directly* and *varies inversely* are used in the following examples with a specified power of the independent variable.

EXAMPLE D

The statement that the volume V of a sphere varies directly as the cube of its radius r is written as $V = kr^3$. In this case we know that $k = \frac{4}{3}\pi$. As the radius of a sphere increases, the volume increases much more rapidly in value. ■

EXAMPLE E

A company finds that the number n of units of a product which are sold is inversely proportional to the square of the price p of the product. This is written as $n = k/p^2$. As the price of the product is raised, the number of units which are sold decreases much more rapidly. ■

One quantity may vary as the product of two or more other quantities. Such variation is called **joint variation.** *We write*

joint variation

$$y = kxz$$

(17-4)

to show that y varies jointly as x and z.

EXAMPLE F

The cost C of a piece of sheet metal varies jointly as the area A of the piece and the cost c per unit area of the metal. This we write as $C = kAc$. Here C increases if the product Ac increases. ■

Direct, inverse, and joint variations may be combined. A given relationship may be a combination of two or all three of these types of variation.

EXAMPLE G

Newton's universal law of gravitation can be stated as "the force F of gravitation between two objects varies jointly as the masses m_1 and m_2 of the objects and inversely as the square of the distance r between their centers." We write this as

$$F = \frac{Gm_1 m_2}{r^2}$$

⟵ force varies jointly as masses
and
⟵ inversely as the square of the distance

■ where G is the constant of proportionality.

Note the use of the word *and* in Example G. There *and* is used to indicate that F varies in more than one way; ***it is* not *interpreted as addition.***

Once we have used the given statement to set up a general equation in terms of the variables and the constant of proportionality, we may calculate the value of the constant of proportionality if one complete set of values of the variables is known. This value can then be substituted into the general equation to find the specific equation relating the variables. We can then find the value of one of the variables if the others are known.

EXAMPLE H ____ If y varies inversely as x, and $x = 15$ when $y = 4$, find the value of y when $x = 12$.
First we write

$$y = \frac{k}{x} \qquad \text{general equation from statement}$$

to show that y varies inversely as x. Next we substitute $x = 15$ and $y = 4$ into the equation. This leads to

$$4 = \frac{k}{15} \quad \text{or} \quad k = 60 \qquad \text{evaluate } k$$

Thus, for this problem the constant of proportionality is 60, and this may be substituted into $y = k/x$, giving

$$y = \frac{60}{x} \qquad \text{specific equation relating } y \text{ and } x$$

as the equation between y and x. Now, for any given value of x, we may find the value of y. For $x = 12$, we have

$$y = \frac{60}{12} = 5 \qquad \text{evaluating } y \text{ for } x = 12$$

EXAMPLE I ____ The frequency f of vibration of a wire varies directly as the square root of the tension T on the wire. If $f = 420$ Hz when $T = 1.14$ N, find f when $T = 3.40$ N.
The steps in making this evaluation are outlined below.

$$f = k\sqrt{T} \qquad \qquad \text{set up general equation: } f \text{ varies directly as } \sqrt{T}$$
$$420 \text{ Hz} = k\sqrt{1.14 \text{ N}} \qquad \text{substitute given set of values and}$$
$$k = 393 \text{ Hz/N}^{1/2} \qquad \text{evaluate } k$$
$$f = 393\sqrt{T} \qquad \qquad \text{substitute value of } k \text{ to get specific equation}$$
$$f = 393\sqrt{3.40} \qquad \text{evaluate } f \text{ for } T = 3.40 \text{ N}$$
$$= 725 \text{ Hz}$$

We note that k has a set of units associated with it, and this will normally be the case in applied problems. As long as we do not change the units for any of the variables, the units for the final variable evaluated will remain the same.

EXAMPLE J _____ The heat H developed in an electric resistor varies jointly as the time t and the square of the current i in the resistor. If the heat developed in t_0 seconds with a current of i_0 amperes passing through the resistor is H_0 joules, how much heat is developed if both the time and the current are doubled?

The solution proceeds as follows:

$$H = kti^2 \qquad\qquad \text{set up general equation}$$

$$H_0 \text{ J} = k(t_0 \text{ s})(i_0 \text{ A})^2 \qquad \text{substitute given values and}$$

$$k = \frac{H_0}{t_0 i_0^2} \text{ J/s} \cdot \text{A}^2 \qquad \text{evaluate } k$$

$$H = \frac{H_0 t i^2}{t_0 i_0^2} \qquad\qquad \text{substitute for } k \text{ to get specific equation}$$

We are asked to determine H when both the time and the current are doubled. This means we are to substitute $t = 2t_0$ and $i = 2i_0$. Making this substitution, we have

$$H = \frac{H_0(2t_0)(2i_0)^2}{t_0 i_0^2} = \frac{8H_0 t_0 i_0^2}{t_0 i_0^2} = 8H_0$$

This tells us that the heat developed is eight times as much as for the original values of i and t. ∎

EXAMPLE K _____

See the chapter introduction.

In Example G we stated Newton's universal law of gravitation. This law was formulated in the late seventeenth century, but it has numerous modern space-age applications. Use this law to solve the following problem.

A spacecraft is traveling from the earth to the moon, which are 240,000 mi apart. The mass of the moon is 0.0123 that of the earth. How far from the earth is the gravitational force of the earth on the spacecraft equal to the gravitational force of the moon on the spacecraft?

From Example G, we have the gravitational force between two objects as

$$F = \frac{Gm_1 m_2}{r^2}$$

where the constant of proportionality G is the same for any two objects. Since we want the force between the earth and the spacecraft to equal the force between the moon and the spacecraft, we have

$$\frac{Gm_s m_e}{r^2} = \frac{Gm_s m_m}{(240{,}000 - r)^2}$$

where m_s, m_e, and m_m are the masses of the spacecraft, the earth, and the moon, respectively, r is the distance from the earth to the spacecraft, and $240{,}000 - r$ is the distance from the moon to the spacecraft.

Since $m_m = 0.0123m_e$, we have

$$\frac{Gm_s m_e}{r^2} = \frac{Gm_s(0.0123m_e)}{(240{,}000 - r)^2}$$

$$\frac{1}{r^2} = \frac{0.0123}{(240{,}000 - r)^2} \qquad \text{divide each side by } Gm_sm_e$$

$$(240{,}000 - r)^2 = 0.0123r^2 \qquad \text{multiply each side by LCD}$$

$$240{,}000 - r = 0.111r \qquad \text{take square roots}$$

$$1.111r = 240{,}000$$

$$r = 216{,}000 \text{ mi}$$

Therefore, the spacecraft is 216,000 mi from the earth and 24,000 mi from the moon when the gravitational forces are equal.

Exercises 17-2

In Exercises 1 through 8, set up the general equations from the given statements.

1. y varies directly as z.

2. p varies inversely as q.

3. s varies inversely as the square of t.

4. w is proportional to the cube of L.

5. f is proportional to the square root of x.

6. n is inversely proportional to the $\frac{3}{2}$ power of s.

7. w varies jointly as x and the cube of y.

8. q varies as the square of r and inversely as the fourth power of t.

In Exercises 9 through 16, give the specific equation relating the variables after evaluating the constant of proportionality for the given set of values.

9. r varies inversely as y, and $r = 2$ when $y = 8$.

10. y varies directly as x, and $y = 18$ when $x = 2$.

11. y varies directly as the square root of x, and $y = 2$ when $x = 64$.

12. n is inversely proportional to the square of p, and $n = \frac{1}{27}$ when $p = 3$.

13. s is inversely proportional to the square root of t, and $s = \frac{1}{2}$ when $t = 49$.

14. f varies inversely as the product xy, and $f = 50$ when $x = 4$ and $y = 0.5$.

15. p is proportional to q and inversely proportional to the cube of r, and $p = 6$ when $q = 3$ and $r = 2$.

16. v is proportional to t and the square of s, and $v = 80$ when $s = 2$ and $t = 5$.

In Exercises 17 through 24, find the required value by setting up the general equation and then evaluating.

17. Find y when $x = 10$ if y varies directly as x and $y = 20$ when $x = 8$.

18. Find y when $x = 5$ if y varies directly as the square of x and $y = 6$ when $x = 8$.

19. Find s when $t = 10$ if s is inversely proportional to t and $s = 100$ when $t = 5$.

20. Find p for $q = 0.8$ if p is inversely proportional to the square of q and $p = 18$ when $q = 0.2$.

21. Find y for $x = 6$ and $z = 5$ if y varies directly as x and inversely as z and $y = 60$ when $x = 4$ and $z = 10$.

22. Find r when $n = 16$ if r varies directly as the square root of n and $r = 4$ when $n = 25$.

23. Find f when $p = 2$ and $c = 4$ if f varies jointly as p and the cube of c and $f = 8$ when $p = 4$ and $c = 0.1$.

24. Find v when $r = 2$, $s = 3$, and $t = 4$ if v varies jointly as r and s and inversely as the square of t and $v = 8$ when $r = 2$, $s = 6$, and $t = 6$.

In Exercises 25 through 48, solve the given applied problems involving variation.

25. The volume V of carbon dioxide (CO_2) which is exhausted from a room in a given time varies directly as the initial volume V_0 which is present. If 75 ft^3 of CO_2 is removed in an hour from a room with an initial volume of 160 ft^3, how much is removed in an hour if the initial volume is 130 ft^3?

26. The amount of heat H required to melt ice is proportional to the mass m of ice which is melted. If it takes 3.35×10^5 J to melt 1000 g of ice, how much heat is required to melt 625 g?

27. In electroplating, the mass m of the material deposited varies directly as the time t during which the electric current is on. Set up the equation for this relationship if 2.50 g are deposited in 5.25 h.

28. Hooke's law states that the force needed to stretch a spring is proportional to the amount the spring is stretched. If 10.0 lb stretches a certain spring 4.00 in., how much will the spring be stretched by a force of 6.00 lb?

29. The energy output E from an engine is proportional to the input I. The constant of proportionality is the *efficiency* of the engine. Find the efficiency of an engine which has an output of 3600 W for an input of 5400 W.

30. The velocity v of a falling object varies directly as the time t of fall, where the constant of proportionality is the acceleration due to gravity. A brick is dropped from the top of a construction site and has a velocity of 19.6 m/s after 2.00 s. What is the acceleration due to gravity?

31. The time t required to empty a wastewater holding tank is inversely proportional to the cross-sectional area A of the drainage pipe. If it takes 2.0 h to empty a tank with a drainage pipe for which $A = 48$ in.^{2}, how long will it take to empty the tank if $A = 68$ in.^{2}?

32. The time t required to make a particular trip is inversely proportional to the average speed v. If a jet takes 2.75 h at an average speed of 520 km/h, how long will it take at an average speed of 620 km/h? What is the meaning of the constant of proportionality?

33. The number N of insects remaining alive t hours after applying an insecticide varies inversely as $t + 1$. If there are initially 7500 insects in the area, how many will remain after 5.00 h?

34. The power P required to propel a ship varies directly as the cube of the speed s of the ship. If 5200 hp will propel a ship at 12.0 mi/h, what power is required to propel it at 15 mi/h?

35. The lift L of a model airplane wing is directly proportional to the square of its width w. If the lift is 50 N for a wing width of 8.0 cm, what is the lift for a width of 6.0 cm?

36. The f-number lens setting of a camera varies directly as the square root of the time t that the film is exposed. If the f-number is 8 (written as $f/8$) for $t = 0.0200$ s, find the f-number for $t = 0.0098$ s.

37. The force F on the blade of a wind generator varies jointly as the blade area A and the square of the wind velocity v. Find the equation relating F, A, and v if $F = 19.2$ lb when $A = 3.72$ ft^2 and $v = 31.4$ ft/s.

38. When the volume V of a gas changes very rapidly, an approximate relation is that the pressure P varies inversely as the 1.4 power of the volume. Given that $P = P_0$ when $V = V_0$, find P if the volume increases by ten times.

39. The average speed s of oxygen molecules in the air is directly proportional to the square root of the absolute temperature T. If the speed of the molecules is 460 m/s at 273 K, what is the speed at 300 K?

40. The time t required to test a computer memory unit varies directly as the square of the number n of memory cells in the unit. If a unit with 4800 memory cells can be tested in 15.0 s, how long does it take to test a unit with 8400 memory cells?

41. The electric resistance of a wire varies directly as its length and inversely as its cross-sectional area. Find the relation between resistance, length, and area for a wire which has a resistance of 0.200 Ω for a length of 200 ft and cross-sectional area of 0.0500 in.^{2}.

42. The general gas law states that the pressure P of an ideal gas varies directly as the thermodynamic temperature T and inversely as the volume V. If $P = 600$ kPa for $V = 10.0$ cm^3 and $T = 300$ K, find V for $P = 400$ kPa and $T = 400$ K.

43. The power in an electric current varies jointly as the resistance and the square of the current. Given that the power is 10.0 W when the current is 0.500 A and the resistance is 40.0 Ω, find the power if the current is 2.00 A and the resistance is 20.0 Ω.

44. Under certain conditions, the natural logarithm of the ratio of an electric current i at time t to the current i_0 at time $t = 0$ is proportional to the time. Given that the current equals $1/e$ of its initial value after 0.100 s, what is its value after 0.200 s?

45. The power gain G by a parabolic microwave dish varies directly as the square of the diameter d of the opening and inversely as the square of the wavelength λ of the wave carrier. Find the equation relating G, d, and λ if $G = 5.5 \times 10^4$ for $d = 2.9$ m and $\lambda = 0.030$ m.

46. The intensity I of sound varies directly as the power P of the source and inversely as the square of the distance r from the source. Two sound sources are separated by a distance d, and one has twice the power output of the other. Where should an observer be located on a line between them such that the intensity of each sound is the same?

47. The x-component of the acceleration of an object moving around a circle with constant angular velocity ω varies jointly as $\cos \omega t$ and the square of ω. If the x-component of the acceleration is -11.4 ft/s^2 when $t = 1.00$ s for $\omega = 0.524$ rad/s, find the x-component of the acceleration when $t = 2.00$ s.

48. The tangent of the proper banking angle θ of the road for a car making a turn is directly proportional to the square of the car's velocity v and inversely proportional to the radius r of the turn. If $7.75°$ is the proper banking angle for a car traveling at 20.0 m/s around a turn of radius 300 m, what is the proper banking angle for a car traveling at 30.0 m/s around a turn of radius 250 m?

17-3   Chapter Equations, Review Exercises, and Practice Test

Chapter Equations

Proportion	$\dfrac{a}{b} = \dfrac{c}{d}$	(17-1)
Direct variation	$y = kx$	(17-2)
Inverse variation	$y = \dfrac{k}{x}$	(17-3)
Joint variation	$y = kxz$	(17-4)

Review Exercises

In Exercises 1 through 8, find the indicated ratios.

1. 4 Mg to 20 kg   **2.** 300 nm to 6 μm   **3.** 20 mL to 5 cL   **4.** 12 ks to 2 h

5. The mechanical advantage of a lever is the ratio of the output force F_0 to the input force F_i. Find the mechanical advantage if $F_0 = 28$ kN and $F_i = 5000$ N.

6. The number π equals the ratio of the circumference c of a circle to its diameter d. Using computer simulation, the circumference and diameter of a metal cylinder are measured as $c = 4.273604$ in. and $d = 1.360330$ in. Find the value of π using these measurements.

7. The pressure p exerted on a surface is the ratio of the force F on the surface to its area A. Find the pressure on a square patch, 2.25 in. on a side, on a tank if the force on the patch is 37.4 lb.

8. The resistance R of a resistor is the ratio of the voltage V across the resistor to the current i in the resistor. Find R if $V = 0.632$ V and $i = 2.03$ mA.

In Exercises 9 through 20, answer the given questions by setting up and solving the appropriate proportions.

9. On a certain map, 1.00 in. represents 16.0 mi. What distance on the map represents 52.0 mi?

10. Given that 1.000 lb = 453.6 g, what is the weight in pounds of a 14.0-g computer diskette?

11. Given that 1.00 Btu = 1060 J, how much heat in joules is produced by a heating element which produces 2660 Btu?

12. Given that 1.00 L = 61.0 in.^{3}, what capacity in liters has a cubical box that is 3.23 in. along an edge?

13. A microcomputer printer can print 3600 characters in 30 s. How many characters can it print in 5 min?

14. A solar heater with a collector area of 58.0 m^2 is required to heat 2560 kg of water. Under the same conditions, how much water can be heated by a rectangular solar collector 9.50 m by 8.75 m?

15. The dosage of a certain medicine is 25 mL for each 10 lb of the patient's weight. What is the dosage for a person weighing 56 kg?

16. A woman invests $50,000 and a man invests $20,000 in a partnership. If profits are to be shared in the ratio that each invested in the partnership, how much does each receive from $10,500 in profits?

17. On a certain blueprint a measurement of 25.0 ft is represented by 2.00 in. What is the actual distance between two points if they are 5.75 in. apart on the blueprint?

18. The chlorine concentration in a water supply is 0.12 parts per million. How much chlorine is there in a cylindrical holding tank 4.22 m in radius and 5.82 m high filled from the water supply?

19. One fiber-optic cable carries 60% as many messages as another fiber-optic cable. Together they carry 12,000 messages. How many does each carry?

20. Two types of roadbed material, one 50% rock and the other 100% rock, are used in the ratio of 4 to 1 to form a roadbed. If a total of 150 tons are used, how much rock is in the roadbed?

In Exercises 21 through 24, give the specific equation relating the variables after evaluating the constant of proportionality for the given set of values.

21. y varies directly as the square of x, and $y = 27$ when $x = 3$.

22. f varies inversely as ℓ, and $f = 5$ when $\ell = 8$.

23. v is directly proportional to x and inversely proportional to the cube of y, and $v = 10$ when $x = 5$ and $y = 4$.

24. r varies jointly as u, v, and the square of w, and $r = 8$ when $u = 2$, $v = 4$, and $w = 3$.

In Exercises 25 through 48, solve the given applied problems.

25. For a lever balanced at the fulcrum, the relation

$$\frac{F_1}{F_2} = \frac{L_2}{L_1}$$

holds, where F_1 and F_2 are forces on opposite sides of the fulcrum at distances L_1 and L_2, respectively. If $F_1 = 4.50$ lb, $F_2 = 6.75$ lb, and $L_1 = 17.5$ in., find L_2. See Fig. 17-3.

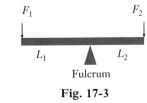

Fig. 17-3

26. A company finds that the volume V of sales of a certain item and the price P of the item are related by

$$\frac{P_1}{P_2} = \frac{V_2}{V_1}$$

Find V_2 if $P_1 = \$8.00$, $P_2 = \$6.00$, and $V_1 = 3000$ per week.

27. The charge C on a capacitor varies directly as the voltage V across it. If the charge is 6.3 μC with a voltage of 220 V across a capacitor, what is the charge on it with a voltage of 150 V across it?

28. The amount of natural gas which is burned is proportional to the amount of oxygen which is consumed. If 24.0 lb of oxygen is consumed in burning 15.0 lb of natural gas, how much air, which is 21.0% oxygen, is consumed to burn 50.0 lb of natural gas?

29. The power of a gas engine is proportional to the area of the piston. If an engine with a piston area of 8.00 in.^{2} can develop 30.0 hp, what power is developed by an engine with a piston area of 6.00 in.^{2}?

30. The volume of hydrogen liberated from hydrochloric acid by zinc is proportional to the amount of zinc used. If 8.28 g of zinc liberates 3.00 L of hydrogen, how much is liberated by 5.36 g of zinc?

31. The distance an object falls under the influence of gravity varies directly as the square of the time of fall. If an object falls 64.0 ft in 2.00 s, how far will it fall in 3.00 s?

32. The kinetic energy of a moving object varies jointly as the mass of the object and the square of its velocity. If a 5.00-kg object, traveling at 10.0 m/s, has a kinetic energy of 250 J, find the kinetic energy of an 8.00-kg object traveling at 50.0 m/s.

33. In a particular computer design, N numbers can be sorted in a time proportional to the square of log N. How many times longer does it take to sort 8000 numbers than to sort 2000 numbers?

34. The period of a pendulum is directly proportional to the square root of its length. Given that a pendulum 2.00 ft long has a period of 1.57 s, what is the period of a pendulum 4.00 ft long?

35. In any given electric circuit containing an inductance and capacitance, the resonant frequency is inversely proportional to the square root of the capacitance. If the resonant frequency in a circuit is 25.0 Hz and the capacitance is 100 μF, what is the resonant frequency of this circuit if the capacitance is 25.0 μF?

36. The rate of emission of radiant energy from the surface of a body is proportional to the fourth power of the thermodynamic temperature. Given that a 25.0-W (the rate of emission) lamp has an operating temperature of 2500 K, what is the operating temperature of a similar 40.0-W lamp?

37. The frequency f of a radio wave is inversely proportional to its wavelength λ. The constant of proportionality is the velocity of the wave, which equals the speed of light. Find this velocity if an FM radio wave has a frequency of 90.9 MHz and a wavelength of 3.29 m.

38. The acceleration of gravity g on a satellite in orbit around the earth varies inversely as the square of its distance from the center of the earth. If $g = 8.7$ m/s^2 for a satellite at an altitude of 400 km above the surface of the earth, find g if it is 1000 km above the surface. The radius of the earth is 6.4×10^6 m.

39. Using *holography* (a method of producing an image without using a lens), an image of concentric circles is formed. The radius r of each circle varies directly as the square root of the wavelength λ of the light used. If $r = 3.56$ cm for $\lambda = 575$ nm, find r if $\lambda = 483$ nm.

40. The velocity of a jet of fluid flowing from an opening in the side of a container is proportional to the square root of the depth of the opening. If the velocity of the jet from an opening at a depth of 4.00 ft is 16.0 ft/s, what is the velocity of a jet from an opening at a depth of 25.0 ft?

41. The heat loss through fiberglass insulation varies directly as the time and inversely as the thickness of the fiberglass. If the loss through 8.0 in. of fiberglass is 1200 Btu in 30 min, what is the loss through 6.0 in. in 1 h 30 min?

42. Kepler's third law of planetary motion states that the square of the period of any planet is proportional to the cube of the mean radius (about the sun) of that planet, with the constant of proportionality being the same for all planets. Using the fact that the period of the earth is one year and its mean radius is 93.0 million miles, calculate the mean radius for Venus, given that its period is 7.38 months.

43. The range R of a projectile varies jointly as the square of its initial velocity v_0 and the sine of twice the angle θ from the horizontal at which it is fired. See Fig. 17-4. A bullet for which $v_0 = 850$ m/s and $\theta = 22.0°$ has a range of 5.12×10^4 m. Find the range if $v_0 = 750$ m/s and $\theta = 43.2°$.

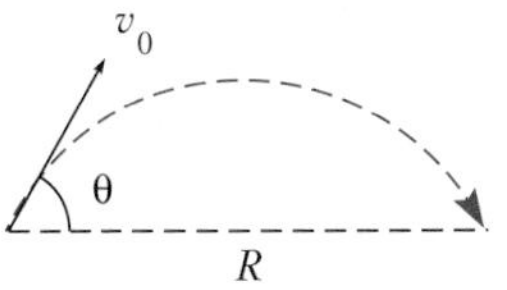

Fig. 17-4

44. The load L which a helical spring can support varies directly as the cube of its wire diameter d and inversely as its coil diameter D. A spring for which $d = 0.120$ in. and $D = 0.953$ in. can support 45.0 lb. What is the coil diameter of a similar spring which supports 78.5 lb and for which $d = 0.156$ in.?

45. The volume rate of flow of blood through an artery varies directly as the fourth power of the radius of the artery and inversely as the distance along the artery. If an operation is successful in effectively increasing the radius of an artery by 25% and decreasing its length by 2%, by how much is the volume rate of flow increased?

46. The safe uniformly distributed load on a horizontal beam, supported at both ends, varies jointly as the width and the square of the depth and inversely as the distance between supports. Given that one beam has double the dimensions of another, how many times heavier is the safe load it can support than the first can support?

47. A bank statement exactly 30 years old is discovered. It states, "This 10-year-old account is now worth $185.03 and pays 4% interest compounded annually." An investment with annual compound interest varies directly as

$1 + r$ to the power n, where r is the interest rate expressed as a decimal and n is the number of years of compounding. What was the value of the original investment, and what is it worth now?

48. The distance s that an object falls due to gravity varies jointly as the acceleration g due to gravity and the square of the time t of fall. The acceleration due to gravity on the moon is 0.172 of that on earth. If a rock falls for t_0 seconds on earth, how many times farther would the rock fall on the moon in $3t_0$ seconds?

Practice Test

1. Express the ratio of 180 s to 4 min in simplest form.

2. The force F between two parallel wires carrying electric currents is inversely proportional to the distance d between the wires. If a force of 0.750 N exists between wires which are 1.25 cm apart, what is the force between them if they are separated by 1.75 cm?

3. Given that 1.00 in. = 2.54 cm, what length in inches is 7.24 cm?

4. The difference p in pressure in a fluid between that at the surface and that at a point below varies jointly as the density d of the fluid and the depth h of the point. The density of water is 1000 kg/m^3, and the density of alcohol is 800 kg/m^3. This difference in pressure at a point 0.200 m below the surface of water is 1.96 kPa. What is the difference in pressure at a point 0.300 m below the surface of alcohol?

5. The perimeter of a rectangular solar panel is 210 in. The ratio of the length to the width is 7 to 3. What are the dimensions of the panel?

6. The crushing load of a pillar varies directly as the fourth power of its radius and inversely as the square of its length. If one pillar has twice the radius and three times the length of a second pillar, what is the ratio of the crushing load of the first pillar to that of the second pillar?

18 Sequences and the Binomial Theorem

In this chapter we will consider briefly the properties of certain sequences of numbers. A *sequence* is a set of numbers arranged in some specified order. We will develop certain basic types of sequences, including those used in the expansion of a binomial to a power.

Applications of sequences can be found in many scientific and technical areas, including certain areas in physics and interest calculations in business. Sequences and binomial expansions are also of importance in developing mathematical topics which in themselves have wide technical applications.

Sequences are basic to many calculations in business, including compound interest. In Section 18-2 we demonstrate such a calculation.

18-1 Arithmetic Sequences

*An **arithmetic sequence** (or **arithmetic progression**) is a set of numbers in which each number after the first can be obtained from the preceding one by adding to it a fixed number called the **common difference.***

EXAMPLE A ————— The sequence 2, 5, 8, 11, 14, . . . is an arithmetic sequence with a common difference of 3. We can obtain any term after the first by adding 3 to the previous term.

The sequence 7, 2, -3, -8, . . . is an arithmetic sequence with a common difference of -5. We can obtain any term after the first by adding -5 to the previous term.

The three dots after the 14 and after the -8 mean that the sequences continue.

If we know the first term of an arithmetic sequence, we can find any other term in the sequence by successively adding the common difference enough times for the desired term to be obtained. This, however, is a very inefficient method, and we can learn more about the sequence if we establish a general way of finding any particular term.

In general, if a_1 is the first term and d is the common difference, the second term is $a_1 + d$, the third term is $a_1 + 2d$, and so forth. If we are looking for the nth term, we note that we need only add d to the first term $n - 1$ times. Thus, *the nth term, a_n, of an arithmetic sequence is given by*

*n*th term

$$a_n = a_1 + (n - 1)d \qquad\qquad (18\text{-}1)$$

Eq. (18-1) can be used to find any given term in any arithmetic sequence. We can refer to a_n as the *last term* of an arithmetic sequence if no terms beyond it are included in the sequence. Such a sequence is called a **finite** sequence. If the terms in a sequence continue without end, the sequence is called an **infinite** sequence.

EXAMPLE B

Find the tenth term of the arithmetic sequence 2, 5, 8, ...

By subtracting any given term from the following term, we find that the common difference is $d = 3$. From the terms given, we know that the first term is $a_1 = 2$. From the statement of the problem, the desired term is the tenth, or $n = 10$. Thus, we may find the tenth term, a_{10}, by

$$\begin{array}{ccc} a_1 & n & d \\ \downarrow & \downarrow & \downarrow \end{array}$$
$$a_{10} = 2 + (10 - 1)3 = 2 + (9)(3)$$
$$= 29$$

Therefore, the tenth term is 29.

The three dots written after the 8 mean that the sequence continues. Since there is no additional information, this would indicate that it is an infinite arithmetic sequence.

EXAMPLE C

Find the common difference between successive terms of the arithmetic sequence for which the first term is 5 and the thirty-second term is -119.

We are to find d given that $a_1 = 5$, $a_{32} = -119$, and $n = 32$. Substitution in Eq. (18-1) gives

$$-119 = 5 + (32 - 1)d$$
$$31d = -124$$
$$d = -4$$

There is no information as to whether this is a finite or an infinite sequence. The solution is the same in either case.

EXAMPLE D _____ How many numbers between 10 and 1000 are divisible by 6?

We must first find the smallest and the largest numbers in this range which are divisible by 6. These numbers are 12 and 996. Obviously the common difference between one multiple of 6 and the next is 6. Thus, we can solve this as an arithmetic sequence with $a_1 = 12$, $a_n = 996$, and $d = 6$. This leads to

$$996 = 12 + (n - 1)6$$
$$6n = 990$$
$$n = 165$$

Thus there are 165 numbers between 10 and 1000 which are divisible by 6.

All of the positive multiples of 6 are included in the infinite arithmetic sequence 6, 12, 18, ..., whereas those between 10 and 1000 are included in the finite arithmetic sequence 12, 18, 24, ..., 996.

EXAMPLE E _____ A block sliding down an inclined plane was given an initial velocity of 8.1 cm/s. If it accelerates such that it gains 3.2 cm/s during each second, after how many seconds is its velocity 36.9 cm/s?

Here we see that the velocity (in cm/s) after each second is

$$11.3, 14.5, 17.7, \ldots, 36.9, \ldots$$

Therefore, $a_1 = 11.3$ (the 8.1 cm/s was at the beginning—after 0 s), $d = 3.2$, $a_n = 36.9$, and we are to find n.

$$36.9 = 11.3 + (n - 1)(3.2)$$
$$25.6 = 3.2n - 3.2$$
$$3.2n = 28.8$$
$$n = 9.0$$

Therefore, the velocity of the block is 36.9 cm/s after 9.0 s.

Another important quantity related to an arithmetic sequence is the sum of the first n terms. We can indicate this sum by either of the two equations

$$S_n = a_1 + (a_1 + d) + (a_1 + 2d) + \cdots + (a_n - d) + a_n$$

or

$$S_n = a_n + (a_n - d) + (a_n - 2d) + \cdots + (a_1 + d) + a_1$$

If we now add these equations, we have

$$2S_n = (a_1 + a_n) + (a_1 + a_n) + (a_1 + a_n) + \cdots + (a_1 + a_n) + (a_1 + a_n)$$

There is one factor $(a_1 + a_n)$ for each term, and there are n terms. Thus, _the sum of the first n terms is given by_

sum

$$S_n = \frac{n}{2}(a_1 + a_n)$$

(18-2)

EXAMPLE F

Find the sum of the first 1000 positive integers.

The first 1000 integers form a finite arithmetic sequence for which $a_1 = 1$, $a_{1000} = 1000$, $n = 1000$, and $d = 1$. Substituting into Eq. (18-2) (in which we do not use the value of d), we have

$$S_{1000} = \frac{1000}{2}(1 + 1000) = 500(1001)$$

$$= 500{,}500$$

EXAMPLE G

Find the sum of the first 10 terms of the arithmetic sequence in which the first term is 4 and the common difference is -5.

We are to find S_n, given that $n = 10$, $a_1 = 4$, and $d = -5$. Since Eq. (18-2) uses the value of a_n but not the value of d, we first find a_{10} by use of Eq. (18-1). This gives us

$$a_{10} = 4 + (10 - 1)(-5) = 4 - 45 = -41$$

Now we can solve for S_{10} by use of Eq. (18-2).

$$S_{10} = \frac{10}{2}(4 - 41) = 5(-37)$$

$$= -185$$

EXAMPLE H

For an arithmetic sequence, given that $a_1 = 2$, $d = \frac{3}{2}$, and $S_n = 72$, find n and a_n.

First we substitute the given values in Eqs. (18-1) and (18-2) in order to identify what is known and how we may proceed. Substituting $a_1 = 2$ and $d = \frac{3}{2}$ in Eq. (18-1), we obtain

$$a_n = 2 + (n - 1)\left(\frac{3}{2}\right)$$

Substituting $S_n = 72$ and $a_1 = 2$ in Eq. (18-2), we obtain

$$72 = \frac{n}{2}(2 + a_n)$$

We note that n and a_n appear in both equations, which means that we must solve them simultaneously. Substituting the expression for a_n from the first equation into the second equation, we proceed with the solution.

$$72 = \frac{n}{2}\left[2 + 2 + (n - 1)\left(\frac{3}{2}\right)\right]$$

$$72 = 2n + \frac{3n(n - 1)}{4}$$

$$288 = 8n + 3n^2 - 3n$$

$$3n^2 + 5n - 288 = 0$$

$$n = \frac{-5 \pm \sqrt{25 - 4(3)(-288)}}{6} = \frac{-5 \pm \sqrt{3481}}{6} = \frac{-5 \pm 59}{6}$$

Since n must be a positive integer, we find that $n = \dfrac{-5 + 59}{6} = 9$. Using this value in the expression for a_n, we find

$$a_9 = 2 + (9 - 1)\left(\frac{3}{2}\right) = 14$$

■ Therefore, $n = 9$ and $a_9 = 14$.

EXAMPLE I — The voltage across a resistor is increased such that during each second the increase is 0.002 mV less than during the previous second. Given that the increase during the first second is 0.350 mV, what is the total voltage increase during the first 10.0 s?

We are asked to find the sum of the voltage increases 0.350 mV, 0.348 mV, 0.346 mV, ... so as to include 10 increases. This means we want the sum of an arithmetic sequence for which $a_1 = 0.350$, $d = -0.002$, and $n = 10$. Since we need a_n to use Eq. (18-2), we first calculate it using Eq. (18-1).

$$a_{10} = 0.350 + (10 - 1)(-0.002) = 0.332 \text{ mV}$$

Now we use Eq. (18-2) to find the sum with $a_1 = 0.350$, $a_{10} = 0.332$, and $n = 10$.

$$S_{10} = \frac{10}{2}(0.350 + 0.332) = 3.410 \text{ mV}$$

■ Thus the total voltage increase is 3.410 mV.

Exercises 18-1

In Exercises 1 through 4, write the first five terms of the arithmetic sequence with the given values.

1. $a_1 = 4$, $d = 2$　　　　　　　　　　**2.** $a_1 = 6$, $d = -\frac{1}{2}$

3. third term $= 5$, fifth term $= -3$　　**4.** second term $= -2$, fifth term $= 7$

In Exercises 5 through 12, find the nth term of the arithmetic sequence with the given values.

5. $1, 4, 7, \ldots n = 8$　　　　**6.** $-6, -4, -2, \ldots n = 10$　　　　**7.** $18, 13, 8, \ldots n = 17$

8. $2, \frac{1}{2}, -1, \ldots n = 25$　　**9.** $a_1 = -7$, $d = 4$, $n = 12$　　　**10.** $a_1 = \frac{3}{2}$, $d = \frac{1}{6}$, $n = 50$

11. $a_1 = b$, $d = 2b$, $n = 25$　**12.** $a_1 = -c$, $d = 3c$, $n = 30$

In Exercises 13 through 16, find the sum of the terms of the indicated arithmetic sequence.

13. $n = 20$, $a_1 = 4$, $a_{20} = 40$　　　**14.** $n = 8$, $a_1 = -12$, $a_8 = -26$

15. $n = 10$, $a_1 = -2$, $d = -\frac{1}{2}$　　**16.** $n = 40$, $a_1 = 3k$, $d = \frac{1}{3}k$

In Exercises 17 through 28, find any of the values of a_1, d, a_n, n, or S_n that are missing for an arithmetic sequence.

17. $a_1 = 5$, $d = 8$, $a_n = 45$　　　**18.** $a_1 = -2$, $n = 60$, $a_n = 28$　　　**19.** $a_1 = \frac{5}{3}$, $n = 20$, $S_{20} = \frac{40}{3}$

20. $a_1 = 0.1$, $a_n = -5.9$, $S_n = -8.7$　**21.** $d = 3$, $n = 30$, $S_{30} = 1875$　　**22.** $d = 9$, $a_n = 86$, $S_n = 455$

23. $a_1 = 74$, $d = -5$, $a_n = -231$　　**24.** $a_1 = -\frac{9}{7}$, $n = 19$, $a_{19} = -\frac{36}{7}$　**25.** $a_1 = -5k$, $d = \frac{1}{2}k$, $S_n = \frac{23}{2}k$

26. $d = -2c$, $n = 50$, $S_{50} = 0$　　　**27.** $a_1 = -c$, $a_n = \dfrac{b}{2}$, $S_n = 2b - 4c$　**28.** $a_1 = 3b$, $n = 7$, $d = \dfrac{b}{3}$

In Exercises 29 through 48, find the indicated quantities for the appropriate arithmetic sequence.

29. Sixth term $= 56$, tenth term $= 72$ (find a_1, d, S_n for $n = 10$).

30. Seventeenth term $= -91$, second term $= -73$ (find a_1, d, S_n for $n = 40$).

31. Fourth term $= 2$, tenth term $= 0$ (find a_1, d, S_n for $n = 10$).

32. Third term $= 1$, sixth term $= -8$ (find a_1, d, S_n for $n = 12$).

33. Find the sum of the first 100 positive integers.

34. Find the sum of the first 100 positive odd integers.

35. Find the sum of the first 200 multiples of 5.

36. Find the number of multiples of 8 between 99 and 999.

37. The temperature in an experiment dropped $16.50°C$ during the first minute and then dropped by $1.25°C$ less each minute than during the previous minute. By how much did the temperature drop during the twelfth minute?

38. A special antenna is made of 9 parallel straight metal rods. The longest rod is 320 cm long, and each rod is 6 cm shorter than the one before it. How long is the shortest rod? What is the combined length of the 9 rods?

39. There are 12 seats in the first row around a semicircular stage. Each row behind the first has 4 more seats than the row in front of it. How many rows of seats are there if there is a total of 300 seats?

40. Each swing of a pendulum is measured to be 3.00 in. shorter than the preceding swing. If the first swing is 10.0 ft, determine the total distance traveled by the pendulum bob in the first five swings.

41. A car depreciates $1800 during the first year after it is purchased. Each year thereafter it depreciates $150 less than the year before. After how many years will it be considered to have no value, and what was the original cost?

42. A clock strikes the number of times of the hour. How many strikes does it make in one day?

43. If a tool dropped from a helicopter falls 4.9 m during the first second, 14.7 m during the second second, 24.5 m during the third second, and so on, how high was the helicopter if the tool takes 10.0 s to reach the ground?

44. In preparing a bid for constructing a new building, a contractor determines that the foundation and basement will cost $600,000 and the first floor will cost $360,000. Each floor above will cost $15,000 more than the one below it. How much will the building cost if it is to be 18 floors high?

45. Derive a formula for S_n in terms of n, a_1, and d.

46. If each term of an arithmetic sequence is multiplied by a constant k, is the resulting sequence an arithmetic sequence?

47. Show that the sum of the first n positive integers is $\frac{1}{2}n(n + 1)$.

48. Show that the sum of the first n positive odd integers is n^2.

18-2 Geometric Sequences

A second type of sequence of numbers is the **geometric sequence** (or **geometric progression**). *In a geometric sequence each number after the first can be obtained from the preceding one by multiplying it by a fixed number **called the common ratio.*** One important application of geometric sequences is in computing interest on savings accounts. Other applications can be found in areas such as biology and physics.

EXAMPLE A — The sequence 2, 4, 8, 16, . . . is a geometric sequence with a common ratio of 2. Any term after the first can be obtained by multiplying the previous term by 2.

The sequence 9, −3, 1, −$\frac{1}{3}$, . . . is a geometric sequence with a common ratio of −$\frac{1}{3}$. We can obtain any term after the first by multiplying the previous term by −$\frac{1}{3}$. ∎

If we know the first term, we can then find any other desired term by multiplying by the common ratio a sufficient number of times. When we do this for a general geometric sequence, we can find the nth term in terms of the first term a_1, the common ratio r, and n. Thus, the second term is $a_1 r$, the third term is $a_1 r^2$, and so forth. In general, the expression for the nth term is

*n*th term

$$a_n = a_1 r^{n-1}$$ (18-3)

EXAMPLE B — Find the eighth term of the geometric sequence 8, 4, 2, ...

By dividing any given term by the previous term, we find the common ratio to be $\frac{1}{2}$. From the terms given, we see that $a_1 = 8$. From the statement of the problem, we know that $n = 8$. Thus we substitute into Eq. (18-3) to find a_8.

$$a_8 = 8\left(\frac{1}{2}\right)^{8-1} = \frac{8}{2^7} = \frac{1}{16}$$

with labels a_1, r, n pointing to the respective values. ∎

EXAMPLE C — Find the tenth term of the geometric sequence for which $a_1 = \frac{8}{625}$ and $r = -\frac{5}{2}$.

Using Eq. (18-3) to find a_{10}, we have

$$a_{10} = \frac{8}{625}\left(-\frac{5}{2}\right)^{10-1} = \frac{8}{625}\left(-\frac{5^9}{2^9}\right) = -\left(\frac{2^3}{5^4}\right)\left(\frac{5^9}{2^9}\right) = -\frac{5^5}{2^6}$$

$$= -\frac{3125}{64}$$

EXAMPLE D — Find the seventh term of a geometric sequence for which the second term is 3, the fourth term is 9, and $r > 0$.

We can find r if we let $a_1 = 3$, $a_3 = 9$, and $n = 3$. (At this point we are considering a sequence made up of 3, the next number, and 9. These are the second, third, and fourth terms of the original sequence.) Thus,

$$9 = 3r^2, \qquad r = \sqrt{3} \qquad \text{(since } r > 0\text{)}$$

We can now find a_1 of the original sequence by considering just the first two terms of the sequence, a_1 and $a_2 = 3$.

$$3 = a_1(\sqrt{3})^{2-1}, \qquad a_1 = \sqrt{3}$$

We can now find the seventh term, using $a_1 = \sqrt{3}$, $r = \sqrt{3}$, and $n = 7$.

$$a_7 = \sqrt{3}(\sqrt{3})^{7-1} = \sqrt{3}(3^3) = 27\sqrt{3}$$

We could have shortened this procedure one step by letting the second term be the first term of a new sequence of six terms. If the first term is of no importance in itself, this is acceptable. ∎

EXAMPLE E

In an experiment, 22.0% of a substance changes chemically each 10.0 min. If there is originally 120 g of the substance, how much will remain after 45.0 min?

Let P = the portion of the substance remaining after each minute. From the statement of the problem we know that $r = 0.780$, since 78.0% remains after each 10.0 min period. We also know that $a_1 = 120$ g, and we let n represent the number of minutes of elapsed time. This means that $P = 120(0.780)^{t/10.0}$. It is necessary to divide by 10.0 because the ratio is given for a 10.0 min period. In order to find P when $n = 45.0$ min, we write

120 ⨯ .78 x^y

4.5 = $\boxed{39.228975}$

$$P = 120(0.780)^{45.0/10.0} = 120(0.780)^{4.50}$$
$$= 39.2 \text{ g}$$

This means that 39.2 g remains after 45.0 min. Note that the power 4.50 represents 4.50 ten-minute periods. The calculator sequence for the calculation is shown at the left. ■

A general expression for the sum S_n of the first n terms of a geometric sequence may be found by directly forming the sum and multiplying this equation by r. By doing this, we have

$$S_n = a_1 + a_1 r + a_1 r^2 + \cdots + a_1 r^{n-1}$$
$$rS_n = a_1 r + a_1 r^2 + a_1 r^3 + \cdots + a_1 r^n$$

If we now subtract the second of these equations from the first, we obtain $S_n - rS_n = a_1 - a_1 r^n$. All other terms cancel by subtraction. Now, factoring S_n from the terms on the left and a_1 from the terms on the right, we solve for S_n. Thus *the sum S_n of the first n terms of a geometric sequence is*

sum

$$\boxed{S_n = \frac{a_1(1 - r^n)}{1 - r} \qquad (r \neq 1)}$$

(18-4)

As with an arithmetic sequence, a geometric sequence can be either a finite sequence or an infinite sequence. We will consider the sum of an infinite sequence in the following section.

EXAMPLE F

Find the sum of the first seven terms of the geometric sequence in which the first term is 2 and the common ratio is $\frac{1}{2}$.

We are to find S_n given that $a_1 = 2$, $r = \frac{1}{2}$, and $n = 7$. Using Eq. (18-4), we have

$$S_7 = \frac{2(1 - (\frac{1}{2})^7)}{1 - \frac{1}{2}} = \frac{2(1 - \frac{1}{128})}{\frac{1}{2}} = 4\left(\frac{127}{128}\right) = \frac{127}{32}$$

EXAMPLE G

See the chapter introduction.

If \$100 is invested each year at 5% interest compounded annually, what would be the total amount of the investment after 10 years (before the eleventh deposit is made)?

After one year the amount invested will have added to it the interest for the year. Thus, for the last \$100 invested, its value will become $\$100(1 + 0.05) =$

$100(1.05) = \$105$. The next to last \$100 will have interest added twice. After one year its value becomes $\$100(1.05)$, and after two years its value becomes $[\$100(1.05)](1.05) = \$100(1.05)^2$. In the same way, the value of the first \$100 becomes $\$100(1.05)^{10}$, since it will have interest added 10 times. This means we are asked to sum the progression

$$100(1.05) + 100(1.05)^2 + 100(1.05)^3 + \cdots + 100(1.05)^{10}$$

1 year	2 years	3 years	10 years
in account	in account	in account	in account

or

$$100[1.05 + (1.05)^2 + (1.05)^3 + \cdots + (1.05)^{10}]$$

For the progression in the brackets we have $a_1 = 1.05$, $r = 1.05$, and $n = 10$. Thus,

$$S_{10} = \frac{1.05[1 - (1.05)^{10}]}{1 - 1.05} = \frac{1.05}{-0.05}(1 - 1.628895) = 13.2068$$

The total value of the \$100 investments is $100(13.2068) = \$1320.68$. We see that \$320.68 in interest has been earned.

The calculation can be performed by using the calculator sequence shown at the left.

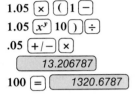

Exercises 18-2

In Exercises 1 through 4, write down the first five terms of the geometric sequence with the given values.

1. $a_1 = 45, r = \frac{1}{3}$ **2.** $a_1 = 9, r = -\frac{2}{3}$ **3.** $a_1 = 2, r = 3$ **4.** $a_1 = -3, r = 2$

In Exercises 5 through 12, find the nth term of the geometric sequence with the given values.

5. $\frac{1}{2}, 1, 2, \ldots$ $(n = 6)$ **6.** $10, 1, 0.1, \ldots$ $(n = 8)$ **7.** $125, -25, 5, \ldots$ $(n = 7)$

8. $0.1, 0.3, 0.9, \ldots$ $(n = 5)$ **9.** $a_1 = -27, r = -\frac{1}{3}, n = 6$ **10.** $a_1 = 48, r = \frac{1}{2}, n = 9$

11. $a_1 = 2, r = 10, n = 7$ **12.** $a_1 = -2, r = 2k, n = 6$

In Exercises 13 through 16, find the sum of the first n terms of the geometric sequence with the given values.

13. $a_1 = \frac{1}{8}, r = 4, n = 5$ **14.** $a_1 = 162, r = -\frac{1}{3}, n = 6$

15. $a_1 = 192, a_n = 6, n = 6$ **16.** $a_1 = 9, a_n = -243, n = 4$

In Exercises 17 through 24, find any of the values of a_1, r, a_n, n, or S_n that are missing.

17. $a_1 = \frac{1}{16}, r = 4, n = 6$ **18.** $r = 0.2, a_n = 0.00032, n = 7$

19. $r = \frac{3}{2}, n = 5, S_5 = 211$ **20.** $r = -\frac{1}{2}, a_n = \frac{1}{8}, n = 7$

21. $a_n = 27, n = 4, S_4 = 40$ **22.** $a_1 = 3, n = 5, a_5 = 48$

23. $a_1 = 75, r = \frac{1}{5}, a_n = \frac{3}{25}$ **24.** $r = -2, n = 6, S_6 = 42$

In Exercises 25 through 40, find the indicated quantities.

25. Find the tenth term of a geometric sequence if the fourth term is 8 and the seventh term is 16.

26. Find the sum of the first 8 terms of the geometric sequence for which the fifth term is 5, the seventh term is 10, and $r > 0$.

27. Each stroke of a pump removes 8.2% of the remaining air from a container. What percent of the air remains after 50 strokes?

28. After each application of a pesticide, 35% of a type of insect remains. What percent of the insects survive after 5 applications of the pesticide?

29. An electric current decreases by 12.5% each 1.00 μs. If the initial current is 3.27 mA, what is the current after 8.20 μs?

30. What is the value of an investment of $10,000 after 10 years if it earns 6% annual interest, compounded semi-annually? (6% annual interest compounded semiannually means that 3% interest is added each six months.)

31. A person invests $250 each three months for 8 years. How much is the investment worth after 8 years if the interest is 8% compounded quarterly?

32. A chemical spill pollutes a stream. A monitoring device finds 620 ppm (parts per million) of the chemical 1.0 mi below the spill, and the readings decrease by 12.5% for each mile farther downstream. How far downstream is the reading 100 ppm?

33. During each oscillation, a pendulum swings through 85% of the distance of the previous oscillation. If the pendulum swings through 80.8 cm in the first oscillation, through what total distance does it move in 12 oscillations?

34. A series of deposits, each of value A and made at equal time intervals, earns an interest rate of i for the time interval. The deposits have a total value of $A(1 + i) + A(1 + i)^2 + A(1 + i)^3 + \cdots + A(1 + i)^n$ after n time intervals (just before the next deposit). Find a formula for this sum.

35. A tank with a 22% acid solution has 25% of its contents removed and replaced with water. How many times must this be done in order to have a 7.0% solution remain?

36. The power on a satellite is supplied by a radioactive isotope. On a given satellite the power decreases by 0.2% each day. What percent of the initial power remains after one year?

37. If you decided to save money by putting away 1¢ on a given day, 2¢ one week later, 4¢ a week later, and so on, how much would you have to put away 6 months (26 weeks) after putting away the 1¢?

38. How many direct ancestors (parents, grandparents, and so on) does a person have in the ten generations which preceded him or her?

39. Derive a formula for S_n in terms of a_1, r, and a_n.

40. Write down several terms of a general geometric sequence. Then take the logarithm of each term. Show that the resulting sequence is an arithmetic sequence.

18-3   Infinite Geometric Series

In the previous sections we developed formulas for the sum of the first n terms of an arithmetic sequence and of a geometric sequence. *The indicated sum of the terms of a sequence is called a* **series.**

EXAMPLE A ───── The infinite arithmetic sequence 2, 5, 8, 11, 14, ... has the series

$$2 + 5 + 8 + 11 + 14 + \cdots$$

associated with it.

   The finite geometric sequence 1, $\frac{1}{2}$, $\frac{1}{4}$, $\frac{1}{8}$ has the series

$$1 + \frac{1}{2} + \frac{1}{4} + \frac{1}{8}$$

associated with it.

The series associated with a finite sequence will sum up to a real number. The series associated with an infinite arithmetic sequence will not sum up to a real number, as the terms being added become larger and larger numerically. The sum is unbounded, as we can see in Example A. The series associated with an infinite geometric sequence may or may not sum up to a real number, as we shall now show.

Let us now consider the sum of the first n terms of the infinite geometric sequence $1, \frac{1}{2}, \frac{1}{4}, \ldots$. This is the sum of the n terms of the associated geometric series

$$1 + \frac{1}{2} + \frac{1}{4} + \cdots + \frac{1}{2^{n-1}}$$

Here $a_1 = 1$ and $r = \frac{1}{2}$, and we find that we get the values of S_n for the given values of n in the following table:

n	2	3	4	5	6	7	8	9	10
S_n	$\frac{3}{2}$	$\frac{7}{4}$	$\frac{15}{8}$	$\frac{31}{16}$	$\frac{63}{32}$	$\frac{127}{64}$	$\frac{255}{128}$	$\frac{511}{256}$	$\frac{1023}{512}$

$1 + \frac{1}{2} + \frac{1}{4} + \frac{1}{8} \rceil$ $\lceil 1 + \frac{1}{2} + \frac{1}{4} + \frac{1}{8} + \frac{1}{16} + \frac{1}{32} + \frac{1}{64}$

The series for $n = 4$ and $n = 7$ are shown. We see that as n gets larger, the numerator of each fraction becomes more nearly twice the denominator. In fact, we find that if we continue to compute S_n as n becomes larger, S_n can be found as close to the value 2 as desired, although it will never actually reach the value 2. For example, if $n = 100$, $S_{100} = 2 - 1.6 \times 10^{-30}$, which could be written as

1.9999999999999999999999999999984

to 32 significant digits. In the formula for the sum of the first n terms of a geometric sequence

$$S_n = a_1 \frac{1 - r^n}{1 - r}$$

the term r^n becomes exceedingly small, and if we consider n as being sufficiently large, we can see that this term is effectively zero. *If this term were exactly zero,* then the sum would be

$$S_n = 1 \frac{1 - 0}{1 - \frac{1}{2}} = 2$$

The only problem is that we cannot find any number large enough for n to make $(\frac{1}{2})^n$ zero. There is, however, an accepted notation for this. This notation is

$$\lim_{n \to \infty} r^n = 0 \qquad (\text{if } |r| < 1)$$

and it is read as "the limit, as n *approaches* infinity, of r to the nth power is zero."

NOTE ▷   The symbol ∞ is read as **infinity,** *but it must not be thought of as a number.* It is simply a symbol which stands for a *process* of considering numbers which become large without bound. The number which is called the **limit** of the sums is simply the number which the sums get closer and closer to, as n is considered to approach infinity. This notation and terminology are of particular importance in the calculus.

If we consider values of r such that $|r| < 1$, and let the values of n become unbounded, we find that $\lim_{n \to \infty} r^n = 0$. The formula for *the sum of the terms of an infinite geometric series then becomes*

sum

$$S = \frac{a_1}{1 - r} \qquad (|r| < 1)$$ (18-5)

where a_1 is the first term and r is the common ratio. If $r \geq 1$, S is unbounded in value.

EXAMPLE B   Find the sum of the infinite geometric series

$$4 - \frac{1}{2} + \frac{1}{16} - \frac{1}{128} + \cdots$$

Here we see that $a_1 = 4$. We find r by dividing any term by the previous term, and we find that $r = -\frac{1}{8}$. Using Eq. (18-5), we have

$$S = \frac{4}{1 - \left(-\frac{1}{8}\right)} = \frac{4}{1 + \frac{1}{8}} = \frac{4}{1} \times \frac{8}{9} = \frac{32}{9}$$ ∎

EXAMPLE C   Find the fraction which has as its decimal form $0.121212\ldots$.
This decimal form can be considered as being

$$0.12 + 0.0012 + 0.000012 + \cdots$$

which means that we have an infinite geometric series in which $a_1 = 0.12$ and $r = 0.01$. Thus,

$$S = \frac{0.12}{1 - 0.01} = \frac{0.12}{0.99} = \frac{4}{33}$$

Therefore, the decimal $0.121212\ldots$ and the fraction $\frac{4}{33}$ represent the same number. ∎

The decimal in Example C is called a **repeating decimal,** because *the numbers in the decimal form appear endlessly in a particular order*. This example verifies the theorem that any repeating decimal represents a rational number. However, not all repeating decimals necessarily start repeating immediately. If numbers never do repeat, the decimal represents an irrational number. For example, there are no repeating decimals which represent π or $\sqrt{2}$.

EXAMPLE D   Find the fraction which has as its decimal form the repeating decimal $0.50345345345\ldots$.
We first separate the decimal into the beginning, nonrepeating part, and the infinite repeating decimal, which follows. Thus we have

$$0.50345345345\ldots = 0.50 + 0.00345345345\ldots$$

This means that we are to add $\frac{50}{100}$ to the fraction which represents the sum of the terms of the infinite geometric series $0.00345 + 0.00000345 + \cdots$. For this series, $a_1 = 0.00345$ and $r = 0.001$. We find this sum to be

$$S = \frac{0.00345}{1 - 0.001} = \frac{0.00345}{0.999} = \frac{115}{33,300} = \frac{23}{6660}$$

Therefore,

$$0.50345345\ldots = \frac{5}{10} + \frac{23}{6660} = \frac{5(666) + 23}{6660} = \frac{3353}{6660}$$

EXAMPLE E
Each swing of a certain pendulum bob is 95% as long as the preceding swing. How far does the bob travel in coming to rest if the first swing is 40.0 in. long?

We are to find the sum of the terms of an infinite geometric series for which $a_1 = 40.0$ and $r = 95\% = \frac{19}{20}$. Substituting these values into Eq. (18-5), we obtain

$$S = \frac{40.0}{1 - \frac{19}{20}} = \frac{40.0}{\frac{1}{20}} = (40.0)(20) = 800 \text{ in.}$$

Therefore, the pendulum bob travels 800 in. (about 67 ft) in coming to rest.

Exercises 18-3

In Exercises 1 through 12, find the sums of the given infinite geometric series.

1. $4 + 2 + 1 + \frac{1}{2} + \cdots$

2. $6 - 2 + \frac{2}{3} - \frac{2}{9} + \cdots$

3. $5 + 1 + 0.2 + 0.04 + \cdots$

4. $2 + \sqrt{2} + 1 + \cdots$

5. $20 - 1 + 0.05 - \cdots$

6. $9 + 8.1 + 7.29 + \cdots$

7. $1 + \frac{7}{8} + \frac{49}{64} + \cdots$

8. $6 - 4 + \frac{8}{3} - \cdots$

9. $1 + 0.0001 + 0.00000001 + \cdots$

10. $30 - 9 + 2.7 - \cdots$

11. $(2 + \sqrt{3}) + 1 + (2 - \sqrt{3}) + \cdots$

12. $(1 + \sqrt{2}) - 1 + (\sqrt{2} - 1) - \cdots$

In Exercises 13 through 24, find the fractions equal to the given decimals.

13. $0.33333\ldots$

14. $0.55555\ldots$

15. $0.404040\ldots$

16. $0.070707\ldots$

17. $0.181818\ldots$

18. $0.336336336\ldots$

19. $0.273273273\ldots$

20. $0.82222\ldots$

21. $0.366666\ldots$

22. $0.66424242\ldots$

23. $0.100841841841\ldots$

24. $0.184561845618456\ldots$

In Exercises 25 through 28, solve the given problems by use of the sum of an infinite geometric series.

25. Liquid is continuously collected in a wastewater holding tank such that during a given hour, only 92.0% as much liquid is collected as in the previous hour. If 28.0 gal are collected in the first hour, what must be the minimum capacity of the tank?

26. In coming to rest, a record turntable makes 0.80 as many revolutions in a second as in the previous second. How many revolutions does the turntable make in coming to rest if it makes 0.50 r in the first second after the power is turned off?

27. The amounts of plutonium 237 which decay each day because of radioactivity form a geometric sequence. Given that the amounts which decay during each of the first four days are 5.882 g, 5.782 g, 5.684 g, and 5.587 g, respectively, what total amount will decay?

28. A square has sides of 20 cm each. Another square is inscribed in the original square by joining the midpoints of the sides. Assuming that such inscribed squares can be formed endlessly, what is the sum of the areas of all of the squares? See Fig. 18-1.

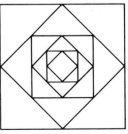

Fig. 18-1

18-4   The Binomial Theorem

If we wish to find the roots of the equation $(x + 2)^5 = 0$, we note that there are five factors of $x + 2$, which in turn tells us that the only distinct root is $x = -2$. However, if we wish to expand the expression $(x + 2)^5$, a number of repeated multiplications would be needed. This would be a relatively tedious operation. In this section we shall develop *the* **binomial theorem,** *by which it is possible to expand binomials to any given power without direct multiplication.* Such direct expansion can be helpful and labor-saving in developing certain mathematical topics. We may also expand certain expressions where direct multiplication is not actually possible. Also, the binomial theorem is used to develop the necessary expressions for use in certain technical applications.

By direct multiplication, we may obtain the following expansions of the binomial $a + b$.

$$(a + b)^0 = 1$$
$$(a + b)^1 = a + b$$
$$(a + b)^2 = a^2 + 2ab + b^2$$
$$(a + b)^3 = a^3 + 3a^2b + 3ab^2 + b^3$$
$$(a + b)^4 = a^4 + 4a^3b + 6a^2b^2 + 4ab^3 + b^4$$
$$(a + b)^5 = a^5 + 5a^4b + 10a^3b^2 + 10a^2b^3 + 5ab^4 + b^5$$

An inspection indicates certain properties which these expansions have, and which we shall assume are valid for the expansion of $(a + b)^n$, where n is any positive integer. These properties are as follows:

1. There are $n + 1$ terms.

2. The first term is a^n and the final term is b^n.

3. Progressing from the first term to the last, the exponent of a decreases by 1 from term to term, the exponent of b increases by 1 from term to term, and the sum of the exponents of a and b in each term is n.

4. If the coefficient of any term is multiplied by the exponent of a in that term, and this product is divided by the number of that term, we obtain the coefficient of the next term.

5. The coefficients of terms equidistant from the ends are equal.

EXAMPLE A — Using the above properties, we shall develop the expansion for $(a + b)^5$.

Since the exponent of the binomial is 5, we have $n = 5$.
From Property 1, we know that there are six terms.
From Property 2, we know that the first term is a^5 and the final term is b^5.
From Property 3, we know that the factors of a and b in terms 2, 3, 4, and 5 are a^4b, a^3b^2, a^2b^3, and ab^4, respectively.

From Property 4, we obtain the coefficients of terms 2, 3, 4, and 5. In the first term, a^5, the coefficient is 1. Multiplying by 5, the power of a, and dividing by 1, the number of the term, we obtain 5, which is the coefficient of the second term. Thus, the second term is $5a^4b$. Again using Property 4, we obtain the coefficient of the third term. The coefficient of the second term is 5. Multiplying by 4, and dividing by 2, we obtain 10. This means that the third term is $10a^3b^2$.

From Property 5, we know that the coefficient of the fifth term is the same as the second, and the coefficient of the fourth term is the same as the third.

These properties are illustrated in the diagram of the expansion shown below.

It is not necessary to use the above properties directly to expand a given binomial. If they are applied to $(a + b)^n$, a general formula for the expansion of a binomial may be obtained. In developing and stating the general formula, it is convenient to use the factorial notation $n!$, where

$$n! = n(n - 1)(n - 2) \cdots (2)(1) \tag{18-6}$$

We see that $n!$, read "n factorial," represents the product of the first n positive integers.

EXAMPLE B

1. $3! = (3)(2)(1) = 6$

2. $5! = (5)(4)(3)(2)(1) = 120$

3. $\dfrac{4!}{2!} = \dfrac{(4)(3)(2)(1)}{(2)(1)} = 12$

We note that $4!/2!$ is *not* $2!$.

Based on the binomial properties, *the* **binomial theorem** *states that the following* **binomial formula** *is valid for all positive integer values of n* (the binomial theorem is proven through advanced methods).

binomial formula

$$(a + b)^n = a^n + na^{n-1}b + \frac{n(n - 1)}{2!} a^{n-2}b^2 + \frac{n(n - 1)(n - 2)}{3!} a^{n-3}b^3 + \cdots + b^n \tag{18-7}$$

The use of the binomial formula is illustrated in the following example.

EXAMPLE C _____ By use of the binomial formula, expand $(2x + 3)^6$.

In using the binomial formula for $(2x + 3)^6$, we use $2x$ for a, 3 for b, and 6 for n. Thus,

$$n = 6 \quad n - 1 \quad n \quad n - 1 \quad n - 2$$

$$(2x + 3)^6 = (2x)^6 + 6(2x)^5(3) + \frac{(6)(5)}{2}(2x)^4(3^2) + \frac{(6)(5)(4)}{(2)(3)}(2x)^3(3^3) + \frac{(6)(5)(4)(3)}{(2)(3)(4)}(2x)^2(3^4) + \frac{(6)(5)(4)(3)(2)}{(2)(3)(4)(5)}(2x)(3^5) + 3^6$$

$$2! \qquad\qquad 3! \qquad\qquad 4! \qquad\qquad 5!$$

$$= 64x^6 + 576x^5 + 2160x^4 + 4320x^3 + 4860x^2 + 2916x + 729$$

For the first few integral powers of a binomial, the coefficients can be obtained by setting them up in the following pattern, known as **Pascal's triangle.**

$n = 0$				1			
$n = 1$			1		1		
$n = 2$		1		2		1	
$n = 3$	1		3		3		1
$n = 4$	1	4		6		4	1
$n = 5$	1	5	10		10	5	1
$n = 6$	1	6	15	20		15	6	1

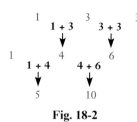

Fig. 18-2

We note that the first and last coefficient shown in each row is 1, and the second and next-to-last coefficients are equal to n. Other coefficients are obtained by adding the two nearest coefficients in the row above, as illustrated in Fig. 18-2 for the indicated section of Pascal's triangle. This pattern may be continued indefinitely, although use of Pascal's triangle is cumbersome for high values of n.

EXAMPLE D _____ By use of Pascal's triangle, expand $(5s - 2t)^4$.

Here we note that $n = 4$. Thus, the coefficients of the five terms are 1, 4, 6, 4, and 1, respectively. Also, here we use $5s$ for a and $-2t$ for b. We are expanding this as $[(5s) + (-2t)]^4$. Therefore,

$$\text{from Pascal's triangle for } n = 4$$

$$(5s - 2t)^4 = (5s)^4 + 4(5s)^3(-2t) + 6(5s)^2(-2t)^2 + 4(5s)(-2t)^3 + (-2t)^4$$

$$= 625s^4 - 1000s^3t + 600s^2t^2 - 160st^3 + 16t^4$$

In certain uses of a binomial expansion it is not necessary to obtain all terms. Only the first few terms are required. The following example illustrates finding the first four terms of an expansion.

EXAMPLE E _____ Find the first four terms of the expansion of $(x + 7)^{12}$.

Here we use x for a, 7 for b, and 12 for n. Thus, from the binomial formula we have

$$(x + 7)^{12} = x^{12} + 12x^{11}(7) + \frac{(12)(11)}{2}x^{10}(7^2) + \frac{(12)(11)(10)}{(2)(3)}x^9(7^3) + \cdots$$

$$= x^{12} + 84x^{11} + 3234x^{10} + 75{,}460x^9 + \cdots$$

If we let $a = 1$ and $b = x$ in the binomial formula, we obtain the **binomial series**

binomial series

$$(1 + x)^n = 1 + nx + \frac{n(n-1)}{2!} x^2 + \frac{n(n-1)(n-2)}{3!} x^3 + \cdots \qquad (18\text{-}8)$$

which through advanced methods can be shown to be valid for any real number n if $|x| < 1$. When n is either negative or a fraction, we obtain an infinite series. In such a case, we calculate as many terms as may be needed although such a series is not obtainable through direct multiplication. The binomial series may be used to develop important expressions which are used in applications and more advanced mathematics topics.

EXAMPLE F

In the analysis of forces on beams, the expression $1/(1 + m^2)^{3/2}$ is used. Use the binomial series to find the first four terms of the expansion of this expression.

Using negative exponents, we have

$$\frac{1}{(1 + m^2)^{3/2}} = (1 + m^2)^{-3/2}$$

Now, in using Eq. (18-8), we have $n = -3/2$ and $x = m^2$.

$$(1 + m^2)^{-3/2} = 1 + \left(-\frac{3}{2}\right)(m^2) + \frac{(-\frac{3}{2})(-\frac{3}{2} - 1)}{2!} (m^2)^2 + \frac{(-\frac{3}{2})(-\frac{3}{2} - 1)(-\frac{3}{2} - 2)}{3!} (m^2)^3 + \cdots$$

Therefore,

$$\frac{1}{(1 + m^2)^{3/2}} = 1 - \frac{3}{2} m^2 + \frac{15}{8} m^4 - \frac{35}{16} m^6 + \cdots$$

Exercises 18-4

In Exercises 1 through 8, expand and simplify the given expressions by use of the binomial formula.

1. $(t + 1)^3$ **2.** $(x - 2)^3$ **3.** $(2x - 1)^4$

4. $(x^2 + 3)^4$ **5.** $(2x + 3)^5$ **6.** $(xy - z)^5$

7. $(2a - b^2)^6$ **8.** $\left(\frac{a}{x} + x\right)^6$

See Appendix E for a computer program for finding coefficients of a binomial expansion.

In Exercises 9 through 12, expand and simplify the given expressions by use of Pascal's triangle.

9. $(5x - 3)^4$ **10.** $(b + 4)^5$ **11.** $(2a + 1)^6$ **12.** $(x - 3)^7$

In Exercises 13 through 20, find the first four terms of the indicated expansions.

13. $(x + 2)^{10}$ **14.** $(x - 3)^8$ **15.** $(2a - 1)^7$ **16.** $(3b + 2)^9$

17. $\left(x^2 - \frac{y}{2}\right)^{12}$ **18.** $\left(2a - \frac{1}{x}\right)^{11}$ **19.** $\left(b^2 + \frac{1}{2b}\right)^{20}$ **20.** $\left(2x^2 + \frac{y}{3}\right)^{15}$

In Exercises 21 through 28, find the first four terms of the indicated expansions by use of the binomial series.

21. $(1 + x)^8$

22. $(1 + x)^{-1/3}$

23. $(1 - x)^{-2}$

24. $(1 - 2x)^9$

25. $\sqrt{1 + x}$

26. $\dfrac{1}{\sqrt{1 + x}}$

27. $\dfrac{1}{\sqrt{9 - 9x}}$

28. $\sqrt{4 + x^2}$

In Exercises 29 through 32, solve the given problems involving factorial notation.

29. Using a calculator (most calculators have an $\boxed{n!}$ key), evaluate (a) $17! + 4!$, (b) $21!$, (c) $17! \times 4!$, (d) $68!$

30. Using a calculator, evaluate (a) $8! - 7!$, (b) $\dfrac{8!}{7!}$, (c) $8! \times 7!$, (d) $56!$

31. Show that $n! = n \times (n - 1)!$ for $n \geq 2$. To use this equation for $n = 1$, show that it is necessary to define $0! = 1$ (this is a standard definition of $0!$).

32. Show that $\dfrac{(n + 1)!}{(n - 2)!} = n^3 - n$ for $n \geq 2$. See Exercise 31.

In Exercises 33 through 36, find the indicated terms by use of the following information. The $r + 1$ term of the expansion of $(a + b)^n$ is given by

$$\frac{n(n - 1)(n - 2) \cdots (n - r + 1)}{r!} a^{n-r} b^r$$

33. the term involving b^5 in $(a + b)^8$

34. the term involving y^6 in $(x + y)^{10}$

35. the fifth term of $(2x - 3b)^{12}$

36. the sixth term of $(a - b)^{14}$

In Exercises 37 through 40, find the indicated expansions.

37. A company purchases a piece of equipment for A dollars, and it depreciates at a rate of r each year. Its value V after n years is given by $V = A(1 - r)^n$. Expand this expression for $n = 5$.

38. In designing a type of tubing, the expression $(D + 0.1)^4$ is used. Expand this expression.

39. In the theory associated with the magnetic field due to an electric current, the expression $1 - \dfrac{x}{\sqrt{a^2 + x^2}}$ is found. By expanding $(a^2 + x^2)^{-1/2}$ find the first three nonzero terms which could be used to approximate the given expression.

40. In the theory related to the dispersion of light, the expression $1 + \dfrac{A}{1 - \dfrac{\lambda_0^2}{\lambda^2}}$ arises. (a) Let $x = \lambda_0^2/\lambda^2$ and find the first four terms of the expansion of $(1 - x)^{-1}$. (b) Find the same expansion by using long division. (c) Write the original expression in expanded form using the results of (a) and (b).

18-5   Chapter Equations, Review Exercises, and Practice Test

Chapter Equations

Arithmetic sequences	nth term	$a_n = a_1 + (n - 1)d$	(18-1)
	sum of n terms	$S_n = \dfrac{n}{2}(a_1 + a_n)$	(18-2)
Geometric sequences	nth term	$a_n = a_1 r^{n-1}$	(18-3)
	sum of n terms	$S_n = \dfrac{a_1(1 - r^n)}{1 - r}$ $(r \neq 1)$	(18-4)

Sum of geometric series $$S = \frac{a_1}{1 - r} \qquad (|r| < 1) \tag{18-5}$$

Factorial notation $$n! = n(n - 1)(n - 2) \cdots (2)(1) \tag{18-6}$$

Binomial formula

$$(a + b)^n = a^n + na^{n-1}b + \frac{n(n - 1)}{2!} a^{n-2}b^2 + \frac{n(n - 1)(n - 2)}{3!} a^{n-3}b^3 + \cdots + b^n \tag{18-7}$$

Binomial series $$(1 + x)^n = 1 + nx + \frac{n(n - 1)}{2!} x^2 + \frac{n(n - 1)(n - 2)}{3!} x^3 + \cdots \tag{18-8}$$

Review Exercises

In Exercises 1 through 8, find the indicated term of each sequence.

1. $1, 6, 11, \ldots$ (17th)

2. $1, -3, -7, \ldots$ (21st)

3. $\frac{1}{2}, 0.1, 0.02, \ldots$ (9th)

4. $0.025, 0.01, 0.004, \ldots$ (7th)

5. $8, \frac{7}{2}, -1, \ldots$ (16th)

6. $-1, -\frac{5}{3}, -\frac{7}{3}, \ldots$ (25th)

7. $\frac{3}{4}, \frac{1}{2}, \frac{1}{3}, \ldots$ (7th)

8. $\frac{2}{3}, 1, \frac{3}{2}, \ldots$ (7th)

In Exercises 9 through 12, find the sum of each sequence with the indicated values.

9. $a_1 = -4, n = 15, a_{15} = 17$ (arith.)

10. $a_1 = 3, d = -\frac{2}{3}, n = 10$

11. $a_1 = 16, r = -\frac{1}{2}, n = 10$

12. $a_1 = 64, a_n = 729, n = 7$ (geom., $r > 0$)

In Exercises 13 through 24, find the indicated quantities for the appropriate sequences.

13. $a_1 = 17, d = -2, n = 9, S_9 = ?$

14. $a = \frac{4}{3}, a_1 = -3, a_n = 17, n = ?$

15. $a_1 = 18, r = \frac{1}{2}, n = 6, a_6 = ?$

16. $a_n = \frac{49}{8}, r = -\frac{2}{7}, S_n = \frac{17199}{288}, a_1 = ?$

17. $a_1 = 8, a_n = -2.5, S_n = 22, d = ?$

18. $a_1 = 2, d = 0.2, n = 11, S_{11} = ?$

19. $n = 6, r = -0.25, S_6 = 204.75, a_6 = ?$

20. $a_1 = 10, r = 0.1, S_n = 11.111, n = ?$

21. $a_1 = -1, a_n = 32, n = 12, S_{12} = ?$ (arith.)

22. $a_1 = 1, a_n = 64, S_n = 325, n = ?$ (arith.)

23. $a_1 = 1, n = 7, a_7 = 64, S_7 = ?$

24. $a_1 = \frac{1}{4}, n = 6, a_6 = 8, S_6 = ?$

In Exercises 25 through 28, find the sums of the given infinite geometric series.

25. $9 + 6 + 4 + \cdots$

26. $80 - 20 + 5 - \cdots$

27. $1 + 1.02^{-1} + 1.02^{-2} + \cdots$

28. $3 - \sqrt{3} + 1 - \cdots$

In Exercises 29 through 32, find the fractions equal to the given decimals.

29. $0.030303\ldots$

30. $0.363363\ldots$

31. $0.0727272\ldots$

32. $0.25399399399\ldots$

In Exercises 33 through 36, expand and simplify the given expressions. In Exercises 37 through 40, find the first four terms of the appropriate expansion.

33. $(x - 2)^4$

34. $(s + 2t)^4$

35. $(x^2 + 1)^5$

36. $(3n - a)^6$

37. $(a + 2b^2)^{10}$

38. $\left(\dfrac{x}{4} - y\right)^{12}$

39. $\left(p^2 - \dfrac{q}{6}\right)^9$

40. $\left(2s^2 - \dfrac{3}{2t}\right)^{14}$

In Exercises 41 through 48, find the first four terms of the indicated expansions by use of the binomial series.

41. $(1 + x)^{12}$

42. $(1 - x)^{10}$

43. $\sqrt{1 + x^2}$

44. $(4 - 4x)^{-1}$

45. $\sqrt{1 - a^2}$

46. $\sqrt{1 + b^4}$

47. $(1 - 2x)^{-3}$

48. $(1 + 4x)^{-1/4}$

In Exercises 49 through 68, solve the given problems by use of an appropriate sequence or expansion.

49. Find the sum of the first 1000 positive even integers.

50. How many numbers divisible by 4 lie between 23 and 121?

51. Each stroke of a pile driver moves a post 2 in. less than the previous stroke. If the first stroke moves the post 24 in., which stroke moves the post 4 in.?

52. During each hour, an exhaust fan removes 15.0% of the carbon dioxide present in the air in a room at the beginning of the hour. What percent of the carbon dioxide remains after 10.0 h?

53. Each 1.0 mm of a filter through which light passes reduces the intensity of the light by 12%. How thick should the filter be to reduce the intensity of the light to 20%?

54. A pile of dirt and 10 holes are in a straight line. It is 20 ft from the dirt pile to the nearest hole, and the holes are 8 ft apart. If a backhoe takes two trips from the dirt pile to fill each hole, how far must it travel in filling all of the holes if it starts and ends at the dirt pile?

55. A roof support with equally spaced vertical pieces is shown in Fig. 18-3. Find the total length of the vertical pieces if the shortest one is 10.0 in. long.

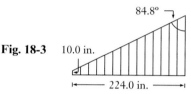

Fig. 18-3

56. During each microsecond the current in an electric circuit decreases by 9.3%. If the initial current is 2.45 mA, how long does it take to reach 0.50 mA?

57. What is the value after 3 years of a $2000 investment which earns 6% annual interest compounded quarterly?

58. A bacterial culture grows 20% less each hour than during the previous hour. If it grows by N bacteria during the first hour, what is the total increase if the growth continues indefinitely?

59. A piece of paper 0.004 in. thick is cut in half. These two pieces are then placed one on the other and cut in half. If this is repeated such that the paper is cut in half 40 times, how high will the pile be?

60. After the power is turned off, an object on a nearly frictionless surface slows down such that it travels 99.9% as far during one second as during the previous second. If it travels 100 cm during the first second after the power is turned off, how far does it travel while stopping?

61. A person invests $1000 each year at the beginning of the year. What is the total value of these investments after 20 years if they earn 7.5% annual interest compounded semiannually?

62. In testing a type of insulation, the temperature in a room was made to fall to $\frac{2}{3}$ of the initial temperature after 1.0 h, to $\frac{2}{5}$ of the initial temperature after 2.0 h, to $\frac{2}{7}$ of the initial temperature after 3.0 h, and so on. If the initial temperature was 50.0°C, what was the temperature after 12.0 h? (This is an illustration of a *harmonic sequence.*)

63. Two competing businesses make the same item and sell it initially for $100. One increases the price by $8 each year for 5 years, and the other increases the price by 8% each year for 5 years. What is the difference in the prices after 5 years?

64. A well driller charges $10.00 for drilling the first meter of a well and for each meter thereafter charges 0.2% more than for the preceding meter. How much is charged for drilling a 150-m well?

65. In hydrodynamics, while studying compressible fluid flow, the expression $\left(1 + \dfrac{a-1}{2}\, m^2\right)^{a/(a-1)}$ arises. Find the first three terms of the expansion of this expression.

66. In finding the partial pressure P_F of fluorine gas under certain conditions, the equation

$$P_F = \frac{(1 + 2 \times 10^{-10}) - \sqrt{1 + 4 \times 10^{-10}}}{2} \text{ atm}$$

is found. By using three terms of the expansion for $\sqrt{1+x}$, approximate the value of this expression.

67. (a) Do the reciprocals of the terms of an arithmetic sequence form an arithmetic sequence? (b) Do the reciprocals of the terms of a geometric sequence form a geometric sequence?

68. The terms a, $a + 12$, $a + 24$ form an arithmetic sequence, and the terms a, $a + 24$, $a + 12$ form a geometric sequence. Find these sequences.

Practice Test

1. Find the sum of the first 7 terms of the sequence 6, -2, $\frac{2}{3}$, ...

2. For a given sequence $a_1 = 6$, $d = 4$ and $S_n = 126$. Find n.

3. Find the fraction equal to the decimal $0.454545\ldots$.

4. Find the first three terms of the expansion of $\sqrt{1 - 4x}$.

5. What is the value after 20 years of an investment of $2500 if it draws 5% annual interest compounded annually?

6. Expand and simplify the expression $(2x - y)^5$.

7. Find the sum of the first 100 even integers.

8. A ball is dropped from a height of 8.00 ft, and on each rebound it rises to $\frac{1}{2}$ of the height it last fell. If it bounces indefinitely, through what total distance will it move?

19 Additional Topics in Trigonometry

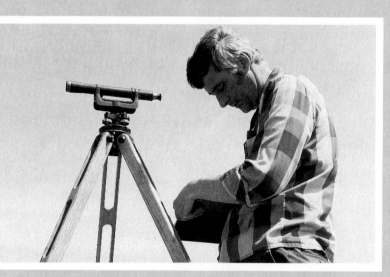

Basic trigonometric relationships can be used to develop useful formulas. In Section 19-3 we derive a formula used in surveying.

The definitions of the trigonometric functions were first introduced in Section 3-2 and were again summarized in Section 7-1. If we take a close look at these definitions, we find that there are many relationships among the various functions. In this chapter we shall study some of these basic relationships, as well as develop certain others. We shall also show how equations with trigonometric functions are solved, and we shall develop the concept of the inverse trigonometric functions, which we introduced in Chapter 3.

The trigonometric relationships we develop in this chapter are important for a number of reasons. We have already actually made limited use of some of them in

Section 9-4 when we graphed certain trigonometric functions. We also used an important *trigonometric identity* in deriving the law of cosines in Chapter 8. When we consider equations with trigonometric functions later in this chapter, we will find that the solution of such equations depends on the proper use of identities. In the study of calculus, there are certain types of problems which require the use of trigonometric identities for solution (even problems in which trigonometric functions do not appear). Also they are used in developing expressions and solving equations in a number of technical areas.

19-1 Fundamental Trigonometric Identities

The definition of the sine of an angle is $\sin \theta = y/r$, and the definition of the cosecant of an angle is $\csc \theta = r/y$ (see Fig. 19-1). But we know that $1/(r/y) = y/r$, which means that $\sin \theta = 1/\csc \theta$. In writing this down we made no reference

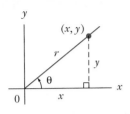

Fig. 19-1

to any particular angle, and since the definitions hold for *any* angle, this relation between the sin θ and csc θ also holds for any angle. *A relation such as this, which holds for any value of the variable, is called an* **identity.** Of course, specific values where division by zero would be indicated are excluded.

Several important identities exist among the six trigonometric functions, and we shall develop these identities in this section. We shall also show how we can use the basic identities to verify other identities among the functions.

From the definitions, we have

$$\sin \theta \csc \theta = \frac{y}{r} \times \frac{r}{y} = 1 \quad \text{or} \quad \sin \theta = \frac{1}{\csc \theta} \quad \text{or} \quad \csc \theta = \frac{1}{\sin \theta}$$

$$\cos \theta \sec \theta = \frac{x}{r} \times \frac{r}{x} = 1 \quad \text{or} \quad \cos \theta = \frac{1}{\sec \theta} \quad \text{or} \quad \sec \theta = \frac{1}{\cos \theta}$$

$$\tan \theta \cot \theta = \frac{y}{x} \times \frac{x}{y} = 1 \quad \text{or} \quad \tan \theta = \frac{1}{\cot \theta} \quad \text{or} \quad \cot \theta = \frac{1}{\tan \theta}$$

$$\frac{\sin \theta}{\cos \theta} = \frac{y/r}{x/r} = \frac{y}{x} = \tan \theta; \qquad \frac{\cos \theta}{\sin \theta} = \frac{x/r}{y/r} = \frac{x}{y} = \cot \theta$$

Also, from the definitions and the Pythagorean theorem in the form of $x^2 + y^2 = r^2$, we arrive at the following identities.

By dividing the Pythagorean relation through by r^2, we have

$$\left(\frac{x}{r}\right)^2 + \left(\frac{y}{r}\right)^2 = 1 \quad \text{which leads us to} \quad \cos^2 \theta + \sin^2 \theta = 1$$

By dividing the Pythagorean relation by x^2, we have

$$1 + \left(\frac{y}{x}\right)^2 = \left(\frac{r}{x}\right)^2 \quad \text{which leads us to} \quad 1 + \tan^2 \theta = \sec^2 \theta$$

By dividing the Pythagorean relation by y^2, we have

$$\left(\frac{x}{y}\right)^2 + 1 = \left(\frac{r}{y}\right)^2 \quad \text{which leads us to} \quad \cot^2 \theta + 1 = \csc^2 \theta$$

The term $\cos^2 \theta$ is the common way of writing $(\cos \theta)^2$, and it means to square the value of the cosine of the angle. Obviously the same holds true for the other functions.

Summarizing these results, we have the following important identities among the trigonometric functions:

basic identities

(19-1) $$\sin \theta = \frac{1}{\csc \theta} \qquad \cos \theta = \frac{1}{\sec \theta}$$ (19-2)

(19-3) $$\tan \theta = \frac{1}{\cot \theta} \qquad \tan \theta = \frac{\sin \theta}{\cos \theta}$$ (19-4)

(19-5) $$\cot \theta = \frac{\cos \theta}{\sin \theta} \qquad \sin^2 \theta + \cos^2 \theta = 1$$ (19-6)

(19-7) $$1 + \tan^2 \theta = \sec^2 \theta \qquad 1 + \cot^2 \theta = \csc^2 \theta$$ (19-8)

In using these identities, θ may stand for any angle or number or expression representing an angle or number.

EXAMPLE A

1. $\sin (x + 1) = \dfrac{1}{\csc (x + 1)}$ using Eq. (19-1)

2. $\tan 157° = \dfrac{\sin 157°}{\cos 157°}$ using Eq. (19-4)

3. $\sin^2 \left(\dfrac{\pi}{4}\right) + \cos^2 \left(\dfrac{\pi}{4}\right) = 1$ using Eq. (19-6)

EXAMPLE B

We shall check the last two illustrations of Example A for the particular values of θ which are used.

Using a calculator, we find that

$$\sin 157° = 0.3907311$$

and

$$\cos 157° = -0.9205049$$

Considering Eq. (19-4) and the second illustration of Example A, and dividing, we find that

$$\frac{\sin 157°}{\cos 157°} = \frac{0.3907311}{-0.9205049} = -0.4244748$$

We also find that $\tan 157° = -0.4244748$, which shows that

$$\tan 157° = \frac{\sin 157°}{\cos 157°}$$

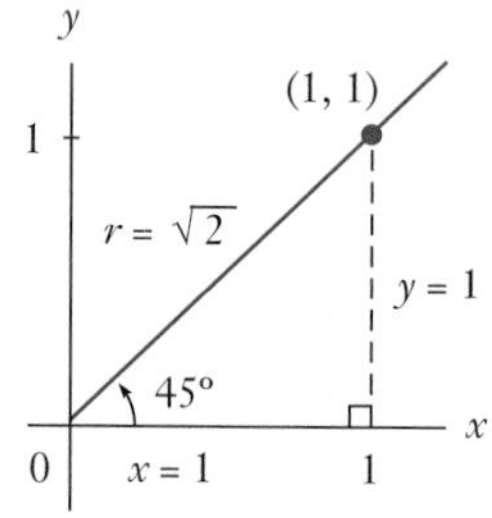

To check the third illustration of Example A, we refer to the values obtained in Example B of Section 3-3, or to Fig. 19-2. These tell us that

$$\sin 45° = \frac{1}{\sqrt{2}} = \frac{\sqrt{2}}{2} \quad \text{and} \quad \cos 45° = \frac{\sqrt{2}}{2}$$

Since $\frac{\pi}{4} = 45°$, by adding the squares of $\sin 45°$ and $\cos 45°$, we have

$$\sin^2 \left(\frac{\pi}{4}\right) + \cos^2 \left(\frac{\pi}{4}\right) = \left(\frac{\sqrt{2}}{2}\right)^2 + \left(\frac{\sqrt{2}}{2}\right)^2 = \frac{1}{2} + \frac{1}{2} = 1$$

Fig. 19-2

We see that this checks with Eq. (19-6) for these values.

A great many identities exist among the trigonometric functions. We are going to use the basic identities already developed in Eqs. (19-1) through (19-8), along with a few additional ones developed in later sections, to prove the validity of still other identities. *The ability to prove such identities depends to a large extent on being very familiar with the basic identities,* so that you can *recognize them in somewhat different forms.* If you do not learn these basic identities and learn them well, you will have difficulty in following the examples and doing the

NOTE ▷

exercises. The more readily you recognize these forms, the more easily you will be able to prove such identities.

In proving identities, we should look for combinations which appear in, or are very similar to, those in the basic identities. Consider the following examples.

EXAMPLE C
In proving the identity

$$\sin x = \frac{\cos x}{\cot x}$$

we know that $\cot x = \frac{\cos x}{\sin x}$. Since $\sin x$ appears on the left, substituting for $\cot x$ on the right will eliminate $\cot x$ and introduce $\sin x$. This should help us proceed in proving the identity. Thus,

$$\sin x = \frac{\cos x}{\cot x} = \frac{\cos x}{\dfrac{\cos x}{\sin x}} = \frac{\cos x}{1} \times \frac{\sin x}{\cos x}$$

Eq. (19-5)   invert

$$= \sin x \qquad \text{cancel } \cos x \text{ factors}$$

By showing that the right side may be changed exactly to $\sin x$, the expression on the left side, we have proved the identity.

Some important points should be made in relation to the proof of the identity of Example C. We must recognize what basic identities may be useful. The proof of an identity requires the use of basic algebraic operations, and these must be done carefully and correctly. Although in Example C we changed the right side to the form on the left, we could have changed the left to the form on the right. From this, and the fact that various substitutions are possible, we see that a variety of procedures can be used to prove any given identity.

EXAMPLE D
Prove that $\tan \theta \csc \theta = \sec \theta$.

In proving this identity, we know that $\tan \theta = \frac{\sin \theta}{\cos \theta}$ and also that $\frac{1}{\cos \theta} = \sec \theta$. Thus, by substituting for $\tan \theta$ we introduce $\cos \theta$ in the denominator, which is equivalent to introducing $\sec \theta$ in the numerator. Therefore, changing only the left side, we have

$$\tan \theta \csc \theta = \frac{\sin \theta}{\cos \theta} \csc \theta = \frac{\sin \theta}{\cos \theta} \frac{1}{\sin \theta}$$

Eq. (19-4)   Eq. (19-1)

$$= \frac{1}{\cos \theta} \qquad \text{cancel } \sin \theta \text{ factors}$$

$$= \sec \theta \qquad \text{using Eq. (19-2)}$$

Having changed the left side into the form of the right side, we have proven the identity.

(Continued on next page)

Many variations of the preceding steps are possible. Also, we could have changed the right side to obtain the form on the left. For example,

$$\tan \theta \csc \theta = \sec \theta = \frac{1}{\cos \theta} \qquad \text{using Eq. (19-2)}$$

$$= \frac{\sin \theta}{\cos \theta \sin \theta} = \frac{\sin \theta}{\cos \theta} \frac{1}{\sin \theta} \qquad \begin{array}{l}\text{multiply numerator and}\\ \text{denominator by } \sin \theta \text{ and rewrite}\end{array}$$

$$= \tan \theta \csc \theta \qquad \text{using Eqs. (19-4) and (19-1)}$$

In proving the identities of Examples C and D we have shown that the expression on one side of the equals sign can be changed into the expression on the other side. Although making the restriction that we change only one side is not entirely necessary, *we shall restrict the method of proof to changing only one side into the same form as the other side.* In this way, we know precisely to which form we are to change, and therefore by looking ahead we are better able to make the proper changes.

NOTE ▷ There is no set procedure which can be stated for working with identities. The most important factors are to be able to **recognize the proper forms,** to be able to *see what effect any change may have* before we actually perform it, and then to *perform it correctly.* Normally *it is easier to change the more complicated side of an identity to the same form as the less complicated side.* If the two sides are of approximately the same complexity, a close look at each side usually suggests steps which will lead to the solution.

EXAMPLE E ——— Prove the identity $\dfrac{\cos x \csc x}{\cot^2 x} = \tan x$.

First, we note that the left-hand side has several factors and the right-hand side has only one. Therefore, let us transform the left-hand side. Next, we note that we want $\tan x$ as the final result. We know that $\cot x = 1/\tan x$. Thus

$$\frac{\cos x \csc x}{\cot^2 x} = \frac{\cos x \csc x}{\dfrac{1}{\tan^2 x}} = \cos x \csc x \tan^2 x$$

At this point, we have two factors of $\tan x$ on the left. Since we want only one, let us factor out one. Therefore,

$$\cos x \csc x \tan^2 x = \tan x (\cos x \csc x \tan x)$$

Now, replacing $\tan x$ within the parentheses by $\sin x / \cos x$, we have

$$\tan x (\cos x \csc x \tan x) = \frac{\tan x (\cos x \csc x \sin x)}{\cos x}$$

Now we may cancel $\cos x$. Also, $\csc x \sin x = 1$ from Eq. (19-1). Finally,

$$\frac{\tan x (\cos x \csc x \sin x)}{\cos x} = \tan x \left(\frac{\cos x}{\cos x}\right)(\csc x \sin x)$$

$$= \tan x (1)(1) = \tan x$$

Since we have transformed the left-hand side into tan x, we have proven the identity. Of course, it is not necessary to rewrite expressions as we did in this example. This was done here only to include the explanations between steps. ■

EXAMPLE F — In determining the rate of radiation by an accelerated electric charge, it is necessary to show that $\sin^3 \theta = \sin \theta - \cos^2 \theta \sin \theta$. Show that this is valid by transforming the left-hand side.

Since each term on the right has a factor of $\sin \theta$, we see that we can proceed by writing $\sin^3 \theta$ as $\sin \theta(\sin^2 \theta)$. Then the factor $\sin^2 \theta$, and the $\cos^2 \theta$ on the right, suggest the use of Eq. (19-6). Thus we have

$$\sin^3 \theta = \sin \theta(\sin^2 \theta) = \sin \theta(1 - \cos^2 \theta)$$

$$= \sin \theta - \sin \theta \cos^2 \theta \quad \text{multiplying}$$

Since we wanted to substitute for $\sin^2 \theta$, we used Eq. (19-6) in the form

$$\sin^2 \theta = 1 - \cos^2 \theta. \quad ■$$

EXAMPLE G — Prove the identity $\dfrac{\sec^2 y}{\cot y} - \tan^3 y = \tan y$.

Here we shall simplify the left side. We note that we can remove cot y from the denominator, since $\cot y = 1/\tan y$. Also, the presence of $\sec^2 y$ suggests the use of Eq. (19-7). Therefore, we have

$$\frac{\sec^2 y}{\cot y} - \tan^3 y = \frac{\sec^2 y}{\dfrac{1}{\tan y}} - \tan^3 y = \sec^2 y \tan y - \tan^3 y$$

$$= \tan y(\sec^2 y - \tan^2 y) = \tan y(1)$$

$$= \tan y$$

Here we have used Eq. (19-7) in the form $\sec^2 y - \tan^2 y = 1$. ■

EXAMPLE H — Prove the identity $\dfrac{1 - \sin x}{\sin x \cot x} = \dfrac{\cos x}{1 + \sin x}$.

The combination $1 - \sin x$ also suggests $1 - \sin^2 x$, since multiplying $(1 - \sin x)$ by $(1 + \sin x)$ gives $1 - \sin^2 x$, which can then be replaced by $\cos^2 x$. Thus, changing only the left side, we have

$$\frac{1 - \sin x}{\sin x \cot x} = \frac{(1 - \sin x)(1 + \sin x)}{\sin x \cot x(1 + \sin x)} \quad \begin{array}{l}\text{multiply numerator and} \\ \text{denominator by } 1 + \sin x\end{array}$$

$$= \frac{1 - \sin^2 x}{\sin x\left(\dfrac{\cos x}{\sin x}\right)(1 + \sin x)} = \frac{\cos^2 x}{\cos x(1 + \sin x)} \longleftarrow \text{Eq. (19-6)}$$

$$\text{cancel } \sin x$$

$$= \frac{\cos x}{1 + \sin x} \quad \text{cancel } \cos x \quad ■$$

EXAMPLE I _____ Prove the identity $\sec^2 x + \csc^2 x = \sec^2 x \csc^2 x$.

Here we note the presence of $\sec^2 x$ and $\csc^2 x$ on each side. This suggests the possible use of the square relationships. By replacing the $\sec^2 x$ on the right-hand side by $1 + \tan^2 x$, we can create $\csc^2 x$ plus another term. The left-hand side is the $\csc^2 x$ plus another term, so this procedure should help. Thus, changing only the right side,

$$\sec^2 x + \csc^2 x = \sec^2 x \csc^2 x$$

$$= (1 + \tan^2 x)(\csc^2 x) \qquad \text{using Eq. (19-7)}$$

$$= \csc^2 x + \tan^2 x \csc^2 x \qquad \text{multiplying}$$

$$= \csc^2 x + \left(\frac{\sin^2 x}{\cos^2 x}\right)\left(\frac{1}{\sin^2 x}\right) \qquad \text{using Eqs. (19-4) and (19-1)}$$

$$= \csc^2 x + \frac{1}{\cos^2 x} \qquad \text{cancel } \sin^2 x$$

$$= \csc^2 x + \sec^2 x \qquad \text{using Eq. (19-2)}$$

We could have used many other variations of this procedure, and they would have been perfectly valid. ■

EXAMPLE J _____ Prove the identity $\dfrac{\csc x}{\tan x + \cot x} = \cos x$.

Here we shall simplify the left-hand side until we have the expression which appears on the right-hand side.

$$\frac{\csc x}{\tan x + \cot x} = \frac{\csc x}{\tan x + \dfrac{1}{\tan x}} = \frac{\csc x}{\dfrac{\tan^2 x + 1}{\tan x}}$$

$$= \frac{\csc x \tan x}{\tan^2 x + 1} = \frac{\csc x \tan x}{\sec^2 x} = \frac{\dfrac{1}{\sin x}\dfrac{\sin x}{\cos x}}{\dfrac{1}{\cos^2 x}}$$

$$= \frac{1}{\sin x}\frac{\sin x}{\cos x}\frac{\cos^2 x}{1}$$

$$= \cos x$$

Therefore, we have shown that $\dfrac{\csc x}{\tan x + \cot x} = \cos x$, which proves the identity. ■

Exercises 19-1

In Exercises 1 through 4, check the indicated basic identities for the given angles.

1. Check Eq. (19-3) for $\theta = 56°$.

2. Check Eq. (19-5) for $\theta = 280°$.

3. Check Eq. (19-6) for $\theta = \dfrac{4\pi}{3}$.

4. Check Eq. (19-7) for $\theta = \dfrac{5\pi}{6}$.

In Exercises 5 through 56, prove the given identities.

5. $\dfrac{\cot \theta}{\cos \theta} = \csc \theta$

6. $\dfrac{\tan y}{\sin y} = \sec y$

7. $\dfrac{\sin x}{\tan x} = \cos x$

See Appendix E for a computer program for checking a trigonometric identity.

8. $\dfrac{\csc \theta}{\sec \theta} = \cot \theta$

9. $\sin y \cot y = \cos y$

10. $\cos x \tan x = \sin x$

11. $\sin x \sec x = \tan x$

12. $\cot \theta \sec \theta = \csc \theta$

13. $\csc^2 x (1 - \cos^2 x) = 1$

14. $\cos^2 x (1 + \tan^2 x) = 1$

15. $\sin x (1 + \cot^2 x) = \csc x$

16. $\sec \theta (1 - \sin^2 \theta) = \cos \theta$

17. $\sin x (\csc x - \sin x) = \cos^2 x$

18. $\cos y (\sec y - \cos y) = \sin^2 y$

19. $\tan y (\cot y + \tan y) = \sec^2 y$

20. $\csc x (\csc x - \sin x) = \cot^2 x$

21. $\sin x \tan x + \cos x = \sec x$

22. $\sec x \csc x - \cot x = \tan x$

23. $\cos \theta \cot \theta + \sin \theta = \csc \theta$

24. $\csc x \sec x - \tan x = \cot x$

25. $\sec \theta \tan \theta \csc \theta = \tan^2 \theta + 1$

26. $\sin x \cos x \tan x = 1 - \cos^2 x$

27. $\cot \theta \sec^2 \theta - \cot \theta = \tan \theta$

28. $\sin y + \sin y \cot^2 y = \csc y$

29. $\tan x + \cot x = \sec x \csc x$

30. $\tan x + \cot x = \tan x \csc^2 x$

31. $\cos^2 x - \sin^2 x = 1 - 2 \sin^2 x$

32. $\tan^2 y \sec^2 y - \tan^4 y = \tan^2 y$

33. $\dfrac{\sin x}{1 - \cos x} = \csc x + \cot x$

34. $\dfrac{1 + \cos x}{\sin x} = \dfrac{\sin x}{1 - \cos x}$

35. $\dfrac{\sec x + \csc x}{1 + \tan x} = \csc x$

36. $\dfrac{\cot x + 1}{\cot x} = 1 + \tan x$

37. $\tan^2 x \cos^2 x + \cot^2 x \sin^2 x = 1$

38. $\dfrac{\sin \theta}{\csc \theta} + \dfrac{\cos \theta}{\sec \theta} = 1$

39. $\dfrac{\sec \theta}{\cos \theta} - \dfrac{\tan \theta}{\cot \theta} = 1$

40. $\dfrac{\csc \theta}{\sin \theta} - \dfrac{\cot \theta}{\tan \theta} = 1$

41. $\dfrac{1 - 2 \cos^2 x}{\sin x \cos x} = \tan x - \cot x$

42. $4 \sin x + \tan x = \dfrac{4 + \sec x}{\csc x}$

43. $\cos^3 x \csc^3 x \tan^3 x = \csc^2 x - \cot^2 x$

44. $\dfrac{1 + \tan x}{\sin x} - \sec x = \csc x$

45. $\sec x + \tan x + \cot x = \dfrac{1 + \sin x}{\cos x \sin x}$

46. $\csc \theta (\csc \theta - \cot \theta) = \dfrac{1}{1 + \cos \theta}$

47. $\dfrac{\cos \theta + \sin \theta}{1 + \tan \theta} = \cos \theta$

48. $\dfrac{\sec x - \cos x}{\tan x} = \sin x$

49. $(\tan x + \cot x) \sin x \cos x = 1$

50. $2 \sin^4 x - 3 \sin^2 x + 1 = \cos^2 x (1 - 2 \sin^2 x)$

51. $\dfrac{\sin^4 x - \cos^4 x}{1 - \cot^4 x} = \sin^4 x$

52. $\dfrac{1}{2} \sin 5y \left(\dfrac{\sin 5y}{1 - \cos 5y} + \dfrac{1 - \cos 5y}{\sin 5y} \right) = 1$

53. $\sec x (\sec x - \cos x) + \dfrac{\cos x - \sin x}{\cos x} + \tan x = \sec^2 x$

54. $\dfrac{\cot 2y}{\sec 2y - \tan 2y} - \dfrac{\cos 2y}{\sec 2y + \tan 2y} = \sin 2y + \csc 2y$

55. $1 + \sin^2 x + \sin^4 x + \cdots = \sec^2 x$

56. $1 - \tan^2 x + \tan^4 x - \cdots = \cos^2 x$

In Exercises 57 through 60, solve the given problems involving trigonometric identities.

57. When designing a solar energy collector, it is necessary to account for the latitude and longitude of the location, the angle of the sun, and the angle of the collector. In doing this, the equation

$$\cos \theta = \cos A \cos B \cos C + \sin A \sin B$$

is used. If $\theta = 90°$, show that $\cos C = -\tan A \tan B$.

58. In determining the fit of two cylindrical pipes joined at right angles, the expression $(r \sin \theta)^2 + (R - r)(R + r)$ is used. Show that this expression can be written as $R^2 - r^2 \cos^2 \theta$.

59. Show that the length ℓ of the straight brace shown in Fig. 19-3 can be found from the equation

$$\ell = \frac{a(1 + \tan \theta)}{\sin \theta}$$

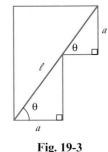

60. In determining the path of least time between two points under certain circumstances, it is necessary to show that

$$\sqrt{\frac{1 + \cos \theta}{1 - \cos \theta}} \sin \theta = 1 + \cos \theta$$

Show this by transforming the left-hand side.

Fig. 19-3

In Exercises 61 through 64, use the given substitutions to show that the given equations are valid. In each,

$$0 < \theta < \frac{\pi}{2}.$$

61. If $x = \cos \theta$, show that $\sqrt{1 - x^2} = \sin \theta$.

62. If $x = 3 \sin \theta$, show that $\sqrt{9 - x^2} = 3 \cos \theta$.

63. If $x = 2 \tan \theta$, show that $\sqrt{4 + x^2} = 2 \sec \theta$.

64. If $x = 4 \sec \theta$, show that $\sqrt{x^2 - 16} = 4 \tan \theta$.

19-2 Sine and Cosine of the Sum and Difference of Two Angles

There are other important relations among the trigonometric functions. The most important and useful relations are those which involve twice an angle and half an angle. To obtain these relations, we shall first derive the expressions for the sine and cosine of the sum and difference of two angles. These expressions will lead directly to the desired relations of double and half angles.

Equation (11-12) gives the polar (or trigonometric) form of the product of two complex numbers. We can use this formula to derive the expressions for the sine and the cosine of the sum of two angles, $\sin (\alpha + \beta)$ and $\cos (\alpha + \beta)$.

Using Eq. (11-12) for the complex numbers $\cos \alpha + j \sin \alpha$ and $\cos \beta + j \sin \beta$, which are represented in Fig. 19-4, we have

$$(\cos \alpha + j \sin \alpha)(\cos \beta + j \sin \beta) = \cos (\alpha + \beta) + j \sin (\alpha + \beta)$$

Expanding the left side, and then switching sides, we have

$$\cos (\alpha + \beta) + j \sin (\alpha + \beta) = (\cos \alpha \cos \beta - \sin \alpha \sin \beta) + j(\sin \alpha \cos \beta + \cos \alpha \sin \beta)$$

Since two complex numbers are equal if their real parts are equal and their imaginary parts are equal, we have

Fig. 19-4

$$\sin (\alpha + \beta) = \sin \alpha \cos \beta + \cos \alpha \sin \beta \tag{19-9}$$

sum identities

and

$$\cos (\alpha + \beta) = \cos \alpha \cos \beta - \sin \alpha \sin \beta \tag{19-10}$$

EXAMPLE A _____ Find $\sin 75°$ from $\sin 75° = \sin (45° + 30°)$.

$\sin 75° = \sin (45° + 30°) = \sin 45° \cos 30° + \cos 45° \sin 30°$ using Eq. (19-9)

$$= \frac{\sqrt{2}}{2} \times \frac{\sqrt{3}}{2} + \frac{\sqrt{2}}{2} \times \frac{1}{2} = \frac{\sqrt{6}}{4} + \frac{\sqrt{2}}{4} = \frac{\sqrt{6} + \sqrt{2}}{4}$$ for values, see Section 3-3

$= 0.9659$

EXAMPLE B _____ Verify that $\sin 90° = 1$, by finding $\sin (60° + 30°)$.

$\sin 90° = \sin (60° + 30°) = \sin 60° \cos 30° + \cos 60° \sin 30°$ using Eq. (19-9)

$$= \frac{\sqrt{3}}{2} \times \frac{\sqrt{3}}{2} + \frac{1}{2} \times \frac{1}{2} = \frac{3}{4} + \frac{1}{4} = 1$$ for values, see Section 3-3

We see that this agrees with the result found in Section 7-2.

NOTE ▷   It should be obvious from this example that **$\sin (\alpha + \beta)$ *is* not *equal to* $\sin \alpha + \sin \beta$,** something which many students assume before they are familiar with the formulas and ideas of this section. If we used such a formula, we would get $\sin 90° = \frac{1}{2}\sqrt{3} + \frac{1}{2} = 1.366$ for the combination $(60° + 30°)$. This is not possible, since the values of the sine never exceed 1 in value. Also, if we used the combination $(45° + 45°)$, we would get 1.414, a different value for the same number, $\sin 90°$.

    From Eqs. (19-9) and (19-10), we can easily find expressions for $\sin (\alpha - \beta)$ and $\cos (\alpha - \beta)$. This is done by finding $\sin (\alpha + (-\beta))$ and $\cos (\alpha + (-\beta))$. Thus, we have

$$\sin (\alpha - \beta) = \sin (\alpha + (-\beta)) = \sin \alpha \cos (-\beta) + \cos \alpha \sin (-\beta)$$

Since $\cos (-\beta) = \cos \beta$ and $\sin (-\beta) = -\sin \beta$ (see Exercise 53 of Section 7-2), we have

$$\boxed{\sin (\alpha - \beta) = \sin \alpha \cos \beta - \cos \alpha \sin \beta} \qquad (19\text{-}11)$$

difference identities   In the same manner, we find that

$$\boxed{\cos (\alpha - \beta) = \cos \alpha \cos \beta + \sin \alpha \sin \beta} \qquad (19\text{-}12)$$

EXAMPLE C _____ Find $\cos 15°$ from $\cos (45° - 30°)$.

$\cos 15° = \cos (45° - 30°) = \cos 45° \cos 30° + \sin 45° \sin 30°$ using Eq. (19-12)

$$= \frac{\sqrt{2}}{2} \times \frac{\sqrt{3}}{2} + \frac{\sqrt{2}}{2} \times \frac{1}{2} = \frac{\sqrt{6} + \sqrt{2}}{4} = 0.9659$$

We get the same result as in Example A, which should be the case since $\sin 75° = \cos 15°$. (See Section 3-4.)

EXAMPLE D _____ Reduce $\sin x \cos (x - y) - \cos x \sin (x - y)$ to a single term.

We could expand $\cos (x - y)$ and $\sin (x - y)$ and simplify. This would be tedious, and it is unnecessary. This expression fits the form of the right side of Eq. (19-11) with $\alpha = x$ and $\beta = x - y$. Thus,

$$\sin x \cos (x - y) - \cos x \sin (x - y) = \sin [x - (x - y)]$$
$$= \sin y$$

∎

EXAMPLE E _____ Evaluate $\cos 23° \cos 67° - \sin 23° \sin 67°$.

We note that this expression fits the form of the right side of Eq. (19-10), so

$$\cos 23° \cos 67° - \sin 23° \sin 67° = \cos (23° + 67°)$$
$$= \cos 90°$$
$$= 0$$

NOTE▷ Again, we are able to evaluate this by **recognizing the form** of the given expres-
∎ sion. Evaluation by a calculator will verify the result.

By using Eqs. (19-9) and (19-10), we can determine expressions for $\tan (\alpha + \beta)$, $\cot (\alpha + \beta)$, $\sec (\alpha + \beta)$, and $\csc (\alpha + \beta)$. These expressions are used less than those for the sine and cosine, and therefore we shall not derive them here, although the expression for $\tan (\alpha + \beta)$ is found in the exercises at the end of this section. By using Eqs. (19-11) and (19-12), we can find similar expressions for the functions of $(\alpha - \beta)$.

Certain trigonometric identities can also be worked out by using the formulas derived in this section. The following examples illustrate the use of these formulas in identities.

EXAMPLE F _____ Prove that $\sin (180° + x) = -\sin x$.

By using Eq. (19-9) we have

$$\sin (180° + x) = \sin 180° \cos x + \cos 180° \sin x$$

Since $\sin 180° = 0$ and $\cos 180° = -1$, we have

$$\sin 180° \cos x + \cos 180° \sin x = (0) \cos x + (-1) \sin x$$

or

$$\sin (180° + x) = -\sin x$$

Although x may or may not be an acute angle, we see that this agrees with the results in Section 7-2 for the sine of a third-quadrant angle if x is acute. See Fig. 19-5.

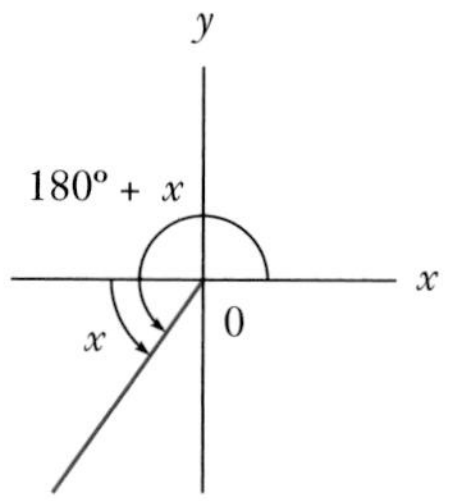

Fig. 19-5

EXAMPLE G _____ Show that $\dfrac{\sin (\alpha - \beta)}{\sin \alpha \sin \beta} = \cot \beta - \cot \alpha$.

$$\frac{\sin (\alpha - \beta)}{\sin \alpha \sin \beta} = \frac{\sin \alpha \cos \beta - \cos \alpha \sin \beta}{\sin \alpha \sin \beta} \qquad \text{using Eq. (19-11)}$$

$$= \frac{\sin \alpha \cos \beta}{\sin \alpha \sin \beta} - \frac{\cos \alpha \sin \beta}{\sin \alpha \sin \beta}$$

$$= \frac{\cos \beta}{\sin \beta} - \frac{\cos \alpha}{\sin \alpha} = \cot \beta - \cot \alpha \qquad \text{using Eq. (19-5)}$$

■

EXAMPLE H ———— Show that $\sin \left(\dfrac{\pi}{4} + x \right) \cos \left(\dfrac{\pi}{4} + x \right) = \dfrac{1}{2} (\cos^2 x - \sin^2 x)$. The solution is as follows:

$$\sin \left(\frac{\pi}{4} + x \right) \cos \left(\frac{\pi}{4} + x \right) = \left(\sin \frac{\pi}{4} \cos x + \cos \frac{\pi}{4} \sin x \right) \left(\cos \frac{\pi}{4} \cos x - \sin \frac{\pi}{4} \sin x \right) \qquad \begin{array}{l}\text{using Eqs.}\\ \text{(19-9) and (19-10)}\end{array}$$

$$= \sin \frac{\pi}{4} \cos \frac{\pi}{4} \cos^2 x - \sin^2 \frac{\pi}{4} \sin x \cos x + \cos^2 \frac{\pi}{4} \sin x \cos x - \sin^2 x \sin \frac{\pi}{4} \cos \frac{\pi}{4} \qquad \text{expanding}$$

$$= \frac{\sqrt{2}}{2} \frac{\sqrt{2}}{2} \cos^2 x - \left(\frac{\sqrt{2}}{2} \right)^2 \sin x \cos x + \left(\frac{\sqrt{2}}{2} \right)^2 \sin x \cos x - \frac{\sqrt{2}}{2} \frac{\sqrt{2}}{2} \sin^2 x \qquad \text{evaluating}$$

$$= \frac{1}{2} \cos^2 x - \frac{1}{2} \sin^2 x = \frac{1}{2} (\cos^2 x - \sin^2 x)$$

EXAMPLE I ———— In analyzing the motion of an object at the end of a spring, it is stated that

$$\sin \omega t = \sin (\omega t + \alpha) \cos \alpha - \cos (\omega t + \alpha) \sin \alpha$$

Show that this is correct.

　　　If we let $x = \omega t + \alpha$, we note that the right side of the equation becomes $\sin x \cos \alpha - \cos x \sin \alpha$, which is the form for $\sin (x - \alpha)$. By replacing x with $\omega t + \alpha$, we obtain $\sin (\omega t + \alpha - \alpha)$, which is $\sin \omega t$. These steps (with $x = \omega t + \alpha$) are shown below.

$$\sin \omega t = \sin x \cos \alpha - \cos x \sin \alpha = \sin (x - \alpha)$$

$$= \sin (\omega t + \alpha - \alpha) = \sin \omega t$$

Therefore, the original equation has been shown to be true.

　　　We again see that the proper recognition of a basic form leads to the solution.   ■

Exercises 19-2

In Exercises 1 through 4, determine the values of the given functions as indicated.

1. Find $\sin 105°$ by using $105° = 60° + 45°$.

2. Find $\cos 75°$ by using $75° = 30° + 45°$.

3. Find $\cos 15°$ by using $15° = 60° - 45°$.

4. Find $\sin 15°$ by using $15° = 45° - 30°$.

In Exercises 5 through 8, evaluate the indicated functions with the following given information: $\sin \alpha = \frac{4}{5}$ (in first quadrant) and $\cos \beta = -\frac{12}{13}$ (in second quadrant).

5. $\sin (\alpha + \beta)$ **6.** $\cos (\beta - \alpha)$ **7.** $\cos (\alpha + \beta)$ **8.** $\sin (\alpha - \beta)$

In Exercises 9 through 16, reduce each of the given expressions to a single term. Expansion of any term is not necessary; proper recognition of the form of the expression leads to the proper result.

9. $\sin x \cos 2x + \sin 2x \cos x$

10. $\sin 3x \cos x - \sin x \cos 3x$

11. $\cos(x + y)\cos y + \sin(x + y)\sin y$

12. $\cos(2x - y)\cos y - \sin(2x - y)\sin y$

13. $\cos 1 \cos(1 - x) - \sin 1 \sin(1 - x)$

14. $\sin x \cos(x + 1) + \cos x \sin(x + 1)$

15. $\sin 3x \cos(3x - \pi) - \cos 3x \sin(3x - \pi)$

16. $\cos(x + \pi)\cos(x - \pi) + \sin(x + \pi)\sin(x - \pi)$

In Exercises 17 through 20, evaluate each of the given expressions. Proper recognition of the given form leads to the result. Verify each result by use of a calculator.

17. $\sin 122° \cos 32° - \cos 122° \sin 32°$

18. $\cos 250° \cos 70° + \sin 250° \sin 70°$

19. $\cos 312° \cos 48° - \sin 312° \sin 48°$

20. $\sin 56° \cos 124° + \cos 56° \sin 124°$

In Exercises 21 through 36, prove the given identities.

21. $\sin(180° - x) = \sin x$

22. $\cos(180° - x) = -\cos x$

23. $\cos(-x) = \cos x$ (*Hint:* $-x = 0 - x$.)

24. $\sin(-x) = -\sin x$

25. $\sin(270° - x) = -\cos x$

26. $\sin(90° + x) = \cos x$

27. $\cos\left(\dfrac{\pi}{2} - x\right) = \sin x$

28. $\cos\left(\dfrac{3\pi}{2} + x\right) = \sin x$

29. $\cos(30° + x) = \dfrac{\sqrt{3}\cos x - \sin x}{2}$

30. $\sin(120° - x) = \dfrac{\sqrt{3}\cos x + \sin x}{2}$

31. $\sin\left(\dfrac{\pi}{4} + x\right) = \dfrac{\sin x + \cos x}{\sqrt{2}}$

32. $\cos\left(\dfrac{\pi}{3} + x\right) = \dfrac{\cos x - \sqrt{3}\sin x}{2}$

33. $\sin(x + y)\sin(x - y) = \sin^2 x - \sin^2 y$

34. $\cos(x + y)\cos(x - y) = \cos^2 x - \sin^2 y$

35. $\cos(\alpha + \beta) + \cos(\alpha - \beta) = 2\cos\alpha\cos\beta$

36. $\cos(x - y) + \sin(x + y) = (\cos x + \sin x)(\cos y + \sin y)$

In Exercises 37 through 40, additional trigonometric identities are shown. Derive these in the indicated manner. Equations (19-14), (19-15), and (19-16) are known as the product formulas.

37. By dividing Eq. (19-9) by Eq. (19-10), show that

$$\tan(\alpha + \beta) = \frac{\tan\alpha + \tan\beta}{1 - \tan\alpha\tan\beta} \tag{19-13}$$

(*Hint:* Divide numerator and denominator by $\cos\alpha\cos\beta$.)

38. By adding Eqs. (19-9) and (19-11), derive the equation

$$\sin\alpha\cos\beta = \tfrac{1}{2}[\sin(\alpha + \beta) + \sin(\alpha - \beta)] \tag{19-14}$$

39. By adding Eqs. (19-10) and (19-12), derive the equation

$$\cos\alpha\cos\beta = \tfrac{1}{2}[\cos(\alpha + \beta) + \cos(\alpha - \beta)] \tag{19-15}$$

40. By subtracting Eq. (19-10) from Eq. (19-12), derive

$$\sin\alpha\sin\beta = \tfrac{1}{2}[\cos(\alpha - \beta) - \cos(\alpha + \beta)] \tag{19-16}$$

In Exercises 41 through 44, additional trigonometric identities are shown. Derive them by letting $\alpha + \beta = x$ and $\alpha - \beta = y$, which leads to $\alpha = \tfrac{1}{2}(x + y)$ and $\beta = \tfrac{1}{2}(x - y)$. The resulting equations are known as the factor formulas.

41. Use Eq. (19-14) and the substitutions above to derive the equation

$$\sin x + \sin y = 2\sin\tfrac{1}{2}(x + y)\cos\tfrac{1}{2}(x - y) \tag{19-17}$$

42. Use Eqs. (19-9) and (19-11) and the substitutions above to derive the equation

$$\sin x - \sin y = 2\sin\tfrac{1}{2}(x - y)\cos\tfrac{1}{2}(x + y) \tag{19-18}$$

43. Use Eq. (19-15) and the substitutions above to derive the equation

$$\cos x + \cos y = 2\cos\tfrac{1}{2}(x + y)\cos\tfrac{1}{2}(x - y) \tag{19-19}$$

44. Use Eq. (19-16) and the substitutions above to derive the equation

$$\cos x - \cos y = -2 \sin \tfrac{1}{2}(x + y) \sin \tfrac{1}{2}(x - y) \qquad (19\text{-}20)$$

In Exercises 45 through 48, use the equations of this section to solve the given problems.

45. An alternating electric current i is given by the equation $i = i_0 \sin (\omega t + \alpha)$. Show that this can be written as $i = i_1 \sin \omega t + i_2 \cos \omega t$, where $i_1 = i_0 \cos \alpha$ and $i_2 = i_0 \sin \alpha$.

46. A weight w is held in equilibrium by forces F and T as shown in Fig. 19-6. Equations relating w, F, and T are

$$F \cos \theta = T \sin \alpha, \qquad w + F \sin \theta = T \cos \alpha$$

Show that $w = \dfrac{T \cos (\theta + \alpha)}{\cos \theta}$.

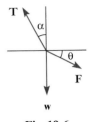

w

Fig. 19-6

47. For the two bevel gears shown in Fig. 19-7, the equation $\tan \alpha = \dfrac{\sin \beta}{R + \cos \beta}$ is

used. Here R is the ratio of gear 1 to gear 2. Show that $R = \dfrac{\sin (\beta - \alpha)}{\sin \alpha}$.

48. In the analysis of the angles of incidence i and reflection r of a light ray subject to certain conditions, the following expression is found:

$$E_2 \left(\frac{\tan r}{\tan i} + 1 \right) = E_1 \left(\frac{\tan r}{\tan i} - 1 \right)$$

Show that an equivalent expression is

$$E_2 = E_1 \frac{\sin (r - i)}{\sin (r + i)}$$

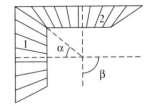

Fig. 19-7

19-3   Double-Angle Formulas

If we let $\beta = \alpha$ in Eqs. (19-9) and (19-10), we can derive the important double-angle formulas. Thus, by making this substitution in Eq. (19-9), we have

$$\sin (\alpha + \alpha) = \sin (2\alpha) = \sin \alpha \cos \alpha + \cos \alpha \sin \alpha = 2 \sin \alpha \cos \alpha$$

Using the same substitution in Eq. (19-10), we have

$$\cos (\alpha + \alpha) = \cos \alpha \cos \alpha - \sin \alpha \sin \alpha = \cos^2 \alpha - \sin^2 \alpha$$

By using the basic identity (19-6), other forms of this last equation may be derived. Thus, summarizing these formulas, we have

double-angle formulas

$$\sin 2\alpha = 2 \sin \alpha \cos \alpha \qquad (19\text{-}21)$$
$$\cos 2\alpha = \cos^2 \alpha - \sin^2 \alpha \qquad (19\text{-}22)$$
$$= 2 \cos^2 \alpha - 1 \qquad (19\text{-}23)$$
$$= 1 - 2 \sin^2 \alpha \qquad (19\text{-}24)$$

We should note carefully that these equations give expressions for the sine and cosine of twice an angle in terms of functions of the angle. They can be used any time we have expressed one angle as twice another. These double-angle formulas are widely used in applications of trigonometry, especially in the calculus. They should be known and recognized quickly in any of the various forms.

EXAMPLE A

1. If $\alpha = 30°$, we have

$$\cos 60° = \cos 2(30°) = \cos^2 30° - \sin^2 30° \qquad \text{using Eq. (19-22)}$$

2. If $\alpha = 3x$, we have

$$\sin 6x = \sin 2(3x) = 2 \sin 3x \cos 3x \qquad \text{using Eq. (19-21)}$$

3. If $2\alpha = x$, we may write $\alpha = x/2$, which means that

$$\sin x = \sin 2\left(\frac{x}{2}\right) = 2 \sin \frac{x}{2} \cos \frac{x}{2} \qquad \text{using Eq. (19-21)}$$

EXAMPLE B

Using the double-angle formulas, simplify the expression $\cos^2 2x - \sin^2 2x$.

Since this is the difference of the square of the cosine of an angle and the square of the sine of the same angle, it fits the right side of Eq. (19-22). Therefore, letting $\alpha = 2x$, we have

$$\cos^2 2x - \sin^2 2x = \cos 2(2x) = \cos 4x$$

EXAMPLE C

See the chapter introduction.

To find the area A of a right triangular piece of land, a surveyor may use the formula $A = \frac{1}{4}c^2 \sin 2\theta$, where c is the hypotenuse and θ is *either* of the acute angles. Derive this formula.

In Fig. 19-8 we see that $\sin \theta = a/c$ and $\cos \theta = b/c$, which gives us

$$a = c \sin \theta \quad \text{and} \quad b = c \cos \theta$$

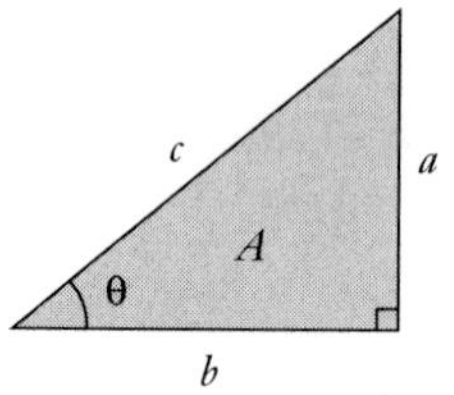

The area is given by $A = \frac{1}{2}ab$, which leads to the solution.

Fig. 19-8

$$A = \frac{1}{2} ab = \frac{1}{2} (c \sin \theta)(c \cos \theta)$$

$$= \frac{1}{2} c^2 \sin \theta \cos \theta = \frac{1}{2} c^2 \left(\frac{1}{2} \sin 2\theta\right) \qquad \text{using Eq. (19-21)}$$

$$= \frac{1}{4} c^2 \sin 2\theta$$

If we had labeled the upper acute angle in Fig. 19-8 as θ, we would have $a = c \cos \theta$ and $b = c \sin \theta$. By substituting these values into the formula for the area, the solution is the same.

EXAMPLE D

Verify the values of $\sin 90°$ and $\cos 90°$ by use of the functions of $45°$.

$$\sin 90° = \sin 2(45°) = 2 \sin 45° \cos 45° = 2\left(\frac{\sqrt{2}}{2}\right)\left(\frac{\sqrt{2}}{2}\right) = 1 \qquad \text{using Eq. (19-21)}$$

$$\cos 90° = \cos 2(45°) = 1 - 2 \sin^2 45° = 1 - 2\left(\frac{\sqrt{2}}{2}\right)^2 = 0 \qquad \text{using Eq. (19-24)}$$

We see that these results agree with those found in Section 7-2.

EXAMPLE E ———— Using a calculator, verify the values of sin 142° and cos 142° by use of the functions of 71°.

$$\sin 142° = 2 \sin 71° \cos 71° = 0.6156615 \qquad \text{using Eq. (19-21)}$$

This value is found by using the calculator sequence

$$2 \boxed{\times} 71 \boxed{\text{SIN}} \boxed{\times} 71 \boxed{\text{COS}} \boxed{=} \boxed{0.6156615}$$

or the sequence $142 \boxed{\text{SIN}}$ $\boxed{0.6156615}$

In the same way,

$$\cos 142° = \cos^2 71° - \sin^2 71° = -0.7880108 \qquad \text{using Eq. (19-22)}$$

is found by the calculator sequence

$$71 \boxed{\text{COS}} \boxed{x^2} \boxed{-} 71 \boxed{\text{SIN}} \boxed{x^2} \boxed{=} \boxed{-0.7880108} \qquad \text{using Eq. (19-22)}$$

or the sequence $142 \boxed{\text{COS}}$ $\boxed{-0.7880108}$

Therefore we see that each of the double-angle formulas is verified. ∎

EXAMPLE F ———— Given that $\cos \alpha = \frac{3}{5}$ (in the fourth quadrant), find sin 2α.

Knowing that $\cos \alpha = \frac{3}{5}$ for an angle in the fourth quadrant, we then determine from Fig. 19-9(a) that

$$\sin \alpha = -\frac{4}{5}$$

Therefore, we have

$$\sin 2\alpha = 2 \sin \alpha \cos \alpha \qquad \text{Eq. (19-21)}$$

$$= 2\left(-\frac{4}{5}\right)\left(\frac{3}{5}\right) = -\frac{24}{25}$$

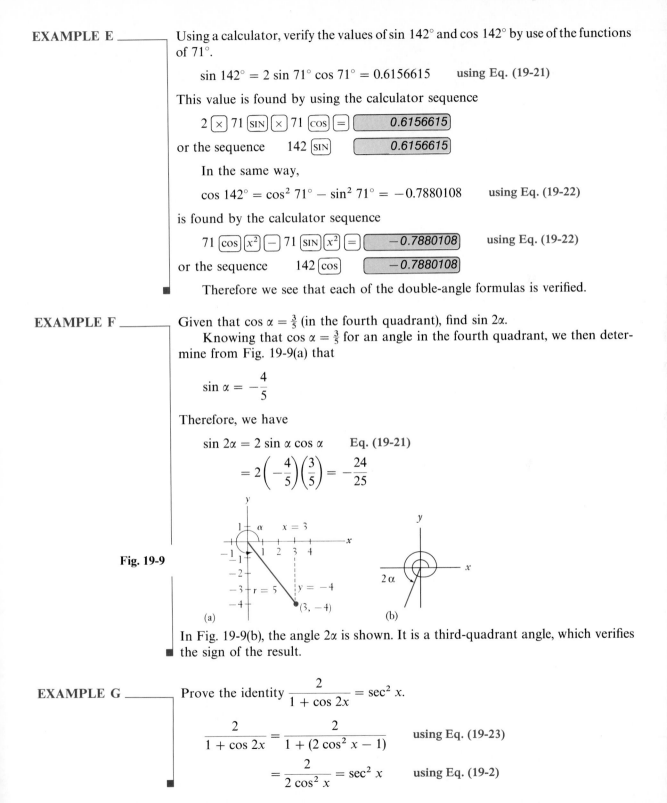

Fig. 19-9

(a)   (b)

In Fig. 19-9(b), the angle 2α is shown. It is a third-quadrant angle, which verifies the sign of the result. ∎

EXAMPLE G ———— Prove the identity $\dfrac{2}{1 + \cos 2x} = \sec^2 x$.

$$\frac{2}{1 + \cos 2x} = \frac{2}{1 + (2 \cos^2 x - 1)} \qquad \text{using Eq. (19-23)}$$

$$= \frac{2}{2 \cos^2 x} = \sec^2 x \qquad \text{using Eq. (19-2)}$$

EXAMPLE H ——— Show that $\dfrac{\sin 3x}{\sin x} + \dfrac{\cos 3x}{\cos x} = 4 \cos 2x$.

Since the left side is the more complex side, we will change it to the form on the right.

$$\frac{\sin 3x}{\sin x} + \frac{\cos 3x}{\cos x} = \frac{\sin 3x \cos x + \cos 3x \sin x}{\sin x \cos x} \qquad \text{combining fractions}$$

$$= \frac{\sin (3x + x)}{\frac{1}{2} \sin 2x} \qquad \begin{array}{l} \longleftarrow \text{ using Eq. (19-9)} \\ \longleftarrow \text{ using Eq. (19-21)} \end{array}$$

$$= \frac{2 \sin 4x}{\sin 2x} = \frac{2(2 \sin 2x \cos 2x)}{\sin 2x} \qquad \text{using Eq. (19-21)}$$

$$= 4 \cos 2x$$

■ Therefore, the expression is shown to be valid.

Exercises 19-3

In Exercises 1 through 4, determine the values of the indicated functions in the given manner.

1. Find $\sin 60°$ by using the functions of $30°$.

2. Find $\sin 120°$ by using the functions of $60°$.

3. Find $\cos 120°$ by using the functions of $60°$.

4. Find $\cos 60°$ by using the functions of $30°$.

In Exercises 5 through 8, use a calculator to verify the values found by use of the double-angle formulas as in Example E.

5. Find $\sin 258°$ directly and by using functions of $129°$.

6. Find $\sin 84°$ directly and by using functions of $42°$.

7. Find $\cos 96°$ directly and by using functions of $48°$.

8. Find $\cos 276°$ directly and by using functions of $138°$.

In Exercises 9 through 12, evaluate the indicated functions with the given information.

9. Find $\sin 2x$ if $\cos x = \frac{4}{5}$ (in first quadrant).

10. Find $\cos 2x$ if $\sin x = -\frac{12}{13}$ (in third quadrant).

11. Find $\cos 2x$ if $\tan x = -\frac{1}{2}$ (in second quadrant).

12. Find $\sin 4x$ if $\sin x = \frac{3}{5}$ (in first quadrant).

In Exercises 13 through 20, simplify the given expressions. Expansion of any term is not necessary; proper recognition of the form of the expression leads to the proper result.

13. $4 \sin 4x \cos 4x$

14. $4 \sin^2 x \cos^2 x$

15. $1 - 2 \sin^2 4x$

16. $\sin^2 4x - \cos^2 4x$

17. $2 \cos^2 \frac{1}{2}x - 1$

18. $2 \sin \frac{1}{2}x \cos \frac{1}{2}x$

19. $4 \sin^2 2x - 2$

20. $\cos 3x \sin 3x$

In Exercises 21 through 36, prove the given identities.

21. $\cos^2 \alpha - \sin^2 \alpha = 2 \cos^2 \alpha - 1$

22. $\cos^2 \alpha - \sin^2 \alpha = 1 - 2 \sin^2 \alpha$

23. $\dfrac{\cos x - \tan x \sin x}{\sec x} = \cos 2x$

24. $2 \tan x \cos x = \sec x \sin 2x$

25. $\cos^4 x - \sin^4 x = \cos 2x$

26. $(\sin x + \cos x)^2 = 1 + \sin 2x$

27. $\dfrac{\sin 4\theta}{\sin 2\theta} = 2 \cos 2\theta$

28. $2 + \dfrac{\cos 2\theta}{\sin^2 \theta} = \csc^2 \theta$

29. $\dfrac{\sin 2\theta}{1 + \cos 2\theta} = \tan \theta$

30. $\dfrac{2 \tan \alpha}{1 + \tan^2 \alpha} = \sin 2\alpha$

31. $\dfrac{1 - \tan^2 x}{\sec^2 x} = \cos 2x$

32. $1 - \cos 2\theta = \dfrac{2}{1 + \cot^2 \theta}$

33. $2 \csc 2x \tan x = \sec^2 x$

34. $2 \sin x + \sin 2x = \dfrac{2 \sin^3 x}{1 - \cos x}$

35. $\dfrac{\sin 3x}{\sin x} - \dfrac{\cos 3x}{\cos x} = 2$

36. $\dfrac{\sin 3x}{\sin x} + \dfrac{\cos 3x}{\cos x} = 4 \cos 2x$

In Exercises 37 and 38, prove the given identities by letting $3x = 2x + x$.

37. $\sin 3x = 3 \cos^2 x \sin x - \sin^3 x$

38. $\cos 3x = \cos^3 x - 3 \sin^2 x \cos x$

In Exercises 39 through 44, solve the given problems.

39. In Exercise 37 of Section 19-2, let $\beta = \alpha$, and show that

$$\tan 2\alpha = \frac{2 \tan \alpha}{1 - \tan^2 \alpha} \tag{19-25}$$

40. Given that $x = \cos 2\theta$ and $y = \sin \theta$, find the relation between x and y by eliminating θ.

41. To find the horizontal range R of a projectile, the equation $R = vt \cos \alpha$ is used, where α is the angle between the line of fire and the horizontal, v is the initial velocity of the projectile, and t is the time of flight. It can be shown that $t = (2v \sin \alpha)/g$, where g is the acceleration due to gravity. Show that $R = (v^2 \sin 2\alpha)/g$.

42. In the theory of reflection of light waves, the expression $\sin(\frac{\pi}{2} - 2\theta)$ arises. Show that this expression can be written as $1 - 2 \sin^2 \theta$.

43. The instantaneous electric power p in an inductor is given by the equation $p = vi \sin \omega t \sin(\omega t - \frac{\pi}{2})$. Show that this equation can be written as $p = -\frac{1}{2} vi \sin 2\omega t$.

44. In the study of the stress at a point in a bar, the equation $s = a \cos^2 \theta + b \sin^2 \theta - 2t \sin \theta \cos \theta$ arises. Show that this equation can be written as $s = \frac{1}{2}(a + b) + \frac{1}{2}(a - b) \cos 2\theta - t \sin 2\theta$.

19-4   Half-Angle Formulas

If we let $\theta = \alpha/2$ in the identity $\cos 2\theta = 1 - 2 \sin^2 \theta$ and then solve for $\sin(\alpha/2)$, we obtain

$$\sin \frac{\alpha}{2} = \pm \sqrt{\frac{1 - \cos \alpha}{2}} \tag{19-26}$$

half-angle formulas

Also, with the same substitution in the identity $\cos 2\theta = 2 \cos^2 \theta - 1$, which is then solved for $\cos(\alpha/2)$, we have

$$\cos \frac{\alpha}{2} = \pm \sqrt{\frac{1 + \cos \alpha}{2}} \tag{19-27}$$

NOTE ▷ In each of Eqs. (19-26) and (19-27), **the sign chosen depends on the quadrant in which $\frac{\alpha}{2}$ lies.**

We can use these half-angle formulas to find values of the functions of angles which are half of those for which the functions are known. The following examples illustrate how these identities are used in evaluations and in identities.

EXAMPLE A We can find $\sin 15°$ by using the relation

$$\sin 15° = \sqrt{\frac{1 - \cos 30°}{2}} \qquad \text{using Eq. (19-26)}$$

$$= \sqrt{\frac{1 - 0.8660}{2}} = 0.2588$$

Here the plus sign is used, since $15°$ is in the first quadrant.

EXAMPLE B

We can find cos 165° by use of the relation

$$\cos 165° = -\sqrt{\frac{1 + \cos 330°}{2}} \qquad \text{using Eq. (19-27)}$$

$$= -\sqrt{\frac{1 + 0.8660}{2}} = -0.9659$$

Here the minus sign is used, since 165° is in the second quadrant, and the cosine of a second-quadrant angle is negative.

EXAMPLE C

Simplify $\sqrt{\dfrac{1 - \cos 114°}{2}}$ by expressing the result in terms of one-half the given angle. Then using a calculator, show that the values are equal.

We note that the given expression fits the form of the right side of Eq. (19-26), which means that

$$\sqrt{\frac{1 - \cos 114°}{2}} = \sin \tfrac{1}{2}(114°) = \sin 57°$$

By using the calculator sequence

$$1 \boxed{-} 114 \boxed{\text{COS}} \boxed{=} \boxed{\div} 2 \boxed{=} \boxed{\sqrt{x}} \boxed{0.8386706}$$

or the sequence $\qquad\qquad$ $57 \boxed{\text{SIN}} \boxed{0.8386706}$

we see that the results are the same, which verifies the equation for these values.

EXAMPLE D

Using the half-angle formulas, simplify the expression $\sqrt{\dfrac{9 + 9 \cos 6x}{2}}$.

$$\sqrt{\frac{9 + 9 \cos 6x}{2}} = \sqrt{\frac{9(1 + \cos 6x)}{2}} = 3\sqrt{\frac{1 + \cos 6x}{2}}$$

$$= 3 \cos \frac{1}{2}(6x) \qquad \text{using Eq. (19-27) with } \alpha = 6x$$

$$= 3 \cos 3x$$

EXAMPLE E

In the kinetic theory of gases, the expression $\sqrt{(1 - \cos \alpha)^2 + \sin^2 \alpha}$ is found. Show that this expression equals $2 \sin \tfrac{1}{2}\alpha$.

$$\sqrt{(1 - \cos \alpha)^2 + \sin^2 \alpha} = \sqrt{1 - 2 \cos \alpha + \cos^2 \alpha + \sin^2 \alpha} \qquad \text{expanding}$$

$$= \sqrt{1 - 2 \cos \alpha + 1} \qquad \text{using Eq. (19-6)}$$

$$= \sqrt{2 - 2 \cos \alpha} = \sqrt{2(1 - \cos \alpha)} \qquad \text{factoring}$$

This last expression is very similar to that for $\sin \tfrac{1}{2}\alpha$, except that no 2 appears in the denominator. Therefore, multiplying the numerator and the denominator under the radical by 2 leads to the solution.

$$\sqrt{2(1 - \cos \alpha)} = \sqrt{\frac{4(1 - \cos \alpha)}{2}} = 2\sqrt{\frac{1 - \cos \alpha}{2}}$$

$$= 2 \sin \frac{1}{2}\alpha \qquad \text{using Eq. (19-26)}$$

EXAMPLE F

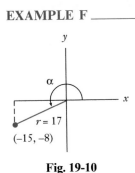

Fig. 19-10

Given that $\tan \alpha = \frac{8}{15} (180° < \alpha < 270°)$, find $\cos \left(\frac{\alpha}{2}\right)$.

Knowing that $\tan \alpha = \frac{8}{15}$ for a third-quadrant angle, we determine from Fig. 19-10 that $\cos \alpha = -\frac{15}{17}$. This means

$$\cos \frac{\alpha}{2} = -\sqrt{\frac{1 + (-15/17)}{2}} = -\sqrt{\frac{2}{34}} \qquad \text{using Eq. (19-27)}$$

$$= -\frac{1}{17}\sqrt{17} = -0.2425$$

Since $180° < \alpha < 270°$, we know that $90° < \frac{\alpha}{2} < 135°$, and therefore $\frac{\alpha}{2}$ is in the second quadrant. Since the cosine is negative for second-quadrant angles, we use the negative value of the radical.

EXAMPLE G

Prove the identity $\sec \frac{\alpha}{2} + \csc \frac{\alpha}{2} = \dfrac{2\left(\sin \dfrac{\alpha}{2} + \cos \dfrac{\alpha}{2}\right)}{\sin \alpha}$.

By expressing $\sin \alpha$ as $2 \sin \dfrac{\alpha}{2} \cos \dfrac{\alpha}{2}$, we have

$$\frac{2\left(\sin \dfrac{\alpha}{2} + \cos \dfrac{\alpha}{2}\right)}{\sin \alpha} = \frac{2\left(\sin \dfrac{\alpha}{2} + \cos \dfrac{\alpha}{2}\right)}{2 \sin \dfrac{\alpha}{2} \cos \dfrac{\alpha}{2}} \qquad \text{using Eq. (19-21)}$$

$$= \frac{\sin \dfrac{\alpha}{2}}{\sin \dfrac{\alpha}{2} \cos \dfrac{\alpha}{2}} + \frac{\cos \dfrac{\alpha}{2}}{\sin \dfrac{\alpha}{2} \cos \dfrac{\alpha}{2}}$$

$$= \frac{1}{\cos \dfrac{\alpha}{2}} + \frac{1}{\sin \dfrac{\alpha}{2}}$$

$$= \sec \frac{\alpha}{2} + \csc \frac{\alpha}{2} \qquad \text{using Eqs. (19-2) and (19-1)}$$

EXAMPLE H

We can find relations for the other functions of $\frac{\alpha}{2}$ by expressing these functions in terms of $\sin \left(\frac{\alpha}{2}\right)$ and $\cos \left(\frac{\alpha}{2}\right)$. For example,

$$\sec \frac{\alpha}{2} = \frac{1}{\cos \dfrac{\alpha}{2}} = \pm \frac{1}{\sqrt{\dfrac{1 + \cos \alpha}{2}}} \qquad \text{using Eq. (19-27)}$$

$$= \pm \sqrt{\frac{2}{1 + \cos \alpha}}$$

EXAMPLE I ———   Show that $2\cos^2 \dfrac{x}{2} - \cos x = 1$.

The first step is to substitute for $\cos \frac{x}{2}$, which will result in each term containing x on the left being in terms of x, and no $\frac{x}{2}$ terms will exist. This might allow us to combine terms. So we perform this operation, and we have for the left side

$$2\cos^2 \frac{x}{2} - \cos x = 2\left(\frac{1 + \cos x}{2}\right) - \cos x \qquad \text{using Eq. (19-27) with both sides squared}$$

$$= 1 + \cos x - \cos x = 1 \qquad \blacksquare$$

Exercises 19-4

In Exercises 1 through 4, use the half-angle formulas to evaluate the given functions.

1. $\cos 15°$ **2.** $\sin 22.5°$ **3.** $\sin 75°$ **4.** $\cos 112.5°$

In Exercises 5 through 8, simplify the given expressions by giving the results in terms of one-half the given angle. Then use a calculator to verify the results, as in Example C.

5. $\sqrt{\dfrac{1 - \cos 236°}{2}}$ **6.** $\sqrt{\dfrac{1 + \cos 98°}{2}}$ **7.** $\sqrt{1 + \cos 164°}$ **8.** $\sqrt{2 - 2\cos 328°}$

In Exercises 9 through 12, use the half-angle formulas to simplify the given expressions.

9. $\sqrt{\dfrac{1 - \cos 6\alpha}{2}}$ **10.** $\sqrt{\dfrac{4 + 4\cos 8\beta}{2}}$ **11.** $\sqrt{8 + 8\cos 4x}$ **12.** $\sqrt{2 - 2\cos 16x}$

In Exercises 13 through 16, evaluate the indicated functions with the information given.

13. Find the value of $\sin \left(\frac{x}{2}\right)$, if $\cos \alpha = \frac{12}{13}$ $(0° < \alpha < 90°)$.

14. Find the value of $\cos \left(\frac{x}{2}\right)$, if $\sin \alpha = -\frac{4}{5}$ $(180° < \alpha < 270°)$.

15. Find the value of $\cos \left(\frac{x}{2}\right)$, if $\tan \alpha = -\frac{7}{24}$ $(90° < \alpha < 180°)$.

16. Find the value of $\sin \left(\frac{x}{2}\right)$, if $\cos \alpha = \frac{8}{17}$ $(270° < \alpha < 360°)$.

In Exercises 17 through 20, derive the required expressions.

17. Derive an expression for $\csc \left(\frac{x}{2}\right)$ in terms of $\cos \alpha$.

18. Derive an expression for $\sec \left(\frac{x}{2}\right)$ in terms of $\sec \alpha$.

19. Derive an expression for $\tan \left(\frac{x}{2}\right)$ in terms of $\sin \alpha$ and $\cos \alpha$.

20. Derive an expression for $\cot \left(\frac{x}{2}\right)$ in terms of $\sin \alpha$ and $\cos \alpha$.

In Exercises 21 through 28, prove the given identities.

21. $\sin \dfrac{\alpha}{2} = \dfrac{1 - \cos \alpha}{2\sin \dfrac{\alpha}{2}}$ **22.** $2\cos \dfrac{x}{2} = (1 + \cos x)\sec \dfrac{x}{2}$ **23.** $2\sin^2 \dfrac{x}{2} + \cos x = 1$

24. $2\cos^2 \dfrac{\theta}{2}\sec \theta = \sec \theta + 1$ **25.** $\cos \dfrac{\theta}{2} = \dfrac{\sin \theta}{2\sin \dfrac{\theta}{2}}$ **26.** $\cos^2 \dfrac{x}{2}\left[1 + \left(\dfrac{\sin x}{1 + \cos x}\right)^2\right] = 1$

27. $2\sin^2 \dfrac{\alpha}{2} - \cos^2 \dfrac{\alpha}{2} = \dfrac{1 - 3\cos \alpha}{2}$ **28.** $\tan \dfrac{\alpha}{2} = \dfrac{\sin \alpha}{1 + \cos \alpha}$

In Exercises 29 through 32, use the half-angle formulas to solve the given problems.

29. In electronics, in order to find the *root-mean-square current* in a circuit, it is necessary to express $\sin^2 \omega t$ in terms of $\cos 2\omega t$. Show how this is done.

30. In the theory of the motion of a pendulum, the expression $2(\sin^2 \frac{1}{2}\alpha - \sin^2 \frac{1}{2}\theta)$ arises. Show that this can be written as $\cos \theta - \cos \alpha$.

31. The index of refraction n, the angle of a prism A, and the minimum angle of deflection ϕ are related by

$$n = \frac{\sin \frac{1}{2}(A + \phi)}{\sin \frac{1}{2}A}$$

See Fig. 19-11. Show that an equivalent expression is

$$n = \sqrt{\frac{1 - \cos A \cos \phi + \sin A \sin \phi}{1 - \cos A}}$$

32. For the structure shown in Fig. 19-12, show that $x = 2\ell \sin^2 \frac{1}{2}\theta$.

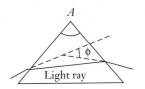

Light ray

Fig. 19-11

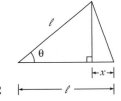

Fig. 19-12

19-5 Trigonometric Equations

One of the most important uses of the trigonometric identities is in the solution of equations involving trigonometric functions. The solution of this type of equation consists of the angles which satisfy the equation. When solving for the angle, we generally first solve for a value of a function of the angle, then find the angle from this value of the function.

When equations are written in terms of more than one function, the identities provide a way of transforming many of them to equations or factors involving only one function of the same angle. Thus *the solution is found by using algebraic methods and trigonometric identities and values.* From Chapter 7 we recall that we must be careful regarding the sign of the value of a trigonometric function in determining the angle. Figure 19-13 shows again the quadrants in which the functions are positive. Functions not listed are negative.

Sin θ Csc θ	All
Tan θ Cot θ	Cos θ Sec θ

Positive functions

Fig. 19-13

EXAMPLE A

Solve the equation $2 \cos \theta - 1 = 0$ for all values of θ such that $0 \le \theta < 2\pi$.

Solving the equation for $\cos \theta$, we obtain $\cos \theta = \frac{1}{2}$. The problem asks for all values of θ from 0 to 2π that satisfy the equation. We know that the cosines of angles in the first and fourth quadrants are positive. Also, we know that $\cos \frac{\pi}{3} = \frac{1}{2}$, which means that $\frac{\pi}{3}$ is the reference angle. Therefore, the solution proceeds as follows:

$$2 \cos \theta - 1 = 0$$

$$2 \cos \theta = 1 \qquad \text{solve for } \cos \theta$$

$$\cos \theta = \frac{1}{2}$$

$$\theta = \frac{\pi}{3}, \frac{5\pi}{3} \qquad \theta \text{ in quadrants I and IV}$$

EXAMPLE B

Solve the equation $5(1 + \sin x) = 2 \sin x + 3$ for $0 \le x < 2\pi$.
    Solving for $\sin x$, we have

$$5(1 + \sin x) = 2 \sin x + 3$$
$$5 + 5 \sin x = 2 \sin x + 3$$
$$3 \sin x = -2$$
$$\sin x = -\frac{2}{3}$$

2 $\div$ 3 $=$ $+/-$
INV SIN
-0.7297277

Since x cannot be found from known values by inspection, we use a calculator. We must remember that the answer is to be expressed in radians. Therefore, with the calculator in radian mode, we use the sequence shown at the left. However, the value $x = -0.7297$ is not in the domain of values $0 \le x < 2\pi$, although it does tell us that the reference angle is 0.7297. Therefore, recalling that $\sin x$ is negative in the third and fourth quadrants, we add 0.7297 to π and subtract it from 2π. This gives us the solutions $x = 3.871, 5.553$. ∎

EXAMPLE C

Solve the equation $2 \cos^2 x - \sin x - 1 = 0$ $(0 \le x < 2\pi)$.
    By use of the identity $\sin^2 x + \cos^2 x = 1$, this equation may be put in terms of $\sin x$ only. Thus, we have

$$2(1 - \sin^2 x) - \sin x - 1 = 0 \qquad \text{use identity}$$
$$-2 \sin^2 x - \sin x + 1 = 0 \qquad \text{solve for } \sin x$$
$$2 \sin^2 x + \sin x - 1 = 0$$

or

$$(2 \sin x - 1)(\sin x + 1) = 0$$

Just as in solving algebraic equations, we can set each factor equal to zero to find valid solutions. Thus, $\sin x = \frac{1}{2}$ or $\sin x = -1$. For the domain from 0 to 2π, the value $\sin x = \frac{1}{2}$ gives values of x as $\frac{\pi}{6}$ and $\frac{5\pi}{6}$, and $\sin x = -1$ gives the value $x = \frac{3\pi}{2}$. Thus, the complete solution is

$$x = \frac{\pi}{6}, \frac{5\pi}{6}, \frac{3\pi}{2}$$

These values check when substituted in the original equation. ∎

EXAMPLE D

Solve the equation $\sec^2 x + 2 \tan x - 6 = 0$ $(0 \le x < 2\pi)$.
    By use of the identity $1 + \tan^2 x = \sec^2 x$ we may express this equation in terms of $\tan x$ only. Therefore,

$$1 + \tan^2 x + 2 \tan x - 6 = 0 \qquad \text{use identity}$$
$$\tan^2 x + 2 \tan x - 5 = 0$$

We note that this expression is not factorable. Therefore, using the quadratic formula, we solve for $\tan x$.

$$\tan x = \frac{-2 \pm \sqrt{4 + 20}}{2} = -1 \pm 2.4495$$

Therefore, $\tan x = 1.4495$ and $\tan x = -3.4495$. In radians, we find that

$\tan x = 1.4495$ for $x = 0.9669$. Since $\tan x$ is also positive in the third quadrant, we have $x = 4.108$ as well. In the same way, using $\tan x = -3.4495$, we obtain $x = 1.853$ and $x = 4.995$. Therefore, the correct solutions are

$$x = 0.9669, 1.853, 4.108, 4.995$$

■ These values check in the original equation.

EXAMPLE E —— The vertical displacement y of an object at the end of a spring, which itself is being moved up and down, is given by $y = 3.50 \sin t + 1.20 \sin 2t$. Find the first two values of t, in seconds, for which $y = 0$.

Using the double-angle formula for $\sin 2t$ leads to the solution.

$$3.50 \sin t + 1.20 \sin 2t = 0 \qquad \text{setting } y = 0$$
$$3.50 \sin t + 2.40 \sin t \cos t = 0 \qquad \text{using identities}$$
$$\sin t (3.50 + 2.40 \cos t) = 0 \qquad \text{factoring}$$
$$\sin t = 0 \quad \text{or} \quad \cos t = -1.46$$
$$t = 0.00, 3.14, \ldots$$

Since $\cos t$ cannot be numerically larger than 1, there are no values of t for which ■ $\cos t = -1.46$. Thus, the required times are $t = 0.00$ s, 3.14 s.

EXAMPLE F —— Solve the equation $\cos \frac{x}{2} = 1 + \cos x$ $(0 \le x < 2\pi)$.

By using the half-angle formula for $\cos(x/2)$ and then squaring both sides of the resulting equation, this equation can be solved.

$$\pm \sqrt{\frac{1 + \cos x}{2}} = 1 + \cos x \qquad\qquad \text{using identity}$$

$$\frac{1 + \cos x}{2} = 1 + 2 \cos x + \cos^2 x \qquad \text{squaring both sides}$$

$$2 \cos^2 x + 3 \cos x + 1 = 0 \qquad\qquad \text{simplifying}$$
$$(2 \cos x + 1)(\cos x + 1) = 0 \qquad\qquad \text{factoring}$$

$$\cos x = -\frac{1}{2}, -1$$

$$x = \frac{2\pi}{3}, \frac{4\pi}{3}, \pi$$

In finding this solution, we squared both sides of the original equation. In doing this we may have introduced extraneous solutions (see Section 13-3). Thus, we must check each solution in the original equation to see if it is valid. Hence,

$$\cos \frac{\pi}{3} \overset{?}{=} 1 + \cos \frac{2\pi}{3} \quad \text{or} \quad \frac{1}{2} \overset{?}{=} 1 + \left(-\frac{1}{2}\right) \quad \text{or} \quad \frac{1}{2} = \frac{1}{2}$$

$$\cos \frac{2\pi}{3} \overset{?}{=} 1 + \cos \frac{4\pi}{3} \quad \text{or} \quad -\frac{1}{2} \overset{?}{=} 1 + \left(-\frac{1}{2}\right) \quad \text{or} \quad -\frac{1}{2} \ne \frac{1}{2}$$

$$\cos \frac{\pi}{2} \overset{?}{=} 1 + \cos \pi \quad \text{or} \quad 0 \overset{?}{=} 1 - 1 \quad \text{or} \quad 0 = 0$$

Thus, the apparent solution $x = \frac{4\pi}{3}$ is not a solution of the original equation. ■ The correct solutions are $x = \frac{2\pi}{3}$ and $x = \pi$.

EXAMPLE G _____ Solve the equation $\tan 3\theta - \cot 3\theta = 0$ $(0 \le \theta < 2\pi)$.

$$\tan 3\theta - \frac{1}{\tan 3\theta} = 0 \qquad \text{using } \cot 3\theta = \frac{1}{\tan 3\theta}$$

$$\tan^2 3\theta = 1 \qquad \text{multiplying by } \tan 3\theta \text{ and adding 1 to each side}$$

$$\tan 3\theta = \pm 1 \qquad \text{taking square roots}$$

Since we require values of θ such that $0 \le \theta < 2\pi$, we must have values of 3θ such that $0 \le 3\theta < 6\pi$. Therefore,

$$3\theta = \frac{\pi}{4}, \frac{3\pi}{4}, \frac{5\pi}{4}, \frac{7\pi}{4}, \frac{9\pi}{4}, \frac{11\pi}{4}, \frac{13\pi}{4}, \frac{15\pi}{4}, \frac{17\pi}{4}, \frac{19\pi}{4}, \frac{21\pi}{4}, \frac{23\pi}{4}$$

This means that the solutions are

$$\theta = \frac{\pi}{12}, \frac{\pi}{4}, \frac{5\pi}{12}, \frac{7\pi}{12}, \frac{3\pi}{4}, \frac{11\pi}{12}, \frac{13\pi}{12}, \frac{5\pi}{4}, \frac{17\pi}{12}, \frac{19\pi}{12}, \frac{7\pi}{4}, \frac{23\pi}{12}$$

It is noted that these values satisfy the original equation. Since we multiplied through by $\tan 3\theta$ in the solution, any value of θ which leads to $\tan 3\theta = 0$ would not be valid, since this would indicate division by zero in the original equation. ■

EXAMPLE H _____ Solve the equation $\cos 3x \cos x + \sin 3x \sin x = 1$ $(0 \le x < 2\pi)$.
 The left side of this equation is of the general form $\cos(A - x)$, where $A = 3x$. Therefore,

$$\cos 3x \cos x + \sin 3x \sin x = \cos(3x - x) = \cos 2x$$

The original equation becomes

$$\cos 2x = 1$$

This equation is satisfied if $2x = 0$ or $2x = 2\pi$. The solutions are $x = 0$ and $x = \pi$. Only through recognition of the proper trigonometric form can we readily solve this equation. ■

Exercises 19-5

In Exercises 1 through 32, solve the given trigonometric equations for values of x so that $0 \le x < 2\pi$.

1. $\sin x - 1 = 0$

2. $2 \sin x + 1 = 0$

3. $\tan x + 1 = 0$

4. $2 \cos x + 1 = 0$

5. $2(2 + \cos x) = 3 + \cos x$

6. $4 \tan x + 2 = 3(1 + \tan x)$

7. $7 \sin x - 2 = 3(2 - \sin x)$

8. $3 - 4 \cos x = 7 - (2 - \cos x)$

9. $4 \cos^2 x - 1 = 0$

10. $\sin^2 x - 1 = 0$

11. $4 \sin^2 x - 3 = 0$

12. $3 \tan^2 x - 1 = 0$

13. $2 \sin^2 x - \sin x = 0$

14. $3 \cos x - 4 \cos^2 x = 0$

15. $\sin 4x - \cos 2x = 0$

16. $\sin 4x - \sin 2x = 0$

17. $\sin 2x \sin x + \cos x = 0$

18. $\cos 2x + \sin^2 x = 0$

19. $2 \sin x - \tan x = 0$

20. $\sin x - \sin \dfrac{x}{2} = 0$

21. $2 \cos^2 x - 2 \cos 2x - 1 = 0$

22. $2 \cos^2 2x + 1 = 3 \cos 2x$

23. $\sin^2 x - 2 \sin x - 1 = 0$

24. $\tan^2 x - 5 \tan x + 6 = 0$

25. $4 \tan x - \sec^2 x = 0$

26. $\tan^2 x - 2 \sec^2 x + 4 = 0$

27. $\tan x + 3 \cot x = 4$

28. $\sin x \sin \frac{1}{2}x = 1 - \cos x$

29. $\sin 2x + \cos 2x = 0$

30. $2 \sin 4x + \csc 4x = 3$

31. $\sin 2x \cos x - \cos 2x \sin x = 0$

32. $\cos 3x \cos x - \sin 3x \sin x = 0$

In Exercises 33 through 36, solve the indicated equations.

33. To find the angle θ subtended by a certain object on a camera film, it is necessary to solve the equation

$$\frac{p^2 \tan \theta}{0.0063 + p \tan \theta} = 1.6$$

where p is the distance from the camera to the object. Find θ if $p = 4.8$ m.

34. In finding the maximum illuminance from a point source of light, it is necessary to solve the equation $\cos \theta \sin 2\theta - \sin^3 \theta = 0$. Find θ if $0 < \theta < 90°$.

35. The vertical displacement y of the end of a robot arm is given by $y = 2.30 \cos 0.1t - 1.35 \sin 0.2t$. Find the first four values of t, in seconds, for which $y = 0$.

36. A building 41.0 ft high is constructed on a vertical 40.0-ft cliff at the edge of a river. How far out on the river from the base of the cliff will the building and the cliff subtend equal angles?

In Exercises 37 through 40, solve the given equations for $0 \le x < 2\pi$. Then solve each graphically and compare results.

37. $3 \sin x - 1 = 0$ **38.** $4 \cos x + 3 = 0$ **39.** $5 \cos 2x + 1 = 0$ **40.** $4 \sin 3x + 1 = 0$

In Exercises 41 through 44, solve the given equations graphically.

41. $\sin 2x = x$ **42.** $\cos 2x = 4 - x^2$

43. In finding the frequencies of vibration of a vibrating wire, the equation $x \tan x = 2.00$ occurs. Find x if $0 < x < \dfrac{\pi}{2}$.

44. An equation used in astronomy is $\theta - e \sin \theta = M$. Solve for θ for $e = 0.25$ and $M = 0.75$.

19-6   The Inverse Trigonometric Functions

When we studied logarithms, we found that we often wished to change a given expression from exponential to logarithmic form, or from logarithmic to exponential form. Each of these forms has its advantages for particular purposes. We found that the exponential function $y = b^x$ can also be written in logarithmic form with x as a function of y, or $x = \log_b y$. We then represented both of these functions as y in terms of x, saying that the letters used for the dependent and independent variables did not matter, when we wished to express a functional relationship. Since it is standard to use y as the dependent variable and x as the independent variable, we wrote the logarithmic function as $y = \log_b x$.

These two functions, the exponential function $y = b^x$ and the logarithmic function $y = \log_b x$, are called **inverse functions.** *This means that if we solve for the independent variable in terms of the dependent variable in one, we will arrive at the functional relationship expressed by the other.* It also means that, for every value of x, there is only one corresponding value of y.

Just as we are able to solve $y = b^x$ for the exponent by writing it in logarithmic form, there are times when it is necessary to solve for the independent variable (the angle) in trigonometric functions. Therefore, we define the **inverse sine function**

$$y = \text{Arcsin } x \qquad \left(-\frac{\pi}{2} \le y \le \frac{\pi}{2} \right) \qquad \text{(19-28)}$$

(It is necessary to designate the range as $-\frac{\pi}{2} \le y \le \frac{\pi}{2}$, as we will see.) In Eq. (19-28), x is the value of the sine of the angle y, or $x = \sin y$. Therefore, the most meaningful way of reading $y = \text{Arcsin } x$ is "y is the angle whose sine is x." The notation $y = \text{Sin}^{-1} x$ is also used, although *the* -1 *is* **not** *an exponent* as used here. We introduced these notations in Chapter 3 when we were finding the angle with a known value of one of the functions.

Similar definitions are used for the other inverse trigonometric functions. They are read as Eq. (19-28).

EXAMPLE A

1. $y = \text{Arccos } x$ is read as "y is the angle whose cosine is x." In this case, $x = \cos y$.

2. $y = \text{Arctan } 2x$ is read as "y is the angle whose tangent is $2x$." In this case, $2x = \tan y$.

We have seen that $y = \text{Arcsin } x$ means that $x = \sin y$. From our previous work with trigonometric functions, we know that there is an unlimited number of possible values of y for a given value of x in $x = \sin y$. Consider the following example.

EXAMPLE B

For $x = \sin y$, we know that $\sin \frac{\pi}{6} = \frac{1}{2}$ and $\sin \frac{5\pi}{6} = \frac{1}{2}$. In fact, $x = \frac{1}{2}$ also for values of y of $-\frac{7\pi}{6}, \frac{13\pi}{6}, \frac{17\pi}{6}$, and so on.

From Chapter 2, we know that *to have a properly defined* **function,** *there must be only one value of the dependent variable for a given value of the independent variable.* (A *relation,* on the other hand, may have more than one such value.) Therefore, as in Eq. (19-28), in order to have only one value of y for each value of x in the domain of the inverse trigonometric functions, it is not possible to include all values of y in the range. For this reason, *the range of each of the* **inverse trigonometric functions** *is defined as follows:*

ranges of the inverse
trigonometric functions

$$-\frac{\pi}{2} \le \text{Arcsin } x \le \frac{\pi}{2}, \; 0 \le \text{Arccos } x \le \pi, \; -\frac{\pi}{2} < \text{Arctan } x < \frac{\pi}{2}, \; 0 < \text{Arccot } x < \pi$$

$$0 \le \text{Arcsec } x \le \pi \quad \left(\text{Arcsec } x \ne \frac{\pi}{2} \right), \; -\frac{\pi}{2} \le \text{Arccsc } x \le \frac{\pi}{2} \quad (\text{Arccsc } x \ne 0)$$

(19-29)

For any of the inverse trigonometric functions $y = f(x)$, we must use a value of y in the range as defined in Eqs. (19-29) which corresponds to a given value of x in the domain. We will discuss the domains and the reasons for these definitions, along with the graphs of the inverse trigonometric functions, following the next three examples.

EXAMPLE C

$$\text{Arcsin}\left(\frac{1}{2}\right) = \frac{\pi}{6} \qquad \text{first-quadrant angle}$$

This is the only value of the function which lies within the defined range. The value $\frac{5\pi}{6}$ is not correct, even though $\sin(\frac{5\pi}{6}) = \frac{1}{2}$, since $\frac{5\pi}{6}$ lies outside the defined range. ■

EXAMPLE D

$$\text{Arccos}\left(-\frac{1}{2}\right) = \frac{2\pi}{3} \qquad \text{second-quadrant angle}$$

Other values such as $\frac{4\pi}{3}$ and $-\frac{2\pi}{3}$ are not correct, since they are not within the defined range of values for the function Arccos x. ■

EXAMPLE E

$$\text{Arctan}(-1) = -\frac{\pi}{4} \qquad \text{fourth-quadrant angle}$$

This is the only value within the defined range for the function Arctan x.

NOTE ▷ We must remember that **when x is negative for Arcsin x and Arctan x, the value of y is a fourth-quadrant angle, expressed as a negative angle.** This is a direct result of the definition. (The single exception is Arcsin $(-1) = -\frac{\pi}{2}$, which is a quadrantal angle and is not *in* the fourth quadrant.) ■

In choosing these values to be the ranges of the inverse trigonometric functions, we first note that the *domain* of $y = \text{Arcsin } x$ and $y = \text{Arccos } x$ is $-1 \le x \le 1$, since the sine and cosine functions take on only these values. Therefore, for each value in this domain we use only one value of y in the range of the function. Although the domain of $y = \text{Arctan } x$ is all real numbers, we still use only one value of y in the range.

The values are so chosen that if x is positive, the resulting value is an angle in the first quadrant. We must, however, account for the possibility that x might be negative. We could not choose second-quadrant angles for Arcsin x. Since the sine of a second-quadrant angle is also positive, this would lead to ambiguity. The sine is negative for fourth-quadrant angles, and to have a continuous range of values we must express the fourth-quadrant angles in the form of negative angles. This range is also chosen for Arctan x, for similar reasons. However, Arccos x cannot be chosen in this way, since the cosines of fourth-quadrant angles are also positive. Thus, again to keep a continuous range of values for Arccos x, the second-quadrant angles are chosen for negative values of x.

As for the values for the other functions, we chose values such that if x is positive, the result is also an angle in the first quadrant. As for negative values of x, it rarely makes any difference, since either positive values of x arise, or we can use one of the other functions. Our definitions, however, are those which are generally used.

The graphs of the inverse trigonometric functions can be used to show the domains and ranges. We can obtain the graph of the inverse sine function by first sketching the sine curve $x = \sin y$ *along the y-axis.* We then mark the specific part of this curve for which $-\frac{\pi}{2} \le y \le \frac{\pi}{2}$ as the graph of the inverse sine function. The graphs of the other inverse functions are found in the same way. In

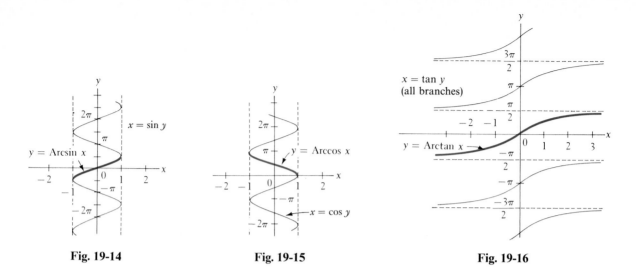

Fig. 19-14   Fig. 19-15   Fig. 19-16

Figs. 19-14, 19-15, and 19-16, the graphs of $x = \sin y$, $x = \cos y$, and $x = \tan y$, respectively, are shown. The heavier, colored portions indicate the graphs of the respective inverse trigonometric functions.

The following examples further illustrate the values and the meanings of the inverse trigonometric functions.

EXAMPLE F

$$\text{Arcsin}\left(-\frac{\sqrt{3}}{2}\right) = -\frac{\pi}{3} \qquad \text{Arccos}\,(-1) = \pi$$

$$\text{Arctan}\,0 = 0 \qquad \text{Arctan}\,(\sqrt{3}) = \frac{\pi}{3}$$

EXAMPLE G

By use of a calculator in radian mode, we find the following values:

$$\text{Arcsin}\,0.6294 = 0.6808 \qquad \text{Arcsin}\,(-0.1568) = -0.1574$$
$$\text{Arccos}\,(-0.8026) = 2.5024 \qquad \text{Arctan}\,(-1.9268) = -1.0921$$

We note that the calculator gives values which are in the defined ranges for each function.

EXAMPLE H

Given that $y = \pi - \text{Arcsec}\,2x$, solve for x.

We first find the expression for Arcsec $2x$ and then use the meaning of the inverse secant. The solution follows.

$$y = \pi - \text{Arcsec}\,2x$$
$$\text{Arcsec}\,2x = \pi - y \qquad \text{solve for Arcsec}\,2x$$
$$2x = \sec(\pi - y) \qquad \text{use meaning of inverse secant}$$
$$x = -\frac{1}{2}\sec y \qquad \sec(\pi - y) = -\sec y$$

We note that as sec $2x$ and 2 sec x are different functions, so are Arcsec $2x$ and 2 Arcsec x. Also, since the values of Arcsec $2x$ are restricted, so are the resulting values of y.

EXAMPLE I _____ The instantaneous power p in an electric inductor is given by $p = vi \sin \omega t \cos \omega t$. Solve for t.

Noting the product $\sin \omega t \cos \omega t$ suggests using the formula

$$\sin 2\alpha = 2 \sin \alpha \cos \alpha$$

Then, using the meaning of the inverse sine, we can complete the solution, which follows.

$$p = vi \sin \omega t \cos \omega t$$

$$= \frac{1}{2} vi \sin 2\omega t \qquad \text{using double-angle formula}$$

$$\sin 2\omega t = \frac{2p}{vi}$$

$$2\omega t = \text{Arcsin}\left(\frac{2p}{vi}\right) \qquad \text{using meaning of inverse sine}$$

$$t = \frac{1}{2\omega} \text{Arcsin}\left(\frac{2p}{vi}\right)$$

If we know the value of x for one of the inverse functions, we can find the trigonometric functions of the angle. If general relations are desired, a representative triangle is very useful. The following examples illustrate these methods.

EXAMPLE J _____ Find $\cos$ (Arcsin 0.5). (Remember again: The inverse functions yield _angles._)

We know Arcsin 0.5 is a first-quadrant angle, since 0.5 is positive. Thus we find Arcsin $0.5 = \frac{\pi}{6}$. The problem now becomes one of finding $\cos\left(\frac{\pi}{6}\right)$. This is, of course, $\sqrt{3}/2$, or 0.8660. Thus,

$$\cos (\text{Arcsin } 0.5) = \cos \frac{\pi}{6} = 0.8660$$

EXAMPLE K _____ 1. $\sin (\text{Arccot } 1) = \sin \dfrac{\pi}{4} \qquad$ first-quadrant angle

$$= \frac{\sqrt{2}}{2} = 0.7071$$

2. $\tan [\text{Arccos} (-1)] = \tan \pi \qquad$ quadrantal angle

$$= 0$$

3. $\cos [\text{Arcsin} (-0.2395)] = 0.9709$

Calculator sequence: .2395 $\boxed{+/-}$ $\boxed{\text{INV}}$ $\boxed{\text{SIN}}$ $\boxed{\text{COS}}$ $\boxed{0.9708964}$

EXAMPLE L

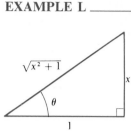

Fig. 19-17

Find sin (Arctan x).

We know that Arctan x is another way of stating "the angle whose tangent is x." Thus, let us draw a right triangle (as in Fig. 19-17) and label one of the acute angles as θ, the side opposite θ as x, and the side adjacent to θ as 1. In this way we see that, by definition, $\tan \theta = \frac{x}{1}$, or $\theta = $ Arctan x, which means θ is the desired angle. By the Pythagorean theorem, the hypotenuse of this triangle is $\sqrt{x^2 + 1}$. Now we find that the $\sin \theta$, which is the same as sin (Arctan x), is $x/\sqrt{x^2 + 1}$, from the definition of the sine. Thus,

$$\sin (\text{Arctan } x) = \frac{x}{\sqrt{x^2 + 1}}$$

EXAMPLE M

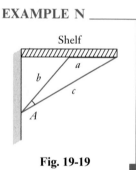

Fig. 19-18

Find cos (2 Arcsin x).

From Fig. 19-18, we see that $\theta = $ Arcsin x. From the double-angle formulas, we have

$$\cos 2\theta = 1 - 2 \sin^2 \theta$$

Thus, since $\sin \theta = x$, we have

$$\cos (2 \text{ Arcsin } x) = 1 - 2x^2$$

EXAMPLE N

Shelf

Fig. 19-19

A triangular brace of sides a, b, and c is made to support a shelf, as shown in Fig. 19-19. Find the expression for the angle between sides b and c.

The law of cosines leads to the solution, which follows.

$$a^2 = b^2 + c^2 - 2bc \cos A \qquad \text{law of cosines}$$
$$2bc \cos A = b^2 + c^2 - a^2 \qquad \text{solving for cos } A$$
$$\cos A = \frac{b^2 + c^2 - a^2}{2bc}$$
$$A = \text{Arccos} \left(\frac{b^2 + c^2 - a^2}{2bc} \right) \qquad \text{using meaning of inverse cosine}$$

Exercises 19-6

In Exercises 1 through 8, write down the meaning of each of the given equations. See Example A.

1. $y = $ Arctan x **2.** $y = $ Arcsec x **3.** $y = $ Arccot $3x$ **4.** $y = $ Arccsc $4x$

5. $y = 2$ Arcsin x **6.** $y = 3$ Arctan x **7.** $y = 5$ Arccos $2x$ **8.** $y = 4$ Arcsin $3x$

In Exercises 9 through 32, evaluate the given expressions.

9. $\text{Arccos} \left(\frac{1}{2} \right)$ **10.** $\text{Arcsin } (1)$ **11.** $\text{Arcsin } 0$ **12.** $\text{Arccos } 0$

13. $\text{Arctan } (-\sqrt{3})$ **14.** $\text{Arcsin} \left(-\frac{1}{2} \right)$ **15.** $\text{Arcsec } 2$ **16.** $\text{Arccot } \sqrt{3}$

17. $\text{Arctan} \left(\frac{\sqrt{3}}{3} \right)$ **18.** $\text{Arctan } 1$ **19.** $\text{Arcsin} \left(-\frac{\sqrt{2}}{2} \right)$ **20.** $\text{Arccos} \left(-\frac{\sqrt{3}}{2} \right)$

21. Arccsc $\sqrt{2}$ **22.** Arccot 1   **23.** Arcsin $\left(-\dfrac{\sqrt{3}}{2}\right)$ **24.** Arccot $(-\sqrt{3})$

25. Arcsec $(-\sqrt{2})$ **26.** Arccsc (-1) **27.** sin (Arctan $\sqrt{3}$)   **28.** tan $\left(\text{Arcsin}\,\dfrac{\sqrt{2}}{2}\right)$

29. cos [Arctan (-1)]   **30.** sec [Arccos $(-\tfrac{1}{2})$]   **31.** cos (2 Arcsin 1)   **32.** sin (2 Arctan 2)

In Exercises 33 through 44, use a calculator to evaluate the given expressions.

33. Arctan (-3.7321) **34.** Arccos (-0.6561) **35.** Arcsin (-0.8326) **36.** Arctan 0.2846

37. Arccos 0.1291   **38.** Arcsin 0.2119   **39.** Arctan 8.2614   **40.** Arcsin (-0.8881)

41. tan [Arccos (-0.6281)]   **42.** cos [Arctan (-1.2256)]

43. sin [Arctan (-0.2297)]   **44.** tan [Arcsin (-0.3019)]

In Exercises 45 through 52, solve the given equations for x.

45. $y = \sin 3x$ **46.** $y = \cos (x - \pi)$ **47.** $y = \text{Arctan}\left(\dfrac{x}{4}\right)$ **48.** $y = 2\,\text{Arcsin}\left(\dfrac{x}{6}\right)$

49. $y = 1 + \sec 3x$ **50.** $4y = 5 - \csc 8x$ **51.** $1 - y = \text{Arccos}\,(1 - x)$ **52.** $2y = \text{Arccot}\,3x - 5$

In Exercises 53 through 60, find an algebraic expression for each of the expressions given.

53. tan (Arcsin x)   **54.** sin (Arccos x)   **55.** cos (Arcsec x)   **56.** cot (Arccot x)

57. sec (Arccsc $3x$)   **58.** tan (Arcsin $2x$)   **59.** sin (2 Arcsin x)   **60.** cos (2 Arctan x)

In Exercises 61 through 64, solve the given problems with the use of the inverse trigonometric functions.

61. In the analysis of ocean tides, the equation $y = A \cos 2(\omega t + \phi)$ is used. Solve for t.

62. For an object of weight w on an inclined plane which is at an angle θ to the horizontal, the equation relating w and θ is $\mu w \cos \theta = w \sin \theta$, where μ is the coefficient of friction between the surfaces in contact. Solve for θ.

63. The electric current in a certain circuit is given by $i = I_m[\sin (\omega t + \alpha) \cos \phi + \cos (\omega t + \alpha) \sin \phi]$. Solve for t.

64. The time t as a function of the displacement d of a piston is given by the equation $t = \dfrac{1}{2\pi f}\,\text{Arccos}\,\dfrac{d}{A}$. Solve for d.

In Exercises 65 and 66, prove that the given expressions are equal. This can be done by use of the relation for sin $(\alpha + \beta)$ *and by showing that the sine of the sum of angles on the left equals the sine of the angle on the right.*

65. Arcsin $\dfrac{3}{5}$ + Arcsin $\dfrac{5}{13}$ = Arcsin $\dfrac{56}{65}$ **66.** Arctan $\dfrac{1}{3}$ + Arctan $\dfrac{1}{2}$ = $\dfrac{\pi}{4}$

In Exercises 67 and 68, verify the given expressions.

67. Arcsin 0.5 + Arccos 0.5 = $\dfrac{\pi}{2}$ **68.** Arctan $\sqrt{3}$ + Arccot $\sqrt{3}$ = $\dfrac{\pi}{2}$

In Exercises 69 and 70, solve for angle A for the triangles in the given figures in terms of the given sides and angles.

69. Fig. 19-20   **70.** Fig. 19-21

Fig. 19-20   **Fig. 19-21**

In Exercises 71 and 72, derive the given expressions.

71. A microwave transmitter of height h is at the top of a vertical cliff. From a point at a horizontal distance d from the base of the cliff, the angles of elevation of the top of the transmitter and the top of the cliff are α and β, respectively. Show that $\alpha = \text{Arctan}\left(\frac{h}{d} + \tan\beta\right)$.

72. Show that the length of the pulley belt indicated in Fig. 19-22 is given by the expression

$$L = 24 + 11\pi + 10 \, \text{Arcsin}\left(\frac{5}{13}\right)$$

Fig. 19-22

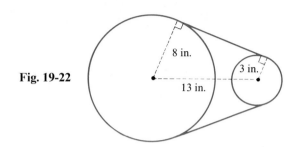

8 in.

3 in.

13 in.

19-7 Chapter Equations, Review Exercises, and Practice Test

Chapter Equations

Basic trigonometric identities	(19-1) $\quad \sin\theta = \dfrac{1}{\csc\theta}$	$\cos\theta = \dfrac{1}{\sec\theta}$ (19-2)
	(19-3) $\quad \tan\theta = \dfrac{1}{\cot\theta}$	$\tan\theta = \dfrac{\sin\theta}{\cos\theta}$ (19-4)
	(19-5) $\quad \cot\theta = \dfrac{\cos\theta}{\sin\theta}$	$\sin^2\theta + \cos^2\theta = 1$ (19-6)
	(19-7) $\quad 1 + \tan^2\theta = \sec^2\theta$	$1 + \cot^2\theta = \csc^2\theta$ (19-8)

Sum and difference identities

$$\sin(\alpha + \beta) = \sin\alpha\cos\beta + \cos\alpha\sin\beta \tag{19-9}$$

$$\cos(\alpha + \beta) = \cos\alpha\cos\beta - \sin\alpha\sin\beta \tag{19-10}$$

$$\sin(\alpha - \beta) = \sin\alpha\cos\beta - \cos\alpha\sin\beta \tag{19-11}$$

$$\cos(\alpha - \beta) = \cos\alpha\cos\beta + \sin\alpha\sin\beta \tag{19-12}$$

Double-angle formulas

$$\sin 2\alpha = 2\sin\alpha\cos\alpha \tag{19-21}$$

$$\cos 2\alpha = \cos^2\alpha - \sin^2\alpha \tag{19-22}$$

$$= 2\cos^2\alpha - 1 \tag{19-23}$$

$$= 1 - 2\sin^2\alpha \tag{19-24}$$

Half-angle formulas

$$\sin\frac{\alpha}{2} = \pm\sqrt{\frac{1 - \cos\alpha}{2}} \tag{19-26}$$

$$\cos\frac{\alpha}{2} = \pm\sqrt{\frac{1 + \cos\alpha}{2}} \tag{19-27}$$

Inverse trigonometric functions

$$y = \text{Arcsin}\, x \qquad \left(-\frac{\pi}{2} \leq y \leq \frac{\pi}{2}\right) \tag{19-28}$$

$$-\frac{\pi}{2} \le \text{Arcsin } x \le \frac{\pi}{2}, 0 \le \text{Arccos } x \le \pi, -\frac{\pi}{2} < \text{Arctan } x < \frac{\pi}{2}, 0 < \text{Arccot } x < \pi,$$

$$0 \le \text{Arcsec } x \le \pi \left(\text{Arcsec } x \ne \frac{\pi}{2} \right), -\frac{\pi}{2} \le \text{Arccsc } x \le \frac{\pi}{2} (\text{Arccsc } x \ne 0) \tag{19-29}$$

Review Exercises

In Exercises 1 through 8, determine the values of the indicated functions in the given manner.

1. Find sin 120° by using 120° = 90° + 30°.

2. Find cos 30° by using 30° = 90° − 60°.

3. Find sin 135° by using 135° = 180° − 45°.

4. Find cos 225° by using 225° = 180° + 45°.

5. Find cos 180° by using 180° = 2(90°).

6. Find sin 180° by using 180° = 2(90°).

7. Find sin 45° by using 45° = $\frac{1}{2}$(90°).

8. Find cos 45° by using 45° = $\frac{1}{2}$(90°).

In Exercises 9 through 16, simplify the given expressions by use of one of the basic formulas of the chapter. Then use a calculator to verify the result by finding the value of the original expression and the value of the simplified expression.

9. sin 14° cos 38° + cos 14° sin 38°

10. cos² 148° − sin² 148°

11. 2 sin 46° cos 46°

12. cos 73° cos 142° + sin 73° sin 142°

13. $\sqrt{\dfrac{1 + \cos 12°}{2}}$

14. cos 3° cos 215° − sin 3° sin 215°

15. 1 − 2 sin² 82°

16. $\sqrt{\dfrac{1 - \cos 166°}{2}}$

In Exercises 17 through 24, simplify each of the given expressions. Expansion of any term is not necessary; proper recognition of the form of the expression leads to the proper result.

17. sin 2x cos 3x + cos 2x sin 3x

18. cos 7x cos 3x + sin 7x sin 3x

19. 8 sin 6x cos 6x

20. 10 sin 5x cos 5x

21. 2 − 4 sin² 6x

22. cos² 2x − sin² 2x

23. $\sqrt{2 + 2 \cos 2x}$

24. $\sqrt{32 - 32 \cos 4x}$

In Exercises 25 through 32, evaluate the given expressions.

25. Arcsin (−1)

26. Arcsec $\sqrt{2}$

27. Arccos (0.9659)

28. Arctan (−0.6249)

29. tan $\left[\text{Arcsin } \left(-\frac{1}{2} \right) \right]$

30. cos $\left[\text{Arctan } (-\sqrt{3}) \right]$

31. Arcsin (tan π)

32. Arccos $\left[\tan \left(-\dfrac{\pi}{4} \right) \right]$

In Exercises 33 through 60, prove the given identities.

33. $\dfrac{\sec y}{\csc y} = \tan y$

34. cos θ csc θ = cot θ

35. sin x(csc x − sin x) = cos² x

36. cos y(sec y − cos y) = sin² y

37. $\dfrac{1}{\sin \theta} - \sin \theta = \cot \theta \cos \theta$

38. sin θ sec θ csc θ cos θ = 1

39. cos θ cot θ + sin θ = csc θ

40. $\dfrac{\sin x \cot x + \cos x}{\cot x} = 2 \sin x$

41. $\dfrac{\sec^4 x - 1}{\tan^2 x} = 2 + \tan^2 x$

42. cos² y − sin² y = $\dfrac{1 - \tan^2 y}{1 + \tan^2 y}$

43. 2 csc 2x cot x = 1 + cot² x

44. sin x cot² x = csc x − sin x

45. $\dfrac{1 - \sin^2 \theta}{1 - \cos^2 \theta} = \cot^2 \theta$

46. $\dfrac{1 + \cos 2\theta}{\cos^2 \theta} = 2$

47. $\dfrac{\cos 2\theta}{\cos^2 \theta} = 1 - \tan^2 \theta$

48. $\dfrac{\sin 2\theta \sec \theta}{2} = \sin \theta$

49. sin $\dfrac{\theta}{2}$ cos $\dfrac{\theta}{2}$ = $\dfrac{\sin \theta}{2}$

50. sin $\dfrac{x}{2}$ = $\dfrac{\sec x - 1}{2 \sec x \sin \left(\dfrac{x}{2} \right)}$

51. $\sec x + \tan x = \dfrac{\cos x}{1 - \sin x}$

52. $\dfrac{\cos \theta - \sin \theta}{\cos \theta + \sin \theta} = \dfrac{\cot \theta - 1}{\cot \theta + 1}$

53. $\cos (x - y) \cos y - \sin (x - y) \sin y = \cos x$

54. $\sin 3y \cos 2y - \cos 3y \sin 2y = \sin y$

55. $\sin 4x(\cos^2 2x - \sin^2 2x) = \dfrac{\sin 8x}{2}$

56. $\csc 2x + \cot 2x = \cot x$

57. $\dfrac{\sin x}{\csc x - \cot x} = 1 + \cos x$

58. $\cos x - \sin \dfrac{x}{2} = \left(1 - 2 \sin \dfrac{x}{2}\right)\left(1 + \sin \dfrac{x}{2}\right)$

59. $\dfrac{\sin (x + y) + \sin (x - y)}{\cos (x + y) + \cos (x - y)} = \tan x$

60. $\sec \dfrac{x}{2} + \csc \dfrac{x}{2} = \dfrac{2\left(\sin \dfrac{x}{2} + \cos \dfrac{x}{2}\right)}{\sin x}$

In Exercises 61 through 64, solve for x.

61. $y = 2 \cos 2x$

62. $y - 2 = 2 \tan \left(x - \dfrac{\pi}{2}\right)$

63. $y = \dfrac{\pi}{4} - 3 \text{ Arcsin } 5x$

64. $2y = \text{Arcsec } 4x - 2$

In Exercises 65 through 76, solve the given equations for x such that $0 \le x < 2\pi$.

65. $3(\tan x - 2) = 1 + \tan x$

66. $5 \sin x = 3 - (\sin x + 2)$

67. $2(1 - 2 \sin^2 x) = 1$

68. $\sec^2 x = 2 \tan x$

69. $\cos^2 2x - 1 = 0$

70. $2 \sin 2x + 1 = 0$

71. $4 \cos^2 x - 3 = 0$

72. $\cos 2x = \sin x$

73. $\sin^2 x - \cos^2 x + 1 = 0$

74. $\cos 3x \cos x + \sin 3x \sin x = 0$

75. $\sin^2 \left(\dfrac{x}{2}\right) - \cos x + 1 = 0$

76. $\sin x + \cos x = 1$

In Exercises 77 through 80, find an algebraic expression for each of the expressions.

77. $\tan (\text{Arccot } x)$

78. $\cos (\text{Arccsc } x)$

79. $\sin (2 \text{ Arccos } x)$

80. $\cos (\pi - \text{Arctan } x)$

In Exercises 81 through 84, use the given substitutions to show that the given equations are valid. (In each, $0 \le \theta < \pi/2$.)

81. If $x = 2 \cos \theta$, show that $\sqrt{4 - x^2} = 2 \sin \theta$

82. If $x = 2 \sec \theta$, show that $\sqrt{x^2 - 4} = 2 \tan \theta$

83. If $x = \tan \theta$, show that $\dfrac{x}{\sqrt{1 + x^2}} = \sin \theta$

84. If $x = \cos \theta$, show that $\dfrac{\sqrt{1 - x^2}}{x} = \tan \theta$

In Exercises 85 through 96, use the formulas and methods of this chapter to solve the given problems.

85. Forces **A** and **B** act on a bolt such that A makes an angle θ with the x-axis and B makes an angle θ with the y-axis as shown in Fig. 19-23. The resultant **R** has components $R_x = A \cos \theta - B \sin \theta$ and $R_y = A \sin \theta + B \cos \theta$. Using these components, show that $R = \sqrt{A^2 + B^2}$.

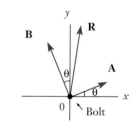

86. In surveying, when determining an azimuth (a measure used for reference purposes), it might be necessary to simplify the expression

$$\dfrac{1}{2 \cos \alpha \cos \beta} - \tan \alpha \tan \beta$$

Fig. 19-23

Perform this operation by expressing it in the simplest possible form when $\alpha = \beta$.

87. The equation for the displacement of a certain object at the end of a spring is $y = A \sin 2t + B \cos 2t$. Show that this equation may be written as $y = C \sin (2t + \alpha)$ where $C = \sqrt{A^2 + B^2}$ and $\tan \alpha = B/A$. (*Hint:* Let $A/C = \cos \alpha$ and $B/C = \sin \alpha$.)

88. In studying the interference of light waves, the identity

$$\dfrac{\sin \frac{3}{2}x}{\sin \frac{1}{2}x} \sin x = \sin x + \sin 2x$$

is used. Prove this identity. [*Hint:* $\sin \frac{3}{2}x = \sin (x + \frac{1}{2}x)$.]

89. In the study of chemical spectroscopy, the equation $\omega t = \text{Arcsin } \dfrac{\theta - \alpha}{R}$ arises. Solve for θ.

90. The power p in a certain electric circuit is given by $p = 2.5[\cos \alpha \sin (\omega t + \phi) - \sin \alpha \cos (\omega t + \phi)]$. Solve for t.

91. In finding the pressure exerted by soil on a retaining wall, the expression $1 - \cos 2\phi - \sin 2\phi \tan \alpha$ is found. Show that this expression can also be written as $2 \sin \phi(\sin \phi - \cos \phi \tan \alpha)$.

92. In analyzing the motion of an automobile universal joint, the equation $\sec^2 A - \sin^2 B \tan^2 A = \sec^2 C$ is used. Show that this equation is true if $\tan A \cos B = \tan C$.

93. Determining the angle between two sections of a robot arm requires the solution of the equation $1.20 \cos \theta + 0.135 \cos 2\theta = 0$. Find the required angle θ if $0 < \theta < 180°$.

94. The angle of elevation of a transmitting tower from a point on level ground 25.0 ft from its base is twice the angle of elevation as from a point 95.0 ft farther from the base. Find the height of the tower.

95. If a plane surface inclined at angle θ moves horizontally, the angle for which the lifting force of the air is a maximum is found by solving the equation $2 \sin \theta \cos^2 \theta - \sin^3 \theta = 0$, where $0 < \theta < 90°$. Solve for θ.

96. The electric current as a function of the time for a particular circuit is given by $i = 8.00e^{-20t}(1.73 \cos 10.0t - \sin 10.0t)$. Find the time, in seconds, when the current is first zero.

Practice Test

1. Prove that $\sec \theta - \dfrac{\tan \theta}{\csc \theta} = \cos \theta$.

2. Solve for x $(0 \le x < 2\pi)$: $\sin 2x + \sin x = 0$.

3. Find an algebraic expression for $\cos (\text{Arcsin } x)$.

4. The angular displacement θ of a certain pendulum in terms of the time t is given by $\theta = e^{-0.1t}(\cos 2t + 3 \sin 2t)$. What is the smallest value of t for which the displacement is zero?

5. Prove that $\dfrac{\tan \alpha + \tan \beta}{\tan \alpha - \tan \beta} = \dfrac{\sin (\alpha + \beta)}{\sin(\alpha - \beta)}$.

6. Prove that $\cot^2 x - \cos^2 x = \cot^2 x \cos^2 x$.

7. Find $\cos \tfrac{1}{2}x$ if $\sin x = -\tfrac{3}{5}$ and $270° < x < 360°$.

8. The intensity of a certain type of polarized light is given by $I = I_0 \sin 2\theta \cos 2\theta$. Solve for θ.

20 Plane Analytic Geometry

We first introduced the graph of a function in Chapter 2, and since that time we have made extensive use of graphs for representing functions. In Chapter 4 we showed the graph of the linear function, and in Chapter 6 we graphed the quadratic function. Also, we have used graphs to represent the trigonometric functions, the exponential and logarithmic functions, and the inverse trigonometric functions. We have also seen how graphs may be used in solving equations and systems of equations. In this chapter we shall consider certain basic principles relating to the graphs of functions. We shall also further show how certain graphs can be constructed by recognizing the form of the equation.

In Section 20-4 we show an important application of analytic geometry in the design of automobile headlights.

The underlying principle of analytic geometry is the relationship of geometry to algebra. A great deal can be learned about a geometric figure if we can find the function which represents its graph. Also, by analyzing certain characteristics of a function, we can obtain useful information as to the nature of its graph.

The concepts developed in analytic geometry are very useful in the study of calculus. Also, there are a great many technical and scientific applications, which include projectile motion, planetary orbits, and fluid motion. Other important applications range from the design of gears, airplane wings, and automobile headlights to the construction of bridges and nuclear cooling towers.

20-1 Basic Definitions

In this section we shall develop certain basic concepts which will be needed for future use in establishing the proper relationships between an equation and a curve.

The first of these concepts involves the distance between any two points in the coordinate plane. If these two points lie on a line parallel to the x-axis, the **directed distance** from the first point $A(x_1, y)$ to the second point $B(x_2, y)$ is denoted as $\overline{AB}$ and is defined as $x_2 - x_1$. We can see that $\overline{AB}$ *is positive if B is to the right of A, and it is negative if B is to the left of A.* Similarly, the directed distance between two points $C(x, y_1)$ and $D(x, y_2)$ on a line parallel to the y-axis is $y_2 - y_1$. We see that $\overline{CD}$ *is positive if D is above C, and it is negative if D is below C.*

EXAMPLE A

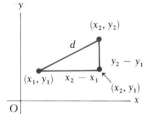
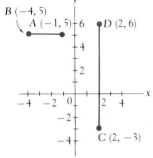
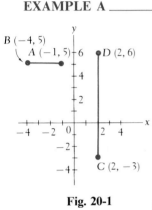

Fig. 20-1

The line segment joining $A(-1, 5)$ and $B(-4, 5)$ in Fig. 20-1 is parallel to the x-axis. Therefore, the directed distance

$$\overline{AB} = -4 - (-1) = -3$$

and the directed distance

$$\overline{BA} = -1 - (-4) = 3$$

Also in Fig. 20-1, the line segment joining $C(2, -3)$ and $D(2, 6)$ is parallel to the y-axis. The directed distance

$$\overline{CD} = 6 - (-3) = 9$$

and the directed distance

$$\overline{DC} = -3 - 6 = -9$$

We now wish to find the length of a line segment joining any two points in the plane. If these points are on a line which is not parallel to either of the axes (Fig. 20-2), we must use the Pythagorean theorem to find the distance between them. By making a right triangle with the line segment joining the two points as the hypotenuse, and line segments parallel to the axes as the legs, we have the formula which gives the distance between any two points in the plane. This formula, called *the* **distance formula,** *is*

Fig. 20-2

distance formula

$$d = \sqrt{(x_2 - x_1)^2 + (y_2 - y_1)^2} \qquad (20\text{-}1)$$

Here we choose the positive square root since we are concerned only with the magnitude of the length of the line segment.

EXAMPLE B _____ The distance between $(3, -1)$ and $(-2, -5)$ is given by

$$d = \sqrt{[(-2) - 3]^2 + [(-5) - (-1)]^2}$$
$$= \sqrt{(-5)^2 + (-4)^2} = \sqrt{25 + 16}$$
$$= \sqrt{41} = 6.403$$

See Fig. 20-3.

NOTE ▷ ***It makes no difference which point is chosen as (x_1, y_1) and which is chosen as (x_2, y_2),*** since the differences in the x-coordinates and the y-coordinates are squared. We obtain the same value for the distance when we calculate it as

$$d = \sqrt{[3 - (-2)]^2 + [(-1) - (-5)]^2}$$
$$= \sqrt{5^2 + 4^2} = \sqrt{41} = 6.403$$

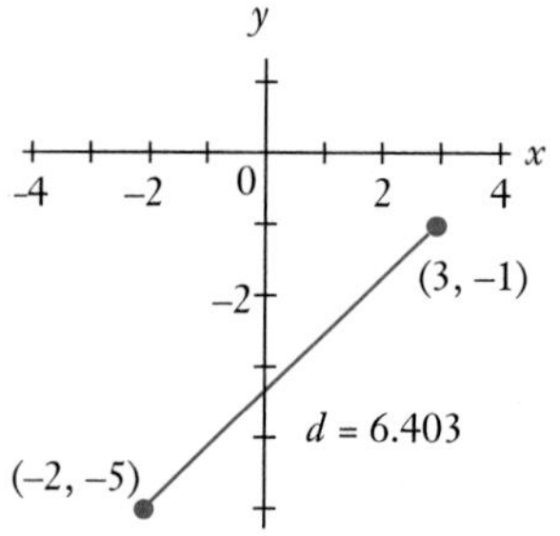

Fig. 20-3

Another important quantity for a line is its *slope,* which we first defined in Chapter 4. Here we shall give it a somewhat more general definition and develop its meaning in more detail, as well as review the basic meaning.

slope

The **slope** *gives a measure of the direction of a line, and is defined as the vertical directed distance from one point to another on the same straight line, divided by the horizontal directed distance from the first point to the second.* Thus, the slope, m, is given by

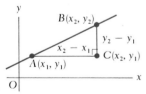

Fig. 20-4

$$m = \frac{y_2 - y_1}{x_2 - x_1} \qquad (20\text{-}2)$$

See Fig. 20-4. When the line is horizontal, $y_2 = y_1$ and $m = 0$. When the line is vertical, $x_2 = x_1$ and the slope is undefined.

EXAMPLE C _____ The slope of the line joining $(3, -5)$ and $(-2, -6)$ is

$$m = \frac{-6 - (-5)}{-2 - 3} = \frac{-6 + 5}{-5} = \frac{1}{5}$$

See Fig. 20-5. Again we may interpret either of the points as (x_1, y_1) and the other as (x_2, y_2). We can also obtain the slope of this same line from

$$m = \frac{-5 - (-6)}{3 - (-2)} = \frac{1}{5}$$

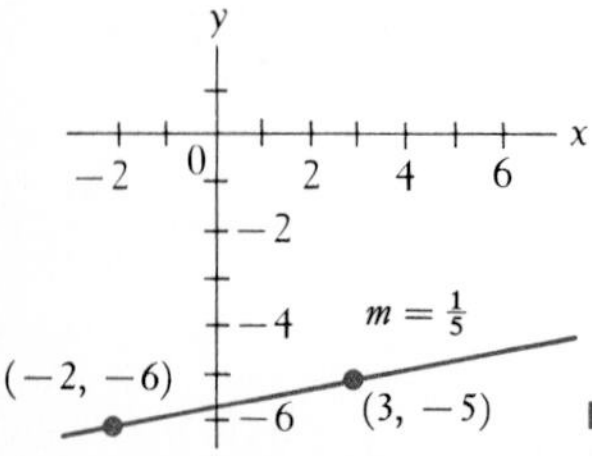

Fig. 20-5

The larger the numerical value of the slope of a line, the more nearly vertical is the line. Also, *a line rising to the right has a positive slope, and a line falling to the right has a negative slope.*

EXAMPLE D

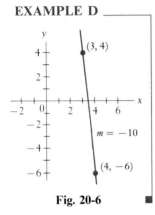

Fig. 20-6

The line through the two points in Example C has a positive slope, which is numerically small. From Fig. 20-5 it can be seen that the line rises slightly to the right.

The line joining (3, 4) and (4, −6) has a slope of

$$m = \frac{4 - (-6)}{3 - 4} = -10$$

This line falls sharply to the right (see Fig. 20-6).

We note again that we get the same value for the slope when we interchange the points as (x_1, y_1) and (x_2, y_2).

$$m = \frac{-6 - 4}{4 - 3} = -10$$

If a given line is extended indefinitely in either direction, it must cross the x-axis at some point unless it is parallel to the x-axis. *The angle measured from the x-axis in a positive direction to the line is called the* **inclination** *of the line* (see Fig. 20-7). The inclination of a line parallel to the x-axis is defined to be zero. *An alternative definition of slope, in terms of the inclination, is*

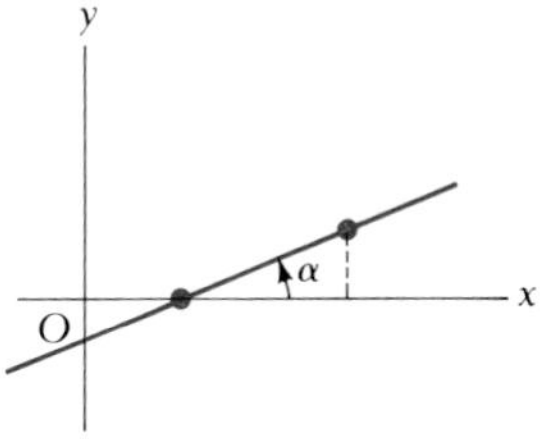

Fig. 20-7

$$m = \tan \alpha \qquad (0° \leq \alpha < 180°) \tag{20-3}$$

where α is the inclination. This can be seen from the fact that the slope can be defined in terms of any two points on the line. Thus, if we choose as one of these points that point where the line crosses the x-axis and any other point, we see from the definition of the tangent of an angle that Eq. (20-3) is in agreement with Eq. (20-2).

EXAMPLE E

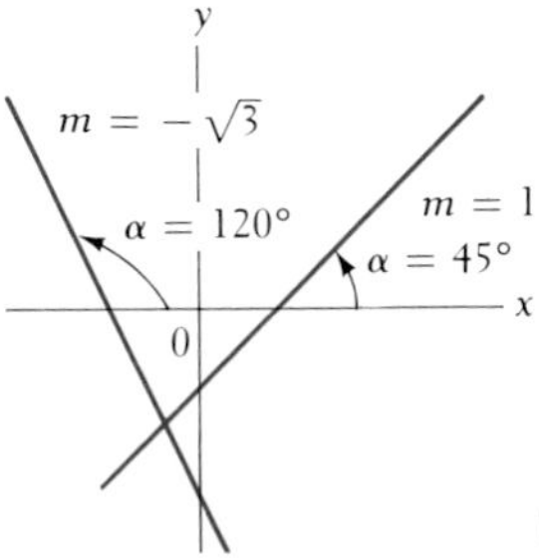

The slope of a line with an inclination of 45° is

$$m = \tan 45° = 1.000$$

The slope of a line having an inclination of 120° is

$$m = \tan 120° = -\sqrt{3} = -1.732$$

See Fig. 20-8.

We see that if the inclination is an acute angle, the slope is positive and the line rises to the right. If the inclination is obtuse, the slope is negative and the line falls to the right.

Fig. 20-8

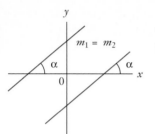

Fig. 20-9

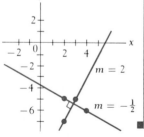

Fig. 20-10

Any two parallel lines crossing the x-axis will have the same inclination. Therefore, as we see in Fig. 20-9, the *slopes of parallel lines are equal*. This can be stated as

$$m_1 = m_2 \qquad \text{(for $\parallel$ lines)} \tag{20-4}$$

If two lines are perpendicular, this means that there must be 90° between their inclinations (Fig. 20-10). The relation between their inclinations is

$$\alpha_2 = \alpha_1 + 90°$$

which can be written as

$$90° - \alpha_2 = -\alpha_1$$

Taking the tangent of each of the angles in this last relation, we have

$$\tan(90° - \alpha_2) = \tan(-\alpha_1)$$

or

$$\cot \alpha_2 = -\tan \alpha_1$$

since a function of the complement of an angle is equal to the cofunction of that angle (see Section 3-4), and since $\tan(-\alpha) = -\tan \alpha$ (see Exercise 53 of Section 7-2). But $\cot \alpha = 1/\tan \alpha$, which means that $1/\tan \alpha_2 = -\tan \alpha_1$. Using the inclination definition of slope, we may write, as *the relation between slopes of perpendicular lines,*

$$m_2 = -\frac{1}{m_1} \quad \text{or} \quad m_1 m_2 = -1 \qquad \text{(for $\perp$ lines)} \tag{20-5}$$

EXAMPLE F

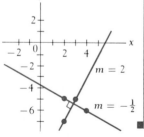

Fig. 20-11

The line through $(3, -5)$ and $(2, -7)$ has a slope of

$$m_1 = \frac{-5 + 7}{3 - 2} = 2$$

The line through $(4, -6)$ and $(2, -5)$ has a slope of

$$m_2 = \frac{-6 - (-5)}{4 - 2} = -\frac{1}{2}$$

Since the slopes of the two lines are negative reciprocals, we know that the lines are perpendicular. See Fig. 20-11. ∎

Using the formulas for distance and slope, we can show certain basic geometric relations. The following examples illustrate the use of the formulas, and thus show the use of algebra in solving problems which are basically geometric. This illustrates the methods of analytic geometry.

EXAMPLE G _____

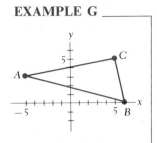

Fig. 20-12

Show that line segments joining $A(-5, 3)$, $B(6, 0)$, and $C(5, 5)$ form a right triangle (see Fig. 20-12).

If these points are vertices of a right triangle, the slopes of two of the sides must be negative reciprocals. This would show perpendicularity. Thus, we find the slopes of the three lines to be

$$m_{AB} = \frac{3 - 0}{-5 - 6} = -\frac{3}{11}, \qquad m_{AC} = \frac{3 - 5}{-5 - 5} = \frac{1}{5}, \qquad m_{BC} = \frac{0 - 5}{6 - 5} = -5$$

We see that the slopes of AC and BC are negative reciprocals, which means that $AC \perp BC$. From this we can conclude that the triangle is a right triangle. ■

EXAMPLE H _____

Fig. 20-13

Find the area of the triangle in Example G. See Fig. 20-13.

Since the right angle is at C, the legs of the triangle are AC and BC. The area is one-half the product of the lengths of the legs of a right triangle. The lengths of the legs are

$$d_{AC} = \sqrt{(-5 - 5)^2 + (3 - 5)^2} = \sqrt{104} = 2\sqrt{26}$$

and

$$d_{BC} = \sqrt{(6 - 5)^2 + (0 - 5)^2} = \sqrt{26}$$

Therefore, the area is

$$A = \frac{1}{2}(2\sqrt{26})(\sqrt{26}) = 26$$

■

Exercises 20-1

In Exercises 1 through 10, find the distances between the given pairs of points.

1. $(3, 8)$ and $(-1, -2)$

2. $(-1, 3)$ and $(-8, -4)$

3. $(4, -5)$ and $(4, -8)$

4. $(-3, 7)$ and $(2, 10)$

5. $(-12, 20)$ and $(32, -13)$

6. $(23, -9)$ and $(-25, 11)$

7. $(-4, -3)$ and $(3, -3)$

8. $(-2, 5)$ and $(-2, -2)$

9. $(1.22, -3.45)$ and $(-1.07, -5.16)$

10. $(-5.6, 2.3)$ and $(8.2, -7.5)$

In Exercises 11 through 20, find the slopes of the lines through the points in Exercises 1 through 10.

In Exercises 21 through 24, find the slopes of the lines with the given inclinations.

21. $30°$

22. $60°$

23. $150°$

24. $135°$

In Exercises 25 through 28, find the inclinations of the lines with the given slopes.

25. 0.364

26. 0.824

27. -6.691

28. -1.428

In Exercises 29 through 32, determine whether the lines through the two pairs of points are parallel or perpendicular.

29. $(6, -1)$ and $(4, 3)$, and $(-5, 2)$ and $(-7, 6)$

30. $(-3, 9)$ and $(4, 4)$, and $(9, -1)$ and $(4, -8)$

31. $(-1, -4)$ and $(2, 3)$, and $(-5, 2)$ and $(-19, 8)$

32. $(-1, -2)$ and $(3, 6)$, and $(2, -6)$ and $(5, 0)$

In Exercises 33 through 36, determine the value of k.

33. The distance between $(-1, 3)$ and $(11, k)$ is 13.

34. The distance between $(k, 0)$ and $(0, 2k)$ is 10.

35. The points $(6, -1)$, $(3, k)$, and $(-3, -7)$ are all on the same line.

36. The points in Exercise 35 are the vertices of a right triangle, with the right angle at $(3, k)$.

In Exercises 37 through 40, show that the given points are vertices of the indicated geometric figures.

37. Show that the points $(2, 3)$, $(4, 9)$, and $(-2, 7)$ are the vertices of an isosceles triangle.

38. Show that $(-1, 3)$, $(3, 5)$, and $(5, 1)$ are the vertices of a right triangle.

39. Show that $(-5, -4)$, $(7, 1)$, $(10, 5)$, and $(-2, 0)$ are the vertices of a parallelogram.

40. Show that $(-5, 6)$, $(0, 8)$, $(-3, 1)$, and $(2, 3)$ are the vertices of a square.

In Exercises 41 through 44, find the indicated areas and perimeters.

41. Find the area of the triangle of Exercise 38.   **42.** Find the area of the square of Exercise 40.

43. Find the perimeter of the triangle of Exercise 37.

44. Find the perimeter of the parallelogram of Exercise 39.

In Exercises 45 through 48, use the following definition to find the midpoints between the given points on a straight line.

The *midpoint* between points (x_1, y_1) and (x_2, y_2) on a straight line is the point

$$\left(\frac{x_1 + x_2}{2}, \frac{y_1 + y_2}{2} \right)$$

45. $(-4, 9)$ and $(6, 1)$ **46.** $(-1, 6)$ and $(-13, -8)$

47. $(-12, 25)$ and $(6, -17)$ **48.** $(2.6, 5.3)$ and $(-4.2, -2.7)$

20-2   The Straight Line

In Chapter 4 we derived the *slope-intercept form* of the equation of the straight line. Here we shall extend the development to include other forms of the equation of the straight line. Also, other methods of finding and applying these equations will be shown. For completeness, we shall review some of the material in Chapter 4.

Using the definition of slope, we can derive the general type of equation which always represents a straight line. This is another basic method of analytic geometry. That is, equations of a particular form can be shown to represent a particular type of curve. When we recognize the form of the equation, we know the kind of curve it represents. This can be of great assistance in sketching the graph.

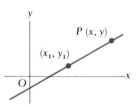

Fig. 20-14

A straight line can be defined as a "curve" with constant slope. By this we mean that for any two different points chosen on a given line, if the slope is calculated, the same value is always found. Thus, if we consider one point (x_1, y_1) on a line to be fixed (Fig. 20-14), and another point $P(x, y)$ which can *represent* any other point on the line, we have

$$m = \frac{y - y_1}{x - x_1}$$

which can be written as

$$\boxed{y - y_1 = m(x - x_1)}$$

(20-6)

point-slope form

Equation (20-6) is known as the **point-slope form** *of the equation of a straight line.* It is useful when we know the slope of a line and some point through which the line passes. Direct substitutions can then give us the equation of the line. Such information is often available about a given line.

EXAMPLE A

Find the equation of the line which passes through $(-4, 1)$ with a slope of -2.
By using Eq. (20-6), we find that

$$y - 1 = (-2)[x - (-4)]$$

slope

coordinates

which can be simplified to

$$y + 2x + 7 = 0$$

■ This line is shown in Fig. 20-15.

Fig. 20-15

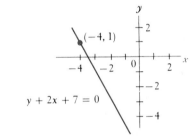

Fig. 20-16

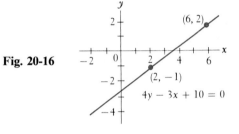

EXAMPLE B

Find the equation of the line through $(2, -1)$ and $(6, 2)$.
We first find the slope of the line through these points:

$$m = \frac{2 + 1}{6 - 2} = \frac{3}{4}$$

Then, by using either of the two known points and Eq. (20-6), we can find the equation of the line:

$$y - (-1) = \frac{3}{4}(x - 2)$$

or

$$4y + 4 = 3x - 6$$

or

$$4y - 3x + 10 = 0$$

■ This line is shown in Fig. 20-16.

Equation (20-6) can be used for any line except one parallel to the y-axis. Such a line has an undefined slope. However, it does have the property that all

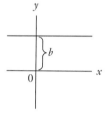

Fig. 20-17

points have the same x-coordinate, regardless of the y-coordinate. *We represent a line parallel to the y-axis (see Fig. 20-17) as*

$$x = a \qquad (20\text{-}7)$$

A line parallel to the x-axis has a slope of 0. From Eq. (20-6), we can find its equation to be $y = y_1$. To keep the same form as Eq. (20-7), we normally write this as

$$y = b \qquad (20\text{-}8)$$

See Fig. 20-18.

Fig. 20-18   **Fig. 20-19**   **Fig. 20-20**

EXAMPLE C ⌐——— The line $x = 2$ is a line parallel to the y-axis and two units to the right of it. This line is shown in Fig. 20-19.

The line $y = -4$ is a line parallel to the x-axis and 4 units below it. This line is shown in Fig. 20-20. ■

If we choose the special point $(0, b)$, which is the y-intercept of the line, as the point to use in Eq. (20-6), we have

$$y - b = m(x - 0)$$

or

$$y = mx + b \qquad (20\text{-}9)$$

slope-intercept form   *Equation (20-9) is the **slope-intercept form** of the equation of a straight line,* and we first derived it in Chapter 4. Its primary usefulness lies in the fact that once we find the equation of a line and then write it in slope-intercept form, we know that the slope of the line is the coefficient of the x-term and that it crosses the y-axis with the coordinate indicated by the constant term. See Fig. 20-21.

Fig. 20-21

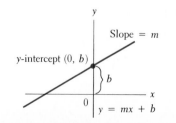

EXAMPLE D

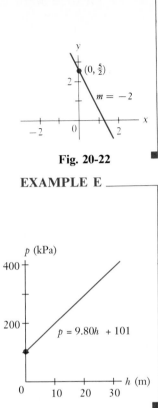

Fig. 20-22

Find the slope and the y-intercept of the straight line whose equation is $2y + 4x - 5 = 0$.

We write this equation in slope-intercept form:

$$2y = -4x + 5$$

slope ┐ ┌ y-coordinate of intercept

$$y = -2x + \frac{5}{2}$$

Since the coefficient of x in this form is -2, the slope of the line is -2. The constant on the right is $\frac{5}{2}$, which means that the y-intercept is $(0, \frac{5}{2})$. See Fig. 20-22. ■

EXAMPLE E

The pressure p_0 at the surface of a body of water (due to the atmosphere) is 101 kPa. The pressure p at a depth of 10.0 m is 199 kPa. In general, the pressure difference $p - p_0$ varies directly as the depth h. Sketch a graph of p as a function of h.

The solution is as follows:

$$p - p_0 = kh \qquad \text{direct variation}$$
$$199 - 101 = k(10.0) \qquad \text{substitute given values}$$
$$k = 9.80 \text{ kPa/m}$$
$$p - 101 = 9.80h \qquad \text{substitute in first equation}$$
$$p = 9.80h + 101$$

We see that this is the equation of a straight line. The slope is 9.80, and the p-intercept is $(0, 101)$. Negative values do not have any physical meaning. The graph is shown in Fig. 20-23. ■

p (kPa)

400

200

$p = 9.80h + 101$

0   10   20   30 h (m)

Fig. 20-23

From Eqs. (20-6) and (20-9), and from the examples of this section, we see that the equation of the straight line has certain characteristics: we have a term in y, a term in x, and a constant term if we simplify as much as possible. *This form is represented by the equation.*

general form

$$\boxed{Ax + By + C = 0}$$

(20-10)

which is known as the **general form** *of the equation of the straight line.* We have seen this form before in Chapter 4. Now we have shown why it represents a straight line.

EXAMPLE F

Find the general form of the equation of the line which is parallel to the line $3x + 2y - 6 = 0$ and which passes through the point $(-1, 2)$.

Since the line whose equation we want is parallel to the line $3x + 2y - 6 = 0$, it must have the same slope. Therefore, writing $3x + 2y - 6 = 0$ in slope-intercept form, we have

$$3x + 2y - 6 = 0$$
$$2y = -3x + 6$$
$$y = -\frac{3}{2}x + 3$$

(Continued on next page)

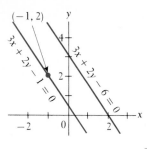

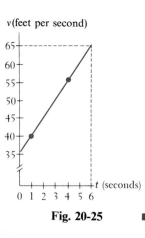

Fig. 20-24

Therefore, the slope of $3x + 2y - 6 = 0$ is $-\frac{3}{2}$, which means that the slope of the required line is also $-\frac{3}{2}$. We use $m = -\frac{3}{2}$, the given point $(-1, 2)$, and the point-slope form to find the equation.

$$y - 2 = -\frac{3}{2}(x + 1)$$

$$2y - 4 = -3(x + 1)$$

$$2y - 4 = -3x - 3$$

$$3x + 2y - 1 = 0$$

This last form is the general form of the equation of the line. This line and the line $3x + 2y - 6 = 0$ are shown in Fig. 20-24.

In many physical situations a linear relationship exists between variables. A few examples of situations where such relationships exist are between (1) the distance traveled by an object and the elapsed time, when the velocity is constant, (2) the amount a spring stretches and the force applied, (3) the change in electric resistance and the change in temperature, (4) the force applied to an object and the resulting acceleration, and (5) the pressure at a certain point within a liquid and the depth of the point. The following example illustrates the use of a straight line in dealing with an applied problem.

EXAMPLE G

Under the condition of constant acceleration, the velocity of an object varies linearly with the time. If after 1 s a certain object has a velocity of 40 ft/s, and 3 s later it has a velocity of 55 ft/s, find the equation relating the velocity and time, and graph this equation. From the graph, determine the initial velocity (the velocity when $t = 0$) and the velocity after 6 s.

If we treat the velocity v as the dependent variable, and the time t as the independent variable, the slope of the straight line is

$$m = \frac{v_2 - v_1}{t_2 - t_1}$$

Using the information given in the problem, we have

$$m = \frac{55 - 40}{4 - 1} = 5$$

Then, using the point-slope form of the equation of a straight line, we have

$$v - 40 = 5(t - 1)$$

$$v = 5t + 35$$

This is the required equation. The graph is shown in Fig. 20-25. For purposes of graphing the line, the values given are sufficient. Of course, there is no need to include negative values of t, since these have no physical meaning. From the graph we see that the line crosses the v-axis at 35. This means that the initial velocity is 35 ft/s. Also, when $t = 6$ we see that $v = 65$ ft/s.

Fig. 20-25

Exercises 20-2

In Exercises 1 through 20, find the equation of each of the lines with the given properties. Sketch the graph of each line.

1. Passes through $(-3, 8)$ with a slope of 4.

2. Passes through $(-2, -1)$ with a slope of -2.

3. Passes through $(2, -5)$ and $(4, 2)$.

4. Passes through $(-3, 5)$ and $(-2, 3)$.

5. Passes through $(1, 3)$ and has an inclination of $45°$.

6. Has a y-intercept $(0, -2)$ and an inclination of $120°$.

7. Passes through $(6, -3)$ and is parallel to the x-axis.

8. Passes through $(-4, -2)$; perpendicular to x-axis.

9. Is parallel to the y-axis and is 3 units to the left of it.

10. Is parallel to the x-axis and is 5 units below it.

11. Has an x-intercept $(4, 0)$ and a y-intercept of $(0, -6)$.

12. Has an x-intercept of $(-3, 0)$ and a slope of 2.

13. Is perpendicular to a line with a slope of 3 and passes through $(1, -2)$.

14. Is perpendicular to a line with a slope of -4 and has a y-intercept of $(0, 3)$.

15. Is parallel to a line with a slope of $-\frac{1}{2}$ and has an x-intercept of $(4, 0)$.

16. Is parallel to a line through $(-1, 2)$ and $(3, 1)$ and passes through $(1, 2)$.

17. Is parallel to a line through $(7, -1)$ and $(4, 3)$ and has a y-intercept of $(0, -2)$.

18. Is perpendicular to the line joining $(4, 2)$ and $(3, -5)$ and passes through $(4, 2)$.

19. Is perpendicular to the line $5x - 2y - 3 = 0$ and passes through $(3, -4)$.

20. Is parallel to the line $2y - 6x - 5 = 0$ and passes through $(-4, -5)$.

In Exercises 21 through 24, draw the lines with the given equations.

21. $4x - y = 8$

22. $2x - 3y - 6 = 0$

23. $3x + 5y - 10 = 0$

24. $4y = 6x - 9$

In Exercises 25 through 28, reduce the equations to slope-intercept form and determine the slope and y-intercept.

25. $3x - 2y - 1 = 0$

26. $4x + 2y - 5 = 0$

27. $5x - 2y + 5 = 0$

28. $6x - 3y - 4 = 0$

In Exercises 29 through 32, determine the value of k.

29. What is the value of k if the lines $4x - ky = 6$ and $6x + 3y + 2 = 0$ are to be parallel?

30. What must k equal in Exercise 29 if the given lines are to be perpendicular?

31. What must be the value of k if the lines $3x - y = 9$ and $kx + 3y = 5$ are to be perpendicular?

32. What must k equal in Exercise 31 if the given lines are to be parallel?

In Exercises 33 and 34, show that the given lines are parallel. In Exercises 35 and 36, show that the given lines are perpendicular.

33. $3x - 2y + 5 = 0$ and $4y = 6x - 1$

34. $3y - 2x = 4$ and $6x - 9y = 5$

35. $6x - 3y - 2 = 0$ and $x + 2y - 4 = 0$

36. $4x - y + 2 = 0$ and $2x + 8y - 1 = 0$

In Exercises 37 through 40, find the equations of the given lines.

37. The line which has an x-intercept of $(4, 0)$ and a y-intercept the same as the line $2y - 3x - 4 = 0$.

38. The line which is perpendicular to the line $8x + 2y - 3 = 0$ and has the same x-intercept as this line.

39. The line with a slope of -3 which also passes through the intersection of the lines $5x - y = 6$ and $x + y = 12$.

40. The line which passes through the point of intersection of $2x + y - 3 = 0$ and $x - y - 3 = 0$ and through the point $(4, -3)$.

In Exercises 41 through 52, some applications involving straight lines are shown.

41. The velocity v of a box sliding down a long ramp is given by $v = v_0 + at$, where v_0 is the initial velocity, a is the acceleration, and t is the time. If $v_0 = 12.2$ ft/s and $v = 35.4$ ft/s when $t = 4.50$ s, find v as a function of t. Sketch the graph.

42. The voltage V across part of an electric circuit is given by $V = E - iR$, where E is a battery voltage, i is the current, and R is the resistance. If $E = 6.00$ V and $V = 4.35$ V for $i = 9.17$ mA, find V as a function of i. Sketch the graph (i and V may be negative).

43. The length ℓ of a pulley belt is 8 cm longer than twice the circumference c of one of the pulley wheels. Express ℓ as a function of c.

44. An acid solution is made from x liters of a 20% solution and y liters of a 30% solution. If the final solution contains 20 L of acid, find the equation relating x and y.

45. One microcomputer printer can print x characters per second and a second printer can print y characters per second. If the first prints for 50 s and the second for 60 s and they print a total of 12,200 characters, find the equation relating x and y.

46. A wall is 15 cm thick. At the outside, the temperature is 3°C and at the inside, it is 23°C. If the temperature changes at a constant rate through the wall, write an equation of the temperature T in the wall as a function of the distance x from the outside to the inside of the wall.

47. One heating unit uses x gallons of fuel at 72% efficiency and another heating unit uses y gallons at 90% efficiency. If 135,000 Btu of heat is delivered by these units together, express y as a function of x.

48. The length of a rectangular solar cell is 10 cm more than the width. Express the perimeter p of the cell as a function of the width w.

49. A light beam is reflected off the edge of an optic fiber at an angle of 0.0032°. The diameter of the fiber is 48 μm. Find the equation of the reflected beam with the x-axis (at the center of the fiber) and the y-axis as shown in Fig. 20-26.

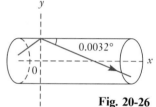

Fig. 20-26

50. A police report stated that a bullet caromed downward off a wall at an angle of 16.5° with the wall, as shown in Fig. 20-27. What is the equation of the path of the bullet after impact?

51. A survey of the traffic on a particular highway showed that the number of cars passing a particular point each minute varied linearly from 6:30 A.M. to 8:30 A.M. on workday mornings. The study showed that an average of 45 cars passed the point in one minute at 7 A.M. and that 115 cars passed in one minute at 8 A.M. If n is the number of cars passing the point in one minute and t is the number of minutes after 6:30 A.M., find the equation relating n and t, and graph the equation. From the graph, determine n at 6:30 A.M. and at 8:30 A.M.

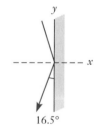

52. In a research project on cancer, a tumor was determined to weigh 30 mg when first discovered. While being treated, it grew smaller by 2 mg each month. Find the equation relating the weight w of the tumor as a function of the time t in months. Graph the equation.

Fig. 20-27

In Exercises 53 through 56, treat the given nonlinear functions as linear functions in order to sketch their graphs. As an example, $y = 2 + 3x^2$ can be sketched as a straight line by graphing y as a function of x^2. A table of values for this graph is shown along with the corresponding graph in Fig. 20-28.

53. The number n of memory cells of a certain computer which can be tested in t seconds is given by $n = 1200\sqrt{t}$. Sketch the graph by graphing n as a function of $\sqrt{t}$.

54. The force F, in pounds, applied to a lever to balance a certain weight on the opposite side of the fulcrum is given by $F = 40/d$, where d is the distance, in feet, of the force from the fulcrum. Sketch the graph by graphing F as a function of $1/d$.

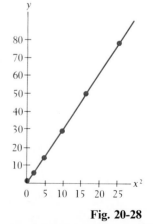

x	0	1	2	3	4	5
x^2	0	1	4	9	16	25
y	2	5	14	29	50	77

Fig. 20-28

55. A spacecraft is launched such that its altitude h, in kilometers, is given by $h = 300 + 2t^{3/2}$ for $0 \le t < 100$ s. Sketch this as a linear function.

56. The current i, in amperes, in a certain electric circuit is given by $i = 6(1 - e^{-t})$. Sketch this as a linear function.

In Exercises 57 through 60, plot the given nonlinear functions on semilogarithmic or logarithmic paper to show that they are linear when plotted. In Section 12-7 we noted that curves plotted on logarithmic and semilogarithmic paper often become straight lines. Since distances corresponding to log y and log x are plotted automatically if the graph is logarithmic in the respective directions, many nonlinear functions are linear when plotted on this paper.

57. A function of the form $y = ax^n$ is straight when plotted on logarithmic paper, since $\log y = \log a + n \log x$ is in the form of a straight line. The variables are $\log y$ and $\log x$; the slope can be found from $(\log y - \log a)/\log x = n$, and the intercept is a. (To get the slope from the graph, it is necessary to measure vertical and horizontal distances between two points. The $\log y$ intercept is found where $\log x = 0$, and this occurs when $x = 1$.) Plot $y = 3x^4$ on logarithmic paper to verify this analysis.

58. A function of the form $y = a(b^x)$ is a straight line on semilogarithmic paper, since $\log y = \log a + x \log b$ is in the form of a straight line. The variables are $\log y$ and x, the slope is $\log b$, and the intercept is a. [To get the slope from the graph, we must calculate $(\log y - \log a)/x$ for some set of values x and y. The intercept is read directly off the graph where $x = 0$.] Plot $y = 3(2^x)$ on semilogarithmic paper to verify this analysis.

59. If experimental data are plotted on logarithmic paper and the points lie on a straight line, it is possible to determine the function (see Exercise 57). The following data come from an experiment to determine the functional relationship between the pressure p and volume V of a gas undergoing an adiabatic (no heat loss) change. From the graph on logarithmic paper, determine p as a function of V.

V (m³)	0.100	0.500	2.00	5.00	10.0
p (kPa)	20.1	2.11	0.303	0.0840	0.0318

60. If experimental data are plotted on semilogarithmic paper, and the points lie on a straight line, it is possible to determine the function (see Exercise 58). The following data come from an experiment designed to determine the relationship between the voltage across an inductor and the time, after the switch is opened. Determine v as a function of t.

v (volts)		40	15	5.6	2.2	0.8
t (milliseconds)	0.0	20	40	60	80

20-3 The Circle

We have found that we can obtain a general equation which represents a straight line by considering a fixed point on the line and then a general point $P(x, y)$ which can represent any other point on the same line. Mathematically we can state this as "the line is the **locus** of a point $P(x, y)$ which *moves* along the line." That is, the point $P(x, y)$ can be considered as a variable point which moves along the line.

In this way we can define a number of important curves. *A **circle** is defined as the locus of a point $P(x, y)$ which moves so that it is always equidistant from a fixed point. We call this fixed distance the **radius**, and we call the fixed point the **center** of the circle.* Thus, using this definition, calling the fixed point (h, k) and the radius r, we have

$$\sqrt{(x - h)^2 + (y - k)^2} = r$$

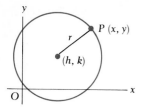

Fig. 20-29

or, by squaring both sides, we have

$$(x - h)^2 + (y - k)^2 = r^2 \qquad (20\text{-}11)$$

Equation (20-11) is called the **standard equation** of a circle with center at (h, k) and radius r (Fig. 20-29).

EXAMPLE A

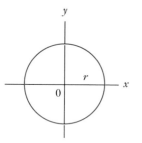

Fig. 20-30

The equation $(x - 1)^2 + (y + 2)^2 = 16$ represents a circle with center at $(1, -2)$ and a radius of 4. We determine these values by considering the equation of this circle to be in the form of Eq. (20-11) as

form requires − signs

$$(x - 1)^2 + [y - (-2)]^2 = 4^2$$

coordinates of center   radius

Note carefully the way in which we find the y-coordinate of the center. This circle is shown in Fig. 20-30. ■

If the center of the circle is at the origin, which means that the coordinates of the center are (0, 0), the equation of the circle (see Fig. 20-31) becomes

$$x^2 + y^2 = r^2 \qquad (20\text{-}12)$$

Fig. 20-31

symmetry

A circle of this type clearly exhibits an important property of the graphs of many equations. *It is **symmetrical** to the x-axis and also to the y-axis.* Symmetry to the x-axis can be thought of as meaning that the lower half of the curve is a reflection of the upper half, and conversely. It can be shown that *if −y can replace y in an equation without changing the equation, the graph of the equation is **symmetrical** to the x-axis.* Symmetry to the y-axis is similar. *If −x can replace x in the equation without changing the equation, the graph is **symmetrical** to the y-axis.*

This type of circle is symmetrical to the origin as well as being symmetrical to both axes. The meaning of symmetry to the origin is that the origin is the midpoint of any two points (x, y) and $(-x, -y)$ which are on the curve. Thus, *if −x can replace x, and −y can replace y at the same time, without changing the equation, the graph of the equation is **symmetrical** to the origin.*

EXAMPLE B

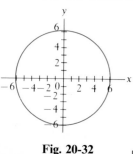

Fig. 20-32

The equation of the circle with its center at the origin and with a radius of 6 is $x^2 + y^2 = 36$.

The symmetry of this circle can be shown analytically by the substitutions mentioned above. Replacing x by $-x$, we obtain $(-x)^2 + y^2 = 36$. Since $(-x)^2 = x^2$, this equation can be rewritten as $x^2 + y^2 = 36$. Since this substitution did not change the equation, the graph is symmetrical to the y-axis.

Replacing y by $-y$, we obtain $x^2 + (-y)^2 = 36$, which is the same as $x^2 + y^2 = 36$. This means that the curve is symmetrical to the x-axis.

Replacing x by $-x$, and simultaneously replacing y by $-y$, we obtain $(-x)^2 + (-y)^2 = 36$, which is the same as $x^2 + y^2 = 36$. This means that the curve is symmetrical to the origin. The circle is shown in Fig. 20-32. ■

EXAMPLE C

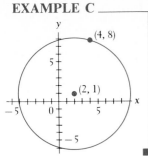

Fig. 20-33

Find the equation of the circle with center at (2, 1) and which passes through (4, 8).

In Eq. (20-11), we can determine the equation if we can find h, k, and r for this circle. From the given information, $h = 2$ and $k = 1$. To find r, we use the fact that *all points on the circle must satisfy the equation of the circle.* The point (4, 8) must satisfy Eq. (20-11), with $h = 2$ and $k = 1$. Thus, $(4 - 2)^2 + (8 - 1)^2 = r^2$. From this relation we find $r^2 = 53$. The equation of the circle is

$$(x - 2)^2 + (y - 1)^2 = 53$$

■ This circle is shown in Fig. 20-33.

EXAMPLE D

A student is drawing a friction drive in which two circular disks are in contact with each other. They are represented by circles in the drawing. The first has a radius of 10.0 cm and the second has a radius of 12.0 cm. What is the equation of each circle if the origin is at the center of the first circle and the positive x-axis passes through the center of the second circle? See Fig. 20-34.

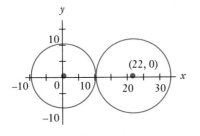

Fig. 20-34

Since the center of the smaller circle is at the origin, we can use Eq. (20-12). Given that the radius is 10.0 cm, we have as its equation

$$x^2 + y^2 = 100$$

The fact that the two disks are in contact tells us that they meet at the point (10.0, 0). Knowing that the radius of the larger circle is 12.0 cm tells us that its center is at (22.0, 0). Thus, using Eq. (20-11) with $h = 22.0$, $k = 0$, and $r = 12.0$, we have

$$(x - 22.0)^2 + (y - 0)^2 = 12.0^2$$

or

$$(x - 22.0)^2 + y^2 = 144$$

as the equation of the larger circle.

If we multiply out each of the terms in Eq. (20-11), we may combine the resulting terms to obtain

$$x^2 - 2hx + h^2 + y^2 - 2ky + k^2 = r^2$$

general equation

$$\boldsymbol{x^2 + y^2 - 2hx - 2ky + (h^2 + k^2 - r^2) = 0} \tag{20-13}$$

Since each of h, k, and r is constant for any given circle, the coefficients of x and y and the term within parentheses in Eq. (20-13) are constants. Equation (20-13) can then be written as

$$\boxed{x^2 + y^2 + Dx + Ey + F = 0} \tag{20-14}$$

Equation (20-14) is called the **general equation** *of the circle. It tells us that any equation which can be written in that form will represent a circle.*

EXAMPLE E

Find the center and radius of the circle

$$x^2 + y^2 - 6x + 8y - 24 = 0$$

NOTE ▷ We can find this information if we write the given equation in standard form. To do so, **we must complete the square in the x-terms and also in the y-terms.** This is done by first writing the equation in the form

$$(x^2 - 6x \quad) + (y^2 + 8y \quad) = 24$$

To complete the square of the x-terms, we take half of −6, which is −3, square it, and add the result, 9, to each side of the equation. In the same way, we complete the square of the y-terms by adding 16 to each side of the equation, which gives

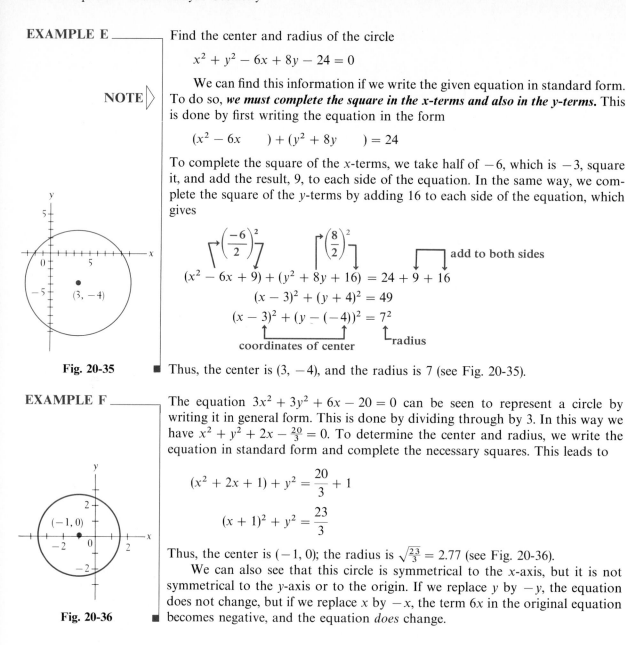

$$\left[\left(\frac{-6}{2}\right)^2\right] \qquad \left[\left(\frac{8}{2}\right)^2\right] \qquad \qquad \text{add to both sides}$$

$$(x^2 - 6x + 9) + (y^2 + 8y + 16) = 24 + 9 + 16$$

$$(x - 3)^2 + (y + 4)^2 = 49$$

$$(x - 3)^2 + (y - (-4))^2 = 7^2$$

coordinates of center   radius

Fig. 20-35

■ Thus, the center is (3, −4), and the radius is 7 (see Fig. 20-35).

EXAMPLE F

The equation $3x^2 + 3y^2 + 6x - 20 = 0$ can be seen to represent a circle by writing it in general form. This is done by dividing through by 3. In this way we have $x^2 + y^2 + 2x - \frac{20}{3} = 0$. To determine the center and radius, we write the equation in standard form and complete the necessary squares. This leads to

$$(x^2 + 2x + 1) + y^2 = \frac{20}{3} + 1$$

$$(x + 1)^2 + y^2 = \frac{23}{3}$$

Thus, the center is (−1, 0); the radius is $\sqrt{\frac{23}{3}} = 2.77$ (see Fig. 20-36).

We can also see that this circle is symmetrical to the x-axis, but it is not symmetrical to the y-axis or to the origin. If we replace y by −y, the equation does not change, but if we replace x by −x, the term 6x in the original equation becomes negative, and the equation *does* change.

Fig. 20-36

Exercises 20-3

In Exercises 1 through 4, determine the center and radius of each circle.

1. $(x - 2)^2 + (y - 1)^2 = 25$

2. $(x - 3)^2 + (y + 4)^2 = 49$

3. $(x + 1)^2 + y^2 = 4$

4. $x^2 + (y - 6)^2 = 64$

In Exercises 5 through 20, find the equation of each of the circles from the given information.

5. Center at (0, 0), radius 3

6. Center at (0, 0), radius 1

7. Center at (2, 2), radius 4

8. Center at (0, 2), radius 2

9. Center at $(-2, 5)$, radius $\sqrt{5}$

10. Center at $(-3, -5)$, radius $2\sqrt{3}$

11. Center at $(12, -15)$, radius 18

12. Center at $(\frac{3}{2}, -2)$, radius $\frac{5}{2}$

13. Center at $(2, 1)$, passes through $(4, -1)$

14. Center at $(-1, 4)$, passes through $(-2, 3)$

15. Center at $(-3, 5)$, tangent to the x-axis

16. Center at $(2, -4)$, tangent to the y-axis

17. Tangent to both axes and the lines $y = 4$ and $x = 4$

18. Tangent to both axes, radius 4, and in the second quadrant

19. Center on the line $5x = 2y$, radius 5, tangent to the x-axis

20. The points $(3, 8)$ and $(-3, 0)$ are the ends of a diameter

In Exercises 21 through 32, determine the center and radius of each circle. Sketch each circle.

21. $x^2 + (y - 3)^2 = 4$

22. $(x - 2)^2 + (y + 3)^2 = 49$

23. $4(x + 1)^2 + 4(y - 5)^2 = 81$

24. $2(x + 4)^2 + 2(y + 3)^2 = 25$

25. $x^2 + y^2 - 25 = 0$

26. $x^2 + y^2 - 9 = 0$

27. $x^2 + y^2 - 2x - 8 = 0$

28. $x^2 + y^2 - 4x - 6y - 12 = 0$

29. $x^2 + y^2 + 8x - 10y - 8 = 0$

30. $x^2 + y^2 + 8x + 6y = 0$

31. $2x^2 + 2y^2 - 4x - 8y - 1 = 0$

32. $3x^2 + 3y^2 - 12x + 4 = 0$

In Exercises 33 through 36, determine whether the circles with the given equations are symmetrical to either axis or the origin.

33. $x^2 + y^2 = 100$

34. $x^2 + y^2 - 4x - 5 = 0$

35. $x^2 + y^2 + 8y - 9 = 0$

36. $x^2 + y^2 - 2x + 4y - 3 = 0$

In Exercises 37 through 44, find the indicated quantities.

37. Determine whether the circle $x^2 - 6x + y^2 - 7 = 0$ crosses the x-axis.

38. Find the points of intersection of the circle $x^2 + y^2 - x - 3y = 0$ and the line $y = x - 1$.

39. Find the locus of a point $P(x, y)$ which moves so that its distance from $(2, 4)$ is twice its distance from $(0, 0)$. Describe the locus.

40. Find the equation of the locus of a point $P(x, y)$ which moves so that the line joining it and $(2, 0)$ is always perpendicular to the line joining it and $(-2, 0)$. Describe the locus.

41. A wire is rotating in a circular path through a magnetic field to induce an electric current in the wire. The wire is rotating at 60.0 Hz with a constant velocity of 37.7 m/s. Taking the origin at the center of rotation, find the equation of the path of the wire.

42. A communications satellite remains stationary at an altitude of 22,500 mi over a point on the earth's equator. It therefore rotates once each day about the earth's center. Its velocity is constant, but the horizontal and vertical components, v_H and v_V, of the velocity constantly change. Show that the equation relating v_H and v_V, measured in mi/h, is that of a circle. The radius of the earth is 3960 mi.

43. In analyzing the stress on a beam, *Mohr's circle* is often used. To form it, normal stress is plotted as the x-coordinate and shear stress is plotted as the y-coordinate. The center of the circle is midway between the minimum and maximum values of normal stress on the x-axis. Find the equation of Mohr's circle if the minimum normal stress is 100×10^{-6} and the maximum normal stress is 900×10^{-6} (stress is unitless). Sketch the graph.

44. A Norman window has the form of a rectangle surmounted by a semicircle. An architect designs a Norman window on a coordinate system as shown in Fig. 20-37. If the circumference of the circular part of the window is on the circle $x^2 + y^2 - 3.00y + 1.25 = 0$, find the area of the window. Measurements are in meters.

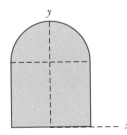

Fig. 20-37

20-4 The Parabola

Another important curve is the parabola. We have come across this curve several times in the earlier chapters. In Chapter 6 we showed that the graph of a quadratic function is a parabola. In this section we define the parabola more generally, and thereby find the general form of its equation.

A **parabola** *is defined as the locus of a point P(x, y) which moves so that it is always equidistant from a given line and a given point. The given line is called the* **directrix,** *and the given point is called the* **focus.** The line through the focus which is perpendicular to the directrix is called the **axis** of the parabola. The point midway between the directrix and focus is the **vertex** of the parabola. Using this definition, we shall find the equation of the parabola for which the focus is the point $(p, 0)$ and the directrix is the line $x = -p$. By choosing the focus and directrix in this manner, we shall be able to find a general representation of the equation of a parabola with its vertex at the origin.

directrix
focus
vertex

According to the definition of the parabola, the distance from a point $P(x, y)$ on the parabola to the focus $(p, 0)$ must equal the distance from $P(x, y)$ to the directrix $x = -p$. The distance from P to the focus can be found by use of the distance formula. The distance from P to the directrix is the perpendicular distance, and this can be found as the distance between two points on a line parallel to the x-axis. These distances are indicated in Fig. 20-38.

Thus, we have

$$\sqrt{(x - p)^2 + (y - 0)^2} = x + p$$

Squaring both sides of this equation, we have

$$(x - p)^2 + y^2 = (x + p)^2$$

or

$$x^2 - 2px + p^2 + y^2 = x^2 + 2px + p^2$$

Simplifying, we obtain

$$y^2 = 4px \qquad (20\text{-}15)$$

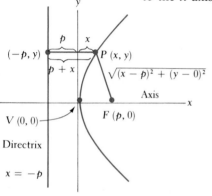

Fig. 20-38

Equation (20-15) is called the **standard form** of the equation of a parabola with its axis along the x-axis and the vertex at the origin. Its symmetry to the x-axis can be proven since $(-y)^2 = 4px$ is the same as $y^2 = 4px$.

EXAMPLE A

Find the coordinates of the focus and the equation of the directrix, and sketch the graph of the parabola $y^2 = 12x$.

Since the equation of this parabola fits the form of Eq. (20-15) we know that the vertex is at the origin. The coefficient of 12 tells us that

$$4p = 12, \qquad p = 3$$

Since $p = 3$, the focus is the point $(3, 0)$ and the directrix is the line $x = -3$, as shown in Fig. 20-39.

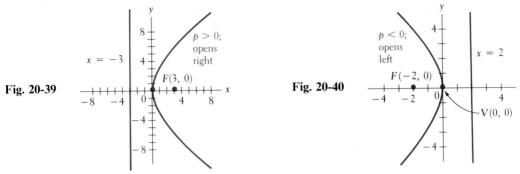

Fig. 20-39

Fig. 20-40

EXAMPLE B

If the focus is to the left of the origin, with the directrix an equal distance to the right, the coefficient of the x-term will be negative. This tells us that the parabola opens to the left, rather than to the right, as is the case when the focus is to the right of the origin. For example, the parabola $y^2 = -8x$ has its vertex at the origin, its focus at $(-2, 0)$, and the line $x = 2$ as its directrix. We determine this from the equation as follows:

$$y^2 = -8x, \qquad 4p = -8 \qquad p = -2$$

Focus $(p, 0)$ is $(-2, 0)$; directrix $x = -p$ is $x = -(-2) = 2$. The parabola opens to the left, as shown in Fig. 20-40. ∎

If we chose the focus as the point $(0, p)$ and the directrix as the line $y = -p$ (see Fig. 20-41), we would find that the resulting equation is

$$\boxed{x^2 = 4py} \tag{20-16}$$

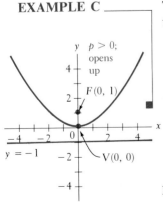

Fig. 20-41

This is the standard form of the equation of a parabola with the y-axis as its axis and the vertex at the origin. Its symmetry to the y-axis can be proven, since $(-x)^2 = 4py$ is the same as $x^2 = 4py$. We note that ***the difference between this equation and Eq. (20-15) is that x is squared and y appears to the first power in Eq. (20-16), rather than the reverse, as in Eq. (20-15).***

NOTE ▷

EXAMPLE C

The parabola $x^2 = 4y$ fits the form of Eq. (20-16). Therefore, its axis is along the y-axis and its vertex is at the origin. From the equation, we have

$$x^2 = 4y, \qquad 4p = 4, \qquad p = 1$$

Focus $(0, p)$ is $(0, 1)$; directrix $y = -p$ is $y = -1$. The parabola is shown in Fig. 20-42, and we see in this case that it opens upward. ∎

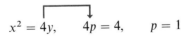

Fig. 20-42

EXAMPLE D

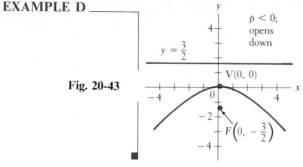

Fig. 20-43

The parabola $x^2 = -6y$ fits the form of Eq. (20-16) with $4p = -6$. Therefore, its axis is along the y-axis and its vertex is at the origin. Since $4p = -6$, we have

$$p = -\frac{3}{2}, \quad \text{focus}\left(0, -\frac{3}{2}\right), \quad \text{directrix } y = \frac{3}{2}$$

The parabola opens downward, as shown in Fig. 20-43.

EXAMPLE E

See the chapter introduction.

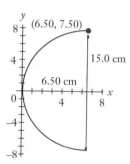

Fig. 20-44

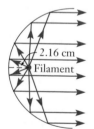

Fig. 20-45

In calculus it can be shown that a light ray coming from the focus of a parabola will be reflected off the parabolic surface parallel to the axis of the parabola. This property of a parabola involving light reflection is very important. One of its important uses is in the design of automobile headlights.

An automobile headlight reflector is designed such that the cross sections of the reflecting surface are equal parabolas. It has an opening of 15.0 cm and is 6.50 cm deep, as shown in Fig. 20-44. Determine where the filament of the bulb should be located so that reflected light rays are parallel in order to create a light beam.

By finding the equation of the parabola we can find the location of the focus, which is the desired location of the filament. Knowing that the parabolic opening is 15.0 cm wide and 6.50 cm deep tells us that the point (6.50, 7.50) is on the parabola.

Since we have shown that the parabola has its vertex at the origin and its axis along the x-axis, its general form is given by Eq. (20-15), or $y^2 = 4px$. We can find the value of p by using the fact that (6.50, 7.50) is a point on the parabola and the fact that the coordinates must satisfy the equation. This means that

$$7.50^2 = 4p(6.50) \quad \text{or} \quad p = 2.16$$

Therefore, the equation of the parabola is $y^2 = 8.64x$. The filament should be located at the focus, which means it should be on the axis of the parabola, 2.16 cm from the vertex. In this way the reflected rays are parallel to the axis, as shown in Fig. 20-45.

Equations (20-15) and (20-16) give us the general form of the equation of a parabola with its vertex at the origin and its focus on one of the coordinate axes. The following example illustrates the use of the definition of the parabola to find the equation of a parabola which has its vertex at a point other than at the origin.

EXAMPLE F

Using the definition of the parabola, determine the equation of the parabola with its focus at (2, 3) and its directrix the line $y = -1$ (see Fig. 20-46).

Choosing a general point $P(x, y)$ on the parabola, and equating the distances from this point to (2, 3) and to the line $y = -1$, we have

$$\sqrt{(x - 2)^2 + (y - 3)^2} = y + 1$$

distance P to $F =$ distance P to $y = -1$

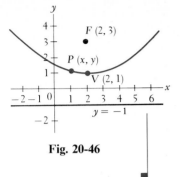

Fig. 20-46

Squaring both sides of this equation and simplifying, we have

$$(x - 2)^2 + (y - 3)^2 = (y + 1)^2$$
$$x^2 - 4x + 4 + y^2 - 6y + 9 = y^2 + 2y + 1$$

or

$$8y = 12 - 4x + x^2$$

When we see it in this form, we note that this type of equation has appeared numerous times in earlier chapters. The x-term and the constant (12 in this case) are characteristic of a parabola which does not have its vertex at the origin if the directrix is parallel to the x-axis.

Thus we can conclude that *a parabola is characterized by the presence of the square of either (but not both) x or y, and a first-power term in the other.* In this way we can recognize a parabola by inspecting the equation.

Exercises 20-4

In Exercises 1 through 12, determine the coordinates of the focus and the equation of the directrix of the given parabolas. Sketch each curve.

1. $y^2 = 4x$ **2.** $y^2 = 16x$ **3.** $y^2 = -4x$ **4.** $y^2 = -16x$

5. $x^2 = 8y$ **6.** $x^2 = 10y$ **7.** $x^2 = -4y$ **8.** $x^2 = -12y$

9. $y^2 = 2x$ **10.** $x^2 = 14y$ **11.** $y = x^2$ **12.** $x = 3y^2$

In Exercises 13 through 20, find the equations of the parabolas satisfying the given conditions.

13. Focus $(3, 0)$, directrix $x = -3$ **14.** Focus $(-2, 0)$, directrix $x = 2$

15. Focus $(0, 4)$, vertex $(0, 0)$ **16.** Focus $(-3, 0)$, vertex $(0, 0)$

17. Vertex $(0, 0)$, directrix $y = -1$ **18.** Vertex $(0, 0)$, directrix $y = \frac{1}{2}$

19. Vertex at $(0, 0)$, axis along the y-axis, passes through $(-1, 8)$

20. Vertex at $(0, 0)$, axis along the x-axis, passes through $(2, -1)$

In Exercises 21 through 24, use the definition of the parabola to find the equations of the parabolas satisfying the given conditions. Sketch each curve.

21. Focus $(6, 1)$, directrix $x = 0$ **22.** Focus $(1, -4)$, directrix $x = 2$

23. Focus $(1, 1)$, vertex $(1, 3)$ **24.** Vertex $(-2, -4)$, directrix $x = 3$

In Exercises 25 through 36, find the indicated quantities.

25. The chord of a parabola that passes through the focus and is parallel to the directrix is called the *latus rectum* of the parabola. Find the length of the latus rectum of the parabola $y^2 = 4px$.

26. Find the equation of the circle which has the focus and the vertex of the parabola $x^2 = 8y$ as the ends of a diameter.

27. The rate of development of heat H (measured in watts) in a resistor of resistance R (measured in ohms) of an electric circuit is given by $H = Ri^2$, where i is the current (measured in amperes) in the resistor. Sketch the graph of H versus i, if $R = 6.0 \, \Omega$.

28. A radio wave reflector has a parabolic cross section, and a wave that comes in parallel to its axis will be reflected through the focus. A particular reflector is designed for the incoming and reflected waves, as shown in Fig. 20-47 (measurements in meters). What is the equation of the parabola?

29. A linear solar reflector has a parabolic cross section, which is shown in Fig. 20-48. From the given dimensions, what is the focal length (vertex to focus) of the reflector?

30. A rocket is fired horizontally from a plane. Its horizontal distance x and vertical distance y from the point at which it was fired are given by $x = v_0 t$ and $y = \frac{1}{2}gt^2$, where v_0 is the initial velocity of the rocket, t is the time, and g is the acceleration due to gravity. Express y as a function of x and show that it is the equation of a parabola.

31. A parabolically shaped window is shown in Fig. 20-49. What is the length of the horizontal bar across the window?

32. A wire is fastened 36.0 ft up on each of two telephone poles which are 200 ft apart. Halfway between the poles the wire is 30.0 ft above the ground. Assuming the wire is parabolic, find the height of the wire 50.0 ft from either pole.

33. The total annual fraction f of energy supplied by solar energy to a home is given by $f = 0.065\sqrt{A}$, where A is the area of the solar collector. Sketch the graph of f as a function of A $(0 < A \le 200 \text{ m}^2)$.

34. The velocity v of a jet of water flowing from an opening in the side of a certain container is given by $v = 8\sqrt{h}$, where h is the depth of the opening. Sketch a graph of v (in feet per second) versus h (in feet).

35. A small island is 4 km from a straight shoreline. A ship channel is equidistant between the island and the shoreline. Write an equation for the channel.

36. Under certain circumstances, the maximum power P in an electric circuit varies as the square of the voltage of the source E_0 and inversely as the internal resistance R_i of the source. If 10 W is the maximum power for a source of 2.0 V and internal resistance of 0.10 Ω, sketch the graph of P versus E_0, if R_i remains constant.

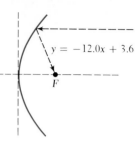

$y = -12.0x + 3.6$

F

Fig. 20-47

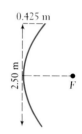

0.425 m

2.50 m

F

Fig. 20-48

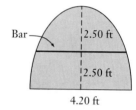

Bar

2.50 ft

2.50 ft

4.20 ft

Fig. 20-49

20-5   The Ellipse

The next important curve we shall discuss is the ellipse. *The **ellipse** is defined as the locus of a point P(x, y) which moves so that the sum of its distances from two fixed points is constant. These two fixed points are called the **foci** of the ellipse.* We will let this sum of distances be $2a$ and the foci be the points $(c, 0)$ and $(-c, 0)$. These choices set up the ellipse with its center at the origin such that c is the length of the line segment from the center to a focus. Also, by choosing $2a$ as the sum of distances, we will see that the length a has a special meaning.

From the definition of an ellipse, we have (see Fig. 20-50)

$$\sqrt{(x - c)^2 + y^2} + \sqrt{(x + c)^2 + y^2} = 2a$$

From the section on solving equations involving radicals (Section 13-4), we find that we should move one of the radicals to the right side and then square each side. Thus we have the following steps:

$$\sqrt{(x + c)^2 + y^2} = 2a - \sqrt{(x - c)^2 + y^2}$$
$$(x + c)^2 + y^2 = 4a^2 - 4a\sqrt{(x - c)^2 + y^2} + (\sqrt{(x - c)^2 + y^2})^2$$

foci

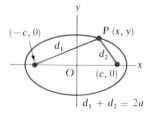

$(-c, 0)$

$P(x, y)$

d_1

d_2

O $(c, 0)$

$d_1 + d_2 = 2a$

Fig. 20-50

$$x^2 + 2cx + c^2 + y^2 = 4a^2 - 4a\sqrt{(x-c)^2 + y^2} + x^2 - 2cx + c^2 + y^2$$
$$4a\sqrt{(x-c)^2 + y^2} = 4a^2 - 4cx$$
$$a\sqrt{(x-c)^2 + y^2} = a^2 - cx$$
$$a^2(x^2 - 2cx + c^2 + y^2) = a^4 - 2a^2cx + c^2x^2$$
$$(a^2 - c^2)x^2 + a^2y^2 = a^2(a^2 - c^2)$$

At this point we define (and the usefulness of this definition will be shown presently) $a^2 - c^2 = b^2$. This substitution gives us

$$b^2x^2 + a^2y^2 = a^2b^2$$

Dividing through by a^2b^2, we have

$$\frac{x^2}{a^2} + \frac{y^2}{b^2} = 1 \qquad (20\text{-}17)$$

If we let $y = 0$, we find that the x-intercepts are $(-a, 0)$ and $(a, 0)$. We see that the distance $2a$, originally chosen as the sum of the distances in the derivation of Eq. (20-17), is the distance between the x-intercepts. *These two points are known as the **vertices** of the ellipse, and the line between them is known as the **major*** **axis** *[Fig. 20-51(a)] of the ellipse. Thus, a is the length of the **semimajor axis.***

vertices
major and minor axes

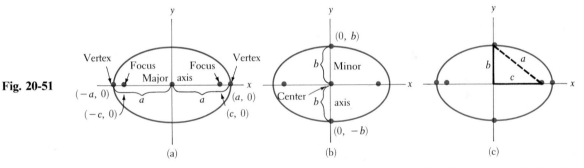

Fig. 20-51

(a)   (b)   (c)

If we now let $x = 0$, we find the y-intercepts to be $(0, -b)$ and $(0, b)$. *The line joining these two intercepts is called the **minor axis** of the ellipse [Fig. 20-51(b)] which means b is the length of the **semiminor axis.*** The point $(0, b)$ is on the ellipse and is also equidistant from $(-c, 0)$ and $(c, 0)$. Since the sum of the distances from these points to $(0, b)$ equals $2a$, the distance from $(c, 0)$ to $(0, b)$ must be a. Thus, we have a right triangle formed by line segments of lengths a, b, and c, with a as the hypotenuse [Fig. 20-51(c)]. From this *we have*

$$a^2 = b^2 + c^2 \qquad (20\text{-}18)$$

as the relation between the distances a, b, and c. This equation also shows us why b was defined as it was in the derivation.

*Equation (20-17) is called the **standard equation** of the ellipse with its major axis along the x-axis and its center at the origin.*

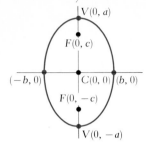

Fig. 20-52

If we choose points on the y-axis as the foci, *the standard equation of the ellipse, with its center at the origin and its major axis along the y-axis, is*

$$\frac{y^2}{a^2} + \frac{x^2}{b^2} = 1 \qquad\qquad (20\text{-}19)$$

In this case the vertices are $(0, a)$ and $(0, -a)$, the foci are $(0, c)$ and $(0, -c)$, and the ends of the minor axis are $(b, 0)$ and $(-b, 0)$. See Fig. 20-52.

The symmetry of the ellipses given by Eqs. (20-17) and (20-19) to each of the axes and the origin can be proven, since each of x and y can be replaced by its negative in these equations without changing the equations.

EXAMPLE A

The ellipse

$$\frac{x^2}{25} + \frac{y^2}{9} = 1$$

NOTE ▷ fits the form of either Eq. (20-17) or Eq. (20-19). Since $a^2 = b^2 + c^2$, we know that *a is always larger than b.* Thus, since the square of the larger number appears under x^2, we know that the equation is in the form of Eq. (20-17). This means that $a^2 = 25$ and $b^2 = 9$, or $a = 5$ and $b = 3$. In turn this tells us that the vertices are $(5, 0)$ and $(-5, 0)$ and that the minor axis extends from $(0, 3)$ to $(0, -3)$. See Fig. 20-53.

We can find c^2 from the relation $c^2 = a^2 - b^2$. Thus, $c^2 = 16$, or $c = 4$. This means that the foci are $(4, 0)$ and $(-4, 0)$. ■

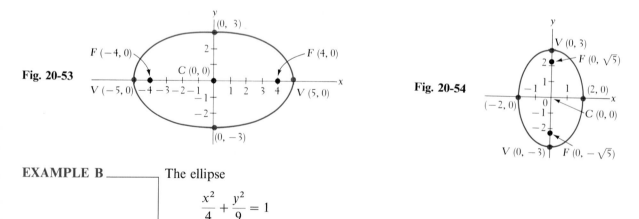

Fig. 20-53

Fig. 20-54

EXAMPLE B

The ellipse

$$\frac{x^2}{4} + \frac{y^2}{9} = 1$$

b^2 ⌐ ⌐ a^2

has vertices at $(0, 3)$ and $(0, -3)$. The minor axis extends from $(2, 0)$ to $(-2, 0)$. This we find directly from the equation, since the square of the larger number appears under y^2. This means the equation fits the form of Eq. (20-19), with $a^2 = 9$ and $b^2 = 4$. Also, $c^2 = 5$ and the foci are $(0, \sqrt{5})$ and $(0, -\sqrt{5})$. See Fig. 20-54. ■

EXAMPLE C _____ Find the coordinates of the vertices, the ends of the minor axis, and the foci of the ellipse $4x^2 + 16y^2 = 64$.

This equation must be put in standard form first, which we do by dividing through by 64. When this is done, we obtain

$$\frac{x^2}{16} + \frac{y^2}{4} = 1 \leftarrow$$

form requires
+ and 1

We see that $a^2 = 16$ and $b^2 = 4$. In turn this tells us that $a = 4$, $b = 2$, and $c = \sqrt{16 - 4} = \sqrt{12} = 2\sqrt{3}$. Since a^2 appears under x^2, the vertices are $(4, 0)$ and $(-4, 0)$. The ends of the minor axis are $(0, 2)$ and $(0, -2)$, and the foci are $(2\sqrt{3}, 0)$ and $(-2\sqrt{3}, 0)$. See Fig. 20-55. ■

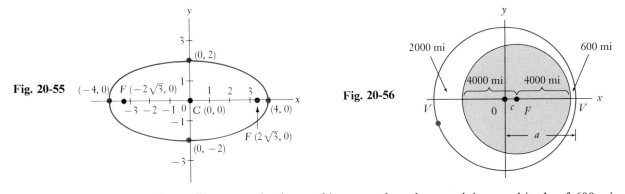

Fig. 20-55 $(-4, 0)$ $F(-2\sqrt{3}, 0)$

Fig. 20-56

EXAMPLE D _____ A satellite to study the earth's atmosphere has a minimum altitude of 600 mi and a maximum altitude of 2000 mi. If the path of the satellite about the earth is an ellipse with the center of the earth at one focus, what is the equation of its path? Assume the radius of the earth is 4000 mi.

We set up the coordinate system such that the center of the ellipse is at the origin and the center of the earth is at the right focus, as shown in Fig. 20-56. We know that the distance between vertices is

$$2a = 2000 + 4000 + 4000 + 600 = 10{,}600 \text{ mi}$$

$$a = 5300 \text{ mi}$$

From the right focus to the right vertex is 4600 mi. This tells us that

$$c = a - 4600 = 5300 - 4600 = 700 \text{ mi}$$

We can now calculate b^2 as

$$b^2 = a^2 - c^2 = 5300^2 - 700^2 = 2.76 \times 10^7 \text{ mi}^2$$

Since $a^2 = 5300^2 = 2.81 \times 10^7 \text{ mi}^2$, the equation is

$$\frac{x^2}{2.81 \times 10^7} + \frac{y^2}{2.76 \times 10^7} = 1$$

or

$$2.76x^2 + 2.81y^2 = 7.76 \times 10^7$$

EXAMPLE E

Find the equation of the ellipse with its center at the origin and an end of its minor axis at $(2, 0)$ and which passes through $(-1, \sqrt{6})$.

Since the center is at the origin and an end of the minor axis is at $(2, 0)$, we know that the ellipse if of the form of Eq. (20-19) and that $b = 2$. Thus, we have

$$\frac{y^2}{a^2} + \frac{x^2}{2^2} = 1$$

In order to find a^2 we use the fact that the ellipse passes through $(-1, \sqrt{6})$. This means that these coordinates satisfy the equation of the ellipse. This gives

$$\frac{(\sqrt{6})^2}{a^2} + \frac{(-1)^2}{4} = 1$$

$$\frac{6}{a^2} = \frac{3}{4}$$

$$a^2 = 8$$

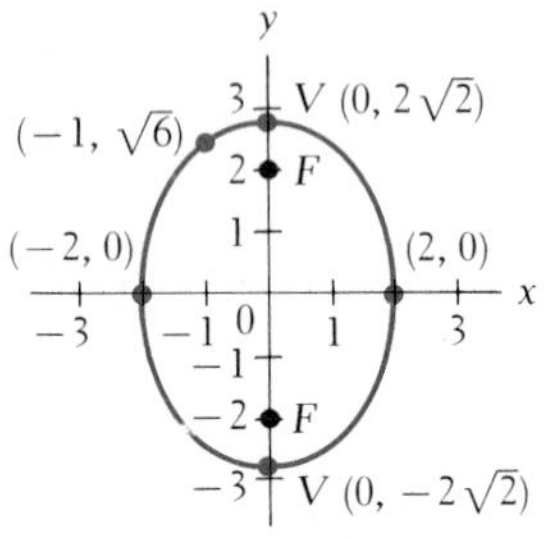

Thus, the equation of the ellipse is

$$\frac{y^2}{8} + \frac{x^2}{4} = 1$$

Fig. 20-57

The ellipse is shown in Fig. 20-57.

Equations (20-17) and (20-19) give us the standard form of the equation of an ellipse with its center at the origin and its foci on one of the coordinate axes. The following example illustrates the use of the definition of the ellipse to find the equation of an ellipse with its center at a point other than the origin.

EXAMPLE F

Using the definition, find the equation of the ellipse with foci at $(1, 3)$ and $(9, 3)$, with a major axis of 10.

Using the same method as in the derivation of Eq. (20-17), we have the following steps. (Remember, the sum of distances in the definition equals the length of the major axis.)

$\sqrt{(x-1)^2 + (y-3)^2} + \sqrt{(x-9)^2 + (y-3)^2} = 10$	use definition of ellipse
$\sqrt{(x-1)^2 + (y-3)^2} = 10 - \sqrt{(x-9)^2 + (y-3)^2}$	isolate a radical
$(x-1)^2 + (y-3)^2 = 100 - 20\sqrt{(x-9)^2 + (y-3)^2}$ $+ (\sqrt{(x-9)^2 + (y-3)^2})^2$	square both sides
$x^2 - 2x + 1 + y^2 - 6y + 9 = 100 - 20\sqrt{(x-9)^2 + (y-3)^2}$ $+ x^2 - 18x + 81 + y^2 - 6y + 9$	simplify
$20\sqrt{(x-9)^2 + (y-3)^2} = 180 - 16x$	isolate radical
$5\sqrt{(x-9)^2 + (y-3)^2} = 45 - 4x$	divide by 4
$25(x^2 - 18x + 81 + y^2 - 6y + 9) = 2025 - 360x + 16x^2$	square both sides
$25x^2 - 450x + 2250 + 25y^2 - 150y = 2025 - 360x + 16x^2$	simplify
$9x^2 - 90x + 25y^2 - 150y + 225 = 0$	

The additional x- and y-terms are characteristic of the equation of an ellipse whose center is not at the origin (see Fig. 20-58).

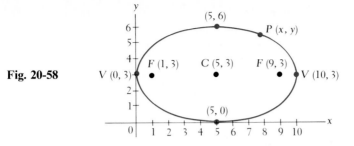

Fig. 20-58

We can conclude that *the equation of an ellipse is characterized by the presence of both an x^2- and a y^2-term, having different coefficients (in value but not in sign)*. The difference between the equation of an ellipse and that of a circle is that the coefficients of the squared terms in the equation of the circle are the same, whereas those of the ellipse differ.

Exercises 20-5

In Exercises 1 through 12, find the coordinates of the vertices and foci of the given ellipses. Sketch each curve.

1. $\dfrac{x^2}{4} + \dfrac{y^2}{1} = 1$

2. $\dfrac{x^2}{100} + \dfrac{y^2}{64} = 1$

3. $\dfrac{x^2}{25} + \dfrac{y^2}{36} = 1$

4. $\dfrac{x^2}{49} + \dfrac{y^2}{81} = 1$

5. $4x^2 + 9y^2 = 36$

6. $x^2 + 36y^2 = 144$

7. $49x^2 + 4y^2 = 196$

8. $25x^2 + y^2 = 25$

9. $8x^2 + y^2 = 16$

10. $2x^2 + 3y^2 = 6$

11. $4x^2 + 25y^2 = 25$

12. $9x^2 + 4y^2 = 9$

In Exercises 13 through 20, find the equations of the ellipses satisfying the given conditions. The center of each is at the origin.

13. Vertex $(15, 0)$, focus $(9, 0)$

14. Minor axis 8, vertex $(0, -5)$

15. Focus $(0, 2)$, major axis 6

16. Semiminor axis 2, focus $(3, 0)$

17. Vertex $(8, 0)$, passes through $(2, 3)$

18. Focus $(0, 2)$, passes through $(-1, \sqrt{3})$

19. Passes through $(2, 2)$ and $(1, 4)$

20. Passes through $(-2, 2)$ and $(1, \sqrt{6})$

In Exercises 21 through 24, find the equations of the ellipses with the given properties by use of the definition of an ellipse.

21. Foci at $(-2, 1)$ and $(4, 1)$, a major axis of 10

22. Foci at $(-3, -2)$ and $(-3, 8)$, major axis 26

23. Vertices at $(1, 5)$ and $(1, -1)$, foci at $(1, 4)$ and $(1, 0)$

24. Vertices at $(-2, 1)$ and $(-2, 5)$, foci at $(-2, 2)$ and $(-2, 4)$

In Exercises 25 through 36, solve the given problems.

25. Show that the ellipse $2x^2 + 3y^2 - 8x - 4 = 0$ is symmetrical to the x-axis.

26. Show that the ellipse $5x^2 + y^2 - 3y - 7 = 0$ is symmetrical to the y-axis.

27. The *eccentricity e* of an ellipse is defined as $e = c/a$. A cam in the shape of an ellipse can be described by the equation $x^2 + 9y^2 = 81$. Find the eccentricity of this elliptical cam.

28. The planet Pluto moves about the sun in an elliptical orbit, with the sun at one focus. The closest that Pluto approaches the sun is 2.8 billion miles, and the farthest it gets from the sun is 4.6 billion miles. Find the eccentricity of Pluto's orbit. (See Exercise 27.)

29. The inside of the top of an arch is a semiellipse with an equation of $x^2 + 4y^2 = 100$. The outside of the arch is also a semiellipse with an equation $2x^2 + 5y^2 = 500$. If measurements are in meters, what is the thickness of the arch (a) at each base and (b) at the center? See Fig. 20-59, in which t_b and t_c represent these thicknesses.

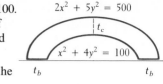

Fig. 20-59

30. The electric power P dissipated in a resistance R is given by $P = Ri^2$, where i is the current in the resistor. Find the equation for the total power of 64 W dissipated in two resistors, with resistances 2.0 Ω and 8.0 Ω, respectively, and with currents i_1 and i_2, respectively. Sketch the graph, assuming that negative values of current are meaningful.

31. An ellipse has a focal property such that a light ray or sound wave emanating from one focus will be reflected through the other focus. Many buildings, such as Statuary Hall in the U.S. Capitol and the Taj Mahal, are built with elliptical ceilings with the property that a sound from one focus is easily heard at the other focus. If a building has a ceiling whose cross sections are part of an ellipse which can be described by the equation $36x^2 + 225y^2 = 8100$ (measurements in meters), how far apart must two persons stand in order to whisper to each other using this focal property?

32. An airplane wing is designed such that a certain cross section is an ellipse 8.40 ft wide and 1.20 ft thick. Find an equation which can be used to describe the perimeter of this cross section.

33. A draftsman draws a series of triangles with a base from $(-3, 0)$ to $(3, 0)$ and a perimeter of 14 cm (all measurements in centimeters). Find the equation of the curve on which all of the third vertices of the triangles are located.

34. An architect designs a window in the shape of an ellipse 4.50 ft wide and 3.20 ft high. Find the perimeter of the window from the formula $p = \pi(a + b)$. This formula gives a good *approximation* for the perimeter when a and b are nearly equal.

35. The ends of a horizontal tank 20.0 ft long are ellipses, which can be described by the equation $9x^2 + 20y^2 = 180$, where x and y are measured in feet. The area of an ellipse is $A = \pi ab$. Find the volume of the tank.

36. A laser beam 6.80 mm in diameter is incident on a plane surface at an angle of 62.0°, as shown in Fig. 20-60. What is the elliptical area that the laser covers on the surface? (See Exercise 35.)

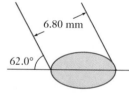

Fig. 20-60

20-6   The Hyperbola

foci

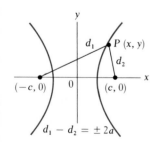

Fig. 20-61

The final curve we shall discuss in detail is the hyperbola. *The **hyperbola** is defined as the locus of a point $P(x, y)$ which moves so that the difference of the distances from two fixed points is a constant. These fixed points are the **foci** of the hyperbola.* We choose the foci of the hyperbola as the points $(c, 0)$ and $(-c, 0)$ (see Fig. 20-61) and the constant difference to be $2a$. As with the ellipse, these choices make c the length of the line segment from the center to a focus and a (as we will see) the length of the line segment from the center to a vertex. Therefore, we have

$$\sqrt{(x + c)^2 + y^2} - \sqrt{(x - c)^2 + y^2} = 2a$$

Following the same procedure as in the preceding section, we find the equation of the hyperbola to be

$$\frac{x^2}{a^2} - \frac{y^2}{b^2} = 1 \qquad\qquad (20\text{-}20)$$

NOTE ▷ When we derive this equation, *we have a definition of the relation between a, b, and c which is different from that for the ellipse.* This relation is

$$c^2 = a^2 + b^2 \qquad (20\text{-}21)$$

An analysis of Eq. (20-20) will reveal the significance of a and b. First, by letting $y = 0$, we find that the x-intercepts are $(a, 0)$ and $(-a, 0)$, just as they are for the ellipse. *These points are called the* **vertices** *of the hyperbola.* By letting $x = 0$, we find that we have imaginary solutions for y, which means that there are no points on the curve which correspond to a value of $x = 0$.

To find the significance of b, we shall solve Eq. (20-20) for y in a particular form. First we have

$$\frac{y^2}{b^2} = \frac{x^2}{a^2} - 1$$

$$= \frac{x^2}{a^2} - \frac{a^2 x^2}{a^2 x^2}$$

$$= \frac{x^2}{a^2}\left(1 - \frac{a^2}{x^2}\right)$$

Multiplying through by b^2 and then taking the square root of each side, we have

$$y^2 = \frac{b^2 x^2}{a^2}\left(1 - \frac{a^2}{x^2}\right)$$

$$y = \pm\frac{bx}{a}\sqrt{1 - \frac{a^2}{x^2}} \qquad (20\text{-}22)$$

We note that, if large values of x are assumed in Eq. (20-22), the quantity under the radical becomes approximately 1. In fact, the larger x becomes, the nearer 1 this expression becomes, since the x^2 in the denominator of a^2/x^2 makes this term nearly zero. Thus, for large values of x, Eq. (20-22) is approximately

$$y = \pm\frac{bx}{a} \qquad (20\text{-}23)$$

Equation (20-23) can be seen to represent the equations for two straight lines, each of which passes through the origin. One has a slope of b/a and the other **NOTE** ▷ has a slope of $-b/a$. *These lines are called the* **asymptotes** *of the hyperbola.* *An asymptote is a line which a curve approaches as one of the variables approaches some particular value.* The tangent curve also has asymptotes, as we can see in Fig. 9-18. We can designate this limiting procedure with notation introduced in Chapter 18 by saying that

$$y \to \frac{bx}{a} \quad \text{as} \quad x \to \infty$$

Since straight lines are easily sketched, the easiest way to sketch a hyperbola is to draw its asymptotes and then to draw the hyperbola so that it comes closer and closer to these lines as x becomes larger numerically. To draw in the asymptotes, the usual procedure is first to draw a small rectangle, $2a$ by $2b$, with the origin in the center. Then straight lines are drawn through opposite vertices. These lines are the asymptotes (see Fig. 20-62). Thus we see that the significance of the value of b lies in the slope of the asymptotes of the hyperbola.

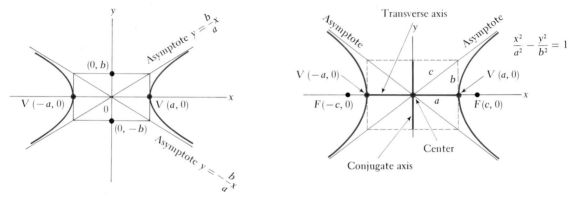

Fig. 20-62   **Fig. 20-63**

transverse and conjugate axes

Equation (20-20) is called the **standard equation** of the hyperbola with its center at the origin. It has a **transverse axis** of length $2a$ along the x-axis and a **conjugate axis** of length $2b$ along the y-axis. This means that a represents the length of the semitransverse axis, and b represents the length of the semiconjugate axis (see Fig. 20-63). From the definition of c we know that c is the length of the line segment from the center to a focus. Also, c is the length of the semidiagonal as shown in Fig. 20-63. This shows us the geometric meaning of the relation among a, b, and c which is given in Eq. (20-21).

If the transverse axis is along the y-axis and the conjugate axis is along the x-axis, the equation of a hyperbola with its center at the origin (see Fig. 20-64) is

Fig. 20-64

$$\frac{y^2}{a^2} - \frac{x^2}{b^2} = 1$$

(20-24)

The symmetry of the hyperbolas given by Eqs. (20-20) and (20-24) to each of the axes and to the origin can be proven since each of x and y can be replaced by its negative in these equations without changing the equations.

EXAMPLE A _____ The hyperbola

$$\frac{x^2}{16} - \frac{y^2}{9} = 1$$

a^2 b^2

fits the form of Eq. (20-20). We know it fits Eq. (20-20) and not Eq. (20-24) since the x^2-term is the positive term with 1 on the right. From the equation we see that $a^2 = 16$ and $b^2 = 9$, or $a = 4$ and $b = 3$. In turn this tells us that the vertices are (4, 0) and (−4, 0) and that the conjugate axis extends from (0, 3) to (0, −3).

Since $c^2 = a^2 + b^2$, we find that $c^2 = 25$, or $c = 5$. The foci are (5, 0) and (−5, 0).

Drawing in the rectangle and then the asymptotes in Fig. 20-65, we draw the hyperbola from each vertex toward the asymptotes. Thus, the curve is sketched.

Fig. 20-65

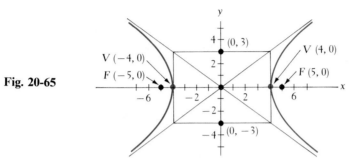

EXAMPLE B

The hyperbola

$$\frac{y^2}{4} - \frac{x^2}{16} = 1$$

has vertices at (0, 2) and (0, −2). Its conjugate axis extends from (4, 0) to (−4, 0). The foci are $(0, 2\sqrt{5})$ and $(0, -2\sqrt{5})$. We find this directly from the equation since the y^2-term is the positive term with 1 on the right. This means the equation fits the form of Eq. (20-24) with $a^2 = 4$ and $b^2 = 16$. Also, $c^2 = 20$, or $c = \sqrt{20} = 2\sqrt{5}$.

Since $2a$ extends along the y axis, we see that the equations of the asymptotes are $y = \pm(a/b)x$. This is not a contradiction of Eq. (20-23) but an extension of it for the case of a hyperbola with its transverse axis along the y-axis. The ratio a/b simply gives the slope of the asymptote. The hyperbola is sketched in Fig. 20-66.

Fig. 20-66

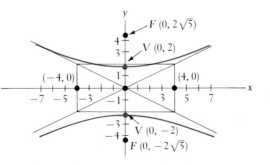

EXAMPLE C _____ Determine the coordinates of the vertices and foci of the hyperbola

$$4x^2 - 9y^2 = 36$$

First, by dividing through by 36, we can put this equation in standard form. Thus, we have

$$\frac{x^2}{9} - \frac{y^2}{4} = 1$$

form requires
—and 1

We see that $a^2 = 9$ and $b^2 = 4$. In turn this tells us that $a = 3$, $b = 2$, and $c = \sqrt{9 + 4} = \sqrt{13}$. Since a^2 appears under x^2, the vertices are $(3, 0)$ and $(-3, 0)$ and the foci are $(\sqrt{13}, 0)$ and $(-\sqrt{13}, 0)$. The hyperbola is sketched in Fig. 20-67.

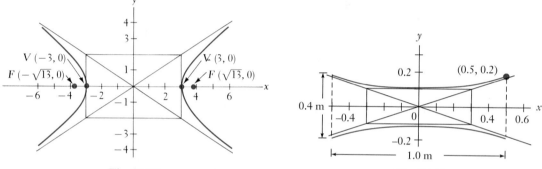

Fig. 20-67   Fig. 20-68

EXAMPLE D _____ In physics it is shown that where the velocity of a fluid is greatest, the pressure is the least. In designing an experiment to study this effect in the flow of water, a pipe is constructed such that its lengthwise cross section is a hyperbola. The pipe is 1.0 m long, 0.2 m in diameter at the narrowest point in the middle, and 0.4 m in diameter at each end. What is the equation of the cross section of the pipe as shown in Fig. 20-68?

As shown, the hyperbola has its transverse axis along the y-axis and its center at the origin. Therefore, its general equation is given by Eq. (20-24). Since the diameter at the middle of the pipe is 0.2 m, we know that $a = 0.1$ m. Also, since it is 1.0 m long and the diameter at the end is 0.2 m, we know that the point $(0.5, 0.2)$ is on the hyperbola. This point must satisfy the equation. This leads to the following solution.

$$\frac{y^2}{a^2} - \frac{x^2}{b^2} = 1 \qquad \text{Eq. (20-24)}$$

point (0.5, 0.2) satisfies equation

$$\frac{0.2^2}{0.1^2} - \frac{0.5^2}{b^2} = 1$$

$a = 0.1$

$$4 - \frac{0.25}{b^2} = 1, \qquad 3b^2 = 0.25, \qquad b^2 = 0.083$$

$$\frac{y^2}{0.1^2} - \frac{x^2}{0.083} = 1 \qquad \text{substituting } a = 0.1, \, b^2 = 0.083 \text{ in Eq. (20-24)}$$

$$100y^2 - 12x^2 = 1 \qquad \text{equation of cross section}$$

Equations (20-20) and (20-24) give us the standard form of the equation of the hyperbola with its center at the origin and its foci on one of the coordinate axes. There is one other important equation form which represents a hyperbola, and that is

$$xy = c \qquad\qquad\qquad (20\text{-}25)$$

The asymptotes of this hyperbola are the coordinate axes, and the foci are on the line $y = x$, or on the line $y = -x$, if c is negative. This hyperbola is symmetrical to the origin, for if $-x$ replaces x, and $-y$ replaces y at the same time, we have $(-x)(-y) = c$, or $xy = c$. The equation remains unchanged. However, either if $-x$ replaces x, or if $-y$ replaces y, but not both, the sign on the left is changed. Therefore, it is not symmetrical to either axis. The c here represents a constant and is not related to the focus. The following example illustrates this type of hyperbola.

EXAMPLE E

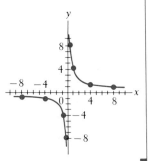

Fig. 20-69

Plot the graph of the equation $xy = 4$.

We find the values of the table below, then plot the appropriate points. Here it is permissible to use a limited number of points, since we know that the equation represents a hyperbola (Fig. 20-69). Thus, using $y = \frac{4}{x}$, we obtain the values in the table.

x	-8	-4	-1	$-\frac{1}{2}$	$\frac{1}{2}$	1	4	8
y	$-\frac{1}{2}$	-1	-4	-8	8	4	1	$\frac{1}{2}$

Note that neither x nor y may equal zero.

If the product xy equals a negative number, the two branches of the hyperbola are in the second and fourth quadrants.

EXAMPLE F

One statement of Boyle's law is that the product of the pressure p and volume V, at a constant temperature, remains constant for a perfect gas. If $p = 300 \, \text{kPa}$ when $V = 8.0 \, \text{L}$ for a certain gas, sketch the graph of p as a function of V.

From the statement of Boyle's law, we know that $pV = c$, where c is a constant. From the given values we have

$$(300 \, \text{kPa})(8.0 \, \text{L}) = c \qquad \text{or} \qquad c = 2400 \, \text{kPa} \cdot \text{L}$$

Thus, we are to sketch the graph of $pV = 2400$.

(Continued on next page)

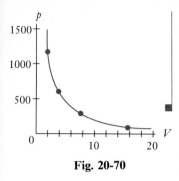

Fig. 20-70

We find the points in the following table by expressing $p = 2400/V$ and noting that only positive values have meaning.

V(L)	2.0	4.0	8.0	16.0
p (kPa)	1200	600	300	150

■ See Fig. 20-70.

We conclude that the equation of a hyperbola is characterized by the presence of both an x^2- and a y^2-term, having different signs, or by the presence of an xy-term with no squared terms.

Exercises 20-6

In Exercises 1 through 12, find the coordinates of the vertices and the foci of the given hyperbolas. Sketch the curve.

1. $\dfrac{x^2}{25} - \dfrac{y^2}{144} = 1$

2. $\dfrac{x^2}{16} - \dfrac{y^2}{4} = 1$

3. $\dfrac{y^2}{9} - \dfrac{x^2}{1} = 1$

4. $\dfrac{y^2}{2} - \dfrac{x^2}{2} = 1$

5. $4x^2 - y^2 = 4$

6. $x^2 - 9y^2 = 81$

7. $2y^2 - 5x^2 = 10$

8. $3y^2 - 2x^2 = 6$

9. $4x^2 - y^2 + 4 = 0$

10. $9x^2 - y^2 - 9 = 0$

11. $4x^2 - 9y^2 = 16$

12. $y^2 - 9x^2 = 25$

See Appendix E for a computer program for graphing a hyperbola.

In Exercises 13 through 20, find the equations of the hyperbolas satisfying the given conditions. The center of each is at the origin.

13. Vertex (3, 0), focus (5, 0)

14. Vertex (0, 1), focus $(0, \sqrt{3})$

15. Conjugate axis = 12, vertex (0, 10)

16. Focus (8, 0), transverse axis = 4

17. Passes through (2, 3), focus (2, 0)

18. Passes through $(8, \sqrt{3})$, vertex (4, 0)

19. Passes through (5, 4) and $(3, \frac{4}{5}\sqrt{5})$

20. Passes through (1, 2) and $(2, 2\sqrt{2})$

In Exercises 21 through 24, sketch the graphs of the hyperbolas given.

21. $xy = 2$

22. $xy = 10$

23. $xy = -2$

24. $xy = -4$

In Exercises 25 through 28, find the equations of the hyperbolas with the given properties by use of the definition of the hyperbola.

25. Foci at (1, 2) and (11, 2), with a transverse axis of 8.

26. Vertices (−2, 4) and (−2, −2), with conjugate axis of 4.

27. Center at (2, 0), vertex at (3, 0), conjugate axis of 6.

28. Center at (1, −1), focus at (1, 4), vertex at (1, 2).

In Exercises 29 through 36, solve the given problems.

29. Two concentric hyperbolas are called conjugate hyperbolas if the transverse and conjugate axes of one are respectively the conjugate and transverse axes of the other. What is the equation of the hyperbola conjugate to the hyperbola of Exercise 14?

30. As with an ellipse, the *eccentricity e* of a hyperbola is defined as $e = c/a$. Find the eccentricity of the hyperbola $2x^2 - 3y^2 = 24$.

31. A plane flying at a constant altitude of 2000 m is observed from the control tower of an airport. Show that the equation relating the horizontal distance x and direct line distance ℓ from the tower to the plane is that of a hyperbola. Sketch the graph of ℓ as a function of x.

32. Two holes of radius r are drilled from a circular area of radius R such that 24 in.^{2} of material remains. Show that the equation relating R and r is that of a hyperbola.

33. Ohm's law in electricity states that the product of the current i and the resistance R equals the voltage V across the resistance. If a battery of 6.00 V is placed across a variable resistor R, find the equation relating i and R and sketch the graph of i as a function of R.

34. The relationship between the frequency f, wavelength λ, and velocity v of a wave is given by $v = f\lambda$. The velocity of light is a constant, being 3.0×10^{10} cm/s. Wavelengths of gamma rays vary from about 10^{-8} cm to about 10^{-13} cm, but the gamma rays have the same velocity as light. On logarithmic paper, plot the graph of frequency versus wavelength for gamma rays. What type of curve results when this type of hyperbola is plotted on logarithmic paper?

35. An electronic instrument located at point P records the sound of a rifle shot and the impact of the bullet striking the target at the same instant. Show that P lies on a branch of a hyperbola.

36. For monochromatic (single-color) light coming from two point sources, curves of maximum intensity occur where the difference in the distances from the sources is an integral number of wavelengths. If a thin translucent film is placed in the plane of the sources, find the equation of the curves of maximum intensity in the film where the difference in paths is two wavelengths and the sources are separated by four wavelengths. Assume the sources are on the x-axis with the origin midway between, and use units of one wavelength for both x and y.

20-7 Translation of Axes

Until now, except by direct use of the definition, the equations considered for the parabola, the ellipse, and the hyperbola have been restricted to the particular cases in which the vertex of the parabola is at the origin and the center of the ellipse or hyperbola is at the origin. In this section we shall consider, without specific use of the definition, the equations of these curves for the cases in which the axis of the curve is parallel to one of the coordinate axes. This is done by **translation of axes.**

We choose a point (h, k) in the xy-coordinate plane and let this point be the origin of another coordinate system, the $x'y'$-coordinate system. The x'-axis is parallel to the x-axis and the y'-axis is parallel to the y-axis. Every point in the plane now has two sets of coordinates associated with it, (x, y) and (x', y'). From Fig. 20-71 we see that

$$x = x' + h \quad \text{and} \quad y = y' + k \qquad (20\text{-}26)$$

Equations (20-26) can also be written in the form

$$x' = x - h \quad \text{and} \quad y' = y - k \qquad (20\text{-}27)$$

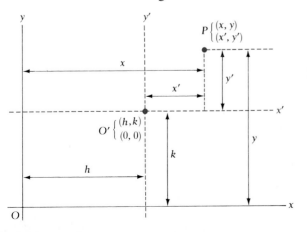

Fig. 20-71

The following examples illustrate the use of Eqs. (20-27) in the analysis of equations of the parabola, ellipse, and hyperbola.

EXAMPLE A

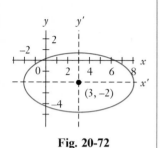

Fig. 20-72

Describe the locus of the equation

$$\frac{(x-3)^2}{25} + \frac{(y+2)^2}{9} = 1$$

In this equation, with $h = 3$ and $k = -2$, we have $x' = x - 3$ and $y' = y + 2$. In terms of x' and y', the equation is

$$\frac{(x')^2}{25} + \frac{(y')^2}{9} = 1$$

We recognize this equation as that of an ellipse (see Fig. 20-72) with a semimajor axis of 5 and a semiminor axis of 3. The center of the ellipse is at $(3, -2)$ since this was the choice of h and k to make the equation fit a standard form.

EXAMPLE B

Find the equation of the parabola with vertex at $(2, 4)$ and focus at $(4, 4)$.

If we let the origin of the $x'y'$-coordinate system be the point $(2, 4)$, the point $(4, 4)$ would be the point $(2, 0)$ in the $x'y'$-system. This means that $p = 2$ and $4p = 8$ (Fig. 20-73). In the $x'y'$-system, the equation is

$$(y')^2 = 8(x')$$

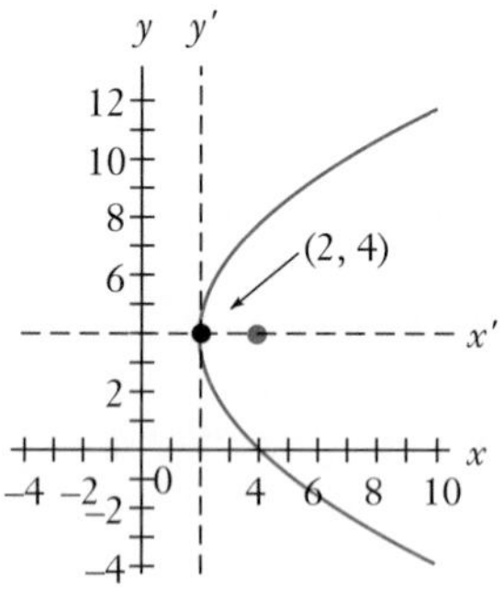

Since $(2, 4)$ is the origin of the $x'y'$-system, this means that $h = 2$ and $k = 4$. Using Eq. (20-27), we have

$$(y - 4)^2 = 8(x - 2)$$

coordinates of vertex $(2, 4)$

as the equation of the parabola in the xy-coordinate system. If this equation is multiplied out, and like terms are combined, we obtain

$$y^2 - 8x - 8y + 32 = 0$$

Fig. 20-73

EXAMPLE C

Find the center of the hyperbola $2x^2 - y^2 - 4x - 4y - 4 = 0$.

To analyze this curve, we first complete the square in the x-terms and in the y-terms. This will allow us to recognize properly the choice of h and k.

$$2x^2 - 4x - y^2 - 4y = 4$$

$$2(x^2 - 2x \quad) - (y^2 + 4y \quad) = 4$$

$$2(x^2 - 2x + 1) - (y^2 + 4y + 4) = 4 + 2 - 4$$

NOTE ▷

We note here that when we added 1 to complete the square of the x-terms within the parentheses, *we were actually adding 2 to the left side.* Thus, we added 2 to the right side. Similarly, when we added 4 to the y-terms within the parentheses, *we were actually subtracting 4 from the left side.* Continuing, we have

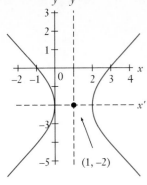

Fig. 20-74

$$2(x - 1)^2 - (y + 2)^2 = 2$$

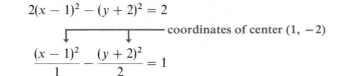

coordinates of center (1, −2)

$$\frac{(x - 1)^2}{1} - \frac{(y + 2)^2}{2} = 1$$

Thus, if we let $h = 1$ and $k = -2$, the equation in the $x'y'$-system becomes

$$\frac{(x')^2}{1} - \frac{(y')^2}{2} = 1$$

This means that the center is at $(1, -2)$, since this point corresponds to the origin of the $x'y'$-coordinate system (see Fig. 20-74).

EXAMPLE D

Find the vertex of the parabola $2x^2 - 12x - 3y + 15 = 0$.

First, we complete the square in the x-terms, placing all other resulting terms on the right. By factoring the resulting expression on the right, we may determine the values of h and k.

$$2x^2 - 12x = 3y - 15$$
$$2(x^2 - 6x \quad) = 3y - 15 \qquad \text{complete the square}$$
$$2(x^2 - 6x + 9) = 3y - 15 + 18$$
$$2(x - 3)^2 = 3(y + 1)$$

coordinates of vertex (3, −1)

$$(x - 3)^2 = \frac{3}{2}(y + 1)$$

$$x'^2 = \frac{3}{2}y'$$

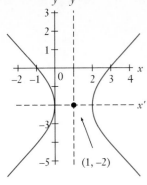

Fig. 20-75

Therefore, the vertex is at $(3, -1)$, since this point corresponds to the origin of the $x'y'$-coordinate system. See Fig. 20-75.

EXAMPLE E

Glass beakers are to be made with a height of 3 in. Express the surface area of the beakers in terms of the radius of the base. Sketch the graph of area versus radius.

The total surface area of a beaker is the sum of the area of the base and the lateral surface area of the side. In general, this surface area S in terms of the radius r of the base and the height h of the side is

$$S = \pi r^2 + 2\pi rh$$

Since h is constant at 3 in., we have

$$S = \pi r^2 + 6\pi r$$

which is the desired relationship. See Fig. 20-76.

For the purposes of sketching the graph of S and r, we complete the square of the r terms. This is done on the following page.

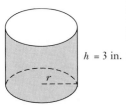

$h = 3$ in.

r

Fig. 20-76

(*Continued on next page*)

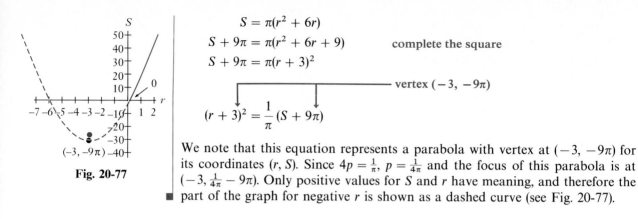

Fig. 20-77

$$S = \pi(r^2 + 6r)$$
$$S + 9\pi = \pi(r^2 + 6r + 9) \qquad \text{complete the square}$$
$$S + 9\pi = \pi(r + 3)^2$$

— vertex $(-3, -9\pi)$

$$(r + 3)^2 = \frac{1}{\pi}(S + 9\pi)$$

We note that this equation represents a parabola with vertex at $(-3, -9\pi)$ for its coordinates (r, S). Since $4p = \frac{1}{\pi}$, $p = \frac{1}{4\pi}$ and the focus of this parabola is at $(-3, \frac{1}{4\pi} - 9\pi)$. Only positive values for S and r have meaning, and therefore the part of the graph for negative r is shown as a dashed curve (see Fig. 20-77). ∎

Exercises 20-7

In Exercises 1 through 8, describe the locus of each of the given equations. Identify the type of curve and its center (or vertex if it is a parabola). Sketch each curve.

1. $(y - 2)^2 = 4(x + 1)$

2. $\dfrac{(x + 4)^2}{4} + \dfrac{(y - 1)^2}{1} = 1$

3. $\dfrac{(x - 1)^2}{4} - \dfrac{(y - 2)^2}{9} = 1$

4. $(y + 5)^2 = -8(x - 2)$

5. $\dfrac{(x + 1)^2}{1} + \dfrac{y^2}{9} = 1$

6. $\dfrac{(y - 4)^2}{16} - \dfrac{(x + 2)^2}{4} = 1$

7. $(x + 3)^2 = -12(y - 1)$

8. $\dfrac{x^2}{16} + \dfrac{(y + 1)^2}{1} = 1$

In Exercises 9 through 20, find the equation of each of the curves described by the given information.

9. Parabola: vertex $(-1, 3)$, focus $(3, 3)$

10. Parabola: vertex $(2, -1)$, directrix $y = 3$

11. Parabola: vertex $(-3, 2)$, focus $(-3, 3)$

12. Parabola: focus $(2, 4)$, directrix $x = 6$

13. Ellipse: center $(-2, 2)$, focus $(-5, 2)$, vertex $(-7, 2)$

14. Ellipse: center $(0, 3)$, focus $(12, 3)$, major axis 26 units

15. Ellipse: vertices $(-2, -3)$ and $(-2, 5)$, end of minor axis $(0, 1)$

16. Ellipse: foci $(1, -2)$ and $(1, 10)$, minor axis 5 units

17. Hyperbola: vertex $(-1, 1)$, focus $(-1, 4)$, center $(-1, 2)$

18. Hyperbola: foci $(2, 1)$ and $(8, 1)$, conjugate axis 6 units

19. Hyperbola: vertices $(2, 1)$ and $(-4, 1)$, focus $(-6, 1)$

20. Hyperbola: center $(1, -4)$, focus $(1, 1)$, transverse axis 8 units

In Exercises 21 through 28, determine the center (or vertex if the curve is a parabola) of the given curves. Sketch each curve.

21. $x^2 + 2x - 4y - 3 = 0$

22. $y^2 - 2x - 2y - 9 = 0$

23. $4x^2 + 9y^2 + 24x = 0$

24. $2x^2 + 9y^2 + 8x - 72y + 134 = 0$

25. $9x^2 - y^2 + 8y - 7 = 0$

26. $5x^2 - 4y^2 + 20x + 8y = 4$

27. $2x^2 - 4x = 9y - 2$

28. $4x^2 + 16x = y - 20$

In Exercises 29 through 32, find the required equations.

29. Find the equation of the hyperbola with asymptotes $x - y = -1$ and $x + y = -3$, and vertex $(3, -1)$.

30. The circle $x^2 + y^2 + 4x - 5 = 0$ passes through the foci and the ends of the minor axis of an ellipse which has its major axis along the x-axis. Find the equation of the ellipse.

31. A first parabola has its vertex at the focus of a second parabola and its focus at the vertex of the second parabola. If the equation of the second parabola is $y^2 = 4x$, find the equation of the first parabola.

32. Verify each of the following equations as being the standard form as indicated.

Parabola, vertex at (h, k), axis parallel to the x-axis: $(y - k)^2 = 4p(x - h)$

Parabola, vertex at (h, k), axis parallel to the y-axis: $(x - h)^2 = 4p(y - k)$

Ellipse, center at (h, k), major axis parallel to the x-axis: $\dfrac{(x - h)^2}{a^2} + \dfrac{(y - k)^2}{b^2} = 1$

Ellipse, center at (h, k), major axis parallel to the y-axis: $\dfrac{(y - k)^2}{a^2} + \dfrac{(x - h)^2}{b^2} = 1$

Hyperbola, center at (h, k), transverse axis parallel to the x-axis: $\dfrac{(x - h)^2}{a^2} - \dfrac{(y - k)^2}{b^2} = 1$

Hyperbola, center at (h, k), transverse axis parallel to the y-axis: $\dfrac{(y - k)^2}{a^2} - \dfrac{(x - h)^2}{b^2} = 1$

In Exercises 33 through 36, solve the given problems.

33. The stream of water from a fire hose follows a parabolic curve. If the stream from a hose nozzle fastened at the ground reaches a maximum height of 60 ft at a horizontal distance of 95 ft from the nozzle, find the equation which describes the stream. Take the origin at the location of the nozzle. Sketch the graph of the stream.

34. For a constant capacitive reactance and a constant resistance, sketch the graph of the impedance and inductive reactance (as abscissas) for an alternating-current circuit. See Section 11-7.

35. Two wheels in a friction drive assembly are equal ellipses, as shown in Fig. 20-78. The wheels are always in contact, with the center of the left wheel fixed in position, and with the right wheel able to move horizontally. Find the equation which can be used to describe the circumference of each wheel in the position shown.

Fig. 20-78

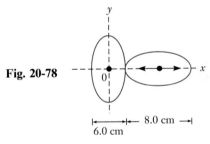

36. A rectangular tract of land is to have a perimeter of 800 m. Express the area in terms of its width and sketch the graph.

20-8 The Second-Degree Equation

The equations of the circle, parabola, ellipse, and hyperbola are all special cases of the same general equation. In this section we discuss this equation and how to identify the particular form it takes when it represents a specific type of curve. *Each of these curves can be represented by a* **second-degree equation** *of the form*

general
second-degree
equation

$$Ax^2 + Bxy + Cy^2 + Dx + Ey + F = 0 \qquad (20\text{-}28)$$

[This equation is the same as Eq. (13-1).] The coefficients of the second-degree terms determine the type of curve which results. Recalling the discussions of the general forms of the equations of the circle, parabola, ellipse, and hyperbola from the previous sections of this chapter, we have the following results.

Equation (20-28), $Ax^2 + Bxy + Cy^2 + Dx + Ey + F = 0$, represents the indicated curve for the given conditions for A, B, and C.

1. If $A = C$, $B = 0$, a circle.
2. If $A \neq C$ (but they have the same sign), $B = 0$, an ellipse.
3. If A and C have different signs, $B = 0$, a hyperbola.
4. If $A = 0$, $C = 0$, $B \neq 0$, a hyperbola.
5. If either $A = 0$ or $C = 0$ (but not both), $B = 0$, a parabola.

(Special cases, such as a single point or no real locus, can also result.)

Another conclusion about Eq. (20-28) is that, if either $D \neq 0$ or $E \neq 0$ (or both), the center (or vertex of a parabola) of the curve is not at the origin. If $B \neq 0$, the axis of the curve has been rotated. We have considered only one such case (the hyperbola $xy = c$) in this chapter.

EXAMPLE A

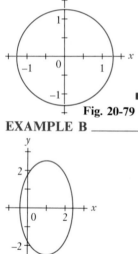

Fig. 20-79

The equation $2x^2 = 3 - 2y^2$ represents a circle. This can be seen by putting it in the form of Eq. (20-28). This form is

$$2x^2 + 2y^2 - 3 = 0$$

$$A = 2 \qquad C = 2$$

We see that $A = C$. Also, since there is no xy-term, we know that $B = 0$. This means that the equation represents a circle. If we write it as $x^2 + y^2 = \frac{3}{2}$ we see that it fits the form of Eq. (20-12). The circle is shown in Fig. 20-79.

EXAMPLE B

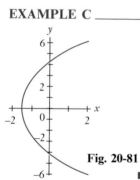

Fig. 20-80

The equation $3x^2 = 6x - y^2 + 3$ represents an ellipse. Before we analyze the equation, we should put it in the form of Eq. (20-28). For the given equation, this form is

$$3x^2 + y^2 - 6x - 3 = 0$$

$$A = 3 \qquad C = 1$$

Here we see that $B = 0$, A and C have the same sign, and $A \neq C$. Therefore, it is an ellipse. The $-6x$ term indicates that the center of the ellipse is not at the origin. The ellipse is shown in Fig. 20-80.

EXAMPLE C

Fig. 20-81

The equation $2(x + 3)^2 = y^2 + 2x^2$ represents a parabola. Putting it in the form of Eq. (20-28), we have

$$2(x^2 + 6x + 9) = y^2 + 2x^2 \qquad \text{expand and simplify}$$

$$2x^2 + 12x + 18 = y^2 + 2x^2$$

$$y^2 - 12x - 18 = 0$$

$$C = 1 \qquad D = -12$$

We now note that $A = 0$, $B = 0$, and $C \neq 0$. This indicates that the equation represents a parabola. Here, the -18 term indicates that the vertex is not at the origin. See Fig. 20-81.

EXAMPLE D _____ Identify the curve represented by the equation $2x^2 + 12x = y^2 - 14$. Determine the appropriate important quantities for the curve, and sketch the graph.

Writing this equation in the form of Eq. (20-28), we have

$$2x^2 - y^2 + 12x + 14 = 0$$

$$A = 2 \uparrow \qquad \uparrow_{C = -1}$$

In this form we identify the equation as representing a hyperbola, since A and C have different signs, and $B = 0$. We now write it in the standard form of a hyperbola.

$$2x^2 + 12x - y^2 = -14$$

$$2(x^2 + 6x \qquad) - y^2 = -14 \qquad \text{complete the square}$$

$$2(x^2 + 6x + 9) - y^2 = -14 + 18$$

$$2(x + 3)^2 - y^2 = 4$$

center is $(-3, 0)$

$$\frac{(x + 3)^2}{2} - \frac{y^2}{4} = 1, \qquad \frac{x'^2}{2} - \frac{y'^2}{4} = 1$$

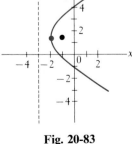

Fig. 20-82

Thus, we see that the center (h, k) of the hyperbola is the point $(-3, 0)$. Also, $a = \sqrt{2}$ and $b = 2$. This means that the vertices are $(-3 + \sqrt{2}, 0)$ and $(-3 - \sqrt{2}, 0)$, and the conjugate axis extends from $(-3, 2)$ to $(-3, -2)$. Also, $c^2 = 2 + 4 = 6$, which means that $c = \sqrt{6}$. The foci are $(-3 + \sqrt{6}, 0)$ and $(-3 - \sqrt{6}, 0)$. The graph is shown in Fig. 20-82. ∎

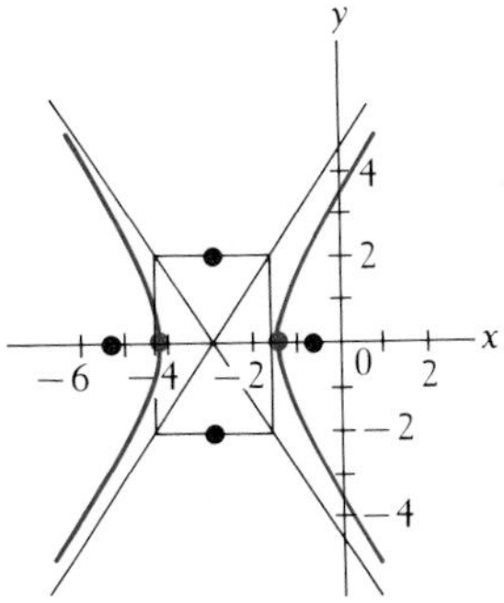

EXAMPLE E _____ Identify the curve represented by the equation $4y^2 - 23 = 4(4x + 3y)$. Determine the appropriate important quantities for the curve, and sketch the graph.

Writing this equation in the form of Eq. (20-28), we have

$$4y^2 - 16x - 12y - 23 = 0$$

Therefore, we recognize the equation as representing a parabola, since $A = 0$ and $B = 0$. Now writing the equation in the standard form of a parabola, we have

$$4y^2 - 12y = 16x + 23$$

$$4(y^2 - 3y \qquad) = 16x + 23 \qquad \text{complete the square}$$

$$4\left(y^2 - 3y + \frac{9}{4}\right) = 16x + 23 + 9$$

$$4\left(y - \frac{3}{2}\right)^2 = 16(x + 2)$$

vertex $\left(-2, \frac{3}{2}\right)$

$$\left(y - \frac{3}{2}\right)^2 = 4(x + 2) \qquad \text{or} \quad y'^2 = 4x'$$

We now note that the vertex is the point $(-2, \frac{3}{2})$ and that $p = 1$. Also, it is symmetric to the x'-axis. Therefore, the focus is $(-1, \frac{3}{2})$ and the directrix is $x = -3$. The graph is shown in Fig. 20-83. ∎

Fig. 20-83

In Chapter 13, when these curves were first introduced, they were referred to as **conic sections.** If a plane is passed through a cone, the intersection of the plane and the cone results in one of these curves; the curve formed depends on the angle of the plane with respect to the axis of the cone. This is indicated in Fig. 20-84.

Fig. 20-84

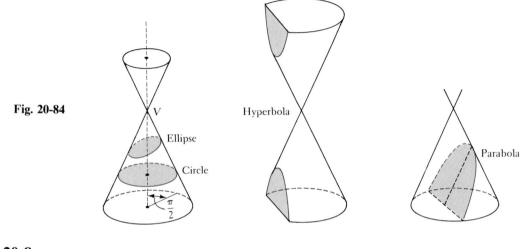

Exercises 20-8

In Exercises 1 through 20, identify each of the equations as representing either a circle, a parabola, an ellipse, or a hyperbola.

1. $x^2 + 2y^2 - 2 = 0$
2. $x^2 - y = 0$
3. $2x^2 - y^2 - 1 = 0$

4. $3x^2 + 3y^2 - 1 = 0$
5. $2x^2 + 2y^2 - 3y - 1 = 0$
6. $x^2 - 2y^2 - 3x - 1 = 0$

7. $2x^2 - x - y = 1$
8. $2x^2 + 4y^2 - y - 2x = 4$
9. $x^2 = y^2 - 1$

10. $3x^2 = 2y - 4y^2$
11. $x^2 = y - y^2$
12. $y = 3 - 6x^2$

13. $y(3 - 2y) = 2(x^2 - y^2)$
14. $x(2 - x) = y^2$
15. $2xy + x - 3y = 6$

16. $(y + 1)^2 = x^2 + y^2 - 1$
17. $2x(x - y) = y(3 - y - 2x)$
18. $2x^2 = x(x - 1) + 4y^2$

19. $x(y + 3x) = x^2 + xy - y^2 + 1$
20. $4x(x - 1) = 2x^2 - 2y^2 + 3$

In Exercises 21 through 28, identify the curve represented by each of the given equations. Determine the appropriate important quantities associated with the curve, and sketch the graph.

21. $x^2 = 8(y - x - 2)$
22. $x^2 = 6x - 4y^2 - 1$
23. $y^2 = 2(x^2 - 2x - 2y)$

24. $4x^2 + 4 = 9 - 8x - 4y^2$
25. $y^2 + 42 = 2x(10 - x)$
26. $x^2 - 4y = y^2 + 4(1 - x)$

27. $4(y^2 - 4x - 2) = 5(4y - 5)$
28. $2(2x^2 - y) = 8 - y^2$

In Exercises 29 through 32, use the given values to determine the type of curve represented.

29. For the equation $x^2 + ky^2 = a^2$, what type of curve is represented if (a) $k = 1$, (b) $k < 0$, and (c) if $k > 0$ ($k \neq 1$)?

30. For the equation $\dfrac{x^2}{4 - C} - \dfrac{y^2}{C} = 1$, what type of curve is represented if (a) $C < 0$, (b) $0 < C < 4$? (For $C > 4$, see Exercise 32.)

31. In Eq. (20-28), if $A = C \neq 0$ and $B = D = E = F = 0$, what locus is described?

32. For the equation in Exercise 30, what type of locus is described if $C > 4$?

In Exercises 33 through 36, determine the type of curve from the given information.

33. The diagonal brace in a rectangular metal frame is 3.0 cm longer than the length of one of the sides. Determine the type of equation relating the lengths of the sides of the frame.

34. One circular solar cell has a radius which is 2.0 in. less than the radius r of a second circular solar cell. Determine the type of curve represented by the equation relating the total area A of both cells and r.

35. A flashlight emits a cone of light onto the floor. What type of curve is the perimeter of the lighted area on the floor, if the floor cuts completely through the cone of light?

36. A supersonic jet emits a conical shock wave behind it. What type of curve is outlined on the surface of a lake by the shock wave if the jet is flying horizontally?

20-9 Polar Coordinates

Thus far we have graphed all curves in one coordinate system. This system, the rectangular coordinate system, is probably the most useful and widely applicable system. However, for certain types of curves, other coordinate systems prove to be better adapted. These coordinate systems are widely used, especially when certain applications of higher mathematics are involved. We shall discuss one of these systems here.

Instead of designating a point by its x- and y-coordinates, we can specify its location by its radius vector and the angle which the radius vector makes with the x-axis. Thus, the r and θ that are used in the definitions of the trigonometric functions can also be used as the coordinates of points in the plane. The important aspect of choosing coordinates is that, for each set of values, there must be only one point which corresponds to this set. We can see that this condition is satisfied by the use of r and θ as coordinates. *In* **polar coordinates** *the origin is called the* **pole,** *and the half-line for which the angle is zero (equivalent to the positive x-axis) is called the* **polar axis.** The coordinates of a point are designated as (r, θ). We shall use radians when measuring the value of θ. See Fig. 20-85.

When using polar coordinates, we generally label the lines for some of the values of θ; namely, those for $\theta = 0$ (the polar axis), $\theta = \frac{\pi}{2}$ (equivalent to the positive y-axis), $\theta = \pi$ (equivalent to the negative x-axis), $\theta = \frac{3\pi}{2}$ (equivalent to the negative y-axis), and possibly others. In Fig. 20-86 these lines and those for the multiples of $\theta = \frac{\pi}{6}$ are shown. The circles for $r = 1, r = 2$, and $r = 3$ are also shown in this figure.

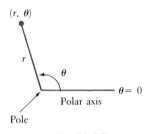

Fig. 20-85

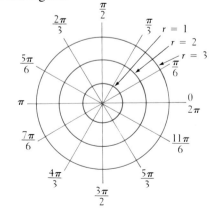

Fig. 20-86

EXAMPLE A ──────

If $r = 2$ and $\theta = \frac{\pi}{6}$, we have the point as indicated in Fig. 20-87. The coordinates (r, θ) of this point are written as $(2, \frac{\pi}{6})$ when polar coordinates are used. This point corresponds to $(\sqrt{3}, 1)$ in rectangular coordinates.

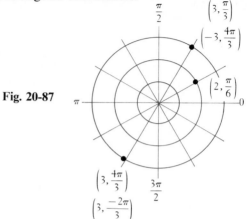

Fig. 20-87

One difference between rectangular coordinates and polar coordinates is that, for each point in the plane, there are limitless possibilities for the polar coordinates of that point. For example, the point $(2, \frac{\pi}{6})$ can also be represented by $(2, \frac{13\pi}{6})$ since the angles $\frac{\pi}{6}$ and $\frac{13\pi}{6}$ are coterminal. We also remove one restriction on r that we imposed in the definition of the trigonometric functions. That is, r is allowed to take on positive and negative values. If r is considered negative, then

NOTE ▷ *the point is found on the opposite side of the pole* from that on which it is positive.

EXAMPLE B ──────

The coordinates $(3, \frac{4\pi}{3})$ and $(3, -\frac{2\pi}{3})$ represent the same point. However, the point $(-3, \frac{4\pi}{3})$ is on the opposite side of the pole, three units from the pole. Another possible set of coordinates for $(-3, \frac{4\pi}{3})$ is $(3, \frac{\pi}{3})$ (see Fig. 20-87).

When plotting a point in polar coordinates it is generally easier to *first locate the terminal side of θ and then measure r along this terminal side.* This is illustrated in the following example.

EXAMPLE C ──────

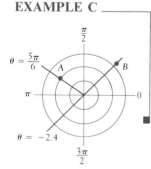

Fig. 20-88

Plot the points $A(2, \frac{5\pi}{6})$ and $B(-3.2, -2.4)$ in the polar coordinate system.

To locate A we determine the terminal side of $\theta = \frac{5\pi}{6}$ and then find $r = 2$. See Fig. 20-83.

To locate B we find the terminal side of $\theta = -2.4$, measuring clockwise from the polar axis (and recalling that $\pi = 3.14 = 180°$). Then we locate $r = -3.2$ on the opposite side of the pole. See Fig. 20-88.

We will find that points with negative values of r occur frequently when plotting curves in polar coordinates.

The relationships between the polar coordinates of a point and the rectangular coordinates of the same point come directly from the definitions of the trigonometric functions. Those most commonly used are (see Fig. 20-89)

$$x = r \cos \theta, \qquad y = r \sin \theta$$

(20-29)

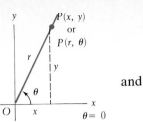

Fig. 20-89

polar and rectangular
coordinates

and $\boxed{\tan \theta = \dfrac{y}{x}, \qquad r = \sqrt{x^2 + y^2}}$ (20-30)

The following examples show the use of Eqs. (20-29) and (20-30) in changing coordinates in one system to coordinates in the other system. Also, they are used to transform equations from one system to the other.

EXAMPLE D

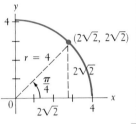

Fig. 20-90

Using Eqs. (20-29), we can transform the polar coordinates of $(4, \frac{\pi}{4})$ into the rectangular coordinates $(2\sqrt{2}, 2\sqrt{2})$, since

$$x = 4 \cos \frac{\pi}{4} = 4\left(\frac{\sqrt{2}}{2}\right) = 2\sqrt{2}$$

and

$$y = 4 \sin \frac{\pi}{4} = 4\left(\frac{\sqrt{2}}{2}\right) = 2\sqrt{2}$$

■ See Fig. 20-90.

EXAMPLE E

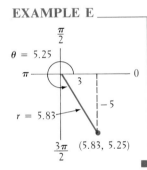

Fig. 20-91

Using Eqs. (20-30), we can transform the rectangular coordinates $(3, -5)$ into polar coordinates.

$$\tan \theta = -\frac{5}{3}, \qquad \theta = 5.25 \qquad (\text{or } -1.03)$$

$$r = \sqrt{3^2 + (-5)^2} = 5.83$$

We know that θ is a fourth-quadrant angle since x is positive and y is negative. Therefore, the point $(3, -5)$ in rectangular coordinates can be expressed as the point $(5.83, 5.25)$ in polar coordinates (see Fig. 20-91). Other polar coordinates for the point are also possible.

EXAMPLE F

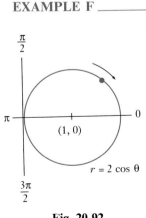

Fig. 20-92

If an electrically charged particle enters a magnetic field at right angles to the field, the particle follows a circular path. This fact is used in the design of nuclear particle accelerators.

A proton (positively charged) enters a magnetic field such that its path may be described by the rectangular equation $x^2 + y^2 = 2x$, where measurements are in meters. Find the polar equation of this circle.

We change this equation expressed in the rectangular coordinates x and y into an equation expressed in the polar coordinates r and θ by using the relations $r^2 = x^2 + y^2$ and $x = r \cos \theta$ as follows:

$$
\begin{array}{ll}
x^2 + y^2 = 2x & \text{rectangular equation} \\
r^2 = 2r \cos \theta & \text{substitute} \\
r = 2 \cos \theta & \text{divide by } r
\end{array}
$$

■ This is the required polar equation of the circle. The circle is shown in Fig. 20-92.

EXAMPLE G

Find the rectangular equation of the *rose* $r = 4 \sin 2\theta$.

Using the trigonometric identity $\sin 2\theta = 2 \sin \theta \cos \theta$ and Eqs. (20-29) and (20-30) leads to the solution.

$$r = 4 \sin 2\theta \qquad \text{polar equation}$$
$$= 4(2 \sin \theta \cos \theta) = 8 \sin \theta \cos \theta \qquad \text{using identity}$$
$$\sqrt{x^2 + y^2} = 8\left(\frac{y}{r}\right)\left(\frac{x}{r}\right) = \frac{8xy}{r^2} = \frac{8xy}{x^2 + y^2} \qquad \text{using Eqs. (20-29) and (20-30)}$$

$$x^2 + y^2 = \frac{64x^2y^2}{(x^2 + y^2)^2} \qquad \text{squaring both sides}$$
$$(x^2 + y^2)^3 = 64x^2y^2 \qquad \text{simplifying}$$

From this example we can see that plotting the graph from the rectangular equation would be complicated. However, as we will see in the following section, plotting this graph in polar coordinates is quite simple. The curve is shown in Fig. 20-93.

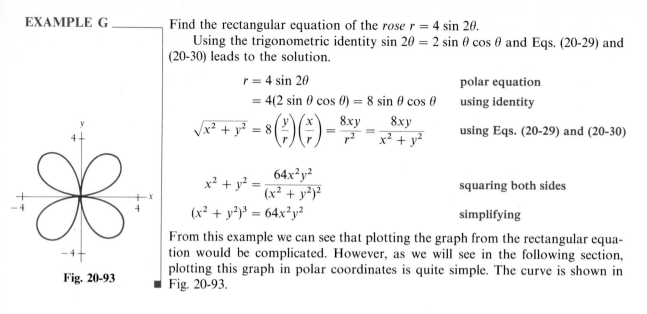

Fig. 20-93

Exercises 20-9

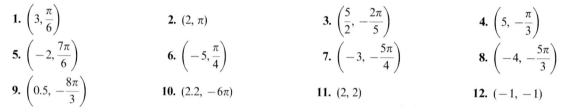

In Exercises 1 through 12, plot the given polar coordinate points on polar coordinate paper.

1. $\left(3, \dfrac{\pi}{6}\right)$

2. $(2, \pi)$

3. $\left(\dfrac{5}{2}, -\dfrac{2\pi}{5}\right)$

4. $\left(5, -\dfrac{\pi}{3}\right)$

5. $\left(-2, \dfrac{7\pi}{6}\right)$

6. $\left(-5, \dfrac{\pi}{4}\right)$

7. $\left(-3, -\dfrac{5\pi}{4}\right)$

8. $\left(-4, -\dfrac{5\pi}{3}\right)$

9. $\left(0.5, -\dfrac{8\pi}{3}\right)$

10. $(2.2, -6\pi)$

11. $(2, 2)$

12. $(-1, -1)$

In Exercises 13 through 16, find a set of polar coordinates for each of the given points expressed in rectangular coordinates.

13. $(\sqrt{3}, 1)$

14. $(-1, -1)$

15. $\left(-\dfrac{\sqrt{3}}{2}, -\dfrac{1}{2}\right)$

16. $(-5, 4)$

In Exercises 17 through 20, find the rectangular coordinates corresponding to the points for which the polar coordinates are given.

17. $\left(8, \dfrac{4\pi}{3}\right)$

18. $(-4, -\pi)$

19. $\left(3, -\dfrac{\pi}{8}\right)$

20. $(-1, 1)$

In Exercises 21 through 28, find the polar equation of the given rectangular equations.

21. $x = 3$

22. $y = 2$

23. $x^2 + y^2 = a^2$

24. $x^2 + y^2 = 4y$

25. $y^2 = 4x$

26. $x^2 - y^2 = a^2$

27. $x^2 + 4y^2 = 4$

28. $y = x^2$

In Exercises 29 through 36, find the rectangular equation of each of the given polar equations.

29. $r = \sin \theta$

30. $r = 4 \cos \theta$

31. $r \cos \theta = 4$

32. $r \sin \theta = -2$

33. $r = 2(1 + \cos \theta)$

34. $r = 1 - \sin \theta$

35. $r^2 = \sin 2\theta$

36. $r^2 = 16 \cos 2\theta$

In Exercises 37 through 40, find the required equations.

37. Under certain conditions, the x- and y-components of a magnetic field B are given by the equations

$$B_x = \frac{-ky}{x^2 + y^2} \quad \text{and} \quad B_y = \frac{kx}{x^2 + y^2}$$

Write these equations in terms of polar coordinates.

38. A cylindrical oil tank 2.60 m in diameter is on its side. Using the center of the end of the tank as the origin, find the equation of the circumference of the end of the tank in rectangular coordinates and in polar coordinates.

39. The shape of a cam can be described by the polar equation $r = 3 - \sin \theta$. Find the rectangular equation for the shape of the cam.

40. The polar equation of the path of a weather satellite of the earth is

$$r = \frac{4800}{1 + 0.14 \cos \theta}$$

where r is measured in miles. Find the rectangular equation of the path of this satellite. The path is an ellipse, with the earth at one of the foci.

20-10   Curves in Polar Coordinates

The basic method for finding a curve in polar coordinates is the same as in rectangular coordinates. We assume values of θ and then find the corresponding values of r. These points are plotted and joined, thus forming the curve which represents the function. However, there are certain basic curves which can be sketched directly from the equation. The following examples illustrate these methods.

EXAMPLE A ———— The graph of the polar equation $r = 3$ is a circle of radius 3, with center at the pole. This can be seen to be the case, since $r = 3$ for all possible values of θ. It is not really necessary to find specific points for this circle, which is shown in Fig. 20-94.

The graph of $\theta = \frac{\pi}{6}$ is a straight line through the pole. It represents all points for which $\theta = \frac{\pi}{6}$ for all possible values of r, positive or negative. This line is also shown in Fig. 20-94.

Fig. 20-94

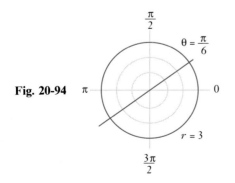

EXAMPLE B

Plot the graph of $r = 1 + \cos \theta$.

We find the following table of values of r corresponding to the assumed values of θ.

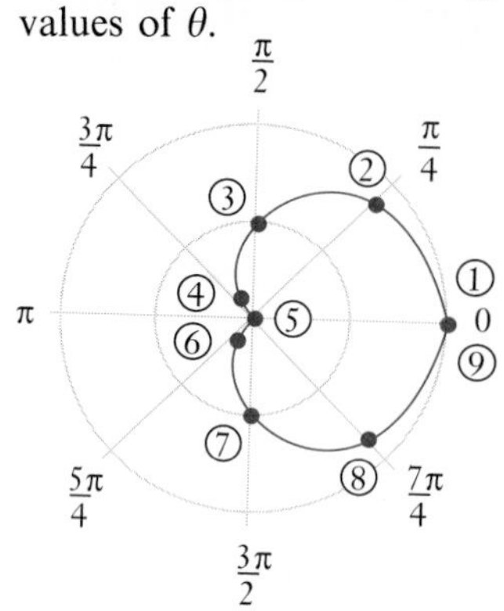

Fig. 20-95

θ	0	$\frac{\pi}{4}$	$\frac{\pi}{2}$	$\frac{3\pi}{4}$	π	$\frac{5\pi}{4}$	$\frac{3\pi}{2}$	$\frac{7\pi}{4}$	2π
r	2	1.7	1	0.3	0	0.3	1	1.7	2
Point number	1	2	3	4	5	6	7	8	9

We now see that the points on the curve start repeating, and it is unnecessary to find additional points. This curve is called a **cardioid** and is shown in Fig. 20-95.

EXAMPLE C

Plot the graph of $r = 1 - 2 \sin \theta$.

θ	0	$\frac{\pi}{4}$	$\frac{\pi}{2}$	$\frac{3\pi}{4}$	π	$\frac{5\pi}{4}$	$\frac{3\pi}{2}$	$\frac{7\pi}{4}$	2π
r	1	-0.4	-1	-0.4	1	2.4	3	2.4	1
Point number	1	2	3	4	5	6	7	8	9

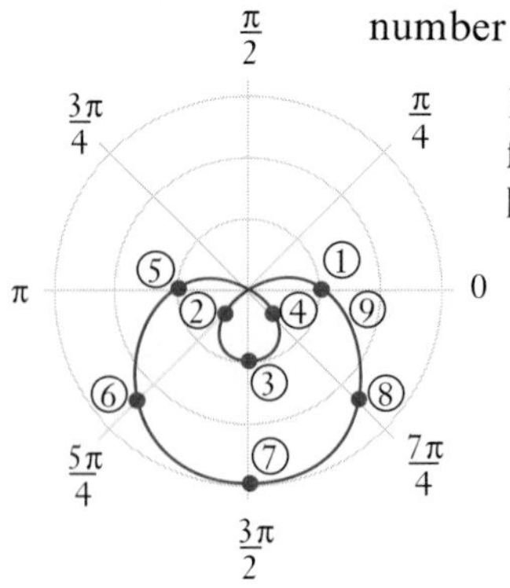

Fig. 20-96

Particular care should be taken in plotting the points for which r is negative. This curve is known as a **limaçon** and is shown in Fig. 20-96.

EXAMPLE D

A cam is shaped such that the edge of the upper "half" can be described by the equation $r = 2.0 + \cos \theta$ and the lower "half" can be described by the equation $r = \dfrac{3.0}{2.0 - \cos \theta}$, where measurements are in inches. Plot the curve which represents the shape of the cam.

We get the points for the edge of the cam by using values of θ from 0 to π for the upper "half" and from π to 2π for the lower half. The table of values follows.

	θ	0	$\frac{\pi}{4}$	$\frac{\pi}{2}$	$\frac{3\pi}{4}$	π
$r = 2.0 + \cos \theta$	r	3.0	2.7	2.0	1.3	1.0
	Point number	1	2	3	4	5

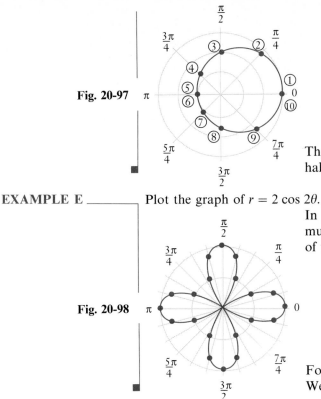

Fig. 20-97

$$r = \frac{3.0}{2.0 - \cos\theta}$$

θ	π	$\frac{5\pi}{4}$	$\frac{3\pi}{2}$	$\frac{7\pi}{4}$	2π
r	1.0	1.1	1.5	2.3	3.0
Point number	6	7	8	9	10

The upper "half" is part of a limaçon, and the lower half is a semiellipse. The cam is shown in Fig. 20-97.

EXAMPLE E _____ Plot the graph of $r = 2\cos 2\theta$.

In finding values of r we must be careful first to multiply the values of θ by 2 before finding the cosine of the angle. The table of values follows.

θ	0	$\frac{\pi}{12}$	$\frac{\pi}{6}$	$\frac{\pi}{4}$	$\frac{\pi}{3}$	$\frac{5\pi}{12}$	$\frac{\pi}{2}$
r	2	1.7	1	0	-1	-1.7	-2

θ	$\frac{7\pi}{12}$	$\frac{2\pi}{3}$	$\frac{3\pi}{4}$	$\frac{5\pi}{6}$	$\frac{11\pi}{12}$	π
r	-1.7	-1	0	1	1.7	2

Fig. 20-98

For values of θ starting with π, the values of θ repeat. We have a four-leaf **rose,** as shown in Fig. 20-98.

EXAMPLE F _____ Plot the graph of $r^2 = 9\cos 2\theta$.

θ	0	$\frac{\pi}{8}$	$\frac{\pi}{4}$	$\cdots$	$\frac{3\pi}{4}$	$\frac{7\pi}{8}$	π
r	± 3	± 2.5	0		0	± 2.5	± 3

There are no values of r corresponding to values of θ in the range $\pi/4 < \theta < 3\pi/4$, since twice these angles are in the second and third quadrants, and the cosine is negative for such angles. The value of r^2 cannot be negative. Also, the values of r repeat for $\theta > \pi$. The figure is called a **lemniscate** (Fig. 20-99).

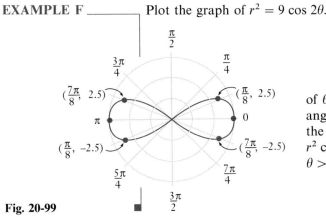

Fig. 20-99

Exercises 20-10

In Exercises 1 through 28, plot the given curves in polar coordinates.

1. $r = 4$ **2.** $r = 2$ **3.** $\theta = \dfrac{3\pi}{4}$ **4.** $\theta = -\dfrac{\pi}{3}$

5. $r = 4\sec\theta$ **6.** $r = 4\csc\theta$ **7.** $r = 2\sin\theta$ **8.** $r = 3\cos\theta$

9. $r = 1 - \cos\theta$ (cardioid)   **10.** $r = \sin\theta - 1$ (cardioid)   **11.** $r = 2 - \cos\theta$ (limaçon)

12. $r = 2 + 3\sin\theta$ (limaçon)   **13.** $r = 4\sin 2\theta$ (rose)   **14.** $r = 2\sin 3\theta$ (rose)

15. $r^2 = 4 \sin 2\theta$ (lemniscate)

16. $r^2 = 2 \sin \theta$

17. $r = 2^\theta$ (spiral)

18. $r = 10^\theta$ (spiral)

19. $r = -4 \sin 4\theta$ (rose)

20. $r = -\cos 3\theta$ (rose)

21. $r = \dfrac{1}{2 - \cos \theta}$ (ellipse)

22. $\dfrac{1}{1 - \cos \theta}$ (parabola)

23. $r = \dfrac{6}{1 - 2 \cos \theta}$ (hyperbola)

24. $r = \dfrac{6}{3 - 2 \sin \theta}$ (ellipse)

25. $r = 4 \cos \frac{1}{2}\theta$

26. $r = 2 + \cos 3\theta$

27. $r = 3 - \sin 3\theta$

28. $r = 4 \tan \theta$

In Exercises 29 through 32, sketch the indicated graphs.

29. An architect designs a patio which is shaped such that it can be described as the area within the polar curve $r = 4.0 - \sin \theta$, where measurements are in meters. Sketch the perimeter of the patio.

30. A missile is fired at an airplane and is always directed toward the airplane. The missile is traveling at twice the speed of the airplane. An equation which describes the distance r between the missile and the airplane is $r = \dfrac{70 \sin \theta}{(1 - \cos \theta)^2}$, where θ is the angle between their directions at any time, and the plane is assumed to be at the pole at all times. See Fig. 20-100. This is a *relative pursuit curve*. Sketch the graph of this equation for $\pi/4 \le \theta \le \pi$.

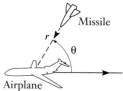

Missile

Airplane

Fig. 20-100

31. In studying the photoelectric effect, an equation used for the rate R at which photoelectrons are ejected at various angles θ is $R = \dfrac{\sin^2 \theta}{(1 - 0.5 \cos \theta)^2}$. Sketch the graph of R and θ.

32. Sketch the graph of the rectangular equation

$$4(x^6 + 3x^4y^2 + 3x^2y^4 + y^6 - x^4 - 2x^2y^2 - y^4) + y^2 = 0$$

[*Hint:* The equation can be written as $4(x^2 + y^2)^3 - 4(x^2 + y^2)^2 + y^2 = 0$. Transform to polar coordinates, and then sketch the curve.]

20-11 Chapter Equations, Review Exercises, and Practice Test

Chapter Equations

Distance formula	Fig. 20-2	$d = \sqrt{(x_2 - x_1)^2 + (y_2 - y_1)^2}$	(20-1)
Slope	Fig. 20-4	$m = \dfrac{y_2 - y_1}{x_2 - x_1}$	(20-2)
	Fig. 20-7	$m = \tan \alpha \quad (0° \le \alpha < 180°)$	(20-3)
	Fig. 20-9	$m_1 = m_2 \quad$ (for $\parallel$ lines)	(20-4)
	Fig. 20-10	$m_2 = -\dfrac{1}{m_1} \quad \text{or} \quad m_1 m_2 = -1 \quad$ (for $\perp$ lines)	(20-5)
Straight line	Fig. 20-14	$y - y_1 = m(x - x_1)$	(20-6)
	Fig. 20-17	$x = a$	(20-7)
	Fig. 20-18	$y = b$	(20-8)
	Fig. 20-21	$y = mx + b$	(20-9)
		$Ax + By + C = 0$	(20-10)

Circle	Fig. 20-29	$(x - h)^2 + (y - k)^2 = r^2$	(20-11)
	Fig. 20-31	$x^2 + y^2 = r^2$	(20-12)
		$x^2 + y^2 + Dx + Ey + F = 0$	(20-14)
Parabola	Fig. 20-38	$y^2 = 4px$	(20-15)
	Fig. 20-41	$x^2 = 4py$	(20-16)
Ellipse	Fig. 20-51	$\dfrac{x^2}{a^2} + \dfrac{y^2}{b^2} = 1$	(20-17)
	Fig. 20-51	$a^2 = b^2 + c^2$	(20-18)
	Fig. 20-52	$\dfrac{y^2}{a^2} + \dfrac{x^2}{b^2} = 1$	(20-19)
Hyperbola	Fig. 20-62	$\dfrac{x^2}{a^2} - \dfrac{y^2}{b^2} = 1$	(20-20)
	Fig. 20-63	$c^2 = a^2 + b^2$	(20-21)
	Fig. 20-62	$y = \pm \dfrac{bx}{a}$ (asymptotes)	(20-23)
	Fig. 20-64	$\dfrac{y^2}{a^2} - \dfrac{x^2}{b^2} = 1$	(20-24)
	Fig. 20-69	$xy = c$	(20-25)
Translation of axes	Fig. 20-71	$x = x' + h$ and $y = y' + k$	(20-26)
		$x' = x - h$ and $y' = y - k$	(20-27)
Second-degree equation		$Ax^2 + Bxy + Cy^2 + Dx + Ey + F = 0$	(20-28)
Polar coordinates	Fig. 20-89	$x = r \cos \theta, \qquad y = r \sin \theta$	(20-29)
		$\tan \theta = \dfrac{y}{x}, \qquad r = \sqrt{x^2 + y^2}$	(20-30)

Review Exercises

In Exercises 1 through 12, find the equation of the indicated curve subject to the given conditions. Sketch each curve.

1. Straight line: passes through $(1, -7)$ with a slope of 4

2. Straight line: passes through $(-1, 5)$ and $(-2, -3)$

3. Straight line: perpendicular to $3x - 2y + 8 = 0$ and has a y-intercept of $(0, -1)$

4. Straight line: parallel to $2x - 5y + 1 = 0$ and has an x-intercept of $(2, 0)$

5. Circle: center at $(1, -2)$, passes through $(4, -3)$

6. Circle: tangent to the line $x = 3$, center at $(5, 1)$

7. Parabola: focus $(3, 0)$, vertex $(0, 0)$

8. Parabola: directrix $y = -5$, vertex $(0, 0)$

9. Ellipse: vertex $(10, 0)$, focus $(8, 0)$, center $(0, 0)$

10. Ellipse: center $(0, 0)$, passes through $(0, 3)$ and $(2, 1)$

11. Hyperbola: $V(0, 13)$, $C(0, 0)$, conj. axis of 24

12. Hyperbola: $F(0, 10)$, $F(0, -10)$, $V(0, 8)$

In Exercises 13 through 24, find the indicated quantities for each of the given equations. Sketch each curve.

13. $x^2 + y^2 + 6x - 7 = 0$, center and radius

14. $x^2 + y^2 - 4x + 2y - 20 = 0$, center and radius

15. $x^2 = -20y$, focus and directrix

16. $y^2 = 24x$, focus and directrix

17. $16x^2 + y^2 = 16$, vertices and foci

18. $2y^2 - 9x^2 = 18$, vertices and foci

19. $2x^2 - 5y^2 = 8$, vertices and foci

20. $2x^2 + 25y^2 = 50$, vertices and foci

21. $x^2 - 8x - 4y - 16 = 0$, vertex and focus

22. $y^2 - 4x + 4y + 24 = 0$, vertex and directrix

23. $4x^2 + y^2 - 16x + 2y + 13 = 0$, center

24. $x^2 - 2y^2 + 4x + 4y + 6 = 0$, center

In Exercises 25 through 32, plot the given curves in polar coordinates.

25. $r = 4(1 + \sin \theta)$

26. $r = 1 - 3 \cos \theta$

27. $r = 4 \cos 3\theta$

28. $r = -3 \sin \theta$

29. $r = \cot \theta$

30. $r = \dfrac{1}{2(\sin \theta - 1)}$

31. $r = 2 \sin\left(\dfrac{\theta}{2}\right)$

32. $r = \theta$

In Exercises 33 through 36, find the polar equation of each of the given rectangular equations.

33. $y = 2x$

34. $2xy = 1$

35. $x^2 - y^2 = 16$

36. $x^2 + y^2 = 7 - 6y$

In Exercises 37 through 40, find the rectangular equation of each of the given polar equations.

37. $r = 2 \sin 2\theta$

38. $r^2 = \sin \theta$

39. $r = \dfrac{4}{2 - \cos \theta}$

40. $r = \dfrac{2}{1 - \sin \theta}$

In Exercises 41 through 44, determine the number of real solutions to the given systems of equations by sketching the curves. (See Section 13-1.)

41. $x^2 + y^2 = 9$
$4x^2 + y^2 = 16$

42. $y = e^x$
$x^2 - y^2 = 1$

43. $x^2 + y^2 - 4y - 5 = 0$
$y^2 - 4x^2 - 4 = 0$

44. $x^2 - 4y^2 + 2x - 3 = 0$
$y^2 - 4x - 4 = 0$

In Exercises 45 through 76, solve the given problems.

45. Find the points of intersection of the ellipses $25x^2 + 4y^2 = 100$ and $4x^2 + 9y^2 = 36$.

46. Find the points of intersection of the hyperbola $y^2 - x^2 = 1$ and the ellipse $x^2 + 25y^2 = 25$.

47. In two ways show that the line segments joining $(-3, 11)$, $(2, -1)$, and $(14, 4)$ form a right triangle.

48. Show that the altitudes of the triangle with vertices $(2, -4)$, $(3, -1)$, and $(-2, 5)$ meet at a single point.

49. By means of the definition of a parabola, find the equation of the parabola with focus at $(3, 1)$ and directrix the line $y = -3$. Find the same equation by the method of translation of axes.

50. Repeat the instructions of Exercise 49 for the parabola with focus at $(0, 2)$ and directrix the line $y = 4$.

51. The total resistance R_T of two resistances in series in an electric circuit is the sum of the resistances. If a variable resistor R is in series with a 2.5-Ω resistor, express R_T as a function of R and sketch the graph.

52. The acceleration of an object is defined as the change in velocity v divided by the corresponding change in time t. Find the equation relating the velocity v and time t for an object for which the acceleration is 20 ft/s² and $v = 5.0$ ft/s when $t = 0$ s.

53. One computer line printer prints 2500 lines/min for x min, and a second printer prints 1500 lines/min for y min. If they print a total of 37,500 lines together, express y as a function of x and sketch the graph.

54. An airplane touches down when landing at 100 mi/h. Its velocity v while coming to a stop is given by $v = 100 - 20,000t$, where t is the time in hours. Sketch the graph of v vs. t.

55. It takes 2.010 kJ of heat to raise the temperature of 1.000 kg of steam by 1.000°C. In a steam generator, a total of y kJ is used to raise the temperature of 50.00 kg of steam from 100°C to T°C. Express y as a function of T and sketch the graph.

56. Let C and F denote corresponding Celsius annd Fahrenheit temperature readings. If the equation relating the two is linear, determine this equation given that $C = 0$ when $F = 32$ and $C = 100$ when $F = 212$.

57. A solar panel reflector is circular and has a circumference of 4.50 m. Taking the origin of a coordinate system at the center of the reflector, what is the equation of the circumference?

58. A friction disk drive wheel is tangent to another wheel, as shown in Fig. 20-101. The radii of the wheels are 2 in. and 6 in. What is the equation of the circumference of each wheel if the coordinate system is located at the center of the drive wheel as shown?

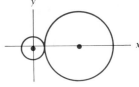

Fig. 20-101

59. The top horizontal cross section of a dam is parabolic. The open area within this cross section is 80 ft across and 50 ft from front to back. Find the equation of the edge of the open area with the vertex at the origin of the coordinate system and the axis along the x-axis.

60. The *quality factor* Q of a series resonant electric circuit with resistance R, inductance L, and capacitance C is given by $Q = \dfrac{1}{R}\sqrt{\dfrac{L}{C}}$. Sketch the graph of Q and L for a circuit in which $R = 1000\ \Omega$ and $C = 4.00\ \mu\text{F}$.

61. A rectangular parking lot is to have a perimeter of 600 m. Express the area A in terms of the width w and sketch the graph.

62. At very low temperatures certain metals have an electric resistance of zero. This phenomenon is called *superconductivity*. A magnetic field also affects the superconductivity. A certain level of magnetic field H_T, the threshold field, is related to the thermodynamic temperature T by $H_T/H_0 = 1 - (T/T_0)^2$, where H_0 and T_0 are specifically defined values of magnetic field and temperature. Sketch H_T/H_0 vs. T/T_0.

63. The electric power P, in watts, supplied by a battery is given by $P = 12.0i - 0.500i^2$, where i is the current. Sketch the graph of P and i.

64. The rate r in grams per second at which a substance is formed in a chemical reaction is given by $r = 10.0 - 0.010t^2$. Sketch the graph of r vs. t.

65. A study indicated that the fraction f of cells destroyed by various dosages d of X rays is given by the graph in Fig. 20-102. Assuming that the curve is a quarter-ellipse, find the equation relating f and d for $0 \le f \le 1$ and $0 < d \le 10$ units.

66. The Colosseum in Rome is in the shape of an ellipse 188 m long and 156 m wide. Find the area of the Colosseum. ($A = \pi ab$ for an ellipse.)

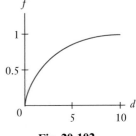

Fig. 20-102

67. Soon after reaching the vicinity of the moon, Apollo 11 (the first spacecraft to land a man on the moon) went into an elliptical lunar orbit. The closest the craft was to the moon in this orbit was 70 mi, and the farthest it was from the moon was 190 mi. What was the equation of the path if the center of the moon was at one of the foci of the ellipse? Assume the major axis to be along the x-axis and that the center of the ellipse is at the origin. The radius of the moon is 1080 mi.

68. A machine-part designer wishes to make a model for an elliptical cam by placing two pins in a design board, putting a loop of string over the pins, and marking off the outline by keeping the string taut. (Note that the definition of the ellipse is being used.) If the cam is to measure 10 cm by 6 cm, how long should the loop of string be and how far apart should the pins be?

69. The vertical cross section of the cooling tower of a nuclear power plant is hyperbolic, as shown in Fig. 20-103. Find the radius r of the smallest circular horizontal cross section.

70. In testing, a rectangular sheet of plastic 2.50 ft wide is stretched in the direction of its length ℓ. Find the equation relating ℓ and the length d of a diagonal, and sketch the graph.

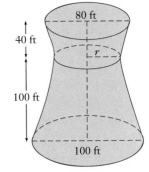

Fig. 20-103

71. An automobile grille is in the shape of an ellipse 90.0 cm wide and 24.0 cm high. Find the width of the grille 10.0 cm from the top.

72. A 60-ft rope passes over a pulley 10 ft above the ground, and a crate on the ground is attached at one end. A man holds the other end at a level of 4 ft above the ground. If the man walks away from the pulley, express the height of the crate above the ground in terms of the distance the man is from directly below the object. Sketch the graph of distance and height. See Fig. 20-104. (Neglect the thickness of the crate.)

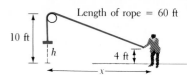

Fig. 20-104

73. A satellite at a height proper to make one revolution per day around the earth will have for an excellent approximation of its projection on the earth of its path the curve $r^2 = R^2 \cos 2(\theta + \frac{\pi}{2})$, where R is the radius of the earth. Sketch the path of the projection.

74. The vertical cross sections of two pipes as drawn on a drawing board are shown in Fig. 20-105. Find the polar equation of each.

75. Under a force which varies inversely as the square of the distance from an attracting object (such as the force the sun exerts on the earth), it can be shown that the equation that an object follows is given in general by

$$\frac{1}{r} = a + b \cos \theta$$

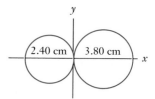

Fig. 20-105

where a and b are constant for a particular path. By transforming this equation to rectangular coordinates, show that this equation represents one of the conic sections, the particular section depending on the values of a and b. It is through this kind of analysis that we know the paths of the planets and comets are conic sections.

76. The sound produced by a jet engine was measured at a distance of 100 m in all directions. The loudness of the sound (in decibels) was found to be $d = 115 + 10 \cos \theta$, where the $0°$ line for the angle θ is directed in front of the engine. Sketch the graph of d vs. θ in polar coordinates (use d as r).

Practice Test

1. Identify the type of curve represented by the equation $2(x^2 + x) = 1 - y^2$.
2. Sketch the graph of the straight line $4x - 2y + 5 = 0$ by finding its slope and y-intercept.
3. Find the polar equation of the curve whose rectangular equation is $x^2 = 2x - y^2$.
4. Find the vertex and the focus of the parabola $x^2 = -12y$. Sketch the graph.
5. Find the equation of the circle with center at $(-1, 2)$ and which passes through $(2, 3)$.
6. Find the equation of the straight line which passes through $(-4, 1)$ and $(2, -2)$.
7. Where is the focus of a parabolic reflector which is 12.0 cm across and 4.00 cm deep?
8. A hallway 16 ft wide has a ceiling whose cross section is a semiellipse. The ceiling is 10 ft high at the walls and 14 ft high at the center. Find the height of the ceiling 4 ft from each wall.
9. Plot the polar curve $r = 3 + \cos \theta$.
10. Find the center and vertices of the conic section $4y^2 - x^2 - 4x - 8y - 4 = 0$. Show completely the sketch of the curve.

21 Introduction to Statistics and Empirical Curve Fitting

Statistical analysis is used extensively in medical research. In Section 21–4, a use of curve fitting in this field is shown.

When a mathematician wishes to state the relation between the area of a circle and its radius, he or she writes $A = \pi r^2$ and knows that this relation is true for all circles. When an engineer wishes to find the safe load which a particular cable is able to support, he or she cannot so simply write an equation for the relation, because the load which a cable may support depends on the diameter of the cable, the material of which it is made, the quality of material in the particular cable, any possible defects in its manufacture, and anything else which could make this cable different from all other cables. Thus, to determine the load which a cable can support, an engineer may test many cables of particular specifications. He or she will find that most of the cables can support approximately the same load, but that there is some variation, and occasionally perhaps a great variation for some reason or other.

The engineer, in testing a number of cables and thereby selecting a certain type of cable, is making use of the basic methods of statistics. That is, the engineer (1) collects data, (2) analyzes these data, and (3) interprets these data. In the first three sections of this chapter, we shall discuss some of the basic methods of tabulating and analyzing this type of statistical information. Applications of statistics occur in nearly all fields of study, including science and technology.

The sections which follow show how to start with a set of points and determine a function for which the graph passes through the given points. The aim is to be able to determine an equation which best "fits" data obtained through experimentation and observation. Such an equation can show a basic relationship between the variables of the experiment and can make it possible to analyze the experiment better. Curve fitting can be particularly useful in research and development leading to new and improved products.

This chapter gives an introduction to some of the basic concepts of statistics. Empirical curve fitting is a method with useful technical applications. In a more complete coverage, a number of other useful methods in statistics with important applications are developed.

21-1 Frequency Distributions

In statistics we deal with various sets of numbers. These could be measurements of length, test scores, weights of objects, or numerous other possibilities. We could deal with the entire set of numbers, but this is often too large to be practical. In such a case we deal with a selected (assumed to be representative) sample.

EXAMPLE A
A company produces a machine part which is designed to be 3.80 cm long. If 10,000 are produced daily, for purposes of quality control it could be impractical to test each one to determine if it meets specifications. Therefore, it might be decided to test every tenth, or every hundredth, part.

If only ten such parts were produced daily, it might be possible to test each one for specifications. ∎

One way of organizing data in order to develop some understanding of it is to *tabulate the number of occurrences for each particular value within the set. This is called a* **frequency distribution.**

EXAMPLE B
A test station measured the loudness of the sound of jet aircraft taking off from a certain airport. The decibel readings of the first twenty jets were as follows: 110, 95, 100, 115, 105, 110, 120, 110, 115, 105, 90, 95, 105, 110, 100, 115, 105, 120, 95, 110.

We can see that it is difficult to determine any pattern to the readings. Therefore, we set up the following table to show the frequency distribution.

Decibel reading	90	95	100	105	110	115	120
Frequency	1	3	2	4	5	3	2

∎ We note that the distribution and pattern of readings is clearer in this form.

Just as graphs are useful in representing algebraic functions, so also are they a very convenient method of representing frequency distributions. There are several useful methods of graphing such distributions, among which *the most important are the* **histogram** *and the* **frequency polygon.** The following examples illustrate these graphical representations.

histogram
frequency polygon

EXAMPLE C  A *histogram represents a particular set of data by use of rectangles.* The width of the base of each rectangle is the same, and is labeled for one of the readings. The height of the rectangle represents the number of values of a given reading. In Fig. 21-1 a histogram representing the data of the frequency distribution of Example B is shown.

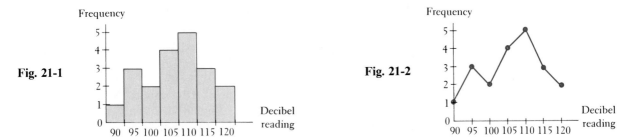

Fig. 21-1

Fig. 21-2

EXAMPLE D *The frequency polygon is used to represent a set of data by plotting as abscissas (x-values) the values of the readings, and as ordinates (y-values) the number of occurrences of each reading (the frequency).* The resulting points are joined by straight-line segments. Figure 21-2 shows a frequency polygon of the data of Example B.

Often the number of measurements is sufficiently large that it is not feasible to tabulate the number of occurrences for a particular value. In other cases such a table does not give a clear idea of the distribution. In these cases we may designate certain intervals, and tabulate the number of values within each interval. Frequency distributions, histograms, and frequency polygons can be used to represent data tabulated in this way.

EXAMPLE E A certain mathematics course had 80 students enrolled in it. After all the tests and exams were recorded, the instructors determined a numerical average (based on 100) for each student. The following is a list of the numerical grades, with the number of students receiving each.

22—1, 37—1, 40—2, 44—1, 47—1, 53—1, 55—3, 56—1, 60—3,
61—1, 63—4, 65—2, 66—1, 67—5, 68—2, 70—2, 71—4, 72—3,
74—5, 75—4, 77—7, 78—4, 79—1, 81—2, 82—4, 84—1, 85—1,
86—3, 87—2, 88—1, 90—2, 92—1, 93—2, 95—1, 97—1

We can observe from this listing that it is difficult to see just how the grades were distributed. Therefore the instructors grouped the grades into intervals, which included five possible grades in each interval. This led to the following table:

Interval	20–24	25–29	30–34	35–39	40–44	45–49	50–54	55–59
Number in interval	1	0	0	1	3	1	1	4

Interval	60–64	65–69	70–74	75–79	80–84	85–89	90–94	95–99
Number in interval	8	10	14	16	7	7	5	2

(Continued on next page)

We can see that the distribution of grades becomes much clearer in the above table. Finally, since the school graded students only with the letters A, B, C, D, and F, the instructors then grouped the grades in intervals below 60, from 60 to 69, from 70 to 79, from 80 to 89, and from 90 to 100, and assigned the appropriate letter grade to each numerical grade within each interval. This led to the following table:

Grade	A	B	C	D	F
Number receiving this grade	7	14	30	18	11

Thus, we can see the distribution according to letter grades. We can also note that, if the number of intervals were reduced much further, it would be difficult to draw any reasonable conclusions regarding the distribution of grades.

Histograms showing the frequency of numerical grades and letter grades are shown in Figs. 21-3 and 21-4. In Fig. 21-3, each interval is represented by one number, the middle number. In Fig. 21-5 a frequency polygon of numerical grades is shown, also with each interval represented by the middle number.

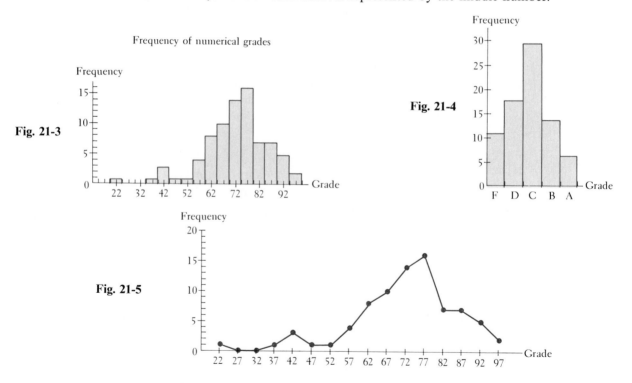

Fig. 21-3

Fig. 21-4

Fig. 21-5

We can see from Example E that care must be taken in grouping data. If the number of intervals is too small, important characteristics of the data may not be apparent. If the number of intervals is too large, it may be difficult to analyze the data to determine the distribution and patterns.

Exercises 21-1

In Exercises 1 through 16, use the following sets of numbers:

    A: 3, 6, 4, 2, 5, 4, 7, 6, 3, 4, 6, 4, 5, 7, 3

    B: 25, 26, 23, 24, 25, 28, 26, 27, 23, 28, 25

    C: 0.48, 0.53, 0.49, 0.45, 0.55, 0.49, 0.47, 0.55, 0.48, 0.57, 0.51, 0.46

    D: 105, 108, 103, 108, 106, 104, 109, 104, 110, 108, 108, 104, 113, 106, 107, 106, 107, 109, 105, 111, 109, 108

In Exercises 1 through 4, set up a frequency distribution table, indicating the frequency of each number given in the indicated set.

1. Set *A*   **2.** Set *B*   **3.** Set *C*   **4.** Set *D*

In Exercises 5 through 8, set up a frequency distribution table, indicating the frequency of numbers for the given intervals of the given sets.

5. Intervals 2–3, 4–5, and 6–7 for set *A*

6. Intervals 22–24, 25–27, and 28–30 for set *B*

7. Intervals 0.43–0.45, 0.46–0.48, 0.49–0.51, 0.52–0.54, 0.55–0.57 for set *C*

8. Intervals 101–105, 106–110, and 111–115 for set *D*

In Exercises 9 through 12, draw histograms for the data in the given exercise.

9. Exercise 1   **10.** Exercise 4   **11.** Exercise 7   **12.** Exercise 8

In Exercises 13 through 16, draw frequency polygons for the data in the given exercise.

13. Exercise 1   **14.** Exercise 4   **15.** Exercise 7   **16.** Exercise 8

In Exercises 17 through 32, find the indicated quantities.

17. In testing a computer system, the number of instructions it could perform in 1 ns was measured at different points in a program. The numbers of instructions recorded were as follows:

      19, 21, 22, 25, 22, 20, 18, 21, 20, 19, 22, 21, 19, 23, 21

Form a frequency distribution table indicating the frequency of each number recorded in this data.

18. For the data of Exercise 17, draw a histogram.

19. For the data of Exercise 17, draw a frequency polygon.

20. For the data of Exercise 17, form a frequency distribution table for intervals 17–19, 20–22, and 23–25. Then draw a histogram to represent these data.

21. A strobe light is designed to flash every 2.25 s at a certain setting. Sample bulbs were tested with the results in the following table:

Number of bulbs	2	7	18	41	56	32	8	3	3
Time between flashes (seconds)	2.21	2.22	2.23	2.24	2.25	2.26	2.27	2.28	2.29

Draw a histogram for these data.

22. Draw a frequency polygon for the data of Exercise 21.

23. In testing a braking system, the distance required to stop a car from 70 mi/h was measured in 120 trials. The results are in the following table:

Stopping distance (feet)	155–159	160–164	165–169	170–174	175–179	180–184	185–189
Number of times car stopped	2	15	32	36	24	10	1

Draw a frequency polygon for these data.

24. Draw a histogram for the data of Exercise 23.

25. The dosage, in milliroentgens (mR), given by a particular X-ray machine was measured 20 times, with the following readings:

   4.25, 4.36, 3.96, 4.21, 4.44, 3.83, 4.37, 4.27, 4.33, 4.34, 4.15, 3.90, 4.41, 4.51, 4.18, 4.26, 4.29, 4.09, 4.36, 4.23

Form a histogram for the intervals 3.80–3.89, 3.90–3.99, and so on.

26. Form a histogram for the data given in Exercise 25 for the intervals 3.71–3.85, 3.86–4.00, and so on.

27. The life of a certain type of battery was measured for a sample of batteries with the following results (in number of hours):

   34, 30, 32, 35, 31, 28, 29, 30, 32, 25, 31, 30, 28, 36, 33, 34, 30, 33, 31, 34, 29, 30, 32

Draw a frequency polygon for these data.

28. Form a histogram for the data given in Exercise 27 for the intervals 25–27, 28–30, 31–33, and 34–36.

29. The diameters of a sample of fiber-optic cables were measured for a sample of cables with the following results:

Diameters (mm)	0.0055	0.0056	0.0057	0.0058	0.0059	0.0060
Number of cables	4	15	32	36	59	64
Diameters (mm)	0.0061	0.0062	0.0063	0.0064	0.0065	0.0066
Number of cables	22	18	10	12	4	4

Draw a histogram for these data.

30. Draw a histogram for the data of Exercise 29 for the intervals 0.0055–0.0057, 0.0058–0.0060, 0.0061–0.0063, and 0.0064–0.0066.

31. Take two dice and toss them 100 times, recording at each toss the sum which appears. Draw a frequency polygon of the sum and the frequency with which it occurred.

32. From the financial section of a newspaper, record (to the nearest dollar) the closing price of the first 50 stocks listed. Form a frequency table with intervals 0–9, 10–19, and so forth. Form a histogram of these data.

21-2   Measures of Central Tendency

Tables and graphical representations give a general description of data. However, it is often profitable and convenient to find representative values for the location of the center of distribution, and other numbers to give a measure of the deviation from this central value. In this way we can obtain an arithmetical description of the data. We shall now discuss the values commonly used to measure the location of the center of the distribution. These are referred to as *measures of central tendency*.

median

The first of these measures of central tendency is the **median.** *The median is the middle number, that number for which there are as many above it as below it in the distribution.* If there is no middle number, the median is that number halfway between the two numbers nearest the middle of the distribution.

EXAMPLE A

Given the numbers 5, 2, 6, 4, 7, 4, 7, 2, 8, 9, 4, 11, 9, 1, 3, we first arrange them in numerical order. This arrangement is

┌ middle number
↓
1, 2, 2, 3, 4, 4, 4, 5, 6, 7, 7, 8, 9, 9, 11

Since there are 15 numbers, the middle number is the eighth. Since the eighth number is 5, the median is 5.

If the number 11 is not included in this set of numbers, and there are only 14 numbers in all, the median is that number halfway between the seventh and eighth numbers. Since the seventh is 4 and the eighth is 5, the median is 4.5.

EXAMPLE B

In the distribution of grades given in Example E of Section 21-1, the median is 74. There are 80 grades in all, and if they are listed in numerical order we find that the 39th through the 43rd grade is 74. This means that the 40th and 41st grades are both 74. The number halfway between the 40th and 41st grades would be the median. The fact that both are 74 means that the median is also 74.

arithmetic mean

Another very widely applied measure of central tendency is the **arithmetic mean.** *The mean is calculated by finding the sum of all the values and then dividing by the number of values.* (The arithmetic mean is the number most people call the "average." However, in statistics the word *average* has the more general meaning of a measure of central tendency.)

EXAMPLE C

The arithmetic mean of the numbers given in Example A is found by finding the sum of all the numbers and dividing by 15. Thus, letting $\bar{x}$ (read as "x bar") represent the mean, we have

$$\bar{x} = \frac{5 + 2 + 6 + 4 + 7 + 4 + 7 + 2 + 8 + 9 + 4 + 11 + 9 + 1 + 3}{15}$$

┌ sum of values
↓
$$= \frac{82}{15} = 5.5$$
↑
└ number of values

Thus, the mean is 5.5.

If we wish to find the arithmetic mean of a large number of values, and if some of them appear more than once, the calculation can be somewhat simplified. The mean can be calculated by multiplying each value by its *frequency* (the number of times it occurs), adding these results, and then dividing by the total number of values considered (the sum of the frequencies). Letting $\bar{x}$ represent the mean of the values $x_1, x_2, \ldots, x_n$, which occur with frequencies $f_1, f_2, \ldots, f_n$, respectively, we have

$$\bar{x} = \frac{x_1 f_1 + x_2 f_2 + \cdots + x_n f_n}{f_1 + f_2 + \cdots + f_n} \tag{21-1}$$

EXAMPLE D ———— Using Eq. (21-1) to find the arithmetic mean of the numbers of Example A, we first set up a table of the values and their respective frequencies.

Value	1	2	3	4	5	6	7	8	9	11
Frequency	1	2	1	3	1	1	2	1	2	1

We now calculate the arithmetic mean $\bar{x}$ by using Eq. (21-1):

multiply each value by its frequency and add results

$$\bar{x} = \frac{1(1) + 2(2) + 3(1) + 4(3) + 5(1) + 6(1) + 7(2) + 8(1) + 9(2) + 11(1)}{1 + 2 + 1 + 3 + 1 + 1 + 2 + 1 + 2 + 1}$$

sum of frequencies

$$= \frac{82}{15} = 5.5$$

Summations such as those in Eq. (21-1) occur frequently in statistics and other branches of mathematics. In order to simplify writing these sums, the symbol Σ is used to indicate the process of summation. (Σ is the Greek capital letter sigma.) In using this symbol, Σx means the sum of the x's.

EXAMPLE E ———— We can show the sum of the numbers $x_1, x_2, x_3, \ldots, x_n$ as

$$\Sigma x = x_1 + x_2 + x_3 + \cdots + x_n$$

If these numbers are 3, 7, 2, 6, 8, 4, and 9, we have

$$\Sigma x = 3 + 7 + 2 + 6 + 8 + 4 + 9 = 39$$

Using the summation symbol Σ, we can write Eq. (21-1) for the arithmetic mean as

$$\bar{x} = \frac{x_1 f_1 + x_2 f_2 + x_3 f_3 + \cdots + x_n f_n}{f_1 + f_2 + f_3 + \cdots + f_n} = \frac{\Sigma xf}{\Sigma f} \tag{21-1}$$

The summation notation Σx is an abbreviated form of the more general notation $\sum_{i=1}^{n} x_i$. This more general form can be used to indicate the sum of the first n numbers of a sequence, or to indicate the sum of a certain set within the sequence. For example, for a set of at least 5 numbers, $\sum_{i=3}^{5} x_i$ indicates the sum of the third through the fifth of these numbers (in Example E, $\sum_{i=3}^{5} x_i = 16$). We will use the abbreviated form Σx to indicate the sum of the numbers being considered.

EXAMPLE F _____ We find the arithmetic mean of the grades in Example E of Section 21-1 by

$$\bar{x} = \frac{\sum xf}{\sum f} = \frac{22(1) + (37)(1) + (40)(2) + \cdots + (67)(5) + \cdots + (97)(1)}{80}$$

$$= \frac{5733}{80} = 71.7$$

mode _____ Another measure of central tendency is *the* **mode,** *which is the value which appears most frequently.* If two or more values appear with the same greatest frequency, each is a mode. If no value is repeated, there is no mode.

EXAMPLE G _____ The mode of the numbers in Example A is 4, since it appears three times and no other value appears more than twice.

The modes of the numbers

1, 2, 2, 4, 5, 5, 6, 7

are 2 and 5, since each appears twice and no other number is repeated.

There is no mode for the values

1, 2, 5, 6, 7, 9

since none of the values is repeated.

EXAMPLE H _____ In order to determine the frictional force between two specially designed surfaces, the force to move a block with one surface along an inclined plane with the other surface is measured 10 times. The results, with forces in newtons, are

2.2, 2.4, 2.1, 2.2, 2.5, 2.2, 2.4, 2.7, 2.1, 2.5

Find the mean, median, and mode of these forces.

To find the mean we sum the values of the forces and divide this total by 10. This gives

$$\bar{F} = \frac{\sum F}{10} = \frac{2.2 + 2.4 + 2.1 + 2.2 + 2.5 + 2.2 + 2.4 + 2.7 + 2.1 + 2.5}{10}$$

$$= \frac{23.3}{10} = 2.3 \text{ N}$$

The median is found by arranging the values in order and finding the middle value. The values in order are

2.1, 2.1, 2.2, 2.2, 2.2, 2.4, 2.4, 2.5, 2.5, 2.7

Since there are 10 values, we see that the fifth value is 2.2 and the sixth is 2.4. The value midway between these is 2.3, which is the median. Therefore, the median force is 2.3 N.

The mode is 2.2 N, since this value appears three times, which is more than any other value.

Exercises 21-2

In Exercises 1 through 12, use the following sets of numbers. They are the same as those used in Exercises 21-1.

A: 3, 6, 4, 2, 5, 4, 7, 6, 3, 4, 6, 4, 5, 7, 3

B: 25, 26, 23, 24, 25, 28, 26, 27, 23, 28, 25

C: 0.48, 0.53, 0.49, 0.45, 0.55, 0.49, 0.47, 0.55, 0.48, 0.57, 0.51, 0.46

D: 105, 108, 103, 108, 106, 104, 109, 104, 110, 108, 108, 104, 113, 106, 107, 106, 107, 109, 105, 111, 109, 108

See Appendix E for a computer program for finding the arithmetic mean.

In Exercises 1 through 4, determine the median of the numbers of the given set.

1. Set *A*   **2.** Set *B*   **3.** Set *C*   **4.** Set *D*

In Exercises 5 through 8, determine the arithmetic mean of the numbers of the given set.

5. Set *A*   **6.** Set *B*   **7.** Set *C*   **8.** Set *D*

In Exercises 9 through 12, determine the mode of the numbers of the given set.

9. Set *A*   **10.** Set *B*   **11.** Set *C*   **12.** Set *D*

In Exercises 13 through 28, the required sets of numbers are those in Section 21-1. Find the indicated measures of central tendency.

13. Median of computer instructions in Exercise 17
14. Mean of computer instructions in Exercise 17
15. Mode of computer instructions in Exercise 17
16. Median of times in Exercise 21
17. Mean of times in Exercise 21
18. Mode of times in Exercise 21
19. Median of stopping distances in Exercise 23. (Use middle value for each interval.)
20. Mean of stopping distances in Exercise 23. (Use middle value for each interval.)
21. Mean of X-ray dosages in Exercise 25
22. Median of X-ray dosages in Exercise 25
23. Mode of X-ray dosages in Exercise 25
24. Mean of battery lives in Exercise 27
25. Median of battery lives in Exercise 27
26. Mode of battery lives in Exercise 27
27. Mean of cable diameters in Exercise 29
28. Median of cable diameters in Exercise 29

In Exercises 29 through 32, find the indicated measures of central tendency.

29. The weekly salaries (in dollars) for the workers in a small factory are as follows:

250, 350, 275, 225, 175, 300, 200, 350, 275, 400, 300, 225, 250, 300

Find the median and the mode of the salaries.

30. Find the mean salary for the salaries in Exercise 29.

31. In a particular month the electrical usage, rounded to the nearest 100 kWh (kilowatt hours), of 1000 homes in a certain city was summarized as follows:

Number of homes	22	80	106	185	380	122	90	15
Usage	500	600	700	800	900	1000	1100	1200

Find the mean of the electrical usage.

32. Find the median and the mode of the electrical usage in Exercise 31.

*In Exercises 33 through 36, use the following information. The **midrange,** another measure of central tendency of a frequency distribution, is found by finding the sum of the highest and the lowest values and dividing this sum by two. Using this definition, find the midrange of the numbers in the indicated sets of numbers at the beginning of this exercise set.*

33. Set *A*   **34.** Set *B*   **35.** Set *C*   **36.** Set *D*

21-3   Standard Deviation

In the preceding section we discussed measures of central tendency of sets of data. However, regardless of the measure which may be used, it does not tell us whether the data are grouped closely together or spread over a large range of values. Therefore, we also need some measure of the deviation, or dispersion, of the values from the median or mean. If the dispersion is small and the numbers are grouped closely together, the measure of central tendency is more reliable and descriptive of the data than the case in which the spread is greater.

standard deviation

There are several measures of dispersion, and we discuss in this section one which is very widely used. It is called the **standard deviation.**

The standard deviation of a set of n numbers is given by the equation

$$s = \sqrt{\frac{\sum(x - \bar{x})^2}{n}} \tag{21-2}$$

The definition of *s* indicates that the following steps are to be taken in computing its value.

1. Find the arithmetic mean $\bar{x}$ of the set of numbers.
2. Subtract the mean from each of the numbers of the set.
3. Square these differences.
4. Find the arithmetic mean of these squares.
5. Find the square root of this last arithmetic mean.

Defined in this way, *s* must be a positive number, and thus indicates a deviation from the mean, regardless of whether or not individual numbers are greater or less than the mean.

In many textbooks, standard deviation is defined such that in Step 4, the sum of the squares found in Step 3 is divided by $n - 1$ rather than n, where n is the number of values. Either definition is correct, and the choice of n or $n - 1$ is determined by considerations beyond the scope of this chapter.

Following the steps listed above, we use Eq. (21-2) for the calculation of standard deviation in the examples on the following page.

EXAMPLE A

Find the standard deviation of the numbers 1, 5, 4, 2, 6, 2, 1, 1, 5, 3.

A table of the necessary values is shown below, and Steps 1–5 are indicated.

	Step 2	Step 3
x	$x - \bar{x}$	$(x - \bar{x})^2$
1	-2	4
5	2	4
4	1	1
2	-1	1
6	3	9
2	-1	1
1	-2	4
1	-2	4
5	2	4
3	0	0
$\overline{30}$		$\overline{32}$

$$\bar{x} = \frac{30}{10} = 3 \qquad \text{Step 1}$$

$$\frac{\sum(x - \bar{x})^2}{n} = \frac{32}{10} = 3.2 \qquad \text{Step 4}$$

$$s = \sqrt{3.2} = 1.8 \qquad \text{Step 5}$$

EXAMPLE B

Find the standard deviation of the numbers given in Example A of Section 21-2. Since several of the numbers appear more than once, it is helpful to use the frequency of each number in the table. Therefore, we have the following table and calculations.

Step 1

x	f	fx	$x - \bar{x}$	$(x - \bar{x})^2$	$f(x - \bar{x})^2$
1	1	1	-4.5	20.25	20.25
2	2	4	-3.5	12.25	24.50
3	1	3	-2.5	6.25	6.25
4	3	12	-1.5	2.25	6.75
5	1	5	-0.5	0.25	0.25
6	1	6	0.5	0.25	0.25
7	2	14	1.5	2.25	4.50
8	1	8	2.5	6.25	6.25
9	2	18	3.5	12.25	24.50
11	1	11	5.5	30.25	30.25
	$\overline{15}$	$\overline{82}$			$\overline{123.75}$

The header for columns "Step 2" is above $(x - \bar{x})$ and "Step 3" is above $(x - \bar{x})^2$.

$$\bar{x} = \frac{82}{15} = 5.5$$

$$\frac{\sum f(x - \bar{x})^2}{n} = \frac{123.75}{15} = 8.25 \qquad \text{Step 4}$$

$$s = \sqrt{8.25} = 2.9 \qquad \text{Step 5}$$

We can see from Examples A and B that there is a great deal of calculational work required to find the standard deviation of a set of numbers. It is possible to reduce this work somewhat, and we now show how this may be done.

Two sets of numbers, x_i and y_i, where each set has n values, have a sum $(x_i + y_i)$. Finding the mean of this sum, we have

$$\overline{x + y} = \frac{\sum(x + y)}{n} = \frac{(x_1 + y_1) + (x_2 + y_2) + \cdots + (x_n + y_n)}{n}$$

$$= \frac{x_1 + x_2 + \cdots + x_n + y_1 + y_2 + \cdots + y_n}{n} = \frac{\sum x}{n} + \frac{\sum y}{n} = \bar{x} + \bar{y}$$

This means that the mean of the sum can be determined by adding the mean of the x's and the mean of the y's.

$$\overline{x + y} = \bar{x} + \bar{y} \tag{21-3}$$

Also, when we find the arithmetic mean, if all the numbers contain a constant factor, this number can be factored before the mean is found. That is,

$$\overline{kx} = k\bar{x} \tag{21-4}$$

We can expand the expression under the radical in Eq. (21-2) and then apply Eq. (21-3). This leads to

$$\frac{\sum(x - \bar{x})^2}{n} = \overline{(x - \bar{x})^2} = \overline{x^2 - 2x\bar{x} + \bar{x}^2} = \overline{x^2} - \overline{2x\bar{x}} + \overline{\bar{x}^2}$$

In this equation the number 2 and $\bar{x}$ are constants for any given problem. Thus, by using Eq. (21-4), we have

$$\overline{x^2} - \overline{2x\bar{x}} + \overline{\bar{x}^2} = \overline{x^2} - 2\bar{x}(\bar{x}) + \bar{x}^2(1) = \overline{x^2} - \bar{x}^2$$

Substituting this last result into Eq. (21-2), we have

$$s = \sqrt{\overline{x^2} - \bar{x}^2} \tag{21-5}$$

Equation (21-5) shows that the standard deviation s of a set of values may be found by following these steps:

1. Calculate the mean of the squares $(\overline{x^2})$.
2. Calculate the square of the mean $(\bar{x}^2)$.
3. Subtract result 2 from result 1 $(\overline{x^2} - \bar{x}^2)$.
4. Take the square root $(\sqrt{\overline{x^2} - \bar{x}^2})$.

Using Eq. (21-5) eliminates the step of subtracting $\bar{x}$ from x, a step required by the original definition in Eq. (21-2). Obviously, a calculator is very useful in making these calculations.

EXAMPLE C ———— By using Eq. (21-5), find s for the numbers in Example A. Again, a table is the most convenient form.

x	x^2
1	1
5	25
4	16
2	4
6	36
2	4
1	1
1	1
5	25
3	9
30	122

$\overline{x^2} = \dfrac{122}{10} = 12.2$ Step 1

$\bar{x} = \dfrac{30}{10} = 3, \quad \bar{x}^2 = 9$ Step 2

$\overline{x^2} - \bar{x}^2 = 12.2 - 9 = 3.2$ Step 3

$s = \sqrt{3.2} = 1.8$ Step 4

EXAMPLE D In an ammeter, two resistances are connected in parallel. Most of the current passing through the meter goes through the one called the shunt. In order to determine the accuracy of the resistance of shunts being made for ammeters, a manufacturer tested a sample of 100 shunts. The measured resistances are indicated in the following table. Calculate the standard deviation of the resistances of the shunts.

R (ohms)	f	fR	fR^2
0.200	1	0.200	0.0400
0.210	3	0.630	0.1323
0.220	5	1.100	0.2420
0.230	10	2.300	0.5290
0.240	17	4.080	0.9792
0.250	40	10.000	2.5000
0.260	13	3.380	0.8788
0.270	6	1.620	0.4374
0.280	3	0.840	0.2352
0.290	2	0.580	0.1682
	100	24.730	6.1421

$$\overline{R^2} = \frac{\sum fR^2}{\sum f} = \frac{6.1421}{100} = 0.061421 \quad \text{Step 1}$$

$$\bar{R} = \frac{\sum fR}{\sum f} = \frac{24.73}{100} = 0.2473 \quad \text{Step 2}$$

$$\bar{R}^2 = 0.06115729$$

$$s = \sqrt{\overline{R^2} - \bar{R}^2} = 0.016 \quad \text{Steps 3, 4}$$

The arithmetic mean of the resistances is 0.247 Ω, with a standard deviation of 0.016 Ω. ■

EXAMPLE E Find the standard deviation of the grades in Example E of Section 21-1. Use the frequency distribution as grouped in intervals 20–24, 25–29, and so forth, and then assume that each value in the interval is equal to the representative value (middle value) of the interval. (This method is not exact, but when a problem involves a large number of values, the method provides a very good approximation and eliminates a great deal of arithmetic work.)

Interval	x	f	fx	fx^2
20–24	22	1	22	484
25–29	27	0	0	0
30–34	32	0	0	0
35–39	37	1	37	1,369
40–44	42	3	126	5,292
45–49	47	1	47	2,209
50–54	52	1	52	2,704
55–59	57	4	228	12,996
60–64	62	8	496	30,752
65–69	67	10	670	44,890
70–74	72	14	1,008	72,576
75–79	77	16	1,232	94,864
80–84	82	7	574	47,068
85–89	87	7	609	52,983
90–94	92	5	460	42,320
95–99	97	2	194	18,818
		80	5,755	429,325

$$\overline{x^2} = \frac{\sum fx^2}{80} = \frac{429,325}{80} = 5,366.5625 \quad \text{Step 1}$$

$$\bar{x} = \frac{\sum fx}{\sum f} = \frac{5,755}{80} = 71.9375 \quad \text{Step 2}$$

$$\bar{x}^2 = 5,175.0039$$

$$s = \sqrt{\overline{x^2} - \bar{x}^2} = 13.8 \quad \text{Steps 3, 4}$$

Many calculators are programmed to give the arithmetic mean and standard deviation when data are entered. Also, these statistical measures are readily programmed to be determined by a computer. Such a program is in Appendix E.

We have seen that the standard deviation is a measure of the dispersion of a set of data. Generally, if we make a great many measurements of a given type, we would expect to find a majority of them near the arithmetic mean. If this is the case, the distribution probably follows, at least to a reasonable extent, the **standard normal distribution curve.** See Fig. 21-6. This curve gives a theoretical distribution about the arithmetic mean, assuming that the number of values in the distribution becomes infinite, and its equation is

$$y = \frac{1}{\sqrt{2\pi}} e^{-x^2/2}$$

(21-6)

Fig. 21-6

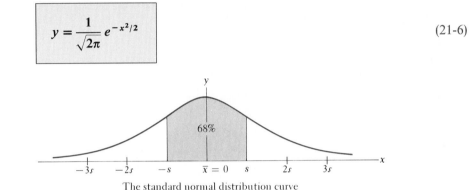

The standard normal distribution curve

An important feature of the standard deviation is that *in a normal distribution, about 68% of the values are found within the interval $\bar{x} - s$ to $\bar{x} + s$.* Also, approximately 95.4% of the numbers are within the interval $\bar{x} - 2s$ to $\bar{x} + 2s$, and 99.7% of the values are in the interval $\bar{x} - 3s$ to $\bar{x} + 3s$.

EXAMPLE F

In Example B, 9 of the 15 values, or 60%, are between 2.6 and 8.4. Here $\bar{x} = 5.5$ and $s = 2.9$. Thus, $\bar{x} - s = 2.6$ and $\bar{x} + s = 8.4$. Therefore, we see that the percentage in this interval is slightly less than in a normal distribution.

In Example E, using $\bar{x} = 72$ and $s = 14$, we have $\bar{x} - s = 58$ and $\bar{x} + s = 86$. We also note that 59 of the 80 values are in the interval from 58 to 86, which means that 74% of the values are in this interval. ∎

We have introduced a few of the measures used in statistics. In using these measures, we have considered sets of numbers for which these measures give a reasonable description of the center and distribution of the numbers. However, care must be used in using and interpreting such measures.

EXAMPLE G

The numbers 1, 2, 3, 4, 5 have a mean of 3 and a median of 3. The standard deviation is 1.4. These values describe fairly well the center and distribution of the numbers in the set.

The numbers 1, 2, 3, 4, 100 have a mean of 22, a median of 3, and a standard deviation of 39. These values do not describe this set of numbers well. Most of the values in the interval $\bar{x} - s$ to $\bar{x} + s$ (-17 to 61) are not even in the set, and the one which is outside the interval is 39 units from it. ∎

Example G illustrates that the statistical measures can be misleading if the numbers in a set are unevenly distributed. Misleading statistics can also come from the source of the data. Consider the probable results of a survey to find the percent of persons in favor of raising income taxes on the wealthy if the survey is taken at the entrance to a welfare office or if it is taken at the entrance to a stock brokerage firm. There are many other considerations in the proper use and interpretation of statistical measures.

Exercises 21-3

In Exercises 1 through 20, use the following sets of numbers. They are the same as those used in Exercises 21-1 and 21-2.

A: 3, 6, 4, 2, 5, 4, 7, 6, 3, 4, 6, 4, 5, 7, 3

B: 25, 26, 23, 24, 25, 28, 26, 27, 23, 28, 25

C: 0.48, 0.53, 0.49, 0.45, 0.55, 0.49, 0.47, 0.55, 0.48, 0.57, 0.51, 0.46

D: 105, 108, 103, 108, 106, 104, 109, 104, 110, 108, 108, 104, 113, 106, 107, 106, 107, 109, 105, 111, 109, 108

See Appendix E for a computer program for finding the standard deviation.

In Exercises 1 through 4, use Eq. (21-2) to find the standard deviation s for the indicated sets of numbers.

1. Set A **2.** Set B **3.** Set C **4.** Set D

In Exercises 5 through 8, use Eq. (21-5) to find the standard deviation s for the indicated sets of numbers.

5. Set A **6.** Set B **7.** Set C **8.** Set D

In Exercises 9 through 16, find the standard deviation s for the indicated sets of numbers from the exercises of Sections 21-1 and 21-2.

9. The computer instructions in Exercise 17 of Section 21-1

10. The X-ray dosages in Exercise 25 of Section 21-1 **11.** The battery lives in Exercise 27 of Section 21-1

12. The salaries of Exercise 29 of Section 21-2 **13.** The strobe light times in Exercise 21 of Section 21-1

14. The stopping distances in Exercise 23 of Section 21-1

15. The fiber-optic cable diameters in Exercise 29 of Section 21-1

16. The electric power usages in Exercise 31 of Section 21-2

In Exercises 17 through 24, find the percent of values in the interval $\bar{x} - s$ to $\bar{x} + s$ for the indicated sets of numbers.

17. Set A **18.** Set B **19.** Set C **20.** Set D

21. The computer instructions in Exercise 17 of Section 21-1

22. The X-ray dosages in Exercise 25 of Section 21-1

23. The strobe light times in Exercise 21 of Section 21-1

24. The stopping distances in Exercise 23 of Section 21-1

21-4 Fitting a Straight Line to a Set of Points

We have considered statistical methods for dealing with one variable. We have discussed methods of tabulating, graphing, and measuring the central tendency and the deviations from this value for one variable. We now shall discuss how to obtain a relationship between two variables for which a set of points is known.

In this section we shall show a method of "fitting" a straight line to a given set of points. In the following section fitting suitable nonlinear curves to given sets of points will be discussed. Some of the reasons for doing this are (1) to express a concise relationship between the variables, (2) to use the equation to predict certain fundamental results, (3) to determine the reliability of certain sets of data, and (4) to use the data for testing theoretical concepts.

We shall assume for the examples and exercises of the remainder of this chapter that there is some relationship between the variables. Often when we are analyzing statistics for variables between which we think a relationship might exist, the points are so scattered as to give no reasonable idea regarding the possible functional relationship. We shall assume here that such combinations of variables have been discarded in the analysis.

EXAMPLE A

All the students enrolled in the mathematics course referred to in Example E of Section 21-1 took an entrance test in mathematics. To study the reliability of this test, an instructor tabulated the test scores of 10 students (selected at random), along with their course averages, and made a graph of these figures (see table below and Fig. 21-7).

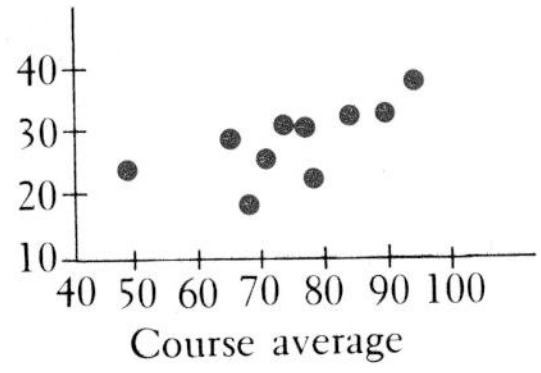

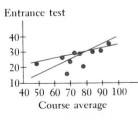

Entrance test

Course average

Fig. 21-7

Student	Entrance test score, based on 40	Course average, based on 100
A	29	63
B	33	88
C	22	77
D	17	67
E	26	70
F	37	93
G	30	72
H	32	81
I	23	47
J	30	74

We now ask whether or not there is a functional relationship between the test scores and the course grades. Certainly no clear-cut relationship exists, but in general we see that the higher the test score, the higher the course grade. This leads to the possibility that there might be some straight line, from which none of the points would vary too significantly. If such a line could be found, then it could be the basis of predictions as to the possible success a student might have in the course, on the basis of his or her grade on the entrance test. Assuming that such a straight line exists, the problem is to find the equation of this line. Figure 21-8 shows two such possible lines.

Entrance test

Course average

Fig. 21-8

There are various methods of determining the line which best fits the given points. We shall employ the one which is most widely used: the **method of least squares.** *The basic principle of this method is that the sum of the squares of the* *deviations of all points from the best line (in accordance with this method) is the least it can be. By* **deviation** *we mean the difference between the y-value of the line and the y-value for the point (of original data) for a particular value of x.*

EXAMPLE B

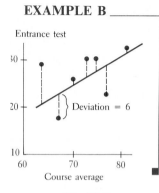

Fig. 21-9

In Fig. 21-9 the deviations of some of the points of Example A are shown. The point (67, 17) (student D of Example A) has a deviation of 6 from the indicated line in the figure. Thus we square the value of this deviation to obtain 36. The method of least squares requires that the sum of all such squares be a minimum in order to determine the line which fits best.

Therefore, in applying this method, it is necessary to use the equation of a straight line and the coordinates of the points of the data. The deviations of these data points are determined and then squared. It is then necessary to determine the constants m and b in the equation of the straight line $y = mx + b$ for which the sum of the squares of the deviations is a minimum. To do this requires certain methods of advanced mathematics.

Using the necessary methods of more advanced mathematics, it can be shown that *the equation of the* **least squares line**

least-squares line

$$y = mx + b \tag{21-7}$$

can be found from the values

$$m = \frac{n \sum xy - (\sum x)(\sum y)}{n \sum x^2 - (\sum x)^2} \tag{21-8}$$

and

$$b = \frac{(\sum x^2)(\sum y) - (\sum xy)(\sum x)}{n \sum x^2 - (\sum x)^2} \tag{21-9}$$

NOTE ▷ In Eqs. (21-8) and (21-9), *the x's and y's are those of the points in the data* and n is the number of points. The calculational work in finding m and b is reduced by noting that the denominators in Eqs. (21-8) and (21-9) are the same.

EXAMPLE C

Find the equation of the least-squares line for the points indicated in the following table. Graph the line and data points on the same graph.

x	1	2	3	4	5
y	3	6	6	8	12

We see from Eqs. (21-8) and (21-9) that we need the sums of x, y, xy, and x^2 in order to find m and b. Thus, we set up a table for these values, along with

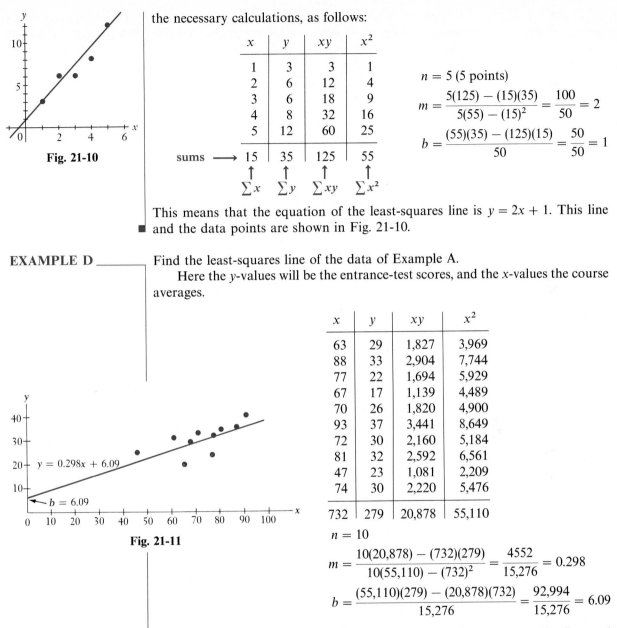

Fig. 21-10

the necessary calculations, as follows:

x	y	xy	x^2
1	3	3	1
2	6	12	4
3	6	18	9
4	8	32	16
5	12	60	25
sums ⟶ 15	35	125	55
$\sum x$	$\sum y$	$\sum xy$	$\sum x^2$

$n = 5$ (5 points)

$$m = \frac{5(125) - (15)(35)}{5(55) - (15)^2} = \frac{100}{50} = 2$$

$$b = \frac{(55)(35) - (125)(15)}{50} = \frac{50}{50} = 1$$

This means that the equation of the least-squares line is $y = 2x + 1$. This line and the data points are shown in Fig. 21-10.

EXAMPLE D _____ Find the least-squares line of the data of Example A.

Here the y-values will be the entrance-test scores, and the x-values the course averages.

x	y	xy	x^2
63	29	1,827	3,969
88	33	2,904	7,744
77	22	1,694	5,929
67	17	1,139	4,489
70	26	1,820	4,900
93	37	3,441	8,649
72	30	2,160	5,184
81	32	2,592	6,561
47	23	1,081	2,209
74	30	2,220	5,476
732	279	20,878	55,110

$n = 10$

$$m = \frac{10(20,878) - (732)(279)}{10(55,110) - (732)^2} = \frac{4552}{15,276} = 0.298$$

$$b = \frac{(55,110)(279) - (20,878)(732)}{15,276} = \frac{92,994}{15,276} = 6.09$$

Fig. 21-11

Thus, the equation of the least-squares line is $y = 0.298x + 6.09$. The line and data points are shown in Fig. 21-11. This is the line which best fits the data, although the fit is obviously approximate. It can be used to predict the approximate course average that a student might be expected to attain, based on the entrance test.

When using the calculator, the value of the denominator can be stored in memory when calculating m and can then be used again when calculating b.

EXAMPLE E

See the chapter introduction.

In a research project to determine the amount of a drug which remains in the bloodstream after a given dosage, the amounts y (in mg of drug/dL of blood) were recorded after t hours.

t (h)	1.0	2.0	4.0	8.0	10.0	12.0
y (mg/dL)	7.6	7.2	6.1	3.8	2.9	2.0

Find the least-squares line for these data, expressing y as a function of t. Sketch the graph of the line and data points.

The table and calculations are shown below.

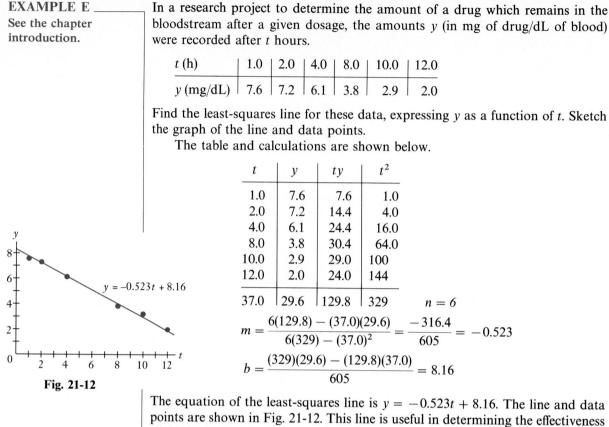

t	y	ty	t^2
1.0	7.6	7.6	1.0
2.0	7.2	14.4	4.0
4.0	6.1	24.4	16.0
8.0	3.8	30.4	64.0
10.0	2.9	29.0	100
12.0	2.0	24.0	144
37.0	29.6	129.8	329

$n = 6$

$$m = \frac{6(129.8) - (37.0)(29.6)}{6(329) - (37.0)^2} = \frac{-316.4}{605} = -0.523$$

$$b = \frac{(329)(29.6) - (129.8)(37.0)}{605} = 8.16$$

$y = -0.523t + 8.16$

Fig. 21-12

The equation of the least-squares line is $y = -0.523t + 8.16$. The line and data points are shown in Fig. 21-12. This line is useful in determining the effectiveness of the drug. It can also be used to determine when additional medication may be administered.

Exercises 21-4

In Exercises 1 through 12, find the equation of the least-squares line for the given data. In each case graph the line and the data points on the same graph.

1. The points in the following table:

x	4	6	8	10	12
y	1	4	5	8	9

2. The points in the following table:

x	1	2	3	4	5	6	7
y	10	17	28	37	49	56	72

3. The points in the following table:

x	20	26	30	38	48	60
y	160	145	135	120	100	90

4. The points in the following table:

x	1	3	6	5	8	10	4	7	3	8
y	15	12	10	8	9	2	11	9	11	7

See Appendix E for a computer program for finding the equation of a least-squares line.

5. In an electrical experiment, the following data were found for the values of current and voltage for a particular element of the circuit. Find the voltage as a function of the current.

Current (milliamperes)	15.0	10.8	9.30	3.55	4.60
Voltage (volts)		3.00	4.10	5.60	8.00	10.50

6. A particular muscle was tested for its speed of shortening as a function of the force which was applied to it. The results were found as tabulated below. Find the speed as a function of the force.

Force (N)	60.0	44.2	37.3	24.2	19.5
Speed (m/s)	1.25	1.67	1.96	2.56	3.05

7. The altitude h of a rocket was measured at several positions at a horizontal distance x from the launch site, as shown in the table. Find the least-squares line for h as a function of x.

x (meters)	0	500	1000	1500	2000	2500
h (meters)	0	1130	2250	3360	4500	5600

8. In testing an air-conditioning system, the temperature T in a building was measured during the afternoon hours with the results shown in the table. Find the least-squares line for T as a function of the time t from noon.

t (hours)	0.0	1.0	2.0	3.0	4.0	5.0
T (degrees Celsius)	20.5	20.6	20.9	21.3	21.7	22.0

9. The pressure p was measured along an oil pipeline at different distances from a reference point, with results as shown. Find the least-squares line for p as a function of x.

x (feet)	0	100	200	300	400
p (lb/in.^{2})	650	630	605	590	570

10. The heat loss L per hour through various thicknesses of a particular type of insulation was measured as shown in the table. Find the least-squares line for L as a function of t.

t (inches)	3.0	4.0	5.0	6.0	7.0
L (Btu)	5900	4800	3900	3100	2450

11. In an experiment on the photoelectric effect, the frequency of the light being used was measured as a function of the stopping potential (the voltage just sufficient to stop the photoelectric current) with the results given below. Find the least-squares line for V as a function of f. The frequency for $V = 0$ is known as the *threshold frequency*. From the graph, determine the threshold frequency.

f (petahertz)	0.550	0.605	0.660	0.735	0.805	0.880
V (volts)		0.350	0.600	0.850	1.10	1.45	1.80

12. If gas is cooled under conditions of constant volume, it is noted that the pressure falls nearly proportionally as the temperature. If this were to happen until there was no pressure, the theoretical temperature for this case is referred to as *absolute zero*. In an elementary experiment, the following data were found for pressure and temperature for a gas under constant volume.

T (degrees Celsius)	0.0	20	40	60	80	100
P (kilopascals)		133	143	153	162	172	183

Find the least-squares line for P as a function of T, and, from the graph, determine the value of absolute zero found in this experiment.

The linear coefficient of correlation, a measure of the relatedness of two variables, is defined by $r = m(s_x/s_y)$, where s_x and s_y are the standard deviations of the x-values and y-values, respectively. Due to its definition, the values of r lie in the range $-1 \leq r \leq 1$. If r is near 1, the correlation is considered good. For values of r between $-\frac{1}{2}$ and $+\frac{1}{2}$, the correlation is poor. If r is near -1, the variables are said to be negatively correlated; that is, one increases while the other decreases.

In Exercises 13 through 16, compute r for the given sets of data.

13. Exercise 1   **14.** Exercise 2   **15.** Exercise 4   **16.** Example A

21-5   Fitting Nonlinear Curves to Data

If the experimental points do not appear to be on a straight line, but we recognize them as being approximately on some other type of curve, the method of least squares can be extended to use on these other curves. For example, if the points are apparently on a parabola, we could use the function $y = a + bx^2$. To use the above method, we shall extend the least-squares line to

$$y = m[f(x)] + b \qquad (21\text{-}10)$$

Here, $f(x)$ must be calculated first, and then the problem can be treated as a least-squares line to find the values of m and b. Some of the functions $f(x)$ which may be considered for use are x^2, $1/x$, and 10^x.

EXAMPLE A ——— Find the least-squares curve $y = mx^2 + b$ for the points in the following table.

x	0	1	2	3	4	5
y	1	5	12	24	53	76

Here $f(x) = x^2$ for use in Eq. (21-10). Therefore, values of x^2 are to be used in place of values of x in order to find m and b. Our first step in making the table will be to calculate values of x^2. From then on we use x^2 as we used x in finding the equation of the least-squares line.

x	$f(x) = x^2$	y	$x^2 y$	$(x^2)^2$
0	0	1	0	0
1	1	5	5	1
2	4	12	48	16
3	9	24	216	81
4	16	53	848	256
5	25	76	1900	625
	55	171	3017	979

$n = 6$

$$m = \frac{6(3017) - (55)(171)}{6(979) - (55)^2} = \frac{8697}{2849} = 3.05$$

$$b = \frac{(979)(171) - (3017)(55)}{2849} = 0.52$$

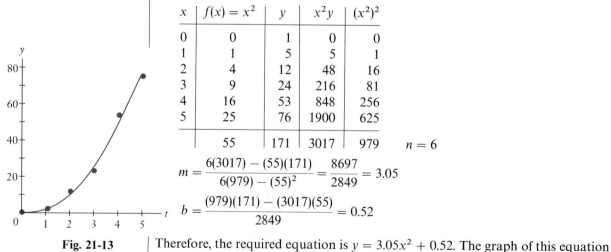

Fig. 21-13

Therefore, the required equation is $y = 3.05x^2 + 0.52$. The graph of this equation and the data points are shown in Fig. 21-13.

EXAMPLE B _____

In a physics experiment designed to measure the pressure and volume of a gas at constant temperature, the following data were found. When the points were plotted, they were seen to approximate the hyperbola $y = \frac{c}{x}$. Find the least-squares approximation to this hyperbola $\left[y = m\left(\frac{1}{x}\right) + b\right]$ (see Fig. 21-14).

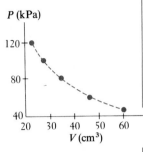

P (kPa)

Fig. 21-14

P (kPa)	V (cm³)	x(= V)	f(x) = 1/x	y(= P)	(1/x)y	(1/x)²
120.0	21.0	21.0	0.0476190	120.0	5.7142857	0.0022676
99.2	25.0	25.0	0.0400000	99.2	3.9680000	0.0016000
81.3	31.8	31.8	0.0314465	81.3	2.5566038	0.0009889
60.6	41.1	41.1	0.0243309	60.6	1.4744526	0.0005920
42.7	60.1	60.1	0.0166389	42.7	0.7104825	0.0002769
			0.1600353	403.8	14.4238246	0.0057254

(*Note on calculator use:* The final digits for the values shown may vary on the type of calculator used and how values are used. Here all individual values are shown assuming that the calculator displays eight digits, but actually carries more. The value of $\frac{1}{x}$ was found from the value of x, with the eight-digit display indicated. However, the values of $y\left(\frac{1}{x}\right)$ and $\left(\frac{1}{x}\right)^2$ were found using the extra digits carried by the calculator for $\frac{1}{x}$. The sums were found using the rounded-off values shown.)

$$m = \frac{5(14.4238246) - (0.1600353)(403.8)}{5(0.0057254) - (0.1600353)^2} = \frac{7.4968689}{0.0030157} = 2490$$

$$b = \frac{(0.0057254)(403.8) - (14.4238246)(0.1600353)}{0.0030157} = 1.2$$

Thus the equation of the hyperbola $y = m\left(\frac{1}{x}\right) + b$ is

$$y = \frac{2490}{x} + 1.2$$

The graph of this hyperbola and the points representing the data are shown in Fig. 21-15.

Fig. 21-15

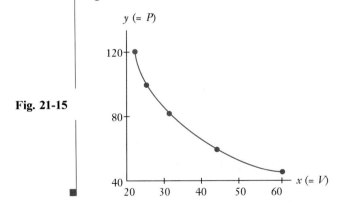

EXAMPLE C

It has been found experimentally that the tensile strength of brass (a copper-zinc alloy) increases (within certain limits) with the percentage of zinc. The following table indicates the values which have been found (also see Fig. 21-16).

Tensile strength (10^5 lb/in.^{2})	0.32	0.36	0.40	0.44	0.48
Percentage of zinc	0	5	13	22	34

Fit a curve of the form $y = m(10^x) + b$ to the data. Let $x =$ tensile strength ($\times 10^5$) and $y =$ percentage of zinc.

x	$f(x) = 10^x$	y	$(10^x)y$	$(10^x)^2$
0.32	2.0892961	0	0.000000	4.3651583
0.36	2.2908677	5	11.454338	5.2480746
0.40	2.5118864	13	32.654524	6.3095734
0.44	2.7542287	22	60.593031	7.5857758
0.48	3.0199517	34	102.67836	9.1201084
	12.6662306	74	207.38025	32.6286905

(See the note on calculator use in Example B.)

$$m = \frac{5(207.38025) - (12.6662306)(74)}{5(32.6286905) - (12.6662306)^2}$$

$$= \frac{99.600186}{2.7100549} = 36.8$$

$$b = \frac{(32.6286905)(74) - (207.38025)(12.6662306)}{2.7100549} = -78.3$$

The equation of the curve is $y = 36.8(10^x) - 78.3$. It must be remembered that for practical purposes, y must be positive. The graph of the equation is shown

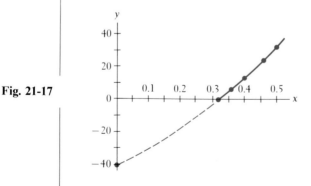

Fig. 21-17

in Fig. 21-17, with the solid portion denoting the meaningful part of the curve. The points of the data are also shown.

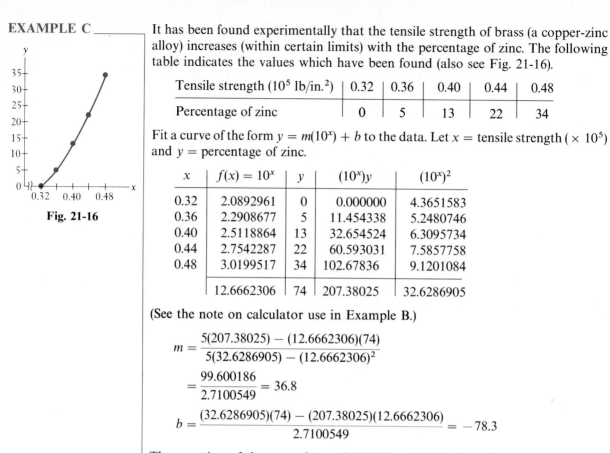

Fig. 21-16

Exercises 21-5

In each of the following exercises, find the indicated least-squares curve. Sketch the curve and plot the data points on the same graph.

1. For the points in the following table, find the least-squares curve $y = mx^2 + b$.

x	2	4	6	8	10
y	12	38	72	135	200

2. For the points in the following table, find the least-squares curve $y = m\sqrt{x} + b$.

x	0	4	8	12	16
y	1	9	11	14	15

3. For the points in the following table, find the least-squares curve $y = m(1/x) + b$.

x	1.10	2.45	4.04	5.86	6.90	8.54
y	9.85	4.50	2.90	1.75	1.48	1.30

4. For the points in the following table, find the least-squares curve $y = m(10^x) + b$.

x	0.00	0.200	0.500	0.950	1.325
y	6.00	6.60	8.20	14.0	26.0

5. The following data were found for the distance y that an object rolled down an inclined plane in time t. Determine the least-squares curve $y = mt^2 + b$.

t (seconds)	1.0	2.0	3.0	4.0	5.0
y (cm)	6.0	23	55	98	148

6. The increase in length of a certain metallic rod was measured in relation to particular increases in temperature. If y represents the increase in length for the corresponding increase in temperature x, find the least-squares curve $y = mx^2 + b$ for these data.

x (degrees Celsius)	50.0	100	150	200	250
y (cm)		1.00	4.40	9.40	16.4	24.0

7. The pressure p at which freon, a refrigerant, vaporizes for temperature T is given in the following table. Find the least-squares curve $p = mT^2 + b$ for these data.

T (°F)	0	20	40	60	80
p (lb/in.²)	23	35	49	68	88

8. A fraction f of annual hot water loads at a certain facility are heated by solar energy. The fractions f for certain values of the collector area A are given in the following table. Find the least-squares curve $f = m\sqrt{A} + b$ for these data.

A (m²)	0	12	27	56	90
f	0.0	0.2	0.4	0.6	0.8

9. The makers of a special blend of coffee found that the demand for the coffee depended on the price charged. The price P per pound and the monthly sales S are shown in the following table. Find the least-squares curve $P = m(\frac{1}{S}) + b$ for these data.

S (thousands)	240	305	420	480	560
P (dollars)	5.60	4.40	3.20	2.80	2.40

10. The resonant frequency of an electric circuit containing a 4-μF capacitor was measured as a function of an inductance in the circuit. The following data were found. Find the least-squares curve $f = m(1/\sqrt{L}) + b$.

L (henrys)	1.0	2.0	4.0	6.0	9.0
f (hertz)	490	360	250	200	170

11. The displacement y of an object at the end of a spring at given times t is given in the following table. Find the least-squares curve $y = me^{-t} + b$.

t (seconds)	0.0	0.5	1.0	1.5	2.0	3.0
y (centimeters)	6.1	3.8	2.3	1.3	0.7	0.3

12. The electric current i in an alternating-current circuit at times t is given in the following table. Find the least-squares curve $y = m \sin 377t + b$.

t (seconds)	0.000	0.002	0.004	0.006	0.008	0.010	0.012
i (milliamperes)	0.00	6.85	9.98	7.70	1.25	−5.88	−9.82

21-6   Chapter Equations, Review Exercises, and Practice Test

Chapter Equations

Arithmetic mean
$$\bar{x} = \frac{x_1 f_1 + x_2 f_2 + \cdots + x_n f_n}{f_1 + f_2 + \cdots + f_n} = \frac{\sum xf}{\sum f} \tag{21-1}$$

Standard deviation
$$s = \sqrt{\frac{\sum (x - \bar{x})^2}{n}} \tag{21-2}$$

$$s = \sqrt{\overline{x^2} - \bar{x}^2} \tag{21-5}$$

Normal distribution
$$y = \frac{1}{\sqrt{2\pi}} e^{-x^2/2} \tag{21-6}$$

Least-squares line
$$y = mx + b \tag{21-7}$$

$$m = \frac{n \sum xy - (\sum x)(\sum y)}{n \sum x^2 - (\sum x)^2} \tag{21-8}$$

$$b = \frac{(\sum x^2)(\sum y) - (\sum xy)(\sum x)}{n \sum x^2 - (\sum x)^2} \tag{21-9}$$

Nonlinear curves
$$y = m[f(x)] + b \tag{21-10}$$

Review Exercises

In Exercises 1 through 4, use the following set of numbers:

  2.3, 2.6, 4.2, 3.6, 3.5, 4.1, 4.8, 2.5, 3.0, 4.1, 3.8

1. Determine the median of the set of numbers.

2. Determine the arithmetic mean of the set of numbers.

3. Determine the standard deviation of the set of numbers.

4. Construct a frequency table with intervals 2.0–2.9, 3.0–3.9, and 4.0–4.9.

In Exercises 5 through 12, use the following set of numbers:

  109, 103, 113, 110, 113, 106, 114, 101, 108, 101, 112, 106, 115, 102, 107, 110, 102, 115, 106, 105

5. Construct a frequency distribution table with intervals 101–103, 104–106, and so on.

6. Determine the median.

7. Determine the mode.

8. Determine the mean.

9. Determine the standard deviation.

10. Draw a frequency polygon for the data of Exercise 5.

11. Draw a histogram for the data of Exercise 5.

12. Construct a frequency distribution table with intervals of 101–105, 106–110, and 111–115. Then draw a histogram for these data.

In Exercises 13 through 18, use the following data:

  An important property of oil is its coefficient of viscosity, which gives a measure of how well it flows. In order to determine the viscosity of a certain motor oil, a refinery took samples from 12 different storage tanks and tested them at 50°C. The results (in pascal-seconds) were 0.24, 0.28, 0.29, 0.26, 0.27, 0.26, 0.25, 0.27, 0.28, 0.26, 0.26, 0.25.

13. Determine the arithmetic mean.

14. Determine the median.

15. Determine the standard deviation.

16. Make a histogram.

17. Make a frequency polygon.

18. Determine the mode.

In Exercises 19 through 24, use the following data:

  A group of wind generators was tested for power output when the wind speed was 30 km/h. The following table gives the number of generators producing the powers shown.

Power (watts)	650	660	670	680	690	700	710	720	730
Number of generators	3	2	7	12	27	34	15	16	5

19. Determine the arithmetic mean.

20. Determine the median.

21. Determine the mode.

22. Make a histogram.

23. Determine the standard deviation.

24. Make a frequency polygon.

In Exercises 25 through 28, use the following data:

  A Geiger counter records the presence of high-energy nuclear particles. Even though no apparent radioactive source is present, a certain number of particles will be recorded. These are primarily cosmic rays, which are caused by very high energy particles from outer space. In an experiment to measure the amount of cosmic radiation, the number of counts were recorded during 200 5-s intervals. The following table gives the number of counts, and the number of 5-s intervals having this number of counts. Draw a frequency curve for these data.

Counts	0	1	2	3	4	5	6	7	8	9	10
Intervals	3	10	25	45	29	39	26	11	7	2	3

25. Determine the median.

26. Determine the arithmetic mean.

27. Make a histogram.

28. Make a frequency polygon.

In Exercises 29 through 34, find the indicated least-squares curves.

29. In a certain experiment, the resistance of a certain resistor was measured as a function of the temperature. The data found were as follows:

T (degrees Celsius)	0.0	20.0	40.0	60.0	80.0	100
R (ohms)	25.0	26.8	28.9	31.2	32.8	34.7

Find the least-squares line for these data, expressing R as a function of T. Sketch the line and data points on the same graph.

30. An air pollution monitoring station took samples of air each hour during the later morning hours and tested each sample for the number n of parts per million (ppm) of carbon monoxide. The results are shown in the table, where t is the number of hours after 6 A.M. Find the least-squares line for n as a function of t.

t (hours)	0.0	1.0	2.0	3.0	4.0	5.0	6.0
n (ppm)	8.0	8.2	8.8	9.5	9.7	10.0	10.7

31. The *Mach number* of a moving object is the ratio of its speed to the speed of sound (740 mi/h). The following table shows the speed s of a jet aircraft, in terms of Mach numbers, and the time t after it starts to accelerate. Find the least-squares line of s as a function of t.

t (minutes)	0.00	0.60	1.20	1.80	2.40	3.00
s (Mach number)	0.88	0.97	1.03	1.11	1.19	1.25

32. In an experiment to determine the relation between the load on a spring and the length of the spring, the following data were found.

Load (pounds)	0.0	1.0	2.0	3.0	4.0	5.0
Length (inches)	10.0	11.2	12.3	13.4	14.6	15.9

Find the least-squares line for these data, which express the length as a function of the load.

33. The distance s of a missile above the ground at time t after being released from a plane is given by the following table. Find the least-squares curve of the form $s = mt^2 + b$ for these data.

t (seconds)	0.0	3.0	6.0	9.0	12.0	15.0	18.0
s (meters)	3000	2960	2820	2600	2290	1900	1410

34. In an elementary experiment which measured the wavelength of sound as a function of the frequency, the following results were obtained.

Frequency (hertz)	240	320	400	480	560
Wavelength (centimeters)	140	107	81.0	70.0	60.0

Find the least-squares curve of the form $y = m(\frac{1}{x}) + b$ for these data, expressing wavelength as y and frequency as x.

In Exercises 35 and 36, solve the given problems.

35. Show that Eqs. (21-8) and (21-9) can be written as

$$m = \frac{\overline{xy} - \bar{x}\bar{y}}{s_x^2} \quad \text{and} \quad b = \frac{\overline{x^2}\bar{y} - \overline{xy}\bar{x}}{s_x^2}$$

where s_x is the standard deviation of the x-values.

36. By using the expressions for m and b from Exercise 35, show that the point $(\bar{x}, \bar{y})$ satisfies Eq. (21-7).

Practice Test

For Problems 1 through 3, use the following set of numbers:

  5, 6, 1, 4, 9, 5, 7, 3, 8, 10, 5, 8, 4, 9, 6

1. Find the median.   **2.** Find the mode.

3. Draw a histogram for the intervals 0–2, 3–5, and so on.

For Problems 4 through 6, use the following data:

  Two machine parts are considered satisfactorily assembled if their total thickness (to the nearest 0.01 in.) is between or equal to 0.92 and 0.94 in. One hundred sample assemblies are tested, and the thicknesses, to the nearest 0.01 in., are given in the following table:

Total thickness	0.90	0.91	0.92	0.93	0.94	0.95	0.96
Number	3	9	31	38	12	5	2

4. Find the mean.   **5.** Find the standard deviation.   **6.** Draw a frequency polygon.

7. Find the equation of the least-squares line for the points indicated in the following table. Graph the line and data points on the same graph.

x	1	3	5	7	9
y	5	11	17	20	27

8. The period of a pendulum as a function of its length was measured, giving the following results:

Length (feet)	1.00	3.00	5.00	7.00	9.00
Period (seconds)	1.10	1.90	2.50	2.90	3.30

Find the equation of the least-squares curve of the form $y = m\sqrt{x} + b$, which expresses the period as a function of the length.

22 The Derivative

Determining how fast an object is moving is important in physics and in many areas of technology. In Section 22-4 we develop the method of finding the instantaneous velocity of a moving object.

The problems which can be solved by the methods of algebra and trigonometry are numerous. There are, however, a great many problems which arise in the various fields of technology which require for their solution methods beyond those available from algebra and trigonometry. These traditional topics remain of definite importance, but it is necessary to develop additional methods of analyzing and solving problems.

In this chapter we shall start developing the methods of **differential calculus.** This subject deals with the important problem involving the *rate of change* of one quantity with respect to another. Examples of rates of change are velocity (the rate of change of distance with respect to time), the rate of change of the length of a metal rod with respect to temperature, the rate of change of light intensity with respect to the distance from the source, the rate of change of electric current with respect to time, and the rate of change of production costs with respect to the number of units a business produces.

Another principal type of problem which calculus allows us to solve is that of finding a function when its rate of change is known. This is **integral calculus.** One of its principal applications comes from electricity, where current is the time rate of change of electric charge. Integral calculus also leads to the solution of a great many apparently unrelated problems, including the determination of plane areas, volumes, and the physical concepts of work and pressure. The study of integral calculus starts in Chapter 24.

Isaac Newton (1642–1727), an English mathematician and physicist, and Gottfried Wilhelm Leibniz (1646–1716), a German mathematician and philosopher, are credited with the creation of calculus. In order to solve problems in fields such as geometry and astronomy, each independently developed the basic methods of calculus.

22-1 Limits

Before dealing directly with the rate of change of a function, we shall first take up the concept of a **limit.** We encountered this concept in our discussion of infinite geometric series and in our discussion of the asymptotes of a hyperbola. It is now necessary to develop this concept further.

continuity

In order to help understand and develop the concept of a limit, we consider briefly the **continuity** of a function. *For a function to be* **continuous at a point,** *the function must exist at the point, and any small change in x produces only a small change in f(x).* In fact, the change in $f(x)$ can be made as small as we wish by restricting the change in x sufficiently, if the function is continuous. Also, *a function is said to be* **continuous over an interval** *if it is continuous at each point in the interval.*

EXAMPLE A

The function $f(x) = 3x^2$ is continuous for all values of x. That is, $f(x)$ is defined for all values of x, and a small change in x for any given value produces only a small change in $f(x)$. If we choose $x = 2$ and then let x change by 0.1, 0.01, and so on, we obtain the values in the following table by use of a calculator.

x	2	2.1	2.01	2.001
$f(x)$	12	13.23	12.1203	12.012003
Change in x		0.1	0.01	0.001
Change in $f(x)$		1.23	0.1203	0.012003

We can see that the change in $f(x)$ is made smaller by the smaller changes in x. This shows that $f(x)$ is continuous at $x = 2$. Since this type of result would be obtained for any other x we may choose, we see that $f(x)$ is continuous for all values, and therefore it is continuous over the interval of all values of x. ∎

EXAMPLE B

The function $f(x) = \dfrac{1}{x - 2}$ is not continuous at $x = 2$. When we substitute 2 for x in the function, we have division by zero. This means that the function is not defined. Therefore, the condition that the function must exist is not satisfied. ∎

Considering continuity from a graphical point of view, a function which is continuous over an interval will have no "breaks" in its graph over that interval. The function is continuous over the interval if we can draw its graph without lifting the marker from the paper. If the function is not continuous (discontinuous), a break occurs because the function is not defined, or the definition of the function leads to an instantaneous "jump" in values.

Several illustrations of continuous functions and discontinuous functions follow on the next page.

EXAMPLE C _____ The graph of the function $f(x) = 3x^2$, which we determined to be continuous at all values of x in Example A, is shown in Fig. 22-1. We see that there are no breaks in the graph.

The graph of $f(x) = \dfrac{1}{x-2}$, which we determined to be not continuous at $x = 2$ in Example B, is shown in Fig. 22-2. We see that there is a break in the graph for $x = 2$, and this shows the fact that $f(x)$ does not exist at $x = 2$. The curve is a hyperbola with an asymptote $x = 2$.

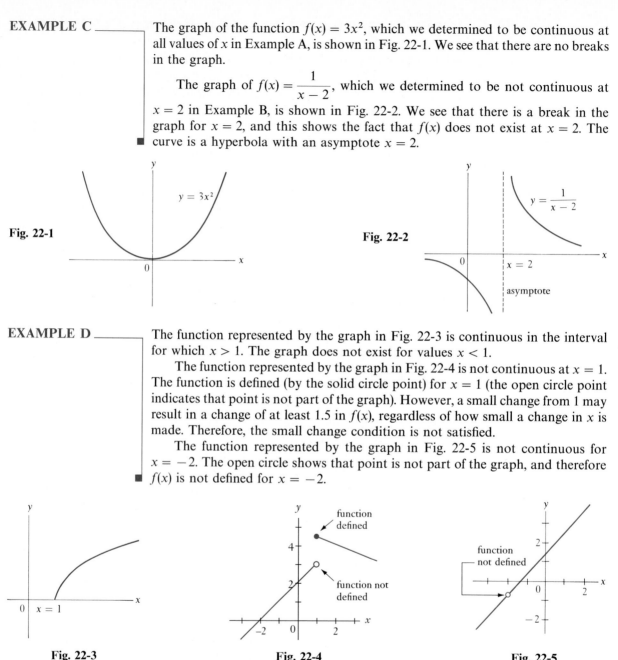

Fig. 22-1

Fig. 22-2

EXAMPLE D _____ The function represented by the graph in Fig. 22-3 is continuous in the interval for which $x > 1$. The graph does not exist for values $x < 1$.

The function represented by the graph in Fig. 22-4 is not continuous at $x = 1$. The function is defined (by the solid circle point) for $x = 1$ (the open circle point indicates that point is not part of the graph). However, a small change from 1 may result in a change of at least 1.5 in $f(x)$, regardless of how small a change in x is made. Therefore, the small change condition is not satisfied.

The function represented by the graph in Fig. 22-5 is not continuous for $x = -2$. The open circle shows that point is not part of the graph, and therefore $f(x)$ is not defined for $x = -2$.

Fig. 22-3

Fig. 22-4

Fig. 22-5

EXAMPLE E _____ We can define the function in Fig. 22-4 as

$$f(x) = \begin{cases} x + 2 & \text{for } x < 1 \\ -\tfrac{1}{2}x + 5 & \text{for } x \geq 1 \end{cases}$$

where we note that the equation differs for different parts of the domain.

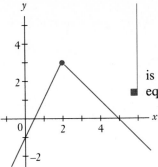

Fig. 22-6

The graph of the function

$$g(x) = \begin{cases} 2x - 1 & \text{for } x \le 2 \\ -x + 5 & \text{for } x > 2 \end{cases}$$

is shown in Fig. 22-6. We see that it is a continuous function even though the equation for $x \le 2$ is different from that for $x > 2$.

In our earlier discussions of infinite geometric series and the asymptotes of a hyperbola, we used the symbol $\rightarrow$, which means "approaches." *When we say that* $x \rightarrow 2$, *we mean that x may take on any value as close to 2 as desired, but it also* **NOTE** ▷ *distinctly means that* **x cannot be set equal to 2.**

EXAMPLE F _____

Let us consider the behavior of the function $f(x) = 2x + 1$ as $x \rightarrow 2$.

Since we are not to use $x = 2$, we use a calculator to set up tables in order to determine values of $f(x)$, as x gets close to 2.

x	1.000	1.500	1.900	1.990	1.999
$f(x)$	3.000	4.000	4.800	4.980	4.998

x	3.000	2.500	2.100	2.010	2.001
$f(x)$	7.000	6.000	5.200	5.020	5.002

values approach 5

We can see that $f(x)$ approaches 5, as x approaches 2, from above 2 and from below 2.

In Example F, since $f(x) \rightarrow 5$ as $x \rightarrow 2$, the number 5 is called the limit of $f(x)$ as $x \rightarrow 2$. This leads us to the meaning of the limit of a function. In general, limit of a function   *the* **limit of a function** $f(x)$ *is that value which the function approaches as x approaches the given value a.* This is written as

$$\lim_{x \to a} f(x) = L \qquad (22\text{-}1)$$

where L is the value of the limit of the function. Remember, in approaching a, x may come as arbitrarily close as desired to a, but it may not equal a.

An important conclusion can be drawn from the limit in Example F. The function $f(x)$ is a continuous function, and $f(2)$ equals the value of the limit as $x \rightarrow 2$. In general, it is true that *the limit of* $f(x)$ *as* $x \rightarrow a$ *is the same as* $f(a)$, *if the function is continuous at* $x = a$.

Although we can evaluate the limit for a continuous function as $x \rightarrow a$ by evaluating $f(a)$, it is possible that a function is not continuous at $x = a$, and that the limit exists and can be determined. Thus, we must be able to determine the value of a limit without finding $f(a)$. The following example illustrates the evaluation of such a limit.

EXAMPLE G ————— Find

$$\lim_{x \to 2} \frac{2x^2 - 3x - 2}{x - 2}$$

We note immediately that the function is not continuous at $x = 2$, for division by zero is indicated. Thus, we cannot evaluate the limit by substituting $x = 2$ into the function. By setting up tables, we determine the value which $f(x)$ approaches, as x approaches 2.

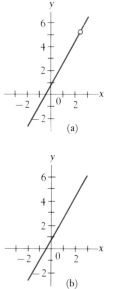

See Appendix E for a computer program for finding the limit for this type of function.

x	1.000	1.500	1.900	1.990	1.999
$f(x)$	3.000	4.000	4.800	4.980	4.998

x	3.000	2.500	2.100	2.010	2.001
$f(x)$	7.000	6.000	5.200	5.020	5.002

values approach 5

We see that the values obtained are identical to those in Example F. Since $f(x) \to 5$ as $x \to 2$, we have

$$\lim_{x \to 2} \frac{2x^2 - 3x - 2}{x - 2} = 5$$

■ Thus, the limit exists as $x \to 2$, although the function does not exist at $x = 2$.

The reason that the functions in Examples F and G have the same limit is shown in the following example.

EXAMPLE H ————— The function $\dfrac{2x^2 - 3x - 2}{x - 2}$ in Example G is the same as the function $2x + 1$ in Example F, except when $x = 2$. By factoring the numerator of the function of Example G, we have

$$\frac{2x^2 - 3x - 2}{x - 2} = \frac{(2x + 1)(x - 2)}{x - 2} = 2x + 1$$

The cancellation here is valid, as long as x does not equal 2, for we have division by zero at $x = 2$. Also, in finding the limit as $x \to 2$, we do not use the value $x = 2$. Therefore,

$$\lim_{x \to 2} \frac{2x^2 - 3x - 2}{x - 2} = \lim_{x \to 2} (2x + 1) = 5$$

The limits of the two functions are equal, since, again, in finding the limit, we do not let $x = 2$. The graphs of the two functions are shown in Fig. 22-7(a) and (b). We can see from the graphs that the limits are the same, although one of the functions is not continuous.

If $f(x) = 5$ for $x = 2$ is added to the definition of the function of Example ■ G, it is then the same as $2x + 1$, and its graph is that in Fig. 22-7(b).

(a)

(b)

Fig. 22-7

The limit of the function in Example G was determined by calculating values near $x = 2$ and by means of an algebraic change in the function. This illustrates that limits may be found through the meaning and definition, and through other procedures when the function is not continuous. The following example illustrates a function for which the limit does not exist as x approaches the indicated value.

EXAMPLE I

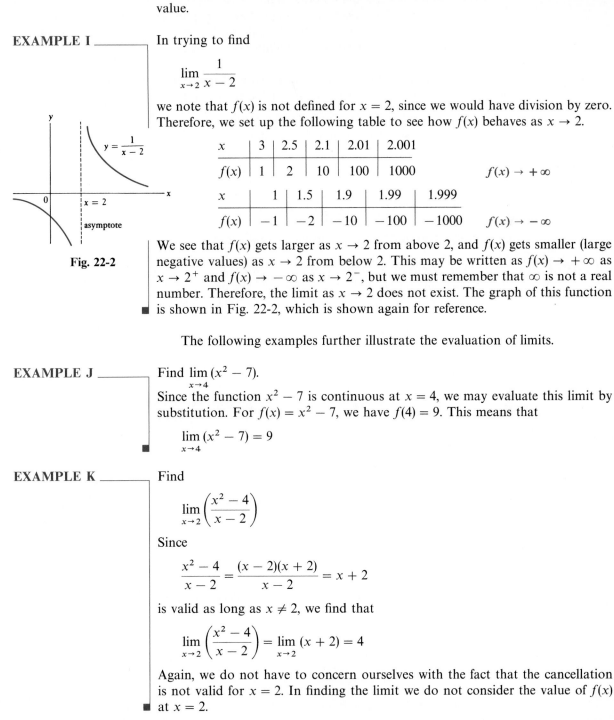

Fig. 22-2

In trying to find

$$\lim_{x \to 2} \frac{1}{x - 2}$$

we note that $f(x)$ is not defined for $x = 2$, since we would have division by zero. Therefore, we set up the following table to see how $f(x)$ behaves as $x \to 2$.

x	3	2.5	2.1	2.01	2.001	
$f(x)$	1	2	10	100	1000	$f(x) \to +\infty$

x	1	1.5	1.9	1.99	1.999	
$f(x)$	-1	-2	-10	-100	-1000	$f(x) \to -\infty$

We see that $f(x)$ gets larger as $x \to 2$ from above 2, and $f(x)$ gets smaller (large negative values) as $x \to 2$ from below 2. This may be written as $f(x) \to +\infty$ as $x \to 2^+$ and $f(x) \to -\infty$ as $x \to 2^-$, but we must remember that ∞ is not a real number. Therefore, the limit as $x \to 2$ does not exist. The graph of this function ■ is shown in Fig. 22-2, which is shown again for reference.

The following examples further illustrate the evaluation of limits.

EXAMPLE J

Find $\lim_{x \to 4} (x^2 - 7)$.

Since the function $x^2 - 7$ is continuous at $x = 4$, we may evaluate this limit by substitution. For $f(x) = x^2 - 7$, we have $f(4) = 9$. This means that

$$\lim_{x \to 4} (x^2 - 7) = 9$$

EXAMPLE K

Find

$$\lim_{x \to 2} \left(\frac{x^2 - 4}{x - 2} \right)$$

Since

$$\frac{x^2 - 4}{x - 2} = \frac{(x - 2)(x + 2)}{x - 2} = x + 2$$

is valid as long as $x \neq 2$, we find that

$$\lim_{x \to 2} \left(\frac{x^2 - 4}{x - 2} \right) = \lim_{x \to 2} (x + 2) = 4$$

Again, we do not have to concern ourselves with the fact that the cancellation is not valid for $x = 2$. In finding the limit we do not consider the value of $f(x)$ ■ at $x = 2$.

EXAMPLE L

Find $\lim_{x \to 0} (x\sqrt{x - 3})$.

We see that $x\sqrt{x - 3} = 0$ if $x = 0$, but this function does not have real values for values of x less than 3 other than $x = 0$. *Since x cannot approach 0, $f(x)$ does not approach 0 and the limit does not exist.* The point of this example is that even if $f(a)$ exists, we cannot evaluate the limit by finding $f(a)$, unless $f(x)$ is continuous at $x = a$. Here, $f(a)$ exists but the limit does not exist. ∎

Returning briefly to the discussion of continuity, we now can see that a function is continuous at $x = a$ if $\lim_{x \to a} f(x) = f(a)$, assuming that both $\lim_{x \to a} f(x)$ and $f(a)$ exist. If $f(a)$ does not exist, the function is discontinuous at $x = a$, and if $f(a)$ does not equal $\lim_{x \to a} f(x)$, a small change in x will not result in a small change in $f(x)$.

NOTE ▷ Limits as x approaches infinity are also of importance. However, when dealing with these limits, we must remember that ∞ *does not represent a real number and that arithmetic operations may not be performed on it.* Therefore, when we write $x \to \infty$, we know that we are to consider values of x which are becoming large without bound. Consider the following examples.

EXAMPLE M

The efficiency E of an engine is given by $E = 1 - Q_2/Q_1$, where Q_1 is the heat taken in and Q_2 is the heat ejected by the engine. ($Q_1 - Q_2$ is the work done by the engine.) If, in an engine cycle, $Q_2 = 500$ kJ, determine E as Q_1 becomes large without bound.

We are to find

$$\lim_{Q_1 \to \infty} \left(1 - \frac{500}{Q_1}\right)$$

As Q_1 becomes larger and larger, $1/Q_1$ becomes smaller and smaller and approaches zero. This means $f(Q_1) \to 1$ as $x \to \infty$. Thus,

$$\lim_{Q_1 \to \infty} \left(1 - \frac{500}{Q_1}\right) = 1$$

We can verify our reasoning and the value of the limit by making a table of values for Q_1 and E as Q_1 becomes large.

Q_1	500	5000	50,000	500,000
E	0	0.9	0.99	0.999

values approach 1

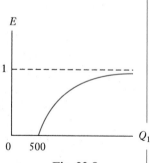

E

1

0   500

Q_1

Fig. 22-8

Again we see that $E \to 1$ as $Q_1 \to \infty$. See Fig. 22-8.

This is primarily a theoretical consideration, as there are obvious practical limitations as to how much heat can be supplied to an engine. An engine for which $E = 1$ would operate at 100% efficiency. ∎

EXAMPLE N ———————— Find

$$\lim_{x \to \infty} \frac{x^2 + 1}{2x^2 + 3}$$

We note that as $x \to \infty$, both the numerator and the denominator become large without bound. Therefore, we use a calculator to make a table to see how $f(x)$ behaves as x becomes very large.

x	1	10	100	1000	
$f(x)$	0.4	0.4975369	0.4999750	0.4999998	values approach 0.5

From this table we see that $f(x) \to 0.5$ as $x \to \infty$. See Fig. 22-9.

This limit can also be found through algebraic operations and an examination of the resulting algebraic form. If we divide both the numerator and the denominator of the function by x^2, **which is the largest power of x which appears in either the numerator or denominator,** we have

NOTE ▷

$$\frac{x^2 + 1}{2x^2 + 3} = \frac{1 + \dfrac{1}{x^2}}{2 + \dfrac{3}{x^2}} \quad \begin{array}{l} \text{terms} \to 0 \\ \text{as } x \to \infty \end{array}$$

Here we see that $1/x^2$ and $3/x^2$ both approach zero as $x \to \infty$. This means that the numerator approaches 1 and the denominator approaches 2. Thus,

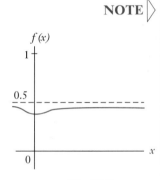

Fig. 22-9

$$\lim_{x \to \infty} \frac{x^2 + 1}{2x^2 + 3} = \lim_{x \to \infty} \frac{1 + \dfrac{1}{x^2}}{2 + \dfrac{3}{x^2}} = \frac{1}{2}$$

which agrees with our previous result.

The definitions and development of continuity and of a limit presented in this section are not mathematically rigorous. However, the development is consistent with a more rigorous development, and the concept of a limit is the principal concern.

Exercises 22-1

In Exercises 1 through 6, determine the values of x for which the function is continuous. If the function is not continuous, determine the reason.

1. $f(x) = 3x - 2$

2. $f(x) = 9 - x^2$

3. $f(x) = \dfrac{1}{x + 3}$

4. $f(x) = \dfrac{2}{x^2 - x}$

5. $f(x) = \dfrac{1}{\sqrt{x}}$

6. $f(x) = \dfrac{\sqrt{x + 2}}{x}$

In Exercises 7 through 12, determine the values of x for which the function, as represented by the graph in Fig. 22-10, is continuous. If the function is not continuous, determine the reason.

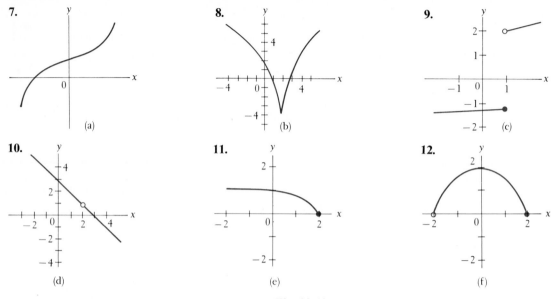

Fig. 22-10

In Exercises 13 through 16, graph the given functions and determine the values of x for which the functions are continuous. If a function is not continuous, determine the reason.

13. $f(x) = \begin{cases} x^2 & \text{for } x < 2 \\ 2 & \text{for } x \geq 2 \end{cases}$

14. $f(x) = \begin{cases} \dfrac{x^3 - x^2}{x - 1} & \text{for } x \neq 1 \\ 1 & \text{for } x = 1 \end{cases}$

15. $f(x) = \begin{cases} \dfrac{2x^2 - 18}{x - 3} & \text{for } x < 3 \text{ or } x > 3 \\ 12 & \text{for } x = 3 \end{cases}$

16. $f(x) = \begin{cases} \dfrac{x + 2}{x^2 - 4} & \text{for } x < -2 \\ \dfrac{x}{8} & \text{for } x > -2 \end{cases}$

In Exercises 17 through 24, use a calculator to evaluate the given function for the values of x shown in the table. Do not change the form of the function. Then, by observing the values obtained, find the indicated limit.

17. Evaluate $f(x) = 3x - 2$. Find $\lim\limits_{x \to 3} (3x - 2)$.

x	2.500	2.900	2.990	2.999	3.001	3.010	3.100	3.500
$f(x)$								

18. Evaluate $f(x) = 2x^2 - 4x$. Find $\lim\limits_{x \to 0.5} (2x^2 - 4x)$.

x	0.4000	0.4900	0.4990	0.4999	0.5001	0.5010	0.5100	0.6000
$f(x)$								

19. Evaluate $f(x) = \dfrac{x^3 - x}{x - 1}$.

Find $\lim\limits_{x \to 1} \dfrac{x^3 - x}{x - 1}$.

x	0.900	0.990	0.999	1.001	1.010	1.100
$f(x)$						

20. Evaluate $f(x) = \dfrac{x^3 + 2x^2 - 2x + 3}{x + 3}$.

Find $\lim\limits_{x \to -3} \dfrac{x^3 + 2x^2 - 2x + 3}{x + 3}$.

x	-3.100	-3.010	-3.001	-2.999	-2.990	-2.900
$f(x)$						

21. Evaluate $f(x) = \dfrac{2 - \sqrt{x + 2}}{x - 2}$.

Find $\displaystyle\lim_{x \to 2} \dfrac{2 - \sqrt{x + 2}}{x - 2}$.

x	1.900	1.990	1.999	2.001	2.010	2.100
$f(x)$						

22. Evaluate $f(x) = \dfrac{2 - \sqrt{x}}{4 - x}$.

Find $\displaystyle\lim_{x \to 4} \dfrac{2 - \sqrt{x}}{4 - x}$.

x	3.900	3.990	3.999	4.001	4.010	4.100
$f(x)$						

23. Evaluate $f(x) = \dfrac{2x + 1}{5x - 3}$.

Find $\displaystyle\lim_{x \to \infty} \dfrac{2x + 1}{5x - 3}$.

x	10	100	1000
$f(x)$			

24. Evaluate $f(x) = \dfrac{1 - x^2}{8x^2 + 5}$.

Find $\displaystyle\lim_{x \to \infty} \dfrac{1 - x^2}{8x^2 + 5}$.

x	10	100	1000
$f(x)$			

In Exercises 25 through 44, evaluate the given limits by direct evaluation as in Examples J through N. Change the form of the function where necessary.

25. $\displaystyle\lim_{x \to 3} (x + 4)$

26. $\displaystyle\lim_{x \to -2} (2x - 1)$

27. $\displaystyle\lim_{x \to 2} \dfrac{x^2 - 1}{x + 1}$

28. $\displaystyle\lim_{x \to 5} \left(\dfrac{3}{x^2 + 2} \right)$

29. $\displaystyle\lim_{x \to 0} \dfrac{x^2 + x}{x}$

30. $\displaystyle\lim_{x \to 2} \dfrac{x^2 - 2x}{x - 2}$

31. $\displaystyle\lim_{x \to -1} \dfrac{x^2 - 1}{x + 1}$

32. $\displaystyle\lim_{x \to 3} \dfrac{x^2 - 2x - 3}{3 - x}$

33. $\displaystyle\lim_{x \to 1} \dfrac{x^3 - x}{x - 1}$

34. $\displaystyle\lim_{x \to 1/3} \dfrac{3x - 1}{3x^2 + 5x - 2}$

35. $\displaystyle\lim_{x \to 1} \dfrac{(2x - 1)^2 - 1}{2x - 2}$

36. $\displaystyle\lim_{x \to 0} \dfrac{(2 + x)^2 - 4}{x}$

37. $\displaystyle\lim_{x \to -1} \sqrt{x}(x + 1)$

38. $\displaystyle\lim_{x \to 1} (x - 1) \sqrt{x^2 - 4}$

39. $\displaystyle\lim_{x \to \infty} \dfrac{\frac{2}{x}}{1 - 2x}$

40. $\displaystyle\lim_{x \to \infty} \dfrac{6}{1 + \frac{2}{x^2}}$

41. $\displaystyle\lim_{x \to \infty} \dfrac{3x^2 + 5}{x^2 - 2}$

42. $\displaystyle\lim_{x \to \infty} \dfrac{x - 1}{7x + 4}$

43. $\displaystyle\lim_{x \to \infty} \dfrac{2x - 6}{x^2 - 9}$

44. $\displaystyle\lim_{x \to \infty} \dfrac{2 - x^2}{3x^3 - 1}$

In Exercises 45 and 46, use a calculator to evaluate the function at 0.1, 0.01, and 0.001 from both sides of the value which x approaches. In Exercises 47 and 48, evaluate the function for the values of x of 10, 100, and 1000. From these values, determine the limit. Then, by using an appropriate change of algebraic form, evaluate the limit directly and compare values.

45. $\displaystyle\lim_{x \to 0} \dfrac{x^2 - 3x}{x}$

46. $\displaystyle\lim_{x \to 3} \dfrac{2x^2 - 6x}{x - 3}$

47. $\displaystyle\lim_{x \to \infty} \dfrac{2x^2 + x}{x^2 - 3}$

48. $\displaystyle\lim_{x \to \infty} \dfrac{x^2 + 6x + 5}{2x^2 + 1}$

In Exercises 49 through 56, solve the given problems involving limits.

49. Velocity can be determined by dividing the displacement of an object by the time elapsed in moving through the displacement. In a certain experiment the following values were determined for the displacements and elapsed times for the motion of an object. Determine the limit of the velocity.

Displacement (cm)	0.480000	0.280000	0.029800	0.002998	0.00029998
Time (seconds)	0.200000	0.100000	0.010000	0.001000	0.00010000

50. A rectangular solar panel is to be designed to have an area of 520 cm². Express the perimeter p as a function of the width w. Find $\displaystyle\lim_{w \to 20} p$ by evaluating the function for the following values of w (in cm): 19.0, 19.9, 19.99, 20.01, 20.1, and 21.

51. A certain object, after being heated, cools at such a rate that its temperature T (in degrees Celsius) decreases 10% each minute. If the object is originally heated to 100°C find $\displaystyle\lim_{t \to 10} T$ and $\displaystyle\lim_{t \to \infty} T$, where t is the time (in minutes).

52. A 5-Ω resistor and a variable resistor of resistance R are placed in parallel. The expression for the resulting resistance R_T is given by $R_T = \dfrac{5R}{5 + R}$. Determine the limiting value of R_T as $R \to \infty$.

53. Using a calculator, find $\lim\limits_{x \to 0} (1 + x)^{1/x}$. Do you recognize the limiting value?

54. Using a calculator in radian mode, find $\lim\limits_{x \to 0} \dfrac{\sin x}{x}$.

55. Using a calculator, find $\lim\limits_{x \to \infty} \dfrac{\sqrt{4x^2 + 100}}{x}$.

56. Show that $\lim\limits_{x \to 0^+} 2^{1/x} \neq \lim\limits_{x \to 0^-} 2^{1/x}$ where $\lim\limits_{x \to 0^+}$ means to find the limit as x approaches zero from the right only, and $\lim\limits_{x \to 0^-}$ means to find the limit as x approaches zero from the left only.

22-2 The Slope of a Tangent to a Curve

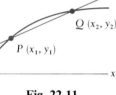

Fig. 22-11

Having developed the basic operations with functions and the concept of a limit, we shall now turn our attention to a graphical interpretation of the rate of change of a function. This interpretation, basic to an understanding of the calculus, deals with the slope of a line tangent to the curve of a function.

Consider the points $P(x_1, y_1)$ and $Q(x_2, y_2)$ in Fig. 22-11. From Chapter 20 we know that the slope of the line through these points is given by

$$m = \frac{y_2 - y_1}{x_2 - x_1}$$

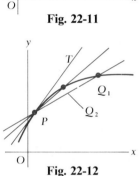

Fig. 22-12

This, however, represents the slope of the line through P and Q and no other line. If we now allow Q to be a point closer to P, the slope of PQ will more closely approximate the slope of a line drawn tangent to the curve at P (see Fig. 22-12). In fact, the closer Q is to P, the better this approximation becomes. It is not possible to allow Q to coincide with P, for then it would not be possible to define the slope of PQ in terms of two points. *The slope of the tangent line, often referred to as the slope of the curve, is the limiting value of the slope of PQ as Q approaches P.*

EXAMPLE A

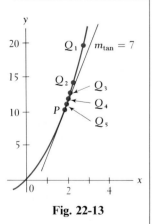

Fig. 22-13

Find the slope of a line tangent to the curve $y = x^2 + 3x$ at the point $P(2, 10)$ by determining the limit of slopes of lines PQ as Q approaches P.

We shall let point Q have the x-values of 3.0, 2.5, 2.1, 2.01, and 2.001. Then, using a calculator, we tabulate the necessary values. Since P is the point $(2, 10)$, $x_1 = 2$ and $y_1 = 10$. Thus, using the values of x_2, we tabulate the values of y_2, $y_2 - 10$, $x_2 - 2$ and thereby the values of the slope m.

Point	Q_1	Q_2	Q_3	Q_4	Q_5	P
x_2	3.0	2.5	2.1	2.01	2.001	2
y_2	18.0	13.75	10.71	10.0701	10.007001	10
$y_2 - 10$	8.0	3.75	0.71	0.0701	0.007001	
$x_2 - 2$	1.0	0.5	0.1	0.01	0.001	
$m = \dfrac{y_2 - 10}{x_2 - 2}$	8.0	7.5	7.1	7.01	7.001	

We can see from the above table that the slope of PQ approaches the value of 7 as Q approaches P. Thus, the slope of the tangent line at $(2, 10)$ is 7. See Fig. 22-13.

With the proper notation, it is possible to express the coordinates of Q in terms of the coordinates of P. If we define the quantity Δx ("delta" x) by the equation

$$x_2 = x_1 + \Delta x \tag{22-2}$$

and Δy by the equation

$$y_2 = y_1 + \Delta y \tag{22-3}$$

NOTE ▷

increments

the coordinates of Q become $(x_1 + \Delta x, y_1 + \Delta y)$. The quantities Δx and Δy represent the difference in the coordinates of P and Q. ***The quantity Δx is not to be thought of as "Δ times x,"*** *for the symbol Δ used here has no meaning by itself. The name* **increment** *is given to the difference of the coordinates of two points, and therefore Δx and Δy are the increments in x and y, respectively.*

Using Eqs. (22-2) and (22-3), along with the definition of slope, we can express the slope of PQ as

$$m_{PQ} = \frac{(y_1 + \Delta y) - y_1}{(x_1 + \Delta x) - x_1} = \frac{\Delta y}{\Delta x} \tag{22-4}$$

By previous discussion, as Q approaches P, the slope of the tangent line is more *nearly* approximated by $\Delta y / \Delta x$.

EXAMPLE B

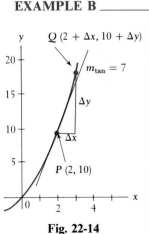

Fig. 22-14

Find the slope of a line tangent to the curve $y = x^2 + 3x$ at the point $(2, 10)$ by the increment method indicated in Eq. (22-4). (This is the same slope as calculated in Example A.)

As in Example A, point P has the coordinates $(2, 10)$. Thus the coordinates of any other point Q can be expressed as $(2 + \Delta x, 10 + \Delta y)$. See Fig. 22-14. The slope of PQ then becomes

$$m_{PQ} = \frac{(10 + \Delta y) - 10}{(2 + \Delta x) - 2} = \frac{\Delta y}{\Delta x}$$

This expression itself does not enable us to find the slope, since values of Δx and Δy are not known. If, however, we can express Δy in terms of Δx, we might derive more information. Both P and Q are on the curve of the function, which means that the coordinates of each must satisfy the function. Using the coordinates of Q, we have

$$(10 + \Delta y) = (2 + \Delta x)^2 + 3(2 + \Delta x)$$
$$= 4 + 4\Delta x + (\Delta x)^2 + 6 + 3\Delta x$$

Subtracting 10 from each side, we have

$$\Delta y = 7\Delta x + (\Delta x)^2$$

(Continued on next page)

As Q approaches P, Δx becomes smaller and smaller. Calculating Δy and the ratio $\Delta y/\Delta x$ for increasingly small Δx, we have the following values.

Δx	0.1	0.01	0.001
Δy	0.71	0.0701	0.007001
$\dfrac{\Delta y}{\Delta x}$	7.1	7.01	7.001

$\Delta y = 7\Delta x + (\Delta x)^2$

We note that the values of $\Delta y/\Delta x$ are the same as those for the slope for points Q_3, Q_4, and Q_5 in Example A. We also see that as $\Delta x \to 0$, $\Delta y/\Delta x \to 7$.

Now, substituting the expression for Δy into the expression for m_{PQ}, we have

$$m_{PQ} = \frac{7\,\Delta x + (\Delta x)^2}{\Delta x} = 7 + \Delta x$$

From this expression we can see directly that as $\Delta x \to 0$, $m_{PQ} \to 7$. Using this fact, we can see that the slope of the tangent line is

$$m_{\text{tan}} = \lim_{\Delta x \to 0} m_{PQ} = 7$$

■ We see that this result agrees with that found in Example A.

EXAMPLE C _____ Find the slope of a line tangent to the curve $y = 4x - x^2$ at the point (x_1, y_1).

The points P and Q are $P(x_1, y_1)$ and $Q(x_1 + \Delta x, y_1 + \Delta y)$. Using the coordinates of Q in the function, we obtain

$$(y_1 + \Delta y) = 4(x_1 + \Delta x) - (x_1 + \Delta x)^2$$
$$= 4x_1 + 4\,\Delta x - x_1^2 - 2x_1\,\Delta x - (\Delta x)^2$$

Using the coordinates of P, we obtain

$$y_1 = 4x_1 - x_1^2$$

Subtracting the second from the first to solve for Δy, we obtain

$$(y_1 + \Delta y) - y_1 = [4x_1 + 4\,\Delta x - x_1^2 - 2x_1\,\Delta x - (\Delta x)^2] - (4x_1 - x_1^2)$$
$$\Delta y = 4\,\Delta x - 2x_1\,\Delta x - (\Delta x)^2$$

Dividing through by Δx to obtain an expression for $\Delta y/\Delta x$, we obtain

$$\frac{\Delta y}{\Delta x} = 4 - 2x_1 - \Delta x$$

In this last equation, the desired expression is on the left, but all we can determine from $\Delta y/\Delta x$ itself is that the ratio will become one very small number divided by another very small number as $\Delta x \to 0$. The right side, however, approaches $4 - 2x_1$ as $\Delta x \to 0$. This indicates that the slope of a tangent at the point (x_1, y_1) is given by

$$m_{\text{tan}} = 4 - 2x_1$$

This method has an advantage over that of Example B—we now have a general expression for the slope of a tangent line for any value x_1. If $x_1 = -1$, $m_{\text{tan}} = 6$, and if $x_1 = 3$, $m_{\text{tan}} = -2$. These tangent lines are indicated in Fig. 22-15.

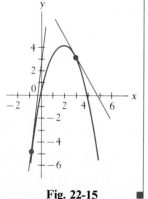

Fig. 22-15

EXAMPLE D _____

Find the expression for the slope of a line tangent to the curve $y = x^3 + 2$ at the general point (x_1, y_1), and use this expression to determine the slope when $x = \frac{1}{2}$.

Using the coordinates of the points $P(x_1, y_1)$ and $Q(x_1 + \Delta x, y_1 + \Delta y)$ in the function, we have the following steps:

$$y_1 + \Delta y = (x_1 + \Delta x)^3 + 2 \qquad \qquad \text{substitute coordinates of } Q$$
$$= x_1^3 + 3x_1^2 \Delta x + 3x_1(\Delta x)^2 + (\Delta x)^3 + 2$$
$$y_1 = x_1^3 + 2 \qquad \qquad \text{substitute coordinates of } P$$
$$\Delta y = 3x_1^2 \Delta x + 3x_1(\Delta x)^2 + (\Delta x)^3 \qquad \text{subtract } y_1 \text{ from } y_1 + \Delta y$$
$$\frac{\Delta y}{\Delta x} = 3x_1^2 + 3x_1 \Delta x + (\Delta x)^2 \qquad \qquad \text{divide by } \Delta x$$

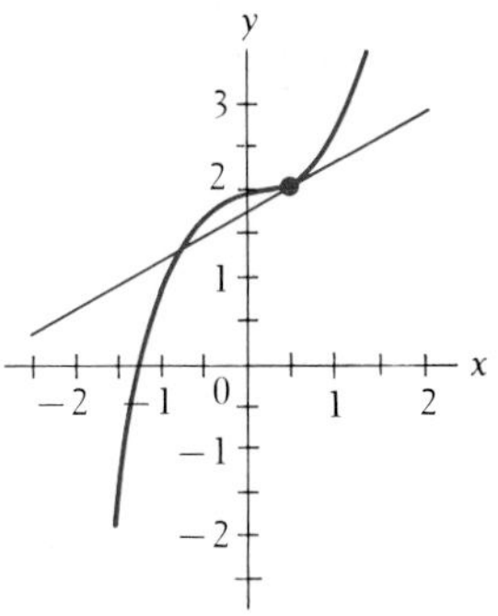

As $\Delta x \to 0$, the right side approaches the value $3x_1^2$. This means that

$$m_{\text{tan}} = 3x_1^2$$

When $x_1 = \frac{1}{2}$, we find that the slope of the tangent is $3(\frac{1}{4}) = \frac{3}{4}$. The curve and this tangent line are indicated in Fig. 22-16.

Fig. 22-16   ■

In interpreting this analysis as the rate of change of a function, we see that if $\Delta x = 1$ and the corresponding value of Δy is found, it can be said that y changes by the amount Δy as x changes one unit. If x changed by some lesser amount, we could still calculate a ratio for the amount of change in y for the given change in x. Therefore, as long as x changes at all, there will be a corresponding change in y. In this way _the ratio_ $\Delta y/\Delta x$ _is the_ **average rate of change** _of y, which is a function of x, with respect to x. As_ $\Delta x \to 0$ _the limit of the ratio_ $\Delta y/\Delta x$ _gives us the_ **instantaneous rate of change** _of y with respect to x._

EXAMPLE E _____

In Example A let us consider points $P(2, 10)$ and $Q_2(2.5, 13.75)$. From P to Q_2, x changes by 0.5 unit and y by 3.75 units. This means that _the average change in y for a 1-unit change in x is_ $3.75/0.5 = 7.5$ _units._ However, this is not the rate at which y is changing with respect to x **at** most points within this interval.

NOTE ▷

_At_ point P, the slope of the tangent line of 7 tells us that y is changing 7 units for a one-unit change in x. However, **this is an instantaneous rate of change at point P,** and tells us the rate at which y is changing with respect to x **at P and only at P.**

Exercises 22-2

In Exercises 1 through 4, use the method of Example A to calculate the slope of a line tangent to the curve of each of the given functions. Let $Q_1, Q_2, Q_3,$ _and_ Q_4 _have the indicated x-values. Sketch the curves and tangent lines._

1. $y = x^2$; P is $(2, 4)$; let Q have x-values of 1.5, 1.9, 1.99, 1.999.

2. $y = 1 - \frac{1}{2}x^2$; P is $(2, -1)$; let Q have x-values of 1.5, 1.9, 1.99, 1.999.

3. $y = x^2 + 2x$; P is $(-3, 3)$; let Q have x-values of $-2.5, -2.9, -2.99, -2.999$.

4. $y = x^3 + 1$; P is $(-1, 0)$; let Q have x-values of $-0.5, -0.9, -0.99, -0.999$.

In Exercises 5 through 8, use the method of Example B (divide the expression for Δy by Δx and use the simplified expression for m_{PQ}) to calculate the slope of a line tangent to the curve of each of the given functions for the given points P. (These are the same functions and points P as for Exercises 1 through 4.)

5. $y = x^2$; P is $(2, 4)$.

6. $y = 1 - \frac{1}{2}x^2$; P is $(2, -1)$.

7. $y = x^2 + 2x$; P is $(-3, 3)$.

8. $y = x^3 + 1$; P is $(-1, 0)$.

In Exercises 9 through 20, use the method of Example C to find a general expression for the slope of a tangent line to each of the indicated curves. Then find the slope for the given values of x. Sketch the curves and tangent lines.

9. $y = x^2$; $x = 2$, $x = -1$

10. $y = 1 - \frac{1}{2}x^2$; $x = 2$, $x = -2$

11. $y = x^2 + 2x$; $x = -3$, $x = 1$

12. $y = 4 - 3x^2$; $x = 0$, $x = 2$

13. $y = x^2 + 4x + 5$; $x = -3$, $x = 2$

14. $y = 2x^2 - 4x$; $x = 1$, $x = 1.5$

15. $y = 6x - x^2$; $x = -2$, $x = 3$

16. $y = 2x - 3x^2$; $x = 0$, $x = 0.5$

17. $y = x^3 - 2x$; $x = -1$, $x = 0$, $x = 1$

18. $y = 3x - x^3$; $x = -2$, $x = 0$, $x = 2$

19. $y = x^4$; $x = 0$, $x = 0.5$, $x = 1$

20. $y = 1 - x^4$; $x = 0$, $x = 1$, $x = 2$

In Exercises 21 through 24, find the average rate of change of y with respect to x from P to Q. Then compare this with the instantaneous rate of change of y with respect to x at P by finding m_{tan} at P.

21. $y = x^2 + 2$, $P(2, 6)$, $Q(2.1, 6.41)$

22. $y = 1 - 2x^2$, $P(1, -1)$, $Q(1.1, -1.42)$

23. $y = 9 - x^3$, $P(2, 1)$, $Q(2.1, -0.261)$

24. $y = x^3 - 6x$, $P(3, 9)$, $Q(3.1, 11.191)$

22-3   The Derivative

Using the terminology and techniques developed in the first section of this chapter, we are now ready to establish one of the fundamental definitions of calculus. Generalizing on the method of Example B of the preceding section, if we consider that the point $P(x, y)$ is held constant while the point $Q(x + \Delta x, y + \Delta y)$ approaches it, then both Δx and Δy approach zero. The points P and Q both lie on the curve of the function $f(x)$, which in turn means that the coordinates of each point must satisfy the function. For the point P this means that $y = f(x)$, and for the point Q this means that $y + \Delta y = f(x + \Delta x)$. Subtracting the first expression from the second, we have

$$(y + \Delta y) - y = f(x + \Delta x) - f(x)$$

$$\Delta y = f(x + \Delta x) - f(x) \tag{22-5}$$

From the discussion of the previous section, we saw that the slope of a line tangent to $f(x)$ at point $P(x, y)$ was found as the limiting value of the ratio $\Delta y / \Delta x$ as Δx approaches zero. Formally, *this limiting value of the ratio of $\Delta y / \Delta x$ is known as the* **derivative** *of the function*. Therefore, the derivative of a function $f(x)$ is defined by the relation

the derivative

$$\lim_{\Delta x \to 0} \frac{\Delta y}{\Delta x} = \lim_{\Delta x \to 0} \frac{f(x + \Delta x) - f(x)}{\Delta x} \tag{22-6}$$

The process of finding the derivative of a function from its definition is called the delta-process. *The general process of finding a derivative is called* **differentiation.**

EXAMPLE A _____ Find the derivative of the function $y = 2x^2 + 3x$ by the delta-process.

To find Δy, we must first derive the quantity $f(x + \Delta x) - f(x)$. Thus, we first find $y + \Delta y = f(x + \Delta x)$ by replacing y by $y + \Delta y$ and x by $x + \Delta x$.

$$y + \Delta y = 2(x + \Delta x)^2 + 3(x + \Delta x)$$

Then we subtract the original function:

$$y + \Delta y - y = 2(x + \Delta x)^2 + 3(x + \Delta x) - (2x^2 + 3x)$$

Simplifying, we find that the result is

$$\Delta y = 4x\,\Delta x + 2(\Delta x)^2 + 3\,\Delta x$$

Next, dividing through by Δx, we obtain

$$\frac{\Delta y}{\Delta x} = 4x + 2\,\Delta x + 3$$

When we find the limit as Δx approaches zero, we see that the $2\,\Delta x$-term on the right approaches zero. Thus,

$$\lim_{\Delta x \to 0} \frac{\Delta y}{\Delta x} = 4x + 3$$

We have found that the derivative of the function $2x^2 + 3x$ is the function $4x + 3$. From the definition of the derivative and from the previous section, this means that we can calculate the slope of a tangent line for any point on the graph of $y = 2x^2 + 3x$ by substituting the x-coordinate into the expression $4x + 3$. For example, the slope of a tangent line is 5 if $x = \frac{1}{2}$ [at the point $(\frac{1}{2}, 2)$]. See Fig. 22-17.

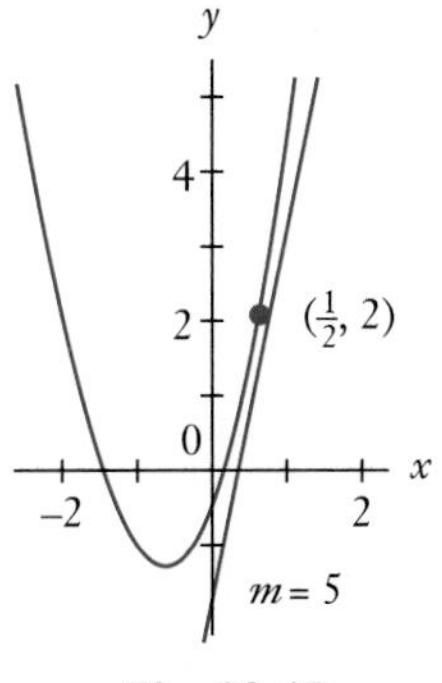

Fig. 22-17

EXAMPLE B _____ Find the derivative of $y = 6x - 2x^3$ by the delta-process.

We have the following steps:

$$y + \Delta y = 6(x + \Delta x) - 2(x + \Delta x)^3 \qquad\qquad y + \Delta y = f(x + \Delta x)$$
$$= 6x + 6(\Delta x) - 2[x^3 + 3x^2(\Delta x) + 3x(\Delta x)^2 + (\Delta x)^3]$$
$$= 6x + 6(\Delta x) - 2x^3 - 6x^2(\Delta x) - 6x(\Delta x)^2 - 2(\Delta x)^3$$
$$y + \Delta y - y = [6x + 6(\Delta x) - 2x^3 - 6x^2(\Delta x) - 6x(\Delta x)^2 - 2(\Delta x)^3] - (6x - 2x^3) \qquad \text{subtract } y = 6x - 2x^3$$
$$\Delta y = 6(\Delta x) - 6x^2(\Delta x) - 6x(\Delta x)^2 - 2(\Delta x)^3$$
$$\frac{\Delta y}{\Delta x} = \frac{6(\Delta x) - 6x^2(\Delta x) - 6x(\Delta x)^2 - 2(\Delta x)^3}{\Delta x} \qquad\qquad \text{divide by } \Delta x$$
$$= 6 - 6x^2 - 6x(\Delta x) - 2(\Delta x)^2$$
$$\lim_{\Delta x \to 0} \frac{\Delta y}{\Delta x} = \lim_{\Delta x \to 0} [6 - 6x^2 - 6x(\Delta x) - 2(\Delta x)^2] \qquad\qquad \text{find limit as } \Delta x \to 0$$
$$= 6 - 6x^2$$

■ Therefore, the derivative of $y = 6x - 2x^3$ is $6 - 6x^2$.

The derivative of a function is itself a function and may not be defined for all values of x. *If x_0 is in the domain of the derivative, then the function is said to be **differentiable** at x_0*. Examples C, D, and F which follow illustrate functions which are not differentiable for all values of x.

EXAMPLE C _____ Find the derivative $y = \dfrac{1}{x}$ by the delta-process.

$$y + \Delta y = \frac{1}{x + \Delta x} \qquad\qquad y + \Delta y = f(x + \Delta x)$$

$$y + \Delta y - y = \frac{1}{x + \Delta x} - \frac{1}{x} \qquad\qquad \text{subtract } y = \frac{1}{x}$$

NOTE $\triangleright$

$$\Delta y = \frac{x - (x + \Delta x)}{x(x + \Delta x)} = \frac{-\Delta x}{x(x + \Delta x)} \qquad \text{combine fractions}$$

$$\frac{\Delta y}{\Delta x} = \frac{-1}{x(x + \Delta x)} \qquad\qquad \text{divide by } \Delta x$$

$$\lim_{\Delta x \to 0} \frac{\Delta y}{\Delta x} = \lim_{\Delta x \to 0} \frac{-1}{x(x + \Delta x)} = \frac{-1}{x^2} \qquad \text{find limit as } \Delta x \to 0$$

We note that neither the function nor the derivative is defined for $x = 0$. This ■ means the function is not differentiable at $x = 0$.

EXAMPLE D _____ Find the derivative of $y = x^2 + \dfrac{1}{x + 1}$ by the delta-process.

$$y + \Delta y = (x + \Delta x)^2 + \frac{1}{x + \Delta x + 1} \qquad\qquad y + \Delta y = f(x + \Delta x)$$

$$y + \Delta y - y = x^2 + 2x\,\Delta x + (\Delta x)^2 + \frac{1}{x + \Delta x + 1} - x^2 - \frac{1}{x + 1} \qquad \text{subtract } y = x^2 + \frac{1}{x + 1}$$

$$\Delta y = 2x\,\Delta x + (\Delta x)^2 + \frac{1}{x + \Delta x + 1} - \frac{1}{x + 1} \qquad\qquad \begin{array}{l}\text{algebra handled most easily}\\\text{if fractions combined separately}\\\text{from other terms}\end{array}$$

$$= 2x\,\Delta x + (\Delta x)^2 + \frac{(x + 1) - (x + \Delta x + 1)}{(x + \Delta x + 1)(x + 1)}$$

$$= 2x\,\Delta x + (\Delta x)^2 - \frac{\Delta x}{(x + \Delta x + 1)(x + 1)}$$

$$\frac{\Delta y}{\Delta x} = 2x + \Delta x - \frac{1}{(x + \Delta x + 1)(x + 1)} \qquad\qquad \text{divide by } \Delta x$$

$$\lim_{\Delta x \to 0} \frac{\Delta y}{\Delta x} = 2x - \frac{1}{(x + 1)^2} \qquad\qquad \text{find limit as } \Delta x \to 0$$

■ We note that this function is not differentiable at $x = -1$.

Several shorter notations are used for the derivative. The notation of the definition is clumsy for general use, so the notations

$$y', \qquad D_x y, \qquad f'(x), \qquad \text{and} \qquad \frac{dy}{dx}$$

are commonly used. The following example illustrates the use of basic notation with derivatives.

EXAMPLE E In Example B, $y = 6x - 2x^3$, and we found that the derivative is $6 - 6x^2$. Thus we may write

$$y' = 6 - 6x^2 \qquad \text{or} \qquad \frac{dy}{dx} = 6 - 6x^2$$

If we had written $f(x) = 6x - 2x^3$, we would then write $f'(x) = 6 - 6x^2$.
If we wish to evaluate the derivative at some point, such as $(-2, 4)$, we write

$$\frac{dy}{dx} = 6 - 6x^2$$

$$\left.\frac{dy}{dx}\right|_{x=-2} = 6 - 6(-2)^2 = 6 - 24 = -18$$

■ We note that only the x-coordinate of the point was needed in the evaluation.

EXAMPLE F Find dy/dx of the function $y = \sqrt{x}$ by the delta-process.
We first square both sides of the equation, thus obtaining $y^2 = x$. (This is valid only for $y \geq 0$, since the original function $y = \sqrt{x}$ is not defined for $y < 0$.) Replacing y by $y + \Delta y$ and x by $x + \Delta x$, we have

$$(y + \Delta y)^2 = x + \Delta x$$

We complete the solution as follows:

$$y^2 + 2y\,\Delta y + (\Delta y)^2 - y^2 = x + \Delta x - x \qquad \text{expand left side and subtract } y^2 = x$$

$$2y\,\Delta y + (\Delta y)^2 = \Delta x$$

$$2y\,\frac{\Delta y}{\Delta x} + \Delta y\,\frac{\Delta y}{\Delta x} = 1 \qquad \text{divide by } \Delta x$$

$$\frac{\Delta y}{\Delta x} = \frac{1}{2y + \Delta y} \qquad \text{solve for } \frac{\Delta y}{\Delta x}$$

$$\lim_{\Delta x \to 0} \frac{\Delta y}{\Delta x} = \lim_{\Delta x \to 0} \frac{1}{2y + \Delta y} = \frac{1}{2y} \qquad \text{find limit as } \Delta x \to 0$$

The Δy is omitted, since $\Delta y \to 0$ as $\Delta x \to 0$. Now, substituting in the original function $y = \sqrt{x}$, we obtain the final result:

$$\frac{dy}{dx} = \frac{1}{2\sqrt{x}}$$

The domain of the function is $x \geq 0$. However, since x appears in the denominator of the derivative, the domain of the derivative is $x > 0$. This means that the function is differentiable for $x > 0$.

NOTE ▷   One might ask why, when we are finding a derivative, we take a limit as Δx approaches zero and do not simply let Δx equal zero. If we did this, the ratio $\Delta y/\Delta x$ would be exactly $0/0$, which would then require division by zero. As we know, this is an undefined operation in mathematics, and therefore *Δx cannot equal zero.* However, it can equal any value as near zero as necessary. This idea is basic in the meaning of the word *limit*.

Exercises 22-3

In Exercises 1 through 24, find the derivative of each of the functions by using the delta-process.

1. $y = 3x - 1$ **2.** $y = 6x + 3$ **3.** $y = 1 - 2x$ **4.** $y = 2 - 5x$

5. $y = x^2 - 1$ **6.** $y = 4 - x^2$ **7.** $y = 5x^2$ **8.** $y = -6x^2$

9. $y = x^2 - 7x$ **10.** $y = x^2 + 4x$ **11.** $y = 8x - 2x^2$ **12.** $y = 3x - \dfrac{1}{2}x^2$

13. $y = x^3 + 4x - 6$ **14.** $y = 2x - 4x^3$ **15.** $y = \dfrac{1}{x + 2}$ **16.** $y = \dfrac{1}{x + 1}$

17. $y = x + \dfrac{1}{x}$ **18.** $y = \dfrac{x}{x - 1}$ **19.** $y = \dfrac{2}{x^2}$ **20.** $y = \dfrac{1}{x^2 + 1}$

21. $y = x^4 + x^3 + x^2 + x$ **22.** $y = \dfrac{1}{3}x^3 + \dfrac{1}{2}x^2 + x$ **23.** $y = x^4 - \dfrac{2}{x}$ **24.** $y = \dfrac{1}{x} + \dfrac{1}{x^2}$

In Exercises 25 through 28, find the derivative of each of the given functions by using the delta-process. Then evaluate the derivative at the given point.

25. $y = 3x^2 - 2x,\ (-1, 5)$ **26.** $y = 9x - x^3,\ (2, 10)$

27. $y = \dfrac{6}{x + 3},\ (3, 1)$ **28.** $y = x^2 - \dfrac{2}{x},\ (-2, 5)$

In Exercises 29 through 32, find dy/dx for the given functions by the method of Example F.

29. $y = \sqrt{x + 1}$ **30.** $y = \sqrt{x - 2}$

31. $y = \sqrt{1 - 3x}$ **32.** $y = \sqrt{x^2 + 3}$

22-4   The Meaning of the Derivative

In Section 22-2 we saw that the slope of a line tangent to a curve at point P was the limiting value of the slope of the line through points P and Q as Q approaches P. In Section 22-3 we defined the limit of the ratio of $\Delta y/\Delta x$ as $\Delta x \to 0$ as the derivative. Thus, the first meaning we have given to the derivative is the slope of a line tangent to a curve, as we noted in Example A of Section 22-3. The following example further illustrates this meaning of the derivative.

EXAMPLE A _____

Find the slope of a line tangent to the curve of $y = 4x - x^2$ at the point $(1, 3)$.

To determine the slope we first find the derivative, then evaluate it at the given point.

$$y + \Delta y = 4(x + \Delta x) - (x + \Delta x)^2$$
$$y + \Delta y - y = 4x + 4\Delta x - x^2 - 2x\Delta x - (\Delta x)^2 - (4x - x^2)$$
$$\Delta y = 4\Delta x - 2x\Delta x - (\Delta x)^2$$
$$\frac{\Delta y}{\Delta x} = 4 - 2x - \Delta x$$
$$\lim_{\Delta x \to 0} \frac{\Delta y}{\Delta x} = 4 - 2x \qquad \text{derivative}$$
$$\left.\frac{dy}{dx}\right|_{(1,3)} = 4 - 2(1) = 2 \qquad \text{evaluate derivative}$$

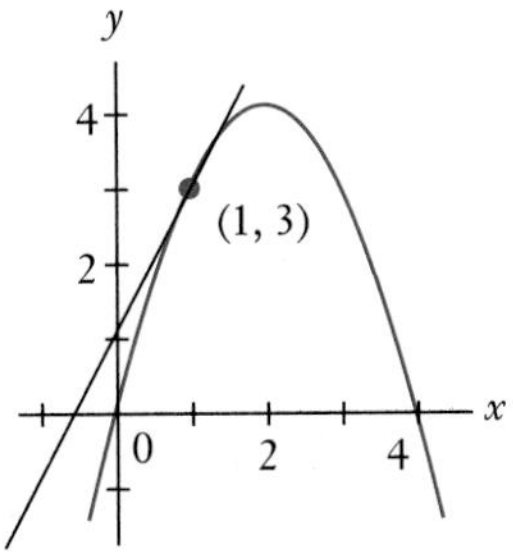

This means that the slope of the tangent line at $(1, 3)$ is 2. We note that only the value $x = 1$ was needed to evaluate the derivative at the point $(1, 3)$. The curve and tangent line are shown in Fig. 22-18.

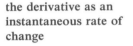

Fig. 22-18

At the end of Section 22-2 we discussed the idea that $\Delta y/\Delta x$ indicates the rate of change of y with respect to x. In defining the derivative as the limit of the ratio of $\Delta y/\Delta x$ as $\Delta x \to 0$, it is a measure of the rate of change of y with respect to x at point P. However, P may represent any point, which means that the value of the derivative changes from one point on a curve to another point.

the derivative as an instantaneous rate of change

*We therefore interpret the derivative as the **instantaneous rate of change** of y with respect to x.*

EXAMPLE B _____

In Examples A and B of Section 22-2, y is changing at the rate of 7 units for every change of 1 unit in x, *when x is 2, and only when x is 2.* In Example C of Section 22-2, y is increasing 6 units for every increase of 1 unit of x *when $x = -1$.* When $x = 3$, y is decreasing 2 units for 1 unit of increase in x.

This gives us a more general meaning of the derivative. If a functional relationship exists between any two variables, then one can be taken to be varying with respect to the other and the derivative gives us the instantaneous rate of change. There are many applications of this principle, one of them being the velocity of an object. We shall consider here the case of *rectilinear motion,* that is, motion along a straight line.

As we have seen, the velocity of an object is found by dividing the change in displacement by the time required for this change. This, however, gives a value only for the **average velocity** for the specified time interval. If the time interval considered becomes smaller and smaller, then the average velocity which is calculated more nearly approximates the **instantaneous velocity** at some particular time. In the limit, the value of the average velocity gives the value of the instantaneous velocity. Using the symbols as defined for the derivative, *the instantaneous* *velocity of an object moving in rectilinear motion at a particular time t is given by*

instantaneous
velocity

$$v = \lim_{\Delta t \to 0} \frac{\Delta s}{\Delta t}$$

(22-7)

where s is the displacement.

In Eq. (22-7) the derivative gives us the instantaneous rate of change of s with respect to t. Therefore, the units of this derivative would be in units of displacement divided by units of time. In general, *units of the derivative of $y = f(x)$ are in units of y divided by units of x.*

EXAMPLE C

See the chapter
introduction.

Find the instantaneous velocity, when $t = 4$ s (exactly), of a falling object for which the displacement s (in feet), which is the distance fallen, is given by $s = 16t^2$ by calculating values of $\Delta s/\Delta t$ and determining the limit as Δt approaches zero.

Here we shall let t take on values of 3.5, 3.9, 3.99, and 3.999 s. When $t = 4$ s, $s = 256$ ft. Therefore, we calculate Δt by subtracting values of t from 4, and Δs is calculated by subtracting values from 256. The values of velocity are then calculated by using $v = \Delta s/\Delta t$.

	t (seconds)	3.5	3.9	3.99	3.999
	s (feet)	196.0	243.36	254.7216	255.872016
$256 - s$	Δs (feet)	60.0	12.64	1.2784	0.127984
$4 - t$	Δt (seconds)	0.5	0.1	0.01	0.001
$\dfrac{\Delta s}{\Delta t}$	v (feet per second)	120.0	126.4	127.84	127.984

We can see that the value of v is approaching 128 ft/s, which is therefore the instantaneous velocity when $t = 4$ s. ∎

EXAMPLE D

See the chapter
introduction.

By use of the delta-process, determine an expression for the instantaneous velocity of the object of Example C, for which the displacement s (in feet) is given by $s = 16t^2$ and t represents the time (in seconds). By use of the derived expression, determine the instantaneous velocity when $t = 2$ s and when $t = 4$ s.

Applying the delta-process to this function, we find the derivative of s with respect to t:

$$s + \Delta s = 16(t + \Delta t)^2 = 16t^2 + 32t\,\Delta t + 16(\Delta t)^2$$

$$\Delta s = 32t\,\Delta t + 16(\Delta t)^2$$

$$\frac{\Delta s}{\Delta t} = 32t + 16\,\Delta t$$

$$v = \lim_{\Delta t \to 0} \frac{\Delta s}{\Delta t} = \frac{ds}{dt} = 32t \qquad \text{expression for instantaneous velocity}$$

Finding the instantaneous velocity for $t = 2$ s and $t = 4$ s, we have

$$\left.\frac{ds}{dt}\right|_{t=2} = 32(2) = 64 \text{ ft/s} \qquad \text{and} \qquad \left.\frac{ds}{dt}\right|_{t=4} = 32(4) = 128 \text{ ft/s}$$

∎ We see that this second result agrees with that obtained in Example C.

By finding $\lim_{\Delta x \to 0} \Delta y / \Delta x$, we can find the instantaneous rate of change of y with respect to x. The expression $\lim_{\Delta t \to 0} \Delta s / \Delta t$ gives the velocity, or instantaneous rate of change of displacement with respect to time. Generalizing, we can say that

the derivative can be interpreted as instantaneous rate of change of the dependent variable with respect to the independent variable.

This is true no matter what the variables represent, so long as a differentiable function is defined between the variables.

EXAMPLE E

A spherical balloon is being inflated. Find an expression for the instantaneous rate of change of the volume with respect to the radius. Evaluate this rate of change for a radius of 2.00 m.

$$V = \frac{4}{3} \pi r^3 \qquad \text{volume of sphere}$$

$$V + \Delta V = \frac{4\pi}{3}(r + \Delta r)^3 \qquad \text{find derivative}$$

$$= \frac{4\pi}{3} \left[r^3 + 3r^2 \, \Delta r + 3r(\Delta r)^2 + (\Delta r)^3 \right]$$

$$\Delta V = \frac{4\pi}{3} \left[3r^2 \, \Delta r + 3r(\Delta r)^2 + (\Delta r)^3 \right]$$

$$\frac{\Delta V}{\Delta r} = \frac{4\pi}{3} \left[3r^2 + 3r \, \Delta r + (\Delta r)^2 \right]$$

$$\lim_{\Delta r \to 0} \frac{\Delta V}{\Delta r} = 4\pi r^2 \qquad \begin{array}{l} \text{expression for instantaneous} \\ \text{rate of change} \end{array}$$

$$\left. \frac{dV}{dr} \right|_{r = 2.00 \text{ m}} = 4\pi(2.00)^2 = 16.0\pi = 50.3 \text{ m}^2 \qquad \begin{array}{l} \text{instantaneous rate of change} \\ \text{when } r = 2.00 \text{ m} \end{array}$$

The instantaneous rate of change of the volume with respect to the radius for $r = 2.00$ m is 50.3 m^3/m (this way of showing the units is actually more meaningful than m^2, the units in simplest form).

We can see that, as the radius of the balloon increases, the instantaneous rate of change of the volume with respect to the radius also increases. This should be expected, since the volume of a sphere varies directly as the cube of the radius.

EXAMPLE F

The power P produced by an electric current i in a resistor varies directly as the square of the current. Given that 1.2 W of power are produced by a current of 0.50 A in a particular resistor, find an expression for the instantaneous rate of change of power with respect to current. Evaluate this rate of change for $i = 2.5$ A.

We must first find the functional relationship between power and current. This we can do by solving the indicated problem in variation:

$$P = ki^2, \qquad 1.2 = k(0.50)^2, \qquad k = 4.8 \text{ W/A}^2, \qquad P = 4.8i^2$$

(*Continued on next page*)

Now, knowing the function, we may determine the expression for the instantaneous rate of change of P with respect to i by use of the delta-process:

$$P + \Delta P = 4.8(i + \Delta i)^2 = 4.8[i^2 + 2i(\Delta i) + (\Delta i)^2]$$
$$\Delta P = 4.8[2i(\Delta i) + (\Delta i)^2]$$
$$\frac{\Delta P}{\Delta i} = 4.8(2i + \Delta i)$$

$$\lim_{\Delta i \to 0} \frac{\Delta P}{\Delta i} = 9.6i \qquad \text{expression for instantaneous rate of change}$$

$$\frac{dP}{di}\bigg|_{i=2.5\,\text{A}} = 9.6(2.5) = 24 \text{ W/A} \qquad \text{instantaneous rate of change when } i = 2.5 \text{ A}$$

This tells us that when $i = 2.5$ A, the rate of change of power with respect to current is 24 W/A. Also, we see that the larger the current is, the greater is the increase in power. This should be expected, since the power varies directly as the square of the current.

Exercises 22-4

In Exercises 1 through 4, find the slope of a line tangent to the curve of the given equation at the given point. Sketch the curve and the tangent line.

1. $y = x^2 - 1$; $(2, 3)$ **2.** $y = 2x - x^2$; $(-1, -3)$ **3.** $y = \dfrac{4}{x+1}$; $(-3, -2)$ **4.** $y = x^3 - 3x$; $(1, -2)$

In Exercises 5 through 8, calculate the instantaneous velocity for the indicated value of the time (in seconds) of an object for which the displacement (in feet) is given by the indicated function. Use the method of Example C, and calculate values of $\Delta s/\Delta t$ for the given values of t and determine the limit as Δt approaches zero.

5. $s = 4t + 10$; when $t = 3$; use values of t of 2.0, 2.5, 2.9, 2.99, 2.999
6. $s = 6 - 3t$; when $t = 4$; use values of t of 3.0, 3.5, 3.9, 3.99, 3.999
7. $s = 3t^2 - 4t$; when $t = 2$; use values of t of 1.0, 1.5, 1.9, 1.99, 1.999
8. $s = 120t - 16t^2$; when $t = 0.5$; use values of t of 0.4, 0.45, 0.49, 0.499, 0.4999

In Exercises 9 through 12, use the delta-process to find an expression for the instantaneous velocity of an object moving with rectilinear motion according to the given functions (the same as those of Exercises 5 through 8) relating s (in feet) and t (in seconds). Then calculate the instantaneous velocity for the given value of t.

9. $s = 4t + 10$; $t = 3$ **10.** $s = 6 - 3t$; $t = 4$ **11.** $s = 3t^2 - 4t$; $t = 2$ **12.** $s = 120t - 16t^2$; $t = 0.5$

In Exercises 13 through 16, use the delta-process to find an expression for the instantaneous velocity of an object moving with rectilinear motion according to the given functions relating s and t.

13. $s = 3t - \dfrac{2}{t}$ **14.** $s = \dfrac{2t}{t+2}$

15. $s = 3t^2 - 2t^3$ **16.** $s = s_0 + v_0 t - \dfrac{1}{2} at^2$ (s_0, v_0, and a are constants.)

In Exercises 17 through 20, use the delta-process to find an expression for the instantaneous acceleration a of an object moving with rectilinear motion according to the given functions. The instantaneous acceleration of an object is defined as the instantaneous rate of change of velocity with respect to time.

17. $v = 6t^2 - 4t + 2$ **18.** $v = \sqrt{2t + 1}$

19. $s = t^3 + 2t$ (Find v, then find a.)

20. $s = s_0 + v_0 t - \dfrac{1}{2} at^2$ (s_0, v_0, and a are constants.) (Find v, then find a.)

In Exercises 21 through 32, find the indicated rates of change.

21. The electric current i at a point in an electric circuit is the instantaneous rate of change of the electric charge q which passes the point, with respect to the time t. Find i in a circuit for which $q = 30 - 2t$.

22. The profit P in selling 100 items at price p, when the cost of each item is \$10, is given by $P = 100(p - 10)$. Find the expression for the instantaneous rate of change of P with respect to p.

23. A rectangular metal plate contracts while cooling. Find the expression for the instantaneous rate of change of the area A of the plate with respect to its width w, if the length of the plate is constantly three times as long as the width.

24. A circular oil spill is increasing in size. Find the instantaneous rate of change of the area A of the spill with respect to its radius r for $r = 240$ m.

25. The total power P, in watts, transmitted by an AM radio station is given by $P = 500 + 250m^2$, where m is the modulation index. Find the instantaneous rate of change of P with respect to m for $m = 0.92$.

26. The number N, in thousands, of bacteria present in a certain culture is given by $N = t^3 - 20t^2 + 100t + 20$, where t is the time in hours. Find the instantaneous rate of change of N with respect to t for $t = 6.0$ h.

27. The total solar radiation H, in watts per square meter, on a particular surface during an average clear day is given by $H = \dfrac{5000}{t^2 + 10}$, where t $(-6 \le t \le 6)$ is the number of hours from noon (6 A.M. is equivalent to $t = -6$ h). Find the instantaneous rate of change of H with respect to t at 3 P.M.

28. The value, in thousands of dollars, of a certain car is given by the function $V = \dfrac{48}{t + 3}$, where t is measured in years. Find a general expression for the instantaneous rate of change of V with respect to t, then evaluate this expression when $t = 3$ years.

29. Oil in a certain machine is stored in a conical reservoir, for which the radius and height are both 4 cm. Find the instantaneous rate of change of the volume V of oil in the reservoir with respect to the depth d of the oil. See Fig. 22-19.

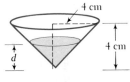

Fig. 22-19

30. The time t required to test a computer memory unit is directly proportional to the square of the number n of memory cells in the unit. For a particular type of unit, $n = 6400$ for $t = 25.0$ s. Find the instantaneous rate of change of t with respect to n for this type of unit for $n = 8000$.

31. A *holograph* (an image formed without using a lens) of concentric circles is formed. The radius r of each circle varies directly as the square root of the wavelength λ of the light used. If $r = 3.72$ cm for $\lambda = 592$ nm, find the expression for the instantaneous rate of change of r with respect to λ.

32. The force F between two electric charges varies inversely as the square of the distance r between them. For two charged particles, $F = 0.12$ N for $r = 0.060$ m. Find the instantaneous rate of change of F with respect to r for $r = 0.120$ m.

22-5   Derivatives of Polynomials

The task of finding the derivative of a function can be considerably shortened from that involved in the direct use of the delta-process. We can use the delta-process to derive certain basic formulas for finding derivatives of particular types of functions. These formulas will then be used to find the derivatives. In this section, we shall derive the formulas for finding the derivatives of polynomial functions of the form $f(x) = a_0 x^n + a_1 x^{n-1} + \cdots + a_n$.

derivative of a
constant

First we shall find the derivative of a constant. By letting $y = c$ and applying the delta-process to this function, we can obtain the desired result:

$$y = c, \qquad y + \Delta y = c, \qquad \Delta y = 0, \qquad \frac{\Delta y}{\Delta x} = 0, \qquad \lim_{\Delta x \to 0} \frac{\Delta y}{\Delta x} = 0$$

From this we conclude that *the derivative of a constant is zero*. No assumption was made as to the value of the constant, and thus this result holds for all constants. Therefore, if $y = c$, $dy/dx = 0$. This is more commonly written as

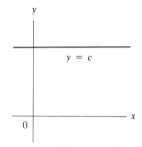

$$\boxed{\frac{dc}{dx} = 0} \tag{22-8}$$

Fig. 22-20

Graphically this means that for any function of the type $y = c$ the slope is always zero. From our knowledge of straight lines, we know that $y = c$ represents a straight line parallel to the x-axis. From the definition of slope, we know that any line parallel to the x-axis has a slope of zero. See Fig. 22-20. We can see that the two results are consistent.

Next we shall find the derivative of any integral power of x. If $y = x^n$, where n is an integer, by use of the binomial theorem we have the following:

$$(y + \Delta y) = (x + \Delta x)^n$$

$$= x^n + nx^{n-1}\,\Delta x + \frac{n(n-1)}{2}\,x^{n-2}(\Delta x)^2 + \cdots + (\Delta x)^n$$

$$\Delta y = nx^{n-1}\,\Delta x + \frac{n(n-1)}{2}\,x^{n-2}(\Delta x)^2 + \cdots + (\Delta x)^n$$

$$\frac{\Delta y}{\Delta x} = nx^{n-1} + \underbrace{\frac{n(n-1)}{2}\,x^{n-2}\,\Delta x + \cdots + (\Delta x)^{n-1}}$$

each term $\to 0$ as $\Delta x \to 0$

$$\lim_{\Delta x \to 0} \frac{\Delta y}{\Delta x} = nx^{n-1}$$

Thus, *the derivative of the nth power of x is found to be*

derivative of x^n

$$\boxed{\frac{dx^n}{dx} = nx^{n-1}} \tag{22-9}$$

EXAMPLE A ———   Find the derivative of the function $y = -5$.

Applying the result of Eq. (22-8), since -5 is a constant, we have

$$\frac{dy}{dx} = \frac{d(-5)}{dx} = 0$$

■

EXAMPLE B _____ Find the derivative of $y = x^3$.

We find that

$$\frac{dy}{dx} = \frac{d(x^3)}{dx} = 3x^{3-1}$$

$$= 3x^2$$

We can see that this result is consistent with the results found in the examples and exercises of the previous sections in this chapter.

EXAMPLE C _____ Find the derivative of the function $y = x$.

In using Eq. (22-9), we have $n = 1$ since $x = x^1$. This means

$$\frac{dy}{dx} = \frac{d(x)}{dx} = (1)x^{1-1} = (1)(x^0)$$

Since $x^0 = 1$, we have

$$\frac{dy}{dx} = 1$$

Thus, the derivative of $y = x$ is 1, which means that the slope of the line $y = x$ is always 1. This is consistent with our previous discussion of the straight line.

EXAMPLE D _____ Find the derivative of $y = x^{10}$.

We find that

$$\frac{dy}{dx} = \frac{d(x^{10})}{dx} = 10x^{10-1}$$

$$= 10x^9$$

Next we shall find the derivative of a constant times a function of x. If $y = cu$, where $u = f(x)$, we have the following result:

$$y + \Delta y = c(u + \Delta u)$$

(As x increases by Δx, u increases by Δu, since u is a function of x.) Then

$$\Delta y = c\,\Delta u, \qquad \frac{\Delta y}{\Delta x} = c\,\frac{\Delta u}{\Delta x}, \qquad \lim_{\Delta x \to 0}\frac{\Delta y}{\Delta x} = c\,\lim_{\Delta x \to 0}\frac{\Delta u}{\Delta x}$$

Therefore, _the derivative of the product of a constant and a differentiable function of x is the product of the constant and the derivative of the function of x._ This is written as

$$\frac{d(cu)}{dx} = c\,\frac{du}{dx}$$

(22-10)

EXAMPLE E

Find the derivative of $y = 3x^2$.

In this case, $c = 3$ and $u = x^2$. Thus, $du/dx = 2x$. Therefore,

$$\frac{dy}{dx} = \frac{d(3x^2)}{dx} = 3\frac{d(x^2)}{dx} = 3(2x)$$

$$= 6x$$

■

Occasionally the derivative of a constant times a function of x is confused with the derivative of a constant (alone). The distinction between a constant multiplying a function and an isolated constant must be made.

Finally, if the types of functions for which we have found derivatives are added, the result is a polynomial function with more than one term. The derivative of such a function is found by letting $y = u + v$, where u and v are functions of x. Applying the delta-process, since u and v are functions of x, each has an increment corresponding to an increment in x. Thus, we have the following result:

$$y + \Delta y = (u + \Delta u) + (v + \Delta v)$$

$$\Delta y = \Delta u + \Delta v$$

$$\frac{\Delta y}{\Delta x} = \frac{\Delta u}{\Delta x} + \frac{\Delta v}{\Delta x}$$

$$\lim_{\Delta x \to 0} \frac{\Delta y}{\Delta x} = \lim_{\Delta x \to 0} \frac{\Delta u}{\Delta x} + \lim_{\Delta x \to 0} \frac{\Delta v}{\Delta x}$$

This tells us that *the derivative of the sum of differentiable functions of x is the sum of the derivatives of the functions.* This is written as

$$\boxed{\frac{d(u + v)}{dx} = \frac{du}{dx} + \frac{dv}{dx}}$$

(22-11)

EXAMPLE F

Find the derivative of $y = 4x^2 + 5$.

Here $u = 4x^2$ and $v = 5$. Thus, $du/dx = 8x$ and $dv/dx = 0$. Hence we have

$$\frac{dy}{dx} = \frac{d(4x^2)}{dx} + \frac{d(5)}{dx}$$

$$= 8x + 0$$

$$= 8x$$

■

EXAMPLE G

Evaluate the derivative of $y = 2x^4 - 6x^2 - 8x - 9$ at the point $(-2, 15)$.

First, finding the derivative, we have

$$\frac{dy}{dx} = \frac{d(2x^4)}{dx} - \frac{d(6x^2)}{dx} - \frac{d(8x)}{dx} - \frac{d(9)}{dx}$$

$$= 8x^3 - 12x - 8$$

Since the derivative is a function only of x, we now evaluate it for $x = -2$.

$$\left.\frac{dy}{dx}\right|_{x=-2} = 8(-2)^3 - 12(-2) - 8$$

$$= -48$$

■

EXAMPLE H

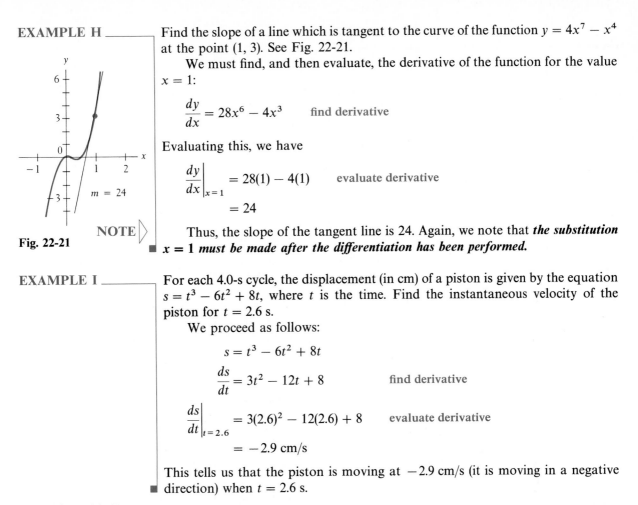

Fig. 22-21

Find the slope of a line which is tangent to the curve of the function $y = 4x^7 - x^4$ at the point (1, 3). See Fig. 22-21.

We must find, and then evaluate, the derivative of the function for the value $x = 1$:

$$\frac{dy}{dx} = 28x^6 - 4x^3 \qquad \text{find derivative}$$

Evaluating this, we have

$$\left.\frac{dy}{dx}\right|_{x=1} = 28(1) - 4(1) \qquad \text{evaluate derivative}$$

$$= 24$$

NOTE ▷   Thus, the slope of the tangent line is 24. Again, we note that **the substitution $x = 1$ must be made after the differentiation has been performed.**

EXAMPLE I

For each 4.0-s cycle, the displacement (in cm) of a piston is given by the equation $s = t^3 - 6t^2 + 8t$, where t is the time. Find the instantaneous velocity of the piston for $t = 2.6$ s.

We proceed as follows:

$$s = t^3 - 6t^2 + 8t$$

$$\frac{ds}{dt} = 3t^2 - 12t + 8 \qquad \text{find derivative}$$

$$\left.\frac{ds}{dt}\right|_{t=2.6} = 3(2.6)^2 - 12(2.6) + 8 \qquad \text{evaluate derivative}$$

$$= -2.9 \text{ cm/s}$$

This tells us that the piston is moving at -2.9 cm/s (it is moving in a negative direction) when $t = 2.6$ s.

Exercises 22-5

In Exercises 1 through 16, find the derivative of each of the given functions.

1. $y = x^5$

2. $y = x^{12}$

3. $y = -4x^9$

4. $y = -7x^6$

5. $y = x^4 - 6$

6. $y = 3x^5 - 1$

7. $y = x^2 + 2x$

8. $y = x^3 - 2x^2$

9. $y = 5x^3 - x - 1$

10. $y = 6x^2 - 6x + 5$

11. $y = x^8 - 4x^7 - x$

12. $y = 4x^4 - 2x + 9$

13. $y = -6x^7 + 5x^3 + 2^3$

14. $y = 13x^4 - 6x^3 - x - 1$

15. $y = \frac{1}{3}x^3 + \frac{1}{2}x^2$

16. $y = -\frac{1}{4}x^8 + \frac{1}{2}x^4 - 3^2$

In Exercises 17 through 20, evaluate the derivatives of the given functions at the given points.

17. $y = 6x^2 - 8x + 1$ (2, 9)

18. $y = x^3 - 5x^2 + 4$ $(-1, -2)$

19. $y = 2x^3 + 9x - 7$ $(-2, -41)$

20. $y = x^4 - 9x^2 - 5x$ $(3, -15)$

In Exercises 21 through 24, find the slope of a tangent line to the curve of each of the given functions for the given values of x.

21. $y = 2x^6 - 6x^2$ $(x = 2)$

22. $y = 3x^3 - 9x$ $(x = 1)$

23. $y = 6x - 2x^4$ $(x = -1)$

24. $y = x^4 - \frac{1}{2}x^2 + 2$ $(x = -2)$

In Exercises 25 through 28, find an expression for the instantaneous velocity of objects moving according to the functions given, if s represents displacement in terms of time t.

25. $s = 6t^5 - 5t + 2$ **26.** $s = 20 + 60t - 4.9t^2$ **27.** $s = 2 - 6t - 2t^3$ **28.** $s = s_0 + v_0 t + \frac{1}{2}at^2$

In Exercises 29 through 32, s represents the displacement and t represents the time for objects moving according to the given functions. Find the value of the instantaneous velocity for the given times.

29. $s = 2t^3 - 4t^2, t = 4$ **30.** $s = 120 + 80t - 16t^2, t = 2.5$

31. $s = 0.5t^4 - 1.5t^2 + 2.5, t = 3$ **32.** $s = 8t^2 - 10t + 6. t = 5$

In Exercises 33 through 44, solve the given problems by finding the appropriate derivative.

33. For what value(s) of x is the tangent to the curve of $y = 3x^2 - 6x$ parallel to the x-axis? (That is, where is the slope zero?)

34. For what value(s) of x is the tangent to the curve of $y = 4x^3 - 12x^2$ parallel to the x-axis? (See Exercise 33.)

35. If the radius of a right circular cylinder always equals its height, find an expression for the instantaneous rate of change of volume with respect to the radius.

36. Find the expression for the instantaneous rate of change of the surface area A of a cube with respect to its edge e.

37. The electric power P, in watts, as a function of the current i, in amperes, in a certain circuit is given by $P = 16i^2 + 60i$. Find the instantaneous rate of change of P with respect to i for $i = 0.75$ A.

38. The torque T on the arm of a robotic control mechanism varies directly as the cube of the diameter d of the arm. If $T = 850$ lb·in. for $d = 0.925$ in., find the expression for the instantaneous rate of change of T with respect to d.

39. The electric polarization P of a light wave for high values of the electric field E is given by $P = a(c_1 E + c_2 E^2 + c_3 E^3)$, where a, c_1, c_2, and c_3 are constants. Find the expression for the instantaneous rate of change of P with respect to E.

40. The ends of a 10-ft beam are supported at different levels. The deflection y of the beam is given by $y = kx^2(x^3 + 450x - 3500)$, where x is the horizontal distance from one end and k is a constant. Determine the expression for the instantaneous rate of change of deflection with respect to x.

41. The force F, in newtons, exerted by a cam on a lever is given by $F = x^4 - 12x^3 + 46x^2 - 60x + 25$, where $x(1 \le x \le 5)$ is the distance (in cm) from the center of rotation of the cam to the edge of the cam in contact with the lever (see Fig. 22-22). Find the instantaneous rate of change of F with respect to x when $x = 4.0$ cm.

42. A company determines that the cost C, in dollars, of producing x machine parts per day is given by $C = x^3 - 300x + 100$. Find the value of the instantaneous rate of change of C with respect to x when $x = 12$ parts.

43. Two ball bearings wear down such that the radius r of one is constantly 1.20 mm less than the radius of the other. Find the instantaneous rate of change of the total volume V_T of the two ball bearings with respect to r for $r = 3.30$ mm.

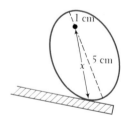

Fig. 22-22

44. An open-top container is to be made from a rectangular piece of cardboard 6.00 in. by 8.00 in. Equal squares of side x are to be cut from each corner, then the sides are to be bent up and taped together. Find the instantaneous rate of change of the volume V of the container with respect to x for $x = 1.75$ in.

22-6   Derivatives of Products and Quotients of Functions ▬▬▬

The formulas developed in the previous section are valid for polynomial functions. However, many functions are not polynomial in form. Some functions can best be expressed as the product of two or more simpler functions, others are the quotient of two simpler functions, and some are expressed as powers of a function. In this section, we shall develop the formula for the derivative of a product of functions and the formula for the derivative of the quotient of two functions.

EXAMPLE A — The functions $f(x) = x^2 + 2$ and $g(x) = 3 - 2x$ can be combined to form new functions which represent the types mentioned above. For example, the function

$$p(x) = f(x)g(x) = (x^2 + 2)(3 - 2x)$$

is an example of a function expressed as the product of two simpler functions. The function

$$q(x) = \frac{g(x)}{f(x)} = \frac{3 - 2x}{x^2 + 2}$$

is an example of a rational function which is the quotient of two other functions. The function

$$F(x) = [g(x)]^3 = (3 - 2x)^3$$

is an example of a power of a function.

If u and v both represent differentiable functions of x, the derivative of the product of u and v is found by applying the delta-process. This leads to the following result:

$$y = uv$$
$$y + \Delta y = (u + \Delta u)(v + \Delta v) = uv + u\,\Delta v + v\,\Delta u + \Delta u\,\Delta v$$

(Since u and v are functions of x, each has an increment corresponding to an increment in x.) Then we have

$$\Delta y = u\,\Delta v + v\,\Delta u + \Delta u\,\Delta v$$
$$\frac{\Delta y}{\Delta x} = u\frac{\Delta v}{\Delta x} + v\frac{\Delta u}{\Delta x} + \Delta u\frac{\Delta v}{\Delta x}$$
$$\lim_{\Delta x \to 0}\frac{\Delta y}{\Delta x} = u\lim_{\Delta x \to 0}\frac{\Delta v}{\Delta x} + v\lim_{\Delta x \to 0}\frac{\Delta u}{\Delta x} + \lim_{\Delta x \to 0}\left(\Delta u\frac{\Delta v}{\Delta x}\right)$$

(The functions u and v are not affected by Δx approaching zero, but Δu and Δv both approach zero as Δx approaches 0.) Thus,

$$\frac{dy}{dx} = u\frac{dv}{dx} + v\frac{du}{dx} + 0\frac{dv}{dx}$$

We conclude that *the derivative of the product of two differentiable functions equals the first function times the derivative of the second function plus the second function times the derivative of the first function.* This is written as

derivative of a product

$$\frac{d(uv)}{dx} = u\frac{dv}{dx} + v\frac{du}{dx} \qquad (22\text{-}12)$$

EXAMPLE B

For the product function in Example A, the derivative is found as follows:

$$y = (x^2 + 2)(3 - 2x), \qquad u = x^2 + 2, \qquad v = 3 - 2x$$

$$\frac{d(uv)}{dx} = u \qquad \frac{dv}{dx} \qquad + \qquad v \qquad \frac{du}{dx}$$

$$\frac{dy}{dx} = (x^2 + 2)(-2) + (3 - 2x)(2x) = -2x^2 - 4 + 6x - 4x^2$$

$$= -6x^2 + 6x - 4$$

EXAMPLE C

Find the derivative of the function $y = (3 - x - 2x^2)(x^4 - x)$.
In this problem, $u = 3 - x - 2x^2$ and $v = x^4 - x$. Hence

$$\frac{dy}{dx} = (3 - x - 2x^2)(4x^3 - 1) + (x^4 - x)(-1 - 4x)$$

$$= 12x^3 - 3 - 4x^4 + x - 8x^5 + 2x^2 - x^4 - 4x^5 + x + 4x^2$$

$$= -12x^5 - 5x^4 + 12x^3 + 6x^2 + 2x - 3$$

In both of these examples, we could have multiplied the functions first and then taken the derivative as a polynomial. However, we shall soon meet functions for which this latter method would not be applicable.

We shall now find the derivative of the quotient of two differentiable functions. We apply the delta-process to the function $y = u/v$ and obtain the result as follows:

$$y + \Delta y = \frac{u + \Delta u}{v + \Delta v}$$

$$\Delta y = \frac{u + \Delta u}{v + \Delta v} - \frac{u}{v} = \frac{vu + v\,\Delta u - uv - u\,\Delta v}{v(v + \Delta v)}$$

$$\frac{\Delta y}{\Delta x} = \frac{v(\Delta u/\Delta x) - u(\Delta v/\Delta x)}{v(v + \Delta v)}$$

$$\lim_{\Delta x \to 0} \frac{\Delta y}{\Delta x} = \frac{v \lim_{\Delta x \to 0} (\Delta u/\Delta x) - u \lim_{\Delta x \to 0} (\Delta v/\Delta x)}{\lim_{\Delta x \to 0} v(v + \Delta v)}$$

$$\frac{dy}{dx} = \frac{v(du/dx) - u(dv/dx)}{v^2}$$

This last result says that *the derivative of the quotient of two differentiable functions equals the denominator times the derivative of the numerator minus the numerator times the derivative of the denominator, all divided by the square of the denominator.* This is written as

derivative of a quotient

$$\boxed{\frac{d\dfrac{u}{v}}{dx} = \frac{v\dfrac{du}{dx} - u\dfrac{dv}{dx}}{v^2}}$$

(22-13)

EXAMPLE D ——— Find the derivative of the quotient indicated in Example A.

$$y = \frac{3 - 2x}{x^2 + 2}, \qquad u = 3 - 2x, \qquad v = x^2 + 2$$

$$\frac{dy}{dx} = \frac{\overset{v}{\downarrow}\overset{\frac{du}{dx}}{\downarrow} - \overset{u}{\downarrow}\overset{\frac{dv}{dx}}{\downarrow}}{}$$

$$\frac{dy}{dx} = \frac{(x^2 + 2)(-2) - (3 - 2x)(2x)}{(x^2 + 2)^2} = \frac{-2x^2 - 4 - 6x + 4x^2}{(x^2 + 2)^2}$$

$$= \frac{2(x^2 - 3x - 2)}{(x^2 + 2)^2} \quad \overset{\nwarrow}{} v^2$$

■

EXAMPLE E ——— Evaluate the derivative of $y = \dfrac{3x^2 + x}{1 - 4x}$ at $(2, -2)$.

We first find dy/dx and then find its value for $x = 2$.

$$\frac{dy}{dx} = \frac{(1 - 4x)(6x + 1) - (3x^2 + x)(-4)}{(1 - 4x)^2}$$

$$= \frac{6x + 1 - 24x^2 - 4x + 12x^2 + 4x}{(1 - 4x)^2}$$

$$= \frac{-12x^2 + 6x + 1}{(1 - 4x)^2}$$

$$\left.\frac{dy}{dx}\right|_{x=2} = \frac{-12(2^2) + 6(2) + 1}{[1 - 4(2)]^2} = \frac{-48 + 12 + 1}{49} = \frac{-35}{49}$$

$$= -\frac{5}{7}$$

■

EXAMPLE F ——— The stress S on a hollow tube is given by

$$S = \frac{16DT}{\pi(D^4 - d^4)}$$

where T is the tension, D is the outer diameter, and d is the inner diameter of the tube. Find the expression for the instantaneous rate of change of S with respect to D, with the other values being constant.

We are to find the derivative of S with respect to D, and it is found as follows:

$$\frac{dS}{dD} = \frac{\pi(D^4 - d^4)(16T) - 16DT(\pi)(4D^3)}{\pi^2(D^4 - d^4)^2} = \frac{16\pi T(D^4 - d^4 - 4D^4)}{\pi^2(D^4 - d^4)^2}$$

$$= \frac{-16T(3D^4 + d^4)}{\pi(D^4 - d^4)^2}$$

■

Exercises 22-6

In Exercises 1 through 8, find the derivative of each function by use of Eq. (22-12). Do not find the product before finding dy/dx.

1. $y = x^2(3x + 2)$ **2.** $y = 3x(x^3 + 1)$ **3.** $y = 6x(3x^2 - 5x)$

4. $y = 2x^3(3x^4 + x)$ **5.** $y = (x + 2)(2x - 5)$ **6.** $y = (3x + 1)(x^2 + 1)$

7. $y = (x^4 - 3x^2 + 3)(1 - 2x^3)$ **8.** $y = (x^3 - 6x)(2 - 4x^3)$

In Exercises 9 through 12, find the derivative of each function by use of Eq. (22-12). Then multiply out each function and find the derivative by treating it as a polynomial. Compare results.

9. $y = (2x - 7)(5 - 2x)$ **10.** $y = (x^2 + 2)(2x^2 - 1)$

11. $y = (x^3 - 1)(2x^2 - x - 1)$ **12.** $y = (3x^2 - 4x + 1)(5 - 6x^2)$

In Exercises 13 through 24, find the derivative of each function by use of Eq. (22-13).

13. $y = \dfrac{x}{2x + 3}$ **14.** $y = \dfrac{2x}{x + 1}$ **15.** $y = \dfrac{1}{x^2 + 1}$ **16.** $y = \dfrac{x + 2}{2x + 3}$

17. $y = \dfrac{x^2}{3 - 2x}$ **18.** $y = \dfrac{2}{3x^2 - 5x}$ **19.** $y = \dfrac{2x - 1}{3x^2 + 2}$ **20.** $y = \dfrac{2x^3}{4 - x}$

21. $y = \dfrac{x + 8}{x^2 + x + 2}$ **22.** $y = \dfrac{3x}{4x^5 - 3x - 4}$ **23.** $y = \dfrac{2x^2 - x - 1}{x^3 + 2x^2}$ **24.** $y = \dfrac{3x^3 - x}{2x^2 - 5x + 4}$

In Exercises 25 through 28, evaluate the derivatives of the given functions for the given values of x. In Exercises 25 and 26, find the derivative by use of Eq. (22-12).

25. $y = (3x - 1)(4 - 7x)$, $x = 3$ **26.** $y = (3x^2 - 5)(2x^2 - 1)$, $x = -1$

27. $y = \dfrac{3x - 5}{2x + 3}$, $x = -2$ **28.** $y = \dfrac{2x^2 - 5x}{3x + 2}$, $x = 2$

In Exercises 29 through 44, solve the given problems by finding the appropriate derivatives.

29. Find the derivative of $y = \dfrac{x^2(1 - 2x)}{3x - 7}$, but do not multiply out the numerator before taking the derivative.

30. Find the derivative of $y = 4x^2 - \dfrac{1}{x - 1}$, but do not combine the terms before finding the derivative.

31. Find the slope of a tangent line to the curve $y = (4x + 1)(x^4 - 1)$ at the point $(-1, 0)$. Do not multiply the factors together before taking the derivative.

32. Find the slope of a line tangent to the curve of the function $y = (3x + 4)(1 - 4x)$ at the point $(2, -70)$. Do not multiply before finding the derivative.

33. For what value(s) of x is the slope of a tangent to the curve $y = \dfrac{x}{x^2 + 1}$ equal to zero?

34. Determine the sign of the derivative of the function $y = \dfrac{2x - 1}{1 - x^2}$ for the following values of x: $-2, -1, 0, 1, 2$.

Is the slope of a tangent to this curve ever negative?

35. During each cycle, the displacement s of the end of a robot arm is given by $s = (t^2 - 8t)(2t^2 + t + 1)$, where t is the time. Find the expression for the instantaneous velocity of the end of the robot arm.

36. The sales S of a product as a function of the time t, in weeks, is given by $S = 5000 - \dfrac{2000}{t}$ $(t \geq 1)$. Find the rate of change of S with respect to t for $t = 10$ weeks. Use the quotient rule.

37. The voltage V at the junction of a 3-Ω resistance and a variable resistance R in a circuit is given by
$V = \dfrac{6R + 25}{R + 3}$. Find the rate of change of V with respect to R for $R = 7\,\Omega$.

38. During a chemical change the number n of grams of a compound being formed is given by $n = \dfrac{6t^2}{2t^2 + 3}$, where
t is measured in seconds. How many grams per second are being formed after 3.0 s?

39. A computer, using data from a refrigeration plant, estimated that in the event of a power failure, the
temperature T (in °C) in the freezers would be given by $T = \dfrac{2t}{0.05t + 1} - 20$, where t is the number of hours
after the power failure. Find the time rate of change of temperature after 6.0 h.

40. The voltage across a resistor in an electric circuit is the product of the resistance and the current. If the current
I (in amperes) varies with time (in seconds) according to the relation $I = 5.00 + 0.01t^2$, and the resistance varies
with time according to the relation $R = 15.00 - 0.10t$, find the time rate of change of the voltage when
$t = 5.00$ s.

41. The frictional radius r_f of a disc clutch is given by $r_f = \dfrac{2(R^2 + Rr + r^2)}{3(R + r)}$, where R and r are the outer radius
and inner radius of the clutch, respectively. Find the derivative of r_f with respect to R with r constant.

42. In thermodynamics, an equation relating the thermodynamic temperature T, pressure p, and volume V of a
gas is $T = \left(p + \dfrac{a}{V^2}\right)\left(\dfrac{V - b}{R}\right)$, where a, b, and R are constants. Find the derivative of T with respect to V,
assuming p is constant.

43. The electric power produced by a certain source is given by $P = \dfrac{E^2 r}{R^2 + 2Rr + r^2}$, where E is the voltage of the
source, R is the resistance of the source, and r is the resistance in the circuit. Find the derivative of P with
respect to r, assuming that the other quantities remain constant.

44. In the theory of lasers, the power P radiated is given by $P = \dfrac{kf^2}{\omega^2 - 2\omega f + f^2 + a^2}$, where f is the field
frequency and a, k, and ω are constants. Find the derivative of P with respect to f.

22-7 The Derivative of a Power of a Function

In Example A of Section 22-6 we illustrated $y = (3 - 2x)^3$ as the power of a
function of x. Here, $3 - 2x$ is the function of x, and it is being raised to the third
power. If we let $u = 3 - 2x$, we can write

$$y = u^3 \quad \text{where } u = 3 - 2x$$

Writing it this way, y is a function of u and u is a function of x. This means that
y is a function of a function of x, referred to as a **composite function.** However, y
is still a function of x, since u is a function of x.

Since we will frequently need to find the derivative of the power of a function, we now develop the necessary formula. Considering $y = f(u)$ and $u = g(x)$,
we can express the derivative dy/dx in terms of dy/du and du/dx. If Δx is the
increment in x, then Δy and Δu are the corresponding increments in y and u,
respectively. We may then write

$$\frac{\Delta y}{\Delta x} = \frac{\Delta y}{\Delta u}\frac{\Delta u}{\Delta x}$$

When Δx approaches zero, Δu and Δy both approach zero, for u and y are functions of x. Thus,

$$\lim_{\Delta x \to 0} \frac{\Delta y}{\Delta x} = \left(\lim_{\Delta u \to 0} \frac{\Delta y}{\Delta u} \right) \left(\lim_{\Delta x \to 0} \frac{\Delta u}{\Delta x} \right)$$

or

chain rule

$$\frac{dy}{dx} = \frac{dy}{du} \frac{du}{dx} \tag{22-14}$$

(Here we have assumed $\Delta u \neq 0$, although it can be shown that this condition is not necessary.) *Equation (22-14) is known as the* **chain rule** *for derivatives.*

Using Eq. (22-14) for $y = u^n$, where u is a differentiable function of x, we have

$$\frac{dy}{dx} = \frac{d(u^n)}{du} \frac{du}{dx}$$

or

derivative of u^n

$$\frac{du^n}{dx} = nu^{n-1} \left(\frac{du}{dx} \right) \tag{22-15}$$

By use of Eq. (22-15) we may find the derivative of a power of a differentiable function of x.

EXAMPLE A ____ Find the derivative of $y = (3 - 2x)^3$.

For this function, we identify $n = 3$ and $u = 3 - 2x$. This means that $du/dx = -2$, and therefore

$$\frac{du^n}{dx} = n \quad u \quad {}^{n-1} \quad \left(\frac{du}{dx} \right)$$

$$\frac{dy}{dx} = 3(3 - 2x)^2(-2)$$

$$= -6(3 - 2x)^2$$

NOTE ▷   A common type of error in finding this type of derivative is to omit the du/dx factor; in this case it is the -2. **The derivative is incomplete, and therefore** ∎ **incorrect, without this factor.**

EXAMPLE B ____ Find the derivative of $y = (1 - 3x^2)^4$.

In this example, $n = 4$ and $u = 1 - 3x^2$. Hence

$$\frac{dy}{dx} = 4(1 - 3x^2)^3(-6x) = -24x(1 - 3x^2)^3$$

NOTE ▷   ∎ *(We must not forget the $-6x$.)*

EXAMPLE C_____ Find the derivative of

$$y = \frac{(3x - 1)^3}{1 - x}$$

NOTE ▷ Here we must *use the quotient rule in combination with the power rule.* We find the derivative of the numerator by using the power rule.

$$\frac{dy}{dx} = \frac{(1 - x)3(3x - 1)^2(3) - (3x - 1)^3(-1)}{(1 - x)^2} \overset{\text{derivative of numerator}}{}$$

$$= \frac{(3x - 1)^2(9 - 9x + 3x - 1)}{(1 - x)^2} = \frac{(3x - 1)^2(8 - 6x)}{(1 - x)^2}$$

$$= \frac{2(3x - 1)^2(4 - 3x)}{(1 - x)^2}$$ ∎

It is normally better to have the derivative in a factored, simplified form, since this is the only form from which useful information may readily be obtained. In this form we can determine where the derivative is undefined (denominator equal to zero) or where the slope is zero (numerator equal to zero); we can also make other required analyses of the derivative. Thus, *all derivatives should be in simplest algebraic form.*

So far, we have derived the formulas for the derivatives of powers of x and for differentiable functions of x for integral powers. We shall now establish that these formulas are also valid for any rational number used as an exponent. If $y = u^{p/q}$, and if each side is raised to the qth power, we have $y^q = u^p$. Applying the power rule to each side of this equation, we have

$$qy^{q-1}\left(\frac{dy}{dx}\right) = pu^{p-1}\left(\frac{du}{dx}\right)$$

Solving for dy/dx, we have

$$\frac{dy}{dx} = \frac{pu^{p-1}(du/dx)}{qy^{q-1}} = \frac{p}{q}\frac{u^{p-1}}{(u^{p/q})^{q-1}}\frac{du}{dx} = \frac{p}{q}\frac{u^{p-1}}{u^{p-p/q}}\frac{du}{dx}$$

$$= \frac{p}{q}u^{p-1-p+(p/q)}\frac{du}{dx}$$

Thus,

$$\boxed{\frac{du^{p/q}}{dx} = \frac{p}{q}u^{(p/q)-1}\frac{du}{dx}}$$ (22-16)

We see that in finding the derivative we multiply the function by the rational exponent and subtract 1 from it to find the exponent of the function in the derivative. This, of course, is the same rule derived for integral exponents in Eq. (22-15).

For reference,
Eq. (22-9) is
$$\frac{dx^n}{dx} = nx^{n-1}$$

In deriving Eqs. (22-15) and (22-16) we made use of Eq. (22-9). In deriving Eq. (22-9) we used the binomial theorem, in which the exponent n can be any positive or negative integer. Together, these tell us that

the power rule for derivatives, Eq. (22-15), can be extended to include all rational exponents, positive or negative.

This, of course, includes all integral exponents, positive and negative. Also, we note that Eq. (22-9) is equivalent to Eq. (22-15) with $u = x$ (since $du/dx = 1$).

EXAMPLE D

We can now find the derivative of $y = \sqrt{x^2 + 1}$.

By use of Eq. (22-16) or Eq. (22-15), and writing the square root as the fractional exponent $\frac{1}{2}$, we can easily derive the result:

$$y = (x^2 + 1)^{1/2}$$

$$\frac{dy}{dx} = \frac{1}{2}(x^2 + 1)^{-1/2}(2x) = \frac{x}{(x^2 + 1)^{1/2}}$$

To avoid introducing apparently significant factors into the numerator, we do not usually rationalize such fractions. ■

Having shown that we may use fractional exponents to find derivatives of roots of functions of x, we may also use them to find derivatives of roots of x itself. Consider the following example.

EXAMPLE E

Find the derivative of $y = 6\sqrt[3]{x}$.

We can write this function as $y = 6x^{1/3}$. In finding the derivative we may use Eq. (22-9) with $n = \frac{1}{3}$. This gives us

$$y = 6x^{1/3}$$

$$\frac{dy}{dx} = 6\left(\frac{1}{3}\right)x^{-2/3} = \frac{2}{x^{2/3}} \qquad \overset{\frac{1}{3} - 1}{}$$

We could also use Eq. (22-15) with $u = x$ and $n = \frac{1}{3}$. This gives us

$$\frac{dy}{dx} = 6\left(\frac{1}{3}\right)x^{-2/3}(1) = \frac{2}{x^{2/3}} \qquad \overset{\frac{du}{dx} = \frac{dx}{dx} = 1}{}$$

This shows us why Eq. (22-9) is equivalent to Eq. (22-15) with $u = x$.

Note that the domain of the function is all real numbers, but it is not differentiable for $x = 0$. ■

We now show the use of Eqs. (22-9) and (22-15) with n being a negative exponent.

EXAMPLE F _____ The electric resistance R of a wire varies inversely as the square of its radius r. For a given wire, $R = 4.66\ \Omega$ for $r = 0.105$ mm. Find the derivative of R with respect to r for this wire.

Since R varies inversely as the square of r, we have $R = k/r^2$. Then, using the fact that $R = 4.66\ \Omega$ for $r = 0.150$ mm, we have

$$4.66 = \frac{k}{(0.150)^2}, \qquad k = 0.105\ \Omega\cdot mm^2$$

which means that

$$R = \frac{0.105}{r^2}$$

We could find the derivative by the quotient rule. However, by using negative exponents the derivative is easily found.

$$R = \frac{0.105}{r^2} = 0.105r^{-2}$$

NOTE ▷ $$\frac{dR}{dr} = 0.105(-2)r^{-3} \longleftarrow -2 - 1 = -3$$

$$= -\frac{0.210}{r^3}$$

■ Here we used Eq. (22-9) directly.

EXAMPLE G _____ Find the derivative of $y = \dfrac{1}{(1 - 4x)^5}$.

The derivative is found as follows:

$$y = \frac{1}{(1 - 4x)^5} = (1 - 4x)^{-5} \qquad \text{use negative exponent}$$

$$\frac{dy}{dx} = (-5)(1 - 4x)^{-6}(-4) \qquad \text{use Eq. (22-15)}$$

$$= \frac{20}{(1 - 4x)^6} \qquad \text{express result with positive exponent}$$

NOTE ▷ ■ Remember: *subtracting* **1** *from* **−5** *gives* **−6.**

We can now see the value of fractional exponents in calculus. They are useful in algebraic operations, but they are almost essential in calculus. Without fractional exponents, it would be necessary to develop more formulas to find derivatives of radicals. To find the derivative of an algebraic function, we need only those equations which we have already developed. Often it is necessary to combine these, as we saw in Example C. Actually, most derivatives are combina-

NOTE ▷ tions. The problem in finding the derivative is *recognizing the form of the function* with which you are dealing. When you have recognized the form, completing the problem is only a matter of mechanics and algebra. You should now see the importance of being able to handle algebraic operations with facility.

EXAMPLE H _____ Evaluate the derivative of

$$y = \frac{x}{\sqrt{1 - 4x}}$$

for $x = -2$.

Here we have a quotient, and in order to find the derivative of this quotient, we must also use the power rule (and a derivative of a polynomial form). With sufficient practice in taking derivatives, we can recognize the rule to use almost automatically. Thus, we find the derivative:

$$\frac{dy}{dx} = \frac{(1 - 4x)^{1/2}(1) - x(\frac{1}{2})(1 - 4x)^{-1/2}(-4)}{1 - 4x}$$

$$= \frac{(1 - 4x)^{1/2} + \dfrac{2x}{(1 - 4x)^{1/2}}}{1 - 4x} = \frac{\dfrac{(1 - 4x)^{1/2}(1 - 4x)^{1/2} + 2x}{(1 - 4x)^{1/2}}}{1 - 4x}$$

$$= \frac{(1 - 4x) + 2x}{(1 - 4x)^{1/2}(1 - 4x)} = \frac{1 - 2x}{(1 - 4x)^{3/2}}$$

Now, evaluating the derivative for $x = -2$, we have

$$\frac{dy}{dx}\bigg|_{x=-2} = \frac{1 - 2(-2)}{[1 - 4(-2)]^{3/2}} = \frac{1 + 4}{(1 + 8)^{3/2}} = \frac{5}{9^{3/2}}$$

$$= \frac{5}{27}$$

■

Exercises 22-7

In Exercises 1 through 24, find the derivative of each of the given functions.

1. $y = \sqrt{x}$

2. $y = \sqrt[4]{x}$

3. $y = \dfrac{1}{x^2}$

4. $y = \dfrac{2}{x^4}$

5. $y = \dfrac{3}{\sqrt[3]{x}}$

6. $y = \dfrac{1}{\sqrt{x}}$

7. $y = x\sqrt{x} - \dfrac{1}{x}$

8. $y = 2x^{-3} - x^{-2}$

9. $y = (x^2 + 1)^5$

10. $y = (1 - 2x)^4$

11. $y = 2(7 - 4x^3)^8$

12. $y = 3(8x^2 - 1)^6$

13. $y = (2x^3 - 3)^{1/3}$

14. $y = (1 - 6x)^{3/2}$

15. $y = \dfrac{1}{(1 - x^2)^4}$

16. $y = \dfrac{4}{\sqrt{1 - 3x}}$

17. $y = 4(2x^4 - 5)^{3/4}$

18. $y = 5(3x^7 - 4)^{2/3}$

19. $y = \sqrt[4]{1 - 8x^2}$

20. $y = \sqrt[3]{4x^6 + 2}$

21. $y = x\sqrt{8x + 5}$

22. $y = x^2(1 - 3x)^5$

23. $y = \dfrac{\sqrt{4x + 3}}{8x + 1}$

24. $y = \dfrac{x\sqrt{x - 1}}{1 - 2x}$

In Exercises 25 through 28, evaluate the derivatives of the given functions at the given values of x.

25. $y = \sqrt{3x + 4}$, $x = 7$

26. $y = (4 - x^2)^{-1}$, $x = -1$

27. $y = \dfrac{\sqrt{x}}{1 - x}$, $x = 4$

28. $y = x^2\sqrt[3]{3x + 2}$, $x = 2$

In Exercises 29 through 44, solve the given problems by finding the appropriate derivatives.

29. Find the derivative of $y = 1/x^3$ as (a) a quotient and (b) a negative power of x, and show that the results are the same.

30. Find the derivative of $y = \dfrac{2}{4x + 3}$ as (a) a quotient and (b) as a negative power of $4x + 3$, and show that the results are the same.

31. Find any values of x for which the derivative of $y = \dfrac{x^2}{\sqrt{x^2 + 1}}$ is zero.

32. Find any values of x for which the derivative of $y = \dfrac{x}{\sqrt{4x - 1}}$ is zero.

33. Find the slope of a line tangent to the parabola $y^2 = 4x$ at the point $(1, 2)$.

34. Find the slope of a tangent to the circle $x^2 + y^2 = 25$ at the point $(4, 3)$.

35. The displacement s (in cm) of a linkage joint of a robot is given by $s = (8t - t^2)^{2/3}$, where t is in seconds. Find the velocity of the joint for $t = 6.25$ s.

36. An analysis of a company's records shows that the profit p, in dollars, of producing x units of a product is $p = 3000\sqrt{x^3 - 12x^2 + 36x} - 2000$. Find dp/dx for $x = 4$ units.

37. When the volume of a gas changes very rapidly, an approximate relation is that the pressure varies inversely as the $\frac{3}{2}$ power of the volume. If P is 300 kPa when $V = 100$ cm³, find the derivative of P with respect to V. Evaluate this derivative for $V = 100$ cm³.

38. The power gain G of a certain antenna is inversely proportional to the square of the wavelength λ, in feet, of the carrier wave. If $G = 5.0 \times 10^4$ for $\lambda = 0.35$ ft, find the derivative of G with respect to λ for $\lambda = 0.35$ ft.

39. The total solar radiation H (in W/m²) on a certain surface during an average clear day is given by $H = \dfrac{4000}{\sqrt{t^6 + 100}}$, where t $(-6 < t < 6)$ is the number of hours from noon. Find the rate at which H is changing with time at 4 P.M.

40. In determining the time for a laser beam to go from S to P (see Fig. 22-23), which are in different mediums, it is necessary to find the derivative of the time

$$t = \frac{\sqrt{a^2 + x^2}}{v_1} + \frac{\sqrt{b^2 + (c - x)^2}}{v_2}$$

with respect to x, where a, b, c, v_1, and v_2 are constants. Here v_1 and v_2 are the velocities of the laser in each medium. Find this derivative.

41. The radio waveguide wavelength λ_r is related to its free-space wavelength λ by

$$\lambda_r = \frac{2a\lambda}{\sqrt{4a^2 - \lambda^2}}$$

where a is a constant. Find $d\lambda_r/d\lambda$.

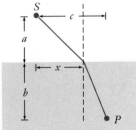

Fig. 22-23

42. The current in a circuit containing a resistance R and an inductance L is found from the expression

$$I = \frac{V}{\sqrt{R^2 + (\omega L)^2}}$$

Find the expression for the rate of change of current with respect to L, assuming that the other quantities remain constant.

43. The length ℓ of a rectangular microprocessor chip is 2 mm longer than its width w. Find the derivative of the diagonal D with respect to w.

44. The trapezoidal structure shown in Fig. 22-24 has an internal support of length ℓ. Find the derivative of ℓ with respect to x.

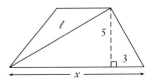

Fig. 22-24

22-8 **Differentiation of Implicit Functions** ▬▬▬▬▬

To this point the functions which we have differentiated have been of the form $y = f(x)$. There are, however, occasions when we need to find the derivative of a function determined by an equation which does not express the dependent variable explicitly in terms of the independent variable.

An equation in which y is not expressed explicitly in terms of x may determine one or more functions. *Any such function, where y is defined implicitly as a function of x, is called an* **implicit function.** Some equations defining implicit functions may be solved to determine the explicit functions, and for others it is not possible to solve for the explicit functions. Also, not all such equations define y as a function of x for real values of x.

EXAMPLE A ───── The equation $3x + 4y = 5$ is an equation which defines a function, although it is not in explicit form. In solving for y as $y = -\frac{3}{4}x + \frac{5}{4}$, we have the explicit form of the function.

The equation $y^2 + x = 3$ is an equation which defines two functions, although we do not have the explicit forms. When we solve for y, we obtain the explicit functions $y = \sqrt{3 - x}$ and $y = -\sqrt{3 - x}$.

The equation $y^5 + xy^2 + 3x^2 = 5$ defines y as a function of x, although we cannot actually solve for the algebraic form of the explicit form of the function.

The equation $x^2 + y^2 + 4 = 0$ is not satisfied by any pair of real values of
■ x and y.

Even when it is possible to determine the explicit form of a function given in implicit form, it is not always desirable to do so. In some cases the implicit form is more convenient than the explicit form.

The derivative of an implicit function may be found directly without having to solve for the explicit function. Thus, *to find dy/dx when y is defined as an implicit function of x, we differentiate each term of the equation with respect to x, regarding y as a differentiable function of x. We then solve for dy/dx, which will usually be in terms of x and y.*

EXAMPLE B ───── Find dy/dx if $y^2 + 2x^2 = 5$.

Here we find the derivative of each term, and then solve for dy/dx. Thus,

For reference, Eq. (22-15) is

$$\frac{du^n}{dx} = nu^{n-1}\frac{du}{dx}$$

$$\frac{d(y^2)}{dx} + \frac{d(2x^2)}{dx} = \frac{d(5)}{dx}$$

$$2y^{2-1}\frac{dy}{dx} + 2\left(2x^{2-1}\frac{dx}{dx}\right) = 0 \qquad \text{see note on next page}$$

$$2y\frac{dy}{dx} + 4x = 0$$

$$\frac{dy}{dx} = -\frac{2x}{y}$$

NOTE▷ *The factor dy/dx arises from the derivative of the first term as a result of using the derivative of a power of a function of x (Eq. 22-15). The factor dy/dx corresponds to the du/dx of the formula.* In the second term, no factor of dy/dx appears, since there are no y factors in the term.

EXAMPLE C ————— Find dy/dx if $3y^4 + xy^2 + 2x^3 - 6 = 0$.

In finding the derivative, we note that the second term is a product, and we must use the product rule for derivatives on it. Thus, we have

$$\frac{d(3y^4)}{dx} + \frac{d(xy^2)}{dx} + \frac{d(2x^3)}{dx} - \frac{d(6)}{dx} = \frac{d(0)}{dx}$$

using product rule

$$12y^3 \frac{dy}{dx} + \left[x\left(2y\frac{dy}{dx}\right) + y^2(1) \right] + 6x^2 - 0 = 0$$

$$12y^3 \frac{dy}{dx} + 2xy\frac{dy}{dx} + y^2 + 6x^2 = 0 \qquad \text{solve for } \frac{dy}{dx}$$

$$(12y^3 + 2xy)\frac{dy}{dx} = -y^2 - 6x^2$$

$$\frac{dy}{dx} = \frac{-y^2 - 6x^2}{12y^3 + 2xy}$$

EXAMPLE D ————— Find dy/dx if $2x^3y + (y^2 + x)^3 = x^4$.

In this case we use the product rule on the first term and the power rule on the second term.

$$\frac{d(2x^3y)}{dx} + \frac{d(y^2 + x)^3}{dx} = \frac{d(x^4)}{dx}$$

product —————— power

$$2x^3\left(\frac{dy}{dx}\right) + y(6x^2) + 3(y^2 + x)^2\left(2y\frac{dy}{dx} + 1\right) = 4x^3$$

$$2x^3\frac{dy}{dx} + 6x^2y + 3(y^2 + x)^2\left(2y\frac{dy}{dx}\right) + 3(y^2 + x)^2 = 4x^3$$

$$[2x^3 + 6y(y^2 + x)^2]\frac{dy}{dx} = 4x^3 - 6x^2y - 3(y^2 + x)^2$$

$$\frac{dy}{dx} = \frac{4x^3 - 6x^2y - 3(y^2 + x)^2}{2x^3 + 6y(y^2 + x)^2}$$

The following example illustrates the evaluation of the derivative of an implicit function.

EXAMPLE E _____   Find the slope of a line tangent to the curve of $2y^3 + xy + 1 = 0$ at the point $(-3, 1)$.

Here we must find dy/dx and evaluate it for $x = -3$ and $y = 1$.

$$\frac{d(2y^3)}{dx} + \frac{d(xy)}{dx} + \frac{d(1)}{dx} = \frac{d(0)}{dx}$$

$$6y^2\frac{dy}{dx} + x\frac{dy}{dx} + y + 0 = 0$$

$$\frac{dy}{dx} = \frac{-y}{6y^2 + x}$$

$$\frac{dy}{dx}\bigg|_{(-3,1)} = \frac{-1}{6(1^2) - 3} = \frac{-1}{6 - 3}$$

$$= -\frac{1}{3}$$

■ Thus, the slope is $-\frac{1}{3}$.

Exercises 22-8

In Exercises 1 through 20, find dy/dx by differentiating implicitly. When applicable, express the result in terms of x and y.

1. $3x + 2y = 5$ **2.** $6x - 3y = 4$ **3.** $4y - 3x^2 = x$ **4.** $x^5 - 5y = 6 - x$

5. $x^2 - y^2 - 9 = 0$ **6.** $x^2 + 2y^2 - 11 = 0$ **7.** $y^5 = x^2 - 1$ **8.** $y^4 = 3x^3 - x$

9. $y^2 + y = x^2 - 4$ **10.** $2y^3 - y = 7 - x^4$ **11.** $y + 3xy - 4 = 0$ **12.** $8y - xy - 7 = 0$

13. $xy^3 + 3y + x^2 = 9$ **14.** $y^2x - \dfrac{y}{x+1} + 3x = 4$ **15.** $\dfrac{3x^2}{y^2+1} + y = 3x + 1$

16. $2x - x^3y^2 = y - x^2 - 1$ **17.** $(2y - x)^4 + x^2 = y + 3$ **18.** $(y^2 + 2)^3 = x^4y + 11$

19. $2(x^2 + 1)^3 + (y^2 + 1)^2 = 17$ **20.** $(2x + 1)(1 - 3y) + y^2 = 13$

In Exercises 21 through 24, evaluate the derivatives of the given functions at the given points.

21. $3x^3y^2 - 2y^3 = -4$, $(1, 2)$ **22.** $2y + 5 - x^2 - y^3 = 0$, $(2, -1)$

23. $5y^4 + 7 = x^4 - 3y$, $(3, -2)$ **24.** $(xy - y^2)^3 = 5y^2 + 22$, $(4, 1)$

In Exercises 25 through 32, solve the given problems by use of implicit differentiation.

25. Find the slope of a line tangent to the curve of $xy + y^2 + 2 = 0$ at the point $(-3, 1)$.

26. Oil moves through a pipeline such that the distance s it moves and the time t are related by $s^3 - t^2 = 7t$. Find the velocity of the oil for $s = 4.01$ m and $t = 5.25$ s.

27. The shelf support shown in Fig. 22-25 is 2.38 ft long. Find the expression for dy/dx in terms of x and y.

28. An open (no top) right circular cylindrical container of radius r and height h has a total surface area of 940 cm^2. Find dr/dh in terms of r and h.

29. Two resistors, with resistances r and $r + 2$, are connected in parallel. Their combined resistance R is related to r by the equation $r^2 = 2rR + 2R - 2r$. Find dR/dr.

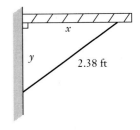

Fig. 22-25

30. The polar moment of inertia I of a rectangular slab of concrete is given by $I = \frac{1}{12}(b^3h + bh^3)$, where b and h are the base and height, respectively, of the slab. If I is constant, find the expression for db/dh.

31. A formula relating the length L and radius of gyration r of a steel column is $24C^3Sr^3 = 40C^3r^3 + 9LC^2r^2 - 3L^3$, where C and S are constants. Find dL/dr.

32. A computer is programmed to draw the graph of $(x^2 + y^2)^3 = 64x^2y^2$ (see Example G of Section 20-9). Find the slope of a line tangent to this curve at $(2.00, 0.56)$ and at $(2.00, 3.07)$. The graph is shown in Fig. 22-26.

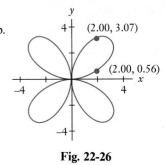

Fig. 22-26

22-9 Higher Derivatives

In the previous sections of this chapter we developed certain important formulas for finding the derivative of a function. The derivative itself is a function, and we may therefore take its derivative. In technology and in mathematics there are many applications which require the use of the derivative of a derivative. Therefore, in this section we will develop the concept and necessary notation, as well as show some of the applications.

The derivative of a function is called the **first derivative** *of the function. The*

the second derivative

derivative of the first derivative is called the **second derivative.** Since the second derivative is a function, we may find its derivative, which is called the third derivative. We may continue to find the fourth derivative, fifth derivative, and so on (provided each derivative is defined). *The second derivative, third derivative, and so on, are known as* **higher derivatives.**

The notations used for higher derivatives follow closely those used for the first derivative. As shown in Section 22-3, the notations for the first derivative are y', D_xy, $f'(x)$, and dy/dx. The notations for the second derivative are y'', D_x^2y, $f''(x)$, and d^2y/dx^2. Similar notations are used for other higher derivatives.

EXAMPLE A

Find the higher derivatives of $y = 5x^3 - 2x$.

First we find the first derivative as

$$\frac{dy}{dx} = 15x^2 - 2 \quad \text{or} \quad y' = 15x^2 - 2$$

Next, we obtain the second derivative by finding the derivative of the first derivative:

$$\frac{d^2y}{dx^2} = 30x \quad \text{or} \quad y'' = 30x$$

Continuing to find the successive derivatives, we have

$$\frac{d^3y}{dx^3} = 30 \quad \text{or} \quad y''' = 30$$

$$\frac{d^4y}{dx^4} = 0 \quad \text{or} \quad y^{(4)} = 0$$

Since the third derivative is a constant, the fourth derivative and all successive derivatives will be zero. This can be shown as $d^ny/dx^n = 0$ for $n \geq 4$.

EXAMPLE B — Find the higher derivatives of $f(x) = x(x^2 - 1)^2$.

Using the product rule, Eq. (22-12), to find the first derivative, we have

For reference,
Eq. (22-12) is

$$\frac{d(uv)}{dx} = u\frac{dv}{dx} + v\frac{du}{dx}$$

$$f'(x) = x(2)(x^2 - 1)(2x) + (x^2 - 1)^2(1)$$
$$= (x^2 - 1)(4x^2 + x^2 - 1) = (x^2 - 1)(5x^2 - 1)$$
$$= 5x^4 - 6x^2 + 1$$

Continuing to find the higher derivatives, we have

$$f''(x) = 20x^3 - 12x$$
$$f'''(x) = 60x^2 - 12$$
$$f^{(4)}(x) = 120x$$
$$f^{(5)}(x) = 120$$
$$f^{(n)}(x) = 0 \quad \text{for } n \geq 6$$

EXAMPLE C — Evaluate the second derivative of $y = \dfrac{2}{1-x}$ for $x = -2$.

We first write the function as $y = 2(1-x)^{-1}$, and then find the first and second derivatives:

$$y = 2(1-x)^{-1}$$
$$\frac{dy}{dx} = 2(-1)(1-x)^{-2}(-1) = 2(1-x)^{-2}$$
$$\frac{d^2y}{dx^2} = 2(-2)(1-x)^{-3}(-1) = 4(1-x)^{-3} = \frac{4}{(1-x)^3}$$

Evaluating the second derivative for $x = -2$, we have

$$\frac{d^2y}{dx^2}\bigg|_{x=-2} = \frac{4}{(1+2)^3} = \frac{4}{27}$$

We note that the function is not differentiable for $x = 1$. Also, we see that if we continue to find higher derivatives, the general expression will not become zero as in Examples A and B.

EXAMPLE D — Find y'' for the implicit function defined by $2x^2 + 3y^2 = 6$.

Differentiating with respect to x, we have

$$2(2x) + 3(2yy') = 0$$
$$4x + 6yy' = 0 \quad \text{or} \quad 2x + 3yy' = 0 \tag{1}$$

NOTE ▷ Before differentiating again we see that **$3yy'$ is a product,** and we note that the derivative of y' is y''. Thus, differentiating again, we have

differentiation of $3yy'$

$$2 + \overline{3yy'' + 3y'(y')} = 0$$
$$2 + 3yy'' + 3(y')^2 = 0 \tag{2}$$

Now, solving Eq. (1) for y' and substituting this into Eq. (2), we have

$$y' = -\frac{2x}{3y}$$

$$2 + 3yy'' + 3\left(-\frac{2x}{3y}\right)^2 = 0$$

$$2 + 3yy'' + \frac{4x^2}{3y^2} = 0$$

$$6y^2 + 9y^3y'' + 4x^2 = 0$$

$$y'' = \frac{-4x^2 - 6y^2}{9y^3} = \frac{-2(2x^2 + 3y^2)}{9y^3}$$

Since $2x^2 + 3y^2 = 6$, we have

$$y'' = \frac{-2(6)}{9y^3} = -\frac{4}{3y^3}$$

■

As we mentioned earlier, higher derivatives are used often in certain applications. This is particularly true of the second derivative. We will use the first and second derivatives in the next chapter in sketching curves, and these and other higher derivatives will be used when we discuss infinite series in Chapter 28. An important technical application of the second derivative is shown in the example which follows, and others are shown in the exercises.

In Section 22-4 we briefly discussed the instantaneous velocity of an object, and in the exercises we mentioned acceleration. From that discussion we recall that the instantaneous velocity is the time rate of change of the displacement, and the **instantaneous acceleration** *is the time rate of change of the instantaneous velocity. Therefore, we see that the acceleration is found from the second derivative of the displacement with respect to time.* Consider the following example.

EXAMPLE E

For the first 12 s after launch, the height s (in meters) of a certain rocket is given by $s = 10\sqrt{t^4 + 25} - 50$. Determine the vertical acceleration of the rocket when $t = 10.0$ s.

Since the velocity is found from the first derivative and the acceleration is found from the second derivative, we must find the second derivative and then evaluate it for $t = 10.0$ s.

$$s = 10\sqrt{t^4 + 25} - 50$$

$$v = \frac{ds}{dt} = 10\left(\frac{1}{2}\right)(t^4 + 25)^{-1/2}(4t^3) = \frac{20t^3}{(t^4 + 25)^{1/2}}$$

$$a = \frac{dv}{dt} = \frac{d^2s}{dt^2} = \frac{(t^4 + 25)^{1/2}(60t^2) - 20t^3(\frac{1}{2})(t^4 + 25)^{-1/2}(4t^3)}{t^4 + 25}$$

$$= \frac{(t^4 + 25)(60t^2) - 40t^6}{(t^4 + 25)^{3/2}} = \frac{20t^6 + 1500t^2}{(t^4 + 25)^{3/2}} = \frac{20t^2(t^4 + 75)}{(t^4 + 25)^{3/2}}$$

Finding the value of the acceleration when $t = 10.0$ s, we have

$$a\Big|_{t=10.0} = \frac{20(10.0^2)(10.0^4 + 75)}{(10.0^4 + 25)^{3/2}} = 20.1 \text{ m/s}^2$$

■

Exercises 22-9

In Exercises 1 through 8, find all the higher derivatives of the given functions.

1. $y = x^3 + x^2$ **2.** $f(x) = 3x - x^4$ **3.** $f(x) = x^3 - 6x^4$ **4.** $y = 2x^5 + 5x^2$

5. $y = (1 - 2x)^4$ **6.** $f(x) = (3x + 2)^3$ **7.** $f(x) = x(2x + 1)^3$ **8.** $y = x(x - 1)^3$

In Exercises 9 through 28, find the second derivative of each of the given functions.

9. $y = 2x^7 - x^6 - 3x$ **10.** $y = 6x - 2x^5$ **11.** $y = 2x + \sqrt{x}$ **12.** $y = x^2 - \dfrac{1}{\sqrt{x}}$

13. $f(x) = \sqrt[4]{8x - 3}$ **14.** $f(x) = \sqrt[3]{6x + 5}$ **15.** $f(x) = \dfrac{4}{\sqrt{1 - 2x}}$ **16.** $f(x) = \dfrac{5}{\sqrt{3 - 4x}}$

17. $y = 2(2 - 5x)^4$ **18.** $y = (4x + 1)^6$ **19.** $y = (3x^2 - 1)^5$ **20.** $y = 3(2x^3 + 3)^4$

21. $f(x) = \dfrac{2x}{1 - x}$ **22.** $f(x) = \dfrac{1 - x}{1 + x}$ **23.** $y = \dfrac{x^2}{x + 1}$ **24.** $y = \dfrac{x}{\sqrt{1 - x^2}}$

25. $x^2 - y^2 = 9$ **26.** $xy + y^2 = 4$ **27.** $x^2 - xy = 1 - y^2$ **28.** $xy = y^2 + 1$

In Exercises 29 through 32, evaluate the second derivative for the given value of x.

29. $f(x) = \sqrt{x^2 + 9}, \ x = 4$ **30.** $f(x) = x - \dfrac{2}{x^3}, \ x = -1$

31. $y = 3x^{2/3} - \dfrac{2}{x}, \ x = -8$ **32.** $y = 3(1 + 2x)^4, \ x = \dfrac{1}{2}$

In Exercises 33 through 36, solve the given problems by finding the appropriate derivatives.

33. A bullet is fired vertically upward. Its distance s, in feet, above the ground is given by $s = 2250t - 16.1t^2$, where t is measured in seconds. Find the acceleration of the bullet.

34. In testing the brakes on a new model automobile it was found that the distance s, in feet, which it traveled under specified conditions after the brakes were applied was given by $s = 57.6t - 1.20t^3$. What were the velocity and acceleration of the automobile for $t = 4.00$ s?

35. The voltage V, in volts, induced in an inductor in an electric circuit is given by $V = L(d^2q/dt^2)$, where L is the inductance in henrys (H). Find the expression for the voltage induced in a 1.60-H inductor if $q = \sqrt{2t + 1} - 1$.

36. How fast is the rate of change of solar radiation changing on the surface in Exercise 27 of Section 22-4 at 3 P.M.?

22-10   Chapter Equations, Review Exercises, and Practice Test

Chapter Equations

Limit of function	$\lim\limits_{x \to a} f(x) = L$	(22-1)
Increments	$x_2 = x_1 + \Delta x$	(22-2)
	$y_2 = y_1 + \Delta y$	(22-3)
Slope	$m_{PQ} = \dfrac{(y_1 + \Delta y) - y_1}{(x_1 + \Delta x) - x_1} = \dfrac{\Delta y}{\Delta x}$	(22-4)
	$\Delta y = f(x + \Delta x) - f(x)$	(22-5)

Definition of derivative $$\lim_{\Delta x \to 0} \frac{\Delta y}{\Delta x} = \lim_{\Delta x \to 0} \frac{f(x + \Delta x) - f(x)}{\Delta x}$$ (22-6)

Instantaneous velocity $$v = \lim_{\Delta t \to 0} \frac{\Delta s}{\Delta t}$$ (22-7)

Derivatives of polynomials $$\frac{dc}{dx} = 0$$ (22-8)

$$\frac{dx^n}{dx} = nx^{n-1}$$ (22-9)

$$\frac{d(cu)}{dx} = c\frac{du}{dx}$$ (22-10)

$$\frac{d(u + v)}{dx} = \frac{du}{dx} + \frac{dv}{dx}$$ (22-11)

Derivative of product $$\frac{d(uv)}{dx} = u\frac{dv}{dx} + v\frac{du}{dx}$$ (22-12)

Derivative of quotient $$\frac{d\frac{u}{v}}{dx} = \frac{v\frac{du}{dx} - u\frac{dv}{dx}}{v^2}$$ (22-13)

Chain rule $$\frac{dy}{dx} = \frac{dy}{du}\frac{du}{dx}$$ (22-14)

Derivative of power $$\frac{du^n}{dx} = nu^{n-1}\left(\frac{du}{dx}\right)$$ (22-15)

$$\frac{du^{p/q}}{dx} = \frac{p}{q}u^{(p/q)-1}\frac{du}{dx}$$ (22-16)

Review Exercises

In Exercises 1 through 12, evaluate the given limits.

1. $\lim_{x \to 4} (8 - 3x)$

2. $\lim_{x \to 3} (2x^2 - 10)$

3. $\lim_{x \to -3} \frac{2x + 5}{x - 1}$

4. $\lim_{x \to 1} \frac{x^2 - 1}{x + 1}$

5. $\lim_{x \to 2} \frac{4x - 8}{x^2 - 4}$

6. $\lim_{x \to 5} \frac{x^2 - 25}{3x - 15}$

7. $\lim_{x \to 2} \frac{x^2 + 3x - 10}{x^2 - x - 2}$

8. $\lim_{x \to 0} \frac{(x - 3)^2 - 9}{x}$

9. $\lim_{x \to \infty} \frac{2 + \dfrac{1}{x + 4}}{3 - \dfrac{1}{x^2}}$

10. $\lim_{x \to \infty} \left(7 - \dfrac{1}{x + 1}\right)$

11. $\lim_{x \to \infty} \frac{x - 2x^3}{1 + x^3}$

12. $\lim_{x \to \infty} \frac{2x + 5}{3x^3 - 2x}$

In Exercises 13 through 20, use the delta-process to find the derivative of each of the given functions.

13. $y = 7 + 5x$

14. $y = 6x - 2$

15. $y = 6 - 2x^2$

16. $y = 2x^2 - x^3$

17. $y = \dfrac{2}{x^2}$

18. $y = \dfrac{1}{1 - 4x}$

19. $y = \sqrt{x + 5}$

20. $y = \dfrac{1}{\sqrt{x}}$

In Exercises 21 through 36, find the derivative of each of the given functions.

21. $y = 2x^7 - 3x^2 + 5$

22. $y = 8x^7 - 2^5 - x$

23. $y = 4\sqrt{x} - \dfrac{3}{x} + \sqrt{3}$

24. $y = \dfrac{3}{x^2} - 8\sqrt[4]{x}$

25. $y = \dfrac{x}{1-x}$

26. $y = \dfrac{2x-1}{x^2+1}$

27. $y = (2 - 3x)^4$

28. $y = (2x^2 - 3)^6$

29. $y = \dfrac{3}{(5 - 2x^2)^{3/4}}$

30. $y = \dfrac{7}{(3x - 1)^3}$

31. $y = x^2\sqrt{1 - 6x}$

32. $y = (x - 1)^3(x^2 - 2)^2$

33. $\dfrac{\sqrt{4x + 3}}{2x}$

34. $y = \dfrac{x}{\sqrt{x^2 + 1}}$

35. $(2x - 3y)^3 = x^2 - y$

36. $x^2y^2 = x^2 + y^2$

In Exercises 37 through 40, evaluate the derivatives of the given functions for the given values of x.

37. $y = \dfrac{4}{x} + 2\sqrt[3]{x},\ x = 8$

38. $y = (3x - 5)^4,\ x = -2$

39. $y = 2x\sqrt{4x + 1},\ x = 6$

40. $y = \dfrac{\sqrt{2x^2 + 1}}{3x},\ x = 2$

In Exercises 41 through 44, find the second derivative of each of the given functions.

41. $y = 3x^4 - \dfrac{1}{x}$

42. $y = \sqrt{1 - 8x}$

43. $y = \dfrac{1 - 3x}{1 + 4x}$

44. $y = 2x(6x + 5)^4$

In Exercises 45 through 68, solve the given problems by finding the appropriate limit or derivative.

45. Two lenses of focal lengths f_1 and f_2, separated by a distance d, are used in the study of lasers. The combined focal length f of this lens combination is $f = \dfrac{f_1 f_2}{f_1 + f_2 - d}$. If f_2 and d remain constant, find the limiting value of f as f_1 continues to increase in value.

46. A company estimates that the sales S, in dollars, of a new product will be $S = \dfrac{8000(2t^2 + 1)}{t^2 + 1}$, where t is the time in months. Find the maximum eventual projected sales in the future.

47. Find the slope of a line tangent to the curve of $y = 7x^4 - x^3$ at $(-1, 8)$.

48. Find the slope of a line tangent to the curve of $y = \sqrt[3]{3 - 8x}$ at $(-3, 3)$.

49. The reliability R of a computer system measures the probability that the system will be operating properly after t hours. For one system $R = 1 - kt + \dfrac{k^2 t^2}{2} - \dfrac{k^3 t^3}{6}$, where k is a constant. Find the expression for the instantaneous rate of change of R with respect to t.

50. The distance s (in feet) traveled by a subway train after the brakes are applied is given by $s = 40t - 5t^2$. How far does it travel, after the brakes are applied, in coming to a stop?

51. The electric field E at a distance r from a point charge is $E = k/r^2$, where k is a constant. Find an expression for the instantaneous rate of change of the electric field with respect to r.

52. The velocity of an object moving with constant acceleration can be found from the equation $v = \sqrt{v_0^2 + 2as}$, where v_0 is the initial velocity, a is the acceleration, and s is the distance traveled. Find dv/ds.

53. The voltage induced in an inductor L is given by $E = L\dfrac{dI}{dt}$, where I is the current in the circuit and t is the time. Find the voltage induced in a 0.4-H inductor if the current I (in amperes) is related to the time (in seconds) by $I = t(0.01t + 1)^3$.

54. The time rate of change of angular velocity is angular acceleration. If a wire is moving through a magnetic field in a circular path such that its angular velocity is given by $\omega = \dfrac{2t^2}{1+t}$, where t is the time (in seconds), find the expression for the angular acceleration (in rad/s²). What is the value of the angular acceleration when $t = 2.00$ s?

55. The frictional radius r_f of a collar used in a braking system is given by $r_f = \dfrac{2(R^3 - r^3)}{3(R^2 - r^2)}$, where R is the outer radius and r is the inner radius. Find dr_f/dR if r is constant.

56. Water is being drained from a pond such that the volume of water in the pond (in m³) after t hours is given by $V = 5000(60 - t)^2$. Find the rate at which the pond is being drained after 4.00 h.

57. The frequency f of a certain electronic oscillator is given by $f = \dfrac{1}{2\pi\sqrt{C(L+2)}}$, where C is a capacitance and L is an inductance. If C is constant, find df/dL.

58. The volume V of fluid produced in the retina of the eye in reaction to exposure to light of intensity I is given by $V = \dfrac{aI^2}{b-I}$, where a and b are constants. Find dV/dI.

59. Under certain conditions, the efficiency of an internal combustion engine is given by

$$\text{eff(in percent)} = 100\left(1 - \frac{1}{(V_1/V_2)^{0.4}}\right)$$

where V_1 and V_2 are the maximum and minimum volumes of air in a cylinder, respectively. Assuming that V_2 is kept constant, find the expression for the rate of change of efficiency with respect to V_1.

60. The temperature T (in °C) in a freezer as a function of the time t (in hours) is given by $T = \dfrac{10(1-t)}{0.5t+1}$. Find dT/dt.

61. The deflection y, in meters, of a 5.00-m beam is given by $y = 0.0001(x^5 - 25x^2)$, where x is the distance, in meters, from one end. Find the value of d^2y/dx^2 (the rate at which the slope of the beam changes) where $x = 3.00$ m.

62. The number n of grams of a compound formed during a certain chemical reaction is given by $n = \dfrac{2t}{t+1}$, where t is the time in minutes. Evaluate d^2n/dt^2 (the rate of increase of the amount of the compound being formed) when $t = 4.00$ min.

63. The area of a rectangular patio is to be 75 m². Express the perimeter p of the patio as a function of its width w and find dp/dw.

64. A water tank is being designed in the shape of a right circular cylinder with a volume of 100 ft³. Find the expression for the instantaneous rate of change of the total surface area A of the tank with respect to the radius r of the base.

65. An arch over a walkway can be described by the first-quadrant part of the parabola $y = 4 - x^2$. In order to determine the size and shape of rectangular objects which can pass under the arch, express the area A of a rectangle inscribed under the parabola in terms of x. Find dA/dx.

66. An airplane flies over an observer with a velocity of 400 mi/h and at an altitude of 2640 ft. If the plane flies horizontally in a straight line, find the rate at which the distance from the observer to the plane is changing 0.600 min after the plane passes over the observer. See Fig. 22-27.

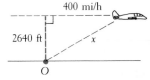

Fig. 22-27

67. A computer analysis showed that a specialized piece of machinery has a value, in dollars, given by $V = 1,500,000/(2t + 10)$, where t is the number of years after purchase. Calculate the value of dV/dt and d^2V/dt^2 for $t = 5$ years. What is the meaning of these values?

68. The *radius of curvature* of $y = f(x)$ at the point (x, y) on a curve is given by

$$R = \frac{[1 + (y')^2]^{3/2}}{|y''|}$$

A certain roadway follows the parabola $y = 1.2x - x^2$ for $0 < x < 1.2$, where x is measured in miles. Find R for $x = 0.2$ mi and $x = 0.6$ mi. See Fig. 22-28.

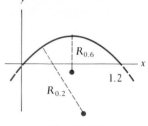

Fig. 22-28

Practice Test

1. Find $\displaystyle\lim_{x \to 1} \frac{x^2 - x}{x^2 - 1}$.

2. Find $\displaystyle\lim_{x \to \infty} \frac{1 - 4x^2}{x + 2x^2}$.

3. Find the slope of a line tangent to the curve of $y = 3x^2 - \dfrac{4}{x^2}$ at (2, 11).

4. The displacement s (in cm) of a pumping machine piston in each cycle is given by $s = t\sqrt{10 - 2t}$. Find its velocity for $t = 4.00$ s.

5. Find dy/dx: $(1 + y^2)^3 - x^2y = 7x$.

6. Under certain conditions, due to the presence of a charge q, the electric potential V along a line is given by

$$V = \frac{kq}{\sqrt{x^2 + b^2}}$$

where k is a constant and b is the minimum distance from the charge to the line. Find the expression for the rate of change of V with respect to x.

7. Find the second derivative of $y = \dfrac{2}{3x + 2}$.

8. By using the delta-process, find the derivative of $y = 5x - 2x^2$ with respect to x.

23 Applications of the Derivative

In Section 23-7 we see how to use the derivative to find the dimensions of a beam of greatest strength.

In Chapter 22 we developed the meaning of the derivative of a function and then went on to find several formulas by which we can differentiate functions. We also established the concept of the derivative as an instantaneous rate of change.

Numerous applications of the derivative were indicated in the examples and exercises throughout Chapter 22. It was not necessary to develop any of these applications in any detail, since finding the derivative was the primary concern. There are, however, certain types of problems in various areas of technology in which the derivative plays a key role in the solution, although other concepts are also involved. In this chapter we shall consider some of these basic types of applications of the derivative.

23-1 Tangents and Normals

The first application of the derivative which we consider involves finding the equations of lines tangent and lines normal (perpendicular) to a given curve.

To find the equation of a line tangent to a curve at a given point, we first find the derivative of the function. The derivative is then evaluated at the point, and this gives us the slope of a line tangent to the curve at the point. Then, by using the point-slope form of the equation of a straight line, we find the equation of the tangent line. The following examples illustrate the method.

EXAMPLE A ———— Find the equation of the line tangent to the parabola $y = x^2 - 1$ at the point $(-2, 3)$.

The derivative of this function is

$$\frac{dy}{dx} = 2x$$

The value of this derivative for the value $x = -2$ (the y-value of 3 is not used since the derivative does not directly contain y) is

$$\frac{dy}{dx} = -4$$

which means that the slope of the tangent line at $(-2, 3)$ is -4. Thus, by using the point-slope form of the equation of the straight line we obtain the desired equation. Thus, we have

$$y - 3 = -4(x + 2)$$
$$y - 3 = -4x - 8$$

or

$$y = -4x - 5$$

Fig. 23-1

■ The parabola and the tangent line $y = -4x - 5$ are shown in Fig. 23-1.

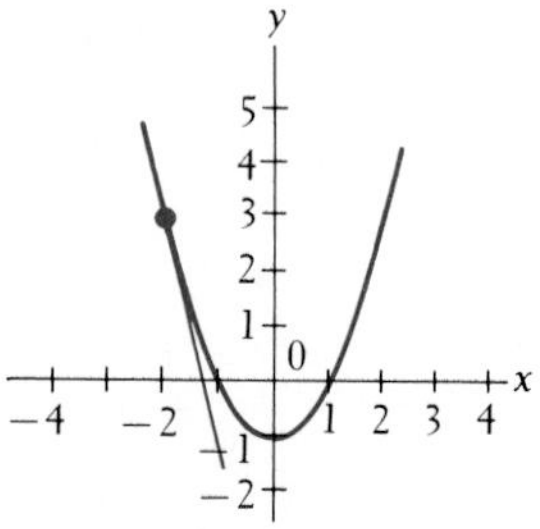

EXAMPLE B ———— Find the equation of the line tangent to the ellipse $4x^2 + 9y^2 = 40$ at the point $(1, 2)$.

The easiest method of finding the derivative of this equation is to treat it as an implicit function. In this way we have the following solution:

$$8x + 18yy' = 0 \qquad \text{find derivative}$$

$$y' = -\frac{4x}{9y}$$

$$y'|_{(1,2)} = -\frac{4}{18} = -\frac{2}{9} \qquad \begin{array}{l}\text{evaluate derivative to find slope} \\ \text{of tangent line}\end{array}$$

$$y - 2 = -\frac{2}{9}(x - 1) \qquad \text{point-slope form of tangent line}$$

$$9y - 18 = -2x + 2$$

$$2x + 9y - 20 = 0 \qquad \text{standard form of tangent line}$$

Fig. 23-2

The ellipse and the tangent line $2x + 9y - 20 = 0$ are shown in Fig. 23-2.

It might be noted that the derivative could also have been found by solving for y. In this way we would have to differentiate $y = \frac{2}{3}\sqrt{10 - x^2}$.

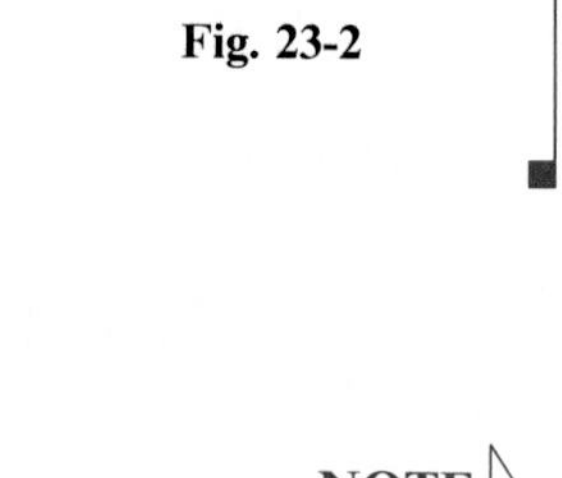

If we wish to obtain the equation of a line normal (perpendicular to a tangent) to a curve, we recall that the slopes of perpendicular lines are negative reciprocals. Thus, the derivative is found and evaluated at the specified point. Since this gives

NOTE ▷ the slope of a tangent line, *we take the negative reciprocal of this number to find*

the slope of the normal line. Then, by using the point-slope form of the equation of a straight line we find the equation of the normal. The following examples illustrate the method.

EXAMPLE C — Find the equation of the line normal to the hyperbola $y = 2/x$ at the point $(2, 1)$.

The derivative of this function is $dy/dx = -2/x^2$, which, evaluated at $x = 2$, gives $dy/dx = -\frac{1}{2}$. Therefore, the slope of a line normal to the curve at the point $(2, 1)$ is 2. The equation of the normal is then

$$y - 1 = 2(x - 2)$$

or

$$y = 2x - 3$$

■ The hyperbola and the normal line are shown in Fig. 23-3.

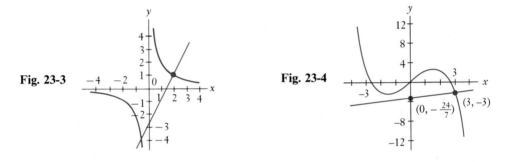

Fig. 23-3

Fig. 23-4

EXAMPLE D — Find the y-intercept of the line normal to the curve of $y = 2x - \frac{1}{3}x^3$, where $x = 3$.

We will first find the equation of the normal line. We will find the y-intercept by writing the equation in slope-intercept form. The solution proceeds as follows:

$$\frac{dy}{dx} = 2 - x^2 \qquad \text{find derivative}$$

$$\frac{dy}{dx}\bigg|_{x=3} = 2 - 3^2 = -7 \qquad \text{evaluate derivative}$$

$$m_{\text{norm}} = \frac{1}{7} \qquad \text{negative reciprocal}$$

$$y|_{x=3} = 2(3) - \frac{1}{3}(3^3) = -3 \qquad \text{find y-coordinate of point}$$

$$y - (-3) = \frac{1}{7}(x - 3) \qquad \text{point-slope form of normal line}$$

$$7y + 21 = x - 3$$

$$y = \frac{1}{7}x - \frac{24}{7} \qquad \text{slope-intercept form}$$

This tells us that the y-intercept is $(0, -\frac{24}{7})$. The curve, normal line, and intercept are shown in Fig. 23-4.

Many of the applications of tangents and normals are geometric. However, there are certain applications in technology, and one of these is shown in the following example. Others are shown in the exercises.

EXAMPLE E

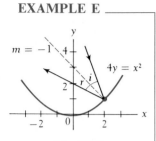

Fig. 23-5

In Fig. 23-5 the cross section of a parabolic solar reflector is shown, along with an incident ray of light and the reflected ray. The angle of incidence i is equal to the angle of reflection r where both angles are measured with respect to the normal to the surface. If the incident ray strikes at the point where the slope of the normal is -1, and the equation of the parabola is $4y = x^2$, what is the equation of the normal line?

If the slope of the normal line is -1, then the slope of a tangent line is $-(\frac{1}{-1}) = 1$. Therefore, we know that the value of the derivative at the point of reflection is 1. This allows us to find the coordinates of the point.

$$4y = x^2$$

$$4\frac{dy}{dx} = 2x \qquad \text{find derivative}$$

$$\frac{dy}{dx} = \frac{1}{2}x$$

$$1 = \frac{1}{2}x \qquad \text{substitute } \frac{dy}{dx} = 1$$

$$x = 2$$

This means that the x-coordinate of the point of reflection is 2. We can find the y-coordinate by substituting $x = 2$ into the equation of the parabola. Thus, the point is (2, 1). Since the slope is -1, the equation is

$$y - 1 = (-1)(x - 2)$$

or

$$y = -x + 3$$

We might note that if the incident ray is vertical, for which $i = 45°$, the reflected ray passes through (0, 1), which is the focus of the parabola. This shows the important reflection property of a parabola that *any incident ray parallel to the axis of a parabola passes through the focus.*

Exercises 23-1

In Exercises 1 through 4, find the equations of the lines tangent to the indicated curves at the given points. In each, sketch the curve and the tangent line.

1. $y = x^2 + 2$ at (2, 6)

2. $y = \frac{1}{3}x^3 - 5x$ at (3, -6)

3. $y = \frac{1}{x^2 + 1}$ at $(1, \frac{1}{2})$

4. $x^2 + y^2 = 25$ at (3, 4)

In Exercises 5 through 8, find the equations of the lines normal to the indicated curves at the given points. In each, sketch the curve and the normal line.

5. $y = 6x - 2x^2$ at $(2, 4)$ **6.** $y = 8 - x^3$ at $(-1, 9)$ **7.** $y = \dfrac{1}{(x^2 + 1)^2}$ at $(1, \frac{1}{4})$ **8.** $x^2 - y^2 = 8$ at $(3, 1)$

In Exercises 9 through 12, find the equations of the tangent lines and the normal lines to the indicated curves.

9. $y = \dfrac{1}{\sqrt{x^2 + 1}}$, where $x = \sqrt{3}$ **10.** $y = \dfrac{4}{(5 - 2x)^2}$, where $x = 2$

11. The parabola with vertex at $(0, 3)$ and focus at $(0, 0)$, where $x = -1$

12. The ellipse with focus at $(4, 0)$, vertex at $(5, 0)$, and center at $(0, 0)$, where $x = 2$

In Exercise 13 through 16, find the equations of the lines tangent or normal to the given curves and with the given slopes.

13. $y = x^2 - 2x$, tangent line with slope 2   **14.** $y = \sqrt{2x - 9}$, tangent line with slope 1

15. $y = (2x - 1)^3$, normal line with slope $-\frac{1}{24}$, $x > 0$ **16.** $y = \frac{1}{2}x^4 + 1$, normal line with slope 4

In Exercises 17 through 24, solve the given problems by finding the equation of the appropriate tangent or normal line.

17. Find the x-intercept of the line tangent to the parabola $y = 4x^2 - 8x$ at $(-1, 12)$.

18. Find the y-intercept of the line normal to the curve $y = x^{3/4}$, where $x = 16$.

19. A certain suspension cable with supports on the same level is closely approximated as being parabolic in shape. If the supports are 200 ft apart and the sag at the center is 30 ft, what is the equation of the line along which the tension acts (tangentially) at the right support? (Choose the origin of the coordinate system at the lowest point of the cable.)

20. A laser source is 2.00 cm from a spherical surface of radius 3.00 cm, and the laser beam is tangent to the spherical surface. By placing the center of the sphere at the origin and the source on the positive x-axis, find the equation of the line along which the beam shown in Fig. 23-6 is directed.

21. In an electric field, the lines of force are perpendicular to the curves of equal electric potential. In a certain electric field, a curve of equal potential is $y = \sqrt{2x^2 + 8}$. If the line along which the force acts on an electron has an inclination of 135°, find its equation.

22. A radio wave reflects from a reflecting surface in the same way as a light wave (see Example E). A certain horizontal radio wave reflects off a parabolic reflector such that the reflected wave is 43.60° below the horizontal, as shown in Fig. 23-7. If the equation of the parabola is $y^2 = 8x$, what is the equation of the normal line through the point of reflection?

23. In designing a flexible tubing system, the supports for the tubing must be perpendicular to the tubing. If a section of the tubing follows the curve $y = \dfrac{4}{x^2 + 1}$ (units in dm, with $-2 < x < 2$), along which lines must the supports be directed if they are located at $x = -1$, $x = 0$, and $x = 1$? See Fig. 23-8.

24. On a particular drawing, a pulley wheel can be described by the equation $x^2 + y^2 = 100$ (units in cm). The pulley belt is directed along the lines $y = -10$ and $4y - 3x - 50 = 0$ when first and last making contact with the wheel. What are the first and last points on the wheel where the belt makes contact?

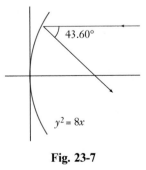

Fig. 23-6

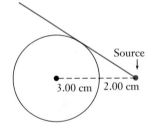

Fig. 23-7

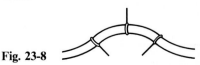

Fig. 23-8

23-2   Newton's Method for Solving Equations ▬▬▬

As we know, finding the roots of an equation $f(x) = 0$ is very important in mathematics and in many types of applications. In algebra and trigonometry some equations can be solved by the use of basic formulas and methods. For example, if $f(x)$ is a linear or quadratic polynomial, we can use the basic methods presented in the early chapters of the book. Also, we developed methods in Chapter 14 for finding the roots if $f(x)$ is a polynomial with rational roots. In Chapter 19 we developed methods for solving certain trigonometric equations. However, for a great many algebraic and nonalgebraic equations there is no method for finding the roots exactly.

We have shown how some equations can be solved graphically, although such solutions are generally of limited accuracy. Therefore, in order to find approximate roots of an equation $f(x) = 0$ with greater accuracy, it is necessary to use other methods. In this section we show a method, known as **Newton's method,** which uses the derivative to locate approximately, but very accurately, the real roots of many kinds of equations. It can be used with polynomial equations of any degree and with nonalgebraic equations.

Newton's method is an example of an **iterative method.** In using such a method, we start with a reasonable guess for a root of the equation. By using the method we obtain a new number, which is a better approximation. This, in turn, gives a still better approximation. Continuing in this way we obtain an approximate answer with the required degree of accuracy. A calculator can be used with Newton's method to find the roots of a great many equations to a very good degree of accuracy. Iterative methods in general are easily programmable for use on a computer.

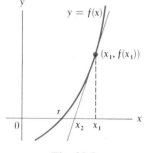

Fig. 23-9

Let us consider a section of a curve $y = f(x)$ which (a) crosses the x-axis, (b) always has either a positive slope or a negative slope, and (c) has a slope which either becomes greater or becomes less as x increases. See Fig. 23-9. The curve in the figure crosses the x-axis at $x = r$, which means that $x = r$ is a root of the equation $f(x) = 0$. If x_1 is sufficiently close to r, a line tangent to the curve at $[x_1, f(x_1)]$ will cross the x-axis at a point $(x_2, 0)$, which is closer to r than is x_1.

We know that the slope of the tangent line is the value of the derivative at x_1, or $m_{tan} = f'(x_1)$. Therefore, the equation of the tangent line is

$$y - f(x_1) = f'(x_1)(x - x_1)$$

For the point $(x_2, 0)$ on this line, we have

$$-f(x_1) = f'(x_1)(x_2 - x_1)$$

Solving for x_2, we have the formula

Newton's method

$$x_2 = x_1 - \frac{f(x_1)}{f'(x_1)} \tag{23-1}$$

Here x_2 is a second approximation to the root. We can then replace x_1 in Eq. (23-1) by x_2 and find a closer approximation, x_3. This process can be repeated as many *times as are needed to find the root to a required accuracy.* We can see that this method lends itself well to the use of a calculator or a computer for finding the root.

EXAMPLE A

Find the root of $x^2 - 3x + 1 = 0$ which lies between $x = 0$ and $x = 1$.

Here $f(x) = x^2 - 3x + 1$. Therefore, $f(0) = 1$ and $f(1) = -1$, which indicates that the root may be near the middle of the interval. Since x_1 must be within the interval, we choose $x_1 = 0.5$.

The derivative is

$$f'(x) = 2x - 3$$

Therefore, $f(0.5) = -0.25$ and $f'(0.5) = -2$, which gives us

$$x_2 = 0.5 - \frac{-0.25}{-2} = 0.375$$

This is a second approximation, which is closer to the actual value of the root. We can get an even better approximation, x_3, by using the method again with $x_2 = 0.375$, $f(0.375) = 0.015625$, and $f'(0.375) = -2.25$. This gives us

$$x_3 = 0.375 - \frac{0.015625}{-2.25} = 0.3819444$$

Since this is a quadratic equation, we can check this result by use of the quadratic formula. Using this formula we find the root is $x = 0.3819660$. Our result using Newton's method is good to three decimal places. Additional accuracy may be obtained by using the method again as many times as needed.

EXAMPLE B

Find the root of $x^4 - 5x^3 + 6x^2 - 5x + 5 = 0$ between 1 and 2 to five decimal places.

Here

$$f(x) = x^4 - 5x^3 + 6x^2 - 5x + 5$$

and

$$f'(x) = 4x^3 - 15x^2 + 12x - 5$$

Since $f(1) = 2$ and $f(2) = -5$ it appears that the root is possibly closer to 1 than to 2. Thus, we let $x_1 = 1.3$. Setting up a table, we have the following values:

n	x_n	$f(x_n)$	$f'(x_n)$	$x_n - \dfrac{f(x_n)}{f'(x_n)}$
1	1.3	0.5111	-5.962	1.3857263
2	1.3857263	-0.0245128	-6.5311513	1.3819731
3	1.3819731	-0.0000461	-6.5066242	1.3819660

Since $x_4 = x_3 = 1.38197$ to five decimal places, this is the value of the required root. (Actually, by finding x_5 we see that x_4 is accurate to the value shown.)

If your calculator can store more than one value in memory, the above results can be more easily obtained by storing at least x_n, $f(x_n)$, and $f'(x_n)$ in memory.

EXAMPLE C

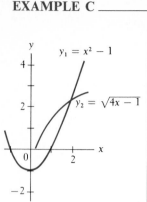

Fig. 23-10

Solve the equation $x^2 - 1 = \sqrt{4x - 1}$.

We can see approximately where the root is by sketching the graphs of $y_1 = x^2 - 1$ and $y_2 = \sqrt{4x - 1}$. We see from Fig. 23-10 that they intersect between $x = 1$ and $x = 2$. Therefore, we choose $x_1 = 1.5$. With

$$f(x) = x^2 - 1 - \sqrt{4x - 1}$$

and

$$f'(x) = 2x - \frac{2}{\sqrt{4x - 1}}$$

we now find the values in the following table:

n	x_n	$f(x_n)$	$f'(x_n)$	$x_n - \dfrac{f(x_n)}{f'(x_n)}$
1	1.5	-0.98606798	2.1055728	1.9683134
2	1.9683134	0.25256859	3.1737598	1.8887332
3	1.8887332	0.00705269	2.9962957	1.8863794
4	1.8863794	0.00000620	2.9910265	1.8863773

Since $x_5 = x_4 = 1.88638$ to five decimal places, this is the required solution. (Here, rounded-off values of x_n are shown, although additional digits were carried and used.)

Exercises 23-2

In Exercises 1 through 4, find the indicated roots of the given quadratic equations by finding x_3 from Newton's method. Then compare this root with that obtained by using the quadratic formula.

1. $x^2 - 2x - 5 = 0$ (between 3 and 4)

2. $2x^2 - x - 2 = 0$ (between 1 and 2)

3. $3x^2 - 5x - 1 = 0$ (between -1 and 0)

4. $x^2 + 4x + 2 = 0$ (between -4 and -3)

See Appendix E for a computer program for Newton's method.

In Exercises 5 through 16, find the indicated roots of the given equations to at least four decimal places.

5. $x^3 - 6x^2 + 10x - 4 = 0$ (between 0 and 1)

6. $x^3 - 3x^2 - 2x + 3 = 0$ (between 0 and 1)

7. $x^3 + 5x^2 + x - 1 = 0$ (the positive root)

8. $2x^3 + 2x^2 - 11x + 3 = 0$ (the larger positive root)

9. $x^4 - x^3 - 3x^2 - x - 4 = 0$ (between 2 and 3)

10. $2x^4 - 2x^3 - 5x^2 - x - 3 = 0$ (between 2 and 3)

11. $x^4 - 2x^3 - 8x - 16 = 0$ (the negative root)

12. $3x^4 - 3x^3 - 11x^2 - x - 4 = 0$ (the negative root)

13. $2x^2 = \sqrt{2x + 1}$ (the positive real solution)

14. $x^3 = \sqrt{x + 1}$ (the real solution)

15. $x = \dfrac{1}{\sqrt{x + 2}}$ (the real solution)

16. $x^{3/2} = \dfrac{1}{2x + 1}$ (the real solution)

In Exercises 17 through 24, determine the required values by use of Newton's method.

17. Find all the real roots of the equation $x^3 - 2x^2 - 5x + 4 = 0$.

18. Find all the real roots of the equation $x^3 - 2x^2 - 2x - 7 = 0$.

19. Find $\sqrt[3]{4}$ by solving the equation $x^3 - 4 = 0$.

20. Find the point of intersection of the curves $y = 1 - x^3$ and $y = \sqrt{2x - 1}$.

21. A dome in the shape of a spherical segment is to be placed over the top of a sports stadium. If the radius r of the dome is to be 60.0 m and the volume V within the dome is 180,000 m³, find the height h of the dome. See Fig. 23-11. $\left[V = \frac{1}{6}\pi h(h^2 + 3r^2). \right]$

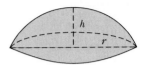

Fig. 23-11

22. The capacitances (in μF) of three capacitors in series are C, $C + 1.00$, and $C + 2.00$. If their combined capacitance is 1.00 μF, their individual values can be found by solving the equation

$$\frac{1}{C} + \frac{1}{C + 1.00} + \frac{1}{C + 2.00} = 1.00$$

Find these capacitances.

23. An oil storage tank has the shape of a right circular cylinder with a hemisphere at each end. See Fig. 23-12. If the volume of the tank is 1500 ft³ and the length ℓ is 12.0 ft, find the radius r.

24. A rectangular block of plastic with edges 2.00 cm, 2.00 cm, and 4.00 cm is heated until its volume doubles. By how much does each edge increase?

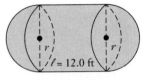

Fig. 23-12

23-3 Curvilinear Motion

A great many phenomena in technology involve the time rate of change of certain quantities. We encountered one of the most fundamental and most important of these when we discussed velocity, the time rate of change of displacement, in Section 22-4. In this section we further develop the concept of velocity and certain other concepts necessary for the discussion. Other time-rate-of-change problems are discussed in the next section.

When velocity was introduced in Section 22-4, the discussion was limited to rectilinear motion, or motion along a straight line. A more general discussion of velocity is necessary when we discuss the motion of an object in a plane. There are many important applications of motion in the plane, a principal one among these being the motion of a projectile.

An important concept in developing this topic is that of a vector. The necessary fundamentals related to vectors are taken up in Chapter 8. Although vectors can be used to represent many physical quantities, we shall restrict our attention to their use in describing the velocity and acceleration of an object moving in a plane along a specified path. Such motion is called **curvilinear motion.**

In describing an object undergoing curvilinear motion, it is very common to express the x- and y-coordinates of its position separately as functions of time. Equations given in this form, that is, *x and y both given in terms of a third variable (in this case t), are* said to be in **parametric form,** which we encountered in Section 9-6. *The third variable, t, is called the* **parameter.**

To find the velocity of an object whose coordinates are given in parametric form, we find its x-component of velocity v_x by determining dx/dt and its y-component of velocity v_y by determining dy/dt. These are then evaluated, and the resultant velocity is found from $v = \sqrt{v_x^2 + v_y^2}$. The direction in which the object is moving is found from $\tan \theta = v_y/v_x$. The following examples illustrate the method.

EXAMPLE A

If the horizontal distance x that an object has moved is given by $x = 3t^2$, and the vertical distance y is given by $y = 1 - t^2$, find the resultant velocity when $t = 2$.

To find the resultant velocity, we must find v and θ, by first finding v_x and v_y as indicated above. After the derivatives are found, they are then evaluated for $t = 2$. Thus,

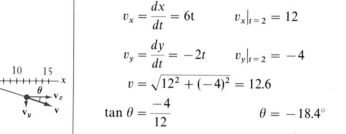

$$v_x = \frac{dx}{dt} = 6t \qquad v_x|_{t=2} = 12 \qquad \text{find velocity components}$$

$$v_y = \frac{dy}{dt} = -2t \qquad v_y|_{t=2} = -4$$

$$v = \sqrt{12^2 + (-4)^2} = 12.6 \qquad \text{magnitude of velocity}$$

$$\tan \theta = \frac{-4}{12} \qquad \theta = -18.4° \qquad \text{direction of motion}$$

Fig. 23-13   ■ The path and velocity vectors are shown in Fig. 23-13.

EXAMPLE B

Find the velocity and direction of motion when $t = 2$ of an object moving such that its x- and y-coordinates of position are given by $x = 1 + 2t$ and $y = t^2 - 3t$.

$$v_x = \frac{dx}{dt} = 2 \qquad v_x|_{t=2} = 2 \qquad \text{find velocity components}$$

$$v_y = \frac{dy}{dt} = 2t - 3 \qquad v_y|_{t=2} = 1$$

$$v|_{t=2} = \sqrt{2^2 + 1^2} = 2.24 \qquad \text{magnitude of velocity}$$

$$\tan \theta = \frac{1}{2} \qquad \theta = 26.6° \qquad \text{direction of motion}$$

Fig. 23-14   ■ These quantities are shown in Fig. 23-14.

NOTE▷   In these examples we note that *we first find the necessary derivatives, and then we evaluate them.* This procedure should always be followed. When a derivative is to be found, it is incorrect to take the derivative of the expression which is the evaluated function.

Acceleration *is the time rate of change of velocity.* Therefore, if the velocity, or its components, is known as a function of time, the acceleration of an object can be found by taking the derivative of the velocity with respect to time. If the displacement is known, the acceleration is found by finding the second derivative with respect to time. Finding the acceleration of an object is illustrated in the following example.

EXAMPLE C

Find the magnitude and direction of the acceleration when $t = 2$ for an object moving such that its x- and y-coordinates of position are given by $x = t^3$ and $y = 1 - t^2$.

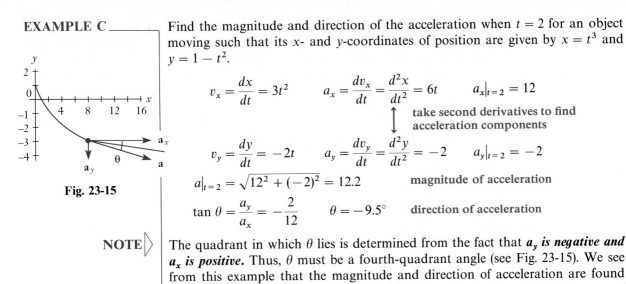

Fig. 23-15

$$v_x = \frac{dx}{dt} = 3t^2 \qquad a_x = \frac{dv_x}{dt} = \frac{d^2x}{dt^2} = 6t \qquad a_x|_{t=2} = 12$$

↑ take second derivatives to find
↓ acceleration components

$$v_y = \frac{dy}{dt} = -2t \qquad a_y = \frac{dv_y}{dt} = \frac{d^2y}{dt^2} = -2 \qquad a_y|_{t=2} = -2$$

$$a|_{t=2} = \sqrt{12^2 + (-2)^2} = 12.2 \qquad \text{magnitude of acceleration}$$

$$\tan \theta = \frac{a_y}{a_x} = -\frac{2}{12} \qquad \theta = -9.5° \qquad \text{direction of acceleration}$$

NOTE ▷ The quadrant in which θ lies is determined from the fact that a_y *is negative and a_x is positive.* Thus, θ must be a fourth-quadrant angle (see Fig. 23-15). We see from this example that the magnitude and direction of acceleration are found ■ from its components just as with velocity.

We now summarize the equations used to find the velocity and acceleration of an object for which the displacement is a function of time. They indicate how to find the components, as well as the magnitude and direction, of each.

$v_x = \dfrac{dx}{dt}$	$v_y = \dfrac{dy}{dt}$	velocity components	(23-2)
$a_x = \dfrac{dv_x}{dt} = \dfrac{d^2x}{dt^2}$	$a_y = \dfrac{dv_y}{dt} = \dfrac{d^2y}{dt^2}$	acceleration components	(23-3)
$v = \sqrt{v_x^2 + v_y^2}$	$a = \sqrt{a_x^2 + a_y^2}$	magnitude	(23-4)
$\tan \theta_v = \dfrac{v_y}{v_x}$	$\tan \theta_a = \dfrac{a_y}{a_x}$	direction	(23-5)

NOTE ▷

For reference,
Eq. (22-15) is

$$\frac{du^n}{dx} = nu^{n-1}\frac{du}{dx}$$

If the curvilinear path which an object follows is given as y as a function of x, *the velocity (and acceleration) is found by taking derivatives of each term of the equation with respect to time.* It is assumed that both x and y are functions of time, although these functions are not stated. When finding derivatives we must be very careful in using the power rule, Eq. (22-15), so that the factor du/dx is not neglected. In the following examples, we illustrate the use of Eqs. (23-2) to (23-5) in applied situations for which we know the equation of the path of the motion. Again, we must be careful to find the direction of the vector as well as its magnitude in order to have a complete solution.

EXAMPLE D

In a physics experiment, a small sphere is constrained to move along a parabolic path described by $y = \frac{1}{3}x^2$. If the horizontal velocity v_x is constant at 6.00 cm/s, find the velocity at the point (2.00, 1.33). See Fig. 23-16.

Since both y and x change with time, both can be considered functions of time. Thus, we can take derivatives of $y = \frac{1}{3}x^2$ with respect to time. When we do this we obtain the following result:

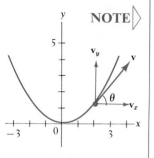

NOTE ▷

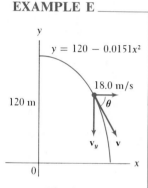

Fig. 23-16

$$\frac{dy}{dt} = \frac{1}{3}\left(2x \boxed{\frac{dx}{dt}}\right) \longleftarrow \frac{dx^2}{dt} = 2x\frac{dx}{dt}$$

$$v_y = \frac{2}{3}xv_x \qquad\qquad \text{using Eqs. (23-2)}$$

$$v_y = \frac{2}{3}(2.00)(6.00) = 8.00 \text{ cm/s} \qquad \text{substituting}$$

$$v = \sqrt{6.00^2 + 8.00^2} = 10.0 \text{ cm/s} \qquad \text{magnitude [Eqs. (23-4)]}$$

$$\tan\theta = \frac{8.00}{6.00}, \qquad \theta = 53.1° \qquad \text{direction [Eqs. (23-5)]}$$

EXAMPLE E

A helicopter is flying at 18.0 m/s and at an altitude of 120 m when a steel ball is released from it. The ball maintains a horizontal velocity and follows a path given by $y = 120 - 0.0151x^2$, as shown in Fig. 23-17. Find the magnitude and direction of the velocity and of the acceleration of the ball 3.00 s after release.

From the given information we know that $v_x = \dfrac{dx}{dt} = 18.0$ m/s. Taking derivatives with respect to time leads to the following solution:

$$y = 120 - 0.0151x^2$$

$$\frac{dy}{dt} = -0.0302x\frac{dx}{dt} \qquad\qquad \text{taking derivatives}$$

$$v_y = -0.0302xv_x \qquad\qquad \text{using Eqs. (23-2)}$$

$$x = (3.00)(18.0) = 54.0 \text{ m} \qquad \text{evaluating at } t = 3.00 \text{ s}$$

$$v_y = -0.0302(54.0)(18.0) = -29.35 \text{ m/s}$$

$$v = \sqrt{18.0^2 + (-29.35)^2} = 34.4 \text{ m/s} \qquad \text{magnitude}$$

$$\tan\theta = \frac{-29.35}{18.0}, \qquad \theta = -58.5° \qquad \text{direction}$$

Therefore, the velocity is 34.4 m/s and is directed at an angle of 58.5° below the horizontal.

To find the acceleration, we return to the equation $v_y = -0.0302xv_x$. Since v_x is constant, we can substitute 18.0 for v_x to get

$$v_y = -0.5436x$$

Again taking derivatives with respect to time, we have

$$\frac{dv_y}{dt} = -0.5436\frac{dx}{dt}$$

Fig. 23-17

$$a_y = -0.5436v_x \qquad\qquad \text{using Eqs. (23-3) and (23-2)}$$

$$a_y = -0.5436(18.0) = -9.78 \text{ m/s}^2 \qquad \text{evaluating}$$

We know that v_x is constant, which means that $a_x = 0$. Therefore, the acceleration is 9.78 m/s^2 and is directed vertically downward. ∎

Exercises 23-3

In Exercises 1 through 4, given that the x- and y-coordinates of a moving particle are given by the indicated parametric equations, find the magnitude and direction of the velocity for the specific value of t. Plot the curves and show the appropriate components of the velocity.

1. $x = 3t$, $y = 1 - t$, $t = 4$

2. $x = \dfrac{5t}{2t + 1}$, $y = 0.1(t^2 + t)$, $t = 2$

3. $x = t(2t + 1)^2$, $y = \dfrac{6}{\sqrt{4t + 3}}$, $t = 0.5$

4. $x = \sqrt{1 + 2t}$, $y = t - t^2$, $t = 4$

In Exercises 5 through 8, use the parametric equations and values of t of Exercises 1 through 4 to find the magnitude and direction of the acceleration in each case.

In Exercises 9 through 24, find the indicated velocities and accelerations.

9. The water from a valve at the bottom of a water tank follows a path described by $y = 4.0 - 0.20x^2$, where units are in meters. If the velocity v_x is constant at 5.0 m/s, find the resultant velocity at the point (4.0, 0.80).

10. A roller mechanism follows a path described by $y = \sqrt{4x + 1}$, where units are in feet. If $v_x = 2x$, find the resultant velocity (in ft/s) at the point (2.0, 3.0).

11. A float is used to test the flow pattern of a stream. It follows a path described by $x = 0.2t^2$, $y = -0.1t^3$, where units of x and y are in feet and t is in minutes. Find the acceleration of the float after 2.0 min.

12. A car on a test track goes into a turn described by $x = 0.2t^3$, $y = 20t - 2t^2$, where x and y are measured in meters and t is in seconds. Find the acceleration of the car at $t = 3.0$ s.

13. A projectile moves according to the equations $x = 120t$ and $y = 160t - 16t^2$, where distances are in feet and time is in seconds. Find the resultant velocity and acceleration for $t = 6.0$ s.

14. A projectile moves according to the parametric equations $x = 30t$ and $y = -4.9t^2$, where distances are in meters and time is in seconds. Find the magnitude and direction of each of the velocity and acceleration when $t = 3.0$ s.

15. A spacecraft moves such that it follows the equation $x = 10(\sqrt{1 + t^4} - 1)$, $y = 40t^{3/2}$ for the first hundred seconds after launch. Here, x and y are measured in meters and t is measured in seconds. Find the magnitude and direction of the velocity of the spacecraft 10.0 s and 100 s after launch.

16. An electron moves in an electric field according to the parametric equations

$$x = \frac{20}{\sqrt{1 + t^2}} \quad \text{and} \quad y = \frac{20t}{\sqrt{1 + t^2}}$$

where distances are in meters and time is in seconds. Find the magnitude and direction of the velocity when $t = 1.0$ s.

17. Find the resultant acceleration for the spacecraft of Exercise 15 for the specified times.

18. Find the resultant acceleration of the electron of Exercise 16 for $t = 1.0$ s.

19. A rocket follows a path given by $y = x - \frac{1}{90}x^3$ (distances in miles). If the horizontal velocity is given by $v_x = x$, find the magnitude and direction of the velocity when the rocket hits the ground (assume level terrain) if time is in minutes.

20. A shipping route around an island is described by $y = 3x^2 - 0.2x^3$. A ship on this route is moving such that $v_x = 1.2$ km/h, where $x = 3.5$ km. Find the velocity of the ship at this point.

21. A personal computer's hard disk is 3.50 in. in diameter and rotates at 3600 r/min. Set up the equation for the circumference of the disk, with its center at the origin. Using this equation, find the components v_x and v_y of a point on the circumference for $x = 1.20$ in., $y > 0$, and $v_x > 0$.

22. A robot arm joint moves in an elliptical path. The horizontal major axis of the ellipse is 8.0 cm long and the minor axis is 4.0 cm long. For $-2 < x < 2$ and $y > 0$, the joint moves such that $v_x = 2.5$ cm/s. On this part of its path, find its velocity if $x = -1.5$ cm. Assume the center of the path is at the origin.

23. An airplane ascends such that its gain h in altitude is proportional to the square root of the change x in horizontal distance traveled. If $h = 280$ m for $x = 400$ m and v_x is constant at 350 m/s, find the velocity at this point.

24. A meteor traveling toward the earth has a velocity inversely proportional to the square root of the distance from the earth's center. Show that its acceleration is inversely proportional to the square of the distance from the center of the earth.

23-4   Related Rates

Any two variables which vary with respect to time and between which a relation is known to exist can have the time rate of change of one expressed in terms of the time rate of change of the other. We do this by taking the derivative with respect to time of the expression which relates the variables, as we did in Examples D and E of Section 23-3. Since the rates of change are related, this type of problem is referred to as a **related-rate** problem. The following examples illustrate the basic method of solution.

EXAMPLE A _____ The voltage of a certain thermocouple as a function of temperature is given by $E = 2.800T + 0.006T^2$. If the temperature is increasing at the rate 1.00°C/min, how fast is the voltage increasing when $T = 100$°C?

Since we are asked to find the time rate of change of voltage, we first take derivatives with respect to time. This gives us

$$\frac{dE}{dt} = 2.800\frac{dT}{dt} + 0.012T\frac{dT}{dt} \longleftarrow \frac{d}{dt}(0.006T^2) = 0.006\left(2T\frac{dT}{dt}\right)$$

NOTE ▷   *again being careful to include the factor dT/dt.* From the given information we know that $dT/dt = 1.00$°C/min and that we wish to know dE/dt when $T = 100$°C. Thus,

$$\frac{dE}{dt}\bigg|_{T=100} = 2.800(1.00) + 0.012(100)(1.00) = 4.00 \text{ V/min}$$

The derivative must be taken before values are substituted. In this problem we are finding the time rate of change of the voltage for a specified value of T. For other values of T, dE/dt would have different values.

EXAMPLE B

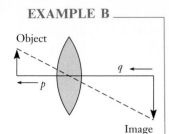

Object

q

p

Image

Fig. 23-18

The distance q that an image is from a certain lens in terms of p, the distance the object is from the lens, is given by

$$q = \frac{10p}{p - 10}$$

If the object distance is increasing at the rate of 0.200 cm/s, how fast is the image distance changing when $p = 15.0$ cm? See Fig. 23-18.

Taking derivatives with respect to time, we have

$$\frac{dq}{dt} = \frac{(p - 10)\left(10\,\dfrac{dp}{dt}\right) - 10p\left(\dfrac{dp}{dt}\right)}{(p - 10)^2} \qquad \text{using the quotient rule}$$

don't forget the $\dfrac{dp}{dt}$

Now, substituting $p = 15.0$ and $dp/dt = 0.200$, we have

$$\left.\frac{dq}{dt}\right|_{p=15} = \frac{5(10)(0.200) - 10(15.0)(0.200)}{25.0}$$

$$= -0.800 \text{ cm/s}$$

Thus, the image distance is decreasing (the significance of the minus sign) at the rate of 0.800 cm/s when $p = 15.0$ cm.

In many related-rate problems the function is not given but must be set up according to the statement of the problem. The following examples illustrate this type of problem.

EXAMPLE C

NOTE ▷

A spherical balloon is being blown up at a constant rate of 2.00 ft³/min. Find the rate at which the radius is increasing when it is 3.00 ft.

We are asked to find the relation between the rate of change of the volume of a sphere with respect to time and the corresponding rate of change of the radius. Thus, we are to **take derivatives of the expression for the volume of a sphere with respect to time.**

$$V = \frac{4}{3}\pi r^3 \qquad\qquad \text{volume of sphere}$$

$$\frac{dV}{dt} = 4\pi r^2 \left(\frac{dr}{dt}\right) \qquad \text{take derivatives with respect to time}$$

$$2.00 = 4\pi(3.00)^2 \left(\frac{dr}{dt}\right) \qquad \text{substitute } \frac{dV}{dt} = 2.00 \text{ ft}^3/\text{min and } r = 3.00 \text{ ft}$$

$$\left.\frac{dr}{dt}\right|_{r=3} = \frac{1}{18.0\pi} \qquad\qquad \text{solve for } \frac{dr}{dt}$$

$$= 0.0177 \text{ ft/min}$$

EXAMPLE D ——— The power supply on a certain earth satellite has an output power P which is inversely proportional to $t + 1000$, where t is the number of days after launch. If the power output is initially 110 W, find the rate at which the power is decreasing after 1000 days (d).

First setting up the equation, we have the following solution:

$$P = \frac{k}{t + 1000}$$ inverse variation

$$110 = \frac{k}{1000}$$ substitute $P = 110$ W, $t = 0$

$$k = 1.10 \times 10^5 \text{ W} \cdot \text{d}$$ solve for k

$$P = \frac{1.10 \times 10^5}{t + 1000}$$ substitute k in equation

$$\frac{dP}{dt} = (1.10 \times 10^5)(-1)(t + 1000)^{-2}(1)$$ take derivatives with respect to time

$$= \frac{-1.10 \times 10^5}{(t + 1000)^2}$$

$$\left.\frac{dP}{dt}\right|_{t=1000} = \frac{-1.10 \times 10^5}{2000^2}$$ evaluate derivative for $t = 1000$ d

$$= -0.0275 \text{ W/d}$$

■ Thus, the power is decreasing at the rate of 0.0275 W/d after 1000 days.

EXAMPLE E ———

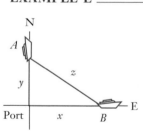

Fig. 23-19

Two ships, A and B, leave a port at noon. Ship A travels north at 6.00 mi/h, and ship B travels east at 8.00 mi/h. How fast are they separating at 2 P.M.?

We are asked to find how fast the length of a line drawn directly from A to B is increasing at a particular time. First a relation must be found between the distances traveled (see Fig. 23-19). If y is the distance traveled by A and x is the distance traveled by B, we can find the distance between them, z, from the Pythagorean theorem. Therefore, we want the value of dz/dt for $t = 2$ h. Even though there are three related variables, each is a function of time. This means we can find dz/dt by taking derivatives of each term with respect to t. This gives us

$$z^2 = x^2 + y^2$$ using Pythagorean theorem

$$2z\frac{dz}{dt} = 2x\frac{dx}{dt} + 2y\frac{dy}{dt}$$ taking derivatives with respect to time

From the given information we know that

$$\frac{dx}{dt} = 8.00 \text{ mi/h} \quad \text{and} \quad \frac{dy}{dt} = 6.00 \text{ mi/h}$$

Also, at 2 P.M. we have the values

$$x = 16.0 \text{ mi}, \quad y = 12.0 \text{ mi}, \quad \text{and} \quad z = 20.0 \text{ mi}$$

With these values we complete the solution.

$$\frac{dz}{dt} = \frac{x(dx/dt) + y(dy/dt)}{z} \qquad\qquad \text{solve for } \frac{dz}{dt}$$

$$\left.\frac{dz}{dt}\right|_{z=20} = \frac{(16.0)(8.00) + (12.0)(6.00)}{20.0} = 10.0 \text{ mi/h} \qquad \text{substitute values}$$

■

Exercises 23-4

In the following exercises, solve the given problems in related rates.

1. The electric resistance of a certain resistor as a function of temperature is given by $R = 4.000 + 0.003T^2$, where R is measured in ohms and T in degrees Celsius. If the temperature is increasing at the rate of 0.100°C/s, find how fast the resistance changes when $T = 150°$C.

2. The kinetic energy of an object is given by $\text{KE} = \frac{1}{2}mv^2$, where m is the mass of the object and v is its velocity. If a 20.0-kg object is accelerating (remember that the time rate of change of velocity is acceleration) at 5.00 m/s^2, how fast is the kinetic energy (in joules) changing when the velocity is 30.0 m/s?

3. A firm found that its profit was p dollars for the production of x tons per week of a product according to the function $p = 30\sqrt{10x - x^2} - 50$. Determine the rate of change of profit if production is increasing such that $dx/dt = 0.200$ tons/week^2, when $x = 4.00$ tons/week.

4. A variable resistor R and an 8-Ω resistor in parallel have a combined resistance R_T given by $R_T = \dfrac{8R}{8 + R}$. If R is changing at 0.3 Ω/min, find the rate at which R_T is changing when $R = 6$ Ω.

5. The radius r of a ring of a certain holograph (an image produced without using a lens) is given by $r = \sqrt{0.4\,\lambda}$, where λ is the wavelength of the light being used. If λ is changing at the rate of 0.10×10^{-7} m/s when $\lambda = 6.0 \times 10^{-7}$ m, find the rate at which r is changing.

6. An earth satellite moves in a path which can be described by $\dfrac{x^2}{28.0} + \dfrac{y^2}{27.6} = 1$, where x and y are in thousands of miles. If $dx/dt = 8000$ mi/h for $x = 2000$ mi and $y > 0$, find dy/dt.

7. The magnetic field B due to a magnet of length ℓ at a distance r is given by $B = \dfrac{k}{[r^2 + (\ell/2)^2]^{3/2}}$, where k is a constant for a given magnet. Find the expression for the time rate of change of B in terms of the time rate of change of r.

8. When air expands so that there is no change in heat (an adiabatic change), the relationship between the pressure and volume is $pv^{1.4} = k$, where k is a constant. At a certain instant, the pressure is 300 kPa and the volume is 10.0 cm^3. The volume is increasing at the rate of 2.00 cm^3/s. What is the time rate of change of pressure at this instant?

9. Fatty deposits have decreased the circular cross-sectional opening of a person's artery. A test drug reduces the fatty deposits such that the radius of the opening increases at the rate of 0.020 mm/month. Find the rate at which the area of the opening increases when $r = 3.0$ mm.

10. A computer program increases the side of a square image on the screen at the rate of 0.25 in./s. Find the rate at which the area of the image increases when the edge is 6.50 in.

11. A metal cube dissolves in acid such that an edge of the cube decreases by 0.50 mm/min. How fast is the volume of the cube changing when the edge is 8.20 mm?

12. A cylindrical water tank is filled with hot water. If the radius equals the height and the radius decreases by 0.02 in./h as the water cools, find the rate at which the volume of the tank decreases when the radius is 38.75 in.

13. One statement of Boyle's law is that the pressure of a gas varies inversely as the volume for constant temperature. If a certain gas occupies 600 cm³ when the pressure is 200 kPa and the volume is increasing at the rate of 20.0 cm³/min, how fast is the pressure changing when the volume is 800 cm³?

14. The tuning frequency f of an electronic tuner is inversely proportional to the square root of the capacitance C in the circuit. If $f = 920$ kHz for $C = 3.5$ pF, find how fast f is changing at this frequency if $dC/dt = 0.3$ pF/s.

15. A spherical metal object is ejected from an earth satellite and reenters the atmosphere. It heats up so that the radius increases at the rate of 5.00 mm/s. What is the rate of change of volume when the radius is 200 mm?

16. The acceleration due to the gravity g on a spacecraft is inversely proportional to its distance from the center of the earth. At the surface of the earth, $g = 32.2$ ft/s². Given that the radius of the earth is 3960 mi, how fast is g changing on a spacecraft approaching the earth at 4500 ft/s at a distance of 25,500 mi from the surface?

17. A tank in the shape of an inverted cone has a height of 3.60 m and a radius at the top of 1.15 m. Water is flowing into the tank at the rate of 0.50 m³/min. How fast is the level rising when it is 1.80 m deep?

18. A ladder is slipping down along a vertical wall. If the ladder is 10.0 ft long, and the top of it is slipping at the constant rate of 10.0 ft/s, how fast is the bottom of the ladder moving along the ground when the bottom is 6.00 ft from the wall?

19. A rope attached to a boat is being pulled in at a rate of 10.0 ft/s. If the water is 20.0 ft below the level at which the rope is being drawn in, how fast is the boat approaching the wharf when 36.0 ft of rope are yet to be pulled in? See Fig. 23-20.

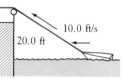

Fig. 23-20

20. A weather balloon leaves the ground 275 ft from an observer and rises vertically at 12.0 ft/s. How fast is the line of sight from the observer to the balloon increasing when the balloon is 450 ft above the ground?

21. A supersonic jet leaves an airfield traveling due east at 1600 mi/h. A second jet leaves the same airfield at the same time and travels at 1800 mi/h along a line north of east such that it remains due north of the first jet. After a half-hour, how fast are the jets separating?

22. A car passes over a bridge at 15.0 m/s at the same time a boat passes under the bridge at a point 10.5 m directly below the car. If the boat is moving perpendicularly to the bridge at 4.0 m/s, how fast are the car and boat separating 5.0 s later?

23. A man 6.00 ft tall approaches a street light 15.0 ft above the ground at the rate of 5.00 ft/s. How fast is the end of the man's shadow moving when he is 10.0 ft from the base of the light?

24. A roller mechanism, as shown in Fig. 23-21, moves such that the right roller is always in contact with the bottom surface and the left roller is always in contact with the left surface. If the right roller is moving to the right at 1.50 cm/s when $x = 10.0$ cm, how fast is the left roller moving?

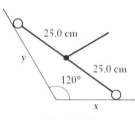

Fig. 23-21

23-5 Using Derivatives in Curve Sketching

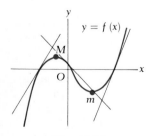

Fig. 23-22

Derivatives can be used effectively in the sketching of curves. An analysis of the first two derivatives can provide useful information as to the graph of a function. This information, possibly along with two or three key points on the curve, is often sufficient to obtain a good, although approximate, graph of the function. Graphs where this information must be supplemented with other analyses are the subject of the next section.

Considering the function $f(x)$ as shown in Fig. 23-22, we see that as x increases (from left to right) the y-values also increase until the point M is reached. From M to m the values of y decrease. To the right of m the values of y again increase. We also note that any tangent line to the left of M or to the right of m will have a positive slope. Any tangent line between M and m will have a negative slope. Since the derivative of a function determines the slope of a tangent line, we can conclude that, *as x increases, y increases if the derivative is positive and decreases if the derivative is negative.* This can be stated as

function increasing
function decreasing

$$f(x) \text{ increases if } f'(x) > 0$$

and

$$f(x) \text{ decreases if } f'(x) < 0$$

NOTE▷ ***It is always assumed that x is increasing.*** *Also, we assume in our present analysis that $f(x)$ and its derivatives are continuous over the indicated interval.*

EXAMPLE A _____ Determine those values of x for which the function $f(x) = x^3 - 3x^2$ is increasing and those values for which it is decreasing.

We find the solution to this problem by determining the values of x for which the derivative is positive and those for which it is negative. Finding the derivative, we have

$$f'(x) = 3x^2 - 6x = 3x(x - 2)$$

This now becomes a problem of solving an inequality. To find the values of x for which $f(x)$ is increasing, we must solve the inequality

$$3x(x - 2) > 0$$

We now recall that the solution of an inequality consists of *all* values of x which may satisfy it. Normally, this consists of certain intervals of values of x. *These intervals are found by first setting the left side of the inequality equal to zero, thus obtaining the* **critical values** *of* x. The function on the left will have the same *sign* for all values of x less than the leftmost critical value. The sign of the function will also be the same within any given interval between critical values and to the right of the rightmost critical value. Those intervals which give the proper sign will satisfy the inequality. In this case the critical values are $x = 0$ and $x = 2$. Therefore, we have the following analysis:

If $x < 0$, $3x(x - 2) > 0$	or $f'(x) > 0$.	$f(x)$ increasing
If $0 < x < 2$, $3x(x - 2) < 0$	or $f'(x) < 0$.	$f(x)$ decreasing
If $x > 2$, $3x(x - 2) > 0$	or $f'(x) > 0$.	$f(x)$ increasing

Therefore, the solution of the above inequality is $x < 0$ or $x > 2$, which means that for these values $f(x)$ is increasing. We can also see that for $0 < x < 2$, $f'(x) < 0$, which means that $f(x)$ is decreasing for these values of x. The solution is now complete. The graph of $f(x) = x^3 - 3x^2$ is shown in Fig. 23-23.

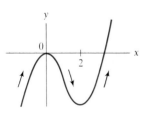

Fig. 23-23

maximum points
minimum points

The points M and m in Fig. 23-22 are called a **relative maximum point** and a **relative minimum point,** respectively. *This means that M has a greater y-value than any other point near it, and that m has a smaller y-value than any point near it.* This does not necessarily mean that M has the greatest y-value of any point on the curve, or that m has the least y-value of any point on the curve. However,

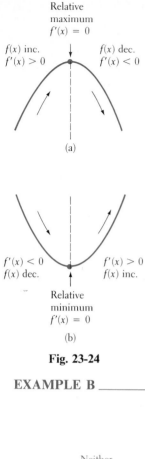

Relative
maximum
$f'(x) = 0$

$f(x)$ inc. $f(x)$ dec.
$f'(x) > 0$ $f'(x) < 0$

(a)

$f'(x) < 0$ $f'(x) > 0$
$f(x)$ dec. $f(x)$ inc.

Relative
minimum
$f'(x) = 0$

(b)

Fig. 23-24

EXAMPLE B

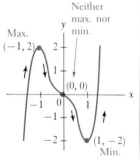

Neither
max. nor
min.

Max.
$(-1, 2)$

$(0, 0)$

$(1, -2)$
Min.

Fig. 23-25

the points M and m are the greatest or least values of y for that part of the curve (that is why we use the word "relative"). Examination of Fig. 23-22 verifies this point. *The characteristic of both M and m is that the derivative is zero at each point.* (We see that this is so since a tangent line would have a slope of zero at each.) *This is how relative maximum and relative minimum points are located. The derivative is found and then set equal to zero. The solutions of the resulting equation give the x-coordinates of the maximum and minimum points.*

It remains now to determine whether a given value of x, for which the derivative is zero, is the coordinate of a maximum or a minimum point (or neither, which is also possible). From the discussion of increasing and decreasing values for y, we can see that *the derivative changes sign from plus to minus when passing through a relative maximum point, and from minus to plus when passing through a relative minimum point.* Thus, we find maximum and minimum points by determining those values of x for which the derivative is zero and by properly analyzing the sign change of the derivative. If the sign of the derivative does not change, it is neither a maximum nor a minimum point. This is known as the **first-derivative test for maxima and minima.**

In Fig. 23-24 a diagram for the first derivative test is shown. The test for a relative maximum is shown in Fig. 23-24(a), and that for a relative minimum is shown in Fig. 23-24(b). For the curves shown in Fig. 23-24, $f(x)$ and $f'(x)$ are continuous throughout the interval shown. [Although $f(x)$ must be continuous, $f'(x)$ may be discontinuous at the maximum point or minimum point, and the sign changes of the first-derivative test remain valid.]

Find any maximum and minimum points on the graph of the function

$$y = 3x^5 - 5x^3$$

Finding the derivative and setting it equal to zero, we have

$$y' = 15x^4 - 15x^2 = 15x^2(x^2 - 1)$$

Therefore,

$$15x^2(x - 1)(x + 1) = 0 \quad \text{for } x = 0, \qquad x = 1, \quad \text{and} \quad x = -1$$

Thus, the sign of the derivative is the same for all points to the left of $x = -1$. For these values $y' > 0$ (thus y is increasing). For values of x between -1 and 0, $y' < 0$. For values of x between 0 and 1, $y' < 0$. For values of x greater than 1, $y' > 0$. Thus, the curve has a maximum at $(-1, 2)$ and a minimum at $(1, -2)$. The point $(0, 0)$ is neither a maximum nor a minimum, since the sign of the derivative did not change for this value of x. The graph of $y = 3x^5 - 5x^3$ is shown in Fig. 23-25.

We now look again at the slope of a tangent drawn to a curve. In Fig. 23-26(a), consider the *change* in the values of the slope of a tangent at a point as the point moves from A to B. At A the slope is positive, and as the point moves toward M, the slope remains positive but becomes smaller until it becomes zero at M. To the right of M the slope is negative and becomes more negative until it reaches I. Therefore, *from A to I the slope continually decreases.* To the right of I the slope remains negative but increases until it becomes zero again at m. To the

(a)

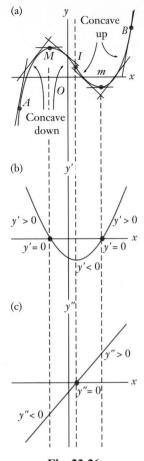

(b)

(c)

Fig. 23-26

concave up
concave down

points of inflection

right of m the slope becomes positive and increases to point B. Therefore, *from I to B, the slope continually increases.* We say that *the curve is* **concave down** *from A to I and* **concave up** *from I to B.*

The curve in Fig. 23-26(b) is that of the derivative, and it therefore indicates the values of the slope of $f(x)$. If the slope changes, we are dealing with the rate of change of slope, or the rate of change of the derivative. This function is the second derivative. The curve in Fig. 23-26(c) is that of the second derivative. We see that *where the second derivative of a function is* **negative,** *the slope is decreasing, or the curve is* **concave down** *(opens down). Where the second derivative is* **positive,** *the slope is increasing, or the curve is* **concave up** *(opens up).* This may be summarized as follows:

If $f''(x) > 0$, the curve is concave up.

If $f''(x) < 0$, the curve is concave down.

We can also now use this information in the determination of maximum and minimum points. By the nature of the definition of maximum and minimum points and of concavity, it is apparent that *a curve is concave down at a maximum point and concave up at a minimum point.* We can see these properties when we make a close analysis of the curve in Fig. 23-26. Therefore, at $x = a$,

if $f'(a) = 0$ and $f''(a) < 0$,

then $f(x)$ has a relative maximum at $x = a$, or

if $f'(a) = 0$ and $f''(a) > 0$,

then $f(x)$ has a relative minimum at $x = a$.

These statements comprise what is known as the **second-derivative test for maxima and minima.** This test is often easier to use than the first-derivative test. However, it can happen that $y'' = 0$ at a maximum or minimum point, and in such cases it is necessary that we use the first-derivative test.

In using the second-derivative test we should note carefully that $f''(x)$ is *neg-ative* at a *maximum* point and *positive* at a *minimum* point. This is contrary to a natural inclination to think of "maximum" and "positive" together or "min-imum" and "negative" together.

The points at which the curve changes from concave up to concave down, or from concave down to concave up, are known as **points of inflection.** Thus point I in the figure is a point of inflection. Inflection points are found by determining those values of x for which the second derivative changes sign. This is analogous to finding maximum and minimum points by the first-derivative test. In Fig. 23-27, various types of points of inflection are illustrated.

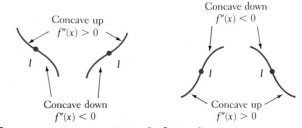

Fig. 23-27   Points of inflection I

EXAMPLE C

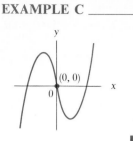

Fig. 23-28

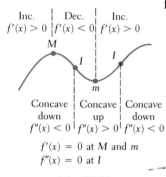

$f'(x) = 0$ at M and m
$f''(x) = 0$ at I

Fig. 23-29

Determine the concavity and find any points of inflection of the function $y = x^3 - 3x$.

This requires inspection and analysis of the second derivative. So we write

$$y' = 3x^2 - 3, \qquad y'' = 6x$$

The second derivative is positive where the function is concave up, and this occurs if $x > 0$. The curve is concave down for $x < 0$, since y'' is negative. Thus, $(0, 0)$ is a point of inflection, since the concavity changes there. The graph of $y = x^3 - 3x$ is shown in Fig. 23-28. ■

At this point we summarize the information regarding the derivatives of a function $f(x)$. See Fig. 23-29.

$f'(x) > 0$ where $f(x)$ increases; $f'(x) < 0$ where $f(x)$ decreases

$f''(x) > 0$ where the graph of $f(x)$ is concave up; $f''(x) < 0$ where the graph of $f(x)$ is concave down

If $f'(x) = 0$ at $x = a$, there is a maximum point if $f'(x)$ changes from $+$ to $-$, or if $f''(a) < 0$

If $f'(x) = 0$ at $x = a$, there is a minimum point if $f'(x)$ changes from $-$ to $+$, or if $f''(a) > 0$

If $f''(x) = 0$ at $x = a$, there is a point of inflection if $f''(x)$ changes from $+$ to $-$ or from $-$ to $+$

The following examples illustrate how the above information is put together to obtain the graph of a function.

EXAMPLE D

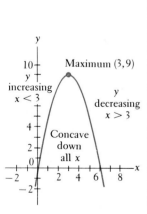

Fig. 23-30

Sketch the graph of $y = 6x - x^2$.

Finding the first two derivatives, we have

$$y' = 6 - 2x = 2(3 - x) \quad \text{and} \quad y'' = -2$$

We now note that $y' = 0$ for $x = 3$. For $x < 3$ we see that $y' > 0$, which means that y is increasing over this interval. Also, for $x > 3$, we note that $y' < 0$, which means that y is decreasing over this interval.

Since y' changes from positive on the left of $x = 3$ to negative on the right of $x = 3$, the curve has a maximum point where $x = 3$. Since $y = 9$ for $x = 3$, this maximum point is $(3, 9)$.

Since $y'' = -2$, this means that its value remains constant for all values of x. Therefore, there are no points of inflection and the curve is concave down for all values of x. This also shows that the point $(3, 9)$ is a maximum point.

Summarizing, we know that y is increasing for $x < 3$, y is decreasing for $x > 3$, there is a maximum point at $(3, 9)$, and the curve is always concave down. Using this information, we sketch the curve shown in Fig. 23-30. ■

EXAMPLE E

Sketch the graph of $y = 2x^3 + 3x^2 - 12x$.

Finding the first two derivatives, we have

$$y' = 6x^2 + 6x - 12 = 6(x + 2)(x - 1)$$

and

$$y'' = 12x + 6 = 6(2x + 1)$$

We note that $y' = 0$ when $x = -2$ and $x = 1$. Using these values in the second derivative, we find that y'' is negative (-18) for $x = -2$ and y'' is positive $(+18)$ when $x = 1$. When $x = -2$, $y = 20$; and when $x = 1$, $y = -7$. Therefore, $(-2, 20)$ is a maximum point and $(1, -7)$ is a minimum point.

Next we see that $y' > 0$ if $x < -2$ or $x > 1$. Also, $y' < 0$ for the interval $-2 < x < 1$. Therefore, y is increasing if $x < -2$ or $x > 1$, and y is decreasing if $-2 < x < 1$.

Now we note that $y'' = 0$ when $x = -\frac{1}{2}$, $y'' < 0$ when $x < -\frac{1}{2}$, and $y'' > 0$ when $x > -\frac{1}{2}$. When $x = -\frac{1}{2}$, $y = \frac{13}{2}$. Therefore, there is a point of inflection at $(-\frac{1}{2}, \frac{13}{2})$, the curve is concave down if $x < -\frac{1}{2}$, and the curve is concave up if $x > -\frac{1}{2}$.

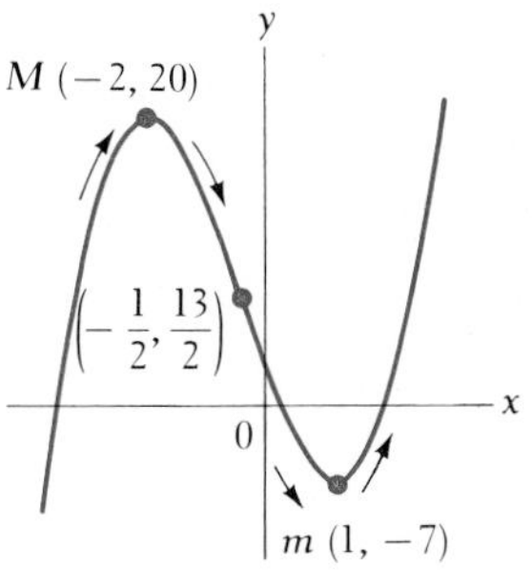

Finally, by locating the points $(-2, 20)$, $(-\frac{1}{2}, \frac{13}{2})$, and $(1, -7)$, we draw the curve *up* to $(-2, 20)$ and then *down* to $(-\frac{1}{2}, \frac{13}{2})$, with the curve *concave down*.

NOTE

***Continuing* down, *but* concave up,** we draw the curve to $(1, -7)$, at which point we start *up* and continue up. This gives us an excellent graph indicating the shape of the curve. If more precision is required, additional points may be used (see Fig. 23-31).

Fig. 23-31

EXAMPLE F

Sketch the graph of $y = x^5 - 5x^4$.

The first two derivatives are

$$y' = 5x^4 - 20x^3 = 5x^3(x - 4)$$
$$y'' = 20x^3 - 60x^2 = 20x^2(x - 3)$$

We now see that $y' = 0$ when $x = 0$ and $x = 4$. For $x = 0$, $y'' = 0$ also, which means ***we cannot use the second-derivative test*** for maximum and minimum points for $x = 0$ in this case. For $x = 4$, $y'' > 0 (+320)$, which means that $(4, -256)$ is a minimum point.

NOTE ▷

Next we note that

$$y' > 0 \quad \text{for } x < 0 \quad \text{or} \quad x > 4, \qquad y' < 0 \quad \text{for } 0 < x < 4$$

Thus, by the first-derivative test, there is a maximum point at $(0, 0)$. Also, y is increasing for $x < 0$ or $x > 4$ and decreasing for $0 < x < 4$.

The second derivative indicates that there is a point of inflection at $(3, -162)$. It also indicates that the curve is concave down for $x < 3$ $(x \neq 0)$ and concave up for $x > 3$. There is no point of inflection at $(0, 0)$ since the second derivative does not change sign at $x = 0$.

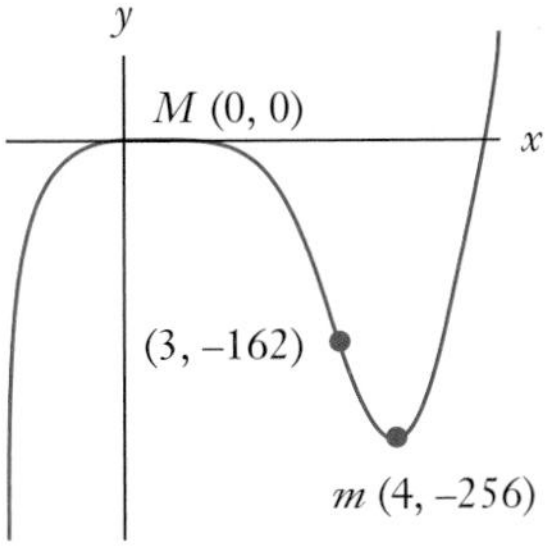

From this information, we sketch the curve in Fig. 23-32.

Fig. 23-32

Exercises 23-5

In Exercises 1 through 4, determine those values of x for which the given functions are increasing and those values of x for which they are decreasing.

1. $y = x^2 + 2x$ **2.** $y = 4 - x^2$ **3.** $y = 12x - x^3$ **4.** $y = x^4 - 6x^2$

In Exercises 5 through 8, find any maximum or minimum points of the given functions. (These are the same functions as in Exercises 1 through 4.)

5. $y = x^2 + 2x$ **6.** $y = 4 - x^2$ **7.** $y = 12x - x^3$ **8.** $y = x^4 - 6x^2$

In Exercises 9 through 12, determine the values of x for which the curve of the given function is concave up, those values for which it is concave down, and any points of inflection. (These are the same functions as in Exercises 1 through 4.)

9. $y = x^2 + 2x$ **10.** $y = 4 - x^2$ **11.** $y = 12x - x^3$ **12.** $y = x^4 - 6x^2$

In Exercises 13 through 16, use the information from Exercises 1 through 12 to sketch the graphs of the given functions.

13. $y = x^2 + 2x$ **14.** $y = 4 - x^2$ **15.** $y = 12x - x^3$ **16.** $y = x^4 - 6x^2$

In Exercises 17 through 26, sketch the graphs of the given functions by determining the appropriate information and points from the first and second derivatives.

17. $y = 12x - 2x^2$ **18.** $y = 3x^2 - 1$ **19.** $y = 2x^3 + 6x^2$

20. $y = x^3 - 9x^2 + 15x + 1$ **21.** $y = x^3 + 3x^2 + 3x + 2$ **22.** $y = x^3 - 12x + 12$

23. $y = x^5 - 5x$ **24.** $y = x^4 + 8x + 2$ **25.** $y = 4x^3 - 3x^4$ **26.** $y = x^5 - 20x^2$

In Exercises 27 through 36, sketch the indicated curves by the methods of this section.

27. A certain projectile follows a path given by $y = x - 0.00025x^2$, where distances are measured in meters. Sketch the graph of the path of the projectile.

28. The angle θ, in degrees, of a robot arm with the horizontal as a function of the time t, in seconds, is given by $\theta = 10 + 12t^2 - 2t^3$. Sketch the graph for $0 \leq t \leq 6$ s.

29. An electric circuit is designed such that the resistance R, in ohms, is a function of the current i, in milliamperes, according to $R = 75 - 18i^2 + 8i^3 - i^4$. Sketch the graph if $R \geq 0$ and i can be positive or negative.

30. The deflection y of a beam at a horizontal distance x from one end is given by $y = k(x^3 - 60x^2)$. Sketch the curve representing the beam if it is 10 ft long and $k = 1/5000$.

31. A rectangular box is made from a piece of cardboard 8 in. by 12 in. by cutting equal squares from each corner and bending up the sides. Express the volume of the box as a function of the side of the square which is cut out, then sketch the curve of the resulting equation.

32. A tool box whose end is square is to be made from 600 in.^{2} of sheet metal. Express the volume of the box as a function of the side of the square of the end. Sketch the curve of the resulting function.

33. Sketch a continuous curve having the following characteristics:

$$f(1) = 0, \quad f'(x) > 0 \quad \text{for all } x, \quad f''(x) < 0 \quad \text{for all } x$$

34. Sketch a continuous curve having the following characteristics:

$$f(0) = 1, \quad f'(x) < 0 \quad \text{for all } x$$
$$f''(x) < 0 \quad \text{for } x < 0, \quad f''(x) > 0 \quad \text{for } x > 0$$

35. Sketch a continuous curve having the following characteristics:

$$f(-1) = 0, \quad f(2) = 2; \quad f'(x) < 0 \quad \text{for } x < -1, \quad f'(x) > 0 \quad \text{for } x > -1$$
$$f''(x) > 0 \quad \text{for } x < 0 \quad \text{or} \quad x > 2, \quad f''(x) < 0 \quad \text{for } 0 < x < 2$$

36. Sketch a continuous curve having the following characteristics:

$f(0) = -1; f'(x) > 0$ for $x < 0$ or $x > 2$, $f'(x) < 0$ for $0 < x < 2$

$f''(x) < 0$ for $x < 1$, $f''(x) > 0$ for $x > 1$

23-6   More on Curve Sketching

We are now in a position to combine the information from the derivative with information obtainable from the function itself to sketch the graph of a function. The information we can obtain from the function includes topics introduced previously. We can obtain intercepts, symmetry, the behavior of the curve as x becomes large, the vertical asymptotes, and the domain and range of the curve. Also, continuity is important in sketching certain functions. The examples which follow demonstrate how these concepts are used along with the information from the derivatives to sketch the graph of a function. We will find that some of these considerations are of much more value than others in graphing any particular curve.

EXAMPLE A _____ Sketch the graph of $y = \dfrac{2}{x^2 + 1}$.

Intercepts: If $x = 0$, $y = 2$, which means that $(0, 2)$ is an intercept. We see that if $y = 0$, there is no corresponding value of x, since $2/(x^2 + 1)$ is a fraction greater than zero for all x. This also indicates that all points on the curve are above the x-axis.

Symmetry: For a review of symmetry, see Section 20-3.

The curve is symmetric to the y-axis, since

$$y = \frac{2}{(-x)^2 + 1} \quad \text{is the same as} \quad y = \frac{2}{x^2 + 1}$$

The curve is not symmetric to the x-axis, since

$$-y = \frac{2}{x^2 + 1} \quad \text{is not the same as} \quad y = \frac{2}{x^2 + 1}$$

The curve is not symmetric to the origin, since

$$-y = \frac{2}{(-x)^2 + 1} \quad \text{is not the same as} \quad y = \frac{2}{x^2 + 1}$$

The value in knowing the symmetry is that we should find those portions of the curve on either side of the y-axis reflections of the other. It is possible to use this fact directly or to use it as a check.

Behavior as x becomes large: We note that as $x \to \infty$, $y \to 0$ since $2/(x^2 + 1)$ is always a fraction which is greater than zero but which becomes smaller as x becomes larger. Therefore, we see that $y = 0$ is an asymptote. From either the symmetry or the function, we also see that $y \to 0$ as $x \to -\infty$.

Vertical asymptotes: From the discussion of the hyperbola we recall that an asymptote is a line which a curve approaches. We have already noted that $y = 0$ is an asymptote for this curve. This asymptote, the x-axis, is a horizontal line.

(*Continued on next page*)

Vertical asymptotes, if any exist, are found by determining those values of x for which the denominator of any term is zero. Such a value of x makes y undefined. Since $x^2 + 1$ cannot be zero, this curve has no vertical asymptotes. The next example illustrates a curve which has a vertical asymptote.

Domain and range: Since the denominator $x^2 + 1$ cannot be zero, x can take on any value. This means the domain of the function is all values of x. Also, we have noted that $2/(x^2 + 1)$ is a fraction greater than zero. Since $x^2 + 1$ is 1 or greater, y is 2 or less. This tells us that the range of the function is $0 < y \leq 2$.

Derivatives: Since

$$y = \frac{2}{x^2 + 1} = 2(x^2 + 1)^{-1}$$

then

$$y' = -2(x^2 + 1)^{-2}(2x) = \frac{-4x}{(x^2 + 1)^2}$$

Since $(x^2 + 1)^2$ is positive for all values of x, the sign of y' is determined by the numerator. Thus, we note that $y' = 0$ for $x = 0$ and that $y' > 0$ for $x < 0$ and $y' < 0$ for $x > 0$. The curve, therefore, is increasing for $x < 0$, is decreasing for $x > 0$, and has a maximum point at $(0, 2)$. Now,

$$y'' = \frac{(x^2 + 1)^2(-4) + 4x(2)(x^2 + 1)(2x)}{(x^2 + 1)^4} = \frac{-4(x^2 + 1) + 16x^2}{(x^2 + 1)^3}$$

$$= \frac{12x^2 - 4}{(x^2 + 1)^3} = \frac{4(3x^2 - 1)}{(x^2 + 1)^3}$$

We note that y'' is negative for $x = 0$, which confirms that $(0, 2)$ is a maximum point. Also, points of inflection are found for the values of x satisfying $3x^2 - 1 = 0$. Thus, $(-\frac{1}{3}\sqrt{3}, \frac{3}{2})$ and $(\frac{1}{3}\sqrt{3}, \frac{3}{2})$ are points of inflection. The curve is concave up if $x < -\frac{1}{3}\sqrt{3}$, or $x > \frac{1}{3}\sqrt{3}$, and it is concave down if $-\frac{1}{3}\sqrt{3} < x < \frac{1}{3}\sqrt{3}$.

Putting this information together, we sketch the curve shown in Fig. 23-33. It might be noted that this curve could have been sketched primarily by use of the fact that $y \to 0$ as $x \to +\infty$ and as $x \to -\infty$, and the fact that a maximum point exists at $(0, 2)$. However, the other parts of the analysis, such as symmetry and concavity, serve as excellent checks and also make the curve more accurate. ■

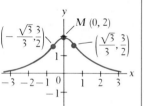

Fig. 23-33

EXAMPLE B _____ Sketch the graph of $y = x + \dfrac{4}{x}$.

Intercepts: If we set $x = 0$, y is undefined. This means that the curve is not *continuous at $x = 0$* and there are no y-intercepts. If we set $y = 0$, $x + 4/x = (x^2 + 4)/x$ cannot be zero since $x^2 + 4$ cannot be zero. Therefore, there are no intercepts. This may seem to be of little value, but we must realize *this curve does not cross either axis.* This, in itself, will be of value when we sketch the graph in Fig. 23-34.

Symmetry: In testing for symmetry, we find that it is not symmetric to either axis. However, this curve does possess symmetry to the origin. This is determined

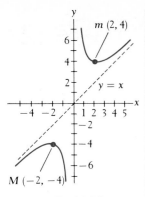

Fig. 23-34

by the fact that when $-x$ replaces x and at the same time $-y$ replaces y, the equation does not change.

Behavior as x becomes large: As $x \to +\infty$, and as $x \to -\infty$, $y \to x$ since $4/x \to 0$. Thus, $y = x$ is an asymptote of the curve.

Vertical asymptotes: As we noted in Example A, vertical asymptotes exist for values of x for which y is undefined. In this equation, $x = 0$ makes the second term on the right undefined and therefore y is undefined. In fact, as $x \to 0$ from the positive side, $y \to +\infty$, and as $x \to 0$ from the negative side, $y \to -\infty$. This is derived from the sign of $4/x$ in each case.

Domain and range: Since x cannot be zero, the domain of the function is all x except zero. As for the range, the analysis from the derivatives will show it to be $y \leq -4$, $y \geq 4$.

Derivatives: Finding the first derivative, we have

$$y' = 1 - \frac{4}{x^2} = \frac{x^2 - 4}{x^2}$$

The x^2 in the denominator indicates that the sign of the first derivative is the same as its numerator. The numerator is zero if $x = -2$ or $x = 2$. If $x < -2$ or $x > 2$, then $y' > 0$; and if $-2 < x < 2$, $x \neq 0$, $y' < 0$. Thus, y is increasing if $x < -2$ or $x > 2$, and also y is decreasing if $-2 < x < 2$, except at $x = 0$ (y is undefined). Also, $(-2, -4)$ is a maximum point and $(2, 4)$ is a minimum point. The second derivative is $y'' = 8/x^3$. This cannot be zero, but it is negative if $x < 0$ and positive if $x > 0$. Thus, the curve is concave down if $x < 0$ and concave up if $x > 0$. Using this information, we have the graph shown in Fig. 23-34.

EXAMPLE C

Sketch the graph of $y = \dfrac{1}{\sqrt{1 - x^2}}$.

Intercepts: If $x = 0$, $y = 1$. If $y = 0$, $1/\sqrt{1 - x^2}$ would have to be zero, but it cannot since it is a fraction with 1 as the numerator for all values of x. Thus, $(0, 1)$ is an intercept.

Symmetry: The curve is symmetric to the y-axis.

Behavior as x becomes large: The values of x cannot be considered beyond 1 or -1, for any value of $x < -1$ or $x > 1$ gives imaginary values for y. Thus, the curve does not exist for values of $x < -1$ or $x > 1$.

Vertical asymptotes: If $x = 1$ or $x = -1$, y is undefined. In each case as $x \to 1$ and as $x \to -1$, $y \to +\infty$.

Domain and range: From the analysis of x becoming large and of the vertical asymptotes, we see that the domain is $-1 < x < 1$. Also, since $\sqrt{1 - x^2}$ is 1 or less, $1/\sqrt{1 - x^2}$ is 1 or more, which means the range is $y \geq 1$.

Derivatives:

$$y' = -\frac{1}{2}(1 - x^2)^{-3/2}(-2x) = \frac{x}{(1 - x^2)^{3/2}}$$

We see that $y' = 0$ if $x = 0$. If $-1 < x < 0$, $y' < 0$ and also if $0 < x < 1$, $y' > 0$. Thus the curve is decreasing if $-1 < x < 0$ and increasing if $0 < x < 1$. There is a minimum point at $(0, 1)$.

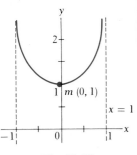

Fig. 23-35

(Continued on next page)

$$y'' = \frac{(1 - x^2)^{3/2} - x(\tfrac{3}{2})(1 - x^2)^{1/2}(-2x)}{(1 - x^2)^3} = \frac{(1 - x^2) + 3x^2}{(1 - x^2)^{5/2}}$$

$$= \frac{2x^2 + 1}{(1 - x^2)^{5/2}}$$

The second derivative cannot be zero since $2x^2 + 1$ is positive for all values of x. The second derivative is also positive for all permissible values of x, which means the curve is concave up for these values.

Using this information, we sketch the graph in Fig. 23-35.

EXAMPLE D

Sketch the graph of $y = \dfrac{x}{x^2 - 4}$.

Intercepts: If $x = 0$, $y = 0$, and if $y = 0$, $x = 0$. The only intercept is $(0, 0)$.

Symmetry: The curve is not symmetric to either axis. However, since $-y = -x/[(-x)^2 - 4]$ is the same as $y = x/(x^2 - 4)$, it is symmetric to the origin.

Behavior as x becomes large: As $x \to +\infty$ and as $x \to -\infty$, $y \to 0$. This means that $y = 0$ is an asymptote.

Vertical asymptotes: If $x = -2$ or $x = 2$, y is undefined. As $x \to -2$, $y \to -\infty$ if $x < -2$ since $x^2 - 4$ is positive, and $y \to +\infty$ if $x > -2$ since $x^2 - 4$ is negative. As $x \to 2$, $y \to -\infty$ if $x < 2$, and $y \to +\infty$ if $x > 2$.

Domain and range: The domain is all real values of x except -2 and 2. As for the range, if $x < -2$, $y < 0$ (the numerator is negative and the denominator is positive). If $x > 2$, $y > 0$ (both numerator and denominator are positive). Since $(0, 0)$ is an intercept, we see that the range is all values of y.

Derivatives:

$$y' = \frac{(x^2 - 4)(1) - x(2x)}{(x^2 - 4)^2} = -\frac{x^2 + 4}{(x^2 - 4)^2}$$

Since $y' < 0$ for all values of x except -2 and 2, the curve is decreasing for all values in the domain.

$$y'' = -\frac{(x^2 - 4)^2(2x) - (x^2 + 4)(2)(x^2 - 4)(2x)}{(x^2 - 4)^4}$$

$$= -\frac{2x(x^2 - 4) - 4x(x^2 + 4)}{(x^2 - 4)^3} = \frac{2x^3 + 24x}{(x^2 - 4)^3} = \frac{2x(x^2 + 12)}{(x^2 - 4)^3}$$

The sign of y'' depends on x and $(x^2 - 4)^3$. If $x < -2$, $y'' < 0$. If $-2 < x < 0$, $y'' > 0$. If $0 < x < 2$, $y'' < 0$. If $x > 2$, $y'' > 0$. This means the curve is concave down for $x < -2$ or $0 < x < 2$ and is concave up for $-2 < x < 0$ or $x > 2$. The curve is sketched in Fig. 23-36.

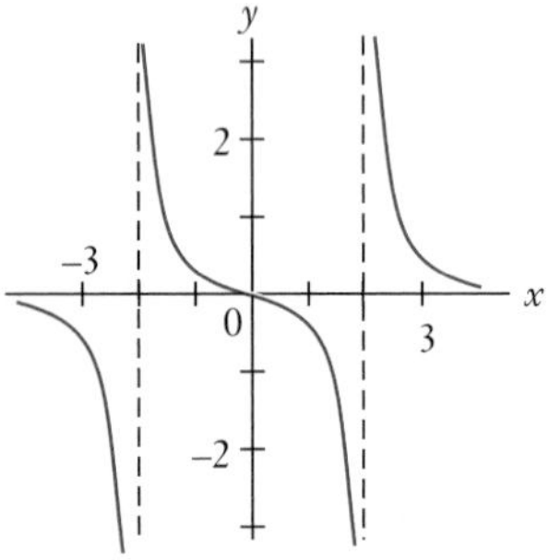

Fig. 23-36

Exercises 23-6

In the following exercises, use the method of the examples of this section to sketch the indicated curves.

1. $y = \dfrac{4}{x^2}$ **2.** $y = \dfrac{2}{x^3}$ **3.** $y = \dfrac{2}{x + 1}$ **4.** $y = \dfrac{x}{x - 2}$

5. $y = x^2 + \dfrac{2}{x}$

6. $y = x + \dfrac{4}{x^2}$

7. $y = x - \dfrac{1}{x}$

8. $y = 3x + \dfrac{1}{x^3}$

9. $y = \dfrac{x^2}{x + 1}$

10. $y = \dfrac{9x}{x^2 + 9}$

11. $y = \dfrac{1}{x^2 - 1}$

12. $y = \dfrac{8}{4 - x^2}$

13. $y = \dfrac{4}{x} - \dfrac{4}{x^2}$

14. $y = 4x + \dfrac{1}{\sqrt{x}}$

15. $y = x\sqrt{1 - x^2}$

16. $y = \dfrac{x - 1}{x^2 - 2x}$

17. $y = \dfrac{9x}{9 - x^2}$

18. $y = \dfrac{x^2 - 4}{x^2 + 4}$

19. The combined capacitance C_T, in microfarads, of a 6-μF capacitance and a variable capacitance C in series is given by $C_T = \dfrac{6C}{6 + C}$. Sketch the graph.

20. A study showed that the percent y of persons surviving burns to x percent of the body is given by $y = \dfrac{300}{0.0005x^2 + 2} - 50$. Sketch the graph.

21. The reliability R of a computer model is found to be $R = \dfrac{200}{\sqrt{t^2 + 40000}}$, where t is the time of operation in hours. ($R = 1$ is perfect reliability and $R = 0.5$ means there is a 50% chance of a malfunction.) Sketch the graph.

22. The electric power P, in watts, produced by a source is given by $P = \dfrac{36R}{R^2 + 2R + 1}$, where R is the resistance in the circuit. Sketch the graph.

23. A cylindrical oil drum is to be made such that it will contain 20 kL. Sketch the area of sheet metal required for construction as a function of the radius of the drum.

24. A fence is to be constructed to enclose a rectangular area of 20,000 m². A previously constructed wall is to be used for one side. Sketch the length of fence to be built as a function of the side of the fence parallel to the wall. See Fig. 23-37.

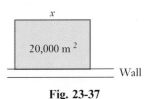

Fig. 23-37

23-7 Applied Maximum and Minimum Problems

Problems from various applied situations frequently occur which require finding a maximum or minimum value of some function. If the function is known, the methods we have already discussed can be used directly. This is discussed in the following example.

EXAMPLE A ⎯⎯⎯⎯ An automobile manufacturer, in testing a new engine on one of its models, found that the efficiency e of the engine as a function of the speed s of the car was given by $e = 0.768s - 0.00004s^3$. Here, e is given in percent and s is given in km/h. What is the maximum efficiency of the engine?

Since we wish to find a maximum value, we find the derivative of e with respect to s.

$$\frac{de}{ds} = 0.768 - 0.00012s^2$$

We then set the derivative equal to zero in order to find the value of s for which a maximum may occur.   *(Continued on next page)*

$$0.768 - 0.00012s^2 = 0$$
$$0.00012s^2 = 0.768$$
$$s^2 = 6400$$
$$s = 80.0 \text{ km/h}$$

We know that s must be positive to have meaning in this problem. Therefore, the apparent solution of $s = -80$ is discarded. The second derivative is

$$\frac{d^2 e}{ds^2} = -0.00024s$$

which is negative for any positive value of s. Thus, a maximum occurs for $s = 80.0$. Substituting $s = 80.0$ in the function for e, we obtain

$$e = 0.768(80.0) - 0.00004(80.0^3)$$
$$= 61.44 - 20.48 = 40.96$$

Therefore, to the nearest tenth, the maximum efficiency is 41.0%, and this occurs for $s = 80.0$ km/h.

In many problems for which a maximum or minimum value is to be determined, the function itself is not given explicitly. It is necessary to find the function from the statement of the problem. **The principal difficulty which arises in these problems is finding this function.** Once we determine this function, we take a derivative with respect to a variable in which it is expressed, and then set the derivative equal to zero. Sufficient information must be given so that a unique solution is possible. Often this information is available, but we must analyze the wording of the problem carefully to find it. Also, many times it is more convenient to set up more than one function, based on the given information, and to find a derivative of each with respect to the same variable. We can then eliminate by substitution any undesired variables or derivatives which appear in each. The following examples illustrate these methods.

NOTE ▷

EXAMPLE B

Find the number which exceeds its square by the greatest amount.

NOTE ▷

We must **set up a function which expresses the difference between a number and its square.** Let the number be x, and let the difference be D. Then $D = x - x^2$. Taking a derivative, we have

$$\frac{dD}{dx} = 1 - 2x$$

Setting the derivative equal to zero and solving for x, we have

$$0 = 1 - 2x, \qquad x = \tfrac{1}{2}$$

The second derivative gives $d^2 D/dx^2 = -2$, which tells us that the second derivative is always negative. This means that whenever the first derivative is zero, it represents a maximum. In many problems it is not necessary to test for maximum or minimum, since the nature of the problem will indicate which must be the case. For example, in this problem we know that numbers greater than 1 do not exceed their squares at all. The same is true for all negative numbers. Thus, the answer must lie between 0 and 1.

EXAMPLE C

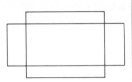

y

x

Fig. 23-38

Find the maximum possible area of a rectangle whose perimeter is 16 units.

When we reflect on this problem, we see that there are limitless possibilities for rectangles of a perimeter of 16 units and differing areas (see Fig. 23-38). For example, if the sides are 7 and 1, the area is 7. If the sides are 6 and 2, the area is 12. Therefore, we set up a function for the area of a rectangle in terms of its sides x and y:

$$A = xy$$

We know, however, one other important fact: that the perimeter is 16. Thus, $2x + 2y = 16$. Solving for y, we have $y = 8 - x$. Substituting this equation into the expression for the area, we have

$$A = x(8 - x) = 8x - x^2$$

We complete the solution as follows:

$$\frac{dA}{dx} = 8 - 2x \qquad \text{take derivative}$$

$$8 - 2x = 0 \qquad \text{set derivative equal to zero}$$

$$x = 4$$

By noting values of the derivative near 4 or by finding the second derivative, we can show that we have a maximum for $x = 4$. Thus, $x = 4$ and $y = 4$ give the maximum area of 16 square units.

EXAMPLE D

See the chapter introduction.

Fig. 23-39

The strength S of a rectangular beam is proportional to the product of its width w and the square of its depth d. Find the dimensions of the strongest beam which can be cut from a circular log 16.0 in. in diameter. See Fig. 23-39.

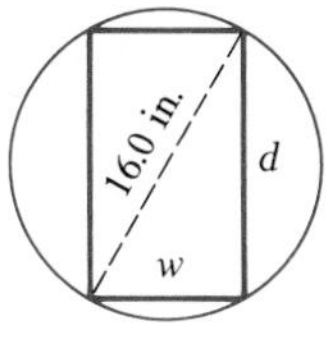

The solution proceeds as follows:

$$S = kwd^2 \qquad \qquad \text{direct variation}$$

$$d^2 = 256 - w^2 \qquad \text{Pythagorean theorem}$$

$$S = kw(256 - w^2) \qquad \text{substituting}$$

$$= k(256w - w^3) \qquad S = f(w)$$

$$\frac{dS}{dw} = k(256 - 3w^2) \qquad \text{take derivative}$$

$$0 = k(256 - 3w^2) \qquad \text{set derivative equal to zero}$$

$$3w^2 = 256 \qquad \text{solve for } w$$

$$w = \frac{16}{\sqrt{3}} = 9.24 \text{ in.}$$

$$d = \sqrt{256 - \frac{256}{3}} = 13.1 \text{ in.} \qquad \text{solve for } d$$

Thus, the strongest beam is 9.24 in. wide and 13.1 in. deep. Since $d^2S/dw^2 = -6kw$ and is negative for $w > 0$, these dimensions give the maximum strength.

EXAMPLE E _____   Find the point on the parabola $y = x^2$ which is nearest the point (6, 3).

In this example we must set up a function for this distance between a general point (x, y) on the parabola and the point (6, 3). This relation is

$$D = \sqrt{(x - 6)^2 + (y - 3)^2}$$

However, to make it easier for us to take derivatives, we shall square both sides of this expression. If a function is a minimum, then so is its square. We shall also use the fact that the point (x, y) is on $y = x^2$ by replacing y by x^2. Thus, we have

$$D^2 = (x - 6)^2 + (x^2 - 3)^2 = x^2 - 12x + 36 + x^4 - 6x^2 + 9$$
$$= x^4 - 5x^2 - 12x + 45$$
$$\frac{dD^2}{dx} = 4x^3 - 10x - 12 \qquad \text{take derivative}$$
$$0 = 2x^3 - 5x - 6 \qquad \text{set derivative equal to zero}$$

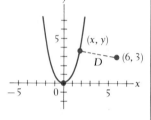

Fig. 23-40

Using synthetic division or some similar method, we find that the solution to this equation is $x = 2$. Thus, the required point on the parabola is (2, 4). (See Fig. 23-40.) We can show that we have a minimum by analyzing the first derivative, by analyzing the second derivative, or by noting that points at much greater distances exist (therefore, it cannot be a maximum). ■

EXAMPLE F _____   A company determines that it can sell 1000 units of a product per month if the price is $5 for each unit. It also estimates that for each 1¢ reduction in unit price, 10 more units can be sold. Under these conditions, what is the maximum possible income and what price per unit gives this income?

If we let $x =$ the number of units over 1000 sold, the total number of units sold is $1000 + x$. The price for each unit is $5 less 1¢ ($0.01) for each block of ten units over 1000 which are sold. Thus, the price for each unit is

$$5 - 0.01\left(\frac{x}{10}\right) \quad \text{or} \quad 5 - 0.001x \text{ dollars}$$

The income I is the number of units sold times the price of each unit. Therefore, we have

$$I = (1000 + x)(5 - 0.001x)$$

Multiplying and finding the first derivative, we have

$$I = 5000 + 4x - 0.001x^2$$
$$\frac{dI}{dx} = 4 - 0.002x \qquad \text{take derivative}$$
$$0 = 4 - 0.002x \qquad \text{set derivative equal to zero}$$
$$x = 2000$$

We note that if $x < 2000$, the derivative is positive, and if $x > 2000$, the derivative is negative. Therefore, if $x = 2000$, I is at a maximum. This means that the maximum income is derived if 2000 units over 1000 are sold, or 3000 units in all. This in turn means that the maximum income is $9000 and the price per unit is $3. These values are found by substituting $x = 2000$ into the expression for I and for the price. ■

EXAMPLE G ———— Find the most economical dimensions for a cylindrical cup which holds a specified volume.

When we analyze the wording of the problem carefully, we see that we are
to minimize the surface area of a right circular cylinder which has a bottom,
but no top. Also the volume is to be considered as constant. Thus, we set up
expressions for the surface area and volume:

$$A = \pi r^2 + 2\pi rh, \qquad V = \pi r^2 h$$

Since the volume is not specified numerically, it is easier to take the derivative
of each of these formulas separately. The derivatives may be taken with respect
to either r or h. Let us choose r, and write

$$\frac{dA}{dr} = 2\pi r + 2\pi r \frac{dh}{dr} + 2\pi h, \qquad \frac{dV}{dr} = 0 = 2\pi rh + \pi r^2 \left(\frac{dh}{dr}\right)$$

Setting dA/dr equal to zero, and at the same time eliminating the derivative
dh/dr by substitution, we have

$$0 = 2\pi r + 2\pi r \left(-\frac{2h}{r}\right) + 2\pi h \quad \text{or} \quad 0 = r - 2h + h$$

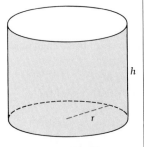

Fig. 23-41

This last equation tells us that $h = r$, which is the relation between dimensions
for the most economical cup (see Fig. 23-41).

EXAMPLE H ———— The illuminance of a light source at any point equals the strength of the source
divided by the square of the distance from the source. Two sources, of strengths
8 units and 1 unit, respectively, are 100 ft apart. Determine at what point between
them the illuminance is the least, assuming that the illuminance at any point is
the sum of the illuminances of the two sources.

Let $I =$ the sum of the illuminances and
$x =$ the distance from the source of strength 8.

Then we find that

$$I = \frac{8}{x^2} + \frac{1}{(100 - x)^2}$$

is the function between the illuminance and the distance from the source of
strength 8. We must now take a derivative of I with respect to x, set it equal
to zero, and solve for x to find the point at which the illuminance is a minimum:

$$\frac{dI}{dx} = -\frac{16}{x^3} + \frac{2}{(100 - x)^3} = \frac{-16(100 - x)^3 + 2x^3}{x^3(100 - x)^3}$$

This function will be zero if the numerator is zero. Therefore, we have

$$2x^3 - 16(100 - x)^3 = 0 \quad \text{or} \quad x^3 = 8(100 - x)^3$$

Taking cube roots of each side, we have

$$x = 2(100 - x) \quad \text{or} \quad x = 66.7 \text{ ft}$$

The point where the illuminance is a minimum is 66.7 ft from the 8-unit source
of illuminance.

Exercises 23-7

In the following exercises, solve the given maximum and minimum problems.

1. The height s (in feet) of an object thrown vertically upward from the ground is given by $s = 112t - 16t^2$, where t is measured in seconds. What is the greatest height to which the object will go?

2. A small oil refinery estimates that its daily profit P, in dollars, from refining x barrels of oil is $P = 8x - 0.02x^2$. How many barrels should be refined for maximum daily profit, and what is the maximum profit?

3. The power output P of a battery of voltage E and internal resistance R is $P = EI - RI^2$, where I is the current. Find the current for which the power is a maximum.

4. The velocity v, in milligrams per second, at which a certain chemical reaction takes place is related to the amount x, in milligrams, of the chemical which is produced by $v = k(100x - x^2)$, where k is a constant. For what value of x is v a maximum?

5. A company projects that its total savings S, in dollars, by converting to a solar heating system with a solar collector area A (in m^2) will be $S = 360A - 0.1A^3$. Find the area which should give the maximum savings, and find the amount of the maximum savings.

6. The altitude h, in feet, of a jet which goes into a dive and then again turns upward is given by $h = 16t^3 - 240t^2 + 10,000$, where t is the time, in seconds, of the dive and turn. What is the altitude of the jet when it turns up out of the dive?

7. The cutting speed s (in feet per minute) of a saw in cutting a particular type of metal piece is given by $s = \sqrt{t - 4t^2}$, where t is the time in seconds. What is the maximum cutting speed in this operation?

8. The electric potential V on the line $3x + 2y = 6$ is given by $V = 3x^2 + 2y^2$. At what point on this line is the potential a minimum?

9. The product of two positive numbers is 64. Find the numbers if their sum is a minimum.

10. The sum of two numbers is 40. Find the numbers such that their product is a maximum.

11. A rectangular microprocessor chip is designed to have an area of 25 mm^2. What must be its dimensions if its perimeter is to be a minimum?

12. A rectangular storage area is to be constructed along the side of a tall building. A security fence is required along the remaining three sides of the area. What is the maximum area which can be enclosed with 800 ft of fencing?

13. A metal plate is to be in the shape of an isosceles triangle with each of the equal sides being 12.0 cm. What should be the length of the base if the plate is to have maximum area?

14. An architect is designing a rectangular building in which the front wall costs twice as much per linear meter as the other three walls. The building is to cover 1350 m^2. What dimensions must it have such that the cost of the walls is a minimum?

15. A computer is programmed to display a slowly changing right triangle with its hypotenuse always equal to 12.0 cm. What are the legs of the triangle when it has its maximum area?

16. A designer must inscribe a rectangle with the greatest area in the drawing of a quarter-circle of radius 2.00 in. What are the dimensions of the rectangle?

17. A culvert designed with a semicircular cross section of diameter 6.00 ft is redesigned to have an isosceles trapezoidal cross section by inscribing the trapezoid in the semicircle. See Fig. 23-42. What is the length of the bottom base b of the trapezoid if its area is to be maximum?

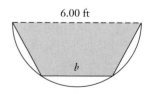

Fig. 23-42

18. A box with a square base is open at the top. If 108 ft^2 of material are used, what is the maximum volume possible for the box?

19. What is the maximum slope of the curve $y = 6x^2 - x^3$?

20. What is the minimum slope of the curve $y = x^5 - 10x^2$?

21. The deflection y of a beam of length L at a horizontal distance x from one end is given by $y = k(2x^4 - 5Lx^3 + 3L^2x^2)$, where k is a constant. For what value of x does the maximum deflection occur?

22. The potential energy E of an electric charge q due to another charge q_1 at a distance r_1 is proportional to qq_1 and inversely proportional to r_1. If charge q is placed directly between two charges of 2.00 nC and 1.00 nC which are separated by 10.0 mm, find the point at which the total potential energy (which is the sum due to the other two charges) of q is a minimum.

23. An open box is to be made from a square piece of cardboard whose sides are 8.00 in. long by cutting equal squares from the corners and bending up the sides. Determine the side of the square which is to be cut out so that the volume of the box may be a maximum. See Fig. 23-43.

24. A cone-shaped paper cup is to hold 100 cm^3 of water. Find the height and radius of the cup which can be made from the least amount of paper.

25. A racetrack 400 m long is to be built around an area which is a rectangle with a semicircle at each end. Find the open side of the rectangle if the area of the rectangle is to be a maximum. See Fig. 23-44.

26. A company finds that there is a net profit of $10 for each of the first 1000 units produced each week. For each unit over 1000 produced, there is 2 cents less profit per unit. How many units should be produced each week to net the greatest profit?

27. A beam of rectangular cross section is to be cut from a log 2.00 ft in diameter. The stiffness varies directly as the width and the cube of the depth. What dimensions will give the beam maximum stiffness? See Fig. 23-45.

28. An alpha particle moves through a magnetic field along the parabolic path $y = x^2 - 4$. Determine the closest that the particle comes to the origin.

29. An oil pipeline is to be built from a refinery to a tanker loading area. The loading area is 10.0 mi downstream from the refinery and on the opposite side of a river 2.5 mi wide. The pipeline is to run along the river and then cross to the loading area. If the pipeline costs $50,000 per mile alongside the river and $80,000 per mile across the river, find the point P (see Fig. 23-46) at which the pipeline should be turned to cross the river if construction costs are to be a minimum.

30. A light ray follows a path of least time. If a ray starts at point A (see Fig. 23-47) and is reflected off a plane mirror to point B, show that the angle of incidence α equals the angle of reflection β. (*Hint:* Set up the appropriate expression in terms of x, which will lead to $\sin \alpha = \sin \beta$.)

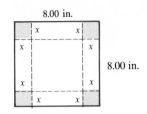

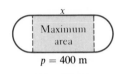

8.00 in.

8.00 in.

Fig. 23-43

x

Maximum area

$p = 400$ m

Fig. 23-44

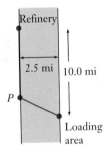

Fig. 23-45

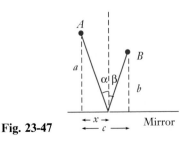

Refinery

2.5 mi   10.0 mi

P

Loading area

Fig. 23-46

A

a

B

α β

b

$\leftarrow x \rightarrow$ Mirror

$\leftarrow c \rightarrow$

Fig. 23-47

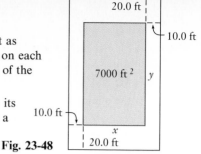

Fig. 23-48

31. A rectangular building covering 7000 ft² is to be built on a rectangular lot as shown in Fig. 23-48. If the building is to be 10.0 ft from the lot boundary on each side and 20.0 ft from the boundary in front and back, find the dimensions of the building if the area of the lot is a minimum.

32. A cylindrical wastewater holding tank has a capacity of 200 kL. What are its radius and height if its total surface area (top, bottom, and lateral area) is a minimum?

23-8 Chapter Equations, Review Exercises, and Practice Test

Chapter Equations

Newton's method
$$x_2 = x_1 - \frac{f(x_1)}{f'(x_1)} \tag{23-1}$$

Curvilinear motion
$$v_x = \frac{dx}{dt} \qquad v_y = \frac{dy}{dt} \tag{23-2}$$

$$a_x = \frac{dv_x}{dt} = \frac{d^2x}{dt^2} \qquad a_y = \frac{dv_y}{dt} = \frac{d^2y}{dt^2} \tag{23-3}$$

$$v = \sqrt{v_x^2 + v_y^2} \qquad a = \sqrt{a_x^2 + a_y^2} \tag{23-4}$$

$$\tan \theta_v = \frac{v_y}{v_x} \qquad \tan \theta_a = \frac{a_y}{a_x} \tag{23-5}$$

Curve sketching and maximum and minimum values

$f'(x) > 0$ where $f(x)$ increases; $f'(x) < 0$ where $f(x)$ decreases

$f''(x) > 0$ where the graph of $f(x)$ is concave up; $f''(x) < 0$ where the graph of $f(x)$ is concave down

If $f'(x) = 0$ at $x = a$, there is a maximum point if $f'(x)$ changes from $+$ to $-$, or if $f''(a) < 0$

If $f'(x) = 0$ at $x = a$, there is a minimum point if $f'(x)$ changes from $-$ to $+$, or if $f''(a) > 0$

If $f''(x) = 0$ at $x = a$, there is a point of inflection if $f''(x)$ changes from $+$ to $-$ or from $-$ to $+$

Review Exercises

In Exercises 1 through 6, find the equations of the tangent and normal lines.

1. Find the equation of the line tangent to the parabola $y = 3x - x^2$ at the point $(-1, -4)$.

2. Find the equation of the line tangent to the curve $y = x^2 - \dfrac{6}{x}$ at the point $(2, 1)$.

3. Find the equation of the line normal to $y = \dfrac{x}{4x - 1}$ at the point $(1, \frac{1}{3})$.

4. Find the equation of the line normal to $y = \dfrac{1}{\sqrt{x - 2}}$ at the point $(6, \frac{1}{2})$.

5. Find the equation of the line tangent to the curve $y = \sqrt{x^2 + 3}$ and which has a slope of $\frac{1}{2}$.

6. Find the equation of the line normal to the curve $y = \dfrac{1}{2x + 1}$ and which has a slope of $\frac{1}{2}$ if $x \geq 0$.

In Exercises 7 through 12, determine the required velocities and accelerations.

7. Given that the x- and y-coordinates of a moving particle are given as a function of time by the parametric equations $x = \sqrt{t} + t$, $y = \frac{1}{12}t^3$, find the magnitude and direction of the velocity when $t = 4$.

8. If the x- and y-coordinates of a moving object as functions of time are given by $x = 0.1t^2 + 1$, $y = \sqrt{4t + 1}$, find the magnitude and direction of the velocity when $t = 6$.

9. An object moves along the curve $y = 0.5x^2 + x$ such that $v_x = 0.5\sqrt{x}$. Find v_y at $(2, 4)$.

10. A particle moves along the curve of $y = \dfrac{1}{x + 2}$ with a constant velocity in the x-direction of 4 cm/s. Find v_y at $(2, \frac{1}{4})$.

11. Find the magnitude and direction of the acceleration for the particle in Exercise 7.

12. Find the magnitude and direction of the acceleration of the particle in Exercise 10.

In Exercises 13 through 16, find the indicated roots of the given equations to at least four decimal places by the use of Newton's method.

13. $x^3 - 3x^2 - x + 2 = 0$ (between 0 and 1)

14. $x^3 + 3x^2 - 6x - 2 = 0$ (between 1 and 2)

15. $3x^3 - x^2 - 8x - 2 = 0$ (between 1 and 2)

16. $x^4 + 3x^3 + 6x + 4 = 0$ (between -1 and 0)

In Exercises 17 through 24, sketch the graphs of the given functions by information obtained from the function as well as information obtained from the derivatives.

17. $y = 4x^2 + 16x$

18. $y = 2x^2 + x - 1$

19. $y = 27x - x^3$

20. $y = x^3 + 2x^2 + x + 1$

21. $y = x^4 - 32x$

22. $y = x^4 + 4x^3 - 16x$

23. $y = \dfrac{x}{x + 1}$

24. $y = x^3 + \dfrac{3}{x}$

In Exercises 25 through 48, solve the given problems.

25. The parabolas $y = x^2 + 2$ and $y = 4x - x^2$ are tangent to each other. Find the equation of the line tangent to them at the point of tangency.

26. Find the equation of the line tangent to $y = x^4 - 8x$ and parallel to $y + 4x + 3 = 0$.

27. The deflection y of a beam at a horizontal distance x from one end is given by $y = k(x^4 - 30x^3 + 1000x)$, where k is a constant. Observing the equation and using Newton's method, find the values of x where the deflection is zero, if the beam is 10.000 m long.

28. The edges of a rectangular water tank are 3.00 ft, 5.00 ft, and 8.00 ft. By Newton's method, determine by how much each edge should be increased equally to double the volume of the tank.

29. A parachutist descends (after the parachute opens) in a path which can be described by $x = 8t$ and $y = -0.15t^2$, where distances are in meters and time is in seconds. Find the parachutist's velocity upon landing if landing occurs when $t = 12$ s.

30. An electron is moving along the path $y = 1/x$ at the constant velocity of 100 m/s. Find the velocity in the x-direction when the electron is at the point $(2, \frac{1}{2})$.

31. In Fig. 23-49, the tension T supports the 40.0-N weight. The relation between the tension T and the deflection d is $d = \dfrac{1000}{\sqrt{T^2 - 400}}$. If the tension is increasing at 2.00 N/s when $T = 28.0$ N, how fast is the deflection, in centimeters, changing?

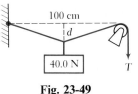

Fig. 23-49

32. The impedance Z, in ohms, in a particular electric circuit is given by $Z = \sqrt{48 + R^2}$, where R is the resistance. If R is increasing at the rate of 0.45 Ω/min for $R = 6.5$ Ω, find the rate at which Z is changing.

33. A calculator manufacturer determines that the monthly profit p, in dollars, from the sale of x calculators is given by $p = -0.25x^2 + 80x - 4000$. What is the maximum monthly profit which can be made?

34. The altitude h, in feet, of a rocket as a function of time after launching is given by $h = 1600t - 16t^2$. What is the maximum altitude which the rocket attains?

35. Sketch a continuous curve having the following characteristics:

$f(0) = 2$ $f'(x) < 0$ for $x < 0$ $f''(x) > 0$ for all x

$f'(x) > 0$ for $x > 0$

36. Sketch a continuous curve having the following characteristics:

$f(0) = 1$ $f'(0) = 0$ $f''(x) < 0$ for $x < 0$

$f'(x) > 0$ for $|x| > 0$ $f''(x) > 0$ for $x > 0$

37. The angle between two equal edges of a triangular plastic sheet is 120°. The plastic sheet is being stretched such that each of these edges is increasing at the rate of 0.750 cm/min. How fast is the third edge changing?

38. The current I through a circuit with a resistance R and a battery whose voltage is E and whose internal resistance is r is given by $I = E/(R + r)$. If R changes at the rate of 0.250 Ω/min, how fast is the current changing when $R = 6.25$ Ω, if $E = 3.10$ V and $r = 0.230$ Ω?

39. The radius of a circular oil spill is increasing at the rate of 15 m/min. How fast is the area of the spill changing when the radius is 400 m?

40. A machine part is to be in the shape of a circular sector of radius r and central angle θ. Find r and θ if the area is one unit and the perimeter is a minimum. See Fig. 23-50.

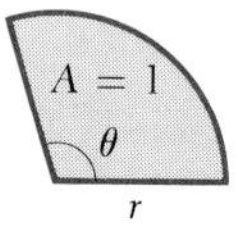

Fig. 23-50

41. A special insulation strip is to be sealed completely around three edges of a rectangular solar panel. If 200 cm of the strip are used, what is the maximum area of the panel?

42. A swimming pool with a rectangular surface of 1200 ft² is to have a cement border area which is 12.0 ft wide at each end and 8.00 ft wide at the sides. Find the surface dimensions of the pool if the total area covered is to be a minimum.

43. Sketch a graph of van der Waals' equation (see Exercise 42 of Section 22-6), assuming the following values: $R = T = a = 1$ and $b = 0$. For many gases the value of a is much greater than that for b. Even though the values of $R = T = 1$ are not realistic, the *shape* of the curve will be correct for the assumed value of $b = 0$.

44. The reciprocal of the total resistance R_T of electric resistances in parallel equals the sum of the reciprocals of the resistances of the individual resistors. If an 8-Ω resistor is in parallel with a variable resistance R, sketch the graph of R_T as a function of R.

45. An airplane flying horizontally at 8000 ft is moving toward a radar installation at 680 mi/h. If the plane is directly over a point on the ground 5.00 mi from the radar installation, what is its actual speed? See Fig. 23-51.

46. The base of a conical machine part is being milled such that the height is decreasing at the rate of 0.05 cm/min. If the part originally had a radius of 1.0 cm and a height of 3.0 cm, how fast is the volume changing when the height is 2.8 cm?

47. An open drawer for small tools is to be made from a rectangular piece of heavy sheet metal 12.0 in. by 10.0 in. by cutting out equal squares from two corners and bending up the three sides, as shown in Fig. 23-52. Determine the side of the square which should be cut out so that the volume of the drawer is a maximum.

48. A Norman window has the form of a rectangle surmounted by a semicircle. Find the dimensions (radius of circular part and height of rectangular part) of the window that will admit the most light if the perimeter of the window is 12 ft. See Fig. 23-53.

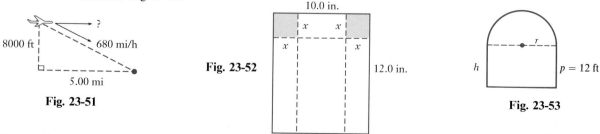

Fig. 23-51

Fig. 23-52

Fig. 23-53

Practice Test

1. Find the equation of the line tangent to the curve $y = x^4 - 3x^2$ at the point $(1, -2)$.

2. If the x- and y-coordinates of a moving object as functions of time are given by the parametric equations $x = 3t^2$, $y = 2t^3 - t^2$, find the magnitude and direction of the acceleration when $t = 2$.

3. The electric power (in watts) produced by a certain source is given by

$$P = \frac{144r}{(r + 0.6)^2}$$

where r is the resistance in the circuit. For what value of r is the power a maximum?

4. Find the root of the equation $x^2 - \sqrt{4x + 1} = 0$ between 1 and 2 to four decimal places by use of Newton's method. Use $x_1 = 1.5$ and find x_3.

5. Sketch the graph of $y = x^3 + 6x^2$ by finding the values of x for which the function is increasing, decreasing, concave up, and concave down, and by finding any maximum points, minimum points, and points of inflection.

6. Sketch the graph of $y = \dfrac{4}{x^2} - x$ by finding the same information as required in Problem 5, as well as intercepts, symmetry, behavior as x becomes large, vertical asymptotes, and the domain and range.

7. Trash is being compacted into a cubical volume. The edge of the cube is decreasing at the rate of 0.50 ft/s. When an edge of the cube is 4.00 ft, how fast is the volume changing?

8. A rectangular field is to be fenced and then divided in half by a fence parallel to two opposite sides. If a total of 6000 m of fencing is used, what is the maximum area which can be fenced?

24 Integration

In the study of physical and technical applications, it is common to find information about the rate of change of a variable. With such information we have to reverse the process of differentiation in order to arrive at the functional relationship between the variables. This leads us to consider the problem of finding the function when we know its derivative, a process known as *integration*.

As we study integration in this chapter and the next, we shall see that it has a great number of applications in science and technology. It can be applied to finding areas and volumes, as well as to physical concepts such as work and pressure. Although these appear to be quite

In Section 24-5 the use of integration is shown in finding the rate of flow of water over an obstacle.

different, we will show that they have a very similar mathematical formulation. A few of these applications are shown in this chapter, although most of them are developed in Chapter 25.

24-1 Differentials

Although the primary concern of this chapter is the development of the inverse process of differentiation, it is necessary to first introduce certain concepts and notation in this section before proceeding to the inverse process. The material presented here is necessary to properly relate differentiation and the methods we shall develop.

We shall now define the **differential** *of a function* $y = f(x)$ *as*

differential
of a function

$$dy = f'(x)\,dx \qquad (24\text{-}1)$$

In Eq. (24-1), the quantity *dy is the differential of y, and dx is the differential of x. The differential dx is defined as equal to* Δx, *the increment in x.* We define it purposely in this way, so that $f'(x) = dy/dx$. In this way we can interpret the derivative as the ratio of the differential of *y* to the differential of *x*. That is, now the derivative can be considered as a fraction. Although we had previously used the notation *dy/dx* for the derivative, we had not interpreted it as a fraction.

EXAMPLE A

Find the differential of $y = 3x^5 - x$.

In this example, $f(x) = 3x^5 - x$, which in turn means that $f'(x) = 15x^4 - 1$. Thus,

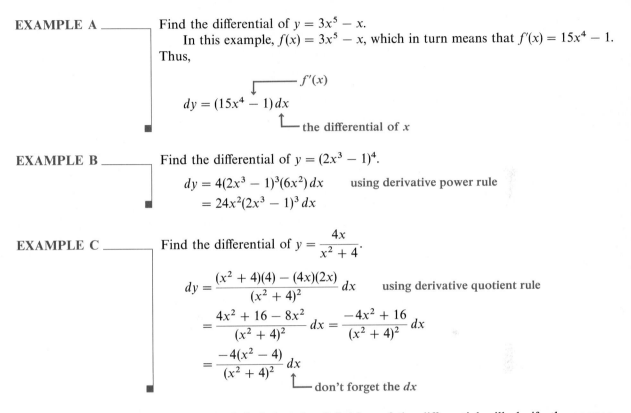

$$dy = (15x^4 - 1)\,dx$$

the differential of *x*

EXAMPLE B

Find the differential of $y = (2x^3 - 1)^4$.

$$dy = 4(2x^3 - 1)^3(6x^2)\,dx \qquad \text{using derivative power rule}$$
$$= 24x^2(2x^3 - 1)^3\,dx$$

EXAMPLE C

Find the differential of $y = \dfrac{4x}{x^2 + 4}$.

$$dy = \frac{(x^2 + 4)(4) - (4x)(2x)}{(x^2 + 4)^2}\,dx \qquad \text{using derivative quotient rule}$$
$$= \frac{4x^2 + 16 - 8x^2}{(x^2 + 4)^2}\,dx = \frac{-4x^2 + 16}{(x^2 + 4)^2}\,dx$$
$$= \frac{-4(x^2 - 4)}{(x^2 + 4)^2}\,dx$$

don't forget the *dx*

We shall find that the definition of the differential will clarify the connection between the various interpretations of the inverse process of differentiation. Although the principal purpose in introducing it here is to be able to use the notation, there are useful direct applications of the differential.

The applications of the differential are based upon the fact that the differential of *y*, *dy*, closely approximates the increment in *y*, Δy, if the differential of *x*, *dx*, is small. To understand this statement, let us look at Fig. 24-1. Recalling the meaning of Δx and Δy, we see that the points $P(x, y)$ and $Q(x + \Delta x, y + \Delta y)$ lie on the curve of $f(x)$. However, $f'(x) = dy/dx$ at *P*, which means if we draw a tangent line at *P*, its slope may be indicated by *dy/dx*. By choosing $\Delta x = dx$, we can see the difference between Δy and *dy*. It can be seen that as *dx* becomes smaller, Δy more nearly equals *dy*.

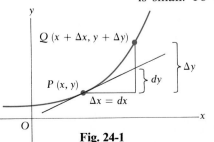

Fig. 24-1

For given changes in x, it is necessary to use the delta-process to find the exact change, Δy, in y. However, *for small values of Δx, dy can be used to approximate Δy closely.* Generally dy is much more easily determined than is Δy.

EXAMPLE D

Calculate Δy and dy for $y = x^3 - 2x$ for $x = 3$ and $\Delta x = 0.1$.

By the delta-process, we find

$$y + \Delta y = (x + \Delta x)^3 - 2(x + \Delta x)$$
$$\Delta y = 3x^2\,\Delta x + 3x(\Delta x)^2 + (\Delta x)^3 - 2\,\Delta x$$

Using the given values, we find

$$\Delta y = 3(9)(0.1) + 3(3)(0.01) + (0.001) - 2(0.1) = 2.591$$

The differential of y is

$$dy = (3x^2 - 2)\,dx$$

Since $dx = \Delta x$, we have

$$dy = [3(9) - 2](0.1) = 2.5$$

Thus, $\Delta y = 2.591$ and $dy = 2.5$. We see that in this case dy is very nearly equal to Δy. ∎

EXAMPLE E

If finding Δy by the delta-process is complicated or lengthy, we can use a calculator to find the difference between Δy and dy. For the function

$$y = f(x) = \sqrt{8x + 3}$$

the differential is

$$dy = \frac{1}{2}(8x + 3)^{-1/2}(8)\,dx = \frac{4\,dx}{\sqrt{8x + 3}}$$

For $x = 2$ and $\Delta x = 0.003$, we have

$$f(x) = \sqrt{8(2) + 3} = \sqrt{19} = 4.3588989$$

and

$$f(x + \Delta x) = \sqrt{8(2.003) + 3} = \sqrt{19.024} = 4.3616511$$

This means that

$$\Delta y = f(x + \Delta x) - f(x)$$
$$= 4.3616511 - 4.3588989 = 0.0027522$$

Now, calculating the value of dy, we have

$$dy = \frac{4(0.003)}{\sqrt{8(2) + 3}} = 0.0027530$$

We can again see that the values of dy and Δy are very nearly equal. ∎

The fact that dy can be used to approximate Δy is useful in determining the error in a result, if the data are in error, or the equivalent problem of finding the change in a result if a change is made in the data. Even though such changes could be found by use of a calculator, the differential can be used to set up a general expression for the change in a particular function. The following example illustrates the method.

EXAMPLE F — The edge of a cube of gold was measured to be 3.850 cm. From this value the volume was found. Later it was discovered that the value of the edge was 0.020 cm too small. By approximately how much was the volume in error?

The volume V of a cube, in terms of an edge, e, is $V = e^3$. Since we wish to find the change in V for a given change in e, this means we want the value of dV for $e = 3.850$ cm and $de = 0.020$ cm.

First, finding the general expression for dV, we have

$$dV = 3e^2\, de$$

Now, evaluating this expression for the given values, we have

$$dV = 3(3.850)^2(0.020) = 0.89 \text{ cm}^3$$

We see that in this case the volume was in error by about 0.89 cm^3. As long as de is small compared with e, we can calculate an error or change in the volume of a cube by calculating the value of $3e^2\, de$. ∎

Often when considering the error of a given value or result, *the actual numerical value of the error, the* **absolute error**, is not as important as its size is in relation to the size of the quantity itself. *The ratio of the absolute error to the size of the quantity itself is known as the* **relative error**. The relative error is commonly expressed as a percent.

EXAMPLE G — Referring to Example F, we see that the absolute error in the edge was 0.020 cm. The relative error in the edge was

$$\frac{de}{e} = \frac{0.020}{3.85} = 0.0052 = 0.52\%$$

The absolute error in the volume was 0.89 cm^3, and the original value of the volume was $3.850^3 = 57.07$ cm^3. This means the relative error in the volume was

$$\frac{dV}{V} = \frac{0.89}{57.07} = 0.016 = 1.6\%$$

∎

Exercises 24-1

In Exercises 1 through 12, find the differential of each of the given functions.

1. $y = x^5 + x$

2. $y = 3x^2 + 6$

3. $y = \dfrac{2}{x^5} + 3$

4. $y = 2x - \dfrac{1}{x}$

5. $y = (x^2 - 1)^4$

6. $y = (4 + 3x)^{1/3}$

7. $y = \dfrac{2}{3x^2 + 1}$

8. $y = \dfrac{1}{\sqrt{1 - x^3}}$

9. $y = x^2(1 - x)^3$ **10.** $y = x\sqrt{1 - 4x}$ **11.** $y = \dfrac{x}{5x + 2}$ **12.** $y = \dfrac{3x + 1}{\sqrt{2x - 1}}$

In Exercises 13 through 16, find the values of Δy and dy for the given values of x and Δx.

13. $y = 7x^2 + 4x$, $x = 4$, $\Delta x = 0.2$

14. $y = 2x^2 - 3x + 1$, $x = 5$, $\Delta x = 0.15$

15. $y = 2x^3 - 4x$, $x = 2.5$, $\Delta x = 0.05$

16. $y = x^4$, $x = 3.2$, $\Delta x = 0.08$

In Exercises 17 through 20, determine the value of dy for the given values of x and Δx. Compare with values of $f(x + \Delta x) - f(x)$ found by use of a calculator.

17. $y = (1 - 3x)^5$, $x = 1$, $\Delta x = 0.01$

18. $y = (x^2 + 2x)^3$, $x = 7$, $\Delta x = 0.02$

19. $y = x\sqrt{1 + 4x}$, $x = 12$, $\Delta x = 0.06$

20. $y = \dfrac{x}{\sqrt{6x - 1}}$, $x = 3.5$, $\Delta x = 0.025$

In Exercises 21 through 32, solve the given problems by finding the appropriate differential.

21. The side of a square microprocessor chip is measured as 0.950 cm, and later it is measured as 0.952 cm. What is the difference in the calculations of the area due to the difference in the measurements of the side? See Fig. 24-2.

22. A circular solar cell is made with a radius measured to be 12.00 ± 0.05 cm. (This means that the radius is 12.00 cm with a possible error of no more than 0.05 cm.) What is the maximum possible relative error in the calculation of the area of the cell?

23. The capacitance C, in microfarads, in an element of an electronic tuner is

$$C = \frac{3.6}{\sqrt{1 + 2V}},$$ where V is the voltage. What is the approximate change in C if V changes from 0.220 V to 0.224 V?

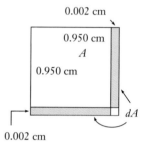

0.002 cm

0.950 cm

A

0.950 cm

dA

0.002 cm

Fig. 24-2

24. The velocity of an object rolling down a certain inclined plane is given by $v = \sqrt{100 + 16h}$, where h is the distance traveled along the plane by the object. What is the increase in velocity (in feet per second) of an object in moving from 20.0 ft to 20.5 ft along the plane? What is the relative change in the velocity?

25. A conical concrete cistern has an inner radius of 6.50 ft and an outer radius of 6.75 ft. If the cistern is 8.20 ft deep, what is the volume of concrete of which it is made?

26. A precisely measured 10.0-Ω resistor (any error in its value will be considered negligible) is put in parallel with a variable resistor of resistance R. The combined resistance of the two resistors is $R_T = \dfrac{10R}{10 + R}$. What is the relative error of the combined resistance if R is measured at 40.0 Ω, with a possible error of 1.5 Ω?

27. The radius r of a holograph is directly proportional to the square root of the wavelength λ of the light used. Show that $dr/r = \frac{1}{2}d\lambda/\lambda$.

28. The gravitational force F of the earth on an object is inversely proportional to the square of the distance r of the object from the center of the earth. Show that $dF/F = -2dr/r$.

29. Show that an error of 2% in the measurement of the side of a square results in an error of approximately 4% in the calculation of the area.

30. Show that the relative error in the calculation of the volume of a sphere is approximately three times the relative error in the measurement of the radius.

31. Using differentials, evaluate 2.03^4. (*Hint:* Let $y = x^4$.)

32. Using differentials, evaluate $\sqrt{9.02}$. (*Hint:* Let $y = \sqrt{x}$.)

24-2 Antiderivatives

Since many kinds of problems in many areas, including science and technology, can be solved by reversing the process of finding a derivative or a differential, we shall introduce the basic technique of this procedure in this section. *This reverse process is known as* **antidifferentiation.** In the next section we shall formalize the process, but it is only the basic idea that is the topic of this section. Many of the applications are found in Chapter 25. The following example illustrates the method.

EXAMPLE A ——— Find a function for which the derivative is $8x^3$. That is, find an antiderivative of $8x^3$.

We know that, when we set out to find the derivative of a polynomial, we reduce the power of x by 1. Also we multiply the coefficient by the power of x. Thus, the power of the function must have been 4, since the power in the derivative is 3. A factor of 4 of the derivative must also be divided out in the process of finding the function. If we write the derivative as $2(4x^3)$, we recognize $4x^3$ as the derivative of x^4. Therefore, the desired function is $2x^4$, which means that an antiderivative of $8x^3$ is $2x^4$. We can verify our result by finding the derivative. ∎

EXAMPLE B ——— Find an antiderivative of $x^2 + 2x$.

As for the x^2, we know that the power of x required in an antiderivative is 3. Also, to make the coefficient correct, we must multiply by $\frac{1}{3}$. The $2x$ should be recognized as the derivative of x^2. Therefore, we have as an antiderivative $\frac{1}{3}x^3 + x^2$. ∎

In Examples A and B, we note that we could add any constant to the antiderivative given as the result and still have a correct antiderivative. This is due to the fact that the derivative of a constant is zero. This is considered further in the following section. For the examples and exercises in this section, we will not include any such constants in the results.

We note that when we find an antiderivative of a given function, we obtain another function. Thus, *we can define an* **antiderivative** *of the function $f(x)$ to be a function $F(x)$ such that $F'(x) = f(x)$.*

EXAMPLE C ——— Find an antiderivative of the function $f(x) = \sqrt{x} - \dfrac{2}{x^3}$.

Since we wish to find an antiderivative of $f(x)$, we know that $f(x)$ is the derivative of the required function.

Considering the term $\sqrt{x}$, we first write it as $x^{1/2}$. To have x to the $\frac{1}{2}$ power in the derivative, we must have x to the $\frac{3}{2}$ power in the antiderivative. Knowing that the derivative of $x^{3/2}$ is $\frac{3}{2}x^{1/2}$, we write $x^{1/2}$ as $\frac{2}{3}(\frac{3}{2}x^{1/2})$. Thus, the first term of the antiderivative is $\frac{2}{3}x^{3/2}$.

As for the term $-2/x^3$, we write it as $-2x^{-3}$. This we recognize as the derivative of x^{-2}, or $1/x^2$.

(Continued on next page)

This means that an antiderivative of the function

$$\text{function} \qquad f(x) = \sqrt{x} - \frac{2}{x^3} \quad \longleftarrow \text{derivative}$$

is the function

$$\text{antiderivative} \longrightarrow F(x) = \frac{2}{3} x^{3/2} + \frac{1}{x^2} \qquad \text{function}$$

A great many functions of which we must find an antiderivative are not polynomials or simple powers of x. It is these functions which may cause more difficulty in the general process of antidifferentiation. The reader is advised to pay special attention to the following examples, for they illustrate a type of problem which *will* be found to be very important.

EXAMPLE D — Find an antiderivative of the function $f(x) = 3(x^3 - 1)^2(3x^2)$.

Noting that we have a power of $x^3 - 1$ in the derivative, it is reasonable that the antiderivative may include a power of $x^3 - 1$. Since, in the derivative, $x^3 - 1$ is raised to the power 2, the antiderivative would then have $x^3 - 1$ raised to the power 3. Noting that the derivative of $(x^3 - 1)^3$ is $3(x^3 - 1)^2(3x^2)$, the desired antiderivative is

$$F(x) = (x^3 - 1)^3$$

NOTE ▷ We note that **the factor of $3x^2$ does not appear in the antiderivative,** for it was included from the process of finding a derivative. Therefore, it must be present for $(x^3 - 1)^3$ to be the proper antiderivative, but it must also be excluded in the process of antidifferentiation.

EXAMPLE E — Find an antiderivative of the function $f(x) = (2x + 1)^{1/2}$.

Here we note a power of $2x + 1$ in the derivative, which infers that the antiderivative has a power of $2x + 1$. Since in finding a derivative 1 is subtracted from the power of $2x + 1$, we should add 1 in finding the antiderivative. Thus, we should have $(2x + 1)^{3/2}$ as part of the antiderivative. Finding a derivative of $(2x + 1)^{3/2}$, we obtain $\frac{3}{2}(2x + 1)^{1/2}(2) = 3(2x + 1)^{1/2}$. This differs from the given derivative by the factor of 3. Thus, if we write $(2x + 1)^{1/2} = \frac{1}{3}[3(2x + 1)^{1/2}]$, we have the required antiderivative as

$$F(x) = \frac{1}{3}(2x + 1)^{3/2}$$

Checking, the derivative of $\frac{1}{3}(2x + 1)^{3/2}$ is $\frac{1}{3}(\frac{3}{2})(2x + 1)^{1/2}(2) = (2x + 1)^{1/2}$.

Exercises 24-2

In the following exercises, find antiderivatives of the given functions.

1. $f(x) = 3x^2$ **2.** $f(x) = 5x^4$ **3.** $f(x) = 6x^5$ **4.** $f(x) = 10x^9$

5. $f(x) = 6x^3 + 1$ **6.** $f(x) = 12x^5 + 2x$ **7.** $f(x) = 2x^2 - x$ **8.** $f(x) = x^2 - 5$

9. $f(x) = \dfrac{5}{2} x^{3/2}$ **10.** $f(x) = \dfrac{4}{3} x^{1/3}$ **11.** $f(x) = 2\sqrt{x} + 3$ **12.** $f(x) = 3\sqrt[3]{x} - 4$

13. $f(x) = -\dfrac{1}{x^2}$ **14.** $f(x) = -\dfrac{7}{x^6}$ **15.** $f(x) = -\dfrac{6}{x^4}$ **16.** $f(x) = -\dfrac{8}{x^5}$

17. $f(x) = 2x^4 + 1$ **18.** $f(x) = 3x^3 - 5x^2 - 3$ **19.** $f(x) = 6x + \dfrac{1}{x^4}$ **20.** $f(x) = \dfrac{1}{2\sqrt{x}} + 4$

21. $f(x) = x^2 + 2 + x^{-2}$ **22.** $f(x) = x + 1 - x^{-3}$ **23.** $f(x) = 6(2x + 1)^5(2)$ **24.** $f(x) = 3(x^2 + 1)^2(2x)$

25. $f(x) = 4(x^2 - 1)^3(2x)$ **26.** $f(x) = 5(2x^4 + 1)^4(8x^3)$ **27.** $f(x) = x^3(2x^4 + 1)^4$ **28.** $f(x) = x(1 - x^2)^7$

29. $f(x) = \dfrac{3}{2}(6x + 1)^{1/2}(6)$ **30.** $f(x) = \dfrac{5}{4}(1 - x)^{1/4}(-1)$ **31.** $f(x) = (3x + 1)^{1/3}$ **32.** $f(x) = (4x + 3)^{1/2}$

24-3 The Indefinite Integral ▰▰▰▰▰▰▰▰▰▰▰

In the previous section, in developing the basic technique of finding an antiderivative, we noted that the results given are not unique. That is, we could have added any constant to the answers and the result would still have been correct. Again, this is the case since the derivative of a constant is zero.

EXAMPLE A ⎤ The derivatives of x^3, $x^3 + 4$, $x^3 - 7$, and $x^3 + 4\pi$ are all $3x^2$. This means that any of the functions listed, as well as others, would be a proper answer to the ▪ problem of finding an antiderivative of $3x^2$.

From Section 24-1, we know that the differential of a function $F(x)$ can be written as $d[F(x)] = F'(x)\,dx$. Therefore, since finding a differential of a function is very closely related to finding the derivative, so is the antiderivative very closely related to the process of finding the function for which the differential is known.

The notation used for finding the general form of the antiderivative, the **indefinite integral,** *is written in terms of the differential.* Thus, *the indefinite integral of a function $f(x)$, for which $dF(x)/dx = f(x)$, or $dF(x) = f(x)\,dx$, is defined as*

the indefinite integral

$$\int f(x)\,dx = F(x) + C \qquad\qquad (24\text{-}2)$$

Here $f(x)$ is called the **integrand,** *$F(x) + C$ is the indefinite integral, and C is an arbitrary constant, called the* **constant of integration.** *It represents any of the constants which may be attached to an antiderivative to have a proper result. We must have additional information beyond a knowledge of the differential to assign a specific value to C. The symbol $\int$ is the* **integral sign,** *and indicates that the inverse of the differential is to be found. Determining the indefinite integral is called* **integration,** *which we can see is essentially the same as finding an antiderivative.*

EXAMPLE B

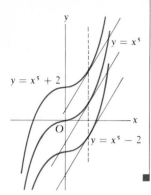

Fig. 24-3

In performing the integration

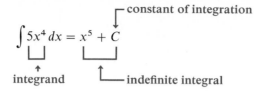

we might think that the inclusion of this constant C would affect the derivative of the function x^5. However, the only effect of the C is to raise or lower the curve. The slope of $x^5 + 2$, $x^5 - 2$, or any function of the form $x^5 + C$ is the same for any given value of x. As Fig. 24-3 shows, tangents drawn to the curves are all parallel for the same value of x. ■

At this point we shall derive some basic formulas for integration. Since

$$\frac{d(cu)}{dx} = c\frac{du}{dx}$$

where u is a function of x and c is a constant, we can write

$$\int c\,du = c\int du = cu + C \tag{24-3}$$

Also, since the derivative of a sum of functions equals the sum of the derivatives, we write

$$\int (du + dv) = u + v + C \tag{24-4}$$

To find the derivative of a power of a function, we multiply by the power and subtract 1 from it. *To find the integral, we reverse this by adding 1 to the power of $f(x)$ in $f(x)\,dx$, and dividing by this new power. The power formula for integration is therefore*

integral of $u^n\,du$

$$\int u^n\,du = \frac{u^{n+1}}{n+1} + C \qquad (n \neq -1) \tag{24-5}$$

When we use the formulas above, we must be able to recognize not only the proper form but also the component parts of the given formula. Unless you can do this and unless you have a good knowledge of differentiation, you will have trouble in applying Eq. (24-5). Most of the difficulty, if it exists, arises from improper identification of du. The following examples illustrate the use of the above formulas.

EXAMPLE C

Integrate: $\int 6x\,dx$.

First we must identify u, n, du, and any multiplying constants. By noting that 6 is a multiplying constant, we identify x as u, which in turn means that dx must be du and $n = 1$. Hence,

$$\int 6x\,dx = 6\int x^1\,dx = 6\left(\frac{x^2}{2}\right) + C = 3x^2 + C$$

do not forget the constant of integration

We see that our result checks since the differential of $3x^2 + C$ is $6x\,dx$.

EXAMPLE D

Integrate: $\int (5x^3 - 6x^2 + 1)\,dx$.

In solving this problem, we have to use a combination of all three of the above formulas. Thus, we have

$$\int (5x^3 - 6x^2 + 1)\,dx = \int 5x^3\,dx + \int(-6x^2)\,dx + \int dx$$

$$= 5\int x^3\,dx - 6\int x^2\,dx + \int dx$$

In the first of these integrals, $u = x$, $n = 3$, and $du = dx$. In the second, $u = x$, $n = 2$, and $du = dx$. The third is a direct application of Eq. (24-3), with $c = 1$ and $du = dx$. Thus, we have

$$5\int x^3\,dx - 6\int x^2\,dx + \int dx = 5\left(\frac{x^4}{4}\right) - 6\left(\frac{x^3}{3}\right) + x + C$$

$$= \frac{5}{4}x^4 - 2x^3 + x + C$$

EXAMPLE E

Integrate: $\int\left(\sqrt{x} - \frac{1}{x^3}\right)dx$.

In order to use Eq. (24-5) we must first write $\sqrt{x} = x^{1/2}$ and $1/x^3 = x^{-3}$. In this way we have

$$\int\left(\sqrt{x} - \frac{1}{x^3}\right)dx = \int x^{1/2}\,dx - \int x^{-3}\,dx$$

$$= \frac{1}{\frac{3}{2}}x^{3/2} - \frac{1}{-2}x^{-2} + C = \frac{2}{3}x^{3/2} + \frac{1}{2}x^{-2} + C$$

$$= \frac{2}{3}x^{3/2} + \frac{1}{2x^2} + C$$

EXAMPLE F

Integrate: $\int (x^2 + 1)^3 (2x\,dx)$.

We first note that $n = 3$, for this is the power involved in the function being integrated. If $n = 3$, then $x^2 + 1$ must be u. If $u = x^2 + 1$, then $du = 2x\,dx$. Thus the integral is in proper form for integration *as it stands*. Using the power formula, we have

$$\int (x^2 + 1)^3 (2x\,dx) = \frac{(x^2 + 1)^4}{4} + C$$

It is in this type of problem that recognition of the proper form of a differential is very important. Only a good knowledge of differential forms, along with sufficient experience, makes ready recognition of this type possible. *It cannot be overemphasized that the entire quantity (2x dx) must be equated to du if we are to integrate properly.* Normally u and n are recognized first, and then the proper form of du is derived from u.

NOTE▷

Showing the use of u directly, we can write the integration as

$$\int (x^2 + 1)^3 (2x\,dx) = \int u^3\,du = \frac{1}{4}u^4 + C = \frac{(x^2 + 1)^4}{4} + C$$

EXAMPLE G

Integrate: $\int x^2 \sqrt{x^3 + 2}\,dx$.

We first note that $n = \frac{1}{2}$ and u is then $x^3 + 2$. Since $u = x^3 + 2$, $du = 3x^2\,dx$. In order to integrate properly, we must group the quantity $3x^2\,dx$ as du, with other factors isolated from this quantity. Since there is no 3 under the integral sign, we introduce one. In order not to change the numerical value, we also introduce $\frac{1}{3}$, normally before the integral sign. In this way we take full advantage of the fact that *a constant (and only a constant) factor may be moved across the integral sign.* The next form we should write is

NOTE▷

$$\int x^2 \sqrt{x^3 + 2}\,dx = \frac{1}{3}\int 3x^2 \sqrt{x^3 + 2}\,dx = \frac{1}{3}\int \sqrt{x^3 + 2}(3x^2\,dx)$$

In this last form we indicate the proper grouping to result in the correct form of Eq. (24-5). Thus,

$$\int x^2 \sqrt{x^3 + 2}\,dx = \frac{1}{3}\int \sqrt{x^3 + 2}(3x^2\,dx) = \frac{1}{3}\left(\frac{2}{3}\right)(x^3 + 2)^{3/2} + C$$

$$= \frac{2}{9}(x^3 + 2)^{3/2} + C$$

The $1/\frac{3}{2}$ was written as $\frac{2}{3}$, since this form is generally more convenient when we are working with fractions.

With $u = x^3 + 3$, and using u directly in the integration, we can write the integral as

$$\int x^2 \sqrt{x^3 + 2}\,dx = \int (x^3 + 2)^{1/2}(x^2\,dx)$$

$$= \int u^{1/2}\left(\frac{1}{3}\,du\right) = \frac{1}{3}\int u^{1/2}\,du \qquad \text{integrating in terms of } u$$

$$= \frac{1}{3}\left(\frac{2}{3}\right)u^{3/2} + C = \frac{2}{9}u^{3/2} + C$$

$$= \frac{2}{9}(x^3 + 2)^{3/2} + C \qquad \text{substituting } x^3 + 2 = u$$

We mentioned earlier that, in addition to knowing the differential, we need more information in order to find the constant of integration. Such information usually consists of a set of values which the function is known to satisfy. That is, a point through which the curve of the function passes would provide the necessary information.

EXAMPLE H

Find y in terms of x, given that $dy/dx = 3x - 1$ and the curve passes through $(1, 4)$.

We write the equation as

$$dy = (3x - 1)\,dx$$

and then indicate and perform the integration:

$$\int dy = \int (3x - 1)\,dx$$

$$y = \frac{3}{2}x^2 - x + C$$

We know that the required curve passes through $(1, 4)$, which means that the coordinates of this point must satisfy the equation. Thus,

$$4 = \frac{3}{2} - 1 + C, \quad \text{or} \quad C = \frac{7}{2}$$

This means that the solution is

$$y = \frac{3}{2}x^2 - x + \frac{7}{2} \quad \text{or} \quad 2y = 3x^2 - 2x + 7$$

The graph of this parabola is shown in Fig. 24-4.

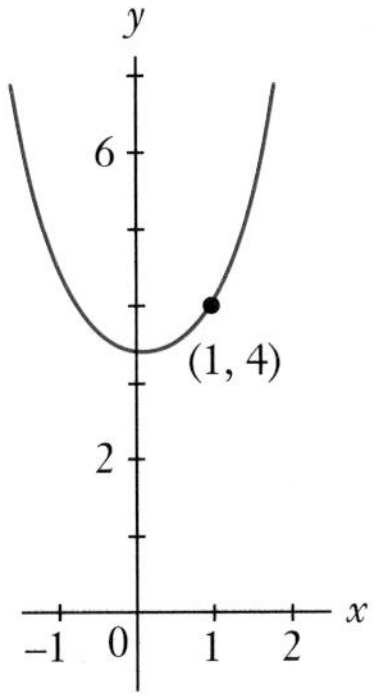

Fig. 24-4

EXAMPLE I

The time rate of change of the displacement (velocity) of a robot arm is $ds/dt = t\sqrt{9 - t^2}$. Find the expression for the displacement as a function of time if $s = 0$ cm when $t = 0$ s.

We start the solution by writing

$$ds = t\sqrt{9 - t^2}\,dt$$

$$\int ds = \int t(9 - t^2)^{1/2}\,dt$$

To integrate the expression on the right, we recognize that $n = \frac{1}{2}$ and $u = 9 - t^2$, and therefore $du = -2t\,dt$. This means we need a -2 with the t and dt to form the proper du, which in turn means we place a $-\frac{1}{2}$ before the integral sign.

$$\int ds = \int t(9 - t^2)^{1/2}\,dt = -\frac{1}{2}\int (9 - t^2)^{1/2}(-2t\,dt)$$

$$s = -\frac{1}{2}\left(\frac{2}{3}\right)(9 - t^2)^{3/2} + C$$

$$= -\frac{1}{3}(9 - t^2)^{3/2} + C$$

(Continued on next page)

Since $s = 0$ cm when $t = 0$ s, we have

$$0 = -\frac{1}{3}(9)^{3/2} + C, \qquad C = 9$$

This means that

$$s = -\frac{1}{3}(9 - t^2)^{3/2} + 9$$

■ is the expression for the displacement.

In this section we have discussed the integration of certain basic types of functions. There are many other methods used to integrate other functions, and some of these methods are discussed in Chapter 27. Also, there are many functions which cannot be integrated.

Exercises 24-3

In Exercises 1 through 32, integrate each of the given expressions.

1. $\int 2x \, dx$

2. $\int 5x^4 \, dx$

3. $\int x^7 \, dx$

4. $\int x^5 \, dx$

5. $\int x^{3/2} \, dx$

6. $\int \sqrt[3]{x} \, dx$

7. $\int x^{-4} \, dx$

8. $\int \frac{1}{\sqrt{x}} \, dx$

9. $\int (x^2 - x^5) \, dx$

10. $\int (1 - 3x) \, dx$

11. $\int (9x^2 + x + 3) \, dx$

12. $\int (4x^2 - 2x + 5) \, dx$

13. $\int \left(\frac{1}{x^3} + \frac{1}{2} \right) dx$

14. $\int \left(4x - \frac{2}{x^3} \right) dx$

15. $\int \sqrt{x}(x^2 - x) \, dx$

16. $\int (x\sqrt{x} - 5x^2) \, dx$

17. $\int (2x^{-2/3} + 3^{-2}) \, dx$

18. $\int (x^{1/3} + x^{1/5} + x^{-1/7}) \, dx$

19. $\int (1 + 2x)^2 \, dx$

20. $\int (1 - x)^2 \, dx$

21. $\int (x^2 - 1)^5 (2x \, dx)$

22. $\int (x^3 - 2)^6 (3x^2 \, dx)$

23. $\int (x^4 + 3)^4 (4x^3 \, dx)$

24. $\int (1 - 2x)^{1/3} (-2 \, dx)$

25. $\int (x^5 + 4)^7 x^4 \, dx$

26. $\int 6x^2 (1 - x^3)^{4/3} \, dx$

27. $\int \sqrt{8x + 1} \, dx$

28. $\int \sqrt[3]{4 - 3x} \, dx$

29. $\int \frac{x \, dx}{\sqrt{6x^2 + 1}}$

30. $\int \frac{2x^2 \, dx}{\sqrt{2x^3 + 1}}$

31. $\int \frac{x - 1}{\sqrt{x^2 - 2x}} \, dx$

32. $\int (x^2 - x) \left(x^3 - \frac{3}{2} x^2 \right)^8 dx$

In Exercises 33 through 36, find y in terms of x.

33. $\frac{dy}{dx} = 6x^2$, curve passes through $(0, 2)$

34. $\frac{dy}{dx} = x + 1$, curve passes through $(-1, 4)$

35. $\frac{dy}{dx} = x^2(1 - x^3)^5$, curve passes through $(1, 5)$

36. $\frac{dy}{dx} = 2x^3(x^4 - 6)^4$, curve passes through $(2, 10)$

In Exercises 37 through 44, find the required equations.

37. Find the equation of the curve whose slope is $-x\sqrt{1 - 4x^2}$ and which passes through $(0, 7)$.

38. Find the equation of the curve whose slope is $\sqrt{6x - 3}$ and which passes through $(2, -1)$.

39. The time rate of change of electric current in a circuit is given by $di/dt = 4t - 0.6t^2$. Find the expression for the current as a function of time if $i = 2$ A when $t = 0$ s.

40. The time rate of change of velocity (acceleration) of a roller mechanism is $dv/dt = 3\sqrt{t} - 2$. Find the expression for the velocity as a function of time if $v = 2$ cm/s when $t = 4$ s.

41. At a given site, the rate of change of the annual fraction f of energy supplied by solar energy with respect to the solar collector area A (in m^2) is $\dfrac{df}{dA} = \dfrac{0.005}{\sqrt{0.01A + 1}}$. Find f as a function of A if $f = 0$ for $A = 0$ m^2.

42. The rate of change of the frequency f of an electronic oscillator with respect to the inductance L is $df/dL = 80(4 + L)^{-3/2}$. Find f as a function of L if $f = 80$ Hz for $L = 0$ H.

43. Find the equation of the curve for which the second derivative is 6. The curve passes through (1, 2) with a slope of 8.

44. Find the equation of the curve for which the second derivative is $12x^2$. The curve passes through (1, 6) with a slope of 4.

24-4   The Area Under a Curve

Another basic problem which can be solved by integration is that of finding the area under a curve. In geometry, there are methods and formulas for finding the area of regular figures. By means of the calculus we will find it possible to find the area between curves for which we know the equations. First let us look at an example which illustrates the basic idea behind the method.

EXAMPLE A ——— Approximate the area in the first quadrant and to the left of the line $x = 4$ under the parabola $y = x^2 + 1$. First make this approximation by inscribing two rectangles of equal width in the area and finding the sum of the areas of these rectangles. Then improve the approximation by repeating the process using eight rectangles.

The area to be approximated is shown in Fig. 24-5(a). The area with two rectangles inscribed under the curve is shown in Fig. 24-5(b). The first approximation, admittedly small, of the area can be found by adding the areas of the two rectangles. Both rectangles have a width of 2. The left rectangle is 1 unit high and the right rectangle is 5 units high. Thus, the area of the two rectangles is

$$A = 2(1 + 5) = 12$$

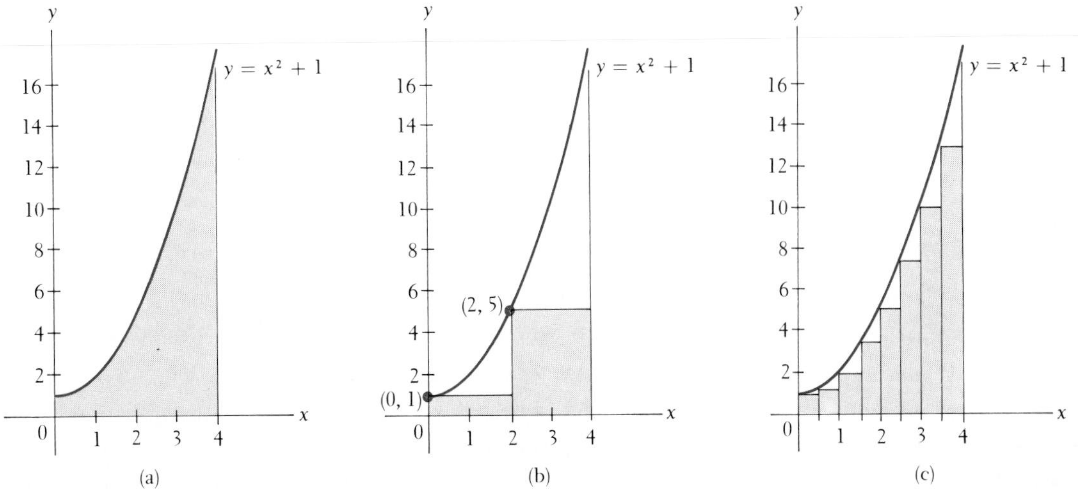

Fig. 24-5

(a)   (b)   (c)

(*Continued on next page*)

TABLE 24-1

Number of rectangles n	Total area of rectangles
8	21.5
100	25.0144
1,000	25.301344
10,000	25.330134

A much better approximation is found by inscribing the eight rectangles as shown in Fig. 24-5(c). Each of these rectangles has a width of $\frac{1}{2}$. The leftmost rectangle has a height of 1. The next has a height of $\frac{5}{4}$, which is determined by finding y for $x = \frac{1}{2}$. The next rectangle has a height of 2, which is found by evaluating y for $x = 1$. Finding the heights of all the rectangles and multiplying their sum by $\frac{1}{2}$ gives the area of the eight rectangles as

$$A = \frac{1}{2}\left(1 + \frac{5}{4} + 2 + \frac{13}{4} + 5 + \frac{29}{4} + 10 + \frac{53}{4}\right) = \frac{43}{2} = 21.5$$

An even better approximation could be obtained by inscribing more rectangles under the curve. The greater the number of rectangles, the more nearly the sum of their areas equals the area under the curve. See Table 24-1. By a method involving integration developed later in this section, we determine the ■ *exact* area to be $\frac{76}{3} = 25\frac{1}{3}$.

We can now develop the basic method involved in finding the area under a curve. In general, the area under a curve is the area bounded by the curve, the x-axis, and the lines $x = a$ and $x = b$. See Fig. 24-6. We will assume here that $f(x)$ is never negative in the interval $a \leq x \leq b$. In Chapter 25 we will extend the method such that $f(x)$ may be negative.

In finding the area under a curve we consider the sum of the areas of inscribed rectangles, as the number of rectangles is assumed to increase without bound. The reason for this last condition is that, as we saw in Example A, as the number of rectangles increases, the approximation of the area is better. Following is a specific example of this technique.

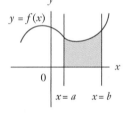

Fig. 24-6

EXAMPLE B

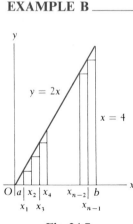

Fig. 24-7

Find the area under the straight line $y = 2x$, above the x-axis, and to the left of the line $x = 4$.

We might mention at the beginning that this area can easily be found, since the figure is a triangle. However, the method we shall use here is the important concept at the moment. We first subdivide the interval from $x = 0$ to $x = 4$ into n inscribed rectangles of Δx in width. Then the extremities of the intervals are labeled $a, x_1, x_2, \ldots, b \ (= x_n)$, as shown in Fig. 24-7, where

$$x_1 = \Delta x$$
$$x_2 = 2\,\Delta x$$
$$\vdots$$
$$x_{n-1} = (n - 1)\,\Delta x$$
$$b = n\,\Delta x$$

We then find the area for each of the rectangles. These areas are (for the respective rectangles):

First $f(a)\,\Delta x$, where $f(a) = f(0) = 2(0) = 0$ is the height.

Second $f(x_1)\,\Delta x$, where $f(x_1) = 2(\Delta x) = 2\,\Delta x$ is the height.

Third $f(x_2)\,\Delta x$, where $f(x_2) = 2(2\,\Delta x) = 4\,\Delta x$ is the height.

Fourth $f(x_3)\,\Delta x$, where $f(x_3) = 2(3\,\Delta x) = 6\,\Delta x$ is the height.

$$\vdots$$

Last $f(x_{n-1})\,\Delta x$, where $f[(n-1)\,\Delta x] = 2(n-1)\,\Delta x$ is the height.

These areas are summed up as follows:

NOTE ▷

$$A_n = \boxed{f(a)\,\Delta x + f(x_1)\,\Delta x + f(x_2)\,\Delta x + \cdots + f(x_{n-1})\,\Delta x}$$

$$= 0 + 2\,\Delta x(\Delta x) + 4\,\Delta x(\Delta x) + \cdots + 2[(n-1)\,\Delta x]\,\Delta x$$
$$= 2(\Delta x)^2[1 + 2 + 3 + \cdots + (n-1)]$$

Now $b = n\,\Delta x$, or $4 = n\,\Delta x$, or $\Delta x = 4/n$. Thus,

$$A_n = 2\left(\frac{4}{n}\right)^2 [1 + 2 + 3 + \cdots + (n-1)]$$

The sum $1 + 2 + 3 + \cdots + n - 1$ is the indicated sum of an arithmetic sequence with the first term equal to 1 and the nth term equal to $n - 1$. This sum is

$$s = \frac{n-1}{2}(1 + n - 1) = \frac{n(n-1)}{2} = \frac{n^2 - n}{2}$$

Now the expression for the sum of the areas can be set forth as

$$A_n = \frac{32}{n^2}\left(\frac{n^2 - n}{2}\right) = 16\left(1 - \frac{1}{n}\right)$$

This expression is an approximation of the actual area under consideration. The larger n becomes, the better the approximation. If we let $n \to \infty$ (which is equivalent to letting $\Delta x \to 0$), the limit of this sum will equal the area in question. Thus,

$$\overset{\displaystyle\ulcorner 1/n \to 0 \text{ as } n \to \infty}{A = \lim_{n \to \infty} 16\left(1 - \frac{1}{n}\right) = 16}$$

(This checks with the geometric result.) The area under the curve can be considered as the limit of the sum of the areas of the inscribed rectangles as the number of rectangles approaches infinity.

The method indicated in Example B illustrates the interpretation of finding an area as a summation process, although it should not be considered as a proof. However, we shall find that integration proves to be a much more useful method for finding an area. Let us now see how integration can be used directly.

Let ΔA represent the area $BCEG$ under the curve, as indicated in Fig. 24-8. We see that the following inequality is true for the indicated areas:

$$A_{BCDG} < \Delta A < A_{BCEF}$$

If the point G is now designated as (x, y) and E as $(x + \Delta x, y + \Delta y)$, we have $y\,\Delta x < \Delta A < (y + \Delta y)\,\Delta x$. Dividing through by Δx, we have

$$y < \frac{\Delta A}{\Delta x} < y + \Delta y$$

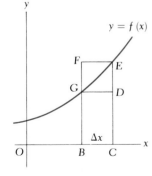

Fig. 24-8

Now we take the limit as $\Delta x \to 0$ (Δy then also approaches 0). This results in

$$\frac{dA}{dx} = y \tag{24-6}$$

This is true since the left member of the inequality is y and the right member approaches y. Also remember that

$$\lim_{\Delta x \to 0} \frac{\Delta A}{\Delta x} = \frac{dA}{dx}$$

We shall now use Eq. (24-6) to show the method of finding the complete area under a curve. We now let $x = a$ be the left boundary of the desired area, and $x = b$ be the right boundary (Fig. 24-9). The area under the curve to the right of $x = a$ and bounded on the right by the line GB is now designated as A_{ax}. From Eq. (24-6) we have

$$dA_{ax} = [y\,dx]_a^x \quad \text{or} \quad A_{ax} = \left[\int y\,dx\right]_a^x$$

where $[\quad]_a^x$ is the notation used to indicate the boundaries of the area. Thus,

$$A_{ax} = \left[\int f(x)\,dx\right]_a^x = [F(x) + C]_a^x \tag{24-7}$$

But we know that if $x = a$, then $A_{aa} = 0$. Thus, $0 = F(a) + C$, or $C = -F(a)$. Therefore,

$$A_{ax} = \left[\int f(x)\,dx\right]_a^x = F(x) - F(a) \tag{24-8}$$

Now, to find the area under the curve that reaches from a to b, we write

$$A_{ab} = F(b) - F(a) \tag{24-9}$$

Thus the area under the curve that reaches from a to b is given by

$$A_{ab} = \left[\int f(x)\,dx\right]_a^b = F(b) - F(a) \tag{24-10}$$

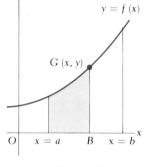

$y = f(x)$

$G(x, y)$

O $x = a$ B $x = b$

Fig. 24-9

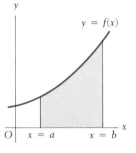

$y = f(x)$

O $x = a$ $x = b$

Fig. 24-10

integration as
summation

This shows that *the area under the curve may be found by integrating the function $f(x)$ to find the function $F(x)$, which is then evaluated at each boundary value. The area is the difference between these values of $F(x)$.* See Fig. 24-10.

In Example B, we found an area under a curve by finding the limit of the sum of the inscribed rectangles as the number of rectangles approaches infinity. Equation (24-10) expresses the area under a curve in terms of integration. We can now see that we have obtained the area by summation and also expressed it in terms of integration. *Thus, we conclude that summations can be evaluated by integration.* Also, we have grasped the connection between the problem of finding the slope of a tangent to a curve (differentiation) and the problem of finding an area (integration). We would not normally suspect that these two problems would have solutions which lead to reverse processes. We have also seen that the definition of integration has much more application than originally anticipated.

EXAMPLE C

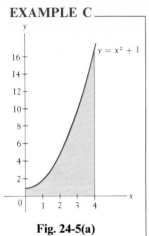

Find the area under the curve $y = x^2 + 1$ between the y-axis and the line $x = 4$. This is the area of Example A shown in Fig. 24-5(a), which is given again here for reference.

Fig. 24-5(a)

In Eq. (24-10), we note that $f(x) = x^2 + 1$. This means that

$$\int (x^2 + 1)\,dx = \frac{1}{3}x^3 + x + C$$

Therefore, with $F(x) = \frac{1}{3}x^3 + x$, the area is given by

$$\begin{aligned} A_{0,4} &= F(4) - F(0) && \text{using Eq. (24-10)} \\ &= \left[\frac{1}{3}(4^3) + 4\right] - \left[\frac{1}{3}(0^3) + 0\right] \\ &= \frac{1}{3}(64) + 4 = \frac{76}{3} \end{aligned}$$

We note that this is about 4 square units greater than the value obtained by the approximation in Example A. This result means that the exact area is $25\frac{1}{3}$, as stated at the end of Example A. ∎

EXAMPLE D

Find the area under the curve $y = x^3$ between the lines $x = 1$ and $x = 2$.

In Eq. (24-10), $f(x) = x^3$. Therefore,

$$\int x^3\,dx = \underbrace{\frac{1}{4}x^4}_{F(x)} + C$$

$$\begin{aligned} A_{1,2} = F(2) - F(1) &= \left[\frac{1}{4}(2^4)\right] - \left[\frac{1}{4}(1^4)\right] && \text{using Eq. (24-10)} \\ &= 4 - \frac{1}{4} = \frac{15}{4} \end{aligned}$$

Fig. 24-11

Therefore, the required area is $\frac{15}{4}$. Again, this is the exact area, not an approximation. See Fig. 24-11. ∎

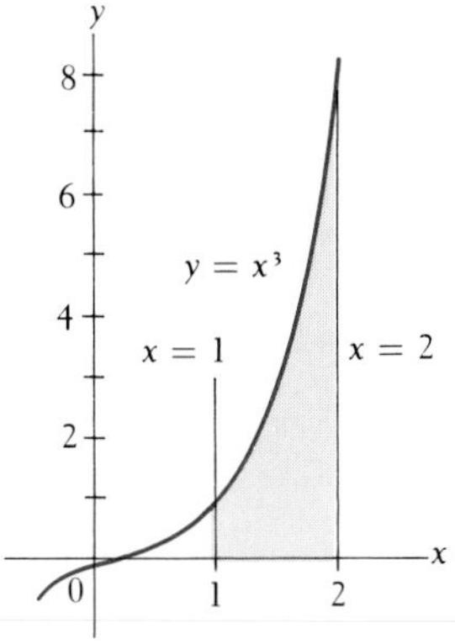

EXAMPLE E

Find the area under the curve $y = 4 - x^2$ which lies in the first quadrant.

By solving the equation $4 - x^2 = 0$, we determine that the area to be found extends from $x = 0$ to $x = 2$ (see Fig. 24-12). Thus,

$$\int (4 - x^2)\,dx = \underbrace{4x - \frac{x^3}{3}}_{F(x)} + C$$

$$A_{0,2} = \left(8 - \frac{8}{3}\right) - (0 - 0) = \frac{16}{3} \qquad \text{using Eq. (24-10)}$$

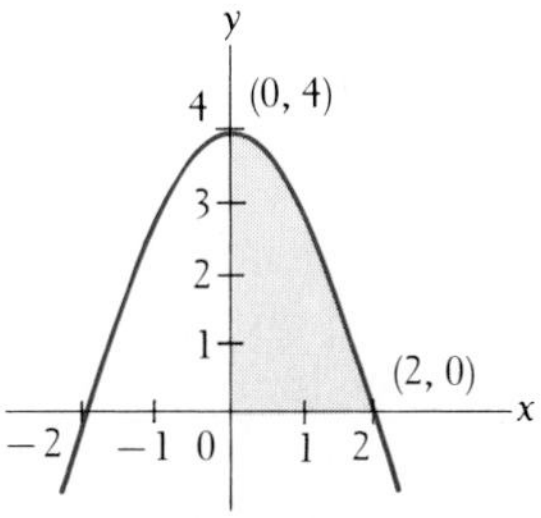

Fig. 24-12

Therefore, the exact area is $\frac{16}{3}$. ∎

We can see from these examples that we do not have to include the constant of integration when we are finding areas. Any constant which might be added to $F(x)$ would cancel out when $F(a)$ is subtracted from $F(b)$.

Exercises 24-4

In each of Exercises 1 through 8, find the approximate area under the given curves by dividing the indicated intervals into n subintervals, and add up the areas of the inscribed rectangles. There are two values of n for each, and therefore two approximations for each area. The height of each rectangle may be found by evaluating the function for the proper value of x. See Example A.

See Appendix E for a computer program for finding the area by summing the areas of n rectangles.

1. $y = 3x$, between $x = 0$ and $x = 3$, for (a) $n = 3$ ($\Delta x = 1$), (b) $n = 10$ ($\Delta x = 0.3$)

2. $y = 2x + 1$, between $x = 0$ and $x = 2$, for (a) $n = 4$ ($\Delta x = 0.5$), (b) $n = 10$ ($\Delta x = 0.2$)

3. $y = x^2$, between $x = 0$ and $x = 2$, for (a) $n = 5$ ($\Delta x = 0.4$), (b) $n = 10$ ($\Delta x = 0.2$)

4. $y = x^2 + 2$, between $x = 0$ and $x = 3$, for (a) $n = 3$ ($\Delta x = 1$), (b) $n = 10$ ($\Delta x = 0.3$)

5. $y = 4x - x^2$, between $x = 1$ and $x = 4$, for (a) $n = 6$, (b) $n = 10$

6. $y = 1 - x^2$, between $x = 0.5$ and $x = 1$, for (a) $n = 5$, (b) $n = 10$

7. $y = \dfrac{1}{x^2}$, between $x = 1$ and $x = 5$, for (a) $n = 4$, (b) $n = 8$

8. $y = \sqrt{x}$, between $x = 1$ and $x = 4$, for (a) $n = 3$, (b) $n = 12$

In Exercises 9 through 16, find the exact area under the given curves between the indicated values of x. The functions are the same as those for which approximate areas were found in Exercises 1 through 8.

9. $y = 3x$, between $x = 0$ and $x = 3$

10. $y = 2x + 1$, between $x = 0$ and $x = 2$

11. $y = x^2$, between $x = 0$ and $x = 2$

12. $y = x^2 + 2$, between $x = 0$ and $x = 3$

13. $y = 4x - x^2$, between $x = 1$ and $x = 4$

14. $y = 1 - x^2$, between $x = 0.5$ and $x = 1$

15. $y = \dfrac{1}{x^2}$, between $x = 1$ and $x = 5$

16. $y = \sqrt{x}$, between $x = 1$ and $x = 4$

24-5   The Definite Integral

Using reasoning similar to that in the preceding section, *we define the* **definite integral** *of a function f(x) as*

the definite integral

$$\int_a^b f(x)\, dx = F(b) - F(a)$$

(24-11)

NOTE ▷

where $F'(x) = f(x)$. *We call this a* **definite integral** *because the final result of integrating and evaluating is* **a number.** (The *indefinite* integral had an arbitrary constant in the result.) *The numbers a and b are called the* **lower limit** *and the*

limits of integration

upper limit, *respectively. We can see that the value of a definite integral is found by evaluating the function (found by integration) at the upper limit and subtracting the value of this function at the lower limit.*

We know, from the analysis in the preceding section, that this definite integral can be interpreted as a summation process, where the size of the subdivision approaches a limit of zero. This fact explains the choice of the $\int$ symbol for integration: It is an elongated S, representing the sum. It is this interpretation of integration which we shall apply to many kinds of problems.

EXAMPLE A _____ Evaluate the integral $\int_0^2 x^4\, dx$.

We write

upper limit

$$\int_0^2 x^4\, dx = \frac{x^5}{5}\Big|_0^2 = \frac{2^5}{5} - 0 = \frac{32}{5}$$

lower limit — $f(x)$ $F(x)$

Note that a vertical line—with the limits written at the top and the bottom—is the way the value is indicated after integration, but before evaluation.

EXAMPLE B _____ Evaluate $\int_1^3 (x^{-2} - 1)\, dx$.

We set this forth as

substitute upper limit   subtract   substitute lower limit

$$\int_1^3 (x^{-2} - 1)\, dx = -\frac{1}{x} - x\Big|_1^3 = \left(-\frac{1}{3} - 3\right) \; - \; (-1 - 1)$$

$$= -\frac{10}{3} + 2 = -\frac{4}{3}$$

EXAMPLE C _____ Evaluate $\int_0^1 5x(x^2 + 1)^5\, dx$.

For purposes of integration, $n = 5$, $u = x^2 + 1$, and $du = 2x\, dx$. Hence,

$$\int_0^1 5x(x^2 + 1)^5\, dx = \frac{5}{2}\int_0^1 (x^2 + 1)^5 (2x\, dx)$$

$$= \frac{5}{2}\left(\frac{1}{6}\right)(x^2 + 1)^6\Big|_0^1 \qquad \text{integrate}$$

$$= \frac{5}{12}(2^6 - 1^6) \qquad\qquad \text{evaluate}$$

$$= \frac{5(63)}{12} = \frac{105}{4}$$

EXAMPLE D

Evaluate $\int_{-1}^{3} x(1 - 3x^2)^{1/3}\, dx$.

For purposes of integration, $n = \frac{1}{3}$, $u = 1 - 3x^2$, and $du = -6x\, dx$. Thus

$$\int_{-1}^{3} x(1 - 3x^2)^{1/3}\, dx = -\frac{1}{6}\int_{-1}^{3} (1 - 3x^2)^{1/3}(-6x\, dx)$$

$$= -\frac{1}{6}\left(\frac{3}{4}\right)(1 - 3x^2)^{4/3}\Big|_{-1}^{3} \qquad \text{integrate}$$

$$= -\frac{1}{8}(-26)^{4/3} + \frac{1}{8}(-2)^{4/3} \qquad \text{evaluate}$$

$$= \frac{1}{8}(2\sqrt[3]{2} - 26\sqrt[3]{26}) = \frac{1}{4}(\sqrt[3]{2} - 13\sqrt[3]{26})$$

The result shown is the exact value. The approximate value to three decimal places (from a calculator) is -9.313. ∎

EXAMPLE E

Evaluate $\int_{0.1}^{2.7} \dfrac{dx}{\sqrt{4x + 1}}$.

In order to integrate this we have $n = -\frac{1}{2}$, $u = 4x + 1$, and $du = 4\, dx$. Therefore,

$$\int_{0.1}^{2.7} \frac{dx}{\sqrt{4x + 1}} = \int_{0.1}^{2.7} (4x + 1)^{-1/2}\, dx = \frac{1}{4}\int_{0.1}^{2.7} (4x + 1)^{-1/2}(4\, dx)$$

$$= \frac{1}{4}\left(\frac{1}{\frac{1}{2}}\right)(4x + 1)^{1/2}\Big|_{0.1}^{2.7} = \frac{1}{2}(4x + 1)^{1/2}\Big|_{0.1}^{2.7} \qquad \text{integrate}$$

$$= \frac{1}{2}(\sqrt{11.8} - \sqrt{1.4}) = 1.126 \qquad \text{evaluate}$$ ∎

EXAMPLE F

Evaluate

$$\int_{0}^{4} \frac{x + 1}{(x^2 + 2x + 2)^3}\, dx$$

For purposes of integration,

$$n = -3, \qquad u = x^2 + 2x + 2, \quad \text{and} \quad du = (2x + 2)\, dx$$

Therefore,

$$\int_{0}^{4} (x^2 + 2x + 2)^{-3}(x + 1)\, dx = \frac{1}{2}\int_{0}^{4} (x^2 + 2x + 2)^{-3}[2(x + 1)\, dx]$$

$$= \frac{1}{2}\left(\frac{1}{-2}\right)(x^2 + 2x + 2)^{-2}\Big|_{0}^{4} \qquad \text{integrate}$$

$$= -\frac{1}{4}(16 + 8 + 2)^{-2} + \frac{1}{4}(0 + 0 + 2)^{-2} \qquad \text{evaluate}$$

$$= \frac{1}{4}\left(-\frac{1}{26^2} + \frac{1}{2^2}\right) = \frac{1}{4}\left(\frac{1}{4} - \frac{1}{676}\right)$$

$$= \frac{1}{4}\left(\frac{168}{676}\right) = \frac{21}{338}$$

It should be pointed out that the definition of the definite integral is valid regardless of the source of $f(x)$. That is, we may apply the definite integral whenever we want to sum a function in a manner similar to that which we use to find an area.

The following example illustrates an application of the definite integral. In Chapter 25 we shall see that the definite integral has applications in many areas of science and technology.

EXAMPLE G

See the chapter introduction.

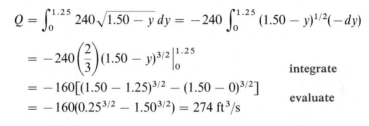

1.25 ft   1.50 ft

Fig. 24-13

The rate of flow Q (in ft^3/s) of water over a certain dam is found by evaluating the definite integral in the equation $Q = \int_0^{1.25} 240\sqrt{1.50 - y}\, dy$. See Fig. 24-13. Find Q.

The solution is as follows:

$$Q = \int_0^{1.25} 240\sqrt{1.50 - y}\, dy = -240 \int_0^{1.25} (1.50 - y)^{1/2}(-dy)$$

$$= -240 \left(\frac{2}{3}\right)(1.50 - y)^{3/2}\Big|_0^{1.25} \qquad \text{integrate}$$

$$= -160[(1.50 - 1.25)^{3/2} - (1.50 - 0)^{3/2}] \qquad \text{evaluate}$$

$$= -160(0.25^{3/2} - 1.50^{3/2}) = 274 \text{ ft}^3/\text{s}$$

Exercises 24-5

In Exercises 1 through 32, evaluate the given definite integrals.

1. $\int_0^1 2x\, dx$

2. $\int_0^2 3x^2\, dx$

3. $\int_1^4 x^{5/2}\, dx$

4. $\int_4^9 (x^{3/2} - 1)\, dx$

5. $\int_3^6 \left(\frac{1}{\sqrt{x}} + 2\right) dx$

6. $\int_{1.2}^{1.6} \left(5 + \frac{6}{x^4}\right) dx$

7. $\int_{-1}^1 (1 - x)^{1/3}\, dx$

8. $\int_1^5 \sqrt{2x - 1}\, dx$

9. $\int_0^3 (x^4 - x^3 + x^2)\, dx$

10. $\int_1^2 (3x^5 - 2x^3)\, dx$

11. $\int_{0.5}^{2.2} (\sqrt[3]{x} - 2)\, dx$

12. $\int_{2.7}^{5.3} \left(\frac{1}{x\sqrt{x}} + 4\right) dx$

13. $\int_0^4 (1 - \sqrt{x})^2\, dx$

14. $\int_1^4 \frac{x^2 + 1}{\sqrt{x}}\, dx$

15. $\int_1^2 2x(4 - x^2)^3\, dx$

16. $\int_0^1 x(3x^2 - 1)^3\, dx$

17. $\int_0^4 \frac{x\, dx}{\sqrt{x^2 + 9}}$

18. $\int_0^3 x^2(x^3 + 2)^{3/2}\, dx$

19. $\int_{2.75}^{3.25} \frac{dx}{\sqrt[3]{6x + 1}}$

20. $\int_2^6 \frac{2\, dx}{\sqrt{4x + 1}}$

21. $\int_1^3 \frac{2x\, dx}{(2x^2 + 1)^3}$

22. $\int_{12.6}^{17.2} \frac{3\, dx}{(6x - 1)^2}$

23. $\int_3^7 3\sqrt{4x - 3}\, dx$

24. $\int_{-5}^1 \sqrt{6 - 2x}\, dx$

25. $\int_0^2 2x(9 - 2x^2)^2\, dx$

26. $\int_{-1}^0 x^3(1 - 2x^4)^3\, dx$

27. $\int_0^1 (x^2 + 3)(x^3 + 9x + 6)\, dx$

28. $\int_2^3 \frac{x^2 + 1}{(x^3 + 3x)^2}\, dx$

29. $\int_{-1}^2 \frac{8x - 2}{(2x^2 - x + 1)^3}\, dx$

30. $\int_{-3}^{-2} (3x^2 - 2)\sqrt[3]{2x^3 - 4x + 1}\, dx$

31. $\int_3^{11} (x + 3 + \sqrt[4]{2x - 6})\, dx$

32. $\int_{-2}^0 (\sqrt{2x + 4} - \sqrt[3]{3x + 8})\, dx$

In Exercises 33 through 36, solve the given problems.

33. The work W (in ft·lb) in winding up an 80-ft cable is $W = \int_0^{80} (1000 - 5x)\, dx$. Evaluate W.

34. The total volume V of liquid flowing through a certain pipe of radius R is $V = k(R^2 \int_0^R r \, dr - \int_0^R r^3 \, dr)$, where k is a constant. Evaluate V.

35. The surface area A (in m^2) of a certain parabolic radio wave reflector is $A = 4\pi \int_0^2 \sqrt{3x + 9} \, dx$. Evaluate A.

36. The total force, in pounds, on the circular end of a water tank is $F = 125 \int_0^5 y\sqrt{25 - y^2} \, dy$. Evaluate F.

24-6 Numerical Integration: The Trapezoidal Rule ▬▬▬▬

For data and functions which cannot be directly integrated by available methods, it is possible to develop numerical methods of integration. These numerical methods are of greater importance today since they are readily adaptable for use on a calculator or computer. There are a great many such numerical techniques for approximating the value of an integral. In this section we shall develop one of these, the trapezoidal rule. In the following section, another numerical method is discussed.

We know from Sections 24-4 and 24-5 that we can interpret a definite integral as the area under a curve. We shall therefore show how we can approximate the value of the integral by approximating the appropriate area by a set of inscribed trapezoids. The basic idea here is very similar to that used when rectangles were inscribed under a curve. However, the use of trapezoids reduces the error and provides a better approximation.

The area to be found is subdivided into n intervals of equal width. Perpendicular lines are then dropped from the curve (or points, if only a given set of numbers is available). If the points on the curve are joined by straight-line segments, the area of successive parts under the curve is then approximated by finding the area of each of the trapezoids formed. However, if these points are not too far apart, the approximation will be very good (see Fig. 24-14). From geometry we recall that the area of a trapezoid equals one-half the product of the sum of the bases times the altitude. For these trapezoids the bases are the y-coordinates and the altitudes are Δx. When we thus indicate the sum of these trapezoidal areas, we have

$$A_T = \frac{1}{2}(y_0 + y_1)\Delta x + \frac{1}{2}(y_1 + y_2)\Delta x + \frac{1}{2}(y_2 + y_3)\Delta x + \cdots$$

$$+ \frac{1}{2}(y_{n-2} + y_{n-1})\Delta x + \frac{1}{2}(y_{n-1} + y_n)\Delta x$$

Fig. 24-14

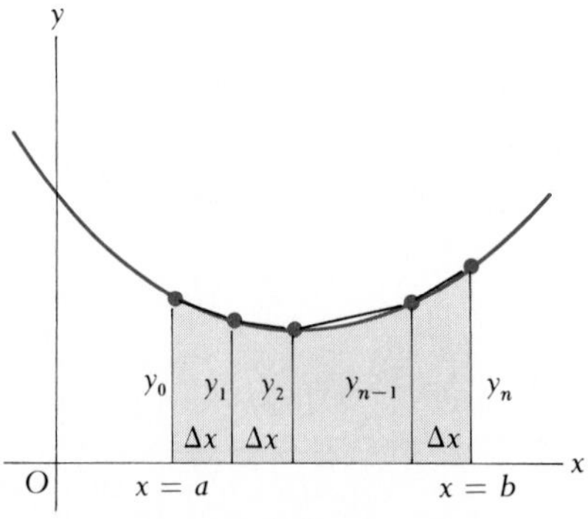

We note, when this addition is performed, that the result is

$$A_T = \left(\frac{1}{2}y_0 + y_1 + y_2 + \cdots + y_{n-1} + \frac{1}{2}y_n\right)\Delta x \qquad (24\text{-}12)$$

The y-values to be used either are derived from the function as $y_1 = f(x_1)$ or are the y-coordinates of a set of data.

Since A_T approximates the area under the curve, it also approximates the value of the definite integral, or

trapezoidal rule

$$\boxed{\int_a^b f(x)\,dx \approx \left(\frac{1}{2}y_0 + y_1 + y_2 + \cdots + y_{n-1} + \frac{1}{2}y_n\right)\Delta x} \qquad (24\text{-}13)$$

Equation (24-13) is known as the **trapezoidal rule.**

EXAMPLE A ——— Approximate the value of $\int_1^3 \frac{1}{x}\,dx$ by the trapezoidal rule. Let $n = 4$.

We are to approximate the area under $y = 1/x$ from $x = 1$ to $x = 3$ by dividing the area into 4 trapezoids. This area is found by applying Eq. (24-12), which is the approximate value of the integral, as shown in Eq. (24-13). Figure 24-15 shows the graph. In this example, $f(x) = 1/x$, and

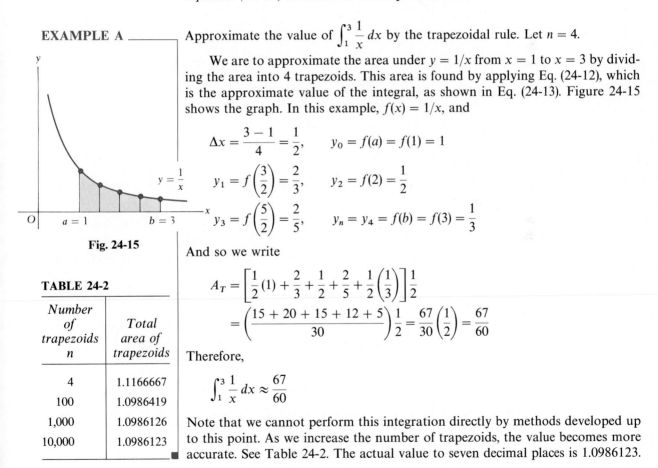

$$\Delta x = \frac{3-1}{4} = \frac{1}{2}, \qquad y_0 = f(a) = f(1) = 1$$

$$y_1 = f\left(\frac{3}{2}\right) = \frac{2}{3}, \qquad y_2 = f(2) = \frac{1}{2}$$

$$y_3 = f\left(\frac{5}{2}\right) = \frac{2}{5}, \qquad y_n = y_4 = f(b) = f(3) = \frac{1}{3}$$

Fig. 24-15

And so we write

$$A_T = \left[\frac{1}{2}(1) + \frac{2}{3} + \frac{1}{2} + \frac{2}{5} + \frac{1}{2}\left(\frac{1}{3}\right)\right]\frac{1}{2}$$

$$= \left(\frac{15 + 20 + 15 + 12 + 5}{30}\right)\frac{1}{2} = \frac{67}{30}\left(\frac{1}{2}\right) = \frac{67}{60}$$

Therefore,

$$\int_1^3 \frac{1}{x}\,dx \approx \frac{67}{60}$$

Note that we cannot perform this integration directly by methods developed up to this point. As we increase the number of trapezoids, the value becomes more accurate. See Table 24-2. The actual value to seven decimal places is 1.0986123. ∎

TABLE 24-2

Number of trapezoids n	Total area of trapezoids
4	1.1166667
100	1.0986419
1,000	1.0986126
10,000	1.0986123

EXAMPLE B

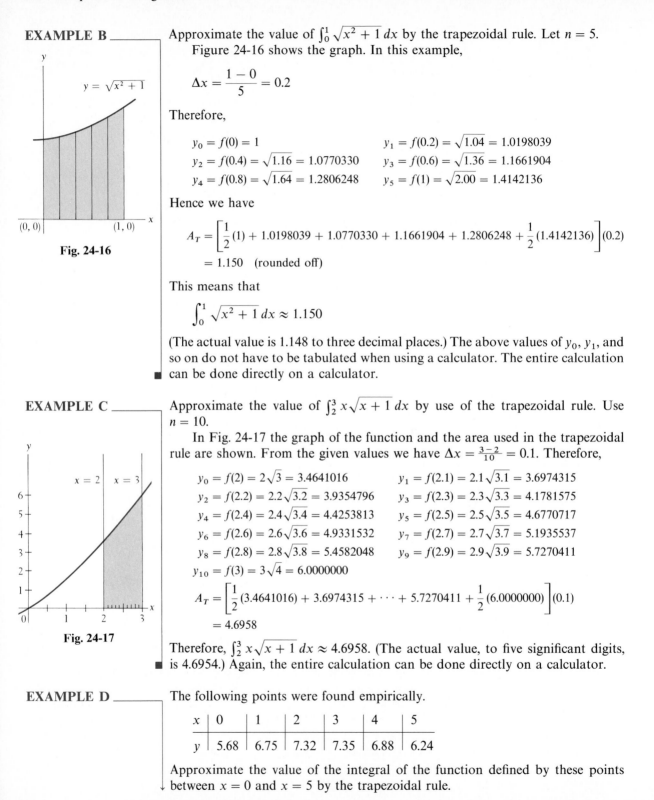

$y = \sqrt{x^2 + 1}$

$(0, 0)$ $(1, 0)$

Fig. 24-16

Approximate the value of $\int_0^1 \sqrt{x^2 + 1} \, dx$ by the trapezoidal rule. Let $n = 5$. Figure 24-16 shows the graph. In this example,

$$\Delta x = \frac{1 - 0}{5} = 0.2$$

Therefore,

$$y_0 = f(0) = 1 \qquad\qquad y_1 = f(0.2) = \sqrt{1.04} = 1.0198039$$
$$y_2 = f(0.4) = \sqrt{1.16} = 1.0770330 \qquad y_3 = f(0.6) = \sqrt{1.36} = 1.1661904$$
$$y_4 = f(0.8) = \sqrt{1.64} = 1.2806248 \qquad y_5 = f(1) = \sqrt{2.00} = 1.4142136$$

Hence we have

$$A_T = \left[\frac{1}{2}(1) + 1.0198039 + 1.0770330 + 1.1661904 + 1.2806248 + \frac{1}{2}(1.4142136) \right](0.2)$$

$$= 1.150 \quad \text{(rounded off)}$$

This means that

$$\int_0^1 \sqrt{x^2 + 1} \, dx \approx 1.150$$

(The actual value is 1.148 to three decimal places.) The above values of y_0, y_1, and so on do not have to be tabulated when using a calculator. The entire calculation can be done directly on a calculator. ∎

EXAMPLE C

$x = 2$ $x = 3$

6
5
4
3
2
1

0 1 2 3

Fig. 24-17

Approximate the value of $\int_2^3 x\sqrt{x + 1} \, dx$ by use of the trapezoidal rule. Use $n = 10$.

In Fig. 24-17 the graph of the function and the area used in the trapezoidal rule are shown. From the given values we have $\Delta x = \frac{3 - 2}{10} = 0.1$. Therefore,

$$y_0 = f(2) = 2\sqrt{3} = 3.4641016 \qquad y_1 = f(2.1) = 2.1\sqrt{3.1} = 3.6974315$$
$$y_2 = f(2.2) = 2.2\sqrt{3.2} = 3.9354796 \qquad y_3 = f(2.3) = 2.3\sqrt{3.3} = 4.1781575$$
$$y_4 = f(2.4) = 2.4\sqrt{3.4} = 4.4253813 \qquad y_5 = f(2.5) = 2.5\sqrt{3.5} = 4.6770717$$
$$y_6 = f(2.6) = 2.6\sqrt{3.6} = 4.9331532 \qquad y_7 = f(2.7) = 2.7\sqrt{3.7} = 5.1935537$$
$$y_8 = f(2.8) = 2.8\sqrt{3.8} = 5.4582048 \qquad y_9 = f(2.9) = 2.9\sqrt{3.9} = 5.7270411$$
$$y_{10} = f(3) = 3\sqrt{4} = 6.0000000$$

$$A_T = \left[\frac{1}{2}(3.4641016) + 3.6974315 + \cdots + 5.7270411 + \frac{1}{2}(6.0000000) \right](0.1)$$

$$= 4.6958$$

Therefore, $\int_2^3 x\sqrt{x + 1} \, dx \approx 4.6958$. (The actual value, to five significant digits, is 4.6954.) Again, the entire calculation can be done directly on a calculator. ∎

EXAMPLE D

The following points were found empirically.

x	0	1	2	3	4	5
y	5.68	6.75	7.32	7.35	6.88	6.24

Approximate the value of the integral of the function defined by these points between $x = 0$ and $x = 5$ by the trapezoidal rule.

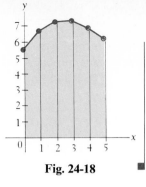

Fig. 24-18

In order to find A_T, we use the values of y_0, y_1, etc., directly from the table. We also note that $\Delta x = 1$. The graph is shown in Fig. 24-18. Therefore, we have

$$A_T = \left[\frac{1}{2}(5.68) + 6.75 + 7.32 + 7.35 + 6.88 + \frac{1}{2}(6.24)\right] = 34.26$$

Although we do not know the algebraic form of the function, we can state that

$$\int_0^5 f(x)\,dx \approx 34.26$$

Exercises 24-6

In Exercises 1 through 4, (a) approximate the value of each of the given integrals by use of the trapezoidal rule, using the given value of n, and (b) check by direct integration.

1. $\int_0^2 2x^2\,dx,\ n = 4$ **2.** $\int_0^1 (1 - x^2)\,dx,\ n = 3$ **3.** $\int_1^4 (1 + \sqrt{x})\,dx,\ n = 6$ **4.** $\int_3^8 \sqrt{1 + x}\,dx,\ n = 5$

In Exercises 5 through 12, approximate each of the given integrals by use of the trapezoidal rule, using the given value of n.

5. $\int_2^3 \frac{1}{2x}\,dx,\ n = 2$ **6.** $\int_2^4 \frac{dx}{x + 3},\ n = 4$ **7.** $\int_0^2 \sqrt{4 - x^2}\,dx,\ n = 4$ **8.** $\int_0^2 \sqrt{x^3 + 1}\,dx,\ n = 4$

9. $\int_1^5 \frac{1}{x^2 + x}\,dx,\ n = 10$ **10.** $\int_2^4 \frac{1}{x^2 + 1}\,dx,\ n = 10$ **11.** $\int_0^4 2^x\,dx,\ n = 12$ **12.** $\int_0^{1.5} 10^x\,dx,\ n = 15$

In Exercises 13 and 14, approximate the values of the integrals defined by the given sets of points.

13. $\int_2^{14} y\,dx$

x	2	4	6	8	10	12	14
y	0.67	2.34	4.56	3.67	3.56	4.78	6.87

14. $\int_{1.4}^{3.2} y\,dx$

x	1.4	1.7	2.0	2.3	2.6	2.9	3.2
y	0.18	7.87	18.23	23.53	24.62	20.93	20.76

In Exercises 15 and 16, solve the given problems using the trapezoidal rule.

15. A force F that a distributed electric charge has on a point charge is $F = k\int_0^2 \frac{dx}{(4 + x^2)^{3/2}}$, where x is the distance along the distributed charge and k is a constant. With $n = 8$, evaluate F in terms of k.

16. The length L of telephone wire needed (considering the sag) between two poles exactly 100 ft apart is $L = 2\int_0^{100} \sqrt{1.6 \times 10^{-7}x^2 + 1}\,dx$. With $n = 10$, evaluate L (to six significant digits).

24-7   Simpson's Rule

The numerical method of integration developed in this section is also readily programmable for use on a computer or easily usable with the necessary calculations done on a calculator. It is obtained by interpreting the definite integral as the area under a curve, as we did in developing the trapezoidal rule, and by approximating the curve by a set of parabolic arcs. The use of parabolic arcs, rather than chords as with the trapezoidal rule, usually gives a better approximation.

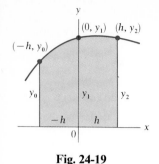

Fig. 24-19

Since we will be using parabolic arcs, we first derive a formula for the area under a parabolic arc. The curve in Fig. 24-19 represents the parabola $y = ax^2 + bx + c$. The indicated points on the curve are $(-h, y_0)$, $(0, y_1)$, and (h, y_2). The area under the parabola is given by

$$A = \int_{-h}^{h} y\,dx = \int_{-h}^{h} (ax^2 + bx + c)\,dx = \frac{ax^3}{3} + \frac{bx^2}{2} + cx\Big|_{-h}^{h}$$

$$= \frac{2}{3}ah^3 + 2ch$$

or

$$A = \frac{h}{3}(2ah^2 + 6c) \tag{24-14}$$

The coordinates of the three points also satisfy the equation $y = ax^2 + bx + c$. This means that

$$y_0 = ah^2 - bh + c$$
$$y_1 = c$$
$$y_2 = ah^2 + bh + c$$

By finding the sum of $y_0 + 4y_1 + y_2$, we have

$$y_0 + 4y_1 + y_2 = 2ah^2 + 6c \tag{24-15}$$

Substituting Eq. (24-15) into Eq. (24-14), we have

$$A = \frac{h}{3}(y_0 + 4y_1 + y_2) \tag{24-16}$$

We note that the area depends only on the distance h and the three y-coordinates.

Now let us consider the area under the curve in Fig. 24-20. If a parabolic arc is passed through the points (x_0, y_0), (x_1, y_1), and (x_2, y_2) we may use Eq. (24-16) to approximate the area under the curve between x_0 and x_2. We also note that the distance h used in finding Eq. (24-16) is the difference in the x-coordinates, or $h = \Delta x$. Therefore, the area under the curve between x_0 and x_2 is

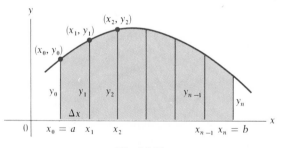

Fig. 24-20

$$A_1 = \frac{\Delta x}{3}(y_0 + 4y_1 + y_2)$$

Similarly, if a parabolic arc is passed through the three points starting with (x_2, y_2), the area between x_2 and x_4 is

$$A_2 = \frac{\Delta x}{3}(y_2 + 4y_3 + y_4)$$

The sum of these areas is

$$A_1 + A_2 = \frac{\Delta x}{3}(y_0 + 4y_1 + 2y_2 + 4y_3 + y_4) \tag{24-17}$$

We can continue this procedure until the approximate value of the entire area has been found. We must note, however, that **the number of intervals n of width Δx must be even.** Therefore, generalizing on Eq. (24-17) and recalling again that the value of the definite integral is the area under the curve, we have the following result:

NOTE ▷

Simpson's rule

$$\int_a^b f(x)\, dx \approx \frac{\Delta x}{3}(y_0 + 4y_1 + 2y_2 + 4y_3 + 2y_4 + \cdots + 4y_{n-1} + y_n) \qquad (24\text{-}18)$$

Equation (24-18) is known as **Simpson's rule.**

EXAMPLE A

Approximate the value of $\int_0^1 \dfrac{dx}{x+1}$ by Simpson's rule. Let $n = 2$.

In Fig. 24-21 the graph of the function and the area used are shown. We are to approximate the integral by use of Eq. (24-18). We therefore note that $f(x) = 1/(x+1)$. Also, $x_0 = a = 0$, $x_1 = 0.5$, and $x_2 = b = 1$. This is due to the fact that $n = 2$ and $\Delta x = 0.5$ since the total interval is 1 unit (from $x = 0$ to $x = 1$). Therefore,

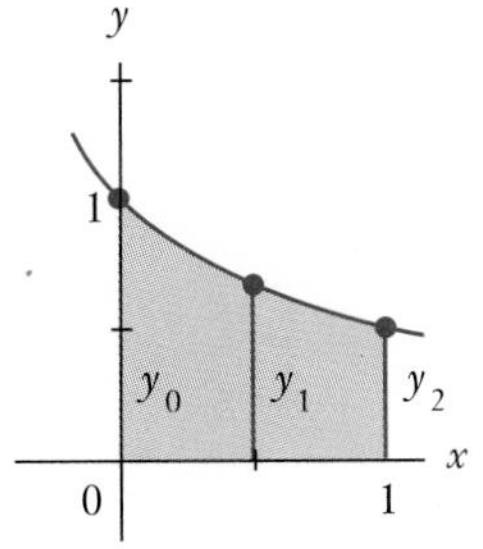

$$y_0 = \frac{1}{0+1} = 1.0000, \qquad y_1 = \frac{1}{0.5+1} = 0.6667, \qquad y_2 = \frac{1}{1+1} = 0.5000$$

Substituting, we have

$$\int_0^1 \frac{dx}{x+1} = \frac{0.5}{3}[1.0000 + 4(0.6667) + 0.5000]$$

$$= 0.694$$

To three decimal places, the actual value of the integral is 0.693. The method of integrating this function will be considered in a later chapter. ∎

Fig. 24-21

EXAMPLE B

Approximate the value of $\int_2^3 x\sqrt{x+1}\, dx$ by Simpson's rule. Let $n = 10$.

Since the necessary values for this function are shown in Example C of Section 24-6, we shall simply tabulate them here. ($\Delta x = 0.1$). See Fig. 24-22.

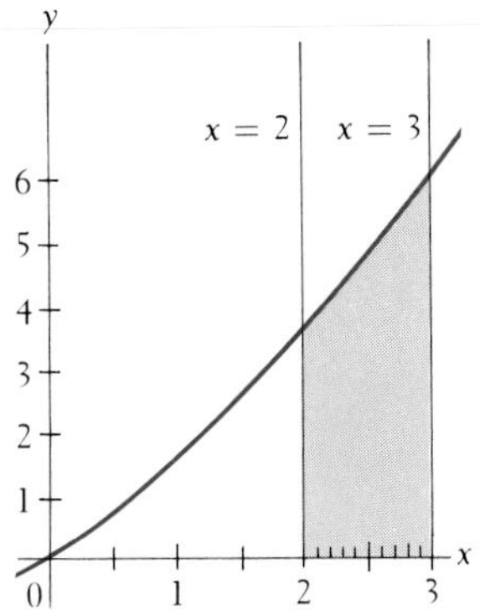

$$y_0 = 3.4641016 \qquad y_1 = 3.6974315 \qquad y_2 = 3.9354796 \qquad y_3 = 4.1781575$$
$$y_4 = 4.4253813 \qquad y_5 = 4.6770717 \qquad y_6 = 4.9331532 \qquad y_7 = 5.1935537$$
$$y_8 = 5.4582048 \qquad y_9 = 5.7270411 \qquad y_{10} = 6.0000000$$

Therefore, we evaluate the integral as follows:

$$\int_2^3 x\sqrt{x+1}\, dx = \frac{0.1}{3}[3.4641016 + 4(3.6974315) + 2(3.9354796) + 4(4.1781575)$$
$$+ 2(4.4253813) + 4(4.6770717) + 2(4.9331532) + 4(5.1935537)$$
$$+ 2(5.4582048) + 4(5.7270411) + 6.0000000]$$

$$= \frac{0.1}{3}(140.86156) = 4.6953854$$

This result agrees with the actual value to the eight significant digits shown. The value we obtained with the trapezoidal rule was 4.6958. ∎

Fig. 24-22

EXAMPLE C _____

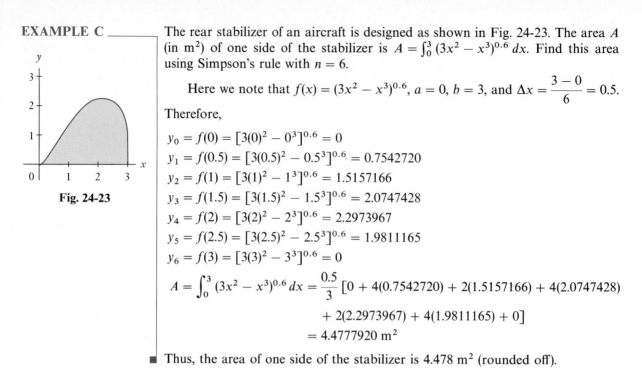

Fig. 24-23

The rear stabilizer of an aircraft is designed as shown in Fig. 24-23. The area A (in m^2) of one side of the stabilizer is $A = \int_0^3 (3x^2 - x^3)^{0.6} \, dx$. Find this area using Simpson's rule with $n = 6$.

Here we note that $f(x) = (3x^2 - x^3)^{0.6}$, $a = 0$, $b = 3$, and $\Delta x = \dfrac{3 - 0}{6} = 0.5$.

Therefore,

$$y_0 = f(0) = [3(0)^2 - 0^3]^{0.6} = 0$$
$$y_1 = f(0.5) = [3(0.5)^2 - 0.5^3]^{0.6} = 0.7542720$$
$$y_2 = f(1) = [3(1)^2 - 1^3]^{0.6} = 1.5157166$$
$$y_3 = f(1.5) = [3(1.5)^2 - 1.5^3]^{0.6} = 2.0747428$$
$$y_4 = f(2) = [3(2)^2 - 2^3]^{0.6} = 2.2973967$$
$$y_5 = f(2.5) = [3(2.5)^2 - 2.5^3]^{0.6} = 1.9811165$$
$$y_6 = f(3) = [3(3)^2 - 3^3]^{0.6} = 0$$

$$A = \int_0^3 (3x^2 - x^3)^{0.6} \, dx = \frac{0.5}{3} [0 + 4(0.7542720) + 2(1.5157166) + 4(2.0747428)$$
$$+ \, 2(2.2973967) + 4(1.9811165) + 0]$$
$$= 4.4777920 \text{ m}^2$$

■ Thus, the area of one side of the stabilizer is 4.478 m^2 (rounded off).

Exercises 24-7

In Exercises 1 through 4, (a) approximate the value of each of the given integrals by use of Simpson's rule, using the given value of n, and (b) check by direct integration.

1. $\int_0^2 (1 + x^3) \, dx$, $n = 2$

2. $\int_0^8 x^{1/3} \, dx$, $n = 2$

3. $\int_1^4 (2x + \sqrt{x}) \, dx$, $n = 6$

4. $\int_0^2 x\sqrt{x^2 + 1} \, dx$, $n = 4$

In Exercises 5 through 12, approximate each of the given integrals by use of Simpson's rule, using the given values of n. Exercises 5 through 10 are the same as Exercises 5 through 10 of Section 24-6.

5. $\int_2^3 \dfrac{1}{2x} \, dx$, $n = 2$

6. $\int_2^4 \dfrac{dx}{x + 3}$, $n = 4$

7. $\int_0^2 \sqrt{4 - x^2} \, dx$, $n = 4$

8. $\int_0^2 \sqrt{x^3 + 1} \, dx$, $n = 4$

9. $\int_1^5 \dfrac{1}{x^2 + x} \, dx$, $n = 10$

10. $\int_2^4 \dfrac{1}{x^2 + 1} \, dx$, $n = 10$

11. $\int_{-4}^5 (2x^4 + 1)^{0.1} \, dx$, $n = 6$

12. $\int_0^{2.4} \dfrac{dx}{(4 + \sqrt{x})^{3/2}}$, $n = 8$

In Exercises 13 and 14, approximate the value of the integrals defined by the given set of points. These are the same as Exercises 13 and 14 of Section 24-6.

13. $\int_2^{14} y \, dx$

x	2	4	6	8	10	12	14
y	0.67	2.34	4.56	3.67	3.56	4.78	6.87

14. $\int_{1.4}^{3.2} y \, dx$

x	1.4	1.7	2.0	2.3	2.6	2.9	3.2
y	0.18	7.87	18.23	23.53	24.62	20.93	20.76

In Exercises 15 and 16, solve the given problems using Simpson's rule.

15. The distance $\bar{x}$, in inches, from one end of a barrel plug (with vertical circular cross sections) to its center of mass, as shown in Fig. 24-24, is
$\bar{x} = 0.9129 \int_0^3 x\sqrt{0.3 - 0.1x}\, dx$.
Find $\bar{x}$ with $n = 12$.

16. The average value of the electric current i_{av}, in amperes, in a circuit for the first 4 s is $i_{av} = \frac{1}{4}\int_0^4 (4t - t^2)^{0.2}\, dt$. Find i_{av} with $n = 10$.

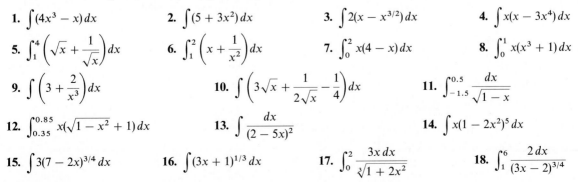

Fig. 24-24

Center of mass

24-8   Chapter Equations, Review Exercises, and Practice Test

Chapter Equations

Differential	$dy = f'(x)\, dx$	(24-1)
Indefinite integral	$\int f(x)\, dx = F(x) + C$	(24-2)
Integrals	$\int c\, du = c \int du = cu + C$	(24-3)
	$\int (du + dv) = u + v + C$	(24-4)
Power formula	$\int u^n\, du = \dfrac{u^{n+1}}{n+1} + C \qquad (n \neq -1)$	(24-5)
Area under a curve	$A_{ab} = \left[\int f(x)\, dx \right]_a^b = F(b) - F(a)$	(24-10)
Definite integral	$\int_a^b f(x)\, dx = F(b) - F(a)$	(24-11)
Trapezoidal rule	$\int_a^b f(x)\, dx \approx \left(\dfrac{1}{2} y_0 + y_1 + y_2 + \cdots + y_{n-1} + \dfrac{1}{2} y_n \right) \Delta x$	(24-13)
Simpson's rule	$\int_a^b f(x)\, dx \approx \dfrac{\Delta x}{3}(y_0 + 4y_1 + 2y_2 + 4y_3 + 2y_4 + \cdots + 4y_{n-1} + y_n)$	(24-18)

Review Exercises

In Exercises 1 through 24, evaluate the given integrals.

1. $\displaystyle\int (4x^3 - x)\, dx$

2. $\displaystyle\int (5 + 3x^2)\, dx$

3. $\displaystyle\int 2(x - x^{3/2})\, dx$

4. $\displaystyle\int x(x - 3x^4)\, dx$

5. $\displaystyle\int_1^4 \left(\sqrt{x} + \frac{1}{\sqrt{x}} \right) dx$

6. $\displaystyle\int_1^2 \left(x + \frac{1}{x^2} \right) dx$

7. $\displaystyle\int_0^2 x(4 - x)\, dx$

8. $\displaystyle\int_0^1 x(x^3 + 1)\, dx$

9. $\displaystyle\int \left(3 + \frac{2}{x^3} \right) dx$

10. $\displaystyle\int \left(3\sqrt{x} + \frac{1}{2\sqrt{x}} - \frac{1}{4} \right) dx$

11. $\displaystyle\int_{-1.5}^{0.5} \frac{dx}{\sqrt{1-x}}$

12. $\displaystyle\int_{0.35}^{0.85} x(\sqrt{1 - x^2} + 1)\, dx$

13. $\displaystyle\int \frac{dx}{(2 - 5x)^2}$

14. $\displaystyle\int x(1 - 2x^2)^5\, dx$

15. $\displaystyle\int 3(7 - 2x)^{3/4}\, dx$

16. $\displaystyle\int (3x + 1)^{1/3}\, dx$

17. $\displaystyle\int_0^2 \frac{3x\, dx}{\sqrt[3]{1 + 2x^2}}$

18. $\displaystyle\int_1^6 \frac{2\, dx}{(3x - 2)^{3/4}}$

19. $\int x^2(1 - 2x^3)^4 \, dx$ **20.** $\int 3x^3(1 - 5x^4)^{1/3} \, dx$ **21.** $\int \dfrac{(2 - 3x^2) \, dx}{(2x - x^3)^2}$ **22.** $\int \dfrac{x^2 - 3}{\sqrt{6 + 9x - x^3}} \, dx$

23. $\int_1^3 (x^2 + x + 2)(2x^3 + 3x^2 + 12x) \, dx$ **24.** $\int_0^2 (4x + 18x^2)(x^2 + 3x^3)^2 \, dx$

In Exercises 25 through 28, find the differential of each of the given functions.

25. $y = \dfrac{1}{(x^2 - 1)^3}$ **26.** $y = \dfrac{1}{(2x - 1)^2}$ **27.** $y = x\sqrt[3]{1 - 3x}$ **28.** $y = \dfrac{3 + x}{4 - x^2}$

In Exercises 29 and 30, evaluate $\Delta y - dy$ for the given functions and values.

29. $y = x^3$, $x = 2$, $\Delta x = 0.1$ **30.** $y = 6x^2 - x$, $x = 3$, $\Delta x = 0.2$

In Exercises 31 and 32, find the required equations.

31. Find the equation of the curve which passes through $(-1, 3)$ for which the slope is given by $3 - x^2$.

32. Find the equation of the curve which passes through $(1, -2)$ for which the slope is $x(x^2 + 1)^2$.

In Exercises 33 and 34, perform the indicated integrations as directed.

33. Perform the integration $\int (1 - 2x) \, dx$ (a) term by term, labeling the constant of integration as C_1, and then (b) by letting $u = 1 - 2x$, using the general power rule, and labeling the constant of integration as C_2. Compare C_1 with C_2.

34. Following the methods (a) and (b) in Exercise 33, perform the integration $\int (3x + 2) \, dx$. In (b) let $u = 3x + 2$. Compare the constants of integration C_1 and C_2.

In Exercises 35 and 36, use Eq. (24-10) to find the indicated areas.

35. The area under $y = 6x - 1$ between $x = 1$ and $x = 3$

36. The first-quadrant area under $y = 8x - x^4$

In Exercises 37 and 38, solve the given problems by the trapezoidal rule.

37. Approximate $\int_1^3 \dfrac{dx}{2x - 1}$ with $n = 4$.

38. Approximate the value of the integral defined by the following set of points:

x	6.0	9.0	12	15	18	21
y	2.0	1.2	0.2	1.0	6.0	12

In Exercises 39 and 40, solve the given problems by Simpson's rule.

39. Approximate $\int_1^3 \dfrac{dx}{2x - 1}$ with $n = 4$ (see Exercise 37).

40. Approximate the value of the integral of the function defined by the following points:

x	1.0	1.4	1.8	2.2	2.6	3.0	3.4
y	1.45	1.89	2.66	3.50	3.22	3.04	2.44

In Exercises 41 and 42, use the function $y = \dfrac{x}{x^2 + 2}$ and approximate the area under the curve in the first quadrant to the left of the line $x = 5$ by the indicated method.

41. Inscribe five rectangles and find the sum of the areas of the rectangles.

42. Use the trapezoidal rule with $n = 5$.

In Exercises 43 and 44, use the function $y = x\sqrt{x^3 + 1}$ and approximate the area under the curve between $x = 1$ and $x = 3$ by the indicated method.

43. Use Simpson's rule with $n = 10$.

44. Inscribe ten rectangles, and find the sum of the areas of the rectangles.

In Exercises 45 through 48, solve the given problems by finding the appropriate differentials.

45. A weather balloon 3.500 m in radius becomes covered with a uniform layer of ice 1.2 cm thick. What is the volume of the ice?

46. The total power P, in watts, transmitted by an AM radio transmitter is $P = 460 + 230m^2$, where m is the modulation index. What is the change in power if m changes from 0.86 to 0.89?

47. The impedance Z of an electric circuit as a function of the resistance R and the reactance X is given by $Z = \sqrt{R^2 + X^2}$. Derive an expression for the relative error in impedance for an error in R and a given value of X.

48. Show that the relative error of the nth root of a given measurement equals approximately $1/n$ of the relative error of the measurement.

In Exercises 49 through 52, solve the given problems using integration.

49. The deflection y of a certain beam at a distance x from one end is $dy/dx = k(2L^3 - 12Lx + 2x^4)$, where k is a constant and L is the length of the beam. Find y as a function of x if $y = 0$ for $x = 0$.

50. The total electric charge Q on a charged sphere is $Q = k \int \left(r^2 - \dfrac{r^3}{R} \right) dr$, where k is a constant, r is the distance from the center of the sphere, and R is the radius of the sphere. Find Q as a function of r if $Q = Q_0$ for $r = R$.

51. Part of the deck of a boat is the parabolic area shown in Fig. 24-25. The area A (in m^2) is $A = 2 \int_0^5 \sqrt{5 - y}\, dy$. Evaluate A.

52. The distance s, in inches, through which a cam follower moves in 4 s is $s = \int_0^4 t\sqrt{4 + 9t^2}\, dt$. Evaluate s.

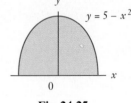

Fig. 24-25

Practice Test

1. For $y = 3x^2 - x$, evaluate (a) Δy, (b) dy, and (c) $\Delta y - dy$ for $x = 3$ and $\Delta x = 0.1$.

2. Integrate: $\int x\sqrt{1 - 2x^2}\, dx$.

3. Evaluate: $\int_2^5 (6 - x)^4\, dx$.

4. Approximate the area under $y = \dfrac{1}{x + 2}$ between $x = 1$ and $x = 4$ (above the x-axis) by inscribing 6 rectangles and finding the sum of their areas.

5. Evaluate $\int_1^4 \dfrac{dx}{x + 2}$ by using the trapezoidal rule with $n = 6$.

6. Evaluate the definite integral of Problem 5 using Simpson's rule with $n = 6$.

7. The inside radius of a pipe is 4.50 cm. By the use of differentials, find the approximate volume of metal in the pipe if it is 20.0 cm long and the metal is 0.30 cm thick.

8. The total electric current (in amperes) to pass a point in the circuit between $t = 1$ s and $t = 3$ s is $i = \int_1^3 \left(t^2 + \dfrac{1}{t^2} \right) dt$. Evaluate i.

25 Applications of Integration

In Section 25-6 integration is used to find the force which water exerts on the floodgate of a dam.

The applications of integration in science, engineering, and technology are numerous. Among the important areas of application are physics, electricity, mechanics, hydrostatics, and geometry. Some of these were indicated in the previous chapter. In this chapter we shall develop the necessary material for setting up the integrals for these types of applications.

In the first section of this chapter we shall present some of the applications of the indefinite integral. In the remaining sections we shall present applications of the definite integral in geometry and certain technical areas.

25-1 Applications of the Indefinite Integral

In the examples of this section we shall show two basic applications of the indefinite integral. Some other applications are shown in the exercises.

The first of these applications deals with velocity and acceleration. The concepts of velocity as a first derivative and acceleration as a second derivative were introduced in Chapter 22. Here we shall apply integration to the problem of finding the displacement and velocity as a function of time, when we know the relationship between acceleration and time, as well as certain specific values of distance and velocity. As we saw in Example I of Section 24-3, these latter values are necessary for determining the values of the constants of integration which are introduced. Recalling now that the acceleration a of an object is given by $a = dv/dt$, we can find the expression for the velocity in terms of a, t, and the constant of integration. We write

$$dv = a\,dt$$

or

velocity

$$v = \int a\,dt \tag{25-1}$$

If the acceleration is constant, we have

$$v = at + C_1 \tag{25-2}$$

Of course, Eq. (25-1) can be used in general to find the velocity as a function of time so long as we know the acceleration as a function of time. However, since the case of constant acceleration is often encountered, Eq. (25-2) is often encountered. If the velocity is known for some specified time, the constant C_1 may be evaluated.

EXAMPLE A _____ Find the expression for the velocity if $a = 12t$, given that $v = 8$ when $t = 1$.
    Using Eq. (25-1), we have

$$v = \int (12t)\,dt = 6t^2 + C_1$$

Substituting the known values, we obtain

$$8 = 6 + C_1 \quad \text{or} \quad C_1 = 2$$

▪ Thus, $v = 6t^2 + 2$.

EXAMPLE B _____ For an object falling under the influence of gravity, the acceleration due to gravity is essentially constant. Its value is -32 ft/s^2. (The minus sign is chosen so
NOTE ▷ that **all quantities directed up are positive and all quantities directed down are negative.**) Find the expression for the velocity of an object under the influence of gravity if $v = v_0$ when $t = 0$.
    We write

$$v = \int (-32)\,dt \qquad \text{substitute } a = -32 \text{ into Eq. (25-1)}$$
$$= -32t + C_1 \qquad \text{integrate}$$
$$v_0 = -32(0) + C_1 \qquad \text{substitute given values}$$
$$C_1 = v_0 \qquad \text{solve for } C_1$$
$$v = v_0 - 32t \qquad \text{substitute}$$

The velocity v_0 is called the *initial velocity*. If the object is given an initial upward velocity of 100 ft/s, $v_0 = 100$ ft/s. If the object is dropped, $v_0 = 0$. If the object
▪ is given an initial downward velocity of 100 ft/s, $v_0 = -100$ ft/s.

Once we obtain the expression for velocity, we can integrate to find the expression for displacement in terms of the time. Since $v = ds/dt$, we can write $ds = v\,dt$, or

displacement

$$s = \int v\,dt$$

(25-3)

Consider the following examples.

EXAMPLE C _____ Find the expression for the distance (in feet) above the ground of an object, given a vertical velocity v_0 from the ground.

From Example B, we know that $v = v_0 - 32t$. In this problem we know that $s = 0$ when $t = 0$ (given velocity v_0 *from the ground*) if distances are measured from ground level. Therefore,

$$s = \int (v_0 - 32t)\,dt \qquad \text{substitute into Eq. (25-3)}$$

$$= v_0 t - 16t^2 + C_2 \qquad \text{integrate}$$

$$0 = 0 - 0 + C_2, \qquad C_2 = 0 \qquad \text{substitute known values; solve for } C_2$$

■ $$s = v_0 t - 16t^2 \qquad \text{substitute}$$

EXAMPLE D _____ During the initial stage of launching a spacecraft vertically, the acceleration a (in m/s^2) of the spacecraft is $a = 6t^2$. Find the height s of the spacecraft after 6.0 s if $s = 12$ m for $t = 0.0$ s and $v = 16$ m/s for $t = 2.0$ s.

First we use Eq. (25-1) to get an expression for the velocity:

$$v = \int 6t^2\,dt = 2t^3 + C_1 \qquad \text{integrate}$$

$$16 = 2(2.0)^3 + C_1, \qquad C_1 = 0 \qquad \text{evaluate } C_1$$

$$v = 2t^3$$

We now use Eq. (25-3) to get an expression for the displacement:

$$s = \int 2t^3\,dt = \frac{1}{2}t^4 + C_2 \qquad \text{integrate}$$

$$12 = \frac{1}{2}(0.0)^4 + C_2, \qquad C_2 = 12 \qquad \text{evaluate } C_2$$

$$s = \frac{1}{2}t^4 + 12$$

Now, finding s for $t = 6.0$ s, we have

■ $$s = \frac{1}{2}(6.0)^4 + 12 = 660 \text{ m}$$

EXAMPLE E

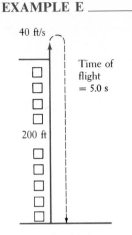

40 ft/s

Time of flight = 5.0 s

200 ft

Fig. 25-1

A ball is thrown vertically from the top of a building 200 ft high and hits the ground 5.0 s later. What initial velocity was the ball given?

Measuring vertical distances from the ground, we know that $s = 200$ ft when $t = 0$. Also, we know that $v = v_0 - 32t$. Thus,

$$s = \int (v_0 - 32t)\, dt = v_0 t - 16t^2 + C \qquad \text{integrate}$$

$$200 = v_0(0) - 16(0) + C, \qquad C = 200 \qquad \text{evaluate } C$$

$$s = v_0 t - 16t^2 + 200$$

We also know that $s = 0$ when $t = 5.0$ s. Thus,

$$0 = v_0(5.0) - 16(5.0)^2 + 200 \qquad \text{substitute given values}$$

$$5.0 v_0 = 200$$

$$v_0 = 40 \text{ ft/s}$$

■ This means that the initial velocity was 40 ft/s upward. See Fig. 25-1.

The second basic application of the indefinite integral which we shall discuss comes from the field of electricity. By definition, *the current i in an electric circuit equals the time rate of change of the charge q (in coulombs) which passes a given point in the circuit, or*

$$i = \frac{dq}{dt} \tag{25-4}$$

Rewriting this expression in differential notation as $dq = i\, dt$ and integrating both sides of the equation, we have

$$q = \int i\, dt \tag{25-5}$$

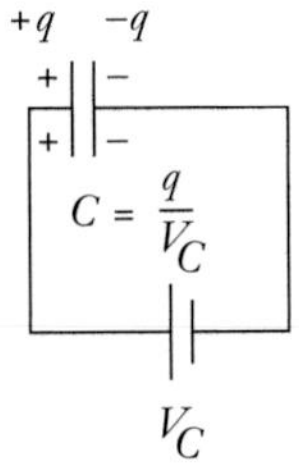

+q -q

$$C = \frac{q}{V_C}$$

V_C

Fig. 25-2

voltage across capacitor

Now, the voltage V_C across a capacitor with capacitance C (see Fig. 25-2) is given by $V_C = q/C$. By combining equations, the voltage V_C is given by

$$V_C = \frac{1}{C} \int i\, dt \tag{25-6}$$

Here, V_C is measured in volts, C in farads, i in amperes, and t in seconds.

EXAMPLE F

The current in a certain electric circuit as a function of time is given by $i = 6t^2 + 4$. Find an expression for the amount of charge which passes a point in the circuit as a function of time. Assuming that $q = 0$ when $t = 0$, determine the total charge which passes the point in 2 s.

Since $q = \int i\, dt$, we have

$$q = \int (6t^2 + 4)\, dt \qquad \text{substitute into Eq. (25-5)}$$

$$= 2t^3 + 4t + C \qquad \text{integrate}$$

(Continued on next page)

This last expression is the desired expression giving charge as a function of time. We note that when $t = 0$, then $q = C$, which means that the constant of integration represents the initial charge, or the charge which passed a given point before we started timing. Using q_0 to represent this charge, we have

$$q = 2t^3 + 4t + q_0$$

Now, returning to the second part of the problem, we see that $q_0 = 0$. Therefore, evaluating q for $t = 2$ s, we have

$$q = 2(8) + 4(2) = 24 \text{ C}$$

[Here, the symbol C represents coulombs and is not the C for capacitance of Eq. (25-6) or the constant of integration.] This is the charge which passes any specified point in the circuit in 2 s. ∎

EXAMPLE G

NOTE▷

The voltage across a 5.0-μF capacitor is zero. What is the voltage after 20 ms if a current of 75 mA charges the capacitor?

Since the current is 75 mA, we know that $i = 0.075$ A $= 7.5 \times 10^{-2}$ A. **We must use the proper power of 10 which corresponds to each prefix.** Since 5.0 μF $= 5.0 \times 10^{-6}$ F, we have

$$V_C = \frac{1}{5.0 \times 10^{-6}} \int 7.5 \times 10^{-2} \, dt \qquad \text{substituting into Eq. (25-6)}$$

$$= (1.5 \times 10^4) \int dt$$

$$= (1.5 \times 10^4)t + C_1 \qquad \text{integrate}$$

From the given information we know that $V_C = 0$ when $t = 0$. Thus,

$$0 = (1.5 \times 10^4)(0) + C_1 \quad \text{or} \quad C_1 = 0 \qquad \text{evaluate } C_1$$

This means that

$$V_C = (1.5 \times 10^4)t$$

Evaluating this expression for $t = 20 \times 10^{-3}$ s, we have

$$V_C = (1.5 \times 10^4)(20 \times 10^{-3})$$

$$= 30 \times 10 = 300 \text{ V}$$

∎

EXAMPLE H

A certain capacitor is measured to have a voltage of 100 V across it. At this instant a current as a function of time given by $i = 0.06\sqrt{t}$ is sent through the circuit. After 0.25 s, the voltage across the capacitor is measured to be 140 V. What is the capacitance of the capacitor?

Substituting $i = 0.06\sqrt{t}$, we find that

$$V_C = \frac{1}{C} \int (0.06\sqrt{t} \, dt) = \frac{0.06}{C} \int t^{1/2} \, dt \qquad \text{using Eq. (25-6)}$$

$$= \frac{0.04}{C} t^{3/2} + C_1 \qquad \text{integrate}$$

From the given information we know that $V_C = 100$ V when $t = 0$. Thus,

$$100 = \frac{0.04}{C}(0) + C_1 \quad \text{or} \quad C_1 = 100 \text{ V} \qquad \text{evaluate } C_1$$

This means that

$$V_C = \frac{0.04}{C} t^{3/2} + 100$$

We also know that $V_C = 140$ V when $t = 0.25$ s. Therefore,

$$140 = \frac{0.04}{C}(0.25)^{3/2} + 100$$

$$40 = \frac{0.04}{C}(0.125)$$

or

$$C = 1.25 \times 10^{-4} \text{ F} = 125 \ \mu\text{F}$$

Exercises 25-1

1. What is the velocity (in ft/s) of a wrench 2.5 s after it is dropped from a building platform?

2. A hoop is started upward along an inclined plane at 16 ft/s. If the acceleration of the hoop is 5.0 ft/s² downward along the plane, find the velocity of the hoop after 6.0 s.

3. A conveyor belt 8.00 m long moves at 0.25 m/s. If a package is placed at one end, find its displacement from the other end as a function of time.

4. During each cycle, the velocity v (in mm/s) of a piston is $v = 6t - 6t^2$, where t is the time in seconds. Find the displacement s of the piston after 0.75 s if the initial displacement is zero.

5. While in the barrel of a tennis ball machine, the acceleration a (in ft/s²) of a ball is $a = 90\sqrt{1 - 4t}$, where t is the time in seconds. If $v = 0$ for $t = 0$, find the velocity of the ball as it leaves the barrel at $t = 0.25$ s.

6. A proton moves in an electric field such that its acceleration (in cm/s²) is $a = -20(1 + 2t)^{-2}$, where t is the time in seconds. Find the velocity as a function of time if $v = 30$ cm/s when $t = 0$ s.

7. A rocket is fired vertically upward. When it reaches an altitude of 16,500 m, the engines cut off and the rocket is moving upward at 450 m/s. What will be its altitude 3.00 s later ($a = -9.80$ m/s²)?

8. A flare is ejected vertically upward from the ground at 50 ft/s. Find the height of the flare after 2.5 s.

9. What must be the nozzle velocity of the water from a fire hose if it is to reach a point 90 ft directly above the nozzle?

10. An arrow is shot upward with a vertical velocity of 120 ft/s from the edge of a cliff. If it hits the ground below after 9.0 s, how high is the cliff?

11. In coming to a stop, the acceleration of a car is $-12t$. If it is traveling at 96.0 ft/s when the brakes are applied, how far does it travel while stopping?

12. A hoist mechanism raises a crate with an acceleration (in m/s²) $a = \sqrt{1 + 0.2t}$, where t is the time in seconds. Find the displacement of the crate as a function of time if $v = 0$ m/s and $s = 2$ m for $t = 0$ s.

13. The electric current in a microprocessor circuit is 0.230 μA. How many coulombs pass a given point in the circuit in 1.50 ms?

14. The electric current (in mA) in a computer circuit as a function of time is $i = 0.3 - 0.2t$. What total charge passes a point in the circuit in 0.050 s?

15. In an amplifier circuit the current (in amperes) changes with time (in seconds) according to $i = 0.06t\sqrt{1 + t^2}$. If 0.015 C of charge have passed a point in the circuit at $t = 0$, find the total charge to have passed the point at $t = 0.25$ s.

16. The current i, in microamperes, in a certain microprocessor circuit is given by $i = 8 - t$, where t is the time in microseconds and $0 \le t \le 20$ μs. If $q_0 = 0$, for what value of t, greater than zero, is $q = 0$? What interpretation can be given to this result?

17. The voltage across a 3.0-μF capacitor is zero. What is the voltage after 10 ms if a current of 0.20 A charges the capacitor?

18. The voltage across an 8.50-nF capacitor in an FM receiver circuit is zero. Find the voltage after 2.00 μs if a current (in mA) $i = 0.042t$ charges the capacitor.

19. The voltage across a 3.75-μF capacitor in a television circuit is 4.50 mV. Find the voltage after 0.565 ms if a current (in μA) $i = \sqrt[3]{1 + 6t}$ further charges the capacitor.

20. A current $i = t/\sqrt{t^2 + 1}$ is sent into a circuit containing a previously uncharged 4.0-μF capacitor. How long does it take for the capacitor voltage to be 100 V?

21. The angular velocity ω is the time rate of change of the angular displacement θ of a rotating object. See Fig. 25-3. If the angular velocity of a pulley wheel is $\omega = 0.40t$, find the angular displacement at $t = 1.50$ s if $\theta = 0.10$ for $t = 0$.

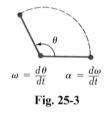

22. The angular acceleration α is the time rate of change of angular velocity ω of a rotating object. See Fig. 25-3. When starting up, the angular acceleration of a helicopter blade is $\alpha = \sqrt{8t + 1}$. Find the expression for θ if $\omega = 0$ and $\theta = 0$ for $t = 0$.

$$\omega = \frac{d\theta}{dt} \qquad \alpha = \frac{d\omega}{dt}$$

Fig. 25-3

23. An inductor in an electric circuit is essentially a coil of wire in which the voltage is affected by a changing current. By definition, the voltage caused by the changing current is given by $V_L = L(di/dt)$, where L is the inductance and is measured in henrys. If $V_L = 12.0 - 0.2t$ for a 3-H inductor, find the current in the circuit after 20 s if the initial current was zero.

24. If the inner and outer walls of a container are at different temperatures, the rate of change of temperature with respect to the distance from one wall is a function of the distance from the wall. Symbolically this is stated as $dT/dx = f(x)$, where T is the temperature. If x is measured from the outer wall, at 20°C, and $f(x) = 72x^2$, find the temperature at the inner wall if the container walls are 0.5 cm thick.

25. Surrounding an electrically charged particle is an electric field. The rate of change of electric potential with respect to the distance from the particle creating the field equals the negative of the value of the electric field. That is, $dV/dx = -E$, where E is the electric field. If $E = k/x^2$, where k is a constant, find the electric potential at a distance x_1 from the particle, if $V \to 0$ as $x \to \infty$.

26. The rate of change of the vertical deflection y with respect to the horizontal distance x from one end of a beam is a function of x. For a particular beam, this function is $k(x^5 + 1350x^3 - 7000x^2)$, where k is a constant. Find y as a function of x.

27. Fresh water is flowing into a brine solution, with an equal volume of mixed solution flowing out. The amount of salt in the solution decreases, but more slowly as time increases. Under certain conditions the time rate of change of mass of salt (in grams per minute) is given by $-1/\sqrt{t + 1}$. Find the mass of salt as a function of time if 1000 g were originally present. Under these conditions, how long would it take for all the salt to be removed?

28. A holograph of a circle is formed. The rate of change of the radius r of the circle with respect to the wavelength λ of the light used is inversely proportional to the square root of λ. If $dr/d\lambda = 3.55 \times 10^4$ and $r = 4.08$ cm for $\lambda = 574$ nm, find r as a function of λ.

25-2 Areas by Integration

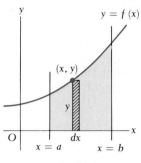

Fig. 25-4

In Section 24-4 we introduced the method of finding the area under a curve by means of integration. In the same section we also showed that the area under a curve can be found by a summation process performed on the rectangles inscribed under the curve. In this way it was shown that integration can be interpreted as a summation process. *The basic applications of the definite integral use this summation interpretation of the integral.* In this section we shall formulate a general procedure for finding the area for which the bounding curves are known. The method is based on the summing of the areas of inscribed rectangles and using integration for the summation.

The first step in finding any area is to make a sketch of the area to be found. Next a representative **element of area** dA (a typical rectangle) should be drawn. In Fig. 25-4 the width of the element is dx. The length of the element is determined by the y-coordinate (of the vertex of the element) of the point on the curve. Thus, we call the length y. The area of this element is $y\,dx$, which in turn means that $dA = y\,dx$, or

$$A = \int_a^b y\,dx = \int_a^b f(x)\,dx \qquad (25\text{-}7)$$

This equation states that the elements are to be summed (this is the meaning of the integral sign) from a (the left boundary) to b (the right boundary).

EXAMPLE A

Find the area bounded by the curves $y = 2x^2$, $y = 0$, $x = 1$, and $x = 2$.

The desired area is shown in Fig. 25-5. The rectangle shown is the representative element. Its area is $y\,dx$. The elements are to be summed from $x = 1$ to $x = 2$. Thus,

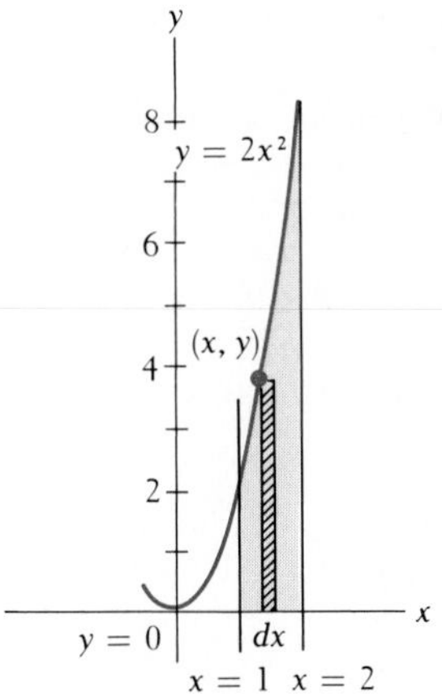

$$A = \overset{\text{sum}}{\underset{\text{left boundary}}{\int_1^2}} \underset{\text{area of element}}{\overset{\text{right boundary}}{y\,dx}}$$

We now complete the solution as follows:

$$A = \int_1^2 y\,dx = \int_1^2 2x^2\,dx \qquad \text{substitute } 2x^2 \text{ for } y$$

$$= \frac{2}{3}x^3 \Big|_1^2 = \frac{2}{3}(8) - \frac{2}{3}(1) \qquad \text{integrate and evaluate}$$

$$= \frac{14}{3}$$

Fig. 25-5

In Figs. 25-4 and 25-5 the elements are vertical. It is also possible to use horizontal elements, and many problems are simplified by using them. The difference is that the length (longest dimension) is now measured in terms of the

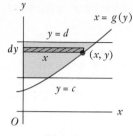

Fig. 25-6

x-coordinate of the point on the curve, and the width becomes dy. In Fig. 25-6 the area of the element is $x\,dy$, which means $dA = x\,dy$, or

$$A = \int_c^d x\,dy = \int_c^d g(y)\,dy \tag{25-8}$$

In using Eq. (25-8) the elements are summed from c (the lower boundary) to d (the upper boundary). In the following example, the area is found by use of both Eq. (25-7) and Eq. (25-8).

EXAMPLE B _____

Find the area in the first quadrant bounded by $y = 9 - x^2$.

The area to be found is shown in Fig. 25-7. Using first the vertical element of length y and width dx, we have

$$A = \int_0^3 y\,dx \qquad \text{sum of areas of elements}$$

$$= \int_0^3 (9 - x^2)\,dx \qquad \text{substitute } 9 - x^2 \text{ for } y$$

$$= \left(9x - \frac{x^3}{3}\right)\Bigg|_0^3 \qquad \text{integrate}$$

$$= (27 - 9) - 0 = 18 \qquad \text{evaluate}$$

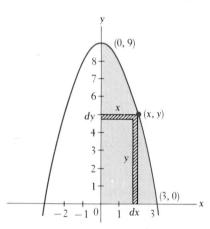

Fig. 25-7

Now, using the horizontal element of length x and width dy, we have

$$A = \int_0^9 x\,dy \qquad \text{sum of areas of elements}$$

$$= \int_0^9 \sqrt{9 - y}\,dy = -\int_0^9 (9 - y)^{1/2}(-dy) \qquad \text{substitute } \sqrt{9 - y} \text{ for } x$$

$$= -\frac{2}{3}(9 - y)^{3/2}\Bigg|_0^9 \qquad \text{integrate}$$

$$= -\frac{2}{3}(9 - 9)^{3/2} + \frac{2}{3}(9 - 0)^{3/2} \qquad \text{evaluate}$$

$$= \frac{2}{3}(27) = 18$$

NOTE ▷

Note that the limits for the vertical elements were 0 and 3, while those for the horizontal elements were 0 and 9. These limits are determined by the direction in which the elements are summed. As we have noted, **vertical elements are summed from left to right, and horizontal elements are summed from bottom to top.** Doing it this way means that the summation will be in a positive direction.

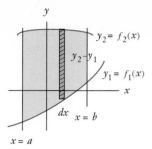

Fig. 25-8

The choice of vertical or horizontal elements is determined by (1) which one leads to the simplest solution, or (2) the form of the resulting integral. In some problems it makes little difference which is chosen. However, our present methods of integration do not include many types of integrals.

It is also possible to find the area between two curves if one is not an axis. In such a case, the length of the element becomes the difference in the y- or x-coordinates, depending on which element is used.

In Fig. 25-8, by using vertical elements, the element of area is bounded on the bottom by $y_1 = f_1(x)$ and on the top by $y_2 = f_2(x)$. The length of the element is $y_2 - y_1$, and its width is dx. Thus, the area is

$$A = \int_a^b (y_2 - y_1)\,dx \qquad (25\text{-}9)$$

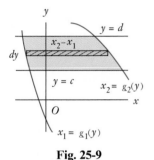

Fig. 25-9

In Fig. 25-9, by using horizontal elements, the element of area is bounded on the left by $x_1 = g_1(y)$ and on the right by $x_2 = g_2(y)$. The length of the element is $x_2 - x_1$, and its width is dy. Thus, the area is

$$A = \int_c^d (x_2 - x_1)\,dy \qquad (25\text{-}10)$$

The following examples show the use of Eqs. (25-9) and (25-10) to find the indicated areas.

EXAMPLE C

$x^2 = x + 2$
$x^2 - x - 2 = 0$
$(x + 1)(x - 2) = 0$
$x = -1, 2$

Find the area bounded by $y = x^2$ and $y = x + 2$.

First, by sketching each curve, we note that the area to be found is that shown in Fig. 25-10. The points of intersection are found by solving the equations simultaneously as shown at the left.

Here we choose vertical elements, since they are all bounded at the top by the line and at the bottom by the parabola. If we choose horizontal elements, the bounding curves are different above $(-1, 1)$ from below it. Choosing horizontal elements would then require two separate integrals for solution. Therefore, using vertical elements, we have

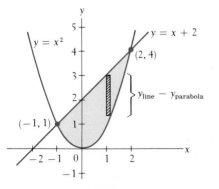

Fig. 25-10

$$A = \int_{-1}^{2} (y_{\text{line}} - y_{\text{parabola}})\,dx \qquad \text{using Eq. (25-9)}$$

$$= \int_{-1}^{2} (x + 2 - x^2)\,dx = \left(\frac{x^2}{2} + 2x - \frac{x^3}{3}\right)\Bigg|_{-1}^{2}$$

$$= \left(2 + 4 - \frac{8}{3}\right) - \left(\frac{1}{2} - 2 + \frac{1}{3}\right)$$

$$= \frac{10}{3} + \frac{7}{6} = \frac{27}{6} = \frac{9}{2}$$

EXAMPLE D

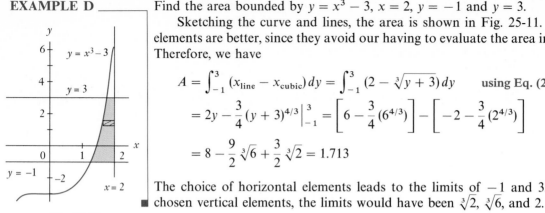

Fig. 25-11

Find the area bounded by $y = x^3 - 3$, $x = 2$, $y = -1$ and $y = 3$.

Sketching the curve and lines, the area is shown in Fig. 25-11. Horizontal elements are better, since they avoid our having to evaluate the area in two parts. Therefore, we have

$$A = \int_{-1}^{3} (x_{\text{line}} - x_{\text{cubic}})\,dy = \int_{-1}^{3} (2 - \sqrt[3]{y + 3})\,dy \qquad \text{using Eq. (25-10)}$$

$$= 2y - \frac{3}{4}(y + 3)^{4/3}\Big|_{-1}^{3} = \left[6 - \frac{3}{4}(6^{4/3})\right] - \left[-2 - \frac{3}{4}(2^{4/3})\right]$$

$$= 8 - \frac{9}{2}\sqrt[3]{6} + \frac{3}{2}\sqrt[3]{2} = 1.713$$

The choice of horizontal elements leads to the limits of -1 and 3. If we had chosen vertical elements, the limits would have been $\sqrt[3]{2}$, $\sqrt[3]{6}$, and 2. ∎

NOTE ▷

It is important that the length of the element be positive. If the difference is taken incorrectly, the result will be negative. ***Getting positive lengths can be assured for vertical elements if we subtract y of the lower curve from y of the upper curve. For horizontal elements we should subtract x of the left curve from x of the right curve.*** Consider the following example.

EXAMPLE E

Fig. 25-12

Find the area bounded by $y = x^3 - 3x - 2$ and the x-axis.

Sketching the graph, we find that $y = x^3 - 3x - 2$ has a maximum point at $(-1, 0)$, a minimum point at $(1, -4)$, and an intercept at $(2, 0)$. The graph is shown in Fig. 25-12, and we see that the area is *below* the x-axis. In using vertical elements, we see that the top is the x-axis ($y = 0$) and the bottom is the curve of $y = x^3 - 3x - 2$. Therefore, we have

$$A = \int_{-1}^{2} [0 - (x^3 - 3x - 2)]\,dx = \int_{-1}^{2} (-x^3 + 3x + 2)\,dx$$

$$= -\frac{1}{4}x^4 + \frac{3}{2}x^2 + 2x\Big|_{-1}^{2} = \left[-\frac{1}{4}(2^4) + \frac{3}{2}(2^2) + 2(2)\right]$$

$$- \left[-\frac{1}{4}(-1)^4 + \frac{3}{2}(-1)^2 + 2(-1)\right]$$

$$= \frac{27}{4} = 6.75$$

If we had simply set up the area as $A = \int_{-1}^{2} (x^3 - 3x - 2)\,dx$, we would have found $A = -6.75$. The negative sign shows that the area is below the x-axis. Again, we avoid any complications with negative areas by making the length of the element positive.

Also note that since

$$0 - (x^3 - 3x - 2) = -(x^3 - 3x - 2)$$

an area bounded on top by the x-axis can be found by setting up the area as being "under" the curve and using the negative of the function. ∎

We must be particularly careful if the bounding curves cross. In such a case, for part of the area one curve is above the area, and for a different part of the area this curve is below the area. When this happens, the area can be found by using two integrals. The following example illustrates the necessity of this procedure.

EXAMPLE F

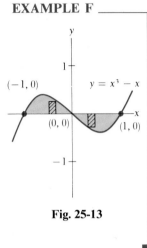

Fig. 25-13

Find the area between $y = x^3 - x$ and the x-axis.

We note from Fig. 25-13 that the area to the left of the origin is above the axis and the area to the right is below. If we find the area from

$$A = \int_{-1}^{1} (x^3 - x)\, dx = \frac{x^4}{4} - \frac{x^2}{2}\Big|_{-1}^{1} = \left(\frac{1}{4} - \frac{1}{2}\right) - \left(\frac{1}{4} - \frac{1}{2}\right) = 0$$

we see that the apparent area is zero. From the figure we know this is not correct. Noting that the y-values (of the area) are negative to the right of the origin, we set up the integrals

$$A = \int_{-1}^{0} (x^3 - x)\, dx + \int_{0}^{1} [-(x^3 - x)]\, dx$$

$$= \left(\frac{x^4}{4} - \frac{x^2}{2}\right)\Big|_{-1}^{0} - \left(\frac{x^4}{4} - \frac{x^2}{2}\right)\Big|_{0}^{1}$$

$$= 0 - \left(\frac{1}{4} - \frac{1}{2}\right) - \left(\frac{1}{4} - \frac{1}{2}\right) + 0 = \frac{1}{2}$$

■

The area under a curve can be applied to various kinds of functions. This is illustrated in the following example and in some of the exercises which follow.

EXAMPLE G

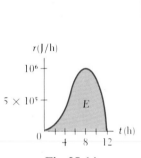

Fig. 25-14

Measurements of solar radiation on a particular surface indicated that the rate r, in joules per hour, at which solar energy is received is given by $r = 3600(12t^2 - t^3)$. Since r is a rate, we may write $r = dE/dt$, where E is the energy, in joules, received. This means that $dE = 3600(12t^2 - t^3)\, dt$, and we can find the total energy received at the surface by evaluating the definite integral

$$E = 3600 \int_{0}^{12} (12t^2 - t^3)\, dt$$

This integral can be interpreted as the area under $f(t) = 3600(12t^2 - t^3)$ from $t = 0$ to $t = 12$, as shown in Fig. 25-14. Evaluating this integral, we have

$$E = 3600 \int_{0}^{12} (12t^2 - t^3)\, dt = 3600\left(4t^3 - \frac{1}{4}t^4\right)\Big|_{0}^{12}$$

$$= 3600\left[4(12^3) - \frac{1}{4}(12^4) - 0\right] = 6.22 \times 10^6 \text{ J}$$

Therefore, 6.22 MJ of energy were received in 12 h. ■

Exercises 25-2

In Exercises 1 through 28, find the areas bounded by the indicated curves.

1. $y = 4x$, $y = 0$, $x = 1$ **2.** $y = 4 - 2x$, $y = 0$, $x = 0$ **3.** $y = x^2$, $y = 0$, $x = 2$

4. $y = 3x^2$, $y = 0$, $x = 3$ **5.** $y = 6 - 4x$, $x = 0$, $y = 0$, $y = 3$ **6.** $y = 8x$, $x = 0$, $y = 4$

7. $y = x^2 + 2$, $x = 0$, $y = 4$, $(x > 0)$ **8.** $y = x^3$, $x = 0$, $y = 3$

9. $y = 2 - x$, $y = 0$, $x = 1$

10. $y = x^2 - 2x$, $y = 0$

11. $y = x^{-2}$, $y = 0$, $x = 2$, $x = 3$

12. $y = 16 - x^2$, $y = 0$, $x = 1$, $x = 2$

13. $y = \sqrt{x}$, $x = 0$, $y = 1$, $y = 3$

14. $y = 2\sqrt{x + 1}$, $x = 0$, $y = 4$

15. $y = 2/\sqrt{x}$, $x = 0$, $y = 1$, $y = 4$

16. $x = y^2 - y$, $x = 0$

17. $y = 4 - 2x$, $x = 0$, $y = 0$, $y = 3$

18. $y = x$, $y = 2 - x$, $x = 0$

19. $y = x^2$, $y = 2 - x$, $x = 0$ $(x \geq 0)$

20. $y = x^2$, $y = 2 - x$, $y = 1$

21. $y = x^4$, $y = 16$

22. $y = x^4 - 8x^2 + 16$, $y = 0$

23. $y = \sqrt{x - 1}$, $y = 3 - x$, $y = 0$

24. $y = x^2 - 3x$, $y = 0$

25. $y = x^3 - 8$, $x = -1$, $y = 0$

26. $y = 4x$, $y = x^3$

27. $y = x$, $x = -1$, $x = 1$, $y = 0$

28. $y = x^2 + 2x - 8$, $y = x + 4$

In Exercises 29 through 36, some applications of areas are shown.

29. Certain physical quantities are often represented as an area under a curve. By definition, power is defined as the time rate of change of performing work. Thus, $p = dw/dt$, or $dw = p\,dt$. Therefore, if $p = 12t - 4t^2$, find the work (in joules) performed in 3 s by finding the area under the curve of p vs. t. See Fig. 25-15.

30. The total electric charge Q, in coulombs, to pass a point in the circuit from time t_1 to t_2 is $Q = \int_{t_1}^{t_2} i\,dt$, where i is the current in amperes. Find Q if $t_1 = 1$ s, $t_2 = 4$ s, and $i = 0.0032t\sqrt{t^2 + 1}$.

31. Since the displacement s, velocity v, and time t of a moving object are related by $s = \int v\,dt$, it is possible to represent the change in displacement as an area. A rocket is launched such that its vertical velocity v (in km/s) as a function of time in seconds is $v = 1 - 0.01\sqrt{2t + 1}$. Find the change in vertical displacement from $t = 10$ s to $t = 100$ s.

32. The total cost of production can be interpreted as an area. If the cost per unit C', in dollars per unit, of producing x units is given by $\dfrac{100}{(0.01x + 1)^2}$, find the total cost C of producing 100 units by finding the area under the curve of C' versus x.

33. A cam is designed such that one face of it is described as being the area between the curves $y = x^3 - 2x^2 - x + 2$ and $y = x^2 - 1$, with units in centimeters. Show that this description does not uniquely describe the face of the cam. Find the area of the face of the cam, if a complete description requires that $x \leq 1$.

34. Using CAD (computer-assisted design), an architect programs a computer to sketch the shape of a swimming pool designed between the curves $y = \dfrac{800x}{(x^2 + 10)^2}$, $y = 0.5x^2 - 4x$, and $x = 8$, with dimensions in meters. Find the area of the surface of the pool.

35. A window is designed to be the area between a parabolic section and a straight base, as shown in Fig. 25-16. What is the area of the window?

36. The vertical ends of a fuel storage tank have a parabolic bottom section and a triangular top section, as shown in Fig. 25-17. What volume does the tank hold?

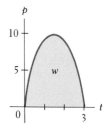

Fig. 25-15

Fig. 25-16

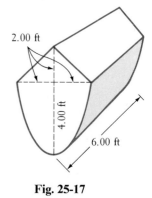

Fig. 25-17

25-3   Volumes by Integration

Consider an area and its representative element [see Fig. 25-18(a)] to be rotated about the x-axis. When an area is rotated in this manner, it is said to generate a volume, which is also indicated in the figure. We shall now show methods of finding volumes which are generated in this manner.

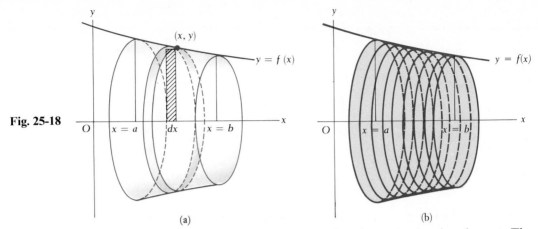

Fig. 25-18

(a)   (b)

As the area rotates about the x-axis, so does its representative element. The element generates a solid for which the volume is known, namely, a thin cylindrical **disk.** We know that the volume of a right circular cylinder is π times the square of the radius times the height (in this case the thickness) of the cylinder. We must now determine the radius and thickness of this disk. Since the element is rotated about the x-axis, the y-coordinate of the point on the curve which touches the element must represent the radius. Also, it can be seen that the thickness is dx (the disk is on its side). This disk, which is the representative **element of volume,** has a volume of $dV = \pi y^2\, dx$. Summing these elements of volume from left to right, as illustrated in Fig. 25-18(b), we have for the total volume V

disk element of volume

$$V = \pi \int_a^b y^2\, dx = \pi \int_a^b [f(x)]^2\, dx \qquad\qquad (25\text{-}11)$$

Here the element of volume is a ***disk,*** and by use of Eq. *(25-11)* we can find the *volume generated by an area bounded by the x-axis, which is rotated about the x-axis.*

EXAMPLE A

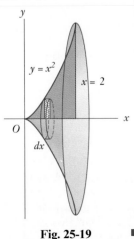

Fig. 25-19

Find the volume generated by rotating the area bounded by $y = x^2$, $x = 2$, and $y = 0$ about the x-axis. See Fig. 25-19.

From the figure we see that the radius of the disk is y and its thickness is dx. The elements are summed from left ($x = 0$) to right ($x = 2$). The solution is as follows:

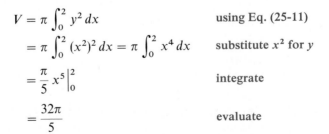

$$V = \pi \int_0^2 y^2\, dx \qquad\qquad \text{using Eq. (25-11)}$$

$$= \pi \int_0^2 (x^2)^2\, dx = \pi \int_0^2 x^4\, dx \qquad \text{substitute } x^2 \text{ for } y$$

$$= \frac{\pi}{5} x^5 \Big|_0^2 \qquad\qquad \text{integrate}$$

$$= \frac{32\pi}{5} \qquad\qquad \text{evaluate}$$

Since π is used in Eq. (25-11), it is common to leave results in terms of π. In ■ applied problems, a decimal result would normally be given.

If an area bounded by the y-axis is rotated about the y-axis, the volume generated is given by

$$V = \pi \int_c^d x^2 \, dy$$

(25-12)

In this case the radius of the element of volume, a **disk,** is the x-coordinate of the point on the curve, and the height of the disk is dy, as shown in Fig. (25-20). One should always be careful to identify the radius and height properly.

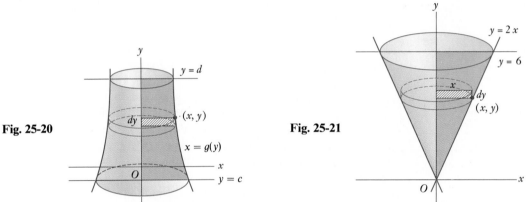

Fig. 25-20

Fig. 25-21

EXAMPLE B

Find the volume generated by rotating the area bounded by $y = 2x$, $y = 6$, and $x = 0$ about the y-axis.

Figure 25-21 shows the volume to be found. Note that the radius of the disk is x and its thickness is dy.

$$V = \pi \int_0^6 x^2 \, dy \qquad\qquad \text{using Eq. (25-12)}$$

$$= \pi \int_0^6 \left(\frac{y}{2}\right)^2 dy = \frac{\pi}{4} \int_0^6 y^2 \, dy \qquad \text{substitute } \frac{y}{2} \text{ for } x$$

$$= \frac{\pi}{12} y^3 \Big|_0^6 = 18\pi \qquad\qquad \text{integrate and evaluate}$$

Since this volume is a right circular cone, it is possible to check the result:

$$V = \frac{1}{3} \pi r^2 h = \frac{1}{3} \pi (3^2)(6) = 18\pi$$

There is another method of finding a volume of a solid of revolution. If the area in Fig. 25-19 is rotated about the y-axis, the element of area $y \, dx$ generates a different element of volume from that generated when it is rotated about the x-axis. We can see in Fig. 25-22 that this element of volume is a **cylindrical shell.** *The total volume is made up of an infinite number of concentric shells.* When the volumes of these shells are summed, we have the total volume generated. Thus we must now find the approximate volume dV of the representative shell. By finding the circumference of the base and multiplying this by the height, we can

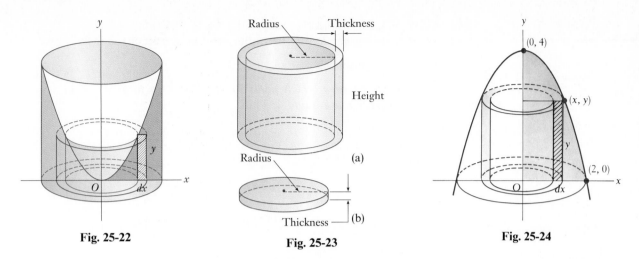

Fig. 25-22   **Fig. 25-23**   **Fig. 25-24**

obtain an expression for the surface area of the shell. Then, by multiplying this by the thickness of the shell, we obtain its volume. The volume of the representative *shell* is [see Fig. 25-23(a)]

$$dV = 2\pi(\text{radius}) \times (\text{height}) \times (\text{thickness}) \tag{25-13}$$

shell element of volume   Similarly, the volume of a *disk* is [see Fig. 25-23(b)]

$$dV = \pi(\text{radius})^2 \times (\text{thickness}) \tag{25-14}$$

It is generally better to remember the formulas for the elements of volume in the general forms given in Eqs. (25-13) and (25-14), and not in specific forms such as Eqs. (25-11) and (25-12) (both of which use *disks*). If we remember them in this manner, we can readily apply these methods to finding any such volume of a solid of revolution.

EXAMPLE C _____   Use the method of cylindrical shells to find the volume generated by rotating the area bounded by $y = 4 - x^2$, $x = 0$, and $y = 0$ about the *y*-axis.

From Fig. 25-24, we identify the radius, height, and thickness of the shell:

$$\text{radius} = x, \qquad \text{height} = y, \qquad \text{thickness} = dx$$

NOTE ▷   *The fact that the elements of area which generate the shells go from $x = 0$ to $x = 2$ determines the limits of integration as **0** and **2**.* Therefore,

$$V = 2\pi \int_0^2 xy\, dx \longleftarrow \text{thickness} \qquad\qquad \text{using Eq. (25-13)}$$
$$\text{radius} \underset{\text{height}}{\text{⎵⎴}}$$

$$= 2\pi \int_0^2 x(4 - x^2)\, dx = 2\pi \int_0^2 (4x - x^3)\, dx \qquad \text{substitute } 4 - x^2 \text{ for } y$$

$$= 2\pi \left(2x^2 - \frac{1}{4}x^4\right)\Big|_0^2 = 8\pi \qquad\qquad \text{integrate and evaluate}$$

∎

EXAMPLE D _____   Use the disk method to find the indicated volume of Example C.

From Fig. 25-25, we identify the radius and thickness of the disk:

radius $= x$,   thickness $= dy$

NOTE ▷ *Since the elements of area which generate the disks go from $y = 0$ to $y = 4$, the limits of integration are* **0** *and* **4.** Thus,

$$V = \pi \int_0^4 x^2 \, dy \qquad\qquad \text{using Eq. (25-14)}$$
$$\text{radius} \rule{0pt}{0pt} \quad \text{thickness}$$

$$= \pi \int_0^4 (4 - y) \, dy \qquad\qquad \text{substitute } \sqrt{4 - y} \text{ for } x$$

$$= \pi \left(4y - \frac{1}{2} y^2 \right) \Big|_0^4 = 8\pi \qquad \text{integrate and evaluate}$$

Fig. 25-26

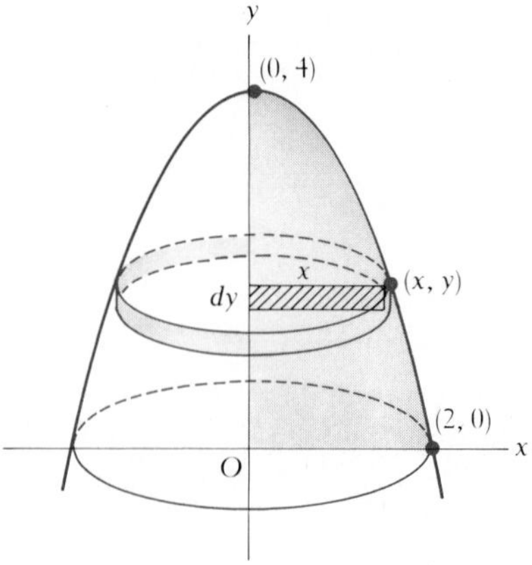

Fig. 25-25

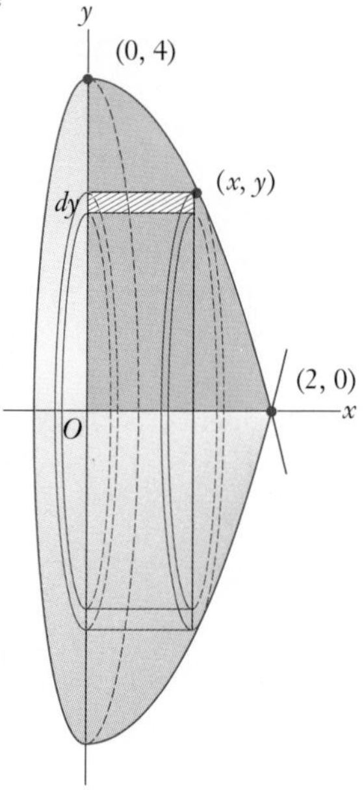

EXAMPLE E _____   Use shells to find the volume if the area of Example C is rotated about the x-axis.

From Fig. 25-26, we see that for the shell we have

radius $= y$,   height $= x$,   thickness $= dy$

Since the elements go from $y = 0$ to $y = 4$, the limits of integration are 0 and 4. Hence

$$V = 2\pi \int_0^4 xy \, dy \qquad\qquad \text{using Eq. (25-13)}$$

$$= 2\pi \int_0^4 \sqrt{4 - y}(y \, dy) \qquad \text{substitute } \sqrt{4 - y} \text{ for } x$$

$$= \frac{256\pi}{15}$$

(The method of integrating this function has not yet been discussed. We present the answer here for the reader's information at this time.)

EXAMPLE F ———— By the use of disks, find the volume indicated in Example E.

For the disk in Fig. 25-27, we have

$$\text{radius} = y, \qquad \text{thickness} = dx$$

and the limits of integration are $x = 0$ and $x = 2$. This gives us

$$V = \pi \int_0^2 y^2 \, dx \qquad\qquad \text{using Eq. (25-14)}$$

$$= \pi \int_0^2 (4 - x^2)^2 \, dx = \pi \int_0^2 (16 - 8x^2 + x^4) \, dx \qquad \text{substitute } 4 - x^2 \text{ for } y$$

$$= \pi \left(16x - \frac{8}{3} x^3 + \frac{1}{5} x^5 \right) \Big|_0^2 = \frac{256\pi}{15} \qquad \text{integrate and evaluate}$$

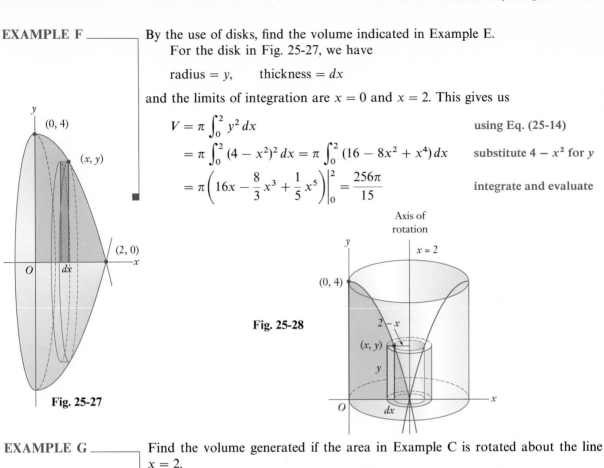

Fig. 25-27

Fig. 25-28

EXAMPLE G ———— Find the volume generated if the area in Example C is rotated about the line $x = 2$.

Shells are convenient, since the volume of a shell can be expressed as a single integral. We can find the radius, height, and thickness of the shell from Fig. 25-28.

NOTE ▷ We carefully note that *the radius is not x but 2 − x,* since the area is rotated about $x = 2$. We see that

$$\text{radius} = 2 - x, \qquad \text{height} = y, \qquad \text{thickness} = dx$$

Since the elements which generate the shells go from $x = 0$ to $x = 2$, the limits of integration are 0 and 2. This means we have

$$V = 2\pi \int_0^2 (2 - x)y \, dx \longleftarrow \text{thickness} \qquad \text{using Eq. (25-13)}$$

$$\text{radius} \rule{0pt}{0pt} \quad \rule{0pt}{0pt} \text{height}$$

$$= 2\pi \int_0^2 (2 - x)(4 - x^2) \, dx \qquad \text{substitute } 4 - x^2 \text{ for } y$$

$$= 2\pi \int_0^2 (8 - 2x^2 - 4x + x^3) \, dx$$

$$= 2\pi \left(8x - \frac{2}{3} x^3 - 2x^2 + \frac{1}{4} x^4 \right) \Big|_0^2 = \frac{40\pi}{3} \qquad \text{integrate and evaluate}$$

(If the area had been rotated about the line $x = 3$, the only difference in the integral would have been that $r = 3 - x$. Everything else, including the limits, would have remained the same.)

Exercises 25-3

In Exercises 1 through 12, find the volume generated by the areas bounded by the given curves if they are rotated about the x-axis. Use the indicated method in each case.

1. $y = 1 - x$, $x = 0$, $y = 0$ (disks)
2. $y = x$, $y = 0$, $x = 2$ (disks)
3. Area of Exercise 1   (shells)
4. $y = \sqrt{x}$, $x = 0$, $y = 2$ (shells)
5. $y = 3\sqrt{x}$, $y = 0$, $x = 4$ (disks)
6. $y = 2x - x^2$, $y = 0$ (disks)
7. $y = x^3$, $y = 8$, $x = 0$ (shells)
8. $y = x^2$, $y = x$ (shells)
9. $y = x^2 + 1$, $x = 0$, $x = 3$, $y = 0$ (disks)
10. $y = 6 - x - x^2$, $x = 0$, $y = 0$ (quadrant I)   (disks)
11. $x = 4y - y^2 - 3$, $x = 0$ (shells)
12. $y = x^4$, $x = 0$, $y = 1$, $y = 2$ (shells)

See Appendix E for a computer program for finding the volume generated by rotating an area about the x-axis.

In Exercises 13 through 24, find the volume generated by the areas bounded by the given curves if they are rotated about the y-axis. Use the indicated method in each case.

13. Area of Exercise 1   (disks)
14. Area of Exercise 4   (disks)
15. Area of Exercise 1   (shells)
16. Area of Exercise 2   (shells)
17. $y = 2\sqrt{x}$, $x = 0$, $y = 2$ (disks)
18. $y^2 = x$, $y = 4$, $x = 0$ (disks)
19. Area of Exercise 6   (shells)
20. Area of Exercise 10   (shells)
21. Area of Exercise 11   (disks)
22. $x^2 + 4y^2 = 4$ (quadrant I)   (disks)
23. $y = \sqrt{4 - x^2}$ (quadrant I) (shells)
24. $y = 8 - x^3$, $x = 0$, $y = 0$ (shells)

In Exercises 25 through 32, find the indicated volumes by integration.

25. Find the volume generated if the area of Exercise 6 is rotated about the line $x = 2$.

26. Find the volume generated if the area of Exercise 8 is rotated about the line $y = 4$.

27. Derive the formula for the volume of a right circular cone of radius r and height h by rotating the area bounded by $y = (r/h)x$, $y = 0$, and $x = h$ about the x-axis.

28. Derive the formula for the volume of a sphere of radius r by rotating the area within the circle $x^2 + y^2 = r^2$ about the x-axis.

29. The capillary tube shown in Fig. 25-29 has circular horizontal cross sections of inner radius 1.1 mm. What is the volume of the liquid in the tube above the level of liquid outside the tube if the top of the liquid in the center vertical cross section is described by the equation $y = x^4 + 1.5$, as shown?

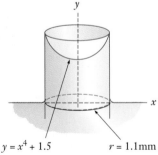

$y = x^4 + 1.5$ $r = 1.1$mm

Fig. 25-29

30. The concrete base for a piece of machinery can be described as being the first-quadrant area bounded by $y = 2$ and a section of $y = 8 - x^3$, as shown in Fig. 25-30, being rotated about the y-axis. Find the volume, in cubic feet, of the machinery base.

31. A hole 2.00 cm in diameter is drilled through the center of a spherical lead weight 6.00 cm in diameter. How much lead is removed?

32. All horizontal cross sections of a keg 4 ft tall are circular, and the sides of the keg are parabolic. The diameter at the top and the bottom is 2 ft, and the diameter in the middle is 3 ft. Find the volume that the keg holds.

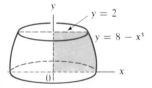

$y = 2$

$y = 8 - x^3$

Fig. 25-30

25-4 Centroids

In the study of mechanics, a very important property of an object is its center of mass. In this section we explain the meaning of center of mass and then show how integration is used to determine the center of mass for areas and solids of rotation.

If a mass m is at a distance d from a specified point O, the **moment** of the mass about O is defined as md. If several masses $m_1, m_2, \ldots, m_n$ are at distances $d_1, d_2, \ldots, d_n$, respectively, from point O, their moment (as a group) about O is defined as $m_1 d_1 + m_2 d_2 + \cdots + m_n d_n$. If all the masses could be concentrated at one point $\bar{d}$ units from O, the moment would be $(m_1 + m_2 + \cdots + m_n)\bar{d}$, and this is what is meant by the previous expression. Therefore, we may write

$$m_1 d_1 + m_2 d_2 + \cdots + m_n d_n = (m_1 + m_2 + \cdots + m_n)\bar{d} \qquad (25\text{-}15)$$

In Eq. (25-15), $\bar{d}$ is the distance from O to the **center of mass.** The moment of a mass is a measure of its tendency to rotate about a point. A weight far from the point of balance of a long rod is more likely to make the rod turn than if the same weight were placed near the point of balance. It is easier to open a door if you push near the doorknob than if you push near the hinges. This is the type of physical property which the moment of mass measures.

EXAMPLE A

One of the simplest and most basic illustrations of moments and center of mass is seen in balancing a long rod with masses of different sizes, one on either side of the balance point.

In Fig. 25-31, a mass of 5.0 kg is hung from the rod 0.8 m to the right of point O. We see that this 5.0-kg mass tends to turn the rod clockwise. A mass placed on the opposite side of O will tend to turn the rod counterclockwise. Neglecting the mass of the rod, in order to balance the rod at O, the moments must be equal in magnitude but opposite in sign. Therefore, a 4.0-kg mass would have to be placed 1.0 m to the left.

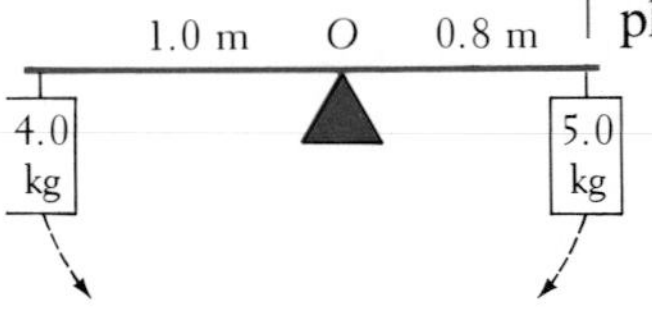

Thus, with $d_1 = 0.8$ m, $d_2 = -1.0$ m, we see that

$$(5.0 + 4.0)\bar{d} = 5.0(0.8) + 4.0(-1.0) = 4.0 - 4.0$$
$$\bar{d} = 0.0 \text{ m}$$

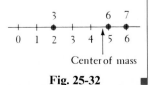

1.0 m O 0.8 m

4.0 kg   5.0 kg

Fig. 25-31

NOTE ▷

This means that the center of mass (of the combination of the 5.0-kg mass and the 4.0-kg mass) is at O. Also, we see that *we **must** use directed **distances in finding moments.***

EXAMPLE B

On the x-axis a mass of 3 units is placed at $(2, 0)$, another of 6 units at $(5, 0)$, and a third, 7 units, at $(6, 0)$. Find the center of mass of the three objects.

Taking the reference point as the origin, we find $d_1 = 2$, $d_2 = 5$, and $d_3 = 6$. Thus, $m_1 d_1 + m_2 d_2 + m_3 d_3 = (m_1 + m_2 + m_3)\bar{d}$ becomes

$$3(2) + 6(5) + 7(6) = (3 + 6 + 7)\bar{d} \quad \text{or} \quad \bar{d} = 4.88$$

This means that the center of mass of the three objects is at $(4.88, 0)$. Therefore a mass of 16 units placed at this point has the same moment as the three masses as a unit (see Fig. 25-32).

Center of mass

Fig. 25-32

EXAMPLE C

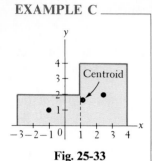

Fig. 25-33

Find the center of mass of the area indicated in Fig. 25-33.

We first note that the center of mass is not *on* either axis. This can be seen from the fact that the major portion of the area is in the first quadrant. *We shall therefore measure the moments with respect to each axis to find the point which is the center of mass. This point is also called the* **centroid** *of the area.*

The easiest method of finding the centroid is to divide the area into rectangles, as indicated by the dashed line in Fig. 25-33, and assume that we may consider the mass of each rectangle to be concentrated at its center. In this way the left rectangle has its center at $(-1, 1)$ and the right rectangle has its center at $(\frac{5}{2}, 2)$. The mass of each rectangle, assumed uniform, is proportional to the area. The area of the left rectangle is 8 units and that of the right rectangle is 12 units. Thus, taking moments with respect to the y-axis, we have

$$8(-1) + 12\left(\frac{5}{2}\right) = (8 + 12)\bar{x}$$

where $\bar{x}$ is the x-coordinate of the centroid. Solving for $\bar{x}$, we have $\bar{x} = \frac{11}{10}$.

Now, taking moments with respect to the x-axis, we have

$$8(1) + 12(2) = (8 + 12)\bar{y}$$

where $\bar{y}$ is the y-coordinate of the centroid. Thus, $\bar{y} = \frac{8}{5}$. This means that the coordinates of the centroid, the center of mass, are $(\frac{11}{10}, \frac{8}{5})$. This may be interpreted as meaning that an area of this shape would balance on a single support under this point. As an approximate check, we note from the figure that this point appears to be a reasonable balance point for the area. ∎

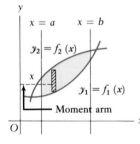

Fig. 25-34

If an area is bounded by the curves of the functions $y_1 = f_1(x)$, $y_2 = f_2(x)$, $x = a$, and $x = b$, as shown in Fig. 25-34, the moment of the element of area about the y-axis is $(k\,dA)x$, where k is the mass per unit area. In this expression, $k\,dA$ is the mass of the element, and x is its distance (moment arm) from the y-axis. The element dA may be written as $(y_2 - y_1)\,dx$, which means that the moment may be written as $kx(y_2 - y_1)\,dx$. If we then sum up the moments of all the elements and express this as an integral (which, of course, means sum), we have $k\int_a^b x(y_2 - y_1)\,dx$. If we consider all the mass of the area to be concentrated at one point $\bar{x}$ units from the y-axis, the moment would be $(kA)\bar{x}$, where kA is the mass of the entire area, and $\bar{x}$ is the distance the center of mass is from the y-axis. By the previous discussion these two expressions should be equal. This means $k\int_a^b x(y_2 - y_1)\,dx = kA\bar{x}$. Since k appears on each side of the equation, we divide it out (we are assuming that the mass per unit area is constant). The area A is found by the integral $\int_a^b (y_2 - y_1)\,dx$. Therefore,

centroid of an area

$$\bar{x} = \frac{\displaystyle\int_a^b x(y_2 - y_1)\,dx}{\displaystyle\int_a^b (y_2 - y_1)\,dx}$$

(25-16)

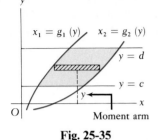

NOTE ▷ *Equation (25–16) gives us the x-coordinate of the centroid of an area if vertical elements are used.* It should be noted that **the two integrals in Eq. (25-16) must be evaluated separately.** We cannot cancel the apparent common factor $y_2 - y_1$, and we cannot combine quantities and perform one integration. The two integrals must be evaluated before cancellations are made.

Following the same reasoning, if an area is bounded by $x_1 = g_1(y)$, $x_2 = g_2(y)$, $y = c$, and $y = d$ (Fig. 25-35), the y-coordinate of the centroid is

Fig. 25-35

$$\bar{y} = \frac{\int_c^d y(x_2 - x_1)\,dy}{\int_c^d (x_2 - x_1)\,dy} \qquad (25\text{-}17)$$

In this equation, horizontal elements are used.

In applying Eqs. (25-16) and (25-17), we should keep in mind that the denominators of the right-hand sides give the area, and once we have found this, we may use it for both $\bar{x}$ and $\bar{y}$. Also, we should utilize any symmetry a curve may have.

EXAMPLE D

Find the coordinates of the centroid of the area bounded by $y = x^2$ and the line $y = 4$.

We sketch a graph indicating the area and an element of area (see Fig. 25-36). The curve is a parabola whose axis is the y-axis. Since the area is symmetrical to the y-axis, the centroid must be on this axis. This means that the x-coordinate of the centroid is zero, or $\bar{x} = 0$. To find the y-coordinate of the centroid, we have

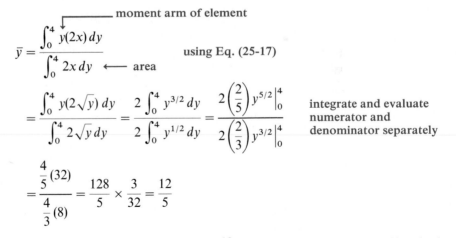

Fig. 25-36

$$\bar{y} = \frac{\int_0^4 y(2x)\,dy}{\int_0^4 2x\,dy} \qquad \text{using Eq. (25-17)}$$

moment arm of element

area

$$= \frac{\int_0^4 y(2\sqrt{y})\,dy}{\int_0^4 2\sqrt{y}\,dy} = \frac{2\int_0^4 y^{3/2}\,dy}{2\int_0^4 y^{1/2}\,dy} = \frac{2\left(\dfrac{2}{5}\right)y^{5/2}\Big|_0^4}{2\left(\dfrac{2}{3}\right)y^{3/2}\Big|_0^4}$$

integrate and evaluate numerator and denominator separately

$$= \frac{\dfrac{4}{5}(32)}{\dfrac{4}{3}(8)} = \frac{128}{5} \times \frac{3}{32} = \frac{12}{5}$$

The coordinates of the centroid are $(0, \frac{12}{5})$. This area would balance if a single pointed support were to be put under this point. ∎

EXAMPLE E

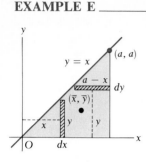

Fig. 25-37

Find the coordinates of the centroid of an isosceles right triangle with side a.

We must first set up this area in the xy-plane. One choice is to place the triangle with one vertex at the origin and the right angle on the x-axis (see Fig. 25-37). Since each side is a, the hypotenuse passes through the point (a, a). The equation of the hypotenuse is $y = x$. The x-coordinate of the centroid is found by using Eq. (25-16):

$$\bar{x} = \frac{\int_0^a xy\,dx}{\int_0^a y\,dx} = \frac{\int_0^a x(x)\,dx}{\int_0^a x\,dx} = \frac{\int_0^a x^2\,dx}{\frac{1}{2}x^2\Big|_0^a} = \frac{\frac{1}{3}x^3\Big|_0^a}{\frac{a^2}{2}} = \frac{\frac{a^3}{3}}{\frac{a^2}{2}} = \frac{2a}{3}$$

The y-coordinate of the centroid is found by using Eq. (25-17):

$$\bar{y} = \frac{\int_0^a y(a - x)\,dy}{\frac{a^2}{2}} = \frac{\int_0^a y(a - y)\,dy}{\frac{a^2}{2}} = \frac{\int_0^a (ay - y^2)\,dy}{\frac{a^2}{2}}$$

$$= \frac{\frac{ay^2}{2} - \frac{y^3}{3}\Big|_0^a}{\frac{a^2}{2}} = \frac{\frac{a^3}{6}}{\frac{a^2}{2}} = \frac{a}{3}$$

Thus, the coordinates of the centroid are $(\frac{2}{3}a, \frac{1}{3}a)$. The results indicate that the center of mass is $\frac{1}{3}a$ units from each of the equal sides.

Another important figure for which we wish to find the centroid is the solid of revolution. If the density of the solid is constant, the coordinates of the centroid will be on the axis of rotation. The problem which remains is to find just where on the axis the centroid is located.

If a given area bounded by the x-axis, as shown in Fig. 25-38, is revolved about the x-axis, a vertical element of area generates a disk element of volume. The center of mass of the disk is at its center, and therefore, for purposes of finding moments, we may consider its mass concentrated there. The moment about the y-axis of a typical element is $x(k)(\pi y^2\,dx)$, where x is the moment arm, k is the density, and $\pi y^2\,dx$ is the volume. The sum of the moments of the elements can be expressed as an integral; it equals the volume times the density times the x-coordinate of the centroid of the volume. Since the density k and the factor of π would appear on each side of the equation, they need not be written. Therefore,

centroid of a volume

$$\bar{x} = \frac{\int_a^b xy^2\,dx}{\int_a^b y^2\,dx} \tag{25-18}$$

is the equation giving the x-coordinate of the centroid of a volume of a solid of revolution about the x-axis.

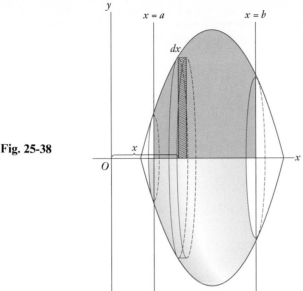

Fig. 25-38

In the same manner we may find *the y-coordinate of the centroid of a volume of a solid of revolution about the y-axis. It is*

$$\bar{y} = \frac{\int_c^d yx^2 \, dy}{\int_c^d x^2 \, dy} \qquad (25\text{-}19)$$

EXAMPLE F

Find the coordinates of the centroid of the volume generated by rotating the first-quadrant area under the curve $y = 4 - x^2$ about the x-axis.

Since the curve (see Fig. 25-39) is rotated about the x-axis, the centroid is on the x-axis, which means that $\bar{y} = 0$. We find the x-coordinate as follows:

$$\bar{x} = \frac{\int_0^2 \overset{\text{moment arm}}{x} y^2 \, dx}{\int_0^2 y^2 \, dx} \qquad \text{using Eq. (25-18)}$$

$$= \frac{\int_0^2 x(4 - x^2)^2 \, dx}{\int_0^2 (4 - x^2)^2 \, dx} = \frac{\int_0^2 (16x - 8x^3 + x^5) \, dx}{\int_0^2 (16 - 8x^2 + x^4) \, dx}$$

$$= \frac{8x^2 - 2x^4 + \dfrac{1}{6}x^6 \Big|_0^2}{16x - \dfrac{8}{3}x^3 + \dfrac{1}{5}x^5 \Big|_0^2} = \frac{32 - 32 + \dfrac{64}{6}}{32 - \dfrac{64}{3} + \dfrac{32}{5}} = \frac{5}{8}$$

■ The coordinates of the centroid are $(\frac{5}{8}, 0)$.

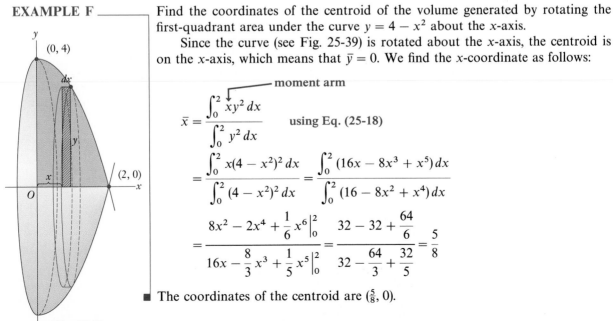

Fig. 25-39

EXAMPLE G ___

Find the coordinates of the centroid of the volume generated by rotating the area of Example F about the y-axis.

Since the curve is rotated about the y-axis, $\bar{x} = 0$ (see Fig. 25-40). The y-coordinate is

$$
\bar{y} = \frac{\int_0^4 \overset{\text{moment arm}}{y} x^2 \, dy}{\int_0^4 x^2 \, dy} \qquad \text{using Eq. (25-19)}
$$

$$
= \frac{\int_0^4 y(4-y)\,dy}{\int_0^4 (4-y)\,dy} = \frac{\int_0^4 (4y-y^2)\,dy}{\int_0^4 (4-y)\,dy} = \frac{2y^2 - \dfrac{1}{3}y^3 \Big|_0^4}{4y - \dfrac{1}{2}y^2 \Big|_0^4}
$$

$$
= \frac{32 - \dfrac{64}{3}}{16 - 8} = \frac{4}{3}
$$

■ The coordinates of the centroid are $(0, \tfrac{4}{3})$.

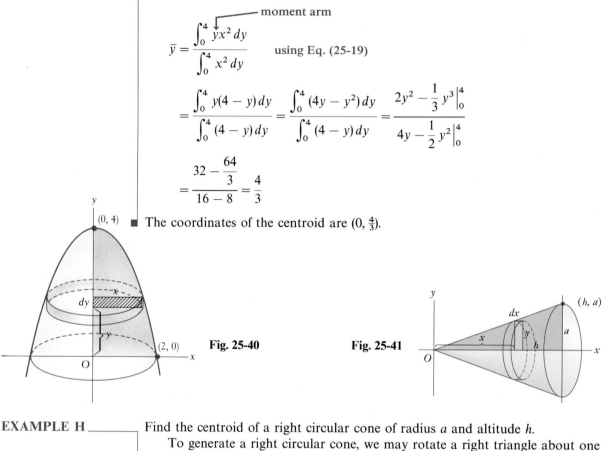

Fig. 25-40   Fig. 25-41

EXAMPLE H ___

Find the centroid of a right circular cone of radius a and altitude h.

To generate a right circular cone, we may rotate a right triangle about one of its legs (Fig. 25-41). Placing the leg of length h along the x-axis, we rotate the right triangle whose hypotenuse is given by $y = (a/h)x$ about the x-axis. The x-coordinate of the centroid is

$$
\bar{x} = \frac{\int_0^h \overset{\text{moment arm}}{x} y^2 \, dx}{\int_0^h y^2 \, dx} \qquad \text{using Eq. (25-18)}
$$

$$
= \frac{\int_0^h x\left[\left(\dfrac{a}{h}\right)x\right]^2 dx}{\int_0^h \left[\left(\dfrac{a}{h}\right)x\right]^2 dx} = \frac{\left(\dfrac{a^2}{h^2}\right)\left(\dfrac{1}{4}x^4\right)\Big|_0^h}{\left(\dfrac{a^2}{h^2}\right)\left(\dfrac{1}{3}x^3\right)\Big|_0^h} = \frac{3}{4}h
$$

Therefore, the centroid is located along the altitude $\frac{3}{4}$ of the way from the vertex, ■ or $\frac{1}{4}$ of the way from the base.

Exercises 25-4

In Exercises 1 through 4, find the center of mass of the particles of the given masses located at the given points.

1. 5 units at $(1, 0)$, 10 units at $(4, 0)$, 3 units at $(5, 0)$

2. 2 units at $(2, 0)$, 9 units at $(3, 0)$, 3 units at $(8, 0)$, 1 unit at $(12, 0)$

3. 4 units at $(-3, 0)$, 2 units at $(0, 0)$, 1 unit at $(2, 0)$, 8 units at $(3, 0)$

4. 2 units at $(-4, 0)$, 1 unit at $(-3, 0)$, 5 units at $(1, 0)$, 4 units at $(4, 0)$

In Exercises 5 through 8, find the coordinates of the centroid of the area shown.

5. Fig. 25-42(a)   **6.** Fig. 25-42(b)   **7.** Fig. 25-42(c)   **8.** Fig. 25-42(d)

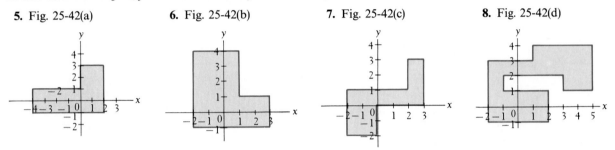

Fig. 25-42

In Exercises 9 through 24, find the coordinates of the centroids of the indicated figures.

9. Find the coordinates of the centroid of the area bounded by $y = x^2$ and $y = 2$.

10. Find the coordinates of the centroid of a semicircular area.

11. Find the coordinates of the centroid of the area bounded by $y = 4 - x$ and the axes.

12. Find the coordinates of the centroid of the area bounded by $y = x^3$, $x = 2$, and the x-axis.

13. Find the coordinates of the centroid of the area bounded by $y = x^2$ and $y = x^3$.

14. Find the coordinates of the centroid of the area bounded by $y^2 = x$, $y = 2$, $x = 0$.

15. Find the coordinates of the centroid of the volume generated by rotating the area in the first quadrant bounded by $y^2 = 4x$, $y = 0$, and $x = 1$ about the y-axis.

16. Find the coordinates of the centroid of the volume generated by rotating the area bounded by $y = x^2$, $x = 2$, and the x-axis about the x-axis.

17. Find the coordinates of the centroid of the volume generated by rotating the area bounded by $y^2 = 4x$ and $x = 1$ about the x-axis.

18. Find the coordinates of the centroid of the volume generated by rotating the area bounded by $x^2 - y^2 = 9$, $y = 4$, and the x-axis about the y-axis.

19. Find the location of the centroid of a right triangle with legs a and b.

20. Find the location of the centroid of a hemisphere of radius a.

21. A lens with semi-elliptical vertical cross sections and circular horizontal cross sections is shown in Fig. 25-43. For proper installation in an optical device, its centroid must be known. Locate its centroid.

22. A sanding machine disc can be described as the volume generated by rotating the area bounded $y^2 = 4/x$, $y = 1$, $y = 2$, and the y-axis about the y-axis, where measurements are in inches. Locate the centroid of the disc.

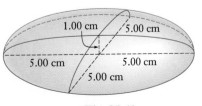

Fig. 25-43

23. A highway marking pylon has the shape of a frustum of a cone. Find its centroid if the radii of its bases are 2.00 in. and 8.00 in. and the height between bases is 24.0 in.

24. A floodgate is in the shape of an isosceles trapezoid. Find the location of the centroid of the floodgate if the upper base is 20 m, the lower base is 12 m, and the height between bases is 6 m.

25-5 Moments of Inertia

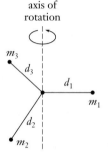

axis of rotation

Fig. 25-44

In applying the laws of physics, we encounter an important quantity when we are discussing the rotation of an object: its **moment of inertia.** The moment of inertia of an object rotating about an axis is analogous to the mass of an object moving through space in reference to some point. *In each case the quantity (moment of inertia or mass) is the measure of the tendency of an object to resist a change in motion.*

Suppose that a particle of mass m is rotating about some point; we define its moment of inertia as md^2, where d is the distance from the particle to the point. If a group of particles of masses $m_1, m_2, \ldots, m_n$ are rotating about some axis, such as is shown in Fig. 25-44, the moment of inertia I with respect to that axis is

$$I = m_1 d_1^2 + m_2 d_2^2 + \cdots + m_n d_n^2$$

where the d's are the respective distances of the particles from the axis. If all of the masses were at the same distance R from the axis of rotation, so that the total moment of inertia were the same, we would have

$$m_1 d_1^2 + m_2 d_2^2 + \cdots + m_n d_n^2 = (m_1 + m_2 + \cdots + m_n)R^2 \tag{25-20}$$

radius of gyration

where R is called the **radius of gyration.**

EXAMPLE A

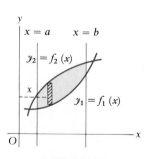

Fig. 25-45

Find the moment of inertia and radius of gyration of three masses, one of 3 units at $(-2, 0)$, the second of 5 units at $(1, 0)$, and the third of 4 units at $(4, 0)$ with respect to the origin (see Fig. 25-45).

The moment of inertia of the group is

$$I = 3(-2)^2 + 5(1)^2 + 4(4)^2 = 81$$

The radius of gyration is found from $I = (m_1 + m_2 + m_3)R^2$. Thus,

$$81 = (3 + 5 + 4)R^2 \qquad R^2 = \frac{81}{12} \quad \text{or} \quad R = 2.60$$

This means that a mass of 12 units placed at $(2.60, 0)$ [or at $(-2.60, 0)$] would have the same rotational inertia about the origin as the three objects as a unit.

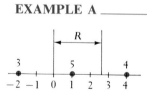

Fig. 25-46

If an area is bounded by the curves of the functions $y_1 = f_1(x)$, $y_2 = f_2(x)$, and $x = a$ and $x = b$, as shown in Fig. 25-46, the moment of inertia of this area with respect to the y-axis, I_y, is given by the sum of the moments of inertia of the individual elements. The mass of each element is $k(y_2 - y_1)dx$, where k is the mass per unit area and $(y_2 - y_1)dx$ is the area of the element. The distance of the element from the y-axis is x. Representing this sum as an integral, we have

$$I_y = k \int_a^b x^2(y_2 - y_1)\,dx \qquad (25\text{-}21)$$

To find the radius of gyration of the area with respect to the y-axis, R_y, we would first find the moment of inertia, divide this by the mass of the area, and take the square root of this result.

moment of inertia
of an area

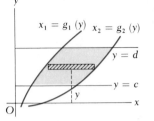

$x_1 = g_1(y)$ $x_2 = g_2(y)$

Fig. 25-47

In the same manner, the moment of inertia of an area, with respect to the x-axis, bounded by $x_1 = g_1(y)$ and $x_2 = g_2(y)$ is given by

$$I_x = k \int_c^d y^2(x_2 - x_1)\,dy \qquad (25\text{-}22)$$

We find the radius of gyration of the area with respect to the x-axis, R_x, in the same manner as we find it with respect to the y-axis (see Fig. 25-47).

EXAMPLE B

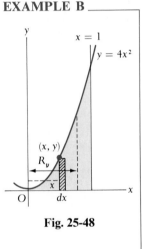

Fig. 25-48

Find the moment of inertia and the radius of gyration of the area bounded by $y = 4x^2$, $x = 1$, and the x-axis with respect to the y-axis.

We find the moment of inertia of the area (see Fig. 25-48) as follows:

distance from element to axis

$$I_y = k \int_0^1 x^2 y\,dx \qquad \text{using Eq. (25-21)}$$

$$= k \int_0^1 x^2(4x^2)\,dx = 4k \int_0^1 x^4\,dx$$

$$= 4k\left(\frac{1}{5}x^5\right)\Bigg|_0^1 = \frac{4k}{5}$$

To find the radius of gyration, we first determine the mass of the area:

$$m = k \int_0^1 y\,dx = k \int_0^1 (4x^2)\,dx = 4k\left(\frac{1}{3}x^3\right)\Bigg|_0^1 = \frac{4k}{3} \qquad m = kA$$

$$R_y^2 = \frac{I_y}{m} = \frac{4k}{5} \times \frac{3}{4k} = \frac{3}{5} \qquad R_y^2 = I_y/m$$

$$R_y = \sqrt{\frac{3}{5}} = \frac{\sqrt{15}}{5}$$

∎

EXAMPLE C

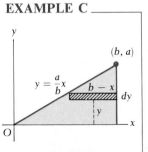

Fig. 25-49

Find the moment of inertia of a right triangle with sides a and b with respect to the side b. Assume that $k = 1$.

Placing the triangle as shown in Fig. 25-49, we see that the equation of the hypotenuse is $y = (a/b)x$. The moment of inertia is

distance from element to axis

$$I_x = \int_0^a y^2(b - x)\,dy \qquad \text{using Eq. (25-22)}$$

$$= \int_0^a y^2\left(b - \frac{b}{a}y\right)dy = b\int_0^a\left(y^2 - \frac{1}{a}y^3\right)dy$$

$$= b\left(\frac{1}{3}y^3 - \frac{1}{4a}y^4\right)\Bigg|_0^a = b\left(\frac{a^3}{3} - \frac{a^3}{4}\right) = \frac{ba^3}{12}$$

Among the most important moments of inertia are those of solids of revolution. Since ***all parts of an element of mass should be at the same distance from the axis***, the most convenient element of volume to choose is the cylindrical shell (see Fig. 25-50). If the area bounded by the curves $y = f(x)$, $x = a$, $x = b$, and the x-axis is rotated about the y-axis, the moment of inertia of the element of volume is $k(2\pi xy\, dx)(x^2)$, where k is the density, $2\pi xy\, dx$ is the volume of the element, and x^2 is the square of its distance from the y-axis. Expressing the sum of the moments of the elements as an integral, *the moment of inertia of the volume with respect to the y-axis*, I_y, is

moment of inertia
of a volume

$$I_y = 2\pi k \int_a^b y x^3\, dx \qquad\qquad (25\text{-}23)$$

The radius of gyration of the volume with respect to the y-axis, R_y, is found by determining (1) the moment of inertia, (2) the mass of the volume, and (3) the square root of the quotient of the moment of inertia divided by the mass.

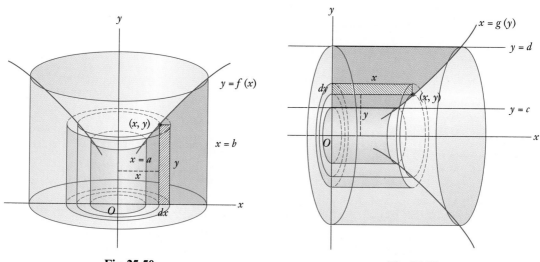

Fig. 25-50   **Fig. 25-51**

The moment of inertia of the volume (see Fig. 25-51) generated by rotating the area bounded by $x = g(y)$, $y = c$, $y = d$, and the y-axis about the x-axis, I_x, is given by

$$I_x = 2\pi k \int_c^d x y^3\, dy \qquad\qquad (25\text{-}24)$$

The radius of gyration of the volume with respect to the x-axis, R_x, is found in the same manner as R_y.

EXAMPLE D — Find the moment of inertia and the radius of gyration with respect to the x-axis of the solid (see Fig. 25-52) generated by rotating the area bounded by $y^3 = x$, $y = 2$, and the y-axis about the x-axis.

distance from element to axis

$$I_x = 2\pi k \int_0^2 xy^3 \, dy \qquad \text{using Eq. (25-24)}$$

$$= 2\pi k \int_0^2 (y^3) y^3 \, dy = 2\pi k \left(\frac{1}{7} y^7\right)\bigg|_0^2 = \frac{256\pi k}{7}$$

$$m = 2\pi k \int_0^2 xy \, dy \qquad \text{mass} = k \times \text{volume}$$

$$= 2\pi k \int_0^2 y^3 y \, dy = 2\pi k \left(\frac{1}{5} y^5\right)\bigg|_0^2 = \frac{64\pi k}{5}$$

$$R_x^2 = \frac{256\pi k}{7} \times \frac{5}{64\pi k} = \frac{20}{7} \qquad R_x^2 = I_x/m$$

$$R_x = \sqrt{\frac{20}{7}} = \frac{2}{7}\sqrt{35}$$

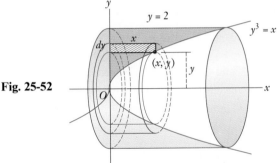

Fig. 25-52

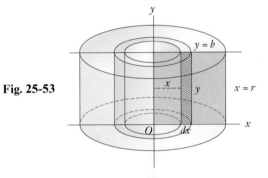

Fig. 25-53

EXAMPLE E — As we noted at the beginning of this section, the moment of inertia is important when studying the rotational motion of an object. For this reason, the moments of inertia of various objects are calculated, and the formulas can be tabulated. Such formulas are usually expressed in terms of the mass of the object.

Among the objects for which the moment of inertia is important is a solid disk. Find the moment of inertia of a disk with respect to its axis and in terms of its mass.

To generate a disk (see Fig. 25-53) we shall rotate the area bounded by the axes, $x = r$, and $y = b$, about the y-axis. We then have

distance from element to axis

$$I_y = 2\pi k \int_0^r x^3 y \, dx \qquad \text{using Eq. (25-23)}$$

$$= 2\pi k \int_0^r x^3 (b) \, dx = 2\pi k b \int_0^r x^3 \, dx$$

$$= 2\pi k b \left(\frac{1}{4} x^4\right)\bigg|_0^r = \frac{\pi k b r^4}{2}$$

(*Continued on next page*)

The mass of the disk is $k(\pi r^2)b$. Rewriting the expresssion for I_y, we have

$$I_y = \frac{(\pi k b r^2)r^2}{2} = \frac{mr^2}{2}$$

∎

Due to the limited methods of integration available at this point, we cannot integrate the expressions for the moments of inertia of circular areas or of a sphere. These will be introduced in Section 27-8 in the exercises, by which point the proper method of integration will have been developed.

Exercises 25-5

In Exercises 1 through 4, find the moment of inertia and the radius of gyration with respect to the origin of the given masses at the given points.

1. 5 units at $(2, 0)$ and 3 units at $(6, 0)$

2. 3 units at $(-1, 0)$, 6 units at $(3, 0)$, 2 units at $(4, 0)$

3. 4 units at $(-4, 0)$, 10 units at $(0, 0)$, 6 units at $(5, 0)$

4. 6 units at $(-3, 0)$, 5 units at $(-2, 0)$, 9 units at $(1, 0)$, 2 units at $(8, 0)$

In Exercises 5 through 24, find the indicated moment of inertia or radius of gyration.

5. Find the moment of inertia of the area bounded by $y^2 = x$, $x = 4$, and the x-axis with respect to the x-axis.

6. Find the moment of inertia of the area bounded by $y = 2x$, $x = 1$, $x = 2$, and the x-axis with respect to the y-axis.

7. Find the radius of gyration of the area bounded by $y = x^3$, $x = 2$, and the x-axis with respect to the y-axis.

8. Find the radius of gyration of the first-quadrant area bounded by the curve $y = 1 - x^2$ with respect to the y-axis.

9. Express the moment of inertia of the triangular area in Example C in terms of the mass of the area.

10. Find the moment of inertia of a rectangular area of sides a and b with respect to side a. Express the result in terms of the mass.

11. Find the radius of gyration of the area bounded by $y = x^2$, $x = 2$, and the x-axis with respect to the x-axis.

12. Find the radius of gyration of the area bounded by $y^2 = x^3$, $y = 8$, and the y-axis with respect to the y-axis.

13. Find the radius of gyration of the area of Exercise 12 with respect to the x-axis.

14. Find the radius of gyration of the first-quadrant area bounded by $x = 1$, $y = 2 - x$, and the y-axis with respect to the y-axis.

15. Find the moment of inertia with respect to its axis of the solid generated by rotating the area bounded by $y^2 = x$, $y = 2$, and the y-axis about the x-axis.

16. Find the radius of gyration with respect to its axis of the volume generated by rotating the first-quadrant area under the curve $y = 4 - x^2$ about the y-axis.

17. Find the radius of gyration with respect to its axis of the volume generated by rotating the area bounded by $y = 2x - x^2$ and the x-axis about the y-axis.

18. Find the radius of gyration with respect to its axis of the volume generated by rotating the area bounded by $y = 2x$ and $y = x^2$ about the y-axis.

19. Find the moment of inertia in terms of its mass of a right circular cone of radius r and height h with respect to its axis. See Fig. 25-54.

20. Find the moment of inertia in terms of its mass of a circular hoop of radius r and of negligible thickness with respect to its center.

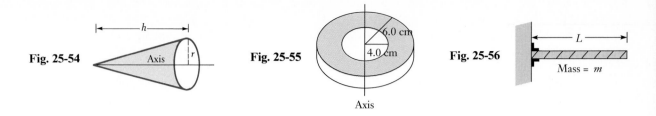

Fig. 25-54   Axis r h

Fig. 25-55   6.0 cm   4.0 cm   Axis

Fig. 25-56 L Mass = m

21. A rotating drill head is in the shape of a right circular cone. Find the moment of inertia of the drill head with respect to its axis if its radius is 0.6 cm, its height is 0.8 cm, and its mass is 3 g. (See Exercise 19.)

22. Find the moment of inertia (in kg·m^2) of a rectangular door 2 m high and 1 m wide with respect to its hinges if $k = 3$ kg/m^2. (See Exercise 10.)

23. Find the moment of inertia of a flywheel with respect to its axis if its inner radius is 4.0 cm, its outer radius is 6.0 cm, and its mass is 1.2 kg. See Fig. 25-55.

24. A cantilever beam is supported only at its left end, as shown in Fig. 25-56. Find the formula for the moment of inertia of this beam with respect to a vertical axis through its left end if its length is L and its mass is m. (Consider the mass to be distributed linearly along the beam. This is not an area or volume type of problem.)

25-6   Other Applications

To illustrate the great variety of applications of integration, we shall demonstrate three more types of applied problems in the examples of this section. Certain other applications are shown in the exercises. The first application is to the physical problem of work done by a variable force.

Work, *in its physical sense, is the product of a constant force times the distance through which it acts.* When we consider the work done in stretching a spring, the first thing we recognize is that the more the spring is stretched, the greater is the force necessary to stretch it. Thus, the force varies. However, if we are stretching the spring a distance Δx, where we are considering the limit as $\Delta x \to 0$, the force can be considered as approaching a constant over Δx. Adding the products of force$_1$ times Δx_1, force$_2$ times Δx_2, and so forth, we see that the total work is the sum of these products. Thus, the work can be expressed as a definite integral in the form

work

$$W = \int_a^b f(x)\,dx$$

(25-25)

NOTE ▷

where $f(x)$ is the force as a function of the distance the spring is stretched. The *limits a and b refer to the initial and final distances the spring is stretched* **from its normal length.**

One problem remains: We must find the function $f(x)$. From physics we learn that the force required to stretch a spring is proportional to the amount it is stretched (Hooke's law). If a spring is stretched x units from its normal length, then $f(x) = kx$. From conditions stated for a particular spring, the value of k may be determined. Thus, $W = \int_a^b kx\,dx$ is the formula for finding the total work done in stretching a spring.

EXAMPLE A

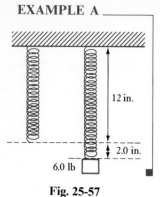

12 in.

2.0 in.

6.0 lb

Fig. 25-57

A spring of natural length 12 in. requires a force of 6.0 lb to stretch it 2.0 in. (see Fig. 25-57). Find the work done in stretching it 6.0 in.

From Hooke's law, we find the constant k for the spring as

$$f(x) = kx, \qquad 6.0 = k(2.0), \qquad k = 3.0 \text{ lb/in.}$$

Since the spring is to be stretched 6.0 in., $a = 0$ (it starts unstretched) and $b = 6.0$ (it is 6.0 in. longer than its normal length). Therefore, the work done in stretching it is

$$W = \int_0^{6.0} 3.0x \, dx = 1.5x^2 \Big|_0^{6.0} \qquad \text{using Eq. (25-25)}$$

$$= 54 \text{ lb} \cdot \text{in.}$$

Problems involving work by a variable force arise in many fields of technology. An illustration from electricity is found in the motion of an electric charge through an electric field created by another electric charge.

Electric charges are of two types, designated as positive and negative. A basic law is that charges of the same sign repel each other and charges of opposite signs attract each other. *The force between charges is proportional to the product of their charges, and inversely proportional to the square of the distance between them.* Thus for this case

$$f(x) = \frac{kq_1q_2}{x^2} \tag{25-26}$$

when q_1 and q_2 are the charges (in coulombs), x is the distance (in meters), the force is in newtons, and $k = 9.0 \times 10^9 \text{ N} \cdot \text{m}^2/\text{C}^2$. For other systems of units, the numerical value of k is different. We can find the work done when electric charges move toward each other or when they separate by use of Eq. (25-26) in Eq. (25-25), which can be used in general for the work done by a variable force $f(x)$ acting through the distance from $x = a$ to $x = b$.

EXAMPLE B

Find the work done when two α-particles, $q = 0.32$ aC each, move until they are 10 nm apart, if they were originally separated by 1.0 m.

From the given information, we have for each α-particle

$$q = 0.32 \text{ aC} = 0.32 \times 10^{-18} \text{ C} = 3.2 \times 10^{-19} \text{ C}$$

Since the particles start 1.0 m apart and are moved to 10 nm apart, $a = 1.0$ m and $b = 10 \times 10^{-9}$ m $= 10^{-8}$ m. The work done is

$f(x)$ from Eq. (25-26)

$$W = \int_{1.0}^{10^{-8}} \frac{9.0 \times 10^9 (3.2 \times 10^{-19})^2}{x^2} \, dx \qquad \text{using Eq. (25-25)}$$

$$= 9.2 \times 10^{-28} \int_{1.0}^{10^{-8}} \frac{dx}{x^2}$$

$$= 9.2 \times 10^{-28} \left(-\frac{1}{x} \right) \Big|_{1.0}^{10^{-8}} = -9.2 \times 10^{-28}(10^8 - 1)$$

$$= -9.2 \times 10^{-20} \text{ J}$$

Since $10^8 \gg 1$, where $\gg$ means "much greater than," the 1 may be neglected in the calculation. The meaning of the minus sign in the result is that work must be done *on* the system to move the particles together. If free to move, they would tend to separate. ∎

The following example illustrates another type of problem in which work by a variable force is involved.

EXAMPLE C

x

$\equiv dx$

$100 - x$

Fig. 25-58

Find the work done in winding up 60.0 ft of a 100-ft cable which weighs 4.00 lb/ft. See Fig. 25-58.

First we let x denote the length of cable which has been wound up at any time. Then the force required to raise the remaining cable equals the weight of the cable which has not yet been wound up. This weight is the product of the unwound cable length, $100 - x$, and its weight per unit length, 4.00 lb/ft, or

$$f(x) = 4.00(100 - x)$$

Since 60.0 ft of cable are to be wound up, $a = 0$ (none is initially wound up) and $b = 60.0$ ft. The work done is

$$W = \int_0^{60.0} 4.00(100 - x)\,dx \qquad \text{using Eq. (25-25)}$$

$$= \int_0^{60.0} (400 - 4.00x)\,dx = 400x - 2.00x^2 \Big|_0^{60.0} = 16{,}800 \text{ ft·lb}$$

∎

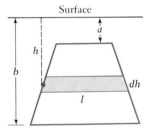

Fig. 25-59

The second application of integration in this section deals with force due to liquid pressure. The force F on an area A at a depth h in a liquid of density w is $F = whA$. Let us assume that the plate shown in Fig. 25-59 is submerged vertically in water. Using integration to sum the forces on the elements of area, *the total force on the plate is given by*

$$\boxed{F = w \int_a^b \ell h\,dh} \qquad\qquad (25\text{-}27)$$

force due to
liquid pressure

Here ℓ is the length of the element of area, h is the depth of the element of area, w is the weight per unit volume of the liquid, a is the depth of the top, and b is the depth of the bottom of the area on which the force is exerted.

EXAMPLE D

See the chapter
introduction.

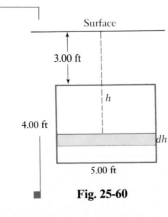

Surface

3.00 ft

h

4.00 ft

dh

5.00 ft

Fig. 25-60

A vertical floodgate of a dam is 5.00 ft wide and 4.00 ft high. Find the force on the gate if its upper edge is 3.00 ft below the surface. See Fig. 25-60.

Each element of area of the floodgate has a length of 5.00 ft, which means that $\ell = 5.00$ ft. Since the top of the gate is 3.00 ft below the surface, $a = 3.00$ ft, and since the gate is 4.00 ft high, $b = 7.00$ ft. Using $w = 62.4$ lb/ft^3, we have the force on the gate as

$$F = 62.4 \int_{3.00}^{7.00} 5.00h\,dh \qquad \text{using Eq. (25-27)}$$

$$= 312 \int_{3.00}^{7.00} h\,dh = 156h^2 \Big|_{3.00}^{7.00} = 156(49.0 - 9.00)$$

$$= 6240 \text{ lb}$$

EXAMPLE E

NOTE ▷

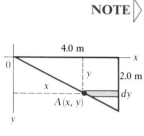

Fig. 25-61

The vertical end of a tank of water is in the shape of a right triangle as shown in Fig. 25-61. What is the force on the end of the tank?

In setting up the figure it is convenient to use coordinate axes. It is also convenient to have the y-axis directed downward, since we integrate from the top to the bottom of the area. The equation of the line OA is $y = \frac{1}{2}x$. Thus, we see that the length of an element of area of the end of the tank is $4.0 - x$, the depth of the element of area is y, the top of the tank is $y = 0$, and the bottom is $y = 2.0$ m. Therefore, the force on the end of the tank is ($w = 9800$ N/m³).

$$
\begin{array}{cc}
\textbf{length} & \textbf{depth} \\
\downarrow & \downarrow
\end{array}
$$

$$F = 9800 \int_0^{2.0} (4.0 - x)(y)(dy) \qquad \text{using Eq. (25-27)}$$

$$= 9800 \int_0^{2.0} (4.0 - 2y)(y\,dy) = 19,600 \int_0^{2.0} (2.0y - y^2)\,dy$$

$$= 19,600 \left(1.0y^2 - \frac{1}{3}y^3 \right) \Bigg|_0^{2.0} = 26,100 \text{ N}$$

∎

The third application of integration in this section is that of the average value of a function. In general, *an average is found by summing up the quantities to be averaged and then dividing by the total number of them.* Generalizing on this and using integration for the summation, *the* **average value** *of a function y with respect to x from $x = a$ to $x = b$ is given by*

average value

$$y_{\text{av}} = \frac{\int_a^b y\,dx}{b - a}$$

(25-28)

The following examples illustrate the applications of the average value of a function.

EXAMPLE F

The velocity of an object falling under the influence of gravity as a function of time is given by $v = 32t$. What is the average velocity (in ft/s) with respect to time for the first 3.0 s?

In this case, we want the average value of the function v from $t = 0$ to $t = 3.0$ s. This gives us

$$v_{\text{av}} = \frac{\int_0^{3.0} v\,dt}{3.0 - 0} \qquad \text{using Eq. (25-28)}$$

$$= \frac{\int_0^{3.0} 32t\,dt}{3.0} = \frac{16t^2}{3.0}\Bigg|_0^{3.0}$$

$$= 48 \text{ ft/s}$$

This result can be interpreted as meaning that an average velocity of 48 ft/s for 3.0 s would result in the same distance, 144 ft, being traveled as that with the variable velocity. Since $s = \int v\,dt$, the numerator represents the distance traveled.

EXAMPLE G ———— The power developed in a certain resistor as a function of current is $P = 6i^2$. What is the average power (in watts) with respect to the current as the current changes from 2.0 A to 5.0 A?

In this case, we are to find the average value of the function P from $i = 2.0$ A to $i = 5.0$ A. This average value of P is

$$P_{av} = \frac{\int_{2.0}^{5.0} P\, di}{5.0 - 2.0} \quad \text{using Eq. (25-28)}$$

$$= \frac{6\int_{2.0}^{5.0} i^2\, di}{3.0} = \frac{2i^3}{3.0}\Big|_{2.0}^{5.0} = \frac{2(125 - 8.0)}{3.0} = 78 \text{ W}$$

∎

In general, it might be noted that the average value of y with respect to x is that value of y which, when multiplied by the length of the interval for x, gives the same area as that under the curve of y as a function of x.

Exercises 25-6

1. The spring of a spring balance is 8.0 in. long when there is no weight on the balance, and it is 9.5 in. long with 6.0 lb hung from the balance. How much work is done in stretching it from 8.0 in. to a length of 10.0 in.?

2. How much work is done in stretching the spring of the spring balance in Exercise 1 from a length of 10.0 in. to a length of 12.0 in.?

3. A force F of 25 N on the spring in the lever-spring mechanism shown in Fig. 25-62 stretches the spring by 16 mm. How much work is done by the 25-N force in stretching the spring?

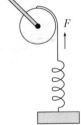

Fig. 25-62

4. An electron has a 1.6×10^{-19} C negative charge. How much work is done in separating two electrons from 1.0 pm to 4.0 pm?

5. How much work is done in separating an electron (see Exercise 4) and an oxygen nucleus, which has a positive charge of 1.3×10^{-18} C, from a distance of 2.0 μm to a distance of 1.0 m?

6. The gravitational force (in lb) of attraction between two objects is given by $F = k/x^2$. If two objects are 10 ft apart, find the work required to separate them until they are 100 ft apart. Express the result in terms of k.

7. Find the work done in winding up 20 m of a 25-m rope on which the force of gravity is 6.00 N/m.

8. Find the work done in winding up a 200-ft cable which weighs 100 lb.

9. At lift-off, a rocket weighs 32.5 tons, including the weight of its fuel. It is fired vertically, and, during the first stage of ascent, the fuel is consumed at the rate of 1.25 tons per 1000 ft of ascent. How much work is done in lifting the rocket to an altitude of 12,000 ft?

10. While descending, a 550-N weather balloon enters a zone of freezing rain in which ice forms on it at the rate of 7.50 N per 100 m of descent. Find the work done on the balloon during the first 1000 m of descent through the freezing rain.

11. Find the work done in pumping the water out of the top of a cylindrical tank 3.0 ft in radius and 10 ft high, if water weighs 62.4 lb/ft³. [*Hint:* If horizontal slices dx ft thick are used, each element weighs $62.4(9\pi\, dx)$ lb, and each element must be raised $10 - x$ ft, if x is the distance from the base to the element. In this way the force, the weight of the slice, and distance are determined. Thus, the products of force and distance are summed by integration.]

12. A hemispherical tank of radius 10 ft is full of water. Find the work done in pumping the water to the top of the tank. [See Exercise 11. This problem is similar, except that the weight of each element is $62.4\pi(\text{radius})^2(\text{thickness})$, where the radius of each element is different. If we let x be the radius of an element and y be the distance the element must be raised, we have $62.4\pi x^2\,dy$, with $x^2 + y^2 = 100$.]

13. One end of a spa is a vertical rectangular wall 12 ft wide. What is the force exerted on this wall by the water if it is 2.5 ft deep?

14. Find the force on one end of a square tank 2.0 m on an edge if the tank is full of water.

15. A rectangular sea aquarium observation window is 10.0 ft wide and 5.00 ft high. What is the force on this window if the upper edge is 4.00 ft below the surface of the water? The density of seawater is 64.0 lb/ft³.

16. A horizontal tank has vertical circular ends, each with a radius of 4.00 ft. It is filled to a depth of 4.00 ft with oil of density 60.0 lb/ft³. Find the force on one end of the tank.

17. A right triangular plate of base 2.0 m and height 1.0 m is submerged vertically, as shown in Fig. 25-63. Find the force on one side of the plate.

18. Find the force on the plate in Exercise 17 if the vertex is 3.0 m below the surface.

19. A small dam is in the shape of the area bounded by $y = x^2$ and $y = 20$. Find the force on the area below $y = 4$ if the surface of the water is at the top of the dam. Assume all distances are in feet.

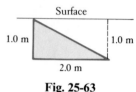

Surface

1.0 m   1.0 m

2.0 m

Fig. 25-63

20. A horizontal cylindrical tank has vertical elliptical ends with the minor axis horizontal. The major axis is 6.0 ft and the minor axis is 4.0 ft. Find the force on one end if the tank is half-filled with fuel oil of density 50 lb/ft³.

21. The electric current as a function of time for a certain circuit is given by $i = 4t - t^2$. Find the average value of the current (in amperes), with respect to time, for the first 4.0 s.

22. A gas is under a pressure of 2.60 kPa. If 2.00 L of this gas undergoes an *adiabatic* (no *heat* gained or lost) change of volume to 6.00 L, find the average value of the pressure p with respect to the volume V. For an adiabatic change, the equation relating p and V is $pV^{1.4} = k$, where k is a constant.

23. The efficiency e (in percent) of an automobile engine is $e = 0.768s - 0.00004s^3$, where s is the speed (in km/h) of the car. Find the average efficiency with respect to the speed for $s = 30.0$ km/h to $s = 90.0$ km/h. (See Example A of Section 23-7.)

24. Find the average value of the volume of a sphere with respect to the radius.

25. The length of arc s of a curve from $x = a$ to $x = b$ is

$$s = \int_a^b \sqrt{1 + \left(\frac{dy}{dx}\right)^2}\,dx$$

Fig. 25-64

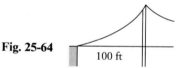

100 ft

The cable of a bridge can be described by the equation $y = 0.04x^{3/2}$ from $x = 0$ to $x = 100$ ft. Find the length of the cable. See Fig. 25-64.

26. A rocket takes off in a path described by the equation $y = \frac{2}{3}(x^2 - 1)^{3/2}$. Find the distance traveled by the rocket for $x = 1.0$ km to $x = 3.0$ km.

27. The area of a surface of revolution generated by rotating an arc from $x = a$ to $x = b$ is given by $S = 2\pi \int_a^b y\sqrt{1 + (dy/dx)^2}\,dx$. Find the formula for the lateral surface area of a right circular cone of radius r and height h.

28. The grinding surface of a grinding machine can be described as the surface generated by rotating the curve $y = 0.2x^3$ from $x = 0$ to $x = 2.0$ cm. Find the grinding surface area.

25-7 Chapter Equations, Review Exercises, and Practice Test

Chapter Equations

Velocity	$v = \int a\,dt$	(25-1)
	$v = at + C_1$	(25-2)
Displacement	$s = \int v\,dt$	(25-3)
Electric current	$i = \dfrac{dq}{dt}$	(25-4)
Electric charge	$q = \int i\,dt$	(25-5)
Voltage across capacitor	$V_C = \dfrac{1}{C} \int i\,dt$	(25-6)
Area	$A = \int_a^b y\,dx = \int_a^b f(x)\,dx$	(25-7)
	$A = \int_c^d x\,dy = \int_c^d g(y)\,dy$	(25-8)
	$A = \int_a^b (y_2 - y_1)\,dx$	(25-9)
	$A = \int_c^d (x_2 - x_1)\,dy$	(25-10)
Volume	$V = \pi \int_a^b y^2\,dx = \pi \int_a^b [f(x)]^2\,dx$	(25-11)
	$V = \pi \int_c^d x^2\,dy$	(25-12)
(Shell)	$dV = 2\pi(\text{radius}) \times (\text{height}) \times (\text{thickness})$	(25-13)
(Disk)	$dV = \pi(\text{radius})^2 \times (\text{thickness})$	(25-14)
Center of mass	$m_1 d_1 + m_2 d_2 + \cdots + m_n d_n = (m_1 + m_2 + \cdots + m_n)\bar{d}$	(25-15)
Centroid of area	$\bar{x} = \dfrac{\int_a^b x(y_2 - y_1)\,dx}{\int_a^b (y_2 - y_1)\,dx}$	(25-16)
	$\bar{y} = \dfrac{\int_c^d y(x_2 - x_1)\,dy}{\int_c^d (x_2 - x_1)\,dy}$	(25-17)
Centroid of volume	$\bar{x} = \dfrac{\int_a^b xy^2\,dx}{\int_a^b y^2\,dx}$	(25-18)
	$\bar{y} = \dfrac{\int_c^d yx^2\,dy}{\int_c^d x^2\,dy}$	(25-19)

Radius of gyration	$m_1 d_1^2 + m_2 d_2^2 + \cdots + m_n d_n^2 = (m_1 + m_2 + \cdots + m_n)R^2$	(25-20)

Moment of inertia of area	$I_y = k \int_a^b x^2 (y_2 - y_1)\, dx$	(25-21)
	$I_x = k \int_c^d y^2 (x_2 - x_1)\, dy$	(25-22)

Moment of inertia of volume	$I_y = 2\pi k \int_a^b yx^3\, dx$	(25-23)
	$I_x = 2\pi k \int_c^d xy^3\, dy$	(25-24)

Work	$W = \int_a^b f(x)\, dx$	(25-25)

Force between electric charges	$f(x) = \dfrac{kq_1 q_2}{x^2}$	(25-26)

Force due to liquid pressure	$F = w \int_a^b \ell h\, dh$	(25-27)

Average value	$y_{av} = \dfrac{\int_a^b y\, dx}{b - a}$	(25-28)

Review Exercises

1. How long after it is dropped does a baseball reach a speed of 140 ft/s (the speed of a good fastball)?

2. If the velocity v (in m/s) of a subway train after the brakes are applied can be expressed as $v = \sqrt{400 - 20t}$, where t is the time in seconds, how far does it travel in coming to a stop?

3. A weather balloon is rising at the rate of 20 ft/s when a small metal part drops off. If the balloon is 200 ft high at this instant, when will the part hit the ground?

4. A float is dropped into a river at a point where it is flowing at 5.0 ft/s. How far does the float travel in 30 s if it accelerates downstream at 0.020 ft/s²?

5. The electric current i, in amperes, in a circuit as a function of the time t, in seconds, is $i = 0.25(2\sqrt{t} - t)$. Find the total charge to pass a point in the circuit in 2.0 s.

6. The current in a certain electric circuit is given by $i = \sqrt{1 + 4t}$, where i is in amperes and t is in seconds. Find the charge which passes a given point during the first two seconds.

7. The voltage across a 55.0-nF capacitor is zero. What is the voltage after 20 ms if a current of 10 mA charges the capacitor?

8. The initial voltage across a capacitor is zero, and $V_C = 2.50$ V after 8.0 ms. If a current $i = t/\sqrt{t^2 + 1}$, where i is in amperes and t is in seconds, charges the capacitor, find the capacitance C of the capacitor.

9. The distribution of weight on a cable is not uniform. If the slope of the cable at any point is given by $dy/dx = 20 + \frac{1}{40}x^2$, and if the origin of the coordinate system is at the lowest point, find the equation which gives the curve described by the cable.

10. The time rate of change of the reliability R, in percent, of a computer system is $dR/dt = -2.5(0.05t + 1)^{-1.5}$, where t is in hours. If $R = 100$ for $t = 0$, find R for $t = 100$ h.

11. Find the area between $y = \sqrt{1 - x}$ and the coordinate axes.

12. Find the area bounded by $y = 3x^2 - x^3$ and the x-axis.

13. Find the area bounded by $y^2 = 2x$ and $y = x - 4$.

14. Find the area bounded by $y = 1/(2x + 1)^2$, $y = 0$, $x = 1$, and $x = 2$.

15. Find the area between $y = x^2$ and $y = x^3 - 2x^2$.

16. Find the area between $y = x^2 + 2$ and $y = 3x^2$.

17. Find the volume generated by rotating the area bounded by $y = 3 + x^2$ and the line $y = 4$ about the x-axis.

18. Find the volume generated by rotating the area of Exercise 12 about the x-axis.

19. Find the volume generated by rotating the area of Exercise 12 about the y-axis.

20. Find the volume generated by rotating the area bounded by $y = x$ and $y = 3x - x^2$ about the y-axis.

21. Find the volume generated by revolving an ellipse about its major axis.

22. A hole of radius 1 cm is bored along the diameter of a sphere of radius 4 cm. Find the volume of the material which is removed from the sphere.

23. Find the centroid of the area bounded by $y^2 = x^3$ and $y = 2x$.

24. Find the centroid of the area bounded by $y = 2x - 4$, $x = 1$, and $y = 0$.

25. Find the centroid of the volume generated by rotating the area bounded by $y = \sqrt{x}$, $x = 1$, $x = 4$, and $y = 0$ about the x-axis.

26. Find the centroid of the volume generated by rotating the area bounded by $yx^4 = 1$, $y = 1$, and $y = 4$ about the y-axis.

27. Find the moment of inertia of the area bounded by $y = 3x - x^2$ and $y = x$ with respect to the y-axis.

28. Find the radius of gyration of the first-quadrant area which is bounded by $y = 8 - x^3$ with respect to the y-axis.

29. Find the moment of inertia with respect to its axis of the solid generated by rotating the area bounded by $y = x^{1/2}$, $y = 0$, and $x = 8$ about the x-axis.

30. Find the radius of gyration with respect to its axis of the solid formed by rotating the area bounded by $xy = 1$, $x = 1$, $x = 3$, and $y = \frac{1}{3}$ about the x-axis.

31. A pail and its contents weigh 80 lb. The pail is attached to the end of a 100-ft rope which weighs 10 lb and is hanging vertically. How much work is done in winding up the rope with the pail attached?

32. The gravitational force, in pounds, of the earth on a satellite (the weight of the satellite) is given by $F = 10^{11}/x^2$, where x is the vertical distance, in miles, from the center of the earth to the satellite. How much work is done in moving the satellite from the earth's surface to an altitude of 2000 mi? The radius of the earth is 3960 mi.

33. The rear stabilizer of a certain aircraft can be described as the area under the curve $y = 3x^2 - x^3$, as shown in Fig. 25-65. Find the x-coordinate, in meters, of the centroid of the stabilizer.

34. The water in a spherical tank 20 m in radius is 15 m deep at the deepest point. How much water is in the tank?

35. The nose cone of a rocket has the shape of a semiellipse rotated about its major axis, as shown in Fig. 25-66. What is the volume of the nose cone?

36. The deck area of a boat is a parabolic section as shown in Fig. 25-67. What is the area of the deck?

37. A chemical waste holding tank 4.50 ft in radius has a depth of 3.25 ft. Find the total force on the circular side of the tank when it is filled with liquid with a density of 68.0 lb/ft³.

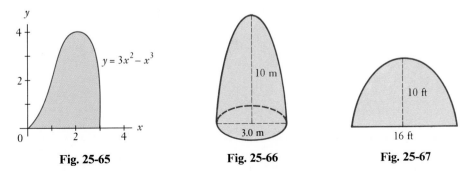

Fig. 25-65   Fig. 25-66   Fig. 25-67

38. A section of a dam is in the shape of a right triangle. The base of the triangle is 6.00 ft and is in the surface of the water. If the triangular section goes to a depth of 4.00 ft, find the force on it. See Fig. 25-68.

39. The electric resistance of a wire is inversely proportional to the square of its radius. If a certain wire has a resistance of 0.30 Ω when its radius is 2.0 mm, find the average value of the resistance with respect to the radius if the radius changes from 2.0 mm to 2.1 mm.

40. A horizontal straight section of pipe is supported at its center by a vertical wire as shown in Fig. 25-69. Find the formula for the moment of inertia of the pipe with respect to an axis along the wire if the pipe is of length L and mass m.

Fig. 25-68

Fig. 25-69

Practice Test

In Problems 1 through 3, use the area bounded by $y = \frac{1}{4}x^2$, $y = 0$, and $x = 2$.

1. Find the area.

2. Find the coordinates of the centroid of the area.

3. Find the volume if the given area is rotated about the x-axis.

In Problems 4 and 5, use the first quadrant area bounded by $y = x^2$, $x = 0$, and $y = 9$.

4. Find the volume if the given area is rotated about the x-axis.

5. Find the moment of inertia of the area with respect to the y-axis.

6. The velocity v of an object as a function of the time t is $v = 60 - 4t$. Find the expression for the displacement s if $s = 10$ for $t = 0$.

7. The natural length of a spring is 8.0 cm. A force of 12 N stretches it to a length of 10.0 cm. How much work is done in stretching it from a length of 10.0 cm to a length of 14.0 cm?

8. A vertical rectangular floodgate is 6.00 ft wide and 2.00 ft high. Find the force on the gate if its upper edge is 1.00 ft below the surface of the water ($w = 62.4 \text{ lb/ft}^3$).

26 Differentiation of Transcendental Functions

In Sections 26-4 and 26-7 we use derivatives of transcendental functions in analyzing the motion of a rocket.

In our development of differentiation and integration in the previous chapters, we have used only algebraic functions. We have not yet used the trigonometric, inverse trigonometric, exponential, or logarithmic functions, for the derivative of each of these is a special form. In this chapter we shall find the formulas for the derivatives of these functions, which are the most important of the **transcendental** (nonalgebraic) **functions.** In the next chapter we shall take up integration involving these functions.

As we have seen in earlier chapters, there are many technical and scientific applications of these functions. These applications are found in alternating current theory, harmonic motion, rocket motion, monetary interest calculations, population growth, and many other types of problems.

26-1 Derivatives of the Sine and Cosine Functions

If we find the derivative of the sine function, it is possible to use this derivative as a basis for finding the derivatives of the other trigonometric and inverse trigonometric functions. Therefore, we shall now find the derivative of the sine function, by using the delta-process.

Let $y = \sin u$, where u is a function of x and is expressed in radians. If u changes by Δu, y then changes by Δy. Thus,

$$y + \Delta y = \sin (u + \Delta u)$$
$$\Delta y = \sin (u + \Delta u) - \sin u$$
$$\frac{\Delta y}{\Delta u} = \frac{\sin (u + \Delta u) - \sin u}{\Delta u}$$

For reference, Eq. (19-18) is

$$\sin x - \sin y = 2 \sin \tfrac{1}{2}(x - y) \cos \tfrac{1}{2}(x + y)$$

Referring now to Eq. (19-18), we have

$$\frac{\Delta y}{\Delta u} = \frac{2 \sin \tfrac{1}{2}(u + \Delta u - u) \cos \tfrac{1}{2}(u + \Delta u + u)}{\Delta u}$$

$$= \frac{\sin (\Delta u/2) \cos [u + (\Delta u/2)]}{\Delta u/2}$$

Looking ahead to the next step of letting $\Delta u \to 0$, we see that the numerator and denominator both approach zero. This situation is precisely the same as that in which we were finding the derivatives of the algebraic functions. To find the limit, we must find

$$\lim_{\Delta u \to 0} \frac{\sin (\Delta u/2)}{\Delta u/2}$$

since these are the factors which cause the numerator and the denominator to approach zero.

In finding this limit, we shall let $\theta = \Delta u/2$ for convenience of notation. This means that we are to determine $\lim\limits_{\theta \to 0} \dfrac{\sin \theta}{\theta}$. Of course, it would be convenient to know before proceeding if this limit does actually exist. Therefore, by using a calculator, we can now develop a table of values of $\dfrac{\sin \theta}{\theta}$ as θ becomes very small.

θ (radians)	0.5	0.1	0.05	0.01	0.001
$\dfrac{\sin \theta}{\theta}$	0.9588511	0.9983342	0.9995834	0.9999833	0.9999998

We see from this table that the limit of $\dfrac{\sin \theta}{\theta}$, as $\theta \to 0$, appears to be 1.

In order to prove that $\lim\limits_{\theta \to 0} \dfrac{\sin \theta}{\theta} = 1$, we use a geometric approach. Considering Fig. 26-1, we can see that the following inequality is true:

Area triangle OBD < area sector OBD < area triangle OBC

$$\frac{1}{2} r(r \sin \theta) < \frac{1}{2} r^2 \theta < \frac{1}{2} r(r \tan \theta) \quad \text{or} \quad \sin \theta < \theta < \tan \theta$$

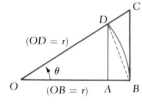

$(OD = r)$

$(OB = r)$

Fig. 26-1

Remembering that we want to find the limit of $(\sin \theta)/\theta$, we next divide through by $\sin \theta$, and then take reciprocals:

$$1 < \frac{\theta}{\sin \theta} < \frac{1}{\cos \theta} \quad \text{or} \quad 1 > \frac{\sin \theta}{\theta} > \cos \theta$$

When we consider the limit as $\theta \to 0$, we see that the left member remains 1 and the right member approaches 1. Thus, $(\sin \theta)/\theta$ must approach 1. This means

$$\lim_{\theta \to 0} \frac{\sin \theta}{\theta} = \lim_{\Delta u \to 0} \frac{\sin (\Delta u/2)}{\Delta u/2} = 1 \tag{26-1}$$

Using the result expressed in Eq. (26-1) in the expression for $\Delta y/\Delta u$, we have

$$\lim_{\Delta u \to 0} \frac{\Delta y}{\Delta u} = \lim_{\Delta u \to 0} \left[\cos \left(u + \frac{\Delta u}{2} \right) \frac{\sin (\Delta u/2)}{\Delta u/2} \right] = \cos u$$

or

$$\frac{dy}{du} = \cos u \qquad (26\text{-}2)$$

However, we want the derivative of y with respect to x. This requires the use of the chain rule, Eq. (22-14), which we repeat here for reference.

$$\frac{dy}{dx} = \frac{dy}{du} \frac{du}{dx} \qquad (26\text{-}3)$$

Combining Eqs. (26-2) and (26-3), we have ($y = \sin u$)

derivative of sin u

$$\frac{d(\sin u)}{dx} = \cos u \frac{du}{dx} \qquad (26\text{-}4)$$

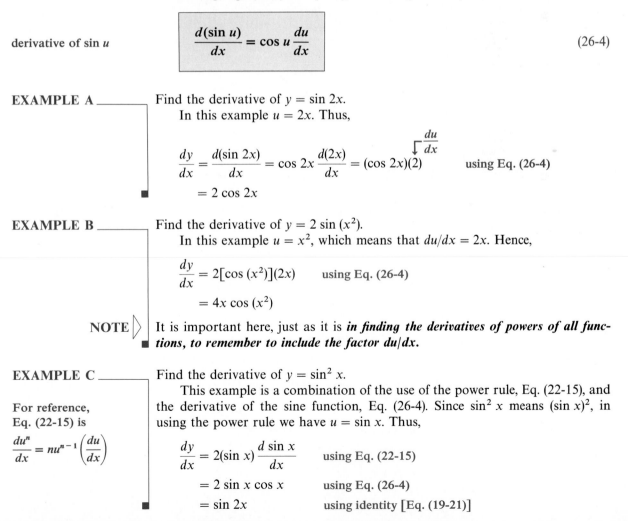

EXAMPLE A Find the derivative of $y = \sin 2x$.
In this example $u = 2x$. Thus,

$$\frac{dy}{dx} = \frac{d(\sin 2x)}{dx} = \cos 2x \frac{d(2x)}{dx} = (\cos 2x)(2) \qquad \text{using Eq. (26-4)}$$
$$= 2 \cos 2x$$

EXAMPLE B Find the derivative of $y = 2 \sin (x^2)$.
In this example $u = x^2$, which means that $du/dx = 2x$. Hence,

$$\frac{dy}{dx} = 2[\cos (x^2)](2x) \qquad \text{using Eq. (26-4)}$$
$$= 4x \cos (x^2)$$

NOTE ▷ It is important here, just as it is *in finding the derivatives of powers of all functions, to remember to include the factor du/dx.*

EXAMPLE C Find the derivative of $y = \sin^2 x$.
This example is a combination of the use of the power rule, Eq. (22-15), and the derivative of the sine function, Eq. (26-4). Since $\sin^2 x$ means $(\sin x)^2$, in using the power rule we have $u = \sin x$. Thus,

For reference,
Eq. (22-15) is
$$\frac{du^n}{dx} = nu^{n-1}\left(\frac{du}{dx}\right)$$

$$\frac{dy}{dx} = 2(\sin x) \frac{d \sin x}{dx} \qquad \text{using Eq. (22-15)}$$
$$= 2 \sin x \cos x \qquad \text{using Eq. (26-4)}$$
$$= \sin 2x \qquad \text{using identity [Eq. (19-21)]}$$

EXAMPLE D

Find the derivative of $y = 2 \sin^3 (2x^4)$.

In the general power rule, $u = \sin (2x^4)$. In the derivative of the sine function, $u = 2x^4$. Thus, we have

$$\frac{dy}{dx} = 2(3) \sin^2 (2x^4) \frac{d(\sin 2x^4)}{dx} \qquad \text{using Eq. (22-15)}$$

$$= 6 \sin^2 (2x^4) \cos (2x^4) \frac{d(2x^4)}{dx} \qquad \text{using Eq. (26-4)}$$

$$= 6 \sin^2 (2x^4) \cos (2x^4)(8x^3)$$

$$= 48x^3 \sin^2 (2x^4) \cos (2x^4)$$

To find the derivative of the cosine function, we write it in the form $\cos u = \sin (\frac{\pi}{2} - u)$. Thus, if $y = \sin (\frac{\pi}{2} - u)$, we have

$$\frac{dy}{dx} = \cos \left(\frac{\pi}{2} - u\right) \frac{d(\frac{\pi}{2} - u)}{dx} = \cos \left(\frac{\pi}{2} - u\right)\left(-\frac{du}{dx}\right)$$

$$= -\cos \left(\frac{\pi}{2} - u\right) \frac{du}{dx}$$

Since $\cos (\frac{\pi}{2} - u) = \sin u$, we have

derivative of cos *u*

$$\boxed{\frac{d(\cos u)}{dx} = -\sin u \frac{du}{dx}} \qquad (26\text{-}5)$$

EXAMPLE E

The electric power p developed in a resistor of an amplifier circuit is $p = 25 \cos^2 120\pi t$, where t is the time. Find the expression for the time rate of change of power.

From Chapter 24, we know that we are to find the derivative dp/dt. To do this we use the power rule and Eq. (26-5). Using these, we have

$$p = 25 \cos^2 120\pi t$$

$$\frac{dp}{dt} = 25(2 \cos 120\pi t) \frac{d \cos 120\pi t}{dt} \qquad \text{using Eq. (22-15)}$$

$$= 50 \cos 120\pi t(-\sin 120\pi t) \frac{d(120\pi t)}{dt} \qquad \text{using Eq. (26-5)}$$

$$= (-50 \cos 120\pi t \sin 120\pi t)(120\pi)$$

$$= -6000\pi \cos 120\pi t \sin 120\pi t$$

Using the double-angle identity [Eq. (19-21)], this result can be written as $-3000\pi \sin 240\pi t$.

EXAMPLE F

Find the derivative of $y = \sqrt{1 + \cos 2x}$.

The derivative is found as follows:

$$y = (1 + \cos 2x)^{1/2}$$

$$\frac{dy}{dx} = \frac{1}{2}(1 + \cos 2x)^{-1/2} \frac{d(1 + \cos 2x)}{dx} \qquad \text{using Eq. (22-15)}$$

$$= \frac{1}{2}(1 + \cos 2x)^{-1/2}(-\sin 2x)(2) \qquad \text{using Eq. (26-5)}$$

$$= -\frac{\sin 2x}{\sqrt{1 + \cos 2x}}$$

EXAMPLE G ————— Find the differential of $y = \sin 2x \cos x^2$.

From Section 24-1, we recall that the differential of a function $y = f(x)$ is $dy = f'(x)\,dx$. Thus, using the derivative product rule and the derivatives of the sine and cosine functions, we arrive at the following result:

$$y = \sin 2x \cos x^2 \qquad y = (\sin 2x)(\cos x^2)$$
$$dy = \left[\sin 2x(-\sin x^2)(2x) + \cos x^2(\cos 2x)(2)\right]dx$$
$$= (-2x \sin 2x \sin x^2 + 2 \cos 2x \cos x^2)\,dx$$

EXAMPLE H ————— Find the slope of a tangent line to the curve of $y = 5 \sin 3x$, where $x = 0.2$.

Here we are to find the derivative of $y = 5 \sin 3x$ and then evaluate the derivative for $x = 0.2$. Therefore, we have the following:

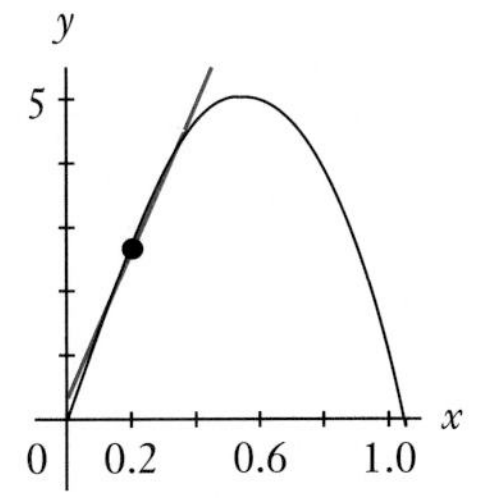

$$y = 5 \sin 3x$$

$$\frac{dy}{dx} = 5(\cos 3x)(3) = 15 \cos 3x \qquad \textbf{find derivative}$$

$$\left.\frac{dy}{dx}\right|_{x=0.2} = 15 \cos 3(0.2) = 15 \cos 0.6 \qquad \textbf{evaluate}$$

$$= 15(0.8253) = 12.38$$

Fig. 26-2

NOTE ▷ In evaluating the slope we must remember that $x = 0.2$ means the *values are in radians.* Therefore, the slope is 12.38. See Fig. 26-2.

Exercises 26-1

In Exercises 1 through 32, find the derivatives of the given functions.

1. $y = \sin(x + 2)$ **2.** $y = 3 \sin 4x$ **3.** $y = 2 \sin(2x - 1)$ **4.** $y = 5 \sin(3 - x)$

5. $y = 6 \cos \frac{1}{2}x$ **6.** $y = \cos(1 - x)$ **7.** $y = 2 \cos(3x - 1)$ **8.** $y = 4 \cos(6x^2 + 5)$

9. $y = \sin^2 4x$ **10.** $y = 3 \sin^3(2x + 1)$ **11.** $y = 3 \cos^3(5x + 2)$ **12.** $y = 4 \cos^2 \sqrt{x}$

13. $y = x \sin 3x$ **14.** $y = x^2 \sin 2x$ **15.** $y = 3x^3 \cos 5x$ **16.** $y = 5x \cos 2x^3$

17. $y = \sin x^2 \cos 2x$ **18.** $y = \sin x \cos 4x$ **19.** $y = \sqrt{1 + \sin 4x}$ **20.** $y = (x - \cos^2 x)^4$

21. $y = \dfrac{\sin 3x}{x}$ **22.** $y = \dfrac{2x + 3}{\sin 4x}$ **23.** $y = \dfrac{2 \cos x^2}{3x - 1}$ **24.** $y = \dfrac{5x}{\cos x}$

25. $y = 2 \sin^2 3x \cos 2x$ **26.** $y = \cos^3 4x \sin^2 2x$ **27.** $y = \dfrac{\cos^2 3x}{1 + 2 \sin^2 2x}$ **28.** $y = \dfrac{\sin 5x}{3 - \cos^2 3x}$

29. $y = \sin^3 x - \cos 2x$ **30.** $y = x \sin x + \cos x$

31. $y = x - \dfrac{1}{3}\sin^3 4x$ **32.** $y = 2x \sin x + 2 \cos x - x^2 \cos x$

In Exercises 33 through 52, solve the given problems.

33. Verify the values of $(\sin \theta)/\theta$ in the table where $(\sin \theta)/\theta$ is calculated for $\theta = 0.5, 0.1, 0.05, 0.01$, and 0.001.

34. Evaluate $\lim\limits_{\theta \to 0} (\tan \theta)/\theta$. (Use the fact that $\lim\limits_{\theta \to 0} (\sin \theta)/\theta = 1$.)

35. On a calculator find the values of (a) cos 1.0000 and (b) $(\sin 1.0001 - \sin 1.0000)/0.0001$. Compare the values and give the meaning of each in relation to the derivative of the sine function, where $x = 1$.

36. On a calculator, find the values of (a) $-\sin 1.0000$ and (b) $(\cos 1.0001 - \cos 1.0000)/0.0001$. Compare the values and give the meaning of each in relation to the derivative of the cosine function, where $x = 1$.

37. On the graph of $y = \sin x$ in Fig. 26-3, draw tangent lines at the indicated points and determine the slopes of these tangent lines. Then plot the values of these slopes for the same values of x and join the points with a smooth curve. Compare the resulting curve with $y = \cos x$.

Fig. 26-3

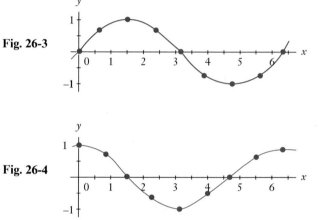

38. Follow the instructions of Exercise 37 for the graph of $y = \cos x$ in Fig. 26-4. Compare the resulting curve with $y = \sin x$.

39. Find the derivative of the implicit function $\sin (xy) + \cos 2y = x^2$.

Fig. 26-4

40. Find the derivative of the implicit function $x \cos 2y + \sin x \cos y = 1$.

41. Show that $\dfrac{d^4 \sin x}{dx^4} = \sin x$.

42. If $y = \cos 2x$, show that $\dfrac{d^2y}{dx^2} = -4y$.

43. Find the derivative of the members of the identity $\cos 2x = 2 \cos^2 x - 1$ and thereby obtain another trigonometric identity.

44. Find the derivative of each of the members of the identity $2 \sin^2 x = 1 - \cos 2x$ and thereby obtain another trigonometric identity.

45. Evaluate the differential of $y = 3 \sin 2x$ for $x = \frac{\pi}{8}$ and $dx = 0.02$.

46. Evaluate the differential of $y = 6 \cos x^2$ for $x = 0.50$ and $dx = 0.04$.

47. Find the slope of a line tangent to the curve of $y = x \cos 2x$ where $x = 1.20$.

48. Find the slope of a line tangent to the curve of $y = \dfrac{2 \sin 3x}{x}$ where $x = 0.15$.

49. The blade of a saber saw moves vertically up and down, and its displacement y (in cm) is given by $y = 1.85 \sin 36\pi t$, where t is the time in seconds. Find the velocity of the blade for $t = 0.0250$ s.

50. The current (in amperes) in a certain electric circuit, as a function of time (in seconds), is given by $i = 10 \cos (120\pi t + \frac{\pi}{6})$. Find the expression for the voltage across a 2-H inductor. (See Exercise 23 of Section 25-1.)

51. The distance r (in km) from an aircraft to a rocket directed at the aircraft is $r = \dfrac{100}{1 - \cos \theta}$, where θ is the angle between their directions. Find $dr/d\theta$ for $\theta = 120°$.

52. The number N of reflections of a light ray passing through an optic fiber of length L and diameter d is $N = \dfrac{L \sin \theta}{d \sqrt{n^2 - \sin^2 \theta}}$. Here n is the index of refraction of the fiber and θ is the angle between the light ray and the fiber's axis. Find $dN/d\theta$.

26-2   **Derivatives of the Other Trigonometric Functions** ▰▰▰▰

We can find the derivatives of the other trigonometric functions by expressing the functions in terms of the sine and cosine. After we perform the differentiation, we use trigonometric relations to put the derivative in a convenient form.

We obtain the derivative of tan u by expressing tan u as sin u/cos u. Therefore, letting $y = \sin u/\cos u$, by employing the quotient rule we have

$$\frac{dy}{dx} = \frac{\cos u[\cos u(du/dx)] - \sin u[-\sin u(du/dx)]}{\cos^2 u}$$

$$= \frac{\cos^2 u + \sin^2 u}{\cos^2 u}\frac{du}{dx} = \frac{1}{\cos^2 u}\frac{du}{dx} = \sec^2 u\frac{du}{dx}$$

$$\boxed{\frac{d(\tan u)}{dx} = \sec^2 u\,\frac{du}{dx}} \tag{26-6}$$

We find the derivative of cot u by letting $y = \cos u/\sin u$ and again using the quotient rule:

$$\frac{dy}{dx} = \frac{\sin u[-\sin u(du/dx)] - \cos u[\cos u(du/dx)]}{\sin^2 u}$$

$$= \frac{-\sin^2 u - \cos^2 u}{\sin^2 u}\frac{du}{dx}$$

$$\boxed{\frac{d(\cot u)}{dx} = -\csc^2 u\,\frac{du}{dx}} \tag{26-7}$$

To obtain the derivative of sec u, we let $y = 1/\cos u$. Then

$$\frac{dy}{dx} = -(\cos u)^{-2}\left[(-\sin u)\left(\frac{du}{dx}\right)\right] = \frac{1}{\cos u}\frac{\sin u}{\cos u}\frac{du}{dx}$$

$$\boxed{\frac{d(\sec u)}{dx} = \sec u \tan u\,\frac{du}{dx}} \tag{26-8}$$

We obtain the derivative of csc u by letting $y = 1/\sin u$. And so

$$\frac{dy}{dx} = -(\sin u)^{-2}\left(\cos u\frac{du}{dx}\right) = -\frac{1}{\sin u}\frac{\cos u}{\sin u}\frac{du}{dx}$$

$$\boxed{\frac{d(\csc u)}{dx} = -\csc u \cot u\,\frac{du}{dx}} \tag{26-9}$$

EXAMPLE A ———— Find the derivative of $y = 2 \tan 8x$.
The derivative is

$$\frac{dy}{dx} = 2(\sec^2 8x)(8) \qquad \text{using Eq. (26-6)}$$

where the factor (8) is $\dfrac{du}{dx}$

$$= 16 \sec^2 8x$$

■

EXAMPLE B ———— Find the derivative of $y = 3 \sec^2 4x$.
Using the power rule and Eq. (26-8), we have

$$\frac{dy}{dx} = 3(2)(\sec 4x)\frac{d(\sec 4x)}{dx} \qquad \text{using } \frac{du^n}{dx} = nu^{n-1}\frac{du}{dx}$$

$$= 6(\sec 4x)(\sec 4x \tan 4x)(4) \qquad \text{using } \frac{d \sec u}{dx} = \sec u \tan u \frac{du}{dx}$$

$$= 24 \sec^2 4x \tan 4x$$

■

EXAMPLE C ———— Find the derivative of $y = x \csc^3 2x$.
Using the power rule, the product rule, and Eq. (26-9), we have

$$\frac{dy}{dx} = x(3 \csc^2 2x)(-\csc 2x \cot 2x)(2) + (\csc^3 2x)(1)$$

$$= \csc^3 2x(-6x \cot 2x + 1)$$

■

EXAMPLE D ———— Find the derivative of $y = (\tan 2x + \sec 2x)^3$.
Using the power rule and Eqs. (26-6) and (26-8), we have

$$\frac{dy}{dx} = 3(\tan 2x + \sec 2x)^2[\sec^2 2x(2) + \sec 2x \tan 2x(2)]$$

$$= 3(\tan 2x + \sec 2x)^2(2 \sec 2x)(\sec 2x + \tan 2x)$$

$$= 6 \sec 2x(\tan 2x + \sec 2x)^3$$

■

EXAMPLE E ———— Find the differential of $y = \sin 2x \tan x^2$.
Here we are to find the derivative of the given function and multiply by dx. Therefore, using the product rule along with Eqs. (26-4) and (26-6), we have the following:

$$dy = [(\sin 2x)(\sec^2 x^2)(2x) + (\tan x^2)(\cos 2x)(2)] dx$$

$$= (2x \sin 2x \sec^2 x^2 + 2 \cos 2x \tan x^2) dx$$

■

EXAMPLE F ———— Find dy/dx if $\cot 2x - 3 \csc xy = y^2$.
In finding the derivative of this implicit function, we must be careful not to forget the factor dy/dx when it occurs. The derivative is found as follows:

$$\cot 2x - 3 \csc xy = y^2$$

$$(-\csc^2 2x)(2) - 3(-\csc xy \cot xy)\left(x\frac{dy}{dx} + y\right) = 2y\frac{dy}{dx}$$

$$3x \csc xy \cot xy \frac{dy}{dx} - 2y \frac{dy}{dx} = 2 \csc^2 2x - 3y \csc xy \cot xy$$

$$\frac{dy}{dx} = \frac{2 \csc^2 2x - 3y \csc xy \cot xy}{3x \csc xy \cot xy - 2y}$$

EXAMPLE G

Evaluate the derivative of $y = \dfrac{2x}{1 - \cot 3x}$, where $x = 0.25$.

Finding the derivative, we have

$$\frac{dy}{dx} = \frac{(1 - \cot 3x)(2) - 2x(\csc^2 3x)(3)}{(1 - \cot 3x)^2}$$

$$= \frac{2 - 2 \cot 3x - 6x \csc^2 3x}{(1 - \cot 3x)^2}$$

Now, substituting $x = 0.25$, we have

$$\frac{dy}{dx}\bigg|_{x=0.25} = \frac{2 - 2 \cot 0.75 - 6(0.25) \csc^2 0.75}{(1 - \cot 0.75)^2}$$

$$= -626.0$$

In using the calculator we recall that we must have it in radian mode. Also, to evaluate cot 0.75 and csc 0.75, we must use reciprocals of tan 0.75 and sin 0.75, respectively.

Exercises 26-2

In Exercises 1 through 32, find the derivatives of the given functions.

1. $y = \tan 5x$ **2.** $y = 3 \tan (3x + 2)$ **3.** $y = \cot (1 - x)^2$ **4.** $y = 3 \cot 6x$

5. $y = 3 \sec 2x$ **6.** $y = \sec \sqrt{1 - x}$ **7.** $y = -3 \csc \sqrt{x}$ **8.** $y = \csc (1 - 2x)$

9. $y = 5 \tan^2 3x$ **10.** $y = 2 \tan^2 (x^2)$ **11.** $y = 2 \cot^4 \frac{1}{2}x$ **12.** $y = \cot^2 (1 - x)$

13. $y = \sqrt{\sec 4x}$ **14.** $y = \sec^3 x$ **15.** $y = 3 \csc^4 7x$ **16.** $y = \csc^2 (2x^2)$

17. $y = x^2 \tan x$ **18.** $y = 3x \sec 4x$ **19.** $y = 4 \cos x \csc x^2$ **20.** $y = \frac{1}{2} \sin 2x \sec x$

21. $y = \dfrac{\csc x}{x}$ **22.** $y = \dfrac{\cot 4x}{2x}$ **23.** $y = \dfrac{\cos 4x}{1 + \cot 3x}$ **24.** $y = \dfrac{\tan^2 3x}{2 + \sin x^2}$

25. $y = \dfrac{1}{3} \tan^3 x - \tan x$ **26.** $y = \csc 2x - 2 \cot 2x$ **27.** $y = \tan 2x - \sec 2x$ **28.** $y = x \tan x + \sec^2 2x$

29. $y = \sqrt{2x + \tan 4x}$ **30.** $y = (1 - \csc^2 3x)^3$ **31.** $x \sec y - 2y = \sin 2x$ **32.** $3 \cot (x + y) = \cos y^2$

In Exercises 33 through 36, find the differentials of the given functions.

33. $y = 4 \tan^2 3x$ **34.** $y = 5 \sec^3 2x$ **35.** $y = \tan 4x \sec 4x$ **36.** $y = 2x \cot 3x$

In Exercises 37 through 48, solve the given problems.

37. On a calculator find the values of (a) $\sec^2 1.0000$ and (b) $(\tan 1.0001 - \tan 1.0000)/0.0001$. Compare the values and give the meaning of each in relation to the derivative of tan x, where $x = 1$.

38. On a calculator find the values of (a) sec 1.0000 tan 1.0000 and (b) (sec 1.0001 − sec 1.0000)/0.0001. Compare the values and give the meaning of each in relation to the derivative of sec x, where $x = 1$.

39. Find the derivative of each member of the identity $1 + \tan^2 x = \sec^2 x$ and show that the results are equal.

40. Find the derivative of each member of the identity $1 + \cot^2 x = \csc^2 x$ and show that the results are equal.

41. Find the slope of a line tangent to the curve of $y = 2 \cot 3x$ where $x = \frac{\pi}{12}$.

42. Find the slope of a line normal to the curve of $y = \csc \sqrt{2x + 1}$ where $x = 0.45$.

43. Show that $y = 2 \tan x - \sec x$ satisfies $\dfrac{dy}{dx} = \dfrac{2 - \sin x}{\cos^2 x}$.

44. Show that $y = \cos^3 x \tan x$ satisfies $\cos x \dfrac{dy}{dx} + 3y \sin x - \cos^2 x = 0$.

45. The vertical displacement y (in cm) of the end of an industrial robot arm for each cycle is $y = 2t^{1.5} - \tan 0.1t$, where t is the time in seconds. Find its vertical velocity for $t = 15$ s.

46. The electric charge q (in coulombs) passing a given point in a circuit is given by $q = t \sec \sqrt{0.2t^2 + 1}$, where t is the time in seconds. Find the current i (in amperes) for $t = 0.80$ s. $(i = dq/dt.)$

47. An observer to a rocket launch was 1000 ft from the take-off position. The observer found the angle of elevation of the rocket as a function of time to be $\theta = 3t/(2t + 10)$. Therefore, the height h of the rocket was

$$h = 1000 \tan \frac{3t}{2t + 10}$$

Find the time rate of change of height after 5.0 s. See Fig. 26-5.

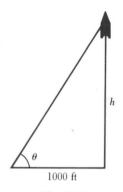

h

1000 ft

Fig. 26-5

48. A surveyor measures the distance between two markers to be 378.00 m. Then, moving along a line equidistant from the markers, the distance d from the surveyor to each marker is $d = 189.00 \csc \frac{1}{2}\theta$, where θ is the angle between the lines of sight to the markers. See Fig. 26-6. By using differentials, find the change in d if θ changes from 98.20° to 98.45°.

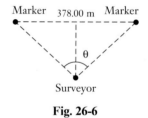

Marker   378.00 m   Marker

θ

Surveyor

Fig. 26-6

26-3   Derivatives of the Inverse Trigonometric Functions

To obtain the derivative of $y = \text{Arcsin } u$, we first solve for u in the form $u = \sin y$ and then take derivatives with respect to x:

$$\frac{du}{dx} = \cos y \frac{dy}{dx}$$

Solving this equation for dy/dx, we obtain

$$\frac{dy}{dx} = \frac{1}{\cos y} \frac{du}{dx} = \frac{1}{\sqrt{1 - \sin^2 y}} \frac{du}{dx} = \frac{1}{\sqrt{1 - u^2}} \frac{du}{dx}$$

We choose the positive square root since $\cos y > 0$ for $-\frac{\pi}{2} < y < \frac{\pi}{2}$, which is the range of the defined values of Arcsin u. Therefore, we obtain the following result:

derivative of
Arcsin u

$$\boxed{\frac{d(\text{Arcsin } u)}{dx} = \frac{1}{\sqrt{1 - u^2}} \frac{du}{dx}}$$

(26-10)

We note here that the derivative of the inverse sine function is an algebraic function.

We find the derivative of the inverse cosine function by letting $y = $ Arccos u and by following the same procedure as that used in finding the derivative of Arcsin u:

$$u = \cos y, \qquad \frac{du}{dx} = -\sin y \frac{dy}{dx}$$

$$\frac{dy}{dx} = -\frac{1}{\sin y}\frac{du}{dx} = -\frac{1}{\sqrt{1 - \cos^2 y}}\frac{du}{dx}$$

Therefore,

derivative of
Arccos u

$$\frac{d(\text{Arccos } u)}{dx} = -\frac{1}{\sqrt{1 - u^2}}\frac{du}{dx} \qquad (26\text{-}11)$$

The positive square root is chosen here since $\sin y > 0$ for $0 < y < \pi$, which is the range of the defined values of Arccos u. We note that the derivative of the inverse cosine is the negative of the derivative of the inverse sine.

By letting $y = $ Arctan u, solving for u, and taking derivatives, we find the derivative of the inverse tangent function:

$$u = \tan y, \qquad \frac{du}{dx} = \sec^2 y \frac{dy}{dx}, \qquad \frac{dy}{dx} = \frac{1}{\sec^2 y}\frac{du}{dx} = \frac{1}{1 + \tan^2 y}\frac{du}{dx}$$

Therefore,

derivative of
Arctan u

$$\frac{d(\text{Arctan } u)}{dx} = \frac{1}{1 + u^2}\frac{du}{dx} \qquad (26\text{-}12)$$

We can see that the derivative of the inverse tangent is an algebraic function also.

The inverse sine, inverse cosine, and inverse tangent prove to be of the greatest importance in applications and in further development of mathematics. Therefore, the formulas for the derivatives of the other inverse functions are not presented here, although they are included in the exercises.

EXAMPLE A

Find the derivative of $y = $ Arcsin $4x$.

Using Eq. (26-10), we have

$$\frac{dy}{dx} = \frac{1}{\sqrt{1 - (4x)^2}}\overset{\displaystyle\frac{du}{dx}}{(4)} = \frac{4}{\sqrt{1 - 16x^2}}$$

$\underset{u}{\uparrow}$

EXAMPLE B

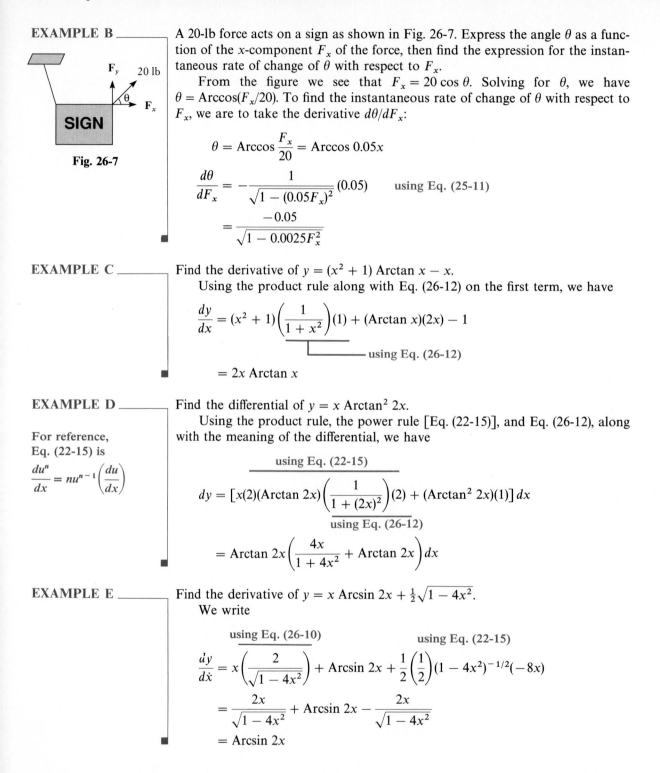

Fig. 26-7

A 20-lb force acts on a sign as shown in Fig. 26-7. Express the angle θ as a function of the x-component F_x of the force, then find the expression for the instantaneous rate of change of θ with respect to F_x.

From the figure we see that $F_x = 20 \cos \theta$. Solving for θ, we have $\theta = \text{Arccos}(F_x/20)$. To find the instantaneous rate of change of θ with respect to F_x, we are to take the derivative $d\theta/dF_x$:

$$\theta = \text{Arccos}\,\frac{F_x}{20} = \text{Arccos}\ 0.05x$$

$$\frac{d\theta}{dF_x} = -\frac{1}{\sqrt{1 - (0.05F_x)^2}}\,(0.05) \qquad \text{using Eq. (25-11)}$$

$$= \frac{-0.05}{\sqrt{1 - 0.0025F_x^2}}$$

EXAMPLE C

Find the derivative of $y = (x^2 + 1)\,\text{Arctan}\ x - x$.

Using the product rule along with Eq. (26-12) on the first term, we have

$$\frac{dy}{dx} = (x^2 + 1)\left(\frac{1}{1 + x^2}\right)(1) + (\text{Arctan}\ x)(2x) - 1$$

$$\underline{\hspace{2cm}}\ \text{using Eq. (26-12)}$$

$$= 2x\ \text{Arctan}\ x$$

EXAMPLE D

For reference,
Eq. (22-15) is

$$\frac{du^n}{dx} = nu^{n-1}\left(\frac{du}{dx}\right)$$

Find the differential of $y = x\,\text{Arctan}^2\ 2x$.

Using the product rule, the power rule [Eq. (22-15)], and Eq. (26-12), along with the meaning of the differential, we have

$$\overline{\text{using Eq. (22-15)}}$$

$$dy = \left[x(2)(\text{Arctan}\ 2x)\left(\frac{1}{1 + (2x)^2}\right)(2) + (\text{Arctan}^2\ 2x)(1)\right]dx$$

$$\underline{\text{using Eq. (26-12)}}$$

$$= \text{Arctan}\ 2x\left(\frac{4x}{1 + 4x^2} + \text{Arctan}\ 2x\right)dx$$

EXAMPLE E

Find the derivative of $y = x\,\text{Arcsin}\ 2x + \frac{1}{2}\sqrt{1 - 4x^2}$.

We write

$$\overline{\text{using Eq. (26-10)}} \qquad\qquad \overline{\text{using Eq. (22-15)}}$$

$$\frac{dy}{dx} = x\left(\frac{2}{\sqrt{1 - 4x^2}}\right) + \text{Arcsin}\ 2x + \frac{1}{2}\left(\frac{1}{2}\right)(1 - 4x^2)^{-1/2}(-8x)$$

$$= \frac{2x}{\sqrt{1 - 4x^2}} + \text{Arcsin}\ 2x - \frac{2x}{\sqrt{1 - 4x^2}}$$

$$= \text{Arcsin}\ 2x$$

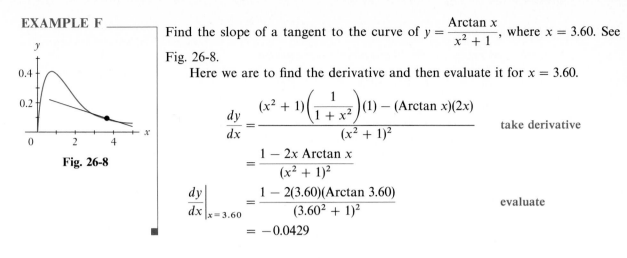

EXAMPLE F

Find the slope of a tangent to the curve of $y = \dfrac{\text{Arctan } x}{x^2 + 1}$, where $x = 3.60$. See Fig. 26-8.

Here we are to find the derivative and then evaluate it for $x = 3.60$.

$$\frac{dy}{dx} = \frac{(x^2 + 1)\left(\dfrac{1}{1 + x^2}\right)(1) - (\text{Arctan } x)(2x)}{(x^2 + 1)^2} \qquad \text{take derivative}$$

$$= \frac{1 - 2x \text{ Arctan } x}{(x^2 + 1)^2}$$

$$\left.\frac{dy}{dx}\right|_{x=3.60} = \frac{1 - 2(3.60)(\text{Arctan } 3.60)}{(3.60^2 + 1)^2} \qquad \text{evaluate}$$

$$= -0.0429 \qquad \blacksquare$$

Fig. 26-8

Exercises 26-3

In Exercises 1 through 32, find the derivatives of the given functions.

1. $y = \text{Arcsin } (x^2)$ **2.** $y = \text{Arcsin } (1 - x^2)$ **3.** $y = 2 \text{ Arcsin } 3x^3$ **4.** $y = \text{Arcsin } \sqrt{1 - 2x}$

5. $y = \text{Arccos } \frac{1}{2}x$ **6.** $y = 3 \text{ Arccos } 5x$ **7.** $y = 2 \text{ Arccos } \sqrt{2 - x}$ **8.** $y = 3 \text{ Arccos } (x^2 + 1)$

9. $y = \text{Arctan } \sqrt{x}$ **10.** $y = \text{Arctan } (1 - x)$ **11.** $y = \text{Arctan } \left(\dfrac{1}{x}\right)$ **12.** $y = 4 \text{ Arctan } 3x^4$

13. $y = x \text{ Arcsin } x$ **14.** $y = x^2 \text{ Arccos } x$ **15.** $y = 2x \text{ Arctan } 2x$ **16.** $y = (x^2 + 1) \text{ Arcsin } 4x$

17. $y = \dfrac{3x - 1}{\text{Arcsin } 2x}$ **18.** $y = \dfrac{\text{Arctan } 2x}{x}$ **19.** $y = \dfrac{\text{Arcsin } 2x}{\text{Arccos } 2x}$ **20.** $y = \dfrac{x^2 + 1}{\text{Arctan } x}$

21. $y = 2 \text{ Arccos}^3 4x$ **22.** $y = 3 \text{ Arcsin}^4 3x$ **23.** $y = \text{Arcsin}^2 4x$ **24.** $y = \sqrt{\text{Arcsin } (x - 1)}$

25. $y = 3 \text{ Arctan}^3 x$ **26.** $y = \dfrac{1}{\text{Arccos } 2x}$ **27.** $y = \dfrac{1}{1 + 4x^2} - \text{Arctan } 2x$

28. $y = \text{Arcsin } x - \sqrt{1 - x^2}$ **29.** $y = 3(4 - \text{Arccos } 2x)^3$ **30.** $\text{Arcsin } (x + y) + y = x^2$

31. $2 \text{ Arctan } xy + x = 3$ **32.** $y = \sqrt{1 - \text{Arcsin } 4x}$

In Exercises 33 through 44, solve the given problems.

33. On a calculator find the values of (a) $1/\sqrt{1 - 0.5^2}$ and (b) $(\text{Arcsin } 0.5001 - \text{Arcsin } 0.5000)/0.0001$. Compare the values and give the meaning of each in relation to the derivative of Arcsin x where $x = 0.5$.

34. On a calculator find the values of (a) $1/(1 + 0.5^2)$ and (b) $(\text{Arctan } 0.5001 - \text{Arctan } 0.5000)/0.0001$. Compare the values and give the meaning of each in relation to the derivative of Arctan x where $x = 0.5$.

35. Find the differential of the function $y = \text{Arcsin}^3 x$.

36. Find the differential of the function $y = x^3 \text{ Arccos } x^2$.

37. Find the second derivative of the function $y = \text{Arctan } 2x$.

38. Show that $\dfrac{d(\text{Arccot } u)}{dx} = -\dfrac{1}{1 + u^2} \dfrac{du}{dx}$.

39. Show that $\dfrac{d(\text{Arcsec } u)}{dx} = \dfrac{1}{\sqrt{u^2(u^2 - 1)}} \dfrac{du}{dx}$.

40. Show that $\dfrac{d(\text{Arccsc } u)}{dx} = -\dfrac{1}{\sqrt{u^2(u^2 - 1)}} \dfrac{du}{dx}$.

41. In analyzing the waveform of an AM radio wave, the equation $t = \dfrac{1}{\omega} \text{Arcsin} \dfrac{A - E}{mE}$
    arises. Find dt/dm, assuming that the other quantities are constant.

42. An equation which arises in the theory of solar collectors is $\alpha = \text{Arccos} \dfrac{2f - r}{r}$.
    Find the expression for $d\alpha/dr$ if f is constant.

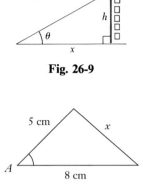

Fig. 26-9

43. As a person approaches a building of height h, the angle of elevation of the top of
    the building is a function of the person's distance from the building. Express the
    angle of elevation θ in terms of h and the distance x from the building and then
    find $d\theta/dx$. Assume the person's height is negligible to that of the building. See
    Fig. 26-9.

44. A triangular metal frame is being designed as shown in Fig. 26-10. Express the
    angle A as a function of x, and evaluate dA/dx for $x = 6$ cm.

Fig. 26-10

26-4   Applications

With our development of the formulas for the derivatives of the trigonometric
and inverse trigonometric functions, it is now possible for us to apply these derivatives in the same manner as we applied the derivatives of algebraic functions.
We can now use trigonometric and inverse trigonometric functions to solve tangent and normal, Newton's method, time rate of change, curve tracing, maximum
and minimum, and differential application problems. The following examples
illustrate the use of these functions in these types of problems.

EXAMPLE A

Sketch the curve $y = \sin^2 x - \dfrac{x}{2} \ (0 \le x \le 2\pi)$.

First, by setting $x = 0$, we see that the only easily obtainable intercept is
$(0, 0)$. Replacing x by $-x$ and y by $-y$, we find that the curve is not symmetrical
to either axis or to the origin. Also, since x does not appear in a denominator,
there are no vertical asymptotes. We are considering only the restricted domain
$0 \le x \le 2\pi$. (Without this restriction, the domain is all x and the range is all y.)
In order to find the information from the derivatives, we write

$$\frac{dy}{dx} = 2 \sin x \cos x - \frac{1}{2} = \sin 2x - \frac{1}{2}, \qquad \frac{d^2y}{dx^2} = 2 \cos 2x$$

Maximum and minimum points will occur for $\sin 2x = \frac{1}{2}$. Thus, we have possible
maximum and minimum points for

$$2x = \frac{\pi}{6}, \frac{5\pi}{6}, \frac{13\pi}{6}, \frac{17\pi}{6} \quad \text{or} \quad x = \frac{\pi}{12}, \frac{5\pi}{12}, \frac{13\pi}{12}, \frac{17\pi}{12}$$

Using the second derivative, we find that d^2y/dx^2 is positive for $x = \frac{\pi}{12}$ and $x =$
$\frac{13\pi}{12}$ and is negative for $x = \frac{5\pi}{12}$ and $x = \frac{17\pi}{12}$. Thus, the maximum points are
$(\frac{5\pi}{12}, 0.279)$ and $(\frac{17\pi}{12}, -1.29)$. Minimum points are $(\frac{\pi}{12}, -0.064)$ and $(\frac{13\pi}{12}, -1.64)$.
Inflection points occur for $\cos 2x = 0$, or

$$2x = \frac{\pi}{2}, \frac{3\pi}{2}, \frac{5\pi}{2}, \frac{7\pi}{2}, \quad \text{or} \quad x = \frac{\pi}{4}, \frac{3\pi}{4}, \frac{5\pi}{4}, \frac{7\pi}{4}$$

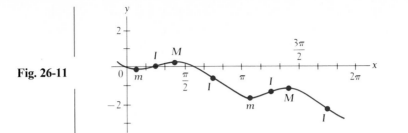

Fig. 26-11

Therefore, the points of inflection are $(\frac{\pi}{4}, 0.11)$, $(\frac{3\pi}{4}, -0.68)$, $(\frac{5\pi}{4}, -1.47)$, and $(\frac{7\pi}{4}, -2.25)$, Using this information, we sketch the curve in Fig. 26-11.

EXAMPLE B

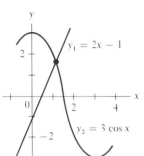

Fig. 26-12

By using Newton's method, solve the equation $2x - 1 = 3 \cos x$.

First we locate the root approximately by sketching $y_1 = 2x - 1$ and $y_2 = 3 \cos x$. From Fig. 26-12 we see that they intersect between $x = 1$ and $x = 2$, nearer to $x = 1$. Therefore, we choose $x_1 = 1.2$. With

$$f(x) = 2x - 1 - 3 \cos x$$

and

$$f'(x) = 2 + 3 \sin x$$

we use Eq. (23-1), which is

$$x_2 = x_1 - \frac{f(x_1)}{f'(x_1)}$$

To find x_2, we have

$$f(x_1) = 2(1.2) - 1 - 3 \cos 1.2 = 0.3129267$$

and

$$f'(x_1) = 2 + 3 \sin 1.2 = 4.7961173$$

$$x_2 = 1.2 - \frac{0.3129267}{4.7961173} = 1.1347542$$

Finding the next approximation, we find $x_3 = 1.1342366$, which is accurate to the value shown. Again, when using the calculator it is not necessary to list the values of $f(x_1)$ and $f'(x_1)$, as the complete calculation can be done directly on the calculator.

EXAMPLE C

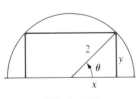

Fig. 26-13

Find the area of the largest rectangle which can be inscribed in a semicircle of radius 2.

First we draw Fig. 26-13 to help set up the necessary equation. From the figure we see that $x = 2 \cos \theta$ and $y = 2 \sin \theta$, which means that the area of the rectangle inscribed within the semicircle is

$$A = (2x)y = 2(2 \cos \theta)(2 \sin \theta) = 8 \cos \theta \sin \theta$$

$$= 4 \sin 2\theta \quad \text{using trigonometric identity [Eq. (19-21)]}$$

(*Continued on next page*)

Now, taking the derivative and setting it equal to zero, we have

$$\frac{dA}{d\theta} = (4 \cos 2\theta)(2) = 8 \cos 2\theta$$

$$8 \cos 2\theta = 0, \qquad 2\theta = \frac{\pi}{2}, \qquad \theta = \frac{\pi}{4} \qquad \cos \frac{\pi}{2} = 0$$

(Using $2\theta = \frac{3\pi}{2}$, $\theta = \frac{3\pi}{4}$ leads to the same solution.) Since the minimum area is zero, we have the maximum area when $\theta = \frac{\pi}{4}$, and this maximum area is

$$A = 4 \sin 2\left(\frac{\pi}{4}\right) = 4 \sin \frac{\pi}{2} = 4 \qquad \sin \frac{\pi}{2} = 1$$

EXAMPLE D

See the chapter
introduction.

A rocket is taking off vertically at a distance of 6500 ft from an observer. If, when the angle of elevation is 38.4°, it is changing at the rate of 5.0°/s, how fast is the rocket ascending?

From Fig. 26-14 we see that

$$\tan \theta = \frac{x}{6500} \qquad \text{or} \qquad \theta = \text{Arctan} \frac{x}{6500}$$

Taking derivatives with respect to time, we have

$$\frac{d\theta}{dt} = \frac{1}{1 + (x/6500)^2} \frac{dx/dt}{6500} = \frac{6500 \, dx/dt}{6500^2 + x^2}$$

When evaluating, we must express angles in radians. Therefore, $d\theta/dt = 5.0°/s = 0.0873$ rad/s. Substituting this value and $x = 6500 \tan 38.4° = 5150$ ft, we have

$$0.0873 = \frac{6500 \, dx/dt}{6500^2 + 5150^2}, \qquad \frac{dx}{dt} = 924 \text{ ft/s}$$

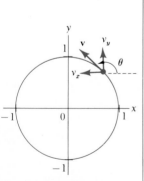

Fig. 26-14

EXAMPLE E

A particle moves so that its x- and y-coordinates are given by $x = \cos 2t$ and $y = \sin 2t$. Find the magnitude and direction of its velocity when $t = \frac{\pi}{8}$.

Taking derivatives with respect to time, we have

$$v_x = \frac{dx}{dt} = -2 \sin 2t, \qquad v_y = \frac{dy}{dt} = 2 \cos 2t$$

$$v_x|_{t=\pi/8} = -2 \sin 2\left(\frac{\pi}{8}\right) = -2\left(\frac{\sqrt{2}}{2}\right) = -\sqrt{2} \qquad \text{evaluating}$$

$$v_y|_{t=\pi/8} = 2 \cos 2\left(\frac{\pi}{8}\right) = 2\left(\frac{\sqrt{2}}{2}\right) = \sqrt{2}$$

$$v = \sqrt{v_x^2 + v_y^2} = \sqrt{2 + 2} = 2 \qquad \text{magnitude}$$

$$\tan \theta = \frac{v_y}{v_x} = \frac{\sqrt{2}}{-\sqrt{2}} = -1$$

Since v_x is negative and v_y is positive, $\theta = 135°$. Note that in this example θ is the angle, in standard position, between the horizontal and the resultant velocity.

By plotting the curve we note that it is a circle (see Fig. 26-15). Thus, the object is moving about a circle in a counterclockwise direction. It can be deter-

Fig. 26-15

mined in another way that the curve is a circle. If we square each of the expressions defining x and y and then add these, we have the equation $x^2 + y^2 = \cos^2 2t + \sin^2 2t = 1$. This is a circle of radius 1.

EXAMPLE F

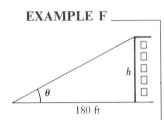

Fig. 26-16

At a point 180 ft from the base of a building on level ground, the angle of elevation of the top of the building is 30.00°. What would be the error in the height of the building due to an error of 0.25° in this angle?

From Fig. 26-16 we see that $h = 180 \tan \theta$. Finding the differential of h, we have

$$dh = 180 \sec^2 \theta \, d\theta$$

Since $\theta = 30.00°$, $\sec \theta = 1.1547005$. Also, the possible error in θ is 0.25°, or 0.0043633 rad, which is the value we must use in the calculation. Thus,

$$dh = (180)(1.1547005)^2(0.0043633) = 1.05 \text{ ft}$$

We see that an error of 0.25° in the angle can result in an error of over 1 ft in the calculation of the height.

Exercises 26-4

1. Show that the slopes of the sine and cosine curves are negatives of each other at the points of intersection.

2. Show that the tangent curve, when defined, is always increasing.

3. Show that $y = \text{Arctan } x$ is always increasing.

4. Sketch the graph of $y = \sin x + \cos x$ $(0 \le x \le 2\pi)$.

5. Sketch the graph of $y = x - \tan x$ $(-\frac{\pi}{2} < x < \frac{\pi}{2})$.

6. Sketch the graph of $y = \sin x + \sin 2x$ $(0 \le x \le 2\pi)$.

7. Find the equation of the line tangent to the curve of $y = \sin 2x$ at $x = \frac{5\pi}{8}$.

8. Find the equation of a line normal to the curve of $y = \text{Arctan } \frac{x}{2}$ at $x = 3$.

9. By Newton's method, find the positive root of the equation $x^2 - 4 \sin x = 0$.

10. By Newton's method, find the smallest positive root of the equation $\tan x = 2x$.

11. Find the maximum value of the function $y = 6 \cos x - 8 \sin x$.

12. Find the minimum value of the function $y = \cos 2x + 2 \sin x$.

13. In studying water waves, the vertical displacement y, in feet, of a wave was determined to be $y = 0.50 \sin 2t + 0.30 \cos t$. Find the velocity and acceleration for $t = 0.40$ s.

14. During each cycle, the vertical displacement y (in cm) of the stamp holder of a stamping machine is $y = t - t \sin t$. Find the velocity of the holder for $t = 7.2$ s.

15. Find the time rate of change of the horizontal component T_x of the constant 46.6-lb tension shown in Fig. 26-17 if $d\theta/dt = 0.36°/s$ for $\theta = 14.2°$.

16. The *apparent power* P_a of an electric circuit whose power is P and whose impedance phase angle is θ is given by $P_a = P \sec \theta$. Given that P is constant at 12 W, find the time rate of change of P_a if θ is changing at the rate of 0.050 rad/min, when $\theta = 40.0°$.

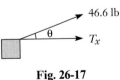

Fig. 26-17

17. A point on the rim of a $5\frac{1}{4}$-in. computer floppy disk can be described by the equations $x = 2.625 \cos 12\pi t$ and $y = 2.625 \sin 12\pi t$. Find the velocity of the point for $t = 1.250$ s.

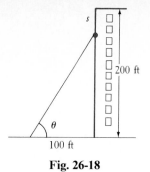

18. A machine is programmed to move an etching tool such that the position of the tool is given by $x = 2 \cos 3t$ and $y = \cos 2t$, where dimensions are in centimeters and time is in seconds. Find the velocity of the tool for $t = 4.1$ s.

19. Find the acceleration of the point on the floppy disk of Exercise 17 for $t = 1.250$ s.

20. Find the acceleration of the tool of Exercise 18 for $t = 4.1$ s.

21. A person observes an object dropped from the top of a building 100 ft away. If the building is 200 ft high, how fast is the angle of elevation of the object changing after 1.0 s? (The distance the object drops is given by $s = 16t^2$.) See Fig. 26-18.

Fig. 26-18

22. An airplane flying horizontally at 500 ft/s passes directly over an observer 4000 ft below. How fast is the angle of elevation changing at this instant?

23. A searchlight is 225 ft from a straight wall. As the beam moves along the wall, the angle between the beam and the perpendicular to the wall is increasing at 1.5°/s. How fast is the length of the beam increasing when it is 315 ft long?

24. In a modern hotel, where the elevators are directly observable from the lobby area (and a person can see from the elevators), a person in the lobby observes one of the elevators rising at the rate of 12.0 ft/s. If the person was 50.0 ft from the elevator when it left the lobby, how fast is the angle of elevation of the line of sight to the elevator increasing 10.0 s later?

25. If a block is placed on a plane inclined with the horizontal at an angle θ such that the block just moves down the plane, the coefficient of friction μ is given by $\mu = \tan \theta$. Use differentials to find the change in μ if θ changes from 20° to 21°.

26. The electric power p, in watts, developed in a resistor in an FM receiver circuit is $p = 0.0307 \cos^2 120\pi t$, where t is the time in seconds. Use differentials to find the approximate change in p between $t = 10.0$ ms and $t = 12.2$ ms.

27. A surveyor measures two sides and the included angle of a triangular parcel of land to be 82.04 m, 75.37 m, and 38.38°. What error is caused in the calculation of the third side by an error of 0.15° in the angle?

28. A firm determined that its total weekly profit from the production of x units is $P = 10,000 \sin 0.1x$, for $0 < x < 30$ and where P is measured in dollars. What is the approximate change in profit when production changes from 20 to 22 units?

29. Use trigonometric functions to find the area of the largest rectangular microprocessor chip with a perimeter of 40 mm.

30. An architect is designing a window in the shape of an isosceles triangle with a perimeter of 60 in. What is the vertex angle of the window of greatest area?

31. A wall is 6.0 ft high and 4.0 ft from a building. What is the length of the shortest pole that can touch the building and the ground beyond the wall? (*Hint:* From Fig. 26-19 it can be shown that $y = 6.0 \csc \theta + 4.0 \sec \theta$.)

32. A painting 8.0 ft high is hung such that the lower edge is 2.0 ft above an observer's eye level. Assuming that the best view is obtained when the angle subtended by the painting at eye level is a maximum, how far from the wall should the observer stand? See Fig. 26-20.

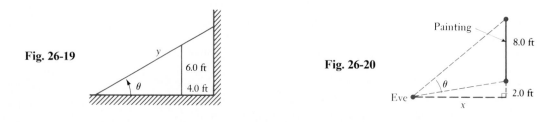

Fig. 26-19

Fig. 26-20

26-5 Derivative of the Logarithmic Function

The next function for which we shall find the derivative is the logarithmic function. Again, we shall use the delta-process.

If we let $y = \log_b u$, where u is a function of x, we have

$$y + \Delta y = \log_b (u + \Delta u)$$

$$\Delta y = \log_b (u + \Delta u) - \log_b u = \log_b \left(\frac{u + \Delta u}{u} \right) = \log_b \left(1 + \frac{\Delta u}{u} \right)$$

$$\frac{\Delta y}{\Delta u} = \frac{\log_b (1 + \Delta u/u)}{\Delta u} = \frac{1}{u} \frac{u}{\Delta u} \log_b \left(1 + \frac{\Delta u}{u} \right)$$

$$= \frac{1}{u} \log_b \left(1 + \frac{\Delta u}{u} \right)^{u/\Delta u}$$

(We multiply and divide by u for purposes of evaluating the limit, as we shall now show.)

Before we can evaluate the $\lim\limits_{\Delta u \to 0} \Delta y/\Delta u$, it is necessary to determine

$$\lim_{\Delta u \to 0} \left(1 + \frac{\Delta u}{u} \right)^{u/\Delta u}$$

We can see that the exponent becomes unbounded, but the number being raised to this exponent approaches 1. Therefore, we shall investigate this limiting value.

To find an approximate value, let us graph the function $y = (1 + x)^{1/x}$ (for purposes of graphing we let $\Delta u/u = x$). We construct a table of values and then graph the function (see Fig. 26-21).

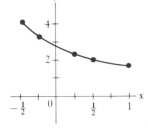

x	-0.5	-0.25	$+0.25$	$+0.50$	$+1.00$
y	4.00	3.16	2.44	2.25	2.00

Fig. 26-21

Only these values are shown, since we are interested in the y-value corresponding to $x = 0$. We see from the graph that this value is approximately 2.7. Choosing very small values of x, we may obtain the following values:

x	0.1	0.01	0.001	0.0001
y	2.5937	2.7048	2.7169	2.71815

By methods developed in Chapter 28, it can be shown that this value is about 2.7182818. *The limiting value is the irrational number e.* This is the same number used in the exponential form of a complex number in Chapter 11 and as the base of natural logarithms in Chapter 12.

Returning to the derivative of the logarithmic function, we have

See Appendix E for a computer program for finding the value for e.

$$\lim_{\Delta u \to 0} \frac{\Delta y}{\Delta u} = \lim_{\Delta u \to 0} \left[\frac{1}{u} \log_b \left(1 + \frac{\Delta u}{u} \right)^{u/\Delta u} \right] = \frac{1}{u} \log_b e$$

Therefore,

$$\frac{dy}{du} = \frac{1}{u} \log_b e$$

For reference,
Eq. (26-3) is

$$\frac{dy}{dx} = \frac{dy}{du}\frac{du}{dx}$$

Combining this equation with Eq. (26-3), we have

$$\frac{d(\log_b u)}{dx} = \frac{1}{u}\log_b e \frac{du}{dx} \qquad (26\text{-}13)$$

At this point we see that if we choose e as the basis of a system of logarithms, the above formula becomes

derivative of ln u

$$\frac{d(\ln u)}{dx} = \frac{1}{u}\frac{du}{dx} \qquad (26\text{-}14)$$

The choice of e as the base b makes $\log_e e = 1$; thus, this factor does not appear in Eq. (26-14). We can now see why the number e is chosen as the base for a system of logarithms, the natural logarithms. The notation $\ln u$ is the same as that used in Chapter 12 for natural logarithms.

EXAMPLE A Find the derivative of $y = \log 4x$.
Using Eq. (26-13), we have

$$\frac{dy}{dx} = \frac{1}{4x}(\log e)(4) \quad \overset{\frac{du}{dx}}{\underset{u}{}}$$

$$= \frac{1}{x}\log e \qquad (\log e = 0.4343)$$

EXAMPLE B Find the derivative of $y = \ln 3x^4$.
Using Eq. (26-14), we have (with $u = 3x^4$)

$$\frac{dy}{dx} = \frac{1}{3x^4}(12x^3) \quad \frac{du}{dx}$$

$$= \frac{4}{x}$$

EXAMPLE C Find the derivative of $y = \ln \tan 4x$.
Using Eq. (26-14), along with the derivative of the tangent, we have

$$\frac{dy}{dx} = \frac{1}{\tan 4x}(\sec^2 4x)(4) \quad \frac{d \tan 4x}{dx}$$

$$= \frac{\cos 4x}{\sin 4x}\frac{4}{\cos^2 4x} \qquad \textbf{using trigonometric relations}$$

$$= \frac{1}{\sin 4x}\frac{4}{\cos 4x} = 4 \csc 4x \sec 4x$$

NOTE ▷

Frequently we can find the derivative of a logarithmic function more simply if we *use the properties of logarithms to simplify the expression before the derivative is found.* The following examples illustrate this.

EXAMPLE D

For reference, Eqs. (12-7) and (12-8) are

$\log_b xy$
$\quad = \log_b x + \log_b y$

$\log_b \left(\dfrac{x}{y}\right)$
$\quad = \log_b x - \log_b y$ ∎

Find the derivative of $y = \ln \dfrac{x - 1}{x + 1}$.

In this example it is easier to write y in the form

$$y = \ln (x - 1) - \ln (x + 1)$$

by using the properties of logarithms [Eq. (12-8)]. Hence,

$$\frac{dy}{dx} = \frac{1}{x - 1} - \frac{1}{x + 1} = \frac{x + 1 - x + 1}{(x - 1)(x + 1)} = \frac{2}{x^2 - 1}$$

EXAMPLE E

For reference, Eq. (12-9) is

$n \log_b x = \log_b x^n$

1. Find the derivative of $y = \ln (1 - 2x)^3$.

First, using Eq. (12-9), we may rewrite the equation as $y = 3 \ln (1 - 2x)$. Then we have

$$\frac{dy}{dx} = 3\left(\frac{1}{1 - 2x}\right)(-2) = \frac{-6}{1 - 2x}$$

2. Find the derivative of $y = \ln^3 (1 - 2x)$.

First we note that

$$y = \ln^3 (1 - 2x) = [\ln (1 - 2x)]^3$$

where $\ln^3 (1 - 2x)$ is usually the preferred notation.

Next, we must be careful to distinguish this function from that in part (1). For $y = \ln^3 (1 - 2x)$, it is the logarithm of $1 - 2x$ which is being cubed, whereas for $y = \ln (1 - 2x)^3$, it is $1 - 2x$ which is being cubed.

Now, finding the derivative of $y = \ln^3 (1 - 2x)$, we have

$$\frac{dy}{dx} = 3[\ln^2 (1 - 2x)]\left(\frac{1}{1 - 2x}\right)(-2)$$

$$= -\frac{6 \ln^2 (1 - 2x)}{1 - 2x} \qquad\underset{\dfrac{d \ln(1 - 2x)}{dx}}{\uparrow}$$

EXAMPLE F

Evaluate the derivative of $y = \ln (\sin 2x)(\sqrt{x^2 + 1})$, where $x = 0.375$.

First, using the properties of logarithms, we rewrite the function as

$$y = \ln \sin 2x + \frac{1}{2} \ln (x^2 + 1)$$

Now we have

$$\frac{dy}{dx} = \frac{1}{\sin 2x} (\cos 2x)(2) + \frac{1}{2}\left(\frac{1}{x^2 + 1}\right)(2x) \qquad \text{take the derivative}$$

$$= 2 \cot 2x + \frac{x}{x^2 + 1}$$

$$\left.\frac{dy}{dx}\right|_{x=0.375} = 2 \cot 0.750 + \frac{0.375}{0.375^2 + 1} = 2.48 \qquad \text{evaluate}$$

Exercises 26-5

In Exercises 1 through 32, find the derivatives of the given functions.

1. $y = \log x^2$

2. $y = \log_2 6x$

3. $y = 2 \log_5 (3x + 1)$

4. $y = 3 \log_7 (x^2 + 1)$

5. $y = \ln(1 - 3x)$

6. $y = 2 \ln (3x^2 - 1)$

7. $y = 2 \ln \tan 2x$

8. $y = \ln \sin^2 x$

9. $y = \ln \sqrt{x}$

10. $y = \ln \sqrt{4x - 3}$

11. $y = \ln (x^2 + 2x)^3$

12. $y = \ln (2x^3 - x)^2$

13. $y = x \ln x^2$

14. $y = x^2 \ln 2x$

15. $y = \dfrac{3x}{\ln (2x + 1)}$

16. $y = \dfrac{\ln x}{x}$

17. $y = \ln (\ln x)$

18. $y = \ln \cos x^2$

19. $y = \ln \dfrac{x}{1 + x}$

20. $y = \ln (x\sqrt{x + 1})$

21. $y = \sin \ln x$

22. $y = \text{Arctan} \ln 2x$

23. $y = 3 \ln^2 2x$

24. $y = x \ln^3 x$

25. $y = \ln (x \tan x)$

26. $y = \ln (x + \sqrt{x^2 - 1})$

27. $y = \ln \dfrac{x^2}{x + 2}$

28. $y = \sqrt{x + \ln 3x}$

29. $y = \sqrt{x^2 + 1} - \ln \dfrac{1 + \sqrt{x^2 + 1}}{x}$

30. $3 \ln xy + \sin y = x^2$

31. $y = x - \ln^2 (x + y)$

32. $y = (x^2 - \ln 2x)^3$

In Exercises 33 through 48, solve the given problems.

33. On a calculator find the value of $(\ln 2.0001 - \ln 2.0000)/0.0001$ and compare it with 0.5. Give the meanings of the value found and 0.5 in relation to the derivative of $\ln x$, where $x = 2$.

34. On a calculator find the value of $(\ln 0.5001 - \ln 0.5000)/0.0001$ and compare it with 2. Give the meanings of the value found and 2 in relation to the derivative of $\ln x$, where $x = 0.5$.

35. Verify the values of $(1 + x)^{1/x}$ in the table where $(1 + x)^{1/x}$ is calculated for $x = 0.1, 0.01, 0.001$, and 0.0001.

36. Find the second derivative of the function $y = x^2 \ln x$.

37. Evaluate the derivative of $y = \text{Arctan } 2x + \ln (4x^2 + 1)$, where $x = 0.625$.

38. Evaluate the derivative of $y = \ln \sqrt{\dfrac{2x + 1}{3x + 1}}$, where $x = 2.75$.

39. Find the differential of the function $y = \ln \cos^2 x - 2 \ln \tan x$.

40. Find the differential of the function $y = \sin 2x \ln 4x$.

41. Find the slope of a line tangent to the curve of $y = \ln \cos x$ at $x = \frac{\pi}{4}$.

42. Find the slope of a line tangent to the curve of $y = x \ln 2x$ at $x = 2$.

43. Find the derivative of $y = x^x$ by first taking logarithms of each side of the equation.

44. Find the derivative of $y = (\sin x)^x$ by first taking logarithms of each side of the equation.

45. If the loudness b (in decibels) of a sound of intensity I is given by $b = 10 \log (I/I_0)$ where I_0 is a constant, find the expression for db/dt in terms of dI/dt.

46. The time t for a particular computer system to process N bits of data is directly proportional to $N \ln N$. Find the expression for dt/dN.

47. When air friction is considered, the time t (in seconds) it takes a certain falling object to attain a velocity v (in ft/s) is given by $t = 5 \ln \dfrac{16}{16 - 0.1v}$. Find dt/dv for $v = 100$ ft/s.

48. The electric potential V at a distance x from an electric charge distributed along a wire of length $2a$ is $V = k \ln \dfrac{\sqrt{a^2 + x^2} + a}{\sqrt{a^2 + x^2} - a}$, where k is a constant. Find the expression for the electric field E, which is defined as $E = -dV/dx$.

26-6   **Derivative of the Exponential Function** ▬▬▬▬▬

To obtain the derivative of the exponential function, we let $y = b^u$ and then take natural logarithms of both sides:

$$\ln y = \ln b^u = u \ln b$$

$$\frac{1}{y} \frac{dy}{dx} = \ln b \frac{du}{dx}$$

$$\frac{dy}{dx} = y \ln b \frac{du}{dx}$$

Thus,

$$\frac{d(b^u)}{dx} = b^u \ln b \left(\frac{du}{dx} \right) \qquad (26\text{-}15)$$

If we let $b = e$, Eq. (26-15) becomes

derivative of e^u

$$\frac{d(e^u)}{dx} = e^u \left(\frac{du}{dx} \right) \qquad (26\text{-}16)$$

The simplicity of Eq. (26-16) compared with Eq. (26-15) again shows the advantage of choosing e as the basis of natural logarithms. It is for this reason that e appears so often in applications of calculus.

EXAMPLE A ——— Find the derivative of $y = e^x$.

Using Eq. (26-16), we have

$$\frac{dy}{dx} = e^x \overset{\overset{\textstyle \frac{du}{dx}}{\downarrow}}{(1)} = e^x$$

We see that the derivative of the function e^x equals itself. This exponential function is widely used in applications of calculus. ■

For reference, Eq. (22-15) is

$$\frac{du^u}{dx} = nu^{n-1}\left(\frac{du}{dx}\right)$$

We should note carefully that Eq. (22-15) is used with a variable raised to a constant exponent, whereas with Eqs. (26-15) and (26-16) we are finding the derivative of a constant raised to a variable exponent.

variable constant   constant variable

$$\overset{\downarrow\,\swarrow}{\frac{du^n}{dx} = nu^{n-1}\left(\frac{du}{dx}\right)} \qquad \overset{\downarrow\,\swarrow}{\frac{db^u}{dx} = b^u \ln b \left(\frac{du}{dx}\right)}$$

EXAMPLE B ____ Find the derivatives of $y = (4x)^2$ and $y = 2^{4x}$.

Using Eq. (22-15), we have

$$y = (4x)^2$$
$$\frac{dy}{dx} = 2(4x)^1(4) \quad \overset{du}{\underset{dx}{\big\downarrow}}$$
$$= 32x$$

Using Eq. (26-15), we have

$$y = 2^{4x}$$
$$\frac{dy}{dx} = 2^{4x}(\ln 2)(4) \quad \overset{du}{\underset{dx}{\big\downarrow}}$$
$$= (4 \ln 2)(2^{4x})$$

We continue now with additional examples of the use of Eq. (26-16).

EXAMPLE C ____ Find the derivative of $y = \ln \cos e^{2x}$.

Using Eq. (26-16), along with the derivatives of the logarithmic and cosine functions, we have

$$\frac{dy}{dx} = \frac{1}{\cos e^{2x}} \frac{d \cos e^{2x}}{dx} \qquad \text{using } \frac{d \ln u}{dx} = \frac{1}{u} \frac{du}{dx}$$

$$= \frac{1}{\cos e^{2x}} (-\sin e^{2x}) \frac{de^{2x}}{dx} \qquad \text{using } \frac{d \cos u}{dx} = -\sin u \frac{du}{dx}$$

$$= -\frac{\sin e^{2x}}{\cos e^{2x}} (e^{2x})(2) \qquad \text{using } \frac{de^u}{dx} = e^u \frac{du}{dx}$$

$$= -2e^{2x} \tan e^{2x} \qquad \text{using } \frac{\sin \theta}{\cos \theta} = \tan \theta$$

EXAMPLE D ____ Find the derivative of $y = xe^{\tan x}$.

Using Eq. (26-16), along with the derivatives of a product and the tangent, we have

$$\frac{dy}{dx} = xe^{\tan x}(\sec^2 x) + e^{\tan x}(1)$$

$$= e^{\tan x}(x \sec^2 x + 1)$$

EXAMPLE E ____ Find the derivative of $y = (e^{1/x})^2$.

In this example we use Eqs. (22-15) and (26-16).

$$\frac{dy}{dx} = 2(e^{1/x})(e^{1/x})\left(-\frac{1}{x^2}\right)$$

$$= \frac{-2(e^{1/x})^2}{x^2} = \frac{-2e^{2/x}}{x^2} \qquad \text{using Eqs. (26-16) and (22-15) to find } \frac{du}{dx} \text{ of Eq. (22-15)}$$

This problem could have also been solved by first writing the function as $y = e^{2/x}$, which is an equivalent form determined by the laws of exponents. When we use this form, the derivative becomes

using Eq. (22-15) to find $\dfrac{du}{dx}$ of Eq. (26-16)

$$\frac{dy}{dx} = e^{2/x}\left(-\frac{2}{x^2}\right)$$

$$= \frac{-2e^{2/x}}{x^2}$$

We can see that this change in form of the function simplifies the steps necessary for finding the derivative.

EXAMPLE F

Find the derivative of $y = (3e^{4x} + 4x^2 \ln x)^3$.

Using the general power rule [Eq. (22-15)] for derivatives, the derivative of the exponential function [Eq. (26-16)], the derivative of a product [Eq. (22-12)], and the derivative of a logarithm [Eq. (26-14)], we have

$$\frac{dy}{dx} = 3(3e^{4x} + 4x^2 \ln x)^2\left[12e^{4x} + 4x^2\left(\frac{1}{x}\right) + (\ln x)(8x)\right]$$

$$= 3(3e^{4x} + 4x^2 \ln x)^2(12e^{4x} + 4x + 8x \ln x)$$

EXAMPLE G

Find the slope of a line tangent to the curve of $y = \dfrac{3e^{2x}}{x^2 + 1}$ where $x = 1.275$. See Fig. 26-22.

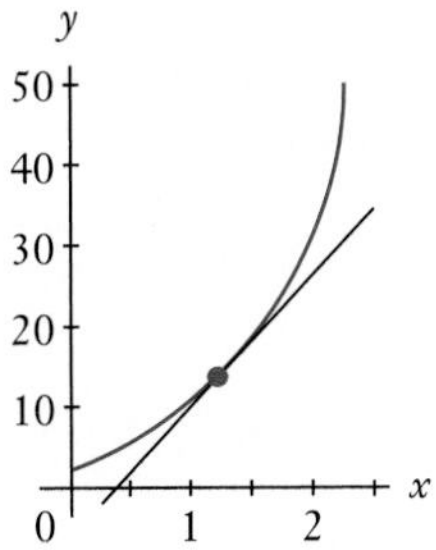

Here we are to find the derivative and then evaluate it for $x = 1.275$. The solution is as follows:

$$\frac{dy}{dx} = \frac{(x^2 + 1)(3e^{2x})(2) - 3e^{2x}(2x)}{(x^2 + 1)^2} \qquad \text{take derivative}$$

$$= \frac{6e^{2x}(x^2 - x + 1)}{(x^2 + 1)^2}$$

$$\left.\frac{dy}{dx}\right|_{x=1.275} = \frac{6e^{2(1.275)}(1.275^2 - 1.275 + 1)}{(1.275^2 + 1)^2} \qquad \text{evaluate}$$

$$= 15.05$$

Fig. 26-22

Exercises 26-6

In Exercises 1 through 32, find the derivatives of the given functions.

1. $y = 3^{2x}$

2. $y = 3^{1-x}$

3. $y = 4^{6x}$

4. $y = 10^{x^2}$

5. $y = e^{6x}$

6. $y = 3e^{x^2}$

7. $y = e^{\sqrt{x}}$

8. $y = e^{2x^4}$

9. $y = xe^{-x}$

10. $y = x^2e^{2x}$

11. $y = xe^{\sin x}$

12. $y = 4e^x \sin \frac{1}{2}x$

13. $y = \dfrac{3e^{2x}}{x + 1}$

14. $y = \dfrac{e^x}{x}$

15. $y = e^{-3x} \sin 4x$

16. $y = (\cos 2x)(e^{x^2 - 1})$

17. $y = \dfrac{2e^{3x}}{4x+3}$

18. $y = \dfrac{\ln 2x}{e^{2x}+2}$

19. $y = \ln(e^{x^2}+4)$

20. $y = (3e^{2x}+x)^3$

21. $y = (2e^{2x})^3 \sin x^2$

22. $y = (e^{3/x} \cos x)^2$

23. $y = (\ln 2x + e^{2x})^2$

24. $y = (2e^{x^2}+x^2)^3$

25. $y = xe^{xy} + \sin y$

26. $y = 4e^{-2/x} \ln y + 1$

27. $y = e^{2x} \ln x$

28. $y = e^{x^2} \ln \cos x$

29. $y = \ln \sin 2e^{6x}$

30. $y = 6 \tan e^{x+1}$

31. $y = 2 \operatorname{Arcsin} e^{2x}$

32. $y = \operatorname{Arctan} e^{3x}$

In Exercises 33 through 44, solve the given problems.

33. On a calculator find the values of (a) e and (b) $(e^{1.0001} - e^{1.0000})/0.0001$. Compare the values and give the meaning of each in relation to the derivative of e^x where $x = 1$.

34. On a calculator find the values of (a) e^2 and (b) $(e^{2.0001} - e^{2.0000})/0.0001$. Compare the values and give the meaning of each in relation to the derivative of e^x where $x = 2$.

35. Find the slope of a tangent to $y = e^{-2x} \cos 2x$ for $x = 0.625$.

36. Find the slope of a tangent to $y = \dfrac{e^{-x}}{1 + \ln 4x}$ for $x = 1.842$.

37. Show that $y = xe^{-x}$ satisfies the equation $(dy/dx) + y = e^{-x}$.

38. Show that $y = e^{-x} \sin x$ satisfies the equation

$$\frac{d^2y}{dx^2} + 2\frac{dy}{dx} + 2y = 0$$

39. Show that the slope of the curve $y = e^x$ at $x = a$ equals the ordinate at $x = a$.

40. If $e^x + e^y = e^{x+y}$, show that $dy/dx = -e^{y-x}$.

41. The electric current i in a certain circuit is given by $i = 2te^{-0.5t}$, where t is the time. Find the expression for the instantaneous time rate of change of i.

42. The Beer-Lambert law of light absorption may be expressed as $I = I_0 e^{-\alpha x}$, where I/I_0 is that fraction of the incident light beam which is transmitted, α is a constant, and x is the distance the light travels through the medium. Find the expression for the instantaneous rate of change of I with respect to x.

43. The reliability R $(0 \le R \le 1)$ of a certain computer system is $R = e^{-0.002t}$, where t is the time of operation in hours. Find dR/dt for $t = 100$ h.

44. An equation used in analyzing biological cells is $n = N(1 - e^{-at})$, where a and N are constants and t is the time. Express dn/dt as a function of n.

In Exercises 45 through 48, use the following information:

*The **hyperbolic sine** of u is defined as*

$$\sinh u = \frac{1}{2}(e^u - e^{-u})$$

Figure 26-23 shows the graph of $y = \sinh x$.

*The **hyperbolic cosine** of u is defined as*

$$\cosh u = \frac{1}{2}(e^u + e^{-u})$$

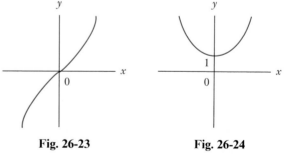

Fig. 26-23   Fig. 26-24

Figure 26-24 shows the graph of $y = \cosh x$.

These functions are called hyperbolic *functions since, if $x = \cosh u$ and $y = \sinh u$, x and y satisfy the equation of the* hyperbola $x^2 - y^2 = 1$.

45. Verify the fact that the expressions for the hyperbolic sine and hyperbolic cosine satisfy the equation of the hyperbola.

46. Show that $\sinh u$ and $\cosh u$ satisfy the identity $\cosh^2 u - \sinh^2 u = 1$.

47. Show that

$$\frac{d}{dx}\sinh u = \cosh u \frac{du}{dx} \quad \text{and} \quad \frac{d}{dx}\cosh u = \sinh u \frac{du}{dx}$$

where u is a function of x.

48. Show that

$$\frac{d^2 \sinh x}{dx^2} = \sinh x \quad \text{and} \quad \frac{d^2 \cosh x}{dx^2} = \cosh x$$

26-7   Applications

The following examples show applications of the logarithmic and exponential functions to curve tracing, Newton's method, and time-rate-of-change problems. Certain other applications are indicated in the exercises.

EXAMPLE A

Sketch the graph of the function $y = x \ln x$.

First we note that x cannot be zero since $\ln x$ is not defined at $x = 0$. Since $\ln 1 = 0$, we have an intercept at $(1, 0)$. There is no symmetry to the axes or origin, and there are no vertical asymptotes. Also, because $\ln x$ is defined only for $x > 0$, the domain is $x > 0$.

Finding the first two derivatives, we have

$$\frac{dy}{dx} = x\left(\frac{1}{x}\right) + \ln x = 1 + \ln x, \qquad \frac{d^2 y}{dx^2} = \frac{1}{x}$$

The first derivative is zero if $\ln x = -1$, or $x = e^{-1}$. The second derivative is positive for this value of x. Thus, there is a minimum point at $(1/e, -1/e)$. Since the domain is $x > 0$, the second derivative indicates that the curve is always concave up. In turn, we can now see that the range of the function is $y \geq -1/e$. The graph is shown in Fig. 26-25.

Although the curve approaches the origin as x approaches zero, the origin is not included on the graph of the function.

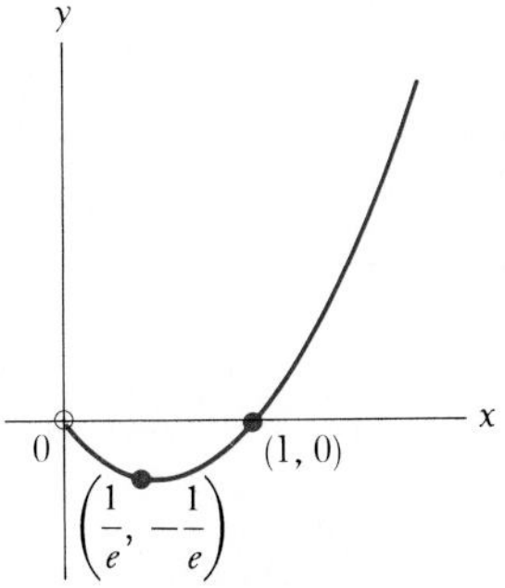

Fig. 26-25

EXAMPLE B

Sketch the graph of the function $y = e^{-x}\cos x$ $(0 \leq x \leq 2\pi)$.

This curve has intercepts for all values for which $\cos x$ is zero. Those values in the domain $0 \leq x \leq 2\pi$ for which $\cos x = 0$ are $x = \frac{\pi}{2}$ and $x = \frac{3\pi}{2}$. The factor e^{-x} is always positive, and $e^{-x} = 1$ for $x = 0$, which means $(0, 1)$ is also an intercept. There is no symmetry to the axes or the origin, and there are no vertical asymptotes.

Next, finding the first derivative, we have

$$\frac{dy}{dx} = -e^{-x}\sin x - e^{-x}\cos x = -e^{-x}(\sin x + \cos x)$$

Setting the derivative equal to zero, since e^{-x} is always positive, we have

$$\sin x + \cos x = 0, \qquad \tan x = -1, \qquad x = \frac{3\pi}{4}, \frac{7\pi}{4}$$

(Continued on next page)

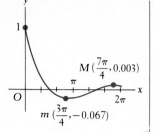

$M\left(\frac{7\pi}{4}, 0.003\right)$

$m\left(\frac{3\pi}{4}, -0.067\right)$

Fig. 26-26

Now, finding the second derivative, we have

$$\frac{d^2y}{dx^2} = -e^{-x}(\cos x - \sin x) - e^{-x}(-1)(\sin x + \cos x) = 2e^{-x}\sin x$$

The sign of the second derivative depends only on $\sin x$. Thus, $d^2y/dx^2 > 0$ for $x = \frac{3\pi}{4}$ and $d^2y/dx^2 < 0$ for $x = \frac{7\pi}{4}$. Hence, $\left(\frac{3\pi}{4}, -0.067\right)$ is a minimum and $\left(\frac{7\pi}{4}, 0.003\right)$ is a maximum. Also, from the second derivative, points of inflection occur for $x = 0$, π, and 2π since $\sin x = 0$ for these values. The graph is shown in Fig. 26-26. ∎

EXAMPLE C By Newton's method, find the root of the equation $e^{2x} - 4\cos x = 0$ which lies between 0 and 1.

Here

$$f(x) = e^{2x} - 4\cos x$$

and

$$f'(x) = 2e^{2x} + 4\sin x$$

This means that $f(0) = -3$ and $f(1) = 5.2$. Therefore, we choose $x_1 = 0.5$. Using Eq. (23-1), which is

$$x_2 = x_1 - \frac{f(x_1)}{f'(x_1)}$$

we have the following values.

$$f(x_1) = e^{2(0.5)} - 4\cos 0.5 = -0.7920484$$
$$f'(x_1) = 2e^{2(0.5)} + 4\sin 0.5 = 7.3542658$$
$$x_2 = 0.5 - \frac{-0.7920484}{7.3542658} = 0.6076992$$

Using the method again, we find $x_3 = 0.5979751$, which is correct to three decimal places. ∎

EXAMPLE D A good model for population growth is that the population P at time t is given by $P = P_0e^{kt}$, where P_0 is the initial population (at $t = 0$, when timing starts for the population being considered) and k is a constant. Show that the time rate of change of population is directly proportional to the population present at time t.

To find the time rate of change, we find the derivative dP/dt:

$$\frac{dP}{dt} = (P_0e^{kt})(k) = kP_0e^{kt}$$

$$= kP \qquad\qquad \text{since } P = P_0e^{kt}$$

∎ Thus we see that population growth increases as the population increases.

EXAMPLE E _____ | A rocket moving vertically, if the only force acting on it is due to gravity and its mass is decreasing at a constant rate r, has its velocity as a function of time given by

See the chapter
introduction.

$$v = v_0 - gt - k \ln\left(1 - \frac{rt}{m_0}\right)$$

where v_0 is the initial velocity, g is the acceleration due to gravity, t is the time, m_0 is the initial mass, and k is a constant. Determine the expression for the acceleration.

Since the acceleration is the time rate of change of the velocity, we must find dv/dt. Therefore,

$$\frac{dv}{dt} = -g - k\,\frac{1}{1 - \dfrac{rt}{m_0}}\left(\frac{-r}{m_0}\right) = -g + \frac{km_0}{m_0 - rt}\left(\frac{r}{m_0}\right)$$

$$= \frac{kr}{m_0 - rt} - g$$

Exercises 26-7

In Exercises 1 through 12, sketch the graphs of the given functions.

1. $y = \ln \cos x$

2. $y = \dfrac{\ln x}{x}$

3. $y = xe^{-x}$

4. $y = \dfrac{e^x}{x}$

5. $y = \ln \dfrac{1}{x^2 + 1}$

6. $y = \ln \dfrac{1}{x}$

7. $y = e^{-x^2}$

8. $y = x - e^x$

9. $y = \ln x - x$

10. $y = e^{-x} \sin x$

11. $y = \frac{1}{2}(e^x - e^{-x})$ (See Exercise 45 of Section 26-6.)

12. $y = \frac{1}{2}(e^x + e^{-x})$ (See Exercise 45 of Section 26-6.)

In Exercises 13 through 32, solve the given problems by finding the appropriate derivatives.

13. Find the equation of the line tangent to the curve $y = x^2 \ln x$ at the point $(1, 0)$.

14. Find the equation of the line tangent to the curve of $y = \text{Arctan } 2x$, where $x = 1$.

15. Find the equation of the line normal to the curve of $y = 2 \sin \frac{1}{2}x$, where $x = \frac{3}{2}\pi$.

16. Find the equation of the line normal to the curve of $y = e^{2x}/x$ at $x = 1$.

17. By Newton's method, solve the equation $x^2 - 2 + \ln x = 0$. See Fig. 26-27.

18. By Newton's method, solve the equation $e^{-2x} - \text{Arctan } x = 0$.

19. The power supply P (in watts) in a satellite is given by $P = 100e^{-0.005t}$, where t is measured in days. Find the time rate of change of power after 100 days.

20. The number N of atoms of radium at any time t is given in terms of the number at $t = 0$, N_0, by $N = N_0 e^{-kt}$. Show that the time rate of change of N is proportional to N.

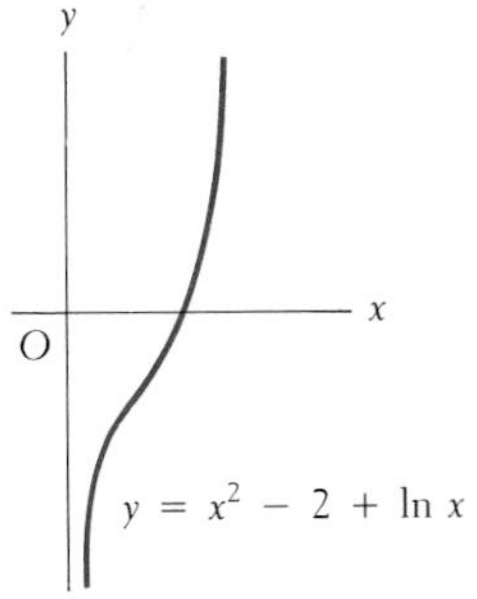

21. The vapor pressure p and thermodynamic temperature T of a gas are related by the equation $\ln p = \dfrac{a}{T} + b \ln T + c$, where a, b, and c are constants. Find the expression for dp/dT.

22. The charge on a capacitor in a circuit containing the capacitor of capacitance C, a resistance R, and a source of voltage E is given by $q = CE(1 - e^{-t/RC})$. Show that this equation satisfies the equation $R\dfrac{dq}{dt} + \dfrac{q}{C} = E$.

Fig. 26-27

23. Assuming that force is proportional to acceleration, show that a particle moving along the x-axis, so that its displacement $x = ae^{kt} + be^{-kt}$, has a force acting on it which is proportional to its displacement.

24. The radius of curvature at any point on a curve is given by

$$R = \frac{[1 + (dy/dx)^2]^{3/2}}{d^2y/dx^2}$$

A roller mechanism moves along the path defined by $y = \ln \sec x$, -1.5 dm $\leq x \leq 1.5$ dm. Find the radius of curvature of this path for $x = 0.85$ dm.

25. Sketch the graph of $y = \ln \sec x$, marking that part which is the path of the roller mechanism of Exercise 24.

26. In an electronic device, the maximum current density i_m as a function of the temperature T is given by $i_m = AT^2e^{k/T}$, where A and k are constants. If the temperature is changing with time, find the expression for the time rate of change of i_m.

27. In developing the theory dealing with the friction between a pulley wheel and the pulley belt, the ratio of the tensions in the belt on either side of the wheel is given by $R_T = e^{k \csc (\theta/2)}$, where k is a constant and θ is the angle of the opening of the pulley wheel. Find the expression for a small change in the ratio of tensions for a small change in the angle θ.

28. The reliability R ($0 \leq R \leq 1$) of a certain computer system for t hours of operation is found from the equation $R = 3e^{-0.004t} - 2e^{-0.006t}$. Use Newton's method to find how long the system operates to have a reliability of 0.8 (80% probability that there will be no system failure).

29. An object on the end of a spring is moving so that its displacement (in cm) from the equilibrium position is given by $y = e^{-0.5t}(0.4 \cos 6t - 0.2 \sin 6t)$. Find the expression for the velocity of the object. What is the velocity when $t = 0.26$ s? The motion described by this equation is called *damped harmonic motion*.

30. A package of weather instruments is propelled into the air to an altitude of about 7 km. A parachute then opens, and the package returns to the surface. The altitude y of the package as a function of the time t, in minutes, is given by
$y = \dfrac{10t}{e^{0.4t} + 1}$. Find the vertical velocity of the package for $t = 8.0$ min.

31. The speed s of signaling by use of a certain communications cable is directly proportional to $x^2 \ln \frac{1}{x}$, where x is the ratio of the radius of the core of the cable to the thickness of the surrounding insulation. For what value of x is s a maximum?

32. A computer is programmed to inscribe a series of rectangles in the first quadrant under the curve of $y = e^{-x}$. What is the area of the largest rectangle which can be inscribed? See Fig. 26-28.

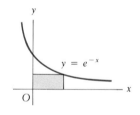

Fig. 26-28

26-8 Chapter Equations, Review Exercises, and Practice Test

Chapter Equations

Limit of $\dfrac{\sin \theta}{\theta}$ as $\theta \to 0$ $\qquad \lim_{\theta \to 0} \dfrac{\sin \theta}{\theta} = \lim_{\Delta u \to 0} \dfrac{\sin (\Delta u/2)}{\Delta u/2} = 1$ $\qquad\qquad$ (26-1)

Chain rule $\qquad\qquad\qquad \dfrac{dy}{dx} = \dfrac{dy}{du}\dfrac{du}{dx}$ $\qquad\qquad$ (26-3)

Derivatives $\qquad\qquad\qquad \dfrac{d(\sin u)}{dx} = \cos u \dfrac{du}{dx}$ $\qquad\qquad$ (26-4)

$$\frac{d(\cos u)}{dx} = -\sin u \frac{du}{dx} \tag{26-5}$$

$$\frac{d(\tan u)}{dx} = \sec^2 u \frac{du}{dx} \tag{26-6}$$

$$\frac{d(\cot u)}{dx} = -\csc^2 u \frac{du}{dx} \tag{26-7}$$

$$\frac{d(\sec u)}{dx} = \sec u \tan u \frac{du}{dx} \tag{26-8}$$

$$\frac{d(\csc u)}{dx} = -\csc u \cot u \frac{du}{dx} \tag{26-9}$$

$$\frac{d(\text{Arcsin } u)}{dx} = \frac{1}{\sqrt{1-u^2}} \frac{du}{dx} \tag{26-10}$$

$$\frac{d(\text{Arccos } u)}{dx} = -\frac{1}{\sqrt{1-u^2}} \frac{du}{dx} \tag{26-11}$$

$$\frac{d(\text{Arctan } u)}{dx} = \frac{1}{1+u^2} \frac{du}{dx} \tag{26-12}$$

$$\frac{d(\log_b u)}{dx} = \frac{1}{u} \log_b e \frac{du}{dx} \tag{26-13}$$

$$\frac{d(\ln u)}{dx} = \frac{1}{u} \frac{du}{dx} \tag{26-14}$$

$$\frac{d(b^u)}{dx} = b^u \ln b \frac{du}{dx} \tag{26-15}$$

$$\frac{d(e^u)}{dx} = e^u \frac{du}{dx} \tag{26-16}$$

Review Exercises

In Exercises 1 through 40, find the derivative of each of the given functions.

1. $y = 3 \cos(4x - 1)$ **2.** $y = 4 \sec(1 - x^3)$ **3.** $y = \tan \sqrt{3 - x}$ **4.** $y = 5 \sin(1 - 6x)$

5. $y = \csc^2(3x + 2)$ **6.** $y = \cot^2 5x$ **7.** $y = 3 \cos^4 x^2$ **8.** $y = 2 \sin^3 \sqrt{x}$

9. $y = (e^{x-3})^2$ **10.** $y = e^{\sin 2x}$ **11.** $y = 3 \ln(x^2 + 1)$ **12.** $y = \ln(3 + \sin x^2)$

13. $y = 3 \text{ Arctan}\left(\dfrac{x}{3}\right)$ **14.** $y = 4 \text{ Arccos}(2x + 3)$ **15.** $y = \ln \text{Arcsin } 4x$ **16.** $y = \sin(\text{Arctan } x)$

17. $y = \sqrt{\csc 4x + \cot 4x}$ **18.** $y = (1 + \sin 2x)^4$ **19.** $y = \ln(x - e^{-x})^2$ **20.** $y = \ln \sqrt{\sin 2x}$

21. $y = \dfrac{\cos^2 x}{e^{3x} + 1}$ **22.** $y = \dfrac{\ln \sqrt{3x + 1}}{3x + 1}$ **23.** $y = \dfrac{x^2}{\text{Arctan } 2x}$ **24.** $y = \dfrac{\text{Arcsin } x}{4x}$

25. $y = \ln(\csc x^2)$ **26.** $y = 2e^{\sqrt{1-x}}$ **27.** $y = \ln^2(3 + \sin x)$ **28.** $y = \ln(3 + \sin x)^2$

29. $y = e^{-2x} \sec x$ **30.** $y = e^{3x} \ln x$ **31.** $y = \sqrt{\sin 2x + e^{4x}}$ **32.** $x + y \ln 2x = y^2$

33. $\text{Arctan} \dfrac{y}{x} = x^2 e^y$ **34.** $x^2 \ln y = y + x$ **35.** $y = x^2(e^{\cos^2 x})^2$ **36.** $y = (\ln 4x - \tan 4x)^3$

37. $\ln xy + ye^{-x} = 1$ **38.** $y = x \, \text{Arcsin}^2 x + 2\sqrt{1 - x^2} \, \text{Arcsin} \, x - 2x$

39. $y = x \, \text{Arccos} \, x - \sqrt{1 - x^2}$ **40.** $y = \ln(4x^2 + 1) + \text{Arctan} \, 2x$

In Exercises 41 through 44, sketch the graphs of the given functions.

41. $y = x - \cos x$ **42.** $y = 4 \sin x + \cos 2x$ **43.** $y = x(\ln x)^2$ **44.** $y = \ln(1 + x)$

In Exercises 45 through 48, find the equations of the indicated tangent or normal lines.

45. Find the equation of the line tangent to the curve $y = 4 \cos^2(x^2)$ at $x = 1$.

46. Find the equation of the line tangent to the curve of $y = \ln \cos x$ at $x = \frac{\pi}{6}$.

47. Find the equation of the line normal to the curve of $y = e^{x^2}$ at $x = \frac{1}{2}$.

48. Find the equation of the line normal to the curve $y = \text{Arctan} \, x$ at $x = 1$.

In Exercises 49 through 80, solve the given problems.

49. Find the derivative of each of the members of the identity $\sin^2 x + \cos^2 x = 1$ and show that the results are equal.

50. Find the derivative of each of the members of the identity

$$\sin(x + 1) = \sin x \cos 1 + \cos x \sin 1$$

and show that the results are equal.

51. By Newton's method, find the negative root of the equation $e^x - x^2 = 0$.

52. By Newton's method, solve the equation $\ln x^2 = \text{Arctan} \, x$.

53. The vertical displacement y (in cm) of an object at the end of a spring is given by $y = 3.5 \sin(0.75\pi t + 0.50)$, where t is the time in seconds. Find the velocity of the object for $t = 1.50$ s.

54. An earth-orbiting satellite is launched such that its altitude y, in miles, is given by $y = 150(1 - e^{-0.05t})$, where t is the time in minutes. Find the vertical velocity of the satellite for $t = 10.0$ min.

55. Power can be defined as the time rate of doing work. If work is being done in an electric circuit according to $W = 10 \cos 2t$, find P as a function of t.

56. The value V of a bank account in which $1000 is deposited and then earns 6% annual interest, compounded continuously (daily compounding approximates this, and some banks actually use continuous compounding), is $V = 1000e^{0.06t}$ after t years. How fast is the account growing after exactly 2 years?

57. In the study of a certain bacterial culture, the time t required for a biological change to occur in x bacteria is found to be $t = a \ln \dfrac{x}{b - x} - c$, where a, b, and c are constants. Find dt/dx.

58. Under certain conditions, the potential V (in volts) due to a magnet is given by $V = -k \ln\left(1 + \dfrac{L}{x}\right)$, where L is the length of the magnet and x is the distance from the point where the potential is measured. Find the expression for dV/dx.

59. In the theory of making images by holography, an expression used for the light intensity distribution is $I = kE_0^2 \cos^2 \frac{1}{2}\theta$, where k and E_0 are constants and θ is the phase angle between two light waves. Find the expression for $dI/d\theta$.

60. Neglecting air resistance, the range R of a bullet fired at an angle θ with the horizontal is $R = \dfrac{v_0^2}{g} \sin 2\theta$, where v_0 is the initial velocity and g is the acceleration due to gravity. Find θ for the maximum range. See Fig. 26–29.

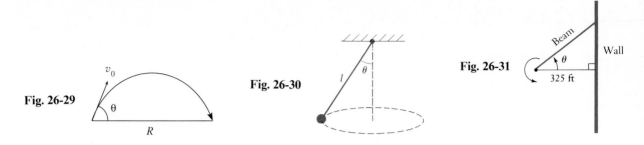

Fig. 26-29

Fig. 26-30

Fig. 26-31

61. In designing a cone-type clutch, an equation relating the cone angle θ and the applied force F is $\theta = \text{Arcsin} \dfrac{Ff}{R}$, where R is the frictional resistance and f is the coefficient of friction. For constant R and f, find $d\theta/dF$.

62. If inflation makes the dollar worth 5% less each year, then the value of $100 in t years will be $V = 100(0.95)^t$. What is the approximate change in the value during the fourth year?

63. An object attached to a cord of length l, as shown in Fig. 26–30, moves in a circular path. The angular velocity ω is given by $\omega = \sqrt{g/l} \cos \theta$. By use of differentials, find the approximate change in ω if θ changes from $32.50°$ to $32.75°$, given that $g = 9.800 \text{ m/s}^2$ and $l = 0.6375$ m.

64. An analysis of samples of air for a city showed that the number of parts per million p of sulfur dioxide on a certain day was $p = 0.05 \ln (2 + 24t - t^2)$, where t is the hour of the day. Using differentials, find the approximate change in the amount of sulfur dioxide between 10 A.M. and noon.

65. According to Newton's law of cooling (Isaac Newton, again), the rate at which a body cools is proportional to the difference in temperature between it and the surrounding medium. Using this law, the temperature T (in °F) of an engine coolant as a function of the time t (in min) is $T = 80 + 120(0.5)^{0.2t}$. The coolant was initially at $200°$F, and the air temperature was $80°$F. Find dT/dt for $t = 5.00$ min.

66. The charge q on a certain capacitor as a function of time is given by $q = e^{-0.1t}(0.2 \sin 100t + \cos 100t)$. The current i in the circuit is the time rate of change of charge. Find the expression for i as a function of t.

67. An object is dropped from a weather balloon. The distance (in feet) it falls, assuming a resisting force of the air on the object, is given by $y = 320(t + 10e^{-0.1t} - 10)$. Find the velocity after 10.0 s.

68. A revolving light 325 ft from a straight wall makes 6.00 r/min. Find the velocity of the light along the wall when it makes an angle of $42.0°$ with the wall. See Fig. 26-31.

69. An architect designs an arch of height y (in m) over a walkway by the curve of the equation $y = 3e^{-0.5x^2}$. What are the dimensions of the largest rectangular passage area under the arch?

70. A force P at an angle θ above the horizontal drags a 50-lb box across a level floor. The coefficient of friction between the floor and box is constant and equals 0.20. The magnitude of the force P is given by

$$P = \frac{(0.20)(50)}{0.20 \sin \theta + \cos \theta}.$$ Find θ such that P is a minimum.

71. A jet is flying at 880 ft/s directly away from the control tower of an airport. If the jet is at a constant altitude of 6800 ft, how fast is the angle of elevation of the jet from the control tower changing when it is $13.0°$?

72. The current i in an electric circuit with a resistance R and an inductance L is $i = i_0 e^{-Rt/L}$, where i_0 is the initial current. Show that the time rate of change of current is directly proportional to the current.

73. When a wheel rolls along a straight line, a point P on the circumference traces a curve called a *cycloid*. See Fig. 26-32. The equations of a cycloid are $x = r(\theta - \sin \theta)$, $y = r(1 - \cos \theta)$. Find the velocity of the point on the rim of a wheel for which $r = 5.500$ cm and $d\theta/dt = 0.12$ rad/s for $\theta = 35.0°$. [An inverted cycloid is the path of least time of descent (the *brachistochrone*) of an object acted on only by gravity.]

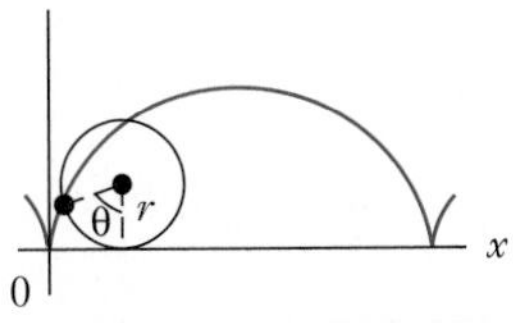

Fig. 26-32

74. In the study of atomic spectra, it is necessary to solve the equation $x = 5(1 - e^{-x})$ for x. Use Newton's method to find the solution.

75. The illuminance from a point source of light varies directly as the cosine of the angle of incidence (measured from the perpendicular) and inversely as the square of the distance r from the source. How high above the center of a circle of radius 10.0 in. should a light be placed so that the illuminance at the circumference will be a maximum? (See Fig. 26-33.)

76. A Y-shaped metal bracket is to be made such that its height is 10.0 cm and its width across the top is 6.00 cm. What shape will require the least amount of material? See Fig. 26-34.

77. A gutter is to be made from a sheet of metal 12 in. wide by turning up strips of width 4 in. along each side to make equal angles θ with the vertical. Sketch a graph of the cross-sectional area A as a function of θ. (See Fig. 26-35.)

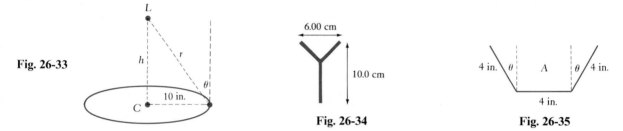

Fig. 26-33   **Fig. 26-34**   **Fig. 26-35**

78. The curve formed by a uniform cable hanging from two points under its own weight (e.g., a telephone wire between two poles) is called the *catenary* (see Fig. 26-36). The equation of the curve is

$$y = \frac{H}{w} \cosh \frac{wx}{H}$$

where w and H are constants. Show that the equation of the catenary satisfies the equation

$$\frac{d^2y}{dx^2} = \frac{w}{H} \sqrt{1 + \left(\frac{dy}{dx}\right)^2}$$

(See Exercise 45 of Section 26-6.)

79. A businessman determines that the gross income I he receives by selling x items per week is $I = 100xe^{-x/10}$. How many items should he sell per week to make I a maximum?

80. A conical filter is made from a circular piece of wire mesh of radius 24.0 cm by cutting out a sector with central angle θ and then taping the cut edges of the remaining piece together (see Fig. 26–37). What is the maximum possible volume which the resulting filter can hold?

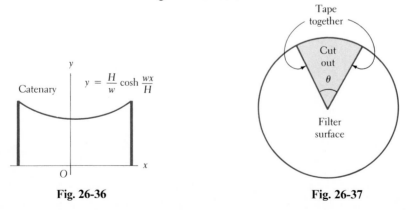

Fig. 26-36   **Fig. 26-37**

Practice Test

In Problems 1 through 3, find the derivative of each of the functions.

1. $y = \tan^3 2x + \text{Arctan } 2x$

2. $y = 2(3 + \cot 4x)^3$

3. $y \sec 2x = \text{Arcsin } 3y$

4. Find the differential of the function $y = \dfrac{\cos^2 (3x + 1)}{x}$.

5. Find the slope of a tangent to the curve of $y = \ln \dfrac{2x - 1}{1 + x^2}$ for $x = 2$.

6. Find the expression for the time rate of change of electric current which is given by the equation $i = 8e^{-t} \sin 10t$, where t is the time.

7. Sketch the graph of the function $y = xe^x$.

8. A balloon leaves the ground 250 ft from an observer and rises at the rate of 5.0 ft/s. How fast is the angle of elevation of the balloon increasing after 8.0 s?

27 Methods of Integration

Having developed the derivatives of the basic transcendental functions, we can now expand considerably the functions which we can integrate. In addition to the transcendental functions, many algebraic functions which we were unable to integrate previously will now lend themselves to integration.

In this chapter we shall expand the use of the general power formula for integration for use with integrands which have transcendental functions. We shall then develop additional standard forms for algebraic and transcendental integrands. Also, certain basic methods of integration, including the use of tables, will be shown.

In Section 27-5 an important application of integration related to the use of electricity is shown.

As we have seen in Chapters 24 and 25, there are numerous applications of integration in geometry, science, and technology. Additional examples of these applications are found in the examples and exercises of this chapter.

27-1 The General Power Formula

The first integration formula we shall discuss is the general power formula. We first used this with basic algebraic functions in Chapter 24.

For reference, we repeat here the general power formula for integration, Eq. (24-5):

integral of $u^n\,du$

$$\int u^n\,du = \frac{u^{n+1}}{n+1} + C \qquad (n \neq -1)$$

(27-1)

NOTE ▷ This formula will now be applied to transcendental functions as well as algebraic functions. When we are applying the general power formula, *we must properly recognize the quantities **u, n,** and **du.*** This requires familiarity with the differential forms presented in Chapters 22 and 26. The following examples illustrate the use of the general power formula.

EXAMPLE A ———— Integrate: $\int \sin^3 x \cos x \, dx$.

Since $d(\sin x) = \cos x \, dx$, we note that this integral fits the form of Eq. (27-1) for $u = \sin x$. Thus, with $u = \sin x$, we have $du = \cos x \, dx$, which means that this integral is of the form $\int u^3 \, du$. Therefore, the integration can now be completed.

$$\int \sin^3 x \cos x \, dx = \int \sin^3 x \overbrace{(\cos x \, dx)}^{du}$$

$$= \frac{1}{4} \sin^4 x + C \longleftarrow \begin{array}{l} \text{do not forget the} \\ \text{constant of integration} \end{array}$$

NOTE ▷ We note here that *the factor* **cos** *x is a necessary part of the du* in order to have the proper form of integration *and therefore does not appear in the final result.*

We also see that our result checks by finding the derivative of $\frac{1}{4} \sin^4 x + C$. This derivative is

$$\frac{d}{dx}\left(\frac{1}{4}\sin^4 x + C\right) = \frac{1}{4}(4)\sin^3 x \cos x = \sin^3 x \cos x \quad ■$$

EXAMPLE B ———— Integrate: $\int 2\sqrt{1 + \tan x} \, \sec^2 x \, dx$.

Here we note that $d(\tan x) = \sec^2 x \, dx$, which means that the integral fits the form of Eq. (27-1) with

$$u = 1 + \tan x, \qquad du = \sec^2 x \, dx, \qquad n = \tfrac{1}{2}$$

The integral is of the form $\int u^{1/2} \, du$. Thus,

$$\int 2\sqrt{1 + \tan x}(\sec^2 x \, dx) = 2 \int \underbrace{(1 + \tan x)^{1/2}}_{u}\overbrace{(\sec^2 x \, dx)}^{du}$$

$$= 2\left(\frac{2}{3}\right)(1 + \tan x)^{3/2} + C = \frac{4}{3}(1 + \tan x)^{3/2} + C \quad ■$$

EXAMPLE C ———— Integrate: $\int \ln x \left(\frac{dx}{x}\right)$.

By noting that $d(\ln x) = \dfrac{dx}{x}$, we have

$$u = \ln x, \qquad du = \frac{dx}{x}, \qquad n = 1$$

This means that the integral is of the form $\int u \, du$. Thus,

$$\int \underset{u}{\ln x}\left(\underset{du}{\frac{dx}{x}}\right) = \frac{1}{2}(\ln x)^2 + C = \frac{1}{2}\ln^2 x + C$$

EXAMPLE D _____ Integrate: $\int (2 + 3e^{2t})^3 e^{2t} \, dt$.

In this case

$$u = 2 + 3e^{2t}, \qquad du = 6e^{2t} \, dt, \qquad n = 3$$

This means that we must introduce a 6 to complete the differential in order that the integral may fit the proper form. Also, a $\frac{1}{6}$ must then be placed before the integral. This means that the integration is as follows:

$$\int (2 + 3e^{2t})^3 e^{2t} \, dt = \frac{1}{6} \int \underbrace{(2 + 3e^{2t})^3}_{u} (\overbrace{6e^{2t} \, dt}^{du})$$

$$= \frac{1}{6}\left(\frac{1}{4}\right)(2 + 3e^{2t})^4 + C$$

$$= \frac{1}{24}(2 + 3e^{2t})^4 + C$$

■

EXAMPLE E _____ Find the value of $\int_0^{0.5} \dfrac{\text{Arcsin } x}{\sqrt{1 - x^2}} \, dx$.

For purposes of integrating, we see that

$$u = \text{Arcsin } x, \qquad du = \frac{dx}{\sqrt{1 - x^2}}, \qquad n = 1$$

Therefore,

$$\int_0^{0.5} \frac{\text{Arcsin } x}{\sqrt{1 - x^2}} \, dx = \int_0^{0.5} \text{Arcsin } x \left(\frac{dx}{\sqrt{1 - x^2}}\right) \qquad \int u \, du$$

$$= \frac{(\text{Arcsin } x)^2}{2}\bigg|_0^{0.5} \qquad\qquad \text{integrate}$$

$$= \frac{\left(\frac{\pi}{6}\right)^2}{2} - 0 = \frac{\pi^2}{72} \qquad\qquad \text{evaluate}$$

■

EXAMPLE F _____ Find the first-quadrant area bounded by $y = \dfrac{e^{2x}}{\sqrt{e^{2x} + 1}}$ and $x = 1.5$.

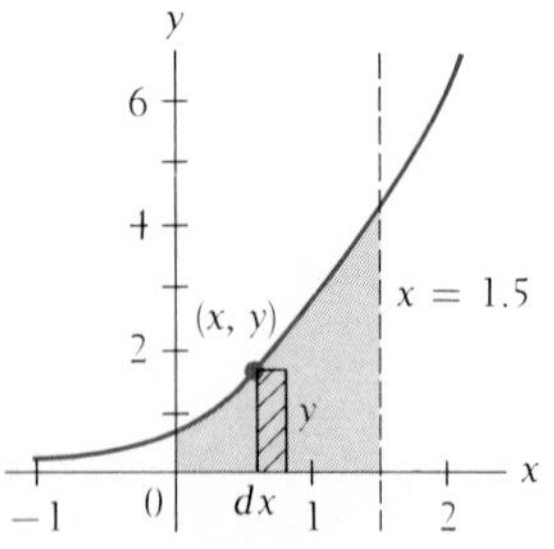

Fig. 27-1

The area to be found is shown in Fig. 27-1. The area of the representative element is $y \, dx$. Therefore, the area is found by evaluating the integral

$$\int_0^{1.5} \frac{e^{2x} \, dx}{\sqrt{e^{2x} + 1}}$$

For the purpose of integration, $n = -\frac{1}{2}$, $u = e^{2x} + 1$, and $du = 2e^{2x} \, dx$. Therefore,

$$\int_0^{1.5} (e^{2x} + 1)^{-1/2} e^{2x} \, dx = \frac{1}{2} \int_0^{1.5} (e^{2x} + 1)^{-1/2}(2e^{2x} \, dx)$$

$$= \frac{1}{2}(2)(e^{2x} + 1)^{1/2}\bigg|_0^{1.5}$$

$$= (e^{2x} + 1)^{1/2} \Big|_0^{1.5} = \sqrt{e^3 + 1} - \sqrt{2}$$

$$= 3.178$$

■ This means that the area is 3.178 square units.

Exercises 27-1

In Exercises 1 through 24, integrate each of the given functions.

1. $\int \sin^4 x \cos x \, dx$

2. $\int \cos^5 x(-\sin x \, dx)$

3. $\int \sqrt{\cos x} \sin x \, dx$

4. $\int \sin^{1/3} x \cos x \, dx$

5. $\int 4 \tan^2 x \sec^2 x \, dx$

6. $\int \sec^3 x(\sec x \tan x) \, dx$

7. $\int_0^{\pi/8} \cos 2x \sin 2x \, dx$

8. $\int_{\pi/6}^{\pi/4} 3\sqrt{\cot x} \csc^2 x \, dx$

9. $\int (\text{Arcsin } x)^3 \left(\dfrac{dx}{\sqrt{1 - x^2}} \right)$

10. $\int \dfrac{(\text{Arccos } 2x)^4 \, dx}{\sqrt{1 - 4x^2}}$

11. $\int \dfrac{\text{Arctan } 5x}{1 + 25x^2} \, dx$

12. $\int \dfrac{\text{Arcsin } 4x \, dx}{\sqrt{1 - 16x^2}}$

13. $\int [\ln (x + 1)]^2 \dfrac{dx}{x + 1}$

14. $\int (3 + \ln 2x)^3 \dfrac{dx}{x}$

15. $\int_0^{1/2} \dfrac{\ln (2x + 3)}{2x + 3} \, dx$

16. $\int_1^e \dfrac{(1 - 2 \ln x) \, dx}{x}$

17. $\int (4 + e^x)^3 e^x \, dx$

18. $\int 2\sqrt{1 - e^{-x}}(e^{-x} dx)$

19. $\int (2e^{2x} - 1)^{1/3} e^{2x} \, dx$

20. $\int \dfrac{(1 + 3e^{-2x})^4 \, dx}{e^{2x}}$

21. $\int (1 + \sec^2 x)^4 (\sec^2 x \tan x \, dx)$

22. $\int (e^x + e^{-x})^{1/4} (e^x - e^{-x}) \, dx$

23. $\int_0^{\pi/6} \dfrac{\tan x}{\cos^2 x} \, dx$

24. $\int_{\pi/3}^{\pi/2} \dfrac{\sin \theta \, d\theta}{\sqrt{1 + \cos \theta}}$

In Exercises 25 through 32, solve the given problems by integration.

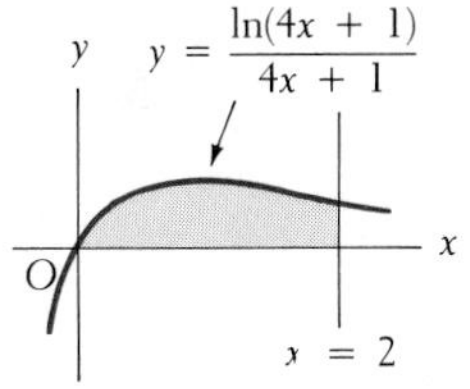

25. Find the area under the curve $y = \dfrac{1 + \text{Arctan } 2x}{1 + 4x^2}$ from $x = 0$ to $x = 2$.

26. Find the first-quadrant area bounded by $y = \dfrac{\ln (4x + 1)}{4x + 1}$ and $x = 2$. See Fig. 27-2.

27. The general expression for the slope of a given curve is $(\ln x)^2/x$. If the curve passes through $(1, 2)$, find its equation.

28. Find the equation of the curve for which $dy/dx = (1 + \tan 2x)^2 \sec^2 2x$ if the curve passes through $(2, 1)$.

29. In developing the expression for the total pressure P on a wall due to molecules with mass m and velocity v striking the wall, the following equation is found: $P = mnv^2 \int_0^{\pi/2} \sin \theta \cos^2 \theta \, d\theta$. The symbol n represents the number of molecules per unit volume and θ represents the angle between a perpendicular to the wall and the direction of the molecule. Find the final expression for P.

30. The solar energy E passing through a hemispherical surface per unit time, per unit area, is given by $E = 2\pi I \int_0^{\pi/2} \cos \theta \sin \theta \, d\theta$, where I is the solar intensity and θ is the angle at which it is directed (from the perpendicular). Evaluate this integral.

31. Following an electric power interruption, the current i in a circuit is given by $i = 3(1 - e^{-t})^2(e^{-t})$, where t is the time. Find the expression for the total electric charge q to pass a point in the circuit if $q = 0$ for $t = 0$.

32. A space vehicle is launched vertically from the ground such that its velocity v (in km/s) is given by $v = [\ln^2 (t^3 + 1)] \dfrac{t^2}{t^3 + 1}$, where t is the time in seconds. Find the altitude of the vehicle after 10.0 s.

$y = \dfrac{\ln(4x + 1)}{4x + 1}$

Fig. 27-2

27-2   The Basic Logarithmic Form

The general power formula for integration, Eq. (27-1), is valid for all values of n except $n = -1$. If n were set equal to -1, this would cause the result to be undefined. When we obtained the derivative of the logarithmic function, we found that

$$\frac{d(\ln u)}{dx} = \frac{1}{u}\frac{du}{dx}$$

which means the differential of the logarithmic form is $d(\ln u) = du/u$. Reversing the process, we then determine that $\int du/u = \ln u + C$. In other words, when the exponent of the expression being integrated is -1, it is a logarithmic form.

Logarithms are defined only for positive numbers. Thus, $\int du/u = \ln u + C$ is valid if $u > 0$. If $u < 0$, then $-u > 0$. In this case $d(-u) = -du$, or $\int (-du)/(-u) = \ln(-u) + C$. However, $\int du/u = \int (-du)/(-u)$. These results can be combined into the single form

integral of $\dfrac{du}{u}$

$$\boxed{\int \frac{du}{u} = \ln|u| + C} \tag{27-2}$$

EXAMPLE A

Integrate: $\displaystyle\int \frac{dx}{x+1}$.

Since $d(x + 1) = dx$, this integral fits the form of Eq. (27-2) with $u = x + 1$ and $du = dx$. Therefore, we have

$$\int \frac{dx}{x+1} = \ln|x+1| + C$$

EXAMPLE B

Newton's law of cooling states that the rate at which an object cools is directly proportional to the difference in its temperature T and the temperature of the surrounding medium. Using this, the time t, in minutes, a certain object takes to cool from 80°C to 50°C in air at 20°C is found to be

$$t = -9.8 \int_{80}^{50} \frac{dT}{T - 20}$$

Find the value of t.

We see that the integral fits Eq. (27-2) with $u = T - 20$ and $du = dT$. Thus,

$$t = -9.8 \int_{80}^{50} \frac{dT}{T - 20} \quad \longleftarrow du \atop \longleftarrow u$$

$$= -9.8 \ln\left|T - 20\right|\Big|_{80}^{50} \qquad \text{integrate}$$

$$= -9.8(\ln 30 - \ln 60) \qquad \text{evaluate}$$

$$= -9.8 \ln \frac{30}{60} = -9.8 \ln (0.50) \qquad \ln x - \ln y = \ln \frac{x}{y}$$

$$= 6.8 \text{ min}$$

EXAMPLE C

Integrate: $\int \dfrac{\cos x}{\sin x} \, dx$.

We note that $d(\sin x) = \cos x \, dx$. This means that this integral fits the form of Eq. (27-2) with $u = \sin x$ and $du = \cos x \, dx$. Thus,

$$\int \frac{\cos x}{\sin x} \, dx = \int \frac{\cos x \, dx}{\sin x} \quad \longleftarrow \quad du$$
$$\qquad\qquad\qquad\qquad\qquad \longleftarrow \quad u$$
$$= \ln |\sin x| + C$$

EXAMPLE D

Integrate: $\int \dfrac{x \, dx}{4 - x^2}$.

This integral fits the form of Eq. (27-2) with $u = 4 - x^2$ and $du = -2x \, dx$. This means that we must introduce a factor of -2 into the numerator and a factor of $-\frac{1}{2}$ before the integral. Therefore,

$$\int \frac{x \, dx}{4 - x^2} = -\frac{1}{2} \int \frac{-2x \, dx}{4 - x^2} \qquad \begin{matrix} du \\ u \end{matrix}$$

$$= -\frac{1}{2} \ln |4 - x^2| + C$$

NOTE ▷

We should note that if the quantity $4 - x^2$ were raised to any power other than that in the example, we would have to employ the general power formula for integration. For example,

$$\int \frac{x \, dx}{(4 - x^2)^2} = -\frac{1}{2} \int \frac{-2x \, dx}{(4 - x^2)^2} \quad \longleftarrow \quad du$$
$$\qquad\qquad\qquad\qquad\qquad\qquad\qquad u^2$$

$$= -\frac{1}{2} \frac{(4 - x^2)^{-1}}{-1} + C = \frac{1}{2(4 - x^2)} + C$$

EXAMPLE E

Integrate: $\int \dfrac{e^{4x} \, dx}{1 + 3e^{4x}}$; $u = 1 + 3e^{4x}$, $du = 12e^{4x} \, dx$.

We write

$$\int \frac{e^{4x} \, dx}{1 + 3e^{4x}} = \frac{1}{12} \int \frac{12e^{4x} \, dx}{1 + 3e^{4x}} \qquad \text{introduce factors of 12}$$

$$= \frac{1}{12} \ln |1 + 3e^{4x}| + C \qquad \text{integrate}$$

$$= \frac{1}{12} \ln (1 + 3e^{4x}) + C \qquad 1 + 3e^{4x} > 0 \text{ for all } x$$

EXAMPLE F

Evaluate: $\int_0^{\pi/8} \dfrac{\sec^2 2x}{1 + \tan 2x}\, dx$; $u = 1 + \tan 2x$, $du = 2 \sec^2 2x\, dx$.

We write

$$\int_0^{\pi/8} \frac{\sec^2 2x}{1 + \tan 2x}\, dx = \frac{1}{2} \int_0^{\pi/8} \frac{2 \sec^2 2x\, dx}{1 + \tan 2x} \qquad \text{introduce factors of 2}$$

$$= \frac{1}{2} \ln \left|1 + \tan 2x\right|\Big|_0^{\pi/8} \qquad \text{integrate}$$

$$= \frac{1}{2} \ln |2| - \frac{1}{2} \ln |1| \qquad \text{evaluate}$$

$$= \frac{1}{2} \ln 2 - \frac{1}{2} \ln 1 = \frac{1}{2} \ln 2 - 0 = \frac{1}{2} \ln 2$$

EXAMPLE G

Find the volume generated by rotating the area bounded by $y = \dfrac{3}{\sqrt{4x + 3}}$, $x = 2.5$, and the axes about the x-axis.

Figure 27-3 shows the volume to be found. As shown, the element of volume is a disk. The volume is found as follows:

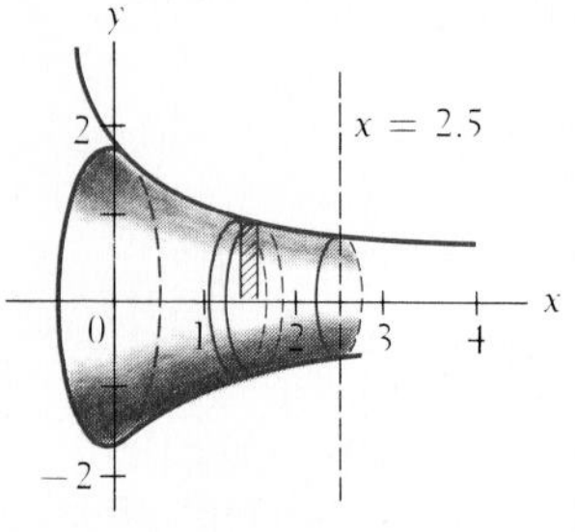

$$V = \pi \int_0^{2.5} y^2\, dx = \pi \int_0^{2.5} \left(\frac{3}{\sqrt{4x + 3}}\right)^2 dx$$

$$= \pi \int_0^{2.5} \frac{9\, dx}{4x + 3} = \frac{9\pi}{4} \int_0^{2.5} \frac{4\, dx}{4x + 3}$$

$$= \frac{9\pi}{4} \ln (4x + 3)\Big|_0^{2.5} = \frac{9\pi}{4} (\ln 13 - \ln 3)$$

$$= \frac{9\pi}{4} \ln \frac{13}{3} = 10.36$$

Fig. 27-3

Exercises 27-2

In Exercises 1 through 28, integrate each of the given functions.

1. $\displaystyle\int \frac{dx}{1 + 4x}$

2. $\displaystyle\int \frac{dx}{1 - 4x}$

3. $\displaystyle\int \frac{2x\, dx}{4 - 3x^2}$

4. $\displaystyle\int \frac{x^2\, dx}{1 - x^3}$

5. $\displaystyle\int_0^2 \frac{dx}{8 - 3x}$

6. $\displaystyle\int_{-1}^3 \frac{2x^3\, dx}{x^4 + 1}$

7. $\displaystyle\int \frac{\csc^2 2x}{\cot 2x}\, dx$

8. $\displaystyle\int \frac{\sin x}{\cos x}\, dx$

9. $\displaystyle\int_0^{\pi/2} \frac{\cos x\, dx}{1 + \sin x}$

10. $\displaystyle\int_0^{\pi/4} \frac{\sec^2 x\, dx}{4 + \tan x}$

11. $\displaystyle\int \frac{e^{-x}}{1 - e^{-x}}\, dx$

12. $\displaystyle\int \frac{5e^{3x}}{1 - e^{3x}}\, dx$

13. $\displaystyle\int \frac{1 + e^x}{x + e^x}\, dx$

14. $\displaystyle\int \frac{e^x - e^{-x}}{e^x + e^{-x}}\, dx$

15. $\displaystyle\int \frac{\sec x \tan x\, dx}{1 + 4 \sec x}$

16. $\displaystyle\int \frac{\sin 2x}{1 - \cos^2 x}\, dx$

17. $\displaystyle\int_1^3 \frac{1 + x}{4x + 2x^2}\, dx$

18. $\displaystyle\int_1^2 \frac{4x + 6x^2}{x^2 + x^3}\, dx$

19. $\displaystyle\int \frac{dx}{x \ln x}$

20. $\displaystyle\int \frac{dx}{x(1 + 2 \ln x)}$

21. $\displaystyle\int \frac{2 + \sec^2 x}{2x + \tan x}\, dx$

22. $\displaystyle\int \frac{x + \cos 2x}{x^2 + \sin 2x}\, dx$

23. $\displaystyle\int \frac{dx}{\sqrt{1 - 2x}}$

24. $\displaystyle\int \frac{4x\, dx}{(1 + x^2)^2}$

25. $\displaystyle\int \frac{x + 2}{x^2}\, dx$

26. $\displaystyle\int \frac{2 - 3x^2}{x^3}\, dx$

27. $\displaystyle\int_0^{\pi/12} \frac{\sec^2 3x}{4 + \tan 3x}\, dx$

28. $\displaystyle\int_1^2 \frac{x^2 + 1}{x^3 + 3x}\, dx$

In Exercises 29 through 40, solve the given problems by integration.

29. Find the area bounded by $y(x + 1) = 1$, $x = 0$, $y = 0$, and $x = 2$. See Fig. 27-4.

30. Find the area bounded by $xy = 1$, $x = 1$, $x = 2$, and $y = 0$.

31. Find the volume generated by rotating the area bounded by $y = 1/(x^2 + 1)$, $x = 0$, $x = 1$, $y = 0$ about the y-axis. (Use shells.)

32. Find the volume of the solid generated by rotating the area bounded by $y = 2/\sqrt{3x + 1}$, $x = 0$, $x = 3.5$, and $y = 0$ about the x-axis.

33. The general expression for the slope of a curve is $\sin x/(3 + \cos x)$. If the curve passes through $(\frac{\pi}{3}, 2)$, find its equation.

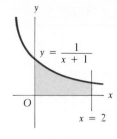

Fig. 27-4

34. Find the average value of the function $xy = 4$ from $x = 1$ to $x = 2$.

35. Under ideal conditions, the natural law of population growth is that it increases at a rate proportional to the population P present at any time t. This leads to the equation $t = \frac{1}{k} \int \frac{dP}{P}$. If $P = P_0$ for $t = 0$, integrate and then solve for P as a function of t.

36. In determining the temperature which is absolute zero (0 K, or about $-273°C$), the equation $\ln T = -\int \frac{dr}{r - 1}$ is used. Here T is the thermodynamic temperature and r is the ratio between certain specific vapor pressures. If $T = 273.16$ K for $r = 1.3361$, find T as a function of r (if $r > 1$ for all T).

37. The time t and electric current i for a circuit with a voltage E, a resistance R, and an inductance L is given by $t = L \int \frac{di}{E - iR}$. If $t = 0$ for $i = 0$, integrate and express i as a function of t.

38. Conditions are often such that a force proportional to the velocity tends to retard the motion of an object. Under such conditions, the acceleration of a certain object moving down an inclined plane is given by $20 - v$. This leads to the equation $t = \int \frac{dv}{20 - v}$. If the object starts from rest, find the expression for the velocity as a function of time.

39. An architect designs a wall panel which can be described as the first-quadrant area bounded by $y = \frac{50}{x^2 + 20}$ and $x = 3.00$. If the area of the panel is 6.61 m^2, find the x-coordinate (in m) of the centroid of the panel.

40. The electric power p developed in a certain resistor is given by $p = 3 \int \frac{\sin \pi t}{2 + \cos \pi t} \, dt$, where t is the time. Express p as a function of t.

27-3 The Exponential Form

In deriving the derivative for the exponential function we obtained the result $de^u/dx = e^u(du/dx)$. This means that the differential of the exponential form is $d(e^u) = e^u \, du$. Reversing this form to find the proper form of the integral for the exponential function, we have

integral of $e^u \, du$

$$\int e^u \, du = e^u + C$$

(27-3)

EXAMPLE A

Integrate: $\int e^{5x}\, dx$.

Since $d(5x) = 5\, dx$, this integral can be put in the form of Eq. (27-3) with $u = 5x$ and $du = 5\, dx$. This means that we place a 5 with the dx and a $\frac{1}{5}$ before the integral. Therefore,

$$\int e^{5x}\, dx = \frac{1}{5} \int \overset{u}{e^{5x}} \underset{du}{(5\, dx)}$$

$$= \frac{1}{5} e^{5x} + C$$

■

EXAMPLE B

Integrate: $\int x e^{x^2}\, dx$.

Since $d(x^2) = 2x\, dx$, we can write this integral in the form of Eq. (27-3) with $u = x^2$ and $du = 2x\, dx$. Thus,

$$\int x e^{x^2}\, dx = \frac{1}{2} \int \overset{u}{e^{x^2}} \underset{du}{(2x\, dx)}$$

$$= \frac{1}{2} e^{x^2} + C$$

■

EXAMPLE C

For an electric circuit containing a voltage source E, a resistance R, and an inductance L, an equation relating the current i and time t is $ie^{Rt/L} = \dfrac{E}{L} \int e^{Rt/L}\, dt$. See Fig. 27-5. If $i = 0$ for $t = 0$, perform the integration and then solve for i as a function of t.

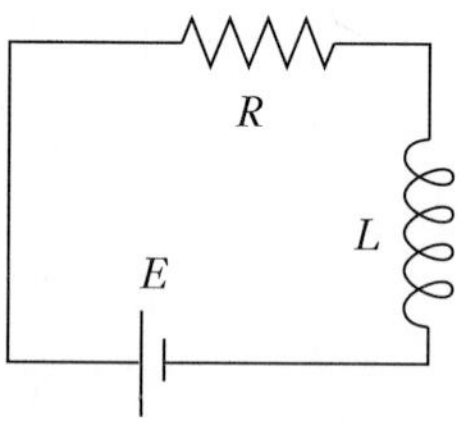

For this integral we see that $u = \dfrac{Rt}{L}$, which means that $du = \dfrac{R\, dt}{L}$. The solution is then as follows:

$$ie^{Rt/L} = \frac{E}{L} \int e^{Rt/L}\, dt = \frac{E}{L}\left(\frac{L}{R}\right) \int e^{Rt/L}\left(\frac{R\, dt}{L}\right) \qquad \text{introduce factor } \frac{R}{L}$$

$$= \frac{E}{R} e^{Rt/L} + C \qquad \text{integrate}$$

$$0(e^0) = \frac{E}{R} e^0 + C, \qquad C = -\frac{E}{R} \qquad i = 0 \text{ for } t = 0; \text{ evaluate } C$$

$$ie^{Rt/L} = \frac{E}{R} e^{Rt/L} - \frac{E}{R} \qquad \text{substitute for } C$$

$$i = \frac{E}{R} - \frac{E}{R} e^{-Rt/L} = \frac{E}{R}\left(1 - e^{-Rt/L}\right) \qquad \text{solve for } i$$

■

Fig. 27-5

EXAMPLE D

Integrate: $\int \dfrac{dx}{e^{3x}}$.

This integral can be put in proper form by writing it as $\int e^{-3x}\, dx$. In this form $u = -3x$, $du = -3\, dx$. Thus,

$$\int \frac{dx}{e^{3x}} = \int e^{-3x}\, dx = -\frac{1}{3} \int e^{-3x}(-3\, dx) = -\frac{1}{3} e^{-3x} + C$$

■

EXAMPLE E

For reference,
Eq. (10-2) is

$$\frac{a^m}{a^n} = a^{m-n}$$

Integrate: $\int \dfrac{4e^{3x} - 3e^x}{e^{x+1}}\, dx.$

This can be put in the proper form for integration, and then integrated, as follows:

$$\int \frac{4e^{3x} - 3e^x}{e^{x+1}}\, dx = \int \frac{4e^{3x}}{e^{x+1}}\, dx - \int \frac{3e^x}{e^{x+1}}\, dx$$

$$= 4 \int e^{3x-(x+1)}\, dx - 3 \int e^{x-(x+1)}\, dx \qquad \text{using Eq. (10-2)}$$

$$= 4 \int e^{2x-1}\, dx - 3 \int e^{-1}\, dx$$

$$= \frac{4}{2} \int e^{2x-1}(2\, dx) - \frac{3}{e} \int dx$$

$$= 2e^{2x-1} - \frac{3}{e} x + C$$

■

EXAMPLE F

Evaluate: $\int_0^{\pi/2} (\sin 2x)(e^{\cos 2x})\, dx.$

With $u = \cos 2x$, $du = -2 \sin 2x\, dx$, we have

$$\int_0^{\pi/2} (\sin 2x)(e^{\cos 2x})\, dx = -\frac{1}{2} \int_0^{\pi/2} (e^{\cos 2x})(-2 \sin 2x\, dx)$$

$$= -\frac{1}{2} e^{\cos 2x}\Big|_0^{\pi/2} \qquad \text{integrate}$$

$$= -\frac{1}{2}\left(\frac{1}{e} - e\right) = 1.175 \qquad \text{evaluate}$$

■

EXAMPLE G

Find the equation of the curve for which $\dfrac{dy}{dx} = \dfrac{e^{\sqrt{x+1}}}{\sqrt{x+1}}$ if the curve passes through $(0, 1)$.

The solution of this problem requires that we integrate the given function and then evaluate the constant of integration. Hence,

$$dy = \frac{e^{\sqrt{x+1}}}{\sqrt{x+1}}\, dx, \qquad \int dy = \int \frac{e^{\sqrt{x+1}}}{\sqrt{x+1}}\, dx$$

For purposes of integrating the right-hand side,

$$u = \sqrt{x+1} \quad \text{and} \quad du = \frac{1}{2\sqrt{x+1}}\, dx$$

$$y = 2 \int e^{\sqrt{x+1}} \left(\frac{1}{2\sqrt{x+1}}\, dx\right) = 2e^{\sqrt{x+1}} + C$$

Letting $x = 0$ and $y = 1$, we have $1 = 2e + C$, or $C = 1 - 2e$. This means that the equation is

$$y = 2e^{\sqrt{x+1}} + 1 - 2e$$

See Fig. 27-6.

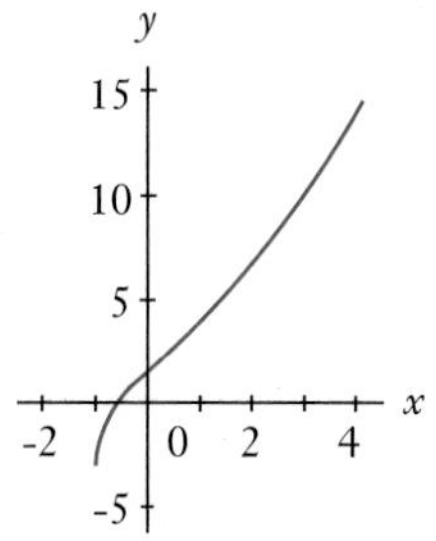

Fig. 27-6

Exercises 27-3

In Exercises 1 through 24, integrate each of the given functions.

1. $\int e^{7x}(7\,dx)$

2. $\int e^{x^4}(4x^3\,dx)$

3. $\int e^{2x+5}\,dx$

4. $\int 2e^{-4x}\,dx$

5. $\int_0^2 e^{x/2}\,dx$

6. $\int_1^2 3e^{4x}\,dx$

7. $\int 6x^2 e^{x^3}\,dx$

8. $\int xe^{-x^2}\,dx$

9. $\int_1^4 \dfrac{e^{\sqrt{x}}}{\sqrt{x}}\,dx$

10. $\int_0^1 4x^3 e^{2x^4}\,dx$

11. $\int (\sec x \tan x)e^{2\sec x}\,dx$

12. $\int (\sec^2 x)e^{\tan x}\,dx$

13. $\int \dfrac{(3 - e^x)\,dx}{e^{2x}}$

14. $\int (e^x - e^{-x})^2\,dx$

15. $\int_1^3 3e^{2x}(e^{-2x} - 1)\,dx$

16. $\int_0^{0.5} \dfrac{3e^{3x+1}}{e^x}\,dx$

17. $\int \dfrac{2\,dx}{\sqrt{x}e^{\sqrt{x}}}$

18. $\int \dfrac{4\,dx}{\sec x\,e^{\sin x}}$

19. $\int \dfrac{e^{\text{Arctan } x}}{x^2 + 1}\,dx$

20. $\int \dfrac{e^{\text{Arcsin } 2x}\,dx}{\sqrt{1 - 4x^2}}$

21. $\int \dfrac{e^{\cos 3x}\,dx}{\csc 3x}$

22. $\int \dfrac{e^{2/x}\,dx}{x^2}$

23. $\int_0^{\pi} (\sin 2x)e^{\cos^2 x}\,dx$

24. $\int_{-1}^1 \dfrac{dx}{e^{2-3x}}$

In Exercises 25 through 36, solve the given problems by integration.

25. Find the area bounded by $y = e^x$, $x = 0$, $y = 0$, and $x = 2$.

26. Prove that the area bounded by $x = a$, $x = b$, $y = 0$, and $y = e^x$ equals the difference of the ordinates at $x = b$ and $x = a$.

27. Find the volume generated by rotating the area bounded by $y = e^{x^2}$, $x = 1$, $y = 0$, and $x = 2$ about the y-axis. See Fig. 27-7.

28. Find the equation of the curve for which $dy/dx = \sqrt{e^{x+3}}$ if the curve passes through $(1, 0)$.

29. Find the average value of the function $y = e^{2x}$ from $x = 0$ to $x = 4$.

30. Find the moment of inertia with respect to the y-axis of the first-quadrant area bounded by $y = e^{x^3}$, $x = 1$, and the axes.

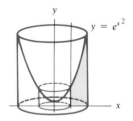

Fig. 27-7

31. Using Eq. (26-15), show that $\int b^u\,du = \dfrac{b^u}{\ln b} + C\ (b > 0, b \neq 1)$.

32. Find the first-quadrant area bounded by $y = 2^x$ and $x = 3$. See Exercise 31.

33. For an electric circuit containing a voltage source E, a resistance R, and a capacitance C, an equation relating the charge q on the capacitor and the time t is $qe^{t/RC} = \dfrac{E}{R}\int e^{t/RC}\,dt$. See Fig. 27-8. If $q = 0$ for $t = 0$, perform the integration and then solve for q as a function of t.

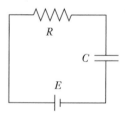

Fig. 27-8

34. In the theory dealing with energy propagation of lasers, the equation $E = a\int_0^{I_0} e^{-Tx}\,dx$ is used. Here a, I_0, and T are constants. Evaluate this integral.

35. An object at the end of a spring is immersed in liquid. Its velocity (in cm/s) is then described by the equation $v = 2e^{-2t} + 3e^{-5t}$, where t is the time in seconds. Such motion is called *overdamped*. Find the displacement s as a function of t if $s = -1.6$ cm for $t = 0$.

36. The force F (in lb) exerted by a robot programmed to staple carton sections together is given by $F = 6\int e^{\sin \pi t} \cos \pi t\,dt$, where t is the time in seconds. Find F as a function of t if $F = 0$ for $t = 1.5$ s.

27-4   Basic Trigonometric Forms

In this section we shall discuss the integrals of the six trigonometric functions and the trigonometric integrals which arise directly from reversing the formulas for differentiation. Other trigonometric forms will be discussed later.

By directly reversing the differentiation formulas, the following integral formulas are obtained:

basic trigonometric
integrals

$$\int \sin u \, du = -\cos u + C \tag{27-4}$$

$$\int \cos u \, du = \sin u + C \tag{27-5}$$

$$\int \sec^2 u \, du = \tan u + C \tag{27-6}$$

$$\int \csc^2 u \, du = -\cot u + C \tag{27-7}$$

$$\int \sec u \tan u \, du = \sec u + C \tag{27-8}$$

$$\int \csc u \cot u \, du = -\csc u + C \tag{27-9}$$

EXAMPLE A

Integrate: $\int x \sec^2 x^2 \, dx$.
   With $u = x^2$, $du = 2x \, dx$, we have

$$\int x \sec^2 x^2 \, dx = \frac{1}{2} \int (\sec^2 x^2)(2x \, dx)$$

$$= \frac{1}{2} \tan x^2 + C \qquad \text{using Eq. (27-6)}$$

EXAMPLE B

Integrate: $\int \dfrac{\tan 2x}{\cos 2x} \, dx$.
   By using the basic identity $\sec \theta = 1/\cos \theta$, we can transform this integral to form $\int \sec 2x \tan 2x \, dx$. In this form $u = 2x$, $du = 2 \, dx$. Therefore,

$$\int \frac{\tan 2x}{\cos 2x} \, dx = \int \sec 2x \tan 2x \, dx = \frac{1}{2} \int \sec 2x \tan 2x (2 \, dx)$$

$$= \frac{1}{2} \sec 2x + C \qquad \text{using Eq. (27-8)}$$

EXAMPLE C

The vertical velocity v (in cm/s) of the end of a vibrating rod is given by $v = 80 \cos 20\pi t$, where t is the time in seconds. Find the vertical displacement y (in cm) as a function of t if $y = 0$ for $t = 0$.
   Since $v = dy/dt$, we have the following solution:

$$\int dy = \int 80 \cos 20\pi t \, dt \qquad \text{set up integration}$$

$$= \frac{80}{20\pi} \int (\cos 20\pi t)(20\pi \, dt)$$

$$y = \frac{4}{\pi} \sin 20\pi t + C \qquad \text{using Eq. (27-5)}$$

$$0 = 1.27 \sin 0 + C, \qquad C = 0 \quad \text{evaluate } C$$

$$y = 1.27 \sin 20\pi t \qquad \text{solution}$$

To find the integrals for the other trigonometric functions, we must change them to a form for which the integral can be determined by methods previously discussed. We can accomplish this by use of the basic trigonometric relations.

The formula for $\int \tan u\,du$ is found by expressing the integral in the form $\int (\sin u/\cos u)\,du$. We recognize this as being a logarithmic form, where the u of the logarithmic form is $\cos u$ in this integral. The differential of $\cos u$ is $-\sin u\,du$. Therefore, we have

$$\int \tan u\,du = \int \frac{\sin u}{\cos u}\,du = -\int \frac{-\sin u\,du}{\cos u} = -\ln|\cos u| + C$$

The formula for $\int \cot u\,du$ is found by writing it in the form $\int (\cos u/\sin u)\,du$. In this manner we obtain the following result:

$$\int \cot u\,du = \int \frac{\cos u}{\sin u}\,du = \int \frac{\cos u\,du}{\sin u} = \ln|\sin u| + C$$

The formula for $\int \sec u\,du$ is found by writing it in the form

$$\int \frac{\sec u(\sec u + \tan u)}{\sec u + \tan u}\,du$$

We see that this form is also a logarithmic form, since

$$d(\sec u + \tan u) = (\sec u \tan u + \sec^2 u)\,du$$

The right side of this equation is the expression appearing in the numerator of the integral. Thus,

$$\int \sec u\,du = \int \frac{\sec u(\sec u + \tan u)\,du}{\sec u + \tan u} = \int \frac{\sec u \tan u + \sec^2 u}{\sec u + \tan u}\,du$$
$$= \ln|\sec u + \tan u| + C$$

To obtain the formula for $\int \csc u\,du$, we write it in the form

$$\int \frac{\csc u(\csc u - \cot u)\,du}{\csc u - \cot u}$$

Thus, we have

$$\int \csc u\,du = \int \frac{\csc u(\csc u - \cot u)}{\csc u - \cot u}\,du$$
$$= \int \frac{(-\csc u \cot u + \csc^2 u)\,du}{\csc u - \cot u}$$
$$= \ln|\csc u - \cot u| + C$$

Summarizing these results, we have the following integrals:

basic trigonometric
integrals

$$\int \tan u \, du = -\ln |\cos u| + C \qquad (27\text{-}10)$$

$$\int \cot u \, du = \ln |\sin u| + C \qquad (27\text{-}11)$$

$$\int \sec u \, du = \ln |\sec u + \tan u| + C \qquad (27\text{-}12)$$

$$\int \csc u \, du = \ln |\csc u - \cot u| + C \qquad (27\text{-}13)$$

EXAMPLE D

Integrate: $\int \tan 4x \, dx$.
Noting that $u = 4x$, $du = 4 \, dx$, we have

$$\int \tan 4x \, dx = \frac{1}{4} \int \tan 4x (4 \, dx) \qquad \text{introducing factors of 4}$$

$$= -\frac{1}{4} \ln |\cos 4x| + C \qquad \text{using Eq. (27-10)}$$

EXAMPLE E

Integrate: $\int \dfrac{\sec e^{-x} \, dx}{e^x}$.
In this integral, $u = e^{-x}$, $du = -e^{-x} \, dx$. Therefore,

$$\int \frac{\sec e^{-x} \, dx}{e^x} = -\int (\sec e^{-x})(-e^{-x} \, dx) \qquad \text{introducing} - \text{sign}$$

$$= -\ln |\sec e^{-x} + \tan e^{-x}| + C \qquad \text{using Eq. (27-12)}$$

EXAMPLE F

Evaluate: $\int_{\pi/6}^{\pi/4} \dfrac{1 + \cos x}{\sin x} \, dx$.
The solution is as follows:

$$\int_{\pi/6}^{\pi/4} \frac{1 + \cos x}{\sin x} \, dx = \int_{\pi/6}^{\pi/4} \csc x \, dx + \int_{\pi/6}^{\pi/4} \cot x \, dx \qquad \text{using Eqs. (19-1) and (19-5)}$$

$$= \ln |\csc x - \cot x| \Big|_{\pi/6}^{\pi/4} + \ln |\sin x| \Big|_{\pi/6}^{\pi/4} \qquad \text{integrating}$$

$$= \ln |\sqrt{2} - 1| - \ln |2 - \sqrt{3}| + \ln \left|\frac{1}{2}\sqrt{2}\right| - \ln \left|\frac{1}{2}\right| \qquad \text{evaluating}$$

$$= \ln \frac{(\frac{1}{2}\sqrt{2})(\sqrt{2} - 1)}{(\frac{1}{2})(2 - \sqrt{3})} = \ln \frac{2 - \sqrt{2}}{2 - \sqrt{3}}$$

$$= 0.782$$

Exercises 27-4

In Exercises 1 through 24, integrate each of the given functions.

1. $\int \cos 2x \, dx$

2. $\int 4 \sin (2 - x) \, dx$

3. $\int \sec^2 3x \, dx$

4. $\int \csc 2x \cot 2x \, dx$

5. $\int \sec \frac{1}{2} x \tan \frac{1}{2} x \, dx$

6. $\int e^x \csc^2 (e^x) \, dx$

7. $\int_{0.5}^{1} x^2 \cot x^3 \, dx$

8. $\int_{0}^{1} \tan \frac{1}{2} x \, dx$

9. $\displaystyle\int 3x \sec x^2 \, dx$

10. $\displaystyle\int 2 \csc 3x \, dx$

11. $\displaystyle\int \frac{\sin (1/x)}{x^2} \, dx$

12. $\displaystyle\int \frac{3 \, dx}{\sin 4x}$

13. $\displaystyle\int_0^{\pi/6} \frac{dx}{\cos^2 2x}$

14. $\displaystyle\int_0^1 2e^x \cos e^x \, dx$

15. $\displaystyle\int \frac{\sec 5x}{\cot 5x} \, dx$

16. $\displaystyle\int \frac{\sin 2x}{\cos^2 x} \, dx$

17. $\displaystyle\int \sqrt{\tan^2 2x + 1} \, dx$

18. $\displaystyle\int (1 + \cot x)^2 \, dx$

19. $\displaystyle\int \frac{1 + \sin 2x}{\tan 2x} \, dx$

20. $\displaystyle\int \frac{1 - \cot^2 x}{\cos^2 x} \, dx$

21. $\displaystyle\int \frac{1 - \sin x}{1 + \cos x} \, dx$

22. $\displaystyle\int \frac{1 + \sec^2 x}{x + \tan x} \, dx$

23. $\displaystyle\int_0^{\pi/9} \sin 3x(\csc 3x + \sec 3x) \, dx$

24. $\displaystyle\int_{\pi/4}^{\pi/3} (1 + \sec x)^2 \, dx$

In Exercises 25 through 32, solve the given problems by integration.

25. Find the area bounded by $y = \tan x$, $x = \frac{\pi}{4}$, and $y = 0$.

26. Find the area under the curve $y = \sin x$ from $x = 0$ to $x = \pi$.

27. Find the volume generated by rotating the area bounded by $y = \sec x$, $x = 0$, $x = \frac{\pi}{3}$, and $y = 0$ about the x-axis.

28. Find the volume generated by rotating the area bounded by $y = \cos x^2$, $x = 0$, $y = 0$, and $x = 1$ about the y-axis.

29. The angular velocity ω (in rad/s) of a pendulum is $\omega = -0.25 \sin 2.5t$. Find the angular displacement θ as a function of t if $\theta = 0.10$ for $t = 0$.

30. If the current in a certain electric circuit is $i = 110 \cos 377t$, find the expression for the voltage across a 500-μF capacitor as a function of time. The initial voltage is zero. Show that the voltage across the capacitor is 90° out of phase with the current.

31. A fin on a wind direction indicator has a shape which can be described as the area bounded by $y = \tan x^2$, $y = 0$, and $x = 1$. Find the x-coordinate (in m) of the fin if its area is 0.3984 m².

32. A force is given as a function of the distance from the origin as $F = \dfrac{2 + \tan x}{\cos x}$. Express the work done by this force as a function of x if $W = 0$ for $x = 0$.

27-5   Other Trigonometric Forms

The basic trigonometric relations developed in Chapter 19 provide the means by which many other integrals involving trigonometric functions may be integrated. By use of the square relations, Eqs. (19-6), (19-7), and (19-8), and the equations for the cosine of the double angle, Eqs. (19-22), (19-23), and (19-24), it is possible to transform integrals involving powers of the trigonometric functions into integrable form. We repeat these equations here for reference:

$$\cos^2 x + \sin^2 x = 1 \qquad (27\text{-}14)$$

$$1 + \tan^2 x = \sec^2 x \qquad (27\text{-}15)$$

$$1 + \cot^2 x = \csc^2 x \qquad (27\text{-}16)$$

$$2 \cos^2 x = 1 + \cos 2x \qquad (27\text{-}17)$$

$$2 \sin^2 x = 1 - \cos 2x \qquad (27\text{-}18)$$

NOTE ▷ *To integrate a product of powers of the sine and cosine, we use Eq. (27-14) **if at least one of the powers is odd.*** The method is based on transforming the integral so that it is made up of powers of either the sine or cosine and the first power of the other. In this way this first power becomes a factor of *du.*

EXAMPLE A ———— Integrate: $\int \sin^3 x \cos^2 x \, dx$.

Since $\sin^3 x = \sin^2 x \sin x = (1 - \cos^2 x) \sin x$, it is possible to write this integral with powers of $\cos x$ along with $\sin x \, dx$. In this way we can have $-\sin x \, dx$ as the necessary *du* of the integral. Therefore,

$$\int \sin^3 x \cos^2 x \, dx = \int (1 - \cos^2 x)(\sin x)(\cos^2 x) \, dx \qquad \text{using Eq. (27-14)}$$

$$= \int (\cos^2 x - \cos^4 x)(\sin x \, dx)$$

$$= \int \cos^2 x (\sin x \, dx) - \int \cos^4 x (\sin x \, dx)$$

$$= -\int \cos^2 x (-\sin x \, dx) + \int \cos^4 x (-\sin x \, dx)$$

$$= -\frac{1}{3} \cos^3 x + \frac{1}{5} \cos^5 x + C$$

EXAMPLE B ———— Integrate: $\int \cos^5 2x \, dx$.

Since $\cos^5 2x = \cos^4 2x \cos 2x = (1 - \sin^2 2x)^2 \cos 2x$, it is possible to write this integral with powers of $\sin 2x$ along with $\cos 2x \, dx$. Thus, with the introduction of a factor of 2, $(\cos 2x)(2 \, dx)$ is the necessary *du* of the integral. Thus,

$$\int \cos^5 2x \, dx = \int (1 - \sin^2 2x)^2 \cos 2x \, dx \qquad \text{using Eq. (27-14)}$$

$$= \int (1 - 2 \sin^2 2x + \sin^4 2x) \cos 2x \, dx$$

$$= \int \cos 2x \, dx - \int 2 \sin^2 2x \cos 2x \, dx + \int \sin^4 2x \cos 2x \, dx$$

$$= \frac{1}{2} \int \cos 2x (2 \, dx) - \int \sin^2 2x (2 \cos 2x \, dx) + \frac{1}{2} \int \sin^4 2x (2 \cos 2x \, dx)$$

$$= \frac{1}{2} \sin 2x - \frac{1}{3} \sin^3 2x + \frac{1}{10} \sin^5 2x + C$$

NOTE ▷ *In products of powers of the sine and cosine, **if the powers to be integrated are even, we use Eqs. (27-17) and (27-18) to transform the integral.*** Those most commonly met are $\int \cos^2 u \, du$ and $\int \sin^2 u \, du$. Consider the examples on the following page.

EXAMPLE C _____   Integrate: $\int \sin^2 2x \, dx$.

NOTE ▷   Using Eq. (27-18) in the form $\sin^2 2x = \frac{1}{2}(1 - \cos 4x)$, this integral can be transformed into a form which can be integrated. (Here we note **the x of Eq. (27-18) is treated as 2x** for this integral.) Therefore, we write

$$\int \sin^2 2x \, dx = \int \left[\frac{1}{2}(1 - \cos 4x) \right] dx \qquad \text{using Eq. (27-18)}$$

$$= \frac{1}{2} \int dx - \frac{1}{8} \int \cos 4x (4 \, dx)$$

$$= \frac{x}{2} - \frac{1}{8} \sin 4x + C$$

∎

NOTE ▷   *To integrate even powers of the secant, powers of the tangent, or products of the secant and tangent, we use Eq. (27-15) to transform the integral. In transform-ing, the forms we look for are **powers of the tangent with sec² x,** which becomes part of du, or **powers of the secant along with sec x tan x,** which becomes part of du in this case. Similar transformations are made when we integrate powers of the cotangent and cosecant, with the use of Eq. (27-16).*

EXAMPLE D _____   Integrate: $\int \tan^5 x \, dx$.

Since $\tan^5 x = \tan^3 x \tan^2 x = \tan^3 x (\sec^2 x - 1)$, we can write this integral with powers of tan x along with $\sec^2 x \, dx$. Thus, $\sec^2 x \, dx$ becomes the necessary du of the integral. It is necessary to replace $\tan^2 x$ with $\sec^2 x - 1$ twice during the integration. Therefore,

$$\int \tan^5 x \, dx = \int \tan^3 x (\sec^2 x - 1) \, dx \qquad \text{using Eq. (27-15)}$$

$$= \int \tan^3 x (\sec^2 x \, dx) - \int \tan^3 x \, dx$$

$$= \frac{1}{4} \tan^4 x - \int \tan x (\sec^2 x - 1) \, dx \qquad \text{using Eq. (27-15) again}$$

$$= \frac{1}{4} \tan^4 x - \int \tan x (\sec^2 x \, dx) + \int \tan x \, dx$$

$$= \frac{1}{4} \tan^4 x - \frac{1}{2} \tan^2 x - \ln |\cos x| + C$$

∎

EXAMPLE E _____   Integrate: $\int \sec^3 x \tan x \, dx$.

By writing $\sec^3 x \tan x$ as $\sec^2 x (\sec x \tan x)$, we can use the $\sec x \tan x \, dx$ as the du of the integral. Thus,

$$\int \sec^3 x \tan x \, dx = \int (\sec^2 x)(\overset{\displaystyle du}{\overbrace{\sec x \tan x \, dx}})$$

$$= \frac{1}{3} \sec^3 x + C$$

∎

EXAMPLE F

Integrate: $\int \csc^4 2x\, dx$.

By writing $\csc^4 2x = \csc^2 2x \csc^2 2x = \csc^2 2x(1 + \cot^2 2x)$, we can write this integral with powers of cot $2x$ along with $\csc^2 2x\, dx$, which becomes part of the necessary du of the integral. Thus,

$$\int \csc^4 2x\, dx = \int \csc^2 2x(1 + \cot^2 2x)\, dx \qquad \text{using Eq. (27-16)}$$

$$= \frac{1}{2} \int \csc^2 2x(2\, dx) - \frac{1}{2} \int \cot^2 2x(-2 \csc^2 2x\, dx)$$

$$\underset{\longmapsto\ du}{}$$

$$= -\frac{1}{2} \cot 2x - \frac{1}{6} \cot^3 2x + C$$

∎

EXAMPLE G

Integrate: $\int_0^{\pi/4} \dfrac{\tan^3 x}{\sec^3 x}\, dx$.

This integral requires the use of several trigonometric relationships to obtain integrable forms.

$$\int_0^{\pi/4} \frac{\tan^3 x}{\sec^3 x}\, dx = \int_0^{\pi/4} \frac{(\sec^2 x - 1)\tan x}{\sec^3 x}\, dx \qquad \text{using Eq. (27-15)}$$

$$= \int_0^{\pi/4} \frac{\tan x}{\sec x}\, dx - \int_0^{\pi/4} \frac{\tan x\, dx}{\sec^3 x} \qquad \frac{\tan x}{\sec x} = \frac{\sin x}{\cos x \sec x} = \sin x$$

$$= \int_0^{\pi/4} \sin x\, dx - \int_0^{\pi/4} \cos^2 x \sin x\, dx \qquad \frac{\tan x}{\sec^3 x} = \frac{\tan x}{\sec x \sec^2 x} = \sin x \cos^2 x$$

$$= -\cos x + \frac{1}{3} \cos^3 x \Big|_0^{\pi/4} \qquad \text{integrate}$$

$$= -\frac{\sqrt{2}}{2} + \frac{1}{3}\left(\frac{\sqrt{2}}{2}\right)^3 - \left(-1 + \frac{1}{3}\right) \qquad \text{evaluate}$$

$$= \frac{8 - 5\sqrt{2}}{12} = 0.0774$$

∎

EXAMPLE H

See the chapter introduction.

The *root-mean-square value of a function* with respect to x is defined by

$$\boxed{\; y_{\text{rms}} = \sqrt{\frac{1}{T} \int_0^T y^2\, dx} \;} \qquad (27\text{-}19)$$

Usually the value of T which is of importance is the period of the function. Find the root-mean-square value of the current in an electric circuit for one period if $i = 3 \cos \pi t$.

The period of the current is $\frac{2\pi}{\pi} = 2$ s. Therefore, we must find the square root of the integral

$$\frac{1}{2} \int_0^2 (3 \cos \pi t)^2\, dt = \frac{9}{2} \int_0^2 \cos^2 \pi t\, dt$$

Hence, we have

$$\frac{9}{2} \int_0^2 \cos^2 \pi t\, dt = \frac{9}{4} \int_0^2 (1 + \cos 2\pi t)\, dt$$

(*Continued on next page*)

after substituting $\frac{1}{2}(1 + \cos 2\pi t)$ for $\cos^2 \pi t$. Evaluating, we have

$$\frac{9}{4} \int_0^2 (1 + \cos 2\pi t)\, dt = \frac{9}{4} t \Big|_0^2 + \frac{9}{8\pi} \int_0^2 \cos 2\pi t (2\pi\, dt)$$

$$= \frac{9}{2} + \frac{9}{8\pi} \sin 2\pi t \Big|_0^2 = \frac{9}{2}$$

Thus, the root-mean-square current is

$$i_{rms} = \sqrt{\frac{9}{2}} = \frac{3\sqrt{2}}{2} = 2.12 \text{ A}$$

This value of the current, often referred to as the *effective current*, is the value of direct current which would produce the same quantity of heat energy in the same time. It is important in the design of electric appliances. ∎

Exercises 27-5

In Exercises 1 through 28, integrate each of the given functions.

1. $\int \sin^2 x \cos x\, dx$

2. $\int \sin x \cos^5 x\, dx$

3. $\int \sin^3 2x\, dx$

4. $\int 3 \cos^3 x\, dx$

5. $\int 2 \sin^2 x \cos^3 x\, dx$

6. $\int \sin^3 x \cos^6 x\, dx$

7. $\int_0^{\pi/4} \sin^5 x\, dx$

8. $\int_{\pi/3}^{\pi/2} \sqrt{\cos x} \sin^3 x\, dx$

9. $\int \sin^2 x\, dx$

10. $\int \cos^2 2x\, dx$

11. $\int \cos^2 3x\, dx$

12. $\int_0^1 \sin^2 4x\, dx$

13. $\int \tan^3 x\, dx$

14. $\int \cot^3 x\, dx$

15. $\int_0^{\pi/4} \tan x \sec^4 x\, dx$

16. $\int \cot 4x \csc^4 4x\, dx$

17. $\int \tan^4 2x\, dx$

18. $\int 4 \cot^4 x\, dx$

19. $\int \tan^3 3x \sec^3 3x\, dx$

20. $\int \sqrt{\tan x} \sec^4 x\, dx$

21. $\int (\sin x + \cos x)^2\, dx$

22. $\int (\tan 2x + \cot 2x)^2\, dx$

23. $\int \frac{1 - \cot x}{\sin^4 x}\, dx$

24. $\int \frac{1 + \sin x}{\cos^4 x}\, dx$

25. $\int_{\pi/6}^{\pi/4} \cot^5 x\, dx$

26. $\int_{\pi/6}^{\pi/3} \frac{2\, dx}{1 + \sin x}$

27. $\int \sec^6 x\, dx$

28. $\int \tan^7 x\, dx$

In Exercises 29 through 40, solve the given problems by integration.

29. Find the volume generated by rotating the area bounded by $y = \sin x$ and $y = 0$, from $x = 0$ to $x = \pi$, about the x-axis.

30. Find the volume generated by rotating the area bounded by $y = \tan^3(x^2)$, $y = 0$, and $x = \frac{\pi}{4}$ about the y-axis.

31. Find the area bounded by $y = \sin x$, $y = \cos x$, and $x = 0$ in the first quadrant.

32. Find the length of the curve $y = \ln \cos x$ from $x = 0$ to $x = \frac{\pi}{3}$. (See Exercise 25 of Section 25-6.)

33. Show that $\int \sin x \cos x\, dx$ can be integrated in two ways. Explain the difference in the answers.

34. Show that $\int \sec^2 x \tan x\, dx$ can be integrated in two ways. Explain the difference in the answers.

35. In determining the rate of radiation by an accelerated charge, the following integral must be evaluated: $\int_0^\pi \sin^3 \theta\, d\theta$. Find the value of the integral.

36. In finding the volume of a special O-ring for a space vehicle, it is necessary to evaluate the integral $\int \frac{\sin^2 \theta}{\cos^2 \theta}\, d\theta$. Perform this integration.

37. Find the root-mean-square value of the voltage for one period of standard house current, which is given by $V = 170 \sin 120\pi t$.

38. For a current $i = i_0 \sin \omega t$, show that the root-mean-square value of the current for one period is $i_0/\sqrt{2}$.

39. In analyzing the intensity of light from a certain source, the equation $I = A \int_{-a/2}^{a/2} \cos^2 [b\pi(c - x)] \, dx$ is used. Here A, a, b, and c are constants. Evaluate this integral. (The simplification of the result is lengthy.)

40. In finding the lifting force L due to a stream of fluid passing around a cylinder, the equation $L = k \int_0^{2\pi} (a \sin \theta + b \sin^2 \theta - b \sin^3 \theta) \, d\theta$ is used. Here, k, a, and b are constants and θ is the angle from the direction of flow. Evaluate the integral.

27-6   Inverse Trigonometric Forms

For reference,
Eq. (26-10) is

$$\frac{d(\text{Arcsin } u)}{dx}$$

$$= \frac{1}{\sqrt{1 - u^2}} \frac{du}{dx}$$

Referring to Eq. (26-10), we can find the differential of Arcsin (u/a), where a is a constant:

$$d\left(\text{Arcsin } \frac{u}{a}\right) = \frac{1}{\sqrt{1 - (u/a)^2}} \frac{du}{a} = \frac{a}{\sqrt{a^2 - u^2}} \frac{du}{a} = \frac{du}{\sqrt{a^2 - u^2}}$$

Reversing this differentiation formula, we have the important integration formula

Integral of

$$\frac{du}{\sqrt{a^2 - u^2}}$$

$$\int \frac{du}{\sqrt{a^2 - u^2}} = \text{Arcsin } \frac{u}{a} + C \qquad\qquad (27\text{-}20)$$

By finding the differential of Arctan (u/a) and then reversing the equation, we derive another important integration formula:

$$d\left(\text{Arctan } \frac{u}{a}\right) = \frac{1}{1 + (u/a)^2} \frac{du}{a} = \frac{a^2}{a^2 + u^2} \frac{du}{a} = \frac{a \, du}{a^2 + u^2}$$

Thus,

Integral of

$$\frac{du}{a^2 + u^2}$$

$$\int \frac{du}{a^2 + u^2} = \frac{1}{a} \text{Arctan } \frac{u}{a} + C \qquad\qquad (27\text{-}21)$$

This shows one of the principal uses of the inverse trigonometric functions: They provide a solution to the integration of important algebraic functions.

EXAMPLE A

Integrate: $\int \dfrac{dx}{\sqrt{9 - x^2}}$.

This integral fits the form of Eq. (27-20) with $u = x$, $du = dx$, and $a = 3$. Thus,

$$\int \frac{dx}{\sqrt{9 - x^2}} = \int \frac{dx}{\sqrt{3^2 - x^2}}$$

$$= \text{Arcsin } \frac{x}{3} + C$$

EXAMPLE B

The volume flow rate Q (in m³/s) of a constantly flowing liquid is given by $Q = 24 \int_0^2 \dfrac{dx}{6 + x^2}$, where x is the distance from the center of flow. Find the value of Q.

For the integral, we see that it fits Eq. (27-21) with $u = x$, $du = dx$, and $a = \sqrt{6}$.

$$Q = 24 \int_0^2 \frac{dx}{6 + x^2} = 24 \int_0^2 \frac{dx}{(\sqrt{6})^2 + x^2}$$

$$= \frac{24}{\sqrt{6}} \operatorname{Arctan} \frac{x}{\sqrt{6}} \Big|_0^2 = \frac{24}{\sqrt{6}} \left(\operatorname{Arctan} \frac{2}{\sqrt{6}} - \operatorname{Arctan} 0 \right)$$

$$= 9.80 \operatorname{Arctan} 0.8165 = 6.71 \text{ m}^3/\text{s}$$

EXAMPLE C

Integrate: $\displaystyle\int \frac{dx}{\sqrt{25 - 4x^2}}$.

This integral fits the form of Eq. (27-20) with $u = 2x$, $du = 2\,dx$, and $a = 5$. Thus, in order to have the proper du, we must include a factor of 2 in the numerator, and therefore we also place a $\frac{1}{2}$ before the integral. This leads to

$$\int \frac{dx}{\sqrt{25 - 4x^2}} = \frac{1}{2} \int \frac{2\,dx \longleftarrow du}{\sqrt{5^2 - (2x)^2}}$$
$$\underset{\uparrow \qquad u}{}$$

$$= \frac{1}{2} \operatorname{Arcsin} \frac{2x}{5} + C$$

EXAMPLE D

Integrate: $\displaystyle\int_{-1}^3 \frac{dx}{x^2 + 6x + 13}$.

At first glance it does not appear that this integral fits any of the forms presented up to this point. However, by writing the denominator in the form $(x^2 + 6x + 9) + 4 = (x + 3)^2 + 2^2$, *we recognize that $u = x + 3$, $du = dx$, and $a = 2$.* Thus,

NOTE ▷

$$\int_{-1}^3 \frac{dx}{x^2 + 6x + 13} = \int_{-1}^3 \frac{dx \longleftarrow du}{(x + 3)^2 + 2^2}$$
$$\underset{\uparrow \qquad u}{}$$

$$= \frac{1}{2} \operatorname{Arctan} \frac{x + 3}{2} \Big|_{-1}^3 \qquad\qquad \text{integrate}$$

$$= \frac{1}{2} (\operatorname{Arctan} 3 - \operatorname{Arctan} 1) \qquad \text{evaluate}$$

$$= 0.2318$$

Now we can see the use of completing the square when we are transforming integrals into proper form.

EXAMPLE E

Integrate: $\int \dfrac{2x + 5}{x^2 + 9}\, dx$.

By writing this integral as the sum of two integrals, we may integrate each of these separately:

$$\int \frac{2x + 5}{x^2 + 9}\, dx = \int \frac{2x\, dx}{x^2 + 9} + \int \frac{5\, dx}{x^2 + 9}$$

The first of these integrals is seen to be a logarithmic form, and the second the inverse tangent form. For the first, $u = x^2 + 9$, $du = 2x\, dx$. For the second, $u = x$, $du = dx$, $a = 3$. Thus

$$\int \frac{2x\, dx}{x^2 + 9} + 5 \int \frac{dx}{x^2 + 9} = \ln |x^2 + 9| + \frac{5}{3} \operatorname{Arctan} \frac{x}{3} + C$$

NOTE ▷   The inverse trigonometric integral forms of this section show very well the importance of *proper recognition of the form of the integral.* It is important that these forms are not confused with those of the general power rule or the logarithmic form. Consider the following examples.

EXAMPLE F

The integral $\int \dfrac{dx}{\sqrt{1 - x^2}}$ is of the inverse sine form, with $u = x$, $du = dx$, and $a = 1$. Thus,

$$\int \frac{dx}{\sqrt{1 - x^2}} = \operatorname{Arcsin} x + C$$

The integral $\int \dfrac{x\, dx}{\sqrt{1 - x^2}}$ is not of the inverse sine form due to the factor of x in the numerator. It is integrated by use of the general power rule, with $u = 1 - x^2$, $du = -2x\, dx$, and $n = -\frac{1}{2}$. Thus,

$$\int \frac{x\, dx}{\sqrt{1 - x^2}} = -\sqrt{1 - x^2} + C$$

The integral $\int \dfrac{x\, dx}{1 - x^2}$ is of logarithmic form with $u = 1 - x^2$ and $du = -2x\, dx$. If $1 - x^2$ were raised to any other power other than 1 in the denominator, we would use the general power rule. To be of the inverse sine form, we would have the square root of $1 - x^2$ and no factor of x, as in the first illustration. Thus,

$$\int \frac{x\, dx}{1 - x^2} = -\frac{1}{2} \ln |1 - x^2| + C$$

EXAMPLE G _____ The following integrals are of the form indicated.

$$\int \frac{dx}{1 + x^2} \qquad \text{Inverse tangent form} \qquad u = x,\ du = dx$$

$$\int \frac{x\,dx}{1 + x^2} \qquad \text{Logarithmic form} \qquad u = 1 + x^2,\ du = 2x\,dx$$

$$\int \frac{x\,dx}{\sqrt{1 + x^2}} \qquad \text{General power form} \qquad u = 1 + x^2,\ du = 2x\,dx$$

$$\int \frac{dx}{1 + x} \qquad \text{Logarithmic form} \qquad u = 1 + x,\ du = dx$$

There are a number of integrals whose forms appear to be similar to those in Examples F and G, but which do not fit the forms which we have discussed. They include

$$\int \frac{dx}{\sqrt{x^2 - 1}}, \qquad \int \frac{dx}{\sqrt{1 + x^2}}, \qquad \int \frac{dx}{1 - x^2}, \quad \text{and} \quad \int \frac{dx}{x\sqrt{1 + x^2}}$$

We will develop methods to integrate some of these, and all of them can be integrated by the tables discussed in Section 27-9.

Exercises 27-6

In Exercises 1 through 24, integrate each of the given functions.

1. $\int \dfrac{dx}{\sqrt{4 - x^2}}$

2. $\int \dfrac{dx}{\sqrt{49 - x^2}}$

3. $\int \dfrac{dx}{64 + x^2}$

4. $\int \dfrac{4\,dx}{4 + x^2}$

5. $\int \dfrac{dx}{\sqrt{1 - 16x^2}}$

6. $\int_0^1 \dfrac{2\,dx}{\sqrt{9 - 4x^2}}$

7. $\int_0^2 \dfrac{3\,dx}{1 + 9x^2}$

8. $\int_1^3 \dfrac{dx}{49 + 4x^2}$

9. $\int_0^{0.4} \dfrac{2\,dx}{\sqrt{4 - 5x^2}}$

10. $\int \dfrac{4x\,dx}{\sqrt{3 - 2x^2}}$

11. $\int \dfrac{8x\,dx}{9x^2 + 16}$

12. $\int \dfrac{3\,dx}{25 + 16x^2}$

13. $\int_1^2 \dfrac{dx}{5x^2 + 7}$

14. $\int_0^1 \dfrac{x\,dx}{1 + x^4}$

15. $\int \dfrac{e^x\,dx}{\sqrt{1 - e^{2x}}}$

16. $\int \dfrac{\sec^2 x\,dx}{\sqrt{1 - \tan^2 x}}$

17. $\int \dfrac{dx}{x^2 + 2x + 2}$

18. $\int \dfrac{2\,dx}{x^2 + 8x + 17}$

19. $\int \dfrac{4\,dx}{\sqrt{-4x - x^2}}$

20. $\int \dfrac{dx}{\sqrt{2x - x^2}}$

21. $\int_{\pi/6}^{\pi/2} \dfrac{\cos 2x}{1 + \sin^2 2x}\,dx$

22. $\int_{-4}^0 \dfrac{dx}{x^2 + 4x + 5}$

23. $\int \dfrac{2 - x}{\sqrt{4 - x^2}}\,dx$

24. $\int \dfrac{3 - 2x}{1 + 4x^2}\,dx$

In Exercises 25 through 28, identify the form of each integral as being inverse sine, inverse tangent, logarithmic, or general power, as in Examples F and G. Do not integrate.

25. (a) $\int \dfrac{2\,dx}{4 + 9x^2}$

(b) $\int \dfrac{2\,dx}{4 + 9x}$

(c) $\int \dfrac{2x\,dx}{\sqrt{4 + 9x^2}}$

26. (a) $\int \dfrac{2x\,dx}{4 - 9x^2}$

(b) $\int \dfrac{2\,dx}{\sqrt{4 - 9x}}$

(c) $\int \dfrac{2x\,dx}{4 + 9x^2}$

27. (a) $\int \dfrac{2x\,dx}{\sqrt{4-9x^2}}$ (b) $\int \dfrac{2\,dx}{\sqrt{4-9x^2}}$ (c) $\int \dfrac{2\,dx}{4-9x}$

28. (a) $\int \dfrac{2\,dx}{9x^2+4}$ (b) $\int \dfrac{2x\,dx}{\sqrt{9x^2-4}}$ (c) $\int \dfrac{2x\,dx}{9x^2+4}$

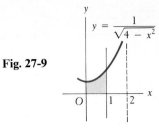

Fig. 27-9

In Exercises 29 through 36, solve the given problems by integration.

29. Find the area bounded by the curve $y(1+x^2)=1$, $x=0$, $y=0$, and $x=2$.

30. Find the area bounded by $y\sqrt{4-x^2}=1$, $x=0$, $y=0$, and $x=1$. See Fig. 27-9.

31. To find the electric field E from an electric charge distributed uniformly over the entire xy-plane at a distance d from the plane, it is necessary to evaluate the integral $kd \int \dfrac{dx}{d^2+x^2}$. Here x is the distance from the origin to the element of charge. Perform the indicated integration.

32. An oil storage tank can be described as the volume generated by rotating the area bounded by $y=24/\sqrt{16+x^2}$, $x=0$, $y=0$, and $x=3$ about the x-axis. Find the volume (in m³) of the tank.

33. In dealing with the theory for simple harmonic motion, it is necessary to solve the following equation:

$$\frac{dx}{\sqrt{A^2-x^2}} = \sqrt{\frac{k}{m}}\,dt \quad \text{(with } k, m, \text{ and } A \text{ as constants)}$$

Determine the solution to this equation if $x=x_0$ when $t=0$.

34. During each cycle, the velocity v (in ft/s) of a robotic welding device is given by $v = 2t - \dfrac{12}{2+t^2}$, where t is the time in seconds. Find the expression for the displacement s (in ft) as a function of t if $s=0$ for $t=0$.

35. Find the moment of inertia with respect to the y-axis for the area bounded by $y=1/(1+x^6)$, the x-axis, $x=1$, and $x=2$.

36. Find the length of arc along the curve $y=\sqrt{1-x^2}$ between $x=0$ and $x=1$. (See Exercise 25 of Section 25-6.)

27-7 Integration by Parts

There are many methods of transforming integrals into a form which can be integrated by one of the basic formulas. In the preceding section we saw that completing the square and trigonometric identities can be used for this purpose. In this section and the following one, we shall develop two general methods. The method of integration by parts is discussed in this section.

Since the derivative of a product of functions is found by use of the formula

$$\frac{d(uv)}{dx} = u\frac{dv}{dx} + v\frac{du}{dx}$$

the differential of a product of functions is given by $d(uv) = u\,dv + v\,du$. Integrating both sides of this equation, we have $uv = \int u\,dv + \int v\,du$. Solving for $\int u\,dv$, we obtain

$$\int u\,dv = uv - \int v\,du \tag{27-22}$$

Integration by use of Eq. (27-22) is called **integration by parts.**

EXAMPLE A

Integrate: $\int x \sin x \, dx$.

This integral does not fit any of the previous forms which have been discussed, since neither x nor $\sin x$ can be made a factor of a proper du. However, by choosing $u = x$ and $dv = \sin x \, dx$, integration by parts may be used. Thus,

$$u = x, \qquad dv = \sin x \, dx$$

By finding the differential of u and integrating dv, we find du and v. This gives us

$$du = dx, \qquad v = -\cos x + C_1$$

Now, substituting in Eq. (27-22), we have

$$
\begin{array}{ccccccc}
\int u & dv & = u & v & - \int & v & du \\
\downarrow & \downarrow & \downarrow & \downarrow & & \downarrow & \downarrow
\end{array}
$$

$$\int (x)(\sin x \, dx) = (x)(-\cos x + C_1) - \int (-\cos x + C_1)(dx)$$

$$= -x \cos x + C_1 x + \int \cos x \, dx - \int C_1 \, dx$$

$$= -x \cos x + C_1 x + \sin x - C_1 x + C$$

$$= -x \cos x + \sin x + C$$

Other choices of u and dv may be made, but they do not prove to be useful. For example, if we choose $u = \sin x$ and $dv = x \, dx$, then $du = \cos x \, dx$ and $v = \frac{1}{2} x^2 + C_2$. This makes $\int v \, du = \int (\frac{1}{2} x^2 + C_2)(\cos x \, dx)$, which is more complex than the integrand of the original problem.

We also note that the constant C_1 which was introduced when we integrated dv does not appear in the final result. This constant will always cancel out and therefore *we will not show any constant of integration when finding* v.

NOTE ▷

As in Example A, there is often more than one choice as to the part of the integrand which is selected to be u and the part which is selected to be dv. There are no set rules which may be stated for the best choice of u and dv, but there are two guidelines which may be stated: **(1) *The quantity u is normally chosen such that du/dx is of simpler form than u.*** (2) *The differential dv is normally chosen such that $\int dv$ is easily obtained.* Working examples, and thereby gaining experience in methods of integrating, is the best way to determine when this method should be used and how to employ it.

EXAMPLE B

Integrate: $\int x \sqrt{1 - x} \, dx$.

We see that this does not fit the general power rule, for $x \, dx$ is not a factor of the differential of $1 - x$. By choosing $u = x$ and $dv = \sqrt{1 - x} \, dx$ we have $du/dx = 1$, and v can readily be determined. Thus,

$$u = x, \qquad dv = \sqrt{1 - x} \, dx = (1 - x)^{1/2} \, dx$$

$$du = dx, \qquad v = -\frac{2}{3}(1 - x)^{3/2}$$

Substituting in Eq. (27-22), we have

$$
\begin{array}{ccccccc}
\int u & dv & = u & v & - \int & v & du \\
\downarrow & \downarrow & \downarrow & \downarrow & & \downarrow & \downarrow
\end{array}
$$

$$\int x[(1 - x)^{1/2} \, dx] = x\left[-\frac{2}{3}(1 - x)^{3/2} \right] - \int \left[-\frac{2}{3}(1 - x)^{3/2} \right] dx$$

At this point we see that we can complete the integration. Thus,

$$\int x(1-x)^{1/2}\,dx = -\frac{2x}{3}(1-x)^{3/2} + \frac{2}{3}\int(1-x)^{3/2}\,dx$$

$$= -\frac{2x}{3}(1-x)^{3/2} + \frac{2}{3}\left(-\frac{2}{5}\right)(1-x)^{5/2} + C$$

$$= -\frac{2}{3}(1-x)^{3/2}\left[x + \frac{2}{5}(1-x)\right] + C$$

$$= -\frac{2}{15}(1-x)^{3/2}(2+3x) + C$$

EXAMPLE C

Integrate: $\int \sqrt{x}\,\ln x\,dx$.

For this integral we have

$$u = \ln x, \qquad dv = x^{1/2}\,dx$$

$$du = \frac{1}{x}\,dx, \qquad v = \frac{2}{3}x^{3/2}$$

This leads to

$$\int \sqrt{x}\,\ln x\,dx = \frac{2}{3}x^{3/2}\ln x - \frac{2}{3}\int x^{1/2}\,dx$$

$$= \frac{2}{3}x^{3/2}\ln x - \frac{4}{9}x^{3/2} + C$$

EXAMPLE D

Integrate: $\int \text{Arcsin } x\,dx$.

We write

$$u = \text{Arcsin } x, \qquad dv = dx$$

$$du = \frac{dx}{\sqrt{1-x^2}}, \qquad v = x$$

$$\int \text{Arcsin } x\,dx = x\,\text{Arcsin } x - \int \frac{x\,dx}{\sqrt{1-x^2}}$$

$$= x\,\text{Arcsin } x + \frac{1}{2}\int \frac{-2x\,dx}{\sqrt{1-x^2}}$$

$$= x\,\text{Arcsin } x + \sqrt{1-x^2} + C$$

EXAMPLE E

Integrate: $\int_0^1 xe^{-x}\,dx$.

We write

$$u = x, \qquad dv = e^{-x}\,dx$$

$$du = dx, \qquad v = -e^{-x}$$

$$\int_0^1 xe^{-x}\,dx = -xe^{-x}\Big|_0^1 + \int_0^1 e^{-x}\,dx = -xe^{-x} - e^{-x}\Big|_0^1$$

$$= -e^{-1} - e^{-1} + 1$$

$$= 1 - \frac{2}{e}$$

EXAMPLE F _____

Integrate: $\int e^x \sin x \, dx$. Let $u = \sin x$, $dv = e^x \, dx$, $du = \cos x \, dx$, $v = e^x$. We write

$$\int e^x \sin x \, dx = e^x \sin x - \int e^x \cos x \, dx$$

NOTE ▷

At first glance it appears that we have made no progress in applying the method of integration by parts. We note, however, that when we integrated $\int e^x \sin x \, dx$, part of the result was a term of $\int e^x \cos x \, dx$. This implies that *if $\int e^x \cos x \, dx$ were integrated, a term of $\int e^x \sin x \, dx$ might result*. Thus, the method of integration by parts is now applied to the integral $\int e^x \cos x \, dx$:

$$u = \cos x, \qquad dv = e^x \, dx, \qquad du = -\sin x \, dx, \qquad v = e^x$$

And so $\int e^x \cos x \, dx = e^x \cos x + \int e^x \sin x \, dx$. Substituting this expression into the expression for $\int e^x \sin x \, dx$, we obtain

$$\int e^x \sin x \, dx = e^x \sin x - \left(e^x \cos x + \int e^x \sin x \, dx \right)$$

$$= e^x \sin x - e^x \cos x - \int e^x \sin x \, dx$$

$$2 \int e^x \sin x \, dx = e^x(\sin x - \cos x) + 2C$$

$$\int e^x \sin x \, dx = \frac{e^x}{2} (\sin x - \cos x) + C$$

■ Thus, by combining integrals of like form, we obtain the desired result.

Exercises 27-7

In Exercises 1 through 16, integrate each of the given functions.

1. $\int x \cos x \, dx$

2. $\int x \sin 2x \, dx$

3. $\int x e^{2x} \, dx$

4. $\int 3x e^x \, dx$

5. $\int x \sec^2 x \, dx$

6. $\int_0^{\pi/4} x \sec x \tan x \, dx$

7. $\int 2 \, \text{Arctan} \, x \, dx$

8. $\int \ln x \, dx$

9. $\int \dfrac{4x \, dx}{\sqrt{1 - x}}$

10. $\int x\sqrt{x + 1} \, dx$

11. $\int x \ln x \, dx$

12. $\int x^2 \ln 4x \, dx$

13. $\int x^2 \sin 2x \, dx$

14. $\int x^2 e^{2x} \, dx$

15. $\int_0^{\pi/2} e^x \cos x \, dx$

16. $\int e^{-x} \sin 2x \, dx$

In Exercises 17 through 24, solve the given problems by integration.

17. Find the area bounded by $y = xe^{-x}$, $y = 0$, and $x = 2$. See Fig. 27-10.

18. Find the volume generated by rotating the area bounded by $y = \sin x$ and $y = 0$ (from $x = 0$ to $x = \pi$) about the y-axis.

19. Find the x-coordinate of the centroid of the area bounded by $y = \cos x$ and $y = 0$ for $0 \le x \le \frac{\pi}{2}$.

20. Find the moment of inertia with respect to its axis of the volume generated by rotating the area bounded by $y = e^x$, $x = 1$, and the coordinate axes about the y-axis.

21. Find the root-mean-square value of the function $y = \sqrt{\text{Arcsin} \, x}$ between $x = 0$ and $x = 1$. (See Example H of Section 27-5.)

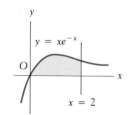

Fig. 27-10

22. The general expression for the slope of a curve is $dy/dx = x^3\sqrt{1 + x^2}$. Find the equation of the curve if it passes through the origin.

23. The current in a given circuit is given by $i = e^{-2t}\cos t$. Find an expression for the amount of charge which passes a given point in the circuit as a function of the time, if $q_0 = 0$.

24. In finding the average length $\bar{x}$ (in nm) of a certain type of large molecule, the equation
$\bar{x} = \lim\limits_{b\to\infty}\left[0.1\int_0^b x^3 e^{-x^2/8}\, dx\right]$ is used. Evaluate the integral and then use a calculator to show that $\bar{x} \to 3.2$ nm as $b \to \infty$.

27-8 Integration by Trigonometric Substitution

Trigonometric relations not only provide a means of transforming trigonometric integrals, they can also be used to transform algebraic integrals. Substitutions based on Eqs. (27-14), (27-15), and (27-16) prove to be particularly useful for integrals involving radicals. The method is illustrated in the following examples.

EXAMPLE A

Integrate: $\displaystyle\int \frac{dx}{x^2\sqrt{1 - x^2}}$.

If we let $x = \sin\theta$, the radical becomes $\sqrt{1 - \sin^2\theta} = \cos\theta$. Therefore, by making this substitution, the integral can be transformed into a trigonometric integral. We must be careful to **replace all factors of the integral by proper expressions in terms of θ**. With $x = \sin\theta$, by finding the differential of x, we have $dx = \cos\theta\, d\theta$.

NOTE ▷

Now, by substituting $x = \sin\theta$ and $dx = \cos\theta\, d\theta$ into the integral, we have

$$\int \frac{dx}{x^2\sqrt{1 - x^2}} = \int \frac{\cos\theta\, d\theta}{\sin^2\theta\sqrt{1 - \sin^2\theta}} \qquad \text{substituting}$$

$$= \int \frac{\cos\theta\, d\theta}{\sin^2\theta\cos\theta} = \int \csc^2\theta\, d\theta \qquad \begin{array}{l}\text{using trigonometric}\\ \text{relations}\end{array}$$

This last integral can be integrated by use of Eq. (27-7). This leads to

$$\int \csc^2\theta\, d\theta = -\cot\theta + C$$

We have now performed the integration, but the answer is expressed in terms of a variable different from the original integral. We must express the result in terms of x. Making a right triangle with an angle θ such that $\sin\theta = x/1$ (see Fig. 27-11), we may express any of the trigonometric functions in terms of x. (This is the method used with inverse trigonometric functions.) Thus,

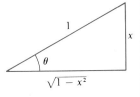

Fig. 27-11

$$\cot\theta = \frac{\sqrt{1 - x^2}}{x}$$

Therefore, the result of the integration becomes

$$\int \frac{dx}{x^2\sqrt{1 - x^2}} = -\cot\theta + C = -\frac{\sqrt{1 - x^2}}{x} + C$$

EXAMPLE B

Integrate: $\int \dfrac{dx}{\sqrt{x^2 + 4}}$.

If we let $x = 2 \tan \theta$, the radical in this integral becomes

$$\sqrt{x^2 + 4} = \sqrt{4 \tan^2 \theta + 4} = 2\sqrt{\tan^2 \theta + 1} = 2\sqrt{\sec^2 \theta} = 2 \sec \theta$$

Therefore, with $x = 2 \tan \theta$ and $dx = 2 \sec^2 \theta \, d\theta$, we have

$$\int \frac{dx}{\sqrt{x^2 + 4}} = \int \frac{2 \sec^2 \theta \, d\theta}{\sqrt{4 \tan^2 \theta + 4}} = \int \frac{2 \sec^2 \theta \, d\theta}{2 \sec \theta} \qquad \text{substituting}$$

$$= \int \sec \theta \, d\theta = \ln |\sec \theta + \tan \theta| + C \qquad \text{using Eq. (27-12)}$$

$$= \ln \left| \frac{\sqrt{x^2 + 4}}{2} + \frac{x}{2} \right| + C \qquad \text{see Fig. (27-12)}$$

$$= \ln \left| \frac{\sqrt{x^2 + 4} + x}{2} \right| + C$$

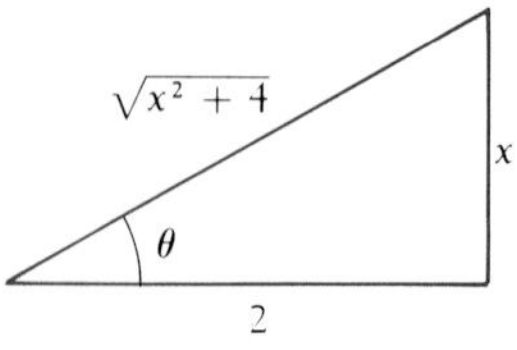

This answer is acceptable, but there is another form. By using the properties of logarithms, we have

$$\ln \left| \frac{\sqrt{x^2 + 4} + x}{2} \right| + C = \ln \left| \sqrt{x^2 + 4} + x \right| + (C - \ln 2)$$

$$= \ln \left| \sqrt{x^2 + 4} + x \right| + C'$$

Here C' is another arbitrary constant, which combines all constants of the previous expression. Combining constants in this manner is a common practice in integration problems. ∎

Fig. 27-12

EXAMPLE C

Integrate: $\int \dfrac{2 \, dx}{x\sqrt{x^2 - 9}}$.

If we let $x = 3 \sec \theta$, the radical in this integral becomes

$$\sqrt{x^2 - 9} = \sqrt{9 \sec^2 \theta - 9} = 3\sqrt{\sec^2 \theta - 1} = 3\sqrt{\tan^2 \theta} = 3 \tan \theta$$

Therefore, with $x = 3 \sec \theta$ and $dx = 3 \sec \theta \tan \theta \, d\theta$, we have

$$\int \frac{2 \, dx}{x\sqrt{x^2 - 9}} = 2 \int \frac{3 \sec \theta \tan \theta \, d\theta}{3 \sec \theta \sqrt{9 \sec^2 \theta - 9}} = 2 \int \frac{\tan \theta \, d\theta}{3 \tan \theta}$$

$$= \frac{2}{3} \int d\theta = \frac{2}{3} \theta + C = \frac{2}{3} \operatorname{Arcsec} \frac{x}{3} + C$$

In this solution it is not necessary to refer to a triangle to express the integral in terms of x. This solution is found by solving $x = 3 \sec \theta$ for θ, as indicated. ∎

From the preceding examples we can see that by making the proper substitution in terms of a trigonometric function, we can solve integrals of algebraic functions by means of simpler equivalent trigonometric forms, due to the properties of the square relations of the trigonometric functions. In summary, for the indicated radical form the following trigonometric substitutions are used:

basic substitutions

$$
\begin{array}{ll}
\text{For} \quad \sqrt{a^2 - x^2} & \text{use} \quad x = a \sin \theta \\
\text{For} \quad \sqrt{a^2 + x^2} & \text{use} \quad x = a \tan \theta \\
\text{For} \quad \sqrt{x^2 - a^2} & \text{use} \quad x = a \sec \theta
\end{array}
\tag{27-23}
$$

EXAMPLE D

Evaluate: $\displaystyle\int_1^4 \frac{\sqrt{9 + x^2}}{x} \, dx$.

Since we have the form $\sqrt{a^2 + x^2}$, where $a = 3$, we make the substitution $x = 3 \tan \theta$. This means that $dx = 3 \sec^2 \theta \, d\theta$. Thus,

$$
\int \frac{\sqrt{9 + x^2}}{x} \, dx = \int \frac{\sqrt{9 + 9 \tan^2 \theta}}{3 \tan \theta} (3 \sec^2 \theta \, d\theta) \qquad \text{substituting}
$$

$$
= 3 \int \frac{\sqrt{1 + \tan^2 \theta}}{\tan \theta} \sec^2 \theta \, d\theta = 3 \int \frac{\sec^3 \theta}{\tan \theta} \, d\theta \qquad \begin{array}{l}\text{using trig.}\\\text{relations}\end{array}
$$

$$
= 3 \int \frac{1 + \tan^2 \theta}{\tan \theta} \sec \theta \, d\theta = 3 \left(\int \frac{\sec \theta}{\tan \theta} \, d\theta + \int \tan \theta \sec \theta \, d\theta \right)
$$

$$
= 3 \left(\int \csc \theta \, d\theta + \int \tan \theta \sec \theta \, d\theta \right)
$$

$$
= 3 [\ln |\csc \theta - \cot \theta| + \sec \theta] + C \qquad \begin{array}{l}\text{using Eqs. (27-13)}\\\text{and (27-8)}\end{array}
$$

$$
= 3 \left[\ln \left| \frac{\sqrt{x^2 + 9}}{x} - \frac{3}{x} \right| + \frac{\sqrt{x^2 + 9}}{3} \right] + C \qquad \text{see Fig. 27-13}
$$

$$
= 3 \left[\ln \left| \frac{\sqrt{x^2 + 9} - 3}{x} \right| + \frac{\sqrt{x^2 + 9}}{3} \right] + C
$$

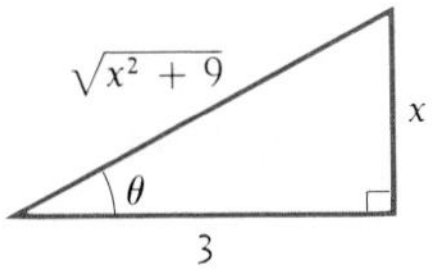

Fig. 27-13

Limits have not been included, due to the changes in variables. The actual evaluation may now be completed:

$$
\int_1^4 \frac{\sqrt{9 + x^2}}{x} \, dx = 3 \left[\ln \left| \frac{\sqrt{x^2 + 9} - 3}{x} \right| + \frac{\sqrt{x^2 + 9}}{3} \right]_1^4
$$

$$
= 3 \left[\left(\ln \frac{1}{2} - \ln \frac{\sqrt{10} - 3}{1} \right) + \left(\frac{5}{3} - \frac{\sqrt{10}}{3} \right) \right]
$$

$$
= 3 \left[\ln \frac{1}{2(\sqrt{10} - 3)} + \frac{5 - \sqrt{10}}{3} \right]
$$

$$
= 3 \ln \frac{\sqrt{10} + 3}{2} + \frac{5 - \sqrt{10}}{1}
$$

$$
= 5.21
$$

Exercises 27-8

In Exercises 1 through 16, integrate each of the given functions.

1. $\displaystyle\int \frac{\sqrt{1-x^2}}{x^2}\,dx$

2. $\displaystyle\int \frac{dx}{(x^2+9)^{3/2}}$

3. $\displaystyle\int \frac{2\,dx}{\sqrt{x^2-4}}$

4. $\displaystyle\int \frac{\sqrt{x^2-25}}{x}\,dx$

5. $\displaystyle\int \frac{dx}{x^2\sqrt{x^2+9}}$

6. $\displaystyle\int \frac{3\,dx}{x\sqrt{4-x^2}}$

7. $\displaystyle\int \frac{4\,dx}{(4-x^2)^{3/2}}$

8. $\displaystyle\int \frac{2x^3\,dx}{\sqrt{9+x^2}}$

9. $\displaystyle\int_0^{0.5} \frac{x^3\,dx}{\sqrt{1-x^2}}$

10. $\displaystyle\int_4^5 \frac{\sqrt{x^2-16}}{x^2}\,dx$

11. $\displaystyle\int \frac{5\,dx}{\sqrt{x^2+2x+2}}$

12. $\displaystyle\int \frac{dx}{\sqrt{x^2+2x}}$

13. $\displaystyle\int \frac{dx}{x\sqrt{4x^2-9}}$

14. $\displaystyle\int \sqrt{16-x^2}\,dx$

15. $\displaystyle\int \frac{dx}{\sqrt{e^{2x}-1}}$

16. $\displaystyle\int \frac{\sec^2 x\,dx}{(4-\tan^2 x)^{3/2}}$

In Exercises 17 through 24, solve the given problems by integration.

17. Find the area of a circle of radius 1 by integration.

18. Find the area bounded by $y = \dfrac{1}{x^2\sqrt{x^2-1}}$, $x = \sqrt{2}$, $x = \sqrt{5}$, and $y = 0$. See Fig. 27-14.

19. Find the moment of inertia with respect to the y-axis of the first-quadrant area under the circle $x^2 + y^2 = a^2$ in terms of its mass.

20. Find the moment of inertia of a sphere of radius a with respect to its axis in terms of its mass.

21. Find the volume generated by rotating the area bounded by $y = \dfrac{\sqrt{x^2-16}}{x^2}$, $y = 0$, and $x = 5$ about the y-axis.

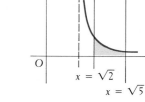

Fig. 27-14

22. Find the length of arc along the curve of $y = \ln x$ from $x = 1$ to $x = 3$. (See Exercise 25 of Section 25-6.)

23. If an electric charge Q is distributed along a straight wire of length $2a$, the electric potential V at a point P, which is at a distance b from the center of the wire, is $V = kQ \displaystyle\int_{-a}^{a} \frac{dx}{\sqrt{b^2+x^2}}$. Here k is a constant and x is the distance along the wire. Evaluate the integral.

24. An electric insulating ring for a machine part can be described as the volume generated by rotating the area bounded by $y = x^2\sqrt{x^2-4}$, $y = 0$, and $x = 2.5$ about the y-axis. Find the volume (in cm^3) of material in the ring.

27-9   Integration by Use of Tables

In this chapter we have introduced certain basic integrals and have also brought in some methods of reducing other integrals to these basic forms. Often this transformation and integration requires a number of steps to be performed, and therefore integrals are tabulated for reference. The integrals found in tables have been derived by using the methods introduced thus far, as well as many other methods which can be used. Therefore, an understanding of the basic forms and some of the basic methods is very useful in finding integrals from tables. Such an understanding forms a basis for proper recognition of the forms which are used in the tables, as well as the types of results which may be expected. Therefore,

NOTE ▷ *the use of the tables depends on proper recognition of the form and the variables and constants of the integral.* The following examples illustrate the use of the table of integrals found in Table 5 in Appendix F. More extensive tables are available in other sources.

EXAMPLE A

Integrate: $\int \dfrac{x\,dx}{\sqrt{2+3x}}$.

We first note that this integral fits the form of Formula 6 of Table 5, with $u = x$, $a = 2$, and $b = 3$. Therefore,

$$\int \frac{x\,dx}{\sqrt{2+3x}} = -\frac{2(4-3x)\sqrt{2+3x}}{27} + C$$

EXAMPLE B

Integrate: $\int \dfrac{\sqrt{4-9x^2}}{x}\,dx$.

This fits the form of Formula 18, with proper identification of constants; $u = 3x$, $du = 3\,dx$, $a = 2$. Hence,

$$\int \frac{\sqrt{4-9x^2}}{x}\,dx = \int \frac{\sqrt{4-9x^2}}{3x}\,3\,dx$$

$$= \sqrt{4-9x^2} - 2\ln\left(\frac{2+\sqrt{4-9x^2}}{3x}\right) + C$$

EXAMPLE C

Integrate: $\int 5\sec^3 2x\,dx$.

This fits the form of Formula 37; $n = 3$, $u = 2x$, $du = 2\,dx$. And so

$$\int 5\sec^3 2x\,dx = 5\left(\frac{1}{2}\right)\int \sec^3 2x(2\,dx)$$

$$= \frac{5}{2}\frac{\sec 2x \tan 2x}{2} + \frac{5}{2}\left(\frac{1}{2}\right)\int \sec 2x(2\,dx)$$

For reference,
Eq. (27-12) is

$\int \sec u\,du$

$= \ln|\sec u + \tan u| + C$

To complete this integral, we must use the basic form of Eq. (27-12). Thus, we complete it by

$$\int 5\sec^3 2x\,dx = \frac{5\sec 2x \tan 2x}{4} + \frac{5}{4}\ln|\sec 2x + \tan 2x| + C$$

EXAMPLE D

Find the area bounded by $y = x^2 \ln 2x$, $y = 0$, and $x = e$.

From Fig. 27-15 we see that the area is

$$A = \int_{0.5}^{e} x^2 \ln 2x\,dx$$

This integral fits the form of Formula 46 if $u = 2x$. Thus, we have

$$A = \frac{1}{8}\int_{0.5}^{e} (2x)^2 \ln 2x(2\,dx) = \frac{1}{8}(2x)^3\left[\frac{\ln 2x}{3} - \frac{1}{9}\right]_{0.5}^{e}$$

$$= e^3\left(\frac{\ln 2e}{3} - \frac{1}{9}\right) - \frac{1}{8}\left(\frac{\ln 1}{3} - \frac{1}{9}\right) = e^3\left(\frac{3\ln 2e - 1}{9}\right) + \frac{1}{72}$$

$$= 9.118$$

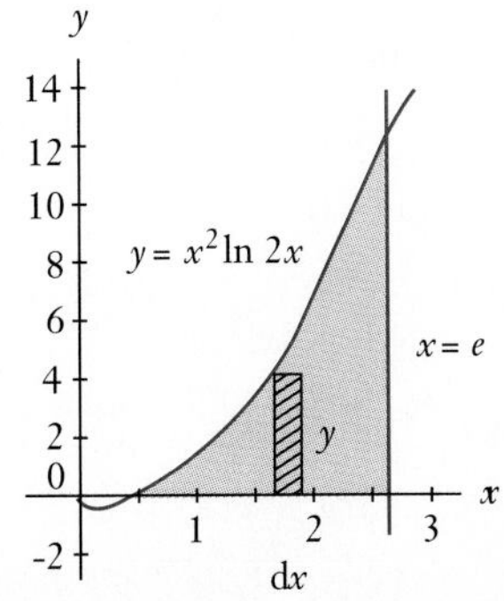

■ **Fig. 27-15**

The proper identification of u and du is the key step in the use of tables. Therefore, for the integrals in the following example the proper u and du, along with the appropriate formula from the table, are identified, but the integrations are not performed.

EXAMPLE E

1. $\int \dfrac{dx}{16x^2 - 9}$ $u = 4x,$ $du = 4\,dx,$ Formula 9

2. $\int x\sqrt{1 - x^4}\,dx$ $u = x^2,$ $du = 2x\,dx,$ Formula 15

3. $\int \dfrac{(4x^6 - 9)^{3/2}}{x}\,dx$ $u = 2x^3,$ $du = 6x^2\,dx,$ Formula 22
(Introduce a factor of x^2 into the numerator and the denominator.)

4. $\int x^3 \sin x^2\,dx$ $u = x^2,$ $du = 2x\,dx,$ Formula 47
$[x^3\,dx = x^2(x\,dx)]$

We summarize briefly here the approach to integrating a function. There are two methods of approaching a problem in order to obtain the exact result for the integral:

1. Write the integral such that it fits an integral form. Either a basic form as developed in this chapter or a form from a table of integrals may be used.

2. Use a technique of transforming the integral such that an integral form may be used. Techniques include those covered in this chapter and others which may be found in other sources.

Also, definite integrals may be approximated by methods such as the trapezoidal rule or Simpson's rule. There are other approximation methods, one of which is developed in Example E of Section 28-3.

Exercises 27-9

In Exercises 1 through 32, integrate each function by use of Table 5 in Appendix F.

1. $\int \dfrac{3x\,dx}{2 + 5x}$

2. $\int \dfrac{4x\,dx}{(1 + x)^2}$

3. $\int_2^7 4x\sqrt{2 + x}\,dx$

4. $\int \dfrac{dx}{x^2 - 4}$

5. $\int \sqrt{4 - x^2}\,dx$

6. $\int_0^{\pi/3} \sin^3 x\,dx$

7. $\int \sin 2x \sin 3x\,dx$

8. $\int \text{Arcsin } 3x\,dx$

9. $\int \dfrac{\sqrt{4x^2 - 9}}{x}\,dx$

10. $\int \dfrac{(9x^2 + 16)^{3/2}}{x}\,dx$

11. $\int \cos^5 4x\,dx$

12. $\int \tan^2 x\,dx$

13. $\int \text{Arctan } x^2(x\,dx)$

14. $\int 5xe^{4x}\,dx$

15. $\int_1^2 (4 - x^2)^{3/2}\,dx$

16. $\int \dfrac{3\,dx}{9 - 16x^2}$

17. $\int \dfrac{dx}{x\sqrt{4x^2 + 1}}$

18. $\int \dfrac{\sqrt{4 + x^2}}{x}\,dx$

19. $\int \dfrac{dx}{x\sqrt{1 - 4x^2}}$

20. $\int \dfrac{dx}{x(1 + 4x)^2}$

21. $\int \sin x \cos 5x\,dx$

22. $\int_0^2 x^2 e^{3x}\,dx$

23. $\int x^5 \cos x^3\,dx$

24. $\int 2 \sin^3 x \cos^2 x\,dx$

25. $\displaystyle\int \frac{2x\,dx}{(1-x^4)^{3/2}}$ **26.** $\displaystyle\int \frac{dx}{x(1-4x)}$ **27.** $\displaystyle\int_1^3 \frac{\sqrt{3+5x^2}\,dx}{x}$ **28.** $\displaystyle\int_0^1 \frac{\sqrt{9-4x^2}}{x}\,dx$

29. $\displaystyle\int x^3 \ln x^2\,dx$ **30.** $\displaystyle\int \frac{x\,dx}{x^2\sqrt{x^4-9}}$ **31.** $\displaystyle\int \frac{x^2\,dx}{(x^6-1)^{3/2}}$ **32.** $\displaystyle\int x^7\sqrt{x^4+4}\,dx$

In Exercises 33 through 40, evaluate the integrals by use of Table 5.

33. Find the length of arc of the curve $y = x^2$ from $x = 0$ to $x = 1$. (See Exercise 25 of Section 25-6.)

34. Find the moment of inertia with respect to its axis of the volume generated by rotating the area bounded by $y = \ln x$, $x = e$, and the x-axis about the y-axis.

35. Find the area bounded by $y = \text{Arctan } 2x$, $x = 2$, and the x-axis.   **Fig. 27-16**

36. Find the area bounded by $y = (x^2 + 4)^{-3/2}$, $x = 3$, and the axes.

37. Find the force (in lb) on the area bounded by $x = 1/\sqrt{1+y}$, $y = 0$, $y = 3$, and the y-axis, if the surface of the water is at the upper edge of the area. See Fig. 27-16.

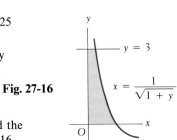

38. If 6.00 g of a chemical are placed in water, the time t (in min) it takes to dissolve half of the chemical is given by $t = 560 \displaystyle\int_3^6 \frac{dx}{x(x+4)}$, where x is the amount of undissolved chemical at any time. Evaluate t.

39. The force F, in pounds, required to move a particular lever mechanism is a function of the displacement s, in feet, according to $F = 4s\sqrt{4s+3}$. Find the work done if the displacement changes from zero to 2.50 ft.

40. If an electric charge Q is distributed along a wire of length $2a$, the force F exerted on an electric charge q placed at point P is $F = kqQ \displaystyle\int \frac{b\,dx}{(b^2+x^2)^{3/2}}$. Integrate to find F as a function of x.

27-10   Chapter Equations, Review Exercises, and Practice Test

Chapter Equations

Integrals

$$\int u^n\,du = \frac{u^{n+1}}{n+1} + C \qquad (n \neq -1) \tag{27-1}$$

$$\int \frac{du}{u} = \ln|u| + C \tag{27-2}$$

$$\int e^u\,du = e^u + C \tag{27-3}$$

$$\int \sin u\,du = -\cos u + C \tag{27-4}$$

$$\int \cos u\,du = \sin u + C \tag{27-5}$$

$$\int \sec^2 u\,du = \tan u + C \tag{27-6}$$

$$\int \csc^2 u\,du = -\cot u + C \tag{27-7}$$

$$\int \sec u \tan u\,du = \sec u + C \tag{27-8}$$

$$\int \csc u \cot u\,du = -\csc u + C \tag{27-9}$$

$$\int \tan u \, du = -\ln |\cos u| + C \tag{27-10}$$

$$\int \cot u \, du = \ln |\sin u| + C \tag{27-11}$$

$$\int \sec u \, du = \ln |\sec u + \tan u| + C \tag{27-12}$$

$$\int \csc u \, du = \ln |\csc u - \cot u| + C \tag{27-13}$$

Trigonometric relations
$$\cos^2 x + \sin^2 x = 1 \tag{27-14}$$

$$1 + \tan^2 x = \sec^2 x \tag{27-15}$$

$$1 + \cot^2 x = \csc^2 x \tag{27-16}$$

$$2 \cos^2 x = 1 + \cos 2x \tag{27-17}$$

$$2 \sin^2 x = 1 - \cos 2x \tag{27-18}$$

Root-mean-square value
$$y_{\text{rms}} = \sqrt{\frac{1}{T} \int_0^T y^2 \, dx} \tag{27-19}$$

Integrals
$$\int \frac{du}{\sqrt{a^2 - u^2}} = \text{Arcsin} \frac{u}{a} + C \tag{27-20}$$

$$\int \frac{du}{a^2 + u^2} = \frac{1}{a} \text{Arctan} \frac{u}{a} + C \tag{27-21}$$

$$\int u \, dv = uv - \int v \, du \tag{27-22}$$

Trigonometric substitutions
For $\sqrt{a^2 - x^2}$ use $x = a \sin \theta$

For $\sqrt{a^2 + x^2}$ use $x = a \tan \theta \tag{27-23}$

For $\sqrt{x^2 - a^2}$ use $x = a \sec \theta$

Review Exercises

In Exercises 1 through 36, integrate the given functions without the use of Table 5.

1. $\int e^{-2x} \, dx$

2. $\int e^{\cos 2x} \sin x \cos x \, dx$

3. $\int \frac{dx}{x(\ln 2x)^2}$

4. $\int x^{1/3} \sqrt{x^{4/3} + 1} \, dx$

5. $\int \frac{4 \cos x \, dx}{1 + \sin x}$

6. $\int \frac{\sec^2 x \, dx}{2 + \tan x}$

7. $\int \frac{2 \, dx}{25 + 49x^2}$

8. $\int \frac{dx}{\sqrt{1 - 4x^2}}$

9. $\int_0^{\pi/2} \cos^3 2x \, dx$

10. $\int_0^{\pi/8} \sec^3 2x \tan 2x \, dx$

11. $\int_0^2 \frac{x \, dx}{4 + x^2}$

12. $\int_1^e \frac{\ln x^2 \, dx}{x}$

13. $\int \sec^4 3x \tan 3x \, dx$

14. $\int \frac{\sin^3 x \, dx}{\sqrt{\cos x}}$

15. $\int \tan 3x \, dx$

16. $\int 3 \sec 4x \, dx$

17. $\int \sec^4 3x \, dx$

18. $\int \frac{\sin^2 2x \, dx}{1 + \cos 2x}$

19. $\int_0^{0.5} \frac{2e^{2x} - 3e^x}{e^{2x}} \, dx$

20. $\int \frac{4 - e^{\sqrt{x}}}{\sqrt{x} e^{\sqrt{x}}} \, dx$

21. $\int \frac{3x \, dx}{4 + x^4}$

22. $\int_1^3 \frac{x \, dx}{2(3 + x^2)}$

23. $\int \frac{dx}{\sqrt{4x^2 - 9}}$

24. $\int \frac{x^2 \, dx}{\sqrt{9 - x^2}}$

25. $\displaystyle\int \frac{e^{2x}\,dx}{\sqrt{e^{2x}+1}}$

26. $\displaystyle\int \frac{(4+\ln 2x)^3\,dx}{x}$

27. $\displaystyle\int 3\sin^2 3x\,dx$

28. $\displaystyle\int \sin^4 x\,dx$

29. $\displaystyle\int x\csc^2 2x\,dx$

30. $\displaystyle\int x\,\text{Arctan}\,x\,dx$

31. $\displaystyle\int e^{2x}\cos e^{2x}\,dx$

32. $\displaystyle\int \frac{dx}{x^2+6x+10}$

33. $\displaystyle\int_1^e \frac{(\ln x)^2\,dx}{x}$

34. $\displaystyle\int_1^3 \frac{2\,dx}{x^2-2x+5}$

35. $\displaystyle\int \frac{x^2-1}{x+2}\,dx$

36. $\displaystyle\int [\cos(\ln x)]\frac{dx}{x}$

In Exercises 37 through 64, solve the given problems by integration.

37. Show that $\int e^x(e^x+1)^2\,dx$ can be integrated in two ways. Explain the difference in the answers.

38. Show that $\int \frac{1}{x}(1+\ln x)\,dx$ can be integrated in two ways. Explain the difference in the answers.

39. Evaluate $\int \sin^2 x\,dx$ (a) by use of Eq. (27-18) and (b) by use of Eq. (27-22). Show that the results are equivalent.

40. Find the equation of the curve for which $dy/dx = e^x(2-e^x)^2$, if the curve passes through $(0, 4)$.

41. Find the equation of the curve for which $dy/dx = \sec^4 x$, if the curve passes through the origin.

42. Find the equation of the curve for which $\dfrac{dy}{dx} = \dfrac{\sqrt{4+x^2}}{x^4}$, if the curves passes through $(2, 1)$.

43. Find the area bounded by $y = 4e^{2x}$, $x = 1.5$, and the axes.

44. Find the area bounded by $y = x/(1+x)^2$, the x-axis, and the line $x = 4$.

45. Find the area inside the circle $x^2+y^2 = 25$ and to the right of the line $x = 3$.

46. Find the area bounded by $y = x\sqrt{x+4}$, $y = 0$, and $x = 5$.

47. Find the volume generated by rotating the area bounded by $y = xe^x$, $y = 0$, and $x = 2$ about the y-axis.

48. Find the volume generated by rotating about the y-axis the area bounded by $y = x+\sqrt{x+1}$, $x = 3$, and the axes.

49. Find the volume of the solid generated by rotating the area bounded by $y = e^x \sin x$ and the x-axis between $x = 0$ and $x = \pi$ about the x-axis.

50. Find the centroid of the area bounded by $y = \ln x$, $x = 2$, and the x-axis.

51. Find the length of arc along the curve of $y = \ln \sin x$ from $x = \frac{1}{3}\pi$ to $x = \frac{2}{3}\pi$. See Exercise 25 of Section 25-6.

52. Find the area of the surface generated by rotating the curve of $y = \sqrt{4-x^2}$ from $x = -2$ to $x = 2$ about the x-axis. See Exercise 27 of Section 25-6.

53. The change in the thermodynamic entity of entropy ΔS may be expressed as $\Delta S = \int (c_v/T)\,dT$, where c_v is the heat capacity at constant volume and T is the temperature. For increased accuracy, c_v is often given by the equation $c_v = a + bT + cT^2$, where a, b, and c are constants. Express ΔS as a function of temperature.

54. A second-order chemical reaction leads to the equation $dt = \dfrac{k_1\,dx}{a-x} + \dfrac{k_2\,dx}{b-x}$, where k_1 and k_2 are constants, a and b are initial concentrations, t is the time, and x is the decrease in concentration. Solve for t as a function of x.

55. Assuming a resisting force, the velocity (in ft/s) of a falling object in terms of the time (in seconds) is given by $\dfrac{dv}{32-0.1v} = dt$. If $v = 0$ when $t = 0$, find v as a function of t.

56. The power delivered to an electric circuit is given by $P = ei$, where e and i are the instantaneous voltage and the instantaneous current in the circuit, respectively. The mean power, averaged over a period $2\pi/\omega$, is given by $P_{av} = \dfrac{\omega}{2\pi}\displaystyle\int_0^{2\pi/\omega} ei\,dt$. If $e = 20\cos 2t$ and $i = 3\sin 2t$, find the average power over a period of $\frac{\pi}{4}$.

57. Find the root-mean-square value of the electric current i if $i = 2\sin t$.

58. In atomic theory, when finding the number n of atoms per unit volume of a substance, the equation $n = A \int_0^{\pi} e^{a \cos \theta} \sin \theta \, d\theta$ is used. Perform the indicated integration.

59. When considering the effects of an electric field on molecular orientation, the integral $\int_0^{\pi} (1 + k \cos \theta) \cos \theta \sin \theta \, d\theta$ is used. Evaluate this integral.

60. In finding the lift of the air flowing around an airplane wing, the integral $\int_{-\pi/2}^{\pi/2} \theta^2 \cos \theta \, d\theta$ is used. Evaluate this integral.

61. Find the volume of the piece of tubing in an oil distribution line shown in Fig. 27-17. All cross sections are circular.

62. A metal plate has the shape shown in Fig. 27-18. Find the x-coordinate of the centroid of the plate.

63. The nose cone of a space vehicle is to be covered with a heat shield. The cone is designed such that a cross section x feet from the tip and perpendicular to its axis is a circle of radius $1.5x^{2/3}$ feet. Find the surface area of the heat shield if the nose cone is 16.0 ft long. See Fig. 27-19. (See Exercise 27 of Section 25-6.)

64. A window has the shape of a semiellipse, as shown in Fig. 27-20. What is the area of the window?

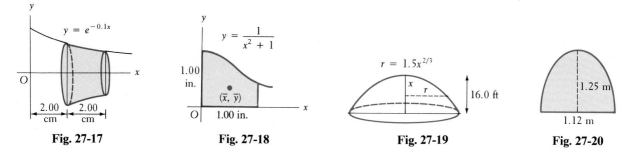

Fig. 27-17   Fig. 27-18   Fig. 27-19   Fig. 27-20

Practice Test

In Problems 1 through 6, evaluate the given integrals.

1. $\displaystyle\int (\sec x - \sec^3 x \tan x) \, dx$

2. $\displaystyle\int \sin^3 x \, dx$

3. $\displaystyle\int \tan^3 2x \, dx$

4. $\displaystyle\int \cos^2 4x \, dx$

5. $\displaystyle\int \frac{dx}{x^2 \sqrt{4 - x^2}}$

6. $\displaystyle\int xe^{-2x} \, dx$

7. The electric current in a certain circuit is given by $i = \displaystyle\int \frac{6t + 1}{4t^2 + 9} \, dt$, where t is the time. Integrate and find the resulting function if $i = 0$ for $t = 0$.

8. Find the first-quadrant area bounded by $y = \dfrac{1}{\sqrt{16 - x^2}}$ and $x = 3$.

28 Expansion of Functions in Series

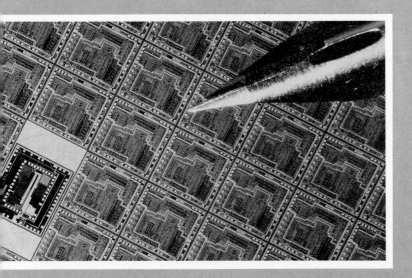

Many types of electronic devices are used to control the current in a circuit. In Section 28-6 we see how series are used to analyze one such device.

The values of the trigonometric functions can be determined exactly only for a few particular angles. The value of *e* can only be approximated in decimal form. The question arises as to how these values may be determined, particularly if a specified degree of accuracy is necessary. In this chapter we shall show how a given function may be expressed in terms of a polynomial. Once the polynomial which approximates a given function has been found, we shall be able to evaluate the function to any desired accuracy. A number of applications and other uses of this polynomial form will be shown as well.

In Section 28-6 we shall show how certain types of functions can be represented by a series of sine and cosine terms. Such series are very useful in the study of functions and applications which are periodic in nature, such as mechanical vibrations and the currents and voltages in an ac electric circuit.

We begin the chapter by reviewing the meanings and properties of sequences and series.

28-1 Infinite Series

In Chapter 18 we discussed arithmetic and geometric sequences. Also, we introduced the concept of an infinite geometric series. In this section we further develop these topics for use in the sections which follow.

As well as arithmetic sequences and geometric sequences, there are many other ways of generating sequences of numbers. The squares of the integers 1,

4, 9, 16, . . . form a sequence. Also, the successive approximations $x_1, x_2, x_3, \ldots$ obtained by using Newton's method in solving a particular equation form a sequence.

In general, *a* **sequence** (*or* **infinite sequence**) *is an infinite succession of numbers. Each of the numbers is a* **term** *of the sequence.* Each term of the sequence is associated with a positive integer, although at times it is convenient to associate the first term with zero (or some specified positive integer). We shall use a_n to designate the term of the sequence corresponding to the integer n.

EXAMPLE A ⎤   Find the first three terms of the sequence for which $a_n = 2n + 1$, $n = 1, 2, 3, \ldots$.

Substituting the values of n, we obtain the values

$$a_1 = 2(1) + 1 = 3, \ a_2 = 2(2) + 1 = 5, \ a_3 = 2(3) + 1 = 7, \ldots$$

Therefore, we have the sequence

3, 5, 7, . . .

If we are given $a_n = 2n + 1$ for $n = 0, 1, 2, \ldots$, the sequence is

1, 3, 5, . . .

As we stated in Chapter 18, *the indicated sum of the terms of a sequence is called an* **infinite series.** Thus, for the sequence

$$a_1, a_2, a_3, \ldots, a_n, \ldots$$

the associated infinite series is

$$a_1 + a_2 + a_3 + \cdots + a_n + \cdots$$

Using the summation sign Σ (see Section 21-2) to indicate the sum, we have

infinite series

$$\sum_{n=1}^{\infty} a_n = a_1 + a_2 + a_3 + \cdots + a_n + \cdots \tag{28-1}$$

We must realize that an infinite series, as shown in Eq. (28-1), does not have a sum in the ordinary sense of the word, for it is not possible actually to carry out the addition of infinitely many terms. Therefore, we define the sum for an infinite series in terms of a limit.

For the infinite series of Eq. (28-1) we let S_n represent the sum of the first n terms. Therefore,

$$S_1 = a_1$$
$$S_2 = a_1 + a_2$$
$$S_3 = a_1 + a_2 + a_3$$
$$S_n = a_1 + a_2 + a_3 + \cdots + a_n$$

partial sum

The numbers $S_1, S_2, S_3, \ldots, S_n, \ldots$ form a sequence. *Each term of this sequence is called a* **partial sum.** *We say that the infinite series, Eq. (28-1), is* **convergent and** *has the sum S given by*

$$S = \lim_{n \to \infty} S_n = \lim_{n \to \infty} \sum_{n=1}^{\infty} a_n$$

(28-2)

if this limit exists. If the limit does not exist, the series is **divergent.**

EXAMPLE B

For the infinite series

$$\sum_{n=0}^{\infty} \frac{1}{5^n} = \frac{1}{5^0} + \frac{1}{5^1} + \frac{1}{5^2} + \cdots + \frac{1}{5^n} + \cdots$$

the first six partial sums are

$$S_0 = 1 \qquad\qquad\qquad \textbf{first term}$$

$$S_1 = 1 + \frac{1}{5} = 1.2 \qquad\qquad \textbf{sum of first two terms}$$

$$S_2 = 1 + \frac{1}{5} + \frac{1}{25} = 1.24 \qquad \textbf{sum of first three terms}$$

$$S_3 = 1 + \frac{1}{5} + \frac{1}{25} + \frac{1}{125} = 1.248$$

$$S_4 = 1 + \frac{1}{5} + \frac{1}{25} + \frac{1}{125} + \frac{1}{625} = 1.2496$$

$$S_5 = 1 + \frac{1}{5} + \frac{1}{25} + \frac{1}{125} + \frac{1}{625} + \frac{1}{3125} = 1.24992$$

where the values are conveniently found by use of a calculator. Here it appears that the sequence of partial sums approaches the value 1.25. We therefore conclude that the infinite series converges and that its sum is approximately 1.25. ■

EXAMPLE C

The infinite series

$$\sum_{n=1}^{\infty} 5^n = 5 + 5^2 + 5^3 + \cdots + 5^n + \cdots$$

is a divergent series. The first four partial sums are

$$S_1 = 5, \qquad S_2 = 30, \qquad S_3 = 155, \quad \text{and} \quad S_4 = 780$$

Obviously they are increasing without bound.
    The infinite series

$$\sum_{n=0}^{\infty} (-1)^n = 1 + (-1) + 1 + (-1) + \cdots + (-1)^n + \cdots$$

has as its first five partial sums

$$S_0 = 1, \qquad S_1 = 0, \qquad S_2 = 1, \qquad S_3 = 0, \quad \text{and} \quad S_4 = 1$$

These values do not approach a limiting value, and therefore the series diverges. ■

Since convergent series are those which have a value attached to them, they are the ones which are of primary use to us. However, it is not generally easy to determine whether a given series is convergent, and many types of tests have been developed for this purpose.

One important series for which we are able to determine the convergence, and its sum if convergent, is the geometric series. For this series the nth partial sum is

$$S_n = a_1 + a_1 r + a_1 r^2 + \cdots + a_1 r^{n-1}$$

where r is the fixed number by which we multiply a given term to get the next term. In Chapter 18 we determined that if $|r| < 1$, the sum S of the infinite geometric series is

$$S = \lim_{n \to \infty} S_n = \frac{a_1}{1 - r} \tag{28-3}$$

If $r = 1$, we see that the series is $a_1 + a_1 + a_1 + \cdots + a_1 + \cdots$ and is therefore divergent. If $r = -1$, the series is $a_1 - a_1 + a_1 - a_1 + \cdots$ and is also divergent.

NOTE $\triangleright$ If $|r| > 1$, $\lim_{n \to \infty} r^n$ is unbounded. Therefore, *the geometric series is convergent only if* $|r| < 1$ *and has the value given by Eq. (28-3).*

EXAMPLE D   Show that the infinite series

$$\sum_{n=0}^{\infty} \frac{1}{5^n} = \frac{1}{5^0} + \frac{1}{5^1} + \frac{1}{5^2} + \cdots + \frac{1}{5^n} + \cdots$$

is convergent and find its sum. This is the same series as in Example B.

We see that this is a geometric series with $r = \frac{1}{5}$. Since $|r| < 1$, the series is convergent. We find the sum to be

$$S = \frac{1}{1 - \frac{1}{5}} = \frac{1}{\frac{4}{5}} = \frac{5}{4} = 1.25 \qquad \text{using Eq. (28-3)}$$

■ We see that this agrees with the conclusion of Example B.

Exercises 28-1

In Exercises 1 through 4, give the first four terms of the sequences for which a_n is given.

1. $a_n = n^2, \quad n = 1, 2, 3, \ldots$

2. $a_n = \frac{2}{3^n}, \quad n = 1, 2, 3, \ldots$

3. $a_n = \frac{1}{n + 2}, \quad n = 0, 1, 2, \ldots$

4. $a_n = \frac{n^2 + 1}{2n + 1}, \quad n = 0, 1, 2, \ldots$

See Appendix E for a computer program for finding the first n partial sums.

In Exercises 5 through 8, give (a) the first four terms of the sequence for which a_n is given and (b) the first four terms of the infinite series associated with the sequence.

5. $a_n = \left(-\frac{2}{5}\right)^n, \quad n = 1, 2, 3, \ldots$

6. $a_n = \frac{1}{n} + \frac{1}{n + 1}, \quad n = 1, 2, 3, \ldots$

7. $a_n = 1 + (-1)^n, \quad n = 0, 1, 2, \ldots$

8. $a_n = \frac{1}{n(n + 1)}, \quad n = 2, 3, 4, \ldots$

In Exercises 9 through 12, find the nth term of the given infinite series for n = 1, 2, 3, ...

9. $\dfrac{1}{2} + \dfrac{1}{3} + \dfrac{1}{4} + \dfrac{1}{5} + \cdots$

10. $\dfrac{1}{2} + \dfrac{1}{4} + \dfrac{1}{8} + \dfrac{1}{16} + \cdots$

11. $\dfrac{1}{2 \times 3} + \dfrac{1}{3 \times 4} + \dfrac{1}{4 \times 5} + \dfrac{1}{5 \times 6} + \cdots$

12. $-\dfrac{2}{3} + \dfrac{4}{9} - \dfrac{8}{27} + \dfrac{16}{81} - \cdots$

In Exercises 13 through 20, find the first five partial sums of the given series and determine whether the series appears to be convergent or divergent. If it is convergent, find its approximate sum.

13. $1 + \dfrac{1}{8} + \dfrac{1}{27} + \dfrac{1}{64} + \dfrac{1}{125} + \cdots$

14. $1 + 2 + 5 + 10 + 17 + \cdots$

15. $1 + \dfrac{1}{2} + \dfrac{2}{3} + \dfrac{3}{4} + \dfrac{4}{5} + \cdots$

16. $\dfrac{1}{3} - \dfrac{1}{9} + \dfrac{1}{27} - \dfrac{1}{81} + \dfrac{1}{243} - \cdots$

17. $\displaystyle\sum_{n=0}^{\infty} (-n)$

18. $\displaystyle\sum_{n=1}^{\infty} \dfrac{2}{n(n+1)}$

19. $\displaystyle\sum_{n=1}^{\infty} \dfrac{2n+1}{n^2(n+1)^2}$

20. $\displaystyle\sum_{n=1}^{\infty} \dfrac{n}{2n+1}$

In Exercises 21 through 28, test each of the given geometric series for convergence or divergence. Find the sum of each convergent series.

21. $1 + 2 + 4 + \cdots + 2^n + \cdots$

22. $1 + \dfrac{1}{2} + \dfrac{1}{4} + \cdots + \dfrac{1}{2^n} + \cdots$

23. $1 - \dfrac{1}{3} + \dfrac{1}{9} - \cdots + \left(-\dfrac{1}{3}\right)^n + \cdots$

24. $1 - \dfrac{3}{2} + \dfrac{9}{4} - \cdots + \left(-\dfrac{3}{2}\right)^n + \cdots$

25. $10 + 9 + 8.1 + 7.29 + 6.561 + \cdots$

26. $4 + 1 + \dfrac{1}{4} + \dfrac{1}{16} + \dfrac{1}{64} + \cdots$

27. $512 - 64 + 8 - 1 + \dfrac{1}{8} - \cdots$

28. $16 + 12 + 9 + \dfrac{27}{4} + \dfrac{81}{16} + \cdots$

In Exercises 29 through 32, solve the given problems as indicated.

29. Using a calculator, take successive square roots of 2 and find at least 20 approximate values for the terms of the sequence $2^{1/2}, 2^{1/4}, 2^{1/8}, \ldots$. From the values obtained, (a) what do you observe about the value of $\lim_{n \to \infty} 2^{1/n}$? (b) Determine whether the infinite series for this sequence converges or diverges. (Also, try 0.01 and 100 in place of 2 and note the results.)

30. If an electric discharge is passed through hydrogen gas, a spectrum of isolated parallel lines, called the Balmer series, is formed. See Fig. 28-1. The wavelengths λ (in nm) of the light for these lines is given by the formula

Fig. 28-1

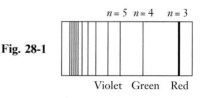

$$\dfrac{1}{\lambda} = 1.097 \times 10^{-2}\left(\dfrac{1}{2^2} - \dfrac{1}{n^2}\right), \qquad n = 3, 4, 5, \ldots$$

Find the wavelengths of the first three lines and the shortest wavelength of all of the lines of the series.

31. Use geometric series to show that $\displaystyle\sum_{n=0}^{\infty} x^n = \dfrac{1}{1-x}$ for $|x| < 1$.

32. Use geometric series to show that $\displaystyle\sum_{n=0}^{\infty} (-1)^n x^n = \dfrac{1}{1+x}$ for $|x| < 1$.

28-2 Maclaurin Series

In this section we shall develop a very important basic polynomial form of a function. Before developing the method using calculus, we shall review how this can be done for some functions algebraically.

EXAMPLE A

$$1 + \frac{x}{2}$$
$$2 - x \overline{\smash)2}$$
$$\underline{2 - x}$$
$$x$$
$$x - \frac{x^2}{2}$$
$$\underline{\phantom{x - {}}\frac{x^2}{2}}$$
$$\frac{x^2}{2}$$

By using long division (as started at left), we have

$$\frac{2}{2 - x} = 1 + \frac{1}{2}x + \frac{1}{4}x^2 + \cdots + \left(\frac{1}{2}x\right)^{n-1} + \cdots \tag{1}$$

where n is the number of the term of the expression on the right. Since x represents a number, the right-hand side of Eq. (1) becomes a geometric series.

From Eq. (28-3) we know that the sum of a geometric series with first term a_1 and common ratio r is

$$S = \frac{a_1}{1 - r}$$

where $|r| < 1$ and the series converges.

If $x = 1$, the right-hand side of Eq. (1) is

$$1 + \frac{1}{2} + \frac{1}{4} + \cdots + \left(\frac{1}{2}\right)^{n-1} + \cdots$$

For this series, $r = \frac{1}{2}$ and $a_1 = 1$, which means that the series converges and $S = 2$. If $x = 3$, the right-hand side of Eq. (1) is

$$1 + \frac{3}{2} + \frac{9}{4} + \cdots + \left(\frac{3}{2}\right)^{n-1} + \cdots$$

which diverges since $r > 1$. Referring to the left side of Eq. (1), we see that it also equals 2 when $x = 1$. Thus we see that the two sides agree for $x = 1$, but that the series diverges for $x = 3$. In fact, as long as $|x| < 2$, the series will converge to the value of the function on the left. From this we can conclude that the series on the right side properly represents the function on the left, as long as $|x| < 2$.

From Example A, we see that an algebraic function may be properly represented by a function of the form

power series

$$\boxed{f(x) = a_0 + a_1 x + a_2 x^2 + \cdots + a_n x^n + \cdots} \tag{28-4}$$

Equation (28-4) is known as a **power-series expansion** *of the function $f(x)$.* The problem now arises as to whether or not functions in general may be represented in this form. If such a representation were possible, it would provide a means of evaluating the transcendental functions for the purpose of making tables of values. Also, since a power-series expansion is in the form of a polynomial, it makes algebraic operations much simpler due to the properties of polynomials. A further study of calculus shows many other uses of power series.

In Example A we saw that the function could be represented by a power series as long as $|x| < 2$. That is, if we substitute any value of x in this interval into the series and also into the function, the series will converge to the value of the function. *This interval of values for which the series converges is called the* **interval of convergence.**

EXAMPLE B ———————— In Example A the interval of convergence for the series

$$1 + \frac{1}{2}x + \frac{1}{4}x^2 + \cdots + \left(\frac{1}{2}x\right)^{n-1} + \cdots$$

is $|x| < 2$. We saw that the series converges for $x = 1$, with $S = 2$, and that the value of the function is 2 for $x = 1$. This verifies that $x = 1$ is in the interval of convergence.

Also, we saw that the series diverges for $x = 3$, which verifies that $x = 3$ is ■ not in the interval of convergence.

At this point we shall assume that unless otherwise noted the functions with which we shall be dealing may be properly represented by a power-series expansion (it takes more advanced methods to prove that this is generally possible), for appropriate intervals of convergence. We shall find that the methods of calculus are very useful in developing the method of general representation. Thus, writing a general power series, along with the first few derivatives, we have

$$f(x) = a_0 + a_1 x + a_2 x^2 + a_3 x^3 + a_4 x^4 + a_5 x^5 + \cdots + a_n x^n + \cdots$$
$$f'(x) = a_1 + 2a_2 x + 3a_3 x^2 + 4a_4 x^3 + 5a_5 x^4 + \cdots + na_n x^{n-1} + \cdots$$
$$f''(x) = 2a_2 + 2(3)a_3 x + 3(4)a_4 x^2 + 4(5)a_5 x^3 + \cdots + (n-1)na_n x^{n-2} + \cdots$$
$$f'''(x) = 2(3)a_3 + 2(3)(4)a_4 x + 3(4)(5)a_5 x^2 + \cdots + (n-2)(n-1)na_n x^{n-3} + \cdots$$
$$f^{iv}(x) = 2(3)(4)a_4 + 2(3)(4)(5)a_5 x + \cdots + (n-3)(n-2)(n-1)na_n x^{n-4} + \cdots$$

NOTE ▷ Regardless of the values of the constants a_n for any power series, *if $x = 0$, the left and right sides must be equal,* and all the terms on the right are zero except the first. Thus, setting $x = 0$ in each of the above equations, we have

$$f(0) = a_0 \qquad f'(0) = a_1 \qquad f''(0) = 2a_2$$
$$f'''(0) = 2(3)a_3 \qquad f^{iv}(0) = 2(3)(4)a_4$$

Solving each of these for the constants a_n, we have

$$a_0 = f(0) \qquad a_1 = f'(0) \qquad a_2 = \frac{f''(0)}{2!} \qquad a_3 = \frac{f'''(0)}{3!} \qquad a_4 = \frac{f^{iv}(0)}{4!}$$

Substituting these into the expression for $f(x)$, we have

Maclaurin series

$$f(x) = f(0) + f'(0)x + \frac{f''(0)x^2}{2!} + \frac{f'''(0)x^3}{3!} + \cdots + \frac{f^{(n)}(0)x^n}{n!} + \cdots \qquad \text{(28-5)}$$

Equation (28-5) is known as the **Maclaurin series expansion** *of a function.* For a function to be represented by a Maclaurin expansion, the function and all its derivatives must exist at $x = 0$. Also, we note that the factorial notation introduced in Section 18-4 is used in writing the Maclaurin series expansion.

As we mentioned earlier, one of the uses we will make of series expansions is that of determining the values of functions for particular values of x. If x is sufficiently small, successive terms become smaller and smaller and the series will converge rapidly. This is considered in the sections which follow.

EXAMPLE C

Find the first four terms of the Maclaurin series expansion of $f(x) = \dfrac{2}{2 - x}$.

This is written as

$$f(x) = \frac{2}{2 - x} \qquad f(0) = 1 \qquad\qquad \text{find derivatives}$$
$$\text{and evaluate}$$
$$f'(x) = \frac{2}{(2 - x)^2} \qquad f'(0) = \frac{1}{2} \qquad \text{each at } x = 0$$

$$f''(x) = \frac{4}{(2 - x)^3} \qquad f''(0) = \frac{1}{2}$$

$$f'''(x) = \frac{12}{(2 - x)^4} \qquad f'''(0) = \frac{3}{4}$$

$$f(x) = 1 + \frac{1}{2}x + \frac{1}{2}\left(\frac{x^2}{2!}\right) + \frac{3}{4}\left(\frac{x^3}{3!}\right) + \cdots \qquad \text{using Eq. (28-5)}$$

or

$$\frac{2}{2 - x} = 1 + \frac{1}{2}x + \frac{1}{4}x^2 + \frac{1}{8}x^3 + \cdots$$

We see that this result agrees with that obtained by direct division. ∎

EXAMPLE D

Find the first four terms of the Maclaurin series expansion of $f(x) = e^{-x}$.

We write

$$f(x) = e^{-x} \qquad f(0) = 1 \qquad\qquad \text{find derivatives}$$
$$\text{and evaluate}$$
$$f'(x) = -e^{-x} \qquad f'(0) = -1 \qquad \text{each at } x = 0$$
$$f''(x) = e^{-x} \qquad f''(0) = 1$$
$$f'''(x) = -e^{-x} \qquad f'''(0) = -1$$

$$f(x) = 1 + (-1)x + 1\left(\frac{x^2}{2!}\right) + (-1)\left(\frac{x^3}{3!}\right) + \cdots \qquad \text{using Eq. (28-5)}$$

or

$$e^{-x} = 1 - x + \frac{x^2}{2!} - \frac{x^3}{3!} + \cdots$$

∎

EXAMPLE E Find the first three nonzero terms of the Maclaurin series expansion of $f(x) = \sin 2x$.

We have

$$\begin{array}{llll} f(x) = \sin 2x & f(0) = 0 & f'''(x) = -8\cos 2x & f'''(0) = -8 \\ f'(x) = 2\cos 2x & f'(0) = 2 & f^{\text{iv}}(x) = 16\sin 2x & f^{\text{iv}}(0) = 0 \\ f''(x) = -4\sin 2x & f''(0) = 0 & f^{\text{v}}(x) = 32\cos 2x & f^{\text{v}}(0) = 32 \end{array}$$

$$f(x) = 0 + 2x + 0 + (-8)\frac{x^3}{3!} + 0 + 32\frac{x^5}{5!} + \cdots$$

or

$$\sin 2x = 2x - \frac{4}{3}x^3 + \frac{4}{15}x^5 - \cdots$$

This series is called an **alternating series,** *since precisely every other term is negative.*

EXAMPLE F A lever is attached to a spring as shown in Fig. 28-2. The frictional forces in the spring are just sufficient so that the lever does not oscillate after being depressed. Such motion is called *critically damped.* The displacement y as a function of the time t for one such case is $y = (1 + t)e^{-t}$. In order to study the motion for small values of t, a polynomial form of $y = f(t)$ is to be used. Find the first four nonzero terms of the expansion.

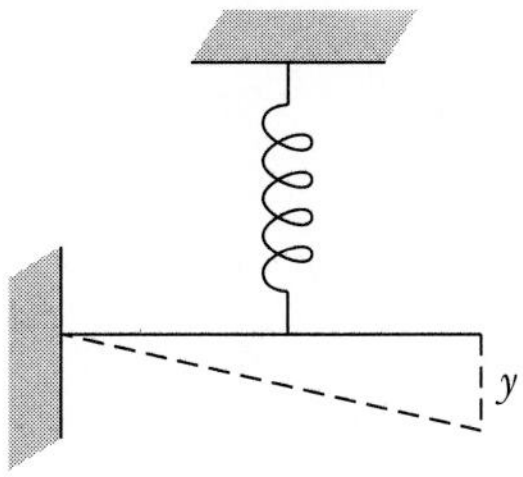

Fig. 28-2

The solution is as follows:

$$\begin{array}{ll} f(t) = (1 + t)e^{-t} & f(0) = 1 \\ f'(t) = (1 + t)e^{-t}(-1) + e^{-t} = -te^{-t} & f'(0) = 0 \\ f''(t) = te^{-t} - e^{-t} & f''(0) = -1 \\ f'''(t) = -te^{-t} + e^{-t} + e^{-t} = 2e^{-t} - te^{-t} & f'''(0) = 2 \\ f^{\text{iv}}(t) = -2e^{-t} + te^{-t} - e^{-t} = te^{-t} - 3e^{-t} & f^{\text{iv}}(0) = -3 \end{array}$$

$$f(t) = 1 + 0 + (-1)\frac{t^2}{2!} + 2\frac{t^3}{3!} + (-3)\frac{t^4}{4!} + \cdots$$

or

$$(1 + t)e^{-t} = 1 - \frac{t^2}{2} + \frac{t^3}{3} - \frac{t^4}{8} + \cdots$$

Exercises 28-2

In Exercises 1 through 16, find in each case the first three nonzero terms of the Maclaurin expansion of the given functions.

1. $f(x) = e^x$ **2.** $f(x) = \sin x$ **3.** $f(x) = \cos x$ **4.** $f(x) = \ln(1 + x)$

5. $f(x) = \sqrt{1 + x}$ **6.** $f(x) = \dfrac{1}{(1 - x)^{1/3}}$ **7.** $f(x) = e^{-2x}$ **8.** $f(x) = \dfrac{1}{2}(e^x + e^{-x})$

9. $f(x) = \cos 4x$ **10.** $f(x) = e^x \sin x$ **11.** $f(x) = \dfrac{1}{1 - x}$ **12.** $f(x) = \dfrac{1}{(1 + x)^2}$

13. $\ln(1 - 2x)$ **14.** $(1 + x)^{3/2}$ **15.** $\cos \frac{1}{2}x$ **16.** $\ln(1 + 4x)$

In Exercises 17 through 24, find the first two nonzero terms of the Maclaurin expansion of the given functions.

17. $f(x) = \text{Arctan } x$ **18.** $f(x) = \cos x^2$ **19.** $f(x) = \tan x$ **20.** $f(x) = \sec x$

21. $f(x) = \ln \cos x$ **22.** $f(x) = xe^{\sin x}$ **23.** $f(x) = \sin^2 x$ **24.** $f(x) = e^{-x^2}$

In Exercises 25 through 32, solve the given problems.

25. Is it possible to find a Maclaurin expansion for (a) $f(x) = \csc x$ or (b) $f(x) = \ln x$? Explain.

26. Is it possible to find a Maclaurin expansion for (a) $f(x) = \sqrt{x}$ or (b) $f(x) = \sqrt{1 + x}$? Explain.

27. Find the first three nonzero terms of the Maclaurin expansion for (a) $f(x) = e^x$ and (b) $f(x) = e^{x^2}$. Compare these expansions.

28. By finding the Maclaurin expansion of $f(x) = (1 + x)^n$, derive the first four terms of the binomial series, which is Eq. (18-7). Its interval of convergence is $|x| < 1$ for all values of n.

29. If $f(x) = x^3$, show that this function is obtained when a Maclaurin expansion is found.

30. If $f(x) = x^4 + 2x^2$, show that this function is obtained when a Maclaurin expansion is found.

31. The reliability R ($0 \le R \le 1$) of a certain computer system is $R = e^{-0.001t}$, where t is the time of operation in minutes. Express $R = f(t)$ in polynomial form by using the first three terms of the Maclaurin expansion.

32. In analyzing the optical paths of light from a narrow slit S to a point P as shown in Fig. 28-3, the law of cosines is used to obtain the equation

$$c^2 = a^2 + (a + b)^2 - 2a(a + b) \cos \frac{s}{a}$$

Fig. 28-3

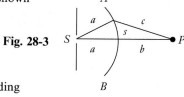

where s is part of the circular arc $\overset{\frown}{AB}$. By using two nonzero terms of the Maclaurin expansion of $\cos \frac{s}{a}$, simplify the right side of the equation. (In finding the expansion, let $x = \frac{s}{a}$ and then substitute back into the expansion.)

28-3   Certain Operations with Series

The series found in the first four exercises and Exercise 28 (the binomial series) of Section 28-2 are of particular importance. They are used to find values of exponential functions, trigonometric functions, logarithms, powers, and roots. Also, they may be used to develop other series. We shall give them here for reference, along with their intervals of convergence.

$$e^x = 1 + x + \frac{x^2}{2!} + \frac{x^3}{3!} + \cdots \qquad \textbf{(all } x\textbf{)} \qquad (28\text{-}6)$$

$$\sin x = x - \frac{x^3}{3!} + \frac{x^5}{5!} - \cdots \qquad \textbf{(all } x\textbf{)} \qquad (28\text{-}7)$$

$$\cos x = 1 - \frac{x^2}{2!} + \frac{x^4}{4!} - \cdots \qquad \textbf{(all } x\textbf{)} \qquad (28\text{-}8)$$

$$\ln(1 + x) = x - \frac{x^2}{2} + \frac{x^3}{3} - \frac{x^4}{4} + \cdots \qquad \textbf{(}|x| < 1\textbf{)} \qquad (28\text{-}9)$$

$$(1 + x)^n = 1 + nx + \frac{n(n - 1)}{2!} x^2 + \cdots \qquad \textbf{(}|x| < 1\textbf{)} \qquad (28\text{-}10)$$

In the next section we shall see how we make use of these series in the development of tables. In this section we shall not only see how new series may be developed by using the above basic series, but also we shall see certain other uses of series.

When we discussed functions in Chapter 2, we briefly mentioned functions such as $f(2x)$, $f(x^2)$, and $f(-x)$. *By using functional notation and the preceding series, we can find the series expansions of a great many other series without the use of direct expansion.* This can often save a great deal of time in finding a desired series. The examples given here illustrate the use of series for this purpose.

EXAMPLE A ——— Find the Maclaurin expansion of e^{2x}.

From Eq. (28-6), we know the expansion of e^x. Hence,

$$f(x) = 1 + x + \frac{x^2}{2!} + \frac{x^3}{3!} + \cdots$$

Since $e^{2x} = f(2x)$, we have

$$f(2x) = 1 + (2x) + \frac{(2x)^2}{2!} + \frac{(2x)^3}{3!} + \cdots \qquad \text{in } f(x), \text{ replace } x \text{ by } 2x$$

or

$$e^{2x} = 1 + 2x + 2x^2 + \frac{4x^3}{3} + \cdots$$

■

EXAMPLE B ——— Find the Maclaurin expansion of $\sin x^2$.

From Eq. (28-7), we know the expansion of $\sin x$. Therefore,

$$f(x) = x - \frac{x^3}{3!} + \frac{x^5}{5!} - \cdots$$

Since $\sin x^2 = f(x^2)$, we have

$$f(x^2) = (x^2) - \frac{(x^2)^3}{3!} + \frac{(x^2)^5}{5!} - \cdots \qquad \text{in } f(x), \text{ replace } x \text{ by } x^2$$

or

$$\sin x^2 = x^2 - \frac{x^6}{3!} + \frac{x^{10}}{5!} - \cdots$$

■ Direct expansion of this series is quite lengthy.

The basic algebraic operations may be applied to series in the same manner they are applied to polynomials. That is, we may add, subtract, multiply, or divide series in order to obtain other series. The interval of convergence for the resulting series is that which is common to those of the series being used. The multiplication of series is illustrated in the following example.

EXAMPLE C ———— Multiply the series for e^x and $\cos x$ in order to obtain the series expansion for $e^x \cos x$.

Using the series expansion for e^x and $\cos x$ as shown in Eqs. (28-6) and (28-8), we have the following indicated multiplication:

$$e^x \cos x = \left(1 + x + \frac{x^2}{2!} + \frac{x^3}{3!} + \frac{x^4}{4!} + \cdots\right)\left(1 - \frac{x^2}{2!} + \frac{x^4}{4!} - \cdots\right)$$

By multiplying the series on the right, we have the following result, considering through the x^4 terms in the product.

$$1\left(1 - \frac{x^2}{2!} + \frac{x^4}{4!}\right) \quad x\left(1 - \frac{x^2}{2!}\right) \quad \frac{x^2}{2!}\left(1 - \frac{x^2}{2!}\right) \quad \left(\frac{x^3}{3!} + \frac{x^4}{4!}\right) \tag{1}$$

$$e^x \cos x = 1 - \frac{x^2}{2} + \frac{x^4}{24} + x - \frac{x^3}{2} + \frac{x^2}{2} - \frac{x^4}{4} + \frac{x^3}{6} + \frac{x^4}{24} + \cdots$$

$$= 1 + x - \frac{1}{3}x^3 - \frac{1}{6}x^4 + \cdots$$

∎

It is also possible to use the operations of differentiation and integration to obtain series expansions, although the proof of this is found in more advanced texts. Consider the following example.

EXAMPLE D ———— Show that by differentiating the expansion for $\ln(1 + x)$ term by term, the result is the same as the expansion for $\dfrac{1}{1 + x}$.

The series for $\ln(1 + x)$ is shown in Eq. (28-9) as

$$\ln(1 + x) = x - \frac{x^2}{2} + \frac{x^3}{3} - \frac{x^4}{4} + \cdots$$

Differentiating, we have

$$\frac{1}{1 + x} = 1 - \frac{2x}{2} + \frac{3x^2}{3} - \frac{4x^3}{4} + \cdots$$

$$= 1 - x + x^2 - x^3 + \cdots$$

Using the binomial expansion for $\dfrac{1}{1 + x} = (1 + x)^{-1}$, we have

$$(1 + x)^{-1} = 1 + (-1)x + \frac{(-1)(-2)}{2!}x^2 + \frac{(-1)(-2)(-3)}{3!}x^3 + \cdots \qquad \text{using Eq. (28-10)}$$
$$\text{with } n = -1$$

$$= 1 - x + x^2 - x^3 + \cdots$$

∎ We see that the results are the same.

We can use algebraic operations on series to verify that the definition of the exponential form of a complex number, as shown in Eq. (11-10), is consistent with other definitions. The only assumption required here is that the Maclaurin expansions for e^x, sin x, and cos x are also valid for complex numbers. This is shown in advanced calculus. Thus,

$$e^{j\theta} = 1 + j\theta + \frac{(j\theta)^2}{2!} + \frac{(j\theta)^3}{3!} + \cdots = 1 + j\theta - \frac{\theta^2}{2!} - j\frac{\theta^3}{3!} + \cdots \qquad (28\text{-}11)$$

$$j \sin \theta = j\theta - j\frac{\theta^3}{3!} + \cdots \qquad (28\text{-}12)$$

$$\cos \theta = 1 - \frac{\theta^2}{2!} + \cdots \qquad (28\text{-}13)$$

When we add the terms of Eq. (28-12) to those of Eq. (28-13), the result is the series given in Eq. (28-11). Thus,

$$\boxed{e^{j\theta} = \cos \theta + j \sin \theta} \qquad (28\text{-}14)$$

A comparison of Eqs. (11-10) and (28-14) indicates the reason for the choice of the definition of the exponential form of a complex number.

An additional use of power series is now shown. Many integrals which occur in practice cannot be integrated by methods given in the preceding chapters. However, power series can be very useful in giving excellent approximations to some definite integrals.

EXAMPLE E

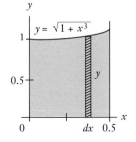

$y = \sqrt{1 + x^3}$

Fig. 28-4

Find the first-quadrant area bounded by $y = \sqrt{1 + x^3}$ and $x = 0.5$.

From Fig. 28-4, we see that the area is

$$A = \int_0^{0.5} \sqrt{1 + x^3}\, dx$$

This integral does not fit any form we have used. However, its value can be closely approximated by using the binomial expansion for $\sqrt{1 + x^3}$ and then integrating.

Using the binomial expansion to find the first three terms of the expansion for $\sqrt{1 + x^3}$, we have

$$\sqrt{1 + x^3} = (1 + x^3)^{0.5} = 1 + 0.5x^3 + \frac{0.5(-0.5)}{2}(x^3)^2 + \cdots$$

$$= 1 + 0.5x^3 - 0.125x^6 + \cdots$$

Substituting in the integral, we have

$$A = \int_0^{0.5} (1 + 0.5x^3 - 0.125x^6 + \cdots)\, dx$$

$$= x + \frac{0.5}{4}x^4 - \frac{0.125}{7}x^7 + \cdots \Big|_0^{0.5}$$

$$= 0.5 + 0.0078125 - 0.0001395 + \cdots = 0.507673 + \cdots$$

We can see that each of the terms omitted was very small. The result shown is correct to four decimal places, or $A = 0.5077$. Additional accuracy can be obtained by using more terms of the expansion. ∎

EXAMPLE F

Evaluate: $\int_0^{0.1} e^{-x^2}\, dx$.

We write

$$e^{-x^2} = 1 + (-x^2) + \frac{(-x^2)^2}{2!} + \cdots \qquad \textbf{using Eq. (28-6)}$$

Thus,

$$\int_0^{0.1} e^{-x^2}\, dx = \int_0^{0.1}\left(1 - x^2 + \frac{x^4}{2} - \cdots\right)dx \qquad \textbf{substitute}$$

$$= \left(x - \frac{x^3}{3} + \frac{x^5}{10} - \cdots\right)\Big|_0^{0.1} \qquad \textbf{integrate}$$

$$= 0.1 - \frac{0.001}{3} + \frac{0.00001}{10} = 0.0996677 \qquad \textbf{evaluate}$$

■ This answer is correct to the indicated accuracy.

The question of accuracy now arises. The integrals just evaluated indicate that the more terms used, the greater the accuracy of the result. As an indication of the accuracy involved, Fig. 28-5 shows the graphs of $y = \sin x$ and the graphs of

$$y = x, \qquad y = x - \frac{x^3}{3!}, \quad \text{and} \quad y = x - \frac{x^3}{3!} + \frac{x^5}{5!}$$

which are the first three approximations. We can see that each term added gives a better fit to the curve of $\sin x$. Also, a graphical representation of the meaning of series expansions is indicated.

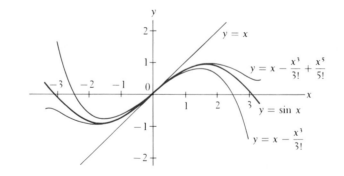

Fig. 28-5

We have just shown that the more terms included, the more accurate the result. For small values of x, a Maclaurin series can give good accuracy with a very few terms. When this happens, we see that the series *converges* rapidly, as we mentioned earlier. For this reason a Maclaurin series is of particular use if small values of x are being considered. If larger values of x are being considered, usually a function is expanded in a Taylor's series (see Section 28-5). Of course, when we omit the later terms in any series, a certain error is inherent in the calculation.

Exercises 28-3

In Exercises 1 through 8, find in each case the first four nonzero terms of the Maclaurin expansions of the given functions by using Eqs. (28-6) to (28-10).

1. $f(x) = e^{3x}$ **2.** $f(x) = e^{-2x}$ **3.** $f(x) = \sin \frac{1}{2}x$ **4.** $f(x) = \sin x^4$

5. $f(x) = \cos 4x$ **6.** $f(x) = \sqrt{1 - x^4}$ **7.** $f(x) = \ln(1 + x^2)$ **8.** $f(x) = \ln(1 - x)$

In Exercises 9 through 12, evaluate the given integrals by use of three terms of the appropriate series.

9. $\int_0^1 \sin x^2\, dx$ **10.** $\int_0^{0.4} \sqrt[4]{1 - 2x^2}\, dx$ **11.** $\int_0^{0.2} \cos \sqrt{x}\, dx$ **12.** $\int_{0.1}^{0.2} \frac{\cos x - 1}{x}\, dx$

In Exercises 13 through 20, determine the indicated series by the given operation.

13. Find the first four nonzero terms of the expansion of the function $f(x) = \frac{1}{2}(e^x + e^{-x})$ by adding the terms of the appropriate series. The result is the series for cosh x. (See Exercise 45 of Section 26-6.)

14. Find the first four nonzero terms of the expansion of the function $f(x) = \frac{1}{2}(e^x - e^{-x})$ by subtracting the terms of the appropriate series. The result is the series for sinh x. (See Exercise 45 of Section 26-6.)

15. Find the first three terms of the expansion for $e^x \sin x$ by multiplying the proper expansions together, term by term.

16. Find the first three nonzero terms of the expansion for $f(x) = \tan x$ by dividing the series for sin x by that for cos x.

17. Show that by differentiating term by term the expansion for sin x, the result is the expansion for cos x.

18. Show that by differentiating term by term the expansion for e^x, the result is also the expansion for e^x.

19. Show that by integrating term by term the expansion for cos x, the result is the expansion for sin x.

20. Show that by integrating term by term the expansion for $-1/(1 - x)$ (see Exercise 11 of Section 28-2), the result is the expansion for $\ln(1 - x)$.

In Exercises 21 through 28, solve the given problems.

21. Evaluate $\int_0^1 e^x\, dx$ directly and compare the result obtained by using four terms of the series for e^x and then integrating.

22. Evaluate $\lim\limits_{x \to 0} \dfrac{\sin x}{x}$ by using the series expansion for sin x. Compare the result with Eq. (26-1).

23. Find the approximate value of the area bounded by $y = x^2 e^x$, $x = 0.2$, and the x-axis by use of three terms of the appropriate Maclaurin series.

24. Find the approximate area bounded by $y = e^{-x^2}$, $x = -1$, $x = 1$, and $y = 0$ by use of three terms of the appropriate series. See Fig. 28-6 and Fig. 21-6.

25. Find the volume generated by rotating the area bounded by $y = \sin x$ from $x = 0$ to $x = \frac{1}{8}\pi$ and the x-axis about the y-axis by use of two terms of the appropriate series.

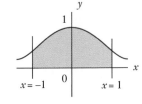

Fig. 28-6

26. Find the volume generated by rotating the area bounded by $y = e^{-x}$, $y = 0$, $x = 0$, and $x = 0.1$ about the y-axis by use of three terms of the appropriate series.

27. The vertical displacement y of a mass at the end of a spring is given by $y = \sin 3t - \cos 2t$, where t is the time. By subtraction of series, find the first four nonzero terms of the series for y.

28. The charge q on a capacitor in a certain electric circuit is given by $q = ce^{-at} \sin 6at$, where t is the time. By multiplication of series, find the first four nonzero terms of the expansion for q.

28-4   Computations by Use of Series Expansions

As mentioned previously, power-series expansions can be used to compute numerical values of transcendental functions. By including a sufficient number of terms, we can calculate the values to any desired degree of accuracy. It is through such calculations that tables of values are made, and values of such numbers as e and π can be found. Also, many of the values determined by a calculator or a computer are found by use of series expansions.

EXAMPLE A   Calculate the value of $e^{0.1}$.

In order to evaluate $e^{0.1}$, we substitute 0.1 for x in the expansion for e^x. The more terms that are used, the more accurate a value we can obtain. The limit of the partial sums would be the actual value. However, since $e^{0.1}$ is irrational, we cannot express the exact value in decimal form.

Therefore, the value is found as follows:

$$e^x = 1 + x + \frac{x^2}{2!} + \cdots \qquad \text{Eq. (28-6)}$$

$$e^{0.1} = 1 + 0.1 + \frac{(0.1)^2}{2} + \cdots \qquad \text{substitute 0.1 for } x$$

$$= 1.105 \qquad \text{using 3 terms}$$

Using a calculator, we find that $e^{0.1} = 1.1051709$, which shows that our answer is valid to the accuracy shown. ■

EXAMPLE B   Calculate the value of sin 2°.

NOTE ▷   In finding trigonometric values, we must be careful to *express the angle in radians.* Thus, the value of sin 2° is found as follows:

$$\sin x = x - \frac{x^3}{3!} + \cdots \qquad \text{Eq. (28-7)}$$

$$\sin 2° = \left(\frac{\pi}{90}\right) - \frac{(\pi/90)^3}{6} + \cdots \qquad 2° = \frac{\pi}{90} \text{ rad}$$

$$= 0.0348995 \qquad \text{using 2 terms}$$

A calculator also gives the value 0.0348995. Here we note that the second term is much smaller than the first. In fact, a good approximation can be found by use of one term. Thus, we see the reason that $\sin \theta = \theta$ (approximately) for small values of θ, as we noted in Section 7-4. ■

EXAMPLE C   Calculate the value of cos 0.5429.

Since the angle is expressed in radians, we have

$$\cos 0.5429 = 1 - \frac{0.5429^2}{2} + \frac{0.5429^4}{4!} - \cdots \qquad \text{using Eq. (28-8)}$$

$$= 0.8562495 \qquad \text{using 3 terms}$$

A calculator shows that cos 0.5429 = 0.8562141. Since the angle is not small, additional terms are needed to obtain this accuracy. With one more term, the value 0.8562139 is obtained. ■

EXAMPLE D

Calculate the value of ln 1.2.

$$\ln (1 + x) = x - \frac{x^2}{2} + \frac{x^3}{3} - \cdots \qquad \text{Eq. (28-9)}$$

$$\ln 1.2 = \ln (1 + 0.2)$$

$$= 0.2 - \frac{(0.2)^2}{2} + \frac{(0.2)^3}{3} - \cdots = 0.1827$$

To four significant digits, ln (1.2) = 0.1823. One more term is required to obtain this accuracy. ■

Series approximations can also be used to determine errors in calculated values due to errors in measured values. We have already discussed this topic as an application of differentials. A series solution of this kind of problem allows as close a calculation of the error as needed. With differentials, only one term can be found. Of course, for many purposes this is sufficient. The following example illustrates the use of series in error calculations.

EXAMPLE E

NOTE▷

The velocity v attained by an object that has fallen h feet is given by $v = 8.00\sqrt{h}$. Find the approximate error in the calculation of the velocity of an object which has fallen 100 ft, with a possible error of 2.0 ft.

If we **let $v = 8.00\sqrt{100 + x}$, where x is the error in h,** we may express v as a Maclaurin series in x:

$$f(x) = 8.00(100 + x)^{1/2} \qquad f(0) = 80.0$$
$$f'(x) = 4.00(100 + x)^{-1/2} \qquad f'(0) = 0.400$$
$$f''(x) = -2.00(100 + x)^{-3/2} \qquad f''(0) = -0.00200$$

Therefore,

$$v = 8.00\sqrt{100 + x} = 80.0 + 0.400x - 0.00100x^2 + \cdots$$

Since the actual calculated value of v for $x = 0$ is 80.0, the error e in the value of v is

$$e = 0.400x - 0.00100x^2 + \cdots$$

Calculating, the error for $x = 2.0$ is

$$e = 0.400(2.0) - 0.00100(4.0) = 0.800 - 0.0040 = 0.796 \text{ ft/s}$$

The value 0.800 is the same as that which would be found by use of differentials, since it is a result of evaluating the first-derivative term of the expansion. The additional terms are corrections to this term. The one additional term in this case indicates that the first term is a good approximation to the error. Although this problem could be done numerically, a series solution, once set up, allows us to calculate the error easily for a given value of x. ■

EXAMPLE F

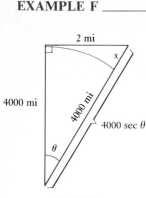

2 mi

x

4000 mi

4000 mi

4000 sec θ

θ

Fig. 28-7

Assuming that the earth is a perfect sphere 4000 mi in radius, find how far the end of a tangent line 2 mi long is from the surface.

From Fig. 28-7 we see that

$$x = 4000 \sec \theta - 4000$$

Finding the series for sec θ, we have

$f(\theta) = \sec \theta$	$f(0) = 1$
$f'(\theta) = \sec \theta \tan \theta$	$f'(0) = 0$
$f''(\theta) = \sec^3 \theta + \sec \theta \tan^2 \theta$	$f''(0) = 1$

Thus, the first two nonzero terms are sec $\theta = 1 + (\theta^2/2)$. Therefore,

$$x = 4000(\sec \theta - 1)$$

$$= 4000 \left(1 + \frac{\theta^2}{2} - 1 \right) = 2000\, \theta^2$$

The first two terms of the expansion for tan θ are $\theta + \theta^3/3$, which means that if θ is small, which it is in this case, then tan $\theta = \theta$. Again, this was noted in Section 7-4. From Fig. 28-7, tan $\theta = 2/4000$. Therefore, $\theta = \frac{1}{2000}$. Substituting this value into the expression for x, we have

$$x = 2000 \left(\frac{1}{2000} \right)^2 = \frac{1}{2000} = 0.0005 \text{ mi}$$

This means that the end of a 2-mi line drawn tangent to the earth would only be about 2.6 ft from the surface!

Exercises 28-4

In Exercises 1 through 16, calculate the value of each of the given functions. Use the indicated number of terms of the appropriate series. Compare with the value found directly on a calculator.

1. $e^{0.2}$ (3)
2. $e^{-0.5}$ (3)
3. $\sin 0.1$ (2)
4. $\cos 0.05$ (2)

5. e (7)
6. $1/\sqrt{e}$ (5)
7. $\cos 3°$ (2)
8. $\sin 4°$ (2)

9. $\ln (1.4)$ (4)
10. $\ln (0.95)$ (4)
11. $\sin 0.3625$ (3)
12. $\cos 0.4072$ (3)

13. $\ln 0.9861$ (3)
14. $\ln 1.0534$ (3)
15. $e^{-0.3165}$ (4)
16. $e^{0.2651}$ (4)

In Exercises 17 through 20, calculate the value of each of the given functions. In Exercises 17 and 18, use the expansion for $\sqrt{1+x}$, and in Exercises 19 and 20, use the expansion for $\sqrt[3]{1+x}$. Use three terms of the appropriate series.

17. $\sqrt{1.1076}$
18. $\sqrt{0.7915}$
19. $\sqrt[3]{0.9628}$
20. $\sqrt[3]{1.1392}$

In Exercises 21 through 24, calculate the maximum error of the values indicated. If a series is alternating (every other term is negative), the maximum possible error in the calculated value is the value of the first term omitted.

21. The value found in Exercise 3
22. The value found in Exercise 2
23. The value found in Exercise 7
24. The value found in Exercise 9

In Exercises 25 through 32, solve the given problems by the use of series expansions.

25. We can evaluate π by use of the fact that $\frac{1}{4}\pi = \text{Arctan } \frac{1}{2} + \text{Arctan } \frac{1}{3}$ (see Exercise 66 of Section 19-6), along with the use of the series expansion for Arctan x. The first three terms are Arctan $x = x - \frac{1}{3}x^3 + \frac{1}{5}x^5$. Using these terms, expand Arctan $\frac{1}{2}$ and Arctan $\frac{1}{3}$, and thereby approximate the value of π.

26. Use the fact that $\frac{1}{4}\pi = \text{Arctan } \frac{1}{7} + 2 \text{ Arctan } \frac{1}{3}$ to approximate the value of π. (See Exercise 25.)

27. The time t, in years, for an investment to increase by 10% when the interest rate is 6% is given by $t = \dfrac{\ln 1.1}{0.06}$.

    Evaluate this expression by using the first four terms of the appropriate series.

28. The period T of a pendulum of length L is given by

$$T = 2\pi \sqrt{\frac{L}{g}} \left(1 + \frac{1}{4}\sin^2 \frac{\theta}{2} + \frac{9}{64}\sin^4 \frac{\theta}{2} + \cdots \right)$$

    where g is the acceleration due to gravity and θ is the maximum angular displacement. If $L = 1.000$ m and $g = 9.800$ m/s², calculate T for $\theta = 10.0°$ (a) if only one term (the 1) of the series is used, and (b) if two terms of the indicated series are used. In the second term, substitute one term of the series for $\sin^2 \dfrac{\theta}{2}$.

29. The electric current in a circuit containing a resistance R, an inductance L, and a battery whose voltage is E is given by $i = \dfrac{E}{R}(1 - e^{-Rt/L})$, where t is the time. Approximate this expression by using the first three terms of the appropriate exponential series. Under what conditions will this approximation be valid?

30. The image distance q from a certain lens as a function of the object distance p is given by $q = 20p/(p - 20)$. Find the first three nonzero terms of the expansion of the right side. From this expression calculate q for $p = 2$ cm and compare with the value found by substituting 2 in the original expression.

31. At what height above sea level must an observer on a ship be in order to see a point 10 mi distant along the surface of the sea?

32. The efficiency (in percent) of an internal combustion engine in terms of its compression ratio c is given by $E = 100(1 - c^{-0.4})$. Determine the possible approximate error in the efficiency for a compression ratio measured to be 6.00 with a possible error of 0.50. [*Hint:* Set up a series for $(6 + x)^{-0.4}$.]

28-5 Taylor's Series

If values of certain functions are determined for values of x which are not close to zero, it is usually necessary to use many terms of a Maclaurin expansion to obtain accurate values. Therefore, another type of series, called a **Taylor's series,** can be used for this purpose. Also, functions for which a Maclaurin series may not be found may have a Taylor's series. Actually, a Taylor's series is a more general expansion than a Maclaurin expansion.

The basic assumption in formulating a Taylor's expansion is that a function may be expanded in a polynomial of the form

$$f(x) = c_0 + c_1(x - a) + c_2(x - a)^2 + \cdots \tag{28-15}$$

By following precisely the same line of reasoning that we did in deriving the general Maclaurin expansion, we may find the constants c_0, c_1, c_2, and so forth. That is, derivatives of Eq. (28-15) are taken, and the function and its derivatives are evaluated for $x = a$. This leads to

Taylor's series

$$f(x) = f(a) + f'(a)(x - a) + \frac{f''(a)(x - a)^2}{2!} + \cdots \tag{28-16}$$

This series converges rapidly for values of x which are close to a, and this is illustrated in Examples C and D. The following examples illustrate the derivation of Taylor's series.

EXAMPLE A _____

Expand $f(x) = e^x$ in a Taylor's series with $a = 1$.

$$f(x) = e^x \qquad f(1) = e \qquad \text{find derivatives and evaluate each at } x = 1$$
$$f'(x) = e^x \qquad f'(1) = e$$
$$f''(x) = e^x \qquad f''(1) = e$$
$$f'''(x) = e^x \qquad f'''(1) = e$$

$$f(x) = e + e(x - 1) + e\frac{(x-1)^2}{2!} + e\frac{(x-1)^3}{3!} + \cdots \qquad \text{using Eq. (28-16)}$$

$$e^x = e\left[1 + (x-1) + \frac{(x-1)^2}{2} + \frac{(x-1)^3}{6} + \cdots\right]$$

■ This series can be used in evaluating e^x for values of x near 1.

EXAMPLE B _____

Expand $f(x) = \sqrt{x}$ in powers of $(x - 4)$.

Another way of stating this is to find the Taylor's series for $f(x) = \sqrt{x}$, with $a = 4$. Thus,

$$f(x) = x^{1/2} \qquad\qquad f(4) = 2 \qquad \text{find derivatives and}$$
$$f'(x) = \frac{1}{2x^{1/2}} \qquad\quad f'(4) = \frac{1}{4} \qquad \text{evaluate each at } x = 4$$

$$f''(x) = -\frac{1}{4x^{3/2}} \qquad f''(4) = -\frac{1}{32}$$

$$f'''(x) = \frac{3}{8x^{5/2}} \qquad f'''(4) = \frac{3}{256}$$

$$f(x) = 2 + \frac{1}{4}(x-4) - \frac{1}{32}\frac{(x-4)^2}{2!} + \frac{3}{256}\frac{(x-4)^3}{3!} - \cdots \qquad \text{using Eq. (28-16)}$$

$$\sqrt{x} = 2 + \frac{(x-4)}{4} - \frac{(x-4)^2}{64} + \frac{(x-4)^3}{512} - \cdots$$

This series would be used to evaluate square roots of numbers near 4.

In Fig. 28-8 we show the graphs of $y = \sqrt{x}$ and $y = 1 + \frac{1}{4}x$, which is the approximation using the first two terms of the Taylor's series. We see that each passes through (4, 2) and they have nearly equal values of y for values of x near 4.

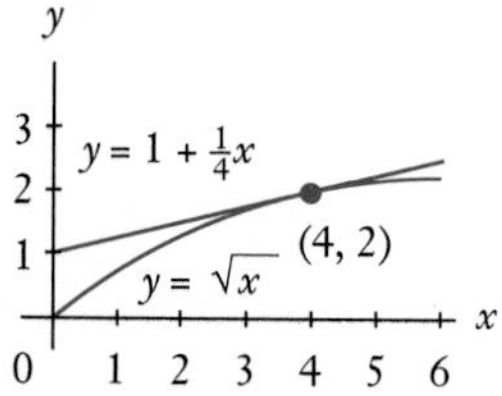

Fig. 28-8

In the last section we evaluated functions by the use of Maclaurin series. In the following examples we shall use Taylor's series to determine certain values of functions.

EXAMPLE C _____ By use of a Taylor's series, evaluate $\sqrt{4.5}$.

Using the four terms of the series found in Example B, we have

$$\sqrt{4.5} = 2 + \frac{(4.5 - 4)}{4} - \frac{(4.5 - 4)^2}{64} + \frac{(4.5 - 4)^3}{512} \qquad \text{substitute} \\ 4.5 \text{ for } x$$

$$= 2 + \frac{(0.5)}{4} - \frac{(0.5)^2}{64} + \frac{(0.5)^3}{512}$$

$$= 2.1213379$$

The value found directly on a calculator is 2.1213203. Therefore, the value found by these terms of the series expansion is correct to four decimal places.

■

NOTE ▷ In Example C we saw that successive terms become small rapidly. If a value of x is chosen such that $x - a$ is larger, the successive terms may not become small rapidly and many terms may be required. Therefore, *we should choose a as conveniently close to the x-values which will be used.* Also, we should note that a Maclaurin expansion for $\sqrt{x}$ cannot be used since the derivatives of $\sqrt{x}$ are not defined for $x = 0$.

EXAMPLE D _____ Calculate the approximate value of $\sin 29°$ by means of the appropriate Taylor expansion.

Since the value of $\sin 30°$ is known to be $\frac{1}{2}$, if we let $a = \frac{\pi}{6}$ (remember, we must use values expressed in radians), when we evaluate the expansion for $x = 29°$ (when expressed in radians) the quantity $(x - a)$ is $-\frac{\pi}{180}$ (equivalent to $-1°$). This means that its numerical values are small and become smaller when it is raised to higher powers.

Therefore,

$$f(x) = \sin x \qquad f\left(\frac{\pi}{6}\right) = \frac{1}{2}$$

$$\qquad\qquad\qquad\qquad\qquad\qquad\qquad\qquad \text{find derivatives and}$$

$$f'(x) = \cos x \qquad f'\left(\frac{\pi}{6}\right) = \frac{\sqrt{3}}{2} \qquad\qquad \text{evaluate each at } x = \frac{\pi}{6}$$

$$f''(x) = -\sin x \qquad f''\left(\frac{\pi}{6}\right) = -\frac{1}{2}$$

$$f(x) = \frac{1}{2} + \frac{\sqrt{3}}{2}\left(x - \frac{\pi}{6}\right) - \frac{1}{4}\left(x - \frac{\pi}{6}\right)^2 - \cdots \qquad \text{using Eq. (28-16)}$$

$$\sin x = \frac{1}{2} + \frac{\sqrt{3}}{2}\left(x - \frac{\pi}{6}\right) - \frac{1}{4}\left(x - \frac{\pi}{6}\right)^2 - \cdots \qquad f(x) = \sin x$$

$$\sin 29° = \sin\left(\frac{\pi}{6} - \frac{\pi}{180}\right) \qquad\qquad 29° = 30° - 1° = \frac{\pi}{6} - \frac{\pi}{180}$$

$$= \frac{1}{2} + \frac{\sqrt{3}}{2}\left(\frac{\pi}{6} - \frac{\pi}{180} - \frac{\pi}{6}\right) - \frac{1}{4}\left(\frac{\pi}{6} - \frac{\pi}{180} - \frac{\pi}{6}\right)^2 - \cdots \qquad \text{substitute } \frac{\pi}{6} - \frac{\pi}{180} \text{ for } x$$

$$= \frac{1}{2} + \frac{\sqrt{3}}{2}\left(-\frac{\pi}{180}\right) - \frac{1}{4}\left(-\frac{\pi}{180}\right)^2 - \cdots$$

$$= 0.4848089$$

■ The value found directly on a calculator is 0.4848096.

Exercises 28-5

In Exercises 1 through 8, evaluate the given functions by use of the series developed in the examples of this section.

1. $e^{1.2}$ (Use $e = 2.7183$.)   **2.** $e^{0.7}$ **3.** $\sqrt{4.2}$ **4.** $\sqrt{3.5}$

5. $\sin 31°$ **6.** $\sin 28°$ **7.** $\sin 29.53°$ **8.** $\sqrt{3.8527}$

In Exercises 9 through 16, find the first three nonzero terms of the Taylor's expansion for the given function and given value of a.

9. e^{-x} $(a = 2)$ **10.** $\cos x$ $(a = \frac{\pi}{4})$ **11.** $\sin x$ $(a = \frac{\pi}{3})$ **12.** $\ln x$ $(a = 3)$

13. $\sqrt[3]{x}$ $(a = 8)$ **14.** $\dfrac{1}{x}$ $(a = 2)$ **15.** $\tan x$ $(a = \frac{\pi}{4})$ **16.** $\ln \sin x$ $(a = \frac{\pi}{2})$

In Exercises 17 through 24, evaluate the given functions by use of three terms of the appropriate Taylor's series.

17. $e^{-2.2}$ (Use $e^{-2} = 0.1353$.)   **18.** $\ln (3.1)$ (Use $\ln 3 = 1.0986$.)

19. $\sqrt{9.3}$ **20.** $\sqrt{15}$

21. $\sqrt[3]{8.3}$ **22.** $\tan 46°$

23. $\sin 61°$ **24.** $\cos 42°$

In Exercises 25 through 28, solve the given problems.

25. By completing the steps indicated before Eq. (28-16) in the text, complete the derivation of Eq. (28-16).

26. Calculate $e^{0.9}$ by use of four terms of the Maclaurin expansion for e^x. Also calculate $e^{0.9}$ by using the first three terms of the Taylor's expansion in Example A, using $e = 2.7183$. Compare the accuracy of the values obtained with that found directly on a calculator.

27. Calculate $\sin 31°$ by use of three terms of the Maclaurin expansion for $\sin x$. Also calculate $\sin 31°$ by using three terms of the Taylor's expansion in Example D. Compare the accuracy of the values obtained with that found directly on a calculator.

28. Calculate $\tan 44°$ by use of two terms of the Maclaurin expansion for $\tan x$ (see Exercise 19 of Section 28-2). Also calculate $\tan 44°$ by using two terms of the Taylor's expansion (see Exercise 15). Compare the accuracy of the values obtained with that found directly on a calculator.

28-6   Fourier Series

Many problems which are encountered in the various fields of science and technology involve functions which are periodic. *A periodic function is one for which* $F(x + P) = F(x)$, *where P is the period*. We noted that the trigonometric functions are periodic when we discussed their graphs in Chapter 9. Illustrations of applied problems which involve periodic functions are alternating-current voltages and mechanical oscillations.

Therefore, in this section we use a series made of terms of sines and cosines. This allows us to represent complicated periodic functions in terms of the simpler sines and cosines. It also provides a good approximation over a greater interval than Maclaurin and Taylor's series, which give good approximations with a few terms only near a specific value. Illustrations of applications of this type of series are given in Example C and in the exercises.

We shall assume that a function $f(x)$ may be represented by the series of sines and cosines as indicated:

$$f(x) = a_0 + a_1 \cos x + a_2 \cos 2x + \cdots + a_n \cos nx + \cdots$$
$$+ b_1 \sin x + b_2 \sin 2x + \cdots + b_n \sin nx + \cdots \qquad (28\text{-}17)$$

Since all the sines and cosines indicated in this expansion have a period of 2π (the period of any given term may be less than 2π, but all do repeat every 2π units—for example, $\sin 2x$ has a period of π, but it also repeats every 2π), the series expansion indicated in Eq. (28-17) will also have a period of 2π. *This series is called a* **Fourier series.**

The principal problem to be solved is that of finding the coefficients a_n and b_n. Derivatives proved to be useful in finding the coefficients for a Maclaurin expansion. We use the properties of certain integrals to find the coefficients of a Fourier series. To utilize these properties, we multiply all terms of Eq. (28-17) by $\cos mx$ and then evaluate from $-\pi$ to π (in this way we take advantage of the period 2π). Thus, we have

$$\int_{-\pi}^{\pi} f(x) \cos mx \, dx = \int_{-\pi}^{\pi} (a_0 + a_1 \cos x + a_2 \cos 2x + \cdots)(\cos mx) \, dx$$
$$+ \int_{-\pi}^{\pi} (b_1 \sin x + b_2 \sin 2x + \cdots)(\cos mx) \, dx \quad (28\text{-}18)$$

Using the methods of integration of Chapter 27, we have the following:

$$\int_{-\pi}^{\pi} a_0 \cos mx \, dx = \frac{a_0}{m} \sin mx \Big|_{-\pi}^{\pi} = \frac{a_0}{m} (0 - 0) = 0 \quad (28\text{-}19)$$

$$\int_{-\pi}^{\pi} a_n \cos nx \cos mx \, dx$$
$$= a_n \left(\frac{\sin (n - m)x}{2(n - m)} + \frac{\sin(n + m)x}{2(n + m)} \right) \Big|_{-\pi}^{\pi} = 0 \quad (n \neq m); \quad (28\text{-}20)$$

(since the sine of any multiple of π is zero),

$$\int_{-\pi}^{\pi} a_n \cos nx \cos nx \, dx = \int_{-\pi}^{\pi} a_n \cos^2 nx \, dx$$
$$= \left(\frac{a_n x}{2} + \frac{1}{2n} \sin nx \cos nx \right) \Big|_{-\pi}^{\pi} = \frac{a_n x}{2} \Big|_{-\pi}^{\pi} = \pi a_n \quad (28\text{-}21)$$

as well as

$$\int_{-\pi}^{\pi} b_n \sin nx \cos mx \, dx = b_n \left(-\frac{\cos (n - m)x}{2(n - m)} - \frac{\cos (n + m)x}{2(n + m)} \right) \Big|_{-\pi}^{\pi}$$
$$= b_n \left(-\frac{\cos (n - m)\pi}{2(n - m)} - \frac{\cos (n + m)\pi}{2(n + m)} \right.$$
$$\left. + \frac{\cos (n - m)(-\pi)}{2(n - m)} + \frac{\cos(n + m)(-\pi)}{2(n + m)} \right)$$
$$= 0 \, [\text{since } \cos \theta = \cos (-\theta)] \quad (n \neq m) \quad (28\text{-}22)$$

$$\int_{-\pi}^{\pi} b_n \sin nx \cos nx \, dx = \frac{b_n}{2n} \sin^2 nx \Big|_{-\pi}^{\pi} = 0 \quad (28\text{-}23)$$

These integrals are seen to be zero, except for the one specific case of $\int_{-\pi}^{\pi} a_n \cos nx \cos mx \, dx$ when $n = m$, for which the result is indicated in Eq. (28-21). Using these results in Eq. (28-18), we have

$$\int_{-\pi}^{\pi} f(x) \cos nx \, dx = a_n \int_{-\pi}^{\pi} \cos^2 nx \, dx = \pi a_n$$

or

$$a_n = \frac{1}{\pi} \int_{-\pi}^{\pi} f(x) \cos nx \, dx \qquad (28\text{-}24)$$

This equation allows us to find the coefficients a_n, except a_0. We find the term a_0 by direct integration of Eq. (28-17) from $-\pi$ to π. When we perform this integration, all the sine and cosine terms integrate to zero, thereby giving the result

$$\int_{-\pi}^{\pi} f(x) \, dx = \int_{-\pi}^{\pi} a_0 \, dx = a_0 x \Big|_{-\pi}^{\pi} = 2\pi a_0$$

or

$$a_0 = \frac{1}{2\pi} \int_{-\pi}^{\pi} f(x) \, dx \qquad (28\text{-}25)$$

By multiplying all terms of Eq. (28-17) by $\sin mx$ and then integrating from $-\pi$ to π, we find the coefficients b_n. We obtain the result

$$b_n = \frac{1}{\pi} \int_{-\pi}^{\pi} f(x) \sin nx \, dx \qquad (28\text{-}26)$$

We can restate our equations for the Fourier series of a function $f(x)$:

Fourier series

$$f(x) = a_0 + a_1 \cos x + a_2 \cos 2x + \cdots + a_n \cos nx + \cdots$$
$$+ b_1 \sin x + b_2 \sin 2x + \cdots + b_n \sin nx + \cdots \qquad (28\text{-}17)$$

where the coefficients are found by

$$a_0 = \frac{1}{2\pi} \int_{-\pi}^{\pi} f(x) \, dx \qquad (28\text{-}25)$$

$$a_n = \frac{1}{\pi} \int_{-\pi}^{\pi} f(x) \cos nx \, dx \qquad (28\text{-}24)$$

$$b_n = \frac{1}{\pi} \int_{-\pi}^{\pi} f(x) \sin nx \, dx \qquad (28\text{-}26)$$

EXAMPLE A ____ Find the Fourier series for the square wave function

$$f(x) = \begin{cases} 0 & \text{for } -\pi \le x < 0 \\ 1 & \text{for } \quad 0 \le x < \pi \end{cases}$$

[Many of the functions which we shall expand in Fourier series are discontinuous (not continuous) like this one. See Section 22-1 for a discussion of continuity.]

NOTE ▷ Since $f(x)$ is defined differently for the intervals of x indicated, *it requires two integrals for each coefficient:*

$$a_0 = \frac{1}{2\pi} \int_{-\pi}^{0} 0 \, dx + \frac{1}{2\pi} \int_{0}^{\pi} (1) \, dx = \frac{1}{2} \qquad \text{using Eq. (28-25)}$$

$$a_n = \frac{1}{\pi} \int_{-\pi}^{0} 0 \, dx + \frac{1}{\pi} \int_{0}^{\pi} (1) \cos nx \, dx = \frac{1}{n\pi} (\sin nx) \Big|_{0}^{\pi} = 0 \qquad \text{using Eq. (28-24)}$$

for all values of n, since $\sin n\pi = 0$;

$$b_1 = \frac{1}{\pi} \int_{-\pi}^{0} 0 \, dx + \frac{1}{\pi} \int_{0}^{\pi} \sin x \, dx = -\frac{1}{\pi} (\cos x) \Big|_{0}^{\pi} \qquad \begin{array}{l} \text{using Eq. (28-26)} \\ \text{with } n = 1 \end{array}$$

$$= -\frac{1}{\pi} (-1 - 1) = \frac{2}{\pi}$$

$$b_2 = \frac{1}{\pi} \int_{-\pi}^{0} 0 \, dx + \frac{1}{\pi} \int_{0}^{\pi} \sin 2x \, dx = -\frac{1}{2\pi} (\cos 2x) \Big|_{0}^{\pi} \qquad \begin{array}{l} \text{using Eq. (28-26)} \\ \text{with } n = 2 \end{array}$$

$$= -\frac{1}{2\pi} (1 - 1) = 0$$

$$b_3 = \frac{1}{\pi} \int_{-\pi}^{0} 0 \, dx + \frac{1}{\pi} \int_{0}^{\pi} \sin 3x \, dx = -\frac{1}{3\pi} (\cos 3x) \Big|_{0}^{\pi} \qquad \begin{array}{l} \text{using Eq. (28-26)} \\ \text{with } n = 3 \end{array}$$

$$= -\frac{1}{3\pi} (-1 - 1) = \frac{2}{3\pi}$$

In general we can see that if n is even, $b_n = 0$, and if n is odd, then $b_n = 2/n\pi$. Thus, we have

$$f(x) = \frac{1}{2} + \frac{2}{\pi} \sin x + \frac{2}{3\pi} \sin 3x + \cdots$$

A graph of the function as defined, and the curve found by using the first three terms of the Fourier series, is shown in Fig. 28-9.

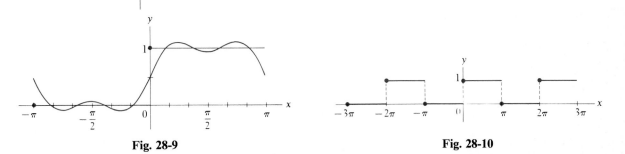

Fig. 28-9   Fig. 28-10

Since functions found by Fourier series have a period of 2π, they can represent functions with this period. If the function $f(x)$ were defined to be periodic with period 2π, with the same definitions as originally indicated, we would graph the function as shown in Fig. 28-10. The Fourier series representation would follow it as in Fig. 28-9. If more terms were used, the fit would be closer.

EXAMPLE B _____   Find the Fourier series for the function

$$f(x) = \begin{cases} 1 & \text{for } -\pi \le x < 0 \\ x & \text{for } \quad 0 \le x < \pi \end{cases}$$

For the periodic function, let $f(x + 2\pi) = f(x)$ for all x.

A graph of three periods of this function is shown in Fig. 28-11. The coefficients are

$$a_0 = \frac{1}{2\pi} \int_{-\pi}^{0} dx + \frac{1}{2\pi} \int_{0}^{\pi} x\, dx = \frac{x}{2\pi}\Big|_{-\pi}^{0} + \frac{x^2}{4\pi}\Big|_{0}^{\pi} \qquad \text{using Eq. (28-25)}$$

$$= \frac{1}{2} + \frac{\pi}{4} = \frac{2+\pi}{4}$$

$$a_1 = \frac{1}{\pi} \int_{-\pi}^{0} \cos x\, dx + \frac{1}{\pi} \int_{0}^{\pi} x \cos x\, dx \qquad \begin{array}{l}\text{using Eq. (28-24)}\\ \text{with } n = 1\end{array}$$

$$= \frac{1}{\pi} \sin x\Big|_{-\pi}^{0} + \frac{1}{\pi} (\cos x + x \sin x)\Big|_{0}^{\pi} = -\frac{2}{\pi}$$

$$a_2 = \frac{1}{\pi} \int_{-\pi}^{0} \cos 2x\, dx + \frac{1}{\pi} \int_{0}^{\pi} x \cos 2x\, dx \qquad \begin{array}{l}\text{using Eq. (28-24)}\\ \text{with } n = 2\end{array}$$

$$= \frac{1}{2\pi} \sin 2x\Big|_{-\pi}^{0} + \frac{1}{4\pi} (\cos 2x + 2x \sin 2x)\Big|_{0}^{\pi} = 0$$

$$a_3 = \frac{1}{\pi} \int_{-\pi}^{0} \cos 3x\, dx + \frac{1}{\pi} \int_{0}^{\pi} x \cos 3x\, dx \qquad \begin{array}{l}\text{using Eq. (28-24)}\\ \text{with } n = 3\end{array}$$

$$= \frac{1}{3\pi} \sin 3x\Big|_{-\pi}^{0} + \frac{1}{9\pi} (\cos 3x + 3x \sin 3x)\Big|_{0}^{\pi} = -\frac{2}{9\pi}$$

$$b_1 = \frac{1}{\pi} \int_{-\pi}^{0} \sin x\, dx + \frac{1}{\pi} \int_{0}^{\pi} x \sin x\, dx \qquad \begin{array}{l}\text{using Eq. (28-26)}\\ \text{with } n = 1\end{array}$$

$$= -\frac{1}{\pi} \cos x\Big|_{-\pi}^{0} + \frac{1}{\pi} (\sin x - x \cos x)\Big|_{0}^{\pi} = \frac{\pi - 2}{\pi}$$

$$b_2 = \frac{1}{\pi} \int_{-\pi}^{0} \sin 2x\, dx + \frac{1}{\pi} \int_{0}^{\pi} x \sin 2x\, dx \qquad \begin{array}{l}\text{using Eq. (28-26)}\\ \text{with } n = 2\end{array}$$

$$= -\frac{\cos 2x}{2\pi}\Big|_{-\pi}^{0} + \frac{\sin 2x - 2x \cos 2x}{4\pi}\Big|_{0}^{\pi} = -\frac{1}{2}$$

Thus,

$$f(x) = \frac{2+\pi}{4} - \frac{2}{\pi} \cos x - \frac{2}{9\pi} \cos 3x - \cdots + \left(\frac{\pi - 2}{\pi}\right) \sin x - \frac{1}{2} \sin 2x + \cdots$$

■

Fig. 28-11

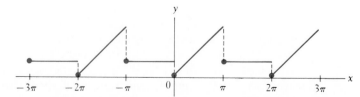

EXAMPLE C

See the chapter
introduction.

Certain electronic devices allow an electric current to pass through in only one direction. The result is that when an alternating current is applied, the current exists for only half the cycle. Figure 28-12 is a representation for such a current as a function of time. A device which causes this effect is called a *half-wave rectifier*. Derive the Fourier series for the half of the rectified wave for which $f(t) = \sin t$ $(0 \le t \le \pi)$, and $f(t) = 0$ for the other half of the cycle.

Fig. 28-12

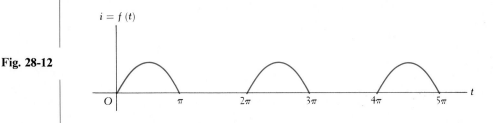

We write

$$a_0 = \frac{1}{2\pi} \int_0^\pi \sin t \, dt = \frac{1}{2\pi}(-\cos t)\Big|_0^\pi = \frac{1}{2\pi}(1 + 1) = \frac{1}{\pi}$$

In the previous example we evaluated each of the coefficients individually. Here we shall show how to set up a general expression for a_n and another for b_n. Once we have determined these, we can substitute values of n in the formula to obtain the individual coefficients:

$$a_n = \frac{1}{\pi} \int_0^\pi \sin t \cos nt \, dt = -\frac{1}{2\pi}\left[\frac{\cos(1-n)t}{1-n} + \frac{\cos(1+n)t}{1+n}\right]_0^\pi$$

$$= -\frac{1}{2\pi}\left[\frac{\cos(1-n)\pi}{1-n} + \frac{\cos(1+n)\pi}{1+n} - \frac{1}{1-n} - \frac{1}{1+n}\right]$$

See Formula 40 of Table 5 in Appendix F. This formula is valid for all values of n except $n = 1$. Now we write

$$a_1 = \frac{1}{\pi} \int_0^\pi \sin t \cos t \, dt = \frac{1}{2\pi} \sin^2 t \Big|_0^\pi = 0$$

$$a_2 = -\frac{1}{2\pi}\left(\frac{-1}{-1} + \frac{-1}{3} - \frac{1}{-1} - \frac{1}{3}\right) = -\frac{2}{3\pi}$$

$$a_3 = -\frac{1}{2\pi}\left(\frac{1}{-2} + \frac{1}{4} - \frac{1}{-2} - \frac{1}{4}\right) = 0$$

$$a_4 = -\frac{1}{2\pi}\left(\frac{-1}{-3} + \frac{-1}{5} - \frac{1}{-3} - \frac{1}{5}\right) = -\frac{2}{15\pi}$$

$$b_n = \frac{1}{\pi} \int_0^\pi \sin t \sin nt \, dt = \frac{1}{2\pi}\left[\frac{\sin(1-n)t}{1-n} - \frac{\sin(1+n)t}{1+n}\right]_0^\pi$$

$$= \frac{1}{2\pi}\left[\frac{\sin(1-n)\pi}{1-n} - \frac{\sin(1+n)\pi}{1+n}\right]$$

See Formula 39 of Table 5. This formula is valid for all values of n except $n = 1$.

(Continued on next page)

Thus,

$$b_1 = \frac{1}{\pi} \int_0^\pi \sin t \sin t \, dt = \frac{1}{\pi} \int_0^\pi \sin^2 t \, dt = \frac{1}{2\pi} (t - \sin t \cos t) \Big|_0^\pi = \frac{1}{2}$$

We see that $b_n = 0$ if $n > 1$, since each is evaluated in terms of the sine of a multiple of π.

Therefore, the Fourier series for the rectified wave is

$$f(t) = \frac{1}{\pi} + \frac{1}{2} \sin t - \frac{2}{\pi} \left(\frac{1}{3} \cos 2t + \frac{1}{15} \cos 4t + \cdots \right)$$

■

The standard form of a Fourier series is the one we have considered to this point. In this form the function is defined in the interval from $x = -\pi$ to $x = \pi$. At times it may be preferable to consider the form of the series of a function which is defined over a different interval.

Noting that

$$\sin \frac{n\pi}{L}(x + 2L) = \sin n \left(\frac{\pi x}{L} + 2\pi \right) = \sin \frac{n\pi x}{L}$$

we see that $\sin (n\pi x/L)$ has a period of $2L$. Thus, by using $\sin (n\pi x/L)$ and $\cos (n\pi x/L)$ and the same method of derivation, the following equations are found for the coefficients for the Fourier series for the interval from $x = -L$ to $x = L$.

$$a_0 = \frac{1}{2L} \int_{-L}^{L} f(x) \, dx \tag{28-27}$$

$$a_n = \frac{1}{L} \int_{-L}^{L} f(x) \cos \frac{n\pi x}{L} \, dx \tag{28-28}$$

$$b_n = \frac{1}{L} \int_{-L}^{L} f(x) \sin \frac{n\pi x}{L} \, dx \tag{28-29}$$

EXAMPLE D ———— Find the Fourier series for the function

$$f(x) = \begin{cases} 0 & \text{for } -4 \le x < 0 \\ 2 & \text{for } \quad 0 \le x < 4 \end{cases}$$

and for which the period is 8.

Here $L = 4$. A graph of the function is shown in Fig. 28-13. The coefficients are

$$a_0 = \frac{1}{2(4)} \int_{-4}^{0} 0 \, dx + \frac{1}{2(4)} \int_0^4 2 \, dx = \frac{1}{8}(2x) \Big|_0^4 = 1 \qquad \text{using Eq. (28-27)}$$

Fig. 28-13

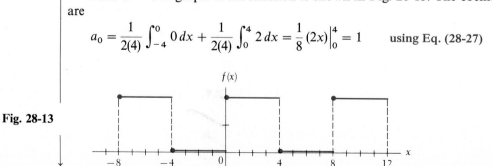

$$a_n = \frac{1}{4} \int_{-4}^{0} 0 \cos \frac{n\pi x}{4} \, dx + \frac{1}{4} \int_{0}^{4} 2 \cos \frac{n\pi x}{4} \, dx \qquad \text{using Eq. (28-28)}$$

$$= \frac{1}{2}\left(\frac{4}{n\pi}\right) \int_{0}^{4} \cos \frac{n\pi x}{4}\left(\frac{n\pi \, dx}{4}\right) = \frac{2}{n\pi} \sin \frac{n\pi x}{4}\Big|_{0}^{4} = 0 \quad \text{for all } n$$

$$b_n = \frac{1}{4} \int_{-4}^{0} 0 \sin \frac{n\pi x}{4} \, dx + \frac{1}{4} \int_{0}^{4} 2 \sin \frac{n\pi x}{4} \, dx \qquad \text{using Eq. (28-29)}$$

$$= \frac{1}{2}\left(\frac{4}{n\pi}\right) \int_{0}^{4} \sin \frac{n\pi x}{4}\left(\frac{n\pi \, dx}{4}\right) = -\frac{2}{n\pi} \cos \frac{n\pi x}{4}\Big|_{0}^{4}$$

$$= -\frac{2}{n\pi}(\cos n\pi - \cos 0) = \frac{2}{n\pi}(1 - \cos n\pi)$$

$$b_1 = \frac{2}{\pi}[1-(-1)] = \frac{4}{\pi}, \qquad b_2 = \frac{2}{2\pi}(1-1) = 0$$

$$b_3 = \frac{2}{3\pi}[1-(-1)] = \frac{4}{3\pi}, \qquad b_4 = \frac{2}{4\pi}(1-1) = 0$$

Therefore,

$$f(x) = 1 + \frac{4}{\pi}\sin\frac{\pi x}{4} + \frac{4}{3\pi}\sin\frac{3\pi x}{4} + \cdots$$

All the types of periodic functions included in this section (as well as many others) may actually be seen on an oscilloscope when the proper signal is sent into it. In this way the oscilloscope may be used to analyze the periodic nature of such phenomena as sound waves and electric currents.

Exercises 28-6

In the following exercises, find a few terms of the Fourier series for the given functions, and sketch at least three periods of the function. In Exercises 1 through 10, 13, and 14, $f(x + 2\pi) = f(x)$.

1. $f(x) = \begin{cases} 1 & -\pi \le x < 0 \\ 0 & 0 \le x < \pi \end{cases}$

2. $f(x) = \begin{cases} -1 & -\pi \le x < 0 \\ 1 & 0 \le x < \pi \end{cases}$

3. $f(x) = \begin{cases} 1 & -\pi \le x < 0 \\ 2 & 0 \le x < \pi \end{cases}$

4. $f(x) = \begin{cases} 0 & -\pi \le x < 0, \dfrac{\pi}{2} < x < \pi \\ 1 & 0 \le x \le \dfrac{\pi}{2} \end{cases}$

5. $f(x) = \begin{cases} 0 & -\pi \le x < 0 \\ x & 0 \le x < \pi \end{cases}$

6. $f(x) = x, \; -\pi \le x < \pi$

7. $f(x) = \begin{cases} -1 & -\pi \le x < 0 \\ 0 & 0 \le x < \dfrac{\pi}{2} \\ 1 & \dfrac{\pi}{2} \le x < \pi \end{cases}$

8. $f(x) = x^2, \; -\pi \le x < \pi$

9. $f(x) = \begin{cases} -x & -\pi \leq x < 0 \\ x & 0 \leq x < \pi \end{cases}$

10. $f(x) = \begin{cases} 0 & -\pi \leq x < 0 \\ x^2 & 0 \leq x < \pi \end{cases}$

11. $f(x) = \begin{cases} 5 & -3 \leq x < 0 \\ 0 & 0 \leq x < 3, \end{cases}$ period = 6

12. $f(x) = \begin{cases} -x & -4 \leq x < 0 \\ x & 0 \leq x < 4, \end{cases}$ period = 8

13. Find the Fourier expansion of the electronic device known as a *full-wave rectifier*. This is found by using as the function for the current $f(t) = -\sin t$ for $-\pi \leq t \leq 0$ and $f(t) = \sin t$ for $0 < t \leq \pi$. See Fig. 28-14. The portion of the curve to the left of the $f(t)$-axis is dashed because from a physical point of view we can give no significance to this part of the wave, although mathematically we can derive the proper form of the expansion by using it.

Fig. 28-14

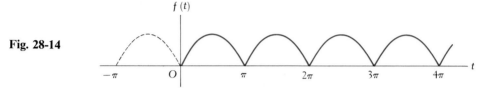

14. The loudness L, in decibels, of a certain siren as a function of time, in seconds, can be described by the function $L = 0$ for $-\pi \leq t < 0$, $L = 120t$ for $0 \leq t < \dfrac{\pi}{2}$, and $L = 120(\pi - t)$ for $\dfrac{\pi}{2} \leq t < \pi$, with a period of 2π seconds (where only positive values of t have physical significance). Find the Fourier expansion for the loudness of the siren.

15. Each pulse of a pulsating force F of a pressing machine is 8 N. The force lasts for 1 s, followed by a 3-s pause. Thus, it can be represented by $F = 0$ for $-2 \leq t < 0$ and $1 \leq t \leq 2$, and $F = 8$ for $0 \leq t < 1$, with a period of 4 s (only positive values of t have physical significance). Find the Fourier expansion for the force.

16. A function for which $f(x) = f(-x)$ is called an **even function,** and a function for which $f(x) = -f(-x)$ is called an **odd function.** From our discussion of symmetry in Section 20-3, we see that an even function is symmetrical to the y-axis, and an odd function is symmetrical to the origin. It can be proven that the Fourier expansion of an even function contains no sine terms, and that of an odd function contains no cosine terms. Show that the function in Exercise 9 is an even function and the function in Exercise 6 is an odd function. Note and compare the Fourier expansions of these functions.

28-7   Chapter Equations, Review Exercises, and Practice Test

Chapter Equations

Infinite series	$\displaystyle\sum_{n=1}^{\infty} a_n = a_1 + a_2 + a_3 + \cdots + a_n + \cdots$	(28-1)

Sum of series
$$S = \lim_{n \to \infty} S_n = \lim_{n \to \infty} \sum_{n=1}^{\infty} a_n \qquad\qquad (28\text{-}2)$$

Sum of geometric series
$$S = \lim_{n \to \infty} S_n = \frac{a_1}{1 - r} \qquad\qquad (28\text{-}3)$$

Power series
$$f(x) = a_0 + a_1 x + a_2 x^2 + \cdots + a_n x^n + \cdots \qquad\qquad (28\text{-}4)$$

Maclaurin series
$$f(x) = f(0) + f'(0)x + \frac{f''(0)x^2}{2!} + \frac{f'''(0)x^3}{3!} + \cdots + \frac{f^{(n)}(0)x^n}{n!} + \cdots \qquad\qquad (28\text{-}5)$$

Special series
$$e^x = 1 + x + \frac{x^2}{2!} + \frac{x^3}{3!} + \cdots \qquad \text{(all } x\text{)} \qquad\qquad (28\text{-}6)$$

$$\sin x = x - \frac{x^3}{3!} + \frac{x^5}{5!} - \cdots \qquad \text{(all } x) \qquad (28\text{-}7)$$

$$\cos x = 1 - \frac{x^2}{2!} + \frac{x^4}{4!} - \cdots \qquad \text{(all } x) \qquad (28\text{-}8)$$

$$\ln(1 + x) = x - \frac{x^2}{2} + \frac{x^3}{3} - \frac{x^4}{4} + \cdots \qquad (|x| < 1) \qquad (28\text{-}9)$$

$$(1 + x)^n = 1 + nx + \frac{n(n - 1)}{2!} x^2 + \cdots \qquad (|x| < 1) \qquad (28\text{-}10)$$

Complex number $\qquad e^{j\theta} = \cos \theta + j \sin \theta \qquad (28\text{-}14)$

Taylor's series $\qquad f(x) = f(a) + f'(a)(x - a) + \dfrac{f''(a)(x - a)^2}{2!} + \cdots \qquad (28\text{-}16)$

Fourier series
$$f(x) = a_0 + a_1 \cos x + a_2 \cos 2x + \cdots + a_n \cos nx + \cdots$$
$$+ b_1 \sin x + b_2 \sin 2x + \cdots + b_n \sin nx + \cdots \qquad (28\text{-}17)$$

Period $= 2\pi \qquad a_0 = \dfrac{1}{2\pi} \displaystyle\int_{-\pi}^{\pi} f(x)\, dx \qquad (28\text{-}25)$

$$a_n = \frac{1}{\pi} \int_{-\pi}^{\pi} f(x) \cos nx\, dx \qquad (28\text{-}24)$$

$$b_n = \frac{1}{\pi} \int_{-\pi}^{\pi} f(x) \sin nx\, dx \qquad (28\text{-}26)$$

Period $= 2L \qquad a_0 = \dfrac{1}{2L} \displaystyle\int_{-L}^{L} f(x)\, dx \qquad (28\text{-}27)$

$$a_n = \frac{1}{L} \int_{-L}^{L} f(x) \cos \frac{n\pi x}{L}\, dx \qquad (28\text{-}28)$$

$$b_n = \frac{1}{L} \int_{-L}^{L} f(x) \sin \frac{n\pi x}{L}\, dx \qquad (28\text{-}29)$$

Review Exercises

In Exercises 1 through 8, find the first three nonzero terms of the Maclaurin expansion of the given functions.

1. $f(x) = \dfrac{1}{1 + e^x}$ **2.** $f(x) = \frac{1}{2}(e^x - e^{-x})$ **3.** $f(x) = \sin 2x^2$ **4.** $f(x) = \dfrac{1}{(1 - x)^2}$

5. $f(x) = (x + 1)^{1/3}$ **6.** $f(x) = \dfrac{2}{2 - x^2}$ **7.** $f(x) = \text{Arcsin } x$ **8.** $f(x) = \dfrac{1}{1 - \sin x}$

In Exercises 9 through 20, calculate the value of each of the given functions. Use three terms of the appropriate series.

9. $e^{-0.2}$ **10.** $\ln(1.10)$ **11.** $\sqrt[3]{1.3}$ **12.** $\sin 3.5°$ **13.** $\sqrt{1.07}$ **14.** $e^{0.4173}$

15. $\ln 0.8172$ **16.** $\cos 0.1376$ **17.** $\tan 43.62°$ **18.** $\sqrt[4]{260}$ **19.** $\sqrt{148}$ **20.** $\cos 47°$

In Exercises 21 and 22, evaluate the given integrals by using three terms of the appropriate series.

21. $\int_{0.1}^{0.2} \frac{\cos x}{\sqrt{x}} dx$

22. $\int_0^{0.1} \sqrt[3]{1 + x^2} \, dx$

In Exercises 23 and 24, find the first three nonzero terms of the Taylor's expansion for the given function and the given value of a.

23. $\cos x \quad (a = \pi/3)$

24. $\ln \cos x \quad (a = \pi/4)$

In Exercises 25 through 28, find a few terms of the Fourier series for the given functions with the given periods. Sketch three periods of the function.

25. $f(x) = \begin{cases} 0 & -\pi \le x < -\dfrac{\pi}{2} \text{ and } \dfrac{\pi}{2} < x < \pi, \\ 1 & -\dfrac{\pi}{2} \le x \le \dfrac{\pi}{2} \quad \text{period} = 2\pi \end{cases}$

26. $f(x) = \begin{cases} -x & -\pi \le x < 0, \quad \text{period} = 2\pi \\ 0 & 0 \le x < \pi \end{cases}$

27. $f(x) = x, \ -2 \le x < 2, \quad \text{period} = 4$

28. $f(x) = \begin{cases} -2 & -3 \le x < 0, \quad \text{period} = 6 \\ 2 & 0 \le x < 3 \end{cases}$

In Exercises 29 and 30, determine the indicated series by the given operation.

29. If h is small, show that $\sin (x + h) - \sin (x - h) = 2h \cos x$.

30. Find the first three nonzero terms of the Maclaurin expansion of the function $\sin x + x \cos x$ by differentiating the expansion term by term for $x \sin x$.

In Exercises 31 through 48, solve the given problems.

31. Find the sum of the series $64 + 48 + 36 + 27 + \cdots$.

32. Find the first five partial sums of the series $\sum_{n=1}^{\infty} \dfrac{n}{3n + 1}$ and determine whether it appears to be convergent or divergent.

33. On the same diagram plot the graphs of $y = e^x$, $y = 1$, $y = 1 + x$, and $y = 1 + x + \frac{1}{2}x^2$.

34. Show that the Maclaurin expansion for the polynomial $f(x) = ax^2 + bx + c$ is the polynomial itself.

35. Find the first four nonzero terms of the expansion for $\cos^2 x$ by using the identity $\cos^2 x = \frac{1}{2}(1 + \cos 2x)$ and the series for $\cos x$.

36. Evaluate the integral $\int_0^1 x \sin x \, dx$ (a) by methods of Chapter 27 and (b) by using three terms of the series for $\sin x$. Compare results.

37. Find the first three terms of the Maclaurin expansion for $\sec x$ by finding the reciprocal of the series for $\cos x$.

38. By simplifying the sum of the squares of the Maclaurin series for $\sin x$ and $\cos x$ through the x^4 terms, verify (to this extent) that $\sin^2 x + \cos^2 x = 1$.

39. From the Maclaurin series for $f(x) = \dfrac{1}{1 - x}$ (see Exercise 11 of Section 28-2), find the series for $\dfrac{1}{1 + x}$.

40. Show that the Maclaurin expansions for $\cos x$ and $\cos (-x)$ are the same.

41. Find the approximate area between the curve $y = \dfrac{x - \sin x}{x^2}$ and the x-axis between $x = 0.1$ and $x = 0.2$.

42. Find the approximate value of the moment of inertia with respect to its axis of the volume generated by rotating the area bounded by $y = \sin x$, $x = 0.3$, and the x-axis about the y-axis. Use two terms of the appropriate series.

43. The number N of radioactive nuclei in a radioactive sample is $N = N_0 e^{-\lambda t}$. Here t is the time, N_0 is the number at $t = 0$, and λ is the *decay constant.* By use of four terms of the appropriate series, express the right side of this equation as a polynomial.

44. If a mass M is hung from a spring of mass m, the ratio of masses is $m/M = k\omega \tan k\omega$, where k is a constant and ω is a measure of the frequency of vibration. By use of two terms of the appropriate series, express m/M as a polynomial in terms of ω.

45. In the study of electromagnetic radiation, the expression $\dfrac{N_0}{1 - e^{-k/T}}$ is used. Here T is the thermodynamic temperature and N_0 and k are constants. Show that this expression can be written as $N_0(1 + e^{-k/T} + e^{-2k/T} + \cdots)$. (*Hint:* Let $x = e^{-k/T}$.)

46. In the analysis of reflection from a spherical mirror, it is necessary to express the x-coordinate on the surface shown in Fig. 28-15 in terms of the y-coordinate and the radius R. Using the equation of the semicircle shown, solve for x (note that $x \le R$). Then express the result as a series. (Note that the first approximation gives a parabolic surface.)

Fig. 28-15

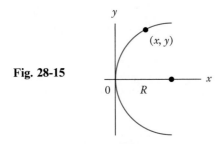

47. A certain electric current is pulsating so that the current as a function of time is given by $f(t) = 0$ if $-\pi \le t < 0$ and $\pi/2 < t < \pi$. If $0 < t < \pi/2$, $f(t) = \sin t$. Find the Fourier expansion for this pulsating current, and plot three periods.

48. By use of series, find the greatest distance that a great circle arc 200 mi long on the earth is from its chord.

Practice Test

1. By direct expansion, find the first four nonzero terms of the Maclaurin expansion for $f(x) = (1 + e^x)^2$.

2. Find the first three nonzero terms of the Taylor's expansion for $f(x) = \cos x$, with $a = \pi/3$.

3. Evaluate $\ln 0.96$ by using four terms of the expansion for $\ln (1 + x)$.

4. Find the first three nonzero terms of the expansion for $f(x) = \dfrac{1}{\sqrt{1 - 2x}}$ by using the binomial series.

5. Evaluate $\int_0^1 x \cos x \, dx$ by using three terms of the appropriate series.

6. An electric current is pulsating such that the current is a function of time with a period of 2π. If $f(t) = 2$ for $0 \le t < \pi$ and $f(t) = 0$ for the other half-cycle, find the first three nonzero terms of the Fourier series for this current.

29 Differential Equations

A great many problems which arise in science, technology, and engineering involve rates of change. For this reason, equations which contain derivatives, called *differential equations,* are of considerable importance in nearly all areas of application. Among the problems using methods of differential equations for solution are those involving velocities, chemical reactions, interest calculations, thermodynamic changes, population growth, forces on beams, electric circuits, radioactivity, and many others.

In this chapter we develop a few methods of solving certain basic types of differential equations. Actually, we have already solved a few simple differential equations in previous chapters when we started a solution with the expression for slope or velocity.

In Section 29-5 we show how differential equations are used in the study of radioactivity, which is important in the development of nuclear energy.

29-1 Solutions of Differential Equations

A **differential equation** *is an equation which contains derivatives or differentials.* In this section it is our purpose to introduce the basic meaning of the solution of a differential equation. In this sections that follow, we consider certain methods of finding such solutions.

The basic types of differential equations we shall consider are those which contain first and second derivatives. *If an equation contains only first derivatives, it is called a* **first-order** *differential equation. If the equation contains second derivatives, and possibly first derivatives, it is called a* **second-order** *differential equation. In general, the* **order** *of a differential equation is that of the highest derivative in the equation. The* **degree** *of a differential equation is the highest power of that derivative.*

order of differential
equation

EXAMPLE A

The equation $\dfrac{dy}{dx} + x = y$ is a first-order differential equation since it contains only a first derivative.

The equations

$$\overbrace{\dfrac{d^2y}{dx^2}}^{\text{order}} + y = 3x^2 \quad \text{and} \quad \dfrac{d^2y}{dx^2} + 2\dfrac{dy}{dx} = x$$

are both second-order equations since each contains a second derivative and no higher derivatives. The presence of dy/dx in the second equation does not affect the order. ∎

EXAMPLE B

The equation

$$\dfrac{d^2y}{dx^2} + \left(\dfrac{dy}{dx}\right)^4 - y = 6$$

is a differential equation of the second order and the first degree. That is, the highest derivative which appears is the second, and it is raised to the first power. Since the second derivative appears, the fourth power of the first derivative does not affect the degree. ∎

In our discussion of differential equations, we shall restrict our attention to equations of the first degree.

A **solution** *of a differential equation is a relation between the variables which satisfies the differential equation. That is, when this relation is substituted into the differential equation, an algebraic identity results. A solution containing a number of independent arbitrary constants equal to the order of the differential equation is called the* **general solution** *of the equation. When specific values are given to at least one of these constants, the solution is called a* **particular solution.**

general solution
particular solution

EXAMPLE C

Any coefficients which are not specified numerically after like terms have been combined are independent arbitrary constants. The expression $c_1x + c_2 + c_3x$ has only two independent constants since the two x terms may be combined with one arbitrary constant; $c_2 + c_4x$ is an equivalent expression, with $c_4 = c_1 + c_3$. ∎

EXAMPLE D

The equation $y = c_1e^{-x} + c_2e^{2x}$ is the general solution of the differential equation

$$\dfrac{d^2y}{dx^2} - \dfrac{dy}{dx} = 2y$$

The order of this differential equation is two, and there are two independent arbitrary constants in the equation representing the solution. The equation $y = 4e^{-x}$ is a particular solution. This particular solution can be derived from the general solution by letting $c_1 = 4$ and $c_2 = 0$. Each of these solutions can be shown to satisfy the differential equation by taking two derivatives and substituting. ∎

To solve a differential equation, we have to find some method of transforming the equation so that the terms may be integrated. This will be the topic of principal concern after this section. The purpose here is to show that a given equation is a solution of a differential equation by taking the required derivatives, and to show that an identity results after substitution.

EXAMPLE E

Show that $y = c_1 \sin x + c_2 \cos x$ is the general solution of the differential equation $y'' + y = 0$.

The function and its first two derivatives are

$$y = c_1 \sin x + c_2 \cos x$$
$$y' = c_1 \cos x - c_2 \sin x$$
$$y'' = -c_1 \sin x - c_2 \cos x$$

Substituting these into the differential equation, we have

$$\begin{array}{ccccc} y'' & + & y & = & 0 \\ \downarrow & & \downarrow & & \end{array}$$

$$(-c_1 \sin x - c_2 \cos x) + (c_1 \sin x + c_2 \cos x) = 0 \quad \text{or} \quad 0 = 0$$

We know that this must be the general solution, since there are two independent arbitrary constants, and the order of the differential equation is two.

EXAMPLE F

Show that $y = cx + x^2$ is a solution of the differential equation $xy' - y = x^2$.

Taking one derivative of the function and substituting into the differential equation, we have

$$y = cx + x^2 \underline{\hspace{3cm}}$$
$$xy' - y = x^2$$
$$y' = c + 2x \underline{\hspace{2cm}}$$
$$x(c + 2x) - (cx + x^2) = x^2 \quad \text{or} \quad x^2 = x^2$$

Exercises 29-1

In Exercises 1 through 4, determine whether the given equation is the general solution or a particular solution of the indicated differential equations.

1. $\dfrac{dy}{dx} + 2xy = 0, \quad y = e^{-x^2}$

2. $y' \ln x - \dfrac{y}{x} = 0, \quad y = c \ln x$

3. $y'' + 3y' - 4y = 3e^x, \quad y = c_1 e^x + c_2 e^{-4x} + \dfrac{3}{5} x e^x$

4. $\dfrac{d^2 y}{dx^2} + 4y = 0, \quad y = c_1 \sin 2x + 3 \cos 2x$

In Exercises 5 through 28, show that the given equation is a solution of the indicated differential equation.

5. $\dfrac{dy}{dx} = 1, \quad y = x + 3$

6. $\dfrac{dy}{dx} = 2x, \quad y = x^2 + 1$

7. $\dfrac{dy}{dx} - y = 1, \quad y = e^x - 1$

8. $\dfrac{dy}{dx} - 3 = 2x, \quad y = x^2 + 3x$

9. $xy' = 2y, \quad y = cx^2$

10. $y' = 2xy^2, \quad y = -\dfrac{1}{x^2 + c}$

11. $y' + 2y = 2x, \quad y = ce^{-2x} + x - \dfrac{1}{2}$

12. $y' - 3x^2 = 1, \quad y = x^3 + x + c$

13. $\dfrac{d^2y}{dx^2} + 4y = 0, \quad y = 3\cos 2x$

14. $y'' + 9y = 4\cos x, \quad 2y = \cos x$

15. $y'' - 4y' + 4y = e^{2x}, \quad y = e^{2x}\left(c_1 + c_2x + \dfrac{x^2}{2}\right)$

16. $\dfrac{d^3y}{dx^3} = \dfrac{d^2y}{dx^2}, \quad y = c_1 + c_2x + c_3e^x$

17. $x^2y' + y^2 = 0, \quad xy = cx + cy$

18. $xy' - 3y = x^2, \quad y = cx^3 - x^2$

19. $x\dfrac{d^2y}{dx^2} + \dfrac{dy}{dx} = 0, \quad y = c_1 \ln x + c_2$

20. $\dfrac{d^3y}{dx^3} + 4\dfrac{d^2y}{dx^2} + 4\dfrac{dy}{dx} = 0, \quad y = c_1 + c_2e^{-2x} + xe^{-2x}$

21. $y' + y = 2\cos x, \quad y = \sin x + \cos x - e^{-x}$

22. $(x + y) - xy' = 0, \quad y = x \ln x - cx$

23. $y'' + y' = 6\sin 2x, \quad y = e^{-x} - \dfrac{3}{5}\cos 2x - \dfrac{6}{5}\sin 2x$

24. $xy''' + 2y'' = 0, \quad y = c_1x + c_2 \ln x$

25. $\cos x\,\dfrac{dy}{dx} + \sin x = 1 - y, \quad y = \dfrac{x + c}{\sec x + \tan x}$

26. $2xyy' + x^2 = y^2, \quad x^2 + y^2 = cx$

27. $(y')^2 + xy' = y, \quad y = cx + c^2$

28. $x^4(y')^2 - xy' = y, \quad y = c^2 + \dfrac{c}{x}$

29-2   Separation of Variables

The first type of differential equation we shall solve is one of the first order and first degree. There are many methods for solving such equations, a few of which are presented in this and the following two sections. The first of these is *the method of* **separation of variables.**

By the definitions of order and degree, a differential equation of the first order and first degree contains the first derivative to the first power. That is, it may be written as $dy/dx = f(x, y)$. This type of equation is more commonly expressed in its differential form,

$$M(x, y)\,dx + N(x, y)\,dy = 0 \qquad\qquad (29\text{-}1)$$

where $M(x, y)$ and $N(x, y)$ may represent constants, functions of either x or y, or functions of x and y.

To find the solution of an equation of the form (29-1), it is necessary to integrate. However, if $M(x, y)$ is a function of x and y, we cannot integrate this term. It is integrable only if $M(x, y)$ is a function of x only. For the same reason, the second term is integrable only if $N(x, y)$ is a function of y only. If it is possible, by some algebraic means, to rewrite Eq. (29-1) in the form

$$A(x)\,dx + B(y)\,dy = 0 \qquad\qquad (29\text{-}2)$$

where $A(x)$ is a function of x alone and $B(y)$ is a function of y alone, then *we may find the solution by integrating each term and adding the constant of integration.* (If division is involved, we must remember that the solution is not valid for values which make the divisor zero.) Many elementary differential equations lend themselves to this type of solution.

EXAMPLE A

Solve the differential equation $dx - 4xy^3\,dy = 0$.

We can write this equation as

$$(1)\ dx + (-4xy^3)\,dy = 0$$

which means that $M(x, y) = 1$ and $N(x, y) = -4xy^3$.

NOTE ▷

We have to remove the x from the coefficient of dy **without introducing any factor of y in the coefficient of dx.** This can be done by dividing each term of the equation by x, which results in the equation

$$\frac{dx}{x} - 4y^3\,dy = 0$$

It is now possible to integrate each term separately. Performing this integration, we have

$$\ln|x| - y^4 = c$$

which is the desired solution. The c is the constant of integration, which becomes the arbitrary constant of the solution. ∎

In Example A we showed the integration of dx/x as $\ln|x|$, which follows our discussion in Section 27-2. We know that $\ln|x| = \ln x$ if $x > 0$ and $\ln|x| = \ln(-x)$ if $x < 0$. Since we know the values being considered when we find a particular solution, we generally will not use the absolute value notation when integrating logarithmic forms. We would show the integration of dx/x as $\ln x$, with the understanding that we know that $x > 0$. If we are using negative values of x, we could express it as $\ln(-x)$.

EXAMPLE B

Solve the differential equation $xy\,dx + (x^2 + 1)\,dy = 0$.

In order to integrate each term, it is necessary to divide each term by $y(x^2 + 1)$. When this is done, we have

$$\frac{x\,dx}{x^2 + 1} + \frac{dy}{y} = 0$$

For reference,
Eq. (12-9) is

$n \log_b x = \log_b x^n$

Eq. (12-7) is

$\log_b x + \log_b y$
$= \log_b xy$

Integrating, we have

$$\frac{1}{2}\ln(x^2 + 1) + \ln y = c$$

which is the desired solution.

It is possible to make use of the properties of logarithms to make the form of this solution much neater. If we write the constant of integration as $\ln c_1$, rather than c, we have $\frac{1}{2}\ln(x^2 + 1) + \ln y = \ln c_1$. Multiplying through by 2 and using the property of logarithms given by Eq. (12-9), we have $\ln(x^2 + 1) + \ln y^2 = \ln c_1^2$. Next, using the property of logarithms given by Eq. (12-7), we have $\ln(x^2 + 1)y^2 = \ln c_1^2$, which means

$$(x^2 + 1)y^2 = c_1^2$$

This form of the solution is much more compact and would be generally preferred.

NOTE ▷ However, it must be pointed out that *any expression representing a constant may be chosen as the constant of integration* and leads to a proper result. In checking answers, we must remember that a different choice of constant will lead to a different form of the answer. Thus, two different-appearing answers may both be correct. *It often happens that there is more than one reasonable choice of a constant, and different forms of the answer may be expected.*

EXAMPLE C

Solve the differential equation $\dfrac{dy}{dx} = \dfrac{y}{x^2 + 4}$.

The solution proceeds as follows:

$$\frac{dy}{y} = \frac{dx}{x^2 + 4} \qquad \text{separate variables by multiplying by } dx \text{ and dividing by } y$$

$$\ln y = \frac{1}{2} \text{Arctan} \frac{x}{2} + \frac{c}{2} \qquad \text{integrate}$$

$$2 \ln y = \text{Arctan} \frac{x}{2} + c$$

or

$$\ln y^2 = \text{Arctan} \frac{x}{2} + c$$

Note the different forms of the result using $c/2$ as the constant of integration. These forms would differ somewhat had we chosen c as the constant.

The choice of $\ln c$ as the constant of integration (on the left) is also reasonable. It would lead to the result $2 \ln cy = \text{Arctan} (x/2)$.

In order to separate the variables of a differential equation with exponential functions, it may be necessary to use the properties of exponents. When trigonometric functions are involved, the basic trigonometric identities may be used.

EXAMPLE D

Solve the differential equation $2e^{3x} \sin y \, dx + e^x \csc y \, dy = 0$.

In the dx-term we want only a function of x, which means that we must divide by $\sin y$. Also, the dy-term indicates that we must divide by e^x. Thus,

$$\frac{2e^{3x} \sin y \, dx}{e^x \sin y} + \frac{e^x \csc y \, dy}{e^x \sin y} = 0 \qquad \text{divide by } e^x \sin y$$

$$2e^{2x} \, dx + \csc^2 y \, dy = 0 \qquad \text{variables separated}$$

$$e^{2x}(2 \, dx) + \csc^2 y \, dy = 0 \qquad \text{form for integrating}$$

$$e^{2x} - \cot y = c \qquad \text{integrate}$$

The following examples show how particular solutions of a differential equation are found. Also, we show graphically the difference between the general solution and the particular solution.

EXAMPLE E _____ Solve the differential equation $(x^2 + 1)^2 dy + 4x\,dx = 0$, subject to the condition that $x = 1$ when $y = 3$.

Separating variables, we have

$$dy + \frac{4x\,dx}{(x^2 + 1)^2} = 0 \qquad\qquad \text{dividing by } (x^2 + 1)^2$$

$$y - \frac{2}{x^2 + 1} = c \qquad\qquad \text{integrating}$$

$$y = \frac{2}{x^2 + 1} + c \qquad \text{general solution}$$

Since a specific set of values is given, we can evaluate the constant of integration and thereby get a particular solution. Using the values $x = 1$ and $y = 3$, we have

$$3 = \frac{2}{1 + 1} + c, \qquad c = 2 \qquad \text{evaluate } c$$

which gives us

$$y = \frac{2}{x^2 + 1} + 2 \qquad \text{particular solution}$$

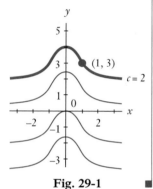

Fig. 29-1

The general solution defines a *family* of curves, one for each value of c. A few of these are shown in Fig. 29-1. When c is specified as in the particular solution, we have the specific (darker) curve shown in Fig. 29-1. ■

EXAMPLE F _____ Find the particular solution in Example B if the function is subject to the condition that $x = 0$ when $y = e$.

Using the solution $\frac{1}{2}\ln(x^2 + 1) + \ln y = c$, we have

$$\frac{1}{2}\ln(0 + 1) + \ln e = c, \qquad \frac{1}{2}\ln 1 + 1 = c \quad \text{or} \quad c = 1$$

The particular solution is then

$$\frac{1}{2}\ln(x^2 + 1) + \ln y = 1 \qquad \text{substitute } c = 1$$

$$\ln(x^2 + 1) + 2\ln y = 2$$

$$\ln y^2(x^2 + 1) = 2 \qquad \text{using properties of logarithms}$$

$$y^2(x^2 + 1) = e^2 \qquad \text{exponential form}$$

Using the general solution $(x^2 + 1)y^2 = c_1^2$, we have

$$(0 + 1)e^2 = c_1^2, \qquad c_1^2 = e^2$$

$$y^2(x^2 + 1) = e^2$$

which is precisely the same solution. We see, therefore, that the choice of the constant does not affect the final result and is truly arbitrary. ■

Exercises 29-2

In Exercises 1 through 24, solve the given differential equations.

1. $2x\,dx + dy = 0$ **2.** $y^2\,dy + x^3\,dx = 0$ **3.** $y^2\,dx + dy = 0$

4. $y\,dx + x\,dy = 0$ **5.** $x^2 + (x^3 + 5)y' = 0$ **6.** $xyy' + \sqrt{1 + y^2} = 0$

7. $e^{x^2}\,dy = x\sqrt{1 - y}\,dx$ **8.** $\sqrt{1 + 4x^2}\,dy = y^3 x\,dx$ **9.** $e^{x+y}\,dx + dy = 0$

10. $e^{2x}\,dy + e^x\,dx = 0$ **11.** $y' - y = 4$ **12.** $y' + 4x = 3$

13. $x\dfrac{dy}{dx} = y^2 + y^2 \ln x$ **14.** $(yx^2 + y)\dfrac{dy}{dx} = \text{Arctan } x$ **15.** $y \tan x\,dx + \cos^2 x\,dy = 0$

16. $\sin x \sec y\,dx = dy$ **17.** $yx^2\,dx = y\,dx - x^2\,dy$ **18.** $e^{\sin x}\,dx + \sec x\,dy = 0$

19. $y\sqrt{1 - x^2}\,dy + 2\,dx = 0$ **20.** $(x^3 + x^2)\,dx + (x + 1)y\,dy = 0$

21. $2 \ln x\,dx + x\,dy = 0$ **22.** $2y(x^3 + 1)\,dy + 3x^2(y^2 - 1)\,dx = 0$

23. $y^2 e^x + (e^x + 1)\dfrac{dy}{dx} = 0$ **24.** $y + 1 + \sec x(\sin x + 1)\dfrac{dy}{dx} = 0$

In Exercises 25 through 32, in each case find the particular solution of the given differential equations for the indicated conditions.

25. $\dfrac{dy}{dx} + yx^2 = 0;$ $x = 0$ when $y = 1$ **26.** $\dfrac{dy}{dx} + 2y = 6;$ $x = 0$ when $y = 1$

27. $(xy^2 + x)\dfrac{dy}{dx} = \ln x;$ $x = 1$ when $y = 0$ **28.** $y' = \sec y;$ $x = 0$ when $y = 0$

29. $y' = (1 - y)\cos x;$ $x = \pi/6$ when $y = 0$ **30.** $x\,dy = y \ln y\,dx;$ $x = 2$ when $y = e$

31. $y^2 e^x\,dx + e^{-x}\,dy = y^2\,dx;$ $x = 0$ when $y = 2$

32. $2y \cos y\,dy - \sin y\,dy = y \sin y\,dx;$ $x = 0$ when $y = \pi/2$

29-3 Integrable Combinations

Many differential equations cannot be solved by the method of separation of variables. Many other methods have been developed for solving such equations. One of these methods is based on the fact that *certain combinations of basic differentials can be integrated together as a unit.* The following differentials suggest some of these combinations which may occur:

$$d(xy) = x\,dy + y\,dx \tag{29-3}$$

$$d(x^2 + y^2) = 2(x\,dx + y\,dy) \tag{29-4}$$

$$d\left(\frac{y}{x}\right) = \frac{x\,dy - y\,dx}{x^2} \tag{29-5}$$

$$d\left(\frac{x}{y}\right) = \frac{y\,dx - x\,dy}{y^2} \tag{29-6}$$

    Equation (29-3) suggests that if the combination of $x\,dy + y\,dx$ occurs in a differential equation, we look for functions of xy as solutions. Equation (29-4) suggests that if the combination $x\,dx + y\,dy$ occurs, we look for functions of $x^2 + y^2$. Equations (29-5) and (29-6) suggest that if either of the combinations

$x\,dy - y\,dx$ or $y\,dx - x\,dy$ occurs, we look for functions of y/x or x/y. The following examples illustrate the use of these combinations.

EXAMPLE A

Solve the differential equation $x\,dy + y\,dx + xy\,dy = 0$.
   By dividing through by xy, we have

$$\frac{x\,dy + y\,dx}{xy} + dy = 0$$

The left term is the differential of xy divided by xy. This means it integrates to $\ln xy$. Thus, we have

$$\frac{d(xy)}{xy} + dy = 0$$

for which the solution is

$$\ln xy + y = c$$

■

EXAMPLE B

NOTE▷

Solve the differential equation $y\,dx - x\,dy + x\,dx = 0$.
   The combination of $y\,dx - x\,dy$ suggests that this equation might make use of either Eq. (29-5) or (29-6). This would require dividing through by x^2 or y^2. If we divide by y^2, the last term cannot be integrated, but *division by x^2 still allows integration of the last term.* Performing this division, we obtain

$$\frac{y\,dx - x\,dy}{x^2} + \frac{dx}{x} = 0$$

This left combination is the negative of Eq. (29-5). Thus we have

$$-d\left(\frac{y}{x}\right) + \frac{dx}{x} = 0$$

for which the solution is $-\dfrac{y}{x} + \ln x = c$.

■

EXAMPLE C

Solve the differential equation $(x^2 + y^2 + x)\,dx + y\,dy = 0$.
   Regrouping the terms of this equation, we have

$$(x^2 + y^2)\,dx + (x\,dx + y\,dy) = 0$$

By dividing through by $x^2 + y^2$, we have

$$dx + \frac{x\,dx + y\,dy}{x^2 + y^2} = 0$$

The right term now can be put in the form of du/u (with $u = x^2 + y^2$) by multiplying each of the terms of the numerator by 2. This leads to

$$d(x^2 + y^2)$$

$$dx + \left(\frac{1}{2}\right)\frac{2x\,dx + 2y\,dy}{x^2 + y^2} = 0$$

$$x + \frac{1}{2}\ln(x^2 + y^2) = \frac{c}{2} \quad \text{or} \quad 2x + \ln(x^2 + y^2) = c$$

■

EXAMPLE D _____ Find the particular solution of the differential equation

$$(x^3 + xy^2 + 2y)\,dx + (y^3 + x^2y + 2x)\,dy = 0$$

which satisfies the condition that $x = 1$ when $y = 0$.

Regrouping the terms of the equation, we have

$$x(x^2 + y^2)\,dx + y(x^2 + y^2)\,dy + 2(y\,dx + x\,dy) = 0$$

Factoring $x^2 + y^2$ from each of the first two terms gives

$$(x^2 + y^2)(x\,dx + y\,dy) + 2(y\,dx + x\,dy) = 0$$

$$\frac{1}{2}(x^2 + y^2)\underset{d(x^2+y^2)}{(2x\,dx + 2y\,dy)} + 2\underset{d(xy)}{(y\,dx + x\,dy)} = 0$$

$$\frac{1}{2}\left(\frac{1}{2}\right)(x^2 + y^2)^2 + 2xy + \frac{c}{4} = 0 \qquad \text{integrating}$$

$$(x^2 + y^2)^2 + 8xy + c = 0$$

Using the given condition gives $(1 + 0)^2 + 0 + c = 0$, or $c = -1$. The particular solution is then

$$(x^2 + y^2)^2 + 8xy = 1$$

NOTE ▷   The use of these integrable combinations depends on proper recognition of the forms. At times it may take two or three arrangements to arrive at the combination which leads to the result. Of course, many problems cannot be so arranged as to give integrable combinations in all terms.

Exercises 29-3

In Exercises 1 through 16, solve the given differential equations.

1. $x\,dy + y\,dx + x\,dx = 0$

2. $(2y + x)\,dy + y\,dx = 0$

3. $y\,dx - x\,dy + x^3\,dx = 2\,dx$

4. $x\,dy - y\,dx + y^2\,dx = 0$

5. $x^3\,dy + x^2y\,dx + y\,dx - x\,dy = 0$

6. $\sec(xy)\,dx + (x\,dy + y\,dx) = 0$

7. $x^3y^4(x\,dy + y\,dx) = 3\,dy$

8. $x\,dy + y\,dx + 4xy^3\,dy = 0$

9. $\sqrt{x^2 + y^2}\,dx - 2y\,dy = 2x\,dx$

10. $x\,dx + (x^2 + y^2 + y)\,dy = 0$

11. $\tan(x^2 + y^2)\,dy + x\,dx + y\,dy = 0$

12. $(x^2 + y^3)^2\,dy + 2x\,dx + 3y^2\,dy = 0$

13. $y\,dy - x\,dx + (y^2 - x^2)\,dx = 0$

14. $e^{x+y}(dx + dy) + 4x\,dx = 0$

15. $10x\,dy + 5y\,dx + 3y\,dy = 0$

16. $x^2\,dy + 3xy\,dx + 2\,dx = 0$

In Exercises 17 through 20, find the particular solutions to the given differential equations which satisfy the given conditions.

17. $2(x\,dy + y\,dx) + 3x^2\,dx = 0;\quad x = 1$ when $y = 2$

18. $x\,dx + y\,dy = 2(x^2 + y^2)\,dx;\quad x = 1$ when $y = 0$

19. $y\,dx - x\,dy = y^3\,dx + y^2x\,dy;\quad x = 2$ when $y = 4$

20. $e^{x/y}(x\,dy - y\,dx) = y^4\,dy;\quad x = 0$ when $y = 2$

29-4   The Linear Differential Equation of the First Order

linear differential equation

There is one type of differential equation of the first order and first degree for which an integrable combination can always be determined. *It is the **linear differential equation** of the first order and is of the form*

$$dy + Py\,dx = Q\,dx \tag{29-7}$$

where P and Q are functions of x only. We shall find that this type of equation occurs widely in applications.

If each side of Eq. (29-7) is multiplied by $e^{\int P\,dx}$ it becomes integrable, since the left side becomes of the form du with $u = ye^{\int P\,dx}$ and the right side is a function of x only. This is shown by finding the differential of $ye^{\int P\,dx}$. Thus,

$$d(ye^{\int P\,dx}) = e^{\int P\,dx}(dy + Py\,dx)$$

[In finding the differential of $\int P\,dx$ we use the fact that, by definition, these are reverse processes. Thus, $d(\int P\,dx) = P\,dx$.] Therefore, if each side is multiplied by $e^{\int P\,dx}$, the left side may be immediately integrated to $ye^{\int P\,dx}$ and the right-side integration may be indicated. The solution becomes

solution of linear differential equation

$$ye^{\int P\,dx} = \int Qe^{\int P\,dx}\,dx + c \tag{29-8}$$

EXAMPLE A

Solve the differential equation $dy + \left(\dfrac{2}{x}\right)y\,dx = 4x\,dx$.

This equation fits the form of Eq. (29-7) with $P = 2/x$ and $Q = 4x$. The first expression to find is $e^{\int P\,dx}$. In this case this is

$$e^{\int(2/x)\,dx} = e^{2\ln x} = e^{\ln x^2} = x^2 \qquad \text{see text comments following example}$$

This means that the left side integrates to yx^2, while the right side becomes $\int 4x(x^2)\,dx$. Thus,

$$y\underbrace{e^{\int P\,dx}} = \int Q\underbrace{e^{\int P\,dx}}\,dx + c$$

$$y(x^2) = \int (4x)(x^2)\,dx + c \qquad \text{using Eq. (29-8)}$$

$$yx^2 = \int 4x^3\,dx + c = x^4 + c \qquad \text{integrating}$$

$$y = x^2 + cx^{-2}$$

As in Example A, in finding the factor $e^{\int P\,dx}$ we often obtain an expression of the form $e^{\ln u}$. Using the properties of logarithms, we now show that $e^{\ln u} = u$.

Let $y = e^{\ln u}$

$$\ln y = \ln e^{\ln u} = \ln u(\ln e) = \ln u$$

$$y = u, \quad \text{or} \quad e^{\ln u} = u$$

Also, in finding $e^{\int P\,dx}$, the constant of integration in the exponent $\int P\,dx$ can always be taken as zero, as we did in Example A. To show why this is so, let $P = 2/x$ as in Example A.

$$e^{\int (2/x)\,dx} = e^{\ln x^2 + c} = (e^{\ln x^2})(e^c) = x^2 e^c$$

The solution to the differential equation, as given in Eq. (29-8), is then

$$y(x^2)(e^c) = \int 4x(x^2)(e^c)\,dx + c_1 e^c$$

Regardless of the value of c, the factor e^c can be divided out. Therefore, it is convenient to let $c = 0$ and have $e^c = 1$.

EXAMPLE B _____ Solve the differential equation $x\,dy - 3y\,dx = x^3\,dx$.

This equation can be put in the form of Eq. (29-7) by dividing through by x. This gives $dy - (3/x)y\,dx = x^2\,dx$. Here, $P = -3/x$, $Q = x^2$, and the factor $e^{\int P\,dx}$ becomes

$$e^{\int (-3/x)\,dx} = e^{-3\ln x} = e^{\ln x^{-3}} = x^{-3}$$

Therefore,

$$y e^{\int P\,dx} = \int Q e^{\int P\,dx}\,dx + c$$

$$yx^{-3} = \int x^2(x^{-3})\,dx + c \qquad \text{using Eq. (29-8)}$$

This means that

$$yx^{-3} = \int x^{-1}\,dx + c = \ln x + c$$

or

$$y = x^3(\ln x + c)$$

EXAMPLE C _____ Solve the differential equation $dy + y\,dx = x\,dx$.

Here, $P = 1$, $Q = x$, and

$$e^{\int P\,dx} = e^{\int (1)\,dx} = e^x$$

Therefore,

$$y e^x = \int x e^x\,dx + c = e^x(x - 1) + c \qquad \text{using Eq. (29-8)}$$

or

$$y = x - 1 + c e^{-x}$$

EXAMPLE D _____ Solve the differential equation $x^2\,dy + 2xy\,dx = \sin x\,dx$.

This equation is first written in the form of Eq. (29-7). This gives us

$$dy + \left(\frac{2}{x}\right)y\,dx = \frac{1}{x^2}\sin x\,dx$$

This shows us that $P = 2/x$ and $e^{\int P\,dx} = e^{\int (2/x)\,dx} = x^2$.

The solution to the equation then becomes

$$yx^2 = \int \sin x\,dx + c = -\cos x + c \qquad \text{using Eq. (29-8)}$$

$$yx^2 + \cos x = c$$

EXAMPLE E _____ Solve the differential equation $\cos x \dfrac{dy}{dx} = 1 - y \sin x$.

Writing this in the form of Eq. (29-7), we have

$$dy + y \tan x \, dx = \sec x \, dx$$

Thus, with $P = \tan x$, we have

$$e^{\int P \, dx} = e^{\int \tan x \, dx} = e^{-\ln \cos x} = \sec x$$

The solution is

$$y \sec x = \int \sec^2 x \, dx = \tan x + c \qquad \text{using Eq. (29-8)}$$

$$y = \sin x + c \cos x$$

∎

EXAMPLE F _____ Find the particular solution of the differential equation

$$dy = (1 - 2y)x \, dx$$

such that $x = 0$ when $y = 2$.

First writing this equation in the form of Eq. (29-7), and then solving, we have

$$dy + 2xy \, dx = x \, dx \qquad\qquad \text{form of Eq. (29-7)}$$

$$e^{\int P \, dx} = e^{\int 2x \, dx} = e^{x^2} \qquad\qquad \text{find } e^{\int P \, dx}$$

$$ye^{x^2} = \int xe^{x^2} \, dx \qquad\qquad \text{using Eq. (29-8)}$$

$$= \frac{1}{2} e^{x^2} + c \qquad\qquad \text{general solution}$$

$$(2)(e^0) = \frac{1}{2}(e^0) + c \qquad\qquad x = 0, y = 2$$

$$2 = \frac{1}{2} + c \qquad c = \frac{3}{2} \qquad \text{evaluate } c$$

$$ye^{x^2} = \frac{1}{2} e^{x^2} + \frac{3}{2} \qquad\qquad \text{substitute } c = \frac{3}{2}$$

$$y = \frac{1}{2}(1 + 3e^{-x^2}) \qquad\qquad \text{particular solution}$$

∎

Exercises 29-4

In Exercises 1 through 24, solve the given differential equations.

1. $dy + y \, dx = e^{-x} \, dx$

2. $dy + 3y \, dx = e^{-3x} \, dx$

3. $dy + 2y \, dx = e^{-4x} \, dx$

4. $dy + y \, dx = e^{-x} \cos x \, dx$

5. $\dfrac{dy}{dx} - 2y = 4$

6. $2\dfrac{dy}{dx} = 5 - 6y$

7. $x \, dy - y \, dx = 3x \, dx$

8. $x \, dy + 3y \, dx = dx$

9. $2x \, dy + y \, dx = 8x^3 \, dx$

10. $3x \, dy - y \, dx = 9x \, dx$

11. $dy + y \cot x \, dx = dx$

12. $y' = x^2 y + 3x^2$

13. $\sin x \dfrac{dy}{dx} = 1 - y \cos x$ 　　**14.** $\dfrac{dy}{dx} - \dfrac{y}{x} = \ln x$ 　　**15.** $y' + y = 3$

16. $y' + 2y = \sin x$ 　　**17.** $\dfrac{dy}{dx} = xe^{4x} + 4y$ 　　**18.** $y' - 2y = 2e^{2x}$

19. $y' = x^3(1 - 4y)$ 　　**20.** $dy + y \tan x\, dx + \sin x\, dx = 0$ 　　**21.** $x \dfrac{dy}{dx} = y + (x^2 - 1)^2$

22. $2x(dy - dx) + y\, dx = 0$ 　　**23.** $x\, dy + (1 - 3x)y\, dx = 3x^2 e^{3x}\, dx$ 　　**24.** $(1 + x^2)\, dy + xy\, dx = x\, dx$

In Exercises 25 through 32, find the indicated particular solutions of the given differential equations.

25. $\dfrac{dy}{dx} + 2y = e^{-x};\quad x = 0$ when $y = 1$ 　　**26.** $\dfrac{dy}{dx} - 4y = 2;\quad y = 2$ when $x = 0$

27. $y' + 2y \cot x = 4 \cos x;\quad x = \dfrac{\pi}{2}$ when $y = \dfrac{1}{3}$ 　　**28.** $y' - 3y = e^{4x};\quad x = 0$ when $y = 2$

29. $y'\sqrt{x} + \dfrac{1}{2}y = e^{\sqrt{x}};\quad x = 1$ when $y = 3$ 　　**30.** $(\sin x)y' + y = \tan x;\quad x = \dfrac{\pi}{4}$ when $y = 0$

31. $x\, dy - 2y\, dx = x^3 \cos x\, dx;\quad y = \pi^2$ when $x = \dfrac{\pi}{2}$

32. $f(x)\, dy + 2yf'(x)\, dx = f(x)f'(x)\, dx;\quad f(x) = -1$ when $y = 3$

29-5 Elementary Applications ▉▉▉▉▉▉

The differential equations of the first order and first degree which we have discussed thus far have a number of applications in geometry and the various fields of technology. In this section we shall illustrate some of these applications, through both the examples and the exercises.

EXAMPLE A ──── The slope of a given curve is given by the expression $6xy$. Find the equation of the curve if it passes through the point $(2, 1)$.

Since the slope is $6xy$, we know that the differential equation for the curve is

$$\frac{dy}{dx} = 6xy$$

We now want to find the particular solution of this equation for which $x = 2$ when $y = 1$. The solution follows:

$$\frac{dy}{y} = 6x\, dx \qquad \qquad \text{separate variables}$$

$$\ln y = 3x^2 + c \qquad \qquad \text{general solution}$$

$$\ln 1 = 3(2^2) + c \qquad \qquad \text{evaluate } c$$

$$0 = 12 + c, \qquad c = -12$$

$$\ln y = 3x^2 - 12 \qquad \qquad \text{particular solution}$$

Fig. 29-2

▉ The graph of this solution is shown in Fig. 29-2.

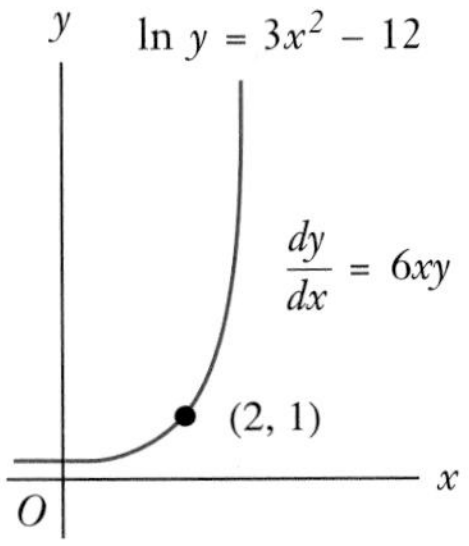

(In margin figure: y, $\ln y = 3x^2 - 12$, $\dfrac{dy}{dx} = 6xy$, $(2, 1)$, O, x)

EXAMPLE B _____ *A curve which intersects all the members of a family of curves at right angles is called an* **orthogonal trajectory** *of the family.* Find the equations of the orthogonal trajectories of the parabolas $x^2 = cy$. As before, each value of c gives us a particular member of the family.

Finding the derivative of the given equation, we have $dy/dx = 2x/c$. This equation contains the constant c, which depends on the point (x, y) on the pa-

NOTE $\triangleright$ rabola. ***Eliminating this constant between the equations of the parabola and derivative,*** we have

$$c = \frac{x^2}{y}; \qquad \frac{dy}{dx} = \frac{2x}{c} = \frac{2x}{x^2/y} \quad \text{or} \quad \frac{dy}{dx} = \frac{2y}{x}$$

This equation gives a general expression for the slope of any of the members of the family. For a curve to be perpendicular, its slope must equal the negative reciprocal of this expression, or the slope of the orthogonal trajectories must be given by

$$\left.\frac{dy}{dx}\right|_{OT} = -\frac{x}{2y} \longleftarrow \text{ this equation must not contain the constant } c$$

Solving this differential equation gives the family of orthogonal trajectories.

Parabola (curve)   y   Ellipse (OT)

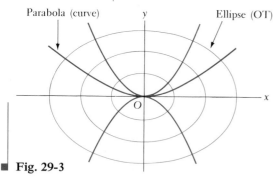

■ **Fig. 29-3**

$$2y\,dy = -x\,dx$$

$$y^2 = -\frac{x^2}{2} + \frac{c}{2}$$

$$2y^2 + x^2 = c \qquad \begin{matrix}\text{orthogonal}\\\text{trajectories}\end{matrix}$$

Thus, the orthogonal trajectories are a family of ellipses. Note in Fig. 29-3 that each parabola intersects each ellipse at right angles.

EXAMPLE C _____ Radioactive elements decay at rates proportional to the amount present. Uranium 231, which is found among the fission products of nuclear reactors, changes form

See the chapter introduction.

such that one-half of an original amount decays in 4.2 days. Determine the equation relating the amount present with the time, and determine the fraction remaining after a week (7.0 days).

Let N_0 be the original amount and N be the amount present at any time t (in days). Since we know information related to the rate of decay, we express this rate as a derivative. Therefore, since the rate of change of N is proportional to N, we have the equation

$$\frac{dN}{dt} = kN$$

Solving this differential equation, we have

$$\frac{dN}{N} = k\,dt \qquad \text{separate variables}$$

$$\ln N = kt + \ln c \qquad \text{general solution}$$

$$\ln N_0 = k(0) + \ln c \qquad N = N_0 \text{ for } t = 0$$

$$c = N_0 \qquad \text{solve for } c$$

$$\ln N = kt + \ln N_0 \qquad \text{substitute } N_0 \text{ for } c$$

$$\ln N - \ln N_0 = kt \qquad \text{use properties of logarithms}$$

$$\ln \frac{N}{N_0} = kt$$

$$N = N_0 e^{kt} \qquad \text{exponential form}$$

Now, using the condition that half this isotope decays in 4.2 days, we have $N = N_0/2$ when $t = 4.2$ days. This gives

$$\frac{N_0}{2} = N_0 e^{4.2k} \quad \text{or} \quad \frac{1}{2} = (e^k)^{4.2}$$

NOTE $\triangleright$

$$e^k = 0.5^{1/4.2} = 0.5^{0.24}$$

Therefore, the equation relating N and t is

$$N = N_0(0.5)^{0.24t}$$

In order to determine the fraction remaining after 7.0 days, we evaluate this equation for $t = 7.0$. This gives

$$N = N_0(0.5)^{0.24(7.0)} = N_0(0.5)^{1.68} = 0.31 N_0$$

■ This means that 31% remains after a week.

EXAMPLE D

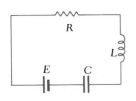

Fig. 29-4

The general equation relating the current i, voltage E, inductance L, capacitance C, and resistance R of a simple electric circuit (see Fig. 29-4) is given by

$$L\frac{di}{dt} + Ri + \frac{q}{C} = E \qquad (29-9)$$

where q is the charge on the capacitor. Find the general expression for the current in a circuit containing an inductance, a resistance, and a voltage source if $i = 0$ when $t = 0$.

The differential equation for this circuit is

$$L\frac{di}{dt} + Ri = E$$

Using the method of the linear differential equation of the first order, we have the equation

$$di + \frac{R}{L} i\, dt = \frac{E}{L}\, dt$$

The factor $e^{\int P\, dt}$ is $e^{\int (R/L)\, dt} = e^{(R/L)t}$. This gives

$$ie^{(R/L)t} = \frac{E}{L} \int e^{(R/L)t}\, dt = \frac{E}{R} e^{(R/L)t} + c$$

(Continued on next page)

15. Under proper conditions, bacteria grow at a rate proportional to the number N present. In a certain culture there were 10^5 bacteria present at a given time, and there were 3.0×10^5 present after 10 h. How many were present after 5.0 h?

16. The marginal profit function gives the change in the total profit P of a business due to a change in the business, such as adding new machinery. A company determines that the marginal profit dP/dx is $e^{-x^2} - 2Px$, where x is the amount invested in new machinery. Determine the total profit (in thousands of dollars) as a function of x, if $P = 0$ for $x = 0$.

17. According to Newton's law of cooling, the rate at which a body cools is proportional to the difference in temperature between it and the surrounding medium. How long will it take a cup of hot water, initially at 200°F, to cool to 100°F if the room temperature is 80°F, if it cools to 140°F in 5 min?

18. An object whose temperature is 100°C is placed in a medium whose temperature is 20°C. The temperature of the object falls to 50°C in 10 min. Express the temperature, T, of the object as a function of time. (See Exercise 17.)

19. If interest in a bank account is compounded continuously, the amount grows at a rate which is proportional to the amount which is present. An account earning daily interest very closely approximates this situation. Determine the amount in an account after one year if $1000 is placed in an account which pays 8% interest per year, compounded continuously.

20. If interest is compounded continuously (see Exercise 19), how long will it take a bank account to double in value if the rate of interest is 5% per year?

21. If the current in an RL circuit with a voltage source E is zero when $t = 0$ (see Example D), show that $\lim_{t \to \infty} i = E/R$. See Fig. 29-5.

22. If a circuit contains only an inductance and a resistance, with $L = 2.0$ H and $R = 30\ \Omega$, find the current i as a function of the time t if $i = 0.020$ A when $t = 0$. See Fig. 29-6.

23. An amplifier circuit contains a resistance R, an inductance L, and a voltage source $E \sin \omega t$. Express the current in the circuit as a function of the time t if the initial current is zero.

24. A radio transmitter circuit contains a resistance of 2.0 Ω, a variable inductor of $100 - t$ henrys, and a voltage source of 4.0 V. Find the current i in the circuit as a function of the time t for $0 \le t \le 100$ s if the initial current is zero.

25. If a circuit contains only a resistance and a capacitance, find the expression relating the charge on the capacitor in terms of the time if $i = dq/dt$ and $q = q_0$ when $t = 0$. See Fig. 29-7.

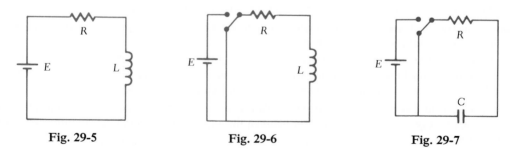

| Fig. 29-5 | Fig. 29-6 | Fig. 29-7 |

26. A circuit contains a 4.0-μF capacitor, a 450-Ω resistor, and a 20.0-mV voltage source. If the charge on the capacitor is 20.0 nC when $t = 0$, find the charge after 0.010 s. $(i = dq/dt)$.

27. One hundred gallons of brine originally containing 30 lb of salt are in a tank into which 5.0 gal of water run each minute. The same amount of mixture from the tank leaves each minute. How much salt is in the tank after 20 min?

28. Repeat Exercise 27 with the change that the water entering the tank contains 1.0 lb of salt per gallon.

29. An object falling under the influence of gravity has a variable acceleration given by $32 - v$, where v represents

the velocity. If the object starts from rest, find an expression for the velocity in terms of the time. Also find the limiting value of the velocity (find $\lim\limits_{t \to \infty} v$).

30. In a ballistics test, a bullet is fired into a sandbag. The acceleration of the bullet is $-40\sqrt{v}$, where v is the velocity in ft/s. When will the bullet stop if it enters the sandbag at 900 ft/s?

31. A boat with a mass of 10 slugs is being towed at 8.0 mi/h. The tow rope is then cut, and a motor which exerts a force of 20 lb on the boat is started. If the water exerts a retarding force which numerically equals twice the velocity, what is the velocity of the boat 3.0 min later?

32. A parachutist is falling at the rate of 200 ft/s when her parachute opens. If the air resists the fall with a force equal to $0.5v^2$, find the velocity as a function of time. The woman and her equipment have a mass of 5.0 slugs (weight is 160 lb).

33. For each cycle, a roller mechanism follows a path described by $y = 2x - x^2$, $y \geq 0$, such that $dx/dt = 6t - 3t^2$. Find x and y (in cm) in terms of t (in seconds) if x and y are zero for $t = 0$.

34. In studying the flow of water in a stream, it is found that an object follows the hyperbolic path $y(x + 1) = 10$ such that $(t + 1)\,dx = (x - 2)\,dt$. Find x and y (in ft) in terms of t (in seconds) if $x = 4$ ft and $y = 2$ ft for $t = 0$.

35. The rate of change of air pressure with respect to height is approximately proportional to the pressure. If the pressure is 15 lb/in.² when $h = 0$, and $p = 10$ lb/in.² when $h = 10{,}000$ ft, find the expression relating pressure and height.

36. Water flows from a vertical cylindrical storage tank through a hole of area A at the bottom of the tank. The rate of flow is $4.8A\sqrt{h}$, where h is the distance from the surface of the water to the hole. If h changes from 9.0 ft to 8.0 ft in 16 min, how long will it take the tank to empty?

37. Assume that the rate of depreciation of an object is proportional to its value at any time t. If a car costs \$8200 new, and its value 3 years later is \$4920, what is its value 11 years after it was purchased?

38. Assume that sugar dissolves at a rate proportional to the undissolved amount. If there are initially 500 g of sugar and 200 g remain after 4.0 min, how long does it take to dissolve 400 g?

39. Fresh air is being circulated into a room whose volume is 4000 ft³. Under specified conditions the number of cubic feet x of carbon dioxide present at any time t is found by solving the differential equation $\dfrac{dx}{dt} = 1 - 0.25x$.

Find x as a function of t if $x = 12$ ft³ when $t = 0$ min.

40. The lines of equal potential in a field of force are all at right angles to the lines of force. In an electric field of force caused by charged particles, the lines of force are given by $x^2 + y^2 = cx$. Find the equation of the lines of equal potential. Sketch a few members of the lines of force and those of equal potential.

29-6 Linear Differential Equations of Higher Order

Until now we have restricted our attention to differential equations of the first order and first degree. Another important type of differential equation is the linear differential equation of the second order with constant coefficients. Before restricting our attention to this specific type, we shall briefly describe the general higher-order equation and the notation we shall use with this type of equation.

The general linear differential equation of the nth order is of the form

$$a_0 \frac{d^n y}{dx^n} + a_1 \frac{d^{n-1} y}{dx^{n-1}} + \cdots + a_{n-1} \frac{dy}{dx} + a_n y = b \qquad (29\text{-}11)$$

where the a's and b are either functions of x or constants.

For convenience of notation, the nth derivative with respect to x will be denoted by D^n. The other derivatives are denoted in a similar manner. In this notation, the general linear differential equation becomes

$$a_0 D^n y + a_1 D^{n-1} y + \cdots + a_{n-1} Dy + a_n y = b \qquad (29\text{-}12)$$

homogeneous equation
nonhomogeneous
equation

If $b = 0$ the general linear equation is called **homogeneous,** *and if $b \neq 0$ it is called* **nonhomogeneous.** Both these types have important applications.

Although the a's may be functions of x, we shall restrict our attention to the case where they are all constants. We shall, however, consider both homogeneous and nonhomogeneous equations. Since second-order equations occur widely in applied problems, the next four sections will be devoted to this type of equation. The methods developed, however, may be applied to equations of higher order.

29-7   Second-Order Homogeneous Equations with Constant Coefficients

In accordance with the notation and terminology introduced in Section 29-6, *a second-order, linear, homogeneous differential equation with constant coefficients is one of the form*

$$a_0 D^2 y + a_1 Dy + a_2 y = 0 \qquad (29\text{-}13)$$

where the a's are now constants. The following example will indicate the kind of solution we should expect for this type of equation.

EXAMPLE A

Solve the differential equation $D^2 y - Dy - 2y = 0$.

First we put this equation in the form $(D^2 - D - 2)y = 0$. This is another way of saying that we are to take the second derivative of y, subtract the first derivative, and finally subtract twice the function. This may now be factored as $(D - 2)(D + 1)y = 0$. (We shall not develop the algebra of the operator D. However, most such algebraic operations can be shown to be valid.) This formula tells us to find the first derivative of the function and add this to the function. Then twice this result is to be subtracted from the derivative of this result. If we let $z = (D + 1)y$, which is valid since $(D + 1)y$ is a function of x, we have $(D - 2)z = 0$. This equation is easily solved by separation of variables. Thus,

$$\frac{dz}{dx} - 2z = 0, \qquad \frac{dz}{z} - 2\,dx = 0, \qquad \ln z - 2x = \ln c_1$$

$$\ln \frac{z}{c_1} = 2x \quad \text{or} \quad z = c_1 e^{2x}$$

Replacing z by $(D + 1)y$, we have

$$(D + 1)y = c_1 e^{2x}$$

This is a linear equation of the first order. Then

$$dy + y\,dx = c_1 e^{2x}\,dx$$

The factor $e^{\int P\,dx}$ is $e^{\int dx} = e^x$. And so

$$ye^x = \int c_1 e^{3x}\,dx = \frac{c_1}{3}e^{3x} + c_2 \qquad \text{using Eq. (29-8)}$$

or

$$y = c_1' e^{2x} + c_2 e^{-x}$$

where $c_1' = \tfrac{1}{3}c_1$. This example indicates that solutions of the form e^{mx} result for this equation. ∎

Basing our reasoning on the result of this example, let us assume that an equation of the form (29-13) has a particular solution e^{mx}. Substituting this into Eq. (29-13) gives

$$a_0 m^2 e^{mx} + a_1 m e^{mx} + a_2 e^{mx} = 0$$

The exponential function e^{mx} is never zero, which means that the polynomial in m is zero, or

auxiliary equation

$$\boxed{a_0 m^2 + a_1 m + a_2 = 0} \tag{29-14}$$

Equation (29-14) is called the **auxiliary equation** *of Eq. (29-13). Note that it may be formed directly by inspection from Eq. (29-13).*

There are two roots of the auxiliary Eq. (29-14) and there are two arbitrary constants in the solution of Eq. (29-13). These factors lead us to *the general solution of Eq. (29-13), which is*

$$\boxed{y = c_1 e^{m_1 x} + c_2 e^{m_2 x}} \tag{29-15}$$

where m_1 and m_2 are the solutions of Eq. (29-14). This is seen to be in agreement with the results of Example A.

EXAMPLE B ⎯⎯⎯ Solve the differential equation $D^2 y - 5\,Dy + 6y = 0$.

From this operator form of the differential equation, we write the auxiliary equation

$$m^2 - 5m + 6 = 0$$

Solving the auxiliary equation, we have

$$(m - 3)(m - 2) = 0$$
$$m_1 = 3, \qquad m_2 = 2$$

Now, using Eq. (29-15), we write the solution of the differential equation as

$$y = c_1 e^{3x} + c_2 e^{2x}$$

Obviously it makes no difference which constant is written with each exponential function. ∎

EXAMPLE C ———— Solve the differential equation $D^2 y - 6\,Dy = 0$.

In this case we have

$$m^2 - 6m = 0 \qquad \text{auxiliary equation}$$

$$m(m - 6) = 0 \qquad \text{solve for } m$$

$$m_1 = 0, \qquad m_2 = 6$$

$$y = c_1 e^{0x} + c_2 e^{6x} \qquad \text{using Eq. (29-15)}$$

Since $e^{0x} = 1$, we have

$$y = c_1 + c_2 e^{6x} \qquad \text{general solution}$$

EXAMPLE D ———— Solve the differential equation $2y'' + 7y' = 4y$.

We first rewrite this equation using the D notation for derivatives. Also, we want to write it in the proper form of a homogeneous equation. This gives us

$$2\,D^2 y + 7\,Dy - 4y = 0$$

Now, writing the auxiliary equation and solving it, we have

$$2m^2 + 7m - 4 = 0 \qquad \text{auxiliary equation}$$

$$(2m - 1)(m + 4) = 0 \qquad \text{solve for } m$$

$$m_1 = \frac{1}{2}, \qquad m_2 = -4$$

$$y = c_1 e^{x/2} + c_2 e^{-4x} \qquad \text{using Eq. (29-15)}$$

EXAMPLE E ———— Solve the differential equation $2\dfrac{d^2 y}{dx^2} + \dfrac{dy}{dx} - 7y = 0$.

First writing this equation with the D notation for derivatives, we have

$$2\,D^2 y + Dy - 7y = 0$$

This means that the auxiliary equation is

$$2m^2 + m - 7 = 0$$

Since this equation is not factorable, we solve it by use of the quadratic formula. This gives us

$$m = \frac{-1 \pm \sqrt{1 + 56}}{4} = \frac{-1 \pm \sqrt{57}}{4}$$

Therefore, using Eq. (29-15), we have

$$y = c_1 e^{\frac{-1+\sqrt{57}}{4} x} + c_2 e^{\frac{-1-\sqrt{57}}{4} x}$$

A somewhat better form can be obtained by observing that $e^{-x/4}$ is a factor of each term. Thus,

$$y = e^{-x/4}(c_1 e^{\sqrt{57}x/4} + c_2 e^{-\sqrt{57}x/4})$$

EXAMPLE F ─────── Solve the differential equation $D^2y - 2Dy - 15y = 0$, and find the particular solution which satisfies the conditions that $Dy = 2$, $y = -1$ when $x = 0$. (It is necessary to give two conditions since there are two constants to evaluate.)

We have

$$m^2 - 2m - 15 = 0, \qquad (m - 5)(m + 3) = 0$$

$$m_1 = 5, \qquad m_2 = -3$$

$$y = c_1 e^{5x} + c_2 e^{-3x}$$

NOTE ▷ This equation is the general solution. In order to evaluate the constants c_1 and c_2, **we use the given conditions to find two simultaneous equations in c_1 and c_2.** These are then solved to determine the particular solution. Thus,

$$y' = 5c_1 e^{5x} - 3c_2 e^{-3x}$$

Using the given conditions in the general solution and its derivative, we have

$$c_1 + c_2 = -1 \qquad y = -1 \text{ when } x = 0$$

$$5c_1 - 3c_2 = 2 \qquad Dy = 2 \text{ when } x = 0$$

The solution to this system of equations is $c_1 = -\frac{1}{8}$ and $c_2 = -\frac{7}{8}$. The particular solution becomes

$$y = -\frac{1}{8}e^{5x} - \frac{7}{8}e^{-3x} \quad \text{or} \quad 8y + e^{5x} + 7e^{-3x} = 0$$

To solve the auxiliary equation for differential equations of higher order, we use methods developed in Chapter 14. With these roots, we form the solutions in the same way as we do those for second-order equations.

Exercises 29-6, 29-7

In Exercises 1 through 20, solve the given differential equations.

1. $\dfrac{d^2y}{dx^2} - \dfrac{dy}{dx} - 6y = 0$ **2.** $\dfrac{d^2y}{dx^2} + \dfrac{dy}{dx} = 0$ **3.** $3\dfrac{d^2y}{dx^2} + 4\dfrac{dy}{dx} + y = 0$ **4.** $\dfrac{d^2y}{dx^2} - 2\dfrac{dy}{dx} - 8y = 0$

5. $D^2y - 3Dy = 0$ **6.** $D^2y + 7Dy + 6y = 0$ **7.** $3D^2y + 12y = 20Dy$ **8.** $4D^2y + 12Dy = 7y$

9. $3y'' + 8y' - 3y = 0$ **10.** $8y'' + 6y' - 9y = 0$ **11.** $3y'' + 2y' - y = 0$ **12.** $2y'' - 7y' + 6y = 0$

13. $2\dfrac{d^2y}{dx^2} - 4\dfrac{dy}{dx} + y = 0$ **14.** $\dfrac{d^2y}{dx^2} + \dfrac{dy}{dx} - 5y = 0$ **15.** $4D^2y - 3Dy - 2y = 0$ **16.** $2D^2y - 3Dy - y = 0$

17. $y'' = 3y' + y$ **18.** $5y'' - y' = 3y$ **19.** $y'' + y' = 8y$ **20.** $8y'' = y' + y$

In Exercises 21 through 24, find the particular solutions of the differential equations which satisfy the given conditions.

21. $D^2y - 4Dy - 21y = 0$; $Dy = 0$ and $y = 2$ when $x = 0$

22. $4D^2y - Dy = 0$; $Dy = 2$ and $y = 4$ when $x = 0$

23. $D^2y - Dy - 12y = 0$; $y = 0$ when $x = 0$ and $y = 1$ when $x = 1$

24. $2D^2y + 5Dy = 0$; $y = 0$ when $x = 0$ and $y = 2$ when $x = 1$

In Exercises 25 through 28, solve the given differential equations. The auxiliary equations will be third degree, and the solutions will have three arbitrary constants.

25. $y''' - 2y'' - 3y' = 0$

26. $2y''' + 3y'' - 2y' = 0$

27. $y''' - 6y'' + 11y' - 6y = 0$

28. $y''' - 2y'' - y' + 2y = 0$

29-8   Auxiliary Equations with Repeated or Complex Roots

The way to solve a second-order homogeneous linear differential equation was shown in the last section. However, we purposely avoided repeated and complex roots of the auxiliary equation. In this section we shall develop the solutions for such equations. The following example indicates the type of solution which results from the case of repeated roots.

EXAMPLE A

Solve the differential equation $D^2y - 4Dy + 4y = 0$.

Using the method of Example A of the previous section, we have the following steps:

$$(D^2 - 4D + 4)y = 0, \qquad (D - 2)(D - 2)y = 0, \qquad (D - 2)z = 0$$

where $z = (D - 2)y$. The solution to $(D - 2)z = 0$ is found by separation of variables. And so

$$\frac{dz}{dx} - 2z = 0, \qquad \frac{dz}{z} - 2\,dx = 0$$

$$\ln z - 2x = \ln c_1 \quad \text{or} \quad z = c_1 e^{2x}$$

Substituting back, we have $(D - 2)y = c_1 e^{2x}$, which is a linear equation of the first order. Then

$$dy - 2y\,dx = c_1 e^{2x}\,dx, \qquad e^{\int -2\,dx} = e^{-2x}$$

This leads to

$$ye^{-2x} = c_1 \int dx = c_1 x + c_2 \quad \text{or} \quad y = c_1 xe^{2x} + c_2 e^{2x}$$

This example indicates the type of solution which results when the auxiliary equation has repeated roots. If the method of the previous section were to be used, the solution of the above example would be $y = c_1 e^{2x} + c_2 e^{2x}$. This would not be the general solution, since both terms are similar, which means that there is only one independent constant. The constants can be combined to give a solution of the form $y = ce^{2x}$, where $c = c_1 + c_2$. This solution would contain only one constant for a second-order equation.

solution with repeated roots ■

For reference, Eq. (29-13) is

$$a_0 D^2 y + a_1 Dy + a_2 y = 0$$

Eq. (29-14) is

$$a_0 m^2 + a_1 m + a_2 = 0$$

Based on the above example, *the solution to Eq. (29-13) when the auxiliary Eq. (29-14) has repeated roots is*

$$\boxed{y = e^{mx}(c_1 + c_2 x)} \tag{29-16}$$

where m is the double root. (In Example A, this double root is 2.)

EXAMPLE B _____ Solve the differential equation $(D + 2)^2 y = 0$.

The auxiliary equation is $(m + 2)^2 = 0$, for which the solutions are $m = -2$, -2. Since we have repeated roots, the solution of the differential equation is

$$y = e^{-2x}(c_1 + c_2 x) \qquad \text{using Eq. (29-16)}$$

■

EXAMPLE C _____ Solve the differential equation $\dfrac{d^2 y}{dx^2} - 10 \dfrac{dy}{dx} + 25y = 0$.

The solution is as follows:

$D^2 y - 10 Dy + 25y = 0$	using operator D notation
$m^2 - 10m + 25 = 0$	auxiliary equation
$(m - 5)^2 = 0$	solve for m
$m = 5, 5$	double root
$y = e^{5x}(c_1 + c_2 x)$	using Eq. (29-16)

■

When the auxiliary equation has complex roots, it can be solved by the method developed in the preceding section. However, the solution can be put in a special, more useful form. Let us assume that the quadratic formula is used to find the roots of the auxiliary equation, and that these roots are complex of the form $m = \alpha \pm j\beta$. This means that the solution is of the form

$$y = c_1 e^{(\alpha + j\beta)x} + c_2 e^{(\alpha - j\beta)x} = e^{\alpha x}(c_1 e^{j\beta x} + c_2 e^{-j\beta x})$$

Using the exponential form of a complex number, Eq. (11-10), we have

$$y = e^{\alpha x}[c_1 \cos \beta x + jc_1 \sin \beta x + c_2 \cos (-\beta x) + jc_2 \sin (-\beta x)]$$
$$= e^{\alpha x}(c_1 \cos \beta x + c_2 \cos \beta x + jc_1 \sin \beta x - jc_2 \sin \beta x)$$
$$= e^{\alpha x}(c_3 \cos \beta x + c_4 \sin \beta x)$$

where $c_3 = c_1 + c_2$ and $c_4 = jc_1 - jc_2$.

Summarizing these results, we see that *if the auxiliary equation has complex roots of the form* $\alpha \pm j\beta$, *the solution to Eq. (29-13) is*

solution with
complex roots

$$\boxed{y = e^{\alpha x}(c_1 \sin \beta x + c_2 \cos \beta x)} \qquad \text{(29-17)}$$

The c_1 and c_2 here are not the same as those above. They are simply the two arbitrary constants of the solution.

EXAMPLE D _____ Solve the differential equation $D^2 y - Dy + y = 0$.

We have the following solution:

$m^2 - m + 1 = 0$	auxiliary equation
$m = \dfrac{1 \pm j\sqrt{3}}{2}$	complex roots
$\alpha = \dfrac{1}{2}, \quad \beta = \dfrac{\sqrt{3}}{2}$	identify α and β
$y = e^{x/2}\left(c_1 \sin \dfrac{\sqrt{3}}{2} x + c_2 \cos \dfrac{\sqrt{3}}{2} x\right)$	using Eq. (29-17)

■

EXAMPLE E

Solve the differential equation $D^2y + 4y = 0$.

In this case we have

$$m^2 + 4 = 0 \qquad\qquad \text{auxiliary equation}$$

$$m_1 = 2j, \qquad m_2 = -2j \qquad\qquad \text{complex roots}$$

$$\alpha = 0, \qquad \beta = 2 \qquad\qquad \text{identify } \alpha \text{ and } \beta$$

$$y = e^{0x}(c_1 \sin 2x + c_2 \cos 2x) \qquad \text{using Eq. (29-17)}$$

$$= c_1 \sin 2x + c_2 \cos 2x \qquad\qquad e^0 = 1$$

EXAMPLE F

Solve the differential equation $y'' - 2y' + 12y = 0$, if $y' = 2$ and $y = 1$ when $x = 0$.

We find the required particular solution as follows:

$$D^2y - 2Dy + 12y = 0 \qquad\qquad \text{use operator } D \text{ notation}$$

$$m^2 - 2m + 12 = 0 \qquad\qquad \text{auxiliary equation}$$

$$m = \frac{2 \pm \sqrt{4 - 48}}{2} = \frac{2 \pm 2j\sqrt{11}}{2} \qquad \text{solve for } m$$

$$= 1 \pm j\sqrt{11} \qquad\qquad \text{complex roots: } \alpha = 1, \beta = \sqrt{11}$$

$$y = e^x(c_1 \cos \sqrt{11}x + c_2 \sin \sqrt{11}x) \qquad \text{general solution}$$

Using the condition that $y = 1$ when $x = 0$, we have

$$1 = e^0(c_1 \cos 0 + c_2 \sin 0) \quad \text{or} \quad c_1 = 1$$

Since the other condition involves the value of y', we find the derivative and then evaluate c_2.

$$y' = e^x(c_1 \cos \sqrt{11}x + c_2 \sin \sqrt{11}x - \sqrt{11}c_1 \sin \sqrt{11}x + \sqrt{11}c_2 \cos \sqrt{11}x)$$

$$2 = e^0(\cos 0 + c_2 \sin 0 - \sqrt{11} \sin 0 + \sqrt{11}c_2 \cos 0) \qquad y' = 2 \text{ when } x = 0$$

$$2 = 1 + \sqrt{11}c_2, \qquad c_2 = \frac{1}{11}\sqrt{11} \qquad\qquad \text{solve for } c_2$$

$$y = e^x\left(\cos \sqrt{11}x + \frac{1}{11}\sqrt{11} \sin \sqrt{11}x\right) \qquad \text{particular solution}$$

Now that we have determined the various types of possible solutions of second-order homogeneous linear differential equations, we can see that it is possible to find the differential equation if the solution is known. Consider the following example.

EXAMPLE G

If we know that the solution of a differential equation is $y = c_1e^x + c_2e^{2x}$, we know that the auxiliary equation is $(m - 1)(m - 2) = 0$, since it gives solutions of $m_1 = 1$ and $m_2 = 2$. Thus, the simplest form of the differential equation is $(D^2 - 3D + 2)y = 0$.

For the solution $y = c_1e^{2x} + c_2xe^{2x}$, the auxiliary equation would be $(m - 2)^2 = 0$, for repeated roots are indicated by the second term of the solution. The differential equation would be $(D^2 - 4D + 4)y = 0$.

For the solution $y = c_1 \sin 2x + c_2 \cos 2x$, the auxiliary equation would be $m^2 + 4 = 0$, for imaginary roots are indicated by the terms $\sin 2x$ and $\cos 2x$. The differential equation would be $(D^2 + 4)y = 0$.

Exercises 29-8

In Exercises 1 through 24, solve the given differential equations.

1. $\dfrac{d^2y}{dx^2} - 2\dfrac{dy}{dx} + y = 0$

2. $\dfrac{d^2y}{dx^2} - 6\dfrac{dy}{dx} + 9y = 0$

3. $D^2y + 12\,Dy + 36y = 0$

4. $16\,D^2y + 8\,Dy + y = 0$

5. $\dfrac{d^2y}{dx^2} + 9y = 0$

6. $\dfrac{d^2y}{dx^2} + y = 0$

7. $D^2y + Dy + 2y = 0$

8. $D^2y - 2\,Dy + 4y = 0$

9. $D^2y = 0$

10. $4\,D^2y - 12\,Dy + 9y = 0$

11. $4\,D^2y + y = 0$

12. $9\,D^2y + 4y = 0$

13. $16y'' - 24y' + 9y = 0$ **14.** $9y'' - 24y' + 16y = 0$ **15.** $25y'' + 2y = 0$ **16.** $y'' - 4y' + 5y = 0$

17. $2\,D^2y + 5y = 4\,Dy$ **18.** $D^2y + 4\,Dy + 6y = 0$ **19.** $25y'' + 16y = 40y'$ **20.** $9y'' + 0.6y' + 0.01y = 0$

21. $2\,D^2y - 3\,Dy - y = 0$ **22.** $D^2y - 5\,Dy - 4y = 0$ **23.** $3\,D^2y + 12\,Dy = 2y$ **24.** $36\,D^2y = 25y$

In Exercises 25 through 28, find the particular solutions of the given differential equations which satisfy the stated conditions.

25. $y'' + 2y' + 10y = 0$; $y = 0$ when $x = 0$ and $y = e^{-1}$ when $x = \pi/6$

26. $9\,D^2y + 16y = 0$; $Dy = 0$ and $y = 2$ when $x = \pi/2$

27. $D^2y - 8\,Dy + 16y = 0$; $Dy = 2$ and $y = 4$ when $x = 0$

28. $4y'' + 20y' + 25y = 0$; $y = 0$ when $x = 0$ and $y = e$ when $x = -\frac{2}{5}$

In Exercises 29 through 32, find the simplest form of the second-order homogeneous linear differential equation which has the given solution.

29. $y = c_1 e^{3x} + c_2 e^{-3x}$

30. $y = c_1 e^{3x} + c_2 x e^{3x}$

31. $y = c_1 \cos 3x + c_2 \sin 3x$

32. $y = c_1 e^{2x} \cos x + c_2 e^{2x} \sin x$

29-9 Solutions of Nonhomogeneous Equations

To solve a nonhomogeneous linear equation of the form

$$a_0\,D^2y + a_1\,Dy + a_2 y = b \tag{29-18}$$

where b is a function of x or is a constant, the solution must be such that if we substitute it into the left side, we obtain the right side. Solutions obtained from the methods of Sections 29-7 and 29-8 will give zero when substituted into the left side, but they do contain the arbitrary constants necessary in the solution. If we could find some particular solution which when substituted into the left side produced the expression of the right, it could be added to the solution containing the arbitrary constants. Thus, *the solution is of the form*

$$y = y_c + y_p \tag{29-19}$$

complementary solution

where y_c, called the **complementary solution,** is obtained by solving the corresponding homogeneous equation and where y_p is the particular solution necessary to produce the expression b of Eq. (29-18).

EXAMPLE A —————— The differential equation $D^2y - Dy - 6y = e^x$ has the solution

$$y = c_1e^{3x} + c_2e^{-2x} - \frac{1}{6}e^x,$$

where the complementary solution y_c and particular solution y_p are

$$y_c = c_1e^{3x} + c_2e^{-2x} \qquad y_p = -\frac{1}{6}e^x$$

The complementary solution y_c is obtained by solving the corresponding homogeneous equation $D^2y - Dy - 6y = 0$, and we shall discuss below the method of finding y_p. We can see that y_p alone will satisfy the differential equation, but since it has no arbitrary constants, it cannot be the general solution.

method of
undetermined
coefficients

NOTE ▷   By inspection of the form of the expression b on the right side of the equation, we can determine the form which the particular solution must have. Since a combination of the derivatives and the function itself must form the function b, *we assume that an expression which contains all possible forms of b and its derivatives will form y_p.* The method used to find the exact form of this expression is called the **method of undetermined coefficients.**

EXAMPLE B —————— If the function b on the right of Eq. (29-18) is $4x$, we would assume that the particular solution is of the form $y_p = A + Bx$. The Bx-term is included to account for the presence of the $4x$. Since the derivative of Bx is a constant, the A-term is included to account for any first derivative of the Bx-term which may be present. Since the derivative of A is zero, no other terms are needed to account for any higher derivatives of the Bx-term.

If the function b on the right is e^{2x}, we assume that the form of the particular solution is $y_p = Ce^{2x}$. Since all derivatives of Ce^{2x} are a constant times e^{2x}, no other forms appear in the derivatives, and no other forms are needed in y_p.

If the function b on the right is $4x + e^{2x}$, we assume that the form of y_p is the sum of all forms of terms needed, or $y_p = A + Bx + Ce^{2x}$.

EXAMPLE C —————— If the expression b is of the form $x^2 + e^{-x}$, we would assume y_p to be of the form $y_p = A + Bx + Cx^2 + Ee^{-x}$.

If b is of the form $xe^{-2x} - 5$, we would assume y_p to be of the form $y_p = Ae^{-2x} + Bxe^{-2x} + C$.

If b is of the form $x \sin x$, we would then assume y_p to be of the form $A \sin x + B \cos x + Cx \sin x + Ex \cos x$. All these terms occur in the derivatives of $x \sin x$.

EXAMPLE D ——————
NOTE ▷   If the expression for b is of the form $e^x + xe^x$, we would assume y_p to be of the form $y_p = Ae^x + Bxe^x$. *These terms occur for xe^x and its derivatives.* Since the form of the e^x-term of b is already included in Ae^x, we do not include another e^x-term in y_p.

In the same way, if b is of the form $2x + 4x^2$, the form of y_p is $y_p = A + Bx + Cx^2$. These are the only forms which occur in either $2x$ or $4x^2$ and their derivatives.

Once we have determined the form of the particular solution, we have to find the numerical values of the coefficients $A, B, \ldots$ *This is done by substituting the form into the differential equation, and equating coefficients of like terms.*

EXAMPLE E

Solve the differential equation $D^2y - Dy - 6y = e^x$.

In this case the solution of the auxiliary equation $m^2 - m - 6 = 0$ gives us the roots $m_1 = 3$ and $m_2 = -2$. Thus,

$$y_c = c_1 e^{3x} + c_2 e^{-2x}$$

The proper form of y_p is $y_p = Ae^x$. This means that $Dy_p = Ae^x$ and $D^2y_p = Ae^x$. Substituting y_p and its derivatives into the differential equation, we have

$$Ae^x - Ae^x - 6Ae^x = e^x$$

To produce equality, the coefficients of e^x must be the same on each side of the equation. Thus,

$$-6A = 1 \quad \text{or} \quad A = -\frac{1}{6}$$

Therefore, $y_p = -\frac{1}{6}e^x$. This gives the complete solution $y = y_c + y_p$,

$$y = c_1 e^{3x} + c_2 e^{-2x} - \frac{1}{6}e^x \qquad \text{see Example A}$$

■ This solution checks when substituted into the original differential equation.

EXAMPLE F

Solve the differential equation $D^2y + 4y = x - 4e^{-x}$.

In this case we have $m^2 + 4 = 0$, which gives us $m_1 = 2j$ and $m_2 = -2j$. Therefore,

$$y_c = c_1 \sin 2x + c_2 \cos 2x$$

The proper form of the particular solution is $y_p = A + Bx + Ce^{-x}$. Finding two derivatives and then substituting into the differential equation gives the following result:

$$y_p = A + Bx + Ce^{-x}, \qquad Dy_p = B - Ce^{-x}, \qquad D^2y_p = Ce^{-x}$$

$$D^2y + 4y \qquad\qquad\qquad = x - 4e^{-x} \qquad \text{differential equation}$$

$$(Ce^{-x}) + 4(A + Bx + Ce^{-x}) = x - 4e^{-x} \qquad \text{substituting}$$

$$Ce^{-x} + 4A + 4Bx + 4Ce^{-x} = x - 4e^{-x}$$

$$(4A) + (4B)x + (5C)e^{-x} = 0 + (1)x + (-4)e^{-x} \qquad \text{note coefficients}$$

(Continued on next page)

NOTE▷ *Equating the constants, and the coefficients* of x and e^{-x}, on either side of the equation gives

$$4A = 0, \quad 4B = 1, \quad 5C = -4, \quad \text{or} \quad A = 0, \quad B = \frac{1}{4}, \quad C = -\frac{4}{5}.$$

This means that the particular solution is

$$y_p = \frac{1}{4}x - \frac{4}{5}e^{-x}$$

In turn this tells us that the complete solution is

$$y = c_1 \sin 2x + c_2 \cos 2x + \frac{1}{4}x - \frac{4}{5}e^{-x}$$

■ Substitution into the original differential equation verifies this solution.

EXAMPLE G — Solve the differential equation $D^2y - 3Dy + 2y = 2 \sin x$.
The solution is as follows:

$$m^2 - 3m + 2 = 0, \quad (m-1)(m-2) = 0, \quad m_1 = 1, \quad m_2 = 2 \qquad \text{auxiliary equation}$$
$$y_c = c_1 e^x + c_2 e^{2x} \qquad \text{complementary solution}$$

We now find the particular solution:

$$y_p = A \sin x + B \cos x \qquad \text{particular solution form}$$
$$Dy_p = A \cos x - B \sin x \qquad \text{find two derivatives}$$
$$D^2y_p = -A \sin x - B \cos x$$
$$(-A \sin x - B \cos x) - 3(A \cos x - B \sin x) + 2(A \sin x + B \cos x) = 2 \sin x \qquad \text{substitute into}$$
$$(A + 3B) \sin x + (B - 3A) \cos x = 2 \sin x \qquad \text{differential equation}$$
$$A + 3B = 2, \qquad -3A + B = 0 \qquad \text{equate coefficients}$$

The solution of this system is $A = \frac{1}{5}$ and $B = \frac{3}{5}$. Thus,

$$y_p = \frac{1}{5}\sin x + \frac{3}{5}\cos x \qquad \text{particular solution}$$

$$y = c_1 e^x + c_2 e^{2x} + \frac{1}{5}\sin x + \frac{3}{5}\cos x \qquad \text{complete general solution}$$

■ Check this solution in the differential equation.

EXAMPLE H — Find the particular solution of the differential equation $y'' + 16y = 2e^{-x}$ if $Dy = -2$ and $y = 1$ when $x = 0$.
In this case we must not only find y_c and y_p, we must also evaluate the constants of y_c from the given conditions. The solution is as follows:

$$D^2y + 16y = 2e^{-x} \qquad \text{operator } D \text{ form}$$
$$m^2 + 16 = 0, \qquad m = \pm 4j \qquad \text{auxiliary equation}$$
$$y_c = c_1 \sin 4x + c_2 \cos 4x \qquad \text{complementary solution}$$
$$y_p = Ae^{-x} \qquad \text{particular solution form}$$

$$Dy_p = -Ae^{-x}, \qquad D^2y_p = Ae^{-x}$$

$$Ae^{-x} + 16Ae^{-x} = 2e^{-x} \qquad \text{substituting}$$

$$17Ae^{-x} = 2e^{-x}, \qquad A = \frac{2}{17} \qquad \text{equate coefficients}$$

$$y_p = \frac{2}{17}e^{-x}$$

$$y = c_1 \sin 4x + c_2 \cos 4x + \frac{2}{17}e^{-x} \qquad \text{complete general solution}$$

We now evaluate c_1 and c_2 from the given conditions:

$$Dy = 4c_1 \cos 4x - 4c_2 \sin 4x - \frac{2}{17}e^{-x}$$

$$1 = c_1(0) + c_2(1) + \frac{2}{17}(1), \qquad c_2 = \frac{15}{17} \qquad y = 1 \text{ when } x = 0$$

$$-2 = 4c_1(1) - 4c_2(0) - \frac{2}{17}(1), \qquad c_1 = -\frac{8}{17} \qquad Dy = -2 \text{ when } x = 0$$

$$y = -\frac{8}{17}\sin 4x + \frac{15}{17}\cos 4x + \frac{2}{17}e^{-x} \qquad \text{required particular solution}$$

■ This solution checks when substituted into the differential equation.

If any terms of the form required for y_p are of the same form as any of the terms of y_c, it is necessary to multiply such terms of y_p by a power of x. We have not included any equations of this type to this point, although we will be able to get solutions by methods presented in Section 29-12.

Exercises 29-9

In Exercises 1 through 12, solve the given differential equations. The form of y_p is indicated.

1. $D^2y - Dy - 2y = 4$ (Let $y_p = A$.)   **2.** $D^2y - Dy - 6y = 4x$ (Let $y_p = A + Bx$.)

3. $4D^2y + y = x^2$ (Let $y_p = A + Bx + Cx^2$.)   **4.** $D^2y - y = 2 + x^2$ (Let $y_p = A + Bx + Cx^2$.)

5. $D^2y + 4Dy + 3y = 2 + e^x$ (Let $y_p = A + Be^x$.)   **6.** $2D^2y + Dy + y = e^{2x}$ (Let $y_p = Ae^{2x}$.)

7. $y'' - 3y' = 2e^x + xe^x$ (Let $y_p = Ae^x + Bxe^x$.)

8. $y'' + y' - 2y = 8 + 4x + 2xe^{2x}$ (Let $y_p = A + Bx + Ce^{2x} + Exe^{2x}$.)

9. $9D^2y - y = \sin x$ (Let $y_p = A \sin x + B \cos x$.)

10. $D^2y + 4y = \sin x + 4$ (Let $y_p = A + B \sin x + C \cos x$.)

11. $\dfrac{d^2y}{dx^2} + 9y = 9x + 5\cos 2x + 10\sin 2x$ (Let $y_p = A + Bx + C \sin 2x + E \cos 2x$.)

12. $\dfrac{d^2y}{dx^2} - 2\dfrac{dy}{dx} + y = 2x + x^2 + \sin 3x$ (Let $y_p = A + Bx + Cx^2 + E \sin 3x + F \cos 3x$.)

In Exercises 13 through 24, solve the given differential equations.

13. $\dfrac{d^2y}{dx^2} - \dfrac{dy}{dx} - 30y = 10$ **14.** $2\dfrac{d^2y}{dx^2} + 11\dfrac{dy}{dx} - 6y = 8x$ **15.** $3\dfrac{d^2y}{dx^2} + 13\dfrac{dy}{dx} - 10y = 14e^{3x}$

16. $\dfrac{d^2y}{dx^2} + 4y = 2 \sin 3x$ **17.** $D^2y - 4y = \sin x + 2 \cos x$ **18.** $2D^2y + 5Dy - 3y = e^x + 4e^{2x}$

19. $D^2y + y = 4 + \sin 2x$ **20.** $D^2y - Dy + y = x + \sin x$ **21.** $D^2y + 5Dy + 4y = xe^x + 4$

22. $3D^2y + Dy - 2y = 4 + 2x + e^x$ **23.** $y'' + 6y' + 9y = e^{2x} - e^{-2x}$ **24.** $y'' + 8y' - y = x^2 + 4e^{-2x}$

In Exercises 25 through 28, find the particular solution of each differential equation for the given conditions.

25. $D^2y - Dy - 6y = 5 - e^x$; $Dy = 4$ and $y = 2$ when $x = 0$

26. $3y'' - 10y' + 3y = xe^{-2x}$; $Dy = 0$ and $y = -1$ when $x = 0$

27. $y'' + y = x + \sin 2x$; $Dy = 1$ and $y = 0$ when $x = \pi$

28. $D^2y - 2Dy + y = xe^{2x} - e^{2x}$; $Dy = 4$ and $y = -2$ when $x = 0$

29-10   Applications of Second-Order Equations

Linear differential equations of the second order have many important applications. We shall restrict our attention to two of these. In this section, we shall apply second-order differential equations to solve problems in simple harmonic motion and simple electric circuits.

EXAMPLE A ⎯⎯⎯⎯⎯⎯⎯

Simple harmonic motion may be defined as motion in a straight line for which the acceleration is proportional to the displacement and in the opposite direction. Examples of this type of motion are a weight on a spring, a simple pendulum, and an object bobbing in water. If x represents the displacement, d^2x/dt^2 is the acceleration.

Using the definition of simple harmonic motion, we have

$$\frac{d^2x}{dt^2} = -k^2x$$

(We chose k^2 for convenience of notation in the solution.) We write this equation in the form

$$D^2x + k^2x = 0$$

where $D = d/dt$. The roots of the auxiliary equation are kj and $-kj$, and the solution is

$$x = c_1 \sin kt + c_2 \cos kt$$

This solution indicates an oscillating motion, which is known to be the case. If, for example, $k = 4$ and we know that $x = 2$ and $Dx = 0$ (which means the velocity is zero) for $t = 0$, we have

$$Dx = 4c_1 \cos 4t - 4c_2 \sin 4t$$
$$2 = c_1(0) + c_2(1) \qquad x = 2 \text{ for } t = 0$$
$$0 = 4c_1(1) - 4c_2(0) \qquad Dx = 0 \text{ for } t = 0$$

which gives $c_1 = 0$ and $c_2 = 2$. Therefore,

$$x = 2 \cos 4t$$

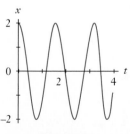

Fig. 29-8

is the equation relating the displacement and time; Dx is the velocity and D^2x is the acceleration. See Fig. 29-8.

EXAMPLE B

In practice an object will in time cease to move due to unavoidable frictional forces. It is found that a "freely" oscillating object has a force retarding it which is approximately proportional to the velocity. The differential equation for this case is $D^2x = -k^2x - b\,Dx$. This results from applying (from physics) Newton's second law of motion, which states that the net force acting on an object equals its mass times its acceleration. The term D^2x represents the acceleration, the term $-k^2x$ is a measure of the restoring force (of the spring, for example), and the term $-b\,Dx$ is the term which represents the retarding (damping) force. This equation can be written as

$$D^2x + b\,Dx + k^2x = 0$$

The auxiliary equation is $m^2 + bm + k^2 = 0$, for which the roots are

$$m = \frac{-b \pm \sqrt{b^2 - 4k^2}}{2}$$

If $k = 3$ and $b = 4$, $m = -2 \pm j\sqrt{5}$, which means the solution is

$$x = e^{-2t}(c_1 \sin \sqrt{5}t + c_2 \cos \sqrt{5}t) \tag{1}$$

Here, $4k^2 > b^2$, and this case is called **underdamped.** In this case the object oscillates as the amplitude becomes smaller.

If $k = 2$ and $b = 5$, $m = -1, -4$, which means the solution is

$$x = c_1e^{-t} + c_2e^{-4t} \tag{2}$$

Here $4k^2 < b^2$, and the case is called **overdamped.** It will be noted that the motion is not oscillatory, since no sine or cosine terms appear. In this case the object returns slowly to equilibrium without oscillating.

If $k = 2$ and $b = 4$, $m = -2, -2$, which means the solution is

$$x = e^{-2t}(c_1 + c_2t) \tag{3}$$

Here $4k^2 = b^2$, and the case is called **critically damped.** Again the motion is not oscillatory. In this case there is just enough damping to prevent any oscillations. The object returns to equilibrium in the minimum time.

See Fig. 29-9, in which Eqs. (1), (2), and (3) are represented in general. Of course, the actual values depend on c_1 and c_2, which in turn depend on the conditions imposed on the motion. ■

Fig. 29-9

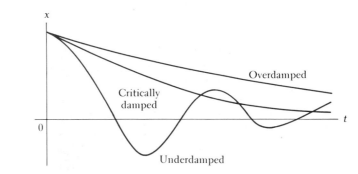

EXAMPLE C

In testing the characteristics of a particular type of spring, it is found that a weight of 16.0 lb stretches the spring 1.60 ft when the weight and spring are placed in a fluid which resists the motion with a force equal to twice the velocity. If the weight is brought to rest and then given a velocity of 12.0 ft/s, find the equation of motion. See Fig. 29-10.

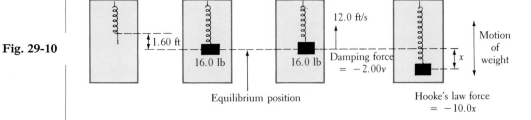

Fig. 29-10

In order to find the equation of motion we use Newton's second law of motion (see Example B). The weight (one force) at the end of the spring is offset by the equilibrium position force exerted by the spring, in accordance with Hooke's law (see Section 25-6). Therefore, the net force acting on the weight is the sum of the Hooke's law force due to the displacement from the equilibrium position and the resisting force. Using Newton's second law, we have

mass × acceleration = resisting force + Hooke's law force

$$m\,D^2x = -2.00\,Dx - kx$$

The mass of an object is its weight divided by the acceleration due to gravity. The weight is 16.0 lb and the acceleration due to gravity is 32.0 ft/s². Thus, the mass m is

$$m = \frac{16.0\ \text{lb}}{32.0\ \text{ft/s}^2} = 0.500\ \text{slug}$$

where the slug is the unit of mass if the weight is in pounds.

The constant k for the Hooke's law force is found from the fact that the spring stretches 1.60 ft for a force of 16.0 lb. Thus, using Hooke's law,

$$16.0 = k(1.60), \qquad k = 10.0\ \text{lb/ft}$$

This means that the differential equation to be solved is

$$0.500\,D^2x + 2.00\,Dx + 10.0x = 0$$

or

$$1.00\,D^2x + 4.00\,Dx + 20.0x = 0$$

Solving this equation, we have

$$1.00m^2 + 4.00m + 20.0 = 0 \qquad \text{auxiliary equation}$$

$$m = \frac{-4.00 \pm \sqrt{16.0 - 4(20.0)(1.00)}}{2.00}$$

$$= -2.00 \pm 4.00j \qquad \textbf{complex roots}$$

$$x = e^{-2.00t}(c_1 \cos 4.00t + c_2 \sin 4.00t) \qquad \textbf{general solution}$$

Since the weight started from the equilibrium position with a velocity of 12.0 ft/s, we know that $x = 0$ and $Dx = 12.0$ for $t = 0$. Thus,

$$0 = e^0(c_1 + 0c_2), \quad \text{or} \quad c_1 = 0 \qquad x = 0 \text{ for } t = 0$$

Thus, since $c_1 = 0$, we have

$$x = c_2 e^{-2.00t} \sin 4.00t$$

$$Dx = c_2 e^{-2.00t}(\cos 4.00t)(4.00) + c_2 \sin 4.00t(e^{-2.00t})(-2.00)$$

$$12.0 = c_2 e^0(1)(4.00) + c_2(0)(e^0)(-2.00) \qquad Dx = 12.0 \text{ for } t = 0$$

$$c_2 = 3.00$$

This means that the equation of motion is

$$x = 3.00 e^{-2.00t} \sin 4.00t$$

■ This motion is underdamped; the graph is shown in Fig. 29-11.

Fig. 29-11

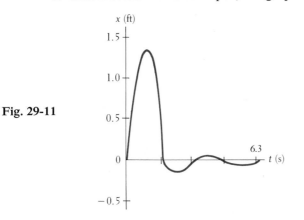

In Example C the force was given in pounds and the mass was therefore expressed in slugs. If metric units are used, it is common to give the mass of the object in kilograms, and its weight is therefore in newtons. The weight, which is needed to determine the constant k in the Hooke's law force, is found by multiplying the mass, in kilograms, by the acceleration of gravity, 9.80 m/s^2.

It is possible to have an additional force acting on a weight such as the one in Example C. For example, a vibratory force may be applied to the support of the spring. In such a case, called *forced vibrations,* this additional external force is added to the other net force. This means that the added force $F(t)$ becomes a nonzero function on the right side of the differential equation, and we must then solve a nonhomogeneous equation.

EXAMPLE D

E or $E_0 \sin \omega t$

Fig. 29-12

A basic relation for electric circuits is that the impressed voltage of a circuit equals the sum of the voltages across the components of the circuit. When this principle is applied to a circuit containing a resistance, an inductance, a capacitance, and a voltage source (see Fig. 29-12), the differential equation is

$$L \frac{d^2q}{dt^2} + R \frac{dq}{dt} + \frac{q}{C} = E \qquad (29\text{-}20)$$

By definition q represents the electric charge, $dq/dt = i$ is the current, and d^2q/dt^2 is the time rate of change of current. This equation may be written as

$$LD^2q + RDq + q/C = E$$

The auxiliary equation is $Lm^2 + Rm + 1/C = 0$. The roots are

$$m = \frac{-R \pm \sqrt{R^2 - 4L/C}}{2L} = -\frac{R}{2L} \pm \sqrt{\frac{R^2}{4L^2} - \frac{1}{LC}}$$

If we let $a = R/2L$ and $\omega = \sqrt{1/LC - R^2/4L^2}$, we have (assuming complex roots, which corresponds to realistic values of R, L, and C)

$$q_c = e^{-at}(c_1 \sin \omega t + c_2 \cos \omega t)$$

This indicates an oscillating charge, or an alternating current. However, the exponential term usually is such that the current dies out rapidly unless there is a source of voltage in the circuit.

If there is no source of voltage in the circuit of Example D, we have a homogeneous differential equation to solve. If we have a constant voltage source, the particular solution is of the form $q_p = A$. If there is an alternating voltage source, the particular solution is of the form $q_p = A \sin \omega_1 t + B \cos \omega_1 t$, where ω_1 is the angular velocity of the source. After a very short time, the exponential factor in the complementary solution makes it negligible. For this reason it is referred to as the **transient** term, and *the particular solution is the steady-state solution*. Therefore, to find the steady-state solution, we need find only the particular solution.

EXAMPLE E

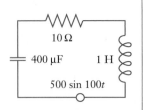

Fig. 29-13

Find the steady-state solution for the current in a circuit containing the following elements: $C = 400 \ \mu F$, $L = 1 \ H$, $R = 10 \ \Omega$, and a voltage source of $500 \sin 100t$. See Fig. 29-13.

This means the differential equation to be solved is

$$\frac{d^2q}{dt^2} + 10 \frac{dq}{dt} + \frac{10^4}{4} q = 500 \sin 100t$$

Since we wish to find the steady-state solution, we must find q_p, from which we may find i_p by finding a derivative. The solution now follows:

$$q_p = A \sin 100t + B \cos 100t \qquad \text{particular solution form}$$

$$\frac{dq_p}{dt} = 100A \cos 100t - 100B \sin 100t$$

$$\left| \frac{d^2 q_p}{dt^2} = -10^4 A \sin 100t - 10^4 B \cos 100t \right.$$

$-10^4 A \sin 100t - 10^4 B \cos 100t + 10^3 A \cos 100t - 10^3 B \sin 100t$ substitute into differential equation

$$+ \frac{10^4}{4} A \sin 100t + \frac{10^4}{4} B \cos 100t = 500 \sin 100t$$

$$(-0.75 \times 10^4 A - 10^3 B) \sin 100t + (-0.75 \times 10^4 B + 10^3 A) \cos 100t = 500 \sin 100t$$

$$\left| \begin{array}{l} -7.5 \times 10^3 A - 10^3 B = 500 \quad \text{equate coefficients of sin } 100t \\ 10^3 A - 7.5 \times 10^3 B = 0 \quad \text{equate coefficients of cos } 100t \end{array} \right.$$

Solving these equations, we obtain

$$B = -8.73 \times 10^{-3} \quad \text{and} \quad A = -65.5 \times 10^{-3}$$

Therefore,

$$q_p = -65.5 \times 10^{-3} \sin 100t - 8.73 \times 10^{-3} \cos 100t$$

$$i_p = \frac{dq_p}{dt} = -6.55 \cos 100t + 0.87 \sin 100t$$

which is the desired solution.

It should be noted that the complementary solutions of the mechanical and electric cases are of identical form. There is also an equivalent mechanical case to that of an impressed sinusoidal voltage source in the electric case. This arises in the case of forced vibrations, when an external force affecting the vibrations is applied to the system. Thus, we may have transient and steady-state solutions to mechanical and other nonelectric situations.

Exercises 29-10

1. When the angular displacement θ of a pendulum is small (less than about $6°$), the pendulum moves with simple harmonic motion closely approximated by $D^2\theta + \frac{g}{\ell} \theta = 0$. Here $D = d/dt$, g is the acceleration due to gravity, and ℓ is the length of the pendulum. Find θ as a function of time if $g = 9.8$ m/s^2, $\ell = 1.0$ m, $\theta = 0.1$, and $D\theta = 0$ when $t = 0$. Sketch the curve.

2. Find θ as a function of time for the pendulum in Exercise 1 if it moves according to the equation $D^2\theta + 0.2 D\theta + \frac{g}{\ell} \theta = 0$. Sketch the curve.

3. A car suspension is depressed from its equilibrium position such that its equation of motion is $D^2 y + b Dy + 25y = 0$, where $D = d/dt$ and y is the displacement. What must be the value of b if the motion is critically damped?

4. A block of wood floating in oil is depressed from its equilibrium position such that its equation of motion is $D^2 y + 8 Dy + 3y = 0$, where y is the displacement (in inches) and $D = d/dt$. Find its displacement after 12 s if $y = 6.0$ in. and $Dy = 0$ when $t = 0$.

5. A 4.00-lb weight stretches a certain spring 0.125 ft. With this weight attached, the spring is pulled 3.00 in. longer than its equilibrium length and released. Find the equation of the resulting motion, assuming no damping.

6. Find the solution for the spring of Exercise 5 if a damping force numerically equal to the velocity is present.

7. Find the solution for the spring of Exercise 5 if no damping is present, but an external force of 4 sin 2t is acting on the spring.

8. Find the solution for the spring of Exercise 5 if the damping force of Exercise 6 and the impressed force of Exercise 7 are both acting.

9. A mass of 0.820 kg stretches a given spring by 0.250 m. The mass is pulled down 0.150 m below the equilibrium position and released. Find the equation of motion of the mass if there is no damping.

10. Find the equation of motion for the mass of Exercise 9 if a damping force numerically equal to three times the velocity is also present.

11. Find the equation relating the charge and time in an electric circuit with the following elements: $L = 0.2$ H, $R = 8\ \Omega$, $C = 1\ \mu$F, $E = 0$. In this circuit $q = 0$ and $i = 0.5$ A when $t = 0$.

12. For a given circuit, $L = 2$ mH, $R = 0$, $C = 50$ nF, and $E = 0$. Find the equation relating the charge and time if $q = 10^5$ C and $i = 0$ when $t = 0$.

13. For a given circuit, $L = 0.1$ H, $R = 0$, $C = 100\ \mu$F, and $E = 100$ V. Find the equation relating the charge and time if $q = 0$ and $i = 0$ when $t = 0$.

14. Find the relation between the current and time for the circuit of Exercise 13.

15. For a given circuit, $L = 0.5$ H, $R = 10\ \Omega$, $C = 200\ \mu$F, and $E = 100 \sin 200t$. Find the expression relating charge and time.

16. Find the steady-state current for the circuit of Exercise 15.

17. In a given circuit $L = 8$ mH, $R = 0$, $C = 0.5\ \mu$F, and $E = 20e^{-200t}$ mV. Find the relation between the current and time if $q = 0$ and $i = 0$ for $t = 0$.

18. Find the current as a function of time for a circuit in which $L = 0.4$ H, $R = 60\ \Omega$, $C = 0.2\ \mu$F, and $E = 0.8e^{-100t}$ V, if $q = 0$ and $i = 5.0$ mA for $t = 0$.

19. Find the steady-state current for a circuit with $L = 1$ H, $R = 5\ \Omega$, $C = 150\ \mu$F, and $E = 120 \sin 100t$ V.

20. Find the steady-state solution for the current in an electric circuit containing the following elements: $C = 20\ \mu$F, $L = 2$ H, $R = 20\ \Omega$, and $E = 200 \sin 10t$.

29-11   Laplace Transforms

The final method of solving differential equations which we shall discuss is by means of **Laplace transforms.** As we shall see, *Laplace transforms provide an algebraic method of obtaining a **particular** solution of a differential equation from stated initial conditions.* Since this is often what is desired in practice, Laplace transforms are often preferred for the solution of differential equations in engineering and electronics. In this section we shall discuss the meaning of the transform and its operations. These methods will be applied to solving differential equations in the following section. The treatment in this text is intended only as an introduction to the topic of Laplace transforms.

The Laplace transform of a function $f(t)$ is defined as the function $F(s)$ by the equation

$$F(s) = \int_0^\infty e^{-st} f(t)\, dt \qquad\qquad (29\text{-}21)$$

By writing the transform as $F(s)$, we show that the result of integrating and evaluating is a function of s. To denote that we are dealing with "the Laplace transform of the function $f(t)$" the notation $L(f)$ is used. Thus,

definition of
Laplace transform

$$F(s) = L(f) = \int_0^\infty e^{-st}f(t)\,dt \qquad (29\text{-}22)$$

We shall see that both notations are quite useful.

A note regarding the form of the integral in Eqs. (29-21) and (29-22) is in order at this point. Since the upper limit is ∞, which means that it is unbounded, this integral is one type of what is known as an **improper integral.** In evaluating this integral at the upper limit, it is necessary to find the limit of the resulting function as the upper limit approaches infinity. This may be denoted by

$$\lim_{c \to \infty} \int_0^c e^{-st}f(t)\,dt$$

where we substitute c for t in the resulting function and determine the limit as $c \to \infty$ to determine the result for the upper limit. This also means that the Laplace transform, $F(s)$, is defined only for those values of s for which the limit is defined.

EXAMPLE A

Find the Laplace transform of the function $f(t) = t$, $t > 0$.

By the definition of the Laplace transform

$$L(f) = L(t) = \int_0^\infty e^{-st}t\,dt$$

This may be integrated by parts or by Formula (44) of Table 5 in Appendix F. Using the formula, we have

$$L(t) = \int_0^\infty te^{-st}\,dt = \lim_{c \to \infty} \int_0^c te^{-st}\,dt = \lim_{c \to \infty} \frac{e^{-st}(-st-1)}{s^2}\Big|_0^c$$

$$= \lim_{c \to \infty} \left[\frac{e^{-sc}(-sc-1)}{s^2}\right] + \frac{1}{s^2}$$

Now, for $s > 0$, as $c \to \infty$, $e^{-sc} \to 0$ and $sc \to \infty$. However, although we cannot prove it here, $e^{-sc} \to 0$ much faster than $sc \to \infty$. We can see that this is reasonable, for $ce^{-c} = 4.5 \times 10^{-4}$ for $c = 10$ and $ce^{-c} = 3.7 \times 10^{-42}$ for $c = 100$. Thus, the value at the upper limit approaches zero, which means the limit is zero. This means that

$$L(t) = \frac{1}{s^2}$$

■ As we noted, this transform is defined for $s > 0$.

EXAMPLE B

Find the Laplace transform of the function $f(t) = \cos at$.
By definition,

$$L(f) = L(\cos at) = \int_0^\infty e^{-st} \cos at\, dt$$

Using Formula (50) of Table 5, we have

$$L(\cos at) = \int_0^\infty e^{-st} \cos at\, dt = \lim_{c \to \infty} \int_0^c e^{-st} \cos at\, dt$$

$$= \lim_{c \to \infty} \left. \frac{e^{-st}(-s \cos at + a \sin at)}{s^2 + a^2} \right|_0^c$$

$$= \lim_{c \to \infty} \frac{e^{-sc}(-s \cos ac + a \sin ac)}{s^2 + a^2} - \left(-\frac{s}{s^2 + a^2} \right)$$

$$= 0 + \frac{s}{s^2 + a^2}$$

$$= \frac{s}{s^2 + a^2} \quad (s > 0)$$

Therefore, the Laplace transform of the function $\cos at$ is

$$L(\cos at) = \frac{s}{s^2 + a^2}$$

■ In both examples the resulting transform was an algebraic function of s.

An important property of transforms is the **linearity property,**

$$L[af(t) + bg(t)] = aL(f) + bL(g) \qquad (29\text{-}23)$$

We state this property here since it determines that the transform of a sum of functions is the sum of the transforms. This is of definite importance when dealing with a sum of functions. This property is a direct result of the definition of the Laplace transform.

We now present a short table of Laplace transforms. The transforms given on the next page are sufficient for our remaining work in this chapter. Much more complete tables are available in many standard reference sources.

Another Laplace transform important to the solution of a differential equation is the transform of the derivative of a function. Let us first find the Laplace transform of the first derivative of a function.

By definition,

$$L(f') = \int_0^\infty e^{-st} f'(t)\, dt$$

To integrate by parts, let $u = e^{-st}$ and $dv = f'(t)\, dt$, so $du = -se^{-st}\, dt$ and $v = f(t)$ (the integral of the derivative of a function is the function). Therefore,

$$L(f') = e^{-st} f(t) \Big|_0^\infty + s \int_0^\infty e^{-st} f(t)\, dt$$

$$= 0 - f(0) + sL(f)$$

TABLE OF LAPLACE TRANSFORMS

	$f(t) = L^{-1}(F)$	$L(f) = F(s)$		$f(t) = L^{-1}(F)$	$L(f) = F(s)$
1.	1	$\dfrac{1}{s}$	11.	te^{-at}	$\dfrac{1}{(s+a)^2}$
2.	$\dfrac{t^{n-1}}{(n-1)!}$	$\dfrac{1}{s^n}\ (n = 1, 2, 3, \ldots)$	12.	$t^{n-1}e^{-at}$	$\dfrac{(n-1)!}{(s+a)^n}$
3.	e^{-at}	$\dfrac{1}{s+a}$	13.	$e^{-at}(1-at)$	$\dfrac{s}{(s+a)^2}$
4.	$1 - e^{-at}$	$\dfrac{a}{s(s+a)}$	14.	$[(b-a)t+1]e^{-at}$	$\dfrac{s+b}{(s+a)^2}$
5.	$\cos at$	$\dfrac{s}{s^2+a^2}$	15.	$\sin at - at\cos at$	$\dfrac{2a^3}{(s^2+a^2)^2}$
6.	$\sin at$	$\dfrac{a}{s^2+a^2}$	16.	$t\sin at$	$\dfrac{2as}{(s^2+a^2)^2}$
7.	$1 - \cos at$	$\dfrac{a^2}{s(s^2+a^2)}$	17.	$\sin at + at\cos at$	$\dfrac{2as^2}{(s^2+a^2)^2}$
8.	$at - \sin at$	$\dfrac{a^3}{s^2(s^2+a^2)}$	18.	$t\cos at$	$\dfrac{s^2-a^2}{(s^2+a^2)^2}$
9.	$e^{-at} - e^{-bt}$	$\dfrac{b-a}{(s+a)(s+b)}$	19.	$e^{-at}\sin bt$	$\dfrac{b}{(s+a)^2+b^2}$
10.	$ae^{-at} - be^{-bt}$	$\dfrac{s(a-b)}{(s+a)(s+b)}$	20.	$e^{-at}\cos bt$	$\dfrac{s+a}{(s+a)^2+b^2}$

It is noted that the integral in the second term on the right is the Laplace transform of $f(t)$ by definition. Therefore, *the Laplace transform of the first derivative of a function is*

Laplace transform of first derivative

$$L(f') = sL(f) - f(0) \tag{29-24}$$

Applying the same analysis, we may find *the Laplace transform of the second derivative of a function. It is*

Laplace transform of second derivative

$$L(f'') = s^2L(f) - sf(0) - f'(0) \tag{29-25}$$

Here it is necessary to integrate by parts twice to derive the result. The transforms of higher derivatives are found in a similar manner.

Equations (29-24) and (29-25) allow us to express the transform of each derivative in terms of s and the transform of the function itself.

EXAMPLE C ——— If $f(0) = 0$ and $f'(0) = 1$, express the transform of $f''(t) - 2f'(t)$ in terms of s and the transform of $f(t)$.

By using the linearity property and the transforms of the derivatives, we have

$$L[f''(t) - 2f'(t)] = L(f'') - 2L(f') \qquad \text{using Eq. (29-23)}$$
$$= [s^2L(f) - sf(0) - f'(0)] - 2[sL(f) - f(0)] \qquad \text{using Eqs. (29-25) and (29-24)}$$
$$= [s^2L(f) - s(0) - 1] - 2[sL(f) - 0] \qquad \text{substitute given values}$$
$$= (s^2 - 2s)L(f) - 1$$

If the Laplace transform of a function is known, it is then possible to find the function by finding the **inverse transform,**

inverse Laplace transform

$$\boxed{L^{-1}(F) = f(t)} \tag{29-26}$$

where L^{-1} denotes the inverse transform.

EXAMPLE D ——— If $F(s) = \dfrac{s}{s^2 + a^2}$, from Transform (5) of the table we see that

$$L^{-1}(F) = L^{-1}\left(\frac{s}{s^2 + a^2}\right) = \cos at$$

or

$$f(t) = \cos at$$

EXAMPLE E ——— If $(s^2 - 2s)L(f) - 1 = 0$, then

$$L(f) = \frac{1}{s^2 - 2s} \quad \text{or} \quad F(s) = \frac{1}{s(s-2)}$$

Therefore, we have

$$f(t) = L^{-1}(F) = L^{-1}\left[\frac{1}{s(s-2)}\right] \qquad \text{inverse transform}$$
$$= -\frac{1}{2}L^{-1}\left[\frac{-2}{s(s-2)}\right] \qquad \text{fit form of Transform (4)}$$
$$= -\frac{1}{2}(1 - e^{2t}) \qquad \text{use Transform (4)}$$

EXAMPLE F ——— If $F(s) = \dfrac{s+5}{s^2 + 6s + 10}$, then

$$L^{-1}(F) = L^{-1}\left[\frac{s+5}{s^2 + 6s + 10}\right]$$

It appears that this function does not fit any of the forms given. However,

$$s^2 + 6s + 10 = (s^2 + 6s + 9) + 1 = (s+3)^2 + 1$$

By writing $F(s)$ as

$$F(s) = \frac{(s+3)+2}{(s+3)^2+1} = \frac{s+3}{(s+3)^2+1} + \frac{2}{(s+3)^2+1}$$

we can find the inverse of each term. Therefore,

$$L^{-1}(F) = e^{-3t}\cos t + 2e^{-3t}\sin t \qquad \text{using Transforms (20) and (19)}$$

or

$$f(t) = e^{-3t}(\cos t + 2\sin t)$$

∎

Exercises 29-11

In Exercises 1 through 4, verify the indicated transforms given in the table.

1. Transform 1   **2.** Transform 3   **3.** Transform 6   **4.** Transform 11

In Exercises 5 through 12, find the transforms of the given functions by use of the table.

5. $f(t) = e^{3t}$ **6.** $f(t) = 1 - \cos 2t$ **7.** $f(t) = t^3 e^{-2t}$

8. $f(t) = 2e^{-3t}\sin 4t$ **9.** $f(t) = \cos 2t - \sin 2t$ **10.** $f(t) = 2t\sin 3t + e^{-3t}\cos t$

11. $f(t) = 3 + 2t\cos 3t$ **12.** $f(t) = t^3 - 3te^{-t}$

In Exercises 13 through 16, express the transforms of the given expressions in terms of s and L(f).

13. $y'' + y'$, $f(0) = 0$, $f'(0) = 0$ **14.** $y'' - 3y'$, $f(0) = 2$, $f'(0) = -1$

15. $2y'' - y' + y$, $f(0) = 1$, $f'(0) = 0$ **16.** $y'' - 3y' + 2y$, $f(0) = -1$, $f'(0) = 2$

In Exercises 17 through 24, find the inverse transform of the given functions of s.

17. $F(s) = \dfrac{2}{s^3}$ **18.** $F(s) = \dfrac{3}{s^2+4}$ **19.** $F(s) = \dfrac{1}{2s+6}$ **20.** $F(s) = \dfrac{3}{s^4+4s^2}$

21. $F(s) = \dfrac{1}{s^3+3s^2+3s+1}$ **22.** $F(s) = \dfrac{s^2-1}{s^4+2s^2+1}$ **23.** $F(s) = \dfrac{s+2}{(s^2+9)^2}$ **24.** $F(s) = \dfrac{s+3}{s^2+4s+13}$

29-12   Solving Differential Equations by Laplace Transforms

We will now show how certain differential equations can be solved by the use of Laplace transforms. *It must be remembered that these solutions are the **particular** solutions of the equations subject to the given conditions.* The necessary operations were developed in the preceding section. The following examples illustrate the method.

EXAMPLE A

Solve the differential equation $2y' - y = 0$ if $y(0) = 1$. (Note that we are using y to denote the function.)

Taking transforms of each term in the equation, we have

$$L(2y') - L(y) = L(0)$$
$$2L(y') - L(y) = 0$$

$L(0) = 0$ by direct use of the definition of the transform. Now, using Eq. (29-24),

(Continued on next page)

$L(y') = sL(y) - y(0)$, we have

$$2[sL(y) - 1] - L(y) = 0 \qquad y(0) = 1$$

Solve for $L(y)$, we obtain

$$2sL(y) - L(y) = 2$$

$$L(y) = \frac{2}{2s - 1} = \frac{1}{s - \frac{1}{2}}$$

Finding the inverse transform, we have

$$y = e^{t/2} \qquad \text{using Transform (3)}$$

NOTE ▷ The reader should check this solution with that obtained by methods developed earlier. Also, it should be noted that the solution was essentially an algebraic one. This points out the power and usefulness of Laplace transforms. *We are able to translate a differential equation into an algebraic form,* which can in turn be translated into the solution of the differential equation. Thus, we are able to solve a differential equation by use of algebra and specific algebraic forms.

EXAMPLE B ———— Solve the differential equation $y'' + 2y' + 2y = 0$, if $y(0) = 0$ and $y'(0) = 1$.

Using the same steps as outlined in Example A, we have the following solution:

$$L(y'') + 2L(y') + 2L(y) = 0 \qquad \text{take transforms}$$

$$[s^2 L(y) - sy(0) - y'(0)] + 2[sL(y) - y(0)] + 2L(y) = 0 \qquad \text{using Eqs. (29-25) and (29-24)}$$

$$[s^2 L(y) - s(0) - 1] + 2[sL(y) - 0] + 2L(y) = 0 \qquad \text{substitute given values}$$

$$s^2 L(y) - 1 + 2sL(y) + 2L(y) = 0$$

$$(s^2 + 2s + 2)L(y) = 1 \qquad \text{solve for } L(y)$$

$$L(y) = \frac{1}{s^2 + 2s + 2} = \frac{1}{(s + 1)^2 + 1} \qquad \text{take inverse transform}$$

$$y = e^{-t} \sin t \qquad \text{using Transform (19)}$$

EXAMPLE C ———— Solve the differential equation $y'' + y = \cos t$ if $y(0) = 1$ and $y'(0) = 2$.

Taking transforms of each term in the equation, we have

$$L(y'') + L(y) = L(\cos t)$$

The transform on the left is found by use of Eq. (29-25) and the transform on the right is found by direct use of Transform (5) in the table. Therefore, we have

$$[s^2 L(y) - s(1) - 2] + L(y) = \frac{s}{s^2 + 1}$$

Solving for $L(y)$, we obtain

$$(s^2 + 1)L(y) = \frac{s}{s^2 + 1} + s + 2$$

$$L(y) = \frac{s}{(s^2 + 1)^2} + \frac{s}{s^2 + 1} + \frac{2}{s^2 + 1}$$

Finding the inverse transform of each expression completes the solution:

$$y = \frac{t}{2} \sin t + \cos t + 2 \sin t \qquad \text{using Transforms (16), (5), (6)}$$

■

EXAMPLE D

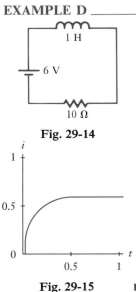

Fig. 29-14

An electric circuit contains a 1-H inductor, a 10-Ω resistor, and a battery whose voltage is 6 V. Find the current as a function of the time, if the initial current is zero. See Fig. 29-14.

The differential equation for this circuit is

$$\frac{di}{dt} + 10i = 6 \qquad \text{using Eq. (29-20)}$$

Following the procedures outlined in the previous examples, the solution is found:

$$L\left(\frac{di}{dt}\right) + 10L(i) = L(6) \qquad \text{take transforms}$$

$$[sL(i) - 0] + 10L(i) = \frac{6}{s} \qquad \begin{array}{l}\text{substitute given values}\\ \text{and find transform on right}\end{array}$$

$$L(i) = \frac{6}{s(s + 10)} \qquad \text{solve for } L(i)$$

$$i = 0.6(1 - e^{-10t}) \qquad \text{take inverse transform}$$

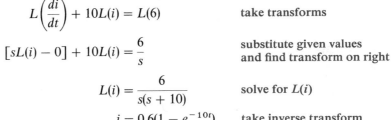

Fig. 29-15 ■ The graph of this solution is shown in Fig. 29-15.

EXAMPLE E

A spring is stretched 1 ft by a weight of 16 lb (mass of $\frac{1}{2}$ slug). The medium resists the motion of the object with a force of $4v$, where v is the velocity. The differential equation describing the motion is

$$\frac{1}{2}\frac{d^2y}{dt^2} + 4\frac{dy}{dt} + 16y = 0$$

Find y, the displacement of the object, as a function of the time if $y(0) = 1$ and $dy/dt = 0$ when $t = 0$.

Clearing fractions and denoting the derivatives by y'' and y', the differential equation becomes

$$y'' + 8y' + 32y = 0$$

Following the procedure of solving this type of an equation by use of Laplace transforms, we have the following steps:

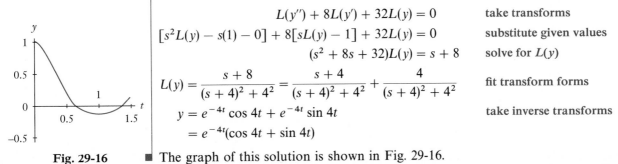

Fig. 29-16 ■ The graph of this solution is shown in Fig. 29-16.

Exercises 29-12

In the following exercises, solve the given differential equations by use of Laplace transforms, where the function is subject to the given conditions.

1. $y' + y = 0$, $y(0) = 1$

2. $y' - 2y = 0$, $y(0) = 2$

3. $2y' - 3y = 0$, $y(0) = -1$

4. $y' + 2y = 1$, $y(0) = 0$

5. $y' + 3y = e^{-3t}$, $y(0) = 1$

6. $y' + 2y = te^{-2t}$, $y(0) = 0$

7. $y'' + 4y = 0$, $y(0) = 0$, $y'(0) = 1$

8. $y'' - 4y = 0$, $y(0) = 2$, $y'(0) = 0$

9. $y'' + 2y' = 0$, $y(0) = 0$, $y'(0) = 2$

10. $y'' + 2y' + y = 0$, $y(0) = 0$, $y'(0) = -2$

11. $y'' - 4y' + 5y = 0$, $y(0) = 1$, $y'(0) = 2$

12. $4y'' + 4y' + y = 0$, $y(0) = 1$, $y'(0) = 0$

13. $y'' + y = 1$, $y(0) = 1$, $y'(0) = 1$

14. $y'' + 4y = 2t$, $y(0) = 0$, $y'(0) = 0$

15. $y'' + 2y' + y = e^{-t}$, $y(0) = 1$, $y'(0) = 2$

16. $y'' + 4y = \sin 2t$, $y(0) = 0$, $y'(0) = 0$

17. A constant force of 6 lb moves a 2-slug mass through a medium which resists the motion with a force equal to v, where v is the velocity. The differential equation relating the velocity and time is $2\dfrac{dv}{dt} = 6 - v$. Find v as a function of time if the object starts from rest.

18. A pendulum moves with simple harmonic motion according to the differential equation $D^2\theta + 20\theta = 0$, where θ is the angular displacement and $D = d/dt$. Find θ as a function of t if $\theta = 0$ and $D\theta = 0.4$ rad/s when $t = 0$.

19. A 50-Ω resistor, a 4-μF capacitor, and a 40-V battery are connected in series. Find the charge as a function of the time if the initial charge on the capacitor is zero.

20. A 2-H inductor, an 80-Ω resistor, and an 8-V battery are connected in series. Find the current in the circuit as a function of time if the initial current is zero.

21. A 10-H inductor, a 40-μF capacitor, and a voltage supply whose voltage is given by $100 \sin 50t$ are connected in series in an electric circuit. Find the current as a function of time if the initial charge on the capacitor and the initial current are zero.

22. A 20-mH inductor, a 40-Ω resistor, a 50-μF capacitor, and a voltage source of $100e^{-1000t}$ are connected in series in a circuit. Find the charge on the capacitor as a function of time if $q = 0$ and $i = 0$ when $t = 0$.

23. The weight on a spring undergoes forced vibrations according to the equation $D^2y + 9y = 18 \sin 3t$. Find its displacement y as a function of the time t, if $y = 0$ and $Dy = 0$ when $t = 0$.

24. A spring is such that it is stretched 6 in. by an 8-lb ($\frac{1}{4}$-slug) weight. The spring is stretched 6 in. below the equilibrium position with the weight on it and is given a velocity of 4 ft/s. If the spring is in a medium which resists the motion with a force equal to $4v$, find the displacement y as a function of time.

29-13   Chapter Equations, Review Exercises, and Practice Test

Chapter Equations

Separation of variables	$M(x, y)\,dx + N(x, y)\,dy = 0$	(29-1)
	$A(x)\,dx + B(y)\,dy = 0$	(29-2)
Integrable combinations	$d(xy) = x\,dy + y\,dx$	(29-3)
	$d(x^2 + y^2) = 2(x\,dx + y\,dy)$	(29-4)
	$d\left(\dfrac{y}{x}\right) = \dfrac{x\,dy - y\,dx}{x^2}$	(29-5)
	$d\left(\dfrac{x}{y}\right) = \dfrac{y\,dx - x\,dy}{y^2}$	(29-6)

Linear differential equation of first order	$dy + Py\,dx = Q\,dx$	(29-7)
	$ye^{\int P\,dx} = \int Qe^{\int P\,dx}\,dx + c$	(29-8)
Electric circuit	$L\dfrac{di}{dt} + Ri + \dfrac{q}{C} = E$	(29-9)
Motion in resisting medium	$m\dfrac{dv}{dt} = F - kv$	(29-10)
General linear differential equation	$a_0\dfrac{d^n y}{dx^n} + a_1\dfrac{d^{n-1}y}{dx^{n-1}} + \cdots + a_{n-1}\dfrac{dy}{dx} + a_n y = b$	(29-11)
	$a_0 D^n y + a_1 D^{n-1}y + \cdots + a_{n-1}Dy + a_n y = b$	(29-12)
Homogeneous linear differential equation	$a_0 D^2 y + a_1 Dy + a_2 y = 0$	(29-13)
Auxiliary equation	$a_0 m^2 + a_1 m + a_2 = 0$	(29-14)
Distinct roots	$y = c_1 e^{m_1 x} + c_2 e^{m_2 x}$	(29-15)
Repeated roots	$y = e^{mx}(c_1 + c_2 x)$	(29-16)
Complex roots	$y = e^{\alpha x}(c_1 \sin \beta x + c_2 \cos \beta x)$	(29-17)
Nonhomogeneous linear differential equation	$a_0 D^2 y + a_1 Dy + a_2 y = b$	(29-18)
	$y = y_c + y_p$	(29-19)
Electric circuit	$L\dfrac{d^2 q}{dt^2} + R\dfrac{dq}{dt} + \dfrac{q}{C} = E$	(29-20)
Laplace transforms	$F(s) = \displaystyle\int_0^\infty e^{-st}f(t)\,dt$	(29-21)
	$F(s) = L(f) = \displaystyle\int_0^\infty e^{-st}f(t)\,dt$	(29-22)
	$L[af(t) + bg(t)] = aL(f) + bL(g)$	(29-23)
	$L(f') = sL(f) - f(0)$	(29-24)
	$L(f'') = s^2 L(f) - sf(0) - f'(0)$	(29-25)
Inverse transform	$L^{-1}(F) = f(t)$	(29-26)

Review Exercises

In Exercises 1 through 28, find the general solution of each of the given differential equations.

1. $4xy^3\,dx + (x^2 + 1)\,dy = 0$

2. $\dfrac{dy}{dx} = e^{x-y}$

3. $\dfrac{dy}{dx} + 2y = e^{-2x}$

4. $x\,dy + y\,dx = y\,dy$

5. $2D^2 y + Dy = 0$

6. $2D^2 y - 5Dy + 2y = 0$

7. $y'' + 2y' + y = 0$

8. $y'' + 2y' + 2y = 0$

9. $(x + y)\,dx + (x + y^3)\,dy = 0$

10. $y \ln x\,dx = x\,dy$

11. $x\dfrac{dy}{dx} - 3y = x^2$

12. $dy - 2y\,dx = (x - 2)e^x\,dx$

13. $dy = 2y\,dx + y^2\,dx$

14. $x^2 y\,dy = (1 + x)\csc y\,dx$

15. $D^2 y + 2Dy + 6y = 0$

16. $4D^2 y - 4Dy + y = 0$

17. $y' + 4y = 2$

18. $2xy\,dx = (2y - \ln y)\,dy$

19. $\sin x \dfrac{dy}{dx} + y \cos x + x = 0$

20. $y\,dy = (x^2 + y^2 - x)\,dx$

21. $2\dfrac{d^2 y}{dx^2} + \dfrac{dy}{dx} - 3y = 6$

22. $\dfrac{d^2 y}{dx^2} + 6\dfrac{dy}{dx} + 9y = 3x$

23. $y'' + y' - y = 2e^x$

24. $4D^2 y + 9y = xe^x$

25. $9D^2 y - 18Dy + 8y = 16 + 4x$

26. $y'' + y = 4\cos 2x$

27. $D^2 y + 9y = \sin x$

28. $y'' + y' = e^x + \cos 2x$

In Exercises 29 through 36, find the particular solution of each of the given differential equations.

29. $3y' = 2y \cot x;$ $x = \dfrac{\pi}{2}$ when $y = 2$

30. $x\,dy - y\,dx = y^3\,dy;$ $x = 1$ when $y = 3$

31. $y' = 4x - 2y;$ $x = 0$ when $y = -2$

32. $xy^2\,dx + e^x\,dy = 0;$ $x = 0$ when $y = 2$

33. $\dfrac{d^2 y}{dx^2} + \dfrac{dy}{dx} + 4y = 0;$ $Dy = \sqrt{15}, y = 0$ when $x = 0$

34. $5y'' + 7y' - 6y = 0;$ $y' = 10, y = 2$ when $x = 0$

35. $(D^2 + 4D + 4)y = 4\cos x;$ $Dy = 1, y = 0$ when $x = 0$

36. $y'' - 2y' + y = e^{2x} + x;$ $Dy = 0, y = 2$ when $x = 0$

In Exercises 37 through 44, solve the given differential equations by use of Laplace transforms where the function is subject to the given conditions.

37. $4y' - y = 0, y(0) = 1$

38. $2y' - y = 4, y(0) = 1$

39. $y' - 3y = e^t, y(0) = 0$

40. $y' + 2y = e^{-2t}, y(0) = 2$

41. $y'' + y = 0, y(0) = 0, y'(0) = -4$

42. $y'' + 4y' + 5y = 0, y(0) = 1, y'(0) = 1$

43. $y'' + 9y = 3t, y(0) = 0, y'(0) = -1$

44. $y'' + 4y' + 4y = 2e^{-2t}, y(0) = 0, y'(0) = 1$

In Exercises 45 through 72, solve the given problems.

45. The time rate of change of volume of an evaporating substance is proportional to the surface area. Express the radius of an evaporating sphere of ice as a function of time. Let $r = r_0$ when $t = 0$. (*Hint:* Express both V and A in terms of the radius r.)

46. An insulated tank is filled with a solution containing radioactive cobalt. Due to the radioactivity, energy is released and the temperature T (in degrees Fahrenheit) of the solution rises with the time t (in hours). The following differential equation expresses the relation between temperature and time for a specific case:

$$56{,}600 = 262(T - 70) + 20{,}200\dfrac{dT}{dt}$$

If the initial temperature is $70°F$, what is the temperature 24 h later?

47. An object with a mass of 1 kg slides down a long inclined plane. The effective force of gravity is 4 N, and the motion is retarded by a force numerically equal to the velocity. If the object starts from rest, what is its velocity (in m/s) 4 s later?

48. A 192-lb object falls from rest under the influence of gravity. Find the equation for the velocity at any time t if the air resists the motion with a force numerically equal to twice the velocity.

49. Initially there are 500 mg of the radioactive hydrogen isotope tritium. After 4.02 years, 400 mg remain. What is the half-life of tritium?

50. When a gas undergoes an adiabatic change (no gain or loss of heat), the rate of change of pressure with respect to volume is directly proportional to the pressure and inversely proportional to the volume. Express the pressure in terms of the volume.

51. Under ideal conditions, the natural law of population change is that it increases at a rate proportional to the population at any time. Under these conditions, project the population of the world in the year 2000 if it reached 4.0 billion in 1976 and 5.0 billion in 1987.

52. In a certain type of chemical reaction, the velocity of the reaction is proportional to the mass of the chemical which remains unchanged. If m is the mass changed in time t, m_0 is the initial mass, and dm/dt is the velocity of the reaction, find m as a function of t.

53. Find the orthogonal trajectories of the family of curves $y = cx^5$.

54. Find the equation of the curves such that their normals at all points are in the direction with the lines connecting the points and the origin.

55. If a circuit contains a resistance R, capacitance C, and a source of voltage E, express the charge q on the capacitor as a function of time.

56. A 2-H inductor, a 40-Ω resistor, and a 20-V battery are connected in series. Find the current in the circuit as a function of time if the initial current is zero.

57. A certain spring stretches 0.5 m by a 40-N weight. With this weight suspended on it the spring is stretched 0.5 m beyond the equilibrium position and released. Find the equation of the resulting motion if the medium in which the weight is suspended retards the motion with a force equal to 16 times the velocity. Classify the motion as underdamped, critically damped, or overdamped.

58. The end of a vibrating rod moves according to the equation $D^2y + 0.2\,Dy + 4000y = 0$, where y is the displacement and $D = d/dt$. Find y as a function of t if $y = 3$ cm and $Dy = -0.3$ cm/s when $t = 0$.

59. A 0.5-H inductor, a 6-Ω resistor, and a 20-mF capacitor are connected in series with a generator for which $E = 24 \sin 10t$. Find the charge on the capacitor as a function of time if the initial charge and current are zero.

60. A 5-mH inductor and a 10-μF capacitor are connected in series with a voltage source of $0.2e^{-200t}$ V. Find the charge on the capacitor as a function of time if $q = 0$ and $i = 4$ mA when $t = 0$.

61. Find the equation for the current as a function of time if a resistance of 20 Ω, an inductor of 4 H, a capacitor of 100 μF, and a battery of 100 V are in series. The initial charge on the capacitor is 10 mC, and the initial current is zero.

62. If a circuit contains an inductor of inductance L, a capacitor of capacitance C, and a sinusoidal source of voltage $E_0 \sin \omega t$, express the charge q on the capacitor as a function of the time. Assume $q = 0$, $i = 0$ when $t = 0$.

63. The differential equation relating current and time for a certain electric circuit is $2\dfrac{di}{dt} + i = 12$. Solve this equation by use of Laplace transforms given that the initial current is zero. Evaluate the current for $t = 0.3$ s.

64. A 6-H inductor and a 30-Ω resistor are connected in series. Find the current as a function of time if the initial current is 15 A. Use Laplace transforms.

65. A 0.25-H inductor, a 4-Ω resistor, and a 100-μF capacitor are connected in series. If the initial charge on the capacitor is 400μC and the initial current is zero, find the charge on the capacitor as a function of time. Use Laplace transforms.

66. An inductor of 0.5 H, a resistor of 6 Ω, and a capacitor of 200 μF are connected in series. If the initial charge on the capacitor is 10 mC and the initial current is zero, find the charge on the capacitor as a function of time after the switch is closed. Use Laplace transforms.

67. An 8-lb weight ($\frac{1}{4}$-slug mass) stretches a spring 6 in. An external force of $\cos 8t$ is applied to the spring. Express the displacement y of the object as a function of time if the initial displacement and velocity are zero. Use Laplace transforms.

68. A spring is stretched 1 m by a mass of 5 kg (assume the weight to be 50 N). Find y, the displacement of the object, as a function of time if $y(0) = 1$ and $dy/dt = 0$ when $t = 0$. Use Laplace transforms.

69. The approximate differential equation relating the displacement y of a beam at
a horizontal distance x from one end is $EI\dfrac{d^2y}{dx^2} = M$, where E is the modulus
of elasticity, I is the moment of inertia of the cross section of the beam perpendicular to its axis, and M is the bending moment at the cross section. If
$M = 2000x - 40x^2$ for a particular beam of length L for which $y = 0$ when $x = 0$
and when $x = L$, express y in terms of x. Consider E and I as constants.

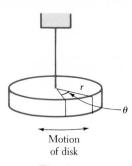

70. When a circular disk of mass m and radius r is suspended by a wire at the center
on one of its flat faces and the disk is twisted through an angle θ, torsion in the
wire tends to turn the disk back in the opposite direction. The differential equa-
tion for this case is $\dfrac{1}{2}mr^2\dfrac{d^2\theta}{dt^2} = -k\theta$, where k is a constant. Determine the equation of motion if $\theta = \theta_0$ and $d\theta/dt = \omega_0$ when $t = 0$. See Fig. 29-17.

Motion
of disk

Fig. 29-17

71. The gravitational acceleration of an object is inversely proportional to the square
of its distance r from the center of the earth. Use the chain rule, Eq. (22-14), to
show that the acceleration is $\dfrac{dv}{dt} = v\dfrac{dv}{dr}$, where $v = \dfrac{dr}{dt}$ is the velocity of the object.
Then solve for v as a function of r if $dv/dt = -g$ and $v = v_0$ for $r = R$, where R is
the radius of the earth. Finally, show that a spacecraft must have a velocity of at least
$v_0 = \sqrt{2gR}$ ($= 7$ mi/s) in order to escape from the earth's gravitation. (Note
the expression for v^2 as $r \to \infty$.)

72. Air containing 20% oxygen passes into a 5.00-L container initially filled with 100% oxygen. A uniform mixture
of the air and oxygen then passes from the container at the same rate. What volume of oxygen is in the container after 5.00 L of air have passed into it?

Practice Test

In Problems 1 through 5, find the general solution of each of the given differential equations.

1. $x\dfrac{dy}{dx} + 2y = 4$

2. $y'' + 2y' + 5y = 0$

3. $x\,dx + y\,dy = x^2\,dx + y^2\,dx$

4. $2D^2y - Dy = 2\cos x$

5. $\dfrac{d^2y}{dx^2} - 4\dfrac{dy}{dx} + 4y = 3x$

6. Find the particular solution of the differential equation $(xy + y)\dfrac{dy}{dx} = 2$, if $y = 2$ when $x = 0$.

7. If interest in a bank account is compounded continuously, the amount grows at a rate which is proportional to
the amount present. Derive the equation for the amount A in an account in which the initial amount is A_0 and
the interest rate is r as a function of the time t after A_0 is deposited.

8. Using Laplace transforms, solve the differential equation $y'' + 9y = 9$, if $y(0) = 0$ and $y'(0) = 1$.

9. Find the equation for the current as a function of the time (in seconds) in a circuit containing a 2-H inductance, an 8-Ω resistance, and a 6-V battery in series, if $i = 0$ for $t = 0$.

10. A 16-lb weight stretches a certain spring 0.5 ft. With this weight attached, the spring is pulled 0.3 ft longer than
its equilibrium length and released. Find the equation of the resulting motion, assuming no damping. (The
acceleration due to gravity is 32 ft/s^2.)

Supplementary Topics

S-1 Gaussian Elimination

In Chapter 4 we solved systems of simultaneous linear equations, dealing primarily with systems containing two equations or three equations. In Chapter 15 we developed methods using determinants which can be used in solving systems with even more equations. Also in Chapter 15, we showed how to use the inverse of a matrix in solving systems of equations.

In this section we show a general method which can be used to solve any system of linear equations. The procedure is very similar to that used in finding the inverse of a matrix in Section 15-5. Also, this procedure, known as **Gaussian elimination,** is commonly used in computer programs to solve systems of equations, particularly when there are several equations in the system.

Gaussian elimination is based on the use of the following two operations on the equations of a system:

basic operations

1. *Both sides of an equation may be multiplied by a constant.*

2. *A multiple of one equation may be added to another equation.*

Using these operations, we obtain an equivalent system of equations, one in which the values of the solution are the same as those of the original system.

Using three linear equations in three unknowns as an example, by the use of the above operations we can change the system

$$a_1x + b_1y + c_1z = d_1$$
$$a_2x + b_2y + c_2z = d_2$$
$$a_3x + b_3y + c_3z = d_3$$

(S1-1)

into the equivalent system

$$x + b_4y + c_4z = d_4$$
$$y + c_5z = d_5$$
$$z = d_6$$

(S1-2)

We can see that the solution is now easily completed by substituting the known value of z into the second equation, then substituting the known values of y and z into the first equation. Consider the following examples.

EXAMPLE A _____ | Solve the following system of equations by Gaussian elimination.

$2x + y = 4$
$3x - 2y = 3$

Given system of equations

$x + \frac{1}{2}y = 2$
$3x - 2y = 3$

We want the coefficient of x in the first equation to be 1. Therefore, divide the first equation by 2, the coefficient of x.

We next eliminate x in the second equation by subtracting 3 times the first equation from the second equation.

$x + \frac{1}{2}y = 2$
$-\frac{7}{2}y = -3$

Now we solve the second equation for y by dividing by $-\frac{7}{2}$.

$x + \frac{1}{2}y = 2$
$y = \frac{6}{7}$

To find the value of x, we substitute the value of $y = \frac{6}{7}$ into the first equation.

$x + \frac{1}{2}(\frac{6}{7}) = 2$
$x = \frac{11}{7}$

The solution is $x = \frac{11}{7}$, $y = \frac{6}{7}$, which checks when substituted into the original equations.

EXAMPLE B _____ | Solve the following system of equations by Gaussian elimination.

$x + 3y - 2z = -5$
$2x - y + 4z = 7$
$-3x + 2y - 3z = -1$

Given system of equations

$x + 3y - 2z = -5$
$-7y + 8z = 17$
$11y - 9z = -16$

Since the coefficient of x in the first equation is 1, we proceed to the next step. We subtract 2 times the first equation from the second equation and add 3 times the first equation to the third equation to eliminate x from the second and third equations.

$x + 3y - 2z = -5$
$y - \frac{8}{7}z = -\frac{17}{7}$
$11y - 9z = -16$

To get the coefficient of y equal to 1 in the second equation, we divide it by -7.

$x + 3y - 2z = -5$
$y - \frac{8}{7}z = -\frac{17}{7}$
$\frac{25}{7}z = \frac{75}{7}$

To eliminate y from the third equation, we subtract 11 times the second equation from the third equation.

$x + 3y - 2z = -5$
$y - \frac{8}{7}z = -\frac{17}{7}$
$z = 3$

We solve for z by multiplying the third equation by $\frac{7}{25}$ (or dividing by $\frac{25}{7}$).

$y - \frac{8}{7}(3) = -\frac{17}{7}$
$y = 1$

The value of y is found by substituting $z = 3$ into the second equation.

$x + 3(1) - 2(3) = -5$
$x = -2$

The value of x is found by substituting $y = 1$ and $z = 3$ back into the first equation.

The solution is $x = -2$, $y = 1$, $z = 3$, which checks when substituted into the original equations.

When we solved systems of equations in Chapters 4 and 15, we found that not all systems have unique solutions, as they have in Examples A and B. The following example illustrates the use of Gaussian elimination on a system of equations for which the solution is not unique.

EXAMPLE C

$$4y + z = 2$$
$$2x + 6y - 2z = 3$$
$$4x + 8y - 5z = 4$$

$$x + 3y - z = \tfrac{3}{2}$$
$$4y + z = 2$$
$$4x + 8y - 5z = 4$$

$$x + 3y - z = \tfrac{3}{2}$$
$$4y + z = 2$$
$$-4y - z = -2$$

$$x + 3y - z = \tfrac{3}{2}$$
$$y + \tfrac{1}{4}z = \tfrac{1}{2}$$
$$-4y - z = -2$$

$$x + 3y - z = \tfrac{3}{2}$$
$$y + \tfrac{1}{4}z = \tfrac{1}{2}$$
$$0 = 0$$

Solve the following system of equations by Gaussian elimination.

Given system of equations

Since the first equation does not contain x, which means that $a_1 = 0$, we cannot divide by a_1. Therefore, we interchange the first and second equations, then divide the new first equation by 2.
We eliminate x from the third equation by subtracting 4 times the first equation from the third equation.

Make the coefficient of y in the second equation equal to 1 by dividing the second equation by 4.

Eliminate y from the third equation by adding 4 times the second equation to the third equation.

Since the third equation, $0 = 0$, is correct, we can continue with the solution. Although there is no specific value for z, it is possible to express both x and y in terms of z.

$$x + 3y - z = \tfrac{3}{2}$$
$$y = \tfrac{1}{2} - \tfrac{1}{4}z$$

$$x + 3(\tfrac{1}{2} - \tfrac{1}{4}z) - z = \tfrac{3}{2}$$
$$x = \tfrac{7}{4}z$$

Solve the second equation for y. In this case it is expressed in terms of z. We no longer need to include the third equation.
Substitute the solution for y in the first equation and solve for x in terms of z.

NOTE ▷

At this point we have expressed the solution as $x = \tfrac{7}{4}z$ and $y = \tfrac{1}{2} - \tfrac{1}{4}z$. Since both x and y are expressed in terms of z, and there is no specific value of z, the value of z can be chosen arbitrarily. This means that **there is an unlimited number of solutions.** For example, if $z = 4$, $x = 7$ and $y = -\tfrac{1}{2}$. If $z = -2$, $x = -\tfrac{7}{2}$ and $y = 1$. ■

In Example C one of the equations became $0 = 0$, and there was an unlimited number of solutions. If any of the equations of a system becomes $0 = a$, $a \neq 0$, then the system is inconsistent and there is no solution.

If a system of equations has more unknowns than equations, or if it can be written in this way, as in Example C, it usually has an unlimited number of solutions. It is possible, however, that such a system is inconsistent.

If a system of equations has more equations than unknowns, it is inconsistent unless enough equations become $0 = 0$ such that at least one solution is determined.

EXAMPLE D ——————— Solve the following systems of equations by Gaussian elimination.

First system		Second system	
$x + 2y =$	5	$x + 2y =$	5
$3x - y =$	1	$3x - y =$	1
$4x + y =$	6	$4x + y =$	2
$x + 2y =$	5	$x + 2y =$	5
$-7y = -14$		$-7y = -14$	
$-7y = -14$		$-7y = -18$	
$x + 2y =$	5	$x + 2y =$	5
$y =$	2	$y =$	2
$-7y = -14$		$-7y = -18$	
$x + 2y =$	5	$x + 2y =$	5
$y =$	2	$y =$	2
$0 =$	0	$0 =$	-4
$x =$	1		

The solutions are shown at the left. We note that each system has three equations and two unknowns.

In the solution of the first system, the third equation becomes $0 = 0$ and only two equations are needed to find the solution $x = 1$, $y = 2$.

In the solution of the second system, the third equation becomes $0 = -4$, which means the system is inconsistent and there is no solution.

The solutions are shown graphically in Figs. S1-1 and S1-2. In Fig. S1-1 each of the three lines passes through the point (1, 2), whereas in Fig. S1-2 there is no point common to the three lines.

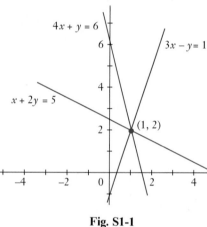

Fig. S1-1

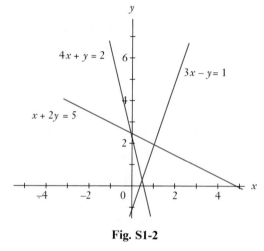

Fig. S1-2

Exercises S-1

In Exercises 1 through 20, solve the given systems of equations by Gaussian elimination. If there is an unlimited number of solutions, find two of them.

1. $x + 2y = 4$
$\quad 3x - y = 5$

2. $2x + y = 1$
$\quad 5x + 2y = 1$

3. $5x - 3y = 2$
$\quad -2x + 4y = 3$

4. $-3x + 2y = 4$
$\quad 4x + y = -5$

5. $2x + y - z = 0$
$\quad 4x + y + z = 2$
$\quad -2x - 2y + 3z = 0$

6. $2y + 2z = -1$
$\quad 3x - 4y + 3z = 1$
$\quad 4x + 2y + 5z = 4$

7. $x + 3y + 3z = -3$
$\quad 2x + 2y + z = -5$
$\quad -2x - y + 4z = 6$

8. $3x - y + 2z = 3$
$\quad 4x - 2y + z = 3$
$\quad 6x + 6y + 3z = 4$

9. $w + 2x - y + 3z = 12$
$\quad 2w \quad - 2y - z = 3$
$\quad 3x - y - z = -1$
$\quad -w + 2x + y + 2z = 3$

10. $2x - 3y + 2z + 2t = 3$
$\quad 4x + 2y - 3z \quad = -4$
$\quad 2x - y + 3z + 2t = 3$
$\quad 6x + 3y - 2z - t = 2$

11. $x - 4y + z = 2$
$3x - y + 4z = -4$

12. $4x + z = 6$
$2x - y - 2z = -2$

13. $2x - y + z = 5$
$3x + 2y - 2z = 4$
$5x + 8y - 8z = 5$

14. $3x + 2y - z = 3$
$2x - y - 3z = 2$
$-x + 4y + 5z = -1$

15. $2x - 4y = 7$
$3x + 5y = -6$
$9x - 7y = 15$

16. $4x - y = 5$
$2x + 2y = 3$
$6x - 4y = 7$
$2x + y = 4$

17. $3x + 5y = -2$
$24x - 18y = 13$
$15x - 33y = 19$
$6x + 68y = -33$

18. $x + 3y - z = 1$
$3x - y + 4z = 4$
$-2x + 2y + 3z = 17$
$3x + 7y + 5z = 23$

19. $x - 2y - 2z = 3$
$2x + y + 3z = 4$
$-2x - y - z = 5$
$3x + 3y - 2z = 2$

20. $2x - y - 2z - t = 4$
$4x + 2y + 3z + 2t = 3$
$-2x - y + 4z = -2$

In Exercises 21 through 24, set up systems of equations and solve by Gaussian elimination.

21. One personal computer can perform x calculations per second, and a second personal computer can perform y calculations per second. If each operates for two seconds, 25.0 million calculations are performed. If the first operates for four seconds and the second for three seconds, 43.2 million calculations are performed. Find x and y.

22. The voltage across an electric resistor equals the current times the resistance. If a current of 3.00 A passes through each of two resistors, the sum of the voltages is 10.5 V. If 2.00 A passes through the first resistor and 4.00 A passes through the second resistor, the sum of the voltages is 13.0 V. Find the resistances (in Ω).

23. Three machines together produce 650 parts each hour. Twice the production of the second machine is 10 parts/h more than the sum of the other two machines. If the first operates for three hours and the others operate for two hours, 1550 parts are produced. Find the production rate of each machine.

24. A total of $12,000 is invested, part at 6.5%, part at 6.0%, and part at 5.5%, yielding a total annual interest of $726. The income from the 6.5% part yields $128 less than the other two combined. How much is invested at each rate?

S-2 Rotation of Axes

In Chapter 20 we discussed the circle, parabola, ellipse, and hyperbola and how these curves are represented by the second-degree equation

$$Ax^2 + Bxy + Cy^2 + Dx + Ey + F = 0 \qquad \text{(S2-1)}$$

Our discussion included the properties of the curves and their equations with center (vertex of a parabola) at the origin. However, except for the special case of the hyperbola $xy = c$, we did not cover what happens when the axes are rotated about the origin.

If a set of axes is rotated about the origin through an angle θ, as shown in Fig. S2-1, we say that there has been a **rotation of axes.** In this case each point P in the plane has two sets of coordinates, (x, y) in the original system and (x', y') in the rotated system.

If we now let r equal the distance from the origin O to point P, and let ϕ be the angle between the x'-axis and the line OP, we have

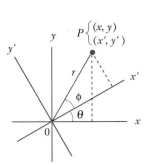

Fig. S2-1

$$x' = r \cos \phi, \qquad y' = r \sin \phi \qquad \text{(S2-2)}$$

$$x = r \cos (\theta + \phi), \qquad y = r \sin (\theta + \phi) \qquad \text{(S2-3)}$$

Using the cosine and sine of the sum of two angles, Eqs. (19-10) and (19-9), we can write Eqs. (S2-3) as

$$x = r \cos \phi \cos \theta - r \sin \phi \sin \theta$$
$$y = r \cos \phi \sin \theta + r \sin \phi \cos \theta$$

(S2-4)

Now, using Eqs. (S2-2), we have

$$x = x' \cos \theta - y' \sin \theta$$
$$y = x' \sin \theta + y' \cos \theta$$

(S2-5)

In our derivation, we have used the special case when θ is acute and P is in the first quadrant of both sets of axes. When simplifying equations of curves using Eqs. (S2-5), we find that a rotation through a positive acute angle θ is sufficient. It can be shown, however, that Eqs. (S2-5) hold for any θ and position of P.

EXAMPLE A ——— Transform the equation $x^2 - y^2 + 8 = 0$ by rotating the axes through 45°. When $\theta = 45°$, the rotation equations (S2-5) become

$$x = x' \cos 45° - y' \sin 45° = \frac{x'}{\sqrt{2}} - \frac{y'}{\sqrt{2}}$$

$$y = x' \sin 45° + y' \cos 45° = \frac{x'}{\sqrt{2}} + \frac{y'}{\sqrt{2}}$$

Substituting into the equation $x^2 - y^2 + 8 = 0$ gives

$$\left(\frac{x'}{\sqrt{2}} - \frac{y'}{\sqrt{2}} \right)^2 - \left(\frac{x'}{\sqrt{2}} + \frac{y'}{\sqrt{2}} \right)^2 + 8 = 0$$

$$\frac{1}{2} x'^2 - x'y' + \frac{1}{2} y'^2 - \frac{1}{2} x'^2 - x'y' - \frac{1}{2} y'^2 + 8 = 0$$

$$x'y' = 4$$

Fig. S2-2

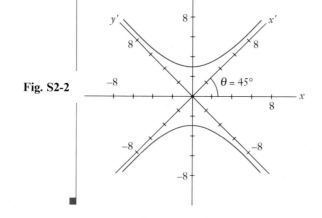

The graph and both sets of axes are shown in Fig. S2-2. Note that the original equation represents a hyperbola and that we have the $xy = c$ form of the hyperbola with rotation through 45°.

When we showed the type of curve represented by the second-degree equation [Eq. (S2-1)] in Section 20-8, we saw that the standard forms of the parabola, ellipse, and hyperbola required that $B = 0$. This means that there is no xy-term in the equation. In fact, in Section 20-8 we considered only one case, $(xy = c)$, for which $B \neq 0$.

If we can remove the xy-term from a second-degree equation, the analysis of the graph is simplified. By a proper rotation of axes we find that Eq. (S2-1) can be transformed into an equation that has no $x'y'$-term.

By substituting Eqs. (S2-5) into Eq. (S2-1) and then simplifying, we have

$$(A \cos^2 \theta + B \sin \theta \cos \theta + C \sin^2 \theta)x'^2 + [B \cos 2\theta - (A - C) \sin 2\theta]x'y'$$
$$+ (A \sin^2 \theta - B \sin \theta \cos \theta + C \cos^2 \theta)y'^2 + (D \cos \theta + E \sin \theta)x' + (E \cos \theta - D \sin \theta)y' + F = 0$$

If there is to be no $x'y'$-term, its coefficient must be zero. This means that $B \cos 2\theta - (A - C) \sin 2\theta = 0$, or

angle of rotation

$$\tan 2\theta = \frac{B}{A - C} \qquad (A \neq C) \tag{S2-6}$$

Eq. (S2-6) gives the angle of rotation except when $A = C$. In this case the coefficient of the $x'y'$-term is $B \cos 2\theta$, which is zero if $2\theta = 90°$. Thus

$$\theta = 45° \qquad (A = C) \tag{S2-7}$$

Consider the following example.

EXAMPLE B _____

By rotation of axes, transform the equation $8x^2 + 4xy + 5y^2 = 9$ into a form without an xy-term. Identify and sketch the curve.

Here $A = 8$, $B = 4$, and $C = 5$. Therefore, using Eq. (S2-6), we have

For reference, Eq. (19-2) is

$$\cos \theta = \frac{1}{\sec \theta}$$

$$\tan 2\theta = \frac{4}{8 - 5} = \frac{4}{3}$$

Eq. (19-7) is

$$1 + \tan^2 \theta = \sec^2 \theta$$

Since $\tan 2\theta$ is positive, we may take 2θ as an acute angle, which in turn means θ is also an acute angle. To make the transformation we need $\sin \theta$ and $\cos \theta$. We find these values by first finding the value of $\cos 2\theta$ and then using the half-angle formulas, Eqs. (19-26) and (19-27).

Eq. (19-26) is

$$\sin \frac{\alpha}{2} = \pm \sqrt{\frac{1 - \cos \alpha}{2}}$$

The value of $\cos 2\theta$ is found by using methods of earlier chapters. For example, using Eqs. (19-2) and (19-7),

Eq. (19-27) is

$$\cos \frac{\alpha}{2} = \pm \sqrt{\frac{1 + \cos \alpha}{2}}$$

$$\cos 2\theta = \frac{1}{\sec 2\theta} = \frac{1}{\sqrt{1 + \tan^2 2\theta}} = \frac{1}{\sqrt{1 + (\frac{4}{3})^2}} = \frac{3}{5}$$

Now, using Eqs. (19-26) and (19-27), we have

$$\sin \theta = \sqrt{\frac{1 - \cos 2\theta}{2}} = \sqrt{\frac{1 - \frac{3}{5}}{2}} = \frac{1}{\sqrt{5}}, \qquad \cos \theta = \sqrt{\frac{1 + \cos 2\theta}{2}} = \sqrt{\frac{1 + \frac{3}{5}}{2}} = \frac{2}{\sqrt{5}}$$

Here, θ is about $26.6°$.

(Continued on next page)

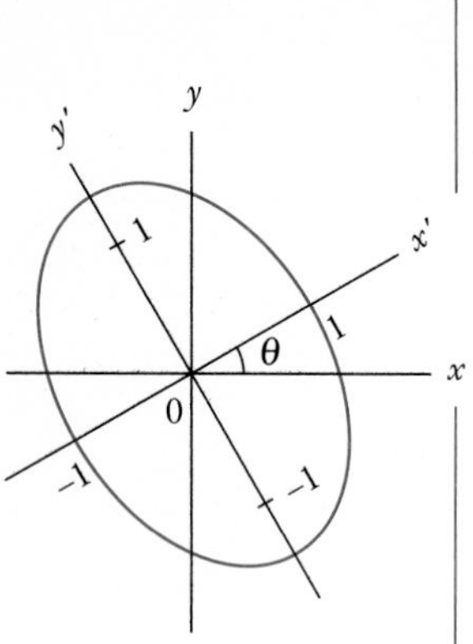

Substituting in Eqs. (S2-5) gives us

$$x = x'\left(\frac{2}{\sqrt{5}}\right) - y'\left(\frac{1}{\sqrt{5}}\right) = \frac{2x' - y'}{\sqrt{5}}, \qquad y = x'\left(\frac{1}{\sqrt{5}}\right) + y'\left(\frac{2}{\sqrt{5}}\right) = \frac{x' + 2y'}{\sqrt{5}}$$

Now, substituting into the equation $8x^2 + 4xy + 5y^2 = 9$ gives

$$8\left(\frac{2x' - y'}{\sqrt{5}}\right)^2 + 4\left(\frac{2x' - y'}{\sqrt{5}}\right)\left(\frac{x' + 2y'}{\sqrt{5}}\right) + 5\left(\frac{x' + 2y'}{\sqrt{5}}\right)^2 = 9$$

$$8(4x'^2 - 4x'y' + y'^2) + 4(2x'^2 + 3x'y' - 2y'^2) + 5(x'^2 + 4x'y' + 4y'^2) = 45$$

$$45x'^2 + 20y'^2 = 45$$

$$\frac{x'^2}{1} + \frac{y'^2}{\frac{9}{4}} = 1$$

Fig. S2-3

We see that this is an ellipse with major axis of $\frac{3}{2}$ and minor axis of 1, as shown in Fig. S2-3. ∎

In Example B, $\tan 2\theta$ was positive, and we made 2θ and θ positive. If, when using Eq. (S2-6), $\tan 2\theta$ is negative, we then make 2θ obtuse ($90° < 2\theta < 180°$). In this case $\cos 2\theta$ will be negative, but θ will be acute ($45° < \theta < 90°$).

In Section 20-7 we showed how to use translation of axes to write an equation in standard form if $B = 0$ in the second-degree equation. In this section we have seen how rotation of axes is used to eliminate the xy-term. It is possible that both a translation of axes and a rotation of axes are needed to write an equation in standard form.

In Section 20-8 we saw how to identify a conic section by inspecting the values of A and C when $B = 0$ in the second-degree equation. If $B \neq 0$, these curves are identified as follows:

1. If $B^2 - 4AC = 0$, a parabola.
2. If $B^2 - 4AC < 0$, an ellipse.
3. If $B^2 - 4AC > 0$, a hyperbola.

It is also possible in special cases that a point, two parallel lines, two intersecting lines, or no curve may result.

EXAMPLE C

Identify the curve of the equation $16x^2 - 24xy + 9y^2 + 20x - 140y - 300 = 0$, simplify the equation to standard form, and sketch the graph.

With $A = 16$, $B = -24$, and $C = 9$, using Eq. (S2-6), we have

$$\tan 2\theta = \frac{-24}{16 - 9} = -\frac{24}{7}$$

In this case $\tan 2\theta$ is negative, and we take 2θ as obtuse. We then find that $\cos 2\theta = -\frac{7}{25}$. In turn we find that $\sin \theta = \frac{4}{5}$ and $\cos \theta = \frac{3}{5}$. Here θ is about 53.1°.

Using these values in Eqs. (S2-5), we find that

$$x = \frac{3x' - 4y'}{5}, \qquad y = \frac{4x' + 3y'}{5}$$

Substituting these into the original equation and simplifying, we get

$$y'^2 - 4x' - 4y' - 12 = 0$$

This equation represents a parabola with its axis parallel to the x' axis. The vertex is found by completing the square:

$$(y' - 2)^2 = 4(x' + 4)$$

This means that the vertex is the point $(-4, 2)$ in the $x'y'$-rotated coordinate system. This can be written as

$$y''^2 = 4x''$$

in the $x''y''$-rotated and then translated system. The graph and coordinate systems are shown in Fig. S2-4.

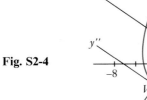

Fig. S2-4

Exercises S-2

In Exercises 1 through 4, transform the given equations by rotating the axes through the given angle. Identify and sketch each resulting curve.

1. $x^2 - y^2 = 25, \theta = 45°$

2. $x^2 + y^2 = 16, \theta = 60°$

3. $8x^2 - 4xy + 5y^2 = 36, \theta = \text{Arctan } 2$

4. $2x^2 + 24xy - 5y^2 = 8, \theta = \text{Arctan } \frac{3}{4}$

In Exercises 5 through 10, transform each equation to a form without an xy-term by a rotation of axes. Identify and sketch each curve.

5. $x^2 + 2xy + y^2 - 2x + 2y = 0$

6. $5x^2 - 6xy + 5y^2 = 32$

7. $3x^2 + 4xy = 4$

8. $9x^2 - 24xy + 16y^2 - 320x - 240y = 0$

9. $11x^2 - 6xy + 19y^2 = 20$

10. $x^2 + 4xy - 2y^2 = 6$

In Exercises 11 and 12, transform each equation to a form without an xy-term by a rotation of axes. Then transform the equation to standard form by a translation of axes. Identify and sketch each curve.

11. $16x^2 - 24xy + 9y^2 - 60x - 80y + 400 = 0$

12. $73x^2 - 72xy + 52y^2 + 100x - 200y + 100 = 0$

S-3   Functions of Two Variables

There are numerous situations in many fields, including science, engineering, and technology, in which functions with more than one independent variable are used. Although some of these applications involve three or more independent variables, many involve only two, and we shall concern ourselves primarily with functions of two variables.

In this section we establish the meaning of a function of two variables, and in the section which follows we discuss how the graph of this type of function is shown.

Many familiar formulas express one variable in terms of two or more other variables. The following example illustrates one from geometry.

EXAMPLE A

Fig. S3-1

The total surface area of a right circular cylinder is a function of the radius and the height of the cylinder. That is, the area will change if either or both of these change. The formula for the total surface area is

$$A = 2\pi r^2 + 2\pi rh$$

We say that A is a function of r and h. See Fig. S3-1. ∎

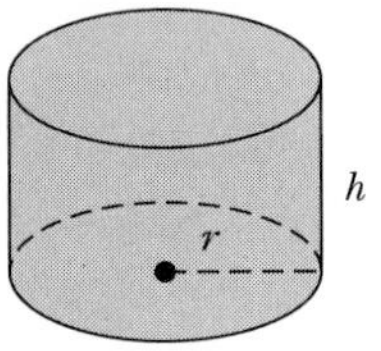

As we have noted, there are numerous applications of functions of two variables in many fields of study. For example, the voltage in a simple electric circuit depends on the resistance and current in the circuit. The moment of a force depends on the magnitude of the force and the distance of the force from the axis. The pressure of a gas depends on its temperature and volume. The temperature at a point on a flat plate depends on the coordinates of the point.

definition

We define a function of two variables as follows. *If z is uniquely determined for given values of x and y, then z is a function of x and y.* Notation similar to that which is used for one independent variable is employed in this case. This notation is $z = f(x, y)$, where both x and y are independent variables. This is the meaning of a function of two variables.

Similar to the case with one independent variable, $f(a, b)$ *means "the value of the function when $x = a$ and $y = b$."*

EXAMPLE B

If $f(x, y) = 3x^2 + 2xy - y^3$, find $f(-1, 2)$. Substituting, we have

$$f(-1, 2) = 3(-1)^2 + 2(-1)(2) - (2)^3$$
$$= 3 - 4 - 8$$
$$= -9$$

∎

EXAMPLE C

If $f(x, y) = 2xy^2 - y$, find $f(x, 2x) - f(x, x^2)$.

We note that in each evaluation the x factor remains as x, but that we are to substitute $2x$ for y and subtract the function for which x^2 is substituted for y.

$$f(x, 2x) - f(x, x^2) = [2x(2x)^2 - (2x)] - [2x(x^2)^2 - x^2]$$
$$= [8x^3 - 2x] - [2x^5 - x^2]$$
$$= 8x^3 - 2x - 2x^5 + x^2$$
$$= -2x^5 + 8x^3 + x^2 - 2x$$

This type of difference of functions is important in Section S-6. ∎

restrictions

Restricting values of the function to real numbers means that certain restrictions may be placed on the independent variables. *Values of either x or y or both which lead to division by zero or to imaginary values for z are not permissible.* The following example illustrates this point.

EXAMPLE D

If $f(x, y) = \dfrac{3xy}{(x - y)(x + 3)}$

all values of x and y are permissible except those for which $x = y$ and $x = -3$. Each of these would indicate division by zero.

If $f(x, y) = \sqrt{4 - x^2 - y^2}$, neither x nor y may be greater than 2 in absolute value, and the sum of their squares may not exceed 4. Otherwise, imaginary values of the function would result. ■

One of the primary difficulties students have with functions is setting them up from stated conditions. Although many examples of this have been encountered in our previous work, an example of setting up a function of two variables should prove helpful. Although no general rules can be given for this procedure, a careful analysis of the statement should lead to the desired function. The following example illustrates setting up such a function.

EXAMPLE E

$V = 2 \text{ ft}^3$

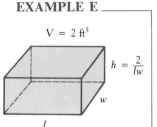

$h = \dfrac{2}{lw}$

w

l

Fig. S3-2

An open rectangular metal box is to be constructed such that it will contain 2 ft³. The cost of sheet metal is c cents per square foot. Express the cost of the sheet metal needed to construct the box as a function of the length and width of the box.

The cost in question depends on the surface area of the sides of the box. An "open" box is one which has no top. Thus, the surface area of the box is

$$S = \ell w + 2\ell h + 2hw$$

where ℓ is the length, w the width, and h the height of the box (see Fig. S3-2). However, this expression contains three independent variables. Using the condition that the volume of the box is 2 ft³, we have $\ell w h = 2$. Since we wish to have only ℓ and w, we solve this expression for h, and find that $h = 2/\ell w$. Substituting this in the expression for the surface area, we have

$$S = \ell w + 2\ell \left(\frac{2}{\ell w}\right) + 2\left(\frac{2}{\ell w}\right) w$$

$$= \ell w + \frac{4}{w} + \frac{4}{\ell}$$

Since the cost C is given by $C = cS$, we have the expression for the cost as

$$C = c\left(\ell w + \frac{4}{w} + \frac{4}{\ell}\right)$$

Exercises S-3

In Exercises 1 through 4, determine the indicated functions.

1. Express the volume of a right circular cone as a function of the radius of the base and the height.

2. A cylindrical can is to be made to contain a volume V. Express the total surface area of the can as a function of V and the radius of the can.

3. The angle between two forces F_1 and F_2 is $30°$. Express the magnitude of the resultant R in terms of F_1 and F_2. See Fig. S3-3.

4. A computer leasing firm charges a monthly fee F based on the length of time a corporation has used the service plus $100 for every hour the computer is used during the month. Express the total monthly charge T as a function of F and the number of hours h the computer is used.

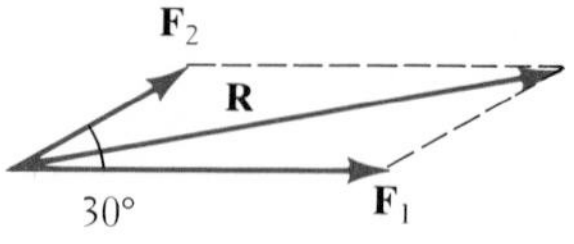

Fig. S3-3

In Exercises 5 through 12, evaluate the given functions.

5. $f(x, y) = 2x - 6y$; find $f(0, -4)$

6. $g(r, s) = r - 2rs - r^2 s$; find $g(-2, 1)$

7. $Y(y, t) = \dfrac{2 - 3y}{t - 1} + 2y^2 t$; find $Y(y, 2)$

8. $X(x, t) = -6xt + xt^2 - t^3$; find $X(x, -t)$

9. $H(p, q) = p - \dfrac{p - 2q^2 - 5q}{p + q}$; find $H(p, q + k)$

10. $f(x, y) = x^2 - 2xy - 4x$; find $f(x + h, y + k) - f(x, y)$

11. $f(x, y) = xy + x^2 - y^2$; find $f(x, x) - f(x, 0)$

12. $g(y, z) = 3y^3 - y^2 z + 5z^2$; find $g(3z^2, z) - g(z, z)$

In Exercises 13 and 14, determine which values of x or y, if any, are not permissible.

13. $f(x, y) = \dfrac{\sqrt{y}}{2x}$

14. $f(x, y) = \sqrt{x^2 + y^2 - x^2 y - y^3}$

In Exercises 15 through 20, solve the given problems.

15. For a certain electric circuit, the current i, in amperes, in terms of the voltage E and resistance R is given by $i = \dfrac{E}{R + 0.25}$. Find the current for $E = 1.50$ V and $R = 1.20$ Ω and for $E = 1.60$ V and $R = 1.05$ Ω.

16. The pressure of a gas as a function of its volume and temperature is given by $p = nRT/V$, where n and R are constants. If $n = 3$ mol and $R = 8.31$ J/mol·K, find the pressure (in pascals) for $T = 300$ K and $V = 50$ m³.

17. A rectangular solar cell panel has a perimeter p and a width w. Express the area A of the panel in terms of p and w and evaluate the area for $p = 250$ cm and $w = 55$ cm. See Fig. S3-4.

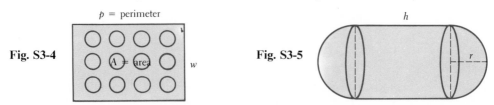

Fig. S3-4 p = perimeter A = area w

Fig. S3-5 h r

18. A gasoline storage tank is in the shape of a right circular cylinder with a hemisphere at each end, as shown in Fig. S3-5. Express the volume V of the tank in terms of r and h and then evaluate the volume for $r = 3.75$ ft and $h = 12.5$ ft.

19. The crushing load L of a pillar varies as the fourth power of its radius and inversely as the square of its length. Express L as a function of r and ℓ for a pillar 20 ft tall and 1 ft in diameter which is crushed by a load of 20 tons.

20. The resonant frequency of an electric circuit containing an inductance L and capacitance C is inversely proportional to the square root of the product of the inductance and capacitance. If the resonant frequency of a circuit containing a 4-H inductor and a 64-μF capacitor is 10 Hz, express f as a function of L and C.

S-4   **Curves and Surfaces in Three Dimensions**

We will now undertake a brief description of the graphical representation of a function of two variables. We shall show first a method of representation in the rectangular coordinate system in two dimensions. The following example illustrates the method.

EXAMPLE A

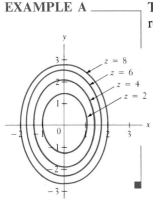

Fig. S4-1

To represent $z = 2x^2 + y^2$, we will assume various values of z and sketch the resulting equation in the xy-plane. For example, if $z = 2$ we have

$$2x^2 + y^2 = 2$$

This we recognize as an ellipse with its major axis along the y-axis and vertices at $(0, \sqrt{2})$ and $(0, -\sqrt{2})$. The ends of the minor axis are $(1, 0)$ and $(-1, 0)$. However, the ellipse $2x^2 + y^2 = 2$ represents the function $z = 2x^2 + y^2$ only if $z = 2$. If $z = 4$, we have $2x^2 + y^2 = 4$, which is another ellipse. In fact, for all positive values of z, an ellipse results. Negative values for z are not possible, since neither x^2 nor y^2 may be negative. Figure S4-1 shows the ellipses obtained by using the indicated values of z.

The method of representation illustrated in Example A is useful if only a few specific values of z are to be used, or at least if the various curves do not intersect in such a way that they cannot be distinguished. If a general representation of z as a function of x and y is desired, it is necessary to use three coordinate axes, one each for x, y, and z. The most widely applicable system of this kind is to place a third coordinate axis at right angles to each of the x- and y-axes. In this way we employ three dimensions for the representation.

The three mutually perpendicular axes, the x-axis, the y-axis, and the z-axis, are the basis of the **rectangular coordinate system in three dimensions.** Together they form three mutually perpendicular planes in space, the xy-plane, the yz-plane, and the xz-plane. To every point in the space of the coordinate system is associated the set of numbers (x, y, z). The point at which the axes meet is the origin. The positive directions of the axes are indicated in Fig. S4-2. *That part of space in which all the values are positive is called the first* **octant.** Numbers are not assigned to the other octants.

Fig. S4-2

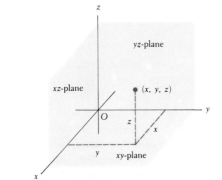

EXAMPLE B

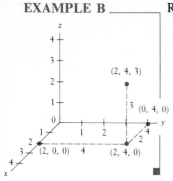

Fig. S4-3

Represent the point (2, 4, 3) in rectangular coordinates.

We first note that a certain distortion is necessary to represent values of x reasonably, since the x-axis "comes out" of the plane of the page. Units $\sqrt{2}/2$ ($=0.7$) as long as those used on the other axes give a good representation. With this in mind, we draw a line 4 units long from the point (2, 0, 0) on the x-axis in the xy-plane. This locates the point (2, 4, 0). From this point a line 3 units long is drawn vertically upward. This locates the desired point, (2, 4, 3). The point may be located by starting from (0, 4, 0), proceeding two units *parallel* to the x-axis to (2, 4, 0), and then proceeding vertically three units to (2, 4, 3). It may also be located by starting from (0, 0, 3) (see Fig. S4-3). ∎

The remainder of this section is devoted to showing certain basic techniques by which three-dimensional figures may be drawn. We start by showing the general equation of a plane.

In Chapter 20 we showed that the graph of the equation $Ax + By + C = 0$ in two dimensions is a straight line. By the following example, we will verify that *the graph of the equation*

equation of a plane

$$Ax + By + Cz + D = 0$$

(S4-1)

is a **plane** *in three dimensions.*

EXAMPLE C

NOTE ▷

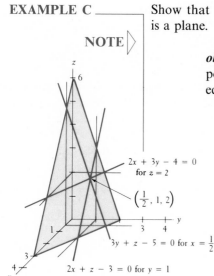

Fig. S4-4

Show that the graph of the equation $2x + 3y + z - 6 = 0$ in three dimensions is a plane.

If we let any of the three variables take on a specific value, we obtain a linear equation in the other two variables. For example, the point ($\frac{1}{2}$, 1, 2) satisfies the equation and therefore lies on the graph of the equation. For $x = \frac{1}{2}$, we have

$$3y + z - 5 = 0$$

which is the equation of a straight line. This means that all pairs of values of y and z which satisfy this equation, along with $x = \frac{1}{2}$, satisfy the given equation. Thus, for $x = \frac{1}{2}$, the straight line $3y + z - 5 = 0$ lies on the graph of the equation.

For $z = 2$, we have

$$2x + 3y - 4 = 0$$

which is also a straight line. By similar reasoning this line lies on the graph of the equation. Since two lines through a point define a plane, these lines through ($\frac{1}{2}$, 1, 2) define a plane. This plane is the graph of the equation (see Fig. S4-4).

To complete the verification, it can be seen that for any point on the graph, there is a straight line parallel to one of the coordinate planes which lies on the graph. Thus, regardless of the point chosen, intersecting straight lines through the point which lies on the graph exist. Thus, the graph is a plane. A similar analysis ∎ could be made for any equation of the same form.

Since we know that the graph of an equation of the form of Eq. (S4-1) is a plane, its graph can be found by determining its three intercepts, and the plane can then be represented by drawing in the lines between these intercepts. If the plane passes through the origin, by letting two of the variables in turn be zero, two straight lines which define the plane are found.

EXAMPLE D

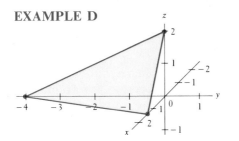

Fig. S4-5

Sketch the graph of $3x - y + 2z - 4 = 0$.

The intercepts of the graph of an equation are those points where it crosses the respective axes. Thus, by letting two of the variables at a time equal zero, we obtain the intercepts. For the given equation the intercepts are $(\frac{4}{3}, 0, 0)$, $(0, -4, 0)$ and $(0, 0, 2)$. These points are located (see Fig. S4-5), and lines are drawn between them to represent the plane.

surface

*In general, the graph of an equation in three variables, which is essentially equivalent to a function with two independent variables, is a **surface** in space. This is seen in the case of the plane, and will be verified for other cases in the examples which follow.*

curve

traces

*The intersection of two surfaces is a **curve** in space. This has been seen in Examples C and D, since the intersection of the given planes and the coordinate planes are lines (which in the general sense are curves). We define the **traces** of a surface to be the curves of intersection of the surface and the coordinate planes.* The traces of a plane are those lines drawn between the intercepts to represent the plane. Many surfaces may be sketched by finding their traces and intercepts.

EXAMPLE E

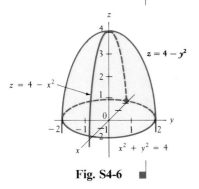

Fig. S4-6 ■

Find the intercepts and traces and sketch the graph of the equation $z = 4 - x^2 - y^2$.

The intercepts of the graph of the equation are $(2, 0, 0)$, $(-2, 0, 0)$. $(0, 2, 0)$, $(0, -2, 0)$, and $(0, 0, 4)$.

Since the traces of a surface lie within the coordinate planes, for each trace one of the variables is zero. Thus, by *letting each variable in turn be zero, we find the trace of the surface in the plane of the other two variables.* Therefore, the traces of this surface are

in the yz-plane: $z = 4 - y^2$ (a parabola)

in the xz-plane: $z = 4 - x^2$ (a parabola)

in the xy-plane: $x^2 + y^2 = 4$ (a circle)

Using the intercepts and sketching the traces (see Fig. S4-6), we obtain the surface of the equation. This figure is called a **circular paraboloid.**

section

There are numerous techniques for analyzing the equation of a surface in order to obtain its graph. Another which we shall discuss here, which is closely associated with a trace, is that of a **section.** *By assuming a specific value of one of the variables, we obtain an equation in two variables, the graph of which lies in a plane parallel to the coordinate plane of the two variables.* The following example illustrates sketching a surface by use of intercepts, traces, and sections.

EXAMPLE F _____ Sketch the graph of $4x^2 + y^2 - z^2 = 4$.

The intercepts are $(1, 0, 0)$, $(-1, 0, 0)$, $(0, 2, 0)$, and $(0, -2, 0)$. We note that there are no intercepts on the z-axis, for this would necessitate $z^2 = -4$.

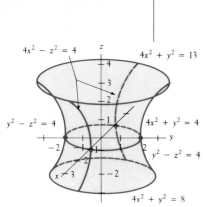

The traces are

      in the yz-plane: $y^2 - z^2 = 4$ (a hyperbola)
      in the xz-plane: $4x^2 - z^2 = 4$ (a hyperbola)
      in the xy-plane: $4x^2 + y^2 = 4$ (an ellipse)

The surface is reasonably defined by these curves, but by assuming suitable values of z we may indicate its shape better. For example, if $z = 3$, we have $4x^2 + y^2 = 13$, which is an ellipse. In using it we must remember that it is valid for $z = 3$ and therefore should be drawn 3 units above the xy-plane. If $z = -2$, we have $4x^2 + y^2 = 8$, which is also an ellipse. Thus, we have the following sections:

      for $z = 3$: $4x^2 + y^2 = 13$ (an ellipse)
      for $z = -2$: $4x^2 + y^2 = 8$ (an ellipse)

Fig. S4-7

Other sections could be found, but these are sufficient to obtain a good sketch of the graph (see Fig. S4-7). The figure is called an **elliptic hyperboloid.**

Now that we have developed the use of the rectangular coordinate system in three dimensions, we can compare the graph of a given function using two dimensions and three dimensions. The following example shows the surface for the function of Example A.

EXAMPLE G _____ Sketch the graph of $z = 2x^2 + y^2$.

The only intercept is $(0, 0, 0)$. The traces are

      in the yz-plane: $z = y^2$ (a parabola)
      in the xz-plane: $z = 2x^2$ (a parabola)
      in the xy-plane: the origin

We can see that the trace in the xy-plane is only the point at the origin, since $2x^2 + y^2 = 0$ may be written as $y^2 = -2x^2$, which is true only for $x = 0$ and $y = 0$.

To obtain a better graph we should use at least one positive value for z. As we noted in Example A, negative values of z cannot be used. Since we used $z = 2$, $z = 4$, $z = 6$, and $z = 8$ in Example A, we shall use these values here. Thus,

      for $z = 2$: $2x^2 + y^2 = 2$,   for $z = 4$: $2x^2 + y^2 = 4$
      for $z = 6$: $2x^2 + y^2 = 6$,   for $z = 8$: $2x^2 + y^2 = 8$

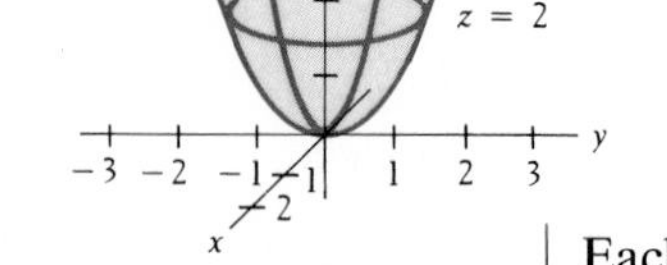

Fig. S4-8

Each of these sections is an ellipse. The surface, called an **elliptic paraboloid,** is shown in Fig. S4-8. Compare with Fig. S4-1.

Example G illustrates how *topographic maps* may be drawn. These maps represent three-dimensional terrain in two dimensions. For example, if Fig. S4-8 represents an excavation in the surface of the earth, then Fig. S4-1 represents the curves of constant elevation, or *contours*, with equally spaced elevations measured from the bottom of the excavation.

An equation with only two variables may represent a surface in space. Since only two variables are included in the equation, the surface is independent of the other variable. Another interpretation is that all sections, for all values of the variable not included, are the same. That is, ***all sections parallel to the co-ordinate plane of the included variables are the same as the trace in that plane.***

NOTE ▷

EXAMPLE H ——————

Sketch the graph of $x + y = 2$ in the rectangular coordinate system in three dimensions and also in two dimensions.

Since z does not appear in the equation, for purposes of graphing in three dimensions we can consider the equation to be $x + y + 0z = 2$. Therefore, for *any value* of z the section is the straight line $x + y = 2$. Thus, the graph is a plane as shown in Fig. S4-9(a). The graph as a straight line in two dimensions is shown in Fig. S4-9(b).

Fig. S4-9

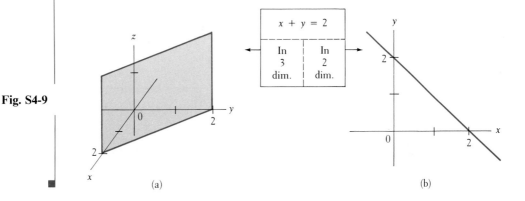

(a)   (b)

EXAMPLE I ——————

The graph of the equation $z = 4 - x^2$ in three dimensions is a surface whose sections, for all values of y, are given by the parabola $z = 4 - x^2$ (see Fig. S4-10).

Fig. S4-10

The surface in Example I is known as a **cylindrical surface.** In general, a cylindrical surface is one which can be generated by a line moving parallel to a fixed line while passing through a plane curve.

It must be realized that most of the figures shown extend beyond the ranges indicated by the traces and sections. However, these traces and sections are convenient for representing and visualizing the surfaces.

A computer can be programmed to draw surfaces by drawing sections which are perpendicular to both the x- and y-axes or by sections perpendicular to only one axis. Such computer-drawn surfaces are of great value for all types of surfaces, especially those of a complex nature. Figure S4-11 shows the graph of $z = \sin(x^2 + y^2)$ drawn by a computer in these two ways.

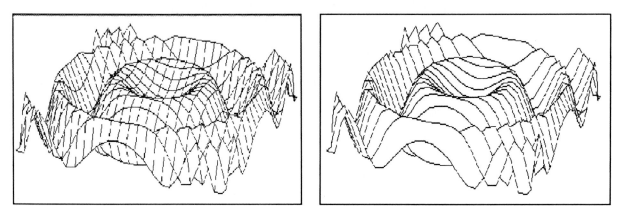

Fig. S4-11

Exercises S-4

In Exercises 1 and 2, use the method of Example A and draw the graphs of the indicated equations for the given values of z.

1. $z = x^2 + y^2, z = 1, z = 4, z = 9$
2. $z = y - x^2, z = 0, z = 2, z = 4$

Sketch the graphs of the equations given in Exercises 3 through 14 in the rectangular coordinate system in three dimensions.

3. $x + y + 2z - 4 = 0$
4. $4x - 2y + z - 8 = 0$
5. $z = y - 2x - 2$
6. $x + 2y = 4$

7. $x^2 + y^2 + z^2 = 4$
8. $z = x^2 + y^2$
9. $z = 4 - 4x^2 - y^2$
10. $x^2 + y^2 - 4z^2 = 4$

11. $z = 2x^2 + y^2 + 2$
12. $x^2 - y^2 - z^2 = 9$
13. $x^2 + y^2 = 16$
14. $y^2 + 9z^2 = 9$

In Exercises 15 through 20, sketch the indicated curves and surfaces.

15. Curves which represent a constant temperature are called *isotherms*. The temperature at a point (x, y) of a flat plate is t degrees Celsius, where $t = 4x - y^2$. In two dimensions draw the isotherms for $t = -4, 0$, and 8.

16. At a point (x, y) in the xy-plane the electric potential V is given by $V = y^2 - x^2$, where V is measured in volts. Draw the lines of equal potential for $V = -9, 0$, and 9.

17. An electric charge is so distributed that the electric potential at all points on an imaginary surface is the same. Such a surface is called an *equipotential surface*. Sketch the graph of the equipotential surface whose equation is $2x^2 + 2y^2 + 3z^2 = 6$.

18. The surface of a small hill can be roughly approximated by the equation $z(2x^2 + y^2 + 100) = 1500$, where the units are meters. Draw the surface of the hill and the contours for $z = 3$ m, $z = 6$ m, $z = 9$ m, $z = 12$ m, and $z = 15$ m.

19. Sketch the line in space defined by the intersection of the planes

$$x + 2y + 3z - 6 = 0 \quad \text{and} \quad 2x + y + z - 4 = 0$$

20. Sketch the graph of $x^2 + y^2 - 2y = 0$ in three dimensions and in two dimensions.

S-5   Partial Derivatives

When we developed the concept of the derivative in Chapter 22, we showed that we were dealing with the rate of change of one variable with respect to another. However, only one independent variable was involved. In order to extend the derivative and its applications to functions of two (or more) variables, we find the derivative of the function with respect to one of the independent variables, while the other is held constant. If $z = f(x, y)$, this is equivalent to saying, for example, that for a specified value of y, we have the derivative of z with respect to x.

definition

Therefore, if $z = f(x, y)$, and y is kept constant, z becomes a function of x alone. The derivative of this function with respect to x is termed the **partial derivative** *of z with respect to x. Similarly, if $z = f(x, y)$ and x is kept constant, the derivative of the resulting function with respect to y is the* **partial derivative** *of z with respect to y.*

For the function $z = f(x, y)$, the notations used for the partial derivative of z with respect to x include

$$\frac{\partial z}{\partial x}, \qquad \frac{\partial f}{\partial x}, \qquad f_x, \qquad \frac{\partial}{\partial x} f(x, y), \qquad f_x(x, y)$$

Similarly, $\partial z/\partial y$ denotes the partial derivative of z with respect to y. In speaking, this is often shortened to "the partial of z with respect to y."

EXAMPLE A

If $z = 4x^2 + xy - y^2$, find $\partial z/\partial x$ and $\partial z/\partial y$.

In finding the partial derivative of z with respect to x, we treat y as a constant. Therefore

$$z = 4x^2 + xy - y^2 \qquad \text{treat as constant}$$

$$\frac{\partial z}{\partial x} = 8x + y$$

Similarly, when finding the partial derivative of z with respect to y, we treat x as a constant. Thus

$$z = 4x^2 + xy - y^2 \qquad \text{treat as constant}$$

$$\frac{\partial z}{\partial y} = x - 2y$$

EXAMPLE B

If $z = \dfrac{x \ln y}{x^2 + 1}$, find $\partial z/\partial x$ and $\partial z/\partial y$.

$$\frac{\partial z}{\partial x} = \frac{(x^2 + 1)(\ln y) - (x \ln y)(2x)}{(x^2 + 1)^2} = \frac{(1 - x^2) \ln y}{(1 + x^2)^2}$$

$$\frac{\partial z}{\partial y} = \left(\frac{x}{x^2 + 1}\right)\left(\frac{1}{y}\right) = \frac{x}{y(x^2 + 1)}$$

We note that in finding $\partial z/\partial x$ it is necessary to use the quotient rule, since x appears in both numerator and denominator. However, when finding $\partial z/\partial y$ the only derivative needed is that of $\ln y$. ∎

EXAMPLE C

For the function $f(x, y) = x^2 y \sqrt{2 + xy^2}$, find $f_y(2, 1)$.

The notation $f_y(2, 1)$ means the partial derivative of f with respect to y, evaluated for $x = 2$ and $y = 1$. Thus, first finding $f_y(x, y)$, we have

$$f(x, y) = x^2 y(2 + xy^2)^{1/2}$$

$$f_y(x, y) = x^2 y\left(\frac{1}{2}\right)(2 + xy^2)^{-1/2}(2xy) + (2 + xy^2)^{1/2}(x^2)$$

$$= \frac{x^3 y^2}{(2 + xy^2)^{1/2}} + x^2(2 + xy^2)^{1/2}$$

$$= \frac{x^3 y^2 + x^2(2 + xy^2)}{(2 + xy^2)^{1/2}} = \frac{2x^2 + 2x^3 y^2}{(2 + xy^2)^{1/2}}$$

$$f_y(2, 1) = \frac{2(4) + 2(8)(1)}{(2 + 2)^{1/2}} = 12$$

∎

To determine the geometric interpretation of a partial derivative, assume that $z = f(x, y)$ is the surface shown in Fig. S5-1. Choosing a point P on the surface, we then draw a plane through P parallel to the xz-plane. On this plane through P, the value of y is constant. The intersection of this plane and the surface is the curve as indicated. *The partial derivative of z with respect to x represents the slope of a line tangent to this curve.* When the values of the coordinates of point P are substituted into the expression, which is the partial derivative, it gives the slope of the tangent line at that point. In the same way, the partial derivative of z with respect to y, evaluated at P, gives the slope of the line tangent to the curve which is found from the intersection of the surface and the plane parallel to the yz-plane through P.

interpretation

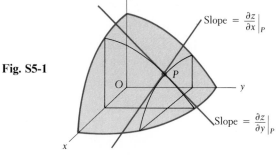

Fig. S5-1

EXAMPLE D —— Find the slope of a line tangent to the surface $2z = x^2 + 2y^2$ and parallel to the xz-plane at the point $(2, 1, 3)$. Also find the slope of a line tangent to this surface and parallel to the yz-plane at the same point.

Finding the partial derivative of z with respect to x, we have

$$2\frac{\partial z}{\partial x} = 2x \quad \text{or} \quad \frac{\partial z}{\partial x} = x$$

NOTE ▷ *This derivative, evaluated at the point (2, 1, 3), will give us the slope of the line tangent which is also parallel to the xz-plane.* Thus, the first required slope is

$$\left.\frac{\partial z}{\partial x}\right|_{(2, 1, 3)} = 2$$

The partial derivative of z with respect to y, evaluated at $(2, 1, 3)$, will give us the second required slope. Thus

$$\frac{\partial z}{\partial y} = 2y, \quad \left.\frac{\partial z}{\partial y}\right|_{(2, 1, 3)} = 2$$

■ Therefore, both slopes are 2 (see Fig. S5-2).

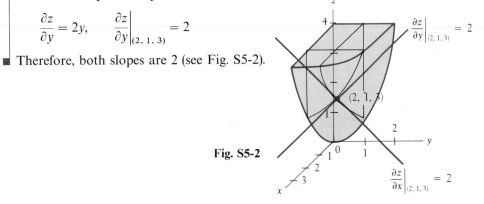

Fig. S5-2

$$\left.\frac{\partial z}{\partial y}\right|_{(2, 1, 3)} = 2$$

$$\left.\frac{\partial z}{\partial x}\right|_{(2, 1, 3)} = 2$$

The general interpretation of the partial derivative follows that of a derivative of a function with one independent variable. *The partial derivative $f_x(x_0, y_0)$ is the instantaneous rate of change of the function $f(x, y)$ with respect to x, with y held constant at the value of y_0.* This holds regardless of what the variables represent.

Applications of partial derivatives are found in various fields of technology. We present here an application from electricity, and others are found in the exercises.

EXAMPLE E —— A parallel electric circuit in a microwave transmitter has resistances r and R. The current through r can be found from

$$i = \frac{IR}{r + R}$$

where I is the total current for the two branches. Assuming that I is constant at 85.4 mA, find $\partial i/\partial r$ and evaluate it for $R = 0.150\ \Omega$ and $r = 0.032\ \Omega$.

Substituting for I and finding the partial derivative, we have the following:

$$i = \frac{85.4R}{r + R} = 85.4R(r + R)^{-1}$$

(Continued on next page)

$$i = \frac{85.4R}{r + R} = 85.4R(r + R)^{-1}$$

$$\frac{\partial i}{\partial r} = (-1)(85.4R)(r + R)^{-2}(1) = \frac{-85.4R}{(r + R)^2}$$

$$\left.\frac{\partial i}{\partial r}\right|_{\substack{r=0.032 \\ R=0.150}} = \frac{-85.4(0.150)}{(0.032 + 0.150)^2} = -387 \text{ mA}/\Omega$$

This result tells us that the current is decreasing at the rate of 387 mA per ohm of change in the smaller resistor at the instant when $r = 0.032$ Ω, for a fixed value of $R = 0.150$ Ω.

■

Since the partial derivatives $\partial f/\partial x$ and $\partial f/\partial y$ are functions of x and y, we can take partial derivatives of each of them. This gives rise to **partial derivatives of higher order,** in a manner similar to the higher derivatives of a function of one independent variable. *The possible* **second-order partial derivatives** *of a function* $f(x, y)$ *are*

$$\frac{\partial^2 f}{\partial x^2} = \frac{\partial}{\partial x}\left(\frac{\partial f}{\partial x}\right) \qquad \frac{\partial^2 f}{\partial y^2} = \frac{\partial}{\partial y}\left(\frac{\partial f}{\partial y}\right)$$

$$\frac{\partial^2 f}{\partial x\, \partial y} = \frac{\partial}{\partial x}\left(\frac{\partial f}{\partial y}\right) \qquad \frac{\partial^2 f}{\partial y\, \partial x} = \frac{\partial}{\partial y}\left(\frac{\partial f}{\partial x}\right)$$

EXAMPLE F

Find the second-order partial derivatives of $z = x^3 y^2 - 3xy^3$.

First we find $\partial z/\partial x$ and $\partial z/\partial y$:

$$\frac{\partial z}{\partial x} = 3x^2 y^2 - 3y^3 \qquad \frac{\partial z}{\partial y} = 2x^3 y - 9xy^2$$

Therefore, we have the following second-order partial derivatives:

$$\frac{\partial^2 z}{\partial x^2} = \frac{\partial}{\partial x}\left(\frac{\partial z}{\partial x}\right) = 6xy^2 \qquad\qquad \frac{\partial^2 z}{\partial y^2} = \frac{\partial}{\partial y}\left(\frac{\partial z}{\partial y}\right) = 2x^3 - 18xy$$

$$\frac{\partial^2 z}{\partial x\, \partial y} = \frac{\partial}{\partial x}\left(\frac{\partial z}{\partial y}\right) = 6x^2 y - 9y^2 \qquad\qquad \frac{\partial^2 z}{\partial y\, \partial x} = \frac{\partial}{\partial y}\left(\frac{\partial z}{\partial x}\right) = 6x^2 y - 9y^2$$

■

In Example F we note that

$$\boxed{\frac{\partial^2 z}{\partial x\, \partial y} = \frac{\partial^2 z}{\partial y\, \partial x}} \tag{S5-1}$$

This is true in general, so long as the function and the partial derivatives are continuous.

EXAMPLE G

For $f(x, y) = \text{Arctan } \dfrac{y}{x^2}$, show that $\dfrac{\partial^2 f}{\partial x\, \partial y} = \dfrac{\partial^2 f}{\partial y\, \partial x}$.

Finding $\partial f/\partial x$ and $\partial f/\partial y$, we have

$$\frac{\partial f}{\partial x} = \frac{1}{1 + \left(\dfrac{y}{x^2}\right)^2}\left(\frac{-2y}{x^3}\right) = \frac{x^4}{x^4 + y^2}\left(-\frac{2y}{x^3}\right) = \frac{-2xy}{x^4 + y^2}$$

$$\frac{\partial f}{\partial y} = \frac{1}{1 + \left(\dfrac{y}{x^2}\right)^2}\left(\frac{1}{x^2}\right) = \frac{x^4}{x^4 + y^2}\left(\frac{1}{x^2}\right) = \frac{x^2}{x^4 + y^2}$$

Now, finding $\partial^2 f/\partial x\, \partial y$ and $\partial^2 f/\partial y\, \partial x$, we have

$$\frac{\partial^2 f}{\partial x\, \partial y} = \frac{(x^4 + y^2)(2x) - x^2(4x^3)}{(x^4 + y^2)^2} = \frac{-2x^5 + 2xy^2}{(x^4 + y^2)^2}$$

$$\frac{\partial^2 f}{\partial y\, \partial x} = \frac{(x^4 + y^2)(-2x) - (-2xy)(2y)}{(x^4 + y^2)^2} = \frac{-2x^5 + 2xy^2}{(x^4 + y^2)^2}$$

■ We see that they are equal.

Exercises S-5

In Exercises 1 through 12, find the partial derivative of the dependent variable or function with respect to each of the independent variables.

1. $z = 5x + 4x^2 y$

2. $z = \dfrac{x^2}{y} - 2xy$

3. $f(x, y) = xe^{2y}$

4. $f(x, y) = \dfrac{2 + \cos x}{1 - \sec 3y}$

5. $\phi = r\sqrt{1 + 2rs}$

6. $z = (x^2 + xy^3)^4$

7. $z = \sin xy$

8. $y = \ln (r^2 + s)$

9. $f(x, y) = \dfrac{2 \sin^3 2x}{1 - 3y}$

10. $z = \dfrac{\text{Arcsin } xy}{3 + x^2}$

11. $z = \sin x + \cos xy - \cos y$

12. $f(x, y) = e^x \cos xy + e^{-2x} \tan y$

In Exercises 13 and 14, evaluate the indicated partial derivatives at the given points.

13. $z = 3xy - x^2 + y$, $\left.\dfrac{\partial z}{\partial x}\right|_{(1, -2, -9)}$

14. $z = e^y \ln xy$, $\left.\dfrac{\partial z}{\partial y}\right|_{(e, 1, e)}$

In Exercises 15 and 16, find all of the second partial derivatives.

15. $z = 2xy^3 - 3x^2 y$

16. $f(x, y) = \dfrac{2 + \cos y}{1 + x^2}$

In Exercises 17 through 24, solve the given problems.

17. Find the slope of a line tangent to the surface $z = 9 - x^2 - y^2$ and parallel to the yz-plane which passes through $(1, 2, 4)$. Repeat the instructions for the line through $(2, 2, 1)$. Draw an appropriate figure.

18. In quality testing, a rectangular sheet of vinyl is stretched. Set up the length of the diagonal d of the sheet as a function of its sides x and y. Find the rate of change of d with respect to x for $x = 6.50$ ft if y remains constant at 4.75 ft.

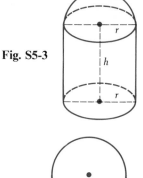

Fig. S5-3

19. A metallic machine part contracts while cooling. It is in the shape of a hemisphere attached to a cylinder, as shown in Fig. S5-3. Find the rate of change of volume with respect to r when $r = 2.65$ cm and $h = 4.20$ cm.

20. Two resistors R_1 and R_2, placed in parallel, have a combined resistance R_T given by $\dfrac{1}{R_T} = \dfrac{1}{R_1} + \dfrac{1}{R_2}$. Find $\partial R_T/\partial R_1$.

21. Two masses M and m are attached as shown in Fig. S5-4. If $M > m$, the downward acceleration a of mass M is given by $a = \dfrac{M - m}{M + m}\,g$, where g is the acceleration due to gravity. Show that $M\dfrac{\partial a}{\partial M} + m\dfrac{\partial a}{\partial m} = 0$.

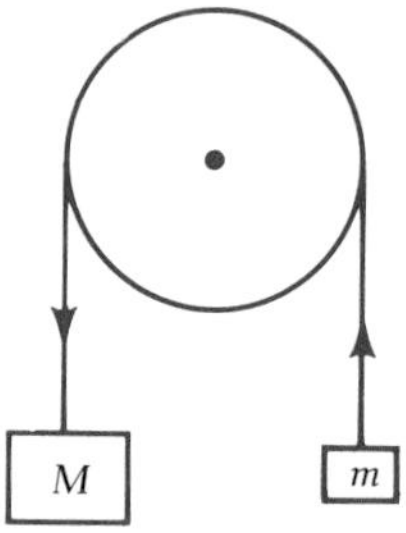

22. If an observer and a source of sound are moving toward or away from each other, the observed frequency of sound is different from that emitted. This is known as the *Doppler effect*. The equation relating the frequency f_0 the observer hears and the frequency f_s emitted by the source (a constant) is $f_0 = f_s\!\left(\dfrac{v + v_0}{v - v_s}\right)$, where v is the velocity of sound in air (a constant), v_0 is the velocity of the observer, and v_s is the velocity of the source. Show that $f_s\dfrac{\partial f_0}{\partial v_s} = f_0\dfrac{\partial f_0}{\partial v_0}$.

Fig. S5-4

23. The *mutual conductance*, measured in reciprocal ohms, of a certain electronic device is defined as $g_m = \partial i_b/\partial e_c$. Under certain circumstances, the current i_b, measured in microamperes, is given by $i_b = 50(e_b + 5e_c)^{1.5}$. Find g_m when $e_b = 200$ V and $e_c = -20$ V.

24. The temperature u in a metal bar depends on the distance x from one end and the time t. Show that $u(x, t) = 5e^{-t}\sin 4x$ satisfies *the one-dimensional heat conduction equation* $\dfrac{\partial u}{\partial t} = k\dfrac{\partial^2 u}{\partial x^2}$, where k is called the *diffusivity*. In this case, $k = \frac{1}{16}$.

S-6 Double Integrals

We now turn our attention to integration in the case of a function of two variables. The analysis has similarities to that of partial differentiation, in that an operation is performed while holding one of the independent variables constant.

If $z = f(x, y)$ and we wish to integrate with respect to x and y, we first consider either x or y constant and integrate with respect to the other. After this integral is evaluated, we then integrate with respect to the variable first held constant. We shall now define this type of integral and then give an appropriate geometric interpretation.

definition

If $z = f(x, y)$, the **double integral** of the function over x and y is defined as

$$\int_a^b \left[\int_{g(x)}^{G(x)} f(x, y)\,dy\right] dx$$

NOTE ▷ It will be noted that the limits on the inner integral are functions of x and those on the outer integral are explicit values of x. In performing the integration, **x is held constant while the inner integral is found and evaluated.** This results in a function of x only. This function is then integrated and evaluated.

It is customary not to include the brackets in stating a double integral. Therefore we write

$$\int_a^b \left[\int_{g(x)}^{G(x)} f(x, y)\, dy \right] dx = \int_a^b \int_{g(x)}^{G(x)} f(x, y)\, dy\, dx \qquad \text{(S6-1)}$$

EXAMPLE A — Evaluate: $\int_0^1 \int_{x^2}^x xy\, dy\, dx$.

Integrating the inner integral with y as the variable and x as a constant, we have

— treat as constant

$$\int_{x^2}^x xy\, dy = \left(x \frac{y^2}{2} \right)\Big|_{x^2}^x = x\left(\frac{x^2}{2} - \frac{x^4}{2} \right) = \frac{1}{2}(x^3 - x^5)$$

This means

$$\int_0^1 \int_{x^2}^x xy\, dy\, dx = \int_0^1 \frac{1}{2}(x^3 - x^5)\, dx = \frac{1}{2}\left(\frac{x^4}{4} - \frac{x^6}{6} \right)\Big|_0^1$$

$$= \frac{1}{2}\left(\frac{1}{4} - \frac{1}{6} \right) - \frac{1}{2}(0) = \frac{1}{24}$$

It is also common to have an integral which is integrated over x first and then over y. The following example illustrates such a case.

EXAMPLE B — Evaluate: $\int_0^{\pi/2} \int_0^{\sin y} e^{2x} \cos y\, dx\, dy$.

NOTE ▷ Since the inner differential is dx *(the inner limits must then be functions of y, which may be constant)*, we first integrate with x as the variable and y as a constant.

$$\int_0^{\pi/2} \int_0^{\sin y} e^{2x} \cos y\, dx\, dy = \int_0^{\pi/2} \left[\frac{1}{2} e^{2x} \cos y \right]_0^{\sin y} dy$$

$$= \frac{1}{2} \int_0^{\pi/2} (e^{2 \sin y} \cos y - \cos y)\, dy$$

$$= \frac{1}{2} \left[\frac{1}{2} e^{2 \sin y} - \sin y \right]_0^{\pi/2}$$

$$= \frac{1}{2}\left(\frac{1}{2} e^2 - 1 \right) - \frac{1}{2}\left(\frac{1}{2} - 0 \right)$$

$$= \frac{1}{4} e^2 - \frac{1}{2} - \frac{1}{4} = \frac{1}{4}(e^2 - 3)$$

$$= 1.097$$

To understand the geometric interpretation of a double integral, consider the surface shown in Fig. S6-1(a). Now consider an **element of volume** (as shown), of dimensions dx, dy, and z, which extends from the xy plane to the surface. We now make x a constant and sum (integrate) these elements of volume from the left boundary, $y = g(x)$, to the right boundary, $y = G(x)$. In this way we have the volume of the vertical slice as a function of x, as shown in Fig. S6-1(b). Now, by summing (integrating) the volumes of these slices from $x = a$ ($x = 0$ in the figure) to $x = b$, we have the complete volume, as shown in Fig. S6-1(c).

Fig. S6-1

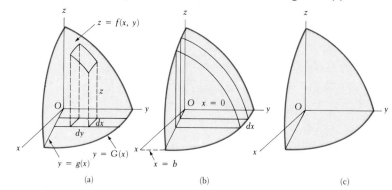

(a)   (b)   (c)

interpretation

Therefore, *we may interpret a double integral as the* **volume under a surface,** in the same way as the integral was interpreted as the area of a plane figure. It will be noted that this is not necessarily a volume of revolution, as discussed in Section 25-3. This means we may find the volume of a more general figure than previously possible. The following examples illustrate the use of double integrals to find volumes.

EXAMPLE C

Find the volume in the first octant under the plane $x + 2y + 4z - 8 = 0$. See Fig. S6-2.

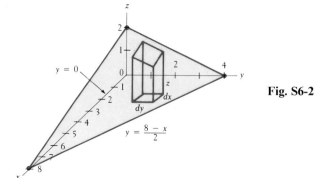

Fig. S6-2

This figure is a tetrahedron, for which $V = \frac{1}{3}Bh$. Assuming the base is in the xy-plane, $B = \frac{1}{2}(4)(8) = 16$, and $h = 2$. Therefore, $V = \frac{1}{3}(16)(2) = \frac{32}{3}$ cubic units. We shall use this value to check the one we find by double integration.

To find $z = f(x, y)$, we solve the given equation for z. Thus

$$z = \frac{8 - x - 2y}{4}$$

Next, we must find the limits on y and x. Choosing to integrate over y first, we
see that y goes from $y = 0$ to $y = (8 - x)/2$. **This last limit is the trace of the
surface in the xy-plane.** Next we note that x goes from $x = 0$ to $x = 8$. Therefore,
we set up and evaluate the integral:

$$
V = \int_0^8 \int_0^{(8-x)/2} \left(\frac{8 - x - 2y}{4} \right) dy\, dx
$$

$$
= \int_0^8 \left[\frac{1}{4} \left(8y - xy - y^2 \right) \Big|_0^{(8-x)/2} \right] dx
$$

$$
= \frac{1}{4} \int_0^8 \left[8 \left(\frac{8 - x}{2} \right) - x \left(\frac{8 - x}{2} \right) - \left(\frac{8 - x}{2} \right)^2 \right] dx
$$

$$
= \frac{1}{4} \int_0^8 \left(32 - 4x - 4x + \frac{x^2}{2} - 16 + 4x - \frac{x^2}{4} \right) dx
$$

$$
= \frac{1}{4} \int_0^8 \left(16 - 4x + \frac{x^2}{4} \right) dx = \frac{1}{4} \left(16x - 2x^2 + \frac{x^3}{12} \right) \Big|_0^8
$$

$$
= \frac{1}{4} \left(128 - 128 + \frac{512}{12} \right) = \frac{1}{4} \left(\frac{128}{3} \right) = \frac{32}{3} \text{ cubic units}
$$

We see that the values obtained by the two different methods agree.

EXAMPLE D _____ Find the volume above the xy-plane, below the surface $z = xy$, and enclosed by
the cylinder $y = x^2$ and the plane $y = x$.

Constructing the figure, shown in Fig. S6-3, we now note that $z = xy$ is the
desired function of x and y. Integrating over y first, the limits on y are $y = x^2$
to $y = x$. The corresponding limits on x are $x = 0$ to $x = 1$. Thus, the integral is

$$
V = \int_0^1 \int_{x^2}^x xy\, dy\, dx
$$

This integral has already been evaluated in Example A of this section, and we
can now see the geometric interpretation of that integral. This means that the
required volume is $\frac{1}{24}$ cubic unit.

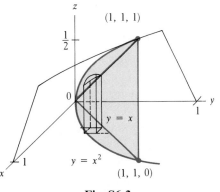

Fig. S6-3

EXAMPLE E

Find the volume in the first octant under the surface

$$z = 4 - x^2 - y^2$$

and between the cylinder $x^2 = 3y$ and the plane $y = 1$ (see Fig. S6-4).
Setting up the integration such that we integrate over x first, we have

$$V = \int_0^1 \int_0^{\sqrt{3y}} (4 - x^2 - y^2)\, dx\, dy$$

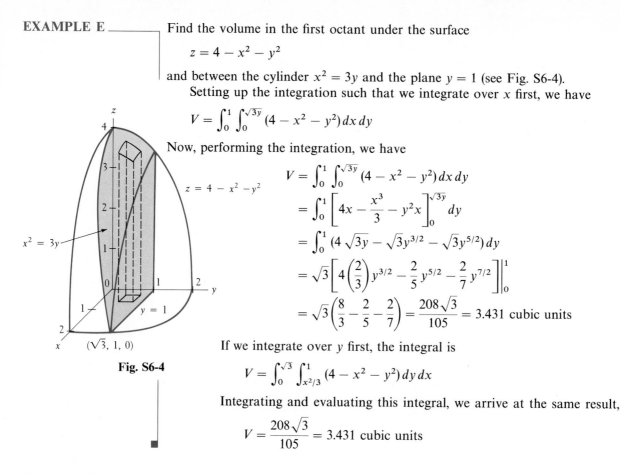

Fig. S6-4

Now, performing the integration, we have

$$V = \int_0^1 \int_0^{\sqrt{3y}} (4 - x^2 - y^2)\, dx\, dy$$

$$= \int_0^1 \left[4x - \frac{x^3}{3} - y^2 x \right]_0^{\sqrt{3y}} dy$$

$$= \int_0^1 (4\sqrt{3y} - \sqrt{3}y^{3/2} - \sqrt{3}y^{5/2})\, dy$$

$$= \sqrt{3} \left[4\left(\frac{2}{3}\right) y^{3/2} - \frac{2}{5} y^{5/2} - \frac{2}{7} y^{7/2} \right]_0^1$$

$$= \sqrt{3}\left(\frac{8}{3} - \frac{2}{5} - \frac{2}{7}\right) = \frac{208\sqrt{3}}{105} = 3.431 \text{ cubic units}$$

If we integrate over y first, the integral is

$$V = \int_0^{\sqrt{3}} \int_{x^2/3}^1 (4 - x^2 - y^2)\, dy\, dx$$

Integrating and evaluating this integral, we arrive at the same result,

$$V = \frac{208\sqrt{3}}{105} = 3.431 \text{ cubic units}$$

Exercises S-6

In Exercises 1 through 8, evaluate the given double integrals.

1. $\int_2^4 \int_0^1 xy^2\, dx\, dy$ **2.** $\int_0^4 \int_1^{\sqrt{y}} (x - y)\, dx\, dy$ **3.** $\int_0^1 \int_0^{\sqrt{1-x^2}} y\, dy\, dx$ **4.** $\int_4^9 \int_0^x \sqrt{x - y}\, dy\, dx$

5. $\int_1^e \int_1^y \frac{1}{x}\, dx\, dy$ **6.** $\int_0^{\pi/6} \int_0^1 y\sin x\, dy\, dx$ **7.** $\int_0^{\ln 3} \int_0^x e^{2x+3y}\, dy\, dx$ **8.** $\int_0^{1/2} \int_y^{y^2} \frac{dx\, dy}{\sqrt{y^2 - x^2}}$

In Exercises 9 through 14, find the indicated volumes by double integration.

9. The first-octant volume under the plane $x + y + z - 4 = 0$

10. The volume above the xy-plane and under the surface $z = 4 - x^2 - y^2$

11. The first-octant volume bounded by the xy-plane, the planes $x = y$, $y = 2$, and $z = 2 + x^2 + y^2$

12. The first-octant volume under the plane $z = x + y$ and inside the cylinder $x^2 + y^2 = 9$

13. A wedge is to be made in the shape shown in Fig. S6-5 (all vertical cross sections are equal right triangles). By double integration, find the volume of the wedge.

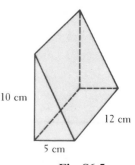

10 cm

12 cm

5 cm

Fig. S6-5

14. A circular piece of pipe is cut as shown in Fig. S6-6. Find the volume within the pipe. (*Hint:* Place the z-axis along the axis of the pipe, with the base of the pipe in the xy-plane.)

In Exercises 15 and 16, draw the appropriate figure.

15. Draw an appropriate figure indicating a volume which is found from the integral

$$\int_0^{1/2} \int_{x^2}^1 (4 - x - 2y)\, dy\, dx$$

Fig. S6-6

16. Repeat Exercise 15 for the integral

$$\int_1^2 \int_0^{2-y} \sqrt{1 + x^2 + y^2}\, dx\, dy$$

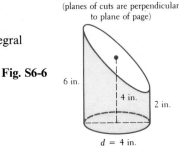

(planes of cuts are perpendicular to plane of page)

6 in.

4 in.

2 in.

d = 4 in.

S-7   Integration by Partial Fractions: Nonrepeated Linear Factors

In Chapter 27 we developed basic integral forms and methods of integration. There are many additional methods by which integrals can be put in a form for integration. In this section and the following section we develop one more useful method of integration.

In algebra we combine fractions into a single fraction by means of addition. However, if we wish to integrate an expression which contains a rational fraction, in which both the numerator and the denominator are polynomials, it is often advantageous to reverse the operation of addition and express the rational fraction as the sum of simpler fractions.

EXAMPLE A

In attempting to integrate $\int \dfrac{7 - x}{x^2 + x - 2}\, dx$ we find that it does not fit any of the standard forms of Chapter 27. However, it is possible to show that

$$\frac{7 - x}{x^2 + x - 2} = \frac{2}{x - 1} - \frac{3}{x + 2}$$

This means that

$$\int \frac{7 - x}{x^2 + x - 2}\, dx = \int \frac{2\, dx}{x - 1} - \int \frac{3\, dx}{x + 2}$$
$$= 2 \ln |x - 1| - 3 \ln |x + 2| + C$$

In Example A we saw that the integral is readily determined once the rational fraction $(7 - x)/(x^2 + x - 2)$ is replaced by the simpler fractions. In this section and the next we will describe how certain rational fractions can be expressed in terms of simpler fractions and thereby be integrated. *This technique is called the* **method of partial fractions.**

partial fractions

In order to express the rational fraction $f(x)/g(x)$ in terms of simpler partial fractions, *the degree of the numerator $f(x)$ must be less than that of the denominator $g(x)$.* If this is not the case, we divide numerator by denominator until the remainder is of the proper form. Then the denominator $g(x)$ is factored into a product of linear and quadratic factors. The method of determining the partial fractions depends on the factors which are obtained. In advanced algebra the form of the partial fractions is shown (but we shall not show the proof here).

There are four cases for the types of factors of the denominator. They are (1) *nonrepeated linear factors,* (2) *repeated linear factors,* (3) *nonrepeated quadratic factors,* and (4) *repeated quadratic factors.* In this section we consider the case of nonrepeated linear factors, and the other cases are discussed in the next section.

For the case of nonrepeated linear factors we use the fact that *corresponding to each linear factor $ax + b$, occurring once in the denominator, there will be a partial fraction of the form*

$$\frac{A}{ax + b}$$

where A is a constant to be determined. The following examples illustrate the method.

EXAMPLE B

Integrate: $\int \dfrac{7 - x}{x^2 + x - 2}\, dx.$

This is the same integral as in Example A. Here we will see how the partial fractions are found.

First we note that the degree of the numerator is 1 (the highest power is x) and that of the denominator is 2 (the highest power is x^2). Since the degree of the denominator is higher, we may proceed to factoring it. Thus,

$$\frac{7 - x}{x^2 + x - 2} = \frac{7 - x}{(x - 1)(x + 2)}$$

There are two linear factors, $(x - 1)$ and $(x + 2)$, in the denominator and they are different. This means that there are two partial fractions. Therefore we write

$$\frac{7 - x}{(x - 1)(x + 2)} = \frac{A}{x - 1} + \frac{B}{x + 2} \tag{1}$$

The constants A and B are to be determined so that the equation is an identity. To assist in finding the values of A and B we clear Eq. (1) of fractions by multiplying both sides by $(x - 1)(x + 2)$. This gives us

$$7 - x = A(x + 2) + B(x - 1) \tag{2}$$

Equation (2) is also an identity, which means that there are two ways of determining the values of A and B.

NOTE ▷

Solution by substitution: Since Eq. (2) is an identity, **it is true for any value of x.** Thus, in turn we pick $x = -2$ and $x = 1$, for each of these values makes a factor on the right equal to zero, and the values of B and A are easily found.

Therefore,

for $x = -2$: $7 - (-2) = A(-2 + 2) + B(-2 - 1)$
$$9 = -3B, \qquad B = -3$$

for $x = 1$: $7 - 1 = A(1 + 2) + B(1 - 1)$
$$6 = 3A, \qquad A = 2$$

Solution by equating coefficients: Since Eq. (2) is an identity, another way of

NOTE▷ finding the constants A and B is to **equate coefficients of like powers of x from each side.** Thus, writing Eq. (2) as

$$7 - x = (2A - B) + (A + B)x$$

we have

$$2A - B = 7 \quad \text{(equating constants: } x^0 \text{ terms)}$$
$$A + B = -1 \quad \text{(equating coefficients of } x)$$

$$2A - B = \ \ 7$$
$$\underline{A + B = -1}$$
$$3A = 6$$
$$A = 2$$
$$2(2) - B = 7$$
$$-B = 3$$
$$B = -3$$

Solving this system, as shown at the left, we have $A = 2$ and $B = -3$.

Now using the values of A and B we have

$$\frac{7 - x}{(x - 1)(x + 2)} = \frac{2}{x - 1} - \frac{3}{x + 2}$$

Therefore, the integral is found as in Example A.

$$\int \frac{7 - x}{x^2 + x - 2}\, dx = \int \frac{7 - x}{(x - 1)(x + 2)}\, dx = \int \frac{2\, dx}{x - 1} - \int \frac{3\, dx}{x + 2}$$
$$= 2 \ln |x - 1| - 3 \ln |x + 2| + C$$

Using the properties of logarithms, we may write this as

$$\int \frac{7 - x}{x^2 + x - 2}\, dx = \ln \left| \frac{(x - 1)^2}{(x + 2)^3} \right| + C$$

In Example B we showed two ways of finding the values of A and B. The method of substitution is generally easier to use for linear factors. However, as we will see in the next section, the method of equating coefficients can also be very useful.

EXAMPLE C ———— Integrate: $\int \dfrac{6x^2 - 14x - 11}{(x + 1)(x - 2)(2x + 1)}\, dx$.

First we note that the denominator is factored and that it is of degree 3 (when the denominator is multiplied out, the highest power of x is x^3). Thus we have three nonrepeated linear factors. This means that

$$\frac{6x^2 - 14x - 11}{(x + 1)(x - 2)(2x + 1)} = \frac{A}{x + 1} + \frac{B}{x - 2} + \frac{C}{2x + 1}$$

Multiplying through by $(x + 1)(x - 2)(2x + 1)$, we have

$$6x^2 - 14x - 11 = A(x - 2)(2x + 1) + B(x + 1)(2x + 1) + C(x + 1)(x - 2)$$

We now substitute the values of 2, $-\frac{1}{2}$, and -1 for x. Again, these are chosen because they make factors of the coefficients of A, B, or C equal to zero, although any values may be chosen. Therefore,

for $x = 2$ $6(4) - 14(2) - 11 = A(0)(5) + B(3)(5) + C(3)(0),$ $B = -1$

for $x = -\dfrac{1}{2}$: $6\left(\dfrac{1}{4}\right) - 14\left(-\dfrac{1}{2}\right) - 11 = A\left(-\dfrac{5}{2}\right)(0) + B\left(\dfrac{1}{2}\right)(0) + C\left(\dfrac{1}{2}\right)\left(-\dfrac{5}{2}\right),$ $C = 2$

for $x = -1$: $6(1) - 14(-1) - 11 = A(-3)(-1) + B(0)(-1) + C(0)(-3),$ $A = 3$

(Continued on next page)

Therefore,

$$\int \frac{6x^2 - 14x - 11}{(x+1)(x-2)(2x+1)} \, dx = \int \frac{3 \, dx}{x+1} - \int \frac{dx}{x-2} + \int \frac{2 \, dx}{2x+1}$$

$$= 3 \ln |x+1| - \ln |x-2| + \ln |2x+1| + C_1 = \ln \left| \frac{(2x+1)(x+1)^3}{x-2} \right| + C_1$$

Here we have let the constant of integration be C_1 since we used C as the numerator of the third partial fraction. ∎

EXAMPLE D

Integrate: $\displaystyle\int \frac{2x^4 - x^3 - 9x^2 + x - 12}{x^3 - x^2 - 6x} \, dx$.

Since the numerator is of a higher degree than the denominator, we must first divide the numerator by the denominator. This gives

$$\frac{2x^4 - x^3 - 9x^2 + x - 12}{x^3 - x^2 - 6x} = 2x + 1 + \frac{4x^2 + 7x - 12}{x^3 - x^2 - 6x}$$

We must now express this rational fraction in terms of its partial fractions.

$$\frac{4x^2 + 7x - 12}{x^3 - x^2 - 6x} = \frac{4x^2 + 7x - 12}{x(x+2)(x-3)} = \frac{A}{x} + \frac{B}{x+2} + \frac{C}{x-3}$$

Clearing fractions, we have

$$4x^2 + 7x - 12 = A(x+2)(x-3) + Bx(x-3) + Cx(x+2)$$

Now, using values of x of -2, 3, and 0 for substitution, we obtain the values of $B = -1$, $C = 3$, and $A = 2$, respectively. Therefore,

$$\int \frac{2x^4 - x^3 - 9x^2 + x - 12}{x^3 - x^2 - 6x} \, dx = \int \left(2x + 1 + \frac{2}{x} - \frac{1}{x+2} + \frac{3}{x-3} \right) dx$$

$$= x^2 + x + 2 \ln |x| - \ln |x+2| + 3 \ln |x-3| + C_1$$

$$= x^2 + x + \ln \left| \frac{x^2(x-3)^3}{x+2} \right| + C_1$$

∎

Exercises S-7

In Exercises 1 through 12, integrate the given functions.

1. $\displaystyle\int \frac{x+3}{(x+1)(x+2)} \, dx$

2. $\displaystyle\int \frac{x+2}{x(x+1)} \, dx$

3. $\displaystyle\int \frac{dx}{x^2 - 4}$

4. $\displaystyle\int \frac{x-9}{2x^2 - 3x + 1} \, dx$

5. $\displaystyle\int \frac{x^2 + 3}{x^2 + 3x} \, dx$

6. $\displaystyle\int \frac{x^3}{x^2 + 3x + 2} \, dx$

7. $\displaystyle\int_0^1 \frac{2x+4}{3x^2 + 5x + 2} \, dx$

8. $\displaystyle\int_1^3 \frac{x-1}{4x^2 + x} \, dx$

9. $\displaystyle\int \frac{4x^2 - 10}{x(x+1)(x-5)} \, dx$

10. $\displaystyle\int \frac{6x^2 - 2x - 1}{4x^3 - x} \, dx$

11. $\displaystyle\int_2^3 \frac{dx}{x^3 - x}$

12. $\displaystyle\int \frac{dx}{(x^2 - 4)(x^2 - 9)}$

In Exercises 13 through 16, solve the given problems by integration.

13. The current i, in amperes, as a function of the time, in seconds, in a certain electric circuit is given by $i = (4t + 3)/(2t^2 + 3t + 1)$. Find the total charge that passes a given point in the circuit during the first second.

14. The force F, in newtons, applied by a stamping machine in making a certain computer part is given by $F = 4x/(x^2 + 3x + 2)$, where x is the distance (in cm) through which the force acts. Find the work done by the force from $x = 0$ to $x = 0.500$ cm.

15. Find the first-quadrant area bounded by $y = 1/(x^3 + 3x^2 + 2x)$, $x = 1$, and $x = 3$. See Fig. S7-1.

16. Find the volume generated if the area of Exercise 15 is rotated about the y-axis.

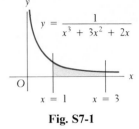

$$y = \frac{1}{x^3 + 3x^2 + 2x}$$

$x = 1$ $x = 3$

Fig. S7-1

S-8   Integration by Partial Fractions: Other Cases

In the previous section we introduced the method of partial fractions and considered the case of nonrepeated linear factors. In this section we develop the use of partial fractions for the cases of repeated linear factors and nonrepeated quadratic factors. We also briefly discuss the case of repeated quadratic factors.

For the case of repeated linear factors we use the fact that *corresponding to each linear factor $ax + b$ that occurs n times in the denominator there will be n partial fractions*

repeated linear factors

$$\frac{A_1}{ax + b} + \frac{A_2}{(ax + b)^2} + \cdots + \frac{A_n}{(ax + b)^n}$$

where $A_1, A_2, \ldots, A_n$ are constants to be determined.

EXAMPLE A

Integrate: $\int \dfrac{dx}{x(x + 3)^2}$.

Here we see that the denominator has a factor of x and two factors of $x + 3$. For the factor of x we use a partial fraction as in the previous section, for it is a nonrepeated factor. For the factor $x + 3$ we need two partial fractions, one with a denominator of $x + 3$ and the other with a denominator of $(x + 3)^2$. Thus, we write

NOTE ▷

$$\frac{1}{x(x + 3)^2} = \boxed{\frac{A}{x} + \frac{B}{x + 3} + \frac{C}{(x + 3)^2}}$$

Multiplying each side by $x(x + 3)^2$, we have

$$1 = A(x + 3)^2 + Bx(x + 3) + Cx \qquad (1)$$

Using the values of x of -3 and 0, we have

for $x = -3$: $1 = A(0^2) + B(-3)(0) + (-3)C$, $C = -\dfrac{1}{3}$

for $x = 0$: $1 = A(3^2) + B(0)(3) + C(0)$, $A = \dfrac{1}{9}$

(Continued on next page)

Since there are no other values which make a factor in Eq. (1) equal to zero, we must either choose some other value of x or equate coefficients of some power of x in Eq. (1). We will let $x = 1$.

Since Eq. (1) is an identity, we may choose any value of x. With $x = 1$, we have

$$1 = A(4^2) + B(1)(4) + C(1)$$
$$1 = 16A + 4B + C$$

Using the known values of A and C, we have

$$1 = 16\left(\frac{1}{9}\right) + 4B - \frac{1}{3}, \qquad B = -\frac{1}{9}$$

This means that

$$\frac{1}{x(x + 3)^2} = \frac{\frac{1}{9}}{x} + \frac{-\frac{1}{9}}{x + 3} + \frac{-\frac{1}{3}}{(x + 3)^2}$$

or

$$\int \frac{dx}{x(x + 3)^2} = \frac{1}{9} \int \frac{dx}{x} - \frac{1}{9} \int \frac{dx}{x + 3} - \frac{1}{3} \int \frac{dx}{(x + 3)^2}$$

$$= \frac{1}{9} \ln |x| - \frac{1}{9} \ln |x + 3| - \frac{1}{3}\left(\frac{1}{-1}\right)(x + 3)^{-1} + C_1$$

$$= \frac{1}{9} \ln \left|\frac{x}{x + 3}\right| + \frac{1}{3(x + 3)} + C_1$$

EXAMPLE B

Integrate: $\int \dfrac{3x^3 + 15x^2 + 21x + 15}{(x - 1)(x + 2)^3}\, dx.$

First we set up the partial fractions as

$$\frac{3x^3 + 15x^2 + 21x + 15}{(x - 1)(x + 2)^3} = \frac{A}{x - 1} + \frac{B}{x + 2} + \frac{C}{(x + 2)^2} + \frac{D}{(x + 2)^3}$$

Next we clear fractions.

$$3x^3 + 15x^2 + 21x + 15 = A(x + 2)^3 + B(x - 1)(x + 2)^2$$
$$+ C(x - 1)(x + 2) + D(x - 1) \qquad (1)$$

For $x = 1$: $\qquad\qquad 3 + 15 + 21 + 15 = 27A, \qquad 54 = 27A, \qquad A = 2$

For $x = -2$: $\quad 3(-8) + 15(4) + 21(-2) + 15 = -3D, \qquad 9 = -3D, \qquad D = -3$

To find B and C we equate coefficients of powers of x. Therefore, we write Eq. (1) as

$$3x^3 + 15x^2 + 21x + 15 = (A + B)x^3 + (6A + 3B + C)x^2$$
$$+ (12A + C + D)x + (8A - 4B - 2C - D)$$

Coefficients of x^3: $3 = A + B$, $3 = 2 + B$, $B = 1$

Coefficients of x^2: $15 = 6A + 3B + C$, $15 = 12 + 3 + C$, $C = 0$

$$\frac{3x^3 + 15x^2 + 21x + 15}{(x-1)(x+2)^3} = \frac{2}{x-1} + \frac{1}{x+2} + \frac{0}{(x+2)^2} + \frac{-3}{(x+2)^3}$$

$$\int \frac{3x^3 + 15x^2 + 21x + 15}{(x-1)(x+2)^3}\, dx = 2 \int \frac{dx}{x-1} + \int \frac{dx}{x+2} - 3 \int \frac{dx}{(x+2)^3}$$

$$= 2\ln|x-1| + \ln|x+2| - 3\left(\frac{1}{-2}\right)(x+2)^{-2} + C_1$$

$$= \ln|(x-1)^2(x+2)| + \frac{3}{2(x+2)^2} + C_1$$

■

If there is one repeated factor in the denominator and it is the only factor present in the denominator, a substitution is easier and more convenient than partial fractions. This is illustrated in the following example.

EXAMPLE C

Integrate: $\displaystyle\int \frac{x\,dx}{(x-2)^3}$.

This could be integrated by first setting up the appropriate partial fractions. However, the solution is more easily found by use of the substitution $u = x - 2$. Using this, we have

$$u = x - 2,\qquad x = u + 2,\qquad dx = du$$

$$\int \frac{x\,dx}{(x-2)^3} = \int \frac{(u+2)(du)}{u^3} = \int \frac{du}{u^2} + 2\int \frac{du}{u^3}$$

$$= \int u^{-2}\,du + 2\int u^{-3}\,du = \frac{1}{-u} + \frac{2}{-2}u^{-2} + C$$

$$= -\frac{1}{u} - \frac{1}{u^2} + C = -\frac{u+1}{u^2} + C$$

$$= -\frac{x-2+1}{(x-2)^2} + C = \frac{1-x}{(x-2)^2} + C$$

■

quadratic factors

For the case of nonrepeated quadratic factors, we use the fact that *corresponding to each irreducible (cannot be further factored) quadratic factor $ax^2 + bx + c$ that occurs once in the denominator there is a partial fraction of the form*

$$\frac{Ax + B}{ax^2 + bx + c}$$

where A and B are constants to be determined.

EXAMPLE D

Integrate: $\int \dfrac{4x + 4}{x^3 + 4x}\, dx$.

In setting up the partial fractions, we note that the denominator factors as $x^3 + 4x = x(x^2 + 4)$. Here the factor $x^2 + 4$ cannot be further factored. Thus,

NOTE ▷

$$\frac{4x + 4}{x^3 + 4x} = \frac{4x + 4}{x(x^2 + 4)} = \boxed{\frac{A}{x} + \frac{Bx + C}{x^2 + 4}}$$

Clearing fractions, we have

$$4x + 4 = A(x^2 + 4) + Bx^2 + Cx = (A + B)x^2 + Cx + 4A$$

Equating coefficients of powers of x gives us

for x^2: $0 = A + B$ for x: $4 = C$

for constants: $4 = 4A$, or $A = 1$

Therefore we easily find that $B = -1$ from the first equation. This means that

$$\frac{4x + 4}{x^3 + 4x} = \frac{1}{x} + \frac{-x + 4}{x^2 + 4}$$

and

$$\int \frac{4x + 4}{x^3 + 4x}\, dx = \int \frac{1}{x}\, dx + \int \frac{-x + 4}{x^2 + 4}\, dx = \int \frac{1}{x}\, dx - \int \frac{x\, dx}{x^2 + 4} + \int \frac{4\, dx}{x^2 + 4}$$

$$= \ln |x| - \frac{1}{2} \ln |x^2 + 4| + 2\, \text{Arctan}\, \frac{x}{2} + C_1$$

EXAMPLE E

Integrate: $\int \dfrac{x^3 + 3x^2 + 2x + 4}{x^2(x^2 + 2x + 2)}\, dx$.

In the denominator we have a repeated linear factor, x^2, and a quadratic factor. Therefore,

$$\frac{x^3 + 3x^2 + 2x + 4}{x^2(x^2 + 2x + 2)} = \frac{A}{x} + \frac{B}{x^2} + \frac{Cx + D}{x^2 + 2x + 2}$$

$$x^3 + 3x^2 + 2x + 4 = Ax(x^2 + 2x + 2) + B(x^2 + 2x + 2) + Cx^3 + Dx^2$$

$$= (A + C)x^3 + (2A + B + D)x^2 + (2A + 2B)x + 2B$$

Equating coefficients, we find that $A = -1$, $B = 2$, $C = 2$, and $D = 3$. Therefore,

$$\frac{x^3 + 3x^2 + 2x + 4}{x^2(x^2 + 2x + 2)} = -\frac{1}{x} + \frac{2}{x^2} + \frac{2x + 3}{x^2 + 2x + 2}$$

$$\int \frac{x^3 + 3x^2 + 2x + 4}{x^2(x^2 + 2x + 2)}\, dx = -\int \frac{dx}{x} + 2\int \frac{dx}{x^2} + \int \frac{2x + 3}{x^2 + 2x + 2}\, dx$$

$$= -\ln |x| - 2\left(\frac{1}{x}\right) + \int \frac{2x + 2 + 1}{x^2 + 2x + 2}\, dx$$

$$= -\ln |x| - \frac{2}{x} + \int \frac{2x + 2}{x^2 + 2x + 2} \, dx + \int \frac{dx}{(x^2 + 2x + 1) + 1}$$

$$= -\ln |x| - \frac{2}{x} + \ln |x^2 + 2x + 2| + \text{Arctan}\,(x + 1) + C_1$$

NOTE ▷ *Note the manner in which the integral for the quadratic denominator was handled. First the numerator, $2x + 3$, was changed to the form $(2x + 2) + 1$ so that we could fit the logarithmic form with the $2x + 2$. Then we completed the square in the denominator of the final integral so that it fitted an inverse tangent form.*

Finally, for the case of repeated quadratic factors we use the fact that *corresponding to each irreducible quadratic factor $ax^2 + bx + c$ that occurs n times in the denominator there will be n partial fractions*

$$\frac{A_1 x + B_1}{ax^2 + bx + c} + \frac{A_2 x + B_2}{(ax^2 + bx + c)^2} + \cdots + \frac{A_n x + B_n}{(ax^2 + bx + c)^n}$$

where $A_1, A_2, \ldots, A_n, B_1, B_2, \ldots, B_n$ are constants to be determined. The procedures which lead to the solution are the same as those for the other cases. Exercises 11 and 12 in the following set are solved by using the above types of partial fractions.

Exercises S-8

In Exercises 1 through 12, integrate each of the given functions.

1. $\displaystyle\int \frac{x - 8}{x^3 - 4x^2 + 4x} \, dx$

2. $\displaystyle\int \frac{dx}{x^3 - x^2}$

3. $\displaystyle\int \frac{2\,dx}{x^2(x^2 - 1)}$

4. $\displaystyle\int_1^3 \frac{3x^3 + 8x^2 + 10x + 2}{x(x + 1)^3} \, dx$

5. $\displaystyle\int_1^2 \frac{2x\,dx}{(x - 3)^3}$

6. $\displaystyle\int \frac{4\,dx}{(x + 1)^2(x - 1)^2}$

7. $\displaystyle\int_0^2 \frac{x^2 + x + 5}{(x + 1)(x^2 + 4)} \, dx$

8. $\displaystyle\int \frac{x^2 + x - 1}{(x^2 + 1)(x - 2)} \, dx$

9. $\displaystyle\int \frac{5x^2 + 8x + 16}{x^2(x^2 + 4x + 8)} \, dx$

10. $\displaystyle\int \frac{10x^3 + 40x^2 + 22x + 7}{(4x^2 + 1)(x^2 + 6x + 10)} \, dx$

11. $\displaystyle\int \frac{2x^3}{(x^2 + 1)^2} \, dx$

12. $\displaystyle\int \frac{-x^3 + x^2 + x + 3}{(x + 1)(x^2 + 1)^2} \, dx$

In Exercises 13 through 16, solve the given problems by integration.

13. Find the area bounded by $y = (x - 3)/(x^3 + x^2)$, $y = 0$, and $x = 1$.

14. Find the volume generated by rotating the first-quadrant area bounded by $y = 4/(x^4 + 6x^2 + 5)$ and $x = 2$ about the y-axis.

15. Under certain conditions the velocity v, in meters per second, of an object moving along a straight line as a function of time, in seconds, is given by $v = \dfrac{t^2 + 14t + 27}{(2t + 1)(t + 5)^2}$. Find the distance traveled by the object during the first 2.00 s.

16. By a computer analysis the electric current i, in amperes, in a certain integrated circuit is given by $i = \dfrac{0.001(7t^2 + 16t + 48)}{(t + 4)(t^2 + 16)}$, where t is the time in seconds. Find the total charge that passes a point in the circuit in the first 0.250 s.

APPENDIX A Study Aids

A-1 Introduction

The primary objective of this text is to give you an understanding of mathematics so that you can use it effectively as a tool in your technology. Without understanding of the basic methods, knowledge is usually short-lived. However, if you do understand, you will find your work much more enjoyable and rewarding. This is true in any course you may take, be it in mathematics or in any other field.

Mathematics is an indispensable tool in almost all scientific fields of study. You will find it used to a greater and greater degree as you work in your chosen field. Generally, in the introductory portions of allied courses, it is enough to have a grasp of elementary concepts in algebra and geometry. However, as you develop in your field, the need for more mathematics will be apparent. This text is designed to develop these necessary tools so they will be available in your allied courses. You cannot derive the full benefit from your mathematics course unless you devote the necessary amount of time to developing a sound understanding of the subject.

It is assumed in this text that you have a background which includes geometry and some algebra. Therefore, many of the topics covered in this book may seem familiar to you, especially in the earlier chapters. However, it is likely that your background in some of these areas is not complete, either because you have not studied mathematics for a year or two or because you did not understand the topics when you first encountered them. If a topic is familiar, do not reason that there is no sense in studying it again, but take the opportunity to clarify any points on which you are not certain. In almost every topic you will probably find certain points which can use further study. If the topic is new, use your time effectively to develop an understanding of the methods involved, and do not simply memorize problems of a certain type.

There is only one good way to develop the understanding and working knowledge necessary in any course, and that is to *work with it*. Many students consider mathematics difficult. They will tell you that it is their lack of mathematical ability and the complexity of the material itself which make it difficult. Some topics in mathematics, especially in the more advanced areas, do require a certain aptitude for full comprehension. However, a large proportion of poor grades in elementary mathematics courses result from the fact that the student is not willing to put in the necessary time to develop the understanding. The student takes a quick glance through the material, tries a few exercises, is largely unsuccessful, and then decides that the material is "impossible." A detailed reading of the text, a careful following of the illustrative examples, and then solving the exercises would lead to more success and therefore would make the work much more enjoyable and rewarding. No matter what text is used, what methods are used in the course, or what other variables may be introduced, if you do not put in an adequate amount of time studying, you will not derive the proper results. More detailed suggestions for study are included in the following section.

If you consider these suggestions carefully, and follow good study habits, you should enjoy a successful learning experience in this course as well as in other courses.

A-2 Suggestions for Study

When you are studying the material presented in this text, the following suggestions may help you to derive full benefit from the time you devote to it.

1. Before attempting to do the exercises, read through the material preceding them.

2. Follow the illustrative examples carefully, being certain that you know how to proceed from step to step. You should then have a good idea of the methods involved.

3. Work through the exercises, spending a reasonable amount of time on each problem. If you cannot solve a certain problem in a reasonable amount of time, leave it and go to the next. Return to this problem later. If you find many problems difficult, you should reread the explanatory material and the examples to determine what point or points you have not understood.

4. When you have completed the exercises, or at least most of them, glance back through the explanatory material to be sure you understand the methods and principles.

5. If you have gone through the first four steps and certain points still elude you, ask to have these points clarified in class. Do not be afraid to ask questions; only be sure that you have made a sincere effort on your own before you ask them.

Some study habits which are useful not only here but in all of your other subjects are the following:

1. Put in the time required to develop the material fully, being certain that you are making effective use of your time. A good place to study helps immeasurably.

2. Learn the *methods and principles* being presented. Memorize as little as possible, for although there are certain basic facts which are more expediently learned by memorization, these should be kept to a minimum.

3. Keep up with the material in all of your courses. Do not let yourself get so behind in your studies that it becomes difficult to make up the time. Usually the time is never really made up. Studying only before tests is a poor way of learning, and is usually rather ineffective.

4. When you are taking examinations, always read each question carefully before attempting the solution. Solve those you find easiest first, and do

not spend too much time on any one problem. Also, use all the time available for the examination. If you finish early, use the remainder of the time to check your work.

A-3   Problem Analysis

Drill-type problems require a working knowledge of the methods presented. However, they do not require, in general, much analysis before being put in proper form for solution. Stated problems, on the other hand, do require proper interpretation before they can be put in a form for solution. The remainder of this section is devoted to some suggestions for solving stated problems.

We have to put stated problems in symbolic form before we attempt to solve them. It is this step which most students find difficult. Because such problems require the student to do more than merely go through a certain routine, they demand more analysis and thus appear more "difficult." There are several reasons for the student's difficulty, some of them being (1) unsuccessful previous attempts at solving such problems, leading the student to believe that all stated problems are "impossible"; (2) failure to read the problem carefully; (3) a poorly organized approach to the solution; and (4) improper and incomplete interpretation of the statements given. The first two of these can be overcome only with the proper attitude and care.

There are over 120 completely worked examples of stated problems (as well as numerous other examples which indicate a similar analysis) throughout this text, illustrating proper interpretations and approaches to these problems. Therefore, we shall not include specific examples here. However, we shall set forth the method of analysis of any stated problem. Such an analysis generally follows these steps:

1. Read the problem carefully.
2. Carefully identify known and unknown quantities.
3. Draw a figure when appropriate (which is quite often the case).
4. Write, in symbols, the relations given in the statements.
5. Solve for the desired quantities.

If you follow this step-by-step method and write out the solution neatly, you should find that stated problems lend themselves to solution more readily than you had previously found.

APPENDIX B Units of Measurement and Approximate Numbers

B-1   Units of Measurement; the Metric System

Most scientific and technical calculations involve numbers that represent a measurement or count of a specific physical quantity. *Such numbers are called* **denominate numbers,** *and associated with these denominate numbers are* **units of measurement.** For calculations and results to be meaningful we must know these units. For example, if we measure the length of an object to be 12, we must know whether it is being measured in feet, yards, or some other specified unit of length.

base units

Certain universally accepted **base units** *are used to measure fundamental quantities.* The units for numerous other quantities are expressed in terms of the base units. Fundamental quantities for which base units are defined are (1) length, (2) mass or force, depending on the system of units being used, (3) time, (4) electric current, (5) temperature, (6) amount of substance, and (7) luminous intensity. *Other units, referred to as* **derived units,** *are expressible in terms of the units for these quantities.*

derived units

Even though all other quantities can be expressed in terms of the fundamental ones, many have units which are given a specified name. This is done primarily for those quantities which are used very commonly, although it is not done for all such quantities. For example, the volt is defined as a meter^2-kilogram/second^3-ampere, which is in terms of (a unit of length)^2(a unit of mass)/(a unit of time)^3(a unit of electric current). The unit for acceleration has no special name and is left in terms of the base units, for example, feet/second^2. For convenience, special symbols are usually used to designate units. The units for acceleration would be written as ft/s^2.

Two basic systems of units, the **metric system** and the **United States Customary** system, are in use today. (The U.S. Customary system is also known as the *British system,* but the metric system is now used in Great Britain.) Nearly every country in the world now uses the metric system. In the United States both systems are used, and international trade has led many major U.S. industrial firms to convert their products to the metric system. Also due to world trade, to further promote conversion to the metric system, Congress has passed legislation which states that the metric system is the preferred system, and which requires Federal agencies to use the metric system in their business-related activities.

Therefore, for the present, both systems are of importance, although the metric system will eventually be used almost universally. For that reason, where units are used, some of the exercises and examples have metric units and others have U.S. Customary units. Technicians and engineers need to have some knowledge of both systems.

Although more than one system has been developed in which metric units are used, the system which is now accepted as the metric system is the **International System of Units (SI).** This was established in 1960 and uses some different definitions for base units from the previously developed metric units. However, the measurement of the base units in the SI system is more accessible, and the differences are very slight. *Therefore, when we refer to the metric system, we are using SI units.*

As we have stated, in each system the base units are specified, and all other units are then expressible in terms of these. Table B-1 lists the fundamental quantities, as well as many other commonly used quantities, along with the symbols and names of units used to represent each quantity shown.

In the U.S. Customary system, the base unit of length is the *foot,* and that of force is the *pound.* In the metric system, the base unit of length is the *meter,* and that of mass is the *kilogram.* Here, we see a difference in the definition of the systems which causes some difficulty when units are converted from one system to the other. That is, a base unit in the U.S. Customary system is a unit of force, and a base unit in the metric system is a unit of mass.

The distinction between mass and force is very significant in physics, and the weight of an object is the force with which it is attracted to the earth. Weight, which is therefore a force, is different from mass, which is a measure of the inertia an object exhibits. Although they are different quantities, mass and weight are, however, very closely related. In fact, the weight of an object equals its mass multiplied by the acceleration due to gravity. Near the surface of the earth the acceleration due to gravity is nearly constant, although it decreases as the distance from the earth increases. Therefore, near the surface of the earth the weight of an object is directly proportional to its mass. However, at great distances from the earth the weight of an object will be zero, whereas its mass does not change.

Since force and mass are different, it is not strictly correct to convert pounds to kilograms. However, since the pound is the base unit in the U.S. Customary system, and the kilogram is the base unit in the metric system, at the earth's surface 1 kg corresponds to 2.21 lb, in the sense that the force of gravity on a 1 kg mass is 2.21 lb.

When designating units for weight, we use pounds in the U.S. Customary system. In the metric system, although kilograms are used for weight, it is preferable to specify the mass of an object in kilograms. The force of gravity on an object is designated in newtons, and the use of the term weight is avoided unless its meaning is completely clear.

As for the other fundamental quantities, both systems use the *second* as the base unit of time. In the SI system the *ampere* is defined as the base unit of electric current, and this can also be used in the U.S. Customary system. As for temperature, *degrees Fahrenheit* are used with the U.S. Customary system, and *degrees Celsius* (formerly centigrade) are used with the metric system (actually, the *kelvin* is defined as the base unit, where the temperature in kelvins is the temperature in degrees Celsius plus 273.16). In the SI system, the base unit for the amount of a substance is the *mole,* and the base unit of luminous intensity is the *candela.* These last two are of limited importance to our use in this text.

SI metric system

TABLE B-1   Quantities and Their Associated Units

Quantity	Quantity symbol	U.S. Customary Name	U.S. Customary Symbol	Metric (SI) Name	Metric (SI) Symbol	In Terms of Other SI Units
Length	s	foot	ft	**meter**	m	
Mass	m	slug		**kilogram**	kg	
Force	F	pound	lb	newton	N	$m \cdot kg/s^2$
Time	t	second	s	**second**	s	
Area	A		ft^2		m^2	
Volume	V		ft^3		m^3	
Capacity	V	gallon	gal	liter	L	$(1\ L = 1\ dm^3)$
Velocity	v		ft/s		m/s	
Acceleration	a		ft/s^2		m/s^2	
Density	d, ρ		lb/ft^3		kg/m^3	
Pressure	p		lb/ft^2	pascal	Pa	N/m^2
Energy, work	E, W		ft·lb	joule	J	$N \cdot m$
Power	P	horsepower	hp	watt	W	J/s
Period	T		s		s	
Frequency	f		1/s	hertz	Hz	1/s
Angle	θ	radian	rad	radian	rad	
Electric current	I, i	ampere	A	**ampere**	A	
Electric charge	q	coulomb	C	coulomb	C	$A \cdot s$
Electric potential	V, E	volt	V	volt	V	$J/A \cdot s$
Capacitance	C	farad	F	farad	F	s/Ω
Inductance	L	henry	H	henry	H	$\Omega \cdot s$
Resistance	R	ohm	Ω	ohm	Ω	V/A
Thermodynamic temperature	T			**kelvin**	K	(temp. interval
Temperature	T	degree Fahrenheit	°F	degree Celsius	°C	1°C = 1 K)
Quantity of heat	Q	British thermal unit	Btu	joule	J	
Amount of substance	n			**mole**	mol	
Luminous intensity	I	candlepower	cp	**candela**	cd	

Special Notes:

1. The SI base units are shown in boldface type.

2. The unit symbols shown above are those which are used in the text. Many of them were adopted with the adoption of the SI system. This means, for example, that we use s rather than sec for seconds, and A rather than amp for amperes. Also, other units, such as volt, are not spelled out, a common practice in the past. When a given unit is used with both systems, we use the SI symbol for the unit.

3. The liter and degree Celsius are not actually SI units. However, they are recognized for use with the SI system due to their practical importance. Also, the symbol for liter has several variations. Presently L is recognized for use in the United States and Canada, l is recognized by the International Committee of Weights and Measures, and ℓ is also recognized for use in several countries.

4. Other units of time, along with their symbols, which are recognized for use with the SI system and are used in this text, are minute, min; hour, h; day, d.

5. There are many additional specialized units which are used with the SI system. However, most of those which appear in this text are shown in the table. A few of the specialized units are noted when used in the text . One which is frequently used is that for revolution, r.

6. Other common U.S. units which are used in the text are inch, in.; yard, yd; mile, mi; ounce, oz.; ton; quart, qt; acre.

7. There are a number of units which were used with the metric system prior to the development of the SI system. However, many of these are not to be used with the SI system. Among those which were commonly used are the dyne, erg, and calorie.

Due to greatly varying sizes of certain quantities, the metric system employs certain prefixes to units to denote different orders of magnitude. These prefixes, with their meanings and symbols, are shown in Table B-2.

TABLE B-2 Metric Prefixes

Prefix	Factor	Symbol	Prefix	Factor	Symbol
exa	10^{18}	E	deci	10^{-1}	d
peta	10^{15}	P	centi	10^{-2}	c
tera	10^{12}	T	milli	10^{-3}	m
giga	10^{9}	G	micro	10^{-6}	μ
mega	10^{6}	M	nano	10^{-9}	n
kilo	10^{3}	k	pico	10^{-12}	p
hecto	10^{2}	h	femto	10^{-15}	f
deca	10^{1}	da	atto	10^{-18}	a

EXAMPLE A

Some of the more commonly used units which use the prefixes in Table B-2, along with their meanings, are shown below.

Unit	Symbol	Meaning	Unit	Symbol	Meaning
megohm	$M\Omega$	10^6 ohms	milligram	mg	10^{-3} gram
kilometer	km	10^3 meters	microfarad	μF	10^{-6} farad
centimeter	cm	10^{-2} meter	nanosecond	ns	10^{-9} second

■ (Mega is shortened to meg when used with "ohm.")

When designating units of area or volume, where square or cubic units are used, we use exponents in the designation. For example, we use m^2 rather than sq m, or in.^{3} rather than cu in.

When we are working with *numbers that represent units of measurement (referred to as* **denominate numbers**), it is sometimes necessary to change from one set of units to another. *A change within a given system is called a* **reduction,** and *a change from one system to another is called a* **conversion.** Table B-3 gives some basic reduction and conversion factors. It should be noted that some calculators are programmed to do conversions.

reduction
conversion

TABLE B-3 Reduction and Conversion Factors

1 in. = 2.54 cm (exact)	1 ft^3 = 28.32 L
1 km = 0.6214 mi	1 L = 1.057 qt
1 lb = 453.6 g	1 Btu = 778.0 ft·lb
1 kg = 2.205 lb	1 hp = 550 ft·lb/s (exact)
1 lb = 4.448 N	1 hp = 746.0 W

The advantages of the metric system should be evident. Reductions within the system are made by use of powers of 10, which amounts to moving the decimal point. Reductions in the U.S. customary system have numerous different multiples that must be used. Thus, comparisons and changes within the metric system are much simpler than in the U.S. customary system.

To change a given number of one set of units into another set of units, *we perform algebraic operations with units in the same manner as we do with any algebraic symbol.* Consider the following example.

EXAMPLE B

If we had a number representing feet per second to be multiplied by another number representing seconds per minute, as far as the units are concerned, we have

$$\frac{ft}{s} \times \frac{s}{min} = \frac{ft \times s}{s \times min} = \frac{ft}{min}$$

■ This means that the final result would be in feet per minute.

In changing a number of one set of units to another set of units, we use reduction and conversion factors and the principle illustrated in Example B. The convenient way to use the values in the tables is in the form of fractions. Since the given values are equal to each other, their quotient is 1. For example, since 1 in. = 2.54 cm,

NOTE ▷

$$\frac{1 \text{ in.}}{2.54 \text{ cm}} = 1 \quad \text{or} \quad \frac{2.54 \text{ cm}}{1 \text{ in.}} = 1$$

since each represents the division of a certain length by itself. Multiplying a quantity by 1 does not change its value. The following examples illustrate reduction and conversion of units.

EXAMPLE C

Reduce 20 kg to milligrams.

$$20 \text{ kg} = 20 \text{ kg}\left(\frac{10^3 \text{ g}}{1 \text{ kg}}\right)\left(\frac{10^3 \text{ mg}}{1 \text{ g}}\right)$$
$$= 20 \times 10^6 \text{ mg} = 2.0 \times 10^7 \text{ mg}$$

We note that this result is found essentially by moving the decimal point three places when changing from kilograms to grams, and another three places when
■ changing from grams to milligrams.

EXAMPLE D

Change 30 mi/h to feet per second.

$$30 \frac{mi}{h} = \left(30 \frac{mi}{h}\right)\left(\frac{5280 \text{ ft}}{1 \text{ mi}}\right)\left(\frac{1 \text{ h}}{60 \text{ min}}\right)\left(\frac{1 \text{ min}}{60 \text{ s}}\right) = \frac{(30)(5280) \text{ ft}}{(60)(60) \text{ s}} = 44 \frac{ft}{s}$$

■ Note that the only units remaining after the division are those required.

EXAMPLE E ——————— Change 575 g/cm³ to kilograms per cubic meter.

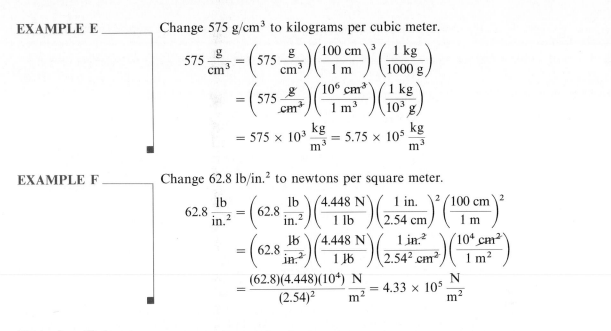

$$575 \, \frac{g}{cm^3} = \left(575 \, \frac{g}{cm^3}\right)\left(\frac{100 \text{ cm}}{1 \text{ m}}\right)^3 \left(\frac{1 \text{ kg}}{1000 \text{ g}}\right)$$

$$= \left(575 \, \frac{g}{cm^3}\right)\left(\frac{10^6 \text{ cm}^3}{1 \text{ m}^3}\right)\left(\frac{1 \text{ kg}}{10^3 \text{ g}}\right)$$

$$= 575 \times 10^3 \, \frac{kg}{m^3} = 5.75 \times 10^5 \, \frac{kg}{m^3}$$

EXAMPLE F ——————— Change 62.8 lb/in.² to newtons per square meter.

$$62.8 \, \frac{lb}{in.^2} = \left(62.8 \, \frac{lb}{in.^2}\right)\left(\frac{4.448 \text{ N}}{1 \text{ lb}}\right)\left(\frac{1 \text{ in.}}{2.54 \text{ cm}}\right)^2\left(\frac{100 \text{ cm}}{1 \text{ m}}\right)^2$$

$$= \left(62.8 \, \frac{lb}{in.^2}\right)\left(\frac{4.448 \text{ N}}{1 \text{ lb}}\right)\left(\frac{1 \text{ in.}^2}{2.54^2 \text{ cm}^2}\right)\left(\frac{10^4 \text{ cm}^2}{1 \text{ m}^2}\right)$$

$$= \frac{(62.8)(4.448)(10^4)}{(2.54)^2} \, \frac{N}{m^2} = 4.33 \times 10^5 \, \frac{N}{m^2}$$

Exercises B-1

In Exercises 1 through 4, give the symbol and meaning for the given unit.

1. megahertz   **2.** kilowatt   **3.** millimeter   **4.** picosecond

In Exercises 5 through 8, give the name and meaning for the units whose symbols are given.

5. kV   **6.** GΩ   **7.** mA   **8.** pF

In Exercises 9 through 40, make the indicated reductions or conversions.

9. Reduce 1 km to centimeters.   **10.** Reduce 1 kg to milligrams.   **11.** Reduce 1 mi to inches.

12. Reduce 1 gal to pints.   **13.** Convert 5.25 in. to centimeters.   **14.** Convert 6.50 kg to pounds.

15. Convert 15.7 qt to liters.   **16.** Convert 185 km to miles.   **17.** Reduce 1 ft² to square inches.

18. Reduce 1 yd³ to cubic feet.   **19.** Reduce 250 mm² to sq. meters.   **20.** Reduce 30.8 kL to milliliters.

21. Convert 4.50 lb to grams.   **22.** Convert 0.360 in. to meters.   **23.** Convert 829 in.³ to liters.

24. Convert 0.0680 kL to cubic feet.   **25.** Convert 2.25 hp to newton centimeters per second.

26. Convert 8.75 Btu to joules.   **27.** Convert 75.0 W to horsepower.   **28.** Convert 326 mL to quarts.

29. A weather satellite orbiting the earth weighs 6500 lb. How many tons is this?

30. An airplane is flying at 37,000 ft. What is its altitude in miles?

31. A foreign car's gasoline tank holds 56 L. Convert this capacity to gallons.

32. A television set weighs 14 kg. Convert this weight to pounds.

33. The speed of sound is about 1130 ft/s. Change this speed to kilometers per hour.

34. A certain pump can pump 72 gal/min from a well. Convert this rate to liters per second.

35. The speedometer of a car is calibrated in kilometers per hour. If the speedometer of such a car reads 60, how fast in miles per hour is the car traveling?

36. The acceleration due to gravity is about 980 cm/s². Convert this to feet per squared second.

37. Fifteen grams of a medication are to be dissolved in 0.060 L of water. Express this concentration in milligrams per deciliter.

38. The earth's surface receives energy from the sun at the rate of 1.35 kW/m². Reduce this to joules per second square centimeter.

39. At sea level, atmospheric pressure is about 14.7 lb/in.². Express this pressure in pascals.

40. The density of water is about 62.4 lb/ft³. Convert this to kilograms per cubic meter.

B-2   Approximate Numbers and Significant Digits

When we perform calculations on numbers, we must consider the accuracy of these numbers, since this affects the accuracy of the results obtained. Most of the numbers involved in technical and scientific work are *approximate*, having been arrived at through some process of measurement. However, certain other numbers are *exact*, having been arrived at through some definition or counting process. We can determine whether or not a number is approximate or exact if we know how the number was determined.

EXAMPLE A   If we measure the length of a rope to be 15.3 ft, we know that the 15.3 is approximate. A more precise measuring device may cause us to determine the length as 15.28 ft. However, regardless of the method of measurement used, we shall not be able to determine this length exactly.

If a voltage shown on a voltmeter is read as 116 V, the 116 is approximate. A more precise voltmeter may show the voltage as 115.7 V. However, this voltage cannot be determined exactly. ■

EXAMPLE B   If a computer counts the names it has processed and prints this number as 837, this 837 is exact. We know the number of names was not 836 or 838. Since 837 was determined through a counting process, it is exact.

When we say that 60 s = 1 min, the 60 is exact, since this is a definition. By this definition there are exactly 60 s in 1 min. ■

When we are writing approximate numbers we often have to include some zeros so that the decimal point will be properly located. However, *except for these zeros, all other digits are considered to be* **significant digits.** When we make computations with approximate numbers, we must know the number of significant digits. The following example illustrates how we determine this.

significant digits

EXAMPLE C   All numbers in this example are assumed to be approximate.

34.7 has three significant digits.

8900 has two significant digits. We assume that the two zeros are place holders, unless we have specific knowledge to the contrary. (Note: The overbar symbol (¯) is sometimes used to show that trailing zeros are significant digits. For example, 89,000 has four significant digits.)

0.039 has two significant digits. The zeros are for proper location of the decimal point.

706.1 has four significant digits. The zero is not used for the location of the decimal point. It shows specifically the number of tens in the number.

NOTE ▷

5.90 has three significant digits. ***The zero is not necessary as a place holder*** and should not be written unless it is significant.

Other approximate numbers with the proper number of significant digits are listed below.

96,000	two	0.0709	three	1.070	four
30,900	three	6.000	four	700.00	five
4.006	four	0.0005	one	20,008	five

■

Note from the example above that *all nonzero digits are significant. Zeros, other than those used as place holders for proper positioning of the decimal point, are also significant.*

precision

accuracy

In computations involving approximate numbers, the position of the decimal point as well as the number of significant digits is important. *The* **precision** *of a number refers directly to the decimal position of the last significant digit, whereas the* **accuracy** *of a number refers to the number of significant digits in the number.* Consider the illustrations in the following example.

EXAMPLE D

Suppose that you are measuring an electric current with two ammeters. One ammeter reads 0.031 A and the second ammeter reads 0.0312 A. The second reading is more precise, in that the last significant digit is the number of ten-thousandths, and the first reading is expressed only to thousandths. The second reading is also more accurate, since it has three significant digits rather than two.

A machine part is measured to be 2.5 cm long. It is coated with a film 0.025 cm thick. The thickness of the film has been measured to a greater precision, although the two measurements have the same accuracy: two significant digits.

A segment of a newly completed highway is 9270 ft long. The concrete surface is 0.8 ft thick. Of these two numbers, 9270 is more accurate, since it contains three

■ significant digits, and 0.8 is more precise, since it is expressed to tenths.

rounding off

The last significant digit of an approximate number is known not to be completely accurate. It has usually been determined by estimation or **rounding off.** However, we do know that it is at most in error by one-half of a unit in its place value.

EXAMPLE E

When we measure the length of the rope referred to in Example A to be 15.3 ft, we are saying that the length is at least 15.25 ft and no longer than 15.35 ft. Any value between these two, rounded off to tenths, would be expressed as 15.3 ft.

In converting the fraction $\frac{2}{3}$ to the decimal form 0.667, we are saying that

■ the value is between 0.6665 and 0.6675.

The principle of rounding off a number is to write the closest approximation, with the last significant digit in a specified position, or with a specified number of significant digits. We shall now formalize the process of rounding off as follows: If we want a certain number of significant digits, we examine the digit in the next place to the right. If this digit is less than 5, we accept the digit in the last place. If the next digit is 5 or greater, we increase the digit in the last place by 1, and this resulting digit becomes the final significant digit of the approximation.

If necessary, we use zeros to replace other digits in order to locate the decimal point properly. Except when the next digit is a 5, and no other nonzero digits are discarded, we have the closest possible approximation with the desired number of significant digits.

EXAMPLE F ⎯⎯⎯⎯ 70,360 rounded off to three significant digits is 70,400.
70,430 rounded off to three significant digits is 70,400.
187.35 rounded off to four significant digits is 187.4.
■ 71,500 rounded off to two significant digits is 72,000.

With the advent of computers and calculators, another method of reducing numbers to a specified number of significant digits is used. This is the process of **truncation,** in which the digits beyond a certain place are discarded. For example, 3.17482 truncated to thousandths is 3.174. For our purposes in this text, when working with approximate numbers, we have used only rounding off.

Exercises B-2

In Exercises 1 through 8, determine whether the numbers given are exact or approximate.

1. An automobile engine has 8 cylinders.
2. The velocity of light is 186,000 mi/s.
3. The 3-stage rocket took 74.6 h to reach the moon.
4. A microprocessor chip 10 mm by 15 mm is priced at $5.50.
5. A cube of gold 1 cm on an edge weighs 19.3 g.
6. The 21 students had an average test grade of 81.6.
7. A 150-acre piece of land is divided into 76 lots.
8. A calculator has 50 keys and its battery lasted for 50 h of use.

In Exercises 9 through 16, determine the number of significant digits in the given approximate numbers.

9. 37.2; 6844	**10.** 3600; 730	**11.** 107; 3004	**12.** 0.8735; 0.0075
13. 6.80; 6.08	**14.** 90,050; 105,040	**15.** 30,000; 30,000.0	**16.** 1.00; 0.01

In Exercises 17 through 24, determine which of each pair of approximate numbers is (a) more precise and (b) more accurate.

17. 3.764; 2.81	**18.** 0.041; 7.673	**19.** 30.8; 0.01	**20.** 70,370; 50,400
21. 0.1; 78.0	**22.** 7040; 37.1	**23.** 7000; 0.004	**24.** 50.060; 8.914

In Exercises 25 through 36, round off each of the given approximate numbers (a) to three significant digits and (b) to two significant digits.

25. 4.933	**26.** 80.53	**27.** 50,893	**28.** 31,490
29. 861.29	**30.** 9555	**31.** 0.30505	**32.** 0.7350
33. 0.9499	**34.** 81.920	**35.** 30.96	**36.** 9.999

In Exercises 37 through 40, solve the given problems.

37. A surveyor measured the road frontage of a parcel of land and obtained a distance of 128.3 ft. Based on this result, find the lowest possible value and the highest possible value.

38. An automobile manufacturer claims that the gasoline tank on a certain car holds approximately 82 L. What are the very least and the very greatest capacities?

39. A chemist has a container of 164.0 mL of sulfuric acid. Convert this volume to quarts and use the same degree of accuracy.

40. A design specification calls for a washer that is 0.0625 in. thick. Convert this thickness to millimeters and use the same degree of accuracy.

B-3 Arithmetic Operations with Approximate Numbers ▬▬▬

NOTE ▷ When performing arithmetic operations on approximate numbers, *we must be careful not to express the result to a precision or accuracy which is not warranted.* The following two examples illustrate how a false indication of the accuracy of a result could be obtained when using approximate numbers.

EXAMPLE A _____ A pipe is made in two sections. The first is measured to be 16.3 ft long and the second is measured to be 0.927 ft long. A plumber wants to know what the total length will be when the two sections are put together.

At first, it appears we might simply add the numbers as follows to obtain the necessary result.

$$\begin{array}{r} 16.3 \ \text{ft} \\ 0.927 \ \text{ft} \\ \hline 17.227 \ \text{ft} \end{array}$$

However, the first length is precise only to tenths, and the digit in this position was obtained by rounding off. It might have been as small as 16.25 ft or as large as 16.35 ft. If we consider only the precision of this first number, the total length might be as small as 17.177 ft or as large as 17.277 ft. These two values agree when rounded off to two significant digits (17). They vary by 0.1 when rounded off to tenths (17.2 and 17.3). When rounded to hundredths, they do not agree at all, since the third significant digit is different (17.18 and 17.28). Therefore, there is no agreement at all in the digits after the third when these two numbers are rounded off to a precision beyond tenths. This may also be deemed reasonable, since the first length is not expressed beyond tenths. The second number does not further change the precision of the result, since it is expressed to thousandths. Therefore we may conclude that the result must be rounded off at least to tenths, ■ the precision of the first number.

EXAMPLE B

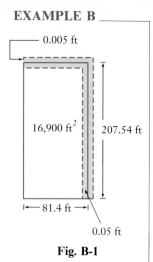

Fig. B-1

We can find the area of a rectangular piece of land by multiplying the length, 207.54 ft, by the width, 81.4 ft. Performing the multiplication, we find the area to be (207.54 ft)(81.4 ft) = 16,893.756 ft². See Fig. B-1.

However, we know this length and width were found by measurement and that the least each could be is 207.535 ft and 81.35 ft. Multiplying these values, we find the least value for the area to be

$$(207.535 \text{ ft})(81.35 \text{ ft}) = 16,882.97225 \text{ ft}^2$$

The greatest possible value for the area is

$$(207.545 \text{ ft})(81.45 \text{ ft}) = 16,904.54025 \text{ ft}^2$$

We now note that the least possible and greatest possible values of the area agree when rounded off to three significant digits (16,900 ft²) and there is no agreement in digits beyond this if the two values are rounded off to a greater accuracy. Therefore we can conclude that the accuracy of the result is good to three significant digits, or certainly no more than four. We also note that the width was accurate to three significant digits, and the length to five significant digits.

The following rules are based on reasoning similar to that in Examples A and B; we go by these rules when we perform the basic arithmetic operations on approximate numbers.

operations with approximate numbers

1. When approximate numbers are added or subtracted, the result is expressed with the precision of the least precise number.

2. When approximate numbers are multiplied or divided, the result is expressed with the accuracy of the least accurate number.

3. When the root of an approximate number is found, the result is accurate to the accuracy of the number.

If a calculator is not being used, there is another procedure which can be helpful. Before and during the calculation all numbers except the least precise or least accurate may be rounded off to one place beyond that of the least precise or least accurate number.

EXAMPLE C

Add the approximate numbers 73.2, 8.0627, 93.57, 66.296.

The least precise of these numbers is 73.2. Therefore, before performing the addition we may round off the other numbers to hundredths. If we were using a calculator, this would not be done. In either case, after the addition we round off the result to tenths. This leads to

Without calculator	With calculator
73.2	73.2
8.06	8.0627
93.57	93.57
66.30	66.296
241.13	241.1287 ⟵ result must be rounded to tenths

Therefore, the final result is 241.1.

EXAMPLE D

If we multiply 2.4832 by 30.5, we obtain 75.7376. However, since 30.5 has only three significant digits, we express the product and result as $(2.4832)(30.5) = 75.7$.

To find the square root of 3.7, we may express the result only to two significant digits. Thus, $\sqrt{3.7} = 1.9$.

NOTE ▷   *If an exact number is included in a calculation, there is no limitation to the number of decimal positions it may take on. The accuracy of the result is limited only by the approximate numbers involved.*

EXAMPLE E

Considering the exact number 144 and the approximate number 2.7, we express the result to tenths if these numbers are added or subtracted. If these numbers are to be multiplied or divided, we express the result to two significant digits:

precision of 2.7

$$144 + 2.7 = 146.7 \qquad 144 - 2.7 = 141.3$$
$$144 \times 2.7 = 390 \qquad 144 \div 2.7 = 53$$

accuracy of 2.7

When using the trigonometric functions, the angle is often an approximate number. Angles of $2.3°$, $92.3°$, and $182.3°$, for example, are considered to be angles with equal accuracy. Therefore, we see that the accuracy of an angle does not depend on the number of digits in the angle. The measurement of an angle and the accuracy of its trigonometric functions are shown in Table B-4.

TABLE B-4   Angles and Accuracy of Trigonometric Functions

Measurement of Angle to Nearest	Accuracy of the Trigonometric Function
1°	2 significant digits
0.1° or 10′	3 significant digits
0.01° or 1′	4 significant digits

EXAMPLE F

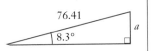

Fig. B-2

In solving for side a in the triangle shown in Fig. B-2, we have

$$a = 76.41 \sin 8.3°$$

The calculator sequence for this solution is

76.41 $\boxed{\times}$ 8.3 $\boxed{\text{SIN}}$ $\boxed{=}$ $\boxed{\textit{11.030257}}$

In rounding off the result, we see that the hypotenuse has four significant digits and that the angle is expressed to tenths of a degree, which is equivalent to three significant digits. Thus we round off the result to three significant digits, or $a = 11.0$.

The rules stated in this section are usually sufficiently valid for the computations encountered in technical work. They are intended only as good, practical rules for working with approximate numbers. It was recognized in Examples A and B that the last significant digit obtained by these rules is subject to some possible error. Therefore it is possible that the most accurate result is not obtained by their use, although this is not often the case.

When it is necessary to round off answers for the examples and exercises in the text, nearly all are expressed with the accuracy of the given values. There are, however, some applied examples and exercises in which more accuracy is given in the result than is shown in the given values. This is done when the process of solution is of primary importance and might be impeded by additional digits.

Exercises B-3

In Exercises 1 through 4, add the given approximate numbers.

1. 3.8
    0.154
   47.26

2. 26
    5.806
  147.29

3. 0.36294
  0.086
  0.5056
  0.74

4. 56.1
   3.0645
 127.38
   0.055

In Exercises 5 through 8, subtract the given approximate numbers.

5. 468.14
  36.7

6. 1.03964
  0.69

7. 57.348
  26.5

8. 8.93
 6.8947

In Exercises 9 through 12, multiply the given approximate numbers.

9. $(3.64)(17.06)$

10. $(0.025)(70.1)$

11. $(704.6)(0.38)$

12. $(0.003040)(6079.52)$

In Exercises 13 through 16, divide the given approximate numbers.

13. $608 \div 3.9$

14. $0.4962 \div 827$

15. $\dfrac{596{,}000}{22}$

16. $\dfrac{53.267}{0.3002}$

In Exercises 17 through 20, find the indicated square roots of the given approximate numbers.

17. $\sqrt{32}$

18. $\sqrt{6.5}$

19. $\sqrt{19.3}$

20. $\sqrt{0.0694}$

In Exercises 21 through 28, evaluate the given expression. All numbers are approximate.

21. $3.862 + 14.7 - 8.3276$

22. $(3.2)(0.386) + 6.842$

23. $\dfrac{8.60}{0.46} + (0.9623)(3.86)$

24. $9.6 - 0.1962(7.30)$

25. $\tan 246.3° + \tan 5.8°$

26. $87.447 \cos 2.09°$

27. $\dfrac{0.0300}{\sin 46°}$

28. $18.182 - 2.3846 \tan 1246.4°$

In Exercises 29 through 32, perform the indicated operations. The first number given is approximate and the second number is exact.

29. $3.62 + 14$

30. $17.382 - 2.5$

31. $(0.3142)(60)$

32. $8.62 \div 1728$

In Exercises 33 through 44, the solution to some of the problems will require the use of reduction and conversion factors.

33. Three adjacent lots have road frontages measured to be 75.4 ft, 39.66 ft, and 81 ft, respectively. What is the total length of the frontages of these three lots?

34. A force of 16.7 lb and another force of 10.258 lb are acting in the same direction on an object. Find the sum of these two forces.

35. Two planes are reported to have flown at speeds of 938 mi/h and 1400 km/h, respectively. Which plane is faster, and by how many miles per hour?

36. The power (in watts) developed in an electric circuit is found by multiplying the current (in amperes) by the voltage. In a certain circuit the current is 0.0125 A and the voltage is 12.68 V. What is the power that is developed?

37. If 1 K of computer memory consists of exactly 1024 bytes, how many bytes are there in a 256 K computer? (The number 256 is also exact.)

38. A certain ore is 5.3% iron. How many tons of ore must be refined to obtain 45,000 lb of iron?

39. A square tabletop has an area of 7026 cm². Find the length of one side.

40. In order to find the velocity (in feet per second) of an object which has fallen a certain height, we calculate the square root of the product of 64.4 (an approximate number) and the height in feet. What is the velocity of an object which has fallen 63 m?

41. Find the angle between the longer side and the diagonal of a rectangular carpet with sides 12.5 ft and 9.5 ft.

42. A jet cruising at 1250 mi/h climbs at a constant angle of 6.5°. What is its gain in altitude in exactly 3 min?

43. A flash of lightning strikes an object 3.25 mi distant from a person. The thunder is heard 15 s later. The person then calculates the speed of sound and reports it to be 1144 ft/s. What is wrong with the conclusion?

44. A student reports the current in a certain experiment to be 0.02 A at one time and later notes that it is 0.023 A. The student then states that the change in current is 0.003 A. What is wrong with this conclusion?

C Review of Geometry

C-1 Basic Geometric Figures and Definitions

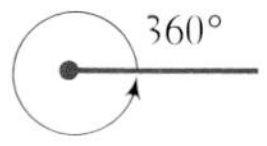

Fig. C-1

Fig. C-2

Fig. C-3

Fig. C-4

Fig. C-5

Fig. C-6

In this appendix we shall present geometric terminology and formulas that are related to the basic geometric figures. It is intended only as a brief summary of basic geometry.

Geometry deals with the properties and measurement of angles, lines, surfaces, and volumes, and the basic figures that are formed. We shall restrict our attention in this appendix to the basic figures and concepts which are related to these figures.

Repeating the definition in Section 3-1, an **angle** is generated by rotating a half-line about its endpoint from an initial position to a terminal position. One complete rotation of a line about a point is defined to be an angle of 360 **degrees** (Fig. C-1), written as 360°. A **straight angle** contains 180° (Fig. C-2), and a **right angle** contains 90° (Fig. C-3) and is shown by the symbol ∟. If two lines meet so that the angle between them is 90°, the lines are said to be **perpendicular.**

An angle less than 90° is an **acute angle** (Fig. C-4). An angle greater than 90° but less than 180° is an **obtuse angle** (Fig. C-5). **Supplementary angles** are two angles whose sum is 180°, and **complementary angles** are two angles whose sum equals 90°.

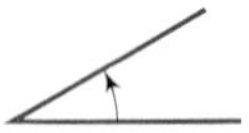

In a plane, if a line crosses two **parallel** or nonparallel lines, it is called a **transversal.** If a transversal crosses a pair of parallel lines, certain pairs of equal angles result. In Fig. C-6, the **corresponding angles** are equal. (That is, $\angle 1 = \angle 5$, $\angle 2 = \angle 6$, $\angle 3 = \angle 7$, $\angle 4 = \angle 8$.) Also, the **alternate interior angles** are equal ($\angle 3 = \angle 6$ and $\angle 4 = \angle 5$).

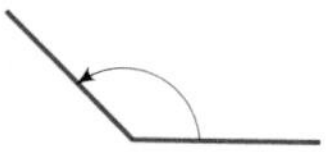

Adjacent angles have a common vertex and a side common to them. For example, in Fig. C-6, ∠s 5 and 6 are adjacent angles. **Vertical angles** are equal angles formed "across" the point of intersection of two intersecting lines. In Fig. C-6, ∠s 5 and 8 are vertical angles.

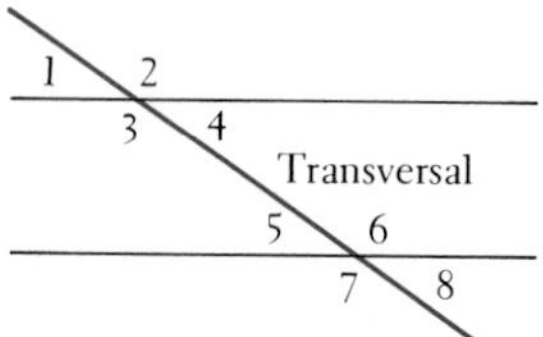

When a part of the plane is bounded and closed by straight line segments, it is called a **polygon.** In general, polygons are named according to the number of sides they contain. A **triangle** has three sides (Fig. C-7), a **quadrilateral** has four sides (Fig. C-8), a **pentagon** has five sides (Fig. C-9), and so on. In a **regular polygon,** all the sides are of equal length (Fig. C-10), and all the interior angles are equal. A **diagonal** of a polygon is a straight line segment joining any two nonadjacent vertices (the dashed lines in Fig. C-9).

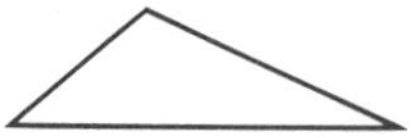

Fig. C-7

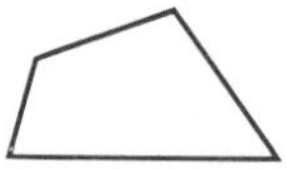

Fig. C-8

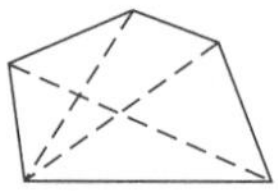

Fig. C-9

Fig. C-10

There are several important types of triangles. In an **equilateral triangle,** the three sides are equal and the three angles are also equal, each being 60° (Fig. C-11). In an **isosceles triangle,** two of the sides are equal, as are the two **base angles** (the angles opposite the equal sides) (Fig. C-12). In a **scalene triangle,** no two sides are equal, and none of the angles is a right angle (Fig. C-13). In a **right triangle,** one of the angles is a right angle (Fig. C-14). The side opposite the right angle is called the **hypotenuse.**

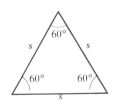 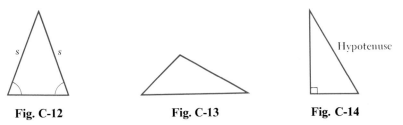

| Fig. C-11 | Fig. C-12 | Fig. C-13 | Fig. C-14 |

There are certain basic properties of triangles which we shall mention here. One very important property is that

the sum of the three angles of any triangle is 180°

Another property is that the three **medians** (line segments drawn from a vertex to the **midpoint** of the opposite side) meet at a single point (Fig. C-15). This point of intersection is called the **centroid** of the triangle. Also, the three **angle bisectors** meet at a common point (Fig. C-16), as do the three **altitudes** (heights), which are drawn from a vertex perpendicular to the opposite side (or the extension of the opposite side) (Fig. C-17).

Of particular importance is the **Pythagorean theorem,** which states that

in a right triangle, the square of the length of the hypotenuse equals the sum of the squares of the lengths of the other two sides

Two triangles are said to be **congruent** if corresponding angles are equal and if corresponding sides are equal. Also,

*two **similar** triangles have the properties that (1) the corresponding angles are equal, and (2) the corresponding sides are proportional*

See Fig. C-18.

$$\triangle ABC \sim \triangle DEF \qquad (\sim \text{ means ``similar to''})$$
$$\angle A = \angle D, \ \angle B = \angle E, \ \angle C = \angle F$$
$$\frac{a}{d} = \frac{b}{e} = \frac{c}{f}$$

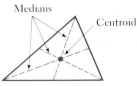

Medians

Centroid

Fig. C-15

Angle bisectors

Fig. C-16

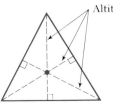

Altitudes

Fig. C-17

Fig. C-18

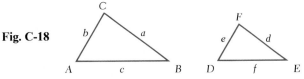

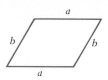

Fig. C-19

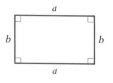

Fig. C-20

There are also several important types of quadrilaterals. A **parallelogram** is a quadrilateral with opposite sides **parallel** (extensions of the sides will not intersect) (Fig. C-19). Also, opposite sides and opposite angles of a parallelogram are equal. A **rectangle** is a parallelogram with intersecting sides perpendicular, which means that all four interior angles are right angles (Fig. C-20). It also means that opposite sides of a rectangle are equal and parallel. A **square** is a rectangle, all sides of which are equal (Fig. C-21). A **trapezoid** is a quadrilateral with two parallel sides (Fig. C-22). These parallel sides are called the **bases** of the trapezoid. A **rhombus** is a parallelogram, all four sides of which are equal (Fig. C-23). One important property of all quadrilaterals is that the sum of the four angles is 360°.

All the points on a **circle** are the same distance from a fixed point in the plane. This point is the **center** of the circle. The distance from the center to a point on the circle is the **radius** of the circle. The distance between two points on the circle and on a line passing through the center of the circle is the **diameter** of the circle. Thus the diameter is twice the radius. See Fig. C-24.

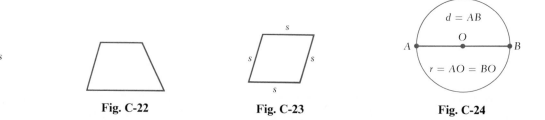

Fig. C-21   **Fig. C-22**   **Fig. C-23**   **Fig. C-24**

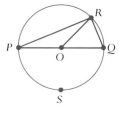

Fig. C-25

Also associated with the circle is the **chord,** which is a line segment having its endpoints on the circle. A **tangent** is a line that touches a circle (does not pass through) at one point. A **secant** is a line that passes through two points of a circle. In Fig. C-25, AB is a chord, TV is a tangent line, and SU is a secant line.

An **arc** is a part of a circle. When two radii form an angle at the center, the angle is called a **central angle.** An **inscribed angle** of an arc is one for which the endpoints of the arc are points on the sides of the angle, and for which the vertex is a point of the arc, although not an endpoint. In Fig. C-26, RQ is an arc and $\angle RPQ$ is an inscribed angle equal to one-half RQ. Angle ROQ is a central angle.

There are two important properties of a circle which we shall mention here.

1. *A tangent to a circle is perpendicular to the radius drawn to the point of contact.*

2. *An inscribed angle is one-half of its intercepted arc. Thus, an angle inscribed in a semicircle is a right angle.*

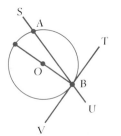

Fig. C-26

There are two important measures associated with the plane figures we have discussed to this point. They are its **perimeter** and its **area.** The perimeter of a plane figure is the distance around it. Formulas for the areas of the important plane figures are found in the following section of this appendix.

In space, the intersection of two planes is a straight line (Fig. C-27). A **polyhedron** is a solid figure which is bounded by planes. The plane surfaces of the polyhedron are called **faces** (Fig. C-28). A **prism** is a polyhedron whose bases are parallel and equal polygons and whose side faces are parallelograms (Fig. C-29).

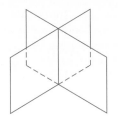

Fig. C-27

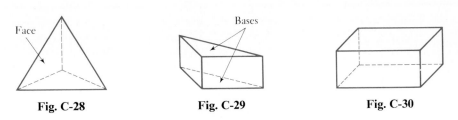

Face

Bases

Fig. C-28   **Fig. C-29**   **Fig. C-30**

A prism of particular importance is the **rectangular solid** (Fig. C-30). The faces of a rectangular solid are all rectangles. The base of a **pyramid** is a polygon (Fig. C-31). The other faces, the **lateral faces,** meet at a common point, the **vertex.**

A **right circular cylinder** is generated by revolving a rectangle about one of its sides (Fig. C-32). A **right circular cone** is generated by revolving a right triangle about one of its legs (Fig. C-33). A **sphere** is generated by revolving a circle about a diameter (Fig. C-34).

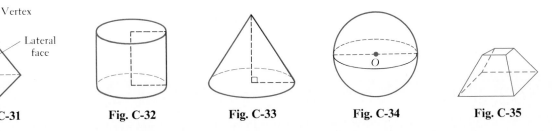

Vertex

Lateral face

Base

Fig. C-31   **Fig. C-32**   **Fig. C-33**   **Fig. C-34**   **Fig. C-35**

Fig. C-36

The **frustum** of a pyramid or cone is the solid figure that remains after the top is cut off by a plane parallel to the base (Fig. C-35 and Fig. C-36).

The **lateral area** of a solid is the area other than that of the bases. The **total area** is the lateral area plus the area of the bases. Formulas for the important areas and those for the **volumes** of solid figures are found in the following section of this appendix.

C-2   Basic Geometric Formulas

For the indicated figures, the following symbols are used: A = area, B = area of base, c = circumference, S = lateral area, V = volume.

    1. Triangle. $A = \frac{1}{2}bh$ (Fig. C-37)

    2. Pythagorean theorem. $c^2 = a^2 + b^2$ (Fig. C-38)

    3. Parallelogram. $A = bh$ (Fig. C-39)

    4. Trapezoid. $A = \frac{1}{2}(a + b)h$ (Fig. C-40)

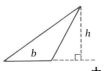

Fig. C-37

Fig. C-38

Fig. C-39

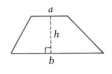

Fig. C-40

5. **Circle.** $A = \pi r^2$, $c = 2\pi r$ (Fig. C-41)

6. **Rectangular solid.** $A = 2(lw + lh + wh)$, $V = lwh$ (Fig. C-42)

7. **Cube.** $A = 6e^2$, $V = e^3$ (Fig. C-43)

8. **Any cylinder or prism with parallel bases.** $V = Bh$ (Fig. C-44)

9. **Right circular cylinder.** $S = 2\pi rh$, $V = \pi r^2 h$ (Fig. C-45)

10. **Any cone or pyramid.** $V = \frac{1}{3}Bh$ (Fig. C-46)

11. **Right circular cone.** $S = \pi rs$, $V = \frac{1}{3}\pi r^2 h$ (Fig. C-47)

12. **Sphere.** $A = 4\pi r^2$, $V = \frac{4}{3}\pi r^3$ (Fig. C-48)

As we noted, the **perimeter** of a plane figure is the distance around it. For example, the perimeter p of the triangle in Fig. C-38 is $p = a + b + c$. The perimeter of a circle is its **circumference.**

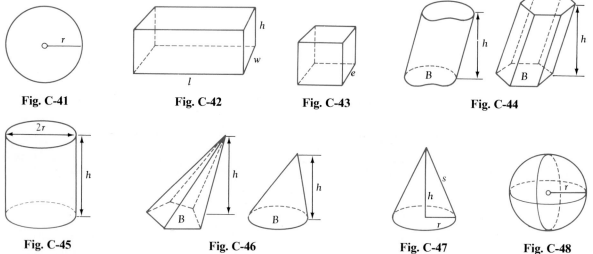

Fig. C-41   Fig. C-42   Fig. C-43   Fig. C-44

Fig. C-45   Fig. C-46   Fig. C-47   Fig. C-48

Exercises for Appendix C

In Exercises 1 through 4, refer to Fig. C-49, in which AOP and KOT are straight lines. $\angle TOP = 40°$, $\angle OBC = 90°$, and $\angle AOL = 55°$. Determine the indicated angles.

1. $\angle LOT$ **2.** $\angle POL$ **3.** $\angle KOP$ **4.** $\angle KCB$

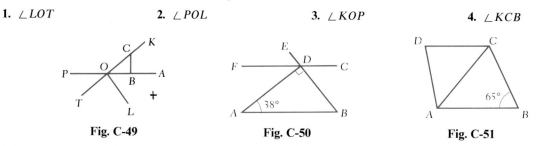

Fig. C-49   Fig. C-50   Fig. C-51

In Exercises 5 through 8, refer to Figs. C-50 and C-51. In Fig. C-50, $AB \parallel FC$, $AD \perp BE$, and $\angle A = 38°$. In Fig. C-51, $AB \parallel CD$, $AB = AC$, and $AD = CD$. (The symbol $\parallel$ means "is parallel to," and the symbol $\perp$ means "is perpendicular to.") Find the indicated angles.

5. In Fig. C-50, $\angle BDC$ **6.** In Fig. C-50, $\angle CDE$ **7.** In Fig. C-51, $\angle DCA$ **8.** In Fig. C-51, $\angle ADC$

In Exercises 9 through 12, refer to Figs. C-52 and C-53. In each, ABCD is a parallelogram. In Fig. C-52, △BEC is isosceles and ∠BCE = 40°. In Fig. C-53, ∠FED = 34°. Determine the indicated angles.

9. In Fig. C-52, ∠CEB **10.** In Fig. C-52, ∠ADC **11.** In Fig. C-53, ∠EDF **12.** In Fig. C-53, ∠DCB

In Exercises 13 through 16, refer to Fig. C-54, where AB is a diameter, line TB is tangent to the circle at B, and ∠ABC = 65°. Determine the indicated angles.

13. ∠CBT **14.** ∠BCT **15.** ∠CAB **16.** ∠BTC

In Exercises 17 through 20, refer to Fig. C-55. Determine the indicated arcs and angles.

17. $\overset{\frown}{BC}$ **18.** $\overset{\frown}{AB}$ **19.** ∠ABC **20.** ∠ACB

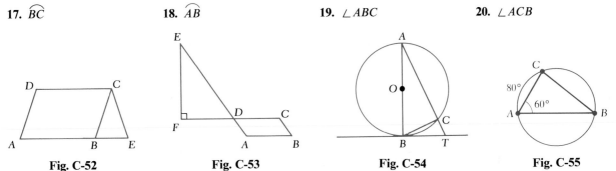

| **Fig. C-52** | **Fig. C-53** | **Fig. C-54** | **Fig. C-55** |

In Exercises 21 through 32, find the indicated sides of the right triangle shown in Fig. C-56. Where necessary, round off the results to four significant digits.

	a	*b*	*c*		*a*	*b*	*c*		*a*	*b*	*c*
21.	3	4	?	**22.**	9	12	?	**23.**	8	15	?
24.	24	10	?	**25.**	6	?	10	**26.**	2	?	4
27.	5.126	?	7.245	**28.**	3.074	?	9.452	**29.**	?	12.32	16.15
30.	?	10.49	18.72	**31.**	?	15.08	32.59	**32.**	?	5.472	36.54

In Exercises 33 through 36, use Fig. C-57. In this figure BD ∥ AE, AE = 18, and DB = 6.

33. If *DC* = 7, find *CE*. **34.** If *CA* = 27, find *CB*. **35.** If *CB* = 10, find *BA*. **36.** If *CE* = 24, find *DE*.

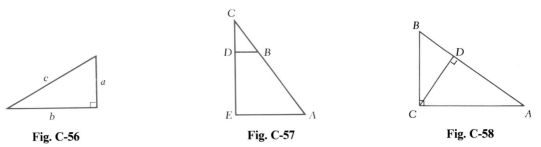

| **Fig. C-56** | **Fig. C-57** | **Fig. C-58** |

In Exercises 37 through 40, determine the required values. In Fig. C-58, AD = 9 and AC = 12.

37. Two triangles are similar, and the sides of the larger triangle are 3.0 in., 5.0 in., and 6.0 in., and the shortest side of the other triangle is 2.0 in. Find the remaining sides of the smaller triangle.

38. Two triangles are similar. The angles of the smaller triangle are 50°, 100°, and 30°, and the sides of the smaller triangle are 7.00 in., 9.00 in., and 4.57 in. The longest side of the larger triangle is 15.00 in. Find the other two sides and the three angles of the larger triangle.

39. In Fig. C-58, find *AB*. **40.** In Fig. C-58, find *BC*.

In Exercises 41 through 52, find the perimeters (or circumferences) of the indicated figures.

41. A triangle with sides 6 ft, 8 ft, and 11 ft

42. A quadrilateral with sides 3 in., 4 in., 6 in., and 9 in.

43. An isosceles triangle whose equal sides are 3 yd long and whose third side is 4 yd long

44. An equilateral triangle whose sides are 7 ft long

45. A rectangle 5.14 in. long and 4.09 in. wide

46. A parallelogram whose longer side is 13.7 mm and whose shorter side is 11.9 mm

47. A square of side 8.18 cm

48. A rhombus of side 15.62 m

49. A trapezoid whose longer base is 74 cm, whose shorter base is 46 cm, and whose other sides are equal to the shorter base

50. A trapezoid whose bases are 17.8 in. and 7.4 in. and whose other sides are each 8.1 in.

51. A circle of radius 10.0 ft

52. A circle of diameter 7.06 cm

In Exercises 53 through 60, find the areas of the indicated figures.

53. A parallelogram of base 7.0 in. and height 4.0 in.

54. A rectangle of base 8.2 ft and height 2.5 ft

55. A triangle of base 6.3 yd and height 4.1 yd

56. A triangle of base 14.25 cm and height 6.838 cm

57. A trapezoid of bases 3.0 m and 9.0 m and height 4.0 m

58. A trapezoid of bases 18.5 in. and 26.3 in. and height 10.5 in.

59. A circle of radius 7.020 in.

60. A semicircle of diameter 8.24 cm

In Exercises 61 through 72, evaluate the volume of the indicated figures.

61. A rectangular solid of length 9.00 ft, width 6.00 ft, and height 4.00 ft

62. A cube of edge 7.15 ft

63. Prism, square base of side 2.0 mm, altitude 5.0 mm

64. Prism, trapezoidal base (bases 4.0 in. and 6.0 in. and height 3.0 in.), altitude 6.0 in.

65. Cylinder, radius of base 7.00 in., altitude 6.00 in.

66. Cylinder, diameter of base 6.36 cm, altitude 18.0 cm

67. Pyramid, rectangular base 12.5 in. by 8.75 in., altitude 4.20 in.

68. Pyramid, equilateral triangular base of side 3.00 in., altitude 4.25 in.

69. Cone, radius of base 2.66 ft, altitude 1.22 yd

70. Cone, diameter of base 16.31 cm, altitude 18.42 cm

71. Sphere, radius 5.483 m

72. Sphere, diameter 15.7 yd

In Exercises 73 through 80, find the total surface area of the indicated figure for the given values.

73. Prism, rectangular base 4.00 in. by 8.00 in., altitude 6.00 in.

74. Prism, parallelogram base (base 16.0 in., height 4.25 in., perimeter 42.0 in.), altitude 6.45 in.

75. Cylinder, radius of base 8.58 m, altitude 1.38 m

76. Cylinder, diameter of base 12.55 ft, altitude 4.600 ft

77. Sphere, radius 16.06 ft

78. Sphere, diameter 15.3 in.

79. Cone, radius of base 7.05 cm, slant height 8.45 cm

80. Cone, diameter of base 76.2 mm, altitude 22.1 mm

In Exercises 81 through 84, solve the given problems.

81. A washer is in the shape of a circular disk with a circular hole in the middle. If the diameter of the disk is 3.40 cm and the diameter of the hole is 1.18 cm, find the volume of the washer if it is 0.250 cm thick. See Fig. C-59.

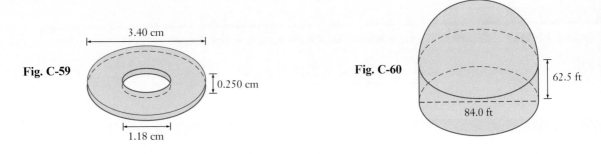

Fig. C-59

3.40 cm

0.250 cm

1.18 cm

Fig. C-60

62.5 ft

84.0 ft

82. Find the total surface area of the washer of Exercise 81.

83. A grain storage container is in the shape of a cylinder surmounted by a hemisphere (half a sphere). See Fig. C-60. If the height of the cylinder is 62.5 ft and it is 84.0 ft in diameter, find the total surface area of the container.

84. Find the volume of the grain storage container of Exercise 83.

D The Scientific Calculator

D-1   Introduction

Since the development of the first electronic calculators in the early 1970s, there has been a wide variety of different types of calculators. This appendix is intended to be a brief reference for the use of scientific calculators. More extensive discussions of specific operations with a calculator are included throughout the text where they are appropriate.

In Section 1-4 the calculator was briefly introduced. We noted there that a scientific calculator can be used for any of the calculations which may be found in this text, but that one which performs only the basic arithmetic operations is not sufficient.

There are many types and models of scientific calculators. Some models now available can display graphs and do symbolic operations. However, the discussions in this appendix are based on the operations which can be performed by use of the basic keys of a scientific calculator and are general enough to apply to most such calculators. Some of the variations and special features which may be found are noted. Nearly all calculators come with a manual that can be used to learn the operations of the calculator or to supplement this material on the variations and special features which a given calculator may have.

In this appendix we discuss the calculator keys and operations which are used for data entry, arithmetic operations, special functions (squares, square roots, reciprocals, powers, and roots), trigonometric and inverse trigonometric functions, exponential and logarithmic functions, and calculator memory. Examples are included here and in many of the examples throughout the text.

Most models of scientific calculators use many of the keys for more than one purpose, and the labeling of some keys varies from one model to another. Some of the variations which are used are noted.

When a number is entered, or a result calculated, it shows on the display at the top of the calculator. *We will use an eight-digit display,* as well as a possible display of the exponent of 10 for scientific notation, in our discussion of the calculator. Many calculators do display ten or more digits, although the operation is essentially the same. If the result contains more significant digits than the display can show, the result shown will be truncated or rounded off. We will use a rounded-off display.

In using a calculator, we must keep in mind that certain operations do not have defined results. If such an operation is attempted, the calculator will show an error display. Operations which can result in an error display include division by zero, square root of a negative number, logarithm of a negative number, and the inverse trigonometric function of a value outside of the defined interval.

As we noted in Section 1-4, the calculator logic determines the order in which entries must be made in order to obtain the desired result. We also noted that *in this text we assume that the calculator uses algebraic logic.* Some discussion of this is found in Section 1-4, but *the user must become acquainted with the logic of the particular calculator being used.*

calculator logic

D-2 Calculator Data Entry and Function Keys ▬▬▬

Following is a listing of certain basic keys which may be found on a scientific calculator, and which can be of use in performing calculations for the exercises in this text. There are other keys and other calculational capabilities which are also found on many calculators. Along with each key is a description and an illustration of its basic use. In the examples shown at the right, to perform the indicated entry or calculation, use the given sequence of entries and operations. The final display is also shown.

It should be emphasized that not all the keys which will probably be found on a given calculator are discussed, but that a scientific calculator will have most of the keys which are described. It should be noted that the use of many of the keys may be beyond the scope of the reader until the appropriate text material has been covered.

Keys	*Examples*
$\boxed{0}$, $\boxed{1}$, ..., $\boxed{9}$ **Digit Keys** These keys are used to enter the digits 0 through 9 in the display, or to enter an exponent of 10 when scientific notation is used.	To enter: 37514 Sequence: $\boxed{3}\boxed{7}\boxed{5}\boxed{1}\boxed{4}$ Display: ⎡ *37514.* ⎤
$\boxed{.}$ **Decimal Point Key** This key is used to enter a decimal point.	To enter: 375.14 Sequence: $\boxed{3}\boxed{7}\boxed{5}\boxed{.}\boxed{1}\boxed{4}$ Display: ⎡ *375.14* ⎤
$\boxed{+/-}$ **Change Sign Key** This key is used to change the sign of the number on the display, or to change the sign of the exponent when scientific notation is used. (May be designated as $\boxed{\text{CHS}}$.	To enter: -375.14 Sequence: $\boxed{3}\boxed{7}\boxed{5}\boxed{.}\boxed{1}\boxed{4}\boxed{+/-}$ Display: ⎡ -375.14 ⎤
	From here on, the entry of a number will be shown as one operation and the display will be shown at the end of the key sequence.

Keys	*Examples*

$\boxed{\pi}$ **Pi Key**
This key is used to enter π to the number of digits of the display. On some calculators a key will have a secondary purpose of providing the value of π. See the second function key.

To enter: π
Sequence: $\boxed{\pi}$ $\boxed{3.1415927}$

$\boxed{\text{EE}}$ **Enter Exponent Key**
This key is used to enter an exponent when scientific notation is used. After the key is pressed, the exponent is entered. For a negative exponent, the $\boxed{+/-}$ key is pressed after the exponent is entered. (May be designated as $\boxed{\text{EEX}}$ or $\boxed{\text{EXP}}$.)

To enter: 2.936×10^8
Sequence: 2.936 $\boxed{\text{EE}}$ 8 $\boxed{2.936 \ \ 08}$

To enter: -2.936×10^{-8}
Sequence: 2.936 $\boxed{+/-}$ $\boxed{\text{EE}}$ 8 $\boxed{+/-}$ $\boxed{-2.936 \ -08}$

$\boxed{=}$ **Equals Key**
This key is used to complete a calculation to give the required result.

See the following examples for illustrations of the use of this key.

$\boxed{+}$ **Add Key**
This key is used to add the next entry to the previous entry or result.

Evaluate: $37.56 + 241.9$
Sequence: 37.56 $\boxed{+}$ 241.9 $\boxed{=}$ $\boxed{279.46}$

$\boxed{-}$ **Subtract Key**
This key is used to subtract the next entry from the previous entry or result.

Evaluate: $37.56 - 241.9$
Sequence: 37.56 $\boxed{-}$ 241.9 $\boxed{=}$ $\boxed{-204.34}$

$\boxed{\times}$ **Multiply Key**
This key is used to multiply the previous entry or result by the next entry.

Evaluate: 8.75×30.92
Sequence: 8.75 $\boxed{\times}$ 30.92 $\boxed{=}$ $\boxed{270.55}$

$\boxed{\div}$ **Divide Key**
This key is used to divide the previous entry or result by the next entry.

Evaluate: $8.75 \div 30.92$
Sequence: 8.75 $\boxed{\div}$ 30.92 $\boxed{=}$ $\boxed{0.2829884}$
(truncated or rounded off)

$\boxed{(}\boxed{)}$ **Parentheses Keys**
These keys are used to group specified numbers in a calculation.

Evaluate: $3.586(30.72 + 47.92)$
Sequence: 3.586 $\boxed{\times}$ $\boxed{(}$ 30.72 $\boxed{+}$ 47.92 $\boxed{)}$ $\boxed{=}$
$\boxed{282.00304}$

$\boxed{\text{CE}}$ **Clear Entry Key**
This key is used to clear the last entry. Its use will not affect any other part of a calculation. On some calculators one press of the $\boxed{\text{C/CE}}$ or $\boxed{\text{CL}}$ key is used for this purpose.

Evaluate: $37.56 + 241.9$, with an improper entry of 242.9
Sequence: 37.56 $\boxed{+}$ 242.9 $\boxed{\text{CE}}$ 241.9 $\boxed{=}$ $\boxed{279.46}$

$\boxed{\text{C}}$ **Clear Key**
This key is used to clear the display and information being calculated (not including memory) so that a new calculation may be started. For calculators with a $\boxed{\text{C/CE}}$ or $\boxed{\text{C}}$ key, and no $\boxed{\text{CE}}$ key, a second press on these keys is used for this purpose.

To clear previous calculation:
Sequence: $\boxed{\text{C}}$ $\boxed{0.}$

Keys	*Examples*
2ND **Second Function Key** This key is used on calculators on which many of the keys serve dual purposes. It is pressed before the second key functions are activated. The INV key may also serve as a second function key.	Evaluate: $(37.4)^2$ on a calculator where x^2 is a second use of a key Sequence: 37.4 2ND x^2 ☐ 1398.76
x^2 **Square Key** This key is used to square the number on the display.	Evaluate: $(37.4)^2$ Sequence: 37.4 x^2 ☐ 1398.76
$\sqrt{x}$ **Square Root Key** This key is used to find the square root of the number on the display.	Evaluate: $\sqrt{37.4}$ Sequence: 37.4 $\sqrt{x}$ ☐ 6.1155539
1/x **Reciprocal Key** This key is used to find the reciprocal of the number on the display.	Evaluate: $\dfrac{1}{37.4}$ Sequence: 37.4 1/x ☐ 0.026738
x^y **x to the y Power Key** This key is used to raise x, the first entry, to the y power, the second entry. (This key may be labeled as y^x.)	Evaluate: $(3.73)^{1.5}$ Sequence: 3.73 x^y 1.5 = ☐ 7.2038266
DRG **Degree-Radian-Grad Key** This key is used to designate a displayed angle as being measured in degrees, radians, or grads. (100 grads = 90°)	This key should show the appropriate angle measurement before the calculation is started. (On some calculators this key is also used to convert from one angle measure to another.)
DMS **Degrees-Minutes-Seconds Key** This key is used to convert an angle in degrees, minutes, and seconds to an angle in decimal degrees. By use of the INV key the conversion is done in reverse.	Convert $52°8'17''$ to decimal degrees. Sequence: 52.0817 DMS ☐ 52.138056 Convert $52.138056°$ to degrees, minutes, and seconds. Sequence: 52.138056 INV DMS ☐ 52.0817 ← $52°8'17''$
☐ **Sine Key** This key is used to find the sine of the angle on the display.	Evaluate: sin 37.4° Sequence: DRG 37.4 SIN ☐ 0.6073758
COS **Cosine Key** This key is used to find the cosine of the angle of the display.	Evaluate: cos 2.475 (rad) Sequence: DRG 2.475 COS ☐ −0.785933
TAN **Tangent Key** This key is used to find the tangent of the angle on the display.	Evaluate: tan $(-24.9°)$ Sequence: DRG 24.9 +/− TAN ☐ −0.4641845

Keys	*Examples*

Cotangent, Secant, Cosecant
These functions are found through their reciprocal relation with the tangent, cosine, and sine functions, respectively.

Evaluate: cot 2.841 (rad)
Sequence: [D®G] 2.841 [TAN] [1/x] [−3.2259549]

[INV] **Inverse Function Key**
On some models this key is used to find the inverse of the primary function on a key. On other models it is the second function key and many of the paired functions are not inverses. For inverse trigonometric functions, some models have ARCSIN or equivalent keys.

Evaluate: Arcsin 0.1758 (the angle whose sine is 0.1758) in degrees.
Sequence: [D RG] 0.1758 [INV] [SIN] [10.125217]

.

[LOG] **Common Logarithm Key**
This key is used to find the logarithm to the base 10 of the number on the display.

Evaluate: log 37.45
Sequence: 37.45 [LOG] [1.5734518]

[LN] **Natural Logarithm Key**
This key is used to find the logarithm to the base e of the number on the display.

Evaluate: ln 0.8421
Sequence: 0.8421 [LN] [−0.1718565]

[10ˣ] **10 to the x Power Key**
This key is used to find antilogarithms (base 10) of the number on the display. (The [xʸ] key can be used for this purpose, if the calculator does not have this key.)

Evaluate: Antilog 0.7265 (or $10^{0.7265}$)
Sequence: 0.7265 [10ˣ] [5.3272122]

[eˣ] **e to the x Power Key**
This key is used to raise e to the power on the display.

Evaluate: $e^{-4.05}$
Sequence: 4.05 [+/−] [eˣ] [0.0174224]

[!] **Factorial Key**
This key is used to calculate the value of $n!$.

Evaluate: 40!
Sequence: 40 [!] [8.1591528 47]

[STO] **Store in Memory Key**
This key is used to store the displayed number in the memory.

Store in memory: 56.02
Sequence: 56.02 [STO]

[RCL] **Recall from Memory Key**
This key is used to recall the number in the memory to the display. (May be designated as [MR].)

Recall from memory: 56.02
Sequence: [RCL] [56.02]

[M] **Other Memory Keys**
Some calculators use an [M] key to store a number in the memory. It also may add the entry to the number in the memory. On such calculators [CM] (Clear Memory) is used to clear the memory. There are also keys for other operations on the number in the memory.

D-3   Combined Operations

There are many types of problems which require a number of calculator steps and keys for their solutions. Throughout the book there are examples of how such calculations are done on a calculator. Therefore, we will not present any examples here.

It can be seen from the examples in the text that a calculator is especially valuable for certain types of problems. These include problems in which special functions such as square roots, exponentials, or logarithms are involved. The calculator is very useful for problems which are primarily arithmetic, especially when there are several significant digits involved. In solving triangles, the calculator is particularly valuable. Also, there are some examples and exercises in which the calculator plays an important role in developing a concept. However, it is still necessary to learn the mathematics required in order to set up and understand a given problem.

The exercises which follow provide an opportunity for additional practice in the use of a calculator.

Exercises for Appendix D

In Exercises 1 through 36, perform all calculations on a scientific calculator.

1. $47.08 + 8.94$
2. $654.1 + 407.7$
3. $4724 - 561.9$
4. $0.9365 - 8.077$
5. 0.0396×471
6. 26.31×0.9393
7. $76.7 \div 194$
8. $52,060 \div 75.09$
9. 3.76^2
10. 0.986^2
11. $\sqrt{0.2757}$
12. $\sqrt{60.36}$
13. $\dfrac{1}{0.0749}$
14. $\dfrac{1}{607.9}$
15. $(19.66)^{2.3}$
16. $(8.455)^{1.75}$
17. $\sin 47.3°$
18. $\sin 1.15$
19. $\cos 3.85$
20. $\cos 119.1°$
21. $\tan 306.8°$
22. $\tan 0.537$
23. $\sec 6.11$
24. $\csc 242.0°$
25. Arcsin 0.6607 (in degrees)
26. Arccos (-0.8311) (in radians)
27. Arctan (-2.441) (in radians)
28. Arcsin 0.0737 (in degrees)
29. $\log 3.857$
30. $\log 0.9012$
31. $\ln 808$
32. $\ln 70.5$
33. $10^{0.545}$
34. $10^{-0.0915}$
35. $e^{-5.17}$
36. $e^{1.672}$

In Exercises 37 through 92, perform all calculations on a scientific calculator. In these exercises some of the combined operations which are encountered in certain types of problems are given. For specific types of applications, solve problems from the appropriate sections of the text.

37. $(4.38 + 9.07) \div 6.55$
38. $(382 + 964) \div 844$
39. $4.38 + (9.07 \div 6.55)$
40. $382 + (964 \div 844)$
41. $\dfrac{5.73 \times 10^{11}}{20.61 - 7.88}$
42. $\dfrac{7.09 \times 10^{23}}{284 + 839}$
43. $50.38\pi^2$
44. $\dfrac{5\pi}{14.6}$
45. $\sqrt{1.65^2 + 6.44^2}$
46. $\sqrt{0.735^2 + 0.409^2}$
47. $3(3.5)^4 - 4(3.5)^2$
48. $\dfrac{3(-1.86)}{(-1.86)^2 + 1}$
49. $29.4 \cos 72.5°$
50. $\dfrac{477}{\sin 58.7°}$
51. $\dfrac{4 + \sqrt{(-4)^2 - 4(3)(-9)}}{2(3)}$
52. $\dfrac{-5 - \sqrt{5^2 - 4(4)(-7)}}{2(4)}$
53. $\dfrac{0.176(180)}{\pi}$
54. $\dfrac{209.6\pi}{180}$

55. $\dfrac{1}{2}\left(\dfrac{51.4\pi}{180}\right)(7.06)^2$

56. $\dfrac{1}{2}\left(\dfrac{148.2\pi}{180}\right)(49.13)^2$

57. Arcsin $\dfrac{27.3 \sin 36.5°}{46.8}$

58. $\dfrac{0.684 \sin 76.1°}{\sin 39.5°}$

59. $\sqrt{3924^2 + 1762^2 - 2(3924)(1762)\cos 106.2°}$

60. Arccos $\dfrac{8.09^2 + 4.91^2 - 9.81^2}{2(8.09)(4.91)}$

61. $\sqrt{5.81 \times 10^8} + \sqrt[3]{7.06 \times 10^{11}}$

62. $(6.074 \times 10^{-7})^{2/5} - (1.447 \times 10^{-5})^{4/9}$

63. $\dfrac{3}{2\sqrt{7} - \sqrt{6}}$

64. $\dfrac{7\sqrt{5}}{4\sqrt{5} - \sqrt{11}}$

65. Arctan $\dfrac{7.37}{5.06}$

66. Arctan $\dfrac{46.3}{-25.5}$

67. $2 + \dfrac{\log 12}{\log 7}$

68. $\dfrac{10^{0.4115}}{\pi}$

69. $\dfrac{26}{2}(-1.450 + 2.075)$

70. $\dfrac{4.55(1 - 1.08^{15})}{1 - 1.08}$

71. $\sin^2\left(\dfrac{\pi}{7}\right) + \cos^2\left(\dfrac{\pi}{7}\right)$

72. $\sec^2\left(\dfrac{2}{9}\pi\right) - \tan^2\left(\dfrac{2}{9}\pi\right)$

73. $\sin 31.6° \cos 58.4° + \sin 58.4° \cos 31.6°$

74. $\cos^2 296.7° - \sin^2 296.7°$

75. $\sqrt{(1.54 - 5.06)^2 + (-4.36 - 8.05)^2}$

76. $\sqrt{(7.03 - 2.94)^2 + (3.51 - 6.44)^2}$

77. $\dfrac{(4.001)^2 - 16}{4.001 - 4}$

78. $\dfrac{(2.001)^2 + 3(2.001) - 10}{2.001 - 2}$

79. $\dfrac{4\pi}{3}(8.01^3 - 8.00^3)$

80. $4\pi(76.3^2 - 76.0^2)$

81. $0.01\left(\dfrac{1}{2}\sqrt{2} + \sqrt{2.01} + \sqrt{2.02} + \dfrac{1}{2}\sqrt{2.03}\right)$

82. $0.2\left[\dfrac{1}{2}(3.5)^2 + 3.7^2 + 3.9^2 + \dfrac{1}{2}(4.1)^2\right]$

83. $\dfrac{e^{0.45} - e^{-0.45}}{e^{0.45} + e^{-0.45}}$

84. $\ln \sin 2e^{-0.055}$

85. $\ln \dfrac{2 - \sqrt{2}}{2 - \sqrt{3}}$

86. $\sqrt{\dfrac{9}{2} + \dfrac{9 \sin 0.2\pi}{8\pi}}$

87. $2 + \dfrac{0.3}{4} - \dfrac{(0.3)^2}{64} + \dfrac{(0.3)^3}{512}$

88. $\dfrac{1}{2} + \dfrac{\pi\sqrt{3}}{360} - \dfrac{1}{4}\left(\dfrac{\pi}{180}\right)^2$

89. $160(1 - e^{-1.50})$

90. $e^{-3.60}(\cos 1.20 + 2 \sin 1.20)$

91. $\dfrac{(10)(9)(8)(7)(6)}{5!}$

92. $\dfrac{20! - 15!}{20! + 15!}$

E Selected Computer Programs in BASIC

E-1 Introduction

In this appendix there are 25 selected computer programs. They are written in the programming language BASIC. This language is chosen because of its simplicity and widespread availability on computer systems.

It is not the intent here to teach how to write a computer program in BASIC. The programs are intended for those who have some experience with computer programs or who have resources available to learn the fundamentals of programming in BASIC. However, only a knowledge of the fundamentals is necessary to use and learn from these programs.

Most of the programs in this appendix are reasonably simple, and when you have some experience you will see how they can be modified and expanded to provide additional information. Also, you will see how many additional programs can be developed. These programs illustrate how a few of the calculations and problems in the text can be solved by use of a computer. The number of programs which could be used for the problems in this text is essentially limitless.

special statements There are many special statements, such as INPUT and PRINT, which are used in BASIC. They provide the computer with the instructions which it must follow. You should have a knowledge of the fundamental BASIC statements before attempting to use these programs. Among those which are necessary are REM, INPUT, LET, PRINT, GOTO, FOR (with NEXT), IF (with THEN), and END. Also, the system commands of NEW, LOAD, LIST, SAVE, and RUN are necessary.

symbols of operation As well as knowing the fundamental statements, it is necessary to know the symbols for the arithmetic operations and the various mathematical functions. The symbols $+$, $-$, $*$, and $/$ are used for addition, subtraction, multiplication, and division, respectively. For exponentiation, $\wedge$, $\uparrow$, or $**$ may be used, depending on the particular computer. A list of the fundamental mathematical functions is shown in Table E-1.

It should be noted that the trigonometric functions use angles expressed in radians and not degrees. Also, many computers include only the Arctangent and do not have the Arcsine or Arccosine. If a problem involves the use of functions or values which the computer does not have, it may be necessary to define them in the particular program.

There are many variations of BASIC. However, except for the three graphics programs (Chapters 2-B, 9, and 20), the programs given in this appendix should run on most computers. Versions of the graphics programs are given for the Apple II, IIe, and IIc and for the IBM PC, PS/2, and compatibles.

TABLE E-1

BASIC Expression	Function	Comment
SIN(X)	sin x	x in radians
COS(X)	cos x	x in radians
TAN(X)	tan x	x in radians
ATN(X)	Arctan x	Result in radians
LOG(X)	ln x	Natural logarithm
EXP(X)	e^x	Exponential function
SQR(X)	$\sqrt{x}$	Square root
ABS(X)	$|x|$	Absolute value
SGN(X)		Result: $+1$ if $x > 0$ -1 if $x < 0$ 0 if $x = 0$

Following are the BASIC programs. There are 25 different programs from 22 different chapters throughout the text. They are identified by the chapter in which the program is appropriate. The program REM statement identifies the type of problem which is involved.

E-2   Programs

In the programs for Chapters 2-B, 9 and 20, lines are generally usable except for those marked with an *a* or an *m* before the line number. For those lines, *a* and *m* have the following meanings:

   a—Applesoft BASIC; use with Apple II, IIe, IIc.

   m—Microsoft BASICA; use with IBM PC, PS/2, and compatibles.

Chapter 2-A

```
10 REM   FIND THE VALUE OF
20 REM   A FUNCTION AT X=A
30 REM
40 REM   ENTER FUNCTION
50 REM   IN LINE 100
60 REM
100 DEF FN F(X) = 2*X^2 - 3*X
110 PRINT "ENTER VALUE OF A"
120 PRINT
130 INPUT "A = ";A
140 PRINT
150 PRINT "F(A) = "; FN F(A)
160 END
```

Chapter 2-B

```
10 REM   PLOT Y = F(X)
20 REM
30 REM   ENTER FUNCTION
40 REM   IN LINE 150
50 REM
60 REM   DOMAIN -10<=X<=10
70 REM   RANGE -20<=Y<=20
80 REM
a100 HGR : HCOLOR = 3
m100 screen 1 : cls
a110 HPLOT 40,80 TO 240,80
m110 line (40,80)-(240,80),1
a120 HPLOT 140,1 TO 140,160
m120 line (140,1)-(140,160),1
130 FOR N = -100 TO 100
140 LET X = .1 * N
150 LET Y = X * X + 5 * X
160 LET B = 4 * Y
170 IF B < -80 OR B > 80
      THEN 190
a180 HPLOT N + 140, 80 - B
m180 line (n+140,80-b) -
      (n+140,80-b),1
190 NEXT N
200 END
```

Chapter 3

```
10 REM   SOLUTION OF RIGHT
20 REM   TRIANGLE GIVEN
30 REM   SIDES A AND B
40 REM
100 PRINT "ENTER SIDES A AND B"
110 PRINT
120 INPUT "SIDE A = ";SA
130 INPUT "SIDE B = ";SB
140 PRINT
150 LET A = ATN (SA/SB)
160 LET SC = SA/SIN(A)
170 LET A = A*180/3.14159265
180 LET B = 90 - A
190 PRINT "ANGLE A = ";A
200 PRINT "ANGLE B = ";B
210 PRINT "SIDE C = "; SC
220 END
```

Chapter 4

```
10 REM   SOLVE A 2X2 SYSTEM
20 REM   OF LINEAR EQUATIONS
30 REM   BY DETERMINANTS
40 REM
100 PRINT "ENTER COEFFICIENTS"
110 PRINT
120 INPUT "A1 = ";A1: INPUT "B1 = ";B1:
    INPUT "C1 = ";C1
130 PRINT
140 INPUT "A2 = ";A2: INPUT "B2 = ";B2:
    INPUT "C2 = ";C2
150 PRINT
160 LET D = A1*B2 - A2*B1
170 IF D=0 THEN 210
180 PRINT "X = ";(C1*B2 - C2*B1)/D
190 PRINT "Y = ";(A1*C2 - A2*C1)/D
200 GOTO 220
210 PRINT "NO UNIQUE SOLUTION"
220 END
```

Chapter 6

```
10 REM   SOLUTION OF A QUADRATIC
20 REM   EQUATION BY USE OF THE
30 REM   QUADRATIC FORMULA
40 REM
100 PRINT "ENTER COEFFICIENTS A,B,C"
110 PRINT
120 INPUT "A = ";A: INPUT "B = ";B:
    INPUT "C = ";C
130 PRINT
140 LET D = B*B - 4*A*C
150 IF D<0 THEN 190
160 PRINT "X1 = ";(-B+SQR(D))/(2*A)
170 PRINT "X2 = ";(-B-SQR(D))/(2*A)
180 GOTO 200
190 PRINT "ROOTS ARE IMAGINARY"
200 END
```

Chapter 7

```
10 REM   CONVERT ANGLE FROM
20 REM   DEGREES TO RADIANS OR
30 REM   RADIANS TO DEGREES
40 REM
100 PRINT "TO CONVERT DEGREES TO RADIANS"
105 PRINT "ENTER THE NUMBER 1":PRINT
110 PRINT "TO CONVERT RADIANS TO DEGREES"
115 PRINT "ENTER A NUMBER OTHER THAN 1"
120 INPUT A:PRINT
130 IF A<>1 THEN 190
140 INPUT "ENTER THE NUMBER OF DEGREES: ";D
150 PRINT
160 LET R = D*3.14159265 /180
170 PRINT D;" DEGREES = ";R;"RADIANS"
180 GOTO 230
190 INPUT "ENTER THE NUMBER OF RADIANS: ";R
200 PRINT
210 LET D = 180*R/3.14159265#
220 PRINT R;" RADIANS = ";D;" DEGREES"
230 END
```

Chapter 8-A

```
10 REM   ADDITION OF TWO VECTORS
20 REM
30 REM   USE ANGLES IN
40 REM   STANDARD POSITION
50 REM
100 PRINT "ENTER MAGNTIUDE AND"
105 PRINT "ANGLE OF VECTOR A"
110 PRINT "AND THEN OF VECTOR B"
120 PRINT
130 INPUT "A = ";A: INPUT "TA = ";TA:
    INPUT "B = ";B: INPUT "TB = ";TB
140 PRINT
150 LET PI = 3.141592654
160 LET TA = TA*PI/180
170 LET TB = TB*PI/180
180 LET RX = A*COS(TA) + B*COS(TB)
190 LET RY = A*SIN(TA) + B*SIN(TB)
195 IF RX = 0 THEN 270
200 LET TR = 180/PI*ATN(RY/RX)
210 LET R =  SQR(RX*RX + RY*RY)
220 PRINT "RESULTANT MAGNITUDE = ";R
230 IF RX>0 AND RY>0 THEN PRINT
    "RESULTANT ANGLE = ";TR
240 IF RX<0 THEN PRINT
    "RESULTANT ANGLE = ";TR+180
250 IF RX>0 AND RY<0 THEN PRINT
    "RESULTANT ANGLE = ";TR+360
260 GOTO 300
270 PRINT "RESULTANT MAGNITUDE = ";
    ABS(RY)
280 IF RY<0 THEN PRINT
    "RESULTANT ANGLE = 270"
290 IF RY>0 THEN PRINT
    "RESULTANT ANGLE = 90"
300 END
```

Chapter 8-B

```
10 REM   SOLUTION OF A TRIANGLE
20 REM   GIVEN THREE SIDES
30 REM
100 PRINT "ENTER THE THREE SIDES"
110 PRINT
120 INPUT "SA = ";SA: INPUT "SB = ";SB:
    INPUT "SC = ";SC
130 PRINT
140 LET PI = 3.14159265
150 DEF FN ACS(X) = ATN(SQR(1-X*X)/X) -
    (SGN(X) - 1)*PI/2
160 LET A = FN ACS((SB*SB + SC*SC -
    SA*SA)/(2*SB*SC))*180/PI
170 LET B = FN ACS((SA*SA + SC*SC -
    SB*SB)/(2*SA*SC))*180/PI
180 LET C = 180 - A - B
190 PRINT
200 PRINT "ANGLE A = ";A
210 PRINT "ANGLE B = ";B
220 PRINT "ANGLE C = ";C
230 END
```

Chapter 9

```
10 REM    PLOT Y = A SIN BX
20 REM      AND Y = SIN X
30 REM
40 REM    VALUE OF A: -3<=A<=3
50 REM    VALUE OF B: -5<=B<=5
60 REM
70 REM    DOMAIN: 0<=X<=4PI
80 REM
100 PRINT "ENTER A AND B"
a110 HGR: HCOLOR = 3
m100 screen 1: cls
120 INPUT "A = ";A:INPUT "B = ";B
130 PRINT
a140 HPLOT 1,80 TO 239,80
m140 line (1,80)-(239,80),1
a150 HPLOT 0,0 TO 0,160
m150 line (0,0)-(0,160),1
160 FOR X = 1 TO 238
a170 HCOLOR = 1
a180 HPLOT X,80 - 26 * SIN (X /
     19)
m180 p= 80 - 26 * sin(x/19):
     line (x,p)-(x,p),1
a190 HCOLOR = 5
a200 HPLOT X, 90 - 26 * A * SIN
     B * X / 19)
m200 p = 80 - 26 * a * sin (b * x
     / 19) : line (x,p)-(x,p),1
210 NEXT X
220 END
```

Chapter 11

```
10 REM   COMPLEX NUMBERS
20 REM   RECTANGULAR FORM
30 REM   TO POLAR FORM
40 REM
100 PRINT "ENTER X AND Y"
110 PRINT
120 INPUT "X = ";X: INPUT "Y = ";Y
130 PRINT
140 LET R = SQR(X*X + Y*Y)
150 IF X = 0 THEN 200
160 LET T = ATN(Y/X)*180/3.14159265
170 IF X>0 AND Y>=0 THEN PRINT R;"/";T
180 IF X<0 THEN PRINT R;"/";T+180
190 IF X>0 AND Y<0 THEN PRINT R;"/";T+360
200 IF X=0 AND Y>0 THEN PRINT R;"/ 90"
210 IF X=0 AND Y<0 THEN PRINT R;"/ 270"
220 END
```

Chapter 12

```
10 REM PROPERTIES OF LOGARITHMS
20 REM
100 PRINT "ENTER TWO POSITIVE NUMBERS"
110 PRINT
120 INPUT "A = ";A: INPUT "B = ";B
130 PRINT
140 PRINT "LOG ";A;" + LOG ";B;" = LOG ";A*B
150 PRINT
160 PRINT "LOG ";A;" - LOG ";B;" = LOG ";A/B
170 PRINT
180 PRINT A;" LOG ";B;" = LOG "; B^A
190 END
```

Chapter 14

```
10 REM   SYNTHETIC DIVISION
20 REM   FOR A POLYNOMIAL
30 REM   OF DEGREE 4 OR LESS
40 REM   OF FORM
50 REM   Y = AX^4+BX^3+CX^2+DX+E
60 REM
100 PRINT "ENTER EACH OF A,B,C,D,E"
110 PRINT "EVEN IF IT IS ZERO"
115 PRINT
120 INPUT "A = ";A:INPUT "B = ";B:
    INPUT "C = ";C:INPUT "D = ";D:
    INPUT "E = ";E
130 PRINT
140 PRINT "ENTER THE VALUE OF THE ROOT"
150 INPUT "R = ";R
160 PRINT
170 LET B1 = B + R*A
180 LET C1 = C + R*B1
190 LET D1 = D + R*C1
200 LET E1 = E + R*D1
210 PRINT "THE BOTTOM ROW "
215 PRINT "AFTER DIVISION IS"
220 PRINT A,B1,C1,D1,E1
230 END
```

Chapter 15

```
10 REM   INVERSE OF A 2X2 MATRIX
20 REM
100 PRINT "ENTER THE ELEMENTS"
110 PRINT
120 INPUT "A11 = ";A: INPUT "A12 = ";B:
    INPUT "A21 = ";C: INPUT "A22 = ";D
130 PRINT
140 PRINT "MATRIX"; TAB(10); A;
    TAB(22); B:PRINT
150 PRINT TAB(10); C; TAB(22); D:PRINT
160 LET E = A*D - B*C
170 IF E=0 THEN 220
180 PRINT "INVERSE"; TAB(10); D/E;
    TAB(22); -B/E
190 PRINT
200 PRINT TAB(10); -C/E; TAB(22); A/E
210 GOTO 230
220 PRINT "NO INVERSE"
230 END
```

Chapter 16

```
10 REM   DETERMINE WHETHER
20 REM   F(X) = AX^3+BX^2+CX+D
30 REM   IS POSITIVE, NEGATIVE, OR
40 REM   ZERO AT X=XO
50 REM
100 PRINT "ENTER EACH OF A,B,C,D"
110 PRINT "EVEN IF IT IS ZERO"
120 PRINT
130 INPUT "A = ";A: INPUT "B = ";B:
    INPUT "C = ";C: INPUT "D = ";D
140 PRINT
150 PRINT "ENTER VALUE OF XO"
155 PRINT "YOU WISH TO TRY"
158 PRINT
160 PRINT "ENTER 999 TO END PROGRAM"
170 PRINT
180 INPUT "XO = ";X
190 IF X=999 THEN 290
200 LET Y = A*X^3 + B*X^2 + C*X + D
210 IF Y>=0 THEN 240
220 PRINT "F(X) IS NEGATIVE FOR X = ";X
230 GOTO 140
240 IF ABS(Y)<.0001 THEN 270
250 PRINT "F(X) IS POSITIVE FOR X = ";X
260 GOTO 140
270 PRINT "F(X) IS ZERO FOR X = ";X
280 GOTO 140
290 END
```

Chapter 18

```
10 REM   BINOMIAL EXPANSION OF
20 REM   FIRST K COEFFICIENTS OF
30 REM   (AX + B)^N
40 REM
100 PRINT "ENTER K,A,B,N
110 PRINT
```

```
120 INPUT "K = ";K: INPUT "A = ";A:
    INPUT "B = ";B: INPUT "N = ";N
130 PRINT
140 LET T = A^N
150 FOR I=1 TO K
160 PRINT "C";I;" = ";T
170 LET T = T*(N-I+1)*B/(A*I)
180 NEXT I
190 END
```

Chapter 19

```
10 REM   CHECK TRIG IDENTITY
20 REM   FOR X=A
30 REM
40 REM   ENTER IDENTITY
50 REM   LEFT SIDE IN LINE 150
60 REM   RIGHT SIDE IN LINE 160
65 REM
70 REM   USE RECIPROCAL FUNCTIONS
80 REM   FOR COT X, SEC X, CSC X
90 REM
100 PRINT "ENTER VALUE OF A"
110 PRINT "IN DEGREES"
120 PRINT
130 INPUT "A = ";X
140 LET X = X*3.14159265 /180
150 LET LS = SIN(X)*TAN(X)+COS(X)
160 LET RS = 1/COS(X)
170 PRINT "VALUE OF LEFT SIDE = ";LS
180 PRINT "VALUE OF RIGHT SIDE = ";RS
190 END
```

Chapter 20

```
10 REM   PLOT HYPERBOLA
20 REM   X^2/A^2 - Y^2/B^2 = 1
30 REM
40 REM   DOMAIN: -10<=X<=10
50 REM   RANGE:  -10<=Y<=10
60 REM
70 REM   VALUES OF A: 1<=A<=8
80 REM   VALUES OF B: 1<=B<=8
90 REM
100 PRINT "ENTER A AND B"
110 PRINT
120 INPUT "A = ";A:INPUT "B = ";B
a130 HGR : HCOLOR = 3
m130 screen 1: cls
a140 HPLOT 40,80 TO 240,80
m140 line (40,80)-(240,80),1
a150 HPLOT 140,1 TO 140,160
m150 line (140,1)-(140,160),1
160 FOR N = -100 TO 100
170 LET X = .1 * N
180 LET Y1 = 8 * B * X / A: LET
    Y2 = - 8 * B * X / A
190 IF Y1 < - 80 OR Y1 > 80 THEN
    210
a200 HCOLOR = 1: HPLOT N + 140,
    80 - Y1: HPLOT N + 140, 80
    - Y2
```

(Program continued on next page)

```
m200 line (n + 140,80 - y1) -
     (n + 140,80 - y1),1
m205 line (n + 140,80 - y2) -
     (n + 140,80 - y2),1
210 NEXT N
220 FOR N = -100 TO 100
230 LET X = .1 * N
240 IF X ^ 2 < A ^ 2 THEN 290
250 LET Y3 = 8 * B * SQR (X ^ 2
    - A ^ 2) / A
260 LET Y4 = - 8 * B * SQR (X ^
    2 - A ^ 2) / A
270 IF Y3 < - 80 OR Y3 > 80 THEN
    290
a280 HCOLOR = 5: HPLOT N + 140,
     80 - Y3: hPLOT N + 140, 80
     - Y4
m280 line (n + 140,80 - y3) -
     (n + 140,80 - y3),1
m285 line (n + 140,80 - y4) -
     (n + 140,80 - y4),1
290 NEXT N
300 END
```

Chapter 21-A

```
10 REM   ARITHMETIC MEAN AND
20 REM   STANDARD DEVIATION
30 REM   OF N NUMBERS
40 REM
50 REM   N AND NUMBERS
60 REM   TO BE ENTERED
70 REM
100 PRINT "ENTER N"
110 LET SQ = 0: LET SUM = 0
120 PRINT
130 INPUT "N = ";N: PRINT
140 PRINT "ENTER ";N;" NUMBERS"
145 PRINT
150 FOR C = 1 TO N
160 INPUT "    ";D
170 LET SUM = SUM + D
180 LET SQ = SQ + D*D
190 NEXT C
200 PRINT
210 LET M = SUM/N: PRINT
220 LET X = SQ/N
230 LET S = SQR(X-M*M)
240 PRINT "ARITHMETIC MEAN = ";M
250 PRINT
260 PRINT "STANDARD DEVIATION = ";S
270 END
```

Chapter 21-B

```
10 REM   FIND EQUATION OF
20 REM   LEAST-SQUARES LINE
30 REM   THRU SEVERAL POINTS
40 REM
50 REM   NUMBER OF POINTS
60 REM   AND COORDINATES
70 REM   TO BE ENTERED
80 REM
```

```
100 PRINT "ENTER NUMBER OF POINTS"
105 PRINT
110 INPUT "N = ";N
120 LET S1 = 0: LET S2 = 0
130 FOR I = 1 TO N
135 PRINT
140 PRINT "ENTER POINT NUMBER";I
150 PRINT "ENTER X-COORDINATE"
155 INPUT X
160 PRINT "ENTER Y-COORDINATE"
165 INPUT Y
170 LET S1 = S1 + X
180 LET S2 = S2 + Y
190 LET S3 = S3 + X*X
200 LET S4 = S4 + X*Y
210 NEXT I
220 LET A1 = S1/N
230 LET A2 = S2/N
240 LET A3 = S3/N
250 LET A4 = S4/N
260 LET D2 = A3 - A1*A1
270 LET M = (A4 - A1*A2)/D2
280 LET B = A2 - M*A1: PRINT
290 PRINT "LEAST SQUARES LINE IS"
295 PRINT
300 PRINT "Y = ";M;"X + ";B
310 END
```

Chapter 22

```
10 REM   LIMIT OF FUNCTION
20 REM
30 REM   LIM    N(X)
40 REM          ----
50 REM   X->A   D(X)
60 REM
70 REM   ENTER N(X) AND D(X)
80 REM   IN LINES 140, 150
90 REM
100 PRINT
110 PRINT "ENTER A"
120 PRINT
130 INPUT "A = ";A: PRINT
140 DEF FN N(X) = X*X - 9
150 DEF FN D(X) = X + 3
160 PRINT "X", "FUNCTION"
170 PRINT
180 FOR I=A-.005 TO A-.0009 STEP .001
190 PRINT I, FN N(I)/FN D(I)
200 NEXT I
210 IF ABS(FN D(A))<.0001 THEN 240
220 PRINT A, FN N(A)/FN D(A)
230 GOTO 250
240 PRINT A, "UNDEFINED"
250 FOR I=A+.001 TO A+.006 STEP .001
260 PRINT I, FN N(I)/FN D(I)
270 NEXT I
280 PRINT
290 PRINT "LIMIT = ";(FN N(A-.001)/
    FN D(A-.001)+FN N(A+.001)/
    FN D(A+.001))/2
300 END
```

Chapter 23

```
10 REM    NEWTON'S METHOD
15 REM
20 REM    SOLVE EQUATION
30 REM    F(X) = 0
40 REM
50 REM    ENTER FUNCTION
55 REM    IN LINE 130
60 REM
70 REM    ENTER DERIVATIVE
75 REM    IN LINE 140
80 REM
100 PRINT "ENTER X1":PRINT
110 INPUT "X1 = ";X
120 FOR A = 1 TO 5
130 LET F = 3*X^2 - 5*X - 2
140 LET D = 6*X - 5
150 LET R = X - F/D
160 PRINT"X";A+1;" = ";R
170 LET X = R
180 NEXT A
190 END
```

Chapter 24

```
10 REM    AREA UNDER A CURVE
20 REM    FROM X=A TO X=B
30 REM    BY SUMMING THE AREAS
40 REM    OF N RECTANGLES
50 REM
60 REM    ENTER FUNCTION
70 REM    IN LINE 100
80 REM
100 DEF FN F(X) = X*X + 1
110 PRINT "ENTER A,B,N":PRINT
120 INPUT "A = ";A: INPUT "B = ";B:
    INPUT "N = ";N: PRINT
130 LET R = 0
140 LET DX = (B-A)/N
150 FOR I = 0 TO N-1
160 LET X = A + I*DX
170 LET R = R + FN F(X)*DX
180 NEXT I
190 PRINT "AREA = ";R
200 END
```

Chapter 25

```
10 REM    VOLUME GENERATED BY
20 REM    ROTATING AREA UNDER
30 REM    Y = C*X^N
40 REM    FROM X=A TO X=B
50 REM    ABOUT THE X-AXIS
60 REM
100 DEF FN I(X) = C^2*X^(2*N+1)/(2*N+1)
110 LET PI = 3.14159265
120 PRINT "ENTER A, B, C, N"
130 PRINT
140 INPUT "A = ";A: INPUT "B = ";B:
    INPUT "C = ";C: INPUT "N = ";N
150 PRINT
160 PRINT "VOLUME = ";PI*FN I(B) -
    PI*FN I(A)
200 END
```

Chapter 26

```
10 REM    LIMITING VALUE FOR E
20 REM
100 PRINT "VALUE OF (1+X)^(1/X)"
110 PRINT "AS X->0":PRINT
120 PRINT "X","(1+X)^(1/X)"
125 PRINT
130 LET X = 1
140 FOR N = 1 TO 5
150 PRINT X, (1+X)^(1/X)
160 LET X = X/10
170 NEXT N
180 PRINT
190 PRINT "ACTUAL VALUE = "; EXP(1)
200 END
```

Chapter 28

```
10 REM    FIND THE FIRST N
20 REM    PARTIAL SUMS
30 REM    OF INFINITE SERIES
40 REM
50 REM    ENTER GENERAL TERM
60 REM    IN LINE 100
70 REM
100 DEF FN F(N) = (2*N+1)/(N*N*(N+1)^2)
110 LET S = 0: PRINT
120 PRINT "ENTER N": PRINT
130 INPUT "N = ";I: PRINT
140 PRINT "N", "TERM", "PARTIAL SUM"
145 PRINT
150 FOR N = 1 TO I
160 LET S = S + FN F(N)
170 LET T = FN F(N)
180 PRINT N,T,S
190 NEXT N
200 END
```

Exercises for Appendix E

In the following exercises, change the given computer programs as indicated. Run each program.

1. Change the function in the program for Chapter 2-A to $f(x) = \dfrac{2x}{x^2 + 1}$.

2. Change the function in the program for Chapter 2-B to $y = 8/x$. Note that $x = 0$ is not in the domain of the function. Therefore, start line 150 with IF X < > 0 THEN LET, or insert an appropriate line between lines 140 and 150.

3. Change the program for Chapter 3 to solve a right triangle given sides a and c.

4. Change the program for Chapter 4 to print out either SYSTEM IS INCONSISTENT or SYSTEM IS DEPENDENT, whichever is correct, rather than NO UNIQUE SOLUTION.

5. Change the program for Chapter 6 to print out the imaginary roots rather than ROOTS ARE IMAGINARY.

6. Indicate how you would use the program for Chapter 8-A to find the sum of three vectors.

7. In the program for Chapter 8-B, if one of the lengths entered is greater than the sum of the other two lengths, the program will not run. These lengths cannot be the sides of a triangle. Change the program to print out NO TRIANGLE HAS THESE THREE SIDES for such a case.

8. Change the program for Chapter 9 to plot the graph of $y = a \sin (bx + c)$ rather than $y = a \sin bx$.

9. Note the number of different combinations of values of x and y which require different print lines in the program for Chapter 11. Write a program of five or fewer lines (not counting REM and PRINT statements for directions and spacing) which converts a complex number in polar form to rectangular form and which prints out the rectangular form.

10. Add a line to the program for Chapter 12 which prints out log $\sqrt[q]{b}$ in the same way it prints out log b^a.

11. In the program for Chapter 16, change the function to $f(x) = \dfrac{ax + b}{cx + d}$.

12. The program for Chapter 18 is generally intended for larger values of n. However, note the results for $n = 0$, 1, and 2.

13. Change the identity in the program for Chapter 19 to $\dfrac{\tan x \sin x}{\cos x} = \sec^2 x - 1$.

14. Change the program for Chapter 20 to plot the graph of the hyperbola $\dfrac{y^2}{a^2} - \dfrac{x^2}{b^2} = 1$ rather than $\dfrac{x^2}{a^2} - \dfrac{y^2}{b^2} = 1$.

15. Change the function in the program for Chapter 22 to $\dfrac{2 - \sqrt{x + 2}}{x - 2}$.

16. Change the function in the program for Chapter 23 to $f(x) = 3x^3 - x^2 - 4x + 1$.

17. Change the function in the program for Chapter 24 to $f(x) = 4x - x^2$. Use $A = 0$, $B = 4$, $N = 40$; $A = 0$, $B = 6$, $N = 60$; $A = 4$, $B = 6$, $N = 20$. Explain the results.

18. In the program for Chapter 25, use $A = 0$, $B = 1$, $C = 1$ and then increasing values of N. Explain the results.

19. In the program for Chapter 26, increase the number of terms to 20. Note the results.

20. In the program for Chapter 28, change the function to $f(n) = 1/5^n$, and start with $n = 0$. Compare with Examples B and D of Section 28-1.

APPENDIX F **Tables**

TABLE 1 Powers and Roots

No.	Sq.	Sq. Root	Cube	Cube Root	No.	Sq.	Sq. Root	Cube	Cube Root
1	1	1.000	1	1.000	51	2 601	7.141	132 651	3.708
2	4	1.414	8	1.260	52	2 704	7.211	140 608	3.733
3	9	1.732	27	1.442	53	2 809	7.280	148 877	3.756
4	16	2.000	64	1.587	54	2 916	7.348	157 464	3.780
5	25	2.236	125	1.710	55	3 025	7.416	166 375	3.803
6	36	2.449	216	1.817	56	3 136	7.483	175 616	3.826
7	49	2.646	343	1.913	57	3 249	7.550	185 193	3.849
8	64	2.828	512	2.000	58	3 364	7.616	195 112	3.871
9	81	3.000	729	2.080	59	3 481	7.681	205 379	3.893
10	100	3.162	1 000	2.154	60	3 600	7.746	216 000	3.915
11	121	3.317	1 331	2.224	61	3 721	7.810	226 981	3.936
12	144	3.464	1 728	2.289	62	3 844	7.874	238 328	3.958
13	169	3.606	2 197	2.351	63	3 969	7.937	250 047	3.979
14	196	3.742	2 744	2.410	64	4 096	8.000	262 144	4.000
15	225	3.873	3 375	2.466	65	4 225	8.062	274 625	4.021
16	256	4.000	4 096	2.520	66	4 356	8.124	287 496	4.041
17	289	4.123	4 913	2.571	67	4 489	8.185	300 763	4.062
18	324	4.243	5 832	2.621	68	4 624	8.246	314 432	4.082
19	361	4.359	6 859	2.668	69	4 761	8.307	328 509	4.102
20	400	4.472	8 000	2.714	70	4 900	8.367	343 000	4.121
21	441	4.583	9 261	2.759	71	5 041	8.426	357 911	4.141
22	484	4.690	10 648	2.802	72	5 184	8.485	373 248	4.160
23	529	4.796	12 167	2.844	73	5 329	8.544	389 017	4.179
24	576	4.899	13 824	2.884	74	5 476	8.602	405 224	4.198
25	625	5.000	15 625	2.924	75	5 625	8.660	421 875	4.217
26	676	5.099	17 576	2.962	76	5 776	8.718	438 976	4.236
27	729	5.196	19 683	3.000	77	5 929	8.775	456 533	4.254
28	784	5.292	21 952	3.037	78	6 084	8.832	474 552	4.273
29	841	5.385	24 389	3.072	79	6 241	8.888	493 039	4.291
30	900	5.477	27 000	3.107	80	6 400	8.944	512 000	4.309
31	961	5.568	29 791	3.141	81	6 561	9.000	531 441	4.327
32	1 024	5.657	32 768	3.175	82	6 724	9.055	551 368	4.344
33	1 089	5.745	35 937	3.208	83	6 889	9.110	571 787	4.362
34	1 156	5.831	39 304	3.240	84	7 056	9.165	592 704	4.380
35	1 225	5.916	42 875	3.271	85	7 225	9.220	614 125	4.397
36	1 296	6.000	46 656	3.302	86	7 396	9.274	636 056	4.414
37	1 369	6.083	50 653	3.332	87	7 569	9.327	658 503	4.431
38	1 444	6.164	54 872	3.362	88	7 744	9.381	681 472	4.448
39	1 521	6.245	59 319	3.391	89	7 921	9.434	704 969	4.465
40	1 600	6.325	64 000	3.420	90	8 100	9.487	729 000	4.481
41	1 681	6.403	68 921	3.448	91	8 281	9.539	753 571	4.498
42	1 764	6.481	74 088	3.476	92	8 464	9.592	778 688	4.514
43	1 849	6.557	79 507	3.503	93	8 649	9.644	804 357	4.531
44	1 936	6.633	85 184	3.530	94	8 836	9.695	830 584	4.547
45	2 025	6.708	91 125	3.557	95	9 025	9.747	857 375	4.563
46	2 116	6.782	97 336	3.583	96	9 216	9.798	884 736	4.579
47	2 209	6.856	103 823	3.609	97	9 409	9.849	912 673	4.595
48	2 304	6.928	110 592	3.634	98	9 604	9.899	941 192	4.610
49	2 401	7.000	117 649	3.659	99	9 801	9.950	970 299	4.626
50	2 500	7.071	125 000	3.684	100	10 000	10.000	1 000 000	4.642

TABLE 2 Four-Place Logarithms

N	0	1	2	3	4	5	6	7	8	9
10	0000	0043	0086	0128	0170	0212	0253	0294	0334	0374
11	0414	0453	0492	0531	0569	0607	0645	0682	0719	0755
12	0792	0828	0864	0899	0934	0969	1004	1038	1072	1106
13	1139	1173	1206	1239	1271	1303	1335	1367	1399	1430
14	1461	1492	1523	1553	1584	1614	1644	1673	1703	1732
15	1761	1790	1818	1847	1875	1903	1931	1959	1987	2014
16	2041	2068	2095	2122	2148	2175	2201	2227	2253	2279
17	2304	2330	2355	2380	2405	2430	2455	2480	2504	2529
18	2553	2577	2601	2625	2648	2672	2695	2718	2742	2765
19	2788	2810	2833	2856	2878	2900	2923	2945	2967	2989
20	3010	3032	3054	3075	3096	3118	3139	3160	3181	3201
21	3222	3243	3263	3284	3304	3324	3345	3365	3385	3404
22	3424	3444	3464	3483	3502	3522	3541	3560	3579	3598
23	3617	3636	3655	3674	3692	3711	3729	3747	3766	3784
24	3802	3820	3838	3856	3874	3892	3909	3927	3945	3962
25	3979	3997	4014	4031	4048	4065	4082	4099	4116	4133
26	4150	4166	4183	4200	4216	4232	4249	4265	4281	4298
27	4314	4330	4346	4362	4378	4393	4409	4425	4440	4456
28	4472	4487	4502	4518	4533	4548	4564	4579	4594	4609
29	4624	4639	4654	4669	4683	4698	4713	4728	4742	4757
30	4771	4786	4800	4814	4829	4843	4857	4871	4886	4900
31	4914	4928	4942	4955	4969	4983	4997	5011	5024	5038
32	5051	5065	5079	5092	5105	5119	5132	5145	5159	5172
33	5185	5198	5211	5224	5237	5250	5263	5276	5289	5302
34	5315	5328	5340	5353	5366	5378	5391	5403	5416	5428
35	5441	5453	5465	5478	5490	5502	5514	5527	5539	5551
36	5563	5575	5587	5599	5611	5623	5635	5647	5658	5670
37	5682	5694	5705	5717	5729	5740	5752	5763	5775	5786
38	5798	5809	5821	5832	5843	5855	5866	5877	5888	5899
39	5911	5922	5933	5944	5955	5966	5977	5988	5999	6010
40	6021	6031	6042	6053	6064	6075	6085	6096	6107	6117
41	6128	6138	6149	6160	6170	6180	6191	6201	6212	6222
42	6232	6243	6253	6263	6274	6284	6294	6304	6314	6325
43	6335	6345	6355	6365	6375	6385	6395	6405	6415	6425
44	6435	6444	6454	6464	6474	6484	6493	6503	6513	6522
45	6532	6542	6551	6561	6571	6580	6590	6599	6609	6618
46	6628	6637	6646	6656	6665	6675	6684	6693	6702	6712
47	6721	6730	6739	6749	6758	6767	6776	6785	6794	6803
48	6812	6821	6830	6839	6848	6857	6866	6875	6884	6893
49	6902	6911	6920	6928	6937	6946	6955	6964	6972	6981
50	6990	6998	7007	7016	7024	7033	7042	7050	7059	7067
51	7076	7084	7093	7101	7110	7118	7126	7135	7143	7152
52	7160	7168	7177	7185	7193	7202	7210	7218	7226	7235
53	7243	7251	7259	7267	7275	7284	7292	7300	7308	7316
54	7324	7332	7340	7348	7356	7364	7372	7380	7388	7396

For a number

$N = P \times 10^k$

the logarithm is

$\log N = k + \log P$

k is the

characteristic

$\log P$ is the

mantissa

The mantissa is found from Table 2.

The characteristic is found from the location of the decimal point.

EXAMPLE

Find log 4870.

$4870 = 4.87 \times 10^3$

$k = 3$

Find 48 (first two significant digits) under N. To the right of 48, find 6875 under 7 (third significant digit).

$\log 4.87 = 0.6875$

$\log 4.87 \times 10^3$
$\quad = 3 + 0.6875$

$\log 4870 = 3.6875$

TABLE 2   (*Continued*)

N	0	1	2	3	4	5	6	7	8	9
55	7404	7412	7419	7427	7435	7443	7451	7459	7466	7474
56	7482	7490	7497	7505	7513	7520	7528	7536	7543	7551
57	7559	7566	7574	7582	7589	7597	7604	7612	7619	7627
58	7634	7642	7649	7657	7664	7672	7679	7686	7694	7701
59	7709	7716	7723	7731	7738	7745	7752	7760	7767	7774
60	7782	7789	7796	7803	7810	7818	7825	7832	7839	7846
61	7853	7860	7868	7875	7882	7889	7896	7903	7910	7917
62	7924	7931	7938	7945	7952	7959	7966	7973	7980	7987
63	7993	8000	8007	8014	8021	8028	8035	8041	8048	8055
64	8062	8069	8075	8082	8089	8096	8102	8109	8116	8122
65	8129	8136	8142	8149	8156	8162	8169	8176	8182	8189
66	8195	8202	8209	8215	8222	8228	8235	8241	8248	8254
67	8261	8267	8274	8280	8287	8293	8299	8306	8312	8319
68	8325	8331	8338	8344	8351	8357	8363	8370	8376	8382
69	8388	8395	8401	8407	8414	8420	8426	8432	8439	8445
70	8451	8457	8463	8470	8476	8482	8488	8494	8500	8506
71	8513	8519	8525	8531	8537	8543	8549	8555	8561	8567
72	8573	8579	8585	8591	8597	8603	8609	8615	8621	8627
73	8633	8639	8645	8651	8657	8663	8669	8675	8681	8686
74	8692	8698	8704	8710	8716	8722	8727	8733	8739	8745
75	8751	8756	8762	8768	8774	8779	8785	8791	8797	8802
76	8808	8814	8820	8825	8831	8837	8842	8848	8854	8859
77	8865	8871	8876	8882	8887	8893	8899	8904	8910	8915
78	8921	8927	8932	8938	8943	8949	8954	8960	8965	8971
79	8976	8982	8987	8993	8998	9004	9009	9015	9020	9025
80	9031	9036	9042	9047	9053	9058	9063	9069	9074	9079
81	9085	9090	9096	9101	9106	9112	9117	9122	9128	9133
82	9138	9143	9149	9154	9159	9165	9170	9175	9180	9186
83	9191	9196	9201	9206	9212	9217	9222	9227	9232	9238
84	9243	9248	9253	9258	9263	9269	9274	9279	9284	9289
85	9294	9299	9304	9309	9315	9320	9325	9330	9335	9340
86	9345	9350	9355	9360	9365	9370	9375	9380	9385	9390
87	9395	9400	9405	9410	9415	9420	9425	9430	9435	9440
88	9445	9450	9455	9460	9465	9469	9474	9479	9484	9489
89	9494	9499	9504	9509	9513	9518	9523	9528	9533	9538
90	9542	9547	9552	9557	9562	9566	9571	9576	9581	9586
91	9590	9595	9600	9605	9609	9614	9619	9624	9628	9633
92	9638	9643	9647	9652	9657	9661	9666	9671	9675	9680
93	9685	9689	9694	9699	9703	9708	9713	9717	9722	9727
94	9731	9736	9741	9745	9750	9754	9759	9763	9768	9773
95	9777	9782	9786	9791	9795	9800	9805	9809	9814	9818
96	9823	9827	9832	9836	9841	9845	9850	9854	9859	9863
97	9868	9872	9877	9881	9886	9890	9894	9899	9903	9908
98	9912	9917	9921	9926	9930	9934	9939	9943	9948	9952
99	9956	9961	9965	9969	9974	9978	9983	9987	9991	9996

EXAMPLE

Find log 0.0624.

$0.0624 = 6.24 \times 10^{-2}$

$k = -2$

$\log 6.24 = 0.7952$
(opposite 62, under 4)

$\log 6.24 \times 10^{-2}$
$= -2 + 0.7952$

$\log 0.0624$
$= 0.7952 - 2$
$= -1.2048$

log 0.0624 is not
equal to -2.7952.

EXAMPLE

**If log $N = 1.9867$,
find N.**

$k = 1$

$\log P = 0.9867$
In the table, the
number nearest to
9867 is 9868, which
is opposite 97 and
under 0.

$P = 9.70$

$N = 9.70 \times 10^1$
$= 97.0$

Interpolation (see
Section 2-5) can be
used to increase the
accuracy of N to
four significant digits.

TABLE 3 Trigonometric Functions

Angle		Sin	Cos	Tan	Cot		
Degree	Radian						
0°	0.000	0.000	1.000	0.000		1.571	90°
1°	0.017	0.017	1.000	0.017	57.29	1.553	89°
2°	0.035	0.035	0.999	0.035	28.64	1.536	88°
3°	0.052	0.052	0.999	0.052	19.08	1.518	87°
4°	0.070	0.070	0.998	0.070	14.30	1.501	86°
5°	0.087	0.087	0.996	0.087	11.43	1.484	85°
6°	0.105	0.105	0.995	0.105	9.514	1.466	84°
7°	0.122	0.122	0.993	0.123	8.144	1.449	83°
8°	0.140	0.139	0.990	0.141	7.115	1.431	82°
9°	0.157	0.156	0.988	0.158	6.314	1.414	81°
10°	0.175	0.174	0.985	0.176	5.671	1.396	80°
11°	0.192	0.191	0.982	0.194	5.145	1.379	79°
12°	0.209	0.208	0.978	0.213	4.705	1.361	78°
13°	0.227	0.225	0.974	0.231	4.332	1.344	77°
14°	0.244	0.242	0.970	0.249	4.011	1.326	76°
15°	0.262	0.259	0.966	0.268	3.732	1.309	75°
16°	0.279	0.276	0.961	0.287	3.487	1.292	74°
17°	0.297	0.292	0.956	0.306	3.271	1.274	73°
18°	0.314	0.309	0.951	0.325	3.078	1.257	72°
19°	0.332	0.326	0.946	0.344	2.904	1.239	71°
20°	0.349	0.342	0.940	0.364	2.748	1.222	70°
21°	0.367	0.358	0.934	0.384	2.605	1.204	69°
22°	0.384	0.375	0.927	0.404	2.475	1.187	68°
23°	0.401	0.391	0.921	0.424	2.356	1.169	67°
24°	0.419	0.407	0.914	0.445	2.246	1.152	66°
25°	0.436	0.423	0.906	0.466	2.145	1.134	65°
26°	0.454	0.438	0.899	0.488	2.050	1.117	64°
27°	0.471	0.454	0.891	0.510	1.963	1.100	63°
28°	0.489	0.469	0.883	0.532	1.881	1.082	62°
29°	0.506	0.485	0.875	0.554	1.804	1.065	61°
30°	0.524	0.500	0.866	0.577	1.732	1.047	60°
31°	0.541	0.515	0.857	0.601	1.664	1.030	59°
32°	0.559	0.530	0.848	0.625	1.600	1.012	58°
33°	0.576	0.545	0.839	0.649	1.540	0.995	57°
34°	0.593	0.559	0.829	0.675	1.483	0.977	56°
35°	0.611	0.574	0.819	0.700	1.428	0.960	55°
36°	0.628	0.588	0.809	0.727	1.376	0.942	54°
37°	0.646	0.602	0.799	0.754	1.327	0.925	53°
38°	0.663	0.616	0.788	0.781	1.280	0.908	52°
39°	0.681	0.629	0.777	0.810	1.235	0.890	51°
40°	0.698	0.643	0.766	0.839	1.192	0.873	50°
41°	0.716	0.656	0.755	0.869	1.150	0.855	49°
42°	0.733	0.669	0.743	0.900	1.111	0.838	48°
43°	0.750	0.682	0.731	0.933	1.072	0.820	47°
44°	0.768	0.695	0.719	0.966	1.036	0.803	46°
45°	0.785	0.707	0.707	1.000	1.000	0.785	45°
		Cos	Sin	Cot	Tan	Radian	Degree
							Angle

TABLE 4   Natural Logarithms and Exponential Functions

x	$\ln x$	e^x	e^{-x}	x	$\ln x$	e^x	e^{-x}	x	$\ln x$	e^x	e^{-x}
0.0		1.0000	1.0000	4.5	1.5041	90.017	0.0111	9.0	2.1972	8,103	0.0001
0.1	−2.3026	1.1052	0.9048	4.6	1.5261	99.484	0.0101	9.1	2.2083	8,955	
0.2	−1.6094	1.2214	0.8187	4.7	1.5476	109.95	0.0091	9.2	2.2192	9,897	
0.3	−1.2040	1.3499	0.7408	4.8	1.5686	121.51	0.0082	9.3	2.2300	10,938	
0.4	−0.9163	1.4918	0.6703	4.9	1.5892	134.29	0.0074	9.4	2.2407	12,088	
0.5	−0.6931	1.6487	0.6065	5.0	1.6094	148.41	0.0067	9.5	2.2513	13,360	0.00007
0.6	−0.5108	1.8221	0.5488	5.1	1.6292	164.02	0.0061	9.6	2.2618	14,765	
0.7	−0.3567	2.0138	0.4966	5.2	1.6487	181.27	0.0055	9.7	2.2721	16,318	
0.8	−0.2231	2.2255	0.4493	5.3	1.6677	200.34	0.0050	9.8	2.2824	18,034	
0.9	−0.1054	2.4596	0.4066	5.4	1.6864	221.41	0.0045	9.9	2.2925	19,930	
1.0	0.0000	2.7183	0.3679	5.5	1.7047	244.69	0.0041	10	2.3026	22,026	4.5×10^{-5}
1.1	0.0953	3.0042	0.3329	5.6	1.7228	270.43	0.0037	15	2.7081		
1.2	0.1823	3.3201	0.3012	5.7	1.7405	298.87	0.0033	20	2.9957	4.9×10^{8}	2.1×10^{-9}
1.3	0.2624	3.6693	0.2725	5.8	1.7579	330.30	0.0030	25	3.2189		
1.4	0.3365	4.0552	0.2466	5.9	1.7750	365.04	0.0027	30	3.4012	1.1×10^{13}	9.4×10^{-14}
1.5	0.4055	4.4817	0.2231	6.0	1.7918	403.43	0.0025	35	3.5553		
1.6	0.4700	4.9530	0.2019	6.1	1.8083	445.86		40	3.6889	2.4×10^{17}	4.2×10^{-18}
1.7	0.5306	5.4739	0.1827	6.2	1.8245	492.75		45	3.8067		
1.8	0.5878	6.0496	0.1653	6.3	1.8405	544.57		50	3.9120	5.2×10^{21}	1.9×10^{-22}
1.9	0.6419	6.6859	0.1496	6.4	1.8563	601.85		55	4.0073		
2.0	0.6931	7.3891	0.1353	6.5	1.8718	665.14	0.0015	60	4.0943	1.1×10^{26}	8.8×10^{-27}
2.1	0.7419	8.1662	0.1225	6.6	1.8871	735.10		70	4.2485		
2.2	0.7885	9.0250	0.1108	6.7	1.9021	812.41		80	4.3820	5.5×10^{34}	1.8×10^{-35}
2.3	0.8329	9.9742	0.1003	6.8	1.9169	897.85		90	4.4998		
2.4	0.8755	11.023	0.0907	6.9	1.9315	992.27		100	4.6052	2.7×10^{43}	3.7×10^{-44}
2.5	0.9163	12.182	0.0821	7.0	1.9459	1097	0.0009				
2.6	0.9555	13.464	0.0743	7.1	1.9601	1212					
2.7	0.9933	14.880	0.0672	7.2	1.9741	1339					
2.8	1.0296	16.445	0.0608	7.3	1.9879	1480					
2.9	1.0647	18.174	0.0550	7.4	2.0015	1636					
3.0	1.0986	20.086	0.0498	7.5	2.0149	1808	0.0006				
3.1	1.1314	22.198	0.0450	7.6	2.0281	1998					
3.2	1.1632	24.553	0.0408	7.7	2.0412	2208					
3.3	1.1939	27.113	0.0369	7.8	2.0541	2441					
3.4	1.2238	29.964	0.0334	7.9	2.0669	2697					
3.5	1.2528	33.115	0.0302	8.0	2.0794	2981	0.0003				
3.6	1.2809	36.598	0.0273	8.1	2.0919	3294					
3.7	1.3083	40.447	0.0247	8.2	2.1041	3641					
3.8	1.3350	44.701	0.0224	8.3	2.1163	4024					
3.9	1.3610	49.402	0.0202	8.4	2.1282	4447					
4.0	1.3863	54.598	0.0183	8.5	2.1401	4915	0.0002				
4.1	1.4110	60.340	0.0166	8.6	2.1518	5432					
4.2	1.4351	66.686	0.0150	8.7	2.1633	6003					
4.3	1.4586	73.700	0.0136	8.8	2.1748	6634					
4.4	1.4816	81.451	0.0123	8.9	2.1861	7332					

$\ln 10$	2.3026
$2 \ln 10$	4.6052
$3 \ln 10$	6.9078
$4 \ln 10$	9.2103
$5 \ln 10$	11.5129

For $N = P \times 10^k$

$\ln N = k \ln 10 + \ln P$

EXAMPLE

$820 = 8.2 \times 10^2$

$\ln 820 = 2 \ln 10 + \ln 8.2$

$\qquad = 4.6052 + 2.1041$

$\qquad = 6.7093$

TABLE 5   A Short Table of Integrals
The basic forms of Chapter 27 are not included. The constant of integration is omitted.

Forms containing $a + bu$ and $\sqrt{a + bu}$

1. $\displaystyle\int \frac{u\,du}{a + bu} = \frac{1}{b^2}[(a + bu) - a\,\ln(a + bu)]$

2. $\displaystyle\int \frac{du}{u(a + bu)} = -\frac{1}{a}\ln\frac{a + bu}{u}$

3. $\displaystyle\int \frac{u\,du}{(a + bu)^2} = \frac{1}{b^2}\left(\frac{a}{a + bu} + \ln(a + bu)\right)$

4. $\displaystyle\int \frac{du}{u(a + bu)^2} = \frac{1}{a(a + bu)} - \frac{1}{a^2}\ln\frac{a + bu}{u}$

5. $\displaystyle\int u\sqrt{a + bu}\,du = -\frac{2(2a - 3bu)(a + bu)^{3/2}}{15b^2}$

6. $\displaystyle\int \frac{u\,du}{\sqrt{a + bu}} = -\frac{2(2a - bu)\sqrt{a + bu}}{3b^2}$

7. $\displaystyle\int \frac{du}{u\sqrt{a + bu}} = \frac{1}{\sqrt{a}}\ln\left(\frac{\sqrt{a + bu} - \sqrt{a}}{\sqrt{a + bu} + \sqrt{a}}\right), \qquad a > 0$

8. $\displaystyle\int \frac{\sqrt{a + bu}}{u}\,du = 2\sqrt{a + bu} + a\int \frac{du}{u\sqrt{a + bu}}$

Forms containing $\sqrt{u^2 \pm a^2}$ and $\sqrt{a^2 - u^2}$

9. $\displaystyle\int \frac{du}{u^2 - a^2} = \frac{1}{2a}\ln\frac{u - a}{u + a}$

10. $\displaystyle\int \frac{du}{\sqrt{u^2 \pm a^2}} = \ln(u + \sqrt{u^2 \pm a^2})$

11. $\displaystyle\int \frac{du}{u\sqrt{u^2 + a^2}} = -\frac{1}{a}\ln\left(\frac{a + \sqrt{u^2 + a^2}}{u}\right)$

12. $\displaystyle\int \frac{du}{u\sqrt{u^2 - a^2}} = \frac{1}{a}\text{Arcsec}\frac{u}{a}$

13. $\displaystyle\int \frac{du}{u\sqrt{a^2 - u^2}} = -\frac{1}{a}\ln\left(\frac{a + \sqrt{a^2 - u^2}}{u}\right)$

14. $\displaystyle\int \sqrt{u^2 \pm a^2}\,du = \frac{u}{2}\sqrt{u^2 \pm a^2} \pm \frac{a^2}{2}\ln(u + \sqrt{u^2 \pm a^2})$

15. $\displaystyle\int \sqrt{a^2 - u^2}\,du = \frac{u}{2}\sqrt{a^2 - u^2} + \frac{a^2}{2}\text{Arcsin}\frac{u}{a}$

16. $\displaystyle\int \frac{\sqrt{u^2 + a^2}}{u}\,du = \sqrt{u^2 + a^2} - a\,\ln\left(\frac{a + \sqrt{u^2 + a^2}}{u}\right)$

17. $\displaystyle\int \frac{\sqrt{u^2 - a^2}}{u}\,du = \sqrt{u^2 - a^2} - a\,\text{Arcsec}\frac{u}{a}$

TABLE 5   (*Continued*)

18. $\int \dfrac{\sqrt{a^2 - u^2}}{u}\,du = \sqrt{a^2 - u^2} - a\,\ln\!\left(\dfrac{a + \sqrt{a^2 - u^2}}{u}\right)$

19. $\int (u^2 \pm a^2)^{3/2}\,du = \dfrac{u}{4}(u^2 \pm a^2)^{3/2} \pm \dfrac{3a^2 u}{8}\sqrt{u^2 \pm a^2} + \dfrac{3a^4}{8}\ln(u + \sqrt{u^2 \pm a^2})$

20. $\int (a^2 - u^2)^{3/2}\,du = \dfrac{u}{4}(a^2 - u^2)^{3/2} + \dfrac{3a^2 u}{8}\sqrt{a^2 - u^2} + \dfrac{3a^4}{8}\text{Arcsin}\dfrac{u}{a}$

21. $\int \dfrac{(u^2 + a^2)^{3/2}}{u}\,du = \dfrac{1}{3}(u^2 + a^2)^{3/2} + a^2\sqrt{u^2 + a^2} - a^3\ln\!\left(\dfrac{a + \sqrt{u^2 + a^2}}{u}\right)$

22. $\int \dfrac{(u^2 - a^2)^{3/2}}{u}\,du = \dfrac{1}{3}(u^2 - a^2)^{3/2} - a^2\sqrt{u^2 - a^2} + a^3\text{Arcsec}\dfrac{u}{a}$

23. $\int \dfrac{(a^2 - u^2)^{3/2}}{u}\,du = \dfrac{1}{3}(a^2 - u^2)^{3/2} - a^2\sqrt{a^2 - u^2} + a^3\ln\!\left(\dfrac{a + \sqrt{a^2 - u^2}}{u}\right)$

24. $\int \dfrac{du}{(u^2 \pm a^2)^{3/2}} = \pm\dfrac{u}{a^2\sqrt{u^2 \pm a^2}}$

25. $\int \dfrac{du}{(a^2 - u^2)^{3/2}} = \dfrac{u}{a^2\sqrt{a^2 - u^2}}$

26. $\int \dfrac{du}{u(u^2 + a^2)^{3/2}} = \dfrac{1}{a^2\sqrt{u^2 + a^2}} - \dfrac{1}{a^3}\ln\!\left(\dfrac{a + \sqrt{u^2 + a^2}}{u}\right)$

27. $\int \dfrac{du}{u(u^2 - a^2)^{3/2}} = -\dfrac{1}{a^2\sqrt{u^2 - a^2}} - \dfrac{1}{a^3}\text{Arcsec}\dfrac{u}{a}$

28. $\int \dfrac{du}{u(a^2 - u^2)^{3/2}} = \dfrac{1}{a^2\sqrt{a^2 - u^2}} - \dfrac{1}{a^3}\ln\!\left(\dfrac{a + \sqrt{a^2 - u^2}}{u}\right)$

Trigonometric forms

29. $\int \sin^2 u\,du = \dfrac{u}{2} - \dfrac{1}{2}\sin u \cos u$

30. $\int \sin^3 u\,du = -\cos u + \dfrac{1}{3}\cos^3 u$

31. $\int \sin^n u\,du = -\dfrac{1}{n}\sin^{n-1} u \cos u + \dfrac{n-1}{n}\int \sin^{n-2} u\,du$

32. $\int \cos^2 u\,du = \dfrac{u}{2} + \dfrac{1}{2}\sin u \cos u$

33. $\int \cos^3 u\,du = \sin u - \dfrac{1}{3}\sin^3 u$

34. $\int \cos^n u\,du = \dfrac{1}{n}\cos^{n-1} u \sin u + \dfrac{n-1}{n}\int \cos^{n-2} u\,du$

35. $\int \tan^n u\,du = \dfrac{\tan^{n-1} u}{n-1} - \int \tan^{n-2} u\,du$

TABLE 5 *(Continued)*

36. $\displaystyle\int \cot^n u \, du = -\frac{\cot^{n-1} u}{n-1} - \int \cot^{n-2} u \, du$

37. $\displaystyle\int \sec^n u \, du = \frac{\sec^{n-2} u \tan u}{n-1} + \frac{n-2}{n-1} \int \sec^{n-2} u \, du$

38. $\displaystyle\int \csc^n u \, du = -\frac{\csc^{n-2} u \cot u}{n-1} + \frac{n-2}{n-1} \int \csc^{n-2} u \, du$

39. $\displaystyle\int \sin au \sin bu \, du = \frac{\sin(a-b)u}{2(a-b)} - \frac{\sin(a+b)u}{2(a+b)}$

40. $\displaystyle\int \sin au \cos bu \, du = -\frac{\cos(a-b)u}{2(a-b)} - \frac{\cos(a+b)u}{2(a+b)}$

41. $\displaystyle\int \cos au \cos bu \, du = \frac{\sin(a-b)u}{2(a-b)} + \frac{\sin(a+b)u}{2(a+b)}$

42. $\displaystyle\int \sin^m u \cos^n u \, du = \frac{\sin^{m+1} u \cos^{n-1} u}{m+n} + \frac{n-1}{m+n} \int \sin^m u \cos^{n-2} u \, du$

43. $\displaystyle\int \sin^m u \cos^n u \, du = -\frac{\sin^{m-1} u \cos^{n+1} u}{m+n} + \frac{m-1}{m+n} \int \sin^{m-2} u \cos^n u \, du$

Other forms

44. $\displaystyle\int u e^{au} du = \frac{e^{au}(au-1)}{a^2}$

45. $\displaystyle\int u^2 e^{au} du = \frac{e^{au}}{a^3}(a^2 u^2 - 2au + 2)$

46. $\displaystyle\int u^n \ln u \, du = u^{n+1}\left(\frac{\ln u}{n+1} - \frac{1}{(n+1)^2}\right)$

47. $\displaystyle\int u \sin u \, du = \sin u - u \cos u$

48. $\displaystyle\int u \cos u \, du = \cos u + u \sin u$

49. $\displaystyle\int e^{au} \sin bu \, du = \frac{e^{au}(a \sin bu - b \cos bu)}{a^2 + b^2}$

50. $\displaystyle\int e^{au} \cos bu \, du = \frac{e^{au}(a \cos bu + b \sin bu)}{a^2 + b^2}$

51. $\displaystyle\int \text{Arcsin } u \, du = u \text{ Arcsin } u + \sqrt{1 - u^2}$

52. $\displaystyle\int \text{Arctan } u \, du = u \text{ Arctan } u - \frac{1}{2}\ln(1 + u^2)$

Answers to Odd-Numbered Exercises

1. Integer, rational, real; irrational, real **3.** Imaginary; irrational, real **5.** $3, \frac{7}{2}$ **7.** $\frac{6}{7}, \sqrt{3}$ **9.** $6 < 8$
11. $\pi > -1$ **13.** $-4 < -3$ **15.** $-\frac{1}{3} > -\frac{1}{2}$ **17.** $\frac{1}{3}, -\frac{1}{2}$ **19.** $-\frac{\pi}{5}, \frac{1}{x}$
21.   **23.**

25. $-18, -|-3|, -1, \sqrt{5}, \pi, |-8|, 9$
27. (a) Positive integer, (b) negative integer, (c) positive rational number less than 1
29. (a) To right of origin, (b) to left of -4 **31.** L, t variables; a constant **33.** $N = 1000an$ **35.** Yes

Exercises 1-2, 1-3, page 11

1. 11 **3.** -11 **5.** 9 **7.** -3 **9.** -24 **11.** 35 **13.** -3 **15.** 20 **17.** 40 **19.** -1
21. 9 **23.** Undefined **25.** 20 **27.** -5 **29.** -9 **31.** 24 **33.** -6 **35.** 3
37. Commutative law of multiplication **39.** Distributive law **41.** Associative law of addition
43. Associative law of multiplication **45.** Positive **47.** $\frac{4}{2} = 2; \frac{2}{4} = \frac{1}{2}; 2 \neq \frac{1}{2}$
49. 100 m + 200 m = 200 m + 100 m; commutative law of addition **51.** 8($2000 + $1000); distributive law

Exercises 1-4, page 13

1. 3.62 **3.** -0.118 **5.** 0.1356 **7.** 6.086 **9.** $43.011 = 43.011$, commutative law of addition
11. $478.30341 = 478.30341$, distributive law **13.** -8.21 **15.** 0.976 **17.** 14.9 **19.** 0.0330
21. (a) 0.36, (b) 0.36 **23.** No, -0.0000073 **25.** 0.2424242, (b) 3.1415927 **27.** Error
29. 2.6 g/day **31.** 59.14%

Exercises 1-5, page 19

1. x^7 **3.** $2b^6$ **5.** m^2 **7.** $\dfrac{1}{n^4}$ **9.** a^8 **11.** t^{20} **13.** $8n^3$ **15.** a^2x^8 **17.** $\dfrac{8}{b^3}$ **19.** $\dfrac{x^8}{16}$ **21.** 1

23. 3 **25.** $\frac{1}{6}$ **27.** s^2 **29.** $-t^{14}$ **31.** $64x^{12}$ **33.** 1 **35.** b^2 **37.** $\frac{1}{8}$ **39.** 1 **41.** $\dfrac{a}{x^2}$

43. $\dfrac{x^3}{64a^3}$ **45.** $64g^2s^6$ **47.** $\dfrac{5a}{n}$ **49.** -53 **51.** -10 **53.** 6230 **55.** -0.421 **57.** $\dfrac{r}{6}$ **59.** 68.9 W

Exercises 1-6, page 23

1. 45,000   **3.** 0.00201   **5.** 3.23   **7.** 18.6   **9.** 4×10^4 **11.** 8.7×10^{-3} **13.** 6×10^0
15. 6.3×10^{-2} **17.** 5.6×10^{13} **19.** 2.2×10^8 **21.** 4.85×10^{10} **23.** 3.67×10^5
25. 9.965×10^{-3} **27.** 5.0350×10^1 **29.** 6.5×10^6 kW   **31.** 0.0000000000016 W   **33.** 200,000
35. 3×10^{-6} W   **37.** 0.000000000000000000000000000000000000000024   **39.** 3.6×10^4 km
41. 2.57×10^{14} cm^2 **43.** 3.433 Ω

Exercises 1-7, page 26

1. 5   **3.** -11 **5.** -7 **7.** 20   **9.** 5   **11.** -6 **13.** 5   **15.** 31   **17.** $3\sqrt{2}$ **19.** $2\sqrt{3}$
21. $2\sqrt{11}$ **23.** $30\sqrt{7}$ **25.** $4\sqrt{21}$ **27.** 7   **29.** 10   **31.** $3\sqrt{10}$ **33.** 9.24   **35.** 0.6877   **37.** 1.47 s
39. 35.4 m   **41.** 19.0 in.   **43.** No, not true if $a < 0$

Exercises 1-8, page 31

1. $8x$ **3.** $4x + y$ **5.** $a + c - 2$ **7.** $-a^2b - a^2b^2$ **9.** $4s + 4$ **11.** $-v + 5x - 4$ **13.** $5a - 5$
15. $-5a + 2$ **17.** $-2t + 5u$ **19.** $7r + 8s$ **21.** $19c - 50$ **23.** $3n - 9$ **25.** $-2t^2 + 18$ **27.** $6a$
29. $2a\sqrt{xy} + 1$ **31.** $4c - 6$ **33.** $8p - 5q$ **35.** $-4x^2 + 22$ **37.** $4a - 3$ **39.** $-6t + 13$
41. $2D + d$ **43.** $22 - 10x$

Exercises 1-9, page 33

1. a^3x **3.** $-a^2c^3x^3$ **5.** $-8a^3x^5$ **7.** $2a^8x^3$ **9.** $a^2x + a^2y$ **11.** $-3s^3 + 15st$ **13.** $5m^3n + 15m^2n$
15. $-3x^2 - 3xy + 6x$ **17.** $a^2b^2c^5 - ab^3c^5 - a^2b^3c^4$ **19.** $acx^4 + acx^3y^3$ **21.** $x^2 + 2x - 15$
23. $2x^2 + 9x - 5$ **25.** $6a^2 - 7ab + 2b^2$ **27.** $6s^2 + 11st - 35t^2$ **29.** $2x^3 + 5x^2 - 2x - 5$
31. $x^3 + 2x^2 - 8x$ **33.** $x^3 - 2x^2 - x + 2$ **35.** $x^5 - x^4 - 6x^3 + 4x^2 + 8x$ **37.** $2a^2 - 16a - 18$
39. $2x^3 + 6x^2 - 8x$ **41.** $4x^2 - 20x + 25$ **43.** $x^2 + 6ax + 9a^2$ **45.** $x^2y^2z^2 - 4xyz + 4$
47. $2x^2 + 32x + 128$ **49.** $-x^3 + 2x^2 + 5x - 6$ **51.** $6x^4 + 21x^3 + 12x^2 - 12x$ **53.** $x^2 + x - 2$ mm^2
55. $R^2 - r^2$

Exercises 1-10, page 37

1. $-4x^2y$ **3.** $4t^4/r^2$ **5.** $4x^2$ **7.** $-6a$ **9.** $a^2 + 4y$ **11.** $t - 2rt^2$ **13.** $q + 2p - 4q^3$
15. $\dfrac{2L}{R} - R$ **17.** $\dfrac{1}{3a} - \dfrac{2b}{3a} + 1$ **19.** $x^2 + a$ **21.** $2x + 1$ **23.** $x - 1$ **25.** $4x^2 - x - 1, R = -3$
27. $x + 5$ **29.** $x^2 + x - 6$ **31.** $2x^2 + 4x + 2, R = 4x + 4$ **33.** $x^2 - 2x + 4$ **35.** $x - y$
37. $A + \dfrac{\mu^2E^2}{2A} - \dfrac{\mu^4E^4}{8A^3}$ **39.** $5x + 16$

Exercises 1-11, page 40

1. 9   **3.** -1 **5.** 10   **7.** -5 **9.** -3 **11.** 1   **13.** $-\frac{7}{2}$ **15.** 8   **17.** $\frac{10}{3}$ **19.** $-\frac{13}{3}$ **21.** 2
23. 0   **25.** 9.5   **27.** -1.5 **29.** True for all values of x, identity   **31.** Not true for any x, contradiction
33. 1.3 mi/h   **35.** 750 gal

Exercises 1-12, page 43

1. $\dfrac{b}{a}$ **3.** $\dfrac{4m - 1}{4}$ **5.** $\dfrac{c + 6}{a}$ **7.** $2a + 8$ **9.** $\theta - kA$ **11.** $\dfrac{E}{I}$ **13.** $\dfrac{P}{2\pi f}$ **15.** $\dfrac{p - p_a}{dg}$ **17.** $\dfrac{PL^2}{\pi^2 I}$
19. $\dfrac{s + 16t^2}{t}$ **21.** $\dfrac{C_0^2 - C_1^2}{2C_1^2}$ **23.** $\dfrac{a + PV^2}{V}$ **25.** $\dfrac{Q_1 + PQ_1}{P}$ **27.** $\dfrac{N + N_2 - N_2T}{T}$
29. $\dfrac{L - \pi r_2 - 2x_1 - x_2}{\pi}$ **31.** $\dfrac{gJP + V_1^2}{V_1}$ **33.** 28.2 in.   **35.** 32.3°C

Exercises 1-13, page 48

1. $138, $252   **3.** 22, 35, 50   **5.** 20 acres at $20,000/acre, 50 acres at $10,000/acre   **7.** 20 girders
9. $-2.3\ \mu A$, $-4.6\ \mu A$, $6.9\ \mu A$ **11.** 36 in., 54 in.   **13.** 6.9 km (main), 9.5 km (others)
15. 360 s (6 min), first car   **17.** 1600 mi/h, 2000 mi/h   **19.** 900 m   **21.** 6 L   **23.** 110 tons, 140 tons

Review Exercises for Chapter 1, page 50

1. -10 **3.** -20 **5.** 2   **7.** -25 **9.** -4 **11.** 5   **13.** $4r^2t^4$ **15.** $-\dfrac{6m^2}{nt^2}$ **17.** $\dfrac{z^6}{y^2}$ **19.** $\dfrac{8t^3}{s^2}$
21. $3\sqrt{5}$ **23.** $2\sqrt{5}$ **25.** 18.12   **27.** -1.440 **29.** $-a-2ab$ **31.** $5xy+3$ **33.** $2x^2+9x-5$
35. $x^2+16x+64$ **37.** $hk-3h^2k^4$ **39.** $7a-6b$ **41.** $13xy-10z$ **43.** $2x^3-x^2-7x-3$
45. $-3x^2y+24xy^2-48y^3$ **47.** $-9p^2+3pq+18p^2q$ **49.** $\dfrac{6q}{p}-2+\dfrac{3q^4}{p^3}$ **51.** $2x-5$ **53.** x^2-2x+3
55. $4x^3-2x^2+6x$, $R=-1$ **57.** $15r-3s-3t$ **59.** y^2+5y-1, $R=4$ **61.** 4   **63.** $-\frac{9}{2}$ **65.** $-\frac{7}{3}$
67. 3   **69.** $-\frac{19}{5}$ **71.** 1   **73.** 2.5×10^4 mi/h   **75.** 1.02×10^9 Hz   **77.** 1.2×10^{-6} cm^2
79. 0.00018 kg/m^3 **81.** $\dfrac{5a-2}{3}$ **83.** $\dfrac{4-2n}{3}$ **85.** $\dfrac{R}{n^2}$ **87.** $\dfrac{I-P}{Pr}$ **89.** $\dfrac{dV-m}{dV}$ **91.** $\dfrac{2C-mN_2}{m}$
93. $J-E+ES$ **95.** $\dfrac{d-3kbx^2+kx^3}{3kx^2}$ **97.** 1.80×10^{-2} ft   **99.** 1.45 N   **101.** $1-4r+3r^2$
103. $9x^2-1600x+80,000$ **105.** 16,384 bytes, 65,536 bytes   **107.** 1900 Ω, 3100 Ω
109. 52 vertical strands, 82 horizontal strands   **111.** After 1.4 h   **113.** 400 L   **115.** 293 ft^2

Exercises 2-1, page 58

1. $A=\pi r^2$ **3.** $c=2\pi r$ **5.** $A=5\ell$ **7.** $A=s^2$; $s=\sqrt{A}$ **9.** 3, -1 **11.** 11, 3.8   **13.** -18, 70
15. $\frac{5}{2}$, $-\frac{1}{2}$ **17.** $\frac{1}{4}a+\frac{1}{2}a^2$, 0   **19.** $3s^2+s+6$, $12s^2-2s+6$ **21.** 62.9, 260   **23.** 0.01988, -0.2998
25. Square the value of the independent variable and add 2.   **27.** Cube the value of the independent variable
and subtract this value from 6 times the value of the independent variable.   **29.** $y=f(x)$, $f(x)=x^2$
31. $P=f(p)$, $f(p)=40(p-24)$ **33.** 3.4 m   **35.** 75 ft, $2v+0.2v^2$, 240 ft, 240 ft

Exercises 2-2, page 62

1. Domain: all real numbers; range: all real numbers   **3.** Domain: all real numbers except 0;
range: all real numbers except 0   **5.** Domain: all real numbers except 0; range: all real numbers $f(s)>0$
7. Domain: all real numbers $h \geq 0$; range: all real numbers $H(h) \geq 1$ **9.** All real numbers $y>2$
11. All real numbers except 2 and -4 **13.** -10, $\frac{8}{9}$ **15.** -16, $\frac{1}{2}$ **17.** -1, $\sqrt{5}$ **19.** $\frac{1}{2}$, 0
21. $d=80+55t$ **23.** $w=5500-2t$ **25.** $w=0.5h-390$ **27.** $C=5\ell+250$ **29.** $y=3000-0.25x$
31. $A=\dfrac{1}{16}p^2+\dfrac{(60-p)^2}{4\pi}$ **33.** $C>0$, with some upper limit depending on circuit
35. $w=\begin{cases}110 & \text{for } h \leq 1000 \text{ m} \quad (w \text{ in kg})\\ 0.5h-390 & \text{for } h > 1000 \text{ m}\end{cases}$

Exercises 2-3, page 66

1. $(2, 1), (-1, 2), (-2, -3)$ **3.**

5. Isosceles triangle   **7.** Rectangle

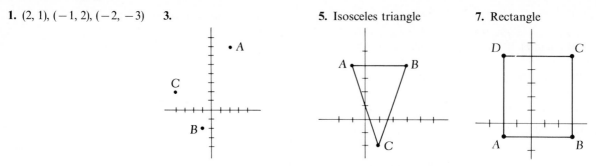

9. $(5, 4)$ **11.** $(3, -2)$ **13.** On a line parallel to the y-axis, one unit to the right
15. On a line parallel to the x-axis, three units above   **17.** On a line bisecting the first and third quadrants   **19.** 0
21. To the right of the y-axis   **23.** To the left of a line which is parallel to the y-axis and one unit to its left
25. On the negative y-axis   **27.** First, third

Exercises 2-4, page 71

1.

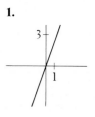

3.

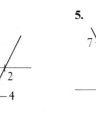

5.

7.

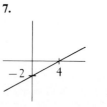

9.

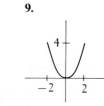

11.

13.

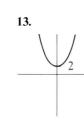

15.

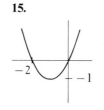

17.

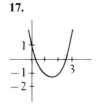

19.

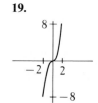

21.

23.

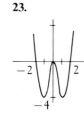

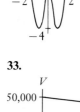

25.

27.

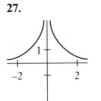

29.

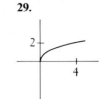

31.

33.

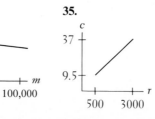

35.

37.

39.

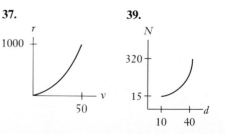

41. $A = 30w - w^2$ **43.**   **45.**   **47.**   **49.** Yes
51. No

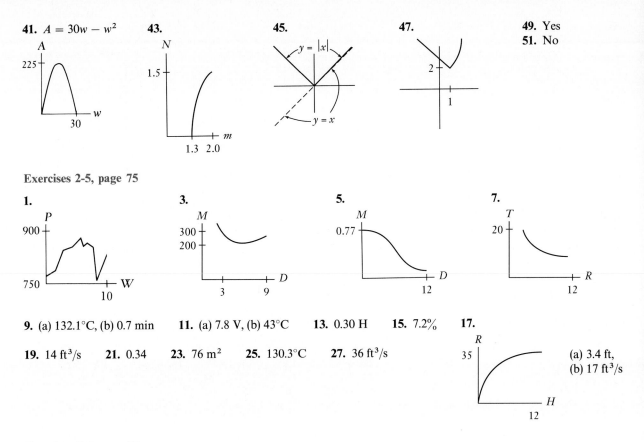

Exercises 2-5, page 75

1.   **3.**   **5.**   **7.**

9. (a) 132.1°C, (b) 0.7 min   **11.** (a) 7.8 V, (b) 43°C   **13.** 0.30 H   **15.** 7.2%   **17.**

19. 14 ft³/s   **21.** 0.34   **23.** 76 m²   **25.** 130.3°C   **27.** 36 ft³/s

(a) 3.4 ft,
(b) 17 ft³/s

Exercises 2-6, page 79

1. 2   **3.** $-\frac{9}{2}$ **5.** 3.5   **7.** 0.5   **9.** 0.0, -1.0 **11.** -1.7 **13.** 0.7   **15.** 2.5   **17.** -2.8, 1.8
19. -1.6, 2.1   **21.** -2.0, 0.0, 2.0   **23.** 0.0, 1.3   **25.** 3.5   **27.** No real zeros   **29.** 6 V   **31.** 5.1 N
33. 18 cm, 30 cm   **35.** 67 ft

Review Exercises for Chapter 2, page 80

1. $V = 8\pi r^2$ **3.** $y = -\frac{10}{9}x + \frac{250}{9}$ **5.** 16, -47 **7.** -5, -1.08 **9.** 3, $\sqrt{1 - 4h}$ **11.** $6hx + 3h^2 - 2h$
13. -3.67, 16.7   **15.** 0.16572, -0.21566 **17.** Domain: all real numbers; range: all real numbers $f(x) \geq 1$
19. Domain: all real numbers $t > -4$; range: all real numbers $g(t) > 0$
21.   **23.**   **25.**   **27.**   **29.**   **31.**

33. -0.5 **35.** 0, 4   **37.** -1.5, 1.0   **39.** -2.4, 0.0, 2.4   **41.** -1.2, 1.2   **43.** 0   **45.** 0.4
47. 0.2, 5.8   **49.** 1.4   **51.** -0.7, 0.7   **53.** Either a or b is positive, the other is negative   **55.** 13.4

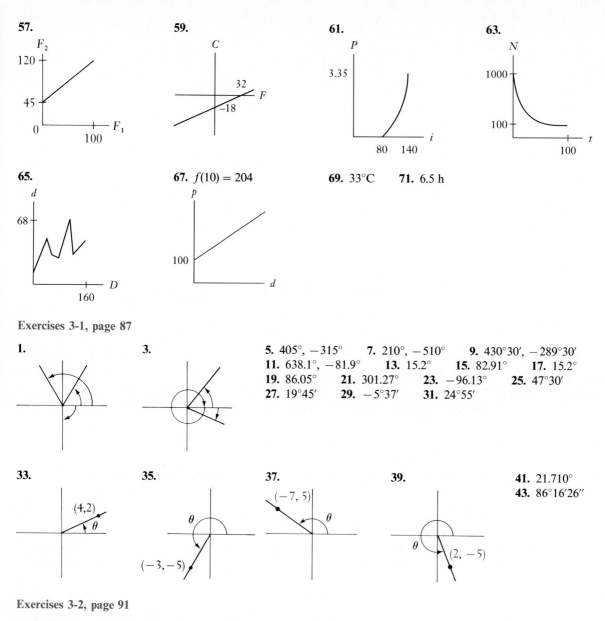

57.

59.

61.

63.

65.

67. $f(10) = 204$

69. 33°C   **71.** 6.5 h

Exercises 3-1, page 87

1.

3.

5. 405°, −315°   **7.** 210°, −510°   **9.** 430°30′, −289°30′
11. 638.1°, −81.9°   **13.** 15.2°   **15.** 82.91°   **17.** 15.2°
19. 86.05°   **21.** 301.27°   **23.** −96.13°   **25.** 47°30′
27. 19°45′   **29.** −5°37′   **31.** 24°55′

33.

35.

37.

39.

41. 21.710°
43. 86°16′26″

Exercises 3-2, page 91

1. $\sin \theta = \frac{4}{5}$, $\cos \theta = \frac{3}{5}$, $\tan \theta = \frac{4}{3}$, $\cot \theta = \frac{3}{4}$, $\sec \theta = \frac{5}{3}$, $\csc \theta = \frac{5}{4}$
3. $\sin \theta = \frac{8}{17}$, $\cos \theta = \frac{15}{17}$, $\tan \theta = \frac{8}{15}$, $\cot \theta = \frac{15}{8}$, $\sec \theta = \frac{17}{15}$, $\csc \theta = \frac{17}{8}$
5. $\sin \theta = \frac{40}{41}$, $\cos \theta = \frac{9}{41}$, $\tan \theta = \frac{40}{9}$, $\cot \theta = \frac{9}{40}$, $\sec \theta = \frac{41}{9}$, $\csc \theta = \frac{41}{40}$
7. $\sin \theta = \frac{\sqrt{15}}{4}$, $\cos \theta = \frac{1}{4}$, $\tan \theta = \sqrt{15}$, $\cot \theta = \frac{1}{\sqrt{15}}$, $\sec \theta = 4$, $\csc \theta = \frac{4}{\sqrt{15}}$
9. $\sin \theta = \frac{1}{\sqrt{2}}$, $\cos \theta = \frac{1}{\sqrt{2}}$, $\tan \theta = 1$, $\cot \theta = 1$, $\sec \theta = \sqrt{2}$, $\csc \theta = \sqrt{2}$
11. $\sin \theta = \frac{2}{\sqrt{29}}$, $\cos \theta = \frac{5}{\sqrt{29}}$, $\tan \theta = \frac{2}{5}$, $\cot \theta = \frac{5}{2}$, $\sec \theta = \frac{\sqrt{29}}{5}$, $\csc \theta = \frac{\sqrt{29}}{2}$
13. $\sin \theta = 0.846$, $\cos \theta = 0.534$, $\tan \theta = 1.58$, $\cot \theta = 0.631$, $\sec \theta = 1.87$, $\csc \theta = 1.18$
15. $\sin \theta = 0.1521$, $\cos \theta = 0.9884$, $\tan \theta = 0.1539$, $\cot \theta = 6.498$, $\sec \theta = 1.012$, $\csc \theta = 6.575$
17. $\frac{5}{13}$, $\frac{12}{5}$ **19.** $\frac{1}{\sqrt{2}}$, $\sqrt{2}$ **21.** 0.882, 1.33   **23.** 0.246, 3.94   **25.** $\sin \theta = \frac{4}{5}$, $\tan \theta = \frac{4}{3}$

27. $\tan \theta = \frac{1}{2}$, $\sec \theta = \frac{\sqrt{5}}{2}$ **29.** $\sec \theta$ **31.** $\dfrac{x}{y} \times \dfrac{y}{r} = \dfrac{x}{r} = \cos \theta$

Exercises 3-3, page 95

1. $\sin 40° = 0.64$, $\cos 40° = 0.77$, $\tan 40° = 0.84$, $\cot 40° = 1.19$, $\sec 40° = 1.30$, $\csc 40° = 1.56$
3. $\sin 15° = 0.26$, $\cos 15° = 0.97$, $\tan 15° = 0.27$, $\cot 15° = 3.73$, $\sec 15° = 1.04$, $\csc 15° = 3.86$
5. 0.381   **7.** 1.58   **9.** 0.9626   **11.** 0.99   **13.** 0.4085   **15.** 1.57   **17.** 1.32   **19.** 0.07063
21. 70.97°   **23.** 65.70°   **25.** 11.7°   **27.** 49.453°   **29.** 53.44°   **31.** 81.79°   **33.** 74.1°   **35.** 17.85°
37. 0.8885   **39.** 0.93614   **41.** 0.326   **43.** 2.356   **45.** 40°   **47.** 83°   **49.** 0.550   **51.** 0.880
53. 72.8°   **55.** 35°10′   **57.** 87 dB   **59.** 70.4°

Exercises 3-4, page 100

1.

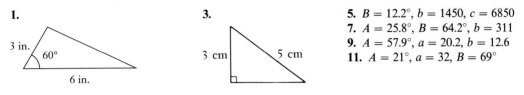

5. $B = 12.2°$, $b = 1450$, $c = 6850$
7. $A = 25.8°$, $B = 64.2°$, $b = 311$
9. $A = 57.9°$, $a = 20.2$, $b = 12.6$
11. $A = 21°$, $a = 32$, $B = 69°$

13. $a = 30.21$, $B = 57.90°$, $b = 48.16$ **15.** $A = 52.15°$, $B = 37.85°$, $c = 71.85$
17. $A = 52.5°$, $b = 0.661$, $c = 1.09$ **19.** $A = 15.82°$, $a = 0.5239$, $c = 1.922$
21. $A = 65.886°$, $B = 24.114°$, $c = 648.46$ **23.** $a = 3.3621$, $B = 77.025°$, $c = 14.974$ **25.** 4.45   **27.** 40.24°
29. $A = 52°20′$, $b = 0.684$, $c = 1.12$ **31.** $A = 58°40′$, $a = 143$, $B = 31°20′$
33. $a = c \sin A$, $b = c \cos A$, $B = 90° - A$ **35.** $A = 90° - B$, $b = a \tan B$, $c = a/\cos B$

Exercises 3-5, page 103

1. 97 ft   **3.** 44.0 ft   **5.** 0.4°   **7.** 850.1 cm   **9.** 25.4 ft   **11.** 26.6°, 63.4°, 90.0°   **13.** 195 ft
15. 1.29 m   **17.** 8.1°   **19.** 1840 km   **21.** 651 ft   **23.** 30.2°   **25.** 47.3 m   **27.** 200 m

Review Exercises for Chapter 3, page 105

1. 377.0°, −343.0°   **3.** 142.5°, −577.5°   **5.** 31.9°   **7.** 38.1°   **9.** 17°30′   **11.** 49°42′
13. $\sin \theta = \frac{7}{25}$, $\cos \theta = \frac{24}{25}$, $\tan \theta = \frac{7}{24}$, $\cot \theta = \frac{24}{7}$, $\sec \theta = \frac{25}{24}$, $\csc \theta = \frac{25}{7}$
15. $\sin \theta = \frac{1}{\sqrt{2}}$, $\cos \theta = \frac{1}{\sqrt{2}}$, $\tan \theta = 1$, $\cot \theta = 1$, $\sec \theta = \sqrt{2}$, $\csc \theta = \sqrt{2}$ **17.** 0.923, 2.40   **19.** 0.447, 1.12
21. 0.952   **23.** 1.853   **25.** 1.05   **27.** 8.074   **29.** 18.2°   **31.** 57.57°   **33.** 12.25°   **35.** 66.8°
37. $a = 1.83$, $B = 73.0°$, $c = 6.27$ **39.** $A = 51.5°$, $B = 38.5°$, $c = 104$ **41.** $B = 52.5°$, $b = 15.6$, $c = 19.7$
43. $A = 31.61°$, $a = 4.006$, $B = 58.39°$ **45.** $a = 0.6292$, $B = 40.33°$, $b = 0.5341$
47. $A = 48.813°$, $B = 41.187°$, $b = 10.196$ **49.** 44.4 N   **51.** 679.2 m²   **53.** 12.0°   **55.** 4.92 km   **57.** 56%
59. 1.4 cm   **61.** 4.43 m   **63.** 0.184 km   **65.** 10.2 in.   **67.** $d = 880 \cot 2.2° \cong 440 \cot 1.1° = 22{,}900$ m
69. 4810 m   **71.** 1.83 km   **73.** 464 m   **75.** 73.3 cm

Exercises 4-1, page 111

1. Yes, no   **3.** Yes, yes   **5.** −1, −16   **7.** $\frac{1}{4}$, −0.6   **9.** Yes   **11.** No   **13.** No   **15.** Yes
17. Yes   **19.** No

Exercises 4-2, page 116

1. 4   **3.** −5   **5.** $\frac{2}{7}$ **7.** $\frac{1}{2}$
9.   **11.**   **13.**   **15.**

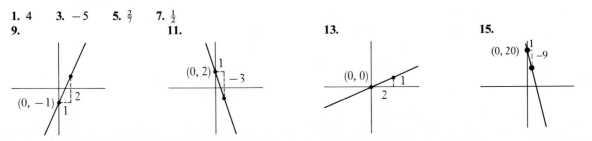

17. $m = -2$, $b = 1$ **19.** $m = 1$, $b = 4$ **21.** $m = \frac{5}{2}$, $b = -2$ **23.** $m = -\frac{1}{3}$, $b = 1$

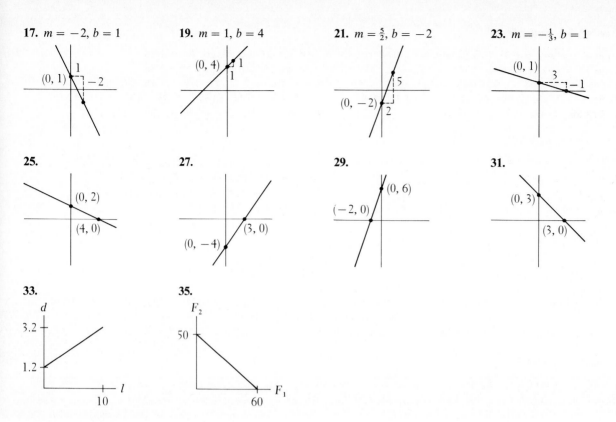

25.   **27.**   **29.**   **31.**

33.   **35.**

Exercises 4-3, page 119

1. $x = 3.0$, $y = 1.0$ **3.** $x = 3.0$, $y = 0.0$ **5.** $x = 2.2$, $y = -0.3$ **7.** $x = -0.9$, $y = -2.3$
9. $x = 6.2$, $y = -0.5$ **11.** $x = 0.0$, $y = 3.0$ **13.** $x = -14.0$, $y = -5.0$ **15.** $x = 4.0$, $y = 7.5$
17. $x = -3.6$, $y = -1.4$ **19.** $x = 1.1$, $y = 0.8$ **21.** Dependent   **23.** $x = -1.2$, $y = -3.6$
25. $x = 1.5$, $y = 4.5$ **27.** Inconsistent   **29.** 50 N, 47 N   **31.** 8 s, 4 s

Exercises 4-4, page 125

1. $x = 1$, $y = -2$ **3.** $x = 7$, $y = 3$ **5.** $x = -1$, $y = -4$ **7.** $x = \frac{1}{2}$, $y = 2$ **9.** $x = -\frac{1}{3}$, $y = 4$
11. $x = \frac{9}{22}$, $y = -\frac{16}{11}$ **13.** $x = 3$, $y = 1$ **15.** $x = -1$, $y = -2$ **17.** $x = 1$, $y = 2$ **19.** Inconsistent
21. $x = -\frac{14}{5}$, $y = -\frac{16}{5}$ **23.** $x = 2.38$, $y = 0.45$ **25.** $x = \frac{1}{2}$, $y = -4$ **27.** $x = -\frac{2}{3}$, $y = 0$
29. Dependent   **31.** $x = -1$, $y = -2$ **33.** $V_1 = 9$ V, $V_2 = 6$ V   **35.** $x = 6250$ L, $y = 3750$ L
37. $t_1 = 32$ s, $t_2 = 20$ s   **39.** 4.0×10^6 calc/s, 2.5×10^6 calc/s   **41.** 750 mL, 250 mL
43. Incorrect conclusion or error in sales figures; system of equations is inconsistent

Exercises 4-5, page 131

1. -10 **3.** 29   **5.** 32   **7.** 93   **9.** 0.9   **11.** 96   **13.** $x = 3$, $y = 1$ **15.** $x = -1$, $y = -2$
17. $x = 1$, $y = 2$ **19.** Inconsistent   **21.** $x = -\frac{14}{5}$, $y = -\frac{16}{5}$ **23.** $x = 2.38$, $y = 0.45$ **25.** $x = \frac{1}{2}$, $y = -4$
27. $x = -\frac{2}{3}$, $y = 0$ **29.** Dependent   **31.** $x = -1$, $y = -2$ **33.** $F_1 = 15$ lb, $F_2 = 6$ lb
35. $A = \$3200$, $B = \$2800$ **37.** 2.5 h, 2.1 h   **39.** 4200, 1400   **41.** 0.8 L, 1.2 L   **43.** $V = 4.5i - 3.2$

Exercises 4-6, page 137

1. $x = 2$, $y = -1$, $z = 1$ **3.** $x = 4$, $y = -3$, $z = 3$ **5.** $x = \frac{1}{2}$, $y = \frac{2}{3}$, $z = \frac{1}{6}$ **7.** $x = \frac{2}{3}$, $y = -\frac{1}{3}$, $z = 1$
9. $x = \frac{4}{15}$, $y = -\frac{3}{5}$, $z = \frac{1}{3}$ **11.** $x = -2$, $y = \frac{2}{3}$, $z = \frac{1}{3}$ **13.** $x = \frac{3}{4}$, $y = 1$, $z = -\frac{1}{2}$
15. $r = 0$, $s = 0$, $t = 0$, $u = -1$ **17.** $P = 800$ h, $M = 125$ h, $I = 225$ h
19. $F_1 = 9.43$ N, $F_2 = 8.33$ N, $F_3 = 1.67$ N   **21.** $A = 22.5°$, $B = 45.0°$, $C = 112.5°$ **23.** 70 lb, 100 lb, 30 lb
25. Unlimited; $x = -10$, $y = -6$, $z = 0$ **27.** No solution

Exercises 4-7, page 143

1. 122   **3.** 651   **5.** -439 **7.** 202   **9.** 128   **11.** 0.128   **13.** $x = -1$, $y = 2$, $z = 0$
15. $x = 2$, $y = -1$, $z = 1$ **17.** $x = 4$, $y = -3$, $z = 3$ **19.** $x = \frac{1}{2}$, $y = \frac{2}{3}$, $z = \frac{1}{6}$ **21.** $x = \frac{2}{3}$, $y = -\frac{1}{3}$, $z = 1$
23. $x = \frac{4}{15}$, $y = -\frac{3}{5}$, $z = \frac{1}{3}$ **25.** $x = -2$, $y = \frac{2}{3}$, $z = \frac{1}{3}$ **27.** $x = \frac{3}{4}$, $y = 1$, $z = -\frac{1}{2}$
29. $A = 125$ N, $B = 60$ N, $F = 75$ N   **31.** 45 mi/h, 540 mi/h, 30 mi/h

Review Exercises for Chapter 4, page 145

1. -17 **3.** -1485 **5.** -4 **7.** $\frac{2}{7}$
9. $m = -2$, $b = 4$ **11.** $m = 4$, $b = -\frac{5}{2}$

13. $x = 2.0$, $y = 0.0$ **15.** $x = 2.2$, $y = 2.7$ **17.** $x = 1.5$, $y = -1.9$ **19.** $x = 1.5$, $y = 0.4$
21. $x = 1$, $y = 2$ **23.** $x = \frac{1}{2}$, $y = -2$ **25.** $x = -\frac{1}{3}$, $y = \frac{7}{4}$ **27.** $x = \frac{11}{19}$, $y = -\frac{26}{19}$
29. $x = -\frac{6}{19}$, $y = \frac{36}{19}$ **31.** $x = 1.10$, $y = 0.54$ **33.** $x = 1$, $y = 2$ **35.** $x = \frac{1}{2}$, $y = -2$
37. $x = -\frac{1}{3}$, $y = \frac{7}{4}$ **39.** $x = \frac{11}{19}$, $y = -\frac{26}{19}$ **41.** $x = -\frac{6}{19}$, $y = \frac{36}{19}$ **43.** $x = 1.10$, $y = 0.54$
45. -115 **47.** 230.08   **49.** $x = 2$, $y = -1$, $z = 1$ **51.** $x = \frac{2}{3}$, $y = -\frac{1}{2}$, $z = 0$ **53.** $r = 3$, $s = -1$, $t = \frac{3}{2}$
55. $x = -\frac{1}{2}$, $y = \frac{1}{2}$, $z = 3$ **57.** $x = 2$, $y = -1$, $z = 1$ **59.** $x = \frac{2}{3}$, $y = -\frac{1}{2}$, $z = 0$ **61.** $r = 3$, $s = -1$, $t = \frac{3}{2}$
63. $x = -\frac{1}{2}$, $y = \frac{1}{2}$, $z = 3$ **65.** $x = \frac{8}{3}$, $y = -8$ **67.** $x = 1$, $y = 3$ **69.** -6 **71.** $-\frac{4}{3}$
73. $F_1 = 232$ lb, $F_2 = 24$ lb, $F_3 = 201$ lb   **75.** $a = 440$ m·°C, $b = 9.6$°C   **77.** 22,800 km/h, 1400 km/h
79. $R_1 = 0.5$ Ω, $R_2 = 1.5$ Ω   **81.** $L = 10$ lb, $w = 40$ lb   **83.** 425 TVs, 475 VCRs, 850 CDs

Exercises 5-1, page 152

1. $40x - 40y$ **3.** $2x^3 - 8x^2$ **5.** $y^2 - 36$ **7.** $9v^2 - 4$ **9.** $16x^2 - 25y^2$ **11.** $144 - 25a^2b^2$
13. $25f^2 + 40f + 16$ **15.** $4x^2 + 28x + 49$ **17.** $x^2 - 2x + 1$ **19.** $16a^2 + 56axy + 49x^2y^2$
21. $16x^2 - 16xy + 4y^2$ **23.** $36s^2 - 12st + t^2$ **25.** $x^2 + 6x + 5$ **27.** $c^2 + 9c + 18$ **29.** $6x^2 + 13x - 5$
31. $20x^2 - 21x - 5$ **33.** $20v^2 + 13v - 15$ **35.** $6x^2 - 13xy - 63y^2$ **37.** $2x^2 - 8$ **39.** $8a^3 - 2a$
41. $6ax^2 + 24abx + 24ab^2$ **43.** $20n^4 + 100n^3 + 125n^2$ **45.** $16a^3 - 48a^2 + 36a$
47. $x^2 + y^2 + 2xy + 2x + 2y + 1$ **49.** $x^2 + y^2 + 2xy - 6x - 6y + 9$ **51.** $125 - 75t + 15t^2 - t^3$
53. $8x^3 + 60x^2t + 150xt^2 + 125t^3$ **55.** $x^2 + 2xy + y^2 - 1$ **57.** $x^3 + 8$ **59.** $64 - 27x^3$

61. $P_0P_1c + P_1G$ **63.** $4p^2 + 8pDA + 4D^2A^2$ **65.** $\frac{1}{2}\pi R^2 - \frac{1}{2}\pi r^2$ **67.** $\dfrac{L}{6}x^3 - \dfrac{L}{2}ax^2 + \dfrac{L}{2}a^2x - \dfrac{L}{6}a^3$

Exercises 5-2, page 156

1. $6(x + y)$ **3.** $5(a - 1)$ **5.** $3x(x - 3)$ **7.** $7b(by - 4)$ **9.** $6n(2n + 1)$ **11.** $2(x + 2y - 4z)$
13. $3ab(b - 2 + 4b^2)$ **15.** $4pq(3q - 2 - 7q^2)$ **17.** $2(a^2 - b^2 + 2c^2 - 3d^2)$ **19.** $(x + 2)(x - 2)$

21. $(10 + y)(10 - y)$ **23.** $(6a + 1)(6a - 1)$ **25.** $(9s + 5t)(9s - 5t)$ **27.** $(12n + 13p^2)(12n - 13p^2)$
29. $(x + y + 3)(x + y - 3)$ **31.** $2(x + 2)(x - 2)$ **33.** $3(x + 3z)(x - 3z)$ **35.** $2(a - 1)(a - 5)$

37. $(x^2 + 4)(x + 2)(x - 2)$ **39.** $(x^4 + 1)(x^2 + 1)(x + 1)(x - 1)$ **41.** $\dfrac{3 + b}{2 - b}$ **43.** $\dfrac{3}{2(t - 1)}$

45. $(3 + b)(x - y)$ **47.** $(a - b)(a + x)$ **49.** $(x + 2)(x - 2)(x + 3)$ **51.** $(x - y)(x + y + 1)$

53. $Rv(1 + v + v^2)$ **55.** $a(D_1 + D_2)(D_1 - D_2)$ **57.** $Pb(L + b)(L - b)$ **59.** $\dfrac{5Y}{9S - 3Y}$

Exercises 5-3, page 162

1. $(x + 1)(x + 4)$ **3.** $(s - 7)(s + 6)$ **5.** $(t + 8)(t - 3)$ **7.** $(x + 1)^2$ **9.** $(x - 2y)^2$ **11.** $(3x + 1)(x - 2)$
13. $(3y + 1)(y - 3)$ **15.** $(2s + 11)(s + 1)$ **17.** $(3f - 1)(f - 5)$ **19.** $(2t - 3)(t + 5)$ **21.** $(3t - 4u)(t - u)$
23. $(4x - 7)(x + 1)$ **25.** $(9x - 2y)(x + y)$ **27.** $(2m + 5)^2$ **29.** $(2x - 3)^2$ **31.** $(3t - 4)(3t - 1)$
33. $(8b - 1)(b + 4)$ **35.** $(4p - q)(p - 6q)$ **37.** $(12x - y)(x + 4y)$ **39.** $2(x - 1)(x - 6)$
41. $2(2x - 1)(x + 4)$ **43.** $ax(x + 6a)(x - 2a)$ **45.** $(a + b + 2)(a + b - 2)$ **47.** $(5a + 5x + y)(5a - 5x - y)$
49. $(x + 1)^3$ **51.** $(2x + 1)(4x^2 - 2x + 1)$ **53.** $4(s + 1)(s + 3)$ **55.** $100(2n + 3)(n - 12)$
57. $wx^2(x - 2L)(x - 3L)$ **59.** $Ad(3u - v)(u - v)$

Exercises 5-4, page 166

1. $\dfrac{14}{21}$ **3.** $\dfrac{2ax^2}{2xy}$ **5.** $\dfrac{2x - 4}{x^2 + x - 6}$ **7.** $\dfrac{ax^2 - ay^2}{x^2 - xy - 2y^2}$ **9.** $\dfrac{7}{11}$ **11.** $\dfrac{2xy}{4y^2}$ **13.** $\dfrac{2}{x + 1}$ **15.** $\dfrac{x - 5}{2x - 1}$

17. $\dfrac{1}{4}$ **19.** $\dfrac{3}{4}x$ **21.** $\dfrac{1}{5a}$ **23.** $\dfrac{3a - 2b}{2a - b}$ **25.** $\dfrac{4x^2 + 1}{(2x + 1)(2x - 1)}$ (cannot be reduced)   **27.** $3x$ **29.** $\dfrac{1}{2y^2}$

31. $\dfrac{x - 4}{x + 4}$ **33.** $\dfrac{2x - 1}{x + 8}$ **35.** $\dfrac{5x + 4}{x(x + 3)}$ **37.** $(x^2 + 4)(x - 2)$ **39.** $\dfrac{x^2y^2(y + x)}{y - x}$ **41.** $\dfrac{x + 3}{x - 3}$ **43.** $-\dfrac{1}{2}$

45. $-\dfrac{2x - 1}{x}$ **47.** $\dfrac{(x + 5)(x - 3)}{(5 - x)(x + 3)}$ **49.** $\dfrac{x^2 - xy + y^2}{2}$ **51.** $\dfrac{(x + 1)^2}{x^2 - x + 1}$ **53.** (a)   **55.** (a)   **57.** $\dfrac{4r}{3\pi}$

59. $\dfrac{E^2(R - r)}{(R + r)^3}$

Exercises 5-5, page 170

1. $\dfrac{3}{28}$ **3.** $6xy$ **5.** $\dfrac{7}{18}$ **7.** $\dfrac{xy^2}{bz^2}$ **9.** $4t$ **11.** $3(u + v)(u - v)$ **13.** $\dfrac{10}{3(a + 4)}$ **15.** $\dfrac{x - 3}{x(x + 3)}$ **17.** $\dfrac{3x}{5a}$

19. $\dfrac{(x + 1)(x - 1)(x - 4)}{4(x + 2)}$ **21.** $\dfrac{x^2}{a + x}$ **23.** $\dfrac{15}{4}$ **25.** $\dfrac{3}{4x + 3}$ **27.** $\dfrac{7x - 1}{3x + 5}$ **29.** $\dfrac{7x^4}{3a^4}$

31. $\dfrac{4t(2t - 1)(t + 5)}{(2t + 1)^2}$ **33.** $\dfrac{1}{2}(x + y)$ **35.** $(x + y)(3p + 7q)$ **37.** $\dfrac{na^2}{V(1 - a)}$ **39.** $\dfrac{\pi}{2}$

Exercises 5-6, page 176

1. $\dfrac{9}{5}$ **3.** $\dfrac{8}{x}$ **5.** $\dfrac{5}{4}$ **7.** $\dfrac{3 + 7ax}{4x}$ **9.** $\dfrac{ax - b}{x^2}$ **11.** $\dfrac{30 + ax^2}{25x^3}$ **13.** $\dfrac{14 - a^2}{10a}$

15. $\dfrac{-x^2 + 4x + xy + y - 2}{xy}$ **17.** $\dfrac{7}{2(2x - 1)}$ **19.** $\dfrac{5 - 3x}{2x(x + 1)}$ **21.** $\dfrac{-3}{4(s - 3)}$ **23.** $\dfrac{7x + 6}{3(x + 3)(x - 3)}$

25. $\dfrac{2x - 5}{(x - 4)^2}$ **27.** $\dfrac{x + 27}{(x - 5)(x + 5)(x - 6)}$ **29.** $\dfrac{9x^2 + x - 2}{(3x - 1)(x - 4)}$ **31.** $\dfrac{13t^2 + 27t}{(t - 3)(t + 2)(t + 3)^2}$ **33.** $\dfrac{x + 1}{x - 1}$

35. $-\dfrac{(x + 1)(x - 1)(2x + 1)}{x^2(x + 2)}$ **37.** $-\dfrac{(3x + 4)(x - 1)}{2x}$ **39.** $\dfrac{2s}{r - s}$ **41.** $\dfrac{h}{(x + 1)(x + h + 1)}$ **43.** $\dfrac{-2hx - h^2}{x^2(x + h)^2}$

45. $\dfrac{y^2 - rx + r^2}{r^2}$ **47.** $\dfrac{2a - 1}{a^2}$ **49.** $\dfrac{3(H - H_0)}{4\pi H}$ **51.** $\dfrac{n(2n + 1)}{2(n + 2)(n - 1)}$

Exercises 5-7, page 180

1. 4   **3.** -3 **5.** $\dfrac{7}{2}$ **7.** $\dfrac{16}{21}$ **9.** -9 **11.** $-\dfrac{2}{13}$ **13.** $\dfrac{5}{3}$ **15.** -2 **17.** $\dfrac{3}{4}$ **19.** 6   **21.** -5

23. $-\dfrac{7}{8}$ **25.** No solution   **27.** $\dfrac{2}{3}$ **29.** $\dfrac{3b}{1 - 2b}$ **31.** $\dfrac{(2b - 1)(b + 6)}{2(b - 1)}$ **33.** $\dfrac{n_1 V - nV}{n_1}$ **35.** $\dfrac{V_r A - V_0}{V_0 A}$

37. $\dfrac{jX}{1 - g_m z}$ **39.** $\dfrac{PV^3 - bPV^2 + aV - ab}{RV^2}$ **41.** $\dfrac{kA_1 A_2 R - A_1 L_2}{A_2}$ **43.** $\dfrac{fnR_2 - fR_2}{R_2 + f - fn}$ **45.** 2.4 h

47. 3.2 min   **49.** 80 km/h   **51.** 2.5 qt

Review Exercises for Chapter 5, page 182

1. $12ax + 15a^2$ **3.** $4a^2 - 49b^2$ **5.** $4a^2 + 4a + 1$ **7.** $b^2 + 3b - 28$ **9.** $2x^2 - 13x - 45$
11. $16c^2 + 6cd - d^2$ **13.** $3(s + 3t)$ **15.** $a^2(x^2 + 1)$ **17.** $(x + 12)(x - 12)$ **19.** $(4x + 8 + t^2)(4x + 8 - t^2)$
21. $(3t - 1)^2$ **23.** $(5t + 1)^2$ **25.** $(x + 8)(x - 7)$ **27.** $(t - 9)(t + 4)$ **29.** $(2x - 9)(x + 4)$
31. $(2x + 5)(2x - 7)$ **33.** $(5b - 1)(2b + 5)$ **35.** $4(x + 4y)(x - 4y)$ **37.** $(x + 3)^3$

39. $(2x + 3)(4x^2 - 6x + 9)$ **41.** $(a - 3)(b^2 + 1)$ **43.** $(x + 5)(n - x + 5)$ **45.** $\dfrac{16x^2}{3a^2}$ **47.** $\dfrac{3x + 1}{2x - 1}$

49. $\dfrac{16}{5x(x - y)}$ **51.** $\dfrac{6}{5 - x}$ **53.** $\dfrac{x + 2}{2x(7x - 1)}$ **55.** $\dfrac{1}{x - 1}$ **57.** $\dfrac{16x - 15}{36x^2}$ **59.** $\dfrac{5y + 6}{2xy}$ **61.** $\dfrac{-2(2a + 3)}{a(a + 2)}$

63. $\dfrac{2x^2 - x + 1}{x(x + 3)(x - 1)}$ **65.** $\dfrac{12x^2 - 7x - 4}{2(x - 1)(x + 1)(4x - 1)}$ **67.** $\dfrac{x^3 + 6x^2 - 2x + 2}{x(x - 1)(x + 3)}$ **69.** 2   **71.** $\dfrac{7}{2c + 4}$

73. $-\dfrac{(a - 1)^2}{2a}$ **75.** 6   **77.** $\frac{1}{4}[(x + y)^2 - (x - y)^2] = \frac{1}{4}(x^2 + 2xy + y^2 - x^2 + 2xy - y^2) = \frac{1}{4}(4xy) = xy$

79. $2zS^2 + 2zS$ **81.** $\pi\ell(r_1 + r_2)(r_1 - r_2)$ **83.** $(t + 1)(1 - t)$ **85.** $W^4 - 4L^4 + 4k^2 L^2$

87. $4(3x^2 + 12x + 16)$ **89.** $\dfrac{12\pi^2 wv^2 D}{gn^2 t}$ **91.** $\dfrac{60t + 9t^2 + 2t^3}{6}$ **93.** $\dfrac{120 - 60d^2 + 5d^4 - d^6}{120}$

95. $\dfrac{2\pi^2 CN + 2\pi^2 Cn + N^2 - 2nN + n^2}{4\pi^2 C}$ **97.** $\dfrac{4r^3 - 3ar^2 - a^3}{4r^3}$ **99.** $\dfrac{q_2 D - fd}{D + d}$ **101.** $\dfrac{RHw}{w - RH}$

103. $\dfrac{-mb^2 s^2 - kL^2}{b^2 s}$ **105.** 3.4 h   **107.** 15 s   **109.** 11.3   **111.** 18 Ω

Exercises 6-1, page 189

1. $a = 1, b = -8, c = 5$ **3.** $a = 1, b = -2, c = -4$ **5.** Not quadratic   **7.** $a = 1, b = -1, c = 0$
9. 2, -2 **11.** $\frac{3}{2}, -\frac{3}{2}$ **13.** $-1, 9$ **15.** 3, 4   **17.** 0, -2 **19.** $\frac{1}{3}, -\frac{1}{3}$ **21.** $\frac{1}{3}, 4$ **23.** $-4, -4$
25. $\frac{2}{3}, \frac{3}{2}$ **27.** $\frac{1}{2}, -\frac{3}{2}$ **29.** 2, -1 **31.** $2b, -2b$ **33.** $0, \frac{5}{2}$ **35.** $\frac{5}{2}, -\frac{9}{2}$ **37.** 0, -2 **39.** $b - a, -b - a$
41. 10 Pa   **43.** 2 A.M., 10 A.M.   **45.** $\frac{3}{2}, 4$ **47.** $-2, \frac{1}{2}$ **49.** 3 N/cm, 6 N/cm   **51.** 8 mi/gal, 12 mi/gal

Exercises 6-2, page 193

1. $-5, 5$ **3.** $-\sqrt{7}, \sqrt{7}$ **5.** $-3, 7$ **7.** $-3 \pm \sqrt{7}$ **9.** 2, -4 **11.** $-2, -1$ **13.** $2 \pm \sqrt{2}$
15. $-5, 3$ **17.** $-3, \frac{1}{2}$ **19.** $\frac{1}{6}(3 \pm \sqrt{33})$ **21.** $\frac{1}{4}(1 \pm \sqrt{17})$ **23.** $-b \pm \sqrt{b^2 - c}$

Exercises 6-3, page 197

1. $2, -4$ **3.** $-2, -1$ **5.** $2 \pm \sqrt{2}$ **7.** $-5, 3$ **9.** $-3, \frac{1}{2}$ **11.** $\frac{1}{6}(3 \pm \sqrt{33})$ **13.** $\frac{1}{4}(1 \pm \sqrt{17})$
15. $-\frac{8}{5}, \frac{5}{6}$ **17.** $\frac{1}{2}(-5 \pm \sqrt{-5})$ **19.** $\frac{1}{6}(1 \pm \sqrt{109})$ **21.** $\frac{3}{2}, -\frac{3}{2}$ **23.** $\frac{3}{4}, -\frac{5}{8}$ **25.** $-0.54, 0.74$

27. $-0.26, 2.43$ **29.** $-c \pm \sqrt{c^2 + 1}$ **31.** $\dfrac{b + 1 \pm \sqrt{-3b^2 + 2b + 4b^2a + 1}}{2b^2}$ **33.** 4.376 cm

35. $\dfrac{-R \pm \sqrt{R^2 - 4L/C}}{2L}$ **37.** 4.5 ft, 6.5 ft   **39.** 173 ft by 116 ft, or 77 ft by 260 ft

Exercises 6-4, page 202

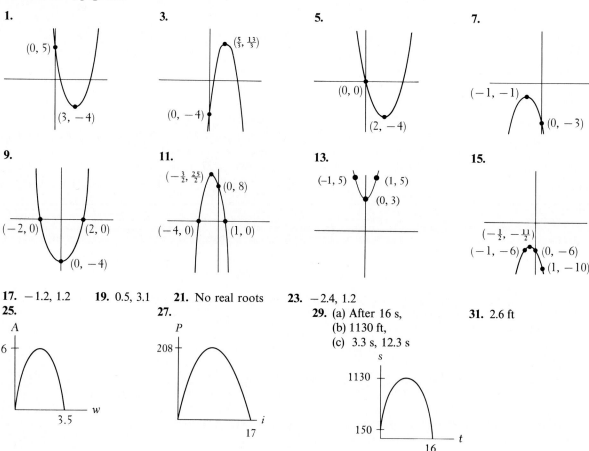

1. $(0, 5)$ $(3, -4)$

3. $\left(\frac{5}{3}, \frac{13}{3}\right)$ $(0, -4)$

5. $(0, 0)$ $(2, -4)$

7. $(-1, -1)$ $(0, -3)$

9. $(-2, 0)$ $(2, 0)$ $(0, -4)$

11. $\left(-\frac{3}{2}, \frac{25}{2}\right)$ $(0, 8)$ $(-4, 0)$ $(1, 0)$

13. $(-1, 5)$ $(1, 5)$ $(0, 3)$

15. $\left(-\frac{1}{2}, -\frac{11}{2}\right)$ $(-1, -6)$ $(0, -6)$ $(1, -10)$

17. $-1.2, 1.2$ **19.** $0.5, 3.1$ **21.** No real roots   **23.** $-2.4, 1.2$

25. [graph: A vs w, 6, 3.5]

27. [graph: P vs i, 208, 17]

29. (a) After 16 s, (b) 1130 ft, (c) 3.3 s, 12.3 s   [graph: s vs t, 1130, 150, 16]

31. 2.6 ft

Review Exercises for Chapter 6, page 203

1. $-4, 1$ **3.** $2, 8$ **5.** $\frac{1}{3}, -4$ **7.** $\frac{1}{2}, \frac{5}{3}$ **9.** $0, \frac{25}{6}$ **11.** $-\frac{3}{2}, \frac{7}{2}$ **13.** $-10, 11$ **15.** $-1 \pm \sqrt{6}$
17. $-4, \frac{9}{2}$ **19.** $\frac{1}{8}(3 \pm \sqrt{41})$ **21.** $\frac{1}{42}(-23 \pm \sqrt{-4091})$ **23.** $\frac{1}{6}(-2 \pm \sqrt{58})$ **25.** $-2 \pm 2\sqrt{2}$

27. $\frac{1}{3}(-4 \pm \sqrt{10})$ **29.** $-1, \frac{5}{4}$ **31.** $\frac{1}{4}(-3 \pm \sqrt{-47})$ **33.** $\dfrac{-1 \pm \sqrt{-1}}{a}$ **35.** $\dfrac{-3 \pm \sqrt{9 + 4a^3}}{2a}$

37. $-5, 6$ **39.** $\frac{1}{4}(1 \pm \sqrt{33})$ **41.** $3 \pm \sqrt{7}$ **43.** 0 (3 is not a solution)

45.

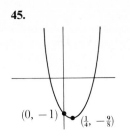

$(0, -1)$ $(\frac{1}{4}, -\frac{9}{8})$

47.

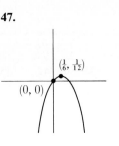

$(\frac{1}{6}, \frac{1}{12})$ $(0, 0)$

49. $-1.7, 1.2$ **51.** No real roots

53. 1.2 cm, 4.0 cm   **55.** 0.6 s, 1.9 s

57. 8000   **59.** $\dfrac{-k \pm \sqrt{k^2 + 4ck}}{2c}$

61.

p

0.135

6 t

63. 61 m, 81 m   **65.** 3.06 mm   **67.** 3.4 ft   **69.** 25

71.

p 6 h, 18 h

0.268

0.017

24 h

Exercises 7-1, page 207

1. $+, -, -$ **3.** $+, +, -$ **5.** $+, +, +$ **7.** $+, -, +$

9. $\sin \theta = \dfrac{1}{\sqrt{5}}$, $\cos \theta = \dfrac{2}{\sqrt{5}}$, $\tan \theta = \frac{1}{2}$, $\cot \theta = 2$, $\sec \theta = \frac{1}{2}\sqrt{5}$, $\csc \theta = \sqrt{5}$

11. $\sin \theta = -\dfrac{3}{\sqrt{13}}$, $\cos \theta = -\dfrac{2}{\sqrt{13}}$, $\tan \theta = \frac{3}{2}$, $\cot \theta = \frac{2}{3}$, $\sec \theta = -\frac{1}{2}\sqrt{13}$, $\csc \theta = -\frac{1}{3}\sqrt{13}$

13. $\sin \theta = \frac{12}{13}$, $\cos \theta = -\frac{5}{13}$, $\tan \theta = -\frac{12}{5}$, $\cot \theta = -\frac{5}{12}$, $\sec \theta = -\frac{13}{5}$, $\csc \theta = \frac{13}{12}$

15. $\sin \theta = -\dfrac{2}{\sqrt{29}}$, $\cos \theta = \dfrac{5}{\sqrt{29}}$, $\tan \theta = -\frac{2}{5}$, $\cot \theta = -\frac{5}{2}$, $\sec \theta = \frac{1}{5}\sqrt{29}$, $\csc \theta = -\frac{1}{2}\sqrt{29}$ **17.** II   **19.** II

21. IV   **23.** III

Exercises 7-2, page 214

1. $\sin 20°$; $-\cos 40°$ **3.** $-\tan 75°$, $-\csc 58°$ **5.** $-\sin 57°$; $-\cot 6°$ **7.** $\cos 40°$; $-\tan 40°$

9. $-\sin 15° = -0.26$ **11.** $-\cos 73.7° = -0.281$ **13.** $\tan 39.15° = 0.8141$ **15.** $\sec 31.67° = 1.175$

17. -0.523 **19.** -0.7620 **21.** -0.34 **23.** -3.910 **25.** 237.99°, 302.01°   **27.** 66.40°, 293.60°

29. 15.8°, 195.8°   **31.** 102.0°, 282.0°   **33.** 119.5°   **35.** 263°   **37.** 306.21°   **39.** 299.24°   **41.** -0.7003

43. -0.777 **45.** $<$ **47.** $=$ **49.** 0.0183 A   **51.** 16.7 in.

53. $\cos(-\theta) = \dfrac{x}{r}$, $\tan(-\theta) = \dfrac{-y}{x}$, $\cot(-\theta) = \dfrac{x}{-y}$, $\sec(-\theta) = \dfrac{r}{x}$, $\csc(-\theta) = \dfrac{r}{-y}$ **55.** (a) 5.7, (b) -1.4

Exercises 7-3, page 219

1. $\dfrac{\pi}{12}, \dfrac{5\pi}{6}$ **3.** $\dfrac{5\pi}{12}, \dfrac{11\pi}{6}$ **5.** $\dfrac{7\pi}{6}, \dfrac{3\pi}{2}$ **7.** $\dfrac{8\pi}{9}, \dfrac{13\pi}{9}$ **9.** 72°, 270°   **11.** 10°, 315°   **13.** 170°, 300°

15. 15°, 27°   **17.** 0.401   **19.** 4.40   **21.** 5.821   **23.** 3.115   **25.** 43.0°   **27.** 195.2°   **29.** 140°

31. 940.8°   **33.** 0.7071   **35.** 3.732   **37.** -0.8660 **39.** -8.327 **41.** 0.9056   **43.** -0.89

45. -0.48 **47.** -0.15 **49.** 0.3141, 2.827   **51.** 2.932, 6.074   **53.** 0.8309, 5.452   **55.** 2.442, 3.841

57. 0.030 ft·lb   **59.** 2400 m

Exercises 7-4, page 224

1. 346 ft   **3.** 5570 cm^2 **5.** $0.382 = 21.9°$ **7.** 4240 ft^2 **9.** 0.52 rad/s   **11.** 34.73 m^2 **13.** 2.30 ft
15. 22.6 m^2 **17.** 369 m^3 **19.** 0.4 km   **21.** 8.17 r/min   **23.** 5940 in./min   **25.** 25.7 ft
27. 1410 in./min   **29.** 0.433 rad/s   **31.** 25,100 ft/min   **33.** 209 rad   **35.** 14.9 m^3
37. 4.848×10^{-6} (all three values)   **39.** 7.13×10^7 mi

Review Exercises for Chapter 7, page 227

1. $\sin\theta = \frac{4}{5}$, $\cos\theta = \frac{3}{5}$, $\tan\theta = \frac{4}{3}$, $\cot\theta = \frac{3}{4}$, $\sec\theta = \frac{5}{3}$, $\csc\theta = \frac{5}{4}$

3. $\sin\theta = -\dfrac{2}{\sqrt{53}}$, $\cos\theta = \dfrac{7}{\sqrt{53}}$, $\tan\theta = -\frac{2}{7}$, $\cot\theta = -\frac{7}{2}$, $\sec\theta = \dfrac{\sqrt{53}}{7}$, $\csc\theta = -\dfrac{\sqrt{53}}{2}$

5. $-\cos 48°$, $\tan 14°$ **7.** $-\sin 71°$, $\sec 15°$ **9.** $\dfrac{2\pi}{9}, \dfrac{17\pi}{20}$ **11.** $\dfrac{4\pi}{15}, \dfrac{9\pi}{8}$ **13.** 252°, 130°   **15.** 12°, 330°

17. 32.1°   **19.** 206.7°   **21.** 1.78   **23.** 0.3534   **25.** 4.5736   **27.** 2.377   **29.** -0.415 **31.** -0.47
33. -1.080 **35.** -0.4264 **37.** -1.64 **39.** 4.140   **41.** -0.5878 **43.** -0.8660 **45.** 0.5569
47. 1.197   **49.** 10.30°, 190.30°   **51.** 118.23°, 241.77°   **53.** 0.5759, 5.707   **55.** 4.187, 5.238   **57.** 223.76°
59. 246.78°   **61.** 0.0562 W   **63.** $0.800 = 45.8°$ **65.** 18.98 cm   **67.** 24.4 ft^2 **69.** 4710 cm/s
71. 1.81×10^6 cm/s   **73.** 170.9 ft   **75.** 3.58×10^5 km

Exercises 8-1, page 234

1. (a) Vector: magnitude and direction are specified; (b) scalar: only magnitude is specified
3. (a) Vector: magnitude and direction are specified; (b) scalar: only magnitude is specified

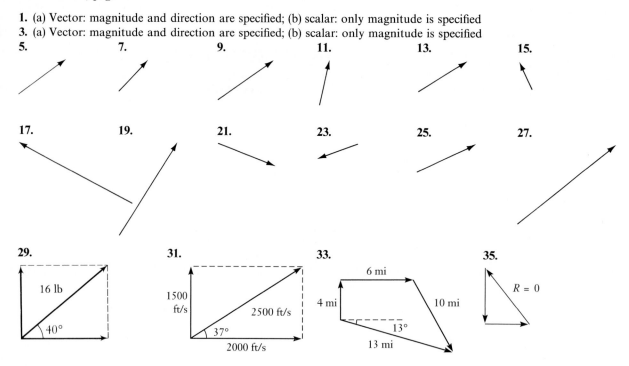

Exercises 8-2, page 237

1. 662, 352   **3.** $-349, -664$ **5.** 3.22, 7.97   **7.** $-62.9, 44.1$ **9.** 2.08, -8.80 **11.** $-2.53, -0.788$
13. $-0.8088, 0.3296$ **15.** 88,920, 12,240   **17.** 23.9 km/h, 7.43 km/h   **19.** 52.2 lb, 17.4 lb
21. 115 km, 88.3 km   **23.** 18,530 mi/h, -825.3 mi/h

Exercises 8-3, page 242

1. $R = 24.2$, $\theta = 52.6°$ with A **3.** $R = 7.781$, $\theta = 66.63°$ with A **5.** $R = 10.0$, $\theta = 58.8°$
7. $R = 2.74$, $\theta = 111.0°$ **9.** $R = 2130$, $\theta = 107.7°$ **11.** $R = 1.426$, $\theta = 299.12°$ **13.** $R = 29.2$, $\theta = 10.8°$
15. $R = 47.0$, $\theta = 101.0°$ **17.** $R = 27.27$, $\theta = 33.14°$ **19.** $R = 12.735$, $\theta = 25.216°$ **21.** $R = 50.2$, $\theta = 50.3°$
23. $R = 235$, $\theta = 121.7°$

Exercises 8-4, page 245

1. 6.60 lb, 29.5° from 5.75-lb force   **3.** 11,400 N, 44.2° above horizontal   **5.** 3070 ft, 17.8° S of W
7. 229.4 ft, 72.82° N of E   **9.** 25.3 km/h, 29.6° S of E   **11.** 175 lb, 9.3° above horizontal
13. 540 km/h, 6.2° from direction of plane   **15.** 18,130 mi/h, 0.03° from direction of shuttle
17. 181,000 in./min^2, $\phi = 89.6°$ **19.** 138 mi, 65.0° N of E
21. 79.0 m/s, 11.3° from direction of plane, 75.6° from vertical   **23.** 4.06 A/m, 11.6° with magnet

Exercises 8-5, page 252

1. $b = 38.1$, $C = 66.0°$, $c = 46.1$ **3.** $a = 2800$, $b = 2620$, $C = 108.0°$ **5.** $B = 12.20°$, $C = 149.57°$, $c = 7.448$
7. $a = 110.5$, $A = 149.70°$, $C = 9.57°$ **9.** $A = 125.6°$, $a = 0.0776$, $c = 0.00566$ **11.** $A = 99.4°$, $b = 55.1$, $c = 24.4$
13. $A = 68.01°$, $a = 5520$, $c = 5376$
15. $A_1 = 61.36°$, $C_1 = 70.51°$, $c_1 = 5.628$; $A_2 = 118.64°$, $C_2 = 13.23°$, $c_2 = 1.366$
17. $A_1 = 107.3°$, $a_1 = 5280$, $C_1 = 41.3°$; $A_2 = 9.9°$, $a_2 = 952$, $C_2 = 138.7°$ **19.** No solution   **21.** 2.47 ft
23. 884 N   **25.** 406 m   **27.** 13.94 cm   **29.** 27,300 km   **31.** 77.3° with bank downstream

Exercises 8-6, page 257

1. $A = 50.3°$, $B = 75.7°$, $c = 6.31$ **3.** $A = 70.9°$, $B = 11.1°$, $c = 4750$ **5.** $A = 34.72°$, $B = 40.67°$, $C = 104.61°$
7. $A = 18.21°$, $B = 22.28°$, $C = 139.51°$ **9.** $A = 6.0°$, $B = 16.0°$, $c = 1150$ **11.** $A = 82.3°$, $b = 21.6$, $C = 11.4°$
13. $A = 36.24°$, $B = 39.09°$, $a = 97.22$ **15.** $A = 46.94°$, $B = 61.82°$, $C = 71.24°$
17. $A = 137.9°$, $B = 33.7°$, $C = 8.4°$ **19.** $b = 37$, $C = 25°$, $c = 24$ **21.** 69.4 mi   **23.** 0.039 km
25. 57.3°, 141.7°   **27.** 18.2 ft   **29.** 5.09 km/h   **31.** 94.7°

Review Exercises for Chapter 8, page 258

1. $A_x = 57.4$, $A_y = 30.5$ **3.** $A_x = -0.7485$, $A_y = -0.5357$ **5.** $R = 602$, $\theta = 57.1°$ with A
7. $R = 5960$, $\theta = 33.60°$ with A **9.** $R = 965$, $\theta = 8.6°$ **11.** $R = 26.12$, $\theta = 146.03°$
13. $R = 71.93$, $\theta = 336.50°$ **15.** $R = 99.42$, $\theta = 359.57°$ **17.** $b = 18.1$, $C = 64.0°$, $c = 17.5$
19. $A = 21.2°$, $b = 34.8$, $c = 51.5$ **21.** $a = 17,340$, $b = 24,660$, $C = 7.99°$ **23.** $A = 39.88°$, $a = 51.94$, $C = 30.03°$
25. $A_1 = 54.8°$, $a_1 = 12.7$, $B_1 = 68.6°$; $A_2 = 12.0°$, $a_2 = 3.24$, $B_2 = 111.4°$ **27.** $A = 32.3°$, $b = 267$, $C = 17.7°$
29. $A = 148.7°$, $B = 9.3°$, $c = 5.66$ **31.** $a = 1782$, $b = 1920$, $C = 16.00°$ **33.** $A = 37°$, $B = 25°$, $C = 118°$
35. $A = 20.6°$, $B = 35.6°$, $C = 123.8°$ **37.** -155.7 lb, 81.14 lb   **39.** 2080 ft/s   **41.** 14.9 mN
43. 12.1 ft/s^2 **45.** 2.30 m, 2.49 m   **47.** 32,900 mi   **49.** 2.65 km   **51.** 186 mi   **53.** 808 N, 36.4° N of E
55. 2510 ft or 3370 ft (ambiguous)

Exercises 9-1, page 264

1. 0, -0.7, -1, -0.7, 0, 0.7, 1, 0.7, 0, -0.7, -1,   **3.** -3, -2.1, 0, 2.1, 3, 2.1, 0, -2.1, -3,
-0.7, 0, 0.7, 1, 0.7, 0 -2.1, 0, 2.1, 3, 2.1, 0, -2.1, -3

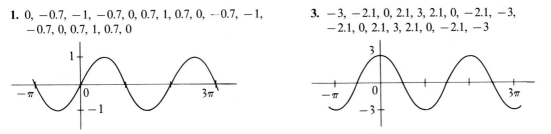

5.

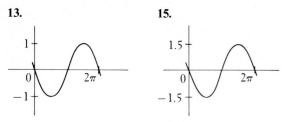

7.

9.

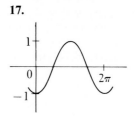

11.

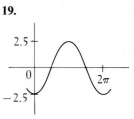

13.

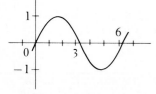

15.

17.

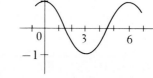

19.

21. 0, 0.84, 0.91, 0.14, −0.76, −0.96, −0.28, 0.66

23. 1, 0.54, −0.42, −0.99, −0.65, 0.28, 0.96, 0.75

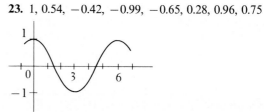

Exercises 9-2, page 268

1. $\dfrac{\pi}{3}$ **3.** $\dfrac{\pi}{4}$ **5.** $\dfrac{\pi}{6}$ **7.** $\dfrac{\pi}{8}$ **9.** 1   **11.** $\dfrac{1}{2}$ **13.** 6π **15.** 3π **17.** 3   **19.** $\dfrac{2}{\pi}$

21.

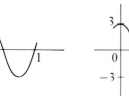

23.

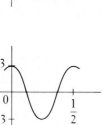

25.

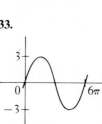

27.

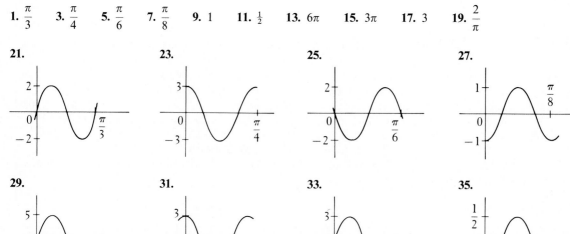

29.

31.

33.

35.

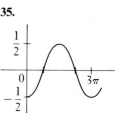

37.

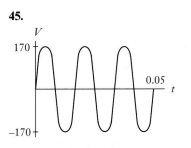

39.

41. $y = \sin 6x$

43. $y = \sin \pi x$

45.

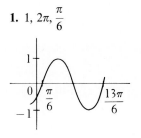

47.

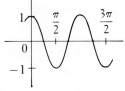

Exercises 9-3, page 271

1. $1, 2\pi, \dfrac{\pi}{6}$

3. $1, 2\pi, -\dfrac{\pi}{6}$

5. $2, \pi, -\dfrac{\pi}{4}$

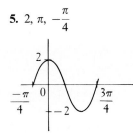

7. $1, \pi, \dfrac{\pi}{2}$

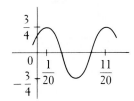

9. $\dfrac{1}{2}, 4\pi, \dfrac{\pi}{2}$

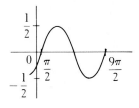

11. $3, 6\pi, -\pi$

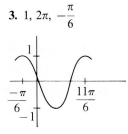

13. $1, 2, -\dfrac{1}{8}$

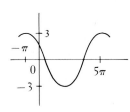

15. $\dfrac{3}{4}, \dfrac{1}{2}, \dfrac{1}{20}$

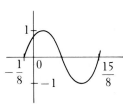

17. $0.6, 1, \dfrac{1}{2\pi}$

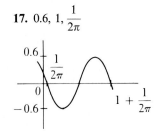

19. $40, \dfrac{2}{3}, -\dfrac{2}{3\pi}$

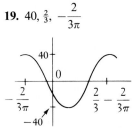

21. $1, \dfrac{2}{\pi}, \dfrac{1}{\pi}$

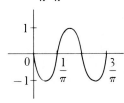

23. $\dfrac{3}{2}, 2, -\dfrac{\pi}{6}$

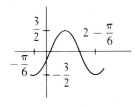

25.

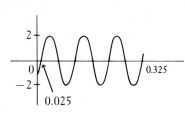

27.

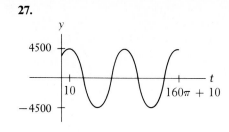

Exercises 9-4, page 275

1. Undef., -1.7, -1, -0.58, 0, 0.58,
1, 1.7, undef., -1.7, -1, -0.58, 0

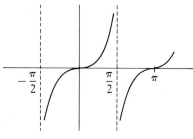

3. Undef., 2, 1.4, 1.2, 1, 1.2, 1.4, 2,
undef., -2, -1.4, -1.2, -1

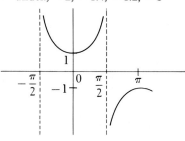

5.

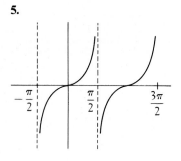

7.

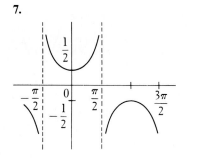

9.

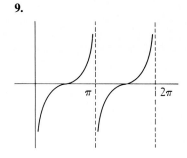

11.

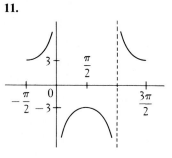

13.

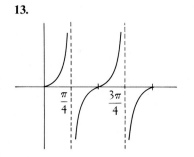

15.

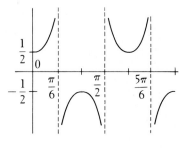

17.
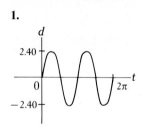

19.
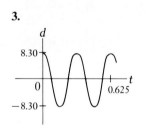

21.
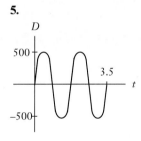

23.
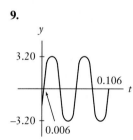

25.
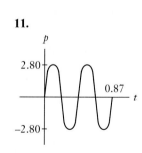

27.
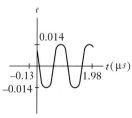

Exercises 9-5, page 279

1.
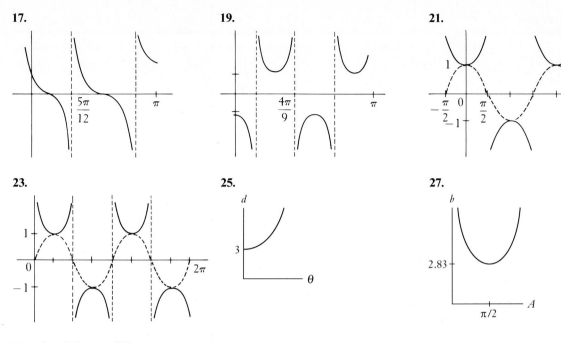

3.

5.

7.

9.

11.

13.

15.

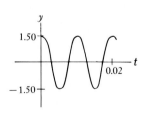
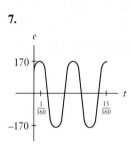

Exercises 9-6, page 284

1.
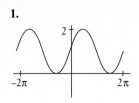

3.
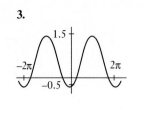

5.
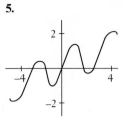

7.
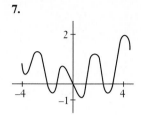

9.
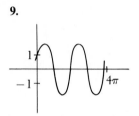

11.
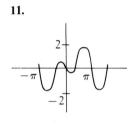

13.
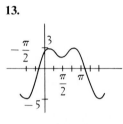

15.
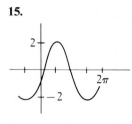

17.
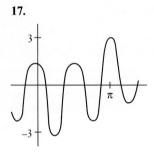

19.
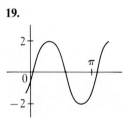

21.
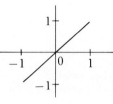

23.
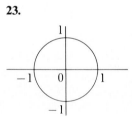

25.
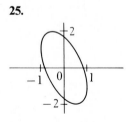

27.
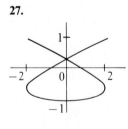

29.
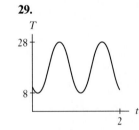

31.

33.
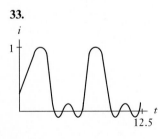

35.
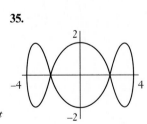

Review Exercises for Chapter 9, page 286

1.
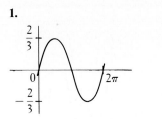

3.
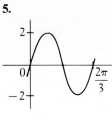

5.
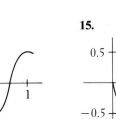

7.
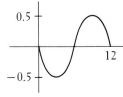

9.
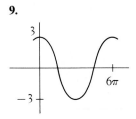

11.
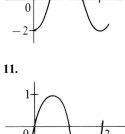

13.
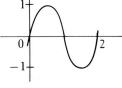

15.
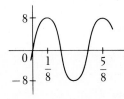

17.
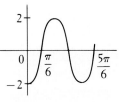

19.
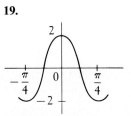

21.
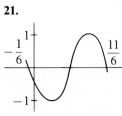

23.
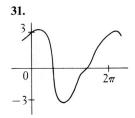

25.
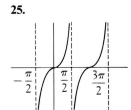

27.
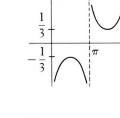

29.
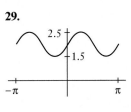

31.
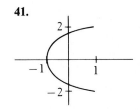

33.
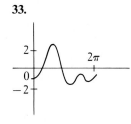

35.
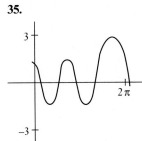

37. $y = 2 \sin\left(2x + \dfrac{\pi}{2}\right)$

39. $y = \cos\left(\dfrac{\pi}{4}x - \dfrac{3\pi}{4}\right)$

41.
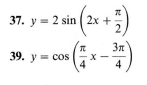

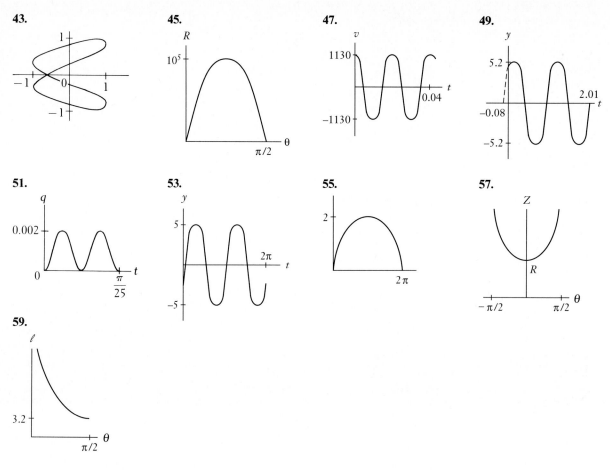

43. **45.** **47.** **49.**

51. **53.** **55.** **57.**

59.

Exercises 10-1, page 292

1. x^3 **3.** $\dfrac{1}{a^4}$ **5.** $\dfrac{1}{25}$ **7.** $\dfrac{25}{4}$ **9.** $\dfrac{4a^2}{x^2}$ **11.** $\dfrac{n^2}{5a}$ **13.** 1 **15.** -7 **17.** $\dfrac{3}{x^2}$ **19.** $\dfrac{1}{7^3a^3x^3}$ **21.** $\dfrac{n^3}{2}$

23. $\dfrac{1}{a^3b^6}$ **25.** $\dfrac{1}{a+b}$ **27.** $\dfrac{2x^2+3y^2}{x^2y^2}$ **29.** $\dfrac{2^3}{3^5}$ **31.** $\dfrac{a^2b^2}{3}$ **33.** $\dfrac{b^3}{432a}$ **35.** $\dfrac{4}{t^4v^4}$ **37.** $\dfrac{x^8-y^2}{x^4y^2}$

39. $\dfrac{2a^6+16}{a^8}$ **41.** $\dfrac{10}{9}$ **43.** $\dfrac{ab}{a+b}$ **45.** $\dfrac{4n^2-4n+1}{n^4}$ **47.** $\dfrac{45}{2}$ **49.** $-\dfrac{x}{y}$ **51.** $\dfrac{a^2-ax+x^2}{ax}$

53. $\dfrac{t^2+t+2}{t^2}$ **55.** $\dfrac{2x}{(x+1)(x-1)}$ **57.** (a) 4^5, (b) 2^{10} **59.** $\left(\dfrac{a}{b}\right)^{-n} = \dfrac{1}{\left(\dfrac{a}{b}\right)^n} = \dfrac{1}{\dfrac{a^n}{b^n}} = \dfrac{b^n}{a^n} = \left(\dfrac{b}{a}\right)^n$ **61.** $\dfrac{\text{Pa·m}^3}{\text{mol}}$

63. $\dfrac{\omega^4-2\omega^2\omega_0^2+\omega_0^4}{\omega^2\omega_0^2}$

Exercises 10-2, page 297

1. 5 **3.** 3 **5.** 16 **7.** 10^{25} **9.** $\dfrac{1}{2}$ **11.** $\dfrac{1}{16}$ **13.** 25 **15.** 4096 **17.** $\dfrac{1}{110}$ **19.** $\dfrac{6}{7}$ **21.** $-\dfrac{1}{2}$

23. 24 **25.** $\dfrac{39}{1000}$ **27.** $\dfrac{3}{5}$ **29.** 2.059 **31.** 0.53891 **33.** $a^{7/6}$ **35.** $\dfrac{1}{y^{9/10}}$ **37.** $s^{23/12}$ **39.** $\dfrac{1}{y^{13/12}}$

41. $2ab^2$ **43.** $\dfrac{1}{8a^3b^{9/4}}$ **45.** $\dfrac{4x}{(4x^2+1)^{1/2}}$ **47.** $\dfrac{27}{64t^3}$ **49.** $\dfrac{b^{11/10}}{2a^{1/12}}$ **51.** $\dfrac{2}{3}x^{1/6}y^{11/12}$ **53.** $\dfrac{x}{(x+2)^{1/2}}$

55. $\dfrac{a^2+1}{a^4}$ **57.** $\dfrac{a+1}{a^{1/2}}$ **59.** $\dfrac{5x^2-2x}{(2x-1)^{1/2}}$ **61.**   **63.**   **65.** 0.84

67. 1.88 mA

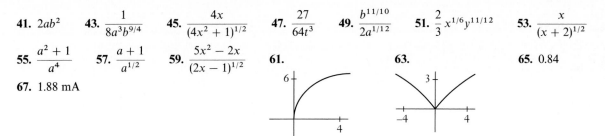

Exercises 10-3, page 301

1. $2\sqrt{6}$ **3.** $3\sqrt{5}$ **5.** $xy^2\sqrt{y}$ **7.** $qr^3\sqrt{pr}$ **9.** $x\sqrt{5}$ **11.** $3ac^2\sqrt{2ab}$ **13.** $2\sqrt[3]{2}$ **15.** $2\sqrt[5]{3}$

17. $2\sqrt[3]{a^2}$ **19.** $2st\sqrt[4]{4r^3t}$ **21.** 2   **23.** $ab\sqrt[3]{b^2}$ **25.** $\frac{1}{2}\sqrt{6}$ **27.** $\dfrac{\sqrt{ab}}{b}$ **29.** $\frac{1}{2}\sqrt[3]{6}$ **31.** $\frac{1}{3}\sqrt[5]{27}$

33. $2\sqrt{5}$ **35.** 2   **37.** 200   **39.** 2000   **41.** $\sqrt{2a}$ **43.** $\frac{1}{2}\sqrt{2}$ **45.** $\sqrt[3]{2}$ **47.** $\sqrt[8]{2}$ **49.** $\frac{1}{6}\sqrt{6}$

51. $\dfrac{\sqrt{b(a^2+b)}}{ab}$ **53.** $\dfrac{\sqrt{2x^2+x}}{2x+1}$ **55.** $a+b$ **57.** $\sqrt{4x^2-1}$ **59.** $\frac{1}{2}\sqrt{4x^2+1}$ **61.** 55.0%

63. $\dfrac{24\sqrt{EIgWL}}{WL^2}$

Exercises 10-4, page 304

1. $7\sqrt{3}$ **3.** $\sqrt{5}-\sqrt{7}$ **5.** $3\sqrt{5}$ **7.** $-4\sqrt{3}$ **9.** $-2\sqrt{2a}$ **11.** $19\sqrt{7}$ **13.** $-4\sqrt{5}$
15. $23\sqrt{3}-6\sqrt{2}$ **17.** $\frac{7}{3}\sqrt{15}$ **19.** 0   **21.** $13\sqrt[3]{3}$ **23.** $\sqrt[4]{2}$ **25.** $(a-2b^2)\sqrt{ab}$ **27.** $(3-2a)\sqrt{10}$

29. $(2b-a)\sqrt[3]{3a^2b}$ **31.** $\dfrac{(a^2-c^3)\sqrt{ac}}{a^2c^3}$ **33.** $\dfrac{(a-2b)\sqrt[3]{ab^2}}{ab}$ **35.** $\dfrac{2b\sqrt{a^2-b^2}}{b^2-a^2}$

37. $15\sqrt{3}-11\sqrt{5}=1.3840144$ **39.** $\frac{1}{6}\sqrt{6}=0.4082483$ **41.** $3\sqrt{3}$ **43.** $18+3\sqrt{2}=22.2$ ft

Exercises 10-5, page 307

1. $\sqrt{30}$ **3.** $2\sqrt{3}$ **5.** 2   **7.** $2\sqrt[5]{2}$ **9.** 50   **11.** 16   **13.** $\frac{1}{3}\sqrt{30}$ **15.** $\frac{1}{33}\sqrt{165}$ **17.** $\sqrt{6}-\sqrt{15}$
19. $8-12\sqrt{3}$ **21.** -1 **23.** $39-12\sqrt{3}$ **25.** $48+9\sqrt{15}$ **27.** $66+13\sqrt{11x}-5x$ **29.** $a\sqrt{b}+c\sqrt{ac}$
31. $5n\sqrt{3}+10\sqrt{mn}$ **33.** $2a-3b+2\sqrt{2ab}$ **35.** $\sqrt{6}-\sqrt{10}-2$ **37.** $\sqrt[6]{72}$ **39.** $c^{12}\sqrt[12]{a^3b^7c^4}$ **41.** 1

43. $2x\sqrt{2x}+x\sqrt[6]{8y^4}-2\sqrt[6]{x^3y^2}-y$ **45.** $\dfrac{2-a-a^2}{a}$ **47.** $4x^2+x-2y-4x\sqrt{x-2y}$ (valid for $x\ge2y$)

49. $-1-\sqrt{66}=-9.1240384$ **51.** $158+9\sqrt{182}=279.41664$ **53.** $\dfrac{2x+1}{\sqrt{x}}$ **55.** $\dfrac{5x^2+2x}{\sqrt{2x+1}}$

57. $(1-\sqrt{2})^2-2(1-\sqrt{2})-1=1-2\sqrt{2}+2-2+2\sqrt{2}-1=0$
59. $[\frac{1}{2}(\sqrt{b^2-4k^2}-b)]^2+b[\frac{1}{2}(\sqrt{b^2-4k^2}-b)]+k^2=\frac{1}{4}(b^2-4k^2)-\frac{b}{2}\sqrt{b^2-4k^2}+\frac{1}{4}b^2+\frac{b}{2}\sqrt{b^2-4k^2}-\frac{1}{2}b^2+k^2=0$

Exercises 10-6, page 310

1. $\sqrt{7}$ **3.** $\dfrac{1}{2}\sqrt{14}$ **5.** $\dfrac{1}{6}\sqrt[3]{9x^2}$ **7.** $\dfrac{\sqrt[6]{200}}{2}$ **9.** $\dfrac{1}{2}\sqrt[6]{4a^3}$ **11.** $\dfrac{2-\sqrt{6}}{2}$ **13.** $\dfrac{a\sqrt{2}-b\sqrt{a}}{a}$

15. $\dfrac{3\sqrt{a}-\sqrt{3b}}{3}$ **17.** $\frac{1}{4}(\sqrt{7}-\sqrt{3})$ **19.** $\frac{1}{23}(5\sqrt{7}+\sqrt{14})$ **21.** $-\frac{3}{8}(\sqrt{5}+3)$ **23.** $\frac{1}{13}(9+\sqrt{3})$

25. $\frac{1}{11}(\sqrt{7}+3\sqrt{2}-6-\sqrt{14})$ **27.** $\frac{1}{13}(4-\sqrt{3})$ **29.** $\frac{1}{17}(-56+9\sqrt{15})$ **31.** $\frac{1}{14}(4-\sqrt{2})$

33. $\dfrac{2x+2\sqrt{xy}}{x-y}$ **35.** $\dfrac{8(3\sqrt{a}+2\sqrt{b})}{9a-4b}$ **37.** $\dfrac{2c+4d\sqrt{2c}+3d^2}{2c-d^2}$ **39.** $-\dfrac{\sqrt{x^2-y^2}+\sqrt{x^2+xy}}{y}$

41. $14 - 2\sqrt{42} = 1.0385186$ **43.** $-\dfrac{16 + 5\sqrt{30}}{26} = -1.6686972$ **45.** $\dfrac{1}{\sqrt{30} - 2\sqrt{3}}$ **47.** $\dfrac{1}{\sqrt{x + h} + \sqrt{x}}$

49. $\dfrac{\sqrt{3gsw}}{6w}$ **51.** $\dfrac{\sqrt{2g}(\sqrt{h_2} + \sqrt{h_1})}{2g(h_2 - h_1)}$

Review Exercises for Chapter 10, page 312

1. $\dfrac{2}{a^2}$ **3.** $\dfrac{2d^3}{c}$ **5.** 375 **7.** $\dfrac{1}{8000}$ **9.** $\dfrac{t^4}{9}$ **11.** -28 **13.** $64a^2b^5$ **15.** $-8m^9n^6$ **17.** $\dfrac{2y - x^2}{x^2 y}$

19. $\dfrac{2y}{x + 2y}$ **21.** $\dfrac{b}{ab - 3}$ **23.** $\dfrac{(x^3y^3 - 1)^{1/3}}{y}$ **25.** $4a(a^2 + 4)^{1/2}$ **27.** $\dfrac{-2(x + 1)}{(x - 1)^3}$ **29.** $2\sqrt{17}$

31. $b^2c\sqrt{ab}$ **33.** $3ab^2\sqrt{a}$ **35.** $2tu\sqrt{21st}$ **37.** $\dfrac{5\sqrt{2s}}{2s}$ **39.** $\dfrac{1}{9}\sqrt{33}$ **41.** $mn^2\sqrt[4]{8m^2n}$ **43.** $\sqrt{2}$

45. $14\sqrt{2}$ **47.** $-7\sqrt{7}$ **49.** $3ax\sqrt{2x}$ **51.** $(2a + b)\sqrt[3]{a}$ **53.** $10 - \sqrt{55}$ **55.** $4\sqrt{3} - 4\sqrt{5}$

57. $-45 - 7\sqrt{17}$ **59.** $42 - 7\sqrt{7a} - 3a$ **61.** $\dfrac{6x + \sqrt{3xy}}{12x - y}$ **63.** $-\dfrac{8 + \sqrt{6}}{29}$ **65.** $\dfrac{13 - 2\sqrt{35}}{29}$

67. $\dfrac{6x - 13a\sqrt{x} + 5a^2}{9x - 25a^2}$ **69.** $\sqrt{4b^2 + 1}$ **71.** $\dfrac{15 - 2\sqrt{15}}{4}$ **73.** $2\sqrt{13} + 5\sqrt{6} = 19.458551$

75. $51 - 7\sqrt{105} = -20.728655$ **77.** 6.0% **79.** (a) $v = k(P/W)^{1/3}$, (b) $v = \dfrac{k\sqrt[3]{PW^2}}{W}$ **81.** $\dfrac{n_1^2 n_2^2 v}{n_1^2 - n_2^2}$

83. $T = \dfrac{2\sqrt{2\pi\mu\ell I}}{\mu a^2}$ **85.** $6\sqrt{2}$ cm   **87.** $\dfrac{\ell\sqrt{2a\ell + a^2}}{2\ell + a}$

Exercises 11-1, page 319

1. $9j$ **3.** $-2j$ **5.** $0.6j$ **7.** $2j\sqrt{2}$ **9.** $\frac{1}{2}j\sqrt{7}$ **11.** $-j\sqrt{\frac{2}{5}} = -\frac{1}{5}j\sqrt{10}$ **13.** $-7; 7$ **15.** $4; -4$
17. $-j$ **19.** 1 **21.** 0 **23.** $-2j$ **25.** $2 + 3j$ **27.** $-7j$ **29.** $6 + 2j$ **31.** $-2 + 3j$
33. $3\sqrt{2} - 2j\sqrt{2}$ **35.** -1 **37.** $6 + 7j$ **39.** $-2j$ **41.** $x = 2, y = -2$ **43.** $x = 10, y = -6$
45. $x = 0, y = -1$ **47.** $x = -2, y = 3$ **49.** Yes   **51.** It is a real number.

Exercises 11-2, page 322

1. $5 - 8j$ **3.** $-9 + 6j$ **5.** $7 - 5j$ **7.** $-5j$ **9.** -1 **11.** $-8 + 21j$ **13.** $7 + 49j$
15. $-22.4 + 6.4j$ **17.** $22 + 3j$ **19.** $-42 - 6j$ **21.** $-18j\sqrt{2}$ **23.** $25j\sqrt{5}$ **25.** $3j\sqrt{3}$
27. $-4\sqrt{3} + 6j\sqrt{7}$ **29.** $-28j$ **31.** $3\sqrt{7} + 3j$ **33.** $-40 - 42j$ **35.** $-2 - 2j$ **37.** $\frac{1}{29}(-30 + 12j)$
39. $0.075 + 0.025j$ **41.** $-\frac{1}{3}(1 + j)$ **43.** $\frac{1}{11}(-13 + 8j\sqrt{2})$ **45.** $\frac{1}{3}(2 + 5j\sqrt{2})$ **47.** $\frac{1}{5}(-1 + 3j)$
49. $(-1 + j)^2 + 2(-1 + j) + 2 = 1 - 2j + j^2 - 2 + 2j + 2 = 0$ **51.** 13 **53.** $-\frac{1}{13}(5 + 12j)$
55. $281 + 35.2j$ volts   **57.** $(a + bj) + (a - bj) = 2a$ **59.** $(a + bj) - (a - bj) = 2bj$

Exercises 11-3, page 324

1.   **3.**   **5.** $5 + 4j$ **7.** $8 + j$

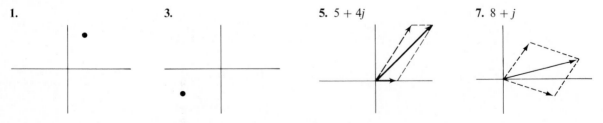

9. $4 + 4j$

11. $-3j$

13. $-1 + 4j$

15. $-2 + 3j$

17. $4.5 + 2.0j$

19. $4 - 11j$

21. $-2j$

23. $-13 + j$

25.

27.

29.

31.

Exercises 11-4, page 328

1. $10(\cos 36.9° + j \sin 36.9°)$

3. $5(\cos 306.9° + j \sin 306.9°)$

5. $3.61(\cos 123.7° + j \sin 123.7°)$

7. $6.00(\cos 203.6° + j \sin 203.6°)$

9. $2(\cos 60° + j \sin 60°)$

11. $8.062(\cos 295.84° + j \sin 295.84°)$

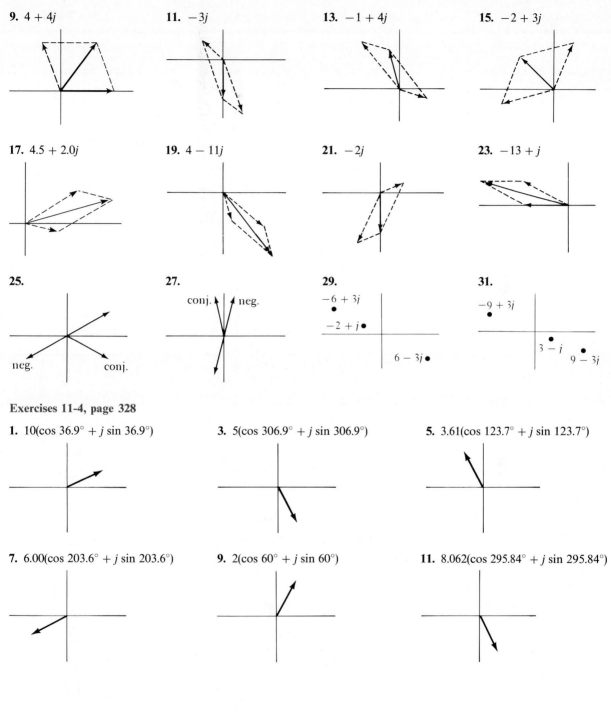

13. $3(\cos 180° + j \sin 180°)$ **15.** $9(\cos 90° + j \sin 90°)$ **17.** $2.94 + 4.05j$ **19.** $-1.39 + 0.800j$

21. -6 **23.** 8 **25.** $11.97 - 3.082j$ **27.** $-0.500 - 0.866j$

29. $-4.71 + 0.595j$ **31.** $-0.6052 - 0.7096j$ **33.** $7.32j$ **35.** $-6.961 + 86.14j$

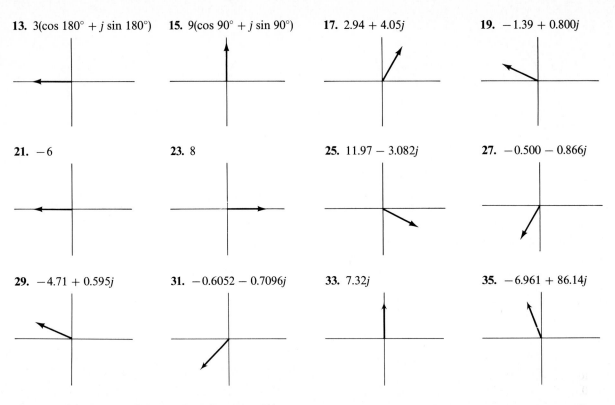

37. $R = 42.7$ N, $\theta = 306.8°$ **39.** $2.51 + 12.1j$ V/m

Exercises 11-5, page 332

1. $3.00e^{1.05j}$ **3.** $4.50e^{4.93j}$ **5.** $375.5e^{1.666j}$ **7.** $0.515e^{3.46j}$ **9.** $4.06e^{-1.07j} = 4.06e^{5.21j}$ **11.** $9245e^{5.172j}$
13. $5.00e^{5.36j}$ **15.** $3.61e^{2.55j}$ **17.** $6.37e^{0.386j}$ **19.** $825.7e^{3.836j}$ **21.** $3.00\underline{/28.6°}$; $2.63 + 1.44j$
23. $4.64\underline{/106.0°}$; $-1.28 + 4.46j$ **25.** $3.20\underline{/310.0°}$; $2.06 - 2.45j$ **27.** $0.1724\underline{/136.99°}$; $-0.1261 + 0.1176j$
29. $391e^{0.285j}$ ohms; 391 Ω **31.** $2.11 - 5.43j$ cm

Exercises 11-6, page 338

1. $8(\cos 80° + j \sin 80°)$ **3.** $3(\cos 250° + j \sin 250°)$ **5.** $2(\cos 35° + j \sin 35°)$ **7.** $2.4(\cos 110° + j \sin 110°)$
9. $8(\cos 105° + j \sin 105°)$ **11.** $256(\cos 0° + j \sin 0°)$ **13.** $0.305 + 1.70j = 1.73\underline{/79.8°}$
15. $-5134 + 10,570j = 11,750 (\cos 115.91° + j \sin 115.91°)$ **17.** $65.0(\cos 345.7° + j \sin 345.7°) = 63 - 16j$
19. $61.4(\cos 343.9° + j \sin 343.9°)$; $59 - 17j$ **21.** $2.21(\cos 71.6° + j \sin 71.6°)$; $\frac{7}{10} + \frac{21}{10}j$
23. $0.385(\cos 120.5° + j \sin 120.5°) = \frac{1}{169}(-33 + 56j)$ **25.** $625(\cos 212.5° + j \sin 212.5°) = -527 - 336j$
27. $609(\cos 281.5° + j \sin 281.5°)$; $122 - 597j$ **29.** $2(\cos 30° + j \sin 30°)$, $2(\cos 210° + j \sin 210°)$
31. $-0.364 + 1.67j$, $-1.26 - 1.15j$, $1.63 - 0.520j$ **33.** $1, -1, j, -j$ **35.** $3j, -\frac{3}{2}(\sqrt{3} + j), \frac{3}{2}(\sqrt{3} - j)$
37. $[\frac{1}{2}(1 - j\sqrt{3})]^3 = \frac{1}{8}[1 - 3(j\sqrt{3}) + 3(j\sqrt{3})^2 - (j\sqrt{3})^3] = \frac{1}{8}[1 - 3j\sqrt{3} - 9 + 3j\sqrt{3}] = \frac{1}{8}(-8) = -1$
39. $p = 47.9\underline{/40.5°}$ watts

Exercises 11-7, page 344

1. 12.9 V   **3.** (a) 2850 Ω, (b) $37.9°$, (c) 16.4 V   **5.** (a) 14.6 Ω, (b) $-90.0°$ **7.** (a) 47.8 Ω, (b) $19.8°$
9. 38.0 V   **11.** 54.5 Ω, $-62.3°$ **13.** 0.682 H   **15.** 376 Hz   **17.** 1.30×10^{-11} F $= 13.0$ pF
19. 1.02 mW

Review Exercises for Chapter 11, page 346

1. $10 - j$ **3.** $6 + 2j$ **5.** $9 + 2j$ **7.** $-12 + 66j$ **9.** $\frac{1}{85}(21 + 18j)$ **11.** $-2 - 3j$ **13.** $\frac{1}{10}(-12 + 9j)$
15. $\frac{1}{5}(13 + 11j)$ **17.** $x = -\frac{2}{3}, y = -2$ **19.** $x = -\frac{1}{2}, y = \frac{1}{2}$
21. $3 + 11j$ **23.** $4 + 8j$ **25.** $1.41(\cos 315° + j \sin 315°) = 1.41e^{5.50j}$
27. $7.28(\cos 254.1° + j \sin 254.1°) = 7.28e^{4.43j}$
29. $4.67(\cos 76.8° + j \sin 76.8°); 4.67e^{1.34j}$
31. $10(\cos 0° + j \sin 0°); 10e^{0j}$ **33.** $-1.41 - 1.41j$
35. $-2.789 + 4.163j$ **37.** $0.19 - 0.59j$

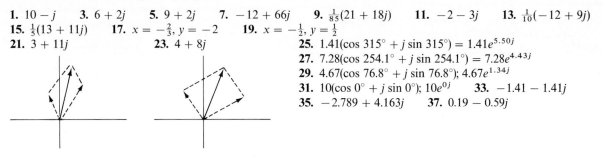

39. $26.31 - 6.427j$ **41.** $1.94 + 0.495j$ **43.** $-8.346 + 23.96j$ **45.** $15(\cos 84° + j \sin 84°)$ **47.** $20\underline{/263°}$
49. $8(\cos 59° + j \sin 59°)$ **51.** $14.29\underline{/133.61°}$ **53.** $1.26\underline{/59.7°}$ **55.** $9682\underline{/249.52°}$
57. $1024(\cos 160° + j \sin 160°)$ **59.** $27\underline{/331.5°}$ **61.** $32(\cos 270° + j \sin 270°) = -32j$
63. $\frac{625}{2}(\cos 270° + j \sin 270°) = -\frac{625}{2}j$ **65.** $1.00 + 1.73j, -2, 1.00 - 1.73j$
67. $\cos 67.5° + j \sin 67.5°, \cos 157.5° + j \sin 157.5°, \cos 247.5° + j \sin 247.5°, \cos 337.5° + j \sin 337.5°$ **69.** 60 V

71. $-21.6°$ **73.** 22.9 Hz   **75.** 814 lb, 317.5°   **77.** $\dfrac{u - j\omega n}{u^2 + \omega^2 n^2}$ **79.** $e^{j\pi} = \cos \pi + j \sin \pi = -1$

Exercises 12-1, page 352

1. 3   **3.** $\frac{1}{81}$ **5.** $\log_3 27 = 3$ **7.** $\log_4 256 = 4$ **9.** $\log_4 \left(\frac{1}{16}\right) = -2$ **11.** $\log_2 \left(\frac{1}{64}\right) = -6$
13. $\log_8 2 = \frac{1}{3}$ **15.** $\log_{1/4} \left(\frac{1}{16}\right) = 2$ **17.** $81 = 3^4$ **19.** $9 = 9^1$ **21.** $5 = 25^{1/2}$ **23.** $3 = 243^{1/5}$
25. $0.1 = 10^{-1}$ **27.** $16 = (0.5)^{-4}$ **29.** 2   **31.** -2 **33.** 343   **35.** $\frac{1}{4}$ **37.** 9   **39.** $\frac{1}{64}$ **41.** 0.2
43. -3 **45.** 3   **47.** $-\frac{1}{2}$ **49.** $t = \log_{1.1} (V/A)$ **51.** $I = I_0 10^{-D}$ **53.** $N = N_0 e^{-kt}$
55. $y = \log_3 x \rightarrow x = 3^y \rightarrow y = 3^x$

Exercises 12-2, page 355

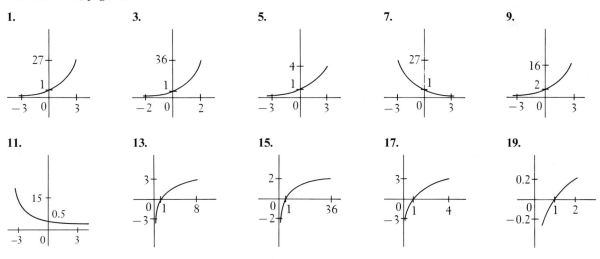

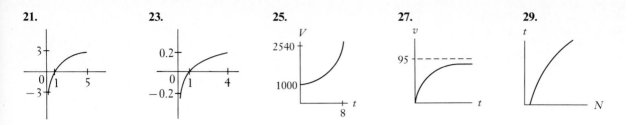

21. **23.** **25.** **27.** **29.**

31. Same graph as in Example C

Exercises 12-3, page 360

1. $\log_5 x + \log_5 y$ **3.** $\log_7 5 - \log_7 a$ **5.** $3 \log_2 a$ **7.** $\log_6 a + \log_6 b + \log_6 c$ **9.** $\frac{1}{4} \log_5 y$
11. $\frac{1}{2} \log_2 x - 2 \log_2 a$ **13.** $\log_b ac$ **15.** $\log_5 3$ **17.** $\log_b x^{3/2}$ **19.** $\log_e 4n^3$ **21.** -5 **23.** 2.5
25. $\frac{1}{2}$ **27.** $\frac{3}{4}$ **29.** $2 + \log_3 2$ **31.** $-1 - \log_2 3$ **33.** $\frac{1}{2}(1 + \log_3 2)$ **35.** $4 + 3 \log_2 3$

37. $3 + \log_{10} 3$ **39.** $-2 + 3 \log_{10} 3$ **41.** $y = 2x$ **43.** $y = \dfrac{3x}{5}$ **45.** $y = \dfrac{49}{x^3}$ **47.** $y = 2(2ax)^{1/5}$

49. $y = \dfrac{2}{x}$ **51.** $y = \dfrac{x^2}{25}$ **53.** 0.602 **55.** -0.301 **57.** **59.** $T = 65e^{-0.41t}$

Exercises 12-4, page 364

1. 2.754 **3.** -1.194 **5.** 6.966 **7.** -3.9311 **9.** 1.8649 **11.** -0.72948 **13.** -0.0104 **15.** 1.219
17. 27,400 **19.** 0.04960 **21.** 2000.4 **23.** 0.7234 **25.** 1.4284 **27.** 0.0057882 **29.** 85.5
31. 94,600 **33.** 1.44×10^{341} **35.** 7.37×10^{101} **37.** 9.0607 **39.** 5.176 **41.** 15.2 dB **43.** 1260 g

Exercises 12-5, page 368

1. 3.258 **3.** 0.4460 **5.** -0.6898 **7.** -4.91623 **9.** 1.92 **11.** 3.418 **13.** 1.933 **15.** 1.795
17. 3.940 **19.** 0.3322 **21.** -0.008335 **23.** -4.34766 **25.** 3.8104 **27.** -0.37793 **29.** 8.94
31. 1.0085 **33.** 0.4757 **35.** 6.20×10^{-11} **37.** $y = 3x$ **39.** 8.155% **41.** 0.384 s **43.** 21.7 s

Exercises 12-6, page 372

1. 4 **3.** -0.748 **5.** 0.587 **7.** 0.6439 **9.** 0.285 **11.** 0.203 **13.** 4.11 **15.** 14.2 **17.** $\frac{1}{4}$
19. 1.649 **21.** 3 **23.** -0.162 **25.** 5 **27.** 0.906 **29.** 4 **31.** 2 **33.** 4 **35.** 1.42 **37.** 28.0
39. 6.84 min **41.** 3.922×10^{-4} **43.** $10^{8.25} = 1.78 \times 10^8$ **45.** $n = 20e^{-0.04t}$ **47.** $P = P_0(0.999)^t$
49. 3.35 **51.** ± 3.7 m

Exercises 12-7, page 376

1. **3.** **5.** **7.** **9.**

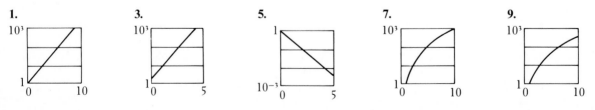

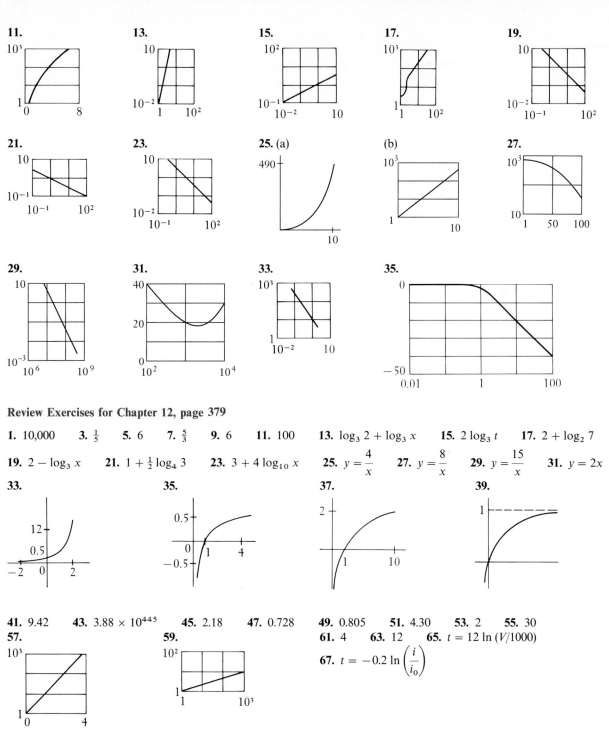

11.   **13.**   **15.**   **17.**   **19.**

21.   **23.**   **25.** (a)   (b)   **27.**

29.   **31.**   **33.**   **35.**

Review Exercises for Chapter 12, page 379

1. 10,000   **3.** $\frac{1}{5}$ **5.** 6   **7.** $\frac{5}{3}$ **9.** 6   **11.** 100   **13.** $\log_3 2 + \log_3 x$ **15.** $2\log_3 t$ **17.** $2 + \log_2 7$

19. $2 - \log_3 x$ **21.** $1 + \frac{1}{2}\log_4 3$ **23.** $3 + 4\log_{10} x$ **25.** $y = \dfrac{4}{x}$ **27.** $y = \dfrac{8}{x}$ **29.** $y = \dfrac{15}{x}$ **31.** $y = 2x$

33.   **35.**   **37.**   **39.**

41. 9.42   **43.** 3.88×10^{445} **45.** 2.18   **47.** 0.728   **49.** 0.805   **51.** 4.30   **53.** 2   **55.** 30
57.   **59.**   **61.** 4   **63.** 12   **65.** $t = 12\ln(V/1000)$

67. $t = -0.2\ln\left(\dfrac{i}{i_0}\right)$

69.

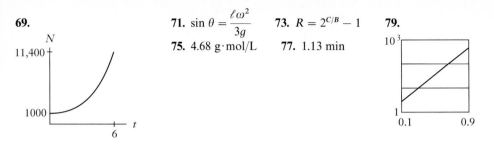

71. $\sin \theta = \dfrac{\ell \omega^2}{3g}$ **73.** $R = 2^{C/B} - 1$ **79.**

75. 4.68 g·mol/L   **77.** 1.13 min

Exercises 13-1, page 386

1. $x = 1.8$, $y = 3.6$; $x = -1.8$, $y = -3.6$ **3.** $x = 0.0$, $y = -2.0$; $x = 2.7$, $y = -0.7$ **5.** $x = 1.5$, $y = 0.2$
7. $x = 1.6$, $y = 2.5$ **9.** $x = 1.1$, $y = 2.8$; $x = -1.1$, $y = 2.8$; $x = 2.4$, $y = -1.8$; $x = -2.4$, $y = -1.8$
11. No solution   **13.** $x = -2.8$, $y = -1.0$; $x = 2.8$, $y = 1.0$; $x = 2.8$, $y = -1.0$; $x = -2.8$, $y = 1.0$
15. $x = 0.7$, $y = 0.7$; $x = -0.7$, $y = -0.7$ **17.** $x = 0.0$, $y = 0.0$; $x = 0.9$, $y = 0.8$
19. $x = -1.1$, $y = 3.0$; $x = 1.8$, $y = 0.2$ **21.** $x = 1.0$, $y = 0.0$ **23.** $x = 3.6$, $y = 1.0$
25. 4.9 mi N, 1.6 mi E   **27.** 2.2 A, 0.9 A

Exercises 13-2, page 391

1. $x = 0$, $y = 1$; $x = 1$, $y = 2$ **3.** $x = -\frac{19}{5}$, $y = \frac{17}{5}$; $x = 5$, $y = -1$ **5.** $x = 1$, $y = 0$
7. $x = \frac{2}{7}(3 + \sqrt{2})$, $y = \frac{2}{7}(-1 + 2\sqrt{2})$; $x = \frac{2}{7}(3 - \sqrt{2})$, $y = \frac{2}{7}(-1 - 2\sqrt{2})$ **9.** $x = 1$, $y = 1$
11. $x = \frac{2}{3}$, $y = \frac{9}{2}$; $x = -3$, $y = -1$ **13.** $x = -2$, $y = 4$; $x = 2$, $y = 4$ **15.** $x = 1$, $y = 2$; $x = -1$, $y = 2$
17. $x = 1$, $y = 0$; $x = -1$, $y = 0$; $x = \frac{1}{2}\sqrt{6}$, $y = \frac{1}{2}$; $x = -\frac{1}{2}\sqrt{6}$, $y = \frac{1}{2}$
19. $x = \sqrt{19}$, $y = \sqrt{6}$; $x = \sqrt{19}$, $y = -\sqrt{6}$; $x = -\sqrt{19}$, $y = \sqrt{6}$; $x = -\sqrt{19}$, $y = -\sqrt{6}$
21. $x = \frac{1}{11}\sqrt{22}$, $y = \frac{1}{11}\sqrt{770}$; $x = \frac{1}{11}\sqrt{22}$, $y = -\frac{1}{11}\sqrt{770}$; $x = -\frac{1}{11}\sqrt{22}$, $y = \frac{1}{11}\sqrt{770}$; $x = -\frac{1}{11}\sqrt{22}$, $y = -\frac{1}{11}\sqrt{770}$
23. $x = -5$, $y = -2$; $x = -5$, $y = 2$; $x = 5$, $y = -2$; $x = 5$, $y = 2$ **25.** $v_1 = -4.0$ m/s, $v_2 = 2.5$ m/s
27. 1.5 cm, 1.4 cm   **29.** 11.8 ft, 38.2 ft   **31.** 80 mi/h

Exercises 13-3, page 395

1. $-3, -2, 2, 3$ **3.** $\frac{3}{2}, -\frac{3}{2}, j, -j$ **5.** $-\frac{1}{2}, \frac{1}{4}$ **7.** $-\frac{1}{2}, \frac{1}{2}, \frac{1}{6}j\sqrt{6}, -\frac{1}{6}j\sqrt{6}$ **9.** $1, \frac{25}{4}$ **11.** $\frac{64}{729}, 1$
13. $-27, 125$ **15.** 256 **17.** 5 **19.** $-2, -1, 3, 4$ **21.** 18
23. $1, -2, 1 + j\sqrt{3}, 1 - j\sqrt{3}, \frac{1}{2}(-1 + j\sqrt{3}), -\frac{1}{2}(1 + j\sqrt{3})$ **25.** $R_1 = 2.62$ Ω, $R_2 = 1.62$ Ω   **27.** 16 ft by 12 ft

Exercises 13-4, page 398

1. 12 **3.** 2 **5.** $\frac{2}{3}$ **7.** $\frac{1}{2}(7 + \sqrt{5})$ **9.** -1 **11.** 32 **13.** 16 **15.** 9 **17.** 12 **19.** 16 **21.** $\frac{1}{2}$
23. $7, -1$ **25.** 0 **27.** 5 **29.** 6 **31.** 258 **33.** $L = \dfrac{1}{4\pi^2 f^2 C}$ **35.** $\dfrac{k^2}{2n(1 - k)}$ **37.** 1.5 ft, 2.0 ft, 2.5 ft
39. 3.4 mi

Review Exercises for Chapter 13, page 399

1. $x = -0.9$, $y = 3.5$; $x = 0.8$, $y = 2.6$ **3.** $x = 2.0$, $y = 0.0$; $x = 1.6$, $y = 0.6$
5. $x = 0.8$, $y = 1.6$; $x = -0.8$, $y = 1.6$ **7.** $x = -1.2$, $y = 2.7$; $x = 1.2$, $y = 2.7$
9. $x = 0.0$, $y = 0.0$; $x = 2.4$, $y = 0.9$ **11.** $x = 0$, $y = 0$; $x = 2$, $y = 16$ **13.** $x = \sqrt{2}$, $y = 1$; $x = -\sqrt{2}$, $y = 1$
15. $x = \frac{1}{12}(1 + \sqrt{97})$, $y = \frac{1}{18}(5 - \sqrt{97})$; $x = \frac{1}{12}(1 - \sqrt{97})$, $y = \frac{1}{18}(5 + \sqrt{97})$
17. $x = 7$, $y = 5$; $x = 7$, $y = -5$; $x = -7$, $y = 5$; $x = -7$, $y = -5$ **19.** $x = -2$, $y = 2$; $x = \frac{2}{3}$, $y = \frac{10}{9}$
21. $-4, -2, 2, 4$ **23.** $1, 16$ **25.** $\frac{1}{3}, -\frac{1}{7}$ **27.** $\frac{25}{4}$ **29.** $-\frac{1}{2}, -2$ **31.** 11 **33.** $1, 4$ **35.** 8 **37.** $\frac{9}{16}$
39. $\frac{1}{3}(11 - 4\sqrt{15})$ **41.** $\ell = \frac{1}{2}(-1 + \sqrt{1 + 16\pi^2 L^2/h^2})$ **43.** $m = \frac{1}{2}(-y \pm \sqrt{2s^2 - y^2})$ **45.** 0.75 s, 1.5 s
47. 230 ft   **49.** $Z = 1.09$ Ω, $X = 0.738$ Ω   **51.** 15.2 in., 11.4 in.   **53.** 3.57 in.   **55.** 26.3 mi/h, 32.3 mi/h

Exercises 14-1, page 403

1. 0   **3.** 0   **5.** -40 **7.** -4 **9.** 8   **11.** 183   **13.** -28 **15.** 14   **17.** Yes   **19.** Yes
21. No   **23.** No   **25.** Yes   **27.** Yes   **29.** $(4x^3 + 8x^2 - x - 2) \div (2x - 1) = 2x^2 + 5x + 2$; No   **31.** -7

Exercises 14-2, page 408

1. $x^2 + 3x + 2$, $R = 0$ **3.** $x^2 - x + 3$, $R = 0$ **5.** $2x^4 - 4x^3 + 8x^2 - 17x + 42$, $R = -40$
7. $3x^3 - x + 2$, $R = -4$ **9.** $x^2 + x - 4$, $R = 8$ **11.** $x^3 - 3x^2 + 10x - 45$, $R = 183$
13. $2x^3 - x^2 - 4x - 12$, $R = -28$ **15.** $x^4 + 2x^3 + x^2 + 7x + 4$, $R = 14$
17. $x^5 + 2x^4 + 4x^3 + 8x^2 + 18x + 36$, $R = 66$ **19.** $x^6 + 2x^5 + 4x^4 + 8x^3 + 16x^2 + 32x + 64$, $R = 0$
21. Yes   **23.** No   **25.** No   **27.** No   **29.** Yes   **31.** No   **33.** Yes   **35.** Yes

Exercises 14-3, page 413

(Note: Unknown roots listed)
1. $-2, -1$ **3.** $-2, 3$ **5.** $-2, -2$ **7.** $-j, -\frac{2}{3}$ **9.** $2j, -2j$ **11.** $-1, -3$ **13.** $-2, 1$
15. $1 - j, \frac{1}{2}, -2$ **17.** $\frac{1}{4}(1 + \sqrt{17}), \frac{1}{4}(1 - \sqrt{17})$ **19.** $2, -3$ **21.** $-j, -1, 1$ **23.** $-2j, 3, -3$

Exercises 14-4, page 419

1. $1, -1, -2$ **3.** $2, -1, -3$ **5.** $\frac{1}{2}, 5, -3$ **7.** $\frac{1}{3}, -3, -1$ **9.** $-2, -2, 2 \pm \sqrt{3}$ **11.** $2, 4, -1, -3$
13. $1, -\frac{1}{2}, 1 \pm \sqrt{3}$ **15.** $\frac{1}{2}, -\frac{2}{3}, -3, -\frac{1}{2}$ **17.** $2, 2, -1, -1, -3$ **19.** $\frac{1}{2}, 1, 1, j, -j$ **21.** 0.59
23. -0.77 **25.** 2 cm^3 **27.** $0, L$ **29.** 1.5 lb, 4.1 lb   **31.** $2 \, \Omega, 3 \, \Omega, 6 \, \Omega$ **33.** 3.0 in., 4.0 in.
35. $-1.86, 0.68, 3.18$

Review Exercises for Chapter 14, page 420

1. 1   **3.** -107 **5.** Yes   **7.** No   **9.** $x^2 + 4x + 10$, $R = 11$ **11.** $2x^2 - 7x + 10$, $R = -17$
13. $x^3 - 3x^2 - 4$, $R = -4$ **15.** $2x^4 + 10x^3 + 4x^2 + 21x + 105$, $R = 516$ **17.** No   **19.** Yes
21. (Unlisted roots) $\frac{1}{2}(-5 \pm \sqrt{17})$ **23.** (Unlisted roots) $\frac{1}{3}(-1 \pm j\sqrt{14})$ **25.** (Unlisted roots) $4, -3$
27. (Unlisted roots) $-j, \frac{1}{2}, -\frac{3}{2}$ **29.** (Unlisted roots) $2, -2$
31. (Unlisted roots) $-2 - j, \frac{1}{2}(-1 + j\sqrt{3}), -\frac{1}{2}(1 + j\sqrt{3})$ **33.** $1, 2, -4$ **35.** $-1, -1, \frac{5}{2}$ **37.** $\frac{5}{3}, -\frac{1}{2}, -1$
39. $\frac{1}{2}, -1, j\sqrt{2}, -j\sqrt{2}$ **41.** 4   **43.** $1, 0.4, 1.5, -0.6$ (last three are irrational)   **45.** 1.91   **47.** April
49. 0.75 cm   **51.** 2 in.   **53.** 2 cm, 3 cm, 4 cm   **55.** 8 ft, 15 ft

Exercises 15-1, page 428

1. 39   **3.** 30   **5.** 50   **7.** $-47,416$ **9.** -6 **11.** -86 **13.** 118   **15.** -2
17. $x = 2, y = -1, z = 3$ **19.** $x = -1, y = \frac{1}{3}, z = -\frac{1}{2}$ **21.** $x = -1, y = 0, z = 2, t = 1$
23. $x = 1, y = 2, z = -1, t = 3$ **25.** $\frac{2}{7}$ A, $\frac{18}{7}$ A, $-\frac{8}{7}$ A, $-\frac{12}{7}$ A   **27.** 500, 300, 700

Exercises 15-2, page 434

1. -60 **3.** -56 **5.** 0   **7.** 0   **9.** 57   **11.** -124 **13.** -13 **15.** -118 **17.** -72 **19.** 0
21. $x = 0, y = -1, z = 4$ **23.** $x = \frac{1}{3}, y = -\frac{1}{2}, z = 1$ **25.** $x = 2, y = -1, z = -1, t = 3$
27. $x = 1, y = 2, z = -1, t = -2$ **29.** $\frac{33}{16}$ A, $\frac{11}{8}$ A, $-\frac{5}{8}$ A, $-\frac{15}{8}$ A, $-\frac{15}{16}$ A
31. ppm of SO_2, NO, NO_2, CO: 0.5, 0.3, 0.2, 5.0

Exercises 15-3, page 439

1. $a = 1, b = -3, c = 4, d = 7$ **3.** $x = 2, y = 3$ **5.** Elements cannot be equated; different number of rows

7. $x = 4, y = 6, z = 9$ **9.** $\begin{pmatrix} 1 & 10 \\ 0 & 2 \end{pmatrix}$ **11.** $\begin{pmatrix} -5 & 0 \\ 11 & 71 \\ 11 & -5 \end{pmatrix}$ **13.** $\begin{pmatrix} 0 & 9 & -13 & 3 \\ 6 & -7 & 7 & 0 \end{pmatrix}$

15. Cannot be added   **17.** $\begin{pmatrix} -1 & 13 & -20 & 3 \\ 8 & -13 & 6 & 2 \end{pmatrix}$ **19.** $\begin{pmatrix} -3 & -6 & 5 & -6 \\ -6 & -4 & -17 & 6 \end{pmatrix}$

21. $A + B = B + A = \begin{pmatrix} 3 & 1 & 0 & 7 \\ 5 & -3 & -2 & 5 \\ 10 & 10 & 8 & 0 \end{pmatrix}$ **23.** $-(A - B) = B - A = \begin{pmatrix} 5 & -3 & -6 & -7 \\ 5 & 3 & 0 & -3 \\ -8 & 12 & 8 & 4 \end{pmatrix}$

25. $v_1 = 58.2$ m/s, $v_2 = 46.3$ m/s   **27.** $\begin{pmatrix} 24 & 18 & 0 & 0 \\ 15 & 12 & 9 & 0 \\ 0 & 9 & 15 & 18 \end{pmatrix}$

Exercises 15-4, page 444

1. $(-8 \quad -12)$ **3.** $\begin{pmatrix} -15 & 15 & -26 \\ 8 & 5 & -13 \end{pmatrix}$ **5.** $\begin{pmatrix} 29 \\ -29 \end{pmatrix}$ **7.** $\begin{pmatrix} -13.21 & -9.52 \\ 11.50 & 0.00 \\ 6.89 & 0.68 \end{pmatrix}$

9. $\begin{pmatrix} 33 & -22 & 7 \\ 31 & -12 & 5 \\ 15 & 13 & -1 \\ 50 & -41 & 12 \end{pmatrix}$ **11.** $\begin{pmatrix} -62 & 68 \\ 73 & -27 \end{pmatrix}$ **13.** $AB = (40)$, $BA = \begin{pmatrix} -1 & 3 & -8 \\ 5 & -15 & 40 \\ 7 & -21 & 56 \end{pmatrix}$

15. $AB = \begin{pmatrix} -5 \\ 10 \end{pmatrix}$, BA not defined   **17.** $AI = IA = A$ **19.** $AI = IA = A$ **21.** $B = A^{-1}$ **23.** $B = A^{-1}$

25. Yes   **27.** No   **29.** $\begin{pmatrix} -1 & 0 \\ 0 & -1 \end{pmatrix}\begin{pmatrix} -1 & 0 \\ 0 & -1 \end{pmatrix} = \begin{pmatrix} 1 & 0 \\ 0 & 1 \end{pmatrix}$ **31.** $A^2 - I = (A + I)(A - I) = \begin{pmatrix} 15 & 28 \\ 21 & 36 \end{pmatrix}$

33. $\begin{pmatrix} 0 & -j \\ j & 0 \end{pmatrix}\begin{pmatrix} 0 & -j \\ j & 0 \end{pmatrix} = \begin{pmatrix} 1 & 0 \\ 0 & 1 \end{pmatrix}$ **35.** $v_2 = v_1$, $i_2 = -v_1/R + i_1$

Exercises 15-5, page 449

1. $\begin{pmatrix} -2 & -\frac{5}{2} \\ -1 & -1 \end{pmatrix}$ **3.** $\begin{pmatrix} -\frac{1}{3} & \frac{1}{6} \\ \frac{2}{15} & \frac{1}{30} \end{pmatrix}$ **5.** $\begin{pmatrix} \frac{3}{4} & \frac{1}{2} \\ -\frac{1}{4} & 0 \end{pmatrix}$ **7.** $\begin{pmatrix} -\frac{8}{283} & -\frac{9}{566} \\ \frac{13}{1415} & \frac{5}{283} \end{pmatrix}$ **9.** $\begin{pmatrix} -3 & 2 \\ 2 & -1 \end{pmatrix}$ **11.** $\begin{pmatrix} -\frac{1}{2} & -2 \\ \frac{1}{2} & 1 \end{pmatrix}$

13. $\begin{pmatrix} \frac{2}{9} & -\frac{5}{9} \\ \frac{1}{9} & \frac{2}{9} \end{pmatrix}$ **15.** $\begin{pmatrix} \frac{3}{8} & \frac{1}{16} \\ -\frac{1}{4} & \frac{1}{8} \end{pmatrix}$ **17.** $\begin{pmatrix} -18 & -7 & 5 \\ -3 & -1 & 1 \\ -5 & -2 & 1 \end{pmatrix}$ **19.** $\begin{pmatrix} 3 & -4 & -1 \\ -4 & 5 & 2 \\ 2 & -3 & -1 \end{pmatrix}$

21. $\begin{pmatrix} 2 & 4 & \frac{7}{2} \\ -1 & -2 & -\frac{3}{2} \\ 1 & 1 & \frac{1}{2} \end{pmatrix}$ **23.** $\begin{pmatrix} \frac{5}{2} & -2 & -2 \\ -1 & 1 & 1 \\ \frac{7}{4} & -\frac{3}{2} & -1 \end{pmatrix}$ **25.** $\begin{pmatrix} 2 & 4 & \frac{7}{2} \\ -1 & -2 & -\frac{3}{2} \\ 1 & 1 & \frac{1}{2} \end{pmatrix}$

27. $\begin{pmatrix} \frac{5}{2} & -2 & -2 \\ -1 & 1 & 1 \\ \frac{7}{4} & -\frac{3}{2} & -1 \end{pmatrix}$ **29.** $\dfrac{1}{ad - bc}\begin{pmatrix} ad - bc & -ba + ab \\ cd - dc & -bc + ad \end{pmatrix} = \begin{pmatrix} 1 & 0 \\ 0 & 1 \end{pmatrix}$ **31.** $v_1 = a_{22}i_1 - a_{12}i_2$
$v_2 = -a_{21}i_1 + a_{11}i_2$

Exercises 15-6, page 453

1. $x = \frac{1}{2}$, $y = 3$ **3.** $x = \frac{1}{2}$, $y = -\frac{5}{2}$ **5.** $x = -4$, $y = 2$, $z = -1$ **7.** $x = -1$, $y = 0$, $z = 3$
9. $x = 1$, $y = 2$ **11.** $x = -\frac{3}{2}$, $y = -2$ **13.** $x = -3$, $y = -\frac{1}{2}$ **15.** $x = 1.6$, $y = -2.5$
17. $x = 2$, $y = -4$, $z = 1$ **19.** $x = 2$, $y = -\frac{1}{2}$, $z = 3$ **21.** $A = 118$ N, $B = 186$ N   **23.** 6.4 L, 1.6 L, 2.0 L

Review Exercises for Chapter 15, page 455

1. 6   **3.** 186   **5.** -438 **7.** 44   **9.** 6   **11.** 186   **13.** -438 **15.** 44   **17.** -9 **19.** -44

21. $a = 4, b = -1$ **23.** $x = 2, y = -3, z = \frac{5}{2}, a = -1, b = -\frac{7}{2}, c = \frac{1}{2}$ **25.** $\begin{pmatrix} 1 & -3 \\ 8 & -5 \\ -8 & -2 \\ 3 & -10 \end{pmatrix}$ **27.** $\begin{pmatrix} 3 & 0 \\ -12 & 18 \\ 9 & 6 \\ -3 & 21 \end{pmatrix}$

29. Cannot be subtracted   **31.** $\begin{pmatrix} 7 & -6 \\ -4 & 20 \\ -1 & 6 \\ 1 & 15 \end{pmatrix}$ **33.** $\begin{pmatrix} 13 \\ -13 \end{pmatrix}$ **35.** $\begin{pmatrix} 34 & 11 & -5 \\ 2 & -8 & 10 \\ -1 & -17 & 20 \end{pmatrix}$ **37.** $\begin{pmatrix} -2 & \frac{5}{2} \\ -1 & 1 \end{pmatrix}$

39. $\begin{pmatrix} \frac{2}{15} & \frac{1}{60} \\ -\frac{1}{15} & \frac{7}{60} \end{pmatrix}$ **41.** $\begin{pmatrix} 11 & 10 & 3 \\ -4 & -4 & -1 \\ 3 & 3 & 1 \end{pmatrix}$ **43.** $\begin{pmatrix} \frac{1}{2} & -\frac{1}{2} & -1 \\ -3 & 2 & 1 \\ -4 & 3 & 2 \end{pmatrix}$ **45.** $x = -3, y = 1$ **47.** $x = 10, y = -15$

49. $x = -1, y = -3, z = 0$ **51.** $x = 1, y = \frac{1}{2}, z = -\frac{1}{3}$ **53.** $x = 3, y = 1, z = -1$

55. $x = 1, y = 2, z = -3, t = 1$ **57.** $2\sqrt{2}$ **59.** 0   **61.** $N^{-1} = -N = \begin{pmatrix} 0 & 1 \\ -1 & 0 \end{pmatrix}$

63. $(A + B)(A - B) = \begin{pmatrix} -6 & 2 \\ 4 & 2 \end{pmatrix}, A^2 - B^2 = \begin{pmatrix} -10 & -4 \\ 8 & 6 \end{pmatrix}$ **65.** $\begin{pmatrix} \frac{1}{2} & \frac{1}{3} \\ 0 & \frac{1}{6} \end{pmatrix} = \frac{1}{2}\begin{pmatrix} 1 & \frac{2}{3} \\ 0 & \frac{1}{3} \end{pmatrix}$ **67.** $R_1 = 4\,\Omega, R_2 = 6\,\Omega$

69. $F = 303$ lb, $T = 175$ lb   **71.** 0.22 h after police pass intersection   **73.** 30 g, 50 g, 20 g

75. 2.0 h, 1.5 h, 1.0 h, 1.0 h   **77.** $\begin{pmatrix} 15,000 & 10,000 \\ 20,000 & 18,000 \\ 8,000 & 30,000 \end{pmatrix} + \begin{pmatrix} 18,000 & 12,000 \\ 30,000 & 22,000 \\ 12,000 & 40,000 \end{pmatrix} = \begin{pmatrix} 33,000 & 22,000 \\ 50,000 & 40,000 \\ 20,000 & 70,000 \end{pmatrix}$

79. $(R_1 + R_2)i_1 - R_2 i_2 = 6$
$-R_2 i_1 + (R_1 + R_2)i_2 = 0$

Exercises 16-1, page 465

1. $7 < 12$ **3.** $20 < 45$ **5.** $-4 > -9$ **7.** $16 < 81$ **9.** $x > -2$ **11.** $x \leq 4$ **13.** $1 < x < 7$
15. $x < -9$ or $x \geq -4$ **17.** $x < 1$ or $3 < x \leq 5$ **19.** $-2 < x < 2$ or $3 \leq x < 4$
21. x is greater than 0 and less than or equal to 2
23. x is less than -1, or greater than or equal to 1 and less than 2

25.   **27.**   **29.**   **31.**

33.   **35.**   **37.** $d > 3 \times 10^{12}$ mi   **39.** $18,000 < v < 25,000$ mi/h

41. $0 < n \leq 2565$ steps   **43.** $E = 0$ for $0 \leq r < a$
$E = k/r^2$ for $r \geq a$

Exercises 16-2, page 468

1. $x > -1$ **3.** $x < 6$ **5.** $x \leq -2$ **7.** $x < 2$ **9.** $x \leq \frac{5}{2}$

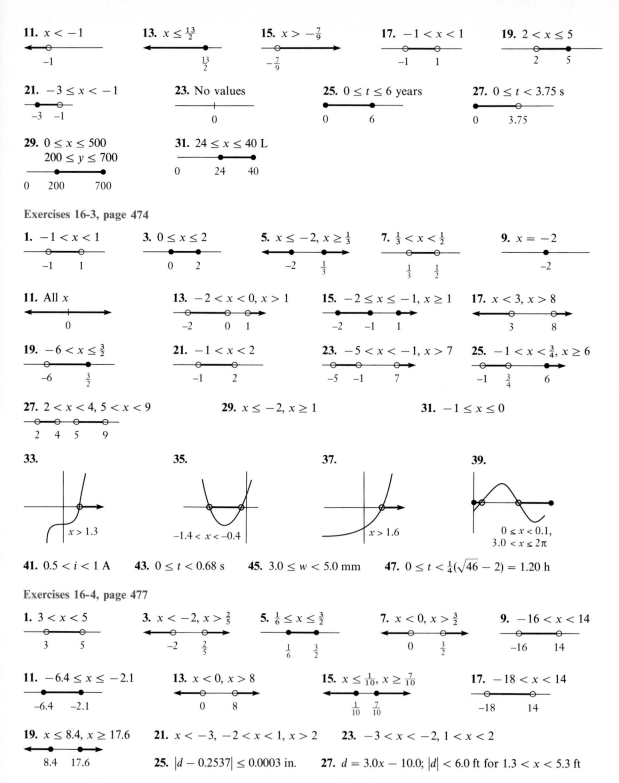

11. $x < -1$

13. $x \leq \frac{13}{2}$

15. $x > -\frac{7}{9}$

17. $-1 < x < 1$

19. $2 < x \leq 5$

21. $-3 \leq x < -1$

23. No values

25. $0 \leq t \leq 6$ years

27. $0 \leq t < 3.75$ s

29. $0 \leq x \leq 500$
$200 \leq y \leq 700$

31. $24 \leq x \leq 40$ L

Exercises 16-3, page 474

1. $-1 < x < 1$

3. $0 \leq x \leq 2$

5. $x \leq -2, x \geq \frac{1}{3}$

7. $\frac{1}{3} < x < \frac{1}{2}$

9. $x = -2$

11. All x

13. $-2 < x < 0, x > 1$

15. $-2 \leq x \leq -1, x \geq 1$

17. $x < 3, x > 8$

19. $-6 < x \leq \frac{3}{2}$

21. $-1 < x < 2$

23. $-5 < x < -1, x > 7$

25. $-1 < x < \frac{3}{4}, x \geq 6$

27. $2 < x < 4, 5 < x < 9$

29. $x \leq -2, x \geq 1$

31. $-1 \leq x \leq 0$

33.

$x > 1.3$

35.

$-1.4 < x < -0.4$

37.

$x > 1.6$

39.

$0 \leq x < 0.1,$
$3.0 < x \leq 2\pi$

41. $0.5 < i < 1$ A

43. $0 \leq t < 0.68$ s

45. $3.0 \leq w < 5.0$ mm

47. $0 \leq t < \frac{1}{4}(\sqrt{46} - 2) = 1.20$ h

Exercises 16-4, page 477

1. $3 < x < 5$

3. $x < -2, x > \frac{2}{5}$

5. $\frac{1}{6} \leq x \leq \frac{3}{2}$

7. $x < 0, x > \frac{3}{2}$

9. $-16 < x < 14$

11. $-6.4 \leq x \leq -2.1$

13. $x < 0, x > 8$

15. $x \leq \frac{1}{10}, x \geq \frac{7}{10}$

17. $-18 < x < 14$

19. $x \leq 8.4, x \geq 17.6$

21. $x < -3, -2 < x < 1, x > 2$

23. $-3 < x < -2, 1 < x < 2$

25. $|d - 0.2537| \leq 0.0003$ in.

27. $d = 3.0x - 10.0; |d| < 6.0$ ft for $1.3 < x < 5.3$ ft

Exercises 16-5, page 482

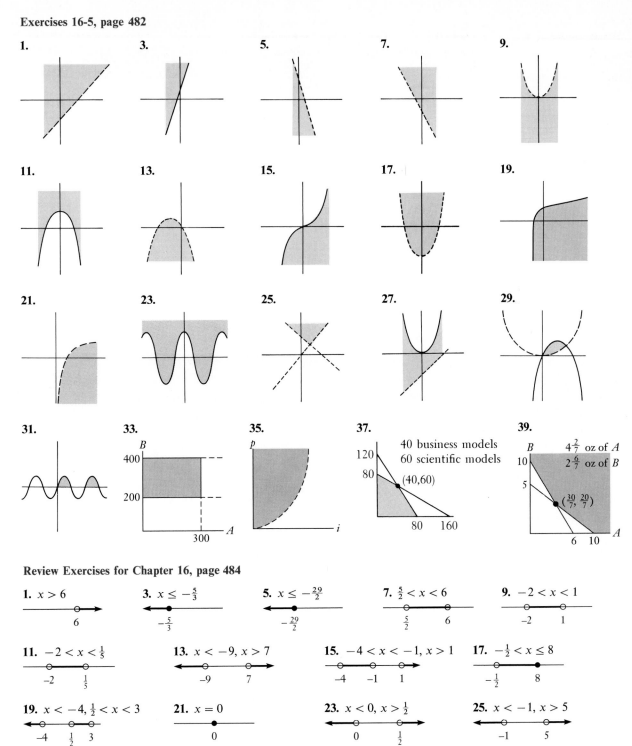

1.

3.

5.

7.

9.

11.

13.

15.

17.

19.

21.

23.

25.

27.

29.

31.

33.

35.

37. 40 business models
60 scientific models
(40,60)

39. $4\frac{2}{7}$ oz of A
$2\frac{6}{7}$ oz of B
$\left(\frac{30}{7}, \frac{20}{7}\right)$

Review Exercises for Chapter 16, page 484

1. $x > 6$

3. $x \le -\frac{5}{3}$

5. $x \le -\frac{29}{2}$

7. $\frac{5}{2} < x < 6$

9. $-2 < x < 1$

11. $-2 < x < \frac{1}{5}$

13. $x < -9, x > 7$

15. $-4 < x < -1, x > 1$

17. $-\frac{1}{2} < x \le 8$

19. $x < -4, \frac{1}{2} < x < 3$

21. $x = 0$

23. $x < 0, x > \frac{1}{2}$

25. $x < -1, x > 5$

27. $-2 \leq x \leq \frac{2}{3}$

29. $x < -\frac{4}{5}, x > 2$

31. $-6 \leq x \leq 14$

33.

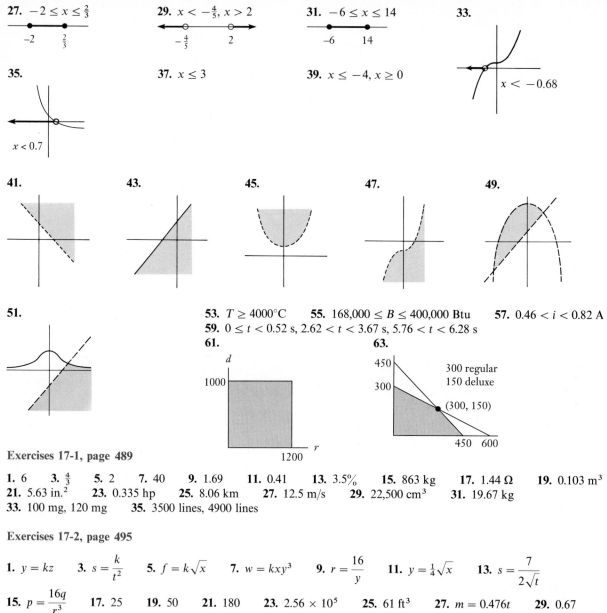

$x < -0.68$

35.

$x < 0.7$

37. $x \leq 3$

39. $x \leq -4, x \geq 0$

41.

43.

45.

47.

49.

51.

53. $T \geq 4000°C$ **55.** $168,000 \leq B \leq 400,000$ Btu   **57.** $0.46 < i < 0.82$ A
59. $0 \leq t < 0.52$ s, $2.62 < t < 3.67$ s, $5.76 < t < 6.28$ s

61.

d

1000

1200 r

63.

450

300

300 regular
150 deluxe

$(300, 150)$

450 600

Exercises 17-1, page 489

1. 6   **3.** $\frac{4}{3}$ **5.** 2   **7.** 40   **9.** 1.69   **11.** 0.41   **13.** 3.5%   **15.** 863 kg   **17.** 1.44 Ω   **19.** 0.103 m³
21. 5.63 in.²   **23.** 0.335 hp   **25.** 8.06 km   **27.** 12.5 m/s   **29.** 22,500 cm³   **31.** 19.67 kg
33. 100 mg, 120 mg   **35.** 3500 lines, 4900 lines

Exercises 17-2, page 495

1. $y = kz$ **3.** $s = \dfrac{k}{t^2}$ **5.** $f = k\sqrt{x}$ **7.** $w = kxy^3$ **9.** $r = \dfrac{16}{y}$ **11.** $y = \frac{1}{4}\sqrt{x}$ **13.** $s = \dfrac{7}{2\sqrt{t}}$

15. $p = \dfrac{16q}{r^3}$ **17.** 25   **19.** 50   **21.** 180   **23.** 2.56×10^5 **25.** 61 ft³   **27.** $m = 0.476t$ **29.** 0.67

31. 1.4 h   **33.** 1250   **35.** 28 N   **37.** $F = 0.00523Av^2$ **39.** 482 m/s   **41.** $R = \dfrac{5.00 \times 10^{-5}\ell}{A}$

43. 80.0 W   **45.** $G = \dfrac{5.9d^2}{\lambda^2}$ **47.** -6.57 ft/s²

Review Exercises for Chapter 17, page 497

1. 200   **3.** $\frac{2}{5}$ **5.** 5.6   **7.** 7.39 lb/in.²   **9.** 3.25 in.   **11.** 2.82×10^6 J = 2.82 MJ   **13.** 36,000 characters

15. 310 mL   **17.** 71.9 ft   **19.** 4500, 7500   **21.** $y = 3x^2$ **23.** $v = \dfrac{128x}{y^3}$ **25.** 11.7 in.   **27.** 4.3 μC

29. 22.5 hp   **31.** 144 ft   **33.** 1.4   **35.** 50.0 Hz   **37.** 2.99×10^8 m/s   **39.** 3.26 cm   **41.** 4800 Btu
43. 5.73×10^4 m   **45.** 149%   **47.** \$125.00, \$600.13

Exercises 18-1, page 505

1. 4, 6, 8, 10, 12   **3.** 13, 9, 5, 1, -3 **5.** 22   **7.** -62 **9.** 37   **11.** $49b$ **13.** 440   **15.** $-\frac{85}{2}$
17. $n = 6$, $S_6 = 150$ **19.** $d = -\frac{2}{19}$, $a_{20} = -\frac{1}{3}$ **21.** $a_1 = 19$, $a_{30} = 106$ **23.** $n = 62$, $S_{62} = -4867$
25. $n = 23$, $a_{23} = 6k$ **27.** $n = 8$, $d = \frac{1}{14}(b + 2c)$ **29.** $a_1 = 36$, $d = 4$, $S_{10} = 540$
31. $a_1 = 3$, $d = -\frac{1}{3}$, $S_{10} = 15$ **33.** 5050   **35.** 100,500   **37.** $2.75°C$ **39.** 10 rows   **41.** 12 years, \$11,700
43. 490 m   **45.** $S_n = \frac{1}{2}n[2a_1 + (n - 1)d]$ **47.** $a_1 = 1$, $a_n = n$

Exercises 18-2, page 509

1. 45, 15, 5, $\frac{5}{3}$, $\frac{5}{9}$ **3.** 2, 6, 18, 54, 162   **5.** 16   **7.** $\frac{1}{125}$ **9.** $\frac{1}{9}$ **11.** 2×10^6 **13.** $\frac{341}{8}$ **15.** 378
17. $a_6 = 64$, $S_6 = \frac{1365}{16}$ **19.** $a_1 = 16$, $a_5 = 81$ **21.** $a_1 = 1$, $r = 3$ **23.** $n = 5$, $S_5 = \frac{2343}{25}$ **25.** 32

27. 1.4%   **29.** 1.09 mA   **31.** \$11,277.89   **33.** 462 cm   **35.** 4   **37.** \$671,088.64   **39.** $S_n = \dfrac{a_1 - ra_n}{1 - r}$

Exercises 18-3, page 513

1. 8   **3.** $\frac{25}{4}$ **5.** $\frac{400}{21}$ **7.** 8   **9.** $\dfrac{10,000}{9999}$ **11.** $\frac{1}{2}(5 + 3\sqrt{3})$ **13.** $\frac{1}{3}$ **15.** $\frac{40}{99}$ **17.** $\frac{2}{11}$ **19.** $\frac{91}{333}$

21. $\frac{11}{30}$ **23.** $\dfrac{100,741}{999,000}$ **25.** 350 gal   **27.** 346 g

Exercises 18-4, page 517

1. $t^3 + 3t^2 + 3t + 1$ **3.** $16x^4 - 32x^3 + 24x^2 - 8x + 1$ **5.** $32x^5 + 240x^4 + 720x^3 + 1080x^2 + 810x + 243$
7. $64a^6 - 192a^5b^2 + 240a^4b^4 - 160a^3b^6 + 60a^2b^8 - 12ab^{10} + b^{12}$ **9.** $625x^4 - 1500x^3 + 1350x^2 - 540x + 81$
11. $64a^6 + 192a^5 + 240a^4 + 160a^3 + 60a^2 + 12a + 1$ **13.** $x^{10} + 20x^9 + 180x^8 + 960x^7 + \cdots$
15. $128a^7 - 448a^6 + 672a^5 - 560a^4 + \cdots$ **17.** $x^{24} - 6x^{22}y + \frac{33}{2}x^{20}y^2 - \frac{55}{2}x^{18}y^3 + \cdots$
19. $b^{40} + 10b^{37} + \frac{95}{2}b^{34} + \frac{285}{2}b^{31} + \cdots$ **21.** $1 + 8x + 28x^2 + 56x^3 + \cdots$ **23.** $1 + 2x + 3x^2 + 4x^3 + \cdots$
25. $1 + \frac{1}{2}x - \frac{1}{8}x^2 + \frac{1}{16}x^3 - \cdots$ **27.** $\frac{1}{3}[1 + \frac{1}{2}x + \frac{3}{8}x^2 + \frac{5}{16}x^3 + \cdots]$
29. (a) 3.557×10^{14}, (b) 5.109×10^{19}, (c) 8.536×10^{15}, (d) 2.480×10^{96}
31. $n! = n(n - 1)(n - 2)(\cdots)(2)(1) = n \times (n - 1)!$; for $n = 1$, $1! = 1 \times 0! = 1 \times 1 = 1$ **33.** $56a^3b^5$

35. $10,264,320x^8b^4$ **37.** $V = A(1 - 5r + 10r^2 - 10r^3 + 5r^4 - r^5)$ **39.** $1 - \dfrac{x}{a} + \dfrac{x^3}{2a^3} - \cdots$

Review Exercises for Chapter 18, page 519

1. 81   **3.** 1.28×10^{-6} **5.** $-\frac{119}{2}$ **7.** $\frac{16}{243}$ **9.** $\frac{195}{2}$ **11.** $\frac{1023}{96}$ **13.** 81   **15.** $\frac{9}{16}$ **17.** -1.5
19. -0.25 **21.** 186   **23.** $\frac{455}{2}$ (as), 127 (gs), or 43 (gs)   **25.** 27   **27.** 51   **29.** $\frac{1}{33}$ **31.** $\frac{8}{110}$
33. $x^4 - 8x^3 + 24x^2 - 32x + 16$ **35.** $x^{10} + 5x^8 + 10x^6 + 10x^4 + 5x^2 + 1$
37. $a^{10} + 20a^9b^2 + 180a^8b^4 + 960a^7b^6 + \cdots$ **39.** $p^{18} - \frac{3}{2}p^{16}q + p^{14}q^2 - \frac{7}{18}p^{12}q^3 + \cdots$
41. $1 + 12x + 66x^2 + 220x^3 + \cdots$ **43.** $1 + \frac{1}{2}x^2 - \frac{1}{8}x^4 + \frac{1}{16}x^6 - \cdots$ **45.** $1 - \frac{1}{2}a^2 - \frac{1}{8}a^4 - \frac{1}{16}a^6 - \cdots$
47. $1 + 6x + 24x^2 + 80x^3 + \cdots$ **49.** 1,001,000   **51.** 11th   **53.** 12.6 mm   **55.** 302.9 in.   **57.** \$2391.24
59. 4.40×10^9 in. = 69,400 mi   **61.** \$47,340.80   **63.** \$6.93   **65.** $1 + \frac{1}{2}am^2 + \frac{1}{8}am^4$ **67.** (a) No, (b) Yes

Exercises 19-1, page 528

(*Note:* "Answers" to trigonometric identities are intermediate steps of suggested reductions of the left member.)

1. $1.483 = \dfrac{1}{0.6745}$ **3.** $\left(-\frac{1}{2}\sqrt{3}\right)^2 + \left(-\frac{1}{2}\right)^2 = \frac{3}{4} + \frac{1}{4} = 1$ **5.** $\dfrac{\cos\theta}{\sin\theta}\left(\dfrac{1}{\cos\theta}\right) = \dfrac{1}{\sin\theta}$ **7.** $\dfrac{\sin x}{\dfrac{\sin x}{\cos x}} = \dfrac{\sin x}{1}\left(\dfrac{\cos x}{\sin x}\right)$

9. $\sin y \left(\dfrac{\cos y}{\sin y} \right)$ **11.** $\sin x \left(\dfrac{1}{\cos x} \right)$ **13.** $\csc^2 x (\sin^2 x)$

15. $\sin x (\csc^2 x) = (\sin x)(\csc x)(\csc x) = \sin x \left(\dfrac{1}{\sin x} \right) \csc x$ **17.** $\sin x \csc x - \sin^2 x = 1 - \sin^2 x$

19. $\tan y \cot y + \tan^2 y = 1 + \tan^2 y$ **21.** $\sin x \left(\dfrac{\sin x}{\cos x} \right) + \cos x = \dfrac{\sin^2 x + \cos^2 x}{\cos x} = \dfrac{1}{\cos x}$

23. $\cos \theta \left(\dfrac{\cos \theta}{\sin \theta} \right) + \sin \theta = \dfrac{\cos^2 \theta + \sin^2 \theta}{\sin \theta} = \dfrac{1}{\sin \theta}$

25. $\sec \theta \left(\dfrac{\sin \theta}{\cos \theta} \right) \csc \theta = \sec \theta \left(\dfrac{1}{\cos \theta} \right)(\sin \theta \csc \theta) = \sec \theta (\sec \theta)(1)$

27. $\cot \theta (\sec^2 \theta - 1) = \cot \theta \tan^2 \theta = (\cot \theta \tan \theta) \tan \theta$ **29.** $\dfrac{\sin x}{\cos x} + \dfrac{\cos x}{\sin x} = \dfrac{\sin^2 x + \cos^2 x}{\cos x \sin x} = \dfrac{1}{\cos x \sin x}$

31. $(1 - \sin^2 x) - \sin^2 x$ **33.** $\dfrac{\sin x (1 + \cos x)}{1 - \cos^2 x} = \dfrac{1 + \cos x}{\sin x}$

35. $\dfrac{(1/\cos x) + (1/\sin x)}{1 + (\sin x/\cos x)} = \dfrac{(\sin x + \cos x)/\cos x \sin x}{(\cos x + \sin x)/\cos x} = \dfrac{\cos x}{\cos x \sin x}$

37. $\dfrac{\sin^2 x}{\cos^2 x} \cos^2 x + \dfrac{\cos^2 x}{\sin^2 x} \sin^2 x = \sin^2 x + \cos^2 x$ **39.** $\dfrac{\sec \theta}{\dfrac{1}{\sec \theta}} - \dfrac{\tan \theta}{\dfrac{1}{\tan \theta}} = \sec^2 \theta - \tan^2 \theta$

41. $\dfrac{\sin^2 x + \cos^2 x - 2 \cos^2 x}{\sin x \cos x} = \dfrac{\sin^2 x - \cos^2 x}{\sin x \cos x} = \dfrac{\sin x}{\cos x} - \dfrac{\cos x}{\sin x}$ **43.** $\cos^3 x \left(\dfrac{1}{\sin^3 x} \right) \left(\dfrac{\sin^3 x}{\cos^3 x} \right) = 1$

45. $\dfrac{1}{\cos x} + \dfrac{\sin x}{\cos x} + \dfrac{\cos x}{\sin x} = \dfrac{\sin x + \cos^2 x + \sin^2 x}{\sin x \cos x}$ **47.** $\dfrac{\cos \theta + \sin \theta}{1 + \dfrac{\sin \theta}{\cos \theta}} = \dfrac{\cos \theta + \sin \theta}{\dfrac{\cos \theta + \sin \theta}{\cos \theta}}$

49. $\left(\dfrac{\sin x}{\cos x} + \dfrac{\cos x}{\sin x} \right) \sin x \cos x = \dfrac{\sin^2 x \cos x}{\cos x} + \dfrac{\sin x \cos^2 x}{\sin x}$

51. $\dfrac{(\sin^2 x - \cos^2 x)(\sin^2 x + \cos^2 x)}{(1 - \cot^2 x)(1 + \cot^2 x)} = \dfrac{\sin^2 x - \cos^2 x}{[1 - (\cos^2 x/\sin^2 x)]\csc^2 x}$ **53.** $\sec^2 x - 1 + 1 - \tan x + \tan x$

55. Infinite series: $\dfrac{1}{1 - \sin^2 x} = \dfrac{1}{\cos^2 x}$ **57.** $0 = \cos A \cos B \cos C + \sin A \sin B$, $\cos C = -\dfrac{\sin A \sin B}{\cos A \cos B}$

59. $\ell = a \csc \theta + a \sec \theta = a \left(\dfrac{1}{\sin \theta} + \dfrac{\tan \theta}{\sin \theta} \right)$ **61.** $\sqrt{1 - \cos^2 \theta} = \sqrt{\sin^2 \theta}$ **63.** $\sqrt{4 + 4 \tan^2 \theta} = 2 \sqrt{1 + \tan^2 \theta}$

Exercises 19-2, page 533

1. $\sin 105° = \sin 60° \cos 45° + \cos 60° \sin 45° = \dfrac{\sqrt{3}}{2} \dfrac{\sqrt{2}}{2} + \dfrac{1}{2} \dfrac{\sqrt{2}}{2} = 0.9659$

3. $\cos 15° = \cos(60° - 45°) = \cos 60° \cos 45° + \sin 60° \sin 45°$
$= (\tfrac{1}{2})(\tfrac{1}{2}\sqrt{2}) + (\tfrac{1}{2}\sqrt{3})(\tfrac{1}{2}\sqrt{2}) = \tfrac{1}{4}\sqrt{2} + \tfrac{1}{4}\sqrt{6} = \tfrac{1}{4}(\sqrt{2} + \sqrt{6}) = 0.9659$
5. $-\tfrac{33}{65}$ **7.** $-\tfrac{56}{65}$ **9.** $\sin 3x$ **11.** $\cos x$ **13.** $\cos (2 - x)$ **15.** 0   **17.** 1   **19.** 1
21. $\sin (180° - x) = \sin 180° \cos x - \cos 180° \sin x = (0) \cos x - (-1) \sin x$
23. $\cos (0 - x) = \cos 0 \cos x + \sin 0 \sin x = (1) \cos x + (0) \sin x$
25. $\sin (270° - x) = \sin 270° \cos x - \cos 270° \sin x = (-1) \cos x - 0(\sin x)$
27. $\cos (\tfrac{1}{2}\pi - x) = \cos \tfrac{1}{2}\pi \cos x + \sin \tfrac{1}{2}\pi \sin x = 0(\cos x) + 1(\sin x)$
29. $\cos (30° + x) = \cos 30° \cos x - \sin 30° \sin x = \tfrac{1}{2}\sqrt{3} \cos x - \tfrac{1}{2} \sin x = \tfrac{1}{2}(\sqrt{3} \cos x - \sin x)$
31. $\sin \left(\dfrac{\pi}{4} + x \right) = \sin \dfrac{\pi}{4} \cos x + \cos \dfrac{\pi}{4} \sin x = \tfrac{1}{2}\sqrt{2} \cos x + \tfrac{1}{2}\sqrt{2} \sin x$

33. $(\sin x \cos y + \cos x \sin y)(\sin x \cos y - \cos x \sin y) = \sin^2 x \cos^2 y - \cos^2 x \sin^2 y$
$$= \sin^2 x\,(1 - \sin^2 y) - (1 - \sin^2 x)\sin^2 y$$

35. $(\cos \alpha \cos \beta - \sin \alpha \sin \beta) + (\cos \alpha \cos \beta + \sin \alpha \sin \beta)$ **37, 39, 41, 43.** Use the indicated method.

45. $i_0 \sin (\omega t + \alpha) = i_0 (\sin \omega t \cos \alpha + \cos \omega t \sin \alpha)$

47. $\tan \alpha\,(R + \cos \beta) = \sin \beta,\ R = \dfrac{\sin \beta - \tan \alpha \cos \beta}{\tan \alpha} = \dfrac{\sin \beta \cos \alpha - \cos \beta \sin \alpha}{\cos \alpha \tan \alpha}$

Exercises 19-3, page 538

1. $\sin 60° = \sin 2(30°) = 2 \sin 30° \cos 30° = 2(\tfrac{1}{2})(\tfrac{1}{2}\sqrt{3}) = \tfrac{1}{2}\sqrt{3}$

3. $\cos 120° = \cos 2(60°) = \cos^2 60° - \sin^2 60° = (\tfrac{1}{2})^2 - (\tfrac{1}{2}\sqrt{3})^2 = -\tfrac{1}{2}$

5. $\sin 258° = 2 \sin 129° \cos 129° = -0.9781476$ **7.** $\cos 96° = \cos^2 48° - \sin^2 48° = -0.1045285$ **9.** $\tfrac{24}{25}$

11. $\tfrac{3}{5}$ **13.** $2 \sin 8x$ **15.** $\cos 8x$ **17.** $\cos x$ **19.** $-2 \cos 4x$ **21.** $\cos^2 \alpha - (1 - \cos^2 \alpha)$

23. $\dfrac{\cos x - (\sin x/\cos x)\sin x}{1/\cos x} = \cos^2 x - \sin^2 x$ **25.** $(\cos^2 x - \sin^2 x)(\cos^2 x + \sin^2 x) = (\cos^2 x - \sin^2 x)(1)$

27. $\dfrac{2 \sin 2\theta \cos 2\theta}{\sin 2\theta}$ **29.** $\dfrac{2 \sin \theta \cos \theta}{1 + 2 \cos^2 \theta - 1} = \dfrac{\sin \theta}{\cos \theta}$ **31.** $\dfrac{1}{\sec^2 x} - \dfrac{\tan^2 x}{\sec^2 x} = \cos^2 x - \dfrac{\sin^2 x}{\cos^2 x \sec^2 x}$

33. $\dfrac{2 \tan x}{\sin 2x} = \dfrac{2(\sin x/\cos x)}{2 \sin x \cos x} = \dfrac{1}{\cos^2 x}$ **35.** $\dfrac{\sin 3x \cos x - \cos 3x \sin x}{\sin x \cos x} = \dfrac{\sin 2x}{\tfrac{1}{2}\sin 2x}$

37. $\sin(2x + x) = \sin 2x \cos x + \cos 2x \sin x = (2 \sin x \cos x)(\cos x) + (\cos^2 x - \sin^2 x)\sin x$

39. Use the indicated method.   **41.** $R = v\left(\dfrac{2v \sin \alpha}{g}\right)\cos \alpha = \dfrac{v^2(2 \sin \alpha \cos \alpha)}{g}$

43. $vi \sin \omega t \sin\left(\omega t - \dfrac{\pi}{2}\right) = vi \sin \omega t\left(\sin \omega t \cos \dfrac{\pi}{2} - \cos \omega t \sin \dfrac{\pi}{2}\right) = vi \sin \omega t[-(\cos \omega t)(1)] = -\dfrac{1}{2} vi(2 \sin \omega t \cos \omega t)$

Exercises 19-4, page 542

1. $\cos 15° = \cos \dfrac{1}{2}(30°) = \sqrt{\dfrac{1 + \cos 30°}{2}} = \sqrt{\dfrac{1.8660}{2}} = 0.9659$

3. $\sin 75° = \sin \dfrac{1}{2}(150°) = \sqrt{\dfrac{1 - \cos 150°}{2}} = \sqrt{\dfrac{1.8660}{2}} = 0.9659$ **5.** $\sin 118° = 0.8829476$

7. $\sqrt{2\left(\dfrac{1 + \cos 164°}{2}\right)} = \sqrt{2} \cos 82° = 0.1968205$ **9.** $\sin 3\alpha$ **11.** $4 \cos 2x$ **13.** $\tfrac{1}{26}\sqrt{26}$ **15.** $\tfrac{1}{10}\sqrt{2}$

17. $\pm\sqrt{\dfrac{2}{1 - \cos \alpha}}$ **19.** $\tan \dfrac{1}{2}\alpha = \dfrac{1 - \cos \alpha}{\sin \alpha} = \dfrac{\sin \alpha}{1 + \cos \alpha}$ **21.** $\dfrac{1 - \cos \alpha}{2 \sin \tfrac{1}{2}\alpha} = \dfrac{1 - \cos \alpha}{2\sqrt{\tfrac{1}{2}(1 - \cos \alpha)}} = \sqrt{\dfrac{1 - \cos \alpha}{2}}$

23. $2\left(\dfrac{1 - \cos x}{2}\right) + \cos x$ **25.** $\sqrt{\dfrac{(1 + \cos \theta)(1 - \cos \theta)}{2(1 - \cos \theta)}} = \dfrac{\sin \theta}{\sqrt{4\left(\dfrac{1 - \cos \theta}{2}\right)}}$ **27.** $2\left(\dfrac{1 - \cos \alpha}{2}\right) - \left(\dfrac{1 + \cos \alpha}{2}\right)$

29. $\sin \omega t = \pm\sqrt{\dfrac{1 - \cos 2\omega t}{2}},\ \sin^2 \omega t = \tfrac{1}{2}(1 - \cos 2\omega t)$

31. $n = \left(\sqrt{\dfrac{1 - \cos (A + \phi)}{2}}\right)\Big/\left(\sqrt{\dfrac{1 - \cos A}{2}}\right) = \sqrt{\dfrac{1 - \cos (A + \phi)}{1 - \cos A}}$

Exercises 19-5, page 546

1. $\dfrac{\pi}{2}$ **3.** $\dfrac{3\pi}{4}, \dfrac{7\pi}{4}$ **5.** π **7.** $0.9273, 2.214$ **9.** $\dfrac{\pi}{3}, \dfrac{2\pi}{3}, \dfrac{4\pi}{3}, \dfrac{5\pi}{3}$ **11.** $\dfrac{\pi}{3}, \dfrac{2\pi}{3}, \dfrac{4\pi}{3}, \dfrac{5\pi}{3}$ **13.** $0, \dfrac{\pi}{6}, \dfrac{5\pi}{6}, \pi$

15. $\dfrac{\pi}{12}, \dfrac{\pi}{4}, \dfrac{5\pi}{12}, \dfrac{3\pi}{4}, \dfrac{13\pi}{12}, \dfrac{5\pi}{4}, \dfrac{17\pi}{12}, \dfrac{7\pi}{4}$ **17.** $\dfrac{\pi}{2}, \dfrac{3\pi}{2}$ **19.** $0, \dfrac{\pi}{3}, \pi, \dfrac{5\pi}{3}$ **21.** $\dfrac{\pi}{4}, \dfrac{3\pi}{4}, \dfrac{5\pi}{4}, \dfrac{7\pi}{4}$ **23.** $3.569, 5.856$

25. 0.2618, 1.309, 3.403, 4.451   **27.** 0.7854, 1.249, 3.927, 4.391   **29.** $\dfrac{3\pi}{8}, \dfrac{7\pi}{8}, \dfrac{11\pi}{8}, \dfrac{15\pi}{8}$ **31.** 0, π

33. 6.56×10^{-4} **35.** 10.2 s, 15.7 s, 21.2 s, 47.1 s   **37.** 0.3398, 2.802   **39.** 0.8861, 2.256, 4.028, 5.397

41. $-0.95, 0.00, 0.95$ **43.** 1.08

Exercises 19-6 page 552

1. y is the angle whose tangent is x.   **3.** y is the angle whose cotangent is $3x$.

5. y is twice the angle whose sine is x.   **7.** y is 5 times the angle whose cosine is $2x$.   **9.** $\dfrac{\pi}{3}$ **11.** 0   **13.** $-\dfrac{\pi}{3}$

15. $\dfrac{\pi}{3}$ **17.** $\dfrac{\pi}{6}$ **19.** $-\dfrac{\pi}{4}$ **21.** $\dfrac{\pi}{4}$ **23.** $-\dfrac{\pi}{3}$ **25.** $\dfrac{3\pi}{4}$ **27.** $\tfrac{1}{2}\sqrt{3}$ **29.** $\tfrac{1}{2}\sqrt{2}$ **31.** -1 **33.** -1.3090

35. -0.9838 **37.** 1.4413   **39.** 1.4503   **41.** -1.2389 **43.** -0.2239 **45.** $x = \tfrac{1}{3}\,\text{Arcsin } y$

47. $x = 4\tan y$ **49.** $x = \tfrac{1}{3}\,\text{Arcsec }(y-1)$ **51.** $x = 1 - \cos(1-y)$ **53.** $\dfrac{x}{\sqrt{1-x^2}}$ **55.** $\dfrac{1}{x}$ **57.** $\dfrac{3x}{\sqrt{9x^2-1}}$

59. $2x\sqrt{1-x^2}$ **61.** $t = \dfrac{1}{2\omega}\,\text{Arccos}\,\dfrac{y}{A} - \dfrac{\phi}{\omega}$ **63.** $t = \dfrac{1}{\omega}\left(\text{Arcsin}\,\dfrac{i}{I_m} - \alpha - \phi\right)$

65. $\sin(\text{Arcsin }\tfrac{3}{5} + \text{Arcsin }\tfrac{5}{13}) = \tfrac{3}{5}\tfrac{12}{13} + \tfrac{4}{5}\tfrac{5}{13} = \tfrac{56}{65}$ **67.** $\dfrac{\pi}{6} + \dfrac{\pi}{3} = \dfrac{\pi}{2}$ **69.** $\text{Arcsin}\left(\dfrac{a}{c}\right)$

71. Let $y = $ height of cliff; $\tan\alpha = \dfrac{h+y}{d}$, $\tan\beta = \dfrac{y}{d}$; $\tan\alpha = \dfrac{h + d\tan\beta}{d}$

Review Exercises for Chapter 19, page 555

1. $\sin(90° + 30°) = \sin 90° \cos 30° + \cos 90° \sin 30° = (1)(\tfrac{1}{2}\sqrt{3}) + (0)(\tfrac{1}{2}) = \tfrac{1}{2}\sqrt{3}$

3. $\sin(180° - 45°) = \sin 180° \cos 45° - \cos 180° \sin 45° = 0(\tfrac{1}{2}\sqrt{2}) - (-1)(\tfrac{1}{2}\sqrt{2}) = \tfrac{1}{2}\sqrt{2}$

5. $\cos 2(90°) = \cos^2 90° - \sin^2 90° = 0 - 1 = -1$ **7.** $\sin\tfrac{1}{2}(90°) = \sqrt{\tfrac{1}{2}(1 - \cos 90°)} = \sqrt{\tfrac{1}{2}(1 - 0)} = \tfrac{1}{2}\sqrt{2}$

9. $\sin 52° = 0.7880108$ **11.** $\sin 92° = 0.9993908$ **13.** $\cos 6° = 0.9945219$ **15.** $\cos 164° = -0.9612617$

17. $\sin 5x$ **19.** $4\sin 12x$ **21.** $2\cos 12x$ **23.** $2\cos x$ **25.** $-\dfrac{\pi}{2}$ **27.** 0.2619   **29.** $-\tfrac{1}{3}\sqrt{3}$ **31.** 0

33. $\dfrac{\dfrac{1}{\cos y}}{\dfrac{1}{\sin y}} = \dfrac{1}{\cos y}\dfrac{\sin y}{1}$ **35.** $\sin x \csc x - \sin^2 x = 1 - \sin^2 x$ **37.** $\dfrac{1 - \sin^2\theta}{\sin\theta} = \dfrac{\cos^2\theta}{\sin\theta}$

39. $\cos\theta\left(\dfrac{\cos\theta}{\sin\theta}\right) + \sin\theta = \dfrac{\cos^2\theta + \sin^2\theta}{\sin\theta}$ **41.** $\dfrac{(\sec^2 x - 1)(\sec^2 x + 1)}{\tan^2 x} = \sec^2 x + 1$

43. $2\left(\dfrac{1}{\sin 2x}\right)\left(\dfrac{\cos x}{\sin x}\right) = 2\left(\dfrac{1}{2\sin x \cos x}\right)\left(\dfrac{\cos x}{\sin x}\right) = \dfrac{1}{\sin^2 x}$ **45.** $\dfrac{\cos^2\theta}{\sin^2\theta}$ **47.** $\dfrac{(\cos^2\theta - \sin^2\theta)}{\cos^2\theta} = 1 - \dfrac{\sin^2\theta}{\cos^2\theta}$

49. $\dfrac{1}{2}\left(2\sin\dfrac{\theta}{2}\cos\dfrac{\theta}{2}\right)$ **51.** $\dfrac{1}{\cos x} + \dfrac{\sin x}{\cos x} = \dfrac{(1 + \sin x)(1 - \sin x)}{\cos x(1 - \sin x)} = \dfrac{1 - \sin^2 x}{\cos x(1 - \sin x)}$ **53.** $\cos[(x - y) + y]$

55. $\sin 4x(\cos 4x)$ **57.** $\dfrac{\sin x}{\dfrac{1}{\sin x} - \dfrac{\cos x}{\sin x}} = \dfrac{\sin^2 x}{1 - \cos x} = \dfrac{1 - \cos^2 x}{1 - \cos x}$

59. $\dfrac{\sin x\cos y + \cos x\sin y + \sin x\cos y - \cos x\sin y}{\cos x\cos y - \sin x\sin y + \cos x\cos y + \sin x\sin y} = \dfrac{2\sin x\cos y}{2\cos x\cos y}$ **61.** $x = \dfrac{1}{2}\,\text{Arccos}\,\dfrac{1}{2}y$

63. $x = \dfrac{1}{5}\sin\dfrac{1}{3}\left(\dfrac{1}{4}\pi - y\right)$ **65.** 1.2925, 4.4341   **67.** $\dfrac{\pi}{6}, \dfrac{5\pi}{6}, \dfrac{7\pi}{6}, \dfrac{11\pi}{6}$ **69.** $0, \dfrac{\pi}{2}, \pi, \dfrac{3\pi}{2}$ **71.** $\dfrac{\pi}{6}, \dfrac{5\pi}{6}, \dfrac{7\pi}{6}, \dfrac{11\pi}{6}$

73. $0, \pi$ **75.** 0 **77.** $\dfrac{1}{x}$ **79.** $2x\sqrt{1-x^2}$ **81.** $2\sqrt{1-\cos^2\theta}$ **83.** $\dfrac{\tan\theta}{\sqrt{1+\tan^2\theta}} = \dfrac{\tan\theta}{\sec\theta}$

85. $R = \sqrt{(A\cos\theta - B\sin\theta)^2 + (A\sin\theta + B\cos\theta)^2} = \sqrt{A^2(\cos^2\theta + \sin^2\theta) + B^2(\sin^2\theta + \cos^2\theta)}$

87. $\cos\alpha = \dfrac{A}{C}, \sin\alpha = \dfrac{B}{C}$:

$$A\sin 2t + B\cos 2t = C\left(\dfrac{A}{C}\sin 2t + \dfrac{B}{C}\cos 2t\right)$$

$$= C(\cos\alpha\sin 2t + \sin\alpha\cos 2t)$$

89. $\theta = \alpha + R\sin\omega t$

91. $1 - (1 - 2\sin^2\phi) - (2\sin\phi\cos\phi)\tan\alpha = 2\sin^2\phi - 2\sin\phi\cos\phi\tan\alpha$ **93.** $83.7°$ **95.** $54.7°$

Exercises 20-1, page 563

1. $2\sqrt{29}$ **3.** 3 **5.** 55 **7.** 7 **9.** 2.86 **11.** $\frac{5}{2}$ **13.** Undefined   **15.** $-\frac{3}{4}$ **17.** 0 **19.** 0.747
21. $\frac{1}{3}\sqrt{3}$ **23.** $-\frac{1}{3}\sqrt{3}$ **25.** $20.0°$ **27.** $98.50°$ **29.** Parallel   **31.** Perpendicular   **33.** $8, -2$
35. -3 **37.** Two sides equal $2\sqrt{10}$ **39.** $m_1 = \frac{5}{12}, m_2 = \frac{4}{3}$ **41.** 10 **43.** $4\sqrt{10} + 4\sqrt{2} = 18.3$ **45.** $(1, 5)$
47. $(-3, 4)$

Exercises 20-2, page 569

1. $4x - y + 20 = 0$ **3.** $7x - 2y - 24 = 0$ **5.** $x - y + 2 = 0$ **7.** $y = -3$ **9.** $x = -3$

11. $3x - 2y - 12 = 0$ **13.** $x + 3y + 5 = 0$ **15.** $x + 2y - 4 = 0$ **17.** $4x + 3y + 6 = 0$

19. $2x + 5y + 14 = 0$ **21.** 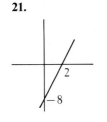   **23.** 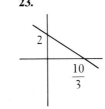   **25.** $y = \frac{3}{2}x - \frac{1}{2}; m = \frac{3}{2}, (0, -\frac{1}{2})$
27. $y = \frac{5}{2}x + \frac{5}{2}; m = \frac{5}{2}, (0, \frac{5}{2})$
29. -2 **31.** 1 **33.** $m_1 = m_2 = \frac{3}{2}$
35. $m_1 = 2, m_2 = -\frac{1}{2}$
37. $x + 2y - 4 = 0$

39. $3x + y - 18 = 0$ **41.** $v = 12.2 + 5.16t$ **43.** $\ell = 2c + 8$ **45.** $5x + 6y = 1220$ **47.** $y = 150{,}000 - 0.80x$

49. $y = 10^{-5}(2.4 - 5.6x)$ **51.** $n = \frac{7}{6}t + 10$; at 6:30, $n = 10$; at 8:30, $n = 150$
53.   **55.**   **57.**   **59.**

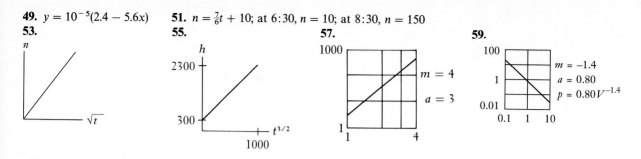

Exercises 20-3, page 574

1. $(2, 1)$, $r = 5$ **3.** $(-1, 0)$, $r = 2$ **5.** $x^2 + y^2 = 9$
7. $(x - 2)^2 + (y - 2)^2 = 16$, or $x^2 + y^2 - 4x - 4y - 8 = 0$
9. $(x + 2)^2 + (y - 5)^2 = 5$, or $x^2 + y^2 + 4x - 10y + 24 = 0$
11. $(x - 12)^2 + (y + 15)^2 = 324$, or $x^2 + y^2 - 24x + 30y + 45 = 0$
13. $(x - 2)^2 + (y - 1)^2 = 8$, or $x^2 + y^2 - 4x - 2y - 3 = 0$
15. $(x + 3)^2 + (y - 5)^2 = 25$, or $x^2 + y^2 + 6x - 10y + 9 = 0$
17. $(x - 2)^2 + (y - 2)^2 = 4$, or $x^2 + y^2 - 4x - 4y + 4 = 0$
19. $(x - 2)^2 + (y - 5)^2 = 25$, or $x^2 + y^2 - 4x - 10y + 4 = 0$; and
$(x + 2)^2 + (y + 5)^2 = 25$, or $x^2 + y^2 + 4x + 10y + 4 = 0$

21. $(0, 3)$,   **23.** $(-1, 5)$,   **25.** $(0, 0)$,   **27.** $(1, 0)$
$r = 2$ $r = \frac{9}{2}$ $r = 5$ $r = 3$

29. $(-4, 5)$,   **31.** $(1, 2)$,
$r = 7$ $r = \frac{1}{2}\sqrt{22}$

33. Symmetrical to both axes and origin
35. Symmetrical to y-axis   **37.** $(7, 0)$, $(-1, 0)$
39. $3x^2 + 3y^2 + 4x + 8y - 20 = 0$, circle

41. $x^2 + y^2 = 0.0100$ **43.** $(x - 500 \times 10^{-6})^2 + y^2 = 0.16 \times 10^{-6}$

Exercises 20-4, page 579

1. $F(1, 0)$, $x = -1$

3. $F(-1, 0)$, $x = 1$

5. $F(0, 2)$, $y = -2$

7. $F(0, -1)$, $y = 1$

9. $F(\frac{1}{2}, 0)$, $x = -\frac{1}{2}$

11. $F(0, \frac{1}{4})$, $y = -\frac{1}{4}$

13. $y^2 = 12x$ **15.** $x^2 = 16y$ **17.** $x^2 = 4y$
19. $x^2 = \frac{1}{8}y$

21. $y^2 - 2y - 12x + 37 = 0$

(3,1)

23. $x^2 - 2x + 8y - 23 = 0$

(1,3)

25. $4p$ **27.** H

i

29. 0.919 m   **31.** 2.97 ft   **33.**

f

0.92

200

A

35. $y^2 = 8x$ or $x^2 = 8y$ with vertex midway between island and shore

Exercises 20-5, page 585

1. $V(2, 0)$, $V(-2, 0)$,
$F(\sqrt{3}, 0)$, $F(-\sqrt{3}, 0)$

3. $V(0, 6)$, $V(0, -6)$,
$F(0, \sqrt{11})$, $F(0, -\sqrt{11})$

5. $V(3, 0)$, $V(-3, 0)$,
$F(\sqrt{5}, 0)$, $F(-\sqrt{5}, 0)$

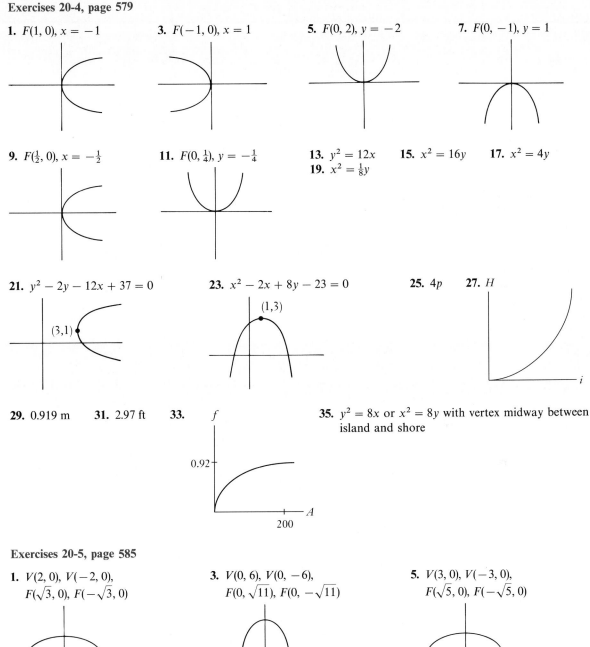

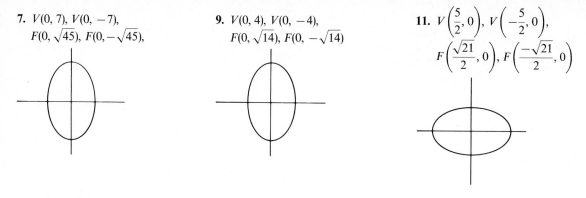

7. $V(0, 7)$, $V(0, -7)$, $F(0, \sqrt{45})$, $F(0, -\sqrt{45})$,

9. $V(0, 4)$, $V(0, -4)$, $F(0, \sqrt{14})$, $F(0, -\sqrt{14})$

11. $V\left(\dfrac{5}{2}, 0\right)$, $V\left(-\dfrac{5}{2}, 0\right)$, $F\left(\dfrac{\sqrt{21}}{2}, 0\right)$, $F\left(\dfrac{-\sqrt{21}}{2}, 0\right)$

13. $\dfrac{x^2}{225} + \dfrac{y^2}{144} = 1$, or $144x^2 + 225y^2 = 32{,}400$ **15.** $\dfrac{y^2}{9} + \dfrac{x^2}{5} = 1$, or $9x^2 + 5y^2 = 45$

17. $\dfrac{x^2}{64} + \dfrac{15y^2}{144} = 1$, or $3x^2 + 20y^2 = 192$ **19.** $\dfrac{x^2}{5} + \dfrac{y^2}{20} = 1$, or $4x^2 + y^2 = 20$

21. $16x^2 + 25y^2 - 32x - 50y - 359 = 0$ **23.** $9x^2 + 5y^2 - 18x - 20y - 16 = 0$

25. $2x^2 + 3y^2 - 8x - 4 = 2x^2 + 3(-y)^2 - 8x - 4$ **27.** $\frac{2}{3}\sqrt{2} = 0.943$ **29.** (a) 5.8 m, (b) 5.0 m

31. 27.5 m **33.** $7x^2 + 16y^2 = 112$ **35.** 843 ft^3

Exercises 20-6, page 592

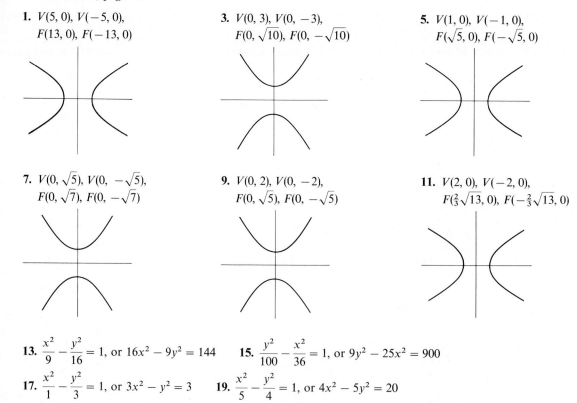

1. $V(5, 0)$, $V(-5, 0)$, $F(13, 0)$, $F(-13, 0)$

3. $V(0, 3)$, $V(0, -3)$, $F(0, \sqrt{10})$, $F(0, -\sqrt{10})$

5. $V(1, 0)$, $V(-1, 0)$, $F(\sqrt{5}, 0)$, $F(-\sqrt{5}, 0)$

7. $V(0, \sqrt{5})$, $V(0, -\sqrt{5})$, $F(0, \sqrt{7})$, $F(0, -\sqrt{7})$

9. $V(0, 2)$, $V(0, -2)$, $F(0, \sqrt{5})$, $F(0, -\sqrt{5})$

11. $V(2, 0)$, $V(-2, 0)$, $F(\frac{2}{3}\sqrt{13}, 0)$, $F(-\frac{2}{3}\sqrt{13}, 0)$

13. $\dfrac{x^2}{9} - \dfrac{y^2}{16} = 1$, or $16x^2 - 9y^2 = 144$ **15.** $\dfrac{y^2}{100} - \dfrac{x^2}{36} = 1$, or $9y^2 - 25x^2 = 900$

17. $\dfrac{x^2}{1} - \dfrac{y^2}{3} = 1$, or $3x^2 - y^2 = 3$ **19.** $\dfrac{x^2}{5} - \dfrac{y^2}{4} = 1$, or $4x^2 - 5y^2 = 20$

21.   **23.**   **25.** $9x^2 - 16y^2 - 108x + 64y + 116 = 0$
27. $9x^2 - y^2 - 36x + 27 = 0$ **29.** $x^2 - 2y^2 = 2$

31. $\ell^2 - x^2 = 2000^2$ **33.** $i = 6.00/R$ **35.** Dist. from rifle to P − dist. from target to P = constant
(related to dist. from rifle to target).

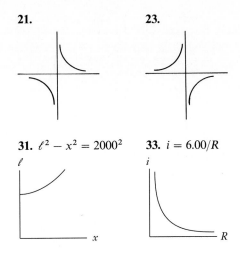

Exercises 20-7, page 596

1. Parabola, $(-1, 2)$ **3.** Hyperbola, $(1, 2)$ **5.** Ellipse, $(-1, 0)$ **7.** Parabola, $(-3, 1)$

9. $(y - 3)^2 = 16(x + 1)$, or $y^2 - 6y - 16x - 7 = 0$ **11.** $(x + 3)^2 = 4(y - 2)$, or $x^2 + 6x - 4y + 17 = 0$

13. $\dfrac{(x + 2)^2}{25} + \dfrac{(y - 2)^2}{16} = 1$, or $16x^2 + 25y^2 + 64x - 100y - 236 = 0$

15. $\dfrac{(y - 1)^2}{16} + \dfrac{(x + 2)^2}{4} = 1$, or $4x^2 + y^2 + 16x - 2y + 1 = 0$

17. $\dfrac{(y - 2)^2}{1} - \dfrac{(x + 1)^2}{3} = 1$, or $x^2 - 3y^2 + 2x + 12y - 8 = 0$

19. $\dfrac{(x + 1)^2}{9} - \dfrac{(y - 1)^2}{16} = 1$, or $16x^2 - 9y^2 + 32x + 18y - 137 = 0$

21. Parabola, $(-1, -1)$ **23.** Ellipse, $(-3, 0)$ **25.** Hyperbola, $(0, 4)$ **27.** Parabola, $(1, 0)$

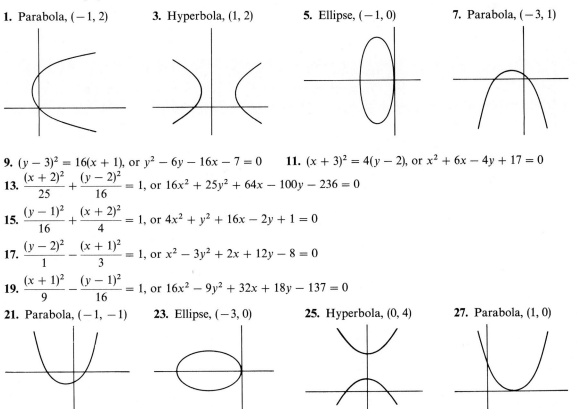

29. $x^2 - y^2 + 4x - 2y - 22 = 0$ **31.** $y^2 + 4x - 4 = 0$ **33.** $(x - 95)^2 = -\dfrac{95^2}{60}(y - 60)$

35. $\dfrac{x^2}{9.0} + \dfrac{y^2}{16} = 1, \dfrac{(x - 7.0)^2}{16} + \dfrac{y^2}{9.0} = 1$

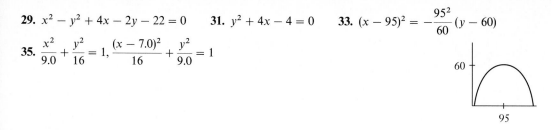

Exercises 20-8, page 600

1. Ellipse   **3.** Hyperbola   **5.** Circle   **7.** Parabola   **9.** Hyperbola   **11.** Circle   **13.** Parabola
15. Hyperbola   **17.** Ellipse   **19.** Ellipse
21. Parabola; $V(-4, 0)$; $F(-4, 2)$ **23.** Hyperbola; $C(1, -2)$;   **25.** Ellipse: $C(5, 0)$; $V(5, \pm 2\sqrt{2})$
$V(1, -2 \pm \sqrt{2})$

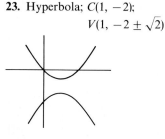

27. Parabola; $V(-\frac{1}{2}, \frac{5}{2})$; $F(\frac{1}{2}, \frac{5}{2})$ **29.** (a) Circle, (b) hyperbola, (c) ellipse
31. The origin   **33.** Parabola
35. Circle if light beam is perpendicular to floor; otherwise an ellipse

Exercise 20-9, page 604

1.   **3.**   **5.**   **7.**   **9.**   **11.**

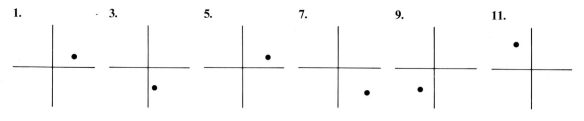

13. $\left(2, \dfrac{\pi}{6}\right)$ **15.** $\left(1, \dfrac{7\pi}{6}\right)$ **17.** $(-4, -4\sqrt{3})$ **19.** $(2.77, -1.15)$ **21.** $r = 3 \sec \theta$ **23.** $r = a$

25. $r = 4 \cot \theta \csc \theta$ **27.** $r^2 = \dfrac{4}{1 + 3 \sin^2 \theta}$ **29.** $x^2 + y^2 - y = 0$ **31.** $x = 4$

33. $x^4 + y^4 - 4x^3 + 2x^2y^2 - 4xy^2 - 4y^2 = 0$ **35.** $(x^2 + y^2)^2 = 2xy$ **37.** $B_x = -\dfrac{k \sin \theta}{r}, B_y = \dfrac{k \cos \theta}{r}$

39. $x^4 + y^4 + 2x^2y^2 + 2x^2y + 2y^3 - 9x^2 - 8y^2 = 0$

Exercises 20-10, page 607

1. 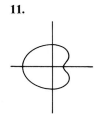 **3.** 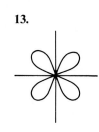 **5.**  **7.** 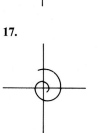 **9.**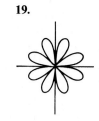

11.   **13.**   **15.** 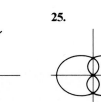 **17.** 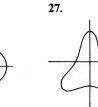 **19.**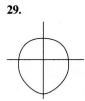

21.   **23.** **25.** **27.** **29.**

31.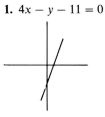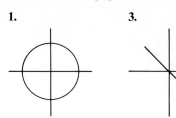

Review Exercises for Chapter 20, page 609

1. $4x - y - 11 = 0$ **3.** $2x + 3y + 3 = 0$ **5.** $x^2 + y^2 - 2x + 4y - 5 = 0$

 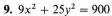

7. $y^2 = 12x$ **9.** $9x^2 + 25y^2 = 900$ **11.** $144y^2 - 169x^2 = 24{,}336$

 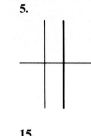 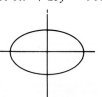 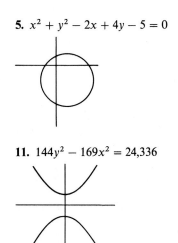

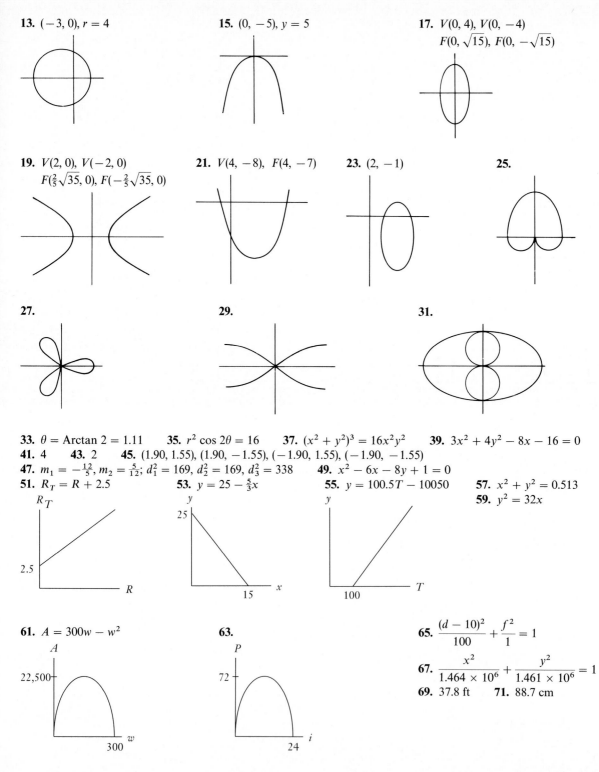

13. $(-3, 0), r = 4$

15. $(0, -5), y = 5$

17. $V(0, 4), V(0, -4)$
$F(0, \sqrt{15}), F(0, -\sqrt{15})$

19. $V(2, 0), V(-2, 0)$
$F(\frac{2}{5}\sqrt{35}, 0), F(-\frac{2}{5}\sqrt{35}, 0)$

21. $V(4, -8), F(4, -7)$

23. $(2, -1)$

25.

27.

29.

31.

33. $\theta = \text{Arctan } 2 = 1.11$ **35.** $r^2 \cos 2\theta = 16$ **37.** $(x^2 + y^2)^3 = 16x^2y^2$ **39.** $3x^2 + 4y^2 - 8x - 16 = 0$

41. 4   **43.** 2   **45.** $(1.90, 1.55), (1.90, -1.55), (-1.90, 1.55), (-1.90, -1.55)$

47. $m_1 = -\frac{12}{5}, m_2 = \frac{5}{12}; d_1^2 = 169, d_2^2 = 169, d_3^2 = 338$ **49.** $x^2 - 6x - 8y + 1 = 0$

51. $R_T = R + 2.5$ **53.** $y = 25 - \frac{5}{3}x$ **55.** $y = 100.5T - 10050$ **57.** $x^2 + y^2 = 0.513$

59. $y^2 = 32x$

61. $A = 300w - w^2$

63.

65. $\dfrac{(d - 10)^2}{100} + \dfrac{f^2}{1} = 1$

67. $\dfrac{x^2}{1.464 \times 10^6} + \dfrac{y^2}{1.461 \times 10^6} = 1$

69. 37.8 ft   **71.** 88.7 cm

73.

75. $(a^2 - b^2)x^2 + a^2y^2 + 2bx - 1 = 0$;
$a = b$, parabola; $a^2 > b^2$, ellipse; $a^2 < b^2$, hyperbola

Exercises 21-1, page 617

1.

No.	2	3	4	5	6	7
Freq.	1	3	4	2	3	2

3.

No.	0.45	0.46	0.47	0.48	0.49	0.50	0.51	0.52	0.53	0.54	0.55	0.56	0.57
Freq.	1	1	1	2	2	0	1	0	1	0	2	0	1

5.

Int.	2–3	4–5	6–7
Freq.	4	6	5

7.

Int.	0.43–0.45	0.46–0.48	0.49–0.51	0.52–0.54	0.55–0.57
Freq.	1	4	3	1	3

9. **11.** **13.** **15.**

17.

No. inst.	18	19	20	21	22	23	24	25
No.	1	3	2	4	3	1	0	1

19. **21.** **23.** **25.**

27. **29.** **31.**

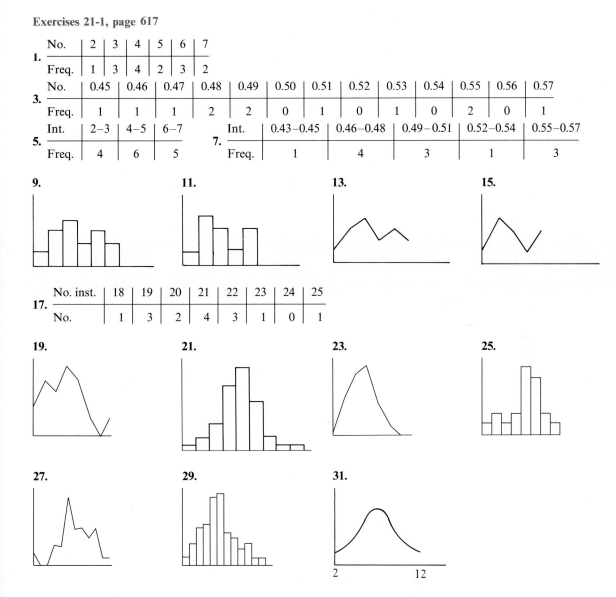

Exercises 21-2, page 622

1. 4   **3.** 0.49   **5.** 4.6   **7.** 0.503   **9.** 4   **11.** 0.48, 0.49, 0.55   **13.** 21   **15.** 21   **17.** 2.248 s
19. 172 ft   **21.** 4.237 mR   **23.** 4.36 mR   **25.** 31 h   **27.** 0.00595 mm   **29.** $275, $300
31. 862 kW·h   **33.** 4.5   **35.** 0.51

Exercises 21-3, page 628

1. 1.50 **3.** 0.037 **5.** 1.50 **7.** 0.037 **9.** 1.7 **11.** 2.5 h **13.** 0.014 s **15.** 0.00022 mm **17.** 60%
19. 58% **21.** 60% **23.** 76%

Exercises 21-4, page 632

1. $y = 1.0x - 2.6$ **3.** $y = -1.77x + 191$ **5.** $V = -0.590i + 11.3$

7. $h = 2.24x + 5.2$ **9.** $p = -0.200x + 649$ **11.** $V = 4.32 \times 10^{-15} f - 2.03$
                                                           $f_0 = 0.470$ PHz

13. 0.985 **15.** -0.901

Exercises 21-5, page 637

1. $y = 1.97x^2 + 4.8$ **3.** $y = 10.9/x$ **5.** $y = 5.97t^2 + 0.38$

7. $p = 0.00967T^2 + 29.4$ **9.** $P = \dfrac{1343}{S}$ **11.** $y = 6.20e^{-t} - 0.05$

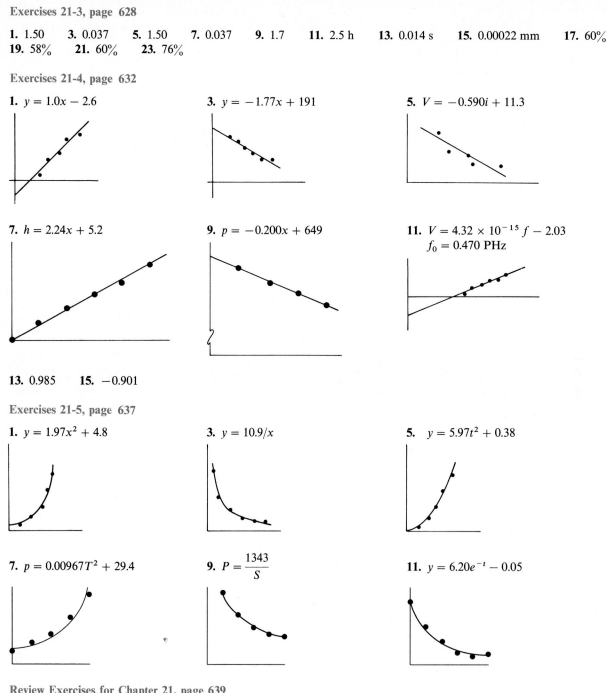

Review Exercises for Chapter 21, page 639

1. 3.6 **3.** 0.77 **5.**

Int.	101–103	104–106	107–109	110–112	113–115
Freq.	5	4	3	3	5

7. 106   **9.** 4.6   **11.**    **13.** 0.264 Pa·s   **15.** 0.014 Pa·s   **17.**

19. 697 W   **21.** 700 W   **23.** 17 W   **25.** 4

27. 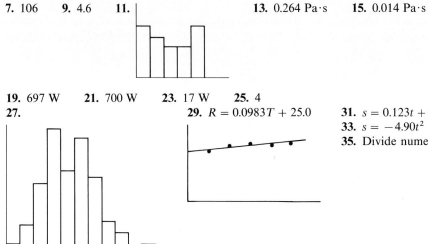   **29.** $R = 0.0983T + 25.0$

31. $s = 0.123t + 0.887$
33. $s = -4.90t^2 + 3000$
35. Divide numerator and denominator by n^2.

Exercises 22-1, page 649

1. Cont. all x **3.** Not cont. $x = -3$, div. by zero   **5.** Cont. $x > 0$ **7.** Cont. all x
9. Not cont. $x = 1$, small change   **11.** Cont. $x < 2$ **13.** Not cont. $x = 2$, small change
15. Cont. all x

17.

x	2.500	2.900	2.990	2.999	3.001	3.010	3.100	3.500
$f(x)$	5.500	6.700	6.970	6.997	7.003	7.030	7.300	8.500

$\lim\limits_{x \to 3} f(x) = 7$

19.

x	0.900	0.990	0.999	1.001	1.010	1.100
$f(x)$	1.7100	1.9701	1.9970	2.0030	2.0301	2.3100

$\lim\limits_{x \to 1} f(x) = 2$

21.

x.	1.900	1.990	1.999	2.001	2.010	2.100
$f(x)$	-0.2516	-0.2502	-0.25002	-0.24998	-0.2498	-0.2485

$\lim\limits_{x \to 2} f(x) = -0.25$

23.

x.	10	100	1000
$f(x)$	0.4468	0.4044	0.4004

$\lim\limits_{x \to \infty} f(x) = 0.4$ **25.** 7   **27.** 1   **29.** 1

31. -2 **33.** 2   **35.** 2   **37.** Does not exist   **39.** 0   **41.** 3   **43.** 0

45.

x	-0.1	-0.01	-0.001	0.001	0.01	0.1
$f(x)$	-3.1	-3.01	-3.001	-2.999	-2.99	-2.9

$\lim\limits_{x \to 0} f(x) = -3$

47.

x	10	100	1000
$f(x)$	2.1649	2.0106	2.0010

$\lim\limits_{x \to \infty} f(x) = 2$

49. 3 cm/s   **51.** 34.9°C, 0°C   **53.** e **55.** 2

Exercises 22-2, page 655

1. (Slopes) 3.5, 3.9, 3.99, 3.999; $m = 4$ **3.** (Slopes) $-3.5, -3.9, -3.99, -3.999$; $m = -4$ **5.** 4   **7.** -4

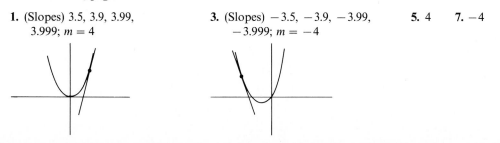

9. $m_{\text{tan}} = 2x_1$; 4, -2

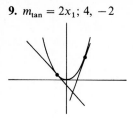

11. $m_{\text{tan}} = 2x_1 + 2$; -4, 4

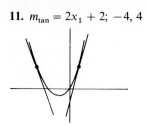

13. $m_{\text{tan}} = 2x_1 + 4$; -2, 8

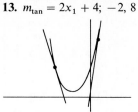

15. $m_{\text{tan}} = 6 - 2x_1$; 10, 0

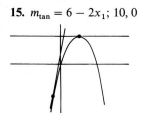

17. $m_{\text{tan}} = 3x_1^2 - 2$; 1, -2, 1

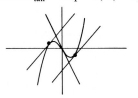

19. $m_{\text{tan}} = 4x_1^3$; 0, 0.5, 4

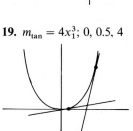

21. $\dfrac{\Delta y}{\Delta x} = 4.1$, $m_{\text{tan}} = 4$

23. $\dfrac{\Delta y}{\Delta x} = -12.61$, $m_{\text{tan}} = -12$

Exercises 22-3, page 660

1. 3   **3.** -2 **5.** $2x$ **7.** $10x$ **9.** $2x - 7$ **11.** $8 - 4x$ **13.** $3x^2 + 4$ **15.** $-\dfrac{1}{(x+2)^2}$

17. $1 - \dfrac{1}{x^2}$ **19.** $-\dfrac{4}{x^3}$ **21.** $4x^3 + 3x^2 + 2x + 1$ **23.** $4x^3 + \dfrac{2}{x^2}$ **25.** $6x - 2$; -8 **27.** $\dfrac{-6}{(x+3)^2}$; $-\dfrac{1}{6}$

29. $\dfrac{1}{2\sqrt{x+1}}$ **31.** $-\dfrac{3}{2\sqrt{1-3x}}$

Exercises 22-4, page 664

1. $m = 4$

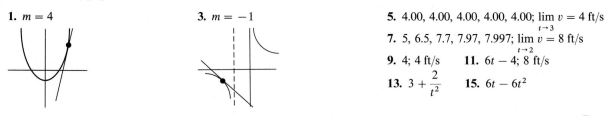

3. $m = -1$

5. 4.00, 4.00, 4.00, 4.00, 4.00; $\lim\limits_{t \to 3} v = 4$ ft/s
7. 5, 6.5, 7.7, 7.97, 7.997; $\lim\limits_{t \to 2} v = 8$ ft/s
9. 4; 4 ft/s   **11.** $6t - 4$; 8 ft/s
13. $3 + \dfrac{2}{t^2}$ **15.** $6t - 6t^2$

17. $12t - 4$ **19.** $6t$ **21.** -2 **23.** $6w$ **25.** 460 W   **27.** -83.1 W/m²·h   **29.** πd^2 **31.** $24.2/\sqrt{\lambda}$

Exercises 22-5, page 669

1. $5x^4$ **3.** $-36x^8$ **5.** $4x^3$ **7.** $2x + 2$ **9.** $15x^2 - 1$ **11.** $8x^7 - 28x^6 - 1$ **13.** $-42x^6 + 15x^2$
15. $x^2 + x$ **17.** 16   **19.** 33   **21.** 360   **23.** 14   **25.** $30t^4 - 5$ **27.** $-6 - 6t^2$ **29.** 64
31. 45   **33.** 1   **35.** $3\pi r^2$ **37.** 84 W/A   **39.** $a(c_1 + 2c_2 E + 3c_3 E^2)$ **41.** -12 N/cm   **43.** 391 mm²

Exercises 22-6, page 674

1. $x^2(3) + (3x + 2)(2x) = 9x^2 + 4x$ **3.** $6x(6x - 5) + (3x^2 - 5x)(6) = 54x^2 - 60x$
5. $(x + 2)(2) + (2x - 5)(1) = 4x - 1$
7. $(x^4 - 3x^2 + 3)(-6x^2) + (1 - 2x^3)(4x^3 - 6x) = -14x^6 + 30x^4 + 4x^3 - 18x^2 - 6x$

9. $(2x - 7)(-2) + (5 - 2x)(2) = -8x + 24$

11. $(x^3 - 1)(4x - 1) + (2x^2 - x - 1)(3x^2) = 10x^4 - 4x^3 - 3x^2 - 4x + 1$

13. $\dfrac{3}{(2x + 3)^2}$ **15.** $\dfrac{-2x}{(x^2 + 1)^2}$ **17.** $\dfrac{6x - 2x^2}{(3 - 2x)^2}$ **19.** $\dfrac{-6x^2 + 6x + 4}{(3x^2 + 2)^2}$ **21.** $\dfrac{-x^2 - 16x - 6}{(x^2 + x + 2)^2}$

23. $\dfrac{-2x^4 + 2x^3 + 5x^2 + 4x}{(x^3 + 2x^2)^2}$ **25.** -107 **27.** 19 **29.** $\dfrac{-12x^3 + 45x^2 - 14x}{(3x - 7)^2}$ **31.** 12 **33.** $1, -1$

35. $8t^3 - 45t^2 - 14t - 8$ **37.** $-0.07 \text{ V}/\Omega$ **39.** $1.2°\text{C/h}$ **41.** $\dfrac{2R(R + 2r)}{3(R + r)^2}$ **43.** $\dfrac{E^2(R - r)}{(R + r)^3}$

Exercises 22-7, page 680

1. $\dfrac{1}{2x^{1/2}}$ **3.** $-\dfrac{2}{x^3}$ **5.** $-\dfrac{1}{x^{4/3}}$ **7.** $\dfrac{3}{2}x^{1/2} + \dfrac{1}{x^2}$ **9.** $10x(x^2 + 1)^4$ **11.** $-192x^2(7 - 4x^3)^7$

13. $\dfrac{2x^2}{(2x^3 - 3)^{2/3}}$ **15.** $\dfrac{8x}{(1 - x^2)^5}$ **17.** $\dfrac{24x^3}{(2x^4 - 5)^{1/4}}$ **19.** $\dfrac{-4x}{(1 - 8x^2)^{3/4}}$ **21.** $\dfrac{12x + 5}{(8x + 5)^{1/2}}$

23. $\dfrac{-16x - 22}{(4x + 3)^{1/2}(8x + 1)^2}$ **25.** $\dfrac{3}{10}$ **27.** $\dfrac{5}{36}$ **29.** $\dfrac{x^3(0) - 1(3x^2)}{x^6} = -3x^{-4}$ **31.** $x = 0$ **33.** 1

35. -1.35 cm/s   **37.** $\dfrac{-450{,}000}{V^{5/2}}$, -4.50 kPa/cm³   **39.** -45.2 W/m² · h   **41.** $\dfrac{8a^3}{(4a^2 - \lambda^2)^{3/2}}$

43. $\dfrac{2(w + 1)}{(2w^2 + 4w + 4)^{1/2}}$

Exercises 22-8, page 684

1. $-\dfrac{3}{2}$ **3.** $\dfrac{6x + 1}{4}$ **5.** $\dfrac{x}{y}$ **7.** $\dfrac{2x}{5y^4}$ **9.** $\dfrac{2x}{2y + 1}$ **11.** $\dfrac{-3y}{3x + 1}$ **13.** $\dfrac{-2x - y^3}{3xy^2 + 3}$

15. $\dfrac{3(y^2 + 1)(y^2 - 2x + 1)}{(y^2 + 1)^2 - 6x^2y}$ **17.** $\dfrac{4(2y - x)^3 - 2x}{8(2y - x)^3 - 1}$ **19.** $\dfrac{-3x(x^2 + 1)^2}{y(y^2 + 1)}$ **21.** 3 **23.** $-\dfrac{108}{157}$ **25.** 1

27. $-\dfrac{x}{y}$ **29.** $\dfrac{r - R + 1}{r + 1}$ **31.** $\dfrac{2C^2r(12CSr - 20Cr - 3L)}{3(C^2r^2 - L^2)}$

Exercises 22-9, page 688

1. $y' = 3x^2 + 2x$, $y'' = 6x + 2$, $y''' = 6$, $y^{(n)} = 0$ $(n \geq 4)$

3. $f'(x) = 3x^2 - 24x^3$, $f''(x) = 6x - 72x^2$, $f'''(x) = 6 - 144x$, $f^{(4)}(x) = -144$, $f^{(n)}(x) = 0$ $(n \geq 5)$

5. $y' = -8(1 - 2x)^3$, $y'' = 48(1 - 2x)^2$, $y''' = -192(1 - 2x)$, $y^{(4)} = 384$, $y^{(n)} = 0$ $(n \geq 5)$

7. $f'(x) = (8x + 1)(2x + 1)^2$, $f''(x) = 12(2x + 1)(4x + 1)$, $f'''(x) = 24(8x + 3)$, $f^{(4)}(x) = 192$, $f^{(n)}(x) = 0$ $(n \geq 5)$

9. $84x^5 - 30x^4$ **11.** $-\dfrac{1}{4x^{3/2}}$ **13.** $-\dfrac{12}{(8x - 3)^{7/4}}$ **15.** $\dfrac{12}{(1 - 2x)^{5/2}}$ **17.** $600(2 - 5x)^2$

19. $30(27x^2 - 1)(3x^2 - 1)^3$ **21.** $\dfrac{4}{(1 - x)^3}$ **23.** $\dfrac{2}{(x + 1)^3}$ **25.** $-\dfrac{9}{y^3}$ **27.** $-\dfrac{6(x^2 - xy + y^2)}{(2y - x)^3}$ **29.** $\dfrac{9}{125}$

31. $-\dfrac{13}{384}$ **33.** -32.2 ft/s²   **35.** $-\dfrac{1.60}{(2t + 1)^{3/2}}$

Review Exercises for Chapter 22, page 689

1. -4 **3.** $\dfrac{1}{4}$ **5.** 1 **7.** $\dfrac{7}{3}$ **9.** $\dfrac{2}{3}$ **11.** -2 **13.** 5 **15.** $-4x$ **17.** $-\dfrac{4}{x^3}$ **19.** $\dfrac{1}{2\sqrt{x + 5}}$

21. $14x^6 - 6x$ **23.** $\dfrac{2}{x^{1/2}} + \dfrac{3}{x^2}$ **25.** $\dfrac{1}{(1-x)^2}$ **27.** $-12(2-3x)^3$ **29.** $\dfrac{9x}{(5-2x^2)^{7/4}}$ **31.** $\dfrac{-15x^2 + 2x}{(1-6x)^{1/2}}$

33. $\dfrac{-2x-3}{2x^2(4x+3)^{1/2}}$ **35.** $\dfrac{2x - 6(2x-3y)^2}{1 - 9(2x-3y)^2}$ **37.** $\dfrac{5}{48}$ **39.** $\dfrac{74}{5}$ **41.** $36x^2 - \dfrac{2}{x^3}$ **43.** $\dfrac{56}{(1+4x)^3}$ **45.** f_2

47. -31 **49.** $-k + k^2 t - \dfrac{1}{2}k^3 t^2$ **51.** $-\dfrac{2k}{r^3}$ **53.** $0.4(0.01t+1)^2(0.04t+1)$ **55.** $\dfrac{2R(R^3 - 3r^2 R + 2r^3)}{3(R^2 - r^2)^2}$

57. $-\dfrac{1}{4\pi\sqrt{C}(L+2)^{3/2}}$ **59.** $\dfrac{40V_2^{0.4}}{V_1^{1.4}}$ **61.** $0.049/\text{m}$ **63.** $p = 2w + \dfrac{150}{w}, \dfrac{dp}{dw} = 2 - \dfrac{150}{w^2}$

65. $A = 4x - x^3, \dfrac{dA}{dx} = 4 - 3x^2$

67. At $t = 5$ years, $dV/dt = -\$7500/\text{year}$ (rate of appreciation is decreasing)
      $d^2V/dt^2 = \$1500/\text{year}^2$ (rate at which appreciation changes is increasing)
   (Machinery is depreciating, but depreciation is lessening.)

Exercises 23-1, page 696

1. $4x - y - 2 = 0$ **3.** $2y + x - 2 = 0$ **5.** $x - 2y + 6 = 0$ **7.** $8x - 4y - 7 = 0$

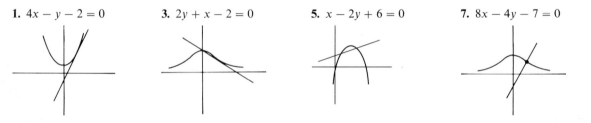

9. $\sqrt{3}x + 8y - 7 = 0, 16x - 2\sqrt{3}y - 15\sqrt{3} = 0$ **11.** $2x - 12y + 37 = 0, 72x + 12y + 37 = 0$ **13.** $y = 2x - 4$
15. $y - 8 = -\frac{1}{24}(x - \frac{3}{2})$, or $2x + 48y - 387 = 0$ **17.** $(-\frac{1}{4}, 0)$ **19.** $3x - 5y - 150 = 0$ **21.** $x + y - 6 = 0$
23. $x + 2y - 3 = 0, x = 0, x - 2y + 3 = 0$

Exercises 23-2, page 700

1. 3.4494897 **3.** -0.1804604 **5.** 0.5857864 **7.** 0.3488942 **9.** 2.5615528 **11.** -1.2360680
13. 0.9175433 **15.** 0.6180340 **17.** $-1.8557725, 0.6783628, 3.1774097$ **19.** 1.5874011 **21.** 29.4 m
23. 5.05 ft

Exercises 23-3, page 705

1. $3.16, 341.6°$ **3.** $8.07, 352.4°$ **5.** $a = 0$ **7.** $20.0, 3.7°$
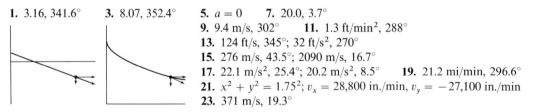
9. 9.4 m/s, $302°$ **11.** 1.3 ft/min^2, $288°$
13. 124 ft/s, $345°$; 32 ft/s^2, $270°$
15. 276 m/s, $43.5°$; 2090 m/s, $16.7°$
17. 22.1 m/s^2, $25.4°$; 20.2 m/s^2, $8.5°$ **19.** 21.2 mi/min, $296.6°$
21. $x^2 + y^2 = 1.75^2$; $v_x = 28,800$ in./min, $v_y = -27,100$ in./min
23. 371 m/s, $19.3°$

Exercises 23-4, page 709

1. $0.0900\ \Omega/\text{s}$ **3.** $\$1.22/\text{week}$ **5.** 4.1×10^{-6} m/s   **7.** $\dfrac{dB}{dt} = \dfrac{-3kr(dr/dt)}{[r^2 + (\ell/2)^2]^{5/2}}$ **9.** 0.38 mm^2/month

11. -101 mm^3/min   **13.** -3.75 kPa/min   **15.** 2.51×10^6 mm^3/s   **17.** 0.48 m/min   **19.** 12.0 ft/s
21. 825 mi/h   **23.** 8.33 ft/s

Exercises 23-5, page 716

1. Inc. $x > -1$, dec. $x < -1$ **3.** Inc. $-2 < x < 2$, dec. $x < -2$, $x > 2$ **5.** Min. $(-1, -1)$
7. Min. $(-2, -16)$, max. $(2, 16)$ **9.** Conc. up all x **11.** Conc. up $x < 0$, conc. down $x > 0$, infl. $(0, 0)$
13.   **15.**   **17.** Max. $(3, 18)$,   **19.** Max. $(-2, 8)$, min. $(0, 0)$,
                                                                                         conc. down all x infl. $(-1, 4)$

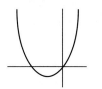  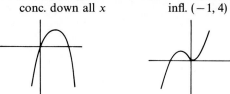

21. No max. or min.,   **23.** Max. $(-1, 4)$, min. $(1, -4)$,   **25.** Max. $(1, 1)$, infl. $(0, 0)$, $\left(\frac{2}{3}, \frac{16}{27}\right)$
    infl. $(-1, 1)$ infl. $(0, 0)$

  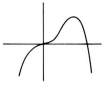

27. Max. $(2000, 1000)$ **29.** Max. $(0, 75)$,   **31.** $V = 4x^3 - 40x^2 + 96x$,
                                   infl. $(1, 64)$, $(3, 48)$ max. $(1.57, 67.6)$

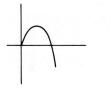 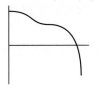 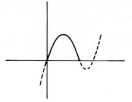

33.   **35.**

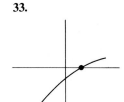 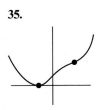

Exercises 23-6, page 720

1. Inc. $x < 0$, dec. $x > 0$,   **3.** Dec. $x < -1$, $x > -1$,   **5.** Int. $(-\sqrt[3]{2}, 0)$, min. $(1, 3)$,
    conc. up $x < 0$, $x > 0$,   conc. up $x > -1$,   infl. $(-\sqrt[3]{2}, 0)$, asym. $x = 0$
    asym. $x = 0$, $y = 0$ conc. down $x < -1$,
                                           int. $(0, 2)$, asym. $x = -1$, $y = 0$

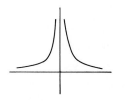 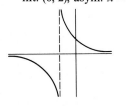 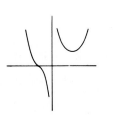

7. Int. $(1, 0)$, $(-1, 0)$,
asym. $x = 0$, $y = x$,
conc. up $x < 0$, conc. down $x > 0$

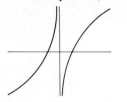

9. Int. $(0, 0)$, max. $(-2, -4)$
min. $(0, 0)$, asym. $x = -1$

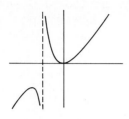

11. Int. $(0, -1)$, max. $(0, -1)$,
asym. $x = 1$, $x = -1$, $y = 0$

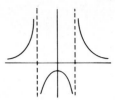

13. Int. $(1, 0)$, max. $(2, 1)$
infl. $(3, \frac{8}{9})$,
asym. $x = 0$, $y = 0$

15. Int. $(0, 0)$, $(1, 0)$, $(-1, 0)$
max. $(\frac{1}{2}\sqrt{2}, \frac{1}{2})$,
min. $(-\frac{1}{2}\sqrt{2}, -\frac{1}{2})$

17. Int. $(0, 0)$, infl. $(0, 0)$,
asym. $x = -3$, $x = 3$, $y = 0$

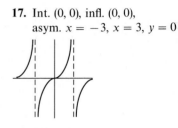

19. Int. $(0, 0)$, asym. $C_T = 6$,
inc. $C \geq 0$,
conc. down $C \geq 0$

21. Int. $(0, 1)$, max. $(0, 1)$,
infl. $(141, 0.82)$,
asym. $R = 0$

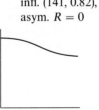

23. $A = 2\pi r^2 + \dfrac{40}{r}$,
min. $(1.47, 40.8)$,
asym. $r = 0$

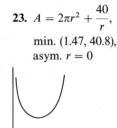

Exercises 23-7, page 726

1. 196 ft   **3.** $\dfrac{E}{2R}$ **5.** 34.6 m^2, \$8310   **7.** 0.250 ft/min   **9.** 8, 8   **11.** 5 mm, 5 mm   **13.** 17.0 cm

15. 8.49 cm, 8.49 cm   **17.** 3.00 ft   **19.** 12   **21.** 0.58L **23.** 1.33 in.   **25.** 100 m

27. $w = 1.00$ ft, $d = 1.73$ ft   **29.** 8.00 mi from refinery   **31.** 59.2 ft, 118 ft

Review Exercises for Chapter 23, page 728

1. $5x - y + 1 = 0$ **3.** $27x - 3y - 26 = 0$ **5.** $x - 2y + 3 = 0$ **7.** 4.19, 72.6°   **9.** 2.12   **11.** 2.00, 90.9°

13. 0.7458983   **15.** 1.9111643

17. Min. $(-2, -16)$,
conc. up all x

19. Int. $(0, 0)$, $(\pm 3\sqrt{3}, 0)$,
max. $(3, 54)$, min. $(-3, -54)$,
infl. $(0, 0)$

21. Min. $(2, -48)$,
conc. up $x < 0$, $x > 0$

23. Int. (0, 0), asym. $x = -1$, $y = 1$

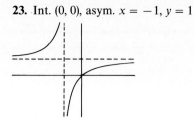

25. $2x - y + 1 = 0$
29. 8.8 m/s, 336°
33. \$2400
35.

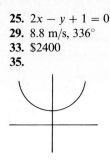

27. 0.0 m, 6.527 m
31. -7.44 cm/s

37. 1.30 cm/min
39. 37,700 m²/min
41. 5000 cm²

43. Int. (1, 0),
max. $(2, \frac{1}{4})$,
infl. $(3, \frac{2}{9})$,
asym. $P = 0$,
$V \geq 1$ (only meaningful values)

45. 711 mi/h

47. 2.2 in.

Exercises 24-1, page 735

1. $(5x^4 + 1)\,dx$ **3.** $\dfrac{-10\,dx}{x^6}$ **5.** $8x(x^2 - 1)^3\,dx$ **7.** $\dfrac{-12x\,dx}{(3x^2 + 1)^2}$ **9.** $x(1 - x)^2(-5x + 2)\,dx$ **11.** $\dfrac{2dx}{(5x + 2)^2}$

13. 12.28, 12   **15.** 1.71275, 1.675   **17.** $-2.4, -2.4730881$ **19.** 0.6257, 0.6264903

21. 0.0038 cm²   **23.** -8.3 nF   **25.** 28 ft³   **27.** $\dfrac{dr}{r} = \dfrac{k\,d\lambda/2\lambda^{1/2}}{k\lambda^{1/2}}$ **29.** $\dfrac{dA}{A} = \dfrac{2\,ds}{s}$ **31.** 16.96

Exercises 24-2, page 738

1. x^3 **3.** x^6 **5.** $\frac{3}{2}x^4 + x$ **7.** $\frac{2}{3}x^3 - \frac{1}{2}x^2$ **9.** $x^{5/2}$ **11.** $\frac{4}{3}x^{3/2} + 3x$ **13.** $\dfrac{1}{x}$ **15.** $\dfrac{2}{x^3}$ **17.** $\frac{2}{5}x^5 + x$

19. $3x^2 - \dfrac{1}{3x^3}$ **21.** $\frac{1}{3}x^3 + 2x - \dfrac{1}{x}$ **23.** $(2x + 1)^6$ **25.** $(x^2 - 1)^4$ **27.** $\frac{1}{40}(2x^4 + 1)^5$ **29.** $(6x + 1)^{3/2}$

31. $\frac{1}{4}(3x + 1)^{4/3}$

Exercises 24-3, page 744

1. $x^2 + C$ **3.** $\frac{1}{8}x^8 + C$ **5.** $\frac{2}{5}x^{5/2} + C$ **7.** $-\dfrac{1}{3x^3} + C$ **9.** $\frac{1}{3}x^3 - \frac{1}{6}x^6 + C$ **11.** $3x^3 + \frac{1}{2}x^2 + 3x + C$

13. $-\dfrac{1}{2x^2} + \dfrac{1}{2}x + C$ **15.** $\frac{2}{7}x^{7/2} - \frac{2}{5}x^{5/2} + C$ **17.** $6x^{1/3} + \frac{1}{9}x + C$ **19.** $\frac{1}{6}(1 + 2x)^3 + C$

21. $\frac{1}{6}(x^2 - 1)^6 + C$ **23.** $\frac{1}{5}(x^4 + 3)^5 + C$ **25.** $\frac{1}{40}(x^5 + 4)^8 + C$ **27.** $\frac{1}{12}(8x + 1)^{3/2} + C$

29. $\frac{1}{6}\sqrt{6x^2 + 1} + C$ **31.** $\sqrt{x^2 - 2x} + C$ **33.** $y = 2x^3 + 2$ **35.** $y = 5 - \frac{1}{18}(1 - x^3)^6$

37. $12y = 83 + (1 - 4x^2)^{3/2}$ **39.** $i = 2t^2 - 0.2t^3 + 2$ **41.** $f = \sqrt{0.01A + 1} - 1$ **43.** $y = 3x^2 + 2x - 3$

Exercises 24-4, page 750

1. 9, 12.15   **3.** 1.92, 2.28   **5.** 7.625, 8.208   **7.** 0.464, 0.5995   **9.** 13.5   **11.** $\frac{8}{3}$ **13.** 9   **15.** 0.8

Exercises 24-5, page 753

1. 1   **3.** $\frac{254}{7}$ **5.** $6 + 2\sqrt{6} - 2\sqrt{3}$ **7.** $\frac{3}{2}\sqrt[3]{2}$ **9.** $\frac{747}{20}$ **11.** $\frac{33}{20}\sqrt[3]{\frac{11}{5}} - \frac{3}{8}\sqrt[3]{\frac{1}{2}} - \frac{17}{5} = -1.552$ **13.** $\frac{4}{3}$
15. $\frac{81}{4}$ **17.** 2   **19.** $\frac{1}{4}(20.5^{2/3} - 17.5^{2/3}) = 0.1875$ **21.** $\frac{88}{3249} = 0.0271$ **23.** 49   **25.** $\frac{364}{3}$ **27.** $\frac{110}{3}$
29. $\frac{33}{784} = 0.0421$ **31.** $\frac{464}{5}$ **33.** 64,000 ft·lb   **35.** 86.8 m²

Exercises 24-6, page 757

1. $\frac{11}{2} = 5.50$, $\frac{16}{3} = 5.33$ **3.** 7.661, $\frac{23}{3} = 7.667$ **5.** 0.2042   **7.** 2.996   **9.** 0.5205   **11.** 21.74   **13.** 45.36
15. $0.177k$

Exercises 24-7, page 760

1. (a) 6, (b) 6   **3.** (a) 19.67, (b) 19.67   **5.** 0.2028   **7.** 3.084   **9.** 0.5114   **11.** 13.085   **13.** 44.63
15. 1.200 in.

Review Exercises for Chapter 24, page 761

1. $x^4 - \frac{1}{2}x^2 + C$ **3.** $x^2 - \frac{4}{5}x^{5/2} + C$ **5.** $\frac{20}{3}$ **7.** $\frac{16}{3}$ **9.** $3x - \frac{1}{x^2} + C$ **11.** $\sqrt{2}$ **13.** $\frac{1}{5(2 - 5x)} + C$

15. $-\frac{6}{7}(7 - 2x)^{7/4} + C$ **17.** $\frac{9}{8}(3\sqrt[3]{3} - 1)$ **19.** $-\frac{1}{30}(1 - 2x^3)^5 + C$ **21.** $-\frac{1}{2x - x^3} + C$ **23.** $\frac{3350}{3}$

25. $\frac{-6x\,dx}{(x^2 - 1)^4}$ **27.** $\frac{(1 - 4x)\,dx}{(1 - 3x)^{2/3}}$ **29.** 0.061   **31.** $y = 3x - \frac{1}{3}x^3 + \frac{17}{3}$
33. (a) $x - x^2 + C_1$, (b) $-\frac{1}{4}(1 - 2x)^2 + C_2 = x - x^2 + C_2 - \frac{1}{4}$; $C_1 = C_2 - \frac{1}{4}$ **35.** 22   **37.** 0.842

39. 0.811   **41.** 1.01   **43.** 13.77   **45.** 1.85 m³   **47.** $\frac{R\,dR}{R^2 + X^2}$ **49.** $y = k(2L^3x - 6Lx^2 + \frac{2}{5}x^5)$

51. 14.9 m²

Exercises 25-1, page 769

1. 80 ft/s   **3.** $s = 8.00 - 0.25t$ **5.** 15 ft/s   **7.** 17,800 m   **9.** 76 ft/s   **11.** 256 ft   **13.** 0.345 nC

15. 0.017 C   **17.** 667 V   **19.** 4.65 mV   **21.** 0.55   **23.** 66.7 A   **25.** $\frac{k}{x_1}$

27. $m = 1002 - 2\sqrt{t + 1}$, 2.51×10^5 min

Exercises 25-2, page 775

1. 2   **3.** $\frac{8}{3}$ **5.** $\frac{27}{8}$ **7.** $\frac{4}{3}\sqrt{2}$ **9.** $\frac{1}{2}$ **11.** $\frac{1}{6}$ **13.** $\frac{26}{3}$ **15.** 3   **17.** $\frac{15}{4}$ **19.** $\frac{7}{6}$ **21.** $\frac{256}{5}$ **23.** $\frac{7}{6}$
25. $\frac{81}{4}$ **27.** 1   **29.** 18.0 J   **31.** 80.8 km   **33.** 4 cm²   **35.** 0.683 m²

Exercises 25-3, page 782

1. $\frac{1}{3}\pi$ **3.** $\frac{1}{3}\pi$ **5.** 72π **7.** $\frac{768}{7}\pi$ **9.** $\frac{348}{5}\pi$ **11.** $\frac{16}{3}\pi$ **13.** $\frac{1}{3}\pi$ **15.** $\frac{1}{3}\pi$ **17.** $\frac{2}{5}\pi$ **19.** $\frac{8}{3}\pi$
21. $\frac{16}{15}\pi$ **23.** $\frac{16}{3}\pi$ **25.** $\frac{8}{3}\pi$ **27.** $\frac{1}{3}\pi r^2 h$ **29.** 7.56 mm³   **31.** 18.3 cm³

Exercises 25-4, page 789

1. $(\frac{10}{3}, 0)$ **3.** $(\frac{14}{15}, 0)$ **5.** $(-\frac{1}{2}, \frac{1}{2})$ **7.** $(\frac{7}{22}, \frac{5}{22})$ **9.** $(0, \frac{6}{5})$ **11.** $(\frac{4}{3}, \frac{4}{3})$ **13.** $(\frac{3}{5}, \frac{12}{35})$ **15.** $(0, \frac{5}{6})$
17. $(\frac{2}{3}, 0)$ **19.** $(\frac{2}{3}a, \frac{1}{3}b)$ **21.** 0.375 cm above center of base   **23.** 7.71 in. from larger base

Exercises 25-5, page 794

1. 128, 4   **3.** 214, 3.27   **5.** $\frac{64}{15}k$ **7.** $\frac{2}{3}\sqrt{6}$ **9.** $\frac{1}{6}ma^2$ **11.** $\frac{4}{7}\sqrt{7}$ **13.** $\frac{8}{11}\sqrt{55}$ **15.** $\frac{64}{3}\pi k$
17. $\frac{2}{5}\sqrt{10}$ **19.** $\frac{3}{10}mr^2$ **21.** 0.324 g·cm²   **23.** 31.2 kg·cm²

Exercises 25-6, page 799

1. 8.0 lb·in.   **3.** 200 N·mm   **5.** 9.4×10^{-22} N·m   **7.** 1800 N·m   **9.** 3.00×10^5 ft·ton
11. 8.82×10^4 ft·lb   **13.** 2340 lb   **15.** 20,800 lb   **17.** 6530 N   **19.** 11,700 lb   **21.** 2.7 A   **23.** 35.3%
25. 109 ft   **27.** $A = \pi r \sqrt{r^2 + h^2}$

Review Exercises for Chapter 25, page 802

1. 4.4 s   **3.** 4.2 s   **5.** 0.44 C   **7.** 3640 V   **9.** $y = 20x + \frac{1}{120}x^3$ **11.** $\frac{2}{3}$ **13.** 18   **15.** $\frac{27}{4}$ **17.** $\frac{48}{5}\pi$
19. $\frac{243}{10}\pi$ **21.** $\frac{4}{3}\pi ab^2$ **23.** $\left(\frac{40}{21}, \frac{10}{3}\right)$ **25.** $\left(\frac{14}{5}, 0\right)$ **27.** $\frac{8}{5}k$ **29.** $\frac{256}{3}\pi k$ **31.** 8500 ft·lb   **33.** 1.8 m
35. 47 m^3 **37.** 10,200 lb   **39.** 0.29 Ω

Exercises 26-1, page 809

1. $\cos(x + 2)$ **3.** $4\cos(2x - 1)$ **5.** $-3\sin\frac{1}{2}x$ **7.** $-6\sin(3x - 1)$ **9.** $8\sin 4x \cos 4x = 4\sin 8x$
11. $-45\cos^2(5x + 2)\sin(5x + 2)$ **13.** $\sin 3x + 3x\cos 3x$ **15.** $9x^2\cos 5x - 15x^3\sin 5x$
17. $2x\cos x^2\cos 2x - 2\sin x^2\sin 2x$ **19.** $\dfrac{2\cos 4x}{\sqrt{1 + \sin 4x}}$ **21.** $\dfrac{3x\cos 3x - \sin 3x}{x^2}$
23. $\dfrac{4x(1 - 3x)\sin x^2 - 6\cos x^2}{(3x - 1)^2}$ **25.** $4\sin 3x(3\cos 3x\cos 2x - \sin 3x\sin 2x)$
27. $\dfrac{-2\cos 3x[3\sin 3x(1 + 2\sin^2 2x) + 4\cos 3x\sin 2x\cos 2x]}{(1 + 2\sin^2 2x)^2}$ **29.** $3\sin^2 x\cos x + 2\sin 2x$
31. $1 - 4\sin^2 4x\cos 4x$ **33.** See the table.
35. (a) 0.5403023, value of derivative, (b) 0.5402602, slope of secant line   **37.** Resulting curve is $y = \cos x$.
39. $\dfrac{2x - y\cos xy}{x\cos xy - 2\sin 2y}$ **41.** $\dfrac{d\sin x}{dx} = \cos x, \dfrac{d^2\sin x}{dx^2} = -\sin x, \dfrac{d^3\sin x}{dx^3} = -\cos x, \dfrac{d^4\sin x}{dx^4} = \sin x$
43. $\sin 2x = 2\sin x\cos x$ **45.** 0.085   **47.** -2.36 **49.** -199 cm/s   **51.** -38.5 km

Exercises 26-2, page 813

1. $5\sec^2 5x$ **3.** $2(1 - x)\csc^2(1 - x)^2$ **5.** $6\sec 2x\tan 2x$ **7.** $\dfrac{3\csc\sqrt{x}\cot\sqrt{x}}{2\sqrt{x}}$ **9.** $30\tan 3x\sec^2 3x$
11. $-4\cot^3\frac{1}{2}x\csc^2\frac{1}{2}x$ **13.** $2\tan 4x\sqrt{\sec 4x}$ **15.** $-84\csc^4 7x\cot 7x$ **17.** $x^2\sec^2 x + 2x\tan x$
19. $-4\csc x^2(2x\cos x\cot x^2 + \sin x)$ **21.** $-\dfrac{\csc x(x\cot x + 1)}{x^2}$
23. $\dfrac{-4\sin 4x - 4\sin 4x\cot 3x + 3\cos 4x\csc^2 3x}{(1 + \cot 3x)^2}$ **25.** $\sec^2 x(\tan^2 x - 1)$ **27.** $2\sec 2x(\sec 2x - \tan 2x)$
29. $\dfrac{1 + 2\sec^2 4x}{\sqrt{2x + \tan 4x}}$ **31.** $\dfrac{2\cos 2x - \sec y}{x\sec y\tan y - 2}$ **33.** $24\tan 3x\sec^2 3x\,dx$ **35.** $4\sec 4x(\tan^2 4x + \sec^2 4x)\,dx$
37. (a) 3.4255188, value of derivative, (b) 3.4260524, slope of secant line   **39.** $2\tan x\sec^2 x = 2\sec x(\sec x\tan x)$
41. -12 **43.** $2\sec^2 x - \sec x\tan x = \dfrac{2}{\cos^2 x} - \dfrac{\sin x}{\cos^2 x}$ **45.** -8.4 cm/s   **47.** 140 ft/s

Exercises 26-3, page 817

1. $\dfrac{2x}{\sqrt{1 - x^4}}$ **3.** $\dfrac{18x^2}{\sqrt{1 - 9x^6}}$ **5.** $-\dfrac{1}{\sqrt{4 - x^2}}$ **7.** $\dfrac{1}{\sqrt{(x - 1)(2 - x)}}$ **9.** $\dfrac{1}{2\sqrt{x}(1 + x)}$ **11.** $-\dfrac{1}{x^2 + 1}$
13. $\dfrac{x}{\sqrt{1 - x^2}} + \text{Arcsin } x$ **15.** $\dfrac{4x}{1 + 4x^2} + 2\text{ Arctan } 2x$ **17.** $\dfrac{3\sqrt{1 - 4x^2}\text{ Arcsin } 2x - 6x + 2}{\sqrt{1 - 4x^2}\text{ Arcsin}^2 2x}$

19. $\dfrac{2(\text{Arccos } 2x + \text{Arcsin } 2x)}{\sqrt{1 - 4x^2}\,\text{Arccos}^2\, 2x}$ **21.** $\dfrac{-24\,\text{Arccos}^2\, 4x}{\sqrt{1 - 16x^2}}$ **23.** $\dfrac{8\,\text{Arcsin } 4x}{\sqrt{1 - 16x^2}}$ **25.** $\dfrac{9\,\text{Arctan}^2\, x}{1 + x^2}$ **27.** $\dfrac{-2(2x + 1)^2}{(1 + 4x^2)^2}$

29. $\dfrac{18(4 - \text{Arccos } 2x)^2}{\sqrt{1 - 4x^2}}$ **31.** $-\dfrac{x^2y^2 + 2y + 1}{2x}$

33. (a) 1.1547005, value of derivative, (b) 1.1547390, slope of secant line   **35.** $\dfrac{3\,\text{Arcsin}^2\, x\, dx}{\sqrt{1 - x^2}}$ **37.** $\dfrac{-16x}{(1 + 4x^2)^2}$

39. Let $y = \text{Arcsec } u$; solve for u; take derivatives; substitute.   **41.** $\dfrac{E - A}{\omega m\sqrt{m^2E^2 - (A - E)^2}}$

43. $\theta = \text{Arctan}\,\dfrac{h}{x}; \dfrac{d\theta}{dx} = \dfrac{-h}{x^2 + h^2}$

Exercises 26-4, page 821

1. $d\sin x/dx = \cos x$ and $d\cos x/dx = -\sin x$, and $\sin x = \cos x$ at points of intersection.

3. $\dfrac{1}{x^2 + 1}$ is always positive.   **5.** Dec. $x > 0$, $x < 0$,   **7.** $8\sqrt{2}x + 8y + 4\sqrt{2} - 5\pi\sqrt{2} = 0$
infl. (0, 0),
asym. $x = \pi/2$, $x = -\pi/2$ **9.** 1.9337538   **11.** 10
13. 0.58 ft/s, -1.7 ft/s^2
15. -0.072 lb/s
17. 98.96 in./s, 270°
19. 3731 in./s^2, 0°

21. -0.073 rad/s   **23.** 8.08 ft/s   **25.** 0.020   **27.** 0.19 m   **29.** 100 mm^2 **31.** 14 ft

Exercises 26-5, page 826

1. $\dfrac{2\log e}{x}$ **3.** $\dfrac{6\log_5 e}{3x + 1}$ **5.** $\dfrac{-3}{1 - 3x}$ **7.** $\dfrac{4\sec^2 2x}{\tan 2x} = 4\sec 2x \csc 2x$ **9.** $\dfrac{1}{2x}$ **11.** $\dfrac{6(x + 1)}{x^2 + 2x}$

13. $2(1 + \ln x)$ **15.** $\dfrac{3(2x + 1)\ln (2x + 1) - 6x}{(2x + 1)[\ln (2x + 1)]^2}$ **17.** $\dfrac{1}{x\ln x}$ **19.** $\dfrac{1}{x^2 + x}$ **21.** $\dfrac{\cos \ln x}{x}$ **23.** $\dfrac{6\ln 2x}{x}$

25. $\dfrac{x\sec^2 x + \tan x}{x\tan x}$ **27.** $\dfrac{x + 4}{x(x + 2)}$ **29.** $\dfrac{\sqrt{x^2 + 1}}{x}$ **31.** $\dfrac{x + y - 2\ln (x + y)}{x + y + 2\ln (x + y)}$

33. 0.5 is value of derivative; 0.4999875 is slope of secant line.   **35.** See the table.   **37.** 2.73

39. $-2(\tan x + \sec x \csc x)\, dx$ **41.** -1 **43.** $x^x(\ln x + 1)$ **45.** $\dfrac{10\log e}{I}\dfrac{dI}{dt}$ **47.** 0.083 s^2/ft

Exercises 26-6, page 829

1. $(2\ln 3)3^{2x}$ **3.** $(6\ln 4)4^{6x}$ **5.** $6e^{6x}$ **7.** $\dfrac{e^{\sqrt{x}}}{2\sqrt{x}}$ **9.** $e^{-x}(1 - x)$ **11.** $e^{\sin x}(x\cos x + 1)$

13. $\dfrac{3e^{2x}(2x + 1)}{(x + 1)^2}$ **15.** $e^{-3x}(4\cos 4x - 3\sin 4x)$ **17.** $\dfrac{2e^{3x}(12x + 5)}{(4x + 3)^2}$ **19.** $\dfrac{2xe^{x^2}}{e^{x^2} + 4}$

21. $16e^{6x}(x\cos x^2 + 3\sin x^2)$ **23.** $2(\ln 2x + e^{2x})\left(\dfrac{1}{x} + 2e^{2x}\right)$ **25.** $\dfrac{e^{xy}(xy + 1)}{1 - x^2e^{xy} - \cos y}$ **27.** $\dfrac{e^{2x}}{x} + 2e^{2x}\ln x$

29. $12e^{6x}\cot 2e^{6x}$ **31.** $\dfrac{4e^{2x}}{\sqrt{1 - e^{4x}}}$ **33.** (a) 2.7182818, value of derivative, (b) 2.7184178, slope of secant line

35. -0.724 **37.** $(-xe^{-x} + e^{-x}) + (xe^{-x}) = e^{-x}$ **39.** $\dfrac{dy}{dx} = e^a; y = e^a$ **41.** $(2 - t)e^{-0.5t}$ **43.** $-0.00164/h$

45. Substitute and simplify.   **47.** $\dfrac{d}{dx}\left[\dfrac{1}{2}(e^u - e^{-u})\right] = \dfrac{1}{2}(e^u + e^{-u})\dfrac{du}{dx}$; $\dfrac{d}{dx}\left[\dfrac{1}{2}(e^u + e^{-u})\right] = \dfrac{1}{2}(e^u - e^{-u})\dfrac{du}{dx}$

Exercises 26-7, page 833

1. Int. (0, 0), max. (0, 0),
not defined for cos $x < 0$,
asym. $x = -\frac{1}{2}\pi, \frac{1}{2}\pi, \ldots$

3. Int. (0, 0), max. $\left(1, \dfrac{1}{e}\right)$,
infl. $\left(2, \dfrac{2}{e^2}\right)$, asym. $y = 0$

5. Int. (0, 0), max. (0, 0),
infl. $(-1, -\ln 2)$, $(1, -\ln 2)$

7. Int. (0, 1), max. (0, 1),
infl. $(\frac{1}{2}\sqrt{2}, \frac{1}{e}\sqrt{e})$, $(-\frac{1}{2}\sqrt{2}, \frac{1}{e}\sqrt{e})$,
asym. $y = 0$

9. Max. $(1, -1)$, asym. $x = 0$

11. Int. (0, 0), infl. (0, 0),
inc. all x

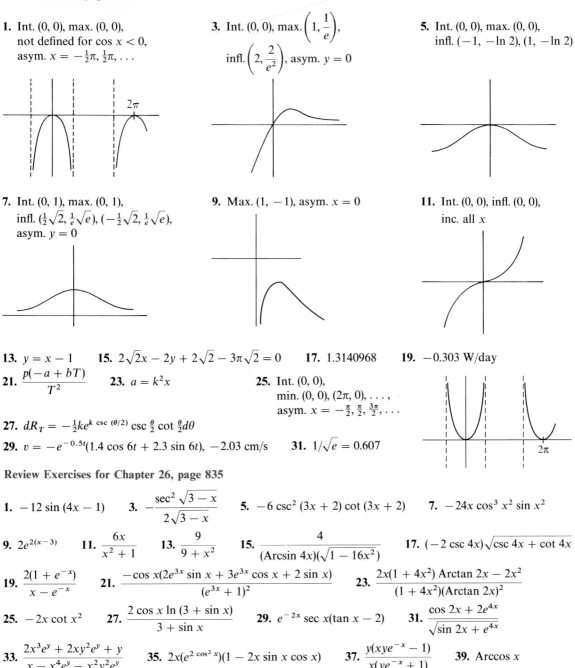

13. $y = x - 1$ **15.** $2\sqrt{2}x - 2y + 2\sqrt{2} - 3\pi\sqrt{2} = 0$ **17.** 1.3140968   **19.** -0.303 W/day

21. $\dfrac{p(-a + bT)}{T^2}$ **23.** $a = k^2 x$

25. Int. (0, 0),
min. (0, 0), $(2\pi, 0), \ldots$,
asym. $x = -\frac{\pi}{2}, \frac{\pi}{2}, \frac{3\pi}{2}, \ldots$

27. $dR_T = -\frac{1}{2}ke^{k\,\csc\,(\theta/2)}\csc\frac{\theta}{2}\cot\frac{\theta}{2}d\theta$

29. $v = -e^{-0.5t}(1.4\cos 6t + 2.3\sin 6t)$, -2.03 cm/s   **31.** $1/\sqrt{e} = 0.607$

Review Exercises for Chapter 26, page 835

1. $-12\sin(4x - 1)$ **3.** $-\dfrac{\sec^2\sqrt{3 - x}}{2\sqrt{3 - x}}$ **5.** $-6\csc^2(3x + 2)\cot(3x + 2)$ **7.** $-24x\cos^3 x^2\sin x^2$

9. $2e^{2(x-3)}$ **11.** $\dfrac{6x}{x^2 + 1}$ **13.** $\dfrac{9}{9 + x^2}$ **15.** $\dfrac{4}{(\text{Arcsin }4x)(\sqrt{1 - 16x^2})}$ **17.** $(-2\csc 4x)\sqrt{\csc 4x + \cot 4x}$

19. $\dfrac{2(1 + e^{-x})}{x - e^{-x}}$ **21.** $\dfrac{-\cos x(2e^{3x}\sin x + 3e^{3x}\cos x + 2\sin x)}{(e^{3x} + 1)^2}$ **23.** $\dfrac{2x(1 + 4x^2)\text{ Arctan }2x - 2x^2}{(1 + 4x^2)(\text{Arctan }2x)^2}$

25. $-2x\cot x^2$ **27.** $\dfrac{2\cos x\ln(3 + \sin x)}{3 + \sin x}$ **29.** $e^{-2x}\sec x(\tan x - 2)$ **31.** $\dfrac{\cos 2x + 2e^{4x}}{\sqrt{\sin 2x + e^{4x}}}$

33. $\dfrac{2x^3e^y + 2xy^2e^y + y}{x - x^4e^y - x^2y^2e^y}$ **35.** $2x(e^{2\cos^2 x})(1 - 2x\sin x\cos x)$ **37.** $\dfrac{y(xye^{-x} - 1)}{x(ye^{-x} + 1)}$ **39.** Arccos x

41. Infl. $(\frac{1}{2}\pi, \frac{1}{2}\pi)$, $(\frac{3}{2}\pi, \frac{3}{2}\pi)$

43. Max. $(e^{-2}, 4e^{-2})$, min. $(1, 0)$, infl. (e^{-1}, e^{-1})

45. $7.27x + y - 8.44 = 0$
47. $2x + 2.57y - 4.30 = 0$
49. $2 \sin x \cos x - 2 \cos x \sin x = 0$
51. -0.7034674
53. -5.17 cm/s
55. $-20 \sin 2t$ **57.** $\dfrac{ab}{x(b - x)}$
59. $-kE_0^2 \cos \frac{1}{2}\theta \sin \frac{1}{2}\theta$

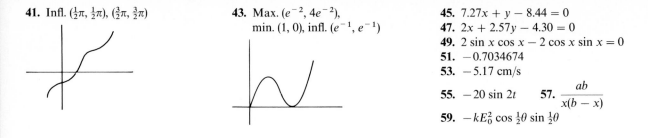

61. $\dfrac{f}{\sqrt{R^2 - F^2 f^2}}$ **63.** 0.005934 rad/s   **65.** $-8.32°$F/min   **67.** 202 ft/s   **69.** 2.00 m wide, 1.82 m high
71. -0.00655 rad/s   **73.** 0.40 cm/s, 72.5°   **75.** 7.07 in.   **77.** $A = 16 \cos \theta(1 + \sin \theta)$ **79.** 10 items

Exercises 27-1, page 843

1. $\frac{1}{5}\sin^5 x + C$ **3.** $-\frac{2}{3}(\cos x)^{3/2} + C$ **5.** $\frac{4}{3}\tan^3 x + C$ **7.** $\frac{1}{8}$ **9.** $\frac{1}{4}(\text{Arcsin } x)^4 + C$
11. $\frac{1}{10}(\text{Arctan } 5x)^2 + C$ **13.** $\frac{1}{3}[\ln(x + 1)]^3 + C$ **15.** 0.179   **17.** $\frac{1}{4}(4 + e^x)^4 + C$ **19.** $\frac{3}{16}(2e^{2x} - 1)^{4/3} + C$
21. $\frac{1}{10}(1 + \sec^2 x)^5 + C$ **23.** $\frac{1}{6}$ **25.** 1.102   **27.** $y = \frac{1}{3}(\ln x)^3 + 2$ **29.** $\frac{1}{3}mnv^2$ **31.** $q = (1 - e^{-t})^3$

Exercises 27-2, page 846

1. $\frac{1}{4}\ln|1 + 4x| + C$ **3.** $-\frac{1}{3}\ln|4 - 3x^2| + C$ **5.** $\frac{1}{3}\ln 4 = 0.462$ **7.** $-\frac{1}{2}\ln|\cot 2x| + C$ **9.** $\ln 2 = 0.693$
11. $\ln|1 - e^{-x}| + C$ **13.** $\ln|x + e^x| + C$ **15.** $\frac{1}{4}\ln|1 + 4\sec x| + C$ **17.** $\frac{1}{4}\ln 5 = 0.402$ **19.** $\ln|\ln x| + C$
21. $\ln|2x + \tan x| + C$ **23.** $-\sqrt{1 - 2x} + C$ **25.** $\ln|x| - \dfrac{2}{x} + C$ **27.** $\frac{1}{3}\ln(\frac{5}{4}) = 0.0744$ **29.** 1.10
31. $\pi \ln 2 = 2.18$ **33.** $y = \ln\dfrac{3.5}{3 + \cos x} + 2$ **35.** $P = P_0 e^{kt}$ **37.** $i = \dfrac{E}{R}(1 - e^{-Rt/L})$ **39.** 1.41 m

Exercises 27-3, page 850

1. $e^{7x} + C$ **3.** $\frac{1}{2}e^{2x+5} + C$ **5.** $2(e - 1) = 3.44$ **7.** $2e^{x^3} + C$ **9.** $2(e^2 - e) = 9.34$ **11.** $\frac{1}{2}e^{2 \sec x} + C$
13. $\dfrac{2e^x - 3}{2e^{2x}} + C$ **15.** $6 - \dfrac{3(e^6 - e^2)}{2} = -588.06$ **17.** $-\dfrac{4}{e^{\sqrt{x}}} + C$ **19.** $e^{\text{Arctan } x} + C$ **21.** $-\frac{1}{3}e^{\cos 3x} + C$
23. 0   **25.** 6.389   **27.** $\pi(e^4 - e) = 163$ **29.** $\frac{1}{8}(e^8 - 1) = 372$ **31.** $\ln b \int b^u \, du = b^u + C_1$
33. $q = EC(1 - e^{-t/RC})$ **35.** $s = -e^{-2t} - 0.6e^{-5t}$

Exercises 27-4, page 853

1. $\frac{1}{2}\sin 2x + C$ **3.** $\frac{1}{3}\tan 3x + C$ **5.** $2 \sec \frac{1}{2}x + C$ **7.** 0.6365   **9.** $\frac{3}{2}\ln|\sec x^2 + \tan x^2| + C$
11. $\cos\left(\dfrac{1}{x}\right) + C$ **13.** $\frac{1}{2}\sqrt{3}$ **15.** $\frac{1}{5}\sec 5x + C$ **17.** $\frac{1}{2}\ln|\sec 2x + \tan 2x| + C$

19. $\frac{1}{2}(\ln|\sin 2x| + \sin 2x) + C$ **21.** $\csc x - \cot x - \ln|\csc x - \cot x| + \ln|\sin x| + C$ **23.** $\frac{1}{9}\pi + \frac{1}{3}\ln 2 = 0.580$
25. 0.347   **27.** $\pi\sqrt{3} = 5.44$ **29.** $\theta = 0.10 \cos 2.5t$ **31.** 0.7726 m

Exercises 27-5, page 858

1. $\frac{1}{3}\sin^3 x + C$ **3.** $-\frac{1}{2}\cos 2x + \frac{1}{6}\cos^3 2x + C$ **5.** $\frac{2}{3}\sin^3 x - \frac{2}{5}\sin^5 x + C$ **7.** $\frac{1}{120}(64 - 43\sqrt{2})$
9. $\frac{1}{2}x - \frac{1}{4}\sin 2x + C$ **11.** $\frac{1}{2}x + \frac{1}{12}\sin 6x + C$ **13.** $\frac{1}{2}\tan^2 x + \ln|\cos x| + C$ **15.** $\frac{3}{4}$

17. $\frac{1}{6} \tan^3 2x - \frac{1}{2} \tan 2x + x + C$ **19.** $\frac{1}{15} \sec^5 3x - \frac{1}{9} \sec^3 3x + C$ **21.** $x - \frac{1}{2} \cos 2x + C$

23. $\frac{1}{4} \cot^4 x - \frac{1}{3} \cot^3 x + \frac{1}{2} \cot^2 x - \cot x + C$ **25.** $1 + \frac{1}{2} \ln 2 = 1.347$ **27.** $\frac{1}{5} \tan^5 x + \frac{2}{3} \tan^3 x + \tan x + C$

29. $\frac{1}{2} \pi^2 = 4.935$ **31.** $\sqrt{2} - 1 = 0.414$ **33.** $\int \sin x \cos x \, dx = \frac{1}{2} \sin^2 x + C_1 = -\frac{1}{2} \cos^2 x + C_2; C_2 = C_1 + \frac{1}{2}$

35. $\frac{4}{3}$ **37.** 120 V   **39.** $\dfrac{aA}{2} + \dfrac{A}{2b\pi} \sin ab\pi \cos 2bc\pi$

Exercises 27-6, page 862

1. $\operatorname{Arcsin} \frac{1}{2} x + C$ **3.** $\frac{1}{8} \operatorname{Arctan} \frac{1}{8} x + C$ **5.** $\frac{1}{4} \operatorname{Arcsin} 4x + C$ **7.** $\operatorname{Arctan} 6 = 1.41$

9. $\frac{2}{5} \sqrt{5} \operatorname{Arcsin} \frac{1}{5} \sqrt{5} = 0.415$ **11.** $\frac{4}{9} \ln |9x^2 + 16| + C$ **13.** $\frac{1}{35} \sqrt{35} (\operatorname{Arctan} \frac{2}{7} \sqrt{35} - \operatorname{Arctan} \frac{1}{7} \sqrt{35}) = 0.057$

15. $\operatorname{Arcsin} e^x + C$ **17.** $\operatorname{Arctan} (x + 1) + C$ **19.** $4 \operatorname{Arcsin} \frac{1}{2} (x + 2) + C$ **21.** -0.357

23. $2 \operatorname{Arcsin} (\frac{1}{2} x) + \sqrt{4 - x^2} + C$ **25.** (a) Inverse tangent, (b) logarithmic, (c) general power

27. (a) General power, (b) inverse sine, (c) logarithmic   **29.** $\operatorname{Arctan} 2 = 1.11$ **31.** $k \operatorname{Arctan} \dfrac{x}{d} + C$

33. $\operatorname{Arcsin} \dfrac{x}{A} = \sqrt{\dfrac{k}{m}} \, t + \operatorname{Arcsin} \dfrac{x_0}{A}$ **35.** $0.22k$

Exercises 27-7, page 866

1. $\cos x + x \sin x + C$ **3.** $\frac{1}{2} x e^{2x} - \frac{1}{4} e^{2x} + C$ **5.** $x \tan x + \ln |\cos x| + C$

7. $2x \operatorname{Arctan} x - 2 \ln \sqrt{1 + x^2} + C$ **9.** $-8x\sqrt{1 - x} - \frac{16}{3} (1 - x)^{3/2} + C$ **11.** $\frac{1}{2} x^2 \ln x - \frac{1}{4} x^2 + C$

13. $\frac{1}{2} x \sin 2x - \frac{1}{4} (2x^2 - 1) \cos 2x + C$ **15.** $\frac{1}{2} (e^{\pi/2} - 1) = 1.91$ **17.** $1 - \dfrac{3}{e^2} = 0.594$ **19.** $\frac{1}{2} \pi - 1 = 0.571$

21. 0.756 **23.** $q = \frac{1}{5} [e^{-2t} (\sin t - 2 \cos t) + 2]$

Exercises 27-8, page 870

1. $-\dfrac{\sqrt{1 - x^2}}{x} - \operatorname{Arcsin} x + C$ **3.** $2 \ln |x + \sqrt{x^2 - 4}| + C$ **5.** $-\dfrac{\sqrt{x^2 + 9}}{9x} + C$ **7.** $\dfrac{x}{\sqrt{4 - x^2}} + C$

9. $\dfrac{16 - 9\sqrt{3}}{24} = 0.017$ **11.** $5 \ln |\sqrt{x^2 + 2x + 2} + x + 1| + C$ **13.** $\frac{1}{3} \operatorname{Arcsec} \frac{2}{3} x + C$ **15.** $\operatorname{Arcsec} e^x + C$

17. π **19.** $\dfrac{1}{4} ma^2$ **21.** 2.68   **23.** $kQ \ln \dfrac{\sqrt{a^2 + b^2} + a}{\sqrt{a^2 + b^2} - a}$

Exercises 27-9, page 872

1. $\frac{3}{25} [2 + 5x - 2 \ln |2 + 5x|] + C$ **3.** $\frac{3544}{15} = 236.3$ **5.** $\frac{1}{2} x\sqrt{4 - x^2} + 2 \operatorname{Arcsin} \frac{1}{2} x + C$

7. $\frac{1}{2} \sin x - \frac{1}{10} \sin 5x + C$ **9.** $\sqrt{4x^2 - 9} - 3 \operatorname{Arcsec} \left(\dfrac{2x}{3} \right) + C$

11. $\frac{1}{20} \cos^4 4x \sin 4x + \frac{1}{5} \sin 4x - \frac{1}{15} \sin^3 4x + C$ **13.** $\frac{1}{2} x^2 \operatorname{Arctan} x^2 - \frac{1}{4} \ln (1 + x^4) + C$

15. $\frac{1}{4} (8\pi - 9\sqrt{3}) = 2.386$ **17.** $-\ln \left(\dfrac{1 + \sqrt{4x^2 + 1}}{2x} \right) + C$ **19.** $-\ln \left(\dfrac{1 + \sqrt{1 - 4x^2}}{2x} \right) + C$

21. $\frac{1}{8} \cos 4x - \frac{1}{12} \cos 6x + C$ **23.** $\frac{1}{3} (\cos x^3 + x^3 \sin x^3) + C$ **25.** $\dfrac{x^2}{\sqrt{1 - x^4}} + C$ **27.** 4.892

29. $\frac{1}{4} x^4 (\ln x^2 - \frac{1}{2}) + C$ **31.** $-\dfrac{x^3}{3\sqrt{x^6 - 1}} + C$ **33.** $\frac{1}{4} [2\sqrt{5} + \ln (2 + \sqrt{5})] = 1.479$

35. $\frac{1}{4} (8 \operatorname{Arctan} 4 - \ln 17) = 1.943$ **37.** 208 lb   **39.** 38.5 ft·lb

Review Exercises for Chapter 27, page 874

1. $-\frac{1}{2}e^{-2x} + C$ **3.** $-\dfrac{1}{\ln 2x} + C$ **5.** $4 \ln (1 + \sin x) + C$ **7.** $\frac{2}{35}$ Arctan $\frac{7}{5}x + C$ **9.** 0

11. $\frac{1}{2} \ln 2 = 0.3466$ **13.** $\frac{1}{12} \sec^4 3x + C$ **15.** $-\frac{1}{3} \ln |\cos 3x| + C$ **17.** $\frac{1}{9} \tan^3 3x + \frac{1}{3} \tan 3x + C$

19. $\dfrac{3}{\sqrt{e}} - 2 = -0.1804$ **21.** $\frac{3}{4}$ Arctan $\dfrac{x^2}{2} + C$ **23.** $\frac{1}{2} \ln |2x + \sqrt{4x^2 - 9}| + C$ **25.** $\sqrt{e^{2x} + 1} + C$

27. $\frac{1}{2}(3x - \sin 3x \cos 3x) + C$ **29.** $-\frac{1}{2}x \cot 2x + \frac{1}{4} \ln |\sin 2x| + C$ **31.** $\frac{1}{2} \sin e^{2x} + C$ **33.** $\frac{1}{3}$

35. $\frac{1}{2}x^2 - 2x + 3 \ln |x + 2| + C$ **37.** $\frac{1}{3}(e^x + 1)^3 + C_1 = \frac{1}{3}e^{3x} + e^{2x} + e^x + C_2; C_2 = C_1 + \frac{1}{3}$

39. (a) $\dfrac{1}{2} \int (1 - \cos 2x)\,dx = \dfrac{x}{2} - \dfrac{1}{4} \sin 2x + C_1,$

(b) $\int \sin x(\sin x\,dx) = -\sin x \cos x + \int \cos^2 x\,dx = -\sin x \cos x + \int (1 - \sin^2 x)\,dx$ **41.** $y = \frac{1}{3} \tan^3 x + \tan x$

43. $2(e^3 - 1) = 38.17$ **45.** 11.18 **47.** $4\pi(e^2 - 1) = 80.29$ **49.** $\frac{1}{8}\pi(e^{2\pi} - 1) = 209.9$ **51.** $\ln 3$

53. $\Delta S = a \ln T + bT + \frac{1}{2}cT^2 + C$ **55.** $v = 320(1 - e^{-t/10})$ **57.** $\sqrt{2}$ **59.** $\dfrac{2}{3} k$

61. 3.47 cm^3 **63.** 644 ft^2

Exercises 28-1, page 880

1. $1, 4, 9, 16$ **3.** $\frac{1}{2}, \frac{1}{3}, \frac{1}{4}, \frac{1}{5}$ **5.** (a) $-\frac{2}{5}, \frac{4}{25}, -\frac{8}{125}, \frac{16}{625}$ (b) $-\frac{2}{5} + \frac{4}{25} - \frac{8}{125} + \frac{16}{625} - \cdots$

7. (a) $2, 0, 2, 0$ (b) $2 + 0 + 2 + 0 + \cdots$ **9.** $a_n = \dfrac{1}{n + 1}$ **11.** $a_n = \dfrac{1}{(n + 1)(n + 2)}$

13. $1, 1.125, 1.1620370, 1.1776620, 1.1856620$; convergent; 1.2

15. $1, 1.5, 2.1666667, 2.9166667, 3.7166667$; divergent   **17.** $0, -1, -3, -6, -10$; divergent

19. $0.75, 0.8888889, 0.9375000, 0.9600000, 0.9722222$; convergent; 1 **21.** Divergent

23. Convergent, $S = \frac{3}{4}$ **25.** Convergent, $S = 100$ **27.** Convergent, $S = \frac{4096}{9}$ **29.** (a) 1, (b) diverges

31. $r = x; S = \dfrac{x^0}{1 - x} = \dfrac{1}{1 - x}$

Exercises 28-2, page 885

1. $1 + x + \frac{1}{2}x^2 + \cdots$ **3.** $1 - \frac{1}{2}x^2 + \frac{1}{24}x^4 - \cdots$ **5.** $1 + \frac{1}{2}x - \frac{1}{8}x^2 + \cdots$ **7.** $1 - 2x + 2x^2 - \cdots$

9. $1 - 8x^2 + \frac{32}{3}x^4 - \cdots$ **11.** $1 + x + x^2 + \cdots$ **13.** $-2x - 2x^2 - \frac{8}{3}x^3 - \cdots$ **15.** $1 - \dfrac{x^2}{8} + \dfrac{x^4}{384} - \cdots$

17. $x - \frac{1}{3}x^3 + \cdots$ **19.** $x + \frac{1}{3}x^3 + \cdots$ **21.** $-\frac{1}{2}x^2 - \frac{1}{12}x^4 - \cdots$ **23.** $x^2 - \frac{1}{3}x^4 + \cdots$

25. No, functions are not defined at $x = 0$.   **27.** $e^x = 1 + x + \dfrac{x^2}{2} + \cdots, e^{x^2} = 1 + x^2 + \dfrac{x^4}{2} + \cdots$

29. $f''(0) = 0$ except $f'''(0) = 6$ **31.** $R = e^{-0.001t} = 1 - 0.001t + (5 \times 10^{-7})t^2 - \cdots$

Exercises 28-3, page 891

1. $1 + 3x + \frac{9}{2}x^2 + \frac{9}{2}x^3 + \cdots$ **3.** $\dfrac{x}{2} - \dfrac{x^3}{2^3 3!} + \dfrac{x^5}{2^5 5!} - \dfrac{x^7}{2^7 7!} + \cdots$ **5.** $1 - 8x^2 + \frac{32}{3}x^4 - \frac{256}{45}x^6 + \cdots$

7. $x^2 - \frac{1}{2}x^4 + \frac{1}{3}x^6 - \frac{1}{4}x^8 + \cdots$ **9.** 0.3103 **11.** 0.1901 **13.** $1 + \frac{1}{2}x^2 + \frac{1}{24}x^4 + \frac{1}{720}x^6 + \cdots$

15. $x + x^2 + \frac{1}{3}x^3 + \cdots$ **17.** $\dfrac{d}{dx}\left(x - \dfrac{1}{6}x^3 + \dfrac{1}{120}x^5 - \cdots\right) = 1 - \dfrac{1}{2}x^2 + \dfrac{1}{24}x^4 - \cdots$

19. $\displaystyle\int \cos x\,dx = x - \dfrac{x^3}{3!} + \cdots$ **21.** $\displaystyle\int_0^1 e^x\,dx = 1.7182818, \int_0^1 \left(1 + x + \dfrac{1}{2}x^2 + \dfrac{1}{6}x^3\right)dx = 1.7083333$

23. 0.003099 **25.** 0.1249 **27.** $y = -1 + 3t + 2t^2 - \dfrac{9}{2}t^3 - \cdots$

Exercises 28-4, page 894

1. 1.22, 1.2214028   **3.** 0.0998333, 0.0998334   **5.** 2.7180556, 2.7182818   **7.** 0.9986292, 0.9986295
9. 0.3349333, 0.3364722   **11.** 0.3546130, 0.3546129   **13.** $-0.0139975, -0.0139975$
15. 0.7283020, 0.7286950   **17.** 1.0523528   **19.** 0.9874462   **21.** 8.3×10^{-8} **23.** 3.1×10^{-7} **25.** 3.146

27. 1.59 years   **29.** $i = \dfrac{E}{L}\left(t - \dfrac{Rt^2}{2L}\right)$; small values of t **31.** 66 ft

Exercises 28-5, page 898

1. 3.32   **3.** 2.049   **5.** 0.5150   **7.** 0.49288   **9.** $e^{-2}\left[1 - (x-2) + \dfrac{(x-2)^2}{2!} - \cdots\right]$

11. $\dfrac{1}{2}\left[\sqrt{3} + \left(x - \dfrac{1}{3}\pi\right) - \dfrac{\sqrt{3}}{2!}\left(x - \dfrac{1}{3}\pi\right)^2 - \cdots\right]$ **13.** $2 + \frac{1}{12}(x-8) - \frac{1}{288}(x-8)^2 + \cdots$

15. $1 + 2(x - \frac{1}{4}\pi) + 2(x - \frac{1}{4}\pi)^2 + \cdots$ **17.** 0.111   **19.** 3.0496   **21.** 2.0247   **23.** 0.87462
25. Use the indicated method.   **27.** 0.5150408, 0.5150388, 0.5150381

Exercises 28-6, page 905

1. $f(x) = \frac{1}{2} - \frac{2}{\pi}\sin x - \frac{2}{3\pi}\sin 3x - \cdots.$ **3.** $f(x) = \frac{3}{2} + \frac{2}{\pi}\sin x + \frac{2}{3\pi}\sin 3x + \cdots$

5. $f(x) = \frac{\pi}{4} - \frac{2}{\pi}(\cos x + \frac{1}{9}\cos 3x + \cdots) + (\sin x - \frac{1}{2}\sin 2x + \cdots)$

7. $f(x) = -\frac{1}{4} - \frac{1}{\pi}\cos x + \frac{1}{3\pi}\cos 3x - \cdots + \frac{3}{\pi}\sin x - \frac{1}{\pi}\sin 2x + \frac{1}{\pi}\sin 3x - \cdots$

9. $f(x) = \frac{\pi}{2} - \frac{4}{\pi}\cos x - \frac{4}{9\pi}\cos 3x - \cdots$ **11.** $f(x) = \frac{5}{2} - \frac{10}{\pi}(\sin \frac{\pi x}{3} + \frac{1}{3}\sin \pi x - \cdots)$

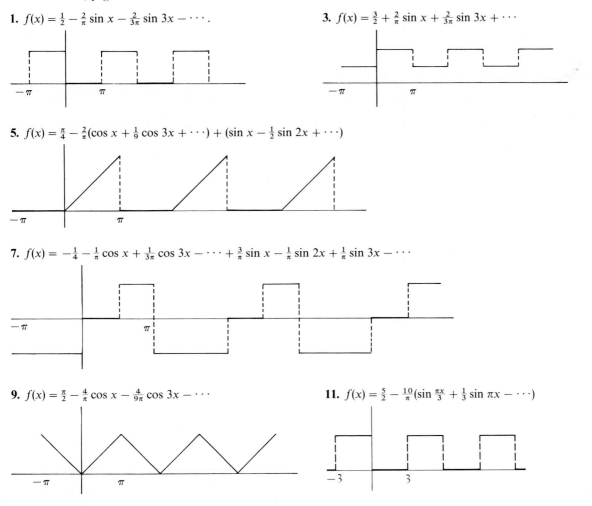

13. $f(t) = \frac{2}{\pi} - \frac{4}{3\pi} \cos 2t - \frac{4}{15\pi} \cos 4t - \cdots$

15. $f(t) = 2 + \frac{8}{\pi}(\cos \frac{\pi}{2}t - \frac{1}{3} \cos \frac{3\pi}{2}t + \cdots + \sin \frac{\pi}{2}t + \sin \pi t + \frac{1}{3} \sin \frac{3\pi}{2}t + \cdots)$

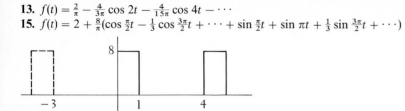

Review Exercises for Chapter 28, page 907

1. $\frac{1}{2} - \frac{1}{4}x + \frac{1}{48}x^3 - \cdots$ **3.** $2x^2 - \frac{4}{3}x^6 + \frac{4}{15}x^{10} - \cdots$ **5.** $1 + \frac{1}{3}x - \frac{1}{9}x^2 + \cdots$ **7.** $x + \frac{1}{6}x^3 + \frac{3}{40}x^5 + \cdots$

9. 0.82   **11.** 1.09   **13.** 1.0344   **15.** -0.2015 **17.** 0.95299   **19.** 12.1655   **21.** 0.259

23. $\frac{1}{2} - \frac{1}{2}\sqrt{3}(x - \frac{1}{3}\pi) - \frac{1}{4}(x - \frac{1}{3}\pi)^2 + \cdots$

25. $f(x) = \frac{1}{2} + \frac{2}{\pi}(\cos x - \frac{1}{3} \cos 3x + \cdots)$ **27.** $f(x) = \frac{4}{\pi}(\sin \frac{\pi x}{2} - \frac{1}{2} \sin \pi x + \frac{1}{3} \sin \frac{3\pi x}{2} - \cdots)$

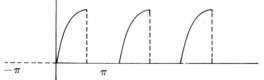

29. $(x + h) - \frac{(x + h)^3}{3!} + \cdots - (x - h) + \frac{(x - h)^3}{3!} - \cdots = 2h - \frac{2hx^2}{2!} + \cdots = 2h\left(1 - \frac{x^2}{2!} + \cdots\right)$

31. 256   **33.**   **35.** $1 - x^2 + \frac{1}{3}x^4 - \frac{2}{45}x^6 + \cdots$

$y = 1 + x + \frac{x^2}{2}$

$y = x + 1$

$y = e^x$ $y = 1$

37. $1 + \frac{x^2}{2} + \frac{5x^4}{24} + \cdots$

39. $1 - x + x^2 - \cdots$

41. 0.00249688   **43.** $N = N_0\left(1 - \lambda t + \frac{\lambda^2 t^2}{2} - \frac{\lambda^3 t^3}{6} + \cdots\right)$

45. $N_0[1 + e^{-k/T} + (e^{-k/T})^2 + \cdots] = N_0(1 + e^{-k/T} + e^{-2k/T} + \cdots)$

47. $f(t) = \frac{1}{2\pi} + \frac{1}{\pi}(\frac{1}{2} \cos t - \frac{1}{3} \cos 2t + \cdots) + \frac{1}{4} \sin t + \frac{2}{3\pi} \sin 2t + \cdots$

Exercises 29-1, page 912

1. Particular solution   **3.** General solution

(The following "answers" are the unsimplified expressions obtained by substituting functions and derivatives.)

5. $1 = 1$ **7.** $e^x - (e^x - 1) = 1$ **9.** $x(2cx) = 2(cx^2)$ **11.** $(-2ce^{-2x} + 1) + 2(ce^{-2x} + x - \frac{1}{2}) = 2x$

13. $(-12 \cos 2x) + 4(3 \cos 2x) = 0$

15. $[e^{2x}(1 + 4c_1 + 4c_2 + 4x + 4c_2x + 2x^2)] - 4[e^{2x}(2c_1 + c_2 + x + 2c_2x + x^2)] + 4[e^{2x}(c_1 + c_2x + x^2/2)] = e^{2x}$

17. $x^2 \left[-\dfrac{c^2}{(x-c)^2} \right] + \left[\dfrac{cx}{(x-c)} \right]^2 = 0$ **19.** $x\left(-\dfrac{c_1}{x^2} \right) + \dfrac{c_1}{x} = 0$

21. $(\cos x - \sin x + e^{-x}) + (\sin x + \cos x - e^{-x}) = 2\cos x$

23. $(e^{-x} + \frac{12}{5}\cos 2x + \frac{24}{5}\sin 2x) + (-e^{-x} + \frac{6}{5}\sin 2x - \frac{12}{5}\cos 2x) = 6\sin 2x$

25. $\cos x \left[\dfrac{(\sec x + \tan x) - (x+c)(\sec x \tan x + \sec^2 x)}{(\sec x + \tan x)^2} \right] + \sin x = 1 - \dfrac{x+c}{\sec x + \tan x}$ **27.** $c^2 + cx = cx + c^2$

Exercises 29-2, page 917

1. $y = c - x^2$ **3.** $x - \dfrac{1}{y} = c$ **5.** $\ln(x^3+5) + 3y = c$ **7.** $4\sqrt{1-y} = e^{-x^2} + c$ **9.** $e^x - e^{-y} = c$

11. $\ln(y+4) = x + c$ **13.** $y(1 + \ln x)^2 + cy + 2 = 0$ **15.** $\tan^2 x + 2\ln y = c$

17. $x^2 + 1 + x\ln y + cx = 0$ **19.** $y^2 + 4\operatorname{Arcsin} x = c$ **21.** $y = c - (\ln x)^2$ **23.** $\ln(e^x+1) - \dfrac{1}{y} = c$

25. $3\ln y + x^3 = 0$ **27.** $\frac{1}{3}y^3 + y = \frac{1}{2}\ln^2 x$ **29.** $2\ln(1-y) = 1 - 2\sin x$ **31.** $e^{2x} - \dfrac{2}{y} = 2(e^x - 1)$

Exercises 29-3, page 919

1. $2xy + x^2 = c$ **3.** $x^3 - 2y = cx - 4$ **5.** $x^2 y - y = cx$ **7.** $(xy)^4 = 12\ln y + c$ **9.** $2\sqrt{x^2+y^2} = x + c$

11. $y = c - \frac{1}{2}\ln\sin(x^2+y^2)$ **13.** $\ln(y^2-x^2) + 2x = c$ **15.** $5xy^2 + y^3 = c$ **17.** $2xy + x^3 = 5$

19. $2x = 2xy^2 - 15y$

Exercises 29-4, page 922

1. $y = e^{-x}(x+c)$ **3.** $y = -\frac{1}{2}e^{-4x} + ce^{-2x}$ **5.** $y = -2 + ce^{2x}$ **7.** $y = x(3\ln x + c)$ **9.** $y = \dfrac{8}{7}x^3 + \dfrac{c}{\sqrt{x}}$

11. $y = -\cot x + c\csc x$ **13.** $y = (x+c)\csc x$ **15.** $y = 3 + ce^{-x}$ **17.** $2y = e^{4x}(x^2 + c)$

19. $y = \frac{1}{4} + ce^{-x^4}$ **21.** $3y = x^4 - 6x^2 - 3 + cx$ **23.** $xy = (x^3 + c)e^{3x}$ **25.** $y = e^{-x}$

27. $y = \frac{4}{3}\sin x - \csc^2 x$ **29.** $y = e^{\sqrt{x}} + (3e - e^2)e^{-\sqrt{x}}$ **31.** $y = x^2(3 + \sin x)$

Exercises 29-5, page 927

1. $y^2 = 2x^2 + 1$ **3.** $y = 2e^x - x - 1$ **5.** $y^2 = c - 2x$ **7.** $y^2 = c - 2\sin x$ **9.** $N = N_0(0.5)^{t/40}$, 35.4%

11. 4130 years   **13.** $S = a + \dfrac{c}{r^2}$ **15.** 1.7×10^5 **17.** 12.9 min   **19.** $1083.29

21. $\displaystyle\lim_{t\to\infty} \dfrac{E}{R}(1 - e^{-Rt/L}) = \dfrac{E}{R}$ **23.** $i = \dfrac{E}{R^2 + \omega^2 L^2}(R\sin\omega t - \omega L\cos\omega t + \omega L e^{-Rt/L})$ **25.** $q = q_0 e^{-t/RC}$

27. 11.04 lb   **29.** $v = 32(1 - e^{-t})$, 32   **31.** 10.0 ft/s   **33.** $x = 3t^2 - t^3$, $y = 6t^2 - 2t^3 - 9t^4 + 6t^5 - t^6$

35. $p = 15.0(0.667)^{10^{-4}h}$ **37.** $1260   **39.** $x = 4(1 + 2e^{-0.25t})$

Exercises 29-6, 29-7, page 933

1. $y = c_1 e^{3x} + c_2 e^{-2x}$ **3.** $y = c_1 e^{-x} + c_2 e^{-x/3}$ **5.** $y = c_1 + c_2 e^{3x}$ **7.** $y = c_1 e^{6x} + c_2 e^{2x/3}$

9. $y = c_1 e^{x/3} + c_2 e^{-3x}$ **11.** $y = c_1 e^{x/3} + c_2 e^{-x}$ **13.** $y = e^x(c_1 e^{x\sqrt{2}/2} + c_2 e^{-x\sqrt{2}/2})$

15. $y = e^{3x/8}(c_1 e^{x\sqrt{41}/8} + c_2 e^{-x\sqrt{41}/8})$ **17.** $y = e^{3x/2}(c_1 e^{x\sqrt{13}/2} + c_2 e^{-x\sqrt{13}/2})$

19. $y = e^{-x/2}(c_1 e^{x\sqrt{33}/2} + c_2 e^{-x\sqrt{33}/2})$ **21.** $y = \frac{1}{5}(3e^{7x} + 7e^{-3x})$ **23.** $y = \dfrac{e^3}{e^7 - 1}(e^{4x} - e^{-3x})$

25. $y = c_1 + c_2 e^{-x} + c_3 e^{3x}$ **27.** $y = c_1 e^x + c_2 e^{2x} + c_3 e^{3x}$

Exercises 29-8, page 937

1. $y = (c_1 + c_2x)e^x$ **3.** $y = (c_1 + c_2x)e^{-6x}$ **5.** $y = c_1 \sin 3x + c_2 \cos 3x$
7. $y = e^{-x/2}(c_1 \sin \frac{1}{2}\sqrt{7}x + c_2 \cos \frac{1}{2}\sqrt{7}x)$ **9.** $y = c_1 + c_2x$ **11.** $y = c_1 \sin \frac{1}{2}x + c_2 \cos \frac{1}{2}x$
13. $y = (c_1 + c_2x)e^{3x/4}$ **15.** $y = c_1 \sin \frac{1}{5}\sqrt{2}x + c_2 \cos \frac{1}{5}\sqrt{2}x$ **17.** $y = e^x(c_1 \cos \frac{1}{2}\sqrt{6}x + c_2 \sin \frac{1}{2}\sqrt{6}x)$
19. $y = (c_1 + c_2x)e^{4x/5}$ **21.** $y = e^{3x/4}(c_1e^{x\sqrt{17}/4} + c_2e^{-x\sqrt{17}/4})$ **23.** $y = c_1e^{x(-6+\sqrt{42})/3} + c_2e^{x(-6-\sqrt{42})/3}$
25. $y = e^{(\pi/6 - 1 - x)} \sin 3x$ **27.** $y = (4 - 14x)e^{4x}$ **29.** $(D^2 - 9)y = 0$ **31.** $(D^2 + 9)y = 0$

Exercises 29-9, page 941

1. $y = c_1e^{2x} + c_2e^{-x} - 2$ **3.** $y = c_1 \sin \frac{1}{2}x + c_2 \cos \frac{1}{2}x + x^2 - 8$ **5.** $y = c_1e^{-x} + c_2e^{-3x} + \frac{1}{8}e^x + \frac{2}{3}$
7. $y = c_1 + c_2e^{3x} - \frac{3}{4}e^x - \frac{1}{2}xe^x$ **9.** $y = c_1e^{x/3} + c_2e^{-x/3} - \frac{1}{10} \sin x$
11. $y = c_1 \sin 3x + c_2 \cos 3x + x + 2 \sin 2x + \cos 2x$ **13.** $y = c_1e^{-5x} + c_2e^{6x} - \frac{1}{3}$
15. $y = c_1e^{2x/3} + c_2e^{-5x} + \frac{1}{4}e^{3x}$ **17.** $y = c_1e^{2x} + c_2e^{-2x} - \frac{1}{5} \sin x - \frac{2}{5} \cos x$
19. $y = c_1 \sin x + c_2 \cos x - \frac{1}{3} \sin 2x + 4$ **21.** $y = c_1e^{-x} + c_2e^{-4x} - \frac{7}{100}e^x + \frac{1}{10}xe^x + 1$
23. $y = (c_1 + c_2x)e^{-3x} + \frac{1}{25}e^{2x} - e^{-2x}$ **25.** $y = \frac{1}{6}(11e^{3x} + 5e^{-2x} + e^x - 5)$
27. $y = -\frac{2}{3} \sin x + \pi \cos x + x - \frac{1}{3} \sin 2x$

Exercises 29-10, page 947

1. $\theta = 0.1 \cos 3.1t$

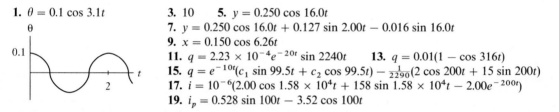

3. 10   **5.** $y = 0.250 \cos 16.0t$
7. $y = 0.250 \cos 16.0t + 0.127 \sin 2.00t - 0.016 \sin 16.0t$
9. $x = 0.150 \cos 6.26t$
11. $q = 2.23 \times 10^{-4}e^{-20t} \sin 2240t$ **13.** $q = 0.01(1 - \cos 316t)$
15. $q = e^{-10t}(c_1 \sin 99.5t + c_2 \cos 99.5t) - \frac{1}{2290}(2 \cos 200t + 15 \sin 200t)$
17. $i = 10^{-6}(2.00 \cos 1.58 \times 10^4t + 158 \sin 1.58 \times 10^4t - 2.00e^{-200t})$
19. $i_p = 0.528 \sin 100t - 3.52 \cos 100t$

Exercises 29-11, page 953

1. $F(s) = \int_0^\infty e^{-st} dt = -\frac{1}{s} e^{-st}\Big|_0^\infty = \frac{1}{s}$ **3.** $F(s) = \int_0^\infty e^{-st} \sin at \, dt = \frac{e^{-st}(-s \sin at - a \cos at)}{s^2 + a^2}\Big|_0^\infty = \frac{a}{s^2 + a^2}$

5. $\frac{1}{s - 3}$ **7.** $\frac{6}{(s + 2)^4}$ **9.** $\frac{s - 2}{s^2 + 4}$ **11.** $\frac{3}{s} + \frac{2(s^2 - 9)}{(s^2 + 9)^2}$ **13.** $s^2L(f) + sL(f)$
15. $(2s^2 - s + 1)L(f) - 2s + 1$ **17.** t^2 **19.** $\frac{1}{2}e^{-3t}$ **21.** $\frac{1}{2}t^2e^{-t}$ **23.** $\frac{1}{54}(9t \sin 3t + 2 \sin 3t - 6t \cos 3t)$

Exercises 29-12, page 956

1. $y = e^{-t}$ **3.** $y = -e^{3t/2}$ **5.** $y = (1 + t)e^{-3t}$ **7.** $y = \frac{1}{2} \sin 2t$ **9.** $y = 1 - e^{-2t}$ **11.** $y = e^{2t} \cos t$
13. $y = 1 + \sin t$ **15.** $y = e^{-t}(\frac{1}{2}t^2 + 3t + 1)$ **17.** $v = 6(1 - e^{-t/2})$ **19.** $q = 1.6 \times 10^{-4}(1 - e^{-5000t})$
21. $i = 5t \sin 50t$ **23.** $y = \sin 3t - 3t \cos 3t$

Review Exercises for Chapter 29, page 957

1. $2 \ln (x^2 + 1) - \frac{1}{2y^2} = c$ **3.** $ye^{2x} = x + c$ **5.** $y = c_1 + c_2e^{-x/2}$ **7.** $y = (c_1 + c_2x)e^{-x}$

9. $2x^2 + 4xy + y^4 = c$ **11.** $y = cx^3 - x^2$ **13.** $y = c(y + 2)e^{2x}$ **15.** $y = e^{-x}(c_1 \sin \sqrt{5}x + c_2 \cos \sqrt{5}x)$
17. $y = \frac{1}{2}(1 + ce^{-4x})$ **19.** $y = \frac{1}{2}(c - x^2) \csc x$ **21.** $y = c_1e^x + c_2e^{-3x/2} - 2$
23. $y = e^{-x/2}(c_1e^{x\sqrt{5}/2} + c_2e^{-x\sqrt{5}/2}) + 2e^x$ **25.** $y = c_1e^{2x/3} + c_2e^{4x/3} + \frac{1}{2}x + \frac{25}{8}$
27. $y = c_1 \sin 3x + c_2 \cos 3x + \frac{1}{8} \sin x$ **29.** $y^3 = 8 \sin^2 x$ **31.** $y = 2x - 1 - e^{-2x}$
33. $y = 2e^{-x/2} \sin (\frac{1}{2}\sqrt{15}x)$ **35.** $y = \frac{1}{25}[16 \sin x + 12 \cos x - 3e^{-2x}(4 + 5x)]$ **37.** $y = e^{t/4}$
39. $y = \frac{1}{2}(e^{3t} - e^t)$ **41.** $y = -4 \sin t$ **43.** $y = \frac{1}{3}t - \frac{4}{9} \sin 3t$ **45.** $r = r_0 + kt$ **47.** 3.93 m/s

49. 12.5 years   **51.** 6.5 billion   **53.** $5y^2 + x^2 = c$ **55.** $q = c_1 e^{-t/RC} + EC$

57. $y = 0.25e^{-2t}(2\cos 4t + \sin 4t)$, underdamped   **59.** $q = e^{-6t}(0.4\cos 8t + 0.3\sin 8t) - 0.4\cos 10t$

61. $i = 0$ **63.** $i = 12(1 - e^{-t/2})$; $i(0.3) = 1.67$ A   **65.** $q = 10^{-4}e^{-8t}(4.0\cos 200t + 0.16\sin 200t)$

67. $y = 0.25t\sin 8t$ **69.** $y = \dfrac{10}{3EI}\left[100x^3 - x^4 + xL^2(L - 100)\right]$

71. $\dfrac{dv}{dt} = \dfrac{dv}{dr}\dfrac{dr}{dt} = v\dfrac{dv}{dr}$; $\dfrac{dv}{dt} = \dfrac{k}{r^2}$, $k = -gR^2$, $\dfrac{-gR^2}{r^2} = v\dfrac{dv}{dr}$, $v^2 = \dfrac{2gR^2}{r} + v_0^2 - 2gR$; $v^2 \to 0$ as $r \to \infty$ if $v_0^2 \ge 2gR$

Exercises S-1, page 964

1. $x = 2$, $y = 1$ **3.** $x = \frac{17}{14}$, $y = \frac{19}{14}$ **5.** $x = -1$, $y = 4$, $z = 2$ **7.** $x = -2$, $y = -\frac{2}{3}$, $z = \frac{1}{3}$

9. $w = 1$, $x = 0$, $y = -2$, $z = 3$ **11.** Unlimited: $x = -3$, $y = -1$, $z = 1$; $x = 12$, $y = 0$, $z = -10$

13. Inconsistent   **15.** $x = \frac{1}{2}$, $y = -\frac{3}{2}$ **17.** $x = \frac{1}{6}$, $y = -\frac{1}{2}$ **19.** Inconsistent   **21.** 5.7 calc/s, 6.8 calc/s

23. 250 parts/h, 220 parts/h, 180 parts/h

Exercises S-2, page 969

1. Hyperbola;
  $2x'y' + 25 = 0$

3. Ellipse;
  $4x'^2 + 9y'^2 = 36$

5. Parabola;
  $x'^2 + \sqrt{2}y' = 0$

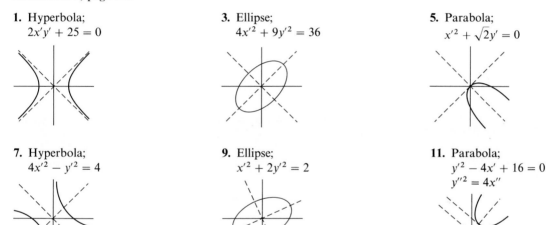

7. Hyperbola;
  $4x'^2 - y'^2 = 4$

9. Ellipse;
  $x'^2 + 2y'^2 = 2$

11. Parabola;
  $y'^2 - 4x' + 16 = 0$
  $y''^2 = 4x''$

Exercises S-3, page 972

1. $V = \frac{1}{3}\pi r^2 h$ **3.** $R = \sqrt{F_1^2 + F_2^2 + 1.732F_1F_2}$ **5.** 24   **7.** $2 - 3y + 4y^2$

9. $\dfrac{p^2 + pq + kp - p + 2q^2 + 4kq + 2k^2 + 5q + 5k}{p + q + k}$ **11.** 0   **13.** $x \ne 0$, $y \ge 0$ **15.** 1.03 A, 1.23 A

17. $A = \dfrac{pw - 2w^2}{2}$, 3850 cm^2 **19.** $L = \dfrac{(1.28 \times 10^5)r^4}{\ell^2}$

Exercises S-4, page 978

1.   **3.**   **5.**   **7.**

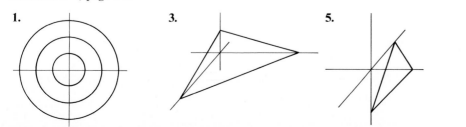
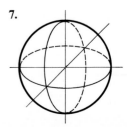

9.

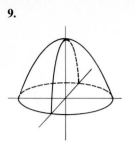

11.

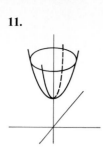

13.

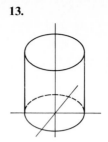

15.

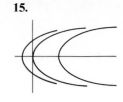

17.

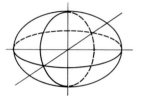

19.

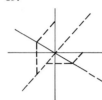

Exercises S-5, page 983

1. $\dfrac{\partial z}{\partial x} = 5 + 8xy, \dfrac{\partial z}{\partial y} = 4x^2$ **3.** $\dfrac{\partial f}{\partial x} = e^{2y}, \dfrac{\partial f}{\partial y} = 2xe^{2y}$ **5.** $\dfrac{\partial \phi}{\partial r} = \dfrac{1 + 3rs}{\sqrt{1 + 2rs}}, \dfrac{\partial \phi}{\partial s} = \dfrac{r^2}{\sqrt{1 + 2rs}}$

7. $\dfrac{\partial z}{\partial x} = y \cos xy, \dfrac{\partial z}{\partial y} = x \cos xy$ **9.** $\dfrac{\partial f}{\partial x} = \dfrac{12 \sin^2 2x \cos 2x}{1 - 3y}, \dfrac{\partial f}{\partial y} = \dfrac{6 \sin^3 2x}{(1 - 3y)^2}$

11. $\dfrac{\partial z}{\partial x} = \cos x - y \sin xy, \dfrac{\partial z}{\partial y} = -x \sin xy + \sin y$ **13.** -8

15. $\dfrac{\partial^2 z}{\partial x^2} = -6y, \dfrac{\partial^2 z}{\partial y^2} = 12xy, \dfrac{\partial^2 z}{\partial x \partial y} = 6y^2 - 6x$ **17.** $-4, -4$

19. 114 cm^2

21. $M\left(\dfrac{2mg}{(M + m)^2}\right) + m\left(\dfrac{-2Mg}{(M + m)^2}\right) = 0$

23. $3.75 \times 10^{-3} \; 1/\Omega$

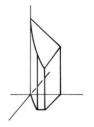

Exercises S-6, page 988

1. $\frac{28}{3}$ **3.** $\frac{1}{3}$ **5.** 1   **7.** $\frac{74}{5}$ **9.** $\frac{32}{3}$ **11.** $\frac{28}{3}$ **13.** 300 cm^3 **15.**

Exercises S-7, page 992

1. $\ln\left|\dfrac{(x+1)^2}{x+2}\right| + C$ **3.** $\dfrac{1}{4}\ln\left|\dfrac{x-2}{x+2}\right| + C$ **5.** $x + \ln\left|\dfrac{x}{(x+3)^4}\right| + C$ **7.** 1.057   **9.** $\ln\left|\dfrac{x^2(x-5)^3}{x+1}\right| + C$
11. 0.08495   **13.** 1.79 C   **15.** $\frac{1}{2}\ln\frac{5}{4} = 0.1116$

Exercises S-8, page 997

1. $\dfrac{3}{x-2} + 2\ln\left|\dfrac{x-2}{x}\right| + C$ **3.** $\dfrac{2}{x} + \ln\left|\dfrac{x-1}{x+1}\right| + C$ **5.** $-\frac{5}{4}$ **7.** $\frac{1}{8}\pi + \ln 3 = 1.491$
9. $-\dfrac{2}{x} + \dfrac{3}{2}\text{Arctan}\dfrac{x+2}{2} + C$ **11.** $\ln(x^2+1) + \dfrac{1}{x^2+1} + C$ **13.** $2 + 4\ln\frac{2}{3} = 0.3781$ **15.** 0.9190 m

Exercises B-1, page A-9

1. MHz, 1 MHz = 10^6 Hz   **3.** mm, 1 mm = 10^{-3} m   **5.** kilovolt, 1 kV = 10^3 V
7. milliampere, 1 mA = 10^{-3} A   **9.** 10^5 cm   **11.** 63,360 in.   **13.** 13.3 cm   **15.** 14.9 L   **17.** 144 in.^{2}
19. 2.50×10^{-4} m^2 **21.** 2040 g   **23.** 13.6 L   **25.** 1.68×10^5 N·cm/s   **27.** 0.101 hp   **29.** 3.25 tons
31. 15 gal   **33.** 1240 km/h   **35.** 37 mi/h   **37.** 25,000 mg/dL   **39.** 101,000 Pa

Exercises B-2, page A-12

1. 8 is exact.   **3.** 3 is exact; 74.6 is approx.   **5.** 1 and 19.3 are approx.
7. 150 is approx; 76 is exact.   **9.** 3, 4   **11.** 3, 4   **13.** 3, 3   **15.** 1, 6   **17.** (a) 3.764, (b) 3.764
19. (a) 0.01, (b) 30.8   **21.** (a) Same, (b) 78.0   **23.** (a) 0.004, (b) same   **25.** (a) 4.93, (b) 4.9
27. (a) 50,900, (b) 51,000   **29.** (a) 861, (b) 860   **31.** (a) 0.305, (b) 0.31   **33.** (a) 0.950, (b) 0.95
35. (a) 31.0, (b) 31   **37.** 128.25 ft, 128.35 ft   **39.** 0.1733 qt

Exercises B-3, page A-16

1. 51.2   **3.** 1.69   **5.** 431.4   **7.** 30.8   **9.** 62.1   **11.** 270   **13.** 160   **15.** 27,000   **17.** 5.7
19. 4.39   **21.** 10.2   **23.** 22   **25.** 2.38   **27.** 0.042   **29.** 17.62   **31.** 18.85   **33.** 196 ft
35. First plane, 70 mi/h   **37.** 262,144 bytes   **39.** 83.82 cm   **41.** 37°
43. Too many sig. digits; time has only two sig. digits

Exercises for Appendix C, page A-22

1. 85°   **3.** 140°   **5.** 52°   **7.** 50°   **9.** 70°   **11.** 56°   **13.** 25°   **15.** 25°   **17.** 120°   **19.** 40°
21. 5   **23.** 17   **25.** 8   **27.** 5.120   **29.** 10.44   **31.** 28.89   **33.** 21   **35.** 20   **37.** 3.3 in., 4.0 in.
39. 16   **41.** 25 ft   **43.** 10 yd   **45.** 18.46 in.   **47.** 32.7 cm   **49.** 212 cm   **51.** 62.8 ft   **53.** 28 in.^{2}
55. 13 yd^2 **57.** 24 m^2 **59.** 154.8 in.^{2} **61.** 216 ft^3 **63.** 20 mm^3 **65.** 924 in.^{3} **67.** 153 in.^{3}
69. 27.1 ft^3 **71.** 690.5 m^3 **73.** 208 in.^{2} **75.** 537 m^2 **77.** 3241 ft^2 **79.** 343 cm^2 **81.** 2.00 cm^3
83. 33,100 ft^2

Exercises for Appendix D, page A-31

(Most answers have been rounded off to four significant digits.)
1. 56.02   **3.** 4162.1   **5.** 18.65   **7.** 0.3954   **9.** 14.14   **11.** 0.5251   **13.** 13.35   **15.** 944.6
17. 0.7349   **19.** -0.7594 **21.** -1.337 **23.** 1.015   **25.** 41.35°   **27.** -1.182 **29.** 0.5862
31. 6.695   **33.** 3.508   **35.** 0.005685   **37.** 2.053   **39.** 5.765   **41.** 4.501×10^{10} **43.** 497.2
45. 6.648   **47.** 401.2   **49.** 8.841   **51.** 2.523   **53.** 10.08   **55.** 22.36   **57.** 20.3°   **59.** 4729
61. 3.301×10^4 **63.** 1.056   **65.** 55.5°   **67.** 3.277   **69.** 8.125   **71.** 1.000   **73.** 1.000   **75.** 12.90
77. 8.001   **79.** 8.053   **81.** 0.04259   **83.** 0.4219   **85.** 0.7822   **87.** 2.0736465   **89.** 124.3   **91.** 252

Exercises for Appendix E, page A-40

1. 100 DEF FN F(X) = 2*X/(X^2 + 1)

3. Change B to C in lines 30 and 100; delete line 160
130 INPUT "SIDE C = ";SC
145 LET SB = SQR(SC^2 − SA^2)
210 PRINT "SIDE B = ";SB

5. 190 PRINT "X1 = (";−B;" + SQR(";D;"))/";2*A
195 PRINT "X2 = (";−B;" − SQR(";D;"))/";2*A

7. 125 IF SA > SB + SC OR SB > SA + SC OR SC > SA + SB THEN GOTO 225
222 GOTO 230
225 PRINT "NO TRIANGLE HAS THESE THREE SIDES"

9. 120 INPUT "R = ";R:INPUT "T = ";T
140 LET T = 3.14159265*T/180
150 PRINT R*COS(T);" + (";R*SIN(T);")J"
160 END

11. Change line 20
195 IF X = −D/C THEN GOTO 285
200 LET Y = (A*X + B)/(C*X + D)
285 PRINT "F(X) IS UNDEFINED"
287 GOTO 140

13. 150 LET LS = TAN(X)*SIN(X)/COS(X)
160 LET RS = 1/(COS(X))^2 − 1

15. 140 DEF FN N(X) = 2 − SQR(X + 2)
150 DEF FN D(X) = X − 2

17. 100 DEF FN F(X) = 4*X − X^2
10.66, 0.59; −10.07; curve is below axis for $x > 4$

19. 140 FOR N = 1 TO 20
On most computers values will increase past the value of e, and then at some point all values become 1. The way in which numbers are calculated on a computer leads to an error when the numbers are extremely small.

Solutions to Practice Test Problems

Chapter 1

1. $\sqrt{9+16} = \sqrt{25} = 5$

2. $\dfrac{(7)(-3)(-2)}{(-6)(0)}$ is undefined (division by zero).

3. $\dfrac{3.372 \times 10^{-3}}{7.526 \times 10^{12}} = 4.480 \times 10^{-16}$ (Calculator sequence: 3.372 $\boxed{\text{EE}}$ 3 $\boxed{+/-}$ $\boxed{\div}$ 7.526 $\boxed{\text{EE}}$ 12 $\boxed{=}$ $\boxed{4.4804677\ -16}$.)

4. $\dfrac{(+6)(-2)-3(-1)}{5-2} = \dfrac{-12-(-3)}{3} = \dfrac{-12+3}{3} = \dfrac{-9}{3} = -3$

5. $\dfrac{523.0}{207.7} - \dfrac{396.4 - 23.5}{249.8} = 1.025$ (Calculator sequence: 523 $\boxed{\div}$ 207.7 $\boxed{-}$

$\boxed{(}$ 396.4 $\boxed{-}$ 23.5 $\boxed{)}$ $\boxed{\div}$ 249.8 $\boxed{=}$ $\boxed{1.0252607}$.)

6. (a) $2x^0 = 2(1) = 2$, (b) $(2a^{-2}b^3)^{-3} = (2^{(1)(-3)})(a^{(-2)(-3)})(b^{3(-3)}) = 2^{-3}a^6b^{-9} = \dfrac{a^6}{2^3b^9} = \dfrac{a^6}{8b^9}$

7. $(3s^2y)^2(-2s) = (3^2s^4y^2)(-2s) = (9)(-2)(s^{4+1})y^2 = -18s^5y^2$

8. $\dfrac{8a^3x^2 - 4a^2x^4}{-2ax^2} = \dfrac{8a^3x^2}{-2ax^2} - \dfrac{4a^2x^4}{-2ax^2} = -4a^2 - (-2ax^2) = -4a^2 + 2ax^2$

9. $3m^2(am - 2m^3) = 3m^2(am) + 3m^2(-2m^3) = 3am^3 - 6m^5$

10. $(2x - 3)(x + 7) = 2x(x) + 2x(7) + (-3)(x) + (-3)(7)$
$$= 2x^2 + 14x - 3x - 21 = 2x^2 + 11x - 21$$

11.

$$
\begin{array}{r}
3x \ - \ 5 \quad \text{(quotient)} \\
2x - 1 \overline{\smash{\big)}\, 6x^2 - 13x + 7} \\
\underline{6x^2 - \ 3x} \\
-10x + 7 \\
\underline{-10x + 5} \\
2 \quad \text{(remainder)}
\end{array}
$$

12. $(2x + 3)^2 = (2x + 3)(2x + 3)$
$$= 2x(2x) + 2x(3) + 3(2x) + 3(3)$$
$$= 4x^2 + 6x + 6x + 9$$
$$= 4x^2 + 12x + 9$$

13.
$$3(x - 3) = x - d$$
$$3x - 9 = x - d$$
$$3x - 9 - x + 9 = x - d - x + 9$$
$$2x = 9 - d$$
$$x = \dfrac{9 - d}{2}$$

14. $5x - 2(x - 4) = 7$
$$5x - 2x + 8 = 7$$
$$3x + 8 - 8 = 7 - 8$$
$$3x = -1$$
$$x = -\dfrac{1}{3}$$

15. $0.0000036 = 3.6 \times 10^{-6}$
(6 places to right)

16.

$-\pi$	-3	0.3	$\sqrt{2}$	$|-4|$	(order)
-3.14	-3	0.3	1.41	4	(value)

17. $3(5 + 8) = 3(5) + 3(8)$ illustrates distributive law.

18. $L = L_0[1 + a(t_2 - t_1)]$
$= L_0[1 + at_2 - at_1]$
$= L_0 + aL_0t_2 - aL_0t_1$
$L - L_0 + aL_0t_1 = aL_0t_2$
$t_2 = \dfrac{L - L_0 + aL_0t_1}{aL_0}$

19. Let x = length of side.
$x + x + x + x + x = 75$
$5x = 75$
$x = 15$ m

20. Let n = number of pounds of second alloy.
$0.3(20) + 0.8n = 0.6(n + 20)$
$6 + 0.8n = 0.6n + 12$
$0.2n = 6$
$n = 30$ lb

Chapter 2

1. $f(x) = 2x - x^2 + \dfrac{8}{x}$

$f(2.385) = 2(2.385) - (2.385)^2 + \dfrac{8}{2.385} = 2.436$

$f(-4) = 2(-4) - (-4)^2 + \dfrac{8}{-4}$

$= -8 - 16 - 2$

$= -26$

(Calculator sequence: $2 \;\boxed{\times}\; 2.385 \;\boxed{-}\; 2.385 \;\boxed{x^2}\; \boxed{+}\; 8$
$\boxed{\div}\; 2.385 \;\boxed{=}\; \boxed{2.4360727}$.)

2. $w = 2000 - 10t$

3. $f(x) = 4 - 2x$
$y = 4 - 2x$

$y = 4 - 2(-1) = 6$
$y = 4 - 2(0) = 4$
$y = 4 - 2(1) = 2$
$y = 4 - 2(2) = 0$
$y = 4 - 2(3) = -2$
$y = 4 - 2(4) = -4$

x	y
-1	6
0	4
1	2
2	0
3	-2
4	-4

4. $f(x) = 2x^2 - 3x - 3$
$y = 2x^2 - 3x - 3$

x	y
-1	2
0	-3
1	-4
2	-1
3	6

$f(x) = 0$ for
$x = -0.7$ and $x = 2.2$ (approx.)

5. $y = \sqrt{4 + 2x}$

$y = \sqrt{4 + 2(-2)} = 0$
$y = \sqrt{4 + 2(-1)} = 1.4$
$y = \sqrt{4 + 2(0)} = 2$
$y = \sqrt{4 + 2(1)} = 2.4$
$y = \sqrt{4 + 2(2)} = 2.8$
$y = \sqrt{4 + 2(4)} = 3.5$

x	y
-2	0
-1	1.4
0	2
1	2.4
2	2.8
4	3.5

6. On negative x-axis

7. Let r = radius of circular part.

square   semicircle

$A = (2r)(2r) + \dfrac{1}{2}(\pi r^2)$

$= 4r^2 + \dfrac{1}{2}\pi r^2$

8. $f(x) = \sqrt{6 - x}$
Domain: $x \le 6$; x cannot be greater than 6 to have real values of $f(x)$.
Range: $f(x) \ge 0$; $\sqrt{6 - x}$ is the principal square root of $6 - x$ and cannot be negative.

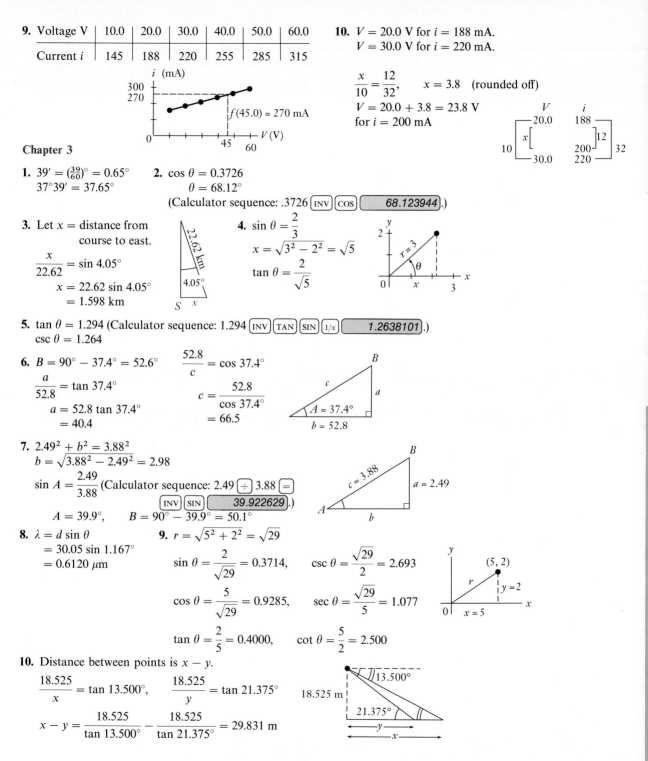

9.

Voltage V	10.0	20.0	30.0	40.0	50.0	60.0
Current i	145	188	220	255	285	315

$f(45.0) = 270$ mA

10. $V = 20.0$ V for $i = 188$ mA.
$V = 30.0$ V for $i = 220$ mA.

$$\frac{x}{10} = \frac{12}{32}, \qquad x = 3.8 \quad \text{(rounded off)}$$

$V = 20.0 + 3.8 = 23.8$ V
for $i = 200$ mA

Chapter 3

1. $39' = \left(\frac{39}{60}\right)^\circ = 0.65^\circ$
$37°39' = 37.65^\circ$

2. $\cos \theta = 0.3726$
$\theta = 68.12^\circ$
(Calculator sequence: .3726 [INV][COS] 68.123944.)

3. Let $x =$ distance from
course to east.
$$\frac{x}{22.62} = \sin 4.05^\circ$$
$x = 22.62 \sin 4.05^\circ$
$= 1.598$ km

4. $\sin \theta = \dfrac{2}{3}$
$x = \sqrt{3^2 - 2^2} = \sqrt{5}$
$\tan \theta = \dfrac{2}{\sqrt{5}}$

5. $\tan \theta = 1.294$ (Calculator sequence: 1.294 [INV][TAN][SIN][1/x] 1.2638101.)
$\csc \theta = 1.264$

6. $B = 90^\circ - 37.4^\circ = 52.6^\circ$
$$\frac{a}{52.8} = \tan 37.4^\circ$$
$a = 52.8 \tan 37.4^\circ$
$= 40.4$

$$\frac{52.8}{c} = \cos 37.4^\circ$$
$$c = \frac{52.8}{\cos 37.4^\circ}$$
$= 66.5$

7. $2.49^2 + b^2 = 3.88^2$
$b = \sqrt{3.88^2 - 2.49^2} = 2.98$
$$\sin A = \frac{2.49}{3.88}$$ (Calculator sequence: 2.49 [÷] 3.88 [=]
[INV][SIN] 39.922629.)
$A = 39.9^\circ, \qquad B = 90^\circ - 39.9^\circ = 50.1^\circ$

8. $\lambda = d \sin \theta$
$= 30.05 \sin 1.167^\circ$
$= 0.6120$ μm

9. $r = \sqrt{5^2 + 2^2} = \sqrt{29}$
$$\sin \theta = \frac{2}{\sqrt{29}} = 0.3714, \qquad \csc \theta = \frac{\sqrt{29}}{2} = 2.693$$
$$\cos \theta = \frac{5}{\sqrt{29}} = 0.9285, \qquad \sec \theta = \frac{\sqrt{29}}{5} = 1.077$$
$$\tan \theta = \frac{2}{5} = 0.4000, \qquad \cot \theta = \frac{5}{2} = 2.500$$

10. Distance between points is $x - y$.
$$\frac{18.525}{x} = \tan 13.500^\circ, \qquad \frac{18.525}{y} = \tan 21.375^\circ$$
$$x - y = \frac{18.525}{\tan 13.500^\circ} - \frac{18.525}{\tan 21.375^\circ} = 29.831 \text{ m}$$

Chapter 4

1. Points $(2, -5)$ and $(-1, 4)$

$$m = \frac{4 - (-5)}{-1 - 2} = \frac{9}{-3} = -3$$

2.

$$x + 2y = 5 \qquad 4y = 3 - 2(5 - 2y)$$
$$4y = 3 - 2x \qquad 4y = 3 - 10 + 4y$$
$$x = 5 - 2y \qquad 0 = -7$$
$$\text{Inconsistent}$$

3. $3x - 2y = 4$

$2x + 5y = -1$

$$x = \frac{\begin{vmatrix} 4 & -2 \\ -1 & 5 \end{vmatrix}}{\begin{vmatrix} 3 & -2 \\ 2 & 5 \end{vmatrix}} = \frac{20 - 2}{15 - (-4)} = \frac{18}{19}$$

$$y = \frac{\begin{vmatrix} 3 & 4 \\ 2 & -1 \end{vmatrix}}{19} = \frac{-3 - 8}{19} = -\frac{11}{19}$$

4. $2x + y = 4$

$$y = -2x + 4$$
$$m = -2, \qquad b = 4$$

5. $2x + 2y = 24$ (perimeter)

$$x = y + 6$$
$$\underline{x + y = 12}$$
$$\underline{x - y = 6}$$
$$2x = 18$$
$$x = 9 \text{ km}$$

$$9 = y + 6$$
$$y = 3 \text{ km}$$

6. $2x - 3y = 6$

$x = 0$: $y = -2$

$y = 0$: $x = 3$

Int: $(0, -2)$, $(3, 0)$

$4x + y = 4$

$x = 0$: $y = 4$

$y = 0$: $x = 1$

Int: $(0, 4)$, $(1, 0)$

$$x = 1.3, \qquad y = -1.1$$

7. Let $x = $ vol. of first alloy,

$y = $ vol. of second alloy.

$z = $ vol. of third alloy.

$$\begin{array}{rl} x + \ \ y + \ \ z = 100 & \text{total vol.} \\ 0.6x + 0.5y + 0.3z = \ \ 40 & \text{copper} \\ 0.3x + 0.3y \qquad\quad = \ \ 15 & \text{zinc} \end{array}$$

$$x + \ y + \ z = 100$$
$$6x + 5y + 3z = 400$$
$$\underline{3x + 3y \qquad = 150}$$
$$3x + 2y \qquad = 100$$
$$\underline{3x + 3y \qquad = 150}$$
$$y = 50 \text{ cm}^3$$
$$x = \ \ 0 \text{ cm}^3$$
$$z = 50 \text{ cm}^3$$

8. $3x + 2y - \ \ z = 4$

$2x - \ \ y + 3z = -2$

$x \qquad + 4z = 5$

$$y = \frac{\begin{vmatrix} 3 & 4 & -1 \\ 2 & -2 & 3 \\ 1 & 5 & 4 \end{vmatrix}}{\begin{vmatrix} 3 & 2 & -1 \\ 2 & -1 & 3 \\ 1 & 0 & 4 \end{vmatrix}}$$

$$= \frac{-24 + 12 - 10 - 2 - 45 - 32}{-12 + 6 + 0 - 1 - 0 - 16}$$

$$= \frac{-101}{-23} = \frac{101}{23}$$

Chapter 5

1. $2x(2x - 3)^2 = 2x[(2x)^2 - 2(2x)(3) + 3^2] = 2x(4x^2 - 12x + 9) = 8x^3 - 24x^2 + 18x$

2. $\dfrac{1}{R} = \dfrac{1}{R_1 + r} + \dfrac{1}{R_2}$

$$\frac{RR_2(R_1 + r)}{R} = \frac{RR_2(R_1 + r)}{R_1 + r} + \frac{RR_2(R_1 + r)}{R_2}$$

$$R_2(R_1 + r) = RR_2 + R(R_1 + r)$$

$$R_1R_2 + rR_2 = RR_2 + RR_1 + rR$$
$$R_1R_2 - RR_1 = RR_2 + rR - rR_2$$
$$R_1(R_2 - R) = RR_2 + rR - rR_2$$

$$R_1 = \frac{RR_2 + rR - rR_2}{R_2 - R}$$

3. $\dfrac{2x^2 + 5x - 3}{2x^2 + 12x + 18} = \dfrac{(2x-1)(x+3)}{2(x+3)^2} = \dfrac{2x-1}{2(x+3)}$

4. $4x^2 - 16y^2 = 4(x^2 - 4y^2) = 4(x+2y)(x-2y)$

6. $\dfrac{x^2 + x}{2 - x} \div \dfrac{x^2}{x^2 - 4x + 4} = \dfrac{x^2 + x}{2 - x} \times \dfrac{x^2 - 4x + 4}{x^2} = \dfrac{x(x+1)}{2-x} \times \dfrac{(x-2)^2}{x^2}$

$= -\dfrac{x(x+1)(x-2)^2}{(x-2)(x^2)} = -\dfrac{(x+1)(x-2)}{x}$

5. $\dfrac{3}{4x^2} - \dfrac{2}{x^2 - x} - \dfrac{x}{2x - 2} = \dfrac{3}{4x^2} - \dfrac{2}{x(x-1)} - \dfrac{x}{2(x-1)}$

$= \dfrac{3(x-1) - 2(4x) - x(2x^2)}{4x^2(x-1)}$

$= \dfrac{3x - 3 - 8x - 2x^3}{4x^2(x-1)}$

$= \dfrac{-2x^3 - 5x - 3}{4x^2(x-1)}$

7. $\dfrac{1 - \dfrac{3}{2x+2}}{\dfrac{x}{5} - \dfrac{1}{2}} = \dfrac{\dfrac{2(x+1) - 3}{2(x+1)}}{\dfrac{2x - 5}{10}} = \dfrac{2x + 2 - 3}{2(x+1)} \times \dfrac{10}{2x - 5} = \dfrac{5(2x - 1)}{(x+1)(2x - 5)}$

8. Let t = time working together.
$\dfrac{t}{12} + \dfrac{t}{16} = 1$
LCD of 12 and 16 is 48.

$\dfrac{48t}{12} + \dfrac{48t}{16} = 48$

$4t + 3t = 48$

$t = \dfrac{48}{7} = 6.9$ days

Chapter 6

1. $x^2 - 3x - 5 = 0$
Not factorable
$a = 1, \quad b = -3, \quad c = -5$
$x = \dfrac{-(-3) \pm \sqrt{(-3)^2 - 4(1)(-5)}}{2(1)}$
$= \dfrac{3 \pm \sqrt{29}}{2}$

2. $2x^2 = 9x - 4$
$2x^2 - 9x + 4 = 0$
$(2x - 1)(x - 4) = 0$
$2x - 1 = 0, \quad x - 4 = 0$
$x = \tfrac{1}{2} \quad$ or $\quad x = 4$

3. $\dfrac{3}{x} - \dfrac{2}{x+2} = 1$
$\dfrac{3x(x+2)}{x} - \dfrac{2x(x+2)}{x+2} = x(x+2)$
$3(x+2) - 2x = x^2 + 2x$
$3x + 6 - 2x = x^2 + 2x$
$0 = x^2 + x - 6$
$(x+3)(x-2) = 0$
$x = -3, 2$

4. $2x^2 - x = 6 - 2x(3 - x)$
$2x^2 - x = 6 - 6x + 2x^2$
$5x = 6 \quad$ (not quadratic)
$x = \tfrac{6}{5}$

5. $y = 2x^2 + 8x + 5$
$\dfrac{-b}{2a} = \dfrac{-8}{2(2)} = -2$
$y = 2(-2)^2 + 8(-2) + 5 = -3$
Min. pt. $(a > 0)$ is $(-2, -3)$,
$c = 5$, y-intercept is $(0, 5)$.

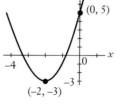

$(0, 5)$
-4
$(-2, -3)$

6. $P = EI - RI^2$
$RI^2 - EI + P = 0$
$I = \dfrac{-(-E) \pm \sqrt{(-E)^2 - 4RP}}{2R}$
$= \dfrac{E \pm \sqrt{E^2 - 4RP}}{2R}$

7. $x^2 - 6x - 9 = 0$
$x^2 - 6x \quad = 9$
$x^2 - 6x + 9 = 9 + 9$
$(x - 3)^2 = 18$
$x - 3 = \pm\sqrt{18}$
$x = 3 \pm 3\sqrt{2}$

8. Let w = width of window,
h = height of window.
$2w + 2h = 8.4 \quad$ (perimeter)
$w + h = 4.2, \quad h = 4.2 - w$
$w(4.2 - w) = 3.8 \quad$ (area)
$-w^2 + 4.2w = 3.8$
$w^2 - 4.2w + 3.8 = 0$
$w = \dfrac{-(-4.2) \pm \sqrt{(-4.2)^2 - 4(3.8)}}{2}$
$= 2.88, 1.32$
$w = 1.3$ m, $h = 2.9$ m   or
$w = 2.9$ m, $h = 1.3$ m

Chapter 7

1. $150° = (\frac{\pi}{180})(150) = \frac{5\pi}{6}$ **2.** $\sin 205° = -\sin(205° - 180°) = -\sin 25°$ **3.** $x = -9, y = 12$

4. $r = \frac{1.38}{2} = 0.690$ ft   **5.** $3.572 = 3.572(\frac{180°}{\pi}) = 204.7°$ $r = \sqrt{(-9)^2 + 12^2} = 15$

   $\omega = 1200$ r/min $= (1200$ r/min$)(2\pi$ rad/r$)$ $\sin\theta = \frac{12}{15} = \frac{4}{5}$

   $= 2400\pi$ rad/min $\sec\theta = \frac{15}{-9} = -\frac{5}{3}$

   $v = \omega r = (2400\pi)(0.690) = 5200$ ft/min

6. $\tan\theta = 0.2396$ **7.** $\cos\theta = -0.8244, \csc\theta < 0;$ **8.** $s = r\theta, 32.0 = 8.50\theta$

   $\theta_{ref} = 13.47°$ $\cos\theta$ negative, $\csc\theta$ negative, $\theta = \frac{32.0}{8.50} = 3.76$ rad

   $\theta = 13.47°$ or θ in third quadrant $A = \frac{1}{2}\theta r^2 = \frac{1}{2}(3.76)(8.50)^2$

   $\theta = 180° + 13.47° = 193.47°$ $\theta_{ref} = 0.6017, \theta = 3.7432$ $= 136$ ft^2

   (Calculator sequence: (Calculator sequence:

   .2396 $\boxed{\text{INV}}$ $\boxed{\text{TAN}}$ $\boxed{13.474061}$ .8244 $\boxed{\text{INV}}$ $\boxed{\text{COS}}$ $\boxed{+}$ $\boxed{\pi}$

   $\boxed{+}$ 180 $\boxed{=}$ $\boxed{193.47406}$.) $\boxed{=}$ $\boxed{3.7432477}$.)

$s = 32.0$ ft

$r = 8.50$ ft

Chapter 8

1.

2.

$c = 30.9$

$78.6°$

$a = 22.5$

$b^2 = a^2 + c^2 - 2ac\cos B$

$b = \sqrt{22.5^2 + 30.9^2 - 2(22.5)(30.9)\cos 78.6°} = 34.4$

3.

Tree

36.50 m   45.0°   21.38 m

Pole

45.0° x

α θ

Set

$x^2 = 36.50^2 + 21.38^2 - 2(36.50)(21.38)\cos 45.00°, x = 26.19$ m

$\dfrac{21.38}{\sin\alpha} = \dfrac{26.19}{\sin 45.00°}, \qquad \sin\alpha = \dfrac{21.38\sin 45.00°}{26.19}, \qquad \alpha = 35.26°$

$\theta = 45.00° - 35.26° = 9.74°$

Displacement is 26.19 m, 9.74° N of E.

4.

C

$a = 426$

$18.9°$ $104.2°$

A B

$C = 180° - (18.9° + 104.2°) = 56.9°$

$\dfrac{c}{\sin C} = \dfrac{a}{\sin A}$

$c = \dfrac{426\sin 56.9°}{\sin 18.9°} = 1100$

5. Since a is longest side,

   find A first.

   $a^2 = b^2 + c^2 - 2bc\cos A$

   $\cos A = \dfrac{b^2 + c^2 - a^2}{2bc}$

   $= \dfrac{3.29^2 + 8.44^2 - 9.84^2}{2(3.29)(8.44)}$

C

$b = 3.29$ $a = 9.84$

A B

$c = 8.44$

   $A = 105.4°$

   $\dfrac{b}{\sin B} = \dfrac{a}{\sin A}$

   $\sin B = \dfrac{b\sin A}{a} = \dfrac{3.29\sin 105.4°}{9.84}$

   $B = 18.8°$

   $C = 180° - (105.4° + 18.8°) = 55.8°$

6. $A_x = 870\cos 284.3° = 215$

   $A_y = 870\sin 284.3° = -843$

$284.3°$ Ax x

Ay

$A = 870$

7. $\dfrac{63.0}{\sin 148.5°} = \dfrac{42.0}{\sin A}$

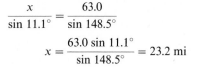

$\dfrac{x}{\sin 11.1°} = \dfrac{63.0}{\sin 148.5°}$

$\sin A = \dfrac{42.0 \sin 148.5°}{63.0}$

$x = \dfrac{63.0 \sin 11.1°}{\sin 148.5°} = 23.2 \text{ mi}$

$A = 20.4°$

$C = 180° - (148.5° + 20.4°) = 11.1°$

8. $A_x = 450 \cos 74.2°, \qquad B_x = 285 \cos 208.9°$

$A_y = 450 \sin 74.2°, \qquad B_y = 285 \sin 208.9°$

$R_x = A_x + B_x = 450 \cos 74.2° + 285 \cos 208.9° = -127.0$

$R_y = A_y + B_y = 450 \sin 74.2° + 285 \sin 208.9° = 295.3$

$R = \sqrt{(-127.0)^2 + 295.3^2} = 321$

$\tan \theta_{\text{ref}} = \dfrac{295.3}{127.0}, \qquad \theta_{\text{ref}} = 66.7°, \qquad \theta = 113.3°$

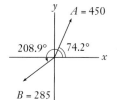

θ is in second quadrant, since R_x is negative and R_y is positive.

Chapter 9

1. $y = 0.5 \cos \frac{\pi}{2}x$

Amp. $= 0.5$, disp. $= 0$,

per. $= \dfrac{2\pi}{\pi/2} = 4$.

x	0	1	2	3	4
y	0.5	0	-0.5	0	0.5

2. $y = 2 + 3 \sin x$

For $y_1 = 3 \sin x$,

amp. $= 3$, per. $= 2\pi$,

disp. $= 0$.

x	0	$\frac{\pi}{2}$	π	$\frac{3\pi}{2}$	2π
$y_1 = 3 \sin x$	0	3	0	-3	0
$y = 2 + 3 \sin x$	2	5	2	-1	2

3. $y = 2 \sin x + \cos 2x$

For $y_1 = 2 \sin x$,

amp. $= 2$, per. $= 2\pi$, disp. $= 0$.

For $y_2 = \cos 2x$,

amp. $= 1$, per. $= \frac{2\pi}{2} = \pi$, disp. $= 0$.

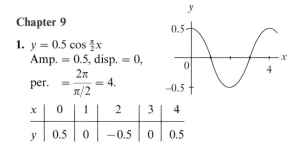

4. $y = 2 \sin (2x - \frac{\pi}{3})$

Amp. $= 2$, per. $= \frac{2\pi}{2} = \pi$,

disp. $= -\dfrac{-\pi/3}{2} = \dfrac{\pi}{6}$.

x	$\frac{\pi}{6}$	$\frac{5\pi}{12}$	$\frac{2\pi}{3}$	$\frac{11\pi}{12}$	$\frac{7\pi}{6}$
y	0	2	0	-2	0

$\frac{\pi}{6} + \frac{\pi}{4} = \frac{5\pi}{12}, \ \frac{\pi}{6} + \frac{\pi}{2} = \frac{2\pi}{3}, \ \frac{\pi}{6} + \frac{3\pi}{4} = \frac{11\pi}{12}$

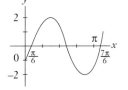

5. $y = A \cos \frac{2\pi}{T} t$

$A = 0.200$ in., $T = 0.100$ s,

$y = 0.200 \cos 20\pi t$,

amp. $= 0.200$ in., per. $= 0.100$ s.

6. $y = 3 \sec x \ (0 \le x \le 2\pi)$

$\sec x = \dfrac{1}{\cos x}$

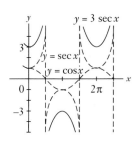

7. $x = \sin \pi t$, $y = 2 \cos 2\pi t$

t	x	y	Point
0	0	2	1
$\frac{1}{4}$	0.7	0	2
$\frac{1}{2}$	1	-2	3
$\frac{3}{4}$	0.7	0	4
1	0	2	5
$\frac{5}{4}$	-0.7	0	6
$\frac{3}{2}$	-1	-2	7
$\frac{7}{4}$	-0.7	0	8
2	0	2	9

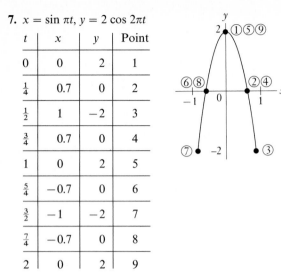

8. $d = R \sin \left(\omega t + \frac{\pi}{6} \right)$
$\omega = 2.00$ rad/s
$d = R \sin \left(2.00t + \frac{\pi}{6} \right)$

Amp. $= R$, per. $= \dfrac{2\pi}{2.00} = \pi$ s $= 3.14$ s,

disp. $= \dfrac{-\pi/6}{2.00} = -\dfrac{\pi}{12}$ s $= -0.26$ s.

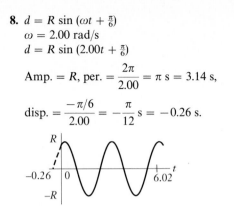

Chapter 10

1. $2\sqrt{20} - \sqrt{125} = 2\sqrt{4 \times 5} - \sqrt{25 \times 5} = 2(2\sqrt{5}) - 5\sqrt{5} = 4\sqrt{5} - 5\sqrt{5} = -\sqrt{5}$

2. $\dfrac{100^{3/2}}{8^{-2/3}} = (100^{3/2})(8^{2/3}) = [(100^{1/2})^3][(8^{1/3})^2] = (10^3)(2^2) = 4000$

3. $\sqrt{x^2 + \dfrac{1}{9}} = \sqrt{\dfrac{9x^2 + 1}{9}} = \dfrac{\sqrt{9x^2 + 1}}{\sqrt{9}} = \dfrac{\sqrt{9x^2 + 1}}{3}$

4. $(2x^{-1} + y^{-2})^{-1} = \dfrac{1}{2x^{-1} + y^{-2}} = \dfrac{1}{\dfrac{2}{x} + \dfrac{1}{y^2}} = \dfrac{1}{\dfrac{2y^2 + x}{xy^2}} = \dfrac{xy^2}{2y^2 + x}$

5. $(\sqrt{2x} - 3\sqrt{y})^2 = (\sqrt{2x})^2 - 2\sqrt{2x}(3\sqrt{y}) + (3\sqrt{y})^2 = 2x - 6\sqrt{2xy} + 9y$

6. $\sqrt[3]{\sqrt[4]{4}} = \sqrt[12]{4} = \sqrt[12]{2^2} = 2^{2/12} = 2^{1/6} = \sqrt[6]{2}$
7. $\dfrac{3 - 2\sqrt{2}}{2\sqrt{x}} = \dfrac{3 - 2\sqrt{2}}{2\sqrt{x}} \times \dfrac{\sqrt{x}}{\sqrt{x}} = \dfrac{3\sqrt{x} - 2\sqrt{2x}}{2x}$

8. $\sqrt{27a^4b^3} = \sqrt{9 \times 3 \times (a^2)^2(b^2)(b)} = 3a^2b\sqrt{3b}$

9. $(2x + 3)^{1/2} + (x + 1)(2x + 3)^{-1/2} = (2x + 3)^{1/2} + \dfrac{x + 1}{(2x + 3)^{1/2}}$

$= \dfrac{(2x + 3)^{1/2}(2x + 3)^{1/2} + x + 1}{(2x + 3)^{1/2}} = \dfrac{2x + 3 + x + 1}{(2x + 3)^{1/2}} = \dfrac{3x + 4}{(2x + 3)^{1/2}}$

10. $2\sqrt{2}(3\sqrt{10} - \sqrt{6}) = 2\sqrt{2}(3\sqrt{10}) - 2\sqrt{2}(\sqrt{6}) = 6\sqrt{20} - 2\sqrt{12} = 6(2\sqrt{5}) - 2(2\sqrt{3}) = 12\sqrt{5} - 4\sqrt{3}$

11. $2\sqrt{\dfrac{5}{3a}} - \sqrt{\dfrac{3}{5a}} = 2\sqrt{\dfrac{5(3a)}{3a(3a)}} - \sqrt{\dfrac{3(5a)}{5a(5a)}} = \dfrac{2\sqrt{15a}}{3a} - \dfrac{\sqrt{15a}}{5a} = \dfrac{10\sqrt{15a} - 3\sqrt{15a}}{15a} = \dfrac{7\sqrt{15a}}{15a}$

12. $\dfrac{2\sqrt{15} + \sqrt{3}}{\sqrt{15} - 2\sqrt{3}} = \dfrac{2\sqrt{15} + \sqrt{3}}{\sqrt{15} - 2\sqrt{3}} \times \dfrac{\sqrt{15} + 2\sqrt{3}}{\sqrt{15} + 2\sqrt{3}} = \dfrac{2(15) + 4\sqrt{45} + \sqrt{45} + 2(3)}{15 - 4(3)} = \dfrac{36 + 15\sqrt{5}}{3} = 12 + 5\sqrt{5}$

13. $\dfrac{3^{-1/2}}{2} = \dfrac{1}{2 \times 3^{1/2}} = \dfrac{1}{2\sqrt{3}}$

$= \dfrac{\sqrt{3}}{2\sqrt{3}\sqrt{3}} = \dfrac{\sqrt{3}}{6}$

14. $0.220N^{-1/6}$, $N = 64 \times 10^6$

$0.220(64 \times 10^6)^{-1/6} = \dfrac{0.220}{(64 \times 10^6)^{1/6}}$

$= \dfrac{0.220}{2 \times 10} = 0.011$

Chapter 11

1. $(3 - \sqrt{-4}) + (5\sqrt{-9} - 1) = (3 - 2j) + [5(3j) - 1] = 3 - 2j + 15j - 1 = 2 + 13j$

2. $(2\underline{/130°})(3\underline{/45°}) = (2)(3)\underline{/130° + 45°} = 6\underline{/175°}$

3. $2 - 7j$: $r = \sqrt{2^2 + (-7)^2} = \sqrt{53} = 7.28$ $2 - 7j = 7.28(\cos 285.9° + j \sin 285.9°)$

   $\tan \theta = \dfrac{-7}{2} = -3.500,$ $\theta_{ref} = 74.1°,$ $\theta = 285.9°$ $= 7.28\underline{/285.9°}$

4. (a) $-\sqrt{-64} = -(8j) = -8j,$ (b) $-j^{15} = -j^{12}j^3 = (-1)(-j) = j$

5. $(4 - 3j) + (-1 + 4j) = 3 + j$ 6. $\dfrac{2 - 4j}{5 + 3j} = \dfrac{(2 - 4j)(5 - 3j)}{(5 + 3j)(5 - 3j)} = \dfrac{10 - 26j + 12j^2}{25 - 9j^2}$

   $= \dfrac{10 - 26j - 12}{25 - 9(-1)} = \dfrac{-2 - 26j}{34} = -\dfrac{1 + 13j}{17}$

7. $2.56(\cos 125.2° + j \sin 125.2°) = 2.56\, e^{2.185j}$

   $125.2° = \dfrac{125.2\pi}{180} = 2.185$ rad

8. $R = 3.50\ \Omega,\ X_L = 6.20\ \Omega,\ X_C = 7.35\ \Omega$ 9. $3.47 - 2.81j = 4.47e^{5.60j}$

   $|Z| = \sqrt{R^2 + (X_L - X_C)^2}$ $R = \sqrt{3.47^2 + (-2.81)^2} = 4.47$

   $= \sqrt{3.50^2 + (6.20 - 7.35)^2} = 3.68\ \Omega$ $\tan \theta = \dfrac{-2.81}{3.47},$ $\theta_{ref} = 39.0°$

   $\tan \theta = \dfrac{X_L - X_C}{R} = \dfrac{6.20 - 7.35}{3.50} = \dfrac{-1.15}{3.50}$ $\theta = 321.0° = 5.60$ rad

   $\theta = -18.2°$

10. $x + 2j - y = yj - 3xj$ 11. $L = 8.75$ mH $= 8.75 \times 10^{-3}$ H

   $x - y + 3xj - yj = -2j$ $f = 600$ kHz $= 6.00 \times 10^5$ Hz

   $(x - y) + (3x - y)j = 0 - 2j$ $2\pi fL = \dfrac{1}{2\pi fC}$

   $x - y = 0$

   $\dfrac{3x - y = -2}{2x = -2}$ $C = \dfrac{1}{(2\pi f)^2 L} = \dfrac{1}{(2\pi)^2(6.00 \times 10^5)^2(8.75 \times 10^{-3})}$

   $x = -1,\qquad y = -1$ $= 8.04 \times 10^{-12} = 8.04$ pF

12. $j = 1(\cos 90° + j \sin 90°)$

   $j^{1/3} = 1^{1/3}\left(\cos \dfrac{90°}{3} + j \sin \dfrac{90°}{3}\right) = \cos 30° + j \sin 30° = 0.8660 + 0.5000j$

   $= 1^{1/3}\left(\cos \dfrac{90° + 360°}{3} + j \sin \dfrac{90° + 360°}{3}\right) = \cos 150° + j \sin 150° = -0.8660 + 0.5000j$

   $= 1^{1/3}\left(\cos \dfrac{90° + 720°}{3} + j \sin \dfrac{90° + 720°}{3}\right) = \cos 270° + j \sin 270° = -j$

   Cube roots of j: $0.8660 + 0.5000j,\ -0.8660 + 0.5000j,\ -j.$

Chapter 12

1. $\log_9 x = -\dfrac{1}{2}$ 2. $\log_3 x - \log_3 2 = 2$ 3. $\log_x 64 = 3$ 4. $3^{3x+1} = 8$

   $x = 9^{-1/2}$ $\log_3 \dfrac{x}{2} = 2$ $64 = x^3$ $(3x + 1)\log 3 = \log 8$

   $= \dfrac{1}{9^{1/2}} = \dfrac{1}{3}$ $\dfrac{x}{2} = 3^2$ $4^3 = x^3$ $3x + 1 = \dfrac{\log 8}{\log 3}$

   $x = 4$

   $x = 2(3^2) = 18$ $x = \dfrac{1}{3}\left(\dfrac{\log 8}{\log 3} - 1\right)$

   $= 0.298$

5. $y = 2 \log_4 x$

x	$\frac{1}{4}$	1	4	16
y	-2	0	2	4

$\log_4 \frac{1}{4} = -1$, $\log_4 16 = 2$

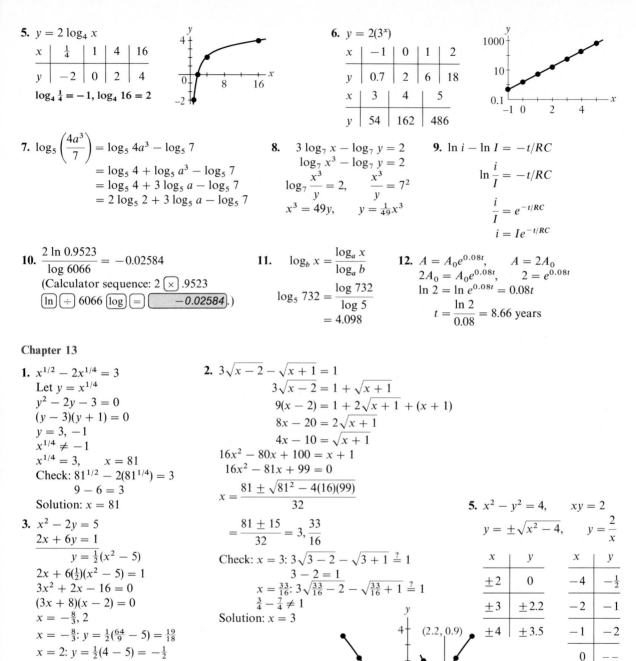

6. $y = 2(3^x)$

x	-1	0	1	2
y	0.7	2	6	18

x	3	4	5
y	54	162	486

7. $\log_5 \left(\dfrac{4a^3}{7} \right) = \log_5 4a^3 - \log_5 7$
$= \log_5 4 + \log_5 a^3 - \log_5 7$
$= \log_5 4 + 3 \log_5 a - \log_5 7$
$= 2 \log_5 2 + 3 \log_5 a - \log_5 7$

8. $3 \log_7 x - \log_7 y = 2$
$\log_7 x^3 - \log_7 y = 2$
$\log_7 \dfrac{x^3}{y} = 2$, $\quad \dfrac{x^3}{y} = 7^2$
$x^3 = 49y$, $\quad y = \frac{1}{49}x^3$

9. $\ln i - \ln I = -t/RC$
$\ln \dfrac{i}{I} = -t/RC$
$\dfrac{i}{I} = e^{-t/RC}$
$i = Ie^{-t/RC}$

10. $\dfrac{2 \ln 0.9523}{\log 6066} = -0.02584$
(Calculator sequence: 2 $\boxed{\times}$ .9523
$\boxed{\ln}$ $\boxed{\div}$ 6066 $\boxed{\log}$ $\boxed{=}$ $\boxed{-0.02584}$.)

11. $\log_b x = \dfrac{\log_a x}{\log_a b}$
$\log_5 732 = \dfrac{\log 732}{\log 5}$
$= 4.098$

12. $A = A_0 e^{0.08t}$, $\qquad A = 2A_0$
$2A_0 = A_0 e^{0.08t}$, $\quad 2 = e^{0.08t}$
$\ln 2 = \ln e^{0.08t} = 0.08t$
$t = \dfrac{\ln 2}{0.08} = 8.66$ years

Chapter 13

1. $x^{1/2} - 2x^{1/4} = 3$
Let $y = x^{1/4}$
$y^2 - 2y - 3 = 0$
$(y - 3)(y + 1) = 0$
$y = 3, -1$
$x^{1/4} \neq -1$
$x^{1/4} = 3$, $\quad x = 81$
Check: $81^{1/2} - 2(81^{1/4}) = 3$
$9 - 6 = 3$
Solution: $x = 81$

2. $3\sqrt{x - 2} - \sqrt{x + 1} = 1$
$3\sqrt{x - 2} = 1 + \sqrt{x + 1}$
$9(x - 2) = 1 + 2\sqrt{x + 1} + (x + 1)$
$8x - 20 = 2\sqrt{x + 1}$
$4x - 10 = \sqrt{x + 1}$
$16x^2 - 80x + 100 = x + 1$
$16x^2 - 81x + 99 = 0$
$x = \dfrac{81 \pm \sqrt{81^2 - 4(16)(99)}}{32}$
$= \dfrac{81 \pm 15}{32} = 3, \dfrac{33}{16}$
Check: $x = 3$: $3\sqrt{3 - 2} - \sqrt{3 + 1} \overset{?}{=} 1$
$3 - 2 = 1$
$x = \frac{33}{16}$: $3\sqrt{\frac{33}{16} - 2} - \sqrt{\frac{33}{16} + 1} \overset{?}{=} 1$
$\frac{3}{4} - \frac{7}{4} \neq 1$
Solution: $x = 3$

3. $x^2 - 2y = 5$
$2x + 6y = 1$
$y = \frac{1}{2}(x^2 - 5)$
$2x + 6(\frac{1}{2})(x^2 - 5) = 1$
$3x^2 + 2x - 16 = 0$
$(3x + 8)(x - 2) = 0$
$x = -\frac{8}{3}, 2$
$x = -\frac{8}{3}$: $y = \frac{1}{2}(\frac{64}{9} - 5) = \frac{19}{18}$
$x = 2$: $y = \frac{1}{2}(4 - 5) = -\frac{1}{2}$
$x = -\frac{8}{3}, y = \frac{19}{18}$
or $x = 2, y = -\frac{1}{2}$

4. $v = \sqrt{v_0^2 + 2gh}$
$v^2 = v_0^2 + 2gh$
$2gh = v^2 - v_0^2$
$h = \dfrac{v^2 - v_0^2}{2g}$

5. $x^2 - y^2 = 4$, $\qquad xy = 2$
$y = \pm\sqrt{x^2 - 4}$, $\qquad y = \dfrac{2}{x}$

x	y
± 2	0
± 3	± 2.2
± 4	± 3.5

x	y
-4	$-\frac{1}{2}$
-2	-1
-1	-2
0	$--$
1	2
2	1
4	$\frac{1}{2}$

$x = 2.2$, $\quad y = 0.9$
$x = -2.2$, $\quad y = -0.9$

6. Let ℓ = length, w = width.

$$2\ell + 2w = 14$$
$$\ell w = 10$$
$$\overline{\ell + \quad w = 7}$$
$$\ell = 7 - w$$
$$(7 - w)w = 10$$

$$7w - w^2 = 10$$
$$w^2 - 7w + 10 = 0$$
$$(w - 5)(w - 2) = 0$$
$$w = 5, 2$$
$$\ell = 5.00 \text{ ft}, \ w = 2.00 \text{ ft}$$
$$(\text{or } \ell = 2.00 \text{ ft}, \ w = 5.00 \text{ ft})$$

Chapter 14

1. $f(x) = 2x^3 + 3x^2 + 7x - 6$

$f(-3) = 2(-3)^3 + 3(-3)^2 + 7(-3) - 6$

$\quad = -54 + 27 - 21 - 6 = -54$

$f(-3) \neq 0, \qquad -3$ is not a zero

3. $(x^3 - 5x^2 + 4x - 9) \div (x - 3)$

$$
\begin{array}{rrrr|l}
1 & -5 & 4 & -9 & \underline{3} \\
  & 3 & -6 & -6 & \\
\hline
1 & -2 & -2 & -15 & \\
\end{array}
$$

Quotient: $x^2 - 2x - 2$.

Remainder $= -15$.

4. $f(x) = 2x^4 + 15x^3 + 23x^2 - 16$

$2x + 1 = 2(x + \frac{1}{2})$

$$
\begin{array}{rrrrr|l}
2 & 15 & 23 & 0 & -16 & \underline{-\frac{1}{2}} \\
  & -1 & -7 & -8 & 4 & \\
\hline
2 & 14 & 16 & -8 & -12 & \\
\end{array}
$$

Remainder is not zero;

$2x + 1$ is not a factor.

6. $2x^4 - x^3 + 5x^2 - 4x - 12 = 0$

$f(x) = 2x^4 - x^3 + 5x^2 - 4x - 12$

$f(-x) = 2x^4 + x^3 + 5x^2 + 4x - 12$

$n = 4$; 4 roots;

no more than 3 positive roots;

one negative root;

rational roots: factors of 12

divided by factors of 2;

possible rational roots: $\pm 1, \pm 2,$

$\pm 3, \pm 4, \pm 6, \pm 12, \pm \frac{1}{2}, \pm \frac{3}{2}$

$$
\begin{array}{rrrrr|l}
2 & -1 & 5 & -4 & -12 & \underline{2} \\
  & 4 & 6 & 22 & 36 & \\
\hline
2 & 3 & 11 & 18 & 24 & \\
\end{array}
$$

2 is too large.

$$
\begin{array}{rrrrr|l}
2 & -1 & 5 & -4 & -12 & \underline{\frac{3}{2}} \\
  & 3 & 3 & 12 & 12 & \\
\hline
2 & 2 & 8 & 8 & 0 & \underline{-1} \\
  & -2 & 0 & -8 & & \\
\hline
2 & 0 & 8 & 0 & & \\
\end{array}
$$

$2x^2 + 8 = 0, \qquad x^2 + 4 = 0, \qquad x = \pm 2j$

roots: $\frac{3}{2}, -1, 2j, -2j$

2. $x^4 - 2x^3 - 7x^2 + 20x - 12 = 0$

$$
\begin{array}{rrrrr|l}
1 & -2 & -7 & 20 & -12 & \underline{2} \\
  & 2 & 0 & -14 & 12 & \\
\hline
1 & 0 & -7 & 6 & & \underline{2} \\
  & 2 & 4 & -6 & & \\
\hline
1 & 2 & -3 & & & \\
\end{array}
$$

$x^2 + 2x - 3 = (x + 3)(x - 1)$

Other roots: $x = -3, 1$.

5. $(x^3 + 4x^2 + 7x - 9) \div (x + 4)$

$f(x) = x^3 + 4x^2 + 7x - 9$

$f(-4) = (-4)^3 + 4(-4)^2 + 7(-4) - 9$

$\quad = -64 + 64 - 28 - 9 = -37$

Remainder $= -37$.

7. $y = kx^2(x^3 + 436x - 4000)$

$y = 0, \ kx^2(x^3 + 436x - 4000) = 0$

$x = 0, \ x^3 + 436x - 4000 = 0$

$$
\begin{array}{rrrr|l}
1 & 0 & 436 & -4000 & \underline{8} \\
  & 8 & 64 & 4000 & \\
\hline
1 & 8 & 500 & 0 & \\
\end{array}
$$

$y = 0 \quad \text{for} \quad x = 0 \text{ ft}, \qquad x = 8 \text{ ft}$

8. Let x = length of edge.

$V = x^3, \qquad 2V = (x + 1)^3$

$2x^3 = x^3 + 3x^2 + 3x + 1$

$x^3 - 3x^2 - 3x - 1 = 0$

$f(x) = x^3 - 3x^2 - 3x - 1$

$f(3) = -10, \qquad f(4) = 3$

$f(3.6) = -4.024, \qquad f(3.8) = -0.848$

$f(3.9) = 0.989, \qquad f(3.85) = 0.049125$

$x = 3.8 \text{ mm} \quad$ (to one decimal place)

Chapter 15

1. $A = \begin{pmatrix} 3 & -1 & 4 \\ 2 & 0 & -2 \end{pmatrix}$, $B = \begin{pmatrix} 1 & 4 & 5 \\ -1 & -2 & 3 \end{pmatrix}$, $2B = \begin{pmatrix} 2 & 8 & 10 \\ -2 & -4 & 6 \end{pmatrix}$

$$A - 2B = \begin{pmatrix} 3-2 & -1-8 & 4-10 \\ 2+2 & 0+4 & -2-6 \end{pmatrix} = \begin{pmatrix} 1 & -9 & -6 \\ 4 & 4 & -8 \end{pmatrix}$$

2. $\begin{vmatrix} 4 & 0 & -2 \\ 3 & -3 & 2 \\ -4 & 1 & -1 \end{vmatrix} = 4 \begin{vmatrix} -3 & 2 \\ 1 & -1 \end{vmatrix} - 0 \begin{vmatrix} 3 & 2 \\ -4 & -1 \end{vmatrix} + (-2) \begin{vmatrix} 3 & -3 \\ -4 & 1 \end{vmatrix}$ using first row

$\qquad = 4(3 - 2) - 0 - 2(3 - 12) = 4 - 2(-9) = 4 + 18 = 22$

$\qquad = -0 \begin{vmatrix} 3 & 2 \\ -4 & -1 \end{vmatrix} + (-3) \begin{vmatrix} 4 & -2 \\ -4 & -1 \end{vmatrix} - (1) \begin{vmatrix} 4 & -2 \\ 3 & 2 \end{vmatrix}$ using second column

$\qquad = 0 - 3(-4 - 8) - (8 + 6) = -3(-12) - 14 = 22$

3. $CD = \begin{pmatrix} 1 & 0 & 4 \\ 2 & -2 & 1 \\ -1 & 3 & 2 \end{pmatrix} \begin{pmatrix} 2 & -2 \\ 4 & -5 \\ 6 & 1 \end{pmatrix} = \begin{pmatrix} 2+0+24 & -2+0+4 \\ 4-8+6 & -4+10+1 \\ -2+12+12 & 2-15+2 \end{pmatrix} = \begin{pmatrix} 26 & 2 \\ 2 & 7 \\ 22 & -11 \end{pmatrix}$

$DC = \begin{pmatrix} 2 & -2 \\ 4 & -5 \\ 6 & 1 \end{pmatrix} \begin{pmatrix} 1 & 0 & 4 \\ 2 & -2 & 1 \\ -1 & 3 & 2 \end{pmatrix}$ **not defined, since D has 2 columns and C has 3 rows**

4. $\begin{vmatrix} 1 & 0 & 4 & -2 \\ -2 & -1 & 3 & 0 \\ 3 & 2 & -1 & 2 \\ 1 & 1 & -1 & -2 \end{vmatrix} = \begin{vmatrix} 1 & 0 & 0 & 0 \\ -2 & -1 & 11 & -4 \\ 3 & 2 & -13 & 8 \\ 1 & 1 & -5 & 0 \end{vmatrix} = 1 \begin{vmatrix} -1 & 11 & -4 \\ 2 & -13 & 8 \\ 1 & -5 & 0 \end{vmatrix} = \begin{vmatrix} -1 & 11 & -4 \\ 0 & 9 & 0 \\ 0 & 6 & -4 \end{vmatrix} = -1 \begin{vmatrix} 9 & 0 \\ 6 & -4 \end{vmatrix}$

$\qquad = -(-36 - 0) = 36$

5. $\begin{pmatrix} 1 & 0 & 4 & | & 1 & 0 & 0 \\ 2 & -2 & 1 & | & 0 & 1 & 0 \\ -1 & 3 & 2 & | & 0 & 0 & 1 \end{pmatrix} \rightarrow \begin{pmatrix} 1 & 0 & 4 & | & 1 & 0 & 0 \\ 0 & -2 & -7 & | & -2 & 1 & 0 \\ 0 & 3 & 6 & | & 1 & 0 & 1 \end{pmatrix} \rightarrow$

$\begin{pmatrix} 1 & 0 & 4 & | & 1 & 0 & 0 \\ 0 & 1 & \frac{7}{2} & | & 1 & -\frac{1}{2} & 0 \\ 0 & 3 & 6 & | & 1 & 0 & 1 \end{pmatrix} \rightarrow \begin{pmatrix} 1 & 0 & 4 & | & 1 & 0 & 0 \\ 0 & 1 & \frac{7}{2} & | & 1 & -\frac{1}{2} & 0 \\ 0 & 0 & -\frac{9}{2} & | & -2 & \frac{3}{2} & 1 \end{pmatrix} \rightarrow$

$\begin{pmatrix} 1 & 0 & 4 & | & 1 & 0 & 0 \\ 0 & 1 & \frac{7}{2} & | & 1 & -\frac{1}{2} & 0 \\ 0 & 0 & 1 & | & \frac{4}{9} & -\frac{1}{3} & -\frac{2}{9} \end{pmatrix} \rightarrow \begin{pmatrix} 1 & 0 & 0 & | & -\frac{7}{9} & \frac{4}{3} & \frac{8}{9} \\ 0 & 1 & 0 & | & -\frac{5}{9} & \frac{2}{3} & \frac{7}{9} \\ 0 & 0 & 1 & | & \frac{4}{9} & -\frac{1}{3} & -\frac{2}{9} \end{pmatrix}$ $C^{-1} = \begin{pmatrix} -\frac{7}{9} & \frac{4}{3} & \frac{8}{9} \\ -\frac{5}{9} & \frac{2}{3} & \frac{7}{9} \\ \frac{4}{9} & -\frac{1}{3} & -\frac{2}{9} \end{pmatrix}$

6. Let A = number of shares of stock A,   Let C = coefficient matrix.
$\qquad B$ = number of shares of stock B.

$\quad 50A + 30B = 2600$
$\quad 30A + 40B = 2000$

$\overline{\quad 5A + \ 3B = 260 \quad}$
$\quad 3A + \ 4B = 200$

$\quad A = 40$ shares, $B = 20$ shares

$C = \begin{pmatrix} 5 & 3 \\ 3 & 4 \end{pmatrix}$, $\begin{vmatrix} 5 & 3 \\ 3 & 4 \end{vmatrix} = 20 - 9 = 11$

$C^{-1} = \frac{1}{11} \begin{pmatrix} 4 & -3 \\ -3 & 5 \end{pmatrix} = \begin{pmatrix} \frac{4}{11} & -\frac{3}{11} \\ -\frac{3}{11} & \frac{5}{11} \end{pmatrix}$

$\begin{pmatrix} A \\ B \end{pmatrix} = \begin{pmatrix} \frac{4}{11} & -\frac{3}{11} \\ -\frac{3}{11} & \frac{5}{11} \end{pmatrix} \begin{pmatrix} 260 \\ 200 \end{pmatrix} = \begin{pmatrix} \frac{4}{11}(260) - \frac{3}{11}(200) \\ -\frac{3}{11}(260) + \frac{5}{11}(200) \end{pmatrix} = \begin{pmatrix} 40 \\ 20 \end{pmatrix}$

Chapter 16

1. $x < 0$, $y > 0$

2. $\dfrac{-x}{2} \geq 3$

$-x \geq 6$

$x \leq -6$

3. $3x + 1 < -5$

$3x < -6$

$x < -2$

4. $-1 < 1 - 2x < 5$

$-2 < -2x < 4$

$-2 < x < 1$

5. $\dfrac{x^2 + x}{x - 2} \leq 0$, $\dfrac{x(x + 1)}{x - 2} \leq 0$

Interval	$\dfrac{x(x + 1)}{x - 2}$	Sign
$x < -1$	$-\quad -$ $-$	$-$
$-1 < x < 0$	$-\quad +$ $-$	$+$
$0 < x < 2$	$+\quad +$ $-$	$-$
$x > 2$	$+\quad +$ $+$	$+$

Solution: $x \leq -1$ or

$0 \leq x < 2$

(*x cannot equal 2*)

6. $|2x + 1| \geq 3$

$2x + 1 \geq 3$, $\qquad 2x + 1 \leq -3$

$2x \geq 2$ $\qquad\qquad 2x \leq -4$

$x \geq 1$ $\quad$ or $\quad x \leq -2$

8. If $\sqrt{x^2 - x - 6}$ is real,

then $x^2 - x - 6 \geq 0$.

$(x - 3)(x + 2) \geq 0$

$x \leq -2$ or $x \geq 3$

9. Let w = width, ℓ = length.

$\ell = w + 20$ $\qquad w^2 + 20w - 4800 \geq 0$

$w\ell \geq 4800$ $\qquad (w + 80)(w - 60) \geq 0$

$w(w + 20) \geq 4800$ $\qquad\qquad w \geq 60$ m

10. Let A = length of type A wire,

B = length of type B wire.

$0.10A + 0.20B < 5.00$

$A + 2B < 50$

7.

Chapter 17

1. $\dfrac{180 \text{ s}}{4 \text{ min}} = \dfrac{180 \text{ s}}{240 \text{ s}} = \dfrac{3}{4}$

2. $F = \dfrac{k}{d}$;

$0.750 = \dfrac{k}{1.25}$, $\qquad k = 0.938 \text{ N} \cdot \text{cm}$

3. $\dfrac{1.00 \text{ in.}}{2.54 \text{ cm}} = \dfrac{x}{7.24 \text{ cm}}$

4. $p = kdh$

Using values for water,

$1.96 = k(1000)(0.200)$

$k = 0.00980 \text{ kPa} \cdot \text{m}^2/\text{kg}$

For alcohol,

$p = 0.00980(800)(0.300)$

$= 2.35 \text{ kPa}$

$F = \dfrac{0.938}{d}$; $\qquad F = \dfrac{0.938}{1.75} = 0.536 \text{ N}$

$x = \dfrac{7.24}{2.54} = 2.85 \text{ in.}$

5. $2\ell + 2w = 210$

$\ell = 105 - w$

$\dfrac{\ell}{w} = \dfrac{7}{3}$

$\dfrac{105 - w}{w} = \dfrac{7}{3}$

$315 - 3w = 7w$

$10w = 315$

$w = 31.5 \text{ in.}$

$\ell = 105.0 - 31.5 = 73.5 \text{ in}$

6. Let L_1 = crushing load of first pillar,

L_2 = crushing load of second pillar.

$L_2 = \dfrac{kr_2^4}{\ell_2^2}$, $\qquad L_1 = \dfrac{k(2r_2)^4}{(3\ell_2)^2}$;

$\dfrac{L_1}{L_2} = \dfrac{\dfrac{k(2r_2)^4}{(3\ell_2)^2}}{\dfrac{kr_2^4}{\ell_2^2}} = \dfrac{16kr_2^4}{9\ell_2^2} \times \dfrac{\ell_2^2}{kr_2^4} = \dfrac{16}{9}$

Chapter 18

1. $6, -2, \frac{2}{3}, \ldots;$
geometric sequence

$$a_1 = 6, \qquad r = \frac{-2}{6} = -\frac{1}{3}$$

$$S_7 = \frac{6[1 - (-\frac{1}{3})^7]}{1 - (-\frac{1}{3})}$$

$$= \frac{6\left(1 + \frac{1}{3^7}\right)}{\frac{4}{3}} = \frac{1094}{243}$$

2. $a_1 = 6, d = 4, s_n = 126;$
arithmetic sequence

$$126 = \frac{n}{2}(6 + a_n)$$
$$a_n = 6 + (n-1)4 = 2 + 4n$$
$$126 = \frac{n}{2}[6 + (2 + 4n)]$$
$$252 = n(8 + 4n) = 4n^2 + 8n$$
$$n^2 + 2n - 63 = 0$$
$$(n + 9)(n - 7) = 0$$
$$n = 7$$

3. $0.454545\ldots = 0.45 + 0.0045 + 0.000045 + \cdots$

$a = 0.45, \qquad r = 0.01$

$$S = \frac{0.45}{1 - 0.01} = \frac{0.45}{0.99} = \frac{5}{11}$$

4. $\sqrt{1 - 4x} = (1 - 4x)^{1/2}$

$$= 1 + (\tfrac{1}{2})(-4x) + \frac{\frac{1}{2}(\frac{1}{2} - 1)}{2}(-4x)^2 + \cdots$$

$$= 1 - 2x - 2x^2 + \cdots$$

5. $5\% = 0.05$
Value after 1 year is
$2500 + 2500(0.05) = 2500(1.05)$
$V_{20} = 2500(1.05)^{20}$
$\quad = \$6633.24$

6. $(2x - y)^5 = (2x)^5 + 5(2x)^4(-y) + \frac{5(4)}{2}(2x)^3(-y)^2$

$$+ \frac{5(4)(3)}{2(3)}(2x)^2(-y)^3 + \frac{5(4)(3)(2)}{2(3)(4)}(2x)(-y)^4 + (-y)^5$$

$$= 32x^5 - 80x^4y + 80x^3y^2 - 40x^2y^3 + 10xy^4 - y^5$$

7. $2 + 4 + \cdots + 200$
$a_1 = 2, \qquad a_{100} = 200, \qquad n = 100$
$S_{100} = \frac{100}{2}(2 + 200)$
$\quad = 10{,}100$

8. Ball falls 8.00 ft, rises 4.00 ft, falls 4.00 ft, etc.
Distance $= 8.00 + (4.00 + 4.00) + (2.00 + 2.00) + \cdots$
$\quad = 8.00 + 8.00 + 4.00 + 2.00 + \cdots$
$\quad = 8.00 + \dfrac{8.00}{1 - 0.5} = 8.00 + 16.0 = 24.0$ ft

Chapter 19

1. $\sec\theta - \dfrac{\tan\theta}{\csc\theta} = \dfrac{1}{\cos\theta} - \dfrac{\dfrac{\sin\theta}{\cos\theta}}{\dfrac{1}{\sin\theta}} = \dfrac{1}{\cos\theta} - \dfrac{\sin^2\theta}{\cos\theta} = \dfrac{1 - \sin^2\theta}{\cos\theta} = \dfrac{\cos^2\theta}{\cos\theta} = \cos\theta; \cos\theta = \cos\theta$

2. $\sin 2x + \sin x = 0$
$2\sin x \cos x + \sin x = 0$
$\sin x(2\cos x + 1) = 0$
$\sin x = 0, \qquad \cos x = -\dfrac{1}{2}$
$x = 0, \pi, \frac{2\pi}{3}, \frac{4\pi}{3}$

3. $\theta = \text{Arcsin } x$
$\cos\theta = \cos(\text{Arcsin } x)$
$\quad = \sqrt{1 - x^2}$

4. $\theta = e^{-0.1t}(\cos 2t + 3\sin 2t)$
$e^{-0.1t}(\cos 2t + 3\sin 2t) = 0$
$\cos 2t + 3\sin 2t = 0$
$3\sin 2t = -\cos 2t$
$3\tan 2t = -1, \qquad \tan 2t = -\dfrac{1}{3}$
$2t = 2.82, \qquad t = 1.41$

5. $\dfrac{\tan\alpha + \tan\beta}{\tan\alpha - \tan\beta} = \dfrac{\dfrac{\sin\alpha}{\cos\alpha} + \dfrac{\sin\beta}{\cos\beta}}{\dfrac{\sin\alpha}{\cos\alpha} - \dfrac{\sin\beta}{\cos\beta}} = \dfrac{\sin\alpha\cos\beta + \sin\beta\cos\alpha}{\sin\alpha\cos\beta - \sin\beta\cos\alpha} = \dfrac{\sin(\alpha + \beta)}{\sin(\alpha - \beta)}$

6. $\cot^2 x - \cos^2 x = \dfrac{\cos^2 x}{\sin^2 x} - \cos^2 x = \cos^2 x(\csc^2 x - 1) = \cos^2 x \cot^2 x; \cos^2 x \cot^2 x = \cos^2 x \cot^2 x$

7. $\sin x = -\dfrac{3}{5}$

$\cos x = \dfrac{4}{5}$

$\cos\dfrac{x}{2} = -\sqrt{\dfrac{1 + (\frac{4}{5})}{2}} = -\sqrt{\dfrac{9}{10}} = -0.9487$

since $270° < x < 360°$, $135° < \frac{x}{2} < 180°$;
$\frac{x}{2}$ is in second quadrant, where $\cos\frac{x}{2}$ is negative

8. $I = I_0 \sin 2\theta \cos 2\theta,$ $\dfrac{2I}{I_0} = 2 \sin 2\theta \cos 2\theta = \sin 4\theta$

$4\theta = \text{Arcsin} \dfrac{2I}{I_0},$ $\theta = \dfrac{1}{4} \text{Arcsin} \dfrac{2I}{I_0}$

Chapter 20

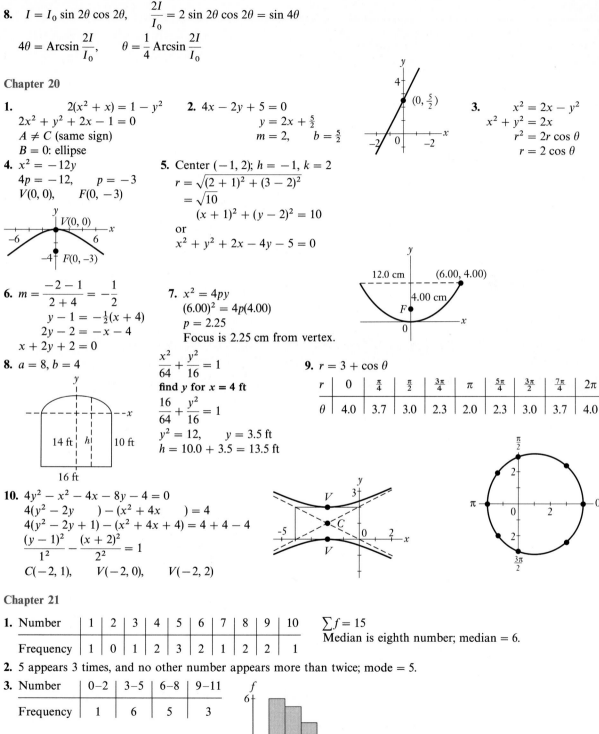

1. $2(x^2 + x) = 1 - y^2$
$2x^2 + y^2 + 2x - 1 = 0$
$A \neq C$ (same sign)
$B = 0$: ellipse

2. $4x - 2y + 5 = 0$
$y = 2x + \frac{5}{2}$
$m = 2,$ $b = \frac{5}{2}$

3. $x^2 = 2x - y^2$
$x^2 + y^2 = 2x$
$r^2 = 2r \cos \theta$
$r = 2 \cos \theta$

4. $x^2 = -12y$
$4p = -12,$ $p = -3$
$V(0, 0),$ $F(0, -3)$

5. Center $(-1, 2)$; $h = -1, k = 2$
$r = \sqrt{(2 + 1)^2 + (3 - 2)^2}$
$= \sqrt{10}$
$(x + 1)^2 + (y - 2)^2 = 10$
or
$x^2 + y^2 + 2x - 4y - 5 = 0$

6. $m = \dfrac{-2 - 1}{2 + 4} = -\dfrac{1}{2}$
$y - 1 = -\frac{1}{2}(x + 4)$
$2y - 2 = -x - 4$
$x + 2y + 2 = 0$

7. $x^2 = 4py$
$(6.00)^2 = 4p(4.00)$
$p = 2.25$
Focus is 2.25 cm from vertex.

8. $a = 8, b = 4$

$\dfrac{x^2}{64} + \dfrac{y^2}{16} = 1$

find y for $x = 4$ ft

$\dfrac{16}{64} + \dfrac{y^2}{16} = 1$

$y^2 = 12,$ $y = 3.5$ ft
$h = 10.0 + 3.5 = 13.5$ ft

9. $r = 3 + \cos \theta$

r	0	$\frac{\pi}{4}$	$\frac{\pi}{2}$	$\frac{3\pi}{4}$	π	$\frac{5\pi}{4}$	$\frac{3\pi}{2}$	$\frac{7\pi}{4}$	2π
θ	4.0	3.7	3.0	2.3	2.0	2.3	3.0	3.7	4.0

10. $4y^2 - x^2 - 4x - 8y - 4 = 0$
$4(y^2 - 2y \quad) - (x^2 + 4x \quad) = 4$
$4(y^2 - 2y + 1) - (x^2 + 4x + 4) = 4 + 4 - 4$
$\dfrac{(y - 1)^2}{1^2} - \dfrac{(x + 2)^2}{2^2} = 1$
$C(-2, 1),$ $V(-2, 0),$ $V(-2, 2)$

Chapter 21

1.

Number	1	2	3	4	5	6	7	8	9	10
Frequency	1	0	1	2	3	2	1	2	2	1

$\sum f = 15$
Median is eighth number; median $= 6$.

2. 5 appears 3 times, and no other number appears more than twice; mode $= 5$.

3.

Number	0–2	3–5	6–8	9–11
Frequency	1	6	5	3

4. Let t = thickness

$$\bar{t} = \frac{3(0.90) + 9(0.91) + 31(0.92) + 38(0.93) + 12(0.94) + 5(0.95) + 2(0.96)}{3 + 9 + 31 + 38 + 12 + 5 + 2}$$

$$= \frac{92.7}{100} = 0.927 \text{ in.}$$

5. $\overline{t^2} = \dfrac{\sum ft^2}{\sum f} = \dfrac{85.9464}{100} = 0.859464$

6.

$s_t = \sqrt{\overline{t^2} - \bar{t}^2} = \sqrt{0.859464 - 0.927^2} = 0.012 \text{ in.}$

7.

x	y	xy	x^2
1	5	5	1
3	11	33	9
5	17	85	25
7	20	140	49
9	27	243	81
25	80	506	165

$n = 5$

$m = \dfrac{5(506) - (25)(80)}{5(165) - 25^2} = 2.65$

$b = \dfrac{165(80) - (506)(25)}{5(165) - 25^2} = 2.75$

$y = 2.65x + 2.75$

8.

x	$\sqrt{x}$	y	$\sqrt{x}\,y$	$(\sqrt{x})^2 = x$
1.00	1.000	1.10	1.1000	1.000
3.00	1.732	1.90	3.2908	3.000
5.00	2.236	2.50	5.5900	5.000
7.00	2.646	2.90	7.6734	7.000
9.00	3.000	3.30	9.9000	9.000
	10.614	11.70	27.5542	25.000

$n = 5$

$m = \dfrac{5(27.5542) - (10.614)(11.70)}{5(25.000) - (10.614)^2} = 1.10$

$b = \dfrac{(25.000)(11.70) - (27.5542)(10.614)}{5(25.000) - (10.614)^2}$

$= 0.00 \quad$ (rounded off)

$y = 1.10\sqrt{x}$

Chapter 22

1. $\displaystyle\lim_{x \to 1} \frac{x^2 - x}{x^2 - 1} = \lim_{x \to 1} \frac{x(x-1)}{(x+1)(x-1)} = \lim_{x \to 1} \frac{x}{x+1} = \frac{1}{2}$

2. $\displaystyle\lim_{x \to \infty} \frac{1 - 4x^2}{x + 2x^2} = \lim_{x \to \infty} \frac{\dfrac{1}{x^2} - 4}{\dfrac{1}{x} + 2} = -2$

3. $y = 3x^2 - \dfrac{4}{x^2}$

$\dfrac{dy}{dx} = 6x + \dfrac{8}{x^3}$

$\dfrac{dy}{dx}\Big|_{x=2} = 6(2) + \dfrac{8}{2^3}$

$= 13$

$m_{\text{tan}} = 13$

4. $s = t\sqrt{10 - 2t}$

$v = \dfrac{ds}{dt} = t(\tfrac{1}{2})(10 - 2t)^{-1/2}(-2) + (10 - 2t)^{1/2}(1)$

$= \dfrac{-t}{(10 - 2t)^{1/2}} + (10 - 2t)^{1/2} = \dfrac{10 - 3t}{(10 - 2t)^{1/2}}$

$v\Big|_{t=4.00} = \dfrac{10 - 3(4.00)}{[10 - 2(4.00)]^{1/2}} = \dfrac{10 - 12.00}{2.00^{1/2}} = -1.41 \text{ cm/s}$

5. $(1 + y^2)^3 - x^2y = 7x$

$3(1 + y^2)^2(2yy') - x^2y' - y(2x) = 7$

$6y(1 + y^2)^2y' - x^2y' = 7 + 2xy$

$y' = \dfrac{7 + 2xy}{6y(1 + y^2)^2 - x^2}$

6. $V = \dfrac{kq}{\sqrt{x^2 + b^2}} = kq(x^2 + b^2)^{-1/2}$

$\dfrac{dV}{dx} = kq\left(-\dfrac{1}{2}\right)(x^2 + b^2)^{-3/2}(2x)$

$= \dfrac{-kqx}{(x^2 + b^2)^{3/2}}$

7. $y = \dfrac{2}{3x + 2} = 2(3x + 2)^{-1}$

$\dfrac{dy}{dx} = 2(-1)(3x + 2)^{-2}(3) = -6(3x + 2)^{-2}$

$\dfrac{d^2y}{dx^2} = -6(-2)(3x + 2)^{-3}(3) = \dfrac{36}{(3x + 2)^3}$

8. $y = 5x - 2x^2$

$y + \Delta y = 5(x + \Delta x) - 2(x + \Delta x)^2$

$\Delta y = 5\,\Delta x - 4x\,\Delta x - 2\,\Delta x^2$

$\dfrac{\Delta y}{\Delta x} = 5 - 4x - 2\,\Delta x$

$\lim\limits_{\Delta x \to 0} \dfrac{\Delta y}{\Delta x} = 5 - 4x$

Chapter 23

1. $y = x^4 - 3x^2$

$\dfrac{dy}{dx} = 4x^3 - 6x$

$\left.\dfrac{dy}{dx}\right|_{x=1} = 4(1^3) - 6(1)$

$= -2$

$y - (-2) = -2(x - 1)$

$y = -2x$

2. $x = 3t^2$

$v_x = \dfrac{dx}{dt} = 6t$

$a_x = \dfrac{dv_x}{dt} = \dfrac{d^2x}{dt^2} = 6$

$a_x|_{t=2} = 6$

$a|_{t=2} = \sqrt{6^2 + 22^2} = 22.8$

$y = 2t^3 - t^2$

$v_y = \dfrac{dy}{dt} = 6t^2 - 2t$

$a_y = \dfrac{dv_y}{dt} = \dfrac{d^2y}{dt^2} = 12t - 2$

$a_y|_{t=2} = 12(2) - 2 = 22$

$\tan \theta = \tfrac{22}{6}, \qquad \theta = 74.7°$

3. $P = \dfrac{144r}{(r + 0.6)^2}$

$\dfrac{dP}{dr} = \dfrac{144[(r + 0.6)^2(1) - r(2)(r + 0.6)(1)]}{(r + 0.6)^4}$

$= \dfrac{144[(r + 0.6) - 2r]}{(r + 0.6)^3} = \dfrac{144(0.6 - r)}{(r + 0.6)^3}$

$\dfrac{dP}{dr} = 0; \qquad 0.6 - r = 0, \qquad r = 0.6\ \Omega$

$\left(r < 0.6, \dfrac{dP}{dr} > 0; r > 0.6, \dfrac{dP}{dr} < 0\right)$

4. $x^2 - \sqrt{4x + 1} = 0; \qquad f(x) = x^2 - \sqrt{4x + 1}$

$f'(x) = 2x - \tfrac{1}{2}(4x + 1)^{-1/2}(4) = 2x - \dfrac{2}{(4x + 1)^{1/2}}$

n	x_n	$f(x_n)$	$f'(x_n)$	$x_n - \dfrac{f(x_n)}{f'(x_n)}$
1	1.5	-0.3957513	2.2440711	1.6763542
2	1.6763542	0.0343001	2.6322118	1.6633233

$x_3 = 1.6633$

5. $y = x^3 + 6x^2$

$y' = 3x^2 + 12x = 3x(x + 4)$

$y'' = 6x + 12 = 6(x + 2)$

$\begin{array}{ll} x < -4 & y \text{ inc.} \\ -4 < x < 0 & y \text{ dec.} \\ x > 0 & y \text{ inc.} \\ x < -2 & y \text{ conc. down} \\ x > -2 & y \text{ conc. up} \end{array}$

Max. $(-4, 32)$
Min. $(0, 0)$
Infl. $(-2, 16)$

6. $y = \dfrac{4}{x^2} - x$

$y' = -\dfrac{8}{x^3} - 1 = -\dfrac{8 + x^3}{x^3}$

$y'' = \dfrac{24}{x^4}$

$\begin{array}{ll} x < -2 & y \text{ dec.} \\ -2 < x < 0 & y \text{ inc.} \\ x > 0 & y \text{ dec., conc. up} \\ x < 0 & y \text{ conc. up} \end{array}$

Min. $(-2, 3)$, no infl., int. $(\sqrt[3]{4}, 0)$, sym. none;
as $x \to \pm\infty$, $y \to -x$, asym. $y = -x$, $x = 0$.
Domain: all real x except 0; range: all real y.

7. Let $V =$ volume of cube,
$e =$ edge of cube.

$V = e^3; \qquad \dfrac{dV}{dt} = 3e^2 \dfrac{de}{dt}$

$\left.\dfrac{dV}{dt}\right|_{e=4.00} = 3(4.00)^2(-0.50)$

$= -24\ \text{ft}^3/\text{s}$

8.

$3x + 2y = 6000$

$y = \dfrac{6000 - 3x}{2}$

$A = xy = x\left(\dfrac{6000 - 3x}{2}\right)$

$= 3000x - \tfrac{3}{2}x^2$

$\dfrac{dA}{dx} = 3000 - 3x$

$3000 - 3x = 0, x = 1000\ \text{m}$

$y = 1500\ \text{m}$

$A_{\max} = (1000)(1500) = 1.5 \times 10^6\ \text{m}^2$

$\left(x < 1000, \dfrac{dA}{dx} > 0; x > 1000, \dfrac{dA}{dx} < 0\right)$

Chapter 24

1. $y = 3x^2 - x$
$y + \Delta y = 3(x + \Delta x)^2 - (x + \Delta x)$
$\Delta y = 6x\,\Delta x + 3\,\Delta x^2 - \Delta x$
$dy = (6x - 1)\,dx$
For $x = 3$, $\Delta x = 0.1$.
$\Delta y = 6(3)(0.1) + 3(0.1)^2 - (0.1)$
$\quad = 1.73$
$dy = [6(3) - 1](0.1) = 1.7$
$\Delta y - dy = 0.03$

2. $\int x\sqrt{1 - 2x^2}\,dx = \int x(1 - 2x^2)^{1/2}\,dx$
$u = 1 - 2x^2, \qquad du = -4x\,dx, \qquad n = \tfrac{1}{2}, \qquad n + 1 = \tfrac{3}{2}$
$\int x(1 - 2x^2)^{1/2}\,dx = -\tfrac{1}{4}\int (1 - 2x^2)^{1/2}(-4x\,dx)$
$\qquad = -\tfrac{1}{4}(\tfrac{2}{3})(1 - 2x^2)^{3/2} + C = -\tfrac{1}{6}(1 - 2x^2)^{3/2} + C$

3. $\int_2^5 (6 - x)^4\,dx = -\int_2^5 (6 - x)^4(-dx) = -\tfrac{1}{5}(6 - x)^5\Big|_2^5$
$\qquad = -\tfrac{1}{5}(1^5 - 4^5) = \tfrac{1023}{5}$

4. $y = \dfrac{1}{x + 2}$, $n = 6$, $\Delta x = \dfrac{4 - 1}{6} = \dfrac{1}{2}$
$A = \tfrac{1}{2}(\tfrac{2}{7} + \tfrac{1}{4} + \tfrac{2}{9} + \tfrac{1}{5} + \tfrac{2}{11} + \tfrac{1}{6}) = 0.6532$

x	1	$\tfrac{3}{2}$	2	$\tfrac{5}{2}$	3	$\tfrac{7}{2}$	4
y	$\tfrac{1}{3}$	$\tfrac{2}{7}$	$\tfrac{1}{4}$	$\tfrac{2}{9}$	$\tfrac{1}{5}$	$\tfrac{2}{11}$	$\tfrac{1}{6}$

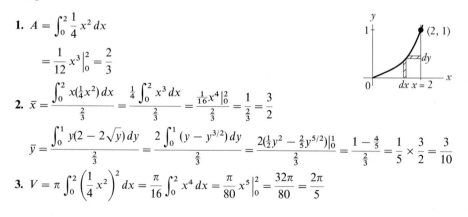

5. (See values for Problem 4.)
$\int_1^4 \dfrac{dx}{x + 2} = \left[\tfrac{1}{2}\left(\tfrac{1}{3}\right) + \tfrac{2}{7} + \tfrac{1}{4} + \tfrac{2}{9} + \tfrac{1}{5} + \tfrac{2}{11} + \tfrac{1}{2}\left(\tfrac{1}{6}\right)\right]\left(\tfrac{1}{2}\right)$
$\qquad = 0.6949$

6. (See values for Problem 4.)
$\int_1^4 \dfrac{dx}{x + 2} = \dfrac{1}{6}\left[\tfrac{1}{3} + 4\left(\tfrac{2}{7}\right) + 2\left(\tfrac{1}{4}\right) + 4\left(\tfrac{2}{9}\right) + 2\left(\tfrac{1}{5}\right) + 4\left(\tfrac{2}{11}\right) + \tfrac{1}{6}\right]$
$\qquad = 0.6932$

7. $A = \pi r^2$
$dA = 2\pi r\,dr$
For $r = 4.50$ cm, $dr = 0.30$ cm.
$dA = 2\pi(4.50)(0.30) = 8.48$ cm^2
$V = 20.0(8.48) = 170$ cm^3

8. $i = \int_1^3 \left(t^2 + \dfrac{1}{t^2}\right)dt$
$\quad = \dfrac{1}{3}t^3 - \dfrac{1}{t}\Big|_1^3$
$\quad = \tfrac{1}{3}(27) - \tfrac{1}{3} - (\tfrac{1}{3} - 1) = 9.3$ A

Chapter 25

1. $A = \int_0^2 \dfrac{1}{4}x^2\,dx$

$\quad = \dfrac{1}{12}x^3\Big|_0^2 = \dfrac{2}{3}$

2. $\bar{x} = \dfrac{\int_0^2 x(\tfrac{1}{4}x^2)\,dx}{\tfrac{2}{3}} = \dfrac{\tfrac{1}{4}\int_0^2 x^3\,dx}{\tfrac{2}{3}} = \dfrac{\tfrac{1}{16}x^4\big|_0^2}{\tfrac{2}{3}} = \dfrac{1}{\tfrac{2}{3}} = \dfrac{3}{2}$

$\bar{y} = \dfrac{\int_0^1 y(2 - 2\sqrt{y})\,dy}{\tfrac{2}{3}} = \dfrac{2\int_0^1 (y - y^{3/2})\,dy}{\tfrac{2}{3}} = \dfrac{2(\tfrac{1}{2}y^2 - \tfrac{2}{5}y^{5/2})\big|_0^1}{\tfrac{2}{3}} = \dfrac{1 - \tfrac{4}{5}}{\tfrac{2}{3}} = \dfrac{1}{5} \times \dfrac{3}{2} = \dfrac{3}{10}$

3. $V = \pi\int_0^2 \left(\dfrac{1}{4}x^2\right)^2 dx = \dfrac{\pi}{16}\int_0^2 x^4\,dx = \dfrac{\pi}{80}x^5\Big|_0^2 = \dfrac{32\pi}{80} = \dfrac{2\pi}{5}$

4. $V = \pi \int_0^3 9^2\,dx - \pi \int_0^3 (x^2)^2\,dx = 81\pi x\big|_0^3 - \dfrac{\pi}{5}x^5\big|_0^3$

$= 243\pi - \dfrac{243\pi}{5} = \dfrac{972\pi}{5}$

or $V = 2\pi \int_0^9 xy\,dy = 2\pi \int_0^9 y^{1/2}y\,dy = \dfrac{4\pi}{5}y^{5/2}\big|_0^9$

$= \dfrac{4\pi}{5}(3^5 - 0) = \dfrac{972\pi}{5}$

5. $I_y = k \int_0^3 x^2(9 - x^2)\,dx = k \int_0^3 (9x^2 - x^4)\,dx$

$= k\left(3x^3 - \dfrac{1}{5}x^5\right)\Big|_0^3 = k\left(81 - \dfrac{243}{5}\right) = \dfrac{162k}{5}$

6. $s = \int (60 - 4t)\,dt = 60t - 2t^2 + C$

$s = 10$ for $t = 0,$ $10 = 60(0) - 2(0^2) + C,$ $C = 10$

$s = 60t - 2t^2 + 10$

7. $F = kx,\ 12 = k(2.0),\ k = 6.0$ N/cm

$W = \int_{2.0}^{6.0} 6.0x\,dx = 3.0x^2\big|_{2.0}^{6.0} = 3.0(36.0 - 4.0) = 96$ N·cm

8. $F = 62.4 \int_{1.00}^{3.00} 6.00h\,dh = (62.4)(6.00)\tfrac{1}{2}h^2\big|_{1.00}^{3.00}$

$= (62.4)(3.00)(9.00 - 1.00) = 1500$ lb

Chapter 26

1. $y = \tan^3 2x + \text{Arctan } 2x$

$\dfrac{dy}{dx} = 3(\tan^2 2x)(\sec^2 2x)(2) + \dfrac{2}{1 + (2x)^2}$

$= 6\tan^2 2x \sec^2 2x + \dfrac{2}{1 + 4x^2}$

2. $y = 2(3 + \cot 4x)^3$

$\dfrac{dy}{dx} = 2(3)(3 + \cot 4x)^2(-\csc^2 4x)(4)$

$= -24\csc^2 4x(3 + \cot 4x)^2$

3. $y\sec 2x = \text{Arcsin } 3y$

$y(\sec 2x \tan 2x)(2) + (\sec 2x)(y') = \dfrac{3y'}{\sqrt{1 - (3y)^2}}$

$\sqrt{1 - 9y^2}\,\sec 2x\,(2y\tan 2x + y') = 3y'$

$y' = \dfrac{2y\sqrt{1 - 9y^2}\,\sec 2x \tan 2x}{3 - \sqrt{1 - 9y^2}\,\sec 2x}$

4. $y = \dfrac{\cos^2(3x + 1)}{x}$

$dy = \dfrac{x\{2\cos(3x + 1)[-\sin(3x + 1)(3)]\} - \cos^2(3x + 1)(1)}{x^2}\,dx$

$= \dfrac{-6x\cos(2x + 1)\sin(3x + 1) - \cos^2(3x + 1)}{x^2}\,dx$

5. $y = \ln\dfrac{2x - 1}{1 + x^2}$

$= \ln(2x - 1) - \ln(1 + x^2)$

$\dfrac{dy}{dx} = \dfrac{2}{2x - 1} - \dfrac{2x}{1 + x^2}$

$m_{\tan} = \dfrac{dy}{dx}\Big|_{x=2} = \dfrac{2}{4 - 1} - \dfrac{4}{1 + 4}$

$= \dfrac{2}{3} - \dfrac{4}{5} = -\dfrac{2}{15}$

6. $i = 8e^{-t}\sin 10t$

$\dfrac{di}{dt} = 8[e^{-t}\cos 10t(10) + \sin 10t(-e^{-t})]$

$= 8e^{-t}(10\cos 10t - \sin 10t)$

7. $y = xe^x$

$y' = xe^x + e^x = e^x(x + 1)$

$y'' = xe^x + e^x + e^x = e^x(x + 2)$

$x < -1$ y dec.

$x > -1$ y inc.

$x < -2$ y conc. down

$x > -2$ y conc. up

Int. $(0, 0)$, min. $(-1, -e^{-1})$, infl. $(-2, -2e^{-2})$, asym. $y = 0$.

8. For $t = 8.0$s, $x = 40$ ft.

$$\theta = \text{Arctan} \frac{x}{250}; \quad \frac{d\theta}{dt} = \frac{1}{1 + \dfrac{x^2}{250^2}} \frac{dx/dt}{250} = \frac{250^2}{250^2 + x^2} \frac{dx/dt}{250}$$

250 ft

$$\frac{d\theta}{dt}\Big|_{t=8.0} = \frac{250}{250^2 + 40^2}(5.0) = 0.020 \text{ rad/s}$$

Chapter 27

1. $\displaystyle\int (\sec x - \sec^3 x \tan x)\,dx = \int \sec x\,dx - \int \sec^2 x(\sec x \tan x)\,dx$

$$= \ln|\sec x + \tan x| - \tfrac{1}{3}\sec^3 x + C$$

2. $\displaystyle\int \sin^3 x\,dx = \int \sin^2 x \sin x\,dx = \int (1 - \cos^2 x)\sin x\,dx = \int \sin x\,dx - \int \cos^2 x \sin x\,dx$

$$= -\cos x + \tfrac{1}{3}\cos^3 x + C$$

3. $\displaystyle\int \tan^3 2x\,dx = \int \tan 2x(\tan^2 2x)\,dx = \int \tan 2x(\sec^2 2x - 1)\,dx$

$$= \tfrac{1}{2}\int \tan 2x \sec^2 2x(2\,dx) - \tfrac{1}{2}\int \tan 2x(2\,dx) = \tfrac{1}{4}\tan^2 2x + \tfrac{1}{2}\ln|\cos 2x| + C$$

4. $\displaystyle\int \cos^2 4x\,dx = \tfrac{1}{2}\int (1 + \cos 8x)\,dx = \tfrac{1}{2}\int dx + \tfrac{1}{16}\int \cos 8x(8\,dx) = \tfrac{1}{2}x + \tfrac{1}{16}\sin 8x + C$

5. $\displaystyle\int \frac{dx}{x^2\sqrt{4 - x^2}} = \int \frac{2\cos\theta\,d\theta}{4\sin^2\theta\sqrt{4 - 4\sin^2\theta}} = \frac{1}{4}\int \frac{\cos\theta\,d\theta}{\sin^2\theta\sqrt{\cos^2\theta}} = \tfrac{1}{4}\int \csc^2\theta\,d\theta = -\tfrac{1}{4}\cot\theta + C = -\frac{\sqrt{4 - x^2}}{4x} + C$

Let $x = 2\sin\theta$,
$\quad dx = 2\cos\theta\,d\theta.$

6. $\displaystyle\int xe^{-2x}\,dx; u = x, du = dx, dv = e^{-2x}\,dx, v = -\tfrac{1}{2}e^{-2x}$

$$\int xe^{-2x}\,dx = x(-\tfrac{1}{2}e^{-2x}) - \int(-\tfrac{1}{2}e^{-2x})\,dx = -\tfrac{1}{2}xe^{-2x} - \tfrac{1}{4}e^{-2x} + C$$

7. $\displaystyle i = \int \frac{6t + 1}{4t^2 + 9}\,dt = \int \frac{6t\,dt}{4t^2 + 9} + \int \frac{dt}{4t^2 + 9} = \frac{6}{8}\int \frac{8t\,dt}{4t^2 + 9} + \frac{1}{2}\int \frac{2\,dt}{9 + (2t)^2}$

$$= \frac{3}{4}\ln(4t^2 + 9) + \frac{1}{2}\left(\frac{1}{3}\right)\text{Arctan}\frac{2t}{3} + C$$

$$i = 0 \text{ for } t = 0: 0 = \frac{3}{4}\ln 9 + \frac{1}{6}\text{Arctan } 0 + C, C = -\frac{3}{4}\ln 9$$

$$i = \frac{3}{4}\ln(4t^2 + 9) + \frac{1}{6}\text{Arctan}\frac{2t}{3} - \frac{3}{4}\ln 9 = \frac{3}{4}\ln\frac{4t^2 + 9}{9} + \frac{1}{6}\text{Arctan}\frac{2t}{3}$$

8. $\displaystyle y = \frac{1}{\sqrt{16 - x^2}}; A = \int_0^3 y\,dx = \int_0^3 \frac{dx}{\sqrt{16 - x^2}} = \text{Arcsin}\frac{x}{4}\Big|_0^3$

$$= \text{Arcsin}\tfrac{3}{4} = 0.8481$$

Chapter 28

1. $\quad f(x) = (1 + e^x)^2 \qquad\qquad f(0) = (1 + 1)^2 = 4$

$\quad f'(x) = 2(1 + e^x)(e^x) \qquad f'(0) = 2(1 + 1)(1) = 4$

$\qquad\quad = 2e^x + 2e^{2x} \qquad\quad f''(0) = 2(1) + 4(1) = 6 \qquad (1 + e^x)^2 = 4 + 4x + \tfrac{6}{2}x^2 + \tfrac{10}{6}x^3 + \cdots$

$\quad f''(x) = 2e^x + 4e^{2x} \qquad f'''(0) = 2(1) + 8(1) = 10 \qquad\qquad = 4 + 4x + 3x^2 + \tfrac{5}{3}x^3 + \cdots$

$\quad f'''(x) = 2e^x + 8e^{2x}$

2. $\quad f(x) = \cos x \qquad\qquad f\left(\dfrac{\pi}{3}\right) = \dfrac{1}{2}$

$\quad f'(x) = -\sin x \qquad\quad f'\left(\dfrac{\pi}{3}\right) = -\dfrac{\sqrt{3}}{2} \qquad \cos x = \dfrac{1}{2} - \dfrac{\sqrt{3}}{2}\left(x - \dfrac{\pi}{3}\right) - \dfrac{\dfrac{1}{2}\left(x - \dfrac{\pi}{3}\right)^2}{2} + \cdots$

$\quad f''(x) = -\cos x \qquad\quad f''\left(\dfrac{\pi}{3}\right) = -\dfrac{1}{2} \qquad\qquad = \dfrac{1}{2}\left[1 - \sqrt{3}\left(x - \dfrac{\pi}{3}\right) - \dfrac{1}{2}\left(x - \dfrac{\pi}{3}\right)^2 + \cdots\right]$

3. $\ln(1+x) = x - \dfrac{x^2}{2} + \dfrac{x^3}{3} - \dfrac{x^4}{4} + \cdots$

$\ln 0.96 = \ln(1 - 0.04)$

$\quad = -0.04 - \dfrac{(-0.04)^2}{2} + \dfrac{(-0.04)^3}{3} - \dfrac{(-0.04)^4}{4}$

$\quad = -0.0408220$

4. $f(x) = \dfrac{1}{\sqrt{1-2x}} = (1-2x)^{-1/2}$

$(1+x)^n = 1 + nx + \dfrac{n(n-1)}{2}x^2 + \cdots$

$(1-2x)^{-1/2} = 1 + \left(-\dfrac{1}{2}\right)(-2x) + \dfrac{-\frac{1}{2}\left(-\frac{3}{2}\right)}{2}(-2x)^2 + \cdots$

$\quad = 1 + x + \tfrac{3}{2}x^2 + \cdots$

5. $\displaystyle\int_0^1 x \cos x\, dx = \int_0^1 x\left(1 - \dfrac{x^2}{2} + \dfrac{x^4}{24}\right) dx = \int_0^1 \left(x - \dfrac{x^3}{2} + \dfrac{x^5}{24}\right) dx = \dfrac{1}{2}x^2 - \dfrac{x^4}{8} + \dfrac{x^6}{144}\bigg|_0^1$

$\quad = \tfrac{1}{2} - \tfrac{1}{8} + \tfrac{1}{144} - 0 = 0.3819$

6. $f(t) = 0 \qquad -\pi \le t < 0$
$f(t) = 2 \qquad\ \ \ 0 \le t < \pi$

$a_0 = \dfrac{1}{2\pi}\displaystyle\int_0^\pi 2\, dt = \dfrac{1}{\pi}t\bigg|_0^\pi = 1$

$a_n = \dfrac{1}{\pi}\displaystyle\int_0^\pi 2\cos nx\, dx = \dfrac{2}{n\pi}\sin nx\bigg|_0^\pi = 0 \quad \text{for all } n$

$b_n = \dfrac{1}{\pi}\displaystyle\int_0^\pi 2\sin nx\, dx = \dfrac{2}{n\pi}(-\cos nx)\bigg|_0^\pi = \dfrac{2}{n\pi}(1 - \cos n\pi) \qquad f(t) = 1 + \dfrac{4}{\pi}\sin x + \dfrac{4}{3\pi}\sin 3x + \cdots$

$b_1 = \dfrac{2}{\pi}(1+1) = \dfrac{4}{\pi}, \qquad b_2 = \dfrac{2}{2\pi}(1-1) = 0, \qquad b_3 = \dfrac{2}{3\pi}(1+1) = \dfrac{4}{3\pi}$

$f(t)$

Chapter 29

1. $x\dfrac{dy}{dx} + 2y = 4$

$dy + \dfrac{2}{x}y\, dx = \dfrac{4}{x}dx$

$e^{\int \frac{2}{x}dx} = e^{2\ln x} = x^2$

$yx^2 = \displaystyle\int \dfrac{4}{x}x^2\, dx = \int 4x\, dx$

$\quad = 2x^2 + c$

$y = 2 + \dfrac{c}{x^2}$

2. $y'' + 2y' + 5y = 0$

$m^2 + 2m + 5 = 0$

$m = \dfrac{-2 \pm \sqrt{4 - 20}}{2}$

$\quad = -1 \pm 2j$

$y = e^{-x}(c_1 \sin 2x + c_2 \cos 2x)$

3. $x\, dx + y\, dy = x^2\, dx + y^2\, dx$

$\quad = (x^2 + y^2)\, dx$

$\dfrac{x\, dx + y\, dy}{x^2 + y^2} = dx$

$\tfrac{1}{2}\ln(x^2 + y^2) = x + \tfrac{1}{2}c$

$\ln(x^2 + y^2) = 2x + c$

4. $2D^2y - Dy = 2\cos x$

$2m^2 - m = 0; \quad m = 0, \tfrac{1}{2}$

$y_c = c_1 + c_2 e^{x/2}$

$y_p = A\sin x + B\cos x$

$Dy_p = A\cos x - B\sin x$

$D^2 y_p = -A\sin x - B\cos x$

$2(-A\sin x - B\cos x) - (A\cos x - B\sin x) = 2\cos x$

$-2A + B = 0, \qquad -2B - A = 2, \qquad A = -\tfrac{2}{5}, B = -\tfrac{4}{5}$

$y = c_1 + c_2 e^{x/2} - \tfrac{2}{5}\sin x - \tfrac{4}{5}\cos x$

5. $\dfrac{d^2 y}{dx^2} - 4\dfrac{dy}{dx} + 4y = 3x$

$m^2 - 4m + 4 = 0; \qquad m = 2, 2$

$y_c = (c_1 + c_2 x)e^{2x}$

$y_p = A + Bx, \qquad Dy_p = B, \qquad D^2 y_p = 0$

$0 - 4B + 4(A + Bx) = 3x$

$4A - 4B = 0, \qquad 4B = 3$

$B = \tfrac{3}{4}, \qquad A = \tfrac{3}{4}$

$y = (c_1 + c_2 x)e^{2x} + \tfrac{3}{4} + \tfrac{3}{4}x$

6. $(xy + y)\dfrac{dy}{dx} = 2$

$y(x + 1)\,dy = 2\,dx$

$y\,dy = \dfrac{2\,dx}{x + 1}$

$\frac{1}{2}y^2 = 2\ln(x + 1) + c$

$y = 2$ when $x = 0$

$\frac{1}{2}(4) = 2\ln(1) + c,$ $c = 2$

$\frac{1}{2}y^2 = 2\ln(x + 1) + 2$

$y^2 = 4\ln(x + 1) + 4$

7. $\dfrac{dA}{dt} = rA$

$\dfrac{dA}{A} = r\,dt$

$\ln A = rt + \ln c$

$\ln\dfrac{A}{c} = rt$

$A = ce^{rt}$

$A_0 = ce^0,$ $c = A_0$

$A = A_0 e^{rt}$

8. $y'' + 9y = 9$

$y(0) = 0,$ $y'(0) = 1$

$L(y'') + 9L(y) = L(9)$

$s^2 L(y) - s(0) - 1 + 9L(y) = \dfrac{9}{s}$

$(s^2 + 9)L(y) = \dfrac{9}{s} + 1$

$L(y) = \dfrac{9}{s(s^2 + 9)} + \dfrac{1}{s^2 + 9}$

$y = 1 - \cos 3t + \frac{1}{3}\sin 3t$

9. $L\dfrac{d^2q}{dt^2} + R\dfrac{dq}{dt} = E$

$L\dfrac{di}{dt} + Ri = E$

$L = 2\text{ H},$ $R = 8\ \Omega,$ $E = 6\text{ V}$

$2\dfrac{di}{dt} + 8i = 6$

$di + 4i\,dt = 3\,dt$

$e^{\int 4\,dt} = e^{4t}$

$ie^{4t} = \int 3e^{4t}\,dt = \frac{3}{4}e^{4t} + c$

$i = 0$ for $t = 0,$ $0 = \frac{3}{4} + c,$ $c = -\frac{3}{4}$

$ie^{4t} = \frac{3}{4}e^{4t} - \frac{3}{4}$

$i = \frac{3}{4}(1 - e^{-4t}) = 0.75(1 - e^{-4t})$

10. $F = kx;\ 16 = k(0.5),\ k = 32\text{ lb/ft}$

$m = \dfrac{16\text{ lb}}{32\text{ ft/s}^2} = 0.5\text{ slug}$

$mD^2 x = -kx,$ $0.5D^2 x + 32x = 0$

$D^2 x + 64x = 0$

$m^2 + 64 = 0;$ $m = \pm 8j$

$x = c_1 \sin 8t + c_2 \cos 8t$

$Dx = 8c_1 \cos 8t - 8c_2 \sin 8t$

$x = 0.3\text{ ft},$ $Dx = 0,$ for $t = 0$

$0.3 = c_1 \sin 0 + c_2 \cos 0,$ $c_2 = 0.3$

$0 = 8c_1 \cos 0 - 8c_2 \sin 0,$ $c_1 = 0$

$x = 0.3\cos 8t$

Index of Applications

Index